Handbook of
Data Structures
and Applications

Second Edition

Chapman & Hall/CRC
Computer and Information Science Series

Series Editor: Sartaj Sahni

Distributed Systems
An Algorithmic Approach, Second Edition
Sukumar Ghosh

Computer-Aided Graphing and Simulation Tools for AutoCAD Users
P. A. Simionescu

Integration of Services into Workflow Applications
Paweł Czarnul

Handbook of Graph Theory, Combinatorial Optimization, and Algorithms
Krishnaiyan "KT" Thulasiraman, Subramanian Arumugam, Andreas Brandstädt, and Takao Nishizeki

From Action Systems to Distributed Systems
The Refinement Approach
Luigia Petre and Emil Sekerinski

Trustworthy Cyber-Physical Systems Engineering
Alexander Romanovsky and Fuyuki Ishikawa

X-Machines for Agent-Based Modeling
FLAME Perspectives
Mariam Kiran

From Internet of Things to Smart Cities
Enabling Technologies
Hongjian Sun, Chao Wang, and Bashar I. Ahmad

Evolutionary Multi-Objective System Design
Theory and Applications
Nadia Nedjah, Luiza De Macedo Mourelle, and Heitor Silverio Lopes

Networks of the Future
Architectures, Technologies, and Implementations
Mahmoud Elkhodr, Qusay F. Hassan, and Seyed Shahrestani

Computer Simulation
A Foundational Approach Using Python
Yahya E. Osais

Internet of Things
Challenges, Advances, and Applications
Qusay F. Hassan, Atta ur Rehman Khan, and Sajjad A. Madani

Handbook of Data Structures and Applications, Second Edition
Dinesh P. Mehta and Sartaj Sahni

For more information about this series please visit:
https://www.crcpress.com/Chapman--HallCRC-Computer-and-Information-Science-Series/book-series/CHCOMINFSCI

Handbook of Data Structures and Applications

Second Edition

Edited by
Dinesh P. Mehta
Sartaj Sahni

CRC Press
Taylor & Francis Group
Boca Raton   London   New York

CRC Press is an imprint of the
Taylor & Francis Group, an **informa** business

A CHAPMAN & HALL BOOK

MATLAB®is a trademark of The MathWorks, Inc. and is used with permission. The MathWorks does not warrant the accuracy of the text or exercises in this book. This book's use or discussion of MATLAB® software or related products does not constitute endorsement or sponsorship by The MathWorks of a particular pedagogical approach or particular use of the MATLAB® software.

CRC Press
Taylor & Francis Group
6000 Broken Sound Parkway NW, Suite 300
Boca Raton, FL 33487-2742

© 2018 by Taylor & Francis Group, LLC
CRC Press is an imprint of Taylor & Francis Group, an Informa business

No claim to original U.S. Government works

Printed on acid-free paper

International Standard Book Number-13: 978-1-4987-0185-3 (Hardback)

This book contains information obtained from authentic and highly regarded sources. Reasonable efforts have been made to publish reliable data and information, but the author and publisher cannot assume responsibility for the validity of all materials or the consequences of their use. The authors and publishers have attempted to trace the copyright holders of all material reproduced in this publication and apologize to copyright holders if permission to publish in this form has not been obtained. If any copyright material has not been acknowledged please write and let us know so we may rectify in any future reprint.

Except as permitted under U.S. Copyright Law, no part of this book may be reprinted, reproduced, transmitted, or utilized in any form by any electronic, mechanical, or other means, now known or hereafter invented, including photocopying, microfilming, and recording, or in any information storage or retrieval system, without written permission from the publishers.

For permission to photocopy or use material electronically from this work, please access www.copyright.com (http://www.copyright.com/) or contact the Copyright Clearance Center, Inc. (CCC), 222 Rosewood Drive, Danvers, MA 01923, 978-750-8400. CCC is a not-for-profit organization that provides licenses and registration for a variety of users. For organizations that have been granted a photocopy license by the CCC, a separate system of payment has been arranged.

Trademark Notice: Product or corporate names may be trademarks or registered trademarks, and are used only for identification and explanation without intent to infringe.

Library of Congress Cataloging-in-Publication Data

Names: Mehta, Dinesh P., editor. | Sahni, Sartaj, editor.
Title: Handbook of data structures and applications / edited by Dinesh P. Mehta and Sartaj Sahni.
Other titles: Data structures and applications
Description: Second edition. | Boca Raton, Florida : CRC Press, [2018] | Includes bibliographical references and index.
Identifiers: LCCN 2017041375| ISBN 9781498701853 (hardback) | ISBN 9781315119335 (e-book)
Subjects: LCSH: System design--Handbooks, manuals, etc. | Data structures (Computer science)--Handbooks, manuals, etc.
Classification: LCC QA76.9.S88 H363 2005 | DDC 005.7/3--dc23 LC record available at https://lccn.loc.gov/2017041375

**Visit the Taylor & Francis Web site at
http://www.taylorandfrancis.com**

**and the CRC Press Web site at
http://www.crcpress.com**

To our wives,
Usha Mehta and Neeta Sahni

Contents

Preface to the Second Edition xi

Preface to the First Edition xiii

Editors xv

Contributors xvii

PART I   Fundamentals

1 Analysis of Algorithms 3
   Sartaj Sahni

2 Basic Structures 23
   Dinesh P. Mehta

3 Trees 35
   Dinesh P. Mehta

4 Graphs 49
   Narsingh Deo

PART II   Priority Queues

5 Leftist Trees 69
   Sartaj Sahni

6 Skew Heaps 77
   C. Pandu Rangan

7 Binomial, Fibonacci, and Pairing Heaps 85
   Michael L. Fredman

8 Double-Ended Priority Queues 97
   Sartaj Sahni

PART III   Dictionary Structures

9 Hash Tables 117
   Pat Morin

10 Bloom Filter and Its Variants 131
    Shigang Chen

11 Balanced Binary Search Trees 151
    Arne Andersson, Rolf Fagerberg, and Kim S. Larsen

12 Finger Search Trees 171
    Gerth Stølting Brodal

13   Splay trees 179
      Sanjeev Saxena

14   Randomized Dictionary Structures 197
      C. Pandu Rangan

15   Trees with Min Weighted Path Length 215
      Wojciech Rytter

16   B Trees 233
      Donghui Zhang

PART IV   Multidimensional/Spatial Structures

17   Multidimensional Spatial Data Structures 251
      Hanan Samet

18   Planar Straight Line Graphs 277
      Siu-Wing Cheng

19   Interval, Segment, Range, Priority Search Trees 291
      D. T. Lee and Hung-I Yu

20   Quadtrees and Octtrees 309
      Srinivas Aluru

21   BSP Trees 329
      Bruce F. Naylor

22   R-Trees 343
      Scott Leutenegger and Mario A. Lopez

23   Managing Spatio-Temporal Data 359
      Sumeet Dua and S. S. Iyengar

24   Kinetic Data Structures 377
      Leonidas Guibas

25   Online Dictionary Structures 389
      Teofilo F. Gonzalez

26   Cuttings 397
      Bernard Chazelle

27   Approximate Geom Query Structures 405
      Christian A. Duncan and Michael T. Goodrich

28   Geometric and Spatial Data Structures in External Memory 419
      Jeffrey Scott Vitter

PART V   Miscellaneous

29   Tries 445
      Sartaj Sahni

30   Suffix Trees and Suffix Arrays 461
      Srinivas Aluru

31   String Searching 477
      Andrzej Ehrenfeucht and Ross M. McConnell

32 Binary Decision Diagrams 495
Shin-ichi Minato

33 Persistent Data Structures 511
Haim Kaplan

34 Data Structures for Sets 529
Rajeev Raman

35 Cache Oblivious Data Structures.. 545
Lars Arge, Gerth Stølting Brodal, and Rolf Fagerberg

36 Dynamic Trees.. 567
Camil Demetrescu, Irene Finocchi, and Giuseppe F. Italiano

37 Dynamic Graphs 581
Camil Demetrescu, Irene Finocchi, and Giuseppe F. Italiano

38 Succinct Representation of Data Structures 595
J. Ian Munro and S. Srinivasa Rao

39 Randomized Graph Data Structures 611
Surender Baswana and Sandeep Sen

40 Searching and Priority Queues in o(log n) Time.................................. 627
Arne Andersson

PART VI Data Structures in Langs and Libraries

41 Functional Data Structures 639
Chris Okasaki

42 LEDA, a Platform for Combinatorial and Geometric Computing............ 653
Stefan Naeher

43 Data Structures in C++... 667
Mark Allen Weiss

44 Data Structures in JDSL.. 679
Michael T. Goodrich, Roberto Tamassia, and Luca Vismara

45 Data Structure Visualization 697
John Stasko

46 Drawing Trees.. 707
Sebastian Leipert

47 Drawing Graphs 723
Peter Eades and Seok-Hee Hong

48 Concurrent Data Structures... 741
Mark Moir and Nir Shavit

PART VII Applications

49 IP Router Tables 765
Sartaj Sahni, Kun Suk Kim, and Haibin Lu

50 Multidimensional Packet Classification.. 783
Pankaj Gupta

51 Data Structures in Web Information Retrieval 799
Monika Henzinger

52 The Web as a Dynamic Graph 805
S. N. Maheshwari

53 Layout Data Structures 817
Dinesh P. Mehta

54 Floorplan Rep in VLSI 833
Zhou Feng, Bo Yao, and Chung-Kuan Cheng

55 Computer Graphics 857
Dale McMullin and Alyn Rockwood

56 Geographic Information Systems 871
Bernhard Seeger and Peter Widmayer

57 Collision Detection 889
Ming C. Lin and Dinesh Manocha

58 Image Data Structures 903
S. S. Iyengar, V. K. Vaishnavi, and S. Gunasekaran

59 Computational Biology 917
Paolo Ferragina, Stefan Kurtz, Stefano Lonardi, and Giovanni Manzini

60 Data Structures for Cheminformatics 935
Dinesh P. Mehta and John D. Crabtree

61 Elimination Structures in Scientific Computing 945
Alex Pothen and Sivan Toledo

62 Data Structures for Databases 967
Joachim Hammer and Markus Schneider

63 Data Structures for Big Data Stores 983
Arun A. Ravindran and Dinesh P. Mehta

64 Data Mining 997
Vipin Kumar, Pang-Ning Tan, and Michael Steinbach

65 Computational Geometry: Fundamental Structures 1013
Mark de Berg and Bettina Speckmann

66 Computational Geometry: Proximity and Location 1027
Sunil Arya and David M. Mount

67 Computational Geometry: Generalized (or Colored) Intersection Searching 1043
Prosenjit Gupta, Ravi Janardan, Saladi Rahul, and Michiel Smid

Index 1059

Preface to the Second Edition

It has been over a decade since the first edition of this handbook was published in 2005. In this edition, we have attempted to capture advances in data structures while retaining the seven-part structure of the first edition. As one would expect, the discipline of data structures has matured with advances not as rapidly forthcoming as in the twentieth century. Nevertheless, there have been areas that have seen significant progress that we have focused on in this edition.

We have added four new chapters on Bloom Filters; Binary Decision Diagrams; Data Structures for Cheminformatics; and Data Structures for Big Data Stores. In addition, we have updated 13 other chapters from the original edition.

Dinesh P. Mehta
Sartaj Sahni
October 2017

Preface to the First Edition

In the late 1960s, Donald Knuth, winner of the 1974 Turing Award, published his landmark book *The Art of Computer Programming: Fundamental Algorithms*. This book brought together a body of knowledge that defined the data structures area. The term data structure, itself, was defined in this book to be *A table of data including structural relationships*. Niklaus Wirth, the inventor of the Pascal language and winner of the 1984 Turing award, stated that "Algorithms + Data Structures = Programs." The importance of algorithms and data structures has been recognized by the community and consequently, every undergraduate Computer Science curriculum has classes on data structures and algorithms. Both of these related areas have seen tremendous advances in the decades since the appearance of the books by Knuth and Wirth. Although there are several advanced and specialized texts and handbooks on algorithms (and related data structures), there is, to the best of our knowledge, no text or handbook that focuses exclusively on the wide variety of data structures that have been reported in the literature. The goal of this handbook is to provide a comprehensive survey of data structures of different types that are in existence today.

To this end, we have subdivided this handbook into seven parts, each of which addresses a different facet of data structures. Part I is a review of introductory material. Although this material is covered in all standard data structures texts, it was included to make the handbook self-contained and in recognition of the fact that there are many practitioners and programmers who may not have had a formal education in computer science. Parts II through IV discuss Priority Queues, Dictionary Structures, and Multidimensional structures, respectively. These are all well-known classes of data structures. Part V is a catch-all used for well-known data structures that eluded easy classification. Parts I through V are largely theoretical in nature: they discuss the data structures, their operations and their complexities. Part VI addresses mechanisms and tools that have been developed to facilitate the *use* of data structures in real programs. Many of the data structures discussed in previous parts are very intricate and take some effort to program. The development of data structure libraries and visualization tools by skilled programmers are of critical importance in reducing the gap between theory and practice. Finally, Part VII examines applications of data structures. The deployment of many data structures from Parts I through V in a variety of applications is discussed. Some of the data structures discussed here have been invented solely in the context of these applications and are not well-known to the broader community. Some of the applications discussed include Internet routing, web search engines, databases, data mining, scientific computing, geographical information systems, computational geometry, computational biology, VLSI floorplanning and layout, computer graphics and image processing.

For data structure and algorithm researchers, we hope that the handbook will suggest new ideas for research in data structures and for an appreciation of the application contexts in which data structures are deployed. For the practitioner who is devising an algorithm, we hope that the handbook will lead to insights in organizing data that make it possible to solve the algorithmic problem more cleanly and efficiently. For researchers in specific application areas, we hope that they will gain some insight from the ways other areas have handled their data structuring problems.

Although we have attempted to make the handbook as complete as possible, it is impossible to undertake a task of this magnitude without some omissions. For this, we apologize in advance and encourage readers to contact us with information about significant data structures or applications that do not appear here. These could be included in future editions of this handbook. We would like to thank the excellent team of authors, who are at the forefront of research in data structures, who have contributed to this handbook. The handbook would not have been possible without their painstaking efforts. We are extremely saddened by the untimely demise of a prominent data structures researcher, Professor Gísli R. Hjaltason, who was to write a chapter for this handbook. He will be missed greatly by the computer science community. Finally, we would like to thank our families for their support during the development of the handbook.

Dinesh P. Mehta
Sartaj Sahni
April 2004

MATLAB® is a registered trademark of The MathWorks, Inc. For product information, please contact:

The MathWorks, Inc.

3 Apple Hill Drive

Natick, MA 01760-2098 USA Tel: 508 647 7000

Fax: 508-647-7001

E-mail: info@mathworks.com

Web: www.mathworks.com

Editors

Dinesh P. Mehta has been on the faculty of the Colorado School of Mines since 2000, where he is currently professor in the Department of Computer Science. He earned degrees (all in computer science) from the Indian Institute of Technology Bombay, the University of Minnesota, and the University of Florida. Before joining Mines in 2000, he was on the faculty of the University of Tennessee Space Institute, where he received the Vice President's Award for Teaching Excellence in 1997. He was a visiting professor at Intel's Strategic CAD Labs for several months in 1996 and 1997 and at the Tata Research Development and Design Center (in Pune, India) in 2007. He has also received graduate teaching awards at Mines in 2007, 2008, and 2009.

He was assistant department head from 2004 to 2008 and interim department head from 2008 to 2010 in the former Department of Mathematical and Computer Sciences and served as president of the Mines Faculty Senate in 2016–2017.

Dr. Mehta is the coauthor of the book, *Fundamentals of Data Structures in C++* and coeditor of the *Handbook of Algorithms for VLSI Physical Design Automation*. He serves as associate director of the ORwE (Operations Research with Engineering) PhD program at Mines and is currently an associate editor of *ACM Computing Surveys*. His current research interests are in cheminformatics, computational materials, and big graph analytics.

Sartaj Sahni is a distinguished professor of computer and information sciences and engineering at the University of Florida. He is also a member of the European Academy of Sciences, a Fellow of IEEE, ACM, AAAS, and Minnesota Supercomputer Institute, and a distinguished alumnus of the Indian Institute of Technology, Kanpur. Dr. Sahni is the recipient of the 1997 IEEE Computer Society Taylor L. Booth Education Award, the 2003 IEEE Computer Society W. Wallace McDowell Award, and the 2003 ACM Karl Karlstrom Outstanding Educator Award. Dr. Sahni earned his BTech (electrical engineering) degree from the Indian Institute of Technology, Kanpur, and MS and PhD in computer science from Cornell University. Dr. Sahni has published over 400 research papers and written 15 books. His research publications are on the design and analysis of efficient algorithms, parallel computing, interconnection networks, design automation, and medical algorithms.

Dr. Sahni is the editor-in-chief of the *ACM Computing Surveys*, a managing editor of the *International Journal of Foundations of Computer Science*, and a member of the editorial boards of 17 other journals. He is a past coeditor-in-chief of the *Journal of Parallel and Distributed Computing*. He has served as program committee chair, general chair, and a keynote speaker at many conferences. Dr. Sahni has served on several NSF (National Science Foundation) and NIH (National Institutes of Health) panels and he has been involved as an external evaluator of several computer science and engineering departments.

Contributors

Srinivas Aluru
Georgia Institute of Technology
Atlanta, Georgia

Arne Andersson
Uppsala University
Uppsala, Sweden

Lars Arge
University of Aarhus
Aarhus, Denmark

Sunil Arya
The Hong Kong University of Science
    and Technology
Kowloon, Hong Kong

Surender Baswana
Indian Institute of Technology, Kanpur
Kanpur, India

Mark de Berg
Eindhoven University of Technology
Eindhoven, The Netherlands

Gerth Stølting Brodal
University of Aarhus
Aarhus, Denmark

Bernard Chazelle
Department of Computer Science
Princeton University
Princeton, New Jersey

John D. Crabtree
Department of Computer Science
    and Information Systems
University of North Alabama
Florence, Alabama

Shigang Chen
Computer and Information Science
    and Engineering
University of Florida
Gainesville, Florida

Chung-Kuan Cheng
University of California
San Diego, California

Siu-Wing Cheng
The Hong Kong University of Science
    and Technology
Kowloon, Hong Kong

Camil Demetrescu
Sapienza University of Rome
Rome, Italy

Narsingh Deo
University of Central Florida
Orlando, Florida

Sumeet Dua
Louisiana Tech University
Ruston, Louisiana

Christian A. Duncan
Quinnipiac University
Hamden, Connecticut

Peter Eades
University of Sydney
Sydney, Australia

Andrzej Ehrenfeucht
University of Colorado
Boulder, Colorado

Rolf Fagerberg
University of Southern Denmark
Odense, Denmark

Zhou Feng
Fudan University
Shanghai, China

Paolo Ferragina
Università di Pisa
Pisa, Italy

Irene Finocchi
Department of Computer Science
Sapienza University of Rome
Rome, Italy

Michael L. Fredman
Rutgers University
New Brunswick, New Jersey

Teofilo F. Gonzalez
Department of Computer Science
University of California
Santa Barbara, California

Michael T. Goodrich
University of California
Irvine, California

Leonidas Guibas
Stanford University
Palo Alto, California

Prosenjit Gupta
Department of Computer Science
    and Engineering
NIIT University
Rajasthan, India

Monika Henzinger
University of Vienna
Vienna, Austria

Seok-Hee Hong
University of Sydney
Sydney, Australia

Giuseppe F. Italiano
Department of Civil Engineering
    and Computer Science Engineering
University of Rome
Tor Vergata Rome, Italy

S. S. Iyengar
Florida International University
Miami, Florida

Ravi Janardan
Department of Computer Science and Engineering
University of Minnesota
Minneapolis, Minnesota

Haim Kaplan
Tel Aviv University
Tel Aviv, Israel

Vipin Kumar
University of Minnesota
Minneapolis, Minnesota

Stefan Kurtz
University of Hamburg
Hamburg, Germany

Kim S. Larsen
University of Southern Denmark
Odense, Denmark

D. T. Lee
Institute of Information Science
Academia Sinica
Taipei, Taiwan

Scott Leutenegger
Department of Computer Science
University of Denver
Denver, Colorado

Ming C. Lin
University of North Carolina
Chapel Hill, North Carolina

Stefano Lonardi
University of California, Riverside
Riverside, California

Mario A. Lopez
Department of Computer Science
University of Denver
Denver, Colorado

S. N. Maheshwari
Indian Institute of Technology Delhi
Delhi, India

Dinesh Manocha
University of North Carolina
Chapel Hill, North Carolina

Giovanni Manzini
Università del Piemonte Orientale
Vercelli VC, Italy

Ross M. McConnell
Colorado State University
Fort Collins, Colorado

Dinesh P. Mehta
Department of Computer Science
Colorado School of Mines
Golden, Colorado

Shin-ichi Minato
Hokkaido University
Sapporo, Japan

Pat Morin
Carleton University
Ottawa, Ontario, Canada

David M. Mount
University of Maryland
College Park, Maryland

J. Ian Munro
University of Waterloo
Waterloo, Ontario, Canada

Stefan Naeher
Department of Computer Science
University of Trier
Trier, Germany

Chris Okasaki
United States Military Academy
West Point, New York

Alex Pothen
Computer Science Department Purdue University
West Lafayette, Indiana

Saladi Rahul
Department of Computer Science
University of Illinois
Urbana, Illinois

Rajeev Raman
University of Leicester
Leicester, United Kingdom

C. Pandu Rangan
Indian Institute of Technology, Madras
Chennai, India

S. Srinivasa Rao
Seoul National University
Seoul, South Korea

Arun A. Ravindran
Department of Electrical and Computer Engineering
University of North Carolina at Charlotte
Charlotte, North Carolina

Wojciech Rytter
Warsaw University
Warsaw, Poland

Sartaj Sahni
University of Florida
Gainesville, Florida

Hanan Samet
Computer Science Department
Center for Automation Research
and

Institute for Advanced Computer Studies
University of Maryland
College Park, Maryland

Sanjeev Saxena
Department of Computer Science & Engineering
Indian Institute of Technology, Kanpur
Kanpur, India

Markus Schneider
University of Florida
Gainesville, Florida

Bernhard Seeger
University of Marburg
Marburg, Germany

Sandeep Sen
Indian Institute of Technology, Delhi
New Delhi, India

Michiel Smid
School of Computer Science
Carleton University
Ottawa, Ontario, Canada

Bettina Speckmann
Eindhoven University of Technology
Eindhoven, The Netherlands

John Stasko
School of Interactive Computing
Georgia Institute of Technology
Atlanta, Georgia

Michael Steinbach
University of Minnesota
Minneapolis, Minnesota

Roberto Tamassia
Brown University
Providence, Rhode Island

Pang-Ning Tan
Michigan State University
East Lansing, Michigan

Sivan Toledo
Tel Aviv University
Tel Aviv, Israel

Luca Vismara
Brown University, Providence
Rhode Island

Jeffrey Scott Vitter
Department of Computer and Information Science
The University of Mississippi
Oxford, Mississippi

Mark Allen Weiss
Florida International University
Miami, Florida

Peter Widmayer
ETH Zürich
Zürich, Switzerland

Bo Yao
University of California
San Diego, California

Hung-I Yu
Institute of Information Science
Academia Sinica
Taipei, Taiwan

I

Fundamentals

1 **Analysis of Algorithms** Sartaj Sahni 3
Introduction • Operation Counts • Step Counts • Counting Cache Misses • Asymptotic
Complexity • Recurrence Equations • Amortized Complexity • Practical Complexities •
Acknowledgments • References

2 **Basic Structures** Dinesh P. Mehta 23
Introduction • Arrays • Linked Lists • Stacks and Queues • Acknowledgments • References

3 **Trees** Dinesh P. Mehta 35
Introduction • Tree Representation • Binary Trees and Properties • Binary Tree Traversals •
Threaded Binary Trees • Binary Search Trees • Heaps • Tournament Trees • Acknowledgments •
References

4 **Graphs** Narsingh Deo 49
Introduction • Graph Representations • Connectivity, Distance, and Spanning Trees • Searching a
Graph • Simple Applications of DFS and BFS • Minimum Spanning Tree • Shortest Paths •
Eulerian and Hamiltonian Graphs • Acknowledgments • References

1

Analysis of Algorithms*

1.1 Introduction.. 3
1.2 Operation Counts... 4
1.3 Step Counts.. 5
1.4 Counting Cache Misses.. 7
    A Simple Computer Model • Effect of Cache Misses on Run Time • Matrix Multiplication
1.5 Asymptotic Complexity 9
    Big Oh Notation (O) • Omega (Ω) and Theta (Θ) Notations • Little Oh Notation (o)
1.6 Recurrence Equations 12
    Substitution Method • Table-Lookup Method
1.7 Amortized Complexity.. 14
    What Is Amortized Complexity? • Maintenance Contract • The McWidget Company •
    Subset Generation
1.8 Practical Complexities.. 20
Acknowledgments 21
References... 21

Sartaj Sahni
University of Florida

1.1 Introduction

The topic "Analysis of Algorithms" is concerned primarily with determining the memory (space) and time requirements (complexity) of an algorithm. Since the techniques used to determine memory requirements are a subset of those used to determine time requirements, in this chapter, we focus on the methods used to determine the time complexity of an algorithm.

The time complexity (or simply, complexity) of an algorithm is measured as a function of the problem size. Some examples are given below.

1. The complexity of an algorithm to sort n elements may be given as a function of n.
2. The complexity of an algorithm to multiply an $m \times n$ matrix and an $n \times p$ matrix may be given as a function of m, n, and p.
3. The complexity of an algorithm to determine whether x is a prime number may be given as a function of the number, n, of bits in x. Note that $n = \lceil \log_2(x + 1) \rceil$.

We partition our discussion of algorithm analysis into the following sections.

1. Operation counts
2. Step counts
3. Counting cache misses
4. Asymptotic complexity
5. Recurrence equations
6. Amortized complexity
7. Practical complexities

See [1–4] for additional material on algorithm analysis.

---

* This chapter has been reprinted from first edition of this Handbook, without any content updates.

1.2 Operation Counts

One way to estimate the time complexity of a program or method is to select one or more operations, such as add, multiply, and compare, and to determine how many of each is done. The success of this method depends on our ability to identify the operations that contribute most to the time complexity.

EXAMPLE 1.1

[Max Element] Figure 1.1 gives an algorithm that returns the position of the largest element in the array a[0:n-1]. When n > 0, the time complexity of this algorithm can be estimated by determining the number of comparisons made between elements of the array a. When n $\leq$ 1, the for loop is not entered. So no comparisons between elements of a are made. When n > 1, each iteration of the for loop makes one comparison between two elements of a, and the total number of element comparisons is n-1. Therefore, the number of element comparisons is max{n-1, 0}. The method max performs other comparisons (e.g., each iteration of the for loop is preceded by a comparison between i and n) that are not included in the estimate. Other operations such as initializing positionOfCurrentMax and incrementing the for loop index i are also not included in the estimate.

```
int max(int [] a, int n)
{
    if (n < 1) return -1; // no max
    int positionOfCurrentMax = 0;
    for (int i = 1; i < n; i++)
        if (a[positionOfCurrentMax] < a[i]) positionOfCurrentMax = i;
    return positionOfCurrentMax;
}
```

FIGURE 1.1   Finding the position of the largest element in a[0:n-1].

The algorithm of Figure 1.1 has the nice property that the operation count is precisely determined by the problem size. For many other problems, however, this is not so. Figure 1.2 gives an algorithm that performs one pass of a bubble sort. In this pass, the largest element in a[0:n-1] relocates to position a[n-1]. The number of swaps performed by this algorithm depends not only on the problem size n but also on the particular values of the elements in the array a. The number of swaps varies from a low of 0 to a high of $n - 1$.

```
void bubble(int [] a, int n)
{
    for (int i = 0; i < n - 1; i++)
        if (a[i] > a[i+1]) swap(a[i], a[i+1]);
}
```

FIGURE 1.2   A bubbling pass.

Since the operation count isn't always uniquely determined by the problem size, we ask for the best, worst, and average counts.

EXAMPLE 1.2

[SEQUENTIAL SEARCH] Figure 1.3 gives an algorithm that searches a[0:n-1] for the first occurrence of x. The number of comparisons between x and the elements of a isn't uniquely determined by the problem size n. For example, if $n = 100$ and x = a[0], then only 1 comparison is made. However, if x isn't equal to any of the a[i]s, then 100 comparisons are made.

A search is *successful* when x is one of the a[i]s. All other searches are *unsuccessful*. Whenever we have an unsuccessful search, the number of comparisons is n. For successful searches the best comparison count is 1, and the worst is n. For the average count assume that all array elements are distinct and that each is searched for with equal frequency. The average count for a successful search is

$$\frac{1}{n}\sum_{i=1}^{n}i = (n+1)/2$$

```
int sequentialSearch(int [] a, int n, int x)
{
    // search a[0:n-1] for x
    int i;
    for (i = 0; i < n && x != a[i]; i++);
    if (i == n) return -1;    // not found
    else return i;
}
```

FIGURE 1.3   Sequential search.

EXAMPLE 1.3

[Insertion into a Sorted Array] Figure 1.4 gives an algorithm to insert an element x into a sorted array a[0:n-1].

We wish to determine the number of comparisons made between x and the elements of a. For the problem size, we use the number n of elements initially in a. Assume that $n \geq 1$. The best or minimum number of comparisons is 1, which happens when the new element x

```
void insert(int [] a, int n, int x)
{
    // find proper place for x
    int i;
    for (i = n - 1; i >= 0 && x < a[i]; i--)
        a[i+1] = a[i];

    a[i+1] = x;    // insert x
}
```

FIGURE 1.4   Inserting into a sorted array.

is to be inserted at the right end. The maximum number of comparisons is n, which happens when x is to be inserted at the left end. For the average assume that x has an equal chance of being inserted into any of the possible n+1 positions. If x is eventually inserted into position i+1 of a, $i \geq 0$, then the number of comparisons is n-i. If x is inserted into a[0], the number of comparisons is n. So the average count is

$$\frac{1}{n+1}\left(\sum_{i=0}^{n-1}(n-i)+n\right) = \frac{1}{n+1}\left(\sum_{j=1}^{n}j+n\right) = \frac{1}{n+1}\left(\frac{n(n+1)}{2}+n\right) = \frac{n}{2}+\frac{n}{n+1}$$

This average count is almost 1 more than half the worst-case count.

1.3 Step Counts

The operation-count method of estimating time complexity omits accounting for the time spent on all but the chosen operations. In the *step-count* method, we attempt to account for the time spent in all parts of the algorithm. As was the case for operation counts, the step count is a function of the problem size.

A *step* is any computation unit that is independent of the problem size. Thus 10 additions can be one step; 100 multiplications can also be one step; but *n* additions, where *n* is the problem size, cannot be one step. The amount of computing represented by one step may be different from that represented by another. For example, the entire statement

```
return a+b+b*c+(a+b-c)/(a+b)+4;
```

TABLE 1.1 Best-Case Step Count for Figure 1.3

Statement	s/e	Frequency	Total Steps
`int sequentialSearch(⋯)`	0	0	0
`{`	0	0	0
`  int i;`	1	1	1
`  for (i = 0; i < n && x != a[i]; i++);`	1	1	1
`  if (i == n) return -1;`	1	1	1
`  else return i;`	1	1	1
`}`	0	0	0
Total			4

TABLE 1.2 Worst-Case Step Count for Figure 1.3

Statement	s/e	Frequency	Total Steps
`int sequentialSearch(⋯)`	0	0	0
`{`	0	0	0
`  int i;`	1	1	1
`  for (i = 0; i < n && x != a[i]; i++);`	1	$n+1$	$n+1$
`  if (i == n) return -1;`	1	1	1
`  else return i;`	1	0	0
`}`	0	0	0
Total			$n+3$

TABLE 1.3 Step Count for Figure 1.3 When `x = a[j]`

Statement	s/e	Frequency	Total Steps
`int sequentialSearch(⋯)`	0	0	0
`{`	0	0	0
`  int i;`	1	1	1
`  for (i = 0; i < n && x != a[i]; i++);`	1	$j+1$	$j+1$
`  if (i == n) return -1;`	1	1	1
`  else return i;`	1	1	1
`}`	0	0	0
Total			$j+4$

can be regarded as a single step if its execution time is independent of the problem size. We may also count a statement such as

```
x = y;
```

as a single step.

To determine the step count of an algorithm, we first determine the number of steps per execution (s/e) of each statement and the total number of times (i.e., frequency) each statement is executed. Combining these two quantities gives us the total contribution of each statement to the total step count. We then add the contributions of all statements to obtain the step count for the entire algorithm.

EXAMPLE 1.4

[`Sequential Search`] Tables 1.1 and 1.2 show the best- and worst-case step-count analyses for `sequentialSearch` (Figure 1.3).

For the average step-count analysis for a successful search, we assume that the n values in a are distinct and that in a successful search, x has an equal probability of being any one of these values. Under these assumptions the average step count for a successful search is the sum of the step counts for the n possible successful searches divided by n. To obtain this average, we first obtain the step count for the case `x = a[j]` where j is in the range $[0, n-1]$ (see Table 1.3).

Now we obtain the average step count for a successful search:

$$\frac{1}{n}\sum_{j=0}^{n-1}(j+4) = (n+7)/2$$

This value is a little more than half the step count for an unsuccessful search.

Now suppose that successful searches occur only 80% of the time and that each a[i] still has the same probability of being searched for. The average step count for sequentialSearch is

0.8 * (average count for successful searches) + 0.2 * (count for an unsuccessful search)
= 0.8(n + 7)/2 + 0.2(n + 3)
= 0.6n + 3.4

1.4 Counting Cache Misses

1.4.1 A Simple Computer Model

Traditionally, the focus of algorithm analysis has been on counting operations and steps. Such a focus was justified when computers took more time to perform an operation than they took to fetch the data needed for that operation. Today, however, the cost of performing an operation is significantly lower than the cost of fetching data from memory. Consequently, the run time of many algorithms is dominated by the number of memory references (equivalently, number of cache misses) rather than by the number of operations. Hence, algorithm designers focus on reducing not only the number of operations but also the number of memory accesses. Algorithm designers focus also on designing algorithms that hide memory latency.

Consider a simple computer model in which the computer's memory consists of an L1 (level 1) cache, an L2 cache, and main memory. Arithmetic and logical operations are performed by the arithmetic and logic unit (ALU) on data resident in registers (R). Figure 1.5 gives a block diagram for our simple computer model.

Typically, the size of main memory is tens or hundreds of megabytes; L2 cache sizes are typically a fraction of a megabyte; L1 cache is usually in the tens of kilobytes; and the number of registers is between 8 and 32. When you start your program, all your data are in main memory.

To perform an arithmetic operation such as an add, in our computer model, the data to be added are first loaded from memory into registers, the data in the registers are added, and the result is written to memory.

Let one cycle be the length of time it takes to add data that are already in registers. The time needed to load data from L1 cache to a register is two cycles in our model. If the required data are not in L1 cache but are in L2 cache, we get an L1 cache miss and the required data are copied from L2 cache to L1 cache and the register in 10 cycles. When the required data are not in L2 cache either, we have an L2 cache miss and the required data are copied from main memory into L2 cache, L1 cache, and the register in 100 cycles. The write operation is counted as one cycle even when the data are written to main memory because we do not wait for the write to complete before proceeding to the next operation. For more details on cache organization, see [5].

1.4.2 Effect of Cache Misses on Run Time

For our simple model, the statement a = b + c is compiled into the computer instructions

```
load a; load b; add; store c;
```

where the load operations load data into registers and the store operation writes the result of the add to memory. The add and the store together take two cycles. The two loads may take anywhere from 4 cycles to 200 cycles depending on whether we get no cache miss, L1 misses, or L2 misses. So the total time for the statement a = b + c varies from 6 cycles to 202 cycles. In practice, the variation in time is not as extreme because we can overlap the time spent on successive cache misses.

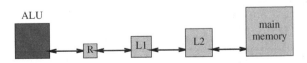

FIGURE 1.5 A simple computer model.

Suppose that we have two algorithms that perform the same task. The first algorithm does 2000 adds that require 4000 `load`, 2000 `add`, and 2000 `store` operations and the second algorithm does 1000 `adds`. The data access pattern for the first algorithm is such that 25% of the `loads` result in an L1 miss and another 25% result in an L2 miss. For our simplistic computer model, the time required by the first algorithm is $2000 * 2$ (for the 50% `loads` that cause no cache miss) $+ 1000 * 10$ (for the 25% `loads` that cause an L1 miss) $+ 1000 * 100$ (for the 25% `loads` that cause an L2 miss) $+ 2000 * 1$ (for the adds) $+ 2000 * 1$ (for the `stores`) $= 118,000$ cycles. If the second algorithm has 100% L2 misses, it will take $2000 * 100$ (L2 misses) $+ 1000 * 1$ (adds) $+ 1000 * 1$ (`stores`) $= 202,000$ cycles. So the second algorithm, which does half the work done by the first, actually takes 76% more time than is taken by the first algorithm.

Computers use a number of strategies (such as preloading data that will be needed in the near future into cache, and when a cache miss occurs, the needed data as well as data in some number of adjacent bytes are loaded into cache) to reduce the number of cache misses and hence reduce the run time of a program. These strategies are most effective when successive computer operations use adjacent bytes of main memory.

Although our discussion has focused on how cache is used for data, computers also use cache to reduce the time needed to access instructions.

1.4.3 Matrix Multiplication

The algorithm of Figure 1.6 multiplies two square matrices that are represented as two-dimensional arrays. It performs the following computation:

$$c[i][j] = \sum_{k=1}^{n} a[i][k] * b[k][j], \quad 1 \leq i \leq n, 1 \leq j \leq n \tag{1.1}$$

Figure 1.7 is an alternative algorithm that produces the same two-dimensional array c as is produced by Figure 1.6. We observe that Figure 1.7 has two nested `for` loops that are not present in Figure 1.6 and does more work than is done by Figure 1.6 with respect to indexing into the array c. The remainder of the work is the same.

You will notice that if you permute the order of the three nested `for` loops in Figure 1.7, you do not affect the result array c. We refer to the loop order in Figure 1.7 as ijk order. When we swap the second and third `for` loops, we get ikj order. In all, there are $3! = 6$ ways in which we can order the three nested `for` loops. All six orderings result in methods that perform exactly

```
void squareMultiply(int [][] a, int [][] b, int [][] c, int n)
{
    for (int i = 0; i < n; i++)
        for (int j = 0; j < n; j++)
        {
            int sum = 0;
            for (int k = 0; k < n; k++)
                sum += a[i][k] * b[k][j];
            c[i][j] = sum;
        }
}
```

FIGURE 1.6   Multiply two n × n matrices.

```
void fastSquareMultiply(int [][] a, int [][] b, int [][] c, int n)
{
    for (int i = 0; i < n; i++)
        for (int j = 0; j < n; j++)
            c[i][j] = 0;

    for (int i = 0; i < n; i++)
        for (int j = 0; j < n; j++)
            for (int k = 0; k < n; k++)
                c[i][j] += a[i][k] * b[k][j];
}
```

FIGURE 1.7   Alternative algorithm to multiply square matrices.

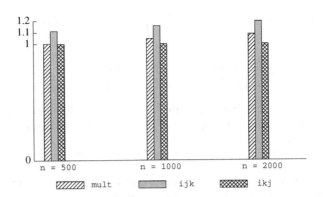

FIGURE 1.8 Normalized run times for matrix multiplication.

the same number of operations of each type. So you might think all six take the same time. Not so. By changing the order of the loops, we change the data access pattern and so change the number of cache misses. This in turn affects the run time.

In `ijk` order, we access the elements of a and c by rows; the elements of b are accessed by column. Since elements in the same row are in adjacent memory and elements in the same column are far apart in memory, the accesses of b are likely to result in many L2 cache misses when the matrix size is too large for the three arrays to fit into L2 cache. In `ikj` order, the elements of a, b, and c are accessed by rows. Therefore, `ikj` order is likely to result in fewer L2 cache misses and so has the potential to take much less time than taken by `ijk` order.

For a crude analysis of the number of cache misses, assume we are interested only in L2 misses; that an L2 cache-line can hold w matrix elements; when an L2 cache-miss occurs, a block of w matrix elements is brought into an L2 cache line; and that L2 cache is small compared to the size of a matrix. Under these assumptions, the accesses to the elements of a, b and c in `ijk` order, respectively, result in n^3/w, n^3, and n^2/w L2 misses. Therefore, the total number of L2 misses in `ijk` order is $n^3(1 + w + 1/n)/w$. In `ikj` order, the number of L2 misses for our three matrices is n^2/w, n^3/w, and n^3/w, respectively. So, in `ikj` order, the total number of L2 misses is $n^3(2 + 1/n)/w$. When n is large, the ration of `ijk` misses to `ikj` misses is approximately $(1 + w)/2$, which is 2.5 when $w = 4$ (e.g., when we have a 32-byte cache line and the data is double precision) and 4.5 when $w = 8$ (e.g., when we have a 64-byte cache line and double-precision data). For a 64-byte cache line and single-precision (i.e., 4 byte) data, $w = 16$ and the ratio is approximately 8.5.

Figure 1.8 shows the normalized run times of a Java version of our matrix multiplication algorithms. In this figure, *mult* refers to the multiplication algorithm of Figure 1.6. The normalized run time of a method is the time taken by the method divided by the time taken by `ikj` order.

Matrix multiplication using `ikj` order takes 10% less time than does `ijk` order when the matrix size is n = 500 and 16% less time when the matrix size is 2000. Equally surprising is that `ikj` order runs faster than the algorithm of Figure 1.6 (by about 5% when $n = 2000$). This despite the fact that `ikj` order does more work than is done by the algorithm of Figure 1.6.

1.5 Asymptotic Complexity

1.5.1 Big Oh Notation (*O*)

Let $p(n)$ and $q(n)$ be two nonnegative functions. $p(n)$ is *asymptotically bigger* ($p(n)$ asymptotically dominates $q(n)$) than the function $q(n)$ iff

$$\lim_{n \to \infty} \frac{q(n)}{p(n)} = 0 \tag{1.2}$$

$q(n)$ is *asymptotically smaller* than $p(n)$ iff $p(n)$ is asymptotically bigger than $q(n)$. $p(n)$ and $q(n)$ are *asymptotically equal* iff neither is asymptotically bigger than the other.

In the following discussion the function $f(n)$ denotes the time or space complexity of an algorithm as a function of the problem size n. Since the time or space requirements of a program are nonnegative quantities, we assume that the function f has a nonnegative value for all values of n. Further, since n denotes an instance characteristic, we assume that $n \geq 0$. The function $f(n)$ will, in general, be a sum of terms. For example, the terms of $f(n) = 9n^2 + 3n + 12$ are $9n^2$, $3n$, and 12. We may compare pairs of terms to determine which is bigger. The biggest term in the example $f(n) = 9n^2$ is $9n^2$.

EXAMPLE 1.5

Since

$$\lim_{n\to\infty} \frac{10n+7}{3n^2+2n+6} = \frac{10/n+7/n^2}{3+2/n+6/n^2} = 0/3 = 0$$

$3n^2+2n+6$ is asymptotically bigger than $10n+7$ and $10n+7$ is asymptotically smaller than $3n^2+2n+6$. A similar derivation shows that $8n^4+9n^2$ is asymptotically bigger than $100n^3-3$, and that $2n^2+3n$ is asymptotically bigger than $83n$. $12n+6$ is asymptotically equal to $6n+2$.

Figure 1.9 gives the terms that occur frequently in a step-count analysis. Although all the terms in Figure 1.9 have a coefficient of 1, in an actual analysis, the coefficients of these terms may have a different value.

We do not associate a logarithmic base with the functions in Figure 1.9 that include $\log n$ because for any constants a and b greater than 1, $\log_a n = \log_b n / \log_b a$. So $\log_a n$ and $\log_b n$ are asymptotically equal.

The definition of asymptotically smaller implies the following ordering for the terms of Figure 1.9 ($<$ is to be read as "is asymptotically smaller than"):

$$1 < \log n < n < n\log n < n^2 < n^3 < 2^n < n!$$

Asymptotic notation describes the behavior of the time or space complexity for large instance characteristics. Although we will develop asymptotic notation with reference to step counts alone, our development also applies to space complexity and operation counts.

The notation $f(n) = O(g(n))$ (read as "$f(n)$ is big oh of $g(n)$") means that $f(n)$ is asymptotically smaller than or equal to $g(n)$. Therefore, in an asymptotic sense $g(n)$ is an upper bound for $f(n)$.

EXAMPLE 1.6

From Example 1.5, it follows that $10n+7 = O(3n^2+2n+6)$; $100n^3-3 = O(8n^4+9n^2)$. We see also that $12n+6 = O(6n+2)$; $3n^2+2n+6 \neq O(10n+7)$; and $8n^4+9n^2 \neq O(100n^3-3)$.

Although Example 1.6 uses the big oh notation in a correct way, it is customary to use $g(n)$ functions that are *unit terms* (i.e., $g(n)$ is a single term whose coefficient is 1) except when $f(n)=0$. In addition, it is customary to use, for $g(n)$, the smallest unit term for which the statement $f(n)=O(g(n))$ is true. When $f(n)=0$, it is customary to use $g(n)=0$.

EXAMPLE 1.7

The customary way to describe the asymptotic behavior of the functions used in Example 1.6 is $10n+7 = O(n)$; $100n^3-3 = O(n^3)$; $12n+6 = O(n)$; $3n^2+2n+6 \neq O(n)$; and $8n^4+9n^2 \neq O(n^3)$.

In asymptotic complexity analysis, we determine the biggest term in the complexity; the coefficient of this biggest term is set to 1. The unit terms of a step-count function are step-count terms with their coefficients changed to 1. For example, the unit

Term	Name
1	constant
$\log n$	logarithmic
n	linear
$n\log n$	$n\log n$
n^2	quadratic
n^3	cubic
2^n	exponential
$n!$	factorial

FIGURE 1.9   Commonly occurring terms.

terms of $3n^2 + 6n \log n + 7n + 5$ are n^2, $n \log n$, n, and 1; the biggest unit term is n^2. So when the step count of a program is $3n^2 + 6n \log n + 7n + 5$, we say that its asymptotic complexity is $O(n^2)$.

Notice that $f(n) = O(g(n))$ is not the same as $O(g(n)) = f(n)$. In fact, saying that $O(g(n)) = f(n)$ is meaningless. The use of the symbol $=$ is unfortunate, as this symbol commonly denotes the equals relation. We can avoid some of the confusion that results from the use of this symbol (which is standard terminology) by reading the symbol $=$ as "is" and not as "equals."

1.5.2 Omega (Ω) and Theta (Θ) Notations

Although the big oh notation is the most frequently used asymptotic notation, the omega and theta notations are sometimes used to describe the asymptotic complexity of a program.

The notation $f(n) = \Omega(g(n))$ (read as "$f(n)$ is omega of $g(n)$") means that $f(n)$ is asymptotically bigger than or equal to $g(n)$. Therefore, in an asymptotic sense, $g(n)$ is a lower bound for $f(n)$. The notation $f(n) = \Theta(g(n))$ (read as "$f(n)$ is theta of $g(n)$") means that $f(n)$ is asymptotically equal to $g(n)$.

EXAMPLE 1.8

$10n + 7 = \Omega(n)$ because $10n + 7$ is asymptotically equal to n; $100n^3 - 3 = \Omega(n^3)$; $12n + 6 = \Omega(n)$; $3n^3 + 2n + 6 = \Omega(n)$; $8n^4 + 9n^2 = \Omega(n^3)$; $3n^3 + 2n + 6 \neq \Omega(n^5)$; and $8n^4 + 9n^2 \neq \Omega(n^5)$.

$10n + 7 = \Theta(n)$ because $10n + 7$ is asymptotically equal to n; $100n^3 - 3 = \Theta(n^3)$; $12n + 6 = \Theta(n)$; $3n^3 + 2n + 6 \neq \Theta(n)$; $8n^4 + 9n^2 \neq \Theta(n^3)$; $3n^3 + 2n + 6 \neq \Theta(n^5)$; and $8n^4 + 9n^2 \neq \Theta(n^5)$.

The best-case step count for sequentialSearch (Figure 1.3) is 4 (Table 1.1), the worst-case step count is $n + 3$, and the average step count is $0.6n + 3.4$. So the best-case asymptotic complexity of sequentialSearch is $\Theta(1)$, and the worst-case and average complexities are $\Theta(n)$. It is also correct to say that the complexity of sequentialSearch is $\Omega(1)$ and $O(n)$ because 1 is a lower bound (in an asymptotic sense) and n is an upper bound (in an asymptotic sense) on the step count.

When using the Ω notation, it is customary to use, for $g(n)$, the largest unit term for which the statement $f(n) = \Omega(g(n))$ is true.

At times it is useful to interpret $O(g(n))$, $\Omega(g(n))$, and $\Theta(g(n))$ as being the following sets:

$$O(g(n)) = \{f(n) | f(n) = O(g(n))\}$$
$$\Omega(g(n)) = \{f(n) | f(n) = \Omega(g(n))\}$$
$$\Theta(g(n)) = \{f(n) | f(n) = \Theta(g(n))\}$$

Under this interpretation, statements such as $O(g_1(n)) = O(g_2(n))$ and $\Theta(g_1(n)) = \Theta(g_2(n))$ are meaningful. When using this interpretation, it is also convenient to read $f(n) = O(g(n))$ as "f of n is in (or is a member of) big oh of g of n" and so on.

1.5.3 Little Oh Notation (o)

The little oh notation describes a strict upper bound on the asymptotic growth rate of the function f. $f(n)$ is little oh of $g(n)$ iff $f(n)$ is asymptotically smaller than $g(n)$. Equivalently, $f(n) = o(g(n))$ (read as "f of n is little oh of g of n") iff $f(n) = O(g(n))$ and $f(n) \neq \Omega(g(n))$.

EXAMPLE 1.9

[LITTLE OH] $3n + 2 = o(n^2)$ as $3n + 2 = O(n^2)$ and $3n + 2 \neq \Omega(n^2)$. However, $3n + 2 \neq o(n)$. Similarly, $10n^2 + 4n + 2 = o(n^3)$, but is not $o(n^2)$.

The little oh notation is often used in step-count analyses. A step count of $3n + o(n)$ would mean that the step count is $3n$ plus terms that are asymptotically smaller than n. When performing such an analysis, one can ignore portions of the program that are known to contribute less than $\Theta(n)$ steps.

1.6 Recurrence Equations

Recurrence equations arise frequently in the analysis of algorithms, particularly in the analysis of recursive as well as divide-and-conquer algorithms.

EXAMPLE 1.10

[BINARY SEARCH] Consider a binary search of the sorted array $a[l:r]$, where $n = r - l + 1 \geq 0$, for the element x. When $n = 0$, the search is unsuccessful and when $n = 1$, we compare x and $a[l]$. When $n > 1$, we compare x with the element $a[m]$ $(m = \lfloor (l+r)/2 \rfloor)$ in the middle of the array. If the compared elements are equal, the search terminates; if $x < a[m]$, we search $a[l:m-1]$; otherwise, we search $a[m+1:r]$. Let $t(n)$ be the worst-case complexity of binary search. Assuming that $t(0) = t(1)$, we obtain the following recurrence.

$$t(n) = \begin{cases} t(1) & n \leq 1 \\ t(\lfloor n/2 \rfloor) + c & n > 1 \end{cases} \tag{1.3}$$

where c is a constant.

EXAMPLE 1.11

[MERGE SORT] In a merge sort of $a[0:n-1]$, $n \geq 1$, we consider two cases. When $n = 1$, no work is to be done as a one-element array is always in sorted order. When $n > 1$, we divide a into two parts of roughly the same size, sort these two parts using the merge sort method recursively, then finally merge the sorted parts to obtain the desired sorted array. Since the time to do the final merge is $\Theta(n)$ and the dividing into two roughly equal parts takes $O(1)$ time, the complexity, $t(n)$, of merge sort is given by the recurrence:

$$t(n) = \begin{cases} t(1) & n = 1 \\ t(\lfloor n/2 \rfloor) + t(\lceil n/2 \rceil) + cn & n > 1 \end{cases} \tag{1.4}$$

where c is a constant.

Solving recurrence equations such as Equations 1.3 and 1.4 for $t(n)$ is complicated by the presence of the floor and ceiling functions. By making an appropriate assumption on the permissible values of n, these functions may be eliminated to obtain a simplified recurrence. In the case of Equations 1.3 and 1.4 an assumption such as n is a power of 2 results in the simplified recurrences:

$$t(n) = \begin{cases} t(1) & n \leq 1 \\ t(n/2) + c & n > 1 \end{cases} \tag{1.5}$$

and

$$t(n) = \begin{cases} t(1) & n = 1 \\ 2t(n/2) + cn & n > 1 \end{cases} \tag{1.6}$$

Several techniques—substitution, table lookup, induction, characteristic roots, and generating functions—are available to solve recurrence equations. We describe only the substitution and table lookup methods.

1.6.1 Substitution Method

In the substitution method, recurrences such as Equations 1.5 and 1.6 are solved by repeatedly substituting right-side occurrences (occurrences to the right of $=$) of $t(x)$, $x > 1$, with expressions involving $t(y)$, $y < x$. The substitution process terminates when the only occurrences of $t(x)$ that remain on the right side have $x = 1$.

Consider the binary search recurrence of Equation 1.5. Repeatedly substituting for $t()$ on the right side, we get

$$
\begin{aligned}
t(n) &= t(n/2) + c \\
&= (t(n/4) + c) + c \\
&= t(n/4) + 2c \\
&= t(n/8) + 3c \\
&\vdots \\
&= t(1) + c \log_2 n \\
&= \Theta(\log n)
\end{aligned}
$$

For the merge sort recurrence of Equation 1.6, we get

$$
\begin{aligned}
t(n) &= 2t(n/2) + cn \\
&= 2(2t(n/4) + cn/2) + cn \\
&= 4t(n/4) + 2cn \\
&= 4(2t(n/8) + cn/4) + 2cn \\
&= 8t(n/8) + 3cn \\
&\vdots \\
&= nt(1) + cn \log_2 n \\
&= \Theta(n \log n)
\end{aligned}
$$

1.6.2 Table-Lookup Method

The complexity of many divide-and-conquer algorithms is given by a recurrence of the form

$$
t(n) = \begin{cases} t(1) & n = 1 \\ a * t(n/b) + g(n) & n > 1 \end{cases} \tag{1.7}
$$

where a and b are known constants. The merge sort recurrence, Equation 1.6, is in this form. Although the recurrence for binary search, Equation 1.5, isn't exactly in this form, the $n \le 1$ may be changed to $n = 1$ by eliminating the case $n = 0$. To solve Equation 1.7, we assume that $t(1)$ is known and that n is a power of b (i.e., $n = b^k$). Using the substitution method, we can show that

$$
t(n) = n^{\log_b a} [t(1) + f(n)] \tag{1.8}
$$

where $f(n) = \sum_{j=1}^{k} h(b^j)$ and $h(n) = g(n)/n^{\log_b a}$.

Table 1.4 tabulates the asymptotic value of $f(n)$ for various values of $h(n)$. This table allows us to easily obtain the asymptotic value of $t(n)$ for many of the recurrences we encounter when analyzing divide-and-conquer algorithms.

Let us solve the binary search and merge sort recurrences using this table. Comparing Equation 1.5 with $n \le 1$ replaced by $n = 1$ with Equation 1.7, we see that $a = 1$, $b = 2$, and $g(n) = c$. Therefore, $\log_b(a) = 0$, and $h(n) = g(n)/n^{\log_b a} = c = c(\log n)^0 = \Theta((\log n)^0)$. From Table 1.4, we obtain $f(n) = \Theta(\log n)$. Therefore, $t(n) = n^{\log_b a}(c + \Theta(\log n)) = \Theta(\log n)$.

For the merge sort recurrence, Equation 1.6, we obtain $a = 2$, $b = 2$, and $g(n) = cn$. So $\log_b a = 1$ and $h(n) = g(n)/n = c = \Theta((\log n)^0)$. Hence $f(n) = \Theta(\log n)$ and $t(n) = n(t(1) + \Theta(\log n)) = \Theta(n \log n)$.

TABLE 1.4 $f(n)$ Values for Various $h(n)$ Values

$h(n)$	$f(n)$
$O(n^r), r < 0$	$O(1)$
$\Theta((\log n)^i), i \ge 0$	$\Theta(((\log n)^{i+1})/(i+1))$
$\Omega(n^r), r > 0$	$\Theta(h(n))$

1.7 Amortized Complexity

1.7.1 What Is Amortized Complexity?

The complexity of an algorithm or of an operation such as an insert, search, or delete, as defined in Section 1.1, is the *actual complexity* of the algorithm or operation. The actual complexity of an operation is determined by the step count for that operation, and the actual complexity of a sequence of operations is determined by the step count for that sequence. The actual complexity of a sequence of operations may be determined by adding together the step counts for the individual operations in the sequence. Typically, determining the step count for each operation in the sequence is quite difficult, and instead, we obtain an upper bound on the step count for the sequence by adding together the worst-case step count for each operation.

When determining the complexity of a sequence of operations, we can, at times, obtain tighter bounds using *amortized complexity* rather than worst-case complexity. Unlike the actual and worst-case complexities of an operation which are closely related to the step count for that operation, the amortized complexity of an operation is an accounting artifact that often bears no direct relationship to the actual complexity of that operation. The amortized complexity of an operation could be anything. *The only requirement is that the sum of the amortized complexities of all operations in the sequence be greater than or equal to the sum of the actual complexities.* That is

$$\sum_{1 \le i \le n} amortized(i) \ge \sum_{1 \le i \le n} actual(i) \tag{1.9}$$

where $amortized(i)$ and $actual(i)$, respectively, denote the amortized and actual complexities of the ith operation in a sequence of n operations. Because of this requirement on the sum of the amortized complexities of the operations in any sequence of operations, we may use the sum of the amortized complexities as an upper bound on the complexity of any sequence of operations.

You may view the amortized cost of an operation as being the amount you charge the operation rather than the amount the operation costs. You can charge an operation any amount you wish so long as the amount charged to all operations in the sequence is at least equal to the actual cost of the operation sequence.

Relative to the actual and amortized costs of each operation in a sequence of n operations, we define a *potential function $P(i)$* as below

$$P(i) = amortized(i) - actual(i) + P(i-1) \tag{1.10}$$

That is, the ith operation causes the potential function to change by the difference between the amortized and actual costs of that operation. If we sum Equation 1.10 for $1 \le i \le n$, we get

$$\sum_{1 \le i \le n} P(i) = \sum_{1 \le i \le n} (amortized(i) - actual(i) + P(i-1))$$

or

$$\sum_{1 \le i \le n} (P(i) - P(i-1)) = \sum_{1 \le i \le n} (amortized(i) - actual(i))$$

or

$$P(n) - P(0) = \sum_{1 \le i \le n} (amortized(i) - actual(i))$$

From Equation 1.9, it follows that

$$P(n) - P(0) \ge 0 \tag{1.11}$$

When $P(0) = 0$, the potential $P(i)$ is the amount by which the first i operations have been overcharged (i.e., they have been charged more than their actual cost).

Generally, when we analyze the complexity of a sequence of n operations, n can be any nonnegative integer. Therefore, Equation 1.11 must hold for all nonnegative integers.

The preceding discussion leads us to the following three methods to arrive at amortized costs for operations:

1. *Aggregate Method*
   In the aggregate method, we determine an upper bound for the sum of the actual costs of the n operations. The amortized cost of each operation is set equal to this upper bound divided by n. You may verify that this assignment of amortized costs satisfies Equation 1.9 and is, therefore, valid.
2. *Accounting Method*
   In this method, we assign amortized costs to the operations (probably by guessing what assignment will work), compute the $P(i)$s using Equation 1.10, and show that $P(n) - P(0) \ge 0$.

3. *Potential Method*

Here, we start with a potential function (probably obtained using good guess work) that satisfies Equation 1.11 and compute the amortized complexities using Equation 1.10.

1.7.2 Maintenance Contract

1.7.2.1 Problem Definition

In January, you buy a new car from a dealer who offers you the following maintenance contract: $50 each month other than March, June, September and December (this covers an oil change and general inspection), $100 every March, June, and September (this covers an oil change, a minor tune-up, and a general inspection), and $200 every December (this covers an oil change, a major tune-up, and a general inspection). We are to obtain an upper bound on the cost of this maintenance contract as a function of the number of months.

1.7.2.2 Worst-Case Method

We can bound the contract cost for the first n months by taking the product of n and the maximum cost incurred in any month (i.e., $200). This would be analogous to the traditional way to estimate the complexity–take the product of the number of operations and the worst-case complexity of an operation. Using this approach, we get $200n$ as an upper bound on the contract cost. The upper bound is correct because the actual cost for n months does not exceed $200n$.

1.7.2.3 Aggregate Method

To use the aggregate method for amortized complexity, we first determine an upper bound on the sum of the costs for the first n months. As tight a bound as is possible is desired. The sum of the actual monthly costs of the contract for the first n months is

$$200 * \lfloor n/12 \rfloor + 100 * (\lfloor n/3 \rfloor - \lfloor n/12 \rfloor) + 50 * (n - \lfloor n/3 \rfloor)$$
$$= 100 * \lfloor n/12 \rfloor + 50 * \lfloor n/3 \rfloor + 50 * n$$
$$\leq 100 * n/12 + 50 * n/3 + 50 * n$$
$$= 50n(1/6 + 1/3 + 1)$$
$$= 50n(3/2)$$
$$= 75n$$

The amortized cost for each month is set to $75. Table 1.5 shows the actual costs, the amortized costs, and the potential function value (assuming $P(0) = 0$) for the first 16 months of the contract.

Notice that some months are charged more than their actual costs and others are charged less than their actual cost. The cumulative difference between what the operations are charged and their actual costs is given by the potential function. The potential function satisfies Equation 1.11 for all values of n. When we use the amortized cost of $75 per month, we get $75n$ as an upper bound on the contract cost for n months. This bound is tighter than the bound of $200n$ obtained using the worst-case monthly cost.

1.7.2.4 Accounting Method

When we use the accounting method, we must first assign an amortized cost for each month and then show that this assignment satisfies Equation 1.11. We have the option to assign a different amortized cost to each month. In our maintenance contract example, we know the actual cost by month and could use this actual cost as the amortized cost. It is, however, easier to work with an equal cost assignment for each month. Later, we shall see examples of operation sequences that consist of two or more types of operations (e.g., when dealing with lists of elements, the operation sequence may be made up of search, insert, and remove operations). When dealing with such sequences we often assign a different amortized cost to operations of different types (however, operations of the same type have the same amortized cost).

TABLE 1.5 Maintenance Contract

Month	1	2	3	4	5	6	7	8	9	10	11	12	13	14	15	16
Actual cost	50	50	100	50	50	100	50	50	100	50	50	200	50	50	100	50
Amortized cost	75	75	75	75	75	75	75	75	75	75	75	75	75	75	75	75
P()	25	50	25	50	75	50	75	100	75	100	125	0	25	50	25	50

To get the best upper bound on the sum of the actual costs, we must set the amortized monthly cost to be the smallest number for which Equation 1.11 is satisfied for all n. From the above table, we see that using any cost less than \$75 will result in $P(n) - P(0) < 0$ for some values of n. Therefore, the smallest assignable amortized cost consistent with Equation 1.11 is \$75.

Generally, when the accounting method is used, we have not computed the aggregate cost. Therefore, we would not know that \$75 is the least assignable amortized cost. So we start by assigning an amortized cost (obtained by making an educated guess) to each of the different operation types and then proceed to show that this assignment of amortized costs satisfies Equation 1.11. Once we have shown this, we can obtain an upper bound on the cost of any operation sequence by computing

$$\sum_{1 \le i \le k} f(i) * amortized(i)$$

where k is the number of different operation types and $f(i)$ is the frequency of operation type i (i.e., the number of times operations of this type occur in the operation sequence).

For our maintenance contract example, we might try an amortized cost of \$70. When we use this amortized cost, we discover that Equation 1.11 is not satisfied for $n = 12$ (e.g.) and so \$70 is an invalid amortized cost assignment. We might next try \$80. By constructing a table such as the one above, we will observe that Equation 1.11 is satisfied for all months in the first 12 month cycle, and then conclude that the equation is satisfied for all n. Now, we can use \$80n as an upper bound on the contract cost for n months.

1.7.2.5 Potential Method

We first define a potential function for the analysis. The only guideline you have in defining this function is that the potential function represents the cumulative difference between the amortized and actual costs. So, if you have an amortized cost in mind, you may be able to use this knowledge to develop a potential function that satisfies Equation 1.11, and then use the potential function and the actual operation costs (or an upper bound on these actual costs) to verify the amortized costs.

If we are extremely experienced, we might start with the potential function

$$t(n) = \begin{cases} 0 & n \bmod 12 = 0 \\ 25 & n \bmod 12 = 1 \text{ or } 3 \\ 50 & n \bmod 12 = 2, 4, \text{ or } 6 \\ 75 & n \bmod 12 = 5, 7, \text{ or } 9 \\ 100 & n \bmod 12 = 8 \text{ or } 10 \\ 125 & n \bmod 12 = 11 \end{cases}$$

Without the aid of the table (Table 1.5) constructed for the aggregate method, it would take quite some ingenuity to come up with this potential function. Having formulated a potential function and verified that this potential function satisfies Equation 1.11 for all n, we proceed to use Equation 1.10 to determine the amortized costs.

From Equation 1.10, we obtain $amortized(i) = actual(i) + P(i) - P(i-1)$. Therefore,

$$amortized(1) = actual(1) + P(1) - P(0) = 50 + 25 - 0 = 75$$
$$amortized(2) = actual(2) + P(2) - P(1) = 50 + 50 - 25 = 75$$
$$amortized(3) = actual(3) + P(3) - P(2) = 100 + 25 - 50 = 75$$

and so on. Therefore, the amortized cost for each month is \$75. So, the actual cost for n months is at most \$75n.

1.7.3 The McWidget Company

1.7.3.1 Problem Definition

The famous McWidget company manufactures widgets. At its headquarters, the company has a large display that shows how many widgets have been manufactured so far. Each time a widget is manufactured, a maintenance person updates this display. The cost for this update is $c + dm$, where c is a fixed trip charge, d is a charge per display digit that is to be changed, and m is the number of digits that are to be changed. For example, when the display is changed from 1399 to 1400, the cost to the company is $c + 3d$ because 3 digits must be changed. The McWidget company wishes to amortize the cost of maintaining the display over the widgets that are manufactured, charging the same amount to each widget. More precisely, we are looking for an amount $e = amortized(i)$ that should levied against each widget so that the sum of these charges equals or exceeds the actual cost of maintaining/updating the display ($e * n \ge$ actual total cost incurred for first n widgets for all $n \ge 1$). To keep the overall selling price of a widget low, we wish to find as small an e as possible. Clearly, $e > c + d$ because each time a widget is made, at least one digit (the least significant one) has to be changed.

TABLE 1.6 Data for Widgets

Widget	1	2	3	4	5	6	7	8	9	10	11	12	13	14
Actual cost	1	1	1	1	1	1	1	1	1	2	1	1	1	1
Amortized cost—	1.12	1.12	1.12	1.12	1.12	1.12	1.12	1.12	1.12	1.12	1.12	1.12	1.12	1.12
P()	0.12	0.24	0.36	0.48	0.60	0.72	0.84	0.96	1.08	0.20	0.32	0.44	0.56	0.68
Widget	15	16	17	18	19	20	21	22	23	24	25	26	27	28
Actual cost	1	1	1	1	1	2	1	1	1	1	1	1	1	1
Amortized cost—	1.12	1.12	1.12	1.12	1.12	1.12	1.12	1.12	1.12	1.12	1.12	1.12	1.12	1.12
P()	0.80	0.92	1.04	1.16	1.28	0.40	0.52	0.64	0.76	0.88	1.00	1.12	1.24	1.36

1.7.3.2 Worst-Case Method

This method does not work well in this application because there is no finite worst-case cost for a single display update. As more and more widgets are manufactured, the number of digits that need to be changed increases. For example, when the 1000th widget is made, 4 digits are to be changed incurring a cost of $c + 4d$, and when the 1,000,000th widget is made, 7 digits are to be changed incurring a cost of $c + 7d$. If we use the worst-case method, the amortized cost to each widget becomes infinity.

1.7.3.3 Aggregate Method

Let n be the number of widgets made so far. As noted earlier, the least significant digit of the display has been changed n times. The digit in the ten's place changes once for every ten widgets made, that in the hundred's place changes once for every hundred widgets made, that in the thousand's place changes once for every thousand widgets made, and so on. Therefore, the aggregate number of digits that have changed is bounded by

$$n(1 + 1/10 + 1/100 + 1/1000 + \cdots) = (1.11111\ldots)n$$

So, the amortized cost of updating the display is $c + d(1.11111\ldots)n/n < c + 1.12d$. If the McWidget company adds $c + 1.12d$ to the selling price of each widget, it will collect enough money to pay for the cost of maintaining the display. Each widget is charged the cost of changing 1.12 digits regardless of the number of digits that are actually changed. Table 1.6 shows the actual cost, as measured by the number of digits that change, of maintaining the display, the amortized cost (i.e., 1.12 digits per widget), and the potential function. The potential function gives the difference between the sum of the amortized costs and the sum of the actual costs. Notice how the potential function builds up so that when it comes time to pay for changing two digits, the previous potential function value plus the current amortized cost exceeds 2. From our derivation of the amortized cost, it follows that the potential function is always nonnegative.

1.7.3.4 Accounting Method

We begin by assigning an amortized cost to the individual operations, and then we show that these assigned costs satisfy Equation 1.11. Having already done an amortized analysis using the aggregate method, we see that Equation 1.11 is satisfied when we assign an amortized cost of $c + 1.12d$ to each display change. Typically, however, the use of the accounting method is not preceded by an application of the aggregate method and we start by guessing an amortized cost and then showing that this guess satisfies Equation 1.11.

Suppose we assign a guessed amortized cost of $c + 2d$ for each display change.

$$
\begin{aligned}
P(n) - P(0) &= \sum_{1 \le i \le n} (amortized(i) - actual(i)) \\
&= (c + 2d)n - \sum_{1 \le i \le n} actual(i) \\
&= (c + 2d)n - (c + (1 + 1/10 + 1/100 + \cdots)d)n \\
&\ge (c + 2d)n - (c + 1.12d)n \\
&\ge 0
\end{aligned}
$$

This analysis also shows us that we can reduce the amortized cost of a widget to $c + 1.12d$.

An alternative proof method that is useful in some analyses involves distributing the excess charge $P(i) - P(0)$ over various accounting entities, and using these stored excess charges (called *credits*) to establish $P(i + 1) - P(0) \ge 0$. For our McWidget example, we use the display digits as the accounting entities. Initially, each digit is 0 and each digit has a credit of 0 dollars.

```
public int [] nextSubset()
{// return next subset; return null if no next subset
    // generate next subset by adding 1 to the binary number x[1:n]
    int i = n;
    while (i > 0 && x[i] == 1)
        {x[i] = 0; i--;}

    if (i == 0) return null;
    else {x[i] = 1; return x;}
}
```

FIGURE 1.10   Subset enumerator.

Suppose we have guessed an amortized cost of $\$c + (1.111\ldots)d$. When the first widget is manufactured, $\$c + d$ of the amortized cost is used to pay for the update of the display and the remaining $\$(0.111\ldots)d$ of the amortized cost is retained as a credit by the least significant digit of the display. Similarly, when the second through ninth widgets are manufactured, $\$c + d$ of the amortized cost is used to pay for the update of the display and the remaining $\$(0.111\ldots)d$ of the amortized cost is retained as a credit by the least significant digit of the display. Following the manufacture of the ninth widget, the least significant digit of the display has a credit of $\$(0.999\ldots)d$ and the remaining digits have no credit. When the tenth widget is manufactured, $\$c + d$ of the amortized cost are used to pay for the trip charge and the cost of changing the least significant digit. The least significant digit now has a credit of $\$(1.111\ldots)d$. Of this credit, $\$d$ are used to pay for the change of the next least significant digit (i.e., the digit in the ten's place), and the remaining $\$(0.111\ldots)d$ are transferred to the ten's digit as a credit. Continuing in this way, we see that when the display shows 99, the credit on the ten's digit is $\$(0.999\ldots)d$ and that on the one's digit (i.e., the least significant digit) is also $\$(0.999\ldots)d$. When the 100th widget is manufactured, $\$c+d$ of the amortized cost are used to pay for the trip charge and the cost of changing the least significant digit, and the credit on the least significant digit becomes $\$(1.111\ldots)d$. Of this credit, $\$d$ are used to pay for the change of the ten's digit from 9 to 0, the remaining $\$(0.111\ldots)d$ credit on the one's digit is transferred to the ten's digit. The credit on the ten's digit now becomes $\$(1.111\ldots)d$. Of this credit, $\$d$ are used to pay for the change of the hundred's digit from 0 to 1, the remaining $\$(0.111\ldots)d$ credit on the ten's digit is transferred to the hundred's digit.

The above accounting scheme ensures that the credit on each digit of the display always equals $\$(0.111\ldots)dv$, where v is the value of the digit (e.g., when the display is 206 the credit on the one's digit is $\$(0.666\ldots)d$, the credit on the ten's digit is $\$0$, and that on the hundred's digit is $\$(0.222\ldots)d$.

From the preceding discussion, it follows that $P(n) - P(0)$ equals the sum of the digit credits and this sum is always nonnegative. Therefore, Equation 1.11 holds for all n.

1.7.3.5 Potential Method

We first postulate a potential function that satisfies Equation 1.11, and then use this function to obtain the amortized costs. From the alternative proof used above for the accounting method, we can see that we should use the potential function $P(n) = (0.111\ldots)d\sum_i v_i$, where v_i is the value of the ith digit of the display. For example, when the display shows 206 (at this time $n = 206$), the potential function value is $(0.888\ldots)d$. This potential function satisfies Equation 1.11.

Let q be the number of 9 s at the right end of j (i.e., when $j = 12903999$, $q = 3$). When the display changes from j to $j+1$, the potential change is $(0.111\ldots)d(1 - 9q)$ and the actual cost of updating the display is $\$c + (q + 1)d$. From Equation 1.10, it follows that the amortized cost for the display change is

$$\text{actual cost} + \text{potential change} = c + (q + 1)d + (0.111\ldots)d(1 - 9q) = c + (1.111\ldots)d$$

1.7.4 Subset Generation

1.7.4.1 Problem Definition

The subsets of a set of n elements are defined by the 2^n vectors $x[1:n]$, where each $x[i]$ is either 0 or 1. $x[i] = 1$ iff the ith element of the set is a member of the subset. The subsets of a set of three elements are given by the eight vectors 000, 001, 010, 011, 100, 101, 110, and 111, for example. Starting with an array $x[1:n]$ has been initialized to zeroes (this represents the empty subset), each invocation of algorithm nextSubset (Figure 1.10) returns the next subset. When all subsets have been generated, this algorithm returns null.

We wish to determine how much time it takes to generate the first m, $1 \le m \le 2^n$ subsets. This is the time for the first m invocations of `nextSubset`.

1.7.4.2 Worst-Case Method

The complexity of `nextSubset` is $\Theta(c)$, where c is the number of $x[i]$s that change. Since all n of the $x[i]$s could change in a single invocation of `nextSubset`, the worst-case complexity of `nextSubset` is $\Theta(n)$. Using the worst-case method, the time required to generate the first m subsets is $O(mn)$.

1.7.4.3 Aggregate Method

The complexity of `nextSubset` equals the number of $x[i]$s that change. When `nextSubset` is invoked m times, $x[n]$ changes m times; $x[n-1]$ changes $\lfloor m/2 \rfloor$ times; $x[n-2]$ changes $\lfloor m/4 \rfloor$ times; $x[n-3]$ changes $\lfloor m/8 \rfloor$ times; and so on. Therefore, the sum of the actual costs of the first m invocations is $\sum_{0 \le i \le \lfloor \log_2 m \rfloor} (m/2^i) < 2m$. So, the complexity of generating the first m subsets is actually $O(m)$, a tighter bound than obtained using the worst-case method.

The amortized complexity of `nextSubset` is (sum of actual costs)$/m < 2m/m = O(1)$.

1.7.4.4 Accounting Method

We first guess the amortized complexity of `nextSubset`, and then show that this amortized complexity satisfies Equation 1.11. Suppose we guess that the amortized complexity is 2. To verify this guess, we must show that $P(m) - P(0) \ge 0$ for all m.

We shall use the alternative proof method used in the McWidget example. In this method, we distribute the excess charge $P(i) - P(0)$ over various accounting entities, and use these stored excess charges to establish $P(i+1) - P(0) \ge 0$. We use the $x[j]$s as the accounting entities. Initially, each $x[j]$ is 0 and has a credit of 0. When the first subset is generated, 1 unit of the amortized cost is used to pay for the single $x[j]$ that changes and the remaining 1 unit of the amortized cost is retained as a credit by $x[n]$, which is the $x[j]$ that has changed to 1. When the second subset is generated, the credit on $x[n]$ is used to pay for changing $x[n]$ to 0 in the while loop, 1 unit of the amortized cost is used to pay for changing $x[n-1]$ to 1, and the remaining 1 unit of the amortized cost is retained as a credit by $x[n-1]$, which is the $x[j]$ that has changed to 1. When the third subset is generated, 1 unit of the amortized cost is used to pay for changing $x[n]$ to 1, and the remaining 1 unit of the amortized cost is retained as a credit by $x[n]$, which is the $x[j]$ that has changed to 1. When the fourth subset is generated, the credit on $x[n]$ is used to pay for changing $x[n]$ to 0 in the while loop, the credit on $x[n-1]$ is used to pay for changing $x[n-1]$ to 0 in the while loop, 1 unit of the amortized cost is used to pay for changing $x[n-2]$ to 1, and the remaining 1 unit of the amortized cost is retained as a credit by $x[n-2]$, which is the $x[j]$ that has changed to 1. Continuing in this way, we see that each $x[j]$ that is 1 has a credit of 1 unit on it. This credit is used to pay the actual cost of changing this $x[j]$ from 1 to 0 in the while loop. One unit of the amortized cost of `nextSubset` is used to pay for the actual cost of changing an $x[j]$ to 1 in the else clause, and the remaining one unit of the amortized cost is retained as a credit by this $x[j]$.

The above accounting scheme ensures that the credit on each $x[j]$ that is 1 is exactly 1, and the credit on each $x[j]$ that is 0 is 0.

From the preceding discussion, it follows that $P(m) - P(0)$ equals the number of $x[j]$s that are 1. Since this number is always nonnegative, Equation 1.11 holds for all m.

Having established that the amortized complexity of `nextSubset` is $2 = O(1)$, we conclude that the complexity of generating the first m subsets equals $m *$ amortized complexity $= O(m)$.

1.7.4.5 Potential Method

We first postulate a potential function that satisfies Equation 1.11, and then use this function to obtain the amortized costs. Let $P(j)$ be the potential just after the jth subset is generated. From the proof used above for the accounting method, we can see that we should define $P(j)$ to be equal to the number of $x[i]$s in the jth subset that are equal to 1.

By definition, the 0th subset has all $x[i]$ equal to 0. Since $P(0) = 0$ and $P(j) \ge 0$ for all j, this potential function P satisfies Equation 1.11. Consider any subset $x[1:n]$. Let q be the number of 1s at the right end of $x[]$ (i.e., $x[n], x[n-1], \ldots, x[n-q+1]$, are all 1s). Assume that there is a next subset. When the next subset is generated, the potential change is $1 - q$ because q 1s are replaced by 0 in the while loop and a 0 is replaced by a 1 in the else clause. The actual cost of generating the next subset is $q + 1$. From Equation 1.10, it follows that, when there is a next subset, the amortized cost for `nextSubset` is

$$\text{actual cost + potential change} = q + 1 + 1 - q = 2$$

When there is no next subset, the potential change is $-q$ and the actual cost of `nextSubset` is q. From Equation 1.10, it follows that, when there is no next subset, the amortized cost for `nextSubset` is

$$\text{actual cost + potential change} = q - q = 0$$

Therefore, we can use 2 as the amortized complexity of `nextSubset`. Consequently, the actual cost of generating the first m subsets is $O(m)$.

1.8 Practical Complexities

We have seen that the time complexity of a program is generally some function of the problem size. This function is very useful in determining how the time requirements vary as the problem size changes. For example, the run time of an algorithm whose complexity is $\Theta(n^2)$ is expected to increase by a factor of 4 when the problem size doubles and by a factor of 9 when the problem size triples.

The complexity function also may be used to compare two algorithms P and Q that perform the same task. Assume that algorithm P has complexity $\Theta(n)$ and that algorithm Q has complexity $\Theta(n^2)$. We can assert that algorithm P is faster than algorithm Q for "sufficiently large" n. To see the validity of this assertion, observe that the actual computing time of P is bounded from above by cn for some constant c and for all n, $n \geq n_1$, while that of Q is bounded from below by dn^2 for some constant d and all n, $n \geq n_2$. Since $cn \leq dn^2$ for $n \geq c/d$, algorithm P is faster than algorithm Q whenever $n \geq \max\{n_1, n_2, c/d\}$.

One should always be cautiously aware of the presence of the phrase *sufficiently large* in the assertion of the preceding discussion. When deciding which of the two algorithms to use, we must know whether the n we are dealing with is, in fact, sufficiently large. If algorithm P actually runs in $10^6 n$ milliseconds while algorithm Q runs in n^2 milliseconds and if we always have $n \leq 10^6$, then algorithm Q is the one to use.

To get a feel for how the various functions grow with n, you should study Figures 1.11 and 1.12 very closely. These figures show that 2^n grows very rapidly with n. In fact, if a algorithm needs 2^n steps for execution, then when $n = 40$, the number of steps

$\log n$	n	$n \log n$	n^2	n^3	2^n
0	1	0	1	1	2
1	2	2	4	8	4
2	4	8	16	64	16
3	8	24	64	512	256
4	16	64	256	4096	65,536
5	32	160	1024	32,768	4,294,967,296

FIGURE 1.11   Value of various functions.

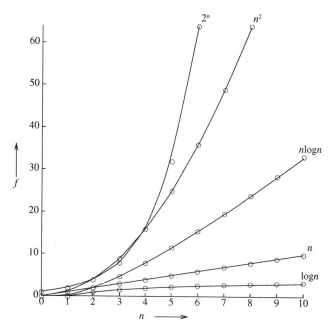

FIGURE 1.12   Plot of various functions.

n	$f(n)$						
	n	$n \log_2 n$	n^2	n^3	n^4	n^{10}	2^n
10	$.01\,\mu s$	$.03\,\mu s$	$.1\,\mu s$	$1\,\mu s$	$10\,\mu s$	$10\,s$	$1\,\mu s$
20	$.02\,\mu s$	$.09\,\mu s$	$.4\,\mu s$	$8\,\mu s$	$160\,\mu s$	$2.84\,h$	$1\,ms$
30	$.03\,\mu s$	$.15\,\mu s$	$.9\,\mu s$	$27\,\mu s$	$810\,\mu s$	$6.83\,d$	$1\,s$
40	$.04\,\mu s$	$.21\,\mu s$	$1.6\,\mu s$	$64\,\mu s$	$2.56\,ms$	$121\,d$	$18\,m$
50	$.05\,\mu s$	$.28\,\mu s$	$2.5\,\mu s$	$125\,\mu s$	$6.25\,ms$	$3.1\,y$	$13\,d$
100	$.10\,\mu s$	$.66\,\mu s$	$10\,\mu s$	$1\,ms$	$100\,ms$	$3171\,y$	$4*10^{13}\,y$
10^3	$1\,\mu s$	$9.96\,\mu s$	$1\,ms$	$1\,s$	$16.67\,m$	$3.17*10^{13}\,y$	$32*10^{283}\,y$
10^4	$10\,\mu s$	$130\,\mu s$	$100\,ms$	$16.67\,m$	$115.7\,d$	$3.17*10^{23}\,y$	
10^5	$100\,\mu s$	$1.66\,ms$	$10\,s$	$11.57\,d$	$3171\,y$	$3.17*10^{33}\,y$	
10^6	$1\,ms$	$19.92\,ms$	$16.67\,m$	$31.71\,y$	$3.17*10^7\,y$	$3.17*10^{43}\,y$	

μs = microsecond = 10^{-6} seconds; ms = milliseconds = 10^{-3} seconds

s = seconds; m = minutes; h = hours; d = days; y = years

FIGURE 1.13 Run times on a 1,000,000,000 instructions per second computer.

needed is approximately $1.1*10^{12}$. On a computer performing 1,000,000,000 steps per second, this algorithm would require about 18.3 minutes. If $n = 50$, the same algorithm would run for about 13 days on this computer. When $n = 60$, about 310.56 years will be required to execute the algorithm, and when $n = 100$, about $4*10^{13}$ years will be needed. We can conclude that the utility of algorithms with exponential complexity is limited to small n (typically $n \le 40$).

Algorithms that have a complexity that is a high-degree polynomial are also of limited utility. For example, if an algorithm needs n^{10} steps, then our 1,000,000,000 steps per second computer needs 10 seconds when $n = 10$; 3171 years when $n = 100$; and $3.17*10^{13}$ years when $n = 1000$. If the algorithm's complexity had been n^3 steps instead, then the computer would need 1 second when $n = 1000$, 110.67 minutes when $n = 10,000$, and 11.57 days when $n = 100,000$.

Figure 1.13 gives the time that a 1,000,000,000 instructions per second computer needs to execute an algorithm of complexity $f(n)$ instructions. One should note that currently only the fastest computers can execute about 1,000,000,000 instructions per second. From a practical standpoint, it is evident that for reasonably large n (say $n > 100$) only algorithms of small complexity (such as n, $n \log n$, n^2, and n^3) are feasible. Further, this is the case even if we could build a computer capable of executing 10^{12} instructions per second. In this case the computing times of Figure 1.13 would decrease by a factor of 1000. Now when $n = 100$, it would take 3.17 years to execute n^{10} instructions and $4*10^{10}$ years to execute 2^n instructions.

Acknowledgments

This work was supported, in part, by the National Science Foundation under grant CCR-9912395.

References

1. T. Cormen, C. Leiserson, and R. Rivest, *Introduction to Algorithms*, McGraw-Hill, New York, NY, 1992.
2. E. Horowitz, S. Sahni, and S. Rajasekaran, *Fundamentals of Computer Algorithms*, W. H. Freeman and Co., New York, NY, 1998.
3. G. Rawlins, *Compared to What: An Introduction to the Analysis of Algorithms*, W. H. Freeman and Co., New York, NY, 1992.
4. S. Sahni, *Data Structures, Algorithms, and Applications in Java*, McGraw-Hill, NY, 2000.
5. J. Hennessey and D. Patterson, *Computer Organization and Design*, Second Edition, Morgan Kaufmann Publishers, Inc., San Francisco, CA, 1998, Chapter 7.

2

Basic Structures*

2.1 Introduction 23
2.2 Arrays 23
Operations on an Array • Sorted Arrays • Array Doubling • Multiple Lists in a Single Array • Heterogeneous Arrays • Multidimensional Arrays • Sparse Matrices
2.3 Linked Lists 28
Chains • Circular Lists • Doubly Linked Circular Lists • Generalized Lists
2.4 Stacks and Queues 31
Stack Implementation • Queue Implementation
Acknowledgments 34
References 34

Dinesh P. Mehta
Colorado School of Mines

2.1 Introduction

In this chapter, we review several basic structures that are usually taught in a first class on data structures. There are several text books that cover this material, some of which are listed here [1–4]. However, we believe that it is valuable to review this material for the following reasons:

1. In practice, these structures are used more often than all of the other data structures discussed in this handbook combined.
2. These structures are used as basic building blocks on which other more complicated structures are based.

Our goal is to provide a crisp and brief review of these structures. For a more detailed explanation, the reader is referred to the text books listed at the end of this chapter. In the following, we assume that the reader is familiar with arrays and pointers.

2.2 Arrays

An array is conceptually defined as a collection of $<$ index,value $>$ pairs. The implementation of the array in modern programming languages such as C++ and Java uses indices starting at 0. Languages such as Pascal permitted one to define an array on an arbitrary range of integer indices. In most applications, the array is the most convenient method to store a collection of objects. In these cases, the index associated with a value is unimportant. For example, if an array `city` is being used to store a list of cities in no particular order, it doesn't really matter whether `city[0]` is "Denver" or "Mumbai". If, on the other hand, an array name is being used to store a list of student names in alphabetical order, then, although the absolute index values don't matter, the ordering of names associated with the ordering of the index does matter: that is, `name[i]` must precede `name[j]` in alphabetical order, if $i < j$. Thus, one may distinguish between sorted arrays and unsorted arrays. Sometimes arrays are used so that the index *does* matter. For example, suppose we are trying to represent a histogram: we want to maintain a count of the number of students that got a certain score on an exam from a scale of 0 to 10. If score[5] $= 7$, this means that 7 students received a score of 5. In this situation, it is possible that the desired indices are not supported by the language. For example, C++ does not directly support using indices such as "blue", "green", and "red". This may be rectified by using enumerated types to assign integer values to the indices. In cases when the objects in the array are large and unwieldy and have to be moved around from one array location to another, it may be advantageous to store *pointers* or *references* to objects in the array rather than the objects themselves.

Programming languages provide a mechanism to retrieve the value associated with a supplied index or to associate a value with an index. Programming languages like C++ do not explicitly maintain the size of the array. Rather, it is the programmer's responsibility to be aware of the array size. Further, C++ does not provide automatic range-checking. Again, it is the

* This chapter has been reprinted from first edition of this Handbook, without any content updates.

programmer's responsibility to ensure that the index being supplied is valid. Arrays are usually allocated contiguous storage in the memory. An array may be allocated statically (i.e., during compile-time) or dynamically (i.e., during program execution). For example, in C++, a static array is defined as:

```
int list[20];
```

while a dynamic one is defined as:

```
int* list;
      .
      .
      .
list = new int[25];
```

An important difference between static and dynamic arrays is that the size of a static array cannot be changed during run time, while that of a dynamic array can (as we will see in Section 2.2.3).

2.2.1 Operations on an Array

1. Retrieval of an element: Given an array index, retrieve the corresponding value. This can be accomplished in $O(1)$ time. This is an important advantage of the array relative to other structures. If the array is sorted, this enables one to compute the minimum, maximum, median (or in general, the ith smallest element) essentially for free in $O(1)$ time.

2. Search: Given an element value, determine whether it is present in the array. If the array is unsorted, there is no good alternative to a sequential search that iterates through all of the elements in the array and stops when the desired element is found:

```
int SequentialSearch(int* array, int n, int x)
// search for x in array[n]
{
    for (int i = 0; i < n; i++)
        if (array[i] == x) return i; // search succeeded
    return -1; // search failed
}
```

In the worst case, this requires $O(n)$ time. If, however, the array is sorted, binary search can be used.

```
int BinarySearch(int* array, int n, int x)
{
    int first = 0, mid, last = n-1;
    while (first < last) {
       mid = (first + last)/2;
       if (array[mid] == x) return mid; // search succeeded
       if (x < array[mid]) last = mid-1;
       else first = mid+1;
    }
    return -1; // search failed
}
```

Binary search only requires $O(\log n)$ time.

3. Insertion and Deletion: These operations can be the array's Achilles heel. First, consider a sorted array. It is usually assumed that the array that results from an insertion or deletion is to be sorted. The worst case scenario presents itself when an element that is smaller than all of the elements currently in the array is to be inserted. This element will be placed in the leftmost location. However, to make room for it, all of the existing elements in the array will have to be shifted one place to the right. This requires $O(n)$ time. Similarly, a deletion from the leftmost element leaves a "vacant" location. Actually, this location can never be vacant because it refers to a word in memory which must contain some value. Thus, if the program accesses a "vacant" location, it doesn't have any way to know that the location is vacant. It may be possible to establish some sort of code based on our knowledge of the values contained in the array. For example, if it is known that an array contains only positive integers, then one could use a zero to denote a vacant location. Because of these and other complications, it is best to eliminate vacant locations that are interspersed among occupied locations by shifting elements to the left so that all vacant locations are placed to the right. In this case, we know which locations are vacant by maintaining an integer variable which contains the number of locations starting at the left that are currently in use. As before, this shifting requires $O(n)$ time. In an unsorted array, the efficiency of insertion and deletion depends on where elements are to be added or removed.

If it is known for example that insertion and deletion will only be performed at the right end of the array, then these operations take $O(1)$ time as we will see later when we discuss stacks.

2.2.2 Sorted Arrays

We have already seen that there are several benefits to using sorted arrays, namely: searching is faster, computing order statistics (the ith smallest element) is $O(1)$, etc. This is the first illustration of a key concept in data structures that will be seen several times in this handbook: *the concept of preprocessing data to make subsequent queries efficient.* The idea is that we are often willing to invest some time at the beginning in setting up a data structure so that subsequent operations on it become faster. Some sorting algorithms such as heap sort and merge sort require $O(n \log n)$ time in the worst case, whereas other simpler sorting algorithms such as insertion sort, bubble sort and selection sort require $O(n^2)$ time in the worst case. Others such as quick sort have a worst case time of $O(n^2)$, but require $O(n \log n)$ on the average. Radix sort requires $\Theta(n)$ time for certain kinds of data. We refer the reader to [5] for a detailed discussion.

However, as we have seen earlier, insertion into and deletion from a sorted array can take $\Theta(n)$ time, which is large. It is possible to merge two sorted arrays into a single sorted array in time linear in the sum of their sizes. However, the usual implementation needs additional $\Theta(n)$ space. See [6] for an $O(1)$-space merge algorithm.

2.2.3 Array Doubling

To increase the length of a (dynamically allocated) one-dimensional array a that contains elements in positions $a[0..n-1]$, we first define an array of the new length (say m), then copy the n elements from a to the new array, and finally change the value of a so that it references the new array. It takes $\Theta(m)$ time to create an array of length m because all elements of the newly created array are initialized to a default value. It then takes an additional $\Theta(n)$ time to copy elements from the source array to the destination array. Thus, the total time required for this operation is $\Theta(m + n)$. This operation is used in practice to increase the array size when it becomes full. The new array is usually twice the length of the original; that is, $m = 2n$. The resulting complexity ($\Theta(n)$) would normally be considered to be expensive. However, when this cost is amortized over the subsequent n insertions, it in fact only adds $\Theta(1)$ time per insertion. Since the cost of an insertion is $\Omega(1)$, this does not result in an asymptotic increase in insertion time. In general, increasing array size by a constant factor every time its size is to be increased does not adversely affect the asymptotic complexity. A similar approach can be used to reduce the array size. Here, the array size would be reduced by a constant factor every time.

2.2.4 Multiple Lists in a Single Array

The array is wasteful of space when it is used to represent a list of objects that changes over time. In this scenario, we usually allocate a size greater than the number of objects that have to be stored, for example by using the array doubling idea discussed above. Consider a completely-filled array of length 8192 into which we are inserting an additional element. This insertion causes the array-doubling algorithm to create a new array of length 16,384 into which the 8192 elements are copied (and the new element inserted) before releasing the original array. This results in a space requirement during the operation which is almost three times the number of elements actually present. When several lists are to be stored, it is more efficient to store them all in a single array rather than allocating one array for each list.

Although this representation of multiple lists is more space-efficient, insertions can be more expensive because it may be necessary to move elements belonging to other lists in addition to elements in one's own list. This representation is also harder to implement (Figure 2.1).

2.2.5 Heterogeneous Arrays

The definition of an array in modern programming languages requires all of its elements to be of the same type. How do we then address the scenario where we wish to use the array to store elements of different types? In earlier languages like C, one could use the `union` facility to artificially coalesce the different types into one type. We could then define an array on this new type.

FIGURE 2.1   Multiple lists in a single array.

The kind of object that an array element actually contains is determined by a tag. The following defines a structure that contains one of three types of data.

```
struct Animal{
  int id;
  union {
     Cat c;
     Dog d;
     Fox f;
  }
}
```

The programmer would have to establish a convention on how the `id` tag is to be used: for example, that `id = 0` means that the animal represented by the struct is actually a cat, etc. The union allocates memory for the largest type among `Cat`, `Dog`, and `Fox`. This is wasteful of memory if there is a great disparity among the sizes of the objects. With the advent of object-oriented languages, it is now possible to define the base type `Animal`. `Cat`, `Dog`, and `Fox` may be implemented using inheritance as derived types of `Animal`. An array of pointers to `Animal` can now be defined. These pointers can be used to refer to any of `Cat`, `Dog`, and `Fox`.

2.2.6 Multidimensional Arrays

2.2.6.1 Row- or Column Major Representation

Earlier representations of multidimensional arrays mapped the location of each element of the multidimensional array into a location of a one- dimensional array. Consider a two-dimensional array with r rows and c columns. The number of elements in the array $n = rc$. The element in location $[i][j]$, $0 \leq i < r$ and $0 \leq j < c$, will be mapped onto an integer in the range $[0, n-1]$. If this is done in row-major order—the elements of row 0 are listed in order from left to right followed by the elements of row 1, then row 2, etc.—the mapping function is $ic + j$. If elements are listed in column-major order, the mapping function is $jr + i$. Observe that we are required to perform a multiplication and an addition to compute the location of an element in an array.

2.2.6.2 Array of Arrays Representation

In Java, a two-dimensional array is represented as a one-dimensional array in which each element is, itself, a one-dimensional array. The array

```
int [][] x = new int[4][5];
```

is actually a one-dimensional array whose length is 4. Each element of x is a one-dimensional array whose length is 5. Figure 2.2 shows an example. This representation can also be used in C++ by defining an array of pointers. Each pointer can then be used to point to a dynamically-created one-dimensional array. The element $x[i][j]$ is found by first retrieving the pointer $x[i]$. This gives the address in memory of $x[i][0]$. Then $x[i][j]$ refers to the element j in row i. Observe that this only requires the addition operator to locate an element in a one-dimensional array.

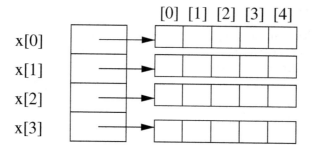

FIGURE 2.2 The array of arrays representation.

2.2.6.3 Irregular Arrays

A two-dimensional array is regular in that every row has the same number of elements. When two or more rows of an array have different number of elements, we call the array irregular. Irregular arrays may also be created and used using the array of arrays representation.

2.2.7 Sparse Matrices

A matrix is sparse if a large number of its elements are 0. Rather than store such a matrix as a two-dimensional array with lots of zeroes, a common strategy is to save space by explicitly storing only the non-zero elements. This topic is of interest to the scientific computing community because of the large sizes of some of the sparse matrices it has to deal with. The specific approach used to store matrices depends on the nature of sparsity of the matrix. Some matrices, such as the tridiagonal matrix have a well-defined sparsity pattern. The tridiagonal matrix is one where all of the nonzero elements lie on one of three diagonals: the main diagonal and the diagonals above and below it. See Figure 2.3a.

There are several ways to represent this matrix as a one-dimensional array. We could order elements by rows giving [2,1,1,3,4,1,1,2,4,7,4,3,5] or by diagonals giving [1,1,4,3,2,3,1,7,5,1,4,2,4]. Figure 2.3 shows other special matrices: the upper and lower triangular matrices which can also be represented using a one-dimensional representation.

Other sparse matrices may have an irregular or unstructured pattern. Consider the matrix in Figure 2.4a. We show two representations. Figure 2.4b shows a one-dimensional array of triples, where each triple represents a nonzero element and consists of the row, column, and value. Figure 2.4c shows an irregular array representation. Each row is represented by a one-dimensional array of pairs, where each pair contains the column number and the corresponding nonzero value.

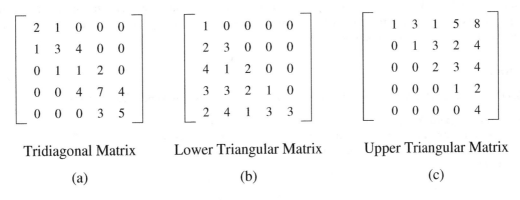

Tridiagonal Matrix   Lower Triangular Matrix   Upper Triangular Matrix

(a)   (b)   (c)

FIGURE 2.3 Matrices with regular structures. (a) Tridiagonal Matrix, (b) Lower Triangular Matrix, (c) Upper Triangular Matrix.

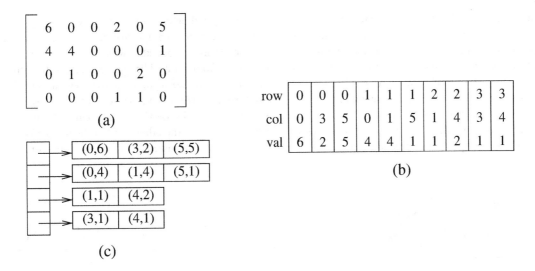

FIGURE 2.4 Unstructured matrices.

2.3 Linked Lists

The linked list is an alternative to the array when a collection of objects is to be stored. The linked list is implemented using pointers. Thus, an element (or node) of a linked list contains the actual data to be stored and a pointer to the next node. Recall that a pointer is simply the address in memory of the next node. Thus, a key difference from arrays is that a linked list does not have to be stored contiguously in memory.

The code fragment below defines a linked list data structure, which is also illustrated in Figure 2.5:

```
class ListNode {
friend class List;
private:
   int data;
   ListNode *link;
}

class List {
public:
  // List manipulation operations go here
  ...
private:
  ListNode *first;
}
```

A chain is a linked list where each node contains a pointer to the next node in the list. The last node in the list contains a null (or zero) pointer. A circular list is identical to a chain *except* that the last node contains a pointer to the first node. A doubly linked circular list differs from the chain and the circular list in that each node contains *two* pointers. One points to the next node (as before), while the other points to the *previous* node.

2.3.1 Chains

The following code searches for a key k in a chain and returns true if the key is found and false, otherwise.

```
bool List::Search(int k) {
  for (ListNode *current = first; current; current = current->next)
    if (current->data == k) then return true;
  return false;
}
```

In the worst case, Search takes $\Theta(n)$ time. In order to insert a node `newnode` in a chain immediately after node `current`, we simply set `newnode`'s pointer to the node following `current` (if any) and `current`'s pointer to `newnode` as shown in the Figure 2.6.

To delete a node `current`, it is *necessary* to have a pointer to the node preceding `current`. This node's pointer is then set to `current->next` and node `current` is freed. Both insertion and deletion can be accomplished in $O(1)$ time *provided* that the required pointers are initially available. Whether this is true or not depends on the context in which these operations are called. For example, if you are required to delete the node with key 50, if it exists, from a linked list, you would first have to search for 50. Your search algorithm would maintain a trailing pointer so that when 50 is found, a pointer to the previous node is available. Even though, deletion takes $\Theta(1)$ time, deletion in this context would require $\Theta(n)$ time in the worst case because of the search. In some cases, the context depends on how the list is organized. For example, if the list is to be sorted, then node insertions should be made so as to maintain the sorted property (which could take $\Theta(n)$ time). On the other hand, if the list is unsorted, then a node insertion can take place anywhere in the list. In particular, the node could be inserted at the front of the list in $\Theta(1)$ time.

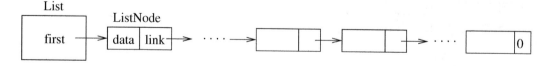

FIGURE 2.5   The structure of a linked list.

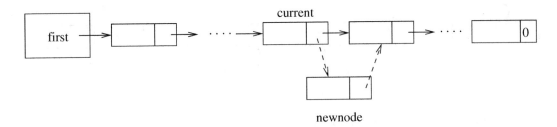

FIGURE 2.6 Insertion into a chain. The dashed links show the pointers after `newnode` has been inserted.

Interestingly, the author has often seen student code in which the insertion algorithm traverses the entire linked list and inserts the new element at the end of the list!

As with arrays, chains can be sorted or unsorted. Unfortunately, however, many of the benefits of a sorted array do not extend to sorted linked lists because arbitrary elements of a linked list cannot be accessed quickly. In particular, it is not possible to carry out binary search in $O(\log n)$ time. Nor is it possible to locate the ith smallest element in $O(1)$ time. On the other hand, merging two sorted lists into one sorted list is more convenient than merging two sorted arrays into one sorted array because the traditional implementation requires space to be allocated for the target array. A code fragment illustrating the merging of two sorted lists is shown below. This is a key operation in mergesort:

```
void Merge(List listOne, List listTwo, List& merged) {
  ListNode* one = listOne.first;
  ListNode* two = listTwo.first;
  ListNode* last = 0;

  if (one == 0) {merged.first = two; return;}
  if (two == 0) {merged.first = one; return;}

  if (one->data < two->data) last = merged.first = one;
  else last = merged.first = two;
  while (one && two)
     if (one->data < two->data) {
       last->next = one; last= one; one = one->next;
     }
     else {
       last->next = two; last = two; two = two->next;
     }
  if (one) last->next = one;
  else last->next = two;
}
```

The merge operation is not defined when lists are unsorted. However, one may need to combine two lists into one. This is the concatenation operation. With chains, the best approach is to attach the second list to the end of the first one. In our implementation of the linked list, this would require one to traverse the first list until the last node is encountered and then set its *next* pointer to point to the first element of the second list. This requires time proportional to the size of the first linked list. This can be improved by maintaining a pointer to the last node in the linked list.

It is possible to traverse a singly linked list in both directions (i.e., left to right and a restricted right-to-left traversal) by reversing links during the left-to-right traversal. Figure 2.7 shows a possible configuration for a list under this scheme.

As with the heterogeneous arrays described earlier, heterogeneous lists can be implemented in object-oriented languages by using inheritance.

2.3.2 Circular Lists

In the previous section, we saw that to concatenate two unsorted chains efficiently, one needs to maintain a rear pointer in addition to the first pointer. With circular lists, it is possible to accomplish this with a single pointer as follows: consider the circular list in Figure 2.8. The second node in the list can be accessed through the first in $O(1)$ time. Now, consider the list that begins at this second node and ends at the first node. This may be viewed as a chain with access pointers to the first and last nodes.

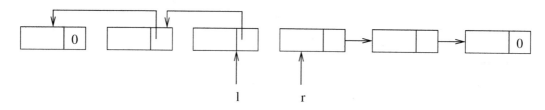

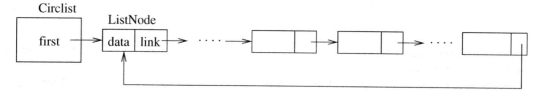

FIGURE 2.7   Illustration of a chain traversed in both directions.

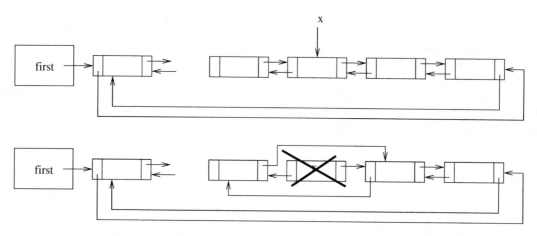

FIGURE 2.8   A circular list.

Concatenation can now be achieved in $O(1)$ time by linking the last node of one chain to the first node of the second chain and vice versa.

2.3.3 Doubly Linked Circular Lists

A node in a doubly linked list differs from that in a chain or a singly linked list in that it has two pointers. One points to the next node as before, while the other points to the previous node. This makes it possible to traverse the list in both directions. We observe that this is possible in a chain as we saw in Figure 2.7. The difference is that with a doubly linked list, one can initiate the traversal from any arbitrary node in the list. Consider the following problem: we are provided a pointer x to a node in a list and are required to delete it as shown in Figure 2.9. To accomplish this, one needs to have a pointer to the previous node. In a chain or a circular list, an expensive list traversal is required to gain access to this previous node. However, this can be done in $O(1)$ time in a doubly linked circular list. The code fragment that accomplishes this is as below:

```
void DblList::Delete(DblListNode* x)
{
 x->prev->next = x->next;
 x->next->prev = x->prev;
 delete x;
}
```

An application of doubly linked lists is to store a list of siblings in a Fibonacci heap (Chapter 7).

FIGURE 2.9   Deletion from a doubly linked list.

2.3.4 Generalized Lists

A generalized list A is a finite sequence of $n \geq 0$ elements, $e_0, e_1, \ldots, e_{n-1}$, where e_i is either an atom or a generalized list. The elements e_i that are not atoms are said to be sublists of A. Consider the generalized list $A = ((a, b, c), ((d, e), f), g)$. This list contains three elements: the sublist (a, b, c), the sublist $((d, e), f)$ and the atom g. The generalized list may be implemented by employing a GenListNode type as follows:

```
private:
  GenListNode* next;
  bool tag;
  union {
    char data;
    GenListNode* down;
  };
```

If tag is true, the element represented by the node is a sublist and down points to the first node in the sublist. If tag is false, the element is an atom whose value is contained in data. In both cases, next simply points to the next element in the list. Figure 2.10 illustrates the representation.

2.4 Stacks and Queues

The stack and the queue are data types that support insertion and deletion operations with well-defined semantics. Stack deletion deletes the element in the stack that was inserted the last, while a queue deletion deletes the element in the queue that was inserted the earliest. For this reason, the stack is often referred to as a LIFO (Last In First Out) data type and the queue as an FIFO (First In First out) data type. A deque (double ended queue) combines the stack and the queue by supporting both types of deletions.

Stacks and queues find a lot of applications in Computer Science. For example, a system stack is used to manage function calls in a program. When a function f is called, the system creates an activation record and places it on top of the system stack. If function f calls function g, the local variables of f are added to its activation record and an activation record is created for g. When g terminates, its activation record is removed and f continues executing with the local variables that were stored in its activation record. A queue is used to schedule jobs at a resource when a first-in first-out policy is to be implemented. Examples could include a queue of print-jobs that are waiting to be printed or a queue of packets waiting to be transmitted over a wire. Stacks and queues are also used routinely to implement higher-level algorithms. For example, a queue is used to implement a breadth-first traversal of a graph. A stack may be used by a compiler to process an expression such as $(a + b) \times (c + d)$.

2.4.1 Stack Implementation

Stacks and queues can be implemented using either arrays or linked lists. Although the burden of a correct stack or queue implementation appears to rest on deletion rather than insertion, it is convenient in actual implementations of these data types to place restrictions on the insertion operation as well. For example, in an array implementation of a stack, elements are inserted in a left-to-right order. A stack deletion simply deletes the rightmost element.

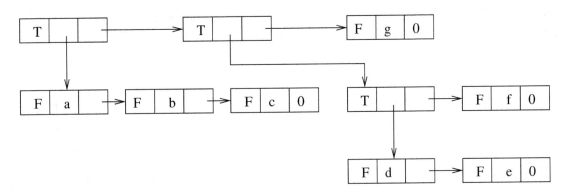

FIGURE 2.10 Generalized list for ((a,b,c),((d,e),f),g).

A simple array implementation of a stack class is shown below:

```
class Stack {
public:
    Stack(int maxSize = 100); // 100 is default size
  void Insert(int);
  int* Delete(int&);
private:
  int *stack;
  int size;
  int top; // highest position in array that contains an element
};
```

The stack operations are implemented as follows:

```
Stack::Stack(int maxSize): size(maxSize)
{
  stack = new int[size];
  top = -1;
}

void Stack::Insert(int x)
{
  if (top == size-1) cerr << "Stack Full" << endl;
  else stack[++top] = x;
}

int* Stack::Delete(int& x)
{
  if (top == -1) return 0; // stack empty
  else {
    x = stack[top--];
    return &x;
  }
}
```

The operation of the following code fragment is illustrated in Figure 2.11.

```
Stack s;
int x;
s.Insert(10);
s.Insert(20);
s.Insert(30);
s.Delete(x);
s.Insert(40);
s.Delete(x);
```

It is easy to see that both stack operations take $O(1)$ time. The stack data type can also be implemented using linked lists by requiring all insertions and deletions to be made at the front of the linked list.

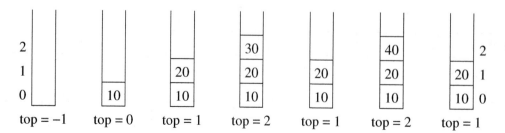

FIGURE 2.11   Stack operations.

2.4.2 Queue Implementation

An array implementation of a queue is a bit trickier than that of a stack. Insertions can be made in a left-to-right fashion as with a stack. However, deletions must now be made from the left. Consider a simple example of an array of size 5 into which the integers 10, 20, 30, 40, and 50 are inserted as shown in Figure 2.12a. Suppose three elements are subsequently deleted (Figure 2.12b).

What if we are now required to insert the integer 60. On one hand, it appears that we are out of room as there is no more place to the right of 50. On the other hand, there are three locations available to the left of 40. This suggests that we use a circular array implementation of a queue, which is described below.

```
class Queue {
public:
     Queue(int maxSize = 100); // 100 is default size
 void Insert(int);
 int* Delete(int&);
private:
 int *queue;
 int size;
 int front, rear;
};
```

The queue operations are implemented below:

```
Queue::Queue(int maxSize): size(maxSize)
{
 queue= new int[size];
 front = rear = 0;
}

void Queue::Insert(int x)
{
 int k = (rear + 1)
 if (front == k) cerr << "Queue Full!" <, endl;
 else queue[rear = k] = x;
}

int* Queue::Delete(int& x)
{
 if (front == rear) return 0; // queue is empty
 x = queue[++front
 return &x;
}
```

Figure 2.13 illustrates the operation of this code on an example. The first figure shows an empty queue with first = rear = 0. The second figure shows the queue after the integer 10 is inserted. The third figure shows the queue when 20, 30, 40, 50, and 60 have been inserted. The fourth figure shows the queue after 70 is inserted. Notice that, although one slot remains empty, the queue is now full because `Queue::Insert` will not permit another element to be inserted. If it did permit an insertion at this stage, rear and front would be the same. This is the condition that `Queue:Delete` checks to determine whether the queue is empty! This would make it impossible to distinguish between the queue being full and being empty. The fifth figure shows the queue after two integers (10 and 20) are deleted. The last figure shows a full queue after the insertion of integers 80 and 90.

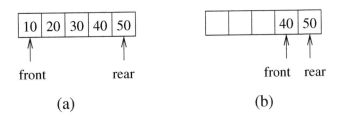

FIGURE 2.12 Pitfalls of a simple array implementation of a queue.

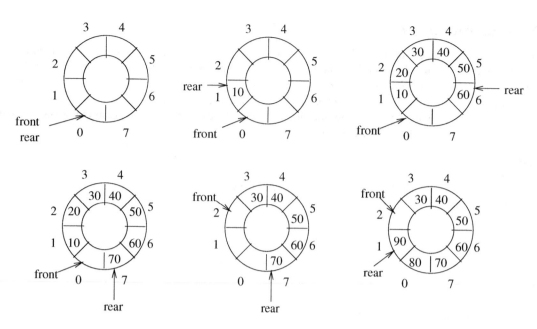

FIGURE 2.13   Implementation of a queue in a circular array.

It is easy to see that both queue operations take $O(1)$ time. The queue data type can also be implemented using linked lists by requiring all insertions and deletions to be made at the front of the linked list.

Acknowledgments

The author would like to thank Kun Gao, Dean Simmons and Mr. U. B. Rao for proofreading this chapter. This work was supported, in part, by the National Science Foundation under grant CCR-9988338.

References

1. D. Knuth, *The Art of Computer Programming*, Vol. 1, Fundamental Algorithms, Third edition. Addison-Wesley, NY, 1997.
2. S. Sahni, *Data Structures, Algorithms, and Applications in C++*, McGraw-Hill, NY, 1998.
3. S. Sahni, *Data Structures, Algorithms, and Applications in Java*, McGraw-Hill, NY, 2000.
4. E. Horowitz, S. Sahni, and D. Mehta, *Fundamentals of Data Structures in C++*, W. H. Freeman, NY, 1995.
5. D. Knuth, *The Art of Computer Programming*, Vol. 3, Sorting and Searching, Second edition. Addison-Wesley, NY, 1997.
6. B. Huang and M. Langston, Practical in-place merging, *Communications of the ACM*, 31:3, 1988, pp. 348–352.

Trees *

3.1	Introduction...	35
3.2	Tree Representation...	37
	List Representation • Left Child-Right Sibling Representation • Binary Tree Representation	
3.3	Binary Trees and Properties..	37
	Properties • Binary Tree Representation	
3.4	Binary Tree Traversals..	39
	Inorder Traversal • Preorder Traversal • Postorder Traversal • Level Order Traversal	
3.5	Threaded Binary Trees	41
	Threads • Inorder Traversal of a Threaded Binary Tree	
3.6	Binary Search Trees	42
	Definition • Search • Insert • Delete • Miscellaneous	
3.7	Heaps..	44
	Priority Queues • Definition of a Max-Heap • Insertion • Deletion	
3.8	Tournament Trees..	46
	Winner Trees • Loser Trees	
	Acknowledgments	48
	References..	48

Dinesh P. Mehta
Colorado School of Mines

3.1 Introduction

The tree is a natural representation for *hierarchical* information. Thus, trees are used to represent genealogical information (e.g., family trees and evolutionary trees), organizational charts in large companies, the directory structure of a file system on a computer, parse trees in compilers and the structure of a knock-out sports tournament. The Dewey decimal notation, which is used to classify books in a library, is also a tree structure. In addition to these and other applications, the tree is used to design fast algorithms in computer science because of its *efficiency* relative to the simpler data structures discussed in Chapter 2. Operations that take linear time on these structures often take logarithmic time on an appropriately organized tree structure. For example, the average time complexity for a search on a key is linear on a linked list and *logarithmic* on a binary search tree. Many of the data structures discussed in succeeding chapters of this handbook are tree structures.

Several kinds of trees have been defined in the literature:

1. *Free or unrooted tree*: this is defined as a graph (a set of vertices and a set of edges that join pairs of vertices) such that there exists a unique path between any two vertices in the graph. The minimum spanning tree of a graph is a well-known example of a free tree. Graphs are discussed in Chapter 4.
2. *Rooted tree*: a finite set of one or more nodes such that
   a. There is a special node called the *root*.
   b. The remaining nodes are partitioned into $n \geq 0$ disjoint sets $T_1, \ldots, T_n$, where each of these sets is a tree. $T_1, \ldots, T_n$ are called the subtrees of the root.
   If the order in which the subtrees are arranged is not important, then the tree is a rooted, unordered (or *oriented*) tree. If the order of the subtrees is important, the tree is rooted and ordered. Figure 3.1 depicts the relationship between the three types of trees. We will henceforth refer to the rooted, ordered tree simply as "tree."
3. *k-ary tree*: a finite set of nodes that is either empty or consists of a root and the elements of k disjoint k-ary trees called the 1st, 2nd, $\ldots$, kth subtrees of the root. The *binary tree* is a k-ary tree with $k = 2$. Here, the first and second subtrees are respectively called the left and right subtrees of the root. Note that binary trees are not trees. One difference is that a binary

---

* This chapter has been reprinted from first edition of this Handbook, without any content updates.

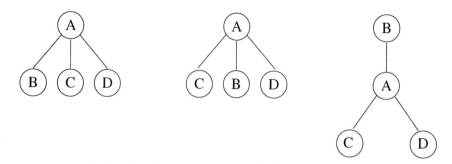

FIGURE 3.1 The three trees shown are distinct if they are viewed as rooted, ordered trees. The first two are identical if viewed as oriented trees. All three are identical if viewed as free trees.

FIGURE 3.2 Different binary trees.

tree can be empty, whereas a tree cannot. Second, the two trees shown in Figure 3.2 are different binary trees but would be different drawings of the same tree.

Figure 3.3 shows a tree with 11 nodes. The number of subtrees of a node is its *degree*. Nodes with degree 0 are called *leaf* nodes. Thus, node *A* has degree 3, nodes *B*, *D*, and *I* have degree 2, node *E* has degree 1, and nodes *C*, *F*, *G*, *H*, *J*, and *K* have degree 0 (and are leaves of the tree). The degree of a tree is the maximum of the degree of the nodes in the tree. The roots of the subtrees of a node *X* are its *children*. *X* is the parent of its children. Children of the same parent are *siblings*. In the example, *B*, *C*, and *D* are each other's siblings and are all children of *A*. The *ancestors* of a node are all the nodes excluding itself along the path from the root to that node. The *level* of a node is defined by letting the root be at level zero. If a node is at level *l*, then its children are at level *l* + 1. The *height* of a tree is the maximum level of any node in the tree. The tree in the example has height 4. These terms are defined in the same way for binary trees. See [1–6] for more information on trees.

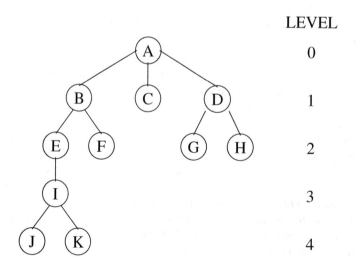

FIGURE 3.3 An example tree.

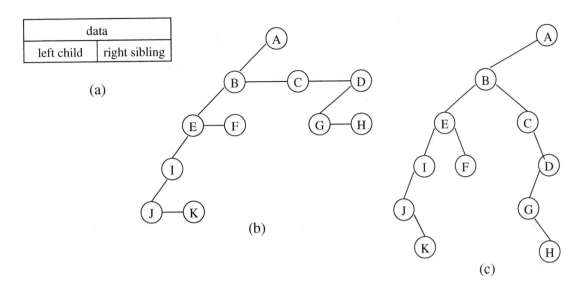

FIGURE 3.4 Tree representations.

3.2 Tree Representation

3.2.1 List Representation

The tree of Figure 3.3 can be written as the generalized list (A (B (E (I (J, K)), F), C, D(G, H))). The information in the root node comes first followed by a list of subtrees of the root. This enables us to represent a tree in memory using generalized lists as discussed in Chapter 2.

3.2.2 Left Child-Right Sibling Representation

Figure 3.4a shows the node structure used in this representation. Each node has a pointer to its leftmost child (if any) and to the sibling on its immediate right (if any). The tree in Figure 3.3 is represented by the tree in Figure 3.4b.

3.2.3 Binary Tree Representation

Observe that the left child-right sibling representation of a tree (Figure 3.4b) may be viewed as a binary tree by rotating it clockwise by 45 degrees. This gives the binary tree representation shown in Figure 3.4c. This representation can be extended to represent a *forest*, which is defined as an ordered set of trees. Here, the roots of the trees are viewed as siblings. Thus, a root's right pointer points to the next tree root in the set. We have

Lemma 3.1

There is a one-to-one correspondence between the set of forests and the set of binary trees.

3.3 Binary Trees and Properties

Binary trees were defined in Section 3.1. For convenience, a binary tree is sometimes extended by adding *external* nodes. External nodes are imaginary nodes that are added wherever an empty subtree was present in the original tree. The original tree nodes are known as *internal* nodes. Figure 3.5a shows a binary tree and (b) the corresponding extended tree. Observe that in an extended binary tree, all internal nodes have degree 2 while all external nodes have degree 0. (Some authors use the term *full* binary tree to denote a binary tree whose nodes have 0 or two children.) The *external path length* of a tree is the sum of the lengths of all root-to-external node paths in the tree. In the example, this is $2 + 2 + 3 + 3 + 2 = 12$. The *internal path length* is similarly defined by adding lengths of all root-to-internal node paths. In the example, this quantity is $0 + 1 + 1 + 2 = 4$.

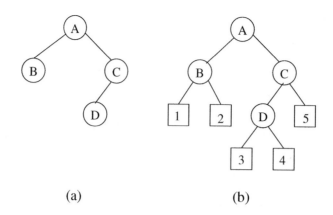

FIGURE 3.5   (b) shows the extended binary tree corresponding to the binary tree of (a). External nodes are depicted by squares.

3.3.1 Properties
Lemma 3.2

A binary tree with n internal nodes has $n+1$ external nodes.

Proof. Each internal node in the extended tree has branches leading to two children. Thus, the total number of branches is $2n$. Only $n-1$ internal nodes have a single incoming branch from a parent (the root does not have a parent). Thus, each of the remaining $n+1$ branches points to an external node.   ∎

Lemma 3.3

For any non-empty binary tree with n_0 leaf nodes and n_2 nodes of degree 2, $n_0 = n_2 + 1$.

Proof. Let n_1 be the number of nodes of degree 1 and $n = n_0 + n_1 + n_2$ (Equation 1) be the total number of nodes. The number of branches in a binary tree is $n-1$ since each non-root node has a branch leading into it. But, all branches stem from nodes of degree 1 and 2. Thus, the number of branches is $n_1 + 2n_2$. Equating the two expressions for number of branches, we get $n = n_1 + 2n_2 + 1$ (Equation 2). From Equations 1 and 2, we get $n_0 = n_2 + 1$.   ∎

Lemma 3.4

The external path length of any binary tree with n internal nodes is $2n$ greater than its internal path length.

Proof. The proof is by induction. The lemma clearly holds for $n=0$ when the internal and external path lengths are both zero. Consider an extended binary tree T with n internal nodes. Let E_T and I_T denote the external and internal path lengths of T. Consider the extended binary tree S that is obtained by deleting an internal node whose children are both external nodes (i.e., a leaf) and replacing it with an external node. Let the deleted internal node be at level l. Thus, the internal path length decreases by l while the external path length decreases by $2(l+1)-l=l+2$. From the induction hypothesis, $E_S = I_S + 2(n-1)$. But, $E_T = E_S + l + 2$ and $I_T = I_S + l$. Thus, $E_T - I_T = 2n$.   ∎

Lemma 3.5

The maximum number of nodes on level i of a binary tree is 2^i, $i \geq 0$.

Proof. This is easily proved by induction on i.   ∎

Lemma 3.6

The maximum number of nodes in a binary tree of height k is $2^{k+1} - 1$.

Proof. Each level i, $0 \leq i \leq k$, has 2^i nodes. Summing over all i results in $\sum_{i=0}^{k} 2^i = 2^{k+1} - 1$.   ∎

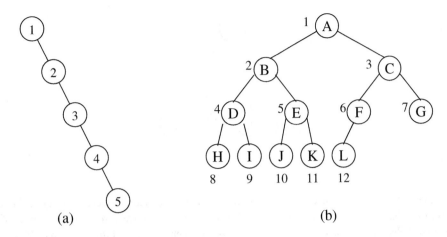

FIGURE 3.6 (a) Skewed and (b) complete binary trees.

Lemma 3.7

The height of a binary tree with n internal nodes is at least $\lceil \log_2(n+1) \rceil$ and at most $n-1$.

Proof. The worst case is a skewed tree (Figure 3.6a) and the best case is a tree with 2^i nodes at every level i except possibly the bottom level (Figure 3.6b). If the height is h, then $n+1 \leq 2^h$, where $n+1$ is the number of external nodes. ∎

Lemma 3.8

The number of distinct binary trees with n nodes is $\frac{1}{n+1}\binom{2n}{n}$.

Proof. For a detailed proof, we refer the reader to [7]. However, we note that $C_n = \frac{1}{n+1}\binom{2n}{n}$ are known as the Catalan numbers, which occur frequently in combinatorial problems. The Catalan number C_n also describes the number of trees with $n+1$ nodes and the number of binary trees with $2n+1$ nodes all of which have 0 or 2 children. ∎

3.3.2 Binary Tree Representation

Binary trees are usually represented using nodes and pointers. A `TreeNode` class may be defined as:

```
class TreeNode {
  TreeNode* LeftChild;
  TreeNode* RightChild;
  KeyType data;
};
```

In some cases, a node might also contain a `parent` pointer which facilitates a "bottom-up" traversal of the tree. The tree is accessed by a pointer *root* of type `TreeNode*` to its root. When the binary tree is *complete* (i.e., there are 2^i nodes at every level i, except possibly the last level which has nodes filled in from left to right), it is convenient to use an array representation. The complete binary tree in Figure 3.6b can be represented by the array

```
 1  2  3  4  5  6  7  8  9 10 11 12
[ A  B  C  D  E  F  G  H  I  J  K  L ]
```

Observe that the children (if any) of a node located at position i of the array can be found at positions $2i$ and $2i+1$ and its parent at $\lfloor i/2 \rfloor$.

3.4 Binary Tree Traversals

Several operations on trees require one to traverse the entire tree: that is, given a pointer to the root of a tree, process every node in the tree systematically. Printing a tree is an example of an operation that requires a tree traversal. Starting at a node, we can

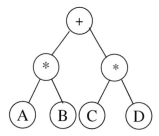

FIGURE 3.7   An expression tree.

do one of three things: visit the node (*V*), traverse the left subtree recursively (*L*), and traverse the right subtree recursively (*R*). If we adopt the convention that the left subtree will be visited before the right subtree, we have three types of traversals *LV R*, *V LR*, and *LRV* which are called *inorder*, *preorder*, and *postorder*, respectively, because of the position of *V* with respect to *L* and *R*. In the following, we will use the expression tree in Figure 3.7 to illustrate the three traversals, which result in infix, prefix, and postfix forms of the expression. A fourth traversal, the level order traversal, is also studied.

3.4.1 Inorder Traversal

The following is a recursive algorithm for an inorder traversal that prints the contents of each node when it is visited. The recursive function is invoked by the call inorder(root). When run on the example expression tree, it returns A*B+C*D.

```
inorder(TreeNode* currentNode)
{
  if (currentNode) {
   inorder(currentNode->LeftChild);
   cout << currentNode->data;
   inorder(currentNode->RightChild);
   }
 }
```

3.4.2 Preorder Traversal

The following is a recursive algorithm for a preorder traversal that prints the contents of each node when it is visited. The recursive function is invoked by the call preorder(root). When run on the example expression tree, it returns +*AB*CD.

```
preorder(TreeNode* currentNode)
{
  if (currentNode) {
    cout << currentNode->data;
    preorder(currentNode->LeftChild);
    preorder(currentNode->RightChild);
   }
 }
```

3.4.3 Postorder Traversal

The following is a recursive algorithm for a postorder traversal that prints the contents of each node when it is visited. The recursive function is invoked by the call postorder(root). When run on the example expression tree, it prints AB*CD*+.

```
postorder(TreeNode* currentNode)
{
  if (currentNode) }
    postorder(currentNode->LeftChild);
    postorder(currentNode->RightChild);
    cout << currentNode->data;
   }
 }
```

The complexity of each of the three algorithms is linear in the number of tree nodes. Non-recursive versions of these algorithms may be found in Reference 6. Both versions require (implicitly or explicitly) a stack.

3.4.4 Level Order Traversal

The level order traversal uses a queue. This traversal visits the nodes in the order suggested in Figure 3.6b. It starts at the root and then visits all nodes in increasing order of their level. Within a level, the nodes are visited in left-to-right order.

```
LevelOrder(TreeNode* root)
{
  Queue q<TreeNode*>;
  TreeNode* currentNode = root;
  while (currentNode) {
    cout << currentNode->data;
    if (currentNode->LeftChild) q.Add(currentNode->LeftChild);
    if (currentNode->RightChild) q.Add(currentNode->RightChild);
    currentNode = q.Delete(); //q.Delete returns a node pointer
  }
}
```

3.5 Threaded Binary Trees

3.5.1 Threads

Lemma 3.2 implies that a binary tree with n nodes has $n + 1$ null links. These null links can be replaced by pointers to nodes called threads. Threads are constructed using the following rules:

1. A null right child pointer in a node is replaced by a pointer to the inorder successor of p (i.e., the node that would be visited after p when traversing the tree inorder).
2. A null left child pointer in a node is replaced by a pointer to the inorder predecessor of p.

Figure 3.8 shows the binary tree of Figure 3.7 with threads drawn as broken lines. In order to distinguish between threads and normal pointers, two boolean fields LeftThread and RightThread are added to the node structure. If p->LeftThread is 1, then p->LeftChild contains a thread; otherwise it contains a pointer to the left child. Additionally, we assume that the tree contains a head node such that the original tree is the left subtree of the head node. The LeftChild pointer of node A and the RightChild pointer of node D point to the head node.

3.5.2 Inorder Traversal of a Threaded Binary Tree

Threads make it possible to perform an inorder traversal without using a stack. For any node p, if p's right thread is 1, then its inorder successor is p->RightChild. Otherwise the inorder successor is obtained by following a path of left-child links from the right child of p until a node with left thread 1 is reached. Function Next below returns the inorder successor of currentNode (assuming that currentNode is not 0). It can be called repeatedly to traverse the entire tree in inorder in $O(n)$ time. The code below assumes that the last node in the inorder traversal has a threaded right pointer to a dummy head node.

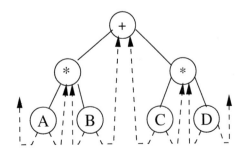

FIGURE 3.8 A threaded binary tree.

```
TreeNode* Next(TreeNode* currentNode)
{
  TreeNode* temp = currentNode->RightChild;
  if (currentNode->RightThread == 0)
    while (temp->LeftThread == 0)
      temp = temp->LeftChild;
  currentNode = temp;
  if (currentNode == headNode)
    return 0;
  else
    return currentNode;
}
```

Threads simplify the algorithms for preorder and postorder traversal. It is also possible to insert a node into a threaded tree in $O(1)$ time [6].

3.6 Binary Search Trees

3.6.1 Definition

A *binary search tree* (BST) is a binary tree that has a key associated with each of its nodes. The keys in the left subtree of a node are smaller than or equal to the key in the node and the keys in the right subtree of a node are greater than or equal to the key in the node. To simplify the discussion, we will assume that the keys in the binary search tree are distinct. Figure 3.9 shows some binary trees to illustrate the definition.

3.6.2 Search

We describe a recursive algorithm to search for a key k in a tree T: first, if T is empty, the search fails. Second, if k is equal to the key in T's root, the search is successful. Otherwise, we search T's left or right subtree recursively for k depending on whether it is less or greater than the key in the root.

```
bool Search(TreeNode* b, KeyType k)
{
  if (b == 0) return 0;
  if (k == b->data) return 1;
  if (k < b->data) return Search(b->LeftChild,k);
  if (k > b->data) return Search(b->RightChild,k);
}
```

3.6.3 Insert

To insert a key k, we first carry out a search for k. If the search fails, we insert a new node with k at the null branch where the search terminated. Thus, inserting the key 17 into the binary search tree in Figure 3.9b creates a new node which is the left child of 18. The resulting tree is shown in Figure 3.10a.

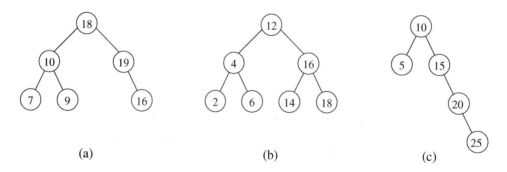

(a)   (b)   (c)

FIGURE 3.9   Binary trees with distinct keys: (a) is not a BST. (b) and (c) are BSTs.

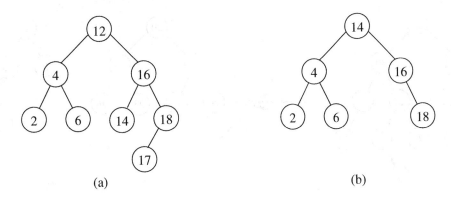

FIGURE 3.10 Tree of Figure 3.9b with (a) 18 inserted and (b) 12 deleted.

```
typedef TreeNode* TreeNodePtr;

Node* Insert(TreeNodePtr& b, KeyType k)
{
  if (b == 0) {b = new TreeNode; b->data= k; return b;}
  if (k == b->data) return 0; // don't permit duplicates
  if (k < b->data) Insert(b->LeftChild, k);
  if (k > b->data) Insert(b->RightChild, k);
}
```

3.6.4 Delete

The procedure for deleting a node x from a binary search tree depends on its degree. If x is a leaf, we simply set the appropriate child pointer of x's parent to 0 and delete x. If x has one child, we set the appropriate pointer of x's parent to point directly to x's child and then delete x. In Figure 3.9c, node 20 is deleted by setting the right child of 15 to 25. If x has two children, we replace its key with the key in its inorder successor y and then delete node y. The inorder successor contains the smallest key greater than x's key. This key is chosen because it can be placed in node x without violating the binary search tree property. Since y is obtained by first following a RightChild pointer and then following LeftChild pointers until a node with a null LeftChild pointer is encountered, it follows that y has degree 0 or 1. Thus, it is easy to delete y using the procedure described above. Consider the deletion of 12 from Figure 3.9b. This is achieved by replacing 12 with 14 in the root and then deleting the leaf node containing 14. The resulting tree is shown in Figure 3.10b.

3.6.5 Miscellaneous

Although Search, Insert, and Delete are the three main operations on a binary search tree, there are others that can be defined which we briefly describe below.

- *Minimum and Maximum* that respectively find the minimum and maximum elements in the binary search tree. The minimum element is found by starting at the root and following LeftChild pointers until a node with a 0 LeftChild pointer is encountered. That node contains the minimum element in the tree.
- Another operation is to find the kth smallest element in the binary search tree. For this, each node must contain a field with the number of nodes in its left subtree. Suppose that the root has m nodes in its left subtree. If $k \leq m$, we recursively search for the kth smallest element in the left subtree. If $k = m + 1$, then the root contains the kth smallest element. If $k > m + 1$, then we recursively search the right subtree for the $k - m - 1$st smallest element.
- The Join operation takes two binary search trees A and B as input such that all the elements in A are smaller than all the elements of B. The objective is to obtain a binary search tree C which contains all the elements originally in A and B. This is accomplished by deleting the node with the largest key in A. This node becomes the root of the new tree C. Its LeftChild pointer is set to A and its RightChild pointer is set to B.
- The Split operation takes a binary search tree C and a key value k as input. The binary search tree is to be split into two binary search trees A and B such that all keys in A are less than or equal to k and all keys in B are greater than k. This is achieved by searching for k in the binary search tree. The trees A and B are created as the search proceeds down the tree as shown in Figure 3.11.

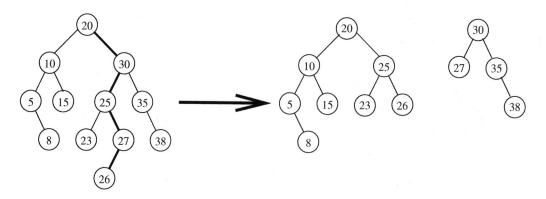

FIGURE 3.11   Splitting a binary search tree with $k = 26$.

- An inorder traversal of a binary search tree produces the elements of the binary search tree in sorted order. Similarly, the inorder successor of a node with key k in the binary search tree yields the smallest key larger than k in the tree. (Note that we used this property in the Delete operation described in the previous section.)

All of the operations described above take $O(h)$ time, where h is the height of the binary search tree. The bounds on the height of a binary tree are derived in Lemma 3.7. It has been shown that when insertions and deletions are made at random, the height of the binary search tree is $O(\log n)$ on the average.

3.7 Heaps

3.7.1 Priority Queues

Heaps are used to implement priority queues. In a priority queue, the element with highest (or lowest) priority is deleted from the queue, while elements with arbitrary priority are inserted. A data structure that supports these operations is called a max(min) priority queue. Henceforth, in this chapter, we restrict our discussion to a max priority queue. A priority queue can be implemented by a simple, unordered linked list. Insertions can be performed in $O(1)$ time. However, a deletion requires a search for the element with the largest priority followed by its removal. The search requires time linear in the length of the linked list. When a max heap is used, both of these operations can be performed in $O(\log n)$ time.

3.7.2 Definition of a Max-Heap

A max heap is a complete binary tree such that for each node, the key value in the node is greater than or equal to the value in its children. Observe that this implies that the root contains the largest value in the tree. Figure 3.12 shows some examples of max heaps.

We define a class Heap with the following data members.

```
private:
  Element *heap;
  int n; // current size of max heap
  int MaxSize; // Maximum allowable size of the heap
```

The heap is represented using an array (a consequence of the complete binary tree property) which is dynamically allocated.

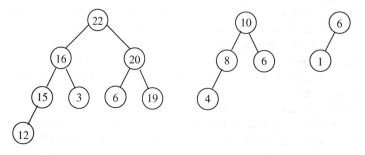

FIGURE 3.12   Max heaps.

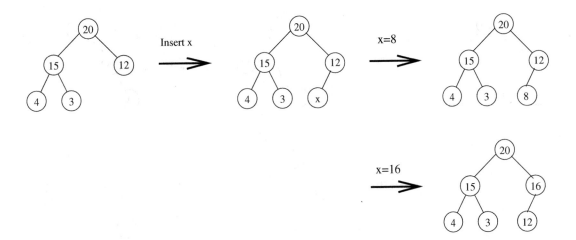

FIGURE 3.13 Insertion into max heaps.

3.7.3 Insertion

Suppose that the max heap initially has n elements. After insertion, it will have $n + 1$ elements. Thus, we need to add a node so that the resulting tree is a complete binary tree with $n + 1$ nodes. The key to be inserted is initially placed in this new node. However, the key may be larger than its parent resulting in a violation of the *max* property with its parent. In this case, we swap keys between the two nodes and then repeat the process at the next level. Figure 3.13 demonstrates two cases of an insertion into a max heap.

The algorithm is described below. In the worst case, the insertion algorithm moves up the heap from leaf to root spending $O(1)$ time at each level. For a heap with n elements, this takes $O(\log n)$ time.

```
void MaxHeap::Insert(Element x)
{
  if (n == MaxSize) {HeapFull(); return;}
  n++;
  for (int i = n; i > 1; i = i/2 ) {
    if (x.key <= heap[i/2].key) break;
    heap[i] = heap[i/2];
  }
  heap[i] = x;
}
```

3.7.4 Deletion

The element to be deleted (i.e., the maximum element in the heap) is removed from the root node. Since the binary tree must be restructured to become a complete binary tree on $n - 1$ elements, the node in position n is deleted. The element in the deleted node is placed in the root. If this element is less than either of the root's (at most) two children, there is a violation of the max property. This is fixed by swapping the value in the root with its larger child. The process is repeated at the other levels until there is no violation. Figure 3.14 illustrates deletion from a max heap.

The deletion algorithm is described below. In the worst case, the deletion algorithm moves down the heap from root to leaf spending $O(1)$ time at each level. For a heap with n elements, this takes $O(\log n)$ time.

```
Element* MaxHeap::DeleteMax(Element& x)
{
  if (n == 0) {HeapEmpty(); return 0;}
  x = heap[1];
  Element last = heap[n];
  n--;
  for (int i = 1, j = 2; j <= n; i = j, j *= 2) {
    if (j < n)
      if (heap[j].key < heap[j+1].key) j++;
    // j points to the larger child
```

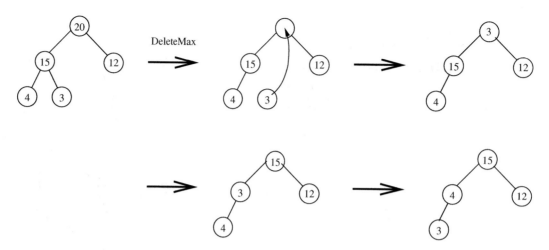

FIGURE 3.14   Deletion from max heaps.

```
    if (last.key >= heap[j].key) break;
    heap[i] = heap[j]; // move child up
  }
  heap[i] = last;
  return &x;
}
```

3.8 Tournament Trees

Consider the following problem: suppose we have k sequences, each of which is sorted in nondecreasing order, that are to be merged into one sequence in nondecreasing order. This can be achieved by repeatedly transferring the element with the smallest key to an output array. The smallest key has to be found from the leading elements in the k sequences. Ordinarily, this would require $k-1$ comparisons for each element transferred. However, with a tournament tree, this can be reduced to $\log_2 k$ comparisons per element.

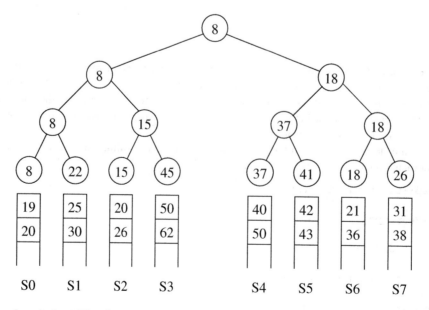

FIGURE 3.15   A winner tree for $k = 8$. Three keys in each of the eight sequences are shown. For example, sequence 2 consists of 15, 20, and 26.

3.8.1 Winner Trees

A winner tree is a complete binary tree in which each node represents the smaller of its two children. The root represents the smallest node in the tree. Figure 3.15 illustrates a winner tree with $k = 8$ sequences. The winner of the tournament is the value 8 from sequence 0. The winner of the tournament is the smallest key from the 8 sequences and is transferred to an output array. The next element from sequence 0 is now brought into play and a tournament is played to determine the next winner. This is illustrated in Figure 3.16. It is easy to see that the tournament winner can be computed in $\Theta(\log n)$ time.

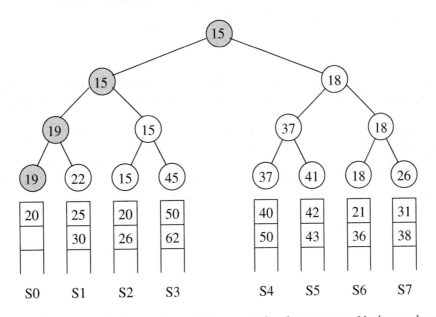

FIGURE 3.16   Winner tree of Figure 3.15 after the next element of sequence 0 plays the tournament. Matches are played at the shaded nodes.

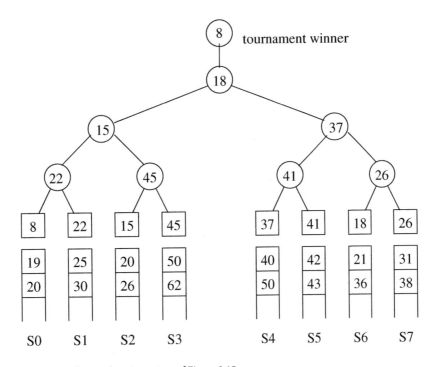

FIGURE 3.17   Loser tree corresponding to the winner tree of Figure 3.15.

3.8.2 Loser Trees

The loser tree is an alternative representation that stores the *loser* of a match at the corresponding node. The loser tree corresponding to Figure 3.15 is shown in Figure 3.17. An advantage of the loser tree is that to restructure the tree after a winner has been output, it is sufficient to examine nodes on the path from the leaf to the root rather than the siblings of nodes on this path.

Acknowledgments

The author would like to thank Kun Gao and Mr. U. B. Rao for proofreading this chapter. This work was supported, in part, by the National Science Foundation under grant CCR-9988338.

References

1. D. Knuth, *The Art of Computer Programming*, Vol. 1, Fundamental Algorithms, Third edition. Addison Wesley, NY, 1997.
2. T. Cormen, C. Leiserson, R. Rivest, and C. Stein, *Introduction to Algorithms*, Second edition. McGraw-Hill, New York, NY, 2001.
3. S. Sahni, *Data Structures, Algorithms, and Applications in C++*, McGraw-Hill, NY, 1998.
4. S. Sahni, *Data Structures, Algorithms, and Applications in Java*, McGraw-Hill, NY, 2000.
5. R. Sedgewick, *Algorithms in C++, Parts 1–4*, Third edition. Addison-Wesley, NY, 1998.
6. E. Horowitz, S. Sahni, and D. Mehta, *Fundamentals of Data Structures in C++*, W. H. Freeman, NY, 1995.
7. D. Stanton and D. White, *Constructive Combinatorics*, Springer-Verlag, NY, 1986.

4

Graphs*

4.1	Introduction	49
4.2	Graph Representations	50
	Weighted Graph Representation	
4.3	Connectivity, Distance, and Spanning Trees	51
	Spanning Trees	
4.4	Searching a Graph	54
	Depth-First Search • Breadth-First Search	
4.5	Simple Applications of DFS and BFS	56
	Depth-First Search on a Digraph • Topological Sorting	
4.6	Minimum Spanning Tree	58
	Kruskal's MST Algorithm • Prim's MST Algorithm • Boruvka's MST Algorithm • Constrained MST	
4.7	Shortest Paths	62
	Single-Source Shortest Paths, Nonnegative Weights • Single-Source Shortest Paths, Arbitrary Weights • All-Pairs Shortest Paths	
4.8	Eulerian and Hamiltonian Graphs	65
	Acknowledgments	66
	References	66

Narsingh Deo
University of Central Florida

4.1 Introduction

Trees, as data structures, are somewhat limited because they can only represent relations of a hierarchical nature, such as that of parent and child. A generalization of a tree so that a binary relation is allowed between any pair of elements would constitute a graph—formally defined as follows:

A *graph* $G = (V, E)$ consists of a finite set of *vertices* $V = \{v_1, v_2, \ldots, v_n\}$ and a finite set E of *edges* $E = \{e_1, e_2, \ldots, e_m\}$ (see Figure 4.1). To each edge e there corresponds a pair of vertices (u, v) which e is said to be *incident* on. While drawing a graph we represent each vertex by a dot and each edge by a line segment joining its two *end vertices*. A graph is said to be a *directed graph* (or *digraph* for short) (see Figure 4.2) if the vertex pair (u, v) associated with each edge e (also called *arc*) is an ordered pair. Edge e is then said to be *directed from* vertex u to vertex v, and the direction is shown by an arrowhead on the edge. A graph is *undirected* if the end vertices of all the edges are unordered (i.e., edges have no direction). Throughout this chapter we use the letters n and m to denote the number of vertices $|V|$ and number of edges $|E|$ respectively, in a graph. A vertex is often referred to as a *node* (a term more popular in applied fields).

Two or more edges having the same pair of end vertices are called *parallel edges* or *multi edges*, and a graph with multi edges is sometimes referred to as a *multigraph*. An edge whose two end vertices are the same is called a *self-loop* (or just *loop*). A graph in which neither parallel edges nor self loops are allowed is often called a *simple graph*. If both self-loops and parallel edges are allowed we have a *general graph* (also referred to as *pseudograph*). Graphs in Figures 4.1 and 4.2 are both simple but the graph in Figure 4.3 is pseudograph. If the graph is simple we can refer to each edge by its end vertices. The number of edges incident on a vertex v, with self-loops counted twice, is called the *degree*, *deg(v)*, of vertex v. In directed graphs a vertex has *in-degree* (number of edges going into it) and *out-degree* (number of edges going out of it).

In a digraph if there is a directed edge (x, y) from x to y, vertex y is called a *successor* of x and vertex x is called a *predecessor* of y. In case of an undirected graph two vertices are said to be *adjacent* or *neighbors* if there is an edge between them.

---

* This chapter has been reprinted from first edition of this Handbook, without any content updates.

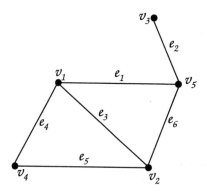

FIGURE 4.1 Undirected graph with 5 vertices and 6 edges.

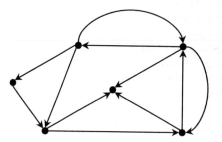

FIGURE 4.2 Digraph with 6 vertices and 11 edges.

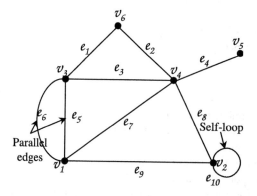

FIGURE 4.3 A pseudograph of 6 vertices and 10 edges.

A *weighted graph* is a (directed or undirected) graph in which a real number is assigned to each edge. This number is referred to as the *weight* of that edge. Weighted directed graphs are often referred to as *networks*. In a practical network this number (weight) may represent the driving distance, the construction cost, the transit time, the reliability, the transition probability, the carrying capacity, or any other such attribute of the edge [1–4].

Graphs are the most general and versatile data structures. Graphs have been used to model and solve a large variety of problems in the discrete domain. In their modeling and problem solving ability graphs are to the discrete world what differential equations are to the world of the continuum.

4.2 Graph Representations

For a given graph a number of different representations are possible. The ease of implementation, as well as the efficiency of a graph algorithm depends on the proper choice of the graph representation. The two most commonly used data structures for representing a graph (directed or undirected) are *adjacency lists* and *adjacency matrix*. In this section we discuss these and other data structures used in representing graphs.

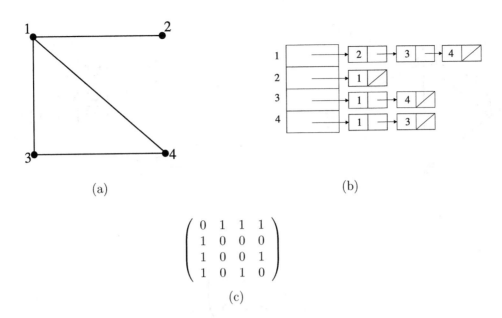

FIGURE 4.4 An undirected graph (a) with four vertices and four edges; (b) its adjacency lists representation, and (c) its adjacency matrix representation.

Adjacency lists: The adjacency lists representation of a graph G consists of an array *Adj* of n linked lists, one for each vertex in G, such that $Adj[\upsilon]$ for vertex υ consists of all vertices adjacent to υ. This list is often implemented as a linked list. (Sometimes it is also represented as a table, in which case it is called the *star representation* [3].)

Adjacency Matrix: The adjacency matrix of a graph $G = (V, E)$ is an $n \times n$ matrix $A = [a_{ij}]$ in which $a_{ij} = 1$ if there is an edge from vertex i to vertex j in G; otherwise $a_{ij} = 0$. Note that in an adjacency matrix a self-loop can be represented by making the corresponding diagonal entry 1. Parallel edges could be represented by allowing an entry to be greater than 1, but doing so is uncommon, since it is usually convenient to represent each element in the matrix by a single bit. The adjacency lists and adjacency matrix of an undirected graph are shown in Figure 4.4, and the corresponding two representations for a digraph are shown in Figure 4.5.

Clearly the memory required to store a graph of n vertices in the form of adjacency matrix is $O(n^2)$, whereas for storing it in the form of its adjacency lists is $O(m + n)$. In general if the graph is sparse, adjacency lists are used but if the graph is dense, adjacency matrix is preferred. The nature of graph processing is an important factor in selecting the data structure.

There are other less frequently used data structures for representing graphs, such as *forward* or *backward star*, the *edge-list*, and *vertex-edge incidence matrix* [1–5].

4.2.1 Weighted Graph Representation

Both adjacency lists and adjacency matrix can be adapted to take into account the weights associated with each edge in the graph. In the former case an additional field is added in the linked list to include the weight of the edge; and in the latter case the graph is represented by a *weight matrix* in which the (i, j)th entry is the weight of edge (i, j) in the weighted graph. These two representations for a weighted graph are shown in Figure 4.6. The boxed numbers next to the edges in Figure 4.6a are the weights of the corresponding edges.

It should be noted that in a weight matrix, W, of a weighted graph, G, if there is no edge (i, j) in G, the corresponding element w_{ij} is usually set to ∞ (in practice, some very large number). The diagonal entries are usually set to ∞ (or to some other value depending on the application and algorithm). It is easy to see that the weight matrix of an undirected graph (like the adjacency matrix) is symmetric.

4.3 Connectivity, Distance, and Spanning Trees

Just as two vertices x and y in a graph are said to be adjacent if there is an edge joining them, two edges are said to be adjacent if they share (i.e., are incident on) a common vertex. A *simple path*, or *path* for short, is a sequence of adjacent edges (υ_1, υ_2),

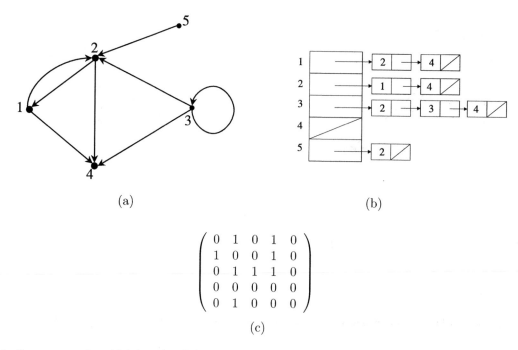

FIGURE 4.5   Two representations: (a) A digraph with five vertices and eight edges; (b) its adjacency lists representation, and (c) its adjacency matrix representation.

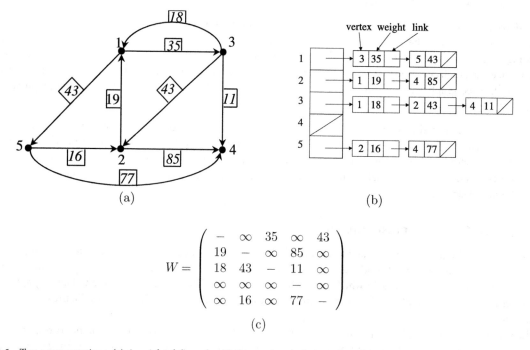

FIGURE 4.6   Two representations: (a) A weighted digraph with five vertices and nine edges; (b) its adjacency lists, and (c) its weight matrix.

$(v_2, v_3), \ldots, (v_{k-2}, v_{k-1}), (v_{k-1}, v_k)$, sometimes written $(v_1, v_2, \ldots, v_k)$, in which all the vertices $v_1, v_2, \ldots, v_k$ are distinct except possibly $v_1 = v_k$. In a digraph this path is said to be *directed from* v_1 to v_k; in an undirected graph this path is said to be *between* v_1 and v_k. The number of edges in a path, in this case, $k - 1$, is called the *length* of the path. In Figure 4.3 sequence $(v_6, v_4), (v_4, v_1), (v_1, v_2) = (v_6, v_4, v_1, v_2)$ is a path of length 3 between v_6 and v_2. In the digraph in Figure 4.6 sequence $(3, 1)$, $(1, 5), (5, 2), (2, 4) = (3, 1, 5, 2, 4)$ is a directed path of length 4 from vertex 3 to vertex 4. A *cycle* or *circuit* is a path in which the

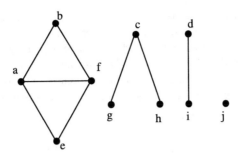

FIGURE 4.7 A disconnected graph of 10 vertices, 8 edges, and 4 components.

first and the last vertices are the same. In Figure 4.3 $(v_3, v_6, v_4, v_1, v_3)$ is a cycle of length 4. In Figure 4.6 (3, 2, 1, 3) is a cycle of length 3. A graph that contains no cycle is called *acyclic*.

A *subgraph* of a graph $G = (V, E)$ is a graph whose vertices and edges are in G. A subgraph g of G is said to be *induced* by a subset of vertices $S \subseteq V$ if g results when the vertices in $V - S$ and all the edges incident on them are removed from G. For example, in Figure 4.3, the subgraph induced by $\{v_1, v_3, v_4\}$ would consists of these three vertices and four edges $\{e_3, e_5, e_6, e_7\}$.

An undirected graph G is said to be *connected* if there is at least one path between every pair of vertices v_i and v_j in G. Graph G is said to be *disconnected* if it has at least one pair of distinct vertices u and v such that there is no path between u and v. Two vertices x and y in an undirected graph $G = (V, E)$ are said to be *connected* if there exists a path between x and y. This relation of being connected is an equivalence relation on the vertex set V, and therefore it partitions the vertices of G into equivalence classes. Each equivalence class of vertices induces a subgraph of G. These subgraphs are called connected components of G. In other words, a connected component is a maximal connected subgraph of G. A connected graph consists of just one component, whereas a disconnected graph consists of several (connected) components. Each of the graphs in Figures 4.1, 4.3, and 4.4 is connected. But the graph given in Figure 4.7 is disconnected, consisting of four components.

Notice that a component may consist of just one vertex such as j in Figure 4.7 with no edges. Such a component (subgraph or graph) is called an *isolated vertex*. Equivalently, a vertex with zero degree is called an isolated vertex. Likewise, a graph (subgraph or component) may consist of just one edge, such as edge (i, d) in Figure 4.7.

One of the simplest and often the most important and useful questions about a given graph G is: Is G connected? And if G is not connected what are its connected components? This question will be taken up in the next section, and an algorithm for determining the connected components will be provided; but first a few more concepts and definitions.

Connectivity in a directed graph G is more involved. A digraph is said to be connected if the undirected graph obtained by ignoring the edge directions in G is connected. A directed graph is said to be *strongly connected* if for every pair of vertices v_i and v_j there exists at least one directed path from v_i to v_j and at least one from v_j to v_i. A digraph which is connected but not strongly connected is called *weakly connected*. A disconnected digraph (like a disconnected undirected graph) consists of connected components; and a weakly-connected digraph consists of strongly-connected components. For example, the connected digraph in Figure 4.5 consists of four strongly-connected components—induced by each of the following subsets of vertices $\{1, 2\}, \{3\}, \{4\}$, and $\{5\}$.

Another important question is that of distance from one vertex to another. The *distance* from vertex a to b is the length of the shortest path (i.e., a path of the smallest length) from a to b, if such a path exists. If no path from a to b exists, the distance is undefined and is often set to ∞. Thus, the distance from a vertex to itself is 0; and the distance from a vertex to an adjacent vertex is 1. In an undirected graph distance from a to b equals the distance from b to a, that is, it is symmetric. It is also not difficult to see that the distances in a connected undirected graph (or a strongly connected digraph) satisfy the triangle inequality. In a connected, undirected (unweighted) graph G, the maximum distance between any pair of vertices is called the diameter of G.

4.3.1 Spanning Trees

A connected, undirected, acyclic (without cycles) graph is called a *tree*, and a set of trees is called a *forest*. We have already seen *rooted* trees and forests of rooted trees in the preceding chapter, but the unrooted trees and forests discussed in this chapter are graphs of a very special kind that play an important role in many applications.

In a connected undirected graph G there is at least one path between every pair of vertices and the absence of a cycle implies that there is at most one such path between any pair of vertices in G. Thus if G is a tree, there is exactly one path between every pair of vertices in G. The argument is easily reversed, and so an undirected graph G is a tree if and only if there is exactly one path between every pair of vertices in G. A tree with n vertices has exactly $(n-1)$ edges. Since $(n-1)$ edges are the fewest possible to connect n points, trees can be thought of as graphs that are *minimally connected*. That is, removing any edge from a tree would disconnect it by destroying the only path between at least one pair of vertices.

A *spanning tree* for a connected graph G is a subgraph of G which is a tree containing every vertex of G. If G is not connected, a set consisting of one spanning tree for each component is called a *spanning forest* of G. To construct a spanning tree (forest) of a given undirected graph G, we examine the edges of G one at a time and retain only those that do not not form a cycle with the edges already selected. Systematic ways of examining the edges of a graph will be discussed in the next section.

4.4 Searching a Graph

It is evident that for answering almost any nontrivial question about a given graph G we must examine every edge (and in the process every vertex) of G at least once. For example, before declaring a graph G to be disconnected we must have looked at every edge in G; for otherwise, it might happen that the one edge we had decided to ignore could have made the graph connected. The same can be said for questions of *separability, planarity*, and other properties [5,6].

There are two natural ways of scanning or searching the edges of a graph as we move from vertex to vertex: (i) once at a vertex v we scan all edges incident on v and then move to an adjacent vertex w, then from w we scan all edges incident on w. This process is continued till all the edges *reachable* from v are scanned. This method of fanning out from a given vertex v and visiting all vertices reachable from v in order of their distances from v (i.e., first visit all vertices at a distance one from v, then all vertices at distances two from v, and so on) is referred to as the *breadth-first search* (BFS) of the graph. (ii) An opposite approach would be, instead of scanning every edge incident on vertex v, we move to an adjacent vertex w (a vertex not visited before) as soon as possible, leaving v with possibly unexplored edges for the time being. In other words, we trace a path through the graph going on to a new vertex whenever possible. This method of traversing the graph is called the *depth-first search* (DFS). Breadth-first and depth-first searches are fundamental methods of graph traversal that form the basis of many graph algorithms [5–8]. The details of these two methods follow.

4.4.1 Depth-First Search

Depth-first search on an undirected graph $G = (V, E)$ explores the graph as follows. When we are "visiting" a vertex $v \in V$, we follow one of the edges (v, w) incident on v. If the vertex w has been previously visited, we return to v and choose another edge. If the vertex w (at the other end of edge (v, w) from v) has not been previously visited, we visit it and apply the process recursively to w. If all the edges incident on v have been thus traversed, we go back along the edge (u, v) that had first led to the current vertex v and continue exploring the edges incident on u. We are finished when we try to back up from the vertex at which the exploration began.

Figure 4.8 illustrates how depth-first search examines an undirected graph G represented as an adjacency lists. We start with a vertex a. From a we traverse the first edge that we encounter, which is (a, b). Since b is a vertex never visited before, we stay at b and traverse the first untraversed edge encountered at b, which is (b, c). Now at vertex c, the first untraversed edge that we find is (c, a). We traverse (c, a) and find that a has been previously visited. So we return to c, marking the edge (c, a) in some way (as a dashed line in Figure 4.8c) to distinguish it from edges like (b, c), which lead to new vertices and shown as the thick lines. Back at vertex c, we look for another untraversed edge and traverse the first one that we encounter, which is (c, d). Once again, since d is a new vertex, we stay at d and look for an untraversed edge. And so on. The numbers next to the vertices in Figure 4.8c show the order in which they were visited; and the numbers next to the edges show the order in which they were traversed.

Depth-first search performed on a connected undirected graph $G = (V, E)$, partitions the edge set into two types: (i) Those that led to new vertices during the search constitute the branches of a spanning tree of G and (ii) the remaining edges in E are called *back edges* because their traversal led to an already visited vertex from which we backed down to the current vertex.

A recursive depth-first search algorithm is given in Figure 4.9. Initially, every vertex x is marked *unvisited* by setting $num[x]$ to 0. Note that in the algorithm shown in Figure 4.9, only the tree edges are kept track of. The time complexity of the depth-first search algorithm is $O(m + n)$, provided the input is in the form of an adjacency matrix.

4.4.2 Breadth-First Search

In breadth-first search we start exploring from a specified vertex s and mark it "visited." All other vertices of the given undirected graph G are marked as "unvisited" by setting $num[] = 0$. Then we visit all vertices adjacent to s (i.e., in the adjacency list of s). Next, we visit all unvisited vertices adjacent to the first vertex in the adjacency list of s. Unlike the depth-first search, in breadth-first search we explore (fan out) from vertices in order in which they themselves were visited. To implement this method of search, we maintain a queue (Q) of visited vertices. As we visit a new vertex for the first time, we place it in (i.e., at the back of) the queue. We take a vertex v from front of the queue and traverse all untraversed edges incident at v—adding to the list of tree edges those

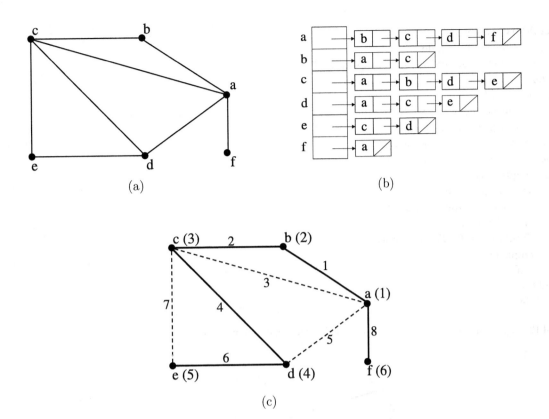

FIGURE 4.8 A graph (a); its adjacency lists (b); and its depth-first traversal (c). The numbers are the order in which vertices were visited and edges traversed. Edges whose traversal led to new vertices are shown with thick lines, and edges that led to vertices that were already visited are shown with dashed lines.

DepthFirstSearch(G)
for each vertex $x \in V$ **do**
  $num[x] \leftarrow 0$
end for
$TreeEdges \leftarrow 0$
$i \leftarrow 0$
for each vertex $x \in V$ **do**
  if $num[x] = 0$ **then**
    DFS-Visit(x)
  end if
end for

DFS-Visit(v)
$i \leftarrow i + 1$
$num[v] \leftarrow i$
for each vertex $w \in Adj[v]$ **do**
  if $num[w] = 0$ **then** {// w is new vertex //}
    $TreeEdges \leftarrow TreeEdges \cup (v, w)$ {// (v, w) is a tree edge //}
    DFS-Visit(w)
  end if
end for

FIGURE 4.9 Algorithm for depth-first search on an undirected graph G.

BreadthFirstSearch(G, s)
for each vertex $x \in V - \{s\}$ **do**
    visited$[x] \leftarrow 0$ {// all vertices unvisited except s //}
end for
TreeEdges $\leftarrow$ *null*
$Q \leftarrow \phi$ {// queue of vertices is initially empty //}
visited$[s] \leftarrow 1$ {// mark s as visited //}
enqueue(Q, s) {// place s in the queue //}
while $Q \neq \phi$ **do** {// queue is not empty //}
    $v \leftarrow$ **dequeue**(Q)
    for each $w \in Adj[v]$ **do**
        if *visited*$[w] = 0$ **then** {// w is a new vertex //}
            visited$[w] \leftarrow 1$
            TreeEdges $\leftarrow$ *TreeEdges* $\cup \{(v, w)\}$
            enqueue(Q, w)
        end if
    end for
end while

FIGURE 4.10   Algorithm for breadth-first search on an undirected graph G from vertex s.

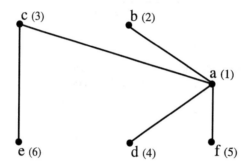

FIGURE 4.11   Spanning tree produced by breadth-first search on graph in Figure 4.8 starting from vertex a. The numbers show the order in which vertices were visited.

edges that lead to unvisited vertices from v ignoring the rest. Once a vertex v has been taken out of the queue, all the neighbors of v have been visited and v is completely explored.

Thus, during the execution of a breadth-first search we have three types of vertices: (i) unvisited, those that have never been in the queue; (ii) completely explored, those that have been in the queue but are not now in the queue; and (iii) visited but not completely explored, that is, those that are currently in the queue.

Since every vertex (reachable from the start vertex s) enters and exits the queue exactly once and every edge in the adjacency list of a vertex is traversed exactly once, the time complexity of the breadth-first search is $O(n + m)$.

An algorithm for performing a breadth-first search on an undirected connected graph G from a specified vertex s is given in Figure 4.10. It produces a breadth-first tree in G rooted at vertex s. For example, the spanning tree produced by BFS conducted on the graph in Figure 4.8 starting at vertex a, is shown in Figure 4.11. The numbers next to the vertices show the order in which the vertices were visited during the BFS.

4.5   Simple Applications of DFS and BFS

In the preceding section we discussed two basic but powerful and efficient techniques for systematically searching a graph such that every edge is traversed exactly once and every vertex is visited once. With proper modifications and embellishments these search techniques can be used to solve a variety of graph problems. Some of the simple ones are discussed in this section.

```
for each vertex v ∈ V do
    compnum[v] ← 0
end for
for each vertex v ∈ V do
    if compnum[v] = 0 then
        c ← c + 1
        COMP(v)
    end if
end for

COMP(x)
compnum[x] ← c
for each w ∈ Adj[x] do
    if compnum[w] = 0 then
        COMP(w)
    end if
end for
```

FIGURE 4.12 Depth-first search algorithm for finding connected components of a graph.

Cycle Detection: The existence of a back edge (i.e., a nontree edge) during a depth-first search indicates the existence of cycle. To test this condition we just add an *else* clause to the **if** $num[w] = 0$ statement in **DFS-Visit**(v) procedure in Figure 4.9. That is, if $num[w] \neq 0$, (v, w) is a back edge, which forms a cycle with tree edges in the path from w to v.

Spanning Tree: If the input graph G for the depth-first (or breadth-first) algorithm is connected, the set TreeEdges at the termination of the algorithm in Figure 4.9 (or in Figure 4.10, for breadth-first) produces a spanning tree of G.

Connected Components: If, on the other hand, the input graph $G = (V, E)$ is disconnected we can use depth-first search to identify each of its connected components by assigning a unique component number $compnum[v]$ to every vertex belonging to one component. The pseudocode of such an algorithm is given below (Figure 4.12).

4.5.1 Depth-First Search on a Digraph

Searching a digraph is somewhat more involved because the direction of the edges is an additional feature that must be taken into account. In fact, a depth-first search on a digraph produces four kinds of edges (rather than just two types for undirected graphs): (i) Tree edges—lead to an unvisited vertex (ii) Back edges—lead to an (visited) ancestor vertex in the tree (iii) Down-edges (also called forward edges) lead to a (visited) descendant vertex in the tree, and (iv) Cross edges, lead to a visited vertex, which is neither ancestor nor descendant in the tree [3,5,6,8,9].

4.5.2 Topological Sorting

The simplest use of the depth-first search technique on digraphs is to determine a labeling of the vertices of an acyclic digraph $G = (V, E)$ with integers $1, 2, \ldots, |V|$, such that if there is a directed edge from vertex i to vertex j, then $i < j$; such a labeling is called *topological sort* of the vertices of G. For example, the vertices of the digraph in Figure 4.13a are topologically sorted but those of Figure 4.13b are not. Topological sorting can be viewed as the process of finding a linear order in which a given partial order can be embedded. It is not difficult to show that it is possible to topologically sort the vertices of a digraph if and only if it is acyclic. Topological sorting is useful in the analysis of activity networks where a large, complex project is represented as a digraph in which the vertices correspond to the goals in the project and the edges correspond to the activities. The topological sort gives an order in which the goals can be achieved [1,3,10].

Topological sorting begins by finding a vertex of $G = (V, E)$ with no outgoing edge (such a vertex must exist if G is acyclic) and assigning this vertex the highest number—namely, $|V|$. This vertex is then deleted from G, along with all its incoming edges. Since the remaining digraph is also acyclic, we can repeat the process and assign the next highest number, namely $|V| - 1$, to a vertex with no outgoing edges, and so on. To keep the algorithm $O(|V| + |E|)$, we must avoid searching the modified digraph for a vertex with no outgoing edges.

We do so by performing a single depth-first search on the given acyclic digraph G. In addition to the usual array *num*, we will need another array, *label*, of size $|V|$ for recording the topologically sorted vertex labels. That is, if there is an edge (u, v) in G,

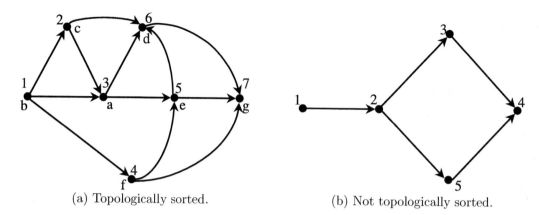

(a) Topologically sorted.   (b) Not topologically sorted.

FIGURE 4.13   Acyclic digraphs.

Topological-Sort(*G*)
for each vertex *x* ∈ *V* **do**
   num[*x*] ← 0
   label[*x*] ← 0
end for
j ← *n* + 1
i ← 0
for each vertex *x* ∈ *V* **do**
   if *num*[*x*] = 0 **then** {// *x* has no labeled ancestor //}
      TOPSORT(*x*)
   end if
end for

TOPSORT(*v*)
i ← *i* + 1
num[*v*] ← *i*
for each *w* ∈ *Adj*[*v*] **do** {// examine all descendants of *w* //}
   if *num*[*w*] = 0 **then**
      TOPSORT(*w*)
   else if *label*[*w*] = 0 **then**
      Error {// cycle detected //}
   end if
   j ← *j* − 1
   label[*v*] ← *j*
end for

FIGURE 4.14   Algorithm for topological sorting.

then *label*[*u*] < *label*[*v*]. The complete search and labeling procedure TOPSORT is given in Figure 4.14. Use the acyclic digraph in Figure 4.13a with vertex set $V = \{a, b, c, d, e, f, g\}$ as the input to the topological sort algorithm in Figure 4.14; and verify that the vertices get relabeled 1 to 7, as shown next to the original names—in a correct topological order.

4.6 Minimum Spanning Tree

How to connect a given set of points with lowest cost is a frequently-encountered problem, which can be modeled as the problem of finding a minimum-weight spanning tree *T* in a weighted, connected, undirected graph $G = (V, E)$. Methods for finding such a spanning tree, called a *minimum spanning tree* (MST), have been investigated in numerous studies and have a long history [11]. In this section we will discuss the bare essentials of the two commonly used MST algorithms—Kruskal's and Prim's—and briefly mention a third one.

4.6.1 Kruskal's MST Algorithm

An algorithm due to J. B. Kruskal, which employs the smallest-edge-first strategy, works as follows: First we sort all the edges in the given network by weight, in nondecreasing order. Then one by one the edges are examined in order, smallest to the largest. If an edge e_i, upon examination, is found to form a cycle (when added to edges already selected) it is discarded. Otherwise, e_i is selected to be included in the minimum spanning tree T. The construction stops when the required $n - 1$ edges have been selected or when all m edges have been examined. If the given network is disconnected, we would get a *minimum spanning forest* (instead of tree). More formally, Kruskal's method may be stated as follows:

$T \leftarrow \phi$
while $|T| < (n - 1)$ and $E \neq \phi$ **do**
  $e \leftarrow$ smallest edge in E
  $E \leftarrow E - \{e\}$
  if $T \cup \{e\}$ has no cycle **then**
    $T \leftarrow T \cup \{e\}$
  end if
end while
if $|T| < (n - 1)$ **then**
  write 'network disconnected'
end if

Although the algorithm just outlined is simple enough, we do need to work out some implementation details and select an appropriate data structure for achieving an efficient execution.

There are two crucial implementations details that we must consider in this algorithm. If we initially sort all m edges in the given network, we may be doing a lot of unnecessary work. All we really need is to be able to to determine the next smallest edge in the network at each iteration. Therefore, in practice, the edges are only partially sorted and kept as a heap with smallest edge at the root of a min heap. In a graph with m edges, the initial construction of the heap would require $O(m)$ computational steps; and the next smallest edge from a heap can be obtained in $O(\log m)$ steps. With this improvement, the sorting cost is $O(m + p \log m)$, where p is the number of edges examined before an MST is constructed. Typically, p is much smaller than m.

The second crucial detail is how to maintain the edges selected (to be included in the MST) so far, such that the next edge to be examined can be efficiently tested for a cycle formation.

As edges are examined and included in T, a forest of disconnected trees (i.e., subtrees of the final spanning tree) is produced. The edge e being examined will form a cycle if and only if both its end vertices belong to the same subtree in T. Thus to ensure that the edge currently being examined does not form a cycle, it is sufficient to check if it connects two different subtrees in T. An efficient way to accomplish this is to group the n vertices of the given network into disjoint subsets defined by the subtrees (formed by the edges included in T so far). Thus if we maintain the partially constructed MST by means of subsets of vertices, we can add a new edge by forming the UNION of two relevant subsets, and we can check for cycle formation by FINDing if the two end vertices of the edge, being examined, are in the same subset. These subsets can themselves be kept as rooted trees. The root is an element of the subset and is used as a name to identify that subset. The FIND subprocedure is called twice—once for each end vertex of edge e—to determine the sets to which the two end vertices belong. If they are different, the UNION subprocedure will merge the two subsets. (If they are the same subset, edge e will be discarded.)

The subsets, kept as rooted trees, are implemented by keeping an array of *parent* pointers for each of the n elements. Parent of a root, of course, is null. (In fact, it is useful to assign $parent[root] = -$ number of vertices in the tree.) While taking the UNION of two subsets, we merge the smaller subset into the larger one by pointing the parent pointer in the root of the smaller subset to the root of the larger subset. Some of these details are shown in Figure 4.15. Note that $r1$ and $r2$ are the roots identifying the sets to which vertices u and v belong.

When algorithm in Figure 4.15 is applied to the weighted graph in Figure 4.16, the order in which edges are included one by one to form the MST are (3, 5), (4, 6), (4, 5), (4, 2), (6, 7), (3, 1). After the first five smallest edges are included in the MST, the 6th and 7th and 8th smallest edges are rejected. Then the 9th smallest edge (1, 3) completes the MST and the last two edges are ignored.

4.6.2 Prim's MST Algorithm

A second algorithm, discovered independently by several people (Jarnik in 1936, Prim in 1957, Dijkstra in 1959) employs the *"nearest neighbor"* strategy and is commonly referred to as Prim's algorithm. In this method one starts with an arbitrary vertex s and joins it to its nearest neighbor, say y. That is, of all edges incident on vertex s, edge (s, y), with the smallest weight, is made

INITIALIZATION:
set parent array to −1 {// n vertices from singleton sets //}
form initial heap of *m* edges
ecount ← 0 {// number of edges examined so far //}
tcount ← 0 {// number of edges in *T* so far //}
T ← φ

ITERATION:
while *tcount* < (*n* − 1) and *ecount* < *m* **do**
  e ← edge(*u*, *v*) from top of heap
  ecount ← *ecount* + 1
  remove *e* from heap
  restore heap
  r1 ← FIND(*u*)
  r2 ← FIND(*v*)
  if *r1* ≠ *r2* **then**
    T ← *T* ∪ {*e*}
    tcount ← *tcount* + 1
    UNION(*r1*, *r2*)
  end if
end while
if *tcount* < (*n* − 1) **then**
  write 'network disconnected'
end if

FIGURE 4.15   Kruskal's minimum spanning tree algorithm.

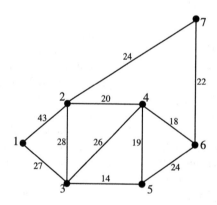

FIGURE 4.16   A connected weighted graph for MST algorithm.

part of the MST. Next, of all the edges incident on *s* or *y* we choose one with minimum weight that leads to some third vertex, and make this edge part of the MST. We continue this process of "reaching out" from the partially constructed tree (so far) and bringing in the "nearest neighbor" until all vertices reachable from *s* have been incorporated into the tree.

As an example, let us use this method to find the minimum spanning tree of the weighted graph given in Figure 4.16. Suppose that we start at vertex 1. The nearest neighbor of vertex 1 is vertex 3. Therefore, edge (1, 3) becomes part of the MST. Next, of all the edges incident on vertices 1 and 3 (and not included in the MST so far) we select the smallest, which is edge (3, 5) with weight 14. Now the partially constructed tree consists of two edges (1, 3) and (3, 5). Among all edges incident at vertices 1, 3, and 5, edge (5, 4) is the smallest, and is therefore included in the MST. The situation at this point is shown in Figure 4.17. Clearly, (4, 6), with weight 18 is the next edge to be included. Finally, edges (4, 2) and (6, 7) will complete the desired MST.

The primary computational task in this algorithm is that of finding the next edge to be included into the MST in each iteration. For each efficient execution of this task we will maintain an array *near*[*u*] for each vertex *u* not yet in the tree (i.e., *u* ∈ *V* − *V*$_T$).

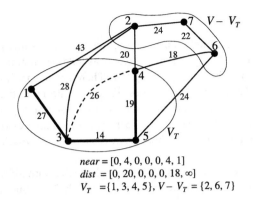

$$near = [0, 4, 0, 0, 0, 4, 1]$$
$$dist = [0, 20, 0, 0, 0, 18, \infty]$$
$$V_T = \{1, 3, 4, 5\}, V - V_T = \{2, 6, 7\}$$

FIGURE 4.17 Partially constructed MST for the network of Figure 4.16.

INITIALIZATION:
choose starting vertex s arbitrarily
for every vertex i other than s **do**
    $near[i] \leftarrow s$
    $dist[i] \leftarrow w_{si}$
end for
$V_T \leftarrow \{s\}$ {// set of vertices in MST so far //}
$E_T \leftarrow \phi$ {// set of edges in MST so far //}

ITERATION:
while $|V_T| < n$ **do**
    $u \leftarrow$ vertex in $(V - V_T)$ with smallest value of dist(u)
    if $dist[u] \geq \infty$ **then**
        write 'graph disconnected' and exit
    end if
    $E_T \leftarrow E_T \cup \{(u, near[u])\}$
    $V_T \leftarrow V_T \cup \{u\}$
    for $x \in (V - V_T)$ **do**
        if $w_{ux} < dist[x]$ **then**
            $dist[x] \leftarrow w_{ux}$
            $near[x] \leftarrow u$
        end if
    end for
end while

FIGURE 4.18 Prim's minimum spanning tree algorithm.

$near[u]$ is that vertex in V_T which is closest to u. (Note that V is the set of all vertices in the network and V_T is the subset of V included in MST thus far.) Initially, we set $near[s] \leftarrow 0$ to indicate that s is in the tree, and for every other vertex v, $near[v] \leftarrow s$.

For convenience, we will maintain another array $dist[u]$ of the actual distance (i.e., edge weight) to that vertex in V_T which is closest to u. In order to determine which vertex is to be added to the set V_T next, we compare all nonzero values in $dist$ array and pick the smallest. Thus $n - i$ comparisons are sufficient to identify the ith vertex to be added. Initially, since s is the only vertex in V_T, $dist[u]$ is set to w_{su}. As the algorithm proceeds, these two arrays are updated in each iteration (see Figure 4.17 for an illustration).

A formal description of the nearest-neighbor algorithm is given in Figure 4.18. It is assumed that the input is given in the form of an $n \times n$ weight matrix W (in which nonexistent edges have ∞ weights). Set $V = \{1, 2, \ldots, n\}$ is the set of vertices of the graph. V_T and E_T are the sets of vertices and edges of the partially formed (minimum spanning) tree. Vertex set V_T is identified by zero entries in array $near$.

4.6.3 Boruvka's MST Algorithm

There is yet a third method for computing a minimum spanning tree, which was first proposed by O. Boruvka in1,926(but rediscovered by G. Chouqet in 1938 and G. Sollin in 1961). It works as follows: First, the smallest edge incident on each vertex is found; these edges form part of the minimum spanning tree. There are at least $\lceil n/2 \rceil$ such edges. The connected components formed by these edges are collapsed into "supernodes" (There are no more than $\lfloor n/2 \rfloor$ such vertices at this point.) The process is repeated on "supernodes" and then on the resulting "supersupernodes," and so on, until only a single vertex remains. This will require at most $\lfloor \log_2 n \rfloor$ steps, because at each step the number of vertices is reduced at least by a factor of 2. Because of its inherent parallelism the nearest-neighbor-from-each-vertex approach is particularly appealing for parallel implementations.

These three "greedy" algorithms and their variations have been implemented with different data structures and their relative performance—both theoretical as well as empirical—have been studied widely. The results of some of these studies can be found in [6,12–14].

4.6.4 Constrained MST

In many applications, the minimum spanning tree is required to satisfy an additional constraint, such as (i) the degree of each vertex in the MST should be equal to or less than a specified value; or (ii) the diameter of the MST should not exceed a specified value; or (iii) the MST must have at least a specified number of leaves (vertices of degree 1 in a tree); and the like. The problem of computing such a *constrained minimum spanning tree* is usually NP-complete. For a discussion of various constrained MST problems and some heuristics solving them see [15].

4.7 Shortest Paths

In the preceding section we dealt with the problem of connecting a set of points with smallest cost. Another commonly encountered and somewhat related problem is that of finding the lowest-cost path (called shortest path) between a given pair of points. There are many types of shortest-path problems. For example, determining the shortest path (i.e., the most economical path or fastest path, or minimum-fuel-consumption path) from one specified vertex to another specified vertex; or shortest paths from a specified vertex to all other vertices; or perhaps shortest path between all pairs of vertices. Sometimes, one wishes to find a shortest path from one given vertex to another given vertex that passes through certain specified intermediate vertices. In some applications, one requires not only the shortest but also the second and third shortest paths. Thus, the shortest-path problems constitute a large class of problems; particularly if we generalize it to include related problems, such as the longest-path problems, the most-reliable-path problems, the largest-capacity-path problems, and various routing problems. Therefore, the number of papers, books, reports, dissertations, and surveys dealing with the subject of shortest paths runs into hundreds [16].

Here we will discuss two very basic and important shortest-path problems: (i) how to determine the shortest distance (and a shortest path) from a specified vertex s to another specified vertex t, and (ii) how to determine shortest distances (and paths) from every vertex to every other vertex in the network. Several other problems can be solved using these two basic algorithms.

4.7.1 Single-Source Shortest Paths, Nonnegative Weights

Let us first consider a classic algorithm due to Dijkstra for finding a shortest path (and its weight) from a specified vertex s (*source* or *origin*) to another specified vertex t (*target* or *sink*) in a network G in which all edge weights are nonnegative. The basic idea behind Dijkstra's algorithm is to fan out from s and proceed toward t (following the directed edges), labeling the vertices with their distances from s obtained so far. The label of a vertex u is made *permanent* once we know that it represents the shortest possible distance from s (to u). All vertices not permanently labeled have *temporary* labels.

We start by giving a permanent label 0 to source vertex s, because zero is the distance of s from itself. All other vertices get labeled ∞, temporarily, because they have not been reached yet. Then we label each immediate successor v of source s, with temporary labels equal to the weight of the edge (s, v). Clearly, the vertex, say x, with smallest temporary label (among all its immediate successors) is the vertex closest to s. Since all edges have nonnegative weights, there can be no shorter path from s to x. Therefore, we make the label of x permanent. Next, we find all immediate successors of vertex x, and shorten their temporary labels if the path from s to any of them is shorter by going through x (than it was without going through x). Now, from among all temporarily labeled vertices we pick the one with the smallest label, say vertex y, and make its label permanent. This vertex y is the second closest vertex from s. Thus, at each iteration, we reduce the values of temporary labels whenever possible (by selecting a shorter path through the most recent permanently labeled vertex), then select the vertex with the smallest temporary label

INITIALIZATION:
for all $v \in V$ **do**
   $dist[v] \leftarrow \infty$
   $final[v] \leftarrow false$
   $pred[v] \leftarrow -1$
end for
$dist[s] \leftarrow 0$
$final[s] \leftarrow true$
$recent \leftarrow s$
{// vertex s is permanently labeled with 0. All other vertices are temporarily labeled with ∞. Vertex s is the most recent vertex to be permanently labeled //}

ITERATION:
while $final[t] = false$ **do**
   for every immediate successor v of $recent$ **do**
      if not $final[v]$ **then** {// update temporary labels //}
         $newlabel \leftarrow dist[recent] + w_{recent,v}$
        if $newlabel < dist[v]$ **then**
           $dist[v] \leftarrow newlabel$
           $pred[v] \leftarrow recent$ {// relabel v if there is a shorter path via vertex $recent$ and make $recent$ the predecessor of v on the
           shortest path from s //}
        end if
      end if
   end for
   let y be the vertex with the smallest temporary label, which is $\neq \infty$
   $final[y] \leftarrow true$
   $recent \leftarrow y$ {// y, the next closest vertex to s gets permanently labeled //}
end while

FIGURE 4.19 Dijkstra's shortest-path algorithm.

and make it permanent. We continue in this fashion until the target vertex t gets permanently labeled. In order to distinguish the permanently labeled vertices from the temporarily labeled ones, we will keep a boolean array *final* of order n. When the ith vertex becomes permanently labeled, the ith element of this array changes from *false* to *true*. Another array, *dist*, of order n will be used to store labels of vertices. A variable *recent* will be used to keep track of most recent vertex to be permanently labeled.

Assuming that the network is given in the form of a weight matrix $W = [w_{ij}]$, with ∞ weights for nonexistent edges, and vertices s and t are specified, this algorithm (which is called *Dijkstra's shortest-path* or the *label-setting algorithm*) may be described as follows (Figure 4.19).

4.7.2 Single-Source Shortest Paths, Arbitrary Weights

In Dijkstra's shortest-path algorithm (Figure 4.19), it was assumed that all edge weights w_{ij} were nonnegative numbers. If some of the edge weights are negative, Dijkstra's algorithm will not work. (Negative weights in a network may represent costs and positive ones, profit.) The reason for the failure is that once the label of a vertex is made permanent, it cannot be changed in future iterations. In order to handle a network that has both positive and negative weights, we must ensure that no label is considered permanent until the program halts. Such an algorithm (called a *label-correcting method*, in contrast to Dijkstra's *label-setting method*) is described as below.

Like Dijkstra's algorithm, the label of the starting vertex s is set to zero and that of every other vertex is set to ∞, a very large number. That is, the initialization consists of

$dist(s) \leftarrow 0$
for all $v \neq s$ **do**
   $dist(v) \leftarrow \infty$
end for

In the iterative step, $dist(v)$ is always updated to the currently known distance from s to v, and the predecessor $pred(v)$ of v is also updated to be the predecessor vertex of v on the currently known shortest path from s to v. More compactly, the iteration may be expressed as follows:

while $\exists$ an edge (u, v) such that $dist(u) + w_{u,v} < dist(v)$ **do**
    $dist(v) \leftarrow dist(u) + w_{u,v}$
    $pred(v) \leftarrow u$
end while

Several implementations of this basic iterative step have been studied, experimented with, and reported in the literature. One very efficient implementation, works as follows.

We maintain a queue of "vertices to be examined." Initially, this queue, Q, contains only the starting vertex s. The vertex u from the front of the queue is "examined" (as follows) and deleted. Examining u consists of considering all edges (u, v) going out of u. If the length of the path to vertex v (from s) is reduced by going through u, that is,

if $dist(u) + w_{u,v} < dist(v)$ **then**
    $dist(v) \leftarrow dist(u) + w_{u,v}$ {// $dist(v)$ is reset to the smaller value //}
    $pred(v) \leftarrow u$
end if

Moreover, this vertex v is added to the queue (if it is not already in the queue) as a vertex to be examined later. Note that v enters the queue only if $dist(v)$ is decremented as above and if v is currently not in the queue. Observe that unlike in Dijkstra's method (the label-setting method) a vertex may enter (and leave) the queue several times—each time a shorter path is discovered. It is easy to see that the label-correcting algorithm will not terminate if the network has a cycle of negative weight.

4.7.3 All-Pairs Shortest Paths

We will now consider the problem of finding a shortest path between every pair of vertices in the network. Clearly, in an n-vertex directed graph there are $n(n-1)$ such paths—one for each ordered pair of distinct vertices—and $n(n-1)/2$ paths in an undirected graph. One could, of course, solve this problem by repeated application of Dijkstra's algorithm, once for each vertex in the network taken as the source vertex s. We will instead consider a different algorithm for finding shortest paths between all pairs of vertices, which is known as *Warshall-Floyd algorithm*. It requires computation time proportional to n^3, and allows some of the edges to have negative weights, as long as no cycles of net negative weight exist.

The algorithm works by inserting one or more vertices into paths, whenever it is advantageous to do so. Starting with $n \times n$ weight matrix $W = [w_{ij}]$ of direct distances between the vertices of the given network G, we construct a sequence of n matrices $W^{(1)}, W^{(2)}, \ldots, W^{(n)}$. Matrix $W^{(l)}$, $1 \leq l \leq n$, may be thought of as the matrix whose (i, j)th entry $w_{ij}^{(l)}$ gives the length of the shortest path among all paths from i to j with vertices $1, 2, \ldots, l$ allowed as intermediate vertices. Matrix $W^{(l)} = w_{ij}^{(l)}$ is constructed as follows:

$$w_{ij}^{(0)} = w_{ij}$$

$$w_{ij}^{(l)} = \min\left\{w_{ij}^{(l-1)}, w_{il}^{(l-1)} + w_{lj}^{(l-1)}\right\} \quad \text{for } l = 1, 2, \ldots, n \tag{4.1}$$

In other words, in iteration 1, vertex 1 is inserted in the path from vertex i to vertex j if $w_{ij} > w_{i1} + w_{1j}$. In iteration 2, vertex 2 can be inserted, and so on.

For example, in Figure 4.6 the shortest path from vertex 2 to 4 is 2–1–3–4; and the following replacements occur:

$$\text{Iteration 1}: \quad w_{23}^{(0)} \text{ is replaced by } \left(w_{21}^{(0)} + w_{13}^{(0)}\right)$$

$$\text{Iteration 2}: \quad w_{24}^{(2)} \text{ is replaced by } \left(w_{23}^{(2)} + w_{34}^{(2)}\right)$$

Once the shortest distance is obtained in $w_{23}^{(3)}$, the value of this entry will not be altered in subsequent operations.

We assume as usual that the weight of a nonexistent edge is ∞, that $x + \infty = \infty$, and that $\min\{x, \infty\} = x$ for all x. It can easily be seen that all distance matrices $W^{(l)}$ calculated from Equation 4.1 can be overwritten on W itself. The algorithm may be stated as follows:

If the network has no negative-weight cycle, the diagonal entries $w_{ii}^{(n)}$ represent the length of shortest cycles passing through vertex i. The off-diagonal entries $w_{ij}^{(n)}$ are the shortest distances. Notice that negative weight of an individual edge has no effect on this algorithm as long as there is no cycle with a net negative weight.

```
for l ← 1 to n do
    for i ← 1 to n do
        if w_il ≠ ∞ then
            for j ← 1 to n do
                w_ij ← min{w_ij, w_il + w_lj}
            end for
        end if
    end for
end for
```

FIGURE 4.20 All-pairs shortest distance algorithm.

Note that the algorithm in Figure 4.20 does not actually list the paths, it only produces their costs or weights. Obtaining paths is slightly more involved than it was in algorithm in Figure 4.19 where a predecessor array *pred* was sufficient. Here the paths can be constructed from a *path matrix* $P = [p_{ij}]$ (also called *optimal policy matrix*), in which p_{ij} is the second to the last vertex along the shortest path from i to j—the last vertex being j. The path matrix P is easily calculated by adding the following steps in Figure 4.20. Initially, we set

$$p_{ij} \leftarrow i, \quad \text{if } w_{ij} \neq \infty, \quad \text{and}$$
$$p_{ij} \leftarrow 0, \quad \text{if } w_{ij} = \infty.$$

In the *l*th iteration if vertex l is inserted between i and j; that is, if $w_{il} + w_{lj} < w_{ij}$, then we set $p_{ij} \leftarrow p_{lj}$. At the termination of the execution, the shortest path $(i, v_1, v_2, \ldots, v_q, j)$ from i to j can be obtained from matrix P as follows:

$$v_q = p_{ij}$$
$$v_{q-1} = p_{i,v_q}$$
$$v_{q-2} = p_{i,v_{q-1}}$$
$$\vdots$$
$$i = p_{i,v_1}$$

The storage requirement is n^2, no more than for storing the weight matrix itself. Since all the intermediate matrices as well as the final distance matrix are overwritten on W itself. Another n^2 storage space would be required if we generated the path matrix P also. The computation time for the algorithm in Figure 4.20 is clearly $O(n^3)$, regardless of the number of edges in the network.

4.8 Eulerian and Hamiltonian Graphs

A path when generalized to include visiting a vertex more than once is called a *trail*. In other words, a trail is a sequence of edges $(v_1, v_2), (v_2, v_3), \ldots, (v_{k-2}, v_{k-1}), (v_{k-1}, v_k)$ in which all the vertices $(v_1, v_2, \ldots, v_k)$ may not be distinct but all the edges are distinct. Sometimes a trail is referred to as a (non-simple) path and path is referred to as a simple path. For example in Figure 4.8a $(b, a), (a, c), (c, d), (d, a), (a, f)$ is a trail (but not a simple path because vertex a is visited twice.

If the first and the last vertex in a trail are the same, it is called a *closed trail*, otherwise an *open trail*. An *Eulerian trail* in a graph $G = (V, E)$ is one that includes every edge in E (exactly once). A graph with a closed Eulerian trail is called a *Eulerian graph*. Equivalently, in an Eulerian graph, G, starting from a vertex one can traverse every edge in G exactly once and return to the starting vertex. According to a theorem proved by Euler in 1736, (considered the beginning of graph theory), a connected graph is Eulerian if and only if the degree of its every vertex is even.

Given a connected graph G it is easy to check if G is Eulerian. Finding an actual Eulerian trail of G is more involved. An efficient algorithm for traversing the edges of G to obtain an Euler trail was given by Fleury. The details can be found in Reference 4.

A cycle in a graph G is said to be *Hamiltonian* if it passes through every vertex of G. Many families of special graphs are known to be Hamiltonian, and a large number of theorems have been proved that give sufficient conditions for a graph to be Hamiltonian. However, the problem of determining if an arbitrary graph is Hamiltonian is NP-complete.

Graph theory, a branch of combinatorial mathematics, has been studied for over two centuries. However, its applications and algorithmic aspects have made enormous advances only in the past fifty years with the growth of computer technology and operations research. Here we have discussed just a few of the better-known problems and algorithms. Additional material is available in the references provided. In particular, for further exploration the Stanford GraphBase [17], the LEDA [18], and the Graph Boost Library [19] provide valuable and interesting platforms with collection of graph-processing programs and benchmark databases.

Acknowledgments

The author gratefully acknowledges the help provided by Hemant Balakrishnan in preparing this chapter.

References

1. R. K. Ahuja, T. L. Magnanti, and J. B. Orlin, *Network Flows: Theory, Algorithms, and Applications*, Prentice Hall, 1993.
2. N. Deo, *Graph Theory with Applications in Engineering and Computer Science*, Prentice-Hall, 1974.
3. M. M. Syslo, N. Deo, and J. S. Kowalik, *Discrete Optimization Algorithms:With Pascal Programs*, Prentice-Hall, 1983.
4. K. Thulasiraman and M. N. S. Swamy, *Graphs: Theory and Algorithms*, Wiley-Interscience, 1992.
5. E. M. Reingold, J. Nievergelt, and N. Deo, *Combinatorial Algorithms: Theory and Practice*, Prentice-Hall, 1977.
6. R. Sedgewick, *Algorithms in C: Part 5 Graph Algorithms*, Addison-Wesley, third edition, 2002.
7. H. N. Gabow, Path-based depth-first search for strong and biconnected components, *Information Processing*, Vol. 74, pp. 107–114, 2000.
8. R. E. Tarjan, *Data Structures and Network Algorithms*, Society for Industrial and Applied Mathematics, 1983.
9. T. H. Cormen, C. L. Leiserson, and R. L. Rivest, *Introduction to Algorithms*, MIT Press and McGraw-Hill, 1990.
10. E. Horowitz, S. Sahni, and B. Rajasekaran, *Computer Algorithms/C++*, Computer Science Press, 1996.
11. R. L. Graham and P. Hell, On the history of minimum spanning tree problem, *Annals of the History of Computing*, Vol. 7, pp. 43–57, 1985.
12. B. Chazelle, A minimum spanning tree algorithm with inverse Ackermann type complexity, *Journal of the ACM*, Vol. 47, pp. 1028–1047, 2000.
13. B. M. E. Moret and H. D. Shapiro, An empirical analysis of algorithms for constructing minimum spanning tree, *Lecture Notes on Computer Science*, Vol. 519, pp. 400–411, 1991.
14. C. H. Papadimitriou and K. Steiglitz, *Combinatorial Optimization: Algorithms and Complexity*, Rentice-Hall, 1982.
15. N. Deo and N. Kumar, Constrained spanning tree problem: Approximate methods and parallel computation, *American Math Society*, Vol. 40, pp. 191–217, 1998.
16. N. Deo and C. Pang, Shortest path algorithms: Taxonomy and annotation, *Networks*, Vol. 14, pp. 275–323, 1984.
17. D. E. Knuth, *The Stanford GraphBase: A Platform for Combinatorial Computing*, Addison-Wesley, 1993.
18. K. Mehlhorn and S. Naher, *LEDA: APlatform forCombinatorial and Geometric Computing*, Cambridge University Press, 1999.
19. J. G. Siek, L. Lee, and A. Lumsdaine, *The Boost Graph Library—User Guide and Reference Manual*, Addison Wesley, 2002.
20. K. Mehlhorn, *Data Structures and Algorithms 2: NP-Completeness and Graph Algorithms*, Springer-Verlag, 1984.

II

Priority Queues

5 **Leftist Trees** Sartaj Sahni 69
   Introduction • Height-Biased Leftist Trees • Weight-Biased Leftist Trees • Acknowledgments •
   References

6 **Skew Heaps** C. Pandu Rangan 77
   Introduction • Basics of Amortized Analysis • Meldable Priority Queues and Skew Heaps •
   Bibliographic Remarks • References

7 **Binomial, Fibonacci, and Pairing Heaps** Michael L. Fredman 85
   Introduction • Binomial Heaps • Fibonacci Heaps • Pairing Heaps • Related Developments •
   References

8 **Double-Ended Priority Queues** Sartaj Sahni 97
   Definition and an Application • Symmetric Min-Max Heaps • Interval Heaps • Min-Max Heaps •
   Deaps • Generic Methods for DEPQs • Meldable DEPQs • Acknowledgment • References

Leftist Trees [*]

5.1 Introduction..69
5.2 Height-Biased Leftist Trees..70
    Definition • Insertion Into a Max HBLT • Deletion of Max Element from a Max HBLT •
    Melding Two Max HBLTs • Initialization • Deletion of Arbitrary Element from a Max HBLT
5.3 Weight-Biased Leftist Trees..74
    Definition • Max WBLT Operations

Sartaj Sahni
University of Florida

Acknowledgments 75
References...75

5.1 Introduction

A single-ended priority queue (or simply, a priority queue) is a collection of elements in which each element has a priority. There are two varieties of priority queues—max and min. The primary operations supported by a max (min) priority queue are (a) find the element with maximum (minimum) priority, (b) insert an element, and (c) delete the element whose priority is maximum (minimum). However, many authors consider additional operations such as (d) delete an arbitrary element (assuming we have a pointer to the element), (e) change the priority of an arbitrary element (again assuming we have a pointer to this element), (f) meld two max (min) priority queues (i.e., combine two max (min) priority queues into one), and (g) initialize a priority queue with a nonzero number of elements.

Several data structures: for example, heaps (Chapter 3), leftist trees [1,2], Fibonacci heaps [3] (Chapter 7), binomial heaps [4] (Chapter 7), skew heaps [5] (Chapter 6), and pairing heaps [6] (Chapter 7) have been proposed for the representation of a priority queue. The different data structures that have been proposed for the representation of a priority queue differ in terms of the performance guarantees they provide. Some guarantee good performance on a per operation basis while others do this only in the amortized sense. Max (min) heaps permit one to delete the max (min) element and insert an arbitrary element into an n element priority queue in $O(\log n)$ time per operation; a find max (min) takes $O(1)$ time. Additionally, a heap is an implicit data structure that has no storage overhead associated with it. All other priority queue structures are pointer-based and so require additional storage for the pointers.

Max (min) leftist trees also support the insert and delete max (min) operations in $O(\log n)$ time per operation and the find max (min) operation in $O(1)$ time. Additionally, they permit us to meld pairs of priority queues in logarithmic time.

The remaining structures do not guarantee good complexity on a per operation basis. They do, however, have good amortized complexity. Using Fibonacci heaps, binomial queues, or skew heaps, find max (min), inserts and melds take $O(1)$ time (actual and amortized) and a delete max (min) takes $O(\log n)$ amortized time. When a pairing heap is used, the amortized complexity is $O(1)$ for find max (min) and insert (provided no decrease key operations are performed) and $O(\log n)$ for delete max (min) operations [7]. Jones [8] gives an empirical evaluation of many priority queue data structures.

In this chapter, we focus on the leftist tree data structure. Two varieties of leftist trees–height-biased leftist trees [2] and weight-biased leftist trees [1] are described. Both varieties of leftist trees are binary trees that are suitable for the representation of a single-ended priority queue. When a max (min) leftist tree is used, the traditional single-ended priority queue operations– find max (min) element, delete/remove max (min) element, and insert an element–take, respectively, $O(1)$, $O(\log n)$ and $O(\log n)$ time each, where n is the number of elements in the priority queue. Additionally, an n-element max (min) leftist tree can be initialized in $O(n)$ time and two max (min) leftist trees that have a total of n elements may be melded into a single max (min) leftist tree in $O(\log n)$ time.

---

[*] This chapter has been reprinted from first edition of this Handbook, without any content updates.

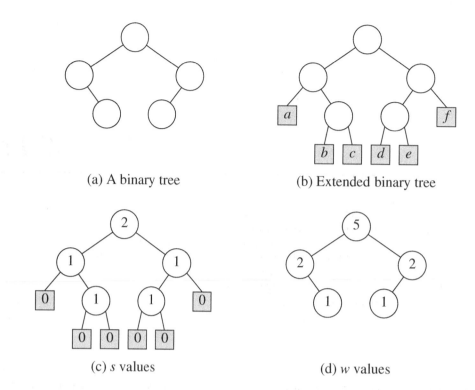

(a) A binary tree

(b) Extended binary tree

(c) *s* values

(d) *w* values

FIGURE 5.1 *s* and *w* values. (a) A binary tree. (b) Extended binary tree.

5.2 Height-Biased Leftist Trees

5.2.1 Definition

Consider a binary tree in which a special node called an **external node** replaces each empty subtree. The remaining nodes are called **internal nodes**. A binary tree with external nodes added is called an **extended binary tree**. Figure 5.1a shows a binary tree. Its corresponding extended binary tree is shown in Figure 5.1b. The external nodes appear as shaded boxes. These nodes have been labeled *a* through *f* for convenience.

Let $s(x)$ be the length of a shortest path from node x to an external node in its subtree. From the definition of $s(x)$, it follows that if x is an external node, its s value is 0. Furthermore, if x is an internal node, its s value is

$$\min\{s(L), s(R)\} + 1$$

where L and R are, respectively, the left and right children of x. The s values for the nodes of the extended binary tree of Figure 5.1b appear in Figure 5.1c.

Definition 5.1: (Crane [2]).

A binary tree is a **height-biased leftist tree (HBLT)** iff at every internal node, the s value of the left child is greater than or equal to the s value of the right child.

The binary tree of Figure 5.1a is not an HBLT. To see this, consider the parent of the external node *a*. The s value of its left child is 0, while that of its right is 1. All other internal nodes satisfy the requirements of the HBLT definition. By swapping the left and right subtrees of the parent of *a*, the binary tree of Figure 5.1a becomes an HBLT.

Theorem 5.1

Let x be any internal node of an HBLT.

 a. The number of nodes in the subtree with root x is at least $2^{s(x)} - 1$.
 b. If the subtree with root x has m nodes, $s(x)$ is at most $\log_2(m+1)$.

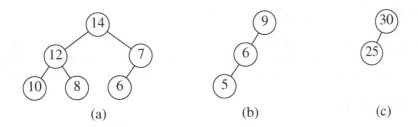

FIGURE 5.2 Some max trees.

c. The length, *rightmost(x)*, of the right-most path from *x* to an external node (i.e., the path obtained by beginning at *x* and making a sequence of right-child moves) is $s(x)$.

Proof. From the definition of $s(x)$, it follows that there are no external nodes on the $s(x) - 1$ levels immediately below node *x* (as otherwise the *s* value of *x* would be less). The subtree with root *x* has exactly one node on the level at which *x* is, two on the next level, four on the next, ..., and $2^{s(x)-1}$ nodes $s(x) - 1$ levels below *x*. The subtree may have additional nodes at levels more than $s(x) - 1$ below *x*. Hence the number of nodes in the subtree *x* is at least $\sum_{i=0}^{s(x)-1} 2^i = 2^{s(x)} - 1$. Part (b) follows from (a). Part (c) follows from the definition of *s* and the fact that, in an HBLT, the *s* value of the left child of a node is always greater than or equal to that of the right child. ■

Definition 5.2

A **max tree (min tree)** is a tree in which the value in each node is greater (less) than or equal to those in its children (if any).

Some max trees appear in Figure 5.2, and some min trees appear in Figure 5.3. Although these examples are all binary trees, it is not necessary for a max tree to be binary. Nodes of a max or min tree may have an arbitrary number of children.

Definition 5.3

A **max HBLT** is an HBLT that is also a max tree. A **min HBLT** is an HBLT that is also a min tree.

The max trees of Figure 5.2 as well as the min trees of Figure 5.3 are also HBLTs; therefore, the trees of Figure 5.2 are max HBLTs, and those of Figure 5.3 are min HBLTs. A max priority queue may be represented as a max HBLT, and a min priority queue may be represented as a min HBLT.

5.2.2 Insertion Into a Max HBLT

The insertion operation for max HBLTs may be performed by using the max HBLT meld operation, which combines two max HBLTs into a single max HBLT. Suppose we are to insert an element *x* into the max HBLT *H*. If we create a max HBLT with the single element *x* and then meld this max HBLT and *H*, the resulting max HBLT will include all elements in *H* as well as the element *x*. Hence an insertion may be performed by creating a new max HBLT with just the element that is to be inserted and then melding this max HBLT and the original.

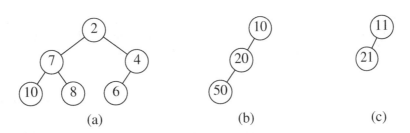

FIGURE 5.3 Some min trees.

5.2.3 Deletion of Max Element from a Max HBLT

The max element is in the root. If the root is deleted, two max HBLTs, the left and right subtrees of the root, remain. By melding together these two max HBLTs, we obtain a max HBLT that contains all elements in the original max HBLT other than the deleted max element. So the delete max operation may be performed by deleting the root and then melding its two subtrees.

5.2.4 Melding Two Max HBLTs

Since the length of the right-most path of an HBLT with n elements is $O(\log n)$, a meld algorithm that traverses only the right-most paths of the HBLTs being melded, spending $O(1)$ time at each node on these two paths, will have complexity logarithmic in the number of elements in the resulting HBLT. With this observation in mind, we develop a meld algorithm that begins at the roots of the two HBLTs and makes right-child moves only.

The meld strategy is best described using recursion. Let A and B be the two max HBLTs that are to be melded. If one is empty, then we may use the other as the result. So assume that neither is empty. To perform the meld, we compare the elements in the two roots. The root with the larger element becomes the root of the melded HBLT. Ties may be broken arbitrarily. Suppose that A has the larger root and that its left subtree is L. Let C be the max HBLT that results from melding the right subtree of A and the max HBLT B. The result of melding A and B is the max HBLT that has A as its root and L and C as its subtrees. If the s value of L is smaller than that of C, then C is the left subtree. Otherwise, L is.

EXAMPLE 5.1

Consider the two max HBLTs of Figure 5.4a. The s value of a node is shown outside the node, while the element value is shown inside. When drawing two max HBLTs that are to be melded, we will always draw the one with larger root value on the left. Ties are broken arbitrarily. Because of this convention, the root of the left HBLT always becomes the root of the final HBLT. Also, we will shade the nodes of the HBLT on the right.

Since the right subtree of 9 is empty, the result of melding this subtree of 9 and the tree with root 7 is just the tree with root 7. We make the tree with root 7 the right subtree of 9 temporarily to get the max tree of Figure 5.4b. Since the s value of the left subtree of 9 is 0 while that of its right subtree is 1, the left and right subtrees are swapped to get the max HBLT of Figure 5.4c.

Next consider melding the two max HBLTs of Figure 5.4d. The root of the left subtree becomes the root of the result. When the right subtree of 10 is melded with the HBLT with root 7, the result is just this latter HBLT. If this HBLT is made the right subtree of 10, we get the max tree of Figure 5.4e. Comparing the s values of the left and right children of 10, we see that a swap is not necessary.

Now consider melding the two max HBLTs of Figure 5.4f. The root of the left subtree is the root of the result. We proceed to meld the right subtree of 18 and the max HBLT with root 10. The two max HBLTs being melded are the same as those melded in Figure 5.4d. The resultant max HBLT (Figure 5.4e) becomes the right subtree of 18, and the max tree of Figure 5.4g results. Comparing the s values of the left and right subtrees of 18, we see that these subtrees must be swapped. Swapping results in the max HBLT of Figure 5.4h.

As a final example, consider melding the two max HBLTs of Figure 5.4i. The root of the left max HBLT becomes the root of the result. We proceed to meld the right subtree of 40 and the max HBLT with root 18. These max HBLTs were melded in Figure 5.4f. The resultant max HBLT (Figure 5.4g) becomes the right subtree of 40. Since the left subtree of 40 has a smaller s value than the right has, the two subtrees are swapped to get the max HBLT of Figure 5.4k. Notice that when melding the max HBLTs of Figure 5.4i, we first move to the right child of 40, then to the right child of 18, and finally to the right child of 10. All moves follow the right-most paths of the initial max HBLTs.

5.2.5 Initialization

It takes $O(n \log n)$ time to initialize a max HBLT with n elements by inserting these elements into an initially empty max HBLT one at a time. To get a linear time initialization algorithm, we begin by creating n max HBLTs with each containing one of the n elements. These n max HBLTs are placed on a FIFO queue. Then max HBLTs are deleted from this queue in pairs, melded, and added to the end of the queue until only one max HBLT remains.

EXAMPLE 5.2

We wish to create a max HBLT with the five elements 7, 1, 9, 11, and 2. Five single-element max HBLTs are created and placed in a FIFO queue. The first two, 7 and 1, are deleted from the queue and melded. The result (Figure 5.5a) is added to the queue. Next the max HBLTs 9 and 11 are deleted from the queue and melded. The result appears in Figure 5.5b. This max HBLT is added to the queue. Now the max HBLT 2 and that of Figure 5.5a are deleted from the queue and melded. The resulting max HBLT (Figure 5.5c) is added to the queue. The next pair to be deleted from the queue consists of the max HBLTs of Figures Figure 5.5b and c. These HBLTs are melded to get the max HBLT of Figure 5.5d. This max HBLT is added to the queue. The queue now has just one max HBLT, and we are done with the initialization.

For the complexity analysis of of the initialization operation, assume, for simplicity, that n is a power of 2. The first $n/2$ melds involve max HBLTs with one element each, the next $n/4$ melds involve max HBLTs with two elements each; the next $n/8$ melds are with trees that have four elements each; and so on. The time needed to meld two leftist trees with 2^i elements each is $O(i+1)$, and so the total time for the initialization is

$$O(n/2 + 2*(n/4) + 3*(n/8) + \cdots) = O\left(n\sum\frac{i}{2^i}\right) = O(n)$$

FIGURE 5.4 Melding max HBLTs.

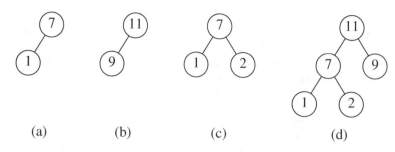

(a)   (b)   (c)   (d)

FIGURE 5.5   Initializing a max HBLT.

5.2.6 Deletion of Arbitrary Element from a Max HBLT

Although deleting an element other than the max (min) element is not a standard operation for a max (min) priority queue, an efficient implementation of this operation is required when one wishes to use the generic methods of Cho and Sahni [9] and Chong and Sahni [10] to derive efficient mergeable double-ended priority queue data structures from efficient single-ended priority queue data structures. From a max or min leftist tree, we may remove the element in any specified node *theNode* in $O(\log n)$ time, making the leftist tree a suitable base structure from which an efficient mergeable double-ended priority queue data structure may be obtained [9,10].

To remove the element in the node *theNode* of a height-biased leftist tree, we must do the following:

1. Detach the subtree rooted at *theNode* from the tree and replace it with the meld of the subtrees of *theNode*.
2. Update *s* values on the path from *theNode* to the root and swap subtrees on this path as necessary to maintain the leftist tree property.

To update *s* on the path from *theNode* to the root, we need parent pointers in each node. This upward updating pass stops as soon as we encounter a node whose *s* value does not change. The changed *s* values (with the exception of possibly $O(\log n)$ values from moves made at the beginning from right children) must form an ascending sequence (actually, each must be one more than the preceding one). Since the maximum *s* value is $O(\log n)$ and since all *s* values are positive integers, at most $O(\log n)$ nodes are encountered in the updating pass. At each of these nodes, we spend $O(1)$ time. Therefore, the overall complexity of removing the element in node *theNode* is $O(\log n)$.

5.3 Weight-Biased Leftist Trees

5.3.1 Definition

We arrive at another variety of leftist tree by considering the number of nodes in a subtree, rather than the length of a shortest root to external node path. Define the weight $w(x)$ of node *x* to be the number of internal nodes in the subtree with root *x*. Notice that if *x* is an external node, its weight is 0. If *x* is an internal node, its weight is 1 more than the sum of the weights of its children. The weights of the nodes of the binary tree of Figure 5.1a appear in Figure 5.1d

Definition 5.4: (Cho and Sahni [1]).

A binary tree is a **weight-biased leftist tree (WBLT)** iff at every internal node the *w* value of the left child is greater than or equal to the *w* value of the right child. A max (min) WBLT is a max (min) tree that is also a WBLT.

Note that the binary tree of Figure 5.1a is not a WBLT. However, all three of the binary trees of Figure 5.2 are WBLTs.

Theorem 5.2

Let *x* be any internal node of a weight-biased leftist tree. The length, *rightmost*(*x*), of the right-most path from *x* to an external node satisfies

$$rightmost(x) \leq \log_2(w(x) + 1).$$

Proof. The proof is by induction on $w(x)$. When $w(x) = 1$, $rightmost(x) = 1$ and $\log_2(w(x) + 1) = \log_2 2 = 1$. For the induction hypothesis, assume that $rightmost(x) \leq \log_2(w(x) + 1)$ whenever $w(x) < n$. Let *RightChild*(*x*) denote the right child of

x (note that this right child may be an external node). When $w(x) = n$, $w(RightChild(x)) \leq (n-1)/2$ and $rightmost(x) = 1 + rightmost(RightChild(x)) \leq 1 + \log_2((n-1)/2 + 1) = 1 + \log_2(n+1) - 1 = \log_2(n+1)$. ∎

5.3.2 Max WBLT Operations

Insert, delete max, and initialization are analogous to the corresponding max HBLT operation. However, the meld operation can be done in a single top-to-bottom pass (recall that the meld operation of an HBLT performs a top-to-bottom pass as the recursion unfolds and then a bottom-to-top pass in which subtrees are possibly swapped and s-values updated). A single-pass meld is possible for WBLTs because we can determine the w values on the way down and so, on the way down, we can update w-values and swap subtrees as necessary. For HBLTs, a node's new s value cannot be determined on the way down the tree.

Since the meld operation of a WBLT may be implemented using a single top-to-bottom pass, inserts and deletes also use a single top-to-bottom pass. Because of this, inserts and deletes are faster, by a constant factor, in a WBLT than in an HBLT [1]. However, from a WBLT, we cannot delete the element in an arbitrarily located node, *theNode*, in $O(\log n)$ time. This is because *theNode* may have $O(n)$ ancestors whose w value is to be updated. So, WBLTs are not suitable for mergeable double-ended priority queue applications [8,9].

C++ and Java codes for HBLTs and WBLTs may be obtained from [11,12], respectively.

Acknowledgments

This work was supported, in part, by the National Science Foundation under grant CCR-9912395.

References

1. S. Cho and S. Sahni, Weight biased leftist trees and modified skip lists, *ACM Journal on Experimental Algorithmics*, Article 2, 1998.
2. C. Crane, Linear Lists and Priority Queues as Balanced Binary Trees, Tech. Rep. CS-72-259, Dept. of Comp. Sci., Stanford University, 1972.
3. M. Fredman and R. Tarjan, Fibonacci Heaps and Their Uses in Improved Network Optimization Algorithms, *JACM*, 34, 3, 1987, 596–615.
4. M. Brown, Implementation and analysis of binomial queue algorithms, *SIAM Journal on Computing*, 7, 3, 1978, 298–319.
5. D. Sleator and R. Tarjan, Self-adjusting heaps, *SIAM Journal on Computing*, 15, 1, 1986, 52–69.
6. M. Fredman, R. Sedgewick, D. Sleator, and R. Tarjan, The pairing heap: A new form of self-adjusting heap, *Algorithmica*, 1, 1986, 111–129.
7. J. Stasko and J. Vitter, Pairing heaps: Experiments and analysis, *Communications of the ACM*, 30, 3, 1987, 234–249.
8. D. Jones, An empirical comparison of priority-queue and event-set implementations, *Communications of the ACM*, 29, 4, 1986, 300–311.
9. S. Cho and S. Sahni, Mergeable double-ended priority queues, *International Journal on Foundations of Computer Science*, 10, 1, 1999, 1–18.
10. K. Chong and S. Sahni, Correspondence based data structures for double ended priority queues, *ACM Journal on Experimental Algorithmics*, 5, 2000, Article 2, 22 pages.
11. S. Sahni, *Data Structures, Algorithms, and Applications in C++*, McGraw-Hill, NY, 1998, 824 pages.
12. S. Sahni, *Data Structures, Algorithms, and Applications in Java*, McGraw-Hill, NY, 2000, 846 pages.

6

Skew Heaps[*]

6.1 Introduction.. 77
6.2 Basics of Amortized Analysis 78
6.3 Meldable Priority Queues and Skew Heaps............................. 80
    Meldable Priority Queue Operations • Amortized Cost of Meld Operation
6.4 Bibliographic Remarks... 83
References.. 83

C. Pandu Rangan
Indian Institute of Technology, Madras

6.1 Introduction

Priority Queue is one of the most extensively studied *Abstract Data Types* (ADT) due to its fundamental importance in the context of resource managing systems, such as operating systems. *Priority Queues* work on finite subsets of a totally ordered universal set U. Without any loss of generality we assume that U is simply the set of all non-negative integers. In its simplest form, a *Priority Queue* supports two operations, namely,

- *insert*(x, S): update S by adding an arbitrary $x \in U$ to S.
- *delete-min*(S): update S by removing from S the minimum element of S.

We will assume for the sake of simplicity, all the items of S are distinct. Thus, we assume that $x \notin S$ at the time of calling *insert*(x, S). This increases the cardinality of S, denoted usually by $|S|$, by one. The well-known data structure *Heaps*, provide an elegant and efficient implementation of *Priority Queues*. In the *Heap* based implementation, both *insert*(x, S) and *delete-min*(S) take $O(\log n)$ time where $n = |S|$.

Several extensions for the basic *Priority Queues* were proposed and studied in response to the needs arising in several applications. For example, if an operating system maintains a set of jobs, say print requests, in a priority queue, then, always, the jobs with "high priority" are serviced irrespective of when the job was queued up. This might mean some kind of "unfairness" for low priority jobs queued up earlier. In order to straighten up the situation, we may extend priority queue to support *delete-max* operation and arbitrarily mix *delete-min* and *delete-max* operations to avoid any undue stagnation in the queue. Such priority queues are called *Double Ended Priority Queues*. It is easy to see that *Heap* is not an appropriate data structure for *Double Ended Priority Queues*. Several interesting alternatives are available in the literature [1–3]. You may also refer Chapter 8 of this handbook for a comprehensive discussion on these structures.

In another interesting extension, we consider adding an operation called *melding*. A *meld* operation takes two disjoint sets, S_1 and S_2, and produces the set $S = S_1 \cup S_2$. In terms of an implementation, this requirement translates to building a data structure for S, given the data structures of S_1 and S_2. A *Priority Queue* with this extension is called a *Meldable Priority Queue*. Consider a scenario where an operating system maintains two different priority queues for two printers and one of the printers is down with some problem during operation. *Meldable Priority Queues* naturally model such a situation.

Again, maintaining the set items in *Heaps* results in very inefficient implementation of *Meldable Priority Queues*. Specifically, designing a data structure with $O(\log n)$ bound for each of the *Meldable Priority Queue* operations calls for more sophisticated ideas and approaches. An interesting data structure called *Leftist Trees*, implements all the operations of *Meldable Priority Queues* in $O(\log n)$ time. *Leftist Trees* are discussed in Chapter 5 of this handbook.

The main objective behind the design of a data structure for an ADT is to implement the ADT operations as efficiently as possible. Typically, efficiency of a structure is judged by its worst-case performance. Thus, when designing a data structure, we seek to minimize the worst case complexity of each operation. While this is a most desirable goal and has been theoretically realized for a number of data structures for key ADTs, the data structures optimizing worst-case costs of ADT operations are

---

[*] This chapter has been reprinted from first edition of this Handbook, without any content updates.

often very complex and pretty tedious to implement. Hence, computer scientists were exploring alternative design criteria that would result in simpler structures without losing much in terms of performance. In Chapter 14 of this handbook, we show that incorporating randomness provides an attractive alternative avenue for designers of the data structures. In this chapter we will explore yet another design goal leading to simpler structural alternatives without any degrading in overall performance.

Since the data structures are used as basic building blocks for implementing algorithms, a typical execution of an algorithm might consist of a sequence of operations using the data structure over and again. In the worst case complexity based design, we seek to reduce the cost of each operation as much as possible. While this leads to an overall reduction in the cost for the sequence of operations, this poses some constraints on the designer of data structure. We may relax the requirement that the cost of each operation be minimized and perhaps design data structures that seek to minimize the total cost of any sequence of operations. Thus, in this new kind of design goal, we will not be terribly concerned with the cost of any individual operations, but worry about the total cost of any sequence of operations. At first thinking, this might look like a formidable goal as we are attempting to minimize the cost of an arbitrary mix of ADT operations and it may not even be entirely clear how this design goal could lead to simpler data structures. Well, it is typical of a novel and deep idea; at first attempt it may puzzle and bamboozle the learner and with practice one tends to get a good intuitive grasp of the intricacies of the idea. This is one of those ideas that requires some getting used to. In this chapter, we discuss about a data structure called *Skew heaps*. For any sequence of a *Meldable Priority Queue* operations, its total cost on *Skew Heaps* is asymptotically same as its total cost on *Leftist Trees*. However, *Skew Heaps* are a bit simpler than *Leftist Trees*.

6.2 Basics of Amortized Analysis

We will now clarify the subtleties involved in the new design goal with an example. Consider a typical implementation of *Dictionary* operations. The so called Balanced Binary Search Tree structure (BBST) implements these operations in $O(m \log n)$ worst case bound. Thus, the total cost of an arbitrary sequence of m dictionary operations, each performed on a tree of size at most n, will be $O(\log n)$. Now we may turn around and ask: Is there a data structure on which the cost of a sequence of m dictionary operations is $O(m \log n)$ but individual operations are not constrained to have $O(\log n)$ bound? Another more pertinent question to our discussion - Is that structure simpler than BBST, at least in principle? An affirmative answer to both the questions is provided by a data structure called *Splay Trees*. *Splay Tree* is the theme of Chapter 13 of this handbook.

Consider for example a sequence of m dictionary operations $S_1, S_2, \ldots, S_m$, performed using a BBST. Assume further that the size of the tree has never exceeded n during the sequence of operations. It is also fairly reasonable to assume that we begin with an empty tree and this would imply $n \leq m$. Let the actual cost of executing S_i be C_i. Then the total cost of the sequence of operations is $C_1 + C_2 + \cdots + C_m$. Since each C_i is $O(\log n)$ we easily conclude that the total cost is $O(m \log n)$. No big arithmetic is needed and the analysis is easily finished. Now, assume that we execute the same sequence of m operations but employ a *Splay Tree* in stead of a BBST. Assuming that c_i is the actual cost of S_i in a *Splay Tree*, the total cost for executing the sequence of operation turns out to be $c_1 + c_2 + \cdots + c_m$. This sum, however, is tricky to compute. This is because a wide range of values are possible for each of c_i and no upper bound other than the trivial bound of $O(n)$ is available for c_i. Thus, a naive, worst case cost analysis would yield only a weak upper bound of $O(nm)$ whereas the actual bound is $O(m \log n)$. But how do we arrive at such improved estimates?

This is where we need yet another powerful tool called *potential function*.

The potential function is purely a conceptual entity and this is introduced only for the sake of computing a sum of widely varying quantities in a convenient way. Suppose there is a function $f:D \rightarrow R^+ \cup \{0\}$, that maps a configuration of the data structure to a non-negative real number. We shall refer to this function as potential function. Since the data type as well as data structures are typically dynamic, an operation may change the configuration of data structure and hence there may be change of potential value due to this change of configuration. Referring back to our sequence of operations $S_1, S_2, \ldots, S_m$, let D_{i-1} denote the configuration of data structure before the executing the operation S_i and D_i denote the configuration after the execution of S_i. The potential difference due to this operation is defined to be the quantity $f(D_i) - f(D_{i-1})$. Let c_i denote the actual cost of S_i. We will now introduce yet another quantity, a_i, defined by

$$a_i = c_i + f(D_i) - f(D_{i-1}).$$

What is the consequence of this definition?

Note that $\sum_{i=1}^m a_i = \sum_{i=1}^m c_i + f(D_m) - f(D_0)$.

Let us introduce one more reasonable assumption that $f(D_0) = f(\phi) = 0$. Since $f(D) \geq 0$ for all non empty structures, we obtain,

$$\sum a_i = \sum c_i + f(D_m) \geq \sum c_i$$

If we are able to choose cleverly a "good" potential function so that a_i's have tight, uniform bound, then we can evaluate the sum $\sum a_i$ easily and this bounds the actual cost sum $\sum c_i$. In other words, we circumvent the difficulties posed by wide variations in c_i by introducing new quantities a_i which have uniform bounds. A very neat idea indeed! However, care must be exercised while defining the potential function. A poor choice of potential function will result in a_is whose sum may be a trivial or useless bound for the sum of actual costs. In fact, arriving at the right potential function is an ingenious task, as you will understand by the end of this chapter or by reading the chapter on *Splay Trees*.

The description of the data structures such as *Splay Trees* will not look any different from the description of a typical data structures—it comprises of a description of the organization of the primitive data items and a bunch of routines implementing ADT operations. The key difference is that the routines implementing the ADT operations will not be analyzed for their individual worst case complexity. We will only be interested in the the cumulative effect of these routines in an arbitrary sequence of operations. Analyzing the average potential contribution of an operation in an arbitrary sequence of operations is called *amortized analysis*. In other words, the routines implementing the ADT operations will be analyzed for their *amortized cost*. Estimating the amortized cost of an operation is rather an intricate task. The major difficulty is in accounting for the wide variations in the costs of an operation performed at different points in an arbitrary sequence of operations. Although our design goal is influenced by the costs of sequence of operations, defining the notion of amortized cost of an operation in terms of the costs of sequences of operations leads one nowhere. As noted before, using a potential function to off set the variations in the actual costs is a neat way of handling the situation.

In the next definition we formalize the notion of amortized cost.

Definition 6.1: (Amortized Cost)

Let A be an ADT with basic operations $O = \{O_1, O_2, \ldots, O_k\}$ and let D be a data structure implementing A. Let f be a potential function defined on the configurations of the data structures to non-negative real number. Assume further that $f(\Phi) = 0$. Let D' denote a configuration we obtain if we perform an operation O_k on a configuration D and let c denote the actual cost of performing O_k on D. Then, the amortized cost of O_k operating on D, denoted as $a(O_k, D)$, is given by

$$a(O_k, D) = c + f(D') - f(D)$$

If $a(O_k, D) \leq c' g(n)$ for all configuration D of size n, then we say that the amortized cost of O_k is $O(g(n))$.

Theorem 6.1

Let D be a data structure implementing an ADT and let $s_1, s_2, \ldots, s_m$ denote an arbitrary sequence of ADT operations on the data structure starting from an empty structure D_0. Let c_i denote actual cost of the operation s_i and D_i denote the configuration obtained which s_i operated on D_{i-1}, for $1 \leq i \leq m$. Let a_i denote the amortized cost of s_i operating on D_{i-1} with respect to an arbitrary potential function. Then,

$$\sum_{i=1}^{m} c_i \leq \sum_{i=1}^{m} a_i.$$

Proof. Since a_i is the amortized cost of s_i working on the configuration D_{i-1}, we have

$$a_i = a(s_i, D_{i-1}) = c_i + f(D_i) - f(D_{i-1})$$

Therefore,

$$\sum_{i=1}^{m} a_i = \sum_{i=1}^{m} c_i + (f(D_m) - f(D_0))$$

$$= f(D_m) + \sum_{i=1}^{m} c_i \text{ (since } f(D_0) = 0)$$

$$\geq \sum_{i=1}^{m} c_i$$

∎

Remark 6.1

The potential function is common to the definition of amortized cost of all the ADT operations. Since $\sum_{i=1}^{m} a_i \geq \sum_{i=1}^{m} c_i$ holds good for any potential function, a clever choice of the potential function will yield tight upper bound for the sum of actual cost of a sequence of operations.

6.3 Meldable Priority Queues and Skew Heaps

Definition 6.2: (Skew Heaps)

A Skew Heap is simply a binary tree. Values are stored in the structure, one per node, satisfying the *heap-order property*: A value stored at a node is larger than the value stored at its parent, except for the root (as root has no parent).

Remark 6.2

Throughout our discussion, we handle sets with distinct items. Thus a set of n items is represented by a skew heap of n nodes. The minimum of the set is always at the root. On any path starting from the root and descending towards a leaf, the values are in increasing order.

6.3.1 Meldable Priority Queue Operations

Recall that a *Meldable Priority queue* supports three key operations: *insert, delete-min* and *meld*. We will first describe the meld operation and then indicate how other two operations can be performed in terms of the *meld* operation.

Let S_1 and S_2 be two sets and H_1 and H_2 be *Skew Heaps* storing S_1 and S_2 respectively. Recall that $S_1 \cap S_2 = \phi$. The *meld* operation should produce a single *Skew Heap* storing the values in $S_1 \cup S_2$. The procedure *meld* (H_1, H_2) consists of two phases. In the first phase, the two right most paths are merged to obtain a single right most path. This phase is pretty much like the merging algorithm working on sorted sequences. In this phase, the left subtrees of nodes in the right most paths are not disturbed. In the second phase, we simply swap the children of every node on the merged path except for the lowest. This completes the process of *melding*.

Figures 6.1–6.3 clarify the phases involved in the *meld* routine.

Figure 6.1 shows two *Skew Heaps* H_1 and H_2. In Figure 6.2 we have shown the scenario after the completion of the first phase. Notice that right most paths are merged to obtain the right most path of a single tree, keeping the respective left subtrees intact. The final *Skew Heap* is obtained in Figure 6.3. Note that left and right child of every node on the right most path of the tree in Figure 6.2 (except the lowest) are swapped to obtain the final *Skew Heap*.

It is easy to implement *delete-min* and *insert* in terms of the *meld* operation. Since minimum is always found at the root, *delete-min* is done by simply removing the root and *melding* its left subtree and right subtree. To *insert* an item x in a *Skew Heap* H_1, we create a Skew Heap H_2 consisting of only one node containing x and then *meld* H_1 and H_2. From the above discussion, it is clear that cost of *meld* essentially determines the cost of *insert* and *delete-min*. In the next section, we analyze the amortized cost of *meld* operation.

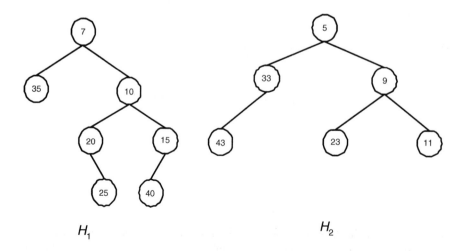

FIGURE 6.1 Skew Heaps for meld operation.

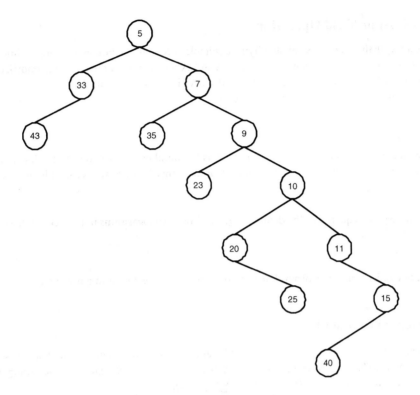

FIGURE 6.2 Rightmost paths are merged. Left subtrees of nodes in the merged path are intact.

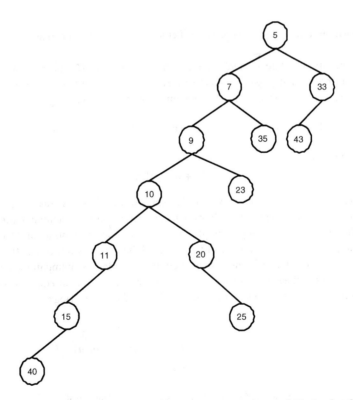

FIGURE 6.3 Left and right children of nodes (5), (7), (9), (10), (11) of Figure 6.2 are swapped. Notice that the children of (15) which is the lowest node in the merged path, are not swapped.

6.3.2 Amortized Cost of Meld Operation

At this juncture we are left with the crucial task of identifying a suitable potential function. Before proceeding further, perhaps one should try the implication of certain simple potential functions and experiment with the resulting amortized cost. For example, you may try the function $f(D) =$ number of nodes in $D($ and discover how ineffective it is!).

We need some definitions to arrive at our potential function.

Definition 6.3

For any node x in a binary tree, the weight of x, denoted $wt(x)$, is the number of descendants of x, including itself. A non-root node x is said to be heavy if $wt(x) > wt(parent(x))/2$. A non-root node that is not heavy is called light. The root is neither light nor heavy.

The next lemma is an easy consequence of the definition given above. All logarithms in this section have base 2.

Lemma 6.1

For any node, at most one of its children is heavy. Furthermore, any root to leaf path in a n-node tree contains at most $\lfloor \log n \rfloor$ light nodes.

Definition 6.4: (Potential Function)

A non-root is called *right* if it is the right child of its parent; it is called *left* otherwise. The potential of a skew heap is the number of right heavy node it contains. That is, $f(H) =$ number of right heavy nodes in H. We extend the definition of potential function to a collection of skew heaps as follows: $f(H_1, H_2, \ldots, H_t) = \sum_{i=1}^{t} f(H_i)$.

Here is the key result of this chapter.

Theorem 6.2

Let H_1 and H_2 be two heaps with n_1 and n_2 nodes respectively. Let $n = n_1 + n_2$. The amortized cost of meld (H_1, H_2) is $O(\log n)$.

Proof. Let h_1 and h_2 denote the number of heavy nodes in the right most paths of H_1 and H_2 respectively. The number of light nodes on them will be at most $\lfloor \log n_1 \rfloor$ and $\lfloor \log n_2 \rfloor$ respectively. Since a node other than root is either heavy or light, and there are two root nodes here that are neither heavy or light, the total number of nodes in the right most paths is at most

$$2 + h_1 + h_2 + \lfloor \log n_1 \rfloor + \lfloor \log n_2 \rfloor \leq 2 + h_1 + h_2 + 2\lfloor \log n \rfloor$$

Thus we get a bound for actual cost c as

$$c \leq 2 + h_1 + h_2 + 2\lfloor \log n \rfloor \tag{6.1}$$

In the process of swapping, the $h_1 + h_2$ nodes that were *right heavy*, will lose their status as *right heavy*. While they remain heavy, they become left children for their parents hence they do not contribute for the potential of the output tree and this means a drop in potential by $h_1 + h_2$. However, the swapping might have created new heavy nodes and let us say, the number of new heavy nodes created in the swapping process is h_3. First, observe that all these h_3 new nodes are attached to the left most path of the output tree. Secondly, by Lemma 6.1, for each one of these right heavy nodes, its sibling in the left most path is a light node. However, the number of light nodes in the left most path of the output tree is less than or equal to $\lfloor \log n \rfloor$ by Lemma 6.1.

Thus $h_3 \leq \lfloor \log n \rfloor$. Consequently, the net change in the potential is $h_3 - h_1 - h_2 \leq \lfloor \log n \rfloor - h_1 - h_2$.

$$\text{The amortized cost} = c + \text{potential difference}$$
$$\leq 2 + h_1 + h_2 + 2\lfloor \log n \rfloor + \lfloor \log n \rfloor - h_1 - h_2$$
$$= 3\lfloor \log n \rfloor + 2.$$

Hence, the amortized cost of meld operation is $O(\log n)$ and this completes the proof. ∎

Since *insert* and *delete-min* are handled as special cases of *meld* operation, we conclude

Theorem 6.3

The amortized cost complexity of all the Meldable Priority Queue operations is $O(\log n)$ where n is the number of nodes in skew heap or heaps involved in the operation.

6.4 Bibliographic Remarks

Skew Heaps were introduced by Sleator and Tarjan [4]. *Leftist Trees* have $O(\log n)$ worst case complexity for all the *Meldable Priority Queue* operations but they require heights of each subtree to be maintained as additional information at each node. *Skew Heaps* are simpler than *Leftist Trees* in the sense that no additional 'balancing' information need to be maintained and the *meld* operation simply swaps the children of the right most path without any constraints and this results in a simpler code. The bound $3 \log_2 n + 2$ for *melding* was significantly improved to $\log_\phi n$ (here ϕ denotes the well-known *golden ratio* $(\sqrt{5}+1)/2$ which is roughly 1.6) by using a different potential function and an intricate analysis in Reference 5. Recently, this bound was shown to be tight in Reference 6. *Pairing Heap*, introduced by Fredman et al. [7], is yet another self-adjusting heap structure and its relation to *Skew Heaps* is explored in Chapter 7 of this handbook.

References

1. A. Aravind and C. Pandu Rangan, Symmetric Min-Max heaps: A simple data structure for double-ended priority queue, *Information Processing Letters*, 69:197–199, 1999.
2. S. Carlson, The Deap—A double ended heap to implement a double ended priority queue, *Information Processing Letters*, 26:33–36, 1987.
3. S. Chang and M. Du, Diamond dequeue: A simple data structure for priority dequeues, *Information Processing Letters*, 46:231–237, 1993.
4. D. D. Sleator and R. E. Tarjan, Self-adjusting heaps, *Journal on Computing*, 15:52–69, 1986.
5. A. Kaldewaij and B. Schoenmakers, The derivation of a tighter bound for top-down skew heaps, *Information Processing Letters*, 37:265–271, 1991.
6. B. Schoenmakers, A tight lower bound for top-down skew heaps, *Information Processing Letters*, 61:279–284, 1997.
7. M. L. Fredman, R. Sedgewick, D. D. Sleator, and R. E. Tarjan, The pairing heap: A new form of self-adjusting heap, *Algorithmica*, 1:111–129, 1986.

7

Binomial, Fibonacci, and Pairing Heaps

7.1	Introduction	85
7.2	Binomial Heaps	85
7.3	Fibonacci Heaps	89
7.4	Pairing Heaps	92
7.5	Related Developments	94
	Other Variations of Pairing Heaps • Adaptive Properties of Pairing Heaps • Soft Heaps	
	References	95

Michael L. Fredman
Rutgers University

7.1 Introduction

This chapter presents three algorithmically related data structures for implementing meldable priority queues: binomial heaps, Fibonacci heaps, and pairing heaps. What these three structures have in common is that (a) they are comprised of heap-ordered trees, (b) the comparisons performed to execute extractmin operations exclusively involve keys stored in the roots of trees, and (c) a common side effect of a comparison between two root keys is the linking of the respective roots: one tree becomes a new subtree joined to the other root.

A tree is considered heap-ordered provided that each node contains one item, and the key of the item stored in the parent $p(x)$ of a node x never exceeds the key of the item stored in x. Thus when two roots get linked, the root storing the larger key becomes a child of the other root. By convention, a linking operation positions the new child of a node as its leftmost child. Figure 7.1 illustrates these notions.

Of the three data structures, the binomial heap structure was the first to be invented [1], designed to efficiently support the operations: insert, extractmin, delete, and meld. The binomial heap has been highly appreciated as an elegant and conceptually simple data structure, particularly given its ability to support the meld operation. The Fibonacci heap data structure [2] was inspired by and can be viewed as a generalization of the binomial heap structure. The raison d'être of the Fibonacci heap structure is its ability to efficiently execute decrease-key operations. A decrease-key operation replaces the key of an item, specified by location, by a smaller value, for example, decrease-key(P,k_{new},H). (The arguments specify that the item is located in node P of the priority queue H and that its new key value is k_{new}.) Decrease-key operations are prevalent in many network optimization algorithms, including minimum spanning tree and shortest path. The pairing heap data structure [3] was devised as a self-adjusting analogue of the Fibonacci heap and has proved to be more efficient in practice [4].

Binomial heaps and Fibonacci heaps are primarily of theoretical and historical interest. The pairing heap is the more efficient and versatile data structure from a practical standpoint. The following three sections describe the respective data structures.

7.2 Binomial Heaps

We begin with an informal overview. A single binomial heap structure consists of a forest of specially structured trees, referred to as binomial trees. The number of nodes in a binomial tree is always a power of two. Defined recursively, the binomial tree B_0 consists of a single node. The binomial tree B_k, for $k > 0$, is obtained by linking two trees B_{k-1} together: one tree becomes the leftmost subtree of the other. In general, B_k has 2^k nodes. Figures 7.2a and b illustrate the recursion and show several trees in the series. An alternative and useful way to view the structure of B_k is depicted in Figure 7.2c: B_k consists of a root and subtrees (in order from left to right) $B_{k-1}, B_{k-2}, \ldots, B_0$. The root of the binomial tree B_k has k children, and the tree is said to have rank k.

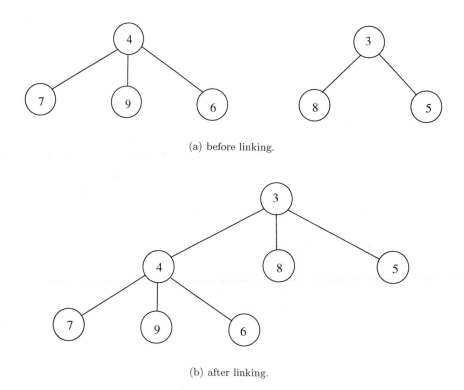

(a) before linking.

(b) after linking.

FIGURE 7.1   Two heap-ordered trees and the result of their linking: (a) before linking and (b) after linking.

We also observe that the height of B_k (maximum number of edges on any path directed away from the root) is k. The name "binomial heap" is inspired by the fact that the root of B_k has $\binom{k}{j}$ descendants at distance j.

Because binomial trees have restricted sizes, a forest of trees is required to represent a priority queue of arbitrary size. A key observation, indeed a motivation for having tree sizes being powers of two, is that a priority queue of arbitrary size can be represented as a union of trees of *distinct* sizes. (In fact, the sizes of the constituent trees are uniquely determined and correspond to the powers of two that define the binary expansion of n, the size of the priority queue.) Moreover, because the tree sizes are unique, the number of trees in the forest of a priority queue of size n is at most $\lg(n+1)$. Thus finding the minimum key in the priority queue, which clearly lies in the root of one of its constituent trees (due to the heap-order condition), requires searching among at most $\lg(n+1)$ tree roots. Figure 7.3 gives an example of binomial heap.

Now let's consider, from a high-level perspective, how the various heap operations are performed. As with leftist heaps (cf. Chapter 5), the various priority queue operations are to a large extent comprised of melding operations, and so we consider first the melding of two heaps.

The melding of two heaps proceeds as follows: (a) the trees of the respective forests are combined into a single forest and then (b) *consolidation* takes place: pairs of trees having common rank are linked together until all remaining trees have distinct ranks. Figure 7.4 illustrates the process. An actual implementation mimics binary addition and proceeds in much the same was as merging two sorted lists in ascending order. We note that insertion is a special case of melding.

The extractmin operation is performed in two stages. First, the *minimum root*, the node containing the minimum key in the data structure, is found by examining the tree roots of the appropriate forest, and this node is removed. Next, the forest consisting of the subtrees of this removed root, whose ranks are distinct (see Figure 7.2c) and thus viewable as constituting a binomial heap, is melded with the forest consisting of the trees that remain from the original forest. Figure 7.5 illustrates the process.

Finally, we consider arbitrary deletion. We assume that the node v containing the item to be deleted is specified. Proceeding up the path to the root of the tree containing v, we permute the items among the nodes on this path, placing in the root the item x originally in v, and shifting each of the other items down one position (away from the root) along the path. This is accomplished through a sequence of exchange operations that move x towards the root. The process is referred to as a *sift-up* operation. Upon reaching the root r, r is then removed from the forest as though an extractmin operation is underway. Observe that the repositioning of items in the ancestors of v serves to maintain the heap-order property among the remaining nodes of the forest. Figure 7.6 illustrates the repositioning of the item being deleted to the root.

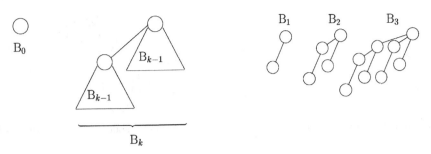

(a) Recursion for binomial trees.   (b) Several binomial trees.

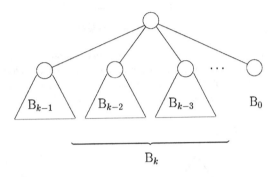

(c) An alternative recursion.

FIGURE 7.2 Binomial trees and their recursions: (a) recursion for binomial trees, (b) several binomial trees, and (c) an alternative recursion.

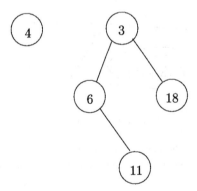

FIGURE 7.3 A binomial heap (showing placement of keys among forest nodes).

This completes our high-level descriptions of the heap operations. For navigational purposes, each node contains a leftmost child pointer and a sibling pointer that points to the next sibling to its right. The children of a node are thus stored in the linked list defined by sibling pointers among these children, and the head of this list can be accessed by the leftmost child pointer of the parent. This provides the required access to the children of a node for the purpose of implementing extractmin operations. Note that when a node obtains a new child as a consequence of a linking operation, the new child is positioned at the head of its list of siblings. To facilitate arbitrary deletions, we need a third pointer in each node pointing to its parent. To facilitate access to the ranks of trees, we maintain in each node the number of children it has, and refer to this quantity as the node rank. Node ranks are readily maintained under linking operations; the node rank of the root gaining a child gets incremented. Figure 7.7 depicts these structural features.

As seen in Figure 7.2c, the ranks of the children of a node form a descending sequence in the children's linked list. However, since the melding operation is implemented by accessing the tree roots in ascending rank order, when deleting a root we first reverse the list order of its children before proceeding with the melding.

Each of the priority queue operations requires in the worst case $O(\log n)$ time, where n is the size of the heap that results from the operation. This follows, for melding, from the fact that its execution time is proportional to the combined lengths of the forest

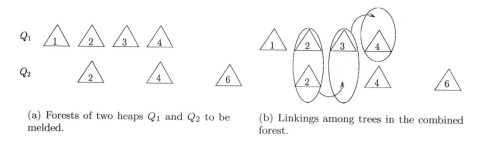

(a) Forests of two heaps Q_1 and Q_2 to be melded.

(b) Linkings among trees in the combined forest.

(c) Forest of meld (Q_1, Q_2).

FIGURE 7.4   Melding of two binomial heaps. The encircled objects reflect trees of common rank being linked. (Ranks are shown as numerals positioned within triangles which in turn represent individual trees.) Once linking takes place, the resulting tree becomes eligible for participation in further linkings, as indicated by the arrows that identify these linking results with participants of other linkings. (a) Forests of two heaps Q_1 and Q_2 to be melded. (b) Linkings among trees in the combined forest. (c) Forest of meld (Q_1, Q_2).

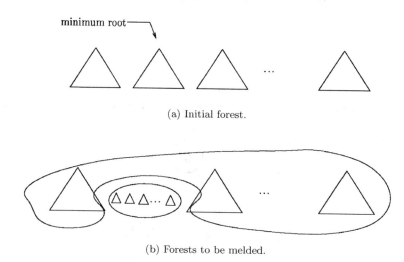

(a) Initial forest.

(b) Forests to be melded.

FIGURE 7.5   Extractmin operation: The location of the minimum key is indicated in (a). The two encircled sets of trees shown in (b) represent forests to be melded. The smaller trees were initially subtrees of the root of the tree referenced in (a).

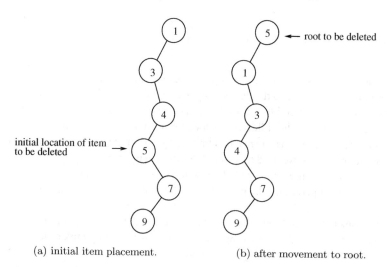

(a) initial item placement.   (b) after movement to root.

FIGURE 7.6   Initial phase of deletion—sift-up operation: (a) initial item placement and (b) after movement to root.

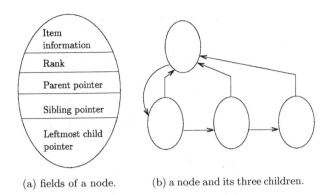

(a) fields of a node.   (b) a node and its three children.

FIGURE 7.7 Structure associated with a binomial heap node: (a) fields of a node and (b) a node and its three children.

lists being merged. For extractmin, this follows from the time for melding, along with the fact that a root node has only $O(\log n)$ children. For arbitrary deletion, the time required for the sift-up operation is bounded by an amount proportional to the height of the tree containing the item. Including the time required for extractmin, it follows that the time required for arbitrary deletion is $O(\log n)$.

A detailed code for manipulating binomial heaps can be found in Weiss [5].

7.3 Fibonacci Heaps

Fibonacci heaps were specifically designed to efficiently support decrease-key operations. For this purpose, the binomial heap can be regarded as a natural starting point. Why? Consider the class of priority queue data structures that are implemented as forests of heap-ordered trees, as will be the case for Fibonacci heaps. One way to immediately execute a decrease-key operation, remaining within the framework of heap-ordered forests, is to simply change the key of the specified data item and sever its link to its parent, inserting the severed subtree as a new tree in the forest. Figure 7.8 illustrates the process. (Observe that the link to the parent only needs to be cut if the new key value is smaller than the key in the parent node, violating heap order.) Fibonacci heaps accomplish this without degrading the asymptotic efficiency with which other priority queue operations can be supported. Observe that to accommodate node cuts, the list of children of a node needs to be doubly linked. Hence, the nodes of a Fibonacci heap require two sibling pointers.

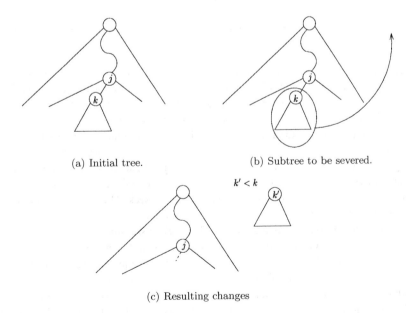

(a) Initial tree.   (b) Subtree to be severed.

(c) Resulting changes

FIGURE 7.8 Immediate decrease-key operation. The subtree severing (b and c) is necessary only when $k' < j$. (a) Initial tree. (b) Subtree to be severed. (c) Resulting changes.

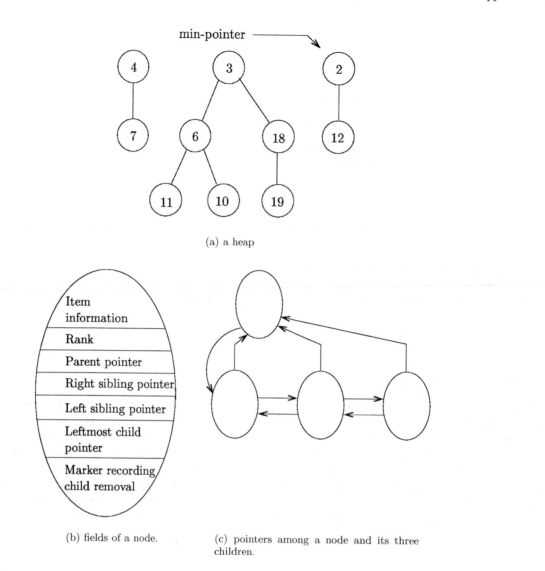

FIGURE 7.9 A Fibonacci heap and associated structure: (a) a heap, (b) fields of a node, and (c) pointers among a node and its three children.

Fibonacci heaps support findmin, insertion, meld, and decrease-key operations in constant amortized time, and deletion operations in $O(\log n)$ amortized time. For many applications, the distinction between worst-case times versus amortized times is of little significance. A Fibonacci heap consists of a forest of heap-ordered trees. As we shall see, Fibonacci heaps differ from binomial heaps in that there may be many trees in a forest of the same rank, and there is no constraint on the ordering of the trees in the forest list. The heap also includes a pointer to the tree root containing the minimum item, referred to as the *min-pointer*, that facilitates findmin operations. Figure 7.9 provides an example of a Fibonacci heap and illustrates certain structural aspects.

The impact of severing subtrees is clearly incompatible with the pristine structure of the binomial tree that is the hallmark of the binomial heap. Nevertheless, the tree structures that can appear in the Fibonacci heap data structure must sufficiently approximate binomial trees in order to satisfy the performance bounds we seek. The linking constraint imposed by binomial heaps, that trees being linked must have the same size, ensures that the number of children a node has (its rank), grows no faster than the logarithm of the size of the subtree rooted at the node. This *rank versus subtree size* relation is the key to obtaining the $O(\log n)$ deletion time bound. Fibonacci heap manipulations are designed with this in mind.

Fibonacci heaps utilize a protocol referred to as *cascading cuts* to enforce the required rank versus subtree size relation. Once a node v has had two of its children removed as a result of cuts, v's contribution to the rank of its parent is then considered suspect in terms of rank versus subtree size. The cascading cut protocol requires that the link to v's parent be cut, with the subtree rooted at v then being inserted into the forest as a new tree. If v's parent has, as a result, had a second child removed, then it in turn needs to be cut, and the cuts may thus cascade. Cascading cuts ensure that no nonroot node has had more than one child removed subsequent to being linked to its parent.

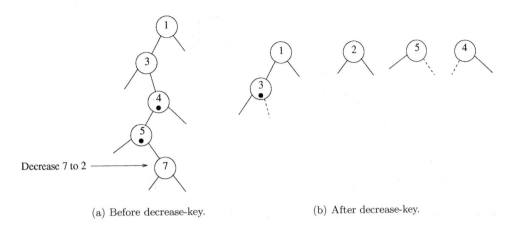

(a) Before decrease-key.   (b) After decrease-key.

FIGURE 7.10 Illustration of cascading cuts. In (b), the dashed lines reflect cuts that have taken place, two nodes marked in (a) get unmarked, and a third node gets marked. (a) Before decrease-key, (b) after decrease-key.

We keep track of the removal of children by marking a node if one of its children has been cut. A marked node that has another child removed is then subject to being cut from its parent. When a marked node becomes linked as a child to another node, or when it gets cut from its parent, it gets unmarked. Figure 7.10 illustrates the protocol of cascading cuts.

Now the induced node cuts under the cascading cuts protocol, in contrast with those primary cuts immediately triggered by decrease-key operations, are bounded in number by the number of primary cuts. (This follows from consideration of a potential function defined to be the total number of marked nodes.) Therefore, the burden imposed by cascading cuts can be viewed as effectively only doubling the number of cuts taking place in the absence of the protocol. One can therefore expect that the performance asymptotics are not degraded as a consequence of proliferating cuts. As with binomial heaps, two trees in a Fibonacci heap can only be linked if they have equal rank. With the cascading cuts protocol in place, we claim that the required rank versus subtree size relation holds, a matter which we address next.

Let's consider how *small* the subtree rooted at a node v having rank k can be. Let ω be the mth child of v from the right. At the time it was linked to v, v had at least $m - 1$ other children (those currently to the right of ω were certainly present). Therefore, ω had rank at least $m - 1$ when it was linked to v. Under the cascading cuts protocol, the rank of ω could have decreased by at most one after its linking to v; otherwise it would have been removed as a child. Therefore, the current rank of ω is at least $m - 2$. We minimize the size of the subtree rooted at v by minimizing the sizes (and ranks) of the subtrees rooted at v's children. Now let F_j be the minimum possible size of the subtree rooted at a node of rank j, so that the size of the subtree rooted at v is F_k. We conclude that (for $k \geq 2$)

$$F_k = \underbrace{F_{k-2} + F_{k-3} + \cdots + F_0 + 1}_{k \text{ terms}} + 1$$

where the final term, 1, reflects the contribution of v to the subtree size. Clearly, $F_0 = 1$ and $F_1 = 2$. See Figure 7.11 for an illustration of this construction. Based on the preceding recurrence, it is readily shown that F_k is given by the $(k + 2)$th Fibonacci number (from whence the name "Fibonacci heap" was inspired). Moreover, since the Fibonacci numbers grow exponentially fast, we conclude that the rank of a node is indeed bounded by the logarithm of the size of the subtree rooted at the node.

We proceed next to describe how the various operations are performed.

Since we are not seeking worst-case bounds, there are economies to be exploited that could also be applied to obtain a variant of Binomial heaps. (In the absence of cuts, the individual trees generated by Fibonacci heap manipulations would all be binomial trees.) In particular, we shall adopt a lazy approach to melding operations: the respective forests are simply combined by concatenating their tree lists and retaining the appropriate min-pointer. This requires only constant time.

An item is deleted from a Fibonacci heap by deleting the node that originally contains it, in contrast with Binomial heaps. This is accomplished by (a) cutting the link to the node's parent (as in decrease-key) if the node is not a tree root and (b) appending the list of children of the node to the forest. Now if the deleted node happens to be referenced by the min-pointer, considerable work is required to restore the min-pointer—the work previously deferred by the lazy approach to the operations. In the course of searching among the roots of the forest to discover the new minimum key, we also link trees together in a *consolidation* process.

Consolidation processes the trees in the forest, linking them in pairs until there are no longer two trees having the same rank, and then places the remaining trees in a new forest list (naturally extending the melding process employed by binomial heaps). This can be accomplished in time proportional to the number of trees in forest plus the maximum possible node rank. Let max-rank denotes the maximum possible node rank. (The preceding discussion implies that max-rank $= O(\log\text{heap-size})$.)

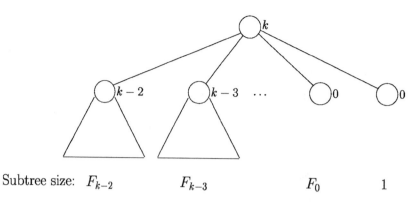

Subtree size: F_{k-2} F_{k-3} F_0 1

FIGURE 7.11   Minimal tree of rank k. Node ranks are shown adjacent to each node.

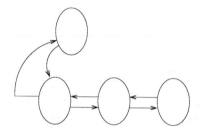

FIGURE 7.12   Pointers among a pairing heap node and its three children.

Consolidation is initialized by setting up an array A of trees (initially empty) indexed by the range [0,max-rank]. A nonempty position A[d] of A contains a tree of rank d. The trees of the forest are then processed using the array A as follows. To process a tree T of rank d, we insert T into A[d] if this position of A is empty, completing the processing of T. However, if A[d] already contains a tree U, then T and U are linked together to form a tree W, and the processing continues as before, but with W in place of T, until eventually an empty location of A is accessed, completing the processing associated with T. After all of the trees have been processed in this manner, the array A is scanned, placing each of its stored trees in a new forest. Apart from the final scanning step, the total time necessary to consolidate a forest is proportional to its number of trees, since the total number of tree pairings that can take place is bounded by this number (each pairing reduces by one the total number of trees present). The time required for the final scanning step is given by max-rank = log(heap-size).

The amortized timing analysis of Fibonacci heaps considers a potential function defined as the total number of trees in the forests of the various heaps being maintained. Ignoring consolidation, each operation takes constant actual time, apart from an amount of time proportional to the number of subtree cuts due to cascading (which, as noted above, is only constant in amortized terms). These cuts also contribute to the potential. The children of a deleted node increase the potential by O(logheap-size). Deletion of a minimum heap node additionally incurs the cost of consolidation. However, consolidation reduces our potential, so that the amortized time it requires is only O(logheap-size). We conclude therefore that all nondeletion operations require constant amortized time, and deletion requires $O(\log n)$ amortized time.

An interesting and unresolved issue concerns the protocol of cascading cuts. How would the performance of Fibonacci heaps be affected by the absence of this protocol?

A detailed code for manipulating Fibonacci heaps can be found in Knuth [6].

7.4 Pairing Heaps

The pairing heap was designed to be a self-adjusting analogue of the Fibonacci heap, in much the same way that the skew heap is a self-adjusting analogue of the leftist heap (see Chapters 5 and 6). The only structure maintained in a pairing heap node, besides item information, consists of three pointers: leftmost child and two sibling pointers. (The leftmost child of a node uses its left sibling pointer to point to its parent, to facilitate updating the leftmost child pointer its parent.) See Figure 7.12 for an illustration of pointer structure.

There are no cascading cuts—only simple cuts for decrease-key and deletion operations. With the absence of parent pointers, decrease-key operations uniformly require a single cut (removal from the sibling list, in actuality), as there is no efficient way

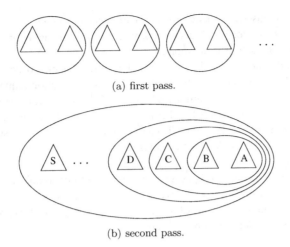

(a) first pass.

(b) second pass.

FIGURE 7.13   Two-pass pairing. The encircled trees get linked. For example, in (b) trees A and B get linked, and the result then gets linked with the tree C, and so on: (a) first pass, (b) second pass.

to check whether heap order would otherwise be violated. Although there are several varieties of pairing heaps, our discussion presents the two-pass version (the simplest), for which a given heap consists of only a single tree. The minimum element is thus uniquely located, and melding requires only a single linking operation. Similarly, a decrease-key operation consists of a subtree cut followed by a linking operation. Extractmin is implemented by removing the tree root and then linking the root's subtrees in a manner described later. Other deletions involve (a) a subtree cut, (b) an extractmin operation on the cut subtree, and (c) linking the remnant of the cut subtree with the original root.

The extractmin operation combines the subtrees of the root using a process referred to as *two-pass pairing*. Let $x_1, \ldots, x_k$ be the subtrees of the root in left-to-right order. The first pass begins by linking x_1 and x_2. Then x_3 and x_4 are linked, followed by x_5 and x_6, and so on, so that the odd positioned trees are linked with neighboring even positioned trees. Let $y_1, \ldots, y_h$, $h = \lceil k/2 \rceil$, be the resulting trees, respecting left-to-right order. (If k is odd, then $y_{\lceil k/2 \rceil}$ is x_k.) The second pass reduces these to a single tree with linkings that proceed from right-to-left. The rightmost pair of trees, y_h and y_{h-1} are linked first, followed by the linking of y_{h-2} with the result of the preceding linking, and so on, until finally we link y_1 with the structure formed from the linkings of $y_2, \ldots, y_h$ (see Figure 7.13).

Since two-pass pairing is not particularly intuitive, a few motivating remarks are offered. The first pass is natural enough, and one might consider simply repeating the process on the remaining trees, until, after logarithmically many such passes, only one tree remains. Indeed, this is known as the multipass variation. Unfortunately, its behavior is less understood than that of the two-pass pairing variation.

The second (right-to-left) pass is also quite natural. Let H be a binomial heap with exactly 2^k items, so that it consists of a single tree. Now suppose that an extractmin followed by an insertion operation are executed. The linkings that take place among the subtrees of the deleted root (after the new node is linked with the rightmost of these subtrees) entail the right-to-left processing that characterizes the second pass. So why not simply rely upon a single right-to-left pass and omit the first? The reason is that although the second pass preserves existing balance within the structure, it doesn't improve upon poorly balanced situations (manifested when most linkings take place between trees of disparate sizes). For example, using a single-right-to-left-pass version of a pairing heap to sort an increasing sequence of length n (n insertions followed by n extractmin operations) would result in an n^2 sorting algorithm. (Each of the extractmin operations yields a tree of height 1 or less.) See Section 7.5, however, for an interesting twist.

In actuality two-pass pairing was inspired [3] by consideration of splay trees (see Chapter 13). If we consider the child, sibling representation that maps a forest of arbitrary trees into a binary tree, then two-pass pairing can be viewed as a splay operation on a search tree path with no bends [3]. The analysis for splay trees then carries over to provide an amortized analysis for pairing heaps.

A detailed code for manipulating pairing heaps can be found in Weiss [5].

The asymptotic behavior of pairing heaps is an interesting and unresolved matter. In the original paper [3], it was demonstrated that the operations require only $O(\log n)$ amortized time. The fact that decrease-key operations require only constant actual time makes seemingly plausible the possibility that the amortized costs of the various operations match those of Fibonacci heaps. This line of thought, however, requires addressing the possibility that the presence of decrease-key operations might degrade the efficiency with which extractmin operations can be performed. On the positive side, Pettie [7] has derived a $2^{O(\sqrt{\log \log n})}$

amortized bound for decrease-key and insertion operations (in the presence of a $O(\log n)$ bound for the extractmin operation), and it is known that for those applications for which decrease-key operations are highly predominant, pairing heaps provably meet the optimal asymptotic bounds characteristic of Fibonacci heaps [8].

Despite the above positive considerations, as a theoretical matter pairing heaps do not match the asymptotic bounds of Fibonacci heaps. More precisely, it has been demonstrated [8] that there exist operation sequences for which the aggregate execution cost via pairing heaps is *not* consistent with amortized costs of $O(\log n)$ for extractmin and insertion operations, and $O(\log \log n)$ for decrease-key operations. (We observe that Pettie's upper bound [7] on the amortized cost of decrease-key operations, cited above, exceeds any fixed power of $\log \log n$.) The derivation of this negative result takes place in an information-theoretic framework, so that it is applicable to all data structures encompassed in a suitably defined model of computation, a matter that we shall return to shortly.

In terms of practice, because the pairing heap design dispenses with the rather complicated, carefully crafted constructs put in place primarily to facilitate proving the time bounds enjoyed by Fibonacci heaps, we can expect efficiency gains at the level of elementary steps such as linking operations. The study of Moret and Shapiro [4] provides experimental evidence in support of the conclusion that pairing heaps are superior to Fibonacci heaps in practice.

7.5 Related Developments

7.5.1 Other Variations of Pairing Heaps

The reader may wonder whether some alternative to two-pass pairing might provably attain the asymptotic performance bounds satisfied by Fibonacci heaps. The strongest result to date is attained by a variation of pairing heap devised by Elmasry [9], for which extractmin is supported with $O(\log n)$ amortized cost and decrease-key with $O(\log \log n)$ amortized cost (and insertion with $O(1)$ amortized cost), thereby matching the known lower bound applicable for two-pass pairing heaps. The information-theoretic framework [8] under which these lower bounds for two-pass pairing heaps have been established, however, does not encompass structures such as Elmasry's, so there remains the possibility that some other alternative to two-pass pairing might further improve the upper bounds attained by Elmasry's structure. We mention that Iacono and Özkan [10] have extended the reach of this information-theoretic framework, developing a "Pure Heap Model" that broadens the class of data structures to which the lower bounds established for two-pass pairing apply. The pure heap model, however, does *not* encompass Elmasry's data structure, and moreover the Elmasry *upper* bounds (matching the lower bounds established for two-pass pairing) have not been demonstrated for any data structure in the pure heap model, though one data structure, the "sort heap" [11], comes close.

7.5.2 Adaptive Properties of Pairing Heaps

Consider the problem of merging k sorted lists of respective lengths $n_1, n_2, \ldots, n_k$, with $\sum n_i = n$. The standard merging strategy that performs $\lg k$ rounds of pairwise list merges requires $n \lg k$ time. However, a merge pattern based upon the binary Huffman tree, having minimal external path length for the weights $n_1, n_2, \ldots, n_k$, is more efficient when the lengths n_i are nonuniform and provides a near-optimal solution. Pairing heaps can be utilized to provide a rather different solution as follows. Treat each sorted list as a linearly structured pairing heap. Then (a) meld these k heaps together and (b) repeatedly execute extractmin operations to retrieve the n items in their sorted order. The number of comparisons that take place is bounded by

$$O\left(\log\binom{n}{n_1, \ldots, n_k}\right)$$

Since the above multinomial coefficient represents the number of possible merge patterns, the information-theoretic bound implies that this result is optimal to within a constant factor. The pairing heap thus self-organizes the sorted list arrangement to approximate an optimal merge pattern. Iacono has derived a "working-set" theorem that quantifies a similar adaptive property satisfied by pairing heaps. Given a sequence of insertion and extractmin operations initiated with an empty heap, at the time a given item x is deleted we can attribute to x a contribution bounded by $O(\log op(x))$ to the total running time of the sequence, where $op(x)$ is the number of heap operations that have taken place since x was inserted (see Reference 12 for a slightly tighter estimate).

7.5.3 Soft Heaps

An interesting development [13] that builds upon and extends binomial heaps in a different direction is a data structure referred to as a *soft heap*. The soft heap departs from the standard notion of priority queue by allowing for a type of error, referred to as *corruption*, which confers enhanced efficiency. When an item becomes corrupted, its key value gets increased. Findmin returns the minimum current key, which might or might not be corrupted. The user has no control over which items become corrupted, this prerogative belonging to the data structure. But the user does control the overall amount of corruption in the following sense.

The user specifies a parameter, $0 < \epsilon \leq 1/2$, referred to as the error rate, that governs the behavior of the data structure as follows. The operations findmin and deletion are supported in constant amortized time, and insertion is supported in $O(\log 1/\epsilon)$ amortized time. Moreover, no more than an ϵ fraction of the items present in the heap are corrupted at any given time.

To illustrate the concept, let x be an item returned by findmin, from a soft heap of size n. Then there are no more than ϵn items in the heap whose original keys are less than the original key of x.

Soft heaps are rather subtle, and we won't attempt to discuss specifics of their design. Soft heaps have been used to construct an optimal comparison-based minimum spanning tree algorithm [14], although its actual running time has not been determined. Soft heaps have also been used to construct a comparison-based algorithm with known running time $m\alpha(m, n)$ on a graph with n vertices and m edges [15], where $\alpha(m, n)$ is a functional inverse of the Ackermann function. Chazelle [13] has also observed that soft heaps can be used to implement median selection in linear time, a significant departure from previous methods.

References

1. J. Vuillemin, A Data Structure for Manipulating Priority Queues, *Communications of the ACM*, 21, 1978, 309–314.
2. M. L. Fredman and R. E. Tarjan, Fibonacci Heaps and Their Uses in Improved Optimization Algorithms, *Journal of the ACM*, 34, 1987, 596–615.
3. M. L. Fredman, R. Sedgewick, D. D. Sleator, and R. E. Tarjan, The Pairing Heap: A New Form of Self-adjusting Heap, *Algorithmica*, 1, 1986, 111–129.
4. B. M. E. Moret and H. D. Shapiro, An Empirical Analysis of Algorithms for Constructing a Minimum Spanning Tree, *Proceedings of the Second Workshop on Algorithms and Data Structures*, 1991, 400–411.
5. M. A. Weiss, *Data Structures and Algorithms in C*, 2nd ed., Addison-Wesley, Reading, MA, 1997.
6. D. E. Knuth, *The Stanford Graph Base*. ACM Press, New York, NY, 1994.
7. S. Pettie, Towards a Final Analysis of Pairing Heaps, *FOCS*, 2005, 2005, 174–183.
8. M. L. Fredman, On the Efficiency of Pairing Heaps and Related Data Structures, *Journal of the ACM*, 46, 1999, 473–501.
9. A. Elmasry, Pairing Heaps with $O(\log \log n)$ Decrease Cost, *SODA*, 2009, 471–476.
10. J. Iacono and O. Özkan, A Tight Lower Bound for Decrease-key in the Pure Heap Model. *CoRR abs/1407.6665*, 2014.
11. J. Iacono and O. Özkan, Why Some Heaps Support Constant-Amortized-Time Decrease-key Operations, and Others Do not, *ICALP*, (1), 2014, 637–649.
12. J. Iacono, Distribution sensitive data structures, *PhD Thesis, Rutgers University*, 2001.
13. B. Chazelle, The Soft Heap: An Approximate Priority Queue with Optimal Error Rate, *Journal of the ACM*, 47, 2000, 1012–1027.
14. S. Pettie and V. Ramachandran, An Optimal Minimum Spanning Tree Algorithm, *Journal of the ACM*, 49, 2002, 16–34.
15. B. Chazelle, A Faster Deterministic Algorithm for Minimum Spanning Trees, *Journal of the ACM*, 47, 2000, 1028–1047.

8

Double-Ended Priority Queues*

8.1	Definition and an Application	97
8.2	Symmetric Min-Max Heaps	98
8.3	Interval Heaps	100
	Inserting an Element • Removing the Min Element • Initializing an Interval Heap • Complexity of Interval Heap Operations • The Complementary Range Search Problem	
8.4	Min-Max Heaps	105
	Inserting an Element • Removing the Min Element	
8.5	Deaps	108
	Inserting an Element • Removing the Min Element	
8.6	Generic Methods for DEPQs	111
	Dual Priority Queues • Total Correspondence • Leaf Correspondence	
8.7	Meldable DEPQs	113
	Acknowledgment	113
	References	113

Sartaj Sahni
University of Florida

8.1 Definition and an Application

A *double-ended priority queue (DEPQ)* is a collection of zero or more elements. Each element has a priority or value. The operations performed on a double-ended priority queue are:

1. *getMin()* ... return element with minimum priority
2. *getMax()* ... return element with maximum priority
3. *put(x)* ... insert the element x into the DEPQ
4. *removeMin()* ... remove an element with minimum priority and return this element
5. *removeMax()* ... remove an element with maximum priority and return this element

One application of a DEPQ is to the adaptation of quick sort, which has the the best expected run time of all known internal sorting methods, to external sorting. The basic idea in quick sort is to partition the elements to be sorted into three groups L, M, and R. The middle group M contains a single element called the *pivot*, all elements in the left group L are $\leq$ the pivot, and all elements in the right group R are $\geq$ the pivot. Following this partitioning, the left and right element groups are sorted recursively.

In an external sort, we have more elements than can be held in the memory of our computer. The elements to be sorted are initially on a disk and the sorted sequence is to be left on the disk. When the internal quick sort method outlined above is extended to an external quick sort, the middle group M is made as large as possible through the use of a DEPQ. The external quick sort strategy is:

1. Read in as many elements as will fit into an internal DEPQ. The elements in the DEPQ will eventually be the middle group of elements.
2. Read in the remaining elements. If the next element is $\leq$ the smallest element in the DEPQ, output this next element as part of the left group. If the next element is $\geq$ the largest element in the DEPQ, output this next element as part of the right group. Otherwise, remove either the max or min element from the DEPQ (the choice may be made randomly or

---

* This chapter has been reprinted from first edition of this Handbook, without any content updates.

alternately); if the max element is removed, output it as part of the right group; otherwise, output the removed element as part of the left group; insert the newly input element into the DEPQ.

3. Output the elements in the DEPQ, in sorted order, as the middle group.
4. Sort the left and right groups recursively.

In this chapter, we describe four implicit data structures—symmetric min-max heaps, interval heaps, min-max heaps, and deaps—for DEPQs. Also, we describe generic methods to obtain efficient DEPQ data structures from efficient data structures for single-ended priority queues (PQ).[*]

8.2 Symmetric Min-Max Heaps

Several simple and efficient implicit data structures have been proposed for the representation of a DEPQ [1–7]. All of these data structures are adaptations of the classical heap data structure (Chapter 2) for a PQ. Further, in all of these DEPQ data structures, *getMax* and *getMin* take $O(1)$ time and the remaining operations take $O(\log n)$ time each (n is the number of elements in the DEPQ). The symmetric min-max heap structure of Arvind and Pandu Rangan [1] is the simplest of the implicit data structures for DEPQs. Therefore, we describe this data structure first.

A *symmetric min-max heap* (SMMH) is a complete binary tree in which each node other than the root has exactly one element. The root of an SMMH is empty and the total number of nodes in the SMMH is $n + 1$, where n is the number of elements. Let x be any node of the SMMH. Let *elements*(x) be the elements in the subtree rooted at x but excluding the element (if any) in x. Assume that *elements*$(x) \neq \emptyset$. x satisfies the following properties:

1. The left child of x has the minimum element in *elements*(x).
2. The right child of x (if any) has the maximum element in *elements*(x).

Figure 8.1 shows an example SMMH that has 12 elements. When x denotes the node with 80, *elements*$(x) = \{6, 14, 30, 40\}$; the left child of x has the minimum element 6 in *elements*(x); and the right child of x has the maximum element 40 in *elements*(x). You may verify that every node x of this SMMH satisfies the stated properties.

Since an SMMH is a complete binary tree, it is stored as an implicit data structure using the standard mapping of a complete binary tree into an array. When $n = 1$, the minimum and maximum elements are the same and are in the left child of the root of the SMMH. When $n > 1$, the minimum element is in the left child of the root and the maximum is in the right child of the root. So the *getMin* and *getMax* operations take $O(1)$ time.

It is easy to see that an $n + 1$-node complete binary tree with an empty root and one element in every other node is an SMMH iff the following are true:

P1: For every node x that has a right sibling, the element in x is less than or equal to that in the right sibling of x.
P2: For every node x that has a grandparent, the element in the left child of the grandparent is less than or equal to that in x.
P3: For every node x that has a grandparent, the element in the right child of the grandparent is greater than or equal to that in x.

Notice that if property P1 is satisfied at node x, then at most one of P2 and P3 may be violated at x. Using properties P1 through P3 we arrive at simple algorithms to insert and remove elements. These algorithms are simple adaptations of the corresponding algorithms for min and max heaps. Their complexity is $O(\log n)$. We describe only the insert operation. Suppose we wish to insert

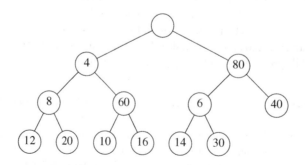

FIGURE 8.1   A symmetric min-max heap.

---

[*] A *minPQ* supports the operations *getmin*(), *put*(x), and *removeMin*() while a *maxPQ* supports the operations *getMax*(), *put*(x), and *removeMax*().

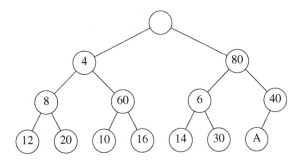

FIGURE 8.2   The SMMH of Figure 8.1 with a node added.

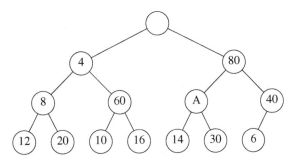

FIGURE 8.3   The SMMH of Figure 8.2 with 6 moved down.

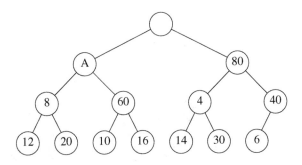

FIGURE 8.4   The SMMH of Figure 8.3 with 4 moved down.

2 into the SMMH of Figure 8.1. Since an SMMH is a complete binary tree, we must add a new node to the SMMH in the position shown in Figure 8.2; the new node is labeled A. In our example, A will denote an empty node.

If the new element 2 is placed in node A, property P2 is violated as the left child of the grandparent of A has 6. So we move the 6 down to A and obtain Figure 8.3.

Now we see if it is safe to insert the 2 into node A. We first notice that property P1 cannot be violated, because the previous occupant of node A was greater than 2. Similarly, property P3 cannot be violated. Only P2 can be violated only when $x = A$. So we check P2 with $x = A$. We see that P2 is violated because the left child of the grandparent of A has the element 4. So we move the 4 down to A and obtain the configuration shown in Figure 8.4.

For the configuration of Figure 8.4 we see that placing 2 into node A cannot violate property P1, because the previous occupant of node A was greater than 2. Also properties P2 and P3 cannot be violated, because node A has no grandparent. So we insert 2 into node A and obtain Figure 8.5.

Let us now insert 50 into the SMMH of Figure 8.5. Since an SMMH is a complete binary tree, the new node must be positioned as in Figure 8.6.

Since A has a right child of its parent, we first check P1 at node A. If the new element (in this case 50) is smaller than that in the left sibling of A, we swap the new element and the element in the left sibling. In our case, no swap is done. Then we check P2 and P3. We see that placing 50 into A would violate P3. So the element 40 in the right child of the grandparent of A is moved down to node A. Figure 8.7 shows the resulting configuration. Placing 50 into node A of Figure 8.7 cannot create a P1 violation

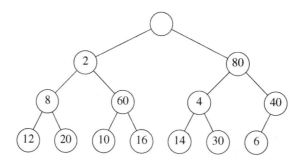

FIGURE 8.5   The SMMH of Figure 8.4 with 2 inserted.

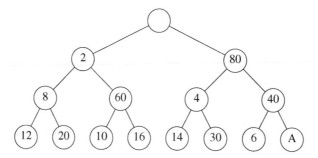

FIGURE 8.6   The SMMH of Figure 8.5 with a node added.

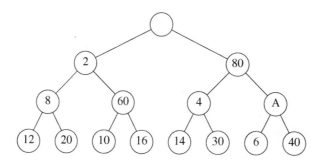

FIGURE 8.7   The SMMH of Figure 8.6 with 40 moved down.

because the previous occupant of node A was smaller. Also, a P2 violation isn't possible either. So only P3 needs to be checked at A. Since there is no P3 violation at A, 50 is placed into A.

The algorithm to remove either the min or max element is a similar adaptation of the trickle-down algorithm used to remove an element from a min or max heap.

8.3 Interval Heaps

The twin heaps of [7], the min-max pair heaps of [6], the interval heaps of [5,8], and the diamond deques of [4] are virtually identical data structures. In each of these structures, an n element DEPQ is represented by a min heap with $\lceil n/2 \rceil$ elements and a max heap with the remaining $\lfloor n/2 \rfloor$ elements. The two heaps satisfy the property that each element in the min heap is $\leq$ the corresponding element (two elements correspond if they occupy the same position in their respective binary trees) in the max heap. When the number of elements in the DEPQ is odd, the min heap has one element (i.e., element $\lceil n/2 \rceil$) that has no corresponding element in the max heap. In the twin heaps of [7], this is handled as a special case and one element is kept outside of the two heaps. In min-max pair heaps, interval heaps, and diamond deques, the case when n is odd is handled by requiring element $\lceil n/2 \rceil$ of the min heap to be $\leq$ element $\lfloor n/4 \rfloor$ of the max heap.

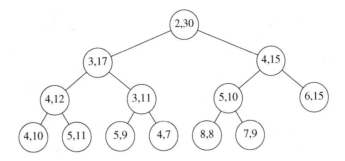

FIGURE 8.8   An interval heap.

In the twin heaps of [7], the min and max heaps are stored in two arrays *min* and *max* using the standard array representation of a complete binary tree[*] [9]. The correspondence property becomes $min[i] \leq max[i]$, $1 \leq i \leq \lfloor n/2 \rfloor$. In the min-max pair heaps of [6] and the interval heaps of [5], the two heaps are stored in a single array *minmax* and we have $minmax[i].min$ being the i'th element of the min heap, $1 \leq i \leq \lceil n/2 \rceil$ and $minmax[i].max$ being the i'th element of the max heap, $1 \leq i \leq \lfloor n/2 \rfloor$. In the diamond deque [4], the two heaps are mapped into a single array with the min heap occupying even positions (beginning with position 0) and the max heap occupying odd positions (beginning with position 1). Since this mapping is slightly more complex than the ones used in twin heaps, min-max pair heaps, and interval heaps, actual implementations of the diamond deque are expected to be slightly slower than implementations of the remaining three structures.

Since the twin heaps of [7], the min-max pair heaps of [6], the interval heaps of [5], and the diamond deques of [4] are virtually identical data structures, we describe only one of these—interval heaps—in detail. An *interval heap* is a complete binary tree in which each node, except possibly the last one (the nodes of the complete binary tree are ordered using a level order traversal), contains two elements. Let the two elements in node *P* be *a* and *b*, where $a \leq b$. We say that the node *P* represents the closed interval $[a, b]$. *a* is the left end point of the interval of *P*, and *b* is its right end point.

The interval $[c, d]$ is contained in the interval $[a, b]$ iff $a \leq c \leq d \leq b$. In an interval heap, the intervals represented by the left and right children (if they exist) of each node *P* are contained in the interval represented by *P*. When the last node contains a single element *c*, then $a \leq c \leq b$, where $[a, b]$ is the interval of the parent (if any) of the last node.

Figure 8.8 shows an interval heap with 26 elements. You may verify that the intervals represented by the children of any node *P* are contained in the interval of *P*.

The following facts are immediate:

1. The left end points of the node intervals define a min heap, and the right end points define a max heap. In case the number of elements is odd, the last node has a single element which may be regarded as a member of either the min or max heap. Figure 8.9 shows the min and max heaps defined by the interval heap of Figure 8.8.
2. When the root has two elements, the left end point of the root is the minimum element in the interval heap and the right end point is the maximum. When the root has only one element, the interval heap contains just one element. This element is both the minimum and maximum element.
3. An interval heap can be represented compactly by mapping into an array as is done for ordinary heaps. However, now, each array position must have space for two elements.
4. The height of an interval heap with *n* elements is $\Theta(\log n)$.

8.3.1 Inserting an Element

Suppose we are to insert an element into the interval heap of Figure 8.8. Since this heap currently has an even number of elements, the heap following the insertion will have an additional node *A* as is shown in Figure 8.10.

The interval for the parent of the new node *A* is [6, 15]. Therefore, if the new element is between 6 and 15, the new element may be inserted into node *A*. When the new element is less than the left end point 6 of the parent interval, the new element is inserted into the min heap embedded in the interval heap. This insertion is done using the min heap insertion procedure starting at node *A*. When the new element is greater than the right end point 15 of the parent interval, the new element is inserted into the max heap embedded in the interval heap. This insertion is done using the max heap insertion procedure starting at node *A*.

---

[*] In a *full* binary tree, every non-empty level has the maximum number of nodes possible for that level. Number the nodes in a full binary tree 1, 2, . . . beginning with the root level and within a level from left to right. The nodes numbered 1 through *n* define the unique *complete binary tree* that has *n* nodes.

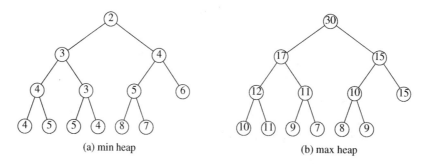

FIGURE 8.9   Min and max heaps embedded in Figure 8.8.

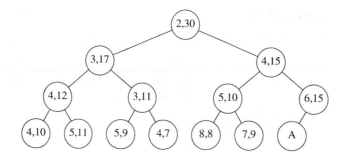

FIGURE 8.10   Interval heap of Figure 8.8 after one node is added.

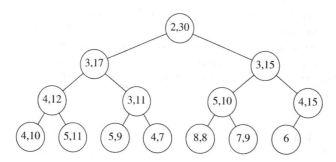

FIGURE 8.11   The interval heap of Figure 8.8 with 3 inserted.

If we are to insert the element 10 into the interval heap of Figure 8.8, this element is put into the node A shown in Figure 8.10. To insert the element 3, we follow a path from node A towards the root, moving left end points down until we either pass the root or reach a node whose left end point is ≤ 3. The new element is inserted into the node that now has no left end point. Figure 8.11 shows the resulting interval heap.

To insert the element 40 into the interval heap of Figure 8.8, we follow a path from node A (see Figure 8.10) towards the root, moving right end points down until we either pass the root or reach a node whose right end point is ≥ 40. The new element is inserted into the node that now has no right end point. Figure 8.12 shows the resulting interval heap.

Now, suppose we wish to insert an element into the interval heap of Figure 8.12. Since this interval heap has an odd number of elements, the insertion of the new element does not increase the number of nodes. The insertion procedure is the same as for the case when we initially have an even number of elements. Let A denote the last node in the heap. If the new element lies within the interval [6, 15] of the parent of A, then the new element is inserted into node A (the new element becomes the left end point of A if it is less than the element currently in A). If the new element is less than the left end point 6 of the parent of A, then the new element is inserted into the embedded min heap; otherwise, the new element is inserted into the embedded max heap. Figure 8.13 shows the result of inserting the element 32 into the interval heap of Figure 8.12.

8.3.2 Removing the Min Element

The removal of the minimum element is handled as several cases:

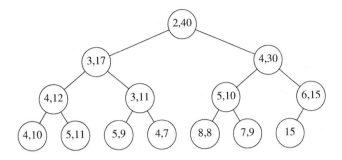

FIGURE 8.12 The interval heap of Figure 8.8 with 40 inserted.

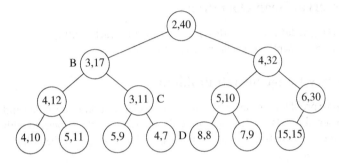

FIGURE 8.13 The interval heap of Figure 8.12 with 32 inserted.

1. When the interval heap is empty, the *removeMin* operation fails.
2. When the interval heap has only one element, this element is the element to be returned. We leave behind an empty interval heap.
3. When there is more than one element, the left end point of the root is to be returned. This point is removed from the root. If the root is the last node of the interval heap, nothing more is to be done. When the last node is not the root node, we remove the left point p from the last node. If this causes the last node to become empty, the last node is no longer part of the heap. The point p removed from the last node is reinserted into the embedded min heap by beginning at the root. As we move down, it may be necessary to swap the current p with the right end point r of the node being examined to ensure that $p \leq r$. The reinsertion is done using the same strategy as used to reinsert into an ordinary heap.

Let us remove the minimum element from the interval heap of Figure 8.13. First, the element 2 is removed from the root. Next, the left end point 15 is removed from the last node and we begin the reinsertion procedure at the root. The smaller of the min heap elements that are the children of the root is 3. Since this element is smaller than 15, we move the 3 into the root (the 3 becomes the left end point of the root) and position ourselves at the left child B of the root. Since, $15 \leq 17$ we do not swap the right end point of B with the current $p = 15$. The smaller of the left end points of the children of B is 3. The 3 is moved from node C into node B as its left end point and we position ourselves at node C. Since $p = 15 > 11$, we swap the two and 15 becomes the right end point of node C. The smaller of left end points of Cs children is 4. Since this is smaller than the current $p = 11$, it is moved into node C as this node's left end point. We now position ourselves at node D. First, we swap $p = 11$ and Ds right end point. Now, since D has no children, the current $p = 7$ is inserted into node D as Ds left end point. Figure 8.14 shows the result.

The max element may removed using an analogous procedure.

8.3.3 Initializing an Interval Heap

Interval heaps may be initialized using a strategy similar to that used to initialize ordinary heaps–work your way from the heap bottom to the root ensuring that each subtree is an interval heap. For each subtree, first order the elements in the root; then reinsert the left end point of this subtree's root using the reinsertion strategy used for the *removeMin* operation, then reinsert the right end point of this subtree's root using the strategy used for the *removeMax* operation.

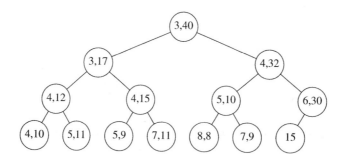

FIGURE 8.14   The interval heap of Figure 8.13 with minimum element removed.

8.3.4 Complexity of Interval Heap Operations

The operations *isEmpty*(), *size*(), *getMin*(), and *getMax*() take $O(1)$ time each; *put(x)*, *removeMin*(), and *removeMax*() take $O(\log n)$ each; and initializing an n element interval heap takes $O(n)$ time.

8.3.5 The Complementary Range Search Problem

In the *complementary range search* problem, we have a dynamic collection (i.e., points are added and removed from the collection as time goes on) of one-dimensional points (i.e., points have only an x-coordinate associated with them) and we are to answer queries of the form: what are the points outside of the interval $[a, b]$? For example, if the point collection is 3, 4, 5, 6, 8, 12, the points outside the range $[5, 7]$ are 3, 4, 8, 12.

When an interval heap is used to represent the point collection, a new point can be inserted or an old one removed in $O(\log n)$ time, where n is the number of points in the collection. Note that given the location of an arbitrary element in an interval heap, this element can be removed from the interval heap in $O(\log n)$ time using an algorithm similar to that used to remove an arbitrary element from a heap.

The complementary range query can be answered in $\Theta(k)$ time, where k is the number of points outside the range $[a, b]$. This is done using the following recursive procedure:

1. If the interval tree is empty, *return.*
2. If the root interval is contained in $[a, b]$, then all points are in the range (therefore, there are no points to report), *return.*
3. Report the end points of the root interval that are not in the range $[a, b]$.
4. Recursively search the left subtree of the root for additional points that are not in the range $[a, b]$.
5. Recursively search the right subtree of the root for additional points that are not in the range $[a, b]$.
6. *return*

Let us try this procedure on the interval heap of Figure 8.13. The query interval is $[4, 32]$. We start at the root. Since the root interval is not contained in the query interval, we reach step 3 of the procedure. Whenever step 3 is reached, we are assured that at least one of the end points of the root interval is outside the query interval. Therefore, each time step 3 is reached, at least one point is reported. In our example, both points 2 and 40 are outside the query interval and so are reported. We then search the left and right subtrees of the root for additional points. When the left subtree is searched, we again determine that the root interval is not contained in the query interval. This time only one of the root interval points (i.e., 3) is to be outside the query range. This point is reported and we proceed to search the left and right subtrees of B for additional points outside the query range. Since the interval of the left child of B is contained in the query range, the left subtree of B contains no points outside the query range. We do not explore the left subtree of B further. When the right subtree of B is searched, we report the left end point 3 of node C and proceed to search the left and right subtrees of C. Since the intervals of the roots of each of these subtrees is contained in the query interval, these subtrees are not explored further. Finally, we examine the root of the right subtree of the overall tree root, that is the node with interval $[4, 32]$. Since this node's interval is contained in the query interval, the right subtree of the overall tree is not searched further.

The complexity of the above six step procedure is $\Theta(\text{number of interval heap nodes visited})$. The nodes visited in the preceding example are the root and its two children, node B and its two children, and node C and its two children. Thus, 7 nodes are visited and a total of 4 points are reported.

We show that the total number of interval heap nodes visited is at most $3k + 1$, where k is the number of points reported. If a visited node reports one or two points, give the node a count of one. If a visited node reports no points, give it a count of zero and add one to the count of its parent (unless the node is the root and so has no parent). The number of nodes with a nonzero count is at most k. Since no node has a count more than 3, the sum of the counts is at most $3k$. Accounting for the possibility that the root reports no point, we see that the number of nodes visited is at most $3k + 1$. Therefore, the complexity of the search is $\Theta(k)$. This complexity is asymptotically optimal because every algorithm that reports k points must spend at least $\Theta(1)$ time per reported point.

In our example search, the root gets a count of 2 (1 because it is visited and reports at least one point and another 1 because its right child is visited but reports no point), node B gets a count of 2 (1 because it is visited and reports at least one point and another 1 because its left child is visited but reports no point), and node C gets a count of 3 (1 because it is visited and reports at least one point and another 2 because its left and right children are visited and neither reports a point). The count for each of the remaining nodes in the interval heap is 0.

8.4 Min-Max Heaps

In the min-max heap structure [2], all n DEPQ elements are stored in an n-node complete binary tree with alternating levels being min levels and max levels (Figure 8.15, nodes at max levels are shaded). The root level of a min-max heap is a min level. Nodes on a min level are called min nodes while those on a max level are max nodes. Every min (max) node has the property that its value is the smallest (largest) value in the subtree of which it is the root. Since 5 is in a min node of Figure 8.15, it is the smallest value in its subtree. Also, since 30 and 26 are in max nodes, these are the largest values in the subtrees of which they are the root.

The following observations are a direct consequence of the definition of a min-max heap.

1. When $n = 0$, there is no min nor max element.
2. When $n = 1$, the element in the root is both the min and the max element.
3. When $n > 1$, the element in the root is the min element; the max element is one of the up to two children of the root.

From these observations, it follows that *getMin*() and *getMax*() can be done in $O(1)$ time each.

8.4.1 Inserting an Element

When inserting an element *newElement* into a min-max heap that has n elements, we go from a complete binary tree that has n nodes to one that has $n + 1$ nodes. So, for example, an insertion into the 12-element min-max heap of Figure 8.15 results in the 13-node complete binary tree of Figure 8.16.

When $n = 0$, the insertion simply creates a min-max heap that has a single node that contains the new element. Assume that $n > 0$ and let the element in the parent, *parentNode*, of the new node j be *parentElement*. If *newElement* is placed in the new node j, the min- and max-node property can be violated only for nodes on the path from the root to *parentNode*. So, the insertion of an element need only be concerned with ensuring that nodes on this path satisfy the required min- and max-node property. There are three cases to consider.

1. *parentElement* = *newElement*
   In this case, we may place *newElement* into node j. With such a placement, the min- and max-node properties of all nodes on the path from the root to *parentNode* are satisfied.
2. *parentNode* > *newElement*

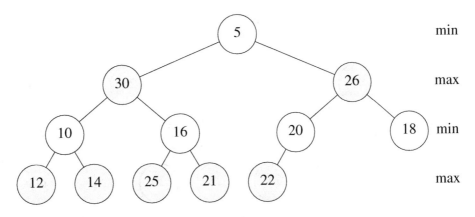

FIGURE 8.15 A 12-element min-max heap.

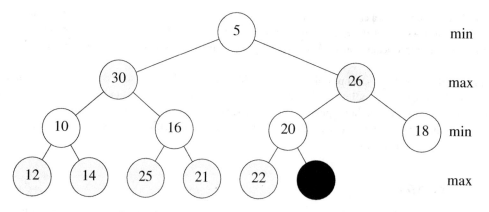

FIGURE 8.16 A 13-node complete binary tree.

If *parentNode* is a min node, we get a min-node violation. When a min-node violation occurs, we know that all max nodes on the path from the root to *parentNode* are greater than *newElement*. So, a min-node violation may be fixed by using the trickle-up process used to insert into a min heap; this trickle-up process involves only the min nodes on the path from the root to *parentNode*. For example, suppose that we are to insert *newElement* = 2 into the min-max heap of Figure 8.15. The min nodes on the path from the root to *parentNode* have values 5 and 20. The 20 and the 5 move down on the path and the 2 trickles up to the root node. Figure 8.17 shows the result. When *newElement* = 15, only the 20 moves down and the sequence of min nodes on the path from the root to *j* have values 5, 15, 20.

The case when *parentNode* is a max node is similar.

3. *parentNode* < *newElement*

When *parentNode* is a min node, we conclude that all min nodes on the path from the root to *parentNode* are smaller than *newElement*. So, we need be concerned only with the max nodes (if any) on the path from the root to *parentNode*. A trickle-up process is used to correctly position *newElement* with respect to the elements in the max nodes of this path. For the example of Figure 8.16, there is only one max node on the path to *parentNode*. This max node has the element 26. If *newElement* > 26, the 26 moves down to *j* and *newElement* trickles up to the former position of 26 (Figure 8.18 shows the case when *newElement* = 32). If *newElement* < 26, *newElement* is placed in *j*.

The case when *parentNode* is a max node is similar.

Since the height of a min-max heap is $\Theta(\log n)$ and a trickle-up examines a single element at at most every other level of the min-max heap, an insert can be done in $O(\log n)$ time.

8.4.2 Removing the Min Element

When $n = 0$, there is no min element to remove. When $n = 1$, the min-max heap becomes empty following the removal of the min element, which is in the root. So assume that $n > 1$. Following the removal of the min element, which is in the root, we need to go from an *n*-element complete binary tree to an $(n-1)$-element complete binary tree. This causes the element in position *n*

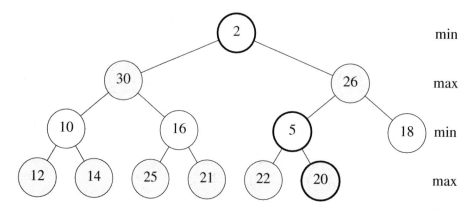

FIGURE 8.17 Min-max heap of Figure 8.15 following insertion of 2.

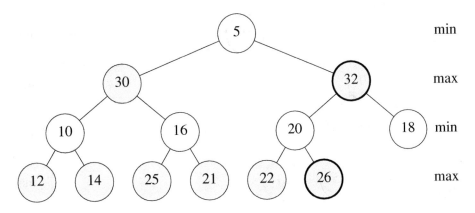

FIGURE 8.18 The min-max heap of Figure 8.15 following the insertion of 32.

of the min-max heap array to drop out of the min-max heap. Figure 8.17 shows the situation following the removal of the min element, 5, from the min-max heap of Figure 8.15. In addition to the 5, which was the min element and which has been removed from the min-max heap, the 22 that was in position $n = 12$ of the min-max heap array has dropped out of the min-max heap. To get the dropped-out element 22 back into the min-max heap, we perform a trickle-down operation that begins at the root of the min-max heap.

The trickle-down operation for min-max heaps is similar to that for min and max heaps. The root is to contain the smallest element. To ensure this, we determine the smallest element in a child or grandchild of the root. If 22 is ≤ the smallest of these children and grandchildren, the 22 is placed in the root. If not, the smallest of the children and grandchildren is moved to the root; the trickle-down process is continued from the position vacated by the just moved smallest element.

In our example, examining the children and grandchildren of the root of Figure 8.19, we determine that the smallest of these is 10. Since $10 < 22$, the 10 moves to the root and the 22 trickles down (Figure 8.20). A special case arises when this trickle down of the 22 by 2 levels causes the 22 to trickle past a smaller element (in our example, we trickle past a larger element 30). When this special case arises, we simply exchange the 22 and the smaller element being trickled past. The trickle-down process applied at the vacant node of Figure 8.20 results in the 22 being placed into the vacant node.

Suppose that *droppedElement* is the element dropped from *minmaxHeap*[n] when a remove min is done from an *n*-element min-max heap. The following describes the trickle-down process used to reinsert the dropped element.

1. *The root has no children.*
   In this case *droppedElement* is inserted into the root and the trickle down terminates.
2. *The root has at least one child.*
   Now the smallest key in the min-max heap is in one of the children or grandchildren of the root. We determine which of these nodes has the smallest key. Let this be node k. The following possibilities need to be considered:
   a. *droppedElement* ≤ *minmaxHeap*[k].
      droppedElement may be inserted into the root, as there is no smaller element in the heap. The trickle down terminates.

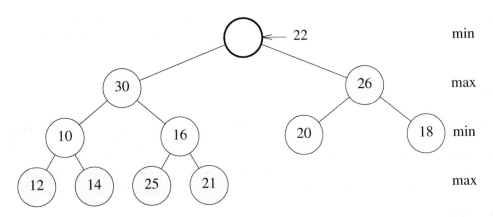

FIGURE 8.19 Situation following a remove min from Figure 8.15.

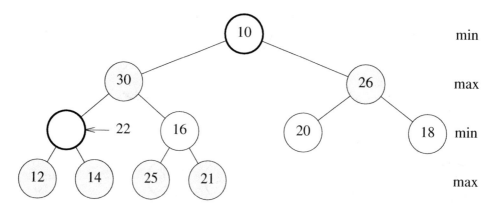

FIGURE 8.20   Situation following one iteration of the trickle-down process.

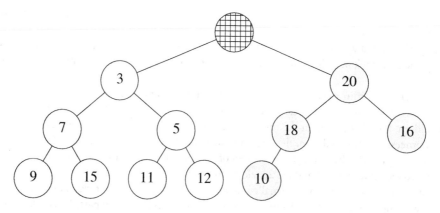

FIGURE 8.21   An 11-element deap.

c. *droppedElement > minmaxHeap[k]* and *k* is a child of the root.
    Since *k* is a max node, it has no descendants larger than *minmaxHeap[k]*. Hence, node *k* has no descendants larger than *droppedElement*. So, the *minmaxHeap[k]* may be moved to the root, and *droppedElement* placed into node *k*. The trickle down terminates.

e. *droppedElement > minmaxHeap[k]* and *k* is a grandchild of the root.
    minmaxHeap[k] is moved to the root. Let *p* be the parent of *k*. If *droppedElement > minmaxHeap[p]*, then *minmax-Heap[p]* and *droppedElement* are interchanged. This interchange ensures that the max node *p* contains the largest key in the subheap with root *p*. The trickle down continues with *k* as the root of a min-max (sub) heap into which an element is to be inserted.

The complexity of the remove-min operation is readily seen to be $O(\log n)$. The remove-max operation is similar to the remove-min operation, and min-max heaps may be initialized in $\Theta(n)$ time using an algorithm similar to that used to initialize min and max heaps [9].

8.5 Deaps

The deap structure of [3] is similar to the two-heap structures of [4–7]. At the conceptual level, we have a min heap and a max heap. However, the distribution of elements between the two is not $\lceil n/2 \rceil$ and $\lfloor n/2 \rfloor$. Rather, we begin with an $(n+1)$-node complete binary tree. Its left subtree is the min heap and its right subtree is the max heap (Figure 8.21, max-heap nodes are shaded). The correspondence property for deaps is slightly more complex than that for the two-heap structures of [4–7].

A *deap* is a complete binary tree that is either empty or satisfies the following conditions:

1. The root is empty.
2. The left subtree is a min heap and the right subtree is a max heap.

3. *Correspondence property.* Suppose that the right subtree is not empty. For every node *x* in the left subtree, define its corresponding node *y* in the right subtree to be the node in the same position as *x*. In case such a *y* doesn't exist, let *y* be the corresponding node for the parent of *x*. The element in *x* is ≤ the element in *y*.

For the example complete binary tree of Figure 8.21, the corresponding nodes for the nodes with 3, 7, 5, 9, 15, 11, and 12, respectively, have 20, 18, 16, 10, 18, 16, and 16.

Notice that every node *y* in the max heap has a unique corresponding node *x* in the min heap. The correspondence property for max-heap nodes is that the element in *y* be ≥ the element in *x*. When the correspondence property is satisfied for all nodes in the min heap, this property is also satisfied for all nodes in the max heap.

We see that when *n* = 0, there is no min or max element, when *n* = 1, the root of the min heap has both the min and the max element, and when *n* > 1, the root of the min heap is the min element and the root of the max heap is the max element. So, both *getMin()* and *getMax()* may be implemented to work in *O*(1) time.

8.5.1 Inserting an Element

When an element is inserted into an *n*-element deap, we go form a complete binary tree that has *n* + 1 nodes to one that has *n* + 2 nodes. So, the shape of the new deap is well defined. Following an insertion, our 11-element deap of Figure 8.21 has the shape shown in Figure 8.22. The new node is node *j* and its corresponding node is node *i*.

To insert *newElement*, temporarily place *newElement* into the new node *j* and check the correspondence property for node *j*. If the property isn't satisfied, swap *newElement* and the element in its corresponding node; use a trickle-up process to correctly position *newElement* in the heap for the corresponding node *i*. If the correspondence property is satisfied, do not swap *newElement*; instead use a trickle-up process to correctly place *newElement* in the heap that contains node *j*.

Consider the insertion of *newElement* = 2 into Figure 8.22. The element in the corresponding node *i* is 15. Since the correspondence property isn't satisfied, we swap 2 and 15. Node *j* now contains 15 and this swap is guaranteed to preserve the max-heap properties of the right subtree of the complete binary tree. To correctly position the 2 in the left subtree, we use the standard min-heap trickle-up process beginning at node *i*. This results in the configuration of Figure 8.23.

To insert *newElement* = 19 into the deap of Figure 8.22, we check the correspondence property between 15 and 19. The property is satisfied. So, we use the trickle-up process for max heaps to correctly position *newElement* in the max heap. Figure 8.24 shows the result.

Since the height of a deap is Θ(log *n*), the time to insert into a deap is *O*(log *n*).

8.5.2 Removing the Min Element

Assume that *n* > 0. The min element is in the root of the min heap. Following its removal, the deap size reduces to *n* − 1 and the element in position *n* + 1 of the deap array is dropped from the deap. In the case of our example of Figure 8.21, the min element 3 is removed and the 10 is dropped. To reinsert the dropped element, we first trickle the vacancy in the root of the min heap down to a leaf of the min heap. This is similar to a standard min-heap trickle down with ∞ as the reinsert element. For our example, this trickle down causes 5 and 11 to, respectively, move to their parent nodes. Then, the dropped element 10 is inserted using a trickle-up process beginning at the vacant leaf of the min heap. The resulting deap is shown in Figure 8.25.

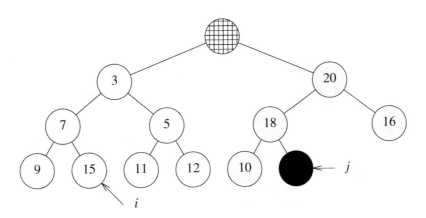

FIGURE 8.22 Shape of a 12-element deap.

Since a *removeMin* requires a trickle-down pass followed by a trickle-up pass and since the height of a deap is $\Theta(\log n)$, the time for a *removeMin* is $O(\log n)$. A *removeMax* is similar. Also, we may initialize a deap in $\Theta(n)$ time using an algorithm similar to that used to initialize a min or max heap [9].

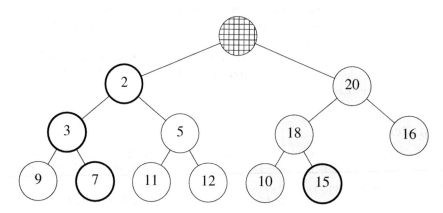

FIGURE 8.23 Deap of Figure 8.21 with 2 inserted.

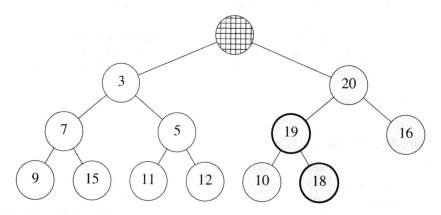

FIGURE 8.24 Deap of Figure 8.21 with 19 inserted.

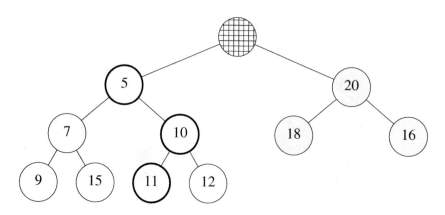

FIGURE 8.25 Deap of Figure 8.21 following a remove min operation.

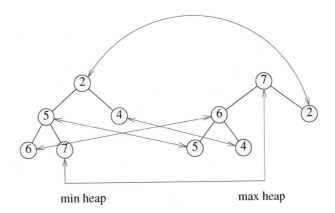

FIGURE 8.26   Dual heap.

8.6 Generic Methods for DEPQs

8.6.1 Dual Priority Queues

General methods [10] exist to arrive at efficient DEPQ data structures from single-ended priority queue data structures that also provide an efficient implementation of the *remove (theNode)* operation (this operation removes the node *theNode* from the PQ). The simplest of these methods, *dual structure method*, maintains both a min PQ (called *minPQ*) and a max PQ (called *maxPQ*) of all the DEPQ elements together with *correspondence pointers* between the nodes of the min PQ and the max PQ that contain the same element. Figure 8.26 shows a dual heap structure for the elements 6, 7, 2, 5, 4. Correspondence pointers are shown as double-headed arrows.

Although Figure 8.26 shows each element stored in both the min and the max heap, it is necessary to store each element in only one of the two heaps.

The minimum element is at the root of the min heap and the maximum element is at the root of the max heap. To insert an element x, we insert x into both the min and the max heaps and then set up correspondence pointers between the locations of x in the min and max heaps. To remove the minimum element, we do a *removeMin* from the min heap and a *remove(theNode)*, where *theNode* is the corresponding node for the removed element, from the max heap. The maximum element is removed in an analogous way.

8.6.2 Total Correspondence

The notion of total correspondence borrows heavily from the ideas used in a twin heap [7]. In the twin heap data structure n elements are stored in a min heap using an array $minHeap[1{:}n]$ and n other elements are stored in a max heap using the array $maxHeap[1:n]$. The min and max heaps satisfy the inequality $minHeap[i] \leq maxHeap[i]$, $1 \leq i \leq n$. In this way, we can represent a DEPQ with $2n$ elements. When we must represent a DEPQ with an odd number of elements, one element is stored in a buffer, and the remaining elements are divided equally between the arrays *minHeap* and *maxHeap*.

In total correspondence, we remove the positional requirement in the relationship between pairs of elements in the min heap and max heap. The requirement becomes: for each element a in *minPQ* there is a distinct element b in *maxPQ* such that $a \leq b$ and vice versa. (a, b) is a corresponding pair of elements. Figure 8.27a shows a twin heap with 11 elements and Figure 8.27b shows a total correspondence heap. The broken arrows connect corresponding pairs of elements.

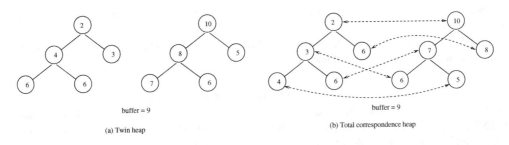

(a) Twin heap   (b) Total correspondence heap

FIGURE 8.27   Twin heap and total correspondence heap.

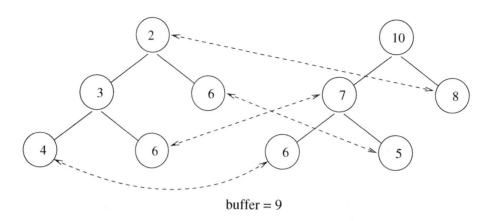

buffer = 9

FIGURE 8.28   Leaf correspondence heap.

In a twin heap the corresponding pairs ($minHeap[i]$, $maxHeap[i]$) are implicit, whereas in a total correspondence heap these pairs are represented using explicit pointers.

In a total correspondence DEPQ, the number of nodes is either n or $n-1$. The space requirement is half that needed by the dual priority queue representation. The time required is also reduced. For example, if we do a sequence of inserts, every other one simply puts the element in the buffer. The remaining inserts put one element in *maxPQ* and one in *minPQ*. So, on average, an insert takes time comparable to an insert in either *maxPQ* or *minPQ*. Recall that when dual priority queues are used the insert time is the sum of the times to insert into *maxPQ* and *minPQ*. Note also that the size of *maxPQ* and *minPQ* together is half that of a dual priority queue.

If we assume that the complexity of the insert operation for priority queues as well as 2 *remove(theNode)* operations is no more than that of the delete max or min operation (this is true for all known priority queue structures other than weight biased leftist trees [11]), then the complexity of *removeMax* and *removeMin* for total correspondence DEPQs is the same as for the *removeMax* and *removeMin* operation of the underlying priority queue data structure.

Using the notion of total correspondence, we trivially obtain efficient DEPQ structures starting with any of the known priority queue structures (other than weight biased leftist trees [11]).

The *removeMax* and *removeMin* operations can generally be programmed to run faster than suggested by our generic algorithms. This is because, for example, a *removeMax()* and *put(x)* into a max priority queue can often be done faster as a single operation *changeMax(x)*. Similarly a *remove(theNode)* and *put(x)* can be programmed as a *change (theNode, x)* operation.

8.6.3 Leaf Correspondence

In leaf correspondence DEPQs, for every leaf element a in *minPQ*, there is a distinct element b in *maxPQ* such that $a \le b$ and for every leaf element c in *maxPQ* there is a distinct element d in *minPQ* such that $d \le c$. Figure 8.28 shows a leaf correspondence heap.

Efficient leaf correspondence DEPQs may be constructed easily from PQs that satisfy the following requirements [10]:

a. The PQ supports the operation *remove(Q, p)* efficiently.

b. When an element is inserted into the PQ, no nonleaf node becomes a leaf node (except possibly the node for the newly inserted item).

c. When an element is deleted (using *remove*, *removeMax* or *removeMin*) from the PQ, no nonleaf node (except possibly the parent of the deleted node) becomes a leaf node.

Some of the PQ structures that satisfy these requirements are height-biased leftist trees (Chapter 5) [9,12, 13], pairing heaps (Chapter 7) [14,15], and Fibonacci heaps [16] (Chapter 7). Requirements (b) and (c) are not satisfied, for example, by ordinary heaps and the FMPQ structure of [17]. Although heaps and Brodal's FMPQ structure do not satisfy the requirements of our generic approach to build a leaf correspondence DEPQ structure from a priority queue, we can nonetheless arrive at leaf correspondence heaps and leaf correspondence FMPQs using a customized approach.

8.7 Meldable DEPQs

A *meldable DEPQ (MDEPQ)* is a DEPQ that, in addition to the DEPQ operations listed above, includes the operation

$$meld(x, y) \ldots \text{meld the DEPQs } x \text{ and } y \text{ into a single DEPQ.}$$

The result of melding the double-ended priority queues x and y is a single double-ended priority queue that contains all elements of x and y. The meld operation is destructive in that following the meld, x and y do not remain as independent DEPQs.

To meld two DEPQs in less than linear time, it is essential that the DEPQs be represented using explicit pointers (rather than implicit ones as in the array representation of a heap) as otherwise a linear number of elements need to be moved from their initial to their final locations. Olariu et al. [6] have shown that when the min-max pair heap is represented in such a way, an n element DEPQ may be melded with a k element one $(k \leq n)$ in $O(\log(n/k) * \log k)$ time. When $k = \sqrt{n}$, this is $O(\log^2 n)$. Hasham and Sack [18] have shown that the complexity of melding two min-max heaps of size n and k, respectively, is $\Omega(n + k)$. Brodal [17] has developed an MDEPQ implementation that allows one to find the min and max elements, insert an element, and meld two priority queues in $O(1)$ time. The time needed to delete the minimum or maximum element is $O(\log n)$. Although the asymptotic complexity provided by this data structure are the best one can hope for [17], the data structure has practical limitations. First, each element is represented twice using a total of 16 fields per element. Second, even though the delete operations have $O(\log n)$ complexity, the constant factors are very high and the data structure will not perform well unless find, insert, and meld are the primary operations.

Cho and Sahni [19] have shown that leftist trees [9, 12, 13] may be adapted to obtain a simple representation for MDEPQs in which *meld* takes logarithmic time and the remaining operations have the same asymptotic complexity as when any of the aforementioned DEPQ representations is used. Chong and Sahni [10] study MDEPQs based on pairing heaps [14, 15], Binomial and Fibonacci heaps [16], and FMPQ [17].

Since leftist heaps, pairing heaps, Binomial and Fibonacci heaps, and FMPQs are meldable priority queues that also support the *remove(theNode)* operation, the MDEPQs of [10, 19] use the generic methods of Section 8.6 to construct an MDEPQ data structure from the corresponding MPQ (meldable PQ) structure.

It is interesting to note that if we use the FMPQ structure of [17] as the base MPQ structure, we obtain a total correspondence MDEPQ structure in which *removeMax* and *removeMin* take logarithmic time, and the remaining operations take constant time. This adaptation is superior to the dual priority queue adaptation proposed in Reference 17 because the space requirements are almost half. Additionally, the total correspondence adaptation is faster. Although Brodal's FMPQ structure does not satisfy the requirements of the generic approach to build a leaf correspondence MDEPQ structure from a priority queue, we can nonetheless arrive at leaf correspondence FMPQs using a customized approach.

Acknowledgment

This work was supported, in part, by the National Science Foundation under grant CCR-9912395.

References

1. A. Arvind and C. Pandu Rangan, Symmetric min-max heap: A simpler data structure for double-ended priority queue, *Information Processing Letters*, 69, 197–199, 1999.
2. M. Atkinson, J. Sack, N. Santoro, and T. Strothotte, Min-max heaps and generalized priority queues, *Communications of the ACM*, 29, 996–1000, 1986.
3. S. Carlsson, The deap—A double ended heap to implement double ended priority queues, *Information Processing Letters*, 26, 33–36, 1987.
4. S. Chang and M. Du, Diamond deque: A simple data structure for priority deques, *Information Processing Letters*, 46, 231–237, 1993.
5. J. van Leeuwen and D. Wood, Interval heaps, *The Computer Journal*, 36, 3, 209–216, 1993.
6. S. Olariu, C. Overstreet, and Z. Wen, A mergeable double-ended priority queue, *The Computer Journal*, 34 5, 423–427, 1991.
7. J. Williams, Algorithm 232, *Communications of the ACM*, 7, 347–348, 1964.
8. Y. Ding and M. Weiss, On the complexity of building an interval heap, *Information Processing Letters*, 50, 143–144, 1994.
9. E. Horowitz, S. Sahni, and D. Mehta, *Fundamentals of Data Structures in C++*, Computer Science Press, NY, 1995.
10. K. Chong and S. Sahni, Correspondence based data structures for double ended priority queues, *ACM Journal on Experimental Algorithmics*, 5, Article 2, 22, 2000.

11. S. Cho and S. Sahni, Weight biased leftist trees and modified skip lists, *ACM Journal on Experimental Algorithms*, Article 2, 1998.

12. C. Crane, *Linear lists and priority queues as balanced binary trees*, Technical Report CS-72-259, Computer Science Department, Stanford University, Standford, CA, 1972.

13. R. Tarjan, *Data Structures and Network Algorithms*, SIAM, Philadelphia, PA, 1983.

14. M. L. Fredman, R. Sedgewick, D. D. Sleator, and R. E. Tarjan, The paring heap: A new form of self-adjusting heap, *Algorithmica*, 1, 111–129, 1986.

15. J. T. Stasko and J. S. Vitter, Pairing heaps: Experiments and analysis, *Communication of the ACM*, 30, 3, 234–249, 1987.

16. M. Fredman and R. Tarjan, Fibonacci heaps and their uses in improved network optimization algorithms, *JACM*, 34, 3, 596–615, 1987.

17. G. Brodal, Fast meldable priority queues, *Workshop on Algorithms and Data Structures*, 1995.

18. A. Hasham and J. Sack, Bounds for min-max heaps, *BIT*, 27, 315–323, 1987.

19. S. Cho and S. Sahni, Mergeable double ended priority queue, *International Journal on Foundation of Computer Sciences*, 10, 1, 1–18, 1999.

20. Y. Ding and M. Weiss, The relaxed min-max heap: A mergeable double-ended priority queue, *Acta Informatica*, 30, 215–231, 1993.

21. D. Sleator and R. Tarjan, Self-adjusting binary search trees, *JACM*, 32, 3, 652–686, 1985.

III

Dictionary Structures

9   Hash Tables Pat Morin.. 117
Introduction • Hash Tables for Integer Keys • Random Probing • Historical Notes • Other
Developments • Acknowledgments • References

10   Bloom Filter and Its Variants Shigang Chen.. 131
Introduction • Bloom Filter • Counting Bloom Filter • Blocked Bloom Filter • Other Bloom Filter
Variants • Summary • References

11   Balanced Binary Search Trees Arne Andersson, Rolf Fagerberg and Kim S. Larsen......... 151
Introduction • Basic Definitions • Generic Discussion of Balancing • Classic Balancing Schemes •
Rebalancing a Tree to Perfect Balance • Schemes with no Balance Information • Low Height
Schemes • Relaxed Balance • References

12   Finger Search Trees Gerth Stølting Brodal.. 171
Finger Searching • Dynamic Finger Search Trees • Level Linked (2,4)-Trees • Randomized Finger
Search Trees • Applications • Acknowledgments • References

13   Splay Trees Sanjeev Saxena.. 179
Introduction • Splay Trees • Analysis • Optimality of Splay Trees • Linking and Cutting Trees •
Case Study: Application to Network Flows • Implementation without Linking and Cutting Trees •
FIFO Dynamic Tree Implementation • Variants of Splay Trees and Top-Down Splaying •
Acknowledgments • References

14   Randomized Dictionary Structures C. Pandu Rangan....................................... 197
Introduction • Preliminaries • Skip Lists • Structural Properties of Skip Lists • Dictionary
Operations • Analysis of Dictionary Operations • Randomized Binary Search Trees • Bibliographic
Remarks • References

15   Trees with Minimum Weighted Path Length Wojciech Rytter........................... 215
Introduction • Huffman Trees • Height Limited Huffman Trees • Optimal Binary Search Trees •
Optimal Alphabetic Tree Problem • Optimal Lopsided Trees • Parallel Algorithms • References

16   B Trees Donghui Zhang.. 233
Introduction • The Disk-Based Environment • The B-tree • The B+-tree • Further Discussions •
References

9

Hash Tables*

9.1 Introduction.. 117
9.2 Hash Tables for Integer Keys.. 118
    Hashing by Division • Hashing by Multiplication • Universal Hashing • Static Perfect
    Hashing • Dynamic Perfect Hashing
9.3 Random Probing 122
    Hashing with Chaining • Hashing with Open Addressing • Linear Probing • Quadratic
    Probing • Double Hashing • Brent's Method • Multiple-Choice Hashing • Asymmetric
    Hashing • LCFS Hashing • Robin-Hood Hashing • Cuckoo Hashing
9.4 Historical Notes 126
9.5 Other Developments 127
    Acknowledgments 127
    References... 127

Pat Morin
Carleton University

9.1 Introduction

A *set abstract data type (set ADT)* is an abstract data type that maintains a set S under the following three operations:

1. INSERT(x): Add the key x to the set.
2. DELETE(x): Remove the key x from the set.
3. SEARCH(x): Determine if x is contained in the set, and if so, return a pointer to x.

One of the most practical and widely used methods of implementing the set ADT is with *hash tables*.

Note that the three set ADT operations can easily be implemented to run in $O(\log n)$ time per operation using balanced binary search trees (See Chapter 11). If we assume that the input data are integers in the set $U = \{0, \ldots, u-1\}$ then they can even be implemented to run in sub-logarithmic time using data structures for integer searching (Chapter 40). However, these data structures actually do more than the three basic operations we require. In particular if we search for an element x that is not present in S then these data structures can report the smallest item in S that is larger than x (the *successor* of x) and/or the largest item in S that is smaller than x (the *predecessor* of x).

Hash tables do away with this extra functionality of finding predecessors and successors and only perform exact searches. If we search for an element x in a hash table and x is not present then the only information we obtain is that $x \notin S$. By dropping this extra functionality hash tables can give better performance bounds. Indeed, any reasonable hash table implementation performs each of the three set ADT operations in $O(1)$ expected time.

The main idea behind all hash table implementations discussed in this chapter is to store a set of $n = |S|$ elements in an array (the hash table) A of length $m \geq n$. In doing this, we require a function that maps any element x to an array location. This function is called a *hash function h* and the value $h(x)$ is called the *hash value* of x. That is, the element x gets stored at the array location $A[h(x)]$. The *occupancy* of a hash table is the ratio $\alpha = n/m$ of stored elements to the length of A.

The study of hash tables follows two very different lines. Many implementations of hash tables are based on the *integer universe assumption*: All elements stored in the hash table come from the universe $U = \{0, \ldots, u-1\}$. In this case, the goal is to design a hash function $h : U \to \{0, \ldots, m-1\}$ so that for each $i \in \{0, \ldots, m-1\}$, the number of elements $x \in S$ such that $h(x) = i$ is as small as possible. Ideally, the hash function h would be such that each element of S is mapped to a unique value in $\{0, \ldots, m-1\}$. Most of the hash functions designed under the integer universe assumption are number-theoretic constructions. Several of these are described in Section 9.2.

---

* This chapter has been reprinted from first edition of this Handbook, without any content updates.

Historically, the integer universe assumption seems to have been justified by the fact that any data item in a computer is represented as a sequence of bits that can be interpreted as a binary number. However, many complicated data items require a large (or variable) number of bits to represent and this make u the size of the universe very large. In many applications u is much larger than the largest integer that can fit into a single word of computer memory. In this case, the computations performed in number-theoretic hash functions become inefficient.

This motivates the second major line of research into hash tables. This research work is based on the random probing assumption: Each element x that is inserted into a hash table is a black box that comes with an infinite random *probe sequence* $x_0, x_1, x_2, \ldots$ where each of the x_i is independently and uniformly distributed in $\{0, \ldots, m-1\}$. Hash table implementations based on the random probing assumption are described in Section 9.3.

Both the integer universe assumption and the random probing assumption have their place in practice. When there is an easily computing mapping of data elements onto machine word sized integers then hash tables for integer universes are the method of choice. When such a mapping is not so easy to compute (variable length strings are an example) it might be better to use the bits of the input items to build a good pseudorandom sequence and use this sequence as the probe sequence for some random probing data structure.

To guarantee good performance, many hash table implementations require that the occupancy α be a constant strictly less than 1. Since the number of elements in a hash table changes over time, this requires that the array A be resized periodically. This is easily done, without increasing the amortized cost of hash table operations by choosing three constants $0 < \alpha_1 < \alpha_2 < \alpha_3 < 1$ so that, whenever n/m is not the interval (α_1, α_3) the array A is resized so that its size is n/α_2. A simple amortization argument (Chapter 1) shows that the amortized cost of this resizing is $O(1)$ per update (Insert/Delete) operation.

9.2 Hash Tables for Integer Keys

In this section we consider hash tables under the integer universe assumption, in which the key values x come from the universe $U = \{0, \ldots, u-1\}$. A *hash function* h is a function whose domain is U and whose range is the set $\{0, \ldots, m-1\}$, $m \leq u$. A hash function h is said to be a *perfect hash function* for a set $S \subseteq U$ if, for every $x \in S$, $h(x)$ is unique. A perfect hash function h for S is *minimal* if $m = |S|$, that is, h is a bijection between S and $\{0, \ldots, m-1\}$. Obviously a minimal perfect hash function for S is desirable since it allows us to store all the elements of S in a single array of length n. Unfortunately, perfect hash functions are rare, even for m much larger than n. If each element of S is mapped independently and uniformly to a random element of $\{0, \ldots, m-1\}$ then the birthday paradox (see, e.g., Feller [1]) states that, if m is much less than n^2 then there will almost surely exist two elements of S that have the same hash value.

We begin our discussion with two commonly used hashing schemes that are heuristic in nature. That is, we can not make any non-trivial statements about the performance of these schemes when storing an arbitrary set S. We then discuss several schemes that have provably good performance.

9.2.1 Hashing by Division

In *hashing by division*, we use the hash function
$$h(x) = x \bmod m.$$

To use this hash function in a data structure, we maintain an array $A[0], \ldots, A[m-1]$ where each element of this array is a pointer to the head of a linked list (Chapter 2). The linked list L_i pointed to by the array element $A[i]$ contains all the elements x such that $h(x) = i$. This technique of maintaining an array of lists is called *hashing with chaining*.

In such a hash table, inserting an element x takes $O(1)$ time; we compute $i = h(x)$ and append (or prepend) x to the list L_i. However, searching for and/or deleting an element x is not so easy. We have to compute $i = h(x)$ and then traverse the list L_i until we either find x or reach the end of the list. The cost of this is proportional to the length of L_i. Obviously, if our set S consists of the elements $0, m, 2m, 3m, \ldots, nm$ then all elements are stored in the list L_0 and searches and deletions take linear time.

However, one hopes that such pathological cases do not occur in practice. For example, if the elements of S are uniformly and independently distributed in U and u is a multiple of m then the expected size of any list L_i is only n/m. In this case, searches and deletions take $O(1 + \alpha)$ expected time. To help avoid pathological cases, the choice of m is important. In particular, m a power of 2 is usually avoided since, in a binary computer, taking the remainder modulo a power of 2 means simply discarding some high-order bits. Taking m to be a prime not too close to a power of 2 is recommended [2].

9.2.2 Hashing by Multiplication

The implementation of a hash table using *hashing by multiplication* is exactly the same as that of hashing by division except that the hash function
$$h(x) = \lfloor mxA \rfloor \bmod m$$

is used. Here A is a real-valued constant whose choice we discuss below. The advantage of the multiplication method is that the value of m is not critical. We can take m to be a power of 2, which makes it convenient for use on binary computers.

Although any value of A gives a hash function, some values of A are better than others. (Setting $A = 0$ is clearly not a good idea.)

Knuth [2] suggests using the *golden ratio* for A, that is, setting

$$A = (\sqrt{5} - 1)/2 = 0.6180339887\ldots$$

This choice of A is motivated by a theorem, first conjectured by Oderfeld and later proven by SŚierczkowski [3]. This theorem states that the sequence

$$mA \bmod m, \quad 2\,mA \bmod m, \quad 3\,mA \bmod m, \ldots, nmA \bmod m$$

partitions the interval $(0, m)$ into $n + 1$ intervals having only three distinct lengths. Furthermore, the next element $(n + 1)mA$ mod m in the sequence is always contained in one of the largest intervals.[*]

Of course, no matter what value of A we select, the pigeonhole principle implies that for $u \geq nm$ then there will always exist some hash value i and some $S \subseteq U$ of size n such that $h(x) = i$ for all $x \in S$. In other words, we can always find a set S all of whose elements get stored in the same list L_i. Thus, the worst case of hashing by multiplication is as bad as hashing by division.

9.2.3 Universal Hashing

The argument used at the end of the previous section applies equally well to any hash function h. That is, if the table size m is much smaller than the universe size u then for any hash function there is some large (of size at least $\lceil u/m \rceil$) subset of U that has the same hash value. To get around this difficulty we need a collection of hash functions from which we can choose one that works well for S. Even better would be a collection of hash functions such that, for any given S, most of the hash functions work well for S. Then we could simply pick one of the functions at random and have a good chance of it working well.

Let $\mathcal{H}$ be a collection of hash functions, that is, functions from U onto $\{0, \ldots, m - 1\}$. We say that $\mathcal{H}$ is *universal* if, for each $x, y \in U$ the number of $h \in \mathcal{H}$ such that $h(x) = h(y)$ is at most $|\mathcal{H}|/m$. Consider any $S \subseteq U$ of size n and suppose we choose a random hash function h from a universal collection of hash functions. Consider some value $x \in U$. The probability that any key $y \in S$ has the same hash value as x is only $1/m$. Therefore, the expected number of keys in S, not equal to x, that have the same hash value as x is only

$$n_{h(x)} = \begin{cases} (n-1)/m & \text{if } x \in S \\ n/m & \text{if } x \notin S \end{cases}$$

Therefore, if we store S in a hash table using the hash function h then the expected time to search for, or delete, x is $O(1 + \alpha)$.

From the preceding discussion, it seems that a universal collection of hash functions from which we could quickly select one at random would be very handy indeed. With such a collection at our disposal we get an implementation of the set ADT that has $O(1)$ insertion time and $O(1)$ expected search and deletion time.

Carter and Wegman [5] describe three different collections of universal hash functions. If the universe size u is a prime number[†] then

$$\mathcal{H} = \{h_{k_1, k_2, m}(x) = ((k_1 x + k_2) \bmod u)) \bmod m : 1 \leq k_1 < u, 0 \leq k_2 < u\}$$

is a collection of universal hash functions. Clearly, choosing a function uniformly at random from $\mathcal{H}$ can be done easily by choosing two random values $k_1 \in \{1, \ldots, u - 1\}$ and $k_2 \in \{0, \ldots, u - 1\}$. Thus, we have an implementation of the set ADT with $O(1)$ expected time per operation.

9.2.4 Static Perfect Hashing

The result of Carter and Wegman on universal hashing is very strong, and from a practical point of view, it is probably the strongest result most people will ever need. The only thing that could be improved about their result is to make it deterministic, so that the running times of all operations are $O(1)$ *worst-case*. Unfortunately, this is not possible, as shown by Dietzfelbinger et al. [6].

Since there is no hope of getting $O(1)$ worst-case time for all three set ADT operations, the next best thing would be to have searches that take $O(1)$ worst-case time. In this section we describe the method of Fredman, Komlós and Szemerédi [7]. This is a static data structure that takes as input a set $S \subseteq U$ and builds a data structure of size $O(n)$ that can test if an element x is in S

---

[*] In fact, any irrational number has this property [4]. The golden ratio is especially good because it is not too close to a whole number.

[†] This is not a major restriction since, for any $u > 1$, there always exists a prime number in the set $\{u, u + 1, \ldots, 2u\}$. Thus we can enforce this assumption by increasing the value of u by a constant factor.

in $O(1)$ worst-case time. Like the universal hash functions from the previous section, this method also requires that u be a prime number. This scheme uses hash functions of the form

$$h_{k,m}(x) = (kx \bmod u)) \bmod m.^*$$

Let $B_{k,m}(S, i)$ be the number of elements $x \in S$ such that $h_{k,m}(x) = i$, that is, the number of elements of S that have hash value i when using the hash function $h_{k,m}$. The function $B_{k,m}$ gives complete information about the distribution of hash values of S. The main lemma used by Fredman et al. is that, if we choose $k \in U$ uniformly at random then

$$E\left[\sum_{i=0}^{m-1} \binom{B_{k,m}(S, i)}{2}\right] < \frac{n^2}{m}. \tag{9.1}$$

There are two important special cases of this result.

In the *sparse case* we take $m = n^2/\alpha$, for some constant $0 < \alpha < 1$. In this case, the expectation in Equation 9.1 is less than α. Therefore, by Markov's inequality, the probability that this sum is greater than or equal to 1 is at most α. But, since this sum is a non-negative integer, then with probability at least $1 - \alpha$ it must be equal to 0. In other words, with probability at least $1 - \alpha$, $B_{k,m}(S, i) \leq 1$ for all $0 \leq i \leq m - 1$, that is, the hash function $h_{k,m}$ is perfect for S. Of course this implies that we can find a perfect hash function very quickly by trying a small number of random elements $k \in U$ and testing if they result in perfect hash functions. (The expected number of elements that we will have to try is only $1/(1 - \alpha)$.) Thus, if we are willing to use quadratic space then we can perform searches in $O(1)$ worst-case time.

In the *dense case* we assume that m is close to n and discover that, for many values of k, the hash values are distributed fairly evenly among the set $1, \ldots, m$. More precisely, if we use a table of size $m = n$, then

$$E\left[\sum_{i=0}^{m-1} B_{k,m}(S, i)^2\right] \leq 3n.$$

By Markov's inequality this means that

$$\Pr\left\{\sum_{i=0}^{m-1} B_{k,m}(S, i)^2 \leq 3n/\alpha\right\} \geq 1 - \alpha. \tag{9.2}$$

Again, we can quickly find a value of k satisfying (9.2) by testing a few randomly chosen values of k.

These two properties are enough to build a two-level data structure that uses linear space and executes searches in worst-case constant time. We call the following data structure the FKS-α data structure, after its inventors Fredman, Komlós and Szemerédi. At the top level, the data structure consists of an array $A[0], \ldots, A[m - 1]$ where $m = n$. The elements of this array are pointers to other arrays $A_0, \ldots, A_{m-1}$, respectively. To decide what will be stored in these other arrays, we build a hash function $h_{k,m}$ that satisfies the conditions of Equation 9.2. This gives us the top-level hash function $h_{k,m}(x) = (kx \bmod u) \bmod m$. Each element $x \in S$ gets stored in the array pointed to by $A[h_{k,m}(x)]$.

What remains is to describe how we use the arrays $A_0, \ldots, A_{m-1}$. Let S_i denote the set of elements $x \in S$ such that $h_{k,m}(s) = i$. The elements of S_i will be stored in A_i. The size of S_i is $n_i = B_{k,m}(S, i)$. To store the elements of S_i we set the size of A_i to $m_i = n_i^2/\alpha = B_{k,n}(S, i)^2/\alpha$. Observe that, by Equation 9.2, all the A_i's take up a total space of $O(n)$, that is, $\sum_{i=0}^{m-1} m_i = O(n)$. Furthermore, by trying a few randomly selected integers we can quickly find a value k_i such that the hash function h_{k_i, m_i} is perfect for S_i. Therefore, we store the element $x \in S_i$ at position $A_i[h_{k_i, m_i}(x)]$ and x is the unique element stored at that location. With this scheme we can search for any value $x \in U$ by computing two hash values $i = h_{k,m}(x)$ and $j = h_{k_i, m_i}(x)$ and checking if x is stored in $A_i[j]$.

Building the array A and computing the values of $n_0, \ldots, n_{m-1}$ takes $O(n)$ expected time since for a given value k we can easily do this in $O(n)$ time and the expected number of values of k that we must try before finding one that satisfies (9.2) is $O(1)$. Similarly, building each subarray A_i takes $O(n_i^2)$ expected time, resulting in an overall expected running time of $O(n)$. Thus, for any constant $0 < \alpha < 1$, an FKS-α data structure can be constructed in $O(n)$ expected time and this data structure can execute a search for any $x \in U$ in $O(1)$ worst-case time.

9.2.5 Dynamic Perfect Hashing

The FKS-α data structure is nice in that it allows for searches in $O(1)$ time, in the worst case. Unfortunately, it is only static; it does not support insertions or deletions of elements. In this section we describe a result of Dietzfelbinger et al. [6] that shows how the FKS-α data structure can be made dynamic with some judicious use of partial rebuilding (Chapter 11).

---

* Actually, it turns out that any universal hash function also works in the FKS scheme [8, Section 11.5].

The main idea behind the scheme is simple: be lazy at both the upper and lower levels of the FKS-α data structure. That is, rebuild parts of the data structure only when things go wrong. At the top level, we relax the condition that the size m of the upper array A is exactly n and allow A to have size anywhere between n and $2n$. Similarly, at the lower level we allow the array A_i to have a size m_i anywhere between n_i^2/α and $2n_i^2/\alpha$.

Periodically, we will perform a *global rebuilding* operation in which we remove all n elements from the hash table. Some elements which have previously been marked as deleted will be discarded, thereby reducing the value of n. We put the remaining elements in a list, and recompute a whole new FKS-$(\alpha/2)$ data structure for the elements in the list. This data structure is identical to the standard FKS-$(\alpha/2)$ data structure except that, at the top level we use an array of size $m = 2n$.

Searching in this data structure is exactly the same as for the static data structure. To search for an element x we compute $i = h_{k,m}(x)$ and $j = h_{k_i, m_i}(x)$ and look for x at location $A_i[j]$. Thus, searches take $O(1)$ worst-case time.

Deleting in this data structure is done in the laziest manner possible. To delete an element we only search for it and then mark it as deleted. We will use the convention that this type of deletion does not change the value of n since it does not change the number of elements actually stored in the data structure. While doing this, we also keep track of the number of elements that are marked as deleted. When this number exceeds $n/2$ we perform a global rebuilding operation. The global rebuilding operation takes $O(n)$ expected time, but only occurs during one out of every $n/2$ deletions. Therefore, the amortized cost of this operation is $O(1)$ per deletion.

The most complicated part of the data structure is the insertion algorithm and its analysis. To insert a key x we know, because of how the search algorithm works, that we must ultimately store x at location $A_i[j]$ where $i = h_{k,m}(x)$ and $j = h_{k_i, m_i}(x)$. However, several things can go wrong during the insertion of x:

1. The value of n increases by 1, so it may be that n now exceeds m. In this case we perform a global rebuilding operation and we are done.
2. We compute $i = h_{k,m}(x)$ and discover that $\sum_{i=0}^{m-1} n_i^2 > 3n/\alpha$. In this case, the hash function $h_{k,m}$ used at the top level is no longer any good since it is producing an overall hash table that is too large. In this case we perform a global rebuilding operation and we are done.
3. We compute $i = h_{k,m}(x)$ and discover that, since the value of n_i just increased by one, $n_i^2/\alpha > m_i$. In this case, the array A_i is too small to guarantee that we can quickly find a perfect hash function. To handle this, we copy the elements of A_i into a list L and allocate a new array A_i with the new size $m_i = 2n_i^2/\alpha$. We then find a new value k_i such that h_{k_i, m_i} is a perfect hash function for the elements of L and we are done.
4. The array location $A_i[j]$ is already occupied by some other element y. But in this case, we know that A_i is large enough to hold all the elements (otherwise we would already be done after Case 3), but the value k_i being used in the hash function h_{k_i, m_i} is the wrong one since it doesn't give a perfect hash function for S_i. Therefore we simply try new values for k_i until we find a find a value k_i that yields a perfect hash function and we are done.

If none of the preceding 4 cases occurs then we can simply place x at location $A_i[j]$ and we are done.

Handling Case 1 takes $O(n)$ expected time since it involves a global rebuild of the entire data structure. However, Case 1 only happens during one out of every $\Theta(n)$ insertions, so the amortized cost of all occurrences of Case 1 is only $O(1)$ per insertion.

Handling Case 2 also takes $O(n)$ expected time. The question is: How often does Case 2 occur? To answer this question, consider the *phase* that occurs between two consecutive occurrences of Case 1. During this phase, the data structure holds at most m distinct elements. Call this set of elements S. With probability at least $(1 - \alpha)$ the hash function $h_{k,m}$ selected at the beginning of the phase satisfies (9.2) so that Case 2 never occurs during the phase. Similarly, the probability that Case 2 occurs exactly once during the phase is at most $\alpha(1 - \alpha)$. In general, the probability that Case 2 occurs exactly i times during a phase is at most $\alpha^i(1 - \alpha)$. Thus, the expected cost of handling all occurrences of Case 2 during the entire phase is at most

$$\sum_{i=0}^{\infty} \alpha^i(1 - \alpha)i \times O(n) = O(n).$$

But since a phase involves $\Theta(n)$ insertions this means that the amortized expected cost of handling Case 2 is $O(1)$ per insertion.

Next we analyze the total cost of handling Case 3. Define a *subphase* as the period of time between two global rebuilding operations triggered either as a result of a deletion, Case 1 or Case 2. We will show that the total cost of handling all occurrences of Case 3 during a subphase is $O(n)$ and since a subphase takes $\Theta(n)$ time anyway this does not contribute to the cost of a subphase by more than a constant factor. When Case 3 occurs at the array A_i it takes $O(m_i)$ time. However, while handling Case 3, m_i increases by a constant factor, so the total cost of handling Case 3 for A_i is dominated by the value of m_i at the end of the subphase. But we maintain the invariant that $\sum_{i=0}^{m-1} m_i = O(n)$ during the entire subphase. Thus, handling all occurrences of Case 3 during a subphase only requires $O(n)$ time.

Finally, we consider the cost of handling Case 4. For a particular array A_i, consider the *subsubphase* between which two occurrences of Case 3 cause A_i to be rebuilt or a global rebuilding operation takes place. During this subsubphase the number of distinct elements that occupy A_i is at most $\alpha\sqrt{m_i}$. Therefore, with probability at least $1 - \alpha$ any randomly chosen value of $k_i \in U$ is a perfect hash function for this set. Just as in the analysis of Case 2, this implies that the expected cost of handling all occurrences of Case 3 at A_i during a subsubphase is only $O(m_i)$. Since a subsubphase ends with rebuilding all of A_i or a global rebuilding, at a cost of $\Omega(m_i)$ all the occurrences of Case 4 during a subsubphase do not contribute to the expected cost of the subsubphase by more than a constant factor.

To summarize, we have shown that the expected cost of handling all occurrences of Case 4 is only a constant factor times the cost of handling all occurrences of Case 3. The cost of handling all occurrences of Case 3 is no more than a constant factor times the expected cost of all global rebuilds. The cost of handling all the global rebuilds that occur as a result of Case 2 is no more than a constant factor times the cost of handling all occurrences of global rebuilds that occur as a consequence of Case 1. And finally, the cost of all global rebuilds that occur as a result of Case 1 or of deletions is $O(n)$ for a sequence of n update operations. Therefore, the total expected cost of n update operation is $O(n)$.

9.3 Random Probing

Next we consider hash table implementations under the random probing assumption: Each element x stored in the hash table comes with a random sequence $x_0, x_1, x_2, \ldots$ where each of the x_i is independently and uniformly distributed in $\{1, \ldots, m\}$.[*] We begin with a discussion of the two basic paradigms: hashing with chaining and open addressing. Both these paradigms attempt to store the key x at array position $A[x_0]$. The difference between these two algorithms is their *collision resolution strategy*, that is, what the algorithms do when a user inserts the key value x but array position $A[x_0]$ already contains some other key.

9.3.1 Hashing with Chaining

In *hashing with chaining*, a collision is resolved by allowing more than one element to live at each position in the table. Each entry in the array A is a pointer to the head of a linked list. To insert the value x, we simply append it to the list $A[x_0]$. To search for the element x, we perform a linear search in the list $A[x_0]$. To delete the element x, we search for x in the list $A[x_0]$ and splice it out.

It is clear that insertions take $O(1)$ time, even in the worst case. For searching and deletion, the running time is proportional to a constant plus the length of the list stored at $A[x_0]$. Notice that each of the at most n elements not equal to x is stored in $A[x_0]$ with probability $1/m$, so the expected length of $A[x_0]$ is either $\alpha = n/m$ (if x is not contained in the table) or $1 + (n - 1)/m$ (if x is contained in the table). Thus, the expected cost of searching for or deleting an element is $O(1 + \alpha)$.

The above analysis shows us that hashing with chaining supports the three set ADT operations in $O(1)$ expected time per operation, as long as the occupancy, α, is a constant. It is worth noting that this does not require that the value of α be less than 1.

If we would like more detailed information about the cost of searching, we might also ask about the *worst-case search time* defined as

$$W = \max\{\text{length of the list stored at } A[i] : 0 \leq i \leq m - 1\}.$$

It is very easy to prove something quite strong about W using only the fact that the length of each list $A[i]$ is a binomial$(n, 1/m)$ random variable. Using Chernoff's bounds on the tail of the binomial distribution [9], this immediately implies that

$$\Pr\{\text{length of } A[i] \geq \alpha c \ln n\} \leq n^{-\Omega(c)}.$$

Combining this with Boole's inequality ($\Pr\{A \text{ or } B\} \leq \Pr\{A\} + \Pr\{B\}$) we obtain

$$\Pr\{W \geq \alpha c \ln n\} \leq n \times n^{-\Omega(c)} = n^{-\Omega(c)}.$$

Thus, with very high probability, the worst-case search time is logarithmic in n. This also implies that $E[W] = O(\log n)$. The distribution of W has been carefully studied and it is known that, *with high probability*, that is, with probability $1 - o(1)$, $W = (1 + o(1))\ln n/\ln\ln n$ [10,11].[†] Gonnet has proven a more accurate result that $W = \Gamma^{-1}(n) - 3/2 + o(1)$ with high probability. Devroye [12] shows that similar results hold even when the distribution of x_0 is not uniform.

---

[*] A variant of the random probing assumption, referred to as the *uniform hashing* assumption, assumes that $x_0, \ldots, x_{m-1}$ is a random permutation of $0, \ldots, m - 1$.

[†] Here, and throughout this chapter, if an asymptotic notation does not contain a variable then the variable that tends to infinity is implicitly n. Thus, for example, $o(1)$ is the set of non-negative functions of n that tend to 0 as $n \to \infty$.

9.3.2 Hashing with Open Addressing

Hashing with open addressing differs from hashing with chaining in that each table position $A[i]$ is allowed to store only one value. When a collision occurs at table position i, one of the two elements involved in the collision must move on to the next element in its probe sequence. In order to implement this efficiently and correctly we require a method of marking elements as deleted. This method could be an auxiliary array that contains one bit for each element of A, but usually the same result can be achieved by using a special key value **del** that does not correspond to any valid key.

To search for an element x in the hash table we look for x at positions $A[x_0]$, $A[x_1]$, $A[x_2]$, and so on until we either (1) find x, in which case we are done or (2) find an empty table position $A[x_i]$ that is not marked as deleted, in which case we can be sure that x is not stored in the table (otherwise it would be stored at position x_i). To delete an element x from the hash table we first search for x. If we find x at table location $A[x_i]$ we then simply mark $A[x_i]$ as deleted. To insert a value x into the hash table we examine table positions $A[x_0]$, $A[x_1]$, $A[x_2]$, and so on until we find a table position $A[x_i]$ that is either empty or marked as deleted and we store the value x in $A[x_i]$.

Consider the cost of inserting an element x using this method. Let i_x denote the smallest value i such that x_{i_x} is either empty or marked as deleted when we insert x. Thus, the cost of inserting x is a constant plus i_x. The probability that the table position x_0 is occupied is at most α so, with probability at least $1 - \alpha$, $i_x = 0$. Using the same reasoning, the probability that we store x at position x_i is at most

$$\Pr\{i_x = i\} \le \alpha^i(1 - \alpha) \tag{9.3}$$

since the table locations $x_0, \ldots, x_{i-1}$ must be occupied, the table location x_i must not be occupied and the x_i are independent. Thus, the expected number of steps taken by the insertion algorithm is

$$\sum_{i=1}^{\infty} i \Pr\{i_x = i\} = (1 - \alpha) \sum_{i=1}^{\infty} i\alpha^{i-1} = 1/(1 - \alpha)$$

for any constant $0 < \alpha < 1$. The cost of searching for x and deleting x are both proportional to the cost of inserting x, so the expected cost of each of these operations is $O(1/(1 - \alpha))$.[*]

We should compare this with the cost of hashing with chaining. In hashing with chaining, the occupancy α has very little effect on the cost of operations. Indeed, any constant α, even greater than 1 results in $O(1)$ time per operation. In contrast, open addressing is very dependent on the value of α. If we take $\alpha > 1$ then the expected cost of insertion using open addressing is infinite since the insertion algorithm never finds an empty table position. Of course, the advantage of hashing with chaining is that it does not require lists at each of the $A[i]$. Therefore, the overhead of list pointers is saved and this extra space can be used instead to maintain the invariant that the occupancy α is a constant strictly less than 1.

Next we consider the worst case search time of hashing with open addressing. That is, we study the value $W = \max\{i_x : x \text{ is stored in the table at location } i_x\}$. Using Equation 9.3 and Boole's inequality it follows almost immediately that

$$\Pr\{W > c \log n\} \le n^{-\Omega(c)}.$$

Thus, with very high probability, W, the worst case search time, is $O(\log n)$. Tighter bounds on W are known when the probe sequences $x_0, \ldots, x_{m-1}$ are random permutations of $0, \ldots, m-1$. In this case, Gonnet [13] shows that

$$E[W] = \log_{1/\alpha} n - \log_{1/\alpha}(\log_{1/\alpha} n) + O(1).$$

Open addressing under the random probing assumption has many nice theoretical properties and is easy to analyze. Unfortunately, it is often criticized as being an unrealistic model because it requires a long random sequences $x_0, x_1, x_2, \ldots$ for each element x that is to be stored or searched for. Several variants of open addressing discussed in the next few sections try to overcome this problem by using only a few random values.

9.3.3 Linear Probing

Linear probing is a variant of open addressing that requires less randomness. To obtain the probe sequence $x_0, x_1, x_2, \ldots$ we start with a random element $x_0 \in \{0, \ldots, m-1\}$. The element x_i, $i > 0$ is given by $x_i = (i + x_0) \bmod m$. That is, one first tries to find x at location x_0 and if that fails then one looks at $(x_0 + 1) \bmod m$, $(x_0 + 2) \bmod m$ and so on.

---

[*] Note that the expected cost of searching for or deleting an element x is proportional to the value of α *at the time x was inserted*. If many deletions have taken place, this may be quite different than the current value of α.

The performance of linear probing is discussed by Knuth [2] who shows that the expected number of probes performed during an unsuccessful search is at most

$$(1 + 1/(1 - \alpha)^2)/2$$

and the expected number of probes performed during a successful search is at most

$$(1 + 1/(1 - \alpha))/2.$$

This is not quite as good as for standard hashing with open addressing, especially in the unsuccessful case.

Linear probing suffers from the problem of *primary clustering*. If j consecutive array entries are occupied then a newly inserted element will have probability j/m of hashing to one of these entries. This results in $j + 1$ consecutive array entries being occupied and increases the probability (to $(j + 1)/m$) of another newly inserted element landing in this cluster. Thus, large clusters of consecutive elements have a tendency to grow larger.

9.3.4 Quadratic Probing

Quadratic probing is similar to linear probing; an element x determines its entire probe sequence based on a single random choice, x_0. Quadratic probing uses the probe sequence $x_0, (x_0 + k_1 + k_2) \bmod m, (x_0 + 2k_1 + 2^2 k_2) \bmod m, \ldots$. In general, the ith element in the probe sequence is $x_i = (x_0 + ik_1 + i^2 k_2) \bmod m$. Thus, the final location of an element depends quadratically on how many steps were required to insert it. This method seems to work much better in practice than linear probing, but requires a careful choice of m, k_1 and k_2 so that the probe sequence contains every element of $\{0, \ldots, m - 1\}$.

The improved performance of quadratic probing is due to the fact that if there are two elements x and y such that $x_i = y_j$ then it is not necessarily true (as it is with linear probing) that $x_{i+1} = y_{j+1}$. However, if $x_0 = y_0$ then x and y will have exactly the same probe sequence. This lesser phenomenon is called *secondary clustering*. Note that this secondary clustering phenomenon implies that neither linear nor quadratic probing can hope to perform any better than hashing with chaining. This is because all the elements that have the same initial hash x_0 are contained in an implicit chain. In the case of linear probing, this chain is defined by the sequence $x_0, x_0 + 1, x_0 + 2, \ldots$ while for quadratic probing it is defined by the sequence $x_0, x_0 + k_1 + k_2, x_0 + 2k_1 + 4k_2, \ldots$

9.3.5 Double Hashing

Double hashing is another method of open addressing that uses two hash values x_0 and x_1. Here x_0 is in the set $\{0, \ldots, m - 1\}$ and x_1 is in the subset of $\{1, \ldots, m - 1\}$ that is relatively prime to m. With double hashing, the probe sequence for element x becomes $x_0, (x_0 + x_1) \bmod m, (x_0 + 2x_1) \bmod m, \ldots$. In general, $x_i = (x_0 + ix_1) \bmod m$, for $i > 0$. The expected number of probes required by double hashing seems difficult to determine exactly. Guibas has proven that, asymptotically, and for occupancy $\alpha \leq 0.31$, the performance of double hashing is asymptotically equivalent to that of uniform hashing. Empirically, the performance of double hashing matches that of open addressing with random probing regardless of the occupancy α [2].

9.3.6 Brent's Method

Brent's method [14] is a heuristic that attempts to minimize the average time for a successful search in a hash table with open addressing. Although originally described in the context of double hashing (Section 9.3.5) Brent's method applies to any open addressing scheme. The *age* of an element x stored in an open addressing hash table is the minimum value i such that x is stored at $A[x_i]$. In other words, the age is one less than the number of locations we will probe when searching for x.

Brent's method attempts to minimize the total age of all elements in the hash table. To insert the element x we proceed as follows: We find the smallest value i such that $A[x_i]$ is empty; this is where standard open-addressing would insert x. Consider the element y stored at location $A[x_{i-2}]$. This element is stored there because $y_j = x_{i-2}$, for some $j \geq 0$. We check if the array location $A[y_{j+1}]$ is empty and, if so, we move y to location $A[y_{j+1}]$ and store x at location $A[x_{i-2}]$. Note that, compared to standard open addressing, this decreases the total age by 1. In general, Brent's method checks, for each $2 \leq k \leq i$ the array entry $A[x_{i-k}]$ to see if the element y stored there can be moved to any of $A[y_{j+1}], A[y_{j+2}], \ldots, A[y_{j+k-1}]$ to make room for x. If so, this represents a decrease in the total age of all elements in the table and is performed.

Although Brent's method seems to work well in practice, it is difficult to analyze theoretically. Some theoretical analysis of Brent's method applied to double hashing is given by Gonnet and Munro [15]. Lyon [16], Munro and Celis [17] and Poblete [18] describe some variants of Brent's method.

9.3.7 Multiple-Choice Hashing

It is worth stepping back at this point and revisiting the comparison between hash tables and binary search trees. For balanced binary search trees, the average cost of searching for an element is $O(\log n)$. Indeed, it easy to see that for at least $n/2$ of the elements, the cost of searching for those elements is $\Omega(\log n)$. In comparison, for both the random probing schemes discussed so far, the expected cost of search for an element is $O(1)$. However, there are a handful of elements whose search cost is $\Theta(\log n/\log \log n)$ or $\Theta(\log n)$ depending on whether hashing with chaining or open addressing is used, respectively. Thus there is an inversion: Most operations on a binary search tree cost $\Theta(\log n)$ but a few elements (close to the root) can be accessed in $O(1)$ time. Most operations on a hash table take $O(1)$ time but a few elements (in long chains or with long probe sequences) require $\Theta(\log n/\log \log n)$ or $\Theta(\log n)$ time to access. In the next few sections we consider variations on hashing with chaining and open addressing that attempt to reduce the worst-case search time W.

Multiple-choice hashing is hashing with chaining in which, during insertion, the element x has the choice of $d \geq 2$ different lists in which it can be stored. In particular, when we insert x we look at the lengths of the lists pointed to by $A[x_0], \ldots, A[x_{d-1}]$ and append x to $A[x_i], 0 \leq i < d$ such that the length of the list pointed to by $A[x_i]$ is minimum. When searching for x, we search for x in each of the lists $A[x_0], \ldots, A[x_{d-1}]$ *in parallel*. That is, we look at the first elements of each list, then the second elements of each list, and so on until we find x. As before, to delete x we first search for it and then delete it from whichever list we find it in.

It is easy to see that the expected cost of searching for an element x is $O(d)$ since the expected length of each the d lists is $O(1)$. More interestingly, the worst case search time is bounded by $O(dW)$ where W is the length of the longest list. Azar et al. [19] show that

$$E[W] = \frac{\ln \ln n}{\ln d} + O(1). \tag{9.4}$$

Thus, the expected worst case search time for multiple-choice hashing is $O(\log \log n)$ for any constant $d \geq 2$.

9.3.8 Asymmetric Hashing

Asymmetric hashing is a variant of multiple-choice hashing in which the hash table is split into d blocks, each of size n/d. (Assume, for simplicity, that n is a multiple of d.) The probe value $x_i, 0 \leq i < d$ is drawn uniformly from $\{in/d, \ldots, (i+1)n/d-1\}$. As with multiple-choice hashing, to insert x the algorithm examines the lengths of the lists $A[x_0], A[x_1], \ldots, A[x_{d-1}]$ and appends x to the shortest of these lists. In the case of ties, it appends x to the list with smallest index. Searching and deletion are done exactly as in multiple-choice hashing.

Vöcking [20] shows that, with asymmetric hashing the expected length of the longest list is

$$E[W] \leq \frac{\ln \ln n}{d \ln \phi_d} + O(1).$$

The function ϕ_d is a generalization of the *golden ratio*, so that $\phi_2 = (1 + \sqrt{5})/2$. Note that this improves significantly on standard multiple-choice hashing (9.4) for larger values of d.

9.3.9 LCFS Hashing

LCFS hashing is a form of open addressing that changes the collision resolution strategy.[*] Reviewing the algorithm for hashing with open addressing reveals that when two elements collide, priority is given to the first element inserted into the hash table and subsequent elements must move on. Thus, hashing with open addressing could also be referred to as *FCFS (first-come first-served) hashing*.

With LCFS (last-come first-served) hashing, collision resolution is done in exactly the opposite way. When we insert an element x, we always place it at location x_0. If position x_0 is already occupied by some element y because $y_j = x_0$ then we place y at location y_{j+1}, possibly displacing some element z, and so on.

Poblete and Munro [22] show that, after inserting n elements into an initially empty table, the expected worst case search time is bounded above by

$$E[W] \leq 1 + \Gamma^{-1}(\alpha n) \left(1 + \frac{\ln \ln(1/(1-\alpha))}{\ln \Gamma^{-1}(\alpha n)} + O\left(\frac{1}{\ln^2 \Gamma^{-1}(\alpha n)} \right) \right),$$

where Γ is the gamma function and

$$\Gamma^{-1}(\alpha n) = \frac{\ln n}{\ln \ln n} \left(1 + \frac{\ln \ln \ln n}{\ln \ln n} + O\left(\frac{1}{\ln \ln n} \right) \right).$$

---

[*] Amble and Knuth [21] were the first to suggest that, with open addressing, any collision resolution strategy could be used.

Historically, LCFS hashing is the first version of open addressing that was shown to have an expected worst-case search time that is $o(\log n)$.

9.3.10 Robin-Hood Hashing

Robin-Hood hashing [23–25] is a form of open addressing that attempts to equalize the search times of elements by using a fairer collision resolution strategy. During insertion, if we are trying to place element x at position x_i and there is already an element y stored at position $y_j = x_i$ then the "younger" of the two elements must move on. More precisely, if $i \leq j$ then we will try to insert x at position x_{i+1}, x_{i+2} and so on. Otherwise, we will store x at position x_i and try to to insert y at positions y_{j+1}, y_{j+2} and so on.

Devroye et al. [26] show that, after performing n insertions on an initially empty table of size $m = \alpha n$ using the Robin-Hood insertion algorithm, the worst case search time has expected value

$$E[W] = \Theta(\log \log n)$$

and this bound is tight. Thus, Robin-Hood hashing is a form of open addressing that has doubly-logarithmic worst-case search time. This makes it competitive with the multiple-choice hashing method of Section 9.3.7.

9.3.11 Cuckoo Hashing

Cuckoo hashing [27] is a form of multiple choice hashing in which each element x lives in one of two tables A or B, each of size $m = n/\alpha$. The element x will either be stored at location $A[x_A]$ or $B[x_B]$. There are no other options. This makes searching for x an $O(1)$ time operation since we need only check two array locations.

The insertion algorithm for cuckoo hashing proceeds as follows[*]: Store x at location $A[x_A]$. If $A[x_A]$ was previously occupied by some element y then store y at location $B[y_B]$. If $B[y_B]$ was previously occupied by some element z then store z at location $A[z_A]$, and so on. This process ends when we place an element into a previously empty table slot or when it has gone on for more than $c \log n$ steps. In the former case, the insertion of x completes successfully. In the latter case the insertion is considered a failure, and the entire hash table is reconstructed from scratch using a new probe sequence for each element in the table. That is, if this reconstruction process has happened i times then the two hash values we use for an element x are $x_A = x_{2i}$ and $x_B = x_{2i+1}$.

Pagh and Rodler [27] (see also Devroye and Morin [28]) show that, during the insertion of n elements, the probability of requiring a reconstruction is $O(1/n)$. This, combined with the fact that the expected insertion time is $O(1)$ shows that the expected cost of n insertions in a Cuckoo hashing table is $O(n)$. Thus, Cuckoo hashing offers a somewhat simpler alternative to the dynamic perfect hashing algorithms of Section 9.2.5.

9.4 Historical Notes

In this section we present some of the history of hash tables. The idea of hashing seems to have been discovered simultaneously by two groups of researchers. Knuth [2] cites an internal IBM memorandum in January 1953 by H. P. Luhn that suggested the use of hashing with chaining. Building on Luhn's work, A. D. Linh suggested a method of open addressing that assigns the probe sequence $x_0, \lfloor x_0/10 \rfloor, \lfloor x_0/100 \rfloor, \lfloor x_0/1000 \rfloor, \ldots$ to the element x.

At approximately the same time, another group of researchers at IBM: G. M. Amdahl, E. M. Boehme, N. Rochester and A. L. Samuel implemented hashing in an assembly program for the IBM 701 computer. Amdahl is credited with the idea of open addressing with linear probing.

The first published work on hash tables was by A. I. Dumey [29], who described hashing with chaining and discussed the idea of using remainder modulo a prime as a hash function. Ershov [30], working in Russia and independently of Amdahl, described open addressing with linear probing.

Peterson [31] wrote the first major article discussing the problem of searching in large files and coined the term "open addressing." Buchholz [32] also gave a survey of the searching problem with a very good discussion of hashing techniques at the time. Theoretical analyses of linear probing were first presented by Konheim and Weiss [33] and Podderjugin. Another, very influential, survey of hashing was given by Morris [34]. Morris' survey is the first published use of the word "hashing" although it was already in common use by practitioners at that time.

---

[*] The algorithm takes its name from the large but lazy cuckoo bird which, rather than building its own nest, steals the nest of another bird forcing the other bird to move on.

9.5 Other Developments

The study of hash tables has a long history and many researchers have proposed methods of implementing hash tables. Because of this, the current chapter is necessarily incomplete. (At the time of writing, the hash.bib bibliography on hashing contains over 800 entries.) We have summarized only a handful of the major results on hash tables in internal memory. In this section we provide a few references to the literature for some of the other results. For more information on hashing, Knuth [2], Vitter and Flajolet [35], Vitter and Chen [36], and Gonnet and Baeza-Yates [37] are useful references.

Brent's method (Section 9.3.6) is a collision resolution strategy for open addressing that reduces the expected search time for a successful search in a hash table with open addressing. Several other methods exist that either reduce the expected or worst-case search time. These include *binary tree hashing* [15,38], *optimal hashing* [15,39,40], Robin-Hood hashing (Section 9.3.10), and *min-max hashing* [13,23]. One interesting method, due to Celis [23], applies to any open addressing scheme. The idea is to study the distribution of the ages of elements in the hash table, that is, the distribution give by

$$D_i = \Pr\{x \text{ is stored at position } x_i\}$$

and start searching for x at the locations at which we are most likely to find it, rather than searching the table positions $x_0, x_1, x_2 \ldots$ in order.

Perfect hash functions seem to have been first studied by Sprugnoli [41] who gave some heuristic number theoretic constructions of minimal perfect hash functions for small data sets. Sprugnoli is responsible for the terms "perfect hash function" and "minimal perfect hash function." A number of other researchers have presented algorithms for discovering minimal and near-minimal perfect hash functions. Examples include Anderson and Anderson [42], Cichelli [43,44], Chang [45,46], Gori and Soda [47], and Sager [48]. Berman et al. [49] and Körner and Marton [50] discuss the theoretical limitations of perfect hash functions. A comprehensive, and recent, survey of perfect hashing and minimal perfect hashing is given by Czech et al. [51].

Tarjan and Yao [52] describe a set ADT implementation that gives $O(\log u / \log n)$ worst-case access time. It is obtained by combining a trie (Chapter 29) of degree n with a compression scheme for arrays of size n^2 that contain only n non-zero elements. (The trie has $O(n)$ nodes each of which has n pointers to children, but there are only a total of $O(n)$ children.) Although their result is superseded by the results of Fredman et al. [7] discussed in Section 9.2.4, they are the first theoretical results on worst-case search time for hash tables.

Dynamic perfect hashing (Section 9.2.5) and cuckoo hashing (Section 9.3.11) are methods of achieving $O(1)$ worst case search time in a dynamic setting. Several other methods have been proposed [53–55].

Yao [56] studies the *membership problem*. Given a set $S \subseteq U$, devise a data structure that can determine for any $x \in U$ whether x is contained in S. Yao shows how, under various conditions, this problem can be solved using a very small number of memory accesses per query. However, Yao's algorithms sometimes derive the fact that an element x is in S without actually finding x. Thus, they don't solve the set ADT problem discussed at the beginning of this chapter since they can not recover a pointer to x.

The "power of two random choices," as used in multiple-choice hashing, (Section 9.3.7) has many applications in computer science. Karp, Luby and Meyer auf der Heide [57,58] were the first to use this paradigm for simulating PRAM computers on computers with fewer processors. The book chapter by Mitzenmacher et al. [59] surveys results and applications of this technique.

A number of table implementations have been proposed that are suitable for managing hash tables in external memory. Here, the goal is to reduce the number of disk blocks that must be accessed during an operation, where a disk block can typically hold a large number of elements. These schemes include *linear hashing* [60], *dynamic hashing* [61], *virtual hashing* [62], *extendible hashing* [63], *cascade hashing* [64], and *spiral storage* [65]. In terms of hashing, the main difference between internal memory and external memory is that, in internal memory, an array is allocated at a specific size and this can not be changed later. In contrast, an external memory file may be appended to or be truncated to increase or decrease its size, respectively. Thus, hash table implementations for external memory can avoid the periodic global rebuilding operations used in internal memory hash table implementations.

Acknowledgments

The author is supported by a grant from the Natural Sciences and Engineering Research Council of Canada (NSERC).

References

1. W. Feller. *An Introduction to Probability Theory and its Applications.* John Wiley & Sons, New York, 1968.
2. D. E. Knuth. *The Art of Computer Programming.* volume 3. 2nd edition, Addison-Wesley, Boston, MA, 1997.

3. S. Świerczkowski. On successive settings of an arc on the circumference of a circle. *Fundamenta Mathematica*, 46:187–189, 1958.

4. V. T. Sós. On the theory of diophantine approximations. i. *Acta Mathematica Budapest*, 8:461–471, 1957.

5. J. L. Carter, and M. N. Wegman. Universal classes of hash functions. *Journal of Computer and System Sciences*, 18(2):143–154, 1979.

6. M. Dietzfelbinger, A. R. Karlin, K. Mehlhorn, F. Meyer auf der Heide, H. Rohnert, and R. E. Tarjan. Dynamic perfect hashing: Upper and lower bounds. *SIAM Journal on Computing*, 23(4):738–761, 1994.

7. M. L. Fredman, J. Komlós, and E. Szemerédi. Storing a sparse table with $O(1)$ worst case access time. *Journal of the ACM*, 31(3):538–544, 1984.

8. T. H. Cormen, C. E. Leiserson, R. L. Rivest, and C. Stein. *Introduction to Algorithms*. MIT Press, Cambridge, Massachussetts, 2nd edition, 2001.

9. H. Chernoff. A measure of the asymptotic efficient of tests of a hypothesis based on the sum of observations. *Annals of Mathematical Statistics*, 23:493–507, 1952.

10. N. L. Johnson, and S. Kotz. *Urn Models and Their Applications*. John Wiley & Sons, New York, 1977.

11. V. F. Kolchin, B. A. Sevastyanov, and V. P. Chistyakov. *Random Allocations*. John Wiley & Sons, New York, 1978.

12. L. Devroye. The expected length of the longest probe sequence when the distribution is not uniform. *Journal of Algorithms*, 6:1–9, 1985.

13. G. H. Gonnet. Expected length of the longest probe sequence in hash code searching. *Journal of the ACM*, 289–304, 1981.

14. R. P. Brent. Reducing the storage time of scatter storage techniques. *Communications of the ACM*, 16(2):105–109, 1973.

15. G. H. Gonnet, and J. I. Munro. Efficient ordering of hash tables. *SIAM Journal on Computing*, 8(3):463–478, 1979.

16. G. E. Lyon. Packed scatter tables. *Communications of the ACM*, 21(10):857–865, 1978.

17. J. I. Munro and P. Celis. Techniques for collision resolution in hash tables with open addressing. In *Proceedings of 1986 Fall Joint Computer Conference*, pages 601–610. ACM Press, 1999.

18. P. V. Poblete. Studies on hash coding with open addressing. M. Math Essay, University of Waterloo, 1977.

19. Y. Azar, A. Z. Broder, A. R. Karlin, and E. Upfal. Balanced allocations. *SIAM Journal on Computing*, 29(1):180–200, 1999.

20. B. Vöcking. How asymmetry helps load balancing. In *Proceedings of the 40th Annual IEEE Symposium on Foundations of Computer Science (FOCS'99)*, pages 131–140. IEEE Press, 1999.

21. O. Amble, and D. E. Knuth. Ordered hash tables. *The Computer Journal*, 17(2):135–142, 1974.

22. P. V. Poblete, and J. Ian Munro. Last-come-first-served hashing. *Journal of Algorithms*, 10:228–248, 1989.

23. P. Celis. Robin Hood hashing. Technical Report CS-86-14, Computer Science Department, University of Waterloo, 1986.

24. P. Celis, P.-Å. Larson, and J. I. Munro. Robin Hood hashing. In *Proceedings of the 26th Annual IEEE Symposium on Foundations of Computer Science (FOCS'85)*, pages 281–288. IEEE Press, 1985.

25. A. Viola, and P. V. Poblete. Analysis of linear probing hashing with buckets. *Algorithmica*, 21:37–71, 1998.

26. L. Devroye, P. Morin, and A. Viola. On worst case Robin-Hood hashing. *SIAM Journal on Computing*, 33(4):923–936, 2004.

27. R. Pagh and F. F. Rodler. Cuckoo hashing. In *Proceedings of the 9th Annual European Symposium on Algorithms (ESA 2001)*, volume 2161 of *Lecture Notes in Computer Science*, pages 121–133. Springer-Verlag, 2001.

28. L. Devroye, and P. Morin. Cuckoo hashing: Further analysis. *Information Processing Letters*, 86(4):215–219, 2002.

29. A. I. Dumey. Indexing for rapid random access memory systems. *Computers and Automation*, 5(12):6–9, 1956.

30. A. P. Ershov. On programming of arithmetic operations. *Doklady Akademii Nauk SSSR*, 118(3):427–430, 1958.

31. W. W. Peterson. Addressing for random-access storage. *IBM Journal of Research and Development*, 1(2):130–146, 1957.

32. W. Buchholz. File organization and addressing. *IBM Systems Journal*, 2(1):86–111, 1963.

33. A. G. Konheim, and B. Weiss. An occupancy discipline and its applications. *SIAM Journal of Applied Mathematics*, 14:1266–1274, 1966.

34. R. Morris. Scatter storage techniques. *Communications of the ACM*, 11(1):38–44, 1968.

35. J. S. Vitter, and P. Flajolet. Analysis of algorithms and data structures. In J. van Leeuwen, editor, *Handbook of Theoretical Computer Science*, volume A: Algorithms and Complexity, Chapter 9, pages 431–524. North Holland, 1990.

36. J. S. Vitter, and W.-C. Chen, *The Design and Analysis of Coalesced Hashing*. Oxford University Press, Oxford, UK, 1987.

37. G. H. Gonnet, and R. Baeza-Yates, *Handbook of Algorithms and Data Structures: In Pascal and C*. Addison-Wesley, Reading, MA, USA, 2nd edition, 1991.

38. E. G. Mallach. Scatter storage techniques: A unifying viewpoint and a method for reducing retrieval times. *The Computer Journal*, 20(2):137–140, 1977.

39. G. Poonan. Optimal placement of entries in hash tables. In *ACM Computer Science Conference (Abstract Only)*, volume 25, 1976. (Also DEC Internal Tech. Rept. LRD-1, Digital Equipment Corp. Maynard Mass).

40. R. L. Rivest. Optimal arrangement of keys in a hash table. *Journal of the ACM*, 25(2):200–209, 1978.

41. R. Sprugnoli. Perfect hashing functions: A single probe retrieving method for static sets. *Communications of the ACM*, 20(11):841–850, 1977.

42. M. R. Anderson, and M. G. Anderson. Comments on perfect hashing functions: A single probe retrieving method for static sets. *Communications of the ACM*, 22(2):104, 1979.

43. R. J. Cichelli. Minimal perfect hash functions made simple. *Communications of the ACM*, 23(1):17–19, 1980.

44. R. J. Cichelli. On Cichelli's minimal perfect hash functions method. *Communications of the ACM*, 23(12):728–729, 1980.

45. C. C. Chang. An ordered minimal perfect hashing scheme based upon Euler's theorem. *Information Sciences*, 32(3):165–172, 1984.

46. C. C. Chang. The study of an ordered minimal perfect hashing scheme. *Communications of the ACM*, 27(4):384–387, 1984.

47. M. Gori, and G. Soda. An algebraic approach to Cichelli's perfect hashing. *Bit*, 29(1):2–13, 1989.

48. T. J. Sager. A polynomial time generator for minimal perfect hash functions. *Communications of the ACM*, 28(5):523–532, 1985.

49. F. Berman, M. E. Bock, E. Dittert, M. J. O'Donnell, and D. Plank. Collections of functions for perfect hashing. *SIAM Journal on Computing*, 15(2):604–618, 1986.

50. J. Körner, and K. Marton. New bounds for perfect hashing via information theory. *European Journal of Combinatorics*, 9(6):523–530, 1988.

51. Z. J. Czech, G. Havas, and B. S. Majewski. Perfect hashing. *Theoretical Computer Science*, 182(1–2):1–143, 1997.

52. R. E. Tarjan, and A. C.-C. Yao. Storing a sparse table. *Communications of the ACM*, 22(11):606–611, 1979.

53. A. Brodnik, and J. I. Munro. Membership in constant time and almost minimum space. *SIAM Journal on Computing*, 28:1627–1640, 1999.

54. M. Dietzfelbinger and F. Meyer auf der Heide. A new universal class of hash functions and dynamic hashing in real time. In *Proceedings of the 17th International Colloquium on Automata, Languages, and Programming (ICALP'90)*, pages 6–19, 1990.

55. M. Dietzfelbinger, J. Gil, Y. Matias, and N. Pippenger. Polynomial hash functions are reliable. In *Proceedings of the 19th International Colloquium on Automata, Languages, and Programming (ICALP'92)*, pages 235–246, 1992.

56. A. C.-C. Yao. Should tables be sorted? *Journal of the ACM*, 28(3):615–628, 1981.

57. R. Karp, M. Luby, and F. Meyer auf der Heide. Efficient PRAM simulation on a distributed memory machine. Technical Report TR-93-040, International Computer Science Institute, Berkeley, CA, USA, 1993.

58. R. M. Karp, M. Luby, and F. Meyer auf der Heide. Efficient PRAM simulation on a distributed memory machine. In *Proceedings of the 24th ACM Symposium on the Theory of Computing (STOC'92)*, pages 318–326. ACM Press, 1992.

59. M. Mitzenmacher, A. W. Richa, and R. Sitaraman. The power of two random choices: A survey of techniques and results. In P. Pardalos, S. Rajasekaran, and J. Rolim, editors, *Handbook of Randomized Computing*, volume 1, chapter 9. Kluwer, 2001.

60. W. Litwin. Linear hashing: A new tool for file and table addressing. In *Proceedings of the 6th International Conference on Very Large Data Bases (VLDB'80)*, pages 212–223. IEEE Computer Society, 1980.

61. P.-Å. Larson. Dynamic hashing. *Bit*, 18(2):184–201, 1978.

62. W. Litwin. Virtual hashing: A dynamically changing hashing. In *Proceedings of the 4th International Conference on Very Large Data Bases (VLDB'80)*, pages 517–523. IEEE Computer Society, 1978.

63. R. Fagin, J. Nievergelt, N. Pippenger, and H. R. Strong. Extendible hashing—a fast access method for dynamic files. *ACM Transactions on Database Systems*, 4(3):315–344, 1979.

64. P. Kjellberg and T. U. Zahle. Cascade hashing. In *Proceedings of the 10th International Conference on Very Large Data Bases (VLDB'80)*, pages 481–492. Morgan Kaufmann, 1984.

65. J. K. Mullin. Spiral storage: Efficient dynamic hashing with constant performance. *The Computer Journal*, 28(3):330–334, 1985.

10

Bloom Filter and Its Variants

10.1 Introduction.. 131
10.2 Bloom Filter 132
     Description of Bloom Filter • Performance Metrics • False-Positive Ratio and Optimal k
10.3 Counting Bloom Filter 135
     Description of CBF • Counter Size and Counter Overflow
10.4 Blocked Bloom Filter.. 136
     Bloom-1 Filter • Impact of Word Size • Bloom-1 Versus Bloom with Small k • Bloom-1
     versus Bloom with Optimal k • Bloom-g: A Generalization of Bloom-1 • Bloom-g versus
     Bloom with Small k • Bloom-g versus Bloom with Optimal k • Discussion • Using
     Bloom-g in a Dynamic Environment
10.5 Other Bloom Filter Variants 144
     Improving Space Efficiency • Reducing False-Positive Ratio • Improving Read/Write
     Efficiency • Reducing Hash Complexity • Bloom Filter for Dynamic Set
10.6 Summary... 147
References... 147

Shigang Chen
University of Florida

10.1 Introduction

The Bloom filter is a compact data structure for fast membership check against a data set. It was first proposed by Burton Howard Bloom in 1970 [1]. As illustrated in Figure 10.1, given an element, it answers such query: *is the element in the set?* The output is either "no" or "maybe." In other words, it may give false positives, but false negatives are not possible.

The Bloom filter has wide applications in routing-table lookup [2–4], online traffic measurement [5,6], peer-to-peer systems [7,8], cooperative caching [9], firewall design [10], intrusion detection [11], bioinformatics [12], database query processing [13,14], stream computing [15], distributed storage system [16], and so on [17,18]. For example, a firewall may be configured with a large watch list of addresses that are collected by an intrusion detection system. If the requirement is to log all packets from those addresses, the firewall must check each arrival packet to see if the source address is a member of the list. Another example is routing-table lookup. The lengths of the prefixes in a routing table range from 8 to 32. A router can extract 25 prefixes of different lengths from the destination address of an incoming packet, and it needs to determine which prefixes are in the routing tables [2]. Some traffic measurement functions require the router to collect the flow labels [6,19], such as source/destination address pairs or address/port tuples that identify TCP flows. Each flow label should be collected only once. When a new packet arrives, the router must check whether the flow label extracted from the packet belongs to the set that has already been collected before. As a last example for the membership check problem, we consider the context-based access control (CBAC) function in Cisco routers [20]. When a router receives a packet, it may want to first determine whether the addresses/ports in the packet have a matching entry in the CBAC table before performing the CBAC lookup.

In all of the previous examples, we face the same fundamental problem. For a large data set, which may be an address list, an address prefix table, a flow label set, a CBAC table, or other types of data, we want to check whether a given element belongs to this set or not. If there is no performance requirement, this problem can be easily solved using conventional exact-checking data structures such as binary search [21] (which stores the set in a sorted array and uses binary search for membership check) or a traditional hash table [22] (which uses linked lists to resolve hash collision). However, these approaches are inadequate if there are stringent speed and memory requirements.

Modern high-end routers and firewalls implement their per-packet operations mostly in hardware. They are able to forward each packet in a couple of clock cycles. To keep up with such high throughput, many network functions that involve per-packet processing also have to be implemented in hardware. However, they cannot store the data structures for membership check in

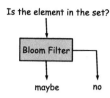

FIGURE 10.1 Black box view of a Bloom filter.

DRAM because the bandwidth and delay of DRAM access cannot match the packet throughput at the line speed. Consequently, the recent research trend is to implement membership check in the high-speed on-die cache memory, which is typically SRAM. The SRAM is however small and must be shared among many online functions. This prevents us from storing a large data set directly in the form of a sorted array or a hash table. On the other hand, a Bloom filter can provide compact membership checking with reasonable precision. For example, it has 1% false-positive probability with fewer than 10 bits per element and 0.1% false-positive probability with fewer than 15 bits per element.

10.2 Bloom Filter

10.2.1 Description of Bloom Filter

A Bloom filter is a space-efficient data structure to encode a set. It includes an array B of m bits, which are initialized to zeros. The array stores the membership information of a set as follows: each member e of the set is mapped to k bits that are randomly selected from B through k hash functions, $H_i(e)$, $1 \leq i \leq k$, whose range is $[0, m-1)$.* To encode the membership information of e, the bits, $B[H_1(e)]$, ..., $B[H_k(e)]$, are set to ones. These are called the "membership bits" in B for the element e. Figure 10.2 illustrates how a set $\{a, b, c\}$ is encoded to a Bloom filter with $m = 16$, $k = 3$.

To check the membership of an arbitrary element e', if the k bits, $B[H_i(e')]$, $1 \leq i \leq k$, are all ones, e' is accepted to be a member of the set. Otherwise, it is denied. Continuing with the Bloom filter shown in Figure 10.2, we illustrate membership checking in the Bloom filter in Figure 10.3. As element a was previously encoded, the three bits $B[H_1(a)]$, $B[H_2(a)]$, $B[H_3(a)]$ were already set to ones. Therefore, a is accepted as a member. In the second example, as element d is not a member of the set $\{a, b, c\}$, chances are it will hit a zero in the array. Therefore, it is denied as a nonmember.

A Bloom filter doesn't have *false negative*, meaning that a member element will always be accepted. The filter however may incur *false positive*, meaning that if it answers that an element is in the set, it may not be really in the set. As illustrated in the third example of Figure 10.3, element e is not a member of set $\{a, b, c\}$, but $B[H_1(e)]$, $B[H_2(e)]$, $B[H_3(e)]$ all happen to be ones, which were individually set by real members.

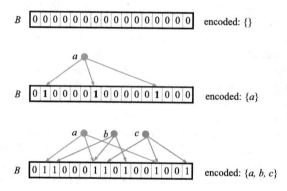

FIGURE 10.2 Encoding a set $\{a, b, c\}$ to a Bloom filter with $k = 3$.

---

* The hash functions can be any random functions as long as the k functions are different, that is, $H_i \neq H_j$, $i \neq j \in [1, k]$. However, hash function outputs are random and two hash results may happen to be the same, that is, $\exists i, j, e, i \neq j, H_i(e) = H_j(e)$.

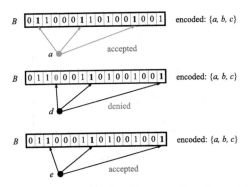

FIGURE 10.3 Membership checking in a Bloom filter with {*a*, *b*, *c*} encoded.

TABLE 10.1 Performance Metrics of the Bloom Filter

Space	Memory Access	Hash Bits	False-Positive Ratio
m	k	$k\lceil \log_2 m \rceil$	$(1 - e^{-nk/m})^k$

m, the size of the bit array; *k*, the number of membership bits (a.k.a. the number of hash functions); *n*, the number of elements encoded.

10.2.2 Performance Metrics

The performance of the Bloom filter is judged based on three criteria. The first one is the processing overhead, which is *k* memory accesses and *k* hash operations for each membership query. The overhead* limits the highest throughput that the filter can support. For example, as both SRAM and the hash function circuit may be shared among different network functions, it is important for them to minimize their processing overhead in order to achieve good system performance. The second performance criterion is the space requirement, which is the size *m* of the array *B*. Minimizing the space requirement to encode each member allows a network function to fit a large set in the limited SRAM space for membership check. The third criterion is the *false-positive ratio* or *false-positive probability*, which is the probability for a Bloom filter to erroneously accept a nonmember. There is a trade-off between the space requirement and the false-positive ratio. We can reduce the latter by allocating more memory.

We can treat the *k* hash functions logically as a single one that produces $k\lceil \log_2 m \rceil$ hash bits. For example, suppose *m* is 2^{20}, $k = 3$, and a hash routine outputs 64 bits. We can extract three 20-bit segments from the first 60 bits of a single hash output and use them to locate three bits in *B*. Hence, from now on, instead of specifying the number of hash functions required by a filter, we will state "the number of hash bits" that are needed, which is denoted as *h*. Table 10.1 summarizes the performance metrics of the Bloom filter. In the next subsection, we will show how the false-positive ratio in Table 10.1 is derived.

10.2.3 False-Positive Ratio and Optimal *k*

We denote the false-positive ratio as f_B. Assume that the hash functions generate fully random locations. The probability that a bit is not set to one by a hash function during the insertion of an element is $1 - (1/m)$, where *m* is the number of bits in the array. The probability that the bit remains zero after inserting *n* elements, each setting *k* bits to ones, is

$$\left(1 - \frac{1}{m}\right)^{kn}.$$

Therefore, the probability that it is one is

$$1 - \left(1 - \frac{1}{m}\right)^{kn}.$$

---

* In the rest of this chapter, we interchangeably use "processing overhead" and "overhead" without further notification.

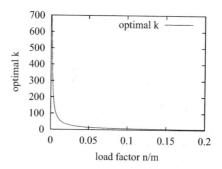

FIGURE 10.4   Optimal number of hash functions (k) with respect to the load factor (n/m).

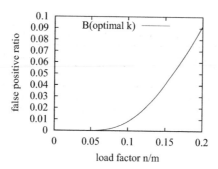

FIGURE 10.5   False-positive ratio of the Bloom filter when optimal k is used.

For an arbitrary nonmember element, each of the k mapped positions has the above probability to be one. The probability of all of them being one, that is, the false-positive ratio, is often given as[*]

$$f_B = \left(1 - \left(1 - \frac{1}{m}\right)^{nk}\right)^k \approx (1 - e^{-nk/m})^k. \tag{10.1}$$

Obviously, the false-positive ratio decreases as the size m of the array increases, and increases as the number n of elements encoded increases. The optimal value of k (denoted as k^*) that minimizes the false-positive ratio can be derived by taking the first-order derivative on Equation 10.1 with respect to k, then letting the right side be zero, and solving the equation.

$$\frac{d}{dk}\left(1 - e^{-nk/m}\right)^k = 0.$$

The result is

$$k^* = \ln 2 \frac{m}{n} \approx 0.7 \frac{m}{n}. \tag{10.2}$$

Figure 10.4 shows the optimal k of the Bloom filter with respect to the load factor, which is defined as n/m. When the load factor is small, the optimal k can be very large. To avoid too many memory accesses, we may also set k as a small constant in practice. When the optimal k is used, the false-positive ratio becomes

$$f_B^* = \left(\frac{1}{2}\right)^k = \left(\frac{1}{2}\right)^{\ln 2 \times m/n} \approx 0.6185^{m/n}. \tag{10.3}$$

Figure 10.5 shows the false-positive ratio of the Bloom filter when the optimal k is used.

If the requirement is to bound the false-positive ratio by a predefined value f_{REQ}, that is, $f_B \le f_{REQ}$, we can determine the length of the Bloom filter by letting $f_B^* \le f_{REQ}$ in Equation 10.3, which gives us

$$m \ge -\frac{n \ln f_{REQ}}{(\ln 2)^2}. \tag{10.4}$$

---

[*] Formula 10.1 is an approximation of the false-positive ratio as bits in the array are not strictly independent. More precise analysis can be found in Reference 23. However, when m is large enough (>1024), the approximation error can be neglected [23].

10.3 Counting Bloom Filter

Removing an element from the traditional Bloom filter is not possible. Each "1" bit in the Bloom filter may be shared by many elements, so we cannot simply reset the membership bits of the element to "0", which may result in the removal of other elements. There are several ways to handle element deletion in the literature (see Section 10.5.5). In this section, we will discuss one of them in detail: the counting Bloom filter (CBF) [9].

10.3.1 Description of CBF

A CBF encodes a set and enables element deletion, which is essentially an array C of l counters. All the counters are initially 0. Similar to the Bloom filter, each member e of the set is mapped to k counters that are randomly selected from C through k hash functions, $H_i(e), 1 \leq i \leq k$, whose range is $[0, l-1]$. To encode the membership information of e, the counters, $C[H_1(e)], \ldots, C[H_k(e)]$, are incremented by one. To remove an element e from the set, the counters, $C[H_1(e)], \ldots, C[H_k(e)]$, are decremented by one. To check the membership of an arbitrary element e', if the k counters, $C[H_i(e')], 1 \leq i \leq k$, are all nonzero, e' is accepted to be a member of the set. Otherwise, it is denied.

Following the same analytical method in Section 10.2.3, we can see that the false-positive ratio of CBF is

$$f_C = \left(1 - \left(1 - \frac{1}{l}\right)^{nk}\right)^k \approx \left(1 - e^{-nk/l}\right)^k,$$ (10.5)

while the optimal k that minimize the false-positive ratio is

$$k_C^* = \ln 2 \frac{l}{n} \approx 0.7 \frac{l}{n},$$ (10.6)

where n is the number of elements encoded. A CBF with l counters has the same false-positive ratio as a traditional Bloom filter with m bits.

10.3.2 Counter Size and Counter Overflow

As counters in CBF are preallocated, they have the same fixed size and hence counter overflow may happen. For example, suppose each counter consists of t bits. It can represent any count between 0 and $2^t - 1$. If a counter with value $2^t - 1$ is about to be incremented for encoding a new element, counter overflow will happen. A smaller value of t makes counter overflow more frequent. However, bigger t means more space: t times the space needed for a Bloom filter!

So our next question is, what is the best choice of t? Or, how small can each counter be to make counter overflow sufficiently rare? To answer this question, we first need to know how big a count value may become. Let c be a random variable which represents the value of an arbitrary counter after inserting n elements, each randomly increments k out of l counters. Obviously, c follows Binomial distribution, $Bino(nk, 1/l)$.

$$\text{Prob}\{c = x\} = \binom{nk}{x} \left(\frac{1}{l}\right)^x \left(1 - \frac{1}{l}\right)^{nk-x}.$$

The probability that the counter is greater or equal to y is

$$\text{Prob}\{c \geq y\} = \sum_{x=y}^{nk} \binom{nk}{x} \left(\frac{1}{l}\right)^x \left(1 - \frac{1}{l}\right)^{nk-x}$$

$$= \binom{nk}{y} \frac{1}{l^y} \sum_{x=0}^{nk-y} \frac{(nk-y)^{\underline{x}}}{(1+x)^{\underline{x}}} \left(\frac{1}{l}\right)^x \left(1 - \frac{1}{l}\right)^{nk-y-x},$$

TABLE 10.2 Probability for Any Counter to Overflow
(Counter Value $\geq 2^t$) in a CBF with l Counters

$t=2$	$t=3$	$t=4$	$t=5$
$0.049 \times l$	$9.5 \times 10^{-6} \times l$	$1.4 \times 10^{-15} \times l$	$4.4 \times 10^{-40} \times l$

Each counter consists of t bits. The optimal k is used.

where $i^{\underline{j}}$ indicates the descending factorial $i^{\underline{j}} = i(i-1)(i-2)\ldots(i-j+1), i \geq j$. Clearly, $(1+x)^{\underline{x}} \geq x!$. As a result, we can bound $\text{Prob}\{c \geq y\}$ by

$$\text{Prob}\{c \geq y\} \leq \binom{nk}{y} \frac{1}{l^y} \sum_{x=0}^{nk-y} \frac{(nk-y)^{\underline{x}}}{x!} \left(\frac{1}{l}\right)^x \left(1 - \frac{1}{l}\right)^{nk-y-x}$$

$$= \binom{nk}{y} \frac{1}{l^y} \sum_{x=0}^{nk-y} \binom{nk-y}{x} \left(\frac{1}{l}\right)^x \left(1 - \frac{1}{l}\right)^{nk-y-x}$$

$$= \binom{nk}{y} \frac{1}{l^y}. \tag{10.7}$$

The probability that any counter is greater or equal to y is (assuming counter values are independent)

$$\text{Prob}\{\max(c) \geq y\} = 1 - (1 - \text{Prob}\{c \geq y\})^l$$

$$\leq l \times \text{Prob}\{c \geq y\} \leq l \binom{nk}{y} \frac{1}{l^y} \leq l \left(\frac{enk}{yl}\right)^y, \tag{10.8}$$

where e is Euler's number ($e = 2.71828\ldots$). If the optimal k is used, that is, $k = k_c^* = \ln 2(l/n)$, we can further bound

$$\text{Prob}\{\max(c) \geq y\} \leq l \left(\frac{e \ln 2}{y}\right)^y. \tag{10.9}$$

From Equation 10.9 and $y = 2^t$, we can calculate the overflow probability of CBF with t-bit counters. The results are shown in Table 10.2. For example, when $t = 4$ and $l = 2^{20}$, the overflow probability is 1.43×10^{-9}, which is extremely small. In other words, 4 bits per counter is enough for most practical values of l.

In practice, even if counter overflow ever happens, we can simply let it stay at the value. Although after many deletions this might lead to a situation where CBF allows a false negative (the count becomes 0 when it shouldn't be), the probability of such a chain of events is usually extremely low.

CBF is a natural extension to the traditional Bloom filter and it is very easy to implement. One issue with CBF is its limited scalability. Because the counter array cannot be expanded, the maximal number of keys to be stored simultaneously in the filter must be known before constructing the filter. Once the designed capacity of the table is exceeded, the false-positive ratio will grow rapidly as more keys are inserted. As a result, even though CBF is designed for dynamic set, it works best when element population has a preknown maximum point.

10.4 Blocked Bloom Filter

Given the fact that Bloom filters have been applied so extensively, any improvement to their performance can potentially have a broad impact. In this section, we study a variant of the Bloom filter, called Bloom-1, which makes just one memory access to perform membership check [24–26]. Therefore, it can be configured to outperform the commonly used Bloom filter with constant k. We will show that when the required false-positive ratio is extremely small, the optimal k of the Bloom filter will be too large to be practical. Next, we generalize Bloom-1 to Bloom-g, which allows g memory accesses. We show that they can achieve the low false-positive ratio of the Bloom filter with optimal k, without incurring the same kind of high overhead. This family of Bloom filter variants is called the *blocked Bloom filter*. We discuss how blocked Bloom filters can be applied for static or dynamic data sets.

10.4.1 Bloom-1 Filter

To check the membership of an element, a Bloom filter requires k memory accesses. We introduce the Bloom-1 filter, which requires one memory access for membership check. The basic idea is that, instead of mapping an element to k bits randomly selected from the entire bit array, we map it to k bits in a word that is randomly selected from the bit array. A word is defined as a continuous block of bits that can be fetched from the memory to the processor in one memory access. In today's computer architectures, most general-purpose processors fetch words of 32 bits or 64 bits. Specifically designed hardware may access words of 72 bits or longer.

A Bloom-1 filter is an array $B1$ of l words, each of which is w bits long. The total number m of bits is $l \times w$. To encode a member e during the filter setup, we first obtain a number of hash bits from e, and use $\log_2 l$ hash bits to map e to a word in $B1$. It is called the "membership word" of e in the Bloom-1 filter. We then use $k \log_2 w$ hash bits to further map e to k membership bits in the word and set them to ones. The total number of hash bits that are needed is $\log_2 l + k \log_2 w$. Suppose $m = 2^{20}$, $k = 3$, $w = 2^6$, and $l = 2^{14}$. Only 32 hash bits are needed, which is smaller than the 60 hash bits required in the previous Bloom filter example under similar parameters.

To check if an element e' is a member in the set that is encoded in a Bloom-1 filter, we first perform hash operations on e' to obtain $\log_2 l + k \log_2 w$ hash bits. We use $\log_2 l$ bits to locate its membership word in $B1$, and then use $k \log_2 w$ bits to identify the membership bits in the word. If all membership bits are ones, it is considered to be a member. Otherwise, it is not.

The change from using k random bits in the array to using k random bits in a word may appear simple, but it is also fundamental. An important question is how it will affect the false-positive ratio and the processing overhead. A more interesting question is how it will open up new design space to configure various new filters with different performance properties. This is what we will investigate in depth.

The false-negative ratio of a Bloom-1 filter is also zero. The false-positive ratio f_{B1} of Bloom-1, which is the probability of mistakenly treating a nonmember as a member, is derived as follows: let F be the false-positive event that a nonmember e' is mistaken for a member. The element e' is hashed to a membership word. Let X be the random variable for the number of members that are mapped to the same membership word. Let x be a constant in the range of $[0, n]$, where n is the number of members in the set. Assume we use fully random hash functions. When $X = x$, the conditional probability for F to occur is

$$\text{Prob}\{F|X = x\} = \left(1 - \left(1 - \frac{1}{w}\right)^{xk}\right)^k. \tag{10.10}$$

Obviously, X follows the binomial distribution, $\text{Bino}(n, 1/l)$, because each of the n elements may be mapped to any of the l words with equal probabilities. Hence,

$$\text{Prob}\{X = x\} = \binom{n}{x}\left(\frac{1}{l}\right)^x\left(1 - \frac{1}{l}\right)^{n-x}, \quad \forall\, 0 \le x \le n. \tag{10.11}$$

Therefore, the false-positive ratio can be written as

$$f_{B1} = \text{Prob}\{F\} = \sum_{x=0}^{n}\left(\text{Prob}\{X = x\} \cdot \text{Prob}\{F|X = x\}\right)$$

$$= \sum_{x=0}^{n}\left(\binom{n}{x}\left(\frac{1}{l}\right)^x\left(1 - \frac{1}{l}\right)^{n-x}\left(1 - \left(1 - \frac{1}{w}\right)^{xk}\right)^k\right). \tag{10.12}$$

10.4.2 Impact of Word Size

We first investigate the impact of word size w on the performance of a Bloom-1 filter. If n, l, and k are known, we can obtain the optimal word size that minimizes Equation 10.12. However, in reality, we can only decide the amount of memory (i.e., m) to be used for a filter, but cannot choose the word size once the hardware is installed. In the left plot of Figure 10.6, we compute the false-positive ratios of Bloom-1 under four word sizes: 32, 64, 72, and 256 bits, when the total amount of memory is fixed at $m = 2^{20}$ and k is set to 3. Note that the number of words, $l = m/w$, is inversely proportional to the word size.

The horizontal axis in the figure is the load factor, n/m, which is the number of members stored by the filter divided by the number of bits in the filter. Since most applications require relatively small false-positive ratios, we zoom in at the load factor range of $[0, 0.2]$ for a detailed look in the right plot of Figure 10.6. The computation results based on Equation 10.12 show that a larger word size helps to reduce the false-positive ratio. In that case, we should simply set $w = m$ for the lowest false-positive ratio. However, in practice, w is given by the hardware, not a configurable parameter. Without losing generality, we choose $w = 64$ in our computations and simulations for the rest of this chapter.

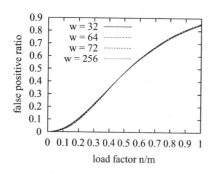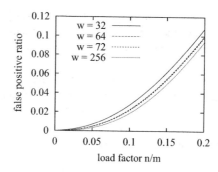

FIGURE 10.6 False-positive ratio with respect to word size *w*. Left plot: False-positive ratios for Bloom-1 under different word sizes. Right plot: Magnified false-positive ratios for Bloom-1 under different word sizes.

10.4.3 Bloom-1 Versus Bloom with Small *k*

Although the optimal *k* always yields the best false positive ratio of the Bloom filter, a small value of *k* is sometimes preferred to bound the processing overhead in terms of memory accesses and hash operations. We compare the performance of the Bloom-1 filter and the Bloom filter in both scenarios. In this section, we use the Bloom filter with $k = 3$ as the benchmark. Using $k = 4, 5, \ldots$ produces quantitatively different results, but the qualitative conclusion will stay the same.

We compare the performance of three types of filters: (1) $B(k = 3)$, which represents a Bloom filter that uses three bits to encode each member; (2) $B1(k = 3)$, which represents a Bloom-1 filter that uses three bits to encode each member; (3) $B1(h = 3\log_2 m)$, which represents a Bloom-1 filter that uses the same number of hash bits as $B(k = 3)$ does.

Let $k1^*$ be the optimal value of *k* that minimizes the false-positive ratio of the Bloom-1 filter in Equation 10.12. For $B1(h = 3\log_2 m)$, we are allowed to use $3\log_2 m$ hash bits, which can encode three or more membership bits in the filter. However, it is not necessary to use more than the optimal number $k1^*$ of membership bits. Hence, $B1(h = 3\log_2 m)$ actually uses the *k* that is closest to $k1^*$ to encode each member.

Table 10.3 presents numerical results of the number of memory accesses and the number of hash bits needed by the three filters for each membership query. They together represent the query overhead and control the query throughput. First, we compare $B(k = 3)$ and $B1(k = 3)$. The Bloom-1 filter saves not only memory accesses but also hash bits. For example, in these examples, $B1(k = 3)$ requires about half of the hash bits needed by $B(k = 3)$. When the hash routine is implemented in hardware, such as CRC [27], the memory access may become the performance bottleneck, particularly when the filter's bit array is located off-chip or, even if it is on-chip, the bandwidth of the cache memory is shared by other system components. In this case, the query throughput of $B1(k = 3)$ will be three times of the throughput of $B(k = 3)$.

Next, we consider $B1(h = 3\log_2 m)$. Even though it still makes one memory access to fetch a word, the processor may check more than 3 bits in the word for a membership query. If the operations of hashing, accessing memory, and checking membership bits are pipelined and the memory access is the performance bottleneck, the throughput of $B1(h = 3\log_2 m)$ will also be three times of the throughput of $B(k = 3)$.

Finally, we compare the false-positive ratios of the three filters in Figure 10.7. The figure shows that the overall performance of $B1(k = 3)$ is comparable to that of $B(k = 3)$, but its false-positive ratio is worse when the load factor is small. The reason is that concentrating the membership bits in one word reduces the randomness. $B1(h = 3\log_2 m)$ is better than $B(k = 3)$ when the load factor is smaller than 0.1. This is because the Bloom-1 filter requires less number of hash bits on average to locate each membership bit than the Bloom filter does. Therefore, when available hash bits are the same, $B1(h = 3\log_2 m)$ is able to use a larger *k* than $B(k = 3)$, as shown in Figure 10.8.

TABLE 10.3 Query Overhead Comparison of Bloom-1 Filters and Bloom Filter with $k = 3$

Data Structure	Number of Memory Access	Number of Hash Bits		
		$m = 2^{16}$	$m = 2^{20}$	$m = 2^{24}$
$B(k = 3)$	3	48	60	72
$B1(k = 3)$	1	28	32	36
$B1(h = 3\log_2 m)$	1	16–46	20–56	24–72

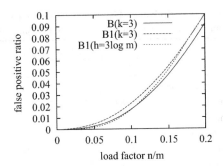

FIGURE 10.7 Performance comparison in terms of false-positive ratio.

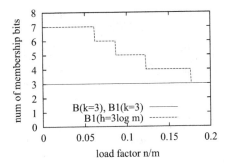

FIGURE 10.8 Number of membership bits used by the filters.

10.4.4 Bloom-1 versus Bloom with Optimal k

We can reduce the false-positive ratio of a Bloom filter or a Bloom-1 filter by choosing the optimal number of membership bits. From Equation 10.2, we find the optimal value k^* that minimizes the false-positive ratio of a Bloom filter. From Equation 10.12, we can find the optimal value $k1^*$ that minimizes the false-positive ratio of a Bloom-1 filter. The values of k^* and $k1^*$ with respect to the load factor are shown in Figure 10.9. When the load factor is less than 0.1, $k1^*$ is significantly smaller than k^*.

We use B(optimal k) to denote a Bloom filter that uses the optimal number k^* of membership bits and $B1$(optimal k) to denote a Bloom-1 filter that uses the optimal number $k1^*$ of membership bits.

To make the comparison more concrete, we present the numerical results of memory access overhead and hashing overhead with respect to the load factor in Table 10.4. For example, when the load factor is 0.04, the Bloom filter requires 17 memory accesses and 340 hash bits to minimize its false-positive ratio, whereas the Bloom-1 filter requires only 1 memory access and 62 hash bits. In practice, the load factor is determined by the application requirement on the false-positive ratio. If an application requires a very small false-positive ratio, it has to choose a small load factor.

Next, we compare the false-positive ratios of B(optimal k) and $B1$(optimal k) with respect to the load factor in Figure 10.10. The Bloom filter has a much lower false-positive ratio than the Bloom-1 filter. On one hand, we must recognize the fact that, as

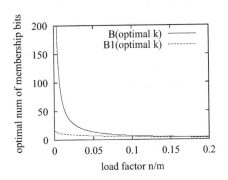

FIGURE 10.9 Optimal number of membership bits with respect to the load factor.

TABLE 10.4   Query Overhead Comparison of Bloom-1 Filter and Bloom Filter with Optimal Number of Membership Bits

	Load Factor n/m				
	0.01	0.02	0.04	0.08	0.16
Memory accesses per query					
B(optimal k)	69	35	17	9	4
$B1$(optimal k)	1	1	1	1	1
Hash bits per query					
B(optimal k)	1380	700	340	180	80
$B1$(optimal k)	80	74	62	50	38

Parameters: $m = 2^{20}$ and $w = 64$.

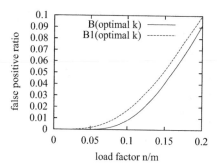

FIGURE 10.10   False-positive ratios of the Bloom filter and the Bloom-1 filter with optimal k.

shown in Table 10.4, the overhead for the Bloom filter to achieve its low false-positive ratio is simply too high to be practical. On the other hand, it raises a challenge for us to improve the design of the Bloom-1 filter so that it can match the performance of the Bloom filter at much lower overhead. In the next section, we generalize the Bloom-1 filter to allow performance-overhead trade-off, which provides flexibility for practitioners to achieve a lower false-positive ratio at the expense of modestly higher query overhead.

10.4.5   Bloom-g: A Generalization of Bloom-1

As a generalization of Bloom-1 filter, a Bloom-g filter maps each member e to g words instead of one and spreads its k membership bits evenly in the g words. More specifically, we use $g \log_2 l$ hash bits derived from e to locate g membership words, and then use $k \log_2 w$ hash bits to locate k membership bits. The first one or multiple words are each assigned $\lceil k/g \rceil$ membership bits, and the remaining words are each assigned $\lfloor k/g \rfloor$ bits, so that the total number of membership bits is k.

To check the membership of an element e', we have to access g words. Hence the query overhead includes g memory accesses and $g \log_2 l + k \log_2 w$ hash bits.

The false-negative ratio of a Bloom-g filter is zero and the false-positive ratio f_{Bg} of the Bloom-g filter is derived as follows: each member encoded in the filter randomly selects g membership words. There are n members. Together they select gn membership words (with replacement). These words are called the "encoded words." In each encoded word, k/g bits are randomly selected to be set as ones during the filter setup. To simplify the analysis, we use k/g instead of taking the ceiling or floor.

Now consider an arbitrary word D in the array. Let X be the number of times this word is selected as an encoded word during the filter setup. Assume we use fully random hash functions. When any member randomly selects a word to encode its membership, the word D has a probability of $1/l$ to be selected. Hence, X is a random number that follows the binomial distribution Bino$(gn, 1/l)$. Let x be a constant in the range $[0, gn]$.

$$\text{Prob}\{X = x\} = \binom{gn}{x} \left(\frac{1}{l}\right)^x \left(1 - \frac{1}{l}\right)^{gn-x}. \tag{10.13}$$

Consider an arbitrary nonmember e'. It is hashed to g membership words. A false positive happens when its membership bits in each of the g words are ones. Consider an arbitrary membership word of e'. Let F be the event that the k/g membership bits of e' in this word are all ones. Suppose this word is selected for x times as an encoded word during the filter setup. We have the following conditional probability:

$$\text{Prob}\{F|X=x\} = \left(1-\left(1-\frac{1}{w}\right)^{x(k/g)}\right)^{k/g}. \tag{10.14}$$

The probability for F to happen is

$$\text{Prob}\{F\} = \sum_{x=0}^{gn}\left(\text{Prob}\{X=x\}\cdot\text{Prob}\{F|X=x\}\right)$$

$$= \sum_{x=0}^{gn}\left(\binom{gn}{x}\cdot\left(\frac{1}{l}\right)^{x}\cdot\left(1-\frac{1}{l}\right)^{gn-x}\cdot\left(1-\left(1-\frac{1}{w}\right)^{x(k/g)}\right)^{k/g}\right). \tag{10.15}$$

Element e' has g membership words. Hence, the false-positive ratio is

$$f_{Bg} = (\text{Prob}\{F\})^{g} = \left[\sum_{x=0}^{gn}\left(\binom{gn}{x}\cdot\left(\frac{1}{l}\right)^{x}\cdot\left(1-\frac{1}{l}\right)^{gn-x}\cdot\left(1-\left(1-\frac{1}{w}\right)^{x(k/g)}\right)^{k/g}\right)\right]^{g}. \tag{10.16}$$

When $g=k$, exactly one bit is set in each membership word. This special Bloom-k is identical to a Bloom filter with k membership bits. (Note that Bloom-k may happen to pick the same membership word more than once. Hence, just like a Bloom filter, Bloom-k allows more than one membership bit in a word.) To prove this, we first let $g=k$, and Equation 10.16 becomes

$$f_{Bk} = \left[\sum_{x=0}^{kn}\left(\binom{kn}{x}\cdot\left(\frac{1}{l}\right)^{x}\cdot\left(1-\frac{1}{l}\right)^{kn-x}\cdot\left(1-\left(1-\frac{1}{w}\right)^{x}\right)\right)\right]^{k}$$

$$= \left[1-\sum_{x=0}^{kn}\left(\binom{kn}{x}\cdot\left(\frac{1}{l}\cdot\left(1-\frac{1}{w}\right)\right)^{x}\cdot\left(1-\frac{1}{l}\right)^{kn-x}\right)\right]^{k}$$

$$= \left(1-\left(1-\frac{1}{lw}\right)^{kn}\right)^{k}. \tag{10.17}$$

As $m=lw$, we have $f_{Bk}=f_{B}$. In other words, a Bloom-k filter is identical to a Bloom filter.

10.4.6 Bloom-g versus Bloom with Small k

We compare the performance and overhead of the Bloom-g filters and the Bloom filter with $k=3$. Because the overhead of Bloom-g increases with g, it is highly desirable to use a small value for g. Hence, we focus on Bloom-2 and Bloom-3 filters as typical examples in the Bloom-g family.

We compare the following filters:

1. $B(k=3)$, the Bloom filter with $k=3$;
2. $B2(k=3)$, the Bloom-2 filter with $k=3$;
3. $B2(h=3\log_2 m)$, the Bloom-2 filter that is allowed to use the same number of hash bits as $B(k=3)$ does.

In this section, we do not consider Bloom-3 because it is equivalent to $B(k=3)$, as we have discussed in Section 10.4.5.

From Equation 10.16, when $g=2$, we can find the optimal value of k, denoted as $k2^{*}$, that minimizes the false-positive ratio. Similar to $B1(h=3\log_2 m)$, the filter $B2(h=3\log_2 m)$ uses the k that is closest to $k2^{*}$ under the constraint that the number of required hash bits is less than or equal to $3\log_2 m$.

Table 10.5 compares the query overhead of three filters. The Bloom filter, $B(k=3)$, needs three memory accesses and $3\log_2 m$ hash bits for each membership query. The Bloom-2 filter, $B2(k=3)$, requires two memory accesses and $2\log_2 l + 3\log_2 w$ hash bits. It is easy to see $2\log_2 l + 3\log_2 w = 3\log_2 m - \log_2 l < 3\log_2 m$. Hence, $B2(k=3)$ incurs fewer memory accesses and fewer hash bits than $B(k=3)$. On the other hand, $B2(h=3\log_2 m)$ uses the same number of hash bits as $B(k=3)$ does, but makes fewer memory accesses.

TABLE 10.5 Query Overhead Comparison of Bloom-2 Filter and Bloom Filter with $k = 3$

Data Structure	Number of Memory Access	Number of Hash Bits		
		$m = 2^{16}$	$m = 2^{20}$	$m = 2^{24}$
$B(k = 3)$	3	48	60	72
$B2(k = 3)$	2	38	46	54
$B2(h = 3 \log_2 m)$	2	32–44	40–58	48–72

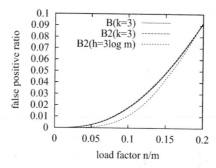

FIGURE 10.11 False-positive ratios of Bloom filter and Bloom-2 filter.

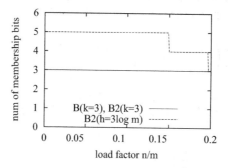

FIGURE 10.12 Number of membership bits used by the filters.

Figure 10.11 presents the false-positive ratios of $B(k = 3)$, $B2(k = 3)$, and $B2(h = 3 \log_2 m)$. Figure 10.12 shows the number of membership bits used by the filters. The figures show that $B(k = 3)$ and $B2(k = 3)$ have comparable false-positive ratios in load factor range of $[0.1, 1]$, whereas $B2(h = 3 \log_2 m)$ performs better in load factor range of $[0.00005, 1]$. For example, when the load factor is 0.04, the false-positive ratio of $B(k = 3)$ is 1.5×10^{-3} and that of $B2(k = 3)$ is 1.6×10^{-3}, while the false-positive ratio of $B2(h = 3 \log_2 m)$ is 3.1×10^{-4}, about one-fifth of the other two. Considering that $B2(h = 3 \log_2 m)$ uses the same number of hash bits as $B(k = 3)$ but only two memory accesses per query, it is a very useful substitute of the Bloom filter to build fast and accurate data structures for membership check.

10.4.7 Bloom-*g* versus Bloom with Optimal *k*

We now compare the Bloom-*g* filters and the Bloom filter when they use the optimal numbers of membership bits determined from Equations 10.1 and 10.16, respectively. We use $B(\text{optimal } k)$ to denote a Bloom filter that uses the optimal number k^* of membership bits to minimize the false-positive ratio. We use $Bg(\text{optimal } k)$ to denote a Bloom-*g* filter that uses the optimal number kg^* of membership bits, where $g = 1, 2$, or 3. Figure 10.13 compares their numbers of membership bits (i.e., k^*, $k1^*$, $k2^*$, and $k3^*$). It shows that the Bloom filter uses many more membership bits when the load factor is small.

Next, we compare the filters in terms of query overhead. For $1 \le g \le 3$, $Bg(\text{optimal } k)$ makes g memory accesses and uses $g \log_2 l + kg^* \log_2 w$ hash bits per membership query. Numerical comparison is provided in Table 10.6. In order to achieve a small false-positive ratio, one has to keep the load factor small, which means that $B(\text{optimal } k)$ will have to make a large number of memory accesses and use a large number of hash bits. For example, when the load factor is 0.08, it makes 9 memory accesses

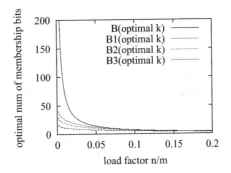

FIGURE 10.13 Optimal number of membership bits with respect to the load factor.

TABLE 10.6 Query Overhead Comparison of Bloom Filter and Bloom-*g* Filter with Optimal *k*

	Load Factor *n/m*				
	0.01	0.02	0.04	0.08	0.16
Memory accesses per query					
B(optimal *k*)	69	35	17	9	4
*B*1(optimal *k*)	1	1	1	1	1
*B*2(optimal *k*)	2	2	2	2	2
*B*3(optimal *k*)	3	3	3	3	3
Hash bits per query					
B(optimal *k*)	1380	700	340	180	80
*B*1(optimal *k*)	80	74	62	50	38
*B*2(optimal *k*)	142	118	94	70	52
*B*3(optimal *k*)	198	162	126	90	66

Parameters: $m = 2^{20}$ and $w = 64$.

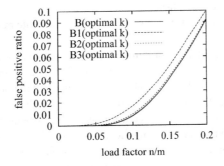

FIGURE 10.14 False-positive ratios of Bloom and Bloom-*g* with optimal *k*.

with 180 hash bits per query, whereas the Bloom-1, Bloom-2, and Bloom-3 filters make 1, 2, and 3 memory accesses with 50, 70, and 90 hash bits, respectively. When the load factor is 0.02, it makes 35 memory accesses with 700 hash bits, whereas the Bloom-1, Bloom-2, and Bloom-3 filters make just 1, 2, and 3 memory accesses with 74, 118 and 162 hash bits, respectively.

Figure 10.14 presents the false-positive ratios of the Bloom and Bloom-*g* filters. As we already showed in Section 10.4.4, *B*1(optimal *k*) performs worse than *B*(optimal *k*). However, the false-positive ratio of *B*2(optimal *k*) is very close to that of *B*(optimal *k*). Furthermore, the curve of *B*3(optimal *k*) is almost entirely overlapped with that of *B*(optimal *k*) for the whole load-factor range. The results indicate that with just two memory accesses per query, *B*2(optimal *k*) works almost as good as *B*(optimal *k*), even though the latter makes many more memory accesses.

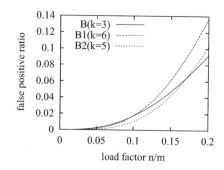

FIGURE 10.15 False-positive ratios of the Bloom filter with $k = 3$, the Bloom-1 filter with $k = 6$, and the Bloom-2 filter with $k = 5$.

10.4.8 Discussion

The mathematical and numerical results demonstrate that Bloom-2 and Bloom-3 have smaller false-positive ratios than Bloom-1 at the expense of larger query overhead. Below we give an intuitive explanation: Bloom-1 uses a single hash to map each member to a word before encoding. It is well known that a single hash cannot achieve an evenly distributed load; some words will have to encode much more members than others, and some words may be empty as no members are mapped to them. This uneven distribution of members to the words is the reason for larger false positives. Bloom-2 maps each member to two words and splits the membership bits among the words. Bloom-3 maps each member to three words. They achieve better load balance such that most words will each encode about the same number of membership bits. This helps them improve their false-positive ratios.

10.4.9 Using Bloom-*g* in a Dynamic Environment

In order to compute the optimal number of membership bits, we must know the values of n, m, w, and l. The values of m, w, and l are known once the amount of memory for the filter is allocated. The value of n is known only when the filter is used to encode a static set of members. In practice, however, the filter may be used for a dynamic set of members. For example, a router may use a Bloom filter to store a watch list of IP addresses, which are identified by the intrusion detection system as potential attackers. The router inspects the arrival packets and logs those packets whose source addresses belong to the list. If the watch list is updated once a week or at the midnight of each day, we can consider it as a static set of addresses during most of the time. However, if the system is allowed to add new addresses to the list continuously during the day, the watch list becomes a dynamic set. In this case, we do not have a fixed optimal value of k^* for the Bloom filter. One approach is to set the number of membership bits to a small constant, such as three, which limits the query overhead. In addition, we should also set the maximum load factor to bound the false-positive ratio. If the actual load factor exceeds the maximum value, we allocate more memory and set up the filter again in a larger bit array.

The same thing is true for the Bloom-*g* filter. For a dynamic set of members, we do not have a fixed optimal number of membership bits, and the Bloom-*g* filter will also have to choose a fixed number of membership bits. The good news for the Bloom-*g* filter is that its number of membership bits is unrelated to its number of memory accesses. The flexible design allows it to use more membership bits while keeping the number of memory accesses small or even a constant one.

Comparing with the Bloom filter, we may configure a Bloom-*g* filter with more membership bits for a smaller false-positive ratio, while in the mean time keeping both the number of memory accesses and the number of hash bits smaller. Imagine a filter of 2^{20} bits is used for a dynamic set of members. Suppose the maximum load factor is set to be 0.1 to ensure a small false-positive ratio. Figure 10.15 compares the Bloom filter with $k = 3$, the Bloom-1 filter with $k = 6$, and the Bloom-2 filter with $k = 5$. As new members are added over time, the load factor increases from 0 to 0.1. In this range of load factors, the Bloom-2 filter has significantly smaller false-positive ratios than the Bloom filter. When the load factor is 0.04, the false-positive ratio of Bloom-2 is just one-fourth of the false positive ratio of Bloom. Moreover, it makes fewer memory accesses per membership query. The Bloom-2 filter uses 58 hash bits per query, and the Bloom filter uses 60 bits. The false-positive ratios of the Bloom-1 filter are close to or slightly better than those of the Bloom filter. It achieves such performance by making just one memory access per query and uses 50 hash bits.

10.5 Other Bloom Filter Variants

Since the Bloom filter was first proposed in 1970 [1], many variants have emerged to meet different requirements, such as supporting set updates, improving the space efficiency, memory access efficiency, false-positive ratio, and hash complexity of the

Bloom filter. A comprehensive survey on Bloom filter variants and their applications can be found in Reference 18. Other Bloom filter variants were proposed to solve specific problems, for example, representing multisets [6,28–34] or sets with multiattribute members [35,36].

We will briefly introduce five categories of Bloom filter variants: the methods that aim to improve the space, hash, accuracy, and memory access efficiency of traditional Bloom filters, and the algorithms to support dynamic set.

10.5.1 Improving Space Efficiency

Some Bloom filter variants aim to improve the space efficiency [37–39]. It has been proven that the Bloom filter is not space-optimal [17,39,40]. Pagh et al. [39] designed a new RAM data structure whose space usage is within a lower order term of the optimal value. Porat [38] designed a dictionary data structure that maps keys to values with optimal space. Mitzenmacher [37] proposed the compressed Bloom filters, which is suitable for message passing, for example, when the filters are transmitted through the Internet. The idea is to use a large, sparse Bloom filter at the sender/receiver, and compress/decompress the filter before/after transmission.

10.5.2 Reducing False-Positive Ratio

Reducing the false-positive ratio of the Bloom filter is another subject of research. Lumetta and Mitzenmacher [41] proposed to use two choices (i.e., two sets of hash functions) for each element encoded in the Bloom filter. The idea is to pick the set of hash functions that produce the least number of ones, so as to reduce the false-positive ratio. This approach generates fewer false positives than the Bloom filter when the load factor is small, but it requires twice the number of hash bits and twice the number of memory access than the Bloom filter does.

Hao et al. [27] use partitioned hashing to divide members into several groups, each group with a different set of hash functions. Heuristic algorithms are used to find the "best" partition. However, this scheme requires the keys of member elements known as a prior to compute the final partition. Therefore, it only works well when members are inserted all at once and membership queries are performed afterwards.

Tabataba and Hashemi [42] proposed the Dual Bloom Filter to improve the false-positive ratio of a single Bloom filter. The idea is to use two equal sized Bloom filters to encode the set with different sets of hash functions. An element is considered a member only when both filters produce positives. In order to reduce the space usage for storing random numbers for different hash functions, the second Bloom filter encodes ones complemented values of the elements.

Lim et al. [43] use two cross-checking Bloom filters together with traditional Bloom filter to reduce the false positives. It first divides the set S to be encoded to two disjoint subsets A and B ($S = A \cup B$, $A \cap B = \emptyset$). Then they use three Bloom filters to encode S, A, B, respectively, denoted as $F(S), F(A), F(B)$. Different hash functions are used to provide cross-checking. All three filters are checked during membership lookup. Only when $F(S)$ gives positive result, while $F(A)$ and $F(B)$ do not both give negative results, it claims the element as a member. Their evaluation shows that it improves the false positive of Bloom filters with the same size by magnitudes.

10.5.3 Improving Read/Write Efficiency

Other works try to improve the read/write efficiency of the Bloom filter. Zhu et al. [44] proposed the hierarchical Bloom filter, which builds a small, less accurate Bloom filter over the large, accurate filter to reduce the number of queries to the large filter. As the small filter has better locality, the total number of memory accesses is reduced. Their method is suitable for the scenarios where large filter access is costive, and element queries are not uniformly distributed [44].

The Partitioned Bloom filter [13,45] can also improve the read/write efficiency. It divides the memory into k segments. k hash functions are used to map an element to one bit in each segment. If the segments reside in different memory banks, reads/writes of the membership bits can proceed in parallel, so that the overall read/write efficiency is improved.

Kim et al. [46] proposed a parallelized Bloom filter design to reduce the power consumption and increase the computation throughput of Bloom filters. The idea is to transform multiple hash functions of Bloom filters into the multiple stages. When the membership bit from earlier hash function is "0," the query string is not progressed to the next stage.

Chen et al. [47] proposed a new Bloom filter variant that allocates two membership bits to each memory block. This reduces the total number of memory accesses by half. Blocked Bloom filters [24–26] generalized [47] and further pushed the problem to extremes: they allocate the membership bits to arbitrary number of blocks. Qiao et al. [26] thoroughly studied the trade-offs between block sizes, number of hash bits, and false-positive ratio.

Canim et al. [48] proposed a two-layer buffered Bloom filter to represent a large set stored in flash memory (a.k.a. solid-state storage, SSD). The design considers several important characteristics of flash memory: (1) data can be read/write by pages. (2) A page write is slower than a page read and data blocks are erased first before they are updated (in-place update problem). (3)

Each cell allows a limited number of erase operations in flash memory life cycle. Therefore, it is important to reduce the number of writes to flash memory. In the design proposed by Canim et al. [48], the filter layer in flash consists of several subfilters, each with the size of a page; while the buffer layer in RAM buffers the read and write operations to each subfilter and applies them in bulk when a buffer is full. The buffered Bloom filter adopts an off-line model, where element insertions and queries are deferred and handled in batch.

Debnath et al. [49] adopted an online model in their design of the BloomFlash, which is also a set representative stored in flash memory. The idea is to reduce random write operations as much as possible. First, bit updates are buffered in RAM so that updates to the same page are handled at once. Second, the Bloom filter is segmented to many subfilters, with each subfilter occupying one flash page. For each element insertion/query, an additional hash function is evoked to choose which subfilter to access. In their design, the size of each subfilter is fixed to the size of a flash page, typically 2 kB or 4 kB [49].

Lu et al. [50] proposed a Forest-structured Bloom filter for dynamic set in flash storage. The proposed data structure resides partially in flash and partially in RAM. It partitions flash space into a collection of flash-page sized subfilters and organizes them into a forest structure. As the set size is increasing, new subfilters are added to the forest.

10.5.4 Reducing Hash Complexity

There are also works that aim to reduce the hash complexity of the Bloom filter. Kirsch and Mitzenmacher [45] have shown that for the Bloom filter, only two hash functions are necessary. Additional hash functions can be produced by a simple linear combination of the output of two hash functions. This gives us an efficient way to produce many hash bits, but it does not reduce the number of hash bits required by the Bloom filter or its variants. There are also other works that design efficient yet well-randomized hash functions that are suitable for Bloom filter-like data structures.

10.5.5 Bloom Filter for Dynamic Set

The Bloom filter does not support element deletion or update well, and it requires the number of elements as a prior to optimize the number of hash functions. Almeida et al. [51] proposed the Scalable Bloom filters for dynamic sets, which create a new Bloom filter when current Bloom filters are "full." Membership query checks all the filters. Each successive bloom filter has a tighter maximum false-positive probability on a geometric progression, so that the compounded probability over the whole series converges to some wanted value. However, this approach uses a lot of space and does not support deletion.

Rottenstreich et al. proposed the Variable-Increment Counting Bloom Filter (VI-CBF) [52] to improve the space efficiency of CBF. For each element insertion, counters are updated by a hashed variable increment instead of a unit increment. Then, during a query, the exact value of a counter is considered, not just its positiveness. Due to possible counter overflow, CBF and VI-CBF both introduce false negatives [53].

The Bloom filter with variable length signatures (VBF) proposed by Lu et al. [34] also enables element deletion. Instead of setting k membership bits, a VBF sets $t(t \leq k)$ bits to one to encode an element. To check the membership of an element, if $q(q \leq t \leq k)$ membership bits are ones, it claims that the element is a member. Deletion is done by setting several membership bits to zero such that remaining ones are less than q. VBF also has false negatives.

Bonomi et al. [54] introduced a data structure based on d-left hashing that is functionally equivalent but uses approximately half as much space as CBF. It has much better scalability than CBF. Once the designed capacity is exceeded, the keys could be reinserted in a new hash table of double size. However, it requires random permutation, which introduces more computational complexity.

Rothenberg et al. [55] achieved false-negative-free deletions by deleting element probabilistically. Their proposed Deletable Bloom filter divides the standard Bloom filter into r regions. It uses a bit map of size r to indicate if there is any "bit collision" (i.e., two members setting the same bit to "1" among member elements in each region. When trying to delete an element, it only resets the bits that are located in collision-free regions. There is a probability that an element can be successfully deleted: when there is at least one membership bit located in a collision-free region. As the load factor increases, the probability of successful deletion decreases. Also, as elements are inserted and deleted, even if the load factor remains the same, the deletion probability still drops. This is because bits the bitmap does not reflect the "real" bit collision when elements previously causing the collision were already deleted. But there is no way to reset bits in the bit map back to "0"s.

In same applications, new data is more meaningful than old data, so some data structure deletes old data in first-in-first-out manner. For example, Chang et al. [28] proposed an aging scheme called double buffering where two large buffers are used alternatively. In this scheme, the memory space is divided into two equal-sized Bloom filters called active filter and warm-up filter. The warm-up filter is a subset of the active filter, which consists only elements that appeared after the active filter is more than half full. When the active filter is full, it is cleared out, and the two filters exchange roles. Each time the active filter is cleared out, a number of old data are deleted, the size of which is half the capacity of the Bloom filter.

Yoon [56] proposed an active–active scheme to better utilize the memory space of double buffering. The idea is to insert new elements to one filter (instead of possibly two in double buffering) and query both filters. When active filter becomes full, the other filter is cleared out and the filters switch roles. It can store more data with the same memory size and tolerable false-positive rate compared to double buffering.

Deng and Rafiei [57] proposed the Stable Bloom filter to detect duplicates in infinite input data stream. Similar to the CBF, a Stable Bloom filter is an array of counters. When an element is inserted, the corresponding counters are set to the maximum value. The false-positive ratio depends on the ratio of zeros, which is a fixed value for a Stable Bloom filter. It achieves this by decrementing some random counters by one whenever a new element is inserted.

Bonomi et al. [58] proposed time-based deletion with a flag associated with each bit in the filter. If a member is accessed, the flags of its corresponding bits are set to "1." At the end of each period, bits with unset flags are reset to zeros, and then all flags are unset to "0"s to prepare for the next period. Elements that are not accessed during a period are deleted consequently. However, it has to wait a time period for the deletion to take effect.

10.6 Summary

This chapter presents the Bloom filters, including many variants designed to realize different performance objectives. We describe the basic form of the Bloom filter that encodes a set for membership lookup, with discussion on performance metrics and minimization of false-positive ratio. We move on to CBFs which allow member deletion. We then discuss a group of blocked Bloom filters which achieve much better performance in terms of memory access overhead at the cost of modestly higher false-positive ratio. Finally, we give a brief overview on many other Bloom filter variants with consideration of space efficiency, false-positive ratio, read/write efficiency, hash complexity, and dynamic sets.

References

1. B. H. Bloom. Space/Time Trade-offs in Hash Coding with Allowable Errors. *Communications of the ACM*, 13(7):422–426, 1970.
2. S. Dharmapurikar, P. Krishnamurthy, and D. Taylor. Longest Prefix Matching Using Bloom Filters. *Proceedings of ACM SIGCOMM*, pages 201–212, August 2003.
3. H. Song, S. Dharmapurikar, J. Turner, and J. Lockwood. Fast Hash Table Lookup Using Extended Bloom Filter: An Aid to Network Processing. *ACM SIGCOMM Computer Communication Review*, 35(4):181–192, 2005.
4. H. Song, F. Hao, M. Kodialam, and T. Lakshman. IPv6 Lookups Using Distributed and Load Balanced Bloom Filters for 100 Gbps Core Router Line Cards. *Proceedings of IEEE INFOCOM*, Rio de Janeiro, Brazil, April 2009.
5. A. Kumar, J. Xu, J. Wang, O. Spatschek, and L. Li. Space-Code Bloom Filter for Efficient Per-flow Traffic Measurement. *IEEE Journal on Selected Areas in Communications*, 24(12):2327–2339, 2006.
6. Y. Lu and B. Prabhakar. Robust Counting Via Counter Braids: An Error-Resilient Network Measurement Architecture. *Proceedings of IEEE INFOCOM*, pages 522–530, April 2009.
7. A. Kumar, J. Xu, and E.W. Zegura. Efficient and Scalable Query Routing for Unstructured Peer-to-Peer Networks. *Proceedings of IEEE INFOCOM*, 2:1162–1173, March 2005.
8. P. Reynolds and A. Vahdat. Efficient Peer-to-Peer Keyword Searching. *Proceedings of the ACM/IFIP/USENIX International Conference on Middleware*, pages 21–40, 2003.
9. F. Li, P. Cao, J. Almeida, and A. Broder. Summary Cache: A Scalable Wide-Area Web Cache Sharing Protocol. *IEEE/ACM Transactions on Networking*, 8(3):281–293, 2000.
10. L. Maccari, R. Fantacci, P. Neira, and R. Gasca. Mesh Network Firewalling with Bloom Filters. *IEEE International Conference on Communications*, pages 1546–1551, June 2007.
11. D. Suresh, Z. Guo, B. Buyukkurt, and W. Najjar. Automatic Compilation Framework for Bloom Filter Based Intrusion Detection. *Reconfigurable Computing: Architectures and Applications*, 3985:413–418, 2006.
12. K. Malde and B. O'Sullivan. Using Bloom Filters for Large Scale Gene Sequence Analysis in Haskell. *Practical Aspects of Declarative Languages*, pages 183–194, 2009.
13. J.K. Mullin. Optimal Semijoins for Distributed Database Systems. *IEEE Transactions on Software Engineering*, 16(5):558–560, 1990.
14. W. Wang, H. Jiang, H. Lu, and J.X. Yu. Bloom Histogram: Path Selectivity Estimation for XML Data with Updates. *Proceedings of Very Large Data Bases (VLDB)*, pages 240–251, 2004.
15. Z. Yuan, J. Miao, Y. Jia, and L. Wang. Counting Data Stream Based on Improved Counting Bloom Filter. *The 9th International Conference on Web-Age Information Management (WAIM)*, pages 512–519, 2008.

16. F. Chang, J. Dean, S. Ghemawat, W.C. Hsieh, D.A. Wallach, M. Burrows, T. Chandra, A. Fikes, and R.E. Gruber. Bigtable: A Distributed Storage System for Structured Data. *ACM Transactions on Computer Systems (TOCS)*, 26(2):4, 2008.

17. A. Broder and M. Mitzenmacher. Network Applications of Bloom Filters: A Survey. *Internet Mathematics*, 1(4):485–509, 2002.

18. S. Tarkoma, C. Rothenberg, and E. Lagerspetz. Theory and Practice of Bloom Filters for Distributed Systems. *Communications Surveys & Tutorials, IEEE*, 14(1):131–155, 2012.

19. Y. Lu, A. Montanari, B. Prabhakar, S. Dharmapurikar, and A. Kabbani. Counter Braids: A Novel Counter Architecture for Per-flow Measurement. *Proceedings of ACM SIGMETRICS*, 36(1):121–132, 2008.

20. Cisco IOS Firewall. Context-Based Access Control (CBAC), Introduction and Configuration. 2008.

21. E. Horowitz, S. Sahni, and S. Rajasekaran, *Computer Algorithms C++ (Chapter 3.2)*. WH Freeman, New York, 1996.

22. M. Dietzfelbinger, A. Karlin, K. Mehlhorn, F. Heide, H. Rohnert, and R. Tarjan. Dynamic Perfect Hashing: Upper and Lower Bounds. *SIAM Journal on Computing*, 23(4):738–761, 1994.

23. K. Christensen, A. Roginsky, and M. Jimeno. A New Analysis of the False Positive Rate of a Bloom Filter. *Information Processing Letters*, 110(21):944–949, 2010.

24. F. Putze, P. Sanders, and J. Singler. Cache-, Hash- and Space-Efficient Bloom Filters. *Experimental Algorithms*, pages 108–121, 2007.

25. Y. Qiao, T. Li, and S. Chen. One Memory Access Bloom Filters and Their Generalization. *Proceedings of IEEE INFOCOM*, pages 1745–1753, 2011.

26. Y. Qiao, T. Li, and S. Chen. Fast Bloom Filters and Their Generalization. *IEEE Transactions on Parallel and Distributed Systems (TPDS)*, 25(1):93–103, 2014.

27. M. Kodialam, F. Hao, and T.V. Lakshman. Building High Accuracy Bloom Filters Using Partitioned Hashing. *Proceedings of ACM SIGMETRICS*, 35(1):277–288, 2007.

28. F. Chang, W. Feng, and K. Li. Approximate Caches for Packet Classification. *Proceedings of IEEE INFOCOM*, 4:2196–2207, March 2004.

29. B. Chazelle, J. Kilian, R. Rubinfeld, and A. Tal. The Bloomier Filter: An Efficient Data Structure for Static Support Lookup Tables. *Proceedings of ACM SODA*, pages 30–39, January 2004.

30. S. Cohen and Y. Matias. Spectral Bloom Filters. *Proceedings of ACM SIGMOD*, pages 241–252, June 2003.

31. M. Goodrich and M. Mitzenmacher. Invertible Bloom Lookup Tables. *Allerton Conference on Communication, Control, and Computing, Allerton*, pages 792–799, 2011.

32. F. Hao, M. S. Kodialam, T. V. Lakshman, and H. Song. Fast Dynamic Multiple-set Membership Testing Using Combinatorial Bloom Filters. *IEEE/ACM Transactions on Networking*, 20(1):295–304, 2012.

33. T. Li, S. Chen, and Y. Ling. Fast and Compact Per-flow Traffic Measurement Through Randomized Counter Sharing. *IEEE INFOCOM*, pages 1799–1807, 2011.

34. Y. Lu, B. Prabhakar, and F. Bonomi. Bloom Filters: Design Innovations and Novel Applications. *Proceedings of Allerton Conference*, 2005.

35. Y. Hua and B. Xiao. A Multi-attribute Data Structure with Parallel Bloom Filters for Network Services. *High Performance Computing (HiPC)*, pages 277–288, 2006.

36. B. Xiao and Y. Hua. Using Parallel Bloom Filters for Multi-attribute Representation on Network Services. *IEEE Transactions on Parallel and Distributed Systems*, 21(1):20–32, 2010.

37. M. Mitzenmacher. Compressed Bloom Filters. *IEEE/ACM Transactions on Networking*, 10(5):604–612, 2002.

38. E. Porat. An Optimal Bloom Filter Replacement Based on Matrix Solving. *Computer Science: Theory and Applications*, pages 263–273, 2009.

39. A. Pagh, R. Pagh, and S.S. Rao. An Optimal Bloom Filter Replacement. *Proceedings of ACM–SIAM Symposium on Discrete Algorithms*, pages 823–829, 2005.

40. S. Lovett and E. Porat. A Lower Bound for Dynamic Approximate Membership Data Structures. *Proceedings of Foundations of Computer Science (FOCS)*, pages 797–804, 2010.

41. S. Lumetta and M. Mitzenmacher. Using the Power of Two Choices to Improve Bloom Filters. *Internet Mathematics*, 4(1):17–33, 2007.

42. F. Tabataba and M. Hashemi. Improving False Positive in Bloom Filter. *Electrical Engineering (ICEE)*, pages 1–5, 2011.

43. H. Lim, N. Lee, J. Lee, and C. Yim. Reducing False Positives of a Bloom Filter Using Cross-checking Bloom Filters. *Applied Mathematics*, 8(4):1865–1877, 2014.

44. Y. Zhu, H. Jiang, and J. Wang. Hierarchical Bloom Filter Arrays (HBA): A Novel, Scalable Metadata Management System for Large Cluster-Based Storage. *IEEE International Conference on Cluster Computing*, pages 165–174, 2004.

45. N. Kamiyama and T. Mori. Simple and Accurate Identification of High-rate Flows by Packet Sampling. *Proceedings of IEEE INFOCOM*, April 2006.

46. D. Kim, D. Oh, and W. Ro. Design of Power-Efficient Parallel Pipelined Bloom Filter. *Electronics Letters*, 48(7):367–369, 2012.

47. Y. Chen, A. Kumar, and J. Xu. A New Design of Bloom Filter for Packet Inspection Speedup. *Proceedings of IEEE GLOBE-COM*, pages 1–5, 2007.

48. M. Canim, G.A. Mihaila, B. Bhattacharhee, C.A. Lang, and K.A. Ross. Buffered Bloom Filters on Solid State Storage. *VLDB ADMS Workshop*, 2010.

49. B. Debnath, S. Sengupta, J. Li, D.J. Lilja, and D.H.C. Du. BloomFlash: Bloom Filter on Flash-Based Storage. *Proceedings of ICDCS*, pages 635–644, 2011.

50. G. Lu, B. Debnath, and D. Du. A Forest-Structured Bloom Filter with Flash Memory. *IEEE Symposium on Mass Storage Systems and Technologies (MSST)*, pages 1–6, 2011.

51. P. Almeida, C. Baquero, N. Preguica, and D. Hutchison. Scalable Bloom Filters. *Information Processing Letters*, 101(6):255–261, 2007.

52. O. Rottenstreich, Y. Kanizo, and I. Keslassy. The Variable-Increment Counting Bloom Filter. *Proceedings of IEEE INFO-COM*, pages 1880–1888, 2012.

53. D. Guo, Y. Liu, X. Li, and P. Yang. False Negative Problem of Counting Bloom Filter. *IEEE Transactions on Knowledge and Data Engineering*, 22(5):651–664, 2010.

54. F. Bonomi, M. Mitzenmacher, R. Panigrah, S. Singh, and G. Varghese. An Improved Construction for Counting Bloom Filters. *Algorithms–ESA*, pages 684–695, 2006.

55. C. Rothenberg, C. Macapuna, F. Verdi, and M. Magalhaes. The Deletable Bloom Filter: A New Member of the Bloom Family. *IEEE Communications Letters*, 14(6)557–559, 2010.

56. M. Yoon. Aging Bloom Filter with Two Active Buffers for Dynamic Sets. *IEEE Transactions on Knowledge and Data Engineering*, 22(1):134–138, 2010.

57. F. Deng and D. Rafiei. Approximately Detecting Duplicates for Streaming Data Using Stable Bloom Filters. *Proceedings of ACM SIGMOD*, pages 25–36, 2006.

58. F. Bonomi, M. Mitzenmacher, R. Panigrah, S. Singh, and G. Varghese. Beyond Bloom Filters: From Approximate Membership Checks to Approximate State Machines. *ACM SIGCOMM Computer Communication Review*, 36(4):315–326, 2006.

59. D. Guo, J. Wu, H. Chen, Y. Yuan, and X. Luo. The Dynamic Bloom Filters. *IEEE Transactions on Knowledge and Data Engineering*, 22(1):120–133, 2010.

11

Balanced Binary Search Trees[*]

11.1 Introduction.. 151
11.2 Basic Definitions... 151
    Trees • Binary Trees as Dictionaries • Implementation of Binary Search Trees
11.3 Generic Discussion of Balancing.. 153
    Balance Definitions • Rebalancing Algorithms • Complexity Results
11.4 Classic Balancing Schemes.. 155
    AVL-Trees • Weight-Balanced Trees • Balanced Binary Trees Based on Multi-Way Trees
11.5 Rebalancing a Tree to Perfect Balance... 158
11.6 Schemes with no Balance Information... 158
    Implicit Representation of Balance Information • General Balanced Trees • Application to
    Multi-Dimensional Search Trees
11.7 Low Height Schemes 162
11.8 Relaxed Balance 164
    Red-Black Trees • AVL-Trees • Multi-Way Trees • Other Results
References.. 167

Arne Andersson
Uppsala University

Rolf Fagerberg
University of Southern Denmark

Kim S. Larsen
University of Southern Denmark

11.1 Introduction

Balanced binary search trees are among the most important data structures in Computer Science. This is because they are efficient, versatile, and extensible in many ways. They are used as a black-box in numerous algorithms and even other data structures.

The main virtue of balanced binary search trees is their ability to maintain a dynamic set in sorted order, while supporting a large range of operations in time logarithmic in the size of the set. The operations include search, insertion, deletion, predecessor/successor search, range search, rank search, batch update, split, meld, and merge. These operations are described in more detail in Section 11.2 below.

Data structures supporting the operations search, insertion, deletion, and predecessor (and/or successor) search are often denoted *ordered dictionaries*. In the comparison based model, the logarithmic performance of balanced binary search trees is optimal for ordered dictionaries, whereas in the RAM model, faster operations are possible [1,2]. If one considers *unordered dictionaries*, that is, only the operations search, insertion, and deletion, expected constant time is possible by hashing.

11.2 Basic Definitions

11.2.1 Trees

There are many ways to define trees. In this section, we define a tree as a hierarchical organization of a collection of nodes. For alternatives to our exposition, see the chapter on trees.

A *tree* can be empty. If it is not empty, it consists of one node, which is referred to as the *root* of the tree, and a collection of trees, referred to as *subtrees*. Thus, a tree consists of many smaller trees, each with their own root. We use r to denote the single node which is the root of the entire tree.

We only consider *finite* trees, that is, every collection of subtrees is finite, and there are no infinite chains of nonempty subtrees. Furthermore, we only consider *ordered* trees, meaning that the collection of subtrees of a node is an ordered sequence rather than

---

[*] This chapter has been reprinted from first edition of this Handbook, without any content updates.

just a set. If every nonempty tree has exactly two subtrees, then the tree is called *binary*. In this case, we refer to the two subtrees as the *left* and *right* subtrees.

We use u, v, w, etc. to denote nodes and T to denote trees, applying apostrophes, index, etc. to increase the name space. For a node u, we use $u.l$ and $u.r$ to denote the left and right subtree, respectively, of the tree rooted by u. However, when no confusion can occur, we do not necessarily distinguish between nodes and subtrees. Thus, by the subtree v, we mean the subtree rooted at the node v and by T we mean the entire tree or the root of the tree.

We use the standard genealogical terminology to denote nodes in the vicinity of a designated node. Thus, if u is the root of a tree and v is the root of a subtree of u, then v is referred to as a *child* of u. By analogy, this defines *grandchildren, parent, grandparent,* and *sibling*.

The set of nodes *belonging* to a nonempty tree is its root, along with all the nodes belonging to its subtrees. For an empty tree, this set is of course empty. If a node v belongs to the subtree of u, then v is a *descendant* of u, and u is an *ancestor* of v. An ancestor or descendant v of a node u is *proper* if $u \neq v$.

Quite often, it is convenient to refer to empty subtrees as real nodes, in which case they are referred to as *external* nodes (or leaves). The remaining nodes are then referred to as *internal* nodes. It is easy to prove by induction that the number of external nodes is always one larger than the number of internal nodes.

The number of nodes belonging to a tree is referred to as its *size* (or its *weight*). In some applications, we define the size of the tree to be the number of internal nodes in the tree, but more often it is convenient to define the size of the tree to be the number of external nodes. We use n to denote the size of the tree rooted by r, and $|u|$ to denote the size of the subtree rooted by u.

A *path* in a tree is a sequence of nodes $u_1, u_2, \ldots, u_k$, $k \geq 1$, such that for $i \in \{1, \ldots, k-1\}$, u_{i+1} is a child of u_i. Note that the length of such a path is $k - 1$. The *depth* of a node u in the tree T is the length of the path from the root of T to u, and the *height* of a tree T is the maximal depth of any external node.

11.2.2 Binary Trees as Dictionaries

When trees are used to implement the abstract data type *dictionary*, nodes have associated values. A dictionary basically organizes a set of *keys*, which must be elements drawn from a total ordering, and must usually supply at least the operations search, insertion, and deletion. There may be additional information associated with each key, but this does not lead to any conceptual complications, so here we simply focus on the keys.

When a tree is used as a dictionary, each node stores one key, and we impose the following ordering invariant (the *in-order* invariant): for each node u in the tree, every key in $u.l$ is strictly smaller than $u.k$, and every key in $u.r$ is strictly larger than $u.k$. A tree organized according to this invariant is referred to as a binary *search* tree.

An important implication of this ordering invariant is that a sorted list of all the keys in the tree can be produced in linear time using an *in-order traversal* defined recursively as follows. On an empty tree, do nothing. Otherwise, recurs on the left subtree, report the root key, and then recurs on the right subtree.

Many different operations can be supported by binary search tree implementations. Here, we discuss the most common. Using the ordering invariant, we can devise a searching procedure of asymptotic time complexity proportional to the height of the tree. Since searching turns out to be at the heart of most of the operations of interest to us, unless we stipulate otherwise, all the operations in the following inherit the same complexity.

1. *Simple Searching*: To *search* for x in a tree rooted by u, we first compare x to $u.k$. If they are equal, a positive response is given. Otherwise, if x is smaller than $u.k$, we search recursively in $u.l$, and if x is larger, we search in $u.r$. If we arrive at an empty tree, a negative response is given. In this description, we have used *ternary* comparisons, in that our decisions regarding how to proceed depend on whether the search key is less than, equal to, or greater than the root key. For implementation purposes, it is possible to use the more efficient *binary* comparisons [3].

   A characteristic feature of search trees is that when a searching fails, a nearest neighbor can be provided efficiently. Dictionaries supporting predecessor/successor queries are referred to as *ordered*. This is in contrast to hashing (described in a chapter of their own) which represents a class of unordered dictionaries. A *predecessor* search for x must return the largest key less than or equal to x. This operation as well as the similar *successor* search are simple generalizations of the search strategy outlined above. The case where x is found on the way is simple, so assume that x is not in the tree. Then the crucial observation is that if the last node encountered during the search is smaller than x, then this node is the predecessor. Otherwise, the predecessor key is the largest key in the left subtree of the last node on the search path containing a key smaller than x. A successor search is similar.

2. *Simple Updates*: An *insertion* takes a tree T and a key x not belonging to T as arguments and adds a node containing x and two empty subtrees to T. The node replaces the empty subtree in T where the search for x terminates.

   A *deletion* takes a tree T and a key x belonging to T as arguments and removes the node u containing x from the tree. If u's children are empty trees, u is simply replaced by an empty tree. If u has exactly one child which is an internal node, then this child is replacing u. Finally, if u has two internal nodes as children, u's predecessor node v is used. First, the key in u is overwritten by the key of v, after which v is deleted. Note that because of the choice of v, the ordering invariant is not violated. Note also that v has at most one child which is an internal node, so one of the simpler replacing strategies described above can be used to remove v.

3. *More Searching Procedures*: A *range* search takes a tree T and two key values $k_1 \leq k_2$ as arguments and returns all keys x for which $k_1 \leq x \leq k_2$. A range search can be viewed as an in-order traversal, where we do not recurs down the left subtree and do not report the root key if k_1 should be in the right subtree; similarly, we do not recurs down the right subtree and do not report the root key if k_2 should be in the left subtree. The complexity is proportional to the height of the tree plus the size of the output.

   A useful technique for providing more complex operations efficiently is to equip the nodes in the tree with additional information which can be exploited in more advanced searching, and which can also be maintained efficiently. A *rank* search takes a tree T and an integer d between one and n as arguments, and returns the dth smallest key in T. In order to provide this functionality efficiently, we store in each node the size of the subtree in which it is the root. Using this information during a search down the tree, we can at each node determine in which subtree the node must be located and we can appropriately adjust the rank that we search for recursively. If the only modifications made to the tree are small local changes, this extra information can be kept up-to-date efficiently, since it can always be recomputed from the information in the children.

4. *Operations Involving More Trees*: The operation *split* takes a key value x and tree T as arguments and returns two trees; one containing all keys from T less than or equal to x and one with the remaining keys. The operations is destructive, meaning that the argument tree T will not be available after the operation. The operation *meld* takes two trees as arguments, where all keys in one tree are smaller than all keys in the other, and combines the trees into one containing all the keys. This operation is also destructive. Finally, *merge* combines the keys from two argument trees, with no restrictions on keys, into one. Also this operation is destructive.

11.2.3 Implementation of Binary Search Trees

In our discussion of time and space complexities, we assume that some standard implementation of trees are used. Thus, in analogy with the recursive definition, we assume that a tree is represented by information associated with its root, primarily the key, along with pointers (references) to its left and right subtrees, and that this information can be accessed in constant time.

In some situations, we may assume that additional pointers are present, such as *parent-pointers*, giving a reference from a node to its parent. We also sometimes use *level-pointers*. A *level* consists of all nodes of the same depth, and a level-pointer to the right from a node with key k points to the node at the same level with the smallest key larger than k. Similar for level-pointers to the left.

11.3 Generic Discussion of Balancing

As seen in Section 11.2, the worst case complexity of almost all operations on a binary search tree is proportional to its height, making the height its most important single characteristic.

Since a binary tree of height h contains at most $2^h - 1$ nodes, a binary tree of n nodes has a height of at least $\lceil \log(n+1) \rceil$. For static trees, this lower bound is achieved by a tree where all but one level is completely filled. Building such a tree can be done in linear time (assuming that the sorted order of the keys is known), as discussed in Section 11.5 below. In the dynamic case, however, insertions and deletions may produce a very unbalanced tree—for instance, inserting elements in sorted order will produce a tree of height linear in the number of elements.

The solution is to rearrange the tree after an insertion or deletion of an element, if the operation has made the tree unbalanced. For this, one needs a *definition of balance* and a *rebalancing algorithm* describing the rearrangement leading to balance after updates. The combined balance definition and rebalancing algorithm we denote a *rebalancing scheme*. In this section, we discuss rebalancing schemes at a generic level.

The trivial rebalancing scheme consists of defining a balanced tree as one having the optimal height $\lceil \log(n+1) \rceil$, and letting the rebalancing algorithm be the rebuilding of the entire tree after each update. This costs linear time per update, which is exponentially larger than the search time of the tree. It is one of the basic results of Computer Science, first proved by Adel'son-Vel'skiĭ and Landis in 1962 [4], that logarithmic update cost can be achieved simultaneously with logarithmic search cost in binary search trees.

Since the appearance of [4], many other rebalancing schemes have been proposed. Almost all reproduce the result of [4] in the sense that they, too, guarantee a height of $c \cdot \log(n)$ for some constant $c > 1$, while handling updates in $O(\log n)$ time. The schemes can be grouped according to the ideas used for definition of balance, the ideas used for rebalancing, and the exact complexity results achieved.

11.3.1 Balance Definitions

The balance definition is a structural constraint on the tree ensuring logarithmic height. Many schemes can viewed as belonging to one of the following three categories: schemes with a constraint based on the *heights of subtrees*, schemes with a constraint based on the *sizes of subtrees*, and schemes which can be seen as *binarizations of multi-way search tree schemes* and which have a constraint inherited from these. The next section will give examples of each.

For most schemes, balance information is stored in the nodes of the tree in the form of single bits or numbers. The structural constraint is often expressed as an invariant on this information, and the task of the rebalancing algorithm is to reestablish this invariant after an update.

11.3.2 Rebalancing Algorithms

The rebalancing algorithm restores the structural constraint of the scheme if it is violated by an update. It uses the balance information stored in the nodes to guide its actions.

The general form of the algorithm is the same in almost all rebalancing schemes—balance violations are removed by working towards the root along the search path from the leaf where the update took place. When removing a violation at one node, another may be introduced at its parent, which is then handled, and so forth. The process stops at the root at the latest.

The violation at a node is removed in $O(1)$ time by a local restructuring of the tree and/or a change of balance information, giving a total worst case update time proportional to the height of the tree. The fundamental restructuring operation is the *rotation*, shown in Figure 11.1. It was introduced in Reference 4. The crucial feature of a rotation is that it preserves the in-order invariant of the search tree while allowing one subtree to be moved upwards in the tree at the expense of another.

A rotation may be seen as substituting a connected subgraph T consisting of two nodes with a new connected subgraph T' on the same number of nodes, redistributing the keys (here x and y) in T' according to in-order, and redistributing the subtrees rooted at leaves of T by attaching them as leaves of T' according to in-order. Described in this manner, it is clear that in-order will be preserved for *any* two subgraphs T and T' having an equal number of nodes. One particular case is the *double rotation* shown in Figure 11.2, so named because it is equivalent to two consecutive rotations.

Actually, any such transformation of a connected subgraph T to another T' on the same number of nodes can be executed through a series of rotations. This can be seen by noting that any connected subgraph can be converted into a right-path, that is, a tree where all left children are empty trees, by repeated rotations (in Figure 11.1, if y but not x is on the rightmost path in the tree, the rotation will enlarge the rightmost path by one node). Using the right-path as an intermediate state and running one of the conversions backwards will transform T into T'. The double rotation is a simple case of this. In a large number of rebalancing schemes, the rebalancing algorithm performs at most one rotation or double rotation per node on the search path.

We note that rebalancing schemes exist [5] where the rebalancing along the search path is done in a top-down fashion instead of the bottom-up fashion described above. This is useful when several processes concurrently access the tree, as discussed in Section 11.8.

In another type of rebalancing schemes, the restructuring primitive used is the rebuilding of an entire subtree to perfect balance, where perfect balance means that any node is the median among the nodes in its subtree. This primitive is illustrated in

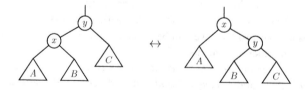

FIGURE 11.1   Rotation.

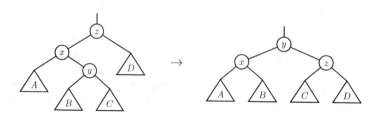

FIGURE 11.2   Double rotation.

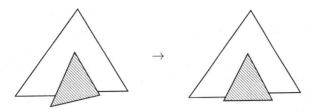

FIGURE 11.3   Rebuilding a subtree.

Figure 11.3. In these rebalancing schemes, the restructuring is only applied to one node on the search path for the update, and this resolves all violations of the balance invariant.

The use of this rebalancing technique is sometimes termed *local* or *partial rebuilding* (in contrast to global rebuilding of data structures, which designates a periodically rebuilding of the entire structure). In Section 11.5, we discuss linear time algorithms for rebalancing a (sub-)tree to perfect balance.

11.3.3   Complexity Results

Rebalancing schemes can be graded according to several complexity measures. One such measure is how much rebalancing work is needed after an update. For this measure, typical values include amortized $O(\log n)$, worst case $O(\log n)$, amortized $O(1)$, and worst case $O(1)$. Values below logarithmic may at first sight seem useless due to the logarithmic search time of balanced search trees, but they are relevant in a number of settings. One setting is finger search trees (described in a chapter of their own in this book), where the search for the update point in the tree does not start at the root and hence may take sub-logarithmic time. Another setting is situations where the nodes of the tree are annotated with information which is expensive to update during restructuring of the tree, such that rotations may take non-constant time. This occurs in Computational Geometry, for instance. A third setting is concurrent access to the tree by several processes. Searching the tree concurrently is not a problem, whereas concurrent updates and restructuring may necessitate lockings of nodes in order to avoid inconsistencies. This makes restructuring more expensive than searches.

Another complexity measure is the exact height maintained. The majority of schemes maintain a height bounded by $c \cdot \log n$ for some constant $c > 1$. Of other results, splay trees [6] have no sub-linear bound on the height, but still perform searches in amortized $O(\log n)$ time. Splay trees are described in a chapter of their own in this book. In the other direction, a series of papers investigate how close c can get to the optimal value one, and at what rebalancing cost. We discuss these results in Section 11.7.

One may also consider the exact amount of balance information stored in each node. Some schemes store an integer, while some only need one or two bits. This may effect the space consumption of nodes, as a single bit may be stored implicitly, for example, as the sign bit of a pointer, or by storing subtrees out of order when the bit is set. Schemes even exist which do not need to store any information at all in nodes. We discuss these schemes in Section 11.6.

Finally, measures such as complexity of implementation and performance in practice can also be considered. However, we will not discuss these here, mainly because these measures are harder to quantify.

11.4   Classic Balancing Schemes

11.4.1   AVL-Trees

AVL-trees where introduced in 1962 in Reference 4, and are named after their inventors Adel'son-Vel'ski and Landis. They proposed the first dictionary structure with logarithmic search and update times, and also introduced the rebalancing technique using rotations.

The balance definition in AVL-trees is based on the height of subtrees. The invariant is that for any node, the heights of its two subtrees differ by at most one. Traditionally, the balance information maintained at each node is $+1$, 0, or -1, giving the difference in heights between the right subtree and the left subtree. This information can be represented by two bits. Another method is to mark a node when its height is larger than its siblings. This requires only one bit per node, but reading the balance of a node now involves visiting its children. In the other direction, storing the height of each node requires $\log \log n$ bits of information per node, but makes the rebalancing algorithms simpler to describe and analyze.

By induction on h, it is easily proved that for an AVL-tree of height h, the minimum number of nodes is $F_{h+2} - 1$, where F_i denotes the i'th Fibonacci number, defined by $F_1 = F_2 = 1$ and $F_{j+2} = F_{j+1} + F_j$. A well-known fact for Fibonacci numbers is that

$F_i \geq \Phi^{i-2}$, where Φ is the golden ratio $(\sqrt{5}+1)/2 \approx 1.618$. This shows that the height of an AVL-tree with n nodes is at most $\log_\Phi (n+1)$, that is, AVL-trees have a height bound of the type $c \cdot \log n$ with $c = 1/\log \Phi \approx 1.440$.

After an update, violations of the balance invariant can only occur at nodes on the search path from the root to the update point, as only these nodes have subtrees changed. The rebalancing algorithm resolves these in a bottom-up fashion. At each node, it either performs a rotation, performs a double rotation, or just updates balance information, with the choice depending on the balance of its child and grandchild on the search path. The algorithm stops when it can guarantee that no ancestor has a balance problem, or when the root is reached.

In AVL-trees, the rebalancing algorithm has the following properties: After an insertion, change of balance information may take place any number of steps towards the root, but as soon as a rotation or double rotation takes place, no further balance problems remain. Hence, only $O(1)$ structural change is made. In contrast, after a deletion it may happen that rotations are performed at all nodes on the search path. If only insertions take place, the amortized amount of rebalancing work, including updating of balance information, can be shown [7] to be $O(1)$. The same is true if only deletions take place [8]. It is not true in the fully dynamic case, as it is easy to find an AVL-tree where alternating insertions and deletions of the same key require rebalancing along the entire search path after each update.

11.4.2 Weight-Balanced Trees

Weight-balanced trees were proposed in 1973 by Nievergelt and Reingold [9], and have a balance definition based on the sizes of subtrees. Here, the size of a subtree is most conveniently defined as the number of external nodes (empty trees) in the subtree, and the size, also denoted the *weight*, of a node is the size of its subtree. The balance invariant of weight-balanced trees states that for any node, the ratio between its own weight and the weight of its right child (or left) is in the interval $[\alpha, 1 - \alpha]$ for some fixed value $\alpha > 0$. This ratio is denoted the *balance* of the node. Since a node of weight three must have subtrees of weight two and one, we must have $\alpha \leq 1/3$. Weight-balanced trees are also called BB$[\alpha]$-trees, which stands for trees of bounded balance with parameter α.

By the balance criterion, for any node v the weight of the parent of v is at least a factor $1/(1 - \alpha)$ larger than the weight of v. A tree of height k therefore has a root of weight at least $1/(1 - \alpha)^k$, which shows that the height of a weight-balanced tree with n nodes is at most $\log_{1/(1-\alpha)} (n+1)$, that is, weight-balanced trees have a height bound of the type $c \cdot \log n$ with $c = -1/\log(1 - \alpha) > 1.709$.

The balance information stored in each node is its weight, for which $\log n$ bits are needed. After an update, this information must be updated for all nodes on the search path from the root to the update point. Some of these nodes may now violate the balance criterion. The rebalancing algorithm proposed in Reference 9 resolves this unbalance in a bottom-up fashion along the search path using either a rotation or a double rotation at each violating node. The choice of rotation depends on the weight of the children and the grandchildren of the node.

In Reference 9, the rebalancing algorithm was claimed to work for α in the interval $[0, 1 - 1/\sqrt{2}]$, but Blum and Mehlhorn [10] later observed that the correct interval is $(2/11, 1 - 1/\sqrt{2}]$. They also showed that for α strictly inside this interval, the rebalancing of an unbalanced node restores its balance to a value in $[(1 + \delta)\alpha, 1 - (1 + \delta)\alpha]$, where δ depends on the choice of α. This implies that when the node becomes unbalanced again, the number of updates which have taken place below it since it was last rebalanced is at least a fraction (depending on α) of its current weight. This feature, unique to weight-balanced trees, has important applications, for example, for data structures in Computational Geometry. A number of these structures are binary search trees where each node has an associated secondary structure built on the elements in the subtree of the node. When a rotation takes place, the structures of the nodes taking part in the rotation will have to be rebuilt. If we attribute the cost of this rebuilding evenly to the updates which have taken place below the node since it was last involved in a rotation, then, as an example, a linear rebuilding cost of the secondary structure will amount to a constant attribution to each of these updates. As the search path for an update contains $O(\log n)$ nodes, any single update can at most receive this many attributions, which implies an amortized $O(\log n)$ update complexity for the entire data structure.

The same analysis allows BB$[\alpha]$-trees to be maintained by local rebuilding instead of rotations in amortized $O(\log n)$ time, as first noted by Overmars and van Leeuwen [11]: After an update, the subtree rooted at the highest unbalanced node (if any) on the search path is rebuilt to perfect balance. Since a rebuilding of a subtree leaves all nodes in it with balance close to 1/2, the number of updates which must have taken place below the node since it was last part of a rebuilding is a constant fraction of its current weight. The rebuilding uses work linear in this weight, which can be covered by attributing a constant amount of work to each of these updates. Again, each update is attributed $O(\log n)$ work. This scheme will work for any $\alpha \leq 1/3$.

For the original rebalancing algorithm using rotations, a better analysis can be made for α chosen strictly inside the interval $(2/11, 1 - 1/\sqrt{2}]$: The total work per rebalancing operation is now $O(1)$, so the work to be attributed to each update below a node is $O(1/w)$, where w is the weight of the node. As noted above in the proof of the height bound of weight-balanced trees, w is exponentially increasing along the search path from the update point to the root. This implies that each update is attributed only $O(1)$ work in total, and also that the number of rotations taking place at a given height decreases exponentially with the height. This result from [10] seems to be the first on $O(1)$ amortized rebalancing in binary search trees. The actual time spent after an update is still logarithmic in weight-balanced trees, though, as the balance information needs to be updated along the entire search path, but this entails no structural changes.

Recently, the idea of balancing by weight has been applied to multi-way search trees [12], leading to trees efficient in external memory which posses the same feature as weight-balanced binary trees, namely that between each rebalancing at a node, the number of updates which have taken place below the node is proportional to the weight of the node.

11.4.3 Balanced Binary Trees Based on Multi-Way Trees

The B-tree [13], which is treated in another chapter of this book, is originally designed to handle data stored on external memory. The basic idea is to associate a physical block with a high-degree node in a multi-way tree. A B-tree is maintained by merging and splitting nodes, and by increasing and decreasing the number of layers of multi-way nodes. The smallest example of a B-tree is the 2-3-tree [14], where the nodes have degree 2 or 3. In a typical B-tree implementation, the degree of a node is much larger, and it varies roughly within a factor of 2.

The concept of multi-way nodes, splitting, and merging, has also proven to be very fruitful in the design of balancing schemes for binary trees. The first such example is the binary B-tree [15], a binary implementation of 2-3-trees. Here, the idea is to organize binary nodes into larger chunks of nodes, here called *pseudo-nodes*. In the binary version of a 2–3-tree, a node of degree 2 is represented by one binary node, while a node of degree 3 is represented as two binary nodes (with the additional constraint that one of the two nodes is the right child of the other). In the terms of binary nodes grouped into pseudo-nodes, it is convenient to say that edges within a pseudo-node are *horizontal* while edges between pseudo-nodes are *vertical*.

As a natural extension of binary B-trees, Bayer invented *Symmetric Binary Trees*, or SBB-trees [16]. The idea was that, instead of only allowing a binary node to have one horizontal outgoing edge to its right child, we can allow both left- and right-edges to be horizontal. For both binary B-trees and Symmetric Binary B-trees, Bayer designed maintenance algorithms, where the original B-tree operations split, merge, and increase/decrease number of levels were implemented for the pseudo-nodes.

Today, SBB-trees mostly appear under the name *red-black trees* [5]. Here, the horizontal and vertical edges are represented by one "color" per node. (Both notations can be represented by one bit per node.) SBB/red-black trees are binary implementations of B-trees where each node has degree between 2 and 4.

One advantage with SBB-trees/red-black trees is that a tree can be updated with only a constant number of rotations per insertion or deletion. This property is important for example when maintaining priority search trees [17] where each rotation requires $\Theta(\log n)$ time.

The first binary search tree with $O(1)$ rotations per update was the half-balanced trees by Olivié [18]. Olivié's idea was to use *path-balancing*, where the quotient between the shortest and longest path from each node is restricted to be at most 1/2, and he showed that this path-balance could be maintained with $O(1)$ rotations per update. It turns out to be the case that half-balanced trees and SBB/red-black trees are structurally equivalent, although their maintenance algorithms are different. It has also been proven by Tarjan [19] that SBB/red-black trees can be maintained by $O(1)$ rotations. These algorithms can also be generalized to maintain pseudo-nodes of higher degree, resulting in binary B-tree implementations with lower height [20], still requiring $O(1)$ rotations per update.

The mechanism behind the constant number of rotations per update can be explained in a simple way by examining three cases of what can happen during insertion and deletion in a binary B-tree representation.

- When a pseudo-node becomes too large, it can be split into two pseudo-nodes without any rotation; we just need to change the balance information.
- Also, when a pseudo-node becomes too small and its sibling has minimal size, these two nodes can be merged without any rotation; we just change balance information.
- In all other cases, when a pseudo-node becomes too small or too large, this will be resolved by moving nodes between the pseudo-node and its sibling and no splitting or merging will take place.

From these three basic facts, it can be shown that as soon as the third case above occurs, no more rebalancing will be done during the same update. Hence, the third case, requiring rotations, will only occur once per update. For details, we refer to the literature [19,20].

Binary B-trees can also be used to design very simple maintenance algorithms that are easy to code. This is illustrated by AA-trees [21,22]. AA-trees are actually the same as Bayer's binary version of 2-3-trees, but with design focused on simplicity. Compared with normal red-black tree implementations, AA-trees require very few different cases in the algorithm and much less code for implementation.

While binary B-trees and SBB/red-black trees deal with small pseudo-nodes, the stratified trees by van Leeuwen and Overmars [23] use large pseudo-nodes arranged in few layers. The concept of stratification does not imply that all pseudo-nodes have similar size; it is mainly a way to conceptually divide the tree into layers, using the notion of merging and splitting.

11.5 Rebalancing a Tree to Perfect Balance

A basic operation is the rebalancing operation, which takes a binary tree as input and produces a balanced tree. This operation is important in itself, but it is also used as a subroutine in balancing schemes (see Section 11.6).

It is quite obvious that one can construct a perfectly balanced tree from an ordered tree, or a sorted list, in linear time. The most straightforward way is to put the elements in sorted order into an array, take the median as the root of the tree, and construct the left and right subtrees recursively from the upper and lower halves of the array. However, this is unnecessarily cumbersome in terms of time, space, and elegance.

A number of restructuring algorithms, from the type mentioned above to more elegant and efficient ones based on rotations, can be found in the literature [24–28]. Of these, the one by Stout and Warren [28] seems to be most efficient. It uses the following principle:

1. *Skew.* Make right rotations at the root until no left child remains. Continue down the right path making right rotations until the entire tree becomes one long rightmost path (a "vine").
2. *Split.* Traverse down the vine a number of times, each time reducing the length of the vine by left rotations.

If we start with a vine of length $2^p - 1$, for some integer p, and make one rotation per visited node, the resulting vine will be of length $2^{p-1} - 1$ after the first pass, $2^{p-2} - 1$ after the second pass, etc., until the vine is reduced to a single node; the resulting tree is a perfectly balanced tree. If the size of the tree is $2^p - 1$, this will work without any problem. If, however, the size is not a power of two, we have to make some special arrangements during the first pass of left rotations. Stout and Warren solved the problem of how to make evenly distributed rotations along the vine in a rather complicated way, but there is a simpler one. It has never before been published in itself, but has been included in demo software and in published code [29,30].

The central operation is a split operation that takes as parameters two numbers p_1 and p_2 and compresses a right-skewed path of p_1 nodes into a path of p_2 nodes ($2p_2 \geq p_1$). The simple idea is to use a counter stepping from $p_1 - p_2$ to $p_2(p_1 - p_2)$ with increment $p_1 - p_2$. Every time this counter reaches or exceeds a multiple of p_2, a rotation is performed. In effect, the operation will make $p_1 - p_2$ evenly distributed left rotations.

With this split operation available, we can do as follows to rebalance a tree of size n (n internal nodes): First, skew the tree. Next, find the largest integer b such that b is an even power of 2 and $b - 1 \leq n$. Then, if $b - 1 < n$, call Split with parameters n and $b - 1$. Now, the vine will have proper length and we can traverse it repeatedly, making a left rotation at each visited node, until only one node remains.

In contrast to the Stout-Warren algorithm, this algorithm is straightforward to implement. We illustrate it in Figure 11.4. We describe the five trees, starting with the topmost:

1. A tree with 12 internal nodes to be balanced.
2. After Skew.
3. With $n = 12$ and $b = 8$, we call split with parameters 12 and 7, which implies that five evenly distributed rotations will be made. As the result, the vine will be of length 7, which fulfills the property of being $2^p - 1$.
4. The next split can be done by traversing the vine, making one left rotation at each node. As a result, we get a vine of length 3 (nodes 3, 6, and 10).
5. After the final split, the tree is perfectly balanced.

11.6 Schemes with no Balance Information

As discussed above, a balanced binary search tree is typically maintained by local constraints on the structure of the tree. By keeping structure information in the nodes, these constraints can be maintained during updates.

In this section, we show that a plain vanilla tree, without any local balance information, can be maintained efficiently. This can be done by coding the balance information implicitly (Section 11.6.1) or by using global instead of local balance criteria, hereby avoiding the need for balance information (Section 11.6.2). Splay trees [6] also have no balance information. They do not have a sub-linear bound on their height, but still perform searches in amortized $O(\log n)$ time. Splay trees are described in a chapter of their own in this book.

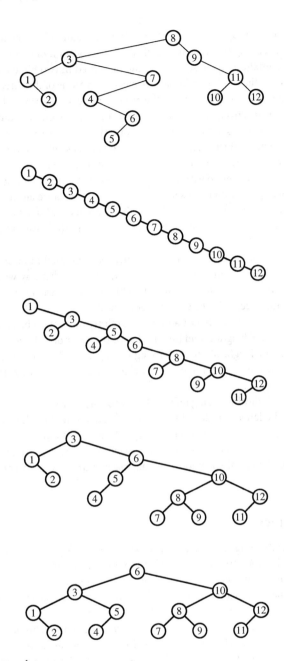

FIGURE 11.4 Rebalancing a binary search tree.

11.6.1 Implicit Representation of Balance Information

One idea of how to remove the need for local balance information is to store the information implicitly. There are two main techniques for this: coding information in the way empty pointers are located or coding information by changing the order between left and right children.

In both cases, we can easily code one bit implicitly at each internal node, but not at external nodes. Therefore, we weed to use balance schemes that can do with only one bit per internal node and no balance information at external nodes.

As an example, we may use the AVL-tree. At each node, we need to keep track of whether the two subtrees have the same height or if one of them is one unit higher than its sibling. We can do this with one bit per internal node by letting the bit be 1 if and only if the node is higher than its sibling. For external nodes we know the height, so no balance information is needed there.

The assumption that we only need one bit per internal node is used in the two constructions below.

1. *Using Empty Pointers*: As pointed out by Brown [31,32], the explicitly stored balance information may in some classes of balanced trees be eliminated by coding the information through the location of empty pointers. We use a tree of pseudo-nodes, where a pseudo-node contains two consecutive elements, stored in two binary nodes. The pseudo-node will have three outgoing pointers, and since the two binary nodes are consecutive, one of the three pointers will be empty. By varying which of the two nodes become parent, we can arrange the pseudo-node in two ways. These two different structures is used to represent bit values 0 and 1, respectively; by checking the position of the empty pointer, we can compute the bit value. In order for this to work, we allow the pseudo-nodes at the bottom of the tree to contain one or two binary nodes.

   During insertion, we traverse down the tree. If the inserted element lies between the two keys in a visited pseudo-node, we replace it by one of the elements in the pseudo-node and continue down the tree with that element instead. At the bottom of the tree, if we find a pseudo-node with only one key, we just add the new key. If, on the other hand, we find a pseudo-node with two keys, we split it into two pseudo-nodes which will cause an insertion in the tree of pseudo-nodes. Rotations etc. can be done with pseudo-nodes instead of ordinary binary nodes. (If a rotation involves the lowest level of the tree of pseudo-nodes, some care has to be taken in order to maintain the invariant that only the lowest pseudo-nodes may contain a single node.)

   Deletions are handled correspondingly. If the deleted element is contained in an internal pseudo-node, we replace it by its predecessor or successor, which resides at the bottom of the tree; in this way we ensure that the deletion occurs at the bottom. If the deletion occurs at a pseudo-node with two binary nodes, we just remove the node, if the pseudo-node contains only one node, a deletion occurs in the tree of pseudo-nodes.

   Despite the pseudo-nodes, the tree is really just a binary search tree where no balance information is explicitly stored. Since each pseudo-node has internal height 2, and the number of pseudo-nodes is less than n, the height of the binary tree is $O(\log n)$. A drawback is that the height of the underlying binary tree will become higher by the use of pseudo-nodes. Instead of n internal nodes we will have roughly $n/2$ pseudo-nodes, each of height 2. In the worst case, the height of the binary tree will be doubled.

2. *Swapping Pointers*: Another possibility for coding information into a structure is to use the ordering of nodes. If we redefine binary search trees, such that the left and right subtree of a node are allowed to change place, we can use this possibility to encode one bit per node implicitly. By comparing the keys of the two children of a node, the one-bit information can be extracted. During search, we have to make one comparison extra at each node. This idea has been used by Munro and Suwanda [33–35] to achieve implicit implementation of binary search trees, but it can of course also be used for traditional pointer-based tree structures.

11.6.2 General Balanced Trees

In the following, we use $|T|$ to denote the weight (number of leaves) in a tree T. We also use $|v|$ to denote the weight of a subtree rooted at node v. It should be noted that for a tree T storing n keys in internal nodes, $|T| = n + 1$.

Instead of coding balance information into the structure of the tree, we can let the tree take any shape, as long as its height is logarithmic. Then, there is no local balance criterion to maintain, and we need no balance information in the nodes, not even implicitly coded. As we show below, the tree can still be maintained efficiently.

When maintaining trees this way, we use the technique of *partial rebuilding*. This technique was first introduced by Overmars and van Leeuwen [11,36] for maintaining weight-balanced trees. By making a partial rebuilding at node v, we mean that the subtree rooted at v is rebuilt into a perfectly balanced tree. The cost of such rebalancing is $\Theta(|v|)$. In Section 11.5, we discuss linear time algorithms for rebalancing a (sub-)tree to perfect balance.

Apart from the advantage of requiring no balance information in the nodes, it can be shown [37] that the constant factor for general balanced trees is lower than what has been shown for the maintenance of weight-balanced trees by partial rebuilding.

The main idea in maintaining a general balanced tree is to let the tree take *any* shape as long as its height does not exceed $\log |T|$ by more than a specified constant factor. The key observation is that whenever the tree gets too high by an insertion, we can find a node where partial rebuilding can be made at a low amortized cost. (Since deletions do not increase the height of the tree, we can handle deletions efficiently by rebuilding the entire tree after a large number of elements have been deleted.)

We use two constants $c > 1$, and $b > 0$, and we maintain a balanced tree T with maximum height $\lceil c \log |T| + b \rceil$.

No balance information is used, except two global integers, containing $|T|$, the number of leaves in T, and $d(T)$, the number of deletions made since the last time the entire tree T was rebalanced.

Updates are performed in the following way:

1. *Insertion*: If the depth of the new leaf exceeds $\lceil c \log(|T| + d(T)) \rceil$, we back up along the insertion path until we find the lowest node v, such that $h(v) > \lceil c \log |v| \rceil$. The subtree v is then rebuilt to perfect balance. The node v is found by explicitly

traversing the subtrees below the nodes on the path from the inserted leaf to v, while counting the number of leaves. The cost for this equals the cost for traversing the subtree below v once, which is $O(|v|)$.

2. *Deletion:* $d(T)$ increases by one. If $d(T) \geq (2^{b/c} - 1)|T|$, we rebuild T to perfect balance and set $d(T) = 0$.

First, we show that the height is low enough. Since deletions do not increase the height of T, we only need to show that the height is not increased too much by an insertion. We prove this by induction. Assume that

$$h(T) \leq \lceil c \log(|T| + d(T)) \rceil \tag{11.1}$$

holds before an insertion. (Note that the height of an empty tree is zero.) During the insertion, the height condition can only be violated by the new node. However, if such a violation occurs, the partial rebuilding will ensure that Inequality (11.1) holds after the insertion. Hence, Inequality (11.1) holds by induction. Combining this with the fact that $d(T) < (2^{b/c} - 1)|T|$, we get that $h(T) \leq \lceil c \log |T| + b \rceil$.

Next, we show that the maintenance cost is low enough. Since the amortized cost for the rebuilding of the entire tree caused by deletions is obviously $O(1)$ per deletion, we only need to consider insertions.

In fact, by the way we choose where to perform rebuilding, we can guarantee that *when a partial rebuilding occurs at node v, $\Omega(v)$ updates have been made below v since the last time v was involved in a partial rebuilding.* Indeed, this observation is the key observation behind general balanced trees.

Let v_H be v's child on the path to the inserted node. By the way v is selected by the algorithm, we know the following about v and v_h:

$$h(v) > \lceil c \log |v| \rceil \tag{11.2}$$

$$h(v_H) \leq \lceil c \log |v_H| \rceil \tag{11.3}$$

$$h(v) = h(v_h) + 1 \tag{11.4}$$

Combining these, we get

$$\lceil c \log |v| \rceil < h(v) = h(v_H) + 1 \leq \lceil c \log |v_H| \rceil + 1 \tag{11.5}$$

and, thus

$$\log |v| < \log |v_H| + 1/c$$
$$|v_H| > 2^{-1/c} |v| \tag{11.6}$$

Since $2^{-1/c} > 1/2$, we conclude that the weight of v_H is $\Theta(v)$ larger than the weight of v's other child. The only way this difference in weight between the two children can occur is by insertions or deletion below v. Hence, $\Omega(v)$ updates must have been made below v since the last time v was involved in a partial rebuilding. In order for the amortized analysis to hold, we need to reserve a constant cost at v for each update below v. At each update, updates are made below $O(\log n)$ nodes, so the total reservation per update is $O(\log n)$.

Since the tree is allowed to take any shape as long as its height is low enough, we call this type of balanced tree *general balanced trees* [37]. We use the notation GB-trees or GB(c)-trees, where c is the height constant above. (The constant b is omitted in this notation.) (The idea of general balanced trees have also been rediscovered under the name scapegoat trees [26].)

Example. The upper tree in Figure 11.5 illustrates a GB(1.2)-tree where five deletions and some insertions have been made since the last global rebuilding. When inserting 10, the height becomes 7, which is too high, since $7 > \lceil c \log(|T| + d(T)) \rceil = \lceil 1.2 \log(20 + 5) \rceil = 6$. We back up along the path until we find the node 14. The height of this node is 5 and the weight is 8. Since $5 > \lceil 1.2 \log 8 \rceil$, we can make a partial rebuilding at that node. The resulting tree is shown as the lower tree in Figure 11.5.

11.6.3 Application to Multi-Dimensional Search Trees

The technique of partial rebuilding is an attractive method in the sense that it is useful not only for ordinary binary search trees, but also for more complicated data structures, such as multi-dimensional search trees, where rotations cannot be used efficiently. For example, partial rebuilding can be used to maintain logarithmic height in k-d trees [38] under updates [11,36, 39]. A detailed study of the use of partial rebuilding can be found in Mark Overmars' Ph.D. thesis [36]. For the sake of completeness, we just mention that if the cost of rebalancing a subtree v is $O(P(|v|))$, the amortized cost of an update will be $O\left(\frac{P(n)}{n} \log n\right)$. For example, applied to k-d trees, we get an amortized update cost of $O(\log^2 n)$.

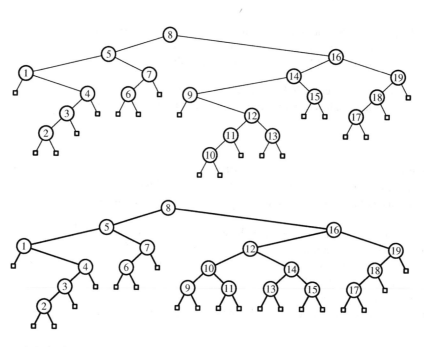

FIGURE 11.5 Upper tree: A GB(1.2)-tree which requires rebalancing. Lower tree: After partial rebuilding.

11.7 Low Height Schemes

Most rebalancing schemes reproduce the result of AVL-trees [4] in the sense that they guarantee a height of $c \cdot \log(n)$ for some constant $c > 1$, while doing updates in $O(\log n)$ time. Since the height determines the worst-case complexity of almost all operations, it may be reasonable to ask exactly how close to the best possible height $\lceil \log(n+1) \rceil$ a tree can be maintained during updates. Presumably, the answer depends on the amount of rebalancing work we are willing to do, so more generally the question is: given a function f, what is the best possible height maintainable with $O(f(n))$ rebalancing work per update?

This question is of practical interest—in situations where many more searches than updates are performed, lowering the height by factor of (say) two will improve overall performance, even if it is obtained at the cost of a larger update time. It is also of theoretical interest, since we are asking about the inherent complexity of maintaining a given height in binary search trees. In this section, we review the existing answers to the question.

Already in 1976, Maurer et al. [40] proposed the k-neighbor trees, which guarantee a height of $c \cdot \log(n)$, where c can be chosen arbitrarily close to one. These are unary-binary trees, with all leaves having the same depth and with the requirement that between any two unary nodes on the same level, at least $k-1$ binary nodes appear. They may be viewed as a type of $(1,2)$-trees where the rebalancing operations exchange children, not only with neighboring nodes (as in standard (a,b)-tree or B-tree rebalancing), but with nodes a horizontal distance k away. Since at each level, at most one out of k nodes is unary, the number of nodes increases by a factor of $(2(k-1)+1)/k = 2 - 1/k$ for each level. This implies a height bound of $\log_{2-1/k} n = \log(n)/\log(2-1/k)$. By first order approximation, $\log(1+x) = \Theta(x)$ and $1/(1+x) = 1 - \Theta(x)$ for x close to zero, so $1/\log(2-1/k) = 1/(1 + \log(1-1/2k)) = 1 + \Theta(1/k)$. Hence, k-trees maintain a height of $(1 + \Theta(1/k)) \log n$ in time $O(k \log n)$ per update.

Another proposal [20] generalizes the red-black method of implementing $(2,4)$-trees as binary trees, and uses it to implement (a,b)-trees as binary trees for $a = 2^k$ and $b = 2^{k+1}$. Each (a,b)-tree node is implemented as a binary tree of perfect balance. If the underlying (a,b)-tree has t levels, the binary tree has height at most $t(k+1)$ and has at least $(2^k)^t = 2^{kt}$ nodes. Hence, $\log n \geq tk$, so the height is at most $(k+1)/k \log n = (1 + 1/k) \log n$. As in red-black trees, a node splitting or fusion in the (a,b)-tree corresponds to a constant amount of recoloring. These operations may propagate along the search path, while the remaining rebalancing necessary takes place at a constant number of (a,b)-tree nodes. In the binary formulation, these operations involve rebuilding subtrees of size $\Theta(2^k)$ to perfect balance. Hence, the rebalancing cost is $O(\log(n)/k + 2^k)$ per update.

Choosing $k = \lfloor \log \log n \rfloor$ gives a tree with height bound $\log n + \log(n)/\log \log(n)$ and update time $O(\log n)$. Note that the constant for the leading term of the height bound is now one. To accommodate a non-constant k, the entire tree is rebuilt when $\lfloor \log \log n \rfloor$ changes. Amortized this is $O(1)$ work, which can be made a worst case bound by using incremental rebuilding [36].

Returning to k-trees, we may use the method of non-constant k also there. One possibility is $k = \Theta(\log n)$, which implies a height bound as low as $\log n + O(1)$, maintained with $O(\log^2 n)$ rebalancing work per update. This height is $O(1)$ from the best possible. A similar result can be achieved using the general balanced trees described in Section 11.6: In the proof of complexity in that section, the main point is that the cost $|v|$ of a rebuilding at a node v can be attributed to at least $(2^{-1/c} - 1/2)|v|$ updates, implying that each update is attributed at most $(1/(2^{-1/c} - 1/2))$ cost at each of the at most $O(\log n)$ nodes on the search path. The rebalancing cost is therefore $O(1/(2^{-1/c} - 1/2) \log n)$ for maintaining height $c \cdot \log n$. Choosing $c = 1 + 1/\log n$ gives a height bound of $\log n + O(1)$, maintained in $O(\log^2 n)$ amortized rebalancing work per update, since $(2^{-1/(1+(1/\log n))} - 1/2)$ can be shown to be $\Theta(1/\log n)$ using the first order approximations $1/(1 + x) = 1 - \Theta(x)$ and $2^x = 1 + \Theta(x)$ for x close to zero.

We note that a binary tree with a height bound of $\log n + O(1)$ in a natural way can be embedded in an array of length $O(n)$: Consider a tree T with a height bound of $\log n + k$ for an integer k, and consider n ranging over the interval $[2^i; 2^{i+1}[$ for an integer i. For n in this interval, the height of T never exceeds $i + k$, so we can think of T as embedded in a virtual binary tree T' with $i + k$ completely full levels. Numbering nodes in T' by an in-order traversal and using these numbers as indexes in an array A of size $2^{i+k} - 1$ gives an embedding of T into A. The keys of T will appear in sorted order in A, but empty array entries may exist between keys. An insertion into T which violates the height bound corresponds to an insertion into the sorted array A at a non-empty position. If T is maintained by the algorithm based on general balanced trees, rebalancing due to the insertion consists of rebuilding some subtree in T to perfect balance, which in A corresponds to an even redistribution of the elements in some consecutive segment of the array. In particular, the redistribution ensures an empty position at the insertion point.

In short, the tree rebalancing algorithm can be used as a maintenance algorithm for a sorted array of keys supporting insertions and deletions in amortized $O(\log^2 n)$ time. The requirement is that the array is never filled to more than some fixed fraction of its capacity (the fraction is $1/2^{k-1}$ in the example above). Such an amortized $O(\log^2 n)$ solution, phrased directly as a maintenance algorithm for sorted arrays, first appeared in Reference 41. By the converse of the embedding just described, [41] implies a rebalancing algorithm for low height trees with bounds as above. This algorithm is similar, but not identical, to the one arising from general balanced trees (the criteria for when to rebuild/redistribute are similar, but differ in the details). A solution to the sorted array maintenance problem with worst case $O(\log^2 n)$ update time was given in Reference 42. Lower bounds for the problem appear in References 43 and 44, with one of the bounds stating that for algorithms using even redistribution of the elements in some consecutive segment of the array, $O(\log^2 n)$ time is best possible when the array is filled up to some constant fraction of its capacity.

We note that the correspondence between the tree formulation and the array formulation only holds when using partial rebuilding to rebalance the tree—only then is the cost of the redistribution the same in the two versions. In contrast, a rotation in the tree will shift entire subtrees up and down at constant cost, which in the array version entails cost proportional to the size of the subtrees. Thus, for pointer based implementation of trees, the above $\Omega(\log^2 n)$ lower bound does not hold, and better complexities can be hoped for.

Indeed, for trees, the rebalancing cost can be reduced further. One method is by applying the idea of *bucketing*: The subtrees on the lowest $\Theta(\log K)$ levels of the tree are changed into buckets holding $\Theta(K)$ keys. This size bound is maintained by treating the buckets as (a, b)-tree nodes, that is, by bucket splitting, fusion, and sharing. Updates in the top tree only happen when a bucket is split or fused, which only happens for every $\Theta(K)$ updates in the bucket. Hence, the amortized update time for the top tree drops by a factor K. The buckets themselves can be implemented as well-balanced binary trees—using the schemes above based on k-trees or general balanced trees for both top tree and buckets, we arrive at a height bound of $\log n + O(1)$, maintained with $O(\log \log^2 n)$ amortized rebalancing work. Applying the idea recursively inside the buckets will improve the time even further. This line of rebalancing schemes was developed in References 45–50, ending in a scheme [48] maintaining height $\lceil \log(n + 1) \rceil + 1$ with $O(1)$ amortized rebalancing work per update.

This rather positive result is in contrast to an observation made in Reference 49 about the cost of maintaining exact optimal height $\lceil \log(n + 1) \rceil$: When $n = 2^i - 1$ for an integer i, there is only one possible tree of height $\lceil \log(n + 1) \rceil$, namely a tree of i completely full levels. By the ordering of keys in a search tree, the keys of even rank are in the lowest level, and the keys of odd rank are in the remaining levels (where the rank of a key k is defined as the number of keys in the tree that are smaller than k). Inserting a new smallest key and removing the largest key leads to a tree of same size, but where all elements previously of odd rank now have even rank, and vice versa. If optimal height is maintained, all keys previously in the lowest level must now reside in the remaining levels, and vice versa—in other words, the entire tree must be rebuilt. Since the process can be repeated, we obtain a lower bound of $\Omega(n)$, even with respect to amortized complexity. Thus, we have the intriguing situation that a height bound of $\lceil \log(n + 1) \rceil$ has amortized complexity $\Theta(n)$ per update, while raising the height bound a trifle to $\lceil \log(n + 1) \rceil + 1$ reduces the complexity to $\Theta(1)$.

Actually, the papers [45–50] consider a more detailed height bound of the form $\lceil \log(n + 1) + \varepsilon \rceil$, where ε is any real number greater than zero. For ε less than one, this expression is optimal for the first integers n above $2^i - 1$ for any i, and optimal plus one for the last integers before $2^{i+1} - 1$. In other words, the smaller an ε, the closer to the next power of two is the height guaranteed to be optimal. Considering tangents to the graph of the logarithm function, it is easily seen that ε is proportional to the fraction of integers n for which the height is non-optimal.

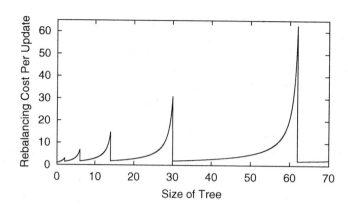

FIGURE 11.6   The cost of maintaining optimal height as a function of tree size.

Hence, an even more detailed formulation of the question about height bound versus rebalancing work is the following: Given a function f, what is the smallest possible ε such that the height bound $\lceil \log(n+1) + \varepsilon \rceil$ is maintainable with $O(f(n))$ rebalancing work per update?

In the case of amortized complexity, the answer is known. In Reference 51, a lower bound is given, stating that no algorithm using $o(f(n))$ amortized rebuilding work per update can guarantee a height of $\lceil \log(n+1) + 1/f(n) \rceil$ for all n. The lower bound is proved by mapping trees to arrays and exploiting a fundamental lemma on density from [43]. In Reference 52, a balancing scheme was given which maintains height $\lceil \log(n+1) + 1/f(n) \rceil$ in amortized $O(f(n))$ time per update, thereby matching the lower bound. The basic idea of the balancing scheme is similar to k-trees, but a more intricate distribution of unary nodes is used. Combined, these results show that for amortized complexity, the answer to the question above is

$$\varepsilon(n) \in \Theta(1/f(n)).$$

We may view this expression as describing the inherent amortized complexity of rebalancing a binary search tree, seen as a function of the height bound maintained. Using the observation above that for any i, $\lceil \log(n+1) + \varepsilon \rceil$ is equal to $\lceil \log(n+1) \rceil$ for n from $2^i - 1$ to $(1 - \Theta(\varepsilon))2^{i+1}$, the result may alternatively be viewed as the cost of maintaining optimal height when n approaches the next power of two: for $n = (1 - \varepsilon)2^{i+1}$, the cost is $\Theta(1/\varepsilon)$. A graph depicting this cost appears in Figure 11.6.

This result holds for the fully dynamic case, where one may keep the size at $(1 - \varepsilon)2^{i+1}$ by alternating between insertions and deletions. In the semi-dynamic case where only insertions take place, the amortized cost is smaller—essentially, it is the integral of the function in Figure 11.6, which gives $\Theta(n \log n)$ for n insertions, or $\Theta(\log n)$ per insertion. More concretely, we may divide the insertions causing n to grow from 2^i to 2^{i+1} into i segments, where segment one is the first 2^{i-1} insertions, segment two is the next 2^{i-2} insertions, and so forth. In segment j, we employ the rebalancing scheme from [52] with $f(n) = \Theta(2^j)$, which will keep optimal height in that segment. The total cost of insertions is $O(2^i)$ inside each of the i segments, for a combined cost of $O(i2^i)$, which is $O(\log n)$ amortized per insertion. By the same reasoning, the lower bound from [51] implies that this is best possible for maintaining optimal height in the semi-dynamic case.

Considering worst case complexity for the fully dynamic case, the amortized lower bound stated above of course still applies. The best existing upper bound is height $\lceil \log(n+1) + \min\{1/\sqrt{f(n)}, \log(n)/f(n)\} \rceil$, maintained in $O(f(n))$ worst case time, by a combination of results in References 46 and 51. For the semi-dynamic case, a worst case cost of $\Theta(n)$ can be enforced when n reaches a power of two, as can be seen by the argument above on odd and even ranks of nodes in a completely full tree.

11.8 Relaxed Balance

In the classic search trees, including AVL-trees [4] and red-black trees [5], balancing is tightly coupled to updating. After an insertion or deletion, the updating procedure checks to see if the structural invariant is violated, and if it is, the problem is handled using the balancing operations before the next operation may be applied to the tree. This work is carried out in a bottom-up fashion by either solving the problem at its current location using rotations and/or adjustments of balance variables, or by carrying out a similar operation which moves the problem closer to the root, where, by design, all problems can be solved.

In relaxed balancing, the tight coupling between updating and balancing is removed. Basically, any restriction on when rebalancing is carried out and how much is done at a time is removed, except that the smallest unit of rebalancing is typically one single or double rotation. The immediate disadvantage is of course that the logarithmic height guarantee disappears, unless other methods are used to monitor the tree height.

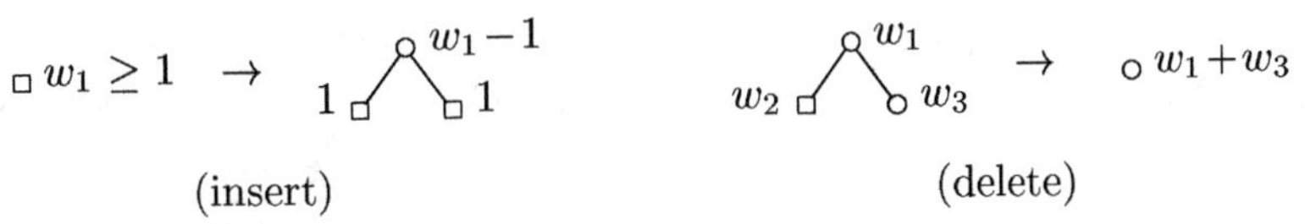

$$\square \; w_1 \geq 1 \quad \rightarrow \quad {}_1\!\!\!\overset{\overset{\textstyle w_1 - 1}{\circ}}{\bigwedge}\!\!\!{}_{\square\, 1} \qquad\qquad {}_{w_2}\square\!\!\!\overset{\overset{\textstyle w_1}{\circ}}{\bigwedge}\!\!\!{}^{\circ}_{w_3} \quad \rightarrow \quad \circ\, w_1 + w_3$$

$$\text{(insert)} \qquad\qquad\qquad\qquad \text{(delete)}$$

FIGURE 11.7 Update operations.

The advantage gained is flexibility in the form of extra control over the combined process of updating and balancing. Balancing can be "turned off" during periods with frequent searching and updating (possibly from an external source). If there is not too much correlation between updates, the tree would likely remain fairly balanced during that time. When the frequency drops, more time can be spend on balancing. Furthermore, in multi-processor environments, balancing immediately after an update is a problem because of the locking strategies with must be employed. Basically, the entire search path must be locked because it may be necessary to rebalance all the way back up to the root. This problem is discussed as early as in Reference 5, where top-down balancing is suggested as a means of avoiding having to traverse the path again bottom-up after an update. However, this method generally leads to much more restructuring than necessary, up to $\Theta(\log n)$ instead of $O(1)$. Additionally, restructuring, especially in the form of a sequence of rotations, is generally significantly more time-consuming than adjustment of balance variables. Thus, it is worth considering alternative solutions to this concurrency control problem.

The advantages outlined above are only fully obtained if balancing is still efficient. That is the challenge: to define balancing constraints which are flexible enough that updating without immediate rebalancing can be allowed, yet at the same time sufficiently constrained that balancing can be handled efficiently at any later time, even if path lengths are constantly super-logarithmic.

The first partial result, dealing with insertions only, is from [53]. Below, we discuss the results which support insertion as well as deletion.

11.8.1 Red-Black Trees

In standard red-black trees, the balance constraints require that no two consecutive nodes are red and that for any node, every path to a leaf has the same number of black nodes. In the relaxed version, the first constraint is abandoned and the second is weakened in the following manner: Instead of a color variable, we use an integer variable, referred to as the *weight* of a node, in such a way that zero can be interpreted as red and one as black. The second constraint is then changed to saying that for any node, every path to a leaf has the same sum of weights. Thus, a standard red-black tree is also a relaxed tree; in fact, it is the ideal state of a relaxed tree. The work on red-black trees with relaxed balance was initiated in References 54 and 55.

Now, the updating operations must be defined so that an update can be performed in such a way that updating will leave the tree in a well-defined state, that is, it must be a relaxed tree, without any subsequent rebalancing. This can be done as shown in Figure 11.7. The operations are from [56].

The trees used here, and depicted in the figure, are assumed to be leaf-oriented. This terminology stems from applications where it is convenient to treat the external nodes differently from the remaining nodes. Thus, in these applications, the external nodes are not empty trees, but real nodes, possibly of another type than the internal nodes. In database applications, for instance, if a sequence of sorted data in the form of a linked list is already present, it is often desirable to build a tree on top of this data to facilitate faster searching. In such cases, it is often convenient to allow copies of keys from the leaves to also appear in the tree structure. To distinguish, we then refer to the key values in the leaves as keys, and refer to the key values in the tree structure as *routers*, since they merely guide the searching procedure. The ordering invariant is then relaxed, allowing keys in the left subtree of a tree rooted by u to be smaller than *or equal to u.k*, and the size of the tree is often defined as the number of leaves. When using the terminology outlined here, we refer to the trees as *leaf-oriented* trees.

The balance problems in a relaxed tree can now be specified as the relations between balance variables which prevent the tree from being a standard red-black tree, that is, consecutive red nodes (nodes of weight zero) and weights greater than one. Thus, the balancing scheme must be targeted at removing these problems. It is an important feature of the design that the global constraint on a standard red-black tree involving the number of black nodes is not lost after an update. Instead, the information is captured in the second requirement and as soon as all weight greater than one has been removed, the standard constraint holds again.

The strategy for the design of balancing operations is the same as for the classical search trees. Problems are removed if this is possible, and otherwise, the problem is moved closer to the root, where all problems can be resolved. In Figure 11.8, examples are shown of how consecutive red nodes and weight greater than one can be eliminated, and in Figure 11.9, examples are given of how these problems may be moved closer to the root, in the case where they cannot be eliminated immediately.

It is possible to show complexity results for relaxed trees which are similar to the ones which can be obtained in the classical case. A logarithmic bound on the number of balancing operations required to balance the tree in response to an update was established in Reference 57. Since balancing operations can be delayed any amount of time, the usual notion of n as the number

FIGURE 11.8 Example operations eliminating balance problems.

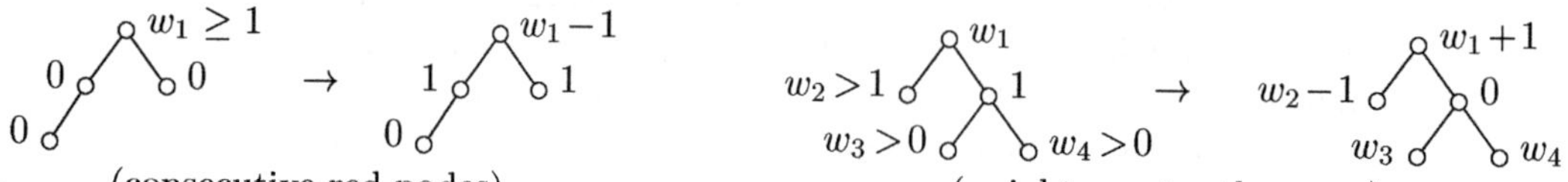

FIGURE 11.9 Example operations moving balance problems closer to the root.

of elements in the tree at the time of balancing after an update is not really meaningful, so the bound is logarithmic in N, which is the maximum number of elements in the tree since it was last in balance. In Reference 58, amortized constant bounds were obtained and in References 59, a version is presented which has fewer and smaller operations, but meets the same bounds. Also, restructuring of the tree is worst-case constant per update. Finally, [56] extends the set of operations with a group insertion, such that an entire search tree can be inserted in between two consecutive keys in amortized time $O(\log m)$, where m is the size of the subtree.

The amortized bounds as well as the worst case bounds are obtained using potential function techniques [60]. For group insertion, the results further depend on the fact that trees with low total potential can build [61], such that the inserted subtree does not increase the potential too dramatically.

11.8.2 AVL-Trees

The first relaxed version of AVL-trees [4] is from [62]. Here, the standard balance constraint of requiring that the heights of any two subtrees differ by at most one is relaxed by introducing a slack parameter, referred to as a *tag* value. The tag value, t_u, of any node u must be an integer greater than or equal to -1, except that the tag value of a leaf must be greater than or equal to zero. The constraint that heights may differ by at most one is then imposed on the *relaxed height* instead. The relaxed height $rh(u)$ of a node u is defined as

$$rh(u) = \begin{cases} t_u, & \text{if } u \text{ is a leaf} \\ \max(rh(u.l), rh(u.r)) + 1 + t_u, & \text{otherwise} \end{cases}$$

As for red-black trees, enough flexibility is introduced by this definition that updates can be made without immediate rebalancing while leaving the tree in a well-defined state. This can be done by adjusting tag values appropriately in the vicinity of the update location. A standard AVL-tree is the ideal state of a relaxed AVL-tree, which is obtained when all tag values are zero. Thus, a balancing scheme aiming at this is designed.

In Reference 63, it is shown that a scheme can be designed such that the complexities from the sequential case are met. Thus, only a logarithmic number of balancing operations must be carried out in response to an update before the tree is again in balance. As opposed to red-black trees, the amortized constant rebalancing result does not hold in full generality for AVL-trees, but only for the semi-dynamic case [7]. This result is matched in Reference 64.

A different AVL-based version was treated in Reference 65. Here, rotations are only performed if the subtrees are balanced. Thus, violations of the balance constraints must be dealt with bottom-up. This is a minimalistic approach to relaxed balance. When a rebalancing operation is carried out at a given node, the children do not violate the balance constraints. This limits the possible cases, and is asymptotically as efficient as the structure described above [66,67].

11.8.3 Multi-Way Trees

Multi-way trees are usually described either as (a, b)-trees or B-trees, which are treated in another chapter of this book. An (a, b)-tree [39,68] consists of nodes with at least a and at most b children. Usually, it is required that $a \geq 2$ to ensure logarithmic height, and in order to make the rebalancing scheme work, b must be at least $2a - 1$. Searching and updating including rebalancing is $O(\log_a n)$. If $b \geq 2a$, then rebalancing becomes amortized $O(1)$. The term B-trees [13] is often used synonymously, but sometimes refers to the variant where $b = 2a - 1$ or the variant where $b = 2a$.

For (a, b)-trees, the standard balance constraints for requiring that the number of children of each node is between a and b and that every leaf is at the same depth are relaxed as follows. First, nodes are allowed to have fewer than a children. This makes it possible to perform a deletion without immediate rebalancing. Second, nodes are equipped with a tag value, which is a non-positive integer value, and leaves are only required to have the same *relaxed depth*, which is the usual depth, except that all tag values encountered from the root to the node in question are added. With this relaxation, it becomes possible to perform an insertion locally and leave the tree in a well-defined state.

Relaxed multi-way trees were first considered in Reference 62, and complexity results matching the standard case were established in Reference 69. Variations with other properties can be found in Reference 70. Finally, a group insertion operation with a complexity of amortized $O(\log_a m)$, where m is the size of the group, can be added while maintaining the already achieved complexities for the other operations [71,72]. The amortized result is a little stronger than usual, where it is normally assumed that the initial structure is empty. Here, except for very small values of a and b, zero-potential trees of any size can be constructed such the amortized results starting from such a tree hold immediately [61].

11.8.4 Other Results

Even though there are significant differences between the results outlined above, it is possible to establish a more general result giving the requirements for when a balanced search tree scheme can be modified to give a relaxed version with corresponding complexity properties [73]. The main requirements are that rebalancing operations in the standard scheme must be local constant-sized operations which are applied bottom-up, but in addition, balancing operation must also move the problems of imbalance towards the root. See [74] for an example of how these general ideas are expressed in the concrete setting of red-black trees.

In Reference 75, it is demonstrated how the ideas of relaxed balance can be combined with methods from search trees of near-optimal height, and [70] contains complexity results made specifically for the reformulation of red-black trees in terms of layers based on black height from [76].

Finally, performance results from experiments with relaxed structures can be found in References [77,78].

References

1. A. A. Andersson and M. Thorup. Tight(er) worst-case bounds on dynamic searching and priority queues. In *Proceedings of the Thirty Second Annual ACM Symposium on Theory of Computing*, pages 335–342. ACM Press, 2000.
2. P. Beame and F. E. Fich. Optimal bounds for the predecessor problem and related problems. *Journal of Computer and System Sciences*, 65(1):38–72, 2002.
3. A. Andersson. A note on searching in a binary search tree. *Software—Practice & Experience*, 21(10):1125–1128, 1991.
4. G. M. Adel'son-Vel'skii and E. M. Landis. An algorithm for the organisation of information. *Doklady Akadamii Nauk SSSR*, 146:263–266, 1962. In Russian. English translation in *Soviet Math. Doklady*, 3:1259–1263, 1962.
5. L. J. Guibas and R. Sedgewick. A dichromatic framework for balanced trees. In *Proceedings of the 19th Annual IEEE Symposium on the Foundations of Computer Science*, pages 8–21, 1978.
6. D. D. Sleator and R. E. Tarjan. Self-adjusting binary search trees. *Journal of the ACM*, 32(3):652–686, July 1985.
7. K. Mehlhorn and A. K. Tsakalidis. An amortized analysis of insertions into AVL-trees. *SIAM Journal on Computing*, 15(1):22–33, 1986.
8. A. K. Tsakalidis. Rebalancing operations for deletions in AVL-trees. *R.A.I.R.O. Informatique Théorique*, 19(4):323–329, 1985.
9. J. Nievergelt and E. M. Reingold. Binary search trees of bounded balance. *SIAM Journal on Computing*, 2(1):33–43, 1973.
10. N. Blum and K. Mehlhorn. On the average number of rebalancing operations in weight-balanced trees. *Theoretical Computer Science*, 11:303–320, 1980.
11. M. H. Overmars and J. van Leeuwen. Dynamic multi-dimensional data structures based on quad- and K-D trees. *Acta Informatica*, 17(3):267–285, 1982.
12. L. Arge and J.S. Vitter. Optimal external memory interval management. *SIAM Journal on Computing*, 32(6):1488–1508, 2003.
13. R. Bayer and E. McCreight. Organization and maintenance of large ordered indexes. *Acta Informatica*, 1:173–189, 1972.
14. A. V. Aho, J. E. Hopcroft, and J. D. Ullman. *Data Structures and Algorithms*. Addison-Wesley, Reading, Massachusetts, 1983.
15. R. Bayer. Binary B-trees for virtual memory. In *Proceedings of the ACM SIGIFIDET Workshop on Data Description, Access and control*, pages 219–235, 1971.
16. R. Bayer. Symmetric binary B-trees: Data structure and maintenance algorithms. *Acta Informatica*, 1(4):290–306, 1972.
17. E. M. McCreight. Priority search trees. *SIAM Journal on Computing*, 14(2):257–276, 1985.

18. H. J. Olivie. A new class of balanced search trees: Half-balanced binary search trees. *R. A. I. R. O. Informatique Theoretique*, 16:51–71, 1982.

19. R. E. Tarjan. Updating a balanced search tree in $O(1)$ rotations. *Information Processing Letters*, 16:253–257, 1983.

20. A. Andersson, C. Icking, R. Klein, and T. Ottmann. Binary search trees of almost optimal height. *Acta Informatica*, 28:165–178, 1990.

21. A. Andersson. Balanced search trees made simple. In *Proceedings of the Third Workshop on Algorithms and Data Structures*, volume 709 of *Lecture Notes in Computer Science*, pages 60–71. Springer-Verlag, 1993.

22. M. A. Weiss. *Data Structures and Algorithm Analysis*. The Benjamin/Cummings Publishing Company, 1992. Several versions during the years since the first version.

23. J. van Leeuwen and M. H. Overmars. Stratified balanced search trees. *Acta Informatica*, 18:345–359, 1983.

24. H. Chang and S. S. Iynegar. Efficient algorithms to globally balance a binary search tree. *Communications of the ACM*, 27(7):695–702, 1984.

25. A. C. Day. Balancing a binary tree. *Computer Journal*, 19(4):360–361, 1976.

26. I. Galperin and R. L. Rivest. Scapegoat trees. In *Proceedings of The Fourth Annual ACM-SIAM Symposium on Discrete Algorithms*, pages 165–174, 1993.

27. W. A. Martin and D. N. Ness. Optimizing binary trees grown with a sorting algorithm. *Communications of the ACM*, 15(2):88–93, 1972.

28. Q. F. Stout and B. L. Warren. Tree rebalancing in optimal time and space. *Communications of the ACM*, 29(9):902–908, 1986.

29. A. Andersson. A demonstration of balanced trees. Algorithm animation program (for macintosh) including user's guide. Can be downloaded from author's homepage, 1993.

30. A. Andersson and S. Nilsson. An efficient list implementation. In *JavaOne Conference*, 1999. Code can be found at Stefan Nilsson's home page.

31. M. R. Brown. A storage scheme for height-balanced trees. *Information Processing Letters*, 7(5):231–232, 1978.

32. M. R. Brown. Addendum to "A storage scheme for height-balanced trees". *Information Processing Letters*, 8(3):154–156, 1979.

33. J. I. Munro. An implicit data structure for the dictionary problem that runs in polylog time. In *Proceedings of the 25th Annual IEEE Symposium on the Foundations of Computer Science*, pages 369–374, 1984.

34. J. I. Munro. An implicit data structure supporting insertion, deletion and search in $O(\log^2 n)$ time. *Journal of Computer and System Sciences*, 33:66–74, 1986.

35. J. I. Munro and H. Suwanda. Implicit data structures for fast search and update. *Journal of Computer and System Sciences*, 21:236–250, 1980.

36. M. H. Overmars. *The Design of Dynamic Data Structures*, volume 156 of *Lecture Notes in Computer Science*. Springer-Verlag, 1983.

37. A. Andersson. General balanced trees. *Journal of Algorithms*, 30:1–28, 1999.

38. J. L. Bentley. Multidimensional binary search trees used for associative searching. *Communications of the ACM*, 18(9):509–517, 1975.

39. K. Mehlhorn. *Sorting and Searching*, volume 1 of *Data Structures and Algorithms*. Springer-Verlag, 1986.

40. H. A. Maurer, T. Ottmann, and H.-W. Six. Implementing dictionaries using binary trees of very small height. *Information Processing Letters*, 5:11–14, 1976.

41. A. Itai, A. G. Konheim, and M. Rodeh. A sparse table implementation of priority queues. In *Proceedings of the 8th International Colloquium on Automata, Languages and Programming*, volume 115 of *Lecture Notes in Computer Science*, pages 417–431. Springer-Verlag, 1981.

42. D. E. Willard. A density control algorithm for doing insertions and deletions in a sequentially ordered file in good worst-case time. *Information and Computation*, 97(2):150–204, 1992.

43. P. F. Dietz, J. I. Seiferas, and J. Zhang. A tight lower bound for on-line monotonic list labeling. In *Proceedings of the Fourth Scandinavian Workshop on Algorithm Theory*, volume 824 of *Lecture Notes in Computer Science*, pages 131–142. Springer-Verlag, 1994.

44. P. F. Dietz and J. Zhang. Lower bounds for monotonic list labeling. In *Proceedings of the Second Scandinavian Workshop on Algorithm Theory*, volume 447 of *Lecture Notes in Computer Science*, pages 173–180. Springer-Verlag, 1990.

45. A. Andersson. Optimal bounds on the dictionary problem. In *Proceeding of the Symposium on Optimal Algorithms*, volume 401 of *Lecture Notes in Computer Science*, pages 106–114. Springer-Verlag, 1989.

46. A. Andersson. *Efficient Search Trees*. PhD thesis, Department of Computer Science, Lund University, Sweden, 1990.

47. A. Andersson and T. W. Lai. Fast updating of well-balanced trees. In *Proceedings of the Second Scandinavian Workshop on Algorithm Theory*, volume 447 of *Lecture Notes in Computer Science*, pages 111–121. Springer-Verlag, 1990.

48. A. Andersson and T. W. Lai. Comparison-efficient and write-optimal searching and sorting. In *Proceedings of the Second International Symposium on Algorithms*, volume 557 of *Lecture Notes in Computer Science*, pages 273–282. Springer-Verlag, 1991.

49. T. Lai. *Efficient Maintenance of Binary Search Trees*. PhD thesis, Department of Computer Science, University of Waterloo, Canada, 1990.

50. T. Lai and D. Wood. Updating almost complete trees or one level makes all the difference. In *Proceedings of the Seventh Annual Symposium on Theoretical Aspects of Computer Science*, volume 415 of *Lecture Notes in Computer Science*, pages 188–194. Springer-Verlag, 1990.

51. R. Fagerberg. Binary search trees: How low can you go? In *Proceedings of the Fifth Scandinavian Workshop on Algorithm Theory*, volume 1097 of *Lecture Notes in Computer Science*, pages 428–439. Springer-Verlag, 1996.

52. R. Fagerberg. The complexity of rebalancing a binary search tree. In *Proceedings of Foundations of Software Technology and Theoretical Computer Science*, volume 1738 of *Lecture Notes in Computer Science*, pages 72–83. Springer-Verlag, 1999.

53. J. L. W. Kessels. On-the-fly optimization of data structures. *Communications of the ACM*, 26:895–901, 1983.

54. O. Nurmi and E. Soisalon-Soininen. Uncoupling updating and rebalancing in chromatic binary search trees. In *Proceedings of the Tenth ACM SIGACT-SIGMOD-SIGART Symposium on Principles of Database Systems*, pages 192–198, 1991.

55. O. Nurmi and E. Soisalon-Soininen. Chromatic binary search trees—a structure for concurrent rebalancing. *Acta Informatica*, 33(6):547–557, 1996.

56. K. S. Larsen. Relaxed red-black trees with group updates. *Acta Informatica*, 38(8):565–586, 2002.

57. J. F. Boyar and K. S. Larsen. Efficient rebalancing of chromatic search trees. *Journal of Computer and System Sciences*, 49(3):667–682, 1994.

58. J. Boyar, R. Fagerberg, and K. S. Larsen. Amortization results for chromatic search trees, with an application to priority queues. *Journal of Computer and System Sciences*, 55(3):504–521, 1997.

59. K. S. Larsen. Amortized constant relaxed rebalancing using standard rotations. *Acta Informatica*, 35(10):859–874, 1998.

60. Robert Endre Tarjan. Amortized computational complexity. *SIAM Journal on Algebraic and Discrete Methods*, 6(2):306–318, 1985.

61. L. Jacobsen, K. S. Larsen, and M. N. Nielsen. On the existence and construction of non-extreme (a, b)-trees. *Information Processing Letters*, 84(2):69–73, 2002.

62. O. Nurmi, E. Soisalon-Soininen, and D. Wood. Concurrency control in database structures with relaxed balance. In *Proceedings of the 6th ACM Symposium on Principles of Database Systems*, pages 170–176, 1987.

63. K. S. Larsen. AVL trees with relaxed balance. In *Proceedings of the 8th International Parallel Processing Symposium*, pages 888–893. IEEE Computer Society Press, 1994.

64. K. S. Larsen. AVL trees with relaxed balance. *Journal of Computer and System Sciences*, 61(3):508–522, 2000.

65. E. Soisalon-Soininen and P. Widmayer. Relaxed balancing in search trees. In *Advances in Algorithms, Languages, and Complexity*, pages 267–283. Kluwer Academic Publishers, 1997.

66. K. S. Larsen, E. Soisalon-Soininen, P. Widmayer. Relaxed balance using standard rotations. *Algorithmica*, 31(4):501–512, 2001.

67. K. S. Larsen, E. Soisalon-Soininen, and P. Widmayer. Relaxed balance through standard rotations. In *Proceedings of the Fifth International Workshop on Algorithms and Data Structures*, volume 1272 of *Lecture Notes in Computer Science*, pages 450–461. Springer-Verlag, 1997.

68. S. Huddleston and K. Mehlhorn. A new data structure for representing sorted lists. *Acta Informatica*, 17:157–184, 1982.

69. K. S. Larsen and R. Fagerberg. Efficient rebalancing of b-trees with relaxed balance. *International Journal of Foundations of Computer Science*, 7(2):169–186, 1996.

70. L. Jacobsen and K. S. Larsen. Complexity of layered binary search trees with relaxed balance. In *Proceedings of the Seventh Italian Conference on Theoretical Computer Science*, volume 2202 of *Lecture Notes in Computer Science*, pages 269–284. Springer-Verlag, 2001.

71. K. S. Larsen. Relaxed multi-way trees with group updates. In *Proceedings of the Twentieth ACM SIGACT-SIGMOD-SIGART Symposium on Principles of Database Systems*, pages 93–101. ACM Press, 2001.

72. K. S. Larsen. Relaxed multi-way trees with group updates. *Journal of Computer and System Sciences*, 66(4):657–670, 2003.

73. K. S. Larsen, T. Ottmann, E. Soisalon-Soininen. Relaxed balance for search trees with local rebalancing. *Acta Informatica*, 37(10):743–763, 2001.

74. S. Hanke, Th. Ottmann, and E. Soisalon-Soininen. Relaxed balanced red-black trees. In *Proceedings of the 3rd Italian Conference on Algorithms and Complexity*, volume 1203 of *Lecture Notes in Computer Science*, pages 193–204. Springer-Verlag, 1997.

75. R. Fagerberg, R. E. Jensen, and K. S. Larsen. Search trees with relaxed balance and near-optimal height. In *Proceedings of the Seventh International Workshop on Algorithms and Data Structures*, volume 2125 of *Lecture Notes in Computer Science*, pages 414–425. Springer-Verlag, 2001.

76. Th. Ottmann and E. Soisalon-Soininen. Relaxed balancing made simple. Technical Report 71, Institut für Informatik, Universität Freiburg, 1995.

77. L. Bougé, J. Gabarró, X. Messeguer, and N. Schabanel. Concurrent rebalancing of AVL trees: A fine-grained approach. In *Proceedings of the Third Annual European Conference on Parallel Processing*, volume 1300 of *Lecture Notes in Computer Science*, pages 421–429. Springer-Verlag, 1997.

78. S. Hanke. The performance of concurrent red-black tree algorithms. In *Proceedings of the 3rd International Workshop on Algorithm Engineering*, volume 1668 of *Lecture Notes in Computer Science*, pages 286–300. Springer-Verlag, 1999.

12

Finger Search Trees*

12.1 Finger Searching.. 171
12.2 Dynamic Finger Search Trees 171
12.3 Level Linked (2,4)-Trees.. 172
12.4 Randomized Finger Search Trees 173
      Treaps • Skip Lists
12.5 Applications 175
      Optimal Merging and Set Operations • Arbitrary Merging Order • List Splitting •
      Adaptive Merging and Sorting
Acknowledgments 177
References... 177

Gerth Stølting Brodal
University of Aarhus

12.1 Finger Searching

One of the most studied problems in computer science is the problem of maintaining a sorted sequence of elements to facilitate efficient searches. The prominent solution to the problem is to organize the sorted sequence as a balanced search tree, enabling insertions, deletions and searches in logarithmic time. Many different search trees have been developed and studied intensively in the literature. A discussion of balanced binary search trees can be found in Chapter 11.

This chapter is devoted to *finger search trees*, which are search trees supporting *fingers*, that is, pointers to elements in the search trees and supporting efficient updates and searches in the vicinity of the fingers.

If the sorted sequence is a *static* set of n elements then a simple and space efficient representation is a sorted array. Searches can be performed by binary search using $1 + \lfloor \log n \rfloor$ comparisons (we throughout this chapter let $\log x$ to denote $\log_2 \max\{2, x\}$). A finger search starting at a particular element of the array can be performed by an *exponential search* by inspecting elements at distance $2^i - 1$ from the finger for increasing i followed by a binary search in a range of $2^{\lfloor \log d \rfloor} - 1$ elements, where d is the rank difference in the sequence between the finger and the search element. In Figure 12.1 is shown an exponential search for the element 42 starting at 5. In the example $d = 20$. An exponential search requires $2 + 2\lfloor \log d \rfloor$ comparisons.

Bentley and Yao [1] gave a close to optimal static finger search algorithm which performs $\sum_{i=1}^{\log^* d - 1} \log^{(i)} d + \mathcal{O}(\log^* d)$ comparisons, where $\log^{(1)} x = \log x$, $\log^{(i+1)} x = \log(\log^{(i)} x)$, and $\log^* x = \min\{i \mid \log^{(i)} x \leq 1\}$.

12.2 Dynamic Finger Search Trees

A dynamic finger search data structure should in addition to finger searches also support the insertion and deletion of elements at a position given by a finger. This section is devoted to an overview of existing dynamic finger search data structures. Sections 12.3 and 12.4 give details concerning how three constructions support efficient finger searches: The level linked (2,4)-trees of Huddleston and Mehlhorn [2], the randomized skip lists of Pugh [3,4] and the randomized binary search trees, treaps, of Seidel and Aragon [5].

Guibas et al. [6] introduced finger search trees as a variant of B-trees [7], supporting finger searches in $\mathcal{O}(\log d)$ time and updates in $\mathcal{O}(1)$ time, assuming that only $\mathcal{O}(1)$ movable fingers are maintained. Moving a finger d positions requires $\mathcal{O}(\log d)$ time. This work was refined by Huddleston and Mehlhorn [2]. Tsakalidis [8] presented a solution based on AVL-trees, and Kosaraju [9] presented a generalized solution. Tarjan and van Wyk [10] presented a solution based on red-black trees.

---

* This chapter has been reprinted from first edition of this Handbook, without any content updates.

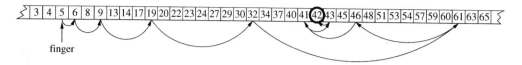

finger

FIGURE 12.1   Exponential search for 42.

The above finger search tree constructions either assume a fixed constant number of fingers or only support updates in amortized constant time. Constructions supporting an arbitrary number of fingers and with worst case update have been developed. Levcopoulos and Overmars [11] presented a search tree that supported updates at an arbitrary position in worst case $\mathcal{O}(1)$ time, but only supports searches in $\mathcal{O}(\log n)$ time. Constructions supporting $\mathcal{O}(\log d)$ time searches and $\mathcal{O}(\log^* n)$ time insertions and deletions were developed by Harel [12,13] and Fleischer [14]. Finger search trees with worst-case constant time insertions and $\mathcal{O}(\log^* n)$ time deletions were presented by Brodal [15], and a construction achieving optimal worst-case constant time insertions and deletions were presented by Brodal et al. [16].

Belloch et al. [17] developed a space efficient alternative solution to the level linked (2,4)-trees of Huddleston and Mehlhorn, see Section 12.3. Their solution allows a single finger, that can be moved by the same performance cost as (2,4)-trees. In the solution no level links and parent pointers are required, instead a special $\mathcal{O}(\log n)$ space data structure, *hand*, is created for the finger that allows the finger to be moved efficiently.

Sleator and Tarjan introduced *splay trees* as a class of self-adjusting binary search trees supporting searches, insertions and deletions in amortized $\mathcal{O}(\log n)$ time [18]. That splay trees can be used as efficient finger search trees was later proved by Cole [19,20]: Given an $\mathcal{O}(n)$ initialization cost, the amortized cost of an access at distance d from the preceding access in a splay tree is $\mathcal{O}(\log d)$ where accesses include searches, insertions, and deletions. Notice that the statement only applies in the presence of one finger, which always points to the last accessed element.

All the above mentioned constructions can be implemented on a pointer machine where the only operation allowed on elements is the comparison of two elements. For the Random Access Machine model of computation (RAM), Dietz and Raman [21,22] developed a finger search tree with constant update time and $\mathcal{O}(\log d)$ search time. This result is achieve by tabulating small tree structures, but only performs the comparison of elements. In the same model of computation, Andersson and Thorup [23] have surpassed the logarithmic bound in the search procedure by achieving $\mathcal{O}\left(\sqrt{\frac{\log d}{\log\log d}}\right)$ query time. This result is achieved by considering elements as bit-patterns/machine words and applying techniques developed for the RAM to surpass lower bounds for comparison based data structures. A survey on RAM dictionaries can be found in Chapter 40.

12.3   Level Linked (2,4)-Trees

In this section we discuss how (2,4)-trees can support efficient finger searches by the introduction of *level links*. The ideas discussed in this section also applies to the more general class of height-balanced trees denoted (a, b)-trees, for $b \geq 2a$. A general discussion of height balanced search trees can be found in Chapter 11. A throughout treatment of level linked (a, b)-trees can be found in the work of Huddleston and Mehlhorn [2,24].

A (2,4)-tree is a height-balanced search tree where all leaves have the same depth and all internal nodes have degree two, three or four. Elements are stored at the leaves, and internal nodes only store search keys to guide searches. Since each internal node has degree at least two, it follows that a (2,4)-tree has height $\mathcal{O}(\log n)$ and supports searches in $\mathcal{O}(\log n)$ time.

An important property of (2,4)-trees is that insertions and deletions given by a finger take amortized $\mathcal{O}(1)$ time (this property is not shared by (2,3)-trees, where there exist sequences of n insertions and deletions requiring $\Theta(n \log n)$ time). Furthermore a (2,4)-tree with n leaves can be split into two trees of size n_1 and n_2 in amortized $\mathcal{O}(\log \min(n_1, n_2))$ time. Similarly two (2,4)-trees of size n_1 and n_2 can be joined (concatenated) in amortized $\mathcal{O}(\log \min(n_1, n_2))$ time.

To support finger searches (2,4)-trees are augmented with level links, such that all nodes with equal depth are linked together in a double linked list. Figure 12.2 shows a (2,4)-tree augmented with level links. Note that all edges represent bidirected links. The additional level links are straightforward to maintain during insertions, deletions, splits and joins of (2,4)-trees.

To perform a finger search from x to y we first check whether y is to the left or right of x. Assume without loss of generality that y is to the right of x. We then traverse the path from x towards the root while examining the nodes v on the path and their right neighbors until it has been established that y is contained within the subtree rooted at v or v's right neighbor. The upwards search is then terminated and at most two downwards searches for y is started at respectively v and/or v's right neighbor. In Figure 12.2 the pointers followed during a finger search from J to T are depicted by thick lines.

The $\mathcal{O}(\log d)$ search time follows from the observation that if we advance the upwards search to the parent of node v then y is to the right of the leftmost subtree of v's right neighbor, that is, d is at least exponential in the height reached so far. In Figure 12.2

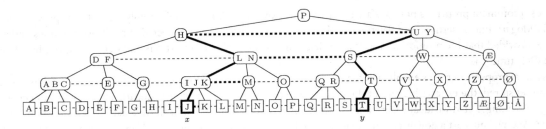

FIGURE 12.2   Level linked (2,4)-trees.

we advance from the internal node labeled "L N" to the node labeled "H" because from "S" we know that y is to the right of the subtree rooted at the node "Q R."

The construction for level linked (2,4)-trees generalizes directly to level linked (a, b)-trees that can be used in external memory. By choosing $a = 2b$ and b such that an internal node fits in a block in external memory, we achieve *external memory* finger search trees supporting insertions and deletions in $\mathcal{O}(1)$ memory transfers, and finger searches with $\mathcal{O}(\log_b n)$ memory transfers.

12.4 Randomized Finger Search Trees

Two randomized alternatives to deterministic search trees are the randomized binary search trees, *treaps*, of Seidel and Aragon [5] and the *skip lists* of Pugh [3,4]. Both treaps and skip lists are elegant data structures, where the randomization facilitates simple and efficient update operations.

In this section we describe how both treaps and skip lists can be used as efficient finger search trees without altering the data structures. Both data structures support finger searches in *expected* $\mathcal{O}(\log d)$ time, where the expectations are taken over the random choices made by the algorithm during the construction of the data structure. For a general introduction to randomized dictionary data structures see Chapter 14.

12.4.1 Treaps

A treap is a rooted binary tree where each node stores an element and where each element has an associated random priority. A treap satisfies that the elements are sorted with respect to an inorder traversal of tree, and that the priorities of the elements satisfy heap order, that is, the priority stored at a node is always smaller than or equal to the priority stored at the parent node. Provided that the priorities are distinct, the shape of a treap is uniquely determined by its set of elements and the associated priorities. Figure 12.3 shows a treap storing the elements A,B,...,T and with random integer priorities between one and hundred.

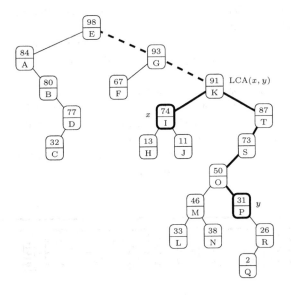

FIGURE 12.3   Performing finger searches on treaps.

The most prominent properties of treaps are that they have expected $\mathcal{O}(\log n)$ height, implying that they provide searches in expected $\mathcal{O}(\log n)$ time. Insertions and deletions of elements can be performed in expected at most two rotations and expected $\mathcal{O}(1)$ time, provided that the position of insertion or deletion is known, that is, insertions and deletions given by a finger take expected $\mathcal{O}(1)$ time [5].

The essential property of treaps enabling expected $\mathcal{O}(\log d)$ finger searches is that for two elements x and y whose ranks differ by d in the set stored, the expected length of the path between x and y in the treap is $\mathcal{O}(\log d)$. To perform a finger search for y starting with a finger at x, we ideally start at x and traverse the ancestor path of x until we reach the *least common ancestor* of x and y, LCA(x, y), and start a downward tree search for y. If we can decide if a node is LCA(x, y), this will traverse exactly the path from x to y. Unfortunately, it is nontrivial to decide if a node is LCA(x, y). In [5] it is assumed that a treap is extended with additional pointers to facilitate finger searches in expected $\mathcal{O}(\log d)$ time. Below an alternative solution is described not requiring any additional pointers than the standard left, right and parent pointers.

Assume without loss of generality that we have a finger at x and have to perform a finger search for $y \geq x$ present in the tree. We start at x and start traversing the ancestor path of x. During this traversal we keep a pointer ℓ to the last visited node that can potentially be LCA(x, y). Whenever we visit a node v on the path from x to the root there are three cases:

1. $v \leq x$, then x is in the right subtree of v and cannot be LCA(x, y); we advance to the parent of v.
2. $x < v \leq y$, then x is in the left subtree of v and LCA(x, y) is either y or an ancestor of y; we reset $\ell = v$ and advance to the parent of v.
3. $x < y < v$, then LCA(x, y) is in the left subtree of v and equals ℓ.

Unfortunately, after LCA(x, y) has been visited case (1) can happen $\omega(\log d)$ times before the search is terminated at the root or by case (3). Seidel and Aragon [5] denote these extra nodes visited above LCA(x, y) the *excess path* of the search, and circumvent this problem by extending treaps with special pointers for this.

To avoid visiting long excess paths we extend the above upward search with a concurrent downward search for y in the subtree rooted at the current candidate ℓ for LCA(x, y). In case (1) we always advance the tree search for y one level down, in case (2) we restart the search at the new ℓ, and in (3) we finalize the search. The concurrent search for y guarantees that the distance between LCA(x, y) and y in the tree is also an upper bound on the nodes visited on the excess path, that is, we visit at most twice the number of nodes as is on the path between x and y, which is expected $\mathcal{O}(\log d)$. It follows that treaps support finger searches in $\mathcal{O}(\log d)$ time. In Figure 12.3 is shown the search for $x = I$, $y = P$, LCA$(x, y) = K$, the path from x to y is drawn with thick lines, and the excess path is drawn with dashed lines.

12.4.2 Skip Lists

A skip list is a randomized dictionary data structure, which can be considered to consists of expected $\mathcal{O}(\log n)$ levels. The lowest level being a single linked list containing the elements in sorted order, and each succeeding level is a random sample of the elements of the previous level, where each element is included in the next level with a fixed probability, for example, 1/2. The pointer representation of a skip is illustrated in Figure 12.4.

The most prominent properties of skip lists are that they require expected linear space, consist of expected $\mathcal{O}(\log n)$ levels, support searches in expected $\mathcal{O}(\log n)$ time, and support insertions and deletions at a given position in expected $\mathcal{O}(1)$ time [3,4].

Pugh in Reference 3 elaborates on the various properties and extensions of skip lists, including pseudo-code for how skip lists support finger searches in expected $\mathcal{O}(\log d)$ time. To facilitate backward finger searches, a finger to a node v is stored as an expected $\mathcal{O}(\log n)$ space finger data structure that for each level i stores a pointer to the node to the left of v where the level i pointer either points to v or a node to the right of v. Moving a finger requires this list of pointers to be updated correspondingly.

A backward finger search is performed by first identifying the lowest node in the finger data structure that is to the left of the search key y, where the nodes in the finger data structure are considered in order of increasing levels. Thereafter the search proceeds downward from the identified node as in a standard skip list search.

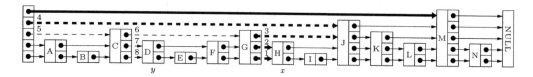

FIGURE 12.4 Performing finger searches on skip list.

Figure 12.4 shows the situation where we have a finger to H, represented by the thick (solid or dashed) lines, and perform a finger search for the element D to the left of H. Dashed (thick and thin) lines are the pointers followed during the finger search. The numbering indicate the other in which the pointers are traversed.

If the level links of a skip list are maintained as double-linked lists, then finger searches can be performed in expected $\mathcal{O}(\log d)$ time by traversing the existing links, without having a separate $\mathcal{O}(\log n)$ space finger data structure.

12.5 Applications

Finger search trees have, for example, been used in algorithms within computational geometry [10,25–29] and string algorithms [30,31]. In the rest of this chapter we give examples of the efficiency that can be obtained by applying finger search trees. These examples typically allow one to save a factor of $\mathcal{O}(\log n)$ in the running time of algorithms compared to using standard balanced search trees supporting $\mathcal{O}(\log n)$ time searches.

12.5.1 Optimal Merging and Set Operations

Consider the problem of merging two sorted sequences X and Y of length respectively n and m, where $n \leq m$, into one sorted sequence of length $n + m$. The canonical solution is to repeatedly insert each $x \in X$ in Y. This requires that Y is searchable and that there can be inserted new elements, that is, a suitable representation of Y is a balanced search tree. This immediately implies an $\mathcal{O}(n \log m)$ time bound for merging. In the following we discuss how finger search trees allow this bound to be improved to $\mathcal{O}\left(n \log \frac{m}{n}\right)$.

Hwang and Lin [32] presented an algorithm for merging two sorted sequence using optimal $\mathcal{O}\left(n \log \frac{m}{n}\right)$ comparisons, but did not discuss how to represent the sets. Brown and Tarjan [33] described how to achieve the same bound for merging two AVL trees [34]. Brown and Tarjan subsequently introduced *level linked* (2,3)-trees and described how to achieve the same merging bound for level linked (2,3)-trees [35].

Optimal merging of two sets also follows as an application of finger search trees [2]. Assume that the two sequences are represented as finger search trees, and that we repeatedly insert the n elements from the shorter sequence into the larger sequence using a finger that moves monotonically from left to right. If the ith insertion advances the finger d_i positions, we have that the total work of performing the n finger searches and insertions is $\mathcal{O}(\sum_{i=1}^{n} \log d_i)$, where $\sum_{i=1}^{n} d_i \leq m$. By convexity of the logarithm the total work becomes bounded by $\mathcal{O}\left(n \log \frac{m}{n}\right)$.

Since sets can be represented as sorted sequences, the above merging algorithm gives immediately raise to optimal, that is, $\mathcal{O}\left(\log \binom{n+m}{n}\right) = \mathcal{O}\left(n \log \frac{m}{n}\right)$ time, algorithms for set union, intersection, and difference operations [2]. For a survey of data structures for set representations see Chapter 34.

12.5.2 Arbitrary Merging Order

A classical $\mathcal{O}(n \log n)$ time sorting algorithm is binary merge sort. The algorithm can be viewed as the merging process described by a balanced binary tree: Each leaf corresponds to an input element and each internal node corresponds to the merging of the two sorted sequences containing respectively the elements in the left and right subtree of the node. If the tree is balanced then each element participates in $\mathcal{O}(\log n)$ merging steps, that is, the $\mathcal{O}(n \log n)$ sorting time follows.

Many divide-and-conquer algorithms proceed as binary merge sort, in the sense that the work performed by the algorithm can be characterized by a treewise merging process. For some of these algorithms the tree determining the merges is unfortunately fixed by the input instance, and the running time using linear merges becomes $\mathcal{O}(n \cdot h)$, where h is the height of the tree. In the following we discuss how finger search trees allow us to achieve $\mathcal{O}(n \log n)$ for unbalanced merging orders to.

Consider an arbitrary binary tree T with n leaves, where each leaf stores an element. We allow T to be arbitrarily unbalanced and that elements are allowed to appear at the leaves in any arbitrary order. Associate to each node v of T the set S_v of elements stored at the leaves of the subtree rooted at v. If we for each node v of T compute S_v by merging the two sets of the children of v using finger search trees, cf. Section 12.5.1, then the total time to compute all the sets S_v is $\mathcal{O}(n \log n)$.

The proof of the total $\mathcal{O}(n \log n)$ bound is by structural induction where we show that in a tree of size n, the total merging cost is $\mathcal{O}(\log(n!)) = \mathcal{O}(n \log n)$. Recall that two sets of size n_1 and n_2 can be merged in $\mathcal{O}\left(\log \binom{n_1+n_2}{n_1}\right)$ time. By induction we get that the total merging in a subtree with a root with two children of size respectively n_1 and n_2 becomes:

$$\log(n_1!) + \log(n_2!) + \log\binom{n_1+n_2}{n_1}$$
$$= \log(n_1!) + \log(n_2!) + \log((n_1+n_2)!) - \log(n_1!) - \log(n_2!)$$
$$= \log((n_1+n_2)!).$$

The above approach of arbitrary merging order was applied in [30,31] to achieve $\mathcal{O}(n \log n)$ time algorithms for finding repeats with gaps and quasiperiodicities in strings. In both these algorithms $\mathcal{T}$ is determined by the suffix-tree of the input string, and the S_v sets denote the set of occurrences (positions) of the substring corresponding to the path label of v.

12.5.3 List Splitting

Hoffmann et al. [36] considered how finger search trees can be used for solving the following *list splitting* problem, that, for example, also is applied in References 26 and 29. Assume we initially have a sorted list of n elements that is repeatedly split into two sequences until we end up with n sequences each containing one element. If the splitting of a list of length k into two lists of length k_1 and k_2 is performed by performing a simultaneous finger search from each end of the list, followed by a split, the searching and splitting can be performed in $\mathcal{O}(\log \min(k_1, k_2))$ time. Here we assume that the splitting order is unknown in advance.

By assigning a list of k elements a potential of $k - \log k \geq 0$, the splitting into two lists of size k_1 and k_2 releases the following amount of potential:

$$(k - \log k) - (k_1 - \log k_1) - (k_2 - \log k_2)$$
$$= -\log k + \log \min(k_1, k_2) + \log \max(k_1, k_2)$$
$$\geq -1 + \log \min(k_1, k_2),$$

since $\max(k_1, k_2) \geq k/2$. The released potential allows each list splitting to be performed in amortized $\mathcal{O}(1)$ time. The initial list is charged $n - \log n$ potential. We conclude that starting with a list of n elements, followed by a sequence of at most $n - 1$ splits requires total $\mathcal{O}(n)$ time.

12.5.4 Adaptive Merging and Sorting

The area of *adaptive sorting* addresses the problem of developing sorting algorithms which perform $\mathcal{O}(n \log n)$ comparisons for inputs with a limited amount of disorder for various definitions of measures of disorder, for example, the measure INV counts the number of pairwise insertions in the input. For a survey of adaptive sorting algorithms see [37].

An adaptive sorting algorithm that is optimal with respect to the disorder measure INV has running time $\mathcal{O}\left(n \log \frac{\text{INV}}{n}\right)$. A simple adaptive sorting algorithm optimal with respect to INV is the *insertion sort* algorithm, where we insert the elements of the input sequence from left to right into a finger search tree. Insertions always start at a finger on the last element inserted. Details on applying finger search trees in insertion sort can be found in References [24,35,38].

Another adaptive sorting algorithm based on applying finger search trees is obtained by replacing the linear merging in binary merge sort by an *adaptive merging* algorithm [39–42]. The classical binary merge sort algorithm alway performs $\Omega(n \log n)$ comparisons, since in each merging step where two lists each of size k is merged the number of comparisons performed is between k and $2k - 1$.

The idea of the adaptive merging algorithm is to identify consecutive blocks from the input sequences which are also consecutive in the output sequence, as illustrated in Figure 12.5. This is done by repeatedly performing a finger search for the smallest element of the two input sequences in the other sequence and deleting the identified block in the other sequence by a split operation. If the blocks in the output sequence are denoted $Z_1, \ldots, Z_k$, it follows from the time bounds of finger search trees that the total time for this adaptive merging operation becomes $\mathcal{O}(\sum_{i=1}^{k} \log |Z_i|)$. From this merging bound it can be argued that

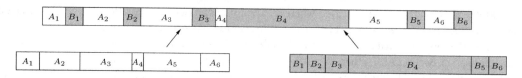

FIGURE 12.5　Adaptive merging.

merge sort with adaptive merging is adaptive with respect to the disorder measure Inv (and several other disorder measures). See [39–41] for further details.

Acknowledgments

This work was supported by the Carlsberg Foundation (contract number ANS-0257/20), BRICS (Basic Research in Computer Science, www.brics.dk, funded by the Danish National Research Foundation), and the Future and Emerging Technologies programme of the EU under contract number IST-1999-14186 (ALCOM-FT).

References

1. J. L. Bentley and A. C.-C. Yao. An almost optimal algorithm for unbounded searching. *Information Processing Letters*, 5(3):82–87, 1976.
2. S. Huddleston and K. Mehlhorn. A new data structure for representing sorted lists. *Acta Informatica*, 17:157–184, 1982.
3. W. Pugh. A skip list cookbook. Technical Report CS-TR-2286.1, Dept. of Computer Science, University of Maryland, College Park, 1989.
4. W. Pugh. Skip lists: A probabilistic alternative to balanced trees. *Communications of the ACM*, 33(6):668–676, 1990.
5. R. Seidel and C. R. Aragon. Randomized search trees. *Algorithmica*, 16(4/5):464–497, 1996.
6. L. J. Guibas, E. M. McCreight, M. F. Plass, and J. R. Roberts. A new representation for linear lists. In *Proc. 9th Ann. ACM Symp. on Theory of Computing*, pages 49–60, 1977.
7. R. Bayer and E. McCreight. Organization and maintenance of large ordered indexes. *Acta Informatica*, 1:173–189, 1972.
8. A. K. Tsakalidis. AVL-trees for localized search. *Information and Control*, 67(1–3):173–194, 1985.
9. S. R. Kosaraju. Localized search in sorted lists. In *Proc. 13th Ann. ACM Symp. on Theory of Computing*, pages 62–69, 1981.
10. R. Tarjan and C. van Wyk. An $o(n \log \log n)$ algorithm for triangulating a simple polygon. *SIAM Journal of Computing*, 17:143–178, 1988.
11. C. Levcopoulos and M. H. Overmars. A balanced search tree with $O(1)$ worst-case update time. *Acta Informatica*, 26:269–277, 1988.
12. D. Harel. Fast updates of balanced search trees with a guaranteed time bound per update. Technical Report 154, University of California, Irvine, 1980.
13. D. Harel and G. S. Lueker. A data structure with movable fingers and deletions. Technical Report 145, University of California, Irvine, 1979.
14. R. Fleischer. A simple balanced search tree with $O(1)$ worst-case update time. *International Journal of Foundations of Computer Science*, 7:137–149, 1996.
15. G. S. Brodal. Finger search trees with constant insertion time. In *Proc. 9th Annual ACM-SIAM Symposium on Discrete Algorithms*, pages 540–549, 1998.
16. G. S. Brodal, G. Lagogiannis, C. Makris, A. Tsakalidis, and K. Tsichlas. Optimal finger search trees in the pointer machine. *Journal of Computer and System Sciences, Special issue on STOC 2002*, 67(2):381–418, 2003.
17. G. E. Blelloch, B. M. Maggs, and S. L. M. Woo. Space-efficient finger search on degree-balanced search trees. In *Proceedings of the Fourteenth Annual ACM-SIAM Symposium on Discrete Algorithms*, pages 374–383. Society for Industrial and Applied Mathematics, 2003.
18. D. D. Sleator and R. E. Tarjan. Self-adjusting binary search trees. *Journal of the ACM*, 32(3):652–686, 1985.
19. R. Cole. On the dynamic finger conjecture for splay trees. part II: The proof. *SIAM Journal of Computing*, 30(1):44–85, 2000.
20. R. Cole, B. Mishra, J. Schmidt, and A. Siegel. On the dynamic finger conjecture for splay trees. Part I. Splay sorting $\log n$-block sequences. *SIAM Journal of Computing*, 30(1):1–43, 2000.
21. P. F. Dietz and R. Raman. A constant update time finger search tree. *Information Processing Letters*, 52:147–154, 1994.
22. R. Raman. *Eliminating Amortization: On Data Structures with Guaranteed Response Time*. PhD thesis, University of Rochester, New York, 1992. Computer Science Dept., U. Rochester, tech report TR-439.
23. A. Anderson and M. Thorup. Tight(er) worst case bounds on dynamic searching and priority queues. In *Proc. 32nd Annual ACM Symposium On Theory of Computing*, pages 335–342, 2000.
24. K. Mehlhorn. *Data Structures and Algorithms 1: Sorting and Searching*. Springer-Verlag, Berlin, 1984.
25. M. Atallah, M. Goodrich, and K. Ramaiyer. Biased finger trees and three-dimensional layers of maxima. In *Proc. 10th ACM Symposium on Computational Geometry*, pages 150–159, 1994.
26. G. S. Brodal and R. Jacob. Dynamic planar convex hull. In *Proc. 43rd Annual Symposium on Foundations of Computer Science*, pages 617–626, 2002.

27. L. Guibas, J. Hershberger, D. Leven, M. Sharir, and R. Tarjan. Linear time algorithms for visibility and shortest path problems inside simple polygons. *Algorithmica*, 2:209–233, 1987.

28. J. Hershberger. Finding the visibility graph of a simple polygon in time proportional to its size. In *Proc. 3rd ACM Symposium on Computational Geometry*, pages 11–20, 1987.

29. R. Jacob. *Dynamic Planar Convex Hull*. PhD thesis, University of Aarhus, Denmark, 2002.

30. G. S. Brodal, R. B. Lyngsø, C. N. S. Pedersen, and J. Stoye. Finding maximal pairs with bounded gap. *Journal of Discrete Algorithms, Special Issue of Matching Patterns*, 1(1):77–104, 2000.

31. G. S. Brodal and C. N. S. Pedersen. Finding maximal quasiperiodicities in strings. In *Proc. 11th Annual Symposium on Combinatorial Pattern Matching, volume 1848 of Lecture Notes in Computer Science*, pages 397–411. Springer-Verlag, 2000.

32. F. K. Hwang and S. Lin. A simple algorithm for merging two disjoint linearly ordered sets. *SIAM Journal of Computing*, 1(1):31–39, 1972.

33. M. R. Brown and R. E. Tarjan. A fast merging algorithm. *Journal of the ACM*, 26(2):211–226, 1979.

34. G. M. Adel'son-Vel'skii and Y. M. Landis. An algorithm for the organization of information. *Doklady Akademii Nauk SSSR*, 146:263–266, 1962. English translation in *Soviet Math. Dokl.*, 3:1259–1262.

35. M. R. Brown and R. E. Tarjan. Design and analysis of a data structure for representing sorted lists. *SIAM Journal of Computing*, 9:594–614, 1980.

36. K. Hoffmann, K. Mehlhorn, P. Rosenstiehl, and R. E. Tarjan. Sorting Jordan sequences in linear time using level/linked search trees. *Information and Control*, 68(1–3):170–184, 1986.

37. V. Estivill-Castro and D. Wood. A survey of adaptive sorting algorithms. *ACM Computing Surveys*, 24:441–476, 1992.

38. H. Mannila. Measures of presortedness and optimal sorting algorithms. *IEEE Transactions on Computers*, C-34:318–325, 1985.

39. S. Carlsson, C. Levcopoulos, and O. Petersson. Sublinear merging and natural mergesort. *Algorithmica*, 9(6):629–648, 1993.

40. A. Moffat. Adaptive merging and a naturally natural merge sort. In *Proceedings of the 14th Australian Computer Science Conference*, pages 08.1–08.8, 1991.

41. A. Moffat, O. Petersson, and N. Wormald. Further analysis of an adaptive sorting algorithm. In *Proceedings of the 15th Australian Computer Science Conference*, pages 603–613, 1992.

42. A. Moffat, O. Petersson, and N. C. Wormald. Sorting and/by merging finger trees. In *Algorithms and Computation: Third International Symposium, ISAAC '92, volume 650 of Lecture Notes in Computer Science*, pages 499–508. Springer-Verlag, 1992.

13

Splay Trees

13.1	Introduction	179
13.2	Splay Trees	180
13.3	Analysis	181
	Access and Update Operations	
13.4	Optimality of Splay Trees	184
	Static Optimality • Static Finger Theorem • Working Set Theorem • Other Properties and Conjectures	
13.5	Linking and Cutting Trees	185
	Data Structure • Solid Trees • Rotation • Splicing • Splay in Virtual Tree • Analysis of Splay in Virtual Tree • Implementation of Primitives for Linking and Cutting Trees	
13.6	Case Study: Application to Network Flows	190
13.7	Implementation without Linking and Cutting Trees	192
13.8	FIFO Dynamic Tree Implementation	193
13.9	Variants of Splay Trees and Top-Down Splaying	195
	Acknowledgments	195
	References	195

Sanjeev Saxena
Indian Institute of Technology at Kanpur

13.1 Introduction

In this chapter, we discuss the following topics:

1. Introduction to splay trees and their applications
2. Splay trees—description, analysis, algorithms, and optimality of splay trees
3. Linking and cutting trees
4. Case study: application to network flows
5. Variants of splay trees.

There are various data structures such as AVL-trees, red-black trees, and 2-3-trees (see Chapter 11), which support operations such as insert, delete (including deleting the minimum item), and search (or membership) in $O(\log n)$ time (for each operation). Splay trees introduced by Sleator and Tarjan [1,2] support all these operations in $O(\log n)$ amortized time, which roughly means that starting from an empty tree, a sequence of m of these operations will take $O(m \log n)$ time (deterministic), an individual operation may take either more time or less time (see Theorem 13.1). We discuss some applications in the rest of this section.

Assume that we are searching for an item in a "large" sorted file, and if the item is in the kth position, then we can search the item in $O(\log k)$ time by exponential and binary search. Similarly, finger search trees (see Chapter 12) can be used to search any item at distance f from a finger in $O(\log f)$ time. Splay trees can search (again in amortized sense) an item from any finger (which need not even be specified) in $O(\log f)$ time, where f is the distance from the finger (see Section 13.4.2). Since the finger is not required to be specified, the time taken will be minimum over all possible fingers (time, again in amortized sense).

If we know the frequency or probability of access of each item, then we can construct an optimum binary search tree (see Chapter 11) for these items; total time for all access will be the smallest for optimal binary search trees. If we do not know the probability (or access frequency), and if we use splay trees, even then the total time taken for all accesses will still be the same as that for a binary search tree, up to a multiplicative constant (see Section 13.4.1).

In addition, splay trees can be used almost as a "black box" in linking and cutting trees (see Section 13.5). Here we need the ability to add (or subtract) a number to key values of all ancestors of a node x.

Moreover, in practice, the rebalancing operations (rotations) are very much simpler than those in height balanced trees. Hence, in practice, we can also use splay trees as an alternative to height balanced trees (such as AVL-trees, red-black trees, and 2-3-trees), if we are interested only in the total time. However, some experimental studies [3] suggest that for random data, splay trees outperform balanced binary trees only for highly skewed data; and for applications like "vocabulary accumulation" of English text [4], even standard binary search trees, which do not have good worst-case performance, outperform both balanced binary trees (AVL-trees) and splay trees. In any case, the constant factor and the algorithms are not simpler than those for the usual heap, hence it will not be practical to use splay trees for sorting (say as in heap sort), even though the resulting algorithm will take $O(n \log n)$ time for sorting, unless the data has some degree of presortedness, in which case splay sort is a practical alternative (see Chapter 11) [5]. Splay trees, however, cannot be used in real-time applications.

Splay trees can also be used for data compression. As splay trees are binary search trees, they can be used directly [6] with guaranteed worst-case performance. They are also used in data compression with some modifications [7]. Routines for data compression can be shown to run in time proportional to the entropy of input sequence [8] for usual splay trees and their variants.

13.2 Splay Trees

Let us assume that for each node x, we store a real number key(x).

In any binary search tree, left subtree of any node x contains items having "key" values less than the value of key(x) and right subtree of the node x contains items with "key" values larger than the value of key(x).

In splay trees, we first search the query item, say x as in the usual binary search trees (see Chapter 3)—compare the query item with the value in the root, if smaller then recursively search in the left subtree else if larger then, recursively search in the right subtree, and if it is equal then we are done. Then, informally speaking, we look at every disjoint pair of consecutive ancestors of x, say $y = $ parent(x) and $z = $ parent(y), and perform certain pair of rotations. As a result of these rotations, x comes in place of z.

In case x has an odd number of proper ancestors, then the ancestor of x (which is child of the root) will also have to be dealt separately, in terminal case—we rotate the edge between x and the root. This step is called *zig* step (see Figure 13.1).

If x and y are both left or are both right children of their respective parents, then we first rotate the edge between y and its parent z and then the edge between x and its parent y. This step is called *zig-zig* step (see Figure 13.2).

If x is a left (respectively right) child and y is a right (respectively left) child, then we first rotate the edge between x and y and then between x and z, this step is called *zig-zag* step (see Figure 13.3).

These rotations (together) not only make x the new root but also roughly speaking halve the depth (length of path to root) of all ancestors of x in the tree. If the node x is at depth "d," splay(x) will take $O(d)$ time, that is, time proportional to access the item in node x.

Formally, **splay**(x) is a sequence of rotations which are performed (as follows) until x becomes a root:

- If parent(x) is root, then we carry out usual rotation (see Figure 13.1).
- If x and parent(x) are both left (or are both right) children of their parents, then we first rotate at $y = $ parent(x) (i.e., the edge between y and its parent) and then rotate at x (see Figure 13.2).

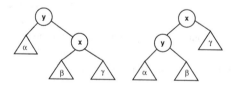

FIGURE 13.1   parent(x) is the root—edge xy is rotated (zig case).

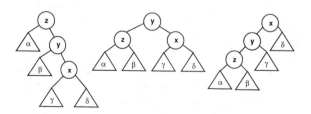

FIGURE 13.2 x and parent(x) are both right children (zig-zig case)—first edge yz is rotated then edge xy.

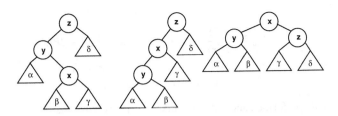

FIGURE 13.3 x is a right child while parent(x) is a left child (zig-zag case)—first edge xy is rotated then edge xz.

- If x is left (or respectively right) child but parent(x) is right (respectively left) child of its parent, then first rotate at x and then again rotate at x (see Figure 13.3).

13.3 Analysis

We will next like to look at the "amortized" time taken by splay operations. Amortized time is the average time taken over a worst-case sequence of operations.

For the purpose of analysis, we give a positive weight $w(x)$ to (any) item x in the tree. The weight function can be chosen completely arbitrarily (as long it is strictly positive). For analysis of splay trees, we need some definitions (or nomenclature) and have to fix some parameters.

Weight of item x: For each item x, an arbitrary positive weight $w(x)$ is associated (see Section 13.4 for some examples of function $w(x)$).

Size of node x: Size(x) is the sum of the individual weights of all items in the subtree rooted at the node x.

Rank of node x: Rank of a node x is $\log_2(\text{size}(x))$.

Potential of a tree: Let α be some positive constant (we will discuss choice of α later), then the potential of a tree T is taken to be

$$\alpha(\text{Sum of rank}(x) \text{ for all nodes } x \in T) = \alpha \sum_{x \in T} \text{rank}(x).$$

Amortized time: As always,

Amortized time = Actual Time + New Potential − Old Potential.

Running time of splaying: Let β be some positive constant, choice of β is also discussed later

but $\beta \leq \alpha$, then the running time for splaying is

$\beta \times$ Number of rotations.

If there are no rotations, then we charge one unit for splaying.

We also need a simple result from algebra. Observe that $4xy = (x+y)^2 - (x-y)^2$. Now if $x + y \leq 1$, then $4xy \leq 1 - (x-y)^2 \leq 1$ or taking logarithms,[*] $\log x + \log y \leq -2$. Note that the maximum value occurs when $x = y = 1/2$.

Fact 13.1

[Result from Algebra] If $x + y \leq 1$, then $\log x + \log y \leq -2$. The maximum value occurs when $x = y = 1/2$.

Lemma 13.1

[Access Lemma] The amortized time to splay a tree (with root "t") at a node "x" is at most

$$3\alpha(\text{rank}(t) - \text{rank}(x)) + \beta = O\left(\log\left(\frac{\text{Size}(t)}{\text{Size}(x)}\right)\right)$$

Proof. We will calculate the change in potential, and hence the amortized time taken in each of the three cases.

Let $s(\ )$ denote the sizes before rotation(s) and $s'(\ )$ be the sizes after rotation(s). Let $r(\ )$ denote the ranks before rotation(s) and $r'(\ )$ be the ranks after rotation(s).

Case 1: x and parent(x) are both left (or both right) children

Please refer to Figure 13.2. Here, $s(x) + s'(z) \leq s'(x)$ or $(s(x)/s'(x)) + (s'(z)/s'(x)) \leq 1$.

---

[*] All logarithms in this chapter are to base 2.

Thus, by Fact 13.1,

$$-2 \geq \log \frac{s(x)}{s'(x)} + \log \frac{s'(z)}{s'(x)} = r(x) + r'(z) - 2r'(x)$$

or

$$r'(z) \leq 2r'(x) - r(x) - 2.$$

Observe that two rotations are performed and only the ranks of $x, y,$ and z are changed.

Further, as $r'(x) = r(z)$, the amortized time is

$= 2\beta + \alpha((r'(x) + r'(y) + r'(z)) - (r(x) + r(y) + r(z)))$

$= 2\beta + \alpha((r'(y) + r'(z)) - (r(x) + r(y)))$

$\leq 2\beta + \alpha((r'(y) + r'(z)) - 2r(x))$, (as $r(y) \geq r(x)$).

As $r'(x) \geq r'(y)$, amortized time

$\leq 2\beta + \alpha((r'(x) + r'(z)) - 2r(x))$

$\leq 2\beta + \alpha((r'(x) + (2r'(x) - r(x) - 2) - 2r(x)))$

$\leq 3\alpha(r'(x) - r(x)) - 2\alpha + 2\beta$

$\leq 3\alpha(r'(x) - r(x))$ (as $\alpha \geq \beta$).

Case 2: x is a left child, parent(x) is a right child

Please refer to Figure 13.3. $s'(y) + s'(z) \leq s'(x)$ or $(s'(y)/s'(x)) + (s'(z)/s'(x)) \leq 1$.

Thus, by Fact 13.1,

$-2 \geq \log(s'(y)/s'(x)) + \log(s'(z)/s'(x)) = r'(y) + r'(z) - 2r'(x)$ or

$r'(y) + r'(z) \leq 2r'(x) - 2$.

Now Amortized time $= 2\beta + \alpha((r'(x) + r'(y) + r'(z)) - (r(x) + r(y) + r(z)))$. But, as $r'(x) = r(z)$, Amortized time

$= 2\beta + \alpha((r'(y) + r'(z)) - (r(x) + r(y)))$. Using $r(y) \geq r(x)$, Amortized time

$\leq 2\beta + \alpha((r'(y) + r'(z)) - 2r(x))$

$\leq 2\alpha(r'(x) - r(x)) - 2\alpha + 2\beta$

$\leq 3\alpha(r'(x) - r(x)) - 2(\alpha - \beta) \leq 3\alpha(r'(x) - r(x))$.

Case 3: parent(x) is a root

Please refer to Figure 13.1. There is only one rotation, Amortized time

$= \beta + \alpha((r'(x) + r'(y)) - (r(x) + r(y)))$.

But as $r'(x) = r(y)$, Amortized time is

$\beta + \alpha(r'(y) - r(x))$

$\leq \beta + \alpha(r'(x) - r(x))$

$\leq \beta + 3\alpha(r'(x) - r(x))$.

As case 3, occurs only once, and other terms vanish by telescopic cancellation, the lemma follows. ∎

Theorem 13.1

Time for m accesses on a tree having at most n nodes is $O((m + n) \log n)$

Proof. Let the weight of each node x be fixed as $1/n$. As there are n nodes in the entire tree, the total weight of all nodes in the tree is 1.

If t is the root of the tree, then size(t) = 1 and as each node x has at least one node (x itself) present in the subtree rooted at x (when x is a leaf, exactly one node will be present), for any node x, size(x) $\geq$ ($1/n$). Thus we have following bounds for the ranks—$r(t) \leq 0$ and $r(x) \geq -\log n$.

Or, from Lemma 13.1, amortized time per splay is at most $1 + 3 \log n$. As maximum possible value of the potential is $n \log n$, maximum possible potential change is also $O(n \log n)$, the theorem follows. ∎

We will generalize the result of Theorem 13.1, where we will be choosing some other weight functions, to discuss other optimality properties of Splay trees.

13.3.1 Access and Update Operations

We are interested in performing the following operations:

1. Access(x): x is a key value which is to be searched.
2. Insert(x): a node with key value x is to be inserted, if a node with this key value is not already present.
3. Delete(x): node containing key value x is to be deleted.

4. Join(t_1, t_2): t_1 and t_2 are two trees. We assume that all items in tree t_1 have smaller key values than the key value of any item in the tree t_2. The two trees are to be combined or joined into a single tree, as a result, the original trees t_1 and t_2 get "destroyed."

5. Split(x, t): the tree t is split into two trees (say) t_1 and t_2 (the original tree is "lost"). The tree t_1 should contain all nodes having key values less than (or equal to) x and tree t_2 should contain all nodes having key values strictly larger than x.

We next discuss implementation of these operations, using a single primitive operation—splay. We will show that each of these operations for splay trees can be implemented using $O(1)$ time and with one or two "splay" operations.

Access(x, t): Search the tree t for key value x, using the routines for searching in a "binary search tree" and splay at the last node—the node containing value x, in case the search is successful, or the parent of "failure" node in case the search is unsuccessful.

Join(t_1, t_2): Here we assume that all items in splay tree t_1 have key values which are smaller than key values of items in splay tree t_2, and we are required to combine these two splay trees into a single splay tree.

Access largest item in t_1, formally, by searching for "$+\infty$," that is, a call to Access($+\infty, t_1$). As a result, the node containing the largest item (say r) will become the root of the tree t_1. Clearly, now the root r of the splay tree t_1 will not have any right child. Make the root of the splay tree t_2 the right child of r, the root of t_1, as a result, t_2 will become the right subtree of the root r and r will be the root of the resulting tree.

Split(x, t): We are required to split the tree t into two trees: t_1 containing all items with key values less than (or equal to) x and t_2 containing items with key values greater than x.

If we carry out Access(x, t), and if a node with key value x is present, then the node containing the value x will become the root. We then remove the link from node containing the value x to its right child (say node containing value y); the resulting tree with root, containing the value x, will be t_1, and the tree with root, containing the value y, will be the required tree t_2.

And if the item with key value x is not present, then the search will end at a node (say) containing key value z. Again, as a result of splay, the node with value z will become the root. If $z > x$, then t_1 will be the left subtree of the root and the tree t_2 will be obtained by removing the edge between the root and its left child.

Otherwise, $z < x$, and t_2 will be the right subtree of the root and t_1 will be the resulting tree obtained by removing the edge between the root and its right child.

Insert(x, t): We are required to insert a new node with key value x in a splay tree t. We can implement insert by searching for x, the key value of the item to be inserted in tree t using the usual routine for searching in a binary search tree. If the item containing the value x is already present, then we splay at the node containing x and return. Otherwise, assume that we reach a leaf (say) containing key y, $y \neq x$. Then if $x < y$, then add the new node containing value x as a left child of node containing value y, and if $x > y$, then the new node containing the value x is made the right child of the node containing the value y, in either case we splay at the new node (containing the value x) and return.

Delete(x, t): We are required to delete the node containing the key value x from the splay tree t. We first access the node containing the key value x in the tree t—Access(x, t). If there is a node in the tree containing the key value x, then that node becomes the root, otherwise, after the access the root will be containing a value different from x and we return(-1)—value not found. If the root contains value x, then let t_1 be the left subtree and t_2 be the right subtree of the root. Clearly, all items in t_1 will have key values less than x and all items in t_2 will have key values greater than x. We delete the links from roots of t_1 and t_2 to their parents (the root of t, the node containing the value x). Then, we join these two subtrees— Join(t_1, t_2) and return.

Observe that in both "Access" and "Insert," after searching, a splay is carried out. Clearly, the time for splay will dominate the time for searching. Moreover, except for splay, everything else in "Insert" can be easily done in $O(1)$ time. Hence the time taken for "Access" and "Insert" will be of the same order as the time for a splay. Again, in "Join," "Split," and "Delete," the time for "Access" will dominate, and everything else in these operations can again be done in $O(1)$ time, hence "Join," "Split," and "Delete" can also be implemented in same order of time as for an "Access" operation, which we just saw is, in turn, of same order as the time for a splay. Thus each of above operations will take same order of time as for a splay. Hence, from Theorem 13.1, we have

Theorem 13.2

Time for m update or access operations on a tree having at most n nodes is $O((m + n) \log n)$.

Observe that, at least in amortized sense, the time taken for first m operations on a tree which never has more than n nodes is the same as the time taken for balanced binary search trees (see Chapter 11) such as AVL-trees, 2-3-trees, and so on.

13.4 Optimality of Splay Trees

If $w(i)$ the weight of node i is independent of the number of descendants of node i, then the maximum value of size(i) will be $W = \sum w(i)$ and minimum value of size(i) will be $w(i)$.

As size of the root t will be W, and hence rank $\log W$, so by Lemma 13.1, the amortized time to splay at a node "x" will be $O(\log(\text{Size}(t)/\text{Size}(x))) = O(\log(W/\text{Size}(x))) = O(\log W/w(x))$.

Also observe that the maximum possible change in the rank (for just node i) will be $\log W - \log w(i) = \log(W/w(i))$ or the total maximum change in all ranks (the potential of the tree, with $\alpha = 1$) will be bounded by $\sum \log(W/w(i))$.

Note that, as $\sum w(i)/W = 1$, $\sum |\log(W/w(i))| \le n \log n$ (the maximum occurs when all $(w(i)/W)$ are equal to $1/n$), hence maximum change in potential is always bounded by $O(n \log n)$.

As a special case, in Theorem 13.1, we had fixed $w(i) = 1/n$, and as a result, the amortized time per operation is bounded by $O(\log n)$ or time for m operations become $O((m + n) \log n)$. We next fix $w(i)$'s in some other cases.

13.4.1 Static Optimality

On any sequence of accesses, a splay tree is as efficient as the optimum binary search tree, up to a constant multiplicative factor. This can be very easily shown.

Let $q(i)$ be the number of times the ith node is accessed, we assume that each item is accessed at least once or $q(i) \ge 1$. Let $m = \sum q(i)$ be the total number of times we access any item in the splay tree. Assign a weight of $q(i)/m$ to item i. We call $q(i)/m$ the *access frequency* of the ith item. Observe that the total (or maximum) weight is 1 and hence the rank of the root $r(t) = 0$.

Thus,

$$r(t) - r(x) = 0 - r(x) = -\log\left(\sum_{i \in T_x} \frac{q(i)}{m}\right) \le -\log\left(\frac{q(x)}{m}\right).$$

Hence, from Lemma 13.1, with $\alpha = \beta = 1$, the amortized time per splay (say at node "x") is at most

$3\alpha(r(t) - r(x)) + \beta$
$= 1 + 3(-\log(q(x)/m))$
$= 1 + 3\log(m/q(x))$.

As ith item is accessed $q(i)$ times, amortized total time for all accesses of the ith item is $O(q(i) + q(i)\log(m/q(i)))$, hence total amortized time will be $O(m + \sum q(i)\log(m/q(i)))$. Moreover, as the maximum value of potential of the tree is $\sum \max\{r(x)\} \le \sum \log(m/q(i)) = O(\sum \log(m/q(i)))$, the total time will be $O(m + \sum q(i)\log(m/q(i)))$.

Theorem 13.3

Time for m update or access operations on an n-node tree is $O(m + \sum q(i)\log(m/q(i)))$, where $q(i)$ is the total number of times item i is accessed, here $m = \sum q(i)$.

Remark 13.1

The total time for this analysis is the same as that for the (static) optimal binary search tree.

13.4.2 Static Finger Theorem

We first need a result from mathematics. Observe that, in the interval $k - 1 \le x \le k$, $1/x \ge (1/k)$ or $(1/x^2) \ge (1/k^2)$. Hence, in this interval, we have $1/k^2 \le \int_{k-1}^{k}(dx/x^2)$ summing from $k = 2$ to n, $\sum_{2}^{n}(1/k^2) \le \int_{1}^{n}(dx/x^2) = 1 - 1/n$ or $\sum_{k=1}^{n}(1/k^2) < 2$.

If f is an integer between 0 and n, then we assign a weight of $1/(|i - f| + 1)^2$ to item i. Then $W \le 2\sum_{k=1}^{\infty}(1/k^2) < 4 = O(1)$.

Consider a particular access pattern (i.e., a snapshot or history or a run). Let the sequence of accessed items be $i_1, \ldots, i_m$, some i_j's may occur more than once.

Then, by the discussion at the beginning of this section, amortized time for the jth access is $O(\log(|i_j - f| + 1))$. Or the total amortized time for all access will be $O(m + \sum_{j=1}^{m} \log(|i_j - f| + 1))$. As weight of any item is at least $1/n^2$, the maximum value of potential is $n \log n$. Thus total time is at most $O(n \log n + m + \sum_{j=1}^{m} \log(|i_j - f| + 1))$.

Remark 13.2

f can be chosen as any fixed item (finger). Thus this outperforms finger search trees, if any fixed point is used as a finger, but here the finger need not be specified.

13.4.3 Working Set Theorem

Splay trees also have the working set property, that is, if only t different items are being repeatedly accessed, then the time for access is actually $O(\log t)$ instead of $O(\log n)$. In fact, if t_j different items were accessed since the last access of i_jth item, then the amortized time for access of i_jth item is only $O(\log (t_j + 1))$.

This time, we number the accesses from 1 to m in the order in which they occur. Assign weights of $1, 1/4, 1/9, \ldots, 1/n^2$ to items in the order of the first access. Item accessed earliest gets the largest weight and those never accessed get the smallest weight. Total weight $W = \sum (1/k^2) < 2 = O(1)$.

It is useful to think of item having weight $1/k^2$ as being in the kth position in a (some abstract) queue. After an item is accessed, we will be putting it in front of the queue, that is, making its weight 1 and "pushing back" items which were originally ahead of it, that is, the weights of items having old weight $1/s^2$ (i.e., items in sth place in the queue) will have a new weight of $1/(s + 1)^2$ (i.e., they are now in place $s + 1$ instead of place s). The position in the queue will actually be the position in the "move to front" heuristic.

Less informally, we will be changing the weights of items after each access. If the weight of item i_j during access j is $1/k^2$, then after access j, assign a weight 1 to item i_j. And an item having weight $1/s^2$, $s < k$ gets weight changed to $1/(s + 1)^2$.

Effectively, item i_j has been placed at the head of queue (weight becomes $1/1^2$) and weights have been permuted. The value of W, the sum of all weights remains unchanged.

If t_j items were accessed after last access of item i_j, then the weight of item i_j would have been $1/t_j^2$, or the amortized time for jth access is $O(\log (t_j + 1))$.

After the access, as a result of splay, the i_jth item becomes the root, thus the new size of i_jth item is the sum of all weights W—this remains unchanged even after changing weights. As weights of all other items, either remain the same or decrease (from $1/s^2$ to $1/(s + 1)^2$), size of all other items also decreases or remains unchanged due to permutation of weights. In other words, as a result of weight reassignment, size of nonroot nodes can decrease and size of the root remains unchanged. Thus weight reassignment can only decrease the potential, or amortized time for weight reassignment is either zero or negative.

Hence, by discussions at the beginning of this section, total time for m accesses on a tree of size at most n is $O(n \log n + \sum \log(t_j + 1))$ where t_j is the number of different items which were accessed since the last access of i_jth item (or from start, if this is the first access).

13.4.4 Other Properties and Conjectures

Splay trees are conjectured [1] to obey "Dynamic Optimality Conjecture" which roughly states that cost for any access pattern for splay trees is of the same order as that of the best possible algorithm. Thus, in amortized sense, the splay trees are the best possible dynamic binary search trees up to a constant multiplicative factor. This conjecture is still open.

However, dynamic finger conjecture for splay trees which says that access which is close to previous access is fast has been proved by Cole [9]. Dynamic finger theorem states that the amortized cost of an access at a distance d from the preceding access is $O(\log (d + 1))$; there is, however, $O(n)$ initialization cost. The accesses include searches, insertions, and deletions (but the algorithm for deletions is different) [9]. Thus splay trees can be used in place of finger search trees to sort a partially sorted file by repeated insertions (see Chapter 11).

Splay trees also obey several other optimality properties [10].

13.5 Linking and Cutting Trees

Tarjan [2] and Sleator and Tarjan [1] have shown that splay trees can be used to implement linking and cutting trees.

We are given a collection of rooted trees. Each node will store a value, which can be any real number. These trees can "grow" by combining with another tree *link* and can shrink by losing an edge *cut*. Less informally, the trees are "dynamic" and grow or shrink by following operations (we assume that we are dealing with a forest of rooted trees).

link: If x is root of a tree, and y is any node, not in the tree rooted at x, then make y the parent of x.

cut: Cut or remove the edge between a nonroot node x and its parent.

Let us assume that we want to perform operations like

- Add (or subtract) a value to all ancestors of a node.
- Find the minimum value stored at ancestors of a query node x.

More formally, following operations are to be supported:

find_cost(v): Return the value stored in the node v.

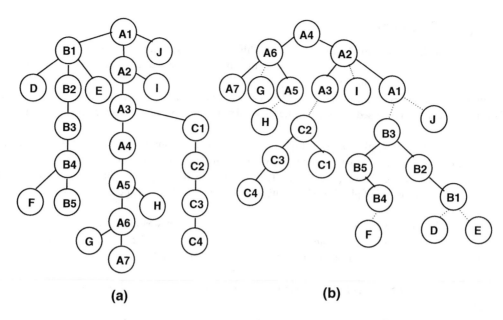

FIGURE 13.4   (a) Original tree and (b) virtual trees: solid and dashed children.

find_root(v): Return the root of the tree containing the node v.

find_min(v): Return the node having the minimum value, on the path from v till find_root(v), the root of the tree containing v. In case of ties, choose the node closest to the root.

add_cost(v, δ): Add a real number δ to the value stored in every node on the path from v to the root (i.e., till find_root(v)).

find_size(v): Find the number of nodes in the tree containing the node v.

link(v, w): Here v is a root of a tree. Make the tree rooted at v a child of node w. This operation does nothing if both vertices v and w are in the same tree or v is not a root.

cut(v): Delete the edge from node v to its parent, thus making v a root. This operation does nothing if v is a root.

13.5.1 Data Structure

For the given forest, we make some of the given edges "dashed" and the rest of them are kept solid. Each nonleaf node will have only one "solid" edge to one of its children. All other children will be connected by a dashed edge. To be more concrete, in any given tree, the rightmost link (to its child) is kept solid, and all other links to its other children are made "dashed."

As a result, the tree will be decomposed into a collection of solid paths. The roots of solid paths will be connected to some other solid path by a dashed edge. A new data structure called a "virtual tree" is constructed. Each linking and cutting tree T is represented by a virtual tree V, containing the same set of nodes. But each solid path of the original tree is modified or converted into a binary tree in the virtual tree; binary trees are as balanced as possible. Thus a virtual tree has a (solid) left child, a (solid) right child, and zero or more (dashed) middle children.

In other words, a virtual tree consists of a hierarchy of solid binary trees connected by dashed edges. Each node has a pointer to its parent, and to its left and right children (see Figure 13.4).

13.5.2 Solid Trees

Recall that each path is converted into a binary tree. Parent (say y) of a node (say x) in the path is the inorder (symmetric order) successor of that node (x) in the solid tree. However, if x is the last node (in symmetric order) in the solid subtree then its parent path will be the parent of the root of the solid subtree containing it (see Figure 13.4). Formally, $\text{Parent}_{\text{path}}(v) = \text{Node}(\text{Inorder}(v) + 1)$.

Note that for any node v, all nodes in the left subtree will have smaller inorder numbers and those in the right subtree will have larger inorder numbers. This ensures that all nodes in the left subtree are descendants and all nodes in the right subtree are ancestors. Thus the parent (in the binary tree) of a left child will be an ancestor (in the original tree). But, parent (in the binary tree) of a right child is a descendant (in the original tree). This order helps us to carry out add_cost effectively.

We need some definitions or notation to proceed.

Let mincost(x) be the cost of the node having the minimum key value among all descendants of x in the same solid subtree. Then in each node, we store two fields δcost(x) and δmin(x). We define,
δmin(x) = cost(x) − mincost(x). And,

$$\delta\text{cost}(x) = \begin{cases} \text{cost}(x) - \text{cost}(\text{parent}(x)) & \text{if } x \text{ has a solid parent} \\ \text{cost}(x) & \text{otherwise } (x \text{ is a solid tree root}). \end{cases}$$

Fact 13.2

δmin(x) − δcost(x) = cost(parent(x)) − mincost(x).

Thus if u and v are solid children of node z, then
mincost(z) = min{cost(z), mincost(v), mincost(w)} or
δmin(z) = cost(z) − mincost(z) = max{0, cost(z) − mincost(v), cost(z) − mincost(w)}.
Using Fact 13.2, and the fact z = parent(u) = parent(v), we have

Fact 13.3

If u and v are children of z, then
δmin(z) = max{0, δmin(u) − δcost(u), δmin(v) − δcost(v)}.
For linking and cutting trees, we need two primitive operations—rotation and splicing.

13.5.3 Rotation

Let us discuss rotation first (see Figure 13.5).

Let w be the parent of v in the solid tree, then rotation of the solid edge $(v, p(v)) \equiv (v, w)$ will make $w = p(v)$ a child of v. Rotation does not have any effect on the middle children. Let a be the left solid child of w and v be the right solid child of w.

Let "nonprimes" denote the values before the rotation and "primes" the values after the rotation of the solid edge (v, w). We next show that the new values δcost', δmin', and δsize' can be calculated in terms of old known values.

We assume that b is the left solid child of v and c is the right solid child of v.

First we calculate the new δcost' values in terms of old δcost values. From Figure 13.5,
$\delta\text{cost}'(v) = \text{cost}(v) - \text{cost}(\text{parent}'(v))$
$= \text{cost}(v) - \text{cost}(\text{parent}(w))$
$= \text{cost}(v) - \text{cost}(w) + \text{cost}(w) - \text{cost}(\text{parent}(w))$
$= \delta\text{cost}(v) + \delta\text{cost}(w)$.
$\delta\text{cost}'(w) = \text{cost}(w) - \text{cost}(v)$
$= -\delta\text{cost}'(v)$.
$\delta\text{cost}'(b) = \text{cost}(b) - \text{cost}(w)$
$= \text{cost}(b) - \text{cost}(v) + \text{cost}(v) - \text{cost}(w)$
$= \delta\text{cost}(b) + \delta\text{cost}(v)$.

Finally,
$\delta\text{cost}'(a) = \delta\text{cost}(a)$ and $\delta\text{cost}'(c) = \delta\text{cost}(c)$.

We next compute δmin' values in terms of δmin and δcost.
$\delta\text{min}'(v) = \text{cost}(v) - \text{mincost}'(v)$
$= \text{cost}(v) - \text{mincost}(w)$

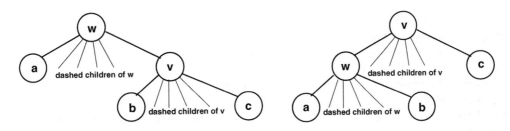

FIGURE 13.5 Rotation in solid trees—rotation of edge (v, w).

$$= \text{cost}(v) - \text{cost}(w) + \text{cost}(w) - \text{mincost}(w)$$
$$= \delta\text{cost}(v) + \delta\text{min}(w).$$

$\delta\text{min}(\ )$ of all nodes other than w will remain same, and for w, from Fact 13.3, we have
$$\delta\text{min}'(w) = \max\{0, \delta\text{min}'(a) -$$
$$\delta\text{cost}'(a), \delta\text{min}'(b) - \delta\text{cost}'(b)\}$$
$$= \max\{0, \delta\text{min}(a) - \delta\text{cost}(a), \delta\text{min}(b) - \delta\text{cost}(b) - \delta\text{cost}(v)\}.$$

13.5.4 Splicing

Let us next look at the other operation, splicing. Let w be the root of a solid tree. And let v be a child of w connected by a dashed edge. If u is the leftmost child of w, then splicing at a dashed child v, of a solid root w, makes v the left child of w. Moreover, the previous left child u now becomes a dashed child of w. Thus informally speaking splicing makes a node the leftmost child of its parent (if the parent is root) and makes the previous leftmost child of parent as dashed.

We next analyze the changes in "cost" and "size" of various nodes after splicing at a dashed child v of solid root w (whose leftmost child is u). As before, "nonprimes" denote the values before the splice and "primes" the values after the splice.

As v was a dashed child of its parent, it was a root earlier (in some solid tree). And as w is also a root,
$$\delta\text{cost}'(v) = \text{cost}(v) - \text{cost}(w)$$
$$= \delta\text{cost}(v) - \delta\text{cost}(w).$$

And as u is now the root of a solid tree,
$$\delta\text{cost}'(u) = \text{cost}(u)$$
$$= \delta\text{cost}(u) + \text{cost}(w)$$
$$= \delta\text{cost}(u) + \delta\text{cost}(w).$$

Finally, $\delta\text{min}'(w) = \max\{0, \delta\text{min}(v) - \delta\text{cost}'(v), \delta\text{min}(\text{right}(w)) - \delta\text{cost}(\text{right}(w))\}$.
All other values are clearly unaffected.

13.5.5 Splay in Virtual Tree

In virtual tree, some edges are solid and some are dashed. Usual splaying is carried out only in the solid trees. To splay at a node x in the virtual tree, following method is used. The algorithm looks at the tree three times, once in each pass, and modifies it. In first pass, by splaying only in the solid trees, starting from the node x, the path from x to the root of the overall tree becomes dashed. This path is made solid by splicing. A final splay at node x will now make x the root of the tree. Less informally, the algorithm is as follows:

Algorithm for Splay(x)

Pass 1: Walk up the virtual tree, but splaying is done only within solid subtree. At the end of this pass, the path from x to root becomes dashed.

Pass 2: Walk up from node x, splicing at each proper ancestor of x. After this step, the path from x to the root becomes solid. Moreover, the node x and all its children in the original tree (the one before pass 1) now become left children.

Pass 3: Walk up from node x to the root, splaying in the normal fashion.

13.5.6 Analysis of Splay in Virtual Tree

Weight of each node in the tree is taken to be the same (say) 1. Size of a node is the total number of descendants—both solid and dashed. And the rank of a node as before is $\text{rank}(x) = \log(\text{size}(x))$. We choose $\alpha = 2$, and hence the potential becomes potential $= 2\sum_x \text{rank}(x)$. We still have to fix β. Let us analyze the complexity of each pass.

Pass 1: We fix $\beta = 1$. Thus, from Lemma 13.1, the amortized cost of single splaying is at most $6(r(t) - r(x)) + 1$. Hence, the total cost of all splays in this pass will be
$$\leq 6(r(t_1) - r(x)) + 1 + 6(r(t_2) - r(p(t_1))) + 1 + \cdots + 6(r(t_k) - r(p(t_{k-1}))) + 1.$$
$$\leq (6(r(t_1) - r(x)) + \cdots + 6(r(t_k) - r(p(t_{k-1})))) + k.$$
Here, k is number of solid trees in path from x to root. Or the total cost
$$\leq k + (6(r(\text{root}) - r(x))) - 6(r(p(t_{k-1})) - r(t_{k-1}) + \cdots + r(p(t_1)) - r(t_1))).$$
Recall that the size includes those of virtual descendants, hence each term in the bracket is nonnegative. Or the total cost
$$\leq k + 6(r(\text{root}) - r(x)).$$
Note that the depth of node x at end of the first pass will be k.

Pass 2: As no rotations are performed, actual time is zero. Moreover as there are no rotations, there is no change in potential. Hence, amortized time is also zero. Alternatively, time taken to traverse k-virtual edges can be accounted by incorporating that in β in pass 3.

Remark 13.3

This means, that in effect, this pass can be done together with Pass 1.

Pass 3: In Pass 1, k extra rotations are performed, (there is a $+k$ factor), thus, we can take this into account, by charging, 2 units for each of the k rotation in Pass 3, hence we set $\beta = 2$. Clearly, the number of rotations is exactly "k." Cost will be $6 \log n + 2$. Thus in effect we can now neglect the $+k$ term of pass 1.

Thus total cost for all three passes is $12 \log n + 2$.

13.5.7 Implementation of Primitives for Linking and Cutting Trees

We next show that various primitives for linking and cutting trees described in the beginning of this section can be implemented in terms of one or two calls to a single basic operation—"splay." We will discuss implementation of each primitive, one by one.

find_cost(v): We are required to find the value stored in the node v. If we splay at node v, then node v becomes the root and $\delta\mathrm{cost}(v)$ will give the required value. Thus the implementation is

> splay(v) and return the value at node v.

find_root(v): We have to find the root of the tree containing the node v. Again, if we splay at v, then v will become the tree root. The ancestors of v will be in the right subtree, hence we follow right pointers till root is reached. The implementation is

> splay(v), follow right pointers till last node of solid tree, say w is reached, splay(w) and return(w).

find_min(v): We have to find the node having the minimum value, on the path from v till the root of the tree containing v; in case of ties, we have to choose the node closest to the root. We again splay at v to make v the root, but, this time, we also keep track of the node having the minimum value. As these values are stored in incremental manner, we have to compute the value by an "addition" at each step.

> splay(v), use $\delta\mathrm{cost}(\ )$ and $\delta\mathrm{min}(\ )$ fields to walk down to the last minimum cost node after v, in the solid tree, say w, splay(w) and return(w).

add_cost($v, \delta x$): We have to add a real number δx to the values stored in each and every ancestors of node v. If we splay at node v, then v will become the root and all ancestors of v will be in the right subtree. Thus, if we add δx to $\delta\mathrm{cost}(v)$, then in effect, we are adding this value not only to all ancestors (in right subtree) but also to the nodes in the left subtree. Hence, we subtract δx from $\delta\mathrm{cost}(\ )$ value of left child of v. Implementation is

> splay(v), add δx to $\delta\mathrm{cost}(v)$, subtract δx from $\delta\mathrm{cost}(\mathrm{LCHILD}(v))$ and return.

For implementing find_size( ), we find the number of descendants of each node in the original tree. This can be done by looking at nodes in postorder (from leaves upwards). These values are stored (say) in a field "size" which is maintained just like the field "cost."

find_size(v): We have to find the number of nodes in the tree containing the node v. If we splay at the node v, then v will become the root and by definition of $\delta\mathrm{size}$, $\delta\mathrm{size}(v)$ will give the required number.

> splay(v) and return($\delta\mathrm{size}(v)$).

link(v, w): If v is a root of a tree, then we have to make the tree rooted at v a child of node w.

> Splay(w), and make v a middle (dashed) child of w. Update the number of descendants of w and its ancestors by add_size(w, size(v)).

cut(v): If v is not a root, then we have to delete the edge from node v to its parent, thus making v a root. The implementation of this is also obvious:

> splay(v), add $\delta\mathrm{cost}(v)$ to $\delta\mathrm{cost}(\mathrm{RCHILD}(v))$, and break link between $\mathrm{RCHILD}(v)$ and v. Update the number of descendants of v and its ancestors by add_size(w, $-\mathrm{size}(v)$).

13.6 Case Study: Application to Network Flows

We next discuss application of linking and cutting trees to the problem of finding maximum flow in a network. Input is a directed graph $G = (V, E)$. There are two distinguished vertices s (source) and t (sink). We need a few definitions and some notations [11,12]. Most of the results in this case study are from [11,12].

PreFlow: $g(*, *)$ is a real-valued function having the following properties:

    Skew-Symmetry: $g(u, v) = -g(v, u)$

    Capacity Constraint: $g(u, v) \leq c(u, v)$

    Positive-Flow Excess: $e(v) \equiv \sum_{w=1}^{n} g(v, w) \geq 0$ for $v \neq s$.

Flow-Excess: Observe that flow-excess at node v is $e(v) = \sum_{w=1}^{n} g(w, v)$, if $v \neq s$, and flow excess at source s is $e(s) = \infty$.

Flow: $f(*, *)$ is a real-valued function having the following additional property:

    Flow Conservation: $\sum_{w=1}^{n} f(v, w) = 0$ for $v \notin \{s, t\}$

    PreFlow: f is a preflow.

Value of flow: $|f| = \sum_{w=1}^{n} f(s, w)$, the net flow out of source.

Remark 13.4

If $(u, v) \notin E$, then $c(u, v) = c(v, u) = 0$. Thus, $f(u, v) \leq c(u, v) = 0$ and $f(v, u) \leq 0$. By skew-symmetry, $f(u, v) = 0$.

Cut: Cut $(S, \bar{S})$ is a partition of vertex set, such that $s \in S$ and $t \in \bar{S}$

Capacity of Cut: $c(S, \bar{S}) = \sum_{v \in S, w \in \bar{S}} c(v, w)$.

Preflow across a Cut: $g(S, \bar{S}) = \sum_{v \in S, w \notin S} g(v, w)$.

Residual Capacity: If g is a flow or preflow, then the residual capacity of an edge (v, w) is $r_g(v, w) = c(v, w) - g(v, w)$.

Residual Graph: G_g contains same set of vertices as the original graph G, but only those edges for which residual capacity is positive; these are either the edges of the original graph or their reverse edges.

Valid Labeling: A valid labeling $d(\ )$ satisfies the following properties:

1. $d(t) = 0$
2. $d(v) > 0$ if $v \neq t$
3. if (v, w) is an edge in residual graph, then $d(w) \geq d(v) - 1$.

A trivial labeling is $d(t) = 0$ and $d(v) = 1$ if $v \neq t$.

Remark 13.5

As for each edge (v, w), $d(v) \leq d(w) + 1$, $\text{dist}(u, t) \geq d(u)$. Thus, label of every vertex from which t is reachable is at most $n - 1$.

Active Vertex: A vertex $v \neq s$ is said to be active if $e(v) > 0$.

The initial preflow is taken to be $g(s, v) = c(s, v)$ and $g(u, v) = 0$ if $u \neq s$.

Flow across a Cut: Please refer to Figure 13.6. Observe that flow conservation is true for all vertices except s and t. In particular sum of flow (total flow) into vertices in set $S - \{s\}$ (set shown between s and cut) is equal to $|f|$ which must be the flow going out of these vertices (into the cut). And this is the flow into vertices (from cut) in set $\bar{S} - \{t\}$ (set after cut before t) which must be equal to the flow out of these vertices into t. Thus the flow into t is $|f|$ which is also the flow through the cut.

Fact 13.4

As $|f| = f(S, \bar{S}) = \sum_{v \in S, w \notin S} f(v, w) \leq \sum_{v \in S, w \notin S} c(v, w) = c(S, \bar{S})$.

Thus maximum value of flow is less than minimum capacity of any cut.

Theorem 13.4

[Max-Flow Min-Cut Theorem] max $|f|$ = minimum cut

Proof. Consider a flow f for which $|f|$ is maximum. Delete all edges for which $(f(u, v) == c(u, v))$ to get the residual graph. Let S be the set of vertices reachable from s in the residual graph. Now, $t \notin S$, otherwise there is a path along which flow can be

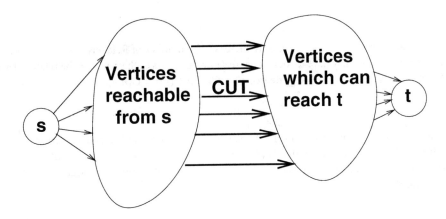

FIGURE 13.6 *s − t* Cut.

increased, contradicting the assumption that flow is maximum. Let $\overline{S}$ be set of vertices not reachable from *s*. $\overline{S}$ is not empty as $t \in S$. Thus $(S, \overline{S})$ is an *s − t* cut and as all edges (v, w) of cut have been deleted, $c(v, w) = f(v, w)$ for edges of cut.

$$|f| = \sum_{v \in S, w \notin S} f(v, w) = \sum_{v \in S, w \notin S} c(v, w) = c(S, \overline{S}).$$

■

Push(v, w)

1. /* v is an active vertex and (v, w) an edge in residual graph with $d(w) = d(v) - 1$ */
   Try to move excess from v to w, subject to capacity constraints, that is, send $\delta = \min\{e(v), r_g(v, w)\}$ units of flow from v to w.
2. /* $g(v, w) = g(v, w) + \delta$; $e(v) = e(v) - \delta$ and $e(w) = e(w) + \delta$ */

   If $\delta = r_g(v, w)$, then the push is said to be *saturating*.

Relabel(v)

For $v \neq s$, the new distance label is
$$d(v) = \min\{d(w) + 1 | (v, w) \text{ is a residual edge}\}.$$

Preflow-Push Algorithms

Following are some of the properties of preflow-push algorithms:

1. If relabel v results in a new label, $d(v) = d(w^*) + 1$, then as initial labeling was valid, $d_{\text{old}}(v) \leq d_{\text{old}}(w^*) + 1$. Thus labels can only increase. Moreover, the new labeling is clearly valid.
2. If push is saturating, edge (v, w) may get deleted from the graph and edge (w, v) will get added to the residual graph, as $d(w) = d(v) - 1$, $d(v) = d(w) + 1 \geq d(w) - 1$, thus even after addition to the residual graph, conditions for labeling to be valid are satisfied.
3. As a result of initialization, each node adjacent to s gets a positive excess. Moreover all arcs out of s are saturated. In other words in residual graph there is no path from s to t. As distances cannot decrease, there can never be a path from s to t. Thus there will be no need to push flow again out of s.
4. By definition of preflow, flow coming into a node is more than flow going out. This flow must come from source. Thus all vertices with positive excess are reachable from s (in the original network). Thus as s is initially the only node, at any stage of the algorithm, there is a path P_v to a vertex v (in the original network) along which preflow has come from s to v. Thus, in the residual graph, there is reverse path from v to s.
5. Consider a vertex v from which there is a path till a vertex X. As we trace back this path from X, then distance label $d(\ )$ increases by at most one. Thus $d(v)$ can be at most $\text{dist}(v, X)$ larger than $d(X)$. That is $d(v) \leq d(X) + \text{dist}(v, X)$.
6. As for vertices from which t is not reachable, s is reachable, $d(v) \leq d(s) + \text{dist}(s, v) = n + (n - 1) = 2n - 1$ (as $d(s) = n$). Thus maximum label of any node is $2n - 1$.

Fact 13.5

As label of t remains zero, and label of other vertices only increase, the number of Relabels which result in change of labels is $(n-1)^2$. In each relabel operation, we may have to look at degree(v) vertices. As each vertex can be relabeled at most $O(n)$ times, time for relabels is $\sum O(n) \times \text{degree}(v) = O(n) \times \sum \text{degree}(v) = O(n) \times O(m) = O(nm)$.

Fact 13.6

If a saturating push occurs from u to v, then $d(u) = d(v) + 1$ and edge (u, v) gets deleted, but edge (v, u) gets added. Edge (u, v) can be added again only if edge (v, u) gets saturated, that is, $d_{\text{now}}(v) = d_{\text{now}}(u) + 1 \geq d(u) + 1 = d(v) + 2$. Thus the edge gets added only if label increases by 2. Thus, for each edge, number of times saturating push can occur is $O(n)$. So the total number of saturating pushes is $O(nm)$.

Remark 13.6

Increase in label of $d(u)$ can make a reverse flow along all arcs (x, u) possible, and not just (v, u); in fact there are at most degree(u) such arcs. Thus number of saturating pushes are $O(nm)$ and not $O(n^2)$.

Fact 13.7

Consider the point in time when the algorithm terminates, that is, when pushes or relabels can no longer be applied. As excess at s is ∞, excess at s could not have been exhausted. The fact that push/relabels cannot be applied means that there is no path from s to t. Thus, $\overline{S_g}$, the set of vertices from which t is reachable, and S_g, set of vertices from which s is reachable, form an $s - t$ cut.

Consider an edge (u, v) with $u \in S_g$ and $v \in \overline{S_g}$. As t is reachable from v, there is no excess at v. Moreover, by definition of cut, the edge is not present in residual graph, or in other words, flow in this edge is equal to capacity. By Theorem 13.4, the flow is the maximum possible.

13.7 Implementation without Linking and Cutting Trees

Each vertex will have a list of edges incident at it. It also has a pointer to *current edge* (candidate for pushing flow out of that node). Each edge (u, v) will have three values associated with it $c(u, v)$, $c(v, u)$, and $g(u, v)$.

Push/Relabel(v)

Here we assume that v is an active vertex and (v, w) is current edge of v.

1. If $(d(w) == d(v) - 1)$ && $(r_g(v, w) > 0)$ then send $\delta = \min\{e(v), r_g(v, w)\}$ units of flow from v to w.
2. Else if v has no next edge, make first edge on edge list the current edge and Relabel(v): $d(v) = \min\{d(w) + 1 | (v, w) \text{ is a residual edge}\}$ /* this causes $d(v)$ to increase by at least one */
3. Else make the next edge out of v, the current edge.

Relabeling v requires a single scan of v's edge list. As each relabeling of v causes $d(v)$ to go up by one, the number of relabeling steps (for v) are at most $O(n)$, each step takes $O(\text{degree}(v))$ time. Thus total time for all relabelings will be:
$O(\sum n\text{degree}(v)) = O(n \sum \text{degree}) = O(n \times 2m) = O(nm)$. Each nonsaturating push clearly takes $O(1)$ time, thus time for algorithm will be $O(nm) + O(\#\text{nonsaturating pushes})$.

Discharge(v)

Keep on applying Push/Relabel(v) until either

1. Entire excess at v is pushed out or
2. Label(v) increases.

FIFO/Queue

Initialize a queue "Queue" to contain s.

Let v be the vertex in front of Queue. Discharge(v), if a push causes excess of a vertex w to become nonzero, add w to the rear of the Queue.

Let phase 1 consists of discharge operations applied to vertices added to the queue by initialization of preflow.

Phase $(i + 1)$ consists of discharge operations applied to vertices added to the queue during phase i.

Let $\Phi = \max\{d(v)|v \text{ is active}\}$, with maximum as zero, if there are no active vertices. If in a phase, no relabeling is done, then the excess of all vertices which were in the queue has been moved. If v is any vertex which was in the queue, then excess has been moved to a node w, with $d(w) = d(v) - 1$. Thus $\max\{d(w)|w \text{ has now become active}\} \leq \max\{d(v) - 1|v \text{ was active}\} = \Phi - 1$.

Thus, if in a phase, no relabeling is done, Φ decreases by at least one. Moreover, as number of relabeling steps are bounded by $2n^2$, number of passes in which relabeling takes place is at most $2n^2$.

Only way in which Φ can increase is by relabeling. Since the maximum value of a label of any active vertex is $n - 1$, and as a label never decreases, the total of all increases in Φ is $(n - 1)^2$.

As Φ decreases by at least one in a pass in which there is no relabeling, number of passes in which there is no relabeling is $(n - 1)^2 + 2n^2 \leq 3n^2$.

Fact 13.8

Number of passes in FIFO algorithm is $O(n^2)$.

13.8 FIFO Dynamic Tree Implementation

Time for nonsaturating push is reduced by performing a succession of pushes along a single path in one operation [12]. After a nonsaturating push, the edge continues to be admissible and we know its residual capacity. Initially each vertex is made into a one vertex node. Arc of dynamic trees are a subset of admissible arcs. Value of an arc is its admissible capacity (if $(u, \text{parent}(u))$ is an arc, value of arc will be stored at u). Each active vertex is a tree root.

Vertices will be kept in a queue as in FIFO algorithm, but instead of discharge(v), Tree-Push(v) will be used. We will further ensure that tree size does not exceed k (k is a parameter to be chosen later). The Tree-Push procedure is as follows.

Tree-Push(v)

/* v is active vertex and (v, w) is an admissible arc */

1. /* link trees rooted at v and the tree containing w by making w the parent of v, if the tree size doesn't exceed k */.
   If v is root and (find_size(v) + find_size(w))$\leq k$, then link v and w. Arc (v, w) gets the value equal to the residual capacity of edge (v, w).
2. If v is root but find_size(v) + find_size(w) > k, then push flow from v to w.
3. If v is not a tree root, then send $\delta = \min\{e(v), \text{find\_cost(find\_min}(v))\}$ units of flow from v, by add_cost($v, -\delta$) /* decrease residual capacity of all arcs */ and while v is not a root and find_cost(find_min(v))$= = 0$ do

   $\{z := \text{find\_min}(v); \text{cut}(z); /* \text{ delete saturated edge } */$
   $f(z, \text{parent}(z)) := c(z, \text{parent}(z));$
   /* in saturated edge, flow=capacity */
   $f(\text{parent}(z), z) := -c(z, \text{parent}(z));$
   $\}$

4. But, if arc(v, w) is not admissible, replace (v, w), as current edge by next edge on v's list. If v has no next edge, then make the first edge, the current edge and cut-off all children of v, and relabel(v).

Analysis

1. Total time for relabeling is $O(nm)$.
2. Only admissible edges are present in the tree, and hence if an edge (u, v) is cut in step (3) or in step (4) then it must be admissible, that is, $d(u) = d(v) + 1$. Edge (v, u) can become admissible and get cut, iff, $d_{\text{then}}(v) = d_{\text{then}}(u) + 1 \geq d(u) + 1 = d(v) + 2$. Thus the edge gets cut again only if label increases by 2. Thus, for each edge, number of times it can get cut is $O(n)$. So total number of cuts are $O(nm)$.
3. As initially, there are at most n-single node trees, number of links are at most $n + \#\text{no\_of\_cuts} = n + O(nm) = O(nm)$.

Moreover, there is at most one tree operation for each relabeling, cut or link. Further, for each item in queue, one operation is performed. Thus,

Lemma 13.2

The time taken by the algorithm is $O(\log k \times (nm + \#\text{No\_of\_times\_an\_item\_is\_added\_to\_the\_queue}))$.

Root-Nodes: Let T_v denote the tree containing node v. Let r be a tree root whose excess has become positive. It can become positive either due to:

1. Push from a nonroot vertex w in Step 3 of the tree-push algorithm
2. Push from a root w in Step 2 /* find_size(w) + find_size(r) > k */

Remark 13.7

Push in Step 3 is accompanied by a cut (unless first push is nonsaturating). As the number of cuts is $O(nm)$, number of times Step 3 (when first push is saturating) can occur is $O(nm)$. Thus, we need to consider only the times when first push was nonsaturating, and the excess has moved to the root as far as push in Step 3 is concerned.

In either case let i be the pass in which this happens (i.e., w was added to the queue in pass $(i-1)$). Let I be the interval from beginning of pass $(i-1)$ to the time when $e(r)$ becomes positive.

Case 1: (T_w changes during I) T_w can change either due to link or cut. But number of times a link or a cut can occur is $O(nm)$. Thus this case occurs at most $O(nm)$ time. Thus we may assume that T_w *does not change during interval I.*
Vertex w is added to the queue either because of relabeling of w or because of a push in Step 2 from (say) a root v to w.

Case 2: (w is added because of relabeling) Number of relabeling steps are $O(n^2)$. Thus number of times this case occurs is $O(n^2)$. Thus we may assume that w *was added to queue because of push* from root v to w in Step 2.

Case 3: (push from w was saturating) As the number of saturating pushes is $O(nm)$, this case occurs $O(nm)$ times. Thus we may assume that *push from w was nonsaturating.*

Case 4: (edge (v, w) was not the current edge at beginning of pass $(i-1)$). Edge (v, w) will become the current edge, only because either the previous current edge (v, x) got saturated, or because of relabel(v), or relabel(x). Note that if entire excess out of v was moved, then (v, w) will remain the current edge.
As number of saturating pushes are $O(nm)$ and number of relabeling are $O(n^2)$, this case can occur at most $O(nm)$ times. Thus we may assume that *(v, w) was the current edge at beginning of pass $(i-1)$.*

Case 5: (T_v changes during interval I) T_v can change either due to link or cut. But the number of times a link or a cut can occur is $O(nm)$. Thus this case occurs at most $O(nm)$ time. Thus we may assume that T_v *has not changed during interval I.*

Remaining Case: Vertex w was added to the queue because of a nonsaturating push from v to w in Step 2 and (v, w) is still the current edge of v. Moreover, T_v and T_w do not change during the interval I.
A tree at the beginning of the pass $(i-1)$ can participate in only one pair (T_w, T_v) as T_w, because this push was responsible for adding w to the queue. Observe that vertex w is uniquely determined by r.
And, a tree at the beginning of the pass $(i-1)$ can participate in only one pair (T_w, T_v) as T_v, because (v, w) was the current edge out of root v, at the beginning of the pass $(i-1)$ (and is still the current edge). Thus, choice of T_v will uniquely determine T_w (and conversely).
Thus as a tree T_x can participate once in a pair as T_v, and once as T_w, and the two trees are unchanged, we have $\sum_{(v,w)} |T_v| + |T_w| \leq 2n$ (a vertex is in at most one tree). As push from v to w was in Step 2, find_size(v) + find_size(w) > k or $|T_v| + |T_w| > k$. Thus the number of such pairs is at most $2n/k$.
But from Fact 13.8, as there are at most $O(n^2)$ passes, the number of such pairs are $O(n^3/k)$.

Nonroot Nodes: Let us count the number of times a nonroot can have its excess made positive. Its excess can only be made positive as a result of push in Step 2. As the number of saturating pushes is $O(nm)$, clearly $O(nm)$ pushes in Step 2 are saturating.
If the push is nonsaturating, then entire excess at that node is moved out, hence it can happen only once after a vertex is removed from Queue. If v was not a root when it was added to the queue, then it has now become a root only because of a cut. But number of cuts is $O(nm)$. Thus we only need to consider the case when v was a root when it was added to the queue. The root was not earlier in queue, because either its excess was then zero, or because its distance label was low. Thus now either

1. Distance label has gone up—this can happen at most $O(n^2)$ times or
2. Now its excess has become positive. This by previous case can happen at most $O(nm + (n^3/k))$ times.

Summary: If k is chosen such that $nm = n^3/k$, or $k = n^2/m$, time taken by the algorithm is $O(nm \log (n^2/m))$.

13.9 Variants of Splay Trees and Top-Down Splaying

Various variants, modifications, and generalization of Splay trees have been studied [13–16]. Two of the most popular "variants" suggested by Sleator and Tarjan [1] are "semi-splay" and "simple-splay" trees. In simple splay trees, the second rotation in the "zig-zag" case is done away with (i.e., we stop at the middle figure in Figure 13.3). Simple splaying can be shown to have a larger constant factor both theoretically [1] and experimentally [14]. In semi-splay [1], in the zig-zig case (see Figure 13.2) we do only the first rotation (i.e., stop at the middle figure) and continue splaying from node y instead of x. Sleator and Tarjan observe that for some access sequences "semi-splaying" may be better but for some others the usual splay is better.

"Top-down" splay trees [1] are another way of implementing splay trees. Both the trees coincide if the node being searched is at an even depth [14], but if the item being searched is at an odd depth, then the top-down and bottom-up trees may differ [14, Theorem 13.2].

Some experimental evidence suggests [3] that top-down splay trees [1,14] are faster in practice as compared to the normal splay trees, but some evidence suggests otherwise [4].

In splay trees as described, we first search for an item, and then restructure the tree. These are called "bottom-up" splay trees. In "top-down" splay trees, we look at two nodes at a time, while searching for the item, and also keep restructuring the tree until the item we are looking for has been located.

Basically, the current tree is divided into three trees, while we move down two nodes at a time searching for the query item.

Left tree: Left tree consists of items known to be smaller than the item we are searching.
Right tree: Similarly, the right tree consists of items known to be larger than the item we are searching.
Middle tree: This is the subtree of the original tree rooted at the current node.

Basically, the links on the access path are broken and the node(s) which we just saw are joined to the bottom right (respectively left) of the left (respectively right) tree if they contain item greater (respectively smaller) than the item being searched. If both nodes are left children or if both are right children, then we make a rotation before breaking the link. Finally, the item at which the search ends is the only item in the middle tree and it is made the root. And roots of left and right trees are made the left and right children of the root.

Acknowledgments

I wish to thank Anjana Singh for pointing an error in the previous version.

References

1. D. Sleator and R.E. Tarjan, Self-adjusting binary search trees, *JACM*, vol 32, 1985, 652–686.
2. R.E. Tarjan, *Data Structures and Network Algorithms*, SIAM, 1983.
3. J. Bell and G. Gupta, An evaluation of self-adjusting binary search tree techniques, *Software: Practice and Experience*, vol 23, 1993, 369–382.
4. H.E. Williams, J. Zobel and S. Heinz, Self-adjusting trees in practice for large text collections, *Software: Practice and Experience*, vol 31, 2001, 925–939.
5. A. Moffat, G. Eddy and O. Petersson, Splaysort, fast, versatile, practical, *Software: Practice and Experience*, vol 26, 1996, 781–797.
6. T. Bell and D. Kulp, Longest-match string searching for Ziv-Lempel compression, *Software: Practice and Experience*, vol 23, 1993, 757–771.
7. D.W. Jones, Application of splay trees to data compression, *CACM*, vol 31, 1988, 996–1007.
8. D. Grinberg, S. Rajagopalan, R. Venkatesh and V.K. Wei, Splay trees for data compression, *Sixth Annual ACM-SIAM Symp. Discrete Algorithms (SODA)*, vol 6, 1995, 522–530.
9. R. Cole, On the dynamic finger conjecture for splay trees. Part II: The Proof, *SIAM Journal on Computing*, vol 30, no.1, 2000, 44–5.
10. J. Iacono. Key independent optimality, *ISAAC 2002*, vol LNCS 2518, 2002, pp. 25–31.
11. R.K. Ahuja, T.L. Magnanti and J.B Orlin, *Network Flows (Theory, Algorithms and Applications)*, Prentice Hall, Inc., Englewood Cliffs, NJ, 1993.
12. A.V. Goldberg and R.E. Tarjan, A new approach to the maximum-flow problem, *JACM*, vol 35, no.4, October 1988, pp 921–940.

13. S. Albers and M. Karpinkski, Randomized splay trees: Theoretical and experimental results, *Information Processing Letters*, vol 81, 2002, 213–221.

14. E. Mäkinen, On top-down splaying, *BIT*, vol 27, 1987, 330–339.

15. M. Sherk, Self-adjusting k-ary search trees, *Journal. of Algorithms*, vol 19, 1995, 25–44.

16. A. Subramanian, An explanation of splaying, *Journal of Algorithms*, vol 20, 1996, 512–525.

14

Randomized Dictionary Structures[*]

14.1 Introduction.. 197
14.2 Preliminaries... 198
    Randomized Algorithms • Basics of Probability Theory • Conditional Probability • Some
    Basic Distributions • Tail Estimates
14.3 Skip Lists 204
14.4 Structural Properties of Skip Lists 205
    Number of Levels in Skip List • Space Complexity
14.5 Dictionary Operations 207
14.6 Analysis of Dictionary Operations.. 207
14.7 Randomized Binary Search Trees.. 209
    Insertion in RBST • Deletion in RBST
14.8 Bibliographic Remarks... 212
References.. 213

C. Pandu Rangan
*Indian Institute of Technology,
Madras*

14.1 Introduction

In the last couple of decades, there has been a tremendous growth in using randomness as a powerful source of computation. Incorporating randomness in computation often results in a much simpler and more easily implementable algorithms. A number of problem domains, ranging from sorting to stringology, from graph theory to computational geometry, from parallel processing system to ubiquitous internet, have benefited from randomization in terms of newer and elegant algorithms. In this chapter we shall see how randomness can be used as a powerful tool for designing simple and efficient data structures. Solving a real-life problem often involves manipulating complex data objects by variety of operations. We use abstraction to arrive at a mathematical model that represents the real-life objects and convert the real-life problem into a computational problem working on the mathematical entities specified by the model. Specifically, we define *Abstract Data Type (ADT)* as a mathematical model together with a set of operations defined on the entities of the model. Thus, an algorithm for a computational problem will be expressed in terms of the steps involving the corresponding ADT operations. In order to arrive at a computer based implementation of the algorithm, we need to proceed further taking a closer look at the possibilities of implementing the ADTs. As programming languages support only a very small number of built-in types, any ADT that is not a built-in type must be represented in terms of the elements from built-in type and this is where the data structure plays a critical role. One major goal in the design of data structure is to render the operations of the ADT as efficient as possible. Traditionally, data structures were designed to minimize the worst-case costs of the ADT operations. When the worst-case efficient data structures turn out to be too complex and cumbersome to implement, we naturally explore alternative design goals. In one of such design goals, we seek to minimize the total cost of a sequence of operations as opposed to the cost of individual operations. Such data structures are said to be designed for minimizing the amortized costs of operations. Randomization provides yet another avenue for exploration. Here, the goal will be to limit the expected costs of operations and ensure that costs do not exceed certain threshold limits with overwhelming probability.

    In this chapter we discuss about the *Dictionary* ADT which deals with sets whose elements are drawn from a fixed universe U and supports operations such as *insert*, *delete* and *search*. Formally, we assume a linearly ordered universal set U and for the sake of concreteness we assume U to be the set of all integers. At any time of computation, the *Dictionary* deals only with a finite subset of U. We shall further make a simplifying assumption that we deal only with sets with distinct values. That is, we

---
[*] This chapter has been reprinted from first edition of this Handbook, without any content updates.

never handle a multiset in our structure, though, with minor modifications, our structures can be adjusted to handle multisets containing multiple copies of some elements. With these remarks, we are ready for the specification of the *Dictionary* ADT.

Definition 14.1: (Dictionary ADT)

Let U be a linearly ordered universal set and S denote a finite subset of U. The Dictionary ADT, defined on the class of finite subsets of U, supports the following operations.

Insert (x, S) : For an $x \in U, S \subset U$, generate the set $S \cup \{x\}$.
Delete (x, S) : For an $x \in U, S \subset U$, generate the set $S - \{x\}$.
Search (x, S) : For an $x \in U, S \subset U$, return TRUE if $x \in S$ and return FALSE if $x \notin S$.

Remark : When the universal set is evident in a context, we will not explicitly mention it in the discussions. Notice that we work with sets and not multisets. Thus, *Insert (x,S)* does not produce new set when x is in the set already. Similarly *Delete (x, S)* does not produce a new set when $x \notin S$.

Due to its fundamental importance in a host of applications ranging from compiler design to data bases, extensive studies have been done in the design of data structures for dictionaries. Refer to Chapters 3 and 11 for data structures for dictionaries designed with the worst-case costs in mind, and Chapter 13 of this handbook for a data structure designed with amortized cost in mind. In Chapter 16 of this book, you will find an account of *B-Trees* which aim to minimize the disk access. All these structures, however, are deterministic. In this sequel, we discuss two of the interesting randomized data structures for Dictionaries. Specifically

- We describe a data structure called *Skip Lists* and present a comprehensive probabilistic analysis of its performance.
- We discuss an interesting randomized variation of a search tree called *Randomized Binary Search Tree* and compare and contrast the same with other competing structures.

14.2 Preliminaries

In this section we collect some basic definitions, concepts and the results on randomized computations and probability theory. We have collected only the materials needed for the topics discussed in this chapter. For a more comprehensive treatment of randomized algorithms, refer to the book by Motwani and Raghavan [1].

14.2.1 Randomized Algorithms

Every computational step in an execution of a *deterministic algorithm* is uniquely determined by the set of all steps executed prior to this step. However, in a *randomized algorithm*, the choice of the next step may not be entirely determined by steps executed previously; the choice of next step might depend on the outcome of a random number generator. Thus, several execution sequences are possible even for the same input. Specifically, when a randomized algorithm is executed several times, even on the same input, the running time may vary from one execution to another. In fact, the running time is a random variable depending on the random choices made during the execution of the algorithm. When the running time of an algorithm is a random variable, the traditional worst case complexity measure becomes inappropriate. In fact, the quality of a randomized algorithm is judged from the statistical properties of the random variable representing the running time. Specifically, we might ask for bounds for the expected running time and bounds beyond which the running time may exceed only with negligible probability. In other words, for the randomized algorithms, there is no bad input; we may perhaps have an *unlucky* execution.

The type of randomized algorithms that we discuss in this chapter is called *Las Vegas* type algorithms. A *Las Vegas* algorithm always terminates with a correct answer although the running time may be a random variable exhibiting wide variations. There is another important class of randomized algorithms, called *Monte Carlo* algorithms, which have fixed running time but the output may be erroneous. We will not deal with *Monte Carlo* algorithms as they are not really appropriate for basic building blocks such as data structures. We shall now define the notion of efficiency and complexity measures for *Las Vegas* type randomized algorithms.

Since the running time of a *Las Vegas* randomized algorithm on any given input is a random variable, besides determining the expected running time it is desirable to show that the running time does not exceed certain threshold value with very high probability. Such threshold values are called *high probability bounds* or *high confidence bounds*. As is customary in algorithmics, we express the estimation of the expected bound or the high-probability bound as a function of the size of the input. We interpret an execution of a *Las Vegas* algorithm as a *failure* if the running time of the execution exceeds the expected running time or the high-confidence bound.

Definition 14.2: (Confidence Bounds)

Let α, β and c be positive constants. A randomized algorithm A requires resource bound $f(n)$ with

1. $n - exponential$ probability or very high probability, if for any input of size n, the amount of the resource used by A is at most $\alpha f(n)$ with probability $1 - O(\beta^{-n})$, $\beta > 1$. In this case $f(n)$ is called a *very high confidence bound*.
2. $n - polynomial$ probability or high probability, if for any input of size n, the amount of the resource used by A is at most $\alpha f(n)$ with probability $1 - O(n^{-c})$. In this case $f(n)$ is called a *high confidence bound*.
3. $n - log$ probability or very good probability, if for any input of size n, the amount of the resource used by A is at most $\alpha f(n)$ with probability $1 - O((\log n)^{-c})$. In this case $f(n)$ is called a *very good confidence bound*.
4. $high - constant$ probability, if for any input of size n, the amount of the resource used by A is at most $\alpha f(n)$ with probability $1 - O(\beta^{-\alpha})$, $\beta > 1$.

The practical significance of this definition can be understood from the following discussions. For instance, let A be a *Las Vegas* type algorithm with $f(n)$ as a high confidence bound for its running time. As noted before, the actual running time $T(n)$ may vary from one execution to another but the definition above implies that, for any execution, on any input, $Pr(T(n) > f(n)) = O(n^{-c})$. Even for modest values of n and c, this bound implies an extreme rarity of failure. For instance, if $n = 1000$ and $c = 4$, we may conclude that the chance that the running time of the algorithm A exceeding the threshold value is one in zillion.

14.2.2 Basics of Probability Theory

We assume that the reader is familiar with basic notions such as *sample space, event* and basic *axioms of probability*. We denote as $Pr(E)$ the probability of the event E. Several results follow immediately from the basic axioms, and some of them are listed in Lemma 14.1.

Lemma 14.1

The following laws of probability must hold:

1. $Pr(\phi) = 0$
2. $Pr(E^c) = 1 - Pr(E)$
3. $Pr(E_1) \leq Pr(E_2)$ if $E_1 \subseteq E_2$
4. $Pr(E_1 \cup E_2) = Pr(E_1) + Pr(E_2) - Pr(E_1 \cap E_2) \leq Pr(E_1) + Pr(E_2)$

Extending item 4 in Lemma 14.1 to countable unions yields the property known as *sub additivity*. Also known as *Boole's Inequality*, it is stated in Theorem 14.1.

Theorem 14.1: (Boole's Inequality)

$Pr(\cup_{i=1}^{\infty} E_i) \leq \sum_{i=1}^{\infty} Pr(E_i)$

A probability distribution is said to be *discrete* if the sample space S is finite or countable. If $E = \{e_1, e_2, \ldots, e_k\}$ is an event, $Pr(E) = \sum_{i=1}^{k} Pr(\{e_i\})$ because all elementary events are mutually exclusive. If $|S| = n$ and $Pr(\{e\}) = \frac{1}{n}$ for every elementary event e in S, we call the distribution a *uniform distribution* of S. In this case,

$$Pr(E) = \sum_{e \in E} Pr(e)$$
$$= \sum_{e \in E} \frac{1}{n}$$
$$= |E|/|S|$$

which agrees with our intuitive and a well-known definition that probability is the ratio of the favorable number of cases to the total number of cases, when all the elementary events are equally likely to occur.

14.2.3 Conditional Probability

In several situations, the occurrence of an event may change the uncertainties in the occurrence of other events. For instance, insurance companies charge higher rates to various categories of drivers, such as those who have been involved in traffic accidents, because the probabilities of these drivers filing a claim is altered based on these additional factors.

Definition 14.3: (Conditional Probability)

The *conditional probability* of an event E_1 given that another event E_2 has occurred is defined by $Pr(E_1/E_2)$ ("$Pr(E_1/E_2)$" is read as "the probability of E_1 given E_2.").

Lemma 14.2

$Pr(E_1/E_2) = \frac{Pr(E_1 \cap E_2)}{Pr(E_2)}$, provided $Pr(E_2) \neq 0$.

Lemma 14.2 shows that the conditional probability of two events is easy to compute. When two or more events do not influence each other, they are said to be independent. There are several notions of independence when more than two events are involved. Formally,

Definition 14.4: (Independence of two events)

Two events are *independent* if $Pr(E_1 \cap E_2) = Pr(E_1)Pr(E_2)$, or equivalently, $Pr(E_1/E_2) = Pr(E_1)$.

Definition 14.5: (Pairwise independence)

Events $E_1, E_2, \ldots E_k$ are said to be *pairwise independent* if $Pr(E_i \cap E_j) = Pr(E_i)Pr(E_j)$, $1 \leq i \neq j \leq n$.

Given a partition $S_1, \ldots, S_k$ of the sample space S, the probability of an event E may be expressed in terms of mutually exclusive events by using conditional probabilities. This is known as the *law of total probability in the conditional form*.

Lemma 14.3: (Law of total probability in the conditional form)

For any partition $S_1, \ldots, S_k$ of the sample space S, $Pr(E) = \sum_{i=1}^{k} Pr(E/S_i) Pr(S_i)$.

The law of total probability in the conditional form is an extremely useful tool for calculating the probabilities of events. In general, to calculate the probability of a complex event E, we may attempt to find a partition $S_1, S_2, \ldots, S_k$ of S such that both $Pr(E/S_i)$ and $Pr(S_i)$ are easy to calculate and then apply Lemma 14.3. Another important tool is *Bayes' Rule*.

Theorem 14.2: (Bayes' Rule)

For events with non-zero probabilities,

1. $Pr(E_1/E_2) = \frac{Pr(E_2/E_1)Pr(E_1)}{Pr(E_2)}$
2. If $S_1, S_2, \ldots, S_k$ is a partition, $Pr(S_i/E) = \frac{Pr(E/S_i)Pr(S_i)}{\sum_{j=1} Pr(E/S_j)Pr(S_j)}$

Proof. Part (1) is immediate by applying the definition of conditional probability; Part (2) is immediate from Lemma 14.3. ∎

14.2.3.1 Random Variables and Expectation

Most of the random phenomena are so complex that it is very difficult to obtain detailed information about the outcome. Thus, we typically study one or two numerical parameters that we associate with the outcomes. In other words, we focus our attention on certain real-valued functions defined on the sample space.

Definition 14.6

A *random variable* is a function from a sample space into the set of real numbers. For a random variable X, $R(X)$ denotes the *range* of the function X.

Having defined a random variable over a sample space, an event of interest may be studied through the values taken by the random variables on the outcomes belonging to the event. In order to facilitate such a study, we supplement the definition of a random variable by specifying how the probability is assigned to (or distributed over) the values that the random variable may assume. Although a rigorous study of random variables will require a more subtle definition, we restrict our attention to the following simpler definitions that are sufficient for our purposes.

A random variable X is a *discrete random variable* if its range $R(X)$ is a finite or countable set (of real numbers). This immediately implies that any random variable that is defined over a finite or countable sample space is necessarily discrete. However, discrete random variables may also be defined on uncountable sample spaces. For a random variable X, we define its *probability mass function (pmf)* as follows:

Definition 14.7: (Probability mass function)

For a random variable X, the *probability mass function* $p(x)$ is defined as $p(x) = Pr(X = x)$, $\forall x \in R(X)$.

The probability mass function is also known as the *probability density function*. Certain trivial properties are immediate, and are given in Lemma 14.4.

Lemma 14.4

The probability mass function $p(x)$ must satisfy

1. $p(x) \geq 0, \forall x \in R(X)$
2. $\sum_{x \in R(X)} p(x) = 1$

Let X be a discrete random variable with probability mass function $p(x)$ and range $R(X)$. The *expectation* of X (also known as the *expected value* or *mean* of X) is its average value. Formally,

Definition 14.8: (Expected value of a discrete random variable)

The *expected value* of a discrete random variable X with probability mass function $p(x)$ is given by $E(X) = \mu_X = \sum_{x \in R(X)} x p(x)$.

Lemma 14.5

The expected value has the following properties:

1. $E(cX) = cE(X)$ if c is a constant
2. (Linearity of expectation) $E(X + Y) = E(X) + E(Y)$, provided the expectations of X and Y exist

Finally, a useful way of computing the expected value is given by Theorem 14.3.

Theorem 14.3

If $R(X) = \{0, 1, 2, \ldots\}$, then $E(X) = \sum_{i=1}^{\infty} Pr(X \geq i)$.

Proof.

$$E(X) = \sum_{i=0}^{\infty} i Pr(X = i)$$

$$= \sum_{i=0}^{\infty} i(Pr(X \geq i) - Pr(X \geq i+1))$$

$$= \sum_{i=1}^{\infty} Pr(X \geq i)$$

∎

14.2.4 Some Basic Distributions

14.2.4.1 Bernoulli Distribution

We have seen that a coin flip is an example of a random experiment and it has two possible outcomes, called *success* and *failure*. Assume that success occurs with probability p and that failure occurs with probability $q = 1 - p$. Such a coin is called p-biased coin. A coin flip is also known as *Bernoulli Trial*, in honor of the mathematician who investigated extensively the distributions that arise in a sequence of coin flips.

Definition 14.9

A random variable X with range $R(X) = \{0, 1\}$ and probability mass function $Pr(X = 1) = p$, $Pr(X = 0) = 1 - p$ is said to follow the Bernoulli Distribution. We also say that X is a Bernoulli random variable with parameter p.

14.2.4.2 Binomial Distribution

Let $\begin{pmatrix} n \\ k \end{pmatrix}$ denote the number of k-combinations of elements chosen from a set of n elements. Recall that $\begin{pmatrix} n \\ k \end{pmatrix} = \frac{n!}{k!(n-k)!}$ and $\begin{pmatrix} n \\ 0 \end{pmatrix} = 1$ since $0! = 1$. $\begin{pmatrix} n \\ k \end{pmatrix}$ denotes the *binomial coefficients* because they arise in the expansion of $(a + b)^n$.

Define the random variable X to be the number of successes in n flips of a p-biased coin. The variable X satisfies the *binomial distribution*. Specifically,

Definition 14.10: (Binomial distribution)

A random variable with range $R(X) = \{0, 1, 2, \ldots, n\}$ and probability mass function

$$Pr(X = k) = b(k, n, p) = \begin{pmatrix} n \\ k \end{pmatrix} p^k q^{n-k}, \ for \ k = 0, 1, \ldots, n$$

satisfies the *binomial distribution*. The random variable X is called a binomial random variable with parameters n and p.

Theorem 14.4

For a binomial random variable X, with parameters n and p, $E(X) = np$ and $Var(X) = npq$.

14.2.4.3 Geometric Distribution

Let X be a random variable X denoting the number of times we toss a p-biased coin until we get a success. Then, X satisfies the *geometric distribution*. Specifically,

Definition 14.11: (Geometric distribution)

A random variable with range $R(X) = \{1, 2, \ldots, \infty\}$ and probability mass function $Pr(X = k) = q^{k-1}p$, for $k = 1, 2, \ldots, \infty$ satisfies the *geometric distribution*. We also say that X is a geometric random variable with parameter p.

The probability mass function is based on $k - 1$ failures followed by a success in a sequence of k independent trials. The mean and variance of a geometric distribution are easy to compute.

Theorem 14.5

For a geometrically distributed random variable X, $E(X) = \frac{1}{p}$ and $Var(X) = \frac{q}{p^2}$.

14.2.4.4 Negative Binomial distribution

Fix an integer n and define a random variable X denoting the number of flips of a p-biased coin to obtain n successes. The variable X satisfies a negative binomial distribution. Specifically,

Definition 14.12

A random variable X with $R(X) = \{0, 1, 2, \ldots\}$ and probability mass function defined by

$$Pr(X = k) = \binom{k-1}{n-1} p^n q^{k-n} \quad \text{if } k \geq n$$
$$= 0 \quad \text{if } 0 \leq k < n \qquad (14.1)$$

is said to be a *negative binomial random variable* with parameters n and p.

Equation 14.1 follows because, in order for the nth success to occur in the kth flip there should be $n-1$ successes in the first $k-1$ flips and the kth flip should also result in a success.

Definition 14.13

Given n identically distributed independent random variables $X_1, X_2, \ldots, X_n$, the sum

$$S_n = X_1 + X_2 + \cdots + X_n$$

defines a new random variable. If n is a finite, fixed constant then S_n is known as the *deterministic sum* of n random variables. On the other hand, if n itself is a random variable, S_n is called a *random sum*.

Theorem 14.6

Let $X = X_1 + X_2 + \cdots + X_n$ be a deterministic sum of n identical independent random variables. Then

1. If X_i is a Bernoulli random variable with parameter p then X is a binomial random variable with parameters n and p.
2. If X_i is a geometric random variable with parameter p, then X is a negative binomial with parameters n and p.
3. If X_i is a (negative) binomial random variable with parameters r and p then X is a (negative) binomial random variable with parameters nr and p.

Deterministic sums and random sums may have entirely different characteristics as the following theorem shows.

Theorem 14.7

Let $X = X_1 + \cdots + X_N$ be a random sum of N geometric random variables with parameter p. Let N be a geometric random variable with parameter α. Then X is a geometric random variable with parameter αp.

14.2.5 Tail Estimates

Recall that the running time of a *Las Vegas* type randomized algorithm is a random variable and thus we are interested in the probability of the running time exceeding a certain threshold value.

Typically we would like this probability to be very small so that the threshold value may be taken as the figure of merit with high degree of confidence. Thus we often compute or estimate quantities of the form $Pr(X \geq k)$ or $Pr(X \leq k)$ during the analysis of randomized algorithms. Estimates for the quantities of the form $Pr(X \geq k)$ are known as *tail estimates*. The next two theorems state some very useful tail estimates derived by Chernoff. These bounds are popularly known as *Chernoff bounds*. For simple and elegant proofs of these and other related bounds you may refer [2].

Theorem 14.8

Let X be a sum of n independent random variables X_i with $R(X_i) \subseteq [0, 1]$. Let $E(X) = \mu$. Then,

$$Pr(X \geq k) \leq \left(\frac{\mu}{k}\right)^k \left(\frac{n-\mu}{n-k}\right)^{n-k} \quad \text{for } k > \mu \qquad (14.2)$$

$$\leq \left(\frac{\mu}{k}\right)^k e^{k-\mu} \quad \text{for } k > \mu \qquad (14.3)$$

$$Pr(X \geq (1+\epsilon)\mu) \leq \left[\frac{e^\epsilon}{(1+\epsilon)^{(1+\epsilon)}}\right]^\mu \quad \text{for } \epsilon \geq 0 \qquad (14.4)$$

Theorem 14.9

Let X be a sum of n independent random variables X_i with $R(X_i) \subseteq [0,1]$. Let $E(X) = \mu$. Then,

$$Pr(X \le k) \le \left(\frac{\mu}{k}\right)^k \left(\frac{n-\mu}{n-k}\right)^{n-k} \quad k < \mu \tag{14.5}$$

$$\le \left(\frac{\mu}{k}\right)^k e^{k-\mu} \quad k < \mu \tag{14.6}$$

$$Pr(X \le (1-\epsilon)\mu) \le e^{-\frac{\mu\epsilon^2}{2}}, \text{ for } \epsilon \in (0,1) \tag{14.7}$$

Recall that a deterministic sum of several geometric variables results in a negative binomial random variable. Hence, intuitively, we may note that only the upper tail is meaningful for this distribution. The following well-known result relates the upper tail value of a negative binomial distribution to a lower tail value of a suitably defined binomial distribution. Hence all the results derived for lower tail estimates of the binomial distribution can be used to derive upper tail estimates for negative binomial distribution. This is a very important result because finding bounds for the right tail of a negative binomial distribution directly from its definition is very difficult.

Theorem 14.10

Let X be a negative binomial random variable with parameters r and p. Then, $Pr(X > n) = Pr(Y < r)$ where Y is a binomial random variable with parameters n and p.

14.3 Skip Lists

Linked list is the simplest of all dynamic data structures implementing a *Dictionary*. However, the complexity of *Search* operation is $O(n)$ in a *Linked list*. Even the *Insert* and *Delete* operations require $O(n)$ time if we do not specify the exact position of the item in the list. *Skip List* is a novel structure, where using randomness, we construct a number of progressively smaller lists and maintain this collection in a clever way to provide a data structure that is competitive to balanced tree structures. The main advantage offered by skip list is that the codes implementing the dictionary operations are very simple and resemble list operations. No complicated structural transformations such as *rotations* are done and yet the expected time complexity of *Dictionary* operations on *Skip Lists* are quite comparable to that of AVL trees or splay trees. *Skip Lists* are introduced by Pugh [3].

Throughout this section let $S = \{k_1, k_2, \ldots, k_n\}$ be the set of keys and assume that $k_1 < k_2 < \cdots < k_n$.

Definition 14.14

Let $S_0, S_1, S_2, \ldots, S_r$ be a collection of sets satisfying

$$S = S_0 \supseteq S_1 \supseteq S_2 \supseteq \cdots \supset S_r = \phi$$

Then, we say that the collection $S_0, S_1, S_2, \ldots, S_r$ defines a *leveling with r levels on S*. The keys in S_i are said to be in level i, $0 \le i \le r$. The set S_0 is called the base level for the leveling scheme. Notice that there is exactly one empty level, namely S_r. The level number $l(k)$ of a key $k \in S$ is defined by

$$l(k) = max\{i \mid k \in L_i\}.$$

In other words, $k \in S_0, S_1, S_2, \ldots, S_{l(k)}$ but $k \notin S_{l(k)+1} \ldots S_r$.

For an efficient implementation of the dictionary, instead of working with the current set S of keys, we would rather work with a leveling of S. The items of S_i will be put in the *increasing order* in a linked list denoted by L_i. We attach the special keys $-\infty$ at the beginning and $+\infty$ at the end of each list L_i as sentinels. In practice, $-\infty$ is a key value that is smaller than any key we consider in the application and $+\infty$ denotes a key value larger than all the possible keys. A leveling of S is implemented by maintaining all the lists $L_0, L_1, L_2, \ldots, L_r$ with some more additional links as shown in Figure 14.1. Specifically, the box containing a key k in L_i will have a pointer to the box containing k in L_{i-1}. We call such pointers *descent pointers*. The links connecting items of the same list are called *horizontal pointers*. Let B be a pointer to a box in the skip list. We use the notations *Hnext*[B], and *Dnext*[B], for the horizontal and descent pointers of the box pointed by B respectively. The notation *key*[B] is used to denote the key stored in the box pointed by B. The name of the skip list is nothing but a pointer to the box containing $-\infty$ in the rth level as shown

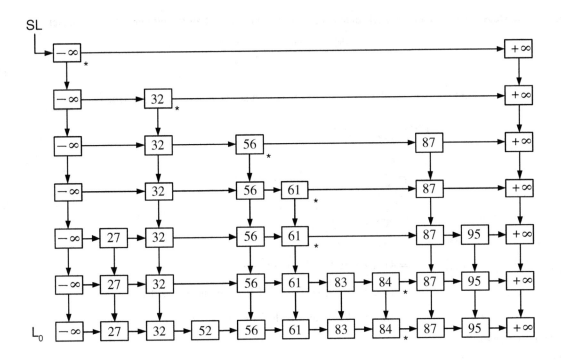

FIGURE 14.1 A Skip List. The starred nodes are marked by Mark(86,SL).

in the figure. From the Figure 14.1 it is clear that L_i has horizontal pointers that skip over several intermediate elements of L_{i-1}. That is why this data structure is called the *Skip List*.

How do we arrive at a leveling for a given set? We may construct S_{i+1} from S_i in a systematic, deterministic way. While this may not pose any problem for a static search problem, it will be extremely cumbersome to implement the dynamic dictionary operations. This is where randomness is helpful. To get S_{i+1} from S_i, we do the following. For each element k of S_i toss a coin and include k in S_{i+1} iff we get *success* in the coin toss. If the coin is p-biased, we expect $|S_{i+1}|$ to be roughly equal to $p|S_i|$. Starting from S, we may repeatedly obtain sets corresponding to successive levels. Since the coin tosses are independent, there is another useful, and equivalent way to look at our construction. For each key k in S, keep tossing a coin until we get a *failure*. Let h be the number of successes before the first failure. Then, we include k in h further levels treating $S = S_0$ as the base level. In other words, we include k in the sets $S_1, S_2, \ldots, S_h$. This suggests us to define a random variable Z_i for each key $k_i \in S_0$ to be the number of times we toss a p-biased coin before we obtain a *failure*. Since Z_i denotes the number of additional copies of k_i in the skip list, the value $maximum\{Z_i : 1 \leq i \leq n\}$ defines the highest nonempty level. Hence, we have,

$$r = 1 + maximum\{Z_i : 1 \leq i \leq n\} \tag{14.8}$$

$$|SL| = n + Z_1 + Z_2 + \cdots + Z_n + 2r + 2 \tag{14.9}$$

where r is the number of levels and $|SL|$ is the number of boxes or the space complexity measure of the *Skip List*. In the expression for $|SL|$ we have added n to count the keys in the base level and $2r + 2$ counts the sentinel boxes containing $+\infty$ and $-\infty$ in each level.

14.4 Structural Properties of Skip Lists

14.4.1 Number of Levels in Skip List

Recall that $r = 1 + max_i\{Z_i\}$. Notice that $Z_i \geq k$ iff the coin tossed for k_i gets a run of at least k successes right from the beginning and this happens with probability p^k. Since $r \geq k$ iff at least one of $Z_i \geq k - 1$, we easily get the following fact from Boole's inequality

$$Pr(r \geq k) \leq np^{k-1}$$

Choosing $k = 4\log_{1/p} n + 1$, we immediately obtain a high confidence bound for the number of levels. In fact,

$$Pr(r \geq 4\log n + 1) \leq nn^{-4}$$

$$= \frac{1}{n^3} \qquad (14.10)$$

We obtain an estimation for the expected value of r, using the formula stated in theorem (14.3) as follows:

$$E(r) = \sum_{i=1}^{\infty} Pr(r \geq i)$$

$$= \sum_{i=1}^{4\log n} Pr(r \geq i) + \sum_{i>4\log n} Pr(r \geq i)$$

$$\leq \sum_{i=1}^{4\log n} 1 + \sum_{i>4\log n}^{\infty} np^{i-1}$$

$$= 4\log n + np^{4\log n}(1 + p + p^2 + \cdots)$$

$$= 4\log n + n \cdot \frac{1}{n^4} \cdot (1-p)^{-1} \text{ if base of the log is } 1/p$$

$$\leq 4\log n + 1 \text{ for sufficiently large } n$$

Thus $E(r) = O(\log n)$
Hence,

Theorem 14.11

The expected number of levels in a skip list of n elements is $O(\log n)$. In fact, the number is $O(\log n)$ with high probability.

14.4.2 Space Complexity

Recall that the space complexity, $|SL|$ is given by

$$|SL| = Z_1 + Z_2 + \cdots + Z_n + n + 2r + 2.$$

As we know that $r = O(\log n)$ with high probability, let us focus on the sum

$$Z = Z_1 + Z_2 + \cdots + Z_n.$$

Since Z_i is a geometric random variable with parameter p, Z is a negative binomial random variable by theorem 14.6.
Thus $E(Z) = \frac{n}{p} = O(n)$.
We can in fact show that Z is $O(n)$ with very high probability.
Now, from theorem (14.10) $Pr(Z > 4n) = Pr(X < n)$ where X is a binomial distribution with parameters $4n$ and p. We now assume that $p = 1/2$ just for the sake of simplicity in arithmetic.
In the first Chernoff bound mentioned in theorem (14.9), replacing n, μ and k respectively with $4n$, $2n$ and n, we obtain,

$$Pr(X < n) \leq \left(\frac{2n}{n}\right)^n \left(\frac{2n}{3n}\right)^{3n}$$

$$= \left(2 \cdot \frac{2^3}{3^3}\right)^n$$

$$= \left(\frac{16}{27}\right)^n$$

$$= \left(\frac{27}{16}\right)^{-n}$$

This implies that $4n$ is in fact a very high confidence bound for Z. Since $|SL| = Z + n + 2r + 2$, we easily conclude that

Theorem 14.12

The space complexity of a skip list for a set of size n is $O(n)$ with very high probability.

14.5 Dictionary Operations

We shall use a simple procedure called *Mark* in all the dictionary operations. The procedure *Mark* takes an arbitrary value x as input and marks in each level the box containing the largest key that is less than x. This property implies that insertion, deletion or search should all be done next to the marked boxes. Let $M_i(x)$ denote the box in the ith level that is marked by the procedure call *Mark(x, SL)*. Recall the convention used in the linked structure that name of a box in a linked list is nothing but a pointer to that box. The keys in the marked boxes $M_i(x)$ satisfy the following condition :

$$key[M_i(x)] < x \le key[Hnext[M_i(x)]] \quad \text{for all } 0 \le i \le r. \tag{14.11}$$

The procedure *Mark* begins its computation at the box containing $-\infty$ in level r. At any current level, we traverse along the level using horizontal pointers until we find that the next box is containing a key larger or equal to x. When this happens, we mark the current box and use the *descent pointer* to reach the next lower level and work at this level in the same way. The procedure stops after marking a box in level 0. See Figure 14.1 for the nodes marked by the call *Mark(86,SL)*.

```
Algorithm Mark(x,SL)
-- x is an arbitrary value.
-- r is the number of levels in the skip list SL.
1. Temp = SL.
2. For i = r down to 0 do
   While (key[Hnext[Temp]] < x)
       Temp = Hnext[Temp];
   Mark the box pointed by Temp;
   Temp = Dnext[Temp];
3. End.
```

We shall now outline how to use marked nodes for *Dictionary* operations. Let $M_0(x)$ be the box in level 0 marked by *Mark(x)*. It is clear that the box next to $M_0(x)$ will have x iff $x \in S$. Hence our algorithm for *Search* is rather straight forward.

To insert an item in the *Skip List*, we begin by marking the *Skip List* with respect to the value x to be inserted. We assume that x is not already in the *Skip List*. Once the marking is done, inserting x is very easy. Obviously x is to be inserted next to the marked boxes. But in how many levels we insert x? We determine the number of levels by tossing a coin on behalf of x until we get a *failure*. If we get h successes, we would want x to be inserted in all levels from 0 to h. If $h \ge r$, we simply insert x at all the existing levels and create a new level consisting of only $-\infty$ and $+\infty$ that corresponds to the empty set. The insertion will be done starting from the base level. However, the marked boxes are identified starting from the highest level. This means, the *Insert* procedure needs the marked boxes in the reverse order of its generation. An obvious strategy is to push the marked boxes in a stack as and when they are marked in the *Mark* procedure. We may pop from the stack as many marked boxes as needed for insertion and then insert x next to each of the popped boxes.

The deletion is even simpler. To delete a key x from S, simply mark the *Skip List* with respect to x. Note that if x is in L_i, then it will be found next to the marked box in L_i. Hence, we can use the *horizontal pointers* in the marked boxes to delete x from each level where x is present. If the deletion creates one or more empty levels, we ignore all of them and retain only one level corresponding to the empty set. In other words, the number of levels in the *Skip List* is reduced in this case. In fact, we may delete an item "on the fly" during the execution of the *Mark* procedure itself. As the details are simple we omit pseudo codes for these operations.

14.6 Analysis of Dictionary Operations

It is easy to see that the cost of *Search*, *Insert* and *Delete* operations are dominated by the cost of *Mark* procedure. Hence we shall analyze only the *Mark* procedure. The *Mark* procedure starts at the 'r'th level and proceeds like a downward walk on a staircase and ends at level 0. The complexity of *Mark* procedure is clearly proportional to the number of edges it traversed. It is advantageous to view the same path as if it is built from level 0 to level r. In other words, we analyze the building of the path in a direction opposite to the direction in which it is actually built by the procedure. Such an analysis is known as *backward analysis*.

Henceforth let P denote the path from level 0 to level r traversed by *Mark(x)* for the given fixed x. The path P will have several vertical edges and horizontal edges. (Note that at every box either P moves vertically above or moves horizontally to the left). Clearly, P has r vertical edges. To estimate number of horizontal edges, we need the following lemmas.

Lemma 14.6

Let the box b containing k at level i be marked by Mark(x). Let a box w containing k be present at level $i+1$. Then, w is also marked.

Proof. Since b is marked, from Equation 14.11, we get that there is no value between k and x in level i. This fact holds good for L_{i+1} too because $S_{i+1} \subseteq S_i$. Hence the lemma.   ∎

Lemma 14.7

Let the box b containing k at level i be marked by Mark(x). Let $k \notin L_{i+1}$. Let u be the first box to the left of b in L_i having a "vertical neighbor" w. Then w is marked.

Proof. Let $w.key = u.key = y$. Since b is marked, k satisfies condition (14.11). Since u is the first node in the left of b having a vertical neighbor, none of the keys with values in between y and x will be in L_{i+1}. Also, $k \notin L_{i+1}$ according to our assumption in the lemma. Thus y is the element in L_{i+1} that is just less than x. That is, y satisfies the condition (14.11) at level $i+1$. Hence the w at level $i+1$ will be marked.   ∎

Lemmas (14.6 and 14.7) characterize the segment of P between two successive marked boxes. This allows us to give an incremental description of P in the following way.

P starts from the marked box at level 0. It proceeds vertically as far as possible (lemma 14.6) and when it cannot proceed vertically it proceeds horizontally. At the "first" opportunity to move vertically, it does so (lemma 14.7), and continues its vertical journey. Since for any box a vertical neighbor exists only with a probability p, we see that P proceeds from the current box vertically with probability p and horizontally with probability $(1-p)$.

Hence, the number of horizontal edges of P in level i is a geometric random variable, say, Y_i, with parameter $(1-p)$. Since the number of vertical edges in P is exactly r, we conclude,

Theorem 14.13

The number of edges traversed by Mark(x) for any fixed x is given by $|P| = r + (Y_0 + Y_1 + Y_2 + \cdots + Y_{r-1})$ where Y_i is a geometric random variable with parameters $1-p$ and r is the random variable denoting the number of levels in the Skip List.

Our next mission is to obtain a high confidence bound for the size of P. As we have already derived high confidence bound for r, let focus on the sum $H_r = Y_0 + Y_1 + \cdots + Y_{r-1}$ for a while. Since r is a random variable H_r is not a deterministic sum of random variables but a *random sum* of random variables.

Hence we cannot directly apply theorem (14.6) and the bounds for negative binomial distributions.

Let X be the event of the random variable r taking a value less than or equal to $4 \log n$. Note that $P(\overline{X}) < \frac{1}{n^3}$ by (14.10).

From the law of total probability, Boole's inequality and equation (14.10) we get,

$$
\begin{aligned}
Pr(H_r > 16 \log n) &= Pr([H_r > 16 \log n] \cap X) + Pr([H_r > 16 \log n] \cap \overline{X}) \\
&= Pr([H_r > 16 \log n] \cap r \leq 4 \log n) + Pr([H_r > 16 \log n] \cap \overline{X}) \\
&\leq \sum_{k=0}^{4 \log n} Pr(H_k > 16 \log n) + Pr(\overline{X}) \\
&\leq (1 + 4 \log n) Pr(H_{4 \log n} > 16 \log n) + \frac{1}{n^3}
\end{aligned}
$$

Now $Pr(H_{4 \log n} > 16 \log n)$ can be computed in a manner identical to the one we carried out in the space complexity analysis. Notice that $H_{4 \log n}$ is a deterministic sum of geometric random variables. Hence we can apply theorems (14.9) and (14.10) to derive a high confidence bound. Specifically, by theorem (14.10),

$$
Pr(H_{4 \log n} > 16 \log n) = Pr(X < 4 \log n),
$$

where X is a binomial random variable with parameters $16 \log n$ and p. Choosing $p = 1/2$ allows us to set $\mu = 8 \log n$, $k = 4 \log n$ and replace n by $16 \log n$ in the first inequality of theorem (14.9). Putting all these together we obtain,

$$
\begin{aligned}
Pr(H_{4\log n} > 16 \log n) &= Pr(X < 4 \log n) \\
&\leq \left(\frac{8 \log n}{4 \log n} \right)^{4 \log n} \left(\frac{8 \log n}{12 \log n} \right)^{12 \log n} \\
&= \left(2 \cdot \frac{2^3}{3^3} \right)^{4 \log n} \\
&= \left(\frac{16}{27} \right)^{4 \log n} \\
&< \left(\frac{1}{8} \right)^{\log n} \\
&= \frac{1}{n^3}
\end{aligned}
$$

Therefore

$$
Pr(H_r > 16 \log n) < (1 + 4 \log n) \left(\frac{1}{n^3} \right) + \frac{1}{n^3}
$$

$$
< \frac{1}{n^2} \quad \text{if } n \geq 32
$$

This completes the derivation of a high confidence bound for H_r. From this, we can easily obtain a bound for expected value of H_r. We use theorem (14.3) and write the expression for $E(H_r)$ as

$$
E(H_r) = \sum_{i=1}^{16 \log n} Pr(H_r \geq i) + \sum_{i > 16 \log n} Pr(H_r \geq i).
$$

The first sum is bounded above by $16 \log n$ as each probability value is less than 1 and by the high confidence bound that we have established just now, we see that the second sum is dominated by $\sum_{i=1}^{\infty} 1/i^2$ which is a constant. Thus we obtain,

$$
E(H_r) \leq 16 \log n + c = O(\log n).
$$

Since $P = r + H_r$ we easily get that $E(|P|) = O(\log n)$ and $|P| = O(\log n)$ with probability greater than $1 - O(\frac{1}{n^2})$. Observe that we have analyzed $Mark(x)$ for a given fixed x. To show that the high confidence bound for any x, we need to proceed little further. Note that there are only $n + 1$ distinct paths possible with respect to the set $\{ -\infty = k_0, k_1, \ldots, k_n, k_{n+1} = +\infty \}$, each corresponding to the x lying in the internal $[k_i, k_{i+1}), i = 0, 1, \ldots, n$.

Therefore, for any x, $Mark(x)$ walks along a path P satisfying $E(|P|) = O(\log n)$ and $|P| = O(\log n)$ with probability greater than $1 - O(\frac{1}{n})$.

Summarizing,

Theorem 14.14

The Dictionary operations Insert, Delete, and Search take $O(\log n)$ expected time when implemented using Skip Lists. Moreover, the running time for Dictionary operations in a Skip List is $O(\log n)$ with high probability.

14.7 Randomized Binary Search Trees

A *Binary Search Tree* (BST) for a set S of keys is a binary tree satisfying the following properties.

a. Each node has exactly one key of S. We use the notation $v.key$ to denote the key stored at node v.
b. For all node v and for all nodes u in the left subtree of v and for all nodes w in the right subtree of v, the keys stored in them satisfy the so called *search tree property*:

$$
u.key < v.key < w.key
$$

The complexity of the dictionary operations are bounded by the height of the binary search tree. Hence, ways and means were explored for controlling the height of the tree during the course of execution. Several clever balancing schemes were discovered with varying degrees of complexities. In general, the implementation of balancing schemes are tedious and we may have to perform a number of rotations which are complicated operations. Another line of research explored the potential of possible randomness in the input data. The idea was to completely avoid balancing schemes and hope to have "*short*" trees due to randomness in input. When only random insertion are done, we obtain so called *Randomly Built Binary Tree* (RBBT). RBBTs have been shown to have $O(\log n)$ expected height.

What is the meaning of random insertion? Suppose we have already inserted the values $a_1, a_2, a_3, \ldots, a_{k-1}$. These values, when considered in sorted order, define k intervals on the real line and the new value to be inserted, say x, is equally likely to be in any of the k intervals.

The first drawback of RBBT is that this assumption may not be valid in practice and when this assumption is not satisfied, the resulting tree structure could be highly skewed and the complexity of search as well as insertion could be as high as $O(n)$. The second major drawback is when deletion is also done on these structures, there is a tremendous degradation in the performance. There is no theoretical results available and extensive empirical studies show that the height of an RBBT could grow to $O(\sqrt{n})$ when we have arbitrary mix of insertion and deletion, even if the randomness assumption is satisfied for inserting elements. Thus, we did not have a satisfactory solution for nearly three decades.

In short, the randomness was not preserved by the deletion operation and randomness preserving binary tree structures for the dictionary operations was one of the outstanding open problems, until an elegant affirmative answer is provided by Martinez and Roura in their landmark paper [4].

In this section, we shall briefly discuss about structure proposed by Martinez and Roura.

Definition 14.15: (Randomized Binary Search Trees)

Let T be a binary search tree of size n. If $n = 0$, then $T = NULL$ and it is a random binary search tree. If $n > 0$, T is a random binary search tree iff both its left subtree L and right subtree R are independent random binary search trees and

$$\Pr\{Size(L) = i | Size(T) = n\} = \frac{1}{n}, 0 \le i \le n.$$

The above definition implies that every key has the same probability of $\frac{1}{n}$ for becoming the root of the tree. It is easy to prove that the expected height of a RBST with n nodes is $O(\log n)$. The RBSTs possess a number of interesting structural properties and the classic book by Mahmoud [5] discusses them in detail. In view of the above fact, it is enough if we prove that when insert or delete operation is performed on a RBST, the resulting tree is also an RBST.

14.7.1 Insertion in RBST

When a key x is inserted in a tree T of size n, we obtain a tree T' of size $n + 1$. For T', as we observed earlier, x should be in the root with probability $\frac{1}{n+1}$. This is our starting point.

```
Algorithm Insert(x, T)
- L is the left subtree of the root
- R is the right subtree of the root
1. n = size(T);
2. r = random(0, n);
3. If (r = n) then
     Insert_at_root(x, T);
4. If (x < key at root of T) then
     Insert(x, L);
   Else
     Insert(x, R);
```

To insert x as a root of the resulting tree, we first split the current tree into two trees labeled $T_<$ and $T_>$, where $T_<$ contains all the keys in T that are smaller than x and $T_>$ contains all the keys in T that are larger than x. The output tree T' is constructed by placing x at the root and attaching $T_<$ as its left subtree and $T_>$ as its right subtree. The algorithm for splitting a tree T into $T_<$ and $T_>$ with respect to a value x is similar to the partitioning algorithm done in *quicksort*. Specifically, the algorithm *split(x, T)* works as follows. If T is empty, nothing needs to be done; both $T_<$ and $T_>$ are empty. When T is non-empty we compare x with Root(T).key. If $x <$ Root(T).key, then root of T as well as the right subtree of T belong to $T_>$. To compute $T_<$ and the remaining part of $T_>$ we recursively call *split(x, L)*, where L is the left subtree for the root of T. If $x >$ Root(T).key, $T_<$ is built first and recursion proceeds with *split(x, R)*. The details are left as easy exercise.

We shall first prove that $T_<$ and $T_>$ are independent *Random Binary Search Trees*. Formally,

Theorem 14.15

Let $T_<$ and $T_>$ be the BSTs produced by split(x, T). If T is a random BST containing the set of keys S, then $T_<$ and $T_>$ are RBBTs containing the keys $S < = \{y \in S | y < x\}$ and $S > = \{y \in S | y > x\}$, respectively.

Proof. Let size(T) $= n > 0$, $x >$ Root(T).key, we will show that for any $z \in S<$, the probability that z is the root of $T_<$ is $1/m$ where $m =$ size($T_<$). In fact,

$$\Pr(z \text{ is root of } T_< \mid \text{ root of } T \text{ is less than } x)$$
$$= \frac{\Pr(z \text{ is root of } T_< \text{ and root of } T \text{ is less than } x)}{\Pr(\text{root of } T \text{ is less than } x)}$$
$$= \frac{1/n}{m/n} = \frac{1}{m}.$$

The independence of $T_<$ and $T_>$ follows from the independence of L and R of T and by induction. ∎

We are now ready to prove that randomness is preserved by insertion. Specifically,

Theorem 14.16

Let T be a RBST for the set of keys S and $x \notin S$, and assume that insert(s, T) produces a tree, say T' for $S \cup \{x\}$. Then, T' is a RBST.

Proof. A key $y \in S$ will be at the root of T' iff

1. y is at root of T.
2. x is not inserted at the root of T'

As (1) and (2) are independent events with probability $\frac{1}{n}$ and $\frac{n}{n+1}$, respectively, it follows that Prob(y is at root of T') $= \frac{1}{n} \cdot \frac{n}{n+1} = \frac{1}{n+1}$. The key x can be at the root of T' only when *insert(x, T)* invokes *insert-at-root(x, T)* and this happens with probability $\frac{1}{n+1}$. Thus, any key in $S \cup \{x\}$ has the probability of $\frac{1}{n+1}$ for becoming the root of T'. The independence of left and right subtrees of T' follows from independence of left and right subtrees of T, induction, and the previous theorem. ∎

14.7.2 Deletion in RBST

Suppose $x \in T$ and let T_x denote the subtree of T rooted at x. Assume that L and R are the left and right subtrees of T_x. To delete x, we build a new BST $T_x =$ Join(L, R) containing the keys in L and R and replace T_x by T_x. Since pseudocode for deletion is easy to write, we omit the details.

We shall take a closer look at the details of the *Join* routine. We need couple of more notations to describe the procedure *Join*. Let L_l and L_r denote the left and right subtrees of L and R_l and R_r denote the left and right subtrees of R, respectively. Let a denote the root of L and b denote the root of R. We select either a or b to serve as the root of T_x with appropriate probabilities. Specifically, the probability for a becoming the root of T_x is $\frac{m}{m+n}$ and for b it is $\frac{n}{n+m}$ where $m =$ size(L) and $n =$ size(R).

If a is chosen as the root, the its left subtree L_l is not modified while its right subtree L_r is replaced with *Join(L_r, R)*. If b is chosen as the root, then its right subtree R_r is left intact but its left subtree R_l is replaced with *Join(L, R_l)*. The join of an empty tree with another tree T is T it self and this is the condition used to terminate the recursion.

```
Algorithm Join(L,R)
-- L and R are RBSTs with roots a and b and size m and n respectively.
-- All keys in L are strictly smaller than all keys in R.
-- $L_1$ and $L_r$ respectively denote the left and right subtree of L.
-- $R_1$ and $R_r$ are similarly defined for R.

1. If ( L is NULL) return R.
2. If ( R is NULL) return L.
3. Generate a random integer i in the range [0, n+m-1].
```

```
4. If ( i < m ) {* the probability for this event is m/(n+m).*}
     L_r = Join(L_l,R);
     return L;
   else    {* the probability for this event is n/(n+m).*}
     R_l = Join(L,R_l);
     return R;
```

It remains to show that *Join* of two RBSTs produces RBST and deletion preserves randomness in RBST.

Theorem 14.17

The Algorithm Join(L,R) produces a RBST under the conditions stated in the algorithm.

Proof. We show that any key has the probability of $1/(n+m)$ for becoming the root of the tree output by *Join(L,R)*. Let x be a key in L. Note that x will be at the root of *Join(L,R)* iff

- x was at the root of L before the *Join* operation, and,
- The root of L is selected to serve as the root of the output tree during the *Join* operation.

The probability for the first event is $1/m$ and the second event occurs with probability $m/(n+m)$. As these events are independent, it follows that the probability of x at the root of the output tree is $\frac{1}{m} \cdot \frac{m}{n+m} = \frac{1}{n+m}$. A similar reasoning holds good for the keys in R. ∎

Finally,

Theorem 14.18

If T is a RBST for a set K of keys, then Delete(x,T) outputs a RBST for $K - \{x\}$.

Proof. We sketch only the chance counting argument. The independence of the subtrees follows from the properties of Join operation and induction. Let $T' = Delete(x, T)$. Assume that $x \in K$ and size of K is n. We have to prove that for any $y \in K, y \neq x$, the probability for y in root of T' is $1/(n-1)$. Now, y will be at the root of T' iff either x was at the root of T and its deletion from T brought y to the root of T' or y was at the root of T (so that deletion of x from T did not dislocate y). The former happens with a probability $\frac{1}{n} \cdot \frac{1}{n-1}$ and the probability of the later is $\frac{1}{n}$. As these events are independent, we add the probabilities to obtain the desired result. ∎

14.8 Bibliographic Remarks

In this chapter we have discussed two randomized data structures for Dictionary ADT. Skip Lists are introduced by Pugh in 1990 [3]. A large number of implementations of this structure by a number of people available in the literature, including the one by the inventor himself. Sedgewick gives an account of the comparison of the performances of Skip Lists with the performances of other balanced tree structures [6]. See [7] for a discussion on the implementation of other typical operations such as merge. Pugh argues how Skip Lists are more efficient in practice than balanced trees with respect to Dictionary operations as well as several other operations. Pugh has further explored the possibilities of performing concurrent operations on Skip Lists in Reference [8]. For a more elaborate and deeper analysis of Skip Lists, refer the PhD thesis of Papadakis [9]. In his thesis, he has introduced deterministic skip lists and compared and contrasted the same with a number of other implementations of Dictionaries. Sandeep Sen [10] provides a crisp analysis of the structural properties of Skip Lists. Our analysis presented in this chapter is somewhat simpler than Sen's analysis and our bounds are derived based on different tail estimates of the random variables involved.

The randomized binary search trees are introduced by Martinez and Roura in their classic paper [4] which contains many more details than discussed in this chapter. In fact, we would rate this paper as one of the best written papers in data structures. Seidel and Aragon have proposed a structure called probabilistic priority queues [11] and it has a comparable performance. However, the randomness in their structure is achieved through randomly generated real numbers (called priorities) while the randomness in Martinez and Roura's structure is inherent in the tree itself. Besides this being simpler and more elegant, it solves one of the outstanding open problems in the area of search trees.

References

1. P. Raghavan and R. Motwani, *Randomized Algorithms*, Cambridge University Press, USA, 1995.
2. T. Hagerup and C. Rub, A guided tour of Chernoff bounds, *Information Processing Letters*, 33:305–308, 1990.
3. W. Pugh, Skip Lists: A probabilistic alternative to balanced trees, *Communications of the ACM*, 33:668–676, 1990.
4. C. Martinez and S. Roura, Randomized binary search trees, *Journal of the ACM*, 45:288–323, 1998.
5. H. M. Mahmoud, *Evolution of Random Search Trees*, Wiley Interscience, USA, 1992.
6. R. Sedgewick, *Algorithms*, Third edition, Addison-Wesley, USA, 1998.
7. W. Pugh, *A Skip List cook book*, Technical report, CS-TR-2286.1, University of Maryland, USA, 1990.
8. W. Pugh, *Concurrent maintenance of Skip Lists*, Technical report, CS-TR-2222, University of Maryland, USA, 1990.
9. Th. Papadakis, *Skip List and probabilistic analysis of algorithms*, PhD Thesis, University of Waterloo, Canada, 1993. (Available as Technical Report CS-93-28).
10. S. Sen, Some observations on Skip Lists, *Information Processing Letters*, 39:173–176, 1991.
11. R. Seidel and C. Aragon, Randomized Search Trees, *Algorithmica*, 16:464–497, 1996.

15

Trees with Minimum Weighted Path Length

15.1	Introduction	215
15.2	Huffman Trees	216
	$O(n \log n)$ Time Algorithm • Linear Time Algorithm for Presorted Sequence of Items • Relation between General Uniquely Decipherable Codes and Prefix-free Codes • Huffman Codes and Entropy • Huffman Algorithm for t-ary Trees	
15.3	Height Limited Huffman Trees	220
	Reduction to the Coin Collector Problem • The Algorithm for the Coin Collector Problem	
15.4	Optimal Binary Search Trees	222
	Approximately Optimal Binary Search Trees	
15.5	Optimal Alphabetic Tree Problem	224
	Computing the Cost of Optimal Alphabetic Tree • Construction of Optimal Alphabetic Tree • Optimal Alphabetic Trees for Presorted Items	
15.6	Optimal Lopsided Trees	226
15.7	Parallel Algorithms	230
	Basic Property of Left-Justified Trees	
	References	231

Wojciech Rytter
Warsaw University

15.1 Introduction

The concept of the "weighted path length" is important in data compression and searching. In case of data compression lengths of paths correspond to lengths of code-words. In case of searching they correspond to the number of elementary searching steps. By a *length of a path* we mean usually its number of edges.

Assume we have n weighted items, where w_i is the non-negative *weight* of the ith item. We denote the sequence of weights by $S = (w_1 \ldots w_n)$. We adopt the convention that the items have unique names. When convenient to do so, we will assume that those names are the positions of items in the list, namely integers in $[1 \ldots n]$.

We consider a binary tree T, where the items are placed in vertices of the trees (in leaves only or in every node, depending on the specific problem). We define the minimum weighted path length (cost) of the tree T as follows:

$$cost(T) = \sum_{i=1}^{n} w_i level_T(i)$$

where $level_T$ is the *level function* of T, that is, $level_T(i)$ is the level (or depth) of i in T, defined to be the length of the path in T from the root to i.

In some special cases (lopsided trees) the edges can have different lengths and the path length in this case is the sum of individual lengths of edges on the path.

In this chapter we concentrate on several interesting algorithms in the area:

- Huffman algorithm constructing optimal prefix-free codes in time $O(n \log n)$, in this case the items are placed in leaves of the tree, the original order of items can be different from their order as leaves;
- A version of Huffman algorithm which works in $O(n)$ time if the weights of items are sorted
- Larmore-Hirschberg algorithm for optimal height-limited Huffman trees working in time $O(n \times L)$, where L is the upper bound on the height, it is an interesting algorithm transforming the problem to so called "coin-collector," see [1].

- Construction of optimal binary search trees (OBST) in $O(n^2)$ time using certain property of monotonicity of "splitting points" of subtrees. In case of OBST every node (also internal) contains exactly one item. (Binary search trees are defined in Chapter 3.)
- Construction of optimal alphabetic trees (OAT) in $O(n \log n)$ time: the Garsia-Wachs algorithm [2]. It is a version of an earlier algorithm of Hu-Tucker [3,4] for this problem. The correctness of this algorithm is nontrivial and this algorithm (as well as Hu-Tucker) and these are the most interesting algorithms in the area.
- Construction of optimal lopsided trees, these are the trees similar to Huffman trees except that the edges can have some lengths specified.
- Short discussion of parallel algorithms

Many of these algorithms look "mysterious," in particular the Garsia-Wachs algorithm for optimal alphabetic trees. This is the version of the Hu-Tucker algorithm. Both algorithms are rather simple to understand in how they work and their complexity, but correctness is a complicated issue.

Similarly one can observe a mysterious behavior of the Larmore-Hirschberg algorithm for height-limited Huffman trees. Its "mysterious" behavior is due to the strange reduction to the seemingly unrelated problem of the *coin collector*.

The algorithms relating the cost of binary trees to shortest paths in certain graphs are also not intuitive, for example the algorithm for lopsided trees, see [5], and parallel algorithm for alphabetic trees, see [6]. The efficiency of these algorithms relies on the *Monge property* of related matrices. Both sequential and parallel algorithms for Monge matrices are complicated and interesting.

The area of weighted paths in trees is especially interesting due to its applications (compression, searching) as well as to their relation to many other interesting problems in combinatorial algorithmics.

15.2 Huffman Trees

Assume we have a text x of length N consisting of n different letters with repetitions. The alphabet is a finite set Σ. Usually we identify the i-th letter with its number i. The letter i appears w_i times in x. We need to encode each letter in binary, as $h(a)$, where h is a morphism of alphabet Σ into binary words, in a way to minimize the total length of encoding and guarantee that it is uniquely decipherable, this means that the extension of the morphism h to all words over Σ is one-to-one. The words $h(a)$, where $a \in \Sigma$, are called codewords or codes.

The special case of uniquely decipherable codes are *prefix-free* codes: none of the code is a prefix of another one. The prefix-free code can be represented as a binary tree, with left edges corresponding to zeros, and right edge corresponding to ones.

Let $S = \{w_1, w2, \ldots, w_n\}$ be the sequence of weights of items. Denote by *HuffmanCost(S)* the total cost of minimal encoding (weighted sum of lengths of code-words) and by HT(S) the tree representing an optimal encoding. Observe that several different optimal trees are possible. The basic algorithm is a greedy algorithm designed by Huffman, the corresponding trees and codes are called Huffman trees and Huffman codes.

Example Let *text* = *abracadabra*. The number of occurrences of letters are

$$w_a = 5, w_b = 2, w_c = 1, w_d = 1, w_r = 2.$$

We treat letters as items, and the sequence of weights is:

$$S = (5, 2, 1, 1, 2)$$

An optimal tree of a prefix code is shown in Figure 15.1. We have, according to the definition of weighted path length:

$$HuffmanCost(S) = 5*1 + 2*2 + 1*4 + 1*4 + 2*3 = 23$$

The corresponding prefix code is:

$$h(a) = 0, h(b) = 10, h(c) = 1100, h(d) = 1101, h(r) = 111.$$

We can encode the original text *abracadabra* using the codes given by paths in the prefix tree. The coded text is then 01011101100011010101110, that is a word of length 23.

If for example the initial code words of letters have length 5, we get the compression ratio $55/23 \approx 2.4$.

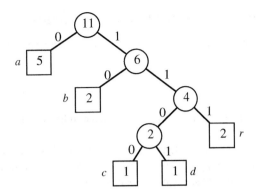

FIGURE 15.1   A Huffman tree T for the items a, b, c, d, r and the weight sequence $S = (5, 2, 1, 1, 2)$. The numbers in internal nodes are sums of weights of leaves in corresponding subtrees. Observe that weighted path length of the tree is the total sum of values in internal nodes. Hence *HuffmanCost*$(S) = 2 + 4 + 6 + 11 = 23$.

15.2.1 $O(n \log n)$ Time Algorithm

The basic algorithm is the *greedy* algorithm given by Huffman. In this algorithm we can assume that two items of smallest weight are at the bottom of the tree and they are sons of a same node. The *greedy* approach in this case is to minimize the cost *locally*.

Two smallest weight items are *combined* into a single *package* with weight equal to the sum of weights of these small items. Then we proceed recursively. The following observation is used.

Observation Assume that the numbers in internal nodes are sums of weights of leaves in corresponding subtrees. Then the total weighted path length of the tree is the total sum of values in internal nodes.

Due to the observation we have for $|S| > 1$:

$$HuffmanCost(S) = HuffmanCost(S - \{u, w\}) + u + w,$$

where u, w are two minimal elements of S. This implies the following algorithm, in which we assume that initially S is stored in a min-priority queue. The algorithm is presented below as a recursive function *HuffmanCost*(S) but it can be easily written without recursion. The algorithm computes only the total cost.

Theorem 15.1

Huffman algorithm constructs optimal tree in $O(n \log n)$ time

Proof. In an optimal tree we can exchange two items which are sons of a same father at a bottom level with two smallest weight items. This will not increase the cost of the tree. Hence there is an optimal tree with two smallest weight items being sons of a same node. This implies correctness of the algorithm.

The complexity is $O(n \log n)$ since each operation in the priority queue takes $O(\log n)$ time and there are $O(n)$ operations of Extract-Min. ∎

---

function *HuffmanCost*(S)
{Huffman algorithm: recursive version}
{computes only the cost of minimum weighted tree}

1. if S contains only one element u then return 0;
2. $u = $ Extract Min(S); $w = $ ExtractMin(S);
3. insert$(u+w, S)$;
4. **return** HuffmanCost$(S) + u + w$

The algorithm in fact computes only the cost of Huffman tree, but the tree can be created on-line in the algorithm. Each time we combine two items we create their father and create links son-father. In the course of the algorithm we keep a forest (collection of trees). Eventually this becomes a single Huffman tree at the end.

15.2.2 Linear Time Algorithm for Presorted Sequence of Items

There is possible an algorithm using "simple" queues with constant time operations of inserting and deleting elements from the queue if the items are *presorted*.

Theorem 15.2

If the weights are already sorted then the Huffman tree can be constructed in linear time.

Proof. If we have sorted queues of remaining original items and remaining newly created items (packages) then it is easy to see that two smallest items can be chosen from among 4 items, two first elements of each queue. This proves correctness.

Linear time follows from constant time costs of single queue operations.   ∎

Linear-Time Algorithm {linear time computation of the cost for presorted items}

1. initialize empty queues Q, S;
   total_cost = 0;
2. place original n weights into nondecreasing order into S;
   the smallest elements at the front of S;
3. **while** $|Q| + |S| > 2$ **do** {
         let u, w be the smallest elements chosen from the
         first two elements of Q and of S;
         remove u, w from $Q \cup S$; insert($u + w$, Q);
         $total\_cost = total\_cost + (u + w)$;}
4. **return** total_cost

15.2.3 Relation between General Uniquely Decipherable Codes and Prefix-free Codes

It would seem that, for some weight sequences, in the class of uniquely decipherable codes there are possible codes which beat every Huffman (prefix-free) code. However it happens that prefix-free codes are optimal within the whole class of uniquely decipherable codes. It follows immediately from the next three lemmas.

Lemma 15.1

For each full (each internal node having exactly two sons) binary tree T with leaves $1 \ldots n$ we have:

$$\sum_{i=1}^{n} 2^{-level_T(i)} = 1$$

Proof. Simple induction on n.   ∎

Lemma 15.2

For each uniquely decipherable code S with word lengths $\ell_1, \ell_2, \ldots, \ell_k$ on the alphabet $\{0, 1\}$ we have :

$$\sum_{i=1}^{k} 2^{-\ell_i} \leq 1$$

Proof. For a set W of words on the alphabet $\{0, 1\}$ define:

$$C(W) = \sum_{x \in W} 2^{-|x|}$$

We have to show that $C(S) \leq 1$. Let us first observe the following simple fact.

Observation.

If S is uniquely decipherable then $C(S)^n = C(S^n)$ for all $n \geq 1$.

The proof that $C(S) \leq 1$ is now by contradiction, assume $C(S) > 1$. Let c be the length of the longest word in S. Observe that

$$C(\Sigma^k) = 1 \text{ for each } k, \ C(S^n) \leq C(\{x \in \Sigma^* : 1 \leq |x| \leq cn \}) = cn$$

Denote $q = C(S)$. Then we have:

$$C(S)^n = q^n \leq cn$$

For $q > 1$ this inequality is not true for all n, since

$$\lim \ q^n/(cn) = +\infty \text{ if } q > 1$$

Therefore it should be $q \leq 1$ and $C(S) \leq 1$. This completes the proof. ∎

Lemma 15.3

[Kraft's inequality] There is a prefix code with word lengths $\ell_1, \ell_2, \ldots, \ell_k$ on the alphabet $\{0, 1\}$ iff

$$\sum_{i=1}^{k} 2^{-\ell_i} \leq 1 \tag{15.1}$$

Proof. It is enough to show how to construct such a code if the inequality holds. Assume the lengths are sorted in the increasing order. Assume we have a (potentially) infinite full binary tree. We construct the next codeword by going top-down, starting from the root. We assign to the i-th codeword, for $i = 1, 2, 3, \ldots, k$, the lexicographically first path of length ℓ_i, such that the bottom node of this path has not been visited before. It is illustrated in Figure 15.2. If the path does not exist then this means that in this moment we covered with paths a full binary tree, and the actual sum equals 1. But some other lengths remained, so it would be :

$$\sum_{i=1}^{k} 2^{-\ell_i} > 1$$

a contradiction. This proves that the construction of a prefix code works, so the corresponding prefix-free code covering all lengths exists. This completes the proof. ∎

The lemmas imply directly the following theorem.

Theorem 15.3

A uniquely decipherable code with prescribed word lengths exists iff a prefix code with the same word lengths exists.

We remark that the problem of testing unique decipherability of a set of codewords is complete in nondeterministic logarithmic space, see [7].

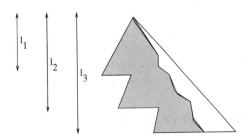

FIGURE 15.2 Graphical illustration of constructing prefix-free code with prescribed lengths sequence satisfying Kraft inequality.

15.2.4 Huffman Codes and Entropy

The performance of Huffman codes is related to a measure of information of the source text, called the *entropy* (denoted by $\mathcal{E}$) of the alphabet. Let w_a be n_a/N, where n_a is the number of occurrences of a in a given text of length N. In this case the sequence S of weights w_a is normalized $\sum_{i=1}^{n} w_i = 1$.

The quantity w_a can be now viewed as the probability that letter a occurs at a given position in the text. This probability is assumed to be independent of the position. Then, the entropy of the alphabet according to w_a's is defined as

$$\mathcal{E}(A) = -\sum_{a \in A} w_a \log w_a.$$

The entropy is expressed in bits (log is a base-two logarithm). It is a lower bound of the average length of the code words $h(a)$,

$$m(A) = \sum_{a \in A} w_a.|h(a)|.$$

Moreover, Huffman codes give the best possible approximation of the entropy (among methods based on coding of the alphabet). This is summarized in the following theorem whose proof relies on the inequalities from Lemma 15.2.

Theorem 15.4

Assume the weight sequence A of weights is normalized. The total cost $m(A)$ of any uniquely decipherable code for A is at least $\mathcal{E}(A)$, and we have

$$\mathcal{E}(A) \leq HuffmanCost(A) \leq \mathcal{E}(A) + 1$$

15.2.5 Huffman Algorithm for *t*-ary Trees

An important generalization of Huffman algorithm is to t-ary trees. Huffman algorithm generalizes to the construction of prefix codes on alphabet of size $t > 2$. The trie of the code is then an *almost full* t-ary tree.

We say that t-ary tree is almost full if all internal nodes have exactly t sons, except possibly one node, which has less sons, in these case all of them should be leaves (let us call this one node a *defect* node).

We perform similar algorithm to Huffman method for binary trees, except that each time we select t items (original or combined) of smallest weight.

There is one technical difficulty. Possibly we start by selecting a smaller number of items in the first step. If we know t and the number of items then it is easy to calculate number q of sons of the defect node, for example if $t = 3$ and $n = 8$ then the defect node has two sons. It is easy to compute the number q of sons of the defect node due to the following simple fact.

Lemma 15.4

If T is a full t-ary tree with m leaves then m *modulo* $(t-1) = 1$.

We start the algorithm by combining q smallest items. Later each time we combine exactly t values. To simplify the algorithm we can add the smallest possible number of dummy items of weigh zero to make the tree full t-ary tree.

15.3 Height Limited Huffman Trees

In this section only, for technical reason, we assume that the length of the path is the number of its vertices. For a sequence S of weights the total cost is changed by adding the sum of weights in S.

Assume we have the same problem as in the case of Huffman trees with additional restriction that the height of the tree is limited by a given parameter L. A beautiful algorithm for this problem has been given by Larmore and Hirschberg, see [8].

15.3.1 Reduction to the Coin Collector Problem

The main component of the algorithm is the reduction to the following problem in which the crucial property play powers of two. We call a real number dyadic iff it has a finite binary representation.

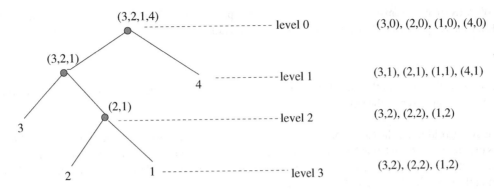

FIGURE 15.3   A Huffman tree T for the items $1, 2, 3, 4$ of height limited by 4 and the corresponding solution to the Coin Collector problem. Each node of the tree can be treated as a *package* consisting of leaves in the corresponding subtree. Assume $weight(i) = i$. Then in the corresponding Coin Collector problem we have $weight(i, h) = i$, $width(i, h) = 2^{-h}$.

Coin Collector problem:

Input: A set I of m items and dyadic number X, each element of I has a *width* and a *weight*, where each width is a (possibly negative) power of two, and each weight is a positive real number.

Output: $CoinColl(I, X)$ - the minimum total weight of a subset $S \subseteq I$ whose widths sum to X.

The following trivial lemma plays important role in the reduction of height limited tree problem to the Coin Collector problem.

Lemma 15.5

Assume T is a full binary tree with n leaves, then

$$\sum_{v \in T} 2^{-level_T(v)} = n + 1$$

Assume we have Huffman coding problem with n items with the sequence of weights weights $W = w_1, w_2, \ldots, w_n$. We define an instance of the Coin Collector problem as follows:

- $I_W = \{(i, l) : i \in [1 \ldots n], l \in [1, \ldots L],$
- $width(i, l) = 2^{-l}$, $weight(i, l) = w_i$ for each i, l
- $X_W = n + 1$.

The intuition behind this strange construction is to view nodes as packages consisting of of original items (elements in the leaves). The internal node v which is a *package* consisting of (leaves in its subtree) items $i_1, i_2, \ldots, i_k$ can be treated as a set of coins $(i_1, h), (i_2, h), \ldots (i_k, h)$, where h is the level of v, and $weight(i_j, h) = weight(i_j)$. The total weight of the set of coins is the cost of the Huffman tree.

Example Figure 15.3 shows optimal Huffman tree for $S = (1, 2, 3, 4)$ with height limited by 4, and the optimal solution to the corresponding Coin Collector problem. The sum of widths of the coins is $4 + 1$, and the total weight is minimal. It is the same as the cost of the Huffman tree on the left, assuming that leaves are also contributing their weights (we scale cost by the sum of weights of S).

Hirschberg and Larmore, see [8], have shown the following fact.

Lemma 15.6

The solution $CoinColl(I_W, X_W)$ to the Coin Collector Problem is the cost of the optimal L-height restricted Huffman tree for the sequence W of weights.

15.3.2 The Algorithm for the Coin Collector Problem

The height limited Huffman trees problem is thus reduced to the Coin Collector Problem. The crucial point in the solution of the latter problem is the fact that weights are powers of two.

Denote *MinWidth(X)* to be the smallest power of two in binary representation of number X. For example $MinWidth(12) = 4$ since $12 = 8 + 4$. Denote by *MinItem(I)* the item with the smallest width in I.

Lemma 15.7

If the items are sorted with respect to the weight then the Coin Collector problem can be solved in linear time (with respect to the total number $|I|$ of coins given in the input).

Proof. The recursive algorithm for the Coin Collector problem is presented below as a recursive function *CoinColl(I, X)*. There are several cases depending on the relation between the smallest width of the item and minimum power of two which constitutes the binary representation of X. In the course of the algorithm the set of weights shrinks as well as the size of X. The linear time implementation of the algorithm is given in References 1.   ∎

```
function CC(I, X); {Coin Collector Problem}
    {compute nnimal weight of a subset of I of total width X
    x := MinItem(X); r := width(x);
    if r > MinWidth(X) then
            no solution exists else
    if r = MinWidth(X) then
            return CC(I − {x}, X − r) + weight(x) else
    if r < MinWidth(X) and there is only one item of width r then
            return CC(I − {x}, X) else
            let x, x′ be two items of smallest weight and width r,
            create new item y such that
                width(y) = 2r, weight(y) = weight(x) + weight(x′);
            return CC(I − {x, x′}∪{y}, X)
```

The last two lemmas imply the following fact.

Theorem 15.5

The problem of the Huffman tree for n items with height limited by L can be solved in $O(n \cdot L)$ time.

Using complicated approach of the Least Weight Concave Subsequence the complexity has been reduced to $n\sqrt{L \log n} + n \log n$ in Reference 9. Another small improvement is by Schieber [10]. An efficient approximation algorithm is given in References 11–13. The dynamic algorithm for Huffman trees is given in Reference 14.

15.4 Optimal Binary Search Trees

Assume we have a sequence of $2n + 1$ weights (nonnegative reals)

$$\alpha_0, \beta_1, \alpha_1, \beta_2, \ldots, \alpha_{n-1}, \beta_n, \alpha_n$$

Let $\text{Tree}(\alpha_0, \beta_1, \alpha_1, \beta_2, \ldots, \alpha_{n-1}, \beta_n, \alpha_n)$ be the set of all full binary weighted trees with n internal nodes, where the i-th internal node (in the in-order) has the weight β_i, and the i-th external node (the leaf, in the left-to-right order) has the weight α_i. The in-order traversal results if we visit all nodes in a recursive way, first the left subtree, then the root, and afterwards the right subtree.

If T is a binary search tree then define the *cost* of T as follows:

$$cost(T) = \sum_{v \in T} level_T(v) \cdot weight(v).$$

Let $OPT(\alpha_0, \beta_1, \ldots, \alpha_{n-1}, \beta_n, \alpha_n)$ be the set of trees $\text{Tree}(\alpha_0, \beta_1, \ldots, \alpha_{n-1}, \beta_n, \alpha_n)$ whose cost is minimal.

We use also terminology from Reference 15. Let $K_1, \ldots K_n$ be a sequence of n weighted items (keys), which are to be placed in a binary search tree. We are given $2n + 1$ weights (probabilities): $q_0, p_1, q_1, p_2, q_2, p_3, \ldots, q_{n-1}, p_n, q_n$ where

- p_i is the probability that K_i is the search argument;
- q_i is the probability that the search argument lies between K_i and K_{i+1}.

The OBST problem is to construct an optimal binary search tree, the keys K_i's are to be stored in internal nodes of this binary search tree and in its external nodes special items are to be stored. The *i*-th special item K'_i corresponds to all keys which are strictly between K_i and K_{i+1}. The binary search tree is a full binary tree whose nodes are labeled by the keys. Using the abstract terminology introduced above the OBST problem consists of finding a tree $T \in OPT(q_0, p_1, q_1, p_2, \ldots, q_{n-1}, p_n, q_n)$, see an example tree in Figure 15.4.

Denote by $obst(i,j)$ the set $OPT(q_i, p_{i+1}, q_{i+1}, \ldots, q_{j-1}, p_j, q_j)$. Let $cost(i,j)$ be the cost of a tree in $obst(i,j)$, for $i < j$, and $cost(i,i) = q_i$. The sequence $q_i, p_{i+1}, q_{i+1}, \ldots, q_{j-1}, p_j, q_j$ is here the subsequence of $q_0, p_1, q_1, p_2, \ldots, q_{n-1}, p_n, q_n$, consisting of some number of consecutive elements which starts with q_i and ends with q_j. Let

$$w(i,j) = q_i + p_{i+1} + q_{i+1} + \cdots + q_{j-1} + p_j + q_j$$

The dynamic programming approach to the computation of the OBST problem relies on the fact that the subtrees of an optimal tree are also optimal. If a tree $T \in obst(i,j)$ contains in the root an item K_k then its left subtree is in $obst(i, k-1)$ and its right subtree is in $obst(k,j)$. Moreover, when we join these two subtrees then the contribution of each node increases by one (as one level is added), so the increase is $w(i,j)$. Hence the costs obey the following *dynamic programming recurrences* for $i < j$:

$$cost(i,j) = \min\{cost(i, k-1) + cost(k,j) + w(i,j) : i < k \leq j\}.$$

Denote the smallest value of k which minimizes the above equation by $cut(i,j)$. This is the first point giving an optimal decomposition of $obst(i,j)$ into two smaller (son) subtrees. Optimal binary search trees have the following crucial property (proved in Reference 16, see Figure 15.5 for graphical illustration)

$$monotonicity\ property: \quad i \leq i' \leq j \leq j' \implies cut(i,j) \leq cut(i',j')$$

The property of monotonicity, the cuts and the quadratic algorithm for the OBST were first given by Knuth. The general dynamic programming recurrences were treated by Yao [17], in the context of reducing cubic time to quadratic.

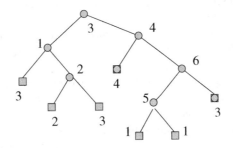

FIGURE 15.4 A binary search tree for the sequences: $\beta = (\beta_1, \beta_2, \ldots, \beta_6) = (1, 2, 3, 4, 5, 6)$, $\alpha = (\alpha_0, \alpha_1, \ldots \alpha_6) = (3, 2, 3, 4, 1, 1, 3)$. We have $cut(0, 6) = 3$.

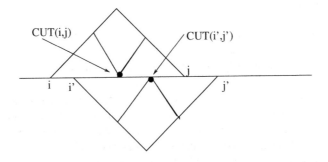

FIGURE 15.5 Graphical illustration of the monotonicity property of cuts.

Theorem 15.6

Optimal binary search trees can be computed in $O(n^2)$ time.

Proof. The values of $cost(i,j)$ are computed by tabulating them in an array. Such tabulation of costs of smaller subproblems is the basis of the *dynamic programming* technique. We use the same name *cost* for this array. It can be computed in $O(n^3)$ time, by processing diagonal after diagonal, starting with the central diagonal.

In case of optimal binary search trees this can be reduced to $O(n^2)$ using additional tabulated values of the cuts in table *cut*. The k-th diagonal consists of entries i,j such that $j - i = k$. If we have computed cuts for k-th diagonal then for (i,j) on the $(k+1)$-th diagonal we know that

$$cut(i,j-1) \le cut(i,j) \le cut(i+1,j)$$

Hence the total work on the $(k+1)$-th diagonal is proportional to the sum of telescoping series:

$$cut(1,k+1) - cut(0,k) + cut(2,k+2) - cut(1,k+1)+$$
$$cut(3,k+3) - cut(2,k+2) + \cdots cut(n-k,k) - cut(n-k-1,k-1)$$

which is $O(n)$. Summing over all diagonals gives quadratic total time to compute the tables of cuts and costs. Once the table $cost(i,j)$ is computed then the construction of an optimal tree can be done in quadratic time by tracing back the values of *cuts*. ∎

15.4.1 Approximately Optimal Binary Search Trees

We can attempt to reduce time complexity at the cost of slightly increased cost of the constructed tree. A common-sense approach would be to insert the keys in the order of decreasing frequencies. However this method occasionally can give quite bad trees.

Another approach would be to choose the root so that the total weights of items in the left and right trees are as close as possible. However it is not so good in pessimistic sense.

The combination of this methods can give quite satisfactory solutions and the resulting algorithm can be linear time, see [18]. *Average* subquadratic time has been given in Reference 19.

15.5 Optimal Alphabetic Tree Problem

The alphabetic tree problem looks very similar to the Huffman problem, except that the leaves of the alphabetic tree (read left-to-right) should be in the same order as in the original input sequence. Similarly as in Huffman coding the binary tree must be *full*, that is, each internal node must have exactly two sons.

The main difficulty is that we cannot localize easily two items which are to be combined.

Assume we have a sequence of n weighted items, where w_i is the non-negative *weight* of the ith item. We write $\alpha = w_1 \ldots w_n$. The sequence will be changing in the course of the algorithm.

An alphabetic tree over α is an ordered binary tree T with n leaves, where the ith leaf (in left-to-right order) corresponds to the ith item of The optimal alphabetic tree problem (OAT problem) is to find an alphabetic tree of minimum cost.

The Garsia-Wachs algorithm solves the alphabetic tree problem, it is a version of an earlier algorithm by Hu and Tucker, see [4]. The strangest part of the algorithm is that it permutes α, though the final tree should have the order of leaves the same as the order of items in the original sequence. We adopt the convention that the items of α have unique names, and that these names are preserved when items are moved. When convenient to do so, we will assume that those names are the positions of items in the list, namely integers in $[1 \ldots n]$.

15.5.1 Computing the Cost of Optimal Alphabetic Tree

First we show how to compute only the cost of the whole tree, however this computation does not give automatically an optimal alphabetic tree, since we will be permuting the sequence of items. Each time we combine two adjacent items in the current permutation, however these items are not necessarily adjacent in the original sequence, so in any legal alphabetic tree they cannot be sons of a same father.

The alphabetic tree is constructed by reducing the initial sequence of items to a shorter sequence in a manner similar to that of the Huffman algorithm, with one important difference. In the Huffman algorithm, the minimum pair of items are combined, because it can be shown that they are siblings in the optimal tree. If we could identify two adjacent items that are siblings in the optimal alphabetic tree, we could combine them and then proceed recursively. Unfortunately, there is no known way to identify such a pair. Even a minimal pair may not be siblings. Consider the weight sequence (8 7 7 8). The second and the third items are not siblings in any optimal alphabetic tree.

Instead, the HT and GW algorithms, as well as the algorithms of [6,20–22], operate by identifying a pair of items that have the same level in the optimal tree. These items are then combined into a single "package," reducing the number of items by one. The details on how this process proceeds differ in the different algorithms. Define, for $1 \leq i < n$, the ith two-sum:

$$TwoSum(i) = w_i + w_{i+1}$$

A pair of adjacent items $(i, i+1)$ is a *locally minimal pair* (or *lmp* for short) if

$$
\begin{array}{lll}
TwoSum(i-1) & \geq & TwoSum(i) \qquad \text{if} \quad i > 1 \\
TwoSum(i) & < & TwoSum(i+1) \quad \text{if} \quad i \leq n-2
\end{array}
$$

A locally minimal pair which is currently being processed is called the *active pair*.

The Operator *Move.* If w is any item in a list π of weighted items, define $RightPos(w)$ to be the predecessor of the nearest right larger or equal neighbor of w. If w has no right larger or equal neighbor, define $RightPos(w)$ to be $|\pi| + 1$.

Let $Move(w, \pi)$ be the operator that changes π by moving w w is inserted between positions $RightPos(w) - 1$ and $RightPos(w)$. For example

$$Move(7, (2, 5, 7, 2, 4, 9, 3, 4) = (2, 5, 2, 4, 7, 9, 3, 4)$$

---

function GW(π); {π is a sequence of names of items}
  {restricted version of the Garsia-Wachs algorithm}
  { computing only the cost of optimal alphabetic tree }
  if $\pi = (v)$ (π consists of a single item)
  then return 0
    else
      find **any** *locally minimal pair* (u, w) of π
      create a new item x whose weight is $weight(u) + weight(w)$;
      replace the pair u, v by the single item x;
      { the items u, v are removed }
      $Move(v, \pi)$;
      return GW(π) $+ weight(v)$;

Correctness of the algorithm is a complicated issue. There are two simplified proofs, see [23,24] and we refer to these papers for detailed proof. In Reference 23 only the rightmost minimal pair can be processed each time, while [24] gives correctness of general algorithm when any minimal pair is processed, this is important in parallel computation, when we process simultaneously many such pairs. The proof in Reference 24 shows that correctness is due to *well-shaped* bottom segments of optimal alphabetic trees, this is expressed as a *structural theorem* in Reference 24 which gives more insight into the global structure of optimal alphabetic trees.

For $j > i + 1$ denote by $\pi_{i,j}$ the sequence π in which elements $i, i+1$ are moved just before left of j.

Theorem 15.7

[Correctness of the GW algorithm]
Let $(i, i+1)$ be a locally minimal pair and $RightPos(i, i+1) = j$, and let T' be a tree over the sequence $\pi_{i,j}$, optimal among all trees over $\pi_{i,j}$ in which $i, i+1$ are siblings. Then there is an optimal alphabetic tree T over the original sequence $\pi = (1, \ldots n)$ such that $T \cong T'$.

Correctness can be also expressed as equivalence between some classes of trees.

Two binary trees T_1 and T_2 are said to be **level equivalent** (we write $T_1 \cong T_2$) if T_1, and T_2 have the same set of leaves (possibly in a different order) and $level_{T_1} = level_{T_2}$.

Denote by OPT(i) the set of all alphabetic trees over the leaf-sequence $(1, \ldots n)$ which are optimal among trees in which i and $i + 1$ are at the same level. Assume the pair $(i, i+1)$ is locally minimal. Let $OPT_{moved}(i)$ be the set of all alphabetic trees over the leaf-sequence $\pi_{i,j}$ which are optimal among all trees in which leaves i and $i + 1$ are at the same level, where $j = RightPos(i, i+1)$.

Two sets of trees OPT and OPT$'$ are said to be *level equivalent*, written OPT $\cong$ OPT$'$, if, for each tree $T \in$ OPT, there is a tree $T' \in$ OPT$'$ such that $T' \cong T$, and vice versa.

Theorem 15.8

Let $(i, i+1)$ be a locally minimal pair. Then

1. $OPT(i) \cong OPT_{moved}(i)$.
2. $OPT(i)$ contains an optimal alphabetic tree T.
3. $OPT_{moved}(i)$ contains a tree T' with $i, i+1$ as siblings.

15.5.2 Construction of Optimal Alphabetic Tree

The full Garsia-Wachs algorithm first computes the level tree. This tree can be easily constructed in the function $GW(\pi)$ when computing the cost of alphabetic tree. Each time we sum weights of two items (original or newly created) then we create new item which is their father with the weight being the sum of weights of sons.

Once we have a level tree, the optimal alphabetic tree can be constructed easily in linear time. Figures 15.6 through 15.8 show the process of construction the level tree and construction an optimal alphabetic tree knowing the levels of original items.

Lemma 15.8

Assume we know level of each leaf in an optimal alphabetic tree. Then the tree can be constructed in linear time.

Proof. The levels give the *"shape"* of the tree, see Figure 15.8.

Assume $l_1, l_2, l_3, \ldots, l_n$ is the sequence of levels. We scan this sequence from left-to-right until we find *the first* two levels l_i, l_{i+1} which are the same. Then we know that the leaves i and $i+1$ are sons of a same father, hence we link these leaves to a newly created father and we remove these leaves, in the level sequence the pair l_i, l_{i+1} is replaced by a single level $l_i - 1$. Next we check if $l_{i-1} = l_i - 1$, if not we search to the right. We keep the scanned and newly created levels on the stack. The total time is linear.   ∎

There are possible many different optimal alphabetic trees for the same sequence, Figure 15.9 shows an alternative optimal alphabetic tree for the same example sequence.

Theorem 15.9

Optimal alphabetic tree can be constructed in $O(n \log n)$ time.

Proof. We keep the array of levels of items. The array *level* is global of size $(2n - 1)$. Its indices are the names of the nodes, that is, the original n items and the $(n - 1)$ nodes ("packages") created during execution of the algorithm. The algorithm works in quadratic time, if implemented in a naive way. Using priority queues, it works in $O(n \log n)$ time. Correctness follows directly from Theorem 15.7.   ∎

15.5.3 Optimal Alphabetic Trees for Presorted Items

We have seen that Huffman trees can be constructed in linear time if the weights are presorted. Larmore and Przytycka, see [21] have shown that slightly weaker similar result holds for alphabetic trees as well:

assume that weights of items are sortable in linear time, then the alphabetic tree problem can be solved in $O(n \log \log n)$ time.

Open problem Is it possible to construct alphabetic trees in linear time in the case when the weights are sortable in linear time?

15.6 Optimal Lopsided Trees

The problem of finding optimal prefix-free codes for unequal letter costs consists of finding a minimal cost prefix-free code in which the encoding alphabet consists of unequal cost (length) letters, of lengths α and β, $\alpha \leq \beta$. We restrict ourselves here only to binary trees. The code is represented by a *lopsided tree*, in the same way as a Huffman tree represents the solution of the Huffman coding problem. Despite the similarity, the case of unequal letter costs is much harder then the classical Huffman problem; no polynomial time algorithm is known for general letter costs, despite a rich literature on the problem, for example, [25,26]. However there are known polynomial time algorithms when α and β are integer constants [26].

The problem of finding the minimum cost tree in this case was first studied by Karp [27] in 1961 who solved the problem by reduction to integer linear programming, yielding an algorithm exponential in both n and β. Since that time there has been much work on various aspects of the problem such as; bounding the cost of the optimal tree, Altenkamp and Mehlhorn [28], Kapoor

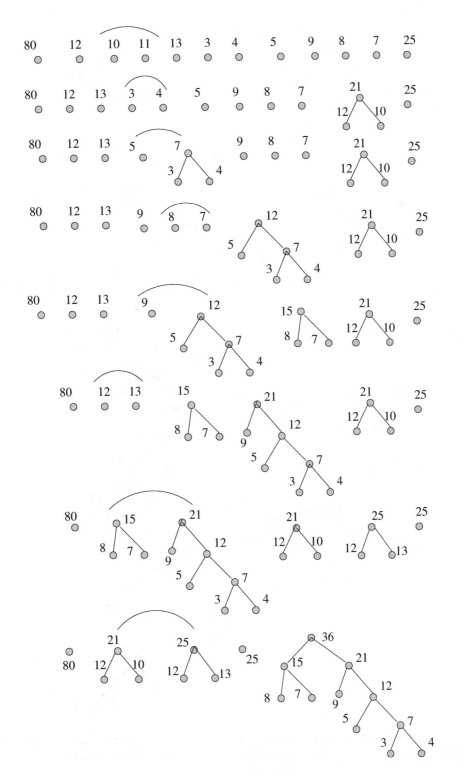

FIGURE 15.6 The first 7 phases of Garsia-Wachs algorithm.

and Reingold [29] and Savari [30]; the restriction to the special case when all of the weights are equal, Cot [31], Perl Gary and Even [32], and Choi and Golin [33]; and approximating the optimal solution, Gilbert [34]. Despite all of these efforts it is still, surprisingly, not even known whether the basic problem is polynomial-time solvable or in *NP*-complete.

Golin and Rote [26] describe an $O(n^{\beta+2})$-time dynamic programming algorithm that constructs the tree in a top-down fashion.

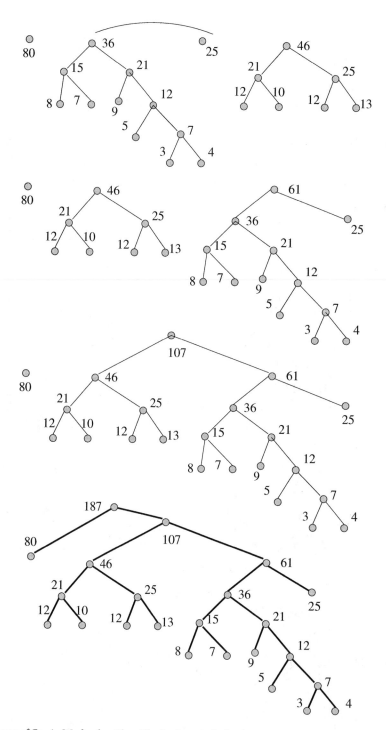

FIGURE 15.7 The last phases of Garsia-Wachs algorithm. The final tree is the level tree (but not alphabetic tree). The total cost (sum of values in internal nodes equal to 538) and the levels of original items are the same as in optimal alphabetic tree. The level sequence for the original sequence (80, 12, 10, 11, 13, 3, 4, 5, 9, 8, 7, 25) is: $\mathscr{L} = (1, 4, 4, 4, 4, 7, 7, 6, 5, 5, 3)$.

This has been improved using a different approach (monotone-matrix concepts, for example, the *Monge property* and the SMAWK algorithm [35].

Theorem 15.10: [5]

Optimal lopsided trees can be constructed in $O(n^\beta)$ time.

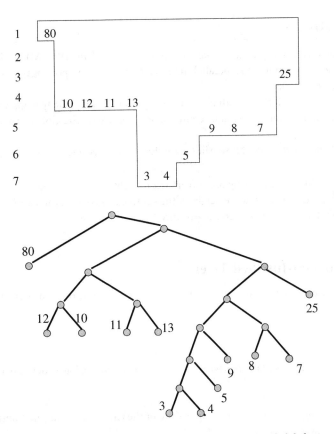

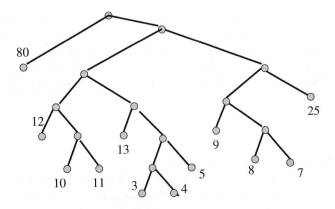

FIGURE 15.8   The shape of the optimal tree given by the level sequence and the final optimal alphabetic tree (cost = 538) corresponding to this shape and to the weight sequence (80, 12, 10, 11, 13, 3, 4, 5, 9, 8, 7, 25).

FIGURE 15.9   An alternative optimal alphabetic tree (cost = 538) for the same weight sequence.

This is the the most efficient known algorithm for the case of small β; in practice the letter costs are typically small (*e.g.,* Morse codes).

Recently a scheme of an efficient approximating algorithm has been given.

Theorem 15.11: [36]

There is a polynomial time approximation scheme for optimal lopsided trees.

15.7 Parallel Algorithms

As a model of parallel computations we choose the *Parallel Random Access Machines* (PRAM), see [37]. From the point of view of parallel complexity two parameters are of interest: parallel time (usually we require polylogarithmic time) and total work (time multiplied by the number of processors).

The sequential greedy algorithm for Huffman coding is quite simple, but unfortunately it appears to be inherently sequential. Its parallel counterpart is much more complicated, and requires a new approach. The global structure of Huffman trees must be explored in depth.

A full binary tree T is said to be *left-justified* if it satisfies the following properties:

1. the depths of the leaves are in non-increasing order from left to right,
2. let u be a left brother of v, and assume that the height of the subtree rooted at v is at least l. Then the tree rooted at u is full at level l, which means that u has 2^l descendants at distance l.

15.7.1 Basic Property of Left-Justified Trees

Let T be a left-justified binary tree. Then, T consists of one leftmost branch and the subtrees hanging from this branch have logarithmic height.

Lemma 15.9

Assume that the weights $w_1, w_2, \ldots, w_n$ are pairwise distinct and in increasing order. Then, there is Huffman tree for $(w_1, w_2, \ldots, w_n)$ that is left-justified.

The left-justified trees are used together with efficient algorithm for the CLWS problem (the *Concave Least Weight Subsequence* problem, to be defined below) to show the following fact.

Theorem 15.12: [38]

The parallel Huffman coding problem can be solved in polylogarithmic time with quadratic work.

Hirschberg and Larmore [8] define the *Least Weight Subsequence* (LWS) problem as follows: Given an integer n, and a real-valued *weight* function $w(i, j)$ defined for integers $0 \le i < j \le n$, find a sequence of integers $\overline{\alpha} = (0 = \alpha_0 < \alpha_1 < \cdots < \alpha_{k-1} < \alpha_k = n)$ such that $w(\overline{\alpha}) = \sum_{i=0}^{k-1} w(\alpha_i, \alpha_{i+1})$ is minimized. Thus, the LWS problem reduces trivially to the minimum path problem on a weighted directed acyclic graph. The *Single Source* LWS problem is to find such a minimal sequence $0 = \alpha_0 < \alpha_1 < \cdots < \alpha_{k-1} < \alpha_k = m$ for all $m \le n$. The weight function is said to be *concave* if for all $0 \le i_0 \le i_1 < j_0 \le j_1 \le n$,

$$w(i_0, j_0) + w(i_1, j_1) \le w(i_0, j_1) + w(i_1, j_0). \tag{15.2}$$

The inequality (15.2) is also called the *quadrangle inequality* [17].

The LWS problem with the restriction that the weight function is concave is called the *Concave Least Weight Subsequence* (CLWS) problem. Hirschberg and Larmore [8] show that the LWS problem can be solved in $O(n^2)$ sequential time, while the CLWS problem can be solved in $O(n \log n)$ time. Wilber [39] gives an $O(n)$-time algorithm for the CLWS problem.

In the parallel setting, the CLWS problem seems to be more difficult. The best current polylogarithmic time algorithm for the CLWS problem uses concave matrix multiplication techniques and requires $O(\log^2 n)$ time with $n^2 / \log^2 n$ processors.

Larmore and Przytycka [40] have shown how to compute efficiently CLWS in sublinear time with the total work smaller than quadratic. Using this approach they showed the following fact (which has been later slightly improved [41,42].

Theorem 15.13

Optimal Huffman tree can be computed in $O(\sqrt{n} \log n)$ time with linear number of processors.

Karpinski and Nekrich have shown an efficient parallel algorithm which *approximates* optimal Huffman code, see [43]. Similar, but much more complicated algorithm works for alphabetic trees. Again the CLWS algorithm is the main tool.

Theorem 15.14: [6]

Optimal alphabetic tree can be constructed in polylogarithmic time with quadratic number of processors.

In case of general binary search trees the situation is more difficult. Polylogarithmic time algorithms need huge number of processors. However sublinear parallel time is easier.

Theorem 15.15: [44,45]

The OBST problem can be solved in **(a)** polylogarithmic time with $O(n^6)$ processors, **(b)** in sublinear time and quadratic total work.

References

1. L. L. Larmore and D. S. Hirschberg, A fast algorithm for optimal length–limited Huffman codes, *Journal of the ACM* 37, 1990, 464–473.
2. A. M. Garsia and M. L. Wachs, A New algorithm for minimal binary search trees, *SIAM Journal of Computing* 6, 1977, 622–642.
3. T. C. Hu, A new proof of the T-C algorithm, *SIAM Journal of Applied Mathematics* 25, 1973, 83–94.
4. T. C. Hu and A. C. Tucker, Optimal computer search trees and variable length alphabetic codes, *SIAM Journal of Applied Mathematics* 21, 1971, 514–532.
5. P. Bradford, M. Golin, L. Larmore, W. Rytter, Optimal prefix-free codes for unequal letter costs and dynamic programming with the monge property, *Journal of Algorithms,* 42 (2), February 2002, 277–303.
6. L. L. Larmore, T. M. Przytycka, W. Rytter, "Parallel Construction of Optimal Alphabetic Trees," in *Proceedings of the 5th ACM Symposium on Parallel Algorithms and Architectures,* 1993, pp. 214–223.
7. W. Rytter, The space complexity of the unique decipherability problem, *IPL* 16 (4), 1983.
8. D. S. Hirschberg and L. L. Larmore, "The Least Weight Subsequence Problem," in *Proc. 26th IEEE Symp. on Foundations of Computer Science, Portland Oregon,* Oct. 1985, pp. 137–143. Reprinted in *SIAM Journal on Computing* 16, 1987, 628–638.
9. A. Aggarwal, B. Schieber, T. Tokuyama, Finding a Minimum-weight k-link path graphs with the concave Monge property and applications, *Discrete & Computational Geometry* 12, 1994, 263–280.
10. B. Schieber, Computing a Minimum Weight k-Link Path in Graphs with the Concave Monge Property. 204–222.
11. R. L. Milidiu and E. Laber, The warm-up algorithm: A Lagrangian construction of length limited Huffman codes, *SIAM J. Comput.* 30 (5), 2000, 1405–1426.
12. R. L. Milidi and E. S. Laber, Linear time recognition of optimal L-restricted prefix codes (extended abstract), *LATIN* 2000, 227–236.
13. R. L. Milidi and E. S. Laber, Bounding the inefficiency of length-restricted prefix codes, *Algorithmica* 31 (4), 2001, 513–529.
14. J. S. Vitter, Dynamic huffman coding, *ACM Trans. Math. Software* 15, June 1989, 158–167.
15. D. E. Knuth, *The Art of computer programming,* Addison–Wesley, 1973.
16. D. E. Knuth, Optimum binary search trees, *Acta Informatica* 1, 1971, 14–25.
17. F. F. Yao, "Efficient Dynamic Programming Using Quadrangle Inequalities," in *Proceedings of the 12th ACM Symposium on Theory of Computing,* 1980, pp. 429–435.
18. K. Mehlhorn, *Data structures and algorithms,* vol. 1, Springer, 1984.
19. M. Karpinski, L. Larmore, W. Rytter, "Sequential and parallel subquadratic work constructions of approximately optimal binary search trees," in *the 7th ACM Symposium on Discrete Algorithms, SODA'96.*
20. M. M. Klawe and B. Mumey, "Upper and Lower Bounds on Constructing Alphabetic Binary Trees," in *Proceedings of the 4th ACM-SIAM Symposium on Discrete Algorithms,* 1993, pp. 185–193.
21. L. L. Larmore and T. M. Przytycka, "The Optimal Alphabetic Tree Problem Revisited," in *Proceedings of the 21st International Colloquium, ICALP'94, Jerusalem, LNCS 820,* Springer-Verlag, 1994, pp. 251–262.
22. P. Ramanan, Testing the optimality of alphabetic trees, *Theoretical Computer Science* 93, 1992, 279–301.
23. J. H. Kingston, A new proof of the Garsia-Wachs algorithm, *Journal of Algorithms* 9, 1988, 129–136.
24. M. Karpinski, L. l. Larmore, W. Rytter, Correctness of constructing optimal alphabetic trees revisited, *Theoretical Computer Science,* 180 (1–2), 10 June 1997, 309–324.
25. J. Abrahams, "Code and Parse Trees for Lossless Source Encoding," in *Sequences '97,* 1997.

26. M. Golin and G. Rote, "A Dynamic Programming Algorithm for Constructing Optimal Prefix-Free Codes for Unequal Letter Costs," in *Proceedings of the 22nd International Colloquium on Automata Languages and Programming (ICALP '95)*, July 1995, pp. 256–267.

27. R. M. Karp, Minimum-redundancy coding for the discrete noiseless channel, *IRE Transactions on Information Theory 7*, 1961, 27–39.

28. D. Altenkamp and K. Mehlhorn, Codes: Unequal probabilities unequal letter costs, *Journal of the Association for Computing Machinery* 27 (3), July 1980, 412–427.

29. S. Kapoor and E. Reingold, Optimum lopsided binary trees, *Journal of the Association for Computing Machinery* 36 (3), July 1989, 573–590.

30. S. A. Savari, Some notes on varn coding, *IEEE Transactions on Information Theory*, 40 (1), Jan. 1994, 181–186.

31. N. Cot, "A Linear-Time Ordering Procedure with Applications to Variable Length Encoding," in *Proc. 8th Annual Princeton Conference on Information Sciences and Systems*, 1974, pp. 460–463.

32. Y. Perl, M. R. Garey and S. Even, Efficient generation of optimal prefix code: Equiprobable words using unequal cost letters, *Journal of the Association for Computing Machinery* 22 (2), April 1975, 202–214.

33. S.-N. Choi, M. Golin, "Lopsided trees: Algorithms, analyses and applications," Automata, Languages and Programming, *Proceedings of the 23rd International Colloquium on Automata, Languages, and Programming (ICALP 96)*.

34. E. N. Gilbert, Coding with digits of unequal costs, *IEEE Transactions on Information Theory*, 41, 1995,

35. A. Aggarwal, M. Klawe, S. Moran, P. Shor, R. Wilber, Geometric applications of a matrix-searching algorithm, *Algorithmica* 2, 1987, 195–208.

36. M. Golin and G. Rote, "A Dynamic Programming Algorithm for Constructing Optimal Prefix-Free Codes for Uneq ual Letter Costs," in *Proceedings of the 22nd International Colloquium on Automata Languages and Programming (ICALP '95)*, July 1995, pp. 256–267.Expanded version to appear in *IEEE Transactions on Information Theory*.

37. A. Gibbons, W. Rytter, *Efficient parallel algorithms*, Cambridge Univ. Press, 1997.

38. M. J. Atallah, S. R. Kosaraju, L. L. Larmore, G. L. Miller, S-H. Teng, "Constructing Trees in Parallel," in *Proc. 1st ACM Symposium on Parallel Algorithms and Architectures*, 1989, pp. 499–533.

39. R. Wilber, The Concave least weight subsequence problem revisited, *Journal of Algorithms* 9, 1988, 418–425.

40. L. L. Larmore and T. M. Przytycka, Constructing Huffman trees in parallel, *SIAM Journal on Computing* 24 (6), 1995, 1163–1169.

41. M. Karpinski, L. Larmore, Y. Nekrich. A work efficient algorithm for the construction of length-limited Huffman codes, to appear in Parallel Processing Letters.

42. C. Levcopulos and T. Przytycka, A work-time trade-off in parallel computation of Huffman trees and concave least weight subsequence problem, *Parallel Processing Letters* 4 (1–2), 1994, 37–43.

43. P. Berman, M. Karpinski, M. Nekrich, "Approximating Huffman codes in parallel," in *Proc. 29th ICALP, LNCS vol. 2380*, Springer, 2002, pp. 845–855.

44. M. Karpinski and W. Rytter, On a Sublinear time parallel construction of optimal binary search trees, *Parallel Processing Letters* 8 (3), 1998.

45. W. Rytter, On efficient parallel computations for some dynamic programming problems, *Theoretical Computer Science*, 59, 1988, 297–307.

16

B Trees*

16.1 Introduction.. 233
16.2 The Disk-Based Environment 233
16.3 The B-tree 234
    B-tree Definition • B-tree Query • B-tree Insertion • B-tree Deletion
16.4 The B+-tree.. 239
    Copy-up and Push-up • B+-tree Query • B+-tree Insertion • B+-tree Deletion
16.5 Further Discussions.. 244
    Efficiency Analysis • Why Is the B+-tree Widely Accepted? • Bulk-Loading a B+-tree •
    Aggregation Query in a B+-tree
References... 247

Donghui Zhang[†]
Northeastern University

16.1 Introduction

We have seen binary search trees in Chapters 3 and 11. When data volume is large and does not fit in memory, an extension of the binary search tree to disk-based environment is the B-tree, originally invented by Bayer and McCreight [1]. In fact, since the B-tree is always balanced (all leaf nodes appear at the same level), it is an extension of the *balanced* binary search tree. Since each disk access exchanges a whole block of information between memory and disk rather than a few bytes, a node of the B-tree is expanded to hold more than two child pointers, up to the block capacity. To guarantee worst-case performance, the B-tree requires that every node (except the root) has to be at least half full. An exact match query, insertion or deletion need to access $O(\log_B n)$ nodes, where B is the page capacity in number of child pointers, and n is the number of objects.

Nowadays, every database management system (see Chapter 62 for more on applications of data structures to database management systems) has implemented the B-tree or its variants. Since the invention of the B-tree, there have been many variations proposed. In particular, Knuth [2] defined the B*-tree as a B-tree in which every node has to be at least 2/3 full (instead of just 1/2 full). If a page overflows during insertion, the B*-tree applies a local redistribution scheme to delay splitting the node till two another sibling node is also full. At this time, the two nodes are split into three. Perhaps the best variation of the B-tree is the B+-tree, whose idea was originally suggested by Knuth [2], but whose name was given by Comer [3]. (Before Comer, Knuth used the name B*-tree to represent both B*-tree and B+-tree.) In a B+-tree, every object stays at the leaf level. Update and query algorithms need to be modified from those of the original B-tree accordingly.

The idea of the B-tree also motivates the design of many other disk-based index structures like the R-tree [4], the state-of-art spatial index structure (Chapter 22).

In this chapter, we describe the B-tree and B+-tree in more detail. In Section 16.2, we briefly describe the disk-based environment and we introduce some notations. The B-tree is described in section 16.3, while the B+-tree is described in Section 16.4. Finally, in Section 16.5 we further discuss some related issues.

16.2 The Disk-Based Environment

Most application software deal with data. For instance, a registration application may keep the name, address and social security number of all students. The data has to be stored somewhere. There are three levels of storage. The computer CPU deals directly with the **primary storage**, which means the main memory (as well as cache). While data stored at this level can be access quickly,

---

[*] This chapter has been reprinted from first edition of this Handbook, without any content updates.
[†] We have used this author's affiliation from the first edition of this handbook, but note that this may have changed since then.

we cannot store everything in memory for two reasons. First, memory is expensive. Second, memory is volatile, that is, if there is a power failure, information stored in memory gets lost.

The **secondary storage** stands for magnetic disks. Although it has slower access, it is less expensive and it is non-volatile. This satisfies most needs. For data which do not need to be accessed often, they can also be stored in the **tertiary storage**, for example, tapes.

Since the CPU does not deal with disk directly, in order for any piece of data to be accessed, it has to be read from disk to memory first. Data is stored on disk in units called **blocks** or **pages**. Every disk access has to read/write one or multiple blocks. That is, even if we need to access a single integer stored in a disk block which contains thousands of integers, we need to read the whole block in. This tells us why internal memory data structures cannot be directly implemented as external-memory index structures.

Consider the binary search tree as an example. Suppose we implement every node as a disk block. The storage would be very inefficient. If a disk page is 8 KB (=8192 bytes), while a node in the binary search tree is 16 bytes (four integers: a key, a value, and two child pointers), we know every page is only 0.2% full. To improve space efficiency, we should store multiple tree nodes in one disk page. However, the query and update will still be inefficient. The query and update need to access $O(\log_2 n)$ nodes, where n is the number of objects. Since it is possible that every node accessed is stored in a different disk page, we need to access $O(\log_2 n)$ disk pages. On the other hand, the B-tree query/update needs to access only $O(\log_B n)$ disk pages, which is a big improvement. A typical value of B is 100. Even if there are as many as billions of objects, the height of a B-tree, $\log_B n$, will be at most 4 or 5.

A fundamental question that the database research addresses is how to reduce the gap between memory and disk. That is, given a large amount of data, how to organize them on disk in such a way that they can efficiently be updated and retrieved. Here we measure efficiency by counting the total number of disk accesses we make. A disk access can be either a read or a write operation. Without going into details on how the data is organized on disk, let's make a few assumptions. First, assume each disk page is identified by a number called its *pageID*. Second, given a *pageID*, there is a function *DiskRead* which reads the page into memory. Correspondingly, there is a *DiskWrite* function which writes the in-memory page onto disk. Third, we are provided two functions which allocate a new disk page and deallocate an existing disk page.

The four functions are listed below.

- **DiskRead**: given a *pageID*, read the corresponding disk page into memory and return the corresponding memory location.
- **DiskWrite**: given the location of an in-memory page, write the page to disk.
- **AllocatePage**: find an unoccupied pageID, allocate space for the page in memory and return its memory location.
- **DeallocatePage**: given a *pageID*, mark the corresponding disk page as being unoccupied.

In the actual implementation, we should utilize a memory buffer pool. When we need to access a page, we should first check if it is already in the buffer pool, and we access the disk only when there is a buffer miss. Similarly, when we want to write a page out, we should write it to the buffer pool. An actual DiskWrite is performed under two circumstances: (a) The buffer pool is full and one page needs to be switched out of buffer. (b) The application program terminates. However, for our purposes we do not differentiate disk access and buffer pool access.

16.3 The B-tree

The problem which the B-tree aims to solve is: given a large collection of objects, each having a *key* and an *value*, design a disk-based index structure which efficiently supports query and update.

Here the query that is of interest is the *exact-match query*: given a key k, locate the value of the object with key= k. The update can be either an insertion or a deletion. That is, insert a new object into the index, or delete from the index an object with a given key.

16.3.1 B-tree Definition

A B-tree is a tree structure where every node corresponds to a disk page and which satisfies the following properties:

- A node (leaf or index) x has a value $x.num$ as the number of objects stored in x. It also stores the list of $x.num$ objects in increasing key order. The key and value of the ith object ($1 \leq i \leq x.num$) are represented as $x.key[i]$ and $x.value[i]$, respectively.
- Every leaf node has the same depth.
- An index node x stores, besides $x.num$ objects, $x.num+1$ child pointers. Here each child pointer is a pageID of the corresponding child node. The ith child pointer is denoted as $x.child[i]$. It corresponds to a key range $(x.key[i-1], x.key[i])$. This means that in the ith sub-tree, any object key must be larger than $x.key[i-1]$ and smaller than $x.key[i]$. For instance,

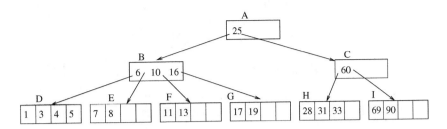

FIGURE 16.1   An example of a B-tree.

in the sub-tree referenced by *x.child*[1], the object keys are smaller than *x.key*[1]. In the sub-tree referenced by *x.child*[2], the objects keys are between *x.key*[1] and *x.key*[2], and so on.

- Every node except the root node has to be at least half full. That is, suppose an index node can hold up to 2*B* child pointers (besides, of course, 2*B* − 1 objects), then any index node except the root must have at least *B* child pointers. A leaf node can hold more objects, since no child pointer needs to be stored. However, for simplicity we assume a leaf node holds between *B* and 2*B* objects.
- If the root node is an index node, it must have at least two children.

A special case of the B-tree is when *B* = 2. Here every index node must have 2 or 3 or 4 child pointers. This special case is called the **2-3-4 tree**.

Figure 16.1 shows an example of a B-tree. In particular, it's a 2-3-4 tree.

In the figure, every index node contains between 2 and 4 child pointers, and every leaf node contains between 2 and 4 objects. The root node *A* is an index node. Currently it has one object with key=25 and two child pointers. In the left sub-tree, every object has key<25. In the right sub-tree, every object has key>25. Every leaf node (*D* through *I*) are located at the same depth: their distance to *A* is 2. Currently, there are two pages which are full: an index node *B* and a leaf node *D*.

16.3.2 B-tree Query

To find the value of an object with key=*k*, we call the *Query* algorithm given below. The parameters are the tree root pageID and the search key *k*. The algorithm works as follows. It follows (at most) a single path from root to leaf. At each index node along the path, there can be at most one sub-tree whose key range contains *k*. A recursive call on that sub-tree is performed (step 2c). Eventually, we reach a leaf node (step 3a). If there exists an object in the node with key=*k*, the algorithm returns the value of the object. Otherwise, the object does not exist in the tree and *NULL* is returned. Since the index nodes of the B-tree also stores objects, it is possible that the object with key=*k* is found in an index node. In this case, the algorithm returns the object value without going down to the next level (step 2a).

Algorithm *Query*(*pageID*, *k*)
Input: *pageID* of a B-tree node, a key *k* to be searched.
Output: *value* of the object with key= *k*; *NULL* if non-exist.

1. *x* = *DiskRead*(*pageID*).
2. **if** *x* is an index node
    a.   If there is an object *o* in *x* s.t. *o.key* = *k*, return *o.value*.
    b.   Find the child pointer *x.child*[*i*] whose key range contains *k*.
    c.   **return** Query(*x.child*[*i*], *k*).
3. **else**
    a.   If there is an object *o* in *x* s.t. *o.key* = *k*, return *o.value*. Otherwise, return *NULL*.
4. **end if**

As an example, Figure 16.2 shows how to perform a search query for *k* = 13. At node *A*, we should follow the left sub-tree since *k* < 25. At node *B*, we should follow the third sub-tree since 10 < *k* < 16. Now we reach a leaf node *F*. An object with key=13 is found in the node.

If the query wants to search for *k* = 12, we still examine the three nodes *A*, *B*, *F*. This time, no object with key=12 is found in *F*, and thus the algorithm returns *NULL*. If the search key is 10 instead, the algorithm only examines node *A* and *B*. Since in node *B* such an object is found, the algorithm stops there.

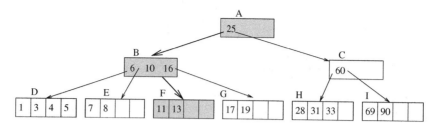

FIGURE 16.2   Query processing in a B-tree.

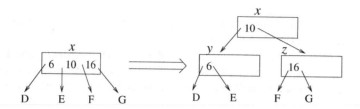

FIGURE 16.3   Splitting the root node increases the height of the tree.

Notice that in the Query algorithm, only *DiskRead* function is called. The other three functions, for example, *DiskWrite* are not needed as the algorithm does not modify the B-tree. Since the query algorithm examines a single path from root to leaf, the complexity of the algorithm in number of I/Os is $O(\log_B n)$, where n is the number of objects.

16.3.3 B-tree Insertion

To insert a new object with key k and value v into the index, we call the *Insert* algorithm given below.

Algorithm *Insert*(*root, k, v*)
Input: *root* pageID of a B-tree, the key k and the value v of a new object.
Prerequisite: The object does not exist in the tree.
Action: Insert the new object into the B-tree.

  1. $x = DiskRead(root)$.
  2. **if** x is full
    a. $y = AllocatePage()$, $z = AllocatePage()$.
    b. Locate the middle object o_i stored in x. Move the objects to the left of o_i into y. Move the objects to the right of o_i into z. If x is an index page, also move the child pointers accordingly.
    c. $x.child[1] = y.pageID$, $x.child[2] = z.pageID$.
    d. $DiskWrite(x), DiskWrite(y), DiskWrite(z)$.
  3. **end if**
  4. $InsertNotFull(x, k, v)$.

Basically, the algorithm makes sure that root page is not currently full, and then it calls the *InsertNotFull* function to insert the object into the tree. If the root page x is full, the algorithm will split it into two nodes y and z, and node x will be promoted to a higher level, thus increasing the height of the tree.

This scenario is illustrated in Figure 16.3. Node x is a full root page. It contains three objects and four child pointers. If we try to insert some record into the tree, the root node is split into two nodes y and z. Originally, x contains $x.num = 3$ objects. The left object (key=6) is moved to a new node y. The right object (key=16) is moved to a new node z. The middle object (key= 10) remains in x. Correspondingly, the child pointers D, E, F, G are also moved. Now, x contains only one object (key= 10). We make it as the new root, and make y and z be the two children of it.

To insert an object into a sub-tree rooted by a non-full node x, the following algorithm *InsertNotFull* is used.

Algorithm *InsertNotFull*(*x, k, v*)
Input: an in-memory page x of a B-tree, the key k and the value v of a new object.
Prerequisite: page x is not full.
Action: Insert the new object into the sub-tree rooted by x.

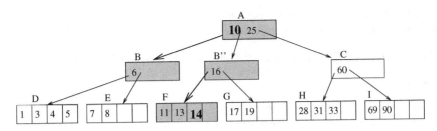

FIGURE 16.4 Inserting an object with key=14 into the B-tree of Figure 16.2 Since node B is full, it is split into two (B and B''). The object is recursively inserted into the sub-tree rooted by B''. At the lowest level, it is stored in node F.

1. **if** x is a leaf page
   a. Insert the new object into x, keeping objects in sorted order.
   b. *DiskWrite*(x).
2. **else**
   a. Find the child pointer $x.child[i]$ whose key range contains k.
   b. $w = DiskRead(x.child[i])$.
   c. **if** w is full
      i. $y = AllocatePage()$.
      ii. Locate the middle object o_j stored in w. Move the objects to the right of o_j into y. If w is an index page, also move the child pointers accordingly.
      iii. Move o_j into x. Accordingly, add a child pointer in x (to the right of o_j) pointing to y.
      iv. *DiskWrite*(x), *DiskWrite*(y), *DiskWrite*(w).
      v. If $k < o_j.key$, call *InsertNotFull*(w, k, v); otherwise, call *InsertNotFull*(y, k, v).
   d. **else**
      InsertNotFull(w, k, v).
   e. **end if**
3. **end if**

Algorithm *InsertNotFull* examines a single path from root to leaf, and eventually insert the object into some leaf page. At each level, the algorithm follows the child pointer whose key range contains the key of the new object (step 2a). If no node along the path is full, the algorithm recursively calls itself on each of these nodes (step 2d) till the leaf level, where the object is inserted into the leaf node (step 1).

Consider the other case when some node w along the path is full (step 2c). The node is first split into two (w and y). The right half of the objects from w are moved to y, while the middle object is pushed into the parent node. After the split, the key range of either w or y, but not both, contains the key of the new object. A recursive call is performed on the correct node.

As an example, consider inserting an object with key=14 into the B-tree of Figure 16.2. The result is shown in Figure 16.4. The child pointers that are followed are thick. When we examine the root node A, we follow the child pointer to B. Since B is full, we first split it into two, by moving the right half of the objects (only one object in our case, with key=16) into a new node B''. The child pointers to F and G are moved as well. Further, the previous middle object in B (key=10) is moved to the parent node A. A new child pointer to B'' is also generated in A. Now, since the key of the new object is 14, which is bigger than 10, we recursively call the algorithm on B''. At this node, since 14 < 16, we recursively call the algorithm on node F. Since F is a leaf node, the algorithm finishes by inserting the new object into F. The accessed disk pages are shown as shadowed.

16.3.4 B-tree Deletion

This section describes the *Delete* algorithm which is used to delete an object with key= k from the B-tree. It is a recursive algorithm. It takes (besides k) as parameter a tree node, and it will perform deletion in the sub-tree rooted by that node.

We know that there is a single path from the root node to the node x that contains k. The *Delete* algorithm examines this path. Along the path, at each level when we examine node x, we first make sure that x has at least one more element than half full (except the case when x is the root). The reasoning behind this is that in order to delete an element from the sub-tree rooted by x, the number of element stored in x can be reduced at most by one. If x has one more element than half full (minimum occupancy), it can be guaranteed that x will not underflow. We distinguish three cases:

1. x is a leaf node;

2. x is an index node which contains an object with key= k;

3. x is an index node which does not contain an object with key = k.

We first describe the *Delete* algorithm and then discuss the three cases in more detail.

Algorithm *Delete*(x, k)

Input: an in-memory node x of a B-tree, the key k to be deleted.

Prerequisite: an object with key= k exists in the sub-tree rooted by x.

Action: Delete the object from the sub-tree rooted by x.

1. **if** x is a leaf page
   a. Delete the object with key= k from x.
   b. *DiskWrite*(x).
2. **else if** x does not contain the object with key= k
   a. Locate the child $x.child[i]$ whose key range contains k.
   b. $y = DiskRead(x.child[i])$.
   c. **if** y is exactly half full
       i. If the sibling node z immediate to the left (right) of y has at least one more object than minimally required, add one more object to y by moving $x.key[i]$ from x to y and move that last (first) object from z to x. If y is an index node, the last (first) child pointer in z is also moved to y.
       ii. Otherwise, any immediate sibling of y is exactly half full. Merge y with an immediate sibling.
       end if
   d. *Delete*(y, k).
3. **else**
   a. If the child y that precedes k in x has at least one more object than minimally required, find the predecessor k' of k in the sub-tree rooted by y, recursively delete k' from the sub-tree and replace k with k' in x.
   b. Otherwise, y is exactly half full. We check the child z that immediately follows k in x. If z has at least one more object than minimally required, find the successor k' of k in the sub-tree rooted by z, recursively delete k' from the sub-tree and replace k with k' in x.
   c. Otherwise, both y and z are half full. Merge them into one node and push k down to the new node as well. Recursively delete k from this new node.
4. **end if**

Along the search path from the root to the node containing the object to be deleted, for each node x we encounter, there are three cases. The simplest scenario is when x is a leaf node (step 1 of the algorithm). In this case, the object is deleted from the node and the algorithm returns. Note that there is no need to handle underflow. The reason is: if the leaf node is root, there is only one node in the tree and it is fine if it has only a few objects; otherwise, the previous recursive step has already guaranteed that x has at least one more object than minimally required.

Steps 2 and 3 of the algorithm correspond to two different cases of dealing with an index node.

For step 2, the index node x does not contain the object with key= k. Thus there exists a child node y whose key range contains k. After we read the child node into memory (step 2b), we will recursively call the *Delete* algorithm on the sub-tree rooted by y (step 2d). However, before we do that, step 2(c) of the algorithm makes sure that y contains at least one more object than half full.

Suppose we want to delete 5 from the B-tree shown in Figure 16.2. When we are examining the root node A, we see that child node B should be followed next. Since B has two more objects than half full, the recursion goes to node B. In turn, since D has two more objects than minimum occupancy, the recursion goes to node D, where the object can be removed.

Let's examine another example. Still from the B+-tree shown in Figure 16.2, suppose we want to delete 33. The algorithm finds that the child node $y = C$ is half full. One more object needs to be incorporated into node C before a recursive call on C is performed. There are two sub-cases. The first sub-case is when one immediate sibling z of node y has at least one more object than minimally required. This case corresponds to step 2(c)i of the algorithm. To handle this case, we drag one object down from x to y, and we push one object from the sibling node up to x. As an example, the deletion of object 33 is shown in Figure 16.5.

Another sub-case is when all immediate siblings of y are exactly half full. In this case, we merge y with one sibling. In our 2-3-4-tree example, an index node which is half full contains one object. If we merge two such nodes together, we also drag an object from the parent node of them down to the merged node. The node will then contain three objects, which is full but does not overflow.

For instance, suppose we want to delete object 31 from Figure 16.5. When we are examining node $x = C$, we see that we need to recursively delete in the child node $y = H$. Now, both immediate siblings of H are exactly half full. So we need to merge H with

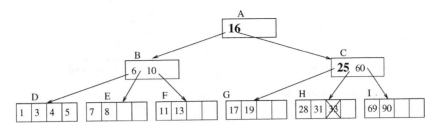

FIGURE 16.5 Illustration of step 2(c)i of the *Delete* algorithm. Deleting an object with key=33 from the B-tree of Figure 16.2. At node *A*, we examine the right child. Since node *C* only had one object before, a new object was added to it in the following way: the object with key=25 is moved from *A* to *C*, and the object with key=16 is moved from *B* to *A*. Also, the child pointer pointing to *G* is moved from *B* to *C*.

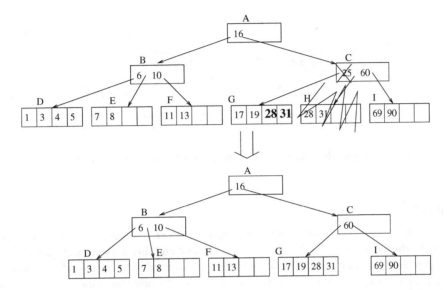

FIGURE 16.6 Illustration of step 3(c) of the *Delete* algorithm. Deleting an object with key=25 from the B-tree of Figure 16.5. At node *A*, we examine the right child. We see that node *C* contains the object with key=25. We cannot move an object up from a child node of *C*, since both children *G* and *H* (around key 25) are exactly half full. The algorithm merges these two nodes into one, by moving objects 28 and 31 from *H* to *G* and then deleting *H*. Node *C* loses an object (key=25) and a child pointer (to *H*).

a sibling, say *G*. Besides moving the remaining object 28 from *H* to *G*, we also should drag object 25 from the parent node *C* to *G*. The figure is omitted for this case.

The third case is that node *x* is an index node which contains the object to be deleted. Step 3 of algorithm *Delete* corresponds to this scenario. We cannot simply delete the object from *x*, because we also need to decrement the number of child pointers by one. In Figure 16.5, suppose we want to delete object with key= 25, which is stored in index node *C*. We cannot simply remove the object, since *C* would have one object but three child pointers left. Now, if child node *G* immediately to the left of key 25 had three or more objects, the algorithm would execute step 3(a) and move the last object from *G* into *C* to fill in the space of the deleted object. Step 3(b) is a symmetric step which shows that we can move an object from the right sub-tree.

However, in our case, both child nodes *G* and *H* are half full and thus cannot contribute an object. Step 3(c) of the algorithm corresponds to this case. As shown in Figure 16.6, the two nodes are merged into one.

16.4 The B+-tree

The most well-know variation of the B-tree is the B+-tree. There are two major differences from the B-tree. First, all objects in the B+-tree are kept in leaf nodes. Second, all leaf nodes are linked together as a double-linked list.

The structure of the B+-tree looks quite similar to the B-tree. Thus we omit the details. We do point out that in an index node of a B+-tree, different from the B-tree, we do not store object values. We still store object keys, though. However, since all objects are stored in the leaf level, the keys stored in index nodes act as *routers*, as they direct the search algorithm to go to the correct child node at each level.

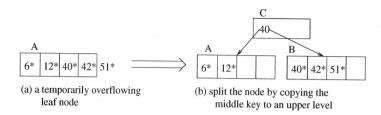

FIGURE 16.7 Illustration of a leaf-node split in the B+-tree. The middle key 40 (same as the first key in the right node) is copied up to the parent node.

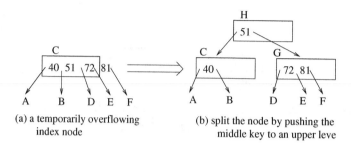

FIGURE 16.8 Illustration of an index-node split in the B+-tree. The middle key 51 is pushed up to the parent node.

16.4.1 Copy-up and Push-up

One may wonder where the routers in the index nodes come from. To understand this, let's look at an example. Initially, the B+-tree has a single node which is a leaf node. After $2B$ insertions, the root node becomes full. In Figure 16.7a, if we try to insert an object to the node A when it is already full, it temporarily overflows. To handle the overflow, the B+-tree will split the node into two nodes A and B. Furthermore, a new node C is generated, which is the new root of the tree. The first key in leaf node B is **copied** up to C. The result B+-tree is shown in Figure 16.7b.

We point out that a key in an index node may be validly replaced by some other keys, unlike in a leaf node. For instance, in node C of Figure 16.7b, we can replace the key 40–35. As long as it is smaller than all keys in the left sub-tree and bigger than or equal to all keys in the right sub-tree, it is fine.

To emphasize the fact that the keys in a index node are different from the keys in a leaf node (a key in an index node is not a real object), in the B+-tree figures we will attach a (*) to each key in a leaf node.

As a comparison, consider the split of an index node. In Figure 16.8a, the index node C temporarily overflows. It is split into two, C and G. Since before the split, C was the tree root, a new root node H is generated. See Figure 16.8b. Here the middle key 51 in the original node C is **pushed** up to the parent node.

16.4.2 B+-tree Query

As in the B-tree, the B+-tree supports the *exact-match query* which finds the object with a given key. Furthermore, the B+-tree can efficiently support the *range query*, which finds the objects whose keys are in a given range.

To perform the exact-match query, the B+-tree follows a single path from root to leaf. In the root node, there is a single child pointer whose key range contains the key to be searched for. If we follow the child pointer to the corresponding child node, inside the child node there is also a single child pointer whose key range contains the object to be searched for. Eventually, we reach a leaf node. The object to be searched, if it exists, must be located in this node. As an example, Figure 16.9 shows the search path if we search key= 42.

Beside the exact-match query, the B+-tree also supports the **range query**. That is, find all objects whose keys belong to a range R. In order to do so, all the leaf nodes of a B+-tree are linked together. If we want to search for all objects whose keys are in the range $R = [low, high]$, we perform an exact match query for key= low. This leads us to a leaf node l. We examine all objects in l, and then we follow the sibling link to the next leaf node, and so on. The algorithm stops when an object with key> $high$ is met. An example is shown in Figure 16.10.

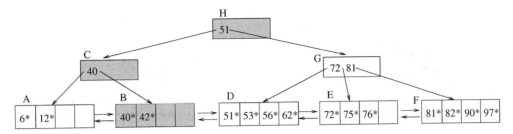

FIGURE 16.9 Illustration of the exact-match query algorithm in the B+-tree. To search for an object with key= 42, nodes *H*, *C* and *B* are examined.

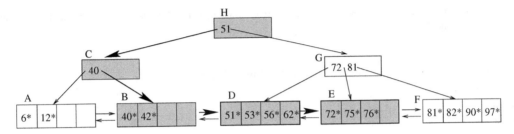

FIGURE 16.10 Illustration of the range query algorithm in the B+-tree. To search for all objects with keys in the range [42,75], the first step is to follow a path from root to leaf to find key 42 (*H*, *C* and *B* are examined). The second step is to follow the right-sibling pointers between leaf nodes and examine *D*, *E*. The algorithm stops at *E* as an object with key= 76 is found.

16.4.3 B+-tree Insertion

Since all objects in the B+-tree are located at the leaf level, the insertion algorithm of the B+-tree is actually easier than that in the B-tree. We basically follow the exact-match query to find the leaf node which should contain the object if it were in the tree. Then we insert the object into the leaf node.

What needs to be taken care of is when the leaf node overflows and is split into two. In this case, a key and a child pointer are inserted into the parent node. This may in turn cause the parent node to overflow, and so on. In the worst case, all nodes along the insertion path are split. If the root node splits into two, the height of the tree increases by one. The insertion algorithm is given below.

Algorithm *Insert*(*root*, *k*, *v*)
Input: the *root* pageID of a B+-tree, the key *k* and the value *v* of a new object.
Prerequisite: the object with key=*k* does not exist in the tree.
Action: Insert the new object into the B+-tree.

1. Starting with the root node, perform an exact-match for key= *k* till a leaf node. Let the search path be x_1, x_2, ..., x_h, where x_1 is the root node, x_h is the leaf node where the new object should be inserted into, and x_i is the parent node of x_{i+1} where $1 \leq i \leq h$-1.
2. Insert the new object with key= *k* and value = *v* into x_h.
3. Let $i = h$.
4. **while** x_i overflows
   a. Split x_i into two nodes, by moving the larger half of the keys into a new node x_i'. If x_i is a leaf node, link x_i' into the double-linked list among leaf nodes.
   b. Identify a key *kk* to be inserted into the parent level along with the child pointer pointing to x_i'. The choice of *kk* depends on the type of node x_i. If x_i is a leaf node, we need to perform *Copy-up*. That is, the smallest key in x_i' is copied as *kk* to the parent level. On the other hand, if x_i is an index node, we need to perform *Push-up*. This means the smallest key in x_i' is removed from x_i' and then stored as *kk* in the parent node.
   c. **if** $i == 1$ /* the root node overflows */
      i. Create a new index node as the new root. In the new root, store one key=*kk* and two child pointers to x_i and x_i'.
      ii. **return**
   d. **else**

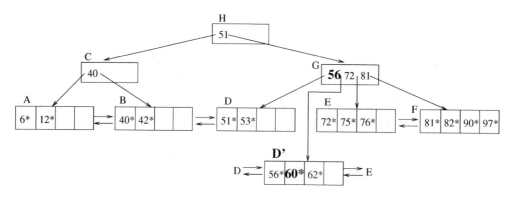

FIGURE 16.11   After inserting an object with key= 60 into the B+-tree shown in Figure 16.9. Leaf node *D* splits into two. The middle key 56 is copied up to the parent node *G*.

>   i.   Insert a key *kk* and a child pointer pointing to x_i' into node x_{i-1}.
>   ii. $i = i - 1$.
>   e.   **end if**
> 5.   **end while**

As an example, Figure 16.11 shows how to insert an object with key= 60 into the B+-tree shown in Figure 16.9.

16.4.4 B+-tree Deletion

To delete an object from the B+-tree, we first examine a single path from root to the leaf node containing the object. Then we remove the object from the node. At this point, if the node is at least half full, the algorithm returns. Otherwise, the algorithm tries to re-distribute objects between a sibling node and the underflowing node. If redistribution is not possible, the underflowing node is merged with a sibling.

Algorithm *Delete*(*root*, *k*)
Input: the *root* pageID of a B+-tree, the key *k* of the object to be deleted.
Prerequisite: the object with key= *k* exists in the tree.
Action: Delete the object with key= *k* from the B+-tree.

1.   Starting with the root node, perform an exact-match for key=*k*. Let the search path be $x_1, x_2, \ldots, x_h$, where x_1 is the root node, x_h is the leaf node that contains the object with key=*k*, and x_i is the parent node of x_{i+1}. ($1 \leq i \leq h - 1$)
2.   Delete the object with key=*k* from x_h.
3.   If $h == 1$, **return**. This is because the tree has only one node which is the root node, and we do not care whether a root node underflows or not.
4.   Let $i = h$.
   while x_i underflows
   a.   **if** an immediate sibling node *s* of x_i has at least one more entry than minimum occupancy
      i.   Re-distribute entries evenly between *s* and x_i.
      ii.   Corresponding to the re-distribution, a key *kk* in the parent node x_{i-1} needs to be modified. If x_i is an index node, *kk* is dragged down to x_i and a key from *s* is pushed up to fill in the place of *kk*. Otherwise, *kk* is simply replaced by a key in *s*.
      iii.   **return**
   b.   **else**
      i.   Merge x_i with a sibling node *s*. Delete the corresponding child pointer in x_{i-1}.
      ii.   If x_i is an index node, drag the key in x_{i-1}, which previously divides x_i and *s*, into the new node x_i. Otherwise, delete that key in x_{i-1}.
      iii. $i = i - 1$.
   c.   **end if**
   end while

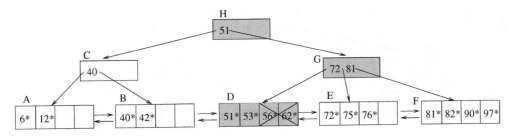

FIGURE 16.12 After deleting keys 56 and 62 from the B+-tree of Figure 16.9. Both keys are deleted from leaf node *D*, which still satisfies the minimum occupancy.

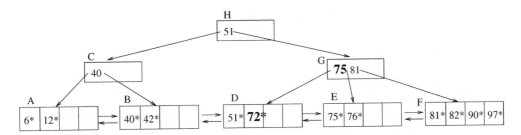

FIGURE 16.13 After deleting keys 53 from Figure 16.12. Objects in *D* and *E* are redistributed. A key in *G* is modified.

Step 1 of the algorithm follows a single path from root to leaf to find the object to be deleted. Step 2 deletes the object. The algorithm will finish at this point if any of the following two conditions hold. One, if the tree has a single node (step 3). Two, if the leaf node is at least half full after the deletion (the while loop of step 4 is skipped).

As an example, suppose we delete object 56 and then 62 from the B+-tree shown in Figure 16.9. The deletions go to the same leaf node *D*, where no underflow occurs. The result is shown in Figure 16.12.

Now, let's try to delete key 53 from Figure 16.12. This time *D* underflows. Step 4 of the *Delete* algorithm handles underflows. In general, when a node x_i underflows, the algorithm tries to borrow some entries from a sibling node *s*, as described in step 4(a). Note that we could borrow just one entry to avoid underflow in x_i. However, this is not good because next time we delete something from x_i, it will underflow again. Instead, the algorithm **redistribute** entries evenly between x_i and *s*. Assume x_i has $B - 1$ objects and *s* has $B + k$ objects, where $k \in [1..B]$. After redistribution, both x_i and *s* will have $B + (k - 1)/2$ objects. Thus x_i can take another $(k - 1)/2$ deletions before another underflow occurs.

In our example, to delete key 53 from node *D*, we re-distribute objects in *D* and *E*, by moving 72* into *D*. As discussed in step 4(a)ii of the algorithm, we also needs to modify a key in the parent node *G*. In our case, since *D* is a leaf node, we simply replace the key 72 by 75 in node *G*. Here 75 is the smallest key in *E*. The result after the redistribution is shown in Figure 16.13. As a comparison, consider the hypothetical case when *D* were an index node. In this case, we would drag down the key 72 from *G* to *D* and push up a key from *E* to *G*.

Let's proceed the example further by deleting object 72 from the tree in Figure 16.13. Now, the node *D* underflows, and redistribution is not possible (since *E*, the only immediate sibling of *D*, is exactly half full). Step 4(b) of the *Delete* algorithm tells us to merge *D* and *E* together. Correspondingly, a key and a child pointer need to be deleted from the parent node *G*. Since *D* is a leaf node, we simply delete the key 75 and the child pointer from *G*. The result is shown in Figure 16.14. As a comparison, imagine *D* were an index node. We would still remove key 75 and the child pointer from *G*, but we would keep the key 75 in node *D*.

One may wonder why in the redistribution and the merge algorithms, the leaf node and the index node are treated differently. The reason is because when we generated an index entry, we had treated two cases differently: the case when the entry points to a leaf node and the case when the entry points to a index node. This is discussed at the beginning of Section 16.4. To generate a new entry pointing to a leaf node, we copied the smallest key from the leaf node. But to generate a new entry pointing to an index node, we pushed a key from the child node up. A key which was copied up can be safely deleted later (when merge occurs). But a key which was pushed up must be kept somewhere. If we delete it from a parent node, we should drag it down to a child node.

As a running example of merging index nodes, consider deleting object 42 from the B+-tree of Figure 16.14. Node *B* underflows, and it is merged with *A*. Correspondingly, in the parent node *C*, the key 40 and the child pointer to *B* are deleted. The temporary result is shown in Figure 16.15. It's temporary since node *C* underflows.

To handle the underflow of node *C*, it is merged with *G*, its sole sibling node. As a consequence, the root node *H* now has only one child. Thus, *H* is removed and *C* becomes the new root. We point out that to merge two index nodes *C* and *G*, a key is

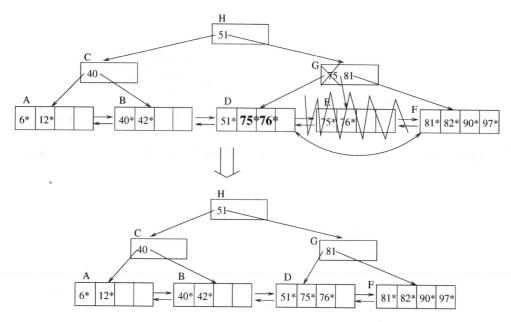

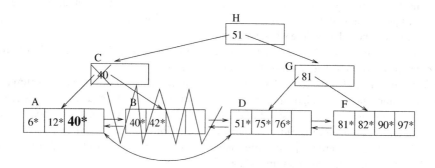

FIGURE 16.14   After deleting keys 72 from Figure 16.13. This figure corresponds to the scenario described in step 4(b) of the *Delete* algorithm. In particular, the example illustrates the merge of two leaf nodes (*D* and *E*). Node *D* underflows, but redistribution is not possible. From the parent node *G*, key 75 and child pointer to *E* are removed.

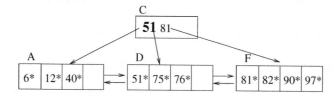

FIGURE 16.15   Temporary tree in the middle of deleting object 42 from Figure 16.14. Nodes *A* and *B* are merged. Key 40 and child pointer to *B* are removed from *C*.

FIGURE 16.16   After the deletion of object 42 is finished. This figure illustrates an example of merging two index nodes. In particular, index nodes *C* and *G* are merged. The key 51 is dragged down from the parent node *H*. Since that is the only key in root *H*, node *C* becomes the new root and the height of the tree is decreased by one.

dragged down from the parent node (versus being deleted in the case of merging leaf nodes). The final result after completing the deletion of 42* is shown in Figure 16.16.

16.5 Further Discussions

In this section we discuss various issues of the B-tree and the B+-tree.

16.5.1 Efficiency Analysis

Theorem: *In the B-tree or the B+-tree, the I/O cost of insertion, deletion and exact-match query is $O(\log_B n)$. In the B+-tree, the I/O cost of a range search is $O(\log_B n + t/B)$. Here B is the minimum page capacity in number of records, n is the total number of objects in the tree, and t is the number of objects in the range query result.*

The correctness of the theorem can be seen from the discussion of the algorithms. Basically, for both the B-tree and the B+-tree, all the insertion, deletion and exact-match query algorithms examine a single path from root to leaf. At each node, the algorithm might examine up to two other nodes. However, asymptotically the complexity of these algorithms are equal to the height of the tree. Since there are n objects, and the minimum fan-out of the tree is B, the height of the tree is $O(\log_B n)$. So the complexity of the algorithms is $O(\log_B n)$ as well.

For the range query in the B+-tree, $\log_B n$ nodes are examined to find the leaf node that contains the low value of the query range. By following the sibling pointers in the leaf nodes, the other leaf nodes that contain objects in the query range are also found. Among all the leaf nodes examined, except for the first and the last, every node contains at least B objects in the query result. Thus if there are t objects in the query result, the range query complexity is $O(\log_B n + t/B)$.

16.5.2 Why Is the B+-tree Widely Accepted?

One can safely claim that the B+-tree has been included in at least 99%, if not all, of the database management systems (DBMS). No other index structure has received so much attention. Why is that?

Let's do some calculation. First, we point out that a practical number of minimum occupancy of a B+-tree is $B = 100$. Thus the fan-out of the tree is between 100 and 200. Analysis has shown that in a real-world B+-tree, the average page capacity is about 66.7% full. Or, a page typically contains 200*66.7%=133 entries. Here is the relationship between the height of the tree and the number of objects that can hold in a typical B+-tree:

- *height=0*: B+-tree holds 133 objects on average. There is a single node, which is 66.7% full.
- *height=1*: B+-tree holds $133^2 = 17,689$ objects. There are 133 leaf nodes, each holds 133 objects.
- *height=2*: B+-tree holds $133^3 = 2,352,637$ objects.
- *height=3*: B+-tree holds $133^4 = 312,900,721$ (over 0.3 billion) objects.

The first two levels of the B+-tree contains $1 + 133 = 134$ disk pages. This is very small. If a disk page is 4 KB large, 134 disk pages occupy 134*4 KB=536 KB disk space. It's quite reasonable to assume that the first two levels of the B+-tree always stays in memory.

The calculations lead to this discovery: in a large database with 0.3 billion objects, to find one object we only need to access two disk pages! This is unbelievably good.

16.5.3 Bulk-Loading a B+-tree

In some cases, we are given a large set of records and we are asked to build a B+-tree index. Of course, we can start with an empty B+-tree and insert one record at a time using the Insert algorithm. However, this approach is not efficient, as the I/O cost is $O(n \cdot \log_B n)$.

Many systems have implemented the *bulk-loading* utility. The idea is as follows. First, sort the objects. Use the objects to fill in leaf nodes in sequential order. For instance, if a leaf node holds up to $2B$ objects, the $2B$ smallest objects are stored in page 1, the next $2B$ objects are stored in page 2, etc. Next, build the index nodes at one level up. Assume an index node holds up to $2B$ child pointers. Create the first index node as the parent of the first $2B$ leaf nodes. Create the second index node as the parent of the next $2B$ leaf nodes, etc. Then, build the index nodes at two levels above the leaf level, and so on. The process stops when there is only one node at a level. This node is the tree root.

If the objects are sorted already, the bulk-loading algorithm has an I/O cost of $O(n/B)$. Otherwise, the bulk-loading algorithm has asymptotically the same I/O cost as external sort, which is $O(n/B \cdot \log_B n)$. Notice that even if the bulk-loading algorithm performs a sorting first, it is still B times faster than inserting objects one at a time into the structure.

16.5.4 Aggregation Query in a B+-tree

The B+-tree can also be used to answer the *aggregation query*: "given a key range R, find the aggregate value of objects whose keys are in R." The standard SQL supports the following aggregation operators: COUNT, SUM, AVG, MIN, MAX. For instance, the COUNT operator returns the number of objects in the query range. Here AVG can be computed as SUM/AVG. Thus we focus on the other four aggregate operators.

Since the B+-tree efficiently supports the range query, it makes sense to utilize it to answer the aggregation query as well. Let's first look at some concepts.

Associated with each aggregate operator, there exists a init_value and an *aggregate* function. The *init_value* is the aggregate for an empty set of objects. For instance, the *init_value* for the COUNT operator is 0. The *aggregate* function computes the aggregate value. There are two versions. One version takes two aggregate values of object set S_1 and S_2, and computes the aggregate value of set $S_1 \cup S_2$. Another version takes one aggregate value of set S_1 and an object o and computes the aggregate value of $S_1 \cup \{o\}$. For instance, if we know $COUNT_1$ and $COUNT_2$ of two sets, the COUNT for the whole set is $COUNT_1 + COUNT_2$. The COUNT of subset 1 added with an object o is $COUNT_1 + 1$. The *init_value* and the *aggregate* functions for COUNT, SUM, MIN, and MAX are shown below.

- COUNT operator:
  - $init\_value = 0$
  - $aggregate(COUNT_1, COUNT_2) = COUNT_1 + COUNT_2$
  - $aggregate(COUNT_1, object) = COUNT_1 + 1$
- SUM operator:
  - $init\_value = 0$
  - $aggregate(SUM_1, SUM_2) = SUM_1 + SUM_2$
  - $aggregate(SUM_1, object) = SUM_1 + object.value$
- MIN operator:
  - $init\_value = +\infty$
  - $aggregate(MIN_1, MIN_2) = min\{MIN_1, MIN_2\}$
  - $aggregate(MIN_1, object) = min\{MIN_1, object.value\}$
- MAX operator:
  - $init\_value = -\infty$
  - $aggregate(MAX_1, MAX_2) = max\{MAX_1, MAX_2\}$
  - $aggregate(MAX_1, object) = max\{MAX_1, object.value\}$

The B+-tree can support the aggregation query in the following way. We keep a temporary aggregate value, which is initially set to be *init_value*. A range search is performed on the B+-tree. For each object found, its value is aggregated with the temporary aggregate value on-the-fly. When all objects whose keys are in the query range are processed, this temporary aggregate value is returned.

However, this approach is not efficient, as the I/O cost is $O(\log_B n + t/B)$, which is linear to the number of objects divided by B. If the query range is large, the algorithm needs to access too many disk pages. It is ideal to find some approach whose query performance is independent to the size of the objects in the query range.

A better way is to store the local aggregate values in the tree. In more detail, along with each child pointer, we store the aggregate value of all objects in the corresponding sub-tree. By doing so, if the query range fully contains the key range of a sub-tree, we take the associated local aggregate value and avoid browsing the sub-tree. We call such a B+-tree with extra aggregate information the **aggregation B+-tree**. The algorithm to perform a aggregation query using the **aggregation B+-tree** is shown below.

Algorithm *Aggregation*(x, R)
Input: a node x of an aggregation B+-tree, the query key range R.
Action: Among objects in the sub-tree rooted by x, compute the aggregate value of objects whose keys belong to R.

1. Initialize the temporary aggregation value v as *init_value*.
2. **if** x is a leaf node
   a. For every object o in x where $o.value \in R$, $v = aggr(v, o)$.
3. **else**
   a. **for** every child pointer $x.child[i]$
      i. **if** the key range of $x.child[i]$ is contained in R
         $v = aggregate(v, x.child[i].aggr)$
      ii. **else if** the key range of $x.child[i]$ intersects R
         $y = DiskRead(x.child[i])$
         $v = aggregate(v, Aggregation(y, R))$
      iii. **end if**
   b. **end for**
4. **return** v.

The algorithm starts with examining the root node. Here the child pointers are divided into three groups. (1) There are at most two child pointers whose key ranges intersect the query range R. (2) The child pointers between them have key ranges fully contained in R. (3) The child pointers outside of them have key ranges non-intersecting with R.

For child pointers in group (2), the local aggregate stored at the child pointer (represented by *x.child*[*i*]*.aggr*) is aggregated to the temporary aggregate value and the examination of the sub-tree is avoided. This is shown in step 3(a)i of the algorithm. For child pointers in group (3), no object in the sub-trees will contribute to the query and the examination of the sub-trees are also avoided.

For each of the two child pointers whose key ranges intersect R, a recursive call to *Aggregation* is performed. This is shown in step 3(a)ii of the algorithm. If we go one level down, in each of the two child nodes, there can be at most one child pointer whose key range intersects R. Take the left child node as an example. If there is a child pointer whose key range intersect R, all child pointers to the left of it will be outside of R and all child pointers to the right of it will be fully contained in R. Thus the algorithm examines two paths from root to leaf.

Theorem: *The I/O cost of the Aggregation query algorithm is $O(\log_B n)$.*

The above theorem shows that the aggregation query performance is independent of the number of objects in the query range.

References

1. R. Bayer and E. M. McCreight: Organization and maintenance of large ordered indices. *Acta Informatica*, 1, 1972.
2. D. Knuth: *The Art of Computer Programming, Vol. 3: Sorting and Searching.* Reading, MA: Addison Wesley, 1973.
3. D. Comer: The Ubiquitous B-Tree. *ACM Computing Surveys*, 11(2), 1979.
4. A. Guttman: R-trees: A dynamic index structure for spatial searching. In *Proceedings of ACM/SIGMOD Annual Conference on Management of Data (SIGMOD)*, 1984.

IV

Multidimensional/
Spatial Structures

17 **Multidimensional Spatial Data Structures** Hanan Samet 251
   Introduction • Point Data • Bucketing Methods • Region Data • Rectangle Data • Line Data and
   Boundaries of Regions • Research Issues and Summary • Acknowledgment • References

18 **Planar Straight Line Graphs** Siu-Wing Cheng 277
   Introduction • Features of PSLGs • Operations on PSLGs • Winged-Edge • Halfedge • Quadedge
   • Further Remarks • Glossary • Acknowledgments • References

19 **Interval, Segment, Range, and Priority Search Trees** D. T. Lee and Hung-I Yu 291
   Introduction • Interval Trees • Segment Trees • Range Trees • Priority Search Trees •
   Acknowledgments • References

20 **Quadtrees and Octrees** Srinivas Aluru 309
   Introduction • Quadtrees for Point Data • Spatial Queries with Region Quadtrees • Image
   Processing Applications • Scientific Computing Applications • Acknowledgments • References

21 **Binary Space Partitioning Trees** Bruce F. Naylor 329
   Introduction • BSP Trees as a Multi-Dimensional Search Structure • Visibility Orderings • BSP
   Tree as a Hierarchy of Regions • Bibliography

22 **R-Trees** Scott Leutenegger and Mario A. Lopez 343
   Introduction • Basic Concepts • Improving Performance • Advanced Operations • Analytical
   Models • Acknowledgment • References

23 **Managing Spatiotemporal Data** Sumeet Dua and S. S. Iyengar 359
   Introduction and Background • Overlapping Linear Quadtree • 3D R-Tree • 2+3 R-Tree •
   HR-Trees • MV3R-Tree • Indexing Structures for Continuously Moving Objects • References

24 **Kinetic Data Structures** Leonidas Guibas 377
   Introduction • Motion in Computational Geometry • Motion Models • Kinetic Data Structures •
   A KDS Application Survey • Querying Moving Objects • Sources and Related Materials •
   References

25 **Online Dictionary Structures** Teofilo F. Gonzalez 389
   Introduction • Trie Implementations • Binary Search Tree Implementations • Balanced BST
   Implementation • Additional Operations • Discussion • References

26 **Cuttings** Bernard Chazelle 397
   Introduction • The Cutting Construction • Applications • Acknowledgments • References

27   Approximate Geometric Query Structures Christian A. Duncan and Michael T. Goodrich............................ 405
Introduction • General Terminology • Approximate Queries • Quasi-BAR Bounds • BBD Trees •
BAR Trees • Maximum-Spread *k*-d Trees • Acknowledgments • References

28   Geometric and Spatial Data Structures in External Memory Jeffrey Scott Vitter 419
Introduction • EM Algorithms for Batched Geometric Problems • EM Tree Data Structures •
Spatial Data Structures and Range Search • Related Problems • Dynamic and Kinetic Data
Structures • Conclusion • Acknowledgments • References

17

Multidimensional Spatial Data Structures

17.1 Introduction..251
17.2 Point Data 252
17.3 Bucketing Methods..256
17.4 Region Data 259
17.5 Rectangle Data...263
17.6 Line Data and Boundaries of Regions..265
17.7 Research Issues and Summary...268
Acknowledgment...269
References...269

Hanan Samet*
University of Maryland at College
Park

17.1 Introduction

The representation of multidimensional data is an important issue in applications in diverse fields that include database management systems (see Chapter 62), computer graphics (see Chapter 55), computer vision, computational geometry (see Chapters 65, 66 and 67), image processing (see Chapter 58), geographic information systems (GIS) (see Chapter 56), pattern recognition, VLSI design (Chapter 53), and others. The most common definition of multidimensional data is a collection of points in a higher dimensional space. These points can represent locations and objects in space as well as more general records where only some, or even none, of the attributes are locational. As an example of nonlocational point data, consider an employee record which has attributes corresponding to the employee's name, address, sex, age, height, weight, and social security number. Such records arise in database management systems and can be treated as points in, for this example, a seven-dimensional space (i.e., there is one dimension for each attribute) albeit the different dimensions have different type units (i.e., name and address are strings of characters, sex is binary; while age, height, weight, and social security number are numbers).

When multidimensional data corresponds to locational data, we have the additional property that all of the attributes have the same unit which is distance in space. In this case, we can combine the distance-denominated attributes and pose queries that involve proximity. For example, we may wish to find the closest city to Chicago within the two-dimensional space from which the locations of the cities are drawn. Another query seeks to find all cities within 50 miles of Chicago. In contrast, such queries are not very meaningful when the attributes do not have the same type.

When multidimensional data spans a continuous physical space (i.e., an infinite collection of locations), the issues become more interesting. In particular, we are no longer just interested in the locations of objects, but, in addition, we are also interested in the space that they occupy (i.e., their extent). Some example objects include lines (e.g., roads, rivers), regions (e.g., lakes, counties, buildings, crop maps, polygons, polyhedra), rectangles, and surfaces. The objects may be disjoint or could even overlap. One way to deal with such data is to store it explicitly by parameterizing it and thereby reduce it to a point in a higher dimensional space. For example, a line in two-dimensional space can be represented by the coordinate values of its endpoints (i.e., a pair of x and a pair of y coordinate values) and then stored as a point in a four-dimensional space [1]. Thus, in effect, we have constructed a transformation (i.e., mapping) from a two-dimensional space (i.e., the space from which the lines are drawn) to a four-dimensional space (i.e., the space containing the representative point corresponding to the line).

The transformation (also known as parameterization) approach is fine if we are just interested in retrieving the data. It is appropriate for queries about the objects (e.g., determining all lines that pass through a given point or that share an endpoint, etc.) and the immediate space that they occupy. However, the drawback of the transformation approach is that it ignores the

---

* All figures © 2017 by Hanan Samet.

geometry inherent in the data (e.g., the fact that a line passes through a particular region) and its relationship to the space in which it is embedded.

For example, suppose that we want to detect if two lines are near each other, or, alternatively, to find the nearest line to a given line. This is difficult to do in the four-dimensional space, regardless of how the data in it is organized, since proximity in the two-dimensional space from which the lines are drawn is not necessarily preserved in the four-dimensional space into which the lines have been transformed. In other words, although the two lines may be very close to each other, the Euclidean distance between their representative points may be quite large, unless the lines are approximately the same size, in which case proximity is preserved [2].

Of course, we could overcome these problems by projecting the lines back to the original space from which they were drawn, but in such a case, we may ask what was the point of using the transformation in the first place? In other words, at the least, the representation that we choose for the data should facilitate the performance of operations on the data. Thus when the multidimensional spatial data is nondiscrete, we need representations besides those that are designed for point data. The most common solution, and the one that we focus on in the rest of this chapter, is to use data structures that are based on spatial occupancy. Such methods decompose the space from which the spatial data is drawn (e.g., the two-dimensional space containing the lines) into regions that are usually called *buckets* because they often contain more than just one element. They are also commonly known as *bucketing methods*.

In this chapter, we explore a number of different representations of multidimensional data bearing the above issues in mind. While we cannot give exhaustive details of all of the data structures, we try to explain the intuition behind their development as well as to give literature pointers to where more information can be found. Many of these representations are described in greater detail in References 3–5 including an extensive bibliography. Our approach is primarily a descriptive one. Most of our examples are of two-dimensional spatial data although the representations are applicable to higher dimensional spaces as well.

At times, we discuss bounds on execution time and space requirements. Nevertheless, this information is presented in an inconsistent manner. The problem is that such analyses are very difficult to perform for many of the data structures that we present. This is especially true for the data structures that are based on spatial occupancy (e.g., quadtree [see Chapter 20 for more details] and R-tree [see Chapter 22 for more details] variants). In particular, such methods have good observable average-case behavior but may have very bad worst cases which may only arise rarely in practice. Their analysis is beyond the scope of this chapter and usually we do not say anything about it. Nevertheless, these representations find frequent use in applications where their behavior is deemed acceptable, and is often found to be better than that of solutions whose theoretical behavior would appear to be superior. The problem is primarily attributed to the presence of large constant factors which are usually ignored in the *big O* and Ω analyses [6].

The rest of this chapter is organized as follows. Section 17.2 reviews a number of representations of point data of arbitrary dimensionality. Section 17.3 describes bucketing methods that organize collections of spatial objects (as well as multidimensional point data) by aggregating the space that they occupy. The remaining sections focus on representations of nonpoint objects of different types. Section 17.4 covers representations of region data, while Section 17.5 discusses a subcase of region data which consists of collections of rectangles. Section 17.6 deals with curvilinear data which also includes polygonal subdivisions and collections of line segments. Section 17.7 contains a summary and a brief indication of some research issues and directions for future work.

17.2 Point Data

The simplest way to store point data of arbitrary dimension is in a sequential list. Accesses to the list can be sped up by forming sorted lists for the various attributes which are known as *inverted lists* [7]. There is one list for each attribute. This enables pruning the search with respect to the value of one of the attributes. It should be clear that the inverted list is not particularly useful for multidimensional range searches. The problem is that it can only speed up the search for one of the attributes (termed the *primary* attribute). A widely used solution is exemplified by the *fixed-grid* method [7,8]. It partitions the space from which the data is drawn into rectangular cells by overlaying it with a grid. Each grid cell c contains a pointer to another structure (e.g., a list) which contains the set of points that lie in c. Associated with the grid is an access structure to enable the determination of the grid cell associated with a particular point p. This access structure acts like a directory and is usually in the form of a d-dimensional array with one entry per grid cell or a tree with one leaf node per grid cell.

There are two ways to build a fixed grid. We can either subdivide the space into equal-sized intervals along each of the attributes (resulting in congruent grid cells) or place the subdivision lines at arbitrary positions that are dependent on the underlying data. In essence, the distinction is between organizing the data to be stored and organizing the embedding space from which the data is drawn [9]. In particular, when the grid cells are congruent (i.e., equal-sized when all of the attributes are locational with the same range and termed a *uniform grid*), use of an array access structure is quite simple and has the desirable property that the grid

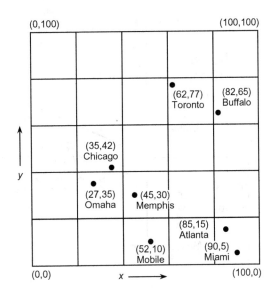

FIGURE 17.1 Uniform-grid representation corresponding to a set of points with a search radius of 20.

cell associated with point p can be determined in constant time. Moreover, in this case, if the width of each grid cell is twice the search radius for a rectangular range query, then the average search time is $O(F \cdot 2^d)$ where F is the number of points that have been found [10]. Figure 17.1 is an example of a uniform-grid representation for a search radius equal to 10 (i.e., a square of size 20×20).

The use of an array access structure when the grid cells are not congruent requires us to have a way of keeping track of their size so that we can determine the entry of the array access structure corresponding to the grid cell associated with point p. One way to do this is to make use of what are termed *linear scales* which indicate the positions of the grid lines (or partitioning hyperplanes in $d > 2$ dimensions). Given a point p, we determine the grid cell in which p lies by finding the "coordinate values" of the appropriate grid cell. The linear scales are usually implemented as one-dimensional trees containing ranges of values.

The array access structure is fine as long as the data is static. When the data is dynamic, it is likely that some of the grid cells become too full while other grid cells are empty. This means that we need to rebuild the grid (i.e., further partition the grid or reposition the grid partition lines or hyperplanes) so that the various grid cells are not too full. However, this creates many more empty grid cells as a result of repartitioning the grid (i.e., empty grid cells are split into more empty grid cells). The number of empty grid cells can be reduced by merging spatially adjacent empty grid cells into larger empty grid cells, while splitting grid cells that are too full, thereby making the grid adaptive. The result is that we can no longer make use of an array access structure to retrieve the grid cell that contains query point p. Instead, we make use of a tree access structure in the form of a k-ary tree where k is usually 2^d. Thus what we have done is marry a k-ary tree with the fixed-grid method. This is the basis of the point quadtree [11] and the PR quadtree [4,12] which are multidimensional generalizations of binary trees.

The difference between the point quadtree and the PR quadtree is the same as the difference between *trees* and *tries* [13], respectively. The binary search tree [7] is an example of the former since the boundaries of different regions in the search space are determined by the data being stored. Address computation methods such as radix searching [7] (also known as digital searching) are examples of the latter, since region boundaries are chosen from among locations that are fixed regardless of the content of the data set. The process is usually a recursive halving process in one dimension, recursive quartering in two dimensions, and so on, and is known as *regular decomposition*.

In two dimensions, a point quadtree is just a two-dimensional binary search tree. The first point that is inserted serves as the root, while the second point is inserted into the relevant quadrant of the tree rooted at the first point. Clearly, the shape of the tree depends on the order in which the points were inserted. For example, Figure 17.2 is the point quadtree corresponding to the data of Figure 17.1 inserted in the order `Chicago`, `Mobile`, `Toronto`, `Buffalo`, `Memphis`, `Omaha`, `Atlanta`, and `Miami`.

In two dimensions, the PR quadtree is based on a recursive decomposition of the underlying space into four congruent (usually square in the case of locational attributes) cells until each cell contains no more than one point. For example, Figure 17.3a is the partition of the underlying space induced by the PR quadtree corresponding to the data of Figure 17.1, while Figure 17.3b is its tree representation. The shape of the PR quadtree is independent of the order in which data points are inserted into it. The disadvantage of the PR quadtree is that the maximum level of decomposition depends on the minimum separation between two points. In particular, if two points are very close, then the decomposition can be very deep. This can be overcome by viewing the cells or nodes as buckets with capacity c and only decomposing a cell when it contains more than c points.

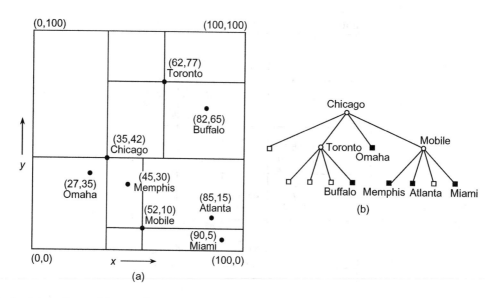

FIGURE 17.2 A point quadtree and the records it represents corresponding to Figure 17.1: (a) the resulting partition of space and (b) the tree representation.

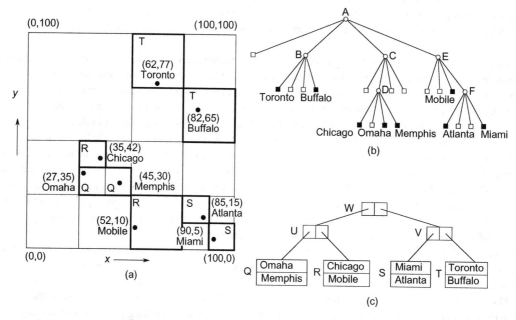

FIGURE 17.3 A PR quadtree and the points it represents corresponding to Figure 17.1: (a) the resulting partition of space, (b) the tree representation, and (c) one possible B$^+$-tree for the nonempty grid cells where each node has a minimum of 2 and a maximum of 3 entries. The nonempty grid cells in (a) have been labeled with the name of the B$^+$-tree leaf node in which they are a member.

As the dimensionality of the space increases, each level of decomposition of the quadtree results in many new cells as the fanout value of the tree is high (i.e., 2^d). This is alleviated by making use of a k-d tree [14]. The k-d tree is a binary tree where at each level of the tree, we subdivide along a different attribute so that, assuming d locational attributes, if the first split is along the x-axis, then after d levels, we cycle back and again split along the x-axis. It is applicable to both the point quadtree and the PR quadtree (in which case we have a *PR k-d tree*, or a bintree in the case of region data).

At times, in the dynamic situation, the data volume becomes so large that a tree access structure such as the one used in the point and PR quadtrees is inefficient. In particular, the grid cells can become so numerous that they cannot all fit into memory thereby causing them to be grouped into sets (termed *buckets*) corresponding to physical storage units (i.e., pages) in secondary storage. The problem is that, depending on the implementation of the tree access structure, each time we must follow a pointer, we may need to make a disk access. Below, we discuss two possible solutions: one making use of an array access structure and one

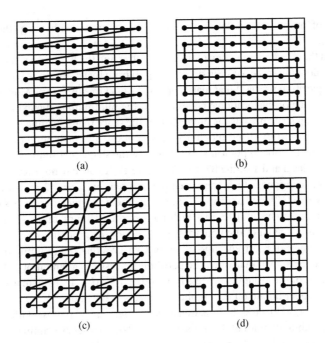

FIGURE 17.4 The result of applying four common different space-ordering methods to an 8 × 8 collection of pixels whose first element is in the upper left corner: (a) row order, (b) row-prime order, (c) Morton order, and (d) Peano–Hilbert order.

making use of an alternative tree access structure with a much larger fanout. We assume that the original decomposition process is such that the data is only associated with the leaf nodes of the original tree structure.

The difference from the array access structure used with the static fixed-grid method described earlier is that the array access structure (termed *grid directory*) may be so large (e.g., when d gets large) that it resides on disk as well, and the fact that the structure of the grid directory can change as the data volume grows or contracts. Each grid cell (i.e., an element of the grid directory) stores the address of a bucket (i.e., page) that contains the points associated with the grid cell. Notice that a bucket can correspond to more than one grid cell. Thus any page can be accessed by two disk operations: one to access the grid cell and one more to access the actual bucket.

This results in *EXCELL* [15] when the grid cells are congruent (i.e., equal-sized for locational data), and *grid file* [9] when the grid cells need not be congruent. The difference between these methods is most evident when a grid partition is necessary (i.e., when a bucket becomes too full and the bucket is not shared among several grid cells). In particular, a grid partition in the grid file only splits one interval in two thereby resulting in the insertion of a $(d-1)$-dimensional cross-section. On the other hand, a grid partition in EXCELL means that all intervals must be split in two thereby doubling the size of the grid directory.

An alternative to the array access structure is to assign an ordering to the grid cells resulting from the adaptive grid, and then to impose a tree access structure on the elements of the ordering that correspond to the nonempty grid cells. The ordering is analogous to using a mapping from d dimensions to one dimension. There are many possible orderings (e.g., see chapter 2 in Reference 5) with the most popular shown in Figure 17.4.

The domain of these mappings is the set of locations of the smallest possible grid cells (termed *pixels*) in the underlying space and thus we need to use some easily identifiable pixel in each grid cell such as the one in the grid cell's lower left corner. Of course, we also need to know the size of each grid cell. One mapping simply concatenates the result of interleaving the binary representations of the coordinate values of the lower left corner (e.g., (a, b) in two dimensions) and i of each grid cell of size 2^i so that i is at the right. The resulting number is termed a *locational code* and is a variant of the Morton order (Figure 17.4c). Assuming such a mapping and sorting the locational codes in increasing order yields an ordering equivalent to that which would be obtained by traversing the leaf nodes (i.e., grid cells) of the tree representation (e.g., Figure 17.8b) in the order SW, SE, NW, NE. It is also known as the Z-order on account of the Z-like pattern formed by the ordering. The Morton order (as well as the Peano–Hilbert order shown in Figure 17.4d) is particularly attractive for quadtree-like decompositions because all pixels within a grid cell appear in consecutive positions in the order. Alternatively, these two orders exhaust a grid cell before exiting it.

For example, Figure 17.3c shows the result of imposing a B$^+$-tree [16] access structure on the nonempty grid cells of the PR quadtree given in Figure 17.3b. Each node of the B$^+$-tree in our example has a minimum of 2 and a maximum of 3 entries. Figure 17.3c does not contain the values resulting from applying the mapping to the individual grid cells nor does it show the discriminator values that are stored in the nonleaf nodes of the B$^+$-tree. The nonempty grid cells of the PR quadtree in Figure

17.3a are marked with the label of the leaf node of the B^+-tree of which they are a member (e.g., the grid cell containing `Chicago` is in leaf node Q of the B^+-tree).

It is important to observe that the above combination of the PR quadtree and the B^+-tree has the property that the tree structure of the partition process of the underlying space has been decoupled [17] from that of the node hierarchy (i.e., the grouping process of the nodes resulting from the partition process) that makes up the original tree directory. More precisely, the grouping process is based on proximity in the ordering of the locational codes and on the minimum and maximum capacity of the nodes of the B^+-tree. Unfortunately, the resulting structure has the property that the space that is spanned by a leaf node of the B^+-tree (i.e., the grid cells spanned by it) has an arbitrary shape and, in fact, does not usually correspond to a k-dimensional hyperrectangle. In particular, the space spanned by the nonempty leaf node may have the shape of a staircase (e.g., the nonempty grid cells in Figure 17.3a that comprise leaf nodes S and T of the B^+-tree in Figure 17.3c) or may not even be connected in the sense that it corresponds to regions that are not contiguous (e.g., the nonempty grid cells in Figure 17.3a that comprise the leaf node R of the B^+-tree in Figure 17.3c). The PK-tree [18] is an alternative decoupling method which overcomes these drawbacks by basing the grouping process on k-instantiation which stipulates that each node of the grouping process contains a minimum of k objects or nonempty grid cells. The result is that all of the nonempty grid cells of the grouping process are congruent at the cost that the result is not balanced although use of relatively large values of k ensures that the resulting trees are relatively shallow or of a limited number of shapes. It can be shown that when the partition process has a fanout of f, then k-instantiation means that the number of objects in each node of the grouping process is bounded by $f \cdot (k - 1)$. Note that k-instantiation is different from bucketing where we only have an upper bound on the number of objects in the node.

Fixed-grids, quadtrees, k-d trees, grid file, EXCELL, as well as other hierarchical representations are good for range searching queries such as finding all cities within 80 miles of St. Louis. In particular, they act as pruning devices on the amount of search that will be performed as many points will not be examined since their containing cells lie outside the query range. These representations are generally very easy to implement and have good expected execution times, although they are quite difficult to analyze from a mathematical standpoint. However, their worst cases, despite being rare, can be quite bad. These worst cases can be avoided by making use of variants of range trees [19] and priority search trees [20]. For more details about these data structures, see Chapter 19.

17.3 Bucketing Methods

There are four principal approaches to decomposing the space from which the objects are drawn. The first approach makes use of an object hierarchy and the space decomposition is obtained in an indirect manner as the method propagates the space occupied by the objects up the hierarchy with the identity of the propagated objects being implicit to the hierarchy. In particular, associated with each object is an object description (e.g., for region data, it is the set of locations in space corresponding to the cells that make up the object). Actually, since this information may be rather voluminous, it is often the case that an approximation of the space occupied by the object is propagated up the hierarchy rather than the collection of individual cells that are spanned by the object. For spatial data, the approximation is usually the minimum bounding rectangle formally known as the minimum axis-aligned bounding box (AABB) for the object, while for nonspatial data it is simply the hyperrectangle whose sides have lengths equal to the ranges of the values of the attributes. Therefore, associated with each element in the hierarchy is a bounding rectangle corresponding to the union of the bounding rectangles associated with the elements immediately below it.

The R-tree [21,22] (also sometimes referenced as an AABB tree) is an example of an object hierarchy which finds use especially in database applications. Note that other bounding objects can be used such as spheres (e.g., the SS-tree [23]), and the intersection of the minimum bounding sphere and the minimum bounding rectangle (e.g., the SR-tree [24]). Moreover, the minimum bounding rectangles need not be axis-aligned (e.g., the OBB-tree [25]). The number of objects or bounding rectangles that are aggregated in each node is permitted to range between a minimum of m where $m \le \lceil M/2 \rceil$ and a maximum of M. The root node in an R-tree has at least two entries unless it is a leaf node in which case it has just one entry corresponding to the bounding rectangle of an object. The R-tree is usually built dynamically as the objects are encountered rather than waiting until all objects have been input (dynamically). The hierarchy is implemented as a tree structure with grouping being based, in part, on proximity of the objects or bounding rectangles.

For example, consider the collection of line segment objects given in Figure 17.5 shown embedded in a 4×4 grid. Figure 17.6a is an example R-tree for this collection with $m = 2$ and $M = 3$. Figure 17.6b shows the spatial extent of the bounding rectangles of the nodes in Figure 17.6a, with heavy lines denoting the bounding rectangles corresponding to the leaf nodes, and broken lines denoting the bounding rectangles corresponding to the subtrees rooted at the nonleaf nodes. Note that the R-tree is not unique. Its structure depends heavily on the order in which the individual objects were inserted into (and possibly deleted from) the tree.

Given that each R-tree node can contain a varying number of objects or bounding rectangles, it is not surprising that the R-tree was inspired by the B-tree [26]. Therefore, nodes are viewed as analogous to disk pages. Thus the parameters defining the tree

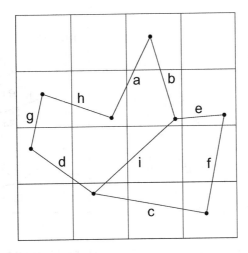

FIGURE 17.5 Example collection of line segments embedded in a 4 × 4 grid.

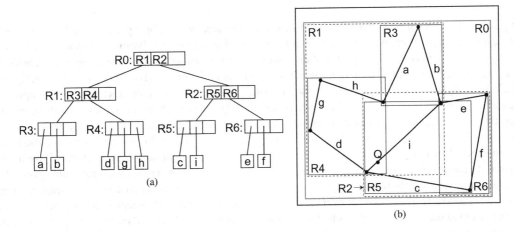

FIGURE 17.6 (a) R-tree for the collection of line segments with $m = 2$ and $M = 3$, in Figure 17.5, and (b) the spatial extents of the bounding rectangles. Notice that the leaf nodes in the index also store bounding rectangles although this is only shown for the nonleaf nodes.

(i.e., m and M) are chosen so that a small number of nodes is visited during a spatial query (i.e., point and range queries), which means that m and M are usually quite large. The actual implementation of the R-tree is really a B$^+$-tree [16] as the objects are restricted to the leaf nodes.

The efficiency of the R-tree for search operations depends on its ability to distinguish between occupied space and unoccupied space (i.e., coverage), and to prevent a node from being examined needlessly due to a false overlap with other nodes. In other words, we want to minimize coverage and overlap. These goals guide the initial R-tree creation process as well, subject to the previously mentioned constraint that the R-tree is usually built as the objects are encountered rather than waiting until all objects have been input (i.e., dynamically).

The drawback of the R-tree (and any representation based on an object hierarchy) is that it does not result in a disjoint decomposition of space. The problem is that an object is only associated with one bounding rectangle (e.g., line segment i in Figure 17.6 is associated with bounding rectangle R5, yet it passes through R1, R2, R4, and R5, as well as through R0 as do all the line segments). In the worst case, this means that if we wish to determine which object (e.g., an intersecting line in a collection of line segment objects or a containing rectangle in a collection of rectangle objects) is associated with a particular point in the two-dimensional space from which the objects are drawn, then we may have to search the entire collection. For example, in Figure 17.6, when searching for the line segment that passes through point Q, we need to examine bounding rectangles R0, R1, R4, R2, and R5, rather than just R0, R2, and R5.

This drawback can be overcome by using one of three other approaches which are based on a decomposition of space into disjoint cells. Their common property is that the objects are decomposed into disjoint subobjects such that each of the subobjects is associated with a different cell. They differ in the degree of regularity imposed by their underlying decomposition rules and by the way in which the cells are aggregated into buckets.

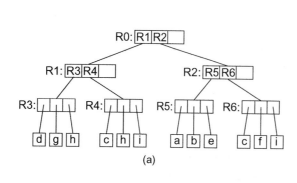
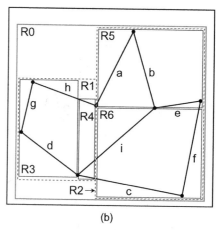

FIGURE 17.7   (a) R$^+$-tree for the collection of line segments in Figure 17.5 with $m = 2$ and $M = 3$ and (b) the spatial extents of the bounding rectangles. Notice that the leaf nodes in the index also store bounding rectangles although this is only shown for the nonleaf nodes.

The price paid for the disjointness is that in order to determine the area covered by a particular object, we have to retrieve all the cells that it occupies. This price is also paid when we want to delete an object. Fortunately, deletion is not so common in such applications. A related costly consequence of disjointness is that when we wish to determine all the objects that occur in a particular region, we often need to retrieve some of the objects more than once [27–29]. This is particularly troublesome when the result of the operation serves as input to another operation via composition of functions. For example, suppose we wish to compute the perimeter of all the objects in a given region. Clearly, each object's perimeter should only be computed once. Avoiding reporting some of the objects more than once is a serious issue (see Reference 27 for a discussion of how to deal with this problem for a collection of line segment objects and Reference 28 for a collection of rectangle objects; see also Reference 29).

The first method based on disjointness partitions the embedding space into disjoint subspaces, and hence the individual objects into subobjects, so that each subspace consists of disjoint subobjects. The subspaces are then aggregated and grouped in another structure, such as a B-tree, so that all subsequent groupings are disjoint at each level of the structure. The result is termed as k-d–B-tree [30]. The R$^+$-tree [31,32] is a modification of the *k*-d–B-tree where at each level we replace the subspace by the minimum bounding rectangle of the subobjects or subtrees that it contains. The cell tree [33] is based on the same principle as the R$^+$-tree except that the collections of objects are bounded by minimum convex polyhedra instead of minimum axis-aligned bounding rectangles.

The R$^+$-tree (as well as the other related representations) is motivated by a desire to avoid overlap among the bounding rectangles. Each object is associated with all the bounding rectangles that it intersects. All bounding rectangles in the tree (with the exception of the bounding rectangles for the objects at the leaf nodes) are nonoverlapping.[*] The result is that there may be several paths starting at the root to the same object. This may lead to an increase in the height of the tree. However, retrieval time is sped up.

Figure 17.7 is an example of one possible R$^+$-tree for the collection of line segments in Figure 17.5. This particular tree is of order (2,3) although in general it is not possible to guarantee that all nodes will always have a minimum of two entries. In particular, the expected B-tree performance guarantees are not valid (i.e., pages are not guaranteed to be *m/M* full) unless we are willing to perform very complicated record insertion and deletion procedures. Notice that line segment objects c, h, and i appear in two different nodes. Of course, other variants are possible since the R$^+$-tree is not unique.

The problem with representations such as the k-d–B-tree and the R$^+$-tree is that overflow in a leaf node may cause overflow of nodes at shallower depths in the tree whose subsequent partitioning may cause repartitioning at deeper levels in the tree. There are several ways of overcoming the repartitioning problem. One approach is to use the LSD-tree [34] at the cost of poorer storage utilization. An alternative approach is to use representations such as the hB-tree [35] and the BANG file [36] which remove the requirement that each block be a hyperrectangle at the cost of multiple postings. This has a similar effect as that obtained when decomposing an object into several subobjects in order to overcome the nondisjoint decomposition problem when using an object hierarchy. The multiple posting problem is overcome by the BV-tree [37] which decouples the partitioning and grouping processes at the cost that the resulting tree is no longer balanced although as in the PK-tree [18] (which we point

---

[*] From a theoretical viewpoint, the bounding rectangles for the objects at the leaf nodes should also be disjoint. However, this may be impossible (e.g., when the objects are line segments and if many of the line segments intersect at a point).

out in Section 17.2 is also based on decoupling), use of relatively large fanout values ensure that the resulting trees are relatively shallow.

Methods such as the R$^+$-tree (as well as the R-tree) also have the drawback that the decomposition is data dependent. This means that it is difficult to perform tasks that require composition of different operations and data sets (e.g., set-theoretic operations such as overlay). This drawback means that although these methods are good are distinguishing between occupied and unoccupied space in the underlying space (termed *image* in much of the subsequent discussion) under consideration, they are unable to correlate occupied space in two distinct images, and likewise for unoccupied space in the two images.

In contrast, the remaining two approaches to the decomposition of space into disjoint cells have a greater degree of data independence. They are based on a regular decomposition. The space can be decomposed either into blocks of uniform size (e.g., the uniform grid [38]) or adapt the decomposition to the distribution of the data (e.g., a quadtree-based approach [39]). In the former case, all the blocks are congruent (e.g., the 4 × 4 grid in Figure 17.5). In the latter case, the widths of the blocks are restricted to be powers of two and their positions are also restricted. Since the positions of the subdivision lines are restricted, and essentially the same for all images of the same size, it is easy to correlate occupied and unoccupied space in different images.

The uniform grid is ideal for uniformly distributed data, while quadtree-based approaches are suited for arbitrarily distributed data. In the case of uniformly distributed data, quadtree-based approaches degenerate to a uniform grid, albeit they have a higher overhead. Both the uniform grid and the quadtree-based approaches lend themselves to set-theoretic operations and thus they are ideal for tasks which require the composition of different operations and data sets. In general, since spatial data is not usually uniformly distributed, the quadtree-based regular decomposition approach is more flexible. The drawback of quadtree-like methods is their sensitivity to positioning in the sense that the placement of the objects relative to the decomposition lines of the space in which they are embedded effects their storage costs and the amount of decomposition that takes place. This is overcome to a large extent by using a bucketing adaptation that decomposes a block only if it contains more than b objects.

17.4 Region Data

There are many ways of representing region data. We can represent a region either by its boundary (termed a *boundary-based* representation) or by its interior (termed an *interior-based* representation). In this section, we focus on representations of collections of regions by their interior. In some applications, regions are really objects that are composed of smaller primitive objects by use of geometric transformations and Boolean set operations. *Constructive Solid Geometry (CSG)* [40] is a term usually used to describe such representations. They are beyond the scope of this chapter although algorithms do exist for converting between them and the quadtree and octree representations discussed in this chapter [41,42]. Instead, unless noted otherwise, our discussion is restricted to regions consisting of congruent cells of unit area (volume) with sides (faces) of unit size that are orthogonal to the coordinate axes.

Regions with arbitrary boundaries are usually represented by either using approximating bounding rectangles or more general boundary-based representations that are applicable to collections of line segments that do not necessarily form regions. In that case, we do not restrict the line segments to be parallel to the coordinate axes. Such representations are discussed in Section 17.6. It should be clear that although our presentation and examples in this section deal primarily with two-dimensional data, they are valid for regions of any dimensionality.

The region data is assumed to be uniform in the sense that all the cells that comprise each region are of the same type. In other words, each region is homogeneous. Of course, an image may consist of several distinct regions. Perhaps the best definition of a region is as a set of four-connected cells (i.e., in two dimensions, the cells are adjacent along an edge rather than a vertex) each of which is of the same type. For example, we may have a crop map where the regions correspond to the four-connected cells on which the same crop is grown. Each region is represented by the collection of cells that comprise it. The set of collections of cells that make up all of the regions is often termed an *image array* because of the nature in which they are accessed when performing operations on them. In particular, the array serves as an access structure in determining the region associated with a location of a cell as well as all remaining cells that comprise the region.

When the region is represented by its interior, then often we can reduce the storage requirements by aggregating identically valued cells into blocks. In the rest of this section, we discuss different methods of aggregating the cells that comprise each region into blocks as well as the methods used to represent the collections of blocks that comprise each region in the image.

The collection of blocks is usually a result of a space decomposition process with a set of rules that guide it. There are many possible decompositions. When the decomposition is recursive, we have the situation that the decomposition occurs in stages and often, although not always, the results of the stages form a containment hierarchy. This means that a block b obtained in stage i is decomposed into a set of blocks b_j that span the same space. Blocks b_j are, in turn, decomposed in stage $i + 1$ using the same decomposition rule. Some decomposition rules restrict the possible sizes and shapes of the blocks as well as their placement in space. Some examples include:

- Congruent blocks at each stage
- Similar blocks at all stages
- All sides of a block are of equal size
- All sides of each block are powers of 2 and so on.

Other decomposition rules dispense with the requirement that the blocks be rectangular (i.e., there exist decompositions using other shapes like triangles, etc.), while still others do not require that they be orthogonal, although, as stated before, we do make these assumptions here. In addition, the blocks may be disjoint or be allowed to overlap. Moreover, we can also constrain the centers of the blocks while letting them span a larger area so long as they cover all of the regions but they need not be disjoint (e.g., the quadtree medial axis transform, QMAT) [43,44]. Clearly, the choice is large. In the following, we briefly explore some of these decomposition processes. We restrict ourselves to disjoint decompositions, although this need not be the case (e.g., the field tree [45]).

The most general decomposition permits aggregation along all dimensions. In other words, the decomposition is arbitrary. The blocks need not be uniform or similar. The only requirement is that the blocks span the space of the environment. The drawback of arbitrary decompositions is that there is little structure associated with them. This means that it is difficult to answer queries such as determining the region associated with a given point, besides exhaustive search through the blocks. Thus we need an additional data structure known as an index or an access structure. A very simple decomposition rule that lends itself to such an index in the form of an array is one that partitions a d-dimensional space having coordinate axes x_i into d-dimensional blocks by use of h_i hyperplanes that are parallel to the hyperplane formed by $x_i = 0$ ($1 \leq i \leq d$). The result is a collection of $\prod_{i=1}^{d}(h_i + 1)$ blocks. These blocks form a grid of irregular-sized blocks rather than congruent blocks. There is no recursion involved in the decomposition process. We term the resulting decomposition an *irregular grid* as the partition lines are at arbitrary positions in contrast to a *uniform grid* [38] where the partition lines are positioned so that all of the resulting grid cells are congruent.

Although the blocks in the irregular grid are not congruent, we can still impose an array access structure by adding d access structures termed *linear scales*. The linear scales indicate the position of the partitioning hyperplanes that are parallel to the hyperplane formed by $x_i = 0$ ($1 \leq i \leq d$). Thus given a location l in space, say (a,b) in two-dimensional space, the linear scales for the x and y coordinate values indicate the column and row, respectively, of the array access structure entry which corresponds to the block that contains l. The linear scales are usually represented as one-dimensional arrays although they can be implemented using tree access structures such as binary search trees, range trees, segment trees, and so on.

Perhaps the most widely known decompositions into blocks are those referred to by the general terms *quadtree* and *octree* [3–5]. They are usually used to describe a class of representations for two- and three-dimensional data (and higher as well), respectively, that are the result of a recursive decomposition of the environment (i.e., space) containing the regions into blocks (not necessarily rectangular) until the data in each block satisfies some condition (e.g., with respect to its size, the nature of the regions that comprise it, the number of regions in it, etc.). The positions and/or sizes of the blocks may be restricted or arbitrary. It is interesting to note that quadtrees and octrees may be used with both interior-based and boundary-based representations although only the former are discussed in this section.

There are many variants of quadtrees and octrees (see also Sections 17.2, 17.5, and 17.6), and they are used in numerous application areas including high energy physics, VLSI, finite element analysis, and many others. Below, we focus on *region quadtrees* [46] and to a lesser extent on *region octrees* [47,48]. They are specific examples of interior-based representations for two- and three-dimensional region data (variants for data of higher dimension also exist), respectively, that permit further aggregation of identically valued cells.

Region quadtrees and region octrees are instances of a restricted decomposition rule where the environment containing the regions is recursively decomposed into four or eight, respectively, rectangular congruent blocks until each block is either completely occupied by a region or is empty (such a decomposition process is termed *regular*). For example, Figure 17.8a is the block decomposition for the region quadtree corresponding to three regions A, B, and C. Notice that in this case, all the blocks are square, have sides whose size is a power of 2, and are located at specific positions. In particular, assuming an origin at the lower left corner of the image containing the regions, then the coordinate values of the lower left corner of each block (e.g., (a,b) in two dimensions) of size $2^i \times 2^i$ satisfy the property that $a \bmod 2^i = 0$ and $b \bmod 2^i = 0$.

The traditional, and most natural, access structure for a region quadtree corresponding to a d-dimensional image is a tree with a fanout of 2^d (e.g., Figure 17.8b). Each leaf node in the tree corresponds to a different block b and contains the identity of the region associated with b. Each nonleaf node f corresponds to a block whose volume is the union of the blocks corresponding to the 2^d sons of f. In this case, the tree is a containment hierarchy and closely parallels the decomposition in the sense that they are both recursive processes and the blocks corresponding to nodes at different depths of the tree are similar in shape. Of course, the region quadtree could also be represented by using a mapping from the domain of the blocks to a subset of the integers and then imposing a tree access structure such as a B$^+$-tree on the result of the mapping as was described in Section 17.2 for point data stored in a PR quadtree.

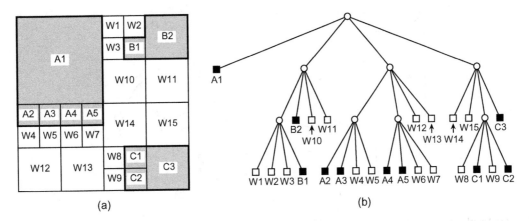

FIGURE 17.8 (a) Block decomposition and (b) its tree representation for the region quadtree corresponding to a collection of three regions A, B, and C.

As the dimensionality of the space (i.e., d) increases, each level of decomposition in the region quadtree results in many new blocks as the fanout value 2^d is high. In particular, it is too large for a practical implementation of the tree access structure. In this case, an access structure termed a *bintree* [49–51] with a fanout value of 2 is used. The bintree is defined in a manner analogous to the region quadtree except that at each subdivision stage, the space is decomposed into two equal-sized parts. In two dimensions, at odd stages we partition along the y-axis and at even stages we partition along the x-axis. In general, in the case of d dimensions, we cycle through the different axes every d levels in the bintree.

The region quadtree, as well as the bintree, is a regular decomposition. This means that the blocks are congruent—that is, at each level of decomposition, all of the resulting blocks are of the same shape and size. We can also use decompositions where the sizes of the blocks are not restricted in the sense that the only restriction is that they be rectangular and be a result of a recursive decomposition process. In this case, the representations that we described must be modified so that the sizes of the individual blocks can be obtained. An example of such a structure is an adaptation of the point quadtree [11] to regions. Although the point quadtree was designed to represent points in a higher dimensional space, the blocks resulting from its use to decompose space do correspond to regions. The difference from the region quadtree is that in the point quadtree, the positions of the partitions are arbitrary, whereas they are a result of a partitioning process into 2^d congruent blocks (e.g., quartering in two dimensions) in the case of the region quadtree.

As in the case of the region quadtree, as the dimensionality d of the space increases, each level of decomposition in the point quadtree results in many new blocks since the fanout value 2^d is high. In particular, it is too large for a practical implementation of the tree access structure. In this case, we can adapt the k-d tree [14], which has a fanout value of 2, to regions. As in the point quadtree, although the k-d tree was designed to represent points in a higher dimensional space, the blocks resulting from its use to decompose space do correspond to regions. Thus the relationship of the k-d tree to the point quadtree is the same as the relationship of the bintree to the region quadtree. In fact, the k-d tree is the precursor of the bintree and its adaptation to regions is defined in a similar manner in the sense that for d-dimensional data we cycle through the d-axes every d levels in the k-d tree. The difference is that in the k-d tree, the positions of the partitions are arbitrary, whereas they are a result of a halving process in the case of the bintree.

The k-d tree can be further generalized so that the partitions take place on the various axes at an arbitrary order, and, in fact, the partitions need not be made on every coordinate axis. The k-d tree is a special case of the *BSP tree* (denoting *Binary Space Partitioning*) [52] where the partitioning hyperplanes are restricted to be parallel to the axes, whereas in the BSP tree they have an arbitrary orientation. The BSP tree is a binary tree. In order to be able to assign regions to the left and right subtrees, we need to associate a direction with each subdivision line. In particular, the subdivision lines are treated as separators between two half-spaces.[*] Let the subdivision line have the equation $a \cdot x + b \cdot y + c = 0$. We say that the right subtree is the "positive" side and contains all subdivision lines formed by separators that satisfy $a \cdot x + b \cdot y + c \geq 0$. Similarly, we say that the left subtree is "negative" and contains all subdivision lines formed by separators that satisfy $a \cdot x + b \cdot y + c < 0$. As an example, consider Figure 17.9a which is an arbitrary space decomposition whose BSP tree is given in Figure 17.9b. Notice the use of arrows to indicate

---

[*] A (linear) *half-space* in d-dimensional space is defined by the inequality $\sum_{i=0}^{d} a_i \cdot x_i \geq 0$ on the $d + 1$ homogeneous coordinates ($x_0 = 1$). The half-space is represented by a column vector a. In vector notation, the inequality is written as $a \cdot x \geq 0$. In the case of equality, it defines a hyperplane with a as its normal. It is important to note that half-spaces are volume elements; they are not boundary elements.

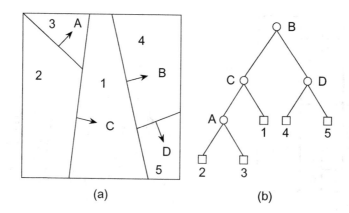

FIGURE 17.9   (a) An arbitrary space decomposition and (b) its BSP tree. The arrows indicate the direction of the positive half-spaces.

the direction of the positive half-spaces. The BSP tree is used in computer graphics to facilitate viewing. It is discussed in greater detail in Chapter 21.

As mentioned before, the various hierarchical data structures that we described can also be used to represent regions in three dimensions and higher. As an example, we briefly describe the region octree which is the three-dimensional analog of the region quadtree. It is constructed in the following manner. We start with an image in the form of a cubical volume and recursively subdivide it into eight congruent disjoint cubes (called octants) until blocks are obtained of a uniform color or a predetermined level of decomposition is reached. Figure 17.10a is an example of a simple three-dimensional object whose region octree block decomposition is given in Figure 17.10b and whose tree representation is given in Figure 17.10c.

The aggregation of cells into blocks in region quadtrees and region octrees is motivated, in part, by a desire to save space. Some of the decompositions have quite a bit of structure thereby leading to inflexibility in choosing partition lines, and so on. In fact, at times, maintaining the original image with an array access structure may be more effective from the standpoint of storage requirements. In the following, we point out some important implications of the use of these aggregations. In particular, we focus on the region quadtree and region octree. Similar results could also be obtained for the remaining block decompositions.

The aggregation of similarly valued cells into blocks has an important effect on the execution time of the algorithms that make use of the region quadtree. In particular, most algorithms that operate on images represented by a region quadtree are implemented by a preorder traversal of the quadtree and, thus, their execution time is generally a linear function of the number of nodes in the quadtree. A key to the analysis of the execution time of quadtree algorithms is the *Quadtree Complexity Theorem* [47] which states that the number of nodes in a region quadtree representation for a simple polygon (i.e., with nonintersecting edges and without holes) is $O(p + q)$ for a $2^q \times 2^q$ image with perimeter p measured in terms of the width of unit-sized cells (i.e., pixels). In all but the most pathological cases (e.g., a small square of unit width centered in a large image), the q factor is negligible and thus the number of nodes is $O(p)$.

The Quadtree Complexity Theorem also holds for three-dimensional data [53] (i.e., represented by a region octree) where perimeter is replaced by surface area, as well as for objects of higher dimensions d for which it is proportional to the size of the $(d - 1)$-dimensional interfaces between these objects. The most important consequence of the Quadtree Complexity Theorem is that it means that most algorithms that execute on a region quadtree representation of an image, instead of one that simply

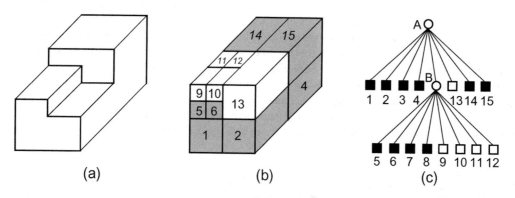

FIGURE 17.10   (a) Example three-dimensional object, (b) its region octree block decomposition, and (c) its tree representation.

imposes an array access structure on the original collection of cells, usually have an execution time that is proportional to the number of blocks in the image rather than the number of unit-sized cells. Some examples of such algorithms include perimeter computation [54], connected component labeling [55], Euler number [56], and quadtree to raster conversion [57]. Other related ramifications are illustrated by algorithms to compute spatial range queries which related to image dilation (i.e., region expansion and corridor or buffer computation in GIS) [58] which have been observed to execute in time that is independent of the radius of expansion when the regions to be expanded are represented using region quadtrees [59].

In its most general case, the above means that the use of the region quadtree, with an appropriate access structure, in solving a problem in d-dimensional space will lead to a solution whose execution time is proportional to the $(d-1)$-dimensional space of the surface of the original d-dimensional image. On the other hand, use of the array access structure on the original collection of cells results in a solution whose execution time is proportional to the number of cells that comprise the image. Therefore, region quadtrees and region octrees act like dimension-reducing devices.

Using a tree access structure captures the characterization of the region quadtree as being of variable resolution in that the underlying space is subdivided recursively until the underlying region satisfies some type of a homogeneity criterion. This is in contrast to a pyramid structure [60,61], which is a family of representations that make use of multiple resolution. This means that the underlying space is subdivided up to the smallest unit (i.e., a pixel) thereby resulting in a complete quadtree, where each leaf node is at the maximum depth and the nonleaf nodes contain information summarizing the contents of their subtrees. The pyramid facilitates feature-based queries [60], which given a feature, identify where it is in space, while the quadtree is more suited for location-based queries, which given a location, identify the features that are associated with it. Feature-based queries have become known as spatial data mining and the pyramid is used to implement them [60,62,63].

17.5 Rectangle Data

The rectangle data type lies somewhere between the point and region data types. It can also be viewed as a special case of the region data type in the sense that it is a region with only four sides. Rectangles are often used to approximate other objects in an image for which they serve as the minimum rectilinear enclosing object. For example, bounding rectangles are used in cartographic applications to approximate objects such as lakes, forests, hills, and so on. In such a case, the approximation gives an indication of the existence of an object. Of course, the exact boundaries of the object are also stored, but they are only accessed if greater precision is needed. For such applications, the number of elements in the collection is usually small, and most often the sizes of the rectangles are of the same order of magnitude as the space from which they are drawn.

Rectangles are also used in VLSI design rule checking as a model of chip components for the analysis of their proper placement. Again, the rectangles serve as minimum enclosing objects. In this application, the size of the collection is quite large (e.g., millions of components) and the sizes of the rectangles are several orders of magnitude smaller than the space from which they are drawn.

It should be clear that the actual representation that is used depends heavily on the problem environment. At times, the rectangle is treated as the Cartesian product of two one-dimensional intervals with the horizontal intervals being treated in a different manner than the vertical intervals. In fact, the representation issue is often reduced to one of representing intervals. For example, this is the case in the use of the plane-sweep paradigm [64] in the solution of rectangle problems such as determining all pairs of intersecting rectangles. In this case, each interval is represented by its left and right endpoints. The solution makes use of two passes.

The first pass sorts the rectangles in ascending order on the basis of their left and right sides (i.e., x coordinate values) and forms a list. The second pass sweeps a vertical scan line through the sorted list from left to right halting at each one of these points, say p. At any instant, all rectangles that intersect the scan line are considered *active* and are the only ones whose intersection needs to be checked with the rectangle associated with p. This means that each time the sweep line halts, a rectangle either becomes active (causing it to be inserted in the set of active rectangles) or ceases to be active (causing it to be deleted from the set of active rectangles). Thus the key to the algorithm is its ability to keep track of the active rectangles (actually just their vertical sides) as well as to perform the actual one-dimensional intersection test.

Data structures such as the segment tree [65], interval tree [66], and the priority search tree [20] can be used to organize the vertical sides of the active rectangles so that, for N rectangles and F intersecting pairs of rectangles, the problem can be solved in $O(N \cdot \log_2 N + F)$ time. All three data structures enable intersection detection, insertion, and deletion to be executed in $O(\log_2 N)$ time. The difference between them is that the segment tree requires $O(N \cdot \log_2 N)$ space while the interval tree and the priority search tree only need $O(N)$ space. These algorithms require that the set of rectangles be known in advance. However, they work even when the size of the set of active rectangles exceeds the amount of available memory, in which case multiple passes are made over the data [67]. For more details about these data structures, see Chapter 19.

In this chapter, we are primarily interested in dynamic problems (i.e., the set of rectangles is constantly changing). The data structures that are chosen for the collection of the rectangles are differentiated by the way in which each rectangle is represented.

One representation discussed in Section 17.1 reduces each rectangle to a point in a higher dimensional space, and then treats the problem as if we have a collection of points [1]. Again, each rectangle is a Cartesian product of two one-dimensional intervals where the difference from its use with the plane-sweep paradigm is that each interval is represented by its centroid and extent. Each set of intervals in a particular dimension is, in turn, represented by a grid file [9] which is described in Section 17.2.

The second representation is region-based in the sense that the subdivision of the space from which the rectangles are drawn depends on the physical extent of the rectangle—not just one point. Representing the collection of rectangles, in turn, with a tree-like data structure has the advantage for some of the representations that there is a relation between the depth of node in the tree and the size of the rectangle(s) that is (are) associated with it. Interestingly, some of the region-based solutions make use of the same data structures that are used in the solutions based on the plane-sweep paradigm.

There are three types of region-based solutions currently in use. The first two solutions use the R-tree and the R$^+$-tree (discussed in Section 17.3) to store rectangle data (in this case the objects are rectangles instead of arbitrary objects). The third is a quadtree-based approach and uses the MX-CIF quadtree [68] (see also Reference 69 for a related variant).

In the *MX-CIF quadtree*, each rectangle is associated with the quadtree node corresponding to the smallest block which contains it in its entirety. Subdivision ceases whenever a node's block contains no rectangles. Alternatively, subdivision can also cease once a quadtree block is smaller than a predetermined threshold size. This threshold is often chosen to be equal to the expected size of the rectangle [68]. For example, Figure 17.11b is the MX-CIF quadtree for a collection of rectangles given in Figure 17.11a. Rectangles can be associated with both leaf and nonleaf nodes.

It should be clear that more than one rectangle can be associated with a given enclosing block and, thus, often we find it useful to be able to differentiate between them. This is done in the following manner [68]. Let P be a quadtree node with centroid (CX,CY), and let S be the set of rectangles that are associated with P. Members of S are organized into two sets according to their intersection (or collinearity of their sides) with the lines passing through the centroid of P's block—that is, all members of S that intersect the line $x = CX$ form one set and all members of S that intersect the line $y = CY$ form the other set.

If a rectangle intersects both lines (i.e., it contains the centroid of P's block), then we adopt the convention that it is stored with the set associated with the line through $x = CX$. These subsets are implemented as binary trees (really tries), which in actuality are one-dimensional analogs of the MX-CIF quadtree. For example, Figures 17.11c and d illustrate the binary trees associated with the y-axes passing through the root and the NE son of the root, respectively, of the MX-CIF quadtree of Figure 17.11b. Interestingly, the MX-CIF quadtree is a two-dimensional analog of the interval tree described earlier. More precisely, the MX-CIF quadtree is a two-dimensional analog of the tile tree [70] which is a regular decomposition version of the interval tree. In fact, the tile tree and the one-dimensional MX-CIF quadtree are identical when rectangles are not allowed to overlap.

The drawback of the MX-CIF quadtree is that the size of the minimum enclosing quadtree cells depends on the position of the centroids of the objects and is independent of the size of the objects, subject to a minimum which is the size of the object. In fact, it may be as large as the entire space from which the objects are drawn. This has bad ramifications for applications where the objects move including games, traffic monitoring, and streaming. In particular, if the objects are moved even slightly, then they usually need to be reinserted in the structure.

The cover fieldtree [45,71] and the more commonly known loose quadtree (octree) [72] are designed to overcome this independence of the size of the minimum enclosing quadtree cell and the size of the object (see also the expanded MX-CIF quadtree [73], multiple shifted quadtree methods [74–76], and the partition fieldtree [45,71]). This is done by expanding the size of the space that is spanned by each quadtree cell c of width w by a cell expansion factor p ($p > 0$) so that the expanded cell is of width $(1 + p) \cdot w$ and an object is associated with its minimum enclosing expanded quadtree (octree) cell. The notion of an expanded quadtree cell can also be seen in the QMAT [43,44]. For example, letting $p = 1$, Figure 17.12 is the loose quadtree corresponding to the

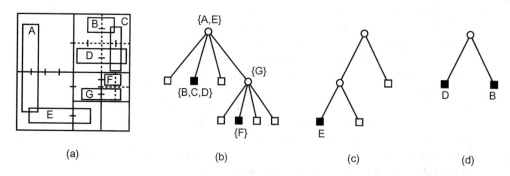

(a)   (b)   (c)   (d)

FIGURE 17.11   (a) Collection of rectangles and the block decomposition induced by the MX-CIF quadtree; (b) the tree representation of (a); the binary trees for the y-axes passing through the root of the tree in (b), and (d) the NE son of the root of the tree in (b).

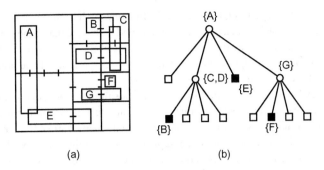

FIGURE 17.12 (a) Cell decomposition induced by the loose quadtree for a collection of rectangle objects identical to those in Figure 17.11 and (b) its tree representation.

collection of objects in Figure 17.11a and its MX-CIF quadtree in Figure 17.11b. In this example, there are only two differences between the loose and MX-CIF quadtrees:

1. Rectangle object E is associated with the SW child of the root of the loose quadtree instead of with the root of the MX-CIF quadtree.
2. Rectangle object B is associated with the NW child of the NE child of the root of the loose quadtree instead of with the NE child of the root of the MX-CIF quadtree.

Ulrich [72] has shown that given a loose quadtree cell c of width w and cell expansion factor p, the radius r of the minimum bounding hypercube box b of the smallest object o that could possibly be associated with c must be greater than $pw/4$. Samet et al. [77] point out that the real utility of the loose quadtree is best evaluated in terms of the inverse of the above relation as we are interested in minimizing the maximum possible width w of c given an object o with minimum bounding hypercube box b of radius r (i.e., half the length of a side of the hypercube). This is because reducing w is the real motivation and goal for the development of the loose quadtree as an alternative to the MX-CIF quadtree for which, as we pointed out, w can be as large as the width of the underlying space. This was done by examining the range of the relative widths of c and b as this provides a way of taking into account the constraints imposed by the fact that the range of values of w is limited to powers of 2. In particular, they showed that this range is just a function of p, and hence independent of the position of o. Moreover, they prove that for $p \geq 0.5$, the relative widths of c and b take on at most two values, and usually just one value for $p \geq 1$. This makes updating the index very simple when objects are moving as there are at most two possible new cells associated with a moved object, instead of $\log_2$ of the width of the space in which the objects are embedded (e.g., as large as 16 assuming a $2^{16} \times 2^{16}$ embedding space). In other words, they have shown how to update in $O(1)$ time for $p \geq 0.5$ which is of great importance as there is no longer a need to perform a search for the appropriate quadtree cell.

17.6 Line Data and Boundaries of Regions

Section 17.4 was devoted to variations on hierarchical decompositions of regions into blocks, an approach to region representation that is based on a description of the region's interior. In this section, we focus on representations that enable the specification of the boundaries of regions, as well as curvilinear data and collections of line segments. The representations are usually based on a series of approximations which provide successively closer fits to the data, often with the aid of bounding rectangles. When the boundaries or line segments have a constant slope (i.e., linear and termed *line segments* in the rest of this discussion), then an exact representation is possible.

There are several ways of approximating a curvilinear line segment. The first is by digitizing it and then marking the unit-sized cells (i.e., pixels) through which it passes. The second is to approximate it by a set of straight line segments termed a *polyline*. Assuming a boundary consisting of straight lines (or polylines after the first stage of approximation), the simplest representation of the boundary of a region is the polygon. It consists of vectors which are usually specified in the form of lists of pairs of x and y coordinate values corresponding to their start and end points. The vectors are usually ordered according to their connectivity. One of the most common representations is the chain code [78] which is an approximation of a polygon's boundary by use of a sequence of unit vectors in the four (and sometimes eight) principal directions.

Chain codes, and other polygon representations, break down for data in three dimensions and higher. This is primarily due to the difficulty in ordering their boundaries by connectivity. The problem is that in two dimensions connectivity is determined by ordering the boundary elements $e_{i,j}$ of boundary b_i of object o so that the end vertex of the vector v_j corresponding to $e_{i,j}$

is the start vertex of the vector v_{j+1} corresponding to $e_{i,j+1}$. Unfortunately, such an implicit ordering does not exist in higher dimensions as the relationship between the boundary elements associated with a particular object is more complex.

Instead, we must make use of data structures which capture the topology of the object in terms of its faces, edges, and vertices. The winged-edge data structure is one such representation which serves as the basis of the boundary model (also known as *BRep* [79]). For more details about these data structures, see Chapter 18.

Polygon representations are very local. In particular, if we are at one position on the boundary, we don't know anything about the rest of the boundary without traversing it element by element. Thus, using such representations, given a random point in space, it is very difficult to find the nearest line to it as the lines are not sorted. This is in contrast to hierarchical representations which are global in nature. They are primarily based on rectangular approximations to the data as well as on a regular decomposition in two dimensions. In the rest of this section, we discuss a number of such representations.

In Section 17.3, we already examined two hierarchical representations (i.e., the R-tree and the R^+-tree) that propagate object approximations in the form of bounding rectangles. In this case, the sides of the bounding rectangles had to be parallel to the coordinate axes of the space from which the objects are drawn. In contrast, the *strip tree* [80] is a hierarchical representation of a single curve that successively approximates segments of it with bounding rectangles that does not require that the sides be parallel to the coordinate axes. The only requirement is that the curve be continuous; it need not be differentiable.

The strip tree data structure consists of a binary tree whose root represents the bounding rectangle of the entire curve. The rectangle associated with the root corresponds to a rectangular strip, that encloses the curve, whose sides are parallel to the line joining the endpoints of the curve. The curve is then partitioned in two at one of the locations where it touches the bounding rectangle (these are not tangent points as the curve only needs to be continuous; it need not be differentiable). Each subcurve is then surrounded by a bounding rectangle and the partitioning process is applied recursively. This process stops when the width of each strip is less than a predetermined value.

In order to be able to cope with more complex curves such as those that arise in the case of object boundaries, the notion of a strip tree must be extended. In particular, closed curves and curves that extend past their endpoints require some special treatment. The general idea is that these curves are enclosed by rectangles which are split into two rectangular strips, and from now on the strip tree is used as before.

The strip tree is similar to the point quadtree in the sense that the points at which the curve is decomposed depend on the data. In contrast, a representation based on the region quadtree has fixed decomposition points. Similarly, strip tree methods approximate curvilinear data with rectangles of arbitrary orientation, while methods based on the region quadtree achieve analogous results by use of a collection of disjoint squares having sides of length power of two. In the following, we discuss a number of adaptations of the region quadtree for representing curvilinear data.

The simplest adaptation of the region quadtree is the MX quadtree [47,81]. It is built by digitizing the line segments and labeling each unit-sized cell (i.e., pixel) through which the line segments pass as of type `boundary`. The remaining pixels are marked `WHITE` and are merged, if possible, into larger and larger quadtree blocks. Figure 17.13a is the MX quadtree for the collection of line segment objects in Figure 17.5. A drawback of the MX quadtree is that it associates a thickness with a line. Also, it is difficult to detect the presence of a vertex whenever five or more line segments meet.

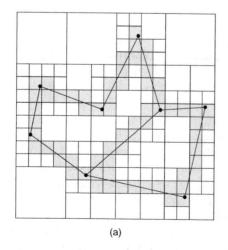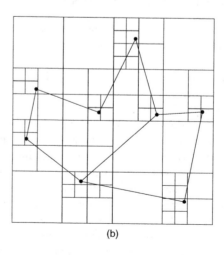

| (a) | (b) |

FIGURE 17.13   (a) MX quadtree and (b) edge quadtree for the collection of line segments of Figure 17.5.

The edge quadtree [82,83] is a refinement of the MX quadtree based on the observation that the number of squares in the decomposition can be reduced by terminating the subdivision whenever the square contains a single curve that can be approximated by a single straight line. For example, Figure 17.13b is the edge quadtree for the collection of line segment objects in Figure 17.5. Applying this process leads to quadtrees in which long edges are represented by large blocks or a sequence of large blocks. However, small blocks are required in the vicinity of corners or intersecting edges. Of course, many blocks will contain no edge information at all.

The PM quadtree family [39,84] (see also edge-EXCELL [15]) represents an attempt to overcome some of the problems associated with the edge quadtree in the representation of collections of polygons (termed *polygonal maps*). In particular, the edge quadtree is an approximation because vertices are represented by pixels. There are a number of variants of the PM quadtree. These variants are either vertex-based or edge-based. They are all built by applying the principle of repeatedly breaking up the collection of vertices and edges (forming the polygonal map) until obtaining a subset that is sufficiently simple so that it can be organized by some other data structure.

The PM_1 quadtree [39] is an example of a vertex-based PM quadtree. Its decomposition rule stipulates that partitioning occurs as long as a block contains more than one line segment unless the line segments are all incident at the same vertex which is also in the same block (e.g., Figure 17.14a). Given a polygonal map whose vertices are drawn from a grid (say $2^m \times 2^m$), and where edges are not permitted to intersect at points other than the grid points (i.e., vertices), it can be shown that the maximum depth of any leaf node in the PM_1 quadtree is bounded from above by $4m + 1$ [85]. This enables a determination of the maximum amount of storage that will be necessary for each node. The PM_2 quadtree is another vertex-based member of the PM quadtree family that is similar to the PM_1 quadtree with the difference that the partitioning of a block b is halted when the line segments in b are all incident at the same vertex v but v need not be in the same block b as is the case for the PM_1 quadtree. For example, this means that in the PM_2 quadtree there is no need to partition the block corresponding to the NE child of the SW quadrant as in the PM_1 quadtree in Figure 17.14a that corresponds to the polygonal map formed by the collection of line segments of Figure 17.5. The PM_2 quadtree is particularly useful for polygonal maps that correspond to triangulations where it has been empirically observed to be relatively insensitive to movement of vertices and edges (L. De Floriani, personal communication, 2004).

A similar representation to the PM_1 quadtree has been devised for polyhedral (i.e., three-dimensional) objects (e.g., see Reference 86 and the references cited in Reference 4). The decomposition criteria are such that no node contains more than one face, edge, or vertex unless the faces all meet at the same vertex or are adjacent to the same edge. This representation is quite useful since its space requirements for polyhedral objects are significantly smaller than those of a region octree.

The PMR quadtree [84] is an edge-based variant of the PM quadtree. It makes use of a probabilistic splitting rule. A node is permitted to contain a variable number of line segments. A line segment is stored in a PMR quadtree by inserting it into the nodes corresponding to all the blocks that it intersects. During this process, the occupancy of each node that is intersected by the line segment is checked to see if the insertion causes it to exceed a predetermined *splitting threshold*. If the splitting threshold is exceeded, then the node's block is split *once*, and only once, into four equal quadrants.

For example, Figure 17.14b is the PMR quadtree for the collection of line segment objects in Figure 17.5 with a splitting threshold value of 2. The line segments are inserted in alphabetic order (i.e., a–i). It should be clear that the shape of the PMR quadtree depends on the order in which the line segments are inserted. Note the difference from the PM_1 quadtree in Figure

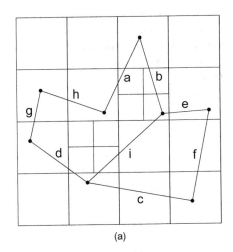
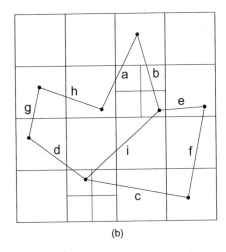

(a)   (b)

FIGURE 17.14 (a) PM_1 quadtree and (b) PMR quadtree for the collection of line segments of Figure 17.5.

17.14a—that is, the NE block of the SW quadrant is decomposed in the PM_1 quadtree while the SE block of the SW quadrant is not decomposed in the PM_1 quadtree.

On the other hand, a line segment is deleted from a PMR quadtree by removing it from the nodes corresponding to all the blocks that it intersects. During this process, the occupancy of the node and its siblings is checked to see if the deletion causes the total number of line segments in them to be less than the predetermined splitting threshold. If the splitting threshold exceeds the occupancy of the node and its siblings, then they are merged and the merging process is reapplied to the resulting node and its siblings. Notice the asymmetry between the splitting and merging rules.

The PMR quadtree is very good for answering queries such as finding the nearest line to a given point [87–90] (see Reference 91 for an empirical comparison with hierarchical object representations such as the R-tree and R^+-tree). It is preferred over the PM_1 quadtree (as well as the MX and edge quadtrees) as it results in far fewer subdivisions. In particular, in the PMR quadtree there is no need to subdivide in order to separate line segments that are very "close" or whose vertices are very "close," which is the case for the PM_1 quadtree. This is important since four blocks are created at each subdivision step. Thus when many subdivision steps that occur in the PM_1 quadtree result in creating many empty blocks, the storage requirements of the PM_1 quadtree will be considerably higher than those of the PMR quadtree. Generally, as the splitting threshold is increased, the storage requirements of the PMR quadtree decrease while the time necessary to perform operations on it will increase.

Using a random image model and geometric probability, it has been shown [92], theoretically and empirically using both random and real map data, that for sufficiently high values of the splitting threshold (i.e., ≥ 4), the number of nodes in a PMR quadtree is asymptotically proportional to the number of line segments and is independent of the maximum depth of the tree. In contrast, using the same model, the number of nodes in the PM_1 quadtree is a product of the number of lines and the maximal depth of the tree (i.e., n for a $2^n \times 2^n$ image). The same experiments and analysis for the MX quadtree confirmed the results predicted by the Quadtree Complexity Theorem (see Section 17.4) which is that the number of nodes is proportional to the total length of the line segments. Observe that although a bucket in the PMR quadtree can contain more line segments than the splitting threshold, this is not a problem. In fact, it can be shown [4] that the maximum number of line segments in a bucket is bounded by the sum of the splitting threshold and the depth of the block (i.e., the number of times the original space has been decomposed to yield this block).

When the line segments that make up the polygonal map are orthogonal and parallel to the coordinate axes, then we can make use of the line quadtree [93,94] which is an enhancement of the region quadtree where the boundaries of the regions are coincident with the quadtree blocks and are also represented in a hierarchical manner. In particular, the line quadtree recursively partitions the underlying space until obtaining blocks (possibly single pixels) that have no boundary elements passing through their interior. Associated with each block is a four bit code which indicates which of its four sides forms a boundary (not a partial boundary) of any single object (including the underlying space). Thus instead of distinguishing blocks on the basis of whether they are part of the background, and of the identity of the object in which they lie, we use the nature of the boundary information. Note the distinction from the region quadtree where no information is associated with the nonleaf nodes. In this respect, it is more like the pyramid (see Section 17.4).

17.7 Research Issues and Summary

A review has been presented of a number of representations of multidimensional data. Our focus has been on multidimensional spatial data with extent rather than just multidimensional point data. There has been a particular emphasis on hierarchical representations. Such representations are based on the "divide-and-conquer" problem-solving paradigm. They are of interest because they enable focusing computational resources on the interesting subsets of data. Thus there is no need to expend work where the payoff is small. Although many of the operations for which they are used can often be performed equally as efficiently, or more so, with other data structures, hierarchical data structures are attractive because of their conceptual clarity and ease of implementation.

When the hierarchical data structures are based on the principle of regular decomposition, we have the added benefit that different data sets (often of differing types) are in registration. This means that they are partitioned in known positions which are often the same or subsets of one another for the different data sets. This is true for all the features including regions, points, rectangles, lines, volumes, and so on. The result is that a query such as "finding all cities with more than 20,000 inhabitants in wheat-growing regions within 30 miles of the Mississippi River" can be executed by simply overlaying the region (crops), point (i.e., cities), and river maps even though they represent data of different types. Alternatively, we may extract regions such as those within 30 miles of the Mississippi River. Such operations find use in applications involving spatial data such as GIS.

Current research in multidimensional representations is highly application dependent in the sense that the work is driven by the application. Many of the recent developments have been motivated by the interaction with databases. This is facilitated by the existence of browsing (e.g., [95–98]) or manipulating (e.g., by what is termed a spatial spreadsheet [99]) capabilities. The choice of a proper representation plays a key role in the speed with which responses are provided to queries. Knowledge of the underlying data distribution is also a factor and research is ongoing to make use of this information in the process of making a choice. Most of the initial applications in which the representation of multidimensional data has been important have involved spatial data of the kind described in this chapter. Such data is intrinsically of low dimensionality (i.e., two and three).

Future applications involve higher dimensional data for applications such as image databases where the data are often points in feature space. Unfortunately, for such applications, the performance of most indexing methods that rely on a decomposition of the underlying space is often unsatisfactory when compared with not using an index at all [100]. The problem is that for uniformly distributed data, most of the data is found to be at or near the boundary of the space in which it lies [101]. The result means that the query region usually overlaps all of the leaf node regions that are created by the decomposition process and thus a sequential scan is preferable. This has led to a number of alternative representations that try to speed up the scan (e.g., VA-file [102], VA$^+$-file [103], IQ-tree [104], etc.). Nevertheless, representations such as the pyramid technique [105] are based on the principle that most of the data lies near the surface and therefore subdivide the data space as if it is an onion by peeling off hypervolumes that are close to its boundary. This is achieved by first dividing the hypercube corresponding to the d-dimensional data space into $2d$ pyramids having the center of the data space as their top point and one of the faces of the hypercube as its base. These pyramids are subsequently cut into slices that are parallel to their base. Of course, the high-dimensional data is not necessarily uniformly distributed which has led to other data structures with good performance (e.g., the hybrid tree [106]). Clearly, more work needs to be done in this area.

It is interesting to observe that the rise of the use of the world wide web has made spatial multidimensional data such as locations much more common. The explicit specification of location has traditionally been geometric (e.g., as latitude–longitude pairs of numbers). Unfortunately, this is not easy as most people don't usually know of a location in this way. Even more important is the fact that they don't know where to obtain this information and are not used to communicate or receive it from others in this way. Instead, they are used to specify a location textually which includes verbally.

The principal drawback of the textual specification of location data is that it is potentially ambiguous. First of all, we must be sure that the textual entity actually corresponds to a location (termed a *toponym*) rather than the name of a person, an organization, or an object, to name a few of many possible alternative interpretations. The process of determining whether a textual entity is a toponym is termed *toponym recognition* [107]. Moreover, having identified the textual entity as a toponym such as "Alexandria," there are many possible locations named "Alexandria" and they must be resolved. This process is known as *toponym resolution* [108,109]. Moreover, in some cases we are not even sure that the term "Lincoln" denotes a location as it could be a reference to the name of a person. Both of these issues can arise when processing documents such as newspaper articles [110–114], tweets [115–117], blogs, and so on. The process of understanding and converting location text to its geometric specification, known as *geotagging* [118–126], serves as a prerequisite to indexing the data and is the subject of much research.

Acknowledgment

This work was supported, in part, by the National Science Foundation under grants IRI-97-12715, EIA-99-00268, IIS-00-86162, EIA-00-91474, IIS-12-19023, and IIS-13-20791.

References

1. K. Hinrichs and J. Nievergelt. The grid file: A data structure designed to support proximity queries on spatial objects. In *Proceedings of WG'83, International Workshop on Graphheoretic Concepts in Computer Science*, M. Nagl and J. Perl, eds., pages 100–113, Trauner Verlag, Haus Ohrbeck (near Osnabrück), West Germany, 1983.

2. J.-W. Song, K.-Y. Whang, Y.-K. Lee, M.-J. Lee, and S.-W. Kim. Spatial join processing using corner transformation. *IEEE Transactions on Knowledge and Data Engineering*, 11(4):688–695, July/August 1999.

3. H. Samet. *Applications of Spatial Data Structures: Computer Graphics, Image Processing, and GIS*. Addison-Wesley, Reading, MA, 1990.

4. H. Samet. *The Design and Analysis of Spatial Data Structures*. Addison-Wesley, Reading, MA, 1990.

5. H. Samet. *Foundations of Multidimensional and Metric Data Structures*. Morgan-Kaufmann, San Francisco, 2006. (Translated to Chinese ISBN 978-7-302-22784-7).

6. D. E. Knuth. Big omicron and big omega and big theta. *SIGACT News*, 8(2):18–24, April–June 1976.

7. D. E. Knuth. *The Art of Computer Programming: Sorting and Searching*, vol. 3. Addison-Wesley, Reading, MA, 1973.

8. J. L. Bentley and J. H. Friedman. Data structures for range searching. *ACM Computing Surveys*, 11(4):397–409, December 1979.

9. J. Nievergelt, H. Hinterberger, and K. C. Sevcik. The grid file: An adaptable, symmetric multikey file structure. *ACM Transactions on Database Systems*, 9(1):38–71, March 1984.

10. J. L. Bentley, D. F. Stanat, and E. H. Williams, Jr. The complexity of finding fixed-radius near neighbors. *Information Processing Letters*, 6(6):209–212, December 1977.

11. R. A. Finkel and J. L. Bentley. Quad trees: A data structure for retrieval on composite keys. *Acta Informatica*, 4(1):1–9, 1974.

12. J. A. Orenstein. Multidimensional tries used for associative searching. *Information Processing Letters*, 14(4):150–157, June 1982.

13. E. Fredkin. Trie memory. *Communications of the ACM*, 3(9):490–499, September 1960.

14. J. L. Bentley. Multidimensional binary search trees used for associative searching. *Communications of the ACM*, 18(9):509–517, September 1975.

15. M. Tamminen. The EXCELL method for efficient geometric access to data. Series Acta Polytechnica Scandinavica, 1981. Also Mathematics and Computer Science Series No. 34.

16. D. Comer. The ubiquitous B-tree. *ACM Computing Surveys*, 11(2):121–137, June 1979.

17. H. Samet. Decoupling: A spatial indexing solution. Computer Science Technical Report TR-4523, University of Maryland, College Park, MD, August 2003.

18. W. Wang, J. Yang, and R. Muntz. PK-tree: A spatial index structure for high dimensional point data. In *Proceedings of the 5th International Conference on Foundations of Data Organization and Algorithms (FODO)*, K. Tanaka and S. Ghandeharizadeh, eds., pages 27–36, Kobe, Japan, November 1998. Also University of California at Los Angeles Computer Science Technical Report 980032, September 1998.

19. J. L. Bentley and H. A. Maurer. Efficient worst-case data structures for range searching. *Acta Informatica*, 13:155–168, 1980.

20. E. M. McCreight. Priority search trees. *SIAM Journal on Computing*, 14(2):257–276, May 1985. Also Xerox Palo Alto Research Center Technical Report CSL-81-5, January 1982.

21. N. Beckmann, H.-P. Kriegel, R. Schneider, and B. Seeger. The R*-tree: An efficient and robust access method for points and rectangles. In *Proceedings of the ACM SIGMOD Conference*, pages 322–331, Atlantic City, NJ, June 1990.

22. A. Guttman. R-trees: A dynamic index structure for spatial searching. In *Proceedings of the ACM SIGMOD Conference*, pages 47–57, Boston, June 1984.

23. D. A. White and R. Jain. Similarity indexing with the SS-tree. In *Proceedings of the 12th IEEE International Conference on Data Engineering*, S. Y. W. Su, ed., pages 516–523, New Orleans, LA, February 1996.

24. N. Katayama and S. Satoh. The SR-tree: An index structure for high-dimensional nearest neighbor queries. In *Proceedings of the ACM SIGMOD Conference*, J. Peckham, ed., pages 369–380, Tucson, AZ, May 1997.

25. S. Gottschalk, M. C. Lin, and D. Manocha. OBBTree: A hierarchical structure for rapid interference detection. In *Proceedings of the SIGGRAPH'96 Conference*, pages 171–180, New Orleans, LA, August 1996.

26. R. Bayer and E. M. McCreight. Organization and maintenance of large ordered indexes. *Acta Informatica*, 1(3):173–189, 1972.

27. W. G. Aref and H. Samet. Uniquely reporting spatial objects: Yet another operation for comparing spatial data structures. In *Proceedings of the 5th International Symposium on Spatial Data Handling*, pages 178–189, Charleston, SC, August 1992.

28. W. G. Aref and H. Samet. Hashing by proximity to process duplicates in spatial databases. In *Proceedings of the 3rd International Conference on Information and Knowledge Management (CIKM)*, pages 347–354, Gaithersburg, MD, December 1994.

29. J.-P. Dittrich and B. Seeger. Data redundancy and duplicate detection in spatial join processing. In *Proceedings of the 16th IEEE International Conference on Data Engineering*, pages 535–546, San Diego, CA, February 2000.

30. J. T. Robinson. The K-D-B-tree: A search structure for large multidimensional dynamic indexes. In *Proceedings of the ACM SIGMOD Conference*, pages 10–18, Ann Arbor, MI, April 1981.

31. T. Sellis, N. Roussopoulos, and C. Faloutsos. The R+-tree: A dynamic index for multi-dimensional objects. In *Proceedings of the 13th International Conference on Very Large Databases (VLDB)*, P. M. Stocker and W. Kent, eds., pages 71–79, Brighton, UK, September 1987. Also University of Maryland Computer Science Technical Report TR-1795, 1987.

32. M. Stonebraker, T. Sellis, and E. Hanson. An analysis of rule indexing implementations in data base systems. In *Proceedings of the 1st International Conference on Expert Database Systems*, pages 353–364, Charleston, SC, April 1986.

33. O. Günther. Efficient structures for geometric data management. PhD thesis, Computer Science Division, University of California at Berkeley, Berkeley, CA, 1987. Also vol. 37 of Lecture Notes in Computer Science, Springer-Verlag, Berlin, West Germany, 1988 and Electronics Research Laboratory Memorandum UCB/ERL M87/77.

34. A. Henrich, H.-W. Six, and P. Widmayer. The LSD tree: Spatial access to multidimensional point and non-point data. In *Proceedings of the 15th International Conference on Very Large Databases (VLDB)*, P. M. G. Apers and G. Wiederhold, eds., pages 45–53, Amsterdam, The Netherlands, August 1989.

35. D. Lomet and B. Salzberg. The hB-tree: A multi-attribute indexing method with good guaranteed performance. *ACM Transactions on Database Systems*, 15(4):625–658, December 1990. Also see *Proceedings of the 5th IEEE International Conference on Data Engineering*, pages 296–304, Los Angeles, February 1989 and Northeastern University Computer Science Technical Report NU-CCS-87-24, 1987.

36. M. Freeston. The BANG file: A new kind of grid file. In *Proceedings of the ACM SIGMOD Conference*, pages 260–269, *San Francisco*, May 1987.

37. M. Freeston. A general solution of the *n*-dimensional B-tree problem. In *Proceedings of the ACM SIGMOD Conference*, pages 80–91, San Jose, CA, May 1995.

38. W. R. Franklin. Adaptive grids for geometric operations. *Cartographica*, 21(2&3):160–167, Summer & Autumn 1984.

39. H. Samet and R. E. Webber. Storing a collection of polygons using quadtrees. *ACM Transactions on Graphics*, 4(3):182–222, July 1985. Also see *Proceedings of Computer Vision and Pattern Recognition'83*, pages 127–132, Washington, DC, June 1983 and University of Maryland Computer Science Technical Report TR-1372, February 1984.

40. A. A. G. Requicha. Representations of rigid solids: Theory, methods, and systems. *ACM Computing Surveys*, 12(4):437–464, December 1980.

41. H. Samet and M. Tamminen. Bintrees, CSG trees, and time. *Computer Graphics*, 19(3):121–130, July 1985. Also in *Proceedings of the SIGGRAPH'85 Conference*, San Francisco, July 1985.

42. H. Samet and M. Tamminen. Approximating CSG trees of moving objects. *Visual Computer*, 6(4):182–209, August 1990. Also University of Maryland Computer Science Technical Report TR-1472, January 1985.

43. H. Samet. A quadtree medial axis transform. *Communications of the ACM*, 26(9):680–693, September 1983. Also see CORRIGENDUM, *Communications of the ACM*, 27(2):151, February 1984 and University of Maryland Computer Science Technical Report TR-803, August 1979.

44. H. Samet. Reconstruction of quadtrees from quadtree medial axis transforms. *Computer Vision, Graphics, and Image Processing*, 29(3):311–328, March 1985. Also University of Maryland Computer Science Technical Report TR-1224, October 1982.

45. A. U. Frank and R. Barrera. The Fieldtree: A data structure for geographic information systems. In *Design and Implementation of Large Spatial Databases—1st Symposium, SSD'89*, A. Buchmann, O. Günther, T. R. Smith, and Y.-F. Wang, eds., vol. 409 of Springer-Verlag Lecture Notes in Computer Science, pages 29–44, Santa Barbara, CA, July 1989.

46. A. Klinger. Patterns and search statistics. In *Optimizing Methods in Statistics*, J. S. Rustagi, ed., pages 303–337. Academic Press, New York, 1971.

47. G. M. Hunter. *Efficient computation and data structures for graphics.* PhD thesis, Department of Electrical Engineering and Computer Science, Princeton University, Princeton, NJ, 1978.

48. D. Meagher. Geometric modeling using octree encoding. *Computer Graphics and Image Processing*, 19(2):129–147, June 1982.

49. K. Knowlton. Progressive transmission of grey-scale and binary pictures by simple efficient, and lossless encoding schemes. *Proceedings of the IEEE*, 68(7):885–896, July 1980.

50. H. Samet and M. Tamminen. Efficient component labeling of images of arbitrary dimension represented by linear bintrees. *IEEE Transactions on Pattern Analysis and Machine Intelligence*, 10(4):579–586, July 1988.

51. M. Tamminen. Comment on quad- and octrees. *Communications of the ACM*, 27(3):248–249, March 1984.

52. H. Fuchs, Z. M. Kedem, and B. F. Naylor. On visible surface generation by a priori tree structures. *Computer Graphics*, 14(3):124–133, July 1980. Also in *Proceedings of the SIGGRAPH'80 Conference*, Seattle, WA, July 1980.

53. D. Meagher. Octree encoding: A new technique for the representation, manipulation, and display of arbitrary 3-D objects by computer. Electrical and Systems Engineering Technical Report IPL-TR-80-111, Rensselaer Polytechnic Institute, Troy, NY, October 1980.

54. H. Samet. Computing perimeters of images represented by quadtrees. *IEEE Transactions on Pattern Analysis and Machine Intelligence*, 3(6):683–687, November 1981. Also University of Maryland Computer Science Technical Report TR-755, April 1979.

55. H. Samet and M. Tamminen. An improved approach to connected component labeling of images. In *Proceedings of Computer Vision and Pattern Recognition'86*, pages 312–318, Miami Beach, FL, June 1986. Also see University of Maryland Computer Science Technical Report TR-1649, August 1986.

56. C. R. Dyer. Computing the Euler number of an image from its quadtree. *Computer Graphics and Image Processing*, 13(3):270–276, July 1980. Also University of Maryland Computer Science Technical Report TR-769, May 1979.

57. H. Samet. Algorithms for the conversion of quadtrees to rasters. *Computer Vision, Graphics, and Image Processing*, 26(1):1–16, April 1984. Also University of Maryland Computer Science Technical Report TR-979, November 1980.

58. A. Amir, A. Efrat, P. Indyk, and H. Samet. Efficient algorithms and regular data structures for dilation, location and proximity problems. *Algorithmica*, 30(2):164–187, 2001. Also see *Proceedings of the 40th IEEE Annual Symposium on Foundations of Computer Science*, pages 160–170, New York, October 1999 and University of Maryland Computer Science Technical Report TR-4201, November 2000.

59. C.-H. Ang, H. Samet, and C. A. Shaffer. A new region expansion for quadtrees. *IEEE Transactions on Pattern Analysis and Machine Intelligence*, 12(7):682–686, July 1990. Also see *Proceedings of the Third International Symposium on Spatial Data Handling*, pages 19–37, Sydney, Australia, August 1988.

60. W. G. Aref and H. Samet. Efficient processing of window queries in the pyramid data structure. In *Proceedings of the 9th ACM SIGACT–SIGMOD–SIGART Symposium on Principles of Database Systems (PODS)*, pages 265–272, Nashville, TN, April 1990. Also in *Proceedings of the Fifth Brazilian Symposium on Databases*, pages 15–26, Rio de Janeiro, Brazil, April 1990.

61. S. L. Tanimoto and T. Pavlidis. A hierarchical data structure for picture processing. *Computer Graphics and Image Processing*, 4(2):104–119, June 1975.

62. J. Han and M. Kamber. *Data Mining: Concepts and Techniques*. Morgan Kaufmann, San Francisco, 2000.

63. W. Wang, J. Yang, and R. R. Muntz. STING: A statistical information grid approach to spatial data mining. In *Proceedings of the 23rd International Conference on Very Large Data Bases (VLDB)*, M. Jarke, M. J. Carey, K. R. Dittrich, F. H. Lochovsky, P. Loucopoulos, and M. A. Jeusfeld, eds., pages 186–195, Athens, Greece, August 1997.

64. F. P. Preparata and M. I. Shamos. *Computational Geometry: An Introduction*. Springer-Verlag, New York, 1985.

65. J. L. Bentley. *Algorithms for Klee's Rectangle Problems (Unpublished)*. Computer Science Department, Carnegie-Mellon University, Pittsburgh, PA, 1977.

66. H. Edelsbrunner. Dynamic rectangle intersection searching. Institute for Information Processing Technical Report 47, Technical University of Graz, Graz, Austria, February 1980.

67. E. Jacox and H. Samet. Iterative spatial join. *ACM Transactions on Database Systems*, 28(3):268–294, September 2003.

68. G. Kedem. The quad-CIF tree: A data structure for hierarchical on-line algorithms. In *Proceedings of the 19th Design Automation Conference*, pages 352–357, Las Vegas, NV, June 1982. Also University of Rochester Computer Science Technical Report TR-91, September 1981.

69. G. G. Lai, D. Fussell, and D. F. Wong. HV/VH trees: A new spatial data structure for fast region queries. In *Proceedings of the 30th ACM/IEEE Design Automation Conference*, pages 43–47, Dallas, TX, June 1993.

70. E. M. McCreight. Efficient algorithms for enumerating intersecting intervals and rectangles. Technical Report CSL-80-09, Xerox Palo Alto Research Center, Palo Alto, CA, June 1980.

71. A. Frank. Problems of realizing LIS: Storage methods for space related data: The fieldtree. Technical Report 71, Institute for Geodesy and Photogrammetry, ETH, Zurich, Switzerland, June 1983.

72. T. Ulrich. Loose octrees. In *Game Programming Gems*, M. A. DeLoura, ed., pages 444–453. Charles River Media, Rockland, MA, 2000.

73. D. J. Abel and J. L. Smith. A data structure and query algorithm for a database of areal entities. *Australian Computer Journal*, 16(4):147–154, November 1984.

74. T. M. Chan. Approximate nearest neighbor queries revisited. In *Proceedings of the 13th Annual Symposium on Computational Geometry*, pages 352–358, Nice, France, June 1997.

75. S. Liao, M. A. Lopez, and S. T. Leutenegger. High dimensional similarity search with space filling curves. In *Proceedings of the 17th IEEE International Conference on Data Engineering*, pages 615–622, Heidelberg, Germany, April 2001.

76. M. D. Lieberman, J. Sankaranarayanan, and H. Samet. A fast similarity join algorithm using graphics processing units. In *Proceedings of the 24th IEEE International Conference on Data Engineering*, pages 1111–1120, Cancun, Mexico, April 2008.

77. H. Samet, J. Sankaranarayanan, and M. Auerbach. Indexing methods for moving object databases: Games and other applications. In *Proceedings of the ACM SIGMOD Conference*, pages 169–180, New York, June 2013.

78. H. Freeman. Computer processing of line-drawing images. *ACM Computing Surveys*, 6(1):57–97, March 1974.

79. B. G. Baumgart. A polyhedron representation for computer vision. In *Proceedings of the 1975 National Computer Conference*, vol. 44, pages 589–596, Anaheim, CA, May 1975.

80. D. H. Ballard. Strip trees: A hierarchical representation for curves. *Communications of the ACM*, 24(5):310–321, May 1981. Also see corrigendum, *Communications of the ACM*, 25(3):213, March 1982.

81. G. M. Hunter and K. Steiglitz. Operations on images using quad trees. *IEEE Transactions on Pattern Analysis and Machine Intelligence*, 1(2):145–153, April 1979.

82. M. Shneier. Two hierarchical linear feature representations: Edge pyramids and edge quadtrees. *Computer Graphics and Image Processing*, 17(3):211–224, November 1981. Also University of Maryland Computer Science Technical Report TR-961, October 1980.

83. J. E. Warnock. A hidden surface algorithm for computer generated half tone pictures. Computer Science Technical Report TR 4–15, University of Utah, Salt Lake City, UT, June 1969.

84. R. C. Nelson and H. Samet. A consistent hierarchical representation for vector data. *Computer Graphics*, 20(4):197–206, August 1986. Also in *Proceedings of the SIGGRAPH'86 Conference*, Dallas, TX, August 1986.

85. H. Samet, C. A. Shaffer, and R. E. Webber. Digitizing the plane with cells of non-uniform size. *Information Processing Letters*, 24(6):369–375, April 1987. Also an expanded version in University of Maryland Computer Science Technical Report TR-1619, January 1986.

86. D. Ayala, P. Brunet, R. Juan, and I. Navazo. Object representation by means of nonminimal division quadtrees and octrees. *ACM Transactions on Graphics*, 4(1):41–59, January 1985.

87. G. R. Hjaltason and H. Samet. Ranking in spatial databases. In *Advances in Spatial Databases—4th International Symposium, SSD'95*, M. J. Egenhofer and J. R. Herring, eds., vol. 951 of Springer-Verlag Lecture Notes in Computer Science, pages 83–95, Portland, ME, August 1995.

88. G. R. Hjaltason and H. Samet. Distance browsing in spatial databases. *ACM Transactions on Database Systems*, 24(2):265–318, June 1999. Also University of Maryland Computer Science Technical Report TR-3919, July 1998.

89. G. R. Hjaltason and H. Samet. Index-driven similarity search in metric spaces. *ACM Transactions on Database Systems*, 28(4):517–580, December 2003.

90. E. G. Hoel and H. Samet. Efficient processing of spatial queries in line segment databases. In *Advances in Spatial Databases—2nd Symposium, SSD'91*, O. Günther and H.-J. Schek, eds., vol. 525 of Springer-Verlag Lecture Notes in Computer Science, pages 237–256, Zurich, Switzerland, August 1991.

91. E. G. Hoel and H. Samet. A qualitative comparison study of data structures for large line segment databases. In *Proceedings of the ACM SIGMOD Conference*, M. Stonebraker, ed., pages 205–214, San Diego, CA, June 1992.

92. M. Lindenbaum and H. Samet. A probabilistic analysis of trie-based sorting of large collections of line segments. Computer Science Technical Report TR-3455, University of Maryland, College Park, MD, April 1995. Also an expanded version in *SIAM Journal on Computing*, 35(1):22–58, September 2005 and TR-3455.1, February 2000.

93. H. Samet and R. E. Webber. On encoding boundaries with quadtrees. Computer Science Technical Report TR-1162, University of Maryland, College Park, MD, February 1982.

94. H. Samet and R. E. Webber. On encoding boundaries with quadtrees. *IEEE Transactions on Pattern Analysis and Machine Intelligence*, 6(3):365–369, May 1984.

95. C. Esperança and H. Samet. Experience with SAND/Tcl: A scripting tool for spatial databases. *Journal of Visual Languages and Computing*, 13(2):229–255, April 2002.

96. H. Samet, H. Alborzi, F. Brabec, C. Esperança, G. R. Hjaltason, F. Morgan, and E. Tanin. Use of the SAND spatial browser for digital government applications. *Communications of the ACM*, 46(1):63–66, January 2003.

97. H. Samet, A. Rosenfeld, C. A. Shaffer, and R. E. Webber. A geographic information system using quadtrees. *Pattern Recognition*, 17(6):647–656, November/December 1984.

98. C. A. Shaffer, H. Samet, R. C. Nelson. QUILT: A geographic information system based on quadtrees. *International Journal of Geographical Information Systems*, 4(2):103–131, April–June 1990. Also University of Maryland Computer Science Technical Report TR-1885.1, July 1987.

99. G. S. Iwerks and H. Samet. The spatial spreadsheet. In *Proceedings of the 3rd International Conference on Visual Information Systems (VISUAL99)*, D. P. Huijsmans and A. W. M. Smeulders, eds., pages 317–324, Amsterdam, The Netherlands, June 1999.

100. K. S. Beyer, J. Goldstein, R. Ramakrishnan, and U. Shaft. When is "nearest neighbor" meaningful? In *Proceedings of the 7th International Conference on Database Theory (ICDT'99)*, C. Beeri and P. Buneman, eds., vol. 1540 of Springer-Verlag Lecture Notes in Computer Science, pages 217–235, Berlin, Germany, January 1999.

101. S. Berchtold, C. Böhm, and H.-P. Kriegel. Improving the query performance of high-dimensional index structures by bulk-load operations. In *Advances in Database Technology—EDBT'98, Proceedings of the 6th International Conference on Extending Database Technology*, H.-J Schek, F. Saltor, I. Ramos, and G. Alonso, eds., vol. 1377 of Springer-Verlag Lecture Notes in Computer Science, pages 216–230, Valencia, Spain, March 1998.

102. R. Weber, H.-J. Schek, and S. Blott. A quantitative analysis and performance study for similarity-search methods in high-dimensional spaces. In *Proceedings of the 24th International Conference on Very Large Data Bases (VLDB)*, A. Gupta, O. Shmueli, and J. Widom, eds., pages 194–205, New York, August 1998.

103. H. Ferhatosmanoglu, E. Tuncel, D. Agrawal, and A. El Abbadi. Vector approximation based indexing for non-uniform high dimensional data sets. In *Proceedings of the 9th International Conference on Information and Knowledge Management (CIKM)*, pages 202–209, McLean, VA, November 2000.

104. S. Berchtold, C. Böhm, H. V. Jagadish, H.-P. Kriegel, and J. Sander. Independent quantization: An index compression technique for high-dimensional data spaces. In *Proceedings of the 16th IEEE International Conference on Data Engineering*, pages 577–588, San Diego, CA, February 2000.

105. S. Berchtold, C. Böhm, and H.-P. Kriegel. The pyramid-technique: Towards breaking the curse of dimensionality. In *Proceedings of the ACM SIGMOD Conference*, L. Hass and A. Tiwary, eds., pages 142–153, Seattle, WA, June 1998.

106. K. Chakrabarti and S. Mehrotra. The hybrid tree: An index structure for high dimensional feature spaces. In *Proceedings of the 15th IEEE International Conference on Data Engineering*, pages 440–447, Sydney, Australia, March 1999. Also University of California at Irvine Information and Computer Science Technical Report TR-MARS-98-14, July 1998.

107. M. D. Lieberman and H. Samet. Multifaceted toponym recognition for streaming news. In *Proceedings of the 34th International Conference on Research and Development in Information Retrieval (SIGIR'11)*, pages 843–852, Beijing, China, July 2011.

108. M. D. Lieberman and H. Samet. Adaptive context features for toponym resolution in streaming news. In *Proceedings of the 35th International Conference on Research and Development in Information Retrieval (SIGIR'12)*, pages 731–740, Portland, OR, August 2012.

109. H. Samet. Using minimaps to enable toponym resolution with an effective 100% rate of recall. In *Proceedings of 8th ACM SIGSPATIAL Workshop on Geographic Information Retrieval (GIR'14)*, R. Purves and C. Jones, eds., pages 9:1–9:8, Dallas, TX, November 2014.

110. M. D. Lieberman and H. Samet. Supporting rapid processing and interactive map-based exploration of streaming news. In *Proceedings of the 20th ACM SIGSPATIAL International Conference on Advances in Geographic Information Systems*, I. Cruz, C. A. Knoblock, P. Kröger, E. Tanin, and P. Widmayer, eds., pages 179–188, Redondo Beach, CA, November 2012.

111. H. Samet, M. D. Adelfio, B. C. Fruin, M. D. Lieberman, and B. E. Teitler. Porting a web-based mapping application to a smartphone app. In *Proceedings of the 19th ACM SIGSPATIAL International Conference on Advances in Geographic Information Systems*, D. Agrawal, I. Cruz, C. S. Jensen, E. Ofek, and E. Tanin, eds., pages 525–528, Chicago, November 2011.

112. H. Samet, J. Sankaranarayanan, M. D. Lieberman, M. D. Adelfio, B. C. Fruin, J. M. Lotkowski, D. Panozzo, J. Sperling, and B. E. Teitler. Reading news with maps by exploiting spatial synonyms. *Communications of the ACM*, 57(10):64–77, October 2014.

113. H. Samet, B. E. Teitler, M. D. Adelfio, and M. D. Lieberman. Adapting a map query interface for a gesturing touch screen interface. In *Proceedings of the Twentieth International Word Wide Web Conference (Companion Volume)*, S. Srinivasan, K. Ramamritham, A. Kumar, M. P. Ravindra, E. Bertino, and R. Kumar, eds., pages 257–260, Hyderabad, India, March-April 2011.

114. B. Teitler, M. D. Lieberman, D. Panozzo, J. Sankaranarayanan, H. Samet, and J. Sperling. News Stand: A new view on news. In *Proceedings of the 16th ACM SIGSPATIAL International Conference on Advances in Geographic Information Systems*, W. G. Aref, M. F. Mokbel, H. Samet, M. Schneider, C. Shahabi, and O. Wolfson, eds., pages 144–153, Irvine, CA, November 2008.

115. N. Gramsky and H. Samet. Seeder finder—Identifying additional needles in the Twitter haystack. In *Proceedings of the 6th ACM SIGSPATIAL International Workshop on Location-Based Social Networks (LBSN'13)*, A. Pozdnukhov, ed., pages 44–53, Orlando, FL, November 2013.

116. A. Jackoway, H. Samet, and J. Sankaranarayanan. Identification of live news events using Twitter. In *Proceedings of the 3rd ACM SIGSPATIAL International Workshop on Location-Based Social Networks (LBSN'11)*, Y. Zheng and M. F. Mokbel, eds., pages 25–32, Chicago, November 2011.

117. J. Sankaranarayanan, H. Samet, B. Teitler, M. D. Lieberman, and J. Sperling. TwitterStand: News in tweets. In *Proceedings of the 17th ACM SIGSPATIAL International Conference on Advances in Geographic Information Systems*, D. Agrawal, W. G. Aref, C.-T. Lu, M. F. Mokbel, P. Scheuermann, C. Shahabi, and O. Wolfson, eds., pages 42–51, Seattle, WA, November 2009.

118. M. D. Adelfio and H. Samet. Structured toponym resolution using combined hierarchical place categories. In *Proceedings of 7th ACM SIGSPATIAL Workshop on Geographic Information Retrieval (GIR'13)*, R. Purves and C. Jones, eds., pages 49–56, Orlando, FL, November 2013.

119. E. Amitay, N. Har'El, R. Sivan, and A. Soffer. Web-a-Where: Geotagging web content. In *Proceedings of the 27th International Conference on Research and Development in Information Retrieval (SIGIR'04)*, M. Sanderson, K. Järvelin, J. Allan, and P. Bruza, eds., pages 273–280, Sheffield, UK, July 2004.

120. C. B. Jones, R. S. Purves, P. D. Clough, and H. Joho. Modelling vague places with knowledge from the Web. *International Journal of Geographical Information Systems*, 22(10):1045–1065, 2008.

121. J. L. Leidner and M. D. Lieberman. Detecting geographical references in the form of place names and associated spatial natural language. *SIGSPATIAL Special*, 3(2):5–11, 2011.

122. M. D. Lieberman, H. Samet, and J. Sankaranarayanan. Geotagging: Using proximity, sibling, and prominence clues to understand comma groups. In *Proceedings of 6th Workshop on Geographic Information Retrieval*, R. Purves, C. Jones, and P. Clough, eds., Zurich, Switzerland, February 2010. Article 6.

123. M. D. Lieberman, H. Samet, and J. Sankaranarayanan. Geotagging with local lexicons to build indexes for textually-specified spatial data. In *Proceedings of the 26th IEEE International Conference on Data Engineering*, pages 201–212, Long Beach, CA, March 2010.

124. B. Martins, H. Manguinhas, and J. Borbinha. Extracting and exploring the geo-temporal semantics of textual resources. In *Proceedings of the 2th IEEE International Conference on Semantic Computing (ICSC'08)*, pages 1–9, Santa Clara, CA, August 2008.

125. R. S. Purves, P. Clough, C. B. Jones, A. Arampatzis, B. Bucher, D. Finch, G. Fu, H. Joho, A. K Syed, S. Vaid, and B. Yang. The design and implementation of SPIRIT: A spatially aware search engine for information retrieval on the internet. *International Journal of Geographical Information Systems*, 21(7):717–745, 2007.

126. G. Quercini, H. Samet, J. Sankaranarayanan, and M. D. Lieberman. Determining the spatial reader scopes of news sources using local lexicons. In *Proceedings of the 18th ACM SIGSPATIAL International Conference on Advances in Geographic Information Systems*, A. El Abbadi, D. Agrawal, M. Mokbel, and P. Zhang, eds., pages 43–52, San Jose, CA, November 2010.

18

Planar Straight Line Graphs*

18.1 Introduction.. 277
18.2 Features of PSLGs.. 277
18.3 Operations on PSLGs.. 279
    Edge Insertion and Deletion • Vertex Split and Edge Contraction
18.4 Winged-Edge... 280
18.5 Halfedge.. 281
    Access Functions • Edge Insertion and Deletion • Vertex Split and Edge Contraction
18.6 Quadedge.. 287
18.7 Further Remarks... 288
18.8 Glossary... 288
Acknowledgments 289
References.. 289

Siu-Wing Cheng
Hong Kong University of Science and Technology

18.1 Introduction

Graphs (Chapter 4) have found extensive applications in computer science as a modeling tool. In mathematical terms, a graph is simply a collection of vertices and edges. Indeed, a popular graph data structure is the adjacency lists representation [1] in which each vertex keeps a list of vertices connected to it by edges. In a typical application, the vertices model entities and an edge models a relation between the entities corresponding to the edge endpoints. For example, the transportation problem calls for a minimum cost shipping pattern from a set of origins to a set of destinations [2]. This can be modeled as a complete directed bipartite graph. The origins and destinations are represented by two columns of vertices. Each origin vertex is labeled with the amount of supply stored there. Each destination vertex is labeled with the amount of demand required there. The edges are directed from the origin vertices to the destination vertices and each edge is labeled with the unit cost of transportation. Only the adjacency information between vertices and edges are useful and captured, apart from the application dependent information.

In geometric computing, graphs are also useful for representing various diagrams. We restrict our attention to diagrams that are planar graphs embedded in the plane using straight edges without edge crossings. Such diagrams are called *planar straight line graphs* and denoted by PSLGs for short. Examples include Voronoi diagrams, arrangements, and triangulations. Their definitions can be found in standard computational geometry texts such as the book by de Berg et al. [3]. See also Chapters 65, 66, and 67. For completeness, we also provide their definitions in Section 18.8. The straight edges in a PSLG partition the plane into regions with disjoint interior. We call these regions *faces*. The adjacency lists representation is usually inadequate for applications that manipulate PSLGs. Consider the problem of locating the face containing a query point in a Delaunay triangulation. One practical algorithm is to walk towards the query point from a randomly chosen starting vertex [4], see Figure 18.1. To support this algorithm, one needs to know the first face that we enter as well as the next face that we step into whenever we cross an edge. Such information is not readily provided by an adjacency lists representation.

There are three well-known data structures for representing PSLGs: the winged-edge, halfedge, and quadedge data structures. In Sections 18.2 and 18.3, we discuss the PSLGs that we deal with in more details and the operations on PSLGs. Afterwards, we introduce the three data structures in Sections 18.4 through 18.6. We conclude in Section 18.7 with some further remarks.

18.2 Features of PSLGs

We assume that each face has exactly one boundary and we allow dangling edges on a face boundary. These assumptions are valid for many important classes of PSLGs such as triangulations, Voronoi diagrams, planar subdivisions with no holes, arrangements of lines, and some special arrangements of line segments (see Figure 18.2).

---

* This chapter has been reprinted from first edition of this Handbook, without any content updates.

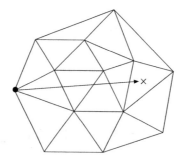

FIGURE 18.1   Locate the face containing the cross by walking from a randomly chosen vertex.

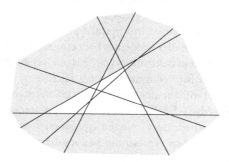

FIGURE 18.2   Dangling edges.

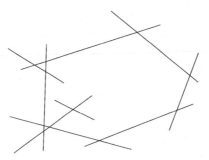

FIGURE 18.3   The shaded faces are the unbounded faces of the arrangement.

There is at least one unbounded face in a PSLG but there could be more than one, for example, in the arrangement of lines shown in Figure 18.3.

The example also shows that there may be some infinite edges. To handle infinite edges like halflines and lines, we need a special vertex v_{inf} at infinity. One can imagine that the PSLG is placed in a small almost flat disk D at the north pole of a giant sphere S and v_{inf} is placed at the south pole. If an edge e is a halfline originating from a vertex u, then the endpoints of e are u and v_{inf}. One can view e as a curve on S from u near the north pole to v_{inf} at the south pole, but e behaves as a halfline inside the disk D. If an edge e is a line, then v_{inf} is the only endpoint of e. One can view e as a loop from v_{inf} to the north pole and back, but e behaves as a line inside the disk D.

We do not allow isolated vertices, except for v_{inf}. Planarity implies that the incident edges of each vertex are circularly ordered around that vertex. This applies to v_{inf} as well.

A PSLG data structure keeps three kinds of attributes: *vertex attributes*, *edge attributes*, and *face attributes*. The attributes of a vertex include its coordinates except for v_{inf} (we assume that v_{inf} is tagged to distinguish it from other vertices). The attributes of an edge include the equation of the support line of the edge (in the form of $Ax + By + C = 0$). The face attributes are useful for auxiliary information, for example, color.

18.3 Operations on PSLGs

The operations on a PSLG can be classified into *access functions* and *structural operations*. The access functions retrieve information without modifying the PSLG. Since the access functions partly depend on the data structure, we discuss them later when we introduce the data structures. In this section, we discuss four structural operations on PSLGs: *edge insertion*, *edge deletion*, *vertex split*, and *edge contraction*. We concentrate on the semantics of these four operations and discuss the implementation details later when we introduce the data structures. For vertex split and edge contraction, we assume further that each face in the PSLG is a simple polygon as these two operations are usually used under such circumstances.

18.3.1 Edge Insertion and Deletion

When a new edge e with endpoints u and v is inserted, we assume that e does not cross any existing edge. If u or v is not an existing vertex, the vertex will be created. If both u and v are new vertices, e is an isolated edge inside a face f. Since each face is assumed to have exactly one boundary, this case happens only when the PSLG is empty and f is the entire plane. Note that e becomes a new boundary of f. If either u or v is a new vertex, then the boundary of exactly one face gains the edge e. If both u and v already exist, then u and v lie on the boundary of a face which is split into two new faces by the insertion of e. These cases are illustrated in Figure 18.4.

The deletion of an edge e has the opposite effects. After the deletion of e, if any of its endpoint becomes an isolated vertex, it will be removed. The vertex v_{inf} is an exception and it is the only possible isolated vertex. The edge insertion is clearly needed to create a PSLG from scratch. Other effects can be achieved by combining edge insertions and deletions appropriately. For example, one can use the two operations to overlay two PSLGs in a plane sweep algorithm, see Figure 18.5.

18.3.2 Vertex Split and Edge Contraction

The splitting of a vertex v is best visualized as the continuous morphing of v into an edge e. Depending on the specification of the splitting, an incident face of v gains e on its boundary or an incident edge of v is split into a triangular face, see Figure 18.6. The incident edges of v are displaced and it is assumed that no self-intersection occurs within the PSLG during the splitting. The contraction of an edge e is the inverse of the vertex split. We also assume that no self-intersection occurs during the edge contraction. If e is incident on a triangular face, that face will disappear after the contraction of e.

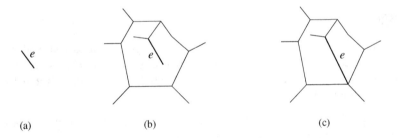

(a)   (b)   (c)

FIGURE 18.4   Cases in edge insertion.

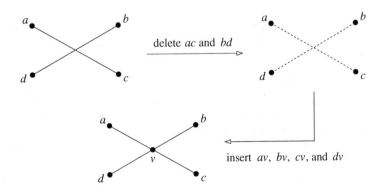

FIGURE 18.5   Intersecting two edges.

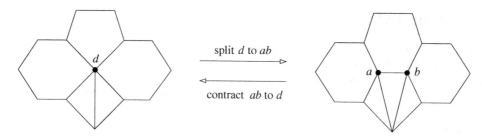

FIGURE 18.6   Vertex split and edge contraction.

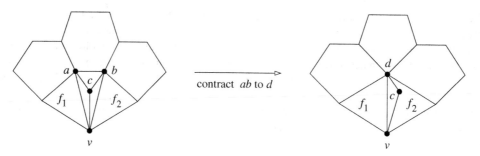

FIGURE 18.7   Non-contractible edge.

Not every edge can be contracted. Consider an edge ab. If the PSLG contains a cycle abv that is not the boundary of any face incident to ab, we call the edge ab *non-contractible* because its contraction is not cleanly defined. Figure 18.7 shows an example. In the figure, after the contraction, there is an ambiguity whether dv should be incident on the face f_1 or the face f_2. In fact, one would expect the edge dv to behave like av and bv and be incident on both f_1 and f_2 after the contraction. However, this is impossible.

The vertex split and edge contraction have been used in clustering and hierarchical drawing of maximal planar graphs [5].

18.4 Winged-Edge

The winged-edge data structure was introduced by Baumgart [6] and it predates the halfedge and quadedge data structures. There are three kinds of records: *vertex records*, *edge records*, and *face records*. Each vertex record keeps a reference to one incident edge of the vertex. Each face record keeps a reference to one boundary edge of the face. Each edge e is stored as an oriented edge with the following references (see Figure 18.8):

- The *origin* endpoint $e.org$ and the *destination* endpoint $e.dest$ of e. The convention is that e is directed from $e.org$ to $e.dest$.
- The faces $e.left$ and $e.right$ on the left and right of e, respectively.
- The two edges $e.lcw$ and $e.lccw$ adjacent to e that bound the face $e.left$. The edge $e.lcw$ is incident to $e.org$ and the edge $e.lccw$ is incident to $e.dest$. Note that $e.lcw$ (resp. $e.lccw$) succeeds e if the boundary of $e.left$ is traversed in the clockwise (resp. anti-clockwise) direction from e.
- The two edges $e.rcw$ and $e.rccw$ adjacent to e that bound the face $e.right$. The edge $e.rcw$ is incident to $e.dest$ and the edge $e.rccw$ is incident to $e.org$. Note that $e.rcw$ (resp. $e.rccw$) succeeds e if the boundary of $e.right$ is traversed in the clockwise (resp. anti-clockwise) direction from e.

The information in each edge record can be retrieved in constant time. Given a vertex v, an edge e, and a face f, we can thus answer in constant time whether v is incident on e and e is incident on f. Given a vertex v, we can traverse the edges incident to v in clockwise order as follows. We output the edge e kept at the vertex record for v. We perform $e := e.rccw$ if $v = e.org$ and $e := e.lccw$ otherwise. Then we output e and repeat the above. Given a face f, we can traverse its boundary edges in clockwise order as follows. We output the edge e kept at the face record for f. We perform $e := e.lcw$ if $f = e.left$ and $e := e.rcw$ otherwise. Then we output e and repeat the above.

Note that an edge reference does not carry information about the orientation of the edge. Also, the orientations of the boundary edges of a face need not be consistent with either the clockwise or anti-clockwise traversal. Thus, the manipulation of the data structure is often complicated by case distinctions. We illustrate this with the insertion of an edge e. Assume that $e.org = u$,

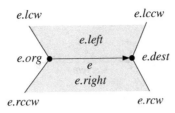

FIGURE 18.8 Winged-edge data structure.

$e.dest = v$, and both u and v already exist. The input also specifies two edges e_1 and e_2 incident to u and v, respectively. The new edge e is supposed to immediately succeed e_1 (resp. e_2) in the anti-clockwise ordering of edges around u (resp. v). The insertion routine works as follows.

1. If $u = v_{inf}$ and it is isolated, we need to store the reference to e in the vertex record for u. We update the vertex record for v similarly.

2. Let e_3 be the incident edge of u following e_1 such that e is to be inserted between e_1 and e_3. Note that e_3 succeeds e_1 in anti-clockwise order. We insert e between e_1 and e_3 as follows.

    $e.rccw := e_1$; $e.lcw := e_3$;
    if $e.org = e_1.org$ **then** $e_1.lcw := e$;   **else** $e_1.rcw := e$;
    if $e.org = e_3.org$ **then** $e_3.rccw := e$;   **else** $e_3.lccw := e$;

3. Let e_4 be the incident edge of v following e_2 such that e is to be inserted between e_2 and e_4. Note that e_4 succeeds e_2 in anti-clockwise order. We insert e between e_2 and e_4 as follows.

    $e.lccw := e_2$; $e.rcw := e_4$;
    if $e.dest = e_2.dest$ **then** $e_2.rcw := e$;   **else** $e_2.lcw := e$;
    if $e.dest = e_4.dest$ **then** $e_4.lccw := e$; **else** $e_4.rccw := e$;

4. The insertion of e has split a face into two. So we create a new face f and make $e.left$ reference it. Also, we store a reference to e in the face record for f. There are further ramifications. First, we make $e.right$ reference the old face.

    if $e.org = e_1.org$ **then** $e.right := e_1.left$; **else** $e.right := e_1.right$;

    Second, we make the *left* or *right* fields of the boundary edges of f reference f.

    $e' := e$; $w := e.org$;
    repeat
       if $e'.org = w$ **then** $e'.left := f$; $w := e'.dest$; $e' := e'.lccw$
            else $e'.right := f$; $w := e'.org$; $e' := e'.rccw$
    until $e' = e$;

Notice the inconvenient case distinctions needed in steps 2, 3, and 4. The halfedge data structure is designed to keep both orientations of the edges and link them properly. This eliminates most of these case distinctions as well as simplifies the storage scheme.

18.5 Halfedge

In the halfedge data structure, for each edge in the PSLG, there are two symmetric edge records for the two possible orientations of the edge [7]. This solves the orientation problem in the winged-edge data structure. The halfedge data structure is also known as the *doubly connected edge list* [3]. We remark that the name doubly connected edge list was first used to denote the variant of the winged-edge data structure in which the *lccw* and *rccw* fields are omitted [8,9].

There are three kinds of records: *vertex records*, *halfedge records*, and *face records*. Let e be a halfedge. The following information is kept at the record for e (see Figure 18.9).

- The reference $e.sym$ to the symmetric version of e.
- The *origin* endpoint $e.org$ of e. We do not need to store the destination endpoint of e since it can be accessed as $e.sym.org$. The convention is that e is directed from $e.org$ to $e.sym.org$.
- The face $e.left$ on the left of e.
- The next edge $e.succ$ and the previous edge $e.pred$ in the anti-clockwise traversal around the face $e.left$.

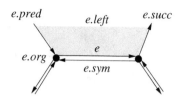

FIGURE 18.9   Halfedge data structure.

For each vertex v, its record keeps a reference to one halfedge $v.edge$ such that $v = v.edge.org$. For each face f, its record keeps a reference to one halfedge $f.edge$ such that $f = f.edge.left$.

We introduce two basic operations `make_halfedges` and `half_splice` which will be needed for implementing the operations on PSLGs. These two operations are motivated by the operations `make_edge` and `splice` introduced by Guibas and Stolfi [10] for the quadedge data structure. We can also do without `make_halfedges` and `half_splice`, but they make things simpler.

- `make_halfedges(u, v)`: Return two halfedges e and $e.sym$ connecting the points u and v. The halfedges e and $e.sym$ are initialized such that they represent a new PSLG with e and $e.sym$ as the only halfedges. That is, $e.succ = e.sym = e.pred$ and $e.sym.succ = e = e.sym.pred$. Also, e is the halfedge directed from u to v. If u and v are omitted, it means that the actual coordinates of $e.org$ and $e.sym.org$ are unimportant.
- `half_splice(e_1, e_2)`: Given two halfedges e_1 and e_2, `half_splice` swaps the contents of $e_1.pred$ and $e_2.pred$ and the contents of $e_1.pred.succ$ and $e_2.pred.succ$. The effects are:
  - Let $v = e_2.org$. If $e_1.org \neq v$, the incident halfedges of $e_1.org$ and $e_2.org$ are merged into one circular list (see Figure 18.10a). The vertex v is now redundant and we finish the merging as follows.

    $e' := e_2$;
    repeat
       $e'.org := e_1.org;\ e' := e'.sym.succ$;
    until $e' = e_2$;
    delete the vertex record for v;

  - Let $v = e_2.org$. If $e_1.org = v$, the incident halfedges of v are separated into two circular lists (see Figure 18.10b). We create a new vertex u for $e_2.org$ with the coordinates of u left uninitialized. Then we finish the separation as follows.

    $u.edge := e_2;\ e' := e_2$;
    repeat
       $e'.org := u;\ e' := e'.sym.succ$;
    until $e' = e_2$.

The behavior of `half_splice` is somewhat complex even in the following special cases. If e is an isolated halfedge, `half_splice(e_1, e)` deletes the vertex record for $e.org$ and makes e a halfedge incident to $e_1.org$ following e_1 in anti-clockwise order. If $e_1 = e.sym.succ$, `half_splice(e_1, e)` detaches e from the vertex $e_1.org$ and creates a new vertex record for $e.org$. If $e_1 = e$, `half_splice(e, e)` has no effect at all.

18.5.1   Access Functions

The information in each halfedge record can be retrieved in constant time. Given a vertex v, a halfedge e, and a face f, we can thus answer the following adjacency queries:

1. Is v incident on e? This is done by checking if $v = e.org$ or $e.sym.org$.
2. Is e incident on f? This is done by checking if $f = e.left$.
3. List the halfedges with origin v in clockwise order. Let $e = v.edge$. Output e, perform $e := e.sym.succ$, and then repeat until we return to $v.edge$.
4. List the boundary halfedges of f in anti-clockwise order. Let $e = f.edge$. Output e, perform $e := e.succ$, and then repeat until we return to $f.edge$.

Other adjacency queries (e.g., listing the boundary vertices of a face) can be answered similarly.

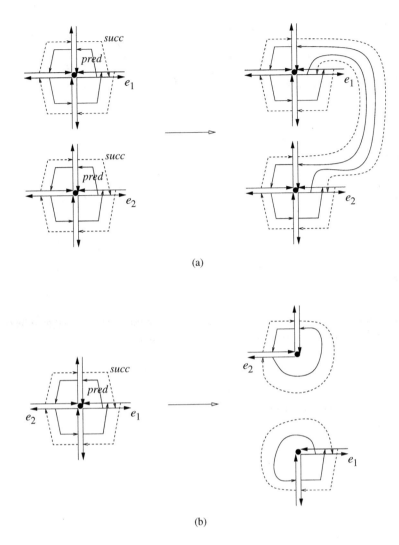

(a)

(b)

FIGURE 18.10   The effects of `half_splice`.

18.5.2 Edge Insertion and Deletion

The edge insertion routine takes two vertices u and v and two halfedges e_1 and e_2. If u is a new vertex, e_1 is ignored; otherwise, we assume that $e_1.org = u$. Similarly, if v is a new vertex, e_2 is ignored; otherwise, we assume that $e_2.org = v$. The general case is that an edge connecting u and v is inserted between e_1 and $e_1.pred.sym$ and between e_2 and $e_2.pred.sym$. The two new halfedges e and $e.sym$ are returned with the convention that e is directed from u to v.

Algorithm $\mathrm{insert}(u, v, e_1, e_2)$
1. $(e, e.sym) := \mathrm{make\_halfedges}(u, v)$;
2. **if** u is not new
3.   **then** $\mathrm{half\_splice}(e_1, e)$;
4. $e.left := e_1.left$;
5. $e.sym.left := e_1.left$;
6. **if** v is not new
7.   **then** $\mathrm{half\_splice}(e_2, e.sym)$;
8. $e.left := e_2.left$;
9. $e.sym.left := e_2.left$;
10. **if** neither u nor v is new
11.   **then** /* A face has been split */
12. $e_2.left.edge := e$;

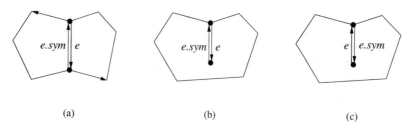

(a)   (b)   (c)

FIGURE 18.11   Cases in deletion.

13.   create a new face f;
14. $f.edge := e.sym$;
15. $e' := e.sym$;
16.   **repeat**
17. $e'.left := f$;
18. $e' := e'.succ$;
19.   **until** $e' = e.sym$;
20.   **return** $(e, e.sym)$;

The following deletion algorithm takes the two halfedges e and $e.sym$ corresponding to the edge to be deleted. If the edge to be deleted borders two adjacent faces, they have to be merged after the deletion.

Algorithm delete$(e, e.sym)$
1.   **if** $e.left \neq e.sym.left$
2.   **then** /* Figure 18.11a */
3.   /* the faces adjacent to e and $e.sym$ are to be merged */
4.   delete the face record for $e.sym.left$;
5. $e' := e.sym$;
6.   **repeat**
7. $e'.left := e.left$;
8. $e' := e'.succ$;
9.   **until** $e' = e.sym$;
10. $e.left.edge := e.succ$;
11.   half_splice$(e.sym.succ, e)$;
12.   half_splice$(e.succ, e.sym)$;
13.   **else if** $e.succ = e.sym$
14.   **then** /* Figure 18.11b */
15. $e.left.edge := e.pred$;
16.   half_splice$(e.sym.succ, e)$;
17.   **else** /* Figure 18.11c */
18. $e.left.edge := e.succ$;
19.   half_splice$(e.succ, e.sym)$;
20.   /* e becomes an isolated edge */
21.   delete the vertex record for $e.org$ if $e.org \neq v_{inf}$;
22.   delete the vertex record for $e.sym.org$ if $e.sym.org \neq v_{inf}$;
23.   delete the halfedges e and $e.sym$;

18.5.3 Vertex Split and Edge Contraction

Recall that each face is assumed to be a simple polygon for the vertex split and edge contraction operations. The vertex split routine takes two points (p, q) and (x, y) and four halfedges e_1, e_2, e_3, and e_4 in anti-clockwise order around the common origin v. It is required that either $e_1 = e_2$ or $e_1.pred = e_2.sym$ and either $e_3 = e_4$ or $e_3.pred = e_4.sym$. The routine splits v into an edge e connecting the points (p, q) and (x, y). Also, e borders the faces bounded by e_1 and e_2 and by e_3 and e_4. Note that if $e_1 = e_2$, we create a new face bounded by e_1, e_2, and e. Similarly, a new face is created if $e_3 = e_4$. The following is the vertex split algorithm.

Algorithm split$(p, q, x, y, e_1, e_2, e_3, e_4)$
1.   **if** $e_1 \neq e_2$ and $e_3 \neq e_4$

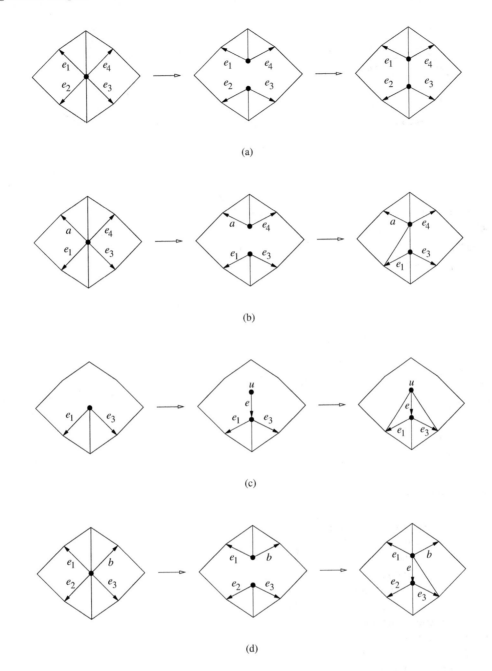

FIGURE 18.12   Cases for split.

```
2.      then /* Figure 18.12a */
3.          half_splice(e₁, e₃);
4.          insert(e₁.org, e₃.org, e₁, e₃);
5.          set the coordinates of e₃.org to (x, y);
6.          set the coordinates of e₁.org to (p, q);
7.      else  if e₁ = e₂
8.          then a := e₁.sym.succ;
9.              if a ≠ e₃
10.                 then /* Figure 18.12b */
11.                     half_splice(a, e₃);
12.                     insert(a.org, e₃.org, a, e₃);
13.                     insert(a.org, e₁.sym.org, a, e₁.sym);
```

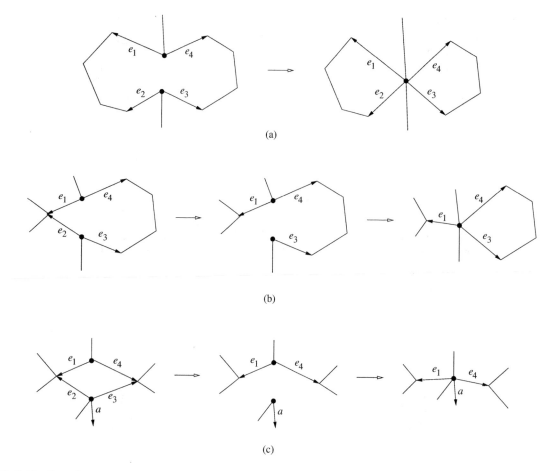

FIGURE 18.13   Cases for contract.

14.   set the coordinates of $a.org$ to (x, y);

15.   **else** /* Figure 18.12c */

16.   let u be a new vertex at (x, y);

17. $(e, e.sym) := \texttt{insert}(u, e_1.org, \cdot, e_3)$;

18. $\texttt{insert}(u, e_1.sym.org, e, e_1.sym)$;

19. $\texttt{insert}(u, e_3.sym.org, e, e_3.succ)$;

20.   set the coordinates of $e_1.org$ to (p, q);

21.   **else** $b := e_3.pred.sym$;

22.   /* since $e_1 \neq e_2$, $b \neq e_2$ */

23.   /* Figure 18.12d */

24. $\texttt{half\_splice}(e_1, e_3)$;

25. $(e, e.sym) := \texttt{insert}(b.org, e_3.org, e_1, e_3)$;

26. $\texttt{insert}(b.org, e_3.sym.org, e, e_3.succ)$;

27.   set the coordinates of $b.org$ to (x, y);

28.   set the coordinates of $e_3.org$ to (p, q);

The following algorithm contracts an edge to a point (x, y), assuming that the edge contractibility has been checked.

Algorithm $\texttt{contract}(e, e.sym, x, y)$

1. $e_1 := e.succ$;

2. $e_2 := e.pred.sym$;

3. $e_3 := e.sym.succ$;

4. $e_4 := e.sym.pred.sym$;

5. $\texttt{delete}(e, e.sym)$;

6.   **if** $e_1.succ \neq e_2.sym$ and $e_3.succ \neq e_4.sym$

7.   **then** /* Figure 18.13a */
8.   half_splice(e_1, e_3);
9.   **else if** $e_1.succ = e_2.sym$ and $e_3.succ \neq e_4.sym$
10.   **then** /* Figure 18.13b */
11.   delete($e_2, e_2.sym$);
12.   half_splice(e_1, e_3);
13.   **else if** $e_1.succ \neq e_2.sym$ and $e_3.succ = e_4.sym$
14.   **then** /* symmetric to Figure 18.13b */
15.   delete($e_4, e_4.sym$);
16.   half_splice(e_1, e_3);
17.   **else** /* Figure 18.13c */
18. $a := e_3.sym.succ$;
19.   delete($e_3, e_3.sym$);
20.   **if** $a \neq e_2$
21.   **then** delete($e_2, e_2.sym$);
22.   half_splice(e_1, a);
23.   **else** delete($e_2, e_2.sym$);
24.   set the coordinates of $e_1.org$ to (x, y);

18.6 Quadedge

The quadedge data structure was introduced by Guibas and Stolfi [10]. It represents the planar subdivision and its dual simultaneously. The dual S^* of a PSLG S is constructed as follows. For each face of S, put a dual vertex inside the face. For each edge of S bordering the faces f and f', put a dual edge connecting the dual vertices of f and f'. The dual of a vertex v in S is a face and this face is bounded by the dual of the incident edges of v. Figure 18.14 shows an example. The dual may have loops and two vertices may be connected by more than one edge, so the dual may not be a PSLG. Nevertheless, the quadedge data structure is expressive enough to represent the dual. In fact, it is powerful enough to represent subdivisions of both orientable and non-orientable surfaces. We describe a simplified version sufficient for our purposes.

Each edge e in the PSLG is represented by four quadedges $e[i]$, where $i \in \{0, 1, 2, 3\}$. The quadedges $e[0]$ and $e[2]$ are the two oriented versions of e. The quadedges $e[1]$ and $e[3]$ are the two oriented versions of the dual of e. These four quadedges are best viewed as a cross such as $e[i + 1]$ is obtained by rotating $e[i]$ for $\pi/2$ in the anti-clockwise direction. This is illustrated in Figure 18.15. The quadedge $e[i]$ has a *next* field referencing the quadedge that has the same origin as $e[i]$ and follows $e[i]$ in anti-clockwise order. In effect, the *next* fields form a circular linked list of quadedges with a common origin. This is called an *edge ring*. The following primitives are needed.

- rot(e, i): Return $e[(i + 1) \bmod 4]$.
- rot$^{-1}(e, i)$: Return $e[(i + 3) \bmod 4]$.
- sym(e, i): This function returns the quadedge with the opposite orientation of $e[i]$. This is done by returning rot(rot(e, i)).
- onext(e, i): Return $e[i].next$.
- oprev(e, i): This function gives the quadedge that has the same origin as $e[i]$ and follows $e[i]$ in clockwise order. This is done by returning rot($e[(i + 1) \bmod 4].next$).

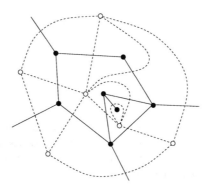

FIGURE 18.14   The solid lines and black dots show a PSLG and the dashed lines and the white dots denote the dual.

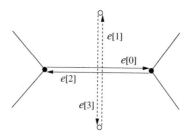

FIGURE 18.15   Quadedges.

The quadedge data structure is entirely edge based and there are no explicit vertex and face records.

The following two basic operations `make_edge` and `splice` are central to the operations on PSLGs supported by the quadedge data structure. Our presentation is slightly different from that in the original paper [10].

- `make_edge`(u, v): Return an edge e connecting the points u and v. The quadedges $e[i]$ where $0 \leq i \leq 3$ are initialized such that they represent a new PSLG with e as the only edge. Also, $e[0]$ is the quadedge directed from u to v. If u and v are omitted, it means that the actual coordinates of the endpoints of are unimportant.
- `splice`(a, i, b, j): Given two quadedges $a[i]$ and $b[j]$, let $(c, k) = rot(a[i].next)$ and $(d, l) = rot(b[j].next)$, `splice` swaps the contents of $a[i].next$ and $b[j].next$ and the contents of $c[k].next$ and $d[l].next$. The effects on the edge rings of the origins of $a[i]$ and $b[j]$ and the edge rings of the origins of $c[k]$ and $d[l]$ are:
  - If the two rings are different, they are merged into one (see Figure 18.16a).
  - If the two rings are the same, it will be split into two separate rings (see Figure 18.16b).

Notice that `make_edge` and `splice` are similar to the operations `make_halfedges` and `half_splice` introduced for the halfedge data structure in the previous section. As mentioned before, they inspire the definitions of `make_halfedges` and `half_splice`. Due to this similarity, one can easily adapt the edge insertion, edge deletion, vertex split, and edge contraction algorithms in the previous section for the quadedge data structure.

18.7 Further Remarks

We have assumed that each face in the PSLG has exactly one boundary. This requirement can be relaxed for the winged-edge and the halfedge data structures. One method works as follows. For each face f, pick one edge from each boundary and keep a list of references to these edges at the face record for f. Also, the edge that belongs to outer boundary of f is specially tagged. With this modification, one can traverse the boundaries of a face f consistently (e.g., keeping f on the left of traversal direction). The edge insertion and deletion algorithms also need to be enhanced. Since a face f may have several boundaries, inserting an edge may combine two boundaries without splitting f. If the insertion indeed splits f, one needs to distribute the other boundaries of f into the two faces resulting from the split. The reverse effects of edge deletion should be taken care of similarly.

The halfedge data structure has also been used for representing orientable polyhedral surfaces [11]. The full power of the quadedge data structure is only realized when one deals with both subdivisions of orientable and non-orientable surfaces. To this end, one needs to introduce a *flip bit* to allow viewing the surface from the above or below. The primitives need to be enhanced for this purpose. The correctness of the data structure is proven formally using *edge algebra*. The details are in the Guibas and Stolfi's original paper [10].

The vertex split and edge contraction are also applicable for polyhedral surfaces. The edge contractibility criteria carries over straightforwardly. Edge contraction is a popular primitive for surface simplification algorithms [12–14]. The edge contractibility criteria for non-manifolds has also been studied [15].

18.8 Glossary

Arrangements. Given a collection of lines, we split each line into edges by inserting a vertex at every intersection on the line. The resulting PSLG is called the *arrangement of lines*. The *arrangement of line segments* is similarly defined.

Voronoi diagram. Let S be a set of points in the plane. For each point $p \in S$, the Voronoi region of p is defined to be $\{x \in R^2 : \|p - x\| \leq \|q - x\|, \forall q \in S\}$. The *Voronoi diagram* of S is the collection of all Voronoi regions (including their boundaries).

Triangulation. Let S be a set of points in the plane. Any maximal PSLG with the points in S as vertices is a *triangulation* of S.

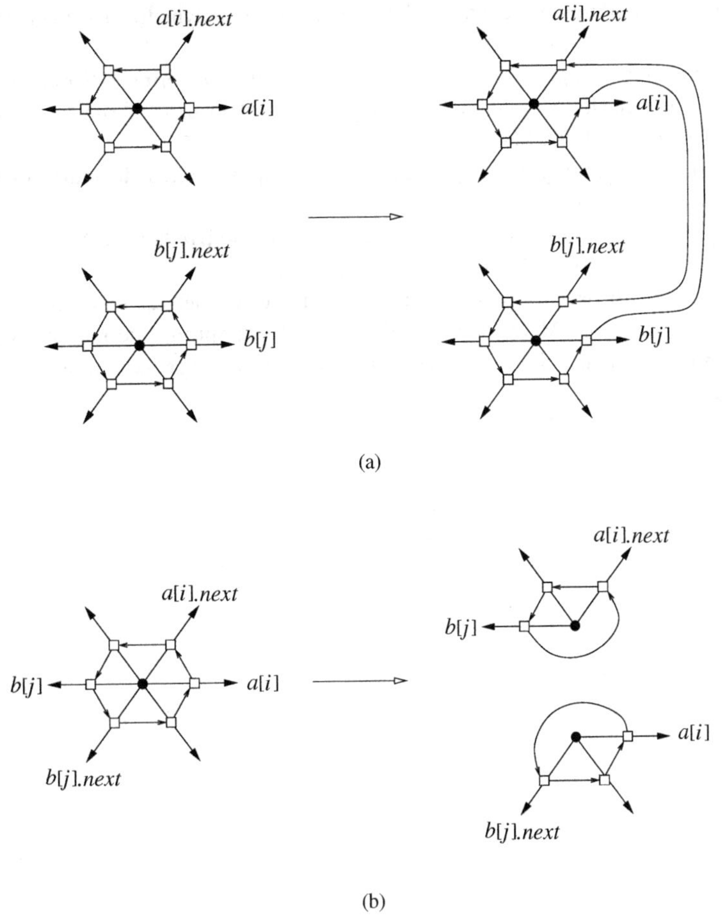

(a)

(b)

FIGURE 18.16 The effects of `splice`.

Delaunay triangulation. Let *S* be a set of points in the plane. For any three points *p*, *q*, and *r* in *S*, if the circumcircle of the triangle *pqr* does not strictly enclose any point in *S*, we call *pqr* a *Delaunay triangle*. The *Delaunay triangulation* of *S* is the collection of all Delaunay triangles (including their boundaries). The Delaunay triangulation of *S* is the dual of the Voronoi diagram of *S*.

Acknowledgments

This work was supported, in part, by the Research Grant Council, Hong Kong, China (HKUST 6190/02E).

References

1. S. Sahni, *Data Structures, Algorithms, and Applications in Java*, McGraw Hill, NY, 2000.
2. M.S. Bazaraa, J.J. Jarvis, and H.D. Sherali, *Linear Programming and Network Flows*, Wiley, 1990.
3. M. deBerg, M. van Kreveld, M. Overmars, and O. Schwarzkopf, *Computational Geometry—Algorithms and Applications*, Springer, 2000.
4. E. Mücke, I. Saias, and B. Zhu, Fast randomized point location without preprocessing in two and three-dimensional Delaunay triangulations, *Computational Geometry: Theory and Applications*, 12(1999), 63–83, 1999.
5. C.A. Duncan, M.T. Goodrich, and S.G. Kobourov, Planarity-preserving clustering and embedding for large graphs, *Proc. Graph Drawing, Lecture Notes Comput. Sci.*, Springer-Verlag, vol. 1731, 1999, 186–196.
6. B.G. Baumgart, A polyhedron representation for computer vision, *National Computer Conference*, 589–596, Anaheim, CA, 1975, AFIPS.
7. K. Weiler, Edge-based data structures for solid modeling in curved-surface environments, *IEEE Computer Graphics and Application*, 5, 1985, 21–40.

8. D.E. Muller and F.P. Preparata, Finding the intersection of two convex polyhedra, *Theoretical Computer Science*, 7, 1978, 217–236.

9. F.P. Preparata and M.I. Shamos, *Computational Geometry: An Introduction*, Springer-Verlag, New York, 1985.

10. L. Guibas and J. Stolfi, Primitives for the manipulation of general subdivisions and the computation of Voronoi diagrams, *ACM Transactions on Graphics*, 4, 1985, 74–123.

11. L. Kettner, Using generic programming for designing a data structure for polyhedral surfaces, *Computational Geometry - Theory and Applications*, 13, 1999, 65–90.

12. S.-W. Cheng, T. K. Dey, and S.-H. Poon, Hierarchy of Surface Models and Irreducible Triangulation, *Computational Geometry: Theory and Applications*, 27, 2004, 135–150.

13. M. Garland and P.S. Heckbert: Surface simplification using quadric error metrics. *Proc. SIGGRAPH '97*, 209–216.

14. H. Hoppe, T. DeRose, T. Duchamp, J. McDonald and W. Stuetzle, Mesh optimization, *Proc. SIGGRAPH '93*, 19–26.

15. T.K. Dey, H. Edelsbrunner, S. Guha, and D.V. Nekhayev, Topology preserving edge contraction, *Publ. Inst. Math. (Beograd) (N.S.)*, 66, 1999, 23–45.

19

Interval, Segment, Range, and Priority Search Trees

19.1 Introduction.. 291
19.2 Interval Trees 292
    Construction of Interval Trees • Example and Its Applications
19.3 Segment Trees.. 294
    Construction of Segment Trees • Examples and Its Applications
19.4 Range Trees.. 298
    Construction of Range Trees • Examples and Its Applications
19.5 Priority Search Trees 304
    Construction of Priority Search Trees • Examples and Its Applications
Acknowledgments 306
References.. 306

D. T. Lee
Academia Sinica

Hung-I Yu
Academia Sinica

19.1 Introduction

In this chapter, we introduce four basic data structures that are of fundamental importance and have many applications as we will briefly cover them in later sections. They are *interval trees*, *segment trees*, *range trees*, and *priority search trees*. Consider for example the following problems. Suppose we have a set of *iso-oriented rectangles* in the plane. A set of rectangles are said to be *iso-oriented* if their edges are parallel to the coordinate axes. The subset of iso-oriented rectangles defines a *clique*, if their common intersection is nonempty. The *largest* subset of rectangles whose common intersection is non-empty is called a *maximum clique*. The problem of finding this largest subset with a non-empty common intersection is referred to as the *maximum clique problem* for a rectangle intersection graph [1,2].[*] The *k*-dimensional, $k \geq 1$, analog of this problem is defined similarly. In one-dimensional (1D) case, we will have a set of *intervals* on the real line, and an *interval intersection graph*, or simply *interval graph*. The maximum clique problem for interval graphs is to find a largest subset of intervals whose common intersection is non-empty. The cardinality of the maximum clique is sometimes referred to as the *density* of the set of intervals.

The problem of finding a subset of objects that satisfy a certain property is often referred to as the *searching problem*. For instance, given a set of numbers $S = \{x_1, x_2, \ldots, x_n\}$, where $x_i \in \Re$, $i = 1, 2, \ldots, n$, the problem of finding the subset of numbers that lie between a range $[\ell, r]$, that is, $F = \{x \in S | \ell \leq x \leq r\}$, is called an (1D) *range search* problem [3,4].

To deal with this kind of geometric searching problem, we need to have appropriate data structures to support efficient searching algorithms. The data structure is assumed to be *static*, that is, the input set of objects is given *a priori*, and no insertions or deletions of the objects are allowed. If the searching problem satisfies *decomposability property*, that is, if they are *decomposable*,[†] then there are general *dynamization* schemes available [7], that can be used to convert static data structures into *dynamic* ones, where *insertions* and *deletions* of objects are permitted. Examples of decomposable searching problems include the *membership* problem in which one queries if a point p lies in S. Let S be partitioned into two subsets S_1 and S_2, and Member(p, S) returns *yes*, if $p \in S$, and *no* otherwise. It is easy to see that Member$(p, S) = $ OR(Member(p, S_1), Member(p, S_2)), where OR is a Boolean operator.

---

[*] A rectangle intersection graph is a graph $G = (V, E)$, in which each vertex in V corresponds to a rectangle, and two vertices are connected by an edge in E, if the corresponding rectangles intersect.

[†] A searching problem is said to be *decomposable* if and only if $\forall x \in T_1, A, B \in 2^{T_2}, Q(x, A \cup B) = \bigcirc(Q(x, A), Q(x, B))$ for some efficiently computable associative operator $\bigcirc$ on the elements of T_3, where Q is a mapping from $T_1 \times 2^{T_2}$ to T_3 [5,6].

19.2 Interval Trees

Consider a set S of intervals, $S = \{I_i | i = 1, 2, \dots, n\}$, each of which is specified by an ordered pair, $I_i = [\ell_i, r_i], \ell_i, r_i \in \Re$, $\ell_i \leq r_i, i = 1, 2, \dots, n$.

An *interval tree* [8,9], *Interval_Tree(S)*, for S is a rooted augmented binary search tree, in which each node v has a key value, $v.key$, two tree pointers $v.left$ and $v.right$ to the left and right subtrees, respectively, and an auxiliary pointer, $v.aux$ to an augmented data structure, and is recursively defined as follows:

- The root node v associated with the set S, denoted Interval_Tree_root(S), has key value $v.key$ equal to the median of the $2 \times |S|$ endpoints. This key value $v.key$ divides S into three subsets S_ℓ, S_r, and S_m, consisting of sets of intervals lying totally to the *left* of $v.key$, lying totally to the *right* of $v.key$, and containing $v.key$, respectively. That is, $S_\ell = \{I_i | r_i < v.key\}$, $S_r = \{I_j | v.key < \ell_j\}$, and $S_m = \{I_k | \ell_k \leq v.key \leq r_k\}$.
- Tree pointer $v.left$ points to the left subtree rooted at Interval_Tree_root(S_ℓ), and tree pointer $v.right$ points to the right subtree rooted at Interval_Tree_root(S_r).
- Auxiliary pointer $v.aux$ points to an augmented data structure consisting of two sorted arrays, $SA(S_m.left)$ and $SA(S_m.right)$ of the set of left endpoints of the intervals in S_m and the set of right endpoints of the intervals in S_m, respectively. That is, $S_m.left = \{\ell_i | I_i \in S_m\}$ and $S_m.right = \{r_i | I_i \in S_m\}$.

19.2.1 Construction of Interval Trees

The following is a pseudo code for the recursive construction of the interval tree of a set S of n intervals. Without loss of generality, we shall assume that the endpoints of these n intervals are all distinct. See Figure 19.1a for an illustration.

function Interval_Tree(S)
/* It returns a pointer v to the root, Interval_Tree_root(S), of the interval tree for a set S of intervals. */
Input: A set S of n intervals, $S = \{I_i | i = 1, 2, \dots, n\}$ and each interval $I_i = [\ell_i, r_i]$, where ℓ_i and r_i are the left and right endpoints, respectively of I_i, $\ell_i, r_i \in \Re$, and $\ell_i \leq r_i, i = 1, 2, \dots, n$.
Output: An interval tree, rooted at Interval_Tree_root(S).
Method:

1. **if** $S = \emptyset$ **return nil**.
2. Create a node v such that $v.key$ equals x, where x is the middle point of the set of endpoints so that there are exactly $|S|/2$ endpoints less than or equal to x and greater than x, respectively. Let $S_\ell = \{I_i | r_i < x\}$, $S_r = \{I_j | x < \ell_j\}$, and $S_m = \{I_k | \ell_k \leq x \leq r_k\}$.

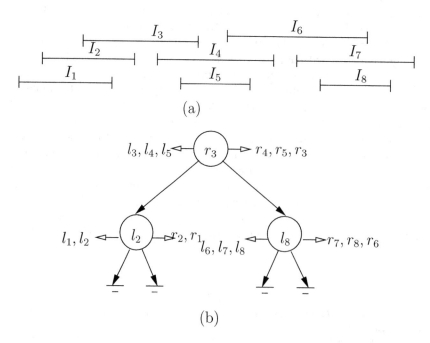

FIGURE 19.1   Interval tree for $S = \{I_1, I_2, \dots, I_8\}$ and its interval models.

3. Set *v.left* equal to Interval_Tree(S_ℓ).
4. Set *v.right* equal to Interval_Tree(S_r).
5. Create a node *w* which is the root node of an auxiliary data structure associated with the set S_m of intervals, such that *w.left* and *w.right* point to two sorted arrays, $SA(S_m.left)$ and $SA(S_m.right)$, respectively. $SA(S_m.left)$ denotes an array of left endpoints of intervals in S_m in *ascending* order, and $SA(S_m.right)$ an array of right endpoints of intervals in S_m in *descending* order.
6. Set *v.aux* equal to node *w*.

Note that this recursively built interval tree structure requires $O(n)$ space, where n is the cardinality of S, since each interval is either in the left subtree, the right subtree, or the middle augmented data structure.

19.2.2 Example and Its Applications

Figure 19.1b illustrates an example of an interval tree of a set of intervals, spread out as shown in Figure 19.1a.

The interval trees can be used to handle quickly queries of the following forms.

Enclosing Interval Searching Problem [10,11]: Given a set S of n intervals and a query point q, report all those intervals containing q, that is, find a subset $F \subseteq S$ such that $F = \{I_i | \ell_i \le q \le r_i\}$.

Overlapping Interval Searching Problem [8,9,12]: Given a set S of n intervals and a query interval Q, report all those intervals in S overlapping Q, that is, find a subset $F \subseteq S$ such that $F = \{I_i | I_i \cap Q \ne \emptyset\}$.

The following pseudo code solves the **Overlapping Interval Searching Problem** in $O(\log n) + |F|$ time. It is invoked by a call to Overlapping_Interval_Search(v, Q, F), where v is Interval_Tree_root(S), and F, initially set to be $\emptyset$, will contain the set of intervals overlapping query interval Q.

procedure Overlapping_Interval_Search(v, Q, F)
Input: A set S of n intervals, $S = \{I_i | i = 1, 2, \ldots, n\}$ and each interval $I_i = [\ell_i, r_i]$, where ℓ_i and r_i are the left and right endpoints, respectively of $I_i, \ell_i, r_i \in \Re$, and $\ell_i \le r_i, i = 1, 2, \ldots, n$ and a query interval $Q = [\ell, r], \ell, r \in \Re$, and $l \le r$.
Output: A subset $F = \{I_i | I_i \cap Q \ne \emptyset\}$.
Method:

1. Set $F = \emptyset$ initially.
2. **if** *v* is **nil return**.
3. **if** ($v.key \in Q$) **then**
      for each interval I_i in the augmented data structure pointed to by *v.aux*
      do $F = F \cup \{I_i\}$.
      Overlapping_Interval_Search($v.left, Q, F$)
      Overlapping_Interval_Search($v.right, Q, F$)
4. **if** ($r < v.key$) **then**
      for each left endpoint ℓ_i in the sorted array pointed to by *v.aux.left* such that $\ell_i < r$
      do $F = F \cup \{I_i\}$.
      Overlapping_Interval_Search($v.left, Q, F$)
5. **if** ($\ell > v.key$) **then**
      for each right endpoint r_i in the sorted array pointed to by *v.aux.right* such that $r_i > \ell$
      do $F = F \cup \{I_i\}$.
      Overlapping_Interval_Search($v.right, Q, F$)

It is obvious to see that an interval I in S overlaps a query interval $Q = [\ell, r]$ if (i) Q contains the left endpoint of I, (ii) Q contains the right endpoint of I, or (iii) Q is totally contained in I. Step 3 reports those intervals I that contain a point *v.key* which is also contained in Q. The for-loop of Step 4 reports intervals in either case (i) or (iii) and the for-loop of Step 5 reports intervals in either case (ii) or (iii).

Note the special case of **procedure** Overlapping_Interval_Search(v, Q, F) when we set the query interval $Q = [\ell, r]$ so that its left and right endpoints coincide, that is, $\ell = r$ will report all the intervals in S containing a query point, solving the **Enclosing Interval Searching Problem**.

However, if one is interested in the problem of finding a special type of overlapping intervals, for example, all intervals containing or contained in a given query interval [10,11], the interval tree data structure does not necessarily yield an efficient solution. Similarly, the interval tree does not provide an effective method to handle queries about the set of intervals, for example, the maximum clique, or the *measure*, the total length of the union of the intervals [13,14].

We conclude with the following theorem.

Theorem 19.1

The **Enclosing Interval Searching Problem** and **Overlapping Interval Searching Problem** for a set S of n intervals can both be solved in $O(\log n)$ time (plus time for output) and in linear space.

19.3 Segment Trees

The segment tree structure, originally introduced by Bentley [3,4], is a data structure for intervals whose endpoints are *fixed* or *known a priori*. The set $S = \{I_1, I_2, \ldots, I_n\}$ of n intervals, each of which is represented by $I_i = [\ell_i, r_i]$, $\ell_i, r_i \in \Re$, $\ell_i \leq r_i$, is represented by a data array, Data_Array(S), whose entries correspond to the endpoints, ℓ_i or r_i, and are sorted in non-decreasing order. This sorted array is denoted $SA[1 \ldots N]$, $N = 2n$. That is, $SA[1] \leq SA[2] \leq \cdots \leq SA[N]$. We will in this section use the indexes in the range $[1, N]$ to refer to the entries in the sorted array $SA[1 \ldots N]$. For convenience, we will be working in the transformed domain using indexes, and a comparison involving a point $q \in \Re$ and an index $i \in \aleph$, unless otherwise specified, is performed in the original domain in $\Re$. For instance, $q < i$ is interpreted as $q < SA[i]$.

The segment tree structure, as will be demonstrated later, can be useful in finding the *measure* of a set of intervals. That is, the length of the union of a set of intervals. It can also be used to find the maximum clique of a set of intervals. This structure can be generalized to higher dimensions.

19.3.1 Construction of Segment Trees

The segment tree, as the interval tree discussed in Section 19.2 is a rooted augmented binary search tree, in which each node v is associated with a range of integers $v.range = [v.B, v.E]$, $v.B, v.E \in \aleph$, $v.B < v.E$, representing a range of indexes from $v.B$ to $v.E$, a key $v.key$ that splits $v.range$ into two subranges, each of which is associated with each child of v, two tree pointers $v.left$ and $v.right$ to the left and right subtrees, respectively, and an auxiliary pointer, $v.aux$ to an augmented data structure. Given integers s and t, with $1 \leq s < t \leq N$, the segment tree, denoted Segment_Tree(s, t), is recursively described as follows:

- The root node v, denoted Segment_Tree_root(s, t), is associated with the range $[s, t]$, and $v.B = s$ and $v.E = t$.
- If $s + 1 = t$, then we have a leaf node v with $v.B = s$, $v.E = t$, and $v.key =$ **nil**.
- Otherwise (i.e., $s + 1 < t$), let m be the mid-point of s and t, or $m = \lfloor (v.B + v.E)/2 \rfloor$. Set $v.key = m$.
- Tree pointer $v.left$ points to the left subtree rooted at Segment_Tree_root(s, m), and tree pointer $v.right$ points to the right subtree rooted at Segment_Tree_root(m, t).
- Auxiliary pointer $v.aux$ points to an augmented data structure, associated with the range $[s, t]$, whose content depends on the usage of the segment tree.

The following is a pseudo code for the construction of a segment tree for a range $[s, t]$ $s < t, s, t \in \aleph$, and the construction of a set of n intervals whose endpoints are indexed by an array of integers in the range $[1, N], N = 2n$ can be done by a call to Segment_Tree($1, N$). See Figure 19.2b for an illustration.

function Segment_Tree(s, t)
/* It returns a pointer v to the root, Segment_Tree_root(s, t), of the segment tree for the range $[s, t]$.*/
Input: A set $\mathcal{N}$ of integers, $\{s, s + 1, \ldots, t\}$ representing the indexes of the endpoints of a subset of intervals.
Output: A segment tree, rooted at Segment_Tree_root(s, t).
Method:

1. Let v be a node, $v.B = s$, $v.E = t$, $v.left = v.right = $ **nil**, and $v.aux$ to be determined.
2. **if** $s + 1 = t$ **then return**.
3. Let $v.key = m = \lfloor (v.B + v.E)/2 \rfloor$.
4. $v.left = $ Segment_Tree(s, m)
5. $v.right = $ Segment_Tree(m, t)

The parameters $v.B$ and $v.E$ associated with node v define a range $[v.B, v.E]$, called a *standard range* associated with v. The standard range associated with a leaf node is also called an *elementary range*. It is straightforward to see that Segment_Tree(s, t) constructed in **function** Segment_Tree(s, t) described above is balanced, and has height, denoted Segment_Tree(s, t).height, equal to $\lceil \log_2(t - s) \rceil$.

We now introduce the notion of *canonical covering* of a range $[b, e]$, where $b, e \in \aleph$ and $1 \leq b < e \leq N$. A node v in Segment_Tree($1, N$) is said to be in the *canonical covering* of $[b, e]$ if its associated standard range satisfies this property $[v.B, v.E] \subseteq [b, e]$, while that of its parent node does not. It is obvious that if a node v is in the canonical covering, then its *sibling*

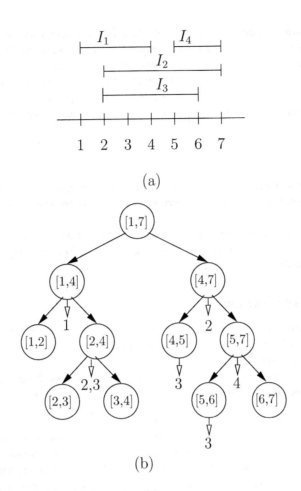

FIGURE 19.2 Segment tree of $S = \{I_1, I_2, I_3, I_4\}$ and its interval models.

node, that is, the node with the same parent node as the present one, is not, for otherwise the common parent node would have been in the canonical covering. Thus, at each level, there are *at most* two nodes that belong to the canonical covering of $[b, e]$.

Thus, for each range $[b, e]$, the number of nodes in its canonical covering is at most $\lceil \log_2(e - b) \rceil + \lfloor \log_2(e - b) \rfloor - 2$. In other words, a range $[b, e]$ (or respectively an interval $[b, e]$) can be decomposed into at most $\lceil \log_2(e - b) \rceil + \lfloor \log_2(e - b) \rfloor - 2$ standard ranges (or respectively subintervals) [3,4].

To identify the nodes in a segment tree T that are in the canonical covering of an interval $I = [b, e]$, representing a range $[b, e]$, we perform a call to Interval_Insertion(v, b, e, Q), where v is Segment_Tree_root$(1, N)$. The procedure Interval_Insertion(v, b, e, Q) is defined below.

procedure Interval_Insertion(v, b, e, Q)
/* It returns a queue Q of nodes $q \in T$ such that $[v.B, v.E] \subseteq [b, e]$ but $[u.B, u.E] \subsetneq [b, e]$, where u is the parent node of v. */
Input: A segment tree T pointed to by its root node, $v =$ Segment_Tree_root$(1, N)$, for a set S of intervals.
Output: A queue Q of nodes in T that are in the canonical covering of $[b, e]$.
Method:

1. Initialize an output queue Q, which supports insertion ($\Rightarrow Q$) and deletion ($\Leftarrow Q$) in constant time.
2. **if** $([v.B, v.E] \subseteq [b, e])$ **then append** $[b, e]$ to v, $v \Rightarrow Q$ and **return.**
3. **if** $(b < v.key)$ **then** Interval_Insertion$(v.left, b, e, Q)$
4. **if** $(v.key < e)$ **then** Interval_Insertion$(v.right, b, e, Q)$

To **append** $[b, e]$ to a node v means to insert interval $I = [b, e]$ into the auxiliary structure associated with node v to indicate that node v whose standard range is totally contained in I is in the canonical covering of I. If the auxiliary structure $v.aux$ associated with node v is an array, the operation **append** $[b, e]$ to a node v can be implemented as $v.aux[j++] = I$.

procedure Interval_Insertion(v, b, e, Q) described above can be used to represent a set S of n intervals in a segment tree by performing the insertion operation n times, one for each interval. As each interval I can have at most $O(\log n)$ nodes in its canonical covering, and hence we perform at most $O(\log n)$ **append** operations for each insertion, the total amount of space required in the auxiliary data structures reflecting all the nodes in the canonical covering is $O(n \log n)$.

Deletion of an interval represented by a range $[b, e]$ can be done similarly, except that the **append** operation will be replaced by its corresponding inverse operation **remove** that removes the interval from the auxiliary structure associated with some canonical covering node.

Theorem 19.2

The segment tree for a set S of n intervals can be constructed in $O(n \log n)$ time, and if the auxiliary structure for each node v contains a list of intervals containing v in the canonical covering, then the space required is $O(n \log n)$.

To simplify the description, given a set S of intervals, we assume that the endpoints of the intervals are represented by the indexes of a sorted array $SA[1 \ldots N], N = 2|S|$, containing these endpoints, and the segment tree representing this set is abbreviated as Segment_Tree(S) in the remaining of this section.

19.3.2 Examples and Its Applications

Figure 19.2b illustrates an example of a segment tree of the set of intervals, as shown in Figure 19.2a. The integers, if any, under each node v represent the indexes of intervals that contain the node in its canonical covering. For example, Interval I_2 contains nodes labeled by standard ranges [2,4] and [4,7].

We now describe how segment trees can be used to solve the **Enclosing Interval Searching Problem** defined before, the *Maximum Clique Problem* of a set of intervals, and that of a set of rectangles, which are defined below.

Maximum Clique Size of a set of Intervals [2,15,16]: Given a set $S = \{I_1, I_2, \ldots, I_n\}$ of n intervals, a *clique* is a subset $C_I \subseteq S$ such that the common intersection of intervals in C_I is non-empty, and a *maximum clique* is a clique of maximum size. That is, $\bigcap_{I_i \in C_I \subseteq S} I_i \neq \emptyset$ and $|C_I|$ is maximized. The problem is to find the maximum size $|C_I|$.

Maximum Clique Size of a set of Rectangles [2]: Given a set $S = \{R_1, R_2, \ldots, R_n\}$ of n rectangles, a *clique* is a subset $C_R \subseteq S$ such that the common intersection of rectangles in C_R is non-empty, and a *maximum clique* is a clique of maximum size. That is, $\bigcap_{R_i \in C_R \subseteq S} R_i \neq \emptyset$ and $|C_R|$ is maximized. The problem is to find the maximum size $|C_R|$.

The following pseudo code solves the **Enclosing Interval Searching Problem** in $O(\log n) + |F|$ time, where F is the output set. It is invoked by a call Point_in_Interval_Search(v, q, F), where v is Segment_Tree_root(S), and F, initially set to be $\emptyset$, will contain the set of intervals containing a query point q.

procedure Point_in_Interval_Search(v, q, F)
/* It returns in F the list of intervals stored in the segment tree pointed to by v and containing query point q. */
Input: A segment tree representing a set S of n intervals, $S = \{I_i | i = 1, 2, \ldots, n\}$ and a query point $q \in \Re$. The auxiliary structure $v.aux$ associated with each node v is a list of intervals $I \in S$ that contain v in their canonical coverings.
Output: A subset $F = \{I_i | \ell_i \leq q \leq r_i\}$.
Method:

1. **if** v is **nil** or ($q < v.B$ **or** $q > v.E$) **then return**.
2. **if** ($v.B \leq q \leq v.E$) **then**
   for each interval I_i in the auxiliary data structure pointed to by $v.aux$ **do** $F = F \cup \{I_i\}$.
3. **if** ($q \leq v.key$) **then** Point_in_Interval_Search($v.left, q, F$)
4. **else** (/* i.e., $q > v.key$ */) Point_in_Interval_Search($v.right, q, F$)

We now address the problem of finding the size of the maximum clique of a set S of intervals, $S = \{I_1, I_2, \ldots, I_n\}$, where each interval $I_i = [\ell_i, r_i]$, and $\ell_i \leq r_i, \ell_i, r_i \in \Re, i = 1, 2, \ldots, n$. There are other approaches, such as plane-sweep [2,4,15,16], that solve this problem within the same complexity.

For this problem, we introduce an auxiliary data structure to be stored at each node v. $v.aux$ will contain two pieces of information: one is the *number* of intervals containing v in the canonical covering, denoted $v.\sharp$, and the other is the *clique size*, denoted $v.clq$. The *clique size* of a node v is the size of the maximum clique whose common intersection is contained in but not equal to the standard range associated with v. It is defined to be equal to the *larger* of the two numbers: $v.left.\sharp + v.left.clq$ and $v.right.\sharp + v.right.clq$. For a leaf node v, $v.clq = 0$. The size of the maximum clique for the set of intervals will then be stored at the root node $r = $ Segment_Tree_root(S) and is equal to the sum of $r.\sharp$ and $r.clq$. It is obvious that the space needed for this segment tree is linear.

As this data structure supports insertion of intervals incrementally, it can be used to answer the maximum clique of the current set of intervals as the intervals are inserted into (or deleted from) the segment tree T. The following pseudo code finds the size of the maximum clique of a set of intervals.

function Interval_Maximum_Clique_Size(S)
/* It returns the size of the maximum clique of a set S of intervals. */
Input: A set S of n intervals and the segment tree T rooted at $r = $ Segment_Tree_root(S).
Output: An integer, which is the size of the maximum clique of S.
Method:

1. Initialize $v.clq = v.\sharp = 0$ for all nodes v in T.
2. **for** each interval $I_i = [\ell_i, r_i] \in S, i = 1, 2, \ldots, n$ **do**
   /* Insert I_i into the tree and update $v.\sharp$ and $v.clq$ for all visited nodes and those nodes in the canonical covering of I_i. */
   $clq_S_i = $ Update_Segment_Tree_by_Weight($r, \ell_i, r_i, +1$) (see below).
3. **return** clq_S_n.

function Update_Segment_Tree_by_Weight(v, b, e, w)
/* Insert into/delete from the segment tree pointed to by v the interval $I = [b, e]$ and update $v.\sharp$ and $v.clq$ by w for all visited nodes and those nodes in the canonical covering of I. */
Input: A segment tree rooted at v containing a set of intervals, an interval $I = [b, e]$ to be inserted/deleted, $b, e \in \mathfrak{R}, b \leq e$, and a weight $w \in \{+1, -1\}$, where $+1$ corresponds to insertion and -1 to deletion.
Output: An integer, which is the size of the maximum clique of the set of intervals after update.
Method:

1. **if** $([v.B, v.E] \subseteq [b, e])$ **then**
   $v.\sharp = v.\sharp + w$
   return $v.\sharp + v.clq$
2. **if** $(b < v.key)$ **then** Update_Segment_Tree_by_Weight($v.left, b, e, w$)
3. **if** $(v.key < e)$ **then** Update_Segment_Tree_by_Weight($v.right, b, e, w$)
4. $v.clq = \max\{v.left.\sharp + v.left.clq, v.right.\sharp + v.right.clq\}$
5. **return** $v.\sharp + v.clq$

Note that **function** Update_Segment_Tree_by_Weight(v, b, e, w) visits at most $O(\log n)$ nodes in the canonical covering of I and takes $O(\log n)$ time. Since **function** Interval_Maximum_Clique_Size(S) invokes the function n times, we obtain the following theorem.

Theorem 19.3

Given a set $S = \{I_1, I_2, \ldots, I_n\}$ of n intervals, the size of the maximum clique of $S_i = \{I_1, I_2, \ldots, I_i\}$ can be found in $O(i \log n)$ time and linear space, for each $i = 1, 2, \ldots, n$, by using a segment tree.

Next, we extend the study of the maximum clique size problem to its 2D version, finding the size of the maximum clique of a set S of rectangles, $S = \{R_1, R_2, \ldots, R_n\}$, where each rectangle $R_i = [x_{\ell,i}, x_{r,i}] \times [y_{\ell,i}, y_{r,i}], x_{\ell,i}, x_{r,i}, y_{\ell,i}, y_{r,i} \in \mathfrak{R}^2, x_{\ell,i} \leq x_{r,i}, y_{\ell,i} \leq y_{r,i}, i = 1, 2, \ldots, n$. We will describe a reduction of this problem to **Maximum Clique Size of a set of Intervals**, which can be easily generalized to higher dimensions.

For each rectangle R_j in S, let L_j denote the horizontal line $y = y_{\ell,j}$ through the bottom of R_j, and S_j denote the set of rectangles in S intersecting L_j, that is, $S_j = \{R_i | R_i \cap L_j \neq \emptyset, 1 \leq i \leq n\}$. It is not difficult to see that any non-empty common intersection of rectangles in S has to intersect a horizontal line L_j for some j, so does the common intersection of rectangles in the maximum clique of S. This implies that the maximum clique of S is equivalent to the maximum clique of some set S_j. Then, for any fixed S_j, finding the size of the maximum clique of S_j can be reduced to the problem **Maximum Clique Size of a set of Intervals** as follows. Consider L_j as a 1D real line. Each rectangle R_i intersecting L_j, that is, $y_{\ell,i} \leq y_{\ell,j} \leq y_{r,i}$, can be mapped to an interval $I_i = [x_{\ell,i}, x_{r,i}]$ on L_j. Let S_j^I denote the set of intervals I_i whose corresponding rectangles R_i intersect L_j. Obviously, the maximum clique of S_j is exactly the same (in terms of size) as that of S_j^I. Thus, the size of the maximum clique of S_j can be obtained by constructing S_j^I in $O(n)$ time and performing a call to Interval_Maximum_Clique_Size(S_j^I) in $O(n \log n)$ time. It follows that we can solve **Maximum Clique Size of a Set of Rectangles** in $O(n^2 \log n)$ time by obtaining the size of the maximum clique for each S_j and taking the maximum.

However, by applying the plane-sweep technique, the above algorithm can be optimally implemented in $O(n \log n)$ time. Let $S^I = \{I_1, I_2, \ldots, I_n\}$ denote the set of intervals $I_i = [x_{\ell,i}, x_{r,i}]$ corresponding to the x-ranges of all rectangles R_i in S. We create a sorted array $SA(S^I)$ containing the endpoints of intervals in S^I sorted in nondecreasing order, so that these endpoints can be represented by the indexes of $SA(S^I)$, and construct an empty segment tree T rooted at $r = \text{Segment\_Tree\_root}(S^I)$, in which each node contains two auxiliary values: interval number and clique size. Then, we scan the y-coordinates of all rectangles in S from bottom to top. If an *entry coordinate* $y_{\ell,j}$ of some rectangle R_j is met, the corresponding x-range I_j is inserted into T. Otherwise if an *exit coordinate* $y_{r,j}$ of R_j is met, I_j is removed from T. We can observe that, when I_j is inserted, the set of intervals stored in T is equivalent to S_j^I, and the size of the maximum clique of S_j^I (and the corresponding S_j) is equal to the sum of $r.\sharp$ and $r.clq$. Thus, for each S_j, the scanning checks its maximum clique and obtains its size in a more efficient way. The following pseudo code finds the size of the maximum clique of S in optimal $O(n \log n)$ time.

function Rectangle_Maximum_Clique_Size(S)
/* It returns the size of the maximum clique of a set S of rectangles. */
Input: A set S of n rectangles.
Output: An integer, which is the size of the maximum clique of S.
Method:

1. Create the set S^I, the sorted array $SA(S^I)$, and the segment tree T, rooted at $r = \text{Segment\_Tree\_root}(S^I)$.
2. Initialize $v.clq = v.\sharp = 0$ for all nodes v in T.
3. Create a sorted array SY containing the y-coordinates of all rectangles in S, that is, SY has $\{y_{\ell,j}, y_{r,j}|R_j \in S\}$ in nondecreasing order (ties broken in favor of entry coordinates).
4. $max\_clq = 0$
5. **for** each coordinate $SY[k], k = 1, 2, \ldots, N, N = 2n$ **do**
   /* Suppose that $SY[k]$ corresponds to $R_j = [x_{\ell,j}, x_{r,j}] \times [y_{\ell,j}, y_{r,j}]$. */
   if $SY[k]$ is the entry coordinate $y_{\ell,j}$ of R_j **then**
       /* Insert I_j into T and update $v.\sharp$ and $v.clq$ for all visited nodes and those nodes in the canonical covering of I_j. */
       $clq\_S_j = \text{Update\_Segment\_Tree\_by\_Weight}(r, x_{\ell,j}, x_{r,j}, +1)$
       if $clq\_S_j > max\_clq$ **then** $max\_clq = clq\_S_j$
   else /* $SY[k]$ is the exit coordinate $y_{r,j}$ of R_j */
       /* Delete I_j from T and update $v.\sharp$ and $v.clq$ for all visited nodes and those nodes in the canonical covering of I_j. */
       $\text{Update\_Segment\_Tree\_by\_Weight}(r, x_{\ell,j}, x_{r,j}, -1)$
6. **return** max_clq.

Similarly, each insertion or deletion of an interval updates at most $O(\log n)$ nodes in T and the scanning of y-coordinates takes $O(n \log n)$ time. We conclude that

Theorem 19.4

Given a set $S = \{R_1, R_2, \ldots, R_n\}$ of n rectangles, the size of the maximum clique of S can be found in $O(n \log n)$ time and linear space by using a segment tree.

We note that the above reduction can be adapted to find the maximum clique of a set of *hyperrectangles* in k-dimensions for $k > 2$ in $O(n^{k-1} \log n)$ time [2].

19.4 Range Trees

Consider a set S of points in k-dimensional space $\Re^k$. A range tree for this set S of points is a data structure that supports general range queries of the form $[x_\ell^1, x_r^1] \times [x_\ell^2, x_r^2] \times \cdots \times [x_\ell^k, x_r^k]$, where each range $[x_\ell^i, x_r^i]$, $x_\ell^i, x_r^i \in \Re$, $x_\ell^i \leq x_r^i$ for all $i = 1, 2, \ldots, k$, denotes an interval in $\Re$. The Cartesian product of these k ranges is referred to as a kD-range. In two-dimensional space, a 2D-range is simply an axes-parallel rectangle in $\Re^2$. The *range search problem* is to find all the points in S that satisfy any range query. In one-dimension, the range search problem can be easily solved in logarithmic time using a sorted array or a balanced binary search tree. The 1D-range is simply an interval $[x_\ell, x_r]$. We first do a binary search using x_ℓ as the search key to find the first node v whose key is no less than x_ℓ. Once v is located, the rest is simply to retrieve the nodes, one at a time, until the node u whose key is greater than x_r. We shall in this section describe an augmented binary search tree which is easily generalized to higher dimensions.

19.4.1 Construction of Range Trees

A range tree is primarily a binary search tree augmented with an auxiliary data structure. The root node v, denoted Range_Tree_root(S), of a kD-range tree [3,4,17,18] for a set S of points in k-dimensional space $\Re^k$, that is, $S = \{p_i = (x_i^1, x_i^2, \ldots, x_i^k), i = 1, 2, \ldots, n\}$, where $p_i.x^j = x_i^j \in \Re$ is the jth coordinate value of point p_i, for $j = 1, 2, \ldots, k$, is associated with the entire set S. The key stored in $v.key$ is to partition S into two approximately equal subsets S_ℓ and S_r, such that all the points in S_ℓ and in S_r lie to the left and to the right, respectively of the hyperplane $H^k{:}x^k = v.key$. That is, we will store the median of the kth coordinate values of all the points in S in $v.key$ of the root node v, that is, $v.key = p_j.x^k$ for some point p_j such that S_ℓ contains points $p_\ell, p_\ell.x^k \leq v.key$, and S_r contains points $p_r, p_r.x^k > v.key$. Each node v in the kD-range tree, as before, has two tree pointers, $v.left$ and $v.right$, to the roots of its left and right subtrees, respectively. The node pointed to by $v.left$ will be associated with the set S_ℓ and the node pointed to by $v.right$ will be associated with the set S_r. The auxiliary pointer $v.aux$ will point to an augmented data structure, in our case a $(k{-}1)$D-range tree.

A 1D-range tree is a sorted array of all the points $p_i \in S$ such that the entries are drawn from the set $\{x_i^1 | i = 1, 2, \ldots, n\}$ sorted in nondecreasing order. This 1D-range tree supports the 1D-range search in logarithmic time.

The following is a pseudo code for a kD-range tree for a set S of n points in k-dimensional space. See Figure 19.3a and b for an illustration. Figure 19.4c is a schematic representation of a kD-range tree.

function kD_Range_Tree(k, S)
/* It returns a pointer v to the root, kD_Range_Tree_root(k, S), of the kD-range tree for a set $S \subseteq \Re^k$ of points, $k \geq 1$. */
Input: A set S of n points in $\Re^k$, $S = \{p_i = (x_i^1, x_i^2, \ldots, x_i^k), i = 1, 2, \ldots, n\}$, where $x_i^j \in \Re$ is the jth coordinate value of point p_i, for $j = 1, 2, \ldots, k$.
Output: A kD-range tree, rooted at kD_Range_Tree_root(k, S).

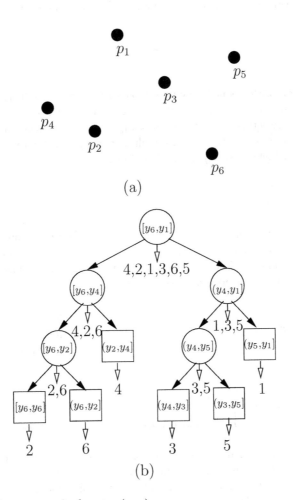

FIGURE 19.3 2D-range tree of $S = \{p_1, p_2, \ldots, p_6\}$, where $p_i = (x_i, y_i)$.

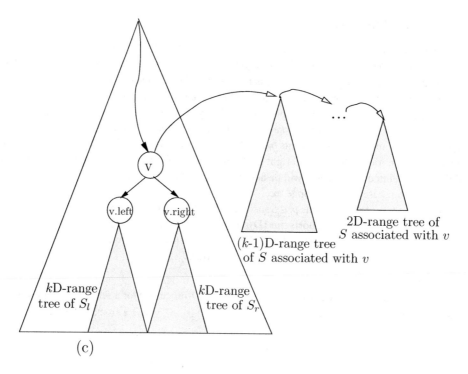

FIGURE 19.4   A schematic representation of a (layered) kD-range tree, where S is the set associated with node v.

Method:

1. **if** $S = \emptyset$, **return nil**.

2. **if** $(k = 1)$ create a sorted array $SA(S)$ pointed to by a node v containing the set of the 1st coordinate values of all the points in S, that is, $SA(S)$ has $\{p_i.x^1 | i = 1, 2, \ldots, n\}$ in nondecreasing order. **return** (v).

3. Create a node v such that $v.key$ equals the *median* of the set $\{p_i.x^k | k$th coordinate value of $p_i \in S, i = 1, 2, \ldots, n\}$. Let S_ℓ and S_r denote the subset of points whose kth coordinate values are not greater than and are greater than $v.key$, respectively. That is, $S_\ell = \{p_i \in S | p_i.x^k \leq v.key\}$ and $S_r = \{p_j \in S | p_j.x^k > v.key\}$.

4. $v.left = k$D_Range_Tree(k, S_ℓ)

5. $v.right = k$D_Range_Tree(k, S_r)

6. $v.aux = k$D_Range_Tree$(k - 1, S)$

As this is a recursive algorithm with two parameters, k and $|S|$, that determine the recursion depth, it is not immediately obvious how much time and how much space are needed to construct a kD-range tree for a set of n points in k-dimensional space.

Let $T(n, k)$ denote the time taken and $S(n, k)$ denote the space required to build a kD-range tree of a set of n points in $\mathfrak{R}^k$. The following are recurrence relations for $T(n, k)$ and $S(n, k)$, respectively.

$$T(n, k) = \begin{cases} O(1) & \text{if } n = 1 \\ O(n \log n) & \text{if } k = 1 \\ 2T(n/2, k) + T(n, k - 1) + O(n) & \text{otherwise.} \end{cases}$$

$$S(n, k) = \begin{cases} O(1) & \text{if } n = 1 \\ O(n) & \text{if } k = 1 \\ 2S(n/2, k) + S(n, k - 1) + O(1) & \text{otherwise.} \end{cases}$$

Note that in one dimension, we need to have the points sorted and stored in a sorted array, and thus $T(n, 1) = O(n \log n)$ and $S(n, 1) = O(n)$. The solutions of $T(n, k)$ and $S(n, k)$ to the above recurrence relations are $T(n, k) = O(n \log^{k-1} n + n \log n)$ for $k \geq 1$ and $S(n, k) = O(n \log^{k-1} n)$ for $k \geq 1$. For a general multidimensional divide-and-conquer scheme, and solutions to the recurrence relation, please refer to Bentley [19] and Monier [20], respectively.

We conclude that

Theorem 19.5

The kD-range tree for a set of n points in k-dimensional space can be constructed in $O(n \log^{k-1} n + n \log n)$ time and $O(n \log^{k-1} n)$ space for $k \geq 1$.

19.4.2 Examples and Its Applications

Figure 19.3b illustrates an example of a range tree for a set of points in two dimensions, shown in Figure 19.3a. This list of integers under each node represents the indexes of points in ascending x-coordinates. Figure 19.4 illustrates a general schematic representation of a kD-range tree, which is a *layered* structure [3,4].

We now discuss how we make use of a range tree to solve the range search problem. We shall use 2D-range tree as an example for illustration purposes. It is rather obvious to generalize it to higher dimensions. Recall we have a set S of n points in the plane $\Re^2$ and 2D range query $Q = [x_\ell, x_r] \times [y_\ell, y_r]$. Let us assume that a 2D-range tree rooted at 2D_Range_Tree_root(S) is available. Besides, for each node v in the range tree, it is associated with a *standard range* for the set S_v of points represented in the subtree rooted at node v, in this case $[v.B, v.E]$ where $v.B = \min\{p_i.y\}$ and $v.E = \max\{p_i.y\}$ for all $p_i \in S_v$. $v.key$ will split the standard range into two standard subranges $[v.B, v.key]$ and $[v.key, v.E]$, each associated with the root nodes $v.left$ and $v.right$ of the left and right subtrees of v, respectively.

The following pseudo code reports in F the set of points in S that lie in the range $Q = [x_\ell, x_r] \times [y_\ell, y_r]$. It is invoked by 2D_Range_Search($v, x_\ell, x_r, y_\ell, y_r, F$), where v is the root, 2D_Range_Tree_root(S), and F, initially empty, will return all the points in S that lie in the range $Q = [x_\ell, x_r] \times [y_\ell, y_r]$.

procedure 2D_Range_Search($v, x_\ell, x_r, y_\ell, y_r, F$)
/* It returns F containing all the points in the range tree rooted at node v that lie in $[x_\ell, x_r] \times [y_\ell, y_r]$. */
Input: A set S of n points in $\Re^2$, $S = \{p_i | i = 1, 2, \ldots, n\}$ and each point $p_i = (p_i.x, p_i.y)$, where $p_i.x$ and $p_i.y$ are the x- and y-coordinates of p_i, $p_i.x, p_i.y \in \Re, i = 1, 2, \ldots, n$.
Output: A set F of points in S that lie in $[x_\ell, x_r] \times [y_\ell, y_r]$.
Method:

1. Start from the root node v to find the split node s, $s = $ Find_split-node(v, y_ℓ, y_r) (see below), such that $s.key$ lies in $[y_\ell, y_r]$.
2. **if** s is a leaf **then** 1D_Range_Search($s.aux, x_\ell, x_r, F$) that checks in the sorted array pointed to by $s.aux$, which contains just a point p, to see if its x-coordinate $p.x$ lies in the x-range $[x_\ell, x_r]$.
3. $v = s.left$.
4. **while** v is not a leaf **do**
   if ($y_\ell \leq v.key$) **then**
       1D_Range_Search($v.right.aux, x_\ell, x_r, F$)
       $v = v.left$
   else $v = v.right$
5. (/* v is a leaf, and check node $v.aux$ directly */)
   1D_Range_Search($v.aux, x_\ell, x_r, F$)
6. $v = s.right$
7. **while** v is not a leaf **do**
   if ($y_r > v.key$) **then**
       1D_Range_Search($v.left.aux, x_\ell, x_r, F$)
       $v = v.right$
   else $v = v.left$
8. (/* v is a leaf, and check node $v.aux$ directly */)
   1D_Range_Search($v.aux, x_\ell, x_r, F$).

function Find_split-node(v, b, e)
/* Given a kD-range tree T rooted at v and an interval $I = [b, e] \subseteq [v.B, v.E]$, this procedure returns the *split-node* s such that either $[s.B, s.E] = [b, e]$ or $[s_\ell.B, s_\ell.E] \cap [b, e] \neq \emptyset$ and $[s_r.B, s_r.E] \cap [b, e] \neq \emptyset$, where s_ℓ and s_r are the left child and right child of s, respectively. */

1. **if** $[v.B, v.E] = [b, e]$ **then return** v
2. **if** ($b < v.key$) **and** ($e > v.key$) **then return** v
3. **if** ($e \leq v.key$) **then return** Find_split-node($v.left, b, e$)
4. **if** ($b \geq v.key$) **then return** Find_split-node($v.right, b, e$)

procedure 1D_Range_Search(v, x_ℓ, x_r, F) is very straightforward. v is a pointer to a sorted array SA. We first do a binary search in SA looking for the first element no less than x_ℓ and then start to report in F those elements no greater than x_r. It is obvious that **procedure** 2D_Range_Search finds all the points in Q in $O(\log^2 n)$ time. Note that there are $O(\log n)$ nodes for which we need to invoke 1D_Range_Search in their auxiliary sorted arrays. These nodes v are in the canonical covering* of the y-range $[y_\ell, y_r]$, since its associated standard range $[v.B, v.E]$ is totally contained in $[y_\ell, y_r]$, and the 2D-range search problem is now reduced to the 1D-range search problem.

This is not difficult to see that the 2D-range search problem can be answered in time $O(\log^2 n)$ plus time for output, as there are $O(\log n)$ nodes in the canonical covering of a given y-range and for each node in the canonical covering we spend $O(\log n)$ time for dealing with the 1D-range search problem.

However, with a modification to the auxiliary data structure, one can achieve an optimal query time of $O(\log n)$, instead of $O(\log^2 n)$ [18,21,22]. This is based on the observation that in each of the 1D-range search subproblem associated with each node in the canonical covering, we perform the same query, reporting points whose x-coordinates lie in the x-range $[x_\ell, x_r]$. More specifically, we are searching for the smallest element no less than x_ℓ.

The modification is performed on the sorted array associated with each of the node in the 2D_Range_Tree(S).

Consider the root node v. As it is associated with the entire set of points, $v.aux$ points to the sorted array containing the x-coordinates of *all* the points in S. Let this sorted array be denoted $SA(v)$ and the entries, $SA(v)_i, i = 1, 2, \ldots, |S|$, are sorted in nondecreasing order of the x-coordinate values. In addition to the x-coordinate value, each entry also contains the index of the corresponding point. That is, $SA(v)_i.key$ and $SA(v)_i.index$ contain the x-coordinate and index of p_j, respectively, where $SA(v)_i.index = j$ and $SA(v)_i.key = p_j.x$.

We shall augment each entry $SA(v)_i$ with two pointers, $SA(v)_i.left$ and $SA(v)_i.right$. They are defined as follows. Let v_ℓ and v_r denote the roots of the left and right subtrees of v, that is, $v.left = v_\ell$ and $v.right = v_r$. $SA(v)_i.left$ points to the entry $SA(v_\ell)_j$ such that entry $SA(v_\ell)_j.key$ is the smallest among all key values $SA(v_\ell)_j.key \geq SA(v)_i.key$. Similarly, $SA(v)_i.right$ points to the entry $SA(v_r)_k$ such that entry $SA(v_r)_k.key$ is the smallest among all key values $SA(v_r)_k.key \geq SA(v)_i.key$.

These two augmented pointers, $SA(v)_i.left$ and $SA(v)_i.right$, possess the following property: if $SA(v)_i.key$ is the smallest key such that $SA(v)_i.key \geq x_\ell$, then $SA(v_\ell)_j.key$ is also the smallest key such that $SA(v_\ell)_j.key \geq x_\ell$. Similarly $SA(v_r)_k.key$ is the smallest key such that $SA(v_r)_k.key \geq x_\ell$.

Thus if we have performed a binary search in the auxiliary sorted array $SA(v)$ associated with node v locating the entry $SA(v)_i$ whose key $SA(v)_i.key$ is the smallest key such that $SA(v)_i.key \geq x_\ell$, then following the left (respectively right) pointer $SA(v)_i.left$ (respectively $SA(v)_i.right$) to $SA(v_\ell)_j$ (respectively $SA(v_r)_k$), the entry $SA(v_\ell)_j.key$ (respectively $SA(v_r)_k.key$) is also the smallest key such that $SA(v_\ell)_j.key \geq x_\ell$ (respectively $SA(v_r)_k.key \geq x_\ell$). Thus there is no need to perform an additional binary search in the auxiliary sorted array $SA(v.left)$ (respectively $SA(v.right)$).

With this additional modification, we obtain an *augmented* 2D-range tree and the following theorem.

Theorem 19.6

The 2D-range search problem for a set of n points in the two-dimensional space can be solved in time $O(\log n)$ plus time for output, using an augmented 2D-range tree that requires $O(n \log n)$ space.

The following procedure is generalized from **procedure** 2D_Range_Search($v, x_\ell, x_r, y_\ell, y_r, F$) discussed in Section 19.4.2 taken into account the augmented auxiliary data structure. It is invoked by kD_Range_Search(k, v, Q, F), where v is the root kD_Range_Tree_root(S) of the range tree, Q is the kD-range, $[x_\ell^1, x_r^1] \times [x_\ell^2, x_r^2] \times \cdots \times [x_\ell^k, x_r^k]$, represented by a two dimensional array, such that $Q_i.\ell = x_\ell^i$ and $Q_i.r = x_r^i$, and F, initially empty, will contain all the points that lie in Q.

procedure kD_Range_Search(k, v, Q, F).

/* It returns F containing all the points in the range tree rooted at node v that lie in kD-range, $[x_\ell^1, x_r^1] \times [x_\ell^2, x_r^2] \times \cdots \times [x_\ell^k, x_r^k]$, where each range $[x_\ell^i, x_r^i]$, $x_\ell^i = Q_i.\ell$, $x_r^i = Q_i.r \in \Re$, $x_\ell^i \leq x_r^i$ for all $i = 1, 2, \ldots, k$ denotes an interval in $\Re$. */
Input: A set S of n points in $\Re^k$, $S = \{p_i | i = 1, 2, \ldots, n\}$ and each point $p_i = (p_i.x^1, p_i.x^2, \ldots, p_i.x^k)$, where $p_i.x^j \in \Re$, are the jth coordinates of p_i, $j = 1, 2, \ldots, k$.
Output: A set F of points in S that lie in $[x_\ell^1, x_r^1] \times [x_\ell^2, x_r^2] \times \cdots \times [x_\ell^k, x_r^k]$.
Method:

  1. **if** $(k > 2)$ **then**
     - Start from the root node v to find the split node s, $s = $ Find_split-node($v, Q_k.\ell, Q_k.r$), such that $s.key$ lies in $[Q_k.\ell, Q_k.r]$.
     - **if** the split node is not found **then return**

---

* See the definition of the canonical covering defined in Section 19.3.1.

- **if** s is a leaf **then** check in the sorted array pointed to by $s.aux$, which contains just a point p. $p \Rightarrow F$ if its coordinate values lie in Q. **return**
- $v = s.left$
- **while** v is not a leaf **do**

  if $(Q_k.\ell \leq v.key)$

  then $kD\_Range\_Search(k - 1, v.right.aux, Q, F)$

      $v = v.left$

  else $v = v.right$
- (/* v is a leaf, and check node $v.aux$ directly */)
  Check in the sorted array pointed to by $v.aux$, which contains just a point p. $p \Rightarrow F$ if its coordinate values lie in Q.
- $v = s.right$
- **while** v is not a leaf **do**

  if $(Q_k.r > v.key)$

  then $kD\_Range\_Search(k - 1, v.left.aux, Q, F)$

      $v = v.right$

  else $v = v.left$
- (/* v is a leaf, and check node $v.aux$ directly */)
  Check in the sorted array pointed to by $v.aux$, which contains just a point p. $p \Rightarrow F$ if its coordinate values lie in Q.

2. **else** /* $k \leq 2$*/

3. **if** $k = 2$ **then**
   - Do binary search in sorted array $SA(v)$ associated with node v, using $Q_1.\ell$ (x_ℓ^1) as key to find entry o_v such that $SA(v)_{o_v}$'s key, $SA(v)_{o_v}.key$ is the smallest such that $SA(v)_{o_v}.key \geq Q_1.\ell$.
   - Find the split node s, $s = Find\_split\text{-}node(v, Q_2.\ell, Q_2.r)$, such that $s.key$ lies in $[Q_2.\ell, Q_2.r]$. Record the root-to-split-node path from v to s, following *left* or *right* tree pointers.
   - **if** the split node is not found **then return**
   - Starting from entry o_v ($SA(v)_i$) follow pointers $SA(v)_{o_v}.left$ or $SA(v)_{o_v}.right$ according to the v-to-s path to point to entry $SA(s)_{o_s}$ associated with $SA(s)$.
   - **if** s is a leaf **then** check in the sorted array pointed to by $s.aux$, which contains just a point p. $p \Rightarrow F$ if its coordinate values lie in Q. **return**
   - $v = s.left, o_v = SA(s)_{o_s}.left$
   - **while** v is not a leaf **do**

     if $(Q_2.\ell \leq v.key)$

     then $\ell = SA(v)_{o_v}.right$

         while $(SA(v.right)_\ell.key \leq Q_1.r)$ **do**

             point $p_m \Rightarrow F$, where $m = SA(v.right)_\ell.index$

             $\ell++$

         $v = v.left, o_v = SA(v)_{o_v}.left$

     else $v = v.right, o_v = SA(v)_{o_v}.right$
   - (/* v is a leaf, and check node $v.aux$ directly */)
     Check in the sorted array pointed to by $v.aux$, which contains just a point p. $p \Rightarrow F$ if its coordinate values lie in Q.
   - $v = s.right, o_v = SA(s)_{o_s}.right$
   - **while** v is not a leaf **do**

     if $(Q_2.r > v.key)$

     then $\ell = SA(v)_{o_v}.left$

         while $(SA(v.left)_\ell.key \leq Q_1.r)$ **do**

             point $p_m \Rightarrow F$, where $m = SA(v.left)_\ell.index$

             $\ell++$

         $v = v.right, o_v = SA(v)_{o_v}.right$

     else $v = v.left, o_v = SA(v)_{o_v}.left$
   - (/* v is a leaf, and check node $v.aux$ directly */)
     Check in the sorted array pointed to by $v.aux$, which contains just a point p. $p \Rightarrow F$ if its coordinate values lie in Q.

The following recurrence relation for the query time $Q(n, k)$ of the kD-range search problem can be easily obtained:

$$Q(n, k) = \begin{cases} O(1) & \text{if } n = 1 \\ O(\log n) + \mathcal{F} & \text{if } k = 2 \\ \Sigma_{v \in CC} Q(n_v, k - 1) + O(\log n) & \text{otherwise} \end{cases}$$

where $\mathcal{F}$ denotes the output size and n_v denotes the size of the subtree rooted at node v that belongs to the canonical covering CC of the query. The solution is $Q(n,k) = O(\log^{k-1} n) + \mathcal{F}$ [3,4].

We conclude with the following theorem.

Theorem 19.7

The kD-range search problem for a set of n points in the k-dimensional space can be solved in time $O(\log^{k-1} n)$ plus time for output, using an augmented kD-range tree that requires $O(n \log^{k-1} n)$ space for $k \geq 1$.

19.5 Priority Search Trees

The priority search tree was originally introduced by McCreight [23]. It is a hybrid of two data structures, binary search tree and a priority queue [24]. A *priority queue* is a queue and supports the following operations: insertion of an item and deletion of the minimum (highest priority) item, so called *delete_min* operation. Normally the delete_min operation takes constant time, while updating the queue so that the minimum element is readily accessible takes logarithmic time. However, searching for an element in a priority queue will normally take linear time. To support efficient searching, the priority queue is modified to be a priority search tree. We will give a formal definition and its construction later. As the priority search tree represents a set S of elements, each of which has two pieces of information, one being a key from a totally ordered set, say the set $\Re$ of real numbers, and the other being a notion of priority, also from a totally ordered set, for each element, we can model this set S as a set of points in two-dimensional space. The x- and y-coordinates of a point p represent the key and the priority, respectively. For instance, consider a set of jobs $S = \{J_1, J_2, \ldots, J_n\}$, each of which has a release time $r_i \in \Re$ and a priority $p_i \in \Re, i = 1, 2, \ldots, n$. Then each job J_i can be represented as a point q such that $q.x = r_i, q.y = p_i$.

The priority search tree can be used to support queries of the form, find, among a set S of n points, the point p with minimum $p.y$ such that its x-coordinate lies in a given range $[\ell, r]$, that is, $\ell \leq p.x \leq r$. As can be shown later, this query can be answered in $O(\log n)$ time.

19.5.1 Construction of Priority Search Trees

As before, the root node, Priority_Search_Tree_root(S), represents the entire set S of points. Each node v in the tree will have a key $v.key$, an auxiliary data $v.aux$ containing the index of the point, and two pointers $v.left$ and $v.right$ to its left and right subtrees, respectively, such that all the key values stored in the left subtree are less than $v.key$ and all the key values stored in the right subtree are greater than $v.key$. The following is a pseudo code for the recursive construction of the priority search tree of a set S of n points in $\Re^2$. See Figure 19.5a for an illustration.

function Priority_Search_Tree(S)
/* It returns a pointer v to the root, Priority_Search_Tree_root(S), of the priority search tree for a set S of points. */
Input: A set S of n points in $\Re^2$, $S = \{p_i | i = 1, 2, \ldots, n\}$ and each point $p_i = (p_i.x, p_i.y)$, where $p_i.x$ and $p_i.y$ are the x- and y-coordinates of p_i, $p_i.x, p_i.y \in \Re, i = 1, 2, \ldots, n$.
Output: A priority search tree, rooted at Priority_Search_Tree_root(S).
Method:

1. **if** $S = \emptyset$, **return nil**.
2. Create a node v such that $v.key$ equals the *median* of the set $\{p.x | p \in S\}$, and $v.aux$ contains the index i of the point p_i whose y-coordinate is the minimum among all the y-coordinates of the set S of points, that is, $p_i.y = \min\{p.y | p \in S\}$.
3. Let $S_\ell = \{p \in S \setminus \{p_{v.aux}\} | p.x \leq v.key\}$ and $S_r = \{p \in S \setminus \{p_{v.aux}\} | p.x > v.key\}$ denote the set of points whose x-coordinates are less than or equal to $v.key$ and greater than $v.key$, respectively.
4. $v.left =$ Priority_Search_Tree(S_ℓ).
5. $v.right =$ Priority_Search_Tree(S_r).
6. **return** v.

Thus, Priority_Search_Tree_root(S) is a minimum heap data structure with respect to the y-coordinates, that is, the point with minimum y-coordinate can be accessed in constant time, and is a balanced binary search tree for the x-coordinates. Implicitly the root node v is associated with an x-range $[x_\ell, x_r]$ representing the span of the x-coordinate values of all the points in the whole set S. The root of the left subtree pointed to by $v.left$ is associated with the x-range $[x_\ell, v.key]$ representing the span of the x-coordinate values of all the points in the set S_ℓ and the root of the right subtree pointed to by $v.right$ is associated with the x-range $(v.key, x_r]$ representing the span of the x-coordinate values of all the points in the set S_r. It is obvious that this algorithm takes $O(n \log n)$ time and linear space. We summarize this in the following.

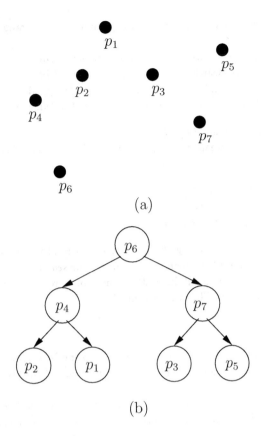

FIGURE 19.5   Priority search tree of $S = \{p_1, p_2, \ldots, p_7\}$.

Theorem 19.8

The priority search tree for a set S of n points in $\Re^2$ can be constructed in $O(n \log n)$ time and linear space.

19.5.2 Examples and Its Applications

Figure 19.5 illustrates an example of a priority search tree of the set of points. Note that the root node contains p_6 since its y-coordinate value is the minimum.

   We now illustrate a usage of the priority search tree by an example. Consider a so-called *grounded 2D-range search problem* for a set of n points in the plane. As defined in Section 19.4.2, a 2D-range search problem is to find all the points $p \in S$ such that $p.x$ lies in an x-range $[x_\ell, x_r], x_\ell \le x_r$ and $p.y$ lies in a y-range $[y_\ell, y_r]$. When the y-range is of the form $[-\infty, y_r]$ then the 2D-range is referred to as *grounded* 2D-range or sometimes as 1.5D-range, and the 2D-range search problem as *grounded* 2D-range search or 1.5D-range search problem.

   Grounded 2D-Range Search Problem: Given a set S of n points in the plane $\Re^2$, with preprocessing allowed, find the subset F of points whose x- and y-coordinates satisfy a grounded 2D-range query of the form $[x_\ell, x_r] \times [-\infty, y_r], x_\ell, x_r, y_r \in \Re, x_\ell \le x_r$.

   The following pseudo code solves this problem optimally. We assume that a priority search tree for S has been constructed via procedure Priority_Search_Tree(S). The answer will be obtained in F via an invocation to Priority_Search_Tree_Range_Search(v, x_ℓ, x_r, y_r, F), where v is Priority_Search_Tree_root(S).

procedure Priority_Search_Tree_Range_Search(v, x_ℓ, x_r, y_r, F)

/* v points to the root of the tree, F is a queue and set to **nil** initially. */

Input: A set S of n points, $\{p_1, p_2, \ldots, p_n\}$, in $\Re^2$, stored in a priority search tree, Priority_Search_Tree(S) pointed to by Priority_Search_Tree_root(S), and a grounded 2D-range $[x_\ell, x_r] \times [-\infty, y_r], x_\ell, x_r, y_r \in \Re, x_\ell \le x_r$.

Output: A subset $F \subseteq S$ of points that lie in the grounded 2D-range, that is, $F = \{p \in S | x_\ell \le p.x \le x_r \text{ and } p.y \le y_r\}$.

Method:

   1. Start from the root v finding the first split node v_{split} such that $v_{split}.key$ lies in the x-range $[x_\ell, x_r]$.

2. For each node u on the path from node v to node v_{split}, **if** the point $p_{u.aux}$ lies in range $[x_\ell, x_r] \times [-\infty, y_r]$ **then** report it by $(p_{u.aux} \Rightarrow F)$.

3. For each node u on the path from $v_{split}.left$ to x_ℓ in the left subtree of v_{split}, **do**
   if the path goes left at u **then** Priority_Search_Tree_1dRange_Search($u.right, y_r, F$)

4. For each node u on the path from $v_{split}.right$ to x_r in the right subtree of v_{split}, **do**
   if the path goes right at u **then** Priority_Search_Tree_1dRange_Search($u.left, y_r, F$)

procedure Priority_Search_Tree_1dRange_Search(v, y_r, F)
/* Report in F all the points p_i, whose y-coordinate values are no greater than y_r, where $i = v.aux$. */

1. **if** v is **nil, return.**
2. **if** $p_{v.aux}.y \leq y_r$ **then** report it by $(p_{v.aux} \Rightarrow F)$.
3. Priority_Search_Tree_1dRange_Search($v.left, y_r, F$)
4. Priority_Search_Tree_1dRange_Search($v.right, y_r, F$)

procedure Priority_Search_Tree_1dRange_Search(v, y_r, F) basically retrieves all the points stored in the priority search tree rooted at v such that their y-coordinates are all less than or equal to y_r. The search terminates at the node u whose associated point has a y-coordinate greater than y_r, implying **all** the nodes in the subtree rooted at u satisfy this property. The amount of time required is proportional to the output size. Thus we conclude that

Theorem 19.9

The **Grounded 2D-Range Search Problem** for a set S of n points in the plane $\mathfrak{R}^2$ can be solved in time $O(\log n)$ plus time for output, with a priority search tree structure for S that requires $O(n \log n)$ time and $O(n)$ space.

Note that the space requirement for the priority search tree is linear, compared to that of a 2D-range tree, which requires $O(n \log n)$ space. That is, the **Grounded 2D-Range Search Problem** for a set S of n points can be solved optimally using priority search tree structure.

Acknowledgments

This work was supported in part by the National Science Council under the grants NSC-91-2219-E-001-001, NSC91-2219-E-001-002, and by the Ministry of Science and Technology under the grants MOST 104–2221-E-001-031 and MOST 104–2221-E-001-032.

References

1. H. Imai and Ta. Asano, Finding the connected components and a maximum clique of an intersection graph of rectangles in the plane, *J. Algorithms*, vol. 4, 1983, pp. 310–323.

2. D. T. Lee, Maximum clique problem of rectangle graphs, in *Advances in Computing Research*, Vol. 1, ed. F.P. Preparata, JAI Press Inc., Greenwich, CT, 1983, 91–107.

3. M. de Berg, M. van Kreveld, M. Overmars and O. Schwarzkopf, *Computational Geometry: Algorithms and Applications*, Springer-Verlag, Berlin, 1997.

4. F. P. Preparata and M. I. Shamos, *Computational Geometry: An Introduction*, 3rd edition, Springer-Verlag, 1990.

5. J. L. Bentley, Decomposable searching problems, *Inform. Process Lett.*, vol. 8, 1979, pp. 244–251.

6. J. L. Bentley and J. B. Saxe, Decomposable searching problems. I: Static-to-dynamic transformation, *J. Algorithms*, vol. 1, 1980, pp. 301–358.

7. M. H. Overmars and J. van Leeuwen, Two general methods for dynamizing decomposable searching problems, *Computing*, vol. 26, 1981, pp. 155–166.

8. H. Edelsbrunner, A new approach to rectangle intersections. Part I, *Intl. J. Comput. Math.*, vol. 13, 1983, pp. 209–219.

9. H. Edelsbrunner, A new approach to rectangle intersections. Part II, *Intl. J. Comput. Math.*, vol. 13, 1983, pp. 221–229.

10. P. Gupta, R. Janardan, M. Smid and B. Dasgupta, The rectangle enclosure and point-dominance problems revisited, *Intl. J. Comput. Geom. Appl.*, vol. 7, 1997, pp. 437–455.

11. D. T. Lee and F. P. Preparata, An improved algorithm for the rectangle enclosure problem, *J. Algorithms*, vol. 3, 1982, pp. 218–224.

12. J.-D. Boissonnat and F. P. Preparata, Robust plane sweep for intersecting segments, *SIAM J. Comput.*, vol. 295, 2000, pp. 1401–1421.

13. M. L. Fredman and B. Weide, On the complexity of computing the measure of $\cup[a_i, b_i]$, *Commun. ACM*, vol. 21, 1978, pp. 540–544.

14. J. van Leeuwen and D. Wood, The measure problem for rectangular ranges in d-space, *J. Algorithms*, vol. 2, 1981, pp. 282–300.

15. U. Gupta, D. T. Lee and J. Y.-T. Leung, An optimal solution for the channel-assignment problem, *IEEE Trans. Comput.*, vol. 28(11), 1979, pp. 807–810.

16. M. Sarrafzadeh and D. T. Lee, Restricted track assignment with applications, *Intl. J. Comput. Geom. Appl.*, vol. 4, 1994, pp. 53–68.

17. G. S. Lueker and D. E. Willard, A data structure for dynamic range queries, *Inform. Process. Lett.*, vol. 15, 1982, pp. 209–213.

18. D. E. Willard, New data structures for orthogonal range queries, *SIAM J. Comput.*, vol. 14, 1985, pp. 232–253.

19. J. L. Bentley, Multidimensional divide-and-conquer, *Commun. ACM*, vol. 23,4, 1980, pp. 214–229.

20. L. Monier, Combinatorial solutions of multidimensional divide-and-conquer recurrences, *J. Algorithms*, vol. 1, 1980, pp. 60–74.

21. B. Chazelle and L. J. Guibas, Fractional cascading. I: A data structuring technique, *Algorithmica*, vol. 13, 1986, pp. 133–162.

22. B. Chazelle and L. J. Guibas, Fractional cascading. II: Applications, *Algorithmica*, vol. 1, 1986, pp. 163–191.

23. E. M. McCreight, Priority search trees, *SIAM J. Comput.*, vol. 14, 1985, pp. 257–276.

24. E. Horowitz, S. Sahni and S. Anderson-Freed, *Fundamentals of Data Structures in C*, Computer Science Press, 1993.

20

Quadtrees and Octrees*

20.1	Introduction	309
20.2	Quadtrees for Point Data	310
	Point Quadtrees • Region Quadtrees • Compressed Quadtrees and Octrees • Cell Orderings and Space-Filling Curves • Construction of Compressed Quadtrees • Basic Operations • Practical Considerations	
20.3	Spatial Queries with Region Quadtrees	318
	Range Query • Spherical Region Queries • k-Nearest Neighbors	
20.4	Image Processing Applications	320
	Construction of Image Quadtrees • Union and Intersection of Images • Rotation and Scaling • Connected Component Labeling	
20.5	Scientific Computing Applications	322
	The N-Body Problem	

Srinivas Aluru
Georgia Institute of Technology

Acknowledgments 325

References 326

20.1 Introduction

Quadtrees are hierarchical spatial tree data structures that are based on the principle of recursive decomposition of space. The term *quadtree* originated from representation of two dimensional data by recursive decomposition of space using separators parallel to the coordinate axis. The resulting split of a region into four regions corresponding to southwest, northwest, southeast and northeast quadrants is represented as four children of the node corresponding to the region, hence the term "quad" tree. In a three dimensional analogue, a region is split into eight regions using planes parallel to the coordinate planes. As each internal node can have eight children corresponding to the 8-way split of the region associated with it, the term *octree* is used to describe the resulting tree structure. Analogous data structures for representing spatial data in higher than three dimensions are called *hyperoctrees*. It is also common practice to use the term *quadtrees* in a generic way irrespective of the dimensionality of the spatial data. This is especially useful when describing algorithms that are applicable regardless of the specific dimensionality of the underlying data.

Several related spatial data structures are described under the common rubric of quadtrees. Common to these data structures is the representation of spatial data at various levels of granularity using a hierarchy of regular, geometrically similar regions (such as cubes, hyperrectangles etc.). The tree structure allows quick focusing on regions of interest, which facilitates the design of fast algorithms. As an example, consider the problem of finding all points in a data set that lie within a given distance from a query point, commonly known as the *spherical region query*. In the absence of any data organization, this requires checking the distance from the query point to each point in the data set. If a quadtree of the data is available, large regions that lie outside the spherical region of interest can be quickly discarded from consideration, resulting in great savings in execution time. Furthermore, the unit aspect ratio employed in most quadtree data structures allows geometric arguments useful in designing fast algorithms for certain classes of applications.

In constructing a quadtree, one starts with a square, cubic or hypercubic region (depending on the dimensionality) that encloses the spatial data under consideration. The different variants of the quadtree data structure are differentiated by the principle used in the recursive decomposition process. One important aspect of the decomposition process is if the decomposition is guided by input data or is based on the principle of equal subdivision of the space itself. The former results in a tree size proportional to the size of the input. If all the input data is available a priori, it is possible to make the data structure height balanced. These

---

* This chapter has been reprinted from first edition of this Handbook, without any content updates.

attractive properties come at the expense of difficulty in making the data structure dynamic, typically in accommodating deletion of data. If the decomposition is based on equal subdivision of space, the resulting tree depends on the distribution of spatial data. As a result, the tree is height balanced and is linear in the size of input only when the distribution of the spatial data is uniform, and the height and size properties deteriorate with increase in nonuniformity of the distribution. The beneficial aspect is that the tree structure facilitates easy update operations and the regularity in the hierarchical representation of the regions facilitates geometric arguments helpful in designing algorithms.

Another important aspect of the decomposition process is the termination condition to stop the subdivision process. This identifies regions that will not be subdivided further, which will be represented by leaves in the quadtree. Quadtrees have been used as fixed resolution data structures, where the decomposition stops when a preset resolution is reached, or as variable resolution data structures, where the decomposition stops when a property based on input data present in the region is satisfied. They are also used in a hybrid manner, where the decomposition is stopped when either a resolution level is reached or when a property is satisfied.

Quadtrees are used to represent many types of spatial data including points, line segments, rectangles, polygons, curvilinear objects, surfaces, volumes and cartographic data. Their use is pervasive spanning many application areas including computational geometry, computer aided design (Chapter 53), computer graphics (Chapter 55), databases (Chapter 62), geographic information systems (Chapter 56), image processing (Chapter 58), pattern recognition, robotics and scientific computing. Introduction of the quadtree data structure and its use in applications involving spatial data dates back to the early 1970s and can be attributed to the work of Klinger [1], Finkel and Bentley [2], and Hunter [3]. Due to extensive research over the last three decades, a large body of literature is available on quadtrees and its myriad applications. For a detailed study on this topic, the reader is referred to the classic textbooks by Samet [4,5]. Development of quadtree like data structures, algorithms and applications continues to be an active research area with significant research developments in recent years. In this chapter, we attempt a coverage of some of the classical results together with some of the more recent developments in the design and analysis of algorithms using quadtrees and octrees.

20.2 Quadtrees for Point Data

We first explore quadtrees in the context of the simplest type of spatial data—multidimensional points. Consider a set of n points in d dimensional space. The principal reason a spatial data structure is used to organize multidimensional data is to facilitate queries requiring spatial information. A number of such queries can be identified for point data. For example:

1. *Range query:* Given a range of values for each dimension, find all the points that lie within the range. This is equivalent to retrieving the input points that lie within a specified hyperrectangular region. Such a query is often useful in database information retrieval.

2. *Spherical region query:* Given a query point p and a radius r, find all the points that lie within a distance of r from p. In a typical molecular dynamics application, spherical region queries centered around each of the input points is required.

3. *All nearest neighbor query:* Given n points, find the nearest neighbor of each point within the input set.

While quadtrees are used for efficient execution of such spatial queries, one must also design algorithms for the operations required of almost any data structure such as constructing the data structure itself, and accommodating searches, insertions and deletions. Though such algorithms will be covered first, it should be kept in mind that the motivation behind the data structure is its use in spatial queries. If all that were required was search, insertion and deletion operations, any one dimensional organization of the data using a data structure such as a binary search tree would be sufficient.

20.2.1 Point Quadtrees

The point quadtree is a natural generalization of the binary search tree data structure to multiple dimensions. For convenience, first consider the two dimensional case. Start with a square region that contains all of the input points. Each node in the point quadtree corresponds to an input point. To construct the tree, pick an arbitrary point and make it the root of the tree. Using lines parallel to the coordinate axis that intersect at the selected point (see Figure 20.1), divide the region into four subregions corresponding to the southwest, northwest, southeast and northeast quadrants, respectively. Each of the subregions is recursively decomposed in a similar manner to yield the point quadtree. For points that lie at the boundary of two adjacent regions, a convention can be adopted to treat the points as belonging to one of the regions. For instance, points lying on the left and bottom edges of a region may be considered included in the region, while points lying on the top and right edges are not. When a region corresponding to a node in the tree contains a single point, it is considered a leaf node. Note that point quadtrees are not unique and their structure depends on the selection of points used in region subdivisions. Irrespective of the choices made, the resulting tree will have n nodes, one corresponding to each input point.

If all the input points are known in advance, it is easy to choose the points for subdivision so as to construct a height balanced tree. A simple way to do this is to sort the points with one of the coordinates, say x, as the primary key and the other coordinate,

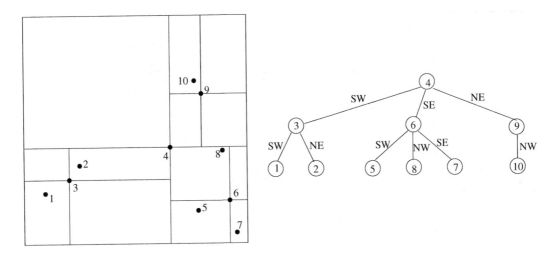

FIGURE 20.1   A two dimensional set of points and a corresponding point quadtree.

say y, as the secondary key. The first subdivision point is chosen to be the median of this sorted data. This will ensure that none of the children of the root node receives more than half the points. In $O(n)$ time, such a sorted list can be created for each of the four resulting subregions. As the total work at every level of the tree is bounded by $O(n)$, and there are at most $O(\log n)$ levels in the tree, a height balanced point quadtree can be built in $O(n \log n)$ time. Generalization to d dimensions is immediate, with $O(dn \log n)$ run time.

The recursive structure of a point quadtree immediately suggests an algorithm for searching. To search for a point, compare it with the point stored at the root. If they are different, the comparison immediately suggests the subregion containing the point. The search is directed to the corresponding child of the root node. Thus, search follows a path in the quadtree until either the point is discovered, or a leaf node is reached. The run time is bounded by $O(dh)$, where h is the height of the tree.

To insert a new point not already in the tree, first conduct a search for it which ends in a leaf node. The leaf node now corresponds to a region containing two points. One of them is chosen for subdividing the region and the other is inserted as a child of the node corresponding to the subregion it falls in. The run time for point insertion is also bounded by $O(dh)$, where h is the height of the tree after insertion. One can also construct the tree itself by repeated insertions using this procedure. Similar to binary search trees, the run time under a random sequence of insertions is expected to be $O(n \log n)$ [6]. Overmars and van Leeuwen [7] present algorithms for constructing and maintaining optimized point quadtrees irrespective of the order of insertions.

Deletion in point quadtrees is much more complex. The point to be deleted is easily identified by a search for it. The difficulty lies in identifying a point in its subtree to take the place of the deleted point. This may require nontrivial readjustments in the subtree underneath. The reader interested in deletion in point quadtrees is referred to [8]. An analysis of the expected cost of various types of searches in point quadtrees is presented by Flajolet *et al.* [9].

For the remainder of the chapter, we will focus on quadtree data structures that use equal subdivision of the underlying space, called *region quadtrees*. This is because we regard Bentley's multidimensional binary search trees [2], also called *k-d* trees, to be superior to point quadtrees. The *k-d* tree is a binary tree where a region is subdivided into two based only on one of the dimensions. If the dimension used for subdivision is cyclically rotated at consecutive levels in the tree, and the subdivision is chosen to be consistent with the point quadtree, then the resulting tree would be equivalent to the point quadtree but without the drawback of large degree (2^d in d dimensions). Thus, it can be argued that point quadtrees are contained in *k-d* trees. Furthermore, recent results on compressed region quadtrees indicate that it is possible to simultaneously achieve the advantages of both region and point quadtrees. In fact, region quadtrees are the most widely used form of quadtrees despite their dependence on the spatial distribution of the underlying data. While their use posed theoretical inconvenience—it is possible to create as large a worst-case tree as desired with as little as three points—they are widely acknowledged as the data structure of choice for practical applications. We will outline some of these recent developments and outline how good practical performance and theoretical performance guarantees can both be achieved using region quadtrees.

20.2.2 Region Quadtrees

The region quadtree for n points in d dimensions is defined as follows: Consider a hypercube large enough to enclose all the points. This region is represented by the root of the d-dimensional quadtree. The region is subdivided into 2^d subregions of equal

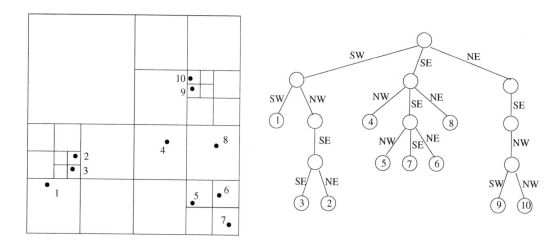

FIGURE 20.2 A two dimensional set of points and the corresponding region quadtree.

size by bisecting along each dimension. Each of these regions containing at least one point is represented as a child of the root node. The same procedure is recursively applied to each child of the root node. The process is terminated when a region contains only a single point. This data structure is also known as the *point region quadtree*, or *PR-quadtree* for short [10]. At times, we will simply use the term *quadtree* when the tree implied is clear from the context. The region quadtree corresponding to a two dimensional set of points is shown in Figure 20.2. Once the enclosing cube is specified, the region quadtree is unique. The manner in which a region is subdivided is independent of the specific location of the points within the region. This makes the size of the quadtree sensitive to the spatial distribution of the points.

Before proceeding further, it is useful to establish a terminology to describe the type of regions that correspond to nodes in the quadtree. Call a hypercubic region containing all the points the *root cell*. Define a hierarchy of cells by the following: The root cell is in the hierarchy. If a cell is in the hierarchy, then the 2^d equal-sized cubic subregions obtained by bisecting along each dimension of the cell are also called *cells* and belong to the hierarchy (see Figure 20.3 for an illustration of the cell hierarchy in two dimensions). We use the term *subcell* to describe a cell that is completely contained in another. A cell containing the subcell is called a *supercell*. The subcells obtained by bisecting a cell along each dimension are called the *immediate subcells* with respect to the bisected cell. Also, a cell is the *immediate supercell* of any of its immediate subcells. We can treat a cell as a set of all points in space contained in the cell. Thus, we use $C \subseteq D$ to indicate that the cell C is a subcell of the cell D and $C \subset D$ to indicate that C is a subcell of D but $C \neq D$. Define the *length of a cell* C, denoted *length*(C), to be the span of C along any dimension. An important property of the cell hierarchy is that, given two arbitrary cells, either one is completely contained in the other or they are disjoint. cells are considered disjoint if they are adjacent to each other and share a boundary.

Each node in a quadtree corresponds to a subcell of the root cell. Leaf nodes correspond to largest cells that contain a single point. There are as many leaf nodes as the number of points, n. The size of the quadtree cannot be bounded as a function of n, as it depends on the spatial distribution. For example, consider a data set of 3 points consisting of two points very close to each other and a faraway point located such that the first subdivision of the root cell will separate the faraway point from the other two. Then, depending on the specific location and proximity of the other two points, a number of subdivisions may be necessary to separate them. In principle, the location and proximity of the two points can be adjusted to create as large a worst-case tree as desired. In practice, this is an unlikely scenario due to limits imposed by computer precision.

From this example, it is intuitively clear that a large number of recursive subdivisions may be required to separate points that are very close to each other. In the worst case, the recursive subdivision continues until the cell sizes are so small that a single cell cannot contain both the points irrespective of their location. Subdivision is never required beyond this point, but the points may be separated sooner depending on their actual location. Let s be the smallest distance between any pair of points and D be the length of the root cell. An upper bound on the height of the quadtree is obtained by considering the worst-case path needed to separate a pair of points which have the smallest pairwise distance. The length of the smallest cell that can contain two points s apart in d dimensions is $\frac{s}{\sqrt{d}}$ (see Figure 20.4 for a two and three dimensional illustration). The paths separating the closest points may contain recursive subdivisions until a cell of length smaller than $\frac{s}{\sqrt{d}}$ is reached. Since each subdivision halves the length of the cells, the maximum path length is given by the smallest k for which $\frac{D}{2^k} < \frac{s}{\sqrt{d}}$, or $k = \left\lceil \log \frac{\sqrt{d}D}{s} \right\rceil$. For a fixed number of dimensions, the worst-case path length is $O\left(\log \frac{D}{s}\right)$. Since the tree has n leaves, the number of nodes in the tree is bounded by $O\left(n \log \frac{D}{s}\right)$. In the worst case, D is proportional to the largest distance between any pair of points. Thus, the height of a quadtree

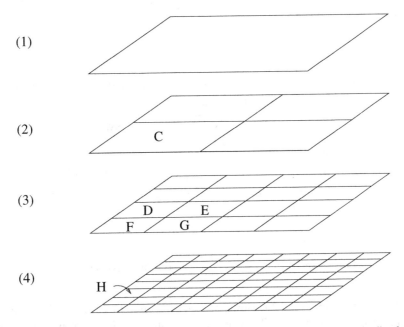

(1)

(2)

C

(3)

D   E

F   G

(4)

H

FIGURE 20.3 Illustration of hierarchy of cells in two dimensions. Cells *D, E, F* and *G* are immediate subcells of *C*. Cell *H* is an immediate subcell of *D*, and is a subcell of *C*.

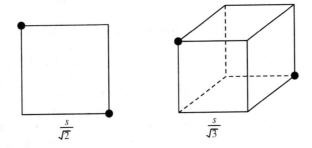

$\frac{s}{\sqrt{2}}$ $\frac{s}{\sqrt{3}}$

FIGURE 20.4 Smallest cells that could possibly contain two points that are a distance *s* apart in two and three dimensions.

is bounded by the logarithm of the ratio of the largest pairwise distance to the smallest pairwise distance. This ratio is a measure of the degree of nonuniformity of the distribution.

Search, insert and delete operations in region quadtrees are rather straightforward. To search for a point, traverse a path from root to a leaf such that each cell on the path encloses the point. If the leaf contains the point, it is in the quadtree. Otherwise, it is not. To insert a point not already in the tree, search for the point which terminates in a leaf. The leaf node corresponds to a region which originally had one point. To insert a new point which also falls within the region, the region is subdivided as many times as necessary until the two points are separated. This may create a chain of zero or more length below the leaf node followed by a branching to separate the two points. To delete a point present in the tree, conduct a search for the point which terminates in a leaf. Delete the leaf node. If deleting the node leaves its parent with a single child, traverse the path from the leaf to the root until a node with at least two children is encountered. Delete the path below the level of the child of this node. Each of the search, insert and delete operations takes $O(h)$ time, where h is the height of the tree. Construction of a quadtree can be done either through repeated insertions, or by constructing the tree level by level starting from the root. In either case, the worst case run time is $O\left(n \log \frac{D}{s}\right)$. We will not explore these algorithms further in favor of superior algorithms to be described later.

20.2.3 Compressed Quadtrees and Octrees

In an *n*-leaf tree where each internal node has at least two children, the number of nodes is bounded by $2n - 1$. The size of quadtrees is distribution dependent because there can be internal nodes with only one child. In terms of the cell hierarchy, a cell may contain all its points in a small volume so that, recursively subdividing it may result in just one of the immediate subcells containing the points for an arbitrarily large number of steps. Note that the cells represented by nodes along such a path have

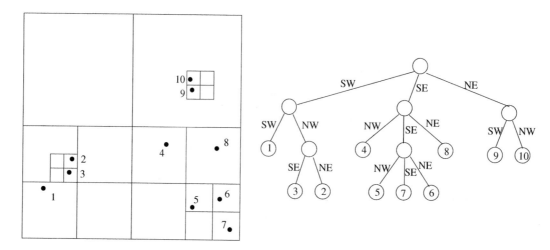

FIGURE 20.5   The two-dimensional set of points from Figure 20.2, and the corresponding compressed quadtree.

different sizes but they all enclose the same points. In many applications, all these nodes essentially contain the same information as the information depends only on the points the cell contains. This prompted the development of compressed quadtrees, which are obtained by compressing each such path into a single node. Therefore, each node in a compressed quadtree is either a leaf or has at least two children. The compressed quadtree corresponding to the quadtree of Figure 20.2 is depicted in Figure 20.5. Compressed quadtrees originated from the work of Clarkson [11] in the context of the all nearest neighbors problem and further studied by Aluru and Sevilgen [12].

A node v in the compressed quadtree is associated with two cells, *large cell of v* ($L(v)$) and *small cell of v* ($S(v)$). They are the largest and smallest cells that enclose the points in the subtree of the node, respectively. When $S(v)$ is subdivided, it results in at least two non-empty immediate subcells. For each such subcell C resulting from the subdivision, there is a child u such that $L(u) = C$. Therefore, $L(u)$ at a node u is an immediate subcell of $S(v)$ at its parent v. A node is a leaf if it contains a single point and the small cell of a leaf node is the hypothetical cell with zero length containing the point.

The size of a compressed quadtree is bounded by $O(n)$. The height of a compressed quadtree has a lower bound of $\Omega(\log n)$ and an upper bound of $O(n)$. Search, insert and delete operations on compressed quadtrees take $O(h)$ time. In practice, the height of a compressed quadtree is significantly smaller than suggested by the upper bound because (a) computer precision limits restrict the ratio of largest pairwise distance to smallest pairwise distance that can be represented, and (b) the ratio of length scales represented by a compressed quadtree of height h is at least $2^h{:}1$. In most practical applications, the height of the tree is so small that practitioners use representation schemes that allow only trees of constant height [13,14] or even assume that the height is constant in algorithm analysis [15]. For instance, a compressed octree of height 20 allows potentially $8^{20} = 2^{60}$ leaf nodes and a length scale of $2^{20} : 1 \approx 10^6 : 1$.

Though compressed quadtrees are described as resulting from collapsing chains in quadtrees, such a procedure is not intended for compressed quadtree construction. Instead, algorithms for direct construction of compressed quadtrees in $O(dn \log n)$ time will be presented, which can be used to construct quadtrees efficiently if necessary. To obtain a quadtree from its compressed version, identify each node whose small cell is not identical to its large cell and replace it by a chain of nodes corresponding to the hierarchy of cells that lead from the large cell to the small cell.

20.2.4 Cell Orderings and Space-Filling Curves

We explore a suitable one dimensional ordering of cells and use it in conjunction with spatial ordering to develop efficient algorithms for compressed quadtrees. First, define an ordering for the immediate subcells of a cell. In two dimensions, we use the order SW, NW, SE and NE. The same ordering has been used to order the children of a node in a two dimensional quadtree (Figures 20.2 and 20.5). Now consider ordering two arbitrary cells. If one of the cells is contained in the other, the subcell precedes the supercell. If the two cells are disjoint, the smallest supercell enclosing both the cells contains them in different immediate subcells of it. Order the cells according to the order of the immediate subcells containing them. This defines a total order on any collection of cells with a common root cell. It follows that the order of leaf regions in a quadtree corresponds to the left-or-right order in which the regions appear in our drawing scheme. Similarly, the ordering of all regions in a quadtree corresponds to the postorder traversal of the quadtree. These concepts naturally extend to higher dimensions. Note that any ordering of the immediate subcells of a cell can be used as foundation for cell orderings.

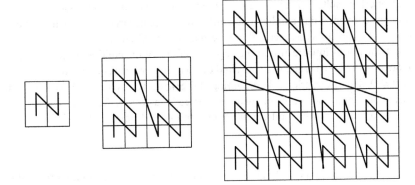

FIGURE 20.6 *Z*-curve for 2 × 2, 4 × 4 and 8 × 8 cell decompositions.

Ordering of cells at a particular resolution in the manner described above can be related to space filling curves. Space filling curves are proximity preserving mappings from a multidimensional uniform cell decomposition to a one dimensional ordering. The path implied in the multidimensional space by the linear ordering, that is, the sequence in which the multidimensional cells are visited according to the linear ordering, forms a non-intersecting curve. Of particular interest is Morton ordering, also known as the *Z*-space filling curve [16]. The *Z*-curves for 2 × 2, 4 × 4 and 8 × 8 cell decompositions are shown in Figure 20.6. Consider a square two dimensional region and its $2^k \times 2^k$ cell decomposition. The curve is considered to originate in the lower left corner and terminate in the upper right corner. The curve for a $2^k \times 2^k$ grid is composed of four $2^{k-1} \times 2^{k-1}$ grid curves one in each quadrant of the $2^k \times 2^k$ grid and the tail of one curve is connected to the head of the next as shown in the figure. The order in which the curves are connected is the same as the order of traversal of the 2 × 2 curve. Note that Morton ordering of cells is consistent with the cell ordering specified above. Other space filling curves orderings such as graycode [17] and Hilbert [18] curve can be used and quadtree ordering schemes consistent with these can also be utilized. We will continue to utilize the *Z*-curve ordering as it permits a simpler bit interleaving scheme which will be presented and exploited later.

Algorithms on compressed quadtree rely on the following operation due to Clarkson [11]:

Lemma 20.1

Let R be the product of d intervals $I_1 \times I_2 \times \cdots \times I_d$, that is, R is a hyperrectangular region in d dimensional space. The smallest cell containing R can be found in $O(d)$ time, which is constant for any fixed d.

The procedure for computing the smallest cell uses floor, logarithm and bitwise exclusive-or operations. An extended RAM model is assumed in which these are considered constant time operations. The reader interested in proof of Lemma 20.1 is referred to [11,19]. The operation is useful in several ways. For example, the order in which two points appear in the quadtree as per our ordering scheme is independent of the location of other points. To determine the order for two points, say $(x_1, x_2, \ldots, x_d)$ and $(y_1, y_2, \ldots, y_d)$, find the smallest cell that contains $[x_1, y_1] \times [x_2, y_2] \times \cdots \times [x_d, y_d]$. The points can then be ordered according to its immediate subcells that contain the respective points. Similarly, the smallest cell containing a pair of other cells, or a point and a cell, can be determined in $O(d)$ time.

20.2.5 Construction of Compressed Quadtrees

20.2.5.1 A Divide-and-Conquer Construction Algorithm

Let T_1 and T_2 be two compressed quadtrees representing two distinct sets S_1 and S_2 of points. Let r_1 (respectively, r_2) be the root node of T_1 (respectively, T_2). Suppose that $L(r_1) = L(r_2)$, that is, both T_1 and T_2 are constructed starting from a cell large enough to contain $S_1 \cup S_2$. A compressed quadtree T for $S_1 \cup S_2$ can be constructed in $O(|S_1| + |S_2|)$ time by merging T_1 and T_2.

To merge T_1 and T_2, start at their roots and merge the two trees recursively. Suppose that at some stage during the execution of the algorithm, node u in T_1 and node v in T_2 are being considered. An invariant of the merging algorithm is that $L(u)$ and $L(v)$ cannot be disjoint. Furthermore, it can be asserted that $S(u) \cup S(v) \subseteq L(u) \cap L(v)$. In merging two nodes, only the small cell information is relevant because the rest of the large cell $(L(v) - S(v))$ is empty. For convenience, assume that a node may be empty. If a node has less than 2^d children, we may assume empty nodes in place of the absent children. Four distinct cases arise:

- *Case I:* If a node is empty, the result of merging is simply the tree rooted at the other node.
- *Case II:* If $S(u) = S(v)$, the corresponding children of u and v have the same large cells which are the the immediate subcells of $S(u)$ (or equivalently, $S(v)$). In this case, merge the corresponding children one by one.
- *Case III:* If $S(v) \subset S(u)$, the points in v are also in the large cell of one of the children of u. Thus, this child of u is merged with v. Similarly, if $S(u) \subset S(v)$, u is merged with the child of v whose large cell contains $S(u)$.
- *Case IV:* If $S(u) \cap S(v) = \emptyset$, then the smallest cell containing both $S(u)$ and $S(v)$ contains $S(u)$ and $S(v)$ in different immediate subcells. In this case, create a new node in the merged tree with its small cell as the smallest cell that contains $S(u)$ and $S(v)$ and make u and v the two children of this node. Subtrees of u and v are disjoint and need not be merged.

Lemma 20.2

Two compressed quadtrees with a common root cell can be merged in time proportional to the sum of their sizes.

Proof. The merging algorithm presented above performs a preorder traversal of each compressed quadtree. The whole tree may not need to be traversed because in merging a node, it may be determined that the whole subtree under the node directly becomes a subtree of the resulting tree. In every step of the merging algorithm, we advance on one of the trees after performing at most $O(d)$ work. Thus, the run time is proportional to the sum of the sizes of the trees to be merged. ∎

To construct a compressed quadtree for n points, scan the points and find the smallest and largest coordinate along each dimension. Find a region that contains all the points and use this as the root cell of every compressed quadtree constructed in the process. Recursively construct compressed quadtrees for $\lfloor \frac{n}{2} \rfloor$ points and the remaining $\lceil \frac{n}{2} \rceil$ points and merge them in $O(dn)$ time. The compressed quadtree for a single point is a single node v with the root cell as $L(v)$. The run time satisfies the recurrence

$$T(n) = T\left(\left\lfloor \frac{n}{2} \right\rfloor\right) + T\left(\left\lceil \frac{n}{2} \right\rceil\right) + O(dn)$$

resulting in $O(dn \log n)$ run time.

20.2.5.2 Bottom-up Construction

To perform a bottom-up construction, first compute the order of the points in $O(dn \log n)$ time using any optimal sorting algorithm and the ordering scheme described previously. The compressed quadtree is then incrementally constructed starting from the single node tree for the first point and inserting the remaining points as per the sorted list. During the insertion process, keep track of the most recently inserted leaf. Let p be the next point to be inserted. Starting from the most recently inserted leaf, traverse the path from the leaf to the root until the first node v such that $p \in L(v)$ is encountered. Two possibilities arise:

- *Case I:* If $p \notin S(v)$, then p is in the region $L(v) - S(v)$, which was empty previously. The smallest cell containing p and $S(v)$ is a subcell of $L(v)$ and contains p and $S(v)$ in different immediate subcells. Create a new node u between v and its parent and insert p as a child of u.
- *Case II:* If $p \in S(v)$, v is not a leaf node. The compressed quadtree presently does not contain a node that corresponds to the immediate subcell of $S(v)$ that contains p, that is, this immediate subcell does not contain any of the points previously inserted. Therefore, it is enough to insert p as a child of v corresponding to this subcell.

Once the points are sorted, the rest of the algorithm is identical to a post-order walk on the final compressed quadtree with $O(d)$ work per node. The number of nodes visited per insertion is not bounded by a constant but the number of nodes visited over all insertions is $O(n)$, giving $O(dn)$ run time. Combined with the initial sorting of the points, the tree can be constructed in $O(dn \log n)$ time.

20.2.6 Basic Operations

Fast algorithms for operations on quadtrees can be designed by simultaneously keeping track of spatial ordering and one dimensional ordering of cells in the compressed quadtree. The spatial ordering is given by the compressed quadtree itself. In addition, a balanced binary search tree (BBST) is maintained on the large cells of the nodes to enable fast cell searches. Both the trees consist of the same nodes and this can be achieved by allowing each node to have pointers corresponding to compressed quadtree structure and pointers corresponding to BBST structure.

20.2.6.1 Point and Cell Queries

Point and cell queries are similar since a point can be considered to be a zero length cell. A node v is considered to represent cell C if $S(v) \subseteq C \subseteq L(v)$. The node in the compressed quadtree representing the given cell is located using the BBST. Traverse the path in the BBST from the root to the node that is being searched in the following manner: To decide which child to visit next on the path, compare the query cell with the large and small cells at the node. If the query cell precedes the small cell in cell ordering, continue the search with the left child. If it succeeds the large cell in cell ordering, continue with the right child. If it lies between the small cell and large cell in cell ordering, the node represents the query cell. As the height of a BBST is $O(\log n)$, the time taken for a point or cell query is $O(d \log n)$.

20.2.6.2 Insertions and Deletions

As points can be treated as cells of zero length, insertion and deletion algorithms will be discussed in the context of cells. These operations are meaningful only if a cell is inserted as a leaf node or deleted if it is a leaf node. Note that a cell cannot be deleted unless all its subcells are previously deleted from the compressed quadtree.

20.2.6.3 Cell Insertion

To insert a given cell C, first check whether it is represented in the compressed quadtree. If not, it should be inserted as a leaf node. Create a node v with $S(v) = C$ and first insert v in the BBST using a standard binary search tree insertion algorithm. To insert v in the compressed quadtree, first find the BBST successor of v, say u. Find the smallest cell D containing C and the $S(u)$. Search for cell D in the BBST and identify the corresponding node w. If w is not a leaf, insert v as a child of w in compressed quadtree. If w is a leaf, create a new node w' such that $S(w') = D$. Nodes w and v become the children of w' in the compressed quadtree. The new node w' should be inserted in the BBST. The overall algorithm requires a constant number of insertions and searches in the BBST, and takes $O(d \log n)$ time.

20.2.6.4 Cell Deletion

As in insertion, the cell should be deleted from the BBST and the compressed quadtree. To delete the cell from BBST, the standard deletion algorithm is used. During the execution of this algorithm, the node representing the cell is found. The node is deleted from the BBST only if it is present as a leaf node in the compressed quadtree. If the removal of this node from the compressed quadtree leaves its parent with only one child, the parent is deleted as well. Since each internal node has at least two children, the delete operation cannot propagate to higher levels in the compressed quadtree.

20.2.7 Practical Considerations

In most practical applications, the height of a region quadtree is rather small because the spatial resolution provided by a quadtree is exponential in its height. This can be used to design schemes that will greatly simplify and speedup quadtree algorithms.

Consider numbering the 2^{2k} cells of a $2^k \times 2^k$ two dimensional cell decomposition in the order specified by the Z-curve using integers $0 \ldots 4^k - 1$ (Figure 20.6). Represent each cell in the cell space using a coordinate system with k bits for each coordinate. From the definition of the Z-curve, it follows that the number of a cell can be obtained by interleaving the bits representing the x and y coordinates of the cell, starting from the x coordinate. For example, $(3, 5) = (011, 101)$ translates to $011011 = 27$. The procedure can be extended to higher dimensions. If $(x_1, x_2, \ldots x_d)$ represents the location of a d dimensional cell, the corresponding number can be obtained by taking bit representations of $x_1, x_2, \ldots x_d$, and interleaving them.

The same procedure can be described using uncompressed region quadtrees. Label the 2^d edges connecting a node to its children with bit strings of length d. This is equivalent to describing the immediate subcells of a cell using one bit for each coordinate followed by bit interleaving. A cell can then be described using concatenation of the bits along the path from the root to the cell. This mechanism can be used to simultaneously describe cells at various length scales. The bit strings are conveniently stored in groups of 32 or 64 bits using integer data types. This creates a problem in distinguishing certain cells. For instance, consider distinguishing cell "000" from cell "000000" in an octree. To ensure each bit string translates to a unique integer when interpreted as a binary number, it is prefixed by a 1. It also helps in easy determination of the length of a bit string by locating the position of the leftmost 1. This leads to a representation that requires $dk + 1$ bits for a d dimensional quadtree with a height of at most k. Such a representation is very beneficial because primitive operations on cells can be implemented using bit operations such as and, or, exclusive-or etc. For example, 128 bits (4 integers on a 32 bit computer and 2 integers on a 64 bit computer) are sufficient to describe an octree of height 42, which allows representation of length scales $2^{42} : 1 > 4 \times 10^{12} : 1$.

The bit string based cell representation greatly simplifies primitive cell operations. In the following, the name of a cell is also used to refer to its bit representation:

- *Check if* $C_1 \subseteq C_2$. If C_2 is a prefix of C_1, then C_1 is contained in C_2, otherwise not.
- *Find the smallest cell enclosing* C_1 *and* C_2. This is obtained by finding the longest common prefix of C_1 and C_2 whose length is $1 \bmod d$.

- *Find the immediate subcell of C_1 that contains C_2.* If $dl + 1$ is the number of bits representing C_1, the required immediate subcell is given by the first $(d + 1)l + 1$ bits of C_2.

Consider n points in a root cell and let k denote the largest resolution to be used. Cells are not subdivided further even if they contain multiple points. From the coordinates of a point, it is easy to compute the leaf cell containing it. Because of the encoding scheme used, if cell C should precede cell D in the cell ordering, the number corresponding to the binary interpretation of the bit string representation of C is smaller than the corresponding number for D. Thus, cells can be sorted by simply treating them as numbers and ordering them accordingly.

Finally, binary search trees and the attendant operations on them can be completely avoided by using hashing to directly access a cell. An n leaf compressed quadtree has at most $2n - 1$ nodes. Hence, an array of that size can be conveniently used for hashing. If all cells at the highest resolution are contained in the quadtree, that is, $n = d^k$, then an array of size $2n - 1$ can be used to directly index cells. Further details of such representation are left as an exercise to the reader.

20.3 Spatial Queries with Region Quadtrees

In this section, we consider a number of spatial queries involving point data, and algorithms for them using compressed region quadtrees.

20.3.1 Range Query

Range queries are commonly used in database applications. Database records with d keys can be represented as points in d-dimensional space. In a range query, ranges of values are specified for all or a subset of keys with the objective of retrieving all the records that satisfy the range criteria. Under the mapping of records to points in multidimensional space, the ranges define a (possibly open-ended) hyperrectangular region. The objective is to retrieve all the points that lie in this query region.

As region quadtrees organize points using a hierarchy of cells, the range query can be answered by finding a collection C of cells that are both fully contained in the query region and completely encompass the points in it. This can be achieved by a top-down traversal of the compressed region quadtree starting from the root. To begin with, C is empty. Consider a node v and its small cell $S(v)$. If $S(v)$ is outside the query region, the subtree underneath it can be safely discarded. If $S(v)$ is completely inside the query region, it is added to C. All points in the subtree of $S(v)$ are within the query region and reported as part of the output. If $S(v)$ overlaps with the query region but is not contained in it, each child of v is examined in turn.

If the query region is small compared to the size of the root cell, it is likely that a path from the root is traversed where each cell on the path completely contains the query region. To avoid this problem, first compute the smallest cell encompassing the query region. This cell can be searched in $O(d \log n)$ time using the cell search algorithm described before, or perhaps in $O(d)$ time if the hashing/indexing technique is applicable. The top-down traversal can start from the identified cell. When the cell is subdivided, at least two of its children represent cells that overlap with the query region. Consider a child cell and the part of the query region that is contained in it. The same idea can be recursively applied by finding the smallest cell that encloses this part of the query region and directly finding this cell rather than walking down a path in the tree to reach there. This ensures that the number of cells examined during the algorithm is $O(|C|)$. To see why, consider the cells examined as organized into a tree based on subcell-supercell relationships. The leaves of the tree are the collection of cells C. Each cell in the tree is the smallest cell that encloses a subregion of the query region. Therefore, each internal node has at least two children. Consequently, the size of the tree, or the number of cells examined in answering the range query is $O(|C|)$. For further study of range queries and related topics, see [17, 20, 21].

Next, we turn our attention to a number of spatial queries which we categorize as group queries. In group queries, a query to retrieve points that bear a certain relation to a query point is specified. The objective of the group query is to simultaneously answer n queries with each input point treated as a query point. For example, given n points, finding the nearest neighbor of each point is a group query. While the run time of performing the query on an individual point may be large, group queries can be answered more efficiently by intelligently combining the work required in answering queries for the individual points. Instead of presenting a different algorithm for each query, we show that the same generic framework can be used to solve a number of such queries. The central idea behind the group query algorithm is to realize run time savings by processing queries together for nearby points. Consider a cell C in the compressed quadtree and the (as yet uncomputed) set of points that result from answering the query for each point in C. The algorithm keeps track of a collection of cells of size as close to C as possible that is guaranteed to contain these points. The algorithm proceeds in a hierarchical manner by computing this information for cells of decreasing sizes (see Figure 20.7).

A node u is said to be *resolved* with respect to node v if, either all points in $S(u)$ are in the result of the query for all points in $S(v)$ or none of the points in $S(u)$ is in the result of the query for any point in $S(v)$. Define the *active set* of a node v to be the set of

Algorithm 1

Group-query (*v*)

> P = active set at v's parent
> A = active set at $v = \emptyset$
> While $P \neq \emptyset$ do
>
> > $u = Select\ (P)$
> > $P = P - \{u\}$
> > decision = $Status\ (v, u)$
> > If decision = PROCESS
> >
> > > $Process\ (v, u)$
> >
> > If decision = UNKNOWN
> >
> > > If $S(u) \subseteq S(v)$
> > >
> > > > $A = A \cup \{u\}$
> > >
> > > Else $P = P \cup children(u)$
>
> For each child u of v
>
> > **Group-query (*u*)**

FIGURE 20.7 Unified algorithm for the group queries.

nodes u that cannot be resolved with respect to v such that $S(u) \subseteq S(v) \subseteq L(u)$. The algorithm uses a depth first search traversal of the compressed quadtree. The active set of the root node contains itself. The active set of a node v is calculated by traversing portions of the subtrees rooted at the nodes in the active set of its parent. The functions *Select, Status* and *Process* used in the algorithm are designed based on the specific group query. When considering the status of a node u with respect to the node v, the function Status(v,u) returns one of the following three values:

- PROCESS – If $S(u)$ is in the result of the query for all points in $S(v)$
- DISCARD – If $S(u)$ is not in the result of the query for all points in $S(v)$
- UNKNOWN – If neither of the above is true

If the result is either PROCESS or DISCARD, the children of u are not explored. Otherwise, the size of $S(u)$ is compared with the size of $S(v)$. If $S(u) \subseteq S(v)$, then u is added to the set of active nodes at v for consideration by v's children. If $S(u)$ is larger, then u's children are considered with respect to v. It follows that the active set of a leaf node is empty, as the length of its small cell is zero. Therefore, entire subtrees rooted under the active set of the parent of a leaf node are explored and the operation is completed for the point inhabiting that leaf node. The function *Process(v,u)* reports all points in $S(u)$ as part of the result of query for each point in $S(v)$.

The order in which nodes are considered is important for proving the run time of some operations. The function *Select* is used to accommodate this.

20.3.2 Spherical Region Queries

Given a query point and a distance $r > 0$, the spherical region query is to find all points that lie within a distance of r from the query point. The group version of the spherical region query is to take n points and a distance $r > 0$ as input, and answer spherical region queries with respect to each of the input points.

A cell D may contain points in the spherical region corresponding to some points in cell C only if the smallest distance between D and C is less than r. If the largest distance between D and C is less than r, then all points in D are in the spherical region of every point in C. Thus, the function $Status(v,u)$ is defined as follows: If the largest distance between $S(u)$ and $S(v)$ is less than r, return PROCESS. If the smallest distance between $S(u)$ and $S(v)$ is greater than r, return DISCARD. Otherwise, return UNKNOWN.

Processing u means including all the points in u in the query result for each point in v. For this query, no special selection strategy is needed.

20.3.3 *k*-Nearest Neighbors

For computing the k-nearest neighbors of each point, some modifications to the algorithm presented in Figure 20.7 are necessary. For each node w in P, the algorithm keeps track of the largest distance between $S(v)$ and $S(w)$. Let d_k be the kth smallest of these distances. If the number of nodes in P is less than k, then d_k is set to ∞. The function $Status(v,u)$ returns DISCARD if the smallest distance between $S(v)$ and $S(u)$ is greater than d_k. The option PROCESS is never used. Instead, for a leaf node, all the points in the nodes in its active set are examined to select the k nearest neighbors. The function Select picks the largest cell in P, breaking ties arbitrarily.

Computing k-nearest neighbors is a well-studied problem [11,22]. The algorithm presented here is equivalent to Vaidya's algorithm [22,23], even though the algorithms appear to be very different on the surface. Though Vaidya does not consider compressed quadtrees, the computations performed by his algorithm can be related to traversal on compressed quadtrees and a proof of the run time of the presented algorithm can be established by a correspondence with Vaidya's algorithm. The algorithm runs in $O(kn)$ time. The proof is quite elaborate, and omitted for lack of space. For details, see [22,23]. The special case of $n = 1$ is called the all nearest neighbor query, which can be computed in $O(n)$ time.

20.4 Image Processing Applications

Quadtrees are ubiquitously used in image processing applications. Consider a two dimensional square array of pixels representing a binary image with black foreground and white background (Figure 20.8). As with region quadtrees used to represent point data, a hierarchical cell decomposition is used to represent the image. The pixels are the smallest cells used in the decomposition and the entire image is the root cell. The root cell is decomposed into its four immediate subcells and represented by the four children of the root node. If a subcell is completely composed of black pixels or white pixels, it becomes a leaf in the quadtree. Otherwise, it is recursively decomposed. If the resolution of the image is $2^k \times 2^k$, the height of the resulting region quadtree is at most k. A two dimensional image and its corresponding region quadtree are shown in Figure 20.8. In drawing the quadtree, the same ordering of the immediate subcells of a cell into SW, NW, SE and NE quadrants is followed. Each node in the tree is colored black, white, or gray, depending on if the cell consists of all black pixels, all white pixels, or a mixture of both black and white pixels, respectively. Thus, internal nodes are colored gray and leaf nodes are colored either black or white.

Each internal node in the image quadtree has four children. For an image with n leaf nodes, the number of internal nodes is $\frac{(n-1)}{3}$. For large images, the space required for storing the internal nodes and the associated pointers may be expensive and several space-efficient storage schemes have been investigated. These include storing the quadtree as a collection of leaf cells using the Morton numbering of cells [24], or as a collection of black leaf cells only [25,26], and storing the quadtree as its preorder traversal [27]. Iyengar *et al.* introduced a number of space-efficient representations of image quadtrees including forests of quadtrees [28,29], translation invariant data structures [30–32] and virtual quadtrees [33].

The use of quadtrees can be easily extended to grayscale images and color images. Significant space savings can be realized by choosing an appropriate scheme. For example, 2^r gray levels can be encoded using r bits. The image can be represented using r binary valued quadtrees. Because adjacent pixels are likely to have gray levels that are closer, a gray encoding of the 2^r levels is

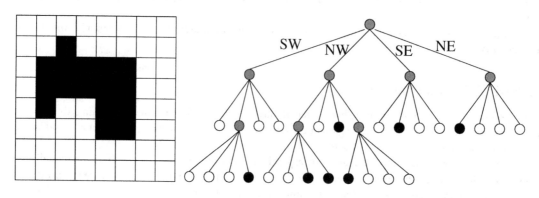

FIGURE 20.8   A two dimensional array of pixels, and the corresponding region quadtree.

advantageous over a binary encoding [27]. The gray encoding has the property that adjacent levels differ by one bit in the gray code representation. This should lead to larger blocks having the same value for a given bit position, leading to shallow trees.

20.4.1 Construction of Image Quadtrees

Region quadtrees for image data can be constructed using an algorithm similar to the bottom-up construction algorithm for point region quadtrees described in Section 20.2.5. In fact, constructing quadtrees for image data is easier because the smallest cells in the hierarchical decomposition are given by the pixels and all the pixels can be read by the algorithm as the image size is proportional to the number of pixels. Thus, quadtree for region data can be built in time linear in the size of the image, which is optimal. The pixels of the image are scanned in Morton order which eliminates the need for the sorting step in the bottom-up construction algorithm. It is wasteful to build the complete quadtree with all pixels as leaf nodes and then compact the tree by replacing maximal subtrees having all black or all white pixels with single leaf nodes, though such a method would still run in linear time and space. By reading the pixels in Morton order, a maximal black or white cell can be readily identified and the tree can be constructed using only such maximal cells as leaf nodes [8]. This will limit the space used by the algorithm to the final size of the quadtree.

20.4.2 Union and Intersection of Images

The union of two images is the overlay of one image over another. In terms of the array of pixels, a pixel in the union is black if the pixel is black in at least one of the images. In the region quadtree representation of images, the quadtree corresponding to the union of two images should be computed from the quadtrees of the constituent images. Let I_1 and I_2 denote the two images and T_1 and T_2 denote the corresponding region quadtrees. Let T denote the region quadtree of the union of I_1 and I_2. It is computed by a preorder traversal of T_1 and T_2 and examining the corresponding nodes/cells. Let v_1 in T_1 and v_2 in T_2 be nodes corresponding to the same region. There are three possible cases:

- Case I: If v_1 or v_2 is black, the corresponding node is created in T and is colored black. If only one of them is black and the other is gray, the gray node will contain a subtree underneath. This subtree need not be traversed.
- Case II: If v_1 (respectively, v_2) is white, v_2 (respectively, v_1) and the subtree underneath it (if any) is copied to T.
- Case III: If both v_1 and v_2 are gray, then the corresponding children of v_1 and v_2 are considered.

The tree resulting from the above merging algorithm may consist of unnecessary subdivisions of a cell consisting of completely black pixels. For example, if a region is marked gray in both T_1 and T_2 but each of the four quadrants of the region is black in at least one of T_1 and T_2, then the node corresponding to the region in T will have four children colored black. This is unnecessary as the node itself can be colored black and should be considered a leaf node. Such adjustments need not be local and may percolate up the tree. For instance, consider a checker board image of $2^k \times 2^k$ pixels with half black and half white pixels such that the north, south, east and west neighbors of a black pixel are white pixels and vice versa. Consider another checker board image with black and white interchanged. The quadtree for each of these images is a full 4-ary tree of level k. But overlaying one image over another produces one black square of size $2^k \times 2^k$. The corresponding quadtree should be a single black node. However, merging initially creates a full 4-ary tree of depth k. To compact the tree resulting from merging to create the correct region quadtree, a bottom-up traversal is performed. If all children of a node are black, then the children are removed and the node is colored black.

The intersection of two images can be computed similarly. A pixel in the intersection of two images is black only if the corresponding pixels in both the images are black. For instance, the intersection of the two complementary checkerboard images described above is a single white cell of size $2^k \times 2^k$. The algorithm for intersection can be obtained by interchanging the roles of black and white in the union algorithm. Similarly, a bottom-up traversal is used to detect nodes all of whose children are white, remove the children, and color the node white. Union and intersection algorithms can be easily generalized to multiple images.

20.4.3 Rotation and Scaling

Quadtree representation of images facilitates certain types of rotation and scaling operations. Rotations on images are often performed in multiples of 90 degrees. Such a rotation has the effect of changing the quadrants of a region. For example, a 90 degree clockwise rotation changes the SW, NW, SE, NE quadrants to NW, NE, SW, SE quadrants, respectively. This change should be recursively carried out for each region. Hence, a rotation that is a multiple of 90 degrees can be effected by simply reordering the child pointers of each node. Similarly, scaling by a power of two is trivial using quadtrees since it is simply a loss of resolution. For other linear transformations, see the work of Hunter [34,35].

20.4.4 Connected Component Labeling

Two black pixels of a binary image are considered adjacent if they share a horizontal or vertical boundary. A pair of black pixels is said to be connected if there is a sequence of adjacent pixels leading from one to the other. A connected component is a maximal set of black pixels where every pair of pixels is connected. The connected component labeling problem is to find the connected components of a given image and give each connected component a unique label. Identifying the connected components of an image is useful in object counting and image understanding. Samet developed algorithms for connected component labeling using quadtrees [36].

Let B and W denote the number of black nodes and white nodes, respectively, in the quadtree. Samet's connected component labeling algorithm works in three stages:

1. Establish the adjacency relationships between pairs of black pixels.
2. Identify a unique label for each connected component. This can be thought of as computing the transitive closure of the adjacency relationships.
3. Label each black cell with the corresponding connected component.

The first stage is carried out using a postorder traversal of the quadtree during which adjacent black pixels are discovered and identified by giving them the same label. To begin with, all the black nodes are unlabeled. When a black region is considered, black pixels adjacent to it in two of the four directions, say north and east, are searched. The post order traversal of the tree traverses the regions in Morton order. Thus when a region is encountered, its south and west adjacent black regions (if any) would have already been processed. At that time, the region would have been identified as a neighbor. Thus, it is not necessary to detect adjacency relationships between a region and its south or west adjacent neighbors.

Suppose the postorder traversal is currently at a black node or black region R. Adjacent black regions in a particular direction, say north, are identified as follows. First, identify the adjacent region of the same size as R that lies to the north of R. Traverse the quadtree upwards to the root to identify the first node on the path that contains both the regions, or equivalently, find the lowest common ancestor of both the regions in the quadtree. If such a node does not exist, R has no north neighbor and it is at the boundary of the image. Otherwise, a child of this lowest common ancestor will be a region adjacent to the north boundary of R and of the same size as or larger than R. If this region is black, it is part of the same connected component as R. If neither the region nor R is currently labeled, create a new label and use it for both. If only one of them is labeled, use the same label for the other. If they are both already labeled with the same label, do nothing. If they are both already labeled with different labels, the label pair is recorded as corresponding to an adjacency relationship. If the region is gray, it is recursively subdivided and adjacent neighbors examined to find all black regions that are adjacent to R. For each black neighbor, the labeling procedure described above is carried out.

The run time of this stage of the algorithm depends on the time required to find the north and east neighbors. This process can be improved using the smallest-cell primitive described earlier. However, Samet proved that the average of the maximum number of nodes visited by the above algorithm for a given black region and a given direction is smaller than or equal to 5, under the assumption that a black region is equally likely to occur at any position and level in the quadtree [36]. Thus, the algorithm should take time proportional to the size of the quadtree, that is, $O(W + B)$ time.

Stage two of the algorithm is best performed using the union-find data structure [37]. Consider all the labels that are used in stage one and treat them as singleton sets. For every pair of labels recorded in stage one, perform a find operation to find the two sets that contain the labels currently, and do a union operation on the sets. At the end of stage two, all labels used for the same connected component will be grouped together in one set. This can be used to provide a unique label for each set (called the set label), and subsequently identify the set label for each label. The number of labels used is bounded by B. The amortized run time per operation using the union-find data structure is given by the inverse Ackermann's function [37], a constant (≤ 4) for all practical purposes. Therefore, the run time of stage two can be considered to be $O(B)$. Stage three of the algorithm can be carried out by another postorder traversal of the quadtree to replace the label of each black node with the corresponding set label in $O(B + W)$ time. Thus, the entire algorithm runs in $O(B + W)$ time.

20.5 Scientific Computing Applications

Region quadtrees are widely used in scientific computing applications. Most of the problems are three dimensional, prompting the development of octree based methods. In fact, some of the practical techniques and shortcuts presented earlier for storing region quadtrees and bit string representation of cells owe their origin to applications in scientific computing.

Octree are used in many ways in scientific computing applications. In many scientific computing applications, the behavior of a physical system is captured through either (1) a discretization of the space using a finite grid, followed by determining the quantities of interest at each of the grid cells, or (2) computing the interactions between a system of (real or virtual) particles which can be related to the behavior of the system. The former are known as grid-based methods and typically correspond to the solution of differential equations. The latter are known as particle-based methods and often correspond to the solution of integral equations. Both methods are typically iterative and the spatial information, such as the hierarchy of grid cells or the spatial distribution of the particles, often changes from iteration to iteration.

Examples of grid-based methods include finite element methods, finite difference methods, multigrid methods and adaptive mesh refinement methods. Many applications use a cell decomposition of space as the grid or grid hierarchy and the relevance of octrees for such applications is immediate. Algorithms for construction of octrees and methods for cell insertions and deletions are all directly applicable to such problems. It is also quite common to use other decompositions, especially for the finite-element method. For example, a decomposition of a surface into triangular elements is often used. In such cases, each basic element can be associated with a point (for example, the centroid of a triangle), and an octree can be built using the set of points. Information on the neighboring elements required by the application is then related to neighborhood information on the set of points.

Particle based methods include such simulation techniques as molecular dynamics, smoothed particle hydrodynamics, and N-body simulations. These methods require many of the spatial queries discussed in this chapter. For example, van der Waal forces between atoms in molecular dynamics fall off so rapidly with increasing distance (inversely proportional to the sixth power of the distance between atoms) that a cutoff radius is employed. In computing the forces acting on an individual atom, van der Waal forces are taken into account only for atoms that lie within the cutoff radius. The list of such atoms is typically referred to as the neighbor list. Computing neighbor lists for all atoms is just a group spherical region query, already discussed before. Similarly, k-nearest neighbors are useful in applications such as smoothed particle hydrodynamics. In this section, we further explore the use of octrees in scientific computing by presenting an optimal algorithm for the N-body problem.

20.5.1 The N-Body Problem

The N-body problem is defined as follows: Given n bodies and their positions, where each pair of bodies interact with a force inversely proportional to the square of the distance between them, compute the force on each body due to all other bodies. A direct algorithm for computing all pairwise interactions requires $O(n^2)$ time. Greengard's fast multipole method [15], which uses an octree data structure, reduces this complexity by approximating the interaction between clusters of bodies instead of computing individual interactions. For each cell in the octree, the algorithm computes a multipole expansion and a local expansion. The multipole expansion at a cell C, denoted $\phi(C)$, is the effect of the bodies within C on distant bodies. The local expansion at C, denoted $\psi(C)$, is the effect of all distant bodies on bodies within C.

The N-body problem can be solved in $O(n)$ time using a compressed octree [12]. For two cells C and D which are not necessarily of the same size, define a predicate *well-separated*(C, D) to be true if D's multipole expansion converges at any point in C, and false otherwise. If two cells are not well-separated, they are *proximate*. Similarly, two nodes v_1 and v_2 in the compressed octree are said to be *well-separated* if and only if $S(v_1)$ and $S(v_2)$ are well-separated. Otherwise, we say that v_1 and v_2 are proximate.

For each node v in the compressed octree, the multipole expansion $\phi(v)$ and the local expansion $\psi(v)$ need to be computed. Both $\phi(v)$ and $\psi(v)$ are with respect to the cell $S(v)$. The multipole expansions can be computed by a simple bottom-up traversal in $O(n)$ time. For a leaf node, its multipole expansion is computed directly. At an internal node v, $\phi(v)$ is computed by aggregating the multipole expansions of the children of v.

The algorithm to compute the local expansions is given in Figure 20.9. The astute reader will notice that this is the same group query algorithm described in Section 20.3, reproduced here again with slightly different notation for convenience. The computations are done using a top-down traversal of the tree. To compute local expansion at node v, consider the set of nodes that are proximate to its parent, which is the *proximity set*, $P(parent(v))$. The proximity set of the root node contains only itself. Recursively decompose these nodes until each node is either (1) well-separated from v or (2) proximate to v and the length of the small cell of the node is no greater than the length of $S(v)$. The nodes satisfying the first condition form the *interaction set* of v, $I(v)$ and the nodes satisfying the second condition are in the proximity set of v, $P(v)$. In the algorithm, the set $E(v)$ contains the nodes that are yet to be processed. Local expansions are computed by combining parent's local expansion and the multipole expansions of the nodes in $I(v)$. For the leaf nodes, potential calculation is completed by using the direct method.

The following terminology is used in analyzing the run time of the algorithm. The set of cells that are proximate to C and having same length as C is called *the proximity set of C* and is defined by $P^=(C) = \{D \mid length(C) = length(D), \neg well\text{-}separated(C, D)\}$. The superscript "$=$" is used to indicate that cells of the same length are being considered. For node v, define the *proximity set $P(v)$* as the set of all nodes proximate to v and having the small cell no greater than and large cell no smaller than $S(v)$. More precisely, $P(v) = \{w \mid \neg well\text{-}separated(S(v), S(w)), length(S(w)) \le length(S(v)) \le length(L(w))\}$. The *interaction set $I(v)$* of v is defined as $I(v) = \{w \mid well\text{-}separated(S(v), S(w)), [w \in P(parent(v)) \vee \{\exists u \in P(parent(v)), w \text{ is a descendant of } u, \neg well\text{-}sep(v, parent(w)), length(S(v)) < length(S(parent(w)))\}]\}$. We use $parent(w)$ to denote the parent of the node w.

The algorithm achieves $O(n)$ run time for any predicate *well-separated* that satisfies the following three conditions:

Algorithm 2

Compute-Local-Exp (v)

I. Find the proximity set $P(v)$ and interaction set $I(v)$ for v

$E(v) = P(parent(v))$
$I(v) = \emptyset; P(v) = \emptyset$
While $E(v) \neq \emptyset$ do

> Pick some $u \in E(v)$
> $E(v) = E(v) - \{u\}$
> If *well-separated*$(S(v), S(u))$
>
> > $I(v) = I(v) \cup \{u\}$
>
> Else if $S(u) \subseteq S(v)$
>
> > $P(v) = P(v) \cup \{u\}$
>
> Else $E(v) = E(v) \cup children(u)$

II. Calculate the local expansion at v

> Assign shifted $\psi(parent(v))$ to $\psi(v)$
> For each node $u \in I(v)$
>
> > Add shifted $\phi(u)$ to $\psi(v)$

III. Calculate the local expansions at the children of v with recursive calls

> For each child u of v
>
> > **Compute-Local-Exp** (u)

FIGURE 20.9   Algorithm for calculating local expansions of all nodes in the tree rooted at v.

---

1. The relation *well-separated* is symmetric for equal length cells, that is, $length(C) = length(D) \Rightarrow$ *well-separated*$(C, D) =$ *well-separated*(D, C).
2. For any cell C, $|P^=(C)|$ is bounded by a constant.
3. If two cells C and D are not well-separated, any two cells C' and D' such that $C \subseteq C'$ and $D \subseteq D'$ are not well-separated as well.

These three conditions are respected by the various well-separatedness criteria used in N-body algorithms and in particular, Greengard's algorithm. In N-body methods, the well-separatedness decision is solely based on the geometry of the cells and their relative distance and is oblivious to the number of bodies or their distribution within the cells. Given two cells C and D of the same length, if D can be approximated with respect to C, then C can be approximated with respect to D as well, as stipulated by Condition C1. The size of the proximity sets of cells of the same length should be $O(1)$ as prescribed by Condition C2 in order that an $O(n)$ algorithm is possible. Otherwise, an input that requires processing the proximity sets of $\Omega(n)$ such cells can be constructed, making an $O(n)$ algorithm impossible. Condition C3 merely states that two cells C' and D' are not well-separated unless every subcell of C' is well-separated from every subcell of D'.

Lemma 20.3

For any node v in the compressed octree, $|P(v)| = O(1)$.

Proof. Consider any node v. Each $u \in P(v)$ can be associated with a unique cell $C \in P^=(S(v))$ such that $S(u) \subseteq C$. This is because any subcell of C which is not a subcell of $S(u)$ is not represented in the compressed octree. It follows that $|P(v)| \leq |P^=(S(v))| = O(1)$ (by Condition C2).   ∎

Lemma 20.4

The sum of interaction set sizes over all nodes in the compressed octree is linear in the number of nodes in the compressed octree, that is, $\sum_v |I(v)| = O(n)$.

Proof. Let v be a node in the compressed octree. Consider any $w \in I(v)$, either $w \in P(parent(v))$ or w is in the subtree rooted by a node $u \in P(parent(v))$. Thus,

$$\sum_v |I(v)| = \sum_v |\{w \mid w \in I(v), w \in P(parent(v))\}|$$
$$+ \sum_v |\{w \mid w \in I(v), w \notin P(parent(v))\}|.$$

Consider these two summations separately. The bound for the first summation is easy; From Lemma 20.3, $|P(parent(v))| = O(1)$. So,

$$\sum_v |\{w \mid w \in I(v), w \in P(parent(v))\}| = \sum_v O(1) = O(n).$$

The second summation should be explored more carefully.

$$\sum_v |\{w \mid w \in I(v), w \notin P(parent(v))\}| = \sum_w |\{v \mid w \in I(v), w \notin P(parent(v))\}|$$

In what follows, we bound the size of the set $M(w) = \{v \mid w \in I(v), w \notin P(parent(v))\}$ for any node.

Since $w \notin P(parent(v))$, there exists a node $u \in P(parent(v))$ such that w is in the subtree rooted by u. Consider $parent(w)$: The node $parent(w)$ is either u or a node in the subtree rooted at u. In either case, $length(S(parent(w))) \leq length(S(parent(v)))$. Thus, for each $v \in M(w)$, there exists a cell C such that $S(v) \subseteq C \subseteq S(parent(v))$ and $length(S(parent(w))) = length(C)$. Further, since v and $parent(w)$ are not well-separated, C and $S(parent(w))$ are not well-separated as well by Condition C3. That is to say $S(parent(w)) \in P^=(C)$ and $C \in P^=(S(parent(w)))$ by Condition C1. By Condition C2, we know that $|P^=(S(parent(w)))| = O(1)$. Moreover, for each cell $C \in P^=(S(parent(w)))$, there are at most 2^d choices of v because $length(C) \leq length(S(parent(v)))$. As a result, $|M(w)| \leq 2^d \times O(1) = O(1)$ for any node w. Thus, $\sum_v |I(v)| = \sum_w |\{v \mid w \in I(v), w \notin P(parent(v))\}| = \sum_w O(1) = O(n)$. ∎

Theorem 20.1

Given a compressed octree for n bodies, the N-body problem can be solved in $O(n)$ time.

Proof. Computing the multipole expansion at a node takes constant time and the number of nodes in the compressed octree is $O(n)$. Thus, total time required for the multipole expansion calculation is $O(n)$. The nodes explored during the local expansion calculation at a node v are either in $P(v)$ or $I(v)$. In both cases, it takes constant time to process a node. By Lemmas 20.3 and 20.4, the total size of both sets for all nodes in the compressed octree is bounded by $O(n)$. Thus, local expansion calculation takes $O(n)$ time. As a conclusion, the running time of the fast multipole algorithm on the compressed octree takes $O(n)$ time irrespective of the distribution of the bodies. ∎

It is interesting to note that the same generic algorithmic framework is used for spherical region queries, all nearest neighbors, k-nearest neighbors and solving the N-body problem. While the proofs are different, the algorithm also provides optimal solution for k-nearest neighbors and the N-body problem.

While this chapter is focused on applications of quadtrees and octrees in image processing and scientific computing applications, they are used for many other types of spatial data and in the context of numerous other application domains. A detailed study of the design and analysis of quadtree based data structures and their applications can be found in [4,5].

Acknowledgments

This work was supported, in part, by the National Science Foundation under grant ACI-0306512.

References

1. A. Klinger. *Optimizing methods in statistics*, chapter Patterns and search statistics, pages 303–337. 1971.

2. J. L. Bentley. Multidimensional binary search trees used for associative searching. *Communications of the ACM*, 18(9):509–517, 1975.

3. G. M. Hunter. *Efficient computation and data structures for graphics*. PhD thesis, Princeton University, Princeton, NJ, USA, 1978.

4. H. Samet. *The Design and Analysis of Spatial Data Structures*. Addison-Wesley, Reading, MA, 1989.

5. H. Samet. *Applications of Spatial Data Structures: Computer Graphics, Image Processing, and GIS*. Addison-Wesley, Reading, MA, 1989.

6. R. A. Finkel and J. L. Bentley. Quad trees: a data structure for retrieval on composite key. *Acta Informatica*, 4(1):1–9, 1974.

7. M. H. Overmars and J. van Leeuwen. Dynamic multi-dimentional data structures based on quad- and *K-D* trees. *Acta Informatica*, 17:267–285, 1982.

8. H. Samet. Deletion in two-dimentional quad trees. *Communications of the ACM*, 23(12):703–710, 1980.

9. P. Flajolet, G. Gonnet, C. Puech, and J. M. Robson. The analysis of multidimensional searching in quad-trees. In Alok Aggarwal, editor, *Proceedings of the 2nd Annual ACM-SIAM Symposium on Discrete Algorithms (SODA '90)*, pages 100–109, San Francisco, CA, USA, 1990.

10. H. Samet, A. Rosenfeld, C. A. Shaffer, and R. E. Webber. Processing geographic data with quadtrees. In *Seventh Int. Conference on Pattern Recognition*, pages 212–215. IEEE Computer Society Press, 1984.

11. K. L. Clarkson. Fast algorithms for the all nearest neighbors problem. In *24th Annual Symposium on Foundations of Computer Science (FOCS '83)*, pages 226–232, 1982.

12. S. Aluru and F. Sevilgen. Dynamic compressed hyperoctrees with application to the N-body problem. In *Springer-Verlag Lecture Notes in Computer Science*, volume 19, 1999.

13. B. Hariharan, S. Aluru, and B. Shanker. A scalable parallel fast multipole method for analysis of scattering from perfect electrically conducting surfaces. In *SC'2002 Conference CD*, http://www.supercomp.org, 2002.

14. M. Warren and J. Salmon. A parallel hashed-octree N-body algorithm. In *Proceedings of Supercomputing '93*, 1993.

15. L. F. Greengard. *The Rapid Evaluation of Potential Fields in Particle Systems*. MIT Press, 1988.

16. G. M. Morton. A computer oriented geodetic data base and a new technique in file sequencing. Technical report, IBM, Ottawa, Canada, 1966.

17. C. Faloutsos. Gray codes for partial match and range queries. *IEEE Transactions on Software Engineering*, 14(10):1381–1393, 1988.

18. D. Hilbert. Uber die stegie abbildung einer linie auf flachenstuck. 38:459–460, 1891.

19. S. Aluru, J. L. Gustafson, G. M. Prabhu, and F. E. Sevilgen. Distribution-independent hierarchical algorithms for the *N*-body problem. *The Journal of Supercomputing*, 12(4):303–323, 1998.

20. J. A. Orenstein and T. H. Merrett. A class of data structures for associative searching. In *Proceedings of the Fourth ACM SIGACT-SIGMOD Symposium on Principles of Database Systems*, pages 181–190, 1984.

21. B. Pagel, H. Six, H. Toben, and P. Widmayer. Toward an analysis of range query performance in spatial data structures. In *Proceedings of the Twelfth ACM SIGACT-SIGMOD-SIGART Symposium on Principles of Database Systems*, pages 214–221, 1993.

22. P. M. Vaidya. An optimal algorithm for the All-Nearest-Neighbors problem. In *27th Annual Symposium on Foundations of Computer Science*, pages 117–122. IEEE, 1986.

23. P. M. Vaidya. An $O(n \log n)$ algorithm for the All-Nearest-Neighbors problem. *Discrete and Computational Geometry*, 4:101–115, 1989.

24. A. Klinger and M. L. Rhodes. Organization and access of image data by areas. *IEEE Transactions on Pattern Analysis and Machine Intelligence*, 1(1):50–60, 1979.

25. I. Gargantini. An effective way to represent quadtrees. *Communications of the ACM*, 25(12):905–910, 1982.

26. I. Gargantini. Translation, rotation and superposition of linear quadtrees. *International Journal of Man-Machine Studies*, 18(3):253–263, 1983.

27. E. Kawaguchi, T. Endo, and J. Matsunaga. Depth-first expression viewed from digital picture processing. *IEEE Transactions on Pattern Analysis and Machine Intelligence*, 5(4):373–384, 1983.

28. N. K. Gautier, S. S. Iyengar, N. B. Lakhani, and M. Manohar. Space and time efficiency of the forest-of-quadtrees representation. *Image and Vision Computing*, 3:63–70, 1985.

29. V. Raman and S. S. Iyengar. Properties and applications of forests of quadtrees for pictorial data representation. *BIT*, 23(4):472–486, 1983.

30. S. S. Iyengar and H. Gadagkar. Translation invariant data structure for 3-D binary images. *Pattern Recognition Letters*, 7:313–318, 1988.

31. D. S. Scott and S. S. Iyengar. A new data structure for efficient storing of images. *Pattern Recognition Letters*, 3:211–214, 1985.

32. D. S. Scott and S. S. Iyengar. Tid – a translation invariant data structure for storing images. *Communications of the ACM*, 29(5):418–429, 1986.

33. L. P. Jones and S. S. Iyengar. Space and time efficient virtual quadtrees. *IEEE Transactions on Pattern Analysis and Machine Intelligence*, 6:244–247, 1984.

34. G. M. Hunter and K. Steiglitz. Linear transformation of pictures represented by quad trees. *Computer Graphics and Image Processing*, 10(3):289–296, 1979.

35. G. M. Hunter and K. Steiglitz. Operations on images using quad trees. *IEEE Transactions on Pattern Analysis and Machine Intelligence*, 1(2):143–153, 1979.

36. H. Samet. Connected component labeling using quadtrees. *Journal of the ACM*, 28(3):487–501, 1981.

37. R. E. Tarjan. Efficiency of a good but not linear disjoint set union algorithm. *Journal of the ACM*, 22:215–225, 1975.

21

Binary Space Partitioning Trees*

21.1 Introduction.. 329
21.2 BSP Trees as a Multi-Dimensional Search Structure............................... 331
21.3 Visibility Orderings 332
    Total Ordering of a Collection of Objects • Visibility Ordering as Tree Traversal •
    Intra-Object Visibility
21.4 BSP Tree as a Hierarchy of Regions.. 334
    Tree Merging • Good BSP Trees • Converting B-Reps to Trees • Boundary
    Representations vs. BSP Trees
Bibliography 340

Bruce F. Naylor[†]
University of Texas, Austin

21.1 Introduction

In most applications involving computation with 3D geometric models, manipulating objects and generating images of objects are crucial operations. Performing these operations requires determining for every frame of an animation the spatial relations between objects: how they might intersect each other, and how they may occlude each other. However, the objects, rather than being monolithic, are most often comprised of many pieces, such as by many polygons forming the faces of polyhedra. The number of pieces may be anywhere from the 100's to the 1,000,000's. To compute spatial relations between n polygons by brute force entails comparing every pair of polygons, and so would require $O(n^2)$. For large scenes comprised of 10^5 polygons, this would mean 10^{10} operations, which is much more than necessary.

The number of operations can be substantially reduced to anywhere from $O(n \log_2 n)$ when the objects interpenetrate (and so in our example reduced to 10^6), to as little as constant time, $O(1)$, when they are somewhat separated from each other. This can be accomplished by using Binary Space Partitioning Trees, also called BSP Trees. They provide a computational representation of space that simultaneously provides a search structure and a representation of geometry. The reduction in number of operations occurs because BSP Trees provide a kind of "spatial sorting." In fact, they are a generalization to dimensions > 1 of binary search trees, which have been widely used for representing sorted lists. Figure 21.1 below gives an introductory example showing how a binary tree of lines, instead of points, can be used to "sort" four geometric objects, as opposed to sorting symbolic objects such as names.

Constructing a BSP Tree representation of one or more polyhedral objects involves computing the spatial relations between polygonal faces once and encoding these relations in a binary tree (Figure 21.2). This tree can then be transformed and merged with other trees to very quickly compute the spatial relations (for visibility and intersections) between the polygons of two moving objects.

As long as the relations encoded by a tree remain valid, which for a rigid body is forever, one can reap the benefits of having generated this tree structure every time the tree is used in subsequent operations. The return on investment manifests itself as substantially faster algorithms for computing intersections and visibility orderings. And for animation and interactive applications, these savings can accrue over hundreds of thousands of frames. BSP Trees achieve an elegant solution to a number of important problems in geometric computation by exploiting two very simple properties occurring whenever a single plane separates (lies between) two or more objects: (1) any object on one side of the plane cannot intersect any object on the other side, (2) given a viewing position, objects on the same side as the viewer can have their images drawn on top of the images of objects on the opposite side (Painter's Algorithm). See Figure 21.3.

These properties can be made dimension independent if we use the term "hyperplane" to refer to planes in 3D, lines in 2D, and in general for *d*-space, to a $(d-1)$-dimensional subspace defined by a single linear equation. The only operation we will need for

---

[*] This chapter has been reprinted from first edition of this Handbook, without any content updates.
[†] We have used this author's affiliation from the first edition of this handbook, but note that this may have changed since then.

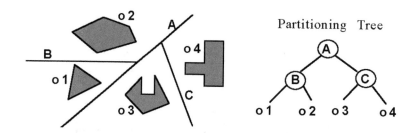

FIGURE 21.1   BSP Tree representation of inter-object spatial relations.

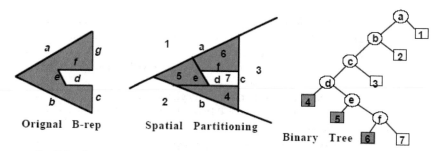

FIGURE 21.2   Partitioning Tree representation of intra-object spatial relations.

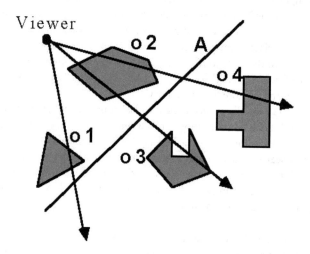

FIGURE 21.3   Plane power: sorting objects w.r.t a hyperplane.

constructing BSP Trees is the partitioning of a convex region by a singe hyperplane into two child regions, both of which are also convex as a result (Figure 21.4).

BSP Trees exploit the properties of separating planes by using one very simple but powerful technique to represent any object or collection of objects: recursive subdivision by hyperplanes. A BSP Tree is the recording of this process of recursive subdivision in the form of a binary tree of hyperplanes. Since there is no restriction on what hyperplanes are used, polytopes (polyhedra, polygons, etc.) can be represented exactly. A BSP Tree is a program for performing intersections between the hyperplane's halfspaces and any other geometric entity. Since subdivision generates increasingly smaller regions of space, the order of the hyperplanes is chosen so that following a path deeper into the tree corresponds to adding more detail, yielding a multi-resolution representation. This leads to efficient intersection computations. To determine visibility, all that is required is choosing at each tree node which of the two branches to draw first based solely on which branch contains the viewer. No other single representation of geometry inherently answers questions of intersection and visibility for a scene of 3D moving objects. And this is accomplished in a computationally efficient and parallelizable manner.

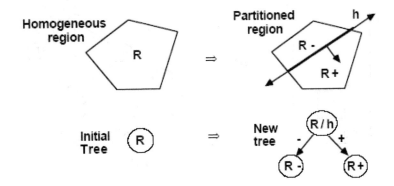

FIGURE 21.4 Elementary operation used to construct BSP Trees.

21.2 BSP Trees as a Multi-Dimensional Search Structure

Spatial search structures are based on the same ideas that were developed in Computer Science during the 60's and 70's to solve the problem of quickly processing large sets of symbolic data, as opposed to geometric data, such as lists of people's names. It was discovered that by first sorting a list of names alphabetically, and storing the sorted list in an array, one can find out whether some new name is already in the list in $\log_2 n$ operations using a binary search algorithm, instead of $n/2$ expected operations required by a sequential search. This is a good example of extracting structure (alphabetical order) existing in the list of names and exploiting that structure in subsequent operations (looking up a name) to reduce computation. However, if one wishes to permit additions and deletions of names while maintaining a sorted list, then a dynamic data structure is needed, that is, one using pointers. One of the most common examples of such a data structure is the binary search tree.

A binary search tree (See also Chapter 3) is illustrated in Figure 21.5, where it is being used to represent a set of integers $S = \{0, 1, 4, 5, 6, 8\}$ lying on the real line. We have included both the binary tree and the hierarchy of intervals represented by this tree. To find out whether a number/point is already in the tree, one inserts the point into the tree and follows the path corresponding to the sequence of nested intervals that contain the point. For a balanced tree, this process will take no more than $O(\log n)$ steps; for in fact, we have performed a binary search, but one using a tree instead of an array. Indeed, the tree itself encodes a portion of the search algorithm since it prescribes the order in which the search proceeds.

This now brings us back to BSP Trees, for as we said earlier, they are a generalization of binary search trees to dimensions > 1 (in 1D, they are essentially identical). In fact, constructing a BSP Tree can be thought of as a geometric version of Quick Sort. Modifications (insertions and deletions) are achieved by merging trees, analogous to merging sorted lists in Merge Sort. However, since points do not divide space for any dimension > 1, we must use hyperplanes instead of points by which to subdivide. Hyperplanes always partition a region into two halfspaces regardless of the dimension. In 1D, they look like points since they are also 0D sets; the one difference being the addition of a normal denoting the "greater than" side. In Figure 21.6, we show a restricted variety of BSP Trees that most clearly illustrates the generalization of binary search trees to higher dimensions. (You may want to call this a k-d tree, but the standard semantics of k-d trees does not include representing continuous sets of points, but rather

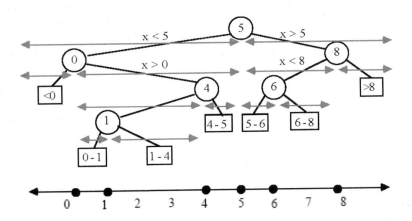

FIGURE 21.5 A binary search tree.

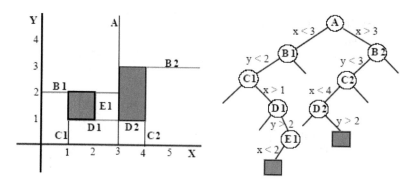

FIGURE 21.6   Extension of binary search trees to 2D as a BSP Tree.

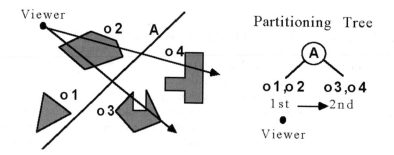

FIGURE 21.7   Left side has priority over right side.

finite sets of points.) BSP Trees are also a geometric variety of Decision Trees, which are commonly used for classification (e.g., biological taxonomies), and are widely used in machine learning. Decision trees have also been used for proving lower bounds, the most famous showing that sorting is $\Omega(n \log n)$. They are also the model of the popular "20 questions" game (I'm thinking of something and you have 20 yes/no question to guess what it is). For BSP Trees, the questions become "what side of a particular hyperplane does some piece of geometry lie."

21.3 Visibility Orderings

Visibility orderings are used in image synthesis for visible surface determination (hidden surface removal), shadow computations, ray tracing, beam tracing, and radiosity. For a given center of projection, such as the position of a viewer or of a light source, they provide an ordering of geometric entities, such as objects or faces of objects, consistent with the order in which any ray originating at the center might intersect the entities. Loosely speaking, a visibility ordering assigns a priority to each object or face so that closer objects have priority over objects further away. Any ray emanating from the center or projection that intersects two objects or faces, will always intersect the surface with higher priority first. The simplest use of visibility orderings is with the "Painter's Algorithm" for solving the hidden surface problem. Faces are drawn into a frame-buffer in far-to-near order (low-to-high priority), so that the image of nearer objects/polygons over-writes those of more distant ones.

A visibility ordering can be generated using a single hyperplane; however, each geometric entity or "object" (polyhedron, polygon, line, point) must lie completely on one side of the hyperplane, that is, no objects are allowed to cross the hyperplane. This requirement can always be induced by partitioning objects by the desired hyperplane into two "halves". The objects on the side containing the viewer are said to have visibility priority over objects on the opposite side; that is, any ray emanating from the viewer that intersects two objects on opposite sides of the hyperplane will always intersect the near side object before it intersects the far side object. See Figures 21.7 and 21.8.

21.3.1 Total Ordering of a Collection of Objects

A single hyperplane cannot order objects lying on the same side, and so cannot provide a total visibility ordering. Consequently, in order to exploit this idea, we must extend it somehow so that a visibility ordering for the entire set of objects can be generated. One way to do this would be to create a unique separating hyperplane for every pair of objects. However, for n objects this would require n^2 hyperplanes, which is too many.

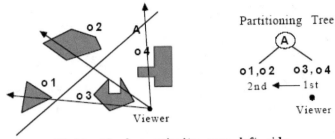

FIGURE 21.8   Right side has priority over left side.

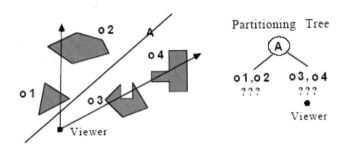

FIGURE 21.9   Separating objects with a hyperplane.

The required number of separating hyperplanes can be reduced to as little as n by using the geometric version of recursive subdivision (divide and conquer). If the subdivision is performed using hyperplanes whose position and orientation is unrestricted, then the result is a BSP Tree. The objects are first separated into two groups by some appropriately chosen hyperplane (Figure 21.9). Then each of the two groups is independently partitioned into two sub-groups (for a total now of 4 sub-groups). The recursive subdivision continues in a similar fashion until each object, or piece of an object, is in a separate cell of the partitioning. This process of partitioning space by hyperplanes is naturally represented as a binary tree (Figure 21.10).

21.3.2 Visibility Ordering as Tree Traversal

How can this tree be used to generate a visibility ordering on the collection of objects? For any given viewing position, we first determine on which side of the root hyperplane the viewer lies. From this we know that all objects in the near-side subtree have higher priority than all objects in the far-side subtree; and we have made this determination with only a constant amount of computation (in fact, only a dot product). We now need to order the near-side objects, followed by an ordering of the far-side objects. Since we have a recursively defined structure, any subtree has the same form computationally as the whole tree. Therefore, we simply apply this technique for ordering subtrees recursively, going left or right first at each node, depending upon which side of the node's hyperplane the viewer lies. This results in a traversal of the entire tree, in near-to-far order, using only $O(n)$ operations, which is optimal (this analysis is correct only if no objects have been split; otherwise it is $> n$).

21.3.3 Intra-Object Visibility

The schema we have just described is only for inter-object visibility, that is, between individual objects. And only when the objects are both convex and separable by a hyperplane is the schema a complete method for determining visibility. To address the general unrestricted case, we need to solve intra-object visibility, that is, correctly ordering the faces of a single object. BSP Trees can solve this problem as well. To accomplish this, we need to change our focus from convex cells containing objects to the idea of hyperplanes containing faces. Let us return to the analysis of visibility with respect to a hyperplane. If instead of ordering objects, we wish to order faces, we can exploit the fact that not only can faces lie on each side of a hyperplane as objects do, but they can also lie on the hyperplane itself. This gives us a 3-way ordering of: near → on → far (Figure 21.11).

If we choose hyperplanes by which to partition space that always contain a face of an object, then we can build a BSP Tree by applying this schema recursively as before, until every face lies in some partitioning hyperplane contained in the tree. To generate a visibility ordering of the faces in this intra-object tree, we use the method above with one extension: faces lying on hyperplanes are included in the ordering, that is, at each node, we generate the visibility ordering of near-subtree → on-faces →

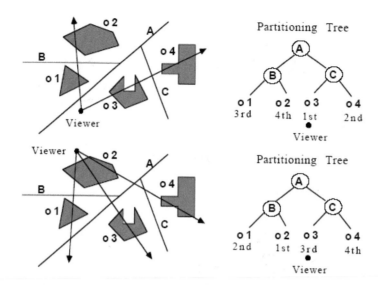

FIGURE 21.10 Binary tree representation of space partitioning.

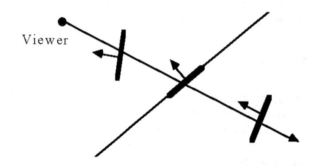

FIGURE 21.11 Ordering of polygons: near → on → far.

far-subtree. Using visibility orderings provides an alternative to z-buffer based algorithms. They obviate the need for computing and comparing z-values, which is very susceptible to numerical error because of the perspective projection. In addition, they eliminate the need for z-buffer memory itself, which can be substantial (80 Mbytes) if used at a sub-pixel resolution of 4×4 to provide anti-aliasing. More importantly, visibility orderings permit unlimited use of transparency (non-refractive) with no additional computational effort, since the visibility ordering gives the correct order for compositing faces using alpha blending. And in addition, if a near-to-far ordering is used, then rendering completely occluded objects/faces can be eliminated, such as when a wall occludes the rest of a building, using a beam-tracing based algorithm.

21.4 BSP Tree as a Hierarchy of Regions

Another way to look at BSP Trees is to focus on the hierarchy of regions created by the recursive partitioning, instead of focusing on the hyperplanes themselves. This view helps us to see more easily how intersections are efficiently computed. The key idea is to think of a BSP Tree region as serving as a bounding volume: each node v corresponds to a convex volume that completely contains all the geometry represented by the subtree rooted at v. Therefore, if some other geometric entity, such as a point, ray, object, etc., is found to not intersect the bounding volume, then no intersection computations need be performed with any geometry within that volume.

Consider as an example a situation in which we are given some test point and we want to find which object if any this point lies in. Initially, we know only that the point lies somewhere space (Figure 21.12).

By comparing the location of the point with respect to the first partitioning hyperplane, we can find in which of the two regions (a.k.a. bounding volumes) the point lies. This eliminates half of the objects (Figure 21.13).

By continuing this process recursively, we are in effect using the regions as a hierarchy of bounding volumes, each bounding volume being a rough approximation of the geometry it bounds, to quickly narrow our search (Figure 21.14).

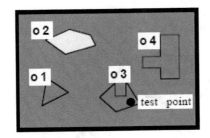

FIGURE 21.12 Point can lie anywhere.

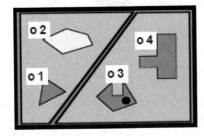
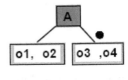

FIGURE 21.13 Point must lie to the right of the hyperplane.

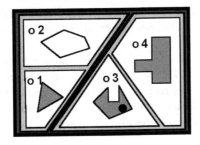
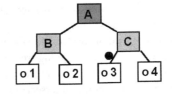

FIGURE 21.14 Point's location is narrowed down to one object.

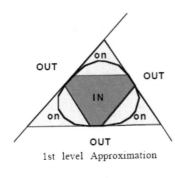
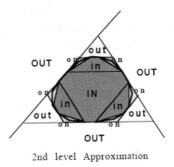

FIGURE 21.15 Multiresolution representation provided by BSP tree.

For a BSP Tree of a single object, this region-based (volumetric) view reveals how BSP Trees can provide a multi-resolution representation. As one descends a path of the tree, the regions decrease in size monotonically. For curved objects, the regions converge in the limit to the curve/surface. Truncating the tree produces an approximation, ala the Taylor series of approximations for functions (Figure 21.15).

21.4.1 Tree Merging

The spatial relations between two objects, each represented by a separate tree, can be determined efficiently by merging two trees. This is a fundamental operation that can be used to solve a number of geometric problems. These include set operations for CSG modeling as well as collision detection for dynamics. For rendering, merging all object-trees into a single model-tree determines

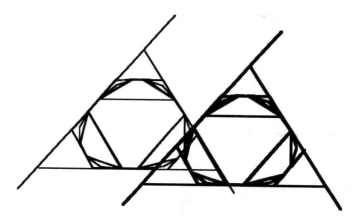

FIGURE 21.16 Merging BSP Trees.

inter-object visibility orderings; and the model-tree can be intersected with the view-volume to efficiently cull away off-screen portions of the scene and provide solid cutaways with the near clipping plane. In the case where objects are both transparent and interpenetrate, tree merging acts as a view independent geometric sorting of the object faces; each tree is used in a manner analogous to the way Merge Sort merges previously sorted lists to quickly created a new sorted list (in our case, a new tree). The model-tree can be rendered using ray tracing, radiosity, or polygon drawing using a far-to-near ordering with alpha blending for transparency. An even better alternative is multi-resolution beam-tracing, since entire occluded subtrees can be eliminated without visiting the contents of the subtree, and distance subtrees can be pruned to the desired resolution. Beam tracing can also be used to efficiently compute shadows.

All of this requires as a basic operation an algorithm for merging two trees. Tree merging is a recursive process, which proceeds down the trees in a multi-resolution fashion, going from low-res to high-res. It is easiest to understand in terms of merging a hierarchy of bounding volumes. As the process proceeds, pairs of tree regions, a.k.a. convex bounding volumes, one from each tree, are compared to determine whether they intersect or not. If they do not, the contents of the corresponding subtrees are never compared. This has the effect of "zooming in" on those regions of space where the surfaces of the two objects intersect (Figure 21.16).

The algorithm for tree merging is quite simple once you have a routine for partitioning a tree by a hyperplane into two trees. The process can be thought of in terms of inserting one tree into the other in a recursive manner. Given trees T1 and T2, at each node of T1 the hyperplane at that node is used to partition T2 into two "halves". Then each half is merged with the subtree of T1 lying on the same side of the hyperplane. (In actuality, the algorithm is symmetric w.r.t. the role of T1 and T2 so that at each recursive call, T1 can split T2 or T2 can split T1.)

```
Merge_Bspts : ( T1, T2 : Bspt ) -> Bspt
    Types
        BinaryPartitioner : { hyperplane, sub-hyperplane}
        PartitionedBspt : ( inNegHs, inPosHs : Bspt )

    Imports
        Merge_Tree_With_Cell : ( T1, T2 : Bspt ) -> Bspt User defined semantics.
        Partition_Bspt : ( Bspt, BinaryPartitioner ) -> PartitionedBspt

    Definition
        IF T1.is_a_cell OR T2.is_a_cell
        THEN
          VAL := Merge_Tree_With_Cell( T1, T2 )
        ELSE
          Partition_Bspt( T2, T1.binary_partitioner ) -> T2_partitioned
          VAL.neg_subtree := Merge_Bspts(T1.neg_subtree, T2_partitioned.inNegHs)
          VAL.pos_subtree:= Merge_Bspts(T1.pos_subtree, T2_partitioned.inPosHs )
        END
    RETURN} VAL
    END Merge_Bspts
```

FIGURE 21.17 Three cases (intersecting, non-intersecting, coincident) when comparing sub-hyperplanes during tree merging.

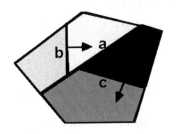
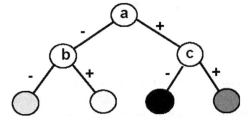

FIGURE 21.18 Before pruning.

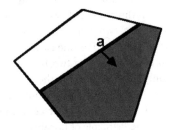
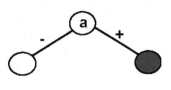

FIGURE 21.19 After pruning.

While tree merging is easiest to understand in terms of comparing bounding volumes, the actual mechanism uses *sub-hyperplanes*, which is more efficient. A sub-hyperplane is created whenever a region is partitioned by a hyperplane, and it is just the subset of the hyperplane lying within that region. In fact, all of the illustrations of trees we have used are drawings of sub-hyperplanes. In 3D, these are convex polygons, and they separate the two child regions of an internal node. Tree merging uses sub-hyperplanes to simultaneously determine the spatial relations of four regions, two from each tree, by comparing the two sub-hyperplanes at the root of each tree. For 3D, this is computed using two applications of convex-polygon clipping to a plane, and there are three possible outcomes: intersecting, non-intersecting and coincident (Figure 21.17). This is the only overtly geometric computation in tree merging; everything else is data structure manipulation.

21.4.2 Good BSP Trees

For any given set, there exist an arbitrary number of different BSP Trees that can represent that set. This is analogous to there being many different programs for computing the same function, since a BSP Tree may in fact be interpreted as a computation graph specifying a particular search of space. Similarly, not all programs/algorithms are equally efficient, and not all searches/trees are equally efficient. Thus the question arises as to what constitutes a good BSP Tree. The answer is a tree that represents the set as a sequence of approximations. This provides a multi-resolution representation. By pruning the tree at various depths, different approximations of the set can be created. Each pruned subtree is replaced with a cell containing a low degree polynomial approximation of the set represented by the subtree (Figures 21.18 and 21.19).

In Figure 21.20, we show two quite different ways to represent a convex polygon, only the second of which employs the sequence of approximations idea. The tree on the left subdivides space using lines radiating from the polygonal center, splitting the number of faces in half at each step of the recursive subdivision. The hyperplanes containing the polygonal edges are chosen only when the number of faces equals one, and so are last along any path. If the number of polygonal edges is n, then the tree is of size $O(n)$ and of depth $O(\log n)$. In contrast, the tree on the right uses the idea of a sequence of approximations. The first three partitioning hyperplanes form a first approximation to the exterior while the next three form a first approximation to the interior. This divides the set of edges into three sets. For each of these, we choose the hyperplane of the middle face by which to partition, and by doing

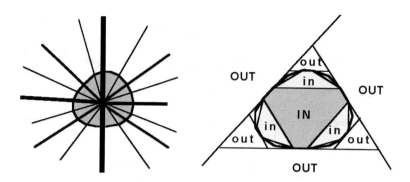

FIGURE 21.20 Illustration of bad vs. good trees.

so refine our representation of the exterior. Two additional hyperplanes refine the interior and divide the remaining set of edges into two nearly equal sized sets. This process precedes recursively until all edges are in partitioning hyperplanes. Now, this tree is also of size $O(n)$ and depth $O(\log n)$, and thus the worst case, say for point classification, is the same for both trees. Yet they appear to be quite different.

This apparent qualitative difference can be made quantitative by, for example, considering the expected case for point classification. With the first tree, all cells are at depth $\log n$, so the expected case is the same as the worst case regardless of the sample space from which a point is chosen. However, with the second tree, the top three out-cells would typically constitute most of the sample space, and so a point would often be classified as OUT by, on average, two point-hyperplane tests. Thus the expected case would converge to $O(1)$ as the ratio of polygon-area/sample-area approaches 0. For line classification, the two trees differ not only in the expected case but also in the worst case: $O(n)$ vs. $O(\log n)$. For merging two trees the difference is $O(n^2)$ vs. $O(n \log n)$. This reduces even further to $O(\log n)$ when the objects are only contacting each other, rather overlapping, as is the case for collision detection. However, there are worst case "basket weaving" examples that do require $O(n^2)$ operations. These are geometric versions of the Cartesian product, as for example when a checkerboard is constructed from n horizontal strips and n vertical strips to produce $n \times n$ squares. These examples, however, violate the Principle of Locality: that geometric features are local not global features. For almost all geometric models of physical objects, the geometric features are local features. Spatial partitioning schemes can accelerate computations only when the features are in fact local, otherwise there is no significant subset of space that can be eliminated from consideration. The key to a quantitative evaluation, and also generation, of BSP Trees is to use expected case models, instead of worst-case analysis. Good trees are ones that have low expected cost for the operations and distributions of input of interest. This means, roughly, that high probability regions can be reached with low cost, that is, they have short paths from the root to the corresponding node, and similarly low probability regions should have longer paths. This is exactly the same idea used in Huffman codes. For geometric computation, the probability of some geometric entity, such as a point, line segment, plane, etc., lying in some arbitrary region is typically correlated positively to the size of the region: the larger the region the greater the probability that a randomly chosen geometric entity will intersect that region. To compute the expected cost of a particular operation for a given tree, we need to know at each branch in the tree the probability of taking the left branch, p, and the probability of taking the right branch p^+. If we assign a unit cost to the partitioning operation, then we can compute the expected cost exactly, given the branch probabilities, using the following recurrence relation:

$E_{cost}[T] =$

    IF T is a cell

    THEN 0

    ELSE $1 + p^- * E_{cost}[T^-] + p^+ * E_{cost}[T^+]$

This formula does not directly express any dependency upon a particular operation; those characteristics are encoded in the two probabilities p^- and p^+. Once a model for these is specified, the expected cost for a particular operation can be computed for any tree.

As an example, consider point classification in which a random point is chosen from a uniform distribution over some initial region R. For a tree region of r with child regions r^+ and r^-, we need the conditional probability of the point lying in r^+ and r^-, given that it lies in r. For a uniform distribution, this is determined by the sizes of the two child-regions relative to their parent:

$$p^+ = \text{vol}(r^+)/\text{vol}(r)$$

$$p^- = \text{vol}(r^-)/\text{vol}(r)$$

Similar models have been developed for line, ray and plane classification. Below we describe how to use these to build good trees.

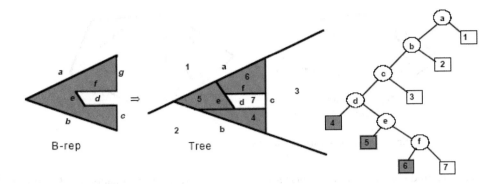

FIGURE 21.21   B-rep and Tree representation of a polygon.

21.4.3 Converting B-Reps to Trees

Since humans do not see physical objects in terms of binary trees, it is important to know how such a tree be constructed from something that is more intuitive. The most common method is to convert a boundary representation, which corresponds more closely to how humans see the world, into a tree. In order for a BSP Tree to represent a solid object, each cell of the tree must be classified as being either entirely inside or outside of the object; thus, each leaf node corresponds to either an in-cell or an out-cell. The boundary of the set then lies between in-cells and out-cells; and since the cells are bounded by the partitioning hyperplanes, it is necessary for the entire boundary to lie in the partitioning hyperplanes (Figure 21.21).

Therefore, we can convert from a b-rep to a tree simply by using all of the face hyperplanes as partitioning hyperplanes. The face hyperplanes can be chosen in any order and the resulting tree will always generate a convex decomposition of the interior and the exterior. If the hyperplane normals of the b-rep faces are consistently oriented to point to the exterior, then all left leaves will be in-cells and all right leaves will be out-cells. The following algorithm summarizes the process.

```
Brep_to_Bspt: Brep b -> Bspt T
IF b == NULL
THEN
 T = if a left-leaf then an in-cell else an out-cell
ELSE
 h = Choose_Hyperplane(b)
 {b+, b-, b0} = Partition_Brep(b,h)
 T.faces = b0
 T.pos_subtree = Brep_to_Bspt(b+)
 T.neg_subtree = Brep_to_Bspt(b-)
END
```

However, this does not tell us in what order to choose the hyperplanes so as to produce the best trees. Since the only known method for finding the optimal tree is by exhaustive enumeration, and there are at least n! trees given n unique face hyperplanes, we must employ heuristics. In 3D, we use both the face planes as candidate partitioning hyperplanes, as well as planes that go through face vertices and have predetermined directions, such as aligned with the coordinates axes. Given any candidate hyperplane, we can try to predict how effective it will be using expected case models; that is, we can estimate the expected cost of a subtree should we choose this candidate to be at its root. We will then choose the least cost candidate. Given a region r containing boundary b which we are going to partition by a candidate h, we can compute exactly p^+ and p^- for a given operation, as well as the size of b^+ and b^-. However, we can only estimate $E_{cost}[T^+]$ and $E_{cost}[T^-]$. The estimators for these values can depend only upon a few simple properties such as number of faces in each halfspace, how many faces would be split by this hyperplane, and how many faces lie on the hyperplane (or area of such faces). Currently, we use $|b^+|^n$ for $E_{cost}[T^+]$, where n typically varies between .8 and .95, and similarly for $E_{cost}[T^-]$. We also include a small penalty for splitting a face by increasing its contribution to b^+ and b^- from 1.0 to somewhere between 1.25 and 1.75, depending upon the object. We also favor candidates containing larger surface area, both in our heuristic evaluation and by first sorting the faces by surface area and considering only the planes of the top k faces as candidates. One interesting consequence of using expected case models is that choosing the candidate that attempts to balance the tree is usually not the best; instead the model prefers candidates that place small amounts of geometry in large regions, since this will result in high probability and low cost subtrees, and similarly large amounts of geometry in small regions. Balanced is optimal only when the geometry is uniformly distributed, which is rarely the case (Figure 21.22). More importantly,

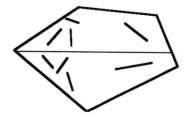 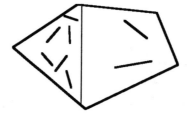

FIGURE 21.22 Balanced is not optimal for non-uniform distributions.

minimizing expected costs produces trees that represents the object as a sequence of approximations, and so in a multi-resolution fashion.

21.4.4 Boundary Representations vs. BSP Trees

Boundary Representations and BSP Trees may be viewed as complementary representations expressing difference aspects of geometry, the former being topological, the later expressing hierarchical set membership. B-reps are well suited for interactive specification of geometry, expressing topological deformations, and scan-conversion. BSP Trees are well suited for intersection and visibility calculations. Their relationship is probably more akin to the capacitor vs. inductor, than the tube vs. transistor.

The most often asked question is what is the size of a BSP Tree representation of a polyhedron vs. the size of its boundary representation. This, of course, ignores the fact that expected cost, measured over the suite of operations for which the representation will be used, is the appropriate metric. Also, boundary representations must be supplemented by other devices, such as octrees, bounding volumes hierarchies, and z-buffers, in order to achieve an efficient system; and so the cost of creating and maintaining these structure should be brought into the equation. However, given the intrinsic methodological difficulties in performing a compelling empirical comparison, we will close with a few examples giving the original number of b-rep faces and the resulting tree using our currently implemented tree construction machinery. The first ratio is number-of-tree-faces/number-of-brep-faces. The second ratio is number-of-tree-nodes/number-of-brep-faces, where number-of-tree-nodes is the number of internal nodes. The last column is the expected cost in terms of point, line and plane classification, respectively, in percentage of the total number of internal nodes, and where the sample space was a bounding box 1.1 times the minimum axis-aligned bounding box. These numbers are pessimistic since typical sample spaces would be much larger than an object's bounding box. Also, the heuristics are controlled by 4 parameters, and these numbers were generated, with some exceptions, without a search of the parameter space but rather using default parameters. There are also quite a number of ways to improve the conversion process, so it should be possible to do even better.

Data Set	# Brep faces	# Tree faces	Faces ratio	# Tree nodes	Faces/Nodes ratio	Point E[T]	Line E[T]	Plane E[T]
Hang glider man	189	406	2.14	390	2.06	1.7	3.4	21.4
Space shuttle	575	1,006	1.75	927	1.61	1.2	2.5	13.2
Human head 1	927	1,095	1.21	1,156	1.24	1.4	4.4	25.0
Human head 2	2,566	5,180	2.01	5,104	1.99	0.2	0.8	9.1
Allosauros	4,072	9,725	2.38	9,914	2.43	NA	NA	NA
Lower Manhattan	4,532	5,510	1.22	4,273	0.94	0.3	0.6	10.5
Berkeley CS Bldg.	9,129	9,874	1.08	4,148	0.45	0.4	1.3	14.6
Dungeon	14,061	20,328	1.44	15,732	1.12	0.1	0.1	1.7
Honda Accord	26,033	51,730	1.98	42,965	1.65	NA	NA	NA
West Point Terrain	29,400	9,208	0.31	7,636	0.26	0.1	0.3	4.2
US Destroyer	45,802	91.928	2.00	65,846	1.43	NA	NA	NA

Bibliography

1. S.H. Bloomberg, A Representation of Solid Objects for Performing Boolean Operations, U.N.C. Computer Science Technical Report 86–006 1986.

2. W. Thibault and B. Naylor, Set operations on polyhedra using binary space BSP Trees, *Computer Graphics*, Vol. 21(4), pp. 153–162, July 1987.

3. B. F. Naylor, J. Amanatides, and W. C. Thibault, Merging BSP Trees yields polyhedral set operations, *Computer Graphics*, Vol. 24(4), pp. 115–124, August 1990.

4. B. F. Naylor, Binary space BSP Trees as an alternative representation of polytopes, *Computer Aided Design*, Vol. 22(4), May 1990.

5. B. F. Naylor, SCULPT: An interactive solid modeling tool, *Proceedings of Graphics Interface*, May 1990.

6. E. Torres, Optimization of the binary space partition algorithm (BSP) for the visualization of dynamic scenes, *Eurographics '90*, September 1990.

7. I. Ihm and B. Naylor, Piecewise linear approximations of curves with applications, *Proceedings of Computer Graphics International '91*, Springer-Verlag, June 1991.

8. G. Vanecek, Brep-index: A multi-dimensional space partitioning tree, *Proceedings of the First ACM symposium on Solid Modeling Foundations and CAD/CAM Applications*, May 1991.

9. B. F. Naylor, Interactive solid modeling via BSP Trees, *Proceeding of Graphics Interface*, pp. 11–18, May 1992.

10. Y. Chrysanthou and M. Slater, Computing dynamic changes to BSP Trees, *Eurographics '92*, Vol. 11(3), pp. 321–332.

11. S. Ar, G. Montag, A. Tal, C. Detection, and A. Reality, Deferred, self-organizing BSP Trees, *Computer Graphics Forum*, Vol. 21(3), September 2002.

12. R. A. Schumacker, R. Brand, M. Gilliland, and W. Sharp, Study for Applying Computer-Generated Images to Visual Simulation, AFHRL-TR-69-14, U.S. Air Force Human Resources Laboratory, 1969.

13. I. E. Sutherland, Polygon Sorting by Subdivision: A Solution to the Hidden-Surface Problem, unpublished manuscript, October 1973.

14. H. Fuchs, Z. Kedem, and B. Naylor, On visible surface generation by a priori tree structures, *Computer Graphics*, Vol. 14(3), pp. 124–133, June 1980.

15. H. Fuchs, G. Abrams, and E. Grant, Near real-time shaded display of rigid objects, *Computer Graphics*, Vol. 17(3), pp. 65–72, July 1983.

16. B. F. Naylor and W. C. Thibault, Application of BSP Trees to Ray-Tracing and CSG Evaluation, Technical Report GIT-ICS 86/03, School of Information and Computer Science, Georgia Institute of Technology, Atlanta, Georgia 30332, February 1986.

17. N. Chin and S. Feiner, Near real-time shadow generation using BSP Trees, *Computer Graphics*, Vol. 23(3), pp. 99–106, July 1989.

18. A. T. Campbell and D.S. Fussell, Adaptive mesh generation for global diffuse illumination, *Computer Graphics*, Vol. 24(4), pp. 155–164, August 1990.

19. A. T. Campbell, Modeling Global Diffuse for Image Synthesis, PhD Dissertation, Department of Computer Science, University of Texas at Austin, 1991.

20. D. Gordon and S. Chen, Front-to-back display of BSP Trees, *IEEE Computer Graphics & Applications*, pp. 79–85, September 1991.

21. N. Chin and S. Feiner, Fast object-precision shadow generation for area light sources using BSP Trees, *Symp. on 3D Interactive Graphics*, March 1992.

22. B. F. Naylor, BSP Tree image representation and generation from 3D geometric models, *Proceedings of Graphics Interface*, May 1992.

23. D. Lischinski, Filippo tampieri and donald greenburg, discontinuity meshing for accurate radiosity, *IEEE Computer Graphics & Applications*, Vol. 12(6), pp. 25–39, November 1992.

24. S. Teller and P. Hanrahan, Global visibility algorithms for illumination computations, *Computer Graphics*, Vol. 27, pp. 239–246, August 1993.

25. Y. Chrysanthou, Shadow Computation for 3D Interaction and Animation, Ph.D. Thesis, Computer Science, University of London, 1996.

26. M. Slater and Y. Chrysanthou, View volume culling using a probabilistic caching scheme, *Proceedings of the ACM Symposium on Virtual Reality Software and Technology*, pp. 71–77, September 1997.

27. Z. Pan, Z. Tao, C. Cheng, and J. Shi, A new BSP tree framework incorporating dynamic LoD models, *Proceedings of the ACM symposium on Virtual reality software and technology*, pp. 134–141, 2000.

28. H. Rahda, R. Leonardi, M. Vetterli, and B. Naylor, Binary space BSP Tree representation of images, *Visual Communications and Image Representation*, Vol. 2(3), pp. 201–221, September 1991.

29. K. R. Subramanian and B. Naylor, Representing medical images with BSP Trees, *Proceeding of Visualization '92*, October 1992.

30. H. M. Sadik Radha, Efficient Image Representation Using Binary Space BSP Trees, Ph.D. dissertation, CU/CTR/TR 343-93-23, Columbia University, 1993.

31. K. R. Subramanian and B.F. Naylor, Converting discrete images to partitioning trees, *IEEE Transactions on Visualization & Computer Graphics*, Vol. 3(3), 1997.

32. A. Rajkumar, F. Fesulin, B. Naylor, and L. Rogers, Prediction RF coverage in large environments using ray-beam tracing and partitioning tree represented geometry, *ACM WINET*, Vol. 2(2), 1996.

33. T. Funkhouser, I. Carlbom, G. Elko, G. Pingali, M. Sondhi, and J. West, A beam tracing approach to acoustic modeling for interactive virtual environments, *Siggraph '98*, July 1998.

34. M. O. Rabin, Proving simultaneous positivity of linear forms, *Journal of Computer and Systems Science*, Vol. 6, pp. 639–650, 1991.

35. E. M. Reingold, On the optimality of some set operations, *Journal of the ACM*, Vol. 19, pp. 649–659, 1972.

36. B. F. Naylor, A Priori Based Techniques for Determining Visibility Priority for 3-D Scenes, Ph.D. Thesis, University of Texas at Dallas, May 1981.

37. M. S. Paterson and F. F. Yao, Efficient binary space partitions for hidden-surface removal and solid modeling, *Discrete & Computational Geometry*, Vol. 5, pp. 485–503, 1990.

38. M. S. Paterson and F. F. Yao, Optimal binary space partitions for orthogonal objects, *Journal of Algorithms*, Vol. 5, pp. 99–113, 1992.

39. B. F. Naylor, Constructing good BSP Trees, *Graphics Interface '93*, Toronto CA, pp. 181–191, May 1993.

40. M. de Berg, M. M. de Groot, and M. Overmars, Perfect binary space partitions, *Canadian Conference on Computational Geometry*, 1993.

41. P. K. Agarwal, L. J. Guibas, T. M. Murali, and J. S. Vitter, Cylindrical and kinetic binary space partitions, *Proceedings of the Thirteenth Annual Symposium on Computational Geometry*, August 1997.

42. J. Comba, Kinetic Vertical Decomposition Trees , Ph.D. Thesis, Computer Science Department, Stanford University, 1999.

43. M. de Berg, J. Comba, and L.J. Guibas, A segment-tree based kinetic BSP, *Proceedings of the Seventeenth Annual Symposium on Computational Geometry*, June 2001.

44. S. Arya, Binary space partitions for axis-parallel line segments: Size-height tradeoffs, *Information Processing Letters*, Vol. 84(4), pp. 201–206, Nov. 2002.

45. C. D. Tóth, Binary space partitions for line segments with a limited number of directions, *SIAM J. on Computing*, Vol. 32(2), pp. 307–325, 2003.

22

R-Trees*

22.1 Introduction.. 343
22.2 Basic Concepts.. 343
    Intersection Queries • Updating the Tree
22.3 Improving Performance 346
    R* Tree • Hilbert Tree • Bulk Loading
22.4 Advanced Operations.. 350
    Nearest Neighbor Queries • Spatial Joins
22.5 Analytical Models.. 353
Acknowledgment.. 356
References.. 356

Scott Leutenegger
University of Denver

Mario A. Lopez
University of Denver

22.1 Introduction

Spatial database management systems must be able to store and process large amounts of disk-resident spatial data. Multidimensional data support is needed in many fields including geographic information systems (GIS), computer aided design (CAD), and medical, multimedia, and scientific databases. Spatial data operations that need to be supported include spatial joins and various types of queries such as intersection, containment, topological and proximity queries. The challenge, and primary performance objective, for applications dealing with disk-resident data is to minimize the number of disk retrievals, or I/Os, needed to answer a query. Main memory data structures are designed to reduce computation time rather than I/O, and hence are not directly applicable to a disk based environment.

Just as the B-tree [1] (Chapter 16) and its variants were proposed to optimize I/O while processing single dimensional disk resident data, the original R-tree [2] and later variants have been proposed to index disk resident multidimensional data efficiently.

R-trees are very versatile, in the sense that they can be used with little or no modification to tackle many different problems related to spatial data. Asymptotically better data structures exist for specific instances of these problems, but the solution of choice is different for different types of inputs or queries. Thus, one of the main advantages of R-trees is that the same data structure can be used to handle different problems on arbitrary types of multidimensional data.

In this chapter we describe the original R-tree proposal and some of its variants. We analyze the efficiency of various operations and examine various performance models. We also describe R-tree based algorithms for additional operations, such as proximity queries and spatial joins. There are many more applications, variants and issues related to R-trees than we can possibly cover in one chapter. Some of the ones we do not cover include parallel and distributed R-trees [3–6], variants for high dimensional [7–9] and for spatio-temporal data [10–16], and concurrency [17,18]. See Chapter 48 for more on concurrent data structures. Other data structures with similar functionality, such as range, quad and k-d trees, are covered in other chapters of Part IV of this handbook. Some of these (see Chapter 28) are specifically designed for disk-resident data.

22.2 Basic Concepts

R-trees were first introduced in Reference 2. An R-tree is a hierarchical data structure derived from the B^+-tree and originally designed to perform intersection queries efficiently. The tree stores a collection of d-dimensional points or rectangles which can change in time via insertions and deletions. Other object types, such as polygons or polyhedra, can be handled by storing their minimum bounding rectangles (MBRs) and performing additional tests to eliminate false hits. A false hit happens when the query object intersects the MBR of a data object but does not intersect the object itself. In the sequel we talk about rectangles only,

---

* This chapter has been reprinted from first edition of this Handbook, without any content updates.

with the understanding that a point is simply a degenerate rectangle. We use the terms MBR and bounding box, interchangeably. In our context, the *d*-dimensional rectangles are "upright," that is, each rectangle is the Cartesian product of *d* one-dimensional intervals: $[l_1, h_1] \times \cdots \times [l_d, h_d]$. Thus, $2d$ values are used to specify a rectangle.

Each node of the tree stores a maximum of *B* entries. With the exception of the root, each node also stores a minimum of $b \leq B/2$ entries. This constraint guarantees a space utilization of at least b/B. Each entry *E* consists of a rectangle *r* and a pointer p_r. As with B^+-trees all input values are stored at the leaves. Thus, at the leaf level, *r* is the bounding box of an actual object pointed to by p_r. At internal nodes, *r* is the bounding box of all rectangles stored in the subtree pointed to by p_r.

A downward path in the tree corresponds to a sequence of nested rectangles. All leaf nodes occur at the same level (i.e., have the same depth), even after arbitrary sequences of updates. This guarantees that the height of the tree is $O(\log_b n)$, where *n* is the number of input rectangles. Notice that MBRs at the same level may overlap, even if the input rectangles are disjoint.

Figure 22.1 illustrates an R-tree with 3 levels (the root is at level 0) and a maximum of $B = 4$ rectangles per node. The 64 small dark data rectangles are grouped into 16 leaf level nodes, numbered 1–16. The bounding box of the set of rectangles stored at the same node is one of the rectangles stored at the parent of the node. In our example, the MBRs of leaf level nodes 1 through 4 are placed in node 17, in level 1. The root node contains the MBRs of the four level 1 nodes: 17, 18, 19, and 20.

22.2.1 Intersection Queries

To perform an intersection query *Q*, all rectangles that intersect the query region must be retrieved and examined (regardless of whether they are stored in an internal node or a leaf node). This retrieval is accomplished by using a simple recursive procedure that starts at the root node and which may follow *multiple* paths down the tree. In the worst case, all nodes may need to be retrieved, even though some of the input data need not be reported. A node is processed by first identifying all the rectangles stored at that node which intersect *Q*. If the node is an internal node, the subtrees corresponding to the identified rectangles are

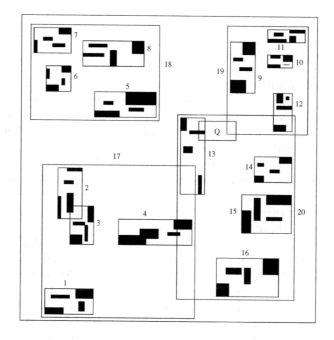

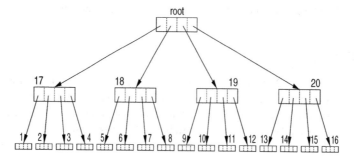

FIGURE 22.1 A sample R-tree using $B = 4$. Input rectangles are shown solid.

searched recursively. Otherwise, the node is a leaf node and the retrieved rectangles (or the data objects themselves) are simply reported.

For illustration, consider the query Q in the example of Figure 22.1. After examining the root node, we determine that nodes 19 and 20 of level 1 must be searched. The search then proceeds with each of these nodes. Since the query region does not intersect any of the MBRs stored in node 19, this sub-query is terminated. While processing the other sub-query, it is determined that Q intersects the MBR corresponding to node 13 and this node is retrieved. Upon checking the rectangles in node 13, the one data rectangle intersected by Q is returned.

Other type of queries, such as arbitrarily shaped queries (e.g., point or polygonal queries) or retrieving all rectangles contained or containing Q, can be handled using a straightforward modification of the above procedure.

22.2.2 Updating the Tree

Many applications require support for update operations, such as insertion and deletion of rectangles. A tree that can change over time via such operations is said to be *dynamic*.

New rectangles can be inserted using a procedure similar to that used to insert a new key in a B^+-tree. In other words, the new rectangle r is first added to a leaf node v and, if the node overflows, a split is performed that requires updating one rectangle and inserting another one in the parent of v. This procedure continues until either a node with fewer than B entries is found or the root is split, in which case a new node is created and the height of the tree grows by one. Independent of whether a split is performed or not, the bounding rectangles of all ancestors of r may need to be updated.

One important difference between R-trees and B^+-trees is that, in our case, there is no incorrect leaf node to which the new rectangle can be added. Choosing a leaf node impacts performance, not correctness, and performance depends on our ability to cluster rectangles so as to minimize the expected number of rectangles intersected by a query of a given size. In practice this means that clustering should attempt to minimize the areas and perimeters of the resulting MBRs. Models for predicting the expected number of nodes that need to be visited while performing an intersection query are discussed in Section 22.5.

The problem of partitioning a set of rectangles into buckets with capacity $B > 2$ such that the expected number of rectangles intersected by a random query is minimized is NP-hard [19]. Hence, it is unlikely we will ever know how to build optimal R-trees efficiently. As a result, many heuristic algorithms for building the tree have been proposed. It is worth noting that, in practice, some of the proposed heuristics result in well tuned R-trees and near optimal I/O for 2-dimensional data. We start with Guttman's original heuristics, which are of two types: leaf selection and node splitting.

To identify a leaf for insertion, Guttman proposes proceeding down the tree, always choosing the rectangle in the current node of the path whose area would increase by the smallest amount were we to insert the new rectangle in the subtree that corresponds to that rectangle. The reasoning behind this approach is that rectangles with small areas are less likely to be chosen for further exploration during a search procedure.

When a node overflows, a split is performed. Ideally, when this happens, one would like to partition a set S of $B + 1$ rectangles into two sets S_1 and S_2 such that the sum of their areas is minimized and each set contains at least b entries. Guttman proposes three different strategies, only the first of which is guaranteed to yield an optimal partition. The first strategy is a brute force algorithm that chooses the best split by checking all candidate partitions of the overflowed set S. This strategy is not practical, as the number of candidates is exponential in the node capacity, which can easily exceed 50 or so rectangles. The second strategy, quadratic split, starts by selecting the two rectangles $r_1, r_2 \in S$ which maximize the quantity $\text{area}(r') - \text{area}(r_1) - \text{area}(r_2)$, where r' is the MBR of $r_1 \cup r_2$. These two rectangles act as seeds which are assigned to different sides of the partition, that is, one is assigned to S_1 and the other to S_2. The remaining entries of S are then assigned to the set (S_1 or S_2) whose MBR area increases the least when including the new rectangle in that set. The entries are not considered in arbitrary order. Rather, the next entry to be allocated is the one with the strongest preference for a group, that is, the entry r that maximizes $|A_1 - A_2|$, where $A_i = \text{area}(\text{MBR}(S_i \cup \{r\})) - \text{area}(\text{MBR}(S_i))$. This heuristic runs in quadratic time and attempts to assign a priority to the unallocated entries according to the performance penalty that the wrong assignment could cause. If at any time during the allocation procedure the size of the smaller set plus the number of unallocated entries is equal to b, then all remaining entries are allocated to that set.

The third and final strategy, linear split, also assigns entries to the group whose MBR area increases the least, but differs from quadratic split in the method used to pick the seeds and in the order in which the remaining entries are allocated. The seeds are the two rectangles r_1 and r_2 whose separation is largest along (at least) one of the dimensions. We elaborate on this. Let $l_j(r)$ and $h_j(r)$ denote the low and high endpoints, respectively, of the j-th interval of r. The width of S along dimension j is simply $w_j = \max_r\{h_j(r)\} - \min_r\{l_j(r)\}$. The normalized separation of S along dimension j is $s_j = (\max_r\{l_j(r)\} - \min_r\{h_j(r)\})/w_j$. The seeds r_1 and r_2 are the two rectangles that yield the largest normalized separation considering all dimensions j. Once the seeds are chosen, the remaining entries are allocated to one set or the other in random order. This last heuristic runs in linear time.

A different linear split algorithm is described in Reference 20. Other efforts [21,22] include polynomial time algorithms to partition a set of rectangles so as to minimize the sum of areas of the two resulting bounding rectangles.

In order to delete a rectangle r we first find the node v containing r and remove the entry, adjusting the bounding rectangles in all ancestors, as necessary. If the node occupancy goes below b the tree needs to be readjusted so as to keep the height in $O(\log_b n)$. There are different ways in which one can readjust the tree. One possibility is to redistribute the remaining entries of v among the siblings of v, in a manner similar to how underflowed nodes are treated in some B-tree algorithms. Instead, Guttman suggests reinserting the remaining entries of v, as this helps the global structure of the tree by considering non-sibling nodes during reinsertion. Of course, this procedure needs to be applied recursively, as internal nodes may underflow as well. Finally, if after deletion the root has exactly one child, the tree height shrinks by one and the only child becomes the new root.

22.3 Improving Performance

Since R-trees were first proposed in Reference 2, many variants and methods to improve the structure and performance of the tree have been proposed. We discuss a few of the more common ones: R*-trees [23], Hilbert R-trees [24] and several bulk loading algorithms [25–27]. Other proposals for improving performance include [28–31].

22.3.1 R* Tree

Given that the known R-tree insertion algorithms are based on heuristic optimization, it is reasonable to assess their merit experimentally. Beckmann et al. [23] conducted an extensive experimental study to explore the impact of alternative approaches for leaf selection and node splitting. Based on their experiments, they proposed the R* tree which has become the most commonly implemented R-tree variant.

The R* tree differs from the original Guttman R-tree in three ways.

First, the leaf where a new object is inserted is chosen differently. The path selection algorithm makes use of the concept of *overlap* of entry E_i in node v_j, defined as $\text{overlap}(E_i) = \sum_{j=1, j \neq i}^{m} \text{area}(r_i \cap r_j)$, where m is the number of entries in node v_j and r_i is the rectangle associated with E_i. When descending from the root, if the next node to be selected is a leaf, the algorithm chooses the node that requires the least increase in overlap, and resolves ties as least area enlargement. If the next node is not a leaf, the entry with the least area enlargement is chosen.

The second difference is the use of *forced reinserts*. The authors discovered that the initial order of inserts significantly impacts tree quality. They also observed that query performance of an existing R-tree can be improved by removing half of the entries and then re-inserting them. Of course, the authors do not recommend performing a restructuring of this magnitude frequently. Rather, they used this insight to modify the policy for dealing with overflowed nodes. If an insertion causes an overflow, calculate the distance from the center of each of the $B + 1$ entries to the center of the MBR enclosing all $B + 1$ entries. Then sort the entries in decreasing order of this distance. Remove the p furthest entries, where p is set to 30% of B, and re-insert the p removed entries into the tree. Some subset of the p re-inserts may be inserted into nodes other than the initial node that overflowed. For each of the p re-inserts, if they do not cause an overflow, do nothing; otherwise, split the node using the algorithm below.

The third difference is in the node splitting algorithm. When a split is needed, the node entries are first sorted twice along each of the d dimensions. The two sorts are based on the low and on the high MBR endpoint values, respectively. Remember that nodes must have a minimum of b and a maximum of B entries. Thus, using one of the sorted lists, the $B + 1$ entries can be partitioned into two groups, S_1 and S_2, by splitting anyplace after the i-th entry, $b \leq i \leq B - b + 1$, of the sorted list. S_1 and S_2 consist of the entries before and after the split position, respectively. In order to choose the best split, the following three objective functions were considered (for 2-d data) and tested using different combinations:

1. Area-value = $\text{area}(\text{MBR}(S_1)) + \text{area}(\text{MBR}(S_2))$
2. Perimeter-value = $\text{perimeter}(\text{MBR}(S_1)) + \text{perimeter}(\text{MBR}(S_2))$
3. Overlap-value = $\text{area}(\text{MBR}(S_1) \cap \text{MBR}(S_2))$

Notice that for a fixed area, the MBR with smallest perimeter is the square.

Based on experiments, the following split policy is adopted. The R* tree computes the perimeter-values for each possible grouping (S_1, S_2) over both sorted lists of all dimensions and chooses the dimension that yields the minimum perimeter-value. Once the dimension has been chosen, the algorithm then chooses the grouping for that dimension that minimizes the overlap-value.

These three changes were shown to substantially improve the I/O performance for all data sets studied.

22.3.2 Hilbert Tree

The Hilbert R-tree [24] further improves performance by imposing a linear order on the input rectangles that results in MBRs of small area and perimeter. The tree is actually an R-tree augmented with order information. Intersection queries are performed as before, using the standard R-tree algorithm; but as a consequence of the ordering constraints, insertion and deletion can proceed as in B$^+$-trees and there is no longer a need to consider various leaf selection heuristics. Additionally, the linear order allows for effective use of deferred splitting, a technique which improves node utilization and performance as trees require fewer nodes for a given input set.

To define an ordering of the input values, Kamel and Faloutsos [24] propose the use of a space-filling curve, such as the Hilbert curve. The power of these curve lies in its ability to linearly order multidimensional points such that nearby points in this order are also close in multidimensional space. Hilbert curves are not the only reasonable choice. Other curves, such as the Peano or Z-order curve, may also be used. See [32] for a discussion on space-filling curves.

A d-dimensional Hilbert curve of order k is a curve H_k^d that visits every vertex of a finite d dimensional grid of size $2^k \times \cdots \times 2^k = 2^{kd}$. Its construction can best be viewed as a sequence of stages. At each stage, an instance of the curve of the previous stage is rotated and placed in each of 2^d equal-sized sub-quadrants. Endpoints of the 2^d sub-curves are then connected to produce the curve at the next stage. The first three stages of the Hilbert curve for two and three dimensions are illustrated in Figures 22.2 and 22.3, respectively.

Each grid vertex is assigned a Hilbert value, which is an integer that corresponds to its position along the curve. For instance, in H_2^2, $(0,0)$ and $(1,2)$ have Hilbert values 0 and 7, respectively. This assignment is easily extended to rectangles, in which case the Hilbert value of the grid point closest to the rectangle center is assigned. Algorithms for computing the position of a point along a space filling curve are given in References 33–35.

The structure of the R-tree is modified as follows. Leaf nodes remain the same. Each entry of an internal node now has the form (r, p, v), where r and p have the same interpretation as before, and v is the largest Hilbert value of all data items stored in the subtree with root p. This assignment results in update algorithms that are similar to those used for B$^+$-trees. In particular, it is straightforward to implement an effective policy for *deferred splitting* which reduces the number of splits needed while performing insertions. The authors propose the following policy, which they call s-to-$(s+1)$ splitting. When a node overflows, an attempt is first made to shift entries laterally to $s-1$ sibling nodes at the same level. An actual split occurs only if the additional entry cannot be accommodated by shifting, because the $s-1$ siblings are already full. When this happens a new node is created and the $sB+1$ entries are distributed among the $s+1$ nodes. Because entries at the leaves are sorted by Hilbert value, bounding boxes of leaves tend to have small area and perimeter, and node utilization is high. Notice that there is a clear trade-off between node

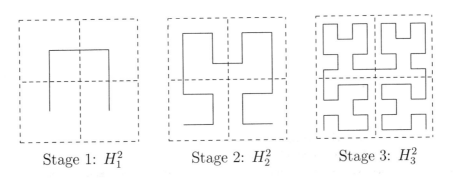

Stage 1: H_1^2 Stage 2: H_2^2 Stage 3: H_3^2

FIGURE 22.2 The first three stages of a 2-dimensional Hilbert curve.

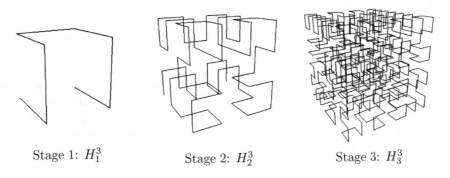

Stage 1: H_1^3 Stage 2: H_2^3 Stage 3: H_3^3

FIGURE 22.3 The first three stages of a 3-dimensional Hilbert curve.

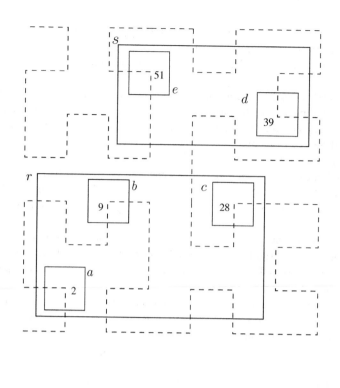

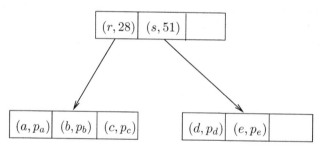

FIGURE 22.4   A Hilbert R-tree with $B = 3$.

utilization and insertion complexity (which increases as s increases). The case $s = 1$ corresponds to the regular split policy used in the original R-tree, that is, split whenever a node overflows.

A sample tree with 5 data rectangles is shown in Figure 22.4. There are two leaf nodes and one internal node (the root). The pointer entries of the internal node are represented by arrows. Each input rectangle has been annotated with its Hilbert value. In reality, the corners of the rectangles would fall on grid vertices. They have been made smaller in order to make the figure more readable. Inserting a new rectangle whose center has Hilbert value 17 would cause an overflow in r. With deferred splitting, a split is not necessary. Instead, the new rectangle would be accommodated in r after shifting rectangle c (with Hilbert value 28) to its sibling s.

The authors report improvements of up to 28% in performance over R*-trees and recommend using 2-to-3 splitting which results in an average node utilization of 82.2%.

In practice, one does not store or compute all bit values on the hypothetical grid. Let β be the number of bits required to describe one coordinate. Storing the Hilbert value of a d-dimensional point requires $d\beta$ bits of storage, which may be larger than the size of native machine integers. It is possible to compare the Hilbert values of two points without storing the values explicitly. Conceptually, the process computes bit positions, one at a time, until discrimination is possible. Consider the case of 2-d and notice that the first bit of the x- and y-coordinates of a point determine which quadrant contains it. Successive bits determine which successively smaller sub-quadrants contain the point. When two center points (x_1, y_1) and (x_2, y_2) need to be compared, the bits of each coordinate are examined until it can be determined that one of the points lies in a different sub-quadrant than the other (one can use the sense and rotation tables described in Reference 25 to accomplish this task). The information gathered is used to decide which point is closer to the origin (along the Hilbert Curve).

22.3.3 Bulk Loading

There are applications where the data is static, or does not change very frequently. Even if the data is dynamic, it may happen that an index needs to be constructed for a large data set which is available a priori. In these circumstances, building an R-tree by inserting one object at a time has several disadvantages: (a) high load time, (b) sub-optimal space utilization, and, most important, (c) poor R-tree structure requiring the retrieval of a large number of nodes in order to satisfy a query. As discussed in the previous section, other dynamic algorithms [23,31] improve the quality of the R-tree, but still are not competitive with regard to query time when compared to loading algorithms that are allowed to pre-process the data to be stored. When done properly, preprocessing results in R-trees with nearly 100% space utilization and improved query times (due to the fact that fewer nodes need to be accessed while performing a query). Such *packing algorithms* were first proposed by Roussopoulos [27] and later by Kamel and Faloutsos [25], and Leutenegger et al. [26]. An approach that is intermediary between inserting a tuple at a time and constructing the entire tree by bulk loading is followed by [36], where an entire batch of input values is processed by partitioning the input into clusters and then inserting R-trees for the clusters into the existing R-tree.

The general approach to bulk loading an R-tree is similar to building a B-tree from a collection of keys by creating the leaf level first and then creating each successively higher level until the root node is created. The general approach is outlined below.

General Algorithm:

1. Sort the n rectangles and partition them into $\lceil n/B \rceil$ consecutive groups of B rectangles. Each group of B rectangles is eventually placed in the same leaf level node. Note that the last group may contain fewer than B rectangles.
2. Load the $\lceil n/B \rceil$ groups of rectangles into nodes and output the (MBR, address) for each leaf level node into a temporary file. The addresses are used as the child pointer fields for the nodes of the next higher level.
3. Recursively pack these MBRs into nodes at the next level, proceeding upwards, until the root node is created.

The three algorithms differ only in how the rectangles are sorted at each level. These differences are described below.

1. *Nearest-X (NX):*
   This algorithm was proposed in Reference 27. The rectangles are sorted by the x-coordinate of a designated point such as the center. Once sorted, the rectangles are packed into nodes, in groups of size B, using this ordering. While our description is in terms of x, a different coordinate can clearly be used.
2. *Hilbert Sort (HS):*
   The algorithm of [25] orders the rectangles using the Hilbert space filling curve. The center points of the rectangles are sorted based on their distance from the origin, measured along the curve. This process determines the order in which the rectangles are placed into the nodes of the R-Tree.
3. *Sort-Tile-Recursive (STR):*
   STR [26] is best described recursively with $d = 2$ providing the base case. (The case $d = 1$ is already handled well by regular B-trees.) Accordingly, we first consider a set of rectangles in the plane. The basic idea is to "tile" the data space using $\sqrt{n/B}$ vertical slices so that each slice contains enough rectangles to pack roughly $\sqrt{n/B}$ nodes. Once again we assume coordinates are for the center points of the rectangles. Determine the number of leaf level pages $P = \lceil n/B \rceil$ and let $S = \lceil \sqrt{P} \rceil$. Sort the rectangles by x-coordinate and partition them into S vertical slices. A slice consists of a run of $S \cdot B$ consecutive rectangles from the sorted list. Note that the last slice may contain fewer than $S \cdot B$ rectangles. Now sort the rectangles of each slice by y-coordinate and pack them into nodes by grouping them into runs of length B (the first B rectangles into the first node, the next n into the second node, and so on).

   The case $d > 2$ is a simple generalization of the approach described above. First, sort the hyper-rectangles according to the first coordinate of their center. Then divide the input set into $S = \lceil P^{\frac{1}{d}} \rceil$ slabs, where a slab consists of a run of $B \cdot \lceil P^{\frac{d-1}{d}} \rceil$ consecutive hyper-rectangles from the sorted list. Each slab is now processed recursively using the remaining $d - 1$ coordinates (i.e., treated as a $(d - 1)$-dimensional data set).

Figure 22.5 illustrates the results from packing a set of segments from a Tiger file corresponding to the city of Long Beach. The figure shows the resultant leaf level MBRs for the same data set for each of the three algorithms using a value of $B = 100$ to bulk load the trees.

As reported in Reference 26, both Hilbert and STR significantly outperform NX packing on all types of data except point data, where STR and NX perform similarly. For tests conducted with both synthetic and actual data sets, STR outperformed Hilbert on all but one set, by factors of up to 40%. In one instance (VLSI data), Hilbert packing performed up to 10% faster. As expected, these differences decrease rapidly as the query size increases.

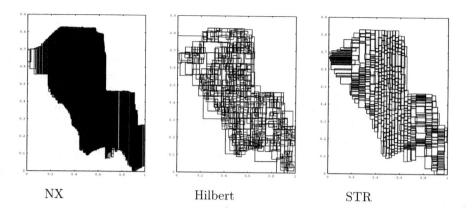

NX   Hilbert   STR

FIGURE 22.5 Leaf level nodes for three packing algorithms.

22.4 Advanced Operations

Even though R-trees were originally designed to perform intersections queries, it did not take long before R-trees were being used for other types of operations, such as nearest neighbor, both simple [37–39] and constrained to a range [40], reverse nearest neighbor [41–43], regular and bichromatic closest pairs [44–46], incremental nearest neighbors [38], topological queries [47], spatial joins [48–56] and distance joins [45,57]. Some of the proposals work on an ordinary R-tree while others require that the tree be modified or augmented in various ways. Because of space limitations we concentrate our attention on two problems: nearest neighbors and spatial joins.

22.4.1 Nearest Neighbor Queries

We discuss the problem of finding the input object closest to an arbitrary query point Q. We assume the standard Euclidean metric, that is, if $a = (a_1, \ldots, a_d)$ and $b = (b_1, \ldots, b_d)$ are arbitrary points then $\text{dist}(a, b) = (\sum_{i=1}^{d} (a_i - b_i)^2)^{1/2}$. For a general object O, such as a rectangle or polygon, we define $\text{dist}(Q, O) = \min_{p \in O} \text{dist}(Q, p)$ (Figure 22.6).

The first proposal to solve this problem using R-trees is from [39]. Roussopoulos et al. define two bounding functions of Q and an arbitrary rectangle r:

$\text{mindist}(Q, r) = \text{dist}(Q, r)$, the distance from Q to the closest point in r.
$\text{minmaxdist}(Q, r) = \min_f \max_{p \in f} (\text{dist}(Q, p))$, where f ranges over all $(d-1)$-dimensional facets of r.

Notice that $\text{mindist}(Q, r) = 0$ if Q is inside r and that for any object or rectangle s that is a descendant of r, $\text{mindist}(Q, r) \leq \text{dist}(Q, s) \leq \text{minmaxdist}(Q, r)$. This last fact follows from the fact that each of the facets of r must share a point with at least one input object, but this object can be as far as possible within an incident face. Thus, the bounding functions serve as optimistic and pessimistic estimates of the distance from Q to the nearest object inside r.

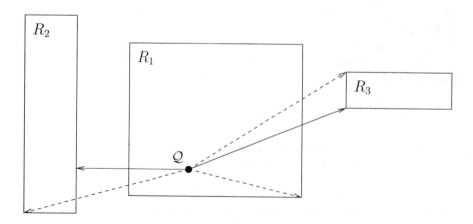

FIGURE 22.6 Illustration of mindist (solid) and minmaxdist (dashed) for three rectangles. Notice that $\text{mindist}(Q, R_1) = 0$ because $Q \in R_1$.

```
findNearestNeighbor(Q,v)
    if v is a leaf then
        foreach rectangle r in v do
            if dist(Q, r) < bestDistance then
                bestDistance ← dist(Q, r)
    else
        produce a list L of all entries (of the form (r, w)) in v
        sort L according to sorting criterion
        prune L using property P1
        while L is not empty do
            retrieve and remove next entry (r, w) from L
            findNearestNeighbor(Q,w)
            prune L using property P2
end
```

FIGURE 22.7   Nearest neighbor algorithm of [39].

The following properties of the bounding functions readily follow:

P1 For any object or MBR r and MBR s, if $\text{mindist}(Q, r) > \text{minmaxdist}(Q, s)$ then r cannot be or contain the nearest neighbor of Q.

P2 For any MBR r and object s, if $\text{mindist}(Q, r) > \text{dist}(Q, s)$ then r cannot contain the nearest neighbor of Q.

The authors describe a branch-and-bound algorithm that performs a depth-first traversal of the tree and keeps track of the best distance so far. The two properties above are used to identify and prune branches that are guaranteed not to contain the answer. For each call the algorithm keeps a list of active nodes, that is, nodes that tentatively need to be explored in search of a better estimate. No node of the list is explored until the subtrees corresponding to nodes appearing earlier in the active list have been processed or pruned. Thus, a sorting policy to determine the order in which rectangles stored in a node are processed is also required. In practice one would like to examine first those nodes that lower the best distance estimate as quickly as possible, but this order is difficult to determine a priori. Two criteria considered include sorting by mindist and by minmaxdist. Experimental results suggest that sorting by mindist results in slightly better performance. The algorithm is summarized in Figure 22.7. Global variable bestDistance stores the estimate of the distance to the nearest neighbor. The initial call uses the query and root of the tree as arguments.

A simple modification to the above algorithm allows [39] to report the $k > 1$ nearest neighbors of Q. All is needed is to keep track of the best k distances encountered so far and to perform the pruning with respect to the k-th best.

Cheung and Fu [37] show that pruning based on P1 is not necessary and do away with computing minmaxdist altogether. They simplify the algorithm by pruning with P2 exclusively, and by rearranging the code so that the pruning step occurs before, and not after, each recursive call.

Hjaltason and Samet [45] also describe an algorithm that avoids using P1. Furthermore, unlike [39] which keeps a local list of active entries for each recursive call, their algorithm uses a global priority queue of active entries sorted by the optimistic distance from Q to that entry. This modification minimizes the number of R-tree nodes retrieved and results in an incremental algorithm, that is, one that reports answers in increasing order of distance, a desirable characteristic when the value of k is not known a priori.

22.4.2 Spatial Joins

We consider the problem of calculating the intersection of two sets $R = \{r_1, r_2, \ldots, r_n\}$ and $S = \{s_1, s_2, \ldots, s_m\}$ of spatial objects. The spatial join, $R \bowtie S$, is the set of all pairs (r_i, s_j) such that $r_i \cap s_j \neq \emptyset$. The join of two spatial data sets is a useful and common operation. Consider, for example, the rivers and roads of a region, both represented using line segments. The spatial join of the river and road data sets yields the set of likely bridge locations. If the subset of river segments whose level is above a given threshold has been previously selected, the same spatial join now computes likely locations for road flooding.

Methods to compute spatial joins without R-trees exist but are not covered in this chapter. We consider the case where R-trees have been constructed for one or both data sets and hence can be used to facilitate the join computation. Unless otherwise stated, we will assume that the r_i's and s_j's refer to the MBRs enclosing the actual data objects.

SpatialJoin(v_1,v_2)
    foreach entry $E_{1j} \in v_1$ **do**
        foreach entry $E_{2j} \in v_2$ such that $R_{1j} \cap R_{2j} \neq \emptyset$ **do**
            if v_1 and v_2 are leaf nodes **then**
                output entries E_{1j} and E_{2j}
            else
                Read the nodes pointed to by p_{1j} and p_{2j}
                SpatialJoin(p_{1j}, p_{2j})
End

FIGURE 22.8   Spatial join algorithm of [49] for two R-trees.

In Reference 49, Brinkhoff et al. proposed the "canonical" spatial join algorithm based on R-trees. A similar join algorithm was proposed at the same time by Gunther [50]. Gunther's algorithm is applicable for general trees and includes R-trees as a special case. In Figure 22.8 we paraphrase the algorithm of [49]. Let v_1 and v_2 be nodes of R-trees T_1 and T_2, respectively. Let E_{ij}, be an entry of v_i of the form (r_{ij}, p_{ij}), where r_{ij} and p_{ij} denote the MBR and child pointer, respectively. To join data sets R and S, indexed by R-trees T_1 and T_2, invoke the SpatialJoin algorithm in Figure 22.8, passing as arguments the roots of T_1 and T_2.

In Reference 49 the authors generalize the algorithm for the case where the two R-trees have different heights. They also improve upon the basic algorithm by reducing CPU computation and disk accesses. The CPU time is improved by reducing the number of rectangle intersection checks. One method to accomplish this is to first calculate the subset of entries within nodes v_1 and v_2 that intersect $\text{MBR}(v_1) \cap \text{MBR}(v_2)$, and then do full intersection tests among this subset. A second approach is to sort the entries and then use a sweep-line algorithm. In addition, the paper considers reduction of page I/O by ordering the accesses to pages so as to improve the buffer hit ratio. This is done by accessing the data in sweep-line order or Z-order.

In Reference 48 Brinkhoff et al. suggest the following 3-step approach for joining complex polygonal data sets: (1) use the R-trees and MBRs of the data sets to reduce the set to a list of potential hits; (2) use a lightweight approximation algorithm to further reduce the set, leaving in some "false positives"; (3) conduct the exact intersection tests on the remaining polygons using techniques from computational geometry (see Chapter 2 of [58], for example).

Often only one of the data sets to be joined will be indexed by an R-tree. This is the case when the DBA decides that maintaining an index is too expensive, when the data set being joined is the result of a series of non-spatial attribute selections, or for multi-step joins where intermediate results are not indexed.

To join two data sets when only one is indexed by an R-tree, Lo and Ravishankar [52] propose the idea of *seeded tree join*. The main idea is to use the existing R-tree to "seed" the creation of a new index, called the seeded tree, for the non-indexed data set, and then perform a join using the method of [49].

Consider the example shown in Figure 22.9 (similar to the example in Reference 52). Assume that the existing tree has four entries in the root node. The four rectangles, R1 to R4, in the left hand side represent the bounding rectangles of these entries. The dark filled squares do not belong to the existing R-tree but, rather, correspond to some of the data items in the non-indexed data set. Assuming a node capacity of four (i.e., $B = 4$) and using the normal insertion or loading algorithms which minimize area and perimeter, the dark rectangles would likely be grouped as shown by the dark MBRs in Figure 22.9b. A spatial join would then require that each node of each tree be joined with two nodes from the other tree, for a total of eight pairs. On the other hand, if the second R-tree was structured as show in Figure 22.9c, then each node would be joined with only one node from the other tree, for a total of four pairs. Hence, when performing a spatial join, it might be better to structure the top levels of the tree in a fashion that is sub-optimal for general window queries.

The general algorithm for seeded tree creation is to copy the top few levels of the existing tree and use them as the top levels for the new tree. These top levels are called the seed levels. The entries at the lowest seed level are called "slots". Non-indexed data items are then inserted into the seeded tree by inserting them into an R-tree that is built under the appropriate slot. In Reference 52 the authors experimentally compare three join algorithms: (1) R-tree join, where an R-tree is fist built on the non-indexed data set and then joined; (2) brute force, where the non-indexed data set is read sequentially and a region query is run against the existing R-tree for each data item in the non-indexed data set; (3) seeded tree join, where a seeded tree is built and then joined. The authors consider several variants of the seeded tree creation algorithm and compare the performance. Experimental studies show that the seeded tree method significantly reduces I/O. Note that if the entire seeded tree does not fit in memory significant I/O can occur during the building process. The authors propose to minimize this I/O by buffering runs for the slots and then building the tree for each slot during a second pass.

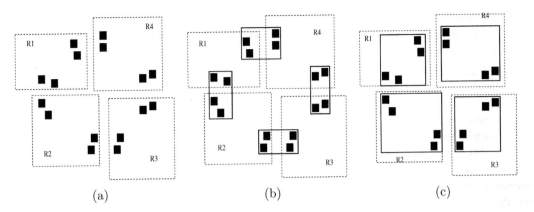

FIGURE 22.9   Seeded Tree Example. (a) Existing R-tree and non-indexed data (dark squares); (b) Normal R-tree structure for non-indexed data; (c) Seeded structure.

Another approach [51] for the one-index case is to sort the non-indexed data using the MBR lower endpoints for one of the dimensions, sort the leaf level MBRs from the existing R-tree on the same dimension, and finally join the two sorted data sets using a sweep-line algorithm. The authors analytically demonstrate that as long as the buffer size is sufficient to merge-sort efficiently, their algorithm results in less I/O than creating any type of index followed by a join. Experimental results also show a significant I/O reduction.

In Reference 56 a relational hash-based join approach is used. Although mostly hash-based, the method does need to default to R-trees in some cases. A sampled subset of R is partitioned into N buckets, $R_1 \ldots R_N$, for some N. Each non-sampled object from R is then added to the bucket that requires the least enlargement. Set S is then partitioned in N corresponding buckets, $S_1 \ldots S_N$ by testing each object o_S of S and placing a copy of it into S_i, for each bucket R_i such that $o_S \cap R_i \neq \emptyset$. If an object in S does not intersect any of the R_i buckets the object is discarded. The bucket pairs (R_i, S_i) are then read into memory and pairwise joined. If a bucket pair is too large to fit in memory, an R-tree index is built for one of the two buckets and an index-loop join is used.

In References 53 and 55 a method that combines the seeded tree and the hash-join approach is proposed. The proposed method, *slot index spatial join*, chooses a level from the existing tree and uses the MBRs of the entries as the buckets for the hashing. The chosen level is determined by the number of entries at that level and the number of buffer pages available for the bucket joining algorithm. Since R-trees have a wide fan out, the optimal number of buckets to use usually falls between level sizes. The paper considers several heuristics for combining MBRs from the next level down to tune the number of buckets. Overall performance is shown to be better than all previously existing methods in all but a few cases.

A method to join multiple spatial relations together is discussed in Reference 53 and 54. The authors propose a multi-way join algorithm called *synchronous traversal* and develop cost models to use for query optimization. They consider using only pairwise joins, only the synchronous traversal method, and combinations of the two approaches. They show that, in general, a combination of the two results in the best performance.

The concept of spatial join has been further generalized to *distance joins*, where a distance based ordering of a subset of the Cartesian product of the two sets is returned. In References 45 and 57, distance join algorithms using R-trees are proposed.

22.5 Analytical Models

Analytical models of R-trees provide performance estimates based on the assumptions of the model. Such models can be useful for gaining insight, comparing algorithms, and for query optimization.

Early analytical models for range queries were proposed by Kamel and Faloutsos [25], and Pagel et al. [59]. The models are similar and use the MBRs of all nodes in the tree as inputs to the model. They derive the probability that Q intersects a given MBR, and use this estimate to compute the expected number of MBRs intersected by Q. In References 60 and 61 the basic model was modified to correct for an error that may arise for MBRs near the boundary of the query sample space. In order to do this, [61] assumes that all queries fall completely within the data space. The changes necessary to handle different sample spaces are straightforward.

The models provide good insight into the problem, especially by establishing a quantitative relationship between performance and the total area and perimeter of MBRs of the tree nodes. We describe the model as presented in Reference 61.

Consider a 2-dimensional data set consisting of rectangles to be stored in an R-tree T with $h + 1$ levels, labeled 0 through h. Assume all input rectangles have been normalized to fit within the unit square $U = [0, 1] \times [0, 1]$. Queries are rectangles Q of size

$q_x \times q_y$. (A point query corresponds to the case $q_x = q_y = 0$.) Initially assume that queries are uniformly distributed over the unit square. Although this description concentrates on 2-d, generalizations to higher dimensions are straightforward.

Assume the following notation:

m_i = number of nodes at the i th level of T
m = Total number nodes in T, that is, $\sum_{i=0}^{h} m_i$
R_{ij} = jth rectangle at the ith level of T
X_{ij} = x-extent (width) of R_{ij}
Y_{ij} = y-extent (height) of R_{ij}
A_{ij} = area of R_{ij}, that is, $A_{ij} = X_{ij} \cdot Y_{ij}$
A_{ij}^Q = probability that R_{ij} is accessed by query Q
B_{ij} = number of accesses to R_{ij}
$\mathcal{A}$ = Sum of the areas of all MBRs in T
L_x = Sum of the x-extents of all MBRs in T
L_y = Sum of the y-extents of all MBRs in T
N = number of queries performed so far
N^* = expected number of queries required to fill the buffer
β = buffer size
$D(N)$ = number of distinct nodes (at all levels) accessed in N queries
$E_T^P(q_x, q_y)$ = expected number of nodes (buffer resident or not) of T accessed while performing a query of size $q_x \times q_y$
$E_T^D(q_x, q_y)$ = expected number of disk accesses while performing a query of size $q_x \times q_y$

The authors of [25,59] assume that performance is measured by the number of nodes accessed (independent of buffering). They observe that for uniform point queries the probability of accessing R_{ij} is just the area of R_{ij}, namely, A_{ij}. They point out that the level of T in which R_{ij} resides is immaterial as all rectangles containing Q (and only those) need to be retrieved. Accordingly, for a point query, the expected number of nodes retrieved as derived in Reference 25 is the sum of node areas[*]:

$$E_T^P(0,0) = \sum_{i=0}^{h} \sum_{j=1}^{m_i} A_{ij} = \mathcal{A} \tag{22.1}$$

which is the sum of the areas of all rectangles (both leaf level MBRs as well as MBRs of internal nodes).

We now turn our attention to region queries. Let $\langle (a,b), (c,d) \rangle$ denote an axis-parallel rectangle with bottom left and top right corners (a,b) and (c,d), respectively. Consider a rectangular query $Q = \langle Q_{bl}, Q_{tr} \rangle$ of size $q_x \times q_y$. Q intersects $R = \langle (a,b), (c,d) \rangle$ if and only if Q_{tr} (the top right corner of Q) is inside the *extended rectangle* $R' = \langle (a,b), (c+q_x, d+q_y) \rangle$, as illustrated in Figure 22.10.

Kamel and Faloutsos infer that the probability of accessing R while performing Q is the area of R', as the region query Q is equivalent to a point query Q_{tr} where *all* rectangles in T have been extended as outlined above. Thus, the expected number of

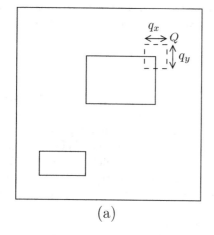

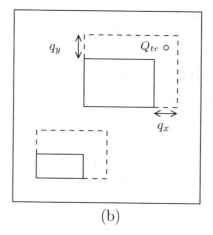

(a)   (b)

FIGURE 22.10 (a) Two data rectangles and region query Q. (b) Corresponding extended rectangles and equivalent point query Q_{tr}.

---

[*] We have modified the notation of [25] to make it consistent with the notation used here.

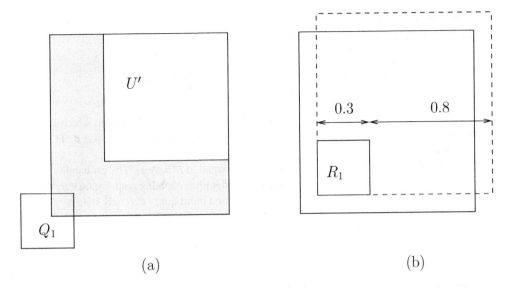

FIGURE 22.11 The domain of Q_{tr} for a query of size 0.3×0.3 is U' (area not shaded).

nodes retrieved (as derived in Reference 25) is:

$$E_T^P(q_x, q_y) = \sum_{i=0}^{h} \sum_{j=1}^{m_i} (X_{ij} + q_x)(Y_{ij} + q_y)$$

$$= \sum_{i=0}^{h} \sum_{j=1}^{m_i} X_{ij} Y_{ij} + q_x \sum_{i=0}^{h} \sum_{j=1}^{m_i} Y_{ij} + q_y \sum_{i=0}^{h} \sum_{j=1}^{m_i} X_{ij} + m q_x q_y$$

$$= \mathcal{A} + q_x L_y + q_y L_x + m q_x q_y \tag{22.2}$$

Equation 22.3 illustrates the fact that a good insertion/loading algorithm should cluster rectangles so as to minimize both the total area and total perimeter of the MBRs of all nodes. For point queries, on the other hand, $q_x = q_y = 0$, and minimizing the total area is enough.

In Reference 61 the model of [25] was modified to handle query windows that fall partially outside the data space as well as data rectangles close to the boundary of the data space, as suggested by Six et al. [59]. Specifically:

1. For uniformly distributed rectangular queries of size $q_x \times q_y$ the top right corner of the query region Q, cannot be an arbitrary point inside the unit square if the entire query region is to fit within the unit square. For example, if $q_x = q_y = 0.3$, a query such as Q_1 in Figure 22.11a should not be allowed. Rather, Q_{tr} must be inside the box $U' = [q_x, 1] \times [q_y, 1]$.
2. The probability of accessing a rectangle $R = \langle (a, b), (c, d) \rangle$ is *not always* the area of $R' = \langle (a, b), (c + q_x, d + q_y) \rangle$ as this value can be bigger than one. For example, in Figure 22.11b, the probability that a query of size 0.8×0.8 accesses rectangle R_1 should not be $1.1 \cdot 1.1 = 1.21$, which is the area of the extended rectangle R'_1, obtained by applying the original formula. Rather, the access probability is the percentage of U' covered by the rectangle $R' \cap U'$.

Thus, we change the probability of accessing rectangle i of level j to:

$$A_{i,j}^Q = \frac{\text{area}(R' \cap U')}{\text{area}(U')}$$

$$= \frac{C \cdot D}{(1 - q_x)(1 - q_y)} \tag{22.3}$$

where $C = [\min(1, c + q_x) - \max(a, q_x)]$ and $D = [\min(1, d + q_y) - \max(b, q_y)]$.

In Reference 62 the R-tree model was expanded to take into account the distribution of the input data. Specifically, rather than being uniformly distributed, the query regions were assumed to be distributed according to the input data distribution.

The above models do not consider the impact of the buffer. In Reference 62 a buffer model is integrated with the query model. Specifically, under uniformly distributed point queries, the probability of accessing rectangle R_{ij} while performing a query is

A_{ij}. Accordingly, the probability that R_{ij} is not accessed during the next N queries is $P[B_{ij} = 0|N] = (1 - A_{ij})^N$. Thus, $P[B_{ij} \geq 1|N] = 1 - (1 - A_{ij})^N$ and the expected number of *distinct* nodes accessed in N queries is,

$$D(N) = \sum_{i=0}^{h} \sum_{j=1}^{m_i} P[B_{ij} \geq 1|N] = m - \sum_{i=0}^{h} \sum_{j=1}^{m_i} (1 - A_{ij})^N \tag{22.4}$$

Note that $D(0) = 0 < \beta$ and $D(1) = \mathcal{A}$ (which may or may not be bigger than β). The buffer, which is initially empty, first becomes full after performing N^* queries, where N^* is the smallest integer that satisfies $D(N^*) \geq \beta$. The value of N^* can be determined by a simple binary search.

While the buffer is not full the probability that R_{ij} is in the buffer is equal to $P[B_{ij} \geq 1]$. The probability that a random query requires a disk access for R_{ij} is $A_{ij} \cdot P[B_{ij} = 0]$. Since the steady state buffer hit probability is approximately the same as the buffer hit probability after N^* queries, the expected number of disk accesses for a point query at steady state is

$$\sum_{i=0}^{h} \sum_{j=1}^{m_i} A_{ij} \cdot P[B_{ij} = 0|N^*] = \sum_{i=0}^{h} \sum_{j=1}^{m_i} A_{ij} \cdot (1 - A_{ij})^{N^*} \tag{22.5}$$

The above derivation also holds for region queries provided that A_{ij}^Q is used instead of A_{ij}, that is:

$$E_T^D(q_x, q_y) = \sum_{i=0}^{h} \sum_{j=1}^{m_i} A_{ij}^Q \cdot \left(1 - A_{ij}^Q\right)^{N^*}$$

In Reference 61 the authors compare the model to simulation and explore the I/O impact of pinning the top levels of an R-tree into the buffer.

Other analytical models include the following. Theodoridis and Sellis [63] provide a fully analytical model that does not require the R-tree MBRs as input. In Reference 64, a technique is developed for analyzing R-tree performance with skewed data distributions. The technique uses the concept of fractal dimension to characterize the data set and resultant R-tree performance. Analysis has also been used for estimating the performance of R-tree based spatial joins [65,66] and nearest neighbor queries [67,68].

Acknowledgment

This work was supported in part by the National Science Foundation under Grants IRI-9610240 and DMS-0107628.

References

1. D. Comer. The ubiquitous B-tree. *ACM Computing Surveys*, 11(2):121–137, 1979.
2. A. Guttman. R-trees: A dynamic index structure for spatial searching. In *Proc. ACM SIGMOD*, pages 47–57, June 1984.
3. T. Brinkhoff, H.-P. Kriegel, and B. Seeger. Parallel processing of spatial joins using R-trees. In *Proc. 12th International Conference on Data Engineering (ICDE)*, pages 258–265, 1996.
4. I. Kamel and C. Faloutsos. Parallel R-trees. In *Proc. ACM SIGMOD*, pages 195–204, April 1992.
5. A. Papadopoulos and Y. Manolopoulos. Similarity query processing using disk arrays. In *Proc. ACM SIGMOD*, pages 225–236, 1998. parallel NN using R-trees.
6. B. Schnitzer and S.T. Leutenegger. Master-client R-trees: A new parallel R-tree architecture. In *Proc. Conference of Scientific and Statistical Database Systems (SSDBM)*, pages 68–77, 1999.
7. S. Berchtold, D. Keim, and H.-P. Kriegel. The X-tree: An index structure for high-dimensional data. In *Proc. 22nd International Conference on Very Large Databases (VLDB)*, pages 28–39, 1996.
8. N. Katayama and S. Satoh. The SR-tree: An index structure for high-dimensional nearest neighbor queries. In *Proc. ACM SIGMOD*, pages 369–380, 1997.
9. D. White and R. Jain. Similarity indexing: Algorithms and performance. In *Proc. SPIE Storage and Retrieval for Still Image and Video Databases IV*, volume 2670, pages 62–73, 1996.
10. G. Kollios, V.J. Tsotras, D. Gunopulos, A. Delis, and M. Hadjieleftheriou. Indexing animated objects using spatiotemporal access methods. *IEEE Transactions on Knowledge and Data Engineering*, 13(5):758–777, 2001.
11. D. Pfoser, C.S. Jensen, and Y. Theodoridis. Novel approaches in query processing for moving object trajectories. In *Proc. 26th International Conference on Very Large Databases (VLDB)*, pages 395–406, 2000.

12. C.M. Procopiuc, P.K. Agarwal, and S. Har-Peled. Star-tree: An efficient self-adjusting index for moving objects. In *Proc. 4th Workshop on Algorithm Engineering and Experiments (ALENEX)*, pages 178–193, 2002.

13. S. Saltenis and C. Jensen. Indexing of now-relative spatio-bitemporal data. *The VLDB Journal*, 11(1):1–16, 2002.

14. S. Saltenis, C. Jensen, S. Leutenegger, and M.A. Lopez. Indexing the positions of continuously moving points. In *Proc. ACM SIGMOD*, pages 331–342, May 2000.

15. Y. Tao and D. Papadias. MV3R-tree: A spatio-temporal access method for timestamp and interval queries. In *Proc. 27th International Conference on Very Large Databases (VLDB)*, pages 431–440, 2001.

16. Y. Tao, D. Papadias, and J. Sun. The TPR*-tree: An optimized spatio-temporal access method for predictive queries. In *Proc. International Conference on Very Large Databases (VLDB)*, pages 790–801, 2003.

17. M. Kornacker and D. Banks. High-concurrency locking in R-trees. In *Proc. International Conference on Very Large DataBases (VLDB)*, pages 134–145, 1995.

18. V. Ng and T. Kameda. Concurrent access to R-trees. In *Proc. SSD*, pages 142–161, 1993.

19. B.-U. Pagel, H.-W. Six, and M. Winter. Window query-optimal clustering of spatial objects. In *Proc. 14th ACM Symposium on Principles of Database Systems (PODS)*, pages 86–94, 1995.

20. C.H. Ang and T.C. Tan. New linear node splitting algorithm for R-trees. In *Advances in Spatial Databases, 5th International Symposium (SSD)*, volume 1262 of *Lecture Notes in Computer Science*, pages 339–349. Springer, 1997.

21. B. Becker, P. G. Franciosa, S. Gschwind, S. Leonardi, T. Ohler, and P. Widmayer. Enclosing a set of objects by two minimum area rectangles. *Journal of Algorithms*, 21(3):520–541, 1996.

22. Y. Garcia, M.A. Lopez, and S.T. Leutenegger. On optimal node splitting for R-trees. In *Proc. International Conference on Very Large Databases (VLDB)*, pages 334–344, August 1998.

23. N. Beckmann, H.P. Kriegel, R. Schneider, and B. Seeger. The R*-tree: An efficient and robust access method for points and rectangles. In *Proc. ACM SIGMOD*, pages 323–331, May 1990.

24. I. Kamel and C. Faloutsos. Hilbert R-tree: An improved R-tree using fractals. In *Proc. International Conference on Very Large Databases (VLDB)*, pages 500–509, September 1994.

25. I. Kamel and C. Faloutsos. On packing R-trees. In *Proc. 2nd International Conference on Information and Knowledge Management (CIKM)*, pages 490–499, Washington D.C., November 1993.

26. S.T. Leutenegger, M.A. Lopez, and J.M. Edgington. STR: A simple and efficient algorithm for R-tree packing. In *Proc. 14th International Conference on Data Engineering (ICDE)*, pages 497–506, 1997.

27. N. Roussopoulos and D. Leifker. Direct spatial search on pictorial databases using packed R-trees. In *Proc. ACM SIGMOD*, pages 17–31, Austin, Texas, May 1985.

28. M. de Berg, J. Gudmundsson, M. Hammar, and M. Overmars. On R-trees with low query complexity. *Computational geometry: Theory of Applications*, 24(3):179–195, 2003.

29. Y. Garcia, M.A. Lopez, and S.T. Leutenegger. Post-optimization and incremental refinement of R-trees. In *Proc. 7th International Symposium on Advances in Geographic Information Systems (ACM GIS)*, pages 91–96, ACM, 1999.

30. T. Schreck and Z. Chen. Branch grafting method for R-tree implementation. *Journal of Systems and Software*, 53:83–93, 2000.

31. T. Sellis, N. Roussopoulos, and C. Faloutsos. The R+ tree: A dynamic index for multidimensional objects. In *Proc. 13th International Conference on Very Large Databases (VLDB)*, pages 507–518, September 1987.

32. H. Sagan. *Space-filling curves*. Springer-Verlag, New York, 1994.

33. T. Bially. Space-filling curves: Their generation and their application to bandwidth reduction. *IEEE Transactions on Information Theory*, IT-15(6):658–664, 1969.

34. A.R. Butz. Alternative algorithm for Hilbert's space-filling curve. *IEEE Transactions on Computers*, C-20:424–426, 1971.

35. K. Sevcik and N. Koudas. Filter trees for managing spatial data over a range of size granularities. Technical Report CSRI-TR-333, Computer Systems Research Institute, University of Toronto, October 1995.

36. R. Choubey, L. Chen, and E.A. Rundensteiner. GBI: A generalized R-tree bulk-insertion strategy. In *Symposium on Large Spatial Databases*, pages 91–108, 1999.

37. K.L. Cheung and A.W.-C. Fu. Enhanced nearest neighbour search on the R-tree. *SIGMOD Record*, 27(3):16–21, 1998.

38. G.R. Hjaltason and H. Samet. Distance browsing in spatial databases. *ACM Transactions on Database Systems*, 24(2):265–318, 1999.

39. N. Roussopoulos, S. Kelley, and F. Vincent. Nearest neighbor queries. In *Proc. ACM SIGMOD*, pages 71–79, May 1995.

40. H. Ferhatosmanoglu, I. Stanoi, D. Agrawal, and A. El Abbadi. Constrained nearest neighbor queries. *Lecture Notes in Computer Science*, 2121:257–278, 2001.

41. F. Korn and S. Muthukrishnan. Influence sets based on reverse nearest neighbor queries. In *Proc. ACM SIGMOD*, pages 201–212, 2000.

42. I. Stanoi, D. Agrawal, and A. El Abbadi. Reverse nearest neighbor queries for dynamic databases. In *ACM SIGMOD Workshop on Research Issues in Data Mining and Knowledge Discovery*, pages 44–53, 2000.

43. C. Yang and K.-I. Lin. An index structure for efficient reverse nearest neighbor queries. In *Proc. International Conference on Data Engineering (ICDE)*, pages 485–492, 2001.

44. A. Corral, Y. Manolopoulos, Y. Theodoridis, and M. Vassilakopoulos. Closest pair queries in spatial databases. In *Proc. ACM SIGMOD*, pages 189–200, 2000.

45. G.R. Hjaltason, and H. Samet. Incremental distance join algorithms for spatial databases. In *Proc. ACM SIGMOD*, pages 237–248, 1998.

46. J. Shan, D. Zhang, and B. Salzberg. On spatia-range closest-pair queries. In *Proc. 8th International Symposium on Spatial and Temporal Databases (SSTD)*, pages 252–269, 2003.

47. D. Papadias, T. Sellis, Y. Theodoridis, and M. Egenhofer. Topological relations in the world of minimum bounding rectangles: A study with R-trees. In *Proc. ACM SIGMOD*, pages 92–103, 1995.

48. T. Brinkhoff, H.-P. Kriegel, R. Schneider, and B. Seeger. Multi-step processing of spatial joins. In *Proc. ACM SIGMOD*, pages 197–208, 1994.

49. T. Brinkhoff, H.-P. Kriegel, and B. Seeger. Efficient processing of spatial joins using R-trees. In *Proc. ACM SIGMOD*, pages 237–246, 1993.

50. O. Gunther. Efficient computation of spatial joins. In *Proc. International Conference on Data Engineering (ICDE)*, pages 50–59, 1993.

51. C. Gurret and P. Rigaux. The sort/sweep algorithm: A new method for R-tree based spatial joins. In *Proc. SSDBM*, pages 153–165, 2002.

52. M-L. Lo and C.V. Ravishankar. Spatial joins using seeded trees. In *Proc. ACM SIGMOD*, pages 209–220, 1994.

53. N. Mamoulis and D. Papadias. Integration of spatial join algorithms for processing multiple inputs. In *Proc. ACM SIGMOD*, pages 1–12, Philadelphia, Pennsylvania, June 1999.

54. N. Mamoulis and D. Papadias. Multiway spatial joins. *ACM Transactions on Database systems*, 26(4):424–475, 2001.

55. N. Mamoulis and D. Papadias. Slot index spatial join. *IEEE Transactions on Knowledge and Data Engineering*, 15(1):1–21, 2003.

56. J.M. Patel and D.J. DeWitt. Partition based spatial-merge join. In *Proc. ACM SIGMOD*, pages 259–270, Montreal, Quebec, June 1996.

57. H. Shin, B. Moon, and S. Lee. Adaptive multi-stage distance join processing. In *Proc. ACM SIGMOD*, pages 343–354, 2000.

58. M. de Berg, O. Cheong, M. van Kreveld, and M. Overmars. *Computational Geometry: Algorithms and Applications*. Springer-Verlag, Berlin, Heidelberg, 3rd edition, 2008.

59. B-U. Pagel, H-W. Six, H. Toben, and P. Widmayer. Towards an analysis of range query performance in spatial data structures. In *Proc. ACM Symposium on Principles of Database Systems (PODS)*, pages 214–221, May 1993.

60. S. Berchtold, C. Bohm, and H.-P. Kriegel. Improving the query performance of high-dimensional index structures by bulk-load operations. In *Proc. 6th International Conference on Extending Database Technology*, pages 216–230, 1998.

61. S.T. Leutenegger and M.A. Lopez. The effect of buffering on the performance of R-trees. In *Proc. 15th International Conference on Data Engineering (ICDE)*, pages 164–171, 1998.

62. S.T. Leutenegger and M.A. Lopez. The effect of buffering on the performance of R-trees. *IEEE Transactions on Knowledge and Data Engineering*, 12(1):33–44, 2000.

63. Y. Theodoridis and T. Sellis. A model for the prediction of R-tree performance. In *Proc. 8th ACM Symposium on Principles of Database Systems (PODS)*, pages 161–171, Montreal, Quebec, Jun 1996.

64. C. Faloutsos and I. Kamel. Beyond uniformity and independence: Analysis of R-trees using the concept of fractal dimension. In *Proc. ACM Symposium on Principles of Database Systems (PODS)*, pages 4–13, 1994.

65. Y.-W. Huang, N. Jing, and E. A. Rundensteiner. A cost model for estimating the performance of spatial joins using R-trees. In *Statistical and Scientific Database Management*, pages 30–38, 1997.

66. Y. Theodoridis, E. Stefanakis, and T. K. Sellis. Efficient cost models for spatial queries using R-trees. *IEEE Transactions on Knowledge and Data Engineering*, 12(1):19–32, 2000.

67. A. Papadopoulos and Y. Manolopoulos. Performance of nearest neighbor queries in R-trees. In *Proc. 6th International Conference on Database Theory*, pages 394–408, 1997.

68. Y. Tao, J. Zhang, D. Papadias, and N. Mamoulis. An efficient cost model for optimization of nearest neighbor search in low and medium dimensional spaces. *IEEE Transactions on Knowledge and Data Engineering*, 16(10):1169–1184, 2004.

23

Managing Spatiotemporal Data

23.1 Introduction and Background 359
    Spatiotemporal Indexing Techniques
23.2 Overlapping Linear Quadtree 360
    Insertion of an Object in MVLQ • Deletion of an Object in MVLQ • Updating an
    Object in MVLQ
23.3 3D R-Tree 363
    Answering Spatiotemporal Queries Using the Unified Schema • Spatiotemporal Query Types
    • Performance Analysis of 3D R-Trees • Handling Queries with Open Transaction Times
23.4 2+3 R-Tree 368
23.5 HR-Trees 368
23.6 MV3R-Tree 368
23.7 Indexing Structures for Continuously Moving Objects 370
    TPR-Tree • R^{EXP}-tree • STAR-Tree • TPR*-Tree
References 373

Sumeet Dua
Louisiana Tech University

S. S. Iyengar
Florida International University

23.1 Introduction and Background

The term spatiotemporal incorporates the two indispensable phenomena of space and time that characterize many objects in the real world. Spatial databases represent, store, and manipulate spatial data in the form of points, lines, areas, surfaces, and hyper-volumes in multidimensional space. Most of these databases suffer from, what is commonly called, the "Curse of Dimensionality" [1]. In the literature, curse of dimensionality refers to a performance degradation of similarity queries with increasing dimensionality of these databases. One way to reduce this curse is to develop data structures for indexing such databases for efficient similarity query handling. Specialized data structures such as R-trees and its variants (see Chapter 22) have been proposed for this purpose which have demonstrated multifold performance gains in access time on this data over sequential search. On the other hand, temporal databases store time-variant data.

Traditional spatial data structures can store only one "copy" of the data, latest at the "present time," while we frequently encounter applications where we need to store more than one copy of the data. While spatial data structures typically handle objects having spatial components and temporal data structures handle time-variant objects, research in spatiotemporal data structures (STDS) and data models have concentrated around storing moving points and moving objects, or objects with geometries changing over time.

Some examples of domains generating spatiotemporal data include but are not limited to the following: geographical information systems, urban planning systems, communication systems, multimedia systems, and traffic planning systems. Moreover [2], advanced spatiotemporal analyses based on trajectory indexing techniques are used in such areas as behavioral pattern analysis and intelligent transportation decisions. While the assemblage of spatiotemporal data is growing, the development of efficient data structures for storage and retrieval of these data types has not kept pace with this increase. Research in spatiotemporal data abstraction is strewn but there are some important research results in this area that have laid elemental foundation for the development of novel structures for these data types. In this chapter, we will present an assortment of key modeling strategies for spatiotemporal data types.

In spatiotemporal databases (STB), two concepts of times are usually considered: *transaction* and *valid* time. According to Jensen et al. [3], transaction time is the time during which a piece of data is recorded in a relation and may be retrieved. Valid time is a time during which a fact is true in the modeled reality. The valid time can be in the future or in the past.

There are two major directions [4] in the development of STDS. The first direction is space-driven structures, which have indexing based upon the partitioning of the embedding multidimensional space into cells, independent of the distribution of the data in this space. An example of such a space-driven data structure is multiversion linear quadtree (MVLQ) for storing spatiotemporal data. The other direction for storing spatiotemporal data is data-driven structures. These structures partition the set of the data objects rather than the embedding space. Examples of data-driven structures include those based upon R-trees and its variants.

23.1.1 Spatiotemporal Indexing Techniques

A key concept that is worth mentioning earlier in our discussion is indexing in STDS. Generally, indexing methods are influenced by a series of factors such as index generation efficiency, storage utilization, query performance, query type, and caching mechanisms. Theoretically, traditional indexing techniques can be used to access spatiotemporal data, for example, a one-dimensional compound indexes based on B-tree variants can be built for multidimensional data like spatiotemporal data where multidimensional spatiotemporal data are transformed into one-dimensional sorting codes. R-tree and Octree can be extended into spatiotemporal indexes, in which time is viewed as another dimension in addition to spatial dimensions. However, concerns over such issues as index generation performance degradation and overall query processing efficiency have led to the quest for hybrid type of indexing methods, such as HBSTR-tree [2], which combines Hash table, B*-tree, and spatiotemporal R-tree to handle trajectory data processing. Currently, three groups of indexes can be identified. The first group is based on multiversion structures in which each timestamp corresponds to a spatial index structure and unchanged nodes are shared between versions, such as HR-tree and MV3R-tree [5]; next are those based on spatial partition methods, such as SETI and CSE, in which trajectory points are firstly divided into respective spatial partitions and then a temporal index is generated for points in each partition [6]; and lastly, those indexes such as STR-tree and TB-tree that are extensions from traditional spatial indexes such as R-tree with the unique ability to adaptively adjust index structures according to generation performance [7]. A more recently proposed technique in Reference 8 utilizes a parallel indexing technique to extend a two-dimensional spatial interval representation of intervals to a multidimensional parallel space and uses a set of formulas to transform spatiotemporal queries into parallel interval set operations. This transformation results in reducing the problem of multidimensional object relationships to simpler two-dimensional spatial intersection problems. In their survey work on predictive spatiotemporal queries, Hendawi and Mokbel [9] mention a data structure-based spatiotemporal indexing as the most popular technique in the literature, with the following categories:

1. *R-tree based*: These include time parameterized R-tree, TPR-tree, and TPR*-tree which are formed by the addition of a time parameter that supports querying current and projected future positions of moving objects, and others based on convex hull property for indexing objects with nonlinear trajectories using a traditional index structure.

2. *B-tree based*: Examples here include B^x-tree that uses a linear technique to index changes in the underlying data values such as moving objects locations and utilized in algorithms for predictive range queries and KNN queries on near-future positions of the indexed objects; others in this category include the self-tunable spatiotemporal B + -tree (SP^2 B-tree) for handling frequent updates for objects locations by allowing automatic online rebuilding of its subtrees using a different set of reference points and different grid size without significant overhead.

3. *kd-tree based*: This category is based on the kd-tree data structure presented in Reference 7 and its main idea to create new indexes for the most updated pieces of the main index and then throw them away from the main memory after some short time period. An example of this is the MOVIES, for (MOVing objects Indexing using frequent Snapshops) technique [10,11] that is characterized as providing support for time-parameterized predictive queries with the qualities of been space-, query-, update-, and multi-CPU efficient.

4. *Quadtree based*: This is a dual scheme that is used to index the predicted trajectories of moving objects in a dual transformed space such that trajectories for objects in d-dimensional space become points in a higher 2D-dimensional space, where this dual transformed space is then indexed using a regular hierarchical grid decomposition indexing structure.

In this chapter, we will discuss both space-driven and data-driven data structures for storing spatiotemporal data. We will initiate our discussion with MVLQ space-driven data structure.

23.2 Overlapping Linear Quadtree

Tzouramanis et al. [12] proposed multiversion linear quadtrees (MVLQ), also called overlapping linear quadtrees, which are analogous to multiversion B-trees (MVBT) [13], but with significant differences. Instead of storing transaction time for each individual object in MVBT, an MVLQ consolidates object descriptors that share a common transaction time. As it will be evident from the following discussion, these object descriptors are code words derived from a linear representation of multiversion quadtrees.

The idea of storing temporal information about the objects is based upon including a parameter for transaction time for these objects. Each object is given a unique timestamp T_i (transaction time), where $i \in [1 \ldots n]$ and n is the number of objects in the

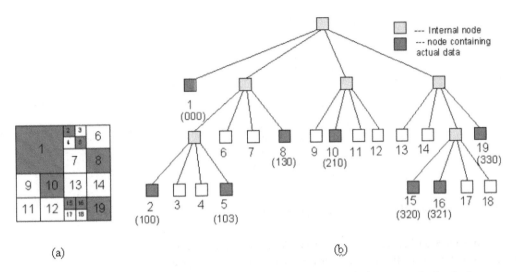

FIGURE 23.1 (a) A binary $2^3 \times 2^3$ image and (b) its corresponding region quadtree. The location codes for the homogeneous codes are indicated in brackets.

database. This timestamp implicitly associates a time value to each record representing the object. Initially, when the object is created in the database, this time interval is set equal to $[T_i,{}^*)$. Here, "*" refers to *as-of-now*, a usage indicating that the current object is valid until a unspecified time in the future, when it would be terminated (at that time, * will be replaced by the T_j, $j \in [1 \ldots n]$ and $j > i$, the timestamp for when the object is terminated).

Before we proceed further, let us gather some brief background in regional Quadtrees, commonly called quadtrees, a spatial data structure. In the following discussion, it is assumed that an STDS is required to be developed for a sequence of evolving regional images, which are stored in an image database. Each image is assumed to be represented as a $2^N \times 2^N$ matrix of pixel values, where N is a positive integer.

Quadtree is a modification of T-pyramid. Every node of the tree, except the leaves, has four children (NW: north-western, NE: north-eastern, SW: south-western, SE: south-eastern). The image is divided into four equal quadrants at each hierarchical level, but it is not necessary to store nodes at all levels. If a parent node has four children of the same homogeneous (e.g., intensity) value, it is not necessary to record them. Figure 23.1 represents an image and its corresponding quadtree representation. Quadtree is a memory-resident data structure, with each node being stored as a record pointing to its children. However, when the represented object becomes too large it may not be possible to store the entire quadtree in the main memory. The strategy that is followed at this point is that the homogeneous components of the quadtrees are stored in a B+ tree, a secondary memory data structure, eliminating the need of pointers in quadtree representation. In this case, addresses representing the location of the homogeneous node and its size constitute a record. Such a version of Quadtree is called linear region quadtree [14].

Several linear representations of regional quadtrees have been proposed, but fixed length (FL), fixed length-depth (FD), and variable length (VL) versions have attracted most attention. We refer the interested reader to Reference 14 for details on these representations, while we concentrate on the FD representation. In this representation, each homogeneous node is represented by a pair of two codes, *location code* and *level number*. Location node C denotes the correct path to this node when traversing the Quadtree from its root till the appropriate leaf is reached and a level number L refers to the level at which the node is located. This makes up the FD linear implementation of Quadtree. The quadrants NW, NE, SW, and SE are represented as 0, 1, 2, and 3, respectively. The location codes for the homogeneous nodes of a quadtree are presented in Figure 23.1.

A persistent data structure (see Chapter 33) [15] is one in which a change to the structure can be made without eliminating the old version, so that all versions of the structure persist and can at least be accessed (the structure is said to be partially persistent) or even modified (the structure is said to be fully persistent). MVLQ is an example of a persistent data structure, in contrast to Linear Quadtree, a transient data structure. In MVLQ, each object in the database is labeled with a time for maintaining the past states. MVLQ couples time intervals with spatial objects in each node.

The leaf of the MVLQ contains the data records of the format:

$$< (C, L), T, EndTime >$$

where

C is the location code of the homogeneous node of the region quadtree,

L is the level of the region quadtree at which the homogeneous node is present,

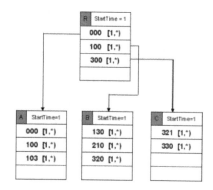

FIGURE 23.2   MVLQ structure after the insertion of image given in Figure 23.1a.

T represents the time interval when the homogeneous node appears in the image sequence, *EndTime* is transaction time when the homogeneous leaf node becomes historical.

The nonleaf nodes contain entries of the form [12]:

$$< C', P', ptr, StartTime >$$

where
 C' is the smallest C recorded in that descendant node,
 P' is the time interval that expresses the lifespan of the latter node,
 Ptr is a pointer to a descendant node,
 StartTime is the time instant when the node was created.

The MVLQ structure after the insertion of image given in Figure 23.1a is given in Figure 23.2.

In addition to the main memory structure of the MVLQ described earlier, it also contains two additional main memory substructures: *Root*table* and Depth First expression (*DF-expression*) of the last inserted object.

Root*table: MVLQ has one tree for each version of the data. Consequently, each version can be reached through its root which can be identified by the time interval of the version that it represents and a pointer to its location. If T'' is the time interval of the root, given by $T'' = [T_i, T_j)$, where $i, j \in [1 \ldots n]$, $i < j$, and *ptr'* is a pointer to the physical address of the root, then each record in the *Root*table* identifying a root is represented in the following form:

$$< T'', ptr' >$$

DF-expression of the last inserted object [12,16]: The purpose of Depth first expression (DF-expression) is to contain the prefetched location of the homogeneous nodes of last insert data object. The advantage of this storage is that when a new data object is being inserted, the stored information can be used to calculate the cost of deletions, insertions, and/or updates. If the DF-expression is not stored, then there is an input/output cost associated with locating the last insert object's homogeneous nodes locations. The DF-expression is an array representation of preorder traversal of the last inserted object's Quadtree.

23.2.1 Insertion of an Object in MVLQ

The first step in insertion of a quadcode of a new object involves identifying the corresponding leaf node. If the corresponding leaf node is full, then a node overflow [12] occurs. Two possibilities may arise at this stage, and depending on the *StartTime* field of the leaf, a split may be initiated. If *NC* is the node capacity, *k* and *c* are integer constants (greater than zero), then the insertion is performed based on the algorithm presented in Figure 23.3.

23.2.2 Deletion of an Object in MVLQ

The algorithm for the deletion of an object from MVLQ is straightforward. If $T_i = StartTime$, then *physical deletion* occurs and the appropriate entry of the object is deleted from the leaf. If number of entries in the leaf $< \lceil NC/k \rceil$ (threshold), then a *node-underflow* is handled as in B+ tree with an additional step of checking for a sibling's *StartTime* for key redistribution. If $T_i > StartTime$, then *logical deletion* occurs and the temporal information of an entry between range $([T_i, *), [T_i, T_j))$ is updated. If an entry is logically deleted in a leaf with exactly $\lceil NC/k \rceil$ present quadcode versions, then a *version underflow* [13] occurs that causes a version split of the node, copying the present versions of its quadcodes into a new node. After version split, the number of present versions of quadcodes is below $(1 + e) \lceil NC/k \rceil$ and a merge is then attempted with a sibling or a copy of that sibling.

If $T_i = StartTime$

    // *Key-split* Old-leaf into Old-leaf and New-leaf

    Old-leaf = First $\lceil NC/2 \rceil$ entries of Old-leaf

    New-Leaf = Remaining entries for the corresponding leaf

If $T_i > StartTime$

    // *Version-Split* the node: Make copy of the old leaf and remove

    //all past entries from the copy leaf. Number of present versions

    //of the quadcodes after version split is within range

    //$[(1+c)*(NC/k),(k-c)*(NC/k)]$.

    If Number of versions after version-split $< (1+c)*(NC/k)$

        Attempt merging with a sibling or a copy of a sibling

        containing *present* version of the quadcode.

    If Number of versions after version-split $> (k-e)*(NC/k)$

        Key-Split the node.

FIGURE 23.3 Algorithm for an insertion of an object in MVLQ.

23.2.3 Updating an Object in MVLQ

Updating a leaf entry refers to update of the field L (level) of the object's code. This is implemented in a two-step fashion. First, a logical deletion of the entry is performed. Second, the new version is inserted in place of that entry through the steps outlined above.

23.3 3D R-Tree

In the previous section, a space-driven, MVLQ-based STDS was presented. In this section, we discuss a data-driven data structure that partitions the set of objects for efficient spatiotemporal representation.

Theodoridis et al. [17] have proposed a data structure termed 3D R-tree for indexing spatiotemporal information for large multimedia applications. The data structure is a derivation of R-trees (see Chapter 22) whose variants have been demonstrated to be an efficient indexing schema for high-dimensional spatial objects.

Theodoridis et al. [17] have adopted a set of operators defined by Papadias and Theodoridis [18] to represent possible topological–directional spatial relationships between two 2-dimensional objects. An anthology of 169 relationships R_{i_j} ($i \in [1 \ldots 13], j \in [1 \ldots 13]$) can represent a complete set of spatial operators. Figure 23.4 represents these relations. An interested reader can find a complete illustration of these topographical relationships in Reference 18. To represent the temporal relationships, a set of temporal operators defined in Reference 19 are employed. Any spatiotemporal relationships among objects can be found using these operators. For example, object Q to appear 7 seconds after the object P, 14 cm to the right and 2 cm down the right bottom vertex of object P can be represented as the following composition tuple:

$$R_t = P[(r_{13\_13}, v3, v4, 14, 2), (-7->)]Q$$

where r_{13_13} is the corresponding spatial relationship, $(-7->)$ is the temporal relationship between the objects, $v3$ and $v4$ are the named vertices of the objects while $(14, 2)$ are their spatial distances on the two axes.

Theodoridis et al. have employed the following typical spatiotemporal relationships to illustrate their indexing schema [17]. These relationships can be defined as *spatiotemporal operators*.

- *overlap_during*(a, b): returns the catalogue of objects a that spatially overlap object b during its execution.
- *overlap_before*(a, b): returns the catalogue of objects a that spatially overlap object b and their execution terminates before the start of execution of b.
- *above_during*(a, b): returns the catalogue of objects a that spatially lie above object b during the course of its execution.

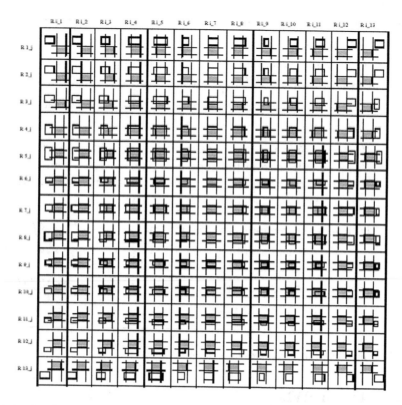

FIGURE 23.4 Spatial relationships between two objects covering directional–topological information. (From M. Abdelguerfi et al., *ACM-GIS*, pp. 29–34, 2002.)

- *above_before(a, b)*: returns the catalogue of objects *a* that spatially lie above object *b* and their execution terminates before the start of execution of *b*.

Spatial and temporal features of objects are typically identified by six dimensions (each spatiotemporal object can be perceived as a point in a six-dimensional space):

$(x_1, x_2, y_1, y_2, t_1, t_2)$, where
(x_1, x_2): Projection of the object on the horizontal plane
(y_1, y_2): Projection of the object on the vertical plane
(t_1, t_2): Projection of the object on the time plane

In a naïve approach, these object descriptors coupled by the object *id* (unique for the object that they represent) can be stored sequentially in a database. An illustration of such an organization is demonstrated in Figure 23.5.

Such a sequential schema has obvious demerits. Answering of spatial–temporal queries, such as one described earlier, would require a full scan of the data organization, at least once. As indicated before, most STB suffer from the curse of dimensionality and sequential organization of data exhibits this curse through depreciated performance, rather than reducing it.

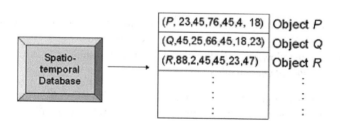

FIGURE 23.5 Schema for sequential organization of spatiotemporal data.

In another schema, two indices can be maintained to store spatial and temporal components separately. Specifically, they can be organized as follows.

1. Spatial index: an index to store the size and coordinates of the objects in two dimensions;
2. Temporal index: a one-dimensional index for storing the duration and start/stop time for objects in one dimension.

R-trees and their variants have been demonstrated to be efficient for storing *n*-dimensional data. Generally speaking, they could be used to store the spatial space components in a 2D R-tree and temporal components in a 1D R-tree. This schema is better than sequential search, since a tree structure provides a hierarchical organization of data leading to a logarithmic time performance. More details on R-trees and its variants can be found in Chapter 22.

Although this schema is better than sequential searching, it still suffers from a limitation [17]. Consider the query *overlap_during*, which would require that both the indices (spatial 2D R-tree and temporal 1D R-tree) are searched individually (in the first phase) and then the intersection of the recovered answer sets from each of the indices is reported as the index's response to the query. Access to both the indices individually and then postintersection can cumulatively be a computationally expensive procedure, especially when each of these indices is dense. Spatial joins [17,20] have been proposed to handle queries on two indexes, provided these indexing schemas adopt the same spatial data structure. It is not straightforward to extend these to handle joins in two varieties of spatial data structures. Additionally, there might be possible inherent relationships between the spatial descriptors and the temporal descriptors resident in two different indexes which can be learned and exploited for enhanced performance. Arranging and searching these descriptors separately may not be able to exploit these relationships. Hence, a unified framework is needed to present spatiotemporal components preferably in a same indexing schema.

Before we proceed further, let us briefly discuss the similarity search procedure in R-trees. In R-trees, minimum bounding boxes (MBB or minimum bounding rectangles in two dimensions) are used to assign geometric descriptors to objects for similarity applications, especially in data mining [21]. The idea behind the usage of MBB in R-trees is the following. If two objects are disjoint, then their MBBs should be disjoint, and if two objects overlap, then their MBB should definitely overlap. Typically, a spatial query on an MBB-based index involves the following steps.

(1) Searching the index: This step is used to select the answer and some possible false alarms from a given data set, ignoring those records that cannot possibly satisfy the query criterion. (2) Dismissals of false alarms: The input to this step is the resultant of the index searching step. In this step, the false alarms are eliminated to identify correct answers, which are reported as the query response. Designing an indexing schema for a multimedia application requires design of a spatiotemporal indexing structure to support spatiotemporal queries. Consider the following scenario.

EXAMPLE 23.1

An application starts with a video clip A located at point (1,7) relative to the application origin Θ. After 2 minutes, an image B appears inside A with 1 unit above its lower horizontal edge and 3 units after its left vertical edge. B disappears after 2 minutes of its presence, while A continues. After 2 minutes of B's disappearance, an text window C appears 3 units below the lower edge of A and 4 units to the right of left edge of it. A then disappears after 1 minute of C's appearance. The moment A disappears, a small image D appears 2 units to the right of right edge of C and 3 units above the top edge of C. C disappears after 2 minutes of D's appearance. As soon as C disappears, a text box E appears 2 units below the lower edge of D and left aligned with it. E lasts for 4 minutes after which it disappears. D disappears 1 minute after E's disappearance. The application ends with D's disappearance. The spatial layout of the above scenario is presented in Figure 23.6a and the temporal layout is presented in Figure 23.6b.

A typical query that can be posed on such a database of objects described earlier is "Which objects overlap the object A during its presentation?" In Reference 17, the authors have proposed a unified schema to handle queries on an STB, such as the one stated above. The schema amalgamates the spatial and temporal components in a single data structure such as R-trees to exhibit advantages and performance gains over the other schemas described earlier. This schema amalgamates need of spatial joins on two spatial data structures besides aggregating both attributes under a unified framework. The idea for representation of spatiotemporal attributes is as follows. If an object which initially lies at point (x_a, y_a) during time $[t_a, t_b)$ and at (x_b, y_b) during $[t_b, t_c)$, it can be modeled by two lines $\overline{((x_a, y_a, t_a), (x_a, y_a, t_b))}$ and $\overline{((x_b, y_b, t_b), (x_b, y_b, t_c))}$. These lines can be presented in a hierarchical R-tree index in three dimensions. Figure 23.7 presents the unified spatial–temporal schema for the spatial and temporal layouts of the example presented above.

23.3.1 Answering Spatiotemporal Queries Using the Unified Schema

Answering queries on the above presented unified schema is similar to handling similarity queries using R-trees. Consider the following queries [17]:

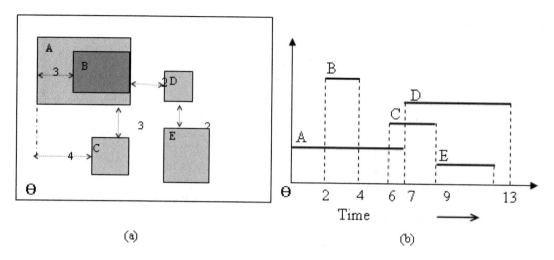

FIGURE 23.6   (a) Spatial layout of the multimedia database and (b) temporal layout of the multimedia database.

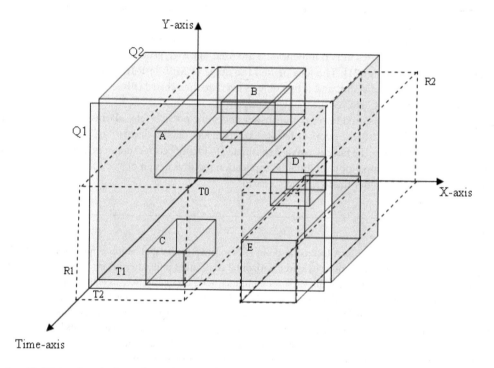

FIGURE 23.7   A unified R-tree-based schema for storing spatial and temporal components.

Query 1: "Find all the objects on the screen at time T_2" (spatial layout query). This query can be answered by considering a rectangle Q1 (Figure 23.7) intersecting the time axis at exactly one point, T_2.

Query 2: "Find all the objects and their corresponding temporal duration between the time interval (T_0, T_1)" (temporal layout query). This query can be answered by considering a box Q2 (Figure 23.7) intersecting the time axis between the intervals (T_0, T_1).

After we have obtained the objects enclosed by the rectangle Q1 and box Q2, these objects are filtered (in main memory) to obtain the answer set.

Query 3: "Find all the objects and their corresponding spatial layout at time $T = 3$ minutes." This query can be answered by looking at the screenshot of spatial layout of objects that were present at the given instant of time. The resultant would be the list of objects and their corresponding spatial descriptors. The response to this query is given in Figure 23.8a.

Query 4: Find the temporal layout of all the objects between the time interval $(T_1 = 2, T_2 = 9)$. This query can be answered by drawing a rectangle on the unified index with dimensions $(X_{max} - 0) \times (Y_{max} - 0) \times (T_2 - T_1)$. The response of the query is shown in Figure 23.8b.

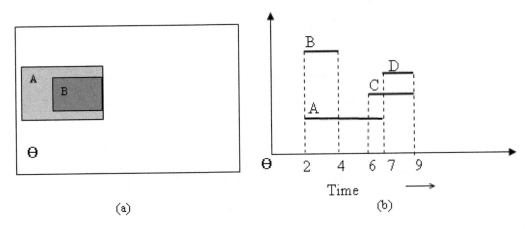

FIGURE 23.8　(a) Response to Query 3 and (b) response to Query 4.

23.3.2　Spatiotemporal Query Types

Here, we consider some major query types that are commonly performed on spatiotemporal data. Existing algorithms for these are mostly predictive in nature [9].

- *Range queries* [22,23]: This has a query range R and a future time t, and requests about the objects expected to be inside R after time t. This utilizes a mobility model that is based on a time parameterized bounding rectangle as a moving region that intersects or otherwise, a moving circle that represents a range query.
- *K-nearest-neighbor queries* [23,24]: This has a location point P, a future time t, and requests about the K objects expected to be closest to P after time t. Two algorithms, RangeSearch and KNNSearchBF, work by traversing spatiotemporal index tree (TPR/TPR*-tree) to find the nodes that intersect with the query circular region for Range and KNN queries, respectively.
- *Reverse-nearest-neighbor queries* [23,25]: Contrary to the predictive KNN query, predictive reverse-nearest-neighbor query finds out the objects expected to have the query region as their nearest neighbor.
- *Aggregate queries* [26]: This type has a query region R, future time t, and requests about the number of objects N predicted to be inside R after time t. Sun et al. [26] present a comprehensive technique that employs adaptive multidimensional histogram (AMH), historical synopsis, and stochastic method to provide approximate answers for aggregate spatiotemporal queries for the future, in addition to the past, and the present.
- *Continuous queries* [27,28]: Continuous query Q, mostly server-side algorithm that is stored until the end of its life, could be any of the above discussed types but with the additional condition that it needs to be continuously reevaluated many times throughout its life in the underlying system. The rate of reevaluation depends on the time gap, t_{gap} between each two consecutive reported answers specified in the received query Q.

23.3.3　Performance Analysis of 3D R-Trees

Theodoridis et al. [17] analyzed the performance of the proposed R-trees using the expected retrieval cost metric that they presented in Reference 15. An interested reader is referred to References 17 and 29 for details on this metric. Based on the analytical model of this metric, it is asserted that one can estimate the retrieval cost of an *overlap query* based on the information attainable from the query window and data set only. In the performance analysis, it was demonstrated that since the expected retrieval cost metric expresses the expected performance of R-trees on overlapping queries, the retrieval of spatiotemporal operators using R-trees is cost equivalent to the cost of retrieval of an overlap query using an appropriate query window Q. Rigorous analysis with 10,000 objects asserted the following conclusions as described in Reference 17:

1. For the operators with high selectivity (*overlap, during, overlap_during*), the proposed 3D R-trees outperformed sequential search at a level of 1 to 2 orders of magnitude.
2. For operators with low selectivity (*above, before, above_before*), the proposed 3D R-trees outperformed sequential search by factors ranging between 0.25 and 0.50 fraction of the sequential cost.

23.3.4　Handling Queries with Open Transaction Times

In the previous section although 3D R-trees were demonstrated to be very efficient compared to sequential search, it suffers from a limitation. The transaction times presented as creation and termination time of an object are expected to be known *a priori*

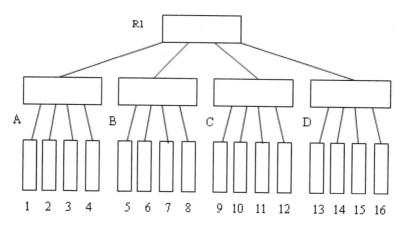

FIGURE 23.9   An R-tree at time *T*0.

before they can be stored and queried from the index. However, in most practical circumstances, the duration of existence of an object is not known. In other words, when the object is created, all we know is that it would remain valid *until changed*. The concept of *until changes* is a well-discussed issue [30,31]. Data structures like R-trees and also its modified form of 3D R-tree are typically not capable of handling such queries having open transaction times. In the following section, we discuss two STDS capable of handling queries with open transaction times.

23.4   2+3 R-Tree

Nascimento et al. [31] have proposed a solution to the problem of handling objects with open transaction times. The idea is to split the indexing of these objects into two parts: one for two-dimensional points and other for three-dimensional lines. Two-dimensional points store the spatial information about the objects and their corresponding start time. Three-dimensional lines store the historical information about the data. Once the "started" point stored in 2D R-tree becomes closed, a corresponding line is constructed to be stored in the 3D R-tree and the point entry is deleted from the 2D R-tree. It should be understood that both trees are expected to be searched depending on the timestamp at which the query is posed. If the end times for each of the spatial points is known *a priori*, then the need of a 2D can be completely eliminated reducing the structure to a 3D R-tree.

23.5   HR-Trees

One possible way to index spatiotemporal data is to build one R-tree for each timestamp, which is certainly space inefficient. The demerit of 2 + 3 R-tree is that it requires two stores to be searched to answer a query. The HR-tree [32] or Historical R-tree "compresses" multiple R-trees, so that R-tree nodes which are common in two timestamps are shared by the corresponding R-trees. The main idea is based on the expectation that a vast majority of indexed points do not change their positions at every timestamp. In other words, it is reasonable to expect that a vast majority of spatial points will "share" common nodes in respective R-trees. The HR-tree exploits this property by keeping a "logical" linkage to a previously present spatial point if it is referenced again in the new R-tree. Consider two R-trees in Figures 23.9 and 23.10 at two different timestamps (*T*0, *T*1). An HR-tree for these R-trees is given in Figure 23.11.

Recently, Ni and Ravishankar [33] proposed the PA-tree as an indexing scheme for historical trajectory data. PA-tree is a parametric space indexing method that uses approximating sequence of movement functions with single continuous polynomial approximations, and demonstrated how this could support both offline and online query processing of historical trajectories. Their work showed that although existing schemes such as MVR-trees and SETI are faster for index construction and timestamp queries, PA-trees are an order of magnitude faster to construct and incur lower I/O cost for spatiotemporal range queries which makes it an excellent choice for both offline and online processing of historical trajectories.

23.6   MV3R-Tree

The MV3R-tree was proposed by Tao and Papadias [34] and is demonstrated to process timestamp and interval queries efficiently. Timestamp queries discover all objects that intersect a window query at a particular timestamp. On the other hand, interval

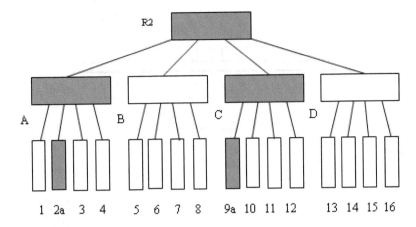

FIGURE 23.10 An R-tree at time *T*1. The modified nodes are in dark boxes.

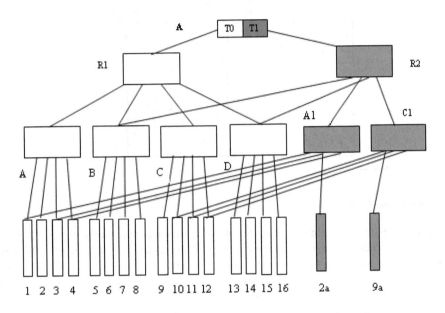

FIGURE 23.11 The HR-tree.

queries include multiple timestamps in the discovered response. While the spatial data structures proposed before have been demonstrated [17,34] to be suitable for either of these queries (or a subset of it), none of them have been established to process both timestamp and interval queries efficiently. MV3R-tree has addressed limitations in terms of efficiency of the above trees in handling both of these queries. MV3R-tree consists of a multiversion R-tree (MVR-tree) and an auxiliary 3D R-tree built on the leaves of the MVR-tree. Although the primary motivation behind the development of MV3R-tree is MVBT [13] (which are extended from B+ trees), they are significantly different from these versions. Before we proceed further, let us understand the working of an MVBT.

A typical entry of an MVBT takes the following form: $< id, time_{start}, time_{end}, P >$. For nonleaf nodes, $time_{start}$ and $time_{end}$ are the minimum and maximum values, respectively, in this node and P is a pointer to the next level node. For a leaf node, the timestamps $time_{start}$ and $time_{end}$ indicate when the object was inserted and deleted from the index and pointer P points to the actual record with a corresponding *id* value. At time $time_{current}$, the entry is said to be *alive* if $time_{start} < time_{current}$, otherwise *dead* [34]. There can be multiple roots in an MVBT, where each root has a distinguishing time range it represents. A search on the tree begins at identifying the root within which the timestamp of the query belongs. The search is continued based on the *id*, $time_{start}$, and $time_{end}$. A *weak version condition* specifies that for each node, except the root, at least $K.T$ entries are alive at time t, where K is the node capacity and T is the tree parameter. This condition ensures that entries alive at the identical timestamps are in a majority of the cases assembled together to allow easy timestamp queries.

3D R-trees are very space efficient and can handle long interval queries efficiently. However, timestamp and short-interval queries using 3D R-trees are expensive. In addition to this, 3D R-trees do not include a methodology such as the weak version

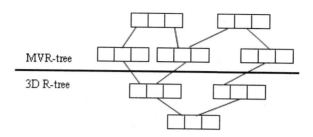

FIGURE 23.12   An MV3R-tree.

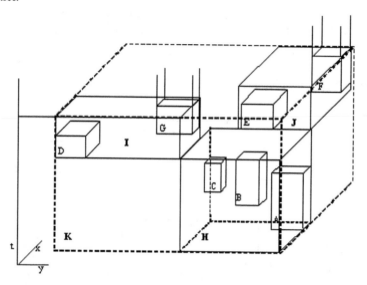

FIGURE 23.13   A 3D visualization of MVR-tree.

condition to ensure that each node has a minimum number of live entries at a given timestamp. HR-trees [32], on the other hand, maintain an R-tree (or its derivative) for each timestamp and the timestamp query is directed to the corresponding R-tree to be searched within it. In other words, the query disintegrates into an ordinary window query and is handled very efficiently. However, in case of an interval query, several timestamps should search the corresponding trees of all the timestamps constituting the interval. The original work in HR-tree did not present a schema for handling interval queries; however, Tao and Papadias [34] have proposed a solution to this problem by the use of negative and positive pointers. The performance of this schema is then compared with MV3R-tree in Reference 34. It is demonstrated that MV3R-trees outperform HR-trees and 3D R-trees even in extreme cases of only timestamp and interval-based queries.

Multiversion 3D R-trees (MV3R-trees) combine a multiversion R-tree (MVR-tree) and a small auxiliary 3D R-tree built on the leaf nodes of the MVR-tree as shown in Figure 23.12. MVR-trees maintain multiple R-trees and have entries of the form $< MBR, time_{start}, time_{end}, P >$, where MBR refers to minimum bounding rectangle and other entries are similar to B+ trees. An example of a 3D visualization of an MV3R-tree of height 2 is shown in Figure 23.13. The tree consists of Object cubes (A–G), leaf nodes (H–J) and root of the tree K. Detailed algorithms for insertion and deletion in these trees are provided in Reference 34. Timestamp queries can be answered efficiently using MVR-trees. An auxiliary 3D R-tree is built on the leaves of the MVR tree in order to process interval queries. It is suggested that for a moderate node capacity, the number of leaf nodes in an MVR-tree is much lower than the actual number of objects, hence this tree is expected to be small compared to a complete 3D R-tree.

Performance analysis has shown that the MV3R-tree offers better trade-off between query performance and structure size than the HR-tree and 3D R-tree. For typical situations where workloads contain both timestamp and interval queries, MV3R-trees outperform HR-trees and 3D R-trees significantly. The incorporation of the auxiliary 3D R-tree not only accelerates interval queries but also provides flexibilities toward other query processing, such as spatiotemporal joins.

23.7   Indexing Structures for Continuously Moving Objects

Continuously moving objects pose new challenges to indexing technology for large databases. Sources for such data include GPS systems, wireless networks, air-traffic controls, and so on. In the previous sections, we have outlined some indexing schemas that

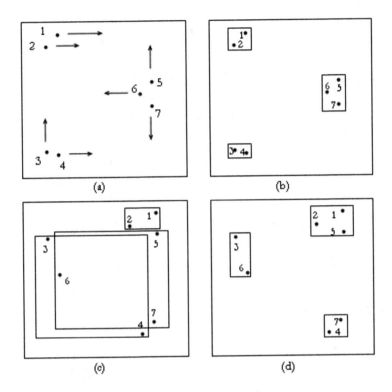

FIGURE 23.14 Moving data points and their leaf-level MBBs at subsequent points.

could efficiently index spatiotemporal data types but some other schemas have been developed specifically for answering predictive queries in a database of continuously moving objects. In this section, we discuss an assortment of such indexing schemas.

Databases of continuously moving objects have two kinds of indexing issues: storing the historical movements in time of objects and predicting the movement of objects based on previous positional and directional information. Such predictions can be made more reliably for a future time T_f, not far from the current timestamp T_f. As T_f increases, the predictions become less and less reliable since the change of trajectory by a moving object results in inaccurate prediction. Traditional indexing schemas such as R*-trees are successful in storing multidimensional data points but are not directly useful for storing moving objects. An example [35] of such kind of system is shown in Figure 23.14. For simplicity, a two-dimensional space is illustrated, but practical systems can have larger dimensions. First part of the figure shows multiple objects moving in different directions. If traditional R*-trees are used for indexing these data, the MBB for the leaf level of the tree are demonstrated in Figure 23.14b. But these objects might be following different trajectories (as shown by arrows in Figure 23.14a) and at subsequent timestamps, the leaf level MBB might change in size and position, as demonstrated in Figure 23.14c and d.

Since traditional indexing methods are not designed for such kind of applications, some novel indexing schemas are described in References 35–38 to handle such data types and queries imposed on them.

23.7.1 TPR-Tree

Saltenis et al. [35] proposed TPR-tree, an acronym for Time-Parameterized R*-tree, based on the underlying principles of R-tree. TPR-tree indexes the current and future anticipated positions of moving objects in one, two, and three dimensions. The basic algorithms of R*-trees are employed for TPR-tree with a modification that the leaf and nonleaf minimum bounding rectangles are now augmented with velocity vectors for these rectangles. The velocity vector for an edge of the rectangle is chosen so that the object remains inside the moving rectangle. TPR-tree can typically handle the following three types of queries:

- Timeslice Query: A query Q specified by hyperrectangle R located at time point t.
- Window Query: A query Q specified by hyperrectangle R covering an interval from $[T_a, T_b]$.
- Moving Query: A query Q specified by hyperrectangles R_a and R_b at different times T_a and T_b, forming a trapezoid.

Figure 23.15 shows objects o1, o2, o3, and o4 moving in time. The trajectories of these objects are shifting, as shown in the figure. The three types of queries as described earlier are illustrated in this figure. Q0 and Q1 are timeslice queries, Q2 and Q3 are window queries, and Q4 is a moving query.

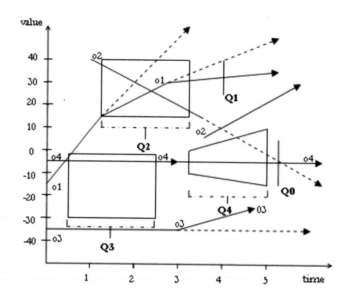

FIGURE 23.15   Types of queries on one-dimensional data.

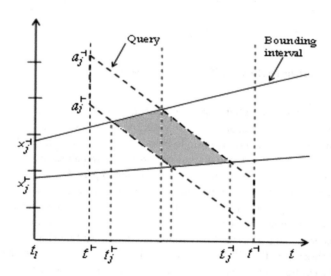

FIGURE 23.16   A bounding interval and a query imposed on the TPR-tree.

The structure of TPR-tree is very similar to R*-tree with leaves consisting of position and pointer of the moving object. The nodes of the tree consist of pointers to subtree and bounding rectangles for the entries in subtree. TPR-trees store the moving objects as linear function of time with time-parameterized bounding rectangles. The index does not consist of points and rectangles for timestamp older than current time. TPR-tree differs from the R*-trees in how its insertion algorithms group points into nodes. While in R*-trees, the heuristics of the minimized area, overlap, and margin of bounding rectangles are used to assign points to the nodes of the tree, in case of TPR-trees these heuristics are replaced by their respective integrals, which are representative of their temporal components. Given an objective function $F(t)$, the following integral is expected to be minimized [35].
$\int_{t_c}^{t_c+H} F(t)dt$, where t_c is the current time and H is the time horizon.

The objective function can be area or perimeters of the bounding rectangles, or could represent the overlap between these rectangles. Figure 23.16 represents a bounding interval and a query in the TPR-tree. The area of the shaded region in Figure 23.16 represents the time integral of the length of the bounding interval.

Saltenis et al. [35] compared the performance of TPR-trees with load-time bounding rectangles, TPR-tree with update-time bounding rectangles, and R-tree with a set of experiments with varying workloads. The results demonstrated that TPR-tree outperforms other approaches by considerable improvement. It was also demonstrated that TPR-tree does not degrade severely in performance with increasing time and it can be tuned to take advantage of a specific update rate.

23.7.2 R^{EXP}-tree

Saltenis and Jensen [36] proposed R^{EXP}-tree as a balanced, multiway tree with a structure of R*-tree. R^{EXP}-tree is an improvement over TPR-tree, assuming that some objects used in indexing expires after a certain period. These trees can handle realistic scenario where certain objects are no longer required, that is when they expire. By removing the expired entries and recomputing bounding rectangles, the index organizes itself to handle subsequent queries efficiently. This tree structure finds its application where the objects do not report their positions for a certain period, possibly implying that they are no more interested in the service.

The index structure of R^{EXP}-tree differs from TPR-tree in insertion and deletion algorithms for disposing the expired nodes. R^{EXP}-tree uses a "lazy strategy" for deleting the expired entries. Another possible strategy is scheduled deletion of entries in TPR-trees. During the search, insertion, and deletion operations, only the live entries are searched and expired entries are physically removed when the content of the node is modified and is written to the disk. Whenever an entry in internal node is deleted, the entire subtree is reallocated. The performance results demonstrated in Reference 36 show that choosing the right bounding rectangles and corresponding algorithms for grouping entries is not straightforward and depends on the characteristics of the workloads.

23.7.3 STAR-Tree

Procopiuc et al. [37] propose a Spatio-Temporal Self-Adjusting R-tree or STAR-tree. STAR-tree indexing schema is similar to TPR trees with few differences. Specifically, STAR-tree groups point according to their current locations and may result in points moving with different velocities being included in the same rectangle. Scheduled events are used to regroup points to control the growth of such bounding rectangles. It improves the structure of TPR-tree by self-adjusting the index, whenever index performance degrades. Intervention of user is not needed for adjustment of the index and the query time is kept low even without continuously updating the index by positions of the objects. STAR-tree does not need periodic rebuilding of indexing and estimation of time horizon. It provides trade-offs between storage and query performance and between time spent in updating the index and in answering queries. STAR-tree can handle not only the timeslice and range queries as those handled by TPR-trees but also nearest neighbor queries for continuously moving objects.

23.7.4 TPR*-Tree

TPR*-tree proposed by Tao et al. [38] is an optimized spatiotemporal indexing method for predictive queries. TPR-tree, described in the previous section, does not propose an analytical model for cost estimation and query optimization and quantification of its performance. TPR*-tree assumes a probabilistic model that accurately estimates the number of disk accesses in answering a window query in a spatiotemporal index. Tao et al. [38] investigate the optimal performance of any data partition index using the proposed model.

The TPR*-tree improves the performance of TPR-tree by employing a new set of insertion and deletion algorithms that minimize the average number of node accesses for answering a window query, whose MBB uniformly distributes in the data space. The static point interval query with the following constraints has been optimized [38] using the TPR*-tree:

- MBB has a length $|Q^R| = 0$ on each axis.
- Velocity bounding rectangle is {0,0,0,0}.
- Query interval $Q_I = [0, H]$, where H is the horizon parameter.

It is demonstrated that the above choice of parameters leads to nearly optimal performance independently of the query parameters. The experiments have also shown that TPR*-trees significantly outperform the conventional TPR-tree under all conditions.

References

1. B.-U. Pagel, F. Korn, C. Faloutsos, "Deflating the Dimensionality Curse using Multiple Fractal Dimensions," In *16th International Conference on Data Engineering (ICDE)*, San Diego, CA, 2000.
2. S. Ke, J. Gong, S. Li, Q. Zhu, X. Liu, Y. Zhang, "A Hybrid Spatio-Temporal Data Indexing Method for Trajectory Databases," *Sensors*, Vol. 14, 2014, pp. 12990–13005.
3. C. S. Jensen, J. Clifford, R. Elmarsi et al., "A Consensus Glossary of Temporal Database Concepts," *SIGMOD Record*, 23 (1), pp. 52–64, 1994.
4. M. Abdelguerfi, J. Givaudan, K. Shaw, R. Ladner, "The 2–3TR-Tree, a Trajectory-Oriented Index Structure for Fully Evolving Valid-Time Spatio-Temporal Datasets," *Proceedings of the 10th ACM international Symposium on Advances in Geographic Information Systems (GIS '02)*. ACM, New York, NY, USA, pp. 29–34, 2002. https://dl.acm.org/ citation.cfm?id=585155

5. M. Mokbel, T. Ghanem, W. Aref, "Spatio-Temporal Access Methods," *IEEE Data Engineering Bulletin*, Vol. 26, 2003, pp. 40–49.

6. V. Chakka, A. Everspaugh, J. Patel, "Indexing Large Trajectory Data Sets with SETI," In *Proceedings of the 2003 CIDR Conference*, Asiloma, USA, 2003.

7. L.-V. Nguyen-Dinh, W. G. Aref, M. Mokbel, "Spatio-Temporal Access Methods: Part 2 (2003–2010)," *Purdue University Libraries—Purdue e-Pubs: Cyber Center Publications*, Paper 19, 2010.

8. Z. He, M. J. Kraak, O. Huisman, X. Ma, J. Xiao, "Parallel Indexing Technique for Spatio-Temporal Data," *ISPRS Journal of Photogrammetry and Remote Sensing*, Vol. 78, 2013, pp. 116–128.

9. A. M. Hendawi, M. F. Mokbel, "Predictive Spatio-Temporal Queries: A Comprehensive Survey and Future Directions," In *ACM SIGSPATIAL MobiGIS'12*, Redondo Beach, CA, 2012.

10. J. Dittrich, L. Blunschi, M. A. V. Salles, "MOVIES: Indexing Moving Objects by Shooting Index Images," *International Journal on Advances of Computer Science for Geographic Information Systems, GeoInformatica*, Vol. 15, No. 4, pp. 727–767, 2011.

11. J. D. L. Blunschi, M. A. V. Salles, "Indexing Moving Objects Using Short-Lived Throwaway Indexes," In *Proceedings of the 11th International Symposium on Advances in Spatial and Temporal Databases*, July 08–10, Aalborg, Denmark, 2007. Doi: 10.1007/978-3-642-02982-0_14.

12. T. Tzouramanis, M. Vassilakopoulos, Y. Manolopoulos, "Multiversion linear quadtree for spatiotemporal Data," In *Proceedings 4th East-European Conference on Advanced Databases and Information Systems (ADBIS-DASFAA'00)*, 279–292, Prague, Czech Republic, 2000.

13. B. Becker, S. Gschwind, T. Ohler, B. Seeger, P. Widmayer, "An Asymptotically Optimal Multiversion B-tree," *The VLDB Journal*, Vol. 5, pp. 264–275, 1996.

14. H. Samet, *The Design and Analysis of Spatial Data Structures*, Addison-Wesley, Reading, MA, 1990.

15. J. R. Driscoll, N. Sarnak, D. D. Sleator, R. E. Tarjan, "Making Data Structures Persistent," *Journal of Computer and System Sciences*, Vol. 38, pp. 86–124, 1989.

16. E. Kawaguchi, T. Endo, "On a Method of Binary Picture Representation and its Application to Data Compression," *IEEE Transactions on Pattern Analysis and Machine Intelligence*, Vol. 2, No. 1, pp. 27–35, 1980.

17. Y. Theodoridis, M. Vazirgiannis, T. Sellis, "Spatio-Temporal Indexing for Large Multimedia Applications," In *Proceedings of the Third IEEE International Conference on Multimedia Computing and Systems*, Hiroshima, Japan, pp. 441–448, 1996.

18. D. Papadias, Y. Theodoridis, "Spatial Relations, Minimum Bounding Rectangles, and Spatial Data Structures," *International Journal of Geographic Information Systems*, Vol. 11, No. 2, 1997.

19. M. Vazirgiannis, Y. Theodoridis, T. Sellis, "Spatio-Temporal Composition in Multimedia Applications," Technical Report KDBSLAB-TR-95-09, Knowledge, Database Systems Laboratory, National Technical University of Athens, 1995.

20. M. G. Martynov, "Spatial Joins and R-trees." In J. Eder, L.A. Kalinichenko (eds.), *Advances in Databases and Information Systems. Workshops in Computing*. Springer, London, 1996.

21. Y. Theodoridis, D. Papadias, "Range Queries Involving Spatial Relations: A Performance Analysis," In A.U. Frank, W. Kuhn (eds.), *Spatial Information Theory A Theoretical Basis for GIS. COSIT 1995. Lecture Notes in Computer Science*, vol 988. Springer, Berlin, Heidelberg, 1995.

22. H. Jeung, M. L. Yiu, X. Zhou, C. S. Jensen, "Path Prediction, Predictive Range Querying in Road Network Databases," *VLDB Journal*, Vol. 19, pp. 585–602, 2010.

23. R. Zhang, H. V. Jagadish, B. T. Dai, K. Ramamohanarao, "Optimized Algorithms for Predictive Range and KNN Queries on Moving Objects," *Information Systems*, Vol. 35, pp. 911–932, 2010.

24. R. Benetis, C. S. Jensen, G. Karciauskas, S. Saltenis, "Nearest and Reverse Nearest Neighbor Queries for Moving Objects," *VLDB Journal*, Vol. 15, pp. 229–249, 2006.

25. T. Xia, D. Zhang, "Continuous Reverse Nearest Neighbor Monitoring," In *Proceedings of the International Conference on Data Engineering, ICDE*, Georgia, USA, 2006.

26. J. Sun, D. Papadias, Y. Tao, B. Liu, "Querying about the Past, the Present, and the Future in Spatio-Temporal," In *Proceedings of the International Conference on Data Engineering, ICDE*, Massachusetts, USA, 2004.

27. J. Kang, M. F. Mokbel, S. Shekhar, T. Xia, D. Zhang, "Continuous Evaluation of Monochromatic and Bichromatic Reverse Nearest Neighbors," In *Proceedings of the International Conference on Data Engineering, ICDE*, Istanbul, Turkey, 2007.

28. H. Wang, R. Zimmermann, W.-S. Ku, "Distributed Continuous Range Query Processing on Moving Objects," In *Proceedings of the International Conference on Database and Expert Systems Applications, DEXA*, Krakow, Poland, 2006.

29. Y. Theodoridis, T. Sellis, "On the Performance Analysis of Multidimensional R-tree-based Data Structures," Technical Report KDBSLABTR-95-03, Knowledge, Database Systems Laboratory, National Technical University of Athens, Greece, 1995.

30. J. Clifford, C. E. Dyreson, T. Isakowitz, C. S. Jensen, R. T. Snodgrass, "On the Semantics of "Now" in Databases," *ACM Transactions on Database Systems*, 22, pp. 171–214, 1997.

31. M. Nascimento, R. Silva, Y. Theodoridis, "Evaluation of Access Structures for Discretely Moving Points," In M.H. Böhlen, C.S. Jensen, and M. Scholl (eds.), *Proceedings of the International Workshop on Spatio-Temporal Database Management* (STDBM '99), Springer-Verlag, London, UK, pp. 171–188, 1999.

32. M. Nascimento, J. Silvia, "Towards Historical Rtrees," In *Proceedings of the 1998 ACM symposium on Applied Computing* (SAC '98). ACM, New York, NY, USA, pp. 235–240, 1998.

33. J. Ni, C. V. Ravishankar, "Indexing Spatio-Temporal Trajectories with Efficient Polynomial Approximations," *IEEE Transactions on Knowledge and Data Engineering*, Vol. 19, pp. 663–678, 2007.

34. Y. Tao, D. Papadias, "MV3R-Tree: A Spatio-Temporal Access Method for Timestamp and Interval Queries," In *Proceedings of the 27th International Conference on Very Large Data Bases*, September 11–14, pp. 431–440, 2001.

35. S. Saltenis, C. Jensen, S. Leutenegger, M. Lopez, "Indexing the Positions of Continuously Moving Objects," In *Proceedings of the 19th ACM-SIGMOD International Conference on Management of Data*, Dallas, TX, 2000.

36. S. Saltenis, C. S. Jensen, "Indexing of Moving Objects for Location-Based Services," *TimeCenter* (2001), TR-63, 24 pages.

37. C.M. Procopiuc, P. K. Agarwal, S. Har-Peled, "STAR-Tree: An Efficient Self-Adjusting Index for Moving Objects," In D.M. Mount, C. Stein (eds.), *Revised Papers from the 4th International Workshop on Algorithm Engineering and Experiments* (ALENEX '02). Springer-Verlag, London, UK, pp. 178–193, 2002.

38. Y. Tao, D. Papadias, J. Sun, "The TPR*-Tree: An Optimized Spatio-Temporal Access Method for Predictive Queries," *Proceedings of the 29th International Conference on Very Large Data Bases, September 09–12*, Berlin, Germany, pp. 790–801, 2003.

24

Kinetic Data Structures*

24.1 Introduction.. 377
24.2 Motion in Computational Geometry 377
24.3 Motion Models 377
24.4 Kinetic Data Structures.. 378
 Convex Hull Example • Performance Measures for KDS • The Convex Hull, Revisited
24.5 A KDS Application Survey.. 381
 Extent Problems • Proximity Problems • Triangulations and Tilings • Collision Detection • Connectivity and Clustering • Visibility • Result Summary • Open Problems
24.6 Querying Moving Objects.. 387
24.7 Sources and Related Materials 387
 References.. 387

Leonidas Guibas
Stanford University

24.1 Introduction

Motion is ubiquitous in the physical world, yet its study is much less developed than that of another common physical modality, namely shape. While we have several standardized mathematical shape descriptions, and even entire disciplines devoted to that area—such as *Computer-Aided Geometric Design* (CAGD)—the state of formal motion descriptions is still in flux. This in part because motion descriptions span many levels of detail; they also tend to be intimately coupled to an underlying physical process generating the motion (dynamics). Thus, until recently, proper abstractions were lacking and there was only limited work on algorithmic descriptions of motion and their associated complexity measures. This chapter aims to show how an algorithmic study of motion is intimately tied via appropriate data structures to more classical theoretical disciplines, such as discrete and computational geometry. After a quick survey of earlier work (Sections 24.2 and 24.3), we devote the bulk of this chapter to discussing the framework of *Kinetic Data Structures* (Section 24.4) [1,2] and its many applications (Section 24.5). We also briefly discuss methods for querying moving objects (Section 24.6).

24.2 Motion in Computational Geometry

Motion has not been studied extensively within the context of theoretical computer science. Until recently, there were only sporadic investigations of moving objects in the computational geometry literature. *Dynamic computational geometry* refers to the study of combinatorial changes in a geometric structure, as its defining objects undergo prescribed motions. For example, we may have n points moving linearly with constant velocities in $\mathcal{R}^2$, and may want to know the time intervals during which a particular point appears on their convex hull, the steady-state form of the hull (after all changes have occurred), or get an upper bound on how many times the convex hull changes during this motion. Such problems were introduced and studied in Reference 3.

A number of other authors have dealt with geometric problems arising from motion, such as collision detection or minimum separation determination [4–6]. See also Chapter 57. For instance, [6] shows how to check in subquadratic time whether two collections of simple geometric objects (spheres, triangles) collide with each other under specified polynomial motions.

24.3 Motion Models

An issue with the above research is that object motion(s) are assumed to be known in advance, sometimes in explicit form (e.g., points moving as polynomial functions of time). Indeed, the proposed methods reduce questions about moving objects to other questions about derived static objects.

---

* This chapter has been reprinted from first edition of this Handbook, without any content updates. A version of this chapter has also appeared in the *CRC Handbook of Discrete and Computational Geometry* (Chapter 50: Modeling Motion), 2nd Edition.

While most evolving physical systems follow known physical laws, it is also frequently the case that discrete events occur (such as collisions) that alter the motion law of one or more of the objects. Thus motion may be predictable in the short term, but becomes less so further into the future. Because of such discrete events, algorithms for modeling motion must be able to adapt in a dynamic way to motion model modifications. Furthermore, the occurrence of these events must be either predicted or detected, incurring further computational costs. Nevertheless, any truly useful model of motion must accommodate this *on-line* aspect of the temporal dimension, differentiating it from spatial dimensions, where all information is typically given at once.

In real-world settings, the motion of objects may be imperfectly known and better information may only be obtainable at considerable expense. The model of *data in motion* of [7] assumes that upper bounds on the rates of change are known, and focuses on being selective in using sensing to obtain additional information about the objects, in order to answer a series of queries.

24.4 Kinetic Data Structures

Suppose we are interested in tracking high-level attributes of a geometric system of objects in motion such as, for example, the convex hull of a set on n points moving in $\mathcal{R}^2$. Note that as the points move continuously, their convex hull will be a continuously evolving convex polygon. At certain discrete moments, however, the combinatorial structure of the convex hull will change (i.e., the circular sequence of a subset of the points that appear on the hull will change). In between such moments, tracking the hull is straightforward: its geometry is determined by the positions of the sequence of points forming the hull. How can we know when the combinatorial structure of the hull changes? The idea is that we can focus on certain elementary geometric relations among the n points, a set of *cached assertions*, which altogether certify the correctness of the current combinatorial structure of the hull. If we have short-term information about the motion of the points, then we can predict failures of these assertions in the near future. Furthermore, we can hope to choose these certifying relations in such a way so that when one of them fails because of point motion, both the hull and its set of certifying relations can be updated locally and incrementally, so that the whole process can continue.

- *Kinetic data structure:* A kinetic data structure (KDS) for a geometric attribute is a collection of simple geometric relations that certifies the combinatorial structure of the attribute, as well as a set of rules for repairing the attribute and its certifying relations when one relation fails.
- *Certificate:* A certificate is one of the elementary geometric relations used in a KDS.
- *Motion plan:* An explicit description of the motion of an object in the near future.
- *Event:* An event is the failure of a KDS certificate during motion. If motion plans are available for all objects in a certificate, then the future time of failure for this certificate can be predicted. Events are classified as *external* when the combinatorial structure of the attribute changes, and *internal*, when the structure of the attribute remains the same, but its certification needs to change.
- *Event queue:* In a KDS, all certificates are placed in an event queue, according to their earliest failure time.

The inner loop of a KDS consists of repeated certificate failures and certification repairs, as depicted in Figure 24.1.

We remark that in the KDS framework, objects are allowed to change their motions at will, with appropriate notification to the data structure. When this happens all certificates involving the object whose motion has changed must re-evaluate their failure times.

24.4.1 Convex Hull Example

Suppose we have four points a, b, c, and d in $\mathcal{R}^2$, and wish to track their convex hull. For the convex hull problem, the most important geometric relation is the CCW predicate: CCW(a, b, c) asserts that the triangle abc is oriented counterclockwise. Figure 24.2 shows a configuration of four points and four CCW relations that hold among them. It turns out that these four relations

FIGURE 24.1   The inner loop of a kinetic data structure.

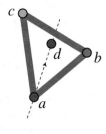

Proof of correctness:

- CCW(a, b, c)
- CCW(d, b, c)
- CCW(d, c, a)
- CCW(d, a, b)

FIGURE 24.2 Determining the convex hull of the points.

Old proof	New proof
CCW(a, b, c)	CCW(a, b, c)
CCW(d, b, c)	CCW(c, b, d)
CCW(d, c, a)	CCW(d, c, a)
CCW(d, a, b)	CCW(d, a, b)

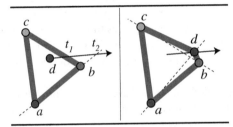

FIGURE 24.3 Updating the convex hull of the points.

are sufficient to prove that the convex hull of the four points is the triangle *abc*. Indeed the points can move and form different configurations, but as long as the four certificates shown remain valid, the convex hull must be *abc*.

Now suppose that points *a*, *b*, and *c* are stationary and only point *d* is moving with a known plan, as shown in Figure 24.3. At some time t_1 the certificate CCW(d, b, c) will fail, and at a later time t_2 CCW(d, a, b) will also fail. Note that the certificate CCW(d, c, a) will never fail in the configuration shown even though *d* is moving. So the certificates CCW(d, b, c) and CCW(d, a, b) schedule events that go into the event queue. At time t_1, CCW(d, b, c) ceases to be true and its negation, CCW(c, b, d), becomes true. In this simple case the three old certificates, plus the new certificate CCW(c, b, d) allow us to conclude that convex hull has now changed to *abdc*.

If the certificate set is chosen wisely, the KDS repair can be a local, incremental process—a small number of certificates may leave the cache, a small number may be added, and the new attribute certification will be closely related to the old one. A good KDS exploits the continuity or coherence of motion and change in the world to maintain certifications that themselves change only incrementally and locally as assertions in the cache fail.

24.4.2 Performance Measures for KDS

Because a KDS is not intended to facilitate a terminating computation but rather an on-going process, we need to use somewhat different measures to assess its complexity. In classical data structures there is usually a tradeoff between operations that interrogate a set of data and operations that update the data. We commonly seek a compromise by building indices that make queries fast, but such that updates to the set of indexed data are not that costly as well. Similarly in the KDS setting, we must at the same time have access to information that facilitates or trivializes the computation of the attribute of interest, yet we want information that is relatively stable and not so costly to maintain. Thus, in the same way that classical data structures need to balance the efficiency of access to the data with the ease of its update, kinetic data structures must tread a delicate path between "knowing too little" and "knowing too much" about the world. A good KDS will select a certificate set that is at once economical and stable, but also allows a quick repair of itself and the attribute computation when one of its certificates fails.

- *Responsiveness:* A KDS is *responsive* if the cost, when a certificate fails, of repairing the certificate set and updating the attribute computation is small. By "small" we mean polylogarithmic in the problem size—in general we consider small quantities that are polylogarithmic or $O(n^\epsilon)$ in the problem size.
- *Efficiency:* A KDS is *efficient* if the number of certificate failures (total number of events) it needs to process is comparable to the number of required changes in the combinatorial attribute description (external events), over some class of allowed motions. Technically, we require that the ratio of total events to external events is small. The class of allowed motions is usually specified as the class of *pseudo-algebraic* motions, in which each KDS certificate can flip between true and false at most a bounded number of times.
- *Compactness:* A KDS is *compact* if the size of the certificate set it needs is close to linear in the degrees of freedom of the moving system.
- *Locality:* A KDS is *local* if no object participates in too many certificates; this condition makes it easier to re-estimate certificate failure times when an object changes its motion law. (The existence of local KDSs is an intriguing theoretical question for several geometric attribute functions.)

24.4.3 The Convex Hull, Revisited

We now briefly describe a KDS for maintaining the convex hull of n points moving around in the plane [1].

The key goal in designing a KDS is to produce a *repairable certification* of the geometric object we want to track. In the convex hull case it turns out that it is a bit more intuitive to look at the dual problem, that of maintaining the upper (and lower) envelope of a set of moving lines in the plane, instead of the convex hull of the primal points. Such dualities represent a powerful toolkit in computational geometry; readers are referred to any standard computational geometry textbook for details, for example [8]. For simplicity we focus only on the upper envelope of the moving lines from now on; the lower envelope case is entirely symmetric. Using a standard divide-and-conquer approach, we partition our lines into two groups of size roughly $n/2$ each, and assume that recursive invocations of the algorithm maintain the upper envelopes of these groups. For convenience call the groups red and blue.

In order to produce the upper envelope of all the lines, we have to merge the upper envelopes of the red and blue groups and also certify this merge, so we can detect when it ceases to be valid as the lines move; see Figure 24.4.

Conceptually, we can approach this problem by sweeping the envelopes with a vertical line from left to right. We advance to the next red (blue) vertex and determine if it is above or below the corresponding blue (red) edge, and so on. In this process we determine when red is above blue or vice versa, as well as when the two envelopes cross. By stitching together all the upper pieces, whether red or blue, we get a representation of the upper envelope of all the lines.

The certificates used in certifying the above merge are of three flavors:

- x-certificates ($<_x$) are used to certify to x-ordering among the red and blue vertices; these involve four original lines.
- y-certificates ($<_y$) are used to certify that a vertex is above or below an edge of the opposite color; these involve three original lines and are exactly the duals of the ccw certificates discussed earlier.
- s-certificates ($<_s$) are slope comparisons between pairs of original lines; though these did not arise in our sweep description above, they are needed to make the KDS local [1].

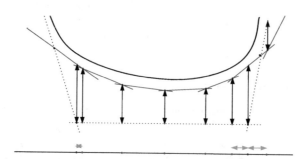

FIGURE 24.4 Merging the red and blue upper envelopes. In this example, the red envelope (solid line) is above the blue (dotted line), except at the extreme left and right areas. Vertical double-ended arrows represent y-certificates and horizontal double-ended arrows represent x-certificates, as described below.

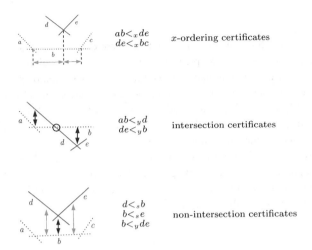

FIGURE 24.5   Using the different types of certificates to certify the red-blue envelope merge.

FIGURE 24.6   Envelope repair after a certificate failure. In the event shown lines b, d, and e become concurrent, producing a red-blue envelope intersection

Figure 24.5 shows examples of how these types of certificates can be used to specify x-ordering constraints and to establish intersection or non-intersection of the envelopes.

A total of $O(n)$ such certificates suffices to verify the correctness of the upper envelope merge.

Whenever the motion of the lines causes one of these certificates to fail, a local, constant-time process suffices to update the envelope and repair the certification. Figure 24.6 shows an example where an y-certificate fails, allowing the blue envelope to poke up above the red.

It is straightforward to prove that this kinetic upper envelope algorithm is responsive, local, and compact, using the logarithmic depth of the hierarchical structure of the certification. In order to bound the number of events processed, however, we must assume that the line motions are polynomial or at least pseudo-algebraic. A proof of efficiency can be developed by extruding the moving lines into space-time surfaces. Using certain well-known theorems about the complexity of upper envelopes of surfaces [9] and the overlays of such envelopes [10] it can be shown that in the worst case the number of events processed by this algorithm is near quadratic ($O(n^{2+\epsilon})$). Since the convex hull of even linearly moving points can change $\Omega(n^2)$ times [11], the efficiency result follows.

No comparable structure is known for the convex hull of points in dimensions $d \geq 3$.

24.5 A KDS Application Survey

Even though we have presented kinetic data structures in a geometric setting, there is nothing intrinsically geometric about KDS. The idea of cached assertions that help track an attribute of interest can be applied to many other settings where there is continuous evolution over time punctuated by discrete events, beyond motion in the physical world. For example, consider a graph whose edge weights or capacities are functions of time, as might arise in an evolving communications network. Then the problem of tracking various substructures of interest, such as the minimum spanning tree (MST) of the graph, or a shortest path tree form a source node, can be formulated and studied within the KDS framework.

We present below a quick summary of some of the areas to which kinetic data structures have been applied so far. The are mostly geometric in nature, but several non-geometric examples appear as well.

24.5.1 Extent Problems

A number of the original problems for which kinetic data structures were developed are aimed at different measures of how "spread out" a moving set of points in $\mathcal{R}^2$ is—one example is the convex hull, whose maintenance was discussed in the previous subsection. Other measures of interest include the diameter, width, and smallest area or perimeter bounding rectangle for a moving set S of n points. All these problems can be solved using the kinetic convex hull algorithm above; the efficiency of the algorithms is $O(n^{2+\epsilon})$, for any $\epsilon > 0$. There are also corresponding $\Omega(n^2)$ lower bounds for the number of combinatorial changes in these measures. Surprisingly, the best known upper bound for maintaining the smallest enclosing disk containing S is still near-cubic. Extensions of these results to dimensions higher than two are also lacking.

These costs can be dramatically reduced if we consider approximate extent measures. If we are content with $(1 + \epsilon)$ approximations to the measures, then an approximate smallest orthogonal rectangle, diameter, and smallest enclosing disk can be maintained with a number of events that is a function ϵ only and not of n [12]. For example, the bound of the number of approximate diameter updates in $\mathcal{R}^2$ under linear motion of the points is $O(1/\epsilon)$.

24.5.2 Proximity Problems

The fundamental proximity structures in computational geometry are the Voronoi Diagram and the Delaunay triangulation (Chapters 65 and 66). The edges of the Delaunay triangulation contain the closest pair of points, the closest neighbor to each point, as well as a wealth of other proximity information among the points. From the kinetic point of view, these are nice structures, because they admit completely local certifications. Delaunay's 1934 theorem [13] states that if a local empty sphere condition is valid for each $(d - 1)$-simplex in a triangulation of points in $\mathcal{R}^d$, then that triangulation must be Delaunay. This makes it simple to maintain a Delaunay triangulation under point motion: an update is necessary only when one of these empty sphere conditions fails. Furthermore, whenever that happens, a local retiling of space (of which the classic "edge-flip" in R^2 is a special case) easily restores Delaunayhood. Thus the KDS for Delaunay (and Voronoi) that follows from this theorem is both responsive and efficient—in fact, each KDS event is an external event in which the structure changes. Though no redundant events happen, an exact upper bound for the total number of such events in the worst-case is still elusive even in R^2, where the best known upper bound is nearly cubic, while the best lower bound only quadratic [14].

This principle of a set of easily checked local conditions that implies a global property has been used in kinetizing other proximity structures as well. For instance, in the *power diagram* [15] of a set of disjoint balls, the two closest balls must be neighbors [16]—and this diagram can be kinetized by a similar approach. Voronoi diagrams of more general objects, such as convex polytopes, have also been investigated. For example, in R^2 [17] shows how to maintain a compact Voronoi-like diagram among moving disjoint convex polygons; again, a set of local conditions is derived which implies the global correctness of this diagram. As the polygons move, the structure of this diagram allows one to know the nearest pair of polygons at all times.

In many applications the exact L_2-distance between objects is not needed and more relaxed notions of proximity suffice. Polyhedral metrics (such as L_1 or L_∞) are widely used, and the normal unit ball in L_2 can be approximated arbitrarily closely by polyhedral approximants. It is more surprising, however, that if we partition the space around each point into a set of polyhedral cones and maintain a number of directional nearest neighbors to each point in each cone, then we can still capture the globally closest pair of points in the L_2 metric. By directional neighbors here we mean that we measure distance only along a given direction in that cone. This geometric fact follows from a packing argument and is exploited in Reference 18 to give a different method for maintaining the closest pair of points in $\mathcal{R}^d$. The advantage of this method is that the kinetic events are changes of the sorted order of the points along a set of directions fixed *a priori*, and therefore the total number of events is provably quadratic.

24.5.3 Triangulations and Tilings

Many areas in scientific computation and physical modeling require the maintenance of a triangulation (or more generally a simplicial complex) that approximates a manifold undergoing deformation. The problem of maintaining the Delaunay triangulation of moving points in the plane mentioned above is a special case. More generally, local re-triangulations are necessitated by collapsing triangles, and sometimes required in order to avoid undesirably "thin" triangles. In certain cases the number of nodes (points) may also have to change in order to stay sufficiently faithful to the underlying physical process; see, for example, [19]. Because in general a triangulation meeting certain criteria is not unique or canonical, it becomes more difficult to assess the efficiency of kinetic algorithms for solving such problems. The lower-bound results in Reference 20 indicate that one cannot hope for a subquadratic bound on the number of events in the worst case in the maintenance an *any* triangulation, even if a linear number of additional Steiner points is allowed.

There is large gap between the desired quadratic upper bound and the current state of art. Even for maintaining an arbitrary triangulation of a set of n points moving linearly in the plane, the best-known algorithm processes $O(n^{7/3})$ events [21] in the worst case. The algorithm actually maintains a pseudotriangulation of the convex hull of the point set and then a triangulation of each pseudotriangle. Although there are only $O(n^2)$ events in the pseudotriangulation, some of the events change too many triangles because of high-degree vertices. Unless additional Steiner points are allowed, there are point configurations for which high-degree vertices are inevitable and therefore some of the events will be expensive. A more clever, global argument is needed to prove a near-quadratic upper bound on the total number of events in the above algorithm. Methods that choose to add additional points, on the other hand, have the burden of defining appropriate trajectories for these Steiner points as well. Finally, today no triangulation that guarantees certain quality on the shapes of triangles as well as a subcubic bound on the number of retiling events is known.

24.5.4 Collision Detection

Kinetic methods are naturally applicable to the problem of collision detection between moving geometric objects. Typically collisions occur at irregular intervals, so that fixed-time stepping methods have difficulty selecting an appropriate sampling rate to fit both the numerical requirements of the integrator as well as those of collision detection. A kinetic method based on the discrete events that are the failures of relevant geometric conditions can avoid the pitfalls of both oversampling and undersampling the system. For two moving convex polygons in the plane, a kinetic algorithm where the number of events is a function of the relative separation of the two polygons is given in Reference 22. The algorithm is based on constructing certain outer hierarchies on the two polygons. Analogous methods for 3D polytopes were presented in Reference 23, together with implementation data.

A tiling of the free space around objects can serve as a proof of non-intersection of the objects. If such a tiling can be efficiently maintained under object motion, then it can be the basis of a kinetic algorithm for collision detection. Several papers have developed techniques along these lines, including the case of two moving simple polygons in the plane [24], or multiple moving polygons [21,25]. These developments all exploit deformable pseudotriangulations of the free space—tilings which undergo fewer combinatorial changes than, for example, triangulations. An example from Reference 21 is shown in Figure 24.7. The figure shows how the pseudotriangulation adjusts by local retiling to the motion of the inner quadrilateral. The approach of [21] maintains a canonical pseudotriangulation, while others are based on letting a pseudotriangulation evolve according to the history of the motion. It is unclear at this point which is best. An advantage of all these methods is that the number of certificates needed is close to size of the min-link separating subdivision of the objects, and thus sensitive to how intertwined the objects are.

Deformable objects are more challenging to handle. Classical methods, such as bounding volume hierarchies [26], become expensive, as the fixed object hierarchies have to be rebuilt frequently. One possibility for mitigating this cost is to let the hierarchies themselves deform continuously, by having the bounding volumes defined implicitly in terms of object features. Such an approach was developed for flexible linear objects (such as rope or macromolecules), using combinatorially defined sphere hierarchies in Reference 27. In that work a bounding sphere is defined not in the usual way, via its center and radius, but in an implicit combinatorial way, in terms of four feature points of the enclosed object geometry. As the object deforms these implicitly defined spheres automatically track their assigned features, and therefore the deformation. Of course the validity of the hierarchy has to be checked at each time step and repaired if necessary. What helps here is that the implicitly defined spheres change their

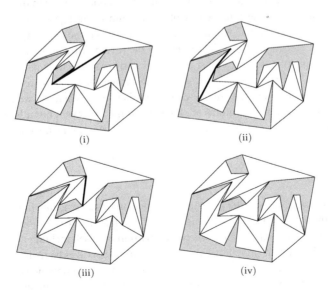

(i)   (ii)

(iii)   (iv)

FIGURE 24.7 Snapshots of the mixed pseudotriangulation of [21]. As the center trapezoid-like polygon moves to the right, the edges corresponding to the next about-to-fail certificate are highlighted.

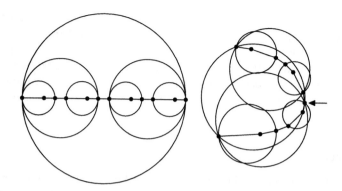

FIGURE 24.8 A thin rod bending from a straight configuration, and a portion of its associated bounding sphere hierarchy. The combinatorially defined sphere hierarchy is stable under deformation. Only the top level sphere differs between the two conformations.

combinatorial description rather infrequently, even under extreme deformation. An example is shown in Figure 24.8 where the rod shown is bent substantially, yet only the top-level sphere needs to update its description.

The pseudotriangulation-based methods above can also be adapted to deal with object deformation.

24.5.5 Connectivity and Clustering

Closely related to proximity problems is the issue of maintaining structures encoding connectivity among moving geometric objects. Connectivity problems arise frequently in *ad hoc* mobile communication and sensor networks, where the viability of links may depend on proximity or direct line-of-sight visibility among the stations desiring to communicate. With some assumptions, the communication range of each station can be modeled by a geometric region, so that two stations can establish a link if and only if their respective regions overlap. There has been work on kinetically maintaining the connected components of the union of a set of moving geometric regions for the case of rectangles [28] and unit disks [29].

Clustering mobile nodes is an essential step in many algorithms for establishing communication hierarchies, or otherwise structuring *ad hoc* networks. Nodes in close proximity can communicate directly, using simpler protocols; correspondingly, well-separated clusters can reuse scarce resources, such the same frequency or time-division multiplexing communication scheme, without interference. Maintaining clusters of mobile nodes requires a tradeoff between the tightness, or optimality of the clustering, and its stability under motion. In Reference 30 a randomized clustering scheme is discussed based on an iterated leader-election algorithm that produces a number of clusters within a constant factor of the optimum, and in which the number of cluster changes is also asymptotically optimal. This scheme was used in Reference 31 to maintain a routing graph on mobile nodes that is always sparse and in which communication paths exist that are nearly as good as those in the full communication graph.

Another fundamental kinetic question is the maintenance of a minimum spanning tree (MST) among n mobile points in the plane, closely related to earlier work on parametric spanning trees [32] in a graph whose edge weights are functions of a parameter λ (λ is time in the kinetic setting). Since the MST is determined by the sorted order of the edge weights in the graph, a simple algorithm can be obtained by maintaining the sorted list of weights and some auxiliary data structures (such an algorithm is quadratic in the graph size, or $O(n^4)$ in our case). This was improved when the weights are linear functions of time to nearly $O(n^{11/6})$ (subquadratic) for planar graphs or other minor-closed families [33]. When the weights are the Euclidean distances between moving points, only approximation algorithms are known and the best event bounds are nearly cubic [18]. For many other optimization problems on geometric graphs, such as shortest paths for example, the corresponding kinetic questions are wide open.

24.5.6 Visibility

The problem of maintaining the visible parts of the environment when an observer is moving is one of the classic questions in computer graphics and has motivated significant developments, such as binary space partition trees, the hardware depth buffer, etc. The difficulty of the question increases significantly when the environment itself includes moving objects; whatever visibility structures accelerate occlusion culling for the moving observer, must now themselves be maintained under object motion.

Binary space partitions (BSP) are hierarchical partitions of space into convex tiles obtained by performing planar cuts (Chapter 21). Tiles are refined by further cuts until the interior of each tile is free of objects or contains geometry of limited complexity. Once a BSP tree is available, a correct visibility ordering for all geometry fragments in the tiles can be easily determined and incrementally maintained as the observer moves. A kinetic algorithm for visibility can be devised by maintaining a BSP tree as

STRUCTURE	BOUNDS ON EVENTS	SOURCE
Convex hull	$\Omega(n^{2+\epsilon})$	[16]
Pseudotriangulation	$O(n^2)$	[5]
Triangulation (arb.)	$\Omega(n^{7/3})$	[5]
MST	$O(n^{11/6} \log^{3/2} n)$	[6]
BSP	$\tilde{O}(n^2)$	[7, 11]

FIGURE 24.9   Bounds on the number of combinatorial changes.

the objects move. The key insight is to certify the correctness of the BSP tree through certain combinatorial conditions, whose failure triggers localized tree rearrangements—most of the classical BSP construction algorithms do not have this property. In $\mathcal{R}^2$, a randomized algorithm for maintaining a BSP of moving disjoint line segments is given in Reference 34. The algorithm processes $O(n^2)$ events, the expected cost per tree update is $O(\log n)$, and the expected tree size is $O(n \log n)$. The maintenance cost increases to $O(n\lambda_{s+2}(n) \log^2 n)$ [35] for disjoint moving triangles in $\mathcal{R}^3$ (s is a constant depending on the triangle motion). Both of these algorithms are based on variants on vertical decompositions (many of the cuts are parallel to a given direction). It turns out that in practice these generate "sliver-like" BSP tiles that lead to robustness issues [36].

As the pioneering work on the visibility complex has shown [37], another structure that is well suited to visibility queries in $\mathcal{R}^2$ is an appropriate pseudotriangulation. Given a moving observer and convex moving obstacles, a full radial decomposition of the free space around the observer is quite expensive to maintain. One can build pseudotriangulations of the free space that become more and more like the radial decomposition as we get closer to the observer. Thus one can have a structure that compactly encodes the changing visibility polygon around the observer, while being quite stable in regions of the free space well-occluded from the observer [38].

24.5.7 Result Summary

We summarize in Figure 24.9 the efficiency bounds on the main KDSs discussed above.

24.5.8 Open Problems

As mentioned above, we still lack efficient kinetic data structures for many fundamental geometric questions. Here is a short list of such open problems:

1. Find an efficient (and responsive, local, and compact) KDS for maintaining the convex hull of points moving in dimensions $d \geq 3$.
2. Find an efficient KDS for maintaining the smallest enclosing disk in $d \geq 2$. For $d = 2$, a goal would be an $O(n^{2+\epsilon})$ algorithm.
3. Establish tighter bounds on the number of Voronoi diagram events, narrowing the gap between quadratic and near-cubic.
4. Obtain a near-quadratic bound on the number of events maintaining an arbitrary triangulation of linearly moving points.[*]
5. Maintain a kinetic triangulation with a guarantee on the shape of the triangles, in subcubic time.
6. Find a KDS to maintain the MST of moving points under the Euclidean metric achieving subquadratic bounds.

Beyond specific problems, there are also several important structural issues that require further research in the KDS framework. These include:

24.5.8.1 Recovery after Multiple Certificate Failures

We have assumed up to now that the KDS assertion cache is repaired after each certificate failure. In many realistic scenarios, however, it is impossible to predict exactly when certificates will fail because explicit motion descriptions may not be available. In such settings we may need to sample the system and thus we must be prepared to deal with multiple (but hopefully few) certificate failures at each time step. A general area of research that this suggests is the study of how to efficiently update common geometric structures, such as convex hulls, Voronoi and Delaunay diagrams, arrangements, etc., after "small motions" of the defining geometric objects.

There is also a related subtlety in the way that a KDS assertion cache can certify the value, or a computation yielding the value, of the attribute of interest. Suppose our goal is to certify that a set of moving points in the plane, in a given circular order, always form a convex polygon. A plausible certificate set for convexity is that all interior angles of the polygon are convex. See Figure

---

[*]   While this handbook was going to print, Agarwal, Wang and Yu, gave a near-quadratic such algorithm [39].

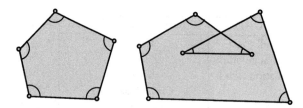

FIGURE 24.10 Certifying the convexity of a polygon.

24.10. In the normal KDS setting where we can always predict accurately the next certificate failure, it turns out that the above certificate set is sufficient, *as long as at the beginning of the motion the polygon was convex*. One can draw, however, nonconvex self-intersecting polygons all of whose interior angles are convex, as also shown in the same figure. The point here is that a standard KDS can offer a *historical* proof of the convexity of the polygon by relying on the fact that the certificate set is valid *and* that the polygon was convex during the prior history of the motion. Indeed the counterexample shown cannot arise under continuous motion without one of the angle certificates failing first. On the other hand, if an oracle can move the points when "we are not looking," we can wake up and find all the angle certificates to be valid, yet our polygon need not be convex. Thus in this oracle setting, since we cannot be sure that no certificates failed during the time step, we must insist on *absolute* proofs—certificate sets that in any state of the world fully validate the attribute computation or value.

24.5.8.2 Hierarchical Motion Descriptions

Objects in the world are often organized into groups and hierarchies and the motions of objects in the same group are highly correlated. For example, though not all points in an elastic bouncing ball follow exactly the same rigid motion, the trajectories of nearby points are very similar and the overall motion is best described as the superposition of a global rigid motion with a small local deformation. Similarly, the motion of an articulated figure, such as a man walking, is most succinctly described as a set of relative motions, say that of the upper right arm relative to the torso, rather than by giving the trajectory of each body part in world coordinates.

What both of these examples suggest is that there can be economies in motion description, if the motion of objects in the environment can be described as a superposition of terms, some of which can be shared among several objects. Such hierarchical motion descriptions can simplify certificate evaluations, as certificates are often local assertions concerning nearby objects, and nearby objects tend to share motion components. For example, in a simple articulated figure, we may wish to assert CCW (A, B, C) to indicate that an arm is not fully extended, where $\overline{AB}$ and $\overline{BC}$ are the upper and lower parts of the arm respectively. Evaluating this certificate is clearly better done in the local coordinate frame of the upper arm than in a world frame—the redundant motions of the legs and torso have already been factored out.

24.5.8.3 Motion Sensitivity

As already mentioned, the motions of objects in the world are often highly correlated and it behooves us to find representations and data structures that exploit such motion coherence. It is also important to find mathematical measures that capture the degree of coherence of a motion and then use this as a parameter to quantify the performance of motion algorithms. If we do not do this, our algorithm design may be aimed at unrealistic worst-case behavior, without capturing solutions that exploit the special structure of the motion data that actually arise in practice—as already discussed in a related setting in Reference 40. Thus it is important to develop a class of kinetic *motion-sensitive* algorithms, whose performance can be expressed a function of how coherent the motions of the underlying objects are.

24.5.8.4 Non-Canonical Structures

The complexity measures for KDSs mentioned earlier are more suitable for maintaining *canonical* geometric structures, which are uniquely defined by the position of the data, for example, convex hull, closest pair, and Delaunay triangulation. In these cases the notion of external events is well defined and is independent of the algorithm used to maintain the structure. On the other hand, as we already discussed, suppose we want to maintain a triangulation of a moving point set. Since the triangulation of a point set is not unique, the external events depend on the triangulation being maintained, and thus depend on the algorithm. This makes it difficult to analyze the efficiency of a kinetic triangulation algorithm. Most of the current approaches for maintaining noncanonical structures artificially impose canonicality and maintain the resulting canonical structure. But this typically increases the number of events. So it is entirely possible that methods in which the current form of the structure may depend on its past history can be more efficient. Unfortunately, we lack mathematical techniques for analyzing such history-dependent structures.

24.6 Querying Moving Objects

Continuous tracking of a geometric attribute may be more than is needed for some applications. There may be time intervals during which the value of the attribute is of no interest; in other scenarios we may be just interested to know the attribute value at certain discrete query times. For example, given n moving points in $\mathcal{R}^2$, we may want to pose queries asking for all points inside a rectangle R at time t, for various values of R and t, or for an interval of time Δt, etc. Such problems can be handled by a mixture of kinetic and static techniques, including standard range-searching tools such as partition trees and range trees [8]. They typically involve tradeoffs between evolving indices kinetically, or prebuilding indices for static snapshots. An especially interesting special case is when we want to be able answer queries about the near future faster than those about the distant future—a natural desideratum in many real-time applications.

A number of other classical range-searching structures, such as k-d-trees and R-trees have recently been investigated for moving objects [41,42].

24.7 Sources and Related Materials

Results not given an explicit reference above may be traced in these surveys.

[2]: An early, and by now somewhat dated, survey of KDS work.

[43]: A report based on an NSF-ARO workshop, addressing several issues on modeling motion from the perspective of a variety of disciplines.

[44]: A "popular-science" type article containing material related to the costs of sensing and communication for tracking motion in the real world.

References

1. J. Basch, L. J. Guibas, and J. Hershberger. Data structures for mobile data. *J. Algorithms*, 31:1–28, 1999.
2. L. J. Guibas. Kinetic data structures—A state of the art report. In P. K. Agarwal, L. E. Kavraki, and M. Mason, editors. *Proc. Workshop Algorithmic Found. Robot.*, pages 191–209. A. K. Peters, Wellesley, MA, 1998.
3. M. J. Atallah. Some dynamic computational geometry problems. *Comput. Math. Appl.*, 11(12):1171–1181, 1985.
4. P. Gupta, R. Janardan, and M. Smid. Fast algorithms for collision and proximity problems involving moving geometric objects. *Comput. Geom. Theory Appl.*, 6:371–391, 1996.
5. E. Schömer and C. Thiel. Efficient collision detection for moving polyhedra. In *Proc. 11th Annu. ACM Sympos. Comput. Geom.*, pages 51–60, 1995.
6. E. Schömer and C. Thiel. Subquadratic algorithms for the general collision detection problem. In *Abstracts 12th European Workshop Comput. Geom.*, pages 95–101. Universität Münster, 1996.
7. S. Kahan. A model for data in motion. In *Proc. 23th Annu. ACM Sympos. Theory Comput.*, pages 267–277, 1991.
8. M. de Berg, M. van Kreveld, M. Overmars, and O. Schwarzkopf. *Comput. Geom.: Algorithms Appl.* Springer-Verlag, Berlin, Germany, 2nd edition, 2000.
9. M. Sharir. Almost tight upper bounds for lower envelopes in higher dimensions. *Discrete Comput. Geom.*, 12:327–345, 1994.
10. P. K. Agarwal, O. Schwarzkopf, and M. Sharir. The overlay of lower envelopes and its applications. *Discrete Comput. Geom.*, 15:1–13, 1996.
11. P. K. Agarwal, L. J. Guibas, J. Hershberger, and E. Veach. Maintaining the extent of a moving point set. *Discrete Comput. Geom.*, 26(3):353–374, 2001.
12. P. K. Agarwal and S. Har-Peled. Maintaining approximate extent measures of moving points. In *Proc.12th ACM-SIAM Sympos. Discrete Algorithms (SODA)*, pages 148–157, 2001.
13. B. Delaunay. Sur la sphère vide. A la memoire de Georges Voronoi. *Izv. Akad. Nauk SSSR, Otdelenie Matematicheskih i Estestvennyh Nauk*, 7:793–800, 1934.
14. G. Albers, L. J. Guibas, J. S. B. Mitchell, and T. Roos. Voronoi diagrams of moving points. *Int. J. Comput. Geom. Appl.*, 8:365–380, 1998.

15. F. Aurenhammer. Power diagrams: Properties, algorithms and applications. *SIAM J. Comput.*, 16:78–96, 1987.

16. L. Guibas and L. Zhang. Euclidean proximity and power diagrams. In *Proc. 10-th Canadian Conf. Comput. Geom.*, pages 90–91, 1998.

17. L. Guibas, J. Snoeyink, and L. Zhang. Compact Voronoi diagrams for moving convex polygons. In *Proc. Scand. Workshop on Alg. and Data Structures (SWAT)*, volume 1851 of *Lecture Notes Comput. Sci.*, pages 339–352. Springer-Verlag, 2000.

18. J. Basch, L. J. Guibas, and L. Zhang. Proximity problems on moving points. In *Proc. 13th Annu. ACM Sympos. Comput. Geom.*, pages 344–351, 1997.

19. H. L. Cheng, T. K. Dey, H. Edelsbrunner, and J. Sullivan. Dynamic skin triangulation. In *Proc. 12-th SIAM Symposium on Discrete Algorithms (SODA)*, pages 47–56, 2001.

20. P. K. Agarwal, J. Basch, M. de Berg, L. J. Guibas, and J. Hershberger. Lower bounds for kinetic planar subdivisions. In *Proc. 15-th Annu. ACM Sympos. Comput. Geom. (SoCG)*, pages 247–254, 1999.

21. P. K. Agarwal, J. Basch, L. J. Guibas, J. Hershberger, and L. Zhang. Deformable free space tiling for kinetic collision detection. In *Fourth Workshop Algorithmic Found. Robot., (WAFR)*, pages 83–96, 2000.

22. J. Erickson, L. J. Guibas, J. Stolfi, and L. Zhang. Separation-sensitive collision detection for convex objects. In *Proc. 10-th ACM-SIAM Symp. Discrete Algorithms (SODA)*, pages 102–111, 1999.

23. L. Guibas, F. Xie, and L. Zhang. Kinetic collision detection: Algorithms and experiments. In *Proceedings of the International Conference on Robotics and Automation (ICRA)*, pages 2903–2910, 2001.

24. J. Basch, J. Erickson, L. J. Guibas, J. Hershberger, and L. Zhang. Kinetic collision detection between two simple polygons. In *Proc. 10-th ACM-SIAM Sympos. Discrete Algorithms (SODA)*, pages 327–336, 1999.

25. D. Kirkpatrick, J. Snoeyink, and B. Speckmann. Kinetic collision detection for simple polygons. In *Proc. 16th Annu. ACM Sympos. Comput. Geom.*, pages 322–330, 2000.

26. S. Gottschalk, M. Lin, and D. Manocha. OBB-Tree: A hierarchical structure for rapid interference detection. In *SIGGRAPH 96 Conference Proceedings*, pages 171–180, 1996.

27. L. Guibas, A. Nguyen, D. Russell, and L. Zhang. Collision detection for deforming necklaces. In *Proc. 18-th Annu. ACM Sympos. Comput. Geom. (SoCG)*, pages 33–42, 2002.

28. J. Hershberger and S. Suri. Kinetic connectivity of rectangles. In *Proc. 15-th Annu. ACM Sympos. Comput. Geom. (SoCG)*, pages 237–246, 1999.

29. L. Guibas, J. Hershberger, and S. Suri, L. Zhang. Kinetic connectivity for unit disks. In *Proc. 16th Annu. ACM Sympos. Comput. Geom.*, pages 331–340, 2000.

30. J. Gao, L. J. Guibas, J. Hershberger, L. Zhang, and A. Zhu. Discrete mobile centers. In *Proc. 17-th ACM Symp. Comput. Geom. (SoCG)*, pages 190–198, June 2001.

31. J. Gao, L. J. Guibas, J. Hershberger, L. Zhang, and A. Zhu. Geometric spanner for routing in mobile networks. In *Proc. 2-nd ACM Symp. on Ad-Hoc Networking and Computing MobiHoc)*, pages 45–55, October 2001.

32. D. Fernàndez-Baca, G. Slutzki, and D. Eppstein. Using sparsification for parametric minimum spanning tree problems. *Nord. J. Comput.*, 3:352–366, 1996.

33. P. K. Agarwal, D. Eppstein, L. J. Guibas, and M. Henzinger. Parametric and kinetic minimum spanning trees. In *Proc. 39th Annu. IEEE Sympos. Found. Comput. Sci.*, pages 596–605, 1998.

34. P. K. Agarwal, L. J. Guibas, T. M. Murali, and J. S. Vitter. Cylindrical static and kinetic binary space partitions. *Comp. Geom.: Theor. Appl.*, 16:103–127, 2000.

35. P. K. Agarwal, J. Erickson, and L. J. Guibas.Kinetic BSPs for intersecting segments and disjoint triangles. In *Proc. 9th ACM-SIAM Sympos. Discrete Algorithms*, pages 107–116, 1998.

36. J. L. D. Comba. *Kinetic vertical decomposition trees*. PhD thesis, Stanford University, 1999.

37. M. Pocchiola and G. Vegter. The visibility complex. *Int. J. Comput. Geom. Ap.*, 6(3):279–308, 1996.

38. O. Hall-Holt. *Kinetic visibility*. PhD thesis, Stanford University, 2002.

39. P. Agarwal, Y. Wang, and H. Yu. A 2d kinetic triangulation with near quadratic topological changes. In *Proc. 20th ACM Symp. Comput. Geom., (SoCG)*, pages 180–189, 2004.

40. M. de Berg, M. J. Katz, A. F. van der Stappen, and J. Vleugels. Realistic input models for geometric algorithms. In *Proc. 13th Annu. ACM Sympos. Comput. Geom.*, pages 294–303, 1997.

41. P. Agarwal, J. Gao, and L. Guibas. Kinetic medians and kd-trees. In *Proc. 10-th Europ. Symp. on Algorithms (ESA)*, pages 5–16, 2002.

42. P. Agarwal, S. Har-Peled, and M. Procopiuc. Star-tree: An efficent self-adjusting index for moving points. In *Workshop on Algorithms Eng.*, 2002.

43. P. K. Agarwal, L. J. Guibas et al. Algorithmic issues in modeling motion. *ACM Comput. Surv.*, 34(4):550–572, 2003.

44. L. J. Guibas. Sensing, tracking, and reasoning with relations. In *IEEE Signal Proc. Magazine*, pages 73–85, 2002.

25

Online Dictionary Structures

25.1 Introduction.. 389
25.2 Trie Implementations 390
25.3 Binary Search Tree Implementations 391
25.4 Balanced BST Implementation 393
25.5 Additional Operations... 395
25.6 Discussion... 396
References.. 396

Teofilo F. Gonzalez
University of California, Santa Barbara

25.1 Introduction

Given an initially empty set S, the dictionary problem consists of executing online any sequence of operations of the form $S.membership(s)$, $S.insert(s)$, and $S.delete(s)$, where each element s is an object (or point in one-dimensional space). Each object can be stored in a single word, and it takes $O(1)$ time to store or retrieve it. It is well known that the set may be represented by using arrays (sorted or unsorted), linked lists (sorted or unsorted), hash tables, binary search trees, AVL-trees, B-trees, 2-3 trees, weighted balanced trees, or balanced binary search trees (i.e., 2-3-4 trees, symmetric B-trees, half balanced trees, or red-black trees).

The worst-case time complexity for performing each of these operations is $O(\log n)$, where n is the maximum number of elements in the set being represented, when using AVL-trees, B-trees (fixed order), 2-3 trees, weighted balanced trees, or balanced binary search trees. See Chapters 2, 3, 9, 11 and 16 for details on these structures. The insertion or deletion of elements in these structures requires a set of additional operations to preserve certain properties of the resulting trees. For binary search trees, these operations are called *rotations*, and for m-way search trees they are called *splitting or combining* nodes. The balanced binary search trees are the only trees that require a constant number of rotations for both the *insert* and *delete* operations [1,2].

This chapter discusses several algorithms [3,4] for the generalized dictionary problem for the case when the data is multidimensional, rather than one dimensional. Each data element consists of d-ordered components which we call *ordered d-tuple* or simply d-tuple. Each component contains a value which can be stored in a memory word and which can be compared against another value to determine whether the values are identical, the first value is larger than the second one, or the first value is smaller than the second one. The comparison operation between components takes constant time. We show that the basic multidimensional dictionary operations can be implemented to take $O(d + \log n)$, where n is the number of d-tuples in the set. We also show that other common operations can also be executed within the same time complexity bounds.

Let us now discuss one of the many applications of multidimensional dictionaries. Given a set of n points in multidimensional space and an integer D, the problem is to find the least number of orthogonal hypersquares (or d-boxes) of size D to cover all the points, that is, each of the points must be in at least one of the d-boxes. This covering problem arises in image processing, as well as when locating emergency facilities so that all users are within a reasonable distance of one of the facilities [5]. The problem has been shown to be NP-hard and several polynomial time approximation schemes for its solution have been developed [5]. The simplest approximation algorithm defines d-boxes along a multidimensional grid with grid d-boxes of length D. The approximation algorithm takes every d-dimensional point and by using simple arithmetic operations, including the floor function, finds its appropriate (grid) d-box. The d-box, which is characterized by d integer components, is inserted into a multidimensional dictionary and the operation is repeated for each d-dimensional point. Then one just visits all the d-tuples in the set and the d-boxes they represent is the solution generated by this simple approximation algorithm. Note that when a d-tuple is inserted into a multidimensional dictionary that already contains it, the dictionary will not change because dictionaries store sets, that is, multiple copies of the d-tuples are not allowed. Other applications of multidimensional dictionaries are when accessing multiattribute data by value. These applications include the management of geometrical objects and the solution of geometry search problems.

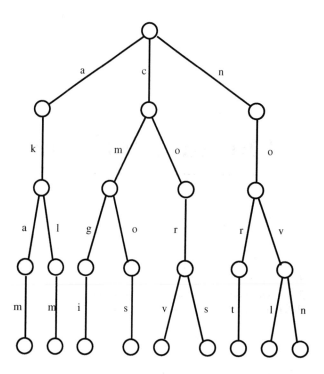

FIGURE 25.1   Trie for set T.

Given a d-tuple s in set S, one may access in constant time the ith element in the d-tuple by using the function $s.x(i)$. In other words, the d-tuple s is simply $(s.x(1), s.x(2), \ldots, s.x(d))$. In this chapter, we examine the methods given in References 3 and 4 to represent the data set and their algorithms to perform online any sequence of the multidimensional dictionary operations. The most efficient of the implementations performs each of the three dictionary operations in $O(d + \log n)$ time, where n is the number of d-tuples and d is the number of dimensions. Each of the insert and delete operations requires no more than a constant number of rotations. The best of the algorithms requires dn words to represent the d-tuples, plus $O(n)$ additional space is required to keep additional pointers and data. Because we are using balanced binary search trees, we can also perform other operations efficiently. For example, find the (lexicographically) smallest or largest d-tuple ($O(\log n)$ time), print in lexicographic order ($O(dn)$ time), and concatenation ($O(d + \log n)$ time). By modifying slightly the representation and introducing additional information, one can also find the (lexicographically) kth smallest or largest d-tuple efficiently ($O(\log n)$ time). In Section 25.5, we show that the structure given in Reference 4 may also be used to implement the split operation in $O(d + \log n)$ time and that the approach can also be used with other balanced structures, such as AVL-trees, weight-balanced trees, and so on.

25.2 Trie Implementations

It is interesting to note that three decades ago balanced structures were written off for this type of applications. As noted in Reference 6, "balanced tree schemes based on key comparisons (e.g., AVL-trees, B-trees, etc.) lose some of their usefulness in this more general context." At that time the approach was to use *tries* in conjunction with balanced tree schemes to represent multidimensional data. We will elaborate on these structures after defining some terms.

Given a set of strings over an alphabet Σ, the tree of all their prefixes is called a *trie* (see Chapter 29). In Figure 25.1, we depict the trie for the set of strings over the English alphabet $T = \{akam, aklm, cmgi, cmos, cors, corv, nort, novl, novn\}$. In this case, all the elements in the trie have the same number of symbols, but in general a trie may contain string with different number of symbols. Each node x in a trie is the destination of all the strings with the same prefix (say p_x) and node x consists of a set of element–pointer pairs. The first component of the pair is an alphabet element and the second one is a pointer to a subtrie that contains all the strings with prefix p_x followed by the alphabet element. In Figure 25.1, the pairs for a trie node are represented by the edges emanating from the lower portion of the circle that represents a node. Note that no two element-pointer pairs for a trie node have the same first component or the same second component. The set of element–pointer pairs in a trie node may be represented in different ways. For example, one may store the element–pointer pairs for each internal node as follows:

1. An array of m pointers, where m is the number of elements in the alphabet Σ. In this case, one needs to define a function to translate each element in the alphabet to an integer in the range $[0, m-1]$. The function can normally be implemented to take constant time;
2. A sorted linked list with all the symbols and corresponding pointers of the branches emanating from the node [7];
3. A binary search tree with all the symbols and corresponding pointers of the branches emanating from the node [8].

We shall refer to the resulting structures as trie-array, trie-list, and trie-bst, respectively. These are by no means all the possibilities. For example, later on we discuss hybrid representations that have some interesting properties.

For multidimensional dictionaries defined over the set of integers $[0, m)$, the trie method treats a point in d-space as a string with d elements defined over the alphabet $\Sigma = \{0, 1, \ldots, m-1\}$ (see Figure 25.1). For the trie-array representation, the element–pointer pairs in a trie node are represented by an m-element sequential array of pointers. The ith pointer corresponds to the ith alphabet symbol, and a null pointer is used when the corresponding alphabet element is not part of any element–pointer pair. The space required to represent n d-tuples in this structure is at most dnm pointers, but the total amount of information can be represented using $dn \log m$ bits. The insert and delete operations take $O(md)$ time, because either of these operations may add or delete $d-1$ trie nodes and each node includes an m-component array of pointers. On the other hand, the membership operation takes only $O(d)$ time. This is the fastest way of implementing the membership operation no matter what structure is used to represent the set. The trie-array implementation is not possible when m is large because there will be too much space wasted. The trie-list representation mentioned above is much better for this scenario. In this case, the list of element–pointer pairs is stored in a sorted linked list. The list is sorted with respect to the alphabet elements, that is, the first component in the element–pointer pairs. In the trie-bst representation, the element–pointers are stored in a binary search tree. The ordering is with respect to the alphabet elements. In the former case, we use dn memory locations for the d-tuples, plus $2dn$ pointers, and in the latter case one needs $3dn$ pointers. The time complexity for insert, delete, and membership in both of these representations is $O(d + n)$. It is important to note that there are two types of pointers: the trie pointers and the linked list or binary search tree pointers.

In practice, one may use hybrid structures in which some nodes in the trie are trie-arrays and others are trie-lists, and depending on the number of element–pointer pairs one transforms from one representation to the other. For example, if the number of element–pointer pairs becomes more than some bound b_u one uses trie-array nodes, but if it is less than b_l then one uses the trie-list nodes. If it is some number between these two bounds, then either representation is fine to use. By using appropriate values for b_l and b_u in this approach, we can avoid "trashing" which means that the algorithm spends most of the time changing back and forth from one representation to the other.

Bentley and Saxe [9] used a modified version of the trie-bst implementation. In this case, the binary search tree is replaced by a completely balanced binary tree. That is, each binary search tree or subtree for trie node x with k element–pointer pairs has as root an element–pointer pair (y, ptr) such that the median of the ordered strings (with prefix equal to the prefix of the trie node x (denoted by p_x) plus any of the k alphabet elements in the k element–pointer pairs) has as prefix p_x followed by y. For example, the root of the trie in Figure 25.1 has the pair with alphabet element c, and the root of the subtrie for the prefix no is the pair with alphabet element v. This balanced structure is the best possible when the trie does not change or it changes very slowly like in multikey sorting or restricted searching [10,11]. Since the overall structure is rigid, it can be shown that rebalancing after inserting or deleting an element can be very expensive and therefore not appropriate in a dynamic environment.

Research into using different balanced strategies for the trie-bst structures has lead into structures that use fixed order B-trees [12], weight-balanced trees [6], AVL-trees [13], and balanced binary search trees [14]. It has been shown that insert, delete, and membership for multidimensional dictionaries for all of the above structures take $O(d + \log n)$ time in the worst case, except for weight-balanced trees in which case it is an expected time complexity bound. The above procedures are complex and require balancing criteria in addition to the obvious ones. Also, the number of rotations needed for these insert and delete operations is not bounded by a constant. Vaishnavi [14] used balanced binary search trees that require only a constant number of rotations, however, since one may encounter d trees, the total number of rotations are no longer bounded by a constant. He left it as an open problem to design a structure such that multidimensional dictionaries can be implemented so that only a constant number of rotations are performed after each insert and delete operation.

25.3 Binary Search Tree Implementations

As pointed out by Gonzalez [3,4], using a balanced binary search tree (without a trie) and storing each tuple at each node lead to membership (and also insert and delete) algorithms that take $O(d \log n)$ time, where n is the number of elements in the tree, because one needs to compare the element being searched with the d-tuples at each node. One can go further and claim that for some problem instances it actually requires $\Theta(d \log n)$ time. Gonzalez [4] also points out that simple shortcuts to the search process do not work. For example, if we reach a node in the search such that the first k components are identical, one may be tempted to conclude that in the subtree rooted at that node one needs to search only from position $k+1$ to d in the d-tuples. This is false because the k-component prefixes of all the d-tuples in a subtree may vary considerable in a binary search tree. One can easily show that even the membership operation cannot be implemented this way. However, this variation is more predictable when comparing against the smallest or largest d-tuple in a subtree. This is a key idea exploited in Reference 3.

Manber and Myers [15] developed an efficient algorithm that given an N symbol text it finds all the occurrences of any input word q. The scenario is that the text is static, but there will be many word searches. Their approach is to preprocess the text and generate a structure where the searching can be performed efficiently. In their preprocessing stage, they construct a sorted list with all the N suffixes of the text. Locating all the occurrences of a string q reduces to performing two binary search operations in the list of suffixes, the first for the first suffix that contains as prefix q and the second search is for the last suffix that contains as prefix q. Both searches are similar, so let's discuss the first one. This operation is similar to the membership operation discussed in this chapter. Manber and Myers [15] binary search process begins by letting L (R) be the largest (smallest) string in the list. Then l (r), the index of the first element where L (R) differs from q, is computed. Note that if two strings are identical the index of the first component where they differ is set to the length of string plus 1. We use this convention throughout this chapter. The middle entry M in the list is located and then they compute the index of the first component where M and q differ. If this value is computed the obvious way, then the procedure will not be efficient. To do this efficiently they compute it with l, r as well as the index of the first component where M and L, and M and R differ. These last two values are precomputed in the preprocessing stage. This indirect computation may take $O(|q|)$ time; however, overall the phases of the computation the process takes at most $O(|q|)$ time. The advantage of this approach is that it requires only $O(N)$ space, and the preprocessing can be done in $O(N)$ expected time [15]. The disadvantage is that it is not dynamic. Updating the text requires expensive recomputation of the precomputed data, that is, one needs to find the first component where many pairs in the list differ in order to carry out efficiently the binary search process. For their application [15], the text is static. The time required to do the search for q is $O(|q| + \log N)$ time. This approach results in a structure that is similar to the fully balanced tree strategy in Reference 9.

Gonzalez [3] solved Vaishnavi's open problem by designing a binary tree data structure where the multidimensional dictionary operations can be performed in $O(d + \log n)$ time while performing only a constant number of rotations. To achieve this goal the set of d-tuples S is represented in a balanced binary search tree that contains additional information at each node. This additional information is similar to the one in the lists of Manber and Myers [15], but it can be recomputed efficiently as we insert and delete elements. The disadvantage is that the membership operation is more complex. But all the procedures developed in Reference 3 are simpler than the ones given in References 6,12–14. One just needs to add a few instructions in addition to the normal code required to manipulate balanced binary search trees [2]. The additional information in Reference 3 includes for every node v in the tree the index of the first element where v and the smallest (largest) d-tuple in the subtree rooted at v differ as well as a pointer to this d-tuple. As mentioned above, testing for membership is more complex. At each iteration in the search process [3], we are at the tree node t and we know that if q is in the tree then it is in the subtree rooted at t. The algorithm knows either the index of a component where q and the smallest d-tuple in the subtree t differ, or the index of a component where q and the largest d-tuple in the subtree t differ. Then the algorithm determines that q is in node t or it advances to the left or right subtrees of node t. In either case, it maintains the above invariant. It is important to point out that the invariant is: "the index of a component where q and the smallest d-tuple in the subtree differ" rather than the "the first index of a ..."It does not seem possible to find "the first index ..." in this structure efficiently with the information stored in the tree. This is not a big problem when q is in the tree since it will be found quickly, but if it is not in the tree then in order to avoid reporting that it is in the tree one must perform an additional verification step at the end that takes $O(d)$ time. Gonzalez [3] calls this search strategy "assume, verify, and conquer" (AVC). That is, in order to avoid multiple expensive verification steps one assumes that some prefixes of strings match. The outcome of the search depends on whether or not these assumptions were valid. This can be determined by performing one simple verification step that takes $O(d)$ time. The elimination of multiple verification steps is very important because in the worst case there are $\Omega(\log n)$ of such steps, and each one could take $\Omega(d)$ time. The difference between this approach and the one in Reference 15 is that Manber and Myers compute the first element where M and q differ, whereas Gonzalez [3] computes an element where M and q differ. As we said before, in Gonzalez [3] structure one cannot compute efficiently the first element where M and q differ, whereas in Reference 15 this is possible because it is a static structure.

Gonzalez [4] modified the structure in Reference 3 to one that follows the search process in Reference 15. The new structure, which we discuss in Section 25.4, is in general faster to update because for every node t one keeps the index of the first component where the d-tuple stored at node t and the smallest (largest) d-tuple greater (smaller) than all the d-tuples in the subtree rooted at t (if any) differ, rather than the one between node t and the smallest (largest) d-tuple in its subtree rooted at t as in Reference 3. In this structure, only several nodes have to be modified when inserting a node or deleting a leaf node, but in the structure in Reference 3 one may need to update $O(\log n)$ nodes. Deleting a nonleaf node from the tree requires more work in this structure than in Reference 3, but membership testing is simpler in the structure in Reference 4. To summarize, the membership algorithm in Reference 4 mimics the search procedure in Reference 15, but follows the update approach developed in Reference 3. Gonzalez [4] established that the dictionary operations can be implemented to take $O(d + \log n)$ time while performing only a constant number of rotations for both insert and delete. Other operations which can be performed efficiently in this multidimensional balanced binary search trees are: finding the (lexicographically) smallest (largest) d-tuple ($O(\log n)$ time), print in lexicographic order ($O(dn)$ time), and concatenation ($O(d + \log n)$ time). Finding the (lexicographically) kth smallest (largest) d-tuple can also be implemented efficiently ($O(\log n)$ time) by adding to each node the number of nodes in its left subtree. The asymptotic time complexity bound for the procedures in Reference 4 is identical to the ones in Reference 3, but the procedures in Reference 4 are simpler. To distinguish this new type of balanced binary search trees from the classic ones and the ones in Reference 3, we refer to these trees as *multidimensional balanced binary search trees*. In this article, we follow the notation in Reference 4.

In Section 25.5, we show that the rotation operation in Reference 4 can be implemented to take only constant time by making some rather simple operations that were first discussed in Reference 16. The implication of this, as pointed out in Reference 16, is that the split operation can also be implemented to take $O(d + \log n)$. Also, the efficient implementation of the rotation operation allows us to use the technique in Reference 4 on many other balanced structures, such as AVL, weight balanced, and so on, since performing $O(\log n)$ rotations does not limit the applicability of the techniques in Reference 4. These observations were first reported in Reference 16, where they present a similar approach, but in a more general setting.

25.4 Balanced BST Implementation

Let us now discuss the data structure and algorithms for the multidimensional dictionaries given in Reference 4. The representation is based on balanced binary search trees, without external nodes, that is, each node represents one d-tuple. Balanced binary search trees and their algorithms [2] are like the typical "bottom-up" algorithms for the Red-Black trees. The ordering of the d-tuples is lexicographic. Each node t in the tree rooted at r has the following information in addition to the color bit required to manipulate balanced binary search trees [2]. Note that if two d-tuples are identical the index of the first component where they differ is set to $d + 1$. We use this convention throughout this chapter.

s:	The d-tuple represented by the node. The individual components may be accessed as $s.x(1), s.x(2), \ldots, s.x(d)$
lchild:	Pointer to the left subtree of t
rchild:	Pointer to the right subtree of t
lptr:	Pointer to the node with largest d-tuple in r with value smaller than all the d-tuples in the subtree rooted at t, or null if no such d-tuple exists
hptr:	Pointer to the node with smallest d-tuple in r with value larger than all the d-tuples in the subtree rooted at t, or null if no such d-tuple exists
lj:	Index of first component where s and the d-tuple at the node pointed at by *lptr* differ, or one if *lptr* = *null*
hj:	Index of first component where s and the d-tuple at the node pointed at by *lptr* differ, or one if *hptr* = *null*

The insert, delete, and membership procedures perform the operations required to manipulate balanced binary search trees, and some new operations to update the structure. The basic operations to manipulate balanced binary search trees are well known [1,2]; so we only discuss in detail the new operations.

To show that membership, insert, and delete can be implemented to take $O(d + \log n)$ time, we only need to show that the following (new) operations can be performed $O(d + \log n)$ time.

- A. Given the d-tuple q determine whether or not it is stored in the tree
- B. Update the structure after adding a node (just before rotation(s), if any)
- C. Update the structure after performing a rotation
- D. Update the structure after deleting a leaf node (just before rotation(s), if any)
- E. Transform the deletion problem to deletion of a leaf node.

The membership operation that tests whether or not the d-tuple q given by $(q.x(1), q.x(2), \ldots, q.x(d))$ is in the multidimensional binary search tree (or subtree) rooted at r appears in [4] and implements (A). The basic steps are as follows. Let t be any node encountered in the search process in the multidimensional balanced binary search tree rooted at r. Let $prev(t)$ to be the d-tuple in r with largest value but whose value is smaller than all the d-tuples stored in the subtree rooted at t, unless no such tuple exists in which case its value is $(-\infty, -\infty, \ldots, -\infty)$, and let $next(t)$ be the d-tuple in r with smallest value but whose value is larger than all the d-tuples stored in the subtree rooted by t, unless no such tuple exists in which case its value is $(+\infty, +\infty, \ldots, +\infty)$. The following invariant is maintained throughout the search process. During the search we will be visiting node t which is initially the root of the tree. The value of d_{low} is the index of the first component where q and $prev(t)$ differ, and variable d_{high} is the index of the first component where q and $next(t)$ differ. The d-tuple being search for, q, has value (lexicographically) greater than $prev(t)$ and (lexicographically) smaller than $next(t)$.

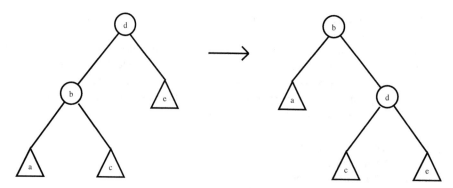

FIGURE 25.2   Rotation.

The computation of j, the index of the first component where $t.s$ and q differ is performed as follows: when $d_{low} \geq d_{high}$, then either (1) $d_{low} = t.lj$ in which case j is just $\min\{d_{low}, t.lj\}$, or (2) $d_{low} = t.lj$ in which case j is set to the index of the first component starting at position d_{low} where q and $t.s$ differ. The case when $d_{low} < d_{high}$ is similar and appears in Reference 4. Gonzalez [4] proved that by setting j by the above procedure will set it to the index of the first component where $t.s$ and q differ. When j is equal to $d + 1$, then q is the d-tuple stored in t and the procedure returns the value of **true**. Otherwise by comparing the jth element of q and t, the procedure decides whether to search in the left or right subtrees of t. In either case d_{high} or d_{low} is set appropriately and the invariant holds at the next iteration. The time complexity at each level is not bounded by a constant; however, it is bounded by 1 plus the difference between the new and old values of $max\{d_{low}, d_{high}\}$. Since $max\{d_{low}, d_{high}\}$ does not decrease and it is at most $d + 1$ at the end of each operation, it follows that the total number of operations performed is of order d plus the height of the tree ($O(\log n)$). Correctness and efficiency are established in the following lemma whose proof appears in Reference 4.

Lemma 25.1

Given a d-tuple q, the formal procedure membership (q, r) given in Reference 4 determines whether or not q is in the multidimensional balanced binary search tree rooted at r in $O(d + \log n)$ time [4].

For operation (B), a node is added as a leaf node, the information that needs to be set for that node are the pointers $lptr$ and $hptr$ which can be copied from the parent node, unless there is no parent in which case they are set to null. The values lj and hj can be computed directly in $O(d)$ time. A predecessor of the node added also needs its $lptr$ and lj, or $hptr$ and hj values changed. This can be easily done in $O(d)$, by remembering the node in question during the search process, that is, the last node where one makes a left turn (moving to the leftchild) or the last one where one makes a right turn (moving to the rightchild).

Lemma 25.2

After inserting a node q in a multidimensional balanced binary search tree and just before rotation, the structure can be updated as mentioned above in $O(d + \log n)$ time [4].

The implementation of operation (D) is similar to the one for (B), therefore, it will not be discussed further. The rotation operation (C) can be reduced to a simple rotation, as a double or triple rotation can reduced to two and three simple rotations, respectively. A simple rotation is shown in Figure 25.2. It is simpler to implement the rotation operation by moving the nodes rather than just their values, because this reduces the number of updates that need to be performed. Clearly, the only nodes whose information needs to be updated are b, d and the parent of d. This can be implemented to take $O(d)$ time.

Lemma 25.3

After a rotation in a multidimensional balanced binary search tree, the structure can be updated as mentioned above in $O(d)$ time [4].

Operation (E) is more complex to implement. It is well known [17] that the problem of deleting an arbitrary node from a balanced binary search tree can be reduced to deleting a leaf node by applying a simple transformation. Since all cases are similar, let's just discuss the case shown in Figure 25.3. The node to delete is a which is not a leaf node. In this case, node b contents are moved to node a, and now we delete the old node b which is labeled x. As pointed out in Reference 4, updating the resulting

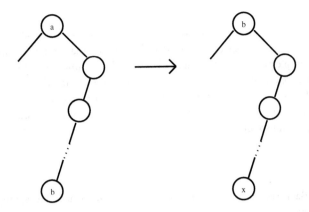

FIGURE 25.3 Transforming deletion of an arbitrary node to deletion of a leaf node.

structure takes $O(d + \log n)$ time. Since the node labeled x will be deleted, we do not need to update it. For the new root (the one labeled b), we need to update the lj and hj values. Since we can use directly the old $lptr$ and $hptr$ pointers, the update can be done in $O(d)$ time. The lj (hj) value of the nodes (if any) in the path that starts at the right (left) child of the new root that continues through the leftchild (rightchild) pointers until the null pointer is reached needs to be updated. There are at most $O(\log n)$ of such nodes. However, the d-tuples stored at each of these nodes are decreasing (increasing) in lexicographic order when traversing the path top down. Therefore, the lj (hj) values appear in increasing (decreasing) order. The correct values can be easily computed in $O(d + \log n)$ time by reusing previously computed lj (hj) values while traversing the path top down. The following lemma, whose proof is omitted, summarizes these observations. Then we state the main result in Reference 4 which is based on the above discussions and the lemmas.

Lemma 25.4

Transforming the deletion problem to deleting a leaf node can be performed, as mentioned above, in $O(d + \log n)$ time [4].

Theorem 25.1

Any online sequence of operations of the form insert, delete, and membership, for any d-tuple can be carried out on a multidimensional balanced binary search tree in $O(d + \log n)$ time, where n is the current number of points, and each insert and delete operation requires no more than a constant number of rotations [4].

25.5 Additional Operations

With respect to other operations, one can easily find the smallest or largest d-tuple in $O(\log n)$ time by just taking all the leftchild or rightchild pointers. By traversing the tree in inorder one can print all the d-tuples in increasing in $O(dn)$ time. An $O(d + \log n)$ time algorithm to concatenate two sets represented by the structure can be easily obtained by using standard procedures provided that all the d-tuples in one set are in lexicographic order smaller than the ones in the other set. The kth smallest or largest d-tuple can be found in $O(\log n)$ time after adding to each node in the tree the number of nodes in its left subtree.

The split operation is given a d-tuple q and a set represented by a multidimensional balanced binary search tree t, split t into two multidimensional balanced binary search trees: one containing all the d-tuples in lexicographic order smaller than or equal to q and the other one containing the remaining elements. At first glance, it seems that the split operation cannot be implemented within the $O(d + \log n)$ time complexity bound. The main reason is that there could be $\Omega(\log n)$ rotations and each rotation takes time proportional to d. However, the analysis in Reference 4 for the rotation operation which is shown in Section 25.4 can be improved and one can show that it can be easily implemented to take constant time. The reason for this is that for a simple rotation, see Figure 25.2, the lj or hj value needs to be updated for the node labeled b, and we know the $lptr$ and $hptr$ pointers. This value can be computed from previous values before the rotation, that is, the lj and hj values for node b and the fact that the $rptr$ value for b is node d before the rotation. The other value to be updated is the lj value for node d after the rotation, but this is simply the rj value for node d before the rotation.

The efficient implementation of the rotation operation allows us to use the same technique on many other balanced structures, such as AVL-trees, weight-balanced trees, and so on, since having $O(\log n)$ rotations does not limit its applicability.

25.6 Discussion

On average, the trie-bst approach requires less space to represent the d-tuples than our structures. However, multidimensional balanced binary search trees have simple procedures that take only $O(d + \log n)$ time, and only a constant number of rotations are required after each insert and delete operations. Furthermore, operations like finding the (lexicographically) smallest or largest d-tuple ($O(\log n)$ time), print in lexicographic order ($O(dn)$ time), concatenation ($O(d + \log n)$ time), and split ($O(d + \log n)$ time) can also be performed efficiently in this new structure. This approach can also be used in other balanced structures, such as AVL-trees, weight-balanced trees, and so on.

References

1. H. J. Olivie, A New Class of Balanced Search Trees: Half-Balanced Binary Search Trees, PhD thesis, University of Antwerp, U.I.A., Wilrijk, Belgium, 1980.
2. R. E. Tarjan, Updating a balanced search tree in $O(1)$ rotations, *Information Processing Letters*, Vol. 16, 1983, pp. 253–257.
3. T. Gonzalez, The on-line D-dimensional dictionary problem, *Proceedings of the Third Symposium on Discrete Algorithms*, Orlando, Florida, January 1992, pp. 376–385.
4. T. Gonzalez, Simple algorithms for the on-line multidimensional problem and related problems, *Algorithmica*, Vol. 28, 2000, pp. 255–267.
5. T. Gonzalez, Covering a set of points with fixed size hypersquares and related problems, *Information Processing Letters*, Vol. 40, 1991, pp. 181–188.
6. K. Mehlhorn, Dynamic binary search, *SIAM Journal of Computing*, Vol. 8, 1979, pp. 175–198.
7. E. H. Sussenguth, Use of tree structures for processing files, *Communications of the ACM*, Vol. 6, 1963, pp. 272–279.
8. H. A. Clampett, Randomized binary searching with the tree structures, *Communications of the ACM*, Vol. 7, 1964, pp. 163–165.
9. J. L. Bentley, and J. B. Saxe, Algorithms on vector sets, *SIGACT News*, Fall 1979, pp. 36–39.
10. D. S. Hirschberg, On the complexity of searching a set of vectors, *SIAM Journal on Computing*, Vol. 9, 1980, pp. 126–129.
11. S. R. Kosaraju, On a Multidimensional Search Problem, *1979 ACM Symposium on the Theory of Computing*, Atlanta, Georgia, pp. 7–73.
12. R. H. Gueting, and H. P. Kriegel, Multidimensional B-tree: An efficient dynamic file structure for exact match queries, *Proceedings 10th GI Annual Conference, Informatik Fachberichte*, Springer-Verlag, Berlin-Heidelberg-New York, 1980, pp. 375–388.
13. V. Vaishnavi, Multidimensional height-balanced trees, *IEEE Transactions on Computers*, Vol. c-33, 1984, pp. 334–343.
14. V. Vaishnavi, Multidimensional balanced binary trees, *IEEE Transactions on Computers*, Vol. 38, 1989, pp. 968–985.
15. U. Manber, and G. Myers, Suffix arrays: A new method for on-line string searches, *SIAM Journal of Computing*, Vol. 22, 1993, pp. 935–948.
16. R. Grossi, and G. F. Italiano, International colloquium on automata, languages and programming (ICALP 99), *Lecture Notes in Computer Science*, Springer-Verlag, Berlin-Heidelberg-New York, Vol. 1644, 1999, pp. 372–381.
17. L. J. Guibas, and R. Sedgewick, A dichromatic framework for balanced trees, *Proceedings of the 19th Annual IEEE Symposium on Foundations of Computer Science*, University of Michigan, Ann Arbor, 1978, pp. 8–21.

26

Cuttings*

26.1	Introduction	397
26.2	The Cutting Construction	397
	Geometric Sampling • Optimal Cuttings	
26.3	Applications	401
	Point Location • Hopcroft's Problem • Convex Hulls and Voronoi Diagrams • Range Searching	
	Acknowledgments	403
	References	403

Bernard Chazelle
Princeton University

26.1 Introduction

For divide-and-conquer purposes, it is often desirable to organize a set S of n numbers into a sorted list, or perhaps to partition it into two equal-sized groups with no element in one group exceeding any element in the other one. More generally, we might wish to break up S into k groups of size roughly n/k, with again a total ordering among the distinct groups. In the first case we sort; in the second one we compute the median; in the third one we compute quantiles. This is all well known and classical. Is it possible to generalize these ideas to higher dimension? Surprisingly the answer is yes. A geometric construction, known as an ε-cutting, provides a space partitioning technique that extends the classical notion of selection to any finite dimension. It is a powerful, versatile data structure with countless applications in computational geometry.

Let H be a set n hyperplanes in $\mathbf{R}^d$. Our goal is to divide up $\mathbf{R}^d$ into simplices, none of which is cut by too many of the n hyperplanes. By necessity, of course, some of the simplices need to be unbounded. We choose a parameter $\varepsilon > 0$ to specify the coarseness of the subdivision. A set C of closed full-dimensional simplices is called an ε-cutting for H (Figure 26.1) if:

1. The union of the simplices is $\mathbf{R}^d$, and their interiors are mutually disjoint;
2. The interior of any simplex is intersected by at most εn hyperplanes of H.

Historically, the idea of using sparsely intersected simplices for divide and conquer goes back to Clarkson [1] and Haussler and Welzl [2], among others. The definition of an ε-cutting given above is essentially due to Matoušek [3]. Efficient but suboptimal constructions were given by Agarwal [4,5] for the two-dimensional case and Matoušek [3,6,7] for arbitrary dimension. The optimal ε-cutting construction cited in the theorem below, due to Chazelle [8], is a simplification of an earlier design by Chazelle and Friedman [9].

Theorem 26.1

Given a set H of n hyperplanes in $\mathbf{R}^d$, for any $0 < \varepsilon < 1$, there exists an ε-cutting for H of size $O(\varepsilon^{-d})$, which is optimal. The cutting, together with the list of hyperplanes intersecting the interior of each simplex, can be found deterministically in $O(n\varepsilon^{1-d})$ time.

26.2 The Cutting Construction

This section explains the main ideas behind the proof of Theorem 26.1. We begin with a quick overview of geometric sampling theory. For a comprehensive treatment of the subject, see References 10 and 11.

---

* This chapter has been reprinted from first edition of this Handbook, without any content updates.

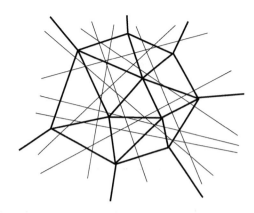

FIGURE 26.1   A two-dimensional cutting.

26.2.1 Geometric Sampling

A *set system* is a pair $\Sigma = (X, \mathcal{R})$, where X is a set and $\mathcal{R}$ is a collection of subsets of X. In our applications, $X \subset \mathbf{R}^d$ and each $R \in \mathcal{R}$ is of the form $X \cap f(K)$, where K is a fixed region of $\mathbf{R}^d$ and f is any member of a fixed group F of transformations. For example, we might consider n points in the plane, together with the subsets lying inside any triangle congruent to a fixed triangle.

Given $Y \subseteq X$, we define the set system "induced by Y" to be $(Y, \mathcal{R}|_Y)$, with $\mathcal{R}|_Y = \{ Y \cap R \mid R \in \mathcal{R} \}$. The *VC-dimension* (named for Vapnik and Chervonenkis [12]) of Σ is defined as the maximum size of any Y such that $\mathcal{R}|_Y = 2^Y$. For example, the VC-dimension of the infinite geometric set system formed by points in the plane and halfplanes is 3. The *shatter function* $\pi_{\mathcal{R}}(m)$ of the set system $\Sigma = (X, \mathcal{R})$ is the maximum number of subsets in the set system $(Y, \mathcal{R}|_Y)$ induced by any $Y \subseteq X$ of size m. If $\pi_{\mathcal{R}}(m)$ is bounded by cm^d, for some constants $c, d > 0$, then the set system is said to have a *shatter function exponent* of at most d. It was shown in References 12–14 that, if the shatter function exponent is $O(1)$, then so is the VC-dimension. Conversely, if the VC-dimension is $d \geq 1$ then, for any $m \geq d$, $\pi_{\mathcal{R}(m)} < (em/d)^d$.

We now introduce two fundamental notions: ε-nets and ε-approximations. For any $0 < \varepsilon < 1$, a set $N \subseteq X$ is called an ε-net for a finite set system $(X, \mathcal{R})$ if $N \cap R \neq \emptyset$ for any $R \in \mathcal{R}$ with $|R|/|X| > \varepsilon$. A finer (but more costly) sampling mechanism is provided by an ε-*approximation* for $(X, \mathcal{R})$, which is a set $A \subseteq X$ such that, given any $R \in \mathcal{R}$,

$$\left| \frac{|R|}{|X|} - \frac{|A \cap R|}{|A|} \right| \leq \varepsilon.$$

Some simple structural facts about nets and approximations:

Lemma 26.1

If X_1, X_2 are disjoint subsets of X of the same size, and A_1, A_2 are same-size ε-approximations for the subsystems induced by X_1, X_2 (respectively), then $A_1 \cup A_2$ is an ε-approximation for the subsystem induced by $X_1 \cup X_2$.

Lemma 26.2

If A is an ε-approximation for $(X, \mathcal{R})$, then any ε'-approximation (resp. -net) for $(A, \mathcal{R}|_A)$ is also an $(\varepsilon + \varepsilon')$-approximation (resp. -net) for $(X, \mathcal{R})$.

In the absence of any restrictive assumption on the set system, it is natural to expect the sample size to depend on both the desired accuracy and the size of the set system itself.

Theorem 26.2

Given a set system $(X, \mathcal{R})$, where $|X| = n$ and $|\mathcal{R}| = m$, for any $1/n \leq \varepsilon < 1$, it is possible to find, in time $O(nm)$, an ε-net for $(X, \mathcal{R})$ of size $O(\varepsilon^{-1} \log m)$ and an ε-approximation for $(X, \mathcal{R})$ of size $O(\varepsilon^{-2} \log m)$.

If we assume bounded VC-dimension, everything changes. In fact the key result in geometric sampling theory is that, for any given level of accuracy, the sample size need not depend on the size of the set system.

In practice, geometric set systems often are "accessible" via an *oracle* function that takes any $Y \subseteq X$ as input and returns the list of sets in $\mathcal{R}|_Y$ (each set represented explicitly). We assume that the time to do that is $O(|Y|^{d+1})$, which is linear in the maximum possible size of the oracle's output, where d is the shatter function exponent. For example, in the case of points and disks in the plane, we have $d = 3$, and so this assumes that, given n points, we can enumerate all subsets enclosed by a disk in time $O(n^4)$. To do this, enumerate all k-tuples of points ($k \leq 3$) and, for each tuple, find which points lie inside the smallest disk enclosing the k points. The main result below is stated in terms of the shatter function exponent d, but the same results hold if d denotes the VC-dimension.

Theorem 26.3

Given a set system $(X, \mathcal{R})$ of shatter function exponent d, for any $\varepsilon \leq 1/2$, an ε-approximation for $(X, \mathcal{R})$ of size $O(d\varepsilon^{-2} \log d\varepsilon^{-1})$ and an ε-net for $(X, \mathcal{R})$ of size $O(d\varepsilon^{-1} \log d\varepsilon^{-1})$ can be computed in time $O(d)^{3d}(\varepsilon^{-2} \log d\varepsilon^{-1})^d |X|$.

Vapnik and Chervonenkis [12] described a probabilistic construction of ε-approximations in bounded VC-dimension. The deterministic construction stated above is due to Chazelle and Matoušek [15], and builds on earlier work [3,6,7,9]. Haussler and Welzl [2] proved the upper bound on the size of ε-nets. The running time for computing an ε-net was improved to $O(d)^{3d}(\varepsilon^{-1} \log d\varepsilon^{-1})^d |X|$ by Brönnimann, Chazelle, and Matoušek [16], using the concept of a *sensitive ε-approximation*. Komlós, Pach, and Woeginger [17] showed that, for any fixed d, the bound of $O(\varepsilon^{-1} \log \varepsilon^{-1})$ for ε-nets is optimal in the worst case (see also Reference 18). The situation is different with ε-approximations: if $d > 1$ is the VC dimension, then there exists an ε-approximation for $(X, \mathcal{R})$ of size $O(\varepsilon^{-2+2/(d+1)})$ [19,20].

An important application of ε-approximations is for estimating how many vertices in an arrangement of hyperplanes in $\mathbf{R}^d$ lie within a given convex region. Let $\Sigma = (H, \mathcal{R})$ be the set system formed by a set H of hyperplanes in $\mathbf{R}^d$, where each $R \in \mathcal{R}$ is the subset of H intersected by an arbitrary line segment. Let σ be a convex body (not necessarily full-dimensional). In the arrangement formed by H within the affine span of σ, let $V(H, \sigma)$ be the set of vertices that lie inside σ. The following was proven in References 8 and 16.

Theorem 26.4

Given a set H of hyperplanes in $\mathbf{R}^d$ in general position, let A be an ε-approximation for $\Sigma = (H, \mathcal{R})$. Given any convex body σ of dimension $k \leq d$,

$$\left| \frac{|V(H, \sigma)|}{|H|^k} - \frac{|V(A, \sigma)|}{|A|^k} \right| \leq \varepsilon.$$

26.2.2 Optimal Cuttings

For convenience of exposition, we may assume that the set H of n hyperplanes in $\mathbf{R}^d$ is in general position. Let $\mathcal{A}(H)$ denote the arrangement formed by H. Obviously, no simplex of an ε-cutting can enclose more than $O(\varepsilon n)^d$ vertices. Since $\mathcal{A}(H)$ itself has exactly $\binom{n}{d}$ vertices, we should expect to need at least on the order of ε^{-d} simplices. But this is precisely the upper bound claimed in Theorem 26.1, which therefore is asymptotically tight.

Our starting point is an ε-net N for H, where the underlying set system $(X, \mathcal{R})$ is formed by a set X of hyperplanes and the collection $\mathcal{R}$ of subsets obtained by intersecting X with all possible open d-simplices. Its VC-dimension is bounded, and so by Theorem 26.3 an ε-net N of size $O(\varepsilon^{-1} \log \varepsilon^{-1})$ can be found in $n\varepsilon^{-O(1)}$ time.

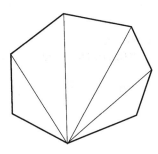

FIGURE 26.2 A canonical triangulation.

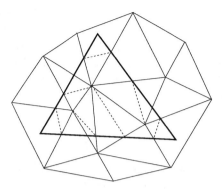

FIGURE 26.3   Clip and retriangulate.

We need to use a systematic way to triangulate the arrangement formed by the ε-net. We build a *canonical triangulation* of $\mathcal{A}(N)$ by induction on the dimension d (Figure 26.2). The case $d = 1$ is trivial, so we assume that $d > 1$.

1. Rank the vertices of $\mathcal{A}(N)$ by the lexicographic order of their coordinate sequences.
2. By induction, form a canonical triangulation of the $(d-1)$-dimensional arrangement made by each hyperplane with respect to the $n - 1$ others.
3. For each cell (i.e., full-dimensional face) σ of $\mathcal{A}(N)$, lift toward its lowest-ranked vertex v each k-simplex ($k = 0, \ldots, d - 2$) on the triangulated boundary of σ that does not lie in a $(d-1)$-face of $\mathcal{A}(N)$ that is incident to v.

It is not hard to see that the combinatorial complexity (i.e., number of all faces of all dimensions) of the canonical triangulation of $\mathcal{A}(N)$ is asymptotically the same as that of $\mathcal{A}(N)$, which is $O(\varepsilon^{-1} \log \varepsilon^{-1})^d$. Therefore, the closures of its cells constitute an ε-cutting for H of size $O(\varepsilon^{-1} \log \varepsilon^{-1})^d$, which is good but not perfect. For optimality we must remove the log factor.

Assume that we have at our disposal an optimal method for building an ε_0-cutting of size $O(\varepsilon_0^{-d})$, for some suitably small constant ε_0. To bootstrap this into an optimal ε-cutting construction for any ε, we might proceed as follows: Beginning with a constant-size cutting, we progressively refine it by producing several generations of finer and finer cuttings, C_1, C_2, etc, where C_k is an ε_0^k-cutting for H of size $O(\varepsilon^{-dk})$. Specifically, assume that we have recursively computed the cutting C_k for H. For each $\sigma \in C_k$, we have the incidence list H_σ of the hyperplanes intersecting the interior of σ. To compute the next-generation cutting C_{k+1}, consider refining each σ in turn as follows:

1. Construct an ε_0-cutting for H_σ, using the algorithm whose existence is assumed.
2. Retain only those simplices that intersect σ and clip them outside of σ.
3. In case the clipping produces nonsimplicial cells within σ, retriangulate them "canonically" (Figure 26.3).

Let C_{k+1} denote the collection of new simplices. A simplex of C_{k+1} in σ is cut (in its interior) by at most $\varepsilon_0 |H_\sigma|$ hyperplanes of H_σ, and hence of H. By induction, this produces at most $n\varepsilon_0^{k+1}$ cuts; therefore, C_{k+1} is an ε_0^{k+1}-cutting. The only problem is that C_{k+1} might be a little too big. The reason is that excess in size builds up from generation to generation. We circumvent this difficulty by using a global parameter that is independent of the construction; namely, the total number of vertices.

Note that we may assume that $|H_\sigma| > n\varepsilon_0^{k+1}$, since σ would otherwise already satisfy the requirement of the next generation. We distinguish between *full* and *sparse* simplices. Given a set X of hyperplanes and a d-dimensional (closed) simplex σ, let $v(X, \sigma)$ be the number of vertices of $\mathcal{A}(X)$ in the interior of σ.

- The simplex $\sigma \in C_k$ is *full* if $v(H, \sigma) \geq c_0 |H_\sigma|^d$, where $c_0 = \varepsilon_0^2$. If so, we compute an ε_0-net for H_σ, and triangulate the portion of the net's arrangement within σ to form an ε_0-cutting of size $O(\varepsilon_0^{-1} \log \varepsilon_0^{-1})^d$. Its simplices form the elements of C_{k+1} that lie within σ.
- A simplex σ that is not full is *sparse*. If so, we find a subset H_σ^o of H_σ that satisfies two conditions:
  i. The canonically triangulated portion of $\mathcal{A}(H_\sigma^o)$ that lies inside σ consists of a set C_σ^o of at most $\frac{1}{2}\varepsilon_0^{-d}$ full-dimensional (closed) simplices.
  ii. Each simplex of C_σ^o is intersected in its interior by at most $\varepsilon_0 |H_\sigma|$ hyperplanes of H.

The elements of C_{k+1} within σ are precisely the simplices of C_σ^o.

Lemma 26.3

C_{k+1} is an ε_0^{k+1}-cutting of size $O(\varepsilon_0^{-d(k+1)})$.

Next, we explain how to enforce conditions (i) and (ii) for sparse simplices. To be able to distinguish between full and sparse simplices, we use a $c_0/2$-approximation A_σ for H_σ of constant size, which we can build in $O(|H_\sigma|)$ time (Theorem 26.3). It follows from Theorem 26.4 that

$$\left| \frac{v(H,\sigma)}{|H_\sigma|^d} - \frac{v(A_\sigma,\sigma)|}{|A_\sigma|^d} \right| \leq \frac{c_0}{2}; \tag{26.1}$$

therefore, we can estimate $v(H,\sigma)$ in constant time with an error of at most $\frac{c_0}{2}|H_\sigma|^d$, which for our purposes here is inconsequential.

How do we go about refining σ and how costly is it? If σ is a full simplex, then by Theorem 26.3, we can compute the required ε_0-net in $O(|H_\sigma|)$ time. Within the same amount of time, we can also find the new set of simplices in σ, together with all of their incidence lists.

The refinement of a sparse simplex σ is a little more involved. We begin with a randomized construction, from which we then remove all the randomness. We compute H_σ^o by choosing a random sample from A_σ of size $c_1\varepsilon_0^{-1}\log\varepsilon_0^{-1}$, for some constant c_1 large enough (independent of ε_0). It can be shown that, with probability at least 2/3, the sample forms an $(\varepsilon_0/2)$-net for A_σ. By Lemma 26.2, H_σ^o is a $(c_0/2 + \varepsilon_0/2)$-net for H_σ; therefore, we ensure that (ii) holds with probability at least 2/3. A slightly more complex analysis shows that (i) also holds with probability at least 2/3; therefore (i,ii) are both true with probability at least 1/3. We derandomize the construction in a trivial manner by trying out all possible samples, which takes constant time; therefore, the running time for refining σ is $O(|H_\sigma|)$.

Putting everything together, we see that refining any simplex takes time proportional to the total size of the incidence lists produced. By Lemma 26.3, the time needed for building generation $k+1$ is $O(n\varepsilon_0^{-(d-1)(k+1)})$. The construction goes on until we reach the first generation such that $\varepsilon_0^k \leq \varepsilon$. This establishes Theorem 26.1.

From the proof above it is not difficult to derive a rough estimate on the constant factor in the $O(\varepsilon^{-d})$ bound on the size of an ε-cutting. A thorough investigation into the smallest possible constant was undertaken by Har-Peled [21] for the two-dimensional case.

26.3 Applications

Cuttings have numerous uses in computational geometry. We mention just a handful: point location, Hopcroft's problem, convex hulls, Voronoi diagrams, and range searching. In many cases, cuttings allow us to derandomize existing probabilistic solutions, that is, to remove any need for random bits and thus produce deterministic algorithms. Many other applications are described in the survey [5].

26.3.1 Point Location

How do we preprocess n hyperplanes in $\mathbf{R}^d$, so that, given a query point q, we can quickly find the face of the arrangement formed by the hyperplanes that contains the point? For an answer, simply set $\varepsilon = 1/n$ in Theorem 26.1, and use the nesting structure of C_1, C_2, etc, to locate q in C_k. Note that this can be done in constant time once we know the location in C_{k-1}.

Theorem 26.5

Point location among n hyperplanes can be done in $O(\log n)$ query time, using $O(n^d)$ preprocessing.

Observe that if we only wish to determine whether the point q lies on one of the hyperplanes, it is possible to cut down the storage requirement a little. To do that, we use an ε-cutting for $\varepsilon = (\log n)/n$. The cells associated with the bottom of the hierarchy are each cut by $O(\log n)$ hyperplanes, which we can therefore check one by one. This reduces the storage to $O(n^d/(\log n)^{d-1})$.

26.3.2 Hopcroft's Problem

Given n points and n lines in $\mathbf{R}^2$, is there any incidence between points and lines? This is *Hopcroft's problem*. It is self-dual; therefore dualizing it won't help. A classical arrangement of n lines due to Erdös has the property that its n highest-degree vertices are each incident to $\Omega(n^{1/3})$ edges. By picking these n lines as input to Hopcroft's problem and positioning the n points in the near vicinity of these high-degree vertices, we get a sense (not a proof) that to solve the problem should require checking each point against the $\Omega(n^{1/3})$ lines incident to their nearby vertex. This leads to an $\Omega(n^{4/3})$ running time, which under some realistic (though restrictive) conditions, can be made into a rigorous lower bound [22]. At the very least this line of reasoning suggests that to beat $\Omega(n^{4/3})$ is unlikely to be easy. This bound has almost been achieved by an algorithm of Matoušek [23] with, at its heart, a highly intricate and subtle use of cuttings.

Theorem 26.6

To decide whether n points and n lines in the plane are free of any incidence can be done in $n^{4/3} 2^{O(\log^* n)}$ time.

26.3.3 Convex Hulls and Voronoi Diagrams

Cuttings play a key role in computing convex hulls in higher dimension. Given n points in $\mathbf{R}^d$, their convex hull is a bounded convex polytope with $O(n^{\lfloor d/2 \rfloor})$ vertices. Of course, it may have much fewer of them: for example, $d + 1$, if $n - d - 1$ points lie strictly inside the convex hull of the $d + 1$ others. It is notoriously difficult to design *output-sensitive* algorithms, the term designating algorithms whose running time is a function of both input and output sizes. In the "worst case" approach our goal is a simpler one: to design an optimal convex hull algorithm that runs in $O(n \log n + n^{\lfloor d/2 \rfloor})$ time. (The extra term $n \log n$ is unavoidable because sorting is easily embedded as a convex hull problem.)

Computing the convex hull of n points is equivalent by duality to computing the intersection of n halfspaces. A naive approach to this problem is to insert each halfspace one after the other while maintaining the intersection of previously inserted halfspaces incrementally. This can be done without difficulty if we maintain a canonical triangulation of the current intersection polyhedron and update a bipartite graph indicating which hyperplane intersects which cell of the triangulation. A surprising fact, first proven by Clarkson and Shor [24], is that if the halfspaces are inserted in random order, then the expected running time of the algorithm can be made optimal. By using an elaborate mix of ε-nets, ε-approximations, and ε-cuttings, Chazelle [25] showed how to compute the intersection deterministically in optimal time; his algorithm was subsequently simplified by Brönnimann, Chazelle, and Matoušek [16]; a complete description is also given in the book [10]. This implies the two theorems below.

Theorem 26.7

The polyhedron formed by the intersection of n halfspaces in $\mathbf{R}^d$ can be computed in $O(n \log n + n^{\lfloor d/2 \rfloor})$ time.

Not only does this result give us an optimal deterministic solution for convex hulls, but it also solves the Voronoi diagram problem. Indeed, recall [26,27] that a Voronoi diagram of n points in $\mathbf{R}^d$ can be "read off" from the facial structure of the convex hull of a lift of the n points into $\mathbf{R}^{d+1}$.

Theorem 26.8

The convex hull of a set of n points in $\mathbf{R}^d$ can be computed deterministically in $O(n \log n + n^{\lfloor d/2 \rfloor})$ time. By duality, the Voronoi diagram (or Delaunay triangulation) of a set of n points in $\mathbf{E}^d$ can be computed deterministically in $O(n \log n + n^{\lceil d/2 \rceil})$ time.

26.3.4 Range Searching

Simplex range searching refers to the problem of preprocessing a set P of n points in $\mathbf{R}^d$ so that, given a query (closed) simplex σ, the size of $P \cap \sigma$ can be quickly evaluated. Variants of the problem include reporting the points of $P \cap \sigma$ explicitly or, assuming that each point p has a weight $w(p) \in \mathbf{R}$, computing $\sum \{ w(p) \mid p \in P \cap \sigma \}$. The most powerful data structure for solving simplex range searching, the *simplicial partition*, vividly illustrates the power of ε-cuttings. A collection $\{(P_i, R_i)\}$ is called a *simplicial partition* if

- The collection $\{P_i\}$ forms a partition of P; and
- Each R_i is a relatively open simplex that contains P_i.

The simplices R_i can be of any dimension and, in fact, need not even be disjoint; furthermore the P_i's need not be equal to $P \cap R_i$. A hyperplane is said to *cut* R_i if it intersects, but does not contain, R_i. The *cutting number* of the simplicial partition refers to the maximum number of R_i's that can be cut by a single hyperplane. Matoušek [28] designed an optimal construction, which happens to be crucially based on ε-cuttings.

Lemma 26.4

Given a set P of n points in $\mathbf{R}^d$ ($d > 1$), for any integer $1 < r \leq n/2$, there exists a simplicial partition of cutting number $O(r^{1-1/d})$ such that $n/r \leq |P_i| < 2n/r$ for each (P_i, R_i) in the partition.

To understand the usefulness of simplicial partitions for range searching, one needs to learn about *partition trees*. A partition tree for P is a tree $\mathcal{T}$ whose root is associated with the point set P. The set P is partitioned into subsets $P_1, \ldots, P_m$, with each P_i

associated with a distinct child v_i of the root. To each v_i corresponds a convex open set R_i, called the *region* of v_i, that contains P_i. The regions R_i are not necessarily disjoint. If $|P_i| > 1$, the subtree rooted at v_i is defined recursively with respect to P_i.

Armed with a partition tree, it is a simple matter to handle range search queries. In preprocessing, at each node we store the sum of the weights of the points associated with the corresponding region. To answer a query σ, we visit all the children v_i of the root and check whether σ intersects the region R_i of v_i: (i) if the answer is yes, but σ does not completely enclose the region R_i of v_i, then we visit v_i and recurse; (ii) if the answer is yes, but σ completely encloses R_i, we add to our current weight count the sum of the weights within P_i, which happens to be stored at v_i; (iii) if the answer is no, then we do not recurse at v_i.

It is straightforward to see that Lemma 26.4 can be used to construct partition trees. It remains for us to choose the branching factor. If we choose a large enough constant r, we end up with a partition tree that lets us answer simplex range search queries in $O(n^{1-1/d+\varepsilon})$ time for any fixed $\varepsilon > 0$, using only $O(n)$ storage. A more complex argument by Matoušek [28] removes the ε term from the exponent.

With superlinear storage, various space-time tradeoffs can be achieved. For example, as shown by Chazelle, Sharir, and Welzl [29], simplex range searching with respect to n points in $\mathbf{R}^d$ can be done in $O(n^{1+\varepsilon}/m^{1/d})$ query time, using a data structure of size m, for any $n \leq m \leq n^d$. Matoušek [23] slightly improved the query time to $O(n(\log m/n)^{d+1}/m^{1/d})$, for m/n large enough. These bounds are essentially optimal under highly general computational models [10].

Acknowledgments

This work was supported in part by NSF grant CCR-998817, and ARO Grant DAAH04-96-1-0181.

References

1. Clarkson, K.L. New applications of random sampling in computational geometry, *Disc. Comput. Geom.* 2, 1987, 195–222.
2. Haussler, D., Welzl, E. ε-nets and simplex range queries, *Disc. Comput. Geom.* 2, 1987, 127–151.
3. Matoušek, J. Cutting hyperplane arrangements, *Disc. Comput. Geom.* 6, 1991, 385–406.
4. Agarwal, P.K. Partitioning arrangements of lines II: Applications, *Disc. Comput. Geom.* 5, 1990, 533–573.
5. Agarwal, P.K. Geometric partitioning and its applications, in *Computational Geometry: Papers from the DIMACS Special Year*, eds., Goodman, J.E., Pollack, R., Steiger, W., Amer. Math. Soc., 1991.
6. Matoušek, J. Construction of ε-nets, *Disc. Comput. Geom.* 5, 1990, 427–448.
7. Matoušek, J. Approximations and optimal geometric divide-and-conquer, *J. Comput. Syst. Sci.* 50, 1995, 203–208.
8. Chazelle, B. Cutting hyperplanes for divide-and-conquer, *Disc. Comput. Geom.* 9, 1993, 145–158.
9. Chazelle, B., Friedman, J. A deterministic view of random sampling and its use in geometry, *Combinatorica* 10, 1990, 229–249.
10. Chazelle, B. *The Discrepancy Method: Randomness and Complexity*, Cambridge University Press, hardcover 2000, paperback 2001.
11. Matoušek, J. *Geometric Discrepancy: An Illustrated Guide, Algorithms and Combinatorics*, 18, Springer, 1999.
12. Vapnik, V.N., Chervonenkis, A.Ya. On the uniform convergence of relative frequencies of events to their probabilities, *Theory of Probability and its Applications* 16, 1971, 264–280.
13. Sauer, N. On the density of families of sets, *J. Combinatorial Theory A* 13, 1972, 145–147.
14. Shelah, S. A combinatorial problem; stability and order for models and theories in infinitary languages, *Pac. J. Math.* 41, 1972, 247–261.
15. Chazelle, B., Matoušek, J. On linear-time deterministic algorithms for optimization problems in fixed dimension, *J. Algorithms* 21, 1996, 579–597.
16. Brönnimann, H., Chazelle, B., Matoušek, J. Product range spaces, sensitive sampling, and derandomization, *SIAM J. Comput.* 28, 1999, 1552–1575.
17. Komlós, J., Pach, J., Woeginger, G. Almost tight bounds for ε-nets, *Disc. Comput. Geom.* 7, 1992, 163–173.
18. Pach, J., Agarwal, P.K. *Combinatorial Geometry, Wiley-Interscience Series in Discrete Mathematics and Optimization*, John Wiley & Sons, Inc., 1995.
19. Matoušek, J. Tight upper bounds for the discrepancy of halfspaces, *Disc. Comput. Geom.* 13, 1995, 593–601.
20. Matoušek, J., Welzl, E., Wernisch, L. Discrepancy and ε-approximations for bounded VC-dimension, *Combinatorica* 13, 1993, 455–466.
21. Har-Peled, S. Constructing planar cuttings in theory and practice, *SIAM J. Comput.* 29, 2000, 2016–2039.
22. Erickson, J. New lower bounds for Hopcroft's problem, *Disc. Comput. Geom.* 16, 1996, 389–418.
23. Matoušek, J. Range searching with efficient hierarchical cuttings, *Disc. Comput. Geom.* 10, 1993, 157–182.

24. Clarkson, K.L., Shor, P.W. Applications of random sampling in computational geometry, II, *Disc. Comput. Geom.* 4, 1989, 387–421.

25. Chazelle, B. An optimal convex hull algorithm in any fixed dimension, *Disc. Comput. Geom.* 10, 1993, 377–409.

26. Edelsbrunner, H. *Algorithms in Combinatorial Geometry*, Springer-Verlag, Berlin, 1987.

27. Ziegler, G.M. *Lectures on Polytopes, Graduate Texts in Mathematics*, 152, Springer, New York, 1995.

28. Matoušek, J. Efficient partition trees, *Disc. Comput. Geom.* 8, 1992, 315–334.

29. Chazelle, B., Sharir, M., Welzl, E. Quasi-optimal upper bounds for simplex range searching and new zone theorems, *Algorithmica* 8, 1992, 407–429.

27

Approximate Geometric Query Structures[*]

27.1 Introduction.. 405
27.2 General Terminology 406
27.3 Approximate Queries 406
27.4 Quasi-BAR Bounds 409
27.5 BBD Trees 410
27.6 BAR Trees... 412
27.7 Maximum-Spread k-d Trees 416
Acknowledgments 416
References.. 417

Christian A. Duncan
Quinnipiac University

Michael T. Goodrich
University of California, Irvine

27.1 Introduction

Specializeddata structures are useful for answering specific kinds of geometric queries. Such structures are tailor-made for the kinds of queries that are anticipated and even then there are cases when producing an exact answer is only slightly better than an exhaustive search. For example, Chazelle and Welzl [1] showed that triangle range queries can be solved in $O(\sqrt{n} \log n)$ time using linear space but this holds only in the plane. In higher dimensions, the running times go up dramatically, so that, in general, the time needed to perform an exact simplex range query and still use small linear space is roughly $\Omega(n^{1-1/d})$, ignoring logarithmic factors [2]. For orthogonal range queries, efficient query processing is possible if superlinear space is allowed. For example, range trees (Chapter 19) can answer orthogonal range queries in $O(\log^{d-1} n)$ time but use $O(n \log^{d-1} n)$ space [3].

In this chapter, we focus instead on general-purpose data structures that can answer nearest-neighbor queries and range queries using linear space. Since the lower-bound of Chazelle [2] applies in this context, in order to get query bounds that are significantly faster than exhaustive search, we need to compromise somewhat on the exactness of our answers. That is, we will answer all queries *approximately*, giving responses that are within an arbitrarily small constant factor of the exact solution. As we discuss, such responses can typically be produced in logarithmic or polylogarithmic time, using linear space. Moreover, in many practical situations, a good approximate solution is often sufficient.

In recent years several interesting data structures have emerged that efficiently solve several general kinds of geometric queries approximately. We review three major classes of such structures in this chapter. The first one we discuss is a structure introduced by Arya et al. [4] for efficiently approximating nearest-neighbor queries in low-dimensional space. Their work developed a new structure known as the balanced box decomposition (BBD) tree. The BBD tree is a variant of the quadtree and octree [5] but is most closely related to the fair-split tree of Callahan and Kosaraju [6]. In Reference 7, Arya and Mount extend the structure to show that it can also answer approximate range queries. Their structure is based on the decomposition of space into "boxes" that may have a smaller box "cut out;" hence, the boxes may not be convex. The second general purpose data structure we discuss is the balanced aspect ratio (BAR) tree of Duncan et al. [8–10], which is a structure that has similar performance as the BBD tree but decomposes space into convex regions. Finally, we discuss an analysis of a type of k-d tree [11] that helps to explain why k-d trees have long been known to exhibit excellent performance bounds in practice for general geometric queries. In particular, we review a result of Dickerson et al. [8,12], which shows that one of the more common variants, the maximum-spread k-d tree, exhibits properties similar to BBD trees and BAR trees; we present efficient bounds on approximate geometric queries for this variant. Unfortunately, the bounds are not as efficient as the BBD tree or BAR tree but are comparable.

---

[*] This chapter has been reprinted from first edition of this Handbook, without any content updates.

27.2 General Terminology

In order to discuss approximate geometric queries and the efficient structures on them without confusion, we must cover a few fundamental terms. We distinguish between general points in $\mathbb{R}^d$ and points given as input to the structures.

For a given metric space $\mathbb{R}^d$, the coordinates of any point $p \in \mathbb{R}^d$ are $(p_1, p_2, \ldots, p_d)$. When necessary to avoid confusion, we refer to points given as input in a set S as *data points* and general points in $\mathbb{R}^d$ as *real points*. For two points $p, q \in \mathbb{R}^d$, the L_m *metric distance* between p and q is

$$\delta(p,q) = \left(\sum_{i=1}^{d} |p_i - q_i|^m \right)^{\frac{1}{m}}.$$

Although our analysis will concentrate on the Euclidean L_2 metric space, the data structures mentioned in this chapter work in all of the L_m metric spaces.

In addition, we use the standard notions of (convex) regions R, rectangular boxes, hyperplanes H, and hyperspheres B. For each of these objects we define two distance values. Let P and Q be any two regions in $\mathbb{R}^d$, the *minimum* and *maximum* metric distances between P and Q are

$$\delta(P,Q) = \min_{p \in P, q \in Q} \delta(p,q) \text{ and}$$

$$\Delta(P,Q) = \max_{p \in P, q \in Q} \delta(p,q) \text{ respectively.}$$

Notice that this definition holds even if one or both regions are simply points.

Let S be a finite data set $S \subset \mathbb{R}^d$. For a subset $S_1 \subseteq S$, the *size* of S_1, written $|S_1|$, is the number of distinct data points in S_1. More importantly, for any region $R \subset \mathbb{R}^d$, the size is $|R| = |R \cap S|$. That is, the *size* of a region identifies the number of data points in it. To refer to the physical size of a region, we define the *outer radius* as $O_R = \min_{R \subseteq B_r} r$, where B_r is defined as the hypersphere with radius r. The *inner radius* of a region is $I_R = \max_{B_r \subseteq R} r$. The outer radius, therefore, identifies the smallest ball that contains the region R whereas the inner radius identifies the largest ball contained in R.

In order to discuss balanced aspect ratio, we need to define the term.

Definition 27.1

A convex region R in $\mathbb{R}^d$ has aspect ratio $\text{asp}(R) = O_R/I_R$ with respect to some underlying metric. For a given balancing factor α, if $\text{asp}(R) \leq \alpha$, R has *balanced aspect ratio* and is called an α-*balanced* region. Similarly, a collection of regions $\mathcal{R}$ has balanced aspect ratio for a given factor α if each region $R \in \mathcal{R}$ is an α-balanced region.

For simplicity, when referring to rectangular boxes, we consider the aspect ratio as simply the ratio of the longest side to the shortest side. It is fairly easy to verify that the two definitions are equal within a constant factor. As is commonly used, we refer to regions as being either *fat* or *skinny* depending on whether their aspect ratios are balanced or not.

The class of structures that we discuss in this chapter are all derivatives of binary space partition (BSP) trees, see for example [13]. See also Chapter 21. Each node u in a BSP tree T represents both a *region* R_u in space and the *data subset* $S_u \subseteq S$ of objects, points, lying inside R_u. For simplicity, regions are considered closed and points falling on the boundary of two regions can be in either of the two regions but not both. Each leaf node in T represents a region with a constant number of data objects, points, from S. Each internal node in T has an associated cut partitioning the region into two subregions, each a child node. The root of T is associated with some bounding (rectangular box) region containing S. In general, BSP trees can store any type of object, points, lines, solids, but in our case we focus on points. Typically, the partitioning cuts used are hyperplanes resulting in convex regions. However, the BBD tree presented in Section 27.5 is slightly different and can introduce regions with a single interior hole. Therefore, we have generalized slightly to accommodate this in our definition.

27.3 Approximate Queries

Before elaborating on the structures and search algorithms used to answer certain geometric queries, let us first introduce the basic definitions of approximate nearest-neighbor, farthest-neighbor, and range queries, see [4,7,8,10,14].

Definition 27.2

Given a set S of points in $\mathbb{R}^d$, a query point $q \in \mathbb{R}^d$, a (connected) query region $Q \subset \mathbb{R}^d$, and $\epsilon > 0$, we define the following queries (see Figure 27.1):

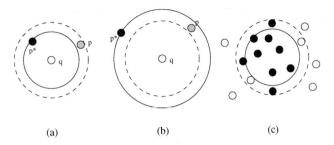

FIGURE 27.1 Examples of (a) an approximate nearest-neighbor p to q with the exact neighbor p^*, (b) an approximate farthest-neighbor p to q with the exact neighbor p^*, and (c) an approximate range query; here the dark points are reported or counted and the lighter points are not.

- A point $p^* \in S$ is a *nearest neighbor of* q if $\delta(p^*, q) \leq \delta(p, q)$ for all $p \in S$.
- A point $p^* \in S$ is a *farthest neighbor of* q if $\delta(p^*, q) \geq \delta(p, q)$ for all $p \in S$.
- A point $p \in S$ is a $(1 + \epsilon)$-*nearest neighbor of* q if $\delta(p, q) \leq (1 + \epsilon)\delta(p^*, q)$, where p^* is the true nearest neighbor of q.
- A point $p \in S$ is a $(1 - \epsilon)$-*farthest neighbor of* q if $\delta(p, q) \geq \delta(p^*, q) - \epsilon O_S$, where p^* is the true farthest neighbor of q.
- An ϵ-*approximate range query* returns or counts a subset $S' \subseteq S$ such that $S \cap Q \subseteq S'$ and for every point $p \in S'$, $\delta(p, Q) \leq \epsilon O_Q$.

To clarify further, a point p is a $(1 + \epsilon)$-approximate nearest neighbor if its distance is within a constant error factor of the true nearest distance. Although we do not discuss it here, we can extend the definitions to report a sequence of k $(1 + \epsilon)$-nearest (or $(1 - \epsilon)$-farthest) neighbors. One may also wonder why the approximate farthest neighbor is defined in absolute terms instead of relative terms as with the nearest version. By observing that the distance from any query point to its farthest neighbor is always at least the radius of the point set O_S, one can see that the approximation is as good as the relative approximation. Moreover, if the query point is extremely far from the point set, then a relative approximation could return any point in S, whereas an absolute approximation would require a much more accurate distance.

Although we do not modify our definition here, one can extend this notion and our later theorems to compensate for the problem of query points distant from the point set in nearest-neighbor queries as well. In other words, when the query point is relatively close to the entire data set, we can use the relative error bound, and when it is relatively far away from the entire data set, we can use the absolute error bound.

The approximate range searching problem described above has one-sided false-positive errors. We do not miss any valid points, but we may include erroneous points near the range boundary. It is a simple modification of the query regions to instead get false-negative errors. That is, we could instead require including only points that are inside of Q but allow missing some points that are inside of Q if they are near the border. In fact, for any region Q one could define two epsilon ranges ϵ_i and ϵ_o for both interior and exterior error bounds and then treat the approximation factor as $\epsilon = \epsilon_i + \epsilon_o$.

There are numerous approaches one may take to solving these problems. Arya et al. [4] introduced a priority search algorithm for visiting nodes in a partition tree to solve nearest-neighbor queries. Using their BBD tree structure, they were able to prove efficient query times. Duncan et al. [10] show how this priority search could also solve farthest-neighbor queries. The nearest and farthest neighbor priority searching algorithms shown in Figures 27.2 and 27.3 come from [8]. In the approximate nearest-neighbor, respectively farthest-neighbor, search, nodes are visited in order of closest node, respectively farthest node. Nodes are extracted via an efficient priority queue, such as the Fibonacci heap [15,16].

Introduced by [7], the search technique used for the approximate range query is a modification to the standard range searching algorithm for regular partition trees. We present the algorithm from [8] in Figure 27.4. In this algorithm, we have two different query regions, the inner region Q and the outer region $Q' \supseteq Q$. The goal is to return all points in S that lie inside Q, allowing some points to lie inside Q' but none outside of Q'. That is $Q' - Q$ defines a buffer zone that is the only place allowed for erroneous points. Whenever a node u is visited, if u is a leaf node, we simply check all of u's associated data points. Otherwise, if R_u does not intersect Q, we know that none of its points can lie in Q, and we ignore u and its subtree. If R_u lies completely inside Q' then all of the data points in its subtree must lie inside Q', and we report all points. Otherwise, we repeat the process on u's two child nodes. For an ϵ-approximate range search, we define $Q' = \{p \in \mathbb{R}^d | \delta(p, Q) \leq \epsilon O_Q\}$. We note that this search algorithm can also be modified to return the count or sum of the weights of the points inside the approximate range rather than explicitly reporting the points.

In all of these search algorithms, the essential criteria behind the running time is the observation that a non-terminating node in the search, one that requires expansion of its child nodes, is a node that must cross certain size boundaries. For example, in the approximate range searching algorithm, the only nodes expanded are those whose region lies partially inside Q, else it

```
APPROXIMATENEARESTNEIGHBOR(T, q, ε)
    Arguments: BSP tree, T, query point q, and error factor ε
    Returns: A (1 + ε)-nearest neighbor p
    Q ← root(T)
    p ← ∞
    do u ← Q.extractMin()
        if δ(u, q) > δ(p, q)/(1 + ε)
            return p
        while u is not a leaf
            u₁ ← leftChild(u)
            u₂ ← rightChild(u)
            if δ(u₁, q) ≤ δ(u₂, q)
                Q.insert(δ(u₂, q), u₂)
                u ← u₁
            else
                Q.insert(δ(u₁, q), u₁)
                u ← u₂
        end while
        // u is now a leaf
        for all p' in dataSet(u)
            if δ(p', q) < δ(p, q)
                p ← p'
    repeat
```

FIGURE 27.2   The basic algorithm to perform nearest-neighbor priority searching.

```
APPROXIMATEFARTHESTNEIGHBOR(T, q, ε)
    Arguments: BSP tree, T, query point q, and error factor ε
    Returns: A (1 − ε)-farthest neighbor p
    Q ← root(T)
    p ← q
    do u ← Q.extractMax()
        if Δ(u, q) ≤ δ(p, q) + εD
            return p
        while u is not a leaf
            u₁ ← leftChild(u)
            u₂ ← rightChild(u)
            if Δ(u₁, q) ≥ Δ(u₂, q)
                Q.insert(Δ(u₂, q), u₂)
                u ← u₁
            else
                Q.insert(Δ(u₁, q), u₁)
                u ← u₂
        end while
        // u is now a leaf
        for all p' in dataSet(u)
            if δ(p', q) > δ(p, q)
                p ← p'
    repeat
```

FIGURE 27.3   The basic algorithm to perform farthest-neighbor priority searching.

```
APPROXIMATERANGESEARCH(u, Q, Q')
    Arguments: Node u in a BSP tree, inner region Q, outer region Q'
        Initially, u ← root(T).
    Reports: All points in the approximate range defined by Q and Q'
    if u is a leaf node
        for all p in dataSet(u)
            if p ∈ Q
                output p
    else if R_u ⊆ Q'
        // The region lies completely inside Q'
        output all points p in the subtree of u
    else if R_u ∩ Q ≠ ∅
        // The region lies partially inside Q
        call APPROXIMATERANGESEARCH(leftChild(u), Q, Q')
        call APPROXIMATERANGESEARCH(rightChild(u), Q, Q')
```

FIGURE 27.4   The basic range search algorithm.

would be discarded, and partially outside Q', else it would be completely counted in the output size. A slightly more complex but similar argument applies for nearest and farthest neighbor algorithms. In the next section, we discuss a general theorem providing provable running time bounds for partition trees satisfying a fundamental packing argument.

27.4 Quasi-BAR Bounds

We are now ready to examine closely a sufficient condition for a data structure to guarantee efficient performance on the aforementioned searches. Before we can proceed, we must first discuss a few more basic definitions presented in Dickerson et al. [12].

Definition 27.3

For any region R, the *region annulus with radius* r, denoted $A_{R,r}$ is the set of all points $p \in \mathbb{R}^d$ such that $p \notin R$ and $\delta(p, R) < r$. A region R' *pierces* an annulus $A_{R,r}$ if and only if there exist two points $p, q \in R'$ such that $p \in R$ and $q \notin R \cup A_{R,r}$.

In other words, an annulus $A_{R,r}$ contains all points outside but near the region R. If R were a sphere of radius r', this would be the standard definition of an annulus with inner radius r' and outer radius $r' + r$. For convenience, when the region and radius of an annulus are understood, we use A. Figure 27.5 illustrates the basic idea of a spherical annulus with multiple piercing regions.

The core of the performance analysis for the searches lies in a critical packing argument. The packing lemmas work by bounding the number of disjoint regions that can pierce an annulus and hence simultaneously fit inside the annulus, see Figure 27.5b. When this packing size is small, the searches are efficient. Rather than cover each structure's search analysis separately, we use the following more generalized notion from Dickerson et al. [12].

Definition 27.4

Given a BSP tree T and a region annulus A, let $\mathcal{P}(A)$ denote the largest set of disjoint nodes in T whose associated regions pierce A. A class of BSP trees is a $\rho(n)$-*quasi-BAR tree* if, for any tree T in the class constructed on a set S of n points in $\mathbb{R}^d$ and any region annulus $A_{R,r}$, $|\mathcal{P}(A_{R,r})| \leq \rho(n)V_A/r^d$, where V_A is the volume of $A_{R,r}$. The function $\rho(n)$ is called the *packing* function.

Basically, the packing function $\rho(n)$ represents the maximum number of regions that can pierce any query annulus. By proving that a class of BSP trees is a $\rho(n)$-quasi-BAR tree, we can automatically inherit the following theorems proven in [4,7,10] and generalized in Reference 12:

Theorem 27.1

Suppose we are given a $\rho(n)$-quasi-BAR tree T with depth $D_T = \Omega(\log n)$ constructed on a set S of n points in $\mathbb{R}^d$. For any query point q, the priority search algorithms in Figures 27.2 and 27.3 find respectively a $(1 + \epsilon)$-nearest and a $(1 - \epsilon)$-farthest neighbor to q in $O(\epsilon^{1-d}\rho(n)D_T)$ time.

Theorem 27.2

Suppose we are given a $\rho(n)$-quasi-BAR tree T with depth D_T constructed on a set S of n points in $\mathbb{R}^d$. For any convex query region Q, the search algorithm in Figure 27.4 solves an ϵ-approximate range searching query in T in $O(\epsilon^{1-d}\rho(n)D_T)$ time (plus

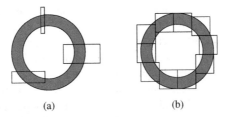

(a)   (b)

FIGURE 27.5 An example of a simple annulus region (a) with three other regions which pierce this annulus and (b) with several "fat" square regions. Observe that only a limited number of such "fat" squares can pierce the annulus.

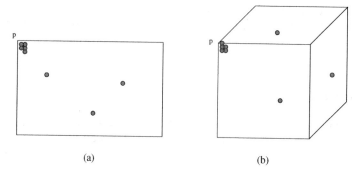

                                    (a)   (b)

FIGURE 27.6   (a) A bad corner of a simple rectangular region with nearly all of the points clustered near a corner. Notice a cut in either the x or y direction dividing the points located inside p would cause a "skinny" region. (b) The same situation in $\mathbb{R}^3$.

output size in the reporting case). For any general non-convex query region Q, the time required is $O(\epsilon^{-d}\rho(n)D_T)$ (plus output size).[*]

Trivially, $\rho(n)$ is always less than n but this accomplishes little in the means of getting good bounds. Having a class of trees with both a good packing function and low depth helps guarantee good asymptotic performance in answering geometric queries. One approach to finding such classes is to require that all regions produced by the tree be fat. The idea behind this is that there is a limit to the number of disjoint fat regions that pierce an annulus dependent upon the aspect ratio of the regions and the thickness of the annulus. Unfortunately, guaranteeing fat regions and good depth is not readily possible using the standard BSP trees like k-d trees and octrees. Imagine building a k-d tree using only axis-parallel partitioning cuts. Figure 27.6 illustrates such an example. Here the majority of the points are concentrated at a particular corner of a rectangular region. Now, any axis-parallel cut is either too close to the opposing face or does not partition the points in the region well, resulting in large tree depth.

Fortunately, there are structures that can provably be shown to be $\rho(n)$-quasi-BAR trees with small values for $\rho(n)$ and D_T. The next few sections discuss some of these structures.

27.5   BBD Trees

Arya et al. [4] introduced the first BSP tree structure to guarantee both balanced aspect ratio and $O(\log n)$ depth. In addition, the aspect ratio achieved allowed them to prove a packing constraint. From this, one can verify that the BBD tree has a packing function of $\rho(n) = O(1)$ where the constant factor depends on the dimension d. In the following section, we describe the basic construction of the BBD tree using terminology from [7].

Every region R_u associated with a node u in a BBD tree is either an *outer* rectangular box or the set theoretic difference between an *outer* rectangular box and an *inner* rectangular box. The *size* of a box is the length of its longest side and the *size* of R_u is the size of the outer box. In order to guarantee balanced aspect ratio for these cells, Arya et al. [4] introduced a *stickiness* restriction on the inner box. Briefly described, an inner box is *sticky* if the distance between the inner box and every face on the outer box is either 0 or not less than the size of the inner box. Although not necessary to the structure, we shall assume that the aspect ratio of the outer box is no more than two.

The construction of the BBD tree is done by a sequence of alternating splitting and shrinking operations. In the *(midpoint) split* operation, a region is bisected by a hyperplane cut orthogonal to one of the longest sides. This is essentially the standard type of cut used in a quadtree or octree. Its simplicity, speed of computation, and effectiveness are major reasons for preferring these operations.

The *shrink* operation partitions a region by a box lying inside the region, creating an inner region. The shrink operation is actually part of a sequence of up to three operations called a *centroid shrink*. The centroid shrink attempts to partition the region into a small number of subregions R_i such that $|R_i| \le 2|R_u|/3$.

When R_u is simply an outer box, with no inner box, a centroid operation is performed with one shrink operation. The inner partitioning box is found by conceptually applying midpoint split operations recursively on the subregion with the larger number of points. The process stops when the subregion contains no more than $2|R_u|/3$ points. The outer box of this subregion is the inner partitioning box for the shrink operation. The other merely conceptual midpoint splits are simply ignored. Choosing this inner box guarantees that both subregions produced by the split have no more than $2|R_u|/3$ points. This can be seen by observing

---
[*]   Actually, BBD trees and BAR trees have a slightly better running time for these searches and we mention this in the respective sections.

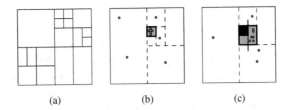

(a)　　　　　(b)　　　　　(c)

FIGURE 27.7 Examples of (a) multiple iterations of the midpoint split rule, (b) centroid shrinking, with dashed lines representing the conceptual midpoint splits and the highlighted inner box being the actual partition cut, (c) centroid shrinking with an inner box. In the final example, the original inner box is solid, the final midpoint split is shown with slightly extended dotted lines, and the new inner box partition cut is shown shaded in gray.

that the inner box has no more than $2|R_u|/3$ points and also must contain at least $|R_u|/3$ points. The technique as stated is not theoretically ideal because the number of midpoint split operations computed cannot be bounded. Each midpoint split may not partition even a single point. Arya et al. [4,7] describe a simple solution to this problem, which repeatedly computes the smallest bounding midpoint box using a technique due to Clarkson [17].

When R_u has an inner box associated with it, we cannot simply find another inner box as this would violate the restriction on having only one inner box. Let b_i represent the original inner box. The solution is to proceed as in the previous centroid shrink operation, repeatedly applying midpoint split operations on the subregion with the larger number of points. However, we now stop in one of two situations; when the size of the larger subregion either has no more than $2|R_u|/3$ points or no longer contains b_i. In the former case, let b be the outer box of this subregion. In the latter case, or in the event both cases happen, let b represent the outer box of the subregion prior to this final split. We perform a shrink operation using b as the partitioning box. Since b clearly contains b_i, the subregion associated with the original outer box continues to have one inner box, albeit a slightly larger one than its parent. The subregion R_1, whose outer box is b, also has one inner box, b_i. If $|R_1| > 2|R_u|/3$, we perform a midpoint split on this subregion. Let R_2 be the larger subregion formed by this last split, which we know does not contain b_i. Since R_2 does not contain an inner box, if R_2 contains more than $2|R_u|/3$ points, we simply perform the basic shrink operation thus dividing R_2 into two smaller subregions as well. Clearly, all the subregions produced by this centroid shrink have no more than $2|R_u|/3$ points. Figure 27.7 shows the three main operations, splitting, shrinking, and the three-step shrinking process.

In addition to this simple version of the BBD tree, there are more flexible variations on this approach. The reader should refer to [4,7] for details on an efficient $O(dn \log n)$ construction algorithm and for discussions on some of the BBD variations. To highlight a few options, at any stage in the construction, rather than alternate between shrinking and splitting operations, it is preferable to perform split operations whenever possible, so long as the point set is divided evenly after every few levels, and to use the more costly shrinking operations only when necessary. Another approach is to use a more flexible split operation, a *fair split*, which attempts to partition the points in the region more evenly. In this case, more care has to be taken to avoid producing skinny regions and to avoid violating the stickiness property; however, as was shown experimentally, the flexibility provides for better experimental performance.

The following theorem summarizes the basic result [4,7]:

Theorem 27.3

Given a set S of n data points in $\mathbb{R}^d$, in $O(dn \log n)$ time it is possible to construct a BBD tree such that

1. The tree has $O(n)$ nodes and depth $O(\log n)$,
2. The regions have outer boxes with balanced aspect ratio and inner boxes that are sticky to the outer box,
3. The sizes of the regions are halved after every $2d$ levels in the tree.

The above conditions imply that the BBD tree is an $O(1)$-quasi-BAR tree.

The size reduction constraint above helps guarantee a slightly better performance for geometric queries than given for general quasi-BAR trees. In particular, Arya and Mount [7] show that the size reduction allows range queries on BBD trees to be solved in $O(2^d \log n + d(3\sqrt{d}/\epsilon)^d)$ time, or $O(2^d \log n + d^3(3\sqrt{d}/\epsilon)^{d-1})$ for convex queries plus output size. Duncan [8] later extended the separation of the n and ϵ dependencies to nearest and farthest neighbor queries showing that the running time for both is $O(\log n + \epsilon^{1-d} \log(1/\epsilon))$ for fixed dimension d.

27.6 BAR Trees

The balanced aspect ratio tree introduced in Reference 9 for the basic two-dimensional case and subsequently revised to higher dimensions in References 8 and 10 can be shown to have a packing function of $\rho(n) = O(1)$ where the constant factor depends on the dimension d and a user-specified aspect ratio parameter α. In the following section, we borrow terminology from [8,10].

Unlike BBD trees, k-d trees, and octrees, BAR trees do not exclusively use axis-orthogonal hyperplane cuts. Instead, to achieve simultaneously the goals of good aspect ratio, balanced depth, and convex regions, cuts in several different directions are used. These directions are called canonical cuts, and the particular choice and size of canonical cuts is essential in creating good BAR trees.

Definition 27.5

The following terms relate to specific cutting directions:

- A *canonical cut set*, $C = \{\vec{v}_1, \vec{v}_2, \ldots, \vec{v}_\gamma\}$, is a collection of γ not necessarily independent vectors that span $\mathbb{R}^d$ (thus, $\gamma \geq d$).
- A *canonical cut direction* is any vector $\vec{v}_i \in C$.
- A *canonical cut* is any hyperplane, H, in $\mathbb{R}^d$ with a normal in C.
- A *canonical region* is any region formed by the intersection of a set of hyperspaces defined by canonical cuts, that is, a convex polyhedron in $\mathbb{R}^d$ with every facet having a normal in C.

Figure 27.8a shows a region composed of three cut directions $(1, 0)$, $(0, 1)$, and $(1, -1)$, or simply cuts along the x, y, and $x - y$ directions. After cutting the region at the dashed line c, we have two regions R_1 and R_2. In R_2 notice the left side is replaced by the new cut c, and more importantly the diagonal cut is no longer tangential to R_2. The following definition describes this property more specifically.

Definition 27.6

A canonical cut c *defines* a canonical region R, written $c \in R$, if and only if c is tangential to R. In other words, c intersects the border of R. For a canonical region R, any two parallel canonical cuts $b, c \in R$ are *opposing canonical cuts*. For any canonical region R, we define the *canonical bounding cuts* with respect to a direction $\vec{v}_i \in C$ to be the two unique opposing canonical cuts normal to $\vec{v}_i$ and tangent to R. We often refer to these cuts as b_i and c_i or simply b and c when i is understood from the context. Intuitively, R is "sandwiched" between b_i and c_i. To avoid confusion, when referring to a canonical cut of a region R, we *always* mean a canonical bounding cut.

For any canonical bounding cut, c, the *facet* of $c \in R$, $\texttt{facet}_c(R)$, is defined as the region formed by the intersection of R with c.

The canonical set used to define a partition tree can vary from method to method. For example, the standard k-d tree algorithm [18] uses a canonical set composed of all axis-orthogonal directions.

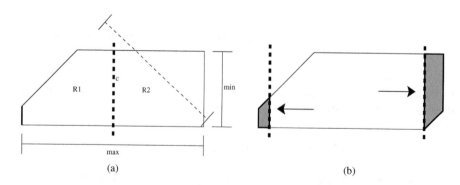

(a)   (b)

FIGURE 27.8 (a) An example of a canonical region of three cut directions, x,y, and $x - y$. Observe the three widths highlighted with lines and the min and max widths of the region. The left facet associated with the x direction is drawn in bold. The bold dashed line in the center represents a cut c and the respective subregions R_1 and R_2. (b) An example of a similar region highlighting the two shield regions associated with the x-direction for $\alpha \approx 4$. Notice the size difference between the two shield regions corresponding to the associated facet sizes.

Definition 27.7

For a canonical set C and a canonical region R, we define the following terms (see Figure 27.8a):

- For a canonical direction $\vec{v}_i \in C$, the *width of R in the direction* $\vec{v}_i$, written $\texttt{width}_i(R)$, is the distance between b_i and c_i, that is, $\texttt{width}_i(R) = \delta(b_i, c_i)$.
- The *maximum side of R* is $\max(R) = \max_{i \in C}(\texttt{width}_i(R))$.
- The *minimum side of R* is $\min(R) = \min_{i \in C}(\texttt{width}_i(R))$.

For simplicity, we also refer to the facets of a region in the same manner. We define the following terms for a facet of a region $R, f = \texttt{facet}_c(R)$:

- The *width of f in the direction* $\vec{v}_i$ is $\texttt{width}_i(f) = \delta(b_i, c_i)$ where b_i and c_i are the opposing bounding cuts of f in the direction $\vec{v}_i$.
- The *maximum side of f* is $\max(f) = \max_{i \in C}(\texttt{width}_i(R))$.
- In addition, for any canonical cut $c \in R$, the *length* of c, $\texttt{len}_c(R)$, is defined as $\max(\texttt{facet}_c(R))$.

When using a canonical cut c_i to partition a region R into two pieces R_1 and R_2 as the cut gets closer to a side of R, one of the two respective regions gets increasingly skinnier. At some point, the region is no longer α-balanced, see Figure 27.8b. This threshold region is referred to as a shield region and is defined in References 8 as the following:

Definition 27.8

Given an α-balanced canonical region R and a canonical cut direction $\vec{v}_i$, sweep a cut c' from the opposing cut b_i toward c_i. Let P be the region of R between c' and c_i. Sweep c' until either region P is empty or just before $\text{asp}(P) > \alpha$. If P is not empty, then P has maximum aspect ratio. Call the region P the *shield region of* c_i *in* R, $\texttt{shield}_{c_i}(R)$. Let the *maximal outer shield*, $\texttt{mos}_i(R)$, be the shield region $\texttt{shield}_{b_i}(R)$ or $\texttt{shield}_{c_i}(R)$ such that $|\texttt{mos}_i(R)| = \max(|\texttt{shield}_{b_i}(R)|, |\texttt{shield}_{c_i}(R)|)$, that is, the maximal outer shield is the shield region with the greater number of points.

Definition 27.9

An α-balanced canonical region, R, is *one-cuttable* with reduction factor β, where $1/2 \leq \beta < 1$, if there exists a cut $c^1 \in C$, called a *one-cut*, dividing R into two subregions R_1 and R_2 such that the following conditions hold:

1. R_1 and R_2 are α-balanced canonical regions,
2. $|R_1| \leq \beta|R|$ and $|R_2| \leq \beta|R|$.

Definition 27.10

An α-balanced canonical region, R, is *k-cuttable* with reduction factor β, for $k > 1$, if there exists a cut $c^k \in C$, called a *k-cut*, dividing R into two subregions R_1 and R_2 such that the following conditions hold:

1. R_1 and R_2 are α-balanced canonical regions,
2. $|R_2| \leq \beta|R|$,
3. Either $|R_1| \leq \beta|R|$ or R_1 is $(k-1)$-cuttable with reduction factor β.

In other words, the sequence of cuts, $c^k, c^{k-1}, \ldots, c^1$, results in $k+1$ α-balanced canonical regions each containing no more than $\beta|R|$ points. If the reduction factor β is understood, we simply say R is k-cuttable.

Definition 27.11

For a canonical cut set, C, a binary space partition tree T constructed on a set S is a *BAR tree with maximum aspect ratio* α if every region $R \in T$ is α-balanced.

Figure 27.9 illustrates an algorithm to construct a BAR tree from a sequence of k-cuttable regions.

Theorem 27.4

For a canonical cut set, C, if every possible α-balanced canonical region is k-cuttable with reduction factor β, then a BAR tree with maximum aspect ratio α can be constructed with depth $O(k \log_{1/\beta} n)$, for any set S of n points in $\mathbb{R}^d$.

```
CREATEBARTREE(u, S_u, R_u, α, β)
    Arguments: Current node u to build (initially the root),
               S_u is the current point set (initially S)
               R_u is the α-balanced region containing S_u
                   (initially a bounding hypercube of S)
               (Optional) node u can contain any of the following:
                   region R_u, sample point p ∈ S_u, size |S_u|
    if |S_u| ≤ leafSize then
        (leaf) node u stores the set S_u
        return
    find c^i, an i-cut for R_u, for smallest value of i
    (internal) node u stores c^i
    create two child nodes of u, v and w
    partition S_u into S_v and S_w by the cut s^i
    partition R_u into R_v and R_w by the cut s^i
    call CREATEBARTREE(v, S_v, R_v, α, β)
    call CREATEBARTREE(w, S_w, R_w, α, β)
```

FIGURE 27.9   General BAR tree construction algorithm.

The main challenge in creating a specific instance of a BAR tree is in defining a canonical set C such that every possible α-balanced canonical region is k-cuttable with reduction factor β for reasonable choices of α, β, and k. The α-balanced regions produced help BAR trees have the following packing function.

Theorem 27.5

For a canonical cut set, C, if every possible α-balanced canonical region is k-cuttable with reduction factor β, then the class of BAR trees with maximum aspect ratio α has a packing function $\rho(n) = O(\alpha^d)$ where the hidden constant factor depends on the angles between the various cut directions. For fixed α, this is constant.

Theorems 27.4 and 27.5 immediately show us that approximate geometric nearest-neighbor and farthest-neighbor queries can be solved in $O(\epsilon^{1-d} \log n)$ time and approximate geometric range searches for convex and non-convex regions take, respectively, $O(\epsilon^{1-d} \log n)$ and $O(\epsilon^{-d} \log n)$ plus the output size. As with the BBD tree, in fact, these structures can also be shown to have running times of $O(\log n + \epsilon^{1-d} \log \frac{1}{\epsilon})$ for nearest-neighbor and farthest-neighbor queries [8] and $O(\log n + \epsilon^{1-d})$ and $O(\log n + \epsilon^{-d})$ for convex and non-convex range queries [7].

Another examination of Figure 27.6 shows why simple axis-orthogonal cuts cannot guarantee k-cuttability. By concentrating a large number of points at an actual corner of the rectangular region, no sequence of axis-orthogonal cuts will divide the points and maintain balanced aspect ratio regions. We can further extend this notion of a bad corner to a general κ-corner associated with a canonical region R.

Definition 27.12

For a canonical cut set C and a canonical region R, a κ-corner $B \in R$ is a ball with center ρ and radius κ such that, for every cut direction $\vec{v}_i \in C$ with bounding cuts b_i and c_i, either b_i or c_i intersects B, that is, $\min(\delta(\rho, b_i), \delta(\rho, c_i)) \leq \kappa$.

When $\kappa = 0$, we are not only defining a vertex of a region but a vertex which is tangential to one of every cut direction's bounding planes. As described in References 8, these corners represent the worst-case placement of points in the region. These corners can always exist in regions. However, if one of the facets associated with this corner has size proportional to (or smaller than) the κ ball, then we can still get close enough to this facet and properly divide the point set without introducing unbalanced regions. The following property formalizes this notion more:

Property 27.13

A canonical cut set C satisfies the κ-Corner Property if for any $\kappa \geq 0$ and any canonical region R containing a κ-corner $B \in R$, there exists a canonical cut $c \in R$ intersecting B such that $\text{len}_c(R) \leq \mathcal{F}_\kappa \kappa$ for some constant $\mathcal{F}_\kappa$.

In particular, notice that if $\kappa = 0$, one of the bounding cutting planes must intersect at a single point. The advantage to this can be seen in the two-dimensional case. Construct any canonical region using any three cutting directions, for simplicity use

```
COMPUTETWOCUT(u)
        Arguments: An α-balanced node u in a BAR tree
        Returns: A one or two-cut for u
        for all cᵢ ∈ C
            if cᵢ is a one-cut, return cᵢ
        let P be the smallest maximal outer shield of R
        let c = cᵢ be the bounding cut associated with P
        let c′ be the cut parallel to c intersecting R such that
            δ(c,c′) = widthᵢ(P) + lenᵢ(R)/σ
        return c′
        // c′ partitions R into two α-balanced regions R₁ and R₂
        // |R₂| ≤ β|R|
        // R₁ incident to cᵢ is one-cuttable
```

FIGURE 27.10   An algorithm to find either a one or two cut in a region.

the two axis-orthogonal cuts and one cut with slope $+1$. It is impossible to find a κ-corner without having at least one of the three bounding sides be small with respect to the corner. This small side has a matching very small shield region. Unfortunately, having a small shield region does not mean that the initial region is one-cuttable. The points may all still be concentrated within this small shield region. However, it is possible that this small shield region is one-cuttable. In fact, in References 8, it is shown that there exist canonical cut sets that guarantee *two-cuttability* for sufficient values of α, β, and σ, where the σ parameter is used in the construction. The sufficiency requirements depend only on certain constant properties associated with the angles of the canonical cut set labeled here as $\mathcal{F}_{\min}$, $\mathcal{F}_{\max}$, $\mathcal{F}_{\text{box}}$, and $\mathcal{F}_\kappa$. For specific values of these constants, see Reference 8. Figure 27.10 describes an algorithm to find an appropriate cut.

The general idea is to find the smallest (in physical size) shield region containing a majority of the points. If none exist, the region must be one-cuttable. Otherwise, we take this shield region and pull the partitioning cut back slightly,[*] increasing the size of this particular shield region. Given an appropriate cut set and various constant bounds, we can guarantee that this new region is one-cuttable. The following theorems summarize this result [8]:

Theorem 27.6: (Two-Cuttable Theorem)

Suppose we are given a canonical cut set, $\mathcal{C}$, which satisfies the κ-Corner Property 27.13. Any α-balanced canonical region R is two-cuttable if the following three conditions are met:

$$\beta \geq (d+1)/(d+2), \tag{27.1}$$

$$\alpha \mathcal{F}_{\min}/4(\mathcal{F}_{\text{box}}+1) > \sigma > (2\mathcal{F}_{\max}+\mathcal{F}_\kappa), \text{ and} \tag{27.2}$$

$$\alpha > 4(\mathcal{F}_{\text{box}}+1)(2\mathcal{F}_{\max}+\mathcal{F}_\kappa)/\mathcal{F}_{\min} + \mathcal{F}_{\max}/\mathcal{F}_{\min}. \tag{27.3}$$

Theorems 27.6 and 27.4 can be combined to yield the following theorem:

Theorem 27.7

Suppose we are given a canonical cut set $\mathcal{C}$ that satisfies the κ-Corner Property and an $\alpha > f(\mathcal{C})$. A BAR tree with depth $O(d \log n)$ and balancing factor α can be constructed in $O(g(\mathcal{C})dn \log n)$ time, where f and g are constant functions depending on properties of the canonical set. In particular, the running time of the algorithm is $O(n \log n)$ for fixed dimensions and fixed canonical sets.

Let us now present two cut sets that do satisfy the κ-Corner Property. The two cut sets we present below are composed of axis-orthogonal cuts and one other set of cuts. Let us give specific names to a few vector directions.

Definition 27.14

A vector $v = (x_0, x_1, x_2, \ldots, x_d)$ is

- an *axis-orthogonal cut* if $x_i = 0$ for all values except one where $x_j = 1$, for example, $(0,0,1,0)$,

---

[*] This is actually only necessary in dimensions greater than 2.

- a *corner cut* if $x_i = \pm 1$ for all values of i, for example, $(1, 1, -1, -1)$,
- a *wedge cut* if $x_i = 0$ for all values except two where $x_j, x_i = \pm 1$, for example, $(0, 1, -1, 0)$.

The *Corner Cut Canonical Set* $\mathcal{C}_c$ is the set of all axis-orthogonal cuts and corner cuts. The *Wedge Cut Canonical Set* $\mathcal{C}_w$ is the set of all axis-orthogonal cuts and wedge cuts.

Notice that $|\mathcal{C}_c|$ is $\Theta(2^d)$ and $|\mathcal{C}_w|$ is $\Theta(d^2)$. Although the corner cut canonical set does not necessarily have to be as large as this, the complexity of the corner cut itself means sidedness tests take longer than axis-orthogonal and wedge cuts, namely d computations instead of 1 or 2. The above two canonical sets satisfy the κ-Corner Property 27.13 and from Theorem 27.7, we get the following two corollaries [8]:

Corollary 27.1

For the Corner Cut Canonical set $\mathcal{C}_c$, a BAR tree with depth $O(d \log n)$ and balancing factor $\alpha = \Omega(d^2)$ can be constructed in $O(n \log n)$ time.

Corollary 27.2

For the Wedge Cut Canonical set $\mathcal{C}_w$, a BAR tree with depth $O(d \log n)$ and balancing factor $\alpha = \Omega(\sqrt{d})$ can be constructed in $O(n \log n)$ time.

To get the exact values needed, see [8]. However, it is important to note that the α bounds above are overestimates of the minimum value needed. In practice, one should try an initially small value of α, say 6, and when that fails to provide two-cuttability double the value for the lower subtree levels. In this manner, one can arrive at the true minimum value in $O(\log \alpha)$ such iterations, if necessary, without having to calculate it. Since the minimum α needed in both cut sets is $O(d^2)$, this adds only an $O(\log(d))$ factor to the depth.

27.7 Maximum-Spread k-d Trees

One very popular class of BSP tree is the k-d tree, see Chapter 17. Although there are very few theoretical bounds known on these structures, there is a lot of empirical evidence that shows them to be extremely efficient for numerous geometric applications. In particular, one variant the maximum-spread k-d tree has long been considered an ideal k-d tree. Given a set of points S and a particular axis dimension x_d, define the *spread* of S in x_d to be the difference between the minimum and maximum coordinates of the points in that dimension. The maximum-spread k-d tree is formed by choosing at each internal node a cutting plane orthogonal to the axis of maximum spread placed at the median point in this direction, see for example [11]. Arya et al. [4] applied the maximum-spread k-d tree to their approximate nearest-neighbor searching algorithm and experimentally showed that they were comparable to the theoretically efficient BBD tree. Later Dickerson et al. [8,12] proved the following theorem regarding maximum-spread k-d trees, referred to there as longest-side k-d trees:

Theorem 27.8

Suppose we are given a maximum-spread k-d tree T constructed on a set S of n points in $\mathbb{R}^d$. Then the packing function $\rho(n)$ of T for a region annulus A is $O(\log^{d-1} n)$. That is, the class of maximum-spread k-d trees is an $O(\log^{d-1} n)$-quasi-BAR tree.

Although the bound is not as good as for BBD trees and BAR trees, the simplicity of the structure yields low constant factors and explains why in practice these trees perform so well. Experimental comparisons to BBD trees and BAR trees verified this result and showed that only for very highly clustered data did the dependency on $\log^{d-1} n$ become prominent [4,8]. In practice, unless data is highly clustered and the dimension is moderately large, the maximal-spread k-d tree is an ideal structure to use. However, for such data sets both the BBD tree and the BAR tree revert to the same behavior as the maximal-spread tree, and they perform well even with highly clustered data. Because of its simpler structure, the BBD tree is potentially more practical than the BAR tree.

Acknowledgments

This work was supported, in part, by the National Science Foundation under grant CCR-0098068.

References

1. B. Chazelle and E. Welzl. Quasi-optimal range searching in spaces of finite VC-dimension. *Discrete Comput. Geom.*, 4: 467–489, 1989.

2. B. Chazelle. Lower bounds on the complexity of polytope range searching. *J. Amer. Math. Soc.*, 2:637–666, 1989.

3. F. P. Preparata and M. I. Shamos. *Computational Geometry: An Introduction.* Springer-Verlag, New York, NY, 1985.

4. S. Arya, D. M. Mount, N. S. Netanyahu, R. Silverman, A. Wu. An optimal algorithm for approximate nearest neighbor searching in fixed dimensions. *J. ACM*, 45(6):891–923, 1998.

5. R. A. Finkel, and J. L. Bentley. Quad trees: a data structure for retrieval on composite keys. *Acta Inform.*, 4, pp. 1–9, 1974.

6. P. B. Callahan, and S. R. Kosaraju. A decomposition of multidimensional point sets with applications to k-nearest-neighbors and n-body potential fields. *J. ACM*, 42:67–90, 1995.

7. S. Arya, and D. M. Mount. Approximate range searching. *Comput. Geom.*, 17(3–4):135–152, 2000.

8. C. A. Duncan. Balanced Aspect Ratio Trees. Ph.D. thesis, Dept. of Computer Science, Johns Hopkins Univ., 1999.

9. C. A. Duncan, M. T. Goodrich, and S. G. Kobourov. Balanced aspect ratio trees and their use for drawing very large graphs. *J. Graph Algorithms Appl.*, 4:19–46, 2000.

10. C. A. Duncan, M. T. Goodrich, and S. G. Kobourov. Balanced aspect ratio trees: Combining the advantages of k-d trees and octrees. *J. Algorithms*, 38:303–333, 2001.

11. J. H. Friedman, J. L. Bentley, and R. A. Finkel. An algorithm for finding best matches in logarithmic expected time. *ACM Trans. Math. Softw.*, 3:209–226, 1977.

12. M. Dickerson, C. A. Duncan, and M. T. Goodrich. K-D trees are better when cut on the longest side. In *ESA: Annual European Symposium on Algorithms*, volume 1879 of *Lecture Notes Comput. Sci.*, pages 179–190, 2000.

13. H. Samet. *The Design and Analysis of Spatial Data Structures.* Addison-Wesley, Reading, MA, 1990.

14. S. Arya and D. M. Mount. Approximate nearest neighbor queries in fixed dimensions. In *Proc. 4th ACM-SIAM Sympos. Discrete Algorithms*, pages 271–280, 1993.

15. J. R. Driscoll, H. N. Gabow, R. Shrairaman, and R. E. Tarjan. Relaxed heaps: An alternative to Fibonacci heaps with applications to parallel computation. *Commun. ACM*, 31:1343–1354, 1988.

16. M. Fredman, and R. E. Tarjan. Fibonacci heaps and their uses in improved network optimization problems. *J. ACM*, 34: 596–615, 1987.

17. K. L. Clarkson. Fast algorithms for the all nearest neighbors problem. In *Proc. 24th Annu. IEEE Sympos. Found. Comput. Sci.*, pages 226–232, 1983.

18. J. L. Bentley. Multidimensional binary search trees used for associative searching. *Commun. ACM*, 18(9):509–517, 1975.

28

Geometric and Spatial Data Structures in External Memory

28.1 Introduction.. 419
    Disk Model • Design Criteria for EM Data Structures • Overview of Chapter
28.2 EM Algorithms for Batched Geometric Problems 421
28.3 EM Tree Data Structures.. 424
    B-trees and Variants • Weight-Balanced B-trees • Parent Pointers and Level-Balanced
    B-trees • Buffer Trees
28.4 Spatial Data Structures and Range Search 428
    Linear-Space Spatial Structures • R-trees • Bootstrapping for 2-D Diagonal Corner and
    Stabbing Queries • Bootstrapping for Three-Sided Orthogonal 2-D Range Search • General
    Orthogonal 2-D Range Search • Lower Bounds for Orthogonal Range Search
28.5 Related Problems.. 434
28.6 Dynamic and Kinetic Data Structures 435
    Logarithmic Method for Decomposable Search Problems • Continuously Moving Items
28.7 Conclusion... 437
Acknowledgments 437
References.. 437

Jeffrey Scott Vitter
The University of Mississippi

28.1 Introduction

Input/Output communication (or simply *I/O*) between the fast internal memory and the slow external memory (EM, such as disk) can be a bottleneck when processing massive amounts of data, as is the case in many spatial and geometric applications [1]. Problems involving massive amounts of geometric data are ubiquitous in spatial databases, geographic information systems (GIS), constraint logic programming, object-oriented databases, statistics, virtual reality systems, and computer graphics. E-Business and space applications produce petabytes (10^{15} bytes) of data per year. A major challenge is to develop mechanisms for processing the data, or else much of the data will be useless.

One promising approach for efficient I/O is to design algorithms and data structures that bypass the virtual memory system and explicitly manage their own I/O. We refer to such algorithms and data structures as *external memory* (or *EM*) *algorithms and data structures*. (The terms *out-of-core algorithms* and *I/O algorithms* are also sometimes used.)

We concentrate in this chapter on the design and analysis of EM memory data structures for batched and online problems involving geometric and spatial data. Luckily, many problems on geometric objects can be reduced to a small core of problems, such as computing intersections, convex hulls, multidimensional search, range search, stabbing queries, point location, and nearest-neighbor search. In this chapter, we discuss useful paradigms for solving these problems in EM.

28.1.1 Disk Model

The three primary measures of performance of an algorithm or data structure are *the number of I/O operations performed, the amount of disk space used, and the internal (parallel) computation time*. For reasons of brevity, we shall focus in this chapter on only the first two measures. Most of the algorithms we mention run in optimal CPU time, at least for the single processor case.

We can capture the main properties of magnetic disks and multiple disk systems by the commonly used *parallel disk model* (PDM) (Figure 28.1) introduced by Vitter and Shriver [2]. Data is transferred in large units of *blocks* of size B so as to amortize the latency of moving the read–write head and waiting for the disk to spin into position. Storage systems such as RAID use multiple

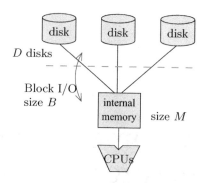

FIGURE 28.1 Parallel disk model.

disks to get more bandwidth [3,4]. The principal parameters of PDM are the following:

$$N = \text{problem input data size (in terms of items)},$$
$$M = \text{size of internal memory (in terms of items)},$$
$$B = \text{size of disk block (in terms of items), and}$$
$$D = \text{\# independent disks},$$

where $M < N$ and $1 \le DB \le M$.

Queries are naturally associated with online computations, but they can also be done in batched mode. For example, in the batched orthogonal 2-D range searching problem discussed in Section 28.2, we are given a set of N points in the plane and a set of Q queries in the form of rectangles, and the problem is to report the points lying in each of the Q query rectangles. In both the batched and online settings, the number of items reported in response to each query may vary. We thus define two more performance parameters:

$$Q = \text{number of input queries (for a batched problem) and}$$
$$Z = \text{query output size (in terms of items)}.$$

If the problem does not involve queries, we set $Q = 0$.

It is convenient to refer to some of the above PDM parameters in units of disk blocks rather than in units of data items; the resulting formulas are often simplified. We define the lowercase notation

$$n = \frac{N}{B}, \quad m = \frac{M}{B}, \quad z = \frac{Z}{B}, \quad q = \frac{Q}{B},$$

to be the problem input size, internal memory size, query output size, and number of queries, respectively, in units of disk blocks.

For simplicity, we restrict our attention in this chapter to the single disk case $D = 1$. The batched algorithms we discuss can generally be sped up by using multiple disks in an optimal manner using the load balancing techniques discussed in Reference 1. Online data structures that use a single disk can generally be transformed automatically by the technique of disk striping to make optimal use of multiple disks [1].

Programs that perform well in terms of PDM will generally perform well when implemented on real systems [1]. More complex and precise models have been formulated [5–7]. Hierarchical (multilevel) memory models are discussed in [1] and its references. Cache-oblivious models are discussed in Chapter 35. Many of the algorithm design techniques we discuss in this chapter, which exploit data locality so as to minimize I/O communication steps, form the basis for algorithms in the other models.

28.1.2 Design Criteria for EM Data Structures

The data structures we examine in this chapter are used in batched algorithms and in online settings. In batched problems, no preprocessing is done, and the entire file of data items must be processed, often by streaming the data through the internal memory in one or more passes. Queries are done in a batched manner during the processing. The goal is generally twofold:

- To solve the problem in $O\big((n + q) \log_m n + z\big)$ I/Os
- To use only a linear number $O(n + q)$ of blocks of disk storage

Most nontrivial problems require the same number of I/Os as does sorting. In particular, criterion B1 is related to the I/O complexity of sorting N items in the PDM model, which is $O(n \log_m n)$ [1].

Online data structures support query operations in a continuous manner. When the data items do not change and the data structure can be preprocessed before any queries are made, the data structure is known as *static*. When the data structure supports insertions and deletions of items, intermixed with the queries, the data structure is called *dynamic*. The primary theoretical challenges in the design and analysis of online EM data structures are threefold:

- To answer queries in $O(\log_B N + z)$ I/Os
- To use only a linear number $O(n)$ of blocks of disk storage
- To do updates (in the case of dynamic data structures) in $O(\log_B N)$ I/Os

Criteria O1–O3 correspond to the natural lower bounds for online search in the comparison model. The three criteria are problem dependent, and for some problems they cannot be met. For dictionary queries, we can do better using hashing, achieving $O(1)$ I/Os per query on the average.

Criterion O1 combines together the $O(\log_B N)$ I/O cost for the search with the $O(\lceil z \rceil)$ I/O cost for reporting the output. When one cost is much larger than the other, the query algorithm has the extra freedom to follow a *filtering* paradigm [8], in which both the search component and the output reporting are allowed to use the larger number of I/Os. For example, when the output size Z is large, the search component can afford to be somewhat sloppy as long as it doesn't use more than $O(z)$ I/Os; when Z is small, the Z output items do not have to reside compactly in only $O(\lceil z \rceil)$ blocks. Filtering is an important design paradigm in online EM data structures.

For many of the batched and online problems we consider, there is a data structure (such as a scanline structure or binary search tree) for the internal memory version of the problem that can answer each query in $O(\log N + Z)$ CPU time, but if we use the same data structure naively in an EM setting (using virtual memory to handle page management), a query may require $\Omega(\log N + Z)$ I/Os, which is excessive.[*]

The goal for online algorithms is to build locality directly into the data structure and explicitly manage I/O so that the $\log N$ and Z terms in the I/O bounds of the naive approach are replaced by $\log_B N$ and z, respectively. The relative speedup in I/O performance, namely $(\log N + Z)/(\log_B N + z)$, is at least $(\log N)/\log_B N = \log B$, which is significant in practice, and it can be as much as $Z/z = B$ for large Z.

For batched problems, the I/O performance can be improved further, since the answers to the queries do not need to be provided immediately but can be reported as a group at the end of the computation. For the batched problems we consider in this chapter, the $Q = qB$ queries collectively use $O(q \log_m n + z)$ I/Os, which is about B times less than a naive approach.

28.1.3 Overview of Chapter

In the next section, we discuss batched versions of geometric search problems. One of the primary methods used for batched geometric problems is distribution sweeping, which uses a data structure reminiscent of the distribution paradigm in external sorting. Other useful batched techniques include persistent B-trees, batched filtering, external fractional cascading, external marriage-before-conquest, and batched incremental construction.

The most popular EM online data structure is the B-tree structure, which provides excellent performance for dictionary operations and one-dimensional range searching. We give several variants and applications of B-trees in Section 28.3. We look at several aspects of multidimensional range search in Section 28.4 and related problems such as stabbing queries and point location. Data structures for other variants and related problems such as nearest-neighbor search are discussed in Section 28.5. Dynamic and kinetic data structures are discussed in Section 28.6. The content of this chapter comes largely from portions of a survey article by Vitter [9]. A more comprehensive manuscript on EM data structures appears in Reference 1.

28.2 EM Algorithms for Batched Geometric Problems

Advances in recent years have allowed us to solve a variety of batched geometric problems optimally, meeting both optimality Criteria B1 and B2 of Section 28.1.2. These problems include

1. Computing the pairwise intersections of N segments in the plane and their trapezoidal decomposition
2. Finding all intersections between N nonintersecting red line segments and N nonintersecting blue line segments in the plane

---

[*] We use the notation $\log N$ to denote the binary (base 2) logarithm $\log_2 N$. For bases other than 2, the base will be specified explicitly, as in the base-B logarithm $\log_B N$.

3. Answering Q orthogonal 2-D range queries on N points in the plane (i.e., finding all the points within the Q query rectangles)

4. Constructing the 2-D and 3-D convex hull of N points

5. Voronoi diagram and Triangulation of N points in the plane

6. Performing Q point location queries in a planar subdivision of size N

7. Finding all nearest neighbors for a set of N points in the plane

8. Finding the pairwise intersections of N orthogonal rectangles in the plane

9. Computing the measure of the union of N orthogonal rectangles in the plane

10. Computing the visibility of N segments in the plane from a point

11. Performing Q ray-shooting queries in 2-D Constructive Solid Geometry (CSG) models of size N

Goodrich et al. [10], Zhu [11], Arge et al. [12,13], and Crauser et al. [14,15] developed EM algorithms for those problems using these EM paradigms for batched problems:

Distribution sweeping, a generalization of the sorting distribution paradigm [1] for "externalizing" plane sweep algorithms.

Persistent B-trees: An offline method for constructing an optimal space persistent version of the B-tree data structure (see Section 28.3.1), yielding a factor of B improvement over the generic persistence techniques of Driscoll et al. [16].

Batched filtering: A general method for performing simultaneous EM searches in data structures that can be modeled as planar layered directed acyclic graphs; it is useful for 3-D convex hulls and batched point location. Multisearch on parallel computers is considered in Reference 17.

External fractional cascading: An EM analogue to fractional cascading on a segment tree, in which the degree of the segment tree is $O(m^\alpha)$ for some constant $0 < \alpha \leq 1$. Batched queries can be performed efficiently using batched filtering; online queries can be supported efficiently by adapting the parallel algorithms of work of Tamassia and Vitter [18] to the I/O setting.

External marriage-before-conquest: An EM analogue to the technique of Kirkpatrick and Seidel [19] for performing output-sensitive convex hull constructions.

Batched incremental construction: A localized version of the randomized incremental construction paradigm of Clarkson and Shor [20], in which the updates to a simple dynamic data structure are done in a random order, with the goal of fast overall performance on the average. The data structure itself may have bad worst-case performance, but the randomization of the update order makes worst-case behavior unlikely. The key for the EM version so as to gain the factor of B I/O speedup is to batch together the incremental modifications.

For illustrative purposes, we focus in the remainder of this section primarily on the distribution sweep paradigm [10], which is a combination of the distribution paradigm for sorting [1] and the well-known sweeping paradigm from computational geometry [21,22]. As an example, let us consider how to achieve optimality Criteria B1 and B2 for computing the pairwise intersections of N orthogonal segments in the plane, making use of the following recursive distribution sweep. At each level of recursion, the region under consideration is partitioned into $\Theta(m)$ vertical *slabs*, each containing $\Theta(N/m)$ of the segments' endpoints.

We sweep a horizontal line from top to bottom to process the N segments. When the sweep line encounters a vertical segment, we insert the segment into the appropriate slab. When the sweep line encounters a horizontal segment h, as pictured in Figure 28.2, we report h's intersections with all the "active" vertical segments in the slabs that are spanned *completely* by h. (A vertical segment is "active" if it intersects the current sweep line; vertical segments that are found to be no longer active are deleted from the slabs.) The remaining two end portions of h (which "stick out" past a slab boundary) are passed recursively to the next level, along with the vertical segments. The downward sweep then proceeds. After the initial sorting (to get the segments with respect to the y-dimension), the sweep at each of the $O(\log_m n)$ levels of recursion requires $O(n)$ I/Os, yielding the desired bound in B1 of $O((n + q) \log_m n + z)$. Some timing experiments on distribution sweeping appear in Reference 23. Arge et al. [13] develop a unified approach to distribution sweep in higher dimensions.

A central operation in spatial databases is spatial join. A common preprocessing step is to find the pairwise intersections of the bounding boxes of the objects involved in the spatial join. The problem of intersecting orthogonal rectangles can be solved by combining the previous sweep line algorithm for orthogonal segments with one for range searching. Arge et al. [13] take a more unified approach using distribution sweep, which is extendible to higher dimensions. The active objects that are stored in the data structure in this case are rectangles, not vertical segments. The authors choose the branching factor to be $\Theta(\sqrt{m})$. Each rectangle is associated with the largest contiguous range of vertical slabs that it spans. Each of the possible $\Theta\left(\binom{\sqrt{m}}{2}\right) = \Theta(m)$ contiguous ranges of slabs is called a *multislab*. The reason why the authors choose the branching factor to be $\Theta(\sqrt{m})$ rather than $\Theta(m)$ is so that the number of multislabs is $\Theta(m)$, and thus there is room in internal memory for a buffer for each multislab. The height of the tree remains $O(\log_m n)$.

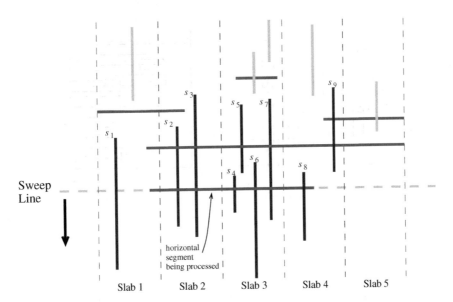

FIGURE 28.2 Distribution sweep used for finding intersections among N orthogonal segments. The vertical segments currently stored in the slabs are indicated in bold (namely, $s_1, s_2, \ldots, s_9$). Segments s_5 and s_9 are not active but have not yet been deleted from the slabs. The sweep line has just advanced to a new horizontal segment that completely spans slabs 2 and 3, so slabs 2 and 3 are scanned and all the active vertical segments in slabs 2 and 3 (namely, s_2, s_3, s_4, s_6, s_7) are reported as intersecting the horizontal segment. In the process of scanning slab 3, segment s_5 is discovered to be no longer active and can be deleted from slab 3. The end portions of the horizontal segment that "stick out" into slabs 1 and 4 are handled by the lower levels of recursion, where the intersection with s_8 is eventually discovered.

The algorithm proceeds by sweeping a horizontal line from top to bottom to process the N rectangles. When the sweep line first encounters a rectangle R, we consider the multislab lists for all the multislabs that R intersects. We report all the active rectangles in those multislab lists, since they are guaranteed to intersect R. (Rectangles no longer active are discarded from the lists.) We then extract the left and right end portions of R that partially "stick out" past slab boundaries, and we pass them down to process in the next lower level of recursion. We insert the remaining portion of R, which spans complete slabs, into the list for the appropriate multislab. The downward sweep then continues. After the initial sorting preprocessing, each of the $O(\log_m n)$ sweeps (one per level of recursion) takes $O(n)$ I/Os, yielding the desired bound $O\big((n + q)\log_m n + z\big)$.

The resulting algorithm, called Scalable Sweeping-Based Spatial Join (SSSJ) [13,24], outperforms other techniques for rectangle intersection. It was tested against two other sweep line algorithms: the Partition-Based Spatial Merge (QPBSM) used in Paradise [25] and a faster version called MPBSM that uses an improved dynamic data structure for intervals [24]. The TPIE (Transparent Parallel I/O Environment) system [26–28] served as the common implementation platform. The algorithms were tested on several data sets. The timing results for the two data sets in Figure 28.3a and b are given in Figure 28.3c and d, respectively. The first data set is the worst case for sweep line algorithms; a large fraction of the line segments in the file are active (i.e., they intersect the current sweep line). The second data set is a best case for sweep line algorithms, but the two PBSM algorithms have the disadvantage of making extra copies of the rectangles. In both cases, SSSJ shows considerable improvement over the PBSM-based methods. In other experiments done on more typical data, such as TIGER/line road data sets [29], SSSJ and MPBSM perform about 30% faster than does QPBSM. The conclusion we draw is that SSSJ is as fast as other known methods on typical data, but unlike other methods, it scales well even for worst-case data. If the rectangles are already stored in an index structure, such as the R-tree index structure we consider in Section 28.4.2, hybrid methods that combine distribution sweep with inorder traversal often perform best [30].

For the problem of finding all intersections among N line segments, Arge et al. [12] give an efficient algorithm based upon distribution sort, but the output component of the I/O bound is slightly nonoptimal: $z \log_m n$ rather than z. Crauser et al. [14,15] attain the optimal I/O bound of criterion B1, namely $O\big((n + q)\log_m n + z\big)$, by constructing the trapezoidal decomposition for the intersecting segments using an incremental randomized construction. For I/O efficiency, they do the incremental updates in a series of batches, in which the batch size is geometrically increasing by a factor of m.

Online issues also arise in the analysis of batched EM algorithms. In practice, batched algorithms must adapt in a robust and online way when the memory allocation changes, and online techniques can play an important role. Some initial work has been done on memory-adaptive EM algorithms in a competitive framework [31].

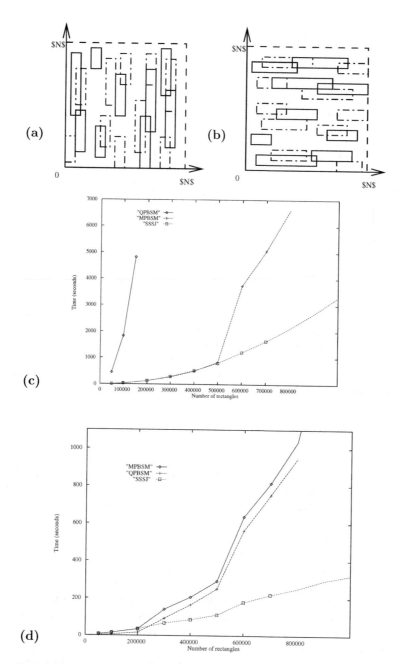

FIGURE 28.3 Comparison of SSSJ with the original PBSM (QPBSM) and a new variant (MPBSM). Each data set contains $N/2$ red rectangles (designated by solid sides) and $N/2$ blue rectangles (designated by dashed sides), and the goal is to find all intersections between red rectangles and blue rectangles. In each data set shown, the number of intersections is $O(N)$: (a) data set 1 consists of tall, skinny (vertically aligned) rectangles; (b) data set 2 consists of short, wide (horizontally aligned) rectangles; (c) running times on data set 1; and (d) running times on data set 2.

28.3 EM Tree Data Structures

In this section, we consider the basic online EM data structures for storing and querying spatial data in one dimension. The dictionary problem is an important special case, which can be solved efficiently in the average case by use of hashing. However, hashing does not support sequential search in a natural way, such as retrieving all the items with key value in a specified range. Some clever work has been done on order-preserving hash functions, in which items with sequential keys are stored in the same block or in adjacent blocks, but the search performance is less robust and tends to deteriorate because of unwanted collisions (see References 1 and 32 for surveys).

A more effective EM approach for geometric queries is to use multiway trees, which we explore in this section. For illustration, we use orthogonal range search as our canonical problem. It is a fundamental database primitive in spatial databases and GIS, and it includes dictionary lookup as a special case. A range query, for a given d-dimensional rectangle, returns all the points in the interior of the rectangle. Other types of spatial queries include point location queries, ray shooting queries, nearest-neighbor queries, and intersection queries, but for brevity we restrict our attention primarily to range searching.

Spatial data structures tend to be of two types: space-driven or data-driven. Quad trees and grid files are space-driven since they are based upon a partitioning of the embedding space, somewhat akin to using order-preserving hash functions, whereas methods like R-trees and kd-trees are organized by partitioning the data items themselves. We shall discuss primarily the latter type in this chapter.

28.3.1 B-trees and Variants

Tree-based data structures arise naturally in the online setting, in which the data can be updated and queries must be processed immediately. Binary trees have a host of applications in the (internal memory) RAM model. In order to exploit block transfer, trees in EM generally use a block for each node, which can store $\Theta(B)$ pointers and data values.

The well-known balanced multiway *B-tree* due to Bayer and McCreight [33–35] is the most widely used nontrivial EM data structure. The degree of each node in the B-tree (with the exception of the root) is required to be $\Theta(B)$, which guarantees that the height of a B-tree storing N items is roughly $\log_B N$. B-trees support dynamic dictionary operations and one-dimensional range search optimally in linear space, $O(\log_B N)$ I/Os per insert or delete, and $O(\log_B N + z)$ I/Os per query, where $Z = zB$ is the number of items output. When a node overflows during an insertion, it splits into two half-full nodes, and if the splitting causes the parent node to overflow, the parent node splits, and so on. Splittings can thus propagate up to the root, which is how the tree grows in height. Deletions are handled in a symmetric way by merging nodes.

In the B^+-*tree* variant, pictured in Figure 28.4, all the items are stored in the leaves, and the leaves are linked together in symmetric order to facilitate range queries and sequential access.

The internal nodes store only key values and pointers and thus can have a higher branching factor. In the most popular variant of B^+-trees, called B^*-*trees*, splitting can usually be postponed when a node overflows by "sharing" the node's data with one of its adjacent siblings. The node needs to be split only if the sibling is also full; when that happens, the node splits into two, and its data and those of its full sibling are evenly redistributed, making each of the three nodes about two-thirds full. This local optimization reduces the number of times new nodes must be created and thus increases the storage utilization. And since there are fewer nodes in the tree, search I/O costs are lower. When no sharing is done (as in B^+-trees), Yao [36] shows that nodes are roughly $\ln 2 \approx 69\%$ full on the average, assuming random insertions. With sharing (as in B^*-trees), the average storage utilization increases to about $2\ln(3/2) \approx 81\%$ [37,38]. Storage utilization can be increased further by sharing among several siblings, at the cost of more complicated insertions and deletions. Some helpful space-saving techniques borrowed from hashing are partial expansions [39] and use of overflow nodes [40].

A cross between B-trees and hashing, where each subtree rooted at a certain level of the B-tree is instead organized as an external hash table, was developed by Litwin and Lomet [41] and further studied in References 42 and 43. O'Neil [44] proposed a B-tree variant called the SB-tree that clusters together on the disk symmetrically ordered nodes from the same level so as to optimize range queries and sequential access. Rao and Ross [45,46] use similar ideas to exploit locality and optimize search tree performance in internal memory. Reducing the number of pointers allows a higher branching factor and thus faster search.

Partially persistent versions of B-trees have been developed by Becker et al. [47] and Varman and Verma [48]. By persistent data structure, we mean that searches can be done with respect to any timestamp y [16,49]. In a partially persistent data structure,

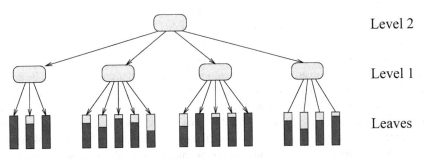

FIGURE 28.4 B^+-tree multiway search tree. Each internal and leaf node corresponds to a disk block. All the items are stored in the leaves; the darker portion of each leaf block indicates its relative fullness. The internal nodes store only key values and pointers, $\Theta(B)$ of them per node. Although not indicated here, the leaf blocks are linked sequentially.

only the most recent version of the data structure can be updated. In a fully persistent data structure, any update done with timestamp y affects all future queries for any time after y. An interesting open problem is whether B-trees can be made fully persistent. Salzberg and Tsotras [50] survey work done on persistent access methods and other techniques for time-evolving data. Lehman and Yao [51], Mohan [52], and Lomet and Salzberg [53] explore mechanisms to add concurrency and recovery to B-trees. Other variants are discussed in Chapter 16.

28.3.2 Weight-Balanced B-trees

Arge and Vitter [54] introduce a powerful variant of B-trees called *weight-balanced B-trees*, with the property that the weight of any subtree at level h (i.e., the number of nodes in the subtree rooted at a node of height h) is $\Theta(a^h)$, for some fixed parameter a of order B. By contrast, the sizes of subtrees at level h in a regular B-tree can differ by a multiplicative factor that is exponential in h. When a node on level h of a weight-balanced B-tree gets rebalanced, no further rebalancing is needed until its subtree is updated $\Omega(a^h)$ times. Weight-balanced B-trees support a wide array of applications in which the I/O cost to rebalance a node of weight w is $O(w)$; the rebalancings can be scheduled in an amortized (and often worst-case) way with only $O(1)$ I/Os. Such applications are very common when the nodes have secondary structures, as in multidimensional search trees, or when rebuilding is expensive. Agarwal et al. [55] apply weight-balanced B-trees to convert partition trees such as kd-trees, BBD trees, and BAR trees, which were designed for internal memory, into efficient EM data structures.

Weight-balanced trees called BB[α]-trees [56,57] have been designed for internal memory; they maintain balance via rotations, which is appropriate for binary trees, but not for the multiway trees needed for EM. In contrast, weight-balanced B-trees maintain balance via splits and merges.

Weight-balanced B-trees were originally conceived as part of an optimal dynamic EM interval tree structure for stabbing queries and a related EM segment tree structure. We discuss their use for stabbing queries and other types of range queries in Sections 28.4.3 through 28.4.5. They also have applications in the (internal memory) RAM model [54,58], where they offer a simpler alternative to BB[α]-trees. For example, by setting a to a constant in the EM interval tree based upon weight-balanced B-trees, we get a simple worst-case implementation of interval trees [59,60] in internal memory. Weight-balanced B-trees are also preferable to BB[α]-trees for purposes of augmenting one-dimensional data structures with range restriction capabilities [61].

28.3.3 Parent Pointers and Level-Balanced B-trees

It is sometimes useful to augment B-trees with parent pointers. For example, if we represent a total order via the leaves in a B-tree, we can answer order queries such as "Is $x < y$ in the total order?" by walking upwards in the B-tree from the leaves for x and y until we reach their common ancestor. Order queries arise in online algorithms for planar point location and for determining reachability in monotone subdivisions [62]. If we augment a conventional B-tree with parent pointers, then each split operation costs $\Theta(B)$ I/Os to update parent pointers, although the I/O cost is only $O(1)$ when amortized over the updates to the node. However, this amortized bound does not apply if the B-tree needs to support cut and concatenate operations, in which case the B-tree is cut into contiguous pieces and the pieces are rearranged arbitrarily. For example, reachability queries in a monotone subdivision are processed by maintaining two total orders, called the leftist and rightist orders, each of which is represented by a B-tree. When an edge is inserted or deleted, the tree representing each order is cut into four consecutive pieces, and the four pieces are rearranged via concatenate operations into a new total order. Doing cuts and concatenation via conventional B-trees augmented with parent pointers will require $\Theta(B)$ I/Os per level in the worst case. Node splits can occur with each operation (unlike the case where there are only inserts and deletes), and thus there is no convenient amortization argument that can be applied.

Agarwal et al. [62] describe an interesting variant of B-trees called *level-balanced B-trees* for handling parent pointers and operations such as cut and concatenate. The balancing condition is "global." The data structure represents a forest of B-trees in which the number of nodes on level h in the forest is allowed to be at most $N_h = 2N/(b/3)^h$, where b is some fixed parameter in the range $4 < b < B/2$. It immediately follows that the total height of the forest is roughly $\log_b N$.

Unlike previous variants of B-trees, the degrees of individual nodes of level-balanced B-trees can be arbitrarily small, and for storage purposes, nodes are packed together into disk blocks. Each node in the forest is stored as a node record (which points to the parent's node record) and a doubly linked list of child records (which point to the node records of the children). There are also pointers between the node record and the list of child records. Every disk block stores only node records or only child records, but all the child records for a given node must be stored in the same block (possibly with child records for other nodes). The advantage of this extra level of indirection is that cuts and concatenates can usually be done in only $O(1)$ I/Os per level of the forest. For example, during a cut, a node record gets split into two, and its list of child nodes is chopped into two separate lists. The parent node must therefore get a new child record to point to the new node. These updates require $O(1)$ I/Os except when there is not enough space in the disk block of the parent's child records, in which case the block must be split into two, and extra

I/Os are needed to update the pointers to the moved child records. The amortized I/O cost, however, is only $O(1)$ per level, since each update creates at most one node record and child record at each level. The other dynamic update operations can be handled similarly.

All that remains is to reestablish the global level invariant when a level gets too many nodes as a result of an update. If level h is the lowest such level out of balance, then level h and all the levels above it are reconstructed via a postorder traversal in $O(N_h)$ I/Os so that the new nodes get degree $\Theta(b)$ and the invariant is restored. The final trick is to construct the new parent pointers that point from the $\Theta(N_{h-1}) = \Theta(bN_h)$ node records on level $h-1$ to the $\Theta(N_h)$ level-h nodes. The parent pointers can be accessed in a blocked manner with respect to the new ordering of the nodes on level h. By sorting, the pointers can be rearranged to correspond to the ordering of the nodes on level $h-1$, after which the parent pointer values can be written via a linear scan. The resulting I/O cost is $O((bN_h/B)\log_m(bN_h/B))$, which can be amortized against the $\Theta(N_h)$ updates that occurred since the last time the level-h invariant was violated, yielding an amortized update cost of $O(1 + (b/B)\log_m n)$ I/Os per level.

Order queries such as "Does leaf x precede leaf y in the total order represented by the tree?" can be answered using $O(\log_B N)$ I/Os by following parent pointers starting at x and y. The update operations insert, delete, cut, and concatenate can be done in $O((1 + (b/B)\log_m n)\log_b N)$ I/Os amortized, for any $2 \leq b \leq B/2$, which is never worse than $O((\log_B N)^2)$ by appropriate choice of b.

Using the multislab decomposition we discuss in Section 28.4.3, Agarwal et al. [62] apply level-balanced B-trees in a data structure for point location in monotone subdivisions, which supports queries and (amortized) updates in $O((\log_B N)^2)$ I/Os. They also use it to dynamically maintain planar st-graphs using $O((1 + (b/B)\log_m n)\log_b N)$ I/Os (amortized) per update, so that reachability queries can be answered in $O(\log_B N)$ I/Os (worst-case). (Planar st-graphs are planar directed acyclic graphs with a single source and a single sink.) An interesting open question is whether level-balanced B-trees can be implemented in $O(\log_B N)$ I/Os per update. Such an improvement would immediately give an optimal dynamic structure for reachability queries in planar st-graphs.

28.3.4 Buffer Trees

An important paradigm for constructing algorithms for batched problems in an internal memory setting is to use a dynamic data structure to process a sequence of updates. For example, we can sort N items by inserting them one by one into a priority queue, followed by a sequence of N *delete_min* operations. Similarly, many batched problems in computational geometry can be solved by dynamic plane sweep techniques. For example, in Section 28.2 we showed how to compute orthogonal segment intersections by dynamically keeping track of the active vertical segments (i.e., those hit by the horizontal sweep line); we mentioned a similar algorithm for orthogonal rectangle intersections.

However, if we use this paradigm naively in an EM setting, with a B-tree as the dynamic data structure, the resulting I/O performance will be highly nonoptimal. For example, if we use a B-tree as the priority queue in sorting or to store the active vertical segments hit by the sweep line, each update and query operation will take $O(\log_B N)$ I/Os, resulting in a total of $O(N \log_B N)$ I/Os, which is larger than the optimal bound $O(n \log_m n)$ by a substantial factor of roughly B. One solution suggested in Reference 63 is to use a binary tree data structure in which items are pushed lazily down the tree in blocks of B items at a time. The binary nature of the tree results in a data structure of height $O(\log n)$, yielding a total I/O bound of $O(n \log n)$, which is still nonoptimal by a significant $\log m$ factor.

Arge [64] developed the elegant *buffer tree* data structure to support *batched dynamic* operations, as in the sweep line example, where the queries do not have to be answered right away or in any particular order. The buffer tree is a balanced multiway tree, but with degree $\Theta(m)$ rather than degree $\Theta(B)$, except possibly for the root. Its key distinguishing feature is that each node has a buffer that can store $\Theta(M)$ items (i.e., $\Theta(m)$ blocks of items). Items in a node are pushed down to the children when the buffer fills. Emptying a full buffer requires $\Theta(m)$ I/Os, which amortizes the cost of distributing the M items to the $\Theta(m)$ children. Each item thus incurs an amortized cost of $O(m/M) = O(1/B)$ I/Os per level, and the resulting cost for queries and updates is $O((1/B)\log_m n)$ I/Os amortized.

Buffer trees have an ever-expanding list of applications. They can be used as a subroutine in the standard sweep line algorithm in order to get an optimal EM algorithm for orthogonal segment intersection. Arge showed how to extend buffer trees to implement segment trees [65] in EM in a batched dynamic setting by reducing the node degrees to $\Theta(\sqrt{m})$ and by introducing *multislabs* in each node, which were explained in Section 28.2 for the related batched problem of intersecting rectangles. Buffer trees provide a natural amortized implementation of priority queues for *time-forward processing* applications such as discrete event simulation, sweeping, and list ranking [66]. Govindarajan et al. [67] use time-forward processing to construct a well-separated pair decomposition of N points in d dimensions in $O(n \log_m n)$ I/Os, and they apply it to the problems of finding the K nearest neighbors for each point and the K closest pairs. Brodal and Katajainen [68] provide a worst-case optimal priority queue, in the sense that every sequence of B insert and *delete_min* operations requires only $O(\log_m n)$ I/Os. Practical implementations of priority queues based upon these ideas are examined in References 69 and 70. In Section 28.4.2, we report on some timing experiments involving buffer trees for use in bulk loading of R-trees. Further experiments on buffer trees appear in Reference 71.

28.4 Spatial Data Structures and Range Search

In this section, we consider online EM data structures for storing and querying spatial data. A fundamental database primitive in spatial databases and GIS is range search, which includes dictionary lookup as a special case. An orthogonal range query, for a given d-dimensional rectangle, returns all the points in the interior of the rectangle. In this section, we use range searching (especially for the orthogonal 2-D case when $d = 2$) as the canonical query operation on spatial data. Other types of spatial queries include point location, ray shooting, nearest neighbor, and intersection queries, but for brevity we restrict our attention primarily to range searching.

There are two types of spatial data structures: data-driven and space-driven. R-trees and kd-trees are data-driven since they are based upon a partitioning of the data items themselves, whereas space-driven methods such as quad trees and grid files are organized by a partitioning of the embedding space, akin to order-preserving hash functions. In this section, we discuss primarily data-driven data structures.

Multidimensional range search is a fundamental primitive in several online geometric applications, and it provides indexing support for constraint and object-oriented data models (see Reference 72 for background). We have already discussed multidimensional range searching in a batched setting in Section 28.2. In this section, we concentrate on data structures for the online case.

For many types of range searching problems, it is very difficult to develop theoretically optimal algorithms and data structures. Many open problems remain. The goal for online data structures is typically to achieve the three optimality Criteria O1–O3 of Section 28.1.2.

We explain in Section 28.4.6 that under a fairly general computational model for general 2-D orthogonal queries, as pictured in Figure 28.5d, it is impossible to satisfy Criteria O1 and O2 simultaneously. At least $\Omega\big(n(\log n)/\log(\log_B N + 1)\big)$ blocks of disk space must be used to achieve a query bound of $O\big((\log_B N)^c + z\big)$ I/Os per query, for any constant c [73]. Three natural questions arise:

- What sort of performance can be achieved when using only a linear amount of disk space? In Sections 28.4.1 and 28.4.2, we discuss some of the linear-space data structures used extensively in practice. None of them come close to satisfying Criteria O1 and O3 for range search in the worst case, but in typical case scenarios they often perform well. We devote Section 28.4.2 to R-trees and their variants, which are the most popular general-purpose spatial structures developed to date.

- Since the lower bound applies only to general 2-D rectangular queries, are there any data structures that meet Criteria O1–O3 for the important special cases of 2-D range searching pictured in Figure 28.5a–c? Fortunately the answer is yes. We show in Sections 28.4.3 and 28.4.4 how to use a "bootstrapping" paradigm to achieve optimal search and update performance.

- Can we meet Criteria O1 and O2 for general four-sided range searching if the disk space allowance is increased to $O\big(n(\log n)/\log(\log_B N + 1)\big)$ blocks? Yes again! In Section 28.4.5, we show how to adapt the optimal structure for three-sided searching in order to handle general four-sided searching in optimal search cost. The update cost, however, is not known to be optimal.

In Section 28.5, we discuss other scenarios of range search dealing with three dimensions and nonorthogonal queries. We discuss the lower bounds for 2-D range searching in Section 28.4.6.

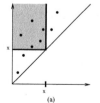 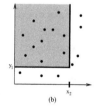 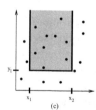 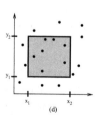

FIGURE 28.5 Different types of 2-D orthogonal range queries: (a) diagonal corner two-sided 2-D query (equivalent to a stabbing query, cf. Section 28.4.3), (b) two-sided 2-D query, (c) three-sided 2-D query, and (d) general four-sided 2-D query.

28.4.1 Linear-Space Spatial Structures

Grossi and Italiano [74] construct an elegant multidimensional version of the B-tree called the *cross tree*. Using linear space, it combines the data-driven partitioning of weight-balanced B-trees (cf. Section 28.3.2) at the upper levels of the tree with the space-driven partitioning of methods such as quad trees at the lower levels of the tree. For $d > 1$, d-dimensional orthogonal range queries can be done in $O(n^{1-1/d} + z)$ I/Os, and inserts and deletes take $O(\log_B N)$ I/Os. The O-tree of Kanth and Singh [75] provides similar bounds. Cross trees also support the dynamic operations of cut and concatenate in $O(n^{1-1/d})$ I/Os. In some restricted models for linear-space data structures, the 2-D range search query performance of cross trees and O-trees can be considered to be optimal, although it is much larger than the logarithmic bound of Criterion O1.

One way to get multidimensional EM data structures is to augment known internal memory structures, such as quad trees and *kd*-trees, with block-access capabilities. Examples include *kd–B-trees* [76], *buddy trees* [77], and *hB-trees* [78,79]. *Grid files* [80–82] are a flattened data structure for storing the cells of a two-dimensional grid in disk blocks. Another technique is to "linearize" the multidimensional space by imposing a total ordering on it (a so-called space-filling curve), and then the total order is used to organize the points into a B-tree [83–85]. Linearization can also be used to represent nonpoint data, in which the data items are partitioned into one or more multidimensional rectangular regions [86,87]. All the methods described in this paragraph use linear space, and they work well in certain situations; however, their worst-case range query performance is no better than that of cross trees, and for some methods, such as grid files, queries can require $\Theta(n)$ I/Os, even if there are no points satisfying the query. We refer the reader to References 32, 88, and 89 for a broad survey of these and other interesting methods. Space-filling curves arise again in connection with R-trees, which we describe next.

28.4.2 R-trees

The *R-tree* of Guttman [90] and its many variants are a practical multidimensional generalization of the B-tree for storing a variety of geometric objects, such as points, segments, polygons, and polyhedra, using linear disk space. Internal nodes have degree $\Theta(B)$ (except possibly the root), and leaves store $\Theta(B)$ items. Each node in the tree has associated with it a bounding box (or bounding polygon) of all the items in its subtree. A big difference between R-trees and B-trees is that in R-trees the bounding boxes of sibling nodes are allowed to overlap. If an R-tree is being used for point location, for example, a point may lie within the bounding box of several children of the current node in the search. In that case the search must proceed to all such children.

In the dynamic setting, there are several popular heuristics for where to insert new items into an R-tree and how to rebalance it (see Chapter 22 and References 32, 88, and 91 for a survey). The *R*-tree* variant of Beckmann et al. [92] seems to give best overall query performance. To insert an item, we start at the root and recursively insert the item into the subtree whose bounding box would expand the least in order to accommodate the item. In case of a tie (e.g., if the item already fits inside the bounding boxes of two or more subtrees), we choose the subtree with the smallest resulting bounding box. In the normal R-tree algorithm, if a leaf node gets too many items or if an internal node gets too many children, we split it, as in B-trees. Instead, in the R*-tree algorithm, we remove a certain percentage of the items from the overflowing node and reinsert them into the tree. The items we choose to reinsert are the ones whose centroids are furthest from the center of the node's bounding box. This *forced reinsertion* tends to improve global organization and reduce query time. If the node still overflows after the forced reinsertion, we split it. The splitting heuristics try to partition the items into nodes so as to minimize intuitive measures such as coverage, overlap, or perimeter. During deletion, in both the normal R-tree and R*-tree algorithms, if a leaf node has too few items or if an internal node has too few children, we delete the node and reinsert all its items back into the tree by forced reinsertion.

The rebalancing heuristics perform well in many practical scenarios, especially in low dimensions, but they result in poor worst-case query bounds. An interesting open problem is whether nontrivial query bounds can be proven for the "typical case" behavior of R-trees for problems such as range searching and point location. Similar questions apply to the methods discussed in Section 28.4.1. New R-tree partitioning methods by de Berg et al. [93] and Agarwal et al. [94] provide some provable bounds on overlap and query performance.

In the static setting, in which there are no updates, constructing the R*-tree by repeated insertions, one by one, is extremely slow. A faster alternative to the dynamic R-tree construction algorithms mentioned above is to bulk-load the R-tree in a bottom-up fashion [86,87,95]. Such methods use some heuristic for grouping the items into leaf nodes of the R-tree, and then recursively build the nonleaf nodes from bottom to top. As an example, in the so-called Hilbert R-tree of Kamel and Faloutsos [95], each item is labeled with the position of its centroid on the Peano–Hilbert space-filling curve, and a B$^+$-tree is built upon the totally ordered labels in a bottom-up manner. Bulk loading a Hilbert R-tree is therefore easy to do once the centroid points are presorted. These static construction methods algorithms are very different in spirit from the dynamic insertion methods. The dynamic methods explicitly try to reduce the coverage, overlap, or perimeter of the bounding boxes of the R-tree nodes, and as a result, they usually achieve good query performance. The static construction methods do not consider the bounding box information at all. Instead, the hope is that the improved storage utilization (up to 100%) of these packing methods compensates for a higher degree of node

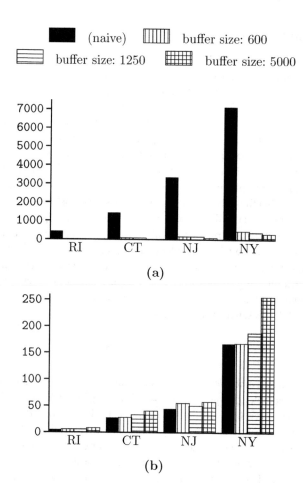

FIGURE 28.6 Costs for R-tree processing (in units of 1000 I/Os) using the naive repeated insertion method and the buffer R-tree for various buffer sizes: (a) cost for bulk-loading the R-tree, (b) query cost.

overlap. A dynamic insertion method related to Reference 95 was presented in Reference 84. The quality of the Hilbert R-tree in terms of query performance is generally not as good as that of an R*-tree, especially for higher dimensional data [96,97].

In order to get the best of both worlds—the query performance of R*-trees and the bulk construction efficiency of Hilbert R-trees—Arge et al. [26] and van den Bercken et al. [98] independently devised fast bulk loading methods based upon buffer trees that do top-down construction in $O(n \log_m n)$ I/Os, which matches the performance of the bottom-up methods within a constant factor. The former method is especially efficient and supports dynamic batched updates and queries. In Figure 28.6 and Table 28.1, we report on some experiments that test the construction, update, and query performance of various R-tree methods. The experimental data came from TIGER/line data sets from four U.S. states [29]; the implementations were done using the TPIE system.

Figure 28.6 compares the construction cost for building R-trees and the resulting query performance in terms of I/Os for the naive sequential method for construction into R*-trees (labeled "naive") and the newly developed buffer R*-tree method [26] (labeled "buffer"). An R-tree was constructed on the TIGER road data for each state and for each of four possible buffer sizes. The four buffer sizes were capable of storing 0, 600, 1250, and 5000 rectangles, respectively; buffer size 0 corresponds to the naive method, and the larger buffers correspond to the buffer method. The query performance of each resulting R-tree was measured by posing rectangle intersection queries using rectangles taken from TIGER hydrographic data. The results, depicted in Figure 28.6, show that buffer R*-trees, even with relatively small buffers, achieve a tremendous speedup in number of I/Os for construction without any worsening in query performance, compared with the naive method. The CPU costs of the two methods are comparable. The storage utilization of buffer R*-trees tends to be in the 90% range, as opposed to roughly 70% for the naive method.

Bottom-up methods can build R-trees even more quickly and more compactly, but they generally do not support bulk dynamic operations, which are a big advantage of the buffer tree approach. Kamel et al. [97] develop a way to do bulk updates with Hilbert R-trees, but at a cost in terms of query performance. Table 28.1 compares dynamic update methods for the naive method, for

TABLE 28.1 Summary of the Costs (in Number of I/Os) for R-tree Updates and Queries

| Data Set | Update Method | Update with 50% of the Data | | |
		Building	Querying	Packing
	Naive	259, 263	6670	64
RI	Hilbert	15, 865	7262	92
	Buffer	13, 484	5485	90
	Naive	805, 749	40, 910	66
CT	Hilbert	51, 086	40, 593	92
	Buffer	42, 774	37, 798	90
	Naive	1, 777, 570	70, 830	66
NJ	Hilbert	120, 034	69, 798	92
	Buffer	101, 017	65, 898	91
	Naive	3, 736, 601	224, 039	66
NY	Hilbert	246, 466	230, 990	92
	Buffer	206, 921	227, 559	90

Note: Packing refers to the percentage storage utilization.

buffer R-trees, and for Hilbert R-trees [97] (labeled "Hilbert"). A single R-tree was built for each of the four U.S. states, containing 50% of the road data objects for that state. Using each of the three algorithms, the remaining 50% of the objects were inserted into the R-tree, and the construction time was measured. Query performance was then tested as before. The results in Table 28.1 indicate that the buffer R*-tree and the Hilbert R-tree achieve a similar degree of packing, but the buffer R*-tree provides better update and query performance.

Arge et al. [99] introduce the priority R-tree, which provides guaranteed query performance of $O(n^{1-1/d} + z)$ I/Os, similarly to the search structures in Section 28.4.1. This query performance is optimal in the worst case for linear-sized data structures. The priority R-tree uses a bulk loading to construct the tree, and it supports updates efficiently in the amortized sense.

28.4.3 Bootstrapping for 2-D Diagonal Corner and Stabbing Queries

An obvious paradigm for developing an efficient dynamic EM data structure, given an existing data structure that works well when the problem fits into internal memory, is to "externalize" the internal memory data structure. If the internal memory data structure uses a binary tree, then a multiway tree such as a B-tree must be used instead. However, when searching a B-tree, it can be difficult to report the outputs in an output-sensitive manner. For example, in certain searching applications, each of the $\Theta(B)$ subtrees of a given node in a B-tree may contribute one item to the query output, and as a result each subtree may need to be explored (costing several I/Os) just to report a single output item.

Fortunately, we can sometimes achieve output-sensitive reporting by augmenting the data structure with a set of filtering substructures, each of which is a data structure for a smaller version of the same problem. We refer to this approach, which we explain shortly in more detail, as the *bootstrapping* paradigm. Each substructure typically needs to store only $O(B^2)$ items and to answer queries in $O(\log_B B^2 + Z'/B) = O(\lceil Z'/B \rceil)$ I/Os, where Z' is the number of items reported. A substructure can even be static if it can be constructed in $O(B)$ I/Os, since we can keep updates in a separate buffer and do a global rebuilding in $O(B)$ I/Os whenever there are $\Theta(B)$ updates. Such a rebuilding costs $O(1)$ I/Os (amortized) per update. We can often remove the amortization and make it worst case using the weight-balanced B-trees of Section 28.3.2 as the underlying B-tree structure.

Arge and Vitter [54] first uncovered the bootstrapping paradigm while designing an optimal dynamic EM data structure for diagonal corner two-sided 2-D queries (see Figure 28.5a) that meets all three design criteria for online data structures listed in Section 28.1.2. Diagonal corner two-sided queries are equivalent to stabbing queries, which have the following form: "Given a set of one-dimensional intervals, report all the intervals 'stabbed' by the query value x" (i.e., report all intervals that contain x). A diagonal corner query x on a set of 2-D points $\{(a_1, b_2), (a_2, b_2), \ldots\}$ is equivalent to a stabbing query x on the set of closed intervals $\{[a_1, b_2], [a_2, b_2], \ldots\}$.

The EM data structure for stabbing queries is a multiway version of the well-known interval tree data structure [59,60] for internal memory, which supports stabbing queries in $O(\log N + Z)$ CPU time and updates in $O(\log N)$ CPU time and uses $O(N)$ space. We can externalize it by using a weight-balanced B-tree as the underlying base tree, where the nodes have degree $\Theta(\sqrt{B})$. Each node in the base tree corresponds in a natural way to a one-dimensional range of x-values; its $\Theta(\sqrt{B})$ children correspond to subranges called slabs, and the $\Theta(\sqrt{B}^2) = \Theta(B)$ contiguous sets of slabs are called *multislabs*, as in Section 28.2 for a similar batched problem. Each input interval is stored in the lowest node v in the base tree whose range completely contains the interval.

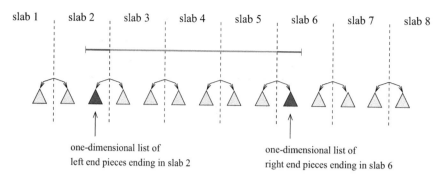

FIGURE 28.7   Internal node v of the EM priority search tree, for $B = 64$ with $\sqrt{B} = 8$ slabs. Node v is the lowest node in the tree completely containing the indicated interval. The middle piece of the interval is stored in the multislab list corresponding to slabs 3–5. (The multislab lists are not pictured.) The left and right end pieces of the interval are stored in the left-ordered list of slab 2 and the right-ordered list of slab 6, respectively.

The interval is decomposed by v's $\Theta(\sqrt{B})$ slabs into at most three pieces: the middle piece that completely spans one or more slabs of v, the left end piece that partially protrudes into a slab of v, and the right end piece that partially protrudes into another slab of v, as shown in Figure 28.7. The three pieces are stored in substructures of v. In the example in Figure 28.7, the middle piece is stored in a list associated with the multislab it spans (corresponding to the contiguous range of slabs 3–5), the left end piece is stored in a one-dimensional list for slab 2 ordered by left endpoint, and the right end piece is stored in a one-dimensional list for slab 6 ordered by right endpoint.

Given a query value x, the intervals stabbed by x reside in the substructures of the nodes of the base tree along the search path from the root to the leaf for x. For each such node v, we consider each of v's multislabs that contains x and report all the intervals in the multislab list. We also walk sequentially through the right-ordered list and left-ordered list for the slab of v that contains x, reporting intervals in an output-sensitive way.

The big problem with this approach is that we have to spend at least one I/O per multislab containing x, regardless of how many intervals are in the multislab lists. For example, there may be $\Theta(B)$ such multislab lists, with each list containing only a few stabbed intervals (or worse yet, none at all). The resulting query performance will be highly nonoptimal. The solution, according to the bootstrapping paradigm, is to use a substructure in each node consisting of an optimal static data structure for a smaller version of the same problem; a good choice is the corner data structure developed by Kanellakis et al. [72]. The corner substructure in this case is used to store all the intervals from the "sparse" multislab lists, namely those that contain fewer than B intervals, and thus the substructure contains only $O(B^2)$ intervals. When visiting node v, we access only v's nonsparse multislab lists, each of which contributes $Z' \geq B$ intervals to the output, at an output-sensitive cost of $O(Z'/B)$ I/Os, for some Z'. The remaining Z'' stabbed intervals stored in v can be found by a single query to v's corner substructure, at a cost of $O(\log_B B^2 + Z''/B) = O(\lceil Z''/B \rceil)$ I/Os. Since there are $O(\log_B N)$ nodes along the search path in the base tree, the total collection of Z stabbed intervals is reported in $O(\log_B N + z)$ I/Os, which is optimal. Using a weight-balanced B-tree as the underlying base tree allows the static substructures to be rebuilt in worst-case optimal I/O bounds.

Stabbing queries are important because, when combined with one-dimensional range queries, they provide a solution to *dynamic interval management*, in which one-dimensional intervals can be inserted and deleted, and intersection queries can be performed. These operations support indexing of one-dimensional constraints in constraint databases. Other applications of stabbing queries arise in graphics and GIS. For example, Chiang and Silva [100] apply the EM interval tree structure to extract at query time the boundary components of the isosurface (or contour) of a surface. A data structure for a related problem, which in addition has optimal output complexity, appears in Reference 101. The above bootstrapping approach also yields dynamic EM segment trees with optimal query and update bound and $O(n \log_B N)$-block space usage.

28.4.4 Bootstrapping for Three-Sided Orthogonal 2-D Range Search

Arge et al. [102] provide another example of the bootstrapping paradigm by developing an optimal dynamic EM data structure for three-sided orthogonal 2-D range searching (see Figure 28.5c) that meets all three design Criteria O1–O3. In internal memory, the optimal structure is the priority search tree [103], which answers three-sided range queries in $O(\log N + Z)$ CPU time, does updates in $O(\log N)$ CPU time, and uses $O(N)$ space. The EM structure of Arge et al. [102] is an externalization of the priority search tree, using a weight-balanced B-tree as the underlying base tree. Each node in the base tree corresponds to a one-dimensional range of x-values, and its $\Theta(B)$ children correspond to subranges consisting of vertical slabs. Each node v contains a small substructure called a *child cache* that supports three-sided queries. Its child cache stores the "Y-set" $Y(w)$ for

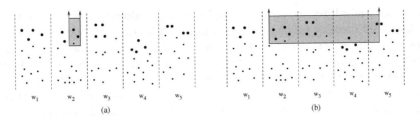

FIGURE 28.8 Internal node v of the EM priority search tree, with slabs (children) $w_1, w_2, \ldots, w_5$. The Y-sets of each child, which are stored collectively in v's child cache, are indicated by the bold points. (a) The three-sided query is completely contained in the x-range of w_2. The relevant (bold) points are reported from v's child cache, and the query is recursively answered in w_2. (b) The three-sided query spans several slabs. The relevant (bold) points are reported from v's child cache, and the query is recursively answered in w_2, w_3, and w_5. The query is *not* extended to w_4 in this case because not all of its Y-set $Y(w_4)$ (stored in v's child cache) satisfies the query, and as a result, none of the points stored in w_4's subtree can satisfy the query.

each of the $\Theta(B)$ children w of v. The Y-set $Y(w)$ for child w consists of the highest $\Theta(B)$ points in w's slab that are not already stored in the child cache of some ancestor of v. There are thus a total of $\Theta(B^2)$ points stored in v's child cache.

We can answer a three-sided query of the form $[x_1, x_2] \times [y_1, +\infty)$ by visiting a set of nodes in the base tree, starting with the root. For each visited node v, we pose the query $[x_1, x_2] \times [y_1, +\infty)$ to v's child cache and output the results. The following rules are used to determine which of v's children to visit. We visit v's child w (see Figure 28.8) if either

1. w is along the leftmost search path for x_1 or the rightmost search path for x_2 in the base tree, or
2. the entire Y-set $Y(w)$ is reported when v's child cache is queried.

There are $O(\log_B N)$ nodes w that are visited because of rule 1. When rule 1 is not satisfied, rule 2 provides an effective filtering mechanism to guarantee output-sensitive reporting. The I/O cost for initially accessing a child node w can be charged to the $\Theta(B)$ points of $Y(w)$ reported from v's child cache; conversely, if not all of $Y(w)$ is reported, then the points stored in w's subtree will be too low to satisfy the query, and there is no need to visit w (see Figure 28.8b). Provided that each child cache can be queried in $O(1)$ I/Os plus the output-sensitive cost to output the points satisfying the query, the resulting overall query time is $O(\log_B N + z)$, as desired.

All that remains is to show how to query a child cache in a constant number of I/Os, plus the output-sensitive cost. Arge et al. [102] provide an elegant and optimal static data structure for three-sided range search, which can be used in the EM priority search tree described earlier to implement the child caches of size $O(B^2)$. The static structure is a persistent B-tree optimized for batched construction. When used for $O(B^2)$ points, it occupies $O(B)$ blocks, can be built in $O(B)$ I/Os, and supports three-sided queries in $O(\lceil Z'/B \rceil)$ I/Os per query, where Z' is the number of points reported. The static structure is so simple that it may be useful in practice on its own.

Both the three-sided structure developed by Arge et al. [102] and the structure for two-sided diagonal queries discussed in Section 28.4.3 satisfy Criteria O1–O3 of Section 28.1.2. So in a sense, the three-sided query structure subsumes the diagonal two-sided structure, since three-sided queries are more general. However, diagonal two-sided structure may prove to be faster in practice, because in each of its corner substructures, the data accessed during a query are always in contiguous blocks, whereas the static substructures used for three-sided search do not guarantee block contiguity. Empirical work is ongoing to evaluate the performance of these data structures.

On a historical note, earlier work on two-sided and three-sided queries was done by Ramaswamy and Subramanian [104] using the notion of *path caching*; their structure met Criterion O1 but had higher storage overheads and amortized and/or nonoptimal update bounds. Subramanian and Ramaswamy [73] subsequently developed the *p-range tree* data structure for three-sided queries, with optimal linear disk space and nearly optimal query and amortized update bounds.

28.4.5 General Orthogonal 2-D Range Search

The dynamic data structure for three-sided range searching can be generalized using the filtering technique of Chazelle [8] to handle general four-sided queries with optimal I/O query bound $O(\log_B N + z)$ and optimal disk space usage $O\big(n (\log n) / \log(\log_B N + 1)\big)$ [102]. The update bound becomes $O\big((\log_B N)(\log n)/\log(\log_B N + 1)\big)$, which may not be optimal.

The outer level of the structure is a balanced $(\log_B N + 1)$-way 1-D search tree with $\Theta(n)$ leaves, oriented, say, along the x-dimension. It therefore has about $(\log n) / \log(\log_B N + 1)$ levels. At each level of the tree, each input point is stored in four substructures (described later) that are associated with the particular tree node at that level that spans the x-value of the point. The space and update bounds quoted above follow from the fact that the substructures use linear space and can be updated in $O(\log_B N)$ I/Os.

To search for the points in a four-sided query rectangle $[x_1, x_2] \times [y_1, y_2]$, we decompose the four-sided query in the following natural way into two three-sided queries, a stabbing query, and $\log_B N - 1$ list traversals. We find the lowest node v in the tree whose x-range contains $[x_1, x_2]$. If v is a leaf, we can answer the query in a single I/O. Otherwise we query the substructures stored in those children of v whose x-ranges intersect $[x_1, x_2]$. Let $2 \le k \le \log_B N + 1$ be the number of such children. The range query when restricted to the leftmost such child of v is a three-sided query of the form $[x_1, +\infty] \times [y_1, y_2]$, and when restricted to the rightmost such child of v, the range query is a three-sided query of the form $[-\infty, x_2] \times [y_1, y_2]$. Two of the substructures at each node are devoted for three-sided queries of these types; using the linear-sized data structures of Arge et al. [102] in Section 28.4.4, each such query can be done in $O(\log_B N + z)$ I/Os.

For the $k - 2$ intermediate children of v, their x-ranges are completely contained inside the x-range of the query rectangle, and thus we need only do $k - 2$ list traversals in y-order and retrieve the points whose y-values are in the range $[y_1, y_2]$. If we store the points in each node in y-order (in the third type of substructure), the Z' output points from a node can be found in $O(\lceil Z'/B \rceil)$ I/Os, once a starting point in the linear list is found. We can find all $k - 2$ starting points via a single query to a stabbing query substructure S associated with v. (This structure is the fourth type of substructure.) For each two y-consecutive points (a_i, b_i) and (a_{i+1}, b_{i+1}) associated with a child of v, we store the y-interval $[b_i, b_{i+1}]$ in S. Note that S contains intervals contributed by each of the $\log_B N + 1$ children of v. By a single stabbing query with query value y_1, we can thus identify the $k - 2$ starting points in only $O(\log_B N)$ I/Os [54], as described in Section 28.4.3. (We actually get starting points for all the children of v, not just the $k - 2$ ones of interest, but we can discard the starting points we don't need.) The total number of I/Os to answer the range query is thus $O(\log_B N + z)$, which is optimal.

28.4.6 Lower Bounds for Orthogonal Range Search

We mentioned in Section 28.4 that Subramanian and Ramaswamy [73] prove that no EM data structure for 2-D range searching can achieve design Criterion O1 of Section 28.1.2 using less than $O(n(\log n)/\log(\log_B N + 1))$ disk blocks, even if we relax the criterion and allow $O((\log_B N)^c + z)$ I/Os per query, for any constant c. The result holds for an EM version of the pointer machine model, based upon the approach of Chazelle [105] for the internal memory model.

Hellerstein et al. [106] consider a generalization of the layout-based lower bound argument of Kanellakis et al. [72] for studying the trade-off between disk space usage and query performance. They develop a model for *indexability*, in which an "efficient" data structure is expected to contain the Z output points to a query compactly within $O(\lceil Z/B \rceil) = O(\lceil z \rceil)$ blocks. One shortcoming of the model is that it considers only data layout and ignores the search component of queries, and thus it rules out the important filtering paradigm discussed earlier in Section 28.4. For example, it is reasonable for any query to perform at least $\log_B N$ I/Os, so if the output size Z is at most B, a data structure may still be able to satisfy Criterion O1 even if the output is contained within $O(\log_B N)$ blocks rather than $O(z) = O(1)$ blocks. Arge et al. [102] modify the model to rederive the same nonlinear space lower bound $O(n(\log n)/\log(\log_B N + 1))$ of Subramanian and Ramaswamy [73] for 2-D range searching by considering only output sizes Z larger than $(\log_B N)^c B$, for which the number of blocks allowed to hold the outputs is $Z/B = O((\log_B N)^c + z)$. This approach ignores the complexity of how to find the relevant blocks, but as mentioned in Section 28.4.5, the authors separately provide an optimal 2-D range search data structure that uses the same amount of disk space and does queries in the optimal $O(\log_B N + z)$ I/Os. Thus, despite its shortcomings, the indexability model is elegant and can provide much insight into the complexity of blocking data in EM. Further results in this model appear in References 107 and 108.

One intuition from the indexability model is that less disk space is needed to efficiently answer 2-D queries when the queries have bounded aspect ratio (i.e., when the ratio of the longest side length to the shortest side length of the query rectangle is bounded). An interesting question is whether R-trees and the linear-space structures of Sections 28.4.1 and 28.4.2 can be shown to perform provably well for such queries. Another interesting scenario is where the queries correspond to snapshots of the continuous movement of a sliding rectangle.

When the data structure is restricted to contain only a single copy of each point, Kanth and Singh [75] show for a restricted class of index-based trees that d-dimensional range queries in the worst case require $\Omega(n^{1 - 1/d} + z)$ I/Os, and they provide a data structure with a matching bound. Another approach to achieve the same bound is the cross tree data structure [74] mentioned in Section 28.4.1, which in addition supports the operations of cut and concatenate.

28.5 Related Problems

For other types of range searching, such as in higher dimensions and for nonorthogonal queries, different filtering techniques are needed. So far, relatively little work has been done, and many open problems remain.

Vengroff and Vitter [109] developed the first theoretically near-optimal EM data structure for static three-dimensional orthogonal range searching. They create a hierarchical partitioning in which all the points that dominate a query point are densely contained in a set of blocks. Compression techniques are needed to minimize disk storage. If we use $(B \log n)$-approximate boundaries rather than B-approximate boundaries, queries can be done in $O(\log_B N + z)$ I/Os, which is optimal, and the space usage is $O\big(n(\log n)^{k+1}\big/(\log(\log_B N + 1))^k\big)$ disk blocks to support $(3 + k)$-sided 3-D range queries, in which k of the dimensions $(0 \leq k \leq 3)$ have finite ranges. The result also provides optimal $O(\log N + Z)$-time query performance for three-sided 3-D queries in the (internal memory) RAM model, but using $O(N \log N)$ space.

Afshani [110] introduced the first optimal 3-D structure for dominance queries, and Afshani et al. [111] developed structures for range queries in higher dimensions.

By the reduction in Reference 112, a data structure for three-sided 3-D queries also applies to *2-D homothetic range search*, in which the queries correspond to scaled and translated (but not rotated) transformations of an arbitrary fixed polygon. An interesting special case is "fat" orthogonal 2-D range search, where the query rectangles are required to have bounded aspect ratio. For example, every rectangle with bounded aspect ratio can be covered by two overlapping squares. An interesting open problem is to develop linear-sized optimal data structures for fat orthogonal 2-D range search. By the reduction, one possible approach would be to develop optimal linear-sized data structures for three-sided 3-D range search.

Agarwal et al. [113] consider halfspace range searching, in which a query is specified by a hyperplane and a bit indicating one of its two sides, and the output of the query consists of all the points on that side of the hyperplane. They give various data structures for half-space range searching in two, three, and higher dimensions, including one that works for simplex (polygon) queries in two dimensions, but with a higher query I/O cost. They have subsequently improved the storage bounds for half-space range queries in two dimensions to obtain an optimal static data structure satisfying Criteria O1 and O2 of Section 28.1.2.

The number of I/Os needed to build the data structures for 3-D orthogonal range search and half-space range search is rather large (more than $\Omega(N)$). Still, the structures shed useful light on the complexity of range searching and may open the way to improved solutions. An open problem is to design efficient construction and update algorithms and to improve upon the constant factors.

Cheng et al. [114,115] look at a model of uncertainty, in which the location of each data point is given by a probability distribution, and give efficient indexes for range queries and join queries.

An interesting model related to range searching involves pattern matching in string data, and there has been great interest recently in EM algorithms for string processing. Chien et al. [116] derive a duality relation that links the number of I/O steps and the space bound for range searching to the corresponding measures for text indexing. Hon et al. [117] provide a formal framework for efficiently retrieving the most relevant occurrences satisfying a query pattern. Their index takes linear space and can be reduced to sublinear space by exploiting the compressibility of the data. Their work is an example of a compressed data structure, in which the goal is to construct a search structure whose size is roughly comparable to the size of the input in compressed format, but which can still be queried directly in an efficient manner. Other work on compressed data structures appears in Vitter [1].

Callahan et al. [118] develop dynamic EM data structures for several online problems in d dimensions. For any fixed $\epsilon > 0$, they can find an approximate nearest neighbor of a query point (within a $1 + \epsilon$ factor of optimal) in $O(\log_B N)$ I/Os; insertions and deletions can also be done in $O(\log_B N)$ I/Os. They use a related approach to maintain the closest pair of points; each update costs $O(\log_B N)$ I/Os. Govindarajan et al. [67] achieve the same bounds for closest pair by maintaining a well-separated pair decomposition. For finding nearest neighbors and approximate nearest neighbors, two other approaches are partition trees [113,119] and locality-sensitive hashing [120]. Numerous other data structures have been developed for range queries and related problems on spatial data. We refer to References 32, 88, and 89 for a broad survey.

28.6 Dynamic and Kinetic Data Structures

In this section, we consider two scenarios where data items change: *dynamic* (in which items are inserted and deleted) and *kinetic* (in which the data items move continuously along specified trajectories). In both cases, queries can be done at any time. It is often useful for kinetic data structures to allow insertions and deletions; for example, if the trajectory of an item changes, we must delete the old trajectory and insert the new one.

28.6.1 Logarithmic Method for Decomposable Search Problems

In previous sections, we've already encountered several dynamic data structures for the problems of dictionary lookup and range search. In Section 28.4, we saw how to develop optimal EM range search data structures by externalizing some known internal memory data structures. The key idea was to use the bootstrapping paradigm, together with weight-balanced B-trees as the underlying data structure, in order to consolidate several static data structures for small instances of range searching into one dynamic data structure for the full problem. The bootstrapping technique is specific to the particular data structures involved. In this section, we look at another technique that is based upon the properties of the problem itself rather than upon that of the data structure.

We call a problem *decomposable* if we can answer a query by querying individual subsets of the problem data and then computing the final result from the solutions to each subset. Dictionary search and range searching are obvious examples of decomposable problems. Bentley developed the *logarithmic method* [121,122] to convert efficient static data structures for decomposable problems into general dynamic ones. In the internal memory setting, the logarithmic method consists of maintaining a series of static substructures, at most one each of size 1, 2, 4, 8, ... When a new item is inserted, it is initialized in a substructure of size 1. If a substructure of size 1 already exists, the two substructures are combined into a single substructure of size 2. If there is already a substructure of size 2, they in turn are combined into a single substructure of size 4, and so on. For the current value of N, it is easy to see that the kth substructure (i.e., of size 2^k) is present exactly when the kth bit in the binary representation of N is 1. Since there are at most $\log N$ substructures, the search time bound is $\log N$ times the search time per substructure. As the number of items increases from 1 to N, the kth structure is built a total of $N/2^k$ times (assuming N is a power of 2). If it can be built in $O(2^k)$ time, the total time for all insertions and all substructures is thus $O(N \log N)$, making the amortized insertion time $O(\log N)$. If we use up to three substructures of size 2^k at a time, we can do the reconstructions in advance and convert the amortized update bounds to worst case [122].

In the EM setting, in order to eliminate the dependence upon the binary logarithm in the I/O bounds, the number of substructures must be reduced from $\log N$ to $\log_B N$, and thus the maximum size of the kth substructure must be increased from 2^k to B^k. As the number of items increases from 1 to N, the kth substructure has to be built NB/B^k times (when N is a power of B), each time taking $O(B^k(\log_B N)/B)$ I/Os. The key point is that the extra factor of B in the numerator of the first term is cancelled by the factor of B in the denominator of the second term, and thus the resulting total insertion time over all N insertions and all $\log_B N$ structures is $O(N(\log_B N)^2)$ I/Os, which is $O((\log_B N)^2)$ I/Os amortized per insertion. By global rebuilding, we can do deletions in $O(\log_B N)$ I/Os amortized. As in the internal memory case, the amortized updates can typically be made worst case.

Arge and Vahrenhold [123] obtain I/O bounds for dynamic point location in general planar subdivisions similar to those of Reference 62, but without use of level-balanced trees. Their method uses a weight-balanced base structure at the outer level and a multislab structure for storing segments similar to that of Arge and Vitter [54] described in Section 28.4.3. They use the logarithmic method to construct a data structure to answer vertical rayshooting queries in the multislab structures. The method relies upon a total ordering of the segments, but such an ordering can be changed drastically by a single insertion. However, each substructure in the logarithmic method is static (until it is combined with another substructure), and thus a static total ordering can be used for each substructure. The authors show by a type of fractional cascading that the queries in the $\log_B N$ substructures do not have to be done independently, which saves a factor of $\log_B N$ in the I/O cost, but at the cost of making the data structures amortized instead of worst case.

Agarwal et al. [55] apply the logarithmic method (in both the binary form and B-way variant) to get EM versions of kd-trees, BBD trees, and BAR trees.

28.6.2 Continuously Moving Items

Early work on temporal data generally concentrated on time-series or multiversion data [50]. A question of growing interest in this mobile age is how to store and index continuously moving items, such as mobile telephones, cars, and airplanes, using kinetic data structures [124–126]. There are two main approaches to storing moving items. The first technique is to use the same sort of data structure as for nonmoving data, but to update it whenever items move sufficiently far so as to trigger important combinatorial *events* that are relevant to the application at hand [127]. For example, an event relevant for range search might be triggered when two items move to the same horizontal displacement (which happens when the x-ordering of the two items is about to switch). A different approach is to store each item's location and speed trajectory, so that no updating is needed as long as the item's trajectory plan does not change. Such an approach requires fewer updates, but the representation for each item generally has higher dimension, and the search strategies are therefore less efficient.

Kollios et al. [128] developed a linear-space indexing scheme for moving points along an 1-D line, based upon the notion of partition trees. Their structure supports a variety of range search and approximate nearest-neighbor queries. For example, given a range and time, the points in that range at the indicated time can be retrieved in $O(n^{1/2+\epsilon} + k)$ I/Os, for arbitrarily small $\epsilon > 0$. Updates require $O((\log n)^2)$ I/Os. Agarwal et al. [119] extend the approach to handle range searches in two dimensions, and they improve the update bound to $O((\log_B n)^2)$ I/Os. They also propose an event-driven data structure with the same query times as the range search data structure of Arge and Vitter [102] discussed in Section 28.4.5, but with the potential need to do many updates. A hybrid data structure combining the two approaches permits a trade-off between query performance and update frequency.

R-trees offer a practical generic mechanism for storing multidimensional points and are thus a natural alternative for storing mobile items. One approach is to represent time as a separate dimension and to cluster trajectories using the R-tree heuristics. However, the orthogonal nature of the R-tree does not lend itself well to diagonal trajectories. For the case of points moving along linear trajectories, Šaltenis et al. [125] build the R-tree upon only the spatial dimensions, but parameterize the bounding box coordinates to account for the movement of the items stored within. They maintain an outer approximation of the true bounding box, which they periodically update to refine the approximation. Agarwal and Har-Peled [129] show how to maintain a provably good approximation of the minimum bounding box with need for only a constant number of refinement events. Further discussion of kinetic data structures, primarily for internal memory, appears in Chapter 24.

28.7 Conclusion

In this chapter, we have surveyed several useful paradigms and techniques for the design and implementation of efficient data structures for EM. A variety of interesting challenges remain in geometric search applications, such as methods for high-dimensional and nonorthogonal range searches as well as the analysis of R-trees and linear-space methods for typical case scenarios. A continuing goal is to translate theoretical gains into observable improvements in practice. For some of the problems that can be solved optimally up to a constant factor, the constant overhead is too large for the algorithm to be of practical use, and simpler approaches are needed.

Acknowledgments

The author thanks the members of the Center for Geometric Computing at Duke University for helpful comments and suggestions. This work was supported in part by the Army Research Office through grant DAAD19-03-1-0321, and by the National Science Foundation through research grants CCR-9877133, CCR-0082986, CCF-0328013, IIS-0415097, CCF-0621457, and CCF-1017623.

References

1. J. S. Vitter. Algorithms and data structures for external memory. *Foundations and Trends in Theoretical Computer Science*. Now Publishers, Hanover, MA, 2008.
2. J. S. Vitter and E. A. M. Shriver. Algorithms for parallel memory I: Two-level memories. *Algorithmica*, 12(2–3):110–147, 1994.
3. P. M. Chen, E. K. Lee, G. A. Gibson, R. H. Katz, and D. A. Patterson. RAID: High-performance, reliable secondary storage. *ACM Computing Surveys*, 26(2):145–185, 1994.
4. L. Hellerstein, G. Gibson, R. M. Karp, R. H. Katz, and D. A. Patterson. Coding techniques for handling failures in large disk arrays. *Algorithmica*, 12(2–3):182–208, 1994.
5. R. D. Barve, E. A. M. Shriver, P. B. Gibbons, B. K. Hillyer, Y. Matias, and J. S. Vitter. Modeling and optimizing I/O throughput of multiple disks on a bus. In *Proceedings of the ACM SIGMETRICS Joint International Conference on Measurement and Modeling of Computer Systems*, pages 83–92, Atlanta, GA, May 1999.
6. C. Ruemmler and J. Wilkes. An introduction to disk drive modeling. *IEEE Computer*, 23(3): 17–28, 1994.
7. E. A. M. Shriver, A. Merchant, and J. Wilkes. An analytic behavior model for disk drives with readahead caches and request reordering. In *Proceedings of ACM SIGMETRICS Joint International Conference on Measurement and Modeling of Computer Systems*, pages 182–191, June 1998.
8. B. Chazelle. Filtering search: A new approach to query-answering. *SIAM Journal on Computing*, 15:703–724, 1986.
9. J. S. Vitter. External memory algorithms and data structures: Dealing with massive data. *ACM Computing Surveys*, 33(2):209–271, 2001.
10. M. T. Goodrich, J.-J. Tsay, D. E. Vengroff, and J. S. Vitter. External-memory computational geometry. In *Proceedings of the IEEE Symposium on Foundations of Computer Science*, volume 34, pages 714–723, Palo Alto, IEEE Computer Society Press, November 1993.
11. B. Zhu. Further computational geometry in secondary memory. In *Proceedings of the International Symposium on Algorithms and Computation*, volume 834 of *LNCS*, pages 514–522, Springer, 1994.
12. L. Arge, D. E. Vengroff, and J. S. Vitter. External-memory algorithms for processing line segments in geographic information systems. *Algorithmica*, 47(1):1–25, 2007.
13. L. Arge, O. Procopiuc, S. Ramaswamy, T. Suel, and J. S. Vitter. Theory and practice of I/O-efficient algorithms for multidimensional batched searching problems. In *Proceedings of the ACM-SIAM Symposium on Discrete Algorithms*, volume 9, pages 685–694, 1998.
14. A. Crauser, P. Ferragina, K. Mehlhorn, U. Meyer, and E. A. Ramos. I/O-optimal computation of segment intersections. In J. Abello and J. S. Vitter, editors, *External Memory Algorithms and Visualization*, DIMACS Series in Discrete Mathematics and Theoretical Computer Science, pages 131–138. American Mathematical Society Press, Providence, RI, 1999.

15. A. Crauser, P. Ferragina, K. Mehlhorn, U. Meyer, and E. A. Ramos. Randomized external-memory algorithms for line segment intersection and other geometric problems. *International Journal of Computational Geometry & Applications*, 11(3):305–337, 2001.

16. J. R. Driscoll, N. Sarnak, D. D. Sleator, and R. E. Tarjan. Making data structures persistent. *Journal of Computer and System Sciences*, 38:86–124, 1989.

17. W. Dittrich, D. A. Hutchinson, and A. Maheshwari. Blocking in parallel multisearch problems. *Theoretical Computer Science*, 34(2):145–189, 2001.

18. R. Tamassia and J. S. Vitter. Optimal cooperative search in fractional cascaded data structures. *Algorithmica*, 15(2):154–171, 1996.

19. D. G. Kirkpatrick and R. Seidel. The ultimate planar convex hull algorithm? *SIAM Journal on Computing*, 15:287–299, 1986.

20. K. L. Clarkson and P. W. Shor. Applications of random sampling in computational geometry, II. *Discrete and Computational Geometry*, 4:387–421, 1989.

21. M. de Berg, M. van Kreveld, M. Overmars, and O. Schwarzkopf. *Computational Geometry Algorithms and Applications*. Springer, Berlin, 1997.

22. F. P. Preparata and M. I. Shamos. *Computational Geometry*. Springer, Berlin, 1985.

23. Y.-J. Chiang. Experiments on the practical I/O efficiency of geometric algorithms: Distribution sweep vs. plane sweep. *Computational Geometry: Theory and Applications*, 8(4):211–236, 1998.

24. L. Arge, O. Procopiuc, S. Ramaswamy, T. Suel, and J. S. Vitter. Scalable sweeping-based spatial join. In *Proceedings of the International Conference on Very Large Databases*, volume 24, pages 570–581, New York, Morgan Kaufmann, August 1998.

25. J. M. Patel and D. J. DeWitt. Partition based spatial-merge join. In *Proceedings of the ACM SIGMOD International Conference on Management of Data*, pages 259–270, ACM Press, June 1996.

26. L. Arge, K. H. Hinrichs, J. Vahrenhold, and J. S. Vitter. Efficient bulk operations on dynamic R-trees. *Algorithmica*, 33(1):104–128, 2002.

27. TPIE User Manual and Reference, 1999. The manual and software distribution are available on the web at http://www.cs.duke.edu/TPIE/.

28. D. E. Vengroff and J. S. Vitter. I/O-efficient scientific computation using TPIE. In *Proceedings of the NASA Goddard Conference on Mass Storage Systems*, volume 5, pages II, 553–570, September 1996.

29. Topologically Integrated Geographic Encoding and Referencing system, TIGER/line1992datafiles, 1992. Available on the World-Wide Web at http://www.census.gov/geo/www/tiger/.

30. L. Arge, O. Procopiuc, S. Ramaswamy, T. Suel, J. Vahrenhold, and J. S. Vitter. A unified approach for indexed and non-indexed spatial joins. In *Proceedings of the International Conference on Extending Database Technology*, volume 7, Konstanz, Germany, ACM Press, March 2000.

31. R. D. Barve and J. S. Vitter. A theoretical framework for memory-adaptive algorithms. In *Proceedings of the IEEE Symposium on Foundations of Computer Science*, volume 40, pages 273–284, New York, IEEE Computer Society Press, October 1999.

32. V. Gaede and O. Günther. Multidimensional access methods. *ACM Computing Surveys*, 30(2):170–231, 1998.

33. R. Bayer and E. McCreight. Organization of large ordered indexes. *Acta Informatica*, 1:173–189, 1972.

34. D. Comer. The ubiquitous B-tree. *ACM Computing Surveys*, 11(2):121–137, 1979.

35. D. E. Knuth. *Sorting and Searching, volume 3 of The Art of Computer Programming*. Addison-Wesley, Reading, MA, 2nd edition, 1998.

36. A. C. Yao. On random 2–3 trees. *Acta Informatica*, 9:159–170, 1978.

37. R. A. Baeza-Yates. Expected behaviour of B$^+$-trees under random insertions. *Acta Informatica*, 26(5):439–472, 1989.

38. K. Küspert. Storage utilization in B*-trees with a generalized overflow technique. *Acta Informatica*, 19:35–55, 1983.

39. R. A. Baeza-Yates and P.-A. Larson. Performance of B$^+$-trees with partial expansions. *IEEE Transactions on Knowledge and Data Engineering*, 1(2):248–257, 1989.

40. B. Srinivasan. An adaptive overflow technique to defer splitting in B-trees. *The Computer Journal*, 34(5):397–405, 1991.

41. W. Litwin and D. Lomet. A new method for fast data searches with keys. *IEEE Software*, 4(2):16–24, 1987.

42. R. Baeza-Yates. Bounded disorder: The effect of the index. *Theoretical Computer Science*, 168:21–38, 1996.

43. D. Lomet. A simple bounded disorder file organization with good performance. *ACM Transactions on Database Systems*, 13(4):525–551, 1988.

44. P. E. O'Neil. The SB-tree. An index-sequential structure for high-performance sequential access. *Acta Informatica*, 29(3):241–265, 1992.

45. J. Rao and K. Ross. Cache conscious indexing for decision-support in main memory. In M. Atkinson et al., editors, *Proceedings of the International Conference on Very Large Databases*, volume 25, pages 78–89, Los Altos, CA, Morgan Kaufmann, 1999.

46. J. Rao and K. A. Ross. Making B^+-trees cache conscious in main memory. In W. Chen, J. Naughton, and P. A. Bernstein, editors, *Proceedings of the ACM SIGMOD International Conference on Management of Data*, pages 475–486, Dallas, TX, 2000.

47. B. Becker, S. Gschwind, T. Ohler, B. Seeger, and P. Widmayer. An asymptotically optimal multiversion B-tree. *VLDB Journal*, 5(4):264–275, 1996.

48. P. J. Varman and R. M. Verma. An efficient multiversion access structure. *IEEE Transactions on Knowledge and Data Engineering*, 9(3):391–409, 1997.

49. M. C. Easton. Key-sequence data sets on indelible storage. *IBM Journal of Research and Development*, 30:230–241, 1986.

50. B. Salzberg and V. J. Tsotras. Comparison of access methods for time-evolving data. *ACM Computing Surveys*, 31:158–221, 1999.

51. P. L. Lehman and S. Bing Yao. Efficient locking for concurrent operations on B-trees. *ACM Transactions on Database Systems*, 6(4):650–670, 1981.

52. C. Mohan. ARIES/KVL: A key-value locking method for concurrency control of multiaction transactions on B-tree indices. In *Proceedings of the International Conference on Very Large Databases*, volume 16, pages 392–405, Brisbane, Australia, Morgan Kaufmann, August 1990.

53. D. B. Lomet and B. Salzberg. Concurrency and recovery for index trees. *VLDB Journal*, 6(3):224–240, 1997.

54. L. Arge and J. S. Vitter. Optimal external memory interval management. *SIAM Journal on Computing*, 32(6):1488–1508, 2003.

55. P. K. Agarwal, L. Arge, O. Procopiuc, and J. S. Vitter. A framework for index bulk loading and dynamization. In *Proceedings of the International Colloquium on Automata, Languages, and Programming*, volume 2076 of *LNCS*, pages 115–127, Crete, Greece, Springer, 2001.

56. N. Blum and K. Mehlhorn. On the average number of rebalancing operations in weight-balanced trees. *Theoretical Computer Science*, 11(3):303–320, July 1980.

57. J. Nievergelt and E. M. Reingold. Binary search tree of bounded balance. *SIAM Journal on Computing*, 2(1):33–43, 1973.

58. R. Grossi and G. F. Italiano. Efficient splitting and merging algorithms for order decomposable problems. *Information and Computation*, 154(1):1–33, 1999.

59. H. Edelsbrunner. A new approach to rectangle intersections, Part I. *International Journal of Computer Mathematics*, 13:209–219, 1983.

60. H. Edelsbrunner. A new approach to rectangle intersections, Part II. *International Journal of Computer Mathematics*, 13:221–229, 1983.

61. D.E. Willard and G.S. Lueker. Adding range restriction capability to dynamic data structures. *Journal of the ACM*, 32(3):597–617, 1985.

62. P. K. Agarwal, L. Arge, G. S. Brodal, and J. S. Vitter. I/O-efficient dynamic point location in monotone planar subdivisions. In *Proceedings of the ACM-SIAM Symposium on Discrete Algorithms*, volume 10, pages 11–20, ACM Press, 1999.

63. J. S. Vitter. Efficient memory access in large-scale computation. In *Proceedings of the Symposium on Theoretical Aspects of Computer Science*, volume 480 of *LNCS*, pages 26–41, Springer, Invited keynote paper, 1991.

64. L. Arge. The Buffer Tree: A technique for designing batched external data structures. *Algorithmica*, 37(1):1–24, 2003.

65. J. L. Bentley. Multidimensional divide and conquer. *Communications of the ACM*, 23(6):214–229, 1980.

66. Y.-J. Chiang, M. T. Goodrich, E. F. Grove, R. Tamassia, D. E. Vengroff, and J. S. Vitter. External-memory graph algorithms. In *Proceedings of the ACM-SIAM Symposium on Discrete Algorithms*, volume 6, pages 139–149, ACM Press, January 1995.

67. S. Govindarajan, T. Lukovszki, A. Maheshwari, and N. Zeh. I/O-efficient well-separated pair decomposition and its applications. *Algorithmica*, 45(4):385–614, 2006.

68. G. S. Brodal and J. Katajainen. Worst-case efficient external-memory priority queues. In *Proceedings of the Scandinavian Workshop on Algorithmic Theory*, volume 1432 of LNCS, pages 107–118, Stockholm, Sweden, Springer, July 1998.

69. K. Brengel, A. Crauser, P. Ferragina, and U. Meyer. An experimental study of priority queues in external memory. *ACM Journal of Experimental Algorithmics*, 5:17, 2000.

70. P. Sanders. Fast priority queues for cached memory. *ACM Journal of Experimental Algorithmics*, 5(7):1–25, 2000.

71. D. Hutchinson, A. Maheshwari, J-R. Sack, and R. Velicescu. Early experiences in implementing the buffer tree. In *Proceedings of the Workshop on Algorithm Engineering*, LNCS, Springer, 1997.

72. P. C. Kanellakis, S. Ramaswamy, D. E. Vengroff, and J. S. Vitter. Indexing for data models with constraints and classes. *Journal of Computer and System Sciences*, 52(3):589–612, 1996.

73. S. Subramanian and S. Ramaswamy. The P-range tree: A new data structure for range searching in secondary memory. In *Proceedings of the ACM-SIAM Symposium on Discrete Algorithms*, volume 6, pages 378–387, PUB, ACM Press, 1995.

74. R. Grossi and G. F. Italiano. Efficient cross-trees for external memory. In J. Abello and J. S. Vitter, editors, *External Memory Algorithms and Visualization*, DIMACS Series in Discrete Mathematics and Theoretical Computer Science, pages 87–106. American Mathematical Society Press, Providence, RI, 1999.

75. K. V. R. Kanth and A. K. Singh. Optimal dynamic range searching in non-replicating index structures. In *Proceedings of the International Conference on Database Theory*, volume 1540 of LNCS, pages 257–276, Springer, January 1999.

76. J. T. Robinson. The k-d-B-tree: A search structure for large multidimensional dynamic indexes. In *Proceedings of the ACM Conference on Principles of Database Systems*, volume 1, pages 10–18, ACM Press, 1981.

77. B. Seeger and H.-P. Kriegel. The buddy-tree: An efficient and robust access method for spatial data base systems. In *Proceedings of the International Conference on Very Large Databases*, volume 16, pages 590–601, Morgan Kaufmann, 1990.

78. G. Evangelidis, D. B. Lomet, and B. Salzberg. The hB$^\Pi$-tree: A multi-attribute index supporting concurrency, recovery and node consolidation. *VLDB Journal*, 6:1–25, 1997.

79. D. B. Lomet and B. Salzberg. The hB-tree: A multiattribute indexing method with good guaranteed performance. *ACM Transactions on Database Systems*, 15(4):625–658, 1990.

80. K. Hinrichs. Implementation of the grid file: Design concepts and experience. *BIT*, 25(4):569–592, 1985.

81. J. Nievergelt, H. Hinterberger, and K. C. Sevcik. The grid file: An adaptable, symmetric multi-key file structure. *ACM Transactions on Database Systems*, 9:38–71, 1984.

82. K.-Y. Whang and R. Krishnamurthy. Multilevel grid files—A dynamic hierarchical multidimensional file structure. In *Proceedings of the International Symposium on Database Systems for Advanced Applications*, pages 449–459, World Scientific Press, 1992.

83. I. Gargantini. An effective way to represent quadtrees. *Communications of the ACM*, 25(12):905–910, 1982.

84. I. Kamel, C. Faloutsos. Hilbert R-tree: An improved R-tree using fractals. In *Proceedings of the International Conference on Very Large Databases*, volume 20, pages 500–509, Morgan Kaufmann, 1994.

85. J. A. Orenstein and T. H. Merrett. A class of data structures for associative searching. In *Proceedings of the ACM Conference on Principles of Database Systems*, volume 3, pages 181–190, ACM Press, 1984.

86. D. J. Abel. A B$^+$-tree structure for large quadtrees. *Computer Vision, Graphics, and Image Processing*, 27(1):19–31, 1984.

87. J. A. Orenstein. Redundancy in spatial databases. In *Proceedings of the ACM SIGMOD International Conference on Management of Data*, pages 294–305, Portland, OR, ACM Press, June 1989.

88. P. K. Agarwal and J. Erickson. Geometric range searching and its relatives. In B. Chazelle, J. E. Goodman, and R. Pollack, editors, *Advances in Discrete and Computational Geometry, Volume 23 of Contemporary Mathematics*, pages 1–56, American Mathematical Society Press, Providence, RI, 1999.

89. J. Nievergelt and P. Widmayer. Spatial data structures: Concepts and design choices. In M. van Kreveld, J. Nievergelt, T. Roos, and P. Widmayer, editors, *Algorithmic Foundations of GIS, Volume 1340 of LNCS*, pages 153–197, Springer, 1997.

90. A. Guttman. R-trees: A dynamic index structure for spatial searching. In *Proceedings of the ACM SIGMOD International Conference on Management of Data*, pages 47–57, ACM Press, 1984.

91. D. Greene. An implementation and performance analysis of spatial data access methods. In *Proceedings of the IEEE International Conference on Data Engineering*, volume 5, pages 606–615, IEEE Press, 1989.

92. N. Beckmann, H.-P. Kriegel, R. Schneider, and B. Seeger. The R*-tree: An efficient and robust access method for points and rectangles. In *Proceedings of the ACM SIGMOD International Conference on Management of Data*, pages 322–331, ACM Press, 1990.

93. M. de Berg, J. Gudmundsson, M. Hammar, and M. H. Overmars. On R-trees with low query complexity. *Computational Geometry*, 24(3):179–195, 2003.

94. P. K. Agarwal, M. de Berg, J. Gudmundsson, M. Hammar, and H. J. Haverkort. Box-trees and R-trees with near-optimal query time. *Discrete and Computational Geometry*, 28(3):291–312, 2002.

95. I. Kamel and C. Faloutsos. On packing R-trees. In *Proceedings of the International ACM Conference on Information and Knowledge Management*, volume 2, pages 490–499, ACM Press, 1993.

96. S. Berchtold, C. Böhm, and H-P. Kriegel. Improving the query performance of high-dimensional index structures by bulk load operations. In *Proceedings of the International Conference on Extending Database Technology*, volume 1377 of LNCS, pages 216–230, Springer, 1998.

97. I. Kamel, M. Khalil, and V. Kouramajian. Bulk insertion in dynamic R-trees. In *Proceedings of the International Symposium on Spatial Data Handling*, volume 4, pages 3B,31–42, Zurich: International Geographical Union, 1996.

98. J. van den Bercken, B. Seeger, and P. Widmayer. A generic approach to bulk loading multidimensional index structures. In *Proceedings of the International Conference on Very Large Databases*, volume 23, pages 406–415, 1997.

99. L. Arge, M. de Berg, H. J. Haverkort, and K. Yi. The priority R-tree: A practically efficient and worst-case optimal R-tree. *ACM Transactions on Algorithms*, 4(1):9, 2008.

100. Y.-J. Chiang and C. T. Silva. External memory techniques for isosurface extraction in scientific visualization. In J. Abello and J. S. Vitter, editors, *External Memory Algorithms and Visualization*, DIMACS Series in Discrete Mathematics and Theoretical Computer Science, pages 247–277, American Mathematical Society Press, Providence, RI, 1999.

101. P. K. Agarwal, L. Arge, T. M. Murali, K. Varadarajan, and J. S. Vitter. I/O-efficient algorithms for contour line extraction and planar graph blocking. In *Proceedings of the ACM-SIAM Symposium on Discrete Algorithms*, volume 9, pages 117–126, ACM Press, 1998.

102. L. Arge, V. Samoladas, and J. S. Vitter. Two-dimensional indexability and optimal range search indexing. In *Proceedings of the ACM Conference on Principles of Database Systems*, volume 18, pages 346–357, Philadelphia, PA, ACM Press, May–June 1999.

103. E. M. McCreight. Priority search trees. *SIAM Journal on Computing*, 14(2):257–276, 1985.

104. S. Ramaswamy and S. Subramanian. Path caching: A technique for optimal external searching. In *Proceedings of the ACM Conference on Principles of Database Systems*, volume 13, pages 25–35, Minneapolis, MN, ACM Press, 1994.

105. B. Chazelle. Lower bounds for orthogonal range searching: I. The reporting case. *Journal of the ACM*, 37(2):200–212, 1990.

106. J. M. Hellerstein, E. Koutsoupias, and C. H. Papadimitriou. On the analysis of indexing schemes. In *Proceedings of the ACM Symposium on Principles of Database Systems*, volume 16, pages 249–256, Tucson, AZ, ACM Press, May 1997.

107. E. Koutsoupias and D. S. Taylor. Tight bounds for 2-dimensional indexing schemes. In *Proceedings of the ACM Symposium on Principles of Database Systems*, volume 17, pages 52–58, Seattle, WA, ACM Press, June 1998.

108. V. Samoladas and D. Miranker. A lower bound theorem for indexing schemes and its application to multidimensional range queries. In *Proceedings of the ACM Symposium on Principles of Database Systems*, volume 17, pages 44–51, Seattle, WA, ACM Press, June 1998.

109. D. E. Vengroff and J. S. Vitter. Efficient 3-d range searching in external memory. In *Proceedings of the ACM Symposium on Theory of Computing*, volume 28, pages 192–201, Philadelphia, PA, ACM Press, May 1996.

110. P. Afshani. On dominance reporting in 3d. In *Proceedings of the European Symposium on Algorithms*, LNCS, pages 41–51. Springer, 2008.

111. P. Afshani, L. Arge, and K. D. Larsen. Orthogonal range reporting in three and higher dimensions. In *Proceedings of the IEEE Symposium on Foundations of Computer Science*, pages 149–158, IEEE Press, October 2009.

112. B. Chazelle and H. Edelsbrunner. Linear space data structures for two types of range search. *Discrete and Computational Geometry*, 2:113–126, 1987.

113. P. K. Agarwal, L. Arge, J. Erickson, P. G. Franciosa, and J. S. Vitter. Efficient searching with linear constraints. *Journal of Computer and System Sciences*, 61(2):194–216, October 2000.

114. R. Cheng, Y. Xia, S. Prabhakar, R. Shah, and J. S. Vitter. Efficient indexing methods for probabilistic threshold queries over uncertain data. In *Proceedings of the International Conference on Very Large Databases*, Toronto, CA, Morgan Kaufmann, August 2004.

115. R. Cheng, Y. Xia, S. Prabhakar, R. Shah, and J. S. Vitter. Efficient join processing over uncertain-valued attributes. In *Proceedings of the International ACM Conference on Information and Knowledge Management*, Arlington, VA, ACM Press, November 2006.

116. Y.-F. Chien, W.-K. Hon, R. Shah, and J. S. Vitter. Geometric Burrows-Wheeler transform: Linking range searching and text indexing. In *Proceedings of the IEEE Data Compression Conference*, pages 252–261, IEEE Press, 2008.

117. W.-K. Hon, R. Shah, S. V. Thankachan, and J. S. Vitter. Space-efficient frameworks for top-*k* string retrieval. *Journal of the ACM*, 61(2):9.1–9.36, 2014.

118. P. Callahan, M. T. Goodrich, and K. Ramaiyer. Topology B-trees and their applications. In *Proceedings of the Workshop on Algorithms and Data Structures*, volume 955 of *LNCS*, pages 381–392, Springer, 1995.

119. P. K. Agarwal, L. Arge, and J. Erickson. Indexing moving points. *Journal of Computer and System Sciences*, 66(1):207–243, 2003.

120. A. Gionis, P. Indyk, and R. Motwani. Similarity search in high dimensions via hashing. In *Proceedings of the International Conference on Very Large Databases*, volume 25, pages 78–89, Edinburgh, Scotland, Morgan Kaufmann, 1999.

121. J. L. Bentley and J. B. Saxe. Decomposable searching problems I: Static-to-dynamic transformations. *Journal of Algorithms*, 1(4):301–358, December 1980.

122. M. H. Overmars, *The Design of Dynamic Data Structures*. LNCS, Berlin: Springer, 1983.

123. L. Arge and J. Vahrenhold. I/O-efficient dynamic planar point location. *Computational Geometry*, 29(2):147–162, 2004.

124. D. Pfoser, C. S. Jensen, and Y. Theodoridis. Novel approaches to the indexing of moving object trajectories. In *Proceedings of the International Conference on Very Large Databases*, volume 26, pages 395–406, Cairo, Morgan Kaufmann, 2000.

125. S. Šaltenis, C. S. Jensen, S. T. Leutenegger, and M. A. Lopez. Indexing the positions of continuously moving objects. In W. Chen, J. Naughton, and P. A. Bernstein, editors, *Proceedings of the ACM SIGMOD International Conference on Management of Data*, pages 331–342, Dallas, Texas, ACM Press, 2000.

126. O. Wolfson. Moving objects information management: The database challenge. In A. Halevy and A. Gal, editors, *Next Generation Information Technologies and Systems, volume 2382 of Lecture Notes in Computer Science*, pages 75–89, Berlin Heidelberg, Springer, 2002.

127. J. Basch, L. J. Guibas, and J. Hershberger. Data structures for mobile data. *Journal of Algorithms*, 31:1–28, 1999.

128. G. Kollios, D. Gunopulos, and V. J. Tsotras. On indexing mobile objects. In *Proceedings of the ACM Symposium on Principles of Database Systems*, volume 18, pages 261–272, Morgan Kaufmann, 1999.

129. P. K. Agarwal and S. Har-Peled. Maintaining the approximate extent measures of moving points. In *Proceedings of the ACM-SIAM Symposium on Discrete Algorithms*, volume 12, pages 148–157, Washington, Morgan Kaufmann, January 2001.

V

Miscellaneous

29 Tries Sartaj Sahni 445
What Is a Trie? • Searching a Trie • Keys with Different Length • Height of a Trie • Space
Required and Alternative Node Structures • Inserting into a Trie • Removing an Element • Prefix
Search and Applications • Compressed Tries • Patricia • Acknowledgments • References

30 Suffix Trees and Suffix Arrays Srinivas Aluru 461
Basic Definitions and Properties • Linear Time Construction Algorithms • Applications • Lowest
Common Ancestors • Advanced Applications • Acknowledgments • References

31 String Searching Andrzej Ehrenfeucht and Ross M. McConnell.. 477
Introduction • Preliminaries • The DAWG • The Compact DAWG • The Position Heap •
References

32 Binary Decision Diagrams Shin-ichi Minato.. 495
Introduction • Basic Concepts • Data Structure • Construction of BDDs from Boolean Expressions
• Variable Ordering for BDDs • Zero-Suppressed BDDs • Related Research Activities •
Acknowledgments • References

33 Persistent Data Structures Haim Kaplan 511
Introduction • Algorithmic Applications of Persistent Data Structures • General Techniques for
Making Data Structures Persistent • Making Specific Data Structures More Efficient • Concluding
Remarks and Open Questions • Acknowledgments • References

34 Data Structures for Sets Rajeev Raman... 529
Introduction • Simple Randomized Set Representations • Equality Testing • Extremal Sets and
Subset Testing • The Disjoint Set Union-Find Problem • Partition Maintenance Algorithms •
Conclusions • References

35 Cache-Oblivious Data Structures Lars Arge, Gerth Stølting Brodal and Rolf Fagerberg..................................... 545
The Cache-Oblivious Model • Fundamental Primitives • Dynamic B-Trees • Priority Queues •
2d Orthogonal Range Searching • Acknowledgments • References

36 Dynamic Trees Camil Demetrescu, Irene Finocchi and Giuseppe F. Italiano 567
Introduction • Linking and Cutting Trees • Topology Trees • Top Trees • ET Trees • Reachability
Trees • Conclusions • Acknowledgments • References

37 Dynamic Graphs Camil Demetrescu, Irene Finocchi and Giuseppe F. Italiano.................... 581
Introduction • Techniques for Undirected Graphs • Techniques for Directed Graphs • Connectivity
• Minimum Spanning Tree • Transitive Closure • All-Pairs Shortest Paths • Conclusions •
Acknowledgments • References

38 Succinct Representation of Data Structures J. Ian Munro and S. Srinivasa Rao 595
Introduction • Bitvector • Succinct Dictionaries • Tree Representations • Graph Representations
• Succinct Structures for Indexing • Permutations and Functions • Partial Sums • Arrays •
Conclusions • References

39 Randomized Graph Data-Structures for Approximate Shortest Paths Surender Baswana and
Sandeep Sen 611
Introduction • A Randomized Data-Structure for Static APASP: Approximate Distance Oracles •
A Randomized Data-Structure for Decremental APASP • Further Reading and Bibliography •
Acknowledgments • References

40 Searching and Priority Queues in o(log n) Time Arne Andersson 627
Introduction • Model of Computation • Overview • Achieving Sub-Logarithmic Time per Element
by Simple Means • Deterministic Algorithms and Linear Space • From Amortized Update Cost to
Worst-Case • Sorting and Priority Queues • References

29

Tries*

29.1	What Is a Trie?	445
29.2	Searching a Trie	446
29.3	Keys with Different Length	446
29.4	Height of a Trie	447
29.5	Space Required and Alternative Node Structures	447
29.6	Inserting into a Trie	449
29.7	Removing an Element	449
29.8	Prefix Search and Applications	451
29.9	Compressed Tries	452
	Compressed Tries with Digit Numbers • Compressed Tries with Skip Fields • Compressed Tries with Edge Information • Space Required by a Compressed Trie	
29.10	Patricia	456
	Searching • Inserting an Element • Removing an Element	
	Acknowledgments	459
	References	460

Sartaj Sahni
University of Florida

29.1 What Is a Trie?

A *trie* (pronounced "try" and derived from the word re*trie*val) is a data structure that uses the digits in the keys to organize and search the dictionary. Although, in practice, we can use any radix to decompose the keys into digits, in our examples, we shall choose our radixes so that the digits are natural entities such as decimal digits (0, 1, 2, 3, 4, 5, 6, 7, 8, 9) and letters of the English alphabet (a–z, A–Z).

Suppose that the elements in our dictionary are student records that contain fields such as student name, major, date of birth, and social security number (SS#). The key field is the social security number, which is a nine digit decimal number. To keep the example manageable, assume that the dictionary has only five elements. Table 29.1 shows the name and SS# fields for each of the five elements in our dictionary.

To obtain a trie representation for these five elements, we first select a radix that will be used to decompose each key into digits. If we use the radix 10, the decomposed digits are just the decimal digits shown in Table 29.1. We shall examine the digits of the key field (i.e., SS#) from left to right. Using the first digit of the SS#, we partition the elements into three groups–elements whose SS# begins with 2 (i.e., Bill and Kathy), those that begin with 5 (i.e., Jill), and those that begin with 9 (i.e., April and Jack). Groups with more than one element are partitioned using the next digit in the key. This partitioning process is continued until every group has exactly one element in it.

The partitioning process described above naturally results in a tree structure that has 10-way branching as is shown in Figure 29.1. The tree employs two types of nodes–*branch nodes* and *element nodes*. Each branch node has 10 children (or pointer/reference) fields. These fields, *child*[0:9], have been labeled 0,1, . . . , 9 for the root node of Figure 29.1. *root.child*[i] points to the root of a subtrie that contains all elements whose first digit is i. In Figure 29.1, nodes A, B, D, E, F, and I are branch nodes. The remaining nodes, nodes C, G, H, J, and K are element nodes. Each element node contains exactly one element of the dictionary. In Figure 29.1, only the key field of each element is shown in the element nodes.

---

* This chapter has been reprinted from first edition of this Handbook, without any content updates.

TABLE 29.1 Five Elements
(Student Records) in a Dictionary

Name	Social Security Number (SS#)
Jack	951-94-1654
Jill	562-44-2169
Bill	271-16-3624
Kathy	278-49-1515
April	951-23-7625

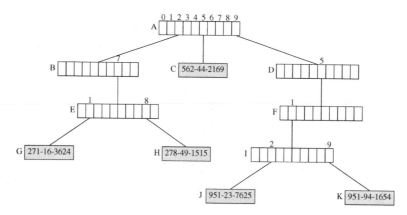

FIGURE 29.1 Trie for the elements of Table 29.1.

29.2 Searching a Trie

To search a trie for an element with a given key, we start at the root and follow a path down the trie until we either fall off the trie (i.e., we follow a *null* pointer in a branch node) or we reach an element node. The path we follow is determined by the digits of the search key. Consider the trie of Figure 29.1. Suppose we are to search for an element with key 951-23-7625. We use the first digit, 9, in the key to move from the root node *A* to the node *A.child*[9] = *D*. Since *D* is a branch node, we use the next digit, 5, of the key to move further down the trie. The node we reach is *D.child*[5] = *F*. To move to the next level of the trie, we use the next digit, 1, of the key. This move gets us to the node *F.child*[1] = *I*. Once again, we are at a branch node and must move further down the trie. For this move, we use the next digit, 2, of the key, and we reach the element node *I.child*[2] = *J*. When an element node is reached, we compare the search key and the key of the element in the reached element node. Performing this comparison at node *J*, we get a match. The element in node *J*, is to be returned as the result of the search.

When searching the trie of Figure 29.1 for an element with key 951-23-1669, we follow the same path as for the key 951-23-7625. The key comparison made at node *J* tells us that the trie has no element with key 951-23-1669, and the search returns the value *null*.

To search for the element with key 562-44-2169, we begin at the root *A* and use the first digit, 5, of the search key to reach the element node *A.child*[5] = *C*. The key of the element in node *C* is compared with the search key. Since the two keys agree, the element in node *C* is returned.

When searching for an element with key 273-11-1341, we follow the path *A*, *A.child*[2] = *B*, *B.child*[7] = *E*, *E.child*[3] = *null*. Since we fall off the trie, we know that the trie contains no element whose key is 273-11-1341.

When analyzing the complexity of trie operations, we make the assumption that we can obtain the next digit of a key in *O*(1) time. Under this assumption, we can search a trie for an element with a *d* digit key in *O*(*d*) time.

29.3 Keys with Different Length

In the example of Figure 29.1, all keys have the same number of digits (i.e., 9). In many applications, however, different keys have different length. This does not pose a problem unless one key is a prefix of another (e.g., 27 is a prefix of 276). For applications in which one key may be a prefix of another, we normally add a special digit (say #) at the end of each key. Doing this ensures that no key is a prefix of another.

To see why we cannot permit a key that is a prefix of another key, consider the example of Figure 29.1. Suppose we are to search for an element with the key 27. Using the search strategy just described, we reach the branch node E. What do we do now? There is no next digit in the search key that can be used to reach the terminating condition (i.e., you either fall off the trie or reach an element node) for downward moves. To resolve this problem, we add the special digit # at the end of each key and also increase the number of children fields in an element node by one. The additional child field is used when the next digit equals #.

An alternative to adding a special digit at the end of each key is to give each node a *data* field that is used to store the element (if any) whose key exhausts at that node. So, for example, the element whose key is 27 can be stored in node E of Figure 29.1. When this alternative is used, the search strategy is modified so that when the digits of the search key are exhausted, we examine the *data* field of the reached node. If this *data* field is empty, we have no element whose key equals the search key. Otherwise, the desired element is in this *data* field.

It is important to note that in applications that have different length keys with the property that no key is a prefix of another, neither of just mentioned strategies is needed; the scheme described in Section 29.2 works as is.

29.4 Height of a Trie

In the worst case, a root-node to element-node path has a branch node for every digit in a key. Therefore, the height[*] of a trie is at most *number of digits* $+ 1$.

A trie for social security numbers has a height that is at most 10. If we assume that it takes the same time to move down one level of a trie as it does to move down one level of a binary search tree, then with at most 10 moves we can search a social-security trie. With this many moves, we can search a binary search tree that has at most $2^{10} - 1 = 1023$ elements. This means that, we expect searches in the social security trie to be faster than searches in a binary search tree (for student records) whenever the number of student records is more than 1023. The breakeven point will actually be less than 1023 because we will normally not be able to construct full or complete binary search trees for our element collection.

Since a SS# is nine digits, a social security trie can have up to 10^9 elements in it. An AVL tree with 10^9 elements can have a height that is as much as (approximately) $1.44 \log_2(10^9 + 2) = 44$. Therefore, it could take us four times as much time to search for elements when we organize our student-record dictionary as an AVL tree than when this dictionary is organized as a trie!

29.5 Space Required and Alternative Node Structures

The use of branch nodes that have as many child fields as the radix of the digits (or one more than this radix when different keys may have different length) results in a fast search algorithm. However, this node structure is often wasteful of space because many of the child fields are *null*. A radix *r* trie for *d* digit keys requires $O(rdn)$ child fields, where *n* is the number of elements in the trie. To see this, notice that in a *d* digit trie with *n* information nodes, each information node may have at most *d* ancestors, each of which is a branch node. Therefore, the number of branch nodes is at most *dn*. (Actually, we cannot have this many branch nodes, because the information nodes have common ancestors like the root node.)

We can reduce the space requirements, at the expense of increased search time, by changing the node structure. For example, each branch node of a trie could be replaced by any of the following:

1. A chain of nodes, each node having the three fields *digit Value, child, next*. Node A of Figure 29.1, for example, would be replaced by the chain shown in Figure 29.2.

   The space required by a branch node changes from that required for *r* children/pointer/reference fields to that required for 2*p* pointer fields and *p* digit value fields, where *p* is the number of children fields in the branch node that are not *null*. Under the assumption that pointer fields and digit value fields are of the same size, a reduction in space is realized when more than two-thirds of the children fields in branch nodes are *null*. In the worst case, almost all the branch nodes have only 1 field that is not *null* and the space savings become almost $(1 - 3/r) * 100\%$.

2. A (balanced) binary search tree in which each node has a digit value and a pointer to the subtrie for that digit value. Figure 29.3 shows the binary search tree for node A of Figure 29.1.

   Under the assumption that digit values and pointers take the same amount of space, the binary search tree representation requires space for 4*p* fields per branch node, because each search tree node has fields for a digit value, a subtrie pointer, a left child pointer, and a right child pointer. The binary search tree representation of a branch node saves us space when more than three-fourths of the children fields in branch nodes are *null*. Note that for large *r*, the binary search tree is faster to search than the chain described above.

---

[*] The definition of height used in this chapter is: the height of a trie equals the number of levels in that trie.

firstNode $\longrightarrow$ | 2 | B | | $\longrightarrow$ | 9 | D | null |

FIGURE 29.2   Chain for node A of Figure 29.1.

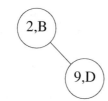

FIGURE 29.3   Binary search tree for node A of Figure 29.1.

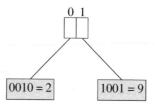

FIGURE 29.4   Binary trie for node A of Figure 29.1.

3. A binary trie (i.e., a trie with radix 2). Figure 29.4 shows the binary trie for node *A* of Figure 29.1. The space required by a branch node represented as a binary trie is at most $(2 * \lceil \log_2 r \rceil + 1)p$.

4. A hash table. When a hash table with a sufficiently small loading density is used, the expected time performance is about the same as when the node structure of Figure 29.1 is used. Since we expect the fraction of *null* child fields in a branch node to vary from node to node and also to increase as we go down the trie, maximum space efficiency is obtained by consolidating all of the branch nodes into a single hash table. To accomplish this, each node in the trie is assigned a number, and each parent to child pointer is replaced by a triple of the form (*currentNode, digitValue, childNode*). The numbering scheme for nodes is chosen so as to easily distinguish between branch and information nodes. For example, if we expect to have at most 100 elements in the trie at any time, the numbers 0 through 99 are reserved for information nodes and the numbers 100 on up are used for branch nodes. The information nodes are themselves represented as an array *information* [100]. (An alternative scheme is to represent pointers as tuples of the form (*currentNode, digitValue, childNode, childNodeIsBranchNode*), where *childNodeIsBranchNode* = *true* iff the child node is a branch node.)

Suppose that the nodes of the trie of Figure 29.1 are assigned numbers as given in Figure 29.5. This number assignment assumes that the trie will have no more than 10 elements.

The pointers in node *A* are represented by the tuples (10, 2, 11), (10, 5, 0), and (10, 9, 12). The pointers in node *E* are represented by the tuples (13, 1, 1) and (13, 8, 2).

The pointer triples are stored in a hash table using the first two fields (i.e., the *currentNode* and *digitValue*) as the key. For this purpose, we may transform the two field key into an integer using the formula *currentNode* $* r + digitValue$, where *r* is the trie radix, and use the division method to hash the transformed key into a home bucket. The data presently in information node *i* is stored in *information*[*i*].

To see how all this works, suppose we have set up the trie of Figure 29.1 using the hash table scheme just described. Consider searching for an element with key 278-49-1515. We begin with the knowledge that the root node is assigned the number 10. Since the first digit of the search key is 2, we query our hash table for a pointer triple with key (10, 2). The hash table search is successful and the triple (10, 2, 11) is retrieved. The *childNode* component of this triple is 11, and since all information nodes have a number 9 or less, the child node is determined to be a branch node. We make a move to the branch node 11. To move to the next level of the trie, we use the second digit 7 of the search key. For the move,

Node	A	B	C	D	E	F	G	H	I	J	K
Number	10	11	0	12	13	14	1	2	15	3	4

FIGURE 29.5   Number assignment to nodes of trie of Figure 29.1.

we query the hash table for a pointer with key (11, 7). Once again, the search is successful and the triple (11, 7, 13) is retrieved. The next query to the hash table is for a triple with key (13, 8). This time, we obtain the triple (13, 8, 2). Since *childNode* = 2 < 10, we know that the pointer gets us to an information node. So, we compare the search key with the key of the element *information* [2]. The keys match, and we have found the element we were looking for.

When searching for an element with key 322-167-8976, the first query is for a triple with key (10, 3). The hash table has no triple with this key, and we conclude that the trie has no element whose key equals the search key.

The space needed for each pointer triple is about the same as that needed for each node in the chain of nodes representation of a trie node. Therefore, if we use a linear open addressed hash table with a loading density of α, the hash table scheme will take approximately $(1/\alpha - 1) * 100\%$ more space than required by the chain of nodes scheme. However, when the hash table scheme is used, we can retrieve a pointer in $O(1)$ expected time, whereas the time to retrieve a pointer using the chain of nodes scheme is $O(r)$. When the (balanced) binary search tree or binary trie schemes are used, it takes $O(\log r)$ time to retrieve a pointer. For large radixes, the hash table scheme provides significant space saving over the scheme of Figure 29.1 and results in a small constant factor degradation in the expected time required to perform a search.

The hash table scheme actually reduces the expected time to insert elements into a trie, because when the node structure of Figure 29.1 is used, we must spend $O(r)$ time to initialize each new branch node (see the description of the insert operation below). However, when a hash table is used, the insertion time is independent of the trie radix.

To support the removal of elements from a trie represented as a hash table, we must be able to reuse information nodes. This reuse is accomplished by setting up an available space list of information nodes that are currently not in use.

Andersson and Nilsson [1] propose a trie representation in which nodes have a variable degree. Their data structure, called LC-tries (level-compressed tries), is obtained from a binary trie by replacing full subtries of the binary trie by single node whose degree is 2^i, where i is the number of levels in the replaced full subtrie. This replacement is done by examining the binary trie from top to bottom (i.e., from root to leaves).

29.6 Inserting into a Trie

To insert an element *theElement* whose key is *theKey*, we first search the trie for an existing element with this key. If the trie contains such an element, then we replace the existing element with *theElement*. When the trie contains no element whose key equals *theKey*, *theElement* is inserted into the trie using the following procedure.

Case 1 **For Insert Procedure**

If the search for *theKey* ended at an element node X, then the key of the element in X and *theKey* are used to construct a subtrie to replace X.

Suppose we are to insert an element with key 271-10-2529 into the trie of Figure 29.1. The search for the key 271-10-2529 terminates at node G and we determine that the key, 271-16-3624, of the element in node G is not equal to the key of the element to be inserted. Since the first three digits of the keys are used to get as far as node E of the trie, we set up branch nodes for the fourth digit (from left) onwards until we reach the first digit at which the two keys differ. This results in branch nodes for the fourth and fifth digits followed by element nodes for each of the two elements. Figure 29.6 shows the resulting trie.

Case 2 **For Insert Procedure**

If the search for *theKey* ends by falling off the trie from the branch node X, then we simply add a child (which is an element node) to the node X. The added element node contains *theElement*.

Suppose we are to insert an element with key 987-33-1122 to the trie of Figure 29.1. The search for an element with key equal to 987-33-1122 ends when we fall off the trie while following the pointer D.child[8]. We replace the *null* pointer D.child[8] with a pointer to a new element node that contains *theElement*, as is shown in Figure 29.7.

The time required to insert an element with a d digit key into a radix r trie is $O(dr)$ because the insertion may require us to create $O(d)$ branch nodes and it takes $O(r)$ time to initialize the children pointers in a branch node.

29.7 Removing an Element

To remove the element whose key is *theKey*, we first search for the element with this key. If there is no matching element in the trie, nothing is to be done. So, assume that the trie contains an element *theElement* whose key is *theKey*. The element node X that contains *theElement* is discarded, and we retrace the path from X to the root discarding branch nodes that are roots of subtries that have only 1 element in them. This path retracing stops when we either reach a branch node that is not discarded or we discard the root.

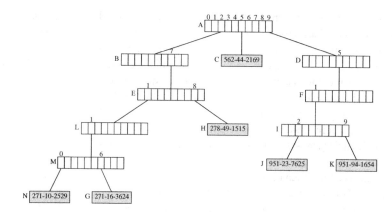

FIGURE 29.6   Trie of Figure 29.1 with 271-10-2529 inserted.

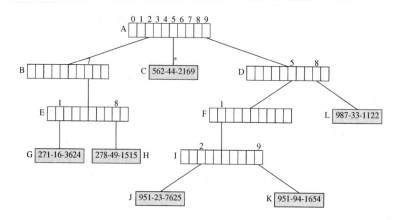

FIGURE 29.7   Trie of Figure 29.1 with 987-33-1122 inserted.

Consider the trie of Figure 29.7. When the element with key 951-23-7625 is removed, the element node *J* is discarded and we follow the path from node *J* to the root node *A*. The branch node *I* is discarded because the subtrie with root *I* contains the single element node *K*. We next reach the branch node *F*. This node is also discarded, and we proceed to the branch node *D*. Since the subtrie rooted at *D* has 2 element nodes (*K* and *L*), this branch node is not discarded. Instead, node *K* is made a child of this branch node, as is shown in Figure 29.8.

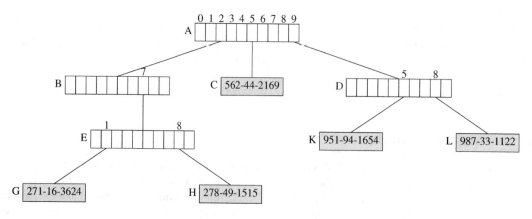

FIGURE 29.8   Trie of Figure 29.7 with 951-23-7635 removed.

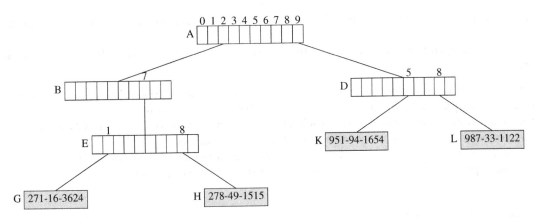

FIGURE 29.9 Trie of Figure 29.8 with 562-44-2169 removed.

To remove the element with key 562-44-2169 from the trie of Figure 29.8, we discard the element node *C*. Since its parent node remains the root of a subtrie that has more than one element, the parent node is not discarded and the removal operation is complete. Figure 29.9 show the resulting trie.

The time required to remove an element with a *d* digit key from a radix *r* trie is $O(dr)$ because the removal may require us to discard $O(d)$ branch nodes and it takes $O(r)$ time to determine whether a branch node is to be discarded. The complexity of the remove operation can be reduced to $O(r + d)$ by adding a *numberOfElementsInSubtrie* field to each branch node.

29.8 Prefix Search and Applications

You have probably realized that to search a trie we do not need the entire key. Most of the time, only the first few digits (i.e., a prefix) of the key is needed. For example, our search of the trie of Figure 29.1 for an element with key 951-23-7625 used only the first four digits of the key. The ability to search a trie using only the prefix of a key enables us to use tries in applications where only the prefix might be known or where we might desire the user to provide only a prefix. Some of these applications are described below.

Criminology

Suppose that you are at the scene of a crime and observe the first few characters *CRX* on the registration plate of the getaway car. If we have a trie of registration numbers, we can use the characters *CRX* to reach a subtrie that contains all registration numbers that begin with *CRX*. The elements in this subtrie can then be examined to see which cars satisfy other properties that might have been observed.

Automatic Command Completion

When using an operating system such as Unix or DOS, we type in system commands to accomplish certain tasks. For example, the Unix and DOS command *cd* may be used to change the current directory. Table 29.2 gives a list of commands that have the prefix *ps* (this list was obtained by executing the command *ls/usr/local/bin/ps** on a Unix system).

We can simply the task of typing in commands by providing a command completion facility which automatically types in the command suffix once the user has typed in a long enough prefix to uniquely identify the command. For instance, once the letters *psi* have been entered, we know that the command must be *psidtopgm* because there is only one command that has the prefix *psi*. In this case, we replace the need to type in a 9 character command name by the need to type in just the first 3 characters of the command!

A command completion system is easily implemented when the commands are stored in a trie using ASCII characters as the digits. As the user types the command digits from left to right, we move down the trie. The command may be completed as soon

TABLE 29.2 Commands that Begin with "ps"

ps2ascii	ps2pdf	psbook	psmandup	psselect
ps2epsi	ps2pk	pscal	psmerge	pstopnm
ps2frag	ps2ps	psidtopgm	psnup	pstops
ps2gif	psbb	pslatex	psresize	pstruct

as we reach an element node. If we fall off the trie in the process, the user can be informed that no command with the typed prefix exists.

Although we have described command completion in the context of operating system commands, the facilty is useful in other environments:

1. A network browser keeps a history of the URLs of sites that you have visited. By organizing this history as a trie, the user need only type the prefix of a previously used URL and the browser can complete the URL.
2. A word processor can maintain a dictionary of words and can complete words as you type the text. Words can be completed as soon as you have typed a long enough prefix to identify the word uniquely.
3. An automatic phone dialer can maintain a list of frequently called telephone numbers as a trie. Once you have punched in a long enough prefix to uniquely identify the phone number, the dialer can complete the call for you.

29.9 Compressed Tries

Take a close look at the trie of Figure 29.1. This trie has a few branch nodes (nodes *B*, *D*, and *F*) that do not partition the elements in their subtrie into two or more nonempty groups. We often can improve both the time and space performance metrics of a trie by eliminating all branch nodes that have only one child. The resulting trie is called a *compressed trie*.

When branch nodes with a single child are removed from a trie, we need to keep additional information so that dictionary operations may be performed correctly. The additional information stored in three compressed trie structures is described below.

29.9.1 Compressed Tries with Digit Numbers

In a *compressed trie with digit numbers*, each branch node has an additional field *digitNumber* that tells us which digit of the key is used to branch at this node. Figure 29.11 shows the compressed trie with digit numbers that corresponds to the trie of Figure 29.1. The leftmost field of each branch node of Figure 29.10 is the *digitNumber* field.

29.9.1.1 Searching a Compressed Trie with Digit Numbers

A compressed trie with digit numbers may be searched by following a path from the root. At each branch node, the digit, of the search key, given in the branch node's *digitNumber* field is used to determine which subtrie to move to. For example, when searching the trie of Figure 29.10 for an element with key 951-23-7625, we start at the root of the trie. Since the root node is a branch node with *digitNumber* = 1, we use the first digit 9 of the search key to determine which subtrie to move to. A move to node *A.child*[9] = *I* is made. Since *I.digitNumber* = 4, the fourth digit, 2, of the search key tells us which subtrie to move to. A move is now made to node *I.child*[2] = *J*. We are now at an element node, and the search key is compared with the key of the element in node *J*. Since the keys match, we have found the desired element.

Notice that a search for an element with key 913-23-7625 also terminates at node *J*. However, the search key and the element key at node *J* do not match and we conclude that the trie contains no element with key 913-23-7625.

29.9.1.2 Inserting into a Compressed Trie with Digit Numbers

To insert an element with key 987-26-1615 into the trie of Figure 29.10, we first search for an element with this key. The search ends at node *J*. Since the search key and the key, 951-23-7625, of the element in this node do not match, we conclude that the trie has no element whose key matches the search key. To insert the new element, we find the first digit where the search key differs from the key in node *J* and create a branch node for this digit. Since the first digit where the search key 987-26-1615 and the

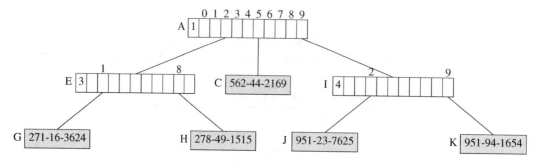

FIGURE 29.10 Compressed trie with digit numbers.

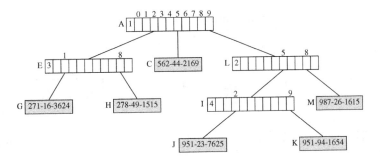

FIGURE 29.11 Compressed trie following the insertion of 987-26-1615 into the compressed trie of Figure 29.10.

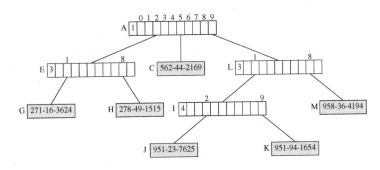

FIGURE 29.12 Compressed trie following the insertion of 958-36-4194 into the compressed trie of Figure 29.10.

element key 951-23-7625 differ is the second digit, we create a branch node with *digitNumber* = 2. Since digit values increase as we go down the trie, the proper place to insert the new branch node can be determined by retracing the path from the root to node *J* and stopping as soon as either a node with digit value greater than 2 or the node *J* is reached. In the trie of Figure 29.10, this path retracing stops at node *I*. The new branch node is made the parent of node *I*, and we get the trie of Figure 29.11.

Consider inserting an element with key 958-36-4194 into the compressed trie of Figure 29.10. The search for an element with this key terminates when we fall of the trie by following the pointer *I.child*[3] = *null*. To complete the insertion, we must first find an element in the subtrie rooted at node *I*. This element is found by following a downward path from node *I* using (say) the first non *null* link in each branch node encountered. Doing this on the compressed trie of Figure 29.10, leads us to node *J*. Having reached an element node, we find the first digit where the element key and the search key differ and complete the insertion as in the previous example. Figure 29.12 shows the resulting compressed trie.

Because of the possible need to search for the first non *null* child pointer in each branch node, the time required to insert an element into a compressed tries with digit numbers is $O(rd)$, where r is the trie radix and d is the maximum number of digits in any key.

29.9.1.3 Removing an Element from a Compressed Trie with Digit Numbers

To remove an element whose key is *theKey*, we do the following:

1. Find the element node *X* that contains the element whose key is *theKey*.
2. Discard node *X*.
3. If the parent of *X* is left with only one child, discard the parent node also. When the parent of *X* is discarded, the sole remaining child of the parent of *X* becomes a child of the grandparent (if any) of *X*.

To remove the element with key 951-94-1654 from the compressed trie of Figure 29.12, we first locate the node *K* that contains the element that is to be removed. When this node is discarded, the parent *I* of *K* is left with only one child. Consequently, node *I* is also discarded, and the only remaining child *J* of node *I* is the made a child of the grandparent of *K*. Figure 29.13 shows the resulting compressed trie.

Because of the need to determine whether a branch node is left with two or more children, removing a d digit element from a radix r trie takes $O(d + r)$ time.

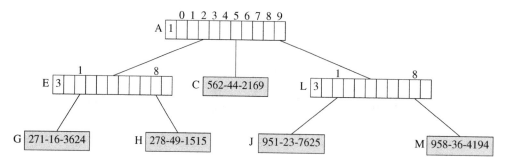

FIGURE 29.13   Compressed trie following the removal of 951-94-1654 from the compressed trie of Figure 29.12.

29.9.2 Compressed Tries with Skip Fields

In a *compressed trie with skip fields*, each branch node has an additional field *skip* which tells us the number of branch nodes that were originally between the current branch node and its parent. Figure 29.15 shows the compressed trie with skip fields that corresponds to the trie of Figure 29.1. The leftmost field of each branch node of Figure 29.14 is the skip field.

The algorithms to search, insert, and remove are very similar to those used for a compressed trie with digit numbers.

29.9.3 Compressed Tries with Edge Information

In a *compressed trie with edge information*, each branch node has the following additional information associated with it: a pointer/reference *element* to an element (or element node) in the subtrie, and an integer *skip* which equals the number of branch nodes eliminated between this branch node and its parent. Figure 29.15 shows the compressed trie with edge information that corresponds to the trie of Figure 29.1. The first field of each branch node is its *element* field, and the second field is the *skip* field.

Even though we store the "edge information" with branch nodes, it is convenient to think of this information as being associated with the edge that comes into the branch node from its parent (when the branch node is not the root). When moving down a trie, we follow edges, and when an edge is followed. we skip over the number of digits given by the *skip* field of the edge information. The value of the digits that are skipped over may be determined by using the *element* field.

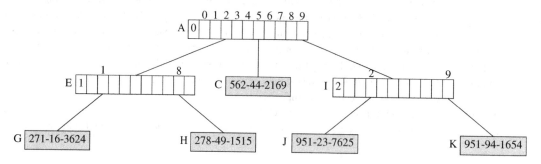

FIGURE 29.14   Compressed trie with skip fields.

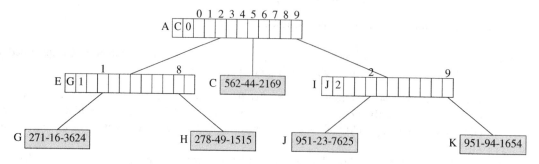

FIGURE 29.15   Compressed trie with edge information.

When moving from node A to node I of the compressed trie of Figure 29.15, we use digit 1 of the key to determine which child field of A is to be used. Also, we skip over the next 2 digits, that is, digits 2 and 3, of the keys of the elements in the subtrie rooted at I. Since all elements in the subtrie I have the same value for the digits that are skipped over, we can determine the value of these skipped over digits from any of the elements in the subtrie. Using the *element* field of the edge information, we access the element node J, and determine that the digits that are skipped over are 5 and 1.

29.9.3.1 Searching a Compressed Trie with Edge Information

When searching a compressed trie with edge information, we can use the edge information to terminate unsuccessful searches (possibly) before we reach an element node or fall off the trie. As in the other compressed trie variants, the search is done by following a path from the root. Suppose we are searching the compressed trie of Figure 29.15 for an element with key 921-23-1234. Since the *skip* value for the root node is 0, we use the first digit 9 of the search key to determine which subtrie to move to. A move to node $A.child[9] = I$ is made. By examining the edge information (stored in node I), we determine that, in making the move from node A to node I, the digits 5 and 1 are skipped. Since these digits do not agree with the next two digits of the search key, the search terminates with the conclusion that the trie contains no element whose key equals the search key.

29.9.3.2 Inserting into a Compressed Trie with Edge Information

To insert an element with key 987-26-1615 into the compressed trie of Figure 29.15, we first search for an element with this key. The search terminates unsuccessfully when we move from node A to node I because of a mismatch between the skipped over digits and the corresponding digits of the search key. The first mismatch is at the first skipped over digit. Therefore, we insert a branch node L between nodes A and I. The *skip* value for this branch node is 0, and its *element* field is set to reference the element node for the newly inserted element. We must also change the *skip* value of I to 1. Figure 29.16 shows the resulting compressed trie.

Suppose we are to insert an element with key 958-36-4194 into the compressed trie of Figure 16. The search for an element with this key terminates when we move to node I because of a mismatch between the digits that are skipped over and the corresponding digits of the search key. A new branch node is inserted between nodes A and I and we get the compressed trie that is shown in Figure 29.17.

The time required to insert a d digit element into a radix r compressed trie with edge information is $O(r + d)$.

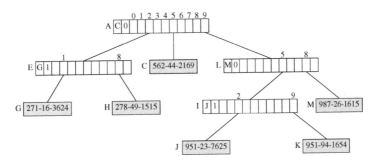

FIGURE 29.16 Compressed trie following the insertion of 987-26-1615 into the compressed trie of Figure 29.15.

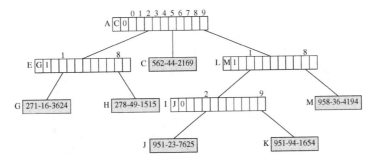

FIGURE 29.17 Compressed trie following the insertion of 958-36-4194 into the compressed trie of Figure 29.15.

29.9.3.3 Removing an Element from a Compressed Trie with Edge Information

This is similar to removal from a compressed trie with digit numbers except for the need to update the *element* fields of branch nodes whose *element* field references the removed element.

29.9.4 Space Required by a Compressed Trie

Since each branch node partitions the elements in its subtrie into two or more nonempty groups, an n element compressed trie has at most $n - 1$ branch nodes. Therefore, the space required by each of the compressed trie variants described by us is $O(nr)$, where r is the trie radix.

When compressed tries are represented as hash tables, we need an additional data structure to store the nonpointer fields of branch nodes. We may use an array (much like we use the array *information*) for this purpose.

29.10 Patricia

The data structure Patricia (**P**ractical **A**lgorithm **T**o **R**etrieve **I**nformation **C**oded **I**n **A**lphanumeric) is a compressed binary trie in which the branch and element nodes have been melded into a single node type. Consider the compressed binary trie of Figure 29.18. Circular nodes are branch nodes and rectangular nodes are element nodes. The number inside a branch node is its bit number field; the left child of a branch node corresponds to the case when the appropriate key bit is 0 and the right child to the case when this bit is 1. The melding of branch and element nodes is done by moving each element from its element node to an ancestor branch node. Since the number of branch nodes is one less than the number of element nodes, we introduce a header node and make the compressed binary trie the left subtree of the header. Pointers that originally went from a branch node to an element node now go from that branch node to the branch node into which the corresponding element has been melded. Figure 29.19 shows a possible result of melding the nodes of Figure 29.18. The number outside a node is its bit number values. The thick pointers are *backward* pointers that replace branch-node to element-node pointers in Figure 29.18. A backward pointer has the property that the bit number value at the start of the pointer is $\geq$ the bit number value at its end. For original branch-node to branch-node pointers (also called *downward* pointers), the bit number value at the pointer end is always greater than the bit number value at the pointer start.

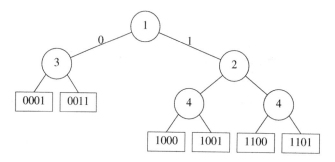

FIGURE 29.18   A compressed binary trie.

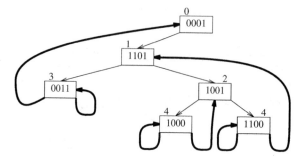

FIGURE 29.19   A Patricia instance that corresponds to Figure 29.18.

29.10.1 Searching

To search for an element with key *theKey*, we use the bits of *theKey* from left to right to move down the Patricia instance just as we would search a compressed binary trie. When a backward pointer is followed, we compare *theKey* with the key in the reached Patricia node. For example, to search for *theKey* = 1101, we start by moving into the left subtree of the header node. The pointer that is followed is a downward pointer (start bit number is 0 and end bit number is 1). We branch using bit 1 of *theKey*. Since this bit is 1, we follow the right child pointer. The start bit number for this pointer is 1 and the end bit number is 2. So, again, a downward pointer has been followed. The bit number for the reached node is 2. So, we use bit 2 of *theKey* to move further. Bit 2 is a 1. So, we use the right child pointer to get to the node that contains 1100. Again, a downward pointer was used. From this node, a move is made using bit 4 of *theKey*. This gets us to the node that contains 1101. This time, a backward pointer was followed (start bit of pointer is 4 and end bit is 1). When a backward pointer is followed, we compare *theKey* with the key in the reached node. In this case, the keys match and we have found the desired element. Notice that the same search path is taken when *theKey* = 1111. In this case, the final compare fails and we conclude that we have no element whose key is 1111.

29.10.2 Inserting an Element

We use an example to illustrate the insert algorithm. We start with an empty instance. Such an instance has no node; not even the header. For our example, we will consider keys that have 7 bits. For the first insert, we use the key 0000101. When inserting into an empty instance, we create a header node whose left child pointer points to the header node; the new element is inserted into the header; and the bit number field of the header set to 0. The configuration of Figure 29.20a results. Note that the right child field of the header node is not used.

The key for the next insert is 0000000. The search for this key terminates at the header. We compare the insert key and the header key and determine that the first bit at which they differ is bit 5. So, we create a new node with bit number field 5 and insert the new element into this node. Since bit 5 of the insert key is 0, the left child pointer of the new node points to the new node and the right child pointer to the header node (Figure 29.20b).

For the next insertion, assume that the key is 0000010. The search for this key terminates at the node with 0000000. Comparing the two keys, we determine that the first bit at which the two keys differ is bit 6. So, we create a new node with bit number field 6 and put the new element into this new node. The new node is inserted into the search path in such a way that bit number fields increase along this path. For our example, the new node is inserted as the left child of the node with 0000000. Since bit 6 of the insert key is 1, the right child pointer of the new node is a self pointer and the left child pointer points to the node with 0000000. Figure 29.20c shows the result.

The general strategy to insert an element other than the first one is given below. The key of the element to be inserted is *theKey*.

1. Search for *theKey*. Let *reachedKey* be the key in the node *endNode* where the search terminates.
2. Determine the leftmost bit position *lBitPos* at which *theKey* and *reachedKey* differ. *lBitPos* is well defined so long as one of the two keys isn't a prefix of the other.
3. Create a new node with bit number field *lBitPos*. Insert this node into the search path from the header to *endNode* so that bit numbers increase along this path. This insertion breaks a pointer from node *p* to node *q*. The pointer now goes from *p* to the new node.
4. If bit *lBitPos* of *theKey* is 1, the right child pointer of the new node becomes a self pointer (i.e., it points to the new node); otherwise, the left child pointer of the new node becomes a self pointer. The remaining child pointer of the new node points to *q*.

For our next insert, the insert key is 0001000. The search for this key terminates at the node with 0000000. We determine that the first bit at which the insert key and *reachedKey* = 0000000 differ is bit 4. We create a new node with bit number 4 and put the new element into this node. The new node is inserted on the search path so as to ensure that bit number fields increase along this path. So, the new node is inserted as the left child of the header node (Figure 29.18d). This breaks the pointer from the header to the node *q* with 0000000. Since bit 4 of the insert key is 1, the right child pointer of the new node is a self pointer and the left child pointer goes to node *q*.

We consider two more inserts. Consider inserting an element whose key is 0000100. The reached key is 0000101 (in the header). We see that the first bit at which the insert and reached keys differ is bit 7. So, we create a new node with bit number 7; the new element is put into the new node; the new node is inserted into the search path so as to ensure that bit numbers increase along this path (this requires the new node to be made a right child of the node with 0000000, breaking the child pointer from 0000000 to the header); for the broken pointer, *p* is the node with 0000000 and *q* is the header; the left child pointer of the new node is a self pointer (because bit 7 of the insert key is 0); and remaining child pointer (in this case the right child) of the new node points to *q* (see Figure 29.21a).

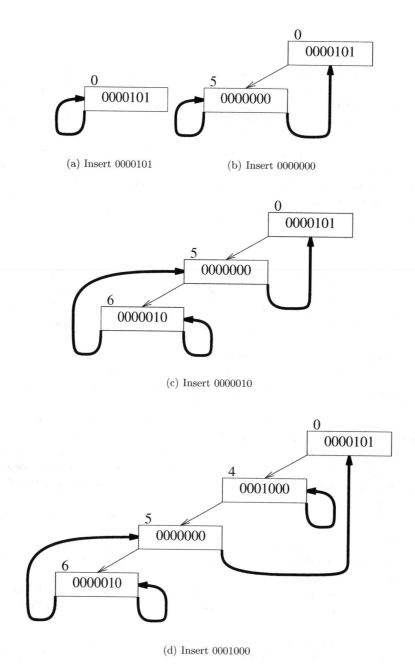

FIGURE 29.20   Insertion example.

For our final insert, the insert key is 0001010. A search for this key terminates at the node with 0000100. The first bit at which the insert and reached keys differ is bit 6. So, we create a new node with bit number 6; the new element is put into the new node; the new node is inserted into the search path so as to ensure that bit numbers increase along this path (this requires the new node to be made a right child of the node with 0001000; so, $p = q =$ node with 001000); the right child pointer of the new node is a self pointer (because bit 6 of the insert key is 1); and the remaining child (in this case the left child) of the new node is q (see Figure 29.21b).

29.10.3 Removing an Element

Let p be the node that contains the element that is to be removed. We consider two cases for p—(a) p has a self pointer and (b) p has no self pointer. When p has a self pointer and p is the header, the Patricia instance becomes empty following the element removal (Figure 29.20a). In this case, we simply dispose of the header. When p has a self pointer and p is not the header, we set

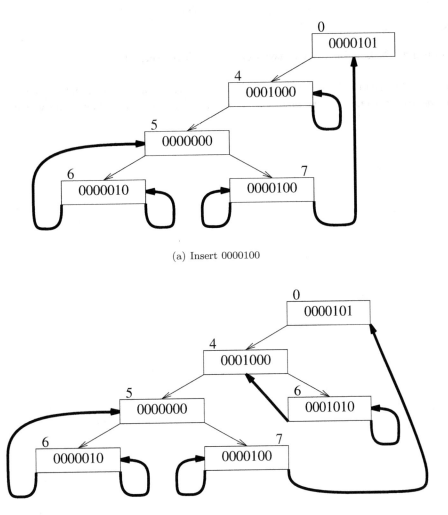

(a) Insert 0000100

(b) Insert 0001010

FIGURE 29.21   Insertion example.

the pointer from the parent of *p* to the value of *p*'s non-self pointer. Following this pointer change, the node *p* is disposed. For example, to remove the element with key 0000010 from Figure 29.21a, we change the left child pointer in the node with 0000000 to point to the node pointed at by *p*'s non-self pointer (i.e., the node with 0000000). This causes the left child pointer of 0000000 to become a self pointer. The node with 0000010 is then disposed.

For the second case, we first find the node *q* that has a backward pointer to node *p*. For example, when we are to remove 0001000 from Figure 29.21b, the node *q* is the node with 0001010. The element *qElement* in *q* (in our example, 0001010) is moved to node *p* and we proceed to delete node *q* instead of node *p*. Notice that node *q* is the node from which we reached node *p* in the search for the element that is to be removed. To delete node *q*, we first find the node *r* that has a back pointer to *q* (for our example *r* = *q*). The node *r* is found by using the key of *qElement*. Once *r* is found, the back pointer to *q* that is in *r* is changed from *q* to *p* to properly account for the fact that we have moved *qElement* to *p*. Finally, the downward pointer from the parent of *q* to *q* is changed to *q*'s child pointer that was used to locate *r*. In our example *p* is the parent of *q* and the right child pointer of *p* is changed from *q* to the right child of *q*, which itself was just changed to *p*.

We see that the time for each of the Patricia operations search, insert, and delete is $O(h)$, where *h* is the height of the Patricia instance. For more information on Patricia and general tries, see [2,3].

Acknowledgments

This work was supported, in part, by the National Science Foundation under grant CCR-9912395.

References

1. A. Andersson and S. Nillson, Improved behavior of tries by adaptive branching, *Information Processing Letters*, 46, 1993, 295–300.
2. E. Horowitz, S. Sahni, and D. Mehta, *Fundamentals of Data Structures in C++*, Computer Science Press, 1995.
3. D. Knuth, *The Art of Computer Programming: Sorting and Searching*, Second Edition, Addison-Wesley, 1998.

30

Suffix Trees and Suffix Arrays*

30.1	Basic Definitions and Properties	461
30.2	Linear Time Construction Algorithms	463
	Suffix Trees vs. Suffix Arrays • Linear Time Construction of Suffix Trees • Linear Time Construction of Suffix Arrays • Space Issues	
30.3	Applications	468
	Pattern Matching • Longest Common Substrings • Text Compression • String Containment • Suffix-Prefix Overlaps	
30.4	Lowest Common Ancestors	472
	Bender and Farach's *lca* algorithm	
30.5	Advanced Applications	473
	Suffix Links from Lowest Common Ancestors • Approximate Pattern Matching • Maximal Palindromes	

Srinivas Aluru
Georgia Institute of Technology

| Acknowledgments | 474 |
| References | 474 |

30.1 Basic Definitions and Properties

Suffix trees and suffix arrays are versatile data structures fundamental to string processing applications. Let s' denote a string over the alphabet Σ. Let $\$ \notin \Sigma$ be a unique termination character, and $s = s'\$$ be the string resulting from appending $\$$ to s'. We use the following notation: $|s|$ denotes the size of s, $s[i]$ denotes the ith character of s, and $s[i..j]$ denotes the substring $s[i]s[i+1]\ldots s[j]$. Let $suff_i = s[i]s[i+1]\ldots s[|s|]$ be the suffix of s starting at ith position.

The suffix tree of s, denoted $ST(s)$ or simply ST, is a compacted trie (See Chapter 28) of all suffixes of string s. Let $|s| = n$. It has the following properties:

1. The tree has n leaves, labeled $1 \ldots n$, one corresponding to each suffix of s.
2. Each internal node has at least 2 children.
3. Each edge in the tree is labeled with a substring of s.
4. The concatenation of edge labels from the root to the leaf labeled i is $suff_i$.
5. The labels of the edges connecting a node with its children start with different characters.

The paths from root to the suffixes labeled i and j coincide up to their longest common prefix, at which point they bifurcate. If a suffix of the string is a prefix of another longer suffix, the shorter suffix must end in an internal node instead of a leaf, as desired. It is to avoid this possibility that the unique termination character is added to the end of the string. Keeping this in mind, we use the notation $ST(s')$ to denote the suffix tree of the string obtained by appending $\$$ to s'.

As each internal node has at least 2 children, an n-leaf suffix tree has at most $n-1$ internal nodes. Because of property (5), the maximum number of children per node is bounded by $|\Sigma| + 1$. Except for the edge labels, the size of the tree is $O(n)$. In order to allow a linear space representation of the tree, each edge label is represented by a pair of integers denoting the starting and ending positions, respectively, of the substring describing the edge label. If the edge label corresponds to a repeat substring, the indices corresponding to any occurrence of the substring may be used. The suffix tree of the string *mississippi* is shown in Figure 30.1. For convenience of understanding, we show the actual edge labels.

The suffix array of $s = s'\$$, denoted $SA(s)$ or simply SA, is a lexicographically sorted array of all suffixes of s. Each suffix is represented by its starting position in s. $SA[i] = j$ iff $Suff_j$ is the ith lexicographically smallest suffix of s. The suffix array is often used in conjunction with an array termed Lcp array, containing the lengths of the longest common prefixes between every consecutive

---

* This chapter has been reprinted from first edition of this Handbook, without any content updates.

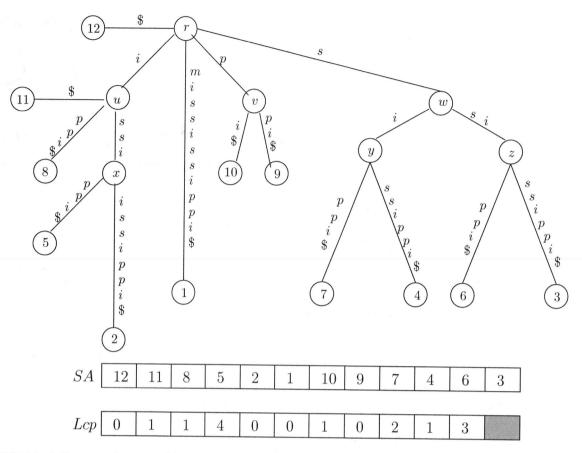

FIGURE 30.1 Suffix tree, suffix array and *Lcp* array of the string *mississippi*. The suffix links in the tree are given by $x \rightarrow z \rightarrow y \rightarrow u \rightarrow r$, $v \rightarrow r$, and $w \rightarrow r$.

pair of suffixes in *SA*. We use $lcp(\alpha, \beta)$ to denote the longest common prefix between strings α and β. We also use the term *lcp* as an abbreviation for the term *longest common prefix*. *Lcp*[*i*] contains the length of the *lcp* between $suff_{SA[i]}$ and $suff_{SA[i+1]}$, that is, $Lcp[i] = lcp(suff_{SA[i]}, suff_{SA[i+1]})$. As with suffix trees, we use the notation $SA(s')$ to denote the suffix array of the string obtained by appending $ to *s'*. The suffix and *Lcp* arrays of the string *mississippi* are shown in Figure 30.1.

Let *v* be a node in the suffix tree. Let *path-label*(*v*) denote the concatenation of edge labels along the path from root to node *v*. Let *string-depth*(*v*) denote the length of *path-label*(*v*). To differentiate this with the usual notion of depth, we use the term *tree-depth* of a node to denote the number of edges on the path from root to the node. Note that the length of the longest common prefix between two suffixes is the string depth of the lowest common ancestor of the leaf nodes corresponding to the suffixes. A repeat substring of string *S* is *right-maximal* if there are two occurrences of the substring that are succeeded by different characters in the string. The path label of each internal node in the suffix tree corresponds to a right-maximal repeat substring and vice versa.

Let *v* be an internal node in the suffix tree with path-label *cα* where *c* is a character and *α* is a (possibly empty) string. Therefore, *cα* is a right-maximal repeat, which also implies that *α* is also a right maximal repeat. Let *u* be the internal node with path label *α*. A pointer from node *v* to node *u* is called a *suffix link*; we denote this by *SL*(*v*) = *u*. Each suffix *suff_i* in the subtree of *v* shares the common prefix *cα*. The corresponding suffix *suff_{i+1}* with prefix *α* will be present in the subtree of *u*. The concatenation of edge labels along the path from *v* to leaf labeled *i*, and along the path from *u* to leaf labeled *i* + 1 will be the same. Similarly, each internal node in the subtree of *v* will have a corresponding internal node in the subtree of *u*. In this sense, the entire subtree under *v* is contained in the subtree under *u*.

Every internal node in the suffix tree other than the root has a suffix link from it. Let *v* be an internal node with *SL*(*v*) = *u*. Let *v'* be an ancestor of *v* other than the root and let *u'* = *SL*(*v'*). As *path-label*(*v'*) is a prefix of *path-label*(*v*), *path-label*(*u'*) is also a prefix of *path-label*(*u*). Thus, *u'* is an ancestor of *u*. Each proper ancestor of *v* except the root will have a suffix link to a distinct proper ancestor of *u*. It follows that *tree-depth*(*u*) ≥ *tree-depth*(*v*) − 1.

Suffix trees and suffix arrays can be generalized to multiple strings. The generalized suffix tree of a set of strings $S = \{s_1, s_2, \ldots, s_k\}$, denoted *GST*(*S*) or simply *GST*, is a compacted trie of all suffixes of each string in *S*. We assume that the unique termination character $ is appended to the end of each string. A leaf label now consists of a pair of integers (i, j), where

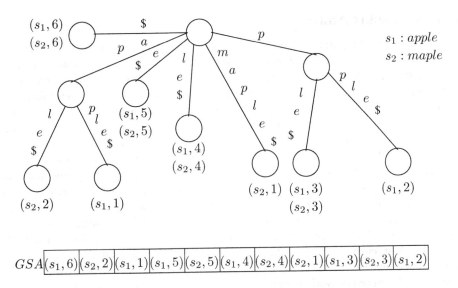

FIGURE 30.2 Generalized suffix tree and generalized suffix array of strings *apple* and *maple*.

i denotes the suffix is from string s_i and j denotes the starting position of the suffix in s_i. Similarly, an edge label in a *GST* is a substring of one of the strings. An edge label is represented by a triplet of integers (i, j, l), where i denotes the string number, and j and l denote the starting and ending positions of the substring in s_i. For convenience of understanding, we will continue to show the actual edge labels. Note that two strings may have identical suffixes. This is compensated by allowing leaves in the tree to have multiple labels. If a leaf is multiply labeled, each suffix should come from a different string. If N is the total number of characters (including the $ in each string) of all strings in $\mathcal{S}$, the *GST* has at most N leaf nodes and takes up $O(N)$ space. The generalized suffix array of $\mathcal{S}$, denoted $GSA(\mathcal{S})$ or simply GSA, is a lexicographically sorted array of all suffixes of each string in $\mathcal{S}$. Each suffix is represented by an integer pair (i, j) denoting suffix starting from position j in s_i. If suffixes from different strings are identical, they occupy consecutive positions in the GSA. For convenience, we make an exception for the suffix $ by listing it only once, though it occurs in each string. The *GST* and *GSA* of strings *apple* and *maple* are shown in Figure 30.2.

Suffix trees and suffix arrays can be constructed in time linear to the size of the input. Suffix trees are very useful in solving a plethora of string problems in optimal run-time bounds. Moreover, in many cases, the algorithms are very simple to design and understand. For example, consider the classic pattern matching problem of determining if a pattern P occurs in text T over a constant sized alphabet. Note that P occurs starting from position i in T iff P is a prefix of $suff_i$ in T. Thus, whether P occurs in T or not can be determined by checking if P matches an initial part of a path from root to a leaf in $ST(T)$. Traversing from the root matching characters in P, this can be determined in $O(|P|)$ time, independent of the size of T. As another application, consider the problem of finding a longest common substring of a pair of strings. Once the *GST* of the two strings is constructed, all that is needed is to identify an internal node with the largest string depth that contains at least one leaf from each string. These and many other applications are explored in great detail in subsequent sections. Suffix arrays are of interest because they require much less space than suffix trees, and can be used to solve many of the same problems. We first concentrate on linear time construction algorithms for suffix trees and suffix arrays. The reader interested in applications can safely skip to Section 30.3.

30.2 Linear Time Construction Algorithms

In this section, we explore linear time construction algorithms for suffix trees and suffix arrays. We also show how suffix trees and suffix arrays can be derived from each other in linear time. In suffix tree and suffix array construction algorithms, three different types of alphabets are considered: a constant or fixed size alphabet ($|\Sigma| = O(1)$), integer alphabet ($\Sigma = \{1, 2, \ldots, n\}$), and arbitrary alphabet. Suffix trees and suffix arrays can be constructed in linear time for both constant size and integer alphabets. The constant alphabet size case covers many interesting application areas, such as English text, or DNA or protein sequences in molecular biology. The integer alphabet case is interesting because a string of length n can have at most n distinct characters. Furthermore, some algorithms use a recursive technique that would generate and require operating on strings over integer alphabet, even when applied to strings over a fixed alphabet.

30.2.1 Suffix Trees vs. Suffix Arrays

We first show that the suffix array and *Lcp* array of a string can be obtained from its suffix tree in linear time. Define lexicographic ordering of the children of a node to be the order based on the first character of the edge labels connecting the node to its children. Define lexicographic depth first search to be a depth first search of the tree where the children of each node are visited in lexicographic order. The order in which the leaves of a suffix tree are visited in a lexicographic depth first search gives the suffix array of the corresponding string. In order to obtain *lcp* information, the string - depth of the current node during the search is remembered. This can be easily updated in $O(1)$ time per edge as the search progresses. The length of the *lcp* between two consecutive suffixes is given by the smallest string - depth of a node visited between the two suffixes.

Given the suffix array and the *Lcp* array of a string s ($|s\$| = n$), its suffix tree can be constructed in $O(n)$ time. This is done by starting with a partial suffix tree for the lexicographically smallest suffix, and repeatedly inserting subsequent suffixes in the suffix array into the tree until the suffix tree is complete. Let T_i denote the compacted trie of the first i suffixes in lexicographic order. The first tree T_1 consists of a single leaf labeled $SA[1]$ connected to the root with an edge labeled $suff_{SA[1]} = \$$.

To insert $SA[i+1]$ into T_i, start with the most recently inserted leaf $SA[i]$ and walk up ($|suff_{SA[i]}| - |lcp$ ($suff_{SA[i]}$, $suff_{SA[i+1]})|) = ((n - SA[i] + 1) - Lcp[i])$ characters along the path to the root. This walk can be done in $O(1)$ time per edge by calculating the lengths of the respective edge labels. If the walk does not end at an internal node, create an internal node. Create a new leaf labeled $SA[i+1]$ and connect it to this internal node with an edge. Set the label on this edge to $s[SA[i+1] + Lcp[i]..n]$. This creates the tree T_{i+1}. The procedure works because $suff_{SA[i+1]}$ shares a longer prefix with $suff_{SA[i]}$ than any other suffix inserted so far. To see that the entire algorithm runs in $O(n)$ time, note that inserting a new suffix into T_i requires walking up the rightmost path in T_i. Each edge that is traversed ceases to be on the rightmost path in T_{i+1}, and thus is never traversed again. An edge in an intermediate tree T_i corresponds to a path in the suffix tree ST. When a new internal node is created along an edge in an intermediate tree, the edge is split into two edges, and the edge below the newly created internal node corresponds to an edge in the suffix tree. Once again, this edge ceases to be on the rightmost path and is never traversed again. The cost of creating an edge in an intermediate tree can be charged to the lowest edge on the corresponding path in the suffix tree. As each edge is charged once for creating and once for traversing, the total run - time of this procedure is $O(n)$.

Finally, the *Lcp* array itself can be constructed from the suffix array and the string in linear time [1]. Let R be an array of size n such that $R[i]$ contains the position in SA of $suff_i$. R can be constructed by a linear scan of SA in $O(n)$ time. The *Lcp* array is computed in n iterations. In iteration i of the algorithm, the longest common prefix between $suff_i$ and its respective right neighbor in the suffix array is computed. The array R facilitates locating an arbitrary suffix $suff_i$ and finding its right neighbor in the suffix array in constant time. Initially, the length of the longest common prefix between $suff_1$ and its suffix array neighbor is computed directly and recorded. Let $suff_j$ be the right neighbor of $suff_i$ in SA. Let l be the length of the longest common prefix between them. Suppose $l \geq 1$. As $suff_j$ is lexicographically greater than $suff_i$ and $s[i] = s[j]$, $suff_{j+1}$ is lexicographically greater than $suff_{i+1}$. The length of the longest common prefix between them is $l - 1$. It follows that the length of the longest common prefix between $suff_{i+1}$ and its right neighbor in the suffix array is $\geq l - 1$. To determine its correct length, the comparisons need only start from the lth characters of the suffixes.

To prove that the run time of the above algorithm is linear, charge a comparison between the rth character in suffix $suff_i$ and the corresponding character in its right neighbor suffix in SA to the position in the string of the rth character of $suff_i$, that is, $i + r - 1$. A comparison made in an iteration is termed successful if the characters compared are identical, contributing to the longest common prefix being computed. Because there is one failed comparison in each iteration, the total number of failed comparisons is $O(n)$. As for successful comparisons, each position in the string is charged only once for a successful comparison. Thus, the total number of comparisons over all iterations is linear in n.

In light of the above discussion, a suffix tree and a suffix array can be constructed from each other in linear time. Thus, a linear time construction algorithm for one can be used to construct the other in linear time. In the following subsections, we explore such algorithms. Each algorithm is interesting in its own right, and exploits interesting properties that could be useful in designing algorithms using suffix trees and suffix arrays.

30.2.2 Linear Time Construction of Suffix Trees

Let s be a string of length n including the termination character $\$$. Suffix tree construction algorithms start with an empty tree and iteratively insert suffixes while maintaining the property that each intermediate tree represents a compacted trie of the suffixes inserted so far. When all the suffixes are inserted, the resulting tree will be the suffix tree. Suffix links are typically used to speedup the insertion of suffixes. While the algorithms are identified by the names of their respective inventors, the exposition presented does not necessarily follow the original algorithms and we take the liberty to comprehensively present the material in a way we feel contributes to ease of understanding.

30.2.2.1 McCreight's Algorithm

McCreight's algorithm inserts suffixes in the order $suff_1, suff_2, \ldots, suff_n$. Let T_i denote the compacted trie after $suff_i$ is inserted. T_1 is the tree consisting of a single leaf labeled 1 that is connected to the root by an edge with label $s[1..n]$. In iteration i of the algorithm, $suff_i$ is inserted into tree T_{i-1} to form tree T_i. An easy way to do this is by starting from the root and following the unique path matching characters in $suff_i$ one by one until no more matches are possible. If the traversal does not end at an internal node, create an internal node there. Then, attach a leaf labeled i to this internal node and use the unmatched portion of $suff_i$ for the edge label. The run-time for inserting $suff_i$ is proportional to $|suff_i| = n - i + 1$. The total run-time of the algorithm is $\Sigma_{i=1}^{n}(n - i + 1) = O(n^2)$.

In order to achieve an $O(n)$ run-time, suffix links are used to significantly speedup the insertion of a new suffix. Suffix links are useful in the following way—Suppose we are inserting $suff_i$ in T_{i-1} and let v be an internal node in T_{i-1} on the path from root to leaf labeled $(i-1)$. Then, $path\text{-}label(v) = c\alpha$ is a prefix of $suff_{i-1}$. Since v is an internal node, there must be another suffix $suff_j$ $(j < i - 1)$ that also has $c\alpha$ as prefix. Because $suff_{j+1}$ is previously inserted, there is already a path from the root in T_{i-1} labeled α. To insert $suff_i$ faster, if the end of path labeled α is quickly found, comparison of characters in $suff_i$ can start beyond the prefix α. This is where suffix links will be useful. The algorithm must also construct suffix links prior to using them.

Lemma 30.1

Let v be an internal node in $ST(s)$ that is created in iteration $i - 1$. Let $path\text{-}label(v) = c\alpha$, where c is a character and α is a (possibly empty) string. Then, either there exists an internal node u with $path\text{-}label(u) = \alpha$ or it will be created in iteration i.

Proof. As v is created when inserting $suff_{i-1}$ in T_{i-2}, there exists another suffix $suff_j$ $(j < i - 1)$ such that $lcp(suff_{i-1}, suff_j) = c\alpha$. It follows that $lcp(suff_i, suff_{j+1}) = \alpha$. The tree T_i already contains $suff_{j+1}$. When $suff_i$ is inserted during iteration i, internal node u with path-label α is created if it does not already exist.   ∎

The above lemma establishes that the suffix link of a newly created internal node can be established in the next iteration.

The following procedure is used when inserting $suff_i$ in T_{i-1}. Let v be the internal node to which $suff_{i-1}$ is attached as a leaf. If v is the root, insert $suff_i$ using character comparisons starting with the first character of $suff_i$. Otherwise, let $path\text{-}label(v) = c\alpha$. If v has a suffix link from it, follow it to internal node u with path-label α. This allows skipping the comparison of the first $|\alpha|$ characters of $suff_i$. If v is newly created in iteration $i - 1$, it would not have a suffix link yet. In that case, walk up to parent v' of v. Let β denote the label of the edge connecting v' and v. Let $u' = SL(v')$ unless v' is the root, in which case let u' be the root itself. It follows that $path\text{-}label(u')$ is a prefix of $suff_i$. Furthermore, it is guaranteed that there is a path below u' that matches the next $|\beta|$ characters of $suff_i$. Traverse $|\beta|$ characters along this path and either find an internal node u or insert an internal node u if one does not already exist. In either case, set $SL(v) = u$. Continue by starting character comparisons skipping the first $|\alpha|$ characters of $suff_i$.

The above procedure requires two different types of traversals—one in which it is known that there exists a path below that matches the next $|\beta|$ characters of $suff_i$ (type I), and the other in which it is unknown how many subsequent characters of $suff_i$ match a path below (type II). In the latter case, the comparison must proceed character by character until a mismatch occurs. In the former case, however, the traversal can be done by spending only $O(1)$ time per edge irrespective of the length of the edge label. At an internal node during such a traversal, the decision of which edge to follow next is made by comparing the next character of $suff_i$ with the first characters of the edge labels connecting the node to its children. However, once the edge is selected, the entire label or the remaining length of β must match, whichever is shorter. Thus, the traversal can be done in constant time per edge, and if the traversal stops within an edge label, the stopping position can also be determined in constant time.

The insertion procedure during iteration i can now be described as follows: Start with the internal node v to which $suff_{i-1}$ is attached as a leaf. If v has a suffix link, follow it and perform a type II traversal to insert $suff_i$. Otherwise, walk up to v's parent, take the suffix link from it unless it is the root, and perform a type I traversal to either find or create the node u which will be linked from v by a suffix link. Continue with a type II traversal below u to insert $suff_i$.

Lemma 30.2

The total time spent in type I traversals over all iterations is $O(n)$.

Proof. A type I traversal is performed by walking down along a path from root to a leaf in $O(1)$ time per edge. Each iteration consists of walking up at most one edge, following a suffix link, and then performing downward traversals (either type II or both type I and type II). Recall that if $SL(v) = u$, then $tree\text{-}depth(u) \geq tree\text{-}depth(v) - 1$. Thus, following a suffix link may reduce the depth in the tree by at most one. It follows that the operations that may cause moving to a higher level in the tree cause a decrease in depth of at most 2 per iteration. As both type I and type II traversals increase the depth in the tree and there are at most n levels in ST, the total number of edges traversed by type I traversals over all the iterations is bounded by $3n$. ■

Lemma 30.3

The total time spent in type II traversals over all iterations is $O(n)$.

Proof. In a type II traversal, a suffix of the string $suff_i$ is matched along a path in T_{i-1} until there is a mismatch. When a mismatch occurs, an internal node is created if there does not exist one already. Then, the remaining part of $suff_i$ becomes the edge label connecting leaf labeled i to the internal node. Charge each successful comparison of a character in $suff_i$ to the corresponding character in the original string s. Note that a character that is charged with a successful comparison is never charged again as part of a type II traversal. Thus, the total time spent in type II traversals is $O(n)$. ■

The above lemmas prove that the total run-time of McCreight's algorithm is $O(n)$. McCreight's algorithm is suitable for constant sized alphabets. The dependence of the run-time and space for storing suffix trees on the size of the alphabet $|\Sigma|$ is as follows: A simple way to allocate space for internal nodes in a suffix tree is to allocate $|\Sigma| + 1$ pointers for children, one for each distinct character with which an edge label may begin. With this approach, the edge label beginning with a given character, or whether an edge label exists with a given character, can be determined in $O(\log |\Sigma|)$ time. However, as all $|\Sigma| + 1$ pointers are kept irrespective of how many children actually exist, the total space is $O(|\Sigma| n)$. If the tree is stored such that each internal node points only to its leftmost child and each node also points to its next sibling, if any, the space can be reduced to $O(n)$, irrespective of $|\Sigma|$. With this, searching for a child connected by an edge label with the appropriate character takes $O(|\Sigma|)$ time. Thus, McCreight's algorithm can be run in $O(n \log |\Sigma|)$ time using $O(n|\Sigma|)$ space, or in $O(n|\Sigma|)$ time using $O(n)$ space.

30.2.2.2 Generalized Suffix Trees

McCreight's algorithm can be easily adapted to build the generalized suffix tree for a set $S = \{s_1, s_2, \ldots, s_k\}$ of strings of total length N in $O(N)$ time. A simple way to do this is to construct the string $S = s_1 \$_1 s_2 \$_2 \ldots s_k \$_k$, where each $\$_i$ is a unique string termination character that does not occur in any string in S. Using McCreight's algorithm, $ST(S)$ can be computed in $O(N)$ time. This differs from $GST(S)$ in the following way: Consider a suffix $suff_j$ of string s_i in $GST(S)$. The corresponding suffix in $ST(S)$ is $s_i[j..|s_i|]\$_i s_{i+1} \$_{i+1} \ldots s_k \$_k$. Let v be the last internal node on the path from root to leaf representing this suffix in $ST(S)$. As each $\$_i$ is unique and $path\text{-}label(v)$ must be a common prefix of at least two suffixes in S, $path\text{-}label(v)$ must be a prefix of $s_i[j..|s_i|]$. Thus, by simply shortening the edge label below v to terminate at the end of the string s_i and attaching a common termination character $\$$ to it, the corresponding suffix in $GST(S)$ can be generated in $O(1)$ time. Additionally, all suffixes in $ST(S)$ that start with some $\$_i$ should be removed and replaced by a single suffix $\$$ in $GST(S)$. Note that the suffixes to be removed are all directly connected to the root in $ST(S)$, allowing easy $O(1)$ time removal per suffix. Thus, $GST(S)$ can be derived from $ST(S)$ in $O(N)$ time.

Instead of first constructing $ST(S)$ and shortening edge labels of edges connecting to leaves to construct $GST(S)$, the process can be integrated into the tree construction itself to directly compute $GST(S)$. When inserting the suffix of a string, directly set the edge label connecting to the newly created leaf to terminate at the end of the string, appended by $\$$. As each suffix that begins with $\$_i$ in $ST(S)$ is directly attached to the root, execution of McCreight's algorithm on S will always result in a downward traversal starting from the root when a suffix starting from the first character of a string is being inserted. Thus, we can simply start with an empty tree, insert all the suffixes of one string using McCreight's algorithm, insert all the suffixes of the next string, and continue this procedure until all strings are inserted. To insert the first suffix of a string, start by matching the unique path in the current tree that matches with a prefix of the string until no more matches are possible, and insert the suffix by branching at this point. To insert the remaining suffixes, continue as described in constructing the tree for one string.

This procedure immediately gives an algorithm to maintain the generalized suffix tree of a set of strings in the presence of insertions and deletions of strings. Insertion of a string is the same as executing McCreight's algorithm on the current tree, and takes time proportional to the length of the string being inserted. To delete a string, we must locate the leaves corresponding to all the suffixes of the string. By mimicking the process of inserting the string in GST using McCreight's algorithm, all the corresponding leaf nodes can be reached in time linear in the size of the string to be deleted. To delete a suffix, examine the corresponding leaf. If it is multiply labeled, it is enough to remove the label corresponding to the suffix. It it has only one label, the leaf and edge leading to it must be deleted. If the parent of the leaf is left with only one child after deletion, the parent and its two incident edges are deleted by connecting the surviving child directly to its grandparent with an edge labeled with the concatenation of the labels of the two edges deleted. As the adjustment at each leaf takes $O(1)$ time, the string can be deleted in time proportional to its length.

Suffix trees were invented by Weiner [2], who also presented the first linear time algorithm to construct them for a constant sized alphabet. McCreight's algorithm is a more space - economical linear time construction algorithm [3]. A linear time on - line construction algorithm for suffix trees is invented by Ukkonen [4]. In fact, our presentation of McCreight's algorithm also draws from ideas developed by Ukkonen. A unified view of these three suffix tree construction algorithms is studied by Giegerich and Kurtz [5]. Farach [6] presented the first linear time algorithm for strings over integer alphabets. The algorithm recursively constructs suffix trees for all odd and all even suffixes, respectively, and uses a clever strategy for merging them. The complexity of suffix tree construction algorithms for various types of alphabets is explored in [7].

30.2.3 Linear Time Construction of Suffix Arrays

Suffix arrays were proposed by Manber and Myers [8] as a space - efficient alternative to suffix trees. While suffix arrays can be deduced from suffix trees, which immediately implies any of the linear time suffix tree construction algorithms can be used for suffix arrays, it would not achieve the purpose of economy of space. Until recently, the fastest known direct construction algorithms for suffix arrays all required $O(n \log n)$ time, leaving a frustrating gap between asymptotically faster construction algorithms for suffix trees, and asymptotically slower construction algorithms for suffix arrays, despite the fact that suffix trees contain all the information in suffix arrays. This gap is successfully closed by a number of researchers in 2003, including Käräkkanen and Sanders [9], Kim et al. [10], and Ko and Aluru [11]. All three algorithms work for the case of integer alphabet. Given the simplicity and/or space efficiency of some of these algorithms, it is now preferable to construct suffix trees via the construction of suffix arrays.

30.2.3.1 Käräkkanen and Sanders' Algorithm

Käräkkanen and Sanders' algorithm is the simplest and most elegant algorithm to date to construct suffix arrays , and by implication suffix trees, in linear time. The algorithm also works for the case of an integer alphabet. Let s be a string of length n over the alphabet $\Sigma = \{1, 2, \ldots, n\}$. For convenience, assume n is a multiple of three and $s[n+1] = s[n+2] = 0$. The algorithm has the following steps:

1. Recursively sort the $(2/3)n$ suffixes $suff_i$ with $i \bmod 3 \neq 0$.
2. Sort the $(1/3)n$ suffixes $suff_i$ with $i \bmod 3 = 0$ using the result of step (1).
3. Merge the two sorted arrays.

To execute step (1), first perform a radix sort of the $(2/3)n$ triples $(s[i], s[i+1], s[i+2])$ for each $i \bmod 3 \neq 0$ and associate with each distinct triple its rank $\in \{1, 2, \ldots, (2/3)n\}$ in sorted order. If all triples are distinct, the suffixes are already sorted. Otherwise, let $suff'_i$ denote the string obtained by taking $suff_i$ and replacing each consecutive triplet with its corresponding rank. Create a new string s' by concatenating $suff'_1$ with $suff'_2$. Note that all $suff_i$ with $i \bmod 3 = 1$ ($i \bmod 3 = 2$, respectively) are suffixes of $suff'_1$ ($suff'_2$, respectively). A lexicographic comparison of two suffixes in s' never crosses the boundary between $suff'_1$ and $suff'_2$ because the corresponding suffixes in the original string can be lexicographically distinguished. Thus, sorting s' recursively gives the sorted order of $suff_i$ with $i \bmod 3 \neq 0$.

Step (2) can be accomplished by performing a radix sort on tuples $(s[i], \text{rank}(suff_{i+1}))$ for all $i \bmod 3 = 0$, where $\text{rank}(suff_{i+1})$ denotes the rank of $suff_{i+1}$ in sorted order obtained in step (1).

Merging of the sorted arrays created in steps (1) and (2) is done in linear time, aided by the fact that the lexicographic order of a pair of suffixes, one from each array, can be determined in constant time. To compare $suff_i$ ($i \bmod 3 = 1$) with $suff_j$ ($i \bmod 3 = 0$), compare $s[i]$ with $s[j]$. If they are unequal, the answer is clear. If they are identical, the ranks of $suff_{i+1}$ and $suff_{j+1}$ in the sorted order obtained in step (1) determines the answer. To compare $suff_i$ ($i \bmod 3 = 2$) with $suff_j$ ($i \bmod 3 = 0$), compare the first two characters of the two suffixes. If they are both identical, the ranks of $suff_{i+2}$ and $suff_{j+2}$ in the sorted order obtained in step (1) determines the answer.

The run - time of this algorithm is given by the recurrence $T(n) = T(\lceil 2n/3 \rceil) + O(n)$, which results in $O(n)$ run - time. Note that the 2/3:1/3 split is designed to make the merging step easy. A 1/2:1/2 split does not allow easy merging because when comparing two suffixes for merging, no matter how many characters are compared, the remaining suffixes will not fall in the same sorted array, where ranking determines the result without need for further comparisons. Kim et al.'s linear time suffix array construction algorithm is based on a 1/2:1/2 split, and the merging phase is handled in a clever way so as to run in linear time. This is much like Farach's algorithm for constructing suffix trees [6] by constructing suffix trees for even and odd positions separately and merging them. Both the above linear time suffix array construction algorithms partition the suffixes based on their starting positions in the string. A completely different way of partitioning suffixes based on the lexicographic ordering of a suffix with its right neighboring suffix in the string is used by Ko and Aluru to derive a linear time algorithm [11]. This reduces solving a problem of size n to that of solving a problem of size no more than $\lceil n/2 \rceil$, while eliminating the complex merging step. The algorithm can be made to run in only $2n$ words plus $1.25n$ bits for strings over constant alphabet. Algorithmically, Käräkkanen and Sanders' algorithm is akin to mergesort and Ko and Aluru's algorithm is akin to quicksort. Algorithms for constructing suffix arrays in external memory are investigated by Crauser and Ferragina [12].

It may be more space efficient to construct a suffix tree by first constructing the corresponding suffix array, deriving the Lcp array from it, and using both to construct the suffix tree. For example, while all direct linear time suffix tree construction algorithms depend on constructing and using suffix links, these are completely avoided in the indirect approach. Furthermore, the resulting algorithms have an alphabet independent run - time of $O(n)$ while using only the $O(n)$ space representation of suffix trees. This should be contrasted with the $O(|\Sigma|n)$ run - time of either McCreight's or Ukkonen's algorithms.

30.2.4 Space Issues

Suffix trees and suffix arrays are space efficient in an asymptotic sense because the memory required grows linearly with input size. However, the actual space usage is of significant concern, especially for very large strings. For example, the human genome can be represented as a large string over the alphabet $\Sigma = \{A, C, G, T\}$ of length over 3×10^9. Because of linear dependence of space on the length of the string, the exact space requirement is easily characterized by specifying it in terms of the number of bytes per character. Depending on the number of bytes per character required, a data structure for the human genome may fit in main memory, may need a moderate sized disk, or might need a large amount of secondary storage. This has significant influence on the run - time of an application as access to secondary storage is considerably slower. It may also become impossible to run an application for large data sizes unless careful attention is paid to space efficiency.

Consider a naive implementation of suffix trees. For a string of length n, the tree has n leaves, at most $n - 1$ internal nodes, and at most $2n - 2$ edges. For simplicity, count the space required for each integer or a pointer to be one word, equal to 4 bytes on most current computers. For each leaf node, we may store a pointer to its parent, and store the starting index of the suffix represented by the leaf, for $2n$ words of storage. Storage for each internal node may consist of 4 pointers, one each for parent, leftmost child, right sibling and suffix link, respectively. This will require approximately $4n$ words of storage. Each edge label consists of a pair of integers, for a total of at most $4n$ words of storage. Putting this all together, a naive implementation of suffix trees takes $10n$ words or $40n$ bytes of storage.

Several techniques can be used to considerably reduce the naive space requirement of 40 bytes per character. Many applications of interest do not need to use suffix links. Similarly, a pointer to the parent may not be required for applications that use traversals down from the root. Even otherwise, note that a depth first search traversal of the suffix tree starting from the root can be conducted even in the absence of parent links, and this can be utilized in applications where a bottom - up traversal is needed. Another technique is to store the internal nodes of the tree in an array in the order of their first occurrence in a depth first search traversal. With this, the leftmost child of an internal node is found right next to it in the array, which removes the need to store a child pointer. Instead of storing the starting and ending positions of a substring corresponding to an edge label, an edge label can be stored with the starting position and length of the substring. The advantage of doing so is that the length of the edge label is likely to be small. Hence, one byte can be used to store edge labels with lengths <255 and the number 255 can be used to denote edge labels with length at least 255. The actual values of such labels can be stored in an exceptions list, which is expected to be fairly small. Using several such techniques, the space required per character can be roughly cut in half to about 20 bytes [13].

A suffix array can be stored in just one word per character, or 4 bytes. Most applications using suffix arrays also need the Lcp array. Similar to the technique employed in storing edge labels on suffix trees, the entries in Lcp array can also be stored using one byte, with exceptions handled using an ordered exceptions list. Provided most of the lcp values fit in a byte, we only need 5 bytes per character, significantly smaller than what is required for suffix trees. Further space reduction can be achieved by the use of compressed suffix trees and suffix arrays and other data structures [14,15]. However, this often comes at the expense of increased run - time complexity.

30.3 Applications

In this section, we present algorithms for several string problems using suffix trees and suffix arrays. While the same run - time bounds can be achieved for many interesting applications with either a suffix tree or a suffix array, there are others which involve a space vs. time trade off. Even in cases where the same run - time bound can be achieved, it is often easier to design the algorithm first for a suffix tree, and then think if the implementation can be done using a suffix array. For this reason, we largely concentrate on suffix trees. The reader interested in reading more on applications of suffix arrays is referred to [16,17].

30.3.1 Pattern Matching

Given a pattern P and a text T, the pattern matching problem is to find all occurrences of P in T. Let $|P| = m$ and $|T| = n$. Typically, $n \gg m$. Moreover, T remains fixed in many applications and the query is repeated for many different patterns. For example,

T could be a text document and P could represent a word search. Or, T could be an entire database of DNA sequences and P denote a substring of a query sequence for homology (similarity) search. Thus, it is beneficial to preprocess the text T so that queries can be answered as efficiently as possible.

30.3.1.1 Pattern Matching using Suffix Trees

The pattern matching problem can be solved in optimal $O(m + k)$ time using $ST(T)$, where k is the number of occurrences of P in T. Suppose P occurs in T starting from position i. Then, P is a prefix of $suff_i$ in T. It follows that P matches the path from root to leaf labeled i in ST. This property results in the following simple algorithm: Start from the root of ST and follow the path matching characters in P, until P is completely matched or a mismatch occurs. If P is not fully matched, it does not occur in T. Otherwise, each leaf in the subtree below the matching position gives an occurrence of P. The positions can be enumerated by traversing the subtree in time proportional to the size of the subtree. As the number of leaves in the subtree is k, this takes $O(k)$ time. If only one occurrence is of interest, the suffix tree can be preprocessed in $O(n)$ time such that each internal node contains the label of one of the leaves in its subtree. Thus, the problem of whether P occurs in T or the problem of finding one occurrence can be answered in $O(m)$ time.

30.3.1.2 Pattern Matching using Suffix Arrays

Consider the problem of pattern matching when the suffix array of the text, $SA(T)$, is available. As before, we need to find all the suffixes that have P as a prefix. As SA is a lexicographically sorted order of the suffixes of T, all such suffixes will appear in consecutive positions in it. The sorted order in SA allows easy identification of these suffixes using binary search. Using a binary search, find the smallest index i in SA such that $suff_{SA[i]}$ contains P as a prefix, or determine that no such suffix is present. If no suffix is found, P does not occur in T. Otherwise, find the largest index $j(\geq i)$ such that $suff_{SA[j]}$ contains P as a prefix. All the elements in the range $SA[i..j]$ give the starting positions of the occurrences of P in T.

A binary search in SA takes $O(\log n)$ comparisons. In each comparison, P is compared with a suffix to determine their lexicographic order. This requires comparing at most $|P| = m$ characters. Thus, the run-time of this algorithm is $O(m \log n)$. Note that while this run-time is inferior to the run-time using suffix trees, the space required by this algorithm is only n words for SA apart from the space required to store the string. Note that the Lcp array is not required. Assuming 4 bytes per suffix array entry and one byte per character in the string, the total space required is only $5n$ bytes.

The run-time can be improved to $O(m + \log n)$, by using slightly more space and keeping track of appropriate lcp information. Consider an iteration of the binary search. Let $SA[L..R]$ denote the range in the suffix array where the binary search is focused. To begin with, $L = 1$ and $R = n$. At the beginning of an iteration, the pattern P is known to be lexicographically greater than or equal to $suff_{SA[L]}$ and lexicographically smaller than or equal to $suff_{SA[R]}$. Let $M = \lceil L + R/2 \rceil$. During the iteration, a lexicographic comparison between P and $suff_{SA[M]}$ is made. Depending on the result, the search range is narrowed to either $SA[L..M]$ or $SA[M..R]$. Assume that $l = |lcp(P, suff_{SA[L]})|$ and $r = |lcp(P, suff_{SA[R]})|$ are known at the beginning of the iteration. Also, assume that $|lcp(suff_{SA[L]}, suff_{SA[M]})|$ and $|lcp(suff_{SA[M]}, suff_{SA[R]})|$ are known. From these values, we wish to determine $|lcp(P, suff_{SA[M]})|$ for use in next iteration, and consequently determine the relative lexicographic order between P and $suff_{SA[M]}$. As SA is a lexicographically sorted array, P and $suff_{SA[M]}$ must agree on at least $min(l, r)$ characters. If l and r are equal, then comparison between P and $suff_{SA[M]}$ is done by starting from the $(l + 1)$th character. If l and r are unequal, consider the case when $l > r$.

Case I: $l < |lcp(suff_{SA[L]}, suff_{SA[M]})|$. In this case, P is lexicographically greater than $suff_{SA[M]}$ and $|lcp(P, suff_{SA[M]})| = |lcp(P, suff_{SA[L]})|$. Change the search range to $SA[M..R]$. No character comparisons are needed.

Case II: $l > |lcp(suff_{SA[L]}, suff_{SA[M]})|$. In this case, P is lexicographically smaller than $suff_{SA[M]}$ and $|lcp(P, suff_{SA[M]})| = |lcp(Suff_{SA[L]}, suff_{SA[M]})|$. Change the search range to $SA[L..M]$. Again, no character comparisons are needed.

Case III: $l = |lcp(suff_{SA[L]}, suff_{SA[M]})|$. In this case, P agrees with the first l characters of $suff_{SA[M]}$. Compare P and $suff_{SA[M]}$ starting from $(l + 1)$th character to determine $|lcp(P, suff_{SA[M]})|$ and the relative lexicographic order of P and $suff_{SA[M]}$.

Similarly, the case when $r > l$ can be handled such that comparisons between P and $suff_{SA[M]}$, if at all needed, start from $(r + 1)$th character. To start the execution of the algorithm, $lcp(P, suff_{SA[1]})$ and $lcp(P, suff_{SA[n]})$ are computed directly using at most $2|P|$ character comparisons. This ensures $|lcp(P, suff_{SA[L]})|$ and $|lcp(P, suff_{SA[R]})|$ are known at the beginning of the first iteration. This property is maintained for each iteration as L or R is shifted to M but $|lcp(P, suff_{SA[M]})|$ is computed. For now, assume that the required $|lcp(suff_{SA[L]}, suff_{SA[M]})|$ and $|lcp(suff_{SA[R]}, suff_{SA[M]})|$ values are available.

Lemma 30.4

The total number of character comparisons made by the algorithm is $O(m + \log n)$.

Proof. The algorithm makes at most $2m$ comparisons in determining the longest common prefixes between P and $suff_{SA[1]}$ and between P and $suff_{SA[n]}$. Classify the comparisons made in each iteration to determine the longest common prefix between P and $suff_{SA[M]}$ into *successful* and *failed* comparisons. A comparison is considered successful if it contributes the longest common prefix. There is at most one failed comparison per iteration, for a total of at most $\log n$ such comparisons over all iterations. As for successful comparisons, note that the comparisons start with $(max(l,r)+1)$th character of P, and each successful comparison increases the value of $max(l,r)$ for next iteration. Thus, each character of P is involved only once in a successful comparison. The total number of character comparisons is at most $3m + \log n = O(m + \log n)$. ∎

It remains to be described how the $|lcp(suff_{SA[L]}, suff_{SA[M]})|$ and $|lcp(suff_{SA[R]}, suff_{SA[M]})|$ values required in each iteration are computed. Suppose the Lcp array of T is known. For any $1 \leq i < j \leq n$

$$|lcp(suff_{SA[i]}, suff_{SA[j]})| = \min_{k=i}^{j-1} Lcp[k]$$

The lcp of two suffixes can be computed in time proportional to the distance between them in the suffix array. In order to find the lcp values required by the algorithm in constant time, consider the binary tree corresponding to all possible search intervals used by any execution of the binary search algorithm. The root of the tree denotes the interval $[1..n]$. If $[i..j]$ $(j - i \geq 2)$ is the interval at an internal node of the tree, its left child is given by $[i..\lceil i+j/2 \rceil]$ and its right child is given by $[\lceil i+j/2 \rceil..j]$. The execution of the binary search tree algorithm can be visualized as traversing a path in the binary tree from root to a leaf. If lcp value for each interval in the tree is precomputed and recorded, any required lcp value during the execution of the algorithm can be retrieved in constant time. The leaf level in the binary tree consists of intervals of the type $[i..i+1]$. The lcp values for these $n-1$ intervals is already given by the Lcp array. The lcp value corresponding to an interval at an internal node is given by the smaller of the lcp values at the children. Using a bottom-up traversal, the lcp values can be computed in $O(n)$ time. In addition to the Lcp array, $n-2$ additional lcp values are required to be stored. Assuming approximately 1 byte per lcp value, the algorithm requires approximately $2n$ bytes of additional space. As usual, lcp values larger than or equal to 255, if any, are stored in an exceptions list and the size of such list should be very small in practical applications.

Thus, pattern matching can be solved in $O(m \log n)$ time using $5n$ bytes of space, or in $O(m + \log n)$ time using $7n$ bytes of space. Abouelhoda et al. [17] reduce this time further to $O(m)$ time by mimicking the suffix tree algorithm on a suffix array with some auxiliary information. Using clever implementation techniques, the space is reduced to approximately $6n$ bytes. An interesting feature of their algorithm is that it can be used in other applications based on a top-down traversal of suffix tree.

30.3.2 Longest Common Substrings

Consider the problem of finding a longest substring common to two given strings s_1 of size m and s_2 of size n. To solve this problem, first construct the GST of strings s_1 and s_2. A longest substring common to s_1 and s_2 will be the path-label of an internal node with the greatest string depth in the suffix tree which has leaves labeled with suffixes from both the strings. Using a traversal of the GST, record the string-depth of each node, and mark each node if it has suffixes from both the strings. Find the largest string-depth of any marked node. Each marked internal node at that depth gives a longest common substring. The total run-time of this algorithm is $O(m + n)$.

The problem can also be solved by using the suffix tree of one of the strings and suffix links. Without loss of generality, suppose the suffix tree of s_2 is given. For each position i in s_1, we find the largest substring of s_1 starting at that position that is also a substring of s_2. For position 1, this is directly computed by matching $suff_1$ of s_1 starting from the root of the suffix tree until no more matches are possible. To determine the longest substring match from position 2, simply walk up to the first internal node, follow the suffix link, and walk down as done in McCreight's algorithm. A similar proof shows that this algorithm runs in $O(m + n)$ time.

Now consider solving the longest common substring problem using the GSA and Lcp array for strings s_1 and s_2. First, consider a one string variant of this problem—that of computing the longest repeat in a string. This is given by the string depth of the deepest internal node in the corresponding suffix tree. All children of such a node must be leaves. Any consecutive pair of such leaves have the longest repeat as their longest common prefix. Thus, each largest value in the Lcp array reveals a longest repeat in the string. The number of occurrences of a repeat is one more than the number of consecutive occurrences of the corresponding largest value in the Lcp array. Thus, all distinct longest repeats, and the number and positions of their occurrences can be determined by a linear scan of the Lcp array.

To solve the longest common substring problem, let v denote an internal node with the greatest string depth that contains a suffix from each of the strings. Because such a pair of suffixes need not be consecutive in the suffix array, it might appear that one has to look at nonconsecutive entries in the Lcp array. However, the subtree of any internal node that is a child of v can only consist of suffixes from one of the strings. Thus, there will be two consecutive suffixes in the subtree under v, one from each string. Therefore, it is enough to look at consecutive entries in the *GSA*. In a linear scan of the *GSA* and *Lcp* arrays, find the largest *lcp* value that corresponds to two consecutive suffixes, one from each string. This gives the length of a longest common substring. The starting positions of the suffixes reveals the positions in the strings where the longest common substring occurs. The algorithm runs in $O(m + n)$ time.

30.3.3 Text Compression

Compression of text data is useful for data transmission and for compact storage. A simple, not necessarily optimal, data compression method is the Ziv - Lempel compression [18,19]. In this method, the text to be compressed is considered a large string, and a compact representation is obtained by identifying repeats in the string. A simple algorithm following this strategy is as follows: Let T denote the text to be compressed and let $|T| = n$. At some stage during the execution of the compression algorithm, suppose that the string $T[1..i-1]$ is already compressed. The compression is extended by finding the length l_i of a largest prefix of $suff_i$ that is a substring of $T[1..i-1]$. Two cases arise:

1. $l_i = 0$. In this case, a compressed representation of $T[1..i]$ is obtained by appending $T[i]$ to the compressed representation of $T[1..i-1]$.
2. $l_i > 0$. In this case, a compressed representation of $T[1..i+l_i-1]$ is obtained by appending (i, l_i) to the compressed representation of $T[1..i-1]$.

The algorithm is initiated by setting $T[1]$ to be the compressed representation of $T[1..1]$, and continuing the iterations until the entire string is compressed. For example, executing the above algorithm on the string *mississippi* yields the compressed string $mis(3,1)(2,3)(2,1)p(9,1)(2,1)$. The decompression method for such a compressed string is immediate.

Suffix trees can be used to carry out the compression in $O(n)$ time [20]. They can be used in obtaining l_i, the length of the longest prefix of $suff_i$ that is a substring of the portion of the string already seen, $T[1..i-1]$. If j is the starting position of such a substring, then $T[j..j+l_i-1] = T[i..i+l_i-1]$ and $i \geq j + l_i$. It follows that $|lcp(suff_j, suff_i)| \geq l_i$. Let $v = lca(i, j)$, where i and j are leaves corresponding to $suff_i$ and $suff_j$, respectively. It follows that $T[i..i+l_i-1]$ is a prefix of *path-label*(v). Consider the unique path from the root of $ST(T)$ that matches $T[i..i+l_i-1]$. Node v is an internal node in the subtree below, and hence j is a leaf in the subtree below. Thus, l_i is the largest number of characters along the path $T[i..n]$ such that $\exists$ leaf j in the subtree below with $j + l_i \leq i$. Note that any j in the subtree below that satisfies the property $j + l_i \leq i$ is acceptable. If such a j exists, the smallest leaf number in the subtree below certainly satisfies this property, and hence can be chosen as the starting position j.

This strategy results in the following algorithm for finding l_i: First, build the suffix tree of T. Using an appropriate linear time tree traversal method, record the string depth of each node and mark each internal node with the smallest leaf label in its subtree. Let $min(v)$ denote the smallest leaf label under internal node v. To find l_i, walk along the path $T[i..n]$ to identify two consecutive internal nodes u and v such that $min(u) + string\text{-}depth(u) < i$ and $min(v) + string\text{-}depth(v) \geq i$. If $min(v) + string\text{-}depth(u) > i$, then set $l_i = string\text{-}depth(u)$ and set the starting position to be $min(u)$. Otherwise, set $l_i = i - min(v)$ and set the starting position to be $min(v)$.

To obtain $O(n)$ run - time, it is enough to find l_i in $O(l_i)$ time as the next l_i characters of the string are compressed into an $O(1)$ space representation of an already seen substring. Therefore, it is enough to traverse the path matching $T[i..n]$ using individual character comparisons. However, as the path is guaranteed to exist, it can be traversed in $O(1)$ time per edge, irrespective of the length of the edge label.

30.3.4 String Containment

Given a set of strings $S = \{s_1, s_2, \ldots, s_k\}$ of total length N, the string containment problem is to identify each string that is a substring of some other string. An example application could be that the strings represent DNA sequence fragments, and we wish to remove redundancy. This problem can be easily solved using suffix trees in $O(N)$ time. First, construct the $GST(S)$ in $O(N)$ time. To find if a string s_i is contained in another, locate the leaf labeled $(s_i, 1)$. If the label of the edge connecting the leaf to its parent is labeled with the string $, s_i$ is contained in another string. Otherwise, it is not. This can be determined in $O(1)$ time per string.

30.3.5 Suffix-Prefix Overlaps

Suppose we are given a set of strings $S = \{s_1, s_2, \ldots, s_k\}$ of total length N. The suffix-prefix overlap problem is to identify, for each pair of strings (s_i, s_j), the longest suffix of s_i that is a prefix of s_j. Suffix-prefix overlaps are useful in algorithms for finding the shortest common superstring of a given set of strings. They are also useful in applications such as genome assembly where significant suffix-prefix overlaps between pairs of fragments are used to assemble fragments into much larger sequences.

The suffix-prefix overlap problem can be solved using $GST(S)$ in optimal $O(N + k^2)$ time. Consider the longest suffix α of s_i that is a prefix of s_j. In $GST(S)$, α is an initial part of the path from the root to leaf labeled $(j, 1)$ that culminates in an internal node. A leaf that corresponds to a suffix from s_i should be a child of the internal node, with the edge label $. Moreover, it must be the deepest internal node on the path from root to leaf $(j, 1)$ that has a suffix from s_i attached in this way. The length of the corresponding suffix - prefix overlap is given by the string depth of the internal node.

Let M be a $k \times k$ output matrix such that $M[i, j]$ should contain the length of the longest suffix of s_i that overlaps a prefix of s_j. The matrix is computed using a depth first search (DFS) traversal of $GST(S)$. The GST is preprocessed to record the string depth of every node. During the DFS traversal, k stacks $A_1, A_2, \ldots, A_k$ are maintained, one for each string. The top of the stack A_i contains the string depth of the deepest node along the current DFS path that is connected with edge label $ to a leaf corresponding to a suffix from s_i. If no such node exists, the top of the stack contains zero. Each stack A_i is initialized by pushing zero onto an empty stack, and is maintained during the DFS as follows: When the DFS traversal visits a node v from its parent, check to see if v is attached to a leaf with edge label $. If so, for each i such that string s_i contributes a suffix labeling the leaf, *push string-depth(v)* on to stack A_i. The string depth of the current node can be easily maintained during the DFS traversal. When the DFS traversal leaves the node v to return back to its parent, again identify each i that has the above property and *pop* the top element from the corresponding stack A_i.

The output matrix M is built one column at a time. When the DFS traversal reaches a leaf labeled $(j, 1)$, the top of stack A_i contains the longest suffix of s_i that matches a prefix of s_j. Thus, column j of matrix M is obtained by setting $M[i, j]$ to the top element of stack S_i. To analyze the run - time of the algorithm, note that each *push* (similarly, *pop*) operation on a stack corresponds to a distinct suffix of one of the input strings. Thus, the total number of *push* and *pop* operations is bounded by $O(N)$. The matrix M is filled in $O(1)$ time per element, taking $O(k^2)$ time. Hence, all suffix - prefix overlaps can be identified in optimal $O(N + k^2)$ time.

30.4 Lowest Common Ancestors

Consider a string s and two of its suffixes $suff_i$ and $suff_j$. The longest common prefix of the two suffixes is given by the path label of their lowest common ancestor. If the string - depth of each node is recorded in it, the length of the longest common prefix can be retrieved from the lowest common ancestor. Thus, an algorithm to find the lowest common ancestors quickly can be used to determine longest common prefixes without a single character comparison. In this section, we describe how to preprocess the suffix tree in linear time and be able to answer lowest common ancestor queries in constant time [21].

30.4.1 Bender and Farach's *lca* algorithm

Let T be a tree of n nodes. Without loss of generality, assume the nodes are numbered $1 \ldots n$. Let $lca(i, j)$ denote the lowest common ancestor of nodes i and j. Bender and Farach's algorithm performs a linear time preprocessing of the tree and can answer *lca* queries in constant time.

Let E be an Euler tour of the tree obtained by listing the nodes visited in a depth first search of T starting from the root. Let L be an array of level numbers such that $L[i]$ contains the tree - depth of the node $E[i]$. Both E and L contain $2n - 1$ elements and can be constructed by a depth first search of T in linear time. Let R be an array of size n such that $R[i]$ contains the index of the first occurrence of node i in E. Let $RMQ_A(i, j)$ denote the position of an occurrence of the smallest element in array A between indices i and j (inclusive). For nodes i and j, their lowest common ancestor is the node at the smallest tree - depth that is visited between an occurrence of i and an occurrence of j in the Euler tour. It follows that

$$lca(i, j) = E[RMQ_L(R[i], R[j])]$$

Thus, the problem of answering *lca* queries transforms into answering range minimum queries in arrays. Without loss of generality, we henceforth restrict our attention to answering range minimum queries in an array A of size n.

To answer range minimum queries in A, do the following preprocessing: Create $\lfloor \log n \rfloor + 1$ arrays $B_0, B_1, \ldots, B_{\lfloor \log n \rfloor}$ such that $B_j[i]$ contains $RMQ_A(i, i + 2^j)$, provided $i + 2^j \leq n$. B_0 can be computed directly from A in linear time. To compute $B_l[i]$, use $B_{l-1}[i]$ and $B_{l-1}[i + 2^{l-1}]$ to find $RMQ_A(i, i + 2^{l-1})$ and $RMQ_A(i + 2^{l-1}, i + 2^l)$, respectively. By comparing the elements in A at these locations, the smallest element in the range $A[i..i + 2^l]$ can be determined in constant time. Using this method, all the $\lfloor \log n \rfloor + 1$ arrays are computed in $O(n \log n)$ time.

Given an arbitrary range minimum query $RMQ_A(i, j)$, let k be the largest integer such that $2^k \leq (j - i)$. Split the range $[i..j]$ into two overlapping ranges $[i..i + 2^k]$ and $[j - 2^k..j]$. Using $B_k[i]$ and $B_k[j - 2^k]$, a smallest element in each of these overlapping ranges can be located in constant time. This will allow determination of $RMQ_A(i, j)$ in constant time. To avoid a direct computation of k, the largest power of 2 that is smaller than or equal to each integer in the range $[1..n]$ can be precomputed and stored in $O(n)$

time. Putting all of this together, range minimum queries can be answered with $O(n \log n)$ preprocessing time and $O(1)$ query time.

The preprocessing time is reduced to $O(n)$ as follows: Divide the array A into $2n/\log n$ blocks of size $1/2\log n$ each. Preprocess each block such that for every pair (i, j) that falls within a block, $RMQ_A(i, j)$ can be answered directly. Form an array B of size $2n/\log n$ that contains the minimum element from each of the blocks in A, in the order of the blocks in A, and record the locations of the minimum in each block in another array C. An arbitrary query $RMQ_A(i, j)$ where i and j do not fall in the same block is answered as follows: Directly find the location of the minimum in the range from i to the end of the block containing it, and also in the range from the beginning of the block containing j to index j. All that remains is to find the location of the minimum in the range of blocks completely contained between i and j. This is done by the corresponding range minimum query in B and using C to find the location in A of the resulting smallest element. To answer range queries in B, B is preprocessed as outlined before. Because the size of B is only $O(n \log n)$, preprocessing B takes $O(n/\log n/\log n/\log n) = O(n)$ time and space.

It remains to be described how each of the blocks in A is preprocessed to answer range minimum queries that fall within a block. For each pair (i, j) of indices that fall in a block, the corresponding range minimum query is precomputed and stored. This requires computing $O(\log^2 n)$ values per block and can be done in $O(\log^2 n)$ time per block. The total run-time over all blocks is $2n/\log n \times O(\log^2 n) = O(n \log n)$, which is unacceptable. The run-time can be reduced for the special case where the array A contains level numbers of nodes visited in an Euler Tour, by exploiting its special properties. Note that the level numbers of consecutive entries differ by $+1$ or -1. Consider the $2n/\log n$ blocks of size $1/2\log n$. Normalize each block by subtracting the first element of the block from each element of the block. This does not affect the range minimum query. As the first element of each block is 0 and any other element differs from the previous one by $+1$ or -1, the number of distinct blocks is $2^{1/2\log n - 1} = 1/2\sqrt{n}$. Direct preprocessing of the distinct blocks takes $1/2\sqrt{n} \times O(\log^2 n) = O(n)$ time. The mapping of each block to its corresponding distinct normalized block can be done in time proportional to the length of the block, taking $O(n)$ time over all blocks.

Putting it all together, a tree T of n nodes can be preprocessed in $O(n)$ time such that lca queries for any two nodes can be answered in constant time. We are interested in an application of this general algorithm to suffix trees. Consider a suffix tree for a string of length n. After linear time preprocessing, lca queries on the tree can be answered in constant time. For a given pair of suffixes in the string, the string-depth of their lowest common ancestor gives the length of their longest common prefix. Thus, the longest common prefix can be determined in constant time, without resorting to a single character comparison! This feature is exploited in many suffix tree algorithms.

30.5 Advanced Applications

30.5.1 Suffix Links from Lowest Common Ancestors

Suppose we are given a suffix tree and it is required to establish suffix links for each internal node. This may become necessary if the suffix tree creation algorithm does not construct suffix links but they are needed for an application of interest. For example, the suffix tree may be constructed via suffix arrays, completely avoiding the construction and use of suffix links for building the tree. The links can be easily established if the tree is preprocessed for lca queries.

Mark each internal node v of the suffix tree with a pair of leaves (i, j) such that leaves labeled i and j are in the subtrees of different children of v. The marking can be done in linear time by a bottom-up traversal of the tree. To find the suffix link from an internal node v (other than the root) marked with (i, j), note that $v = lca(i, j)$ and $lcp(suff_i, suff_j) = path\text{-}label(v)$. Let *path-label*$(v) = c\alpha$, where c is the first character and α is a string. To establish a suffix link from v, node u with path label α is needed. As $lcp(suff_{i+1}, suff_{j+1}) = \alpha$, node u is given by $lca(i + 1, j + 1)$, which can be determined in constant time. Thus, all suffix links can be determined in $O(n)$ time. This method trivially extends to the case of a generalized suffix tree.

30.5.2 Approximate Pattern Matching

The simpler version of approximate pattern matching problem is as follows: Given a pattern P ($|P| = m$) and a text T ($|T| = n$), find all substrings of length $|P|$ in T that match P with at most k mismatches. To solve this problem, first construct the GST of P and T. Preprocess the GST to record the string-depth of each node, and to answer lca queries in constant time. For each position i in T, we will determine if $T[i..i + m - 1]$ matches P with at most k mismatches. First, use an lca query $lca((P, 1), (T, i))$ to find the largest substring from position i of T that matches a substring from position 1 and P. Suppose the length of this longest exact match is l. Thus, $P[1..l] = T[i..i + l - 1]$, and $P[l + 1] \neq T[i + l]$. Count this as a mismatch and continue by finding $lca((P, l + 2), (T, i + l + 1))$. This procedure is continued until either the end of P is reached or the number of mismatches crosses k. As each lca query takes constant time, the entire procedures takes $O(k)$ time. This is repeated for each position i in T for a total run-time of $O(kn)$.

Now, consider the more general problem of finding the substrings of T that can be derived from P by using at most k character insertions, deletions or substitutions. To solve this problem, we proceed as before by determining the possibility of such a match for every starting position i in T. Let $l = string\text{-}depth(lca((P,1),(T,i)))$. At this stage, we consider three possibilities:

1. Substitution—$P[l+1]$ and $T[i+l]$ are considered a mismatch. Continue by finding $lca((P, l+2), (T, i+l+1))$.
2. Insertion—$T[i+l]$ is considered an insertion in P after $P[l]$. Continue by finding $lca((P, l+1), (T, i+l+1))$.
3. Deletion—$P[l+1]$ is considered a deletion. Continue by finding $lca((P, l+2), (T, i+l))$.

After each lca computation, we have three possibilities corresponding to substitution, insertion and deletion, respectively. All possibilities are enumerated to find if there is a sequence of k or less operations that will transform P into a substring starting from position i in T. This takes $O(3^k)$ time. Repeating this algorithm for each position i in T takes $O(3^k n)$ time.

The above algorithm always uses the longest exact match possible from a given pair of positions in P and T before considering the possibility of an insertion or deletion. To prove the correctness of this algorithm, we show that if there is an approximate match of P starting from position i in T that does not use such a longest exact match, then there exists another approximate match that uses only longest exact matches. Consider an approximate match that does not use longest exact matches. Consider the leftmost position j in P and the corresponding position $i+k$ in T where the longest exact match is violated. That is, $P[j] = T[i+k]$ but this is not used as part of an exact match. Instead, an insertion or deletion is used. Suppose that an exact match of length r is used after the insertion or deletion. We can come up with a corresponding approximate match where the longest match is used and the insertion/deletion is taken after that. This will either keep the number of insertions/deletions the same or reduce the count. If the value of k is small, the above algorithms provide a quick and easy way to solve the approximate pattern matching problem. For sophisticated algorithms with better run-times, see [22,23].

30.5.3 Maximal Palindromes

A string is called a palindrome if it reads the same forwards or backwards. A substring $s[i..j]$ of a string s is called a maximal palindrome of s, if $s[i..j]$ is a palindrome and $s[i-1] \neq s[j+1]$ (unless $i=1$ or $j=n$). The maximal palindrome problem is to find all maximal palindromes of a string s.

For a palindrome of odd length, say $2k+1$, define the center of the palindrome to be the $(k+1)$th character. For a palindrome of even length, say $2k$, define the center to be the position between characters k and $k+1$ of the palindrome. In either case, the palindrome is said to be of radius k. Starting from the center, a palindrome is a string that reads the same in both directions. Observe that each maximal palindrome in a string must have a distinct center. As the number of possible centers for a string of length n is $2n-1$, the total number of maximal palindromes of a string is $2n-1$. All such palindromes can be identified in linear time using the following algorithm.

Let s^r denote the reverse of string s. Construct a *GST* of the strings s and s^r and preprocess the *GST* to record string depths of internal nodes and for answering lca queries. Now, consider a character $s[i]$ in the string. The maximal odd length palindrome centered at $s[i]$ is given by the length of the longest common prefix between $suff_{i+1}$ of s and $suff_{n-i+2}$ of s^r. This is easily computed as the string-depth of $lca((s, i+1), (s^r, n-i+2))$ in constant time. Similarly, the maximal even length palindrome centered between $s[i]$ and $s[i+1]$ is given by the length of the longest common prefix between $suff_{i+1}$ of s and $suff_{n-i+1}$ of s^r. This is computed as the string-depth of $lca((s, i+1), (s^r, n-i+1))$ in constant time.

These and many other applications involving strings can be solved efficiently using suffix trees and suffix arrays. A comprehensive treatise of suffix trees, suffix arrays and string algorithms can be found in the textbooks by Gusfield [24], and Crochemore and Rytter [25].

Acknowledgments

This work was supported, in part, by the NSF under grant ACI-0203782.

References

1. T. Kasai, G. Lee, H. Arimura, S. Arikawa, and K. Park. Linear-time longest-common-prefix computation in suffix arrays and its applications. In *12th Annual Symposium, Combinatorial Pattern Matching*, pages 181–192, 2001.
2. P. Weiner. Linear pattern matching algorithms. In *14th Symposium on Switching and Automata Theory*, pages 1–11. IEEE, 1973.
3. E. M. McCreight. A space-economical suffix tree construction algorithm. *Journal of the ACM*, 23:262–272, 1976.
4. E. Ukkonen. On-line construction of suffix-trees. *Algorithmica*, 14:249–260, 1995.

5. R. Giegerich and S. Kurtz. From Ukkonen to McCreight and Weiner: A unifying view of linear-time suffix tree construction. *Algorithmica*, 19:331–353, 1997.

6. M. Farach. Optimal suffix tree construction with large alphabets. In *38th Annual Symposium on Foundations of Computer Science*, pages 137–143. IEEE, 1997.

7. M. Farach-Colton, P. Ferragina, and S. Muthukrishnan. On the sorting-complexity of suffix tree construction. *Journal of the ACM*, 47:987–1011, 2000.

8. U. Manber and G. Myers. Suffix arrays: A new method for on-line search. *SIAM Journal on Computing*, 22:935–948, 1993.

9. J. Kärkkänen and P. Sanders. Simpler linear work suffix array construction. In *International Colloquium on Automata, Languages and Programming*, 2003.

10. D. K. Kim, J. S. Sim, H. Park, and K. Park. Linear-time construction of suffix arrays. In *14th Annual Symposium, Combinatorial Pattern Matching*, 2003.

11. P. Ko and S. Aluru. Space-efficient linear-time construction of suffix arrays. In *14th Annual Symposium, Combinatorial Pattern Matching*, 2003.

12. A. Crauser and P. Ferragina. A theoretical and experimental study on the construction of suffix arrays in external memory. *Algorithmica*, 32(1):1–35, 2002.

13. S. Kurtz. Reducing the space requirement of suffix trees. *Software - Practice and Experience*, 29(13):1149–1171, 1999.

14. P. Ferragina and G. Manzini. Opportunistic data structures with applications. In *41th Annual Symposium on Foundations of Computer Science*, pages 390–398. IEEE, 2000.

15. R. Grossi and J. S. Vitter. Compressed suffix arrays and suffix trees with applications to text indexing and string matching. In *Symposium on the Theory of Computing*, pages 397–406. ACM, 2000.

16. M. I. Abouelhoda, S. Kurtz, and E. Ohlebusch. The enhanced suffix array and its applications to genome analysis. In *2nd Workshop on Algorithms in Bioinformatics*, pages 449–463, 2002.

17. M. I. Abouelhoda, E. Ohlebusch, and S. Kurtz. Optimal exact string matching based on suffix arrays. In *International Symposium on String Processing and Information Retrieval*, pages 31–43. IEEE, 2002.

18. J. Ziv and A. Lempel. A universal algorithm for sequential data compression. *IEEE Transactions on Information Theory*, 23:337–343, 1977.

19. J. Ziv and A. Lempel. Compression of individual sequences via variable length coding. *IEEE Transactions on Information Theory*, 24:530–536, 1978.

20. M. Rodeh, V. R. Pratt, and S. Even. A linear algorithm for data compression via string matching. *Journal of the ACM*, 28:16–24, 1981.

21. M. A. Bender and M. Farach-Colton. The LCA problem revisited. In *Latin American Theoretical Informatics Symposium*, pages 88–94, 2000.

22. R. Cole and R. Hariharan. Approximate string matching: A simpler faster algorithm. *SIAM Journal on Computing*, 31:1761–1782, 2002.

23. E. Ukkonen. Approximate string-matching over suffix trees. In A. Apostolico, M. Crochemore, Z. Galil, and U. Manber, editors, *Combinatorial Pattern Matching, 4th Annual Symposium*, volume 684 of *Lecture Notes in Computer Science*, pages 228–242, Springer, Padova, Italy, 1993.

24. D. Gusfield. *Algorithms on Strings Trees and Sequences*. Cambridge University Press, New York,NY, 1997.

25. M. Crochemore and W. Rytter. *Jewels of Stringology*. World Scientific Publishing Company, Singapore, 2002.

31

String Searching

31.1 Introduction.. 477
31.2 Preliminaries.. 479
31.3 The DAWG.. 479
     A Simple Algorithm for Constructing the DAWG • Constructing the DAWG in Linear Time
31.4 The Compact DAWG.. 486
     Using the Compact DAWG to Find the Locations of a String in the Text • Variations and
     Applications
31.5 The Position Heap... 489
     Building the Position Heap • Querying the Position Heap • Time Bounds • Improvements
     to the Time Bounds
References.. 494

Andrzej Ehrenfeucht
University of Colorado

Ross M. McConnell
Colorado State University

31.1 Introduction

Searching for occurrences of a substring in a text is a common operation familiar to anyone who uses a text editor, word processor, or web browser. It is also the case that algorithms for analyzing textual databases can generate a large number of searches. If a text, such as a portion of the genome of an organism, is to be searched repeatedly, it is sometimes the case that it pays to preprocess the text to create a data structure that facilitates the searches. The suffix tree [1] and suffix array [2] discussed in Chapter 30 are examples.

In this chapter, we give some alternatives to these data structures that have advantages over them in some circumstances, depending on what type of searches or analysis of the text are desired, the amount of memory available, and the amount of effort to be invested in an implementation.

In particular, we focus on the problem of finding the locations of all occurrences of a string x in a text t, where the letters of t are drawn from a fixed alphabet Σ, such as the ASCII letter codes.

The *length* of a string x, denoted $|x|$, is the number of characters in it. The *empty string*, denoted λ is the string of length 0 that has no characters in it. If $t = a_1a_2, \ldots, a_n$ is a text and $p = a_ia_{i+1} \ldots a_j$ is a substring of it, then i is a *starting position* of p in t and j is an *ending position* of p in t. For instance, the starting positions of *abc* in *aabcabcaac* are $\{2, 5\}$, and its ending positions are $\{5, 8\}$. We consider the empty string to have starting and ending positions at $\{0, 1, 2, \ldots, n\}$, once at each position in the text, and once at position 0, preceding the first character of the text. Let *EndPositions*(p, t) denote the ending positions of p in t; when t is understood, we may denote it *EndPositions*(p).

A *deterministic finite automaton* on Σ is a directed graph where each directed edge is labeled with a letter from Σ, and where, for each node, there is at most one edge directed out of the node that is labeled with any given letter. Exactly one of the nodes is designated as a *start node*, and some of the nodes are designated as *accept nodes*. The *label* of a directed path is the word given by the sequence of letters on the path. A deterministic finite automaton is used for representing a set of words, namely, the set of labels of paths from the start node to an accept node.

The first data structure that we examine is the *directed acyclic word graph (DAWG)*. The DAWG is just a deterministic finite automaton representing the language of subwords of a text t, which is finite, hence regular. All of its states except for one are accept states. There is no edge from the nonaccepting state to any accepting state, so it is convenient to omit the nonaccept state when representing the DAWG. In this representation, a string p is a substring of t iff it is the label of a directed path originating at the start node.

There exists a labeling of each node of the DAWG with a set of positions so that the DAWG has the following property:

- Whenever p is a substring of t, its ending positions in t are given by the label of the last node of the path of label p that originates at the start node.

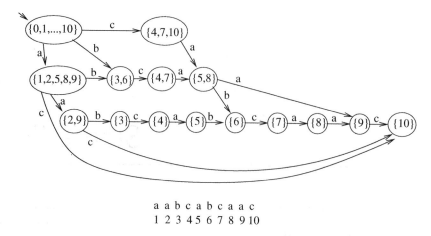

aabcabcaac
1 2 3 4 5 6 7 8 9 10

FIGURE 31.1 The DAWG of the text *aabcabcaac*. The starting node is at the upper left. A string *p* is a substring of the text if and only if it is the label of a path originating at the start node. The nodes can be labeled so that whenever *p* is the label of such a path, the last node of the path gives *EndPositions(p)*. For instance, the strings that lead to the state labeled {5, 8} are *ca*, *bca*, and *abca*, and these have occurrences in the text with their last letter at positions 5 and 8.

To find the locations where *p* occurs, one need only begin at the start node, follow edges that match the letters of *p* in order, and retrieve the set of positions at the node where this process halts.

In view of the fact that there are $\Theta(|t|^2)$ intervals on *t*, each of which represents a substring that is contained in the interval, it is surprising that the number of nodes and edges of the DAWG of *t* is $O(|t|)$. The reason for this is that all possible query strings fall naturally into *equivalence classes*, which are sets of strings such that two strings are in the same set if they have the same set of ending positions. The size of an equivalence class can be large, and this economy makes the $O(|t|)$ bound possible.

In an application such as a search engine, one may be interested not in the locations of a string in a text, but the number of occurrences of a string in the text. This is one criterion for deciding which texts are most relevant to a query. Since all strings in an equivalence class have the same number of occurrences, each state can be labeled not with the position set, but with the cardinality of its position set. The label of the node reached on the path labeled *p* originating at the start node tells the number of occurrences of *p* in *t* in $O(|p|)$ time. This variant requires $O(|t|)$ space and can be constructed in $O(|t|)$ time.

Unfortunately, the sum of cardinalities of the position sets of the nodes of the DAWG of *t* is not $O(|t|)$. However, a second data structure that we describe, called the *compact DAWG* does use $O(|t|)$ space. If a string *p* has *k* occurrences in *t*, then it takes $O(|p| + k)$ time to return the set of occurrences where *p* occurs in *t*, given the compact DAWG of *t*. It can be built in $O(|t|)$ time. These bounds are the same as that for the suffix tree and suffix array, but the compact DAWG requires substantially less space in most cases. An example is illustrated in Figure 31.2.

Another important issue is the ease with which a programmer can understand and program the construction algorithm. Like the computer time required for queries, the time spent by a programmer understanding, writing, and maintaining a program is also a resource that must be considered. The third data structure that we present, called the *position heap*, has worse worst-case bounds for construction and queries, but has the advantage of being as easy to understand and construct as elementary data structures such as unbalanced binary search trees and heaps. One trade-off is that the worst-case bounds for a query is $O(|p|^2 + k)$, rather than $O(|p| + k)$. However, on randomly generated strings, the expected time for a query is $O(|p| + k)$, and on most practical applications, the query time can be expected not to differ greatly from this. Like the other structures, it can be constructed in linear time. However, an extremely simple implementation takes $O(|t| \log |t|)$ expected time on randomly generated strings and does not depart much from this in most practical applications. Those who wish to expend minimal programming effort may wish to consider this simple variant of the construction algorithm.

The position heap for the string of Figure 31.1 is illustrated in Figure 31.3.

More recently, an alternative to the construction algorithm given here has been given in Reference 3. That algorithm can be considered online, since it can construct the position heap incrementally as characters are appended to the text. The algorithm we give here can construct it as characters are prepended to the text. The case where they are appended can be reduced to the case where they are prepended, by working with the reverse of the text and reversing query strings before each query. The algorithm of Reference 3 is more natural for this and contributes additional algorithmic insights.

This analysis assumes that the size of the alphabet is a constant, which is a reasonable assumption for genomic sequences, for example. Recently, an algorithm that takes linear time, independently of the size of the alphabet, has been given in Reference 4. However, a step of that algorithm is construction of the suffix tree, hence it is conceptually more difficult than the one given here.

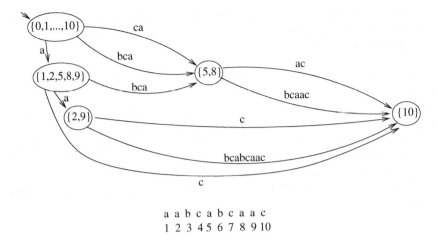

a a b c a b c a a c
1 2 3 4 5 6 7 8 9 10

FIGURE 31.2 The compact DAWG of the text *aabcabcaac* (cf. Figure 31.1) The labels depicted in the nodes are the ending positions of the corresponding principal nodes of the DAWG. The compact DAWG is obtained from the DAWG by deleting nodes that have only one outgoing edge, and representing deleted paths between the remaining nodes with edges that are labeled with the path's label.

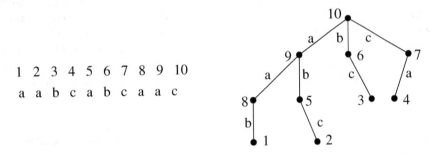

FIGURE 31.3 The position heap of *aabcabcaa*.

Linear-time construction of the position heap for a set of strings stored as a trie is given in Reference 5.

31.2 Preliminaries

The infinite set of all strings that can be formed from letters of an alphabet Σ is denoted as Σ^*. If $a \in \Sigma$, let a^n denote the string that consists of n repetitions of a.

If x is a string, then for $1 \leq j \leq |x|$, let x_j denote the character in position j. Thus x can be written as $x_1 x_2, \ldots, x_{|x|}$. The *reversal* x^R of x is the string $x_{|x|} x_{|x|-1} \ldots x_1$. Let $x[i:j]$ denote the substring $x_i x_{i+1}, \ldots, x_j$.

The *prefixes* of a string $x = x_1 x_2, \ldots, x_k$ are those with a starting position at the leftmost position of x, namely, the empty string and those strings of the form $x[1:j]$ for $1 \leq j \leq k$. Its *suffixes* are those with an ending position at the rightmost position of x, namely, the empty string and those of the form $x[j:k]$.

A *trie* on Σ is a deterministic finite automaton that is a rooted tree whose start node is the root.

Given a family $\mathcal{F}$ of subsets of a domain $\mathcal{V}$, the *transitive reduction* of the subset relation can be viewed as a pointer from each $X \in \mathcal{F}$ to each $Y \in \mathcal{F}$ such that $X \subset Y$ and there exists no Z such that $X \subset Z \subset Y$. This is sometimes referred to as the *Hasse diagram* of the subset relation on the family. The Hasse diagram is a tree if $\mathcal{V} \in \mathcal{F}$, $\emptyset \notin \mathcal{F}$, and for each $X, Y \in \mathcal{F}$, either $X \subseteq Y$, $Y \subset X$, or $X \cap Y = \emptyset$.

31.3 The DAWG

Lemma 31.1

Let x and y be two strings such that $EndPositions(x) \cap EndPositions(y) \neq \emptyset$. One of x and y must be a suffix of the other, and either $EndPositions(x) = EndPositions(y)$, $EndPositions(x) \subset EndPositions(y)$, or $EndPositions(y) \subset EndPositions(x)$.

Proof. If x and y have a common ending position i, then the two occurrences coincide in a way that forces one to be a suffix of the other. Suppose without loss of generality that y is a suffix of x. Then every occurrence of x contains an occurrence of y inside of it that ends at the same position, so $Endpositions(x) \subseteq Endpositions(y)$.   ∎

For instance, in the string $aabcabcaac$, the string ca has ending positions $\{5, 8\}$, while the string $aabca$ has ending positions $\{5\}$, and ca is a suffix of $aabca$.

Let x's *right-equivalence class* in t be the set $\{y | EndPositions(y) = EndPositions(x)\}$. The only infinite class is *degenerate class* of strings with the empty set as ending positions, namely, those elements of Σ^* that are not substrings of t.

The right-equivalence classes on t are a partition of Σ^*: each member of Σ^* is in one and only one right-equivalence class. By Lemma 31.1, whenever two strings are in the same nondegenerate right-equivalence class, then one of them is a suffix of the other. It is easily seen that if y is the shortest string in the class and x is the longest, then the class consists of the suffixes of x whose length is at least $|y|$. For instance, in Figure 31.1, the class of strings with end positions $\{5, 8\}$ consists of $y = ca$, $x = abca$, and since bca is a longer suffix of x than y is.

Lemma 31.2

A text t of length n has at most $2n$ right-equivalence classes.

Proof. The degenerate class is one right equivalence class. All others have nonempty ending positions, and we must show that there are at most $2n - 1$ of them. The set $V = \{0, 1, 2, \ldots, n\}$ is the set of ending positions of the empty string. If X and Y are sets of ending positions of two right-equivalence classes, then $X \subseteq Y$, $Y \subseteq X$, or $Y \cap X = \emptyset$, by Lemma 31.1. Therefore, the transitive reduction (Hasse diagram) of the subset relation on the nonempty position sets is a tree rooted at V. For any i such that $\{i\}$ is not a leaf, we can add $\{i\}$ as a child of the lowest set that contains i as a member. The leaves are now a partition of $\{1, 2, \ldots, n\}$ so it has at most n leaves. Since each node of the tree has at least two children, there are at most $2n - 1$ nodes.   ∎

Definition 31.1

The DAWG is defined as follows. The states of the DAWG are the nondegenerate right-equivalence classes that t induces on its substrings. For each $a \in \Sigma$ and $x \in \Sigma^*$ such that xa is a substring of t, there is an edge labeled a from x's right-equivalence class to xa's right-equivalence class.

Figure 31.1 depicts the DAWG by labeling each right-equivalence class with its set of ending positions. The set of words in a class is just the set of path labels of paths leading from the source to a class. For instance, the right-equivalence class represented by the node labeled $\{5, 8\}$ is $\{ca, bca, abca\}$.

It would be natural to include the infinite degenerate class of strings that do not occur in t. This would ensure that every state had an outgoing edge for every letter of Σ. However, it is convenient to omit this state when representing the DAWG: for each $a \in \Sigma$, there is an edge from the degenerate class to itself, and this does not need to be represented explicitly. An edge labeled a from a nondegenerate class to the degenerate class is implied by the absence of an edge out of the state labeled a in the representation.

For each node X and each $a \in \Sigma$, there is at most one transition out of X that is labeled a. Therefore, the DAWG is a deterministic finite automaton. Any word p such that $EndPositions(p) \neq \emptyset$ spells out the labels of a path to the state corresponding to $EndPositions(p)$. Therefore, all states of the DAWG are reachable from the start state. The DAWG cannot have a directed cycle, as this would allow an infinite set of words to spell out a path, and the set of subwords of t is finite. Therefore, it can be represented by a directed acyclic graph.

A state is a *sink* if it has no outgoing edges. A sink must be the right-equivalence class containing position n, so there is exactly one sink.

Theorem 31.1

The DAWG for a text of length n has at most $2n - 1$ nodes and $3n - 3$ edges.

Proof. The number of nodes follows from Lemma 31.2. There is a single sink, namely, the one that has position set $\{|t|\}$, this represents the equivalence class containing those suffixes of t that have a unique occurrence in t. Let T be a directed spanning tree of the DAWG rooted at the start state. T has one fewer edges than the number of states, hence $2n - 2$ edges. For every $e \notin T$, let $P_1(e)$ denote the path in T from the start state to the tail of e, let $P_2(e)$ denote an arbitrary path in the DAWG from the head of e to the sink, and let $P(e)$ denote the concatenation of $(P_1(e), e, P_2(e))$. Since $P(e)$ ends at the sink, the labels of its edges yield a suffix of t. For $e_1, e_2 \notin T$ with $e_1 \neq e_2$, $P(e_1) \neq P(e_2)$, since they differ in their first edge that is not in T. One suffix is given by the labels of the path in T to the sink. Each of the remaining $n - 1$ suffixes is the sequence of labels of $P(e)$ for at most one edge $e \notin T$, so there are at most $n - 1$ edges not in T.

The total number of edges of the DAWG is bounded by $2n - 2$ tree edges and $n - 1$ nontree edges.   ■

To determine whether a string p occurs as a substring of t, one may begin at the start state and either find the path that spells out the letters of p, thereby accepting p, or else reject p if there is no such path. This requires finding, at each node x, the transition labeled a leaving x, where a is the next letter of p. If $|\Sigma| = O(1)$, this takes $O(1)$ time, so it takes $O(|p|)$ time to determine whether p is a subword of t. Note that, in contrast to naive approaches to this problem, this time bound is independent of the length of t.

If the nodes of the DAWG are explicitly labeled with the corresponding end positions, as in Figure 31.1, then it is easy to find the positions where a substring occurs: it is the label of the state reached on the substring. However, doing this is infeasible if one wishes to build the DAWG in $O(|t|)$ time and use $O(|t|)$ storage, since the sum of cardinalities of the position sets can be greater than this. For this problem, it is preferable to use the compact DAWG that is described below.

For the problem of finding the number of occurrences of a substring in t, it suffices to label each node with the *number* of positions in its position set. This may be done in postorder in a depth-first search, starting at the start node, and applying the following rule: the label of a node v is the sum of labels of its outneighbors, which have already been labeled by the time one must label v. Handling v takes time proportional to the number of edges originating at v, which we have already shown is $O(|t|)$.

31.3.1 A Simple Algorithm for Constructing the DAWG

Definition 31.2

If x is a substring of t, let us say that x's *redundancy* in t is the number of ending (or beginning) positions it has in t. If i is a position in t, let $h(i)$ be the longest substring x of t with an ending position at i whose redundancy is at least as great as its length, $|x|$. Let $h(t)$ be the average of $h(i)$ over all i, namely $\left(\sum_{i=1}^{|t|} h(i) \right) / |t|$.

Clearly, $h(t)$ is a measure of how redundant t is; the greater the value of $h(t)$, the less information it can contain.

In this section, we given an $O(|t| h(t))$ algorithm for constructing the DAWG of a string t. This is quadratic in the worst case, which is illustrated by the string $t = a^n$, consisting of n copies of one letter. However, we claim that the algorithm is a practical one for most applications, where $h(t)$ is rarely large even when t has a long repeated substring. In most applications, $h(t)$ can be expected to behave like an $O(\log |t|)$ function.

The algorithm explicitly labels the nodes of the DAWG with their ending positions, as illustrated in Figure 31.1. Each set of ending positions is represented with a list, where the positions appear in ascending order. It begins by creating a start node, and then iteratively *processes* an unprocessed node by creating its neighbors. To identify an unprocessed node, it is convenient to keep a list of the unprocessed nodes, insert a node in this list, and remove a node from the front of the list when it is time to process a new node.

Algorithm 31.1

```
DAWGBuild(t)
        Create a start node with position set {0, 1, ..., n}
        While there is an unprocessed node v
                Create a copy of v's position set
                Add 1 to every element of this set
                Remove n + 1 from this copy if it occurs
                Partition the copy into sets of positions that have a common letter
                For each partition class W
                        If W is already the position set of a node, then let w denote that node
                        Else create a new node w with position set W
                        Let a be the letter that occurs at the positions in W
                        Install an edge labeled a from v to w
```

Figure 31.4 gives an illustration. For the correctness, it is easy to see by induction on k that every substring w of the text that has length k leads to a node whose position set is the ending positions of w.

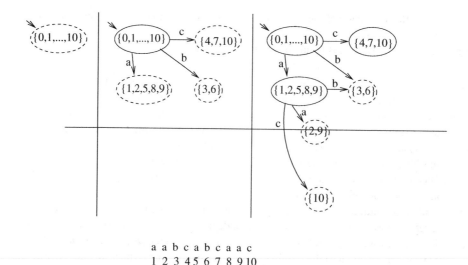

a a b c a b c a a c
1 2 3 4 5 6 7 8 9 10

FIGURE 31.4 Illustration of the first three iterations of Algorithm31.1 on *aabcabcaac*. Unprocessed nodes are drawn with dashed outlines. The algorithm initially creates a start state with position set $\{0, 1, \ldots, n\}$ (left figure). To process the start node, it creates a copy of this position set and adds 1 to each element, yielding $\{1, 2, \ldots, n+1\}$. It discards $n+1$, yielding $\{1, 2, \ldots, n\}$. It partitions this into the set $\{1, 2, 5, 8, 9\}$ of positions that contain *a*, the set $\{3, 6\}$ of positions that contain *b*, and the set $\{4, 7, 10\}$ of positions that contain *c*, creates a node for each, and installs edges labeled with the corresponding letters to the new nodes (middle figure). To process the node *v* labeled $\{1, 2, 5, 8, 9\}$, it adds 1 to each element of this set to obtain $\{2, 3, 6, 9, 10\}$, and partitions them into $\{2, 9\}$, $\{3, 6\}$, and $\{10\}$. Of these, $\{2, 9\}$ and $\{10\}$ are new position sets, so a new node is created for each. It then installs edges from *v* to the nodes with these three position sets.

Lemma 31.3

The sum of cardinalities of the position sets of the nodes of the DAWG is $O(|t|h(t))$.

Proof. For a position *i*, let $N(i)$ be the number of ending position sets in which position *i* appears. By Lemma 31.1, position sets that contain *i* form a chain $\{X_1, X_2, \ldots, X_{N(i)}\}$, where for each *i* from 1 to $N(i) - 1$, $|X_i| > |X_{i+1}|$, and a string with X_i as its ending positions must be shorter than one with X_{i+1} as its ending positions. Therefore, $|X_{\lfloor N(i)/2 \rfloor}| \geq N(i)/2$, and any string with this set as its ending position set must have length at least $\lfloor (N(i)/2 \rfloor - 1$. This is a string whose set of ending positions is at least as large as its length, so $N(i) = O(h(t))$.

The sum of cardinalities of the position sets is given by $\sum_{i=0}^{|t|} N(i)$, since each appearance of *i* in a position set contributes 1 to the sum, and this sum is $O(|t|h(T))$. ∎

It is easy to keep the classes as sorted linked lists. When a class *X* is partitioned into smaller classes, these fall naturally into smaller sorted lists in time linear in the size of *X*. A variety of data structures, such as tries, are suitable for looking up whether the sorted representation of a class *W* already occurs as the position set of a node. The time is therefore linear in the sum of cardinalities of the position sets, which is $O(|t|h(t))$ by Lemma 31.3.

31.3.2 Constructing the DAWG in Linear Time

The linear-time algorithm given in Reference 6 to construct the DAWG works incrementally by induction on the length of the string. The DAWG of a string of length 0 (the null string) is just a single start node. For $k = 0$ to $n - 1$, it iteratively performs an *induction step* that modifies the DAWG of $t[1:k]$ to obtain the DAWG of $t[1:k+c1]$.

To gain insight into how the induction step must be performed, consider Figure 31.5. An occurrence of a substring of *t* can be specified by giving its ending position and its length. For each occurrence of a substring, it gives the number of times the substring occurs up to that point in the text, indexed by length and position. For instance, the string that has length 3 and ends at position 5 is *bca*. The entry in row 3, column 5 indicates that there is one occurrence of it up through position 5 of the text. There is another at position 8, and the entry at row 3 column 8 indicates that it has two occurrences up through position 8.

The lower figure, which we may call the *incremental landscape*, gives a simplified representation of the table, by giving an entry only if it differs from the entry immediately above it. Let $L[i, j]$ denote the entry in row *i*, column *k* of the incremental landscape. Some of these entries are blank; the implicit value of such an entry is the value of the first nonblank entry above it.

String Searching

Ending position

	0	1	2	3	4	5	6	7	8	9	10
10											1
9										1	1
8			Number of						1	1	1
7			occurrences					1	1	1	1
6							1	1	1	1	1
5						1	1	1	1	1	1
4					1	1	1	1	2	1	1
3				1	1	1	1	2	2	1	1
2			1	1	1	1	2	2	2	2	1
1		1	2	1	1	3	2	2	4	5	3
0	1	2	3	4	5	6	7	8	9	10	11
	a	a	b	c	a	b	c	a	a	c	

Length (row label, left of table)

FIGURE 31.5 Displaying the number of occurrences of substrings in a text. In the upper figure, the entry in row i column j corresponds to the substring of length j that ends at position i in the text t, and gives the number of occurrences of the substring at position i or before. That is, it gives the number of occurrences of the substring in $t[1:i]$. Row 0 is included to reflect occurrences of the null substring, which has occurrences at $\{0, 1, \ldots, n\}$.

Column k has one entry for each right-equivalence class of $t[1:k]$ that has k as an ending position. For instance, in column 8, we see the following:

1. $L[0,8]$: A right-equivalence class for the suffix of $t[1:8]$ of length 0, namely, the empty string, which has nine occurrences $(\{0, 1, \ldots, 8\})$ in $t[1:8]$.
2. $L[1,8]$: A right-equivalence class for the suffix of $t[1:8]$ of length 1, namely, the suffix a, which has four occurrences $(\{1, 2, 5, 8\})$ in $t[1:8]$.
3. $L[4,8]$: A right-equivalence class for suffixes of $t[1:8]$ of lengths 2 through 4, namely, $\{ca, bca, abca\}$, which have two occurrences $(\{5, 8\})$ in $t[1:8]$. The longest of these, $abca$, is given by the nonblank entry at $L[4,8]$, and membership of the others in the class is given implicitly by the blank entries immediately below it.
4. $L[8,8]$: A right-equivalence class for suffixes of $t[1:k]$ of lengths 5 through 8, namely, $\{cabca, bcabca, abcabca, abcabca\}$ that have one occurrence in $t[1:8]$.

We may therefore treat nonblank entries in the incremental landscape as nodes of the DAWG. Let the *height* of a node denote the length of the longest substring in its right-equivalence class; this is the height (row number) where it appears in the incremental landscape.

When modifying the DAWG of $t[1:k]$ to obtain the DAWG of $t[1:k+1]$, all new nodes that must be added to the DAWG appear in column $k+1$. However, not every node in column $k+1$ is a new node, as some of the entries reflect nodes that were created earlier.

For instance, consider Figure 31.7, which shows the incremental step from $t[1:6]$ to $t[1:7]$. One of the nodes, which represents the class $\{cabc, bcabc, abcabc, aabcabc\}$ of substrings of $t[1:7]$ that are not substrings of $t[1:6]$. It is the top circled node of column 7 in Figure 31.6. Another represents the class $Z_2 = \{c, bc, abc\}$. This appears in $L[3,7]$. To see why this is new, look at the previous occurrence of its longest substring, abc, which is represented by $L[3,4]$, which is blank. Therefore, in the DAWG of $t[1:6]$, it is part of a right-equivalence Z, which appears at $L[4,4]$, and which contains a longer word, $aabc$. Since $\{c, bc, abc\}$ are suffixes of $t[1:7]$ and $aabc$ is not, they cease to be right-equivalent in $t[1:7]$. Therefore, Z must be split into two right-equivalence classes, $Z_2 = \{c, bc, abc\}$ and $Z_1 = Z - Z_2 = \{aabc\}$. Let us call this operation a *split*.

Let us say that a node is *new* in column k if it is created when the DAWG of $t[1:k]$ is modified to obtain the DAWG of $t[1:k+1]$. In Figure 31.6, a node in a column is circled if it is new in that column. In general, a node is new in column k iff it is the top node of the column or the previous occurrence of its longest member corresponds to a blank space in the incremental landscape.

An important point is that only the top two nodes of a column can be new.

Lemma 31.4

If a new node is the result of a split, only one node lies above it in its column.

Proof. Let a be the character that causes the split, and let xa be the largest string in Z_2, and let bxa be the smallest string in $Z_1 = Z - Z_2$. Since bxa previously had the same set of ending positions as xa and now it does not, it must be that xa occurs as a suffix

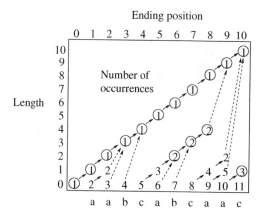

FIGURE 31.6 The incremental landscape is a simplification of the table of Figure 31.5, where an entry is displayed only if it differs from the entry above it. The entries in column i are right-equivalence classes of $t[1:i]$. These are right-equivalence classes that may be affected during the induction step, when the DAWG of $t[1:i-1]$ is modified to obtain the DAWG of $t[1:i]$. Equivalence classes of $t[i:1]$ that are not right-equivalence classes in $t[1:i]$ are circled; these correspond to nodes of the DAWG that must be created during the induction step. Edges of the DAWG of $t[1:i]$ from classes in column $i-1$ are depicted as arrows. (The distinction between solid and dashed arrows is used in the proof of Theorem 31.2.)

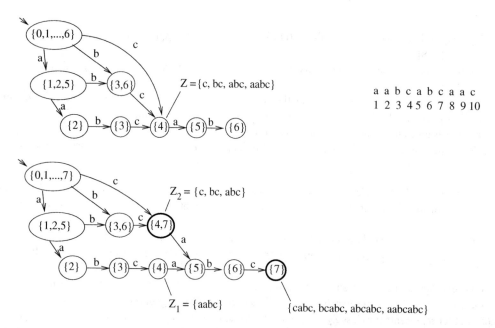

FIGURE 31.7 Modifying the DAWG of $t[1:6] = aabcab$ to obtain the DAWG of $t[1:7] = aabcabc$. New nodes are shown with bold outlines. The sixth column of the incremental landscape, from top to bottom, consists of the nodes $\{6\}$, $\{3,6\}$, and the start node. The seventh column, from top to bottom, consists of $\{7\}$, $\{4,7\}$, and the start node. The node $\{4,7\}$ is *split* from $\{4\}$; of the strings $\{c, bc, abc, aabc\}$ that end at node $\{4\}$, only $\{c, bc, abc\}$ also occur at position 7, so these must now be handled by a different node from the one that handles $aabc$. All edges from nodes in the previous column to $\{4\}$ are redirected to the new node $\{4,7\}$.

of t_k, but bxa does not. Let cxa be the smallest string in the next higher class C in column $k+1$. On all previous occurrences of xa, it was inside bxa, so the first occurrence of cxa is at position $k+1$. The frequency of the strings in C must be 1, so C is the top class of the column.   ∎

The foregoing shows how nodes must be added to the DAWG in the inductive step. In order to understand how the edges of the DAWG must be modified, note that every edge directed into a node in column $k+1$ comes from a node in column k. These edges are given by the following rule:

Lemma 31.5

In the DAWG of $t[1:k+1]$, a node of height i in column k has an edge labeled t_{k+1} to the lowest node of column $k+1$ that has height greater than i.

These edges are drawn in as solid and dashed arrows in Figure 31.6. According to the figure, when the DAWG of $t[1:7]$ is obtained from $t[1:6]$, the new top node in the column must have an incoming edge from the top node of column 6, which is labeled {6} in Figure 31.7. The second new node in column 7, which is labeled {4,7} in the figure, must have edges from the nodes at $L[0,6]$ and $L[2,6]$, which are the source and the node labeled {3,6}. These are obtained by diverting edges into Z.

The induction step consists of implicitly marching in parallel down columns k and $k+1$, creating the required nodes in column t_{k+1} and installing the appropriate edges from right-equivalence classes in column k to right-equivalence classes of column $k+1$, as well as the appropriate outgoing edges and parent pointers on each right-equivalence class in column $k+1$ that is new in t_{k+1}. The new nodes in column $k+1$ are the one with frequency one, and possibly one other, Z_2, that results from a split. By Lemma 31.4, this requires marching down through at most two nodes of column $k+1$, but possibly through many nodes of column k.

To facilitate marching down a column k efficiently, the algorithm needs a *suffix pointer* Suffix(x) on each node x of column k to the next lower node in column k. If $y = y_1y_2y_3 \ldots y_j$ is the shortest string in the x's right-equivalence class, then Suffix(x) points to the right-equivalence class that contains the longest proper suffix $y_2y_3 \ldots y_j$ of y. The suffix pointer for each node is uniquely defined, so the algorithm ensures that suffix pointers are available on nodes of column k by keeping suffix pointers current on all nodes of the DAWG.

The induction step is given in Algorithm 31.2. The algorithm does not build the incremental landscape. However, we may identify the nodes by where they would go in the incremental landscape. The meanings of the variables can be summarized as follows. Top$_k$ is the top node of column k, and Top$_{k+1}$ is the top node of column $k+1$. Mid denotes the highest node of column k that already has an outgoing labeled with the $(k+1)^{th}$ letter. The variable curNode is a looping variable that travels down through nodes of column k, becoming undefined if it travels past the bottom node of the column.

Algorithm 31.2

Update(Top$_k$): Update the DAWG of $t[1:k]$ to obtain the DAWG of $t[1:k+1]$.

> Create a node Top$_{k+1}$ of frequency 1 and height $k+1$
> Let curNode = Top$_k$
> While curNode is defined and has no outgoing edge labeled t_{k+1}
>> Install an edge labeled t_{k+1} from curNode to Top$_{k+1}$
>> curNode := Suffix(curNode)
> If curNode is defined
>> Mid := curNode
>> Let Z be the neighbor of Mid on t_{k+1}
>> Define Suffix(Top$_{k+1}$) to be Z
>> If height(Z) > height(Mid) + 1
>>> Split(k, Mid, Z); Create a second new node in Column $k+1$
> Else define Suffix(Top$_{k+1}$) to be the start node
> Return Top$_{k+1}$

Procedure Split(k, Mid, Z)

> Create a copy Z_2 of the node representing Z, together with its outgoing edges
> Let the height of Z_2 be one plus the height of Mid
> Let curNode = Mid
>
> While curNode is defined and Z is its neighbor on t_{k+1}
>> Divert curNode's edge labeled t_{k+1} so that it points to Z_2
>> curNode := Suffix(curNode)
> Redefine Suffix(Z_2) to be Suffix(Z)
> Redefine Suffix(Z) to be Z_2

Theorem 31.2

It takes $O(|t|)$ time to build the DAWG of a text t of length n.

Proof. No node is ever discarded once it is created, and the final DAWG has $O(|t|)$ nodes. Therefore, the cost of creating nodes is $O(|t|)$. Once an edge is created it remains in the DAWG, though it might be diverted in calls to Split. No edge is ever discarded and the final DAWG has $O(|t|)$ edges, so the cost of creating edges is $O(|t|)$.

It remains to bound the cost of diverting edges in calls to Split. Let an edge that appears in the incremental landscape be *solid* if it goes from a node of height i to one of height $i + 1$, and *dashed* otherwise (see Figure 31.6). We may partition the edges in the landscape into *terminating paths*, each of which starts in row 0, and contains zero or more solid edges, and either followed by a dashed edge or ending in the last column. At most one terminating path begins in any column, and every dashed edge terminates a path. Thus there are at most n dashed edges.

When Z_2 is created in Split, at most one of the edges diverted into it is solid. The cost of diverting this edge is subsumed in the cost of creating Z_2. The cost of diverting other edges is $O(|t|)$ over all calls to Split, since each of them is one of the at most n dashed edges that appear in the incremental landscape. ∎

31.4 The Compact DAWG

By Theorem 31.1 and Lemma 31.3, we cannot assume that the DAWG requires linear space if the nodes are explicitly labeled with their position sets. The algorithm for building the DAWG in linear time does not label the nodes with their position sets. However, without the labels, it is not possible to use the DAWG to find the k locations where a substring p occurs in t in $O(|p| + k)$ time.

One remedy for this problem is to label a node with a position i if it represents the smallest position set that contains i as a member. The total number of these labels is n. We can reverse the directions of the suffix pointers that are installed during the DAWG construction algorithm, yielding a tree on the position sets. If a node represents a set X of positions, the members of X can be returned in $O(|X|)$ time by traversing the subtree rooted at X, assembling a list of these labels. (This tree is isomorphic to the suffix tree of the reverse of the text, but there is no need to adopt the common practice of labeling each of its edges with a string.)

Another alternative, which has considerable advantage in space requirements over the suffix tree, is to "compact" the DAWG, yielding a smaller data structure that still supports a query about the positions of a substring $O(|p| + k)$ time. The algorithm for compacting it runs in $O(|t|)$ time.

If x is a substring of t, let $\alpha(x)$ denote the longest string y such every ending position of x is also an ending position of yx. That is, y is the maximal string that precedes every occurrence of x in t. Note that $\alpha(x)$ may be the null string. Similarly, let $\beta(x)$ denote the longest string z such that every starting position of x is a starting position of xz. This is the longest string that follows every occurrence of x.

For instance, if $t = aabcabcaac$ and $x = b$, then $\alpha(x) = a$ and $\beta(x) = ca$.

Lemma 31.6

1. For x and y in a right-equivalence class, $\alpha(x)x = \alpha(y)y$ is the longest string in the class.
2. For x and y in a right-equivalence class, $\beta(x) = \beta(y)$.

Let a substring x of t be *prime* if $\alpha(x)$ and $\beta(x)$ are both the empty string. For any substring x of t, $\alpha(x)x\beta(x)$ is prime; this is the *prime implicant* of x. If x is prime, it is its own prime implicant.

Definition 31.3

The *compact DAWG* of a text t is defined as follows. The nodes are the prime substrings of t. If x is a prime substring, then for each $a \in \Sigma$ such that xa is a substring of t, let $y = \alpha(xa)$ and $z = a\beta(xa)$. There is an edge labeled z from x to the prime implicant yxz of xa.

If a right-equivalence class contains a prime substring x, then x is the longest member of the class. Stretching the terminology slightly, let us call a class *prime* if it contains a prime substring. If C is a right-equivalence class in t, we may define $\beta(C) = \beta(x)$ such that $x \in C$. By Part 2 of Lemma 31.6, $\beta(C)$ is uniquely defined. We may define C's *prime implicant* to be the right-equivalence class D that contains $x\beta(x)$ for $x \in C$. D is also uniquely defined and contains the prime implicant of the members of C.

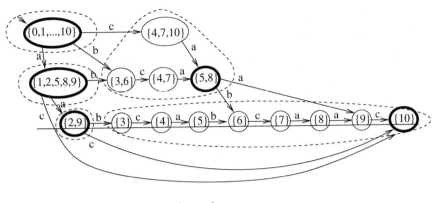

a a b c a b c a a c
1 2 3 4 5 6 7 8 9 10

FIGURE 31.8   Partition of nodes into groups with the same prime implicant.

The nodes of the DAWG may therefore be partitioned into groups that have the same prime implicant. This is illustrated in Figure 31.8.

Lemma 31.7

A right-equivalence class is nonprime if and only if it has exactly one outgoing edge in the DAWG.

We now describe how to obtain the compact DAWG from the DAWG in linear time. For ease of presentation, we describe how to carry it out in four depth-first traversals of the DAWG. However, in practice, only two depth-first traversals are required, since the operations of the first three traversals can be carried out during a single depth-first traversal.

In the first depth-first traversal, we may label each class with a single position from its set of ending positions. This is done in postorder: when retreating from a node, copy its label from the label of any of its successors, which have already been labeled, and subtract 1 from it.

By Lemma 31.7, the prime implicant of a class is the class itself if it is prime; otherwise, it is the unique successor that is prime. Let the *distance* to its prime implicant be the length of this unique path.

In postorder during the second traversal, we may label each node with a pointer to its prime implicant and label this pointer with the distance to the prime implicant. If the class C is a sink or has more than one outgoing edge, this is just a pointer from C to itself with distance label 0. Otherwise, C has a unique successor D, which is already labeled with a pointer to D's prime implicant A with distance label i. Label C with a pointer to A with distance label $i + 1$.

In the third traversal, we install the compact DAWG edges. If we label the edges explicitly with their string labels, we will exceed linear time and storage. Instead, we may take advantage of the fact that the label of every edge is a substring of t. We label each edge with the length of its label (see Figure 31.9). When retreating from a prime node B during the traversal, for each DAWG edge (BC) out of B, let D be C's prime implicant, let i be the distance of D from C. Install a compact DAWG edge from B to D that has length label $i + 1$.

On the final traversal, we may remove the DAWG nodes, DAWG edges, and the prime implication pointers.

31.4.1 Using the Compact DAWG to Find the Locations of a String in the Text

Let v be a node of the compact DAWG, and let x be the corresponding prime implicant. Let the *total length* of a path from v to the sink be the sum of the length labels of the edges on the path. Observe that there is a path of total length i from v to the sink iff x has an ending position at $n - i + 1$.

Lemma 31.8

Let x be a prime substring of t, and let k be the number of occurrences of x in t. Given x's node in the compact DAWG of t, it takes $O(k)$ time to retrieve the ending positions of x in t.

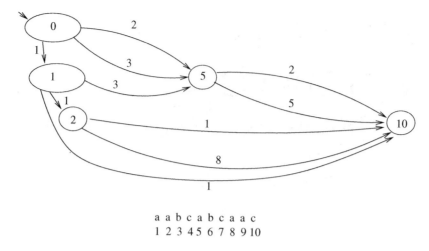

a a b c a b c a a c
1 2 3 4 5 6 7 8 9 10

FIGURE 31.9 Representing the edge labels of the compact DAWG (cf. Figure 31.2) Each edge label is a substring of t with end positions at the end position labels of the principal nodes. The label of the edge can therefore be represented implicitly by labeling each node with one member of its position set, and labeling each edge with the length of its label. For instance, the edge labeled 3 from the source to the node labeled "5" is labeled with the substring of length 3 that ends at position 5, hence, the one occupying positions 3, 4, and 5 of the text. Since the text can be randomly accessed, the text can be used to look up the label of the edge. This ensures that the edge labels take $O(|t|)$ space, since they take $O(1)$ for each node and edge.

Proof. Recursively explore all paths out of the node, and whenever the sink is encountered, subtract the total length of the current path from $n + 1$ and return it.

The running time follows from the following observations. One position is returned for each leaf of the recursion tree; the sets of positions returned by recursive calls are disjoint; and every internal node of the recursion tree has at least two children since every node of the compact DAWG has at least two outgoing edges. ∎

The representation of Figure 31.9 is therefore just as powerful as that of Figure 31.2: The edge labels are implied by accessing t using the numbers on edges and nodes, while the position labels of the vertices can be retrieved in linear time by the algorithm of Lemma 31.8.

The representation of Figure 31.9 now gives an $O(|p| + k)$ algorithm for finding the k occurrences of a substring p in a text t. One must index into the compact DAWG from the source, matching letters of p with letters of the implicit word labels of the compact edges. If a letter of p cannot be matched, then p does not occur as a subword of t. Otherwise, p is the concatenation of a set of word labels on a path, followed by part of the word label of a final edge (u, v). This takes $O(|p|)$ time. Let i be the number of remaining unmatched letters of the word label of (u, v). The k ending positions of p are given by subtracting i from the k ending positions of v, which can be retrieved in $O(k)$ time using the algorithm of Lemma 31.8.

For instance, using the compact DAWG of Figure 31.2 to find the locations where abc occurs, we match a to the label a of an edge out of the source to the node with position set $\{1, 2, 5, 8, 9\}$, then bc to the word label of the edge to the node labeled $\{5, 8\}$. Though the node is labeled with the position set in the picture, this position set is not available in the linear-space data structure. Instead, we find two paths of length 2 and 5 from this node to the sink, and subtracting 2 and 5 from $n = 10$ yields the position set $\{5, 8\}$. Then, since one letter in the word label bca remains unmatched, we subtract 1 from each member of $\{5, 8\}$ to obtain $\{4, 7\}$, which is the desired answer.

31.4.2 Variations and Applications

In Reference 6, a variation of the compact DAWG is given for a collection $\{t_1, t_2, \ldots, t_k\}$ of texts and can be used to find the k occurrences of a string p in the texts in $O(|p| + k)$ time.

That paper also gives a symmetric version of the compact DAWG. By the symmetry in the definition of the prime subwords of t, the set of prime subwords of the reversal of t are given by reversing the set of prime subwords of t. The compact DAWG of t and of the reversal of t therefore have essentially the same set of nodes; only the edges are affected by the reversal. The symmetric version has a single set of nodes and two sets of edges, one corresponding to the edges of the compact DAWG of t and one corresponding to the edges of the reversal of t. The utility of this structure as a tool for exploring the subword structure of t is described in the paper.

Another variant occurs when t is a cyclic ordering of characters, rather than a linear one. A string p has an occurrence anywhere where it matches the subword contained in an interval on this cycle. A variant of the DAWG, compact DAWG, and compact symmetric DAWG for retrieving occurrences of subwords for t in this case is given in Reference 6. The paper gives algorithms that have time bounds analogous to those given here.

Variations of landscapes, such as that of Figure 31.6, are explored in Reference 7. They give a graphical display of the structure of repetitions in a text. The suffix tree can be used to find the longest common substring of two texts t_1 and t_2 efficiently. The paper gives $O(|t|h(t))$ algorithms that use the DAWG to generate the landscape of t (see Definition 31.2), which can be used to help identify functional units in a genomic sequence. One variation of the landscape explored in the paper inputs two texts t_1 and t_2, and gives a graphical display of the number of occurrences of every substring of t_1 in t_2, which has obvious applications to the study of evolutionary relationships among organisms.

Mehta and Sahni give a generalization of the compact DAWG and the compact symmetric DAWG to circular sequences is given in Reference 8, and give techniques for analyzing and displaying the structure of strings using the compact symmetric DAWG in References 9 and 10.

31.5 The Position Heap

We now give a data structure that gives much simpler algorithms, at a cost of slightly increasing the worst-case time required for a query. The algorithms can easily be programmed by undergraduate data structure students.

The data structure is a trie and has one node for each position in the text. The data structures and algorithms can be modified to give the same bounds for construction and searching, but this undermines the principal advantages, which are simplicity and low memory requirements.

The data structure is closely related to trees that are used for storing hash keys in Reference 11.

31.5.1 Building the Position Heap

Let a string be *represented* by a trie if it is the label of a path from the root in the trie.

For analyzing the position heap we adopt the convention of indexing the characters of t in descending order, so $t = t_n t_{n-1} \ldots t_1$. In this case, we let $t[i:j]$ denote $t_i t_{i-1} \ldots t_j$.

The algorithm for constructing the position heap can be described informally as follows. The positions of t are visited from right to left as a trie is built. At each position i, a new substring z is added to the set of words represented by the trie. To do this, the longest prefix $t[i:j]$ of $t[i:1]$ that is already represented in the trie is found by indexing into the trie from the root, using the leading letters of $t[i:1]$, until one can advance no further. A leaf child of the last node of this path is added, and the edge to it is labeled t_{i+1}.

The procedure, PHBuild, is given in Algorithm 31.3. Figure 31.10 gives an illustration.

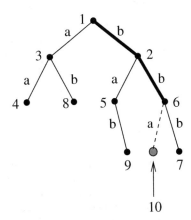

FIGURE 31.10 Construction of the position heap with PHBuild (see Algorithm 31.3). The solid edges reflect the recursively constructed position heap for the suffix $t[9:1]$ of t. To get the position heap for $t[10:1]$, we use the heap to find the largest prefix bb of $t[10:1]$ that is the label of a path in the tree, and add a new child at this spot to record the next larger prefix bba.

Algorithm 31.3

Constructing the Position Heap for a String $t = t_i t_{i-1} \ldots t_1$

PHBuild(t, i)

> If $i = 1$ return a single tree node labeled 1
> Else
>> Recursively construct the position heap H' for the suffix $t[i - 1, 1]$
>> Let $t' = t[i : k]$ be the maximal prefix of t that is the
>>> label of a path originating at the root in the tree
>> Let u be the last node of this path
>> Add a child of u to the tree on edge labeled t_{k-1}, and give it label i

31.5.2 Querying the Position Heap

Algorithm 31.4 gives a procedure, PHFind, to find all starting positions of p in t, and Figure 31.11 gives an illustration. The worst-case running time of $O(|p|^2 + k)$ to find the k occurrences of p is worse than the $O(|p| + k)$ bound for the suffix tree or DAWG.

Algorithm 31.4

Find All Places in a Text t Where a Substring p Occurs, Given the Position Heap H for t

PHFind(p, t, H)

> Let p' be the maximal prefix of p that is the label of a path P' from the root of H
> S_1 be the set of position labels in P'
> Let S_2 be the subset of S_1 that are the positions of occurrences of p in t
> If $p' \neq p$ then let S_3 be the empty set
> Else let S_3 be the position labels of descendants of the last node of P'
> Return $S_2 \cup S_3$

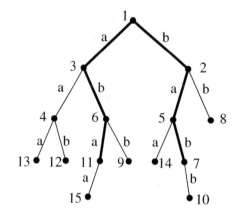

15 14 13 12 11 10 9 8 7 6 5 4 3 2 1
T: a b a a a b a b b a b a a b a

Search strings: aba and baba

FIGURE 31.11 Searching for occurrences of a string in the text t in $O(|p|^2 + k)$ time with PHFind (see Algorithm 31.4). How the search is conducted depends on whether the search string is the path label of a node in the position heap. One case is illustrated by search string *aba*, which is the path label of position 11. The only places where *aba* may match t are at positions given by ancestors and descendants of t. The descendants $\{11, 15\}$ do not need to be verified, but the proper ancestors $\{1, 3, 6\}$ must be verified against t at those positions. Of these, only 3 and 6 are matches. The algorithm returns $\{3, 6, 11, 15\}$. The other case is illustrated by *baba*, which is not the path label of a node. Indexing on it yields position 7 and path label *bab* $\neq$ *baba*. Only the ancestors $\{2, 5, 7\}$ are candidates, and they must be verified against t. Of these, only 7 is a match. The algorithm returns $\{7\}$. Since the ancestor positions occur on the search path, there are $O(|p|)$ of them, and each takes $O(|p|)$ time to verify each of them against t. Descendants can be more numerous, but take $O(1)$ time apiece to retrieve and return, since they do not need to be verified.

Lemma 31.9

`PHFind` returns all positions where p occurs in t.

Proof. Let $p = p_1 p_2 \ldots p_m$ and let $t = t_n t_{n-1} \ldots t_1$. Suppose that i is a position in t where p does not occur. Then $i \notin S_2$. Any node u with position label i has a path label that is a prefix of $t[i:1]$. Since p is not a prefix of this string, it is not a prefix of the path label of u, so $i \notin S_3$. We conclude that i is not among the positions returned by `PHFind`.

Next, let h be the position of an occurrence of p. Let $x = p[1:j]$ be the maximal prefix of p that is represented in the position heap of $t[h-1:1]$, where $j = 0$ if $x = \lambda$. If $x \neq p$, then `PHBuild` created a node with position label h and path label $x p_{j+1}$. This is a prefix of p, so $h \in S_1$, and, since p occurs at position h, $h \in S_2$. If $x = p$, let $y = t[h:k]$ be the largest prefix of $t[h:1]$ that is active in $t[h-1:1]$. Then `PHBuild` created a node with position label h and path label $y t_{k-1}$, and $h \in S_3$. In either case, h is returned as a position where P occurs. ∎

31.5.3 Time Bounds

Lemma 31.10

`PHFind` takes $O(|p|^2 + k)$ worst-case time to return the k occurrences of p in t.

Proof. The members of S_3 can be retrieved in $O(1)$ time apiece using a depth-first traversal of the subtree rooted at the last node on path P'. Since all nodes of S_1 occur on a path whose label is a prefix of p, there are at most $m + 1$ members of S_1. Checking them against t to see which are members of S_2 takes $O(|p|)$ time apiece, for a total of $O(|p|^2)$ time in the worst case. ∎

This time bound overstates what can be expected in practice, since, in most cases, the string is known to match on a prefix, but there is no reason to expect that it will be similar to the position that it is supposed to match in the region beyond this prefix. A good heuristic is to match the string from the end, rather than from the beginning, since the string has a prefix that is already known to match at the position. Checking to see whether a string matches at a given position will usually require examining one or two characters, discovering a mismatch, and rejecting the string.

Lemma 31.11

`PHBuild` takes $O(|t| h(t^R))$ time.

Proof. If $P = (v_0, v_1, \ldots, v_k)$ be a path from the root v_0 in the position heap, let $P_1 = (v_0, v_1, \ldots, v_{\lfloor k/2 \rfloor})$, and let $P_2 = (v_{\lfloor k/2 \rfloor + 1}, v_{\lfloor k/2 \rfloor + 2}, \ldots, v_k)$ be the remainder of the path. Let i be the position label of v_k, and let $h'(i)$ denote the length of the maximum prefix x of $t[i:1]$ that occurs at least $|x|$ times in t. The path label y of P_1 has an occurrence at the position labels of each of its descendants, including those on P_2, of which there are at least $|y|$. Therefore, $|y| = O(h'(i))$. The time spend by the algorithm at position i of t is proportional to the length of P, which is $O(|y|)$. Therefore, the time spent by the algorithm adding the node for position i is $O(h'(i))$, hence the time to build the whole heap is $O(\sum_{i=1}^{|t|} h'(i)) = O(|t| h(t^R))$ by Definition 31.2. ∎

As with the $O(|t| h(t))$ algorithm for building the DAWG, this time bound is a practical one in most settings, since $h(t)$ is relatively insensitive to long repeated strings or localized areas of the string with many repetitions. Only strings where most areas of the string are repeated many times elsewhere have high values of $h(t)$, and $h(t)$ can be expected to behave like an $O(\log n)$ function in most settings.

31.5.4 Improvements to the Time Bounds

In this section, we given an algorithm for constructing the position heap to $O(|t|)$. We also sketch an approach for finding the occurrences of a string p in t to $O(|p| + k)$ using position heaps. Each of these has trade-off costs, such as having greater space requirements and being harder to understand.

The position heap has a dual, which we may call the **dual heap** (see Figure 31.12). They have the same node set: a node has path label x in the heap iff its path label in the dual is the reversal x^R of x. We will refer to the position heap as the **primal heap** when we wish to contrast it to the dual.

It is tempting to think that the dual is just the position heap of the reversal t^R of t, but this is not the case. As in the primal heap, the rightmost positions of t are near the root of the dual, but in the primal heap of t^R, the leftmost positions of t are near the root. In the primal heap of t^R the heap order is inverted, which affects the shape of the tree. Neither the primal nor the dual heap of t is necessarily isomorphic to the primal or dual heap of t^R.

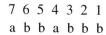

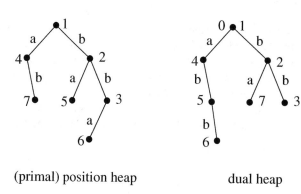

(primal) position heap   dual heap

FIGURE 31.12   The position heap and the dual heap of the string *abbabbb*. The node set of both heaps is the same, but the label of the path leading to a node in the dual heap is the reverse of the label of the path leading to it in the position heap.

Algorithm 31.5

Construct the Position Heap H and its Dual D for a String $t[i:1]$. Return (H, D, y), Where y Is a Pointer to the Node with Position Label i.

```
FastPHBuild (T, i)
```
> If $i = 1$, return a single tree node labeled 1
> Let $(H', D', v) =$ `FastPHBuild`$(t, i-1)$
> Search upward from v in H' to find the lowest ancestor v' of v that has
>   a child u on edge labeled t_i in the dual
> Let w be the penultimate node on this path
> Let $d = d_w$ be the depth of w in the heap
> Create a new child y of u in the position heap on edge labeled t_{i-d}
> Make y be the child of w in the dual on edge labeled t_i
> Give y position label i
> Give y depth label $d_y = d + 1$
> Return the modified position heap, the modified dual, and y

For `PHBuild`, the bottleneck is finding the node u whose path label is $t' = t_i t_{i+1} \ldots t_k$. The dual heap allows us to carry out this step more efficiently. We get an $O(|t|)$ time bound for constructing the position heap by simultaneously constructing the position heap and its dual. It is also necessary to label each node v with its depth d_v during construction, in addition to its position label, p_v. This gives a compact representation of the path label of v if v is not the root: it is $t[p_v : p_v - d_v + 1]$.

During construction, the primal edges are directed from child to parent, while the dual edges are directed from parent to child. The modified procedure, `FastPHBuild`, is given in Algorithm 31.5.

Lemma 31.12

`FastPHBuild` is correct.

Proof. The path label of v is $t[i-1:i-1-d_v+1] = t[i-1:i-d_v]$. Let $d = d_w$ be the depth of w. Since w is an ancestor of v, its path label is a prefix of this, so w's path label is $t[i-1:i-d]$. Since v' is the parent of w, the path label of v' is the next shorter prefix, $t[i-1:i-d+1]$. The path label of v' in the dual is the reversal of this, and since u is reachable on the dual edge out of v' that is labeled t_i, the path label of u is the reversal of $t[i:i-d+1]$ in the dual, hence $t[i:i-d+1]$ in the primal heap. Since w has no child labeled t_i in the dual, there is no node whose path label in the dual is the reversal of $t[i:i-d]$, hence no node whose path label is $t[i:i-d]$ in the primal heap.

Therefore, u has path label $t[i:i-d+1]$ and has no child in the primal graph on t_{i-d}. It follows that updating the primal heap to reflect $t[i:1]$ requires adding a new child y labeled t_{i-d_2} to u in the primal heap. Since w's path label is the longest proper suffix of y's path label, w must be the parent of y in the dual. Since its depth is one greater than w's, $d_y = d + 1$.   ∎

Lemma 31.13

`FastPHBuild` takes $O(|t|)$ time.

Proof. The inductive step takes $O(1)$ time, except for the cost of searching upward from v to find v'. Let k be the distance from v' to v and let $k' = k - 1$. The cost of searching upward is $O(k)$. The depth of the new node y is $d_v + 2$, so it is $d_v - k + 2 \leq d_v + 1$. Since v is the node added just before y, the depth of each successive node added increases by at most one and decreases by $\Theta(k)$. The total increases are $O(|t|)$, so the sum of k's overall recursive calls is bounded by this, hence also $O(|t|)$. ∎

On tests we have run on several megabyte texts, `FastPHBuild` is noticeably faster than `PHBuild`. This advantage must be weighed against the fact that the algorithm is slightly more difficult to understand and uses more memory during construction, to store the dual edges.

By contrast, the algorithm we describe next for finding the positions of p in t in $O(|p| + k)$ time is unlikely to compete in practice with `PHFind`, since the worst-case bound of $O(|p|^2 + k)$ for `PHFind` overstates the typical case. However, it is interesting from a theoretical standpoint.

Let # be a character that is not in Σ. Let $t\#t$ denote the concatenation of two copies of t with the special character # in between. To obtain the time bound for `PHFind`, we may build the position heap of $t\#t$ in $O(|t|)$ time using `FastPHBuild`. Index the positions from $|t|$ to $-|t|$ in descending order. This gives 0 as the position of the # character (see Figure 31.13).

To find the starting positions of p in t, it suffices to find only its positive starting positions in $t\#t$. Suppose that there is a path labeled p that has at most one node with a positive position number. Finding the last node v of the path takes $O(|p|)$ time, and all k positive starting positions are descendants. We can retrieve them in $O(k)$ time. Since we are not required to find negative position numbers where p occurs, we do not have the $\Theta(|p|^2)$ cost of finding which ancestors of v are actual matches. This gives an $O(|p| + k)$ bound in this case.

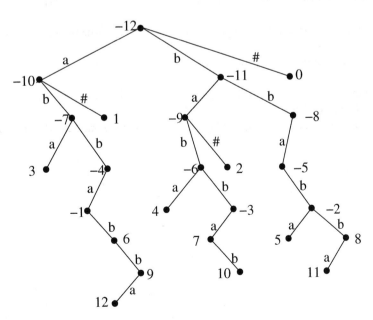

FIGURE 31.13 Finding occurrences of p in t in $O(|p| + k)$ time, using a position heap. Because of the extra memory requirements and the good expected performance of the $O(|p|^2 + k)$ approach, the algorithm is of theoretical interest only. The trick is to build the position heap of $t\#t$, indexing so that positions in the second occurrence are indexed with negative numbers. To find the occurrences of p in t, it suffices to return only its positive positions in $t\#t$. Indexing into the heap is organized so that positive positions are descendants of nodes that are indexed to. Negative occurrences, which are ancestors, do not need to be verified against the text, eliminating the $\Theta(|p|^2)$ step of the simpler algorithm.

Otherwise, the problem can be solved by chopping p into segments $\{x_1, x_2, \ldots, x_k\}$ such that each x_i is the label of a path from the root in the heap that has exactly one node v_i with a positive position number, namely, the last node of the path. Every positive position of x_i is matched by a negative position, which must correspond to an ancestor of v_i. Since there are at most $|x_i|$ ancestors of v_i, v_i has at most $|x_i|$ (positive) descendants, which can be retrieved in $O(|x_i|)$ time.

To see that this implies an $O(|p|)$ time bound to return all occurrences of p in t, the reader should first note that a family $\mathcal{F}$ of k sets $X_1, X_2, \ldots, X_k$ of integers are represented with sorted lists, it takes $O(|X_1| + |X_2| + \cdots |X_k|)$ time to find their intersection. The key to this insight is that when two sets in $\mathcal{F}$ are merged, replacing them with their intersection, the sum of cardinalities of sets in $\mathcal{F}$ drops by an amount proportional to the time to perform the intersection. Therefore, the bound for all intersections is proportional to the sum of cardinalities of the initial lists. The problem of finding the occurrences of p reduces to this one as follows. Let X_i denote the positive positions of segment x_i of p. Shift these positions to the left by $|x_1 x_2 \ldots x_{i-1}|$ to find the candidate positions they imply for the left endpoint of p in t. Intersecting the sets of candidates gives the locations of p in t.

To find the substrings $\{x_1, x_2, \ldots, x_k\}$ of p, index from the root of the position heap on the leading characters of p until a positive node is reached. The label of this path is x_1, and recursing on the remaining suffix of p gives $\{x_2, x_3, \ldots, x_{k-1}\}$. It doesn't give x_k, since an attempt to produce x_k in this way it may run out of characters of p before a node with a positive position number is reached. Instead, find x_k by indexing from the right end of p using the dual heap until a positive position number is reached. Therefore, $\{x_1, x_2, \ldots, x_{k-1}\}$ represent disjoint intervals p, while x_{k-1} and x_k can represent overlapping intervals of p. The sum of their lengths is therefore $O(|p|)$, giving an $O(|p|)$ bound to find all occurrences of p in t in this case.

References

1. E. M. McCreight. A space-economical suffix tree construction algorithm. *Journal of the ACM*, 23:262–272, 1976.
2. U. Manber and E. Myers. Suffix arrays a new method for on-line search. *SIAM Journal of Computing*, 22:935–948, 1993.
3. G. Kucherov. On-line construction of position heaps. *Journal of Discrete Algorithms*, 20:3–11, 2013.
4. Y. Nakamura, I. Tomohiro, S. Inenaga, and H. Bannai. Constructing LZ78 tries and position heaps in linear time for large alphabets. *Information Processing Letters*, 115:655–659, 2015.
5. Y. Nakamura, S. Inenaga, H. Bannai, and M. Takaheda. The position heap of a trie. *Lecture Notes in Computer Science*, 7608:360–371, 2012.
6. A. Blumer, J. Blumer, D. Ehrenfeucht, D. Haussler, and R. McConnell. Complete inverted files for efficient text retrieval and analysis. *Journal of the ACM*, 34:578–595, 1987.
7. B. Clift, D. Haussler, R. McConnell, T. D. Schneider, and G. D. Stormo. Sequence landscapes. *Nucleic Acids Research*, 14:141–158, 1986.
8. D. P. Mehta and S. Sahni. A data structure for circular string analysis and visualization. *IEEE Transactions on Computers*, 42:992–997, 1993.
9. D. P. Mehta and S. Sahni. Computing display conflicts in string visualization. *IEEE Transactions on Computers*, 43:350–361, 1994.
10. D. P. Mehta and S. Sahni. Models, techniques, and algorithms for finding, selecting and displaying patterns in strings and other discrete objects. *Journal of Systems and Software*, 39:201–221, 1997.
11. E. G. Coffman and J. Eve. File structures using hashing functions. *Communications of the ACM*, 11:13–21, 1981.

32

Binary Decision Diagrams

32.1   Introduction.. 495
32.2   Basic Concepts.. 495
32.3   Data Structure.. 498
32.4   Construction of BDDs from Boolean Expressions....................................... 499
       Binary Logic Operation • Complement Edges
32.5   Variable Ordering for BDDs 502
32.6   Zero-Suppressed BDDs... 504
32.7   Related Research Activities 506
Acknowledgments 507
References.. 507

Shin-ichi Minato
Hokkaido University

32.1 Introduction

Binary decision diagrams (BDDs) are a powerful means for computer processing of Boolean functions because in many cases, with BDDs, smaller memory space is required for storing Boolean functions, and values of functions can be calculated faster than with truth tables or logic expressions. The BDD-based techniques are now used in many application areas in computer science, for example, hardware/software system design, fault analysis of large-scale systems, constraint satisfaction problems, data mining/knowledge discovery, machine learning/classification, bioinformatics, and web data analysis.

Systematic methods for Boolean function manipulation were first studied by Shannon in 1938 [40], who applied Boolean algebra to logic network design. The concept of BDD was devised by Lee in 1959 [22]. Binary decision programs that Lee discussed are essentially BDDs. Then, in 1978, its usefulness for expression Boolean functions was shown by Akers [1]. But since the epoch-making paper by Bryant in 1986 [6], BDD-based methods have been extensively used for many practical applications.

BDD was originally developed for the efficient Boolean function manipulation required in VLSI logic design, but Boolean functions are also used for modeling many kinds of combinatorial problems. Zero-suppressed BDD (ZDD) [29] is a variant of BDD, customized for manipulating "sets of combinations." ZDDs have been successfully applied not only for VLSI design but also for solving various combinatorial problems, such as constraint satisfaction, frequent pattern mining, and graph enumeration. Recently, BDD and ZDD become more widely known, since Knuth intensively discussed BDD/ZDD-based algorithms in a volume of his book series [19].

Although a quarter of a century has passed since Bryant first put forth his idea, there are still many interesting research topics related to BDD and ZDD. For example, Knuth presented a very fast algorithm "Simpath" [19] to construct a ZDD which represents all the paths connecting two points in a given graph structure. This work is important since it may have many useful applications. Another example of recent activity is to extend BDDs to represent other kinds of discrete structures, such as sequences and permutations. In this context, new variants of BDDs called sequence BDD [22] and πDD [32] have recently been proposed and the scope of BDD-based techniques is now increasing.

In this chapter, we discuss the basic data structures and algorithms for manipulating BDDs. Then we describe the variable ordering problem, which is important for the effective use of BDDs. Finally we give a brief overview of the recent research activities related to BDDs.

32.2 Basic Concepts

From a truth table, we can easily derive the corresponding BDD. For example, the truth table in Figure. 32.1a can be converted into the BDD in Figure. 32.1b. But there are generally many BDDs for a given truth table, that is, the Boolean function expressed by

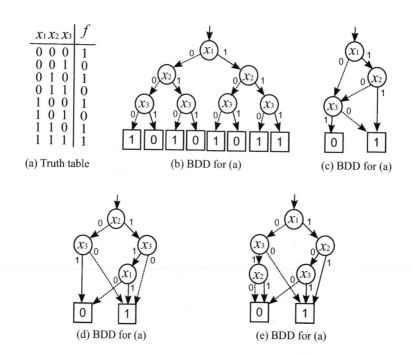

FIGURE 32.1   Truth table and BDDs for $(x_1 \wedge x_2) \vee \overline{x_3}$.

this truth table. For example, all BDDs shown in Figure 32.1c–e represent the Boolean function that the truth table in Figure 32.1a expresses. Here, note that in each of Figures 32.1b–d, the variables appear in the same order and none of them appears more than once in every path from the top node. But in (e), they appear in different orders in different paths. BDDs in Figures 32.1b–d are called **ordered BDDs** (or **OBDDs**). But the BDD in Figure 32.1(e) is called an **unordered BDD**. These BDDs can be reduced into a simple BDD by the following procedure.

In a BDD, the top node is called the **root** that represents the given function $f(x_1, x_2, \ldots, x_n)$. Rectangles that have 1 or 0 inside are called **terminal nodes**. They are also called **1-terminal** and **0-terminal**. Other nodes are called **nonterminal nodes** (or **decision nodes**) denoted by circles with variables inside. They are also called simply **nodes**, differentiating themselves from the 0- and 1-terminals. From a node with x_i inside, two lines go down. The solid line is called **1-edge**, representing $x_i = 1$; and the dotted line is called **0-edge**, representing $x_i = 0$.

In an OBDD, the value of the function f can be evaluated by following a path of edges from the root node to one of the terminal nodes. If the nodes in every path from the root node to a terminal node are assigned with variables, $x_1, x_2, \ldots,$ and x_n, in this order, then f can be expressed as follows, according to the Shannon expansion. By branching with respect to x_1 from the root node, $f(x_1, x_2, \ldots, x_n)$ can be expanded as follows, where $f(0, x_2, \ldots, x_n)$ and $f(1, x_2, \ldots, x_n)$ are functions that the nodes at the low ends of 0-edge and 1-edge from the root represent, respectively:

$$f(x_1, x_2, \ldots, x_n) = (\overline{x_1} \wedge f(0, x_2, \ldots, x_n)) \vee (x_1 \wedge f(1, x_2, \ldots, x_n))$$

Then, by branching with respect to x_2 from each of the two nodes that $f(0, x_2, \ldots, x_n)$ and $f(1, x_2, \ldots, x_n)$ represent, each $f(0, x_2, \ldots, x_n)$ and $f(1, x_2, \ldots, x_n)$ can be expanded as follows:

$$f(0, x_2, \ldots, x_n) = (\overline{x_2} \wedge f(0, 0, x_2, \ldots, x_n)) \vee (x_2 \wedge f(0, 1, x_2, \ldots, x_n))$$

and

$$f(1, x_2, \ldots, x_n) = (\overline{x_2} \wedge f(1, 0, x_2, \ldots, x_n)) \vee (x_2 \wedge f(1, 1, x_2, \ldots, x_n))$$

and so on.

As we go down from the root node toward the 0- or 1-terminal, more variables of f are set to 0 or 1. Each term excluding x_i or $\overline{x_i}$ in each of these expansions (i.e., $f(0, x_2, \ldots, x_n)$ and $f(1, x_2, \ldots, x_n)$ in the first expansion, $f(0, 0, \ldots, x_n)$ and $f(0, 1, \ldots, x_n)$ in the second expansion, etc.), are called **cofactors**. Each node at the low ends of 0-edge and 1-edge from a node in an OBDD represents cofactors of the **Shannon expansion** of the Boolean function at the node, from which these 0-edge and 1-edge come down.

Procedure 32.1

Reduction of a BDD

1. For the given BDD, apply the following steps in any order.
   a. Sharing equivalent nodes: If two nodes v_a and v_b, that represent the same variable x_i, branch to the same nodes in a lower level for each of $x_i = 0$ and $x_i = 1$, then combine them into one node that still represents variable x_i (see Figure 32.2). Similarly, two terminal nodes with the same value, 0 or 1, are merged into one terminal node with the same original value.
   b. Deleting redundant nodes: If a node that represents a variable x_i branches to the same node in a lower level for both $x_i = 0$ and $x_i = 1$, then that node is deleted, and the edges that come down to the former are extended to the latter (see Figure 32.3).
2. When we cannot apply any step after repeatedly applying these steps (a) and (b), the **reduced ordered BDD** (i.e., **ROBDD**) or simply called the **reduced BDD** is obtained for the given function.

In Figure 32.4, we show an example of applying the BDD reduction rules, starting from the same BDD shown in Figure 32.1b.

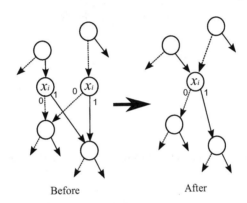

Before   After

FIGURE 32.2   Step 1(a) of Procedure 32.1 (node sharing rule).

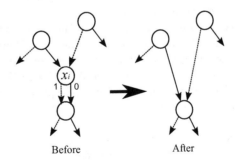

Before   After

FIGURE 32.3   Step 1(b) of Procedure 32.1 (node deletion rule).

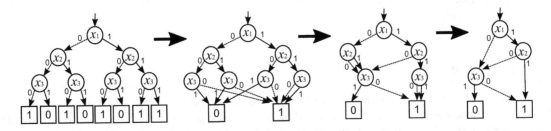

FIGURE 32.4   An example of applying node reduction rules.

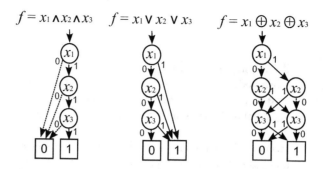

FIGURE 32.5 BDDs for typical Boolean functions.

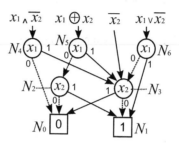

FIGURE 32.6 A shared BDD.

Theorem 32.1

Any Boolean function has a unique reduced ordered BDD and any other ordered BDD for the function in the same variable order (i.e., not reduced) has more nodes.

According to this theorem, the ROBDD is unique for a given Boolean function when the order of the variables is fixed. Thus ROBDDs give canonical forms for Boolean functions. This property is very important to practical applications, as we can easily check the equivalence of two Boolean functions by only checking the isomorphism of their ROBDDs. **Henceforth in this chapter, ROBDDs will be referred to as BDDs for the sake of simplicity.**

It is known that a BDD for an n-variable function requires an exponentially increasing memory space in the worst cases, as n increases [6]. However, the size of memory space for the BDD varies with the types of functions, unlike truth tables which always require memory space proportional to 2^n. But many Boolean functions that we encounter in design practice can be represented in BDDs without large memory space. This is an attractive feature of BDDs. Figure 32.5 shows the BDDs representing typical functions, AND, OR, and the parity function with three input variables. The parity function for n variables, $x_1 \oplus x_2 \oplus \cdots \oplus x_n$, can be represented in the BDD of $2(n-1)$ + decision nodes and 2 terminals, whereas, if a truth table or a logic expression is used, the size increases exponentially as n increases.

A set of BDDs representing many functions can be obtained into a graph that consists of BDDs sharing subgraphs among them, as shown in Figure 32.6. This idea saves time and space for duplicating isomorphic BDDs. By sharing all the isomorphic subgraphs completely, no two nodes that express the same function coexist in the graph. We call it **shared BDDs** [29] or **multirooted BDDs**. In a shared BDD, no two root nodes express the same function.

Shared BDDs are now widely used and those algorithms are more concise than ordinary BDDs. In the remainder of this chapter, we deal with shared BDD only.

32.3 Data Structure

In a typical realization of a BDD manipulator, all the nodes are stored in a single table in the main memory of the computer. Figure 32.7 is a simple example of realization for the shared BDD shown in Figure 32.6. Each node has three basic attributes: input variable and the next nodes accessed by 0- and 1-edges. Also, 0- and 1-terminals are first allocated in the table as special nodes. The other nonterminal nodes are created one by one during the execution of logic operations.

(nodeID)	variable	0-edge	1-edge	(function)
N_0	–	–	–	$\leftarrow \boxed{0}$
N_1	–	–	–	$\leftarrow \boxed{1}$
N_2	x_2	N_0	N_1	
N_3	x_2	N_1	N_0	$\leftarrow \overline{x_2}$
N_4	x_1	N_0	N_3	$\leftarrow x_1 \wedge \overline{x_2}$
N_5	x_1	N_2	N_3	$\leftarrow x_1 \oplus x_2$
N_6	x_1	N_3	N_1	$\leftarrow x_1 \vee \overline{x_2}$

FIGURE 32.7 Table-based realization of a shared BDD.

Before creating a new node, we check the reduction rules of Procedure 32.1. If the 0- and 1-edges go to the same next node (Step 1(b) of Procedure 32.1) or if an equivalent node already exists (Step 1(a) of Procedure 32.1), then we do not create a new node but simply copy the pointer to the existing node. To find an equivalent node, we check a table which displays all the existing nodes. The hash table technique is very effective to accelerate this checking. (It can be done in a constant time for any large-scale BDDs, unless the table overflows in main memories.) The procedure to get a decision node is summarized as follows.

Procedure 32.2

GetNode(variable x, 0-edge to f_0, 1-edge to f_1)

1. If ($f_0 = f_1$), just return f_0.
2. Check the hash table, and if an equivalent node f is found, then return f.
3. If no more available space, return by memory overflow error, otherwise, create a new node f with the attribute (x, f_0, f_1) and return f.

In manipulating BDDs for practical applications, we sometimes generate many intermediate (or temporary) BDDs. It is important for memory efficiency to delete such unnecessary BDDs. In order to determine the necessity of the nodes, a **reference counter** is attached to each node, which shows the number of incoming edges to the node.

In a typical implementation, the BDD manipulator consumes 30–40 bytes of memory for each node. Today, there are personal computers with 32 or 64 Gbytes of memory, and those facilitate us to generate BDDs containing as many as billions of nodes. However, the BDDs still grow beyond the memory capacity in some practical applications.

32.4 Construction of BDDs from Boolean Expressions

Procedure 32.1 shows a way of constructing compact BDDs from the truth table for a function f of n variables. This procedure, however, is not efficient because the size of its initial BDD is always of the order of 2^n, even for a very simple function. In order to avoid this problem, Bryant [6] presented a method of constructing BDDs by applying a sequence of logic operations in a Boolean expression.

Figure 32.8 summarizes the list of primitive operations for constructing and manipulating BDDs. Using these operations, we can construct BDDs for an arbitrary Boolean expression. This method is generally far more efficient than that based on a truth table.

$\emptyset$	Returns the constant-0 function. (0-terminal node)
1	Returns the constant-1 function. (1-terminal node)
$var(x)$	Returns a primitive function for the variable x. (a single decision node)
$f \wedge g$	Logical AND operation.
$f \vee g$	Logical OR operation.
$f \oplus g$	EXOR operation.
$\overline{f}$	Logical NOT operation.
$f_{(x=0)}, f_{(x=1)}$	Returns a sub-function of f when variable $x = 0$ or $x = 1$, respectively.
$f.top$	Returns the variable of the root node of f.
$f.count$	Returns the number of input assignments that satisfy $f = 1$.

FIGURE 32.8 Primitive operations for BDD manipulation.

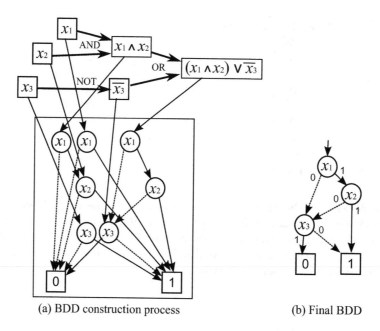

(a) BDD construction process   (b) Final BDD

FIGURE 32.9   Constructing BDDs for $f = (x_1 \wedge x_2) \vee \overline{x_3}$.

Figure 32.9 shows a simple example of constructing a BDD for $f = (x_1 \wedge x_2) \vee \overline{x_3}$. First, trivial BDDs for $x_1, x_2,$ and x_3 are created in Figure 32.9a. Next, applying the AND operation between x_1 and x_2, the BDD for $(x_1 \wedge x_2)$ is then generated. The BDD for $\overline{x_3}$ can be derived by applying NOT operation to the BDD for x_3. Then the BDD for the entire expression is obtained as the result of the OR operation between $(x_1 \wedge x_2)$ and $\overline{x_3}$. After deleting the nodes that are not on the paths from the root node for f toward the 0- or 1-terminal, the final BDD is obtained as shown in Figure 32.9b.

In the following, we describe a formal procedure for generating a BDD by applying binary logic operation between two BDDs.

32.4.1 Binary Logic Operation

Suppose we perform a binary operation between two functions f and g, and this operation is denoted by $f \circ g$, where $\circ$ is one of OR, AND, EXOR, and others. Then by the Shannon expansion of a function explained previously, $f \circ g$ can be expanded as follows:

$$f \circ g = \overline{x_i}\,(f_{(x_i=0)} \circ g_{(x_i=0)}) \vee x_i(f_{(x_i=1)} \circ g_{(x_i=1)})$$

with respect to variable x_i, which is the variable of the node that is in the highest level among all the nodes in f and g. This expansion creates a new decision node with the variable x_i having 0- and 1-edges pointing to the two nodes generated by suboperations $(f_{(x_i=0)} \circ g_{(x_i=0)})$ and $(f_{(x_i=1)} \circ g_{(x_i=1)})$. This formula means that the operation can be expanded to the two sub-operations $(f_{(x_i=0)} \circ g_{(x_i=0)})$ and $(f_{(x_i=1)} \circ g_{(x_i=1)})$. Repeating this expansion recursively for all the variables, as we go down toward the 0- or 1-terminal, we will have eventually trivial operations, such as $f \cdot 1 = f, f \oplus f = 0$, and $0 \vee g = g$, and the result is obtained.

The procedure of constructing $h \circ f \circ g$ is summarized as follows. Here f_0 and f_1 are the BDDs pointed to by the 0- and 1-edges from the root node of f.

Procedure 32.3

Construction of a BDD for Function $h \circ f \circ g$, given BDDs for f and g

1. When f or g is a constant or when $f = g$:
   return a result according to the type of the operation
   (e.g., $f \cdot 1 = f, f \oplus f = 0, 0 \vee g = g$)
2. If $f.top$ and $g.top$ are identical:
   $h_0 \leftarrow f_0 \circ g_0$; $h_1 \leftarrow f_1 \circ g_1$;
   $h \leftarrow \text{GetNode}(f.top, h_0, h_1)$
3. If $f.top$ is higher than $g.top$:
   $h_0 \leftarrow f_0 \circ g$; $h_1 \leftarrow f_1 \circ g$;
   $h \leftarrow \text{GetNode}(f.top, h_0, h_1)$

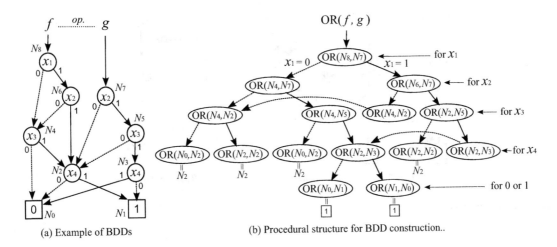

(a) Example of BDDs　　　　　(b) Procedural structure for BDD construction..

FIGURE 32.10　Procedural structure of generating BDDs for $h = f \vee g$.

4. If $f.top$ is lower than $g.top$:
   compute similarly to 3 by exchanging f and g.

As mentioned in the previous section, we check the BDD reduction rules before creating a new nodes to avoid duplication of the equivalent nodes.

Figure 32.10a shows an example of execution. When we perform the operation between the node N_8 and N_7, the procedure is broken down into the binary tree, as shown in Figure 32.10b. We usually execute the procedure calls according to this tree structure in a depth-first manner, and it may require an exponential time for the number of variables. However, some of these subtrees may be redundant; for instance, the operations of $N_4 - N_2$ and $N_2 - N_3$ appear more than once. We can accelerate the procedure by using a hash-based cache that memorizes the results of recent operations. We call it **operation cache**. By referring to the operation cache before every recursive call, we can avoid duplicate executions for equivalent suboperations. This technique enables the binary logic operations to be executed in a time almost proportional to the total size of BDDs for f, g, and h. (For detailed discussions, refer the article [45].)

It is too memory-consuming to keep all the operation results for any different pairs of BDD nodes, therefore, we usually prepare a limited size of the operation cache. If some of operation results have been lost, then some number of redundant suboperations may be called, but anyway, the correct BDD can be constructed by applying BDD reduction rules. The size of operation cache has a significant influence to the performance of BDD manipulation. If the cache size is insufficient, the more redundant procedure calls increase, the more results are overwritten, and the total execution time grows rapidly. Usually we set up the cache size empirically, for instance, it is proportional to the total number of BDD nodes created in the system.

32.4.2 Complement Edges

A BDD for $\overline{f}$, the complement of f, has a form similar to that of the BDD for f: just the 0- and the 1-terminals are exchanged. Complemental BDDs contain the same number of nodes, in contrast to the conjunctive/disjunctive normal forms (CNFs/DNFs), which sometimes suffer an exponential increase of the data size.

By using binary logic operation ($f \oplus 1$), we can generate a BDD for $\overline{f}$ in a linear time for the BDD size. However, the operation is improved to a constant time by using **complemented edges**. The complemented edge is a technique to reduce computation time and space for BDDs by using attributed edges that indicate to complement the function of the subgraph pointed to by the edge, as shown in Figure 32.11a. This idea was first shown by Akers [1] and later discussed by Madre et al. [23]. Today, this technique is commonly used in most of BDD packages. The use of complement edges brings the following outstanding merits:

- The BDD size is reduced by up to half.
- NOT operation can be performed in a constant time.
- Binary logic operations are sped up by applying the rules, such as $f \wedge \overline{f} = 0$, $f \vee \overline{f} = 1$, and $f \oplus \overline{f} = 1$.

Use of complement edges may break the uniqueness of BDDs. Therefore, we have to put the two rules as illustrated in Figure 32.11b:

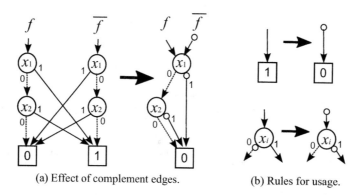

(a) Effect of complement edges.   (b) Rules for usage.

FIGURE 32.11   Complement edges.

- Using the 0-terminal only
- Not using a complement edges at the 0-edge of any node (i.e., it can be used at 1-edge only). If necessary, the complement edges can be carried over to higher nodes.

32.5 Variable Ordering for BDDs

BDDs are a canonical representation of Boolean functions under a fixed order of input variables. A change of the order of variables, however, may yield different BDDs of significantly different sizes for the same function. The effect of variable ordering depends on Boolean functions, sometimes dramatically changing the size of BDDs. Variable ordering is an important problem in using BDDs.

It is generally very time-consuming to find the best order. Friedman et al. presented an algorithm [15] of $O(n^2 3^n)$ time based on dynamic programming, where n is the number of variables. Although this algorithm has been improved [12], existing methods are limited to run on the small size of BDDs with up to about 20 variables. This problem is known as an NP-complete problem [44], and it would be difficult to find the best order for larger problems in reasonably short processing time. However, if we can find a fairly good order, it is useful for practical applications. There are many research works on heuristic methods of variable ordering.

Empirically, the following properties are known on the variable ordering.

1. Closely-related variables:
   Variables that are in close relationship in a Boolean expression should be close in variable order. For example, the Boolean function of AND–OR two-level logic $(x_1 \wedge x_2) \vee (x_3 \wedge x_4) \vee \ldots \vee (x_{2n-1} \wedge x_{2n})$ has a BDD with $2n$ decision nodes in the best order as shown for $n = 4$ in Figure 32.12a, while it needs $(2^{(n+1)} - 2)$ nodes in the worst order, as shown in Figure 32.12b.
2. Influential variables:
   The variables that greatly influence the nature of a function should be at higher position. For example, the 8-to-1 data selector shown in Figure 32.13a can be represented by a linear size of BDD when the three control inputs are ordered high, but when the order is reversed, it becomes of exponentially increasing size as the number of variables increases, as shown in Figure 32.13b.

Based on empirical rules like this, Fujita et al. [13] and Malik et al. [25] presented methods; in these methods, an output of the given logic networks is reached, traversing in a depth-first manner, then an input variable that can be reached by going back toward the inputs of the network is placed at highest position in variable ordering. Minato [33] devised another heuristic method based on a weight propagation procedure for the given logic network. Butler et al. [5] proposed another heuristic based on a measure which uses not only the connection configuration of the network but also the output functions of the network. These methods probably find a good order before generating BDDs. They find good orders in many cases, but there is no method that is always effective to a given network.

Another approach reduces the size of BDDs by reordering input variables. A greedy local exchange (swapping adjacent variables) method was developed by Fujita et al. [14]. Minato [28] presented another reordering method which measures the width of BDDs as a cost function. In many cases, these methods find a fairly good order using no additional information. A drawback of the approach of these methods is that they cannot start if an initial BDD is too large.

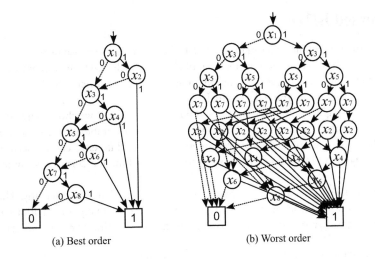

(a) Best order   (b) Worst order

FIGURE 32.12   BDDs for $(x_1 \wedge x_2) \vee (x_3 \wedge x_4) \vee (x_5 \wedge x_6) \vee (x_7 \wedge x_8)$.

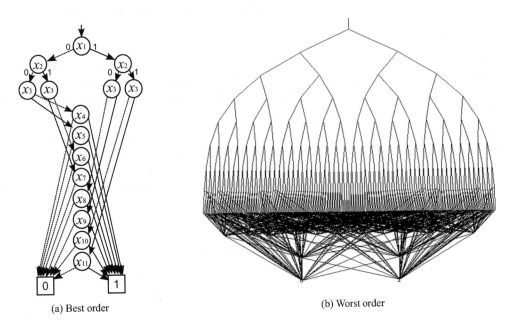

(a) Best order   (b) Worst order

FIGURE 32.13   BDDs for 8-to-1 data selector.

One remarkable work is **dynamic variable ordering**, presented by Rudell [39]. In this technique, the BDD manipulation system itself determines and maintains the variable order. Every time the BDD grow to a certain size, the reordering process is invoked automatically. This method is very effective in terms of the reduction of BDD size, although it sometimes takes a long computation time.

Unfortunately, there are some hard examples where variable ordering is powerless. For example, Bryant [6] proved that an n-bit multiplier function requires an exponentially increasing number of BDD nodes for any variable order, as n increases. From a theoretical viewpoint, the total number of different n-input Boolean functions is 2^{2^n}, and the number of permutation of n variables is only $n!$, which is much less than 2^{2^n}. This means that after variable ordering, the variation of Boolean functions is still a double exponential number, and we still need exponential size of data structure in the worst case to distinguish all the functions. Thus we expect that the variable ordering methods would not be effective for generally randomized functions. However, empirically we can say that, for many practical functions, the variable ordering methods are very useful for generating a small size of BDDs in a reasonably short time.

32.6 Zero-Suppressed BDDs

BDDs are originally developed for handling Boolean function data, however, they can also be used for implicit representation of combinatorial sets. Here we call **set of combinations** for a set of elements each of which is a combination out of n items. This data model often appears in real-life problems, such as combinations of switching devices, Boolean item sets in the database, and combinatorial sets of edges or nodes in the graph data model.

A set of combinations can be mapped into Boolean space of n input variables. If we choose any one combination of items, a Boolean function determines whether the combination is included in the set of combinations. Such Boolean functions are called *characteristic functions*. The set operations such as union, intersection, and difference can be performed by logic operations on characteristic functions. Figure 32.14 shows an example of the Boolean function corresponding to the set of combinations $F = \{ac, b\}$.

By using BDDs for characteristic functions, we can manipulate sets of combinations efficiently. They can be generated and manipulated within a time roughly proportional to the BDD size. When we handle many combinations including similar patterns (subcombinations), BDDs are greatly reduced by node sharing effect, and sometimes an exponential reduction benefit can be obtained.

Zero-suppressed BDD (ZDD or ZBDD) [29,30] is a special type of BDD for efficient manipulation of sets of combinations. Figure 32.15 shows an example of BDD and ZDD for the same set of combinations $F = \{ac, b\}$.

ZDDs are based on the following special reduction rules:

- Delete all nodes whose 1-edge directly points to the 0-terminal node, and jump through to the 0-edge's destination, as shown in Figure 32.16
- Share equivalent nodes as well as ordinary BDDs.

Notice that we do not delete the nodes whose two edges point to the same node, which used to be deleted by the original rule. The zero-suppressed deletion rule is asymmetric for the two edges, as we do not delete the nodes whose 0-edge points to a terminal node. It is proved that ZDDs also give canonical forms as well as ordinary BDDs under a fixed variable ordering.

Here we summarize the features of ZDDs.

- In ZDDs, the nodes of irrelevant items (never chosen in any combination) are automatically deleted by ZDD reduction rule. In ordinary BDDs, irrelevant nodes still remain and they may cause unnecessary computation.
- ZDDs are especially effective for representing sparse combinations. If the average appearance ratio of each item is 1%, ZDDs are possibly up to 100 times more compact than ordinary BDDs. Such situations often appear in real-life problems,

a	b	c	F
0	0	0	0
0	0	1	0
0	1	0	1
0	1	1	0
1	0	0	0
1	0	1	1
1	1	0	0
1	1	1	0

As a Boolean function:
$$F = (a \wedge \overline{b} \wedge c) \vee (\overline{a} \wedge b \wedge \overline{c})$$

As a set of combinations:
$$F = \{ac, \ b\}$$

FIGURE 32.14   Correspondence of Boolean function and set of combinations.

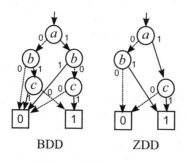

BDD   ZDD

FIGURE 32.15   BDD and ZDD for the same set of combinations $F = \{ac, b\}$.

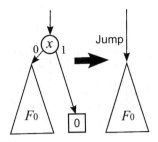

FIGURE 32.16 Reduction rule for ZDDs.

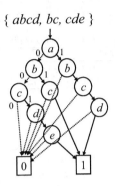

FIGURE 32.17 Explicit representation with ZDD.

for example, in a supermarket, the number of items in a customer's basket is usually much less than the number of all the items displayed.

- Each path from the root node to the 1-terminal node corresponds to each combination in the set. Namely, the number of such paths in the ZDD equals to the number of combinations in the set. In ordinary BDDs, this property does not always hold.
- When no equivalent nodes exist in a ZDD, that is the worst case, the ZDD structure explicitly stores all items in all combinations, as well as using an explicit linear linked list data structure. Namely, (the order of) ZDD size never exceeds the explicit representation. An example is shown in Figure 32.17. If more nodes are shared, the ZDD is more compact than linear list. Ordinary BDDs have larger overhead to represent sparser combinations while ZDDs have no such overhead.

Figure 32.18 shows an example of constructing ZDDs. As well as ordinary BDDs, any set of combinations can be generated by a sequence of primitive operations. Figure 32.19 summarizes most of the primitive operations of the ZDDs. In these operations,

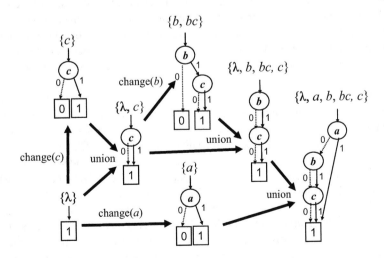

FIGURE 32.18 Example of constructing ZDDs.

Primitive operations corresponding to Boolean algebra

$\emptyset$	Returns empty set. (0-terminal node)
$\{\lambda\}$	Returns the set of a null-combination. (1-terminal node)
$F.offset(x)$	Subset of combinations not including item x.
$F.onset(x)$	Gets $F \setminus F.offset(x)$ and then deletes x from each combination.
$F.change(x)$	Inverts the existence of x (add/delete) on each combination.
$F \cup G$	Returns the union set.
$F \cap G$	Returns the intersection set.
$F \setminus G$	Returns the difference set. (in F but not in G.)
$F.top$	Returns the item of the root node of F.
$F.count$	Counts number of combinations.

Extended operations for sets of combinations [29]

$F * G$	Cartesian product of F and G.
F/G	Quotient of F divided by G.
$F\%G$	Remainder of F divided by G.

FIGURE 32.19 Primitive ZDD operations.

$\emptyset$, $\{\lambda\}$, and $F.top$ can be obtained in constant time. Here λ means a null combination. $F.offset(x)$, $F.onset(x)$, and $F.change(x)$ operations require a constant time if x is the top variable of F, otherwise they require a time linear in the number of ZDD nodes located at a higher position than x. The union, intersection, and difference operations can be performed in a time that is linear in the size of the ZDDs in many practical cases.

The last three operations in the table constitute an interesting algebra for sets of combinations with multiplication and division. Knuth has been inspired by this idea and has developed more various algebraic operations, such as $F \sqcap G$, $F \sqcup G$, $F \boxplus G$, $F \nearrow G$, $F \searrow G$, $F^{\uparrow}$, and $F^{\downarrow}$. He called these ZDD-based operations a "Family Algebra" in the recent fascicle of his book series [19].

The original BDD was developed for VLSI logic design, but ZDD is now recognized as the most important variant of BDD, and is widely used in various kinds of problems in computer science [10,17,21,35].

32.7 Related Research Activities

In the last section of this chapter, we present an overview of research activities related to BDDs and ZDDs. Of course, it is impossible to cover all the research activities in this limited space, so there may be many other interesting topics which are not shown here.

In 1986, Bryant proposed the algorithm called "Apply" [6]. It was the origin of much work on the use of BDDs as modern data structures and algorithms for efficient Boolean function manipulation. Just after Bryant's paper, implementation techniques for the BDD manipulation, such as hash-table operations and memory management techniques, emerged [4,33]. A BDD is a unique representation of a given Boolean function. However, if the order of input variables is changed, a different BDD is obtained for what is essentially the same Boolean function. Since the size of BDDs greatly depends on the order of input variables, variable ordering methods were intensively developed in the early days [12,14,39,44]. Those practical techniques were implemented in a software library called "BDD package." Currently, several academic groups provide such packages as open software (e.g., CUDD package [42]).

At first, BDDs were applied to equivalence checking of logic circuits [7,13,25] and logic optimization [26] in VLSI logic design. Next, BDD-based symbolic manipulation techniques were combined with the already known theory of model checking. It was really a breakthrough that formal verification becomes possible for some practical sizes of sequential machines [3,27]. After that, many researchers became involved with formal hardware verification using BDDs, and Clarke received the Turing Award in 2008 for this work. In addition, the BDD-based symbolic model checking method led to the idea of bounded model checking using SAT solvers [2,37,42]. These research produced many practical applications of SAT solvers that are widely utilized today.

ZDDs [29] deal with sets of combinations, representing a model that is different from Boolean functions. However, the original motivation of developing ZDDs was also for VLSI logic design. ZDDs were first used for manipulating very large-scale logic expressions with an AND–OR two-level structure (also called DNF or CNF), namely, representing a set including very large number of combinations of input variables. Sets of combinations often appear not only in VLSI logic design area but also in various areas of computer science. It is known that ZDDs are effective for handling many kinds of constraint satisfaction problems in graph theory and combinatorics [10,38].

After the enthusiastic research activities in 1990s, one can observe a relatively inactive period between 1999 and 2005. We may call these years the "Winter of BDDs." On that time, most basic implementation techniques had been matured, and BDD applications to VLSI design tools seemed to be almost exhausted. So, many researchers moved from BDD-related work to other research areas, such as actual VLSI chip design issues or SAT-based problem solving.

However, after about 2005, many people understood that BDDs/ZDDs are useful not only for VLSI logic design but also in various other areas, and then BDD-related research activity was revived. For example, we can see some applications to data mining [21,31,36], Bayesian network and probabilistic inference models [17,35], and game theory [43]. More recently, new types of BDD variants, which have not been considered before, have been proposed. Sequence BDDs [22] represent sets of strings or sequences, and πDD [32] represent sets of permutations.

Synchronizing with this new movement, the BDD section of Knuth's book was published [19]. As Knuth presented the potential for wide-ranging applications of BDDs and ZDDs, these data structures and algorithms were recognized as fundamental techniques for whole fields of information science. In particular, his book includes the "Simpath" algorithm which constructs a ZDD representing all the connecting paths between two points in a given graph structure, and a surprisingly fast program is provided for public use on his web page. Experimental results suggest that this algorithm is not just an exercise but is the most efficient method using current technology. Based on this method, the author's research group is now developing extended and generalized algorithms, called "Frontier-based methods," for efficiently enumerating and indexing various kinds of discrete structures [16,18,46].

As the background of this resurgence and new generation applications, we should note the great progress of the computer hardware system, especially the increase of main memory capacity. Actually, in the early days of using BDDs in 1990s, there was some literature on applications for intelligent information processing. Madre and Coudert proposed a TMS (Truth Maintenance System) using BDDs for automatic logic inference and reasoning [24]. A method of probabilistic risk analysis for large industrial plants was also considered [9]. Coudert also proposed a fast method of constructing BDDs to represent prime sets (minimal support sets) for satisfying Boolean functions [8], which is a basic operation for logic inference. However, at that time, the main memory capacities of high specification computers were only 10–100 megabytes, about 10,000 times smaller than those available today, and thus, only small BDDs could be generated.

In the VLSI design process, the usual approach was that the whole circuit was divided into a number of small submodules, and each submodule was designed individually by hand. So, it was natural that the BDD-based design tools are used for the sufficiently small submodules which could be handled with a limited main memory capacity. With the progress of computer hardware performance, BDD-based methods could gradually be applied to larger submodule. On the other hand, in the applications of data mining or knowledge processing, the input data were stored on a very large hard disks. The processor loaded a small fragment of data from the hard disk into the main memory and executed some meaningful operations on the data in the main memory, then the processed data was saved with the original data on the hard disk. Such procedures were common, but it was very difficult to apply BDD-based methods to hard disk data. After 2000, computers' main memory capacity grew rapidly, and in many practical cases, all the input data can be stored in the main memory. Thus many kinds of in-memory algorithms could be actively studied for data processing applications. The BDD/ZDD algorithm is a typical instance of such in-memory techniques.

In view of the above technical background, most research on BDDs/ZDDs after the winter period is not just about remaking old technologies. It also compares classical and well-known efficient methods (such as suffix trees, string matching, and frequent pattern mining) with the BDD/ZDD-based methods, in order to propose combined or improved techniques to obtain the current best performance. For example, Darwiche, who is a well-known researcher in data structures of probabilistic inference models, is very interested in the techniques of BDDs, and he recently proposed a new data structure, the SDD (Sentential Decision Diagram) [11], to combine BDDs with the classical data structures of knowledge databases. As shown in this example, we had better collaborate with many researchers in various fields of information science to import the current best known techniques into BDD/ZDD related work, and then we may develop more efficient new data structures and algorithms for discrete structure manipulation.

Acknowledgments

This work was supported, in part, by JST ERATO Minato Discrete Structure Manipulation System Project and by JSPS KAKENHI Grant Number 15H05711. A part of this article is based on my old work [34] coauthored with the late Professor Saburo Muroga.

References

1. S. B. Akers. Binary decision diagrams. *IEEE Transactions on Computers*, C-27(6): 509–516, 1978.
2. A. Biere, A. Cimatti, E. M. Clarke, M. Fujita, and Y. Zhu. Symbolic model checking using sat procedures instead of BDDs. In *Proceedings of the 36th Annual ACM/IEEE Design Automation Conference*, DAC'99, pages 317–320, New Orleans, LA, 1999. ACM.

3. J. R. Burch, E. M. Clarke, K. L. McMillan, and David L. Dill. Sequential circuit verification using symbolic model checking. In *Proceedings of the 27th ACM/IEEE Design Automation Conference*, DAC'90, pages 46–51, Orland, FL, 1990. IEEE Computer Society Press.

4. K. S. Brace, R. L. Rudell, and R. E. Bryant. Efficient implementation of a BDD package. In *Proceedings of the 27th ACM/IEEE Design Automation Conference*, DAC'90, pages 40–45, Orland, FL, 1990. IEEE Computer Society Press.

5. K. M. Butler, D. E. Ross, R. Kapur, and M. R. Mercer. Heuristics to compute variable orderings for efficient manipulation of ordered binary decision diagrams, In *Proceedings of 28th ACM/IEEE Design Automation Conference (DAC'88)*, pages 417–420, San Francisco, CA, June 1991.

6. R. E. Bryant. Graph-based algorithms for Boolean function manipulation. *IEEE Transactions on Computers*, C-35(8): 677–691, 1986.

7. O. Coudert and J. C Madre. A unified framework for the formal verification of sequential circuits. In *Proceedings of IEEE International Conference on Computer-Aided Design (ICCAD'90)*, pages 126–129, Santa Clara, CA, 1990. IEEE Computer Society Press.

8. O. Coudert and J. C. Madre. Implicit and incremental computation of primes and essential primes of Boolean functions. In *Proceedings of the 29th ACM/IEEE Design Automation Conference*, DAC'92, pages 36–39, Anaheim, CA, 1992. IEEE Computer Society Press.

9. O. Coudert and J. C. Madre. Towards an interactive fault tree analyser. In *Proceedings of IASTED International Conference on Reliability, Quality Control and Risk Assessment*, Washington D.C., 1992.

10. O. Coudert. Solving graph optimization problems with ZBDDs. In *Proceedings of ACM/IEEE European Design and Test Conference (ED&TC'97)*, pages 224–228, Munchen, Germany, 1997. IEEE Computer Society Press.

11. A. Darwiche. SDD: A new canonical representation of propositional knowledge bases. In *Proceedings of 22nd International Joint Conference of Artificial Intelligence (IJCAI'2011)*, pages 819–826, Barcelona, Spain, 2011. AAAI Press.

12. R. Drechsler, N. Drechsler, and W. Günther. Fast exact minimization of BDDs. In *Proceedings of the 35th Annual Design Automation Conference*, DAC'98, pages 200–205, San Francisco, CA, 1998. ACM.

13. M. Fujita, H. Fujisawa, and N. Kawato. Evaluation and implementation of boolean comparison method based on binary decision diagrams. In *Proceedings of IEEE International Conference on Computer-Aided Design (ICCAD'88)*, pages 2–5, Santa Clara, CA, 1988. IEEE Computer Society Press.

14. M. Fujita, Y. Matsunaga, and T. Kakuda. On variable ordering of binary decision diagrams for the application of multi-level logic synthesis. In *Proceedings of the Conference on European Design Automation*, EURO-DAC'91, pages 50–54, Amsterdam, Netherlands, 1991. IEEE Computer Society Press.

15. S. J. Friedman and K. J. Supowit. Finding the optimal variable ordering for binary decision diagrams, In *Proceedings of the 24th ACM/IEEE Design Automation Conference*, DAC'87, Miami Beach, FL, pages 348–356, 1987. ACM.

16. H. Iwashita, J. Kawahara, and S. Minato. ZDD-based computation of the number of paths in a graph. *Hokkaido University, Division of Computer Science, TCS Technical Reports*, TCS-TR-A-10-60, 2012.

17. M. Ishihata, Y. Kameya, T. Sato, and S. Minato. Propositionalizing the EM algorithm by BDDs. In *Proceedings of 18th International Conference on Inductive Logic Programming (ILP'2008)*, Prague, Czech Republic, 2008. Springer.

18. T. Inoue, K. Takano, T. Watanabe, J. Kawahara, R. Yoshinaka, A. Kishimoto, K. Tsuda, S. Minato, and Y. Hayashi. Loss minimization of power distribution networks with guaranteed error. Hokkaido University, Division of Computer Science, TCS Technical Reports, TCS-TR-A-10-59, 2012.

19. D.E. Knuth. *Bitwise Tricks & Techniques; Binary Decision Diagrams*, volume 4, fascicle 1. Addison-Wesley, Boston, MA, 2009.

20. C. Y. Lee. Representation of switching circuits by binary-decision programs. *Bell System Technical Journal*, 38(4):985–999, 7 1959.

21. E. Loekit, J Bailey. Fast mining of high dimensional expressive contrast patterns using zero-suppressed binary decision diagrams. In *Proceedings of the Twelfth ACM SIGKDD International Conference on Knowledge Discovery and Data Mining (KDD'2006)*, pages 307–316, Philadelphia, PA, 2006. ACM.

22. E. Loekito, J. Bailey, and J. Pei. A binary decision diagram based approach for mining frequent subsequences. *Knowledge and Information Systems*, 24(2):235–268, 2010.

23. J. C. Madre and J. P Billon. Proving circuit correctness using formal comparison between expected and extracted behaviour. In *Proceedings of the 25th ACM/IEEE Design Automation Conference (DAC'88)*, pages 205–210, Anaheim, CA, 1988. IEEE Computer Sciety Press.

24. J. C. Madre and O. Coudert A logically complete reasoning maintenance system based on a logical constraint solver. In *Proceedings of the 12th International Joint Conference on Artificial Intelligence—Volume 1*, pages 294–299, San Francisco, CA, 1991. Morgan Kaufmann Publishers Inc.

25. S. Malik, A. Wang, R. K. Brayton, and A. S.- Vincentelli. Logic verification using binary decision diagrams in a logic synthesis environment. In *Proceedings of the IEEE International Conference on Computer-Aided Design (ICCAD'88)*, pages 6–9, Santa Clara, CA, 1988. IEEE Computer Society Press.

26. Y. Matsunaga, M Fujita. Multi-level logic optimization using binary decision diagrams. In *Proceedings of the IEEE International Conference on Computer-Aided Design (ICCAD'89)*, pages 556–559, Santa Clara, CA, 1989. IEEE Computer Society Press.

27. K. L. McMillan. *Symbolic Model Checking*. Kluwer Academic Publishers, 1993.

28. S. Minato. Minimum-width method of variable ordering for binary decision diagrams. *IEICE Transactions on Fundamentals*, E75-A(3): 392–399, 1992.

29. S. Minato. Zero-suppressed BDDs for set manipulation in combinatorial problems. In *Proceedings of the 30th ACM/IEEE Design Automation Conference (DAC'93)*, pages 272–277, Dallas, TX, 1993. ACM.

30. S. Minato. Zero-suppressed BDDs and their applications. *International Journal on Software Tools for Technology Transfer*, 3:156–170, Springer, Berlin/Heidelberg.

31. S. Minato. VSOP (valued-sum-of-products) calculator for knowledge processing based on zero-suppressed BDDs. In K. P.SEP Jantke, A. Lunzer, N. Spyratos, and Y. Tanaka, editors, *Federation over the Web*, volume 3847 of *Lecture Notes in Computer Science*, pages 40–58. Springer, Berlin/Heidelberg, 2006.

32. S. Minato. π DD: A new decision diagram for efficient problem solving in permutation space. In K. A. Sakallah and L. Simon, editors, *Theory and Applications of Satisfiability Testing—SAT 2011*, volume 6695 of *Lecture Notes in Computer Science*, pages 90–104. Springer, Berlin/Heidelberg, 2011.

33. S. Minato, N. Ishiura, and S. Yajima. Shared binary decision diagram with attributed edges for efficient Boolean function manipulation. In *Proceedings of 27th ACM/IEEE Design Automation Conference*, pages 52–57, Orlando, FL, 1990. IEEE Computer Society Press.

34. S. Minato and S. Muroga. Binary decision diagrams. In W.-K. Chen, editor, *The VLSI Handbook*, chapter 26, pages 26.1–26.14. CRC/IEEE Press, 1999.

35. S. Minato, K. Satoh, and T. Sato. Compiling Bayesian networks by symbolic probability calculation based on zero-suppressed BDDs, In *Proceedings of 20th International Joint Conference of Artificial Intelligence (IJCAI'2007)*, pages 2550–2555, Hyderabad, India, 2007, AAAI Press. 2007.

36. S. Minato, T. Uno, and H. Arimura. LCM over ZBDDs: Fast generation of very large-scale frequent itemsets using a compact graph-based representation. In T. Washio, E. Suzuki, K. Ting, and A. Inokuchi, editors, *Advances in Knowledge Discovery and Data Mining*, volume 5012 of *Lecture Notes in Computer Science*, pages 234–246. Springer, Berlin/Heidelberg, 2008.

37. M. W. Moskewicz, C. F. Madigan, Y. Zhao, L. Zhang, and S. Malik. Chaff: Engineering an efficient sat solver, In *Proceedings of the 38th Annual Design Automation Conference*, DAC'01, pages 530–535, Las Vegas, NV, 2001. ACM.

38. H. Okuno, S. Minato, and H. Isozaki. On the properties of combination set operations. *Information Processing Letters*, 66:195–199, 1998. Elsevier.

39. R. Rudell. Dynamic variable ordering for ordered binary decision diagrams. In Proceedings of the 1993 IEEE International Conference on Computer-Aided Design, ICCAD'93, pages 42–47, Santa Clara, CA, 1993. IEEE Computer Society Press.

40. C.E. Shannon. A symbolic analysis of relay and switching circuits. *Transactions of the American Institute of Electrical Engineers*, 57(12):713–723, 1938.

41. F. Somenzi. Cudd: Cu decision diagram package, 1997. http://vlsi.colorado.edu/ fabio/CUDD/.

42. J. P. M. Silva, and K. A Sakallah. Grasp—A new search algorithm for satisfiability, In *Proceedings of the IEEE/ACM International Conference on Computer-Aided Design (ICCAD'96)*, pages 220–227, San Jose, CA, 1996. IEEE Computer Society Press/ACM.

43. Y. Sakurai, S. Ueda, A. Iwasaki, S. Minato, and M. Yokoo. A compact representation scheme of coalitional games based on multi-terminal zero-suppressed binary decision diagrams. In D. Kinny, J-j. Hsu, G. Governatori, and A. K. Ghose, editors, *Agents in Principle, Agents in Practice*, volume 7047 of *Lecture Notes in Computer Science*, pages 4–18. Springer, Berlin/Heidelberg, 2011.

44. S. Tani, K. Hamaguchi, and S. Yajima. The complexity of the optimal variable ordering problems of shared binary decision diagrams. In K. W. Ng, P. Raghavan, N. V. Balasubramanian, and F. Y. L. Chin, editors, *Algorithms and Computation*, volume 762 of *Lecture Notes in Computer Science*, pages 389–398. Springer, Berlin/Heidelberg, 1993.

45. R. Yoshinaka, J. Kawahara, S. Denzumi, H.SEP Arimura, and S. Minato. Counterexamples to the long-standing conjecture on the complexity of BDD binary operations. *Information Processing Letters*, 112(16): 636–640, 2012. Elsevier.

46. R. Yoshinaka, T. Saitoh, J. Kawahara, K. Tsuruma, H.SEP Iwashita, and S. Minato. Finding all solutions and instances of numberlink and slitherlink by ZDDs. *Algorithms*, 5(2):176–213, 2012. Elsevier.

33

Persistent Data Structures*

33.1	Introduction	511
33.2	Algorithmic Applications of Persistent Data Structures	513
33.3	General Techniques for Making Data Structures Persistent	516
	The Fat Node Method • Node Copying and Node Splitting • Handling Arrays • Making Data Structures Confluently Persistent	
33.4	Making Specific Data Structures More Efficient	522
	Redundant Binary Counters • Persistent Deques	
33.5	Concluding Remarks and Open Questions	526
	Acknowledgments	526
	References	526

Haim Kaplan
Tel Aviv University

33.1 Introduction

Think of the initial configuration of a data structure as version zero, and of every subsequent update operation as generating a new version of the data structure. Then a data structure is called *persistent* if it supports access to all versions and it is called *ephemeral* otherwise. The data structure is *partially persistent* if all versions can be accessed but only the newest version can be modified. The structure is *fully persistent* if every version can be both accessed and modified. The data structure is *confluently persistent* if it is fully persistent and has an update operation which combines more than one version. Let the version graph be a directed graph where each node corresponds to a version and there is an edge from node V_1 to a node V_2 if and only of V_2 was created by an update operation to V_1. For partially persistent data structure the version graph is a path; for fully persistent data structure the version graph is a tree; and for confluently persistent data structure the version graph is a directed acyclic graph (DAG).

A notion related to persistence is that of purely functional data structures. (See Chapter 41 by Okasaki in this handbook.) A purely functional data structure is a data structure that can be implemented without using an assignment operation at all (say using just the functions car, cdr, and cons, of pure lisp). Such a data structure is automatically persistent. The converse, however, is not true. There are data structures which are persistent and perform assignments.

Since the seminal paper of Driscoll, Sarnak, Sleator, and Tarjan (DSST) [1], and over the past fifteen years, there has been considerable development of *persistent* data structures. Persistent data structures have important applications in various areas such as functional programming, computational geometry and other algorithmic application areas.

The research on persistent data structures splits into two main tracks. The first track is of designing general transformations that would make any ephemeral data structure persistent while introducing low overhead in space and time. The second track is on how to make specific data structures, such as lists and search trees, persistent. The seminal work of DSST mainly addresses the question of finding a general transformation to make any data structure persistent. In addition DSST also address the special case of making search trees persistent in particular. For search trees they obtain a result which is better than what one gets by simply applying their general transformation to, say, red-black trees.

There is a *naive scheme* to make any data structure persistent. This scheme performs the operations exactly as they would have been performed in an ephemeral setting but before each update operation it makes new copies of all input versions. Then it performs the update on the new copies. This scheme is obviously inefficient as it takes time and space which is at least linear in the size of the input versions.

When designing an efficient general transformation to make a data structure persistent DSST get started with the so called *fat node method*. In this method you allow each field in the data structure to store more than one value, and you tag each value by

---

* This chapter has been reprinted from first edition of this Handbook, without any content updates.

the version which assigned it to the field. This method is easy to apply when we are interested only in a partially persistent data structure. But when the target is a fully persistent data structure, the lack of linear order on the versions already makes navigation in a naive implementation of the fat node data structure inefficient. DSST manage to limit the overhead by linearizing the version tree using a data structure of Dietz and Sleator so we can determine fast whether one version precedes another in this linear order.

Even when implemented carefully the fat node method has logarithmic (in the number of versions) time overhead to access or modify a field of a particular node in a particular version. To reduce this overhead DSST described two other methods to make data structures persistent. The simpler one is the *node copying method* which is good to obtain partially persistent data structures. For obtaining fully persistent data structures they suggest the *node splitting method*. These methods simulate the fat node method using nodes of constant size. They show that if nodes are large enough (but still of constant size) then the amount of overhead is constant per access or update of a field in the ephemeral data structure.

These general techniques suggested by DSST have some limitations. First, all these methods, including even the fat node method, fail to work when the data structure has an update operation which combines more than one version, and confluent persistence is desired. Furthermore, the node splitting and node copying methods apply only to pointer based data structures (no arrays) where each node is of constant size. Since the simulation has to add reverse pointers to the data structure the methods require nodes to be of bounded indegree as well. Last, the node coping and the node splitting techniques have $O(1)$ amortized overhead per update or access of a field in the ephemeral data structure. DSST left open the question of how to make this overhead $O(1)$ in the worst case.

These limitations of the transformations of DSST were addressed by subsequent work. Dietz and Raman [2] and Brodal [3] addressed the question of bounding the worst case overhead of an access or an update of a field. For partial persistence Brodal gives a way to implement node coping such that the overhead is $O(1)$ in the worst case. For fully persistence, the question of whether there is a transformation with $O(1)$ worst case overhead is still unresolved.

The question of making data structures that use arrays persistent with less than logarithmic overhead per step has been addressed by Dietz [4]. Dietz shows how to augment the fat node method with a data structure of van Emde Boaz, Kaas, and Zijlstra [5,6] to make an efficient fully persistent implementation of an array. With this implementation, if we denote by m the number of updates, then each access takes $O(\log \log m)$ time, an update takes $O(\log \log m)$ expected amortized time and the space is linear in m. Since we can model the memory of a RAM by an array, this transformation of Dietz can make any data structure persistent with slowdown double logarithmic in the number of updates to memory.

The question of how to make a data structure with an operation that combines versions confluently persistent has been recently addressed by Fiat and Kaplan [7]. Fiat and Kaplan point out the fundamental difference between fully persistent and confluently persistent data structures. Consider the naive scheme described above and assume that each update operation creates constantly many new nodes. Then, as long as no update operation combines more than one version, the size of any version created by the naive scheme is linear in the number of versions. However when updates combine versions the size of a single version can be exponential in the number of versions. This happens in the simple case where we update a linked list by concatenating it to itself n times. If the initial list is of size one then the final list after n concatenations is of size 2^n.

Fiat and Kaplan prove by simple information theoretic argument that for any general reduction to make a data structure confluently persistent there is a DAG of versions which cannot be represented using only constant space per assignment. Specifically, Fiat and Kaplan define the *effective depth of the DAG* which is the logarithm of the maximum number of different paths from the root of the DAG to any particular vertex. They show that the number of bits that may be required for assignment is at least as large as the *effective depth of the DAG*. Fiat and Kaplan also give several methods to make a data structure confluently persistent. The simplest method has time and space overhead proportional to the depth of the DAG. Another method has overhead proportional to the effective depth of the DAG and degenerate to the fat node method when the DAG is a tree. The last method reduces the time overhead to be polylogarithmic in either the depth of the DAG or the effective depth of the DAG at the cost of using randomization and somewhat more space.

The work on making specific data structures persistent has started even prior to the work of DSST. Dobkin and Munro [8] considered a persistent data structure for computing the rank of an object in an ordered set of elements subject to insertions and deletions. Overmars [9] improved the time bounds of Dobkin and Munro and further reduced the storage for the case where we just want to determine whether an element is in the current set or not. Chazelle [10] considered finding the predecessor of a new element in the set. As we already mentioned DSST suggest two different ways to make search trees persistent. The more efficient of their methods has $O(\log n)$ worst case time bound and $O(1)$ worst case space bound for an update.

A considerable amount of work has been devoted to the question of how to make concatenable double ended queues (deques) confluently persistent. Without catenation, one can make deques fully persistent either by the general techniques of DSST or via real-time simulation of the deque using stacks (see [11] and the references there). Once catenation is added, the problem of making stacks or deques persistent becomes much harder, and the methods mentioned above fail. A straightforward use of balanced trees gives a representation of persistent catenable deques in which an operation on a deque or deques of total size n takes $O(\log n)$ time. Driscoll, Sleator, and Tarjan [12] combined a tree representation with several additional ideas to obtain an implementation of persistent catenable stacks in which the kth operation takes $O(\log \log k)$ time. Buchsbaum and Tarjan [13] used a recursive decomposition of trees to obtain two implementations of persistent catenable deques. The first has a time bound of $2^{O(\log^* k)}$ and the second a time bound of $O(\log^* k)$ for the kth operation, where $\log^* k$ is the iterated logarithm, defined by $\log^{(1)} k = \log_2 k, \log^{(i)} k = \log \log^{(i-1)} k$ for $i > 1$, and $\log^* k = \min\{i \mid \log^{(i)} k \leq 1\}$.

Finally, Kaplan and Tarjan [11] gave a real-time, purely functional (and hence confluently persistent) implementation of deques with catenation in which each operation takes $O(1)$ time in the worst case. A related structure which is simpler but not purely functional and has only amortized constant time bound on each operation has been given by Kaplan, Okasaki, and Tarjan [14]. A key ingredient in the results of Kaplan and Tarjan and the result of Kaplan, Okasaki, and Tarjan is an algorithmic technique related to the redundant digital representations devised to avoid carry propagation in binary counting [15]. If removing elements from one side of the deque is disallowed. Okasaki [16] suggested another confluently persistent implementation with $O(1)$ time bound for every operation. This technique is related to path reversal technique which is used in some union-find data structures [17].

Search trees also support catenation and split operations [18] and therefore confluently persistent implementation of search trees is natural to ask for. Search trees can be made persistent and even confluently persistent using the path copying technique [1]. In path copying you copy every node that changes while updating the search tree and its ancestors. Since updates to search trees affect only a single path, this technique results in copying at most one path and thereby costs logarithmic time and space per update. Making finger search trees confluently persistent is more of a challenge, as we want to prevent the update operation to propagate up on the leftmost and rightmost spines of the tree. This allows an update to be made at distance d from the beginning or end of the list in $O(\log d)$ time. Kaplan and Tarjan [19] used the redundant counting technique to make finger search tree confluently persistent. Using the same technique they also managed to reduce the time (and space) overhead of catenation to be $O(\log \log n)$ where n is the number of elements in the larger tree.

The structure of the rest of this chapter is as follows. Section 33.2 describes few algorithms that use persistent data structures to achieve their best time or space bounds. Section 33.3 surveys the general methods to make data structures persistent. Section 33.4 gives the highlights underlying persistent concatenable deques. We conclude in Section 33.5.

33.2 Algorithmic Applications of Persistent Data Structures

The basic concept of persistence is general and may arise in any context where one maintains a record of history for backup and recovery, or for any other purpose. However, the most remarkable consequences of persistent data structures are specific algorithms that achieve their best time or space complexities by using a persistent data structure. Most such algorithms solve geometric problems but there are also examples from other fields. In this section we describe few of these algorithms.

The most famous geometric application is the algorithm for planar point location by Sarnak and Tarjan [20] that triggered the development of the whole area. In the planar point location problem we are given a subdivision of the Euclidean plane into polygons by n line segments that intersect only at their endpoints. The goal is to preprocess these line segments and build a data structure such that given a query point we can efficiently determine which polygon contains it. As common in this kind of computational geometry problem, we measure a solution by three parameters: The space occupied by the data structure, the preprocessing time, which is the time it takes to build the data structure, and the query time.

Sarnak and Tarjan suggested the following solution (which builds upon previous ideas of Dobkin and Lipton [21] and Cole [22]). We partition the plane into vertical slabs by drawing a vertical line through each vertex (intersection of line segments) in the planar subdivision. Notice that the line segments of the subdivision intersecting a slab are totally ordered. Now it is possible to answer a query by two binary searches. One binary search locates the slab that contains the query, and another binary search locates the segment preceding the query point within the slab. If we associate with each segment within a slab, the polygon just above it, then we have located the answer to the query. If we represent the slabs by a binary search tree from left to right, and the segments within each slab by a binary search tree sorted from bottom to top, we can answer a query in $O(\log n)$ time.* However if we build a separate search tree for each slab then the worst case space requirement is $\Omega(n^2)$, when $\Omega(n)$ lines intersect $\Omega(n)$ slabs.

The key observation is that the sets of line segments intersecting adjacent slabs are similar. If we have the set of one particular slab we can obtain the set of the slab to its right by deleting segments that end at the boundary between these slabs, and inserting segments that start at that boundary. As we sweep all the slabs from left to right we get that in total there are n deletions and n insertions; one deletion and one insertion for every line segment. This observation reduces the planar point location to the problem of maintaining partially persistent search trees. Sarnak and Tarjan [20] suggested a simple implementation of partially persistent search tree where each update takes $O(\log n)$ amortized time and consumes $O(1)$ amortized space. Using these search

---

* Note that testing whether a point is above or below a line takes $O(1)$ time.

trees they obtained a data structure for planar point location that requires $O(n)$ space, takes $O(n \log n)$ time to build, and can answer each query in $O(\log n)$ time.

The algorithm of Sarnak and Tarjan for planar point location in fact suggests a general technique for transforming a 2-dimensional geometric search problem into a persistent data structure problem. Indeed several applications of this technique have emerged since Sarnak and Tarjan published their work [23]. As another example consider the problem of 3-sided range searching in the plane. In this problem we preprocess a set of n points in the plane so given a triple (a, b, c) with $a \leq b$ we can efficiently reports all points $(x, y) \in S$ such that $a \leq x \leq b$, and $y \leq c$. The priority search tree of McCreight [24] yields a solution to this problem with $O(n)$ space, $O(n \log n)$ preprocessing time, and $O(\log n)$ time per query. Using persistent data structure, Boroujerdi and Moret [23] suggest the following alternative. Let $y_1 \leq y_2 \leq \cdots \leq y_n$ be the y-coordinates of the points in S in sorted order. For each i, $1 \leq i \leq n$ we build a search tree containing all i points $(x, y) \in S$ where $y \leq y_i$, and associate that tree with y_i. Given this collection of search tree we can answer a query (a, b, c) in $O(\log n)$ time by two binary searches. One search uses the y coordinate of the query point to find the largest i such that $y_i \leq c$. Then we use the search tree associated with y_i to find all points (x, y) in it with $a \leq x \leq b$. If we use partially persistent search trees then we can build the trees using n insertions so the space requirement is $O(n)$, and the preprocessing time is $O(n \log n)$.

This technique of transforming a 2-dimensional geometric search problem into a persistent data structure problem requires only a partially persistent data structure. This is since we only need to modify the last version while doing the sweep. Applications of fully persistent data structures are less common. However, few interesting ones do exist.

One such algorithm that uses a fully persistent data structure is the algorithm of Alstrup et al. for the binary dispatching problem [25]. In object oriented languages there is a hierarchy of classes (types) and method names are overloaded (i.e., a method may have different implementations for different types of its arguments). At run time when a method is invoked, the most specific implementation which is appropriate for the arguments has to be activated. This is a critical component of execution performance in object oriented languages. Here is a more formal specification of the problem.

We model the class hierarchy by a tree T with n nodes, each representing a class. A class A which is a descendant of B is *more specific* than B and we denote this relation by $A \leq B$ or $A < B$ if we know that $A \neq B$. In addition we have m different implementations of methods, where each such implementation is specified by a name, number of arguments, and the type of each argument. We shall assume that $m > n$, as if that is not the case we can map nodes that do not participate in any method to their closest ancestor that does participate in $O(n)$ time. A method invocation is a query of the form $s(A_1, \ldots, A_d)$ where s is a method name that has d arguments with types $A_1, \ldots, A_d$, respectively. An implementation $s(B_1, \ldots, B_d)$ is applicable for $s(A_1, \ldots, A_d)$ if $A_i \leq B_i$ for every $1 \leq i \leq d$. The *most specific* method which is applicable for $s(A_1, \ldots, A_d)$ is the method $s(B_1, \ldots, B_d)$ such that $A_i \leq B_i$ for $1 \leq i \leq d$, and for any other implementation $s(C_1, \ldots, C_d)$ which is applicable for $s(A_1, \ldots, A_d)$ we have $B_i \leq C_i$ for $1 \leq i \leq d$. Note that for $d > 1$ this may be ambiguous, that is, we might have two applicable methods $s(B_1, \ldots, B_d)$ and $s(C_1, \ldots, C_d)$ where $B_i \neq C_i$, $B_j \neq C_j$, $B_i \leq C_i$ and $C_j \leq B_j$. The dispatching problem is to find for each invocation the most specific applicable method if it exists. If it does not exist or in case of ambiguity, "no applicable method" or "ambiguity" has to be reported, respectively. In the binary dispatching problem, $d \neq 2$, that is, we assume that all implementations and invocations have two arguments.

Alstrup et al. describe a data structure for the binary dispatching problem that use $O(m)$ space, $O(m(\log \log m)^2)$ preprocessing time and $O(\log m)$ query time. They obtain this data structure by reducing the problem to what they call the *bridge color problem*. In the bridge color problem the input consists of two trees T_1 and T_2 with edges, called bridges, connecting vertices in T_1 to vertices in T_2. Each bridge is colored by a subset of colors from C. The goal is to construct a data structure which allows queries of the following form. Given a triple (v_1, v_2, c) where $v_1 \in T_1$, $v_2 \in T_2$, and $c \in C$ finds the bridge (w_1, w_2) such that

1. $v_1 \leq w_1$ in T_1, and $v_2 \leq w_2$ in T_2, and c is one of the colors associated with (w_1, w_2).
2. There is no other such bridge (w', w'') with $v_2 \leq w'' < w_2$ or $v_1 \leq w' < w_1$.

If there is no bridge satisfying the first condition the query just returns nothing and if there is a bridge satisfying the first condition but not the second we report "ambiguity." We reduce the binary dispatching problem to the bridge color problem by taking T_1 and T_2 to be copies of the class hierarchy T of the dispatching problem. The set of colors is the set of different method names. (Recall that each method name may have many implementations for different pairs of types.) We make a bridge (v_1, v_2) between $v_1 \in T_1$ and $v_2 \in T_2$ whenever there is an implementation of some method for classes v_1 and v_2. We color the bridge by all names of methods for which there is an implementation specific to the pair of type (v_1, v_2). It is easy to see now that when we invoke a method $s(A_1, A_2)$ the most specific implementation of s to activate corresponds to the bridge colored s connecting an ancestor of v_1 to an ancestor of v_2 which also satisfies Condition (2) above.

In a way which is somewhat similar to the reduction between static two dimensional problem to a dynamic one dimensional problem in the plane sweep technique above, Alstrup et al. reduce the static bridge color problem to a similar dynamic problem on a single tree which they call the *tree color problem*. In the tree color problem you are given a tree T, and a set of colors C. At any time each vertex of T has a set of colors associated with it. We want a data structure which supports the updates, color(v,c):

which add the color c to the set associated with v; and uncolor(v,c) which deletes the color c from the set associated with v. The query we support is given a vertex v and a color c, find the closest ancestor of v that has color c.

The reduction between the bridge color problem and the tree color problem is as follows. For each node $v \in T_1$ we associate an instance ℓ_v of the tree color problem where the underlying tree is T_2 and the set of colors C is the same as for the bridge color problem. The label of a node $w \in T_2$ in ℓ_v contains color c if w is an endpoint of a bridge with color c whose endpoint in T_1 is an ancestor of v. For each pair (w, c) where $w \in T_2$ and c is a color associated with w in ℓ_v we also keep the closest ancestor v' to v in T_1 such that there is a bridge (v', w) colored c. We can use a large (sparse) array indexed by pairs (w, c) to map each such pair to its associated vertex. We denote this additional data structure associated with v by a_v. Similarly for each vertex $u \in T_2$ we define an instance ℓ_u of the tree color problem when the underlying tree is T_1, and the associated array a_u.

We can answer a query (v_1, v_2, c) to the bridge color data structure as follows. We query the data structure ℓ_{v_1} with v_2 to see if there is an ancestor of v_2 colored c in the coloring of T_2 defined by ℓ_{v_1}. If so we use the array a_{v_1} to find the bridge (w_1, w_2) colored c where $v_1 \leq w_1$ and $v_2 \leq w_2$, and w_1 is as close as possible to v_1. Similarly we use the data structures ℓ_{v_2} and a_{v_2} to find the bridge (w_1, w_2) colored c where $v_1 \leq w_1$ and $v_2 \leq w_2$, and w_2 is as close as possible to v_2, if it exists. Finally if both bridges are identical then we have the answer to the query (v_1, v_2, c) to the bridge color data structure. Otherwise, either there is no such bridge or there is an ambiguity (when the two bridges are different).

The problem of this reduction is its large space requirement if we represent each data structure ℓ_v, and a_v for $v \in T_1 \cup T_2$ independently.* The crucial observation though is that these data structures are strongly related. Thus if we use a dynamic data structure for the tree color problem we can obtain the data structure corresponding to w from the data structure corresponding to its parent using a small number of modifications. Specifically, suppose we have generated the data structures ℓ_v and a_v for some $v \in T_1$. Let w be a child of v in T_1. We can construct ℓ_w by traversing all bridges whose one endpoint is w. For each such bridge (w, u) colored c, we perform color(u,c), and update the entry of (u, c) in a_v to contain w.

So if we were using fully persistent arrays and a fully persistent data structure for the tree color problem we can construct all data structures mentioned above while doing only $O(m)$ updates to these persistent data structures. Alstrup et al. [25] describe a data structure for the tree color problem where each update takes $O(\log \log m)$ expected time and query time is $O(\log m / \log \log m)$. The space is linear in the sum of the sizes of the color-sets of the vertices. To make it persistent without consuming too much space Alstrup et al. [25] suggest how to modify the data structure so that each update makes $O(1)$ memory modifications in the worst case (while using somewhat more space). Then by applying the technique of Dietz [4] (see also Section 33.3.3) to this data structure we can make it fully persistent. The time bounds for updates and queries increase by a factor of $O(\log \log m)$, and the total space is $O(|C|m)$. Similarly, we can make the associated arrays a_v fully persistent. The resulting solution to the binary dispatching problem takes $O(m(\log \log m)^2)$ time to construct, requires $O(|C|m)$ space and support a query in $O(\log m)$ time. Since the number of memory modifications while constructing the data structure is only $O(m)$ Alstrup et al. also suggest that the space can be further reduces to $O(m)$ by maintaining the entire memory as a dynamic perfect hashing data structure.

Fully persistent lists proved useful in reducing the space requirements of few three dimensional geometric algorithms based on the sweep line technique, where the items on the sweep line have secondary lists associated with them. Kitsios and Tsakalidis [26] considered hidden line elimination and hidden surface removal. The input is a collection of (non intersecting) polygons in three dimensions. The hidden line problem asks for the parts of the edges of the polygons that are visible from a given viewing position. The hidden surface removal problem asks to compute the parts of the polygons that are visible from the viewing position.

An algorithm of Nurmi [27] solves these problems by projecting all polygons into a collection of possible intersecting polygons in the plane and then sweeping this plane, stopping at any vertex of a projected polygon, or crossing point of a pair of projected edges. When the sweep stops at such point, the visibility status of its incident edges is determined. The algorithm maintain a binary balanced tree which stores the edges cut by the sweep line in sorted order along the sweep line. With each such edge it also maintains another balanced binary tree over the faces that cover the interval between the edge and its successor edge on the sweep line. These faces are ordered in increasing depth order along the line of sight. An active edge is visible if the topmost face in its list is different from the topmost face in the list of its predecessor. If n is the number of vertices of the input polygons and I is the number of intersections of edges on the projection plane then the sweep line stops at $n + I$ points. Looking more carefully at the updates one has to perform during the sweep, we observe that a constant number of update operations on balanced binary search trees has to be performed non destructively at each point. Thus, using fully persistent balanced search trees one can implement the algorithm in $O((n + I) \log n)$ time and $O(n + I)$ space. Kitsios and Tsakalidis also show that by rebuilding the data structure from scratch every $O(n)$ updates we can reduce the space requirement to $O(n)$ while retaining the same asymptotic running time.

Similar technique has been used by Bozanis et al. [28] to reduce the space requirement of an algorithm of Gupta et al. [29] for the rectangular enclosure reporting problem. In this problem the input is a set S of n rectangles in the plane whose sides are parallel to the axes. The algorithm has to report all pairs (R, R') of rectangles where $R, R' \in S$ and R encloses R'. The algorithm uses the equivalence between the rectangle enclosure reporting problem and the 4-dimensional dominance problem. In the

---

* We can compress the sparse arrays using hashing but even if we do that the space requirement may be quadratic in m.

4-dimensional dominance problem the input is a set of n points P in four dimensional space. A point $p \neq (p_1, p_2, p_3, p_4)$ dominates $p' \neq (p_1', p_2', p_3', p_4')$ if and only if $p_i \geq p_i'$ for $i \neq 1,2,3,4$. We ask for an algorithm to report all *dominating pairs* of points, (p, p'), where $p, p' \in P$, and p dominates p'. The algorithm of Gupta et al. first sorts the points by all coordinates and translates the coordinates to ranks so that they become points in U^4 where $U \neq \{0,1,2,\ldots,n\}$. It then divides the sets into two equal halves R and B according to the forth coordinate (R contains the points with smaller forth coordinate). Using recurrence on B and on R it finds all dominating pairs (p, p') where p and p' are either both in B or both in R. Finally it finds all dominating pairs (r, b) where $r \in R$ and $b \in B$ by iterating a plane sweeping algorithm on the three dimensional projections of the points in R and B. During the sweep, for each point in B, a list of points that it dominates in R is maintained. The size of these lists may potentially be as large as the output size which in turn may be quadratic. Bozanis et al. suggest to reduce the space by making these lists fully persistent, which are periodically being rebuilt.

33.3 General Techniques for Making Data Structures Persistent

We start in Section 33.3.1 describing the fat node simulation. This simulation allows us to obtain fully persistent data structures and has an optimal space expansion but time slowdown logarithmic in the number of versions. Section 33.3.2 describes the node copying and the node splitting methods that reduce the time slowdown to be constant while increasing the space expansion only by a constant factor. In Section 33.3.3 we address the question of making arrays persistent. Finally in Section 33.3.4 we describe simulation that makes data structures confluently persistent.

33.3.1 The Fat Node Method

DSST first considered the *fat node method*. The fat node method works by allowing a field in a node of the data structure to contain a list of values. In a partial persistent setting we associate field value x with version number i, if x was assigned to the field in the update operation that created version i.[*] We keep this list of values sorted by increasing version number in a search tree. In this method simulating an assignment takes $O(1)$ space, and $O(1)$ time if we maintain a pointer to the end of the list. An access step takes $O(\log m)$ time where m is the number of versions.

The difficulty with making the fat node method work in a fully persistent setting is the lack of total order on the versions. To eliminate this difficulty, DSST impose a total order on the versions consistent with the partial order defined by the version tree. They call this total order the *version list*. When a version i is created it is inserted into the version list immediately after its parent (in the version tree). This implies that the version list defines a preorder on the version tree where for any version i, the descendants of i in the version tree occur consecutively in the version list, starting with i.

The version list is maintained in a data structure that given two versions x and y allows to determine efficiently whether x precedes y. Such a data structure has been suggested by Dietz and Sleator [30]. (See also a simpler related data structure by [31].) The main idea underlying these data structures is to assign an integer label to each version so that these labels monotonically increase as we go along the list. Some difficulty arises since in order to use integers from a polynomial range we occasionally have to relabel some versions. For efficient implementation we need to control the amount of relabeling being done. We denote such a data structure that maintains a linear order subject to the operation $insert(x, y)$ which inserts x after y, and $order(x, y)$ which returns "yes" if x precedes y, an *Order Maintenance* (OM) data structure.

As in the partial persistence case we keep a list of version-value pairs in each field. This list contains a pair for each value assigned to the field in any version. These pairs are ordered according to the total order imposed on the versions as described above. We maintain these lists such that the value corresponding to field f in version i is the value associated with the largest version in the list of f that is not larger than i. We can find this version by carrying out a binary search on the list associated with the field using the OM data structure to do comparisons.

To maintain these lists such that the value corresponding to field f in version i is the value associated with the largest version in the list of f that is not larger than i, the simulation of an update in the fully persistent setting differ slightly from what happens in the partially persistent case. Assume we assign a value x to field f in an update that creates version i. (Assume for simplicity that this is the only assignment to f during this update.) First we add the pair (i, x) to the list of pairs associated with field f. Let i' be the version following i in the version list (i.e., in the total order of all versions) and let i'' be the version following i in the list associated with f. (If there is no version following i in one of these lists we are done.) If $i'' > i'$ then the addition of the pair (i, x) to the list of pairs associated with f may change the value of f in all versions between i' and the version preceding i'' in the version list, to be x. To fix that we add another pair (i', y) to the list associated with f, where y is the value of f before the assignment of x to f. The overhead of the fat node method in a fully persistent settings is $O(\log m)$ time and $O(1)$ space per assignment, and $O(\log m)$ time per access step, where m is the number of versions. Next, DSST suggested two methods to reduce the logarithmic

---

[*] If the update operation that created version i assigned to a particular field more than once we keep only the value that was assigned last.

time overhead of the fat node method. The simpler one obtains a partially persistent data structure and is called *node copying*. To obtain a fully persistent data structure DSST suggested the *node splitting* method.

33.3.2 Node Copying and Node Splitting

The node-copying and the node splitting methods simulate the fat node method using nodes of constant size. Here we assume that the data structure is a pointer based data structure where each node contains a constant number of fields. For reasons that will become clear shortly we also assume that the nodes are of constant bounded in-degree, that is, the number of pointer fields that contains the address of any particular node is bounded by a constant.

In the node copying method we allow nodes in the persistent data structure to hold only a fixed number of field values. When we run out of space in a node, we create a new copy of the node, containing only the newest value of each field. Let d be the number of pointer fields in an ephemeral node and let p be the maximum in-degree of an ephemeral node. Each persistent node contains d fields which corresponds to the fields in the ephemeral node, p *predecessor fields*, e *extra fields*, where e is a sufficiently large constant that we specify later, and one field for a *copy pointer*.

All persistent nodes which correspond to the same ephemeral node are linked together in a single linked list using the copy pointer. Each field in a persistent node has a version stamp associated with it. As we go along the chain of persistent nodes corresponding to one ephemeral node then the version stamps of the fields in one node are no smaller than version stamps of the fields in the preceding nodes. The last persistent node in the chain is called *live*. This is the persistent node representing the ephemeral node in the most recent version which we can still update. In each live node we maintain *predecessor pointers*. If x is a live node and node z points to x then we maintain in x a pointer to z.

We update field f in node v, while simulating the update operation creating version i, as follows.[*] Let x be the *live* persistent node corresponding to v in the data structure. If there is an empty extra field in x then we assign the new value to this extra field, stamp it with version i, and mark it as a value associated with original field f. If f is a pointer field which now points to a node z, we update the corresponding predecessor pointer in z to point to x. In case all extra fields in x are used (and none of them is stamped with version i) we copy x as follows.

We create a new persistent node y, make the copy pointer of x point to y, store in each original field in y the most recent value assigned to it, and stamp these values with version stamp i. In particular, field f in node y stores its new value marked with version i. For each pointer field in y we also update the corresponding predecessor pointer to point to y rather than to x.

Then we have to update each field pointing to x in version $i-1$ to point to y in version i. We follow, in turn, each predecessor pointer in x. Let z be a node pointed to by such a predecessor pointer. We identify the field pointing to x in z and update its value in version i to be y. We also update a predecessor pointer in y to point to z. If the old value of the pointer to x in z is not tagged with version i (in particular this means that z has not been copied) then we try to use an extra field to store the new version-value pair. If there is no free extra pointer in z we copy z as above. Then we update the field that points to x to point to y in the new copy of z. This sequence of node copying may cascade, but since each node is copied at most once, the simulation of the update step must terminate. In version i, y is the *live* node corresponding to v.

A simple analysis shows that if we use at least as many extra fields as predecessor fields at each node (i.e., $e \geq p$) then the amortized number of nodes that are copied due to a single update is constant. Intuitively, each time we copy a node we gain e empty extra fields in the live version that "pay" for the assignments that had to be made to redirect pointers to the new copy.

A similar simulation called the *node splitting method* makes a data structure fully persistent with $O(1)$ amortized overhead in time and space. The details however are somewhat more involved so we only sketch the main ideas. Here, since we need predecessor pointers for any version[†] it is convenient to think of the predecessor pointers as part of the ephemeral data structure, and to apply the simulation to the so called *augmented* ephemeral data structure.

We represent each fat node by a list of persistent nodes each of constant size, with twice as many extra pointers as original fields in the corresponding node of the augmented ephemeral data structure. The values in the fields of the persistent nodes are ordered by the version list. Thus each persistent node x is associated with an interval of versions in the version lists, called the *valid interval* of x, and it stores all values of its fields that fall within this interval. The first among these values is stored in an original field and the following ones occupy extra fields.

The key idea underlying this simulation is to maintain the pointers in the persistent structure *consistent* such that when we traverse a pointer valid in version i we arrive at a persistent node whose valid interval contains version i. More precisely, a value c of a pointer field must indicate a persistent node whose valid interval contains the valid interval of c.

---

[*] We assume that each field has only one value in any particular version. When we update a field in version i that already has a value stamped with version i then we overwrite its previous value.

[†] So we cannot simply overwrite a value in a predecessor pointer.

We simulate an update step to field f, while creating version i from version $p(i)$, as follows. If there is already a persistent node x containing f stamped with version i then we merely change the value of f in x. Otherwise, let x be the persistent node whose valid interval contains version i. Let $i+$ be the version following i in the version list. Assume the node following x does not have version stamp of $i+$. We create two new persistent node x', and x'', and insert them into the list of persistent nodes of x, such that x' follows x, and x'' follows x'. We give node x' version stamp of i and fill all its original fields with their values at version i. The extra fields in x' are left empty. We give x'' version stamp of $i+$. We fill the original fields of x'' with their values at version $i+$. We move from the extra fields of x all values with version stamps following $i+$ in the version list to x''. In case the node which follows x in its list has version stamp $i+$ then x'' is not needed.

After this first stage of the update step, values of pointer fields previously indicating x may be inconsistent. The simulation then continues to restore consistency. We locate all nodes containing inconsistent values and insert them into a set S. Then we pull out one node at the time from S and fix its values. To fix a value we may have to replace it with two or more values each valid in a subinterval of the valid interval of the original value. This increases the number of values that has to be stored at the node so we may have to split the node. This splitting may cause more values to become inconsistent. So node splitting and consistency fixing cascades until consistency is completely restored. The analysis is based on the fact that each node splitting produce a node with sufficiently many empty extra fields. For further details see [1].

33.3.3 Handling Arrays

Dietz [4] describes a general technique for making arrays persistent. In his method, it takes $O(\log \log m)$ time to access the array and $O(\log \log m)$ expected amortized time to change the content of an entry, where m is the total number of updates. The space is linear in m. We denote the size of the array by n and assume that $n < m$.

Dietz essentially suggests to think of the array as one big fat node with n fields. The list of versions-values pairs describing the assignments to each entry of the array is represented in a data structure of van Emde Boas et al. [5,6]. This data structure is made to consume space linear in the number of items using dynamic perfect hashing [32]. Each version is encoded in this data structure by its label in the associated Order Maintenance (OM) data structure. (See Section 33.3.1.)

A problem arises with the solution above since we refer to the labels not solely via order queries on pairs of versions. Therefore when a label of a version changes by the OM data structure the old label has to be deleted from the corresponding van Emde Boaz data structure and the new label has to be inserted instead. We recall that any one of the known OM data structures consists of two levels. The versions are partitioned into sublists of size $O(\log m)$. Each sublist gets a label and each version within a sublist gets a label. The final label of a version is the concatenation of these two labels. Now this data structure supports an insertion in $O(1)$ time. However, this insertion may change the labels of a constant number of sublists and thereby implicitly change the labels of $O(\log m)$ versions. Reinserting all these labels into the van Emde Boaz structures containing them may take $\Omega(\log m \log \log m)$ time.

Dietz suggests to solve this problem by bucketizing the van Emde Boaz data structure. Consider a list of versions stored in such a data structure. We split the list into buckets of size $O(\log m)$. We maintain the versions in each bucket in a regular balanced search tree and we maintain the smallest version from each bucket in a van Emde Boaz data structure. This way we need to delete and reinsert a label of a version into the van Emde Boaz data structure only when the minimum label in a bucket gets relabeled.

Although there are only $O(m/\log m)$ elements now in the van Emde Boaz data structures, it could still be the case that we relabel these particular elements too often. This can happen if sublists that get split in the OM data structure contains a particular large number of buckets' minima. To prevent that from happening we modify slightly the OM data structure as follows.

We associate a potential to each version which equals 1 if the version is currently not a minimum in its bucket of its van Emde Boaz data structure and equals $\log \log m$ if it is a minimum in its bucket. Notice that since there are only $O(m/\log m)$ buckets' minima the total potential assigned to all versions throughout the process is $O(m)$. We partition the versions into sublists according to their potentials where the sum of the potentials of the elements in each sublist is $O(\log m)$. We assign labels to the sublists and within each sublists as in the original OM data structure. When we have to split a sublist the work associated with the split, including the required updates on the associated van Emde Boaz data structures, is proportional to the increase in the potential of this sublist since it had last split.

Since we can model the memory of a Random Access Machine (RAM) as a large array. This technique of Dietz is in fact general enough to make any data structure on a RAM persistent with double logarithmic overhead on each access or update to memory.

33.3.4 Making Data Structures Confluently Persistent

Finding a general simulation to make a pointer based data structure confluently persistent is a considerably harder task. In a fully persistent setting we can construct any version by carrying out a particular sequence of updates ephemerally. This seemingly innocent fact is already problematic in a confluently persistent setting. In a confluently persistent setting when an update applies to two versions, one has to produce these two versions to perform the update. Note that these two versions may originate from the same ancestral version so we need some form of persistence even to produce a single version. In particular, methods that achieve persistence typically create versions that share nodes. Semantically however, when an update applied to versions that share nodes we would like the result to be as if we perform the update on two completely independent copies of the input versions.

In a fully persistent setting if each operation takes time polynomial in the number of versions then the size of each version is also polynomial in the number of versions. This breaks down in a confluently persistent setting where even when each operation takes constant time the size of a single version could be exponential in the number of versions. Recall the example of the linked list mentioned in Section 33.1. It is initialized to contain a single node and then concatenated with itself n time. The size of the last versions is 2^n. It follows that any polynomial simulation of a data structure to make it confluently persistent must in some cases represent versions is a compressed form.

Consider the naive scheme to make a data structure persistent which copies the input versions before each update. This method is polynomial in a fully persistent setting when we know that each update operation allocates a polynomial (in the number of versions) number of new nodes. This is not true in a confluently persistent setting as the linked list example given above shows. Thus there is no easy polynomial method to obtain confluently persistence at all.

What precisely causes this difficulty in obtaining a confluently persistent simulation? Lets assume first a fully persistent setting and the naive scheme mentioned above. Consider a single node x created during the update that constructed version v. Node x exists in version v and copies of it may also exist in descendant versions of v. Notice however that each version derived from v contains only a *single* node which is either x or a copy of it. In contrast if we are in a confluently persistent setting a descendant version of v may contain more than a single copy of x. For example, consider the linked list being concatenated to itself as described above. Let x be the node allocated when creating the first version. Then after one catenation we obtain a version which contains two copies of x, after 2 catenations we obtain a version containing 4 copies of x, and in version n we have 2^n copies of x.

Now, if we get back to the fat node method, then we can observe that it identifies a node in a specific version using a pointer to a fat node and a version number. This works since in each version there is only one copy of any node, and thus breaks down in the confluently persistent setting. In a confluently persistent setting we need more than a version number and an address of a fat node to identify a particular node in a particular version.

To address this identification problem Fiat and Kaplan [7] used the notion of *pedigree*. To define pedigree we need the following notation. We denote the version DAG by D, and the version corresponding to vertex $v \in D$ by D_v. Consider the naive scheme defined above. Let w be some node in the data structure D_v. We say that node w in version v was *derived from* node y in version u if version u was one of the versions on which the update producing v had been performed, and furthermore node $w \in D_v$ was formed by a (possibly empty) set of assignments to a copy of node $y \in D_u$.

Let w be a node in some version D_u where D_u is produced by the naive scheme. We associate a *pedigree* with w, and denote it by $p(w)$. The pedigree, $p(w)$, is a path $p \neq \langle v_0, v_1, \ldots, v_k \neq u \rangle$ in the version DAG such that there exist nodes $w_0, w_1, \ldots, w_{k-1}$, $w_k \neq w$, where w_i is a node of D_{v_i}, w_0 was allocated in v_0, and w_i is derived from w_{i-1} for $1 \leq i \leq k$. We also call w_0 the *seminal node* of w, and denote it by $s(w)$. Note that $p(w)$ and $s(w)$ uniquely identify w among all nodes of the naive scheme.

As an example consider Figure 33.1. We see that version v_4 has three nodes (the 1st, 3rd, and 5th nodes of the linked list) with the same seminal node w_0'. The pedigree of the 1st node in D_{v_4} is $\langle v_0, v_1, v_3, v_4 \rangle$. The pedigree of the 2nd node in D_{v_4} is also $\langle v_0, v_1, v_3, v_4 \rangle$ but its seminal node is w_0. Similarly, we can see that the pedigrees of the 3rd, and the 5th nodes of D_{v_4} are $\langle v_0, v_2, v_3, v_4 \rangle$ and $\langle v_0, v_2, v_4 \rangle$, respectively.

The basic simulation of Fiat and Kaplan is called the *full path method* and it works as follows. The data structure consists of a collection of *fat nodes*. Each fat node corresponds to an explicit allocation of a node by an update operation or in another words, to some seminal node of the naive scheme. For example, the update operations of Figure 33.1 performs 3 allocations (3 seminal nodes) labeled w_0, w_0', and w_0'', so our data structure will have 3 fat nodes, $f(w_0)$, $f(w_0')$ and $f(w_0'')$. The full path method represents a node w of the naive scheme by a pointer to the fat node representing $s(w)$, together with the pedigree $p(w)$. Thus a single fat node f represents all nodes sharing the same seminal node. We denote this set of nodes by $N(f)$. Note that $N(f)$ may contain nodes that co-exist within the same version and nodes that exist in different versions. A *fat node* contains the same fields as the corresponding seminal node. Each of these fields, however, rather than storing a single value as in the original node stores a dynamic table of field values in the fat node. The simulation will be able to find the correct value in node $w \in N(f)$ using $p(w)$. To specify the representation of a set of values we need the following definition of an *assignment pedigree*.

Let $p \neq \langle v_0, \ldots, v_k \neq u \rangle$ be the pedigree of a node $w \in D_u$. Let $w_k \neq w, w_{k-1}, \ldots, w_1$, $w_i \in D_{v_i}$ be the sequence of nodes such that $w_i \in D_{v_i}$ is derived from $w_{i-1} \in D_{v_{i-1}}$. This sequence exists by the definition of node's pedigree. Let A be a field in w and let j be the maximum such that there has been an assignment to field A in w_j during the update that created v_j. We define the *assignment pedigree of a field A in node w*, denoted by $p(A, w)$, to be the pedigree of w_j, that is, $p(A, w) \neq \langle v_0, v_1, \ldots, v_j \rangle$.

In the example of Figure 33.1 the nodes contain one pointer field (named *next*) and one data field (named x). The assignment pedigree of x in the 1st node of D_{v_4} is simply $\langle v_0 \rangle$, the assignment pedigree of x in the 2nd node of D_{v_4} is likewise $\langle v_0 \rangle$, the assignment pedigree of x in the 3rd node of D_{v_4} is $\langle v_0, v_2, v_3 \rangle$. Pointer fields also have assignment pedigrees. The assignment

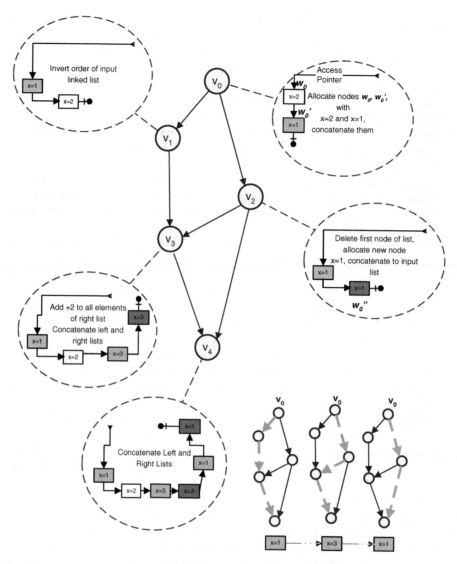

FIGURE 33.1 A DAG of five versions. In each circle we show the corresponding update operation and the resulting version. Nodes with the same color originate from the same seminal node. The three gray nodes in version D_{v_4} all have the same seminal node (w_0'), and are distinguished by their pedigrees $\langle v_0, v_1, v_3, v_4 \rangle$, $\langle v_0, v_2, v_3, v_4 \rangle$, and $\langle v_0, v_2, v_4 \rangle$.

pedigree of the pointer field in the 1st node of D_{v_4} is $\langle v_0, v_1 \rangle$, the assignment pedigree of the pointer field in the 2nd node of D_{v_4} is $\langle v_0, v_1, v_3 \rangle$, the assignment pedigree of the pointer field of the 3rd node of D_{v_4} is $\langle v_0, v_2 \rangle$, finally, the assignment pedigree of the pointer field of the 4th node of D_{v_4} is $\langle v_2, v_3, v_4 \rangle$.

We call the set $\{p(A, w) \mid w \in N(f)\}$ *the set of all assignment pedigrees for field A in a fat note f*, and denote it by $P(A, f)$. The table that represents field A in fat node f contains an entry for each assignment pedigree in $P(A, f)$. The value of a table entry, indexed by an assignment pedigree $p \neq \langle v_0, v_1, \ldots, v_j \rangle$, depends on the type of the field as follows. If A is a data field then the value stored is the value assigned to A in the node $w_j \in D_{v_j}$ whose pedigree is p. If A is a pointer field then let w be the node pointed to by field A after the assignment to A in w_j. We store the pedigree of w and the address of the fat node that represents the seminal node of w.

An access pointer to a node w in version v is represented by a pointer to the fat node representing the seminal node of w and the pedigree of w.

In Figure 33.2 we give the fat nodes of the persistent data structure given in Figure 33.1. For example, the field next has three assignments in nodes of $N(f(w_0'))$. Thus, there are three assignment pedigrees in $P(next, f(w_0'))$:

1. $\langle v_0 \rangle$—allocation of w_0' in version D_{v_0} and default assignment of null to next.
2. $\langle v_0, v_1 \rangle$—inverting the order of the linked list in version D_{v_1} and thus assigning next a new value. The pointer is to a node whose pedigree is $\langle v_0, v_1 \rangle$ and whose seminal node is w_0. So we associate the value $(\langle v_0, v_1 \rangle, f(w_0))$ with $\langle v_0, v_1 \rangle$.

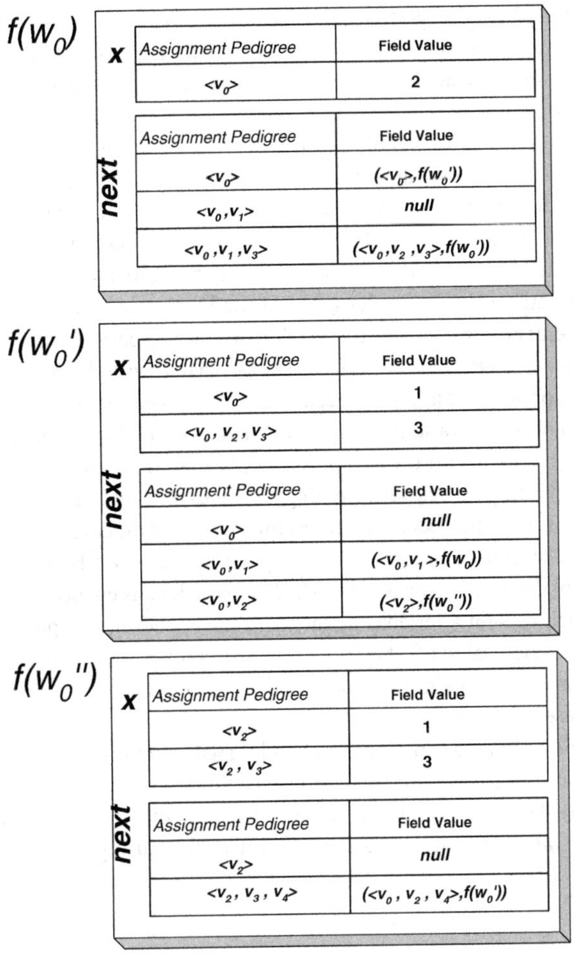

FIGURE 33.2 The fat nodes for the example of Figure 33.1.

3. $\langle v_0, v_2 \rangle$—allocating a new node, w_0'', in version D_{v_2}, and assigning next to point to this new node. The pedigree of w_0'' is $\langle v_2 \rangle$ so we associate the value $(\langle v_2 \rangle, f(w_0''))$ with $\langle v_0, v_2 \rangle$.

You can see all three entries in the table for next in the fat node $f(w_0')$ (Figure 33.2). Similarly, we give the table for field x in $f(w_0')$ as well as the tables for both fields in fat nodes $f(w_0)$ and $f(w_0'')$.

When we traverse the data structure we are pointing to some fat node f and hold a pedigree q of some node w whose seminal node corresponds to f and we would like to retrieve the value of field A in node w from the table representing field A in f. We do that as follows. First we identify the assignment pedigree $p(A, w)$ of field A in node w. This is the longest pedigree which is a prefix of q and has an entry in this table. In case A is a data field, the value we are after is simply the value associated with $p(A, w)$. However if A is a pointer field then the value stored with $p(A, w)$ may not be the value of A in w. This value identifies a node in the version where the assignment occurred, whereas we are interested in a node in the version of w that this pointer field points to.

Let $q \neq \langle q_0, \ldots, q_k \rangle$ and let $p(A, w) \neq \langle q_0, q_1, \ldots, q_j \rangle$. Let the value of $p(A, w)$ be (t, f), where t is the pedigree of the target node in D_{q_j} and f is the fat node representing the seminal node of this target node. The nodes identified by the pedigrees $p(A, w)$ and t were copied in versions $q_{j+1}, \ldots, q_k$ without any assignment made to field A in the nodes derived from the node whose pedigree is $p(A, w)$. Thus the pedigree of the target node of field A of node w in D_{q_k} is $t \| \langle q_{j+1}, \ldots, q_k \rangle$, where $\|$ represents concatenation.

It follows that we need representations for pedigrees and the tables representing field values that support an efficient implementation of the followings.

1. Given a pedigree q find the longest prefix of q stored in a table.
2. Given a pedigree q, replace a prefix of q with another pedigree p.
3. To facilitate updates we also need to be able to add a pedigree to a table representing some field with a corresponding value.

In their simplest simulation Fiat and Kaplan suggested to represent pedigrees as linked lists of version numbers, and to represent tables with field values as tries. Each assignment pedigree contained in the table is represented by a path in the corresponding trie. The last node of the path stores the associated value. Nodes in the trie can have large degrees so for efficiency we represent the children of each node in a trie by a splay tree.

Let U be the total number of assignments the simulation performs and consider the update creating version v. Then with this implementation each assignment performed during this update requires $O(d(v))$ words of size $O(\log U)$ bits, and takes $O(d(v) + \log U)$ time, where $d(v)$ is the depth of v in the DAG. Field retrieval also takes $O(d(v) + \log U)$ time.

The second method suggested by Fiat and Kaplan is the *compressed path method*. The essence of the compressed path method is a particular partition of our DAG into disjoint trees. This partition is defined such that every path enters and leaves any specific tree at most once. The compressed path method encodes paths in the DAG as a sequence of pairs of versions. Each such pair contains a version where the path enters a tree T and the version where the path leaves the tree T. The length of each such representation is $O(e(D))$.[*] Each value of a field in a fat node is now associated with the compressed representation of the path of the node in $N(f)$ in which the corresponding assignment occurred. A key property of these compressed path representations is that they allow easy implementation of the operations we need to perform on pedigree, like replacing a prefix of a pedigree with another pedigree when traversing a pointer. With the compressed path method each assignment requires up to $O(e(D))$ words each of $O(\log \mathcal{U})$ bits. Searching or updating the trie representing all values of a field in a fat node requires $O(e(D) + \log \mathcal{U})$ time. For the case where the DAG is a tree this method degenerates to the fat node simulation of [1].

Fiat and Kaplan also suggested how to use randomization to speed up their two basic methods at the expense of (slightly) larger space expansion and polynomially small error probability. The basic idea is encode each path (or compressed path) in the DAG by an integer. We assign to each version a random integer, and the encoding of a path p is simply the sum of the integers that correspond to the versions on p. Each value of a field in a fat node is now associated with the integer encoding the path of the node in $N(f)$ in which the corresponding assignment occurred. To index the values of each field we use a hash table storing all the integers corresponding to these values.

To deal with values of pointer fields we have to combine this encoding with a representation of paths in the DAG (or compressed paths) as balanced search trees, whose leaves (in left to right order) contain the random integers associated with the vertices along the path (or compressed path). This representation allows us to perform certain operations on these paths in logarithmic (or poly-logarithmic) time whereas the same operations required linear time using the simpler representation of paths in the non-randomized methods.

33.4 Making Specific Data Structures More Efficient

The purely functional deques of Kaplan and Tarjan [11], the confluently persistent deques of Kaplan, Okasaki, and Tarjan [14], the purely functional heaps of Brodal and Okasaki [33], and the purely functional finger search trees of Kaplan and Tarjan [19], are all based on a simple and useful mechanism called redundant counters, which to the best of our knowledge first appeared in lecture notes by Clancy and Knuth [15]. In this section we describe what redundant counters are, and demonstrate how they are used in simple persistent deques data structure.

A persistent implementation of deques support the following operations: $q' \neq push(x, q)$: Inserts an element x to the beginning of the deque q returning a new deque q' in which x is the first element followed by the elements of q. $(x, q') \neq pop(q)$: Returns a pair where x is the first element of q and q' is a deque containing all elements of q but x. $q' \neq Inject(x, q)$: Inserts an element x to the end of the deque q returning a new deque q' in which x is the last element preceded by the elements of q. $(x, q') \neq eject(q)$: Returns a pair where x is the last element of q and q' is a deque containing all elements of q but x. A *stack* supports only push and pop, a *queue* supports only push and eject. Catenable deques also support the operation $q \neq catenate(q_1, q_2)$: Returns a queue q containing all the elements of q_1 followed by the elements of q_2.

Although queues, and in particular catenable queues, are not trivial to make persistent, stacks are easy. The regular representation of a stack by a singly linked list of nodes, each containing an element, ordered from first to last, is in fact purely functional. To push an element onto a stack, we create a new node containing the new element and a pointer to the node containing the previously first element on the stack. To pop a stack, we retrieve the first element and a pointer to the node containing the previously second element.

Direct ways to make queues persistent simulate queues by stacks. One stack holds elements from the beginning of the queue and the other holds elements from its end. If we are interested in fully persistence this simulation should be real time and its details are not trivial. For a detailed discussion see Kaplan and Tarjan [11] and the references there.

---

[*] Recall that $e(D)$ is the logarithm of the maximum number of different paths from the root of the DAG to any particular version.

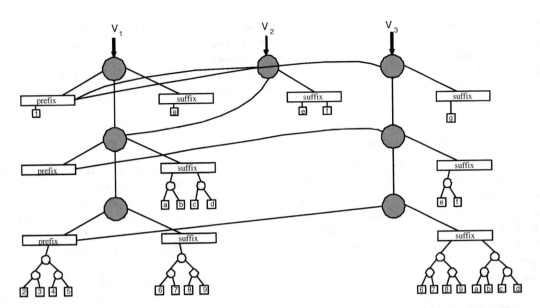

FIGURE 33.3 Representation of a deque of elements over *A*. Each circle denotes a deque and each rectangle denotes a buffer. Squares correspond to elements from *A* which we denote by numbers and letters. Each buffer contains 0, 1, or 2 elements. Three versions are shown V_1, V_2, and V_3. Version V_2 was obtained from V_1 by injecting the element *f*. Version V_3 obtained from version V_2 by injecting the element *g*. The latter inject triggered two recursive injects into the child and grandchild deques of V_2. Note that identical binary trees and elements are represented only once but we draw them multiple times to avoid cluttering of the figure.

Kaplan and Tarjan [11] described a new way to do a simulation of a deque with stacks. They suggest to represent a deque by a recursive structure that is built from bounded-size deques called *buffers*. Buffers are of two kinds: *prefixes* and *suffixes*. A non-empty deque *q* over a set *A* is represented by an ordered triple consisting of a *prefix*, *prefix(q)*, of elements of *A*, a *child deque*, *child(q)*, whose elements are ordered pairs of elements of *A*, and a *suffix*, *suffix(q)*, of elements of *A*. The order of elements within *q* is the one consistent with the orders of all of its component parts. The *child* deque child(q), if non-empty, is represented in the same way. Thus the structure is recursive and unwinds linearly. We define the descendants $\{child^i(q)\}$ of deque *d* in the standard way, namely $child^0(q) \neq q$ and $child^{i+1}(q) \neq child(child^i(q))$ for $i \geq 0$ if $child^i(q)$ is non-empty.

Observe that the elements of *q* are just elements of *A*, the elements of *child(q)* are pairs of elements of *A*, the elements of *child(child(q))* are pairs of pairs of elements of *A*, and so on. One can think of each element of $child^i(q)$ as being a complete binary tree of depth *i*, with elements of *A* at its 2^i leaves. One can also think of the entire structure representing *q* as a stack (of *q* and its descendants), each element of which is prefix-suffix pair. All the elements of *q* are stored in the prefixes and suffixes at the various levels of this structure, grouped into binary trees of the appropriate depths: level *i* contains the prefix and suffix of $child^i(q)$. See Figure 33.3.

Because of the pairing, we can bring *two* elements up to level *i* by doing *one pop* or *eject* at level $i + 1$. Similarly, we can move two elements down from level *i* by doing one *push* or *inject* at level $i + 1$. This two-for-one payoff leads to real-time performance.

Assume that each prefix or suffix is allowed to hold 0, 1, or 2 elements, from the beginning or end of the queue, respectively. We can implement $q' \neq push(x, q)$ as follows. If the prefix of *q* contains 0 or 1 elements we allocate a new node to represent q' make its child deque and its suffix identical to the child and suffix of *q*, respectively. The prefix of q' is a newly allocated prefix containing *x* and the element in the prefix of *q*, if the prefix of *q* contained one element. We return a pointer the new node which represents q'. For an example consider version V_2 shown in Figure 33.3 that was obtained from version V_1 by a case of inject symmetric to the case of the push just described.

The hard case of the push is when the prefix of *q* already contains two elements. In this case we make a pair containing these two elements and push this pair recursively into *child(q)*. Then we allocate a new node to represent q', make its suffix identical to the suffix of *q*, make the deque returned by the recursive push to *child(q)* the child of q', and make the prefix of q' be a newly allocated prefix containing *x*. For an example consider version V_3 shown in Figure 33.3 that was obtained from version V_2 by a recursive case of inject symmetric to the recursive case of the push just described. The implementations of pop and eject is symmetric.

This implementation is clearly purely functional and therefore fully persistent. However the time and space bounds per operation are $O(\log n)$. The same bounds as one gets by using search trees to represent the deques with the path copying technique. These logarithmic time bounds are by far off from the ephemeral $O(1)$ time and space bounds.

Notice that there is a clear correspondence between this data structure and binary counters. If we think of a buffer with two elements as the digit 1, and of any other buffer as the digit 0, then the implementation of *push(q)* is similar to adding one to the binary number defined by the prefixes of the queues *child^i(q)*. It follows that if we are only interested in partially persistent deques then this implementation has $O(1)$ amortized time bound per operation (see the discussion of binary counters in the next section). To make this simulation efficient in a fully persistent setting and even in the worst case, Kaplan and Tarjan suggested to use redundant counters.

33.4.1 Redundant Binary Counters

To simplify the presentation we describe redundant binary counters, but the ideas carry over to any basis. Consider first the regular binary representation of an integer i. To obtain from this representation the representation of $i + 1$ we first flip the rightmost digit. If we flipped a 1 to 0 then we repeat the process on the next digit to the left. Obviously, this process can be long for some integers. But it is straightforward to show that if we carry out a sequence of such increments starting from zero then on average only a constant number of digits change per increment.[*] Redundant binary representations (or counters as we will call them) address the problem of how to represent i so we can obtain a representation of $i + 1$ while changing only a constant number of digits in the worst case.

A *redundant binary representation*, d, of a non-negative integer x is a sequence of digits $d_n, \ldots, d_0$, with $d_i \in \{0, 1, \ldots, 2\}$, such that $x = \sum_{i=0}^{n} d_i 2^i$. We call d *regular* if, between any two digits equal to 2, there is a 0, and there is a 0 between the rightmost 2 and the least significant digit. Notice that the traditional binary representation of each integer (which does not use the digit 2) is *regular*. In the sequel when we refer to a regular representation we mean a regular redundant binary representation, unless we explicitly state otherwise.

Suppose we have a regular representation of i. We can obtain a regular representation of $i + 1$ as follows. First we increment the rightmost digit. Note that since the representation of i is regular, its rightmost digit is either 0 or 1. So after the increment the rightmost digit is either 1 or 2 and we still have a redundant binary representation for $i + 1$. Our concern is that this representation of $i + 1$ may not be regular. However, since the representation of i we started out with was regular the only violation to regularity that we may have in the representation of $i + 1$ is lacking a 0 between the rightmost 2 and the least significant digit. It is easy to check that between any two digits equal to 2, there still is a 0, by the regularity of i.

We can change the representation of $i + 1$ to a representation which is guaranteed to be regular by a simple *fix* operation. A *fix* operation on a digit $d_i \neq 2$ increments d_{i+1} by 1 and sets d_i to 0, producing a new regular representation d' representing the same number as d.[†] If after incrementing the rightmost digit we perform a fix on the rightmost 2 then we switch to another representation of $i + 1$ that must be regular. We omit the proof here which is straightforward.

It is clear that the increment together with the fix that may follow change at most three digits. Therefore redundant binary representations allow to perform an increment while changing constantly many digits. However notice that in any application of this numbering system we will also need a representation that allows to get to the digits which we need to fix efficiently. We show one such representation in the next section.

These redundant representations can be extended so we can also decrement it while changing only a constant number of digits, or even more generally so that we can increment or decrement any digit (add or subtract 2^i) while changing a constant number of other digits. These additional properties of the counters were exploited by other applications (see, e.g., [19,34]).

33.4.2 Persistent Deques

Kaplan and Tarjan use this redundant binary system to improve the deque implementation we described above as follows. We allow each of the prefixes and suffixes to contain between 0 and 5 elements. We label each buffer, and each deque, by one of the digits 0, 1, and 2. We label a buffer 0 if it has two or three elements, we label it 1 if it has one or four elements, and we label it 2 if it has zero or five elements. Observe that we can add one element to or delete one element from a buffer labeled 0 or 1 without violating its size constraint: A buffer labeled 0 may change its label to 1, and a buffer labeled 1 may change its label to 2. (In fact a 1 can also be changed to 0 but this may not violate regularity.) The label of a deque is the larger among the labels of its buffers, unless its child and one of its buffers are empty, in which case the label of the deque is identical to the label of its nonempty buffer.

This coloring of the deques maps each deque to a redundant binary representation. The least significant digit of this representation is the digit of q, the next significant digit is the digit of *child(q)*, and, in general, the ith significant digit is the digit corresponding to *child^i(q)* if the latter is not empty. We impose an additional constraint on the deques and require that the redundant binary representation of any top-level deque is regular.

---

[*] The rightmost digit changes every increment, the digit to it left changes every other operation, and so on.
[†] We use the *fix* only when we know that d_{i+1} is either 0 or 1.

A regular top-level deque is labeled 0 or 1 which implies that both its prefix and its suffix are labeled 0 or 1. This means that any deque operation can be performed by operating on the appropriate top-level buffer. Suppose that the operation is either a push or a pop, the case of inject and eject is symmetric. We can construct the resulting queue q' by setting $child(q') \neq child(q)$ and $suffix(q') \neq suffix(q)$. The prefix of q' is a newly allocated buffer that contains the elements in $prefix(q)$ together with the new element in case of push or without the first element in case of pop. Clearly all these manipulations take constant time.

The label of q', however, may be one larger than the label of q. So the redundant binary representation corresponding to q' is either the same as the redundant binary representation of q in which case it is regular, or it is obtained from the redundant binary representation of q by incrementing the least significant digit. (The least significant digit can also decrease in which case regularity is also preserved.) This corresponds to the first step in the increment procedure for redundant regular representations described in the previous section.

To make the redundant binary representation of q' regular we may have to apply a fix operation. Let i be the minimum such that $child^i(q')$ is labeled 2. If for all $j < i$, $child^j(q')$ is labeled 1 then the fix has to change the label of $child^i(q')$ to 0 and increment the label of $child^{i+1}(q')$.

Fortunately, we have an appropriate interpretation for such a fix. Assume $child^{i+1}(q')$ have a non-empty child. (We omit the discussion of the case where $child^{i+1}(q')$ have an empty child which is similar.) We know that the label of $child^{i+1}(q')$ is either 0 or 1 so neither of its buffers is empty or full. If the prefix of $child^i(q')$ has at least four elements we eject 2 of these elements and push them as a single pair to the prefix of $child^{i+1}(q')$. If the prefix of $child^i(q')$ has at most one element we pop a pair from the prefix of $child^{i+1}(q')$ and inject the components of the pair into the prefix of $child^i(q')$. This makes the prefix of $child^i(q')$ containing either two or three elements. Similarly by popping a pair from or pushing a pair to the suffix of $child^i(q')$, and injecting a pair to or ejecting a pair from the suffix of $child^{i+1}(q')$ we make the suffix of $child^i(q')$ containing two or three elements. As a result the label of $child^i(q')$ and its two buffers becomes 0 while possibly increasing the label of one or both buffers of $child^{i+1}(q')$ and thereby the label of $child^{i+1}(q')$ as well.

There is one missing piece for this simulation to work efficiently. The topmost deque labeled 2 may be arbitrarily deep in the recursive structure of q', since it can be separated from the top level by many deques labeled 1. To implement the fix efficiently we have to be able to find this deque fast and change it in a purely functional way by copying the deques that change without having to copy all their ancestors deques.

For this reason we do not represent a deque in the obvious way, as a stack of prefix-suffix pairs. Instead, we break this stack up into substacks. There is one substack for the top-level deque and one for each descendant deque labeled 0 or 2 not at the top level. Each substack consists of a top-level, or a deque labeled 0, or a deque labeled 2 and all consecutive proper descendant deques labeled 1. We represent the entire deque by a stack of substacks of prefix-suffix pairs using this partition into substacks. This can be realized with four pointers per each node representing a deque at some level. Two of the pointers are to the prefix and suffix of the deque. One pointer is to the node for the child deque if this deque is non-empty and labeled 1. One pointer is to the node of the nearest proper descendant deque not labeled 1, if such a deque exists and q itself is not labeled 1 or top-level. See Figure 4.2.

A single deque operation will require access to at most the top three substacks, and to at most the top two elements in any such substack. The label changes caused by a deque operation produce only minor changes to the stack partition into substacks, changes that can be made in constant time. In particular, changing the label of the top-level deque does not affect the partition into substacks. Changing the label of the topmost deque which is labeled 2 to 0 and the label of its child from 1 to 2 splits one substack into its first element, now a new substack, and the rest. This is just a substack pop operation. Changing the label of the topmost deque which is labeled 2 to 0 and the label of its child from 0 to 1 merges a singleton substack with the substack under it. This is just a substack push operation.

To add catenation, Kaplan and Tarjan had to change the definition of the data structure and allow deques to be stored as components of elements of recursive deques. The redundant binary numbering system, however, still plays a key role. To represent a catenable deque, Kaplan and Tarjan use noncatenable deques as the basic building blocks. They define a *triple* over a set A recursively as a prefix of elements of A, a possibly empty deque of triples over A, and a suffix of elements of A, where each prefix or suffix is a noncatenable deque. Then, they represent a catenable deque of elements from A by either one or two triples over A. The underlying skeleton of this structure is a binary tree (or two binary trees) of triples. The redundant binary number system is extended so that it can distribute work along these trees by adding an extra digit.

Kaplan, Okasaki, and Tarjan [14] simplified these data structures at the expense of making the time bounds amortized rather than worst case and using assignment, thus obtaining a confluently persistent data structure which is not purely functional. The key idea underlying their data structure is to relax the rigid constraint of maintaining regularity. Instead, we "improve" the representation of a deque q with full or empty prefix when we try to push or pop an element from it. Similarly, with full or empty suffix. This improvement in the representation of q is visible to all deques that contain q as a subdeque at some level and prevents from pushing into deques with full prefixes or popping from deques with empty prefixes from happening too often.

More specifically, assume that we push into a deque q with full prefix. First, we eject two elements from this prefix, make a pair containing them, and push the pair recursively into $child(q)$. Let the result of the recursive push be $child'(q)$. Then we change the representation of q so that it has a new prefix which contains all the elements in the prefix of q but the two which we ejected, and its child deque is $child'(q)$. The suffix of q does not change. Finally we perform the push into q by creating a new queue q' that has the same suffix and child deque as q, but has a new prefix that contains the elements in the prefix of q together with the new element. A careful but simple analysis shows that each operation in this implementation takes $O(1)$ amortized time. By extending this idea, Kaplan, Okasaki, and Tarjan managed to construct catenable deques using only constant size buffers as the basic building blocks.

33.5 Concluding Remarks and Open Questions

Much progress has been made on persistent data structures since the seminal paper of Driscoll et al. [1]. This progress has three folds: In developing general techniques to make any data structure persistent, in making specific data structures persistent, and in emerging algorithmic applications. Techniques developed to address these challenges sometimes proved useful for other applications as well.

This algorithmic field still comprise intriguing challenges. In developing general techniques to make data structures persistent, a notable challenge is to find a way to make the time slowdown of the node splitting method worst case. Another interesting research track is how to restrict the operations that combine versions in a confluently persistent setting so that better time bounds, or simpler simulations, are possible. We also believe that the techniques and data structures developed in this field would prove useful for numerous forthcoming applications.

Acknowledgments

This work was supported, in part, by the Israel Science Foundation (ISF) under grant 548/00.

References

1. J. R. Driscoll, N. Sarnak, D. Sleator, and R. Tarjan. Making data structures persistent. *J. of Computer and System Science*, 38:86–124, 1989.

2. P. F. Dietz and R. Raman. Persistence, amortization and randomization. In *Proc. 2nd annual ACM-SIAM symposium on Discrete algorithms*, pages 78–88. Society for Industrial and Applied Mathematics, 1991.

3. G. S. Brodal. Partially persistent data structures of bounded degree with constant update time. *Nordic J. of Computing*, 3(3):238–255, 1996.

4. P. F. Dietz. Fully persistent arrays. In *Proc. 1st Worksh. Algorithms and Data Structures (WADS 1989)*, volume 382 of *Lecture Notes in Computer Science*, pages 67–74. Springer, 1989.

5. P. van Emde Boaz, R. Kaas, and E. Zijlstra. Design and implementation of an efficient priority queue. *Mathematical Systems Theory*, 10:99–127, 1977.

6. P. van Emde Boaz. Preserving order in a forest in less than logarithmic time. *Information Processing Letters*, 6(3):80–82, 1977.

7. A. Fiat and H. Kaplan. Making data structures confluently persistent. *J. of Algorithms*, 48(1):16–58, 2003. Symposium on Discrete Algorithms.

8. D. P. Dobkin and J. I. Munro. Efficient uses of the past. *J. of Algorithms*, 6:455–465, 1985.

9. M. H. Overmars. Searching in the past, I. Technical Report RUU-CS-81-7, Department of Computer Science, University of Utrecht, Utrecht, The Netherlands, 1981.

10. B. Chazelle. How to search in history. *Information and Control*, 64:77–99, 1985.

11. H. Kaplan and R. E. Tarjan. Purely functional, real-time deques with catenation. *J. of the ACM*, 46(5):577–603, 1999.

12. J. Driscoll, D. Sleator, and R. Tarjan. Fully persistent lists with catenation. *J. of the ACM*, 41(5):943–959, 1994.

13. A. L. Buchsbaum and R. E. Tarjan. Confluently persistent deques via data structural bootstrapping. *J. of Algorithms*, 18:513–547, 1995.

14. H. Kaplan, C. Okasaki, and R. E. Tarjan. Simple confluently persistent catenable lists. *Siam J. on Computing*, 30(3):965–977, 2000.

15. M. J. Clancy and D. E. Knuth. A programming and problem-solving seminar. Technical Report STAN-CS-77-606, Department of Computer Science, Stanford University, Palo Alto, 1977.

16. C. Okasaki. Amortization, lazy evaluation, and persistence: Lists with catenation via lazy linking. In *IEEE Symposium on Foundations of Computer Science*, pages 646–654, 1995.

17. R. E. Tarjan and J. Van Leeuwen. Worst case analysis of set union algorithms. *J. of the ACM*, 31:245–281, 1984.

18. R. E. Tarjan. *Data Structures and Network algorithms*. SIAM, Philadelphia, 1982.

19. H. Kaplan and R. E. Tarjan. Purely functional representations of catenable sorted lists. In *Proc. 28th Annual ACM Symposium on Theory of Computing*, pages 202–211. ACM Press, 1996.

20. N. Sarnak and R. E. Tarjan. Planar point location using persistent search trees. *Communications of the ACM*, 29(7):669–679, 1986.

21. D. P. Dobkin and R. J. Lipton. Multidimensional searching problems. *SIAM J. on Computing*, 5(2):181–186, 1976.

22. R. Cole. Searching and storing similar lists. *J. of Algorithms*, 7:202–220, 1986.

23. A. Boroujerdi and B. Moret. Persistence in computational geometry. In *Proc. 7th Canadian Conf. Comp. Geometry (CCCG 95)*, pages 241–246, 1995.

24. E. M. McCreight. Priority search trees. *Siam J. on Computing*, 14:257–276, 1985.

25. S. Alstrup, G. S. Brodal, I. L. Gørtz, and T. Rauhe. Time and space efficient multi-method dispatching. In *Proc. 8th Scandinavian Workshop on Algorithm Theory*, volume 2368 of Lecture Notes in Computer Science, pages 20–29. Springer, 2002.

26. N. Kitsios and A. Tsakalidis. Space reduction and an extension for a hidden line elimination algorithm. *Computational Geometry*, 6(6):397–404, 1996.

27. O. Nurmi. A fast line sweep algorithm for hidden line elimination. *BIT*, 25(3):466–472, 1985.

28. P. Bozanis, N. Kitsios, C. Makris, and A. Tsakalidis. The space-optimal version of a known rectangle enclosure reporting algorithm. *Information Processing Letters*, 61(1):37–41, 1997.

29. P. Gupta, R. Janardan, H. M. Smid, and B. DasGupta. The rectangle enclosure and point-dominance problems revisited. *International J. of Computational Geometry and Applications*, 7(5):437–455, 1997.

30. P. F. Dietz and D. D. Sleator. Two algorithms for maintaining order in a list. In *Proc. 19th Annual ACM Symposium on Theory of Computing*, pages 365–372, 1987.

31. M. A. Bender, R. Cole, E. D. Demaine, M. Farach-Colton, and J. Zito. Two simplified algorithms for maintaining order in a list. In *Proc. 10th Annual European Symposium on Algorithms (ESA 2002)*, volume 2461 of Lecture Notes in Computer Science, pages 152–164. Springer, 2002.

32. M. Dietzfelbinger, A. Karlin, K. Mehlhorn, M. Meyer auf der Heide, H. Rohnert, and R. E. Tarjan. Dynamic perfect hashing: Upper and lower bounds. *SIAM J. on Computing*, 23(4):738–761, 1994.

33. G. S. Brodal and C. Okasaki. Optimal purely functional priority queues. *J. of Functional Programming*, 6(6):839–858, 1996.

34. H. Kaplan, N. Shafrir, and R. E. Tarjan. Meldable heaps and boolean union-find. In *Proc. 34th Annual ACM Symposium on Theory of Computing (STOC)*, pages 573–582, 2002.

34

Data Structures for Sets[*]

34.1	Introduction..	529
	Models of Computation	
34.2	Simple Randomized Set Representations...	530
	The Hash Trie • Some Remarks on Unique Representations	
34.3	Equality Testing	533
34.4	Extremal Sets and Subset Testing	535
	Static Extremal Sets • Dynamic Set Intersections and Subset Testing	
34.5	The Disjoint Set Union-Find Problem...	538
	The Classical Union-Find Problem and Variants	
34.6	Partition Maintenance Algorithms...	540
34.7	Conclusions...	542
	References..	542

Rajeev Raman
University of Leicester

34.1 Introduction

Setsare a fundamental concept in computer science: the study of algorithms and data structures for maintaining them were among the earliest developments in data structures. Our focus will be on problems that involve maintaining a family $\mathcal{F}$ of sets, where all sets are drawn from a *universe U* of elements. We do not assume that there is a particular natural order among the elements of U, and in particular do not focus on data structuring problems on sets where the operations explicitly assume such an order (e.g., priority queues). A *base* repertoire of actions is as follows:

$$
\left.
\begin{array}{ll}
\texttt{create()} & \text{Create a new empty set, add it to } \mathcal{F} \text{ and return the name of the set.} \\
\texttt{destroy}(A) & \text{If } A = \emptyset \text{ then remove } A \text{ from } \mathcal{F}. \text{ If } A \neq \emptyset, \text{ flag an error.} \\
\texttt{insert}(x, A) & \text{Set } A \leftarrow A \cup \{x\}. \\
\texttt{delete}(x, A) & \text{Set } A \leftarrow A - \{x\}.
\end{array}
\right\}
\tag{34.1}
$$

The following operation is fundamental for data structuring problems involving sets in general, but plays only an auxiliary role in this chapter:

$$
\texttt{member}(x, A) \text{ Returns 'true' if } x \in A \text{ and 'false' otherwise.}
\tag{34.2}
$$

If we take only `insert`, `delete` and `member`, we get the *dictionary* problem, covered in Part III. The base repertoire plus member is essentially no more difficult, as it represents the problem of maintaining a collection of independent dictionaries over a common universe. In this chapter, we focus on adding operations to this base repertoire that take two or more sets as arguments. For example, we could consider the standard set-theoretic operations on two sets A and B:

$$
A \text{ op } B, \text{op} \in \{\cup, \cap, -\}.
$$

A data structure may support only an *enumerative* form of these operations, whereby the result of A op B is, in essence, some kind of enumeration of the set A op B. This result may not be in the same representation as A or B, and so one may not be able operate on it (e.g., $((A \text{ op } B) \text{ op } C)$ may not involve just two executions of op). The complexity of the algorithms will generally

---
[*] This chapter has been reprinted from first edition of this Handbook, without any content updates.

be measured in terms of the following parameters:

$$n = \sum_{A \in \mathcal{F}} |A| \text{ (the total size of all sets in the family)}$$

$$m = |U| \text{ (the size of the universe)} \tag{34.3}$$

$$k = |\mathcal{F}| \text{ (the number of sets in the family)}$$

34.1.1 Models of Computation

The problems that we consider in this chapter have been studied on a variety of different computation models. The primary models for proving upper bounds are the *pointer machine* model and the *random-access machine (RAM)* model. The pointer machine [1–3] postulates a storage consisting of an unbounded collection of registers (or records) connected by pointers. Each register can contain an arbitrary amount of additional information, and no arithmetic is allowed to compute the address of a register. The processor has a constant number of (data and pointer) registers that it can manipulate directly, but all other temporary results must be held in the storage records. In particular, this means that to answer a query, the processor must either start from a pointer p into the data structure provided by the "user" or from one of the constant number of pointers it itself holds, and explore the data structure by following pointers starting from p.

The RAM model we use is a standard variant of the original RAM model [4], the *word RAM* model [5]. Briefly, the word RAM model uses the unit-cost criterion, and a word size of $\Theta(\log n)$ bits, where $n = \sum_{S \in \mathcal{F}} |S|$.[*] Clearly the word size should be at least $\log n + O(1)$ bits—otherwise one could not even address the amount of memory required to store $\mathcal{F}$. Nevertheless, there are instances where the solutions that result from the use of this model could be viewed as "cheating." For example, we could have $n = 2^{\Theta(|U|)}$, in which case the word size would be $\Theta(|U|)$ bits, which would allow most set operations to be done in $O(1)$ time by bitwise operations on a single word. The solutions that we discuss, however, do not exploit this fact. In the related *cell probe* model, the storage of the computer is modeled by cells numbered $0, 1, 2, \ldots$, each of which can store a number of $O(\log n)$ bits. The running time of an algorithm in this model is measured as just the number of words (cells) accessed during an operation. All other computations are free.

The *arithmetic* model, used primarily for proving lower bounds, was proposed by Fredman [6] and Yao [7]. We now give a somewhat simplified description of this model which conveys the essential aspects: the interested reader is referred to [8, section 7.2.3] for further details. In many cases of interest, it is useful to assume that the data structure operates on values from a set $\mathcal{M}$, which is a *monoid*. This means that $\mathcal{M} = (\mathcal{M}, +, 0)$ is augmented with an associative and commutative operator $+$ such that $\mathcal{M}$ is closed under $+$ and 0 is the identity element for $+$. The data structure is modeled as an collection of variables $v_0, v_1, \ldots$, each of which can hold a value from $\mathcal{M}$ and initially contains 0. After receiving the input to each operation, the algorithm executes a sequence of operations of the form $v_i \leftarrow \text{INPUT}$, $v_i \leftarrow v_j + v_k$ or $\text{OUTPUT} \leftarrow v_i$. The algorithm must be correct for all choices of $\mathcal{M}$, thus in particular it cannot assume that the operator $+$ is invertible. The cost of an algorithm in processing a sequence of operations is the total number of such instructions executed by it. The restriction that $\mathcal{M}$ is a monoid (rather than say, a group) is partly for ease of proving lower bounds. However, known algorithms in this framework do not gain significant asymptotic speedups by exploiting stronger models (e.g., by using the fact that $+$ is invertible in groups).

34.2 Simple Randomized Set Representations

In this section we cover a simple, but general-purpose set representation due to [9,10]. In addition to the base repertoire (34.1), we wish to support:

$$A \text{ op } B, \text{op} \in \{\cup, \cap, -\},$$

as well as the following boolean operations:

$\text{equal}(A, B)$	Returns 'true' if $A = B$ and 'false' otherwise.	(34.4)
$\text{subset}(A, B)$	Returns 'true' if $A \subseteq B$ and 'false' otherwise.	(34.5)

This representation touches upon a topic of interest in its own right, that of the *unique* representation of sets, which can be defined as follows. We consider a class of representations that suffice to represent all subsets of a given universe U. However, we require that each set should have a a *unique* representation within that class. In particular the representation should be independent of the

---

[*] To some readers, the idea that the wordsize of a machine can change as we update the data structure may appear a little strange, but it is merely a formal device to ensure reasonable usage.

sequence of operations that created the set (e.g., a red-black tree representation is not unique). Unique representations have the desirable property that it is then possible to ensure that all instances of the same set within a family (including sets that are created as intermediate results in a computation) are represented by the same object within the data structure. This allows constant-time equality tests of two sets: just check if they are the same object! The difficulty, of course, is to combine unique representations with rapid updates.

This definition also applies to *randomized* algorithms, which may access a source of random bits while executing a query or update, and the uniqueness requirement here means that the choice of representation for a set depends solely upon the sequence of random bits that are output by the source. (If, as in practice, one uses a pseudo-random generator, then the representation for a given set depends only upon the seed used by the pseudo-random number generator.)

34.2.1 The Hash Trie

We first weaken the notion of a unique representation, and speak about the unique representation of sets from a *labeled* universe. Here, we assume that the elements of the universe U are labeled with (not necessarily unique) b-bit strings, and for $x \in U$ we denote its label by $\ell(x)$. In addition, we also have the notion of a *labeled set* A_y, where y is a sequence of $\leq b$ bits. Any labeled set A_y satisfies the property that for all $x \in A_y$, y is a prefix of $\ell(x)$.

Given a set from a labeled universe, one can of course use the well-known binary trie [11] to represent it. For the sake of precision we now define the binary trie, when used to represent a (labeled) set S_y:

- If $|S_y| = 0$ then the trie is empty.
- If $|y| = b$ then the trie is a leaf node that contains a pointer to a linked list containing all elements of S_y.
- Otherwise, the trie comprises an internal node with an edge labeled 0 pointing to a binary trie for the set

$$S_{y0} = \{x \in S_y \mid y0 \text{ is a prefix of } \ell(x)\}$$

and an edge labeled 1 pointing to a binary trie for the labeled set

$$S_{y1} = \{x \in S_y \mid y1 \text{ is a prefix of } \ell(x)\}$$

A set $S \subseteq U$ is represented as the labeled set S_Λ, where Λ denotes the empty string.

Consider a family $\mathcal{F}$ of sets from a universe U and let each node in the resulting collection of tries represent a labeled set in the natural way. Then if two nodes represent labeled sets A_z and B_z, such that $A_z = B_z$ (note that the label is the same) then the subtrees rooted at these nodes have an identical structure. One can then save space by ensuring that all sets in $\mathcal{F}$ whose representations have instances of the set A_z point to a single instance of A_z. By consistently applying this principle, we will ensure that two sets $S, T, S = T$ point to a single instance of S_Λ.

We now give an example. Let $U = \{a, b, c, d, e, f\}$ and $\mathcal{F}$ contain the sets $X = \{a, c, f\}$, $Y = \{c, d, e\}$ and $Z = \{a, b, c, f\}$. Suppose that the elements of U are labeled as follows:

$$\ell(a) = 001, \ \ell(b) = 011, \ \ell(c) = 010, \ \ell(d) = 101, \ \ell(e) = 110, \ \ell(f) = 010$$

Without exploiting the unique representation property, we would get the representation in Figure 34.1(i), and by exploiting it and storing subtrees uniquely we get the representation in Figure 34.1(ii). Updates now need to be done with a little care: one cannot, for example, add a new child to a node, as the subtree in which the node lies may be shared among a number of sets. The solution is to use *nondestructive* updates, namely, implementing updates by making a *copy* of the path from the leaf to be added/deleted to the name of the sets (cf. *persistent* data structures). Figure 34.1(iii) shows the state of the data structure after executing the commands insert(a, Y) and insert(b, X), which we now explain.

First, let us consider the operation insert(a, Y). We need to insert a as a sibling of c in the trie for Y. Since the path from c up to the root of Y could potentially be shared by other sets, we do not modify the node that is the parent of c, instead making a copy of the entire path from c to the node that is the root of the representation of Y, as shown in the figure. The pointer for Y is then appropriately modified. The nodes that were previously part of the representation of Y (shown in dotted outline) are, in this example, not reachable as part of any set and may be cleared as part of a garbage-collection phase (they are not explicitly freed during the insertion). Now coming to insert(b, X), we proceed as before, and do not insert b directly as a sibling of (c, f) under X. However, before creating a new node with the leaves b and (c, f) as children, we check the data structure to see if such a node exists and find one (this is the node representing the set Z_{01}). Therefore, we avoid creating this node. Continuing, we discover that all the nodes that we would have tried to create as part of this insertion already exist and therefore conclude that the sets X and Z are now the same (and hence their tries should be represented by the same node).

To support the reuse of existing nodes, a dictionary data structure that stores all nodes currently in the data structure is maintained. A node x with left child y and right child z is stored in this dictionary with the key $\langle y, z \rangle$ (either of y or z could be nil).

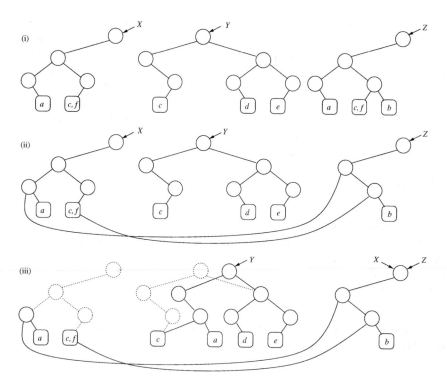

FIGURE 34.1 Unique set representations.

Each insertion requires $\Theta(b)$ lookups or insertions into this dictionary. Avoiding this overhead is important, but in a randomized setting the dictionary can be implemented using dynamic perfect hashing, giving $O(1)$ overhead. This idea is also used in practical implementations of functional programming languages such as LISP, and is referred to as *hashed consing* in that context.[*]

In a randomized setting, the other problem, that of obtaining suitable labels, can also be solved easily. We simply let ℓ be a function chosen at random from a universal class of hash functions (cf. Chapter 9). By choosing $\ell : U \to \{1, \ldots, n^3\}$ (e.g.) we can ensure that the number of *collisions*, or pairs of distinct elements x, y in U with $\ell(x) = \ell(y)$ is $O(1)$ with a probability that tends to 1 as $n \to \infty$. This means that we can essentially ignore the possibility that there are more than a constant number of elements at the leaf of any trie. Clearly, labels are $O(\log n)$ bits long.

Note that we can ensure that both the lengths of the labels and the number of collisions stays under control as n changes, by simply rehashing whenever the value of n doubles or halves since the last rehash. We now analyze some parameters of this data structure. Clearly, equality testing takes $O(1)$ time. Each insertion or deletion takes time and space proportional to the depth of the tries, which is $O(\log n)$. Both insertions and deletions may cause some nodes to become "garbage" (i.e., unreachable from any set)—these need to be compacted periodically, for example, when a rehash occurs. It is easy to verify that the amortized cost of rehashing and garbage collection is negligible. This gives parts (i) and (ii) of the following theorem; for part (iii) we refer the reader to [9, Chapter 8]:

Theorem 34.1

There is a data structure for a family of sets that:

i. Supports `insert` and `delete` in $O(\log n)$ amortized expected time and equality-testing in $O(1)$ worst-case time;
ii. Uses $O(n \log n)$ space, and
iii. Given two sets A and B, supports the constructive version of A op B for op $\in \{\cup, \cap, -, \subseteq\}$ in $O(|S_2|(1 + \log(S_3/S_2))$ amortized expected time, where S_2 and S_3 are the middle and largest sets of $A - B$, $B - A$ and $A \cap B$.

---

[*] The operation for creating a new node is called a Cons.

Remark 34.1

Note that a fairly naive bound for an operation A op B, for op $\in \{\cap, \cup, -, \subseteq\}$ is $O(|A|\log|B|)$, implemented by storing each tree in the family in a binary-tree based dictionary. Thus, Theorem 34.1 is not, in the worst case, a lot better. On the other hand, if we have two large sets S and T that differ in only a constant number k of elements then the expected running time of the above algorithm is $O(k\log\max\{|S|,|T|\})$.

34.2.2 Some Remarks on Unique Representations

There are interesting issues regarding the unique representation problem, in particular for unlabeled universes. We first mention here that both the *randomized search tree* and the *skip list* (cf. Chapter 14) satisfy the unique representation described above, even for unlabeled universes, and support the dictionary operations (insert, delete and member) in $O(\log n)$ expected time. In each of these cases, the algorithm described there should be modified so that instead of choosing a random height for a node in a skip list, or a random priority for a node in a randomized search tree, we should choose a function $U \to [0,1]$ that behaves "randomly" (we refer the reader to [12] for a precise theoretical statement).

By representing each set in a family using either one of these representations we get alternatives to parts (i) and (ii) for Theorem 34.1. However, one may ask about deterministic unique representations that support rapid updates. The reader is directed to [13,14] and the references therein for pointers to the literature on this fascinating area. We only note here that [14] show a $\Theta(n^{1/3})$ time bound for supporting the above dictionary operations, provided that the class of representations is restricted to "graph-like" representations (including, of course, trees). This shows an exponential separation between randomized uniquely represented dictionaries and deterministic (non-unique) dictionaries on the other hand, and deterministic uniquely represented dictionaries on the other. This result is not entirely conclusive, however: by labeling elements in U with integers from $\{0, \ldots, m-1\}$, as observed in [13, Section 2.3], the above trie-based approach gives uniquely represented dictionaries that support all operations in $O(\log m)$ time, where $m = |U|$. Thus, there is a "dichotomy" depending upon whether we are interested in bounds that depend on m alone or n alone (cf. Chapter 40). It is not known how the complexity of unique representation depends on the relationship between m and n. Indeed, the gap in knowledge is quite large, as the $\Omega(n^{1/3})$ applies when $n \sim \log^*|U|$, while the $O(\log m)$ algorithm is optimal only when $n \sim |U|^\epsilon$ for some constant $\epsilon > 0$.

34.3 Equality Testing

We now consider representations that only support the base repertoire (34.1) and the equal operation (34.4). Clearly we would like solutions that are better (at least in some respects) than those given by Theorem 34.1, and we begin by considering deterministic data structures. We first discuss one due to [13], which is based on binary tries (cf. Section 34.2).

The operations applied to our data structure are divided into *epochs*. At the start of an epoch the data structure is rebuilt "from scratch." Without loss of generality we can consider U to consist of all the elements that were present in $\mathcal{F}$ at the start of an epoch, plus all the elements that have appeared as the argument to an insert or delete operation since then (note that U may contain elements no longer present in $\mathcal{F}$). We can then label elements in U with values from $\{0, \ldots, |U|-1\}$, using $\lceil \log m \rceil$-bit integers, where $m = |U|$. Whenever we insert an element that is not in U, we give it the next higher integer. If $\sum_{A \in \mathcal{F}} |A| = n_0$ at the start of an epoch, we start the next epoch after $n_0/2$ updates (inserts or deletes). Rebuilding from scratch involves resetting U to be only those elements that are currently present in $\mathcal{F}$, relabeling all elements with integers from $\{0, \ldots, |U|-1\}$, and recreating the data structure with these labels, as well as any auxiliary data structures. It will be easy to see that the amortized cost of rebuilding is negligible, and that $|U| = O(n)$ at all times.

Each set is represented as a binary trie, with shared subtrees as in the previous section. Updates are also done non-destructively. A key difference is the method used to avoid creating nodes that already exist. Nodes are numbered serially in order of creation. If node p points to node v, we say that p is one of (possibly many) parents of v. Each node v, maintains a set *parents*(v) of all its parents. Each parent $p \in$ *parents*(v) is assigned a key equal to $\langle$*node-number*$(w), b\rangle$, where w is the other node (besides v) pointed to by p, and b equals 0 or 1 depending on whether v is a left child of p or not. When doing an update, before creating a new node with left child x and right child y, we we search set *parents*(x) for the key $\langle$*node-number*$(y), 0\rangle$. If such a key is found, we return the matching parent, which is precisely a node with left child x and right child y. Otherwise, create a new node p with pointers to v and w, set *parents*(p) to empty, insert p into *parents*(v) and *parents*(w), and return p. The issue is how to represent the sets *parents*(v) (cf. the dictionary in the previous section).

Each set *parents*(v) is represented by a binary search tree and a variation of the splay algorithm is used to perform searches. The splay algorithm is especially appropriate as it can be used to exploit patterns in the sequence of accesses. Although a detailed consideration of splay trees is beyond the scope of this chapter, the necessary facts are, roughly, that (a) insertion of a key that is larger than the maximum key currently in the tree (a *passive* insertion) has constant amortized cost while an insertion that is not passive (an *active*) insertion) has logarithmic amortized cost (b) if the frequency of search for an item i is $0 < \alpha_i < 1$ then the amortized cost of all searches for item i is essentially $O(1 + \log(1/\alpha_i))$. This is summarized in the lemma below:

Lemma 34.1

Consider a sequence of insertions and searches performed on an (initially empty) binary search tree using splays. Let:

$$s_i = \text{the number of searches of item } i$$
$$s = \text{the total number of searches,}$$
$$a = \text{the number of active insertions, and}$$
$$p = \text{the number of passive insertions.}$$

The cost of this sequence is $O((p + a) + s + a \log a + \sum_{s_i \geq 1} s_i \log(s/s_i))$.

Remark 34.2

Insertions into an initially non-empty tree are easily handled: pretend to start with an empty tree and initialize the tree with a sequence of passive insertions. By the above lemma, the additional cost is linear in the size of the initial tree.

We now argue that although an update (`insert` or `delete`) may result in $\Theta(\log m)$ node creations, and hence as many insertions into the *parents* dictionaries, most of the insertions are passive. Specifically, let $x \langle y, z \rangle$ denote the creation of a node x with children y and z. Then, an update may result in a sequence of node creations: $y_1 \langle x_0, x_1 \rangle, y_2 \langle y_1, x_2 \rangle, \ldots, y_l \langle y_{l-1}, x_l \rangle$, where $y_1, \ldots, y_l$ are "new" nodes and $x_0, \ldots, x_l$ are nodes that existed previously in the data structure. Note that because the y_i's are new nodes, they have higher *node-numbers* than the x_is. As a result, of the $2l$ insertions into *parents* dictionaries caused by this update, all but two—the insertion of y_i with key $\langle node\text{-}number(x_0), b \rangle$ ($\langle node\text{-}number(x_1), 1 - b \rangle$) into $parents(x_1)$ ($parents(x_0)$)—are passive.

We now need to bound the time to perform searches on the *parents* dictionary. For this, we consider a single epoch, and let $G = (V, E)$ be directed acyclic graph formed by the nodes at the end of the epoch. The sources (nodes of zero in degree) are the roots of (current and past) sets and the sinks (nodes of zero outdegree) are the elements of U. Note that all nodes have outdegree at most 2.

An update to sets in $\mathcal{F}$ results searches in the the *parents* dictionary of nodes that lie along a path π from a source to a sink in G. It should be noted that the path is traversed in reverse, from a sink to a source. Note that each time that a search traverses an edge (v, w), where w is a parent of v, the key that is searched for in $parents(v)$ is the same. Thus we can identify the traversal of an edge in G with a search of a particular key in a particular dictionary. Let a_e denote the number of times an edge $e \in E$ is traversed, and note that we can delete from G all edges with $a_e = 0$. Letting $A_v = \sum_{(w,v) \in E} a_{(w,v)}$, the cost of searches in $parents(v)$ for any vertex $v \in V$, denoted $cost(v)$, is given by $\sum_{(w,v) \in E} a_{(w,v)} \log A_v / a_{w,v}$, by Lemma 34.1. Let V' denote the set of nodes that are neither sinks nor sources, and note that for any $v \in V'$, $\sum_{(w,v) \in E} a_{(w,v)} = \sum_{(v,w) \in E} a_{(v,w)}$. Thus, we have:

$$\sum_{v \in V} cost(v) = \sum_{(w,v) \in E} a_{(w,v)} \log A_v / a_{w,v}$$

$$= \sum_{v \in V'} \sum_{(w,v) \in E} a_{(w,v)} \log A_v + \sum_{v \in V - V'} \sum_{(w,v) \in E} a_{(w,v)} \log A_v + \sum_{e \in E} a_e \log 1/a_e$$

$$= \sum_{v \in V'} \sum_{(v,w) \in E} a_{(v,w)} \log A_v + \sum_{v \in V - V'} \sum_{(w,v) \in E} a_{(w,v)} \log A_v + \sum_{e \in E} a_e \log 1/a_e$$

$$\leq \sum_{v \in V'} \sum_{(v,w) \in E} a_{(v,w)} \log A_v / a_{(v,w)} + \sum_{v \in V - V'} \sum_{(w,v) \in E} a_{(w,v)} \log A_v$$

$$\leq \sum_{v \in V'} \log out(v) + \sum_{v \in V - V'} \sum_{(w,v) \in E} a_{(w,v)} \log A_v$$

where $out(v)$ denotes the out-degree of v. The last step uses the fact that for all $\alpha_1, \ldots, \alpha_d$, $\alpha_i \in [0,1]$ and $\sum_i \alpha_i = 1$, $\sum_{i=1}^d \alpha_i \log(1/\alpha_i) \leq \log d$ (the "entropy" inequality).

Note that $\sum_{v \in V - V'} \sum_{(w,v) \in E} a_{(w,v)}$ is just the number of updates in this epoch and is therefore $\Theta(n)$. Since $A_v = O(n)$ we can bound the latter term by $O(n \log n)$. Since the outdegree of each node is 2, the former term is $O(V)$, which is also $O(n \log n)$. We thus find that all dictionary operations in the sets $parents(v)$ take $O(n \log n)$ time, and so the amortized cost of an update is $O(\log n)$. To conclude:

Theorem 34.2

There is a data structure for a family of sets that supports insert, delete and member in $O(\log n)$ amortized time, equal in $O(1)$ worst-case time, and uses $O(n \log n)$ words of space, where n is the current size of the family of sets.

We now describe the *partition tree* data structure for this problem [15,16]. Since the partition tree is closely related to the above data structure, results based on the partition tree are not significantly different from those of Theorem 34.2. We describe a slightly simplified version of the partition tree by assuming that U is fixed. The partition tree is a *full* binary tree T with $|U|$ leaves, that is, a tree where the leaves are all at depth $\lfloor \log U \rfloor$ or $\lceil \log U \rceil$, and all nodes have either two or zero children. At the leaves of this tree we place the elements of U (in any order). For each internal node v of T, we let $D(v)$ denote the elements of U that are at leaves descended from v. A set $A \in \mathcal{F}$ is stored at an internal node v if $D(v) \cap A \neq \emptyset$. Furthermore, all sets that are stored at an internal node v are grouped into equivalence classes, where the equivalence relation is given by $A \equiv B \Longleftrightarrow (A \cup D(v) = B \cup D(v))$. Clearly, two sets are equal iff they belong to the same equivalence class at the root of the tree and so this representation supports constant-time equality testing. Note that if n_v is the number of sets stored at an internal node v of T, then $\sum_v n_v = O(n \log |U|)$. This is because each set containing x appears once on each node from the path leading from x to the root (and hence $O(\log|U|)$ times in all), and $\sum_{x \in U} |\{A \in \mathcal{F} \mid x \in A\}| = n$. The key issue, of course, is how to update the partition tree when executing $\text{insert}(x, A)$ ($\text{delete}(x, A)$). We traverse T from x to the root, storing (or removing) A from all the nodes along the path. We now show how to maintain the equivalence classes at these nodes.

At the leaf, there is only one equivalence class, consisting of all sets that contain x. We merely need to add (delete) A to (from) this equivalence class. In general, however, at each node we need to determine whether adding/deleting x to/from A causes A to move into a new equivalence class of its own or into an existing equivalence class. This can be done as follows. Suppose γ is a (non-empty) equivalence class at a (non-leaf) node u in T and suppose that v, w are u's children. A little thought shows that must be (non-empty) equivalence classes α, β at v, w respectively such that $\gamma = \alpha \cap \beta$. A global dictionary stores the name of γ with the key $\langle \alpha, \beta \rangle$. Inductively assume that following an operation $\text{insert}(x, A)$ (or $\text{delete}(x, A)$) we have updated the equivalence class of A at all ancestors of x up to and including a node v, and suppose that A is in the equivalence class α at v. Next we determine the equivalence class β of A in v's sibling (this would not have changed by the update). We then look up up the dictionary with the key $\langle \alpha, \beta \rangle$; if we find an equivalence class γ stored with this key then A belongs to γ in v's parent u, otherwise we create a new equivalence class in u and update the dictionary.

Lam and Lee [16] asked whether a solution could be found that performed all operations in good single-operation worst-case time. The main obstacles to achieving this are the amortization in the splay tree and the periodic rebuilding of the partition tree. Again dividing the operation sequence into *epochs*, Lam and Lee noted that at the end of an epoch, the partition tree could be copied and rebuilt incrementally whilst allowing the old partition tree to continue to answer queries for a while (this is an implementation of the *global rebuilding* approach of Overmars [17]). By "naming" the equivalence classes using integers from an appropriate range, they note that the dictionary may be implemented using a two-dimensional array of size $O(n^2 \log n)$ words, which supports dictionary operations in $O(1)$ worst-case time. Alternatively, using standard balanced binary trees or Willard's q-fast tries [18], or the more complex data structure of Beame and Fich [19] gives a worst-case complexity of $O(\log n)$, $O(\sqrt{\log n})$ and $O((\log \log n)^2)$ for the dictionary lookup, respectively. Using these data structures for the dictionary we get:

Theorem 34.3

There is a data structure for a family of sets that supports equal in $O(1)$ worst-case time, and insert and delete in either (i) $O(\log n(\log \log n)^2)$ worst-case time and $O(n \log n)$ space or (ii) $O(\log n)$ worst-case time and $O(n^2 \log n)$ space.

Remark 34.3

The reader may have noticed the similarities between the lookups in partition trees and the lookup needed to avoid creating existing nodes in the solutions of Theorems 34.1 and 34.2; indeed the examples in Figure 34.2 should make it clear that, at least in the case that $|U|$ is a power of 2, there is a mapping from nodes in the DAG of Theorem 34.2 and partitions in the partition tree. The figure illustrates the following example: $U = \{a, b, c, d\}$, $\mathcal{F} = \{A, B, C, D\}$, $A = \{a, c, d\} = C$, $B = \{a, b, d\}$, $D = \{c, d\}$, and an assumed labeling function that labels a, b, c and d with 00,01,10 and 11 respectively. The partitions in (ii) shown circled with a dashed/dotted line correspond to the nodes circled with a dashed/dotted line in (i).

34.4 Extremal Sets and Subset Testing

This section deals with testing sets in $\mathcal{F}$ for the subset containment relationship. We first survey a static version of this problem, and then consider a dynamisation.

(i)

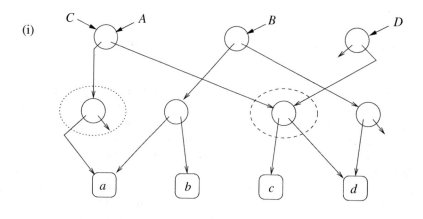

(ii)

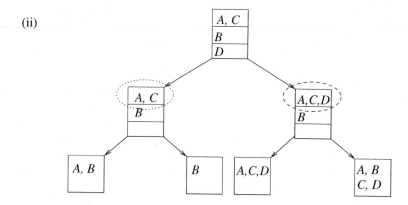

FIGURE 34.2 The partition tree (ii) and its relation to the DAG created by the binary trie representation with shared subtrees.

34.4.1 Static Extremal Sets

Here we assume that we are given a family $\mathcal{F}$ of sets from a common universe as input. A set S is *maximal* in $\mathcal{F}$ if there is no set $T \in \mathcal{F}$ such that $S \subset T$. A set S is *minimal* in $\mathcal{F}$ if there is no set $T \in \mathcal{F}$ such that $S \supset T$. The *extremal sets* problem is that of finding all the maximal (or all the minimal) sets in $\mathcal{F}$. A closely related problem is the computation of the *complete subset graph*, which represents all edges in the partial order induced among members of $\mathcal{F}$ by the subset relation. Specifically, the complete subset graph is a directed graph whose vertices are the members of $\mathcal{F}$ and there is an edge from x to y iff $x \subset y$. This is a problem that arises in a number of practical contexts, for example, it can be used to maximally simplify formulae in restricted conjunctive normal form [20].

This problem has been considered in a number of papers including [20–24]. We now summarize the results in these papers. As before, we let $k = |\mathcal{F}|$ denote the size of the family, and $n = \sum_{S \in \mathcal{F}} |S|$ be the total cardinality of all sets. A number of straightforward approaches can be found with a worst-case complexity of $O(n^2)$. It can be shown that the subset graph has $\Omega(n^2/(\log n)^2)$ edges [21], which gives a lower bound for any algorithm that explicitly lists the complete subset graph. An aim, therefore, is to bridge the gap between these bounds; all results below are in the word RAM model.

Yellin and Jutla gave an algorithm that computes the subset graph using $O(n^2/\log n)$ dictionary operations. Using hashing, all dictionary operations take $O(1)$ expected time, and we get an $O(n^2/\log n)$ expected running time. Using Willard's q-fast tries [18], or the more complex data structure of Beame and Fich [19] gives running times of $O(n^2/\sqrt{\log n})$ or $O((n \log \log n)^2/\log n)$ respectively. Pritchard [22] gave a simple algorithm that ran in $O(n^2/\log n)$ time. In [23] he re-analyzed an earlier algorithm [20] to show that this algorithm, too, ran in $O(n^2/\log n)$ time (the algorithm of [20,23] uses RAM operations less extensively and is a good candidate for porting to the pointer machine model). Finally, Pritchard [24] gave an algorithm that uses the bit-parallelism of the word RAM extensively to achieve an $O(n \log \log n/(\log n)^2)$ running time.

All of the above are static problems—it is very natural to ask about the complexity of this problem when updates may change the sets in $\mathcal{F}$. Again, if one wishes explicitly to maintain the entire subset graph, it is easy to see that $\Omega(n)$ edges may change as the result of a single update. Take $U = \{1, \ldots, u\}$ and $\mathcal{F} = \{A, B_2, \ldots, B_u\}$, where $A = U$ and $B_i = \{1, i\}$ for $i = 2, \ldots, u$. The sum of the sizes of the sets in $\mathcal{F}$ is $2u - 2$, but deleting the element 1 from A removes all $u - 1$ edges from the complete subset graph. It is not very hard to come up with algorithms that can achieve an $O(n)$ running time (see, e.g., [21]). To obtain a better complexity, therefore, we must consider algorithms that do not explicitly store this graph. One way of doing this is to consider the *dynamic subset testing* problem, defined as the problem of supporting the base repertoire (34.1) and the subset operation (34.5). Since the query subset(A, B) tests if there is an edge between A and B in the complete subset graph, this problem seems to be an appropriate dynamisation of the extremal sets problem, and it is dealt with in the next section.

34.4.2 Dynamic Set Intersections and Subset Testing

Rather than consider the dynamic subset testing problem directly, we consider a related problem, the *dynamic set intersection* problem. In addition to the base repertoire (34.1) of actions above, we consider an enumerative version of the following:

$$\text{intersect}(A, B) \quad \text{Report the intersection of sets } A \text{ and } B. \tag{34.6}$$

Other variations could include returning the size of the intersection, or retrieving some values associated with the elements in the intersection. A unifying way to study these problems is as follows: we are given a set $\mathcal{M}$ of *information* items that will be associated with elements of U, a function $I : U \to \mathcal{M}$ that associates values from M with keys in U. We assume that $\mathcal{M} = (\mathcal{M}, +, 0)$ is a monoid. The query intersect(A, B) then returns $\sum_{x \in A \cap B} I(x)$, where the sum is taken with respect to $+$. It is easy to cast the intersection problem and its variants in this framework. The basic problem defined above can be obtained by letting $\mathcal{M} = (2^U, \cup, \{\})$ and $I(x) = \{x\}$ for all x, and the problem where one merely has to report the size of the intersection can be obtained by setting $\mathcal{M} = (\mathbb{N}, +, 0)$ and $I(x) = 1$ for all x, where $+$ here is normal (arithmetic) addition. Note that dynamic subset testing, defined in the previous section, easily reduces to the intersection problem: $A \subseteq B$ iff $|A| = |A \cap B|$.

We now survey the results in this area. It is useful to use the notation $\widetilde{O}(f(n)) = \cup_{c=0}^{\infty} O(f(n) \log^c n)$, which ignores polylogarithmic factors are ignored (a similar convention is used for the Ω notation, with inverse polylogarithmic factors being ignored).

For this problem, Yellin [25] gave an algorithm that processed a sequence of n insertions and deletions and q intersect operations in time $\widetilde{O}(n \cdot n^{1/k} + qn^{(1-1/k)})$ time for any fixed k. The intuition behind this result is as follows. Take $t = n^{1-1/k}$ and say that a set $S \in \mathcal{F}$ is *large* if $|S| \geq t$ and *small* otherwise. All sets are represented as dictionaries (allowing membership testing in logarithmic time). Intersections between two small sets, or between a small set and a large one, are handled by testing each element of one set for membership in the other. Clearly this takes $\widetilde{O}(t)$ time, and insertion into and deletion from a small set takes $\widetilde{O}(1)$ time. For every pair of large sets S, T, the algorithm keeps track of $I(S \cap T)$ explicitly, by storing the elements of $S \cap T$ in a balanced binary search tree, and storing at any internal node x the monoid sums of all the elements under x.* Since there are at most $n/t = n^{1/k}$ large sets, an insertion into, or a deletion from, a large set requires updating at most $n^{1/k}$ of the pairwise intersections with other sets, and takes $\widetilde{O}(n^{1/k})$ time. This proves the time bound, modulo some details such as dealing with changes in the value of t caused by changes in n and so on.

Dietz et al. [26] noticed that if one were to process a sequence of n updates and q queries, where n and q are known in advance, then the overall cost of Yellin's algorithm for processing the sequence is minimized by taking $n^{1/k} = \min\{n, \sqrt{q}\}$, giving an overall time bound of $\widetilde{O}(q + n\sqrt{q})$. They gave an algorithm that achieves this bound even when n and q are not known in advance.

We now show that this bound is essentially tight in the arithmetic model, by giving a lower bound of $\widetilde{\Omega}(q + n\sqrt{q})$. For simplicity, consider the problem where the elements in the intersection are to be reported, that is, take $\mathcal{M} = (2^U, \cup, \{\})$ and $I(x) = \{x\}$. Starting from an empty family $\mathcal{F}$, we build it up by insertions so that the (sums of) sizes of the pairwise intersections of sets in $\mathcal{F}$ are large, and query all possible intersections of the sets. If the answers to all the queries were to be obtained by adding (unioning) together singleton sets, then the lower bound would follow. Unfortunately, this is too simplistic: subsets obtained as temporary values during the computation of one answer may be re-used to answer another query. To get around this, we note that a subset that is used to compute the answer to several intersection queries must lie in the common intersection of all the sets involved, and construct $\mathcal{F}$ so that the common intersection of any of a sufficiently large (yet small) number of sets in $\mathcal{F}$ is small. This means that no large subset can be reused very often.

We assume that $q \leq n^2$ (otherwise the lower bound is trivial). Starting with an empty family, we perform a sequence of $n = \Theta(N)$ insert operations and q queries. We choose $U = \{1, \ldots, 2N/\sqrt{q}\}$ and $k = |\mathcal{F}| = \sqrt{q}$; we let $m = |U|$ and let $\ell = c \log N$ for some sufficiently large constant $c > 0$. We assume the existence of a family $\mathcal{F} = \{S_1, \ldots, S_k\}$ with the following properties:

a. $|S_i \cap S_j| = \Omega(m)$ for all i, j with $1 \leq i < j \leq k$.
b. for any pairwise distinct indices $i_1, \ldots, i_\ell$, $|\cap_{j=1}^\ell S_{i_j}| < \ell$.

The existence of such a family is easily shown by a probabilistic argument (a collection of random sets of the appropriate size suffices). We now represent $\mathcal{F}$ in the data structure by executing the appropriate insertions. Note that this requires at most

---

* If the monoid is $\mathcal{M} = (2^U, \cup, \{\})$, we do not store the monoid sum explicitly, but instead take the monoid sum at an internal node to be implicitly represented by the subtree rooted at it.

$km = N = \Theta(n)$ update operations, since $|S_i| \leq m$ for all i. We then query the pairwise intersections of all the sets; there are $(k\ 2) = \Theta(q)$ queries in all.

Firstly, note that the sizes of all the output sets sum to $\Omega(mk^2)$ by (a) above. The output to each query is conceptually obtained by a binary *union tree* in which each internal node combines the answers from its children; the external nodes represent singleton sets. Each node can be labeled with a set in the obvious way. Consider the entire forest; we wish to count the number of distinct nodes in the forest (that is, nodes labeled with distinct sets). Since each distinct set corresponds to at least one instruction that is not counted elsewhere, counting the number of distinct sets is a lower bound on the number of instructions executed.

We consider only nodes that correspond to sets of size ℓ or larger. Clearly, the number of such sets is $\Omega(mk^2/\ell)$. Furthermore, no such set can be used to answer more than $(\ell\ 2)$ different queries by (b). It follows that there are $\Omega(mk^2/\ell^3) = \widetilde{\Omega}(n\sqrt{q})$ distinct sets, giving us a lower bound of this magnitude (as mentioned above, a lower bound of $\Omega(q)$ is trivial whenever $q = \widetilde{\Omega}(n\sqrt{q})$).

Dietz et al. also considered the relatively high memory usage of the above algorithms. For example, if $q = \Theta(n)$ then both Yellin's algorithm and Dietz et al.'s algorithm use $\Omega(n^{3/2})$ space. Dietz et al. also considered the complexity of the above problem when the algorithms were restricted to use s memory locations, where $n \leq s \leq n^2$, and gave an algorithm that processed n operations (queries and updates) in $\widetilde{O}(n/s^{1/3})$ time. Both the upper bound and the lower bound are more complex than those for the unlimited-memory case. We summarize with the main theorems of Dietz et al.:

Theorem 34.4

For the problem of maintaining a family of sets under create, insert, delete and intersect, we have the following results:

  i. Any algorithm requires $\widetilde{\Omega}(q + n\sqrt{q})$ time to process a sequence of n updates and q queries in the arithmetic model of computation.
  ii. Any intermixed sequence of n updates and q queries can be processed in $\widetilde{O}(n\sqrt{q} + q)$ time using $O(\min\{n\sqrt{q}, n^2\})$ space.
  iii. There is a sequence of $O(n)$ updates and queries that requires $\widetilde{\Omega}(n^2 s^{-1/3})$ operations in the arithmetic model, if the algorithm is restricted to using s memory locations.
  iv. Any intermixed sequence of $O(n)$ updates and queries can be performed in $\widetilde{O}(n^2/s^{1/3})$ time and $O(s)$ space.

As mentioned above, the arithmetic model is a relatively weak one (making it easy to prove lower bounds), but all the algorithms that realise the upper bounds fit into this model. It would be interesting to prove similar lower bounds for stronger models or to design algorithms that improve upon the performance of the above in stronger models. Note also, that the lower bounds are for the intersection problem, not the problem that we originally considered, that of subset testing. Since we get back a boolean answer from a subset testing query, it does not fit into the arithmetic framework.

34.5 The Disjoint Set Union-Find Problem

In this section we cover the best-studied set problem: the *disjoint set union-find* problem. The study of this problem and its variants has generated dozens of papers; indeed, it would not be unreasonable to devote an entire chapter to this problem alone. Fortunately, much of the work on this problem took place from the 1960s through to the early 1990s, and this work is surveyed in an excellent article by Galil and Italiano [27]. In this section, we summarise the main results prior to 1991 but focus on developments since then.

We begin with by formulating the problem in a recent, more general way [28], that fits better in our framework. We start with the base repertoire, but now require that insert(x, A) is only permitted if $x \notin B$, for all sets $B \in \mathcal{F} - \{A\}$. This ensures that all sets in $\mathcal{F}$ are pairwise disjoint. We now add the following operations:

dunion(A, B, C) This sets $C \leftarrow A \cup B$, but destroys (removes from $\mathcal{F}$) A and B.
find(x) For any $x \in \cup_{A \in \mathcal{F}} A$, returns the name of the set that x is in.

This problem has a number of applications, beginning historically with the efficient implementation of EQUIVALENCE and COMMON statements in early FORTRAN compilers in the 1960s, and continues to find applications in diverse areas such as dynamic graph algorithms, meldable data structures and implementations of unification algorithms.

It is difficult to discuss the disjoint-set union-find problem without reference to *Ackermann's function*, which is defined for integers $i, n \geq 0$ as follows:

$$A(i, n) = \begin{cases} 2n & \text{for } i = 0, n \geq 0 \\ 1 & \text{for } i \geq 1, n = 0 \\ A(i - 1, A(i, n - 1)) & \text{for } i, n \geq 1 \end{cases}$$

For a fixed i, the *row* inverse of Ackermann's function, denoted by $a(i, n)$, is defined as:

$$a(i, n) = \min\{j \mid A(i, j) \geq n\}.$$

It can be verified that $a(1, n) = \Theta(\log n)$ and $a(2, n) = \Theta(\log^* n)$, where $\log^* n$ is the *iterated logarithm* function. The functional inverse of Ackermann's function is defined for $m, n \geq 1$ by:

$$\alpha(m, n) = \min\{i \geq 1 \mid A(i, 4\lceil m/n \rceil) \geq n\}$$

Since $A(3, 4) = 2^{2^{\cdot^{\cdot^{\cdot^2}}}} \Big\}\, ^{65,536\ \text{times}}$, for all practical purposes, $\alpha(m, n) \leq 3$ whenever $m, n \geq 1$. Indeed, if m grows sufficiently faster than n (e.g., $m = \Omega(n \log^* n)$) then $\alpha(m, n)$ is indeed a constant.

The union-find problem in the above formalisation was studied by [29], who gave the following results:

Theorem 34.5

   i. An intermixed sequence of f find, m insert, m create, $d \leq m$ delete operations and at most $m - 1$ dunion operations takes $O(m + (f + d)\alpha(f + m, m))$ time. The size of the data structure at any time during the sequence is proportional to the number of live items in it.
   ii. For any parameter $k > 1$, an intermixed sequence of find, dunion, insert, create, and delete operations can be processed in the following worst-case time bounds: find and delete in $O(\log m/\log k)$ time, create and insert in $O(1)$ time and dunion in $O(k)$ time, where m is the current number of elements in the data structure.

In fact, [29] showed that this problem could be reduced to the classical union-find problem (see below). This reduction is such that the time bounds for find and dunion change only by a constant factor, but the time to delete an element x is the same as the time it takes to find the set containing x plus the time it takes to unite a singleton set with this set. The results follow by applying this result, respectively, to the classical union-find data structures given in Theorem 34.6(i) and Theorem 34.6(ii) (in fact they use a variant of the latter result due to Smid [30]).

An interesting variant of the above problem was considered by the same authors in [28]. They replaced the find operation with the seemingly simpler:

bfind(x, A) Return "true" if $x \in A$ and "false" otherwise.

This "boolean" union-find problem was motivated by the problem of implementing *meldable* priority queues. Kaplan et al. showed that the lower bounds established for the classical union-find problem apply to this problem as well (see Theorem 34.7).

34.5.1 The Classical Union-Find Problem and Variants

In the classical version of this problem, we have $U = \{1, \ldots, m\}$ and $\mathcal{F}$ initially equals $\{\{1\}, \{2\}, \ldots, \{m\}\}$; we assume that the name of the set $\{i\}$ is simply i. The operation dunion is modified as follows:

dunion(A, B) This sets $A \leftarrow A \cup B$, but destroys (removes from $\mathcal{F}$) B.

The operation find(x) operates as described above. We now mention the main upper bounds that are known for the classical version of the problem:

Theorem 34.6

A sequence of $m - 1$ dunion and f find operations can be processed as follows:

   i. in $O(m + f\alpha(f + m, m))$ time;
   ii. for any parameter $k > 1$, each find takes $O(\log m/(\log k + \log \log m))$ worst-case time and each dunion takes $O(k)$ worst-case time;
   iii. the ith find operation takes $O(\alpha(i, m))$ worst-case time, and the entire sequence takes $O(m + f\alpha(f + m, m))$ time;
   iv. for any parameter $k > 0$, the entire sequence takes $O(ma(k, m) + kf)$ time;
   v. for any parameter $k > 1$, dunion takes $O(k)$ worst-case time, find takes $t_q = O(\log m/\log k)$ worst-case time, and the entire sequence takes time $O((m + f)(\alpha(m + f, m) + a(t_q, m)))$ time.

We begin by remarking that the bounds (i) and (ii) are in fact implied by Theorem 34.5, but are included here for completeness (see [27] for a more complete history). The upper bound of Theorem 34.5(i) is due to Tarjan [31] (with many interesting refinements due to [32]). Note that since $\alpha(f + m, m)$ is essentially constant, the running time is linear for all practical purposes. In (i), each dunion takes $O(1)$ worst-case time, but a single find may take $\Omega(\log m)$ time. Result (iii) allows us to process each find quickly, in essentially constant time, but an individual dunion could be quite slow and may in fact take $\Omega(m)$ time. The overall cost of the sequence remains as in (i), though. This was explicitly proved in [33], but the main ideas were already present in [34]. In (iv), the time for a dunion is traded off for find, and the overall running time remains essentially linear (at least for $k \geq 2$) [35].

In either (i) or (iii), although the overall cost of the operation sequence is very low, an individual operation could take $\Omega(\log m)$ or $\Omega(m)$ time respectively. If we wish to minimise the maximum cost of a single operation over the sequence, then choosing (say) $k = \Theta(\sqrt{\log m})$ in (ii), we get that no operation takes more than $O(\log m / \log \log m)$ time. However, the cost of the entire sequence increases to $\Omega(f \log m / \log \log m)$. This result is due to [36].

In [37], the authors ask the question whether one can combine the best features of the results (i) (or (iii)) and (ii). Except possibly for a few extreme cases, this is achieved in [37, Theorem 11]. The result presented there has some technical restrictions (e.g., the algorithm needs to know the value of m in advance), but the authors claim that these are not essential.

We now come to lower bounds for this problem, which are either proved in the pointer machine model or the cell probe model. In general the cell probe model is more powerful (since there are no restrictions on the way that memory is accessed), but the lower bounds are proven assuming that all memory locations can store only numbers of $O(\log m)$ bits. By contrast, the pointer machine model does not place any restrictions on the size of the numbers that can be stored at each node. In the pointer machine model, however, the algorithm is required to maintain the name of each set in distinct nodes, and an operation such as find(x) must traverse the graph of nodes from the node for x to the node that contains the name of the set that x is in. Lower bounds for the pointer machine model sometimes make the following *separability* assumption [3]:

> At any time during the computation, the contents of the memory can be partitioned into collections of records such that each collection corresponds to a currently existing set, and no record in one collection contains a pointer to a record in another collection.

We now give some of the known lower bounds:

Theorem 34.7

i. Any algorithm that processes a sequence of $m - 1$ dunion and f find operations must take $\Omega(m + f\alpha(f + m, m))$ steps on either the pointer machine model or the cell probe model with a word size of $O(\log m)$ bits.
ii. Any algorithm that processes a sequence of $m - 1$ dunion and f find operations must take $\Omega(\log m / \log \log m)$ steps for some operation, on either the separable pointer machine model or the cell probe model with a word size of $O(\log m)$ bits.

The lower bound (i) for the pointer machine model is due to [38]; this generalizes an earlier lower bound for separable pointer machine algorithms due to [3] (see also the result due to [39]). The lower bound (ii) is due to [36], while both cell probe lower bounds are due to [40]. Additional lower bounds may be found in [37].

We refer the reader to [27] for variations of the classical union-find problem, including union-find with backtracking and persistent union-find (for the latter, see also [41,42]). Another important variant, which has a similar complexity to that of the union-find problem, is the *split-find* problem. Here we are given an ordered universe U and start with $\mathcal{F}$ containing a single set which contains all of U. We perform an intermixed series of operations $S, T \leftarrow \mathrm{split}(S, x)$ for some $x \in S$, which removes from S all elements $> x$ and places them in a new set T, and find(x), which as before returns the name of the set containing x. Upper and lower bounds of the form given in Theorems 34.6(i) or 34.7(i) have been proven by [33,34,38]. Finally, we mention the very interesting *Union-copy* problem studied in [43]. In addition to the classical union-find operations, they support an additional operation copy, which takes one set as its argument, and adds an exact copy of it to $\mathcal{F}$. Although it is obviously no longer true that all sets in $\mathcal{F}$ are disjoint, they require that two sets that are to be united must be disjoint. In addition, they want the data structure to support *dual* operations, in the sense that the roles of sets and elements are reversed. For example, the operation element-union(x, y) takes two elements x and y that do not both belong to any set in $\mathcal{F}$, and turns them into a single element that is member of all the sets to which x or y previously belonged. Duals of copy and find are also supported. They show applications of this data structure to the (geometric) dynamic segment tree data structure.

34.6 Partition Maintenance Algorithms

A partition P of a given universe $U = \{1, 2, \ldots, m\}$ is a collection of #P disjoint subsets (*parts*), $P^{(1)}, P^{(2)}, \ldots, P^{(\#P)}$, of U such that $\cup_i P_i = U$. A case of special interest is that of *bipartitions*, which is a partition with two parts. A partition P is an equivalence relation on U, which we denote as $\equiv_P$. Given two partitions P and Q, the *induced* partition of P and Q is the partition that represents the equivalence relation $x \equiv y \iff ((x \equiv_{PY}) \vee (x \equiv_{QY}))$ (in words, two elements belong to the same part of the induced partition iff they belong to the same part in both P and Q). The problem we consider is the following. Given a collection $\mathcal{F}$ of partitions (initially empty), to support the following operations:

$$
\left.
\begin{array}{ll}
\texttt{report}(\mathcal{F}) & \text{Report the partition induced by the members of } \mathcal{F}. \\
\texttt{insert}(P) & \text{Add partition } P \text{ to } \mathcal{F}. \\
\texttt{delete}(P) & \text{Remove partition } P \text{ from } \mathcal{F}.
\end{array}
\right\}
\qquad (34.7)
$$

We assume here that each new partition P is given by an array $A[1..m]$ containing integers from 0 to $\#P - 1$. Other operations include asking if two elements belong to the same partition of the global induced partition, reporting the set of elements that belong to the same part of the global induced partition, or, for any two elements, reporting the number of partitions in which these two elements are in different parts.

As noted by Bender et al. [44] and Calinescu [45], this problem has a number of applications. The general issue is supporting a *classification system* that attempts to categorize objects based on a number of tests, where each test realizes a partition of the objects (a bipartition, for instance, could model the outcome of a boolean test). The above data structure would be useful in the pre-processing phase of such a classification system, where, for example, an algorithm may repeatedly insert and delete partitions (tests) until it finds a small set of tests that distinguish all items of the universe. Examples in optical character recognition (OCR) and VLSI are given by the above authors.

We now discuss some solutions to these problems. The model used is the word RAM model with word size $\Theta(\log n)$, where we take $k = |\mathcal{F}|$ and $n = km$. We first consider the case of general partitions, and then that of bipartitions. A simple data structure for general partitions was given by Bender et al. [44], and resembles the partition tree of Section 34.3. A key fact they use is (apply radix sorting on triples of the form $\langle P[i], Q[i], i \rangle$):

Proposition 34.1

Given two partitions P and Q of $U = \{1, \ldots, m\}$ we can calculate the induced partition of P and Q in $O(m)$ time.

This immediately implies that we can maintain the induced partition of $\mathcal{F}$ under inserts alone in $O(m)$ time and also support report in $O(m)$ time—we simply maintain the global induced partition and update it with each insert. deletes are not so easy, however.

We place the partitions at the leaves of a full binary tree with k leaves. At each internal node v of this binary tree we store the partition induced by all the partitions descended from v. Clearly, the root contains the global induced partition. Inserting a partition P involves replacing a leaf containing a partition Q by an internal node that has P and Q as children, and updating the partitions stored at this internal node and all its ancestors. Deleting a partition involves deleting the appropriate leaf, and maintaining fullness by possibly swapping a "rightmost" leaf with the deleted leaf. In either case, we update partitions from at most two leaves up to the root. This gives a data structure that supports insert and delete in $O(m \log k)$ time and report in $O(m)$ time. The space used is $O(mn)$ words of memory.

The time for insert can be improved to amortised $O(m)$, while leaving the time complexity of all other operations the same. The idea is to group all partitions into disjoint groups of $t = \lceil \log k \rceil$ each, leaving perhaps one incomplete group of size smaller than t. For each group we store its induced partition, and also store the induced partitions of all groups, except the incomplete group, at the leaves of a tree T (which may now have $O(k/\log k)$ leaves) as before. In addition, we explicitly store the global induced partition G.

When performing insert(P) we add P to the incomplete group and in $O(m)$ time, update the global partition G. If the incomplete group reaches size t, in addition, we calculate the group's induced partition in $O(mt) = O(m \log k)$ time and insert this induced partition into T, also taking $O(m \log k)$ time, and start a new empty incomplete group. Deleting a partition P is done as follows. We delete P from its group. If P is in the incomplete group we recompute the G in $O(mt)$ time. If P's group is stored in T, we recompute the new partition induced by P's group in $O(mt)$ time and update T in $O(m \log k)$ time (if P is the last remaining partition in its group we delete the corresponding leaf, as before). We then recompute G in $O(mt) = O(m \log k)$ time as well. Note that the amortised cost of an insertion is now $O(m)$, since we spend $O(m \log k)$ time every $\log k$ insertions. Finally, Bender et al. note that a signature-based scheme gives a Monte-Carlo method that performs all operations in $O(m)$ time, but has a small chance of outputting an incorrect result (the algorithm runs correctly with probability $1 - O(m^{-c})$ on all inputs).

We now consider the important special case of bi-partitions, and give a sketch of Calinescu's [45] algorithm for solving this problem in optimal amortised time. Again letting $k = |\mathcal{F}|$, one can associate a k-bit binary string σ_x with each $x \in U = \{1, \ldots, m\}$, which specifies in which part of each of the k partitions x lies. Let π denote a permutation such that $\sigma_{\pi^{-1}(i)} \leq \sigma_{\pi^{-1}(i+1)}$, for $i = 1, \ldots, m - 1$; that is, π represents a sorted order on the (multi-set) of strings $\{\sigma_x | x \in U\}$. Furthermore, let lcp_i denote the most significant position where $\sigma_{\pi^{-1}(i)}$ and $\sigma_{\pi^{-1}(i+1)}$ differ, for $i = 1, \ldots, m-1$ ($lcp_i = k+1$ if $\sigma_{\pi^{-1}(i)} = \sigma_{\pi^{-1}(i+1)}$). We can now clearly support report in $O(m)$ time, as elements in the same part of the global induced partition will be consecutive in sorted order, and the *lcp* values allow us to determine the boundaries of parts of the global induced partition without inspecting the strings. An insert of a partition is also easily supported, as only local re-arrangements of elements lying within the same part of the previous global induced partition are required.

Deleting a partition is a little harder, and requires a few observations. Suppose that we delete the partition that gives the tth most significant bit of $\sigma_1, \ldots, \sigma_m$. Suppose that for two indices $i, j, j > i$, $lcp_i < t$ and $lcp_j < t$, but all lcp's in between are at least t. Then we can conclude that the strings in positions 1 to $i-1$ of the sorted order continue to appear in (possibly a different order in) positions 1 to $i-1$ after the deletion of this partition, and likewise the strings in positions $j+1$ to m also do not "move" except internally. Let Σ denote the strings that appear in positions i through j in sorted order, that is, $\Sigma = \{\sigma_l \mid l \in \{\pi^{-1}(i), \pi^{-1}(i+1), \ldots, \pi^{-1}(j)\}\}$. We now show how to sort Σ, and repeated application of this procedure suffices to re-sort the array. Note that all the strings in Σ that have 0 in the tth position maintain their relative order after the deletion, and likewise those strings with 1 in the tth position. Thus, re-sorting Σ is simply a matter of merging these two sets of strings. At a high level, the merging procedure proceeds like the (standard, trivial) algorithm. However, a naive approach would require $\Theta(k)$ time per comparison, which is excessive. Instead, we note that at each step of the merging, the next candidate can either be determined in $O(1)$ time (when the relative order of the two candidates is implicit from the *lcp* data) or a number, c, of comparisons need to be made. However, if c comparisons are made, then there is at least one *lcp* value in the new array that is c more than its counterpart in the old array. Since *lcp* values are bounded by $O(k)$, no string will be involved in more than $O(k)$ comparisons during its lifetime, and the cost of these comparisons can be charged to the insertions of the partitions. This intuition can be formalized by a potential function argument to show that the amortised cost of a deletion is indeed $O(n)$, thus giving an optimal (amortised) algorithm for this problem.

34.7 Conclusions

We have presented a number of algorithms and data structures for supporting (largely) basic set-theoretic operations. Aside from the union-find problem, which has been extensively studied, relatively little research has been done into these problems. For instance, even the most basic problem, that of finding a general-purpose data structure that supports basic set operations, is not yet satisfactorily solved. The problem becomes more acute if one is concerned about the space usage of the data structures—for example, it is not known whether one can solve set equality testing efficiently in linear space.

Due to the proliferation of unstructured data, set operations are increasingly important. For instance, many search engines return a set of documents that match a given boolean keyword query by means of set operations on the sets of documents that contain each of the keywords in the query. The characteristics of this kind of application also suggest directions for research. For example, given the large data-sets that could be involved, it is a little surprising that work on external-memory algorithms for these problems is somewhat limited. Another issue is that these sets usually have patterns (e.g., the number of sets that contain a given keyword may satisfy a power law; certain sets of keywords may be more likely to be queried together etc.), which should be exploited by efficient algorithms.

With the latter motivation in mind, Demaine et al. [46] have considered the *adaptive* complexity of these problems. They assume that they are given a collection of sorted lists that comprise the sets, and need to compute unions, intersections and differences of these sets. If one is only interested in worst-case complexity (across all instances) then this problem is uninteresting (it essentially boils down to merging). However, some instances can be harder than others: for instance, computing the intersection of two sets when all elements in one set are smaller than the other is much easier than for sets that interleave substantially. Building on this idea, they develop a notion of the complexity of a given instance of a problem and develop algorithms that, for each particular instance, are efficient with respect to the difficulty of that instance.

References

1. A. N. Kolmogorov. On the notion of algorithm. *Uspekhi Matematicheskikh Nauk*, 8, 1953, pp. 175–176.
2. A. Schönage. Storage modification machines. *SIAM Journal on Computing*, 9, 1980, pp. 490–508.
3. R. E. Tarjan. A class of algorithms which require non linear time to maintain disjoint sets. *Journal of Computer and System Sciences*, 18, 1979, pp. 110–127.
4. A. V. Aho, J. E. Hopcroft and J. D. Ullman. *The Design and Analysis of Computer Algorithms*. Addison Wesley, 1974.
5. T. Hagerup. Sorting and searching on the word RAM. In *Proc. 15th Symposium on Theoretical Aspects of Computer Science (STACS 1998)*, Lecture Notes in Computer Science, Springer-Verlag, Berlin, vol. 1373, pp. 366–398, 1998.
6. M. L. Fredman. A lower bound on the complexity of orthogonal range queries. *Journal of the ACM*, 28, 1981, pp. 696–705.

7. A. C. Yao. On the complexity of maintaining partial sums. *SIAM Journal on Computing*, 14, 1985, pp. 277–288.

8. K. Mehlhorn. *Data Structures and Algorithms, Vol. III: Multi-dimensional Searching and Computational Geometry*. Springer-Verlag, Berlin, 1984.

9. W. Pugh. *Incremental computation and the incremental evaluation of function programs*. PhD thesis, Cornell University, 1989.

10. W. Pugh and T. Teitelbaum. Incremental computation via function caching. In *Conference Record of the 16th Annual ACM Symposium on Principles of Programming Languages*, ACM, pp. 315–328, 1989.

11. D. E. Knuth. *The Art of Computer Programming*, vol. 3, 2nd Ed. Addison Wesley, 1998.

12. R. Seidel and C. Aragon. Randomized search trees. *Algorithmica*, 16, 1996, pp. 464–497.

13. R. Sundar and R. E. Tarjan. Unique binary-search-tree representations and equality testing of sets and sequences. *SIAM Journal on Computing*, 23, 1994, pp. 24–44.

14. A. Andersson and T. Ottmann. New tight bounds on uniquely represented dictionaries. *SIAM Journal on Computing*, 24, 1995, pp. 1091–1101.

15. D. M. Yellin. Representing sets with constant time equality testing. *Journal of Algorithms*, 13, 1992, pp. 353–373.

16. T. W. Lam and K. H. Lee. An improved scheme for set equality testing and updating. *Theoretical Computer Science*, 201, 1998, pp. 85–97.

17. M. H. Overmars. *The Design of Dynamic Data Structures*, vol. 156. Lecture Notes in Computer Science. Springer-Verlag, Berlin, 1983.

18. D. E. Willard. New trie data structures which support very fast search operations.

19. P. Beame and F. E. Fich. Optimal bounds for the predecessor problem and related problems. *Journal of Computer and System Sciences*, 65, 2002, pp. 38–72.

20. P. Pritchard. Opportunistic algorithms for eliminating supersets. *Acta Informatica*, 28, 1991, pp. 733–754.

21. D. M. Yellin and C. S. Jutla. Finding extremal sets in less than quadratic time. *Information Processing Letters*, 48, 1993, pp. 29–34.

22. P. Pritchard. A simple sub-quadratic algorithm for computing the subset partial order. *Information Processing Letters*, 56, 1995, pp. 337–341.

23. P. Pritchard. An old sub-quadratic algorithm for finding extremal sets. *Information Processing Letters*, 62, 1997, pp. 329–334.

24. P. Pritchard. A fast bit-parallel algorithm for computing the subset partial order. *Algorithmica*, 24, 1999, pp. 76–86.

25. D. M. Yellin. An algorithm for dynamic subset and intersection testing. *Theoretical Computer Science*, 129, 1994, pp. 397–406.

26. P. F. Dietz, K. Mehlhorn, R. Raman, and C. Uhrig. Lower bounds for set intersection queries. *Algorithmica*, 14, 1995, pp. 154–168.

27. Z. Galil and G. F. Italiano. Data structures and algorithms for disjoint set union problems. *ACM Computing Surveys*, 23, 1991, pp. 319–344.

28. H. Kaplan, N. Shafrir, and R. E. Tarjan. Meldable heaps and boolean union-find. In *Proc. 34th Annual ACM Symposium on Theory of Computing*, ACM, pp. 573–582, 2002.

29. H. Kaplan, N. Shafrir, and R. E. Tarjan. Union-find with deletions. In *Proc. 13th Annual ACM-SIAM Symposium on Discrete Algorithms*, ACM/SIAM, pp. 19–28, 2002.

30. M. Smid. A data structure for the union-find problem having good single-operation complexity. In *Algorithms Review, Newsletter of the ESPRIT II Basic Research Action program project no. 3075*, **1**, ALCOM, 1990.

31. R. E. Tarjan. Efficiency of a good but not linear set union algorithm. *Journal of the ACM*, 22, 1975, pp. 215–225.

32. R. E. Tarjan and J. van Leeuwen. Worst-case analysis of set union algorithms. *Journal of the ACM*, 31, 1984, pp. 245–281.

33. J. A. La Poutré. New techniques for the union-find problem. In *Proc. 1st Annual ACM-SIAM Symposium on Discrete Algorithms*, ACM/SIAM, pp. 54–63, 1990.

34. H. N. Gabow. A scaling algorithm for weighted matching on general graphs. In *Proc. 26th Annual Symposium on Foundations of Computer Science*, IEEE Computer Society, pp. 90–100, 1985.

35. A. M. Ben-Amram and Z. Galil. On data structure tradeoffs and an application to union-find. *Electronic Colloquium on Computational Complexity*, Technical Report TR95-062, 1995.

36. N. Blum. The single-operation worst case.

37. S. Alstrup, A. M. Ben-Amram, and T. Rauhe. Worst-case and amortised optimality in union-find (extended abstract). In *Proc. 31st Annual Symposium on Theory of Computing*, ACM, pp. 499–506, 1999.

38. J. A. La Poutré. Lower bounds for the union-find and the split-find problem on pointer machines. *Journal of Computer and System Sciences*, 52, 1996, pp. 87–99.

39. L. Banachowsky. A complement to Tarjan's result about the lower bound on the complexity of the set union problem. *Information Processing Letters*, 11, 1980, pp. 59–65.

40. M. L. Fredman and M. Saks. The cell probe complexity of dynamic data structures. In *Proc. 21st Annual ACM Symposium on Theory of Computing*, ACM, pp. 345–354, 1989.

41. P. F. Dietz and R. Raman. Persistence, amortisation and randomisation. In *Proc. 2nd Annual ACM-SIAM Symposium on Discrete Algorithms*, ACM/SIAM, pp. 77–87, 1991.

42. P. F. Dietz and R. Raman. Persistence, randomization and parallelization: On some combinatorial games and their applications (Abstract). In *Proc. 3rd Workshop on Algorithms and Data Structures*, Lecture Notes in Computer Science, Springer-Verlag, vol. 709, pp. 289–301, 1993.

43. M. J. van Kreveld, and M. H. Overmars. Union-copy data structures and dynamic segment trees. *Journal of the ACM*, 40, 1993, pp. 635–652.

44. M. A. Bender, S. Sethia, and S. Skiena. Data structures for maintaining set partitions. Manuscript, 2004. Preliminary version. In *Proc. 7th Scandinavian Workshop on Algorithm Theory*, Lecture Notes in Computer Science, Springer-Verlag, Berlin, vol.1851, pp. 83–96, 2000.

45. G. Calinescu. A note on data structures for maintaining partitions. Manuscript, 2003.

46. E. D. Demaine, A. López-Ortiz, and J. I. Munro. Adaptive set intersections, unions and differences. In *Proc. 12th Annual ACM-SIAM Symposium on Discrete Algorithms*, ACM/SIAM, pp. 743–752, 2001.

35

Cache-Oblivious Data Structures*

Lars Arge
University of Aarhus

Gerth Stølting Brodal
University of Aarhus

Rolf Fagerberg
University of Southern Denmark

35.1 The Cache-Oblivious Model 545
35.2 Fundamental Primitives 546
    Van Emde Boas Layout • *k*-Merger
35.3 Dynamic B-Trees 550
    Density Based • Exponential Tree Based
35.4 Priority Queues 554
    Merge Based Priority Queue: Funnel Heap • Exponential Level Based Priority Queue
35.5 2d Orthogonal Range Searching 560
    Cache-Oblivious kd-Tree • Cache-Oblivious Range Tree
Acknowledgments 563
References 564

35.1 The Cache-Oblivious Model

The memory system of most modern computers consists of a hierarchy of memory levels, with each level acting as a cache for the next; for a typical desktop computer the hierarchy consists of registers, level 1 cache, level 2 cache, level 3 cache, main memory, and disk. One of the essential characteristics of the hierarchy is that the memory levels get larger and slower the further they get from the processor, with the access time increasing most dramatically between main memory and disk. Another characteristic is that data is moved between levels in large blocks. As a consequence of this, the memory access pattern of an algorithm has a major influence on its practical running time. Unfortunately, the RAM model (Figure 35.1) traditionally used to design and analyze algorithms is not capable of capturing this, since it assumes that all memory accesses take equal time.

Because of the shortcomings of the RAM model, a number of more realistic models have been proposed in recent years. The most successful of these models is the simple two-level I/O-model introduced by Aggarwal and Vitter [1] (Figure 35.2). In this model the memory hierarchy is assumed to consist of a fast memory of size M and a slower infinite memory, and data is transfered between the levels in blocks of B consecutive elements. Computation can only be performed on data in the fast memory, and it is assumed that algorithms have complete control over transfers of blocks between the two levels. We denote such a transfer a *memory transfer*. The complexity measure is the number of memory transfers needed to solve a problem. The strength of the I/O model is that it captures part of the memory hierarchy, while being sufficiently simple to make design and analysis of algorithms feasible. In particular, it adequately models the situation where the memory transfers between two levels of the memory hierarchy dominate the running time, which is often the case when the size of the data exceeds the size of main memory. Agarwal and Vitter showed that comparison based sorting and searching require $\Theta(\text{Sort}_{M,B}(N)) = \Theta\left(\frac{N}{B}\log_{M/B}\frac{N}{B}\right)$ and $\Theta(\log_B N)$ memory transfers in the I/O-model, respectively [1]. Subsequently a large number of other results have been obtained in the model; see the surveys by Arge [2] and Vitter [3] for references. Also see Chapter 28.

More elaborate models of multi-level memory than the I/O-model have been proposed (see, e.g., [3] for an overview) but these models have been less successful, mainly because of their complexity. A major shortcoming of the proposed models, including the I/O-model, have also been that they assume that the characteristics of the memory hierarchy (the level and block sizes) are known. Very recently however, the *cache-oblivious* model, which assumes no knowledge about the hierarchy, was introduced by Frigo et al. [4]. In essence, a cache-oblivious algorithm is an algorithm formulated in the RAM model but analyzed in the I/O model, with the analysis required to hold for any B and M. Memory transfers are assumed to be performed by an off-line

---

* This chapter has been reprinted from first edition of this Handbook, without any content updates.

optimal replacement strategy. The beauty of the cache-oblivious model is that since the I/O-model analysis holds for any block and memory size, it holds for *all* levels of a multi-level memory hierarchy (see [4] for details). In other words, by optimizing an algorithm to one unknown level of the memory hierarchy, it is optimized on all levels simultaneously. Thus the cache-oblivious model is effectively a way of modeling a complicated multi-level memory hierarchy using the simple two-level I/O-model.

Frigo et al. [4] described optimal $\Theta(\text{Sort}_{M,B}(N))$ memory transfer cache-oblivious algorithms for matrix transposition, fast Fourier transform, and sorting; Prokop also described a static search tree obtaining the optimal $O(\log_B N)$ transfer search bound [5]. Subsequently, Bender et al. [6] described a cache-oblivious dynamic search trees with the same search cost, and simpler and improved cache-oblivious dynamic search trees were then developed by several authors [7–10]. Cache-oblivious algorithms have also been developed for, for example, problems in computational geometry [7,11,12], for scanning dynamic sets [7], for layout of static trees [13], for partial persistence [7], and for a number of fundamental graph problems [14] using cache-oblivious priority queues [14,15]. Most of these results make the so-called *tall cache assumption*, that is, they assume that $M > \Omega(B^2)$; we make the same assumption throughout this chapter.

Empirical investigations of the practical efficiency of cache-oblivious algorithms for sorting [16], searching [9,10,17] and matrix problems [4] have also been performed. The overall conclusion of these investigations is that cache-oblivious methods often outperform RAM algorithms, but not always as much as algorithms tuned to the specific memory hierarchy and problem size. On the other hand, cache-oblivious algorithms perform well on all levels of the memory hierarchy, and seem to be more robust to changing problem sizes than cache-aware algorithms.

In the rest of this chapter we describe some of the most fundamental and representative cache-oblivious data structure results. In Section 35.2 we discuss two fundamental primitives used to design cache-oblivious data structures. In Section 35.3 we describe two cache-oblivious dynamic search trees, and in Section 35.4 two priority queues. Finally, in Section 35.5 we discuss structures for 2-dimensional orthogonal range searching.

35.2 Fundamental Primitives

The most fundamental cache-oblivious primitive is scanning—scanning an array with N elements incurs $\Theta\left(\frac{N}{B}\right)$ memory transfers for any value of B. Thus algorithms such as median finding and data structures such as stacks and queues that only rely on scanning are automatically cache-oblivious. In fact, the examples above are optimal in the cache-oblivious model. Other examples of algorithms that only rely on scanning include Quicksort and Mergesort. However, they are not asymptotically optimal in the cache-oblivious model since they use $O\left(\frac{N}{B} \log \frac{N}{M}\right)$ memory transfers rather than $\Theta(\text{Sort}_{M,B}(N))$.

Apart from algorithms and data structures that only utilize scanning, most cache-oblivious results use recursion to obtain efficiency; in almost all cases, the sizes of the recursive problems decrease double-exponentially. In this section we describe two of the most fundamental such recursive schemes, namely the *van Emde Boas layout* and the *k-merger*.

35.2.1 Van Emde Boas Layout

One of the most fundamental data structures in the I/O-model is the B-tree [18]. A B-tree is basically a fanout $\Theta(B)$ tree with all leaves on the same level. Since it has height $O(\log_B N)$ and each node can be accessed in $O(1)$ memory transfers, it supports

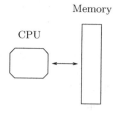

FIGURE 35.1   The RAM model.

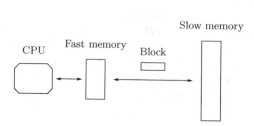

FIGURE 35.2   The I/O model.

searches in $O(\log_B N)$ memory transfers. It also supports range queries, that is, the reporting of all K elements in a given query range, in $O\left(\log_B N + \frac{K}{B}\right)$ memory transfers. Since B is an integral part of the definition of the structure, it seems challenging to develop a cache-oblivious B-tree structure. However, Prokop [5] showed how a binary tree can be laid out in memory in order to obtain a (static) cache-oblivious version of a B-tree. The main idea is to use a recursively defined layout called the *van Emde Boas layout* closely related to the definition of a van Emde Boas tree [19]. The layout has been used as a basic building block of most cache-oblivious search structures (e.g., in [6–11,13]).

35.2.1.1 Layout

For simplicity, we only consider complete binary trees. A binary tree is complete if it has $N = 2^h - 1$ nodes and height h for some integer h. The basic idea in the van Emde Boas layout of a complete binary tree $\mathcal{T}$ with N leaves is to divide $\mathcal{T}$ at the middle level and lay out the pieces recursively (Figure 35.3). More precisely, if $\mathcal{T}$ only has one node it is simply laid out as a single node in memory. Otherwise, we define the *top tree* $\mathcal{T}_0$ to be the subtree consisting of the nodes in the topmost $\lfloor h/2 \rfloor$ levels of $\mathcal{T}$, and the *bottom trees* $\mathcal{T}_1, \ldots, \mathcal{T}_k$ to be the $\Theta(\sqrt{N})$ subtrees rooted in the nodes on level $\lceil h/2 \rceil$ of $\mathcal{T}$; note that all the subtrees have size $\Theta(\sqrt{N})$. The van Emde Boas layout of $\mathcal{T}$ consists of the van Emde Boas layout of $\mathcal{T}_0$ followed by the van Emde Boas layouts of $\mathcal{T}_1, \ldots, \mathcal{T}_k$.

35.2.1.2 Search

To analyze the number of memory transfers needed to perform a search in $\mathcal{T}$, that is, traverse a root-leaf path, we consider the first recursive level of the van Emde Boas layout where the subtrees are smaller than B. As this level $\mathcal{T}$ is divided into a set of *base trees* of size between $\Theta(\sqrt{B})$ and $\Theta(B)$, that is, of height $\Omega(\log B)$ (Figure 35.4). By the definition of the layout, each base tree is stored in $O(B)$ contiguous memory locations and can thus be accessed in $O(1)$ memory transfers. That the search is performed in $O(\log_B N)$ memory transfers then follows since the search path traverses $O((\log N)/\log B) = O(\log_B N)$ different base trees.

35.2.1.3 Range query

To analyze the number of memory transfers needed to answer a range query $[x_1, x_2]$ on $\mathcal{T}$ using the standard recursive algorithm that traverses the relevant parts of $\mathcal{T}$ (starting at the root), we first note that the two paths to x_1 and x_2 are traversed in $O(\log_B N)$

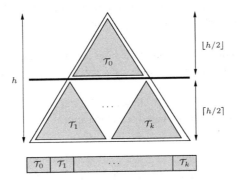

FIGURE 35.3   The van Emde Boas layout.

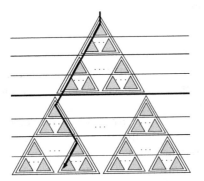

FIGURE 35.4   A search path.

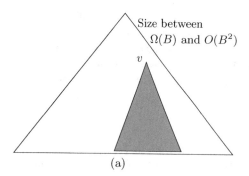
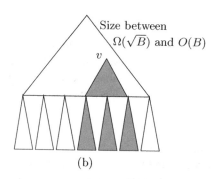

(a)   (b)

FIGURE 35.5 Traversing tree T_v with $O(B)$ leaves; (a) smallest recursive van Emde Boas level containing T_v has size between $\Omega(B)$ and $O(B^2)$; (b) next level in recursive subdivision.

memory transfers. Next we consider traversed nodes v that are not on the two paths to x_1 and x_2. Since all elements in the subtree T_v rooted at such a node v are reported, and since a subtree of height $\log B$ stores $\Theta(B)$ elements, $O\left(\frac{K}{B}\right)$ subtrees T_v of height $\log B$ are visited. This in turn means that the number of visited nodes above the last $\log B$ levels of T is also $O\left(\frac{K}{B}\right)$; thus they can all be accessed in $O\left(\frac{K}{B}\right)$ memory transfers. Consider the smallest recursive level of the van Emde Boas layout that completely contain T_v. This level is of size between $\Omega(B)$ and $O(B^2)$ (Figure 35.5a). On the next level of recursion T_v is broken into a top part and $O(\sqrt{B})$ bottom parts of size between $\Omega(\sqrt{B})$ and $O(\sqrt{B})$ each (Figure 35.5b). The top part is contained in a recursive level of size $O(B)$ and is thus stored within $O(B)$ consecutive memory locations; therefore it can be accessed in $O(1)$ memory accesses. Similarly, the $O(B)$ nodes in the $O(\sqrt{B})$ bottom parts are stored consecutively in memory; therefore they can all be accessed in a total of $O(1)$ memory transfers. Therefore, the optimal paging strategy can ensure that any traversal of T_v is performed in $O(1)$ memory transfers, simply by accessing the relevant $O(1)$ blocks. Thus overall a range query is performed in $O\left(\log_B N + \frac{K}{B}\right)$ memory transfers.

Theorem 35.1

Let T be a complete binary tree with N leaves laid out using the van Emde Boas layout. The number of memory transfers needed to perform a search (traverse a root-to-leaf path) and a range query in T is $O(\log_B N)$ and $O\left(\log_B N + \frac{K}{B}\right)$, respectively.

Note that the navigation from node to node in the van Emde Boas layout is straight-forward if the tree is implemented using pointers. However, navigation using arithmetic on array indexes is also possible [9]. This avoids the use of pointers and hence saves space.

The constant in the $O(\log_B N)$ bound for searching in Theorem 35.1 can be seen to be four. Further investigations of which constants are possible for cache-oblivious comparison based searching appear in [20].

35.2.2 *k*-Merger

In the I/O-model the two basic optimal sorting algorithms are multi-way versions of Mergesort and distribution sorting (Quicksort) [1]. Similarly, Frigo et al. [4] showed how both merge based and distribution based optimal cache-oblivious sorting algorithms can be developed. The merging based algorithm, *Funnelsort*, is based on a so-called *k-merger*. This structure has been used as a basic building block in several cache-oblivious algorithms. Here we describe a simplified version of the *k*-merger due to Brodal and Fagerberg [12].

35.2.2.1 Binary mergers and merge trees

A *binary merger* merges two sorted input streams into a sorted output stream: In one merge step an element is moved from the head of one of the input streams to the tail of the output stream; the heads of the input streams, as well as the tail of the output stream, reside in *buffers* of a limited capacity.

Binary mergers can be combined to form *binary merge trees* by letting the output buffer of one merger be the input buffer of another—in other words, a binary merge tree is a binary tree with mergers at the nodes and buffers at the edges, and it is used to merge a set of sorted input streams (at the leaves) into one sorted output stream (at the root). Refer to Figure 35.6 for an example.

An *invocation* of a binary merger in a binary merge tree is a recursive procedure that performs merge steps until the output buffer is full (or both input streams are exhausted); if an input buffer becomes empty during the invocation (and the corresponding stream is not exhausted), the input buffer is recursively filled by an invocation of the merger having this buffer as output buffer.

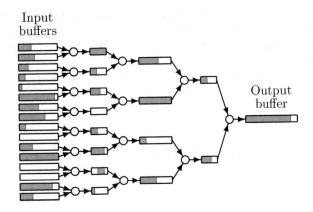

Input
buffers

Output
buffer

FIGURE 35.6   A 16-merger consisting of 15 binary mergers. Shaded parts represent elements in buffers.

Procedure FILL(v)
  while v's output buffer is not full
    if left input buffer empty
      FILL(left child of v)
    if right input buffer empty
      FILL(right child of v)
    perform one merge step

FIGURE 35.7   Invocation of binary merger v.

If both input streams of a merger get exhausted, the corresponding output stream is marked as exhausted. A procedure FILL(v) performing an invocation of a binary merger v is shown in Figure 35.7 (ignoring exhaustion issues). A single invocation FILL(r) on the root r of a merge tree will merge the streams at the leaves of the tree.

35.2.2.2 *k*-merger

A k-merger is a binary merge tree with specific buffer sizes. For simplicity, we assume that k is a power of two, in which case a k-merger is a complete binary tree of $k-1$ binary mergers. The output buffer at the root has size k^3, and the sizes of the rest of the buffers are defined recursively in a manner resembling the definition of the van Emde Boas layout: Let $i = \log k$ be the height of the k-merger. We define the *top tree* to be the subtree consisting of all mergers of depth at most $\lceil i/2 \rceil$, and the *bottom trees* to be the subtrees rooted in nodes at depth $\lceil i/2 \rceil + 1$. We let the edges between the top and bottom trees have buffers of size $k^{3/2}$, and define the sizes of the remaining buffers by recursion on the top and bottom trees. The input buffers at the leaves hold the input streams and are not part of the k-merger definition. The space required by a k-merger, excluding the output buffer at the root, is given by $S(k) = k^{1/2} \cdot k^{3/2} + (k^{1/2} + 1) \cdot S(k^{1/2})$, which has the solution $S(k) = \Theta(k^2)$.

We now analyze the number of memory transfers needed to fill the output buffer of size k^3 at the root of a k-merger. In the recursive definition of the buffer sizes in the k-merger, consider the first level where the subtrees (excluding output buffers) have size less than $M/3$; if $\bar{k}$ is the number of leaves of one such subtree, we by the space usage of k-mergers have $\bar{k}^2 \leq M/3$ and $(\bar{k}^2)^2 = \bar{k}^4 = \Omega(M)$. We call these subtrees of the k-merger *base trees* and the buffers between the base trees *large buffers*. Assuming $B^2 \leq M/3$, a base tree $\mathcal{T}_v$ rooted in v together with one block from each of the large buffers surrounding it (i.e., its single output buffer and $\bar{k}$ input buffers) can be contained in fast memory, since $M/3 + B + \bar{k} \cdot B \leq M/3 + B + (M/3)^{1/2} \cdot (M/3)^{1/2} \leq M$. If the k-merger consists of a single base tree, the number of memory transfers used to fill its output buffer with k^3 elements during an invocation is trivially $O(k^3/B + k)$. Otherwise, consider an invocation of the root v of a base tree $\mathcal{T}_v$, which will fill up the size $\Omega(\bar{k}^3)$ output buffer of v. Loading $\mathcal{T}_v$ and one block for each of the $\bar{k}$ buffers just below it into fast memory will incur $O(\bar{k}^2/B + \bar{k})$ memory transfers. This is $O(1/B)$ memory transfer for each of the $\Omega(\bar{k}^3)$ elements output, since $\bar{k}^4 = \Omega(M)$ implies $\bar{k}^2 = \Omega(M^{1/2}) = \Omega(B)$, from which $\bar{k} = O(\bar{k}^3/B)$ follows. Provided that none of the input buffers just below $\mathcal{T}_v$ become empty, the output buffer can then be filled in $O(\bar{k}^3/B)$ memory transfers since elements can be read from the input buffers in $O(1/B)$ transfers amortized. If a buffer below $\mathcal{T}_v$ becomes empty, a recursive invocation is needed. This invocation may evict $\mathcal{T}_v$ from memory, leading to its reloading when the invocation finishes. We charge this cost to the $\Omega(\bar{k}^3)$ elements in the filled buffer, or $O(1/B)$ memory transfers per element. Finally, the last time an invocation is used to fill a particular buffer, the buffer may not be completely filled (due to exhaustion). However, this happens only once for each buffer, so we can pay the cost by charging $O(1/B)$ memory transfers to each position in each buffer in the k-merger. As the entire k-merger uses $O(k^2)$ space and merges k^3 elements, these charges add up to $O(1/B)$ memory transfers per element.

We charge an element $O(1/B)$ memory transfers each time it is inserted into a large buffer. Since $\bar{k} = \Omega(M^{1/4})$, each element is inserted in $O(\log_{\bar{k}} k) = O(\log_M k^3)$ large buffers. Thus we have the following.

Theorem 35.2

Excluding the output buffers, the size of a k-merger is $O(k^2)$ and it performs $O\left(\frac{k^3}{B}\log_M k^3 + k\right)$ memory transfers during an invocation to fill up its output buffer of size k^3.

35.2.2.3 Funnelsort

The cache-oblivious sorting algorithm Funnelsort is easily obtained once the k-merger structure is defined: Funnelsort breaks the N input elements into $N^{1/3}$ groups of size $N^{2/3}$, sorts them recursively, and then merges the sorted groups using an $N^{1/3}$-merger.

Funnelsort can be analyzed as follows: Since the space usage of a k-merger is sub-linear in its output, the elements in a recursive sort of size $M/3$ only need to be loaded into memory once during the entire following recursive sort. For k-mergers at the remaining higher levels in the recursion tree, we have $k^3 \geq M/3 \geq B^2$, which implies $k^2 \geq B^{4/3} > B$ and hence $k^3/B > k$. By Theorem 35.2, the number of memory transfers during a merge involving N' elements is then $O(\log_M(N')/B)$ per element. Hence, the total number of memory transfers per element is

$$O\left(\frac{1}{B}\left(1 + \sum_{i=0}^{\infty} \log_M N^{(2/3)^i}\right)\right) = O\left((\log_M N)/B\right).$$

Since $\log_M x = \Theta(\log_{M/B} x)$ when $B^2 \leq M/3$, we have the following theorem.

Theorem 35.3

Funnelsort sorts N element using $O(\mathrm{Sort}_{M,B}(N))$ memory transfers.

In the above analysis, the exact (tall cache) assumption on the size of the fast memory is $B^2 \leq M/3$. In [12] it is shown how to generalize Funnelsort such that it works under the weaker assumption $B^{1+\varepsilon} \leq M$, for fixed $\varepsilon > 0$. The resulting algorithm incurs the optimal $O(\mathrm{Sort}_{M,B}(N))$ memory transfers when $B^{1+\varepsilon} = M$, at the price of incurring $O(1/\varepsilon \cdot \mathrm{Sort}_{M,B}(N))$ memory transfers when $B^2 \leq M$. It is shown in [21] that this trade-off is the best possible for comparison based cache-oblivious sorting.

35.3 Dynamic B-Trees

The van Emde Boas layout of a binary tree provides a static cache-oblivious version of B-trees. The first dynamic solution was given Bender et al. [6], and later several simplified structures were developed [7–10]. In this section, we describe two of these structures [7,9].

35.3.1 Density Based

In this section we describe the dynamic cache-oblivious search tree structure of Brodal et al. [9]. A similar proposal was given independently by Bender et al. [8].

The basic idea in the structure is to embed a dynamic binary tree of height $\log N + O(1)$ into a static complete binary tree, that is, in a tree with $2^h - 1$ nodes and height h, which in turn is embedded into an array using the van Emde Boas layout. Refer to Figure 35.8.

To maintain the dynamic tree we use techniques for maintaining small height in a binary tree developed by Andersson and Lai [22]; in a different setting, similar techniques has also been given by Itai et al. [23]. These techniques give an algorithm for maintaining height $\log N + O(1)$ using amortized $O(\log^2 N)$ time per update. If the height bound is violated after performing an update in a leaf l, this algorithm performs rebalancing by rebuilding the subtree rooted at a specific node v on the search path from the root to l. The subtree is rebuilt to perfect balance in time linear in the size of the subtree. In a binary tree of perfect balance the element in any node v is the median of all the elements stored in the subtree T_v rooted in v. This implies that only the lowest level in T_v is not completely filled and the empty positions appearing at this level are evenly distributed across the level. Hence,

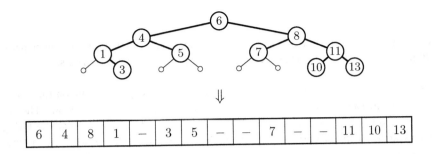

FIGURE 35.8 Illustration of embedding a height H tree into a complete static tree of height H, and the van Emde Boas layout of this tree.

the net effect of the rebuilding is to redistribute the empty positions in T_v. Note that this can lower the cost of future insertions in T_v, and consequently it may in the long run be better to rebuild a subtree larger than strictly necessary for reestablishment of the height bound. The criterion for choosing how large a subtree to rebuild, that is, for choosing the node v, is the crucial part of the algorithms by Andersson and Lai [22] and Itai et al. [23]. Below we give the details of how they can be used in the cache-oblivious setting.

35.3.1.1 Structure

As mentioned, the data structure consists of a dynamic binary tree T embedded into a static complete binary tree T' of height H, which in turn is embedded into an array using the van Emde Boas layout.

In order to present the update and query algorithms, we define the *density* $\rho(u)$ of a node u as $|T_u|/|T'_u|$, where $|T_u|$ and $|T'_u|$ are the number of nodes in the trees rooted in u in T and T', respectively. In Figure 35.8, the node containing the element 4 has balance 4/7. We also define two *density thresholds* τ_i and γ_i for the nodes on each level $i = 1, 2, \ldots, H$ (where the root is at level 1). The upper density thresholds τ_i are evenly space values between 3/4 and 1, and the lower density thresholds γ_i are evenly spaced values between 1/4 and 1/8. More precisely, $\tau_i = 3/4 + (i-1)/(4(H-1))$ and $\gamma_i = 1/4 - (i-1)/(8(H-1))$.

35.3.1.2 Updates

To insert a new element into the structure we first locate the position in T of the new node w. If the insertion of w violates the height bound H, we rebalance T as follows: First we find the lowest ancestor v of w satisfying $\gamma_i \leq \rho(v) \leq \tau_i$, where i is the level of v. If no ancestor v satisfies the requirement, we rebuild the entire structure, that is, T, T' and the layout of T': For k the integer such that $2^k \leq N < 2^{k+1}$ we choose the new height H of the tree T' as $k+1$ if $N \leq 5/4 \cdot 2^k$; otherwise we choose $H = k+2$. On the other hand, if the ancestor v exists we rebuild T_v: We first create a sorted list of all elements in T_v by an in-order traversal of T_v. The $\lceil |T_v|/2 \rceil$th element becomes the element stored at v, the smallest $\lfloor (|T_v| - 1)/2 \rfloor$ elements are recursively distributed in the left subtree of v, and the largest $\lceil (|T_v| - 1)/2 \rceil$ elements are recursively distributed in the right subtree of v.

We can delete an element from the structure in a similar way: We first locate the node w in T containing the element e to be deleted. If w is not a leaf and has a right subtree, we then locate the node w' containing the immediate successor of e (the node reached by following left children in the right subtree of w), swap the elements in w and w', and let $w = w'$. We repeat this until w is a leaf. If on the other hand w is not a leaf but only has a left subtree, we instead repeatedly swap w with the node containing the predecessor of e. Finally, we delete the leaf w from T, and rebalance the tree by rebuilding the subtree rooted at the lowest ancestor v of w satisfying satisfying $\gamma_i \leq \rho(v) \leq \tau_i$, where i is the level of v; if no such node exists we rebuild the entire structure completely.

Similar to the proof of Andersson and Lai [22] and Itai et al. [23] that updates are performed in $O(\log^2 N)$ time, Brodal et al. [9] showed that using the above algorithms, updates can be performed in amortized $O(\log_B N + (\log^2 N)/B)$ memory transfers.

35.3.1.3 Range queries

In Section 35.2, we discussed how a range query can be answered in $O\left(\log_B N + \frac{K}{B}\right)$ memory transfers on a complete tree T' laid out using the van Emde Boas layout. Since it can be shown that the above update algorithm maintains a lower density threshold of 1/8 for all nodes, we can also perform range queries in T efficiently: To answer a range query $[x_1, x_2]$ we traverse the two paths to x_1 and x_2 in T, as well as $O(\log N)$ subtrees rooted in children of nodes on these paths. Traversing one subtree T_v in T incurs at most the number of memory transfers needed to traverse the corresponding (full) subtree T'_v in T'. By the lower density threshold of 1/8 we know that the size of T'_v is at most a factor of eight larger than the size of T_v. Thus a range query is answered in $O\left(\log_B N + \frac{K}{B}\right)$ memory transfers.

Theorem 35.4

There exists a linear size cache-oblivious data structure for storing N elements, such that updates can be performed in amortized $O(\log_B N + (\log^2 N)/B)$ memory transfers and range queries in $O\left(\log_B N + \frac{K}{B}\right)$ memory transfers.

Using the method for moving between nodes in a van Emde Boas layout using arithmetic on the node indices rather than pointers, the above data structure can be implemented as a single size $O(N)$ array of data elements. The amortized complexity of updates can also be lowered to $O(\log_B N)$ by changing leaves into pointers to buckets containing $\Theta(\log N)$ elements each. With this modification a search can still be performed in $O(\log_B N)$ memory transfers. However, then range queries cannot be answered efficiently, since the $O\left(\frac{K}{\log N}\right)$ buckets can reside in arbitrary positions in memory.

35.3.2 Exponential Tree Based

The second dynamic cache-oblivious search tree we consider is based on the so-called *exponential layout* of Bender et al. [7]. For simplicity, we here describe the structure slightly differently than in [7].

35.3.2.1 Structure

Consider a complete balanced binary tree $\mathcal{T}$ with N leaves. Intuitively, the idea in an exponential layout of $\mathcal{T}$ is to recursively decompose $\mathcal{T}$ into a set of *components*, which are each laid out using the van Emde Boas layout. More precisely, we define component C_0 to consist of the first $\frac{1}{2}\log N$ levels of $\mathcal{T}$. The component C_0 contains $\sqrt{N}$ nodes and is called an N-component because its root is the root of a tree with N leaves (that is, $\mathcal{T}$). To obtain the exponential layout of $\mathcal{T}$, we first store C_0 using the van Emde Boas layout, followed immediately by the recursive layout of the $\sqrt{N}$ subtrees, $\mathcal{T}_1, \mathcal{T}_2, \ldots, \mathcal{T}_{\sqrt{N}}$, of size $\sqrt{N}$, beneath C_0 in $\mathcal{T}$, ordered from left to right. Note how the definition of the exponential layout naturally defines a decomposition of $\mathcal{T}$ into $\log \log N + O(1)$ *layers*, with layer i consisting of a number of $N^{1/2^{i-1}}$-components. An X-component is of size $\Theta(\sqrt{X})$ and its $\Theta(\sqrt{X})$ leaves are connected to $\sqrt{X}$-components. Thus the root of an X-component is the root of a tree containing X elements. Refer to Figure 35.9. Since the described layout of $\mathcal{T}$ is really identical to the van Emde Boas layout, it follows immediately that it uses linear space and that a root-to-leaf path can be traversed in $O(\log_B N)$ memory transfers.

By slightly relaxing the requirements on the layout described above, we are able to maintain it dynamically: We define an *exponential layout* of a balanced binary tree $\mathcal{T}$ with N leaves to consist of a composition of $\mathcal{T}$ into $\log \log N + O(1)$ layers, with layer i consisting of a number of $N^{1/2^{i-1}}$-components, each laid out using the van Emde Boas layout (Figure 35.9). An X-component has size $\Theta(\sqrt{X})$ but unlike above we allow its root to be root in a tree containing between X and $2X$ elements. Note how this means that an X-component has between $X/2\sqrt{X} = \frac{1}{2}\sqrt{X}$ and $2X/\sqrt{X} = 2\sqrt{X}$ leaves. We store the layout of $\mathcal{T}$ in memory almost as previously: If the root of $\mathcal{T}$ is root in an X-component C_0, we store C_0 first in $2 \cdot 2\sqrt{X} - 1$ memory locations (the maximal size of an X-component), followed immediately by the layouts of the subtrees ($\sqrt{X}$-components) rooted in the leaves of C_0 (in no particular order). We make room in the layout for the at most $2\sqrt{X}$ such subtrees. This exponential layout for $\mathcal{T}$ uses $S(N) = \Theta(\sqrt{N}) + 2\sqrt{N} \cdot S(\sqrt{N})$ space, which is $\Theta(N \log N)$.

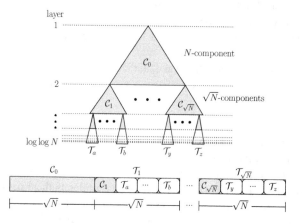

FIGURE 35.9 Components and exponential layout.

35.3.2.2 Search

Even with the modified definition of the exponential layout, we can traverse any root-to-leaf path in $\mathcal{T}$ in $O(\log_B N)$ memory transfers: The path passes through exactly one $N^{1/2^{i-1}}$-component for $1 \le i \le \log \log N + O(1)$. Each X-component is stored in a van Emde Boas layout of size $\Theta(\sqrt{X})$ and can therefore be traversed in $\Theta(\log_B \sqrt{X})$ memory transfers (Theorem 35.1). Thus, if we use at least one memory transfer in each component, we perform a search in $O(\log_B N) + \log \log N$ memory accesses. However, we do not actually use a memory transfer for each of the $\log \log N + O(1)$ components: Consider the traversed X-component with $\sqrt{B} \le X \le B$. This component is of size $O(\sqrt{B})$ and can therefore be loaded in $O(1)$ memory transfers. All smaller traversed components are of total size $O(\sqrt{B} \log \sqrt{B}) = O(B)$, and since they are stored in consecutively memory locations they can also be traversed in $O(1)$ memory transfers. Therefore only $O(1)$ memory transfers are used to traverse the last $\log \log B - O(1)$ components. Thus, the total cost of traversing a root-to-leaf path is $O(\log_B N + \log \log N - \log \log B) = O(\log_B N)$.

35.3.2.3 Updates

To perform an insertion in $\mathcal{T}$ we first search for the leaf l where we want to perform the insertion; inserting the new element below l will increase the number of elements stored below each of the $\log \log N + O(1)$ components on the path to the root, and may thus result in several components needing *rebalancing* (an X-component with $2X$ elements stored below it). We perform the insertion and rebalance the tree in a simple way as follows: We find the topmost X-component C_j on the path to the root with $2X$ elements below it. Then we divide these elements into two groups of X elements and store them separately in the exponential layout (effectively we *split* C_j with $2X$ elements below it into two X-components with X elements each). This can easily be done in $O(X)$ memory transfers. Finally, we update a leaf and insert a new leaf in the X^2-component above C_j (corresponding to the two new X-components); we can easily do so in $O(X)$ memory transfers by rebuilding it. Thus overall we have performed the insertion and rebalancing in $O(X)$ memory transfers. The rebuilding guarantees that after rebuilding an X-component, X inserts have to be performed below it before it needs rebalancing again. Therefore we can charge the $O(X)$ cost to the X insertions that occurred below C_j since it was last rebuilt, and argue that each insertion is charged $O(1)$ memory accesses on each of the $\log \log N + O(1)$ levels. In fact, using the same argument as above for the searching cost, we can argue that we only need to charge an insertion $O(1)$ transfers on the last $\log \log B - O(1)$ levels of $\mathcal{T}$, since rebalancing on any of these levels can always be performed in $O(1)$ memory transfers. Thus overall we perform an insertion in $O(\log_B N)$ memory transfers amortized.

Deletions can easily be handled in $O(\log_B N)$ memory transfers using global rebuilding: To delete the element in a leaf l of $\mathcal{T}$ we simply mark l as deleted. If l's sibling is also marked as deleted, we mark their parent deleted too; we continue this process along one path to the root of $\mathcal{T}$. This way we can still perform searches in $O(\log_B N)$ memory transfers, as long as we have only deleted a fraction of the elements in the tree. After $\frac{N}{2}$ deletes we therefore rebuild the entire structure in $O(N \log_B N)$ memory accesses, or $O(\log_B N)$ accesses per delete operation.

Bender et al. [7] showed how to modify the update algorithms to perform updates "lazily" and obtain worst case $O(\log_B N)$ bounds.

35.3.2.4 Reducing space usage

To reduce the space of the layout of a tree $\mathcal{T}$ to linear we simply make room for $2 \log N$ elements in each leaf, and maintain that a leaf contains between $\log N$ and $2 \log N$ elements. This does not increase the $O(\log_B N)$ search and update costs since the $O(\log N)$ elements in a leaf can be scanned in $O((\log N)/B) = O(\log_B N)$ memory accesses. However, it reduces the number of elements stored in the exponential layout to $O(N/\log N)$.

Theorem 35.5

The exponential layout of a search tree $\mathcal{T}$ on N elements uses linear space and supports updates in $O(\log_B N)$ memory accesses and searches in $O(\log_B N)$ memory accesses.

Note that the analogue of Theorem 35.1 does not hold for the exponential layout, that is, it does not support efficient range queries. The reason is partly that the $\sqrt{X}$-components below an X-component are not located in (sorted) order in memory because components are rebalanced by splitting, and partly because of the leaves containing $\Theta(\log N)$ elements. However, Bender et al. [7] showed how the exponential layout can be used to obtain a number of other important results: The structure as described above can easily be extended such that if two subsequent searched are separated by d elements, then the second search can be performed in $O(\log^* d + \log_B d)$ memory transfers. It can also be extended such that R queries (*batched searching*) can be answered simultaneously in $O(R \log_B \frac{N}{R} + \text{Sort}_{M,B}(R))$ memory transfers. The exponential layout can also be used to develop a *persistent B-tree*, where updates can be performed in the current version of the structure and queries can be performed in the current as well as all previous versions, with both operations incurring $O(\log_B N)$ memory transfers. It can also be used as a basic building block in a linear space *planar point location* structure that answers queries in $O(\log_B N)$ memory transfers.

35.4 Priority Queues

A priority queue maintains a set of elements with a priority (or key) each under the operations INSERT and DELETEMIN, where an INSERT operation inserts a new element in the queue, and a DELETEMIN operation finds and deletes the element with the minimum key in the queue. Frequently we also consider a DELETE operation, which deletes an element with a given key from the priority queue. This operation can easily be supported using INSERT and DELETEMIN: To perform a DELETE we insert a special delete-element in the queue with the relevant key, such that we can detect if an element returned by a DELETEMIN has really been deleted by performing another DELETEMIN.

A balanced search tree can be used to implement a priority queue. Thus the existence of a dynamic cache-oblivious B-tree immediately implies the existence of a cache-oblivious priority queue where all operations can be performed in $O(\log_B N)$ memory transfers, where N is the total number of elements inserted. However, it turns out that one can design a priority queue where all operations can be performed in $\Theta(\text{Sort}_{M,B}(N)/N) = O\left(\frac{1}{B}\log_{M/B}\frac{N}{B}\right)$ memory transfers; for most realistic values of N, M, and B, this bound is less than 1 and we can, therefore, only obtain it in an amortized sense. In this section we describe two different structures that obtain these bounds [14,15].

35.4.1 Merge Based Priority Queue: Funnel Heap

The cache-oblivious priority queue *Funnel Heap* due to Brodal and Fagerberg [15] is inspired by the sorting algorithm Funnelsort [4,12]. The structure only uses binary merging; essentially it is a heap-ordered binary tree with mergers in the nodes and buffers on the edges.

35.4.1.1 Structure

The main part of the Funnel Heap structure is a sequence of k-mergers (Section 35.2.2) with double-exponentially increasing k, linked together in a list using binary mergers; refer to Figure 35.10. This part of the structure constitutes a single binary merge tree. Additionally, there is a single insertion buffer I.

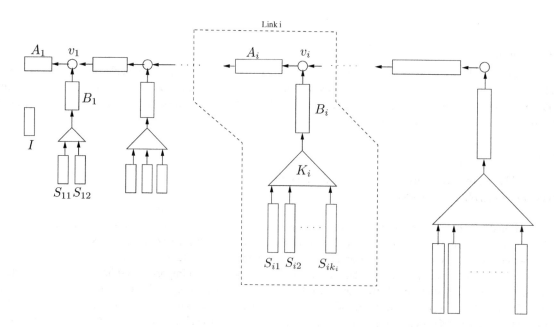

FIGURE 35.10 Funnel Heap: Sequence of k-mergers (triangles) linked together using buffers (rectangles) and binary mergers (circles).

More precisely, let k_i and s_i be values defined inductively by

$$
\begin{aligned}
(k_1, s_1) &= (2, 8), \\
s_{i+1} &= s_i(k_i + 1), \\
k_{i+1} &= \lceil\lceil s_{i+1}^{1/3} \rceil\rceil,
\end{aligned}
\tag{35.1}
$$

where $\lceil\lceil x \rceil\rceil$ denotes the smallest power of two above x, that is, $\lceil\lceil x \rceil\rceil = 2^{\lceil \log x \rceil}$. We note that $s_i^{1/3} \le k_i < 2s_i^{1/3}$, from which $s_i^{4/3} < s_{i+1} < 3s_i^{4/3}$ follows, so both s_i and k_i grow double-exponentially: $s_{i+1} = \Theta(s_i^{4/3})$ and $k_{i+1} = \Theta(k_i^{4/3})$. We also note that by induction on i we have $s_i = s_1 + \sum_{j=1}^{i-1} k_j s_j$ for all i.

A Funnel Heap consists of a linked list with *link* i containing a binary merger v_i, two buffers A_i and B_i, and a k_i-merger K_i having k_i input buffers $S_{i1}, \ldots, S_{ik_i}$. We refer to B_i, K_i, and $S_{i1}, \ldots, S_{ik_i}$ as the *lower part* of the link. The size of both A_i and B_i is k_i^3, and the size of each S_{ij} is s_i. Link i has an associated counter c_i for which $1 \le c_i \le k_i + 1$. The initial value of c_i is one for all i. The structure also has one insertion buffer I of size s_1. We maintain the following invariants:

Invariant 35.1

For link i, $S_{ic_i}, \ldots, S_{ik_i}$ are empty.

Invariant 35.2

On any path in the merge tree from some buffer to the root buffer A_1, elements appear in decreasing order.

Invariant 35.3

Elements in buffer I appear in sorted order.

Invariant 35.2 can be rephrased as the entire merge tree being in heap order. It implies that in all buffers in the merge tree, the elements appear in sorted order, and that the minimum element in the queue will be in A_1 or I, if buffer A_1 is non-empty. Note that an invocation (Figure 35.7) of any binary merger in the tree maintains the invariants.

35.4.1.2 Layout

The Funnel Heap is laid out in consecutive memory locations in the order I, link 1, link 2, ..., with link i being laid out in the order $c_i, A_i, v_i, B_i, K_i, S_{i1}, S_{i2}, \ldots, S_{ik_i}$.

35.4.1.3 Operations

To perform a DELETEMIN operation we compare the smallest element in I with the smallest element in A_1 and remove the smallest of these; if A_1 is empty we first perform an invocation of v_1. The correctness of this procedure follows immediately from Invariant 35.2.

To perform an INSERT operation we insert the new element among the (constant number of) elements in I, maintaining Invariant 35.3. If the number of elements in I is now s_1, we examine the links in order to find the lowest index i for which $c_i \le k_i$. Then we perform the following SWEEP(i) operation.

In SWEEP(i), we first traverse the path p from A_1 to S_{ic_i} and record how many elements are contained in each encountered buffer. Then we traverse the part of p going from A_1 to S_{ic_i}, remove the elements in the encountered buffers, and form a sorted stream σ_1 of the removed elements. Next we form another sorted stream σ_2 of all elements in links $1, \ldots, i-1$ and in buffer I; we do so by marking A_i temporarily as exhausted and calling DELETEMIN repeatedly. We then merge σ_1 and σ_2 into a single stream σ, and traverse p again while inserting the front (smallest) elements of σ in the buffers on p such that they contain the same numbers of elements as before we emptied them. Finally, we insert the remaining elements from σ into S_{ic_i}, reset c_l to one for $l = 1, 2, \ldots, i-1$, and increment c_i.

To see that SWEEP(i) does not insert more than the allowed s_i elements into S_{ic_i}, first note that the lower part of link i is emptied each time c_i is reset to one. This implies that the lower part of link i never contains more than the number of elements inserted into $S_{i1}, S_{i2}, \ldots, S_{ik_i}$ by the at most k_i SWEEP(i) operations occurring since last time c_i was reset. Since $s_i = s_1 + \sum_{j=1}^{i-1} k_j s_j$ for all i, it follows by induction on time that no instance of SWEEP(i) inserts more than s_i elements into S_{ic_i}.

Clearly, SWEEP(i) maintains Invariants 35.1 and 35.3, since I and the lower parts of links $1, \ldots, i-1$ are empty afterwards. Invariant 35.2 is also maintained, since the new elements in the buffers on p are the smallest elements in σ, distributed such that each buffer contains exactly the same number of elements as before the SWEEP(i) operation. After the operation, an element on this path can only be smaller than the element occupying the same location before the operation, and therefore the merge tree is in heap order.

35.4.1.4 Analysis

To analyze the amortized cost of an INSERT or DELETEMIN operation, we first consider the number of memory transfers used to move elements upwards (towards A_1) by invocations of binary mergers in the merge tree. For now we assume that all invocations result in full buffers, that is, that no exhaustions occur. We imagine charging the cost of filling a particular buffer evenly to the elements being brought into the buffer, and will show that this way an element from an input buffer of K_i is charged $O\left(\frac{1}{B}\log_{M/B} s_i\right)$ memory transfers during its ascent to A_1.

Our proof rely on the optimal replacement strategy keeping as many as possible of the first links of the Funnel Heap in fast memory at all times. To analyze the number of links that fit in fast memory, we define Δ_i to be the sum of the space used by links 1 to i and define i_M to be the largest i for which $\Delta_i \leq M$. By the space bound for k-mergers in Theorem 35.2 we see that the space used by link i is dominated by the $\Theta(s_i k_i) = \Theta(k_i^4)$ space use of $S_{i1}, \ldots, S_{ik_i}$. Since $k_{i+1} = \Theta(k_i^{4/3})$, the space used by link i grows double-exponentially with i. Hence, Δ_i is a sum of double-exponentially increasing terms and is therefore dominated by its last term. In other words, $\Delta_i = \Theta(k_i^4) = \Theta(s_i^{4/3})$. By the definition of i_M we have $\Delta_{i_M} \leq M < \Delta_{i_M+1}$. Using $s_{i+1} = \Theta(s_i^{4/3})$ we see that $\log_M(s_{i_M}) = \Theta(1)$.

Now consider an element in an input buffer of K_i. If $i \leq i_M$ the element will not get charged at all in our charging scheme, since no memory transfers are used to fill buffers in the links that fit in fast memory. So assume $i > i_M$. In that case the element will get charged for the ascent through K_i to B_i and then through v_j to A_j for $j = i, i-1, \ldots, i_M$. First consider the cost of ascending through K_i: By Theorem 35.2, an invocation of the root of K_i to fill B_i with k_i^3 elements incurs $O\left(k_i + \frac{k_i^3}{B}\log_{M/B} k_i^3\right)$ memory transfers altogether. Since $M < \Delta_{i_M+1} = \Theta(k_{i_M+1}^4)$ we have $M = O(k_i^4)$. By the tall cache assumption $M = \Omega(B^2)$ we get $B = O(k_i^2)$, which implies $k_i = O(k_i^3/B)$. Under the assumption that no exhaustions occur, that is, that buffers are filled completely, it follows that an element is charged $O\left(\frac{1}{B}\log_{M/B} k_i^3\right) = O\left(\frac{1}{B}\log_{M/B} s_i\right)$ memory transfers to ascend through K_i and into B_i. Next consider the cost of ascending through v_j, that is, insertion into A_j, for $j = i, i-1, \ldots, i_M$: Filling of A_j incurs $O(1 + |A_j|/B)$ memory transfers. Since $B = O(k_{i_M+1}^2) = O(k_{i_M}^{8/3})$ and $|A_j| = k_j^3$, this is $O(|A_j|/B)$ memory transfers, so an element is charged $O(1/B)$ memory transfers for each A_j (under the assumption of no exhaustions). It only remains to bound the number of such buffers A_j, that is, to bound $i - i_M$. From $s_i^{4/3} < s_{i+1}$ we have $s_{i_M}^{(4/3)^{i-i_M}} < s_i$. Using $\log_M(s_{i_M}) = \Theta(1)$ we get $i - i_M = O(\log \log_M s_i)$. From $\log \log_M s_i = O(\log_M s_i)$ and the tall cache assumption $M = \Omega(B^2)$ we get $i - i_M = O(\log_M s_i) = O(\log_{M/B} s_i)$. In total we have proved our claim that, assuming no exhaustions occur, an element in an input buffer of K_i is charged $O\left(\frac{1}{B}\log_{M/B} s_i\right)$ memory transfers during its ascent to A_1.

We imagine maintaining the *credit invariant* that each element in a buffer holds enough credits to be able to pay for the ascent from its current position to A_1, at the cost analyzed above. In particular, an element needs $O\left(\frac{1}{B}\log_{M/B} s_i\right)$ credits when it is inserted in an input buffer of K_i. The cost of these credits we will attribute to the SWEEP(i) operation inserting it, effectively making all invocations of mergers be prepaid by SWEEP(i) operations.

A SWEEP(i) operation also incurs memory transfers by itself; we now bound these. In the SWEEP(i) operation we first form σ_1 by traversing the path p from A_1 to S_{ic_i}. Since the links are laid out sequentially in memory, this traversal at most constitutes a linear scan of the consecutive memory locations containing A_1 through K_i. Such a scan takes $O((\Delta_{i-1} + |A_i| + |B_i| + |K_i|)/B) = O(k_i^3/B) = O(s_i/B)$ memory transfers. Next we form σ_2 using DELETEMIN operations; the cost of which is paid for by the credits placed on the elements. Finally, we merge of σ_1 and σ_2 into σ, and place some of the elements in buffers on p and some of the elements in S_{ic_i}. The number of memory transfers needed for this is bounded by the $O(s_i/B)$ memory transfers needed to traverse p and S_{ic_i}. Hence, the memory transfers incurred by the SWEEP(i) operation itself is $O(s_i/B)$.

After the SWEEP(i) operation, the credit invariant must be reestablished. Each of the $O(s_i)$ elements inserted into S_{ic_i} must receive $O\left(\frac{1}{B}\log_{M/B} s_i\right)$ credits. Additionally, the elements inserted into the part of the path p from A_1 through A_{i-1} must receive enough credits to cover their ascent to A_1, since the credits that resided with elements in the same positions before the operations were used when forming σ_2 by DELETEMIN operations. This constitutes $O(\Delta_{i-1}) = o(s_i)$ elements, which by the analysis above, must receive $O\left(\frac{1}{B}\log_{M/B} s_i\right)$ credits each. Altogether $O(s_i/B) + O\left(\frac{s_i}{B}\log_{M/B} s_i\right) = O\left(\frac{s_i}{B}\log_{M/B} s_i\right)$ memory transfers are attributed to a SWEEP(i) operation, again under the assumption that no exhaustions occur during invocations.

To actually account for exhaustions, that is, the memory transfers incurred when filling buffers that become exhausted, we note that filling a buffer partly incurs at most the same number of memory transfers as filling it entirely. This number was analyzed above to be $O(|A_i|/B)$ for A_i and $O\left(\frac{|B_i|}{B}\log_{M/B} s_i\right)$ for B_i, when $i > i_M$. If B_i become exhausted, only a SWEEP(i) can remove that status. If A_i become exhausted, only a SWEEP(j) for $j \geq i$ can remove that status. As at most a single SWEEP(j) with $j > i$

can take place between one SWEEP(i) and the next, B_i can only become exhausted once for each SWEEP(i), and A_i can only become exhausted twice for each SWEEP(i). From $|A_i| = |B_i| = k_i^3 = \Theta(s_i)$ it follows that charging SWEEP(i) an additional cost of $O\left(\frac{s_i}{B} \log_{M/B} s_i\right)$ memory transfers will cover all costs of filling buffers when exhaustion occurs.

Overall we have shown that we can account for all memory transfers if we attribute $O\left(\frac{s_i}{B} \log_{M/B} s_i\right)$ memory transfers to each SWEEP(i). By induction on i, we can show that at least s_i insertions have to take place between each SWEEP(i). Thus, if we charge the SWEEP(i) cost to the last s_i insertions preceding the SWEEP(i), each insertion is charged $O\left(\frac{1}{B} \log_{M/B} s_i\right)$ memory transfers. Given a sequence of operation on an initial empty priority queue, let i_{max} be the largest i for which SWEEP(i) takes place. We have $s_{i_{max}} \leq N$, where N is the number of insertions in the sequence. An insertion can be charged by at most one SWEEP(i) for $i = 1, \ldots, i_{max}$, so by the double-exponential growth of s_i, the number of memory transfers charged to an insertion is

$$O\left(\sum_{k=0}^{\infty} \frac{1}{B} \log_{M/B} N^{(3/4)^k}\right) = O\left(\frac{1}{B} \log_{M/B} N\right) = O\left(\frac{1}{B} \log_{M/B} \frac{N}{B}\right),$$

where the last equality follows from the tall cache assumption $M = \Omega(B^2)$.

Finally, we bound the space use of the entire structure. To ensure a space usage linear in N, we create a link i when it is first used, that is, when the first SWEEP(i) occurs. At that point in time, c_i, A_i, v_i, B_i, K_i, and S_{i1} are created. These take up $\Theta(s_i)$ space combined. At each subsequent SWEEP(i) operation, we create the next input buffer S_{ic_i} of size s_i. As noted above, each SWEEP(i) is preceded by at least s_i insertions, from which an $O(N)$ space bound follows. To ensure that the entire structure is laid out in consecutive memory locations, the structure is moved to a larger memory area when it has grown by a constant factor. When allocated, the size of the new memory area is chosen such that it will hold the input buffers S_{ij} that will be created before the next move. The amortized cost of this is $O(1/B)$ per insertion.

Theorem 35.6

Using $\Theta(M)$ fast memory, a sequence of N INSERT, DELETEMIN, and DELETE operations can be performed on an initially empty Funnel Heap using $O(N)$ space in $O\left(\frac{1}{B} \log_{M/B} \frac{N}{B}\right)$ amortized memory transfers each.

Brodal and Fagerberg [15] gave a refined analysis for a variant of the Funnel Heap that shows that the structure adapts to different usage profiles. More precisely, they showed that the ith insertion uses amortized $O\left(\frac{1}{B} \log_{M/B} \frac{N_i}{B}\right)$ memory transfers, where N_i can be defined in any of the following three ways: (a) N_i is the number of elements present in the priority queue when the ith insertion is performed, (b) if the ith inserted element is removed by a DELETEMIN operation prior to the jth insertion then $N_i = j - i$, or (c) N_i is the maximum rank of the ith inserted element during its lifetime in the priority queue, where rank denotes the number of smaller elements in the queue.

35.4.2 Exponential Level Based Priority Queue

While the Funnel Heap is inspired by Mergesort and uses k-mergers as the basic building block, the exponential level priority queue of Arge et al. [14] is somewhat inspired by distribution sorting and uses sorting as a basic building block.

35.4.2.1 Structure

The structure consists of $\Theta(\log \log N)$ *levels* whose sizes vary from N to some small size c below a constant threshold c_t; the size of a level corresponds (asymptotically) to the number of elements that can be stored within it. The ith level from above has size $N^{(2/3)^{i-1}}$ and for convenience we refer to the levels by their size. Thus the levels from largest to smallest are level N, level $N^{2/3}$, level $N^{4/9}, \ldots$, level $X^{9/4}$, level $X^{3/2}$, level X, level $X^{2/3}$, level $X^{4/9}, \ldots$, level $c^{9/4}$, level $c^{3/2}$, and level c. In general, a level can contain any number of elements less than or equal to its size, except level N, which always contains $\Theta(N)$ elements. Intuitively, smaller levels store elements with smaller keys or elements that were more recently inserted. In particular, the minimum key element and the most recently inserted element are always in the smallest (lowest) level c. Both insertions and deletions are initially performed on the smallest level and may propagate up through the levels.

Elements are stored in a level in a number of *buffers*, which are also used to transfer elements between levels. Level X consists of one *up buffer* u^X that can store up to X elements, and at most $X^{1/3}$ *down buffers* $d_1^X, \ldots, d_{X^{1/3}}^X$ each containing between $(1/2)X^{2/3}$ and $2X^{2/3}$ elements. Thus level X can store up to $3X$ elements. We refer to the maximum possible number of elements that can be stored in a buffer as the *size* of the buffer. Refer to Figure 35.11. Note that the size of a down buffer at one level matches the size (up to a constant factor) of the up buffer one level down.

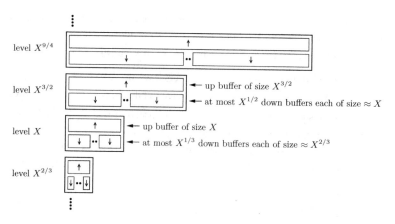

FIGURE 35.11 Levels $X^{2/3}$, X, $X^{3/2}$, and $X^{9/4}$ of the priority queue data structure.

We maintain three invariants about the relationships between the elements in buffers of various levels:

Invariant 35.4

At level X, elements are sorted **among** the down buffers, that is, elements in d_i^X have smaller keys than elements in d_{i+1}^X, but elements within d_i^X are unordered.

The element with largest key in each down buffer d_i^X is called a *pivot element*. Pivot elements mark the boundaries between the ranges of the keys of elements in down buffers.

Invariant 35.5

At level X, the elements in the down buffers have smaller keys than the elements in the up buffer.

Invariant 35.6

The elements in the down buffers at level X have smaller keys than the elements in the down buffers at the next higher level $X^{3/2}$.

The three invariants ensure that the keys of the elements in the down buffers get larger as we go from smaller to larger levels of the structure. Furthermore, an order exists between the buffers on one level: keys of elements in the up buffer are larger than keys of elements in down buffers. Therefore, down buffers are drawn below up buffers on Figure 35.11. However, the keys of the elements in an up buffer are unordered relative to the keys of the elements in down buffers one level up. Intuitively, up buffers store elements that are "on their way up," that is, they have yet to be resolved as belonging to a particular down buffer in the next (or higher) level. Analogously, down buffers store elements that are "on their way down"—these elements are by the down buffers partitioned into several clusters so that we can quickly find the cluster of smallest key elements of size roughly equal to the next level down. In particular, the element with overall smallest key is in the first down buffer at level c.

35.4.2.2 Layout

The priority queue is laid out in memory such that the levels are stored consecutively from smallest to largest with each level occupying a single region of memory. For level X we reserve space for exactly $3X$ elements: X for the up buffer and $2X^{2/3}$ for each possible down buffer. The up buffer is stored first, followed by the down buffers stored in an arbitrary order but linked together to form an ordered linked list. Thus $O\left(\sum_{i=0}^{\log_{3/2} \log_c N} N^{(2/3)^i}\right) = O(N)$ is an upper bound on the total memory used by the priority queue.

35.4.2.3 Operations

To implement the priority queue operations we use two general operations, *push* and *pull*. Push inserts X elements into level $X^{3/2}$, and pull removes the X elements with smallest keys from level $X^{3/2}$ and returns them in sorted order. An INSERT or a DELETEMIN is performed simply by performing a push or pull on the smallest level c.

Push. To push X elements into level $X^{3/2}$, we first sort the X elements cache-obliviously using $O\left(1 + \frac{X}{B} \log_{M/B} \frac{X}{B}\right)$ memory transfers. Next we distribute the elements in the sorted list into the down buffers of level $X^{3/2}$ by scanning through the list and simultaneously visiting the down buffers in (linked) order. More precisely, we append elements to the end of the current down buffer $d_i^{X^{3/2}}$, and advance to the next down buffer $d_{i+1}^{X^{3/2}}$ as soon as we encounter an element with larger key than the pivot of $d_i^{X^{3/2}}$. Elements with larger keys than the pivot of the last down buffer are inserted in the up buffer $u^{X^{3/2}}$. Scanning through the X elements take $O\left(1 + \frac{X}{B}\right)$ memory transfers. Even though we do not scan through every down buffer, we might perform at least one memory transfer for each of the $X^{1/2}$ possible buffers. Thus the total cost of distributing the X elements is $O\left(\frac{X}{B} + X^{1/2}\right)$ memory transfers.

During the distribution of elements a down buffer may run full, that is, contain $2X$ elements. In this case, we split the buffer into two down buffers each containing X elements using $O\left(1 + \frac{X}{B}\right)$ transfers. We place the new buffer in any free down buffer location for the level and update the linked list accordingly. If the level already has the maximum number $X^{1/2}$ of down buffers, we remove the last down buffer $d^X_{X^{1/2}}$ by inserting its no more than $2X$ elements into the up buffer using $O\left(1 + \frac{X}{B}\right)$ memory transfers. Since X elements must have been inserted since the last time the buffer split, the amortized splitting cost per element is $O\left(\frac{1}{X} + \frac{1}{B}\right)$ transfers. In total, the amortized number of memory transfers used on splitting buffers while distributing the X elements is $O\left(1 + \frac{X}{B}\right)$.

If the up buffer runs full during the above process, that is, contains more than $X^{3/2}$ elements, we recursively *push* all of these elements into the next level up. Note that after such a recursive push, $X^{3/2}$ elements have to be inserted (pushed) into the up buffer of level $X^{3/2}$ before another recursive push is needed.

Overall we can perform a push of X elements from level X into level $X^{3/2}$ in $O\left(X^{1/2} + \frac{X}{B} \log_{M/B} \frac{X}{B}\right)$ memory transfers amortized, not counting the cost of any recursive push operations; it is easy to see that a push maintains all three invariants.

Pull. To describe how to pull the X smallest keys elements from level $X^{3/2}$, we first assume that the down buffers contain at least $\frac{3}{2}X$ elements. In this case the first three down buffers $d_1^{X^{3/2}}$, $d_2^{X^{3/2}}$, and $d_3^{X^{3/2}}$ contain the between $\frac{3}{2}X$ and $6X$ smallest elements (Invariants 4 and 5). We find and remove the X smallest elements simply by sorting these elements using $O\left(1 + \frac{X}{B} \log_{M/B} \frac{X}{B}\right)$ memory transfers. The remaining between $X/2$ and $5X$ elements are left in one, two, or three down buffers containing between $X/2$ and $2X$ elements each. These buffers can easily be constructed in $O\left(1 + \frac{X}{B}\right)$ transfers. Thus we use $O\left(1 + \frac{X}{B} \log_{M/B} \frac{X}{B}\right)$ memory transfers in total. It is easy to see that Invariants 4–6 are maintained.

In the case where the down buffers contain fewer than $\frac{3}{2}X$ elements, we first *pull* the $X^{3/2}$ elements with smallest keys from the next level up. Because these elements do not necessarily have smaller keys than the, say U, elements in the up buffer $u^{X^{3/2}}$, we then sort this up buffer and merge the two sorted lists. Then we insert the U elements with largest keys into the up buffer, and distribute the remaining between $X^{3/2}$ and $X^{3/2} + \frac{3}{2}X$ elements into $X^{1/2}$ down buffers containing between X and $X + \frac{3}{2}X^{1/2}$ each (such that the $O\left(\frac{1}{X} + \frac{1}{B}\right)$ amortized down buffer split bound is maintained). It is easy to see that this maintains the three invariants. Afterwards, we can find the X minimal key elements as above. Note that after a recursive pull, $X^{3/2}$ elements have to be deleted (pulled) from the down buffers of level $X^{3/2}$ before another recursive pull is needed. Note also that a pull on level $X^{3/2}$ does not affect the number of elements in the up buffer $u^{X^{3/2}}$. Since we distribute elements into the down and up buffers after a recursive pull using one sort and one scan of $X^{3/2}$ elements, the cost of doing so is dominated by the cost of the recursive pull operation itself. Thus ignoring the cost of recursive pulls, we have shown that a pull of X elements from level $X^{3/2}$ down to level X can be performed in $O\left(1 + \frac{X}{B} \log_{M/B} \frac{X}{B}\right)$ memory transfers amortized, while maintaining Invariants 4–6.

35.4.2.4 Analysis

To analyze the amortized cost of an INSERT or DELETEMIN operation, we consider the total number of memory transfers used to perform push and pull operations during $\frac{N}{2}$ operations; to ensure that the structure always consists of $O(\log \log N)$ levels and use $O(N)$ space we rebuild it using $O\left(\frac{N}{B} \log_{M/B} \frac{N}{B}\right)$ memory transfers (or $O\left(\frac{1}{B} \log_{M/B} \frac{N}{B}\right)$ transfers per operation) after every $\frac{N}{2}$ operations [14].

The total cost of $\frac{N}{2}$ such operations is analyzed as follows: We charge a push of X elements from level X up to level $X^{3/2}$ to level X. Since X elements have to be inserted in the up buffer u^X of level X between such pushes, and as elements can only be inserted in u^X when elements are inserted (pushed) into level X, $O(N/X)$ pushes are charged to level X during the $\frac{N}{2}$ operations. Similarly, we charge a pull of X elements from level $X^{3/2}$ down to level X to level X. Since between such pulls $\Theta(X)$ elements have to be deleted from the down buffers of level X by pulls on X, $O(N/X)$ pulls are charged to level X during the $\frac{N}{2}$ operations.

Above we argued that a push or pull charged to level X uses $O\left(X^{1/2} + \frac{X}{B} \log_{M/B} \frac{X}{B}\right)$ memory transfers. We can reduce this cost to $O\left(\frac{X}{B} \log_{M/B} \frac{X}{B}\right)$ by more carefully examining the costs for differently sized levels. First consider a push or pull of $X \geq B^2$

elements into or from level $X^{3/2} \geq B^3$. In this case $\frac{X}{B} \geq \sqrt{X}$, and we trivially have that $O\left(X^{1/2} + \frac{X}{B} \log_{M/B} \frac{X}{B}\right) = O\left(\frac{X}{B} \log_{M/B} \frac{X}{B}\right)$.
Next, consider the case $B^{4/3} \leq X < B^2$, where the $X^{1/2}$ term in the push bound can dominate and we have to analyze the cost of a push more carefully. In this case we are working on a level $X^{3/2}$ where $B^2 \leq X^{3/2} < B^3$; there is only one such level. Recall that the $X^{1/2}$ cost was from distributing X sorted elements into the less than $X^{1/2}$ down buffers of level $X^{3/2}$. More precisely, a block of each buffer may have to be loaded and written back without transferring a full block of elements into the buffer. Assuming $M = \Omega(B^2)$, we from $X^{1/2} \leq B$ see that a block for each of the buffers can fit into fast memory. Consequently, if a fraction of the fast memory is used to keep a partially filled block of each buffer of level $X^{3/2}$ ($B^2 \leq X^{3/2} \leq B^3$) in fast memory at all times, and full blocks are written to disk, the $X^{1/2}$ cost would be eliminated. In addition, if all of the levels of size less than B^2 (of total size $O(B^2)$) are also kept in fast memory, all transfer costs associated with them would be eliminated. The optimal paging strategy is able to keep the relevant blocks in fast memory at all times and thus eliminates these costs.

Finally, since each of the $O(N/X)$ push and pull operations charged to level X ($X > B^2$) uses $O\left(\frac{X}{B} \log_{M/B} \frac{X}{B}\right)$ amortized memory transfers, the total amortized transfer cost of an INSERT or DELETEMIN operation in the sequence of $\frac{N}{2}$ such operations is

$$O\left(\sum_{i=0}^{\infty} \frac{1}{B} \log_{M/B} \frac{N^{(2/3)^i}}{B}\right) = O\left(\frac{1}{B} \log_{M/B} \frac{N}{B}\right).$$

Theorem 35.7

Using $\Theta(M)$ fast memory, N INSERT, DELETEMIN, and DELETE operations can be performed on an initially empty exponential level priority queue using $O(N)$ space in $O\left(\frac{1}{B} \log_{M/B} \frac{N}{B}\right)$ amortized memory transfers each.

35.5 2d Orthogonal Range Searching

As discussed in Section 35.3, there exist cache-oblivious B-trees that support updates and queries in $O(\log_B N)$ memory transfers (e.g., Theorem 35.5); several cache-oblivious B-tree variants can also support (one-dimensional) range queries in $O\left(\log_B N + \frac{K}{B}\right)$ memory transfers [6,8,9], but at an increased amortized update cost of $O\left(\log_B N + \frac{\log^2 N}{B}\right) = O(\log_B^2 N)$ memory transfers (e.g., Theorem 35.4).

In this section we discuss cache-oblivious data structures for two-dimensional orthogonal range searching, that is, structures for storing a set of N points in the plane such that the points in a axis-parallel query rectangle can be reported efficiently. In Section 35.5.1 we first discuss a cache-oblivious version of a *kd-tree*. This structure uses linear space and answers queries in $O\left(\sqrt{N/B} + \frac{K}{B}\right)$ memory transfers; this is optimal among linear space structures [24]. It supports updates in $O\left(\frac{\log N}{B} \cdot \log_{M/B} N\right) = O(\log_B^2 N)$ transfers. In Section 35.5.2 we then discuss a cache-oblivious version of a two-dimensional *range tree*. The structure answers queries in the optimal $O\left(\log_B N + \frac{K}{B}\right)$ memory transfers but uses $O(N \log^2 N)$ space. Both structures were first described by Agarwal et al. [11].

35.5.1 Cache-Oblivious kd-Tree

35.5.1.1 Structure

The cache-oblivious kd-tree is simply a normal kd-tree laid out in memory using the van Emde Boas layout. This structure, proposed by Bentley [25], is a binary tree of height $O(\log N)$ with the N points stored in the leaves of the tree. The internal nodes represent a recursive decomposition of the plane by means of axis-orthogonal lines that partition the set of points into two subsets of equal size. On even levels of the tree the dividing lines are horizontal, and on odd levels they are vertical. In this way a rectangular region R_v is naturally associated with each node v, and the nodes on any particular level of the tree partition the plane into disjoint regions. In particular, the regions associated with the leaves represent a partition of the plane into rectangular regions containing one point each. Refer to Figure 35.12.

35.5.1.2 Query

An orthogonal range query Q on a kd-tree $\mathcal{T}$ is answered recursively starting at the root: At a node v we advance the query to a child v_c of v if Q intersects the region R_{v_c} associated with v_c. At a leaf w we return the point in w if it is contained in Q. A standard argument shows that the number of nodes in $\mathcal{T}$ visited when answering Q, or equivalently, the number of nodes v where R_v intersects Q, is $O(\sqrt{N} + K)$; $\sqrt{N}$ nodes v are visited where R_v is intersected by the boundary of Q and K nodes u with R_u completely contained in Q [25].

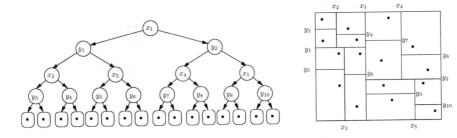

FIGURE 35.12 kd-tree and the corresponding partitioning.

If the kd-tree $\mathcal{T}$ is laid out using the van Emde Boas layout, we can bound the number of memory transfers used to answer a query by considering the nodes $\log B$ levels above the leaves of $\mathcal{T}$. There are $O\left(\frac{N}{B}\right)$ such nodes as the subtree $\mathcal{T}_v$ rooted in one such node v contains B leaves. By the standard query argument, the number of these nodes visited by a query is $O\left(\sqrt{N/B}+\frac{K}{B}\right)$. Thus, the number of memory transfers used to visit nodes more than $\log B$ levels above the leaves is $O\left(\sqrt{N/B}+\frac{K}{B}\right)$. This is also the overall number of memory transfers used to answer a query, since (as argued in Section 35.2.1) the nodes in $\mathcal{T}_v$ are contained in $O(1)$ blocks, that is, any traversal of (any subset of) the nodes in a subtree $\mathcal{T}_v$ can be performed in $O(1)$ memory transfers.

35.5.1.3 Construction

In the RAM model, a kd-tree on N points can be constructed recursively in $O(N \log N)$ time; the root dividing line is found using an $O(N)$ time median algorithm, the points are distributed into two sets according to this line in $O(N)$ time, and the two subtrees are constructed recursively. Since median finding and distribution can be performed cache-obliviously in $O(N/B)$ memory transfers [4,5], a cache-oblivious kd-tree can be constructed in $O\left(\frac{N}{B}\log N\right)$ memory transfers. Agarwal et al. [11] showed how to construct $\log \sqrt{N} = \frac{1}{2}\log N$ levels in $O(\text{Sort}_{M,B}(N))$ memory transfers, leading to a recursive construction algorithms using only $O(\text{Sort}_{M,B}(N))$ memory transfers.

35.5.1.4 Updates

In the RAM model a kd-tree $\mathcal{T}$ can relatively easily be modified to support deletions in $O(\log N)$ time using global rebuilding. To delete a point from $\mathcal{T}$, we simply find the relevant leaf w in $O(\log N)$ time and remove it. We then remove w's parent and connect w's grandparent to w's sibling. The resulting tree is no longer a kd-tree but it still answers queries in $O(\sqrt{N}+T)$ time, since the standard argument still applies. To ensure that N is proportional to the actual number of points in $\mathcal{T}$, the structure is completely rebuilt after $\frac{N}{2}$ deletions. Insertions can be supported in $O(\log^2 N)$ time using the so-called logarithmic method [26], that is, by maintaining $\log N$ kd-trees where the i'th kd-tree is either empty or of size 2^i and then rebuilding a carefully chosen set of these structures when performing an insertion.

Deletes in a cache-oblivious kd-tree is basically done as in the RAM version. However, to still be able to load a subtree $\mathcal{T}_v$ with B leaves in $O(1)$ memory transfers and obtain the $O\left(\sqrt{N/B}+\frac{K}{B}\right)$ query bound, data locality needs to be carefully maintained. By laying out the kd-tree using (a slightly relaxed version of) the exponential layout (Section 35.3.2) rather than the van Emde Boas layout, and by periodically rebuilding parts of this layout, Agarwal et al. [11] showed how to perform a delete in $O(\log_B N)$ memory transfers amortized while maintaining locality. They also showed how a slightly modified version of the logarithmic method and the $O(\text{Sort}_{M,B}(N))$ construction algorithms can be used to perform inserts in $O\left(\frac{\log N}{B}\log_{M/B} N\right) = O(\log_B^2 N)$ memory transfers amortized.

Theorem 35.8

There exists a cache-oblivious (kd-tree) data structure for storing a set of N points in the plane using linear space, such that an orthogonal range query can be answered in $O\left(\sqrt{N/B}+\frac{K}{B}\right)$ memory transfers. The structure can be constructed cache-obliviously in $O(\text{Sort}_{M,B}(N))$ memory transfers and supports updates in $O\left(\frac{\log N}{B}\log_{M/B} N\right) = O(\log_B^2 N)$ memory transfers.

35.5.2 Cache-Oblivious Range Tree

The main part of the cache-oblivious range tree structure for answering (four-sided) orthogonal range queries is a structure for answering three-sided queries $Q = [x_l, x_r] \times [y_b, \infty)$, that is, for finding all points with x-coordinates in the interval $[x_l, x_r]$ and y-coordinates above y_b. Below we discuss the two structures separately.

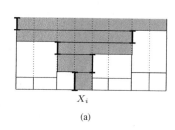
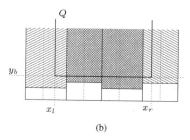

$$(a) \qquad\qquad\qquad\qquad (b)$$

FIGURE 35.13   (a) Active intervals of buckets spanning slab X_i; (b) Buckets active at y_b.

35.5.2.1 Three-Sided Queries

35.5.2.1.1 Structure

Consider dividing the plane into $\sqrt{N}$ vertical *slabs* $X_1, X_2, \ldots, X_{\sqrt{N}}$ containing $\sqrt{N}$ points each. Using these slabs we define $2\sqrt{N} - 1$ *buckets*. A bucket is a rectangular region of the plane that completely spans one or more consecutive slabs and is unbounded in the positive y-direction, like a three-sided query. To define the $2\sqrt{N} - 1$ buckets we start with $\sqrt{N}$ active buckets $b_1, b_2, \ldots, b_{\sqrt{N}}$ corresponding to the $\sqrt{N}$ slabs. The x-range of the slabs define a natural linear ordering on these buckets. We then imagine sweeping a horizontal sweep line from $y = -\infty$ to $y = \infty$. Every time the total number of points above the sweep line in two adjacent active buckets, b_i and b_j, in the linear order falls to $\sqrt{N}$, we mark b_i and b_j as *inactive*. Then we construct a new active bucket spanning the slabs spanned by b_i and b_j with a bottom y-boundary equal to the current position of the sweep line. This bucket replaces b_i and b_j in the linear ordering of active buckets. The total number of buckets defined in this way is $2\sqrt{N} - 1$, since we start with $\sqrt{N}$ buckets and the number of active buckets decreases by one every time a new bucket is constructed. Note that the procedure defines an *active* y-interval for each bucket in a natural way. Buckets overlap but the set of buckets with active y-intervals containing a given y-value (the buckets active when the sweep line was at that value) are non-overlapping and span all the slabs. This means that the active y-intervals of buckets spanning a given slab are non-overlapping. Refer to Figure 35.13a.

After defining the $2\sqrt{N} - 1$ buckets, we are ready to present the three-sided query data structure; it is defined recursively: It consists of a cache-oblivious B-tree $\mathcal{T}$ on the $\sqrt{N}$ boundaries defining the $\sqrt{N}$ slabs, as well as a cache-oblivious B-tree for each of the $\sqrt{N}$ slabs; the tree $\mathcal{T}_i$ for slab i contains the bottom endpoint of the active y-intervals of the $O(\sqrt{N})$ buckets spanning the slab. For each bucket b_i we also store the $\sqrt{N}$ points in b_i in a list $\mathcal{B}_i$ sorted by y-coordinate. Finally, recursive structures $\mathcal{S}_1, \mathcal{S}_2, \ldots, \mathcal{S}_{2\sqrt{N}-1}$ are built on the $\sqrt{N}$ points in each of the $2\sqrt{N} - 1$ buckets.

35.5.2.1.2 Layout

The layout of the structure in memory consists of $O(N)$ memory locations containing $\mathcal{T}$, then $\mathcal{T}_1, \ldots, \mathcal{T}_{\sqrt{N}}$, and $\mathcal{B}_1, \ldots, \mathcal{B}_{2\sqrt{N}-1}$, followed by the recursive structures $\mathcal{S}_1, \ldots, \mathcal{S}_{2\sqrt{N}-1}$. Thus the total space use of the structure is $S(N) \le 2\sqrt{N} \cdot S(\sqrt{N}) + O(N) = O(N \log N)$.

35.5.2.1.3 Query

To answer a three-sided query Q, we consider the buckets whose active y-interval contain y_b. These buckets are non-overlapping and together they contain all points in Q, since they span all slabs and have bottom y-boundary below y_b. We report all points that satisfy Q in each of the buckets with x-range completely between x_l and x_r. At most two other buckets b_l and b_r—the ones containing x_l and x_r—can contain points in Q, and we find these points recursively by advancing the query to $\mathcal{S}_l$ and $\mathcal{S}_r$. Refer to Figure 35.13b.

We find the buckets b_l and b_r that need to be queried recursively and report the points in the completely spanned buckets as follows. We first query $\mathcal{T}$ using $O(\log_B \sqrt{N})$ memory transfers to find the slab X_l containing x_l. Then we query $\mathcal{T}_l$ using another $O(\log_B \sqrt{N})$ memory transfers to find the bucket b_l with active y-interval containing y_b. We can similarly find b_r in $O(\log_B \sqrt{N})$ memory transfers. If b_l spans slabs $X_l, X_{l+1}, \ldots, X_m$ we then query $\mathcal{T}_{m+1}$ with y_b in $O(\log_B \sqrt{N})$ memory transfers to find the active bucket b_i to the right of b_l completely spanned by Q (if it exists). We report the relevant points in b_i by scanning $\mathcal{B}_i$ top-down until we encounter a point not contained in Q. If K' is the number or reported points, a scan of $\mathcal{B}_i$ takes $O\left(1 + \frac{K'}{B}\right)$ memory transfers. We continue this procedure for each of the completely spanned active buckets. By construction, we know that every two adjacent such buckets contain at least $\sqrt{N}$ points above y_b. First consider the part of the query that takes place on recursive levels of size $N \ge B^2$, such that $\sqrt{N}/B \ge \log_B \sqrt{N} \ge 1$. In this case the $O(\log_B \sqrt{N})$ overhead in finding and processing two consecutive completely spanned buckets is smaller than the $O(\sqrt{N}/B)$ memory transfers used to report output points; thus

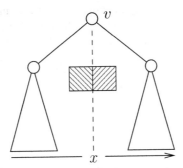

FIGURE 35.14 Answering a four-sided query in v using two three-sided queries in v's children.

we spend $O\left(\log_B \sqrt{N} + \frac{K_i}{B}\right)$ memory transfers altogether to answer a query, not counting the recursive queries. Since we perform at most two queries on each level of the recursion (in the active buckets containing x_l and x_r), the total cost over all levels of size at least B^2 is $O\left(\sum_{i=1}^{\log \log_B N} \log_B N^{1/2^i} + \frac{K_i}{B}\right) = O\left(\log_B N + \frac{K}{B}\right)$ transfers. Next consider the case where $N = B$. In this case the whole level, that is, $\mathcal{T}$, $\mathcal{T}_1, \ldots, \mathcal{T}_{\sqrt{B}}$ and $\mathcal{B}_1, \ldots, \mathcal{B}_{2\sqrt{B}-1}$, is stored in $O(B)$ contiguously memory memory locations and can thus be loaded in $O(1)$ memory transfers. Thus the optimal paging strategy can ensure that we only spend $O(1)$ transfers on answering a query. In the case where $N \le \sqrt{B}$, the level and *all* levels of recursion below it occupies $O(\sqrt{B} \log \sqrt{B}) = O(B)$ space. Thus the optimal paging strategy can load it and all relevant lower levels in $O(1)$ memory transfers. This means that overall we answer a query in $O\left(\log_B N + \frac{K}{B}\right)$ memory transfers, *provided* that N and B are such that we have a level of size B^2 (and thus of size B and $\sqrt{B}$); when answering a query on a level of size between B and B^2 we cannot charge the $O(\log_B \sqrt{N})$ cost of visiting two active consecutive buckets to the ($< B$) points found in the two buckets. Agarwal et al. [11] showed how to guarantee that we have a level of size B^2 by assuming that $B = 2^{2^d}$ for some non-negative integer d. Using a somewhat different construction, Arge et al. [27] showed how to remove this assumption.

Theorem 35.9

There exists a cache-oblivious data structure for storing N points in the plane using $O(N \log N)$ space, such that a three-sided orthogonal range query can be answered in $O\left(\log_B N + \frac{K}{B}\right)$ memory transfers.

35.5.2.2 Four-sided queries

Using the structure for three-sided queries, we can construct a cache-oblivious range tree structure for four-sided orthogonal range queries in a standard way. The structure consists of a cache-oblivious B-tree $\mathcal{T}$ on the N points sorted by x-coordinates. With each internal node v we associate a secondary structure for answering three-sided queries on the points stored in the leaves of the subtree rooted at v: If v is the left child of its parent then we have a three-sided structure for answering queries with the opening to the right, and if v is the right child then we have a three-sided structure for answering queries with the opening to the left. The secondary structures on each level of the tree use $O(N \log N)$ space, for a total space usage of $O(N \log^2 N)$.

To answer an orthogonal range query Q, we search down $\mathcal{T}$ using $O(\log_B N)$ memory transfers to find the first node v where the left and right x-coordinate of Q are contained in different children of v. Then we query the right opening secondary structure of the left child of v, and the left opening secondary structure of the right child of v, using $O\left(\log_B N + \frac{K}{B}\right)$ memory transfers. Refer to Figure 35.14. It is easy to see that this correctly reports all K points in Q.

Theorem 35.10

There exists a cache-oblivious data structure for storing N points in the plane using $O(N \log^2 N)$ space, such that an orthogonal range query can be answered in $O\left(\log_B N + \frac{K}{B}\right)$ memory transfers.

Acknowledgments

Lars Arge was supported in part by the National Science Foundation through ITR grant EIA–0112849, RI grant EIA–9972879, CAREER grant CCR–9984099, and U.S.–Germany Cooperative Research Program grant INT–0129182.

Gerth Stølting Brodal was supported by the Carlsberg Foundation (contract number ANS-0257/20), BRICS (Basic Research in Computer Science, www.brics.dk, funded by the Danish National Research Foundation), and the Future and Emerging Technologies programme of the EU under contract number IST-1999-14186 (ALCOM-FT).

Rolf Fagerberg was supported by BRICS (Basic Research in Computer Science, www.brics.dk, funded by the Danish National Research Foundation), and the Future and Emerging Technologies programme of the EU under contract number IST-1999-14186 (ALCOM-FT). Part of this work was done while at University of Aarhus.

References

1. A. Aggarwal and J. S. Vitter. The input/output complexity of sorting and related problems. *Communications of the ACM*, 31(9):1116–1127, Sept. 1988.

2. L. Arge. External memory data structures. In J. Abello, P. M. Pardalos, and M. G. C. Resende, editors, *Handbook of Massive Data Sets*, pages 313–358. Kluwer Academic Publishers, 2002.

3. J. S. Vitter. External memory algorithms and data structures: Dealing with massive data. *ACM Computing Surveys*, 33(2):209–271, June 2001.

4. M. Frigo, C. E. Leiserson, H. Prokop, and S. Ramachandran. Cache-oblivious algorithms. In *Proc. 40th Annual IEEE Symposium on Foundations of Computer Science*, pages 285–298. IEEE Computer Society Press, 1999.

5. H. Prokop. Cache-oblivious algorithms. Master's thesis, Massachusetts Institute of Technology, Cambridge, MA, June 1999.

6. M. A. Bender, E. D. Demaine, and M. Farach-Colton. Cache-oblivious B-trees. In *Proc. 41st Annual IEEE Symposium on Foundations of Computer Science*, pages 339–409. IEEE Computer Society Press, 2000.

7. M. A. Bender, R. Cole, and R. Raman. Exponential structures for cache-oblivious algorithms. In *Proc. 29th International Colloquium on Automata, Languages, and Programming*, volume 2380 of Lecture Notes in Computer Science, pages 195–207. Springer, 2002.

8. M. A. Bender, Z. Duan, J. Iacono, and J. Wu. A locality-preserving cache-oblivious dynamic dictionary. In *Proc. 13th Annual ACM-SIAM Symposium on Discrete Algorithms*, pages 29–38. SIAM, 2002.

9. G. S. Brodal, R. Fagerberg, and R. Jacob. Cache oblivious search trees via binary trees of small height. In *Proc. 13th Annual ACM-SIAM Symposium on Discrete Algorithms*, pages 39–48. SIAM, 2002.

10. N. Rahman, R. Cole, and R. Raman. Optimized predecessor data structures for internal memory. In *Proc. 3rd Workshop on Algorithm Engineering*, volume 2141 of Lecture Notes in Computer Science, pages 67–78. Springer, 2001.

11. P. K. Agarwal, L. Arge, A. Danner, and B. Holland-Minkley. Cache-oblivious data structures for orthogonal range searching. In *Proc. 19th ACM Symposium on Computational Geometry*, pages 237–245. ACM Press, 2003.

12. G. S. Brodal and R. Fagerberg. Cache oblivious distribution sweeping. In *Proc. 29th International Colloquium on Automata, Languages, and Programming*, volume 2380 of Lecture Notes in Computer Science, pages 426–438. Springer, 2002.

13. M. Bender, E. Demaine, and M. Farach-Colton. Efficient tree layout in a multilevel memory hierarchy. In *Proc. 10th Annual European Symposium on Algorithms*, volume 2461 of Lecture Notes in Computer Science, pages 165–173. Springer, 2002. Full version at http://www.cs.sunysb.edu/ bender/pub/treelayout-full.ps.

14. L. Arge, M. Bender, E. Demaine, B. Holland-Minkley, and J. I. Munro. Cache-oblivious priority-queue and graph algorithms. In *Proc. 34th ACM Symposium on Theory of Computation*, pages 268–276. ACM Press, 2002.

15. G. S. Brodal and R. Fagerberg. Funnel heap - a cache oblivious priority queue. In *Proc. 13th International Symposium on Algorithms and Computation*, volume 2518 of Lecture Notes in Computer Science, pages 219–228. Springer, 2002.

16. G. S. Brodal, R. Fagerberg, and K. Vinther. Engineering a cache-oblivious sorting algorithm. In *Proc. 6th Workshop on Algorithm Engineering and Experiments*, 2004.

17. R. E. Ladner, R. Fortna, and B.-H. Nguyen. A comparison of cache aware and cache oblivious static search trees using program instrumentation. In *Experimental Algorithmics, From Algorithm Design to Robust and Efficient Software (Dagstuhl seminar, September 2000)*, volume 2547 of Lecture Notes in Computer Science, pages 78–92. Springer, 2002.

18. R. Bayer and E. McCreight. Organization and maintenance of large ordered indexes. *Acta Informatica*, 1:173–189, 1972.

19. P. van Emde Boas. Preserving order in a forest in less than logarithmic time and linear space. *Information Processing Letters*, 6:80–82, 1977.

20. M. A. Bender, G. S. Brodal, R. Fagerberg, D. Ge, S. He, H. Hu, J. Iacono, and A. López-Ortiz. The cost of cache-oblivious searching. In *Proc. 44th Annual IEEE Symposium on Foundations of Computer Science*, pages 271–282. IEEE Computer Society Press, 2003.

21. G. S. Brodal and R. Fagerberg. On the limits of cache-obliviousness. In *Proc. 35th ACM Symposium on Theory of Computation*, pages 307–315. ACM Press, 2003.

22. A. Andersson and T. W. Lai. Fast updating of well-balanced trees. In *Proc. 2nd Scandinavian Workshop on Algorithm Theory*, volume 447 of *Lecture Notes in Computer Science*, pages 111–121. Springer, 1990.

23. A. Itai, A. G. Konheim, and M. Rodeh. A sparse table implementation of priority queues. In *Proc. 8th International Colloquium on Automata, Languages, and Programming*, volume 115 of *Lecture Notes in Computer Science*, pages 417–431. Springer, 1981.

24. K. V. R. Kanth and A. K. Singh. Optimal dynamic range searching in non-replicating index structures. In *Proc. International Conference on Database Theory*, volume 1540 of *Lecture Notes in Computer Science*, pages 257–276. Springer, 1999.

25. J. L. Bentley. Multidimensional binary search trees used for associative searching. *Communication of the ACM*, 18:509–517, 1975.

26. J. L. Bentley. Decomposable searching problems. *Information Processing Letters*, 8(5):244–251, 1979.

27. L. Arge, G. S. Brodal, and R. Fagerberg. Improved cache-oblivious two-dimensional orthogonal range searching. Unpublished results, 2004.

36

Dynamic Trees[*]

36.1	Introduction	567
36.2	Linking and Cutting Trees	568

Using Operations on Vertex-Disjoint Paths • Implementing Operations on Vertex-Disjoint Paths

36.3	Topology Trees	571

Construction • Updates • Applications

36.4	Top Trees	574

Updates • Representation and Applications

36.5	ET Trees	577

Updates • Applications

36.6	Reachability Trees	578
36.7	Conclusions	579
	Acknowledgments	579
	References	579

Camil Demetrescu
Sapienza University of Rome

Irene Finocchi
Sapienza University of Rome

Giuseppe F. Italiano
University of Rome Tor Vergata

36.1 Introduction

In this chapter we consider the problem of maintaining properties of a collection of vertex-disjoint trees that change over time as edges are added or deleted. The trees can be rooted or free, and vertices and edges may be associated with real-valued costs that may change as well. A straightforward solution would be to store explicitly with each vertex its parent and cost, if any: with this representation each update would cost only $O(1)$ time, but answering queries would be typically proportional to the size or to the depth of the tree, which may be linear in the worst case. By representing the structure of the trees implicitly, one can reduce the query time while slightly increasing the update time. The typical achieved bounds are logarithmic in the number of vertices of the forest, either in the worst-case or amortized over a sequence of operations.

While the basic tree update operations are edge insertions, edge deletions, and possibly vertex/edge cost changes, many properties of dynamically changing trees have been considered in the literature. The basic query operation is tree membership: while the forest of trees is dynamically changing, we would like to know at any time which tree contains a given vertex, or whether two vertices are in the same tree. Dynamic tree membership is a special case of dynamic connectivity in undirected graphs, and indeed in Chapter 37 we will see that some of the data structures developed here for trees are used to solve the more general problem on graphs. We remark that, if only edge insertions are allowed, the tree membership problem is equivalent to maintaining disjoint sets under union operations and thus the well known set union data structures can solve it [1]. In this chapter we will instead consider the problem in a fully dynamic setting, in which also edge deletions are allowed, and present efficient data structures such as the *linking and cutting trees* of Sleator and Tarjan [2] and the *topology trees* of Frederickson [3].

Other properties that have been considered are finding the least common ancestor of two vertices, the center, the median, or the diameter of a tree [2,4, 5]. When costs are associated either to vertices or to edges, one could also ask what is the minimum or maximum cost in a given path. A variant of topology trees, known as *top trees* [4], are especially well suited at maintaining this kind of path information.

ET trees, first introduced in [6] and much used in [7], allow it to deal easily with forests whose vertices are associated with weighted or unweighted keys, supporting, for example, minkey queries, which require to return a key of minimum weight in the tree that contains a given vertex. *Reachability trees*, introduced by Even and Shiloach in [8], support instead distance and shortest path queries and have been widely used to solve dynamic path problems on directed graphs (see, e.g., [7,9]).

---

[*] This chapter has been reprinted from first edition of this Handbook, without any content updates.

36.2 Linking and Cutting Trees

In this section we present a data structure due to Sleator and Tarjan [2] useful to maintain a collection of rooted trees, each of whose vertices has a real-valued cost, under an arbitrary sequence of the following operations:

- **maketree**(v): initialize a new tree consisting of a single vertex v with cost zero.
- **findroot**(v): return the root of the tree containing vertex v.
- **findcost**(v): return a vertex of minimum cost in the path from v to **findroot**(v).
- **addcost**(v, δ): add the real number δ to the cost of every vertex in the path from v to **findroot**(v).
- **link**(v, w): merge the trees containing vertices v and w by inserting edge (v, w). This operation assumes that v and w are in different trees and that v is a tree root.
- **cut**(v): delete the edge from v to its parent, thus splitting the tree containing vertex v into two trees. This operation assumes that v is not a tree root.

The data structure of Sleator and Tarjan is known as *linking and cutting trees* and supports all the above operations in $O(\log n)$ time, by representing the structure of the trees implicitly. Other operations that can be supported within the same time bound are changing the root of a tree, finding the parent of a vertex, and finding the least common ancestor of two vertices. In particular, the possibility of making a given vertex v root of a tree makes the data structure powerful enough to handle problems requiring linking and cutting of free (i.e., unrooted) trees. Furthermore, the same time bounds can be obtained when real costs are associated with edges rather than with vertices.

The rest of this section is organized as follows. In Section 36.2.1 we show how to implement the operations given above using simpler primitives defined on paths (rather than on trees), and in Section 36.2.2 we describe the implementation of these primitives on paths. For simplicity, we only describe a solution that achieves $O(\log n)$ amortized (rather than worst-case) time per operation. Details of all these results may be found in [2].

36.2.1 Using Operations on Vertex-Disjoint Paths

In this section we show the reduction between the operations on trees and a suitable collection of operations on vertex-disjoint paths. Assume we know how to perform the following operations:

- **makepath**(v): initialize a new path consisting of a single vertex v with cost zero;
- **findpath**(v): return the path containing vertex v;
- **findpathtail**(p): return the tail (last vertex) of path p;
- **findpathcost**(p): return a vertex of minimum cost in path p;
- **addpathcost**(p, δ): add the real value δ to the cost of every vertex in path p;
- **join**(p, v, q): merge path p, vertex v, and path q into a single path by inserting one edge from the tail of p to v, and one edge from v to the head (first vertex) of q, and return the new path. Either p or q can be empty;
- **split**(v): divide the path containing vertex v into at most three paths by deleting the edges incident to v. Return the two new paths p (containing all the vertices before v) and q (containing all the vertices after v). Again, either p or q can be empty.

In order to solve the problem of linking and cutting trees, we partition each tree into a set of vertex disjoint paths. Each tree operation will be defined in terms of one or more path operations. This partition is defined by allowing each tree edge to be either *solid* or *dashed* and by maintaining the invariant that at most one solid edge enters each vertex (we consider an edge oriented from a child to its parent). Removing dashed edges therefore partitions the tree into vertex-disjoint *solid paths*. Dashed edges are represented implicitly: we associate with each path p its *successor*, that is the vertex entered by the dashed edge leaving the tail of p. If the tail of p is a root, *successor*(p) is *null*. Each path will be represented by a vertex on it (an empty path being represented by *null*). In order to convert dashed edges to solid (and vice-versa) we will be using the following operation:

- **expose**(v): make the tree path from v to **findroot**(v) solid. This is done by converting dashed edges in the path to solid, and solid edges incident to the path to dashed. Return the resulting solid path.

Now we describe how to implement tree operations in terms of path operations:

- a **maketree**(v) is done by a **makepath**(v) followed by setting *successor*(v) to *null*;
- a **findroot**(v) is a **findpathtail**(expose(v));
- a **findcost**(v) is a **findpathcost**(expose(v));
- an **addcost**(v, δ) is an **addpathcost**(expose(v), δ);

```
        function expose (v)
   1.   begin
   2.        p ← null
   3.        while v ≠ null do
   4.            w ← successor(findpath(v))
   5.            [q,r] ← split(v)
   6.            if q ≠ null then successor(q) ← v
   7.            p ← join(p,v,r)
   8.            v ← w
   9.        successor(p) ← null
   10.  end
```

FIGURE 36.1   Implementation of **expose**(v).

- a **link**(v, w) is implemented by performing first an **expose**(v) that makes v into a one-vertex solid path, then an **expose**(w) that makes the path from w to its root solid, and then by joining these two solid paths: in short, this means assigning *null* to **successor**(**join**(*null*,**expose**(v),**expose**(w)));
- to perform a **cut**(v), we first perform an **expose**(v), which leaves v with no entering solid edge. We then perform a **split**(v), which returns paths p and q: since v is the head of its solid path and is not a tree root, p will be empty, while q will be non-empty. We now complete the operation by setting both *successor*(v) and *successor*(q) to *null*.

To conclude, we need to show how to perform an **expose**, that is, how to convert all the dashed edges in the path from a given vertex to the tree root to solid maintaining the invariant that at most one solid edge enters each vertex. Let x be a vertex of this path such that the edge from x to its parent w is dashed (and thus $w = successor(x)$). What we would like to do is to convert edge (x, w) into solid, and to convert the solid edge previously entering w (if any) into dashed. We call this operation a *splice*. The pseudocode in Figure 36.1 implements **expose**(v) as a sequence of splices. Path p, initialized to be empty, at the end of the execution will contain the solid path from v to **findroot**(v). Each iteration of the **while** loop performs a splice at v by converting to solid the edge from the tail of p to v (if $p \neq null$) and to dashed the edge from the tail of q to v (if $q \neq null$). A step-by-step execution of **expose** on a running example is shown in Figure 36.2.

From the description above, each tree operation takes $O(1)$ path operations and at most one **expose**. Each splice within an **expose** requires $O(1)$ path operations. Hence, in order to compute the running time, we need first to count the number of splices per **expose** and then to show how to implement the path operations. With respect to the former point, Sleator and Tarjan prove that a sequence of m tree operations causes $O(m \log n)$ splices, and thus $O(\log n)$ splices amortized per **expose**.

Theorem 36.1

[2] Any sequence of m tree operations (including n **maketree**) requires $O(m)$ path operations and at most m **expose**. The exposes can be implemented with $O(m \log n)$ splices, each of which requires $O(1)$ path operations.

36.2.2 Implementing Operations on Vertex-Disjoint Paths

We now describe how to represent solid paths in order to implement efficiently tree operations. Each solid path is represented by a binary tree whose nodes in symmetric order are the vertices in the path; each node x contains pointers to its parent $p(x)$, to its left child $l(x)$, and to its right child $r(x)$. We call the tree representing a solid path a *solid tree*. The vertex representing a solid path is the root of the corresponding solid tree, and thus the root of the solid tree contains a pointer to the successor of the path in the dynamic tree.

Vertex costs are represented as follows. Let $cost(x)$ be the cost of vertex x, and let $mincost(x)$ be the minimum cost among the descendants of x in its solid tree. Rather than storing these two values, in order to implement **addcost** operations efficiently, we store at x the incremental quantities $\Delta cost(x)$ and $\Delta min(x)$ defined as follows:

$$\Delta cost(x) = cost(x) - mincost(x) \tag{36.1}$$

$$\Delta min(x) = \begin{cases} mincost(x) & \text{if } x \text{ is a solid tree root} \\ mincost(x) - mincost(p(x)) & \text{otherwise} \end{cases} \tag{36.2}$$

An example of this representation is given in Figure 36.3. Given $\Delta cost$ and Δmin, we can compute $mincost(x)$ by summing up Δmin for all vertices in the solid tree path from the root to x, and $cost(x)$ as $mincost(x) + \Delta cost(x)$. Moreover, note that $\Delta cost(x) = 0$ if and only if x is a minimum cost node in the subtree rooted at x. If this is not the case and the minimum cost node is

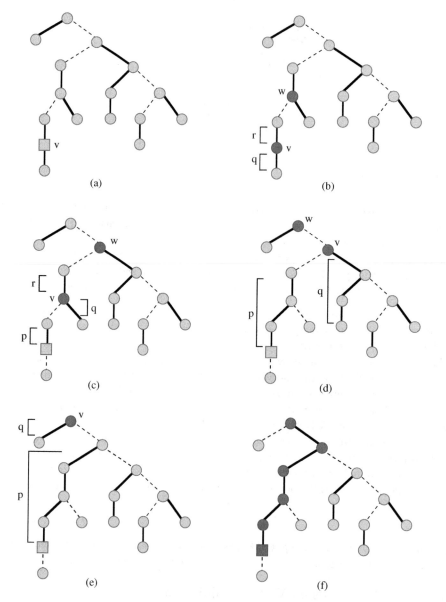

FIGURE 36.2   Effect of **expose**(ν). (a) The original decomposition into solid paths; (b–e) vertices and paths after the execution of line 5 in the four consecutive iterations of the **while** loop; (f) the decomposition into solid paths after **expose**(ν).

in the right subtree, then $\Delta min(r(x)) = 0$; otherwise $\Delta min(l(x)) = 0$. With this representation, rotations can be still implemented in $O(1)$ time. The path operations can be carried out as follows:

- **makepath**(ν): initialize a binary tree of one vertex ν with $\Delta min(ν) = 0$ and $\Delta cost(ν) = 0$.
- **findpath**(ν): starting from ν, follow parent pointers in ν's solid tree until a node with no parent is found. Return this node.
- **findpathtail**(p): assuming that p is the root of a solid tree, follow right pointers and return the rightmost node in the solid tree.
- **findpathcost**(p): initialize ν to p and repeat the following step until $\Delta cost(ν) = 0$: if ν has a right child and $\Delta min(r(ν)) = 0$, replace ν by $r(ν)$; otherwise, replace ν by $l(ν)$. At the end, return ν.
- **addpathcost**(p, δ): add δ to $\Delta min(p)$.
- **join**(p, ν, q): join the solid trees with roots p, ν and q.
- **split**(ν): split the solid tree containing node ν.

We observe that operations **findpath**, **findpathtail** and **findpathcost** are essentially a look up in a search tree, while **split** and **join** are exactly the same operations on search trees. If we represent solid paths by means of balanced search trees,

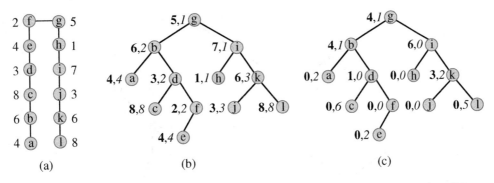

FIGURE 36.3 Representing solid paths with binary trees. (a) A solid path and its vertex costs; (b) solid tree with explicit costs (the bold value is *cost*(*v*) and the italic value is *mincost*(*v*)); (c) corresponding solid tree with incremental costs (the bold value is $\Delta cost(v)$ and the italic value is $\Delta min(v)$).

the time per path operation becomes $O(\log n)$, and by Theorem 36.1 any sequence of *m* tree operations can be supported in $O(m(\log n)^2)$ time. Using self-adjusting binary search trees [10] to represent solid paths, together with a more careful analysis, yields a better bound:

Theorem 36.2

[2] Any sequence of *m* tree operations (including *n* `maketree`) requires $O(m \log n)$ time.

Insights on the use of self-adjusting binary search trees in the implementation of path operations are given in Chapter 13. Using biased search trees [11], the $O(\log n)$ amortized bound given in Theorem 36.2 can be made worst-case. Details can be found in [2].

36.3 Topology Trees

Topology trees have been introduced by Frederickson [3] in order to maintain information about trees subject to insertions and deletions of edges and answer efficiently, for example, tree membership queries. Similarly to the linking and cutting trees of Sleator and Tarjan [2] that we have discussed in Section 36.2, topology trees follow the idea of partitioning a tree into a set of vertex-disjoint paths. However, they are very different in how this partition is chosen, and in the data structures used to represent the paths inside the partition. Indeed, Sleator and Tarjan [2] use a simple partition of the trees based upon a careful choice of sophisticated data structures to represent paths. On the contrary, Frederickson [3] uses a more sophisticated partition that is based upon the topology of the tree; this implies more complicated algorithms but simpler data structures for representing paths.

The basic idea is to partition the tree into a suitable collection of subtrees, called *clusters*, and to implement updates such that only a small number of such clusters is involved. The decomposition defined by the clusters is applied recursively to get faster update and query times.

In order to illustrate how such a recursive decomposition is computed, we assume that *T* has maximum vertex degree 3: this is without loss of generality, since a standard transformation can be applied if this is not the case [12]. Namely, each vertex *v* of degree *d* > 3 is replaced by new vertices $v_0, \ldots, v_{d-1}$; for each neighbor u_i of vertex *v*, $0 \le i \le d-1$, edge (v, u_i) is replaced by (v_i, u_i), and a new edge (v_i, v_{i+1}) is created if $i < d-1$.

Given a tree *T* of maximum degree 3, a *cluster* is any connected subgraph of *T*. The *cardinality* and the *external degree* of a cluster are the number of its vertices and the number of tree edges incident to it, respectively. We now define a partition of the vertices of *T* such that the resulting clusters possess certain useful properties. Let *z* be a positive integer.

Definition 36.1

A *restricted partition* of order *z* w.r.t. *T* is a partition of the vertex set *V* into clusters of degree at most 3 such that:

1. Each cluster of external degree 3 has cardinality 1.
2. Each cluster of external degree <3 has cardinality at most *z*.
3. No two adjacent clusters can be combined and still satisfy the above.

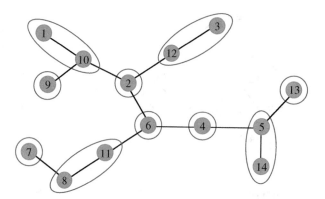

FIGURE 36.4   A restricted partition of order 2 of a tree *T*.

A restricted partition of order 2 of a tree *T* is shown in Figure 36.4. There can be several restricted partitions for a given tree *T*, based upon different choices of the vertices to be unioned. For instance, vertex 8 in Figure 36.4 could be unioned with 7, instead of 11, and the partition would still be valid. It can be proved that a restricted partition of order *z* has $\Theta(n/z)$ clusters [3,13].

We now show that the partition defined above can be applied recursively for $\Theta(\log n)$ levels. Such a recursive application yields a *restricted multilevel partition* [3,13], from which the topology tree can be finally obtained.

Definition 36.2

A *topology tree* is a hierarchical representation of a tree *T* such that each level of the topology tree partitions the vertices of *T* into clusters. Clusters at level 0 contain one vertex each. Clusters at level $\ell \geq 1$ form a restricted partition of order 2 of the vertices of the tree T' obtained by shrinking each cluster at level $\ell - 1$ into a single vertex.

As shown in Figure 36.5, level *l* of the restricted multilevel partition is obtained by computing a restricted partition of order 2 with respect to the tree resulting from viewing each cluster at level $l - 1$ as a single vertex. Figure 36.5 also shows the topology tree corresponding to the restricted multilevel partition. Call any cluster of level $l - 1$ *matched* if it is unioned with another cluster to give a cluster of level *l*: *unmatched* clusters have a unique child in the topology tree. It can be proved that, for any level $l > 0$ of a restricted multilevel partition, the number of matched clusters at level $l - 1$ is at least 1/3 of the total number of vertex clusters at level $l - 1$. Since each pair of matched clusters is replaced by their union at level *l*, the number of clusters at level *l* is at most 5/6 the number of clusters at level $l - 1$. The number of levels of the topology tree is therefore $\Theta(\log n)$.

36.3.1 Construction

It is sufficient to show how to compute a restricted partition: the levels of the topology tree can be then built in a bottom up fashion by repeatedly applying the clustering algorithm as suggested by Definition 36.2. Because of property (3) in Definition 36.1, it is natural to compute a restricted partition according to a locally greedy heuristic, which does not always obtain the minimum number of clusters, but has the advantage of requiring only local adjustments during updates. The tree is first rooted at any vertex of degree 1 and the procedure **cluster** is called with the root as argument. At a generic step, procedure **cluster**(*v*) works as follows. It initializes the cluster $C(v)$ containing vertex *v* as $C(v) = \{v\}$. Then, for each child *w* of *v*, it recursively calls **cluster**(*w*), that computes $C(w)$: if $C(w)$ can be unioned with $C(v)$ without violating the size and degree bounds in Definition 36.1, $C(v)$ is updated as $C(v) \cup C(w)$, otherwise $C(w)$ is output as a cluster. As an example, the restricted partition shown in Figure 36.4 is obtained by running procedure **cluster** on the tree rooted at vertex 7.

36.3.2 Updates

We first describe how to update the clusters of a restricted partition when an edge *e* is inserted in or deleted from the dynamic tree *T*: this operation is the crux of the update of the entire topology tree.

Update of a restricted partition. We start from edge deletion. First, removing an edge *e* splits *T* into two trees, say T_1 and T_2, which inherit all of the clusters of *T*, possibly with the following exceptions.

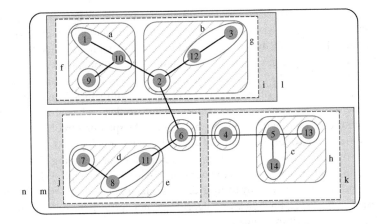

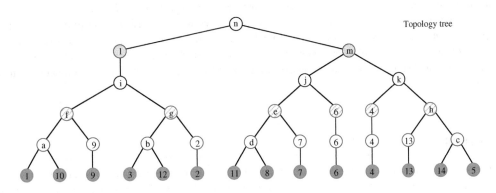

FIGURE 36.5 Restricted multilevel partition and corresponding topology tree.

1. Edge *e* is entirely contained in a cluster: this cluster is no longer connected and therefore must be split. After the split, we must check whether each of the two resulting clusters is adjacent to a cluster of tree degree at most 2, and if these two adjacent clusters together have cardinality ≤ 2. If so, we combine these two clusters in order to maintain condition (3).

2. Edge *e* is between two clusters: in this case no split is needed. However, since the tree degree of the clusters containing the endpoints of *e* has been decreased, we must check if each cluster should be combined with an adjacent cluster, again because of condition (3).

Similar local manipulations can be applied to restore invariants (1)–(3) in Definition 36.1 in case of edge insertions. We now come to the update of the topology tree.

Update of the topology tree. Each level can be updated upon insertions and deletions of edges in tree *T* by applying few locally greedy adjustments similar to the ones described above. In particular, a constant number of basic clusters (corresponding to leaves in the topology tree) are examined: the changes in these basic clusters percolate up in the topology tree, possibly causing vertex clusters to be regrouped in different ways. The fact that only a constant amount of work has to be done on $O(\log n)$ topology tree nodes implies a logarithmic bound on the update time.

Theorem 36.3

[3,13] The update of a topology tree because of an edge insertion or deletion in the dynamic tree *T* can be supported in $O(\log n)$ time, where *n* is the number of vertices of *T*.

36.3.3 Applications

In the *fully dynamic tree membership* problem we would like to maintain a forest of unrooted trees under insertion of edges (which merge two trees into one), deletion of edges (which split one tree into two), and membership queries. Typical queries require to return the name of the tree containing a given vertex, or ask whether two vertices are in a same tree. Most of the solutions presented in the literature root each tree arbitrarily at one of its vertices; by keeping extra information at the root (such as the name of the tree), membership queries are equivalent to finding the tree root a vertex.

The dynamic tree clustering techniques of Frederickson have also found wide application in dynamic graph algorithms. Namely, topology trees have been originally designed in order to solve the *fully dynamic minimum spanning tree* problem [3], in which we wish to maintain a minimum spanning tree of a dynamic weighted undirected graph upon insertions/deletions of edges and edge cost changes. Let $G = (V, E)$ be the dynamic graph and let S be a designated spanning tree of G. As G is updated, edges in the spanning tree S may change: for example, if the cost of an edge e is increased in G and e is in the spanning tree, we need to check for the existence of a replacement edge e' of smaller cost, and swap e' with e in S. The clustering approach proposed in [3,13] consists of partitioning the vertex set V into subtrees connected in S, so that each subtree is only adjacent to a few other subtrees. A topology tree is used for representing this recursive partition of the spanning tree S. A generalization of topology trees, called *2-dimensional topology trees*, is also formed from pairs of nodes in the topology tree in order to maintain information about the edges in $E \setminus S$ [3]. Fully dynamic algorithms based only on a single level of clustering obtain typically time bounds of the order of $O(m^{2/3})$ (see for instance [14,15]), where m is the number of edges of the graph. When the partition is applied recursively, better $O(m^{1/2})$ time bounds can be achieved by using 2-dimensional topology trees: we refer the interested reader to [3,13] for details. As we will see in Section 36.5.2, Frederickson's algorithm is not optimal: the fully dynamic minimum spanning tree problem has been later solved in polylogarithmic time [16].

With the same technique, an $O(m^{1/2})$ time bound can be obtained also for *fully dynamic connectivity* and *2-edge connectivity* [3,13]. For instance, [13] shows that edges and vertices can be inserted to or deleted from an undirected graph in $O(m^{1/2})$ time, and a query as to whether two vertices are in the same 2-edge-connected component can be answered in $O(\log n)$ time, n being the number of vertices. This result is based on the use of *ambivalent data structures* [13], a refinement of the clustering technique in which edges can belong to multiple groups, only one of which is actually selected depending on the topology of the given spanning tree.

36.4 Top Trees

Top trees have been introduced by Alstrup et al. [4] to maintain efficiently information about paths in trees, such as, for example, the maximum weight on the path between any pair of vertices in a tree. The basic idea is taken from Frederickson's topology trees, but instead of partitioning vertices, top trees work by partitioning edges: the same vertex can then appear in more than one cluster. Top trees can be also seen as a natural generalization of standard balanced binary trees over dynamic collections of lists that may be concatenated and split, where each node of the balanced binary tree represents a segment of a list. As we will see, in the terminology of top trees this is just a special case of a cluster.

We follow here the presentation in [5]. Similarly to [3,13], a *cluster* is a connected subtree of the dynamic tree T, with the additional constraint that at most two vertices, called *boundary vertices*, have edges out of the subtree. We will denote the boundary of a cluster C as δC. If the boundary contains two vertices u and v, we call *cluster path* of C the unique path between u and v in T and we denote it as $\pi(C)$. If $|\delta C < 2|$, then $\pi(C) = \emptyset$. Two clusters C_1 and C_2 are called *neighbors* if their intersection contains exactly one vertex: since clusters are connected and have no edges in common, the intersection vertex must be in $\delta C_1 \cap \delta C_2$. It is also possible to define a boundary δT, consisting of one or two vertices, for the entire tree T: we will call such vertices, if any, *external boundary vertices*. If external boundary vertices are defined, we have to extend the notion of boundary of a cluster: namely, if a cluster C contains an external boundary vertex v, then $v \in \delta C$ even if v has no edge out of C.

Definition 36.3

A *top tree* $\mathcal{T}$ over a pair $(T, \delta T)$ is a binary tree such that:

- The leaves of $\mathcal{T}$ are the edges of T.
- The internal nodes of $\mathcal{T}$ are clusters of T.
- The subtree represented by each internal node is the union of the subtrees represented by its two children, which must be neighbors.
- The root of $\mathcal{T}$ represents the entire tree T.
- The height of $\mathcal{T}$ is $O(\log n)$, where n is the number of vertices of T.

A tree with a single node has an empty top tree. Figure 36.6 shows the clusters at different levels of the recursive partition and the corresponding top tree. Note that a vertex can appear in many clusters, as many as $\Theta(n)$ in the worst case. However, it can be a non-boundary vertex only in $O(\log n)$ clusters. Indeed, for each vertex v which is neither an external boundary vertex nor a leaf in T, there exists a unique cluster C with children A and B such that $v \in \delta A$, $v \in \delta B$, and $v \notin \delta C$. Then v is non-boundary vertex only in cluster C and in all its ancestors in the top tree.

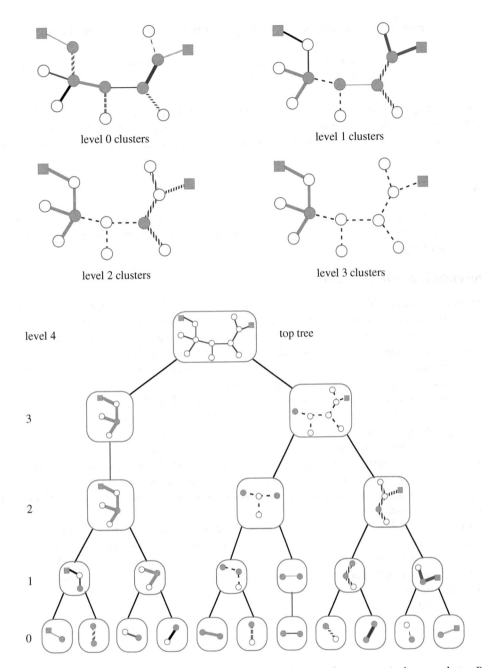

FIGURE 36.6 Clusters and top tree of a tree T. Edges with the same color, thickness, and pattern are in the same cluster. Boundary vertices are grey. External boundary vertices are squared.

A locally greedy approach similar to the one described in Section 36.3.1 for topology trees can be used to build a top tree. The only modifications require to reason in terms of edges, instead of vertices, and to check the condition on the cardinality of the boundary before unioning any two neighboring clusters.

36.4.1 Updates

Given a dynamic forest, top trees over the trees of the forest are maintained under the following operations:

- **link**(u, v), where u and v are in different trees T_u and T_v of the forest: link trees T_u and T_v by adding edge (u, v);
- **cut**(e): remove edge e from the forest;
- **expose**(u, v), where u and v are in the same tree T of the forest: make u and v the external boundary vertices of T and return the new root cluster of the top tree over T.

Top trees can be maintained under these operations by making use of two basic **merge** and **split** primitives:

- **merge:** it takes two top trees whose roots are neighbor clusters and joins them to form a unique top tree;
- **split:** this is the reverse operation, deleting the root of a given top tree.

The implementation of update operations starts with a sequence of **Split** of all ancestor clusters of edges whose boundary changes and finishes with a sequence of **Merge**. In case of insertion and deletions, since an end-point v of an edge has to be already boundary vertex of the edge if v is not a leaf, an update can change the boundary of at most two edges, excluding the edge being inserted/deleted. From [3–5] we have:

Theorem 36.4

[4,5] For a dynamic forest we can maintain top trees of height $O(\log n)$ supporting each **link**, **cut**, and **expose** operation with a sequence of $O(\log n)$ **split** and **merge**. The sequence itself is identified in $O(\log n)$ time. The space usage of top trees is linear in the size of the dynamic forest.

36.4.2 Representation and Applications

Top trees are represented as standard binary trees, with pointers to parent and children for each node. With each leaf is associated the corresponding edge in T and with each internal node the at most two boundary vertices of the corresponding cluster. In addition, each vertex v of T has a pointer to the deepest cluster for which v is a non-boundary vertex, or to the root cluster containing v if v is an external boundary vertex. Given this representation, top trees can be used as a black box for maintaining different kinds of information. Typically, the user needs to attach extra information to the top tree nodes and to show how such information can be maintained upon **merge** and **split** operations.

A careful choice of the extra information makes it possible to maintain easily path properties of trees, such as the minimum weight of an edge on the path between any two vertices. In this example, the extra information w_C associated with a cluster C is the weight of the lightest edge on the cluster path $\pi(C)$. Before showing how to maintain it, note that if cluster A is a child of cluster C in the top tree and A contains an edge from $\pi(C)$, then $\pi(A) \subseteq \pi(C)$: we call A a *path child* of C. When a cluster is created by a **merge**, we store as extra information the minimum weight stored at its path children. In case of a **split**, we just discard the information. Now, in order to find the minimum weight between any two vertices u and v, we compute the root cluster C of the top tree in which u and v are external boundary vertices by calling **expose(u,v)**. Then $\pi(C)$ is the path between u and v and w_C is exactly the value we are looking for.

Top trees can be used quite easily if the property we wish to maintain is a *local* property, that is, being satisfied by a vertex/edge in a tree implies that the property is also satisfied in all the subtrees containing the vertex/edge. Non-local properties appear to be more challenging. For general non-local searching the user has to supply a function **select** that can be used to guide a binary search towards a desired edge: given the root of a top tree, the function selects one of the two children according to the property to be maintained. Since the property is non-local, in general it is not possible to recurse directly on the selected child as is. However, Alstrup et al. [5] show that the top tree can be temporarily modified by means of a few **merge** operations so that **select** can be provided with the "right" input in the recursive call and guide the search to a correct solution.

Lemma 36.1

Given a top tree, after $O(\log n)$ calls to **select**, **merge**, and **split**, there is a unique edge (u, v) contained in all clusters chosen by **select**, and then (u, v) is returned.

We refer the interested reader to [5] for the proof of Lemma 36.1, which shows how to modify the top tree in order to facilitate calls to **select**. We limit here to use the search as a black box in order to show how to maintain dynamically the *center* of a tree (i.e., a vertex which maximizes the distance from any other vertex) using top trees.

The extra information maintained for a cluster C with boundary vertices a and b are: the distance $dist(C)$ between a and b, and the lengths $\ell_a(C)$ and $\ell_b(C)$ of a longest path in C departing from a and b, respectively. Now we show how to compute the extra information for a cluster C obtained by merging two neighboring clusters A and B. Let c be a boundary vertex of cluster C and, w.l.o.g., let $c \in \delta A$. The longest path from c to a vertex in A has length $\ell_c(A)$. Instead, in order to get from c to a vertex in B, we must traverse a vertex, say x, such that $x \in \delta A \cap \delta B$: thus, the longest path from c to a vertex in B has length $dist(c, x) + \ell_x(B)$. This is equal to $\ell_c(B)$ if $c \in \delta B$ (i.e., if $x = c$), or to $dist(A) + \ell_x(B)$ if $c \notin \delta B$. Now, we set $\ell_c(C) = max\{\ell_c(A), dist(c, x) + \ell_x(B)\}$. We can compute $dist(C)$ similarly. Finally, function **select** can be implemented as follows. Given a cluster C with children A and B, let u be the vertex in the intersection of A and B: if $\ell_u(A) \geq \ell_u(B)$ **select** picks A, otherwise it picks B. The correctness of **select**

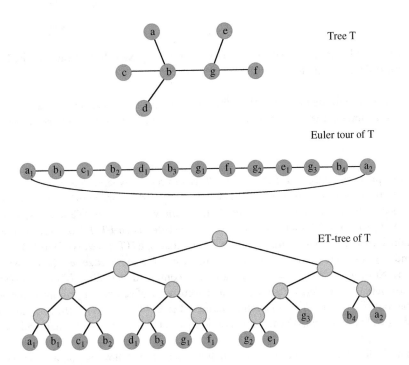

FIGURE 36.7 A tree, an Euler tour, and the corresponding ET tree with $d = 2$.

depends on the fact that if $\ell_u(A) \geq \ell_u(B)$, then A contains all the centers of C. Using Lemma 36.1 we can finally conclude that the center of a tree can be maintained dynamically in $O(\log n)$ time.

We refer the interested reader to [4,5] for other sample applications of top trees, such as maintaining the median or the diameter of dynamic trees. As we will recall in Section 36.5.2, top trees are fundamental in the design of the fastest algorithm for the fully dynamic minimum spanning tree problem [16].

36.5 ET Trees

ET trees, first introduced in [6], have been later used to design algorithms for a variety of problems on dynamic graphs, such as fully dynamic connectivity and bipartiteness (see, e.g., [7,17]). They provide an implicit representation of dynamic forests whose vertices are associated with weighted or unweighted keys. In addition to arbitrary edge insertions and deletions, updates allow it to add or remove the weighted key associated to a vertex. Supported queries are the following:

- **connected**(u, v): tells whether vertices u and v are in the same tree.
- **size**(v): returns the number of vertices in the tree that contains v.
- **minkey**(v): returns a key of minimum weight in the tree that contains v; if keys are unweighted, an arbitrary key is returned.

The main concept for the definition of ET trees is that of Euler tour.

Definition 36.4

An *Euler tour* of a tree T is a maximal closed walk over the graph obtained by replacing each edge of T by two directed edges with opposite direction.

The Euler tour of a tree can be easily computed in linear time by rooting the tree at an arbitrary vertex and running a depth first visit [18]. Each time a vertex v is encountered on the walk, we call this an *occurrence* of v and we denote it by $o(v)$. A vertex of degree Δ has exactly Δ occurrences, expect for the root which is visited $\Delta + 1$ times. Furthermore, the walk traverses each directed edge exactly once; hence, if T has n vertices, the Euler tour has length $2n - 2$. Given an n-vertex tree T, we encode it with a sequence of $2n - 1$ symbols given by an Euler tour. We refer to this encoding as $ET(T)$. For instance, the encoding of the tree given in Figure 36.7 derived from the Euler tour shown below the tree itself is $ET(T) = a\,b\,c\,b\,d\,b\,g\,f\,g\,e\,g\,b\,a$.

Definition 36.5

An *ET tree* is a dynamic balanced d-ary tree over some Euler tour around T. Namely, the leaves of the ET tree are the nodes of the Euler tour, in the same order in which they appear (see Figure 36.7).

An ET tree has $O(n)$ nodes due to the linear length of the Euler tour and to properties of d-ary trees. However, since each vertex of T may occur several times in the Euler tour, an arbitrary occurrence is marked as *representative* of the vertex.

36.5.1 Updates

We first analyze how to update the encoding $ET(T)$ when T is subject to dynamic edge operations. If an edge $e = (u, v)$ is deleted from T, denote by T_u and T_v the two trees obtained after the deletion, with $u \in T_u$ and $v \in T_v$. Let $o_1(u)$, $o_1(v)$, $o_2(u)$ and $o_2(v)$ be the occurrences of u and v encountered during the visit of (u, v). Without loss of generality assume that $o_1(u) < o_1(v) < o_2(v) < o_2(u)$ so that $ET(T) = \alpha o_1(u) \beta o_1(v) \gamma o_2(v) \delta o_2(u) \epsilon$. Then $ET(T_v)$ is given by the interval $o_1(v) \gamma o_2(v)$, and $ET(T_u)$ can be obtained by splicing out the interval from $ET(T)$, that is, $ET(T_u) = \alpha o_1(u) \beta \delta o_2(u) \epsilon$.

If two trees T_1 and T_2 are joined in a new tree T because of a new edge $e = (u, v)$, with $u \in T_1$ and $v \in T_2$, we first reroot T_2 at v. Now, given $ET(T_1) = \alpha o_1(u) \beta$ and the computed encoding $ET(T_2) = o_1(v) \gamma o_2(v)$, we compute $ET(T) = \alpha o_1(u) o_1(v) \gamma o_2(v) o(u) \beta$, where $o(u)$ is a newly created occurrence of vertex u. To complete the description, we need to show how to change the root of a tree T from r to another vertex s. Let $ET(T) = o(r) \alpha o_1(s) \beta$, where $o_1(s)$ is any occurrence of s. Then, the new encoding will be $o_1(s) \beta \alpha o(s)$, where $o(s)$ is a newly created occurrence of s that is added at the end of the new sequence.

In summary, if trees in the forest are linked or cut, a constant number of the following operations is required: (i) splicing out an interval from the encoding, (ii) inserting an interval into the encoding, (iii) inserting or (iv) deleting a single occurrence from the encoding. If the encoding $ET(T)$ is stored in a balanced search tree of degree d, then one may perform each operation in time $O(d \log_d n)$ while maintaining the balance of the tree.

36.5.2 Applications

The query **connected**(u, v) can be easily supported in time $O(\log n/d)$ by finding the roots of the ET trees containing u and v and checking if they coincide. The same time is sufficient to check whether one element precedes another element in the ordering.

To support **size** and **minkey** queries, each node q of the ET tree maintains two additional values: the number $s(q)$ of representatives below it and the minimum weight key $k(q)$ attached to a representative below it. Such values can be easily maintained during updates and allow it to answer queries of the form **size**(v) and **minkey**(v) in $O(\log n/d)$ time for any vertex v of the forest: the root r of the ET tree containing v is found and values $s(r)$ and $k(r)$ are returned, respectively. We refer the interested reader to [7] for additional details of the method.

In Section 36.4 we will see how ET trees have been used [7,16] to design a polylogarithmic algorithm for fully dynamic connectivity. Here we limit to observe that trees in a spanning forest of the dynamic graph are maintained using the Euler tour data structure, and therefore updates and connectivity queries within the forest can be supported in logarithmic time. The use of randomization and of a logarithmic decomposition makes it possible to maintain also non-tree edges in polylogarithmic time upon changes in the graph.

36.6 Reachability Trees

In this section we consider a tree data structure that has been widely used to solve dynamic path problems on directed graphs.

The first appearance of this tool dates back to 1981, when Even and Shiloach showed how to maintain a breadth-first tree of an undirected graph under any sequence of edge deletions [8]; they used this as a kernel for decremental connectivity on undirected graphs. Later on, Henzinger and King [7] showed how to adapt this data structure to fully dynamic transitive closure in directed graphs. King [9] designed an extension of this tree data structure to weighted directed graphs for solving fully dynamic transitive closure (see Section 37.6.1) and all pairs shortest paths (see Section 37.7.1).

In the unweighted directed version, the goal is to maintain information about breadth-first search (BFS) on a directed graph G undergoing deletions of edges. In particular, in the context of dynamic path problems, we are interested in maintaining BFS trees of depth up to d, with $d \leq n$. Given a directed graph $G = (V, E)$ and a vertex $r \in V$, we would like to support any intermixed sequence of the following operations:

Delete(x, y): delete edge (x, y) from G.

Level(u): return the level of vertex u in the BFS tree of depth d rooted at r (return $+\infty$ if u is not reachable from r within distance d).

In [9] it is shown how to maintain efficiently the BFS levels, supporting any **Level** operation in constant time and any sequence of **Delete** operations in $O(md)$ overall time:

Lemma 36.2

(King [9]) Maintaining BFS levels up to depth d from a given root requires $O(md)$ time in the worst case throughout any sequence of edge deletions in a directed graph with m initial edges.

Lemma 36.2 implies that maintaining BFS levels requires d times the time needed for constructing them. Since $d \leq n$, we obtain a total bound of $O(mn)$ if there are no limits on the depth of the BFS levels.

As it was shown in [7,9], it is possible to extend the BFS data structure presented in this section to deal with weighted directed graphs. In this case, a shortest path tree is maintained in place of BFS levels: after each edge deletion or edge weight increase, the tree is reconnected by essentially mimicking Dijkstra's algorithm rather than BFS. Details can be found in [9].

36.7 Conclusions

In this chapter we have surveyed data structures for maintaining properties of dynamically changing trees, focusing our attention on linking and cutting trees [2], topology trees [3], top trees [4], ET trees [6,7], and reachability trees [8]. We have shown that these data structures typically support updates and many different kinds of queries in logarithmic (amortized or worst-case) time. Hence, problems such as tree membership, maintaining the center or diameter of a tree, finding the minimum cost on a given path can be solved in $O(\log n)$ time in a fully dynamic setting.

All the data structures for maintaining properties of dynamically changing trees are not only important and interesting on their own, but are often used as building blocks by many dynamic graph algorithms, as we will see in Chapter 37. Some of these data structures, such as the union find data structures and the linking and cutting trees of Sleator and Tarjan have myriads of applications in other problems, such as implementing property grammars, computational geometry problems, testing equivalence of finite state machines, computing the longest common subsequence of two sequences, performing unification in logic programming and theorem proving, finding minimum spanning trees, and maximum flow algorithms. Since all these problems are outside the scope of this survey, we have not mentioned these applications and have restricted our attention to the applications of these data structures to dynamic tree and graph algorithms only.

Acknowledgments

This work has been supported in part by the IST Programmes of the EU under contract numbers IST-1999-14186 (ALCOM-FT) and IST-2001-33555 (COSIN), and by the Italian Ministry of University and Scientific Research (Project "ALINWEB: Algorithmics for Internet and the Web").

References

1. R. E. Tarjan and J. van Leeuwen. Worst-case analysis of set union algorithms. *J. Assoc. Comput. Mach.*, 31:245–281, 1984.
2. D. D. Sleator and R. E. Tarjan. A data structure for dynamic trees. *J. Comp. Syst. Sci.*, 24:362–381, 1983.
3. G. N. Frederickson. Data structures for on-line updating of minimum spanning trees. *SIAM J. Comput.*, 14:781–798, 1985.
4. S. Alstrup, J. Holm, K. de Lichtenberg, and M. Thorup. Minimizing diameters of dynamic trees. In *Proc. 24th Int. Colloquium on Automata, Languages and Programming (ICALP 97)*, LNCS 1256, pages 270–280, 1997.
5. S. Alstrup, J. Holm, and M. Thorup. Maintaining center and median in dynamic trees. In *Proc. 7th Scandinavian Workshop on Algorithm Theory (SWAT 00)*, pages 46–56, 2000.
6. R. E. Tarjan and U. Vishkin. An efficient parallel biconnectivity algorithm. *SIAM J. on Computing*, 14:862–874, 1985.
7. M. R. Henzinger and V. King. Randomized fully dynamic graph algorithms with polylogarithmic time per operation. *J. Assoc. Comput. Mach.*, 46(4):502–536, 1999.
8. S. Even and Y. Shiloach. An on-line edge deletion problem. *J. Assoc. Comput. Mach.*, 28:1–4, 1981.
9. V. King. Fully dynamic algorithms for maintaining all-pairs shortest paths and transitive closure in digraphs. In *Proc. 40th Symposium on Foundations of Computer Science (FOCS 99)*, 1999.

10. D. D. Sleator and R. E. Tarjan. Self-adjusting binary search trees. *J. Assoc. Comput. Mach.*, 32:652–686, 1985.

11. S. W. Bent, D. D. Sleator, and R. E. Tarjan. Biased search trees. *SIAM J. Comput.*, 14(3):545–568, 1985.

12. F. Harary. *Graph Theory*. Addison-Wesley, Reading, MA, 1969.

13. G. N. Frederickson. Ambivalent data structures for dynamic 2-edge-connectivity and k smallest spanning trees. *SIAM J. Comput.*, 26(2):484--538, 1997.

14. Z. Galil and G. F. Italiano. Fully-dynamic algorithms for 2-edge connectivity. *SIAM J. Comput.*, 21:1047–1069, 1992.

15. M. Rauch. Fully dynamic biconnectivity in graphs. *Algorithmica*, 13:503–538, 1995.

16. J. Holm, K. de Lichtenberg, and M. Thorup. Poly-logarithmic deterministic fully-dynamic algorithms for connectivity, minimum spanning tree, 2-edge, and biconnectivity. *J. Assoc. Comput. Mach.*, 48(4):723–760, 2001.

17. P. B. Miltersen, S. Subramanian, J. S. Vitter, and R. Tamassia. Complexity models for incremental computation. *Theoret. Comput. Science*, 130:203–236, 1994.

18. T. H. Cormen, C. E. Leiserson, R. L. Rivest, and C. Stein. *Introduction to Algorithms*, Second Edition. The MIT Press, Cambridge, MA, 2001.

37

Dynamic Graphs*

37.1 Introduction...581
37.2 Techniques for Undirected Graphs...582
    Clustering • Sparsification • Randomization
37.3 Techniques for Directed Graphs...584
    Kleene Closures • Long Paths • Locality • Matrices
37.4 Connectivity...587
    Updates in $O(\log^2 n)$ Time
37.5 Minimum Spanning Tree..588
    Deletions in $O(\log^2 n)$ Time • Updates in $O(\log^4 n)$ Time
37.6 Transitive Closure..589
    Updates in $O(n^2 \log n)$ Time • Updates in $O(n^2)$ Time
37.7 All-Pairs Shortest Paths..591
    Updates in $O(n^{2.5} \sqrt{C \log n})$ Time • Updates in $O(n^2 \log^3 n)$ Time
37.8 Conclusions..593
Acknowledgments...593
References..593

Camil Demetrescu
Sapienza University of Rome

Irene Finocchi
Sapienza University of Rome

Giuseppe F. Italiano
University of Rome Tor Vergata

37.1 Introduction

In many applications of graph algorithms, including communication networks, VLSI design, graphics, and assembly planning, graphs are subject to discrete changes, such as insertions or deletions of vertices or edges. In the last two decades there has been a growing interest in such dynamically changing graphs, and a whole body of algorithmic techniques and data structures for dynamic graphs has been discovered. This chapter is intended as an overview of this field.

An *update on a graph* is an operation that inserts or deletes edges or vertices of the graph or changes attributes associated with edges or vertices, such as cost or color. Throughout this chapter by *dynamic graph* we denote a graph that is subject to a sequence of updates. In a typical dynamic graph problem one would like to answer queries on dynamic graphs, such as, for instance, whether the graph is connected or which is the shortest path between any two vertices. The goal of a dynamic graph algorithm is to update efficiently the solution of a problem after dynamic changes, rather than having to recompute it from scratch each time. Given their powerful versatility, it is not surprising that dynamic algorithms and dynamic data structures are often more difficult to design and to analyze than their static counterparts.

We can classify dynamic graph problems according to the types of updates allowed. In particular, a dynamic graph problem is said to be *fully dynamic* if the update operations include unrestricted insertions and deletions of edges or vertices. A dynamic graph problem is said to be *partially dynamic* if only one type of update, either insertions or deletions, is allowed. More specifically, a dynamic graph problem is said to be *incremental* if only insertions are allowed, while it is said to be *decremental* if only deletions are allowed.

In the first part of this chapter we will present the main algorithmic techniques used to solve dynamic problems on both *undirected* and *directed* graphs. In the second part of the chapter we will deal with dynamic problems on graphs, and we will investigate as paradigmatic problems the dynamic maintenance of minimum spanning trees, connectivity, transitive closure and shortest paths. Interestingly enough, dynamic problems on directed graphs seem much harder to solve than their counterparts on undirected graphs, and require completely different techniques and tools.

---

* This chapter has been reprinted from first edition of this Handbook, without any content updates.

37.2 Techniques for Undirected Graphs

Many of the algorithms proposed in the literature use the same general techniques, and hence we begin by describing these techniques. In this section we focus on undirected graphs, while techniques for directed graphs will be discussed in Section 37.3. Typically, most of these techniques use some sort of graph decomposition, and partition either the vertices or the edges of the graph to be maintained. Moreover, data structures that maintain properties of dynamically changing trees, such as the ones described in Chapter 36 (linking and cutting trees, topology trees, and Euler tour trees), are often used as building blocks by many dynamic graph algorithms.

37.2.1 Clustering

The clustering technique has been introduced by Frederickson [1] and is based upon partitioning the graph into a suitable collection of connected subgraphs, called *clusters*, such that each update involves only a small number of such clusters. Typically, the decomposition defined by the clusters is applied recursively and the information about the subgraphs is combined with the topology trees described in Section 36.3. A refinement of the clustering technique appears in the idea of *ambivalent data structures* [2], in which edges can belong to multiple groups, only one of which is actually selected depending on the topology of the given spanning tree.

As an example, we briefly describe the application of clustering to the problem of maintaining a minimum spanning forest [1]. Let $G = (V, E)$ be a graph with a designated spanning tree S. Clustering is used for partitioning the vertex set V into subtrees connected in S, so that each subtree is only adjacent to a few other subtrees. A topology tree, as described in Section 36.3, is then used for representing a recursive partition of the tree S. Finally, a generalization of topology trees, called *2-dimensional topology trees*, are formed from pairs of nodes in the topology tree and allow it to maintain information about the edges in $E \setminus S$ [1].

Fully dynamic algorithms based only on a single level of clustering obtain typically time bounds of the order of $O(m^{2/3})$ (see for instance [3,4]). When the partition can be applied recursively, better $O(m^{1/2})$ time bounds can be achieved by using 2-dimensional topology trees (see, for instance, [1,2]).

Theorem 37.1: (Frederickson [1])

The minimum spanning forest of an undirected graph can be maintained in time $O(\sqrt{m})$ per update, where m is the current number of edges in the graph.

We refer the interested reader to [1,2] for details about Frederickson's algorithm. With the same technique, an $O(\sqrt{m})$ time bound can be obtained also for fully dynamic connectivity and 2-edge connectivity [1,2]

The type of clustering used can very problem-dependent, however, and makes this technique difficult to be used as a black box.

37.2.2 Sparsification

Sparsification is a general technique due to Eppstein *et al.* [5] that can be used as a black box (without having to know the internal details) in order to design and dynamize graph algorithms. It is a divide-and-conquer technique that allows it to reduce the dependence on the number of edges in a graph, so that the time bounds for maintaining some property of the graph match the times for computing in sparse graphs. More precisely, when the technique is applicable, it speeds up a $T(n, m)$ time bound for a graph with n vertices and m edges to $T(n, O(n))$, that is, to the time needed if the graph were sparse. For instance, if $T(n, m) = O(\sqrt{m})$, we get a better bound of $O(\sqrt{n})$. The technique itself is quite simple. A key concept is the notion of certificate.

Definition 37.1

For any graph property P and graph G, a *certificate* for G is a graph G' such that G has property P if and only if G' has the property.

Let G be a graph with m edges and n vertices. We partition the edges of G into a collection of $O(m/n)$ sparse subgraphs, that is, subgraphs with n vertices and $O(n)$ edges. The information relevant for each subgraph can be summarized in a sparse certificate. Certificates are then merged in pairs, producing larger subgraphs which are made sparse by again computing their certificate. The result is a balanced binary tree in which each node is represented by a sparse certificate. Each update involves $O(\log (m/n))^*$ graphs with $O(n)$ edges each, instead of one graph with m edges.

---

[*] Throughout this chapter, we assume that $\log x$ stands for $\max\{1, \log_2 x\}$, so $\log (m/n)$ is never smaller than 1 even if $m < 2n$.

There exist two variants of sparsification. The first variant is used in situations where no previous fully dynamic algorithm is known. A static algorithm is used for recomputing a sparse certificate in each tree node affected by an edge update. If the certificates can be found in time $O(m + n)$, this variant gives time bounds of $O(n)$ per update.

In the second variant, certificates are maintained using a dynamic data structure. For this to work, a *stability* property of certificates is needed, to ensure that a small change in the input graph does not lead to a large change in the certificates. We refer the interested reader to [5] for a precise definition of stability. This variant transforms time bounds of the form $O(m^p)$ into $O(n^p)$.

Definition 37.2

A time bound $T(n)$ is *well-behaved* if, for some $c < 1$, $T(n/2) < cT(n)$. Well-behavedness eliminates strange situations in which a time bound fluctuates wildly with n. For instance, all polynomials are well-behaved.

Theorem 37.2: (Eppstein et al. [5])

Let P be a property for which we can find sparse certificates in time $f(n, m)$ for some well-behaved f, and such that we can construct a data structure for testing property P in time $g(n, m)$ which can answer queries in time $q(n, m)$. Then there is a fully dynamic data structure for testing whether a graph has property P, for which edge insertions and deletions can be performed in time $O(f(n, O(n))) + g(n, O(n))$, and for which the query time is $q(n, O(n))$.

Theorem 37.3: (Eppstein et al. [5])

Let P be a property for which stable sparse certificates can be maintained in time $f(n, m)$ per update, where f is well-behaved, and for which there is a data structure for property P with update time $g(n, m)$ and query time $q(n, m)$. Then P can be maintained in time $O(f(n, O(n))) + g(n, O(n))$ per update, with query time $q(n, O(n))$.

Basically, the first version of sparsification (Theorem 37.2) can be used to dynamize static algorithms, in which case we only need to *compute* efficiently *sparse* certificates, while the second version (Theorem 37.3) can be used to speed up existing fully dynamic algorithms, in which case we need to *maintain* efficiently *stable sparse* certificates.

Sparsification applies to a wide variety of dynamic graph problems, including minimum spanning forests, edge and vertex connectivity. As an example, for the fully dynamic minimum spanning tree problem, it reduces the update time from $O(\sqrt{m})$ [1,2] to $O(\sqrt{n})$ [5].

Since sparsification works on top of a given algorithm, we need not to know the internal details of this algorithm. Consequently, it can be applied orthogonally to other data structuring techniques: in a large number of situations both clustering and sparsification have been combined to produce an efficient dynamic graph algorithm.

37.2.3 Randomization

Clustering and sparsification allow one to design efficient deterministic algorithms for fully dynamic problems. The last technique we present in this section is due to Henzinger and King [6], and allows one to achieve faster update times for some problems by exploiting the power of randomization.

We now sketch how the randomization technique works taking the fully dynamic connectivity problem as an example. Let $G = (V, E)$ be a graph to be maintained dynamically, and let F be a spanning forest of G. We call edges in F *tree edges*, and edges in $E \backslash F$ *non-tree edges*. The algorithm by Henzinger and King is based on the following ingredients.

37.2.3.1 Maintaining Spanning Forests

Trees are maintained using the Euler Tours data structure (ET trees) described in Section 36.5: this allows one to obtain logarithmic updates and queries within the forest.

37.2.3.2 Random Sampling

Another key idea is the following: when e is deleted from a tree T, use random sampling among the non-tree edges incident to T, in order to find quickly a replacement edge for e, if any.

37.2.3.3 Graph Decomposition

The last key idea is to combine randomization with a suitable graph decomposition. We maintain an edge decomposition of the current graph G using $O(\log n)$ edge disjoint subgraphs $G_i = (V, E_i)$. These subgraphs are hierarchically ordered. The lower levels contain tightly-connected portions of G (i.e., dense edge cuts), while the higher levels contain loosely-connected portions of G (i.e., sparse cuts). For each level i, a spanning forest for the graph defined by all the edges in levels i or below is also maintained.

Note that the hard operation in this problem is the deletion of a tree edge: indeed, a spanning forest is easily maintained with the help of the linking and cutting trees described in Section 36.2 throughout edge insertions, and deleting a non-tree edge does not change the forest.

The goal is an update time of $O(\log^3 n)$: after an edge deletion, in the quest for a replacement edge, we can afford a number of sampled edges of $O(\log^2 n)$. However, if the candidate set of edge e is a small fraction of all non-tree edges which are adjacent to T, it is unlikely to find a replacement edge for e among this small sample. If we found no candidate among the sampled edges, we must check explicitly all the non-tree edges adjacent to T. After random sampling has failed to produce a replacement edge, we need to perform this check explicitly, otherwise we would not be guaranteed to provide correct answers to the queries. Since there might be a lot of edges which are adjacent to T, this explicit check could be an expensive operation, so it should be made a low probability event for the randomized algorithm. This can produce pathological updates, however, since deleting all edges in a relatively small candidate set, reinserting them, deleting them again, and so on will almost surely produce many of those unfortunate events.

The graph decomposition is used to prevent the undesirable behavior described above. If a spanning forest edge e is deleted from a tree at some level i, random sampling is used to quickly find a replacement for e at that level. If random sampling succeeds, the tree is reconnected at level i. If random sampling fails, the edges that can replace e in level i form with high probability a sparse cut. These edges are moved to level $i+1$ and the same procedure is applied recursively on level $i+1$.

Theorem 37.4: (Henzinger and King [6])

Let G be a graph with m_0 edges and n vertices subject to edge deletions only. A spanning forest of G can be maintained in $O(\log^3 n)$ expected amortized time per deletion, if there are at least $\Omega(m_0)$ deletions. The time per query is $O(\log n)$.

37.3 Techniques for Directed Graphs

In this section we discuss the main techniques used to solve dynamic path problems on directed graphs. We first address combinatorial and algebraic properties, and then we consider some efficient data structures, which are used as building blocks in designing dynamic algorithms for transitive closure and shortest paths. Similarly to the case of undirected graphs seen in Section 37.2, data structures that maintain properties of dynamically changing trees, such as the ones described in Chapter 36 (reachability trees), are often used as building blocks by many dynamic graph algorithms.

37.3.1 Kleene Closures

Path problems such as transitive closure and shortest paths are tightly related to matrix sum and matrix multiplication over a closed semiring (see [7] for more details). In particular, the transitive closure of a directed graph can be obtained from the adjacency matrix of the graph via operations on the semiring of Boolean matrices, that we denote by $\{+, \cdot, 0, 1\}$. In this case, $+$ and $\cdot$ denote the usual sum and multiplication over Boolean matrices.

Lemma 37.1

Let $G = (V, E)$ be a directed graph and let $TC(G)$ be the (reflexive) transitive closure of G. If X is the Boolean adjacency matrix of G, then the Boolean adjacency matrix of $TC(G)$ is the Kleene closure of X on the $\{+, \cdot, 0, 1\}$ Boolean semiring:

$$X^* = \sum_{i=0}^{n-1} X^i.$$

Similarly, shortest path distances in a directed graph with real-valued edge weights can be obtained from the weight matrix of the graph via operations on the semiring of real-valued matrices, that we denote by $\{\oplus, \odot, \mathcal{R}\}$, or more simply by $\{\min, +\}$. Here, $\mathcal{R}$ is the set of real values and $\oplus$ and $\odot$ are defined as follows. Given two real-valued matrices A and B, $C = A \odot B$ is the matrix product such that $C[x, y] = \min_{1 \leq z \leq n}\{A[x, z] + B[z, y]\}$ and $D = A \oplus B$ is the matrix sum such that $D[x, y] = \min\{A[x, y], B[x, y]\}$. We also denote by AB the product $A \odot B$ and by $AB[x, y]$ entry (x, y) of matrix AB.

Lemma 37.2

Let $G = (V, E)$ be a weighted directed graph with no negative-length cycles. If X is a weight matrix such that $X[x, y]$ is the weight of edge (x, y) in G, then the distance matrix of G is the Kleene closure of X on the $\{\oplus, \odot, \mathcal{R}\}$ semiring:

$$X^* = \bigoplus_{i=0}^{n-1} X^i.$$

We now briefly recall two well-known methods for computing the Kleene closure X^* of X. In the following, we assume that X is an $n \times n$ matrix.

37.3.1.1 Logarithmic Decomposition

A simple method to compute X^*, based on repeated squaring, requires $O(n^\mu \cdot \log n)$ worst-case time, where $O(n^\mu)$ is the time required for computing the product of two matrices over a closed semiring. This method performs $\log_2 n$ sums and products of the form $X_{i+1} = X_i + X_i^2$, where $X = X_0$ and $X^* = X_{\log_2 n}$.

37.3.1.2 Recursive Decomposition

Another method, due to Munro [8], is based on a Divide and Conquer strategy and computes X^* in $O(n^\mu)$ worst-case time. Munro observed that, if we partition X in four submatrices A, B, D, C of size $n/2 \times n/2$ (considered in clockwise order), and X^* similarly in four submatrices E, F, H, G of size $n/2 \times n/2$, then X^* is defined recursively according to the following equations:

$E = (A + BD^*C)^*$	$F = EBD^*$	$G = D^*CE$	$H = D^* + D^*CEBD^*$

Surprisingly, using this decomposition the cost of computing X^* starting from X is asymptotically the same as the cost of multiplying two matrices over a closed semiring.

37.3.2 Long Paths

In this section we recall an intuitive combinatorial property of long paths in a graph. Namely, if we pick a subset S of vertices at random from a graph G, then a sufficiently long path will intersect S with high probability. This can be very useful in finding a long path by using short searches.

To the best of our knowledge, the long paths property was first given in [9], and later on it has been used many times in designing efficient algorithms for transitive closure and shortest paths (see e.g., [10–13]). The following theorem is from [12].

Theorem 37.5: (Ullman and Yannakakis [12])

Let $S \subseteq V$ be a set of vertices chosen uniformly at random. Then the probability that a given simple path has a sequence of more than $(cn \log n)/|S|$ vertices, none of which are from S, for any $c > 0$, is, for sufficiently large n, bounded by $2^{-\alpha c}$ for some positive α.

Zwick [13] showed it is possible to choose set S deterministically by a reduction to a hitting set problem [14,15]. King used a similar idea for maintaining fully dynamic shortest paths [11].

37.3.3 Locality

Recently, Demetrescu and Italiano [16] proposed a new approach to dynamic path problems based on maintaining classes of paths characterized by local properties, that is, properties that hold for all proper subpaths, even if they may not hold for the entire paths. They showed that this approach can play a crucial role in the dynamic maintenance of shortest paths. For instance, they considered a class of paths defined as follows:

Definition 37.3

A path π in a graph is *locally shortest* if and only if every proper subpath of π is a shortest path.

This definition is inspired by the optimal-substructure property of shortest paths: all subpaths of a shortest path are shortest. However, a locally shortest path may not be shortest.

The fact that locally shortest paths include shortest paths as a special case makes them an useful tool for computing and maintaining distances in a graph. Indeed, paths defined locally have interesting combinatorial properties in dynamically changing graphs. For example, it is not difficult to prove that the number of locally shortest paths that may change due to an edge weight update is $O(n^2)$ if updates are partially dynamic, that is, increase-only or decrease-only:

Theorem 37.6

Let G be a graph subject to a sequence of increase-only or decrease-only edge weight updates. Then the amortized number of paths that start or stop being locally shortest at each update is $O(n^2)$.

Unfortunately, Theorem 37.6 does not hold if updates are fully dynamic, that is, increases and decreases of edge weights are intermixed. To cope with pathological sequences, a possible solution is to retain information about the history of a dynamic graph, considering the following class of paths:

Definition 37.4

A *historical shortest path* (in short, *historical path*) is a path that has been shortest at least once since it was last updated.

Here, we assume that a path is updated when the weight of one of its edges is changed. Applying the locality technique to historical paths, we derive locally historical paths:

Definition 37.5

A path π in a graph is *locally historical* if and only if every proper subpath of π is historical.

Like locally shortest paths, also locally historical paths include shortest paths, and this makes them another useful tool for maintaining distances in a graph:

Lemma 37.3

If we denote by *SP*, *LSP*, and *LHP* respectively the sets of shortest paths, locally shortest paths, and locally historical paths in a graph, then at any time the following inclusions hold: $SP \subseteq LSP \subseteq LHP$.

Differently from locally shortest paths, locally historical paths exhibit interesting combinatorial properties in graphs subject to fully dynamic updates. In particular, it is possible to prove that the number of paths that become locally historical in a graph at each edge weight update depends on the number of historical paths in the graph.

Theorem 37.7: (Demetrescu and Italiano [16])

Let G be a graph subject to a sequence of update operations. If at any time throughout the sequence of updates there are at most $O(h)$ historical paths in the graph, then the amortized number of paths that become locally historical at each update is $O(h)$.

To keep changes in locally historical paths small, it is then desirable to have as few historical paths as possible. Indeed, it is possible to transform every update sequence into a slightly longer equivalent sequence that generates only a few historical paths. In particular, there exists a simple *smoothing* strategy that, given any update sequence Σ of length k, produces an operationally equivalent sequence $F(\Sigma)$ of length $O(k \log k)$ that yields only $O(\log k)$ historical shortest paths between each pair of vertices in the graph. We refer the interested reader to [16] for a detailed description of this smoothing strategy. According to Theorem 37.7, this technique implies that only $O(n^2 \log k)$ locally historical paths change at each edge weight update in the smoothed sequence $F(\Sigma)$.

As elaborated in [16], locally historical paths can be maintained very efficiently. Since by Lemma 37.3 locally historical paths include shortest paths, this yields the fastest known algorithm for fully dynamic all pairs shortest paths (see Section 37.7.2).

37.3.4 Matrices

Another useful data structure for keeping information about paths in dynamic directed graphs is based on matrices subject to dynamic changes. As we have seen in Section 37.3.1, Kleene closures can be constructed by evaluating polynomials over matrices. It is therefore natural to consider data structures for maintaining polynomials of matrices subject to updates of entries, like the one introduced in [17].

In the case of Boolean matrices, the problem can be stated as follows. Let P be a polynomial over $n \times n$ Boolean matrices with constant degree, constant number of terms, and variables $X_1 \ldots X_k$. We wish to maintain a data structure for P subject to any intermixed sequence of update and query operations of the following kind:

SetRow$(i, \Delta X, X_b)$: sets to one the entries in the ith row of variable X_b of polynomial P corresponding to one-valued entries in the ith row of matrix ΔX.

SetCol$(i, \Delta X, X_b)$: sets to one the entries in the ith column of variable X_b of polynomial P corresponding to one-valued entries in the ith column of matrix ΔX.

Reset$(\Delta X, X_b)$: resets to zero the entries of variable X_b of polynomial P corresponding to one-valued entries in matrix ΔX.

Lookup(): returns the maintained value of P.

We add to the previous four operations a further update operation especially designed for maintaining path problems:

LazySet$(\Delta X, X_b)$: sets to 1 the entries of variable X_b of P corresponding to one-valued entries in matrix ΔX. However, the maintained value of P might not be immediately affected by this operation.

Let C_P be the correct value of P that we would have by recomputing it from scratch after each update, and let M_P be the actual value that we maintain. If no LazySet operation is ever performed, then always $M_P = C_P$. Otherwise, M_P is not necessarily equal to C_P, and we guarantee the following weaker property on M_P: if $C_P[u, v]$ flips from 0 to 1 due to a SetRow/SetCol operation on a variable X_b, then $M_P[u, v]$ flips from 0 to 1 as well. This means that SetRow and SetCol always correctly reveal new 1's in the maintained value of P, possibly taking into account the 1's inserted through previous LazySet operations. This property is crucial for dynamic path problems, since the appearance of new paths in a graph after an edge insertion, which corresponds to setting a bit to one in its adjacency matrix, is always correctly recorded in the data structure.

Lemma 37.4: (Demetrescu and Italiano [17])

Let P be a polynomial with constant degree of matrices over the Boolean semiring. Any SetRow, SetCol, LazySet, and Reset operation on a polynomial P can be supported in $O(n^2)$ amortized time. Lookup queries are answered in optimal time.

Similar data structures can be given for settings different from the semiring of Boolean matrices. In particular, in [10] the problem of maintaining polynomials of matrices over the $\{min, +\}$ semiring is addressed.

The running time of operations for maintaining polynomials in this semiring is given below.

Theorem 37.8: (Demetrescu and Italiano [10])

Let P be a polynomial with constant degree of matrices over the $\{min, +\}$ semiring. Any SetRow, SetCol, LazySet, and Reset operation on variables of P can be supported in $O(D \cdot n^2)$ amortized time, where D is the maximum number of different values assumed by entries of variables during the sequence of operations. Lookup queries are answered in optimal time.

37.4 Connectivity

In this section and in Section 37.5 we consider dynamic problems on undirected graphs, showing how to deploy some of the techniques and data structures presented in Section 37.2 to obtain efficient algorithms. These algorithms maintain efficiently some property of an undirected graph that is undergoing structural changes defined by insertion and deletion of edges, and/or updates of edge costs. To check the graph property throughout a sequence of these updates, the algorithms must be prepared to answer queries on the graph property efficiently. In particular, in this section we address the *fully dynamic connectivity* problem, where we are interested in algorithms that are capable of inserting edges, deleting edges, and answering a query on whether the graph is connected, or whether two vertices are in the same connected component. We recall that the goal of a dynamic algorithm is to minimize the amount of recomputation required after each update. All the dynamic algorithms that we describe are able to maintain dynamically the graph property at a cost (per update operation) which is significantly smaller than the cost of recomputing the graph property from scratch.

37.4.1 Updates in $O(\log^2 n)$ Time

In this section we give a high level description of the fastest deterministic algorithm for the fully dynamic connectivity problem in undirected graphs [18]: the algorithm, due to Holm, de Lichtenberg and Thorup, answers connectivity queries in $O(\log n / \log \log n)$ worst-case running time while supporting edge insertions and deletions in $O(\log^2 n)$ amortized time.

Similarly to the randomized algorithm in [6], the deterministic algorithm in [18] maintains a spanning forest F of the dynamically changing graph G. As above, we will refer to the edges in F as *tree edges*. Let e be a tree edge of forest F, and let T be the tree of F containing it. When e is deleted, the two trees T_1 and T_2 obtained from T after the deletion of e can be reconnected if and only if there is a non-tree edge in G with one endpoint in T_1 and the other endpoint in T_2. We call such an edge a *replacement edge* for e. In other words, if there is a replacement edge for e, T is reconnected via this replacement edge; otherwise, the deletion of e creates a new connected component in G.

To accommodate systematic search for replacement edges, the algorithm associates to each edge e a level $\ell(e)$ and, based on edge levels, maintains a set of sub-forests of the spanning forest F: for each level i, forest F_i is the sub-forest induced by tree edges of level $\geq i$. If we denote by L denotes the maximum edge level, we have that:

$$F = F_0 \supseteq F_1 \supseteq F_2 \supseteq \cdots \supseteq F_L,$$

Initially, all edges have level 0; levels are then progressively increased, but never decreased. The changes of edge levels are accomplished so as to maintain the following invariants, which obviously hold at the beginning.

Invariant (1): F is a maximum spanning forest of G if edge levels are interpreted as weights.
Invariant (2): The number of nodes in each tree of F_i is at most $n/2^i$.

Invariant (1) should be interpreted as follows. Let (u, v) be a non-tree edge of level $\ell(u, v)$ and let $u \cdots v$ be the unique path between u and v in F (such a path exists since F is a spanning forest of G). Let e be any edge in $u \cdots v$ and let $\ell(e)$ be its level. Due to (1), $\ell(e) \geq \ell(u, v)$. Since this holds for each edge in the path, and by construction $F_{\ell(u,v)}$ contains all the tree edges of level $\geq \ell(u, v)$, the entire path is contained in $F_{\ell(u,v)}$, that is, u and v are connected in $F_{\ell(u,v)}$.

Invariant (2) implies that the maximum number of levels is $L \leq \lfloor \log_2 n \rfloor$.

Note that when a new edge is inserted, it is given level 0. Its level can be then increased at most $\lfloor \log_2 n \rfloor$ times as a consequence of edge deletions. When a tree edge $e = (v, w)$ of level $\ell(e)$ is deleted, the algorithm looks for a replacement edge at the highest possible level, if any. Due to invariant (1), such a replacement edge has level $\ell \leq \ell(e)$. Hence, a replacement subroutine `Replace((u, w), ℓ(e))` is called with parameters e and $\ell(e)$. We now sketch the operations performed by this subroutine.

`Replace((u, w), ℓ)` finds a replacement edge of the highest level $\leq \ell$, if any. If such a replacement does not exist in level ℓ, we have two cases: if $\ell > 0$, we recurse on level $\ell - 1$; otherwise, $\ell = 0$, and we can conclude that the deletion of (v, w) disconnects v and w in G.

During the search at level ℓ, suitably chosen tree and non-tree edges may be promoted at higher levels as follows. Let T_v and T_w be the trees of forest F_ℓ obtained after deleting (v, w) and let, w.l.o.g., T_v be smaller than T_w. Then T_v contains at most $n/2^{\ell+1}$ vertices, since $T_v \cup T_w \cup \{(v, w)\}$ was a tree at level ℓ and due to invariant (2). Thus, edges in T_v of level ℓ can be promoted at level $\ell + 1$ by maintaining the invariants. Non-tree edges incident to T_v are finally visited one by one: if an edge does connect T_v and T_w, a replacement edge has been found and the search stops, otherwise its level is increased by 1.

We maintain an ET-tree, as described in Section 36.5, for each tree of each forest. Consequently, all the basic operations needed to implement edge insertions and deletions can be supported in $O(\log n)$ time. In addition to inserting and deleting edges from a forest, ET-trees must also support operations such as finding the tree of a forest that contains a given vertex, computing the size of a tree, and, more importantly, finding tree edges of level ℓ in T_v and non-tree edges of level ℓ incident to T_v. This can be done by augmenting the ET-trees with a constant amount of information per node: we refer the interested reader to [18] for details.

Using an amortization argument based on level changes, the claimed $O(\log^2 n)$ bound on the update time can be finally proved. Namely, inserting an edge costs $O(\log n)$, as well as increasing its level. Since this can happen $O(\log n)$ times, the total amortized insertion cost, inclusive of level increases, is $O(\log^2 n)$. With respect to edge deletions, cutting and linking $O(\log n)$ forest has a total cost $O(\log^2 n)$; moreover, there are $O(\log n)$ recursive calls to `Replace`, each of cost $O(\log n)$ plus the cost amortized over level increases. The ET-trees over $F_0 = F$ allows it to answer connectivity queries in $O(\log n)$ worst-case time. As shown in [18], this can be reduced to $O(\log n / \log \log n)$ by using a $\Theta(\log n)$-ary version of ET-trees.

Theorem 37.9: (Holm et al. [18])

A dynamic graph G with n vertices can be maintained upon insertions and deletions of edges using $O(\log^2 n)$ amortized time per update and answering connectivity queries in $O(\log n / \log \log n)$ worst-case running time.

37.5 Minimum Spanning Tree

One of the most studied dynamic problem on undirected graphs is the *fully dynamic minimum spanning tree* problem, which consists of maintaining a minimum spanning forest of a graph during insertions of edges, deletions of edges, and edge cost changes. In this section, we show that a few simple changes to the connectivity algorithm presented in Section 37.4 are sufficient to maintain a minimum spanning forest of a weighted undirected graph upon deletions of edges [18]. A general reduction from [19] can then be applied to make the deletions-only algorithm fully dynamic.

37.5.1 Deletions in $O(\log^2 n)$ Time

In addition to starting from a *minimum* spanning forest, the only change concerns function `Replace`, that should be implemented so as to consider candidate replacement edges of level ℓ in order of increasing weight, and not in arbitrary order. To do so, the ET-trees described in Section 36.5 can be augmented so that each node maintains the minimum weight of a non-tree edge incident to the Euler tour segment below it. All the operations can still be supported in $O(\log n)$ time, yielding the same time bounds as for connectivity.

We now discuss the correctness of the algorithm. In particular, function `Replace` returns a replacement edge of minimum weight on the highest possible level: it is not immediate that such a replacement edge has the minimum weight among all levels. This can be proved by first showing that the following invariant, proved in [18], is maintained by the algorithm:

Invariant (3): Every cycle C has a non-tree edge of maximum weight and minimum level among all the edges in C.

Invariant (3) can be used to prove that, among all the replacement edges, the lightest edge is on the maximum level. Let e_1 and e_2 be two replacement edges with $w(e_1) < w(e_2)$, and let C_i be the cycle induced by e_i in F, $i = 1, 2$. Since F is a minimum spanning forest, e_i has maximum weight among all the edges in C_i. In particular, since by hypothesis $w(e_1) < w(e_2)$, e_2 is also the heaviest edge in cycle $C = (C_1 \cup C_2) \backslash (C_1 \cap C_2)$. Thanks to Invariant (3), e_2 has minimum level in C, proving that $\ell(e_2) \leq \ell(e_1)$. Thus, considering non-tree edges from higher to lower levels is correct.

Lemma 37.5: (Holm et al. [18])

There exists a deletions-only minimum spanning forest algorithm that can be initialized on a graph with n vertices and m edges and supports any sequence of edge deletions in $O(m \log^2 n)$ total time.

37.5.2 Updates in $O(\log^4 n)$ Time

The reduction used to obtain a fully dynamic algorithm, which involves the top tree data structure discussed in Section 36.4, is a slight generalization of the construction proposed by Henzinger and King [19], and works as follows.

Lemma 37.6: ([18,19])

Suppose we have a deletions-only minimum spanning tree algorithm that, for any k and l, can be initialized on a graph with k vertices and l edges and supports any sequence of $\Omega(l)$ deletions in total time $O(l \cdot t(k, l))$, where t is a non-decreasing function. Then there exists a fully-dynamic minimum spanning tree algorithm for a graph with n nodes starting with no edges, that, for m edges, supports updates in time

$$O\left(\log^3 n + \sum_{i=1}^{3+\log_2 m} \sum_{j=1}^{i} t(min\{n, 2^j\}, 2^j) \right)$$

We refer the interested reader to references [18,19] for the description of the construction that proves Lemma 37.6. From Lemma 37.5 we get $t(k, l) = O(\log^2 k)$. Hence, combining Lemmas 37.5 and 37.6, we get the claimed result.

Theorem 37.10: (Holm et al. [18])

There exists a fully-dynamic minimum spanning forest algorithm that, for a graph with n vertices, starting with no edges, maintains a minimum spanning forest in $O(\log^4 n)$ amortized time per edge insertion or deletion.

37.6 Transitive Closure

In the rest of this chapter we survey the newest results for dynamic problems on directed graphs. In particular, we focus on two of the most fundamental problems: transitive closure and shortest paths. These problems play a crucial role in many applications, including network optimization and routing, traffic information systems, databases, compilers, garbage collection, interactive verification systems, industrial robotics, dataflow analysis, and document formatting. In this section we consider the best known algorithms for fully dynamic transitive closure. Given a directed graph G with n vertices and m edges, the problem consists of supporting any intermixed sequence of operations of the following kind:

Insert(u, v): insert edge (u, v) in G;
Delete(u, v): delete edge (u, v) from G;
Query(x, y): answer a reachability query by returning "yes" if there is a path from vertex x to vertex y in G, and "no" otherwise;

A simple-minded solution to this problem consists of maintaining the graph under insertions and deletions, searching if y is reachable from x at any query operation. This yields $O(1)$ time per update (Insert and Delete), and $O(m)$ time per query, where m is the current number of edges in the maintained graph.

Another simple-minded solution would be to maintain the Kleene closure of the adjacency matrix of the graph, rebuilding it from scratch after each update operation. Using the recursive decomposition of Munro [8] discussed in Section 37.3.1 and fast matrix multiplication [20], this takes constant time per reachability query and $O(n^\omega)$ time per update, where $\omega < 2.38$ is the current best exponent for matrix multiplication.

Despite many years of research in this topic, no better solution to this problem was known until 1995, when Henzinger and King [6] proposed a randomized Monte Carlo algorithm with one-sided error supporting a query time of $O(n/\log n)$ and an amortized update time of $O(n\widehat{m}^{0.58} \log^2 n)$, where $\widehat{m}$ is the average number of edges in the graph throughout the whole update sequence. Since $\widehat{m}$ can be as high as $O(n^2)$, their update time is $O(n^{2.16} \log^2 n)$. Khanna, Motwani and Wilson [21] proved that, when a lookahead of $\Theta(n^{0.18})$ in the updates is permitted, a deterministic update bound of $O(n^{2.18})$ can be achieved.

King and Sagert [22] showed how to support queries in $O(1)$ time and updates in $O(n^{2.26})$ time for general directed graphs and $O(n^2)$ time for directed acyclic graphs; their algorithm is randomized with one-sided error. The bounds of King and Sagert were further improved by King [11], who exhibited a deterministic algorithm on general digraphs with $O(1)$ query time and $O(n^2 \log n)$ amortized time per update operations, where updates are insertions of a set of edges incident to the same vertex and deletions of an arbitrary subset of edges. Using a different framework, in 2000 Demetrescu and Italiano [17] obtained a deterministic fully dynamic algorithm that achieves $O(n^2)$ amortized time per update for general directed graphs. We note that each update might change a portion of the transitive closure as large as $\Omega(n^2)$. Thus, if the transitive closure has to be maintained explicitly after each update so that queries can be answered with one lookup, $O(n^2)$ is the best update bound one could hope for.

If one is willing to pay more for queries, Demetrescu and Italiano [17] showed how to break the $O(n^2)$ barrier on the single-operation complexity of fully dynamic transitive closure: building on a previous path counting technique introduced by King and Sagert [22], they devised a randomized algorithm with one-sided error for directed acyclic graphs that achieves $O(n^{1.58})$ worst-case time per update and $O(n^{0.58})$ worst-case time per query. Other recent results for dynamic transitive closure appear in [23,24].

37.6.1 Updates in $O(n^2 \log n)$ Time

In this section we address the algorithm by King [11], who devised the first deterministic near-quadratic update algorithm for fully dynamic transitive closure. The algorithm is based on the reachability tree data structure considered in Section 36.6 and on the logarithmic decomposition discussed in Section 37.3.1.

It maintains explicitly the transitive closure of a graph G in $O(n^2 \log n)$ amortized time per update, and supports inserting and deleting several edges of the graph with just one operation. Insertion of a bunch of edges incident to a vertex and deletion of any subset of edges in the graph require asymptotically the same time of inserting/deleting just one edge.

The algorithm maintains $\log n + 1$ levels: level i, $0 \le i \le \log n$, maintains a graph G_i whose edges represent paths of length up to 2^i in the original graph G. Thus, $G_0 = G$ and $G_{\log n}$ is the transitive closure of G.

Each level graph G_i is built on top of the previous level graph G_{i-1} by keeping two trees of depth ≤ 2 rooted at each vertex v: an out-tree $OUT_i(v)$ maintaining vertices reachable from v by traversing at most two edges in G_{i-1}, and an in-tree $IN_i(v)$ maintaining vertices that reach v by traversing at most two edges in G_{i-1}. Trees $IN_i(v)$ can be constructed by considering the orientation of edges in G_{i-1} reversed. An edge (x, y) will be in G_i if and only if $x \in IN_i(v)$ and $y \in OUT_i(v)$ for some v. In/out trees are maintained with the deletions-only reachability tree data structure considered in Section 36.6.

To update the levels after an insertion of edges around a vertex v in G, the algorithm simply rebuilds $IN_i(v)$ and $OUT_i(v)$ for each i, $1 \le i \le \log n$, while other trees are not touched. This means that some trees might not be up to date after an insertion operation. Nevertheless, any path in G is represented in at least the in/out trees rooted at the latest updated vertex in the path, so the reachability information is correctly maintained. This idea is the key ingredient of King's algorithm.

When an edge is deleted from G_i, it is also deleted from any data structures $IN_i(v)$ and $OUT_i(v)$ that contain it. The interested reader can find further details in [11].

37.6.2 Updates in $O(n^2)$ Time

In this section we address the algorithm by Demetrescu and Italiano [17]. The algorithm is based on the matrix data structure considered in Section 37.3.4 and on the recursive decomposition discussed in Section 37.3.1.

It maintains explicitly the transitive closure of a graph in $O(n^2)$ amortized time per update, supporting the same generalized update operations of King's algorithm, that is, insertion of a bunch of edges incident to a vertex and deletion of any subset of edges in the graph with just one operation. This is the best known update bound for fully dynamic transitive closure with constant query time.

The algorithm maintains the Kleene closure X^* of the $n \times n$ adjacency matrix X of the graph as the sum of two matrices X_1 and X_2. Let V_1 be the subset of vertices of the graph corresponding to the first half of indices of X, and let V_2 contain the remaining vertices. Both matrices X_1 and X_2 are defined according to Munro's equations of Section 37.3.1, but in such a way that paths appearing due to an insertion of edges around a vertex in V_1 are correctly recorded in X_1, while paths that appear due to an insertion of edges around a vertex in V_2 are correctly recorded in X_2. Thus, neither X_1 nor X_2 encode complete information about X^*, but their sum does. In more detail, assuming that X is decomposed in sub-matrices A, B, C, D as explained in Section 37.3.1, and that X_1, and X_2 are similarly decomposed in sub-matrices E_1, F_1, G_1, H_1 and E_2, F_2, G_2, H_2, the algorithm maintains X_1 and X_2 with the following 8 polynomials using the data structure discussed in Section 37.3.4:

$Q = A + BP^2C$	$E_2 = E_1BH_2^2CE_1$
$F_1 = E_1^2BP$	$F_2 = E_1BH_2^2$
$G_1 = PCE_1^2$	$G_2 = H_2^2CE_1$
$H_1 = PCE_1^2BP$	$R = D + CE_1^2B$

where $P = D^*$, $E_1 = Q^*$, and $H_2 = R^*$ are Kleene closures maintained recursively as smaller instances of the problem of size $n/2 \times n/2$.

To support an insertion of edges around a vertex in V_1, strict updates are performed on polynomials Q, F_1, G_1, and H_1 using `SetRow` and `SetCol`, while E_2, F_2, G_2, and R are updated with `LazySet`. Insertions around V_2 are performed symmetrically, while deletions are supported via `Reset` operations on each polynomial in the recursive decomposition. Finally, P, E_1, and H_2 are updated recursively. The interested reader can find the low-level details of the method in [17].

37.7 All-Pairs Shortest Paths

In this section we survey the best known algorithms for fully dynamic all pairs shortest paths (in short APSP). Given a weighted directed graph G with n vertices and m edges, the problem consists of supporting any intermixed sequence of operations of the following kind:

Update(u, v, w): updates the weight of edge (u, v) in G to the new value w (if $w = +\infty$ this corresponds to edge deletion);

Query(x, y): returns the distance from vertex x to vertex y in G, or $+\infty$ if no path between them exists;

The dynamic maintenance of shortest paths has a remarkably long history, as the first papers date back to 35 years ago [25–27]. After that, many dynamic shortest paths algorithms have been proposed (see, e.g., [28–33]), but their running times in the worst case were comparable to recomputing APSP from scratch.

The first dynamic shortest path algorithms which are provably faster than recomputing APSP from scratch, only worked on graphs with small integer weights. In particular, Ausiello et al. [34] proposed a decrease-only shortest path algorithm for directed graphs having positive integer weights less than C: the amortized running time of their algorithm is $O(Cn \log n)$ per edge insertion. Henzinger et al. [35] designed a fully dynamic algorithm for APSP on planar graphs with integer weights, with a running time of $O(n^{4/3} \log (nC))$ per operation. Fakcharoemphol and Rao in [36] designed a fully dynamic algorithm for single-source shortest paths in planar directed graphs that supports both queries and edge weight updates in $O(n^{4/5} \log^{13/5} n)$ amortized time per edge operation.

The first big step on general graphs and integer weights was made by King [11], who presented a fully dynamic algorithm for maintaining all pairs shortest paths in directed graphs with positive integer weights less than C: the running time of her algorithm is $O(n^{2.5} \sqrt{C \log n})$ per update.

Demetrescu and Italiano [10] gave the first algorithm for fully dynamic APSP on general directed graphs with real weights assuming that each edge weight can attain a limited number S of different *real* values throughout the sequence of updates. In particular, the algorithm supports each update in $O(n^{2.5} \sqrt{S \log^3 n})$ amortized time and each query in $O(1)$ worst-case time. The same authors discovered the first algorithm that solves the fully dynamic all pairs shortest paths problem in its generality [16]. The algorithm maintains explicitly information about shortest paths supporting any edge weight update in $O(n^2 \log^3 n)$ amortized time per operation in directed graphs with non-negative real edge weights. Distance queries are answered with one lookup and actual shortest paths can be reconstructed in optimal time. We note that each update might change a portion of the distance matrix as large as $\Omega(n^2)$. Thus, if the distance matrix has to be maintained explicitly after each update so that queries can be answered with one lookup, $O(n^2)$ is the best update bound one could hope for. Other deletions-only algorithms for APSP, in the simpler case of unweighted graphs, are presented in [37].

37.7.1 Updates in $O(n^{2.5}\sqrt{C \log n})$ Time

In this section we consider the dynamic shortest paths algorithm by King [11]. The algorithm is based on the long paths property discussed in Section 37.3.2 and on the reachability tree data structure Section 36.6.

Similarly to the transitive closure algorithms described in Section 37.6 generalized update operations are supported within the same bounds, that is, insertion (or weight decrease) of a bunch of edges incident to a vertex, and deletion (or weight increase) of any subset of edges in the graph with just one operation.

The main idea of the algorithm is to maintain dynamically all pairs shortest paths up to a distance d, and to recompute longer shortest paths from scratch at each update by stitching together shortest paths of length $\leq d$. For the sake of simplicity, we only consider the case of unweighted graphs: an extension to deal with positive integer weights less than C is described in [11].

To maintain shortest paths up to distance d, similarly to the transitive closure algorithm by King described in Section 37.6, the algorithm keeps a pair of in/out shortest paths trees $IN(v)$ and $OUT(v)$ of depth $\leq d$ rooted at each vertex v. Trees $IN(v)$ and $OUT(v)$ are maintained with the decremental data structure mentioned in Chapter 36. It is easy to prove that, if the distance d_{xy} between any pair of vertices x and y is at most d, then d_{xy} is equal to the minimum of $d_{xv} + d_{vy}$ over all vertices v such that $x \in IN(v)$ and $y \in OUT(v)$. To support updates, insertions of edges around a vertex v are handled by rebuilding only $IN(v)$ and $OUT(v)$, while edge deletions are performed via operations on any trees that contain them. The amortized cost of such updates is $O(n^2 d)$ per operation.

To maintain shortest paths longer than d, the algorithm exploits the long paths property of Theorem 37.5: in particular, it hinges on the observation that, if H is a random subset of $\Theta((n \log n)/d)$ vertices in the graph, then the probability of finding more than d consecutive vertices in a path, none of which are from H, is very small. Thus, if we look at vertices in H as "hubs," then any shortest path from x to y of length $\geq d$ can be obtained by stitching together shortest subpaths of length $\leq d$ that first go from x to a vertex in H, then jump between vertices in H, and eventually reach y from a vertex in H. This can be done by first computing shortest paths only between vertices in H using any cubic-time static all-pairs shortest paths algorithm, and then by extending them at both endpoints with shortest paths of length $\leq d$ to reach all other vertices. This stitching operation requires $O(n^2|H|) = O((n^3 \log n)/d)$ time.

Choosing $d = \sqrt{n \log n}$ yields an $O(n^{2.5}\sqrt{\log n})$ amortized update time. As mentioned in Section 37.3.2, since H can be computed deterministically, the algorithm can be derandomized. The interested reader can find further details on the algorithm in [11].

37.7.2 Updates in $O(n^2 \log^3 n)$ Time

In this section we address the algorithm by Demetrescu and Italiano [16], who devised the first deterministic near-quadratic update algorithm for fully dynamic all-pairs shortest paths. This algorithm is also the first solution to the problem in its generality. The algorithm is based on the notions of historical paths and locally historical paths in a graph subject to a sequence of updates, as discussed in Section 37.3.3.

The main idea is to maintain dynamically the locally historical paths of the graph in a data structure. Since by Lemma 37.3 shortest paths are locally historical, this guarantees that information about shortest paths is maintained as well.

To support an edge weight update operation, the algorithm implements the smoothing strategy mentioned in Section 37.3.3 and works in two phases. It first removes from the data structure all maintained paths that contain the updated edge: this is correct since historical shortest paths, in view of their definition, are immediately invalidated as soon as they are touched by an update. This means that also locally historical paths that contain them are invalidated and have to be removed from the data structure. As a second phase, the algorithm runs an all-pairs modification of Dijkstra's algorithm [38], where at each step a shortest path with minimum weight is extracted from a priority queue and it is combined with existing historical shortest paths to form new locally historical paths. At the end of this phase, paths that become locally historical after the update are correctly inserted in the data structure.

The update algorithm spends constant time for each of the $O(zn^2)$ new locally historical path (see Theorem 37.7). Since the smoothing strategy lets $z = O(\log n)$ and increases the length of the sequence of updates by an additional $O(\log n)$ factor, this yields $O(n^2 \log^3 n)$ amortized time per update. The interested reader can find further details about the algorithm in [16].

37.8 Conclusions

In this chapter we have surveyed the algorithmic techniques underlying the fastest known dynamic graph algorithms for several problems, both on undirected and on directed graphs. Most of the algorithms that we have presented achieve bounds that are close to the best possible. In particular, we have presented fully dynamic algorithms with polylogarithmic amortized time bounds for connectivity and minimum spanning trees [18] on undirected graphs. It remains an interesting open problem to show whether polylogarithmic update bounds can be achieved also in the worst case: we recall that for both problems the current best worst-case bound is $O(\sqrt{n})$ per update, and it is obtained with the sparsification technique [5] described in Section 37.2.

For directed graphs, we have shown how to achieve constant-time query bounds and nearly-quadratic update bounds for transitive closure and all pairs shortest paths. These bounds are close to optimal in the sense that one update can make as many as $\Omega(n^2)$ changes to the transitive closure and to the all pairs shortest paths matrices. If one is willing to pay more for queries, Demetrescu and Italiano [17] have shown how to break the $O(n^2)$ barrier on the single-operation complexity of fully dynamic transitive closure for directed acyclic graphs. This also yields the first efficient update algorithm that maintains reachability in acyclic directed graphs between two fixed vertices s and t, or from s to all other vertices. However, in the case of general graphs or shortest paths, no solution better that the static is known for these problems. In general, it remains an interesting open problem to show whether effective query/update tradeoffs can be achieved for general graphs and for shortest paths problems.

Finally, dynamic algorithms for other fundamental problems such as matching and flow problems deserve further investigation.

Acknowledgments

This work has been supported in part by the IST Programmes of the EU under contract numbers IST-1999-14186 (ALCOM-FT) and IST-2001-33555 (COSIN), and by the Italian Ministry of University and Scientific Research (Project "ALINWEB: Algorithmics for Internet and the Web").

References

1. G. N. Frederickson. Data structures for on-line updating of minimum spanning trees. *SIAM J. Comput.*, 14:781–798, 1985.
2. G. N. Frederickson. Ambivalent data structures for dynamic 2-edge-connectivity and k smallest spanning trees. *SIAM J. Comput.*, 26(2):484–538, 1997.
3. Z. Galil and G. F. Italiano. Fully-dynamic algorithms for 2-edge connectivity. *SIAM J. Comput.*, 21:1047–1069, 1992.
4. M. Rauch. Fully dynamic biconnectivity in graphs. *Algorithmica*, 13:503–538, 1995.
5. D. Eppstein, Z. Galil, G. F. Italiano, and A. Nissenzweig. Sparsification – A technique for speeding up dynamic graph algorithms. *J. Assoc. Comput. Mach.*, 44:669–696, 1997.
6. M. R. Henzinger and V. King. Randomized fully dynamic graph algorithms with polylogarithmic time per operation. *J. Assoc. Comput. Mach.*, 46(4):502–536, 1999.
7. T. H. Cormen, C. E. Leiserson, R. L. Rivest, and C. Stein. *Introduction to Algorithms*, Second Edition. The MIT Press, Cambridge, MA, 2001.
8. I. Munro. Efficient determination of the transitive closure of a directed graph. *Information Processing Letters*, 1(2):56–58, 1971.
9. D. H. Greene and D. E. Knuth. *Mathematics for the Analysis of Algorithms*. Birkhäuser, Boston-Basel-Stuttgart, 1982.
10. C. Demetrescu and G. F. Italiano. Fully dynamic all pairs shortest paths with real edge weights. In *Proc. of the 42nd IEEE Annual Symposium on Foundations of Computer Science (FOCS'01)*, Las Vegas, Nevada, pages 260–267, 2001.
11. V. King. Fully dynamic algorithms for maintaining all-pairs shortest paths and transitive closure in digraphs. In *Proc. 40th Symposium on Foundations of Computer Science (FOCS 99)*, 1999.
12. J. D. Ullman and M. Yannakakis. High-probability parallel transitive-closure algorithms. *SIAM J. Comput.*, 20(1):100–125, 1991.
13. U. Zwick. All pairs shortest paths in weighted directed graphs - exact and almost exact algorithms. In *Proc. of the 39th IEEE Annual Symposium on Foundations of Computer Science (FOCS'98)*, Los Alamitos, CA, November 8–11, pages 310–319, 1998.
14. V. Chvátal. A greedy heuristic for the set-covering problem. *Math. Oper. Res.*, 4(3):233–235, 1979.
15. L. Lovász. On the ratio of optimal integral and fractional covers. *Discrete Math.*, 13:383–390, 1975.
16. C. Demetrescu and G. F. Italiano. A new approach to dynamic all pairs shortest paths. In *Proc. 35th Symp. on Theory of Computing (STOC'03)*, 2003.

17. C. Demetrescu and G. F. Italiano. Fully dynamic transitive closure: Breaking through the $O(n^2)$ barrier. In *Proc. of the 41st IEEE Annual Symposium on Foundations of Computer Science (FOCS'00)*, pages 381–389, 2000.

18. J. Holm, K. de Lichtenberg, and M. Thorup. Poly-logarithmic deterministic fully-dynamic algorithms for connectivity, minimum spanning tree, 2-edge, and biconnectivity. *J. Assoc. Comput. Mach.*, 48(4):723–760, 2001.

19. M. R. Henzinger and V. King. Maintaining minimum spanning trees in dynamic graphs. In *Proc. 24th Int. Colloquium on Automata, Languages and Programming (ICALP 97)*, pages 594–604, 1997.

20. D. Coppersmith and S. Winograd. Matrix multiplication via arithmetic progressions. *J. Symb. Comput.*, 9:251–280, 1990.

21. S. Khanna, R. Motwani, and R. H. Wilson. On certificates and lookahead on dynamic graph problems. *Algorithmica*, 21(4):377–394, 1998.

22. V. King and G. Sagert. A fully dynamic algorithm for maintaining the transitive closure. In *Proc. 31-st ACM Symposium on Theory of Computing (STOC 99)*, pages 492–498, 1999.

23. L. Roditty. A faster and simpler fully dynamic transitive closure. In *Proceedings of the 14th ACM-SIAM Symposium on Discrete Algorithms (SODA'03)*, Baltimore, Maryland, USA, January 12–14, pages 404–412, 2003.

24. L. Roditty and U. Zwick. Improved dynamic reachability algorithms for directed graphs. In *Proceedings of the 43th Annual IEEE Symposium on Foundations of Computer Science (FOCS'02)*, Vancouver, Canada, pages 679, 2002.

25. P. Loubal. A network evaluation procedure. *Highway Research Record 205*, pages 96–109, 1967.

26. J. Murchland. The effect of increasing or decreasing the length of a single arc on all shortest distances in a graph. Technical report, LBS-TNT-26, London Business School, Transport Network Theory Unit, London, UK, 1967.

27. V. Rodionov. The parametric problem of shortest distances. *U.S.S.R. Computational Math. and Math. Phys.*, 8(5):336–343, 1968.

28. S. Even and H. Gazit. Updating distances in dynamic graphs. *Method Oper. Res.*, 49:371–387, 1985.

29. D. Frigioni, A. Marchetti-Spaccamela, and U. Nanni. Semi-dynamic algorithms for maintaining single source shortest paths trees. *Algorithmica*, 22(3):250–274, 1998.

30. D. Frigioni, A. Marchetti-Spaccamela, and U. Nanni. Fully dynamic algorithms for maintaining shortest paths trees. *J. Algorithm*, 34:251–281, 2000.

31. G. Ramalingam and T. Reps. An incremental algorithm for a generalization of the shortest path problem. *J. Algorithm*, 21:267–305, 1996.

32. G. Ramalingam and T. Reps. On the computational complexity of dynamic graph problems. *Theor. Comput. Sci.*, 158:233–277, 1996.

33. H. Rohnert. A dynamization of the all-pairs least cost problem. In *Proc. 2nd Annual Symposium on Theoretical Aspects of Computer Science, (STACS 85)*, LNCS 182, pages 279–286, 1985.

34. G. Ausiello, G. F. Italiano, A. Marchetti-Spaccamela, and U. Nanni. Incremental algorithms for minimal length paths. *J. Algorithm*, 12(4):615–38, 1991.

35. M. R. Henzinger, P. Klein, S. Rao, and S. Subramanian. Faster shortest-path algorithms for planar graphs. *J. Comput. Syst. Sci.*, 55(1):3–23, August 1997.

36. J. Fakcharoemphol and S. Rao. Planar graphs, negative weight edges, shortest paths, and near linear time. In *Proc. of the 42nd IEEE Annual Symposium on Foundations of Computer Science (FOCS'01)*, Las Vegas, Nevada, pages 232–241, 2001.

37. S. Baswana, R. Hariharan, and S. Sen. Improved decremental algorithms for transitive closure and all-pairs shortest paths. In *Proc. 34th ACM Symposium on Theory of Computing (STOC'02)*, pages 117–123, 2002.

38. E. W. Dijkstra. A note on two problems in connexion with graphs. *Numer. Math.*, 1:269–271, 1959.

38

Succinct Representation of Data Structures [*]

38.1	Introduction	595
38.2	Bitvector	596
38.3	Succinct Dictionaries	597
	Indexable Dictionary • Fully Indexable Dictionary • Dynamic Dictionary	
38.4	Tree Representations	599
	Binary Trees • Ordinal Trees • Cardinal Trees • Dynamic Binary Trees	
38.5	Graph Representations	603
38.6	Succinct Structures for Indexing	603
38.7	Permutations and Functions	604
	Permutations • Functions	
38.8	Partial Sums	606
38.9	Arrays	607
	Resizable Arrays • Dynamic Arrays	
38.10	Conclusions	608
	References	608

J. Ian Munro
University of Waterloo

S. Srinivasa Rao
Seoul National University

38.1 Introduction

Although computer memories, at all levels of the hierarchy, have grown dramatically over the past few years, increased problem sizes continues to outstrip this growth. Minimizing space is crucial not only in keeping data in the fastest memory possible, but also in moving it from one level to another, be it from main memory to cache or from a web site around the world. Standard data compression, say Huffman code or grammar based code, applied to a large text file reduces space dramatically, but basic operations on the text require that it be fully decoded.

In this chapter we focus on representations that are not only terse but also permit the basic operations one would expect on the underlying data type to be performed quickly. Jacobson [1] seems to have been the first to apply the term *succinct* to such structures; the goal is to use the information-theoretic minimum number of bits and to support the expected operations on the data type in optimal time. Our archetypical example (discussed in Section 38.4) is the representation of a binary tree. Suppose, we would like to support the operations of navigating through a binary tree moving to either child or the parent of the current node, asking the size of the subtree rooted at the current node or giving the unique "number" of the node so that data can be stored in that position of an array. At $\lg n$ bits per reference, this adds up to at least $5n \lg n$ bits. However, there are only $\binom{2n+1}{n}/(2n+1)$ binary trees, so the information-theoretic minimum space is fewer than $2n$ bits. Our archetypical data structure is a $2n + o(n)$-bit representation that supports the operations noted above, and others, in constant time.

We consider a variety of abstract data types, or combinatorial objects, with the goal of producing such succinct data structures. Most, though not all, of the structures we consider are static. In most cases the construction of a succinct data structure from the standard representation is fairly straightforward in linear time.

Memory Model: We study the problems under the RAM model with word size $\Theta(\lg n)$, where n is the input size of the problem under consideration. This supports arithmetic (addition, subtraction, multiplication and division), indexing and bit-wise boolean operations (AND, OR, NOT, XOR etc.) on words, and reading/writing of words from/to the memory in constant time.

---

[*] This chapter has been reprinted from first edition of this Handbook, without any content updates.

38.2 Bitvector

A bitvector provides a simple way to represent a set from any universe that is easily mapped onto $[m]$.* Membership queries (checking whether a given element from the universe is present in the set) can be answered in constant time (in fact a single bit probe) using a bitvector. Furthermore, one can easily support updates (inserting and deleting elements) in constant time. The most interesting twist on the bitvector came with Jacobson [1] considering two more operations:

- rank(i): return the number of 1s before the position i, and
- select(i): return the position of the ith 1.

As we shall see, these operations are crucial to a number of more complex structures supporting a variety of data types. An immediate use is to support the queries:

- predecessor(x): find the largest element $y \leq x$ in the set S,
- successor(x): find the smallest element $y \geq x$ in the set S.

Given a bitvector of length m, Jacobson [1] gave a structure that takes $o(m)$ bits of additional space and supports rank and select operations by making $O(\lg m)$ bit probes to the structure. On a RAM with word size $\Theta(\lg m)$ bits, the structure given by Munro [2] enhanced this structure and the algorithms to support the operations in $O(1)$ time, without increasing the space bound. We briefly describe the details of this structure (Figure 38.1).

The structure for computing rank, the rank directory, consists of the following:

- Conceptually break the bitvector into blocks of length $\lceil \lg^2 m \rceil$. Keep a table containing the number of 1s up to the last position in each block. This takes $O(m/\lg m)$ bits of space.
- Conceptually break each block into sub-blocks of length $\lceil (1/2) \lg m \rceil$. Keep a table containing the number of 1s within the block up to the last position in each sub-block. This takes $O(m \lg \lg m/\lg m)$ bits.
- Keep a precomputed table giving the number of 1s up to every possible position in every possible distinct sub-block. Since there $O(\sqrt{m})$ distinct possible blocks, and $O(\lg m)$ positions in each, this takes $O(\sqrt{m} \lg m \lg \lg m)$ bits.

Thus, the total space occupied by this auxiliary structure is $o(m)$ bits. The rank of an element is, then, simply the sum of three values, one from each table.

The structure for computing select uses three levels of directories and is more complex. The first one records the position of every $(\lg m \lg \lg m)$th 1 bit in the bitvector. This takes $O(m/\lg \lg m)$ bits. Let r be the subrange between two values in the first directory, and consider the sub-directory for this range. If $r \geq (\lg m \lg \lg m)^2$ then explicitly store the positions of all ones, which requires $O(r/\lg \lg m)$ bits. Otherwise, subdivide the range and store the position (relative to the beginning of this range) of every $(\lg r \lg \lg m)$th one bit in the second level directory. This takes $O(r/\lg \lg m)$ bits for each range of size r, and hence $O(m/\lg \lg m)$ bits over the entire bitvector. After one more level of similar range subdivision, the range size will reduce to at most $(\lg \lg m)^4$.

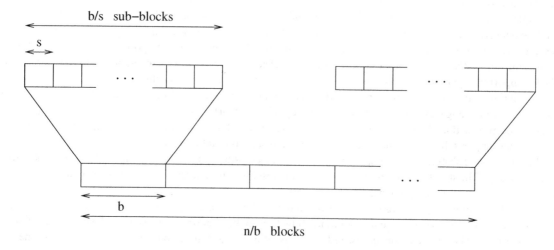

FIGURE 38.1   Two-level rank directory.

---

* for positive integers m, $[m]$ denotes the set $\{0,1,\ldots,m-1\}$

Computing select on these small ranges is performed using a precomputed table. The total space occupied by this auxiliary structure is $o(m)$ bits. The query algorithm is straightforward. See [3,4] for details.

This "indexable bitvector" is used as a substructure in several succinct data structures. To represent a bitvector of length m, it takes $m + o(m)$ bits of space. In general, if nothing is known about the bitvector then any representation needs at least m bits to distinguish between all possible bitvectors, and hence this is close to the optimal space. But if we also know the density (the number of ones) of the bitvector, then the space bound is no longer optimal, in general. The "fully indexable dictionary" described in Section 38.3.2 gives a solution that takes nearly optimal space.

Using the ideas involved in constructing rank and select directories, one can also support the following generalizations of these two operations, using $o(m)$ bits of extra space: Given a bitvector of length m, and a fixed binary pattern p of length up to $(1 - \epsilon) \lg m$, for some fixed constant $0 < \epsilon < 1$

- $\mathrm{rank}_p(i)$: return the number of (possibly overlapping) occurrences of p before the position i, and
- $\mathrm{select}_p(i)$: return the ith occurrence of the pattern p.

One can extend the ideas of rank and select directories to support indexing into a fixed or variable length encoded text (e.g., Huffman coding, prefix-free encoding etc.) in constant time, using negligible extra space. See [1,4] for some examples.

38.3 Succinct Dictionaries

The (static) dictionary problem is to store a subset S of size n so that membership queries can be answered efficiently. In our case, the universe is taken to be the set $[m]$. This problem has been widely studied and various solutions have been proposed to support membership queries in constant time.

As we have seen in the last section, a bitvector is a simple way of representing a set from a given universe. But this requires m bits of space. Since there are $\binom{m}{n}$ sets of size n from a universe of size m, one would require only $\mathcal{B} \equiv \lg \binom{m}{n}$ ($\approx n(\lg m - \lg n + \lg e)$, when $n = o(m)$) bits to store a canonical representation of any such set. Thus a bitvector is quite wasteful of space when the set is sparse. A sorted array is another simple representation, but it requires $\Theta(\lg n)$ time to answer queries. A *fusion tree* (see Chapter 40) also takes linear space and supports membership queries in $\Theta(\lg n / \lg \lg n)$ time. In this section, we consider representations of sets whose space complexity is close to the information theoretic minimum and support queries in constant time. (As all the structures outlined below support membership queries in worst case constant time, we do not mention the query complexity explicitly.)

Fredman, Komlós and Szemerédi [5] gave the first linear space structure for the static dictionary problem. This takes $n \lg m + O(n\sqrt{\lg n} + \lg \lg m)$ bits of space. The lower order term was later improved by Schmidt and Siegel [6] to $O(n + \lg \lg m)$. This structure uses a universe reduction function followed by a two-level hash function to hash the given subset one-to-one onto the set $[n]$, and stores the elements of subset in a hash table (in the order determined by the hash function). The hash table takes $n \lg m$ bits and a clever encoding of the hash function takes $O(n + \lg \lg m)$ bits of space. We refer to this as the FKS hashing scheme. Note that the space required for this structure is $\Theta(n \lg n)$ bits more than the optimal bound of $\mathcal{B}$ bits.

Brodnik and Munro [7] gave a static dictionary representation that takes $\mathcal{B} + o(\mathcal{B})$ bits of space. It uses two different solutions depending on the relative values of n and m. When the set is relatively sparse (namely, when $n \le m/(\lg m)^{\lg \lg m}$), it partitions the elements into buckets based on the first $\lg n - \lg \lg m$ bits of their bit representations, and store explicit pointers to refer to the representations of individual buckets. Each bucket is represented by storing all the elements that fall into it in a perfect hash table for that bucket. Otherwise, when the set is dense, it uses two levels of bucketing (at each level splitting the universe into a number of equal-range buckets, depending only on the universe size) after which the range of these buckets reduces to $\Theta(\lg n)$. These small buckets are stored (almost) optimally by storing pointers into a precomputed table that contains all possible small buckets. In either case the space occupancy can be shown to be $\mathcal{B} + o(\mathcal{B})$ bits.

Pagh [8] observed that each bucket of the hash table may be resolved with respect to the part of the universe hashing to that bucket. Thus, one can save space by compressing the hash table, storing only the *quotient* information, rather than the element itself. From the FKS hash function, one can obtain a quotienting function that takes $\lg(m/n) + O(1)$ bits for each element. Using this idea one can obtain a dictionary structure that takes $n \lg(m/n) + O(n + \lg \lg m)$ bits of space, which is only $\Theta(n)$ bits more than the information-theoretic lower bound (except for the $O(\lg \lg m)$ term). Pagh has also given a dictionary structure that takes only $\mathcal{B} + o(n) + O(\lg \lg m)$ bits of space.

38.3.1 Indexable Dictionary

One useful feature of the sorted array representation of a set is that, given an index i, the ith smallest element in the set can be retrieved in constant time. Furthermore, when we locate an element in the array, we immediately know its rank (the number of elements in the set which are less than the given element). On the other hand, hashing based schemes support membership in constant time, but typically do not maintain the ordering information. In this section we look at a structure that combines the good features of both these approaches.

An *indexable dictionary* is a structure representing a set S of size n from the universe $[m]$ to support the following queries in constant time:

- rank(x, S): Given $x \in [m]$, return -1 if $x \notin S$, and $|\{y \in S | y < x\}|$ otherwise, and
- select(i, S): Given $i \in \{1, \ldots n\}$, return the ith smallest element in S.

Here the rank operation is only supported for the elements present in the set S. Ajtai [9] showed that the more general problem of supporting rank for every element in the universe has a query lower bound of $\Omega(\lg \lg n)$, even if the space used is polynomial in n. As a consequence, we emphasize the need for handling both S and its complement in the next section.

A dictionary that supports rank operation [10], as well as an indexable dictionary is very useful in representing trees [11] (see Section 38.4.3).

Elias [12] considered the indexable dictionary problem and gave a representation that takes $n \lg m - n \lg n + O(n)$ bits and supports the queries in $O(1)$ time, though he only considered the average case time complexity of the queries. Raman et al. [11] have given an indexable dictionary structure that takes $\mathcal{B} + o(n) + O(\lg \lg m)$ bits. The main idea here is, again, to partition the elements into buckets based on their most significant bits, as in the static dictionary structure of Brodnik and Munro [7]. The difference is that instead of storing explicit pointers to the bucket representations, they store the bucket sizes using a succinct representation that supports partial sum queries (see Section 38.8) in constant time. This not only saves a significant amount of space, but also provides the extra functionality needed for supporting rank and select.

Using similar ideas, one can also represent multisets and collections of sets using almost optimal space. See [11] for details.

38.3.2 Fully Indexable Dictionary

Given a set $S \subseteq [m]$, a *fully indexable dictionary* (FID) of S is a representation that supports rank and select operations on both S and its complement $\bar{S} = [m] \setminus S$ in constant time [11].

It is easy to see that the bitvector representation of a set, with auxiliary structures to support rank and select on both the bits as mentioned in Section 38.2, is an FID. But this requires $m + o(m)$ bits, where m is the size of the universe. Here we look at an FID representation that takes $\mathcal{B} + o(m)$ bits of space. Note that when the set is reasonably sparse (namely when $n = m/\omega(\lg m)$) $\mathcal{B} = o(m)$, and hence it improves the space complexity of the bitvector representation.

Let $S \subseteq [m]$ be a given set of size n. Divide $[m]$ into blocks of consecutive elements, with block size $u = \lfloor \frac{1}{2} \lg m \rfloor$. Let S_i be the subset of S that falls into the ith block. Each of the S_i's is represented by storing an index into a table that contains the characteristic bitvectors of all possible subsets of a particular size from the universe $[u]$. As a consequence, the space occupied by these representations together with all the precomputed tables can be shown to be $\mathcal{B} + o(m)$ bits. To enable fast access to the representations of these subsets, we store the partial sums of the sizes of the subsets, and also the partial sums of the lengths of the representations of these subsets, which take $O(m \lg \lg m / \lg m)$ bits. This can be used to support rank in $O(1)$ time.

To support select, we first store the positions of every $(\lg^2 m)$th element explicitly in an array, which takes $O(m/\lg m)$ bits. Call the part the universe that lies between two successive elements in this array a *segment*. If the size of a segment is more than $\lg^4 m$, then we explicitly store all the $\lg^2 m$ elements of S that belong to this segment in sorted order. This takes $\lg^3 m$ bits for every such "sparse" segment, and hence at most $m/\lg m$ bits, over all the sparse segments. Dense segments are handled by constructing a complete tree with branching factor $\sqrt{\lg m}$, and so constant height, whose leaves are the blocks that constitute this segment, and storing some additional information to navigate this tree efficiently (see the *searchable partial sum* structure in Section 38.8).

To support rank and select on $\bar{S}$, first observe that an implicit representation of a set over a given universe is also an implicit representation of its complement. Thus, we need not store the implicit representations of $\bar{S}_i$ again. Except for this, we repeat the above construction with S_i's replaced by $\bar{S}_i$'s.

The overall space requirement of the structure is $\mathcal{B} + O(m \lg \lg m / \lg m)$ bits, and rank and select are supported on both S and $\bar{S}$ in $O(1)$ time. See [11] for details.

38.3.3 Dynamic Dictionary

We have looked at several succinct structures for static dictionaries. We now briefly consider the dynamic dictionary problem where one can add and delete elements from the set while supporting the membership queries.

Model: The model of memory allocation is very important in dynamic data structures. One widely used model [11,13,14] is to assume the existence of a "system" memory manager that would allocate and free memory in variable-sized chunks. In this model, the space complexity of a structure is counted as the total size of all the blocks allocated for that structure, and hence this approach does not account for the space wastage due to *external fragmentation*.

Fundamentally, memory is most easily viewed as a large array. If we are to use the storage, we must manage it. Therefore a simple view is to count all the fragmentation we may cause and count the memory usage as the difference between the addresses of the first and last locations used by the structure. While more complex scenarios may be more realistic in certain cases, we take this simple address difference model as our focus. The methods we discuss are equivalent under either model up to constant factors.

A balanced tree can be used to support all the dynamic dictionary operations in $O(\lg n)$ time using $n \lg m + O(n \lg n)$ bits, where n is the current size of the set. Using the ideas of the FKS dictionary, Dietzfelbinger et al. [15] gave a dynamic dictionary structure that supports membership in $O(1)$ time and updates (insert/delete) in $O(1)$ expected amortized time. This structure takes $O(n \lg m)$ bits of space. There have been several improvements, lowering the space complexity close to the information theoretic-minimum, culminating in a structure that takes $B + o(B)$ bits with the same query complexity as above. See [7,8,16–18] and the references therein.

All these structures also support associating satellite information with the elements, so that whenever an element is found to be in the set, we can also retrieve the satellite information associated with it in constant time.

38.4 Tree Representations

Trees are one of the most fundamental objects in computer science. We consider the problem of representing large trees succinctly. Storing a tree with a pointer per child as well as other structural information can account for the dominant storage cost. For example, standard representations of a binary tree on n nodes, using pointers, take $O(n \lg n)$ bits of space. Since there are only $\binom{2n+1}{n}/(2n+1)$ different binary trees on n nodes, less than $2n$ bits suffice to distinguish between them. We look at some binary tree representations that take $2n + o(n)$ bits and support the basic navigational operations in constant time.

38.4.1 Binary Trees

First, if the tree is a complete binary tree (i.e., a binary tree in which every level, except possibly the deepest, is completely filled, and the last level is filled from the left as far as required), then there is a unique tree of a given size and we require no additional space to store the tree structure. In fact, by numbering the nodes from 1 to n in the "heap order" [19] (left-to-right level-order traversal of the tree), one can support navigational operations on the tree by observing that the parent of a node numbered i is the node numbered $\lfloor i/2 \rfloor$, and the left and right children of node i are $2i$ and $2i + 1$ respectively. But this property does not hold when the tree is not complete.

If the tree is not complete, one could extend it to a complete binary tree with the same height and store a bit vector indicating which nodes are present in the tree (in the heap order of the complete tree) to support the operations efficiently. But this takes space exponential in the number of nodes, in the worst case.

To save space, one can use the following compressed representation due to Jacobson [1]: First, mark all the nodes of the tree with 1 bits. Then add external nodes to the tree, and mark them with 0 bits. Construct a bitvector by reading off the bits that are marking the nodes in left-to-right level-order. (See Figure 38.2.) It is easy to see that the original tree can be reconstructed from this bitvector. For a binary tree with n nodes, this bitvector representation takes $2n + 1$ bits. Moving between parent and child is just a slight twist on the method used in a heap. By storing the rank and select directories for this bitvector, one can support the navigational operations in constant time using the following equations:

$$\text{parent}(i) = \text{select}(\lfloor i/2 \rfloor); \quad \text{leftchild}(i) = 2 \cdot \text{rank}(i); \quad \text{rightchild}(i) = 2 \cdot \text{rank}(i) + 1.$$

38.4.2 Ordinal Trees

Now, consider optimal representations of trees of higher degree, of which there are two different notions.

An *ordinal tree* is a rooted tree of arbitrary degree in which the children of each node are ordered. Ordinal trees on n nodes are in one to one correspondence with binary trees on n nodes. Hence about $2n$ bits are necessary to represent an arbitrary ordinal tree on n nodes. A *cardinal tree* of degree k is a rooted tree in which each node has k positions for an edge to a child. Hence, a binary tree is a cardinal tree of degree 2. There are $C_n^k \equiv \binom{kn+1}{n}/(kn+1)$ cardinal trees of degree k on n nodes [20]. Hence we need roughly $(\lg(k-1) + k \lg \frac{k}{(k-1)})n$ bits to represent an arbitrary such tree.

The basic operations we would like to support on tree representations are: given a node, finding its parent, ith child, degree and the size of the subtree rooted at that node (subtree size). For the cardinal trees we also need to support the additional operation of finding a child with a given label.

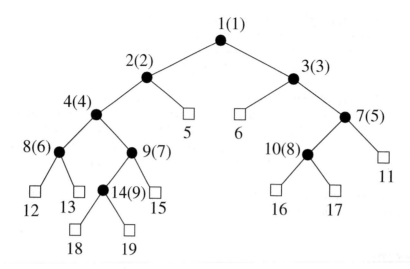

level–order bitmap: 1 1 1 1 0 0 1 1 1 1 0 0 0 1 0 0 0 0 0

FIGURE 38.2 Level-order bitmap representation of a binary tree.

We outline three different representations of an ordinal tree. All the three representations map the n nodes of the tree onto the integers $1, \ldots, n$, and hence all are appropriate for applications in which data is to associated with nodes or leaves.

Level-order unary degree sequence representation: A rooted ordered tree can be represented by storing its degree sequence in any of a number of standard orderings of the nodes. The ordinal tree encoding of Jacobson [1] represents a node of degree d as a string of d **1**s followed by a **0**. Thus the degree of a node is represented by a binary prefix code. These prefix codes are then written in a level-order traversal of the entire tree. Using auxiliary structures to support rank and select operations on this sequence, one can support finding the parent, the ith child and the degree of any node in constant time. Thus, it gives a representation that takes $2n + o(n)$ bits of space and supports the above three operations in constant time, for an ordered tree on n nodes (Figure 38.3).

Balanced parenthesis representation: The tree encoding of Munro and Raman [21] uses a balanced sequence of parentheses to represent an ordinal tree. This balanced representation is derived from the depth-first traversal of the tree, writing an open parenthesis on the way down and a close parenthesis on the way up. Thus, a tree on n nodes can be represented by a balanced parenthesis sequence of length $2n$. Extending the ideas of Jacobson, they showed how to support the following operations in $O(1)$ time, using negligible extra space ($o(n)$ bits):

- findopen/findclose(i): find the position of the open/close parenthesis matching the given close/open parenthesis in position i.
- excess(i): find the difference between the number of open and closing parentheses before the position i.
- enclose(i): given a parenthesis pair whose open parenthesis is in position i, find the open parenthesis corresponding to its closest enclosing matching parenthesis pair.

The parent of a node can be found in constant time using the enclose operation. In the parenthesis representation, the nodes of a subtree are stored together, which enables us to support the operation of finding the size of the subtree rooted at a given node in constant time. The problem with this representation is that finding the ith child takes $\Theta(i)$ time.

Depth-first unary degree sequence representation: Jacobson's representation allows access to the ith child in constant time, whereas Munro and Raman's representation supports subtree size operation in constant time. To combine the virtues of these two representations, Benoit et al. [22] used a representation that writes the unary degree sequence of each node in the depth-first traversal order of the tree. The representation of each node contains essentially the same information as in Jacobson's level-order degree sequence, but written in a different order. Thus, it gives another $2n$ bit encoding of a tree on n nodes. Replacing the **0**'s and **1**'s by open and close parentheses respectively, and adding an extra open parenthesis at the beginning, creates a string of balanced parentheses. Using auxiliary structures to support rank and select operations on this bit string and also the operations on balanced parenthesis sequences defined above, one can support finding the parent, ith child, degree and subtree size of a given node in constant time.

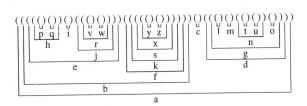

(b) Jacobson's degree sequence representation.

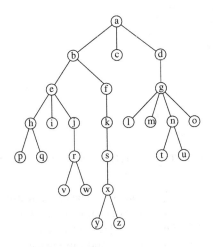

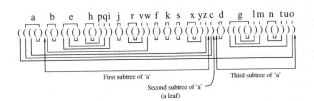

(a) The Ordinal Tree

(c) Munro and Raman's Balanced Parentheses Representation.

(d) Depth-first degree sequence representation.

FIGURE 38.3 Three encodings of an ordinal tree. (a) The Ordinal Tree, (b) Jacobson's degree sequence representation, (c) Munro and Raman's Balanced Parentheses Representation, (c) Depth-first degree sequence representation.

Other operations: Sadakane [23] has shown that the parenthesis representation of an ordinal tree can be used to support least common ancestor queries in $O(1)$ time using a $o(n)$-bit auxiliary structure. Munro and Rao [24] have shown that one can also support the level ancestor queries in $O(1)$ time, using an additional $o(n)$ bit auxiliary structure by storing the parenthesis representation. Geary et al. [25] obtained another structure that takes $2n + o(n)$ bits and supports level-ancestor queries, in addition to all the other navigational operations mentioned above in $O(1)$ time.

38.4.3 Cardinal Trees

A simple cardinal tree encoding can be obtained as follows: Encode each node of a k-ary tree by k bits, where the ith bit specifies whether child i is present. These can be written in any fixed ordering of the tree nodes, such as level order or depth-first order, to obtain the tree encoding. By storing the rank and select directories for this bitvector encoding, one can support parent, ith child and degree queries in constant time. This encoding has the major disadvantage of taking kn bits, far from the lower bound of roughly $(\lg k + \lg e)n$ bits, as there are $C_n^k \equiv \binom{kn+1}{n}/(kn+1)$ k-ary cardinal trees on n nodes.

Using some probabilistic assumptions, Darragh et al. [26] have implemented a structure that takes $\lg k + O(1)$ bits per node, though the implementation treats $\lg \lg n$ as "a constant" (indeed 5). This structure supports the navigational operations in constant expected time and also supports updates "efficiently" (compared with other linear space representations), and was also shown to perform well in practice.

To achieve a better space bound with good worst-case performance, one can use the ordinal tree encoding to store the underlying tree, and store some additional information about which children are present at each node. The ordinal information (using the depth-first unary degree sequence representation) can be used to support the parent, ith child, degree and subtree size queries in constant time.

Let $S_x = \{i_1, i_2, \ldots, i_d\}$ be the child labels of a node x with degree d in the cardinal tree. To find the child labeled j of node x, it suffices to find $i = \text{rank}(j)$ in the set S_x, if $j \in S_x$. If $i = -1$ (i.e., $j \notin S_x$), then there is no child labeled j at node x, otherwise the ith child of x is the child labeled j of node x. The ith child can be found using the ordinal information. Storing each of these sets S_x using the indexable dictionary representation of Section 38.3.1, which takes $d \lg k + o(d) + O(\lg \lg k)$ bits for each S_x, requires $n \lg k + o(n) + O(n \lg \lg k)$ bits in the worst case. Using a representation that stores a collection of indexable dictionaries efficiently [11], one can reduce the space consumption to $n \lg k + o(n) + O(\lg \lg k)$ bits.

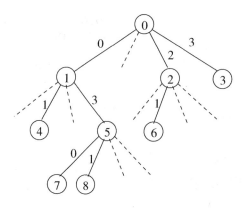

FIGURE 38.4   The tree is represented by storing an indexable dictionary of the set $\{\langle 0,0\rangle, \langle 0,2\rangle, \langle 0,3\rangle, \langle 1,1\rangle, \langle 1,3\rangle, \langle 2,1\rangle, \langle 5,0\rangle, \langle 5,1\rangle\}$.

Thus, this structure uses $2n + o(n)$ bits to represent the underlying ordinal tree, $n \lg k + o(n + \lg k)$ bits to represent the labels of the children at each node, and supports all the navigational operations and the subtree size operation in $O(1)$ time.

Using the succinct indexable dictionary structure mentioned in Section 38.3, Raman et al. [11] obtained an optimal space cardinal tree representation. The main idea is to store the set of all pairs, $\langle i, j\rangle$ such that the ith node, in the level-order of the nodes, has a child labeled j, using an indexable dictionary representation. (See Figure 38.4 for an example.) Since this set is of size n from the universe $[nk]$, it requires $\lg \binom{nk}{n} + o(n + \lg k) = C_n^k + o(n + \lg k)$ bits to store an indexable dictionary for this set. One can easily map the navigational operations on the tree to the operations on this set, to support them in constant time. But this structure does not support the subtree size operation efficiently.

38.4.4 Dynamic Binary Trees

All the tree representations mentioned so far are static. Even to make a minor modification to the tree, such as adding a leaf, the entire structure has to be reconstructed (in the worst case). In this section we look at some representations that are more efficient in supporting updates to the tree.

Munro et al. [14] gave a binary tree representation that takes $2n + o(n)$ bits, and supports parent, left child, right child and subtree size operations in $O(1)$ time. Updating the tree (adding a leaf or adding a node along an edge) requires $O(\lg^c n)$ time, for some constant $c \geq 1$ which depends on the size of the data associated with the nodes. Extending some of the ideas involved in this, Raman and Rao [18] improved the update time to $O((\lg \lg n)^{\epsilon})$, for any fixed $\epsilon > 0$, while maintaining the other time and space bounds (Figure 38.5).

We briefly outline the key issues involved in the construction of these structures. First, we divide the tree into blocks of size $\Theta(\lg^c n)$, for some $c \geq 2$, and each block in turn into sub-blocks of size $\epsilon \lg n$, for some fixed $\epsilon < 1$. The sub-blocks are stored using an implicit representation and are operated upon using precomputed tables. The block structure of the tree is stored using explicit pointers. Since there are only $\Theta(\lg^{c-1} n)$ sub-blocks in each block, we can store the sub-block structure within a block explicitly using $\Theta(\lg \lg n)$ sized pointers. Each block stores its parent block and the size, using a constant number of words. Thus,

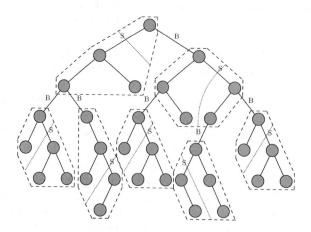

FIGURE 38.5   Dynamic binary tree representation. B denotes an inter-block pointer and S denotes an inter-subblock pointer.

the overall block structure of the tree is easily handled by conventional means (storing explicit pointers) and only takes $O(n/\lg n)$ bits. The blocks and sub-blocks tolerate some slack in their sizes and are moved to appropriate sized areas to avoid wasting space. Ultimately, the key issues boil down to the memory management.

To support subtree size, we maintain the the subtree sizes of the roots of all blocks and sub-blocks. Since each update changes the subtree sizes of several nodes, it is not possible to update all the effected blocks and sub-blocks in constant time, in general. For this reason, we assume that the navigation through the tree begins at the root and may end at any point (or at the root, to achieve worst-case constant time for updates), and navigates the tree by moving from a node only to either its parent or one of its children. Hence, updates to a node higher in the tree regarding the insertions and deletions to descendants are made on return to that node.

38.5 Graph Representations

In this section, we briefly describe some space efficient representations of graphs. In particular, we consider representations that take close to the information theoretic minimum space and support degree and adjacency queries efficiently. A degree query asks for the degree of a given node in the graph, and an adjacency query asks whether two given vertices are adjacent or not in the graph. In addition, we would also like to support listing all the vertices adjacent to a given vertex.

Turán [27] gave a linear time constructible representation of an arbitrary planar graph that takes at most 12 bits per node. Though this gives a space efficient representation of planar graphs, it does not support the queries efficiently. Kannan et al. [28] have given an *implicit* (linear space) graph representation that supports adjacency queries using $O(\lg n)$ bit probes.

Jacobson [1] gave a representation that takes $O(n)$ bits of space to represent a planar graph on n nodes and supports degree and adjacency queries in constant time. It uses a simple mapping of one-page graphs to sequences of balanced parentheses, and the fact that a planar graph always has a 4-page embedding. By storing auxiliary structures to support some natural operations on these sequences (see Section 38.4.2), one can also support the navigational operations in optimal time.

Munro and Raman [21] improved the space to $8n + 2m + o(n)$ bits, for a planar graph on n vertices with m edges, still supporting the queries in constant time. In general, their representation takes $2kn + 2m + o(nk + m)$ bits to store a k page graph on n vertices and m edges and supports degree and adjacency queries in $O(k)$ time.

There have been several improvements [29–33], improving the space close to the information theoretic-lower bound, simultaneously expanding the class of graphs for which the scheme works. In particular, Lu [33] gave an optimal space (within lower-order terms) representation that can be constructed in linear time. This supports degree and adjacency queries in $O(1)$ time for constant-genus graphs.

The main idea is to partition the given graph G on n vertices into $o(n/\lg n)$ disjoint subgraphs of size $O(\lg^6 n)$ by removing a subgraph H of size $o(n/\lg n)$. This is done using a "planarization algorithm" for bounded genus graphs, and an algorithm to construct a "separator decomposition tree" of a planar graph. The representation of G is obtained by storing a rerepresentation of H, and recursing on each of the smaller subgraphs upto a constant number of levels, after which one can use a precomputed table to operate on the small subgraphs. See [33] for details.

38.6 Succinct Structures for Indexing

A *text index* is a data structure storing a *text* (a string or a set of strings) and supporting string matching queries: given a pattern P, find all the occurrences of P in the text. Two well-known and widely used index structures are the suffix trees and suffix arrays. In this section we briefly describe some succinct data structures for these two.

A *suffix tree* for a text is the *compressed digital trie* of all the suffixes of the text [34,35]. A suffix tree for a text of length n has n leaves and at most $n - 1$ internal nodes. The space bound is a consequence of skipping nodes with only one child, hence there are precisely $n - 1$ internal nodes if we use a binary trie. Each leaf points to the position in the text of the corresponding suffix it represents uniquely. The edges are labeled by substrings of the text, which are usually represented by storing a position in the text where the substring starts and its length. Thus, a standard representation of a suffix tree for a text of length n takes $O(n \lg n)$ bits. Searching for an occurrence of a pattern of length m using a suffix tree takes $O(m)$ time.

The *suffix array* of a text is an array storing pointers to the suffixes of the text in their lexicographic order. Thus, a suffix array for a text of length n takes $n\lceil \lg n \rceil$ bits. Note that the leaf labels of a suffix tree written from left to right form the suffix array, if the children of each node are arranged in lexicographic order of their edge labels. Searching for an occurrence of a pattern of length m using a suffix array takes $O(m + \lg n)$ time.

We now briefly sketch the ideas involved in representing a suffix tree (and hence also a suffix array) using $O(n)$ bits. We first convert the trie into binary by using a fixed length encoding of the characters of the alphabet. We then store the parenthesis representation of the underlying tree structure (see Section 38.4.2). The edge labels of a suffix tree can be omitted, as this can be determined by finding the longest common prefix of the leftmost and rightmost leaves of the parent node (of the edge). The parenthesis representation of an ordinal tree can be augmented with $o(n)$-bit additional structure to support finding the leftmost and rightmost leaves of a given node in constant time. Thus, one can use this tree representation to store the tree structure of a suffix tree, and store the leaf pointers (suffix array) explicitly. This gives a suffix tree representation that takes $n \lg n + O(n)$ bits of space and supports indexing queries in optimal time. See [4] for details.

The above structure uses $n \lceil \lg n \rceil$ bits to represent the pointers to the text or the suffix array. Grossi and Vitter [36] obtained a suffix array structure that takes $O(n)$ bits and supports finding the ith element in the suffix array (lookup queries) in $O(\lg^\epsilon n)$ time, for any fixed $\epsilon > 0$. Using this structure they also obtained a suffix tree representation that takes $O(n)$ bits of space and supports finding all the s occurrences of a given pattern of length m in $O(m + s \lg^\epsilon n)$ time. The structure given by Rao [37] generalizes the suffix array structure of Grossi and Vitter, which takes $O(nt(\lg n)^{1/(t+1)})$ bits and supports lookup in $O(t)$ time, for any parameter $1 \le t \le \lg \lg n$. Using this structure, one can get an index structure that takes $o(n \lg n)$ bits and supports finding all the s occurrences of a given pattern of length m in $O(m + s + \lg^\epsilon n)$ time.

Ferragina and Manzini [38] presented an *opportunistic* data structure taking $O(nH_k(n)) + o(n)$ bits of space, where $H_k(n)$ denotes the kth order entropy of the given text of length n. This supports finding all the occurrences of a pattern of length m in $O((m + s) \lg^\epsilon n)$ time, where s is the number of occurrences of the pattern. They also presented its practical performance [39].

Sadakane [23] gave a data structure that takes $O(n \cdot (1 + H_0) + O(|\Sigma| \lg |\Sigma|))$ bits for a text of length n over an alphabet Σ, where $H_0 \le \lg |\Sigma|$ is the order-0 entropy for the text. This supports finding all the s occurrences of a given pattern P in $O(|P| \lg n + s \lg^\epsilon n)$ time, and *decompress* a portion of the text of length l in $O(l + \lg^\epsilon n)$ time, for any fixed $\epsilon > 0$.

Grossi et al. [40] gave another index structure that takes $nH_k(n) + O(n \lg \lg n \lg |\Sigma| / \lg n)$ bits for a text of length n over an alphabet Σ. Finding an occurrence of a pattern of length m using this structure takes $O(m \lg |\Sigma| + polylog(n))$ time. This is also shown to perform well, in terms of space as well as query times, in practice [41].

38.7 Permutations and Functions

38.7.1 Permutations

Permutations are fundamental in computer science and have been the focus of extensive study. Here we consider the problem of representing permutations succinctly to support computing $\pi^k(i)$ for any integer k, where $\pi^0(i) = i$ for all i; $\pi^k(i) = \pi(\pi^{k-1}(i))$ when $k > 0$ and $\pi^k(i) = \pi^{-1}(\pi^{k+1}(i))$ when $k < 0$.

The most obvious way of representing an arbitrary permutation, π, of the integers $\{0, 1, \ldots, n-1\}$ is to store the sequence $\pi(0), \pi(1), \ldots, \pi(n-1)$. This takes $n \lceil \lg n \rceil$ bits, which is $\Theta(n)$ bits more than the information-theoretic lower bound of $\lg(n!) \approx n \lg n - n \lg e$ bits. This representation can be used to find $\pi(i)$ in $O(1)$ time, but finding $\pi^{-1}(i)$ takes $O(n)$ time in the worst case, for $0 \le i \le n-1$. Using this representation, one can easily compute $\pi^k(i)$ in k steps, for $k \ge 1$. To facilitate the computation in constant time, one could store $\pi^k(i)$ for all i and k ($|k| \le n$, along with its cycle length), but that would require $\Theta(n^2 \lg n)$ bits. The most natural compromise is to retain $\pi^k(i)$ with $|k| \le n$ a power of 2. Unfortunately, this $n \lceil \lg n^2 \rceil$ bit representation leaves us with a logarithmic time evaluation scheme and a factor of $\lg n$ from the minimal space representation.

We first show how to augment the standard representation to support π^{-1} queries efficiently, while avoiding most of the extra storage cost one would expect. In addition to storing the standard representation, we trace the cycle structure of the permutation, and for every cycle whose length is at least t, we store a shortcut pointer with the elements which are at a distance of a multiple of t steps from an arbitrary starting point. The shortcut pointer points to the element which is t steps before it in the cycle of the permutation. This *shortcut representation* of a permutation can be stored using $(1 + 1/t)n \lg n + o(n)$ bits, and it supports π queries in $O(1)$ time and π^{-1} queries in $O(t)$ time, for any parameter $1 \le t \le n$.

Consider the cycle representation of a permutation π over $\{0, 1, \ldots, n-1\}$, which is a collection of disjoint cycles of π (where the cycles are ordered arbitrarily). Let σ be this permutation, that is, the standard representation of σ is a cycle representation of π. Let B be a bit vector of length n that has a **1** corresponding to the starting position of each cycle of π and **0** everywhere else, together with its rank and select directories with respect to both bits. Let S be a representation of σ that supports $\sigma(i)$ and $\sigma^{-1}(i)$ queries efficiently. Then to find $\pi^k(i)$, first find the index j of the cycle to which $\sigma^{-1}(i)$ belongs, using B and S. Find the length l of the jth cycle and the number p of elements up to (but not including) the jth cycle. Then, one can verify that $\pi^k(i) = \sigma(p + (i - p + k \mod l))$. Combining this with the shortcut representation, one can get a representation taking $(1 + \epsilon)n \lg n + O(1)$ bits that supports computing arbitrary powers in $O(1)$ time.

Benes network: A *Benes network* [42] is a communication network composed of a number of switches. Each switch has 2 inputs x_0 and x_1 and 2 outputs y_0 and y_1 and can be configured either so that x_0 is connected to y_0 (i.e., a packet that is input along x_0 comes out of y_0) and x_1 is connected to y_1, or the other way around. An r-Benes network has 2^r inputs and outputs, and is defined as follows. For $r = 1$, the Benes network is a single switch with 2 inputs and 2 outputs. An $(r+1)$-Benes network is composed of 2^{r+1} switches and two r-Benes networks, connected as as shown in Figure 38.6a. A particular setting of the switches

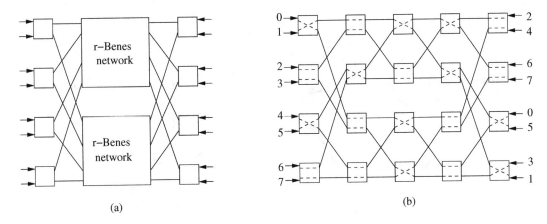

FIGURE 38.6 Benes network: (a) construction of $(r+1)$-Benes network (b) Benes network realizing the permutation (4,7,0,6,1,5,2,3).

of a Benes network *realizes* a permutation π if a packet introduced at input i comes out at output $\pi(i)$, for all i. See Figure 38.6b for an example.

Clearly, a Benes network may be used to represent a permutation. For example, if $n = 2^r$, a representation of a permutation π on $[n]$ may be obtained by configuring an r-Benes network to realize π and then listing the settings of the switches in some canonical order (e.g., level-order). This represents π using $r2^r - 2^{r-1} = n \lg n - n/2$ bits. Given i, one can trace the path taken by a packet at input i by inspecting the appropriate bits in this representation, and thereby calculate $\pi(i)$ in $O(\lg n)$ time (indeed, in $O(\lg n)$ bit-probes). In fact, by tracing the path back from output i we can also compute $\pi^{-1}(i)$ in $O(\lg n)$ time.

One can compress the middle levels of a Benes network by storing an implicit representation of the permutation represented by the middle $O(\lg n/\lg \lg n)$ levels. This reduces the space to $\lg(n!) + o(n)$ bits. One can also group the remaining bits of this Benes network into words of size $\Theta(\lg n)$ bits (by taking $O(\lg \lg n)$ consecutive levels and $O(\lg \lg n)$ appropriate rows). This enables us to traverse $\Theta(\lg \lg n)$ levels in a Benes network in $O(1)$ time. Thus, it gives a representation that takes the optimal $\lceil \lg(n!) \rceil + o(n)$ bits, and supports computing arbitrary powers in $O(\lg n/\lg \lg n)$ time.

One can obtain a structure with same time and space bounds even when n is not a power of 2. See [43] for details.

38.7.2 Functions

Now consider the problem of representing arbitrary functions $f : [n] \to [n]$, so that queries for $f^k(i)$, for any integer k can be answered efficiently. Here $f^0(i) = i$ and for any $k > 0, f^k(i) = f(f^{k-1}(i))$ and $f^{-k}(i) = \{j | f^k(j) = i\}$, for all i. This is a generalization of the problem considered in the previous section. Since there are n^n functions from $[n]$ to $[n]$, any representation scheme takes at least $\lceil n \lg n \rceil$ bits to store an arbitrary function.

A standard way of representing a function is to store the sequence $f(i)$, for $i = 0, \ldots, n-1$. This representation does not support the efficient evaluation of $f^k(i)$ for $k \gg 1$. We look at a representation that takes $(1 + \epsilon)n \lg n + O(1)$ bits of space to store a function $f : [n] \to [n]$ and supports computing arbitrary positive powers in constant time and negative powers $f^{-k}(i)$, in $O(1 + |f^{-k}(i)|)$ time.

Given an arbitrary function $f : [n] \to [n]$, consider the directed graph, $G_f = (V, E)$, obtained from it, where $V = [n]$ and $E = \{(i, j) : f(i) = j\}$. In general this directed graph consists of a disjoint set of subgraphs, each of which is a directed cycle with trees rooted at the nodes on the cycle and edges directed towards the roots. See Figure 38.7 for an example.

The main idea of the solution is as follows: in each directed cycle, we re-order the nodes of each tree such that the leftmost path of any subtree is the longest path in that subtree. This enables finding a node at a given depth from any internal node, if it exists, in constant time using the parenthesis representation. We then preprocess each of the trees and store auxiliary structures to support level-ancestor queries on them in constant time (see Section 38.4.2). Observe that finding $f^k(i)$, for $k > 0$, can be translated to finding the ancestor of node i which is k levels above it, if i is at a depth at least k in its tree T. Otherwise, we have to traverse the cycle to which the root of T belongs, to find the required answer. This can be done by storing these cycles as a permutation.

When i belongs to one of the trees in a subgraph, one can answer $f^k(i)$ queries for $k < 0$ in optimal time by finding all the nodes that are at the kth level in the subtree rooted at i. Otherwise, if i is part of the cycle in the subgraph, we store an auxiliary structure that, for any given i and k, outputs all the trees in the subgraph containing i that have an answer in time proportional to the number of such nodes. From this, one can easily find the required answer in optimal time. The auxiliary structure takes $O(m)$ bits for a subgraph with m nodes, and hence $O(n)$ bits overall. See [24] for details.

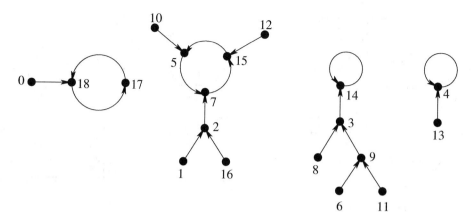

FIGURE 38.7 Graph representation of the function $f(x) = (x^2 + 2x - 1) \bmod 19$, for $0 \le x \le 18$.

For functions from $[n] \to [m]$ one can show the following: If there is a representation of a permutation that takes $P(n)$ space to represent a permutation on $[n]$ and supports forward in t_1 time and inverse in t_2 time, then there is a representation of a function from $[n]$ to $[m]$, $m \le n$ that takes $(n - m) \lg m + P(m) + O(m)$ bits, and supports $f^k(i)$ in $O(t_1 + t_2)$ time, for any positive integer k and for any $i \in [n]$. When $m > n$, larger powers are not defined in general. In this case, we can have a structure that takes $n \lg m + P(n) + O(n)$ bits of space and answers queries for positive powers (returns the power if defined or returns -1 otherwise) in $O(t_1 + t_2)$ time.

38.8 Partial Sums

Let $a_1, a_2, \ldots, a_n$ be a sequence of n non-negative k-bit numbers. The *partial sums* problem maintains the sequence under the following operations:

- sum(i): return $\sum_{j=1}^{i} a_j$,
- update(i, δ): set $a_i \leftarrow a_i + \delta$, for some integer δ such that $0 \le a_i + \delta \le 2^k - 1$. Our later solutions have the additional constraint that $|\delta| \le \lg^{O(1)} n$.

Dietz [44] gave a structure for the partial sum problem that supports sum and update in $O(\lg n / \lg \lg n)$ time using $O(n \lg n)$ bits of extra space, for the case when $k = \Theta(\lg n)$ and no constraints on δ. The time bounds are optimal due to the lower bound of Fredman and Saks [45]. As the information-theoretic space lower bound is kn bits, this structure uses space within a constant factor of the optimal.

The main idea of this structure is to store the elements at the leaves of a complete tree with branching factor $O(\lg^\epsilon n)$ for some $\epsilon < 1$. The operations are performed by traversing a path from a leaf to the root, querying/updating the nodes along the path.

The *searchable partial sums* problem is an extension of the partial sums problem that also supports the following operation:

- search(j): find the smallest i such that sum(i) $\ge j$.

When $k = 1$ (i.e., each element is a bit), the special case is commonly known as the *dynamic bit vector* problem, which maintains a bit vector of length n under rank, select and flip (update) operations.

For the searchable partial sums problem there is a structure that supports all operations in $O(\lg n / \lg \lg n)$ time, and uses $kn + o(kn)$ bits of space [46]. For $k = O(\lg \lg n)$, one can also obtain a structure that again takes $kn + o(kn)$ bits and supports sum and search in $O(\log_b n)$ time and update in $O(b)$ amortized time, for any parameter $b \ge \lg n / \lg \lg n$ [47]. For the partial sums problem, one can support the above trade-off for $k = O(\lg n)$ [46], and the time bounds can be shown to be optimal [45].

For the dynamic bit vector problem, one can support rank and select in $O(\log_b n)$ time and flip in $O(b)$ (worst-case) time, for any parameter $\lg n / \lg \lg n \le b \le n$, using $o(n)$ bits of extra space. One can also extend the above trade-off for $k = O(\lg \lg n)$, using $kn + o(kn)$ bits of space.

See [46,47] for details.

38.9 Arrays

38.9.1 Resizable Arrays

A basic problem that arises in many applications is accumulating elements into a list when the number of elements is unknown ahead of time. The operations needed from such a structure are the ability to append elements to the end of the structure, removing the last element from the structure (in applications such as implementing a stack) and some method of accessing the elements currently in the structure.

One simple solution is a linked list which can easily grow and shrink, and supports sequential access. But this does not support random access to the elements. Moreover, its space overhead is $O(n)$ pointers to store n elements.

Another standard solution is the *doubling technique* [48]. Here the elements are stored in an array. Whenever the array becomes full, an array of double its size allocated and all the elements are copied to it. Similarly, whenever the array shrinks so that it is only one-fourth full, an array of half its size is allocated and all the elements are copied to it. The advantage of this solution over the linked lists is that random access to the elements takes only $O(1)$ time (as opposed to $O(n)$ for linked lists). The amortized update time is $O(1)$, though the worst-case update time is $O(n)$. The space overhead of this solution is $O(n)$.

Sitarski [49] has proposed a solution whose space overhead is only $O(\sqrt{n})$. The idea is to divide the given list of n elements into sublists of size $\lceil \sqrt{n} \rceil$, store them in separate arrays, and store an array (of length $O(\sqrt{n})$) of pointers to these sublists (in order). Whenever $\lceil \sqrt{n} \rceil$ changes, the entire structure is reconstructed with the new size. Thus the amortized update time is $O(1)$ (though the worst-case time is $O(n)$). This also supports random access in $O(1)$ time.

Brodnik et al. [13] gave a structure that takes $O(\sqrt{n})$ extra locations, where n is the current size of the array, and supports the operations in $O(1)$ time. One advantage of this structure is that elements are never re-allocated. They have also shown that any such structure requires $\Omega(\sqrt{n})$ extra locations even if there are no constraints on the access time.

38.9.2 Dynamic Arrays

A resizable array supports adding/deleting elements only at the end of the list, but does not support insertion/deletion of elements at arbitrary positions in the array. A *dynamic array* is data structure that maintains a sequence of records under the following operations:

- access(i): return the ith record in the sequence,
- insert(r, i): insert the record r at position i in the sequence, and
- delete(i): delete the ith record in the sequence.

A standard way of implementing a dynamic array is to store the records in an array and maintain it using the doubling technique. This supports access in $O(1)$ but requires $O(n)$ time to support insert and delete operations.

Goodrich and Kloss [50] gave a structure, the *tiered vector*, that takes $n + O(\sqrt{n})$ words of space to represent a sequence of length n, where each record fits in a word. This structure supports access in $O(1)$ time and updates in $O(\sqrt{n})$ amortized time. The major component of a tiered vector is a set of *indexable circular deques*. A deque is a linear list which provides constant time insert and delete operations at either the head or the tail of the list [51]. A circular deque is a list which is stored in a sequential section of memory of fixed size. An indexable circular deque maintains pointers h and t, which reference the index in memory of the head and tail of this list. A tiered vector is a set of indexable circular deques. Insertions and deletions in an arbitrary indexable circular deque require time linear in its size, but inserting/deleting at either the head or the tail of the list takes $O(1)$ time.

Thus, by maintaining the given sequence of n elements using $O(\sqrt{n})$ indexable circular deques each of size $O(\sqrt{n})$, one can support access in $O(1)$ time and updates in $O(\sqrt{n})$ amortized time. One can easily generalize this structure to one that supports access in $O(1/\epsilon)$ time and updates in $O(n^\epsilon)$ time, for any parameter $0 < \epsilon \le 1$.

Using this structure to represent a block of $O(\lg^{O(1)} n)$ records, Raman et al. [46] gave a structure that supports access and updates in $O(\lg n / \lg \lg n)$ amortized time, using $o(n)$ bits of extra space. The main idea is to divide the given list of length n into sublists of length between $(1/2)\lg^4 n$ and $2\lg^4 n$, and store the sublists using the above dynamic array structure. One can maintain these sublists as the leaves of a weight-balanced B-tree with branching factor $O(\sqrt{\lg n})$, and hence height $O(\lg n / \lg \lg n)$.

By restricting the length of the array, Raman and Rao [18] obtained a dynamic array structure that maintains a sequence of $l = O(w^{O(1)})$ records of $r = O(w)$ bits each, where w is the word size. This structure supports *access* in $O(1)$ time and updates in $O(1 + lr/kw)$ amortized time, and uses $lr + O(k \lg l)$ bits, for any parameter $k \le l$. The data structure also requires a precomputed table of size $O(2^{\epsilon w})$ bits, for any fixed $\epsilon > 0$. The main idea is to store the newly added elements separately from the existing elements, and store a structure to indicate all the positions of the "updated" elements. The structure is rebuilt after every k updates.

38.10 Conclusions

We looked at several succinct data structures that achieve almost optimal space while supporting the required operations efficiently. Apart from being of theoretical interest, succinct data structures will also have many practical applications due to the enormous growth in the amount of data that needs to be stored in a structured fashion.

Most of the succinct data structures we presented here can be constructed in linear time from the standard representation. But this method requires more space than necessary during the construction. Developing algorithms that directly construct the succinct representations without using more space during construction, preferably in optimal time, is an important open problem. See [52] for an example.

Another aspect, that is more of theoretical significance, is to study the cell probe (in particular, bit probe) complexity of succinct data structures [53–55]. For most problems, no bounds other than the straightforward translations from the bounds on the RAM model are known. It is also interesting to consider the time-space trade-offs of these structures.

References

1. G. Jacobson. Succinct Static Data Structures. PhD thesis, Carnegie Mellon University, 1988.
2. J. Ian Munro. Tables. *Proceedings of the Conference on Foundations of Software Technology and Theoretical Computer Science*, LNCS 1180: 37–42, 1996.
3. D. R. Clark. Compact Pat Trees. PhD thesis, University of Waterloo, 1996.
4. J. Ian Munro, V. Raman, and S. Srinivasa Rao. Space Efficient Suffix trees. *Journal of Algorithms*, 39(2): 205–222, 2001.
5. M. L. Fredman, J. Komlós, and E. Szemerédi. Storing a Sparse Table with $O(1)$ Worst Case Access Time. *Journal of the ACM*, 31(3): 538–544, 1984.
6. J. P. Schmidt and A. Siegel. The Spatial Complexity of Oblivious k-probe Hash Functions. *SIAM Journal on Computing*, 19(5): 775–786, 1990.
7. A. Brodnik and J. Ian Munro. Membership in Constant Time and Almost Minimum Space. *SIAM Journal on Computing*, 28(5): 1627–1640, 1999.
8. R. Pagh. Low Redundancy in Static Dictionaries with Constant Query Time. *SIAM Journal on Computing*, 31(2): 353–363, 2001.
9. M. Ajtai. A Lower Bound for Finding Predecessors in Yao's Cell Probe Model. *Combinatorica*, 8(3): 235–247, 1988.
10. V. Raman and S. Srinivasa Rao. Static Dictionaries Supporting Rank. *Proceedings of the International Symposium on Algorithms and Computation*, LNCS 1741: 18–26, 1999.
11. R. Raman, V. Raman, and S. Srinivasa Rao. Succinct Indexable Dictionaries with Applications to Encoding k-ary Trees and Multisets. *Proceedings of the ACM-SIAM Symposium on Discrete Algorithms*, 233–242, 2002.
12. P. Elias. Efficient Storage and Retrieval by Content and Address of Static Files. *Journal of the ACM*, 21(2): 246–260, 1974.
13. A. Brodnik, S. Carlsson, E. D. Demaine, J. Ian Munro, and R. Sedgewick. Resizable Arrays in Optimal Time and Space. *Proceedings of the Workshop on Algorithms and Data Structures*, LNCS 1663: 37–48, 1999.
14. J. Ian Munro, V. Raman, and A. Storm. Representing Dynamic Binary Trees Succinctly. *Proceedings of the ACM-SIAM Symposium on Discrete Algorithms*, 529–536, 2001.
15. M. Dietzfelbinger, A. R. Karlin, K. Mehlhorn, F. Meyer auf der Heide, H. Rohnert, and R. E. Tarjan. Dynamic Perfect Hashing: Upper and Lower Bounds. *SIAM Journal on Computing*, 23(4): 738–761, 1994.
16. R. Pagh and F. F. Rodler. Cuckoo Hashing. *Proceedings of the European Symposium on Algorithms*, LNCS 2161: 121–133, 2001.
17. D. Fotakis, R. Pagh, P. Sanders, and P. G. Spirakis. Space Efficient Hash Tables with Worst Case Constant Access Time. *Proceedings of Symposium on Theoretical Aspects of Computer Science*, LNCS 2607: 271–282, 2003.
18. R. Raman and S. Srinivasa Rao. Dynamic Dictionaries and Binary Trees in Near-minimum Space. *Proceedings of the International Colloquium on Automata, Languages and Programming*, LNCS 2719: 357–368, 2003.
19. J. W. J. Williams. Algorithm 232: Heap Sort. *Communications of the ACM*, 7(6): 347–348, 1964.
20. R. L. Graham, D. E. Knuth, and O. Patashnik. *Concrete Mathematics*. Addison-Wesley, 1989.
21. J. Ian Munro and V. Raman. Succinct Representation of Balanced Parentheses and Static Trees. *SIAM Journal on Computing*, 31(3): 762–776, 2001.
22. D. Benoit, E. D. Demaine, J. Ian Munro, and V. Raman. Representing Trees of Higher Degree. *Proceedings of the Workshop on Algorithms and Data Structures*, LNCS 1663: 169–180, 1999.
23. K. Sadakane. Succinct Representations of lcp Information and Improvements in the Compressed Suffix Arrays. *Proceedings of the ACM-SIAM Symposium on Discrete Algorithms*, 225–232, 2002.

24. J. Ian Munro and S. Srinivasa Rao. Succinct Representations of Functions. *Proceedings of the International Colloquium on Automata, Languages and programming*, LNCS 3142, 1006–1015, 2004.

25. R. F. Geary, R. Raman, and V. Raman. Succinct Ordinal Trees with Level-ancestor Queries. *Proceedings of the ACM-SIAM Symposium on Discrete Algorithms*, 1–10, 2004.

26. J. J. Darragh, J. G. Cleary, and I. H. Witten. Bonsai: A Compact Representation of Trees. *Software Practice and Experience*, 23(3): 277–291, 1993.

27. G. Turán. Succinct Representations of Graphs. *Discrete Applied Mathematics*, 8: 289–294, 1984.

28. S. Kannan, M. Naor, and S. Rudich. Implicit Representation of Graphs. *Proceedings of the ACM Symposium on Theory of Computing*, 334–343, 1988.

29. R. C. Chuang, A. Garg, X. He, M.-Y. Kao, and H.-I. Lu. Compact Encodings of Planar Graphs via Canonical Orderings and Multiple Parentheses. *Proceedings of the International Colloquium on Automata, Languages and Programming*, LNCS 1443: 118–129, 1998.

30. Y.-T. Chiang, C.-C. Lin, and H.-I. Lu. Orderly Spanning Trees with Applications to Graph Drawing and Graph Encoding. *Proceedings of the ACM-SIAM Symposium on Discrete Algorithms*, 506–515, 2001.

31. X. He, M.-Y. Kao, and H.-I. Lu. Linear-time Succinct Encodings of Planar Graphs via Canonical Orderings. *SIAM Journal on Discrete Mathematics*, 12: 317–325, 1999.

32. H.-I. Lu. Linear-time Compression of Bounded-genus Graphs into Information-theoretically Optimal Number of Bits. *Proceedings of the ACM-SIAM Symposium on Discrete Algorithms*, 223–224, 2002.

33. H.-I. Lu. Linear-Time Information-theoretically Optimal Encodings Supporting Constant-time Queries for Constant-genus Graphs. Manuscript, 2002.

34. P. Weiner. Linear Pattern Matching Algorithm. *Proceedings of the IEEE Symposium on Switching and Automata Theory*, 1–11, 1973.

35. E. M. McCreight. A Space-economical Suffix Tree Construction Algorithm. *Journal of the ACM*, 23(2): 262–272, 1976.

36. R. Grossi and J. S. Vitter. Compressed Suffix Arrays and Suffix Trees with Applications to Text Indexing and String Matching. *Proceedings of the ACM Symposium on Theory of Computing*, 397–406, 2000.

37. S. Srinivasa Rao. Time Space Tradeoffs for Compressed Suffix Arrays. *Information Processing Letters*, 82(6): 307–311, 2002.

38. P. Ferragina and G. Manzini. Opportunistic Data Structures with Applications. *Proceedings of the Annual Symposium on Foundations of Computer Science*, 390–398, 2000.

39. P. Ferragina and G. Manzini. An Experimental Study of an Opportunistic Index. *Proceedings of the Annual Symposium on Foundations of Computer Science*, 269–278, 2001.

40. R. Grossi, A. Gupta, and J. S. Vitter. High-order Entropy-compressed Text Indexes. *Proceedings of the ACM-SIAM Symposium on Discrete Algorithms*, 841–850, 2003.

41. R. Grossi, A. Gupta, and J. S. Vitter. When Indexing Equals Compression: Experiments with Compressing Suffix Arrays and Applications. *Proceedings of the ACM-SIAM Symposium on Discrete Algorithms*, 629–638, 2004.

42. F. T. Leighton. *Introduction to Parallel Algorithms and Architectures: Arrays, Trees and Hypercubes*. Computer Science and Information Processing, Morgan Kauffman, 1992.

43. J. Ian Munro, V. Raman, R. Raman, and S. Srinivasa Rao. Succinct Representations of Permutations. *Proceedings of the International Colloquium on Automata, Languages and Programming*, LNCS 2719: 345–356, 2003.

44. P. F. Dietz. Optimal Algorithms for List Indexing and List Ranking. *Proceedings of the Workshop on Algorithms and Data Structures*, LNCS 382: 39–46, 1989.

45. M. L. Fredman and M. E. Saks. The Cell Probe Complexity of Dynamic Data Structures. *Proceedings of the ACM Symposium on Theory of Computing*, 345–354, 1989.

46. R. Raman, V. Raman, and S. Srinivasa Rao. Succinct Dynamic Data Structures. *Proceedings of the Workshop on Algorithms and Data Structures*, LNCS 2125, 426–437, 2001.

47. W.-K. Hon, K. Sadakane, and W.-K. Sung. Succinct Data Structures for Searchable Partial Sums. *Proceedings of the International Symposium on Algorithms and Computation*, LNCS 2906, 505–516, 2003.

48. J. Boyer. Algorithm Alley: Resizable Data Structures. *Dr. Dobb's Journal*, 23(1), January 1998.

49. E. Sitarski. Algorithm Alley: HATs: Hashed Array Tables. *Dr. Dobb's Journal*, 21(11), September 1996.

50. M. T. Goodrich and J. G. Kloss II. Tiered Vectors: Efficient Dynamic Array for JDSL. *Proceedings of the Workshop on Algorithms and Data Structures*, LNCS 1663: 205–216, 1999.

51. D. E. Knuth. *The Art of Computer Programming, Vol. 1: Fundamental Algorithms*. Addison-Wesley, Third edition, 1997.

52. W.-K. Hon, K. Sadakane, and W.-K. Sung. Breaking a Time-and-Space Barrier in Constructing Full-Text Indices. *Proceedings of the Annual Symposium on Foundations of Computer Science*, 251–260, 2003.

53. H. Buhrman, P. B. Miltersen, J. Radhakrishnan, and S. Venkatesh. Are Bitvectors Optimal? *Proceedings of the ACM Symposium on Theory of Computing*, 449–458, 2000.

54. A. Gál and P. B. Miltersen. The Cell Probe Complexity of Succinct Data Structures. *Proceedings of the International Colloquium on Automata, Languages and Programming*, LNCS 2719: 332–344, 2003.

55. J. Radhakrishnan, V. Raman, and S. Srinivasa Rao. Explicit Deterministic Constructions for Membership in the Bitprobe Model. *Proceedings of the European Symposium on Algorithms*, LNCS 2161: 290–299, 2001.

56. D. R. Clark and J. Ian Munro. Efficient Suffix Trees on Secondary Storage. *Proceedings of the ACM-SIAM Symposium on Discrete Algorithms*, 383–391, 1996.

57. E. D. Demaine. Algorithm Alley: Fast and Small Resizable Arrays. *Dr. Dobb's Journal*, 326: 132–134, July 2001.

58. M. L. Fredman and D. E. Willard. Trans-Dichotomous Algorithms for Minimum Spanning Trees and Shortest Paths. *Journal of Computer and System Sciences*, 48(3): 533–551, 1994.

59. G. Jacobson. Space-efficient Static Trees and Graphs. *Proceedings of the Annual Symposium on Foundations of Computer Science*, 549–554, 1989.

39

Randomized Graph Data-Structures for Approximate Shortest Paths[*]

39.1 Introduction...611
39.2 A Randomized Data-Structure for Static APASP: Approximate Distance Oracles.. 611
   3-Approximate Distance Oracle • Preliminaries • (2k‑1)-Approximate Distance Oracle •
   Computing Approximate Distance Oracles
39.3 A Randomized Data-Structure for Decremental APASP...618
   Main Idea • Notations • Hierarchical Distance Maintaining Data-Structure • Bounding
   the Size of $B_u^{d,S}$ under Edge-Deletions • Improved Decremental Algorithm for APASP up to
   Distance d
39.4 Further Reading and Bibliography...624
Acknowledgments 624
References...624

Surender Baswana
Indian Institute of Technology, Kanpur

Sandeep Sen
Indian Institute of Technology, Delhi

39.1 Introduction

Let $G = (V, E)$ be an undirected weighted graph on $n = |V|$ vertices and $m = |E|$ edges. Length of a path between two vertices is the sum of the weights of all the edges of the path. The shortest path between a pair of vertices is the path of least length among all possible paths between the two vertices in the graph. The length of the shortest path between two vertices is also called the distance between the two vertices. An α-approximate shortest path between two vertices is a path of length at-most α times the length of the shortest path.

Computing all-pairs exact or approximate distances in G is one of the most fundamental graph algorithmic problem. In this chapter, we present two randomized graph data-structures for all-pairs approximate shortest paths (APASP) problem in static and dynamic environments. Both the data-structures are hierarchical data-structures and their construction involves random sampling of vertices or edges of the given graph.

The first data-structure is a randomized data-structure designed for efficiently computing APASP in a given static graph. In order to answer a distance query in constant time, most of the existing algorithms for APASP problem output a data-structure which is an $n \times n$ matrix that stores the exact/approximate distance between each pair of vertices explicitly. Recently a remarkable data-structure of $o(n^2)$ size has been designed for reporting all-pairs approximate distances in undirected graph. This data-structure is called *approximate distance oracle* because of its ability to answer a distance query in constant time in spite of its sub-quadratic size. We present the details of this novel data-structure and an efficient algorithm to build it.

The second data-structure is a dynamic data-structure designed for efficiently maintaining APASP in a graph that is undergoing deletion of edges. For a given graph $G = (V, E)$ and a distance parameter $d \le n$, this data-structure provides the first $o(nd)$ update time algorithm for maintaining α-approximate shortest paths for all pairs of vertices separated by distance $\le d$ in the graph.

39.2 A Randomized Data-Structure for Static APASP: Approximate Distance Oracles

There exist classical algorithms that require $O(mn \log n)$ time for solving all-pairs shortest paths (APSP) problem. There also exist algorithms based on fast matrix multiplication that achieve sub-cubic time. However, there is still no combinatorial algorithm

---

[*] This chapter has been reprinted from first edition of this Handbook, without any content updates.

that could achieve $O(n^{3-\epsilon})$ running time for APSP problem. In recent past, many simple combinatorial algorithms have been designed that compute all-pairs approximate shortest paths (APASP) for undirected graphs. These algorithms achieve significant improvement in the running time compared to those designed for APSP, but the distance reported has some additive or/and multiplicative error. An algorithm is said to compute all pairs α-approximate shortest paths, if for each pair of vertices $u, v \in V$, the distance reported is bounded by $\alpha\delta(u,v)$, where $\delta(u,v)$ denotes the actual distance between u and v.

Among all the data-structures and algorithms designed for computing all-pairs approximate shortest paths, the approximate distance oracles are unique in the sense that they achieves simultaneous improvement in running time (sub-cubic) as well as space (sub-quadratic), and still answers any approximate distance query in constant time. For any $k \geq 1$, it takes $O(kmn^{1/k})$ time to compute $(2k-1)$-approximate distance oracle of size $O(kn^{1+1/k})$ that would answer any $(2k-1)$-approximate distance query in $O(k)$ time.

39.2.1 3-Approximate Distance Oracle

For a given undirected graph, storing distance information from each vertex to all the vertices requires $\theta(n^2)$ space. To achieve sub-quadratic space, the following simple idea comes to mind.

> $\mathcal{I}$: *From each vertex, if we store distance information to a small number of vertices, can we still be able to report distance between any pair of vertices?*

The above idea can indeed be realized using a simple random sampling technique, but at the expense of reporting approximate, instead of exact, distance as an answer to a distance query. We describe the construction of 3-approximate distance oracle as follows.

1. Let $R \subset V$ be a subset of vertices formed by picking each vertex randomly independently with probability γ (the value of γ will be fixed later on).
2. For each vertex $u \in V$, store the distances to all the vertices of the sample set R.
3. For each vertex $u \in V$, let $p(u)$ be the vertex nearest to u among all the sampled vertices, and let S_u be the set of all the vertices of the graph G that lie closer to u than the vertex $p(u)$. Store the vertices of set S_u along with their distance from u.

For each vertex $u \in V$, storing distance to vertices S_u helps in answering distance query to vertices in locality of u, whereas storing distance from all the vertices of the graph to all the sampled vertices will be required (as shown below) to answer distance query for vertices that are not present in locality of each other. In order to extract distance information in constant time, for each vertex $u \in V$, we use two *hash tables* (see Reference 1, chapter 12), for storing distances from u to vertices of sets S_u and R respectively. The size of each hash-table is of the order of the size of corresponding set (S_u or R). A typical hash table would require $O(1)$ expected time to determine whether $w \in S_u$, and if so, report the distance $\delta(u,w)$. In order to achieve $O(1)$ worst case time, the $2-level\ hash\ table$ (see Fredman, Komlos, Szemeredi, [2]) of optimal size is employed.

The collection of these hash-tables (two tables per vertex) constitute a data-structure that we call approximate distance oracle. Let $u, v \in V$ be any two vertices whose intermediate distance is to be determined approximately. If either u or v belong to set R, we can report exact distance between the two. Otherwise also exact distance $\delta(u,v)$ will be reported if v lies in S_u or vice versa. The only case, that is left, is when neither $v \in S_u$ nor $u \in S_v$. In this case, we report $\delta(u,p(u)) + \delta(v,p(u))$ as approximate distance between u and v. This distance is bounded by $3\delta(u,v)$ as shown below.

$$\delta(u,p(u)) + \delta(v,p(u)) \leq \delta(u,p(u)) + (\delta(v,u) + \delta(u,p(u))) \quad \{\text{using triangle inequality}\}$$
$$= 2\delta(u,p(u)) + \delta(u,v) \quad \{\text{since graph is undirected}\}$$
$$\leq 2\delta(u,v) + \delta(u,v) \quad \{\text{since } v \text{ lies farther to } u \text{ than } p(u), \text{ see Figure 39.1}\}$$
$$= 3\delta(u,v)$$

Hence distance reported by the approximate distance oracle described above is no more than three times the actual distance between the two vertices. In other words, the oracle is a 3-approximate distance oracle. Now, we shall bound the expected size of the oracle. Using linearity of expectation, the expected size of the sample set R is $n\gamma$. Hence storing the distance from each vertex to all the vertices of sample set will take a total of $O(n^2\gamma)$ space. The following lemma gives a bound on the expected size of the sets $S_u, u \in V$.

Lemma 39.1

Given a graph $G = (V, E)$, let $R \subset V$ be a set formed by picking each vertex independently with probability γ. For a vertex $u \in V$, the expected number of vertices in the set S_u is bounded by $1/\gamma$.

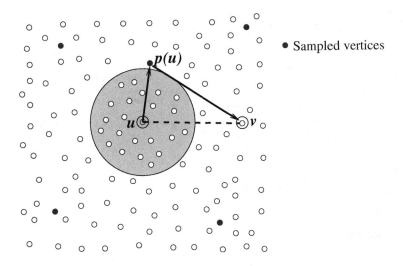

FIGURE 39.1 v is farther to u than $p(u)$, bounding $\delta(p(u),v)$ using triangle inequality.

Proof. Let $\{v_1, v_2, \ldots, v_{n-1}\}$ be the sequence of vertices of set $V\backslash\{u\}$ arranged in non-decreasing order of their distance from u. The set S_u consists of all those vertices of the set $V\backslash\{u\}$ that lie closer to u than any vertex of set R. Note that the vertex v_i belongs to S_u if none of the vertices of set $\{v_1, v_2, \ldots, v_{i-1}\}$ (i.e., the vertices preceding v_i in the sequence above) is picked in the sample R. Since each vertex is picked independently with probability p, therefore the probability that vertex v_i belongs to set S_u is $(1 - \gamma)^{i-1}$. Using linearity of expectation, the expected number of vertices lying closer to u than any sampled vertex is

$$\sum_{i=1}^{n-1}(1-\gamma)^{i-1} \leq \frac{1}{\gamma}$$

Hence the expected number of vertices in the set S_u is no more than $1/\gamma$. ∎

So the total expected size of the 3-approximate distance oracle is $O(n^2\gamma + n/\gamma)$. Choosing $\gamma = 1/\sqrt{n}$ to minimize the size, we conclude that there is a 3-approximate distance oracle of expected size $n^{3/2}$.

39.2.2 Preliminaries

In the previous subsection, 3-approximate distance oracle was presented based on the idea $\mathcal{I}$. The $(2k - 1)$-approximate distance oracle is a k-level hierarchical data-structure. An important construct of the data-structure is $Ball(\cdot)$ defined as follows.

Definition 39.1

For a vertex u, and subsets $X, Y \subset V$, the set $Ball(u, X, Y)$ is the set consisting of all those vertices of the set X that lie closer to u than any vertex from set Y. (see Figure 39.2)

It follows from the definition given above that $Ball(u, X, \emptyset)$ is the set X itself, whereas $Ball(u, X, X) = \emptyset$. It can also be seen that the 3-approximate distance oracle described in the previous subsection stores $Ball(u, V, R)$ and $Ball(u, R, \emptyset)$ for each vertex $u \in V$.

If the set Y is formed by picking each vertex of set X independently with probability γ, it follows from Lemma 39.1 that the expected size of $Ball(u, X, Y)$ is bounded by $1/\gamma$.

Lemma 39.2

Let $G = (V, E)$ be a weighted graph, and $X \subset V$ be a set of vertices. If $Y \subset X$ is formed by selecting each vertex independently with probability γ, the expected number of vertices in $Ball(u, X, Y)$ for any vertex $u \in V$ is at-most $1/\gamma$.

39.2.3 (2k-1)-Approximate Distance Oracle

In this subsection we shall give the construction of a $(2k - 1)$-approximate distance oracle which is also based on the idea $\mathcal{I}$, and can be viewed as a generalization of 3-approximate distance oracle.

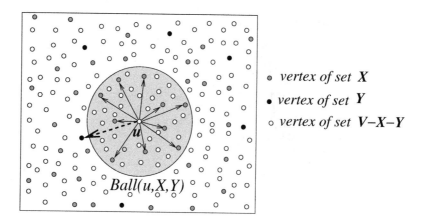

FIGURE 39.2 The vertices pointed by solid-arrows constitute $Ball(u, X, Y)$.

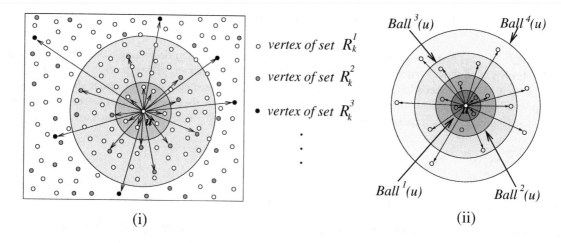

FIGURE 39.3 (i) Close description of $Ball^i(u), i < k$, (ii) hierarchy of balls around u.

The $(2k - 1)$-approximate distance oracle is obtained as follows.

---

1. Let $\mathcal{R}_k^1 \supset \mathcal{R}_k^2 \supset \cdots \mathcal{R}_k^k$ be a a hierarchy of subsets of vertices with $\mathcal{R}_k^1 = V$, and $\mathcal{R}_k^i, i > 1$ is formed by selecting each vertex of set $\mathcal{R}_k^{i-1}$ independently with probability $n^{-1/k}$.

2. For each vertex $u \in V$, store the distance from u to all the vertices of $Ball(u, \mathcal{R}_k^k, \emptyset)$ in a hash table.

3. For each $u \in V$ and each $i < k$, store the vertices of $Ball(u, R_k^i, R_k^{i+1})$ along with their distance from u in a hash table.

    For sake of conciseness and without causing any ambiguity in notations, henceforth we shall use $Ball^i(u)$ to denote $Ball(u, R_k^i, R_k^{i+1})$ or the corresponding hash-table storing $Ball(u, R_k^i, R_k^{i+1})$ for $i < k$.

The collection of the hash-tables $Ball^i(u) : u \in V, i \leq k$ constitute the data-structure that will facilitate answering of any approximate distance query in constant time. To provide a better insight into the data-structure, Figure 39.3 depicts the set of vertices constituting $\{Ball^i(u) \mid i \leq k\}$.

39.2.3.1 Reporting Distance with Stretch At-most (2k − 1)

Given any two vertices $u, v \in V$ whose intermediate distance has to be determined approximately. We shall now present the procedure to find approximate distance between the two vertices using the k-level data-structure described above.

Let $p^1(u) = u$ and let $p^i(u), i > 1$ be the vertex from the set $\mathcal{R}_k^i$ nearest to u. Since $p^i(u) \in Ball^i(u)$ for each $u \in V$, so distance from each u to $p^i(u)$ is known for each $i \leq k$.

The query answering process performs at-most k search steps. In the first step, we search $Ball^1(u)$ for the vertex $p^1(v)$. If $p^1(v)$ is not present in $Ball^1(u)$, we move to the next level and in the second step we search $Ball^2(v)$ for vertex $p^2(u)$. We proceed in this way querying balls of u and v alternatively: In ith step, we search $Ball^i(x)$ for $p^i(y)$, where $(x = u, y = v)$ if i is odd, and $(x = v, y = u)$ otherwise. The search ends at ith step if $p^i(y)$ belongs to $Ball^i(x)$, and then we report $\delta(x, p^i(y)) + \delta(y, p^i(y))$ as an approximate distance between u and v.

Distance_Report(u, v)

Algorithm for reporting $(2k-1)$-approximate distance between $u, v \in V$

$l \leftarrow 1$,
$x \leftarrow u, y \leftarrow v$
While $(p^l(y) \notin Ball^l(x))$ **do**
         swap(x, y),
         $l \leftarrow l + 1$
return $\delta(y, p^l(y)) + \delta(x, p^l(y))$

Note that $p^k(y) \in \mathcal{R}_k^k$, and we store the distance from x to all the vertices of set $\mathcal{R}_k^k$ in $Ball^k(x)$ (which is $Ball(x, \mathcal{R}_k^k, \emptyset)$). Therefore, the "while loop" of the distance reporting algorithm will execute at-most $k - 1$ iterations, spending $O(1)$ time querying a hash table in each iteration.

In order to ensure that the above algorithm reports $(2k-1)$-approximate distance between $u, v \in V$, we first show that the following assertion holds:

$\mathcal{A}_i$: At the end of ith iteration of the "while loop", $\delta(y, p^{i+1}(y)) \leq i\delta(u, v)$.

The assertion $\mathcal{A}_i$ can be proved using induction on i as follows. First note that the variables x and y take the values u and v alternatively during the "while-loop". So $\delta(x, y) = \delta(u, v)$ always.

For the base case $(i = 0)$, $p^1(y)$ is same as y, and y is v. So $\delta(y, p^1(y)) = 0$. Hence $\mathcal{A}_0$ is true. For the rest of the inductive proof, it suffices to show that if $\mathcal{A}_j$ is true, then after $(j+1)$th iteration $\mathcal{A}_{j+1}$ is also true. The proof is as follows.

We consider the case of "even j," the arguments for the case of "odd j" are similar. For even j, at the end of jth iteration, $\{x = u, y = v\}$, Thus $\mathcal{A}_j$ implies that at the end of jth iteration $\delta(v, p^{j+1}(v)) \leq j\delta(u, v)$. Consider the $(j+1)$th iteration. For the execution of $(j+1)$th iteration, the condition in the "while-loop" must have been true. Thus $p^{j+1}(v)$ does not belong to $Ball(u, \mathcal{R}_k^{j+1}, \mathcal{R}_k^{j+2})$. Hence by Definition 39.1, the vertex $p^{j+2}(u)$ must be lying closer to u than the vertex $p^{j+1}(v)$. So at the end of $(j+1)$th iteration, $\delta(y, p^{j+2}(y))$ can be bounded as follows

$$
\begin{aligned}
\delta(y, p^{j+2}(y)) &= \delta(u, p^{j+2}(u)) \\
&\leq \delta(u, p^{j+1}(v)) \\
&\leq \delta(u, v) + \delta(v, p^{j+1}(v)) \quad \{\text{using triangle inequality}\} \\
&\leq \delta(u, v) + j\delta(u, v) \quad \{\text{using } \mathcal{A}_j\} \\
&= (j+1)\delta(u, v)
\end{aligned}
$$

Thus the assertion $\mathcal{A}_{j+1}$ holds.

Theorem 39.1

The algorithm *Distance_Report(u, v)* reports $(2k-1)$-approximate distance between u and v

Proof. As an approximate distance between u and v, note that the algorithm *Distance-Report(u, v)* would output $\delta(y, p^l(y)) + \delta(x, p^l(y))$, which by triangle inequality is no more than $2\delta(y, p^l(y)) + \delta(x, y)$. Since $\delta(x, y) = \delta(u, v)$, and $\delta(y, p^l(y)) \leq (l-1)\delta(u, v)$ as follows from assertion $\mathcal{A}_l$. Therefore, the distance reported is no more than $(2l-1)\delta(u, v)$. Since the "while loop" will execute at-most $k - 1$ iterations, so $l = k$, and therefore the distance reported by the oracle is at-most $(2k-1)\delta(u, v)$. ∎

39.2.3.2 Size of the (2k-1)-approximate Distance Oracle

The expected size of the set $\mathcal{R}_k^k$ is $O(n^{1/k})$, and the expected size of each $Ball^i(u)$ is $n^{1/k}$ using Lemma 39.2. So the expected size of the $(2k-1)$-approximate distance oracle is $O(n^{1/k} \cdot n + (k-1) \cdot n \cdot n^{1/k}) = O(kn^{1+1/k})$.

39.2.4 Computing Approximate Distance Oracles

In this subsection, a sub-cubic running time algorithm is presented for computing $(2k-1)$-approximate distance oracles. It follows from the description of the data-structure associated with approximate distance oracle that after forming the sampled sets of vertices $\mathcal{R}_k^i$, that takes $O(m)$ time, all that is required is the computation of $Ball^i(u)$ along with the distance from u to the vertices belonging to these balls for each u and $i \leq k$.

Since $Ball^i(u)$ is the set of all the vertices of set R_k^i that lie closer to u than the vertex $p^{i+1}(u)$. So, in order to compute $Ball^i(u)$, first we compute $p^i(u)$ for all $u \in V$, $i \leq k$.

39.2.4.1 Computing $p^i(u)$, $\forall u \in V$

Recall from definition itself that $p^i(u)$ is the vertex of the set R_k^i that is nearest to u. Hence, computing $p^i(u)$ for each $u \in V$ requires solving the following problem with $X = R_k^i$, $Y = V \backslash X$.

Given $X, Y \subset V$ in a graph $G = (V, E)$, with $X \cap Y = \emptyset$, compute the nearest vertex of set X for each vertex $y \in Y$.

The above problem can be solved by running a single source shortest path algorithm (Dijkstra's algorithm) on a modified graph as follows. Modify the original graph G by adding a dummy vertex s to the set V, and joining it to each vertex of the set X by an edge of zero weight. Let G' be the modified graph. Running Dijkstra's algorithm from the vertex s as the source, it can be seen that the distance from s to a vertex $y \in Y$ is indeed the distance from y to the nearest vertex of set X. Moreover, if $e(s, x), x \in X$ is the edge leading to the shortest path from s to y, then x is the vertex from the set X that lies nearest to y. The running time of the Dijkstra's algorithm is $O(m \log n)$, we can thus state the following lemma.

Lemma 39.3

Given $X, Y \subset V$ in a graph $G = (V, E)$, with $X \cap Y = \emptyset$, it takes $O(m \log n)$ to compute the nearest vertex of set X for each vertex $y \in Y$.

Corollary 39.1

Given a weighted undirected graph $G = (V, E)$, and a hierarchy of subsets $\{\mathcal{R}_k^i \mid i \leq k\}$, we can compute $p^i(u)$ for all $i \leq k, u \in V$ in $O(km \log n)$ time.

39.2.4.2 Computing $Ball^i(u)$ Efficiently

In order to compute $Ball^i(u)$ for each vertex $u \in V$ efficiently, we first compute clusters $\{C(v, \mathcal{R}_k^{i+1}) \mid v \in \mathcal{R}_k^i\}$ which are defined as follows:

Definition 39.2

For a graph $G = (V, E)$, and a set $X \subset V$, the cluster $C(v, X)$ consists of each vertex $w \in V$ for whom v lies closer than any vertex of set X. That is, $\delta(w, v) < \delta(w, x)$ for each $x \in X$.

It follows from the definition given above that $u \in C(v, \mathcal{R}_k^{i+1})$ if and only if $v \in Ball^i(u)$. So, given clusters $\{C(v, \mathcal{R}_k^{i+1}) \mid v \in \mathcal{R}_k^i\}$, we can compute $\{Ball^i(u) : u \in V\}$ as follows.

For each $v \in \mathcal{R}_k^i$ **do**
    For each $u \in C(v, \mathcal{R}_k^{i+1})$ **do**
        $Ball^i(u) \longleftarrow Ball^i(u) \cup \{v\}$

Hence we can state the following Lemma.

Lemma 39.4

Given the family of clusters $\{C(v, \mathcal{R}_k^{i+1}) \mid v \in \mathcal{R}_k^i\}$, the time required to compute $\{Ball^i(u)\}$ is bounded by $O(\sum_{u \in V} |Ball^i(u)|)$.

The following property of the cluster $C(v, \mathcal{R}_k^{i+1})$ will be used in its efficient computation.

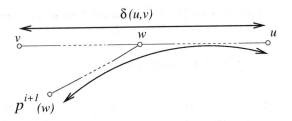

FIGURE 39.4   if w does not lie in $C(v, \mathcal{R}_k^{i+1})$, then $p^{i+1}(w)$ would lie closer to u than v.

Lemma 39.5

If $u \in C(v, \mathcal{R}_k^{i+1})$, then all the vertices on the shortest path from v to u also belong to the set $C(v, \mathcal{R}_k^{i+1})$.

Proof. We give a proof by contradiction. Given that $u \in C(v, \mathcal{R}_k^{i+1})$, let w be any vertex on the shortest path from v to u. If $w \neq C(v, \mathcal{R}_k^{i+1})$, the vertex v doesn't lie closer to w than the vertex $p^{i+1}(w)$. See Figure 39.4. In other words $\delta(w, v) \geq \delta(w, p^{i+1}(w))$. Hence

$$\delta(u, v) = \delta(u, w) + \delta(w, v) \geq \delta(u, w) + \delta(w, p^{i+1}(w)) \geq \delta(u, p^{i+1}(w))$$

Thus v does not lie closer to u than $p^{i+1}(w)$ which is a vertex of set $\mathcal{R}_k^{i+1}$. Hence by definition, $u \neq C(v, \mathcal{R}_k^{i+1})$, thus a contradiction. ∎

From Lemma 39.5, it follows that the graph induced by the vertices of the cluster $C(v, \mathcal{R}_k^{i+1})$ is connected (hence the name cluster). Moreover, the entire cluster $C(v, \mathcal{R}_k^{i+1})$ appears as a sub-tree of the shortest path tree rooted at v in the graph. As follows from the definition, for each vertex $x \in C(v, \mathcal{R}_k^{i+1})$, $\delta(v, x) < \delta(x, p^{i+1}(x))$. Based on these two observations, here follows an efficient algorithm that computes the set $C(v, \mathcal{R}_k^{i+1})$. The algorithm performs a restricted Dijkstra's algorithm from the vertex v, wherein we don't proceed along any vertex that does not belong to the set $C(v, \mathcal{R}_k^{i+1})$.

A restricted Dijkstra's algorithm: Note that the Dijkstra's algorithm starts with singleton tree $\{v\}$ and performs $n - 1$ steps to grow the complete shortest path tree. Each vertex $x \in V \backslash \{v\}$ is assigned a label $L(x)$, which is infinity in the beginning, but eventually becomes the distance from v to x. Let V_i denotes the set of i nearest vertices from v. The algorithm maintains the following invariant at the end of lth step:

$\mathcal{I}(l)$: For all the vertices of the set V_l, the label $L(x) = \delta(v, x)$, and for every other vertex $y \in V \backslash V_l$, the label $L(y)$ is equal to the length of the shortest path from v to y that passes through vertices of V_l only.

During the $(j + 1)$th step, we select the vertex, say w from set $V - V_j$ with least value of $L(\cdot)$. Since all the edge weights are positive, it follows from the invariant $\mathcal{I}(j)$ that $L(w) = \delta(w, v)$. Thus we add w to set V_j to get the set V_{j+1}. Now in order to satisfy the invariant $\mathcal{I}(j + 1)$, we relax each edge $e(w, y)$ incident from w to a vertex $y \in V - V_{j+1}$ as follows: $L(y) \leftarrow \min\{L(y), L(w) + weight(w, y)\}$. It is easy to observe that this ensures the validity of the invariant $\mathcal{I}(j + 1)$.

In the restricted Dijkstra's algorithm, we will put the following restriction on relaxation of an edge $e(w, y)$: we relax the edge $e(w, y)$ only if $L(w) + weight(w, y)$ is less than $\delta(y, p^i(y))$. This will ensure that a vertex $y \neq C(v, \mathcal{R}_k^{i+1})$ will never be visited during the algorithm. The fact that the vertices of the cluster $C(v, \mathcal{R}_k^{i+1})$ form a sub-tree of the shortest path tree rooted at v, ensures that the above restricted Dijkstra's algorithm indeed finds all the vertices (along with their distance from v) that form the cluster $C(v, \mathcal{R}_k^{i+1})$. Since the running time of Dijkstra's algorithm is dominated by the number of edges relaxed, and each edge relaxation takes $\log(n)$ time only, therefore, the restricted Dijkstra's algorithm will run in time of the order of $\sum_{x \in C(v, \mathcal{R}_k^{i+1})} degree(x) \log n$. Thus the total time for computing all the clusters $\{C(v, \mathcal{R}_k^{i+1}) \mid v \in \mathcal{R}_k^i\}$ is given by:

$$\sum_{v \in \mathcal{R}_k^i, x \in C(v, \mathcal{R}_k^{i+1})} degree(x) \log n = \left(\sum_{x \in V, v \in Ball^i(x)} degree(x) \right) \log n$$

$$= \left(\sum_{x \in V} |Ball^i(x)| \cdot degree(x) \right) \log n$$

By Lemma 39.2, the expected size of $Ball^i(x)$ is bounded by $n^{1/k}$, hence using linearity of expectation, the total expected cost of computing $\{C(v, \mathcal{R}_k^{i+1}) \mid v \in \mathcal{R}_k^i\}$ is asymptotically bounded by

$$\sum_{x \in V} n^{1/k} \cdot degree(x) \log n = 2mn^{1/k} \log n$$

Using the above result and Lemma 39.4, we can thus conclude that for a given weighted graph $G = (V, E)$ and an integer k, it takes a total of $\tilde{O}(kmn^{1/k} \log n)$ time for computing $\{Ball^i(u) \mid i < k, u \in V\}$. If we use Fibonacci heaps instead of binary heaps in implementation of the restricted Dijkstra's algorithm, we can get rid of the logarithmic factor in the running time. Hence the total expected running time for building the data-structure is $O(kmn^{1/k})$. As mentioned before, the expected size of the data-structure will be $O(kn^{1+1/k})$. To get $O(kn^{1+1/k})$ bound on the worst case size of the data-structure, we repeat the preprocessing algorithm. The expected number of iterations will be just a constant. Hence, we can state the following theorem.

Theorem 39.2

Given a weighted undirected graph $G = (V, E)$ and an integer k, a data-structure of size $O(kn^{1+1/k})$ can be built in $O(kmn^{1/k})$ expected time so that given any pair of vertices, $(2k - 1)$-approximate distance between them can be reported in $O(k)$ time.

39.3   A Randomized Data-Structure for Decremental APASP

There are a number of applications that require efficient solutions of the APASP problem for a dynamic graph. In these applications, an initial graph is given, followed by an on-line sequence of queries interspersed with updates that can be insertion or deletion of edges. We have to carry out the updates and answer the queries on-line in an efficient manner. The goal of a dynamic graph algorithm is to update the solution efficiently after the dynamic changes, rather than having to re-compute it from scratch each time.

The approximate distance oracles described in the previous section can be used for answering approximate distance query in a static graph. However, there does not seem to be any efficient way to dynamize these oracles in order to answer distance queries in a graph under deletion of edges. In this section we shall describe a hierarchical data structure for efficiently maintaining APASP in an undirected unweighted graph under deletion of edges. In addition to maintaining approximate shortest paths for all-pairs of vertices, this scheme has been used for efficiently maintaining approximate shortest paths for pair of vertices separated by distance in an interval $[a, b]$ for any $1 \le a < b \le n$. However, to avoid giving too much detail in this chapter, we would outline an efficient algorithm for the following problem only.

APASP-d: *Given an undirected unweighted graph $G = (V, E)$ that is undergoing deletion of edges, and a distance parameter $d \le n$, maintain approximate shortest paths for all-pairs of vertices separated by distance at-most d.*

39.3.1 Main Idea

For an undirected unweighted graph $G = (V, E)$, a breadth-first-search (BFS) tree rooted at a vertex $u \in V$ stores distance information with respect to the vertex u. So in order to maintain shortest paths for all-pairs of vertices separated by distance $\le d$, it suffices to maintain a BFS tree of depth d rooted at each vertex under deletion of edges. This is the approach taken by the previously existing algorithms.

The main idea underlying the hierarchical data-structure that would provide efficient update time for maintaining APASP can be summarized as follows: Instead of maintaining exact distance information separately from each vertex, keep *small* BFS trees around each vertex for maintaining distance information within locality of each vertex, and some what *larger* BFS trees around *fewer* vertices for maintaining global distance information.

We now provide the underlying intuition of the above idea and a brief outline of the new techniques used.

Let B_u^d denote the BFS tree of depth d rooted at vertex $u \in V$. There exists a simple algorithm for maintaining a BFS tree B_u^d under deletion of edges that takes a total of $\mu(B_u^d) \cdot d$ time, where $\mu(t)$ is the number of edges in the graph induced by tree t. Thus the total update time for maintaining shortest path for all-pairs separated by distance at-most d is of the order of $\sum_{u \in V} \mu(B_u^d) \cdot d$. Potentially $\mu(B_u^d)$ can be as large as $\theta(m)$, and so the total update time over any sequence of edge deletions will be $O(mnd)$. Dividing this total update cost uniformly over the entire sequence of edge deletions, we can see that it takes $O(nd)$ amortized update time per edge deletion, and $O(1)$ time for reporting exact distance between any pair of vertices separated by distance at-most d.

In order to achieve $o(nd)$ bound on the update time for the problem APASP-d, we closely look at the expression of total update time $\sum_{u \in V} \mu(B_u^d) \cdot d$. There are n terms in this expression each of potential size $\theta(m)$. A decrease in either the total number of

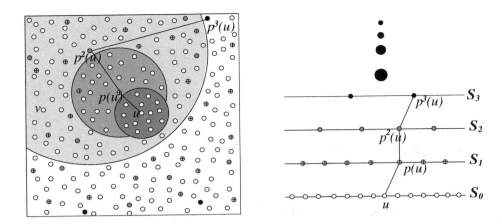

FIGURE 39.5 Hierarchical scheme for maintaining approximate distance.

terms or the size of each term would give an improvement in the total update time. Thus the following simple ideas come to mind.

- Is it possible to solve the problem APASP-d by keeping very *few* depth-d BFS trees?
- Is there some other alternative t for depth bounded BFS tree B_u^d that has $o(m)$ bound on $\mu(t)$?

While it appears difficult for any of the above ideas to succeed individually, they can be combined in the following way: *Build and maintain BFS trees of depth 2d on vertices of a set $S \subset V$ of size $o(n)$, called the set of special vertices, and for each remaining vertex $u \in V \backslash S$, maintain a BFS tree (denoted by B_u^S) rooted at u and containing all the vertices that lie closer to u than the nearest special vertex, say $\mathcal{N}(u, S)$.*

Along the above lines, we present a 2-level data-structure (and its generalization to k-levels) for the problem APASP-d.

It can be seen that unlike the tree B_u^d, the new BFS tree B_u^S might not contain all the vertices lying within distance d from u. In order to ensure that our scheme leads to a solution of problem APASP-d, we use the following observation similar to that of 3-approximate distance oracle in the previous section. If v is a vertex lying within distance d from u but not present in B_u^S, an *approximate* distance from u to v can be extracted from the tree rooted at the nearest special vertex $\mathcal{N}(u, S)$. This is because (by triangle inequality) the distance from $\mathcal{N}(u, S)$ to v is at most twice the distance from u to v.

For our hierarchical scheme to lead to improved update time, it is crucial that we establish sub-linear upper bounds on $\mu(B_u^S)$. We show that if the set S is formed by picking each vertex independently with *suitable* probability, then $\mu(B_u^S) = \tilde{O}(m/|S|)$ with probability arbitrarily close to 1.

39.3.2 Notations

For an undirected unweighted graph $G = (V, E)$, $S \subset V$, and a distance parameter $d \leq n$,

- $\delta(u, v)$: distance between u and v.
- $\mathcal{N}(v, S)$: the vertex of the set $S \subset V$ nearest to v.
- B_v^d: The BFS tree of depth d rooted at $v \in V$.
- B_v^S: The BFS tree of depth $(\delta(u, \mathcal{N}(u, S)) - 1)$ rooted at v.
- $B_v^{d,S}$: The BFS tree of depth $\min\{d, \delta(v, \mathcal{N}(v, S)) - 1\}$ rooted at v.
- $\mu(t)$: the number of edges in the sub-graph (of G) induced by the tree t.
- $\nu(t)$: the number of vertices in tree t.
- For a sequence $\{S_0, S_1, \ldots S_{k-1}\}$, $S_i \subset V$, and a vertex $u \in S_0$, we define

  $p^0(u) = u.$
  $p^{i+1}(u) = $ the vertex from set S_{i+1} nearest to $p^i(u)$.

- $\overline{\alpha}$: the smallest integer of the form 2^i which is greater than α.

39.3.3 Hierarchical Distance Maintaining Data-Structure

Based on the idea of "keeping *many small* trees, *and a few large* trees," we define a k-level hierarchical data-structure for efficiently maintaining approximate distance information as follows. (See Figure 39.5)

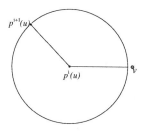

FIGURE 39.6   Bounding the approximate distance between $p^{i+1}(u)$ and v.

Let $\mathcal{S} = \{S_0, S_1, \ldots, S_{k-1} : S_i \subset V, |S_{i+1}| < |S_i|\}$ be a sequence. For a given distance parameter $d \leq n$ and $i < k-1$, let $\mathcal{F}_i$ be the collection $\{B_u^{2^i d, S_{i+1}} : u \in S_i\}$ of BFS trees, and $\mathcal{F}_{k-1}$ be the collection of BFS trees of depth $2^{k-1} d$ rooted at each $u \in S_{k-1}$. We shall denote the set $\{(S_0, \mathcal{F}_0), (S_1, \mathcal{F}_1), \ldots, (S_{k-1}, \mathcal{F}_{k-1})\}$ as the k-level hierarchy $\mathcal{H}_d^k$ induced by the sequence $\mathcal{S}$.

Let v be a vertex within distance d from u. If v is present in B_u^{d, S_1}, we can report exact distance between them. Otherwise, (as will soon become clear) we can extract the approximate distance between u and v from the collection of the BFS trees rooted at the vertices $u, p(u), \ldots, p^{k-1}(u)$ (see Figure 39.5). The following Lemma is the basis for estimating the distance between two vertices using the hierarchy $\mathcal{H}_d^k$.

Lemma 39.6

Given a hierarchy $\mathcal{H}_d^k$, if $j < k-1$ is such that v is not present in any of the BFS trees $\{B_{p^i(u)}^{2^i d, S_{i+1}} \mid 0 \leq i \leq j\}$, then for all $i \leq j$

$$\delta(p^{i+1}(u), p^i(u)) \leq 2^i \delta(u, v) \quad \text{and} \quad \delta(p^{i+1}(u), v) \leq 2^{i+1} \delta(u, v).$$

Proof. We give a proof by induction on j.

Base Case $(j = 0)$: Since v is not present in $B_u^{S_1}$, so the vertex $p(u)$ must be lying equidistant or closer to u than v. Hence $\delta(p(u), u) \leq \delta(u, v)$. Using triangle inequality, it follows that $\delta(p(u), v) \leq \delta(p(u), u) + \delta(u, v) = 2\delta(u, v)$.

Induction Hypothesis:
$\delta(p^{i+1}(u), p^i(u)) \leq 2^i \delta(u, v)$, and
$\delta(p^{i+1}(u), v) \leq 2^{i+1} \delta(u, v)$, for all $i < l$.

Induction Step $(j = l)$: if $v \neq B_{p^l(u)}^{S_{l+1}}$, then the distance between p^{l+1} and $p^l(u)$ must not be longer than $\delta(p^l(u), v)$, which is less than $2^l \delta(u, v)$ (using induction hypothesis).

Now using triangle inequality (see the Figure 39.6) we can bound $\delta(p^{l+1}(u), v)$ as follows.

$$\begin{aligned}
\delta(p^{l+1}(u), v) &\leq \delta(p^{l+1}(u), p^l(u)) + \delta(p^l(u), v) \\
&\leq 2^l \delta(u, v) + \delta(p^l(u), v) \\
&\leq 2^l \delta(u, v) + 2^l \delta(u, v) \quad \{\text{using I.H.}\} \\
&= 2^{l+1} \delta(u, v)
\end{aligned}$$

$\blacksquare$

Since the depth of a BFS tree at $(k-1)$th level of hierarchy $\mathcal{H}_d^k$ is $2^{k-1} d$, therefore the following corollary holds true.

Corollary 39.2

If $\delta(u, v) \leq d$, then there is some $p^i(u), i < k$ such that v is present in the BFS tree rooted at $p^i(u)$ in the hierarchy $\mathcal{H}_d^k$.

Lemma 39.7

Given a hierarchy $\mathcal{H}_d^k$, if $j < k-1$ is such that v is not present in any of the BFS trees $\{B_{p^i(u)}^{2^i d, S_i} \mid 0 \leq i \leq j\}$, then $\delta(p^{i+1}(u), u) \leq (2^{i+1} - 1)\delta(u, v)$, for all $i \leq j$.

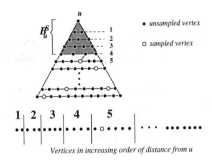

FIGURE 39.7 Bounding the size of BFS tree B_u^S.

Proof. Using simple triangle inequality, it follows that

$$\delta(p^{i+1}(u), u) \leq \sum_{l \leq i} \delta(p^{l+1}(u), p^l(u))$$

$$\leq \sum_{l \leq i} 2^l \delta(u, v) = (2^{i+1} - 1)\delta(u, v)$$

∎

It follows from Lemmas 39.6 and 39.7 that if l is the smallest integer such that v is present in the BFS tree rooted at $p^l(u)$ in the hierarchy $\mathcal{H}_d^k$, then we can report $\delta(p^l(u), u) + \delta(p^l(u), v)$ as an approximate distance between u and v. Along these lines, we shall present an improved decremental algorithms for APASP-d.

39.3.4 Bounding the Size of $B_u^{d,S}$ under Edge-Deletions

We shall now present a scheme based on random sampling to find a set $S \subset V$ of vertices that will establish a sub-linear bound on the number of vertices ($\nu(B_u^S)$) as well as the number of edges ($\mu(B_u^S)$) induced by B_u^S under deletion of edges. Since $B_u^{d,S} \subset B_u^S$, so these upper bounds also hold for $B_u^{d,S}$.

Build the set S of vertices by picking each vertex from V independently with probability $\frac{n^c}{n}$. The expected size of S is $O(n^c)$. Consider an ordering of vertices V according to their levels in the BFS tree B_u^S (see Figure 39.7). The set of vertices lying at higher levels than the nearest sampled vertex in this ordering is what constitutes the BFS tree B_u^S. Along similar lines as that of Lemma 39.1, it follows that the expected size of this set (and hence $\nu(B_u^S)$) is $\frac{n}{n^c}$. Moreover, it can be shown that $\nu(B_u^S)$ is no more than $\frac{4n \ln n}{n^c}$ with probability $> 1 - \frac{1}{n^4}$. Now as the edges are being deleted, the levels of the vertices in the tree B_u^S may fall, and so the ordering of the vertices may change. There will be a total of m such orderings during the entire course of edge deletions. Since the vertices are picked randomly and independently, therefore, the upper bound of $\frac{4n \ln n}{n^c}$ holds for $\nu(B_u^S)$ with probability $(1 - \frac{1}{n^4})$ for any of these orderings. So we can conclude that $\nu(B_u^S)$, the number of vertices of tree B_u^S never exceeds $(\frac{4n \ln n}{n^c})$ during the entire course of edge deletions with probability $> 1 - \frac{1}{n^2}$.

To bound the number of edges induced by B_u^S, consider the following scheme. Pick every edge independently with probability $\frac{n^c}{m}$. The set S consists of the end points of the sampled edges. The expected size of S is $O(n^c)$. Consider an ordering of the edges according to their level in B_u^S (level of an edge is defined as the minimum of the levels of its end points). Along the lines of arguments given above (for bounding the the number of vertices of B_u^S), it can be shown that $\mu(B_u^S)$, the number of edges induced by B_u^S remains $\leq \frac{4m \ln n}{n^c}$ with probability $> 1 - \frac{1}{n^2}$ during the entire course of edge deletions.

Note that in the sampling scheme to bound the number of vertices of tree B_u^S, a vertex v is picked with probability $\frac{n^c}{n}$. Whereas in the sampling scheme for bounding the number of edges in the sub-graph induced by B_u^S, a vertex v is picked with probability $\frac{degree(v) \cdot n^c}{m}$. It can thus be seen that both the bounds can be achieved simultaneously by the following random sampling scheme:

$\mathcal{R}(c)$: Pick each vertex $v \in V$ independently with probability $\frac{n^c}{n} + \frac{degree(v) \cdot n^c}{m}$.

It is easy to see that the expected size of the set formed by the sampling scheme $\mathcal{R}(c)$ will be $O(n^c)$.

Theorem 39.3

Given an undirected unweighted graph $G = (V, E)$, a constant $c < 1$, and a distance parameter d; a set S of size $O(n^c)$ vertices can be found that will ensure the following bound on the number of vertices and number of edges in the sub-graph of G induced by $B_u^{d,S}$.

$$v(B_u^{d,S}) = O\left(\frac{n \ln n}{n^c}\right), \quad \mu(B_u^{d,S}) = O\left(\frac{m \ln n}{n^c}\right)$$

with probability $\Omega(1 - \frac{1}{n^2})$ during the entire sequence of edge deletions.

39.3.4.1 Maintaining the BFS Tree $B_u^{d,S}$ under Edge Deletions

Even and Shiloach [3] design an algorithm for maintaining a depth-d BFS tree in an undirected unweighted graph.

Lemma 39.8: (Even, Shiloach [3])

Given a graph under deletion of edges, a BFS tree B_u^d, $u \in V$ can be maintained in $O(d)$ amortized time per edge deletion.

For maintaining a $B_u^{d,S}$ tree under edge deletions, we shall use the same algorithm of [3] with the modification that whenever the depth of $B_u^{d,S}$ has to be increased (due to recent edge deletion), we grow the tree to its new level min $\{d, \delta(u, \mathcal{N}(u,S)) - 1\}$. We analyze the total update time required for maintaining $B_u^{d,S}$ as follows.

There are two computational tasks: one extending the level of the tree, and another that of maintaining the levels of the vertices in the tree $B_u^{d,S}$ under edge deletions. For the first task, the time required is bounded by the edges of the new level introduced which is $O(\mu(B_u^{d,S}))$. For the second task, we give a variant of the proof of Even and Shiloach [3] (for details, please refer [3]). The running time is dominated by the processing of the edges in this process. In-between two consecutive processing of an edge, level of one of the end-points of the edge falls down by at least one unit. The processing cost of an edge can thus be charged to the level from which it has fallen. Clearly the maximum number of edges passing a level i is bounded by $\mu(B_u^{d,S})$. The number of levels in the tree $B_u^{d,S}$ is $\min\{d, v(B_u^{d,S})\}$. Thus the total cost for maintaining the BFS tree $B_u^{d,S}$ over the entire sequence of edge deletions is $O(\mu(B_u^{d,S}) \cdot \min\{d, v(B_u^{d,S})\})$.

Lemma 39.9

Given an undirected unweighted graph $G = (V, E)$ under edge deletions, a distance parameter d, and a set $S \subset V$; a BFS tree $B_u^{d,S}$ can be maintained in

$$O\left(\frac{\mu(B_u^{d,S})}{m} \cdot \min\left\{d, v\left(B_u^{d,S}\right)\right\}\right)$$

amortized update time per edge deletion.

39.3.4.2 Some Technical Details

As the edges are being deleted, we need an efficient mechanism to detect any increase in the depth of tree $B_u^{d,S}$. We outline one such mechanism as follows.

For every vertex $v \neq S$, we keep a count $C[v]$ of the vertices of the S that are neighbors of v. It is easy to maintain $C[u], \forall u \in V$ under edge-deletions. We use the count $C[v]$ in order to detect any increase in the depth of a tree $B_u^{d,S}$ as follows. Note that when depth of a tree $B_u^{d,S}$ is less than d, there has to be at-least one vertex w at leaf-level in $B_u^{d,S}$ with $C[w] \geq 1$ (as an indicator that the vertex $p(u)$ is at next level). Therefore, after an edge deletion if there is no vertex w at leaf level with $C[w] \geq 1$, we grow the BFS tree $B_u^{d,S}$ beyond its previous level until either depth becomes d or we reach some vertex w' with $C[w'] \geq 1$.

Another technical issue is that when an edge $e(x, y)$ is deleted, we must update only those trees which contain x and y. For this purpose, we maintain for each vertex, a set of roots of all the BFS trees containing it. We maintain this set using any dynamic search tree.

39.3.5 Improved Decremental Algorithm for APASP up to Distance d

Let $\{(S_0, \mathcal{F}_0), (S_1, \mathcal{F}_1), \ldots, (S_{k-1}, \mathcal{F}_{k-1})\}$ be a k-level hierarchy $\mathcal{H}_d^k$ with $S_0 = V$ and $n^{c_i} = |S_i|$, where each $c_i, i < k$ is a fraction to be specified soon. Each set $S_i, i > 0$ is formed by picking the vertices from set V using the random sampling scheme $\mathcal{R}$ mentioned in the previous subsection.

To report distance from u to v, we start form the level 0. We first inquire if v lies in B_u^{d,S_1}. If v does not lie in the tree, we move to the first level and inquire if v lies in $B_{p(u)}^{2d,S_2}$. It follows from the Corollary 39.2 that if $\delta(u,v) \leq d$, then proceeding in this way, we eventually find a vertex $p^l(u), l \leq k-1$ in the hierarchy $\mathcal{H}_d^k$ such that v is present in the BFS tree rooted at $p^l(u)$. (See Figure 39.5). We then report the sum of distances from $p^l(u)$ to both u and v.

Algorithm for reporting approximate distance using $\mathcal{H}_d^k$

$$
\begin{aligned}
&\textbf{Distance}(u,v) \\
&\{ \quad D \leftarrow 0; l \leftarrow 0 \\
&\quad \textbf{While } (v \notin B_{p^l(u)}^{2^l d, S_{l+1}} \wedge l < k-1) \textbf{ do} \\
&\quad \{ \\
&\qquad \textbf{If } u \in B_{p^l(u)}^{2^l d, S_{l+1}}, \textbf{ then } D \leftarrow \delta(p^l(u), u), \\
&\qquad D \leftarrow D + \delta(p^l(u), p^{l+1}(u)) \\
&\qquad l \leftarrow l+1; \\
&\quad \} \\
&\quad \textbf{If } v \notin B_{p^l(u)}^{2^l d, S_{l+1}}, \textbf{ then } \text{``}\delta(u,v) \text{ is greater than } d\text{''}, \\
&\qquad\qquad\qquad \textbf{else return } \delta(p^l(u), v) + D \\
&\}
\end{aligned}
$$

The approximation factor ensured by the above algorithm can be bounded as follows.

It follows from the Lemma 39.7 that the final value of D in the algorithm given above is bounded by $(2^l - 1)\delta(u,v)$, and it follows from Lemma 39.6 that $\delta(p^l(u), v)$ is bounded by $2^l \delta(u,v)$. Since $l \leq k-1$, therefore the distance reported by the algorithm is bounded by $(2^k - 1)\delta(u,v)$ if v is at distance $\leq d$.

Lemma 39.10

Given an undirected unweighted graph $G = (V, E)$, and a distance parameter d. If α is the desired approximation factor, then there exists a hierarchical scheme $\mathcal{H}_d^k$ with $k = \log_2 \bar{\alpha}$, that can report α-approximate shortest distance between any two vertices separated by distance $\leq d$, in time $O(k)$.

Update time for maintaining the hierarchy H_k^d: The update time per edge deletion for maintaining the hierarchy $\mathcal{H}_k^d$ is the sum total of the update time for maintaining the set of BFS trees $\mathcal{F}_i, i \leq k-1$.

Each BFS tree from the set $\mathcal{F}_{k-1}$ has depth $2^{k-1}d$, and edges $O(m)$. Therefore, using Lemma 39.8, each tree from set $\mathcal{F}_{k-1}$ requires $O(2^{k-1}d)$ amortized update time per edge deletion. So, the amortized update time T_{k-1} per edge deletion for maintaining the set $\mathcal{F}_{k-1}$ is

$$T_{k-1} = O(n^{c_{k-1}} 2^{k-1} d)$$

It follows from the Theorem 39.3 that a tree t from a set $\mathcal{F}_i, i < (k-1)$, has $\mu(t) = m \ln n / n^{c_{i+1}}$, and depth $= \min\{2^i d, n \ln n / n^{c_{i+1}}\}$. Therefore, using the Lemma 39.9, each tree $t \in \mathcal{F}_i, i < k-1$ requires $O(\min\{2^i d / n^{c_{i+1}}, n \ln n / n^{2c_{i+1}}\})$ amortized update time per edge deletion. So the amortized update time T_i per edge deletion for maintaining the set $\mathcal{F}_i$ is

$$T_i = O\left(\min\left\{2^i d \frac{n^{c_i}}{n^{c_{i+1}}} \ln n, \frac{n^{1+c_i}}{n^{2c_{i+1}}} \ln^2 n\right\}\right), \quad i < k-1$$

Hence, the amortized update time T per edge deletion for maintaining the hierarchy $\mathcal{H}_k^d$ is

$$T = T_{k-1} + \sum_{i < k-1} T_i$$

$$= O(n^{c_{k-1}} 2^{k-1} d) + \sum_{i=0}^{i=k-2} O\left(\min\left\{2^i d \frac{n^{c_i}}{n^{c_{i+1}}} \ln n, \frac{n^{1+c_i}}{n^{2c_{i+1}}} \ln^2 n\right\}\right)$$

To minimize the sum on right hand side in the above equation, we balance all the terms constituting the sum, and get

$$T = \tilde{O}\left(2^{k-1} \cdot \min\left\{\sqrt[k]{nd}, (nd)^{\frac{2^{(k-1)}}{2^k - 1}}\right\}\right)$$

If α is the desired approximation factor, then it follows from Lemma 39.10 that the number of levels k, in the hierarchy are $\log_2 \bar{\alpha}$. So the amortized update time required is $\tilde{O}(\alpha \cdot \min\{\sqrt[\log_2 \bar{\alpha}]{nd}, (nd)^{\frac{\bar{\alpha}}{2(\bar{\alpha}-1)}}\})$.

TABLE 39.1 Maintaining α-approximate distances for all-pairs of vertices separated by distance $\leq d$.

Data-structure	α (the approximation factor)	Amortized update time per edge deletion
$\dot{\mathcal{D}}_3(1, d)$	3	$\tilde{O}(\min(\sqrt{n}d, (nd)^{2/3}))$
$\dot{\mathcal{D}}_7(1, d)$	7	$\tilde{O}(\min(\sqrt[3]{n}d, (nd)^{4/7}))$
$\dot{\mathcal{D}}_{15}(1, d)$	15	$\tilde{O}(\min(\sqrt[4]{n}d, (nd)^{8/15}))$

Theorem 39.4

Let $G = (V, E)$ be an undirected unweighted graph undergoing edge deletions, d be a distance parameter, and $\alpha > 2$ be the desired approximation factor. There exists a data-structure $\dot{\mathcal{D}}_\alpha(1, d)$ for maintaining α-approximate distances for all-pairs separated by distance $\leq d$ in $\tilde{O}(\alpha \cdot \min\{ \sqrt[\log_2\overline{\alpha}]{n}d, (nd)^{\frac{\overline{\alpha}}{2(\overline{\alpha}-1)}} \})$ amortized update time per edge deletion, and $O(\log\overline{\alpha})$ query time.

Based on the data-structure of [3], the previous best algorithm for maintaining all-pairs exact shortest paths of length $\leq d$ requires $O(nd)$ amortized update time. We have been able to achieve $o(nd)$ update time at the expense of introducing approximation as shown in Table 39.1 on the following page.

39.4 Further Reading and Bibliography

Zwick [4] presents a very recent and comprehensive survey on the existing algorithms for all-pairs approximate/exact shortest paths. Based on the fastest known matrix multiplication algorithms given by Coppersmith and Winograd [5], the best bound for computing all-pairs shortest paths is $O(n^{2.575})$ [6].

Approximate distance oracles are designed by Thorup and Zwick [7]. Based on a 1963 girth conjecture of Erdös [8], they also show that $\Omega(n^{1+1/k})$ space is needed in the worst case for any oracle that achieves stretch strictly smaller than $(2k + 1)$. The space requirement of their approximate distance oracle is, therefore, essentially optimal. Also note that the preprocessing time of $(2k - 1)$-approximate distance oracle is $O(mn^{1/k})$, which is sub-cubic. However, for further improvement in the computation time for approximate distance oracles, Thorup and Zwick pose the following question: *Can $(2k - 1)$-approximate distance oracle be computed in $\tilde{O}(n^2)$ time?* Recently Baswana and Sen [9] answer their question in affirmative for unweighted graphs. However, the question for weighted graphs is still open.

For maintaining fully dynamic all-pairs shortest paths in graphs, the best known algorithm is due to Demetrescu and Italiano [10]. They show that it takes $O(n^2)$ amortized time to maintain all-pairs exact shortest paths after each update in the graph. Baswana et al. [11] present a hierarchical data-structure based on random sampling that provides efficient decremental algorithm for maintaining APASP in undirected unweighted graphs. In addition to achieving $o(nd)$ update time for the problem APASP-d (as described in this chapter), they also employ the same hierarchical scheme for designing efficient data-structures for maintaining approximate distance information for all-pairs of vertices separated by distance in an interval $[a, b], 1 \leq a < b \leq n$.

Acknowledgments

The work of the first author is supported, in part, by a fellowship from Infosys Technologies Limited, Bangalore.

References

1. T.H. Cormen, C.E. Leiserson, and R.L. Rivest, *Introduction to Algorithms*, the MIT Press, 1990, chapter 12.
2. M.L. Fredman, J. Komlós, and E. Szemerédi, Storing a sparse table with $O(1)$ worst case time, *J. ACM*, 31, 538, 1984.
3. S. Even and Y. Shiloach, An on-line edge-deletion problem, *J. ACM*, 28, 1, 1981.
4. U. Zwick, Exact and approximate distances in graphs - a survey, in *Proc. of the 9th European Symposium on Algorithms (ESA)*, 2001, 33.
5. D. Coppersmith and S. Winograd, Matrix multiplication via arithmetic progressions, *J. Symbolic Computation*, 9, 251, 1990.
6. U. Zwick, All-pairs shortest paths in weighted directed graphs - exact and almost exact algorithms, in *Proc. of the 39th IEEE Symposium on Foundations of Computer Science (FOCS)*, 1998, 310.
7. M. Thorup and U. Zwick, Approximate distance oracles, in *Proc. of 33rd ACM Symposium on Theory of Computing (STOC)*, 2001, 183.

8. P. Erdös, Extremal problems in graph theory, in *Theory of Graphs and its Applications* (Proc. Sympos. Smolenice, 1963), Publ. House Czechoslovak Acad. Sci., Prague. 1964, 29.

9. S. Baswana and S. Sen, Approximate distance oracle for unweighted graphs in $\tilde{O}(n^2)$ time, to appear in *Proc. of the 15th Annual ACM-SIAM Symposium on Discrete Algorithms (SODA)*, 2004.

10. C. Demetrescu and G.F. Italiano, A new approach to dynamic all pairs shortest paths, in *Proc. of 35th ACM Symposium on Theory of Computing (STOC)*, 2003, 159.

11. S. Baswana, R. Hariharan, and S. Sen, Maintaining all-pairs approximate shortest paths under deletion of edges, in *Proc. of the 14th Annual ACM-SIAM Symposium on Discrete Algorithms (SODA)*, 2003, 394.

40

Searching and Priority Queues in o(log n) Time[*]

40.1	Introduction	627
40.2	Model of Computation	627
40.3	Overview	628
40.4	Achieving Sub-Logarithmic Time per Element by Simple Means	628
	Range Reduction • Packing Keys • Combining	
40.5	Deterministic Algorithms and Linear Space	630
	Fusion Trees • Exponential Search Trees	
40.6	From Amortized Update Cost to Worst-Case	633
40.7	Sorting and Priority Queues	634
	Range Reduction • Packed Sorting • Combining the Techniques • Further Techniques and Faster Randomized Algorithms	
	References	636

Arne Andersson
Uppsala University

40.1 Introduction

In many cases of algorithm design, the comparison-based model of computation is not the obvious choice. In this chapter, we show how to design data structures with very good complexity on a realistic model of computation where keys are regarded as binary strings, each one contained in one or more machine words (registers). This model is sometimes referred to as the *RAM model,*[†] and it may be argued that it reflects real computers more accurately than the comparison-based model.

In the RAM-model the *word length* becomes a natural part of the model. A comparison does not necessarily take constant time, on the other hand we may use a larger variety of operations on data. This model allows for comparison-based algorithms to be used but also for algorithms like tries, bucket sort, radix sort etc, which are known to be efficient in practice.

40.2 Model of Computation

We use a unit-cost RAM with word size w. In the standard case we assume that the n keys are w-bit keys that can be treated as binary strings or integers, but we may also consider key that occupy multiple words. It should be noted that the assumption that keys can be treated as binary strings or integers also holds for floating-point numbers (*cf.* IEEE 754 floating-point standard [1, p. 228]).

In the RAM-model, we can use other operations than comparisons, for instance indirect addressing, shifting, bitwise logical operations, and multiplication. Without loss of generality, we assume that $w = \Omega(\log n)$, since otherwise we could not even fit the number n, or a pointer, into a machine word. (If we can not fit the number n into a constant number of words, the traditional analysis for comparison-based algorithms would also fail.)

Our complexity analysis has *two* parameters, the number of keys n and the word length w. In cases where the complexity is expressed only in terms of n, it is supposed to hold for any possible value of w, and vice versa.

For the searching problem, we assume that an ordered set is maintained and that operations like range queries and neighbour queries are supported. We say that we study ordered dictionaries, as defined below.

---

[*] This chapter has been reprinted from first edition of this Handbook, without any content updates.
[†] The term RAM is used for many models. There are also RAM models with infinite word length.

Definition 40.1

A dictionary is ordered if neighbour queries and range queries are supported at the same cost as member queries (plus the reporting cost), and if the keys can be reported in sorted order in linear time.

40.3 Overview

The basic purpose of this chapter is to introduce some of the basic techniques and give references to recent development:

- We start by presenting some simple data structures, which allow us to explain how the "information-theoretic $O(\log n)$ barrier" bay be surpassed. These data structures use a two-step approach: First, *range reduction* is used to decrease key length, such that we only need to consider keys that are much shorter than w. Secondly, we treat these short keys efficiently by *packed computing* where many keys are packed together in words.
- Next, we discuss some more elaborate data structures. In particular, we show how to achieve low worst-case complexity in linear space.
  i. The fusion tree, the first data structure presented that achieved sublogarithmic complexity.
  ii. The *exponential search tree*, which achieves tight worst-case bound on dynamic ordered dictionaries.
- We also give references to recent results on efficient priority queue implementations and sorting.

40.4 Achieving Sub-Logarithmic Time per Element by Simple Means

In this section, we show that it is surprisingly simple to achieve a sublogarithmic complexity in n independent of w, which implies sorting and searching asymptotically faster than comparison-based algorithms.

We use indirect addressing and large arrays. As a consequence, the data structures will need much space. However, all algorithms presented here can be fit into linear space with randomization (i.e., with universal hashing [2]).

In some cases, we will consider keys that are shorter than w, we will then use b or k to denote key length.

In this section, we will use $F(n, b)$ to express the complexity of searching, as specified below.

Definition 40.2

Let $F(n, b)$ be the worst-case cost of performing one search or update in an ordered dictionary storing n keys of length b.

Unless we use hashing to obtain linear space, the methods discussed in this section can all be implemented with a simple instruction set. All necessary instructions are standard, they are even in AC^0. (An instruction is in AC^0 if it is implementable by a constant depth, unbounded fan-in (AND,OR,NOT)-circuit of size $w^{O(1)}$. An example of a non-AC^0 instruction is multiplication [3].)

40.4.1 Range Reduction

One way to simplify a computational problem is by *range reduction*. In this case, we reduce the problem of dealing with w-bit keys to that of dealing with k-bits keys, $k < w$.

Assume that we view our w-bit keys as consisting of two $w/2$-bit characters and store these in a trie of height 2. Each internal node in the trie contains

- a reference to the min-element below the node; the min-element is not stored in any subtrie;
- a table of subtries, where each existing subtrie is represented by a $w/2$-bit key;
- a data structure for efficient neighbour search among the $w/2$-bit keys representing the subtries.

Since each node except the root has one incoming edge and each node contains exactly one element (the min-element), the trie has exactly n nodes and $n - 1$ edges.

We make neighbour searches in the following way: Traverse down the trie. If we find a leaf, the search ends, otherwise we end up at an empty entry in the subtrie table of some node. By making a neighbour search in that node, we are done. The cost for traversing the trie is $O(1)$ and the cost for a local neighbour search is $O(F(n, b/2))$ by definition.

The space requirements depend on how the table of subtrie pointers is implemented. If the table is implemented as an array of length $2^{b/2}$, each node in the trie requires $\Theta(2^{b/2})$ space. If we instead represent each table as a hash table, the total space of all hash tables is proportional to the total number of edges in the trie, which is $n - 1$.

We summarize this in the following equation.

$$F(n, w) = O(1) + F\left(\frac{n, w}{2}\right). \tag{40.1}$$

We can use the same construction recursively. That is, the local data structure for neighbour search among $w/2$-bit keys can be a trie of height 2 where $w/4$ bits are used for branching, etc.

In order to apply recursion properly, we have to be a bit careful with the space consumption. First, note that if the number of edges in a trie with n elements was larger than n, for instance $2n$, the total space (number of edges) would grow exponentially with the number of recursive levels. Therefore, we need to ensure that the number of edges in a trie is not just $O(n)$ but actually at most n. This is the reason why each node contains a min-element; in this way we guarantee that the number of edges is $n-1$.

Secondly, even when we can guarantee that the space per recursive level does not increase, we are still faced with $\Theta(n)$ space (with hashing) per level. If we use more than a constant number of levels, this will require superlinear space. This is handled in the following way: When we apply the recursive construction r times, we only keep a small part of the elements in the recursive structure. Instead, the elements are kept in sorted lists of size $\Theta(r)$, and we keep only the smallest element from each list in our recursive trie. When searching for a key, we first search for its list in the recursive trie structure, we then scan the list. Insertions and deletions are made in the lists, and the sizes of the lists are maintained by merging and splitting lists. Now, the total space taken by each level of the recursive trie construction is $\Theta(n/r)$ and the total space for r recursive levels is $\Theta(n)$. Searching, splitting and merging within the lists only takes $O(r)$ time. In summary, setting $r = \log(w/k)$ we get the following lemma.

Lemma 40.1

$$F(n, w) = O(\log(w/k)) + F(n, k).$$

This recursive reduction was first used in van Emde Boas trees [4–7].

40.4.2 Packing Keys

If the word length is small enough—as in today's computers—the range reduction technique discussed above will decrease the key length to a constant at a low cost. However, in order to make a really convincing comparison between comparison-based algorithms and algorithms based on indirect addressing, we must make the complexity independent of the word size. This can be done by combining range reduction with *packed computation*. The basic idea behind packed computation is to exploit the bit-parallelism in a computer; many short keys can be packed in a word and treated simultaneously.

The central observation is due to Paul and Simon [8]; they observed that one subtraction can be used to perform comparisons in parallel. Assume that the keys are of length k. We may then pack $\Theta(w/k)$ keys in a word in the following way: Each key is represented by a $(k+1)$-bit field. The first (leftmost) bit is a *test bit* and the following bits contain the key, cf. Figure 40.1. Let X and Y be two words containing the same number of packed keys, all test bits in X are 0 and all test bits in Y are 1. Let M be a fixed mask in which all test bits are 1 and all other bits are 0. Let

$$R \leftarrow (Y - X) \text{ AND } M. \tag{40.2}$$

Then, the ith test bit in R will be 1 if and only if $y_i > x_i$. All other test bits, as well as all other bits, in R will be 0.

We use packed comparisons to achieve the following result.

Lemma 40.2

$$F(n, k) = O\left(\log(w/k) + \frac{\log n}{\log(w/k)}\right).$$

Y	1 | 00010	1 | 00111	1 | 01001	1 | 01110	1 | 10101	1 | 11000	1 | 11011	1 | 11110
X	0 | 01011	0 | 01011	0 | 01011	0 | 01011	0 | 01011	0 | 01011	0 | 01011	0 | 01011
$Y - X$	0 | 10111	0 | 11100	0 | 11110	1 | 00011	1 | 01010	1 | 01101	1 | 10000	0 | 10011
M	1 | 00000	1 | 00000	1 | 00000	1 | 00000	1 | 00000	1 | 00000	1 | 00000	1 | 00000
$(Y - X)$ AND M	0 | 00000	0 | 00000	0 | 00000	1 | 00000	1 | 00000	1 | 00000	1 | 00000	1 | 00000

FIGURE 40.1 A multiple comparison in a packed B-tree.

Proof. (Sketch) We use a packed B-tree [9].

Th packed B-tree has nodes of degree $\Theta(w/k)$. In each node, the search keys are packed together in a single word, in sorted order from left to right. When searching for a k-bit key x in a packed B-tree, we take the following two steps:

1. We construct a word X containing multiple copies of the query key x. X is created by a simple doubling technique: Starting with a word containing x in the rightmost part, we copy the word, shift the copy $k+1$ steps and unite the words with a bitwise or. The resulting word is copied, shifted $2k+2$ steps and united, etc. Altogether X is generated in $O(\log(w/k))$ time.
2. After the word X has been constructed, we traverse the tree. At each node, we compute the rank of x in constant time with a packed comparison. The cost of the traversal is proportional to the height of the tree, which is $O(\log n/\log(w/k))$.

A packed comparison at a node is done as in Expression 40.2. The keys in the B-tree node are stored in Y and X contains multiple copies of the query key. After subtraction and masking, the rightmost p test bits in R will be 1 if and only if there are p keys in Y which are greater than x. This is illustrated in Figure 40.1. Hence, by finding the position of the leftmost 1-bit in R we can compute the rank of x among the keys in Y. In order to find the leftmost key, we can simply store all possible values of R in a lookup table. Since the number of possible values equals the number of keys in a B-tree node plus one, a hash table implementation of this lookup table would require only $\Theta(w/k)$ space. ∎

Above, we omitted a lot of details, such as how to perform updates and how pointers within a packed B-tree are represented. Details can be found in [9].

40.4.3 Combining

We can now derive our first bounds for searching. First, we state bounds in terms of w. The following bound holds for searching [4–6]:

Theorem 40.1

$F(n, w) = O(\log w)$.

Proof. (Sketch) Apply Lemma 40.1 with $k = 1$. ∎

Next, we show how to remove the dependency of word length [9]:

Theorem 40.2

$F(n, w) = O(\sqrt{\log n})$.

Proof. (Sketch) If $\log w = O(\sqrt{\log n})$, Theorem 40.1 is sufficient. Otherwise, Lemma 40.1 with $k = w/2^{\sqrt{\log n}}$ gives $F(n, w) = O(\sqrt{\log n}) + F(n, w/2^{\sqrt{\log n}})$. Lemma 40.2 gives that $F(n, w/2^{\sqrt{\log n}}) = O(\sqrt{\log n})$. ∎

40.5 Deterministic Algorithms and Linear Space

The data structures in this section are more complicated than the previous ones. They also need more powerful—but standard—instructions, like multiplication. On the other hand, these structures achieves linear space without randomization (i.e., without hashing).

Definition 40.3

Let $D(n)$ be the worst-case search cost and the amortized update cost in an ordered dictionary storing n keys in $O(n)$ space.

40.5.1 Fusion Trees

The fusion tree was the first data structure to surpass the logarithmic barrier for searching. The central part of the fusion tree [10] is a static data structure with the following properties:

Lemma 40.3

For any d, $d = O(w^{1/6})$, a static data structure containing d keys can be constructed in $O(d^4)$ time and space, such that it supports neighbour queries in $O(1)$ worst-case time.

Proof. (Sketch) The main idea behind the fusion tree is to view the keys as stored in an implicit binary trie and concentrate at the branching levels in this trie. We say that branching occurs at significant bit positions. We illustrate this view with an example, shown in Figure 40.2.

In the example, $w = 16$ and $d = 6$. We store a set Y of keys $y_1, \ldots, y_d$. Each key in Y is represented as a path in a binary trie. In the figure, a left edge denotes a 0 and a right edge denotes a 1. For example, y_3 is $\boxed{1010010101011010}$. The significant bit positions correspond to the branching levels in the trie. In this example the levels are 4, 9, 10, and 15, marked by horizontal lines. By extracting the significant bit positions from each key, we create a set Y' of compressed keys $y_1', \ldots, y_d'$. In our example the compressed keys are $\boxed{0000}$, $\boxed{0001}$, $\boxed{0011}$, $\boxed{0110}$, $\boxed{1001}$, and $\boxed{1011}$. Since the trie has exactly d leaves, it contains exactly $d - 1$ binary nodes. Therefore, the number of significant bit positions, and the length of a compressed key, is at most $d - 1$. This implies that we can pack the d keys, including test bits, in d^2 bits. Since $d = O(w^{1/6})$, the packed keys fit in a constant number of words.

This extraction of bits is nontrivial; it can be done with multiplication and masking. However, the extraction is not as perfect as described here; in order to avoid problems with carry bits etc, we need to extract some more bits than just the significant ones. Here, we ignore these problems and assume that we can extract the desired bits properly. For details we refer to Fredman and Willard [10].

The d compressed keys may be used to determine the rank of a query key among the original d keys with packed computation. Assume that we search for $x = \boxed{1010011001110100}$, represented as the fat path in Figure 40.2. First, we extract the proper bits to form a compressed key $x' = \boxed{0010}$. Then, we use packed searching to determine the rank of x' among $y_1', \ldots, y_d'$. In this case, the packed searching will place x' between y_2' and y_3'. as indicated by the arrow in Figure 40.2. This is not the proper rank of the original key x, but nevertheless it is useful. The important information is obtained by finding the position of the first differing bit of x and one of the keys y_2 and y_3. In this example, the 7th bit is the first differing bit. and, since x has a 1 at this bit position, we can conclude that it is greater than all keys in Y with the same 6-bit prefix. Furthermore, the remaining bits in x are insignificant. Therefore, we can replace x by the key $\boxed{1010011111111111}$, where all the last bits are 1s. When compressed, this new key becomes $\boxed{0111}$. Making a second packed searching with this key instead, the proper rank will be found.

Hence, in constant time we can determine the rank of a query key among our d keys. ∎

The original method by Fredman and Willard is slightly different. Instead of filling the query keys with 1s (or 0s) and making a second packed searching, they use a large lookup table in each node. Fusion trees can be implemented without multiplication, using only AC^0 instructions, provided that some simple non-standard instructions are allowed [11].

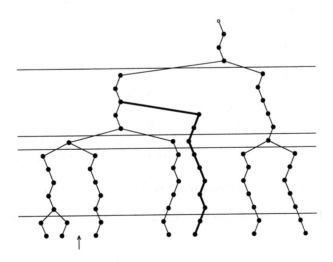

FIGURE 40.2 Searching in the internal trie in a fusion tree node. Horizontal lines represent significant bit positions. The thin paths represent the 6 keys in the trie, while the fat path represents the search key x (which is not present in the trie). The arrow marks the position of the compressed key x' among the keys in the trie.

Theorem 40.3

$D(n) = O(\log n / \log \log n)$.

Proof. (Sketch) Based on Lemma 40.3, we use a B-tree where only the upper levels in the tree contain B-tree nodes, all having the same degree (within a constant factor). At the lower levels, traditional (i.e., comparison-based) weight-balanced trees are used. The reason for using weight-balanced trees is that the B-tree nodes are costly to reconstruct; the trees at the bottom ensure that few updates propagate to the upper levels. In this way, the amortized cost of updating a B-tree node is small.

The amortized cost of searches and updates is $O(\log n / \log d + \log d)$ for any $d = O(w^{1/6})$. The first term corresponds to the number of B-tree levels and the second term corresponds to the height of the weight-balanced trees. Since $w \geq \log n$ (otherwise a pointer would not fit in a word), the cost becomes at most $O(\log n / \log \log n)$. ∎

40.5.2 Exponential Search Trees

The *exponential search tree* [12,13] allows efficient dynamization of static dictionary structures. The key feature is:

> Any static data structure for searching that can be constructed in polynomial time and space can be efficiently used in a dynamic data structure.

The basic data structure is a multiway tree where the degrees of the nodes decrease doubly-exponentially down the tree. In each node, we use a static data structure for navigation. The way the tree is maintained, we can guarantee that, before an update occurs at a certain node, a polynomial number of updates will be made below it. Hence, even if an update requires a costly reconstruction of a static data structure, this will occur with large enough intervals.

Lemma 40.4

Suppose a static data structure containing d keys can be constructed in $O(d^4)$ time and space, such that it supports neighbour queries in $O(S(d))$ worst-case time. Then,

$$D(n) = O(S(n^{1/5})) + D(n^{4/5});$$

Proof. (Sketch) We use an exponential search tree. It has the following properties:

- Its root has degree $\Theta(n^{1/5})$.
- The keys of the root are stored in a local (static) data structure, with the properties stated above. During a search, the local data structure is used to determine in which subtree the search is to be continued.
- The subtrees are exponential search trees of size $\Theta(n^{4/5})$.

First, we show that, given n sorted keys, an exponential search tree can be constructed in linear time and space. The cost of constructing a node of degree d is $O(d^4)$, and the total construction cost $C(n)$ is (essentially) given by

$$C(n) = O((n^{1/5})^4) + n^{1/5} \cdot C(n^{4/5}) \Rightarrow C(n) = O(n). \tag{40.3}$$

Furthermore, with a similar equation, the space required by the data structure can be shown to be $O(n)$.

Balance is maintained by joining and splitting subtrees. The basic idea is the following: A join or split occurs when the size of a subtree has changed significantly, that is, after $\Omega(n^{4/5})$ updates. Then, a constant number of subtrees will be reconstructed; according to Equation 40.3, the cost of this is linear in the size of the subtrees $= O(n^{4/5})$. Also, some keys will be inserted or deleted from the root, causing a reconstruction of the root; the cost of this is by definition $O(n^{4/5})$. Amortizing these two costs over the $\Omega(n^{4/5})$ updates, we get $O(1)$ amortized cost for reconstructing the root. Hence, the restructuring cost is dominated by the search cost.

Finally, the search cost follows immediately from the description of the exponential search tree. ∎

Exponential search trees may be combined with various other data structures, as illustrated by the following two lemmas:

Lemma 40.5

A static data structure containing d keys can be constructed in $O(d^4)$ time and space, such that it supports neighbour queries in $O\left(\frac{\log d}{\log w} + 1\right)$ worst-case time.

Proof. (Sketch) We just construct a static B-tree where each node has the largest possible degree according to Lemma 40.3. That is, it has a degree of $\min(d, w^{1/6})$. This tree satisfies the conditions of the lemma. ∎

Lemma 40.6

A static data structure containing d keys and supporting neighbour queries in $O(\log w)$ worst-case time can be constructed in $O(d^4)$ time and space.

Proof. (Sketch) We study two cases.

Case 1: $w > d^{1/3}$. Lemma 40.5 gives constant query cost.

Case 2: $w \leq d^{1/3}$. The basic idea is to combine a van Emde Boas tree (Theorem 40.1) with perfect hashing. The data structure of Theorem 40.1 uses much space, which can be reduced to $O(d)$ by hash coding. Since we can afford a rather slow construction, we can use the deterministic algorithm by Fredman, Komlós, and Szemerédi [14]. With this algorithm, we can construct a perfectly hashed van Emde Boas tree in $O(d^3 w) = o(d^4)$ time. ∎

Combining these two lemmas, we get a significantly improved upper bound on deterministic sorting and searching in linear space:

Theorem 40.4

$D(n) = O(\sqrt{\log n})$.

Proof. (Sketch) If we combine Lemmas 40.4–40.6, we obtain the following equation

$$D(n) = O\left(\min\left(1 + \frac{\log n}{\log w}, \log w\right)\right) + D(n^{4/5}) \tag{40.4}$$

which, when solved, gives the theorem. ∎

Taking both n and w as parameters, $D(n)$ is $o(\sqrt{\log n})$ in many cases [12]. For example, it can be shown that $D(n) = O(\log w \log \log n)$.

The strongest possible bound is achieved by using the following result by Beame and Fich [3]

Lemma 40.7: (Beame and Fich [3])

In polynomial time and space, we can construct a deterministic data structure over d keys supporting searches in $O(\min\{\sqrt{\log d / \log \log d}, \frac{\log w}{\log \log w}\})$ time.

Combining this with the exponential search tree we get, among others, the following theorem.

Theorem 40.5

$D(n) = O(\sqrt{\log n / \log \log n})$.

Since a matching lower bound was also given by Beame and Fich, this bound is optimal.

40.6 From Amortized Update Cost to Worst-Case

In fact, there are worst-case efficient versions of the data structures above. Willard [15] gives a short sketch on how to make fusion trees worst-case efficient, and as shown by Andersson and Thorup [13], the exponential search tree can be modified into a worst-case data structure. Here, we give a brief description of how exponential search trees are modified.

In the above definition of exponential search trees, the criteria for when a subtree is too large or too small depend on the degree of its parent. Therefore, when a node is joined or split, the requirements on its children will change. Above, we handled that by simply rebuilding the entire subtree at each join or split, but in a worst-case setting, we need to let the children of a node remain unchanged at a join or split. In order to do this, we need to switch from a top-down definition to a bottom-up definition.

Definition 40.4: (Worst-case efficient exponential search trees)

In an exponential search tree all leaves are on the same depth. Let the height of a node to be the distance from the node to the leaves descending from it. For a non-root node v at height $i > 0$, the weight (number of descending leaves) is $|v| = \Theta(n_i)$ where $n_i = \alpha^{(1 + 1/(k-1))^i}$ and $\alpha = \Theta(1)$. If the root has height h, its weight is $O(n_h)$.

With the exception of the root, Definition 40.4 follows our previous definition of exponential search trees (when $k = 5$), that is, if v is a non-root node, it has $\Theta(|v|^{1/k})$ children, each of weight $\Theta(|v|^{1 - 1/k})$.

The worst-case efficiency is mainly based on careful scheduling: the static search structures in the nodes are rebuilt in the background so that they remain sufficiently updated as nodes get joined and split. This scheduling is developed in terms of a general theorem about rebuilding, which has some interesting properties as a tool for other de-amortization applications [13].

40.7 Sorting and Priority Queues

In the comparison-based model of computation, the cost per element is the same ($O(\log n)$) for searching and sorting. However, in the RAM model of computation, the sorting problem can be solved faster than searching. The simple intuitive explanation of this is that the bit-parallelism in packed computation can be utilized more efficiently when a number of keys are treated simultaneously, as in sorting, than when they are treated one-by-one as in searching.

Even more, it turns out that priority queues can be as implemented as efficiently as sorting. (The intuitive reason for this is that a priority queue can use sorting as a subroutine, and only keep a small part of the queue perfectly sorted.) Thorup [16] has shown the following reduction:

Theorem 40.6

If we can sort n keys in time $S(n)$ per key, then we can implement a priority queue supporting find-min in constant time and updates (insert and delete) in $S(n)$ time.

In the following, we will use $T(n, b)$ to denote the cost of sorting n b-bit keys.

40.7.1 Range Reduction

In sorting algorithms, range reduction is an often used technique. For example, we may view traditional radix sort, where we sort long strings by dividing them into shorter parts, as range reduction.

For our purposes, we will use a range reduction technique by Kirkpatrick and Reisch [17], which is similar to the van Emde Boas tree, cf. Section 40.4.1. The only difference from Section 40.4.1 is that instead of letting each trie node contain a data structure for efficient neighbour search among the outgoing edges, we just keep an unsorted list of all outgoing edges (plus the array for constant-time indexing of edges). Then, after all elements have been inserted into the trie, we create sorted lists of edges at all nodes by the following method:

1. Mark each edge with its parent node.
2. Concatenate all edge lists and sort the entire list.
3. Scan the sorted list and put each edge back in the proper node.
4. All edges lists are now sorted. By a recursive traversal of the trie we can report all leafs in sorted order.

Other details, such as the need to store one key per node to avoid space blow up, are handled in the same way as in Section 40.4.1. Altogether, we get the reduction

$$T(n, w) = O(n) + T(n, w/2). \tag{40.5}$$

Applied recursively, this gives

$$T(n, w) = O(n \log(w/k)) + T(n, k). \tag{40.6}$$

40.7.2 Packed Sorting

For the sorting problem, multiple comparisons can be utilized more efficiently than in a packed B-tree. In the packed B-tree, we used the bit-parallelism to compare one key to many keys, in this way we implemented a parallel version of a linear search, which is not the most efficient search method.

For sorting, however, we can utilize the packed computation more efficiently. It turns out that algorithms for *sorting networks* are well suited for implementation by packed computation. A sorting network is a "static" algorithm; it contains a number of compare-and-swap operations, these are the same regardless of the outcome of the comparisons. The merging technique by Batcher [18], originally used to design *odd-even merge sort*, can be efficiently implemented with packed computation. As a sorting network, Batcher's merging algorithm has depth $\Theta(\log n)$ where each level has $O(n)$ compare-and-swap units. Based on the merging, we can sort in $\Theta(\log^2 n)$ time where the total work is $\Theta(n \log^2 n)$.

Batcher's merging technique is well suited for combination with the Paul-Simon technique, as shown by Albers and Hagerup [19].

Lemma 40.8

$T(n, w/\log n) \leq O(n \log \log n)$.

Proof. (Sketch) The key idea is that by packing $\Theta(\log n)$ keys in a machine word, we can combine Batcher's algorithm with packed computation to merge two sorted sequences in $O(\log \log n)$ time. And, if we can merge two sequences of length $\Theta(\log n)$ in $O(\log \log n)$ time (instead of $O(\log n)$ by a comparison-based algorithm), we can use this as a subroutine tom implement a variant of merge sort that sorts n keys in $O(n \log \log n)$ time (instead of $O(n \log n)$). ∎

40.7.3 Combining the Techniques

First, the bound on searching from Theorem 40.1 has a corresponding theorem for sorting [17]:

Theorem 40.7

$T(n, w) = O(n \log(w/\log n))$.

Proof. (Sketch) Apply Eq. 40.6 with $k = \log n$. Keys of length $\log n$ can be sorted in linear time with bucket sort. ∎

Secondly, we combine range reduction with packed computation. We get the following bound [20]:

Theorem 40.8

$T(n, w) = O(n \log \log n)$.

Proof. (Sketch) If $\log w = O(\log \log n)$, Theorem 40.7 is sufficient. Otherwise, Eq. 40.6 with $k = w/\log n$ gives $T(n,w) = O(n \log \log n) + T(n, w/\log n)$. Lemma 40.8 gives the final bound. ∎

40.7.4 Further Techniques and Faster Randomized Algorithms

Apart from these rather simple techniques, there are a number of more elaborate techniques that allows the complexity to be improved further. Examples of such techniques are *signature sort* [20] and *packed bucketing* [21]. Here, we give a short sketch of signature sorting.

Consider a situation where the word length w is very large, and we wish to reduce the problem of sorting w-bit keys to that of sorting k-bit keys, $k \gg \log n$. Instead of treating these k-bit keys directly, we represent each such key by a b-bit *signature*, where the b bits are a hash function of the k bits. In fact, for one w-bit key, we can in constant time replace it by a shorter key, consisting of q b-bit signatures (for details, we refer to the literature [20,21]).

1. Replace each w-bit key by a qb-bit key of concatenated signatures.
2. Sort the qb-bit keys.
3. Compute, for each qb-bit key, its first distinguishing signature. This can be done by constructing a signature-based trie of all keys.
4. If we know the first distinguishing signature in a qb-bit key, we know the first distinguishing k-bit field in the corresponding w-bit key. Finding these k-bit fields, the range reduction is completed and we can continue by sorting these shorter keys.

It should be noted that the sorted set of qb-bit keys does not correspond to a sorted set of w-bit keys. However, the ordering we get is enough to find the proper distinguishing fields. Furthermore, since we use hash coding, we might get collisions, in which case the method will not work. By choosing b large enough, the risk of failure is small enough that we can afford to redo the entire sorting in case of failure: still the expected time for the range reduction step will be linear.

As an important recent result, Han and Thorup presents a linear algorithm for splitting n integers into subsets, where each subset is of size $O(\log n)$. Combining this with techniques like signature sorting, they manage to improve the randomized complexity of sorting to $O(n\sqrt{\log\log n})$. This, in turn, implies that a priority queue can be implemented at $O(\sqrt{\log\log n})$ time per update and find-min in constant time.

Other relevant reading can be found in the cited articles, or in [22–28]

References

1. J. L. Hennessy and D. A. Patterson. *Computer Organization and Design: The Hardware/Software Interface*. Morgan Kaufmann Publ., San Mateo, CA, 1994.

2. J. L. Carter and M. N. Wegman. Universal classes of hash functions. *Journal of Computer and System Sciences*, 18:143–154, 1979.

3. P. Beame and F. Fich. Optimal bounds for the predecessor problem. In *Proc. 31st STOC*, pages 295–304, 1999.

4. P. van Emde Boas. Preserving order in a forest in less than logarithmic time. In *Proceedings of the 16th Annual IEEE Symposium on Foundations of Computer Science*, pages 75–84, 1975.

5. P. van Emde Boas. Preserving order in a forest in less than logarithmic time and linear space. *Inf. Proc. Lett.*, 6(3):80–82, 1977.

6. P. van Emde Boas, R. Kaas, and E. Zijlstra. Design and implementation of an efficient priority queue. *Math. Syst. Theory*, 10:99–127, 1977.

7. K. Mehlhorn and S. Nähler. Bounded ordered dictionaries in $O(\log\log n)$ time and $O(n)$ space. *Inf. Proc. Lett.*, 35(4):183–189, 1990.

8. W. J. Paul and J. Simon. Decision trees and random access machines. In *Logic and Algorithmic: An International Symposium Held in Honour of Ernst Specker*, pages 331–340. L'Enseignement Mathématique, Université de Genevè, 1982.

9. A. Andersson. Sublogarithmic searching without multiplications. In *Proc. 36th FOCS*, pages 655–663, 1995.

10. M. L. Fredman and D. E. Willard. Surpassing the information theoretic bound with fusion trees. *J. Comput. Syst. Sci.*, 47:424–436, 1993. Announced at STOC'90.

11. A. Andersson, P. B. Miltersen, and M. Thorup. Fusion trees can be implemented with AC^0 instructions only. *Theoretical Computer Science*, 215(1-2):337–344, 1999.

12. A. Andersson. Faster deterministic sorting and searching in linear space. In *Proc. 37th FOCS*, pages 135–141, 1996.

13. A. Andersson and M. Thorup. Tight(er) worst-case bounds on dynamic searching and priority queues. In *Proc. 32th STOC*, 2000.

14. M. L. Fredman, J. Komlós, and E. Szemerédi. Storing a sparse table with $O(1)$ worst case access time. *Journal of the ACM*, 31(3):538–544, 1984.

15. D. E. Willard. Examining computational geometry, van Emde Boas trees, and hashing from the perspective of the fusion tree. *SIAM Journal on Computing*, 29(3):1030–1049, 2000. Announced at SODA'92.

16. M. Thorup. Equivalence between priority queues and sorting. In *Proc. FOCS'02*, 2002.

17. D. Kirkpatrick and S. Reisch. Upper bounds for sorting integers on random access machines. *Theor. Comp. Sci.*, 28:263–276, 1984.

18. K. E. Batcher. Sorting networks and their applications. In *Proceedings of the AFIPS Spring Joint Computer Conference*, pages 307–314, 1968. Volume 32.

19. S. Albers and T. Hagerup. Improved parallel integer sorting without concurrent writing. *Inf. Contr.*, 136:25–51, 1997. Announced at SODA '92.

20. A. Andersson, T. Hagerup, S. Nilsson, and R. Raman. Sorting in linear time? *J. Comp. Syst. Sc.*, 57:74–93, 1998. Announced at STOC'95.

21. Y. Han and M. Thorup. Integer sorting in $o(n\sqrt{\log\log n})$ expected time and linear space. In *Proc. FOCS '02*, 2002.

22. A. Andersson and S. Nilsson. A new efficient radix sort. In *Proc. 35th Annual IEEE Symposium on Foundations of Computer Science*, pages 714–721. IEEE Computer Society Press, 1994.

23. A. Brodnik, P. B. Miltersen, and I. Munro. Trans-dichotomous algorithms without multiplication - some upper and lower bounds. In *Proc. 5th WADS*, LNCS 1272, pages 426–439, 1997.

24. M. L. Fredman and D. E. Willard. Trans-dichotomous algorithms for minimum spanning trees and shortest paths. *J. Comput. Syst. Sci.*, 48:533–551, 1994.

25. Y. Han. Fast integer sorting in linear space. In *Proc. 34th STOC*, pages 602–608, 2002.

26. Y. Han. Improved fast integer sorting in linear space. *Inform. Comput.*, 170(8):81–94, 2001. Announced at STACS'00 and SODA'01.

27. R. Raman. Priority queues: small, monotone and trans-dichotomous. In *Proc. 4th ESA*, LNCS 1136, pages 121–137, 1996.

28. M. Thorup. On RAM priority queues. *SIAM J. Comp.*, 30(1):86–109, 2000.

VI

Data Structures in Langs and Libraries

41 Functional Data Structures Chris Okasaki..639
Introduction • Stacks: A Simple Example • Binary Search Trees: Path Copying • Skew Heaps:
Amortization and Lazy Evaluation • Difficulties • Further Reading • Acknowledgments •
References

42 LEDA, a Platform for Combinatorial and Geometric Computing Stefan Naeher.........................653
Introduction • The Structure of LEDA • Data Structures and Data Types • Algorithms •
Visualization • Example Programs • Projects Enabled by LEDA • References

43 Data Structures in C++ Mark Allen Weiss..667
Introduction • Basic Containers • Iterators • Additional Components of the STL •
Acknowledgments • References

44 Data Structures in JDSL Michael T. Goodrich, Roberto Tamassia and Luca Vismara....................679
Introduction • Design Concepts in JDSL • The Architecture of JDSL • A Sample Application •
Acknowledgments • References

45 Data Structure Visualization John Stasko 697
Introduction • Value of Data Structure Rendering • Issues in Data Structure Visualization Systems •
Existing Research and Systems • Summary and Open Problems • References

46 Drawing Trees Sebastian Leipert..707
Introduction • Preliminaries • Level Layout for Binary Trees • Level Layout for n-ary Trees •
Radial Layout • HV-Layout • Acknowledgments • References

47 Drawing Graphs Peter Eades and Seok-Hee Hong..723
Introduction • Preliminaries • Convex Drawing • Symmetric Drawing • Visibility Drawing •
Conclusion • References

48 Concurrent Data Structures Mark Moir and Nir Shavit...741
Designing Concurrent Data Structures • Shared Counters and Fetch-and-ϕ Structures • Stacks and
Queues • Pools • Linked Lists • Hash Tables • Search Trees • Priority Queues • Summary •
References

41

Functional Data Structures*

41.1	Introduction...	639
	Data Structures in Functional Languages • Functional Data Structures in Mainstream Languages	
41.2	Stacks: A Simple Example	640
41.3	Binary Search Trees: Path Copying	642
41.4	Skew Heaps: Amortization and Lazy Evaluation...............................	644
	Analysis of Lazy Skew Heaps	
41.5	Difficulties..	649
41.6	Further Reading..	649
	Acknowledgments	650
	References..	650

Chris Okasaki
United States Military Academy

41.1 Introduction

A *functional data structure* is a data structure that is suitable for implementation in a functional programming language, or for coding in an ordinary language like C or Java using a functional style. Functional data structures are closely related to *persistent data structures* and *immutable data structures*—in fact, the three terms are often used interchangeably. Howevber, there are subtle differences.

- The term *persistent data structures* refers to the general class of data structures in which an update does not destroy the previous version of the data structure, but rather creates a new version that co-exists with the previous version. See Chapter 33 for more details about persistent data structures.
- The term *immutable data structures* emphasizes a particular implementation technique for achieving persistence, in which memory devoted to a particular version of the data structure, once initialized, is never altered.
- The term *functional data structures* emphasizes the language or coding style in which persistent data structures are implemented. Functional data structures are always immutable, except in a technical sense discussed in Section 41.4.

In this chapter, we will discuss the main issues surrounding the implementation of data structures in functional languages, and illustrate these issues with several extended examples. We will also show how to adapt functional data structures to a mainstream language such as Java, for use when a persistent data structure is required. Readers wishing more details about functional data structures should consult Okasaki [1].

41.1.1 Data Structures in Functional Languages

Functional programming languages differ in several important ways from ordinary programming languages like C or Java, and these differences can sometimes have a large effect on how data structures are implemented. The main differences (at least from the perspective of data structures) are *immutability*, *recursion*, *garbage collection*, and *pattern matching*.

41.1.1.1 Immutability

In functional languages, variables and records cannot be modified, or *mutated*, once they have been created.[†] Many textbook data structures depend critically on the ability to mutate variables and records via assignments. Such data structures can be difficult to adapt to a functional setting.

---

* This chapter has been reprinted from first edition of this Handbook, without any content updates.
† Actually, many functional languages do provide mechanisms for mutation, but their use is discouraged.

41.1.1.2 Recursion

Functional languages frequently do not support looping constructs, such as for-loops or while-loops, because such loops depend on being able to mutate the loop control variable. Functional programmers use recursion instead.[*]

41.1.1.3 Garbage Collection

Functional languages almost always depend on automatic garbage collection. Because objects are immutable in functional languages, they are shared much more widely than in ordinary languages, which makes the task of deciding when to deallocate an object very complicated. In functional languages, programmers ignore deallocation issues and allow the garbage collector to deallocate objects when it is safe to do so.

41.1.1.4 Pattern Matching

Pattern matching is a method of defining functions by cases that are essentially textual analogues of the kinds of pictures data-structure designers often draw. Pattern matching is not supported by all functional languages, but, when available, it allows many data structures to be coded very concisely and elegantly.

41.1.2 Functional Data Structures in Mainstream Languages

Even if you are programming in a mainstream language, such as C or Java, you may find it convenient to use a functional data structure. Functional data structures offer three main advantages in such a setting: *fewer bugs*, *increased sharing*, and *decreased synchronization*.

41.1.2.1 Fewer Bugs

A very common kind of bug arises when you observe a data structure in a certain state and shortly thereafter perform some action that assumes the data structure is still in that same state. Frequently, however, something has happened in the meantime to alter the state of the data structure, so that the action is no longer valid. Because functional data structures are immutable, such alterations simply cannot occur. If someone tries to change the data structure, they may produce a new version of it, but they will in no way effect the version that you are using.

41.1.2.2 Increased Sharing

Precisely to avoid the kinds of bugs described above, programmers in mainstream languages are careful to limit access to their internal data structures. When sharing is unavoidable, programmers will often clone their data structures and share the clones rather than granting access to their own internal copies. In contrast, functional data structures can be shared safely without cloning. (Actually, functional data structures typically perform a substantial amount of cloning internally, but this cloning is of individual nodes rather than entire data structures.)

41.1.2.3 Decreased Synchronization

Again, precisely to avoid the kinds of bugs described above, programmers in concurrent settings are careful to synchronize access to their data structures, so that only a single thread can access the data structure at a time. On the other hand, because functional data structures are immutable, they can often be used with little or no synchronization. Even simultaneous writes are not a problem, because each writer thread will get a new version of the data structure that reflects only its own updates. (This assumes, of course, an application where participants do not necessarily want to see changes made by all other participants.)

41.2 Stacks: A Simple Example

Stacks (see Chapter 2) represented as singly-linked lists are perhaps the simplest of all data structures to make persistent. We begin by describing functional stacks supporting four main primitives:

- `empty`: a constant representing the empty stack.
- `push(x, s)`: push the element x onto the stack s and return the new stack.
- `top(s)`: return the top element of s.
- `pop(s)`: remove the top element of s and return the new stack.

---

[*] Or higher-order functions, but we will not discuss higher-order functions further because they have relatively little effect on the implementation of data structures.

We can see right away several differences between this interface and the interface for ordinary stacks. First, for ordinary stacks, push and pop would implicitly change the existing stack *s* rather than returning a new stack. However, the hallmark of functional data structures is that update operations return a new version of the data structure rather than modifying the old version. Second, ordinary stacks would support a function or constructor to *create* a fresh, new stack, rather than offering a single constant to represent all empty stacks. This highlights the increased sharing possible with functional data structures. Because pushing an element onto the empty stack will not change the empty stack, different parts of the program can use the same empty stack without interfering with each other.

Figure 41.1 shows an implementation of stacks in Haskell [2], a popular functional programming language. Figure 41.2 shows a similar implementation in Java. Like all code fragments in this chapter, these implementations are intended only to illustrate the relevant concepts and are not intended to be industrial strength. In particular, all error handling has been omitted. For example, the top and pop operations should check whether the stack is empty. Furthermore, programming conventions in Haskell and Java have been ignored where they would make the code harder for non-fluent readers to understand. For example, the Haskell code makes no use of currying, and the Java code makes no attempt to be object-oriented.

The Haskell code illustrates a simple use of pattern matching. The declaration

```
data Stack = Empty | Push(Element,Stack)
```

states that stacks have two possible shapes, Empty or Push, and that a stack with the Push shape has two fields, an element and another stack. The tags Empty and Push are called *constructors*. Later function declarations can match against these constructors. For example, the declaration

```
top(Push(x,s)) = x
```

says that when top is called on a stack with the Push shape, it returns the contents of the first field. If desired, more clauses can be added to deal with other shapes. For example, a second clause could be added to the definition of top to handle the error case:

```
top(Push(x,s)) = x
top(Empty) = ...signal an error...
```

How do these implementations achieve persistence? First, consider the push operation. Calling push creates a new node containing the new element and a pointer to the old top of stack, but it in no way alters the old stack. For example, if the old stack *s* contains the numbers 3, 2, 1 and we push 4, then the new stack *s′* contains the numbers 4, 3, 2, 1. Figure 41.3 illustrates

```
data Stack = Empty | Push(Element,Stack)

empty = Empty
push(x,s) = Push(x,s)
top(Push(x,s)) = x
pop(Push(x,s)) = s
```

FIGURE 41.1   Stacks in Haskell.

```
public class Stack {
    public static final Stack empty = null;
    public static Stack push(Element x,Stack s) { return new Stack(x,s); }
    public static Element top(Stack s) { return s.element; }
    public static Stack pop(Stack s) { return s.next; }

    private Element element;
    private Stack next;
    private Stack(Element element,Stack next) {
        this.element = element;
        this.next = next;
    }
}
```

FIGURE 41.2   Stacks in Java.

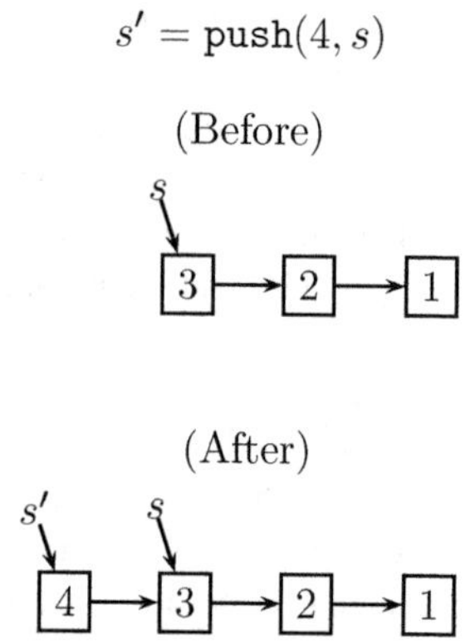

$$s' = \text{push}(4, s)$$

(Before)

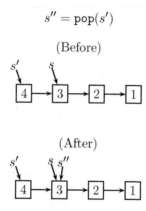

(After)

FIGURE 41.3   The push operation.

$$s'' = \text{pop}(s')$$

(Before)

(After)

FIGURE 41.4   The pop operation.

the relationship between s and s'. Notice how the nodes containing 3, 2, and 1 are shared between both stacks. Because of this sharing, it is crucial that the nodes are immutable. Consider what would happen if nodes could be changed. If we were to change the 3 to a 5, for example, perhaps by calling an operation to update the top element of s, that change would affect not only s (which would now contain 5, 2, 1) but also s' (which would now contain 4, 5, 2, 1). Such unintended consequences would make sharing impossible.

Next, consider the pop operation, which simply returns the *next* pointer of the current node without changing the current node in any way. For example, Figure 41.4 illustrates the result of popping the stack s' to get the stack s'' (which shares its entire representation with the original stack s). Notice that, after popping s', the node containing 4 may or may not be garbage. It depends on whether any part of the program is still using the s' stack. If not, then automatic garbage collection will eventually deallocate that node.

41.3 Binary Search Trees: Path Copying

Stacks are unusual in that there is never a need to update an existing node. However, for most data structures, there is such a need. For example, consider inserting an element into a binary search tree (see Chapter 3). At the very least, we need to update the new node's parent to point to the new node. But how can we do this if nodes are immutable? The solution is a technique called *path copying*. To update an existing node, we copy the node and make the necessary changes in the copy. However, we then have to update the existing node's parent in a similar fashion to point to the copy. In this way, changes propagate all the way from the site of the update to the root, and we end up copying that entire path. That may seem like a lot of copying, but notice that all nodes not on that path are shared between the old and new versions.

To see how path copying works in practice, consider a simple implementation of integer sets as unbalanced binary search trees. Figure 41.5 shows an implementation in Haskell and Figure 41.6 shows the same implementation in Java.

```
data Tree = Empty | Node(Tree,Int,Tree)

empty = Empty
insert(x,Empty) = Node(Empty,x,Empty)
insert(x,Node(t1,y,t2)) = if x < y then Node(insert(x,t1),y,t2)
                          else if x > y then Node(t1,y,insert(x,t2))
                          else Node(t1,y,t2)
search(x,Empty) = False
search(x,Node(t1,y,t2)) = if x < y then search(x,t1)
                          else if x > y then search(x,t2)
                          else True
```

FIGURE 41.5   Binary search trees in Haskell.

```
public class Tree {
  public static final Tree empty = null;
  public static Tree insert(int x,Tree t) {
    if (t == null) return new Tree(null,x,null);
    else if (x < t.element)
      return new Tree(insert(x,t.left),t.element,t.right);
    else if (x > t.element)
      return new Tree(t.left,t.element,insert(x,t.right));
    else return t;
  }
  public static boolean search(int x,Tree t) {
    if (t == null) return false;
    else if (x < t.element) return search(x,t.left);
    else if (x > t.element) return search(x,t.right);
    else return true;
  }

  private int element;
  private Tree left,right;
  private Tree(Tree left,int element,Tree right) {
    this.left = left;
    this.element = element;
    this.right = right;
  }
}
```

FIGURE 41.6   Binary search trees in Java.

The key to understanding path copying lies in the `insert` operation. Consider the case where the element being inserted is larger than the element at the current node. In the Java implementation, this case executes the code

```
return new Tree(t.left,t.element,insert(x,t.right));
```

First, `insert` calls itself recursively on the right subtree, returning a pointer to the new right subtree. It then allocates a new tree node, copying the `left` and `element` fields from the old node, and installing the new pointer in the `right` field. Finally, it returns a pointer to the new node. This process continues until it terminates at the root. Figure 41.7 illustrates a sample insertion. Notice how the parts of the tree not on the path from the root to the site of the update are shared between the old and new trees.

This functional implementation of binary search trees has exactly the same time complexity as an ordinary non-persistent implementation. The running time of `insert` is still proportional to the length of the search path. Of course, the functional implementation allocates more space, but even that issue is not clear cut. If the old tree is no longer needed, then the just-copied nodes can immediately be garbage collected, leaving a net space increase of one node—exactly the same space required by a non-persistent implementation. On the other hand, if the old tree is still needed, then the just-copied nodes cannot be garbage collected, but in that case we are actively taking advantage of functionality not supported by ordinary binary search trees.

$$t' = \texttt{insert}(8, t)$$

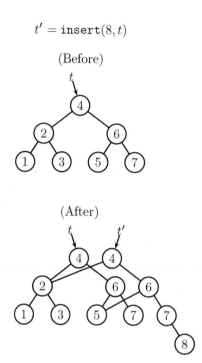

FIGURE 41.7   The insert operation.

Of course, the binary search trees described above suffer from the same limitations as ordinary unbalanced binary search trees, namely a linear time complexity in the worst case. Whether the implementation is functional or not as no effect in this regard. However, we can easily apply the ideas of path copying to most kinds of *balanced* binary search trees (see Chapter 11), such as AVL trees [3,4], red-black trees [5], 2-3 trees [6], and weight-balanced trees [7]. Such a functional implementation retains the logarithmic time complexity of the underlying design, but makes it persistent.

Path copying is sufficient for implementing many tree-based data structures besides binary search trees, including binomial queues [8,9] (Chapter 7), leftist heaps [1,10] (Chapter 5), Patricia tries [11] (Chapter 29), and many others.

41.4 Skew Heaps: Amortization and Lazy Evaluation

Next, we turn to priority queues, or *heaps*, supporting the following primitives:

- `empty`: a constant representing the empty heap.
- `insert(x, h)`: insert the element x into the heap h and return the new heap.
- `findMin(h)`: return the minimum element of h.
- `deleteMin(h)`: delete the minimum element of h and return the new heap.
- `merge(h_1, h_2)`: combine the heaps h_1 and h_2 into a single heap and return the new heap.

Many of the standard heap data structures can easily be adapted to a functional setting, including binomial queues [8,9] and leftist heaps [1,10]. In this section, we describe a simple, yet interesting, design known as *skew heaps* [12]. (Non-persistent skew heaps are described in detail in Chapter 6.)

A skew heap is a heap-ordered binary tree. Each node contains a single element, and the nodes are ordered such that the element at each node is no larger than the elements at the node's children. Because of this ordering, the minimum element in a tree is always at the root. Therefore, the `findMin` operation simply returns the element at the root. The `insert` and `deleteMin` operations are defined in terms of `merge`: `insert` creates a new node and merges it with the existing heap, and `deleteMin` discards the root and merges its children.

The interesting operation is `merge`. Assuming both heaps are non-empty, `merge` compares their roots. The smaller root (that is, the root with the smaller element) becomes the new overall root and its children are swapped. Then the larger root is merged with the new left child of the smaller root (which used to be the right child). The net effect of a `merge` is to interleave the rightmost paths of the two trees in sorted order, swapping the children of nodes along the way. This process is illustrated in Figure 41.8.

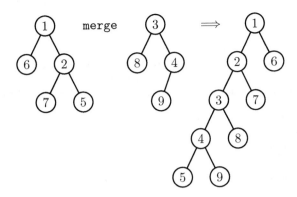

FIGURE 41.8 Merging two skew heaps.

```
data Skew = Empty | Node(Int,Skew,Skew)

empty = Empty
insert(x,s) = merge(Node(x,Empty,Empty),s)
findMin(Node(x,s1,s2)) = x
deleteMin(Node(x,s1,s2)) = merge(s1,s2)

merge(s1,Empty) = s1
merge(Empty,s2) = s2
merge(Node(x,s1,s2),Node(y,t1,t2)) =
   if x < y then Node(x,merge(Node(y,t1,t2),s2),s1)
            else Node(y,merge(Node(x,s1,s2),t2),t1)
```

FIGURE 41.9 Skew heaps in Haskell.

Notice how the nodes on the rightmost paths of the arguments end up on the leftmost path of the result. A Haskell implementation of skew heaps incorporating path copying is shown in Figure 41.9. A naive Java implementation is shown in Figure 41.10.

Skew heaps are not balanced, and individual operations can take linear time in the worst case. For example, Figure 41.11 shows an unbalanced shew heap generated by inserting the elements

$$5, 6, 4, 6, 3, 6, 2, 6, 1, 6$$

into an initially empty heap. Inserting a new element such as 7 into this unbalanced skew heap would take linear time. However, in spite of the fact that any one operation can be inefficient, the way that children are regularly swapped keeps the operations efficient in the amortized sense—insert, deleteMin, and merge run in logarithmic amortized time [12].

Or, at least, those would be the bounds for non-persistent skew heaps. When we analyze skew heaps in a persistent setting, we receive a nasty shock. Making an amortized data structure such as skew heaps persistent using path copying breaks its amortized bounds! In the case of skew heaps, naively incorporating path copying causes the logarithmic amortized bounds to degrade to the linear worst-case bounds.

Consider, for example, the unbalanced skew heap *s* in Figure 41.11, for which insert takes linear time for large elements. The result of the insert is a new skew heap *s'*. Performing another insert on *s'* would actually be quite efficient, but because these structures are persistent, we are free to ignore *s'* and perform the next insert on the old skew heap *s* instead. This insert again takes linear time. We can continue performing operations on the old skew heap as often as we want. The average cost per operation over a sequence of such operations is linear, which means that the amortized cost per operation is now linear, rather than logarithmic. Simple experiments on the Java implementation from Figure 41.10 confirm this analysis.

However, if we repeat those experiments on the Haskell implementation from Figure 41.9, we do not observe linear behavior. Instead, the operations appear to retain their logarithmic amortized bounds, even under persistent usage. This pleasant result is a consequence of a fortuitous interaction between path copying and a property of the Haskell language called *lazy evaluation*. (Many other functional programming languages also support lazy evaluation).

Under lazy evaluation, operations such as merge are not actually executed until their results are needed. Instead, a new kind of node that we might call a *pending merge* is automatically created. The pending merge lays dormant until some other operation

```
public class Skew {
  public static final Skew empty = null;
  public static Skew insert(int x,Skew s) { return merge(new Skew(x,null,null),s); }
  public static int findMin(Skew s) { return s.element; }
  public static Skew deleteMin(Skew s) { return merge(s.left,s.right); }

  public static Skew merge(Skew s,Skew t) {
    if (t == null) return s;
    else if (s == null) return t;
    else if (s.element < t.element)
      return new Skew(s.element,merge(t,s.right),s.left);
    else
      return new Skew(t.element,merge(s,t.right),t.left);
  }

  private int element;
  private Skew left,right;
  private Skew(int element, Skew left, Skew right) {
    this.element = element;
    this.left = left;
    this.right = right;
  }
}
```

FIGURE 41.10 First attempt at skew heaps in Java

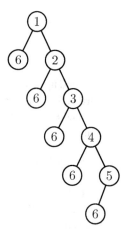

FIGURE 41.11 An unbalanced skew heap.

such as findMin needs to know the result. Then and only then is the pending merge executed. The node representing the pending merge is overwritten with the result so that it cannot be executed twice.

Although Java does not directly support lazy evaluation, it is easy to simulate, as shown in Figure 41.12. A minor difference between the lazy Java implementation and the Haskell implementation is that the Java implementation avoids creating pending merge nodes when one of the arguments is null.

A crucial aspect of lazy evaluation is that a pending computation, once triggered, is executed only far enough to produce the part of the result that is needed. The remaining parts of the computation may be delayed further by creating new pending nodes. In the case of the merge operation, this means that when a pending merge is executed, the two roots are compared and the children of the smaller root are swapped as normal, but the recursive merge of the larger root with the former right child of the smaller root is *not* performed. Instead, a new pending merge is created and installed as the left child of the new root node. This process is illustrated in Figure 41.13, with the pending merges drawn as diamonds.

Figure 41.14 illustrates the propagation of pending merges through a sequence of operations. First, the initial tree is built via a series of inserts. Then findMin executes those pending merges to find the value of the root. Next, deleteMin deletes the

```
public class Skew {
  public static final Skew empty = null;
  public static Skew insert(int x,Skew s) { return merge(new Skew(x,null,null),s); }
  public static int findMin(Skew s) {
    executePendingMerge(s);
    return s.element;
  }
  public static Skew deleteMin(Skew s) {
    executePendingMerge(s);
    return merge(s.left,s.right);
  }

  public static Skew merge(Skew s,Skew t) {
    if (t == null) return s;
    else if (s == null) return t;
    else return new Skew(s,t); // create a pending merge
  }

  private static void executePendingMerge(Skew s) {
    if (s != null && s.pendingMerge) {
      Skew s1 = s.left, s2 = s.right;
      executePendingMerge(s1);
      executePendingMerge(s2);
      if (s2.element < s1.element) { Skew tmp = s1; s1 = s2; s2 = tmp; }
      s.element = s1.element;
      s.left = merge(s2,s1.right);
      s.right = s1.left;
      s.pendingMerge = false;
    }
  }

  private boolean pendingMerge;
  private int element;
  private Skew left,right;
  private Skew(int element, Skew left, Skew right) {
    this.element = element;
    this.left = left;
    this.right = right;
    pendingMerge = false;
  }
  private Skew(Skew left,Skew right) { // create a pending merge
    this.left = left;
    this.right = right;
    pendingMerge = true;
  }
}
```

FIGURE 41.12 Skew heaps with lazy evaluation in Java.

root and creates a new pending merge of the two children. Finally, findMin again executes the pending merges to find the new value of the root.

Notice that pending nodes and lazy evaluation affect *when* the various steps of a merge are carried out, but that they do not affect the end results of those steps. After all the pending merges have been executed, the final tree is identical to the one produced by skew heaps without lazy evaluation.

Strictly speaking, the nodes of a lazy skew heap can no longer be called immutable. In particular, when a pending merge is executed, the node representing the pending merge is updated with the result so that it cannot be executed twice. Functional languages typically allow this kind of mutation, known as *memoization*, because it is invisible to the user, except in terms of efficiency. Suppose that memoization was not performed. Then pending merges might be executed multiple times. However,

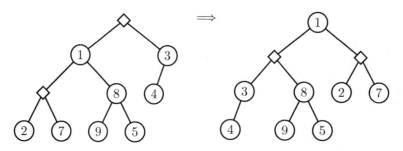

FIGURE 41.13 Executing a pending merge.

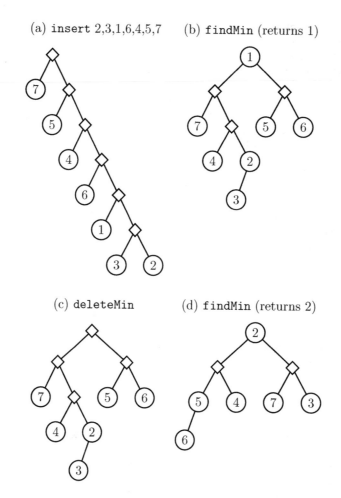

FIGURE 41.14 A sequence of operations on skew heaps.

every time a given pending merge was executed, it would produce the same result. Therefore, memoization is an optimization that makes lazy evaluation run faster, but that has no effect on the output of any computation.

41.4.1 Analysis of Lazy Skew Heaps

Next, we prove that merge on lazy skew heaps runs in logarithmic amortized time. We use the banker's method, associating a certain number of credits with each pending merge.

We begin with a few definitions. The *logical view* of a tree is what the tree would look like if all of its pending merges were executed. The *logical left spine* of a tree is the leftmost path from the root in the logical view. Similarly, the *logical right spine* of a tree is the rightmost path from the root in the logical view. A pending node is called *left-heavy* if its left subtree is at least as large as its right subtree in the logical view, or *right-heavy* if its right subtree is larger than its left subtree in the logical view. The *successor* of a pending node is the new pending node that is created as the new left child of the existing node when the existing node is executed.

Now, we charge one credit to execute a single pending merge. This credit pays for the comparison, the allocation of the successor node, and all necessary pointer manipulations, but it does not pay for executing the child nodes if they happen to be pending merges as well. When a right-heavy pending node is executed, it spends one of its own credits. When a left-heavy pending node is executed, the credit must be supplied either by its parent node, if it has one, or by the heap operation that originated the request, if the node is a root. This adds a single credit to the costs of the `insert` and `deleteMin` operations. After a pending node is executed, it passes any remaining credits to its successor node.

When we create a pending node, we must provide it with all the credits it will ever need. This includes

- One credit for itself, if it is right-heavy,
- One credit for each child that is a left-heavy pending node, and
- Any credits needed by its successors.

Notice that a pending node and its successors all lay on the logical left spine of the resulting tree in the logical view. Similarly, the physical children of a pending node and its successors all lay on the logical right spines of the argument trees to the original `merge`. Therefore, the number of credits that we must create during a `merge` is bounded by the number of right-heavy nodes in the logical left spine of the resulting tree plus the numbers of left-heavy nodes in the logical right spines of the argument trees.

It is easy to see that the number of right-heavy nodes in a logical left spine is at most logarithmic in the size of the logical view. Similarly, the number of left-heavy nodes in a logical right spine is at most logarithmic in the size of the logical view. The total number of credits created by `merge` is therefore bounded by the sum of three logarithmic values, and thus is logarithmic itself.

41.5 Difficulties

As this chapter has shown, many data structures work quite nicely in a functional setting. However, some do not. We close with a description of several warning signs to watch for.

- *Random access*: All of the data structures described in this chapter have been pointer-based. Unfortunately, data structures that depend on arrays—such as hash tables—are more difficult to handle. No entirely satisfactory approach is known for making arrays persistent. The best known approach from a theoretical point of view is that of Dietz [13], in which array accesses run in $O(\log \log n)$ expected amortized time. However, his approach is quite complicated and difficult to implement. Competing approaches, such as [14–16], degrade to logarithmic (or worse!) time per access in common cases.
- *Cycles*: Not all pointer-based data structures are suitable for implementation in a functional setting. The most common problem is the presence of cycles, such as those found in doubly-linked lists or in binary search trees with parent pointers. Path copying requires copying *all* paths from the root to the site of the update. In the presence of cycles, this frequently means copying the entire data structure.
- *Multiple entrypoints*: Even without cycles, a pointer-based data structure can run into difficulties if it has multiple entrypoints. For example, consider a pointer-based implementation of the union-find data structure [17]. All the pointers go from children to parents, so there are no cycles (except sometimes trivial ones at the roots). However, it is common for every node of the union-find data structure to be a potential entrypoint, rather than just the root(s). Path copying requires copying all paths from *any entrypoint* to the site of the update. With multiple entrypoints, this again frequently degrades into copying the entire data structure.
- *Unpredictable access patterns*: Section 41.4 described how to use lazy evaluation to make an amortized data structure persistent. Although this works for many amortized data structures, such as skew heaps [12], it does not work for all amortized data structures. In particular, data structures with highly unpredictable access patterns, such as splay trees [18] (see Chapter 13), are difficult to make persistent in this fashion.

41.6 Further Reading

The most complete general reference on functional data structures is Okasaki [1]. For more information on specific data structures, consult the following sources:

- Queues and deques [19–22]
- Priority queues and priority search queues [23,24]
- Random-access lists and flexible arrays [25–30]
- Catenable lists and deques [31–33]

Acknowledgments

This work was supported, in part, by the National Science Foundation under grant CCR-0098288. The views expressed in this chapter are those of the author and do not reflect the official policy or position of the United States Military Academy, the Department of the Army, the Department of Defense, or the U.S. Government.

References

1. C. Okasaki. *Purely Functional Data Structures.* Cambridge University Press, 1998.

2. S. Peyton Jones et al. Haskell 98: A non-strict, purely functional language. http://haskell.org/onlinereport/, Feb. 1999.

3. E. W. Myers. Efficient applicative data types. In *ACM Symposium on Principles of Programming Languages*, pages 66–75, Jan. 1984.

4. F. Rabhi and G. Lapalme. *Algorithms: A Functional Programming Approach.* Addison-Wesley, 1999.

5. C. Okasaki. Red-black trees in a functional setting. *Journal of Functional Programming*, 9(4):471–477, July 1999.

6. C. M. P. Reade. Balanced trees with removals: an exercise in rewriting and proof. *Sci. Comput. Programming*, 18(2):181–204, Apr. 1992.

7. S. Adams. Efficient sets—a balancing act. *Journal of Functional Programming*, 3(4):553–561, Oct. 1993.

8. R. Hinze. Explaining binomial heaps. *Journal of Functional Programming*, 9(1):93–104, Jan. 1999.

9. D. J. King. Functional binomial queues. In *Glasgow Workshop on Functional Programming*, pages 141–150. Sept. 1994.

10. M. Núñez, P. Palao, and R. Peña. A second year course on data structures based on functional programming. In *Functional Programming Languages in Education*, volume 1022 of *LNCS*, pages 65–84. Springer-Verlag, Dec. 1995.

11. C. Okasaki and A. Gill. Fast mergeable integer maps. In *Workshop on ML*, pages 77–86, Sept. 1998.

12. D. D. K. Sleator and R. E. Tarjan Self-adjusting heaps. *SIAM Journal on Computing*, 15(1):52–69, Feb. 1986.

13. P. F. Dietz. Fully persistent arrays. In *Workshop on Algorithms and Data Structures*, volume 382 of *LNCS*, pages 67–74. Springer-Verlag, Aug. 1989.

14. A. Aasa, S. Holmström, and C. Nilsson. An efficiency comparison of some representations of purely functional arrays. *BIT*, 28(3):490–503, 1988.

15. T. Chuang. A randomized implementation of multiple functional arrays. In *ACM Conference on LISP and Functional Programming*, pages 173–184, June 1994.

16. E. O'Neill and F. W. Burton A new method for functional arrays. *Journal of Functional Programming*, 7(5):487–513, Sept. 1997.

17. R. E. Tarjan. Data Structures and Network Algorithms, volume 44 of *CBMS Regional Conference Series in Applied Mathematics*. Society for Industrial and Applied Mathematics, Philadelphia, 1983.

18. D. D. K. Sleator and R. E. Tarjan. Self-adjusting binary search trees. *Journal of the ACM*, 32(3):652–686, July 1985.

19. T. Chuang and B. Goldberg. Real-time deques, multihead Turing machines, and purely functional programming. In *Conference on Functional Programming Languages and Computer Architecture*, pages 289–298, June 1993.

20. R. Hood. *The Efficient Implementation of Very-High-Level Programming Language Constructs.* PhD thesis, Department of Computer Science, Cornell University, Aug. 1982. (Cornell TR 82–503).

21. R. Hood and R. Melville. Real-time queue operations in pure Lisp. *Information Processing Letters*, 13(2):50–53, Nov. 1981.

22. C. Okasaki. Simple and efficient purely functional queues and deques. *Journal of Functional Programming*, 5(4):583–592, Oct. 1995.

23. G. S. Brodal and C. Okasaki. Optimal purely functional priority queues. *Journal of Functional Programming*, 6(6):839–857, Nov. 1996.

24. R. Hinze. A simple implementation technique for priority search queues. In *ACM SIGPLAN International Conference on Functional Programming*, pages 110–121, Sept. 2001.

25. R. Hinze. Bootstrapping one-sided flexible arrays. In *ACM SIGPLAN International Conference on Functional Programming*, pages 2–13, Sept. 2002.

26. R. R. Hoogerwoord. A logarithmic implementation of flexible arrays. In *Conference on Mathematics of Program Construction*, volume 669 of *LNCS*, pages 191–207. Springer-Verlag, July 1992.

27. A. Kaldewaij and V. J. Dielissen. Leaf trees. *Science of Computer Programming*, 26(1–3):149–165, May 1996.

28. E. W. Myers. An applicative random-access stack. *Information Processing Letters*, 17(5):241–248, Dec. 1983.

29. C. Okasaki. Purely functional random-access lists. In *Conference on Functional Programming Languages and Computer Architecture*, pages 86–95, June 1995.

30. C. Okasaki. Three algorithms on Braun trees. *Journal of Functional Programming*, 7(6):661–666, Nov. 1997.

31. H. Kaplan and R. E. Tarjan. Purely functional, real-time deques with catenation. *Journal of the ACM*, 46(5):577–603, Sept. 1999.

32. C. Okasaki. Amortization, lazy evaluation, and persistence: Lists with catenation via lazy linking. In *IEEE Symposium on Foundations of Computer Science*, pages 646–654, Oct. 1995.

33. C. Okasaki. Catenable double-ended queues. In *ACM SIGPLAN International Conference on Functional Programming*, pages 66–74, June 1997.

42

LEDA, a Platform for Combinatorial and Geometric Computing*

42.1	Introduction	653
	Ease of Use • Extensibility • Correctness • Availability and Usage	
42.2	The Structure of LEDA	654
42.3	Data Structures and Data Types	655
	Basic Data Types • Numbers and Matrices • Advanced Data Types • Graph Data Structures • Geometry Kernels • Advanced Geometric Data Structures	
42.4	Algorithms	656
42.5	Visualization	657
	GraphWin • GeoWin	
42.6	Example Programs	659
	Word Count • Shortest Paths • Curve Reconstruction • Upper Convex Hull • Delaunay Flipping • Discussion	
42.7	Projects Enabled by LEDA	665
	References	665

Stefan Naeher
University of Trier

42.1 Introduction

LEDA, the Library of Efficient Data Types and Algorithms, aims at being a comprehensive software platform for the area of combinatorial and geometric computing. It provides a sizable collection of data types and algorithms in C++. This collection includes most of the data types and algorithms described in the text books of the area [1–10]. LEDA supports a broad range of applications. It has already been used in such diverse areas as code optimization, VLSI design, graph drawing, graphics, robot motion planning, traffic scheduling, geographic information systems, machine learning and computational biology.

The LEDA project was started in 1988 by Kurt Mehlhorn and Stefan Näher. The first months they spent on the specification of different data types and on selecting the implementation language. At that time the item concept came up as an abstraction of the notion "pointer into a data structure." Items provide direct and efficient access to data and are similar to iterators in the standard template library. The item concept worked successfully for all test cases and is now used for most data types in LEDA. Concurrently with searching for the correct specifications several languages were investigated for their suitability as an implementation platform. Among the candidates were Smalltalk, Modula, Ada, Eiffel, and C++. The language had to support abstract data types and type parameters (genericity) and should be widely available. Based on the experiences with different example programs C++ was selected because of its flexibility, expressive power, and availability.

We discuss some of the general aspects of the LEDA system.

42.1.1 Ease of Use

The library is easy to use. In fact, only a small fraction of the users are algorithms experts and many users are not even computer scientists. For these users the broad scope of the library, its ease of use, and the correctness and efficiency of the algorithms in the library are crucial. The LEDA manual [11] gives precise and readable specifications for the data types and algorithms mentioned

---

* This chapter has been reprinted from first edition of this Handbook, without any content updates.

above. The specifications are short (typically not more than a page), general (so as to allow several implementations) and abstract (so as to hide all details of the implementation).

42.1.2 Extensibility

Combinatorial and geometric computing is a diverse area and hence it is impossible for a library to provide ready-made solutions for all application problems. For this reason it is important that LEDA is easily extendible and can be used as a platform for further software development. In many cases LEDA programs are very close to the typical text book presentation of the underlying algorithms. The goal is the equation *Algorithm* + LEDA = *Program*.

LEDA *extension packages* (LEPs) extend LEDA into particular application domains and areas of algorithmics not covered by the core system. LEDA extension packages satisfy requirements, which guarantee compatibility with the LEDA philosophy. LEPs have a LEDA-style documentation, they are implemented as platform independent as possible and the installation process allows a close integration into the LEDA core library. Currently, the following LEPs are available: PQ-trees, dynamic graph algorithms, a homogeneous *d*-dimensional geometry kernel, and a library for graph drawing.

42.1.3 Correctness

Programming is a notoriously error-prone task; this is even true when programming is interpreted in a narrow sense: going from a (correct) algorithm to a program. The standard way to guard against coding errors is program testing. The program is exercised on inputs for which the output is known by other means, typically as the output of an alternative program for the same task. Program testing has severe limitations. It is usually only done during the testing phase of a program. Also, it is difficult to determine the "correct" suite of test inputs. Even if appropriate test inputs are known it is usually difficult to determine the correct outputs for these inputs: alternative programs may have different input and output conventions or may be too inefficient to solve the test cases.

Given that program verification, that is, formal proof of correctness of an implementation, will not be available on a practical scale for some years to come, *program checking* has been proposed as an extension to testing [12,13]. The cited papers explored program checking in the area of algebraic, numerical, and combinatorial computing. In References 14–16 program checkers are presented for planarity testing and a variety of geometric tasks. LEDA uses program checkers for many of its implementations.

In computational geometry the correctness problem is even more difficult because geometric algorithms are frequently formulated under two unrealistic assumptions: computers are assumed to use exact real arithmetic (in the sense of mathematics) and inputs are assumed to be in general position. The naive use of floating point arithmetic as an approximation to exact real arithmetic very rarely leads to correct implementations. In a sequence of papers [17–21] the degeneracy and precision issues have been investigated and LEDA was extended based on this theoretical work. It now provides exact geometric kernels for two-dimensional and higher dimensional computational geometry [22] and also correct implementations for basic geometric tasks, for example, two-dimensional convex hulls, Delaunay diagrams, Voronoi diagrams, point location, line segment intersection, and higher-dimensional convex hulls and Delaunay triangulations.

An elegant (theoretical) approach to the degeneracy problem is *symbolic perturbation*. However, this method of forcing input data into general position can cause some serious problems in practice. In many cases, it increases the complexity of (intermediate) results considerably and furthermore, the final limit process turns out to be very difficult in particular in the presence of combinatorial structures. For this reason, LEDA follows a different approach. It copes with degeneracies directly by treating the degenerate case as the "normal" case. This approach proved to be very effective for many geometric problems.

42.1.4 Availability and Usage

LEDA is realized in C++ and can be used on many different platforms with many different compilers. LEDA is now used at many academic sites. A commercial version of LEDA is marketed by Algorithmic Solutions Software GmbH (<www.algorithmic-solutions.com>).

42.2 The Structure of LEDA

LEDA uses templates for the implementation of parameterized data types and for generic algorithms. However, it is not a pure template library and therefore is based on a number of object code libraries of precompiled code. Programs using LEDA data types or algorithms have to include the appropriate LEDA header files into their source code and have to be linked with one or more of these libraries. The four object code libraries are built on top of each other. Here, we only give a short overview. Consult the LEDA user manual [11] or the LEDA book [23] for a detailed description.

- The Basic Library (*libL*) contains system dependent code, basic data structures, numbers and types for linear algebra, dictionaries, priority queues, partitions, and many more basic data structures and algorithms.

- The Graph Library (*libG*) contains different types of graphs and a large collection of graph and network algorithms
- The 2D Geometry Library (*libP*) contains the two-dimensional geometric kernels advanced geometric data structures, and a large number of algorithms for two-dimensional geometric problems.
- The 3D Geometry Library (*libP*) contains the three-dimensional kernels and some algorithms for three-dimensional problems.
- The Window Library(*libW*) supports graphical output and user interaction for both the X11 platform (Unix) and Microsoft Windows systems. It also contains animation support: GraphWin, a powerful graph editor, and GeoWin, a interactive tool for the visualization of geometric algorithms. See Section 42.5 for details.

42.3 Data Structures and Data Types

LEDA contains a large number of data structures from the different areas of combinatorial and geometric computing. However, as indicated in the name of the library, LEDA was not designed as a collection of *data structures* but as a library of (parameterized) *data types*. For each (abstract) data type in the library there is at least one data structure which implements this type. This separation of specification (by an abstract data type) and implementation (by an actual data structure) is crucial for a software component library. It allows to change the underlying implementation or to choose between different implementations without having to change the application program. In general, there is one default data structure for each of the advanced data types, for example, avl-trees for dictionaries, skiplists for sorted sequences, binary heaps for priority queues, and a union-find structure for partitions. For most of these data types a number of additional data structures are available and can be specified by an optional template argument. For instance `dictionary<string,int,skiplist>` specifies a dictionary type with key type *string* and information type *int* which is implemented by the *skiplist* data structure (see Reference 23 for more details and Section 42.6.1 for an example).

42.3.1 Basic Data Types

Of course, LEDA contains a complete collection of all basic data types, such as strings, stacks, queues, lists, arrays, tuples . . . which are ubiquitous in all areas of computing.

42.3.2 Numbers and Matrices

Numbers are at the origin of computing. We all learn about integers, rationals, and real numbers during our education. Unfortunately, the number types *int*, *float*, and *double* provided by C++ are only crude approximations of their mathematical counterparts: there are only finitely many numbers of each type and for floats and doubles the arithmetic incurs rounding errors. LEDA offers the additional number types *integer*, *rational*, *bigfloat*, and *real*. The first two are the exact realization of the corresponding mathematical types and the latter two are better approximations of the real numbers. Vectors and matrices are one- and two-dimensional arrays of numbers, respectively. They provide the basic operations of linear algebra.

42.3.3 Advanced Data Types

A collection of parameterized data types representing sets, partitions, dictionaries, priority queues, sorted sequences, partitions and sparse arrays (maps, dictionary and hashing arrays). Type parameters include the key, information, priority, or element type and (optional) the data structure used for the implementation of the type. The list of data structures includes skiplists (see Chapter 14), (a, b)-trees, avl-trees, red-black trees, $bb[\alpha]$-trees (see Chapter 11), randomized search trees (see Chapter 14), Fibonacci-heaps, binary heaps, pairing heaps (see Chapter 7), redistributive heaps, union find with path compression (Chapter 34), dynamic perfect hashing, cuckoo hashing (see Chapter 9), Emde-Boas trees, ... This list of data structures is continuously growing and adapted to new results from the area for data structures and algorithms.

42.3.4 Graph Data Structures

The graph data type (Chapter 4) is one of the central data types in LEDA. It offers the standard iterations such as "for all nodes v of a graph G do" (written *forall_nodes*(v, G)) or "for all edges e incident to a node v do" (written *forall_out_edges*(e, v)), it allows to add and delete vertices and edges and it offers arrays and matrices indexed by nodes and edges, (see Reference 23 Chapter 6 for details and Section 42.6.2 for an example program). The data type *graph* allows to write programs for graph problems in a form very close to the typical text book presentation.

42.3.5 Geometry Kernels

LEDA offers kernels for two- and three-dimensional geometry, a kernel of arbitrary dimension is available as an extension package. In either case there exists a version of the kernel based on floating point Cartesian coordinates (called *float-kernel*) as well as a kernel based on rational homogeneous coordinates (called *rat-kernel*). All kernels provide a complete collection of geometric objects (points, segments, rays, lines, circles, simplices, polygons, planes, etc.) together with a large set of geometric primitives and predicates (orientation of points, side-of-circle tests, side-of-hyperplane, intersection tests and computation, etc.). For a detailed discussion and the precise specification see Chapter 9 of the LEDA book [23]. Note that only for the rational kernel, which is based on exact arithmetic and floating-point filters, all operations and primitives are guaranteed to compute the correct result.

42.3.6 Advanced Geometric Data Structures

In addition to the basic kernel data structures LEDA provides many advanced data types for computational geometry. Examples are

- A general polygon type (**gen_polygon** or **rat_gen_polygon**) with a complete set of boolean operations. Its implementation is based on an efficient and robust plane sweep algorithms for the construction of the arrangement of a set of straight line segments (see References 23 and 24 Chapter 11.7 for a detailed description).
- Two- and higher-dimensional geometric tree structures, such as range, segment, interval and priority search trees (see Chapter 19).
- Partially and fully persistent search trees (Chapter 33).
- Different kinds of geometric graphs (triangulations, Voronoi diagrams, and arrangements)
- A dynamic **point_set** data type supporting update, search, closest point, and different types of range query operations on one single representation which is based on a dynamic Delaunay triangulation (see Reference 23 Chapter 11.6).

42.4 Algorithms

The LEDA project had never the goal to provide a complete collection of all algorithms. LEDA was designed and implemented to establish a *platform* for combinatorial and geometric computing enabling programmers to implement these algorithms themselves more easily and customized to their particular needs. But of course the library already contains a considerable number of standard algorithms.

Here we give a brief overview and refer the reader to the user manual for precise specifications and to Chapter 10 of the LEDA-book [23] for a detailed description and analysis of the corresponding implementations. In the current version LEDA offers different implementation of algorithms for the following problems:

- Sorting and searching
- Basic graph properties
- Graph traversals
- Different kinds of connected components
- Planarity test and embeddings
- Minimum spanning trees
- Shortest paths
- Network flows
- Maximum weight and cardinality matchings
- Graph drawing
- Convex hulls (also three-dimensional)
- Half-plane intersections
- (Constraint) triangulations
- Closest and farthest Delaunay and Voronoi diagrams
- Euclidean minimum spanning trees
- Closest pairs
- Boolean operations on generalized polygons
- Segment intersection and construction of line arrangements
- Minkowski sums and differences
- Nearest neighbors and closest points
- Minimum enclosing circles and annuli
- Curve reconstruction

42.5 Visualization

Visualization and animation of programs is very important for the understanding, presentation, and debugging of algorithms (see Chapter 45). LEDA provides two powerful tools for interactive visualization and animation of data structures and algorithms:

42.5.1 GraphWin

The *GraphWin* data type (see Reference 23, Chapter 12) combines the *graph* and the *window* data type. An object of type *GraphWin* (short: a GraphWin) is a window, a graph, and a drawing of the graph, all at once. The graph and its drawing can be modified either by mouse operations or by running a graph algorithm on the graph. The GraphWin data type can be used to:

- Construct and display graphs,
- Visualize graphs and the results of graph algorithms,
- Write interactive demos for graph algorithms,
- Animate graph algorithms.

All demos and animations of graph algorithms in LEDA are based on GraphWin, many of the drawings in the LEDA book [23] have been made with GraphWin, and many of the geometry demos in LEDA have a GraphWin button that allows us to view the graph structure underlying a geometric object.

GraphWin can easily be used in programs for constructing, displaying and manipulating graphs and for animating and debugging graph algorithms. It offers both a programming and an interactive interface, most applications use both of them. Figure 42.1 shows a screenshot of a typical GraphWin window.

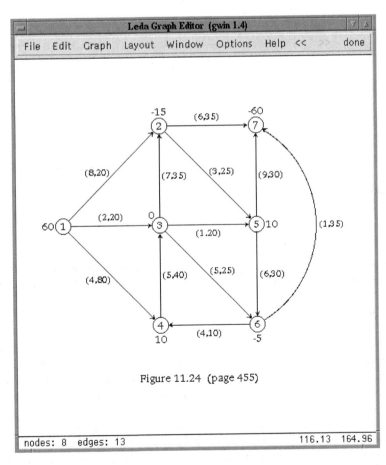

FIGURE 42.1 A GraphWin. The display part of the window shows a graph *G* and the panel part of the window features the default menu of a GraphWin. *G* can be edited interactively, e.g., nodes and edges can be added, deleted, and moved around. It is also possible to run graph algorithms on *G* and to display their result or to animate their execution.

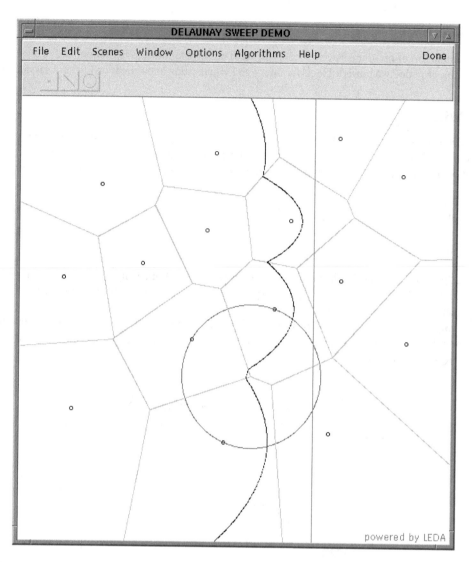

FIGURE 42.2   GeoWin animating Fortune's Sweep Algorithm.

42.5.2 GeoWin

GeoWin [25] is a generic tool for the interactive visualization of geometric algorithms. *GeoWin* is implemented as a C++ data type. It provides support for a number of programming techniques which have shown to be very useful for the visualization and animation of algorithms. The animations use *smooth transitions* to show the result of geometric algorithms on dynamic user-manipulated input objects, for example, the Voronoi diagram of a set of moving points or the result of a sweep algorithm that is controlled by dragging the sweep line with the mouse (see Figure 42.2).

A GeoWin maintains one or more geometric scenes. A geometric *scene* is a collection of geometric objects of the same type. A collection is simply either a standard C++-list (STL-list) or a LEDA-list of objects. GeoWin requires that the objects provide a certain functionality, such as stream input and output, basic geometric transformations, drawing and input in a LEDA window. A precise definition of the required operations can be found in the manual pages [11]. GeoWin can be used for any collection of basic geometric objects (geometry kernel) fulfilling these requirements.

The visualization of a scene is controlled by a number of attributes, such as color, line width, line style, etc. A scene can be subject to user interaction and it may be defined from other scenes by means of an algorithm (a C++ function). In the latter case the scene (also called *result scene*) may be recomputed whenever one of the scenes on which it depends is modified. There are three main modes for re-computation: user-driven, continuous, and event-driven.

GeoWin has both an interactive and a programming interface. The interactive interface supports the interactive manipulation of input scenes, the change of geometric attributes, and the selection of scenes for visualization.

42.6 Example Programs

In this section we give several programming examples showing how LEDA can be used to implement combinatorial and geometric algorithms in a very elegant and readable way. In each case we will first state the algorithm and then show the program. It is not essential to understand the algorithms in full detail; our goal is to show:

- How easily the algorithms are transferred into programs and
- How natural and elegant the programs are.

In other words, Algorithm + LEDA = Program.

42.6.1 Word Count

We start with a very simple program. The task is to read a sequence of strings from standard input, to count the number of occurrences of each string in the input, and to print a list of all occurring strings together with their frequencies on standard output.

The program uses the LEDA types *string* and *d_array* (dictionary arrays). The parameterized data type *d_array* $<I, E>$ realizes arrays with index type *I* and element type *E*. We use it with index type *string* and element type *int*.

```
#include <LEDA/d_array.h>
#include <LEDA/impl/skiplist.h>
int main()
{ d_array<string,int,skiplist> N(0);
  string s;
  while (cin >> s) N[s]++;
  forall_defined(s,N) cout << s << " " << N[s] << endl;
}
```

We give some more explanations. The program starts with the include statement for dictionary arrays and skiplists. In the first line of the main program we define a dictionary array *N* with index type *string*, element type *int* and implementation *skiplist* and initialize all entries of the array to zero. Conceptually, this creates an infinite array with one entry for each conceivable string and sets all entries to zero; the implementation of d_arrays stores the non-zero entries in a balanced search tree with key type string. In the second line we define a string *s*. The while-loop does most of the work. We read strings from the standard input until the end of the input stream is reached. For every string *s* we increment the entry *N[s]* of the array *N* by one. The iteration *forall_defined(s, N)* in the last line successively assigns all strings to *s* for which the corresponding entry of *N* was touched during execution. For each such string, the string and its frequency are printed on the standard output.

42.6.2 Shortest Paths

Dijkstra's shortest path algorithm [26] takes a directed graph $G = (V, E)$, a node $s \in V$, called the source, and a non-negative cost function on the edges $cost : E \rightarrow R_{\geq 0}$. It computes for each node $v \in V$ the distance from *s*, see Figure 42.3. A typical text book presentation of the algorithm is as follows:

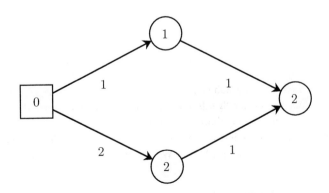

FIGURE 42.3 A shortest path in a graph. Each edge has a non-negative cost. The cost of a path is the sum of the cost of its edges. The source node *s* is indicated as a square. For each node the length of the shortest path from *s* is shown.

forall nodes *v* **do**
    set *dist*(*v*) to infinity.
    declare *v* unreached.
od
set *dist*(*s*) to 0.

while there is an unreached node **do**
    let *u* be an unreached node with minimal dist-value.   (*)
    declare *u* reached.
    forall edges *e* = (*u*, *v*) out of *u* **do**
        set *dist*(*v*) = min(*dist*(*v*), *dist*(*u*) + *cost*(*e*)).
    od
od

The text book presentation will then continue to discuss the implementation of line (*). It will state that the pairs {(*v*, *dist*(*v*));*v* unreached} should be stored in a priority queue, for example, a Fibonacci heap, because this will allow the selection of an unreached node with minimal distance value in logarithmic time. It will probably refer to some other chapter of the book for a discussion of priority queues.

We next give the corresponding LEDA program; it is very similar to the pseudo-code above. In fact, after some experience with LEDA you should be able to turn the pseudo-code into code within a few minutes.

```
#include <LEDA/graph.h>
#include <LEDA/node_pq.h>
void DIJKSTRA(const graph& G, node s, const edge_array<double>& cost,
                                      node_array<double>& dist)
{ node_pq<double> PQ(G);
  node v;
  dist[s] = 0;
  forall_nodes(v,G)
  { if (v != s) dist[v] = MAXDOUBLE;
    PQ.insert(v,dist[v]);
  }

while (!PQ.empty()) {
  node u = PQ.del_min();
  edge e;
  forall_out_edges(e,u) {
    v = target(e);
    double c = dist[u] + cost[e];
    if (c < dist[v]) {
      PQ.decrease_p(v,c);
      dist[v] = c;
      }
    }
  }
}
```

We start by including the graph and the node priority queue data types. The function *DIJKSTRA* takes a graph *G*, a node *s*, an *edge_array* cost, and a *node_array* dist. Edge arrays and node arrays are arrays indexed by edges and nodes, respectively. We declare a priority queue *PQ* for the nodes of graph *G*. It stores pairs (*v*, *dist*[*v*]) and is initially empty. The *forall_nodes*-loop initializes *dist* and *PQ*. In the main loop we repeatedly select a pair (*u*, *dist*[*u*]) with minimal distance value and then iterate over all out-going edges to update distance values of neighboring vertices.

We next incorporate the shortest path program into a small demo. We generate a random graph with *n* nodes and *m* edges and choose the edge costs as random number in the range [0, 100]. We call the function above and report the running time.

```
int main()
{ int n = read_int("number of nodes = ");
  int m = read_int("number of edges = ");
  graph G;
```

```
        random_graph(G,n,m);
        edge_array<double> cost(G);
        node_array<double> dist(G);
        edge e;
        forall_edges(e,G) cost[e] = rand_int(0,100);

        float T = used_time();
        DIJKSTRA(G,G.first_node(),cost,dist);
        cout << "The computation took" << used_time(T) << "seconds." << endl;
}
```

On a graph with 10,000 nodes and 100,000 edges the computation takes less than a second.

42.6.3 Curve Reconstruction

The reconstruction of a curve from a set of sample points is an important problem in computer vision. Amenta, Bern, and Eppstein [27] introduced a reconstruction algorithm which they called CRUST. Figure 42.4 shows a point set and the curves reconstructed by their algorithm. The algorithm *CRUST* takes a list *S* of points and returns a graph *G*. CRUST makes use of Delaunay diagrams and Voronoi diagrams and proceeds in three steps:

- It first constructs the Voronoi diagram *VD* of the points in *S*.
- It then constructs a set $L = S \cup V$, where *V* is the set of vertices of *VD*.
- Finally, it constructs the Delaunay triangulation *DT* of *L* and makes *G* the graph of all edges of *DT* that connect points in *S*.

The algorithm is very simple to implement

```
#include <LEDA/graph.h>
#include <LEDA/map.h>
#include <LEDA/float_kernel.h>
#include <LEDA/geo_alg.h>
void CRUST(const list<point>& S, GRAPH<point,int>& G)
{
list<point> L = S;
GRAPH<circle,point> VD;
VORONOI(L,VD);
// add Voronoi vertices and mark them
map<point,bool> voronoi_vertex(false);
node v;
forall_nodes(v,VD)
{ if (VD.outdeg(v) < 2) continue;
  point p = VD[v].center();
  voronoi_vertex[p] = true;
  L.append(p);
}
DELAUNAY_TRIANG(L,G);
```

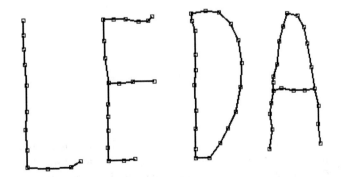

FIGURE 42.4   A set of points in the plane and the curve reconstructed by CRUST. The figure was generated by the program presented in Section 42.6.3.

```
forall_nodes(v,G)
 if (voronoi_vertex[G[v]]) G.del_node(v);
}
```

We give some explanations. We start by including graphs, maps, the floating point geometry kernel, and the geometry algorithms. In CRUST we first make a copy of *S* in *L*. Next we compute the Voronoi diagram *VD* of the points in *L*. In LEDA we represent Voronoi diagrams by graphs whose nodes are labeled with circles. A node *v* is labeled by a circle passing through the defining sites of the vertex. In particular, *VD*[*v*].*center*() is the position of the node *v* in the plane. Having computed *VD* we iterate over all nodes of *VD* and add all finite vertices (a Voronoi diagram also has nodes at infinity, they have degree one in our graph representation of Voronoi diagrams) to *L*. We also mark all added points as vertices of the Voronoi diagram. Next we compute the Delaunay triangulation of the extended point set in *G*. Having computed the Delaunay triangulation, we collect all nodes of *G* that correspond to vertices of the Voronoi diagram in a list *vlist* and delete all nodes in *vlist* from *G*. The resulting graph is the result of the reconstruction.

We next incorporate CRUST into a small demo which illustrates its speed. We generate *n* random points in the plane and construct their crust. We are aware that it does really make sense to apply CRUST to a random set of points, but the goal of the demo is to illustrate the running time.

```
int main()
{ int n = read_int("number of points = ");
  list<point> S;
  random_points_in_unit_square(n,S);
  GRAPH<point,int> G;

  float T = used_time();
  CRUST(S,G);
  cout << "The crust computation took" << used_time(T) << "seconds.";
  cout << endl;
}
```

For 3000 points the computation takes less than a second. We can now use the preceding program for a small interactive demo.

```
#include <LEDA/window.h>
int main()
{ window W;
  W.display();
  W.set_node_width(2);
  W.set_line_width(2);
  point p;
  list<point> S;
  GRAPH<point,int> G;
while (W >> p)
{ S.append(p);
  CRUST(S,G);
  W.clear();
  node v;
  forall_nodes(v,G) W.draw_node(G[v]);
  edge e;
  forall_edges(e,G) W.draw_segment(G[source(e)], G[target(e)]);
  }
}
```

We give some more explanations. We start by including the window type. In the main program we define a window and open its display. A window will pop up. We state that we want nodes and edges to be drawn with width two. We define the list *S* and the graph *G* required for CRUST. In each iteration of the while-loop we read a point in *W* (each click of the left mouse button enters a point), append it to *S* and compute the crust of *S* in *G*. We then draw *G* by drawing its vertices and its edges. Each edge is drawn as a line segment connecting its endpoints. Figure 42.4 was generated with the program above.

42.6.4 Upper Convex Hull

In the next example we show how to use LEDA for computing the upper convex hull of a given set of points. The following function UPPER_HULL takes a list *L* of rational points (type `rat_point`) as input and returns the list of points of the upper convex hull of *L* in clockwise ordering. The algorithm is a variant of Graham's Scan [28]. First we sort *L* according to the lexicographic ordering of the Cartesian coordinates and remove multiple points. If the list contains not more than two points after this step we stop. Before starting the actual Graham Scan we first skip all initial points lying on or below the line connecting the two extreme points. Then we scan the remaining points from left to right and maintain the upper hull of all points seen so far in a list called *hull*. Note however that the last point of the hull is not stored in this list but in a separate variable *p*. This makes it easier to access the last two hull points as required by the algorithm. Note also that we use the rightmost point as a sentinel avoiding the special case that *hull* becomes empty.

```
using namespace leda;
list<rat_point> UPPER_HULL(list<rat_point> L)
{ L.sort();
  L.unique();
  if (L.length() <= 2) return L;

  rat_point p_min = L.front(); // leftmost point
  rat_point p_max = L.back();  // rightmost point

  list<rat_point> hull;
  hull.append(p_max); // use rightmost point as sentinel
  hull.append(p_min); // first hull point

  // goto first point p above (p_min,p_max)
  while (!L.empty() && !left_turn(p_min,p_max,L.front()) L.pop();
  if (L.empty()) return hull;

  rat_point p = L.pop(); // second (potential) hull point
  rat_point q;
  forall(q,L)
{ while (!right_turn(hull.back(),p,q)) p = hull.pop_back();
  hull.append(p);
  p = q;
}

  hull.append(p); // add last hull point
  hull.pop();     // remove sentinel
  return hull;
}
```

42.6.5 Delaunay Flipping

LEDA represents triangulations by bidirected plane graphs (from the graph library) whose nodes are labeled with points and whose edges may carry additional information, for example, integer flags indicating the type of edge (hull edge, triangulation edge, etc.). All edges incident to a node *v* are ordered in counter-clockwise ordering and every edge has a reversal edge. In this way the faces of the graph represent the triangles of the triangulation. The graph type offers methods for iterating over the nodes, edges, and adjacency lists of the graph. In the case of plane graphs there are also operations for retrieving the reverse edge and for iterating over the edges of a face. Furthermore, edges can be moved to new nodes. This graph operation is used in the following program to implement edge flips.

Function DELAUNAY_FLIPPING takes as input an arbitrary triangulation and turns into a Delaunay triangulation by the well-known flipping algorithm. This algorithm performs a sequence of local transformations as shown in Figure 42.5 to establish the Delaunay property: for every triangle the circumscribing sphere does not contain any vertex of the triangulation in its interior. The test whether an edge has to be flipped or not can be realized by a so-called *side_of_circle* test. This test takes four points *a*, *b*, *c*, *d* and decides on which side of the oriented circle through the first three points *a*, *b*, and *c* the last point *d* lies. The result is positive or negative if *d* lies on the left or on the right side of the circle, respectively, and the result is zero if all four points lie on one common circle. The algorithms uses a list of candidates which might have to be flipped (initially all edges). After a flip the

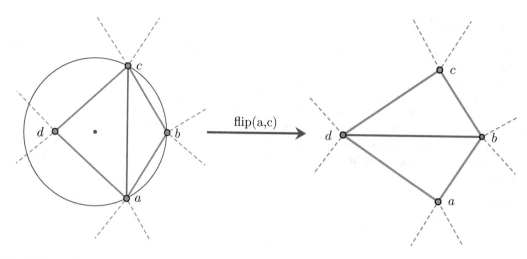

FIGURE 42.5 Delaunay Flipping.

four edges of the corresponding quadrilateral are pushed onto this candidate list. Note that G[v] returns the position of node *v* in the triangulation graph G. A detailed description of the algorithm and its implementation can be found in the LEDA book ([23]).

```
void DELAUNAY_FLIPPING(GRAPH<POINT,int>& G)
{
 list<edge> S = G.all_edges();

 while ( !S.empty() )
 { edge e = S.pop();
   edge r = G.rev_edge(e);

   // e1,e2,e3,e4: edges of quadrilateral with diagonal e
   edge e1 = G.face_cycle_succ(r);
   edge e2 = G.face_cycle_succ(e1);
   edge e3 = G.face_cycle_succ(e);
   edge e4 = G.face_cycle_succ(e3);

   // a,b,c,d: corners of quadrilateral
   POINT a = G[G.source(e1)];
   POINT b = G[G.target(e1)];
   POINT c = G[G.source(e3)];
   POINT d = G[G.target(e3)];

   if (side_of_circle(a,b,c,d) > 0)
   { S.push(e1); S.push(e2); S.push(e3); S.push(e4);
     // flip diagonal
     G.move_edge(e,e2,source(e4));
     G.move_edge(r,e4,source(e2));
   }
  }
}
```

42.6.6 Discussion

In each of the above examples only a few lines of code were necessary to achieve complex functionality and, moreover, the code is elegant and readable. The data structures and algorithms in LEDA are efficient. For example, the computation of shortest paths in a graph with 10,000 nodes and 100,000 edges and the computation of the crust of 5000 points takes less than a second each. We conclude that LEDA is ideally suited for rapid prototyping as summarized in the equation

$$\text{Algorithm} + \text{LEDA} = \text{Efficient Program}.$$

42.7 Projects Enabled by LEDA

A large number academic and industrial projects from almost every area of combinatorial and geometric computing have been enabled by LEDA. Examples are graph drawing, algorithm visualization, geographic information systems, location problems, visibility algorithms, DNA sequencing, dynamic graph algorithms, map labeling, covering problems, railway optimization, route planning and many more. A detailed list of academic LEDA projects can be found on <http://www.mpi-sb.mpg.de/LEDA/friends> and a selection of industrial projects is shown on <http://www.algorithmic-solutions.com/enreferenzen.htm>.

References

1. A. V. Aho, J. E. Hopcroft and J. D. Ullman, *The Design and Analysis of Computer Algorithms*, Addison-Wesley, Longman Publishing Co., Inc. Boston, MA, 1974.
2. T. H. Cormen and C. E. Leiserson and R. L. Rivest, *Introduction to Algorithms*, MIT Press/McGraw-Hill Book Company, 1990.
3. J. H. Kingston, *Algorithms and Data Structures*, Addison-Wesley, 1990.
4. K. Mehlhorn, *Data Structures and Algorithms 1,2, and 3*, Springer, 1984.
5. J. Nievergelt and K.H. Hinrichs, *Algorithms and Data Structures*, Prentice Hall, 1993.
6. J. O'Rourke, *Computational Geometry in C*, Cambridge University Press, 1994.
7. R. Sedgewick, *Algorithms*, Addison-Wesley, 1991.
8. R. E. Tarjan, Data Structures and Network Algorithms, *CBMS-NSF Regional Conference Series in Applied Mathematics* 44, 1983.
9. C. J. van Wyk, *Data Structures and C Programs*, Addison-Wesley, 1988.
10. D. Wood, *Data Structures, Algorithms, and Performance*, Addison-Wesley, 1993.
11. K. Mehlhorn, S. Näher, M. Seel and C. Uhrig, *The LEDA User Manual*, Technical Report, Max-Planck-Institut für Informatik, 1999.
12. M. Blum, M. Luby and R. Rubinfeld, Self-Testing/Correcting with Applications to Numerical Problems, *Proceedings of the 22nd Annual ACM Symposium on Theory of Computing (STOC90)*, 73–83, 1990.
13. M. Blum and S. Kannan, Designing Programs That Check Their Work, *Proceedings of the 21th Annual ACM Symposium on Theory of Computing (STOC89)*, 86–97, 1989.
14. C. Hundack, K. Mehlhorn and S. Näher, *A simple linear time algorithm for identifying Kuratowski subgraphs of non-planar graphs*, unpublished manuscript, 1996.
15. K. Mehlhorn and P. Mutzel, On the Embedding Phase of the Hopcroft and Tarjan Planarity Testing Algorithm, *Algorithmica*, Vol. 16, No. 2, 233–242, 1995.
16. K. Mehlhorn, S. Näher, M. Seel, R. Seidel, T. Schilz, S. Schirra and C. Uhrig, Checking Geometric Programs or Verification of Geometric Structures, *Computational Geometry: Theory and Applications*, Vol. 12, No. 1–2, 85–103, 1999.
17. C. Burnikel, K. Mehlhorn and S. Schirra, How to compute the Voronoi diagram of line Segments: Theoretical and experimental results, *Proceedings of the 2nd Annual European Symposium on Algorithms (ESA94), Lecture Notes in Computer Science* 855, 227–239, 1994.
18. K. Mehlhorn and S. Näher, The Implementation of Geometric Algorithms, *Proceedings of the 13th IFIP World Computer Congress*, 223–231, Elsevier Science B.V. North-Holland, Amsterdam, 1994.
19. M. Seel, *Eine Implementierung abstrakter Voronoidiagramme*, Diplomarbeit, Fachbereich Informatik, Universität des Saarlandes, Saarbrücken, 1994.
20. C. Burnikel, K. Mehlhorn and S. Schirra, On Degeneracy in Geometric Computations, *Proceedings of the 5th ACM-SIAM Sympos. Discrete Algorithms*, 16–23, 1994.
21. A. Fabri, G.-J. Giezeman, L. Kettner, S. Schirra and S. Schönherr, On the Design of CGAL a Computational Geometry Algorithms Library, *Software – Practice and Experience*, Vol. 30, No. 11, 1167–1202, 2000.
22. Mehlhorn K., M. Müller, S. Näher, S. Schirra, M. Seel, C. Uhrig and J. Ziegler, A Computational Basis for Higher-Dimensional Computational Geometry and Applications, *Computational Geometry: Theory and Applications*, Vol. 10, No. 4, 289–303, 1998.
23. K. Mehlhorn and S. Näher, *LEDA: A Platform for Combinatorial and Geometric Computing*, Cambridge University Press, 2000.
24. K. Mehlhorn and S. Näher, *Implementation of a sweep line algorithm for the straight line segment intersection problem*, Technical Report, Max-Planck-Institut für Informatik, MPI-I-94-160, 1994.

25. M. Bäsken and S. Näher, GeoWin—A Generic Tool for Interactive Visualization of Geometric Algorithms, *Software Visualization, LNCS*, Vol. 2269, 88–100, 2002.

26. E. W. Dijkstra, A Note on Two Problems in Connection With Graphs, *Num. Math.*, Vol. 1, 269–271, 1959.

27. N. Amenta, M. Bern and D. Eppstein, The Crust and the beta-Skeleton: Combinatorial Curve Reconstruction, *Graphical Models and Image Processing*, Vol. 60, 125–135, 1998.

28. R. L. Graham, An Efficient Algorithm for Determining the Convex Hulls of a Finite Point Set, *Information Processing Letters*, Vol. 1, 132–133, 1972.

43

Data Structures in C++

43.1	Introduction	667
43.2	Basic Containers	667
	Sequence Containers • Associative Containers • Container Adapters	
43.3	Iterators	674
	Basics • Reverse Iterators	
43.4	Additional Components of the STL	676
	Sorting, Searching, and Selection	
	Acknowledgments	677
	References	677

Mark Allen Weiss
Florida International University

43.1 Introduction

In C++, several classic data structures are implemented as a part of the standard library, commonly known as the *Standard Template Library*. The Standard Template Library (or simply, the STL) consists of a set of *container classes*, such as lists, sets, and maps; *iterator classes* that are used to traverse the container classes; and *generic algorithms*, such as sorting and searching. As its name implies, the library consists of (both function and class) templates. The STL is very powerful and makes some use of advanced C++ features. Our discussion is focused on the most basic uses of the STL, but does include some features found only in C++11 or later.

We partition our discussion of the Standard Template Library into the following sections:

1. Containers
2. Iterators
3. Generic Algorithms

See References 1–7 for additional material on the Standard Template Library.

43.2 Basic Containers

The STL defines several container templates that store collections of objects. Some collections are unordered, while others are ordered. Some collections allow duplicates, while others do not. All containers support the following operations:

- `int size( ) const`: returns the number of elements in the container.
- `void clear( )`: removes all elements from the container.
- `bool empty( ) const`: returns `true` if the container contains no elements and `false` otherwise.

There is no universal `add` or `insert` member function; different containers use different names.
The container classes can be split into three broad categories:

1. *Sequence containers* maintain items with a notion of a position in the collection. Examples include `vector`, `list`, and `deque`.
2. *Associative containers* maintain items in a manner that allows efficient searching, insertion, and deletion (using balanced search trees, and starting in C++11, hash tables). Examples include `set`, `multiset`, `map`, and `multimap`.
3. *Container adapters* built on top of other containers to yield classic data structures. Examples include `stack`, `queue`, and `priority_queue`.

Associated with all containers is an *iterator*. An iterator represents the notion of a position in the container and is used to step through the container. All containers support the following operations:

- `iterator begin( )`: returns an appropriate iterator representing the first item in the container.
- `iterator end( )`: returns an appropriate iterator representing the end marker in the container (i.e., the position after the last item in the container).

We defer the discussion of iterators to Section 43.3.

43.2.1 Sequence Containers

The three basic sequence containers in the STL are the `vector`, `list`, and `deque`.

`vector` is a growable array. The `vector` wraps an internal array, maintaining a notion of its size, and internally its current capacity. If a sequence of additions would cause the size to exceed capacity, the capacity is automatically doubled. This makes additions at the end of the `vector` take constant amortized time. `list` is a doubly linked list, in which pointers to both ends are maintained. `deque` is, in effect, a growable array that grows at both ends. There are several well-known ways to implement `deque` efficiently, but the standard is silent about which implementation must be used.

For `vector`, an insertion or deletion takes *amortized time* that is proportional to the distance from the back, while for a `deque`, these operations take amortized time that is proportional to the smaller of the distances from the front and back. For a `list`, these operations take worst-case time that is proportional to the smaller of the distances from the front and back if an index is specified, but constant worst-case time if the position is specified by an existing iterator. `vector` and `deque` support indexing in constant worst-case time; `list` does not.

The basic operations that are supported by both containers are:

- `void push_back( const Object & x )`: adds x to the end of the container.
- `Object & back( )`: returns the object at the end of the container (an accessor that returns a constant reference is also provided).
- `void pop_back( )`: removes the object at the end of the container.
- `Object & front( )`: returns the object at the front of the container (an accessor that returns a constant reference is also provided).
- `iterator insert( iterator pos, const Object & x )`: adds x into the container, prior to the position given by the iterator pos. This is a constant-time operation for `list`, but not for `vector` or `deque`. The return value is an iterator representing the position of the inserted item.
- `iterator erase( iterator pos )`: removes the object at the position given by the iterator. This is a constant-time operation for `list`, but not `vector` or `deque`. The return value is the position of the element that followed pos prior to the call. This operation invalidates pos, which is now stale.
- `iterator erase( iterator start, iterator end )`: removes all items beginning at position start, up to but not including end. Observe that an entire container can be erased by the call: `c.erase( c.begin( ), c.end( ) )`.

For `deque` and `list`, two additional operations are available with expected semantics:

- `void push_front( const Object & x )`: adds x to the front of the container.
- `void pop_front( )`: removes the object at the front of the container.

The `list` also provides a `splice` operation that allows the transfer of a sublist to somewhere else.

- `void splice( iterator pos, list & l, iterator start, iterator end )`: moves all items beginning at position start, up to but not including end to after pos. The moved items are assumed to have come from list l. The running time of this operation is proportional to the number of items moved, unless the items are moved within the same list (i.e., `&l==this`), in which case the running time is constant time.

For `vector` and `deque`, additional operations include:

- `Object & operator[] ( int idx )`: returns the object at index idx in the container, with no bounds-checking (an accessor that returns a constant reference is also provided).
- `Object & at( int idx )`: returns the object at index idx in the container, with bounds-checking (an accessor that returns a constant reference is also provided).
- `int capacity( ) const`: returns the internal capacity.
- `void reserve( int newCapacity )`: for `vector` only, sets the new capacity. If a good estimate is available, it can be used to avoid array expansion.

C++11 adds an `array` container that represents a fixed size array with no space overhead, and which is instantiated with a type and size (e.g., `array<int,6>`). `array` is useful for embedded systems as a safer replacement for built-in arrays, but does not support insertion or deletion. Generally speaking, the non-modifying operations discussed earlier, such as `at`, `begin` and `end`, and `size` work for `array` also.

C++11 also adds a `forward_list`, which is a singly linked list. Because in a standard `list`, `insert` and `erase` add and remove prior to the position given in the iterator, those operations would be linear time in a singly linked list. Accordingly, in a `forward_list` you'll find additional member functions `insert_after` and `erase_after` that provide constant-time performance.

43.2.2 Associative Containers

The STL provides two basic types of associative containers. Sets are used to store items. The `multiset` allows duplicates, while the `set` does not. Maps are used to store key/value pairs. The `multimap` allows duplicate keys, while the `map` does not. In classic C++, for sets, the items are logically maintained in sorted order, so an iterator to the "beginning item" represents the smallest item in the collection. For maps, the keys are logically maintained in sorted order. As a result, the most natural implementation of sets and maps makes use of balanced binary search trees; typically a top-down red black tree (see Chapter 11) is used. The significant liability is that basic operations take logarithmic worst-case time.

Historically, STL implementations (such as SGI's version) provided `hash_set` and `hash_map`, which make use of hash tables (see Chapter 9) and provided constant average time per basic operation, but were non-standard. Finally, C++11 added four additional classes that maintain the sets and maps without order, using separate chaining hash tables. These classes are `unordered_set`, `unordered_multiset`, `unordered_map`, and `unordered_multimap`; the class names are chosen to avoid conflicting with various preexisting nonstandard classes.

C++ ordered sets and maps use either a default ordering (`operator<`) or one provided by a function object. In C++, a function object is implemented by providing a class that contains an overloaded `operator()` and then instantiating a template with the class name as a template parameter; similarly, unordered sets and maps use either a default hash function and equality operator (for instance, with strings), or ones provided by function objects (an example of this is shown later in Figures 43.1 and 43.2).

Both sets and maps make use of the `pair` class template. The `pair` class template stores two data members `first` and `second`, which can be directly accessed, without invoking member functions. Internally, maps store items of type `pair`. Additionally, several `set` member functions need to return two things; this is done by returning an appropriately instantiated `pair`.

43.2.2.1 Sets and Multisets

The `set` and `unordered_set` provide several member functions in addition to the usual suspects, including:

- `pair<iterator,bool> insert( const Object & x )`: adds x to the set. If x is not already present, the returned `pair` will contain the iterator representing the already contained x, and `false`. Otherwise, it will contain the iterator representing the newly inserted x, and `true`.
- `pair<iterator,bool> insert( iterator hint, const Object & x )`: same behavior as the one-parameter `insert`, but allows specification of a hint, representing the position where x should go. If the underlying implementation is a finger search tree (see Chapter 12), or equivalent, the performance could be better than using the one-parameter `insert`.
- `iterator find( const Object & x ) const`: returns an iterator representing the position of x in the set. If x is not found, the endmarker is returned.
- `int erase( const Object & x )`: removes x (if present) and returns the number of items removed, which is either 0 or 1 in a `set`, and perhaps larger in a multiset.
- `iterator erase( iterator pos )`: same behavior as the sequence container version.
- `iterator erase( iterator start, iterator end )`: same behavior as the sequence container version.
- `iterator lower_bound( const Object & x )`: returns an iterator to the first element in the set with a key that is greater than or equal to x (only for `set`).
- `iterator upper_bound( const Object & x )`: returns an iterator to the first element in the set with a key that is greater than x (only for `set`).
- `pair<iterator,iterator> equal_range( const Object & x )`: returns a pair of iterators representing `lower_bound` and `upper_bound`.

A `multiset` is like a set except that duplicates are allowed. The return type of `insert` is modified to indicate that the insert always succeeds:

```
#include <set>
#include <string>
#include <cstring>
#include <iostream>
using namespace std;

class CaseInsensitive
{
  public:
    bool operator() ( const string & lhs, const string & rhs ) const
      { return strcasecmp( lhs.c_str( ), rhs.c_str( ) ) > 0; }
};

template <typename Set1, typename Set2>
void testSets( Set1 & s1, Set2 & s2 )
{
    s1.clear( ); s2.clear( );
    s1.insert( "hello" ); s1.insert( "world" ); s1.insert( "HELLO" );
    s2.insert( "hello" ); s2.insert( "world" ); s2.insert( "HELLO" );

    cout << s1.size( ) << " " << s2.size( ) << endl;      // prints 3 2
    cout << (s1.find( "HeLlO" ) == s1.end( )) << endl;  // returns true
    cout << (s2.find( "HeLlO" ) == s2.end( )) << endl;  // returns false
}

int main( )
{
    set<string>                   s1;
    set<string,CaseInsensitive> s2;
    testSets( s1, s2 );

    return 0;
}
```

FIGURE 43.1 Using the set class template.

- iterator insert(const Object & x): adds x to the multiset; returns an iterator representing the newly inserted x.
- iterator insert(iterator hint, const Object & x): adds x to the multiset; returns an iterator representing the newly inserted x. Performance might be enhanced if x is inserted close to the position given by hint.

For the multiset, the erase member function that takes an Object x removes all occurrences of x. To simply remove one occurrence, first call find to obtain an iterator, and then use the erase member function that takes an iterator.

In a multiset, to find all occurrences of x, we cannot simply call find; that returns an iterator referencing one occurrence (if there is one), but which specific occurrence is returned is not guaranteed. Instead, the range returned by a call to equal_range is used.

Figure 43.1 illustrates two sets: s1 which stores strings using the normal case-sensitive ordering and s2 which stores strings using case-insensitive comparisons. s2 is instantiated by providing the class of a function object as a template parameter. As a result, s1 contains three strings, but s2 considers "hello" and "HELLO" to be identical, and thus only stores one of them. Figure 43.2 shows a main that implements identical logic for unordered_sets.

43.2.2.2 Maps and Multimaps

A map behaves like a set instantiated with a pair representing a key and value, with a comparison function that refers only to the key. Thus it supports all of the set operations. Similarly, an unordered_map behaves like a map, with hash function and equality test refers only to the key.

Data Structures in C++

```
#include <unordered_set>
#include <string>
#include <cstring>
using namespace std;

class CaseInsensitiveHash
{
  public:
    bool operator() ( const string & lhs, const string & rhs ) const
      { return strcasecmp( lhs.c_str( ), rhs.c_str( ) ) == 0; }

    size_t operator() ( const string & str ) const
    {
        size_t hashVal = 0;

        for( char ch : str )
            hashVal = 37 * hashVal + tolower( ch );

        return hashVal;
    }
};

int main( )
{
    unordered_set<string>                                           h1;
    unordered_set<string,CaseInsensitiveHash,CaseInsensitiveHash> h2;
    testSets( h1, h2 );

    return 0;
}
```

FIGURE 43.2 Using the set class template.

The find operation for maps requires only a key, but the iterator that it returns references a pair. Similarly, erase requires only a key and otherwise behaves like the set's erase. insert is supported, but to use insert, we must insert a properly instantiated pair, which is cumbersome, and thus rarely done. Instead, the map overloads the array indexing operator[]:

- ValueType & operator[] (const KeyType & key): if the key is present in the map, a reference to the value is returned. If the key is not present in the map, it is inserted with a default value into the map, and then a reference to the inserted default value is returned. The default value is obtained by applying a zero-parameter constructor or is zero for the primitive types.

The semantics of operator[] do not allow an accessor version of operator[], and so operator[] cannot be used on an immutable map. For instance, if a map is passed by constant reference, inside the routine, operator[] is unusable. In such a case, we can use find to obtain an iterator, test if the iterator represents the endmarker (in which case the find has failed), and if the iterator is valid, we can use it to access the pair's second component, which is the value.

A multimap is a map in which duplicate keys are allowed. multimaps behave like maps but do not support operator[].

Figure 43.3 illustrates a combination of map and vector in which the keys are email aliases, and the values are the email addresses corresponding to each alias. Since there can be several addresses for each alias, the values are themselves vectors of strings. The trickiest part (other than the various parameter-passing modes) concerns the one-line implementation of addAlias. There, we see that aliasMap[name] refers to the vector value in the map for the name key. If name is not in the map, then the call to operator[] causes it to be placed into the map, with a default value being a zero-sized vector. Thus whether or not name was in the map prior to calling addAlias, after the access of the map, the call to push_back updates the vector value correctly.

```
#include <unordered_map>
#include <string>
#include <vector>
#include <iostream>
using namespace std;

template <typename Map, typename NameType, typename AliasType>
void addAlias( Map & aliasMap, const NameType & name, const AliasType & alias )
{
    aliasMap[ name ].push_back( alias );
}


template <typename Map, typename NameType>
auto getAliases( Map & aliasMap, const NameType & name ) -> decltype( aliasMap[ name ] )
{
    return aliasMap[ name ];
}

int main( )
{
    unordered_map<string,vector<string>> aliasMap;

    addAlias( aliasMap, "john", "john.doe@sillymail.com" );
    addAlias( aliasMap, "authors", "sahni@cise.ufl.edu" );
    addAlias( aliasMap, "authors", "dmehta@mines.edu" );
    addAlias( aliasMap, "authors", "weiss@fiu.edu" );

    cout << "authors is aliased to: ";
    auto & authors = getAliases( aliasMap, "authors" );
    for( auto email : authors )
        cout << email << " ";
    cout << endl;

    return 0;
}
```

FIGURE 43.3 Using the map class template.

43.2.3 Container Adapters

The STL provides adapter class templates to implement stacks, queues, and priority queues. These class templates are instantiated with the type of object to be stored and (optionally, if the default is unacceptable) the container class (such as vector, list, or deque) that is used to store the objects. Thus we can specify a linked list or array implementation for a stack, though in reality, this feature does little more than add nicer names than push_back to the (existing list or vector, respectively) interface.

43.2.3.1 stack and queue

For stack, the member functions are push, pop, and top. For queue, we get push, front, and pop. By default, a vector is used for stack and a deque for queue. Figure 43.4 shows how to use the queue adapter, implemented with a doubly linked list.

43.2.3.2 priority_queue

The priority_queue template contains member functions named push, top, and pop, that mirror insert, findmax, and deletemax in a classic max-heap.

The priority queue template is instantiated with an item type, the container type (as in stack and queue), and the comparator, with defaults allowed for the last two parameters. Sometimes priority queues are set up to remove and access the smallest item instead of the largest item. In such a case, the priority queue can be instantiated with an appropriate greater function object to reverse the comparison order and allow access to the smallest item.

Figure 43.5 illustrates both a max-heap and a min-heap.

```
#include <queue>
#include <iostream>
#include <list>
using namespace std;

int main( )
{
    queue<int,list<int>> q;
    q.push( 3 ); q.push( 7 );

    for( ; !q.empty( ); q.pop( ) )
        cout << q.front( ) << endl;

    return 0;
}
```

FIGURE 43.4 Using the queue adapter.

```
#include <vector>
#include <queue>
#include <functional>
#include <string>
#include <iostream>
using namespace std;

// Empty the priority queue and print its contents.
template <typename PQueue>
void dumpPQ( const string & msg, PQueue & pq )
{
    if( pq.empty( ) )
        cout << msg << " is empty" << endl;
    else
    {
        cout << msg << ": " << pq.top( );
        pq.pop( );
        while( !pq.empty( ) )
        {
            cout << " " <<  pq.top( );
            pq.pop( );
        }
        cout << endl;
    }
}

int main( )
{
    priority_queue<int>                           maxpq;
    priority_queue<int,vector<int>,greater<int>> minpq;

    minpq.push( 3 ); minpq.push( 7 ); minpq.push( 3 );
    maxpq.push( 3 ); maxpq.push( 7 ); maxpq.push( 3 );

    dumpPQ( "minpq", minpq );      // minpq: 3 3 7
    dumpPQ( "maxpq", maxpq );      // maxpq: 7 3 3

    return 0;
}
```

FIGURE 43.5 Using the priority_queue adapter.

43.3 Iterators

The *iterator* in C++ abstracts the notion of a position in the container. As we have already seen, there are several ways to obtain such a position. All containers provide `begin` and `end` member functions that return iterators representing the first item and end-marker, respectively. Many containers provide a searching member function (e.g., `set::find`) that returns an iterator viewing the found item or the endmarker if the item is not found.

43.3.1 Basics

Throughout our discussion, we have used `iterator` as a type. In reality, in C++, each container defines several iterator types, nested in the scope of the container, and these specific iterator types are used by the programmer instead of simply using the word `iterator`. For instance, if we have a `vector<int>`, the basic iterator type is `vector<int>::iterator`. The basic iterator can be used to traverse and change the contents of the container.

Another iterator type, `vector<int>::const_iterator`, does not allow changes to the container on which the iterator is operating. All iterators are guaranteed to have at least the following set of operations:

- `++itr` and `itr++`: advance the iterator `itr` to the next location. Both the prefix and postfix forms are available. This does not cause any change to the container.
- `*itr`: returns a reference to the container object that `itr` is currently representing. The reference that is returned is modifiable for basic iterators, but is not modifiable (i.e., a constant reference) for `const_iterators`.
- `itr1==itr2`: returns `true` if iterators `itr1` and `itr2` refer to the same position in the same container and `false` otherwise.
- `itr1!=itr2`: returns `true` if iterators `itr1` and `itr2` refer to different positions or different containers and `false` otherwise.

Some iterators efficiently support `--itr` and `itr--`. Those iterators are called *bidirectional iterators*. All of the iterators for the common containers `vector`, `list`, `deque`, `set`, and `map` are bidirectional.

Some iterators efficiently support both `itr+=k` and `itr+k`. Those iterators are called *random-access iterators*. `itr+=k` advances the iterator `k` positions. `itr+k` returns a new iterator that is `k` positions ahead of `itr`. Also supported by random access iterators are `itr-=k` and `itr-k`, with obvious semantics, and `itr1-itr2` which yields a separation distance as an integer. The iterators for `vector` and `deque` support random access.

All C++ containers have two member functions: `begin` and `end` that return iterators. Each collection defines four member functions:

- `iterator begin( )`
- `const_iterator begin( ) const`
- `iterator end( )`
- `const_iterator end( ) const`

`begin` returns an iterator that is positioned at the first item in the container. `end` returns an iterator that is positioned at the endmarker, which represents a position one past the last element in the container. On an empty container, `begin` and `end` return the same position. For random access iterators, the result of subtracting the return values of `end` and `begin` is always the size of the container.

A common use of an iterator is to sequentially scan through a container. In classic C++, this is done by initializing a local iterator to be a copy of the `begin` iterator, and have it step through the container, stopping as soon as it hits the endmarker.

As an example, Figure 43.6 shows a `print` function that prints the elements of any container, provided that the elements in the container have implemented an `operator<<`. Note that we must use a `const_iterator` to traverse the container, because the container is itself immutable in the scope of `print`. The test program illustrates five different containers that invoke the `print` function, along with the expected output (in comments). Observe that both `set` and `multiset` output in sorted order, with `multiset` allowing the second insertion of `beta`. Additionally, for `map` to be compatible with the `print` routine, we must provide an `operator<<` that works for the elements of the `map`, namely the appropriate `pairs`.

In C++11, the *range for* loop can be used to simplify this idiom, for any container that provides an appropriate iteration mechanism. The code to implement `print` in Figure 43.7 is mechanically converted by the compiler into the equivalent code, shown in Figure 43.6. Note that since C++11 allows the compiler to deduce the needed type when `auto` is appropriately used, in many instances the details of whether `iterator` or `const_iterator` is needed are no longer something the programmer needs to worry about.

```cpp
#include <iostream>
#include <vector>
#include <list>
#include <set>
#include <string>
#include <map>
using namespace std;

template <typename Container>
void print( const Container & c, ostream & out = cout )
{
    typename Container::const_iterator itr;

    for( itr = c.begin( ); itr != c.end( ); ++itr )
        out << *itr << " ";
    out << endl;
}

template <typename Type1, typename Type2>
ostream & operator<<( ostream & out, const pair<Type1,Type2> & p )
{
    return out << "[" << p.first << "," << p.second << "]";
}

int main( )
{
    vector<int> vec;
    vec.push_back( 3 ); vec.push_back( 7 );

    list<double> lst;
    lst.push_back( 3.14 ); lst.push_front( 6.28 );

    set<string> s;
    s.insert( "beta" ); s.insert( "alpha" ); s.insert( "beta" );

    multiset<string> ms;
    ms.insert( "beta" ); ms.insert( "alpha" ); ms.insert( "beta" );

    map<string,string> zip;
    zip.insert( pair<string,string>( "Miami", "33199" ) );
    zip.insert( pair<string,string>( "Gainesville", "32611" ) );
    zip[ "Golden" ] = "80401";

    print( vec );       // 3 7
    print( lst );       // 6.28 3.14
    print( s );         // alpha beta
    print( ms );        // alpha beta beta
    print( zip );       // [Gainesville,32611] [Golden,80401] [Miami,33199]

    return 0;
}
```

FIGURE 43.6   Generic printing routine that works with five different containers.

```
template <typename Container>
void print( const Container & c, ostream & out = cout )
{
    for( auto item : c )
        out << item << " ";
    out << endl;
}
```

FIGURE 43.7 Range for loop version of code in Figure 43.6.

```
#include <iostream>
using namespace std;

template <typename Container>
void printReverse( const Container & c, ostream & out = cout )
{
    for( auto itr = c.rbegin( ); itr != c.rend( ); ++itr )
        out << *itr << " ";
    out << endl;
}
```

FIGURE 43.8 Printing a container in reverse.

43.3.2 Reverse Iterators

Sometimes it is important to be able to traverse a container in the reverse direction. Because of the asymmetric nature of `begin` and `end` (representing the first element and the endmarker, rather than the last element, respectively) this is cumbersome to do, even for bidirectional iterators. As a result, containers that support bidirectional iterators typically also provide a *reverse iterator*. The reverse iterator comes in two flavors: `reverse_iterator` and `const_reverse_iterator`. For reverse iterators, `++` retreats one position toward the beginning, while `--` advances one position toward the end. Container member functions `rbegin` and `rend` are used to obtain iterators representing the last element and the beginmarker, respectively. The code in Figure 43.8 prints any container in reverse order. Once again, the use of `auto` in C++11 removes from the programmer the burden of deciding whether `reverse_iterator` or `const_reverse_iterator` is needed.

43.4 Additional Components of the STL

43.4.1 Sorting, Searching, and Selection

The Standard Library includes a rich set of functions that can be applied to the standard containers. Some of the algorithms include routines for sorting, searching, copying (possibly with substitutions), shuffling, reversing, rotating, merging, and so on. In all there are over 60 generic algorithms. In this section, we highlight those related to efficient sorting, selection, and binary search.

43.4.1.1 Sorting

Sorting in C++ is accomplished by use of function template `sort`. The parameters to `sort` represent the start and endmarker of a (half-open range in a) container and an optional comparator:

- `void sort( iterator begin, iterator end )`
- `void sort( iterator begin, iterator end, Comparator cmp )`

The iterators must support random access. As an example, in

```
sort( v.begin( ), v.end( ) );
sort( v.begin( ), v.end( ), greater<int>( ) );
sort( v.begin( ), v.begin( ) + min( v.size ( ), 5 ) );
```

the first call sorts the entire container, v, in nondecreasing order. The second call sorts the entire container in nonincreasing order. The third call sorts the first five elements of the container in nondecreasing order. C++11 requires that `sort` provides

$O(N \log N)$ worst-case performance; recent implementations have used a quicksort/heapsort hybird algorithm (*introsort*) [8], but no specific implementation is mandated.

The C++ library inherits the sorting routine `qsort` from C. `qsort` uses pointers to functions to perform its comparison making it significantly slower on average than the `sort` routine. Furthermore, many implementations use a version of quicksort that has been shown to provide quadratic behavior on some commonly occurring inputs [9]. Avoid using `qsort`.

43.4.1.2 Selection

The function template `nth_element` is used for selection, and as expected has $O(N)$ average-case running time. The parameters include a pair of iterators, and *k*:

- `void nth_element( iterator begin, int k, iterator end )`
- `void nth_element( iterator begin, int k, iterator end, Comparator c)`

As a result of calling `nth_element`, the item in the *k*th position is the *k*th smallest element, where counting as usual in C++ begins at 0.

Thus, in

```
nth_element( v.begin( ), 0, v.end( ) );
nth_element( v.begin( ), 0, v.end( ), greater<int>( ) );
nth_element( v.begin( ), v.begin( ) + ( v.end( ) - vbegin( ) / 2 ), v.end( ) );
```

the first call places the smallest element in the position given by `v.begin( )`, the second call places the largest element in that position, and the third call places the median in the middle position.

43.4.1.3 Searching

Several generic searching algorithms are available for containers. The most basic is:

- `iterator find( iterator begin, iterator end, const Object & x )`: returns an iterator representing the first occurrence of `x` in the half-open range specified by `begin` and `end`, or end if `x` is not found.

`find` is simply a sequential search. If the range is sorted, `binary_search` can be used to find an object in logarithmic time. A comparator can be provided or the default ordering can be used. The signatures for `binary_search` are:

- `iterator binary_search( iterator begin, iterator end, const Object & x )`
- `iterator binary_search( iterator begin, iterator end, const Object & x, Comparator cmp )`

`equal_range`, `lower_bound`, and `upper_bound` search sorted ranges and behave with the same semantics as the identically named member functions in `set`.

Acknowledgments

Parts of this chapter are based upon the author's work in References 6 and 7.

References

1. J. Lajoie, S. Lippman, and B.E. Moo, *C++ Primer*, Fifth Edition, Addison-Wesley, Reading, MA, 2013.
2. S. Meyers, *Effective STL*, Addison-Wesley, Reading, MA, 2001.
3. D.R. Musser and A. Saini, *C++ Programming with the Standard Template Library*, Addison-Wesley, Reading, MA, 1996.
4. B. Stroustrop, *The C++ Programming Language*, Fourth Edition, Addison-Wesley, Boston, MA, 2013.
5. B. Stroustrop, *The Design and Evolution of C++*, Addison-Wesley, Reading, MA, 1994.
6. M.A. Weiss, *C++ for Java Programmers*, Prentice-Hall, Englewood Cliffs, NJ, 2004.
7. M.A. Weiss, *Data Structures and Algorithm Analysis in C++*, Fourth Edition, Pearson, Boston, MA, 2014.
8. D.R. Musser, Introspective Sorting and Selection Algorithms, *Software: Practice and Experience*, 27, 1997, 983–993.
9. J.L. Bentley and M.D. McIlroy, Engineering: A Sort Function, *Software: Practice and Experience*, 23, 1993, 1249–1265.

44

Data Structures in JDSL[*]

	44.1	Introduction	679
	44.2	Design Concepts in JDSL	681
		Containers and Accessors • Iterators • Decorations • Comparators • Algorithms	
Michael T. Goodrich	44.3	The Architecture of JDSL	684
University of California, Irvine		Packages • Positional Containers • Key-Based Containers • Algorithms	
	44.4	A Sample Application	689
Roberto Tamassia		Minimum-Time Flight Itineraries • Class IntegerDijkstraTemplate • Class	
Brown University		IntegerDijkstraPathfinder • Class FlightDijkstra	
Luca Vismara[†]		Acknowledgments	695
Brown University		References	695

44.1 Introduction

In the last four decades the role of computers has dramatically changed: once mainly used as *number processors* to perform fast numerical computations, they have gradually evolved into *information processors*, used to store, analyze, search, transfer, and update large collections of structured information. In order for computer programs to perform these tasks effectively, the data they manipulate must be well organized, and the methods for accessing and maintaining those data must be reliable and efficient. In other words, computer programs need advanced *data structures* and *algorithms*. Implementing advanced data structures and algorithms, however, is not an easy task and presents some risks:

Complexity Advanced data structures and algorithms are often difficult to understand thoroughly and to implement.

Unreliability Because of their complexity, the implementation of advanced data structures and algorithms is prone to subtle errors in boundary cases, which may require a considerable effort to identify and correct.

Long development time As a consequence, implementing and testing advanced data structures and algorithms is usually a time consuming process.

As a result, programmers tend to ignore advanced data structures and algorithms and to resort to simple ones, which are easier to implement and test but that are usually not as efficient. It is thus clear how the development of complex software applications, in particular their rapid prototyping, can greatly benefit from the availability of libraries of reliable and efficient data structures and algorithms.

Various libraries are available for the C++ programming language. They include the *Standard Template Library* (STL, see Chapter 43) [1], now part of the C++ standard, the extensive *Library of Efficient Data Structures and Algorithms* (LEDA, see Chapter 42) [2], and the *Computational Geometry Algorithms Library* (CGAL) [3].

The situation for the Java programming language is the following: a small library of data structures, usually referred to as *Java Collections* (JC), is included in the `java.util` package of the Java 2 Platform.[‡] An alternative to the Java Collections are the *Java Generic Libraries* (JGL) by Recursion Software, Inc.,[§] which are patterned after STL. Both the Java Collections and JGL provide implementations of basic data structures such as sequences, sets, maps, and dictionaries. JGL also provides a considerable number of generic programming algorithms for transforming, permuting, and filtering data.

None of the above libraries for Java, however, seems to provide a coherent framework, capable of accommodating both elementary and advanced data structures and algorithms, as required in the development of complex software applications. This

---

[*] This chapter has been reprinted from first edition of this Handbook, without any content updates.

[†] We have used this author's affiliation from the first edition of this handbook, but note that this may have changed since then.

[‡] http://java.sun.com/j2se/

[§] http://www.recursionsw.com/jgl.htm

circumstance motivated the development of the *Data Structures Library in Java* (JDSL) [4], a collection of Java interfaces and classes implementing fundamental data structures and algorithms, such as:

- Sequences and trees;
- Priority queues, binary search trees, and hash tables;
- Graphs;
- Sorting and traversal algorithms;
- Topological numbering, shortest path, and minimum spanning tree.

JDSL is suitable for use by researchers, professional programmers, educators, and students. It comes with extensive documentation, including detailed Javadoc,[*] an overview, a tutorial with seven lessons, and several associated research papers. It is available free of charge for noncommercial use at `http://www.jdsl.org/`.

The development of JDSL began in September 1996 at the Center for Geometric Computing at Brown University and culminated with the release of version 1.0 in 1998. A major part of the project in the first year was the experimentation with different models for data structures and algorithms, and the construction of prototypes. A significant reimplementation, documentation, and testing [5] effort was carried out in 1999 leading to version 2.0, which was officially released in 2000. Starting with version 2.1, released in 2003 under a new license, the source code has been included in the distribution. During its life cycle JDSL 2.0 was downloaded by more than 5700 users, while JDSL 2.1 has been downloaded by more than 3500 users as of this writing. During these seven years a total of 25 people[†] have been involved, at various levels, in the design and development of the library. JDSL has been used in data structures and algorithms courses worldwide as well as in two data structures and algorithms textbooks[‡] written by the first two authors [6,7].

In the development of JDSL we tried to learn from other approaches and to progress on them in terms of ease of use and modern design. The library was designed with the following goals in mind:

Functionality: The library should provide a significant collection of existing data structures and algorithms.

Reliability: Data structures and algorithms should be correctly implemented, with particular attention to boundary cases and degeneracies. All input data should be validated and, where necessary, rejected by means of exceptions.

Efficiency: The implementations of the data structures and algorithms should match their theoretical asymptotic time and space complexity; constant factors, however, should also be considered when evaluating efficiency.

Flexibility: Multiple implementations of data structures and algorithms should be provided, so that the user can experiment and choose the most appropriate implementation, in terms of time or space complexity, for the application at hand. It should also be possible for the user to easily extend the library with additional data structures and algorithms, potentially based on the existing ones.

Observe that there exist some trade-offs between these design goals, for example, between efficiency and reliability, or between efficiency and flexibility.

In JDSL each data structure is specified by an interface and each algorithm uses data structures only via the interface methods. Actual classes need only be specified when objects are instantiated. Programming through interfaces, rather than through actual classes, creates more general code. It allows different implementations of the same interface to be used interchangeably, without having to modify the algorithm code.

A comparison of the key features of the Java Collections, JGL, and JDSL is shown in Table 44.1. The main advantages of JDSL are the definition of a large set of data structure APIs (including binary tree, general tree, priority queue and graph) in terms of Java interfaces, the availability of reliable and efficient implementations of those APIs, and the presence of some fundamental graph algorithms. Note, in particular, that the Java Collections do not include trees, priority queues and graphs, and provide only sorting algorithms.

A good library of data structures and algorithms should be able to integrate smoothly with other existing libraries. In particular, we have pursued compatibility with the Java Collections. JDSL supplements the Java Collections and is not meant to replace them. No conflicts arise when using data structures from JDSL and from the Java Collections in the same program. To facilitate the use of JDSL data structures in existing programs, adapter classes are provided to translate a Java Collections data structure into a JDSL one and vice versa, whenever such a translation is applicable.

---

[*] `http://java.sun.com/j2se/javadoc/`

[†] `http://www.jdsl.org/team.html`

[‡] `http://java.datastructures.net/` and `http://algorithmdesign.net/`

TABLE 44.1 A Comparison of the Java Collections (JC), the Generic Library for Java (JGL), and the Data Structures Library in Java (JDSL)

	JC	JGL	JDSL
Sequences (lists, vectors)	✓	✓	✓
Trees			✓
Priority queues (heaps)		✓	✓
Dictionaries (hash tables, red-black trees)	✓		✓
Sets	✓		
Graphs			✓
Templated algorithms			✓
Sorting	✓	✓	✓
Data transforming, permuting, and filtering		✓	
Graph traversals			✓
Topological numbering			✓
Shortest path			✓
Minimum spanning tree			✓
Accessors (positions and locators)			✓
Iterators	✓	✓	✓
Range views	✓	✓	
Decorations (attributes)			✓
Thread-safety	✓		
Serializability	✓	✓	

44.2 Design Concepts in JDSL

In this section we examine some data organization concepts and algorithmic patterns that are particularly important in the design of JDSL.

44.2.1 Containers and Accessors

In JDSL each data structure is viewed as a *container*, that is, an organized collection of objects, called the *elements* of the container. An element can be stored in many containers at the same time and multiple times in the same container. JDSL containers can store heterogeneous elements, that is, instances of different classes.[*]

JDSL provides two general and implementation-independent ways to access (but not modify) the elements stored in a container: individually, by means of accessors, and globally, by means of iterators (see Section 44.2.2). An *accessor* [8] abstracts the notion of membership of an element into a container, hiding the details of the implementation. It provides constant-time access to an element stored in a container, independently from its implementation. Every time an element is inserted in a container, an accessor associated with it is returned. Most operations on JDSL containers take one or more accessors as their operands.

We distinguish between two kinds of containers and, accordingly, of accessors (see Figure 44.1 for a diagram of the accessor interface hierarchy):

Positional containers: Typical examples are sequences, trees, and graphs. In a positional container, some topological relation is established among the "placeholders" that store the elements, such as the predecessor-successor relation in a sequence, the parent-child relation in a tree, and the incidence relation in a graph. It is the user who decides, when inserting an element in the container, what the relationship is between the new "placeholder" and the existing ones (in a sequence, for instance, the user may decide to insert an element before a given "placeholder"). A positional container does not change its topology, unless the user requests a change specifically. The implementation of these containers usually involves linked structures or arrays.

Positions: The concept of position is an abstraction of the various types of "placeholders" in the implementation of a positional container (typically the nodes of a linked structure or the cells of an array). Each position stores an element. Position implementations may store the following additional information:

---

[*] This is possible since in Java every class extends (directly or indirectly) `java.lang.Object`.

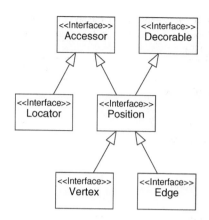

FIGURE 44.1   The accessors interface hierarchy.

- The adjacent positions (e.g., the previous and next positions in a sequence, the right and left child and the parent in a binary tree, the list of incident edges in a graph);
- Consistency information (e.g., what container the position is in, the number of children in a tree).

A position can be directly queried for its element through method element(), which hides the details of where the element is actually stored, be it an instance variable or an array cell. Through the positional container, instead, it is possible to replace the element of a position or to swap the elements between two positions. Note that, as an element moves about in its container, or even from container to container, its position changes. Positions are similar to the concept of items used in LEDA [2].

Key-based containers: Typical examples are dictionaries and priority queues. Every element stored in a key-based container has a *key* associated with it. Keys are used as an indexing mechanism for their associated elements. Typically, a key-based container is internally implemented using a positional container; for example, a possible implementation of a priority queue uses a binary tree (a heap). The details of the internal representation, however, are completely hidden to the user. Thus, the user has no control over the organization of the positions that store the key/element pairs. It is the key-based container itself that modifies its internal representation based on the keys of the key/element pairs inserted or removed.

Locators: The key/element pairs stored in a key-based container may change their positions in the underlying positional container, due to some internal restructuring, say, after the insertion of a new key/element pair. For example, in the binary tree implementation of a priority queue, the key/element pairs move around the tree to preserve the top-down ordering of the keys, and thus their positions change. Hence, a different, more abstract type of accessor, called locator, is provided to access a key/element pair stored in a key-based container. Locators hide the complications of dynamically maintaining the implementation-dependent binding between the key/element pairs and their positions in the underlying positional container.

A locator can be directly queried for its key and element, and through the key-based container it is possible to replace the key and the element of a locator. An example of using locators is given in Section 44.4.

44.2.2 Iterators

While accessors allow users to access single elements or key/element pairs in a container, *iterators* provide a simple mechanism for iteratively listing through a collection of objects. JDSL provides various iterators over the elements, the keys (where present), and the positions or the locators of a container (see Figure 44.2 for a diagram of the iterator interface hierarchy). They are similar to the iterators provided by the Java Collections.

All JDSL containers provide methods that return iterators over the entire container (e.g., all the positions of a tree or all the locators of a dictionary). In addition, some methods return iterators over portions of the container (e.g., the children of a position of a tree or the locators with a given key in a dictionary). JDSL iterators can be traversed only forward; however, they can be reset to start a new traversal.

For simplicity reasons iterators in JDSL have *snapshot* semantics: they refer to the state of the container at the time the iterator was created, regardless of the possible subsequent modifications of the container. For example, if an iterator is created over all the positions of a tree and then a subtree is cut off, the iterator will still include the positions of the removed subtree.

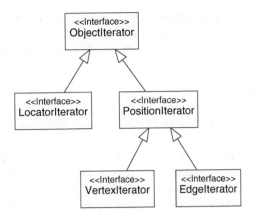

FIGURE 44.2   The iterators interface hierarchy.

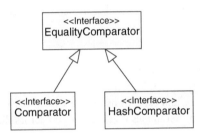

FIGURE 44.3   The comparators interface hierarchy.

44.2.3 Decorations

Another feature of JDSL is the possibility to "decorate" individual positions of a positional container with attributes, that is, with arbitrary objects. This mechanism is more convenient and flexible than either subclassing the position class to add new instance variables or creating global hash tables to store the attributes. Decorations are useful for storing temporary or permanent results of the execution of an algorithm. For example, in a depth-first search (DFS) traversal of a graph, we can use decorations to (temporarily) mark the vertices being visited and to (permanently) store the computed DFS number of each vertex. An example of using decorations is given in Section 44.4.

44.2.4 Comparators

When using a key-based container, the user should be able to specify the comparison relation to be used with the keys. In general, this relation depends on the type of the keys and on the specific application for which the key-based container is used: keys of the same type may be compared differently in different applications. One way to fulfill this requirement is to specify the comparison relation through a *comparator* object, which is passed to the key-based container constructor and is then used by the key-based container every time two keys need to be compared.

Three comparator interfaces are defined in JDSL (see Figure 44.3 for a diagram of the comparators interface hierarchy). The concept of comparator is present also in the `java.util` package of the Java 2 Platform, where a `Comparator` interface is defined.

44.2.5 Algorithms

JDSL views algorithms as objects that receive the input data as arguments of their `execute(.)` method, and provide access to the output during or after the execution via additional methods. Most algorithms in JDSL are implemented following the *template method pattern* [9]. The invariant part of the algorithm is implemented once in an abstract class, deferring the implementation of the steps that can vary to subclasses. These varying steps are defined either as abstract methods (whose implementation must be provided by a subclass) or as "hook" methods (whose default implementation may be overridden in a subclass). In other words, algorithms perform "generic" computations that can be specialized for specific tasks by subclasses.

An example of applying the template method pattern is given in Section 44.4, where we use the JDSL implementation of Dijkstra's single-source shortest path algorithm [10]. The algorithm refers to the edge weights by means of an abstract method that can be specialized depending on how the weights are actually stored or computed in the application at hand.

44.3 The Architecture of JDSL

In this section we describe the interfaces of the data structures present in JDSL, the classes that implement those interfaces, and the algorithms that operate on them. Most containers are described by two interfaces, one (whose name is prefixed with Inspectable) that comprise all the methods to query the container, and the other, extending the first, that comprise all the methods to modify the container. Inspectable interfaces can be used as variable or argument types in order to obtain an immutable view of a container (for instance, to prevent an algorithm from modifying the container it operates on).

As described in Section 44.2.1, we can partition the set of containers present in JDSL into two subsets: the positional containers and the key-based containers. Accordingly, the interfaces for the various containers are organized into two hierarchies (see Figures 44.4 and 44.5), with a common root given by interfaces InspectableContainer and Container. At the same time, container interfaces, their implementations, and algorithms that operate on them are grouped into various Java packages.

In the rest of this section, we denote with n the current number of elements stored in the container being considered.

44.3.1 Packages

JDSL currently consists of eight Java packages, each containing a set of interfaces and/or classes. Interfaces and exceptions for the data structures are defined in packages with the api suffix, while the reference implementations of these interfaces are defined in packages with the ref suffix. Interfaces, classes, and exceptions for the algorithms are instead grouped on a functional basis. As we will see later, the interfaces are arranged in hierarchies that may extend across different packages. The current packages are the following:

jdsl.core.api Interfaces and exceptions that compose the API for the core containers (sequences, trees, priority queues, and dictionaries), for their accessors and comparators, and for the iterators on their elements, positions and locators.

jdsl.core.ref Implementations of the interfaces in jdsl.core.api. Most implementations have names of the form *ImplementationStyleInterfaceName*. For instance, ArraySequence and NodeSequence implement the jdsl.core.api.Sequence interface with a growable array and with a linked structure, respectively. Classes with names of the form Abstract*InterfaceName* implement some methods of the interface for the convenience of developers building alternative implementations.

jdsl.core.algo.sorts Sorting algorithms that operate on the elements stored in a jdsl.core.api.Sequence object. They are parameterized with respect to the comparison rule used to sort the elements, provided as a jdsl.core.api.Comparator object.

jdsl.core.algo.traversals Traversal algorithms that operate on jdsl.core.api.InspectableTree objects. A traversal algorithm performs operations while visiting the nodes of the tree, and can be extended by applying the template method pattern.

jdsl.core.util This package contains a Converter class to convert some JDSL containers to the equivalent data structures of the Java Collections and vice versa.

jdsl.graph.api Interfaces and exceptions that compose the API for the graph container and for the iterators on its vertices and edges.

jdsl.graph.ref Implementations of the interfaces in jdsl.graph.api; in particular, class IncidenceListGraph is an implementation of interface jdsl.graph.api.Graph.

jdsl.graph.algo Basic graph algorithms, including depth-first search, topological numbering, shortest path, and minimum spanning tree, all of which can be extended by applying the template method pattern.

44.3.2 Positional Containers

All positional containers implement interfaces InspectablePositionalContainer and PositionalContainer, which extend InspectableContainer and Container, respectively (see Figure 44.4). Every positional container implements a set of essential operations, including being able to determine its own size (size()), to determine whether it contains a specific position (contains(Accessor)), to replace the element associated with a position (replaceElement(Accessor, Object)), to swap the elements associated with two positions (swapElements(Position, Position)), and to get iterators over the positions (positions()) or the elements (elements()) of the container.

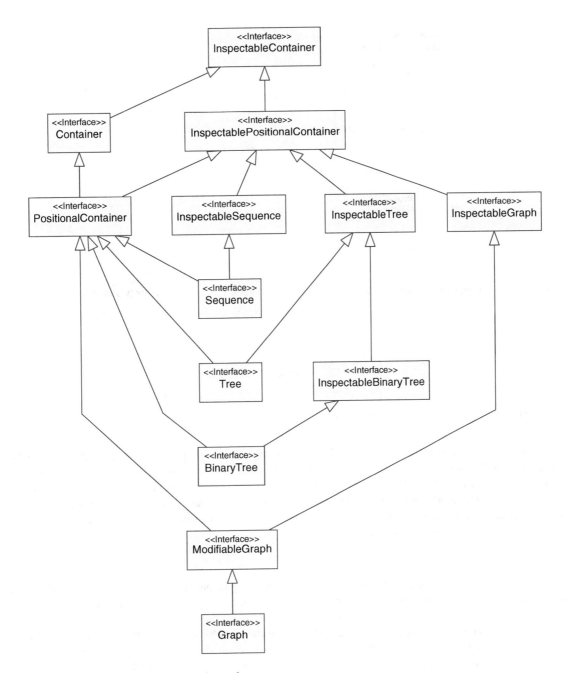

FIGURE 44.4 The positional containers interface hierarchy.

44.3.2.1 Sequences

A sequence is a basic data structure used for storing elements in a linear, ranked fashion (see Chapter 2). Sequences can be implemented in many ways, for example, as a linked list of nodes or on top of an array. In JDSL, sequences are described by interfaces `InspectableSequence` and `Sequence`, which extend `InspectablePositionalContainer` and `PositionalContainer`, respectively. In addition to the basic methods common to all positional containers, the sequence interfaces provide methods to access and modify positions at the sequence ends (methods such as `first()`, `insertLast()` and `removeFirst()`) or specific positions along the sequence (methods such as `after(Position)`, `atRank(int)`, `insertBefore(Position)` and `removeAtRank(int)`).

`NodeSequence` is an implementation of `Sequence` based on a doubly linked list of nodes. The nodes are the positions of the sequence. It takes $O(1)$ time to insert, remove, or access both ends of the sequence or a position before or after a given one, while it takes $O(n)$ time to insert, remove, or access positions at a given rank in the sequence. Thus, `NodeSequence` instances can be suitably used as stacks, queues, or deques.

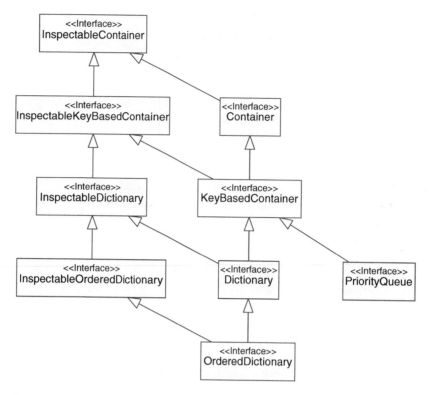

FIGURE 44.5   The key-based containers interface hierarchy.

ArraySequence is an implementation of Sequence based on a growable array of positions. Instances can be created with an initial capacity, and can be told whether or not to reduce this capacity when their size drops below a certain value, depending on whether the user prefers space or time efficiency. It takes $O(1)$ time to access any position in the sequence, $O(1)$ amortized time over a series of operations to insert or remove elements at the end of the sequence, and $O(n)$ time to insert or remove elements at the beginning or middle of the sequence. Hence, ArraySequence instances can be suitably used for quick access to the elements after their initial insertion, when filled only at the end, or as stacks.

44.3.2.2 Trees

Trees allow more sophisticated relationships between elements than is possible with a sequence: they allow relationships between a child and its parent, or between siblings of a parent (see Chapter 3). InspectableTree and Tree are the interfaces describing a general tree; they extend InspectablePositionalContainer and PositionalContainer, respectively. InspectableBinaryTree, which extends InspectableTree, and BinaryTree, which extends PositionalContainer, are the interfaces describing a binary tree. In addition to the basic methods common to all positional containers, the tree interfaces provide methods to determine where in the tree a position lies (methods such as isRoot(Position) and isExternal(Position)), to return the parent (parent(Position)), siblings (siblings(Position)) or children (methods such as children(Position), childAtRank(Position,int) and leftChild(Position)) of a position, and to cut (cut(Position)) or link (link(Position,Tree)) a subtree.

NodeTree is an implementation of Tree based on a linked structure of nodes. The nodes are the positions of the tree. It is the implementation to use when a generic tree is needed or for building more specialized (nonbinary) trees. NodeTree instances always contain at least one node.

NodeBinaryTree is an implementation of BinaryTree based on a linked structure of nodes. The nodes are the positions of the tree. Similarly to NodeTree, NodeBinaryTree instances always contain at least one node; in addition, each node can have either zero or two children. If a more complex tree is not necessary, using NodeBinaryTree instances will be faster and easier than using NodeTree ones.

44.3.2.3 Graphs

A graph is a fundamental data structure describing a binary relationship on a set of elements (see Chapter 4) and it is used in a variety of application areas. Each vertex of the graph may be linked to other vertices through edges. Edges can be either one-way, *directed* edges, or two-way, *undirected* edges. In JDSL, both vertices and edges are positions of the graph. JDSL handles all graph special cases such as self-loops, multiple edges between two vertices, and disconnected graphs.

The main graph interfaces are `InspectableGraph`, which extends `InspectablePositionalContainer`, `ModifiableGraph`, which extends `PositionalContainer`, and `Graph`, which extends both `InspectableGraph` and `ModifiableGraph`. These interfaces provide methods to determine whether two vertices are adjacent (`areAdjacent(Vertex,Vertex)`) or whether a vertex and an edge are incident (`areIncident(Vertex,Edge)`), to determine the degree of a vertex (`degree(Vertex)`), to determine the origin (`origin(Edge)`) or destination (`destination(Edge)`) of an edge, to insert (`insertVertex(Object)`) or remove (`removeVertex(Vertex)`) a vertex, to set the direction of an edge (`setDirectionFrom(Edge,Vertex)` and `setDirectionTo(Edge,Vertex)`), to insert (`insertEdge(Vertex,Vertex,Object)`), remove (`removeEdge(Edge)`), split (`splitEdge(Edge, Object)`), or unsplit (`unsplitEdge(Vertex,Object)`) an edge.

`IncidenceListGraph` is an implementation of `Graph`. As its name suggests, it is based on an incidence list representation of a graph.

44.3.3 Key-Based Containers

All key-based containers implement interfaces `InspectableKeyBasedContainer` and `KeyBasedContainer`, which extend `InspectableContainer` and `Container`, respectively (see Figure 44.5). Every key-based container implements a set of essential operations, including being able to determine its own size (`size()`), to determine whether it contains a specific locator (`contains(Accessor)`), to replace the key (`replaceKey(Locator,Object)`) or the element (`replaceElement(Accessor,Object)`) associated with a locator, to insert (`insert(Object,Object)`) or remove (`remove(Locator)`) a key/element pair, and to get iterators over the locators (`locators()`), the keys (`keys()`) or the elements (`elements()`) of the container.

44.3.3.1 Priority Queues

A priority queue is a data structure used for storing a collection of elements prioritized by keys, where the smallest key value indicates the highest priority (see Part II). It supports arbitrary insertions and deletions of elements and keeps track of the highest-priority key. A priority queue is useful, for instance, in applications where the user wishes to store a queue of tasks of varying priority, and always process the most important task.

Interface `PriorityQueue` extends `KeyBasedContainer`. In addition to the basic methods common to all the key-based containers, it provides methods to access (`min()`) or remove (`removeMin()`) the key/element pair with highest priority, that is, with minimum key. Note that the priority of an element can be changed using method `replaceKey(Locator,Object)`, inherited from `KeyBasedContainer`.

`ArrayHeap` is an efficient implementation of `PriorityQueue` based on a heap. Inserting, removing, or changing the key of a key/element pair takes $O(\log n)$ time, while examining the key/element pair with the minimum key takes $O(1)$ time. The implementation is parameterized with respect to the comparison rule used to order the keys; to this purpose, a `Comparator` object is passed as an argument to the `ArrayHeap` constructors.

44.3.3.2 Dictionaries

A dictionary is a data structure used to store key/element pairs and then quickly search for them using their keys (see Part III). An ordered dictionary is a particular dictionary where a total order on the set of keys is defined. All JDSL dictionaries are *multimaps*, that is, they can store multiple key/element pairs with the same key.

The main dictionary interfaces are `InspectableDictionary` and `Dictionary`, which extend `InspectableKeyBasedContainer` and `KeyBasedContainer`, respectively. In addition to the basic methods common to all the key-based containers, these interfaces provide methods to find key/element pairs by their keys (`find(Object)` and `findAll(Object)`) and to remove all key/element pairs with a specific key (`removeAll(Object)`). Other dictionary interfaces are `InspectableOrderedDictionary` and `OrderedDictionary`, which extend `InspectableDictionary` and `Dictionary`, respectively. They provide additional methods to access the first (`first()`) or last (`last()`) key/element pair in the ordered dictionary, and to access the key/element pair before (`before(Locator)`) or after (`after(Locator)`) a given key/element pair.

`HashtableDictionary` is an implementation of `Dictionary`. As its name suggests, it is based on a hash table. Insertions and removals of key/element pairs usually take $O(1)$ time, although individual insertions and removals may require $O(n)$ time. The implementation is parameterized with respect to the hashing function used to store the key/element pairs; to this purpose, a `HashComparator` object is passed as an argument to the `HashtableDictionary` constructors. `HashtableDictionary` is a good choice when overall speed is necessary.

`RedBlackTree` is an implementation of `OrderedDictionary`. It is a particular type of binary search tree, where insertion, removal, and access to key/element pairs require each $O(\log n)$ time. The implementation is parameterized with respect to the comparison rule used to order the keys; to this purpose, a `Comparator` object is passed as an argument to the `RedBlackTree` constructors.

44.3.4 Algorithms

In addition to the data structures described above, JDSL includes various algorithms that operate on them.

44.3.4.1 Sequence Sorting

JDSL provides a suite of sorting algorithms for different applications. They all implement the `SortObject` interface, whose only method is `sort(Sequence, Comparator)`. Sorting algorithms with the prefix `List` are most efficient when used on instances of `NodeSequence` while those with the prefix `Array` are most efficient when used on instances of `ArraySequence`.

`ArrayQuickSort` is an implementation of the quicksort algorithm. This algorithm runs in $O(n \log n)$ expected time and performs very well in practice. Its performance, however, degrades greatly if the sequence is already very close to being sorted. Also, it is not *stable*, that is, it does not guarantee that elements with the same value will remain in the same order they were in before sorting. In all cases whether neither of these caveats apply, it is the best choice.

`ListMergeSort` and `ArrayMergeSort` are two implementations of the mergesort algorithm. This algorithm is not as fast as quicksort in practice, even though its theoretical time complexity is $O(n \log n)$. There are no cases where its performance will degrade due to peculiarities in the input data, and it is a stable sort.

`HeapSort` is an implementation of the heapsort algorithm, and uses an instance of `ArrayHeap` (see Section 44.3.3) as a sorting device. Its performance, like that of mergesort, will not degrade due to peculiarities in the input data, but it is not a stable sort. Its theoretical time complexity is $O(n \log n)$.

44.3.4.2 Iterator-Based Tree Traversals

JDSL provides two types of tree traversals. The first type is based on iterators: the tree is passed as an argument to the iterator constructor and is then iterated over using methods `hasNext()` and `nextPosition()`. Iterators give a quick traversal of the tree in a specific order, and are the proper traversals to use when this is all that is required. We recall that iterators in JDSL have snapshots semantics (see Section 44.2.2).

A preorder iterator visits the nodes of the tree in preorder, that is, it returns a node before returning any of its children. Preorder iterators work for both binary and general trees; they are implemented in class `PreOrderIterator`.

A postorder iterator visits the nodes of the tree in postorder, that is, it returns a node after returning all of its children. Postorder iterators work for both binary and general trees; they are implemented in class `PostOrderIterator`.

An inorder iterator visits the nodes of the tree in inorder, that is, it returns a node in between its left and right children. Inorder iterators work only for binary trees; they are implemented in class `InOrderIterator`.

44.3.4.3 Euler Tour Tree Traversal

The second type of tree traversals in JDSL is named Euler tour: it is implemented—in class `EulerTour`—as an algorithm object, which can be extended by applying the template method pattern.

The Euler tour visits each node of the tree several times, namely, a first time before traversing any of the subtrees of the node, then between the traversals of any two consecutive subtrees, and a last time after traversing all the subtrees. Each time a node is visited, one of the methods `visitFirstTime(Position)`, `visitBetweenChildren(Position)` and `visitLastTime(Position)`, if the node is internal, or method `visitExternal(Position)`, if the node is a leaf, is automatically called. A particular computation on the visited tree may be performed by suitably overriding those methods in a subclass of `EulerTour`.

The Euler tour is more powerful than the iterators described above as it can be used to implement more general kinds of algorithms. Note that, unlike the iterators, the Euler tour does not have snapshot semantics; this means that any modification of the tree during the execution of the Euler tour will cause undefined behavior.

44.3.4.4 Graph Traversals

The depth-first search (DFS) traversal of a graph is available in JDSL. Depth-first search proceeds by visiting an unvisited vertex adjacent to the current one; if no such vertex exists, then the algorithm backtracks to the previous visited vertex.

Similarly to the Euler tour, depth-first search is implemented in JDSL as an algorithm object, which can be extended by applying the template method pattern. The basic implementation of depth-first search—`DFS`—is designed to work on undirected graphs.

The user can specify actions to occur when a vertex is first or last visited or when different sorts of edges (such as "tree" edges of the DFS tree or "back" edges to previously visited vertices) are traversed by subclassing DFS and suitably overriding some methods.

DFS has two subclasses: FindCycleDFS, an algorithm for determining cycles in an undirected graph, and DirectedDFS, a depth-first search specialized for directed graphs. In turn, DirectedDFS has one subclass: DirectedFindCycleDFS, an algorithm for determining cycles in a directed graph. These subclasses are examples of how to apply the template method pattern to DFS in order to implement a more specific algorithm.

44.3.4.5 Topological Numbering

A topological numbering is a numbering of the vertices of a directed acyclic graph such that, if there is an edge from vertex u to vertex v, then the number associated with v is higher than the number associated with u.

Two algorithms that compute a topological numbering are included in JDSL: TopologicalSort, which decorates each vertex with a unique number, and UnitWeightedTopologicalNumbering, which decorates each vertex with a nonnecessarily unique number based on how far the vertex is from the source of the graph. Both topological numbering algorithms extend abstract class AbstractTopologicalSort.

44.3.4.6 Dijkstra's Algorithm

Dijkstra's algorithm computes the shortest path from a specific vertex to every other vertex of a weighted connected graph. The JDSL implementation of Dijkstra's algorithm—IntegerDijkstraTemplate—follows the template method pattern and can be easily extended to change its functionality. Extending it makes it possible, for instance, to set the function for calculating the weight of an edge, to change the way the results are stored, or to stop the execution of the algorithm after computing the shortest path to a specific vertex (as done in subclass IntegerDijkstraPathfinder). An example of using Dijkstra's algorithm is given in Section 44.4.

44.3.4.7 The Prim-Jarník Algorithm

The Prim-Jarník algorithm computes a *minimum spanning tree* of a weighted connected graph, that is, a tree that contains all the vertices of the graph and has the minimum total weight over all such trees. The JDSL implementation of the Prim-Jarník algorithm—IntegerPrimTemplate—follows the template method pattern and can be easily extended to change its functionality. Extending it makes it possible, for instance, to set the function for calculating the weight of an edge, to change the way the results are stored, or to stop the execution of the algorithm after computing the minimum spanning tree for a limited set of vertices.

44.4 A Sample Application

In this section we explore the implementation of a sample application using JDSL. In particular, we show the use of some of the concepts described above, such as the graph and priority queue data structures, locators, decorations, and the template method pattern.

44.4.1 Minimum-Time Flight Itineraries

We consider the problem of calculating a minimum-time flight itinerary between two airports. The flight network can be modeled using a weighted directed graph: each vertex of the graph represents an airport, each directed edge represents a flight from the origin airport to the destination airport, and the weight of each directed edge is the duration of the flight. The problem we are considering can be solved by computing a shortest path between two vertices of the directed graph or determining that a path does not exists. To this purpose, we can suitably modify the classical algorithm by Dijkstra [10], which takes as input a graph G with nonnegative edge weights and a distinguished source vertex s, and computes a shortest path from s to any reachable vertex of G. Dijkstra's algorithm maintains a priority queue Q of vertices: at any time, the key of a vertex u in the priority queue is the length of the shortest path from s to u thus far. The priority queue is initialized by inserting vertex s with key 0 and all the other vertices with key $+\infty$ (some very large number). The algorithm repeatedly executes the following two steps:

1. Remove a minimum-key vertex u from the priority queue and mark it as *finished*, since a shortest path from s to u has been found.
2. For each edge e connecting u to an unfinished vertex v, if the path formed by extending a shortest path from s to u with edge e is shorter than the shortest known path from s to v, update the key of v (this operation is known as the *relaxation* of edge e).

```
package jdsl.graph.algo;

import jdsl.core.api.*;
import jdsl.core.ref.ArrayHeap;
import jdsl.core.ref.IntegerComparator;
import jdsl.graph.api.*;

public abstract class IntegerDijkstraTemplate {

  // instance variables

  protected PriorityQueue pq ;
  protected InspectableGraph g ;
  protected Vertex source ;
  private final Integer ZERO = new Integer(0);
  private final Integer INFINITY = new Integer(Integer.MAX VALUE);
  private final Object LOCATOR = new Object();
  private final Object DISTANCE = new Object();
  private final Object EDGE TO PARENT = new Object();

  // abstract instance methods

  protected abstract int weight (Edge e);

  // instance methods that may be overridden for special applications

  protected void shortestPathFound (Vertex v, int vDist) {
    v.set(DISTANCE,new Integer(vDist));
  }

  protected void vertexNotReachable (Vertex v) {
    v.set(DISTANCE,INFINITY);
    setEdgeToParent(v,Edge.NONE);
  }

  protected void edgeRelaxed (Vertex u, int uDist, Edge uv, int uvWeight, Vertex v, int vDist) { }

  protected boolean shouldContinue () {
    return true;
  }

  protected boolean isFinished (Vertex v) {
    return v.has(DISTANCE);
  }

  protected void setLocator (Vertex v, Locator vLoc) {
    v.set(LOCATOR,vLoc);
  }

  protected Locator getLocator (Vertex v) {
    return (Locator)v.get(LOCATOR);
  }

  protected void setEdgeToParent (Vertex v, Edge vEdge) {
    v.set(EDGE TO PARENT,vEdge);
  }
```

FIGURE 44.6 Class `IntegerDijkstraTemplate`.

44.4.2 Class IntegerDijkstraTemplate

As seen in Section 44.3.4, JDSL provides an implementation of Dijkstra's algorithm that follows the template method pattern. The abstract class implementing Dijkstra's algorithm is `jdsl.graph.algo.IntegerDijkstraTemplate` (see Figures 44.6; for brevity, the Javadoc comments present in the library code have been removed). The simplest way to run the algorithm is by calling `execute(InspectableGraph,Vertex)`, which first initializes the various auxiliary data structures with `init(g,source)`

```
protected EdgeIterator incidentEdges (Vertex v) {
    return g .incidentEdges(v,EdgeDirection.OUT | EdgeDirection.UNDIR);
}

protected Vertex destination (Vertex origin, Edge e) {
    return g .opposite(origin,e);
}

protected VertexIterator vertices () {
    return g .vertices();
}

protected PriorityQueue newPQ () {
    return new ArrayHeap(new IntegerComparator());
}

// output instance methods

public final boolean isReachable (Vertex v) {
    return v.has(EDGE TO PARENT) && v.get(EDGE TO PARENT) != Edge.NONE;
}

public final int distance (Vertex v) throws InvalidQueryException {
    try {
        return ((Integer)v.get(DISTANCE)).intValue();
    }
    catch (InvalidAttributeException iae) {
        throw new InvalidQueryException(v+" has not been reached yet");
    }
}

public Edge getEdgeToParent (Vertex v) throws InvalidQueryException {
    try {
        return (Edge)v.get(EDGE TO PARENT);
    }
    catch (InvalidAttributeException iae) {
        String s = (v == source ) ? " is the source vertex" : " has not been reached yet";
        throw new InvalidQueryException(v+s);
    }
}

// instance methods composing the core of the algorithm

public void init (InspectableGraph g, Vertex source) {
    g  = g;
    source  = source;
    pq  = newPQ();
    VertexIterator vi = vertices();
    while (vi.hasNext()) {
        Vertex u = vi.nextVertex();
        Integer uKey = (u == source ) ? ZERO : INFINITY;
        Locator uLoc = pq .insert(uKey,u);
        setLocator(u,uLoc);
    }
}
```

FIGURE 44.6 (Continued)   Class `IntegerDijkstraTemplate`.

and then repeatedly calls `doOneIteration()`. Note that the number of times `doOneIteration()` is called is controlled by `should Continue()`. Another possibility, instead of calling `execute(InspectableGraph,Vertex)`, is to call `init(InspectableGraph,Vertex)` directly and then single-step the algorithm by explicitly calling `doOneIteration()`.

For an efficient implementation of the algorithm, it is important to access a vertex stored in the priority queue in constant time, whenever its key has to be modified. This is possible through the locator accessors provided by class `jdsl.core.ref.ArrayHeap` (see Section 44.3.3). In `init(InspectableGraph,Vertex)`, each vertex u of the graph is inserted in the priority queue and a locator `uLoc` for the key/element pair is returned. By calling `setLocator(u,uLoc)`, each

```
protected final void runUntil () {
  while (!pq .isEmpty() && shouldContinue())
    doOneIteration();
}

public final void doOneIteration () throws InvalidEdgeException {
  Integer minKey = (Integer)pq .min().key();
  Vertex u = (Vertex)pq .removeMin(); // remove a vertex with minimum distance from the source
  if (minKey == INFINITY)
    vertexNotReachable(u);
  else {   // the general case
    int uDist = minKey.intValue();
    shortestPathFound(u,uDist);
    int maxEdgeWeight = INFINITY.intValue()−uDist−1;
    EdgeIterator ei = incidentEdges(u);
    while (ei.hasNext()) { // examine all the edges incident with u
      Edge uv = ei.nextEdge();
      int uvWeight = weight(uv);
      if (uvWeight < 0 || uvWeight > maxEdgeWeight)
        throw new InvalidEdgeException
          ("The weight of "+uv+" is either negative or causing overflow");
      Vertex v = destination(u,uv);
      Locator vLoc = getLocator(v);
      if (pq .contains(vLoc)) { // v is not finished yet
        int vDist = ((Integer)vLoc.key()).intValue();
        int vDistViaUV = uDist+uvWeight;
        if (vDistViaUV < vDist) { // relax
          pq .replaceKey(vLoc,new Integer(vDistViaUV));
          setEdgeToParent(v,uv);
        }
        edgeRelaxed(u,uDist,uv,uvWeight,v,vDist);
      }
    }
  }
}

public final void execute (InspectableGraph g, Vertex source) {
  init(g,source);
  runUntil();
}

public void cleanup () {
  VertexIterator vi = vertices();
  while (vi.hasNext()) {
    vi.nextVertex().destroy(LOCATOR);
    try {
      vi.vertex().destroy(EDGE TO PARENT);
      vi.vertex().destroy(DISTANCE);
    }
    catch (InvalidAttributeException iae) { }
  }
}

} // class IntegerDijkstraTemplate
```

FIGURE 44.6 (Continued)   Class `IntegerDijkstraTemplate`.

vertex u is decorated with its locator uLoc; variable LOCATOR is used as the attribute name. Later, in doOneIteration(), the locator of vertex v is retrieved by calling getLocator(v) in order to access and possibly modify the key of v; we recall that the key of v is the shortest known distance from the source vertex source_ to v. In addition to its locator in the priority queue, every unfinished vertex v is also decorated with its last relaxed incident edge uv by calling setEdgeToParent(v,uv); variable EDGE_TO_PARENT is used as the attribute name, in this case. When a vertex is finished, this decoration stores the edge to the parent in the shortest path tree, and can be retrieved with getEdgeToParent(Vertex).

Methods runUntil() and doOneIteration() are declared final and thus cannot be overridden. Following the template method pattern, they call some methods, namely, shouldContinue(), vertexNotReachable(Vertex), shortestPathFound(Vertex,int), and edgeRelaxed(Vertex,int,Edge, int,Vertex,int), that may be overridden in a subclass for special applications. For each vertex u of the graph, either vertexNotReachable(u) or shortestPathFound(u,uDist) is called exactly once, when u is removed from the priority queue and marked as finished. In particular, shortestPathFound(u,uDist) decorates u with uDist, the shortest distance from source_; variable DISTANCE is used as the attribute name. Method edgeRelaxed(u,uDist,uv,uvWeight,v,vDist) is called every time an edge uv from a finished vertex u to an unfinished vertex v is examined. The only method whose implementation must be provided by a subclass is abstract method weight(Edge), which returns the weight of an edge. Other important methods are isFinished(Vertex), which returns whether a given vertex is marked as finished, and distance(Vertex), which returns the shortest distance from source_ to a given finished vertex.

44.4.3 Class IntegerDijkstraPathfinder

JDSL also provides a specialization of Dijkstra's algorithm to the problem of finding a shortest path between two vertices of a graph. This algorithm is implemented in abstract class jdsl.graph.algo.IntegerDijkstraPathfinder (see Figure 44.7;

```java
package jdsl.graph.algo;

import jdsl.core.api.*;
import jdsl.core.ref.NodeSequence;
import jdsl.graph.api.*;
import jdsl.graph.ref.EdgeIteratorAdapter;

public abstract class IntegerDijkstraPathfinder extends IntegerDijkstraTemplate {

    // instance variables

    private Vertex dest ;

    // overridden instance methods from IntegerDijkstraTemplate

    protected boolean shouldContinue () {
      return !isFinished(dest );
    }

    // output instance methods

    public boolean pathExists () {
      return isFinished(dest );
    }

    public EdgeIterator reportPath () throws InvalidQueryException {
      if (!pathExists())
        throw new InvalidQueryException("No path exists between "+source +" and "+dest );
      else {
        Sequence retval = new NodeSequence();
        Vertex currVertex = dest ;
        while (currVertex != source ) {
          Edge currEdge = getEdgeToParent(currVertex);
          retval.insertFirst(currEdge);
          currVertex = g .opposite(currVertex,currEdge);
        }
        return new EdgeIteratorAdapter(retval.elements());
      }
    }

    // instance methods

    public final void execute (InspectableGraph g, Vertex source, Vertex dest) {
      dest  = dest;
      init(g,source);
      if (source  != dest )
        runUntil();
    }

  }   // class IntegerDijkstraPathfinder
```

FIGURE 44.7   Class IntegerDijkstraPathfinder.

for brevity, the Javadoc comments present in the library code have been removed), which extends `IntegerDijkstraTemplate`. The algorithm is run by calling `execute(InspectableGraph,Vertex,Vertex)`. The execution of Dijkstra's algorithm is stopped as soon as the destination vertex is finished. To this purpose, `shouldContinue()` is overridden to return true only if the destination vertex has not been finished yet. Additional methods are provided in `IntegerDijkstraPathfinder` to test, after the execution of the algorithm, whether a path from the source vertex to the destination vertex exists (`pathExists()`), and, in this case, to return it (`reportPath()`).

44.4.4 Class FlightDijkstra

Our application for computing a minimum-time flight itinerary between two airports can be implemented as a specialization of `IntegerDijkstraPathfinder`. The distance of each vertex represents, in this case, the time elapsed from the beginning of the travel to the arrival at the airport represented by that vertex. In Figure 44.8 we show the code of class `FlightDijkstra`; this class is part of the tutorial[*] distributed with JDSL. All it takes to implement our application is to override method `incidentEdges()`, so that only the outgoing edges of a finished vertex are examined, and to define method `weight(Edge)`. As noted before, the weighted graph representing the flight network is a directed graph. Each edge stores, as an element, an instance of auxiliary class `FlightSpecs` providing the departure time and the duration of the corresponding flight. Note that the weights of the edges are not determined before the execution of the algorithm, but rather depend on the computed shortest distance between the source vertex and the origin of each edge. Namely, they are obtained by adding the duration of the flight corresponding to the edge and the connecting time at the origin airport for that flight.[†] Method `TimeTable.diff(int,int)` simply computes the difference between its two arguments modulo 24 hours. The algorithm is run by calling `execute(InspectableGraph,Vertex,Vertex,int)`, where the fourth argument is the earliest time the passenger can begin traveling.

```
import jdsl.graph.api.*;
import jdsl.graph.algo.IntegerDijkstraPathfinder;
import support.*;

public class FlightDijkstra extends IntegerDijkstraPathfinder {

    // instance variables

    private int startTime_;

    // overridden instance methods from IntegerDijkstraPathfinder

    protected int weight (Edge e) {
        FlightSpecs eFS = (FlightSpecs)e.element(); // the flightspecs for the flight along edge e
        int connectingTime = TimeTable.diff(eFS.departureTime(),startTime_+distance(g_.origin(e)));
        return connectingTime+eFS.flightDuration();
    }

    protected EdgeIterator incidentEdges (Vertex v) {
        return g_.incidentEdges(v,EdgeDirection.OUT);
    }

    // instance methods

    public void execute (InspectableGraph g, Vertex source, Vertex dest, int startTime) {
        startTime_ = startTime;
        super.execute(g,source,dest);
    }

}
```

FIGURE 44.8   Class `FlightDijkstra`.

---

[*] `http://www.jdsl.org/tutorial/tutorial.html`

[†] In this sample application we ignore the minimum connecting time requirement, which however could be accommodated with minor code modifications.

As we can see from this example, the availability in JDSL of a set of carefully designed and extensible data structures and algorithms makes it possible to implement moderately complex applications with a small amount of code, thus dramatically reducing the development time.

Acknowledgments

We would like to thank all the members of the JDSL Team for their fundamental role in the design, implementation, and testing of JDSL. It has been a great experience and a real pleasure working together.

This work was supported in part by the National Science Foundation under grants CCR-0098068 and DUE-0231202.

References

1. D. R. Musser, G. J. Derge, and A. Saini. *STL Tutorial and Reference Guide: C++ Programming with the Standard Template Library*. Addison-Wesley, Reading, MA, 2nd edition, 2001.

2. K. Mehlhorn and S. Näher. *LEDA: A Platform for Combinatorial and Geometric Computing*. Cambridge University Press, Cambridge, England, 1999.

3. A. Fabri, G.-J. Giezeman, L. Kettner, S. Schirra, and S. Schönherr. On the design of CGAL a computational geometry algorithms library. *Softw.—Pract. Exp.*, 30(11):1167–1202, 2000.

4. R. Tamassia, M. T. Goodrich, L. Vismara, M. Handy, G. Shubina, R. Cohen, B. Hudson, R. S. Baker, N. Gelfand, and U. Brandes. JDSL: The data structures library in Java. *Dr. Dobb's Journal*, (323):21–31, Apr. 2001.

5. R. S. Baker, M. Boilen, M. T. Goodrich, R. Tamassia, and B. A. Stibel. Testers and visualizers for teaching data structures. In *Proc. 30th ACM SIGCSE Tech. Sympos.*, pages 261–265, 1999.

6. M. T. Goodrich and R. Tamassia, *Data Structures and Algorithms in Java*. John Wiley & Sons, New York, NY, 2nd edition, 2001.

7. M. T. Goodrich and R. Tamassia. *Algorithm Design: Foundations, Analysis, and Internet Examples*. John Wiley & Sons, New York, NY, 2002.

8. M. T. Goodrich, M. Handy, B. Hudson, and R. Tamassia. Accessing the internal organization of data structures in the JDSL library. In M. T. Goodrich and C. C. McGeoch, editors, *Algorithm Engineering and Experimentation (Proc. ALENEX '99)*, volume 1619 of *Lecture Notes Comput. Sci.*, pages 124–139. Springer-Verlag, 1999.

9. E. Gamma, R. Helm, R. Johnson, and J. Vlissides. *Design Patterns*. Addison-Wesley, Reading, MA, 1995.

10. E. W. Dijkstra. A note on two problems in connexion with graphs. *Numerische Mathematik*, 1269–271, 1959.

45

Data Structure Visualization*

45.1	Introduction	697
45.2	Value of Data Structure Rendering	698
45.3	Issues in Data Structure Visualization Systems	698
	Purpose and Environment • Data Structure Views • Interacting with a System	
45.4	Existing Research and Systems	700
	Incense • VIPS • GELO • DDD • Other Systems	
45.5	Summary and Open Problems	704
	References	704

John Stasko
Georgia Institute of Technology

45.1 Introduction

Important advances in programming languages have occurred since the early days of assembly languages and op codes, and modern programming languages have significantly simplified the task of writing computer programs. Unfortunately, tools for *understanding* programs have not achieved the accompanying levels of improvement. Software developers still face difficulties in understanding, analyzing, debugging, and improving code.

As an example of the difficulties still evident, consider a developer who has just implemented a new algorithm and is now ready to examine the program's behavior. After issuing a command to begin program execution, the developer examines output data and/or interactive behavior, attempting to ascertain if the program functioned correctly, and if not, the reasons for the program's failure. This task can be laborious and usually requires the use of a debugger or perhaps even the addition of explicit checking code, usually through output statements. Furthermore, the process can be quite challenging and the average developer may not be that skilled in fault diagnosis and correction. It would be extremely advantageous to have a tool that allows programmers to "look inside" programs as they execute, examining the changes to data and data structures that occur over time. Such a tool also could provide helpful context by identifying the current execution line, the active program unit, and the set of activations that have occurred to reach this configuration.

Tools, such as the one described above, are one form of *software visualization* [1]. Price, Baecker, and Small describe software visualization as, "the use of the crafts of typography, graphic design, animation, and cinematography with modern human-computer interaction and computer graphics technology to facilitate both the human understanding and effective use of computer software [2]." Fundamentally, software visualization seeks to take advantage of the visual perception and pattern matching skills of people to assist software development and software understanding. Human beings have very sophisticated visual perception systems that we constantly use to help us think [3]. Software visualization is a subfield in the larger area of *information visualization* [4,5] that studies how visualizations serve as external cognition aids.

Software visualization research and systems fall into two primary categories, *program visualization* and *algorithm visualization*.

Program visualization systems present data about and attributes of some existing software system. Typically, program visualizations are used in a software engineering context, that is, to help design, implement, optimize, test, or debug software. Thus, program visualization systems often illustrate source code, program data, execution statistics, program analysis information, or program objects such as data structures. The imagery in program visualization systems also is usually displayed with little or no input from the user/viewer. That is, a program (source code or executable) is analyzed by the system, which then automatically produces the graphics when the user requests them.

Algorithm visualization, conversely, strives to capture the abstractions and semantics of an algorithm or program primarily for pedagogical purposes. One often thinks of algorithm visualization as being more "high-level" as compared to program visualization being more "low level." The term *algorithm animation* is used interchangeably for this area simply because the displays are so dynamic and algorithm operations appear to animate over time. More specifically, Brown states, "Algorithm animation displays are, essentially, dynamic displays showing a program's fundamental *operations*—not merely its data or code

---

* This chapter has been reprinted from first edition of this Handbook, without any content updates.

[6]." Algorithm visualizations typically are hand-crafted, user-conceptualized views of what is "important" about a program. These types of views usually require some person, often the system developer or the program's creator, design and implement the graphics that accompany the program. The highly abstract, artificial nature of algorithm animations makes their automatic generation exceptionally difficult.

This chapter focuses on one specific subarea of software visualization, the visualization of data structures. By displaying pictures of data structures, one seeks to help people better understand the characteristics and the use of those structures.

How the data in computer programs are manipulated and organized by groups is a key aspect of successful programming. Understanding how elements of the data relate to other elements is a vital component in being able to comprehend and create sophisticated algorithms and programs. The word "structure" in "data structure" simply reinforces this point.

A person learning a new piece of software and seeking to modify it for some added functionality will likely ask a number of different questions. How are data elements organized? Is some kind of sequential structure used or is a more complicated organization present? Which items reference which other items? Understanding the answers to questions such as these helps a person to understand the utility and value of data structures. Providing pictures to express the relationships of data elements clearly then is one of the most useful ways to answer those questions.

45.2 Value of Data Structure Rendering

The graphical display of data structures can benefit a number of different activities, and one key role is as an aid for computer science education. It is not uncommon for students to have difficulty learning to understand and use non-trivial data structures such as arrays, lists, stacks, queues (Chapter 2), and trees (Chapter 3). One common problem occurs in the mapping back-and-forth between the pseudo code or programming language implementation of a structure and its conceptual model. In current educational methods, to foster understanding, the conceptual model (abstraction) is often presented in some graphical representation or picture—an interesting challenge is to find a book about data structures that does not make liberal use of pictures.

Systems for visualizing data structures typically provide or operate in an environment containing both the structures' programming language implementation and their graphical display, which facilitates students making connections between the two. This allows an instructor to prepare a collection of "interesting" data structures for students to interact with and learn from. By examining the accompanying displays, students can make the connection from a structure's abstract properties to its concrete implementation.

Another activity in which data structure visualization systems can provide significant help is program debugging. When a data structure display tool is used in conjunction with a traditional textual debugger, it allows the active data structures to be displayed and examined in some graphical format as program execution occurs and units of code are examined. When the program's data change, the accompanying visualizations change too, reflecting the new program state.

The graphical display of data structures adds many advantages to strict textual debugging. Because a human's visual perception system is a highly-tuned information processing machine, the sheer amount of information that can be transmitted to the user is much greater using graphics. With a textual debugger, people usually need to issue queries variable-by-variable to discover values. One glance at a computer display showing simultaneous representations of multiple structures may convey as much information as several minutes of the "query-reply" loop in a conventional debugger.

In addition to discovering data values, programmers use debuggers to access the relationships and connections between different pieces of data. This is a particular area in which graphical display is much more beneficial than textual interpreters. For example, to check the structure of a linked list in a language such as C++ or Java, a programmer must print out a list item's values and references to other list items. By carefully examining the resulting values, the programmer can determine if the list is in the desired configuration. A data structure visualization tool that could display a linked list as a series of rectangular nodes with arrows acting as the references or pointers connecting the different nodes would tremendously simplify the programmer's task of determining relationships among the data.

Data structure visualization tools also can aid the acquisition of contextual information about the state of a debugging session. For instance, a display could show a simple persistent view of the call stack and current execution line number in addition to the data pictures. In most debuggers, a user must often make repeated explicit queries to determine this information.

For these reasons and many more, data structure visualization systems remain a relatively untapped source of value to programmers. It is quite surprising that, given these benefits, more attention and research have not been devoted to the area, and data structure visualization systems have yet to achieve widespread use and adoption.

45.3 Issues in Data Structure Visualization Systems

45.3.1 Purpose and Environment

The software environment in which a data structure visualization system resides will influence many of the system's capabilities. Data structure display systems usually target a specific programming language or set of languages for visualization. In particular, many systems display only strongly-typed languages, simply because one is able to infer more information concerning the structure and composition of data objects in a strongly typed language. It also is easier to "customize" a specific graphical look for the different data structure types in a strongly typed language. A weakly typed language such as an assembly language generally forces a more generic looking display of data.

Clearly, the inherent characteristics of a particular programming language will also influence the language's resulting data visualizations. We would expect the node-pointer list data depiction to be an integral part of a LISP visualization system, while imperative languages such as C or Pascal would necessitate repeated use of a flexible structured, tiled image with easy to recognize and identify sub-fields. The display of graphical objects representing class instances and some form of arcs or arrows for messages (method invocations) would be important for illustrating an object-oriented language such as C++ or Java.

The intended audience of a data structure visualization system also affects the system's resulting displays. For example, a system designed for use by introductory programmers would probably stress consistent, simplified data structure views. Conversely, a system targeted to advanced programmers should support sophisticated, customized displays of complex, large data objects. Such a system should also allow a viewer to quickly suppress unwanted display objects and focus on items of special interest.

One common use of data structure visualization systems is in conjunction with other tools in a programming environment or an integrated development environment (IDE). For example, "attaching" a data structure visualization system to a program execution monitor allows the display of graphical data snap-shots at various points in the program's execution. By consistently displaying imagery, a form of program data animation results, showing how a program's data change over time.

Most often, a data structure visualization system functions in conjunction with a program debugger. In this arrangement, the display system provides a graphical window onto the program, permitting a persistent view of the data as opposed to the query-based approach of a textual debugger. While a programmer navigates through her program and its data by using a debugger, a data structure visualization system can map the particular data objects of interest onto a graphical format. As discussed in the previous section, the graphical format provides many benefits to typical debugging activities.

45.3.2 Data Structure Views

A primary distinguishing characteristic of a data structure visualization system is the types of data views it supports. Most systems provide a set of default views for the common data types such as integers, floats, and strings. Some systems also provide default views for more complex composite structures such as arrays, lists, and trees. Visualization systems frequently differ in their ability to handle a display request for a data object of a type other than those in the default view set. Reactions to such a query may range from taking no display action to utilizing a generic view applicable to all possible data types. Some systems also provide users with the ability to tailor a special graphical view for a particular, possibly uncommon data type. The process of defining these views is often tedious, however, discouraging such forms of improvisation.

Specific, relatively common data views that are often handled in varying ways include pointers/references, fields within composite structures, and arrays. Pointers/references, usually represented in a line-arrow manner, present a particularly tricky display problem because they involve issues similar issues to those evident in graph layout, a known difficult problem [7] (see also Chapter 47). Ideally, a view with pointers should minimize pointer overlap, edge crossings, and collisions with other display objects. Both polyline and spline display formats for pointers are common.

Representing fields within a composite data structure is a difficult problem because of spacing and layout concerns. Subfields can be complex structures themselves, complicating efforts to make the data structure view clear and comprehensible.

Arrays present a challenging visualization problem because of the variety of displays that are possible. For instance, a simple one-dimensional array can be presented in a horizontal or vertical format. The ideal view for multi-dimensional arrays can also be a difficult rendering task, particularly for arrays of dimension three or higher.

The ability to display multiple views of a specific data instance is also a valuable capability of a data structure visualization system because certain display contexts may be more informative in particular situations. Allowing varying levels of view abstraction on data is another important feature. For example, consider an array of structures in the C programming language. At one level, we may wish to view the array formation globally, with each structure represented by a small rectangle. At a closer level, we may wish to view a particular array structure element in full detail, with little attention paid to the other array elements.

Some data structure display systems provide visualizations for components of program execution such as program flow of control and the call stack. While these components are not program data structures, their graphical visualizations can be helpful additions to a data structure display system, especially when used in conjunction with a debugger.

45.3.3 Interacting with a System

Data structure visualization systems provide many different ways for users to interact with the system. For instance, consider the manner that users employ to actually display a particular piece of data. Systems that work in conjunction with a debugger may provide a special *display* command in the debugger. Other systems may utilize a specialized user interface with a particular direct manipulation protocol for invoking data display, such as choosing a menu item and supplying a variable name or graphically selecting a variable in a source code view.

Visualization systems vary in the manner of interactions they provide to users as well. For instance, once data is selected for display, the corresponding image(s) must be rendered on the viewing area. One option for this rendering is to allow the viewer to position the images, usually with the mouse. This method has the advantage of giving the user explicit control, but often, such repeated positioning can become tedious. Another rendering option gives the system total placement control. This method has the advantage of requiring less viewer input, but it requires sophisticated layout algorithms to avoid poor layout decisions. Perhaps the most attractive rendering option is a combination of these two: automatic system display with subsequent user repositioning capabilities.

Limits in display window viewing area force data structure visualization systems to confront sizing issues also. One simple solution to the problem of limited viewing space is to provide an infinite, scrollable viewing plane. Another solution, one more closely integrated with the system, is to utilize varying display abstractions dependent upon the amount of space available to view a data object. For example, given no space limitations, an object could be rendered in its default view. With very limited space, the object could be rendered in a space-saving format such as a small rectangle.

Once data has been displayed in the viewing window, the viewer should be able to interact with and control the imagery. Allowing a user to interactively move and delete images is certainly desirable. Even more beneficial, however, is the capability to suppress aspects of the data that are not of interest. For instance, only a small section of an array may be "interesting," so a visualization system could deemphasize other portions as a user dictates. Similarly, only certain sections of linked lists and trees may require attention, and only certain fields with a particular structure type may be of interest. Allowing a viewer to quickly dispose of uninteresting attributes and focus on the matter at hand is a very valuable feature.

If a data structure visualization system works in conjunction with a debugger, display interaction may take on a further role, that of interactive debugging itself. Graphical editing of data imagery by the viewer can be interpreted as corresponding changes to the underlying data. For example, a system may allow a viewer to enter a new data value for a variable by graphically selecting its accompanying image and overwriting the value displayed there. An even more sophisticated action is to allow a viewer to "grab" a pointer variable, usually represented by an arrow on the display, and alter its value by dragging and repositioning the arrow image in order to make the pointer reference some other object.

45.4 Existing Research and Systems

The idea of creating a system that would visualize data structures in computer programs is an old one, dating back even to the 1960's [8]. During this long history, a number of systems for data structure visualization have been developed. Most, however, have only been research prototypes and have never been used by anyone outside the system's creator(s). This relative lack of adoption is surprising in light of the potential benefits of data structure display identified above. A quick survey of widely-used current IDEs notes that none provide anything beyond the most limited abilities to persistently display data values, or more importantly, visualize data structures.

For a good overview of the aesthetic considerations and the general challenges in drawing data structures, see the article by Ding and Mateti [9]. The authors provide a rigorous survey of the different styles of drawings used in data structure diagrams. They enumerate a set of characteristics that influence people's perceptions of data structure illustrations, including visual complexity, regularity, symmetry, consistency, modularity, size, separation, shape, and traditional ways of drawing. Further, they develop a set of guidelines for drawing aesthetically appealing diagrams. The guidelines are based on rules, factors, and objectives, and include the ability to quantitatively assess the aesthetics of different diagrams. Ding and Mateti note that the automatic drawing of data structure representations is difficult because a diagram should be determined not only by a structure's declaration, but also by its intended usage.

Stasko and Patterson echo that thought by identifying the notion of *intention content* in program visualizations [10]. They note that a programmer's intent in creating and using a data structure can significantly influence how the structure should best be visualized. For instance, a simple array of integers may be shown as a row of rectangles with the values inside; a row of rectangles whose height is scaled to the value of each element; a round pie chart with a wedge whose percentage size of the circle corresponds to the value of that array element; or a stack with the array elements sitting on top of each other. To generate the latter two representations, some knowledge of the purpose and goal of the surrounding program code is necessary.

In the remainder of this section, we present brief summaries of a few of the most noteworthy data structure visualization systems that have been developed. Each of these grew from research projects, and the final one, DDD, has been used significantly by outside users. In reviewing each system, we highlight its unique aspects, and we evaluate it with respect to the issues raised in the previous section.

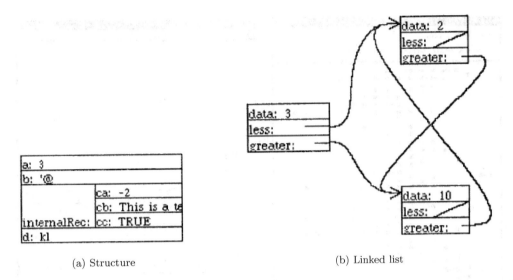

(a) Structure (b) Linked list

FIGURE 45.1 Example data structure views shown in the Incense system. Pictures provided by and reprinted with permission of Brad Myers.

45.4.1 Incense

The Incense system [11] developed by Myers was one of the earliest attempts to build a relatively full-featured data structure display system. Myers stated that Incense allowed a "programmer to interactively investigate data structures in actual programs."

The system operated on a Xerox Alto personal computer and displayed data structures for the Mesa strongly typed programming language. In the system, data structures could be displayed at debug time by supplying a variable's name; Incense examined the symbol table to determine the variable's type. Once a variable was chosen, the user specified via the mouse a rectangular viewing area inside the active window to display the data. Incense automatically chose the appropriate view abstraction given the space requirements. Data with insufficient screen space were displayed as grey boxes.

The system included a large number of sophisticated default views for the language's data types and structures, and it utilized a spline-based method for displaying pointers. Examples of views of different types of data structures are shown in Figure 45.1. Additionally, users could define their own data views using a predefined graphics library in Mesa. The difficulty of that process is unclear.

In order to implement Incense, Myers created the concept of an *artist*, a collection of procedures and data that handled display, erasure, and modification of the data being displayed. An artist had to be associated with a piece of data in the system in order to display it. To display pointers as arrows, Incense utilized *layouts*, special types of artists designed to locate and manage the various pointer and referent components of data.

The system was designed to allow users to also edit variables' values at run-time in the graphical presentation, but it does not appear that this feature was ever implemented. Unfortunately, the host hardware's space and processing limitations made Incense's performance relatively poor which likely restricted it to being a research prototype. The system did, however, illustrate a multitude of valuable capabilities that a data structure visualization system could offer.

45.4.2 VIPS

The VIPS [12] visual debugger by Shimomura and Isoda focused on providing a flexible set of views of lists in programs. The system ran on top of the DBX debugger, with VIPS and DBX communicating through pipes. For instance, when a user would issue a request to display a list, the VIPS system would send necessary messages to DBX to extract information needed for the display. In addition to views of list data structures, VIPS provided views (windows) of program text, I/O, variable displays, as well as control windows for debugger interactions.

VIPS provided two main views of program lists, the whole list view and the partial list view. In the whole view, shown on the left side of Figure 45.2, list nodes were represented as small black rectangles thus allowing more than 100 nodes to be shown at once. The basic layout style used by the system was a classic binary tree node-link layout with the root at the left and the tree growing horizontally to the right. The whole list view assisted users in understanding structure. The partial list view, shown on the right-hand side of Figure 45.2, presented each node in a more magnified manner, allowing the user to see the values of individual fields within the node. By dragging a rectangle around a set of nodes in the whole list view, a user could issue a command to

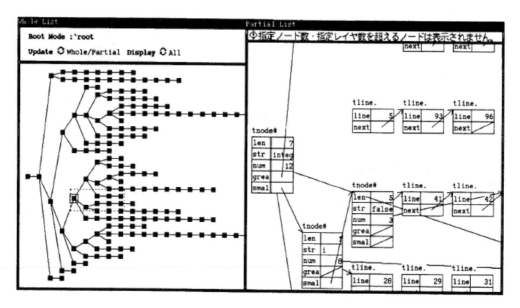

FIGURE 45.2   Data structure views in the VIPS system. (Picture reprinted with permission from T. Shimomura and S. Isoda. *IEEE Software*, 8, 3, 1991, 44–51 page 47, copyright 1991 IEEE.)

generate a new partial list view. VIPS also allowed users to select particular pointer fields in list structures and only have the nodes connected by those pointers to be displayed.

To assist debugging, VIPS also provided a *highlight* feature in which list structures with values that had recently changed were highlighted in order to draw the viewer's attention.

An earlier version of the system [13] provided multiple run-time views of executing Ada programs. Default data views in the system included scalars, linked lists, one-dimensional arrays and records. VIPS also allowed users to design their own data displays or "figures" by writing programs in the Figure Description Language FDL, a subset of Ada. FDL programs contained parameters that could be associated with a variable or variables, promoting view control via data values. Data displays were rendered automatically by the system. When space was tight, smaller representations were utilized.

The VIPS debugger provided one of the most extensive graphically-aided debugging systems onto a particular language. By adding views of the call stack, data flow, and program structure to program data visualizations, the system explored the boundary of visual debugging efforts at that time.

45.4.3   GELO

The GELO system [14] was designed to ease the production and display of complex pictures of programs and data structures. In the system, strongly typed data structures were displayed as picture objects of types data, tile, layout, arc, and empty. GELO differed from many other data structure display systems in that diagrams were not described in a world coordinate space, but by giving a set of topological constraints for their views.

The system was organized as three components: GELO managed the specification and display of classical program and data structures; APPLE allowed a user to design and customize the way that a particular data structure would appear by providing mechanisms for defining the mapping between the data structure and the GELO structures (more than one mapping was allowed); PEAR used these mappings to allow the user to edit the structures' display, thereby modifying the underlying data structures.

GELO allowed users to specify both the data instances to be displayed and the drawing attributes of the display through fairly complex dialog box selections. It provided default displays for common structures such as lists and trees, as well as various heuristics for graph layout. A sample list view is shown in Figure 45.3 and a tree view is shown in Figure 45.4. GELO included view panning and zooming, abstractions on small objects, and scrollable windows.

Reiss noted that the system's topological layout scheme was sometimes overly restrictive; users may have wanted to have a picture "look just so" but GELO did not provide this capability. The sheer amount of dialog and menu choices for designing a display also appeared to be daunting, but the sophistication of automatic layout GELO provided was quite impressive.

45.4.4   DDD

The DDD System [15,16], developed by Zeller, provides some of the most sophisticated graphical layout capabilities ever found in a data structure display system. DDD is technically a front-end to command-line debuggers such as GDB and DBX, and it provides

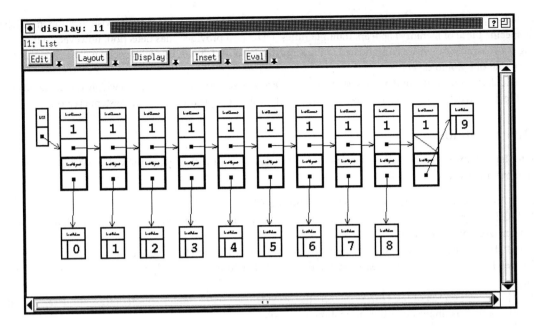

FIGURE 45.3 GELO view of a list. Picture provided courtesy of Steve Reiss.

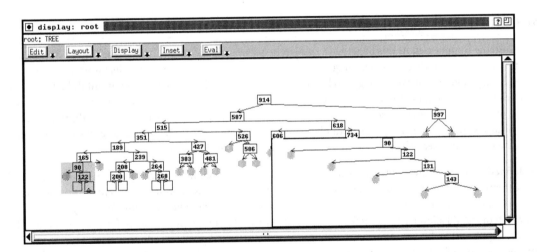

FIGURE 45.4 GELO view of a tree. Picture provided courtesy of Steve Reiss.

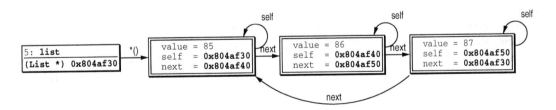

FIGURE 45.5 An example DDD representation of a linked list structure. Picture provided courtesy of Andreas Zeller.

all the capabilities of those debuggers. Additionally, program variables can be visualized in *displays*, rectangular windows that can be placed on a user's canvas.

DDD's sophistication comes in its techniques for visualizing pointer dereferencing. It draws a directed edge from one display to another to indicate the reference. DDD contains simple placement rules with edges pointing right and downward, or if desired, sophisticated graph layout algorithms can be invoked. Figure 45.5 shows an example data structure view from DDD.

In the default display mode, only one edge can point to any display. Thus, it is possible to have a visualization in which a display (variable or specific piece of memory) appears multiple times. DDD also provides alias detection, however, so that all program data residing at the same location are consolidated to the same display object. Thus, lists with cycles will be shown as the circular representation that one would expect.

DDD is particularly noteworthy in that the system is available as free software and it has been widely downloaded and utilized. It is perhaps the data structure visualization that has been most used by people other than the system creators.

45.4.5 Other Systems

In addition to the systems mentioned above, a number of other data structure display tools have been developed over the years. Some of the more noteworthy ones include

- GDBX [17]—A graphical extension to the UNIX debugger DBX.
- PROVIDE [18]—An ambitious process visualization and debugging environment for programs written in a subset/variant of the C programming language.
- Amethyst [19]—A system intended to simplify data structure visualization for novice programmers.
- DS-Viewer [20]—A system that focused on the display of structures in a weakly typed language such as an assembler language.
- Lens [21]—A layer on top of the DBX debugger that provided, via programmer annotations, both data structure display capabilities as well as simple algorithm animation operations.
- SWAN [22]—A system that allowed instructors and students to easily annotate programs to produce data structure views.

45.5 Summary and Open Problems

Although the systems summarized in this paper have shown the promise that data structure visualization holds, data structure display systems still are not widely used in teaching and still are not commonly included in IDEs for programmers to use. Until data structure display systems achieve more general acceptance, the area will continue to exhibit unfulfilled potential. Anyone who has ever programmed will surely agree that a visualization of the complex interactions of data can only help comprehension and debugging of computer programs. Now that software and hardware improvements for graphical displays have made these types of visualizations routine, hopefully, this unfulfilled potential will be reached. Below we describe some specific problems that remain to be solved in order to help foster the growth of data structure visualization.

- **User-defined displays** – Invariably, advanced programmers or programmers in a specific application area will want to build customized views of particular data types and structures. Existing systems have only either provided generic default displays or have required people to do tedious graphical design in order to build custom views. A powerful data structure display system should allow users to *quickly* and *easily* demonstrate a new graphical form for a data structure. Thus, a data structure display system likely must include some form of sophisticated graphical editor or toolkit to facilitate view design.
- **Mapping data to their display** – In addition to designing new views, designers must be able to specify how data is to be interpreted to generate the views. Creating an easy-to-understand and easy-to-use mapping scheme is quite difficult, particularly if multiple pieces of data can "drive" a particular view or one piece of data can be presented in different view abstractions.
- **Complex, large data** – With the possible exception of DDD, prior data structure display systems have been better suited for programming-in-the-small with relatively straightforward, moderate-sized data structures. Sophisticated display imagery becomes significantly more difficult as the complexity and sheer size of the data increases, thereby complicating screen layout issues. Further abstraction and mapping strategies are required to properly address these issues also.

References

1. J. Stasko, J. Domingue, M. Brown, and B. Price, editors. *Software Visualization: Programming as a Multimedia Experience.* MIT Press, Cambridge, MA, 1998, 562 pages.
2. B. Price, R. Baecker, and I. Small. An introduction to software visualization. In J. Stasko, J. Domingue, M. Brown, and B. Price, editors, *Software Visualization: Programming as a Multimedia Experience*, MIT Press, Cambridge, MA, 1998, pp. 3–27.
3. C. Ware. *Information Visualization: Perception for Design.* Morgan Kaufman, San Francisco, 2000, 438 pages.

4. S. Card, J. Mackinlay, and B. Shneiderman, editors. *Readings in Information Visualization: Using Vision to Think*. Morgan Kaufmann, San Francisco, 1999, 686 pages.

5. R. Spence. *Information Visualization*. ACM Press, Pearson Education, Essex, England, 2001, 206 pages.

6. M. Brown. Perspectives on algorithm animation. *ACM CHI*, Washington, DC, 1988, pp. 33–38.

7. G. Di Battista, P. Eades, R. Tamassia, and I. Tollis. *Graph Drawing: Algorithms for the Visualization of Graphs*. Prentice Hall, Upper Saddle River, NJ, 1998, 397 pages.

8. R. Baecker. Experiments in on-line graphical debugging: The interrogation of complex data structures. *Hawaii International Conference on the System Sciences*, Honolulu, HI, 1968, pp. 128–129.

9. C. Ding and P. Mateti. A framework for the automated drawing of data structure diagrams. *IEEE Transactions on Software Engineering*, 16, 5, 1990, 543–557.

10. J. Stasko and C. Patterson. Understanding and characterizing software visualization systems. *IEEE Workshop on Visual Languages*, Seattle, WA, 1992, pp. 3–10.

11. B. Myers. A system for displaying data structures. *SIGGRAPH*, Detroit, MI, 1983, pp. 115–125.

12. T. Shimomura and S. Isoda. Linked-list visualization for debugging. *IEEE Software*, 8, 3, 1991, 44–51.

13. S. Isoda, T. Shimomura, and Y. Ono. VIPS: A visual debugger. *IEEE Software*, 4, 3, 1987, 8–19.

14. S. Reiss, S. Meyers, and C. Duby. Using GELO to visualize software systems. *ACM UIST*, Williamsburg, VA, 1989, pp. 149–157.

15. A. Zeller. Visual debugging with DDD. *Dr. Dobb's Journal*, 322, 2001, 21–28.

16. T. Zimmermann and A. Zeller. Visualizing memory graphs. In S. Diehl, editor, *Software Visualization State-of-the-Art Survey*, LNCS 2269, Springer-Verlag, Berlin, Germany, 2002, pp. 191–204.

17. D. Baskerville. Graphic presentation of data structures in the DBX debugger. Technical Report UCB/CSD 86/260, University of California at Berkeley, 1985.

18. T. Moher. PROVIDE: A process visualization and debugging environment. *IEEE Transactions on Software Engineering*, 14, 6, 1988, 849–857.

19. B. Myers, R. Chandhok, and A. Sareen. Automatic data visualization for novice Pascal programmers. *IEEE Workshop on Visual Languages*, Pittsburgh, PA, 1988, pp. 192–198.

20. D. Pazel. DS-Viewer: An interactive graphical data structure presentation facility. *IBM Systems Journal*, 28, 2, 1989, 307–323.

21. S. Mukherjea and J. Stasko. Toward visual debugging: Integrating algorithm animation capabilities within a source level debugger. *ACM Transactions on Computer-Human Interaction*, 1, 3, 1994, 215–244.

22. C. Shaffer, L. Heath, J. Nielsen, and J. Yang. SWAN: A student-controllable data structure visualization system. *ED-MEDIA*, Boston, MA, 1996, pp. 632–637.

46

Drawing Trees*

46.1	Introduction	707
46.2	Preliminaries	708
46.3	Level Layout for Binary Trees	709
46.4	Level Layout for *n*-ary Trees	710
	PrePosition • Combining a Subtree and its Left Subforest • Ancestor • Apportion • Shifting the Smaller Subtrees	
46.5	Radial Layout	718
46.6	HV-Layout	719
	Acknowledgments	720
	References	720

Sebastian Leipert†
Center of Advanced European Studies and Research

46.1 Introduction

Constructing geometric representations of graphs in a readable and efficient way is crucial for understanding the inherent properties of the structures in many applications. The desire to generate a layout of such representations by algorithms and not by hand meeting certain aesthetics has motivated the research area *Graph Drawing*. Examples of these aesthetics include minimizing the number of edge crossings, minimizing the number of edge bends, minimizing the display area of the graph, visualizing a common direction (flow) in the graph, maximizing the angular resolution at the vertices, and maximizing the display of symmetries. Certainly, two aesthetic criteria cannot be simultaneously optimized in general and it depends on the data which criterion should be preferably optimized. Graph Drawing Software relies on a variety of mathematical results in graph theory, topology, geometry, as well as computer science techniques mainly in the areas algorithms and data structures, software engineering and user interfaces.

A typical graph drawing problem is to create for a graph $G = (V, E)$ a geometric representation where the nodes in V are drawn as geometric objects such as points or two dimensional shapes and edges $(u, v) \in E$ are drawn as simple Jordan curves connecting the geometric objects associated with u and v. Apart from the, in the context of this book obvious, visualization of Data Structures, other application areas are, for example, software engineering (Unified Modelling Language (UML), data flow diagrams, subroutine-call graphs) databases (entity-relationship diagrams), decision support systems for project management (business process management, work flow).

A fundamental issue in Automatic Graph Drawing is to display trees, since trees are a common type of data structures. Thus a good drawing of a tree is often a powerful intuitive guide for analyzing data structures and debugging their implementations. It is a trivial observation that a tree $T = (V, E)$ always admits a planar drawing, that is a drawing in the plane such that no two edges cross. Thus all algorithms that have been developed construct a planar drawing of a tree. Furthermore it is noticed that for trees the condition $|E| = |V| - 1$ holds and therefore the time complexity of the layout algorithms is always given in dependency to the number of nodes $|V|$ of a tree.

In 1979, Wetherell and Shannon [1] presented a linear time algorithm for drawing binary trees satisfying the following aesthetic requirements: the drawing is *strictly upward*, that is, the y-coordinate of a node corresponds to its level, so that the hierarchical structure of the tree is displayed; the left child of a node is placed to the left of the right child, that is, the order of the children is displayed; finally, each parent node is centered over its children. Moreover, edges are drawn *straight line*. Nevertheless, this algorithm showed some deficiencies. In 1981, Reingold and Tilford [2] improved the Wetherell-Shannon algorithm by adding the following feature: each pair of isomorphic subtrees is drawn identically up to translation, that is, the drawing does not depend

---

* This chapter has been reprinted from first edition of this Handbook, without any content updates.
† We have used this author's affiliation from the first edition of this handbook, but note that this may have changed since then.

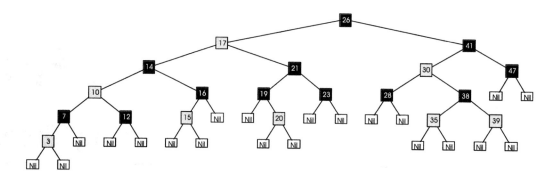

FIGURE 46.1   A typical layout of a binary tree. The tree is a red-black tree as given in Reference 3.

on the position of a subtree within the complete tree. They also made the algorithm symmetrical: if all orders of children in a tree are reversed, the computed drawing is the reflected original one. The width of the drawing is not always minimized subject to these conditions, but it is close to the minimum in general. The algorithm of Reingold and Tilford that runs in linear time, is given in Section 46.3. Figure 46.1 gives an example of a typical layout produced by Reingold and Tilfords Algorithm.

Extending this algorithm to rooted ordered trees of unbounded degree in a straightforward way produces layouts where some subtrees of the tree may get clustered on a small space, even if they could be dispersed much better. This problem was solved in 1990 by the quadratic time algorithm of Walker [4], which spaces out subtrees whenever possible. Very recently Buchheim et al. [5] showed how to improve the algorithm of Walker to linear running time. The algorithm by Buchheim et al. is given in Section 46.4, including a pseudo code that allows the reader a straightforward application of the algorithm. This algorithm for n-ary trees gives similar results on binary trees as the algorithm of Reingold and Tilford.

In Section 46.5 algorithms for straight line circular drawings (see [6–9]) are introduced. This type of layout proved to be useful for free trees that do not have a designated node as a root. Section 46.6 presents *hv*-drawings, an approach for drawing binary and n-ary trees on a grid. For binary trees, edges in an *hv*-drawing are drawn as rightward horizontal or downward vertical segments. This type of drawing is straight line *upward orthogonal*. It is, however, not strictly upward. Such drawings are for example, investigated in References 10–17.

Variations of the *hv*-layout algorithms have been used to obtain results on minimal area requirements of tree layouts. Shiloach and Crescenzi et al. [10,16] showed that any rooted tree admits an upward straight-line drawing with area $O(|V| \log |V|)$. Crescenzi et al. [10] moreover proved that there exists a class of rooted trees that require $\Omega(|V| \log |V|)$ area if drawn strictly upward straight-line. In [10,18] algorithms are given that produce $O(|V|)$-area strictly upward straight line drawings for some classes of balanced trees such as complete binary trees, Fibonacci trees, and AVL trees. These results have been expanded in Reference 19 to a class of balanced trees that include k-balanced trees, red-black trees, BB[α]-trees, and (a, b)-trees. Garg et al. [20] gave an $O(|V| \log |V|)$ area algorithm for ordered trees that produces upward layouts with polyline edges. Moreover they presented an upward orthogonal (not necessarily straight line) algorithm with an asymptotically optimal area of $O(|V| \log \log |V|)$.

Shin et al. [21] showed that bounded degree trees admit upward straight line layouts with an $O(|V| \log \log |V|)$ area. The result can be modified to derive an algorithm that gives an upward polyline grid drawing with an $O(|V| \log \log |V|)$ area, at most one bend per edge, and $O(|V|/ \log |V|)$ bends in total. Moreover in Reference 21 an $O(|V| \log \log |V|)$ area algorithm has been presented for non upward straight line grid layouts with arbitrary aspect ratio. Recently, Garg and Rusu [22] improved this result giving for binary trees an $O(|V|)$ area algorithm for non upward straight line orthogonal layouts with a pre specified aspect ratio in the range of $[1, |V|^{\alpha}]$, with $\alpha \in [0, 1)$ that can be constructed in $O(|V|)$.

46.2 Preliminaries

A (*rooted*) *tree* T is a directed acyclic graph with a single source, called the *root* of the tree, such that there is a unique directed path from the root to any other node. The *level* $l(v)$ of a node v is the length of this path. The largest level of any node in T is the height $h(T)$ of T. For each edge (v, w), v is the *parent* of w and w is a *child* of v. Children of the same node are called *siblings*. Each node w on the path from the root to a node v is called an *ancestor* of v, while v is called a *descendant* of w. A node that has no children is a *leaf*. If v_1 and v_2 are two nodes such that v_1 is not an ancestor of v_2 and vice versa, the *greatest distinct ancestors* of v_1 and v_2 are defined as the unique ancestors w_1 and w_2 of v_1 and v_2, respectively, such that w_1 and w_2 are siblings. Each node v of a rooted tree T induces a unique subtree $T(v)$ of T with root v.

Binary trees are trees with a maximum number of two children per node. In contrast to binary trees, trees that have an arbitrary number of children are called *n-ary trees*. A tree is said to be ordered if for every node the order of its children is fixed. The first (last) child according to this order is called the *leftmost* (*rightmost*) child. The *left* (*right*) sibling of a node v is its predecessor (successor) in the list of children of the parent of v. The *leftmost* (*rightmost*) *descendant* of v on a level l is the leftmost (rightmost) node on the level l belonging to the subtree $T(v)$ induced by v. Finally, if w_1 is the left sibling of w_2, v_1 is the rightmost descendant of w_1 on some level l, and v_2 is the leftmost descendant of w_2 on the same level l, the node v_1 is called the *left neighbor* of v_2 and v_2 is the *right neighbor* of v_1.

To *draw* a tree into the plane means to assign x- and y-coordinates to its nodes and to represent each edge (v, w) by a straight line connecting the points corresponding to v and w. Objects that represent the nodes are centered above the point corresponding to the node. The computation of the coordinates must respect the sizes of the objects. For simplicity however, we assume throughout this paper that all nodes have the same dimensions and that the minimal distance required between neighbors is the same for each pair of neighbors. Both restrictions can be relaxed easily, since we will always compare a single pair of neighbors.

Reingold and Tilford have defined the following aesthetic properties for drawings of trees:

A1. The layout displays the hierarchical structure of the tree, that is, the y-coordinate of a node is given by its level.

A2. A parent is centered above its children.

A3. The drawing of a subtree does not depend on its position in the tree, that is, isomorphic subtrees are drawn identically up to translation.

By their appearance, these drawings are called *level drawings*. If the tree that has to be drawn is ordered, we additionally require the following:

A4. The order of the children of a node is displayed in the drawing, thus a left child is placed to the left with a smaller x-coordinate and a right child is placed to the right with a bigger x-coordinate.

A5. A tree and its mirror image are drawn identically up to reflection.

Here, the *reflection* of an ordered tree is the tree with reversed order of children for each parent node. It is desirable to find a layout satisfying (A1–A5) with a small width. However, it has been shown by Supovit and Reingold [23] that achieving the minimum grid width is NP-hard. Moreover, there is no polynomial time algorithm for achieving a width that is smaller than $\frac{25}{24}$ time the minimum width, unless P = NP. If on the other hand continuous coordinates instead of integral coordinates are allowed, minimum width drawing can be found in linear time by applying linear programming techniques (see [23]).

46.3 Level Layout for Binary Trees

For ordered binary trees, the first linear time algorithm satisfying (A1–A5) was presented by Reingold and Tilford [2]. The algorithm follows the divide and conquer principle implemented in form of a postorder traversal of a tree $T = (V, E)$ and places the nodes on grid units. For each $v \in V$ with left and right child w_{left}, w_{right} the algorithm computes layouts for the trees $T(w_{left})$ and $T(w_{right})$ up to horizontal translation. When v is visited, the drawing of the right subtree $T(w_{right})$ is shifted to the right such that on every level l of the subtrees the rightmost node v_{left} of $T(w_{left})$ and its neighbor, the leftmost node v_{right} of $T(w_{right})$ are separated at least by two or three grid points. The separation value between v_{left} and v_{right} is chosen such that v can be centered above the roots of $T(w_{left})$ and $T(w_{right})$ at an integer grid coordinate.

Shifting $T(w_{right})$ is partitioned into two subtask: first, determining the amount of shift and second, performing the shift of the subtree. To determine the amount of shift, define the *left contour* of a tree T to be the vertices with minimum x-coordinate at each level in the tree. The *right contour* is defined analogously. For an illustration, see Figure 46.2, where nodes belonging to the contours are shaded. To place $T(w_{right})$ as close to $T(w_{left})$ as possible, the right contour of $T(w_{left})$ and the left contour of $T(w_{right})$ are traversed calculating for every level l the amount of shift to separate the two subtrees on that specific level l. The maximum over all shift then gives the displacement for the trees such that they do not overlap. Since each node belongs to the traversed part of the left contour of the right subtree at most for one subtree combination, the total number of such comparisons is linear for the complete tree.

In order to achieve linear running time it must be ensured that the contours are traversed without traversing (too many) nodes not belonging to the contours. This is achieved by introducing a *thread* for each leaf of the subtree that has a successor in the same contour. The thread is a pointer to this successor. See Figure 46.2 for an illustration where the threads are represented by dotted arrows. For every node of the contour, we now have a pointer to its successor in the left (right) contour given either by its leftmost (rightmost) child or by the thread. Finally, to update the threads, a new thread has to be added whenever two subtrees of different height are combined.

Once the shift has been determined for $T(w_{right})$ we omit shifting the subtree by updating the coordinates of its nodes, since this would result in quadratic running time. Instead, the position of each node is only determined preliminary and the shift for $T(w_{right})$ is stored at w_{right} as the *modifier* $mod(w_{right})$ (see [1]). Only $mod(w_{right})$ and the preliminary x-coordinate *prelim*(v) of

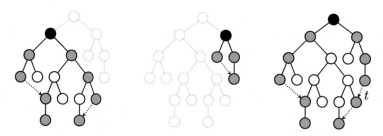

FIGURE 46.2   Combining two subtrees and adding a new thread t.

the parent v are adjusted by the shift of $T(w_{right})$. The modifier of w_{right} is interpreted as a value to be added to all preliminary x-coordinates in the subtree rooted at w_{right}, except for w_{right} itself. Thus, the coordinates of a node v in T in the final layout is its preliminary position plus the aggregated modifier $modsum(v)$ given by the sum of all modifiers on the path from the parent of v to the root. Once all preliminary positions and modifiers have been determined the final coordinates can be easily computed by a top-down sweep.

Theorem 46.1

[Reingold and Tilford [2]] The layout algorithm for binary trees meets the aesthetic requirements (A1–A5) and can be implemented such that the running time is $O(|V|)$.

Proof. By construction of the algorithm it is obvious that the algorithm meets (A1–A5). So it is left to show that the running time is linear in the number of nodes. Every node of T is traversed once during the postorder and the preorder traversal. So it is left to show that the time needed to traverse the contour of the two subtrees $T(w_{left})$ and $T(w_{right})$ for every node v with children w_{left} and w_{right} is linear in the number of nodes of T over all such traversals.

It is obviously necessary to travel down the contours of $T(w_{left})$ and $T(w_{right})$ only as far as the tree with lesser height. Thus the time spent processing a vertex v in the postorder traversal is proportional to the minimum of the height of $h(T(w_{left}))$ and $h(T(w_{right}))$. The running time of the postorder traversal is then given by:

$$\sum_{v \in T}(1 + \min\{h(T(w_{left})), h(T(w_{right}))\}) = |V| + \sum_{v \in T} \min\{h(T(w_{left})), h(T(w_{right}))\}$$

The sum $\sum_{v \in T} \min\{h(T(w_{left})), h(T(w_{right}))\}$ can be estimated as follows. Consider a node w that is part of a contour of a subtree $T(w_{left})$. When comparing the right contour of $T(w_{left})$ and the left contour of $T(w_{right})$ two cases are possible. Either w is not traversed and therefore will be part of the contour of $T(v)$ or it is traversed when comparing $T(w_{left})$ and $T(w_{right})$. In the latter case, w is part of the right contour of $T(w_{left})$ and after merging $T(w_{left})$ and $T(w_{right})$ the node w is not part of the right contour of $T(v)$. Thus every node of T is traversed at most twice and the total number of comparisons is bounded by $|V|$. ∎

We do not give the full algorithm for drawing binary trees here. Instead, the layout algorithm for n-ary trees is given in full length in the next section. The methods presented there are an expansion of Reingold and Tilfords algorithm to the more general case and still proceed for binary trees as described in this section.

46.4 Level Layout for n-ary Trees

A straightforward manner to draw trees of unbounded degree is to adjust the Reingold-Tilford algorithm by traversing the children of each node v from left to right, successively computing the preliminary coordinates and the modifiers.

This however violates property (A5): the subtrees are placed as close to each other as possible and small subtrees between larger ones are piled to the left; see Figure 46.3a. A simple trick to avoid this effect is to add an analogous second traversal from right to left; see Figure 46.3b, and to take average positions after that. This algorithm satisfies (A1–A5), but smaller subtrees are usually clustered then; see Figure 46.3c.

To obtain a layout where smaller subtrees are spaced out evenly between larger subtrees as for example shown in Figure 46.3d, we process the subtrees for each node $v \in V$ from left to right, see Figure 46.4. In a first step, every subtree is then placed as close as possible to the right of its left subtrees. This is done similarly to the algorithm for binary trees as described in Section 46.3 by traversing the left contour of the right subtree $T(w_{right})$ and the right contour of the subforest induced by the left siblings of w_{right}.

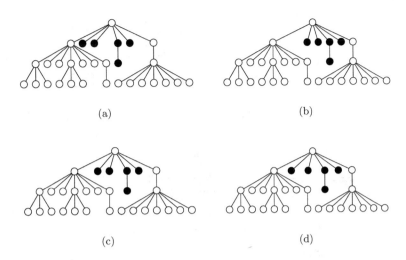

FIGURE 46.3 Extending the Reingold-Tilford algorithm to trees of unbounded degree.

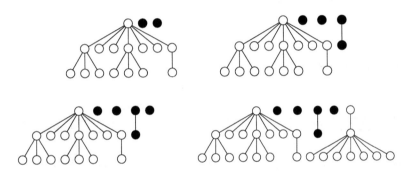

FIGURE 46.4 Spacing out the smaller subtrees.

Whenever two conflicting neighbors v_{left} and v_{right} are detected, forcing v_{right} to be shifted to the right by an amount of σ, we apply an appropriate shift to all smaller subtrees between the subtrees containing v_{left} and v_{right}.

More precisely, let w_{left} and w_{right} be the greatest distinct ancestors of v_{left} and v_{right}. Notice that both w_{left} and w_{right} are children of the node v that is currently processed. Let k be the number of children $w_1, w_2, \ldots, w_k$ of the current root between w_{left} and w_{right} plus 1. The subtrees between w_{left} and w_{right} are spaced out by shifting the subtree $T(w_i)$ to the right of w_{left} by an amount of $i \cdot \sigma / k$, for $i = 1, 2, \ldots, k$. We notice that spacing out the smaller subtrees may result in subtrees that are shifted more than once. Below, the effect on the running time for computing the shifts is determined and it is shown how to obtain a linear time implementation.

It is easy to see that this approach satisfies (A1–A5) and in addition spaces out the smaller subtrees between larger subtrees evenly. In contrast to the algorithm for binary trees (see Section 46.3) nodes are not placed on integer coordinates. Instead real values are used.

Figure 46.5 gives a layout example of a 5-ary tree. The layout algorithm considers different nodes sizes.

Figure 46.6 shows a layout of a *PQ*-tree produced by the algorithm for binary trees. A straightforward modification of the algorithm allows the application of different nodes sizes to the *Q*-nodes and the usage of strictly vertical edges for the children of the *Q*-nodes.

The algorithm TREELAYOUT given below works as a frame that initializes modifiers, threads, and ancestors (see Section 46.4.3), before evoking the methods PREPOSITION and ADJUST. The method PREPOSITION computes the shifts and preliminary positions of the nodes.

Based on the results of the function PREPOSITION the function ADJUST given in Algorithm 46.2 computes the final coordinates by summing up the modifiers recursively.

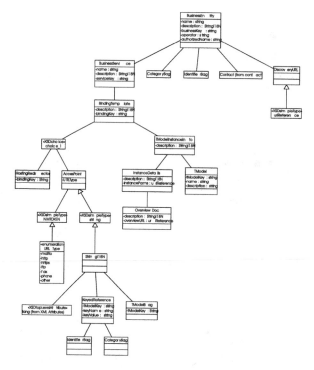

FIGURE 46.5   A level layout of an *n*-ary tree with different node sizes.

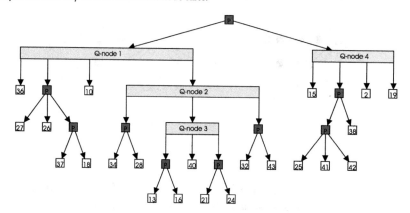

FIGURE 46.6   A level layout of a *PQ*-tree.

Algorithm 46.1: TreeLayout (*T*)

> **forall** *nodes v of T* **do**
> $mod(v) = thread(v) = 0$;
> $ancestor(v) = v$;
> **od**
> let *r* be the root of *T*;
> PREPOSITION(*r*);
> ADJUST(*r*, −*prelim*(*r*));

Algorithm 46.2: Adjust(*v*, *m*)

> $x(v) = prelim(v) + m$;
> let $y(v)$ be the level of *v*;
> **forall** *children w of v* **do**
>   ADJUST(*w*, *m* + *mod*(*v*));
> **od**

46.4.1 PrePosition

Algorithm 46.3 presents the method PREPOSITION(v) that computes a preliminary x-coordinate for a node v. PREPOSITION is applied recursively to the children of v. After each call of PREPOSITION on a child w a function APPORTION is executed on w. The procedure APPORTION is the core of the algorithm and shifts a subtree such that it does not conflict with its left subforest. After spacing out the smaller subtrees by calling EXECUTESHIFTS, the node v is placed to the midpoint of its outermost children. The value *distance* prescribes the minimal distance between two nodes. If objects of different size are considered for the representation of the nodes, or if different minimal distances, that is, between subtrees, are specified, the value *distance* has to be modified appropriately.

Algorithm 46.3: PREPOSITION(v)

> **if** *v is a leaf* **then**
> > *prelim*(v) = 0;
> > **if** *v has a left sibling w* **then**
> > > *prelim*(v) = *prelim*(w) + *distance*;
> >
> > **end**
>
> **else**
> > let *defaultAncestor* be the leftmost child of v;
> > **forall** *children w of v from left to right* **do**
> > > PREPOSITION(w);
> > > APPORTION(w, *defaultAncestor*);
> >
> > **od**
> > EXECUTESHIFTS(v);
> > *midpoint* = $\frac{1}{2}$(*prelim*(leftmost child of v) + *prelim*(rightmost child of v));
> > **if** *v has a left sibling w* **then**
> > > *prelim*(v) = *prelim*(w) + *distance*;
> > > *mod*(v) = *prelim*(v) − *midpoint*;
> >
> > **else**
> > > *prelim*(v) = *midpoint*;
> >
> > **end**
>
> **end**

46.4.2 Combining a Subtree and its Left Subforest

Before presenting APPORTION in detail, we need to consider efficient strategies for the different task that are performed by this function.

Similar to the threads used for binary trees (see Section 46.3), APPORTION uses threads to follow the contours of the right subtree and the left subforest. The fact that the left subforest is no tree in general does not create any additional difficulty.

One major task of APPORTION is that it has to space out the smaller subtrees between the larger ones. More precisely, if APPORTION shifts a subtree to the right to avoid a conflict with its left subforest, APPORTION has to make sure that the shifts of the smaller subtrees of the left subforest are determined.

A straightforward implementation computes the shifts for the intermediate smaller subtrees after the right subtree has been moved. However, as has been shown in Reference 5, this strategy has an aggregated runtime of $\Omega(|V|^{3/2})$. To prove this consider a tree T^k such that the root has k children $v_1, v_2, \ldots, v_k$ (see Figure 46.7 for $k = 3$). The children are numbered from left to right. Except for v_1 let the i-th child v_i be the root of a subtree $T^k(v_i)$ that consist of a chain of i nodes. Between each pair v_i, v_{i+1}, $i = 1, 2, \ldots, k-1$, of these children, add k children as leaves. Moreover, the subtree $T^k(v_1)$ is modified as follows. Its root v_1 has $2k + 5$ children, and up to level $k - 1$, every rightmost child of the $2k + 5$ children again has $2k + 5$ children. The number of nodes of T^k is

$$1 + \sum_{i=1}^{k} i + (k-1)k + (k-1)(2k+5) \in \Theta(k^2).$$

When traversing the nodes $v_1, v_2, \ldots, v_k$ in order to determine their shifts, the subtree chain $T^k(v_i)$ conflicts with the first subtree $T^k(v_1)$ on level i. Hence, all $(i-1)(k+1) - 1$ smaller subtrees between $T^k(v_1)$ and $T^k(v_i)$ are shifted. Thus, the total number of

FIGURE 46.7 T^3.

shifting steps is

$$\sum_{i=2}^{k}((i-1)(k+1)-1) = (k+1)k(k-1)/2 - k + 1 \in \Theta(k^3).$$

Since the number of nodes in T_k is $|V| \in \Theta(k^2)$, this shows that a straightforward implementation needs $\Omega(|V|^{3/2})$ time in total.

Let $T(v_i), i \in \{2, 3, \ldots, k\}$ be the subtree that has to be shifted. Let v_{left} and v_{right} be two neighboring nodes on some level l such that

- v_{left} is in the left subforest $\cup_{h=1}^{i-1} T(v_h)$ and v_{right} is in $T(v_i)$ respectively, and
- v_{left} and v_{right} determine the shift of $T(v_i)$.

In order to develop an efficient method for spacing out the smaller subtrees two tasks have to be solved.

- The tree $T(v_j), j \in \{1, 2, \ldots, i-1\}$, with $v_{left} \in T(v_j)$ has to be maintained in order to determine the smaller subtrees to the left of $T(v_i)$ that have to be spaced out.
- Shifting the smaller subtrees between $T(v_j)$ and $T(v_i)$ has to be done efficiently.

The next Section 46.4.3 shows how to obtain the tree $T(v_j)$ efficiently. Section 46.4.4 gives a detailed description on how to compute the shift of $T(v_i)$. and Section 46.4.5 presents a method that spaces out smaller subtrees.

46.4.3 Ancestor

We first describe how to obtain the subtree $T(v_j)$ that contains the node v_{left}. The problem is equivalent to finding the greatest distinct ancestors w_{left} and w_{right} of the nodes v_{left} and v_{right}, where in this case w_{left} is equal to the root v_j of the subtree that we need to determine and $w_{right} = v_i$. It is possible to apply an algorithm by Schieber and Vishkin [24] that determines for each pair of nodes its greatest distinct ancestors in constant time, after an $O(|V|)$ preprocessing step. Since their algorithm is somewhat tricky, and one of the greatest distinct ancestors, namely v_i, is known anyway, we apply a much simpler algorithm. Furthermore, as v_{right} is always the right neighbor of v_{left}, the left one of the greatest distinct ancestors only depends on v_{left}. Thus we may shortly call it *the ancestor* of v_{left} in the following.

To store the ancestor of a node u a pointer $ancestor(u)$ is introduced and initialize it to u itself. The pointer $ancestor(u)$ for any node u is not updated throughout the algorithm. Instead, a $defaultAncestor$ is used and ancestors are only determined for the nodes u on the right contour of the left subforest $\cup_{h=1}^{i-1} T(v_h)$ during the shift of the subtree $T(v_i)$. This strategy ensures that we obtain linear running time.

We make sure that for every $i = 2, 3, \ldots, k$ the following property for all nodes u on the right contour of $\cup_{h=1}^{i-1} T(v_h)$ holds:

(*) If $ancestor(u)$ is up to date, that is, u is a child of v, then $ancestor(u) = w_{left}$; otherwise, $defaultAncestor$ is the correct $ancestor(u)$.

For $i = 2$ we have that $\cup_{h=1}^{i-1} T(v_h) = T(v_1)$. The $defaultAncestor$ for all nodes on the right contour is obviously v_1 and therefore $defaultAncestor$ is set equal to v_1. It is easy to recognize if a node u on the right contour has a pointer $ancestor(u)$ that is up to date: either $ancestor(u) = v_1$, or the level $l(ancestor(u))$ is greater than $l(v_1)$. This obviously fulfills property (*), see Figure 46.8a for an illustration.

After adding a subtree $T(v_{i-1})$ to $\cup_{h=1}^{i-2} T(v_h)$ for each $i = 3, 4, \ldots, k$ two cases need to be considered. If the height $h(T(v_{i-1}))$ is lesser or equal to the height of the subforest $\cup_{h=1}^{i-2} T(v_h)$, the pointer $ancestor(u)$ is set to v_{i-1} for all nodes u on the right contour of $T(v_{i-1})$. This obviously fulfills property (*); see Figure 46.8b for an illustration. Moreover, the number of update operations is equal to the number of comparisons between the nodes on the left contour of $T(v_{i-1})$ and their neighbors in $\cup_{h=1}^{i-2} T(v_h)$. Hence the total number of all these update operations is in $O(|V|)$.

If the height $h(T(v_{i-1}))$ is greater than the height of $\cup_{h=1}^{i-2} T(v_h)$, we omit updating all nodes on the right contour of $T(v_{i-1})$ in order to obtain linear running time. Instead, it suffices to set $defaultAncestor$ to v_{i-1}, since either $ancestor(u) = v_{i-1}$, or $l(ancestor(u)) > l(v_1)$ holds for any node in the right contour, and all smaller subtrees in the $\cup_{h=1}^{i-1} T(v_h)$ do not contribute to the right contour anymore. Thus property (*) is fulfilled; see Figure 46.8c for an illustration.

In algorithm 46.4 the function ANCESTOR is given. It returns w_{left} of u as described above.

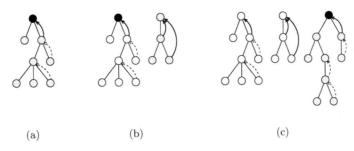

(a)   (b)   (c)

FIGURE 46.8 Adjusting ancestor pointers when adding new subtrees: the pointer *ancestor(u)* is represented by a solid arrow if it is up to date and by a dashed arrow if it is expired. In the latter case, the *defaultAncestor* is used and drawn black. When adding a small subtree, all ancestor pointers *ancestor(u)* of its right contour are updated. When adding a large subtree, only *defaultAncestor* is updated.

Algorithm 46.4: ANCESTOR($u, w_{right}, defaultAncestor$)

> **if** *ancestor(u) is a sibling of w_{right}* **then**
>> return *ancestor(w_{right})*;
> **else**
>> return *defaultAncestor*;
> **end**

46.4.4 Apportion

To give the function APPORTION in Algorithm 46.5 a more readable annotation, we use subscripts f and o to describe the different contours of the left subforest and the right subtree. The subscript f is used for neighboring nodes on the contours that are *facing* each other and thus is used for the nodes on the right contour of the left subforest $\cup_{h=1}^{i-1} T(v_h)$ and the left contour of the right subtree $T(v_i)$, $i = 2, 3, \ldots, k$. We use o for the left contour of $\cup_{h=1}^{i-1} T(v_h)$ and for the right contour of $T(v_i)$, describing the nodes that are on the *outside* of the combined subforest $\cup_{h=1}^{i} T(v_h)$.

The nodes traversing the contours are v^f_{right}, v^f_{left}, v^o_{left}, and v^o_{right}, where the subscript *left* describes nodes of the left subforest and *right* nodes of the right subtree. For summing up the modifiers along the contour (see also Sect. 46.3), respective variables s^f_{right}, s^f_{left}, s^o_{left}, and s^o_{right} are used.

Whenever two nodes of the facing contours conflict, we compute the left one of the greatest distinct ancestors using the function ANCESTOR and call MOVESUBTREE to shift the subtree and prepare the shifts of smaller subtrees.

Finally, a new thread is added (if necessary) as explained in Section 46.3. Observe that we have to adjust *ancestor(v^o_{right})* or *defaultAncestor* to keep property (*). The functions NEXTLEFT and NEXTRIGHT return the next node on the left and right contour, respectively (see Algorithms 46.6 and 46.7).

46.4.5 Shifting the Smaller Subtrees

For spacing out the smaller subtrees evenly, the number of the smaller subtrees between the larger ones has to be maintained. Since simply counting the number of smaller children between two larger subtrees $T(v_j)$ and $T(v_i)$, $1 \le j < i \le k$ would result in $\Omega(n^{3/2})$ time in total, it is determined as follows. The children of v are numbered consecutively. Once the pair of nodes v_{left}, v_{right} that defines the maximum shift on $T(v_i)$ has been determined, the greatest distinct ancestors $w_{left} = v_j$ and $w_{right} = v_i$ are easily determined by the approach described in Section 46.4.3. The number $i - j - 1$ gives the number of in between subtrees in constant time.

In order to obtain a linear runtime, we make sure that each smaller subtree between a pair of larger ones is shifted at most once.

Thus whenever a subtree of $T(v_i)$ is considered to be placed next to the left subforest $\cup_{h=1}^{i-1} T(v_h)$, we do not modify the shift of the smaller subtrees. Such an approach would result in quadratic runtime by the fact that smaller subtrees are shifted every time they are in between a pair of greater subtrees. Instead, a system is installed, that allows to adopt the shift of the smaller subtrees by traversing the nodes $v_1, v_2, \ldots, v_k$ once after the last child v_k of v has been shifted.

For every node v_g, $g = 1, 2, \ldots, k$, real numbers *shift(v_g)* and *change(v_g)* are introduced and initialized by zero. Let $T(v_j)$, $1 \le j < i \le k$, be the subtree that defines the σ on the subtree $T(v_i)$. Let $\theta = i - j$ be the number of subtrees between v_i and v_j, plus 1. Then, for $t = 1, 2, \ldots, i - j - 1$, the t-th subtree $T(v_{j+t})$ has to be moved by $t \cdot \sigma / \theta$. In other words: the tree $T(v_{j+t})$ is shifted by $\sigma - (i - (j + t)) \cdot \sigma / \theta$, for example,

Algorithm 46.5: APPORTION(v,*defaultAncestor*)

> **if** v *has a left sibling w* **then**
>> $v^f_{right} = v^o_{right} = v$;
>>
>> $v^f_{left} = w$;
>>
>> v^o_{left} be the leftmost sibling of v^f_{right};
>>
>> $s^f_{right} = mod(v^f_{right})$;
>>
>> $s^o_{right} = mod(v^o_{right})$;
>>
>> $s^f_{left} = mod(v^f_{left})$;
>>
>> $s^o_{left} = mod(v^o_{left})$;
>>
>> **while** NEXTRIGHT(v^f_{left}) $\neq 0$ *and* NEXTLEFT(v^f_{right}) $\neq 0$ **do**
>>> $v^f_{left} = $ NEXTRIGHT(v^f_{left});
>>>
>>> $v^f_{right} = $ NEXTLEFT(v^f_{right});
>>>
>>> $v^o_{left} = $ NEXTLEFT(v^o_{left});
>>>
>>> $v^o_{right} = $ NEXTRIGHT(v^o_{right});
>>>
>>> $ancestor(v^o_{right}) = v$;
>>>
>>> $\sigma = (prelim(v^f_{left}) + s^f_{left}) - (prelim(v^f_{right}) + s^f_{right}) + distance$;
>>>
>>> **if** $\sigma > 0$ **then**
>>>> MOVESUBTREE(ANCESTOR(v^f_{left},v,*defaultAncestor*),v,σ);
>>>>
>>>> $s^f_{right} = s^f_{right} + \sigma$;
>>>>
>>>> $s^o_{right} = s^o_{right} + \sigma$;
>>>
>>> **end**
>>>
>>> $s^f_{left} = s^f_{left} + mod(v^f_{left})$;
>>>
>>> $s^f_{right} = s^f_{right} + mod(v^f_{right})$;
>>>
>>> $s^o_{left} = s^o_{left} + mod(v^o_{left})$;
>>>
>>> $s^o_{right} = s^o_{right} + mod(v^o_{right})$;
>>
>> **end**
>
> **end**
>
> **if** NEXTRIGHT(v^f_{left}) $\neq 0$ *and* NEXTRIGHT(v^o_{right}) $= 0$ **then**
>> $thread(v^o_{right}) = $ NEXTRIGHT(v^f_{left});
>>
>> $mod(v^o_{right}) = mod(v^o_{right}) + s^f_{left} - s^o_{right}$;
>
> **end**
>
> **if** NEXTLEFT(v^f_{right}) $\neq 0$ *and* NEXTLEFT(v^o_{left}) $= 0$ **then**
>> $thread(v^o_{left}) = $ NEXTLEFT(v^f_{right});
>>
>> $mod(v^o_{left}) = mod(v^o_{left}) + s^f_{right} - s^o_{left}$;
>>
>> $defaultAncestor = v$;
>
> **end**

Algorithm 46.6: NEXTLEFT(v)

> **if** v *has a child* **then**
>> return the leftmost child of v;
>
> **else**
>> return $thread(v)$;
>
> **end**

Algorithm 46.7: NEXTRIGHT(*v*)

if *v has a child* **then**
    return the rightmost child of *v*;
else
    return *thread(v)*;
end

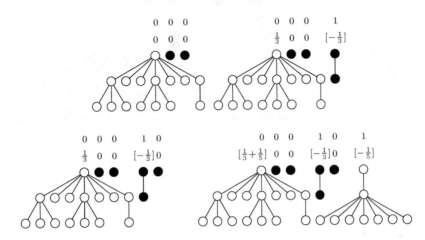

FIGURE 46.9 Aggregating the shifts: the top number at every node *u* indicates the value of *shift(u)*, and the bottom number indicates the value of *change(u)*.

- for $t = i - j - 1$ the subtree $T(v_{i-1})$ is shifted by $\sigma - \sigma/\theta$,
- for $t = i - j - 2$ the subtree $T(v_{i-2})$ is shifted by $\sigma - 2\cdot\sigma/\theta$,
- and for $t = 1$ the subtree $T(v_{j+1})$ is shifted by $\sigma - (i - j - 1)\cdot\sigma/\theta$.

The shift of the subtrees $T(v_{j+t})$, $t = 1, 2, \ldots, i - j - 1$, depends on the shift of $T(v_i)$ and if traversed from right to left, is decreased linear by σ/θ.

Since the decrease σ/θ is linear for every pair of greater subtrees, we use a trick and aggregate the amount of shift for the intermediate smaller subtrees by storing σ in an array *shift()* of size *k* and by storing the σ/θ in an array *change()* of size *k*. The values for shifting the subtrees $T(v_{j+t})$, $t = 1, 2, \ldots, i - j - 1$, are then stored at v_i and v_j:

1. The value $shift(v_i)$ is increased by σ
2. The value $change(v_i)$ is decreased by σ/θ
3. The value $change(v_j)$ is increased by σ/θ

Figure 46.9 shows an example on setting the values of *shift()* and *change()*.

By construction of the arrays *shift()* and *change()* we then obtain the shift of $T(v_g)$, $g = 1, 2, \ldots, k$, (including the "original" shift of the subtree $T(v_i)$) as follows. The children v_g, $g = k, k - 1, \ldots, 1$, are traversed from right to left. Two real values σ and *change* are maintained to store the shifts and the decreases of shift per subtree, respectively. These values are initialized with zero. When visiting child v_g, $g \in \{k, k - 1, \ldots, 1\}$ the subtree $T(v_g)$ is shifted to the right by σ (i.e., we increase *prelim(v_g)* and *mod(v_g)* by σ. Furthermore, *change* ins increased by *change(v_g)*, and σ is increased by *shift(v_g)* and by *change(v_g)* and we continue with v_{g-1}.

It is easy to see that this algorithm shifts each subtree by the correct amount, see Figure 46.10 for an example.

The function MOVESUBTREE($w_{left}, w_{right}, \sigma$) given in algorithm 46.8 performs an update of the arrays *shift()* and *change()*. We recall that w_{left} and w_{right} are the greatest distinct ancestors of v_{left} and v_{right} and correspond to children v_i and v_j, $1 \le i < j \le k$, respectively. MOVESUBTREE shifts the subtree $T(w_{right})$ by increasing *prelim(w_{right})* and *mod(w_{right})* by the amount of σ. The shifts of the intermediate smaller subtrees between $T(w_{left})$ and $T(w_{right})$ are prepared by adjust *change(w_{right})*, *shift(w_{right})*, and *change(w_{left})*.

The function EXECUTESHIFTS(*v*) traverses its children from right to left and determines the total shift of the children based on the arrays *shift()* and *change()*.

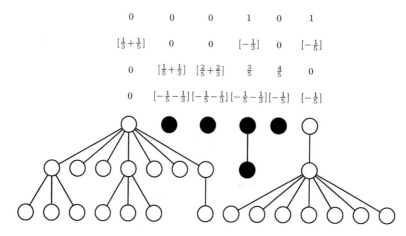

FIGURE 46.10   Executing the shifts: the third and the fourth line of numbers at a node u indicate the values of σ and *change* before shifting u, respectively.

Algorithm 46.8: MOVESUBTREE($w_{left}, w_{right}, \sigma$)

$\theta = number(w_{right}) - number(w_{left})$;
$change(w_{right}) = change(w_{right}) - \sigma / \theta$;
$shift(w_{right}) = shift(w_{right}) + \sigma$;
$change(w_{left}) = change(w_{left}) + \sigma / \theta$;
$prelim(w_{right}) = prelim(w_{right}) + \sigma$;
$mod(w_{right}) = mod(w_{right}) + \sigma$;

Algorithm 46.9: EXECUTESHIFTS

$\sigma = 0$;
$change = 0$;
forall *children w of v from right to left* **do**
    $prelim(w) = prelim(w) + \sigma$;
    $mod(w) = mod(w) + \sigma$;
    $change = change + change(w)$;
    $\sigma = \sigma + shift(w) + change$;
od

Theorem 46.2

[Buchheim et al [5]] The layout algorithm for n-ary trees meets the aesthetic requirements (A1)–(A5), spaces out smaller subtrees evenly, and can be implemented to run in $O(|V|)$.

Proof.   By construction of the algorithm it is obvious that the algorithm meets (A1–A5) and spaces out the smaller subtrees evenly. So it is left to show that the running time is linear in the number of nodes.

    Every node of T is traversed once during the traversals PREPOSITION and ADJUST. Similar reasoning as in the proof of theorem 46.1 for binary trees shows that the time needed to traverse the left contour of the subtree $T(v_i)$ and the right contour of subforest $\cup_{h=1}^{i-1} T(v_h)$, $i = 2, 3, \ldots, k$ for every node v with children v_i, $i = 1, 2, \ldots, k$ is linear in the number of nodes of T over all such traversals. Moreover, we have that by construction the number of extra operations for spacing out the smaller subtrees is linear in $|V|$ plus the number of nodes traversed in the contours. This proofs the theorem.   ■

46.5 Radial Layout

A *radial layout* of a tree is a variation of a level drawing, where the levels are concentric circles $c_1, c_2, \ldots, c_{h(T)}$ around the root placed at the origin c_0. Figure 46.11 shows an example of a radial layout. The radius of a circle c_i, $i = 1, 2, \ldots, h(T)$, is given by an

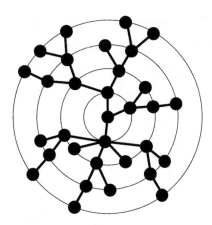

FIGURE 46.11   A radial layout of a binary tree.

increasing function $r(i)$. This type of layout is frequently used for representing a free tree. A free tree is a tree without a specific root. To layout a free tree, a node is chosen as a fictitious root that minimizes the height of the resulting subtrees. A straightforward manner to obtain a radial layout of a tree is to modify the algorithms for level drawings as presented in Sections 46.3 and 46.4.

To guarantee that the resulting drawing is planar, the subtree $T(v)$ of each node v is drawn within an *annulus wedge* $w(v)$. In order to permit edges with endpoints in $w(v)$ to extend outside $w(v)$ and thus to conflict with other edges, the nodes in the subtree $T(v)$ are placed within a convex subset $s(v)$ of $w(v)$. Given the level $l(v)$, the node v is placed on $c_{l(v)}$. Suppose that the tangent to $c_{l(v)}$ through v meets the points a and b on $c_{l(v)+1}$. Then $s(v)$ is chosen to be the unbounded region that is given by the line segment ab and the rays from the root of T and the node a and b. The children $v_j, j = 1, 2, \ldots, k$ of v are then arranged on $c_{l(v)+1}$ within $s(v)$ according to the number of leaves in $T(v_j)$.

If the distance between consecutive circles c_i, c_{i+1} $i = 0, 1, \ldots, h(T) - 1$, is equal, it can be easily shown that the area occupied by the layout is in

$$O(h(T)^2 \max_{v \in T}\{k \mid k \text{ number of children of } v\}).$$

Different algorithms for the radial layout of trees have been presented by [6–8] depending on the choice of the root, the radii of the circles and the angle of the annulus wedge. Symmetry oriented algorithms have also been developed, see for example, [9].

46.6 HV-Layout

An *hv*-layout of a binary tree is an upward (but not strictly upward) straight line orthogonal drawing with edges drawn as rightward horizontal or downward vertical segments. The "hv" stands for horizontal-vertical. For any vertex v in a binary tree T we either have:

1. A child of v is either
   a. Aligned horizontally with v and to the right of v or
   b. Vertically aligned below v.
2. The smallest rectangles that cover the area of the subtrees of the children of v in the layout do not intersect.

Figure 46.12 shows an example of a *hv*-layout

A *hv*-layout is generated by applying a dived and conquer approach. The divide step constructs the *hv*-layout of the left and the right subtrees, while the conquer step either performs a *horizontal* or a *vertical combination*. Figure 46.13 shows the two types of combination.

If the left subtree a node v is placed to the left in a horizontal combination and below in a vertical combination, the layout preserves the ordering of the children of v. The height and width of such a drawing is at most $|V| - 1$. A straightforward way to reduce the size of a *hv*-drawing to a height of at most $\log |V|$ is to use only horizontal combinations and to place the larger subtree (in terms of number of nodes) to the right of the smaller subtree. This *right heavy hv layout* is not order preserving and can be produced in $O(|V|)$ requiring only an area of $O(|V| \log |V|)$. The right heavy *hv* approach can be easily extended to draw n-ary trees as scetched in Figure 46.14.

AS already presented in the introduction, this type of layout has been extensively studied for example, in References 10–21 21 to obtain results on minimal area requirements of tree layouts.

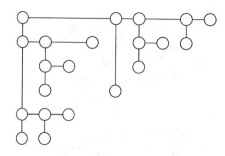

FIGURE 46.12　A hv-layout of a binary tree.

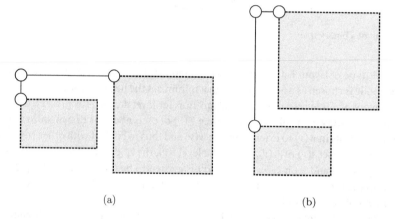

(a)　　　　　　　　　　　　　　　　　　　　(b)

FIGURE 46.13　A horizontal combination (a) and a vertical combination (b).

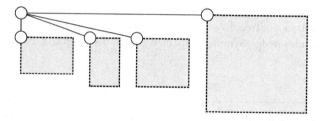

FIGURE 46.14　A hv-layout of a *n*-ary tree.

Recently Garg and Ruse [22] showed by flipping the subtrees rooted at a node v horizontally or vertically, it is possible to obtain tree layouts for binary trees with an $O(|V|)$ area and with a pre specified aspect ratio in the range of $[1, |V|^\alpha]$, with $\alpha \in [0, 1)$. These layouts are non upward straight line orthogonal layouts.

Acknowledgments

I gratefully acknowledge the contributions of Christoph Buchheim, who provided the figures and the implementation from our paper [5] and Merijam Percan for careful proof reading.

References

1. C. Wetherell and A. Shannon. Tidy drawings of trees. *IEEE Trans. Softw. Eng.*, 5(5):514–520, 1979.
2. E. Reingold and J. Tilford. Tidier drawing of trees. *IEEE Trans. Softw. Eng.*, SE-7(2):223–228, 1981.
3. T. Cormen, C. Leiserson, and R. Rivest. *Introduction to Algorithms.* MIT-Press.(1990).
4. Q. Walker II. A node-positioning algorithm for general trees. *Softw. - Pract. Exp.*, 20(7):685–705, 1990.

5. C. Buchheim, M. Jünger, and S. Leipert. Improving walker's algorithm to run in linear time. In M. Goodrich, editor, *Graph Drawing (Proc. GD 2002)*, volume 2528 of *LNCS*, pages 344–353. Springer-Verlag, 2002.

6. M. Bernard. On the automated drawing of graphs. In *3rd Caribbean Conference on Combinatorics and Computing*, pp. 43–55, 1981.

7. P. Eades. Drawing free trees. *Bulletin of the Inst. for Combinatorics and its Applications*, 5:10–36, 1992.

8. C. Esposito. Graph graphics: Theory and practice. *Comput. Math. Appl.*, 15(4):247–253, 1988.

9. J. Manning and M. Atallah. Fast detection and display of symmetry in trees. *Congr. Numer.*, 64:159–169, 1988.

10. P. Crescenzi, G. Di Battista, and A. Piperno. A note on optimal area algorithms for upward drawings of binary trees. *Comput. Geom*, 2:187–200, 1992.

11. P. Crescenzi and A. Piperno. Optimal-area upward drawings of AVL trees. In R. Tamassia and I. G. Tollis, editors, *Graph Drawing (Proc. GD 1994), volume 894 of LNCS*, pages 307–317. Springer-Verlag, 1995.

12. S. K. Kim. Simple algorithms for orthogonal upward drawing of binary and ternary trees. In *Proc. 7th Canad. Conf. Comput. Geometry*, pp. 115–120, 1995.

13. S. K. Kim. H-V drawings of binary trees. In *Software Visualization*, volume 7 of *Software Engineering and Knowledge Engineering*, pages 101–116, 1996.

14. X. Lin, P. Eades, T. Lin. Minimum size h-v drawings. In *Advanced Visual Interfaces*, volume 36 of *World Scientific Series in Computer Science*, pages 386–394, 1992.

15. X. Lin, P. Eades, T. Lin. Two tree drawing conventions. *Int. J. Comput. Geom. Appl.*, 3(2):133–153, 1993.

16. Y. Shiloach. Arrangements of Planar Graphs on the Planar Lattice. PhD thesis, Weizmann Institute of Science, 1976.

17. L. Trevisan. A note on minimum-area upward drawing of complete and Fibonacci trees. *Inform. Process. Lett.*, 57(5):231–236, 1996.

18. P. Crescenzi, P. Penna, and A. Piperno. Linear-area upward drawings of AVL trees. *Comput. Geom. Theory Appl.*, 9(25–42), 1998. (special issue on Graph Drawing, edited by G. Di Battista and R. Tamassia).

19. P. Crescenzi and P. Penna. Strictly-upward drawings of ordered search trees. *Theor. Comput. Sci.*, 203(1):51–67, 1998.

20. A. Garg, M. T. Goodrich, and R. Tamassia. Planar upward tree drawings with optimal area. *Int. J. Comput. Geom. Appl.*, 6:333–356, 1996.

21. C.-S. Shin, S. K. Kim, and K.-Y. Chwa. Area-efficient algorithms for straight-line tree drawings. *Comput. Geom. Theory Appl.*, 15:175–202, 2000.

22. A. Garg and A. Rusu. Straight-line drawings of binary trees with linear area and arbitrary aspect ratio. In M. Goodrich, editor, *Graph Drawing (Proc. GD 2002)*, volume 2528 of LNCS, pages 320–331. Springer-Verlag, 2002.

23. K.R. Supovit and E. Reingold. The complexity of drawing trees nicely. *Act Informatica*, 18:377–392, 1983.

24. B. Schieber and U. Vishkin. On finding lowest common ancestors: Simplification and parallelization. In *Proceedings of the Third Aegean Workshop on Computing*, volume 319 of *Lecture Notes in Computer Science*, pages 111–123. Springer-Verlag, 1998.

47

Drawing Graphs*

47.1 Introduction..723
47.2 Preliminaries...725
47.3 Convex Drawing...726
    Barycenter Algorithm • Divide and Conquer Algorithm • Algorithm Using Canonical Ordering
47.4 Symmetric Drawing 728
    Displaying Rotational Symmetry • Displaying Axial Symmetry • Displaying Dihedral Symmetry
47.5 Visibility Drawing...732
    Planar St-graphs • The Bar Visibility Algorithm • Bar Visibility Representations and Layered Drawings • Bar Visibility Representations for Orthogonal Drawings
47.6 Conclusion 737
References...738

Peter Eades
University of Sydney

Seok-Hee Hong
University of Sydney

47.1 Introduction

Graph Drawing is the art of making pictures of relationships. For example, consider the social network defined in Table 47.1. This table expresses the "has-written-a-joint-paper-with" relation for a small academic community.

It is easier to understand this social network if we draw it. A drawing is in Figure 47.1a; a better drawing is in Figure 47.1b. The challenge for Graph Drawing is to automatically create good drawings of graphs such as in Figure 47.1b, starting with tables such as in Table 47.1.

The *criteria* of a good drawing of a graph have been the subject of a great deal of attention (see, e.g., [1–3]). The four criteria below are among the most commonly used criteria.

- *Edge crossings* should be avoided. Human tests clearly indicate that edge crossings inhibit the readability of the graph [3]. Figure 47.1a has 6 edge crossings, and Figure 47.1b has none; this is the main reason that Figure 47.1b is more readable than Figure 47.1a. The algorithms presented in this chapter deal with *planar drawings*, that is, drawings with no edge crossings.
- The *resolution* of a graph drawing should be as large as possible. There are several ways to define the intuitive concept of resolution; the simplest is the *vertex resolution*, that is, the ratio of the minimum distance between a pair of vertices to the maximum distance between a pair of vertices. High resolution helps readability because it allows a larger font size to be used in the textual labels on vertices. In practice it can be easier to measure the *area* of the drawing, given that the minimum distance between a pair of vertices is one. If the vertices in Figure 47.1 are on an integer grid, then the drawing is 4 units by 2 units, for both (a) and (b). Thus the vertex resolution is $1/2\sqrt{5} \simeq 0.2236$, and the area is 8. One can refine the concept of resolution to define *edge resolution* and *angular resolution*.
- The *symmetry* of the drawing should be maximized. Symmetry conveys the structure of a graph, and Graph Theory textbooks commonly use drawings that are as symmetric as possible. Intuitively, Figure 47.1b displays more symmetry than Figure 47.1a. A refined concept of symmetry display is presented in Section 47.4.
- *Edge bends* should be minimized. There is some empirical evidence to suggest that bends inhibit readability. In Figure 47.1a, three edges contain bends and eleven are straight lines; in total, there are seven edge bends. Figure 47.1b is much better: all edges are straight lines.

---

* This chapter has been reprinted from first edition of this Handbook, without any content updates.

TABLE 47.1   Table Representing the *has-a-joint-paper-with*
Relation

Name	has-a-joint-paper-with
Jane	Harry, Paula, and Sally
Sally	Jane, Paula and Dennis
Dennis	Sally and Monty
Harry	Jane and Paula
Monty	Dennis and Kerry
Ying	Paula
Paula	Jane, Harry, Ying, Tan, Cedric, Chris, Kerry and Sally
Kerry	Paula and Monty
Tan	Paul and Cedric
Cedric	Tan, Paula, and Chris
Chris	Paula and Cedric

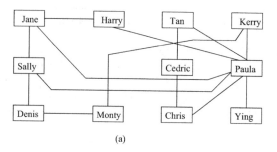

(a)

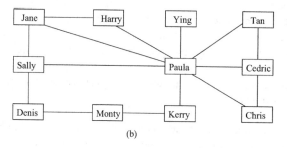

(b)

FIGURE 47.1   Two drawings of a social network.

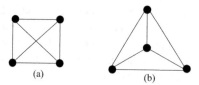

(a)   (b)

FIGURE 47.2   Two drawings of a graph.

Each of these criteria can be measured. Thus Graph Drawing problems are commonly stated as optimization problems: given a graph *G*, we need to find a drawing of *G* for which one or more of these measures are optimal. These optimization problems are usually NP-complete.

In most cases it is not possible to optimize one of the criteria above without compromise on another. For example, the graph in Figure 47.2a has 8 symmetries but one edge crossing. It is possible to draw this graph without edge crossings, but in this case we can only get 6 symmetries, as in Figure 47.2b.

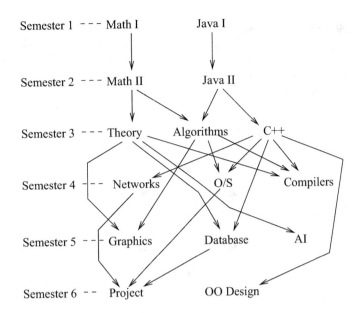

FIGURE 47.3   Prerequisite diagram.

The optimization problem may be constrained. For example, consider a graph representing prerequisite dependencies between the units in a course, as in Figure 47.3; in this case, the y coordinate of a unit depends on the semester in which the unit ought to be taken.

Graph Drawing has a long history in Mathematics, perhaps beginning with the theorem of Wagner that every planar graph can be drawn with straight line edges [4]. The importance of graph drawing algorithms was noted early in the history of Computer Science (note the paper of Knuth on drawing flowcharts, in 1960 [5], and the software developed by Read in the 1970s [6]). In the 1980s, the availability of graphics workstations spawned a broad interest in visualization in general and in graph visualization in particular. The field became mature in the mid 1990s with a conference series (e.g., see the annual proceedings of Symposium on Graph Drawing), some books [7–10], and widely available software (see Reference 11).

In this chapter we briefly describe just a few of the many graph available drawing algorithms. Section 47.3 presents some algorithms to draw planar graphs with straight line edges and convex faces. Section 47.4 describes two algorithms that can be used to construct planar symmetric drawings. Section 47.5 presents a method for constructing "visibility" representations of planar graphs, and shows how this method can be used to construct drawings with few bends and reasonable resolution.

47.2 Preliminaries

Basic concepts of mathematical Graph Theory are described in Reference 12. In this section we define mathematical notions that are used in many Graph Drawing algorithms.

A *drawing* D of a graph $G = (V, E)$ assigns a location $D(u) \in R^2$ for every vertex $u \in V$ and a curve $D(e)$ in R^2 for every edge $e \in E$ such that if $e = (u, v)$ then the curve $D(e)$ has endpoints $D(u)$ and $D(v)$. If $D(u)$ has integer coordinates for each vertex u, then D is a *grid* drawing. If the curve $D(e)$ is a straight line segment for each edge e, then D is a straight line drawing.

For convenience we often identify the vertex u with its location $D(u)$, and the edge e with its corresponding curve $D(e)$; for example, when we say "the edge e crosses the edge f," strictly speaking we should say "the curve $D(e)$ crosses the curve $D(f)$." A graph drawing is *planar* if adjacent edges intersect only at their common endpoint, and no two nonadjacent edges intersect. A graph is *planar* if it has a planar drawing. Planarity is a central concern of Graph Theory, Graph Algorithms, and especially Graph Drawing; see Reference 12.

A planar drawing of a graph divides the plane into regions called *faces*. One face is unbounded; this is the outside face. Two faces are *adjacent* if they share a common edge. The graph G together its faces and the adjacency relationship between the faces is a *plane graph*. The graph whose vertices are the faces of a plane graph G and whose edges are the adjacency relationships between faces is the *planar dual*, or just *dual*, of G. A graph and its planar dual are in Figure 47.4.

The *neighborhood* $N(u)$ of a vertex u is the list of vertices that are adjacent to u. If G is a plane graph, then $N(u)$ is given in the clockwise circular ordering about u.

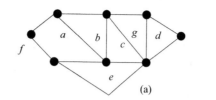
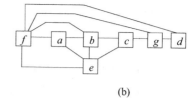

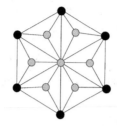

FIGURE 47.4   (a) A graph G and its faces. (b) The dual of G.

FIGURE 47.5   Example output from the algorithm of Tutte.

47.3 Convex Drawing

A straight-line drawing of a planar graph is *convex* if all every face is drawn as convex polygon. This section reviews three different algorithms for constructing such a drawing for planar graphs.

47.3.1 Barycenter Algorithm

Tutte showed that a convex drawing exist for triconnected planar graphs and gave an algorithm for constructing such representations [13,14].

The algorithm divides the vertex set V into two subsets; a set of *fixed* vertices and a set of *free* vertices. The fixed vertices are placed at the vertices of a strictly convex polygon. The positions of free vertices are decided by solving a system of $O(n)$ linear equations, where n is the number of vertices of the graph [14]. In fact, solving the equations is equivalent to placing each free vertex at the barycenter of its neighbors. That is, each position $p(v) = (x(v), y(v))$ of a vertex v is:

$$x(v) = \frac{1}{deg(v)} \sum_{(v,w)\in E} x(w), \quad y(v) = \frac{1}{deg(v)} \sum_{(v,w)\in E} y(w). \tag{47.1}$$

An example of a drawing computed by the barycenter algorithm is illustrated in Figure 47.5. Here the black vertices represent fixed vertices.

The main theorem can be described as follows.

Theorem 47.1: (Tutte [13,14])

Suppose that f is a face in a planar embedding of a triconnected planar graph G, and P is a strictly convex planar drawing of f. Then the barycenter algorithm with the vertices of f fixed and positioned according to P gives a convex planar drawing of G.

The matrix resulting from the equations can be solved in $O(n^{1.5})$ time at best, using a sophisticated sparse matrix elimination method [15]. However, in practice a Newton-Raphson iteration of the equation above converges rapidly.

47.3.2 Divide and Conquer Algorithm

Chiba, Yamanouchi and Nishizeki [16] present a linear time algorithm for constructing convex drawings of planar graphs, using divide and conquer. The algorithm constructs a convex drawing of a planar graph, if it is possible, with a given outside face. The drawing algorithm is based on a classical result by Thomassen [17].

The input to their algorithm is a biconnected plane graph with given outside face and a convex polygon. The output of their algorithm is a convex drawing of the biconnected plane graph with outside face drawn as the input convex polygon, if this is possible. The conditions under which such a drawing is possible come from the following theorem [17].

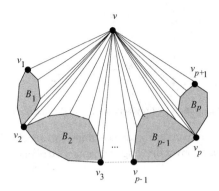

FIGURE 47.6   The algorithm of Chiba et al.

Theorem 47.2: (Thomassen [17])

Let G be a biconnected plane graph with outside facial cycle S, and let S^* be a drawing of S as a convex polygon. Let $P_1, P_2, \ldots, P_k$ be the paths in S, each corresponding to a side of S^*. (Thus S^* is a k-gon. It should be noted that not every vertex of the cycle S is necessarily an apex of the polygon S^*.) Then S^* can be extended to a convex drawing of G if and only if the following three conditions hold.

1. For each vertex v of $G - V(S)$ having degree at least three in G, there are three paths disjoint except at v, each joining v and a vertex of S;
2. The graph $G - V(S)$ has no connected component C such that all the vertices on S adjacent to vertices in C lie on a single path P_i; and no two vertices in each P_i are joined by an edge not in S; and
3. Any cycle of G which has no edge in common with S has at least three vertices of degree at least 3 in G.

The basic idea of the algorithm of Chiba et al. [16] is to reduce the convex drawing of G to those of several subgraphs of G as follows:

> *Algorithm* ConvexDraw(G, S, S^*)
> 1. Delete from G an arbitrary apex v of S^* together with edges incident to v.
> 2. Divide the resulting graph $G' = G - v$ into the biconnected components $B_1, B_2, \ldots, B_p$.
> 3. Determine a convex polygon S_i^* for the outside facial cycle S_i of each B_i so that B_i with S_i^* satisfies the conditions in Theorem 47.2.
> 4. Recursively apply the algorithm to each B_i with S_i^* to determine the positions of vertices not in S_i.

The main idea of the algorithm is illustrated in Figure 47.6; for more details see Reference 16.
It is easy to show that the algorithm runs in linear time, and its correctness can be established using Theorem 47.2.

Theorem 47.3: (Chiba, Yamanouchi and Nishizeki [16])

Algorithm ConvexDraw constructs a convex planar drawing of a biconnected plane graph with given outside face in linear time, if such a drawing is possible.

47.3.3 Algorithm Using Canonical Ordering

Both the Tutte algorithm and the algorithm of Chiba et al. give poor resolution. Kant [18] gives a linear time algorithm for constructing a planar convex grid drawing of a triconnected planar graph on a $(2n - 4) \times (n - 2)$ grid, where n is the number of vertices. This guarantees that the vertex resolution of the drawing is $\Omega(1/n)$.

Kant's algorithm is based on a method of de Fraysseix, Pach and Pollack [19] for drawing planar *triangulated* graphs on a grid of size $(2n - 4) \times (n - 2)$.

The algorithm of de Fraysseix et al. [19] computes a special ordering of vertices called the *canonical ordering* based on fixed embedding. Then the vertices are added, one by one, in order to construct a straight-line drawing combined with shifting method. Note that the outside face of the drawing is always triangle, as shown in Figure 47.7.

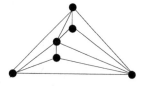

FIGURE 47.7   Planar straight-line grid drawing given by de Fraysseix, Pach and Pollack method.

Kant [18] uses a generalization of the canonical ordering called *leftmost canonical ordering*. The main difference from the algorithm of de Fraysseix et al. [19] is that it gives a convex drawing of *triconnected* planar graph. The following Theorem summarizes his main result.

Theorem 47.4: (Kant [18])

There is a linear time algorithm to construct a planar straight-line convex drawing of a triconnected planar graph with n vertices on a $(2n - 4) \times (n - 2)$ grid.

Details of Kant's algorithm are in Reference [18]. Chrobak and Kant improved the algorithm so that the size of the grid is $(n - 2) \times (n - 2)$ [20].

47.4 Symmetric Drawing

A graph drawing can have two kinds of symmetry: *axial* (or reflectional) symmetry and *rotational* symmetry. For example, see Figure 47.8. The drawing in Figure 47.8a displays four rotational symmetries and Figure 47.8b displays one axial symmetry. Figure 47.8c displays eight symmetries, four rotational symmetries as well as four axial symmetries.

Manning [21] showed that, in general, the problem of determining whether a given graph can be drawn symmetrically is NP-complete. De Fraysseix [22] presents heuristics for symmetric drawings of general graphs. Exact algorithms for finding symmetries in general graphs were presented by Buchheim and Junger [23], and Abelson, Hong and Taylor [24].

There are linear time algorithms available for restricted classes of graphs. Manning and Atallah present a linear time algorithm for detecting symmetries in trees [25] and outerplanar graphs [26]. Hong, Eades and Lee present a linear time algorithm for drawing series-parallel digraphs symmetrically [27]. A linear time algorithm for drawing planar graphs with maximum number of symmetries are given by Hong, McKay and Eades [28] for triconnected planar graphs, by Hong and Eades for biconnected planar graphs [29], and oneconnected planar graphs [30], and disconnected planar graphs [31]. Recently, a linear time algorithm for symmetric convex drawings of internally-triconnected planar graphs are given by Hong and Nagamochi [32]. For a recent survey on symmetric graph drawing, see Reference 33.

In this section, we briefly describe a linear time algorithm to draw triconnected planar graphs with maximum symmetry. Note that the algorithm of Tutte described in Section 47.3.1 can be used to construct symmetric drawings of triconnected planar graphs, but it does not work in linear time. Here we give the linear time algorithm of Hong, McKay and Eades [28]. It has two steps:

1. *Symmetry finding*: this step finds symmetries, or so-called *geometric automorphisms* [24,34], in graphs.
2. *Symmetric drawing*: this step constructs a drawing that displays these automorphisms.

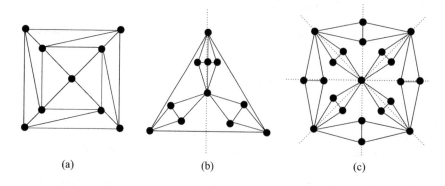

(a)   (b)   (c)

FIGURE 47.8   Three symmetric drawings of graphs.

More specifically, the symmetry finding step finds a plane embedding that displays the maximum number of symmetries. It is based on a classical theorem from group theory called the orbit-stabilizer theorem [35]. It also uses a linear time algorithm of an algorithm of Fontet [36] to compute orbits and generators for groups of geometric automorphisms. For details, see Reference 28. Here we present only the second step, that is, the drawing algorithm. This constructs a straight-line planar drawing of a given triconnected plane graph such that given symmetries are displayed.

The drawing algorithm has three routines, each corresponding to a type of group of symmetries in two dimensions:

1. The *cyclic* case: displaying k rotational symmetries;
2. One *axial* case: displaying one axial symmetry;
3. The *dihedral* case: displaying $2k$ symmetries (k rotational symmetries and k axial symmetries).

The dihedral case is important for displaying *maximum* number of symmetries, however it is the most difficult to achieve. Here, we concentrate on the first two cases; the cyclic case and the one axial case. In Section 47.4.1, we describe an algorithm for constructing a drawing displaying rotational symmetry and in Section 47.4.2 we describe an algorithm for constructing a drawing displaying axial symmetry. We briefly comments on the dihedral case in Section 47.4.3.

The input of the drawing algorithm is a triconnected plane graph; note that the outside face is given. Further, the algorithm needs to compute a triangulation as a preprocessing step. More specifically, given a plane embedding of triconnected planar graphs with a given outside face, we triangulate each internal face by inserting a new vertex in the face and joining it to each vertex of the face. This process is called star-triangulation and clearly it takes only linear time.

A major characteristic of symmetric drawings is the repetition of congruent drawings of isomorphic subgraphs. The algorithms use this property, that is, they compute a drawing for a subgraph and then use copies of it to draw the whole graph. More specifically, each symmetric drawing algorithm consists of three steps. First, it finds a subgraph. Then it draws the subgraph using the algorithm of Chiba et al. [16] as a subroutine. The last step is to replicate the drawing; that is, merge copies of the subgraphs to construct a drawing of the whole graph.

The application of Theorem 47.2 is to subgraphs of a triconnected plane graph defined by a cycle and its interior. In that case Thomassen's conditions can be simplified. Suppose that P is a path in a graph G; a *chord* for P in G is an edge (u, v) of G not in P, but whose endpoints are in P.

Corollary 47.1

Let G be a triconnected plane graph and let S be a cycle of G. Let W be the graph consisting of S and its interior. Let S^* be a drawing of S as a convex k-gon (where S might have more than k vertices). Let $P_1, P_2, \ldots, P_k$ be the paths in S corresponding to the sides of S^*. Then S^* is extendable to a straight-line planar drawing of W if and only if no path P_i has a chord in W.

Proof. We can assume that the interior faces of W are triangles, since otherwise we can star triangulate them.

We need to show that Thomassen's three conditions are met. Condition 1 is a standard implication of the triconnectivity of G, and Condition 3 follows just from the observation that the internal vertices have degree at least 3.

To prove the first part of Condition 2, suppose on the contrary that C is a connected component of $W - S$ which is adjacent to P_i but not to any other of $P_1, \ldots, P_k$. Let u and v be the first and last vertices on P_i which are adjacent to C. If $u = v$, or u and v are adjacent on P_i, then $\{u, v\}$ is a cut (separation pair) in G, which is impossible as G is triconnected. Otherwise, there is an interior face containing u and v and so u and v are adjacent contrary to our hypothesis (since the internal faces are triangles).

The second part of Condition 2 is just the condition that we are imposing. ∎

47.4.1 Displaying Rotational Symmetry

Firstly, we consider the cyclic case, that is, displaying k rotational symmetries. Note that after the star-triangulation, we may assume that there is either an edge or a vertex fixed by the symmetry. The fixed edge case can only occur if $k = 2$ and this case can be transformed into the fixed vertex case by inserting a dummy vertex into the fixed edge with two dummy edges. Thus we assume that there is a fixed vertex c.

The symmetric drawing algorithm for the cyclic case consists of three steps.

Algorithm `Cyclic`
1. `Find_Wedge_Cyclic`.
2. `Draw_Wedge_Cyclic`.
3. `Merge_Wedges_Cyclic`.

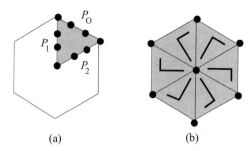

FIGURE 47.9 Example of (a) a wedge and (b) merging.

The first step is to find a subgraph W, called a *wedge*, as follows.

> *Algorithm* `Find_Wedge_Cyclic`
> 1. Find the fixed vertex c.
> 2. Find a shortest path P_1, from c to a vertex v_1 on the outside face. This can be done in linear time by breadth first search.
> 3. Find the path P_2 which is a mapping of P_1 under the rotation.
> 4. Find the induced subgraph of G enclosed by the cycle formed from P_1, P_2 and a path P_0 along the outside face from v_1 to v_2 (including the cycle). This is the wedge W.

A wedge W is illustrated in Figure 47.9a. It is clear that Algorithm `Find_Wedge_Cyclic` runs in linear time.

The second step, `Draw_Wedge_Cyclic`, constructs a drawing D of the wedge W using Algorithm `ConvexDraw`, such that P_1 and P_2 are drawn as straight-lines. This is possible by the next lemma.

Lemma 47.1

Suppose that W is the wedge computed by Algorithm `Find_Wedge_Cyclic`, S is the outside face of W, and S^* is the drawing of the outside face S of W using three straight-lines as in Figure 47.9a. Then S^* is extendable to a planar straight-line drawing of W.

Proof. One of the three sides of S^* corresponds to part of the outside face of G, which cannot have a chord as G is triconnected. The other two sides cannot have chords either, as they are shortest paths in G. Thus W and S^* satisfy the conditions of Corollary 47.1.

∎

The last step, `Merge_Wedges_Cyclic`, constructs a drawing of the whole graph G by replicating the drawing D of W, k times. Note that this merge step relies on the fact that P_1 and P_2 are drawn as straight-lines. This is illustrated in Figure 47.9b.

Clearly, each of these three steps takes linear time. Thus the main result of this section can be stated as follows.

Theorem 47.5

Algorithm `Cyclic` constructs a straight-line drawing of a triconnected plane graph which shows k rotational symmetry in linear time.

47.4.2 Displaying Axial Symmetry

A critical element of the algorithm for displaying one axial symmetry is the subgraph that is fixed by the axial symmetry. Consider a drawing of a star-triangulated planar graph with one axial symmetry. There are fixed vertices, edges and/or fixed faces on the axis. The subgraph formed by these vertices, edges and faces, is called a *fixed string of diamonds*.

The first step of the algorithm is to identify the fixed string of diamonds. Then the second step is to use Algorithm `Symmetric_ConvexDraw`, a modified version of Algorithm `ConvexDraw`. Thus Algorithm `One_Axial` can be described as follows.

> *Algorithm* `One_Axial`
> 1. Find a fixed string of diamonds. Suppose that $\omega_1, \omega_2, \ldots, \omega_k$ are the fixed edges and vertices in the fixed string of diamonds, in order from the outside face (ω_1 is on the outside face). For each ℓ, ω_ℓ may be a vertex or an edge. This is illustrated in Figure 47.10.
> 2. Choose an axially symmetric convex polygon S^* for the outside face of G.
> 3. `Symmetric_ConvexDraw(1, `S^*`, `G`)`.

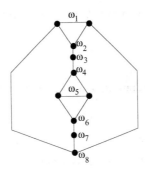

FIGURE 47.10 A string of diamonds.

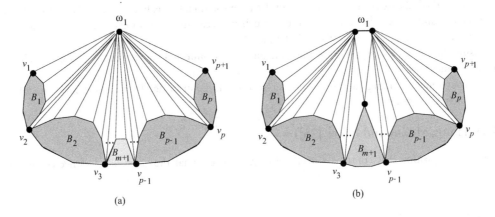

FIGURE 47.11 Symmetric version of ConvexDraw.

The main subroutine in Algorithm `One_Axial` is Algorithm `Symmetric_ConvexDraw`. This algorithm, described below, modifies Algorithm `ConvexDraw` so that the following three conditions are satisfied.

1. Choose v in Step 1 of Algorithm `ConvexDraw` to be ω_1. In Algorithm `ConvexDraw`, v must be a vertex; here we extend Algorithm `ConvexDraw` to deal with an edge or a vertex. The two cases are illustrated in Figure 47.11.
2. Let $D(B_i)$ be the drawing of B_i and α be the axial symmetry. Then, $D(B_i)$ should be a reflection of $D(B_j)$, where $B_j = \alpha(B_i)$, $i = 1, 2, \ldots, m$ and $m = \lfloor p/2 \rfloor$.
3. If p is odd, then $D(B_{m+1})$ should display one axial symmetry.

It is easy to see that satisfying these three conditions ensures the display a single axial symmetry.

Let $B_j = \alpha(B_i)$. The second condition can be achieved as follows. First define S_j^* to be the reflection of S_i^*, $i = 1, 2, \ldots, m$. Then apply Algorithm `ConvexDraw` for B_i, $i = 1, 2, \ldots, m$ and construct $D(B_j)$ using a reflection of $D(B_i)$. If p is odd then we recursively apply Algorithm `Symmetric_ConvexDraw` to B_{m+1}. Thus Algorithm `Symmetric_ConvexDraw` can be described as follows.

> *Algorithm* `Symmetric_ConvexDraw`(ℓ, S^*, G)
> 1. Delete ω_ℓ from G together with edges incident to ω_ℓ. Divide the resulting graph $G' = G - \omega_\ell$ into the blocks $B_1, B_2, \ldots, B_p$, ordered anticlockwise around the outside face. Let $m = \lfloor p/2 \rfloor$.
> 2. For each $i = 1$ to m, determine a convex polygon S_i^* of the outside facial cycle S_i of each B_i so that B_i with S_i^* satisfy the conditions in Theorem 47.2 and chose S_{p-i+1}^* to be a reflection of S_i^*.
> 3. For each $i = 1$ to m,
>   a. Construct a drawing $D(B_i)$ of B_i using Algorithm `ConvexDraw`.
>   b. Construct $D(B_{p-i+1})$ as a reflection of $D(B_i)$.
> 4. If p is odd, then construct a drawing $D(B_{m+1})$ using `Symmetric_ConvexDraw`$(\ell + 1, S_{m+1}^*, B_{m+1})$.
> 5. Merge the $D(B_i)$ to form a drawing of G.

Using the same argument as for the linearity of Algorithm `ConvexDraw` [16], one can show that Algorithm `One_Axial` takes linear time.

Theorem 47.6

Algorithm `One_Axial` constructs a straight-line drawing of a triconnected plane graph which shows one axial symmetry in linear time.

47.4.3 Displaying Dihedral Symmetry

In this section, we briefly review an algorithm for constructing a drawing displaying k axial symmetries and k rotational symmetries.

As with the cyclic case, we assume that there is a vertex fixed by all the symmetries. The algorithm adopts the same general strategy as for the cyclic case: divide the graph into "wedges," draw each wedge, then merge the wedges together to make a symmetric drawing. However, the dihedral case is more difficult than the pure rotational case, because an axial symmetry in the dihedral case can have fixed edges and/or fixed faces. This requires a much more careful applications of the Algorithm `ConvexDraw` and Algorithm `Symmetric_ConvexDraw`; for details see Reference 28.

Nevertheless, using three steps above, a straight-line drawing of a triconnected plane graph which shows dihedral symmetry can be constructed in linear time.

In conclusion, the symmetry finding algorithm together with the three drawing algorithms ensures the following theorem which summarizes their main result.

Theorem 47.7: (Hong, McKay and Eades [28])

There is a linear time algorithm that constructs maximally symmetric planar drawings of triconnected planar graphs, with straight-line edges.

47.5 Visibility Drawing

In general, a *visibility representation* of a graph has a geometric shape for each vertex, and a "line of sight" for each edge. Different visibility representations involve different kinds of geometric shapes and different restrictions on the "line of sight." A three dimensional visibility representation of the complete graph on five vertices is in Figure 47.12. Here the geometric objects are rectangular prisms, and each "line of sight" is parallel to a coordinate axis.

Visibility representations have been extensively investigated in both two and three dimensions.

For two dimensional graph drawing, the simplest and most common visibility representation uses horizontal line segments (called "bars") for vertices and vertical line segments for "lines of sight." More precisely, a *bar visibility representation* of a graph $G = (V, E)$ consists of a horizontal line segment ω_u for each vertex $u \in V$, and a vertical line segment λ_e for each edge $e \in E$, such that the following properties hold:

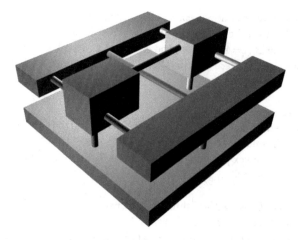

FIGURE 47.12   A visibility representation in three dimensions (courtesy of Nathalie Henry).

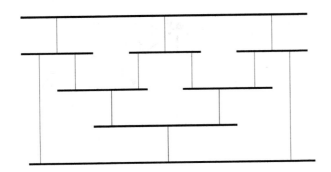

FIGURE 47.13 A bar visibility representation of the 3-cube.

- If u and u' are distinct vertices of G, then ω_u has empty intersection with $\omega_{u'}$.
- If e and e' are distinct edges of G, then λ_e has empty intersection with $\lambda_{e'}$.
- If e is not incident with u then λ_e has empty intersection with ω_u.
- If e is incident with u then the intersection of λ_e and ω_u consists of an endpoint of λ_e on ω_u.

Figure 47.13 shows a bar visibility representation of the 3-cube.

It is clear that if G admits a bar visibility representation, then G is planar. Tamassia and Tollis [37] and Rosenstiehl and Tarjan [38] proved the converse: every planar graph has a bar visibility representation. Further, they showed that the resulting drawing has quadratic area, that is, the resolution is good. The algorithms from [37] and [38] construct such a representation in linear time.

In the Section 47.5.1 below we describe some concepts used to present the algorithm. Section 47.5.2 describes the basic algorithm for constructing bar visibility representations, and Sections 47.5.3 and 47.5.4 show how to apply these representations to obtain layered drawings and orthogonal drawings respectively.

The algorithm in Section 47.5.2 and some of the preceding results are stated in terms of biconnected planar graphs. Note that this restriction can be avoided by augmenting graphs of lower connectivity with dummy edges.

47.5.1 Planar St-graphs

Bar visibility representation algorithms use the theory of planar st-graphs. This is a powerful theory which is useful in a variety of graph drawing applications. In this section we describe enough of this theory to allow presentation of a bar visibility algorithm; for proofs of the results here, and more details about planar st-graphs, see Reference 7.

A directed plane graph is a *planar st-graph* if it has one source s and one sink t, and both s and t are on the outside face. We say that a vertex u on a face f of a planar st-graph is a source for f (respectively *sink for f*) if the edges incident with u on f are out-edges (respectively in-edges).

Lemma 47.2

Every face f of a planar st-graph has one source s_f for f and one sink t_f for f.

We need to extend the concept of the "planar dual," defined in Section 47.2, to directed graphs. For a directed plane graph G, the *directed dual* G^* has a vertex v_f for each internal face f of G, as well as two vertices ℓ_{ext} and r_{ext} for the external face of G.

To define the edges of G^*, we must first define the *left* and *right* side face of each edge of G. Suppose that f is an internal face of G, and the cycle of edges traversing f in a clockwise direction is $(e_1, e_2, \ldots, e_k)$. The clockwise traversal may pass through some edges in the same direction as the edge, and some edges in the opposite direction. If we pass through e_i in the same direction as e_i then we define $right(e_i) = v_f$; if we pass through e_j in the opposite direction to e_j then we define $left(e_j) = v_f$.

Now consider a clockwise traversal $(e'_1, e'_2, \ldots, e'_k)$ of the external face of G. If we pass through e'_i in the same direction as e'_i then we define $left(e_i) = \ell_{ext}$; if we pass through e_j in the opposite direction to e_j then we define $right(e_j) = r_{ext}$.

One can easily show that the definitions of *left* and *right* are well founded, that is, that for each edge e there is precisely one face f of G such that $left(e) = v_f$ and precisely one one face f' of G such that $right(e) = v_{f'}$.

For each edge e of G, G^* has an edge from $left(e)$ to $right(e)$.

An example is in Figure 47.14. Here $\ell_{ext} = 1 = left(e)$, and $r_{ext} = 5 = right(j) = right(g)$, $left(f) = left(i) = 2$, $right(f) = right(h) = 4$, and $left(h) = right(i) = 3$.

The following Lemma (see Reference 7 for a proof) is important for the bar visibility algorithm.

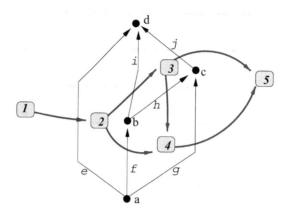

FIGURE 47.14   A directed planar graph and its directed dual.

Lemma 47.3

Suppose that f and f' are faces in a planar st-graph G. Then precisely one of the following statements is true.

    a. There is a directed path in G from the sink of f to the source of f'.
    b. There is a directed path in G from the sink of f' to the source of f.
    c. There is a directed path in G^* from v_f to $v_{f'}$.
    d. There is a directed path in G^* from $v_{f'}$ to v_f.

The next Lemma is, in a sense, dual to Lemma 47.2.

Lemma 47.4

Suppose that u is a vertex in a planar st-graph. The outgoing edges from u are consecutive in the circular ordering of $N(u)$, and the incoming edges are consecutive in the circular ordering of $N(u)$.

Lemma 47.4 leads to a definition of the *left* and *right* of a vertex. Suppose that $(e_1, e_2, \ldots, e_k)$ is the circular list of edges around a non-source non-sink vertex u, in clockwise order. From Lemma 47.4, for some i, edge e_i comes into u and edge e_{i+1} goes out from u. we say that e_i is the *leftmost in-edge* of u and e_{i+1} is the *leftmost out-edge* of u. If f is the face shared by e_i and e_{i+1} then we define $left(u) = f$. Similarly, there is a j such that edge e_j goes out from u and edge e_{j+1} comes into u; we say that e_j is the rightmost out-edge of u and e_{j+1} is the *rightmost in-edge* of u. If f' is shared by e_j and e_{j+1} then we define $right(e_j) = f'$. If u is either the source or the sink, then $left(u) = \ell_{ext}$ and $right(u) = r_{ext}$.

In Figure 47.14, for example, $left(a) = left(d) = 1$, $left(b) = 2$, $left(c) = 3$, $right(b) = 4$, and $right(a) = right(c) = right(d) = 5$.

Finally, we need the following Lemma, which is a kind of dual to Lemma 47.3.

Lemma 47.5

Suppose that u and u' are vertices in G. Then precisely one of the following statements is true.

    a. There is a directed path in G from u to u'.
    b. There is a directed path in G from u' to u.
    c. There is a directed path in G^* from $right(u)$ to $left(u')$.
    d. There is a directed path in G^* from $right(u')$ to $left(u)$.

47.5.2 The Bar Visibility Algorithm

The bar visibility algorithm takes a biconnected plane graph as input, and converts it to a planar st-graph. Then it computes the directed dual, and a "topological number" (defined below) for each vertex in both the original graph and the dual. Using these numbers, it assigns a y coordinate for each vertex and an x coordinate for each edge. The bar representing the vertex extends far enough to touch each incident edge, and the vertical line representing the edge extends far enough to touch each the bars representing its endpoints.

If $G = (V, E)$ is an acyclic directed graph with n vertices, a *topological numbering* Z of G assigns an integer $Z(u) \in \{0, 1, \ldots, n-1\}$ to every vertex $u \in V$ such that $Z(u) < Z(v)$ for all directed edges $(u, v) \in E$. Note that we do not require that the function Z be one-one, that is, it is possible that two vertices are assigned the same number.

The algorithm is described below.

> *Algorithm* `Bar_Visibility`
> *Input: a biconnected plane graph $G = (V, E)$*
> *Output: a bar visibility representation of G*

1. Choose two vertices s and t of G on the same face.
2. Direct the edges of G so that s is the only source, t is the only sink, and the resulting digraph is acyclic.
3. Compute a topological numbering Y for G.
4. Compute the directed planar dual $G^* = (V^*, E^*)$ of G.
5. Compute a topological numbering X for G^*.
6. For each vertex u, let ω_u be the line segment $\big[(X(left(u)), Y(u)), (X(right(u)) - 1, Y(u))\big]$.
7. For each edge $e = (u, v)$, let λ_e be the line segment $\big[(X(left(e)), Y(u)), (X(left(e)), Y(v))\big]$.

Step 2 can be implemented by a simple variation on the depth-first-search method, based on a biconnectivity algorithm.

There are many kinds of topological numberings for acyclic digraphs: for example, one can define $Z(u)$ to be the number of edges in the longest path from the source s to u. Any of these methods can be used in steps 3 and 5.

Theorem 47.8: (Tamassia and Tollis [37]; Rosenstiehl and Tarjan [38])

A visibility representation of a biconnected planar graph with area $O(n) \times O(n)$ can be computed in linear time.

Proof. We need to show that the drawing defined by ω and λ in Algorithm `Bar_Visibility` satisfies the four properties P1, P2, P3 and P4 above.

First consider P1, and suppose that u and u' are two vertices of G. If $Y(u) \neq Y(u')$ then ω_u has empty intersection with $\omega_{u'}$ and so P1 holds. Now suppose that $Y(u) = Y(u')$. Since Y is a topological numbering of G, it follows that there is no directed path in between u and u' (in either direction). Thus, from Lemma 47.5, in G^* either there is a directed path from $right(u)$ to $left(u')$ or a directed path from $right(u')$ to $left(u)$. The first case implies that $X(right(u)) < X(left(u'))$, so that the whole of ω_u is to the left of $\omega_{u'}$; the second case implies that the whole of $\omega_{u'}$ is to the left of ω_u. This implies P1.

Now consider P2, and suppose that $e = (u, v)$ and $e' = (u', v')$ are edges of G, and denote $left(e)$ by f and $left(e')$ by f'. If $X(f) \neq X(f')$ then λ_e cannot intersect $\lambda_{e'}$ and P2 holds. Now suppose that $X(f) = X(f')$; thus in G^* there is no directed path between f and f'. It follows from Lemma 47.3 that in G either there is a directed path from the sink t_f of f to the source $s_{f'}$ of f', or a directed path from the sink $t_{f'}$ of f' to the source s_f of f. In the first case we have $Y(t_f) < Y(s_{f'})$. Also, since e is on f and e' is on f', we have $Y(v) \leq Y(t_f)$ and $Y(s_{f'}) \leq Y(u')$; thus $Y(v) < Y(u')$ and λ_e cannot intersect $\lambda_{e'}$. The second case is similar and so P2 holds.

A similar argument shows that P3 holds.

Property P4 follows form the simple observation that for any edge $e = (u, v)$, $X(left(u)) \leq X(left(e)) \leq X(right(u))$.

The drawing has dimensions $\max_{v_f \in V^*} X(v_f) \times \max_{u \in V} Y(u)$, which is $O(n) \times O(n)$.

It is easy to see that each step can be implemented in linear time. ∎

47.5.3 Bar Visibility Representations and Layered Drawings

A graph in which the nodes are constrained to specified y coordinates, as in Figure 47.3, is called a layered graph. More precisely, a layered graph consists of a directed graph $G = (V, E)$ as well as a topological numbering L of G. We say that $L(u)$ is the *layer* of u. A drawing of G that satisfies the layering constraint, that is, the y coordinate of u is $L(u)$ for each $u \in V$, is called a *layered drawing*. Drawing algorithms for layered graphs have been extensively explored [7].

A layered graph is *planar* (sometimes called *h-planar*) if it can be drawn with no edge crossings, subject to the layering constraint. Note that underlying the prerequisite Figure 47.3 is a planar graph; however, as a layered graph, it is not planar. The theory of planarity for layered graphs has received some attention; see References 39–42.

If a planar layered graph has one source and one sink, then it is a planar st-graph. Clearly the source s has $L(s) = \min_{u \in V} L_u$ and the sink t has $L(t) = \max_{v \in V} L_v$. Since the layers define a topological numbering of the graph, application of Algorithm `Bar_Visibility` yields a visibility representation that satisfies the layering constraints. Further, this can be used to construct a graph drawing by using simple local transformations, illustrated in Figure 47.15.

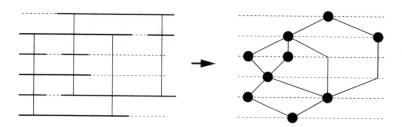

FIGURE 47.15   Transformation from a visibility representation of a layered graph to a layered drawing.

The following theorem is immediate.

Theorem 47.9

If G is a planar layered graph then we can construct a planar layered drawing of G in linear time, with the following properties:

- There are at most two bends in each edge.
- The area is $O(n) \times O(n)$.

Although Algorithm `Bar_Visibility` produces the visibility representation with good resolution, it may not produce minimum area drawings. Lin and Eades [42] give a linear time variation on Algorithm *Bar Visibility* that, for a fixed embedding, maximizes resolution. The algorithm works by using a topological numbering for the dual computed from a dependency relation between the edges.

47.5.4   Bar Visibility Representations for Orthogonal Drawings

An *orthogonal drawing* of a graph G has each vertex represented as a point and each edge represented by a polyline whose line segments are parallel to a coordinate axis. An orthogonal drawing of the 3-cube is in Figure 47.16.

It is clear that a graph that has a planar orthogonal drawing, has maximum degree of G at most 4. Conversely, one can obtain a planar orthogonal drawing of a planar graph with maximum degree at most 4 from a visibility representation by using a simple set of local transformations, described in Figure 47.17.

Figure 47.16 can be obtained from Figure 47.13 by applying transformation (c) in Figure 47.17 to the source and sink, and applying transformation (d) to each other vertex.

Note that the transformations involve the introduction of a few bends in each edge. The worst case is for where a transformation introduces two bends near a vertex; this may occur at each end of an edge, and thus may introduce 4 bends into an edge.

Theorem 47.10

If G is a planar graph then we can construct a planar orthogonal drawing of G in linear time, with the following properties:

- There are at most four bends in each edge.
- The area is $O(n) \times O(n)$.

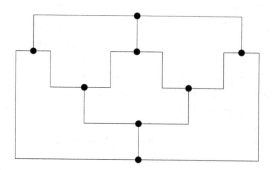

FIGURE 47.16   An orthogonal drawing of the 3-cube.

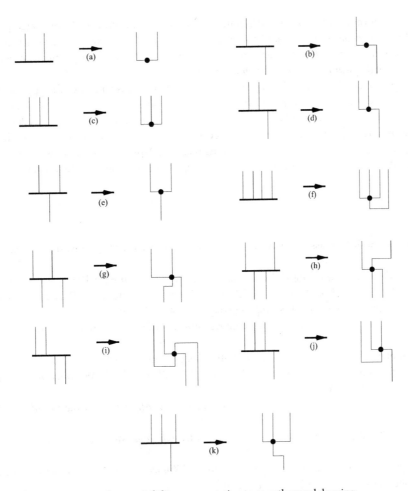

FIGURE 47.17 Local transformations to transform a visibility representation to an orthogonal drawing.

In fact, a variation of Algorithm `Visibility` together with a careful application of some transformations along the lines of those in Figure 47.17 can ensure that the resulting drawing has a total of at most $2n + 4$ bends, where n is the number of vertices; see Reference 7.

47.6 Conclusion

In this chapter we have described just a small sample of graph drawing algorithms. Notable omissions include:

- *Force directed methods*: A graph can be used to define a system of forces. For example, we can define a Hooke's law spring force between two adjacent vertices, and magnetic repulsion between nonadjacent vertices. A minimum energy configuration of the graph can lead to a good drawing. For methods using the force-directed paradigm, see References 7–10, 43.
- *Clustered graph drawing*: in practice, to handle large graphs, one needs to form clusters of the vertices to form "supervertices." Drawing graphs in which some vertices represent graphs is a challenging problem. For algorithms drawing clustered graphs, see References 8,40.
- *Three dimensional graph drawing*: the widespread availability of cheap three dimensional graphics systems has lead to the investigation of graph drawing in three dimensions. For algorithms drawing graphs in three dimensions, see References 8,10, 44.
- *Crossing minimization methods*: in this chapter we have concentrated on planar graphs. In practice, we need to deal with non-planar graphs, by choosing a drawing with a small number of crossings. For algorithms drawing non-planar graphs with crossing minimization methods, see References 7–10, 45,46.
- *Generalization of convex drawing*: Note that graphs admit convex drawings if and only if they are internally-triconnected [16,17]. In practice, we need to relax and generalize convex drawings to "good" non-convex drawings for biconnected graphs. For algorithms drawing planar graphs with non-convex boundary constraints (such as star-shaped

boundary), called *inner convex drawings*, see Reference 47. For algorithms drawing biconnected planar graphs with star-shaped faces, called *star-shaped drawings*, see Reference 48; algorithms for star-shape drawings with minimum number of concave corners, see References 49,50. For algorithms drawing biconnected planar graphs with non-convex polytopes in three dimensions, called *upward star-shaped polytopes*, see Reference 51.

- *Beyond planar graphs*: Recently, the problem of drawing non-planar graphs with specific types of crossing constraints or forbidden crossing constraints, called *beyond planar graphs* and *beyond planarity*, have been studied [52–57]. This include *k-planar graphs*, that is, graphs that can be drawn in the plane with at most k crossings per edge. For algorithms testing beyond planarity and drawing beyond planar graphs, see Reference 56 for drawing 1-planar graphs, see Reference 53 for drawing almost planar graphs, see Reference 54 for testing maximal-1-planar graphs, see Reference 55 for testing outer-1-planar graphs, see Reference 52 for testing maximal-outer-fanplanar graphs, and see Reference 57 for testing full-outer-2-planar graphs.

References

1. C. Batini, E. Nardelli and R. Tamassia, A Layout Algorithm for Data-Flow Diagrams, *IEEE Trans. Softw. Eng.*, SE-12(4), 538–546, 1986.
2. J. Martin and C. McClure. *Diagramming Techniques for Analyst and Programmers*. Prentice Hall Inc., Michigan, 1985.
3. H. Purchase, Which Aesthetic Has the Greatest Effect on Human Understanding, *Proceedings of Graph Drawing 1997*, Lecture Notes in Computer Science 1353, Springer-Verlag, pp. 248–261, 1998.
4. K. Wagner, Bemerkungen zum Vierfarbenproblem, *Jahres-bericht der Deutschen Mathematiker-Vereinigung*, 46, pp. 26–32, 1936.
5. D. E. Knuth, Computer Drawn Flowcharts, *Commun, ACM*, 6, pp. 555–563, 1963.
6. R. Read, Some Applications of Computers in Graph Theory, In L.W. Beineke, R. Wilson, *Selected Topics in Graph Theory*, chapter 15, pp. 415–444. Academic Press, 1978.
7. G. Di Battista, P. Eades, R. Tamassia and I. G. Tollis. *Graph Drawing: Algorithms for the Visualization of Graphs*. Prentice-Hall, 1998.
8. M. Kaufmann and D. Wagner Eds., *Drawing Graphs - Methods and Models*, Lecture Notes in Computer Science 2025, 2001.
9. K. Sugiyama. *Graph Drawing and Applications for Software and Knowledge Engineers*. Series on Software Engineering and Knowledge Engineering - Vol. 11, World Scientific, 2002.
10. R. Tamassia. *Handbook on Graph Drawing and Visualization*. Chapman and Hall, CRC, 2013.
11. M. Junger and P. Mutzel Eds., *Graph Drawing Software*, Series: Mathematics and Visualization, Springer-Verlag, 2003.
12. J. A. Bondy and U. S. R. Murty, *Graph Theory with Applications*, MacMillan Press, London, 1977.
13. W. T. Tutte, Convex Representations of Graphs, *Proc. London Math. Soc.*, 10, pp. 304–320, 1960.
14. W. T. Tutte, How to Draw a Graph, *Proc. London Math. Soc.*, 13, pp. 743–768, 1963.
15. R. J. Lipton, D. J. Rose and R. E. Tarjan, Generalized Nested Dissection, *SIAM J. Numer. Anal.*, 16(2), pp. 346–358, 1979.
16. N. Chiba, T. Yamanouchi and T. Nishizeki, Linear Algorithms for Convex Drawings of Planar Graphs, In J.A. Bondy and U.S.R. Murty, *Progress in Graph Theory*, Academic Press, pp. 153–173, 1984.
17. C. Thomassen, Planarity and Duality of Finite and Infinite Graphs, *J. of Combinatorial Theory,*, Series B , 29, pp. 244–271, 1980.
18. G. Kant, Drawing Planar Graphs Using the Canonical Ordering, *Algorithmica*, 16, pp. 4–32, 1996.
19. H. de Fraysseix, J. Pach and R. Pollack, How to Draw a Planar Graph on a Grid, *Combinatorica*, 10, pp. 41–51, 1990.
20. M. Chrobak and G. Kant, Convex Grid Drawings of 3-connected Planar Graphs, *Int. Journal of Computational Geometry and Applications*, 7(3), 211–224, 1997.
21. J. Manning, Geometric Symmetry in Graphs, *Ph.D. Thesis, Purdue Univ.,*, 1990.
22. H. de Fraysseix, An Heuristic for Graph Symmetry Detection, *Proceedings of Graph Drawing 1999*, Lecture Notes in Computer Science 1731, Springer-Verlag, pp. 276–285, Springer-Verlag, 1999.
23. C. Buchheim and M. Jünger, Detecting Symmetries by Branch and Cut, *Math. Program.*, 98(1–3), 369–384, 2003.
24. D. Abelson, S. Hong and D. E. Taylor, Geometric Automorphism Groups of Graphs, *Discrete Applied Mathematics*, 155(17), 2211–2226, 2007.
25. J. Manning and M. J. Atallah, Fast Detection and Display of Symmetry in Trees, *Congressus Numerantium*, 64, pp. 159–169, 1988.
26. J. Manning and M. J. Atallah, Fast Detection and Display of Symmetry in Outerplanar Graphs, *Discrete Applied Mathematics*, 39, pp. 13–35, 1992.

27. S. Hong, P. Eades and S. Lee, Drawing Series Parallel Digraphs Symmetrically, *Computational Geometry: Theory and Applicatons*, 17(3-4), pp. 165–188, 2000.

28. S. Hong, B. D. McKay and P. Eades, A Linear Time Algorithm for Constructing Maximally Symmetric Straight Line Drawings of Triconnected Planar Graphs, *Discrete and Computational Geometry*, 36(2), pp. 283–311, 2006.

29. S. Hong and P. Eades, Drawing Planar Graphs Symmetrically II: Biconnected Planar Graphs, *Algorithmica*, 42(2), pp. 159–197, 2005.

30. S. Hong and P. Eades, Drawing Planar Graphs Symmetrically III: Oneconnected Planar Graphs, *Algorithmica*, 44(1), pp. 67–100, 2006.

31. S. Hong and P. Eades, Symmetric Layout of Disconnected Graphs, *Proceedings of ISAAC 2003*, LNCS, Kyoto, Japan, pp. 405–414, 2003.

32. S. Hong and H. Nagamochi, A Linear-Time Algorithm for Symmetric Convex Drawings of Internally Triconnected Plane Graphs, *Algorithmica*, 58(2), pp. 433–460, 2010.

33. P. Eades and S. Hong, Symmetric Graph Drawing, In R. Tamassia, *Handbook on Graph Drawing and Visualization*, pp. 87–113, 2013.

34. P. Eades and X. Lin, Spring Algorithms and Symmetry, *Theoretical Computer Science*, 240(2), pp. 379–405, 2000.

35. M. A. Armstrong, *Groups and Symmetry*, Springer-Verlag, 1988.

36. M. Fontet, Linear Algorithms for Testing Isomorphism of Planar Graphs, *Proceedings of Third Colloquium on Automata, Languages, and Programming*, pp. 411–423, 1976.

37. R. Tamassia and I. G. Tollis, A Unified Approach to Visibility Representations of Planar Graphs, *Discrete and Computational Geometry*, 1(4), pp. 321–341, 1986.

38. P. Rosenstiehl and R. E. Tarjan, Rectilinear Planar Layouts and Bipolar Orientations of Planar Graphs, *Discrete and Compututational Geometry*, 1(4), pp. 342–351, 1986.

39. P. Eades, X. Lin and R. Tamassia, An Algorithm for Drawing a Hierarchical Graph, *International Journal of Computational Geometry and Applications*, 6(2), pp. 145–155, 1996.

40. P. Eades, Q. Feng, X. Lin and H. Nagamochi, Straight-Line Drawing Algorithms for Hierarchical Graphs and Clustered Graphs, *Algorithmica*, 44(1), pp. 1–32, 2006.

41. M. Junger, S. Leipert and P. Mutzel, Level Planarity Testing in Linear Time, *Proceedings of Graph Drawing 1998*, Lecture Notes in Computer Science 1547, Stirin Castle, Czech Republic, Springer-Verlag, pp. 167–182, 1999.

42. X. Lin and P. Eades, Towards Area Requirements for Drawing Hierarchically Planar Graphs, *Theoretical Computer Science*, 292(3), pp. 679–695, 2003.

43. S. G. Kobourov, Force-Directed Drawing Algorithms, In R. Tamassia, *Handbook on Graph Drawing and Visualization*, pp. 383–408, 2013.

44. V. Dujmovic, Three-Dimensional Drawings, In R. Tamassia, *Handbook on Graph Drawing, Visualization*, pp. 455–488, 2013.

45. C. Buchheim, M. Chimani, C. Gutwenger, M. Jnger and P. Mutzel, Crossings and Planarization, In R. Tamassia, *Handbook on Graph Drawing and Visualization*, pp. 43–85, 2013.

46. P. Healy and N. S. Nikolov, Hierarchical Drawing Algorithms, In R. Tamassia, *Handbook on Graph Drawing and Visualization*, pp. 409–453, 2013.

47. S. Hong and H. Nagamochi, Convex Drawings of Graphs with Non-convex Boundary Constraints, *Discrete Applied Mathematics*, 156(12), pp. 2368–2380, 2008.

48. S. Hong and H. Nagamochi, An Algorithm for Constructing Star-shaped Drawings of Plane Graphs, *Compututational Geometry: Theory and Application*, 43(2), pp. 191–206, 2010.

49. S. Hong and H. Nagamochi, A Linear-Time Algorithm for Star-Shaped Drawings of Planar Graphs with the Minimum Number of Concave Corners, *Algorithmica*, 62(3-4), pp. 1122–1158, 2012.

50. S. Hong and H. Nagamochi, Minimum Cost Star-shaped Drawings of Plane Graphs with a Fixed Embedding and Concave Corner Constraints, *Theoretical Computer Science*, 445, pp. 36–51, 2012.

51. S. Hong and H. Nagamochi, Extending Steinitz's Theorem to Upward Star-Shaped Polyhedra and Spherical Polyhedra, *Algorithmica*, 61(4), pp. 1022–1076, 2011.

52. M. A. Bekos, S. Cornelsen, L. Grilli, S. Hong and M. Kaufmann, On the Recognition of Fan-Planar and Maximal Outer-Fan-Planar Graphs, *Proceedings of Graph Drawing 2014*, Wuezburg, Germany, pp. 198–209, 2014.

53. P. Eades, S. Hong, G. Liotta, N. Katoh and S. Poon, Straight-Line Drawability of a Planar Graph Plus an Edge, *Proceedings of WADS 2015*, Victoria, Canada, pp. 301–313, 2015.

54. P. Eades, S. Hong, N. Katoh, G. Liotta, P. Schweitzer and Y. Suzuki, A Linear Time Algorithm for Testing Maximal 1-planarity of Graphs with a Rotation System, *Theoretical Computer Science*, 513, pp. 65–76, 2013.

55. P. Eades, S. Hong, N. Katoh, G. Liotta, P. Schweitzer and Y. Suzuki, A Linear-Time Algorithm for Testing Outer-1-Planarity, *Algorithmica*, 72(4), pp. 1033–1054, 2015.

56. S. Hong, P. Eades, G. Liotta and S. Poon, Fry's Theorem for 1-Planar Graphs, *Proceedings of COCOON 2012*, LNCS, Sydney, Australia, pp. 335–346, 2012.

57. S. Hong and H. Nagamochi, Testing Full-outer-2-planarity in Linear Time, In E. Mayr, *Proceedings of WG 2015*, LNCS, Garching, Germany, pp. 406–421, 2015.

48

Concurrent Data Structures [*]

48.1 Designing Concurrent Data Structures... 741
    Performance • Blocking Techniques • Nonblocking Techniques • Complexity Measures •
    Correctness • Verification Techniques • Tools of the Trade

48.2 Shared Counters and Fetch-and-ϕ Structures 749
    Combining • Counting Networks

48.3 Stacks and Queues... 751
    Stacks • Queues • Deques

48.4 Pools 752

48.5 Linked Lists.. 753

48.6 Hash Tables 754

48.7 Search Trees 754

48.8 Priority Queues... 756
    Heap-Based Priority Queues • Tree-Based Priority Pools

48.9 Summary... 757

References... 757

Mark Moir[†]
Sun Microsystems Laboratories

Nir Shavit[†]
Sun Microsystems Laboratories

The proliferation of commercial shared-memory multiprocessor machines has brought about significant changes in the art of concurrent programming. Given current trends towards low-cost chip multithreading (CMT), such machines are bound to become ever more widespread.

Shared-memory multiprocessors are systems that concurrently execute multiple threads of computation which communicate and synchronize through data structures in shared memory. The efficiency of these data structures is crucial to performance, yet designing effective data structures for multiprocessor machines is an art currently mastered by few. By most accounts, concurrent data structures are far more difficult to design than sequential ones because threads executing concurrently may interleave their steps in many ways, each with a different and potentially unexpected outcome. This requires designers to modify the way they think about computation, to understand new design methodologies, and to adopt a new collection of programming tools. Furthermore, new challenges arise in designing *scalable* concurrent data structures that continue to perform well as machines that execute more and more concurrent threads become available. This chapter provides an overview of the challenges involved in designing concurrent data structures, and a summary of relevant work for some important data structure classes. Our summary is by no means comprehensive; instead, we have chosen popular data structures that illustrate key design issues, and hope that we have provided sufficient background and intuition to allow the interested reader to approach the literature we do not survey.

48.1 Designing Concurrent Data Structures

Several features of shared-memory multiprocessors make concurrent data structures significantly more difficult to design and to verify as correct than their sequential counterparts.

The primary source of this additional difficulty is concurrency: Because threads are executed concurrently on different processors, and are subject to operating system scheduling decisions, page faults, interrupts, etc., we must think of the computation as completely asynchronous, so that the steps of different threads can be interleaved arbitrarily. This significantly complicates the task of designing correct concurrent data structures.

---

[*] This chapter has been reprinted from first edition of this Handbook, without any content updates.
[†] We have used this author's affiliation from the first edition of this handbook, but note that this may have changed since then.

```
                                  acquire(Lock);
oldval = X;                       oldval = X;
X = oldval + 1;                   X = oldval + 1;
return oldval;                    release(Lock);
                                  return oldval;
```

FIGURE 48.1   Code fragments for sequential and lock-based fetch-and-inc operations.

Designing concurrent data structures for multiprocessor systems also provides numerous challenges with respect to performance and scalability. On today's machines, the layout of processors and memory, the layout of data in memory, the communication load on the various elements of the multiprocessor architecture all influence performance. Furthermore, the issues of correctness and performance are closely tied to each other: algorithmic enhancements that seek to improve performance often make it more difficult to design and verify a correct data structure implementation.

The following example illustrates various features of multiprocessors that affect concurrent data structure design. Suppose we wish to implement a *shared counter* data structure that supports a `fetch-and-inc` operation that adds one to the counter and returns the value of the counter immediately before the increment. A trivial sequential implementation of the `fetch-and-inc` operation contains code like that shown on the left in Figure 48.1.[*]

If we allow concurrent invocations of the `fetch-and-inc` operation by multiple threads, this implementation does not behave correctly. To see why, observe that most compilers will translate this source code into machine instructions that load X into a register, then add one to that register, then store that register back to X. Suppose that the counter is initially 0, and two `fetch-and-inc` operations execute on different processors concurrently. Then there is a risk that both operations read 0 from X, and therefore both store back 1 and return 0. This is clearly incorrect: one of the operations should return 1.

The incorrect behavior described above results from a "bad" interleaving of the steps of the two `fetch-and-inc` operations. A natural and common way to prevent such interleavings is to use a mutual exclusion lock (also known as a *mutex* or a *lock*). A lock is a construct that, at any point in time, is unowned or is owned by a single thread. If a thread t_1 wishes to acquire ownership of a lock that is already owned by another thread t_2, then t_1 must wait until t_2 releases ownership of the lock.

We can obtain a correct sequential implementation of the `fetch-and-inc` operation by using a lock as shown on the right in Figure 48.1. With this arrangement, we prevent the bad interleavings by preventing *all* interleavings. While it is easy to achieve a correct shared counter this way, this simplicity comes at a price: Locking introduces a host of problems related to both performance and software engineering.

48.1.1 Performance

The *speedup* of an application when run on P processors is the ratio of its execution time on a single processor to its execution time on P processors. It is a measure of how effectively the application is utilizing the machine it is running on. Ideally, we want *linear speedup*: we would like to achieve a speedup of P when using P processors. Data structures whose speedup grows with P are called *scalable*. In designing scalable data structures we must take care: naive approaches to synchronization can severely undermine scalability.

Returning to the lock-based counter, observe that the lock introduces a *sequential bottleneck*: at any point in time, at most one `fetch-and-inc` operation is doing useful work, that is, incrementing the variable X. Such sequential bottlenecks can have a surprising effect on the speedup one can achieve. The effect of the sequentially executed parts of the code on performance is illustrated by a simple formula based on Amdahl's law [1]. Let b be the fraction of the program that is subject to a sequential bottleneck. If the program takes 1 time unit when executed on a single processor, then on a P-way multiprocessor the sequential part takes b time units, and the concurrent part takes $(1-b)/P$ time units in the best case, so the speedup S is at most $1/(b+(1-b)/P)$. This implies that if just 10% of our application is subject to a sequential bottleneck, the best possible speedup we can achieve on a 10-way machine is about 5.3: we are running the application at half of the machine's capacity. Reducing the number and length of sequentially executed code sections is thus crucial to performance. In the context of locking, this means reducing the number of locks acquired, and reducing *lock granularity*, a measure of the number of instructions executed while holding a lock.

A second problem with our simple counter implementation is that it suffers from *memory contention*: an overhead in traffic in the underlying hardware as a result of multiple threads concurrently attempting to access the same locations in memory. Contention can be appreciated only by understanding some aspects of common shared-memory multiprocessor architectures. If the lock protecting our counter is implemented in a single memory location, as many simple locks are, then in order to acquire

---

[*] Throughout our examples, we ignore the fact that, in reality, integers are represented by a fixed number of bits, and will therefore eventually "wrap around" to zero.

the lock, a thread must repeatedly attempt to modify that location. On a *cache-coherent* multiprocessor[*] for example, exclusive ownership of the cache line containing the lock must be repeatedly transferred from one processor to another; this results in long waiting times for each attempt to modify the location, and is further exacerbated by the additional memory traffic associated with unsuccessful attempts to acquire the lock. In Section 48.1.7, we discuss lock implementations that are designed to avoid such problems for various types of shared memory architectures.

A third problem with our lock-based implementation is that, if the thread that currently holds the lock is delayed, then all other threads attempting to access the counter are also delayed. This phenomenon is called *blocking*, and is particularly problematic in *multiprogrammed* systems, in which there are multiple threads per processor and the operating system can preempt a thread while it holds the lock. For many data structures, this difficulty can be overcome by devising *nonblocking* algorithms in which the delay of a thread does not cause the delay of others. By definition, these algorithms cannot use locks.

Below we continue with our shared counter example, discussing blocking and nonblocking techniques separately; we introduce more issues related to performance as they arise.

48.1.2 Blocking Techniques

In many data structures, the undesirable effects of memory contention and sequential bottlenecks discussed above can be reduced by using a *fine-grained* locking scheme. In fine-grained locking, we use multiple locks of small granularity to protect different parts of the data structure. The goal is to allow concurrent operations to proceed in parallel when they do not access the same parts of the data structure. This approach can also help to avoid excessive contention for individual memory locations. For some data structures, this happens naturally; for example, in a hash table, operations concerning values that hash to different buckets naturally access different parts of the data structure.

For other structures, such as the lock-based shared counter, it is less clear how contention and sequential bottlenecks can be reduced because—abstractly—all operations modify the same part of the data structure. One approach to dealing with contention is to spread operations out in time, so that each operation accesses the counter in a separate time interval from the others. One widely used technique for doing so is called *backoff* [2]. However, even with reduced contention, our lock-based shared counter still lacks parallelism, and is therefore not scalable. Fortunately, more sophisticated techniques can improve scalability.

One approach, known as a *combining tree* [3–6], can help implement a scalable counter. This approach employs a binary tree with one leaf per thread. The root of the tree contains the actual counter value, and the other internal nodes of the tree are used to coordinate access to the root. The key idea is that threads climb the tree from the leaves towards the root, attempting to "combine" with other concurrent operations. Every time the operations of two threads are combined in an internal node, one of those threads—call it the loser—simply waits at that node until a return value is delivered to it. The other thread—the winner—proceeds towards the root carrying the sum of all the operations that have combined in the subtree rooted at that node; a winner thread that reaches the root of the tree adds its sum to the counter in a single addition, thereby effecting the increments of all of the combined operations. It then descends the tree distributing a return value to each waiting loser thread with which it previously combined. These return values are distributed so that the effect is as if all of the increment operations were executed one after the other at the moment the root counter was modified.

The technique losers employ while waiting for winners in the combining tree is crucial to its performance. A loser operation waits by repeatedly reading a memory location in a tree node; this is called *spinning*. An important consequence in a cache-coherent multiprocessor is that this location will reside in the local cache of the processor executing the loser operation until the winner operation reports the result. This means that the waiting loser does not generate any unnecessary memory traffic that may slow down the winner's performance. This kind of waiting is called *local spinning*, and has been shown to be crucial for scalable performance [7].

In so-called *non-uniform memory access* (NUMA) architectures, processors can access their local portions of shared memory faster than they can access the shared memory portions of other processors. In such architectures, *data layout*—the way nodes of the combining tree are placed in memory—can have a significant impact on performance. Performance can be improved by locating the leaves of the tree near the processors running the threads that own them. (We assume here that threads are statically bound to processors.)

Data layout issues also affect the design of concurrent data structures for cache-coherent multiprocessors. Recall that one of the goals of the combining tree is to reduce contention for individual memory locations in order to improve performance. However, because cache-coherent multiprocessors manage memory in cache-line-sized chunks, if two threads are accessing different memory locations that happen to fall into the same cache line, performance suffers just as if they had been accessing the same memory location. This phenomenon is known as *false sharing*, and is a common source of perplexing performance problems.

---

[*] A *cache-coherent* multiprocessor is one in which processors have local caches that are updated by hardware in order to keep them consistent with the latest values stored.

By reducing contention for individual memory locations, reducing memory traffic by using local spinning, and allowing operations to proceed in parallel, counters implemented using combining trees scale with the number of concurrent threads much better than the single lock version does [5]. If all threads manage to repeatedly combine, then a tree of width P will allow P threads to return P values after every $O(\log P)$ operations required to ascend and descend the tree, offering a speedup of $O(P/\log P)$ (but see Section 48.1.4).

Despite the advantages of the combining tree approach, it also has several disadvantages. It requires a known bound P on the number of threads that access the counter, and it requires $O(P)$ space. While it provides better throughout under *heavy loads*, that is, when accessed by many concurrent threads, its best-case performance under low loads is poor: It must still traverse $O(\log P)$ nodes in the tree, whereas a `fetch-and-inc` operation of the single-lock-based counter completes in constant time. Moreover, if a thread fails to combine because it arrived at a node immediately after a winner left it on its way up the tree, then it must wait until the winner returns before it can continue its own ascent up the tree. The coordination among ascending winners, losers, and ascending late threads, if handled incorrectly, may lead to *deadlocks*: threads may block each other in a cyclic fashion such that none ever makes progress. Avoiding deadlocks significantly complicates the task of designing correct and efficient implementations of blocking concurrent data structures.

In summary, blocking data structures can provide powerful and efficient implementations if a good balance can be struck between using enough blocking to maintain correctness, while minimizing blocking in order to allow concurrent operations to proceed in parallel.

48.1.3 Nonblocking Techniques

As discussed earlier, nonblocking implementations aim to overcome the various problems associated with the use of locks. To formalize this idea, various nonblocking progress conditions—such as wait-freedom [8,9], lock-freedom [8], and obstruction-freedom [10]—have been considered in the literature. A *wait-free* operation is guaranteed to complete after a finite number of its own steps, regardless of the timing behavior of other operations. A *lock-free* operation guarantees that after a finite number of its own steps, *some* operation completes. An *obstruction-free* operation is guaranteed to complete within a finite number of its own steps after it stops encountering interference from other operations.

Clearly, wait-freedom is a stronger condition than lock-freedom, and lock-freedom in turn is stronger than obstruction-freedom. However, all of these conditions are strong enough to preclude the use of blocking constructs such as locks.* While stronger progress conditions seem desirable, implementations that make weaker guarantees are generally simpler, more efficient in the common case, and easier to design and to verify as correct. In practice, we can compensate for the weaker progress conditions by employing backoff [2] or more sophisticated contention management techniques [11].

Let us return to our shared counter. It follows easily from results of Fischer et al. [12] (extended to shared memory by Herlihy [8] and Loui and Abu-Amara [13]) that such a shared counter cannot be implemented in a lock-free (or wait-free) manner using only `load` and `store` instructions to access memory. These results show that, in any proposed implementation, a bad interleaving of operations can cause incorrect behaviour. These bad interleavings are possible because the `load` and `store` are separate operations. This problem can be overcome by using a hardware operation that atomically combines a load and a store. Indeed, all modern multiprocessors provide such synchronization instructions, the most common of which are *compare-and-swap* (CAS) [14–16] and *load-linked/store-conditional* (LL/SC) [17–19]. The semantics of the CAS instruction is shown in Figure 48.2. For purposes of illustration, we assume an `atomically` keyword which requires the code block it labels to be executed *atomically*, that is, so that that no thread can observe a state in which the code block has been partially executed. The CAS operation atomically loads from a memory location, compares the value read to an expected value, and stores a new value to the location if the comparison succeeds. Herlihy [8] showed that instructions such as CAS and LL/SC are *universal*: there exists a wait-free implementation for *any* concurrent data structure in a system that supports such instructions.

A simple lock-free counter can be implemented using CAS. The idea is to perform the `fetch-and-inc` by loading the counter's value and to then use CAS to atomically change the counter value to a value greater by one than the value read. The CAS instruction fails to increment the counter only if it changes between the load and the CAS. In this case, the operation can retry, as the failing CAS had no effect. Because the CAS fails only as a result of another `fetch-and-inc` operation succeeding, the implementation is lock-free. However, it is not wait-free because a single `fetch-and-inc` operation can repeatedly fail its CAS.

The above example illustrates an *optimistic* approach to synchronization: the `fetch-and-inc` operation completes quickly in the hopefully common case in which it does not encounter interference from a concurrent operation, but must employ more expensive techniques under contention (e.g., backoff).

---

* We use the term "nonblocking" broadly to include all progress conditions requiring that the failure or indefinite delay of a thread cannot prevent other threads from making progress, rather than as a synonym for "lock-free," as some authors prefer.

```
bool CAS(L, E, N) {
  atomically {
    if (*L == E) {
      *L = N;
      return true;
    } else
      return false;
  }
}
```

FIGURE 48.2   The semantics of the CAS operation.

While the lock-free counter described above is simple, it has many of the same disadvantages that the original counter based on a single lock has: a sequential bottleneck and high contention for a single location. It is natural to attempt to apply similar techniques as those described above in order to improve the performance of this simple implementation. However, it is usually more difficult to incorporate such improvements into nonblocking implementations of concurrent data structures than blocking ones. Roughly, the reason for this is that a thread can use a lock to prevent other threads from "interfering" while it performs some sequence of actions. Without locks, we have to design our implementations to be correct despite the actions of concurrent operations; in current architectures, this often leads to the use of complicated and expensive techniques that undermine the improvements we are trying to incorporate. As discussed further in Section 48.1.7, transactional mechanisms make it much easier to design and modify efficient implementations of complex concurrent data structures. However, hardware support for such mechanisms does not yet exist.

48.1.4 Complexity Measures

A wide body of research is directed at analyzing the asymptotic complexity of concurrent data structures and algorithms in idealized models such as *parallel random access machines* [20–22]. However, there is less work on modeling such data structures in a realistic multiprocessor setting. There are many reasons for this, most of which have to do with the interplay of the architectural features of the hardware and the asynchronous execution of threads. Consider the combining tree example. Though we argued a speedup of $O(P/\log P)$ by counting instructions, this is not reflected in empirical studies [5,23]. Real-world behavior is dominated by other features discussed above, such as cost of contention, cache behavior, cost of universal synchronization operations (e.g., CAS), arrival rates of requests, effects of backoff delays, layout of the data structure in memory, and so on. These factors are hard to quantify in a single precise model spanning all current architectures. Complexity measures that capture some of these aspects have been proposed by Dwork et al. [24] and by Anderson and Yang [25]. While these measures provide useful insight into algorithm design, they cannot accurately capture the effects of all of the factors listed above. As a result, concurrent data structure designers compare their designs empirically by running them using micro-benchmarks on real machines and simulated architectures [5,7, 26,27]. In the remainder of this chapter, we generally qualify data structures based on their empirically observed behavior and use simple complexity arguments only to aid intuition.

48.1.5 Correctness

It is easy to see that the behavior of the simple lock-based counter is "the same" as that of the sequential implementation. However, it is significantly more difficult to see this is also true for the combining tree. In general, concurrent data structure implementations are often subtle, and incorrect implementations are not uncommon. Therefore, it is important to be able to state and prove rigorously that a particular design correctly implements the required concurrent data structure. We cannot hope to achieve this without a precise way of specifying what "correct" means.

Data structure specifications are generally easier for sequential data structures. For example, we can specify the semantics of a sequential data structure by choosing a set of states, and providing a *transition function* that takes as arguments a state, an operation name and arguments to the operation, and returns a new state and a return value for the operation. Together with a designated initial state, the transition function specifies all acceptable sequences of operations on the data structure. The sequential semantics of the counter is specified as follows: The set of states for the counter is the set of integers, and the initial state is 0. The transition function for the `fetch-and-inc` operation adds one to the old state to obtain the new state, and the return value is the old state of the counter.

Operations on a sequential data structure are executed one-at-a-time in order, and we simply require that the resulting sequence of operations respects the sequential semantics specified as discussed above. However, with concurrent data structures, operations are not necessarily totally ordered. Correctness conditions for concurrent data structures generally require that *some* total order of the operations exists that respects the sequential semantics. Different conditions are distinguished by their different requirements on this total ordering.

A common condition is Lamport's *sequential consistency* [28], which requires that the total order preserves the order of operations executed by each thread. Sequential consistency has a drawback from the software engineering perspective: a data structure implemented using sequentially consistent components may not be sequentially consistent itself.

A natural and widely used correctness condition that overcomes this problem is Herlihy and Wing's *linearizability* [29], a variation on the *serializability* [30] condition used for database transactions. Linearizability requires two properties: (1) that the data structure be sequentially consistent, and (2) that the total ordering which makes it sequentially consistent respect the *real-time ordering* among the operations in the execution. Respecting the real-time ordering means that if an operation O_1 finishes execution before another operation O_2 begins (so the operations are not concurrent with each other), then O_1 must be ordered before O_2. Another way of thinking of this condition is that it requires us to be able to identify a distinct point within each operation's execution interval, called its *linearization point*, such that if we order the operations according to the order of their linearization points, the resulting order obeys the desired sequential semantics.

It is easy to see that the counter implementation based on a single lock is linearizable. The state of the counter is always stored in the variable x. We define the linearization point of each fetch-and-inc operation as the point at which it stores its incremented value to x. The linearizability argument for the CAS-based lock-free implementation is similarly simple, except that we use the semantics of the CAS instruction, rather than reasoning about the lock, to conclude that the counter is incremented by one each time it is modified.

For the combining tree, it is significantly more difficult to see that the implementation is linearizable because the state of the counter is incremented by more than one at a time, and because the increment for one fetch-and-inc operation may in fact be performed by another operation with which it has combined. We define the linearization points of fetch-and-inc operations on the combining-tree-based counter as follows. When a winner thread reaches the root of the tree and adds its accumulated value to the counter, we linearize each of the operations with which it has combined in sequence immediately after that point. The operations are linearized in the order of the return values that are subsequently distributed to those operations. While a detailed linearizability proof is beyond the scope of our presentation, it should be clear from this discussion that rigorous correctness proofs for even simple concurrent data structures can be quite challenging.

The intuitive appeal and modularity of linearizability makes it a popular correctness condition, and most of the concurrent data structure implementations we discuss in the remainder of this chapter are linearizable. However, in some cases, performance and scalability can be improved by satisfying a weaker correctness condition. For example, the *quiescent consistency* condition [31] drops the requirement that the total ordering of operations respects the real-time order, but requires that every operation executed after a quiescent state—one in which no operations are in progress—must be ordered after every operation executed before that quiescent state. Whether an implementation satisfying such a weak condition is useful is application-dependent. In contrast, a linearizable implementation is always usable, because designers can view it as atomic.

48.1.6 Verification Techniques

In general, to achieve a rigorous correctness proof for a concurrent data structure implementation, we need mathematical machinery for specifying correctness requirements, accurately modeling a concurrent data structure implementation, and ensuring that a proof that the implementation is correct is complete and accurate. Most linearizability arguments in the literature treat at least some of this machinery informally, and as a result, it is easy to miss cases, make incorrect inferences, etc. Rigorous proofs inevitably contain an inordinate amount of mundane detail about trivial properties, making them tedious to write and to read. Therefore, computer-assisted methods for verifying implementations are required. One approach is to use a *theorem prover* to prove a series of assertions which together imply that an implementation is correct. Another approach is to use *model checking* tools, which exhaustively check all possible executions of an implementation to ensure that each one meets specified correctness conditions. The theorem proving approach usually requires significant human insight, while model checking is limited by the number of states of an implementation that can be explored.

48.1.7 Tools of the Trade

Below we discuss three key types of tools one can use in designing concurrent data structures: locks, barriers, and transactional synchronization mechanisms. Locks and barriers are traditional low-level synchronization mechanisms that are used to restrict certain interleavings, making it easier to reason about implementations based on them. Transactional mechanisms seek to hide the tricky details of concurrency from programmers, allowing them to think in a more traditional sequential way.

```
void acquire(Lock *lock) {              void release(Lock *lock) {
  int delay = MIN_DELAY;                  *lock = UNOWNED;
  while (true) {                        }
    if (CAS(lock,UNOWNED,OWNED))
      return;
    sleep(random() % delay);
    if (delay < MAX_DELAY)
      delay = 2 * delay;
  }
}
```

FIGURE 48.3   Exponential backoff lock.

48.1.7.1 Locks

As discussed earlier, locks are used to guarantee mutually exclusive access to (parts of) a data structure, in order to avoid "bad" interleavings. A key issue in designing a lock is what action to take when trying to acquire a lock already held by another thread. On uniprocessors, the only sensible option is to yield the processor to another thread. However, in multiprocessors, it may make sense to repeatedly attempt to acquire the lock, because the lock may soon be released by a thread executing on another processor. Locks based on this technique are called *spinlocks*. The choice between blocking and spinning is often a difficult one because it is hard to predict how long the lock will be held. When locks are supported directly by operating systems or threads packages, information such as whether the lock-holder is currently running can be used in making this decision.

A simple spinlock repeatedly uses a synchronization primitive such as compare-and-swap to atomically change the lock from unowned to owned. As mentioned earlier, if locks are not designed carefully, such spinning can cause heavy contention for the lock, which can have a severe impact on performance. A common way to reduce such contention is to use *exponential backoff* [2]. In this approach, which is illustrated in Figure 48.3, a thread that is unsuccessful in acquiring the lock waits some time before retrying; repeated failures cause longer waiting times, with the idea that threads will "spread themselves out" in time, resulting in lower contention and less memory traffic due to failed attempts.

Exponential backoff has the disadvantage that the lock can be unowned, but all threads attempting to acquire it have backed off too far, so none of them is making progress. One way to overcome this is to have threads form a queue and have each thread that releases the lock pass ownership of the lock to the next queued thread. Locks based on this approach are called *queuelocks*. Anderson [32] and Graunke and Thakkar [33] introduce array-based queuelocks, and these implementations are improved upon by the list-based MCS queue locks of Mellor-Crummey and Scott [7] and the CLH queue locks of Craig and Landin and Hagersten [34,35].

Threads using CLH locks form a virtual linked list of nodes, each containing a done flag; a thread enters the critical section only after the done flag of the node preceding its own node in the list is raised. The lock object has a pointer to the node at the tail of the list, the last one to join it. To acquire the lock, a thread creates a node, sets its done flag to false indicate that it has not yet released the critical section, and uses a synchronization primitive such as CAS to place its own node at the tail of the list while determining the node of its predecessor. It then spins on the done flag of the predecessor node. Note that each thread spins on a different memory location. Thus, in a cache-based architecture, when a thread sets its done flag to inform the next thread in the queue that it can enter the critical section, the done flags on which all other threads are spinning are not modified, so those threads continue to spin on a local cache line, and do not produce additional memory traffic. This significantly reduces contention and improves scalability in such systems. However, if this algorithm is used in a non-coherent NUMA machine, threads will likely have to spin on remote memory locations, again causing excessive memory traffic. The MCS queuelock [7] overcomes this problem by having each thread spin on a done flag in its *own* node. This way, nodes can be allocated in local memory, eliminating the problem.

There are several variations on standard locks that are of interest to the data structure designer in some circumstances. The queuelock algorithms discussed above have more advanced *abortable* versions that allow threads to give up on waiting to acquire the lock, for example, if they are delayed beyond some limit in a real-time application [36,37], or if they need to recover from deadlock. The algorithms of [36] provide an abort that is nonblocking. Finally [27], presents *preemption-safe* locks, which attempt to reduce the negative performance effects of preemption by ensuring that a thread that is in the queue but preempted does not prevent the lock from being granted to another running thread.

In many data structures it is desirable to have locks that allow concurrent readers. Such *reader-writer* locks allow threads that only read data in the critical section (but do not modify it) to access the critical section exclusively from the writers but concurrently with each other. Various algorithms have been suggested for this problem. The reader-writer queuelock algorithms of Mellor-Crummey and Scott [38] are based on MCS queuelocks and use read counters and a special pointer to writer nodes. In Reference 39 a version of these algorithms is presented in which readers remove themselves from the lock's queue. This is done by keeping a doubly-linked list of queued nodes, each having its own simple "mini-lock." Readers remove themselves from the queuelock list by acquiring mini-locks of their neighboring nodes and redirecting the pointers of the doubly-linked list. The above-mentioned real-time queuelock algorithms of Reference 36 provide a similar ability without locking nodes.

The reader-writer lock approach can be extended to arbitrarily many operation types through a construct called *group mutual exclusion* or *room synchronization*. The idea is that operations are partitioned into groups, such that operations in the same group can execute concurrently with each other, but operations in different groups must not. An interesting application of this approach separates push and pop operations on a stack into different groups, allowing significant simplifications to the implementations of those operations because they do not have to deal with concurrent operations of different types [40]. Group mutual exclusion was introduced by Joung [41]. Implementations based on mutual exclusion locks or `fetch-and-inc` counters appear in References 40,42.

More complete and detailed surveys of the literature on locks can be found in References 43,44.

48.1.7.2 Barriers

A barrier is a mechanism that collectively halts threads at a given point in their code, and allows them to proceed only when all threads have arrived at that point. The use of barriers arises whenever access to a data structure or application is layered into phases whose execution should not overlap in time. For example, repeated iterations of a numerical algorithm that converges by iterating a computation on the same data structure or the marking and sweeping phases of a parallel garbage collector.

One simple way to implement a barrier is to use a counter that is initialized to the total number of threads: Each thread decrements the counter upon reaching the barrier, and then spins, waiting for the counter to become zero before proceeding. Even if we use the techniques discussed earlier to implement a scalable counter, this approach still has the problem that waiting threads produce contention. For this reason, specialized barrier implementations have been developed that arrange for each thread to spin on a different location [45–47]. Alternatively, a barrier can be implemented using a *diffusing computation* tree in the style of Dijkstra and Scholten [48]. In this approach, each thread is the owner of one node in a binary tree. Each thread awaits the arrival of its children, then notifies its parent that it has arrived. Once all threads have arrived, the root of the tree releases all threads by disseminating the release information down the tree.

48.1.7.3 Transactional Synchronization Mechanisms

A key use of locks in designing concurrent data structures is to allow threads to modify multiple memory locations atomically, so that no thread can observe partial results of an update to the locations. Transactional synchronization mechanisms are tools that allow the programmer to treat sections of code that access multiple memory locations as a single atomic step. This substantially simplifies the design of correct concurrent data structures because it relieves the programmer of the burden of deciding which locks should be held for which memory accesses and of preventing deadlock.

As an example of the benefits of transactional synchronization, consider Figure 48.4, which shows a concurrent queue implementation achieved by requiring operations of a simple sequential implementation to be executed atomically. Such atomicity could be ensured either by using a global mutual exclusion lock, or via a transactional mechanism. However, the lock-based

```
typedef struct qnode_s { qnode_s *next; valuetype value; } qnode_t;
shared variables:
  // initially null
  qnode_t *head, *tail;

void enqueue(qnode_t *n) {
  atomically {
    if (tail == null)
      tail = head = n;
    else {
      tail->next = n;
      tail = n;
    }
  }
}

qnode_t * dequeue() {
  atomically {
    if (head == null)
      return null;
    else {
      n = head;
      head = head->next;
      if (head == null)
        tail = null;
      return n;
    }
  }
}
```

FIGURE 48.4 A concurrent shared FIFO queue.

approach prevents concurrent `enqueue` and `dequeue` operations from executing in parallel. In contrast, a good transactional mechanism will allow them to do so in all but the empty state because when the queue is not empty, concurrent `enqueue` and `dequeue` operations do not access any memory locations in common.

The use of transactional mechanisms for implementing concurrent data structures is inspired by the widespread use of transactions in database systems. However, the problem of supporting transactions over shared memory locations is different from supporting transactions over databases elements stored on disk. Thus, more lightweight support for transactions is possible in this setting.

Kung and Robinson's *optimistic concurrency control* (OCC) [49] is one example of a transactional mechanism for concurrent data structures. OCC uses a global lock, which is held only for a short time at the end of a transaction. Nonetheless, the lock is a sequential bottleneck, which has a negative impact on scalability. Ideally, transactions should be supported without the use of locks, and transactions that access disjoint sets of memory locations should not synchronize with each other at all.

Transactional support for multiprocessor synchronization was originally suggested by Herlihy and Moss, who also proposed a hardware-based *transactional memory* mechanism for supporting it [50]. Recent extensions to this idea include *lock elision* [51,52], in which the hardware automatically translates critical sections into transactions, with the benefit that two critical sections that do not in fact conflict with each other can be executed in parallel. For example, lock elision could allow concurrent `enqueue` and `dequeue` operations of the above queue implementation to execute in parallel, even if the atomicity is implemented using locks. To date, hardware support for transactional memory has not been built.

Various forms of *software transactional memory* have been proposed by Shavit and Touitou [53], Harris et al. [54], Herlihy et al. [11], and Harris and Fraser [55].

Transactional mechanisms can easily be used to implement most concurrent data structures, and when efficient and robust transactional mechanisms become widespread, this will likely be the preferred method. In the following sections, we mention implementations based on transactional mechanisms only when no direct implementation is known.

48.2 Shared Counters and Fetch-and-ϕ Structures

Counters have been widely studied as part of a broader class of *fetch-and-ϕ* coordination structures, which support operations that fetch the current value of a location and apply some function from an allowable set ϕ to its contents. As discussed earlier, simple lock-based implementations of fetch-and-ϕ structures such as counters suffer from contention and sequential bottlenecks. Below we describe some approaches to overcoming these problems.

48.2.1 Combining

The combining tree technique was originally invented by Gottlieb et al. [4] to be used in the hardware switches of processor-to-memory networks. In Section 48.1.2 we discussed a software version of this technique, first described by Goodman et al. [3] and Yew et al. [6], for implementing a fetch-and-add counter. (The algorithm in Reference 6 has a slight bug; see Reference 5.) This technique can also be used to implement fetch-and-ϕ operations for a variety of sets of combinable operations, including arithmetic and boolean operations, and synchronization operations such as load, store, swap, test-and-set, etc. [56].

As explained earlier, scalability is achieved by sizing the tree such that the there is one leaf node per thread. Under maximal load, the throughput of such a tree is proportional to $O(P/\log P)$ operations per time unit, offering a significant speedup. Though it is possible to construct trees with fan-out greater than two in order to reduce tree depth, that would sacrifice the simplicity of the nodes and, as shown by Shavit and Zemach [57], will most likely result in reduced performance. Moreover, Herlihy et al. [5] have shown that combining trees are extremely sensitive to changes in the arrival rate of requests: as the load decreases, threads must still pay the price of traversing the tree while attempting to combine, but the likelihood of combining is reduced because of the reduced load.

Shavit and Zemach overcome the drawbacks of the static combining tree structures by introducing combining funnels [57]. A *combining funnel* is a linearizable fetch-and-ϕ structure that allows combining trees to form dynamically, adapting its overall size based on load patterns. It is composed of a (typically small) number of *combining layers*. Each such layer is implemented as a *collision array* in memory. Threads pass through the funnel layer by layer, from the first (widest) to the last (narrowest). These layers are used by threads to locate each other and combine their operations. As threads pass through a layer, they read a thread ID from a randomly chosen array element, and write their own in its place. They then attempt to combine with the thread whose ID they read. A successful combination allows threads to exchange information, allowing some to continue to the next layer, and others to await their return with the resulting value. Combining funnels can also support the *elimination* technique (described in Section 48.3) to allow two operations to complete without accessing the central data structure in some cases.

48.2.2 Counting Networks

Combining structures provide scalable and linearizable fetch-and-ϕ operations. However, they are blocking. An alternative approach to parallelizing a counter that overcomes this problem is to have multiple counters instead of a single one, and to use a counting network to coordinate access to the separate counters so as to avoid problems such as duplicated or omitted values. *Counting networks*, introduced by Aspnes et al. [31], are a class of data structures for implementing, in a highly concurrent and nonblocking fashion, a restricted class of fetch-and-ϕ operations, the most important of which is `fetch-and-inc`.

Counting networks, like sorting networks [58], are acyclic networks constructed from simple building blocks called balancers. In its simplest form, a *balancer* is a computing element with two input wires and two output wires. Tokens arrive on the balancer's input wires at arbitrary times, and are output on its output wires in a balanced way. Given a stream of input tokens, a balancer alternates sending one token to the top output wire, and one to the bottom, effectively balancing the number of tokens between the two wires.

We can wire balancers together to form a network. The *width* of a network is its number of output wires (wires that are not connected to an input of any balancer). Let $y_0, \ldots, y_{w-1}$ respectively represent the number of tokens output on each of the output wires of a network of width w. A *counting network* is an acyclic network of balancers whose outputs satisfy the following *step property*:

> In any quiescent state, $0 \leq y_i - y_j \leq 1$ for any $i < j$.

Figure 48.5 shows a sequence of tokens traversing a counting network of width four based on Batcher's Bitonic sorting network structure [59]. The horizontal lines are wires and the vertical lines are balancers, each connected to two input and output wires at the dotted points. Tokens (numbered 1 through 5) traverse the balancers starting on arbitrary input wires and accumulate on specific output wires meeting the desired step-property. Aspnes et al. [31] have shown that every counting network has a layout isomorphic to a sorting network, but not every sorting network layout is isomorphic to a counting network.

On a shared memory multiprocessor, balancers are records, and wires are pointers from one record to another. Threads performing increment operations traverse the data structure from some input wire (either preassigned or chosen at random) to some output wire, each time shepherding a new token through the network.

The counting network distributes input tokens to output wires while maintaining the step property stated above. Counting is done by adding a local counter to each output wire, so that tokens coming out of output wire i are assigned numbers $i, i + w, \ldots, i + (y_i - 1)w$. Because threads are distributed across the counting network, there is little contention on the balancers, and the even distribution on the output wires lowers the load on the shared counters. However, as shown by Shavit and Zemach [23], the dynamic patterns through the networks increase cache miss rates and make optimized layout almost impossible.

There is a significant body of literature on counting networks, much of which is surveyed by Herlihy and Busch [60]. An empirical comparison among various counting techniques can be found in Reference 5. Aharonson and Attiya [61] and Felten et al. [62] study counting networks with arbitrary fan-in and fan-out. Shavit and Touitou [63] show how to perform decrements on counting network counters by introducing the notion of "anti-tokens" and elimination. Busch and Mavronicolas [64] provide a combinatorial classification of the various properties of counting networks. Randomized counting networks are introduced by Aiello et al. [65] and fault-tolerant networks are presented by Riedel and Bruck [66].

The classical counting network structures in the literature are lock-free but not linearizable, they are only quiescently consistent. Herlihy et al. [67] show the tradeoffs involved in making counting networks linearizable.

Klugerman and Plaxton present an optimal $\log w$-depth counting network [68]. However, this construction is not practical, and all practical counting network implementations have $\log^2 w$ depth. Shavit and Zemach introduce *diffracting trees* [23], improved counting networks made of balancers with one input and two output wires laid out as a binary tree. The simple balancers of the counting network are replaced by more sophisticated *diffracting balancers* that can withstand high loads by using a randomized collision array approach, yielding lower depth counting networks with significantly improved throughput. An adaptive diffracting tree that adapts its size to load is presented in References 69.

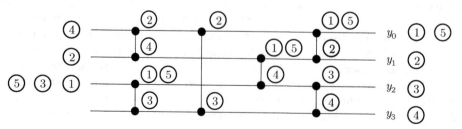

FIGURE 48.5   A bitonic counting network of width four.

48.3 Stacks and Queues

Stacks and queues are among the simplest sequential data structures. Numerous issues arise in designing concurrent versions of these data structures, clearly illustrating the challenges involved in designing data structures for shared-memory multiprocessors.

48.3.1 Stacks

A concurrent *stack* is a data structure linearizable to a sequential stack that provides push and pop operations with the usual *LIFO* semantics. Various alternatives exist for the behavior of these data structures in full or empty states, including returning a special value indicating the condition, raising an exception, or blocking.

Michael and Scott present several linearizable lock-based concurrent stack implementations: they are based on sequential linked lists with a top pointer and a global lock that controls access to the stack [27]. They typically scale poorly because even if one reduces contention on the lock, the top of the stack is a sequential bottleneck. Combining funnels [57] have been used to implement a linearizable stack that provides parallelism under high load. As with all combining structures, it is blocking, and it has a high overhead which makes it unsuitable for low loads.

Treiber [70] was the first to propose a lock-free concurrent stack implementation. He represented the stack as a singly-linked list with a top pointer and used CAS to modify the value of the top pointer atomically. Michael and Scott [27] compare the performance of Treiber's stack to an optimized nonblocking algorithm based on Herlihy's methodology [71], and several lock-based stacks (such as an MCS lock [7]) in low load situations. They concluded that Treiber's algorithm yields the best overall performance, and that this performance gap increases as the degree of multiprogramming grows. However, because the top pointer is a sequential bottleneck, even with an added backoff mechanism to reduce contention, the Treiber stack offers little scalability as concurrency increases [72].

Hendler et al. [72] observe that any stack implementation can be made more scalable using the *elimination* technique of Shavit and Touitou [63]. Elimination allows pairs of operations with reverse semantics—like pushes and pops on a stack—to complete without any central coordination, and therefore substantially aids scalability. The idea is that if a pop operation can find a concurrent push operation to "partner" with, then the pop operation can take the push operation's value, and both operations can return immediately. The net effect of each pair is the same as if the push operation was followed immediately by the pop operation, in other words, they eliminate each other's effect on the state of the stack. Elimination can be achieved by adding a collision array from which each operation chooses a location at random, and then attempts to coordinate with another operation that concurrently chose the same location [63]. The number of eliminations grows with concurrency, resulting in a high degree of parallelism. This approach, especially if the collision array is used as an adaptive backoff mechanism on the shared stack, introduces a high degree of parallelism with little contention [72], and delivers a scalable lock-free linearizable stack.

There is a subtle point in the Treiber stack used in the implementations above that is typical of many CAS-based algorithms. Suppose several concurrent threads all attempt a pop operation that removes the first element, located in some node "A," from the list by using a CAS to redirect the head pointer to point to a previously-second node "B." The problem is that it is possible for the list to change completely just before a particular pop operation attempts its CAS, so that by the time it does attempt it, the list has the node "A" as the first node as before, but the rest of the list including "B" is in a completely different order. This CAS of the head pointer from "A" to "B" may now succeed, but "B" might be anywhere in the list and the stack will behave incorrectly. This is an instance of the "ABA" problem [73], which plagues many CAS-based algorithms. To avoid this problem, Treiber augments the head pointer with a version number that is incremented every time the head pointer is changed. Thus, in the above scenario, the changes to the stack would cause the CAS to fail, thereby eliminating the ABA problem.[*]

48.3.2 Queues

A concurrent *queue* is a data structure linearizable to a sequential queue that provides enqueue and dequeue operations with the usual *FIFO* semantics.

Michael and Scott [27] present a simple lock-based queue implementation that improves on the naive single-lock approach by having separate locks for the head and tail pointers of a linked-list-based queue. This allows an enqueue operation to execute in parallel with a dequeue operation (provided we avoid false sharing by placing the head and tail locks in separate cache lines). This algorithm is quite simple, with one simple trick: a "dummy" node is always in the queue, which allows the implementation to avoid acquiring both the head and tail locks in the case that the queue is empty, and therefore it avoids deadlock.

---

[*] Note that the version number technique does not technically eliminate the ABA problem because the version number can wrap around; see Reference 74 for a discussion of the consequences of this point in practice, and also a "bounded tag" algorithm that eliminates the problem entirely, at some cost in space and time.

It is a matter of folklore that one can implement an array-based lock-free queue for a single enqueuer thread and a single dequeuer thread using only load and store operations [75]. A linked-list-based version of this algorithm appears in References 76. Herlihy and Wing [29] present a lock-free array-based queue that works if one assumes an unbounded size array. A survey in Reference 27 describes numerous flawed attempts at devising general (multiple enqueuers, multiple dequeuers) nonblocking queue implementations. It also discusses some correct implementations that involve much more overhead than the ones discussed below.

Michael and Scott [27] present a linearizable CAS-based lock-free queue with parallel access to both ends. The structure of their algorithm is very simple and is similar to the two-lock algorithm mentioned above: it maintains head and tail pointers, and always keeps a dummy node in the list. To avoid using a lock, the `enqueue` operation adds a new node to the end of the list using CAS, and then uses CAS to update the tail pointer to reflect the addition. If the `enqueue` is delayed between these two steps, another `enqueue` operation can observe the tail pointer "lagging" behind the end of the list. A simple *helping technique* [71] is used to recover from this case, ensuring that the tail pointer is always behind the end of the list by at most one element.

While this implementation is simple and efficient enough to be used in practice, it does have a disadvantage. Operations can access nodes already removed from the list, and therefore the nodes cannot be freed. Instead, they are put into a *freelist*—a list of nodes stored for reuse by future `enqueue` operations—implemented using Treiber's stack. This use of a freelist has the disadvantage that the space consumed by the nodes in the freelist cannot be freed for arbitrary reuse. Herlihy et al. [77] and Michael [78] have presented nonblocking memory management techniques that overcome this disadvantage.

It is interesting to note that the elimination technique is not applicable to queues: we cannot simply pass a value from an `enqueue` operation to a concurrent `dequeue` operation, because this would not respect the FIFO order with respect to other values in the queue.

48.3.3 Deques

A concurrent double-ended queue (*deque*) is a linearizable concurrent data structure that generalizes concurrent stacks and queues by allowing pushes and pops at both ends [79]. (See Chapter 2 for an introduction to stacks, queues and deques.) As with queues, implementations that allow operations on both ends to proceed in parallel without interfering with each other are desirable.

Lock-based deques can be implemented easily using the same two-lock approach used for queues. Given the relatively simple lock-free implementations for stacks and queues, it is somewhat surprising that there is no known lock-free deque implementation that allows concurrent operations on both ends. Martin et al. [80] provide a summary of concurrent deque implementations, showing that, even using nonconventional two-word synchronization primitives such as *double-compare-and-swap* (DCAS) [81], it is difficult to design a lock-free deque. The only known nonblocking deque implementation for current architectures that supports noninterfering operations at opposite ends of the deque is an obstruction-free CAS-based implementation due to Herlihy et al. [10].

48.4 Pools

Much of the difficulty in implementing efficient concurrent stacks and queues arises from the ordering requirements on when an element that has been inserted can be removed. A concurrent *pool* [82] is a data structure that supports `insert` and `delete` operations, and allows a `delete` operation to remove *any* element that has been inserted and not subsequently deleted. This weaker requirement offers opportunities for improving scalability.

A high-performance pool can be built using any quiescently consistent counter implementation [23,31]. Elements are placed in an array, and a `fetch-and-inc` operation is used to determine in which location an `insert` operation stores its value, and similarly from which location a `delete` operation takes its value. Each array element contains a full/empty bit or equivalent mechanism to indicate if the element to be removed has already been placed in the location. Using such a scheme, any one of the combining tree, combining funnel, counting network, or diffracting tree approaches described above can be used to create a high throughput shared pool by parallelizing the main bottlenecks: the shared counters. Alternatively, a "stack like" pool can be implemented by using a counter that allows increments and decrements, and again using one of the above techniques to parallelize it.

Finally, the elimination technique discussed earlier is applicable to pools constructed using combining funnels, counting networks, or diffracting trees: if `insert` and `delete` operations meet in the tree, the `delete` can take the value being inserted by the `insert` operation, and both can leave without continuing to traverse the structure. This technique provides high performance under high load.

The drawback of all these implementations is that they perform rather poorly under low load. Moreover, when used for work-load distribution [26,83,84], they do not allow us to exploit locality information, as pools designed specifically for work-load distribution do.

Workload distribution (or *load balancing*) algorithms involve a collection of pools of units of work to be done; each pool is local to a given processor. Threads create work items and place them in local pools, employing a load balancing algorithm to ensure that the number of items in the pools is balanced. This avoids the possibility that some processors are idle while others still have work in their local pools. There are two general classes of algorithms of this type: *work sharing* [76,84] and *work stealing* [26,83]. In a work sharing scheme, each processor attempts to continuously offload work from its pool to other pools. In work stealing, a thread that has no work items in its local pool steals work from other pools. Both classes of algorithms typically use randomization to select the pool with which to balance or the target pool for stealing.

The classical work stealing algorithm is due to Arora et al. [26]. It is based on a lock-free construction of a *deque* that allows operations by only one thread (the thread to which the pool is local) at one end of the deque, allowing only pop operations at the other end, and allowing concurrent pop operations at that end to "abort" if they interfere. A deque with these restrictions is suitable for work stealing, and the restrictions allow a simple implementation in which the local thread can insert and delete using simple low-cost load and store operations, resorting to a more expensive CAS operation only when it competes with the remote deleters for the last remaining item in the queue.

It has been shown that in some cases it is desirable to steal more than one item at a time [85,86]. A nonblocking multiple-item work-stealing algorithm due to Hendler and Shavit appears in References 87. It has also been shown that in some cases it desirable to use affinity information of work items in deciding which items to steal. A locality-guided work stealing algorithm due to Acar et al. appears in References 88.

48.5 Linked Lists

Consider implementations of concurrent search structures supporting `insert`, `delete`, and `search` operations. If these operations deal only with a key value, then the resulting data structure is a *set*; if a data value is associated with each key, we have a *dictionary* [58]. These are closely related data structures, and a concurrent set implementation can often be adapted to implement a dictionary. In the next three sections, we concentrate on implementing sets using different structures: linked lists, hash tables, and trees.

Suppose we use a linked list to implement a set. Apart from globally locking the linked list to prevent concurrent manipulation, the most popular approach to concurrent lock-based linked lists is *hand-over-hand locking* (sometimes called *lock coupling*) [89,90]. In this approach, each node has an associated lock. A thread traversing the linked list releases a node's lock only after acquiring the lock of the next node in the list, thus preventing overtaking which may cause unnoticed removal of a node. This approach reduces lock granularity but significantly limits concurrency because insertions and deletions at disjoint list locations may delay each other.

One way to overcome this problem is to design lock-free linked lists. The difficulty in implementing a lock-free ordered linked list is ensuring that during an insertion or deletion, the adjacent nodes are still valid, that is, they are still in the list and are still adjacent. As Figure 48.6 shows, designing such lock-free linked lists is not a straightforward matter.

The first CAS-based lock-free linked list is due to Valois [91], who uses a special auxiliary node in front of every regular node to prevent the undesired phenomena depicted in Figure 48.6. Valois's algorithm is correct when combined with a memory management solution due to Michael and Scott [92], but this solution is not practical. Harris [93] presents a lock-free list that

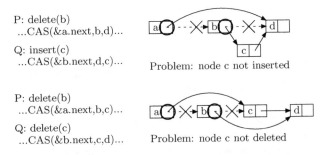

FIGURE 48.6 CAS-based list manipulation is hard. In both examples, P is deleting b from the list (the examples slightly abuse CAS notation). In the upper example, Q is trying to insert c into the list, and in the lower example, Q is trying to delete c from the list. Circled locations indicate the target addresses of the CAS operations; crossed out pointers are the values before the CAS succeeds.

uses a special "deleted" bit that is accessed atomically with node pointers in order to signify that a node has been deleted; this scheme is applicable only in garbage collected environments. Michael [94] overcomes this disadvantage by modifying Harris's algorithm to make it compatible with memory reclamation methods [77,78].

48.6 Hash Tables

A typical extensible hash table is a resizable array of buckets, each holding an expected constant number of elements, and thus requiring on average a constant time for insert, delete and search operations [58]. The principal cost of *resizing*—the redistribution of items between old and new buckets—is amortized over all table operations, thus keeping the cost of operations constant on average. Here resizing means extending the table, as it has been shown that as a practical matter, hash tables need only increase in size [95]. See Chapter 9 for more on hash tables.

Michael [94] shows that a concurrent non-extensible hash table can be achieved by placing a read-write lock on every bucket in the table. However, to guarantee good performance as the number of elements grows, hash tables must be extensible [96].

In the eighties, Ellis [97] and others [95,98] extended the work of Fagin et al. [96] by designing an extensible concurrent hash table for distributed databases based on two-level locking schemes. A recent extensible hash algorithm by Lea [99] is known to be highly efficient in non-multiprogrammed environments [100]. It is based on a version of Litwin's sequential linear hashing algorithm [101]. It uses a locking scheme that involves a small number of high-level locks rather than a lock per bucket, and allows concurrent searches while resizing the table, but not concurrent inserts or deletes. Resizing is performed as a global restructuring of all buckets when the table size needs to be doubled.

Lock-based extensible hash-table algorithms suffer from all of the typical drawbacks of blocking synchronization, as discussed earlier. These problems become more acute because of the elaborate "global" process of redistributing the elements in all the hash table's buckets among newly added buckets. Lock-free extensible hash tables are thus a matter of both practical and theoretical interest.

As described in Section 48.5, Michael [94] builds on the work of Harris [93] to provide an effective CAS-based lock-free linked list implementation. He then uses this as the basis for a lock-free hash structure that performs well in multiprogrammed environments: a fixed-sized array of hash buckets, each implemented as a lock-free list. However, there is a difficulty in making a lock-free array of lists extensible since it is not obvious how to redistribute items in a lock-free manner when the bucket array grows. Moving an item between two bucket lists seemingly requires two CAS operations to be performed together atomically, which is not possible on current architectures.

Greenwald [102] shows how to implement an extensible hash table using his *two-handed emulation* technique. However, this technique employs a DCAS synchronization operation, which is not available on current architectures, and introduces excessive amounts of work during global resizing operations.

Shalev and Shavit [100] introduce a lock-free extensible hash table which works on current architectures. Their key idea is to keep the items in a single lock-free linked list instead of a list per bucket. To allow operations fast access to the appropriate part of the list, the Shalev-Shavit algorithm maintains a resizable array of "hints" (pointers into the list); operations use the hints to find a point in the list that is close to the relevant position, and then follow list pointers to find the position. To ensure a constant number of steps per operation on average, finer grained hints must be added as the number of elements in the list grows. To allow these hints to be installed simply and efficiently, the list is maintained in a special *recursive split ordering*. This technique allows new hints to be installed incrementally, thereby eliminating the need for complicated mechanisms for atomically moving items between buckets, or reordering the list.

48.7 Search Trees

A concurrent implementation of any search tree (See Chapters 3 and 11 for more on sequential search trees) can be achieved by protecting it using a single exclusive lock. Concurrency can be improved somewhat by using a reader-writer lock to allow all read-only (search) operations to execute concurrently with each other while holding the lock in *shared mode*, while update (insert or delete) operations exclude all other operations by acquiring the lock in *exclusive mode*. If update operations are rare, this may be acceptable, but with even a moderate number of updates, the exclusive lock for update operations creates a sequential bottleneck that degrades performance substantially. By using fine-grained locking strategies—for example by using one lock per node, rather than a single lock for the entire tree—we can improve concurrency further.

Kung and Lehman [103] present a concurrent binary search tree implementation in which update operations hold only a constant number of node locks at a time, and these locks only exclude other update operations: search operations are never blocked. However, this implementation makes no attempt to keep the search tree balanced. In the remainder of this section, we focus on balanced search trees, which are considerably more challenging.

As a first step towards more fine-grained synchronization in balanced search tree implementations, we can observe that it is sufficient for an operation to hold an exclusive lock on the subtree in which it causes any modifications. This way, update operations that modify disjoint subtrees can execute in parallel. We briefly describe some techniques in this spirit in the context of B$^+$-trees. Recall that in B$^+$-trees, all keys and data are stored in leaf nodes; internal nodes only maintain routing information to direct operations towards the appropriate leaf nodes. Furthermore, an insertion into a leaf may require the leaf to be split, which may in turn require a new entry to be added to the leaf's parent, which itself may need to be split to accommodate the new entry. Thus, an insertion can potentially result in modifying all nodes along the path from the root to the leaf. However, such behavior is rare, so it does not make sense to exclusively lock the whole path just in case this occurs.

As a first step to avoiding such conservative locking strategies, we can observe that if an `insert` operation passes an internal B$^+$-tree node that is not full, then the modifications it makes to the tree cannot propagate past that node. In this case, we say that the node is *safe* with respect to the `insert` operation. If an update operation encounters a safe node while descending the tree acquiring exclusive locks on each node it traverses, it can safely release the locks on all ancestors of that node, thereby improving concurrency by allowing other operations to traverse those nodes [104,105]. Because `search` operations do not modify the tree, they can descend the tree using lock coupling: as soon as a lock has been acquired on a child node, the lock on its parent can be released. Thus, `search` operations hold at most two locks (in shared mode) at any point in time, and are therefore less likely to prevent progress by other operations.

This approach still requires each update operation to acquire an exclusive lock on the root node, and to hold the lock while reading a child node, potentially from disk, so the root is still a bottleneck. We can improve on this approach by observing that most update operations will not need to split or merge the leaf node they access, and will therefore eventually release the exclusive locks on all of the nodes traversed on the way to the leaf. This observation suggests an "optimistic" approach in which we descend the tree acquiring the locks in shared mode, acquiring only the leaf node exclusively [89]. If the leaf does not need to be split or merged, the update operation can complete immediately; in the rare cases in which changes do need to propagate up the tree, we can release all of the locks and then retry with the more pessimistic approach described above. Alternatively, we can use reader-writer locks that allow locks held in a shared mode to be "upgraded" to exclusive mode. This way, if an update operation discovers that it does need to modify nodes other than the leaf, it can upgrade locks it already holds in shared mode to exclusive mode, and avoid completely restarting the operation from the top of the tree [89]. Various combinations of the above techniques can be used because nodes near the top of the tree are more likely to conflict with other operations and less likely to be modified, while the opposite is true of nodes near the leaves [89].

As we employ some of the more sophisticated techniques described above, the algorithms become more complicated, and it becomes more difficult to avoid deadlock, resulting in even further complications. Nonetheless, all of these techniques maintain the invariant that operations exclusively lock the subtrees that they modify, so operations do not encounter states that they would not encounter in a sequential implementation. Significant improvements in concurrency and performance can be made by relaxing this requirement, at the cost of making it more difficult to reason that the resulting algorithms are correct.

A key difficulty we encounter when we attempt to relax the strict subtree locking schemes is that an operation descending the tree might follow a pointer to a child node that is no longer the correct node because of a modification by a concurrent operation. Various techniques have been developed that allow operations to recover from such "confusion," rather than strictly avoiding it.

An important example in the context of B$^+$-trees is due to Lehman and Yao [106], who define B^{link}-trees: B$^+$-trees with "links" from each node in the tree to its right neighbor at the same level of the tree. These links allow us to "separate" the splitting of a node from modifications to its parent to reflect the splitting. Specifically, in order to split a node n, we can create a new node n' to its right, and install a link from n to n'. If an operation that is descending the tree reaches node n while searching for a key position that is now covered by node n' due to the split, the operation can simply follow the link from n to n' to recover. This allows a node to be split without preventing access by concurrent operations to the node's parent. As a result, update operations do not need to simultaneously lock the entire subtree they (potentially) modify. In fact, in the Lehman-Yao algorithm, update operations as well as `search` operations use the lock coupling technique so that no operation ever holds more than two locks at a time, which significantly improves concurrency. This technique has been further refined, so that operations never hold more than one lock at a time [107].

Lehman and Yao do not address how nodes can be merged, instead allowing `delete` operations to leave nodes underfull. They argue that in many cases `delete` operations are rare, and that if space utilization becomes a problem, the tree can occasionally be reorganized in "batch" mode by exclusively locking the entire tree. Lanin and Shasha [108] incorporate merging into the `delete` operations, similarly to how `insert` operations split overflowed nodes in previous implementations. Similar to the Lehman-Yao link technique, these implementations use links to allow recovery by operations that have mistakenly reached a node that has been evacuated due to node merging.

In all of the algorithms discussed above, the maintenance operations such as node splitting and merging (where applicable) are performed as part of the regular update operations. Without such tight coupling between the maintenance operations and the regular operations that necessitate them, we cannot guarantee strict balancing properties. However, if we relax the balance requirements, we can separate the tree maintenance work from the update operations, resulting in a number of advantages that outweigh the desire to keep search trees strictly balanced. As an example, the B^{link}-tree implementation in References 107 supports a *compression process* that can run concurrently with regular operations to merge nodes that are underfull. By separating this work from the regular update operations, it can be performed concurrently by threads running on different processors, or in the background.

The idea of separating rebalancing work from regular tree operations was first suggested for red-black trees [109], and was first realized in References 110 for AVL trees [111] supporting `insert` and `search` operations. An implementation that also supports `delete` operations is provided in References 112. These implementations improve concurrency by breaking balancing work down into small, local tree transformations that can be performed independently. Analysis in References 113 shows that with some modifications, the scheme of Reference112 guarantees that each update operation causes at most $O(\log N)$ rebalancing operations for an N-node AVL tree. Similar results exist for B-trees [112,114] and red-black trees [115,116].

The only nonblocking implementations of balanced search trees have been achieved using Dynamic Software Transactional Memory mechanisms [11,117]. These implementations use transactions translated from sequential code that performs rebalancing work as part of regular operations.

The above brief survey covers only basic issues and techniques involved with implementing concurrent search trees. To mention just a few of the numerous improvements and extensions in the literature, [118] addresses practical issues for the use of B^+-trees in commercial database products, such as recovery after failures; [119] presents concurrent implementations for *generalized search trees* (GiSTs) that facilitate the design of search trees without repeating the delicate work involved with concurrency control; and [120,121] present several types of trees that support the efficient insertion and/or deletion of a group of values. Pugh [122] presents a concurrent version of his skiplist randomized search structure [123]. *Skiplists* are virtual tree structures consisting of multiple layers of linked lists. The expected search time in a skiplist is logarithmic in the number of elements in it. The main advantage of skiplists is that they do not require rebalancing: insertions are done in a randomized fashion that keeps the search tree balanced.

Empirical and analytical evaluations of concurrent search trees and other data structures can be found in References 124,125.

48.8 Priority Queues

A concurrent *priority queue* is a data structure linearizable to a sequential priority queue that provides `insert` and `delete-min` operations with the usual priority queue semantics. (See Part II of this handbook for more on sequential priority queues.)

48.8.1 Heap-Based Priority Queues

Many of the concurrent priority queue constructions in the literature are linearizable versions of the heap structures described earlier in this book. Again, the basic idea is to use fine-grained locking of the individual heap nodes to allow threads accessing different parts of the data structure to do so in parallel where possible. A key issue in designing such concurrent heaps is that traditionally `insert` operations proceed from the bottom up and `delete-min` operations from the top down, which creates potential for deadlock. Biswas and Brown [126] present such a lock-based heap algorithm assuming specialized "cleanup" threads to overcome deadlocks. Rao and Kumar [127] suggest to overcome the drawbacks of [126] using an algorithm that has both `insert` and `delete-min` operations proceed from the top down. Ayani [128] improved on their algorithm by suggesting a way to have consecutive insertions be performed on opposite sides of the heap. Jones [129] suggests a scheme similar to [127] based on a skew heap.

Hunt et al. [130] present a heap-based algorithm that overcomes many of the limitations of the above schemes, especially the need to acquire multiple locks along the traversal path in the heap. It proceeds by locking for a short duration a variable holding the size of the heap and a lock on either the first or last element of the heap. In order to increase parallelism, insertions traverse the heap bottom-up while deletions proceed top-down, without introducing deadlocks. Insertions also employ a left-right technique as in References 128 to allow them to access opposite sides on the heap and thus minimize interference.

On a different note, Huang and Weihl [131] show a concurrent priority queue based on a concurrent version of Fibonacci Heaps [132].

Nonblocking linearizable heap-based priority queue algorithms have been proposed by Herlihy [71], Barnes [133], and Israeli and Rappoport [134]. Sundell and Tsigas [135] present a lock-free priority queue based on a lock-free version of Pugh's concurrent skiplist [122].

48.8.2 Tree-Based Priority Pools

Huang and Weihl [131] and Johnson [136] describe concurrent *priority pools*: priority queues with relaxed semantics that do not guarantee linearizability of the delete-min operations. Their designs are both based on a modified concurrent B^+-tree implementation. Johnson introduces a "delete bin" that accumulates values to be deleted and thus reduces the load when performing concurrent delete-min operations. Shavit and Zemach [137] show a similar pool based on Pugh's concurrent skiplist [122] with an added "delete bin" mechanism based on [136]. Typically, the weaker pool semantics allows for increased concurrency. In References 137 they further show that if the size of the set of allowable keys is bounded (as is often the case in operating systems) a priority pool based on a binary tree of combining funnel nodes can scale to hundreds (as opposed to tens) of processors.

48.9 Summary

We have given an overview of issues related to the design of concurrent data structures for shared-memory multiprocessors, and have surveyed some of the important contributions in this area. Our overview clearly illustrates that the design of such data structures provides significant challenges, and as of this writing, the maturity of concurrent data structures falls well behind that of sequential data structures. However, significant progress has been made towards understanding key issues and developing new techniques to facilitate the design of effective concurrent data structures; we are particularly encouraged by renewed academic and industry interest in stronger hardware support for synchronization. Given new understanding, new techniques, and stronger hardware support, we believe significant advances in concurrent data structure designs are likely in the coming years.

References

1. D. Patterson and J. Hennessy: *Computer Organization and Design: The Hardware/Software Interface*. Morgan Kaufmann, 2nd edition, 1997.
2. A. Agarwal and M. Cherian: Adaptive backoff synchronization techniques. In *Proceedings of the 16th International Symposium on Computer Architecture*, pages 396–406, May 1989.
3. J. Goodman, M. Vernon, and P. Woest: Efficient synchronization primitives for large-scale cache-coherent shared-memory multiprocessors. In *Proceedings of the 3rd International Conference on Architectural Support for Programming Languages and Operating Systems (ASPLOS-III)*, pages 64–75. ACM Press, April 1989.
4. A. Gottlieb, R. Grishman, C.P. Kruskal, K.P. McAuliffe, L. Rudolph, and M. Snir: The NYU ultracomputer: Designing an MIMD parallel computer. *IEEE Transactions on Computers*, C-32(2):175–189, February 1984.
5. M. Herlihy, B. Lim, and N. Shavit: Scalable concurrent counting. *ACM Transactions on Computer Systems*, 13(4):343–364, 1995.
6. P. Yew, N. Tzeng, and D. Lawrie: Distributing hot-spot addressing in large-scale multiprocessors. *IEEE Transactions on Computers*, C-36(4):388–395, April 1987.
7. J. Mellor-Crummey and M. Scott: Algorithms for scalable synchronization on shared-memory multiprocessors. *ACM Transactions on Computer Systems*, 9(1):21–65, 1991.
8. M. Herlihy: Wait-free synchronization. *ACM Transactions on Programming Languages and Systems*, 13(1):124–149, January 1991.
9. L. Lamport: A new solution of Dijkstra's concurrent programming problem. *Communications of the ACM*, 17(8):453–455, 1974.
10. M. Herlihy, V. Luchangco, and M. Moir: Obstruction-free synchronization: Double-ended queues as an example. In *Proceedings of the 23rd International Conference on Distributed Computing Systems*, pages 522–529. IEEE, 2003.
11. M. Herlihy, V. Luchangco, M. Moir, and W. Scherer: Software transactional memory for dynamic data structures. In *Proceedings of the 22nd Annual ACM Symposium on Principles of Distributed Computing*, 2003.
12. M. Fischer, N. Lynch, and M. Paterson: Impossibility of distributed consensus with one faulty process. *Journal of the ACM*, pages 374–382, 1985.
13. M. Loui and H. Abu-Amara: Memory requirements for agreement among unreliable asynchronous processes. In F. P. Preparata, ed., Advances in Computing Research, Vol. 4, pages 163–183. JAI Press, Greenwich, CT, 1987.
14. IBM. System/370 principles of operation. Order Number GA22-7000.
15. Intel. Pentium Processor Family User's Manual: Vol 3, Architecture and Programming Manual, 1994.
16. D. Weaver and T. Germond (eds): *The SPARC Architecture Manual (Version 9)*. PTR Prentice Hall, Englewood Cliffs, NJ, 1994.

17. IBM. Powerpc microprocessor family: Programming environments manual for 64 and 32-bit microprocessors, version 2.0, 2003.

18. G. Kane: *MIPS RISC Architecture*. Prentice Hall, New York, 1989.

19. R. Sites: Alpha Architecture Reference Manual, 1992.

20. P. Gibbons, Y. Matias, and V. Ramachandran: The queue-read queue-write PRAM model: Accounting for contention in parallel algorithms. *SIAM Journal on Computing*, 28(2):733–769, 1998.

21. W. Savitch and M. Stimson: Time bounded random access machines with parallel processing. *Journal of the ACM*, 26:103–118, 1979.

22. L. Valiant: *Bulk-Synchronous Parallel Computers*. Wiley, New York, NY, 1989.

23. N. Shavit and A. Zemach: Diffracting trees. *ACM Transactions on Computer Systems*, 14(4):385–428, 1996.

24. C. Dwork, M. Herlihy, and O. Waarts: Contention in shared memory algorithms. *Journal of the ACM*, 44(6):779–805, 1997.

25. J. Anderson and J. Yang: Time/contention trade-offs for multiprocessor synchronization. *Information and Computation*, 124(1):68–84, 1996.

26. N. Arora, R. Blumofe, and G. Plaxton: Thread scheduling for multiprogrammed multiprocessors. *Theory of Computing Systems*, 34(2):115–144, 2001.

27. M. Michael and M. Scott: Nonblocking algorithms and preemption-safe locking on multiprogrammed shared: Memory multiprocessors. *Journal of Parallel and Distributed Computing*, 51(1):1–26, 1998.

28. L. Lamport: How to make a multiprocessor computer that correctly executes multiprocess programs. *IEEE Transactions on Computers*, C-28(9):241–248, September 1979.

29. M. Herlihy and J. Wing: Linearizability: A correctness condition for concurrent objects. *ACM Transactions on Programming Languages and Systems*, 12(3):463–492, July 1990.

30. P. Bernstein, V. Hadzilacos, and N. Goodman: *Concurrency Control and Recovery in Database Systems*. Addison-Wesley, 1987.

31. J. Aspnes, M. Herlihy, and N. Shavit: Counting networks. *Journal of the ACM*, 41(5):1020–1048, 1994.

32. T. Anderson: The performance implications of spin lock alternatives for shared-memory multiprocessors. *IEEE Transactions on Parallel and Distributed Systems*, 1(1):6–16, 1990.

33. G. Graunke and S. Thakkar: Synchronization algorithms for shared-memory multiprocessors. *IEEE Computer*, 23(6):60–70, June 1990.

34. T. Craig: Building FIFO and priority-queueing spin locks from atomic swap. Technical Report TR 93-02-02, University of Washington, Department of Computer Science, February 1993.

35. P. Magnussen, A. Landin, and E. Hagersten: Queue locks on cache coherent multiprocessors. In *Proceedings of the 8th International Symposium on Parallel Processing (IPPS)*, pages 165–171. IEEE Computer Society, April 1994.

36. M. Scott: Non-blocking timeout in scalable queue-based spin locks. In *Proceedings of the Twenty-first Annual Symposium on Principles of Distributed Computing*, pages 31–40. ACM Press, 2002.

37. M. Scott and W. Scherer: Scalable queue-based spin locks with timeout. In *Proceedings of the Eighth ACM SIGPLAN Symposium on Principles and Practices of Parallel Programming*, pages 44–52, 2001.

38. J. Mellor-Crummey and M. Scott: Scalable reader-writer synchronization for shared-memory multiprocessors. *SIGPLAN Not.*, 26(7):106–113, 1991.

39. O. Krieger, M. Stumm, R. Unrau, and J. Hanna: A fair fast scalable reader-writer lock. In *Proceedings of the 1993 International Conference on Parallel Processing, Volume II—Software*, pages II-201–II-204, Boca Raton, FL, 1993. CRC Press.

40. G. Blelloch, P. Cheng, and P. Gibbons: Room synchronizations. In *Proceedings of the thirteenth annual ACM symposium on Parallel algorithms and architectures*, pages 122–133. ACM Press, 2001.

41. Y. Joung: The congenial talking philosophers problem in computer networks. *Distrib. Comput.*, 15(3):155–175, 2002.

42. P. Keane and M. Moir: A simple local-spin group mutual exclusion algorithm. In *Proceedings of the Eighteenth Annual ACM Symposium on Principles of Distributed Computing*, pages 23–32. ACM Press, 1999.

43. J. Anderson, Y. Kim, and T. Herman: Shared-memory mutual exclusion: Major research trends since 1986. *Distributed Computing*, 16:75–110, 2003.

44. M. Raynal: *Algorithms for Mutual Exclusion*. The MIT Press, Cambridge, MA, 1986.

45. E.D. Brooks III: The butterfly barrier. *International Journal of Parallel Programming*, 15(4):295–307, October 1986.

46. D. Hensgen, R. Finkel, and U. Manber: Two algorithms for barrier synchronization. *International Journal of Parallel Programming*, 17(1):1–17, 1988.

47. M. Scott and J. Mellor-Crummey: Fast, contention-free combining tree barriers. *International Journal of Parallel Programming*, 22(4), August 1994.

48. E. Dijkstra and C. Sholten: Termination detection for diffusing computations. *Information Processing Letters*, 11(1):1–4, 1980.

49. H. Kung and J. Robinson: On optimistic methods for concurrency control. *ACM Transactions on Database Systems*, 6(2):213–226, 1981.

50. M. Herlihy and E. Moss: Transactional memory: Architectural support for lock-free data structures. In *Proceedings of the Twentieth Annual International Symposium on Computer Architecture*, 1993.

51. R. Rajwar and J. Goodman: Speculative lock elision: Enabling highly concurrent multithreaded execution. In *Proceedings of the 34th Annual International Symposium on Microarchitecture*, pages 294–305, 2001.

52. R. Rajwar and J. Goodman: Transactional lock-free execution of lock-based programs. In *10th Symposium on Architectural Support for Programming Languages and Operating Systems*, October 2003.

53. N. Shavit and D. Touitou: Software transactional memory. *Distributed Computing*, 10(2):99–116, February 1997.

54. T. Harris, K. Fraser, and I. Pratt: A practical multi-word compare-and-swap operation. In *Proceedings of the 16th International Symposium on DIStributed Computing*, pages 265–279, 2002.

55. T. Harris and K. Fraser: Language support for lightweight transactions. In *Proceedings of the 18th ACM SIGPLAN Conference on Object-Oriented Programing, Systems, Languages, and Applications*, pages 388–402. ACM Press, 2003.

56. C. Kruskal, L. Rudolph, and M. Snir: Efficient synchronization on multiprocessors with shared memory. In *Fifth ACM SIGACT-SIGOPS Symposium on Principles of Distributed Computing*, August 1986.

57. N. Shavit and A. Zemach: Combining funnels: A dynamic approach to software combining. *J. Parallel Distrib. Comput.*, 60(11):1355–1387, 2000.

58. T. Cormen, C. Leiserson, R. Rivest, and C. Stein: *Introduction to Algorithms*. MIT Press, Cambridge, MA, second edition, 2001.

59. K. Batcher: Sorting networks and their applications. In *Proceedings of AFIPS Joint Computer Conference*, pages 338–334, 1968.

60. C. Busch and M. Herlihy: A survey on counting networks.

61. E. Aharonson and H. Attiya: Counting networks with arbitrary fan-out. *Distributed Computing*, 8(4):163–169, 1995.

62. E. Felten, A. Lamarca, and R. Ladner: Building counting networks from larger balancers. Technical Report TR-93-04-09, University of Washington, 1993.

63. N. Shavit and D. Touitou: Elimination trees and the construction of pools and stacks. *Theory of Computing Systems*, 30:645–670, 1997.

64. C. Busch and M. Mavronicolas: A combinatorial treatment of balancing networks. *Journal of the ACM*, 43(5):794–839, September 1996.

65. B. Aiello, R. Venkatesan, and M. Yung: Coins, weights and contention in balancing networks. In *Proceedings of the 13th Annual ACM Symposium on Principles of Distributed Computing*, pages 193–214, August 1994.

66. M. Riedel and J. Bruck: Tolerating faults in counting networks. *Poster presented at 10th ACM Symposium on Parallel Algorithms and Architectures*, pages 27–36, 1998.

67. M. Herlihy, N. Shavit, and O. Waarts: Linearizable counting networks. *Distributed Computing*, 9(4):193–203, 1996.

68. M. Klugerman and G. Plaxton: Small-depth counting networks. In *Proceedings of the Twenty-Fourth Annual ACM Symposium on the Theory of Computing*, pages 417–428, 1992.

69. G. Della-Libera and N. Shavit: Reactive diffracting trees. *Journal of Parallel Distributed Computing*, 60(7):853–890, 2000.

70. R. Treiber: Systems programming: Coping with parallelism. Technical Report RJ5118, IBM Almaden Research Center, 1986.

71. M. Herlihy: A methodology for implementing highly concurrent data objects. *ACM Transactions on Programming Languages and Systems*, 15(5):745–770, November 1993.

72. D. Hendler, N. Shavit, and L. Yerushalmi: A scalable lock-free stack algorithm. In *Proceedings of the 16th ACM Symposium on Parallelism in Algorithms and Architectures (SPAA 2004)*, pages 206–215, 2004.

73. S. Prakash, Y. Lee, and T. Johnson: A non-blocking algorithm for shared queues using compare-and-swap. *IEEE Transactions on Computers*, 43(5):548–559, 1994.

74. M. Moir: Practical implementations of non-blocking synchronization primitives. In *Proceedings of the 16th Annual ACM Symposium on Principles of Distributed Computing*, pages 219–228, 1997.

75. L. Lamport: Specifying concurrent program modules. *ACM Transactions on Programming Languages and Systems*, 5(2):190–222, 1983.

76. D. Hendler and N. Shavit: Work dealing. In *Proceedings of the 14th Annual ACM Symposium on Parallel Algorithms and Architectures (SPAA 2002)*, pages 164–172, 2002.

77. M. Herlihy, V. Luchangco, and M. Moir: The repeat offender problem: A mechanism for supporting lock-free dynamic-sized data structures. In *Proceedings of the 16th International Symposium on DIStributed Computing*, volume 2508, pages 339–353. Springer-Verlag Heidelberg, January 2002. A improved version of this paper is in preparation for journal submission; please contact authors.

78. M. Michael: Safe memory reclamation for dynamic lock-free objects using atomic reads and writes. In *The 21st Annual ACM Symposium on Principles of Distributed Computing*, pages 21–30. ACM Press, 2002.

79. D. Knuth: *The Art of Computer Programming: Fundamental Algorithms*. Addison-Wesley, 2nd edition, 1968.

80. P. Martin, M. Moir, and G. Steele: Dcas-based concurrent deques supporting bulk allocation. Technical Report TR-2002-111, Sun Microsystems Laboratories, 2002.

81. Motorola. *MC68020 32-bit microprocessor user's manual*. Prentice-Hall, 1986.

82. U. Manber: On maintaining dynamic information in a concurrent environment. In *Proceedings of the Sixteenth Annual ACM symposium on Theory of Computing*, pages 273–278. ACM Press, 1984.

83. R. Blumofe and C. Leiserson: Scheduling multithreaded computations by work stealing. *Journal of the ACM*, 46(5):720–748, 1999.

84. L. Rudolph, M. Slivkin-Allalouf, and E. Upfal: A simple load balancing scheme for task allocation in parallel machines. In *In Proceedings of the 3rd Annual ACM Symposium on Parallel Algorithms and Architectures*, pages 237–245. ACM Press, July 1991.

85. P. Berenbrink, T. Friedetzky, and L. Goldberg: The natural work-stealing algorithm is stable. *SIAM Journal on Computing*, 32(5):1260–1279, 2003.

86. M. Mitzenmacher: Analyses of load stealing models based on differential equations. In *Proceedings of the 10th ACM Symposium on Parallel Algorithms and Architectures*, pages 212–221, 1998.

87. D. Hendler and N. Shavit: Non-blocking steal-half work queues. In *Proceedings of the 21st Annual ACM Symposium on Principles of Distributed Computing*, 2002.

88. U. Acar, G. Blelloch, and R. Blumofe: The data locality of work stealing. In *ACM Symposium on Parallel Algorithms and Architectures*, pages 1–12, 2000.

89. R. Bayer and M. Schkolnick: Concurrency of operations on B-trees. *Acta Informatica*, 9:1–21, 1979.

90. D. Lea: *Concurrent Programming in Java(TM): Design Principles and Pattern*. Addison-Wesley, 2nd edition, 1999.

91. J. Valois: Lock-free linked lists using compare-and-swap. In *ACM Symposium on Principles of Distributed Computing*, pages 214–222, 1995.

92. M. Michael and M. Scott: Correction of a memory management method for lock-free data structures. Technical Report TR599, University of Rochester, 1995.

93. T. Harris: A pragmatic implementation of non-blocking linked-lists. In *15th International Conference on Distributed Computing, Lecture Notes in Computer Science*, 2180, pages 300–314, 2001.

94. M. Michael: High performance dynamic lock-free hash tables and list-based sets. In *Proceedings of the Fourteenth Snnual ACM symposium on Parallel Algorithms and Architectures*, pages 73–82. ACM Press, 2002.

95. M. Hsu and W. Yang: Concurrent operations in extendible hashing. In *Symposium on Very Large Data Bases*, pages 241–247, 1986.

96. R. Fagin, J. Nievergelt, N. Pippenger, and H. Strong: Extendible hashing: a fast access method for dynamic files. *ACM Transactions on Database Systems*, 4(3):315–355, September 1979.

97. C. Ellis: Concurrency in linear hashing. *ACM Transactions on Database Systems (TODS)*, 12(2):195–217, 1987.

98. V. Kumar: Concurrent operations on extendible hashing and its performance. *Communications of the ACM*, 33(6):681–694, 1990.

99. D. Lea: Concurrent hash map in JSR166 concurrency utilities. http://gee.cs.oswego.edu/dl/concurrency-interest/index.html.

100. O. Shalev and N. Shavit: Split-ordered lists: Lock-free extensible hash tables. In *The 22nd Annual ACM Symposium on Principles of Distributed Computing*, pages 102–111. ACM Press, 2003.

101. W. Litwin: Linear hashing: A new tool for file and table addressing. In *Sixth International Conference on Very Large Data Bases*, pages 212–223. IEEE Computer Society, October 1980.

102. M. Greenwald: Two-handed emulation: How to build non-blocking implementations of complex data structures using DCAS. In *Proceedings of the 21st Annual Symposium on Principles of Distributed Computing*, 2002.

103. H. Kung and P. Lehman: Concurrent manipulation of binary search trees. *ACM Transactions on Programming Languages and Systems*, 5:354–382, September 1980.

104. J. Metzger: Managing simultaneous operations in large ordered indexes. Technical report, Technische Universität München, Institut für Informatik, TUM-Math, 1975.

105. B. Samadi: B-trees in a system with multiple users. *Information Processing Letter*, 5(4):107–112, October 1976.

106. P. Lehman and S. Bing Yao: Efficient locking for concurrent operations on B-trees. *ACM Transactions on Database Systems (TODS)*, 6(4):650–670, December 1981.

107. Y. Sagiv: Concurrent operations on B-trees with overtaking. *Journal of Computer and System Sciences*, 33(2):275–296, October 1986.

108. V. Lanin and D. Shasha: A symmetric concurrent B-tree algorithm. In *Proceedings of the Fall Joint Computer Conference 1986*, pages 380–389. IEEE Computer Society Press, November 1986.

109. L. Guibas and R. Sedgewick: A dichromatic framework for balanced trees. In *In Proceedings of the 19th IEEE Symposium on Foundations of Computer Science*, pages 8–21, October 1978.

110. J. Kessels: On-the-fly optimization of data structures. *Communications of the ACM*, 26(11):895–901, November 1983.

111. G. Adel'son-Vel'skii and E. Landis: An algorithm for the organization of information. *Doklady Akademii Nauk USSR*, 146(2):263–266, 1962.

112. O. Nurmi, E. Soisalon-Soininen, and D. Wood: Concurrency control in database structures with relaxed balance. In *Proceedings of the sixth ACM SIGACT-SIGMOD-SIGART Symposium on Principles of Database Systems*, pages 170–176. ACM Press, 1987.

113. K. Larsen: AVL trees with relaxed balance. In *Eighth International Parallel Processing Symposium*, pages 888–893. IEEE Computer Society Press, April 1994.

114. K. Larsen and R. Fagerberg: B-trees with relaxed balance. In *Ninth International Parallel Processing Symposium*, pages 196–202. IEEE Computer Society Press, April 1995.

115. J. Boyar and K. Larsen: Efficient rebalancing of chromatic search trees. *Journal of Computer and System Sciences*, 49(3):667–682, December 1994.

116. O. Nurmi and E. Soisalon-Soininen: Uncoupling updating and rebalancing in chromatic binary search trees. In *Proceedings of the Tenth ACM SIGACT-SIGMOD-SIGART Symposium on Principles of Database Systems*, pages 192–198. ACM Press, May 1991.

117. K. Fraser: Practical Lock-Freedom. Ph.D. dissertation, Kings College, University of Cambridge, Cambridge, England, September 2003.

118. C. Mohan and F. Levine: Aries/im: An efficient and high concurrency index management method using write-ahead logging. In *Proceedings of the 1992 ACM SIGMOD International Conference on Management of Data*, pages 371–380. ACM Press, 1992.

119. M. Kornacker, C. Mohan, and J. Hellerstein: Concurrency and recovery in generalized search trees. In *Proceedings of the 1997 ACM SIGMOD International Conference on Management of Data*, pages 62–72. ACM Press, 1997.

120. K. Larsen: Relaxed multi-way trees with group updates. In *Proceedings of the Twentieth ACM SIGMOD-SIGACT-SIGART Symposium on Principles of Database Systems*, pages 93–101. ACM Press, 2001.

121. K. Larsen: Relaxed red-black trees with group updates. *Acta Informatica*, 38(8):565–586, 2002.

122. W. Pugh: Concurrent maintenance of skip lists. Technical Report CS-TR-2222.1, Institute for Advanced Computer Studies, Department of Computer Science, University of Maryland, 1989.

123. W. Pugh: Skip lists: A probabilistic alternative to balanced trees. *ACM Transactions on Database Systems*, 33(6):668–676, 1990.

124. S. Hanke: The performance of concurrent red-black tree algorithms. *Lecture Notes in Computer Science*, 1668:286–300, 1999.

125. T. Johnson and D. Sasha: The performance of current B-tree algorithms. *ACM Transactions on Database Systems (TODS)*, 18(1):51–101, 1993.

126. J. Biswas and J. Browne: Simultaneous update of priority structures. In *In Proceedings of the 1987 International Conference on Parallel Processing*, pages 124–131, August 1987.

127. V. Rao and V. Kumar: Concurrent access of priority queues. *IEEE Transactions on Computers*, 37:1657–1665, December 1988.

128. R. Ayani: LR-algorithm: Concurrent operations on priority queues. In *In Proceedings of the 2nd IEEE Symposium on Parallel and Distributed Processing*, pages 22–25, 1991.

129. D. Jones: Concurrent operations on priority queues. *Communications of the ACM*, 32(1):132–137, 1989.

130. G. Hunt, M. Michael, S. Parthasarathy, and M. Scott: An efficient algorithm for concurrent priority queue heaps. *Information Processing Letters*, 60(3):151–157, November 1996.

131. Q. Huang and W. Weihl: An evaluation of concurrent priority queue algorithms. In *IEEE Parallel and Distributed Computing Systems*, pages 518–525, 1991.

132. M. Fredman and R. Tarjan: Fibonacci heaps and their uses in improved network optimization algorithms. *J. ACM*, 34(3):596–615, 1987.

133. G. Barnes: Wait free algorithms for heaps. Technical Report TR-94-12-07, University of Washington, 1994.

134. A. Israeli and L. Rappoport: Efficient wait-free implementation of a concurrent priority queue. In *The 7th International Workshop on Distributed Algorithms*, pages 1–17, 1993.

135. H. Sundell and P. Tsigas: Fast and lock-free concurrent priority queues for multi-thread systems. In *Proceedings of the 17th International Parallel and Distributed Processing Symposium*, pages 84–94. IEEE press, 2003.

136. T. Johnson: A highly concurrent priority queue based on the B-link tree. Technical Report 91-007, University of Florida, August 1991.

137. N. Shavit and A. Zemach: Scalable concurrent priority queue algorithms. In *Proceedings of the Eighteenth Annual ACM Symposium on Principles of Distributed Computing*, pages 113–122. ACM Press, 1999.

VII

Applications

49 IP Router Tables Sartaj Sahni, Kun Suk Kim and Haibin Lu 765
Introduction • Longest Matching-Prefix • Highest-Priority Matching • Most-Specific-Range
Matching • Acknowledgments • References

50 Multi-Dimensional Packet Classification Pankaj Gupta 783
Introduction • Performance Metrics for Classification Algorithms • Classification Algorithms •
Summary • References

51 Data Structures in Web Information Retrieval Monika Henzinger 799
Introduction • Inverted Indices • Fingerprints • Finding Near-Duplicate Documents •
Conclusions • References

52 The Web as a Dynamic Graph S. N. Maheshwari 805
Introduction • Experimental Observations • Theoretical Growth Models • Properties of Web
Graphs and Web Algorithmics • Conclusions • References

53 Layout Data Structures Dinesh P. Mehta 817
Introduction • VLSI Technology • Layout Data Structures: An Overview • Corner Stitching •
Corner Stitching Extensions • Quad Trees and Variants • Concluding Remarks •
Acknowledgments • References

54 Floorplan Representation in VLSI Zhou Feng, Bo Yao and Chung-Kuan Cheng 833
Introduction • Graph Based Representations • Placement Based Representations • Relationships of
the Representations • Rectilinear Shape Handling • Conclusions • Acknowledgments •
References

55 Computer Graphics Dale McMullin and Alyn Rockwood 857
Introduction • Basic Applications • Data Structures • Applications of Previously Discussed
Structures • References

56 Geographic Information Systems Bernhard Seeger and Peter Widmayer 871
Geographic Information Systems: What They Are All About • Space Filling Curves: Order in Many
Dimensions • Spatial Join • Models, Toolboxes, and Systems for Geographic Information •
Acknowledgment • References

57 Collision Detection Ming C. Lin and Dinesh Manocha 889
Introduction • Convex Polytopes • General Polygonal Models • Penetration Depth Computation •
Large Environments • References

58 Image Data Structures S. S. Iyengar, V. K. Vaishnavi and S. Gunasekaran 903
Introduction • What is Image Data? • Quadtrees • Virtual Quadtrees • Quadtrees and R-trees •
Octrees • Translation Invariant Data Structure (TID) • Content-Based Image Retrieval System •
Summary • Acknowledgments • References

59 Computational Biology Paolo Ferragina, Stefan Kurtz, Stefano Lonardi and Giovanni Manzini 917
Introduction • Discovering Unusual Words • Comparing Whole Genomes • The FM-index •
References

60 Data Structures for Cheminformatics Dinesh P. Mehta and John D. Crabtree 935
Introduction • Exact Searches • Chemical Fingerprints and Similarity Search • References

61 Elimination Structures in Scientific Computing Alex Pothen and Sivan Toledo 945
The Elimination Tree • Applications of Etrees • The Clique Tree • Clique Covers and Quotient
Graphs • Column Elimination Trees and Elimination DAGS • Acknowledgments • References

62 Data Structures for Databases Joachim Hammer and Markus Schneider 967
Overview of the Functionality of a Database Management System • Data Structures for Query
Processing • Data Structures for Buffer Management • Data Structures for Disk Space Management
• Conclusion • References

63 Data Structures for Big Data Stores Arun A. Ravindran and Dinesh P. Mehta 983
Introduction • Data Models • Partitioning • Replication and Consistency • Persistence •
Concurrency • Conclusion • References

64 Data Mining Vipin Kumar, Pang-Ning Tan and Michael Steinbach 997
Introduction • Classification • Association Analysis • Clustering • Conclusion •
Acknowledgments • References

65 Computational Geometry: Fundamental Structures Mark de Berg and Bettina Speckmann....................... 1013
Introduction • Arrangements • Convex Hulls • Voronoi Diagrams • Triangulations • References

66 Computational Geometry: Proximity and Location Sunil Arya and David M. Mount 1027
Introduction • Point Location • Proximity Structures • Nearest Neighbor Searching • Sources and
Related Material • Acknowledgments • References

67 Computational Geometry: Generalized (or Colored) Intersection Searching Prosenjit Gupta, Ravi
Janardan, Saladi Rahul and Michiel Smid.. 1043
Geometric Intersection Searching Problems • Summary of Known Results • Techniques •
Conclusion and Future Directions • Acknowledgments • References

49

IP Router Tables[*]

49.1	Introduction	765
49.2	Longest Matching-Prefix	767
	Linear List • End-Point Array • Sets of Equal-Length Prefixes • Tries • Binary Search Trees • Priority Search Trees	
49.3	Highest-Priority Matching	777
	The Data Structure BOB • Search for the Highest-Priority Matching Range	
49.4	Most-Specific-Range Matching	780
	Nonintersecting Ranges • Conflict-Free Ranges	
	Acknowledgments	780
	References	781

Sartaj Sahni
University of Florida

Kun Suk Kim[†]
University of Florida

Haibin Lu[†]
University of Florida

49.1 Introduction

An Internet router classifies incoming packets into flows[‡] utilizing information contained in packet headers and a table of (classification) rules. This table is called the *rule table* (equivalently, *router table*). The packet-header information that is used to perform the classification is some subset of the source and destination addresses, the source and destination ports, the protocol, protocol flags, type of service, and so on. The specific header information used for packet classification is governed by the rules in the rule table. Each rule-table rule is a pair of the form (F, A), where F is a filter and A is an action. The action component of a rule specifies what is to be done when a packet that satisfies the rule filter is received. Sample actions are drop the packet, forward the packet along a certain output link, and reserve a specified amount of bandwidth. A rule filter F is a tuple that is comprised of one or more fields. In the simplest case of destination-based packet forwarding, F has a single field, which is a destination (address) prefix and A is the next hop for packets whose destination address has the specified prefix. For example, the rule $(01*, a)$ states that the next hop for packets whose destination address (in binary) begins with 01 is a. IP (Internet Protocol) multicasting uses rules in which F is comprised of the two fields source prefix and destination prefix; QoS routers may use five-field rule filters (source-address prefix, destination-address prefix, source-port range, destination-port range, and protocol); and firewall filters may have one or more fields.

In the d-dimensional packet classification problem, each rule has a d-field filter. In this chapter, we are concerned solely with 1-dimensional packet classification. It should be noted, that data structures for multidimensional packet classification are usually built on top of data structures for 1-dimensional packet classification. Therefore, the study of data structures for 1-dimensional packet classification is fundamental to the design and development of data structures for d-dimensional, $d > 1$, packet classification.

For the 1-dimensional packet classification problem, we assume that the single field in the filter is the destination field and that the action is the next hop for the packet. With these assumptions, 1-dimensional packet classification is equivalent to the destination-based packet forwarding problem. Henceforth, we shall use the terms rule table and router table to mean tables in which the filters have a single field, which is the destination address. This single field of a filter may be specified in one of two ways:

1. *As a range.* For example, the range [35, 2096] matches all destination addresses d such that $35 \leq d \leq 2096$.
2. *As an address/mask pair.* Let x_i denote the ith bit of x. The address/mask pair a/m matches all destination addresses d for which $d_i = a_i$ for all i for which $m_i = 1$. That is, a 1 in the mask specifies a bit position in which d and a must agree, while a 0

---

[*] This chapter has been reprinted from first edition of this Handbook, without any content updates.

[†] We have used this author's affiliation from the first edition of this handbook, but note that this may have changed since then.

[‡] A **flow** is a set of packets that are to be treated similarly for routing purposes.

Prefix Name	Prefix	Range Start	Range Finish
P1	*	0	31
P2	0101*	10	11
P3	100*	16	19
P4	1001*	18	19
P5	10111	23	23

FIGURE 49.1   Prefixes and their ranges.

in the mask specifies a don't care bit position. For example, the address/mask pair 101100/011101 matches the destination addresses 101100, 101110, 001100, and 001110.

When all the 1-bits of a mask are to the left of all 0-bits, the address/mask pair specifies an address prefix. For example, 101100/110000 matches all destination addresses that have the prefix 10 (i.e., all destination addresses that begin with 10). In this case, the address/mask pair is simply represented as the prefix 10*, where the * denotes a sequence of don't care bits. If W is the length, in bits, of a destination address, then the * in 10* represents all sequences of $W - 2$ bits. In IPv4 the address and mask are both 32 bits, while in IPv6 both of these are 128 bits.

Notice that every prefix may be represented as a range. For example, when $W = 6$, the prefix 10* is equivalent to the range [32,47]. A range that may be specified as a prefix for some W is called a *prefix range*. The specification 101100/011101 may be abbreviated to ?011?0, where ? denotes a don't-care bit. This specification is not equivalent to any single range. Also, the range specification [3,6] isn't equivalent to any single address/mask specification.

Figure 49.1 shows a set of five prefixes together with the start and finish of the range for each. This figure assumes that $W = 5$. The prefix P1 = *, which matches all legal destination addresses, is called the *default* prefix.

Suppose that our router table is comprised of five rules R1–R5 and that the filters for these five rules are P1–P5, respectively. Let N1–N5, respectively, be the next hops for these five rules. The destination address 18 is matched by rules R1, R3, and R5 (equivalently, by prefixes P1, P3, and P5). So, N1, N3, and N5 are candidates for the next hop for incoming packets that are destined for address 18. Which of the matching rules (and associated action) should be selected? When more than one rule matches an incoming packet, a *tie* occurs. To select one of the many rules that may match an incoming packet, we use a *tie breaker*.

Let *RS* be the set of rules in a rule table and let *FS* be the set of filters associated with these rules. *rules(d, RS)* (or simply *rules(d)* when *RS* is implicit) is the subset of rules of *RS* that match/cover the destination address *d*. *filters(d, FS)* and *filters(d)* are defined similarly. A tie occurs whenever $|rules(d)| > 1$ (equivalently, $|filters(d)| > 1$).

Three popular tie breakers are:

1. *First matching rule in table.* The rule table is assumed to be a linear list [1] of rules with the rules indexed 1 through n for an n-rule table. The action corresponding to the first rule in the table that matches the incoming packet is used. In other words, for packets with destination address *d*, the rule of *rules(d)* that has least index is selected.

   For our example router table corresponding to the five prefixes of Figure 49.1, rule R1 is selected for every incoming packet, because P1 matches every destination address. When using the first-matching-rule criteria, we must index the rules carefully. In our example, P1 should correspond to the last rule so that every other rule has a chance to be selected for at least one destination address.

2. *Highest-priority rule.* Each rule in the rule table is assigned a priority. From among the rules that match an incoming packet, the rule that has highest priority wins is selected. To avoid the possibility of a further tie, rules are assigned different priorities (it is actually sufficient to ensure that for every destination address *d*, *rules(d)* does not have two or more highest-priority rules).

   Notice that the first-matching-rule criteria is a special case of the highest-priority criteria (simply assign each rule a priority equal to the negative of its index in the linear list).

3. *Most-specific-rule matching.* The filter F1 is **more specific** than the filter F2 iff F2 matches all packets matched by F1 plus at least one additional packet. So, for example, the range [2, 4] is more specific than [1, 6], and [5, 9] is more specific than [5, 12]. Since [2, 4] and [8, 14] are disjoint (i.e., they have no address in common), neither is more specific than the other. Also, since [4, 14] and [6, 20] intersect,* neither is more specific than the other. The prefix 110* is more specific than the prefix 11*.

---

* Two ranges $[u, v]$ and $[x, y]$ intersect iff $u < x \leq v < y \vee x < u \leq y < v$.

In most-specific-rule matching, ties are broken by selecting the matching rule that has the most specific filter. When the filters are destination prefixes, the most-specific-rule that matches a given destination d is the longest[*] prefix in *filters(d)*. Hence, for prefix filters, the most-specific-rule tie breaker is equivalent to the longest-matching-prefix criteria used in router tables. For our example rule set, when the destination address is 18, the longest matching-prefix is P4. When the filters are ranges, the most-specific-rule tie breaker requires us to select the most specific range in *filters(d)*. Notice also that most-specific-range matching is a special case of the the highest-priority rule. For example, when the filters are prefixes, set the prefix priority equal to the prefix length. For the case of ranges, the range priority equals the negative of the range size.

In a *static* rule table, the rule set does not vary in time. For these tables, we are concerned primarily with the following metrics:

1. *Time required to process an incoming packet.* This is the time required to search the rule table for the rule to use.
2. *Preprocessing time.* This is the time to create the rule-table data structure.
3. *Storage requirement.* That is, how much memory is required by the rule-table data structure?

In practice, rule tables are seldom truly static. At best, rules may be added to or deleted from the rule table infrequently. Typically, in a "static" rule table, inserts/deletes are batched and the rule-table data structure reconstructed as needed.

In a *dynamic* rule table, rules are added/deleted with some frequency. For such tables, inserts/deletes are not batched. Rather, they are performed in real time. For such tables, we are concerned additionally with the time required to insert/delete a rule. For a dynamic rule table, the initial rule-table data structure is constructed by starting with an empty data structures and then inserting the initial set of rules into the data structure one by one. So, typically, in the case of dynamic tables, the preprocessing metric, mentioned above, is very closely related to the insert time.

In this paper, we focus on data structures for static and dynamic router tables (1-dimensional packet classification) in which the filters are either prefixes or ranges.

49.2 Longest Matching-Prefix

49.2.1 Linear List

In this data structure, the rules of the rule table are stored as a linear list[1] L. Let $LMP(d)$ be the longest matching-prefix for address d. $LMP(d)$ is determined by examining the prefixes in L from left to right; for each prefix, we determine whether or not that prefix matches d; and from the set of matching prefixes, the one with longest length is selected. To insert a rule q, we first search the list L from left to right to ensure that L doesn't already have a rule with the same filter as does q. Having verified this, the new rule q is added to the end of the list. Deletion is similar. The time for each of the operations determine $LMP(d)$, insert a rule, delete a rule is $O(n)$, where n is the number of rules in L. The memory required is also $O(n)$.

Note that this data structure may be used regardless of the form of the filter (i.e., ranges, Boolean expressions, etc.) and regardless of the tie breaker in use. The time and memory complexities are unchanged.

Although the linear list is not a suitable data structure for a purely software implementation of a router table with a large number of prefixes, it is leads to a very practical solution using TCAMs (ternary content-addressable memories) [2,3]. Each memory cell of a TCAM may be set to one of three states 0, 1, and don't care. The prefixes of a router table are stored in a TCAM in descending order of prefix length. Assume that each word of the TCAM has 32 cells. The prefix 10* is stored in a TCAM word as 10???...?, where ? denotes a don't care and there are 30 ?s in the given sequence. To do a longest-prefix match, the destination address is matched, in parallel, against every TCAM entry and the first (i.e., longest) matching entry reported by the TCAM arbitration logic. So, using a TCAM and a sorted-by-length linear list, the longest matching-prefix can be determined in $O(1)$ time. A prefix may be inserted or deleted in $O(W)$ time, where W is the length of the longest prefix [2].[†] For example, an insert of a prefix of length 3 (say) can be done by relocating a the first prefix of length 1 to the end of the list; filling the vacated slot with the first prefix of length 2; and finally filling this newly vacated spot with the prefix of length 3 that is to be inserted.

Despite the simplicity and efficiency of using TCAMs, TCAMs present problems in real applications [4]. For example, TCAMs consume a lot of power and board area.

49.2.2 End-Point Array

Lampson, Srinivasan, and Varghese [5] have proposed a data structure in which the end points of the ranges defined by the prefixes are stored in ascending order in an array. $LMP(d)$ is found by performing a binary search on this ordered array of end points.

---

[*] The length of a prefix is the number of bits in that prefix (note that the * is not used in determining prefix length). The length of P1 is 0 and that of P2 is 4.
[†] More precisely, W may be defined to be the number of different prefix lengths in the table.

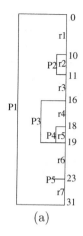

EndPoint	>	=
0	P1	P1
10	P2	P2
11	P1	P2
16	P3	P3
18	P4	P4
19	P1	P4
23	P1	P5
31	–	P1

(a)   (b)

FIGURE 49.2   (a) Pictorial representation of prefixes and ranges (b) Array for binary search.

Prefixes and their ranges may be drawn as nested rectangles as in Figure 49.2a, which gives the pictorial representation of the five prefixes of Figure 49.1.

In the data structure of Lampson et al. [5], the distinct range end-points are stored in ascending order as in Figure 49.2b. The distinct end-points (range start and finish points) for the prefixes of Figure 49.1 are [0, 10, 11, 16, 18, 19, 23, 31]. Let r_i, $1 \leq i \leq q \leq 2n$ be the distinct range end-points for a set of n prefixes. Let $r_{q+1} = \infty$. With each distinct range end-point, r_i, $1 \leq i \leq q$, the array stores $LMP(d)$ for d such that (a) $r_i < d < r_{i+1}$ (this is the column labeled "$>$" in Figure 49.2b and (b) $r_i = d$ (column labeled "$=$"). Now, $LMP(d)$, $r_1 \leq d \leq r_q$ can be determined in $O(\log n)$ time by performing a binary search to find the unique i such that $r_i \leq d < r_{i+1}$. If $r_i = d$, $LMP(d)$ is given by the "$=$" entry; otherwise, it is given by the "$>$" entry. For example, since $d = 20$ satisfies $19 \leq d < 23$ and since $d \neq 19$, the "$>$" entry of the end point 19 is used to determine that $LMP(20)$ is P1.

As noted by Lampson et al. [5], the range end-point table can be built in $O(n)$ time (this assumes that the end points are available in ascending order). Unfortunately, as stated in Reference 5, updating the range end-point table following the insertion or deletion of a prefix also takes $O(n)$ time because $O(n)$ "$>$" and/or "$=$" entries may change. Although Lampson et al. [5] provide ways to reduce the complexity of the search for the LMP by a constant factor, these methods do not result in schemes that permit prefix insertion and deletion in $O(\log n)$ time.

It should be noted that the end-point array may be used even when ties are broken by selecting the first matching rule or the highest-priority matching rule. Further, the method applies to the case when the filters are arbitrary ranges rather than simply prefixes. The complexity of the preprocessing step (i.e., creation of the array of ordered end-points) and the search for the rule to use is unchanged. Further, the memory requirements are the same, $O(n)$ for an n-rule table, regardless of the tie breaker and whether the filters are prefixes or general ranges.

49.2.3 Sets of Equal-Length Prefixes

Waldvogel et al. [6] have proposed a data structure to determine $LMP(d)$ by performing a binary search on prefix length. In this data structure, the prefixes in the router table T are partitioned into the sets S_0, S_1, ... such that S_i contains all prefixes of T whose length is i. For simplicity, we assume that T contains the default prefix. *So, $S_0 = \{*\}$. Next, each S_i is augmented with markers that represent prefixes in S_j such that $j > i$ and i is on the binary search path to S_j. For example, suppose that the length of the longest prefix of T is 32 and that the length of $LMP(d)$ is 22. To find $LMP(d)$ by a binary search on length, we will first search S_{16} for an entry that matches the first 16 bits of d. This search* will need to be successful for us to proceed to a larger length. The next search will be in S_{24}. This search will need to fail. Then, we will search S_{20} followed by S_{22}. So, the path followed by a binary search on length to get to S_{22} is S_{16}, S_{24}, S_{20}, and S_{22}. For this to be followed, the searches in S_{16}, S_{20}, and S_{22} must succeed while that in S_{24} must fail. Since the length of $LMP(d)$ is 22, T has no matching prefix whose length is more than 22. So, the search in S_{24} is guaranteed to fail. Similarly, the search in S_{22} is guaranteed to succeed. However, the searches in S_{16} and S_{20} will succeed iff T has matching prefixes of length 16 and 20. To ensure success, every length 22 prefix P places a *marker* in S_{16} and S_{20}, the marker in S_{16} is the first 16 bits of P and that in S_{20} is the first 20 bits in P. Note that a marker M is placed in S_i only if S_i doesn't contain a prefix equal to M. Notice also that for each i, the binary search path to S_i has $O(\log l_{max}) = O(\log W)$, where l_{max} is the

---
* When searching S_i, only the first i bits of d are used, because all prefixes in S_i have exactly i bits.

length of the longest prefix in T, S_js on it. So, each prefix creates $O(\log W)$ markers. With each marker M in S_i, we record the longest prefix of T that matches M (the length of this longest matching-prefix is necessarily smaller than i).

To determine $LMP(d)$, we begin by setting $leftEnd = 0$ and $rightEnd = l_{max}$. The repetitive step of the binary search requires us to search for an entry in S_m, where $m = \lfloor (leftEnd + rightEnd)/2 \rfloor$, that equals the first m bits of d. If S_m does not have such an entry, set $rightEnd = m - 1$. Otherwise, if the matching entry is the prefix P, P becomes the longest matching-prefix found so far. If the matching entry is the marker M, the prefix recorded with M is the longest matching-prefix found so far. In either case, set $leftEnd = m + 1$. The binary search terminates when $leftEnd > rightEnd$.

One may easily establish the correctness of the described binary search. Since, each prefix creates $O(\log W)$ markers, the memory requirement of the scheme is $O(n \log W)$. When each set S_i is represented as a hash table, the data structure is called SELPH (sets of equal length prefixes using hash tables). The expected time to find $LMP(d)$ is $O(\log W)$ when the router table is represented as an SELPH. When inserting a prefix, $O(\log W)$ markers must also be inserted. With each marker, we must record a longest-matching prefix. The expected time to find these longest matching-prefixes is $O(\log^2 W)$. In addition, we may need to update the longest-matching prefix information stored with the $O(n \log W)$ markers at lengths greater than the length of the newly inserted prefix. This takes $O(n \log^2 W)$ time. So, the expected insert time is $O(n \log^2 W)$. When deleting a prefix P, we must search all hash tables for markers M that have P recorded with them and then update the recorded prefix for each of these markers. For hash tables with a bounded loading density, the expected time for a delete (including marker-prefix updates) is $O(n \log^2 W)$. Waldvogel et al. [6] have shown that by inserting the prefixes in ascending order of length, an n-prefix SELPH may be constructed in $O(n \log^2 W)$ time.

When each set is represented as a balanced search tree (see Chapter 11), the data structure is called SELPT. In an SELPT, the time to find $LMP(d)$ is $O(\log n \log W)$; the insert time is $O(n \log n \log^2 W)$; the delete time is $O(n \log n \log^2 W)$; and the time to construct the data structure for n prefixes is $O(W + n \log n \log^2 W)$.

In the full version of [6], Waldvogel et al. show that by using a technique called marker partitioning, the SELPH data structure may be modified to have a search time of $O(\alpha + \log W)$ and an insert/delete time of $O(\alpha \sqrt[\alpha]{n} W \log W)$, for any $\alpha > 1$.

Because of the excessive insert and delete times, the sets of equal-length prefixes data structure is suitable only for static router tables. Note that in an actual implementation of SELPH or SELPT, we need only keep the non-empty S_is and do a binary search over the collection of non-empty S_is. Srinivasan and Varghese [7] have proposed the use of controlled prefix-expansion to reduce the number of non-empty sets S_i. The details of their algorithm to reduce the number of lengths are given in Reference 8. The complexity of their algorithm is $O(nW^2)$, where n is the number of prefixes, and W is the length of the longest prefix. The algorithm of [8] does not minimize the storage required by the prefixes and markers for the resulting set of prefixes. Kim and Sahni [9] have developed an algorithm that minimizes storage requirement but takes $O(nW^3 + kW^4)$ time, where k is the desired number of non-empty S_is. Additionally, Kim and Sahni [9] propose improvements to the heuristic of [8].

We note that Waldvogel's scheme is very similar to the k-ary search-on-length scheme developed by Berg et al. [10] and the binary search-on-length schemes developed by Willard [11]. Berg et al. [10] use a variant of stratified trees [12] for one-dimensional point location in a set of n disjoint ranges. Willard [11] modified stratified trees and proposed the y-fast trie data structure to search a set of disjoint ranges. By decomposing filter ranges that are not disjoint into disjoint ranges, the schemes of [10,11] may be used for longest-prefix matching in static router tables. The asymptotic complexity for a search using the schemes of [10,11] is the same as that of Waldvogel's scheme. The decomposition of overlapping ranges into disjoint ranges is feasible for static router tables but not for dynamic router tables because a large range may be decomposed into $O(n)$ disjoint small ranges.

49.2.4 Tries

49.2.4.1 1-Bit Tries

A *1-bit trie* is a tree-like structure in which each node has a left child, left data, right child, and right data field. Nodes at level[*] $l - 1$ of the trie store prefixes whose length is l. If the rightmost bit in a prefix whose length is l is 0, the prefix is stored in the left data field of a node that is at level $l - 1$; otherwise, the prefix is stored in the right data field of a node that is at level $l - 1$. At level i of a trie, branching is done by examining bit i (bits are numbered from left to right beginning with the number 0) of a prefix or destination address. When bit i is 0, we move into the left subtree; when the bit is 1, we move into the right subtree. Figure 49.3a gives the prefixes in the 8-prefix example of [7], and Figure 49.3b shows the corresponding 1-bit trie. The prefixes in Figure 49.3a are numbered and ordered as in Reference 7.

The 1-bit tries described here are an extension of the 1-bit tries described in Reference 1. The primary difference being that the 1-bit tries of [1] are for the case when no key is a prefix of another. Since in router-table applications, this condition isn't satisfied, the 1-bit trie representation of [1] is extended so that keys of length l are stored in nodes at level $l - 1$ of the trie. Note that at most

---

[*] Level numbers are assigned beginning with 0 for the root level.

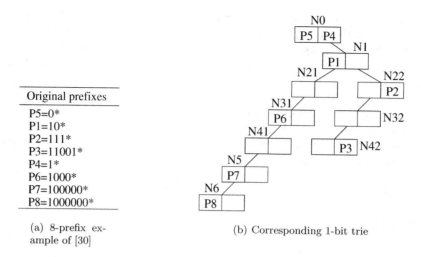

Original prefixes
P5=0*
P1=10*
P2=111*
P3=11001*
P4=1*
P6=1000*
P7=100000*
P8=1000000*

(a) 8-prefix example of [30]

(b) Corresponding 1-bit trie

FIGURE 49.3 Prefixes and corresponding 1-bit trie.

two keys may be stored in a node; one of these has bit l equal to 0 and other has this bit equal to 1. In this extension, every node of the 1-bit trie has 2 child pointers and 2 data fields. The height of a 1-bit trie is $O(W)$.

For any destination address d, all prefixes that match d lie on the search path determined by the bits of d. By following this search path, we may determine the longest matching-prefix, the first prefix in the table that matches d, as well as the highest-priority matching-prefix in $O(W)$ time. Further, prefixes may be inserted/deleted in $O(W)$ time. The memory required by the 1-bit trie is $O(nW)$.

IPv4 backbone routers may have more than 100 thousand prefixes. Even though the prefixes in a backbone router may have any length between 0 and W, there is a concentration of prefixes at lengths 16 and 24, because in the early days of the Internet, Internet address assignment was done by classes. All addresses in a class B network have the same first 16 bits, while addresses in the same class C network agree on the first 24 bits. Addresses in class A networks agree on their first 8 bits. However, there can be at most 256 class A networks (equivalently, there can be at most 256 8-bit prefixes in a router table). For our backbone routers that occur in practice [13], the number of nodes in a 1-bit trie is between $2n$ and $3n$. Hence, in practice, the memory required by the 1-bit-trie representation is $O(n)$.

49.2.4.2 Fixed-Stride Tries

Since the trie of Figure 49.3b has a height of 6, a search into this trie may make up to 7 memory accesses, one access for each node on the path from the root to a node at level 6 of the trie. The total memory required for the 1-bit trie of Figure 49.3b is 20 units (each node requires 2 units, one for each pair of (child, data) fields).

When 1-bit tries are used to represent IPv4 router tables, the trie height may be as much as 31. A lookup in such a trie takes up to 32 memory accesses.

Degermark et al. [14] and Srinivasan and Varghese [7] have proposed the use of fixed-stride tries to enable fast identification of the longest matching prefix in a router table. The *stride* of a node is defined to be the number of bits used at that node to determine which branch to take. A node whose stride is s has 2^s child fields (corresponding to the 2^s possible values for the s bits that are used) and 2^s data fields. Such a node requires 2^s memory units. In a *fixed-stride trie* (FST), all nodes at the same level have the same stride; nodes at different levels may have different strides.

Suppose we wish to represent the prefixes of Figure 49.3a using an FST that has three levels. Assume that the strides are 2, 3, and 2. The root of the trie stores prefixes whose length is 2; the level one nodes store prefixes whose length is 5 (2 + 3); and level three nodes store prefixes whose length is 7 (2 + 3 + 2). This poses a problem for the prefixes of our example, because the length of some of these prefixes is different from the storeable lengths. For instance, the length of P5 is 1. To get around this problem, a prefix with a nonpermissible length is expanded to the next permissible length. For example, P5 = 0* is expanded to P5a = 00* and P5b = 01*. If one of the newly created prefixes is a duplicate, natural dominance rules are used to eliminate all but one occurrence of the prefix. For instance, P4 = 1* is expanded to P4a = 10* and P4b = 11*. However, P1 = 10* is to be chosen over P4a = 10*, because P1 is a longer match than P4. So, P4a is eliminated. Because of the elimination of duplicate prefixes from the expanded prefix set, all prefixes are distinct. Figure 49.4a shows the prefixes that result when we expand the prefixes of Figure 49.3 to lengths 2, 5, and 7. Figure 49.4b shows the corresponding FST whose height is 2 and whose strides are 2, 3, and 2.

Expanded prefixes (3 levels)
00* (P5a)
01* (P5b)
10* (P1)
11* (P4)
11100* (P2a)
11101* (P2b)
11110* (P2c)
11111* (P2d)
11001* (P3)
10000* (P6a)
10001* (P6b)
1000001* (P7)
1000000* (P8)

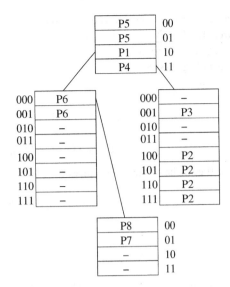

(a) Expanded prefixes

(b) Corresponding fixed-stride trie

FIGURE 49.4 Prefix expansion and fixed-stride trie.

Since the trie of Figure 49.4b can be searched with at most 3 memory references, it represents a time performance improvement over the 1-bit trie of Figure 49.3b, which requires up to 7 memory references to perform a search. However, the space requirements of the FST of Figure 49.4b are more than that of the corresponding 1-bit trie. For the root of the FST, we need 8 fields or 4 units; the two level 1 nodes require 8 units each; and the level 3 node requires 4 units. The total is 24 memory units.

We may represent the prefixes of Figure 49.3a using a one-level trie whose root has a stride of 7. Using such a trie, searches could be performed making a single memory access. However, the one-level trie would require $2^7 = 128$ memory units.

For IPv4 prefix sets, Degermark et al. [14] propose the use of a three-level trie in which the strides are 16, 8, and 8. They propose encoding the nodes in this trie using bit vectors to reduce memory requirements. The resulting data structure requires at most 12 memory accesses. However, inserts and deletes are quite expensive. For example, the insertion of the prefix 1* changes up to 2^{15} entries in the trie's root node. All of these changes may propagate into the compacted storage scheme of [14].

Lampson et al. [5] have proposed the use of hybrid data structures comprised of a stride-16 root and an auxiliary data structure for each of the subtries of the stride-16 root. This auxiliary data structure could be the end-point array of Section 49.2.2 (since each subtrie is expected to contain only a small number of prefixes, the number of end points in each end-point array is also expected to be quite small). An alternative auxiliary data structure suggested by Lampson et al. [5] is a 6-way search tree for IPv4 router tables. In the case of these 6-way trees, the keys are the remaining up to 16 bits of the prefix (recall that the stride-16 root consumes the first 16 bits of a prefix). For IPv6 prefixes, a multicolumn scheme is suggested [5]. None of these proposed structures is suitable for a dynamic table.

In the *fixed-stride trie optimization* (FSTO) problem, we are given a set P of prefixes and an integer k. We are to select the strides for a k-level FST in such a manner that the k-level FST for the given prefixes uses the smallest amount of memory.

For some P, a k-level FST may actually require more space than a $(k-1)$-level FST. For example, when $P = \{00^*, 01^*, 10^*, 11^*\}$, the unique 1-level FST for P requires 4 memory units while the unique 2-level FST (which is actually the 1-bit trie for P) requires 6 memory units. Since the search time for a $(k-1)$-level FST is less than that for a k-level tree, we would actually prefer $(k-1)$-level FSTs that take less (or even equal) memory over k-level FSTs. Therefore, in practice, we are really interested in determining the best FST that uses at most k levels (rather than exactly k levels). The *modified* MSTO problem (MFSTO) is to determine the best FST that uses at most k levels for the given prefix set P.

Let O be the 1-bit trie for the given set of prefixes, and let F be any k-level FST for this prefix set. Let $s_0, \ldots, s_{k-1}$ be the strides for F. We shall say that level 0 of F covers levels $0, \ldots, s_0 - 1$ of O, and that level j, $0 < j < k$ of F covers levels $a, \ldots, b$ of O, where $a = \sum_0^{j-1} s_q$ and $b = \sum_0^j s_q - 1$. So, level 0 of the FST of Figure 49.4b covers levels 0 and 1 of the 1-bit trie of Figure 49.3b. Level 1 of this FST covers levels 2, 3, and 4 of the 1-bit trie of Figure 49.3b; and level 2 of this FST covers levels 5 and 6 of the 1-bit trie. We shall refer to levels $e_u = \sum_0^u s_q$, $0 \le u < k$ as the *expansion levels* of O. The expansion levels defined by the FST of Figure 49.4b are 0, 2, and 5.

Let *nodes(i)* be the number of nodes at level *i* of the 1-bit trie *O*. For the 1-bit trie of Figure 49.3a, $nodes(0:6) = [1,1,2,2,2,1,1]$. The memory required by *F* is $\sum_0^{k-1} nodes(e_q) * 2^{s_q}$. For example, the memory required by the FST of Figure 49.4b is $nodes(0) * 2^2 + nodes(2) * 2^3 + nodes(5) * 2^2 = 24$.

Let $C(j,r)$ be the cost of the best FST that uses *at most r* expansion levels. A simple dynamic programming recurrence for *C* is [13]:

$$C(j,r) = \min_{m\in\{-1..j-1\}} \{C(m,r-1) + nodes(m+1) * 2^{j-m}\}, \quad j \ge 0, r > 1 \tag{49.1}$$

$$C(-1,r) = 0 \text{ and } C(j,1) = 2^{j+1}, \quad j \ge 0 \tag{49.2}$$

Let $M(j,r), r > 1$, be the smallest *m* that minimizes

$$C(m,r-1) + nodes(m+1) * 2^{j-m},$$

in Equation 49.1.

Theorem 49.1

(Sahni and Kim [13])$\forall(j \ge 0, k > 2)[M(j,k) \ge max\{M(j-1,k), M(j,k-1)\}]$.

Theorem 49.1 results in an algorithm to compute $C(W-1,k)$ in $O(kW^2)$. Using the computed *M* values, the strides for the OFST that uses at most *k* expansion levels may be determined in an additional $O(k)$ time. Although the resulting algorithm has the same asymptotic complexity as does the optimization algorithm of Srinivasan and Varghese [7], experiments conducted by Sahni and Kim [13] using real IPv4 prefix-data-sets indicate that the algorithm based on Theorem 49.1 runs 2 to 4 times as fast.

Basu and Narliker [15] consider implementing FST router tables on a pipelined architecture. Each level of the FST is assigned to a unique pipeline stage. The optimization problem to be solved in this application requires an FST that has a number of levels no more than the number of pipeline stages, the memory required per level should not exceed the available per stage memory, and the total memory required is minimum subject to the stated constraints.

49.2.4.3 Variable-Stride Tries

In a *variable-stride trie* (VST) [7], nodes at the same level may have different strides. Figure 49.5 shows a two-level VST for the 1-bit trie of Figure 49.3. The stride for the root is 2; that for the left child of the root is 5; and that for the root's right child is 3. The memory requirement of this VST is 4 (root) + 32 (left child of root) + 8 (right child of root) = 44.

Since FSTs are a special case of VSTs, the memory required by the best VST for a given prefix set *P* and number of expansion levels *k* is less than or equal to that required by the best FST for *P* and *k*. Despite this, FSTs may be preferred in certain router applications "because of their simplicity and slightly faster search time" [7].

Let *r*-VST be a VST that has at most *r* levels. Let $Opt(N,r)$ be the cost (i.e., memory requirement) of the best *r*-VST for a 1-bit trie whose root is *N*. Nilsson and Karlsson [16] propose a greedy heuristic to construct optimal VSTs. The resulting VSTs are known as LC-tries (level-compressed tries) and were first proposed in a more general context by Andersson and Nilsson [17]. An LC-tries obtained from a 1-bit trie by replacing full subtries of the 1-bit trie by single multibit nodes. This replacement is done by examining the 1-bit trie top to bottom (i.e., from root to leaves). Srinivasan and Varghese [7], have obtained the following dynamic programming recurrence for $Opt(N,r)$.

$$Opt(N,r) = \min_{s\in\{1...1+height(N)\}} \left\{ 2^s + \sum_{M\in D_s(N)} Opt(M,r-1) \right\}, \quad r > 1 \tag{49.3}$$

where $D_s(N)$ is the set of all descendants of *N* that are at level *s* of *N*. For example, $D_1(N)$ is the set of children of *N* and $D_2(N)$ is the set of grandchildren of *N*. *height(N)* is the maximum level at which the trie rooted at *N* has a node. For example, in Figure 49.3b, the height of the trie rooted at N1 is 5. When $r=1$,

$$Opt(N,1) = 2^{1+height(N)} \tag{49.4}$$

Srinivasan and Varghese [7], describe a way to determine $Opt(R,k)$ using Equations 49.4 and 49.5. The complexity of their algorithm is $O(p * W * k)$, where *p* is the number of nodes in the 1-bit trie for the prefixes ($p = O(n)$ for realistic router tables). Sahni and Kim [18] provide an alternative way to compute $Opt(R,k)$ in $O(pWk)$ time. The algorithm of [18], however, performs fewer operations and has fewer cache misses. When the cost of operations dominates the run time, the algorithm of [18] is

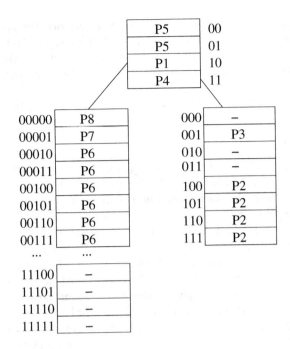

FIGURE 49.5 Two-level VST for prefixes of Figure 49.3a.

expected to be about 6 times as fast as that of [7] (for available router databases). When cache miss time dominates the run time, the algorithm of [18] could be 12 times as fast when $k = 2$ and 42 times as fast when $k = 7$.

We describe the formulation used in Reference 18. Let

$$Opt(N, s, r) = \sum_{M \in D_s(N)} Opt(M, r), \quad s > 0, \quad r > 1,$$

and let $Opt(N, 0, r) = Opt(N, r)$. From Equations 49.4 and 49.5, it follows that:

$$Opt(N, 0, r) = min_{s \in \{1...1+height(N)\}} \{2^s + Opt(N, s, r - 1)\}, \quad r > 1 \tag{49.5}$$

and

$$Opt(N, 0, 1) = 2^{1+height(N)}. \tag{49.6}$$

For $s > 0$ and $r > 1$, we get

$$Opt(N, s, r) = \sum_{M \in D_s(N)} Opt(M, r)$$

$$= Opt(LeftChild(N), s - 1, r)$$

$$+ Opt(RightChild(N), s - 1, r). \tag{49.7}$$

For Equation 49.9, we need the following initial condition:

$$Opt(null, *, *) = 0 \tag{49.8}$$

With the assumption that the number of nodes in the 1-bit trie is $O(n)$, we see that the number of $Opt(*, *, *)$ values is $O(nWk)$. Each $Opt(*, *, *)$ value may be computed in $O(1)$ time using Equations 49.7 through 49.11 provided the Opt values are computed in postorder. Therefore, we may compute $Opt(R, k) = Opt(R, 0, k)$ in $O(pWk)$ time. The algorithm of [18] requires $O(W^2k)$ memory for the $Opt(*, *, *)$ values. To see this, notice that there can be at most $W + 1$ nodes N whose $Opt(N, *, *)$ values must be retained at any given time, and for each of these at most $W + 1$ nodes, $O(Wk)$ $Opt(N, *, *)$ values must be retained. To determine the optimal strides, each node of the 1-bit trie must store the stride s that minimizes the right side of Equation 49.7 for each value of r. For this purpose, each 1-bit trie node needs $O(k)$ space. Since the 1-bit trie has $O(n)$ nodes in practice, the memory requirements of the 1-bit trie are $O(nk)$. The total memory required is, therefore, $O(nk + W^2k)$.

In practice, we may prefer an implementation that uses considerably more memory. If we associate a cost array with each of the p nodes of the 1-bit trie, the memory requirement increases to $O(pWk)$. The advantage of this increased memory implementation is that the optimal strides can be recomputed in $O(W^2k)$ time (rather than $O(pWk)$) following each insert or delete of a prefix. This is so because, the $Opt(N, *, *)$ values need be recomputed only for nodes along the insert/delete path of the 1-bit trie. There are $O(W)$ such nodes.

Faster algorithms to determine optimal 2- and 3-VSTs also are developed in Reference 18.

49.2.5 Binary Search Trees

Sahni and Kim [19] propose the use of a collection of red-black trees (see Chapter 11) to determine $LMP(d)$. The CRBT comprises a front-end data structure that is called the *binary interval tree* (BIT) and a back-end data structure called a *collection of prefix trees* (CPT). For any destination address d, define the *matching basic interval* to be a basic interval with the property that $r_i \le d \le r_{i+1}$ (note that some ds have two matching basic intervals).

The BIT is a binary search tree that is used to search for a matching basic interval for d. The BIT comprises internal and external nodes and there is one internal node for each r_i. Since the BIT has q internal nodes, it has $q+1$ external nodes. The first and last of these, in inorder, have no significance. The remaining $q-1$ external nodes, in inorder, represent the $q-1$ basic intervals of the given prefix set. Figure 49.6a gives a possible (we say possible because, any red-black binary search tree organization for the internal nodes will suffice) BIT for our five-prefix example of Figure 49.2a. Internal nodes are shown as rectangles while circles denote external nodes. Every external node has three pointers: *startPointer*, *finishPointer*, and *basicIntervalPointer*. For an external node that represents the basic interval $[r_i, r_{i+1}]$, *startPointer* (*finishPointer*) points to the header node of the prefix tree (in the back-end structure) for the prefix (if any) whose range start and finish points are r_i (r_{i+1}). Note that only prefixes whose length is W can have this property. *basicIntervalPointer* points to a prefix node in a prefix tree of the back-end structure. In Figure 49.6a, the labels in the external (circular) nodes identify the represented basic interval. The external node with r1 in it, for example, has a *basicIntervalPointer* to the rectangular node labeled r1 in the prefix tree of Figure 49.6b.

For each prefix and basic interval, x, define $next(x)$ to be the smallest range prefix (i.e., the longest prefix) whose range includes the range of x. For the example of Figure 49.2a, the $next()$ values for the basic intervals r1 through r7 are, respectively, P1, P2, P1, P3, P4, P1, and P1. Notice that the next value for the range $[r_i, r_{i+1}]$ is the same as the ">" value for r_i in Figure 49.2b, $1 \le i < q$. The $next()$ values for the nontrivial prefixes P1 through P4 of Figure 49.2a are, respectively, "-," P1, P1, and P3.

The back-end structure, which is a collection of prefix trees (CPT), has one prefix tree for each of the prefixes in the router table. Each prefix tree is a red-black tree. The prefix tree for prefix P comprises a header node plus one node, called a *prefix node*, for every nontrivial prefix or basic interval x such that $next(x) = P$. The header node identifies the prefix P for which this is the prefix tree. The prefix trees for each of the five prefixes of Figure 49.2a are shown in Figure 49.6b–f. Notice that prefix trees do

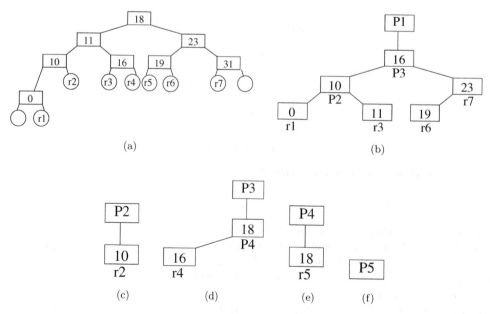

FIGURE 49.6 CBST for Figure 49.2a. (a) base interval tree (b) prefix tree for P1 (c) prefix tree for P2 (d) prefix tree for P3 (e) prefix tree for P4 (f) prefix tree for P5.

not have external nodes and that the prefix nodes of a prefix tree store the start point of the range or prefix represented by that prefix node. In the figures, the start points of the basic intervals and prefixes are shown inside the prefix nodes while the basic interval or prefix name is shown outside the node.

The search for $LMP(d)$ begins with a search of the BIT for the matching basic interval for d. Suppose that external node Q of the BIT represents this matching basic interval. When the destination address equals the left (right) end-point of the matching basic interval and *startPointer* (*finishPointer*) is not null, $LMP(d)$ is pointed to by *startPointer* (*finishPointer*). Otherwise, the back-end CPT is searched for $LMP(d)$. The search of the back-end structure begins at the node $Q.basicIntervalPointer$. By following parent pointers from $Q.basicIntervalPointer$, we reach the header node of the prefix tree that corresponds to $LMP(d)$.

When a CRBT is used, $LMP(d)$ may be found in $O(\log n)$ time. Inserts and deletes also take $O(\log n)$ time when a CRBT is used. In Reference 20, Sahni and Kim propose an alternative BIT structure (ABIT) that has internal nodes only. Although the ABIT structure increases the memory requirements of the router table, the time to search, insert, and delete is reduced by a constant factor [20]. Suri et al. [21] have proposed a B-tree data structure for dynamic router tables. Using their structure, we may find the longest matching prefix in $O(\log n)$ time. However, inserts/deletes take $O(W \log n)$ time. The number of cache misses is $O(\log n)$ for each operation. When W bits fit in $O(1)$ words (as is the case for IPv4 and IPv6 prefixes) logical operations on W-bit vectors can be done in $O(1)$ time each. In this case, the scheme of [21] takes $O(\log W * \log n)$ time for an insert and $O(W + \log n) = O(W)$ time for a delete. An alternative B-tree router-table design has been proposed by Lu and Sahni [22]. Although the asymptotic complexity of each of the router-table operations is the same using either B-tree router-table design, the design of Lu and Sahni [22] has fewer cache misses for inserts and deletes; the number of cache misses when searching for $lmp(d)$ is the same using either design. Consequently, inserts and deletes are faster when the design of Lu and Sahni [22] is used.

Several researchers ([20,23–25], for example), have investigated router table data structures that account for bias in access patterns. Gupta, Prabhakar, and Boyd [25], for example, propose the use of ranges. They assume that access frequencies for the ranges are known, and they construct a bounded-height binary search tree of ranges. This binary search tree accounts for the known range access frequencies to obtain near-optimal IP lookup. Although the scheme of [25] performs IP lookup in near-optimal time, changes in the access frequencies, or the insertion or removal of a prefix require us to reconstruct the data structure, a task that takes $O(n \log n)$ time.

Ergun et al. [24] use ranges to develop a biased skip list structure that performs longest prefix-matching in $O(\log n)$ expected time. Their scheme is designed to give good expected performance for bursty* access patterns."The biased skip list scheme of Ergun et al. [24] permits inserts and deletes in $O(\log n)$ time only in the severely restricted and impractical situation when all prefixes in the router table are of the same length. For the more general, and practical, case when the router table comprises prefixes of different length, their scheme takes $O(n)$ expected time for each insert and delete. Sahni and Kim [20] extend the biased skip lists of Ergun et al. [24] to obtain a biased skip lists structure in which longest prefix-matching as well as inserts and deletes take $O(\log n)$ expected time. They also propose a splay tree scheme (see Chapter 13) for bursty access patterns. In this scheme, longest prefix-matching, insert and delete have an $O(\log n)$ amortized complexity.

49.2.6 Priority Search Trees

A priority-search tree (PST) [28] (see Chapter 19) is a data structure that is used to represent a set of tuples of the form $(key1, key2, data)$, where $key1 \geq 0$, $key2 \geq 0$, and no two tuples have the same $key1$ value. The data structure is simultaneously a min-tree on $key2$ (i.e., the $key2$ value in each node of the tree is $\leq$ the $key2$ value in each descendant node) and a search tree on $key1$. There are two common PST representations [28]:

1. In a **radix priority-search tree** (RPST), the underlying tree is a binary radix tree on $key1$.
2. In a **red-black priority-search tree** (RBPST), the underlying tree is a red-black tree.

McCreight [28] has suggested a PST representation of a collection of ranges with distinct finish points. This representation uses the following mapping of a range r into a PST tuple:

$$(key1, key2, data) = (finish(r), start(r), data) \tag{49.9}$$

where *data* is any information (e.g., next hop) associated with the range. Each range r is, therefore mapped to a point $map1(r) = (x, y) = (key1, key2) = (finish(r), start(r))$ in 2-dimensional space.

Let $ranges(d)$ be the set of ranges that match d. McCreight [28] has observed that when the mapping of Equation 49.12 is used to obtain a point set $P = map1(R)$ from a range set R, then $ranges(d)$ is given by the points that lie in the rectangle (including

---
* In a *bursty* access pattern the number of different destination addresses in any subsequence of q packets is $< < q$. That is, if the destination of the current packet is d, there is a high probability that d is also the destination for one or more of the next few packets. The fact that Internet packets tend to be bursty has been noted in References 26 and 27, for example.

TABLE 49.1   Time Complexity of Data Structures for Longest Matching-Prefix

Data Structure	Search	Update
Linear List	$O(n)$	$O(n)$
End-point Array	$O(\log n)$	$O(n)$
Sets of Equal-Length Prefixes	$O(\alpha + \log W)$ expected	$O(\alpha \sqrt[\alpha]{n} W \log W)$ expected
1-bit tries	$O(W)$	$O(W)$
s-bit Tries	$O(W/s)$	-
CRBT	$O(\log n)$	$O(\log n)$
ACRBT	$O(\log n)$	$O(\log n)$
BSLPT	$O(\log n)$ expected	$O(\log n)$ expected
CST	$O(\log n)$ amortized	$O(\log n)$ amortized
PST	$O(\log n)$	$O(\log n)$

points on the boundary) defined by $x_{left} = d$, $x_{right} = \infty$, $y_{top} = d$, and $y_{bottom} = 0$. These points are obtained using the method $enumerateRectangle(x_{left}, x_{right}, y_{top}) = enumerateRectangle(d, \infty, d)$ of a PST (y_{bottom} is implicit and is always 0).

When an RPST is used to represent the point set P, the complexity of

$$enumerateRectangle(x_{left}, x_{right}, y_{top})$$

is $O(\log maxX + s)$, where $maxX$ is the largest x value in P and s is the number of points in the query rectangle. When the point set is represented as an RBPST, this complexity becomes $O(\log n + s)$, where $n = |P|$. A point (x, y) (and hence a range $[y, x]$) may be inserted into or deleted from an RPST (RBPST) in $O(\log maxX)$ ($O(\log n)$) time [28].

Let R be a set of prefix ranges. For simplicity, assume that R includes the range that corresponds to the prefix *. With this assumption, $LMP(d)$ is defined for every d. One may verify that $LMP(d)$ is the prefix whose range is

$$[maxStart(ranges(d)), minFinish(ranges(d))].$$

Lu and Sahni [29] show that R must contain such range. To find this range easily, we first transform $P = map1(R)$ into a point set $transform1(P)$ so that no two points of $transform1(P)$ have the same x-value. Then, we represent $transform1(P)$ as a PST. For every $(x, y) \in P$, define $transform1(x, y) = (x', y') = (2^W x - y + 2^W - 1, y)$. Then, $transform1(P) = \{transform1(x, y) | (x, y) \in P\}$.

We see that $0 \le x' < 2^{2W}$ for every $(x', y') \in transform1(P)$ and that no two points in $transform1(P)$ have the same x'-value. Let $PST1(P)$ be the PST for $transform1(P)$. The operation

$$enumerateRectangle(2^W d - d + 2^W - 1, \infty, d)$$

performed on $PST1$ yields $ranges(d)$. To find $LMP(d)$, we employ the

$$minXinRectangle(x_{left}, x_{right}, y_{top})$$

operation, which determines the point in the defined rectangle that has the least x-value. It is easy to see that

$$minXinRectangle(2^W d - d + 2^W - 1, \infty, d)$$

performed on $PST1$ yields $LMP(d)$.

To insert the prefix whose range in $[u, v]$, we insert $transform1(map1([u, v]))$ into $PST1$. In case this prefix is already in $PST1$, we simply update the next-hop information for this prefix. To delete the prefix whose range is $[u, v]$, we delete $transform1(map1([u, v]))$ from $PST1$. When deleting a prefix, we must take care not to delete the prefix *. Requests to delete this prefix should simply result in setting the next-hop associated with this prefix to $\emptyset$.

Since, $minXinRectangle$, insert, and delete each take $O(W)$ ($O(\log n)$) time when $PST1$ is an RPST (RBPST), $PST1$ provides a router-table representation in which longest-prefix matching, prefix insertion, and prefix deletion can be done in $O(W)$ time each when an RPST is used and in $O(\log n)$ time each when an RBPST is used.

Tables 49.1 and 49.2 summarize the performance characteristics of various data structures for the longest matching-prefix problem.

TABLE 49.2 Memory Complexity of Data
Structures for Longest Matching-Prefix

Data Structure	Memory Usage
Linear List	$O(n)$
End-point Array	$O(n)$
Sets of Equal-Length Prefixes	$O(n \log W)$
1-bit tries	$O(nW)$
s-bit Tries	$O(2^s nW/s)$
CRBT	$O(n)$
ACRBT	$O(n)$
BSLPT	$O(n)$
CST	$O(n)$
PST	$O(n)$

49.3 Highest-Priority Matching

The trie data structure may be used to represent a dynamic prefix-router-table in which the highest-priority tie-breaker is in use. Using such a structure, each of the dynamic router-table operations may be performed in $O(W)$ time. Lu and Sahni [30] have developed the binary tree on binary tree (BOB) data structure for highest-priority dynamic router-tables. Using BOB, a lookup takes $O(\log^2 n)$ time and cache misses; a new rule may be inserted and an old one deleted in $O(\log n)$ time and cache misses. Although BOB handles filters that are non-intersecting ranges, specialized versions of BOB have been proposed for prefix filters. Using the data structure PBOB (prefix BOB), a lookup, rule insertion and deletion each take $O(W)$ time and cache misses. The data structure LMPBOB (longest matching-prefix BOB) is proposed in Reference 30 for dynamic prefix-router-tables that use the longest matching-prefix rule. Using LMPBOB, the longest matching-prefix may be found in $O(W)$ time and $O(\log n)$ cache misses; rule insertion and deletion each take $O(\log n)$ time and cache misses. On practical rule tables, BOB and PBOB perform each of the three dynamic-table operations in $O(\log n)$ time and with $O(\log n)$ cache misses. Other data structures for maximum-priority matching are developed in References 31 and 32.

49.3.1 The Data Structure BOB

The data structure binary tree on binary tree (BOB) comprises a single balanced binary search tree at the top level. This top-level balanced binary search tree is called the point search tree (PTST). For an n-rule NHRT, the PTST has at most $2n$ nodes (we call this the PTST **size constraint**). With each node z of the PTST, we associate a point, $point(z)$. The PTST is a standard red-black binary search tree (actually, any binary search tree structure that supports efficient search, insert, and delete may be used) on the $point(z)$ values of its node set [1]. That is, for every node z of the PTST, nodes in the left subtree of z have smaller point values than $point(z)$, and nodes in the right subtree of z have larger point values than $point(z)$.

Let R be the set of nonintersecting ranges. Each range of R is stored in exactly one of the nodes of the PTST. More specifically, the root of the PTST stores all ranges $r \in R$ such that $start(r) \leq point(root) \leq finish(r)$; all ranges $r \in R$ such that $finish(r) < point(root)$ are stored in the left subtree of the root; all ranges $r \in R$ such that $point(root) < start(r)$ (i.e., the remaining ranges of R) are stored in the right subtree of the root. The ranges allocated to the left and right subtrees of the root are allocated to nodes in these subtrees using the just stated **range allocation rule** recursively.

For the range allocation rule to successfully allocate all $r \in R$ to exactly one node of the PTST, the PTST must have at least one node z for which $start(r) \leq point(z) \leq finish(r)$. Figure 49.7 gives an example set of nonintersecting ranges and a possible PTST for this set of ranges (we say possible, because we haven't specified how to select the $point(z)$ values and even with specified $point(z)$ values, the corresponding red-black tree isn't unique). The number inside each node is $point(z)$, and outside each node, we give $ranges(z)$.

Let $ranges(z)$ be the subset of ranges of R allocated to node z of the PTST.[*] Since the PTST may have as many as $2n$ nodes and since each range of R is in exactly one of the sets $ranges(z)$, some of the $ranges(z)$ sets may be empty.

The ranges in $ranges(z)$ may be ordered using the $<$ relation for non-intersecting ranges.[†] Using this $<$ relation, we put the ranges of $ranges(z)$ into a red-black tree (any balanced binary search tree structure that supports efficient search, insert,

---

[*] We have overloaded the function *ranges*. When u is a node, $ranges(u)$ refers to the ranges stored in node u of a PTST; when u is a destination address, $ranges(u)$ refers to the ranges that match u.

[†] Let r and s be two ranges. $r < s \iff start(r) < start(s) \vee (start(r) = start(s) \wedge finish(r) > finish(s))$.

range	priority
[2, 100]	4
[2, 4]	33
[2, 3]	34
[8, 68]	10
[8, 50]	9
[10, 50]	20
[10, 35]	3
[15, 33]	5
[16, 30]	30
[54, 66]	18
[60, 65]	7
[69, 72]	10
[80, 80]	12

(a)

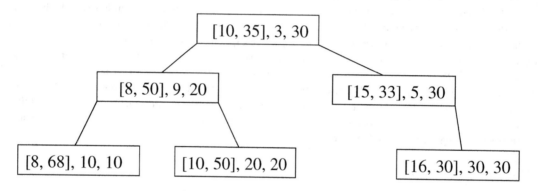

(b)

FIGURE 49.7   (a) A nonintersecting range set and (b) A possible PTST.

FIGURE 49.8   An example RST for *ranges*(30) of Figure 49.7b. Each node shows (*range*(x), *p*(x), *mp*(x)).

delete, join, and split may be used) called the range search-tree or *RST*(z). Each node *x* of *RST*(z) stores exactly one range of *ranges*(z). We refer to this range as *range*(x). Every node *y* in the left (right) subtree of node *x* of *RST*(z) has *range*(y) < *range*(x) (*range*(y) > *range*(x)). In addition, each node *x* stores the quantity *mp*(x), which is the maximum of the priorities of the ranges associated with the nodes in the subtree rooted at *x*. *mp*(x) may be defined recursively as below.

$$mp(x) = \begin{cases} p(x) & \text{if } x \text{ is leaf} \\ \max\{mp(leftChild(x)), mp(rightChild(x)), p(x)\} & \text{otherwise} \end{cases}$$

where *p*(x) = *priority*(*range*(x)). Figure 49.8 gives a possible RST structure for *ranges*(30) of Figure 49.7b.

Lemma 49.1

(Lu and Sahni [30])Let *z* be a node in a PTST and let *x* be a node in *RST*(z). Let *st*(x) = *start*(*range*(x)) and *fn*(x) = *finish*(*range*(x)).

1. For every node *y* in the right subtree of *x*, *st*(y) ≥ *st*(x) and *fn*(y) ≤ *fn*(x).
2. For every node *y* in the left subtree of *x*, *st*(y) ≤ *st*(x) and *fn*(y) ≥ *fn*(x).

Both BOB and BOT (the binary tree on trie structure of Gupta and McKeown [32]) use a range allocation rule identical to that used in an interval tree [33] (See Chapter 19). While BOT may be used with any set of ranges, BOB applies only to a set of non-intersecting ranges. However, BOB reduces the search complexity of BOT from $O(W \log n)$ to $O(\log^2 n)$ and reduces the update complexity from $O(W)$ to $O(\log n)$.

49.3.2 Search for the Highest-Priority Matching Range

The highest-priority range that matches the destination address d may be found by following a path from the root of the PTST toward a leaf of the PTST. Figure 49.9 gives the algorithm. For simplicity, this algorithm finds $hp = priority(hpr(d))$ rather than $hpr(d)$. The algorithm is easily modified to return $hpr(d)$ instead.

We begin by initializing $hp = -1$ and z is set to the root of the PTST. This initialization assumes that all priorities are ≥ 0. The variable z is used to follow a path from the root toward a leaf. When $d > point(z)$, d may be matched only by ranges in $RST(z)$ and those in the right subtree of z. The method $RST(z)$->$hpRight(d,hp)$ (Figure 49.10) updates hp to reflect any matching ranges in $RST(z)$. This method makes use of the fact that $d > point(z)$. Consider a node x of $RST(z)$. If $d > fn(x)$, then d is to the right (i.e., $d > finish(range(x))$) of $range(x)$ and also to the right of all ranges in the right subtree of x. Hence, we may proceed to examine the ranges in the left subtree of x. When $d \leq fn(x)$, $range(x)$ as well as all ranges in the left subtree of x match d. Additional matching ranges may be present in the right subtree of x. $hpLeft(d,hp)$ is the analogous method for the case when $d < point(z)$.

Complexity The complexity of the invocation $RST(z)$->$hpRight(d,hp)$ is readily seen to be $O(height(RST(z))) = O(\log n)$. Consequently, the complexity of $hp(d)$ is $O(\log^2 n)$. To determine $hpr(d)$ we need only add code to the methods $hp(d)$, $hpRight(d,hp)$, and $hpLeft(d,hp)$ so as to keep track of the range whose priority is the current value of hp. So, $hpr(d)$ may be found in $O(\log^2 n)$ time also.

The details of the insert and delete operation as well of the BOB variants PBOB and LMPBOB may be found in Reference 30.

```
Algorithm hp(d) {
    // return the priority of hpr(d)
    // easily extended to return hpr(d)
    hp = -1; // assuming 0 is the smallest priority value
    z  = root; // root of PTST
    while (z != null) {
        if (d > point(z)) {
            RST(z)->hpRight(d, hp);
            z = rightChild(z);
        }
        else if (d < point(z)) {
            RST(z)->hpLeft(d, hp);
            z = leftChild(z);
        }
        else // d == point(z)
            return max{hp, mp(RST(z)->root)};
    }
    return hp;
}
```

FIGURE 49.9 Algorithm to find *priority(hpr(d))*.

```
Algorithm hpRight(d, hp) {
    // update hp to account for any ranges in RST(z) that match d
    // d > point(z)
    x  = root; // root of RST(z)
    while (x != null)
        if (d > fn(x))
            x = leftChild(x);
        else {
            hp = max{hp, p(x), mp(leftChild(x))};
            x = rightChild(x);
        }
}
```

FIGURE 49.10 Algorithm *hpRight(d, hp)*.

49.4 Most-Specific-Range Matching

Let $msr(d)$ be the most-specific range that matches the destination address d. For static tables, we may simply represent the n ranges by the up to $2n-1$ basic intervals they induce. For each basic interval, we determine the most-specific range that matches all points in the interval. These up to $2n-1$ basic intervals may then be represented as up to $4n-2$ prefixes [31] with the property that $msr(d)$ is uniquely determined by $LMP(d)$. Now, we may use any of the earlier discussed data structures for static tables in which the filters are prefixes. In this section, therefore, we discuss only those data structures that are suitable for dynamic tables.

49.4.1 Nonintersecting Ranges

Let R be a set of nonintersecting ranges. For simplicity, assume that R includes the range z that matches all destination addresses ($z = [0, 2^{32} - 1]$ in the case of IPv4). With this assumption, $msr(d)$ is defined for every d. Similar to the case of prefixes, for nonintersecting ranges, $msr(d)$ is the range $[maxStart(ranges(d)), minFinish(ranges(d))]$ (Lu and Sahni [29] show that R must contain such a range). We may use

$$PST1(transform1(map1(R)))$$

to find $msr(d)$ using the same method described in Section 49.2.6 to find $LMP(d)$.

Insertion of a range r is to be permitted only if r does not intersect any of the ranges of R. Once we have verified this, we can insert r into $PST1$ as described in Section 49.2.6. Range intersection may be verified by noting that there are two cases for range intersection. When inserting $r = [u, v]$, we need to determine if $\exists s = [x, y] \in R[u < x \le v < y \lor x < u \le y < v]$. We see that $\exists s \in R[x < u \le y < v]$ iff $map1(R)$ has at least one point in the rectangle defined by $x_{left} = u$, $x_{right} = v - 1$, and $y_{top} = u - 1$ (recall that $y_{bottom} = 0$ by default). Hence, $\exists s \in R[x < u \le y < v]$ iff $minXinRectangle(2^W - (u-1) + 2^W - 1, 2^W(v-1) + 2^W - 1, u - 1)$ exists in PST1.

To verify $\exists s \in R[u < x \le v < y]$, map the ranges of R into 2-dimensional points using the mapping, $map2(r) = (start(r), 2^W - 1 - finish(r))$. Call the resulting set of mapped points $map2(R)$. We see that $\exists s \in R[u < x \le v < y]$ iff $map2(R)$ has at least one point in the rectangle defined by $x_{left} = u + 1$, $x_{right} = v$, and $y_{top} = (2^W - 1) - v - 1$. To verify this, we maintain a second PST, $PST2$ of points in $transform2(map2(R))$, where $transform2(x, y) = (2^W x + y, y)$. Hence, $\exists s \in R[u < x \le v < y]$ iff $minXinRectangle(2^W(u+1), 2^W v + (2^W - 1) - v - 1, (2^W - 1) - v - 1)$ exists.

To delete a range r, we must delete r from both $PST1$ and $PST2$. Deletion of a range from a PST is similar to deletion of a prefix as discussed in Section 49.2.6.

The complexity of the operations to find $msr(d)$, insert a range, and delete a range are the same as that for these operations for the case when R is a set of ranges that correspond to prefixes.

49.4.2 Conflict-Free Ranges

The range set R has a **conflict** iff there exists a destination address d for which $ranges(d) \ne \emptyset \land msr(d) = \emptyset$. R is **conflict free** iff it has no conflict. Notice that sets of prefix ranges and sets of nonintersecting ranges are conflict free. The two-PST data structure of Section 49.4.1 may be extended to the general case when R is an arbitrary conflict-free range set. Once again, we assume that R includes the range z that matches all destination addresses. $PST1$ and $PST2$ are defined for the range set R as in Sections 49.2.6 and 49.4.1.

Lu and Sahni [29] have shown that when R is conflict free, $msr(d)$ is the range

$$[maxStart(ranges(d)), minFinish(ranges(d))]$$

Hence, $msr(d)$ may be obtained by performing the operation

$$minXinRectangle(2^W d - d + 2^W - 1, \infty, d)$$

on PST1. Insertion and deletion are complicated by the need to verify that the addition or deletion of a range to/from a conflict-free range set leaves behind a conflict-free range set. To perform this check efficiently, Lu and Sahni [29] augment $PST1$ and PST with a representation for the chains in the normalized range-set, $norm(R)$, that corresponds to R. This requires the use of several red-black trees. The reader is referred to [29] for a description of this augmentation.

The overall complexity of the augmented data structure of [29] is $O(\log n)$ for each operation when $RBPSTs$ are used for $PST1$ and $PST2$. When $RPSTs$ are used, the search complexity is $O(W)$ and the insert and delete complexity is $O(W + \log n) = O(W)$.

Acknowledgments

This work was supported, in part, by the National Science Foundation under grant CCR-9912395.

References

1. E. Horowitz, S. Sahni, and D. Mehta, *Fundamentals of data structures in C++*, W.H. Freeman, NY, 1995, 653 pages.
2. D. Shah and P. Gupta, Fast updating algorithms for TCAMs, *IEEE Micro*, 21, 1, 36–47, 2002.
3. F. Zane, G. Narlikar, and A. Basu, CoolCAMs: Power-efficient TCAMs for forwarding engines, *IEEE INFOCOM*, 2003.
4. F. Baboescu, S. Singh, and G. Varghese, Packet classification for core routers: Is there an alternative to CAMs?, *IEEE INFOCOM*, 2003.
5. B. Lampson, V. Srinivasan, and G. Varghese, IP lookup using multi-way and multicolumn search, *IEEE INFOCOM*, 1998.
6. M. Waldvogel, G. Varghese, J. Turner, and B. Plattner, Scalable high speed IP routing lookups, *ACM SIGCOMM*, 25–36, 1997.
7. V. Srinivasan and G. Varghese, Faster IP lookups using controlled prefix expansion, *ACM Transactions on Computer Systems*, Feb:1–40, 1999.
8. V. Srinivasan, Fast and efficient Internet lookups, *CS Ph.D Dissertation*, Washington University, Aug., 1999.
9. K. Kim and S. Sahni, IP lookup by binary search on length, *Journal of Interconnection Networks*, 3, 3&4, 105–128, 2002.
10. M. Berg, M. Kreveld, and J. Snoeyink, Two- and three-dimensional point location in rectangular subdivisions, *Journal of Algorithms*, 18, 2, 256–277, 1995.
11. D. E. Willard, Log-logarithmic worst-case range queries are possible in space $\theta(N)$, *Information Processing Letters*, 17, 81–84, 1983.
12. P. V. Emde Boas, R. Kass, and E. Zijlstra, Design and implementation of an efficient priority queue, *Mathematical Systems Theory*, 10, 99–127, 1977.
13. S. Sahni and K. Kim, Efficient construction of fixed-stride multibit tries for IP lookup, *Proceedings 8th IEEE Workshop on Future Trends of Distributed Computing Systems*, 2001.
14. M. Degermark, A. Brodnik, S. Carlsson, and S. Pink, Small forwarding tables for fast routing lookups, *ACM SIGCOMM*, 3–14, 1997.
15. A. Basu and G. Narlikar, Fast incremental updates for pipelined forwarding engines, *IEEE INFOCOM*, 2003.
16. S. Nilsson and G. Karlsson, Fast address look-up for Internet routers, *IEEE Broadband Communications*, 1998.
17. A. Andersson and S. Nillson, Improved behavior of tries by adaptive branching, *Information Processing Letters*, 46, 295–300, 1993.
18. S. Sahni and K. Kim, Efficient construction of variable-stride multibit tries for IP lookup, *Proceedings IEEE Symposium on Applications and the Internet (SAINT)*, 220–227, 2002.
19. S. Sahni and K. Kim, $O(\log n)$ dynamic packet routing, *IEEE Symposium on Computers and Communications*, 443–448, 2002.
20. S. Sahni and K. Kim, Efficient dynamic lookup for bursty access patterns, submitted.
21. S. Suri, G. Varghese, and P. Warkhede, Multiway range trees: Scalable IP lookup with fast updates, *GLOBECOM*, 2001.
22. H. Lu and S. Sahni, A B-tree dynamic router-table design: Submitted.
23. G. Cheung and S. McCanne, Optimal routing table design for IP address lookups under memory constraints, *IEEE INFOCOM*, 1999.
24. F. Ergun, S. Mittra, S. Sahinalp, J. Sharp, and R. Sinha, A dynamic lookup scheme for bursty access patterns, *IEEE INFOCOM*, 2001.
25. P. Gupta, B. Prabhakar, and S. Boyd, Near-optimal routing lookups with bounded worst case performance, *IEEE INFOCOM*, 2000.
26. K. Claffy, H. Braun, and G Polyzos, A parameterizable methodology for internet traffic flow profiling, *IEEE Journal of Selected Areas in Communications*, 1995.
27. S. Lin and N. McKeown, A simulation study of IP switching, *IEEE INFOCOM*, 2000.
28. E. McCreight, Priority search trees, *SIAM Jr. on Computing*, 14, 1, 257–276, 1985.
29. H. Lu and S. Sahni, $O(\log n)$ dynamic router-tables for prefixes and ranges: Submitted.
30. H. Lu and S. Sahni, Dynamic IP router-tables using highest-priority matching: Submitted.
31. A. Feldman and S. Muthukrishnan, Tradeoffs for packet classification, *IEEE INFOCOM*, 2000.
32. P. Gupta and N. McKeown, Dynamic algorithms with worst-case performance for packet classification, *IFIP Networking*, 2000.
33. M. deBerg, M. vanKreveld, and M. Overmars, *Computational Geometry: Algorithms and Applications*, Second Edition, Springer Verlag, 1997.
34. W. Doeringer, G. Karjoth, and M. Nassehi, Routing on longest-matching prefixes, *IEEE/ACM Transactions on Networking*, 4, 1, 86–97, 1996.

50

Multi-Dimensional Packet Classification⋆

50.1 Introduction...783
    Problem Statement
50.2 Performance Metrics for Classification Algorithms......................785
50.3 Classification Algorithms..785
    Background • Taxonomy of Classification Algorithms • Basic Data Structures • Geometric
    Algorithms • Heuristics • Hardware-Based Algorithms
50.4 Summary..797
References..797

Pankaj Gupta[†]

Cypress Semiconductor

50.1 Introduction

Chapter 49 discussed algorithms for 1-d packet classification. In this chapter we consider multi-dimensional classification in detail. First we discuss the motivation and then the algorithms for multi-dimensional classification. As we will see, packet classification on multiple fields is in general a difficult problem. Hence, researchers have proposed a variety of algorithms which, broadly speaking, can be categorized as "basic search algorithms", geometric algorithms, heuristic algorithms, or hardware-specific search algorithms. In this chapter, we will describe algorithms that are representative of each category, and discuss which type of algorithm might be suitable for different applications.

Until recently, Internet routers provided only "best-effort" service, servicing packets in a first-come-first-served manner. Routers are now called upon to provide different qualities of service to different applications which means routers need new mechanisms such as admission control, resource reservation, per-flow queueing, and fair scheduling. All of these mechanisms require the router to distinguish packets belonging to different flows.

Flows are specified by *rules* applied to incoming packets. We call a collection of rules a *classifier*. Each rule specifies a flow that a packet may belong to based on some criteria applied to the packet header, as shown in Figure 50.1. To illustrate the variety of classifiers, consider some examples of how packet classification can be used by an ISP to provide different services. Figure 50.2 shows ISP_1 connected to three different sites: enterprise networks E_1 and E_2 and a Network Access Point[‡] (NAP), which is in turn connected to ISP_2 and ISP_3. ISP_1 provides a number of different services to its customers, as shown in Table 50.1.

Table 50.2 shows the flows that an incoming packet must be classified into by the router at interface X. Note that the flows specified may or may not be mutually exclusive. For example, the first and second flows in Table 50.2 overlap. This is common in practice, and when no explicit priorities are specified, we follow the convention that rules closer to the top of the list take priority (referred to as the "*First matching rule in table*" tie-breaker rule in Chapter 49).

50.1.1 Problem Statement

Each rule of a classifier has d components. $R[i]$ is the ith component of rule R, and is a regular expression on the ith field of the packet header. A packet P is said to match rule R, if $\forall i$, the ith field of the header of P satisfies the regular expression $R[i]$. In practice, a rule component is not a general regular expression but is often limited by syntax to a simple address/mask or

---

⋆ This chapter has been reprinted from first edition of this Handbook, without any content updates.

† We have used this author's affiliation from the first edition of this handbook, but note that this may have changed since then.

‡ A network access point is a network site which acts as an exchange point for Internet traffic. ISPs connect to the NAP to exchange traffic with other ISPs.

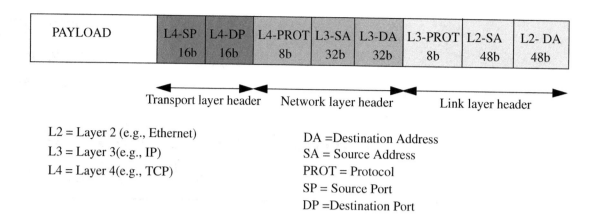

L2 = Layer 2 (e.g., Ethernet)
L3 = Layer 3(e.g., IP)
L4 = Layer 4(e.g., TCP)

DA =Destination Address
SA = Source Address
PROT = Protocol
SP = Source Port
DP =Destination Port

FIGURE 50.1   This figure shows some of the header fields (and their widths) that might be used for classifying the packet. Although not shown in this figure, higher layer (e.g., application-level) headers may also be used.

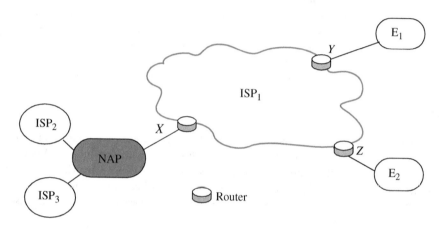

FIGURE 50.2   Example network of an ISP (ISP_1) connected to two enterprise networks (E_1 and E_2) and to two other ISP networks across a network access point (NAP).

TABLE 50.1   Examples of Services Enabled by Packet Classification

Service	Example
Packet Filtering	Deny all traffic from ISP_3 (on interface X) destined to E_2.
Policy Routing	Send all voice-over-IP traffic arriving from E_1 (on interface Y) and destined to E_2 via a separate ATM network.
Accounting & Billing	Treat all video traffic to E_1 (via interface Y) as highest priority and perform accounting for the traffic sent this way.
Traffic Rate Limiting	Ensure that ISP_2 does not inject more than 10 Mbps of email traffic and 50 Mbps of total traffic on interface X.
Traffic Shaping	Ensure that no more than 50 Mbps of web traffic is injected into ISP_2 on interface X.

operator/number(s) specification. In an address/mask specification, a 0 (respectively 1) at bit position x in the mask denotes that the corresponding bit in the address is a don't care (respectively significant) bit. Examples of operator/number(s) specifications are *eq 1232* and *range 34–9339*. Note that a prefix can be specified as an address/mask pair where the mask is contiguous — i.e., all bits with value 1 appear to the left of bits with value 0 in the mask. It can also be specified as a range of width equal to 2^t where $t = 32 - prefixlength$. Most commonly occurring specifications can be represented by ranges. An example real-life classifier in four dimensions is shown in Table 50.3. By convention, the first rule R1 is of highest priority and rule R7 is of lowest priority. Some example classification results are shown in Table 50.4

Longest prefix matching for routing lookups is a special-case of one-dimensional packet classification. All packets destined to the set of addresses described by a common prefix may be considered to be part of the same flow. The address of the next hop where the packet should be forwarded to is the associated action. The length of the prefix defines the priority of the rule.

TABLE 50.2 Flows That an Incoming Packet Must be Classified Into by the Router at Interface X in Figure 50.2

Flow	Relevant Packet Fields:
Email and from ISP_2	Source Link-layer Address, Source Transport port number
From ISP_2	Source Link-layer Address
From ISP_3 and going to E_2	Source Link-layer Address, Destination Network-Layer Address
All other packets	–

TABLE 50.3 An Example 4-Dimensional Real-Life Classifier

Rule	Network-Layer Destination (address/mask)	Network-Layer Source (address/mask)	Transport-Layer Destination	Transport-Layer Protocol	Action
R1	152.163.190.69/ 255.255.255.255	152.163.80.11/ 255.255.255.255	*	*	Deny
R2	152.168.3.0/ 255.255.255.0	152.163.200.157/ 255.255.255.255	eq www	udp	Deny
R5	152.163.198.4/ 255.255.255.255	152.163.160.0/ 255.255.252.0	gt 1023	tcp	Permit
R6	0.0.0.0/0.0.0.0	0.0.0.0/0.0.0.0	*	*	Permit

TABLE 50.4 Example Classification Results from the Real-Life Classifier of Table 50.3

Packet Header	Network-Layer Destination	Network-Layer Source	Transport-Layer Destination	Transport-Layer Protocol	Best Matching Rule, Action
P1	152.163.190.69	152.163.80.11	www	tcp	R1, Deny
P2	152.168.3.21	152.163.200.157	www	udp	R2, Deny
P3	152.163.198.4	152.163.160.10	1024	tcp	R5, Permit

50.2 Performance Metrics for Classification Algorithms

1. *Search speed*—Faster links require faster classification. For example, links running at 10Gbps can bring 31.25 million packets per second (assuming minimum sized 40 byte TCP/IP packets).
2. *Low storage requirements*—Small storage requirements enable the use of fast memory technologies like SRAM (Static Random Access Memory). SRAM can be used as an on-chip cache by a software algorithm and as on-chip SRAM for a hardware algorithm.
3. *Ability to handle large real-life classifiers.*
4. *Fast updates*—As the classifier changes, the data structure needs to be updated. We can categorize data structures into those which can add or delete entries incrementally, and those which need to be reconstructed from scratch each time the classifier changes. When the data structure is reconstructed from scratch, we call it "pre-processing". The update rate differs among different applications: a very low update rate may be sufficient in firewalls where entries are added manually or infrequently, whereas a router with per-flow queues may require very frequent updates.
5. *Scalability in the number of header fields used for classification.*
6. *Flexibility in specification*—A classification algorithm should support general rules, including prefixes, operators (range, less than, greater than, equal to, etc.) and wildcards. In some applications, non-contiguous masks may be required.

50.3 Classification Algorithms

50.3.1 Background

For the next few sections, we will use the example classifier in Table 50.5 repeatedly. The classifier has six rules in two fields labeled $F1$ and $F2$; each specification is a prefix of maximum length 3 bits. We will refer to the classifier as $C = \{R_j\}$ and each rule R_j as a 2-tuple: $< R_{j1}, R_{j2} >$.

TABLE 50.5 Example 2-Dimensional Classifier

Rule	F1	F2
R_1	00 *	00 *
R_2	0 *	01 *
R_3	1 *	0 *
R_4	00 *	0 *
R_5	0 *	1 *
R_6	*	1 *

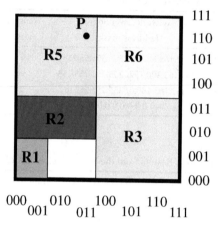

FIGURE 50.3 Geometric representation of the classifier in Table 50.5. A packet represents a point, for instance P(011,110) in two-dimensional space. Note that R4 is hidden by R1 and R2.

50.3.1.1 Bounds from Computational Geometry

There is a simple geometric interpretation of packet classification. While a prefix represents a contiguous interval on the number line, a two-dimensional rule represents a rectangle in two-dimensional euclidean space, and a rule in d dimensions represents a d-dimensional hyper-rectangle. A classifier is therefore a collection of prioritized hyper-rectangles, and a packet header represents a point in d dimensions. For example, Figure 50.3 shows the classifier in Table 50.5 geometrically in which high priority rules overlay lower priority rules. Classifying a packet is equivalent to finding the highest priority rectangle that contains the point representing the packet. For example, point P(011,110) in Figure 3 would be classified by rule $R5$.

There are several standard geometry problems such as *ray shooting, point location* and *rectangle enclosure* that resemble packet classification. Point location involves finding the enclosing region of a point, given a set of non-overlapping regions. The best bounds for point location in N rectangular regions and $d > 3$ dimensions are $O(\log N)$ time with $O(N^d)$ space;[*] or $O((\log N)^{d-1})$ time with $O(N)$ space [1,2]. In packet classification, hyper-rectangles can overlap, making classification at least as hard as point location. Hence, a solution is either impracticably large (with 100 rules and 4 fields, N^d space is about 100MBytes) or too slow ($(\log N)^{d-1}$ is about 350 memory accesses).

We can conclude that: (1) Multi-field classification is considerably more complex than one-dimensional longest prefix matching, and (2) Complexity may require that practical solutions use heuristics.

50.3.1.2 Range Lookups

Packet classification is made yet more complex by the need to match on ranges as well as prefixes. The cases of looking up static arbitrary ranges, and dynamic *conflict-free* ranges in one dimension have been discussed in Chapter 49. One simple way to handle dynamic arbitrary (overlapping) ranges is to convert each W-bit range to a set of $2W-2$ prefixes (see Table 50.6) and then use any of the longest prefix matching algorithms detailed in Chapter 49, thus resulting in $O(NW)$ prefixes for a set consisting of N ranges.

---

[*] The time bound for $d \leq 3$ is $O(\log \log N)$ [1] but has large constant factors.

TABLE 50.6 Example 4-Bit Ranges and Their Constituent Prefixes

Range	Constituent Prefixes
[4, 7]	01**
[3, 8]	0011,01**,1000
[1, 14]	0001,001*,01**,10**,110*,1110

TABLE 50.7 Categories (not Non-overlapping) of Classification Algorithms

Category	Algorithms
Basicdatastructures	Linear search, caching, hierarchical tries, set-pruning tries
Geometry-based	Grid-of-tries, AQT, FIS
Heuristic	RFC, hierarchical cuttings, tuple-space search
Hardware	Ternary CAM, bitmap-intersection (and variants)

50.3.2 Taxonomy of Classification Algorithms

The classification algorithms we will describe here can be categorized into the four classes shown in Table 50.7.

We now proceed to describe representative algorithms from each class.

50.3.3 Basic Data Structures

50.3.3.1 Linear Search

The simplest data structure is a linked-list of rules stored in order of decreasing priority. A packet is compared with each rule sequentially until a rule is found that matches all relevant fields. While simple and storage-efficient, this algorithm clearly has poor scaling properties; the time to classify a packet grows linearly with the number of rules.

50.3.3.2 Hierarchical Tries

A d-dimensional hierarchical radix trie is a simple extension of the one dimensional radix trie data structure, and is constructed recursively as follows. If d is greater than 1, we first construct a 1-dimensional trie, called the $F1$-trie, on the set of prefixes $\{R_{j1}\}$, belonging to dimension $F1$ of all rules in the classifier, $C = \{R_j\} = \{< R_{j1}, R_{j2}, \ldots, R_{jd} >\}$. For each prefix, p, in the $F1$-trie, we recursively construct a $(d-1)$-dimensional hierarchical trie, T_p, on those rules which specify exactly p in dimension $F1$, i.e., on the set of rules $\{R_j : R_{j1} = p\}$. Prefix p is linked to the trie T_p using a next-trie pointer. The storage complexity of the data structure for an N-rule classifier is $O(NdW)$. The data structure for the classifier in Table 50.5 is shown in Figure 50.4. Hierarchical tries are sometimes called "multi-level tries," "backtracking-search tries," or "trie-of-tries".

Classification of an incoming packet $(v_1, v_2, \ldots, v_d)$ proceeds as follows. The query algorithm first traverses the $F1$-trie based on the bits in v_1. At each $F1$-trie node encountered, the algorithm follows the next-trie pointer (if present) and traverses the $(d-1)$-dimensional trie. The query time complexity for d-dimensions is therefore $O(W^d)$. Incremental updates can be carried out similarly in $O(d^2 W)$ time since each component of the updated rule is stored in exactly one location at maximum depth $O(dW)$.

50.3.3.3 Set-Pruning Tries

A set-pruning trie data structure [3] is similar, but with reduced query time obtained by replicating rules to eliminate recursive traversals. The data structure for the classifier in Table 50.5 is shown in Figure 50.5. The query algorithm for an incoming packet $(v_1, v_2, \ldots, v_d)$ need only traverse the $F1$-trie to find the longest matching prefix of v_1, follow its next-trie pointer (if present), traverse the $F2$-trie to find the longest matching prefix of v_1, and so on for all dimensions. The rules are replicated to ensure that every matching rule will be encountered in the path. The query time is reduced to $O(dW)$ at the expense of increased storage of $O(N^d dW)$ since a rule may need to be replicated $O(N^d)$ times. Update complexity is $O(N^d)$, and hence, this data structure works only for relatively static classifiers.

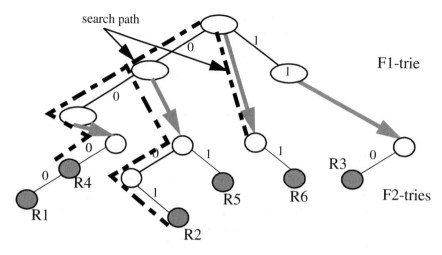

FIGURE 50.4 A hierarchical trie data structure. The gray pointers are the "next-trie" pointers. The path traversed by the query algorithm on an incoming packet (000, 010) is shown.

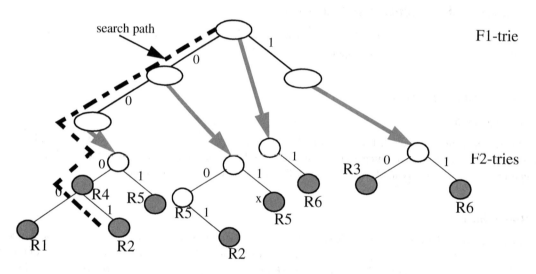

FIGURE 50.5 A set-pruning trie data structure. The gray pointers are the "next-trie" pointers. The path traversed by the query algorithm on an incoming packet (000, 010) is shown.

50.3.4 Geometric Algorithms

50.3.4.1 Grid-of-Tries

The grid-of-tries data structure, proposed by Srinivasan et al. [4] for 2-dimensional classification, reduces storage space by allocating a rule to only one trie node as in a hierarchical trie, and yet achieves $O(W)$ query time by pre-computing and storing a *switch pointer* in some trie nodes. A switch pointer is labeled with "0' or "1" and guides the search process. The conditions which must simultaneously be satisfied for a switch pointer labeled b ($b = $ '0' or "1") to exist from a node w in the trie T_w to a node x of another trie T_x are (see Figure 50.6):

1. T_x and T_w are distinct tries built on the prefix components of dimension $F2$. T_x and T_w are pointed to by two distinct nodes, say r and s respectively of the same trie, T, built on prefix components of dimension $F1$.
2. The bit-string that denotes the path from the root node to node w in trie T_w concatenated with the bit b is identical to the bit-string that denotes the path from the root node to node x in the trie T_x.
3. Node w does not have a child pointer labeled b, and
4. Node s in trie T is the closest ancestor of node r that satisfies the above conditions.

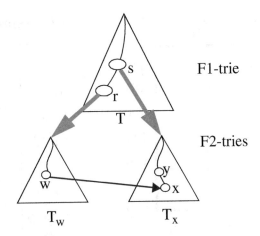

FIGURE 50.6 The conditions under which a switch pointer exists from node *w* to node *x*.

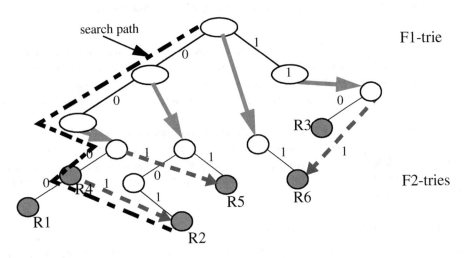

FIGURE 50.7 The grid-of-tries data structure. The switch pointers are shown dashed. The path traversed by the query algorithm on an incoming packet (000, 010) is shown.

If the query algorithm traverses paths $U1(s, root(T_x), y, x)$ and $U2(r, root(T_w), w)$ in a hierarchical trie, it need only traverse the path $(s, r, root(T_w), w, x)$ on a grid-of-tries. This is because paths $U1$ and $U2$ are identical (by condition 2 above) till $U1$ terminates at node w because it has no child branch (by condition 3 above). The switch pointer eliminates the need for backtracking in a hierarchical trie without the storage overhead of a set-pruning trie. Each bit of the packet header is examined at most once, so the time complexity reduces to $O(W)$, while storage complexity $O(NW)$ is the same as a 2-dimensional hierarchical trie. However, the presence of switch pointers makes incremental updates difficult, so the authors recommend rebuilding the data structure (in time $O(NW)$) for each update. An example of the grid-of-tries data structure is shown in Figure 50.7.

Reference 4 reports 2MBytes of storage for a 20,000 two-dimensional classifier with destination and source IP prefixes in a maximum of 9 memory accesses.

Grid-of-tries works well for two dimensional classification, and can be used for the last two dimensions of a multi-dimensional hierarchical trie, decreasing the classification time complexity by a factor of W to $O(NW^{d-1})$. As with hierarchical and set-pruning tries, grid-of-tries handles range specifications by splitting into prefixes.

50.3.4.2 Cross-Producting

Cross-producting [4] is suitable for an arbitrary number of dimensions. Packets are classified by composing the results of separate 1-dimensional range lookups for each dimension as explained below.

Constructing the data structure involves computing a set of ranges (basic intervals), G_k, of size $s_k = |G_k|$, projected by rule specifications in each dimension k, $1 \leq k \leq d$. Let r_k^j, $1 \leq j \leq s_k$, denote the jth range in G_k. r_k^j may be encoded simply as j in the kth

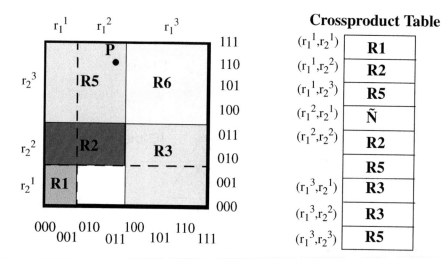

FIGURE 50.8   The table produced by the crossproducting algorithm and its geometric representation.

dimension. A cross-product table C_T of size $\prod_{k=1}^{k=d} s_k$ is constructed, and the best matching rule for each entry $(r_1^{i_1}, r_2^{i_2}, \ldots, r_d^{i_d})$, $1 \le i_k \le s_k, 1 \le k \le d$ is pre-computed and stored. Classifying a packet $(v_1, v_2, \ldots, v_d)$ involves a range lookup in each dimension k to identify the range $r_k^{i_k}$ containing point v_k. The tuple $< r_1^{i_1}, r_2^{i_2}, \ldots, r_d^{i_d} >$ (or, if using the above encoding for r_k^j, the tuple $< i_1, i_2, \ldots, i_d >$) is then found in the cross-product table C_T which contains the pre-computed best matching rule. Figure 50.8 shows an example.

Given that N prefixes lead to at most $2N - 2$ ranges, $s_k \le 2N$ and C_T is of size $O(N^d)$. The lookup time is $O(dt_{RL})$ where t_{RL} is the time complexity of finding a range in one dimension. Because of its high worst case storage complexity, cross-producting is suitable only for very small classifiers. Reference [4] proposes using an on-demand cross-producting scheme together with caching for classifiers bigger than 50 rules in five dimensions. Updates require reconstruction of the cross-product table, and so cross-producting is suitable for relatively static classifiers.

50.3.4.3 A 2-Dimensional Classification Scheme [5]

Lakshman and Stiliadis [5] propose a 2-dimensional classification algorithm where one dimension, say $F1$, is restricted to have prefix specifications while the second dimension, $F2$, is allowed to have arbitrary range specifications. The data structure first builds a trie on the prefixes of dimension $F1$, and then associates a set G_w of non-overlapping ranges to each trie node, w, that represents prefix p. These ranges are created by (possibly overlapping) projections on dimension $F2$ of those rules, S_w, that specify exactly p in dimension $F1$. A range lookup data structure (e.g., an array or a binary search tree) is then constructed on G_w and associated with trie node w. An example is shown in Figure 50.9.

Searching for point $P(v_1, v_2)$ involves a range lookup in data structure G_w for each trie node, w, encountered. The search in G_w returns the range containing v_2, and hence the best matching rule. The highest priority rule is selected from the rules $\{R_w\}$ for all trie nodes encountered during the traversal.

The storage complexity is $O(NW)$ because each rule is stored only once in the data structure. Queries take $O(W \log N)$ time because an $O(\log N)$ range lookup is performed for every node encountered in the $F1$-trie. This can be reduced to $O(W + \log N)$ using fractional cascading [6], but that makes incremental updates impractical.

50.3.4.4 Area-Based Quadtree

The Area-based Quadtree (AQT) was proposed by Buddhikot et al. [7] for two-dimensional classification. AQT allows incremental updates whose complexity can be traded off with query time by a tunable parameter. Each node of a quadtree [6] represents a two dimensional space that is decomposed into four equal sized quadrants, each of which is represented by a child node. The initial two dimensional space is recursively decomposed into four equal-sized quadrants till each quadrant has at most one rule in it (Figure 50.10 shows an example of the decomposition). Rules are allocated to each node as follows. A rule is said to cross a quadrant if it completely spans at least one dimension of the quadrant. For instance, rule R6 spans the quadrant represented by the root node in Figure 50.10, while R5 does not. If we divide the 2-dimensional space into four quadrants, rule R5 crosses the north-west quadrant while rule R3 crosses the south-west quadrant. We call the set of rules crossing the quadrant represented by a node in dimension k, the k-crossing filter set (k-CFS) of that node.

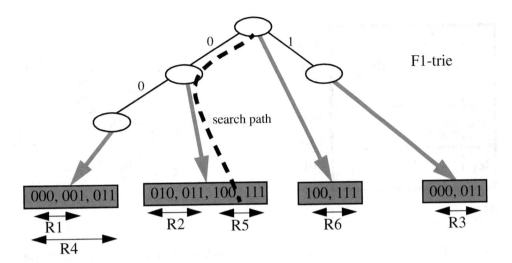

FIGURE 50.9 The data structure of [5] for the example classifier of Table 50.5. The search path for example packet P(011, 110) resulting in R5 is also shown.

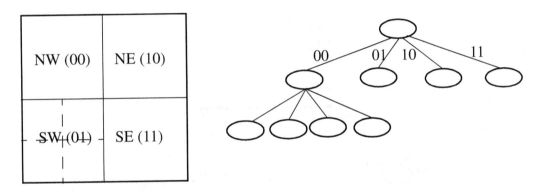

FIGURE 50.10 A quadtree constructed by decomposition of two-dimensional space. Each decomposition results in four quadrants.

Two instances of the same data structure are associated with each quadtree node—each stores the rules in k-CFS ($k = 1,2$). Since rules in crossing filter sets span at least one dimension, only the range specified in the other dimension need be stored. Queries proceed two bits at a time by transposing one bit from each dimension, with two 1-dimensional lookups being performed (one for each dimension on k-CFS) at each node. Figure 50.11 shows an example.

Reference 7 proposes an efficient update algorithm that, for N two-dimensional rules, has $O(NW)$ space complexity, $O(\alpha W)$ search time and $O(\alpha \sqrt[\alpha]{N})$ update time, where α is a tunable integer parameter.

50.3.4.5 Fat Inverted Segment Tree (FIS-tree)

Feldman and Muthukrishnan [8] propose the Fat Inverted Segment tree (FIS-tree) for two dimensional classification as a modification of a segment tree. A segment tree [6] stores a set S of possibly overlapping line segments to answer queries such as finding the highest priority line segment containing a given point. A segment tree is a balanced binary search tree containing the end points of the line segments in S. Each node, w, represents a range G_w, the leaves represent the original line segments in S, and parent nodes represent the union of the ranges represented by their children. A line segment is allocated to a node w if it contains G_w but not $G_{parent(w)}$. The highest priority line segment allocated to a node is pre-computed and stored at the node. A query traverses the segment tree from the root, calculating the highest priority of all the pre-computed segments encountered. Figure 50.12 shows an example segment tree.

An FIS-tree is a segment tree with two modifications: (1) The segment tree is compressed (made "fat" by increasing the degree to more than two) in order to decrease its depth and occupies a given number of levels l, and (2) Up-pointers from child to parent nodes are used. The data structure for 2-dimensions consists of an FIS-tree on dimension $F1$ and a range lookup data associated with each node. An instance of the range lookup data structure associated with node w of the FIS-tree stores the ranges formed by the $F2$-projections of those classifier rules whose $F1$-projections were allocated to w.

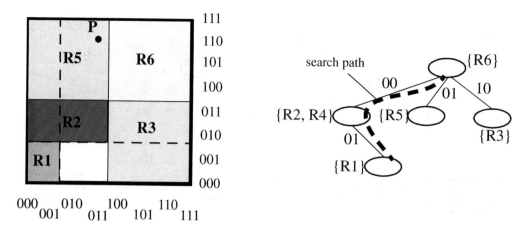

FIGURE 50.11 An AQT data structure. The path traversed by the query algorithm on an incoming packet (000, 010) yields R1 as the best matching rule.

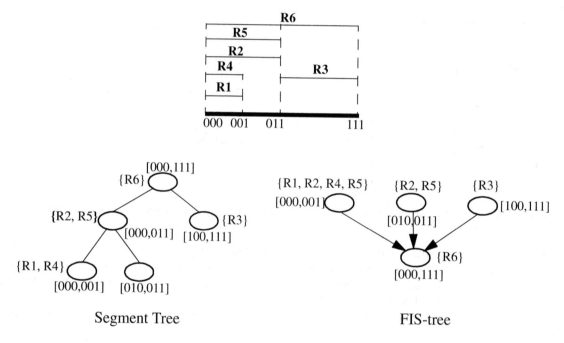

FIGURE 50.12 The segment tree and the 2-level FIS-tree for the classifier of Table 50.5.

A query for point $P(v_1, v_2)$ first solves the range lookup problem on dimension F1. This returns a leaf node of the FIS-tree representing the range containing the point v_1. The query algorithm then follows the up-pointers from this leaf node towards the root node, carrying out 1-dimensional range lookups at each node. The highest priority rule containing the given point is calculated at the end of the traversal.

Queries on an l-level FIS-tree have complexity $O((l+1)t_{RL})$ with storage complexity $O(ln^{1+1/l})$, where t_{RL} is the time for a 1-dimensional range lookup. Storage space can be traded off with search time by varying l. Modifications to the FIS-tree are necessary to support incremental updates—even then, it is easier to support inserts than deletes [8]. The static FIS-tree can be extended to multiple dimensions by building hierarchical FIS-trees, but the bounds are similar to other methods studied earlier [8].

Measurements on real-life 2-dimensional classifiers are reported in Reference 8 using the static FIS-tree data structure. Queries took 15 or fewer memory operations with a two level tree, 4–60 K rules and 5 MBytes of storage. Large classifiers with one million 2-dimensional rules required 3 levels, 18 memory accesses per query and 100MBytes of storage.

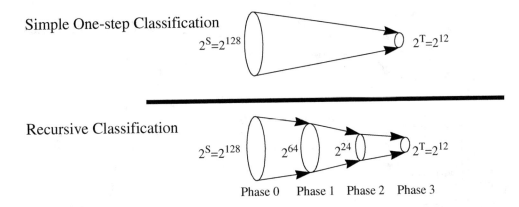

FIGURE 50.13 Showing the basic idea of Recursive Flow Classification. The reduction is carried out in multiple phases, with a reduction in phase I being carried out recursively on the image of the phase $I - 1$. The example shows the mapping of 2^S bits to 2^T bits in three phases.

50.3.4.6 Dynamic Multi-Level Tree Algorithms

Two algorithms, called *Heap-on-Trie (HoT)* and *Binarysearchtree-on-Trie (BoT)* are introduced in Reference 9 that build a heap and binary search tree respectively on the last dimension, and multi-level tries on all the remaining $d - 1$ dimensions. If $W = O(\log N)$: *HoT* has query complexity $O(\log^d N)$, storage complexity $O(N \log^d N)$, and update complexity $O(\log^{d+1} N)$; and *BoT* has query complexity $O(\log^{d+1} N)$, storage complexity $O(N \log^d N)$, and update complexity $O(\log^d N)$. If $W \neq O(\log N)$, each of the above complexity formulae need to be modified to replace a factor $O(\log^{d-1} N)$ with $O(W^{d-1})$.

50.3.5 Heuristics

As we saw in Section 50.3.1.1, the packet classification problem is expensive to solve in the worst-case—theoretical bounds state that solutions to multi-field classification either require storage that is geometric, or a number of memory accesses that is poly-logarithmic, in the number of classification rules. We can expect that classifiers in real networks have considerable structure and redundancy that might be exploited by a heuristic. That is the motivation behind the algorithms described in this section.

50.3.5.1 Recursive Flow Classification (RFC)

RFC [10] is a heuristic for packet classification on multiple fields. Classifying a packet involves mapping S bits in the packet header to a T bit action identifier, where $T = \log N$, $T \ll S$. A simple, but impractical method could pre-compute the action for each of the 2^S different packet headers, yielding the action in one step. RFC attempts to perform the same mapping over several phases, as shown in Figure 50.13; at each stage the algorithm maps one set of values to a smaller set. In each phase a set of memories return a value shorter (i.e., expressed in fewer bits) than the index of the memory access. The algorithm, illustrated in Figure 50.14, operates as follows:

1. In the first phase, d fields of the packet header are split up into multiple chunks that are used to index into multiple memories in parallel. The contents of each memory are chosen so that the result of the lookup is narrower than the index.
2. In subsequent phases, memories are indexed using the results from earlier phases.
3. In the final phase, the memory yields the action.

The algorithm requires construction of the contents of each memory detailed in Reference 10. This paper reports that with real-life four-dimensional classifiers of up to 1700 rules, RFC appears practical for 10Gbps line rates in hardware and 2.5Gbps rates in software. However, the storage space and pre-processing time grow rapidly for classifiers larger than 6000 rules. An optimization described in Reference 10 reduces the storage requirement of a 15,000 four-field classifier to below 4MBytes.

50.3.5.2 Hierarchical Intelligent Cuttings (HiCuts)

HiCuts [11] partitions the multi-dimensional search space guided by heuristics that exploit the structure of the classifier. Each query leads to a leaf node in the HiCuts tree, which stores a small number of rules that can be searched sequentially to find the best match. The characteristics of the decision tree (its depth, degree of each node, and the local search decision to be made at each node) are chosen while pre-processing the classifier based on its characteristics (see Reference 11 for the heuristics used).

Each node, v, of the tree represents a portion of the geometric search space. The root node represents the complete d-dimensional space, which is partitioned into smaller geometric sub-spaces, represented by its child nodes, by cutting across

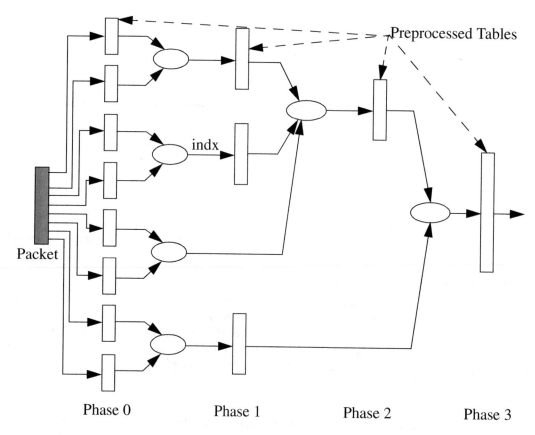

FIGURE 50.14   Packet flow in RFC.

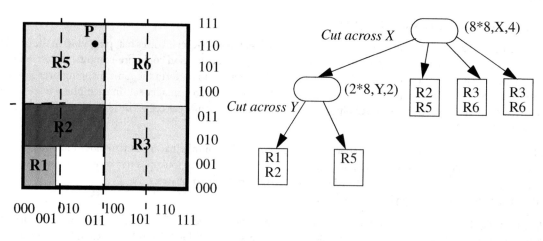

FIGURE 50.15   A possible HiCuts tree for the example classifier in Table 50.5. Each ellipse in the tree denotes an internal node *v* with a tuple <size of 2-dimensional space represented, dimension to cut across, number of children> . Each square is a leaf node which contains the actual classifier rules.

one of the *d* dimensions. Each sub-space is recursively partitioned until no sub-space has more than *B* rules, where *B* is a tunable parameter of the pre-processing algorithm. An example is shown in Figure 50.15 for two dimensions with $B = 2$.

Parameters of the HiCuts algorithm can be tuned to trade- off query time against storage requirements. On 40 real-life four-dimensional classifiers with up to 1,700 rules, HiCuts requires less than 1 MByte of storage with a worst case query time of 20 memory accesses, and supports fast updates.

Rule	Specification	Tuple
R1	(00*,00*)	(2,2)
R2	(0**,01*)	(1,2)
R3	(1**,0**)	(1,1)
R4	(00*,0**)	(2,1)
R5	(0**,1**)	(1,1)
R6	(***,1**)	(0,1)

Tuple	Hash Table Entries
(0,1)	{R6}
(1,1)	{R3,R5}
(1,2)	{R2}
(2,1)	{R4}
(2,2)	{R1}

FIGURE 50.16 The tuples and associated hash tables in the tuple space search scheme for the example classifier of Table 50.5.

50.3.5.3 Tuple Space Search

The basic tuple space search algorithm [12] decomposes a classification query into a number of exact match queries. The algorithm first maps each d-dimensional rule into a d-tuple whose i^{th} component stores the length of the prefix specified in the ith dimension of the rule (the scheme supports only prefix specifications). Hence, the set of rules mapped to the same tuple are of a fixed and known length, and can be stored in a hash table. Queries perform exact match operations on each of the hash tables corresponding to all possible tuples in the classifier. An example is shown in Figure 50.16.

Query time is M hashed memory accesses, where M is the number of tuples in the classifier. Storage complexity is $O(N)$ since each rule is stored in exactly one hash table. Incremental updates are supported and require just one hashed memory access to the hashed table associated with the tuple of the modified rule. In summary, the tuple space search algorithm performs well for multiple dimensions in the average case if the number of tuples is small. However, the use of hashing makes the time complexity of searches and updates non-deterministic. The number of tuples could be very large, up to $O(W^d)$, in the worst case. Furthermore, since the scheme supports only prefixes, the storage complexity increases by a factor of $O(W^d)$ for generic rules as each range could be split into W prefixes in the manner explained in Section 50.3.1. Of course, the algorithm becomes attractive for real-life classifiers that have a small number of tuples.

50.3.6 Hardware-Based Algorithms

50.3.6.1 Ternary CAMs

A TCAM stores each W-bit field as a *(val, mask)* pair; where *val* and *mask* are each W-bit numbers. A *mask* of "0" wildcards the corresponding bit position. For example, if $W = 5$, a prefix 10* will be stored as the pair (10000, 11000). An element matches a given input key by checking if those bits of *val* for which the *mask* bit is "1", match those in the key.

A TCAM is used as shown in Figure 50.17. The TCAM memory array stores rules in decreasing order of priorities, and compares an input key against every element in the array in parallel. The N-bit bit-vector, *matched*, indicates which rules match and so the N-bit priority encoder indicates the address of the highest priority match. The address is used to index into a RAM to find the action associated with this prefix. TCAMs are being increasingly deployed because of their simplicity of use and speed (as they are able to do classification in hardware at the rate of the hardware clock).

Several companies today ship 9Mb TCAMs capable of single and multi-field classification in as little as 10 ns. Both faster and denser TCAMs can be expected in the near future. There are, however, some disadvantages to TCAMs:

1. A TCAM is less dense than a RAM, storing fewer bits in the same chip area. One bit in an SRAM typically requires 4–6 transistors, while one bit in a TCAM requires 11–15 transistors [13]. A 9Mb TCAM running at 100 MHz costs about $200 today, while the same amount of SRAM costs less than $10. Furthermore, range specifications need to be split into multiple masks, reducing the number of entries by up to $(2W - 2)^d$ in the worst case. If only two 16-bit dimensions specify ranges, this is a multiplicative factor of 900. Newer TCAMs, based on DRAM technology, have been proposed and promise higher densities.
2. TCAMs dissipate more power than RAM solutions because an address is compared against every TCAM element in parallel. At the time of writing, a 9Mb TCAM chip running at 100 MHz dissipates about 10–15 watts (the exact number varies with manufacturer). In contrast, a similar amount of SRAM running at the same speed dissipates 1W.
3. A TCAM is more unreliable while being in operational use in a router in the field than a RAM, because a *soft-error* (error caused by alpha particles and package impurities that can flip a bit of memory from 0 to 1, or vice-versa) could go undetected for a long amount of time. In a SRAM, only one location is accessed at any time, thus enabling easy on-the-fly error detection and correction. In a TCAM, wrong results could be given out during the time that the error is undetected – which is particularly problematic in such applications as filtering or security.

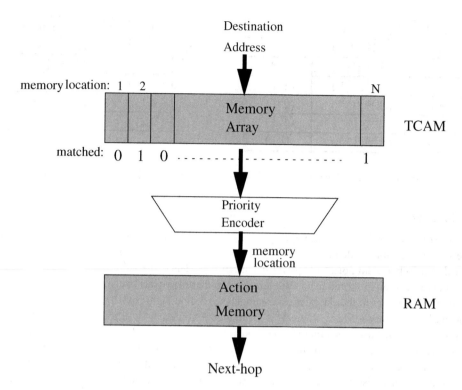

FIGURE 50.17   The classification operation using a ternary CAM. The packet header in this example is assumed to comprise only the destination address.

However, the above disadvantages of a TCAM need to be weighed against its enormous simplicity of use in a hardware platform. Besides, it is the only known "algorithm" that is capable of doing classification at high speeds without explosion in either storage space or time.

Due to their high cost and power dissipation, TCAMs will probably remain unsuitable in the near future for (1) Large classifiers (512K-1M rules) used for microflow recognition at the edge of the network, (2) Large classifiers (256-512 K rules) used at edge routers that manage thousands of subscribers (with a few rules per subscriber). Of course, software-based routers need to look somewhere else.

50.3.6.2 Bitmap-Intersection

The bitmap-intersection classification scheme, proposed in Reference 5, is based on the observation that the set of rules, S, that match a packet is the intersection of d sets, S_i, where S_i is the set of rules that match the packet in the ith dimension alone. While cross-producting pre-computes S and stores the best matching rule in S, this scheme computes S and the best matching rule during each classification operation.

In order to compute intersection of sets in hardware, each set is encoded as an N-bit bitmap where each bit corresponds to a rule. The set of matching rules is the set of rules whose corresponding bits are "1' in the bitmap. A query is similar to cross-producting: First, a range lookup is performed in each of the d dimensions. Each lookup returns a bitmap representing the matching rules (pre-computed for each range) in that dimension. The d sets are intersected (a simple bit-wise AND operation) to give the set of matching rules, from which the best matching rule is found. See Figure 50.18 for an example.

Since each bitmap is N bits wide, and there are $O(N)$ ranges in each of d dimensions, the storage space consumed is $O(dN^2)$. Query time is $O(dt_{RL} + dN/w)$ where t_{RL} is the time to do one range lookup and w is the memory width. Time complexity can be reduced by a factor of d by looking up each dimension independently in parallel. Incremental updates are not supported.

Reference [5] reports that the scheme can support up to 512 rules with a 33 MHz field-programmable gate array and five 1Mbit SRAMs, classifying 1Mpps. The scheme works well for a small number of rules in multiple dimensions, but suffers from a quadratic increase in storage space and linear increase in classification time with the size of the classifier. A variation is described in Reference 5 that decreases storage at the expense of increased query time. This work has been extended significantly by [14] by aggregating bitmaps wherever possible and thus decreasing the time spent in reading the bitmaps. Though the bitmap-intersection scheme is primarily meant for hardware, it is easy to see how it can be used in software, where the aggregated bitmap technique of [14] could be especially useful.

Dimension 1

r_1^1	{R1,R2,R4,R5,R6}	110111
r_1^2	{R2,R5,R6}	010011
r_1^3	{R3,R6}	001001

Dimension 2

r_2^1	{R1,R3,R4}	101100
r_2^2	{R2,R3}	011000
r_2^3	{R5,R6}	000111

Query on P(011,010):

010011 Dimension 1 bitmap
000111 Dimension 2 bitmap
—————
000011

R5 Best matching rule

FIGURE 50.18 Bitmap tables used in the "bitmap-intersection" classification scheme. See Figure 50.8 for a description of the ranges. Also shown is classification query on an example packet P(011, 110).

TABLE 50.8 Complexity of Various Classification Algorithms

Algorithm	Worst-Case Time Complexity	Worst-Case Storage Complexity
Linear Search	N	N
Ternary CAM1	1	N
Hierarchical Tries	W^d	NdW
Set-pruning Tries	dW	N^d
Grid-of-Tries	W^{d-1}	NdW
Cross-producting	dW	N^d
FIS-tree	$(l+1)W$	$l*N^{1+1/l}$
RFC	d	N^d
Bitmap-intersection	$dW + N/memwidth$	dN^2
HiCuts	d	N^d
Tuple Space Search	N	N

Note: N is the Number of Rules in d, W-bit Wide, Dimensions

50.4 Summary

Please see Table 50.8 for a summary of the complexities of classification algorithms described in this chapter.

References

1. M.H. Overmars and A.F. van der Stappen. "Range Searching and Point Location Among Fat Objects," *Journal of Algorithms*, vol. 21, no. 3, pages 629–656, November 1996.
2. F. Preparata and M. I. Shamos. *Computational Geometry: An Introduction*, Springer-Verlag, 1985.
3. P. Tsuchiya. "A Search Algorithm for Table Entries with Non-Contiguous Wildcarding," unpublished report, Bellcore.
4. V. Srinivasan, S. Suri, G. Varghese, and M. Waldvogel. "Fast and Scalable Layer four Switching," *Proceedings of ACM Sigcomm*, pages 203–14, September 1998.
5. T.V. Lakshman and D. Stiliadis. "High-Speed Policy-based Packet Forwarding Using Efficient Multi-dimensional Range Matching", *Proceedings of ACM Sigcomm*, pages 191–202, September 1998.
6. M. de Berg, M. van Kreveld, and M. Overmars. *Computational Geometry: Algorithms and Applications*, Springer-Verlag, 2nd rev. ed. 2000.
7. M.M. Buddhikot, S. Suri, and M. Waldvogel. "Space Decomposition Techniques for Fast Layer-4 Switching," *Proceedings of Conference on Protocols for High Speed Networks*, pages 25–41, August 1999.
8. A. Feldman and S. Muthukrishnan. "Radeoffs for Packet Classification," *Proceedings of Infocom*, vol. 3, pages 1193–202, March 2000.
9. P. Gupta and N. McKeown. "Dynamic Algorithms with Worst-case Performance for Packet Classification," Proc. IFIP Networking, pages 528–39, May 2000, Paris, France.

10. P. Gupta and N. McKeown. "Packet Classification on Multiple Fields," *Proc. Sigcomm, Computer Communication Review,* vol. 29, no. 4, pp 147–60, September 1999, Harvard University.

11. P. Gupta and N. McKeown. "Packet Classification using Hierarchical Intelligent Cuttings," *Proc. Hot Interconnects VII,* August 99, Stanford. This paper is also available in IEEE Micro, pages 34–41, vol. 20, no. 1, January/February 2000.

12. V. Srinivasan, S. Suri, and G. Varghese. "Packet Classification using Tuple Space Search," *Proceedings of ACM Sigcomm,* pages 135–46, September 1999.

13. F. Shafai, K.J. Schultz, G.F. R. Gibson, A.G. Bluschke and D.E. Somppi. "Fully Parallel 30-Mhz, 2.5 Mb CAM," *IEEE Journal of Solid-State Circuits,* vol. 33, no. 11, November 1998.

14. F. Baboescu and G. Varghese. "Scalable Packet Classification," *Proc. Sigcomm,* 2001.

51

Data Structures in Web Information Retrieval*

51.1 Introduction.. 799
51.2 Inverted Indices 799
    Index Compression • Index Granularity
51.3 Fingerprints 801
51.4 Finding Near-Duplicate Documents... 801
51.5 Conclusions.. 803
References.. 803

Monika Henzinger
University of Vienna

51.1 Introduction

Current search engines process thousands of queries per second over a collection of billions of web pages with a sub-second average response time. There are two reasons for this astonishing performance: Massive parallelism and a simple yet efficient data structure, called *inverted index*.

In this chapter we will describe inverted indices. The parallelism deployed by search engines is quite straightforward: Given a collection of documents and a user query the goal of information retrieval is to find all documents that are relevant to a user query and return them in decreasing order of relevance. Since on the Web there are often thousands of matches for a given user query, Web search engines usually return just the top 10 results and retrieve more results only upon request. This can be easily parallelized over m machines: Distribute the documents equally over $m - 1$ machines, find the best up to 10 documents for each machine and return them to the machine without documents, which then merges the lists to determine the overall top 10.

Since the users are only presented with the top 10 results they are usually annoyed if these results contain duplicates or near-duplicates. Thus, it is crucial for a search engine to detect near-duplicate web pages. In Section 51.4 we will describe a technique for doing so based on fingerprints, which we will introduce in Section 51.3.

51.2 Inverted Indices

Given a user query consisting of terms $t_1, \ldots, t_n$, a search engine has to find all documents relevant to the query. Usually Web search engines make the following simplifying assumption: A document is relevant only if it contains *all* query terms. To find all these documents the search engine uses the *inverted (file) index data structure*. It consists of two parts:

- A data structure containing every word t that appears in at least one document together with a pointer p_t for each word and potentially a count f_t of the number of documents containing the word; and
- An *inverted list* for every term t that consists of all the documents containing t and is pointed to by p_t.

One popular implementation is to store the sorted list of words in a search tree whose leaves are connected by a linked list. This imposes an order on the words. The inverted lists are implemented as one large (virtual) array such that t's inverted list consists of all documents between the position in the array pointed to by p_t and by $p_{t'}$, where t' is the term following t in the sorted list. in the web scenario this array does not fit onto a single machine and thus needs to be distributed over multiple machines, either in RAM or on disk.

---

* This chapter has been reprinted from first edition of this Handbook, without any content updates.

Another implementation would be to store the list of words in any search tree or hash function together with the pointers p_t. A special terminator document id (like 0) is appended to the inverted lists and they can be stored either in one large array or in multiple arrays.

To determine all documents containing the query terms $t_1, \ldots, t_n$ (called a *Boolean AND*) the intersection of their inverted lists can be computed by traversing their lists in parallel. If one of the query terms is negated (Boolean NOT), i.e. the user does not want this term to be contained in the query, the negation of the corresponding list is used. To speed up the computation in the case that a very frequent term is combined with a very infrequent term the documents in the inverted list are stored in sorted order and a one-level or two-level search tree is stored for each inverted list. The one-level search tree consists of every $1/f$-th entry entry in the inverted list together with a pointer to the location of this entry in the inverted list, where f is a constant like 100. At query time entries in the inverted list of the frequent term can be skipped quickly to get to the next document that contains the infrequent term. The space requirement is only $1/f$ of the space requirement of the space requirement of the inverted lists.

51.2.1 Index Compression

Of course, data compression is crucial when indexing billions of documents. For that it is usually assumed that the documents are numbered consecutively from 1 to N and that the documents are stored in the inverted list by increasing document id. A simple but powerful technique is called *delta-encoding*: Instead of storing the actual document ids in the inverted list, the document id of the first document is stored together with the difference between the i-th and the $i+1$-st document containing the term. When traversing the inverted list the first document id is summed up with delta-values seen so far to get the document id of the current document. Note that delta-encoding does not interfere with the one-level or two-level search tree stored on top of the inverted list. The advantage of delta-encoding is that it turns large integer (document ids) into mostly small integers (the delta values), depending of course on the value of f_t. A variable length encoding scheme, like Golomb codes, of the delta values then leads to considerable space saving. We sketch Golomb codes in the next paragraph. See Reference 1 for more details on variable compression schemes.

There if a trade-off between the space used for the index and the time penalty encountered at query time for decompressing the index. If the index is read from disk, the time to read the appropriate part of the index dominates the compression time and thus more compression can be used. If the index is in RAM, then less compression should be used, except if sophisticated compression is the only way to make the index fit in RAM.

Let n be the number of unique words in all the documents and let f be the number of words in all the documents, including repetitions. Given an integer x it takes at least $\log x$ bits to represent x in binary. However, the algorithm reading the index needs to know how many bits to read. There are these two possibilities: (1) represent x is in binary by $\log N$ bits in binary with the top $\log N - \log x$ bits set to 0; or (2) represent x in unary by x 1's followed by a 0. Usually the binary representation is chosen, but the unary representation is more space efficient when $x < \log N$. Golomb codes combine both approaches as follows.

Choose $b \approx 0.69 \frac{N \cdot n}{f}$. The *Golumb code* for an integer $x \geq 1$ consists of two parts:

- A unary encoding of q followed by 0, where $q = \lfloor (x-1)/b \rfloor$; and
- A binary encoding of the remainder $r = x - qb - 1$. If b is a power of 2, this requires $\log b$ bits.

Note that $\frac{N \cdot n}{f}$ is the average distance between entries in the inverted index, that is, the average value after the delta-encoding. The value of b was chosen to be roughly 69% of the average. Every entry of value b or less requires 1 bit for the unary part and $\log b$ bits for the binary part. This is a considerable space saving over the binary representation since $\log b + 1 < \log N$. As long as $(x-1)/b < \log N - \log b - 1$, there is a space saving over implementing using the binary encoding. So only entries that are roughly $\log N$ times larger than the average require more space than the average.

If the frequency f_t of a term t is known, then the inverted list of t can be compressed even better using $b_t \approx 0.69 \frac{N}{f_t}$ instead of b. This implies that for frequent terms, b_t is smaller than b and thus fewer bits are used for small integers, which occur frequently in the delta-encoding of frequent terms.

51.2.2 Index Granularity

Another crucial issue is the level of granularity used in the inverted index. So far, we only discussed storing document ids. However to handle some query operators, like quotes, that require the query term to be next to each other in the document, the exact position of the document is needed. There are two possible ways to handle this:

1. One way is to consider all documents to be concatenated into one large document and then store the positions in this large document instead of the document ids. Additionally one special inverted list is constructed that stores the position of the first word of each document. At query time an implicit Boolean AND operation is performed with this special inverted list to determine to which document a given position belongs. Note that the above compression techniques continue to work.

This approach is very space efficient but incurs a run-time overhead at query time for traversing the special inverted list and performing the Boolean AND operation.

2. Another solution is to store (document id,position)-pairs in the inverted list and to delta-encode the document ids and also the position within the same document. This approach uses more space, but does not incur the run-time overhead.

Another data structure that a search engine could use are suffix trees (see Chapter 30). They require, more space and are more complicated, but allow more powerful operations, like searching for syllables or letters.

51.3 Fingerprints

Fingerprints are short strings that represent larger strings and have the following properties:

- If the fingerprints of two strings are different then the strings are guaranteed to be different.
- If the fingerprints of two strings are identical then there is only a small probability that the strings are different.

If two different strings have the same fingerprint, it is called a *collision*. Thus, fingerprints are a type of hash functions where the hash table is populated only very sparsely in exchange for a very low collision probability.

Fingerprints are very useful. For example, search engines can use them to quickly check whether a URL that is contained on a web page is already stored in the index. We will describe how they are useful for finding near-duplicate documents in the next section. One fingerprinting scheme is due to Rabin [2] and is equivalent to cyclic redundancy checks. We follow here the presentation by Broder [3]. Let s be a binary string and let $s' = 1s$, i.e. s' equals s prefixed with a 1. The string $s' = (s_1, s_2, \ldots, s_m)$ induces a polynomial $S(t)$ over Z_2 as follows:

$$S(t) = s_1 t^{m-1} + s_2 t^{m-2} + \cdots + s_m.$$

Let $P(t)$ be an irreducible polynomial of degree k over Z_2. The fingerprint of s is defined to be the polynomial

$$f(s) = S(t) \bmod P(t)$$

over Z_2, which can be simply represented as a string of its coefficients. As shown in Reference 3 the probability that two distinct strings of length m from a set of n strings have the same fingerprint is less than $nm^2/2^k$. Thus, given n and m the probability of collision can be reduced to the required value simply by increasing k, that is, the length of a fingerprint.

Rabin's fingerprints have the property that given two overlapping strings $s = (s_1, s_2, \ldots, s_m)$ and $t = (t_1, t_2, \ldots, t_m)$ such that $t_{i+1} = s_i$ for all $1 \le i < m$ then the fingerprint of t can be computed from the fingerprint of s very efficiently: If

$$f(s) = r_1 t^{k-1} + r_2 t^{k-2} + \cdots + r_k,$$

then

$$f(t) = r_2 t^{k-1} + r_3 t^{k-2} + \cdots + r_k t + t_m + (r_1 t^k) \bmod P(t).$$

Note that $Q(t) = t^k \bmod P(t)$ is equivalent to $P(t)$ with the leading coefficient removed, which means it can be computed easily. Thus

$$f(t) = r_2 t^{k-1} + r_3 t^{k-2} + \cdots + r_k t + t_m + r_1 Q(t).$$

Hence $f(t)$ can be computed from $f(s)$ as follows: Assume $f(s)$ is stored in a shift register and $Q(t)$ is stored in another register. Shift left with t_m as input and if $r_1 = 1$ perform a bit-wise EX-OR operation with $Q(t)$. See [3] for a software implementation for a typical 32-bit word computer.

Other ways of computing fingerprints are cryptographic checksums like Sha-1 or MD5.

51.4 Finding Near-Duplicate Documents

Users of search engine strongly dislike getting near-duplicate results on the same results page. Of course it depends on the user what s/he considers to be near-duplicate. However, most users would agree that pages with the same content except for different ads or a different layout or a different header or footer are near-duplicates. By "near-duplicate" we mean these kinds of "syntactic" duplicates – semantic duplication is even harder detect.

Unfortunately, there are a variety of circumstances that create near-duplicate pages. The main reasons are (1) local copies of public-domain pages (for example the PERL-manual or the preamble of the US constitution) or various databases and (2) multiple visits of the same page by the crawler (the part of the search engine that retrieves web pages to store in the index) without detection of the replication. The latter can happen because there were small changes to the URL and the content of the page. Usually, crawlers fingerprint the URL as well as the content of the page and thus can easily detect exact duplicates.

There are various approaches to detect near-duplicate pages. One approach that works very well in the Web setting was given by Broder [4] and was validated on the Web [5]. The basic idea is to associate a set of fingerprints with each document such that the similarity of two pages is proportional to the size of the intersection of their respective sets.

We describe first how to compute the fingerprints and show then how they can be used to efficiently detect near-duplicates. The set of fingerprints for a document is computed using the *shingling* technique introduced by Brin et al. [6]. Let a *token* be either a letter, a word, a line, or a sentence in a document. Each document consists of a sequence of token. We call a contiguous sequence of q tokens a *q-shingle* of the document. We want to compute fingerprints for all q-shingles in a document. Using Rabin's fingerprints this can be done efficiently if we start from the first shingle and then use a sliding window approach to compute the fingerprint of the shingle currently in the window.

Let S_D be the set of fingerprints generated for document D. The idea is simply to define the similarity or *resemblance* $r(A, B)$ of document A and B as

$$r(A, B) = \frac{|S_A \cap S_B|}{|S_A \cup S_B|}.$$

Experiments have indicated that a resemblance value close to 1 captures well the information notion of "syntactic" near-duplication that we discussed above.

Storing the whole set S_D would take a lot of space. It suffices, however, to keep a fixed number of fingerprints from S_D for each document D. This subset is called a *sketch*. As we will show below the sketches can be computed in time linear in the size of D and will be stored for each document. The resemblance of two documents can be approximated by using the sketches instead of the full sets of fingerprints. Thus, the time is only linear in the size of the sketches.

Sketches are determined as follows: Recall that each fingerprints requires k bits. Thus, $S_D \subseteq \{1, \ldots, n\}$, where $n = 2^k$. Let π be a permutation chosen uniformly at random from the set of all permutations of $[n]$. Let $X = \min(\min\{\pi(S_A)\}, \min\{\pi(S_B)\})$.

The crucial observation is that if $\min\{\pi(S_A)\} = \min\{\pi(S_B)\}$, then $X = \min\{\pi(S_A)\}$ must belong to $\pi(S_A) \cap \pi(S_B)$. If $\min\{\pi(S_A)\} \neq \min\{\pi(S_B)\}$, then X belongs to either $\pi(S_A) - \pi(S_B)$ or to $\pi(S_B) - \pi(S_A)$, and, thus, X does not belong to $\pi(S_A) \cap \pi(S_B)$. It follows that $\min\{\pi(S_A)\} = \min\{\pi(S_B)\}$ if and only if X belongs to $\pi(S_A) \cap \pi(S_B)$. Since π was chosen uniformly at random the probability of the latter is

$$\frac{|S_A \cap S_B|}{|S_A \cup S_B|} = r(A, B).$$

It follows that

$$Pr\left(\min\{\pi(S_A)\} = \min\{\pi(S_B)\}\right) = r(A, B).$$

We choose p independent random permutations. For each document the sketch consists of the permuted fingerprint values $\min\{\pi_1(S_D)\}, \min\{\pi_2(S_D)\}, \ldots, \min\{\pi_p(S_D)\}$. The resemblance of two documents is then estimated by the intersection of their sketches, whose expected value is proportional to the resemblance.

In practice, π cannot be chosen uniformly at random which led to the study of min-wise independent permutations [7].

To detect the near-duplicates in a set of documents we build for each shingle in a sketch the list of documents to which this shingle belongs. Using this list we generate for each shingle and each pair of document containing this shingle an (document id, document id)-pair. Finally we simply sort these pairs to determine the resemblance for each pair. The running time is proportional to the number of (document id, document id)-pairs. If each shingle only belongs to a constant number of documents, it is linear in the number of unique shingles.

In a web search engine it often suffices to remove the near-duplicates at query time. To do this the sketch is stored with each document. The search engine determines the top 50 or so results and then performs a pair-wise near-duplicate detection with some resemblance threshold, removing the near duplicate results, to determine the top 10 results.

The described near-duplicate detection algorithm assumes that the similarity between documents depends on the size of the overlap in their fingerprint sets. This corresponds to the Jaccard coefficient of similarity used in information retrieval. A more standard approach in information retrieval is to assign a weight vector of terms to each document and to define the similarity of two documents as the cosine of their (potentially normalized) term vectors. Charikar [8] gives an efficient near-duplicate detection algorithm based on this measure. He also presents a near-duplicate detection algorithms for another metric, the Earth Mover Distance.

Other similarity detection mechanisms are given in References 6 and 9–11. Many near-duplicate web pages are created because a whole web host is duplicated. Thus, a large percentage of the near-duplicate web pages can be detected if near-duplicate web hosts can be found. Techniques for this problem are presented in References 12 and 13.

51.5 Conclusions

In this chapter we presented the dominant data structure for web search engines, the inverted index. We also described fingerprints, which are useful at multiple places in a search engine, and document sketches for near-duplicate detection. Other useful data structures for web information retrieval are adjacency lists: All web search engines claim to perform some type of analysis of the hyperlink structure. For this, they need to store the list of incoming links for each document, a use of the classic adjacency list representation.

Web search engines also sometimes analyze the log of the user queries issued at the search engine. With hundreds of millions of queries per day these logs are huge and can only be processed with one-pass data stream algorithms.

References

1. Witten, I. H., Moffat, A., and Bell, T. C., 1999. *Managing Gigabytes. Compressing and Indexing Documents and Images.* Morgan Kaufmann, San Francisco.
2. Rabin, M. O., 1981. Fingerprinting by random polynomials. Technical Report TR-15-81, Center for Research in Computing Technology, Harvard University.
3. Broder, A. Z., 1993. Some applications of Rabin's fingerprinting method. In Capocelli, R., De Santis, A., and Vaccaro, U., editors, *Sequences II: Methods in Communications, Security, and Computer Science*, Springer-Verlag, New York, 143–152.
4. Broder, A. Z., 1997. On the resemblance and containment of documents. In *Proceedings of Compression and Complexity of Sequences 1997*, IEEE Computer Society, Washington, DC, 21–29.
5. Broder, A. Z., Glassman, S. C., Manasse, M. S., and Zweig, G., 1997. Syntactic clustering of the Web. In *Proceedings Sixth International World Wide Web Conference*, Elsevier Science, Santa Clara, 391–404.
6. Brin, S., Davis, J., and Garcia-Molina, H., 1995. Copy detection mechanisms for digital documents. In *Proceedings 1995 ACM SIGMOD International conference on Management of Data*, San Jose, 398–409.
7. Broder, A. Z., Charikar, M., Frieze, A., and Mitzenmacher, M., 1998. Min-wise independent permutations. In *Proceedings 30th Annual ACM Symposium on Theory of Computing*, ACM Press, Dallas, 327–336.
8. Charikar, M. S., 2002. Similarity estimation techniques from rounding algorithms. In *Proceedings 34th Annual ACM Symposium on Theory of Computing*, ACM Press, Montreal, 380–388.
9. Heintze, N., 1996. Scalable document fingerprinting. In *Proceedings second USENIX Workshop on Electronic Commerce*, Oakland, 191–200.
10. Manber, U., 1994. Finding similar files in a large file system. In *Proceedings Winter 1994 USENIX Conference*, San Francisco, 1–10.
11. Shivakumar, N. and Garcia-Molina, H., 1995. SCAM: A copy detection mechanism for digital documents. In *Proceedings Second Annual Conference on the Theory and Practice of Digital Libraries*, Austin.
12. Bharat, K., Broder, A. Z., Dean, J., and Henzinger, M. R., 2000. A comparison of techniques to find mirrored hosts on the WWW. *Journal of the American Society for Information Science*, 51(12):114–1122.
13. Cho, J., Shivakumar, N., and Garcia-Molina, H., 2000. Finding replicated web collections. In *Proceedings 2000 ACM International Conference on Management of Data*, ACM Press, Dallas, 355–366.

52

The Web as a Dynamic Graph *

52.1	Introduction	805
52.2	Experimental Observations	806
52.3	Theoretical Growth Models	806
52.4	Properties of Web Graphs and Web Algorithmics	809
	Generating Function Framework • Average Path Length • Emergence of Giant Components • Search on Web Graphs • Crawling and Trawling	
52.5	Conclusions	814
	References	814

S. N. Maheshwari
Indian Institute of Technology Delhi, Delhi

52.1 Introduction

The World Wide Web (the Web) was started as an experimental network in 1991. Its growth since then can only be termed explosive. It has several billion pages today and is growing exponentially with time. This growth is totally distributed. There is no central authority to control the growth. The hyperlinks endow the Web with some structure in the sense that viewing the individual web pages as nodes and the hyperlinks as directed edges between them, the Web can be looked upon as a directed graph. What stands out is that this directed graph is not only dynamic—it is rapidly growing and changing—it has been much too large for some time to even have a complete snapshot. Experimental understanding of its structure is based on large but partial web crawls. What properties are being investigated is itself driven by the requirements of the increasingly sophisticated nature of the applications being developed as well as analogies and insights from fields like bibliometrics involving study of citations in academic literature [1].

Let us briefly consider topic search, discussed in detail in Chapter 51, which involves searching for pages on the Web which correspond closely to a given search topic. The seminal work in this area is Kleinberg's HITS algorithm [2] that assumes that for any topic on the Web there are pages which could be considered to be "authoritative" on that topic, and pages which are "hubs" in the sense that they contain links to relevant pages on that topic. Given a collection of pages and links between them, selected by some sampling method as pertaining to the given topic, HITS algorithm ranks the pages by weights which are representative of the quality of the pages as hubs or authorities. These weights are nothing but principal eigen values, and are, in some sense, a "measure" of the "denseness" of the interconnections between the pages. This model of dense interconnection between hubs and authorities of a given topic gave rise to the notion of "cyber communities" in the Web associated with different topics. Underlying this model of cyber communities was the hypothesis that a subgraph representing a Web community would contain "bipartite cores." Bipartite cores, as the name suggests, are complete bipartite graphs corresponding to the hubs and authorities around which the communities are supposed to have developed. Experimental investigations of the structure of the web graph, taking place on the graphs extracted out of the partial crawls, has confirmed much of the above and more. The structural understanding resulting from the experimental investigations has fueled both, theoretical model building which attempts to explain experimentally observed phenomena, and development of new algorithmic techniques that solve traditional problems of search and information retrieval on the web graph in novel ways. Moreover, the reality of the Web as a structure which is too large and continuously changing, makes the standard off-line and on-line models for algorithm design totally inapplicable. Over the last seven-eight years researchers have attempted to grapple with these unique issues of complexity in the study of the Web. Interestingly, contributions to this study of the Web have come not only from from computer scientists, but also from physicists who have brought to Web model building techniques from statistical mechanics that have been successful in predicting macro level behavior of a variety of natural phenomenon from millions of its constituent parts. In this chapter an attempt is made to put together what to the author are the main strands of this rapidly evolving model

---

* This chapter has been reprinted from first edition of this Handbook, without any content updates.

building. The rest of the chapter is organized as follows: Section 2 surveys the experimental observations and reflects what are the major trends in the findings. Section 3 contains the basic theoretical framework developed to explain the experimental findings. Section 4 contains examples of web algorithmics. Section 5 is crystal gazing and reflects what to the author are the grand challenges.

52.2 Experimental Observations

Recent literature contains reports of a number of experiments conducted to investigate topological properties satisfied by the web graph [3–5]. These experiments were conducted over a period of time, and using Web samples of varying sizes. Albert, Jeong and Barabasi [3] used nd.edu subset of the Web. Kumar et al.[5] used a cleaned up version of a 1997 web crawl carried out by Alexa Inc. Broder et al. [4] based their measurements on an Altavista crawl having about 200 million pages and 1.5 billion hyperlinks. The most fundamental observation that emerges from these experiments conducted at different times, focusing on different subparts of the Web, is that the degree distribution of nodes in the web graph follows a power law. The degree distribution is said to satisfy power law if the fraction of nodes of degree x is proportional to $x^{-\alpha}$ for $\alpha > 0$. Power law distribution is observed for both the indegrees and the outdegrees of the web graph. Broder at al. report that for indegrees the power coefficient —indexexperimental observations!indegree distribution$\alpha \approx 2.1$, and for outdegrees $\alpha \approx 2.72$. There is a very close match in literature in the value of α for indegree distribution. For outdegree distribution the value of α reported varies from 2.4 to 2.72 [3].

Broder et al. [4] also analysed the crawl for connectedness. Viewing the web graph as an undirected graph, it was observed that 91% of the nodes were connected to each other and formed a giant connected component. Interestingly, it was found that the distribution of the number of connected components by their sizes also satisfied power law ($\alpha \approx 2.5$). Power law distribution in the sizes of the components was observed even when the graph was viewed as directed. However, the size of the largest strongly connected component (giant SCC) was only 28% of the total web crawl. The giant SCC was reachable from about 22% of the nodes (the set IN). About similar percentage of nodes were reachable from the giant SCC (the set OUT). A significant portion of the rest of the nodes constituted, in Broder et al's terminology, "tendrils", nodes reachable from IN or from which the set OUT is reachable. All the experiments done so far point to fractal like self similar nature of the Web in the sense that structure described above is likely to be exhibited in any non-trivial crawl carried out at any time.

Kumar et al. [5] also carried out experiments to measure the number of bipartite cores in the Web. In a cleaned up version of the web graph consisting of 24 million nodes, they reported discovering around 135,000 bipartite cliques $K_{i,j}$ with $i \geq 3$ and $j = 3$. The number of $K_{i,j}$'s with $i, j = 3$ was approximately 90,000, and the numbers dropped exponentially with increase in j. Finding such cliques in the web graph is an algorithmically challenging problem which we will further discuss in the section on Web algorithmics.

Measurements have also been made of the "diameter" of the Web [3,4]. If the Web has the structure as asserted in Reference 4, then the probability of a path existing between any two random vertices is approximately 24%, and the average shortest path length is 16. Albert et al. [3] measured the average shortest path length on a directed graph generated having in and outdegree distribution satisfying the power law coefficients of 2.1 and 2.45 respectively. In a directed graph with 8×10^8 vertices the average shortest path length was determined to be approximately 19, and was a linear function of the logarithm of the number of vertices. The Web, therefore, is considered to exhibit the "small worlds" phenomenon [6].

52.3 Theoretical Growth Models

This fractal like structure of an evolving graph, whose growth processes are so organised that the degree distribution of its vertices satisfy power law, which has a large number of bipartite cliques as subgraphs, and exhibits the small world phenomenon has generated a lot of interest among computer scientists recently. Part of the reason is that the web graph does not belong to the much studied $G_{n,p}$ model which consists of all graphs with n vertices having p as the probability that there is an between any two vertices [7]. Graphs in $G_{n,p}$ are essentially sparse and are unlikely to have many bipartite cliques as subgraphs. Moreover, for large n the degree distribution function is Poisson. There is consensus that the primary reason for the web graph to be different from a traditional random graph is that edges in the Web exhibit *preferential attachment*. Hyperlinks from a new web page are more likely to be directed to popular well established website/webpage just as a new manuscript is more likely to cite papers that are already well cited. It is interesting to note that preferential attachment has been used to model evolving phenomena in fields ranging from economics [8], biology [9], to languages [10]. Simon [11] used preferential attachment to explain power law distributions already observed in phenomena like distribution of incomes, distribution of species among genera, distribution of word frequencies in documents etc. A modern exposition of Simon's argument in the framework of its applicability to web graphs is provided by Mitzenmacher [12], and is the basis of Barabasi et al's "mean field theory" based model [13], as well as the "rate equation" approach of Krapivsky and Redner [14]. All three use continuous differential equations to model the dynamics of the evolving web graph. This approach is attractive because of its simplicity and the ease with which it enables one to focus on understanding the issues involved, particularly those relating to power law distributions associated with the Web.

Let us consider the following growth model (we will call this the basic model) which forms the kernel of most of the reported models in literature. At each time step a new node is added to the web graph. This node gets one edge incident at it from one of the existing nodes in the graph, and has one edge pointing out of it. Let us assume that the tail (the node from which the edge emanates) of the edge which is incident at the node just added is chosen uniformly at random from the existing nodes. The head (the node at which the edge is incident) of the edge which emanates from the node just added is chosen with probability proportional to the indegree of the head in keeping with the preferential attachment paradigm that new web pages tend to attach themselves to popular web pages. We assume, to keep matters simple, that whole process starts with a single node with one edge emanating which is incident at the node itself.

Let $I_k(t)$ and $O_k(t)$ denote the number of nodes in the evolved web graph with indegree and outdegree equal to k respectively after t timesteps. $I_k(t)$ and $O_k(t)$ are random variables for all $k, t \geq 1$. We will assume for the purposes of what follows that their expected values are concentrated around their means, and we will use $I_k(t)$ and $O_k(t)$ to denote the expected value also. Note that, at time t, one node and two directed edges are added to the graph. The expected increase in the value of $I_k(t)$ is controlled by two processes. One is the expected increase in the value of $I_{k-1}(t)$, and the other is the expected decrease in the value of $I_k(t)$. The expected increase is $(k-1)I_{k-1}(t)/t$. This is because of our assumption that the probability is proportional to the indegree of the head of the node at which the edge emanating from the node just added is incident and t is the total number of nodes in the graph. Reasoning in the same way we get that the expected decrease is $kI_k(t)/t$. Since this change has has taken place over one unit of time we can write

$$\frac{\Delta(I_k(t))}{\Delta(t)} = \frac{(k-1)I_{k-1}(t) - kI_k(t)}{t}$$

or, in the continuous domain

$$\frac{dI_k(t)}{dt} = \frac{(k-1)I_{k-1}(t) - kI_k(t)}{t}. \tag{52.1}$$

We can solve Equation 52.1 for different values of k starting with 1. For $k=1$ we note that the growth process at each time instance introduces a node of indegree 1. Therefore, Equation 52.1 takes the form

$$\frac{dI_1(t)}{dt} = 1 - \frac{I_1(t)}{t}$$

whose solution has the form $I_1(t) = i_1 t$. Substituting this in the above equation we get $i_1 = 1/2$. Working in the same way we can show that $I_2(t) = t/6$. It is not too difficult to see that $I_k(t)$ are a linear function of t. Therefore, substituting $I_k(t) = i_k t$ in Equation 52.1 we get the recurrence equation

$$i_k = \frac{k-1}{k+1} i_{k-1}$$

whose solution is

$$i_k = \frac{1}{k(k+1)}. \tag{52.2}$$

What interpretation do we put to the solution $I_k(t) = i_k t$? It essentially means that in the steady state (i.e., $t \to \infty$) the number of of nodes in the graph with indegree k is proportional to i_k. Equation 52.2 implies $I_k \sim k^{-2}$, i.e. indegree distribution satisfies power law with $\alpha = 2$.

Let us now develop an estimate for O_k. The counterpart of Equation 52.1 in this case is

$$\frac{dO_k(t)}{dt} = \frac{O_{k-1}(t) - O_k(t)}{t}. \tag{52.3}$$

Assuming that in the steady state $O_k(t) = o_k t$, we can show that $o_1 = 1/2$, and for $k > 1$

$$o_k = \frac{o_{k-1}}{2}.$$

This implies $O_k \sim 2^{-k}$. That the outdegree distribution is exponential and not power law should not come as a surprise if we recall that the process that affected outdegree was uniform random and had no component of preferential attachment associated with it.

It would be instructive to note that very simple changes in the growth model affect the degree distribution functions substantially. Consider the following variation on the growth model analysed above. The edge emanating out of the node added does not just attach itself to another node on the basis of its indegree. With probability β the edge points to a node chosen uniformly at random, and with probability $1 - \beta$ the edge is directed to a node chosen proportionally to its indegree. The Equation 52.1 now takes the form

$$\frac{dI_k(t)}{dt} = \frac{(\beta I_{k-1}(t) + (1 - \beta)(k-1)I_{k-1}(t)) - (\beta I_k(t) + (1 - \beta)kI_k(t))}{t}.$$

Note that the first term of the first part of the r.h.s. corresponds to increase due to the uniform process and the second term increase due to the preferential attachment process. The recurrence equation now takes the form [12]

$$i_k(1 + \beta + k(1 - \beta)) = i_{k-1}(\beta + (k-1)(1 - \beta))$$

or,

$$\frac{i_k}{i_{k-1}} = 1 - \frac{2 - \beta}{1 + \beta + k(1 - \beta)}$$

$$\sim 1 - \left(\frac{2 - \beta}{1 - \beta}\right)\left(\frac{1}{k}\right)$$

for large k. It can be verified by substitution that

$$i_k \sim k^{-\frac{2-\beta}{1-\beta}} \tag{52.4}$$

satisfies the above recurrence. It should be noted that in this case the power law coefficient can be any number larger than 2 depending upon the value of β.

How do we get power law distribution in the outdegree of the nodes of the graph? Simplest modification to the basic model to ensure that would be to choose the tail of the edge incident at the added node to be chosen with probability proportional to the outdegree of the nodes. Aiello et al. [15], and Cooper and Frieze [16] have both given analysis of the version of the basic model in which, at any time epoch, apart from edges incident and emanating out of the added node being chosen randomly and according to the out and in degree distributions of the existing nodes, edges are added between existing nodes of the graph. Both in and out degree distributions show power law behaviour. From Web modeling perspective this is not particularly satisfying because there is no natural analogue of preferential attachment to explain the process that controls the number of hyperlinks in a web page. Never-the-less all models that enable power law distribution in outdegree of nodes in literature resort to it in one form or the other. We will now summarize the other significant variations of the basic model that have been reported in literature.

Kumar et al. [17] categorise evolutionary web graph models according to the rate of growth enabled by them. *Linear growth* models allow one node to be added at one time epoch along with a fixed number of edges to the nodes already in the graph. *Exponential growth* models allow the graph to grow by a fixed fraction of the current size at each time epoch. The models discussed above are linear growth models. Kumar et al. in Reference 17 introduce the notion of *copying* in which the head of the edge emanating out of the added node is chosen to be the head of an edge emanating out of a "designated" node (chosen randomly). The intuition for copying is derived from the insight that links out of a new page are more likely to be directed to pages that deal with the "topic" associated with the page. The designated node represents the choice of the topic and the links out of it are very likely links to "other" pages relating to the topic. Copying from the designated node is done with probability $1 - \beta$. With probability β the head is chosen uniformly at random. The exponential growth model that they have analyzed does not involve the copying rule for distribution of the new edges to be added. The tail of a new edge is chosen to be among the new nodes with some probability factor. If the tail is to be among the old nodes, then the old node is chosen with probability proportional to its out degree. The analysis, as in References 15,16, is carried out totally within the discrete domain using martingale theory to establish that the expected values are sharply clustered. For the linear growth model the power law coefficient is the same as in Equation 52.4. The copying model is particularly interesting because estimates of the distribution of bipartite cliques in graphs generated using copying match those found experimentally.

Another approach that has been used to model evolutionary graphs is the so called *Master Equation Approach* introduced by Dorogovtsev et al. [18] which focuses on the probability that at time t a node introduced at time i, has degree k. If we denote this quantity by $p(k, i, t)$, then the equation controlling this quantity for indegree in the basic model becomes

$$p(k, i, t+1) = \frac{k-1}{t+1}p(k-1, i, t) + \left(1 - \frac{k}{t+1}\right)p(k, i, t). \tag{52.5}$$

First term on the r.h.s. corresponds to the probability with which the node increases its indegree. The second term is the complementary probability with which the node remains in its former state. The over all degree distribution of nodes of indegree k

in the graph is

$$P(k, t) = \frac{1}{t+1} \sum_{i=0}^{i=t} p(k, i, t).$$

Using this definition over Equation 52.5 we get

$$(t+1)P(k, t+1) = (k-1)P(k-1, t) + (t-k)P(k, t).$$

For extremely large networks the stationary form of this equation, i.e. at $t \to \infty$ is

$$P(k) = \frac{k-1}{k+1}P(k-1). \tag{52.6}$$

Notice that the solution to Equation 52.6 with appropriate initial conditions is of the form $P(k) \sim k^{-2}$ which is the same as that obtained by working in the continuous domain.

Rigorous analysis of the stochastic processes done so far in References 15–17 has so far taken into account growth, that is, birth, process only. The combinatorics of a web graph model that involves death processes also has still to be worked out. It must be pointed out that using less rigorous techniques a series of results dealing with issues like non-linear preferential attachment and growth rates, growth rates that change with time (aging and decay), and death processes in the form of edge removals have appeared in literature. This work is primarily being done by physicists. Albert and Barabasi [19], and Dorogovtsev and Mendes [20] are two comprehensive surveys written for physicists which the computer scientists would do well to go through. However, with all this work we are still far away from having a comprehensive model that takes into account all that we understand of the Web. All models view web growth as a global process. But we know that a web page to propagate Esperanto in India is more likely to have hyperlinks to and from pages of Esperanto enthusiasts in rest of the world. From modeling perspective every new node added during the growth process may be associated with the topic of the node chosen, let us say by preferential attachment, to which the added node first points to. This immediately reduces the set of nodes from which links can point to it. In effect the probability for links to be added between two nodes which are associated with some topics will be a function of how related the topics are. That there is some underlying structure to the web graph that is defined by the topics associated with the nodes in the graph has been in the modeling horizon for some time. Search engines that use web directories organised hierarchically as graphs and trees of topics have been designed [21]. Experiments to discover the topic related underlying structures have been carried out [22–24]. A model that takes into account the implicit topic based structure and models web growth as a number of simultaneously taking place local processes [25] is as step in this direction. All that is known about the Web and the models that have been developed so far are pointers to the realisation that development mathematically tractable models that model the phenomena faithfully is a challenging problem which will excite the imagination and the creative energies of researchers for some time.

52.4 Properties of Web Graphs and Web Algorithmics

The properties of web graphs that have been studied analytically are average shortest path lengths between two vertices, size of giant components and their distributions. All these properties have been studied extensively for $G_{n,p}$ class of random graphs. Seminal work in this area for graphs whose degrees were given was done by Malloy and Reed [26] who came up with a precise condition under which phase transition would take place and giant components as large as the graph itself would start to show up. In what follows we will discuss these issues using generating functions as done by Newman, Strogatz, and Watts [27] primarily because of the simplicity with which these reasonably complex issues can be handled in an integrated framework.

52.4.1 Generating Function Framework

Following [27] we define for a large undirected graph with N nodes the generating function

$$G_0(x) = \sum_{k=0}^{k=\infty} p_k x^k, \tag{52.7}$$

where p_k is the probability that a randomly chosen vertex has degree k. We assume that the probability distribution is correctly normalised, that is, $G_0(1) = 1$. $G_0(x)$ can also represent graphs where we know the exact number n_k of vertices of degree k by defining

$$G_0(x) = \frac{\sum_{k=0}^{k=\infty} n_k x^k}{\sum_{k=0}^{k=\infty} n_k}.$$

The denominator is required to ensure that the generating function is properly normalised. Consider the function $[G_0(x)]^2$. Note that coefficient of the power of x^n in $[G_0(x)]^2$ is given by $\sum_{i+j=n} p_i p_j$ which is nothing but the probability of choosing two vertices such that the sum of their degrees is n. The product of two or more generating functions representing different degree distributions can be interpreted in the same way to represent probability distributions reflecting independent choices from the two or more distributions involved.

Consider the problem of estimating the average degree of a vertex chosen at random. The average degree is given by $\sum_k kp_k$ which is also equal to $G_0'(1)$. Interestingly the average degree of a vertex chosen at random and the average degree of a vertex pointed to by a random edge are different. A random edge will point to a vertex with probability proportional to the degree of the vertex which is of the order of kp_k. The appropriately normalised distribution is given by the generating function

$$\frac{\sum_k kp_k x^k}{\sum_k kp_k} = x\frac{G_0'(x)}{G_0'(1)}. \tag{52.8}$$

The generating function given above in Equation 52.8 will have to be divided by x if we wanted to consider the distribution of degree of immediate neighbors excluding the edges by which one reached them. We will denote that generating function by $G_1(x)$. The neighbors of these immediate neighbors are the *second* neighbors of the original node. The probability that any of the second neighbors connect to any of the immediate neighbors or one another is no more than order of N^{-1} and hence can be neglected when N is large. Under this assumption the distribution of the second neighbors of the originally randomly chosen node is

$$\sum_k p_k[G_1(x)]^k = G_0(G_1(x)).$$

The average number of second neighbors, therefore, is

$$z_2 = \left[\frac{d}{dx}G_0(G_1(x))\right]_{x=1} = G_0'(1)G_1'(1), \tag{52.9}$$

using the fact that $G_1(1) = 1$.

52.4.2 Average Path Length

We can extend this reasoning to estimate the distributions for the mth neighbors of a randomly chosen node. The generating function for the distribution for the mth neighbor, denoted by $G^m(x)$, is given by

$$G^m(x) = \begin{cases} G_0(x) & m = 1 \\ G^{(m-1)}(G_1(x)) & m \geq 2. \end{cases}$$

Let z_m denote the average number of mth nearest neighbors. We have

$$z_m = \frac{dG^m(x)}{dx}\bigg|_{x=1} = G_1'(1)G^{(m-1)'}(1) = [G_1'(1)]^{m-1}z_1 = \left[\frac{z_2}{z_1}\right]^{m-1}z_1. \tag{52.10}$$

Let l be the smallest integer such that

$$1 + \sum_{i=1}^{i=l} z_i \geq N.$$

The average shortest path length between two random vertices can be estimated to be of the order l. Using Equation 52.10 we get

$$l \sim \frac{\log[(N-1)(z_2 - z_1) + z_1^2] - \log z_1^2}{\log(z_2/z_1)}.$$

When $N \gg z_1$ and $z_2 \gg z_1$ the above simplifies to

$$l \sim \frac{\log(N/z_1)}{\log(z_2/z_1)} + 1.$$

There do exist more rigorous proofs of this result for special classes of random graphs. The most interesting observation made in Reference 27 is that only estimates of nearest and second nearest neighbors are necessary for calculation of average shortest path length, and that making these purely local measurements one can get a fairly good measure of the average shortest distance which is a global property. One, of course, is assuming that the graph is connected or one is limiting the calculation to the giant connected component.

52.4.3 Emergence of Giant Components

Consider the process of choosing a random edge and determining the component(s) of which one of its end nodes is a part. If there are no other edges incident at that node then that node is a component by itself. Otherwise the end node could be connected to one component, or two components and so on. Therefore, the probability of a component attached to the end of a random edge is the sum of the probability of the end node by itself, the end node connected to one other component, or two other components and so on. If $H_1(x)$ is the generating function for the distribution of the sizes of the components which are attached to one of the ends of the edge, then $H_1(x)$ satisfies the recursive equation

$$H_1(x) = xG_1(H_1(x)). \tag{52.11}$$

Note that each such component is associated with the end of an edge. Therefore, the component associated with a random vertex is a collection of such components associated with the ends of the edges leaving the vertex, and so, $H_0(x)$ the generating function associated with size of the whole component is given by

$$H_0(x) = xG_0(H_1(x)). \tag{52.12}$$

We can use Equations 52.11 and 52.12 to compute the average component size which, in an analogous manner to computing the average degree of a node, is nothing but

$$H_0'(1) = 1 + G_0'(1)H_1'(1). \tag{52.13}$$

Similarly using Equation (52.11) we get

$$H_1'(1) = 1 + G_1'(1)H_1'(1),$$

which gives the average component size as

$$1 + \frac{G_0'(1)}{1 - G_1'(1)}. \tag{52.14}$$

The giant component first appears when $G_1'(1) = 1$. This condition is is equivalent to

$$G_0'(1) = G_0''(1). \tag{52.15}$$

Using Equation 52.7 the condition implied by Equation 52.15 can also written as

$$\sum_k k(k-2)p_k = 0. \tag{52.16}$$

This condition is the same as that obtained by Molloy and Reed in References 26. The sum on the l.h.s. of Equation 52.16 increases monotonically as edges are added to the graph. Therefore, the giant component comes into existence when the sum on the l.h.s. of Equation 52.16 becomes positive. Newman et al. state in Reference 27 that once there is a giant component in the graph, then $H_0(x)$ generates the probability distribution of the sizes of the components excluding the giant component. Molloy and Reed have shown in References 28 that the giant component almost surely has $cN + o(N)$ nodes. It should be remembered that the results in References 26–28 are all for graphs with specified degree sequences. As such they are applicable to graphs with power law degree distributions. However, web graphs are directed and a model that explains adequately the structure consistent with all that is known experimentally is still to be developed.

52.4.4 Search on Web Graphs

Perhaps the most important algorithmic problem on the Web is mining of data. We will, however, have not much to say about it partly because it has been addressed in Chapter 51, but also because we want to focus on issues that become specially relevant in an evolving web graph. Consider the problem of finding a path between two nodes in a graph. On the Web this forms the core of the peer to peer (P2P) search problem defined in the context of locating a particular file on the Web when there is no information available in a global directory about the node on which the file resides. The basic operation available is to pass messages to neighbors in a totally decentralised manner. Normally a distributed message passing flooding technique on an arbitrary random network would be suspect suspect because of the very large number of messages that may be so generated. Under the assumption that a node knows about the identities of its neighbors and perhaps neighbors' neighbors, Adamic et al. [29] claim, experimentally through simulations, that in power law graphs with N nodes the average search time is of the order of $N^{0.79}$ (graphs are undirected and power law coefficients for the generated graphs is 2.1) when the search is totally random. The intuitive explanation is that even in a random search the search process tends to gravitate towards nodes of high degree. When the strategy is to choose to move to the highest degree neighbor the exponent comes down to 0.70. Adamic et al. [29] have come up a with a fairly simple analysis to explain why random as well as degree directed search algorithms need not have time complexity more than rootic in N (reported exponent is 0.1). In what follows we develop the argument along the lines done by Mehta [30] whose analysis, done assuming that the richest node is selected on the basis of looking at the neighbors and neighbors' neighbors, has resulted in a much sharper bound than the one in Reference 29. We will assume that the cut off degree, m, beyond which the the probability distribution is small enough to be ignored is given by $m = N^{1/\alpha}$ [31].

Using

$$p_k = p(k) = \frac{k^{-\alpha}}{\sum k^{-\alpha}}$$

as the probability distribution function, the expected degree z of a random node is

$$z = G_0'(1) \approx \frac{\int_1^m k^{1-\alpha} dk}{\int_1^m k^{-\alpha} dk} = \frac{(\alpha-1)(m^{2-\alpha}-1)}{(\alpha-2)(m^{1-\alpha}-1)} \sim \ln m, \tag{52.17}$$

under the assumption that α tends to 2 and $m^{1-\alpha}$ is small enough to be ignored. Similarly the PDF for $G_1(x)$ is

$$p_1(k) = \frac{p_k k}{\sum p_k k} = \frac{k^{1-\alpha}}{\left(\sum k^{1-\alpha}\right)} = ck^{1-\alpha}, \tag{52.18}$$

where

$$c = \frac{1}{\sum k^{1-\alpha}} = \frac{\alpha-2}{(m^{2-\alpha}-1)} \approx \frac{1}{\ln m} = \frac{\alpha}{\ln N}. \tag{52.19}$$

The CDF for $G_1(x)$ using (52.18) and (52.19) is

$$P(x) = \int_1^x p_1(k)dk = \frac{\ln x}{\ln m}. \tag{52.20}$$

The search proceeds by choosing at each step the highest degree node among the n neighbors of the current node. The total time taken can be estimated to be the sum of the number of nodes traversed at each step. The number of steps itself can be bounded by noting that the sum of the number of steps at each step can not exceed the size of the graph. Let $p_{max}(x, n)$ be the distribution of the degree of the richest node (largest degree node) among the n neighbors of a node. Determining the richest node is equivalent to taking the maximum of n independent random variables and in terms of the CDF $P(x)$,

$$p_{max}(x, n) = \begin{cases} 0 & x = 0 \\ (P(x) - P(x-1))^n & 0 < x \le m \\ 0 & x > m \end{cases} \tag{52.21}$$

This can be approximated as follows

$$p_{max}(x, n) = \frac{d(P(x)^n)}{dx} = nP(x)^{n-1}\frac{dP(x)}{dx} = n(\log_m x)^{n-1}cx^{-1}. \tag{52.22}$$

We can now calculate the expected degree of the richest node among the n nodes as

$$f(n) = E[x_{max}(n)] = \sum_1^m xp_{max}(x, n) = nc\sum_1^m(\log_m x)^{n-1}. \tag{52.23}$$

Note that if the number of neighbors of every node on the average is z, then the number of second neighbors seen when one is at the current richest node is $f(z)(z-1)$. this is because one of the neighbors of every other node is the richest node. At the next step, the expected degree of the node whose current degree is $f(z)$ is given by $E(f(z)) = f(f(z))$ and the number of second neighbors seen at this stage correspondingly is $f(f(z))(z-1)$. The overlap between two successive computations of neighbors takes place only when there is an edge which makes a node a neighbor as well as the second neighbor. However, the probability of this happening is of the order of N^{-1} and can be ignored in the limit for large N. Therefore we can assume that at every stage new nodes are scanned and the number of steps, l, is controlled by

$$(z-1)\sum_{i=0}^{i=l} f^i(z) = N. \tag{52.24}$$

Assuming for the sake of simplicity $f(f(z)) = f(z)$ (simulations indicate that value of $f^i(z)$ increases with i and then stabilises) we can set $E[n] = n$ or

$$E(n) = n = nc\sum_{1}^{m}(\log_m(x))^{n-1},$$

or

$$(\ln m)^n = \sum_{1}^{m}(\ln x)^{n-1} \sim \int_{1}^{m}(\ln x)^{n-1}dx. \tag{52.25}$$

Substituting e^t for x we get

$$(\ln m)^n = \int_{1}^{\ln m} e^t t^{n-1} dt$$

$$\sim \int_{1}^{\ln m} t^{n-1}(1 + \sum_{0}^{\infty}\frac{t^i}{i!})dt$$

$$= \frac{(\ln m)^n}{n} + \left|\sum_{0}^{\infty}\frac{t^{n+i}}{i!(n+i)}\right|_{0}^{\ln m}$$

$$\leq \frac{(\ln m)^n}{n} + \frac{(\ln m)^n}{n+c}e^{\ln m},$$

where c is a constant. The above implies

$$1 = \frac{1}{n} + \frac{m}{n+c},$$

which gives

$$n = O(m).$$

Substituting this in Equation 52.24 we get

$$(\ln m - 1)\sum_{0}^{l} O(m) = N,$$

or

$$l \sim \frac{N}{m\ln m} = \frac{N^{1-1/\alpha}}{\ln m}.$$

Taking α to be 2.1 the number of steps would be of the order of $N^{0.52}/\ln m$. Mehta [30] also reports results of simulations carried out on graphs generated with α equal to 2.1 using the method given in Reference 31. The simulations were carried out on the giant components. He reports the number of steps to be growing proportional to $N^{0.34}$. The simulation results are very preliminary as the giant components were not very large (largest was of the order of 20 thousand nodes).

52.4.5 Crawling and Trawling

Crawling can be looked upon as a process where an agent moves along the nodes and edges of a randomly evolving graph. Off-line versions of the Web which are the basis of much of what is known experimentally about the Web have been obtained essentially through this process. Cooper and Frieze [32] have attempted to study the expected performance of a crawl process where the agent makes a fixed number of moves between the two successive time steps involving addition of nodes and edges to the graph. The results are fairly pessimistic. Expected proportion of unvisited nodes is of order of 0.57 of the graph size when the edges are added uniformly at random. The situation is even worse when the edges are distributed in proportion to the degree of nodes. The proportion of the unvisited vertices increases to 0.59.

Trawling, on the other hand, involves analyzing the web graph obtained through a crawl for subgraphs which satisfy some particular structure. Since the web graph representation may run into Tera bytes of data, the traditional random access model for computing the algorithmic complexity looses relevance. Even the standard external memory models may not apply as data may be on tapes and may not fit totally on the disks available. In these environments it may even be of interest to develop algorithms where the figure of merit may be the number of passes made over the graph represented on tape (S. Rajagopalan. Private Communication). In any case the issue in trawling is to design algorithms that efficiently stream data between secondary and main memory.

Kumar et al. discuss in Reference 5 the design of trawling algorithms to determine bipartite cores in web graphs. We will focus only on the core issue here which is that the size of the web graph is huge in comparison to the cores that are to be detected. This requires pruning of the web graph before algorithms for core detection are run. If (i, j) cores (i represents outdegree and j indegree) have to be detected then all nodes with outdegree less than i and indegree less than j have to be pruned out. This can be done with repeated sorting of the web graph by in and out degrees and keeping track of only those nodes that satisfy the pruning criteria. If an index is kept in memory of all the pruned in vertices then repeated sorting may also be avoided.

The other pruning strategy discussed in Reference 5 is the so called *inclusion exclusion* pruning in which the focus at every step is to either discover an (i, j) core or exclude a node from further contention. Consider a node x with outdegree equal to i (these will be termed fans) and let $\Gamma(x)$ be the set of nodes that are potentially those with which x could form an (i, j) core (called centers). An (i, j) core will be formed if and only if there are $i - 1$ other nodes all pointing to each node in $\Gamma(x)$. While this condition is easy to check if there are two indices in memory, the whole process can be done in two passes. In the first identify all the fans with outdegree i. Output for each such fan the set of i centers adjacent to it. In the second pass use an index on the destination id to generate the set of fans pointing to each of the i centers and compute the intersection of these sets. Kumar et al. [5] mention that this process can be batched with index only being maintained of the centers that result out of the fan pruning process. If we maintain the set of fans corresponding to the centers that have been indexed, then using the dual condition that x is a part of the core if and only if the intersection of the sets $\Gamma^{-1}(x)$ has size at least j. If this process results in identification of a core then a core is outputted other wise the node x is pruned out. Kumar et al. [5] claim that this process does not result in any not yet identified cores to be eliminated.

The area of trawling is in its infancy and techniques that will be developed will depend primarily on the structural properties being discovered. It is likely to be influenced by the underlying web model and the whether any semantic information is also used in defining the structure. This semantic information may be inferred by structural analysis or may be available in some other way. Development of future web models will depend largely on our understanding of the web, and they themselves will influence the algorithmic techniques developed.

52.5 Conclusions

The modeling and analysis of web as a dynamic graph is very much in its infancy. Continuous mathematical models, which has been the focus of this write up provides good intuitive understanding at the expense of rigour. Discrete combinatorial models that do not brush the problems of proving concentration bounds under the rug are available for very simple growth models. These growth models do not incorporate death processes, or the issues relating to aging (newly created pages are more likely to be involved in link generation processes) or for that matter that in and out degree distributions on the web are not independent and may depend upon the underlying community structure. The process of crawling which can visit only those vertices that have at least one edge pointing to it gives a very limited understanding of degree distributions. There are reasons to believe that a significantly large number of pages on the web have only out hyperlinks. All this calls for extensive experimental investigations and development of mathematical models that are tractable and help both in development of new analytical as well as algorithmic techniques.

The author would like to acknowledge Amit Agarwal who provided a willing sounding board and whose insights have significantly influenced the author's approach on these issues.

References

1. L. Egghe and R. Rousseau. *Introduction to Infometrics: Quantitative Methods in Library, Documentation and Information Science*. Elsevier, 1990.
2. J. Kleinberg. Authoritative sources in a hyperlinked environment. *Journal of the ACM*, 46(5):604–632, 1999.
3. R. Albert, H. Jeong, and A. L. Barabasi. Diameter of the World Wide Web. *Nature*, 401:130–131, 1999.
4. A. Broder, R. Kumar, F. Maghoul, P. Raghavan, S. Rajagopalan, R. Stata, A. Tomkins, and J. Wiener. Graph Structure in the Web: Experiments and Models. In *Proceedings of the 9th World Wide Web Conference*, 2000.
5. R. Kumar, P. Raghavan, S. Rajagopalan, and A. Tomkins. Trawling Emerging Cyber-communities Automatically. In *Proceedings of the 8th World Wide Web Conference*, 1999.
6. D. J. Watts. *Small Worlds: The Dynamics of Networks between Order and Randomness*. Princeton University Press, 1999.

7. B. Bollobas. *Modern Graph Theory*. Springer-Verlag, New York, 1998.

8. B. Mandelbrot. *Fractals and Scaling in Finance*. Springer-Verlag, 1997.

9. G. U. Yule. A Mathematical Theory of Evolution based on the Conclusions of Dr J.C. Willis, F.R.S. *Philosophical Transaction of the Royal Society of London (Series B)*, 213:21–87, 1925.

10. G. K. Zipf. *Selective Studies and the Principle of Relative Frequency in Language*. Harvard University Press, 1932.

11. H. A. Simon. On a Class of Skew Distribution Functions. *Biometrika*. 42:425–440, 1955.

12. M. Mitzenmacher. A Brief History of Generative Models for Power Law and Lognormal Distributions. *Internet Mathematics*, 1(2):226–251, 2004.

13. A. L. Barabasi, R. Albert, and H. Jeong. Mean Field Theory for Scale-free Random Networks. *Physica A*, 272:173–189, 1999.

14. P. L. Krapivsky and S. Redner. Organisation of Growing Random Networks. *Physical Review E*, 63:066123001–066123014, 2001.

15. W. Aiello, F. Chung, and L. Lu. Random Evolution in Massive Graphs. In *Handbook on Massive Data Sets*, (Eds. J. Abello et al.), pp. 97–122, 2002.

16. C. Cooper and A. Frieze. A General model of Web Graphs. *Random Structures & Algorithms*, 22(3):311–335, 2003.

17. R. Kumar, P. Raghavan, S. Rajagopalan, D. Sivakumar, A. Tomkins, and E. Upfal. Stochastic Models for the Web Graph. In *Proceedings of the 41st Annual Symposium on Foundations of Computer Science*, pp. 57–65, 2000.

18. S. N. Dorogovtsev, J. F. F. Mendes, and A. N. Samukhin. Structure of Growing Networks with Preferential Linking. *Phys. Rev. Lett.* 85:4633, 2000.

19. R. Albert and A. L. Barabasi. Statistical Mechanics of Complex Networks. *Reviews of Modern Physics*, 74:47–97, 2002.

20. S. N. Dorogovtsev and J. F. F. Mendes. Evolution of Networks. *Advances in Physics*, 51:1079–1187, 2002.

21. S. Chakrabarti, M. Van den Berg, and B. Dom. Focused Crawling: A New Approach to Topic Specific Web Resource Discovery. *Computer Networks: The International Journal of Computer and Telecommunications Networking*, 31:1623–1640, 1999.

22. S. Chakrabarti, M. M. Joshi, K. Punera, and D. V. Pennock. The Structure of Broad Topics on the Web. In *Proceedings of the 11th World Wide Web Conference*, pp. 225–234, 2002.

23. C. Chekuri, M. Goldwasser, P. Rahgavan, and E. Upfal. Web Search using Automatic Classification. In *Proceedings of the 6th World Wide Web Conference*, 1997.

24. D. Gibson, J. Kleinberg, and P. Raghavan. Inferring Web Communities from Link Topology. In *Proceedings of the 9th ACM Conference on Hypertyext and Hypermedia*, pp. 225–234, 1998.

25. A. Agarwal, S. N. Maheshwari, and B. Mehta. An Evolutionary web graph Model. Technical Report, Department of Computer Science and Engineering, IIT Delhi, 2002.

26. M. Malloy and B. Reed. A Critical Point for Random Graphs with a given Degree Sequence. *Random Structures and Algorithms*, 6:161–179, 1995.

27. M. E. J. Newman, S.H. Strogatz, and D.J. Watts. Random Graphs with Arbitrary Degree Distribution and their Applications. *Physical Review E*, 64:026118, 2001.

28. M. Malloy and B. Reed. The Size of a Giant Component of a Random Graph with a given Degree Sequence. *Combinatorics Probability and Computing*, 7:295–305, 1998.

29. L. A. Adamic, R. M. Lukose, A. R. Puniyani, and B. A. Huberman. Search in Power Law Networks. *Physical Review E.*, 64:46135–46143, 2001.

30. B. Mehta, Search in Web Graphs. M. Tech. Thesis, Department of Computer Science and Engineering I.I.T. Delhi, 2002.

31. W. Aiello, F. Chung, and L. Lu. A Random Graph Model for Massive Graphs. In *Proceedings of the 32nd Annual ACM Symposium on Theory of Computing*, pp. 171–180, 2000.

32. C. Cooper and A. Frieze. Crawling on Web Graphs. In *Proceedings of the 43rd Annual Symposium on Theory of Computing*, pp. 419–427, 2002.

53

Layout Data Structures*

53.1	Introduction	817
53.2	VLSI Technology	817
53.3	Layout Data Structures: An Overview	819
53.4	Corner Stitching	819
	Point Finding • Tile Insertion • Storage Requirements of the Corner Stitching Data Structure	
53.5	Corner Stitching Extensions	822
	Expanded Rectangles • Trapezoidal Tiles • Curved Tiles • L-Shaped Tiles	
53.6	Quad Trees and Variants	826
	Bisector List Quad Trees • *k*-d Trees • Multiple Storage Quad Trees • Quad List Quad Trees • Bounded Quad Trees • HV Trees • Hinted Quad Trees	
53.7	Concluding Remarks	830
	Acknowledgments	830
	References	830

Dinesh P. Mehta
Colorado School of Mines

53.1 Introduction

VLSI (Very Large Scale Integration) is a technology that has enabled the manufacture of large circuits in silicon. It is not uncommon to have circuits containing millions of transistors, and this quantity continues to increase very rapidly. Designing a VLSI circuit is itself a very complex task and has spawned the area of VLSI design automation. The purpose of VLSI design automation is to develop software that is used to design VLSI circuits. The VLSI design process is sufficiently complex that it consists of the four steps shown in Figure 53.1. Architectural design is carried out by expert human engineers with some assistance from tools such as simulators. Logic design is concerned with the boolean logic required to implement a circuit. Physical design is concerned with the implementation of logic on a three dimensional physical structure: the VLSI chip. VLSI physical design consists of steps such as floorplanning, partitioning, placement, routing, circuit extraction, etc. Details about VLSI physical design automation may be found in References 1–3. Chapter 54 describes the rich area of data structures for floorplanning. In this chapter, our concern will be with the representation of a circuit in its "physical" form. In order to proceed with this endeavor, it is necessary to first understand the basics of VLSI technology.

53.2 VLSI Technology

We begin with the caveat that our presentation here only seeks to convey the basics of VLSI technology. Detailed knowledge about this area may be obtained from texts such as [4]. The transistor is the fundamental device in VLSI technology and may be viewed as a switch. It consists of a gate, a source, and a drain. The voltage on the gate controls the passage of current between the source and the drain. Thus, the gate can be used to switch the transistor "on" (current flows between the source and the drain) and "off" (no current flows). Basic logic elements such as the inverter (the NOT gate), the NAND gate, and the NOR gate are built using transistors. Transistors and logic gates can be manufactured in layers on a silicon disk called a wafer. Pure silicon is a semiconductor whose electrical resistance is between that of a conductor and an insulator. Its conductivity can be significantly improved by introducing "impurities" called dopants. N-type dopants such as phosphorus supply free electrons, while p-type dopants like boron supply holes. Dopants are diffused into the silicon wafer. This layer of the chip is called the diffusion layer and is further classified into n-type and p-type depending on the type of dopant used. The source and drain of a transistor are

---

* This chapter has been reprinted from first edition of this Handbook, without any content updates.

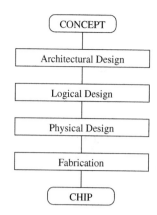

FIGURE 53.1   The VLSI design stages.

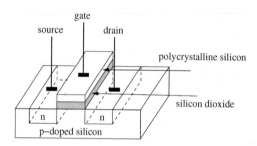

FIGURE 53.2   A transistor.

formed by separating two n-type regions with a p-type region (or vice versa). A gate is formed by sandwiching a silicon dioxide (an insulator) layer between the p-type region and a layer of polycrystalline silicon (a conductor). Figure 53.2 illustrates these concepts. Since polycrystalline silicon (poly) is a conductor, it is also used for short interconnections (wires). Up to this point, we have described the two layers (diff and poly) that are used to make all electronic devices. Although poly conducts electricity, it is not sufficient to complete all the interconnections using one layer. Modern chips usually have several layers of aluminum ("metal"), a conductor, separated from each other by insulators on top of the poly layer. These make it possible for the gates to be interconnected as specified in the design. Note that a layer of material X (e.g., poly) does not mean that there is a monolithic slab of poly over the entire chip area. The poly is only deposited where gates or wires are needed. The remaining areas are filled with insulating materials and for our purposes may be viewed as being empty. In addition to the layers as described above, it is necessary to have a mechanism for signals to pass between layers. This is achieved by contacts (to connect poly with diffusion or metal) and vias (to connect metal on different layers). Figure 53.3 shows the layout and a schematic of an nMOS inverter. We briefly describe the functioning of the inverter. If the input gate voltage is "0," the transistor is switched off and there is no connection between the ground signal and the output. The voltage at the output is identical to that of the power source, which is a "1." If the gate is at "1," the transistor is switched on and there is a connection between the ground signal "0" and the output, making the output "0."

The purpose of a layout data structure is to store and manipulate the rectangles on each layer. Some important high-level operations that a layout data structure must support are design-rule checking, layout compaction, and parasitic extraction.

Design Rule Checking (DRC): Design rules are the interface between the circuit designer and the process engineer. They specify geometric constraints on the layout so that the patterns on the processed wafer preserve the topology of the designs. An example of a design rule is that the width of a wire must be greater than a specified minimum. If this constraint is violated, it is possible that for the wire to be discontinuous because of errors in the fabrication process. Similarly, if two wires are placed too close to each other, they could touch each other. The DRC step verifies that all design rules have been met. Additional design rules for CMOS technology may be found in Reference 4, page 142.

Parasitic Extraction: Each layer of the chip has a resistance and a capacitance that are critical to the estimation of circuit performance. Inductance is usually less important on the chip, but has greater impact on the I/O components of the chip. Capacitance, resistance, and inductance are commonly referred to as "parasitics." After a layout has been created, the parasitics must be computed in order to verify that the circuit will meet its performance goals. (Performance is usually measured by clock cycle times and power dissipation.) The parasitics are computed from the geometry of the layout. For example, the resistance of a

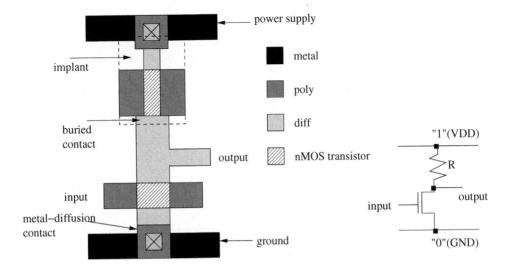

FIGURE 53.3 An inverter.

rectangular slab of metal is $\frac{\rho l}{tw}$, where ρ is the resistivity of the metal and l, w, and t are the slab's length, width, and thickness, respectively. See [4, Chapter 4] for more examples.

Compaction: The compaction step, as its name suggests, tries to make the layout as small as possible without violating any design rules. This reduces the area of the chip, which could result in more chips being manufactured from a single wafer, which significantly reduces cost per chip. Interestingly, the cost of a chip could grow as a *power of five* of its area [5] making it imperative that area be minimized! Two-dimensional compaction is NP-hard, but one-dimensional compaction can be carried out in polynomial time. Heuristics for 2D compaction often iteratively interleave one-dimensional compactions in the x- and y-directions. For more details, see [6].

53.3 Layout Data Structures: An Overview

There are two types of layout data structures that are based on differing philosophies. The first is in the context of a layout editor. Here, the idea is that a user manually designs the layout, for example, by inserting rectangles of the appropriate dimensions at the appropriate layer. This customized approach is used for library cells. The MAGIC system [7] developed at U.C. Berkeley is an example of of a system that included a layout editor. MAGIC was in the public domain and was used to support classes on VLSI design in many universities. The layout editor context is especially important here because it permitted the developers of MAGIC to assume locality of reference; that is, a user is likely to perform several editing operations in the same area of the layout over a short period of time. The second philosophy is that the layout process is completely automated. This has the advantage that some user-interaction operations do not need to be supported and run time is critical. This approach is more common in industrial software, where automatic translation techniques convert electronic circuits into physical layouts. This philosophy is supported by the quad-tree and variants that were designed specifically for VLSI layout.

53.4 Corner Stitching

The corner stitching data structure was proposed by Ousterhout [8] to store non-overlapping rectilinear circuit components in MAGIC. The data structure is obtained by partitioning the layout area into horizontally maximal rectangular tiles. There are two types of tiles: solid and vacant, both of which are explicitly stored in the corner-stitching data structure. Tiles are obtained by extending horizontal lines from corners of all solid tiles until another solid tile or a boundary of the layout region is encountered. The set of solid and vacant tiles so obtained is unique for a given input. The partitioning scheme ensures that no two vacant or solid tiles share a vertical side. Each tile T is stored as a node which contains the coordinates of its bottom left corner, x_1 and y_1, and four pointers N, E, W, and S. N (respectively, E, W, S) points to the rightmost (respectively, topmost, bottommost, leftmost) tile neighboring its north (respectively, east, west, south) boundary. The x and y coordinates of the top right corner of T are $T.E \rightarrow x_1$ and $T.N \rightarrow y_1$, respectively, and are easily obtained in O(1) time. Figure 53.4 illustrates the corner stitching data structure.

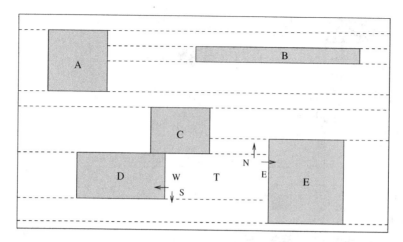

FIGURE 53.4   The corner stitching data structure. Pointers (stitches) are shown for tile *T*.

The corner stitching data structure supports a rich set of operations.

1. Point Finding: given a tile *T* and a point *p*(*x*, *y*), search for the tile containing *p* by following a sequence of stitches starting at *T*.
2. Neighbor Finding: find all solid and vacant tiles that abut a given tile *T*.
3. Area Searches: determine whether any solid tiles intersect a given rectangular area *R*. This operation is used to determine whether a new solid tile can subsequently be inserted into area *R*. (Recall that tiles in a layer are not permitted to overlap.)
4. Directed Area Enumeration: enumerate all the tiles contained in a given rectangular area in a specified order. This is used during the compaction operation which may require tiles to be visited and compacted in a left-to-right order.
5. Tile Creation: insert a solid tile *T* into the data structure at a specified location.
6. Tile Deletion: delete a specified solid tile *T* from the data structure.
7. Plowing: translate a large piece of a design. Move other pieces of the design that lie in its path in the same direction.
8. Compaction: this refers to one-dimensional compaction.

We describe two operations to illustrate corner stitching:

53.4.1 Point Finding

Next, we focus on the point find operation because of its effect on the performance of corner stitching. The algorithm is presented below. Given a pointer to an arbitrary tile *T* in the layout, the algorithm seeks the tile in the layout containing the point *P*.

Tile Point_Find (Tile T, Point P)
1. **begin**
2.   *current* = *T*;
3.   **while** (*P* is not contained in *current*)
4.   **begin**
5.   **while** (*P.y* does not lie in *current*'s y-range)
6.   **if** (*P.y* is above *current*) *current* = *current*→ *N*;
7.   **else** *current* = *current*→ *S*;
8.   **while** (*P.x* does not lie in *current*'s x-range)
9.   **if** (*P.x* is to the right of *current*) *current* = *current*→ *E*;
10.   **else** *current* = *current*→ *W*;
11.   **end**
12.   **return** (*current*);
13. **end**

Figure 53.5 illustrates the execution of the point find operation on a pathological example. From the start tile *T*, the **while** loop of line 5 follows north pointers until tile *A* is reached. We change directions at tile *A* since its *y*-range contains *P*. Next, west pointers are followed until tile *F* is reached (whose *x*-range contains *P*). Notice that the sequence of west moves causes the algorithm to descend in the layout resulting in a vertical position that is similar to that of the start tile! As a result of this

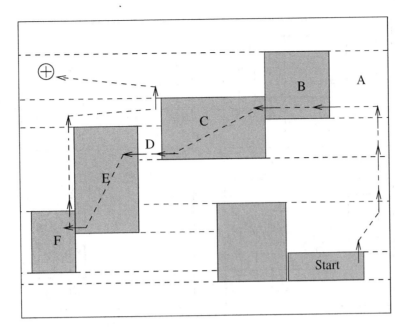

FIGURE 53.5   Illustration of point find operation and misalignment.

misalignment, the outer while loop of the algorithm must execute repeatedly until the point is found (note that the point will eventually be found since the point find algorithm is guaranteed to converge).

53.4.2 Tile Insertion

This operation creates a new solid tile and inserts it into the layout. It accomplishes this by a sequence of split and merge operations. The split operation breaks a tile into two tiles along either a vertical or a horizontal line. The merge operation combines two tiles to form a rectangular tile. Algorithm *Insert* discusses the insertion of a rectangle into the layout.

   Insert(A) // (x_1, y_1) and (x_2, y_2) are the bottom left and top right corners of *A*.

1. **if** (!*AreaSearch(A)*) **return**; //area is not empty, abort.
2. Let $i = 0$; Split Q_i, the tile containing the north edge of *A* into two tiles along the line $y = y_2$; Let *T* be the upper tile and Q_i be the lower tile.
3. **while** (Q_i does not contain the south edge of *A*)
   a.   Split Q_i vertically into three tiles along $x = x_1$ and $x = x_2$; let the resulting tiles from left to right be L_i, Q_i, and R_i.
   b.   **if** ($i > 0$)
         i.   Merge R_{i-1} and L_{i-1} into R_i and L_i, respectively, if possible.
         ii. Merge Q_{i-1} into Q_i;
   c.   Let Q_{i+1} be the tile beneath Q_i.
   d.   Increment i;
4. Split Q_i along $y = y_1$; Let Q_i be the tile above the split and B the tile below the split; Split Q_i into L_i, Q_i, and R_i using the lines $x = x_1$ and $x = x_2$;
5. Merge Q_i and Q_{i-1} to get Q_i. Q_i is the newly inserted solid tile;
6. Merge R_{i-1}, L_{i-1} with neighboring tiles; if R_i (L_i) gets merged, the merged tile is called R_i (L_i). Merge R_i and L_i with neighboring tiles;

   Figure 53.6 shows the various steps involved in the insertion algorithm.

   The rectangle *A* to be inserted is represented by thick, dashed lines in Figure 53.6a. The coordinates of the bottom left and top right corners of *A* are (x_1, y_1) and (x_2, y_2), respectively. First, Step 1 of the algorithm uses *AreaSearch* to ensure that no solid tiles intersect *A*. Step 2 identifies tile Q_0 as the vacant tile containing *A*'s north edge and splits it by the horizontal line $y = y_2$ into two tiles: *T* above the split-line and Q_0 below the split-line. Next, in the while loop of Step 3, Q_0 is split by vertical lines at x_1 and x_2 to form L_0, Q_0, R_0. Tile Q_1 is the vacant tile below Q_0. The resulting configuration is shown in Figure 53.6b. In the next iteration, Q_1 is split to form L_1, Q_1, and R_1. L_0 merges into L_1 and Q_0 merges into Q_1. Tile Q_2 is the vacant tile below Q_1. The resulting configuration is shown in Figure 53.6c. Next, Q_2 is split to form L_2, Q_2, and R_2. R_1 is merged into R_2 and Q_1 merged into Q_2.

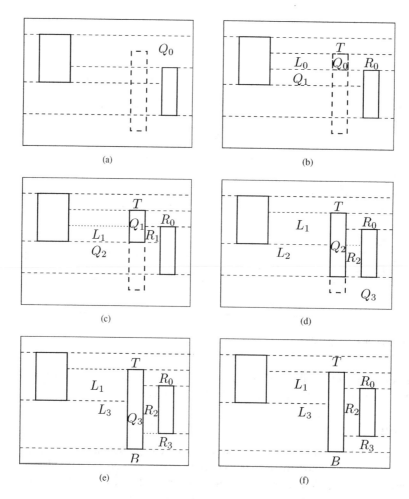

FIGURE 53.6   Illustration of insertion.

Figure 53.6d shows the configuration after Tile Q_2 is processed. The vacant tile Q_3 below Q_2 contains R's bottom edge and the while loop of Step 3 is exited. Steps 4, 5, and 6 of the algorithm result in the configuration of Figure 53.6e. The final layout is shown in Figure 53.6f.

53.4.3 Storage Requirements of the Corner Stitching Data Structure

Unlike simpler data structures such as arrays and linked lists, it is not trivial to manually estimate the storage requirements of a corner stitched layout. For example, if n items are inserted into a linked list, then the amount of storage required is n multiplied by the number of bytes required by a single list node. Because of vacant tiles, the total number of nodes in corner stitching is considerably more than n and depends on the relative positions of the n rectangles. In References 9 and 10, a general formula for the memory requirements of the corner stitching data structure on a given layout. This formula requires knowledge about certain geometric properties of the layout called *violations* of the **CV** property and states that a corner stitching data structure representing a set of N solid, rectangular tiles with k violations contains $3N + 1 - k$ vacant tiles. Since each rectangular tile requires 28 bytes, the memory requirements are $28(4N + 1 - k)$ bytes.

53.5 Corner Stitching Extensions

53.5.1 Expanded Rectangles

Expanded rectangles [11] expands solid tiles in the corner stitching data structure so that each tile contains solid material and the empty space around it. No extra tiles are needed to represent empty space. Thus, there are fewer tiles than in corner stitching. However, each tile now requires 44 rather than 28 bytes because additional fields are needed to store the coordinates of the solid

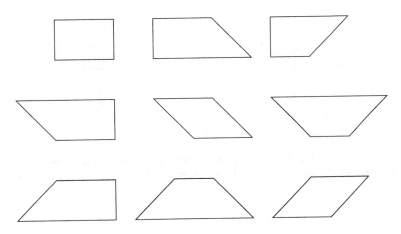

FIGURE 53.7 The different types of trapezoidal tiles.

portion of the tile. It was determined that expanded rectangles required less memory than corner stitching when the ratio of vacant to solid tiles in corner stitching was greater than 0.414. Operations on the expanded rectangles data structure are similar to those in corner stitching.

53.5.2 Trapezoidal Tiles

Marple et al. [12] developed a layout system called "Tailor" that was similar to MAGIC except that it allowed 45° layout. Thus, rectangular tiles were replaced by trapezoidal tiles. There are 9 types of trapezoidal tiles as shown in Figure 53.7. An additional field that stores the type of trapezoidal tile is used. The operations on Tailor are similar to those in MAGIC and are implemented in a similar way. It is possible to extend these techniques to arbitrary angles making it possible to describe arbitrary polygonal shapes.

53.5.3 Curved Tiles

Séquin and Façanha [13] proposed two generalizations to geometries including circles and arbitrary curved shapes, which arise in microelectromechanical systems (MEMS). As with its corner stitching-based predecessors, the layout area is decomposed in a horizontally maximal fashion into tiles. Consequently, tiles have upper and lower horizontal sides. Their left and right sides are represented by parameterized cubic Bezier curves or by composite paths composed of linear, circular, and spline segments. Strategies for reducing storage by minimizing the number of tiles and curve-sharing among tiles are discussed.

53.5.4 L-Shaped Tiles

Mehta and Blust [14] extended Ousterhout's corner stitching data structure to directly represent L- and other simple rectilinear shapes without partitioning them into rectangles. This results in a data structure that is topologically different from the other versions of corner stitching described above. A key property of this L-shaped corner stitching (LCS) data structure is that

1. All vacant tiles are either rectangles or L-shapes.
2. No two vacant tiles in the layout can be merged to form a vacant rectangle or L-shaped tile.

Figure 53.8 shows three possible configurations for the same set of solid tiles.

There are four L-shape types (Figure 53.9), one for each orientation of the L-shape. The L-shapes are numbered according to the quadrant represented by the two lines meeting at the single concave corner of the L-shape. Figure 53.10 describes the contents of L-shapes and rectangles in LCS and rectangles in the original rectangular corner stitching (RCS) data structure. The actual memory requirements of a node in bytes (last column of the table) are obtained by assuming that pointers and coordinates, each, require 4 bytes of storage, and by placing all the remaining bits into a single 4-byte word. Note that the space required by any L-shape is less than the space required by two rectangles in RCS and that the space required by a rectangle in LCS is equal to the space required by a rectangle in RCS. The following theorem has been proved in Reference 9:

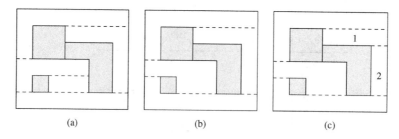

FIGURE 53.8 Layouts represented by the L-shaped corner stitching data structure: Layout (c) is invalid because the vacant tiles 1 and 2 can be merged to form an L-shaped tile. Layouts (b) and (c) are both valid, illustrating that unlike RCS, the LCS data structure does not give a unique partition for a given set of solid tiles.

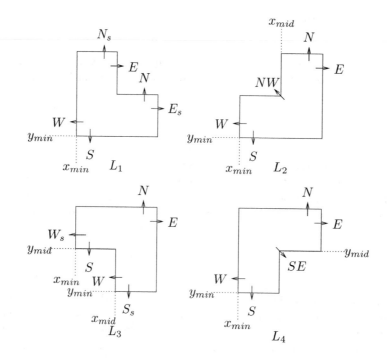

FIGURE 53.9 L-shapes and their pointers.

Theorem 53.1

The LCS data structure never requires more memory than the RCS data structure for the same set of solid rectangular tiles.

Proof. Since all solid tiles are rectangles, and since a rectangle occupies 28 bytes in both LCS and RCS, the total memory required to store solid tiles is the same in each data structure. From the definitions of RCS and LCS, there is a one-to-one correspondence between the set S_1 of vacant rectangles in RCS and the set S_2 consisting of (i) vacant rectangles in LCS and (ii) rectangles obtained by using a horizontal line to split each vacant L-shape in LCS; where each pair of related rectangles (one from S_1, the other from S_2) have identical dimensions and positions in the layout. Vacant rectangles in LCS require the same memory as the corresponding vacant rectangles in RCS. However, a vacant L-shape requires less memory than the two corresponding rectangles in RCS. Therefore, if there is at least one vacant L-shape, LCS requires less memory than RCS.   ∎

Theorem 53.2

The *LCS* data structure requires 8.03%–26.7% less memory than the *RCS* data structure for a set of solid, rectangular tiles that satisfies the **CV** property.

Tile type	Number of Coordinates	Number of Pointers	Number of N/E bits	solid/vacant tile bit	R/L bit	L type bits	Number of Bytes
L_1	2	6	4	1	1	2	36
L_2	3	5	3	1	1	2	36
L_3	4	6	2	1	1	2	44
L_4	3	5	3	1	1	2	36
R in LCS	2	4	2	1	1	0	28
R in RCS	2	4	0	1	0	0	28

FIGURE 53.10 Space requirements of tiles in LCS and RCS.

53.5.4.1 Rectilinear Polygons

Since, in practice, circuit components can be arbitrary rectilinear polygons, it is necessary to partition them into rectangles to enable them to be stored in the corner stitching format. MAGIC handles this by using horizontal lines to partition the polygons. This is not necessary from a theoretical standpoint, but it simplifies the implementation of the various corner stitching operations. Nahar and Sahni [15] studied this problem and presented an $O(n + k_v \log k_v)$ algorithm to decompose a polygon with n vertices and k_v vertical inversions into rectangles using horizontal cuts only. (The number of *vertical inversions* of a polygon is defined as the minimum number of changes in vertical direction during a walk around the polygon divided by 2.) The quantity k_v was observed to be small for polygons encountered in VLSI mask data. Consequently, the Nahar-Sahni algorithm outperforms an $O(n \log n)$ planesweep algorithm on VLSI mask data. We note that this problem is different from the problem of decomposing a rectilinear polygon into a minimum number of rectangles using *both* horizontal and vertical cuts, which has been studied extensively in the literature [16–19].

However, the extension is slower than the original corner stitching, and also harder to implement. Lopez and Mehta [20] presented algorithms for the problem of optimally breaking an arbitrary rectilinear polygon into L-shapes using horizontal cuts.

53.5.4.2 Parallel Corner Stitching

Mehta and Wilson [21] have studied a parallel implementation of corner stitching. Their work focuses on two batched operations (batched insert and delete). Their approach results in a significant speed-up in run times for these operations.

53.5.4.3 Comments about Corner Stitching

1. Corner stitching requires rectangles to be non-overlapping. A single layer of the chip consists of non-overlapping rectangles, but all the layers taken together will consist of overlapping rectangles. So, an instance of the corner stitching data structure can only be used for a single layer. However, corner stitching can be used to store multiple layers in the following way: consider two layers A and B. Superimpose the two layers. This can be thought of as a single layer with four types of rectangles: vacant rectangles, type A rectangles, type B rectangles, and type AB rectangles. Unfortunately, this could greatly increase the number of rectangles to be stored. It also makes it harder to perform insertion and deletion operations. Thus, in MAGIC, the layout is represented by a number of single-layer corner stitching instances and a few multiple-layer instances when the intersection between rectangles in different layers is meaningful; for example, transistors are formed by the intersection of poly and diffusion rectangles.

2. Corner stitching is difficult to implement. While the data structure itself is quite elegant, the author's experience is that its implementation requires one to consider a lot of details that are not considered in a high-level description. This is supported by the following remark attributed to John Ousterhout [13]:

   > Corner-stitching is pretty straightforward at a high level, but it can become much more complicated when you actually sit down to implement things, particularly if you want the implementation to run fast

3. The *Point Find* operation can be slow. For example, *Point Find* could visit all the tiles in the data structure resulting in an $O(n)$ time complexity. On the average, it requires $O(\sqrt{n})$ time. This is expensive when compared with the $O(\log n)$ complexity that may be possible by using a tree type data structure. From a practical standpoint, the slow speed may be tolerable in an interactive environment in which a user performs one operation at a time (e.g., a *Point Find* could be performed by a mouse button click). Here, a slight difference in response time might not be noticeable by the user. Furthermore, in an interactive environment, these operations might actually be fast because they take advantage of locality

of reference (i.e., the high likelihood that two successive points being searched by a user are near each other in the layout). However, in batch mode, where a number of operations are performed without user involvement, one is more likely to experience the average case complexity (unless the order of operations is chosen carefully so as to take advantage of locality of reference). The difference in time between corner stitching and a faster logarithmic technique will be significant.

4. Corner stitching requires more memory to store vacant tiles.

53.6 Quad Trees and Variants

Quad trees have been considered in Chapters 17 and 20. These chapters demonstrate that there are different flavors of quad-trees depending on the type of the data that are to be represented. For example, there are quad trees for regions, points, rectangles, and boundaries. In this chapter, we will be concerned with quad-trees for rectangles. We also note that Chapter 19 describes several data structures that can be used to store rectangles. To the best of my knowledge, the use of these structures has not been reported in the VLSI design automation literature.

The underlying principle of the quad tree is to recursively subdivide the two-dimensional layout area into four "quads" until a stopping criterion is satisfied. The resulting structure is represented by a tree with a node corresponding to each quad, with the entire layout area represented by the root. A node contains children pointers to the four nodes corresponding the quads formed by the subdivision of the node's quad. Quads that are not further subdivided are represented by leaves in the quad tree.

Ideally, each rectangle is the sole occupant of a leaf node. In general, of course, a rectangle does not fit inside any leaf quad, but rather intersects two or more leaf quads. To state this differently, it may intersect one or more of the horizontal and vertical lines (called bisectors) used to subdivide the layout region into quads. Three strategies have been considered in the literature as to where in the quad tree these rectangles should be stored. These strategies, which have given rise to a number of quad tree variants, are listed below and are illustrated in Figure 53.11.

1. SMALLEST: Store a rectangle in the *smallest quad (not necessarily a leaf quad) that contains it*. Such a quad is guaranteed to exist since each rectangle must be contained in the root quad.
2. SINGLE: Store a rectangle in precisely *one of the leaf quads that it intersects*.
3. MULTIPLE: Store a rectangle in *all of the leaf quads that it intersects*.

Obviously, if there is only one rectangle in a quad, there is no need to further subdivide the quad. However, this is an impractical (and sometimes impossible) stopping criterion. Most of the quad tree variants discussed below have auxiliary stopping criteria. Some subdivide a quad until it reaches a specified size related to the typical size of a small rectangle. Others stop if the number of rectangles in a quad is less than some threshold value. Figure 53.12 lists and classifies the quad tree variants.

53.6.1 Bisector List Quad Trees

Bisector List Quad Trees (BLQT) [22], which was the first quad-tree structure proposed for VLSI layouts, used the SMALLEST strategy. Here, a rectangle is associated with the smallest quad (leaf or non-leaf) that contains it. Any non-leaf quad Q is subdivided into four quads by a vertical bisector and a horizontal bisector. Any rectangle associated with this quad *must* intersect one or both of the bisectors (otherwise, it is contained in one of Q's children, and should not be associated with Q). The set of rectangles are partitioned into two sets: V, which consists of rectangles that intersect the vertical bisector and H, which consists of rectangles that intersect the horizontal bisector. Rectangles that intersect both bisectors are arbitrarily assigned to one of V and H. These "lists" were actually implemented using binary trees. The rationale was that since most rectangles in IC layouts were small and uniformly distributed, most rectangles will be at leaf quads. A region search operation identifies all the quads that intersect a query window and checks all the rectangles in each of these quads for intersection with the query window. The BLQT (which is also called the MX-CIF quadtree) is also described in Chapter 17.

53.6.2 *k*-d Trees

Rosenberg [23] compared BLQT with k-d trees and showed experimentally that k-d trees outperformed an implementation of BLQT. Rosenberg's implementation of the BLQT differs from the original in that linked lists rather than binary trees were used to represent bisector lists. It is hard to evaluate the impact of this on the experimental results, which showed that point-find and region-search queries visit fewer nodes when the k-d tree is used instead of BLQT. The experiments also show that k-d trees consume about 60%–80% more space than BLQTs.

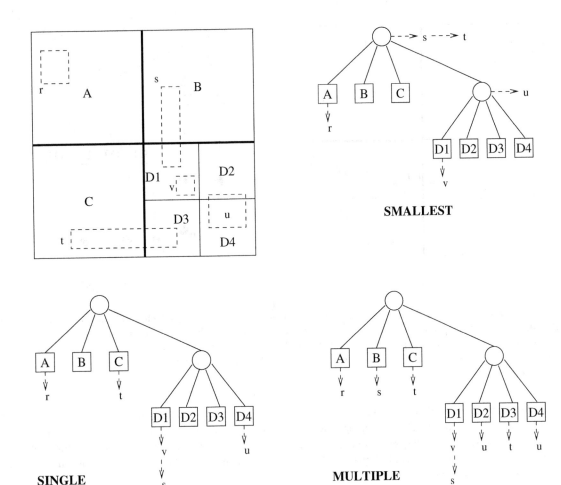

FIGURE 53.11 Quadtree variations.

Author	Abbreviation	Year of Publication	Strategy
Kedem	BLQT	1982	SMALLEST
Rosenberg	k-d	1985	N/A
Brown	MSQT	1986	MULTIPLE
Weyten et al	QLQT	1989	MULTIPLE
Pitaksanonkul et al	BQT	1989	SINGLE
Lai et al	HV	1993	SMALLEST
Lai et al	HQT	1996	MULTIPLE

FIGURE 53.12 Summary of Quad-tree variants.

53.6.3 Multiple Storage Quad Trees

In 1986, Brown proposed a variation [24] called Multiple Storage Quad Trees (MSQT). Each rectangle is stored in every leaf quad it intersects. (See the quad tree labeled "MULTIPLE" in Figure 53.11.) An obvious disadvantage of this approach is that it results in wasted space. This is partly remedied by only storing a rectangle once and having all of the leaf quads that it intersects contain a pointer to the rectangle. Another problem with this approach is that queries such as Region Search may report the same rectangle more than once. This is addressed by marking a rectangle when it is reported for the first time and by not reporting rectangles that have been previously marked. At the end of the Region Search operation, all marked rectangles need to be unmarked in preparation for the next query. Experiments on VLSI mask data were used to evaluate MSQT for different threshold values and for different Region Search queries. A large threshold value results in longer lists of pointers in the leaf quads that have to be

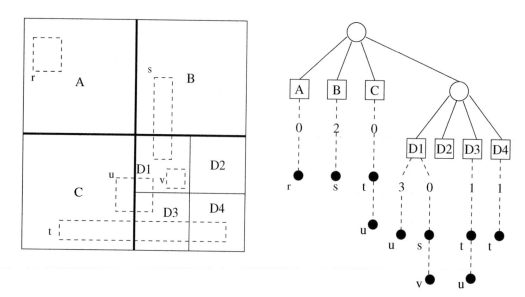

FIGURE 53.13 The leaf quads are *A, B, C, D*1, *D*2, *D*3, and *D*4. The rectangles are *r − v*. Rectangle *t* intersects quads *C, D*3, and *D*4 and must appear in the lists of each of the leaf nodes in the quad tree. Observe that *t* does not cross the lower boundaries of any of the three quads and *x* = 0 in each case. However, *t* does cross the left boundaries of *D*3 and *D*4 and *y* = 1 in these cases. Thus *t* goes into list 1 in *D*3 and *D*4. Since *t* does not cross the left boundary of *C*, it goes into list 0 in *C*. Note that the filled circles represent pointers to the rectangles rather than the rectangles themselves.

searched. On the other hand, a small threshold value results in a quad-tree with greater height and more leaf nodes as quads have to be subdivided more before they meet the stopping criterion. Consequently, a rectangle now intersects and must be pointed at by more leaf nodes. A Region Search query with a small query rectangle (window) benefits from a smaller threshold because it has to search smaller lists in a handful of leaf quads. A large window benefits from a higher threshold value because it has to search fewer quads and encounters fewer duplicates.

53.6.4 Quad List Quad Trees

In 1989, Weyten and De Pauw [25] proposed a more efficient implementation of MSQT called Quad List Quad Trees (QLQT). For Region Searches, experiments on VLSI data showed speedups ranging from 1.85to 4.92 over MSQT, depending on the size of the window. In QLQT, four different lists (numbered 0–3) are associated with each leaf node. If a rectangle intersects the leaf quad, a pointer to it is stored in one of the four lists. The choice of the list is determined by the relative position of this rectangle with respect to the quad. The relative position is encoded by a pair of bits *xy*. *x* is 0 if the rectangle does not cross the lower boundary of the leaf quad and is 1, otherwise. Similarly, *y* is 0 if the rectangle does not cross the left boundary of the leaf quad and is 1, otherwise. The rectangle is stored in the list corresponding to the integer represented by this two bit string. Figure 53.13 illustrates the concept. Notice that each rectangle belongs to exactly one list 0. This corresponds to the quad that contains the bottom left corner of the rectangle. Observe, also, that the combination of the four lists in a leaf quad gives the same pointers as the single list in the same leaf in MSQT. The Region Search of MSQT can now be improved for QLQT by using the following procedure for each quad that intersects the query window: if the query window's left edge crosses the quad, only the quad's lists 0 and 1 need to be searched. If the window's bottom edge crosses the quad, the quad's lists 0 and 2 need to be searched. If the windows bottom left corner belongs to the quad, all four lists must be searched. For all other quads, only list 0 must be searched. Thus the advantages of the QLQT over MSQT are:

1. QLQT has to examine fewer list nodes than MSQT for a Region Search query.
2. Unlike MSQT, QLQT does not require marking and unmarking procedures to identify duplicates.

53.6.5 Bounded Quad Trees

Later, in 1989, Pitaksanonkul et al. proposed a variation of quad trees [26] that we refer to as Bounded Quad Trees (BQT). Here, a rectangle is only stored in the quad that contains its bottom left corner. (See the quad tree labeled "SINGLE" in Figure 53.11.) This may be viewed as a version of QLQT that only uses list 0. Experimental comparisons with *k*-d trees show that for small threshold values, quad trees search fewer nodes than *k*-d trees.

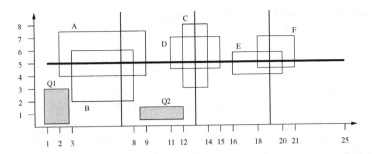
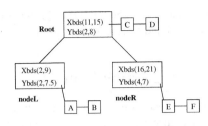

FIGURE 53.14 Bisector list implementation in HVT. All rectangles intersect the thick horizontal bisector line ($y = 5$). The first vertical cutline at $x = 13$ corresponding to the root of the tree intersects rectangles C and D. These rectangles are stored in a linked list at the root. Rectangles A and B are to the left of the vertical cutline and are stored in the left subtree. Similarly, rectangles C and D are stored in the right subtree. The X bounds associated with the root node are obtained by examining the x coordinates of rectangles C and D, while its Y bounds are obtained by examining the y coordinates of all six rectangles stored in the tree. The two shaded rectangles are query rectangles. For Q1, the search will start at **Root**, but will not search the linked list with C and D because Q1's right side is to the left of **Root**'s lower x bound. The search will then examine **nodeL**, but not **nodeR**. For Q2, the search will avoid searching the bisector list entirely because its upper side is below **Root**'s lower y bound.

53.6.6 HV Trees

Next, in 1993, Lai et al. [27] presented a variation that once again uses bisector lists. It overcomes some of the inefficiencies of the original BLQT by a tighter implementation: An HV Tree consists of alternate levels of H-nodes and V-nodes. An H node splits the space assigned to it into two halves with a horizontal bisector while a V-node does the same by using a vertical bisector. A node is not split if the number of rectangles assigned to it is less than some fixed threshold.

Rectangles intersecting an H node's horizontal bisector are stored in the node's bisector list. Bisector lists are implemented using cut trees. A vertical cutline divides the horizontal bisector into two halves. All rectangles that intersect this vertical cutline are stored in the root of the cut tree. All rectangles to the left of the cutline are recursively stored in the left subtree and all rectangles to the right are recursively stored in the right subtree. So far, the data structure is identical to Kedem's binary tree implementation of the bisector list. In addition to maintaining a list of rectangles intersecting a vertical cutline at the corresponding node n, the HV tree also maintains four additional bounds which significantly improve performance of the Region Search operation. The bounds *y_upper_bound* and *y_lower_bound* are the maximum and minimum y coordinates of *any of the rectangles stored in n or in any of n's descendants*. The bounds *x_lower_bound* and *x_upper_bound* are the minimum and maximum x coordinates of the rectangles stored in node n. Figure 53.14 illustrates these concepts. Comprehensive experimental results comparing HVT with BQT, kD, and QLQT showed that the data structures ordered from best to worst in terms of space requirements were HVT, BQT, kD, and QLQT. In terms of speed, the best data structures were HVT and QLQT followed by BQT and finally kD.

53.6.7 Hinted Quad Trees

In 1997, Lai et al. [28] described a variation of the QLQT that was specifically designed for design rule checking. Design-rule checking requires one to check rectangles in the vicinity of the query rectangle for possible violations. Previously, this was achieved by employing a traditional region query whose rectangle was the original query rectangle extended in all directions by a specified amount. Region searches start at the root of the tree and proceed down the tree as discussed previously. The hinted quadtree is based on the philosophy that it is wasteful to begin searching at the root, when, with an appropriate hint, the algorithm can start the search lower down in the tree. Two questions arise here: at which node should the search begin and how does the algorithm get to that node? The node at which the design rule check for rectangle r begins is called the *owner* of r. This is defined as the lowest node in the quad-tree that completely contains r expanded in all four directions. Since the type of r is known (e.g., whether it is n-type diffusion or metal), the amount by which r has to be expanded is also known in advance. Clearly, any rectangle that intersects the expanded r must be referenced by at least one leaf in the owner node's subtree. The owner node may be reached by following parent pointers from the rectangle. However, this could be expensive. Consequently, in HQT, each rectangle maintains a pointer to the owner virtually eliminating the cost of getting to that node. Although this is the main contribution of the HQT, there are additional implementation improvements over the underlying QLQT that are used to speed up the data structure. First, the HQT resolves the situation where the boundary of a rectangle stored in the data structure or a query rectangle coincides with that of a quad. Second, HQT sorts the four lists of rectangles stored in each leaf node with one of their x or y coordinates as keys. This reduces the search time at the leaves and consequently makes it possible to use a higher threshold than that used in QLQT.

Experimental results showed that HQT out-performs QLQT, BQT, HVT, and *k*-d on neighbor-search queries by at least 20%. However, its build-time and space requirements were not as good as some of the other data structures.

53.7 Concluding Remarks

Most of the research in VLSI layout data structures was carried out in the early 80s through the mid-90s. This area has not been very active since then. One reason is that there continue to be several important challenges in VLSI physical design automation that must be addressed quickly so that physical design does not become a bottleneck to the continuing advances in semiconductors predicted by Moore's Law. These challenges require physical design tools to consider "deep sub-micron effects" into account and take priority over data structure improvements. The implication of these effects is that problems in VLSI physical design are no longer purely geometric, but rather need to merge geometry with electronics. For example, in the early 1980s, one could safely use the length of a wire to estimate the delay of a signal traveling along it. This is no longer true for the majority of designs. Thus, the delay of a signal along the wire must now be estimated by factoring in its resistance and capacitance, in addition to its geometry. On the other hand, the more detailed computations of electronic quantities like capacitance, resistance, and inductance require the underlying data structures to support more complex operations. Thus, the author believes that there remain opportunities for developing better layout data structures. Another area that, to the best of our knowledge, has been neglected in the academic literature is that VLSI layout data needs to be efficiently stored in secondary storage because of its sheer size. Thus, although there is potential for academic research in this area, we believe that the design automation industry may well have addressed these issues in their proprietary software.

Unlike corner stitching, quad-trees permit rectangles to overlap. This addresses the problem that corner stitching has with handling multiple layers. On the other hand, corner stitching was designed for use in the context of an interactive layout editor, whereas quad trees are designed in the context of batched operations. We note that, to our knowledge, relatively recent data structures such as R trees have not been used for VLSI layout. It is unclear what the outcome of this comparison will be. On one hand, R-trees are considered to be faster than quad trees. On the other, the quad-trees developed in VLSI have been improved and optimized significantly as we have seen in this chapter. Also, VLSI data has generally been considered to consist of uniformly small rectangles, which may make it particularly suitable for quad trees, although this property may not be true when wires are considered.

Acknowledgments

This work was supported, in part, by the National Science Foundation under grant CCR-9988338.

References

1. N. Sherwani, *Algorithms for VLSI Physical Design Automation*. Boston: Kluwer Academic Publishers, 1992.
2. M. Sarrafzadeh and C. K. Wong, *An Introduction to VLSI Physical Design*. New York: McGraw Hill, 1996.
3. S. Sait and H. Youssef, *VLSI Physical Design Automation:Theory andPractice*. Piscataway, NJ: IEEE Press, 1995.
4. N. H. E. Weste and K. Eshraghian, *Principles of CMOS VLSI Design: A Systems Perspective*, Second Edition. New York: Addison Wesley, 1993.
5. J. L. Hennessy and D. A. Patterson, *Computer Architecture: A Quantitative Approach*, Third Edition. New York: Morgan Kaufmann, 2003.
6. D. G. Boyer, "Symbolic Layout Compaction Review," in *Proceedings of 25th Design Automation Conference*, Anaheim, CA, pp. 383–389, 1988.
7. J. Ousterhout, G. Hamachi, R. Mayo, W. Scott, and G. Taylor, "Magic: A VLSI Layout System," in *Proc. of 21st Design Automation Conf.*, Albuquerque, New Mexico, pp. 152–159, 1984.
8. J. K. Ousterhout, "Corner Stitching: A Data Structuring Technique for VLSI Layout Tools," *IEEE Transactions on Computer-Aided Design*, vol. 3, no. 1, pp. 87–100, 1984.
9. D. P. Mehta, "Estimating the Memory Requirements of the Rectangular and L-Shaped Corner Stitching Data Structures," *ACM Transactions on the Design Automation of Electronic Systems*, vol. 3, no. 2, pp. 272–284, 1998.
10. D. P. Mehta, "CLOTH_MEASURE: A Software Tool for Estimating the Memory Requirements of Corner Stitching Data Structures," *VLSI Design*, vol. 7, no. 4, pp. 425–436, 1998.
11. M. Quayle and J. Solworth, "Expanded Rectangles: A New VLSI Data Structure," in *ICCAD*, Santa Clara, California, pp. 538–541, 1988.

12. D. Marple, M. Smulders, and H. Hegen, "Tailor: A Layout System Based on Trapezoidal Corner Stitching," *IEEE Transactions on Computer-Aided Design*, vol. 9, no. 1, pp. 66–90, 1990.

13. C. H. Séquin and H. da Silva Façanha, "Corner Stitched Tiles with Curved Boundaries," *IEEE Transactions on Computer-Aided Design*, vol. 12, no. 1, pp. 47–58, 1993.

14. D. P. Mehta and G. Blust, "Corner Stitching for Simple Rectilinear Shapes," *IEEE Transactions on Computer-Aided Design of Integrated Circuits and Systems*, vol. 16, 186–198, Feb. 1997.

15. S. Nahar and S. Sahni, "A Fast Algorithm for Polygon Decomposition," *IEEE Transactions on Computer-Aided Design*, vol. 7, 478–483, Apr. 1988.

16. T. Ohtsuki, "Minimum Dissection of Rectilinear Regions," in *Proceedings 1982 International Symposium on Circuits and Systems (ISCAS)*, Rome, Italy, pp. 1210–1213, 1982.

17. H. Imai and T. Asano, "Efficient Algorithms for Geometric Graph Search Algorithms," *SIAM Journal on Computing*, vol. 15, 478–494, May 1986.

18. W. T. Liou, J. J. M. Tan, and R. C. T. Lee, "Minimum Partitioning of Simple Rectilinear Polygons in $O(n \log\log n)$ Time," in *Proceedings of the Fifth Annual Symposium on Computational Geometry*, Saarbruchen, Germany, pp. 344–353, 1989.

19. S. Wu and S. Sahni, "Fast Algorithms to Partition Simple Rectilinear Polygons," *International Journal on Computer Aided VLSI Design*, vol. 3, 241–270, 1991.

20. M. Lopez and D. Mehta, "Efficient Decomposition of Polygons into L-Shapes with Applications to VLSI Layouts," *ACM Transactions on Design Automation of Electronic Systems*, vol. 1, 371–395, 1996.

21. D. Mehta and E. Wilson, "Parallel Algorithms for Corner Stitching," *Concurrency: Practice and Experience*, vol. 10, 1317–1341, 1998.

22. G. Kedem, "The Quad-CIF Tree: A Data Structure for Hierarchical On-Line Algorithms," in *Proceedings of the 19th Design Automation Conference*, Piscataway, NJ, pp. 352–357, 1982.

23. J. B. Rosenberg, "Geographical Data Structures Compared: A Study of Data Structures Supporting Region Queries," *IEEE Transactions on Computer-Aided Design*, vol. 4, no. 1, pp. 53–67, 1985.

24. R. L. Brown, "Multiple Storage Quad Trees: A Simpler Faster Alternative to Bisector List Quad Trees," *IEEE Transactions on Computer-Aided Design*, vol. 5, no. 3, pp. 413–419, 1986.

25. L. Weyten and W. de Pauw, "Quad List Quad Trees: A Geometric Data Structure with Improved Performance for Large Region Queries," *IEEE Transactions on Computer-Aided Design*, vol. 8, no. 3, pp. 229–233, 1989.

26. A. Pitaksanonkul, S. Thanawastien, and C. Lursinsap, "Comparison of Quad Trees and 4-D Trees: New Results," *IEEE Transactions on Computer-Aided Design*, vol. 8, no. 11, pp. 1157–1164, 1989.

27. G. Lai, D. S. Fussell, and D. F. Wong, "HV/VH Trees: A New Spatial Data Structure for Fast Region Queries," in *Proceedings of the 30th Design Automation Conference*, Dallas, TX, pp. 43–47, 1993.

28. G. Lai, D. S. Fussell, and D. F. Wong, "Hinted Quad Trees for VLSI Geometry DRC Based on Efficient Searching for Neighbors," *IEEE Transactions on Computer-Aided Design*, vol. 15, no. 3, pp. 317–324, 1996.

54

Floorplan Representation in VLSI*

54.1 Introduction.. 833
    Statement of Floorplanning Problem • Motivation of the Representation • Combinations
    and Complexities of the Various Representations • Slicing, Mosaic, LB Compact, and
    General Floorplans

54.2 Graph Based Representations 837
    Constraint Graphs • Corner Stitching • Twin Binary Tree • Single Tree Representations

54.3 Placement Based Representations 844
    Sequence-Pair • Bounded-Sliceline Grid • Corner Block List • Slicing Tree

54.4 Relationships of the Representations.. 851
    Summary of the Relationships • A Mosaic Floorplan Example • A General Floorplan
    Example

54.5 Rectilinear Shape Handling 851

54.6 Conclusions.. 853

Acknowledgments 853

References.. 854

Zhou Feng
Fudan University

Bo Yao
University of California, San Diego

Chung-Kuan Cheng
University of California, San Diego

54.1 Introduction

There are two main data models that can be used for representing floorplans: graph-based and placement based [23–29].

The graph-based approach includes constraint graphs, corner stitching, twin binary tree, and O-tree. They utilize constraint graphs or their simplified versions directly for the encoding. Constraint graphs are basic representations. The corner stitching simplifies the constraint graph by recording only the four neighboring blocks to each block. The twin binary tree then reduces the recorded information to only two neighbors of each block, and organizes the neighborhood relations in a pair of binary trees. The O-tree is a further simplification to the twin binary tree. It keeps only one tree for encoding.

The placement-based representations use the relative positions between blocks in a placement for encoding. This category includes sequence pair, bounded-sliceline grid, corner block list and slicing trees. The sequence pair and bounded-sliceline grid can be applied to general floorplan. The corner block list records only the relative position of adjacent blocks, and is available to mosaic floorplan only. The slicing trees are for slicing floorplan, which is a type of mosaic floorplan. The slicing floorplan can be constructed by hierarchical horizontal or vertical merges and thus can be captured by a binary tree structure known as the slicing tree.

The rest of this chapter is organized as follows. In Section 54.1, we give the introduction and the problem statement of floorplanning. In Section 54.2, we discuss the graph-based representations. In Section 54.3, we introduce the placement-based representations. We describe the relationship between different representations in Section 54.4. We illustrate the shape handling of rectilinear blocks in Section 54.5 and summarize the chapter in Section 54.6.

54.1.1 Statement of Floorplanning Problem

Today's complexity in circuitry design wants a hierarchical approach [1]. The entire circuit is partitioned into several sub-circuits, and the sub-circuits are further partitioned into smaller sub-circuits, until they are small enough to be handled. The relationship between the sub-circuits can be represented with a tree as shown in Figure 54.1. Here every sub-circuit is called a block, and hence

---

* This chapter has been reprinted from first edition of this Handbook, without any content updates.

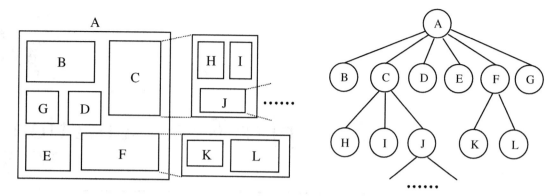

FIGURE 54.1   Hierarchical structure of blocks.

the entire circuit is called the top block. From the layout point of view, every block corresponds to a rectangle, which contains sub-blocks or directly standard cells. Among the decisions to be made is the determination of shape (aspect ratio) and the pin positions on the blocks. In the top-down hierarchical methodology, blocks are designed from the top block (the entire circuit) to the leaf blocks (small modules). The minimizations of chip area and wire length are the basic targets for any layout algorithm. In addition, there are case-dependent constraints that will influence the layouts, such as the performance, the upper or/and lower boundaries of aspect ratio, the directions of pins, etc.

Now we give the definition of floorplanning problem:

Inputs:

1. The net-lists of the sub-circuits;
2. The area estimation of blocks and, if any, the aspect ratio constraints on the blocks;
3. The target area and shape of the entire chip.

Outputs:

1. The shapes and positions of blocks;
2. The pin positions on the blocks.

The objective functions involve: the chip area, the total wire length and, if any, the performances.

54.1.2 Motivation of the Representation

Floorplan representation becomes an important issue in floorplanning for the following reasons.

1. The blocks may have arbitrary shapes and locations, while the size of memory used to represent a two-dimensional floorplan should be $O(n)$.
2. For the general floorplanning problem, iterative improvement is the commonly used approach. The search for the best solution has proven to be NP-complete, so many heuristic optimizing algorithms, such as dynamic programming, simulated annealing, zone refinement, cluster refinement, have been adopted. The representation should also facilitate the operations called by those optimizing algorithms.
3. The storage resources requirement, the redundancy of the representation and the complexity of translating the representation into floorplan are always the concerns in floorplanning. Here redundancy refers to the situation that several different expressions actually correspond to the same physical layout. Essentially, a heuristic algorithm searches part of the solution space to find the local optimal solution, which is hopefully very close to the global optimal solution. Redundancy causes the optimizing algorithm evaluate the same floorplan repeatedly.

54.1.3 Combinations and Complexities of the Various Representations

The number of possible floorplan representations describes how large the searching space is. It also discloses the redundancy of the representation. For general floorplans with n blocks, the combinations of the various representations are listed in Table 54.1. Note that the twin binary tree representation has a one to one relation with the mosaic floorplan, and the slicing tree has a one to one relation with the slicing floorplan [2]. In other words, for these two representations, the number of combinations is equal to the number of possible floorplan configurations and there is no redundancy.

TABLE 54.1 Combinations of the Representations

Representation	Combinations
Twin binary tree	$n!*B(n)$, where $B(n) = \begin{pmatrix} n+1 \\ 1 \end{pmatrix}^{-1} \begin{pmatrix} n+1 \\ 2 \end{pmatrix}^{-1} \sum_{k=1}^{n} \begin{pmatrix} n+1 \\ k-1 \end{pmatrix} \begin{pmatrix} n+1 \\ k \end{pmatrix} \begin{pmatrix} n+1 \\ k+1 \end{pmatrix}$
O-Tree	$O(n!2^{2n-2}/n^{1.5})$
Sequence pair	$(n!)^2$
Bounded-sliceline grid	$n! \begin{pmatrix} n^2 \\ n \end{pmatrix}$
Corner block list	$O(n!2^{3n-3}/n^{1.5})$
Slicing tree	$n!*A(n-1)$, where $A(n)$ is the super-Catalan number with the following definition: $A(0) = A(1) = 1$ and $A(n) = (3(2n-3)A(n-1) - (n-3)A(n-2))/n$

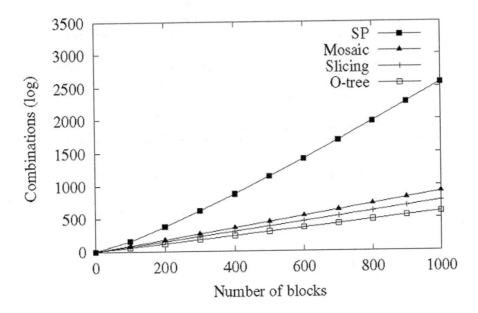

FIGURE 54.2 Combination of floorplans and representations.

The combination numbers of sequence pairs, mosaic floorplans, slicing floorplans, and O-trees are illustrated on a log scale in Figure 54.2. The combination numbers are normalized by $n!$, which is the number of permutations of n blocks. The slopes of the lines for mosaic floorplans, slicing floorplans, and O-tree structures are the constants 0.89, 0.76, and 0.59, respectively. On the other hand, the slope of the line for sequence pair increases with a rate of *log n*. Table 54.2 provides the exact numbers of the combinations for the floorplans or representations with the block number ranging from 1 to 17.

The time complexities of the operations transforming a representation to a floorplan are very important, because they determine the efficiency of the floorplan optimizations. The complexities of the representations covered in this chapter are compared in Table 54.3, where n is the number of blocks in a floorplan.

For sequence-pair, the time complexity to derive a floorplan is *O(nloglogn)* due to a fast algorithm proposed in Reference 3. We will discuss more on this in Section 54.3.1 For bounded slicing grid, there is a trade off between the solution space and the time complexity of deriving a floorplan. To ensure that the solution space covers all the optimal solutions, we need the grid size to be at lease n by n. This results in an $O(n^2)$ time complexity in deriving a floorplan [4]. For the rest of the representations, there are algorithms with $O(n)$ time complexity to convert them into constraint graphs. The time complexity to derive a floorplan is thus $O(n)$.

54.1.4 Slicing, Mosaic, LB Compact, and General Floorplans

A layout can be classified as a slicing floorplan if we can partition the chip with recursive horizontal or vertical cut lines (Figure 54.3a). In a mosaic floorplan, the chip is partitioned by horizontal and vertical segments into rectangular regions and each region corresponds to exactly one block (Figure 54.3b). For a general floorplan, we may find empty space outside rectangular block

TABLE 54.2 Exact Number of Combinations of Different Floorplan Configurations and Representations

Number of Blocks	Combinations of O-Tree	Combinations of Slicing Floorplan	Combinations of Mosaic Floorplan	Combinations of Sequence Pairs
1	1	1	1	1
2	2	2	2	2
3	5	6	6	6
4	14	22	22	24
5	42	90	92	120
6	132	394	422	720
7	429	1806	2074	5040
8	1430	8558	10,754	40,320
9	4862	41,586	58,202	362,880
10	16,796	206,098	326,240	3,628,800
11	58,786	1,037,718	1,882,690	39,916,800
12	208,012	5,293,446	11,140,560	479,001,600
13	742,900	27,297,738	67,329,992	6,227,020,800
14	2,674,440	142,078,746	414,499,438	87,178,291,200
15	9,694,845	745,387,038	2,593,341,586	1,307,674,368,000
16	35,357,670	3,937,603,038	16,458,756,586	20,922,789,888,000
17	129,644,790	20,927,156,706	105,791,986,682	355,687,428,096,000

TABLE 54.3 Time Complexity Comparison of the Representations

Representation	From a Representation to a Floorplan
Constraint graph	$O(n)$
Corner stitching	$O(n)$
Twin binary tree	$O(n)$
O-Tree	$O(n)$
Sequence pair	$O(n \log \log n)$
Bounded-sliceline grid	$O(n^2)$
Corner block list	$O(n)$
Slicing tree	$O(n)$

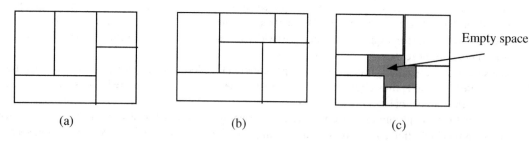

FIGURE 54.3 (a) Slicing floorplan; (b) Mosaic floorplan; (c) General floorplan.

regions (Figure 54.3c). An LB-compact floorplan is a restricted general floorplan in that all blocks are shifted to their left and bottom limits. In summary, the set of general floorplans covers the set of LB-compact floorplans, which covers the set of mosaic floorplans, and which covers the set of slicing floorplans (Figure 54.4).

For slicing and mosaic floorplans, the vertical segments define the left-right relation among the separated regions, and the horizontal segments define the above-below relation. Suppose that we shift the segments to change the sizes of the regions, we view the new floorplan to be equivalent to the original floorplan in terms of their topologies [5–7]. Therefore, we can devise representations to define the topologies of slicing and mosaic floorplans independent of the sizes of the blocks (Figure 54.3a and b).

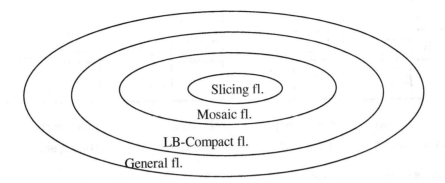

FIGURE 54.4 Set coverage of slicing floorplan, mosaic floorplans, LB-compact floorplans and general floorplans.

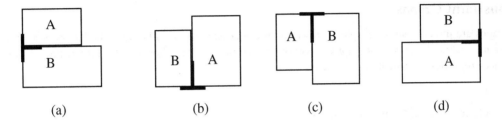

    (a)   (b)   (c)   (d)

FIGURE 54.5 Four directions of T-junctions.

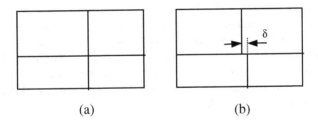

    (a)   (b)

FIGURE 54.6 Degeneration.

In contrast, for a general floorplan, it is rather difficult to draw a meaningful layout (Figure 54.3c) without the knowledge of the block dimensions. One approach is to extend the mosaic floorplans to general floorplans by adding empty blocks [8]. It is shown that to convert a mosaic floorplan with n blocks into a general floorplan, the upper bound on the number of empty blocks inserted is $n - 2\sqrt{n} + 1$. [8]

In a mosaic floorplan, two adjacent blocks meet at a T-junction by sharing the non-crossing edge of the junction. There are four directions of T-junctions as is illustrated in Figure 54.5. In the case of Figure 54.5a and b, block B is termed the C-neighbor at the lower left corner of block A. In Figure 54.5c and d, block B is the C-neighbor at the upper right corner of block A. The C-neighbor is used to construct twin binary tree representation.

The degeneration of a mosaic floorplan refers to the phenomenon that two T-junctions meet together to make up a cross-junction, as illustrated in Figure 54.6a. Some representations forbid the occurrence of degeneration. One scheme to solve the problem is to break one of the intersecting lines and assume a slight shift between the two segments, as shown in Figure 54.6b. Thus the degeneration disappears.

We generate an LB compact floorplan by compacting all blocks toward left and bottom. For a placement, suppose no block can be moved left, the placement is called L-compact. Similarly, if no block can be moved down, the placement is called B-compact. A floorplan is LB-compact if it is both L-compact and B-compact (Figure 54.7).

54.2 Graph Based Representations

Graph based representations include constraint graphs, corner stitching, twin binary trees, and O-tree. They all utilize the constraint graphs or their simplified version for floorplan encoding.

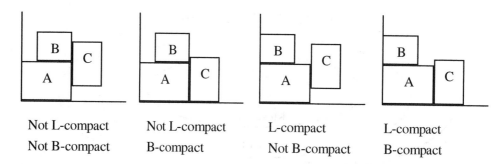

Not L-compact   Not L-compact   L-compact   L-compact

Not B-compact   B-compact   Not B-compact   B-compact

FIGURE 54.7 Examples of L-compact and B-compact.

54.2.1 Constraint Graphs

Constraint graphs are directed acyclic graphs representing the adjacency relations between the blocks in a floorplan. In this subsection, we first define the constraint graphs for general floorplans. We then show that for mosaic floorplan, the constraint graphs have nice properties of triangulation and duality.

54.2.1.1 The Generation of Constraint Graphs

Constraint graphs reflect the relative positions between blocks [9]. Constraint graphs can be applied to general floorplans, as is shown in Figure 54.8. A node in the constraint graph represents a block. A directed edge denotes the location relationship between two blocks. For example, A→B means block A is to the left of B in a horizontal constraint graph (Figure 54.8b). A→E means block A is on top of E in a vertical constraint graph (Figure 54.8c). Here we imply that if A→B and B→C, then A→C. Thus even though block *A* stands to the left of *C*, the edge between *A* and *C* is not necessarily shown. To mark the four sides of the chip, we add the nodes labeled with "left," "right," "top" and "down." A pair of horizontal and vertical constraints graphs can represent a floorplan. Every constraint graph, whether it is a horizontal one or a vertical one, is planar and acyclic. Figure 54.9 shows an example of the constraint graphs to a mosaic floorplan. Figure 54.9a is the mosaic floorplan. Figure 54.9b and c are the corresponding horizontal and vertical constraint graphs, respectively.

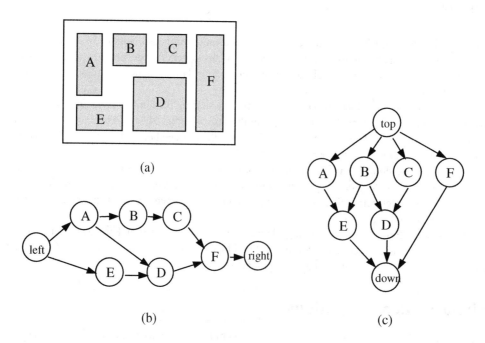

(a)

(b)   (c)

FIGURE 54.8 Constraint graphs for a general floorplan.

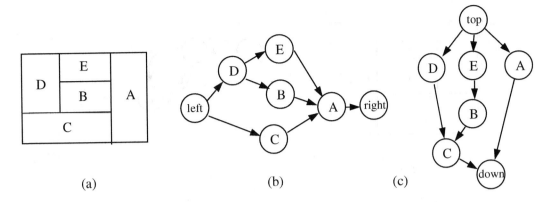

FIGURE 54.9 Constraint graphs for a mosaic floorplan.

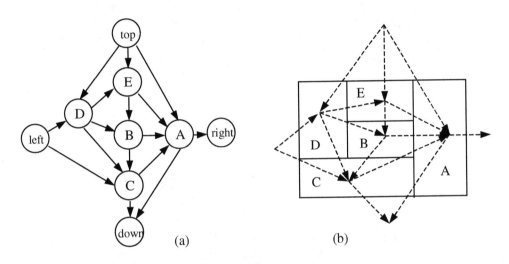

FIGURE 54.10 Triangulation.

54.2.1.2 Triangulation

For a mosaic floorplan without degeneration, if its horizontal and vertical constraint graphs are merged together, then we have the following conclusions [9]:

1. Every face is a triangle.
2. All internal nodes have a degree of at least 4.
3. All cycles, if not faces, have the length of at least 4.

Shown in Figure 54.10a is the result of merging the pair of constraint graphs in Figure 54.9. In fact, the merged constraint graph is a dual graph of its original floorplan (Figure 54.9b).

54.2.1.3 Tutte's Duality [10]

We can also build an edge-based constraint graph for a mosaic floorplan, where the nodes denote the lines separating the blocks while the edges denote the blocks. Labeling the lines with numbers (Figure 54.11a), we build a vertical constraint graph (Figure 54.11b) and a horizontal constraint graph (Figure 54.11c). Figure 54.11d demonstrates the result of merging the vertical and horizontal constraint graphs. Here, to make the merged graph clear, the edges representing horizontal constraints are drawn with dotted lines, and a letter at the intersection of a solid edge and a dotted edge denotes the two edges simultaneously. It is very interesting that, for mosaic floorplans, the vertical and horizontal constraint graphs are dual, as is called Tutte's duality.

Let's see how Tutte's duality is used to solve the sizing problem in floorplanning. We map the constraint graphs into circuit diagrams by replacing the edges in the vertical constraint graph with resistors, as illustrated in Figure 54.12.

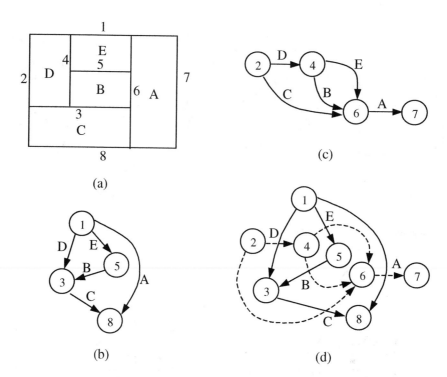

FIGURE 54.11   Line-based constraint graphs.

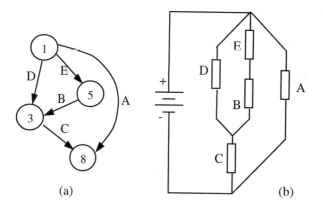

FIGURE 54.12   Mapping a constraint graph to a circuit.

The circuit is subject to Kirchoff voltage law. As a result, taking Figure 54.12 as an example, we have:

$$
\begin{aligned}
V_E + V_B &= V_D & I_E &= I_B \\
V_D + V_C &= V_A & I_E + I_D &= I_C \\
V_A &= V_{source} & I_C + I_A &= I_{source} \\
R_i = \frac{V_i}{I_i}, & \quad i \in \{A, B, C, D, E\}
\end{aligned}
$$

Note that, if we denote the height and the width of block i to be H_i and W_i, we have (refer to Figure 54.12a):

$$
\begin{aligned}
H_E + H_B &= H_D & W_E &= W_B \\
H_D + H_C &= H_A & W_E + W_D &= W_C \\
H_A &= H_{chip} & W_C + W_A &= W_{chip} \\
AspectRatio_i = \frac{H_i}{W_i}, & \, i \in \{A, B, C, D, E\}
\end{aligned}
$$

By comparing the above two equation arrays, we can find that there is a perfect correspondence between the solutions of the circuit and the floorplan. Let us set the following assumptions: (only two of the three equations are independent)

$$R_i = AspectRatio_i$$
$$V_{source} = H_{chip}$$
$$I_{source} = W_{chip}$$

Then we have:

$$V_i = H_i$$
$$I_i = W_i$$

Thus the theories dealing with large circuits can be borrowed to solve the sizing problem in floorplanning.

Constraint graphs have the advantages of being able to cope with any types of floorplans. However, it would be rather difficult to shift to a new floorplan with just a few simple operations on the graph. The efficiency is greatly discounted for the iteratively improving algorithms. Thus in Reference 11, a transitive closure graph (TCG) was proposed to simplify the rules of the operations. The relation of each pair of blocks is prescribed in either horizontal or vertical constraint graph via transitive closure, but not in both graphs.

54.2.2 Corner Stitching

Corner stitching is used to represent the topology of a mosaic floorplan. Simplified from constraint graphs, corner stitching [12] keeps only four pointers at the two opposite corners for every block. All the operations on a constraint graph can also be fulfilled on a corner stitching representation with acceptable increases in time complexity, while the storage for corner stitching becomes easier since the number of pointers attached to every block is fixed (equals to 4). Readers are referred to Chapter 53 for detailed descriptions and analyses of corner stitching.

54.2.3 Twin Binary Tree

Twin binary tree [2] representation applies to mosaic floorplans without degeneration. The twin binary tree is constructed as follows, every block takes its C-neighbor (Figure 54.5) as its parent. In the first tree, only the C-neighbors in lower left corners (Figure 54.5a and b) are taken into account. If the related T-junction is of type (a), then the block is a left child of its parent, and if the T-junction is of type (b), then the block is a right child. The most bottom-left block in the floorplan acts as the root of the tree. Similarly, in the second tree, the C-neighbors in upper right corners (Figure 54.12c and d) are used, and the most upper-right block becomes the tree's root. Figure 54.13 gives an example of a twin binary tree.

The pointers of twin binary trees are in fact a subset of those in corner stitching. Besides, It has been proved that twin binary tree is a non-redundant representation. In other words, every pair of trees corresponds to a unique mosaic floorplan.

The twin properties of binary trees can be illustrated with Figure 54.14. Consider the trees shown in Figure 54.13, we add a left child labeled "0" to every node without left child except the most left node, and a right child labeled "1" to every node without right child except the most right node. The resultant trees are the so-called extended trees (Figure 54.14). The sequences of the in-order traverse of the two extended trees shown in Figure 54.14 are "A0B1C1F0D1E" and "A1B0C0F1D0E" respectively. If we separate them into the label parts and the bit parts, we have $\pi_1 = ABCFDE$, $\alpha_1 = 01101$, $\pi_2 = ABCFDE$ and $\alpha_2 = 10010$. It is interesting to find that $\pi_1 = \pi_2$ and $\alpha_1 = \overline{\alpha_2}$. So rather than store the two binary trees, we only keep the linear lists π and α in memory.

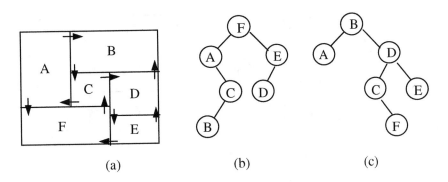

(a)   (b)   (c)

FIGURE 54.13 Twin binary tree.

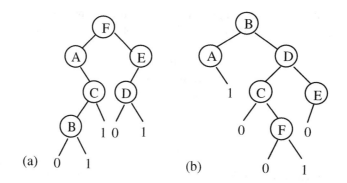

FIGURE 54.14   Extended trees.

However, it is insufficient to recover the two binary trees just from π and α, so we need two more lists, β and β'. If the *i-th* element in π is the left child of its parent in the first tree or the root of the tree, we set the *i-th* element in β to be "0," otherwise we set it "1." In the similar way β' is constructed according to the second tree. Thus we use a group of linear lists $\{\pi, \alpha, \beta, \beta'\}$, called the twin binary sequence [8], to express the twin binary tree, and equivalently the floorplan.

Finally, we give the main properties of twin binary tree representation:

1. The memory required for storing the twin binary sequence is $log_2 n + 3n - 1$, because $|\pi| = log_2 n$, $|\alpha| = n - 1$, and $|\beta| = |\beta'| = n$.
2. Twin binary tree is a non-redundancy floorplan representation.
3. The complexity of the translation between twin binary sequence and floorplan is $O(n)$.

54.2.4 Single Tree Representations

An ordered-tree (O-tree) [13,14], or equivalently the B* tree [15] representation uses a spanning tree of the constraint graph and the dimensions of the blocks. Because the widths and heights of the blocks are given, the representation can describe one kind of general floorplan termed LB-compact. With a proper encoding, a great enhancement on the storage efficiency and the perturbation easiness is obtained.

A horizontal O-tree is derived with the following rules:

1. If block A lies adjacent to the left side of block B, or, $X_a + W_a = X_b$ (here X_a is the coordinate of block A on the X-axis and W_a is the width of block A), then on the O-tree, B is A's child. If there happens to exist more than one block adjacent to the left side of block B (satisfying the requirement $X_i + W_i = X_b$), one of them is assigned to be the parent of block B.
2. If block A lies on top of block B and the two blocks have the same parents on the O-tree, then B is A's elder brother.
3. A virtual block is presumed to have the left most position, and therefore serves as the root of the O-tree.

Figure 54.15 shows an example of an O-tree representation for the same mosaic floorplan shown in Figure 54.13. If we show the pointer of every block pointing to its parent in the O-tree (Figure 54.15b), we can find that the pointers are in fact a subset of those in twin binary tree. Similarly a vertical O-tree can be built up. Without the loss of generality, hereafter we only discuss the horizontal O-tree.

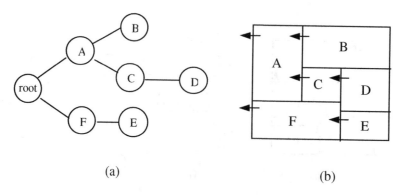

(a)   (b)

FIGURE 54.15   Building an O-tree.

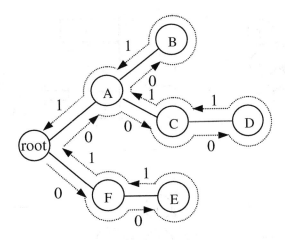

FIGURE 54.16   Derivation of T, π.

Next, let's describe the O-tree with linear lists: a bit string (denoted with T) and a label string (denoted with π). The bit string records the structure of an O-tree. We make a depth-first traverse on the O-tree. A "0" is inserted into T for descending an edge while a "1" for ascending an edge. For example, the O-tree in Figure 54.16 corresponds to the bit string $T =$ "001100011011." Inversely, given the bit string, the structure of the O-tree is solely determined. To get the complete description of an O-tree, we need a second linear list, the label string, to record the labels of the tree's nodes. Not only should the label string include all the labels of the tree's nodes, but also reflect the labels' positions on the tree. A depth-first traverse generates such a linear list. For the example in Figure 54.16, we have $\pi =$ "*FEACDB.*"

Given a horizontal O-tree, or $\{T, \pi\}$, and the dimensions of blocks, the positions of blocks can be derived with the following rules:

1. Each block has to be placed to the left of its children.
2. If two blocks overlap along their x-coordinate projection, then the block having a higher index in π must be placed on top of the other one.

In fact the second rule applies to the situation in which none of the two blocks is a descendant of the other, like blocks "*f*" and "*a*," or "*b*" and "*d*," in Figure 54.16. The way we derive π indicates that the block with higher index always stands "higher." To get to a placement, we need to locate all the blocks. The following operation is executed on blocks according to their orders in π. Here B_i refers to the *i-th* block, and (x_i, y_i) refers to the coordinate of B_i's left-bottom corner.

1. Find B_i's parent, B_j, and then we have $x_i = x_j + w_j$, here w_j is the width of block B_j.
2. Let ψ be a set of blocks who have a lower index than B_i in π and have an projection overlap with B_i in the X-axis, find the maximum $y_k + w_k$ for $B_k \in \psi$, then we have $y_i = max(y_k + w_k)$.

Now we analyze the performance of O-tree:

1. The bit string has a length of $2n$ for an O-tree of n blocks, because each node except the root has an edge towards its parents and each edge is traversed twice during the construction of the bit string. The label string takes $n \log_2 n$ bits for we have to use $\log_2 n$ bits to distinguish the n blocks. So the total memory occupied by an O-tree is $n(2 + \log_2 n)$. By comparison, a sequence pair takes $2n \log_2 n$ bits and a slicing structure takes $n(6 + \log_2 n)$ bits.
2. The time needed to transform $\{T, \pi\}$ to a placement is linear to the number of blocks, or we can say the complexity of transforming $\{T, \pi\}$ to a placement is $O(n)$. For a sequence pair or a slicing structure, the complexity is $O(n \log_2 log_2 n)$ or $O(n)$ respectively. Upon this point, O-tree has the same performance as the slicing structure but is more powerful for representing a placement.
3. The number of combinations is $O(n! \cdot 2^{2n-2}/n^{1.5})$, which is smaller than any other representation that has ever been discussed.

The floorplan of an arbitrarily given O-tree is not necessarily LB-compact. Yet in this case, we can compact the floorplan to reduce the chip area. Notice that an O-tree-to-placement transform tightens the blocks, with one direction having the priority. For example, in a placement transformed from a horizontal O-tree, the blocks are placed tighter along the X-axis than along the Y-axis. It would be undoubted that a placement transformed from a vertical O-tree will give Y-axis the priority. Thus, by

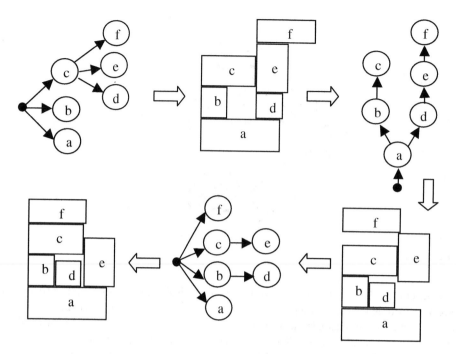

FIGURE 54.17   Transformation between horizontal and vertical O-trees.

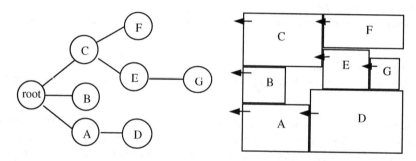

FIGURE 54.18   An O-tree representation of a general floorplan.

iteratively making the transforms between horizontal O-trees and vertical O-trees via placements, we can finally reach an LB-compact floorplan (Figure 54.17).

On the other hand, given an LB-compact floorplan, we can always derive an O-tree representation. For example, Figure 54.18 illustrates a general floorplan with seven blocks and the corresponding O-tree. O-tree is able to cover LB-compact floorplan with the smallest number of combinations in Table 54.1 and Figure 54.2 because the O-tree structure avoids redundancy and screens improper floorplans by taking advantage of the knowledge of the block dimensions. For example, given an O-tree, we can convert it to a binary tree and find many other possible trees to pair up as twin binary trees, which correspond to many different mosaic floorplans. In the perspective of O-tree representation, these floorplans are the variations due to the differences of the block sizes.

The B* tree and the O-tree are equivalent because the transformation between the two is one to one mapping. The merit of the B tree is a different data structure and implementation.

In summary, among the constraint-graphs based representations, from the point of view of the pointers necessary for representing the floorplans, O-tree is a subset of twin binary tree; twin binary is a subset of corner stitching; and corner stitching is a subset of constraint graphs, as demonstrated in Figure 54.19.

54.3 Placement Based Representations

The placement-based representations include sequence pair, bounded-sliceline grid, corner block list and slicing tree. With the dimensions of all the blocks known, the sequence pair and bounded-sliceline grid can be applied to general floorplans. The corner

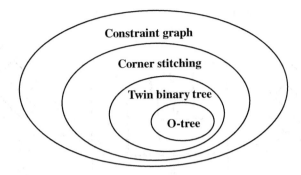

FIGURE 54.19 Relationship between the constraint graph based representation.

block list records the relative position of adjacent blocks, and is applicable to mosaic floorplan only. The slicing tree is for slicing floorplan, which is a special case of mosaic floorplan.

54.3.1 Sequence-Pair

Sequence-pair expresses the topological positions of blocks with a pair of block lists [16,17]. For each pair of blocks, there are four possible relations, left-right, right-left, above-below, and below-above. We use a pair of sequences $\{\pi_1, \pi_2\}$ to record the order of the two blocks. We can record two blocks A, B in four different orders: (AB, AB), (AB, BA), (BA, AB), (BA, BA). Hence, we use the four orders to represent the four placement relations. Figure 54.20 shows the four possible relations between a pair of blocks A and B and the corresponding sequence pair for each relation.

The two sequences can be viewed as two coordinate systems that define a grid placement of the blocks. To see it more clearly, we construct a grid system illustrated in Figure 54.21. The slopes are denoted with the labels of blocks according to their orders in the two sequences. The intersections of two perpendicular lines that have the same label indicate the topological positions of the blocks.

The grid placement defines the floorplan relation between blocks. For each node in the grid placement, we divide the whole space into four quadrants, quadrant I: -45 to $45°$, quadrant II: 45–135, quadrant III: 135–225, quadrant IV: 225–315. Block *B*'s floorplan relation with block *A* can be derived from block B's location in block A's quadrants.

1. Quadrant I: block B is on the right of block A
2. Quadrant II: block B is above block A

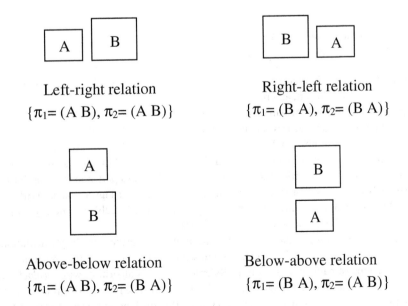

Left-right relation
$\{\pi_1 = (A\ B), \pi_2 = (A\ B)\}$

Right-left relation
$\{\pi_1 = (B\ A), \pi_2 = (B\ A)\}$

Above-below relation
$\{\pi_1 = (A\ B), \pi_2 = (B\ A)\}$

Below-above relation
$\{\pi_1 = (B\ A), \pi_2 = (A\ B)\}$

FIGURE 54.20 Four possible relations between a pair of blocks.

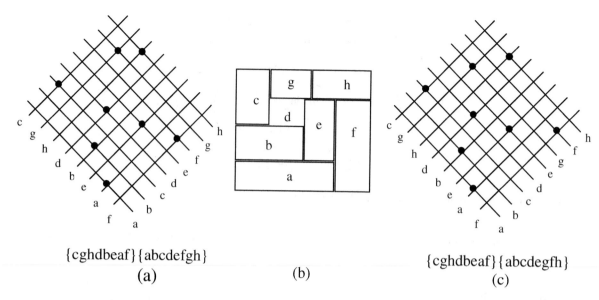

$$\{cghdbeaf\}\{abcdefgh\}$$
(a)   (b) $$\{cghdbeaf\}\{abcdegfh\}$$
(c)

FIGURE 54.21   The grid system and the redundancy.

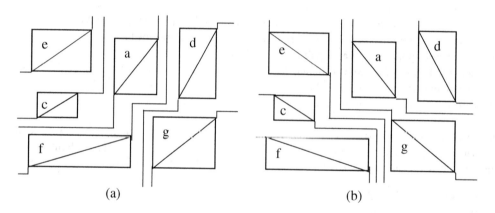

(a)   (b)

FIGURE 54.22   Derivation of sequence-pair.

3. Quadrant III: block B is on the left of block A
4. Quadrant IV: block B is below block A

For example, from the grid placement in Figure 54.22, using node d as a reference, blocks e and f are on the right of block d; blocks g and h are above; blocks c is on the left; and blocks a and b are below.

We can derive the sequence-pair from a floorplan by tracing the blocks through two directions. We first extend the boundary of each block by drawing lines from its four corners (Figure 54.22). In Figure 54.22a, we draw a line from the upper-right corner of each block that goes either rightward or upward. When the line is obstructed with a block or an already existing line, it changes its direction. The extension of the line ends when it reaches the boundary of the chip. We also draw a line from the lower-left corner of each block that goes either downwards or leftwards. The extended lines partition the chip into zones. Following the order of the zones, we get a label sequence. For instance, we have $\pi_1 = {}'ecafdg'$ for Figure 54.22a. With similar operations, the second sequence is derived in the orthogonal direction ($\pi_2 = {}'fcegad'$ in Figure 54.22b). π_1 and π_2 consist of a sequence-pair, which involves all the information of the topological positions of blocks.

For translating a sequence-pair into a topological floorplan with n blocks, a brute force implementation requires time complexity $O(n^2)$, since there are $O(n^2)$ pairs of blocks and we need to scan all the pairs to get the constraint graph and then derive the floorplan by performing a longest path computation. In Reference 3, a fast algorithm with $O(n\log\log n)$ time complexity is proposed to complete the transformation of a sequence pair to the floorplan. The fast algorithm takes advantage of the longest common sequence of the two sequences in a sequence-pair. They show that the longest common sequence calculation is equivalent to the longest path computation in the constraint graph for a sequence-pair. The authors use a priority queue data structure to record the longest common sequence, thus reduce the time for calculating the position of a single block to *O(loglogn)* by amortizing.

The representation of a sequence-pair has the following properties:

1. The number of bits for storing a sequence-pair is $2n \log_2 n$, where n is the number of blocks in the placement.
2. The time complexity of translating a sequence-pair into a topological floorplan is $n \log_2 \log_2 n$ by the newest report.
3. There are totally $(n!)^2$ possible sequence-pairs for n blocks. However, there exists the redundancy. For example, the sequence-pairs in Figure 54.21a and c actually correspond to the same placement (Figure 54.21b).

54.3.2 Bounded-Sliceline Grid

Another method rising from the same idea as sequence-pair is BSG [4], where, instead of the intersections of grid lines, the squares surrounded by the grid lines, called rooms, are used to assign the blocks (Figure 54.23). Although put forward aiming at the packing problem, BSG can also act as a compacting algorithm to determine the accurate positions of blocks.

In BSG, grid lines are separated into vertical and horizontal segments (Figure 54.23). Two constraint graphs are set up with their vertexes corresponding to the grid segments and their directed edges corresponding to the relative positions of the grid segments (Figure 54.24). The vertexes "source" and "sink" are added to make the operation on the graph easy. Every room is crossed by one edge of the constraint graph (respectively in the vertical and horizontal constraint graphs). If the room is not empty, or there is a block assigned into the room, the crossing edge has a weight equal to the width (in the horizontal graph) or the height (in the vertical graph) of the assigned block, otherwise the crossing edge has a weight of zero. Figure 54.24 shows the weights derived from the example in Figure 54.23. By calculating the longest path lengths between the source and the other vertexes in the constraint graphs the real coordinates of the grid segments can be determined and in fact the translation to the placement is implemented. Table 54.4 gives the final results of the example in Figure 54.23 and Figure 54.24. Notice that segment $(1,3)$ and $(2,2)$ have the same position, for the edge between the two vertexes in the horizontal constraint graph has the weight of zero.

Due to the homology, BSG has the similar performance as sequence-pair, except that the complexity of the translation into placement, or finding the longest path lengths for all the vertexes in the constraint graphs, is $O(pq)$, provided that the grid array has p columns and q rows.

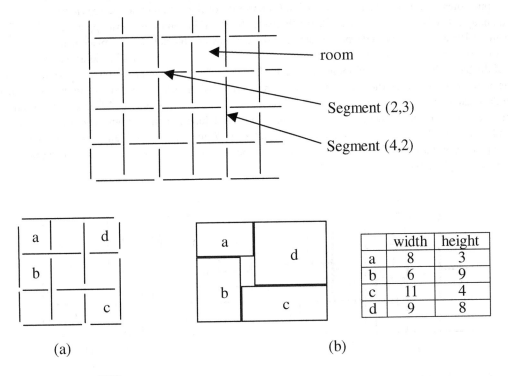

room

Segment (2,3)

Segment (4,2)

	width	height
a	8	3
b	6	9
c	11	4
d	9	8

(a) (b)

FIGURE 54.23   Basic structure of BSG.

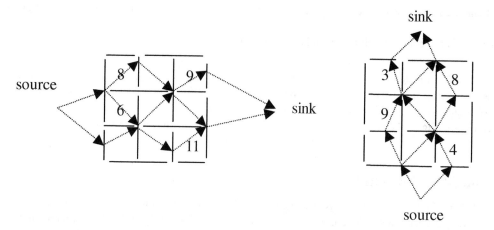

FIGURE 54.24 Constraint graphs.

TABLE 54.4 Physical Positions of Segments

Segment	0,1	1,1	0,2	1,2	1,3	0,3	2,2	2,3	3,3	3,2	2,1	2,0	3,1	3,0
Position (X or Y)	0	6	0	9	8	12	8	12	17	4	4	6	17	0
	Y	X	X	Y	X	Y	X	Y	X	Y	Y	X	X	Y

54.3.3 Corner Block List

Corner block list is a third method of representing non-slicing structures [5,18,19]. Refer to Figure 54.25. The floorplan is required to be mosaic and without degeneration.

For every block, the T-junction at the bottom-left corner can be in either of the two directions. Accordingly we say the orientation of the block is 1 or 0 respectively. Now let's observe the most upper-right blocks in Figure 54.25a and (b), both denoted with "d." If we push the bottom boundary of block "d" in (a) upwards, or the left boundary in (b) rightwards, then block "d" is squeezed and finally deleted (Figure 54.25c and d), and thereafter block "a" becomes the most upper-right block. The operation of squeezing and deleting can be repeatedly performed until there is only one block left in the placement.

According to the order in which the blocks are squeezed out, we get two lists. The labels of blocks are stored in list "S" while the orientations of the blocks in list "L." For example, for Figure 54.25a, $S =$ "fcegbad" and $L =$ "001100." S and L are not sufficient to recover the placement, so we need the third list, T. Each time we squeeze a block, we record the number of T-junctions on the moving boundary, excluding the two at its ends, then add the same number of "1"s and one "0" into T. For example, in Figure 54.25a, when squeezing block d, we have $T =$ "01," for there is only one T-junction on the bottom boundary, which separates the blocks a, d and g. Next, while squeezing block a, only a "0" is added to T, for there is no T-junction on the bottom boundary of block a. Consequently, we have $T =$ "001" after the deletion of block a, and $T =$ "00101001" after the deletion of block c. With an inverse process, the placement can be recovered from S, L and T.

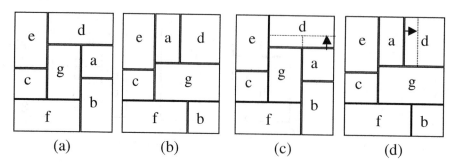

FIGURE 54.25 Corner block list.

The performance of corner block list is:

1. The storage of the 3 lists needs at most $n(3 + logn)$ bits.
2. The number of combinations is *TRIALRESTRICTION*, which is the same as that of a slicing tree. However, slicing structure only covers part of all possible combinations, therefore has larger redundancy than corner block list.
3. The complexity of the translation between placement and lists is *TRIALRESTRICTION*.

54.3.4 Slicing Tree

A floorplan is call a slicing structure [20,21] if it can be iteratively partitioned into two parts with vertical or horizontal cut lines throughout the separated parts of the area, until the individual blocks. Figure 54.26a gives an example. Line 1 splits the whole layout into two parts. The left half involves only one block but the right half can been further split by line 2. Then follows line 3, line 4, and so on. On the contrary, Figure 54.26b is a non-slicing structure. We can't find a cut line throughout the floorplan to split it into two parts.

Now that a slicing floorplan can been iteratively split into two parts with vertical or horizontal cut lines, a binary tree can be structured according to the splitting process [22].

Figure 54.27 gives an example of a slicing floorplan and its binary tree. The inter nodes on the tree denoted with "H" or "V" indicate the direction of the line splitting the areas, while the leaf nodes correspond to the blocks. We call the binary tree a slicing tree.

Just as in twin binary tree, we hope to find a simple data structure to express the slicing tree. Polish Expression is a good solution. It is the result of a post-order traversal on the slicing tree. As an example, the Polish Expression of Figure 54.27 is: *123V56V8H47HVHV*. We regard "H" and "V" as operators and the blocks operands.

Formally, we have the following definition for a Polish expression:

A sequence $b_1 b_2 \ldots b_{2n\_1}$ of elements from $\{1, 2, \ldots, n, V, H\}$ is a Polish expression of length $2n - 1$ iff:

1. Every subscript i appears exactly once in the sequence, $1 \leq i \leq 2n - 1$.
2. The number of operators is less than the number of the components for any prefix sub-string.

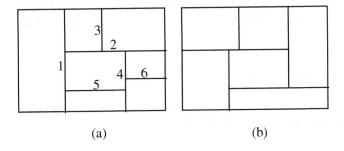

(a)   (b)

FIGURE 54.26 Slicing structure and non-slicing structure.

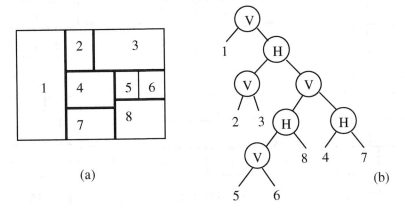

(a)   (b)

FIGURE 54.27 Slicing structure and its binary tree representation.

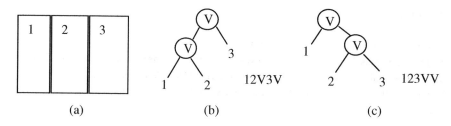

(a)   (b)   (c)

FIGURE 54.28   Redundancy of slicing trees.

So needless to re-construct the slicing tree, we are able to judge a legal Polish Expression. For example, *21H57V43H6VHV* and *215H7V43H6VHV* are legal Polish Expressions while *215HVHV7436HV* and *2H15437VVVHH* are not. An illegal Polish Expression can't be converted to a binary tree.

It is obvious that for a given slicing tree, we have a unique Polish Expression. Inversely, we can easily conduct the slicing tree according to its Polish Expression. So we conclude that there is a one-to-one correspondence between the slicing tree and Polish Expression. However, there may be more than one slicing tree corresponding to one floorplan, as shown in Figure 54.28. Thus there exists redundancy. To compact the solution space, we define the skewed slicing tree to be a slicing tree without a node that has the same operator as its right son. Correspondingly, a normalized Polish expression is defined:

A Polish expression $b_1 b_2 \ldots b_{2n-1}$ is called to be normalized iff there is no consecutive "V" or "H" in the sequence.

In Figure 54.28, (b) is a normalized Polish expression while (c) is not. There is a one-to-one correspondence between the set of normalized Polish expressions and the set of slicing structures.

To perturb the blocks to get new solutions, we define three types of perturbations on polish expressions:

M1: Swap two adjacent operands. For example, *12V3H → 13V2H*;

M2: Complement some chain of operators. For example, *12V3H→ 12V3V* ;

M3: Swap two adjacent operand and operator. For example, *12V3H→ 123VH*;

The operation of M3 may result in a non-normalized Polish Expression. So a check is necessary after an operation of M3 to guarantee that a normalized expression is generated.

Figure 54.29 illustrates two sequences of polish expression transformations as well as their corresponding floorplans. We can see that a slight modification on the Polish Expression can result in a completely different floorplan topology. This is just what we want.

Finally, we analyze the performance of slicing tree.

1. Given a floorplan of n blocks, the slicing tree contains n leaves and $n-1$ internal nodes.
2. The number of possible configurations for the tree is $O(n! \cdot 2^{3n-3} \cdot n^{1.5})$.
3. The storage of a Polish Expression needs a $3n$-bit stream plus a permutation of n blocks.
4. It will take linear time to transform a floorplan into a slicing tree, and vice versa.

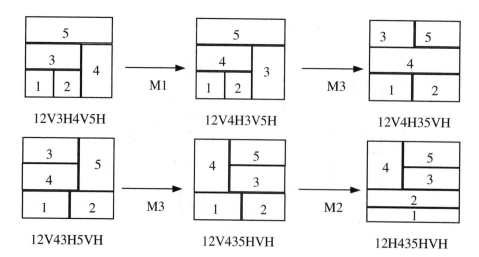

FIGURE 54.29   Slicing structure moves.

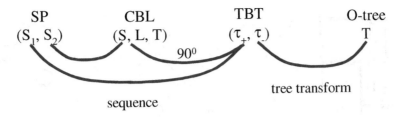

FIGURE 54.30   Summary of the relationships between floorplan representations.

54.4 Relationships of the Representations

We summarize the relationships between the representations in this section. A mosaic floorplan example and a general floorplan example are also discussed in detail to show the relationships.

54.4.1 Summary of the Relationships

According to the definitions of the representations, we have four relationships between the sequence-pair (SP), the corner block list (CBL), the twin binary tree (TBT) and the O-Tree representations.

For further discussions, we first define the 90° rotation of a floorplan as follows:

Definition 90° rotation of a floorplan: We use F^{90} to denote the floorplan obtained by rotating floorplan F, 90 degrees counterclockwise.

The four relationships are as follows and summarized in Figure. 54.30.

1. Given a mosaic floorplan and its corresponding corner block list $CB = (S, L, T)$, there exists a sequence pair $SP = (S_1, S_2)$ corresponding to the same floorplan such that the second sequence of SP is same as the sequence S of the corner block list.
2. Given a mosaic floorplan and its corresponding twin binary trees (τ_+, τ_-), there exists a sequence pair $SP = (S_1, S_2)$ corresponding to the same floorplan such that the first sequence of SP is the same as the node sequence of in-order traversal in tree τ_+.
3. Given a mosaic floorplan and its corresponding twin binary trees $TBT(\tau_+, \tau_-)$, there exists an O-tree corresponding to the same floorplan such that τ_- is identical to the O-tree after the tree conversion from a binary tree to an ordered tree.
4. Given a mosaic floorplan F, and its corresponding twin binary trees $TBT(\tau_+, \tau_-)$, the node sequence of in-order traversal in tree τ_+ is identical to the sequence S in the corner block list $CB = (S, L, T)$ of the floorplan F^{90}.

54.4.2 A Mosaic Floorplan Example

Figure 54.31 describes an example of a mosaic floorplan. We illustrate the four representations. The twin binary trees are marked with circles and crosses as shown in Figure 54.31. Two *SPs*, *SP1* and *SP2*, out of many possible choices, are described in the figure.

The in-order traversal of the τ_+ in the twin binary trees representation produces the sequence $\pi(\tau_+) = ABCDFE$, which is same as the first sequence of *SP1* and *SP2*. In Figure 54.31, an O-tree representation is also given. Its binary tree representation is identical to the τ_- of the twin binary trees after tree conversion. The corner block list representation is next to the O-tree representation in Figure 54.31. Its sequence $S = FADEBC$ is same as the second sequence of *SP1*.

Figure 54.32 shows the 90° rotation of the mosaic floorplan in Figure 54.31. The CB (S,L,T) representation of F^{90} has $S = ABCDFE$ (Figure 54.32.), which is identical to $\pi(\tau)$, the order of the twin binary trees representation (τ_+, τ_-) of the floorplan in Figure 54.31.

54.4.3 A General Floorplan Example

Figure 54.33 illustrates a general floorplan. Only the O-tree and the sequence pair are capable of representing this general floorplan. The O-tree and *SP* representations are shown in the figure.

54.5 Rectilinear Shape Handling

Shape handling makes feasible the treating of blocks with arbitrary shapes. It makes sense especially in deep sub-micron technology, where, with the number of routing layers increasing, wires are placed in layers other than the device layer, and the placement

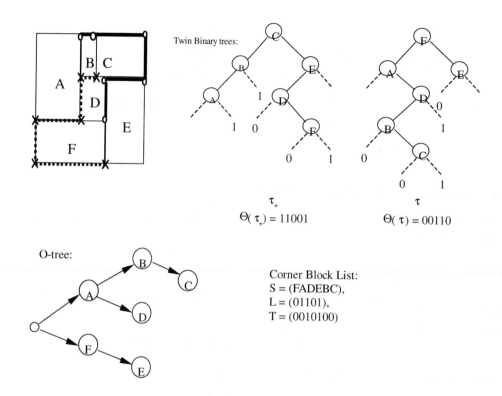

FIGURE 54.31   A mosaic floorplan and its different representations.

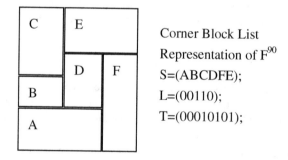

FIGURE 54.32   90° rotation of the floorplan in Figure 54.31.

becomes more like a packing problem: a set of macro blocks are packed together without overlaps, with some objective functions minimized.

The basic idea to handle the arbitrary-shaped blocks is to partition them into rectangles, and then the representations and the optimizing strategies described above can be used. Figure 54.34a and b demonstrate two schemes of partitioning, the horizontal one and the vertical one. However, if the generated rectangle blocks were treated as individual ones, it would be quite impossible to recover the original blocks. The generated rectangle blocks may be shifted (Figure 54.34c) or separated (Figure 54.34d). So extra constraints have to be imposed on the generated rectangle blocks in order to keep the original blocks' shapes.

One proposed method of adding extra constraints is illustrated with Figure 54.35b for horizontal and (c) for vertical. The vertexes in the constraint graph correspond to the generated rectangle blocks, while the direct edges denote the relative positions of the blocks. For example, an edge from "a" to "b" with a weight of 1 means that block "b" lies to the right of "a," the distance

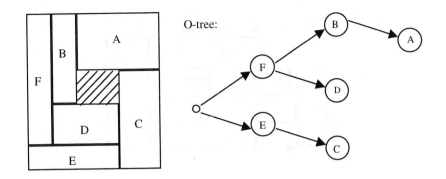

Sequence Pair: SP1= (Γ_+ , Γ_-) = (FBADEC, EFDBCA),
     Also, SP2=(Γ_+, Γ_-) = (FBADEC, EDFBCA)

FIGURE 54.33   A general floorplan with its O-tree and SP representation.

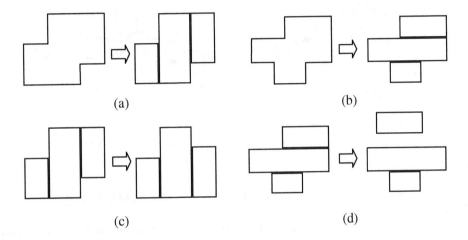

(a)   (b)

(c)   (d)

FIGURE 54.34   Shape handling.

between their left boundaries is 1, which is exactly the width of "a." Because the relative-position constraints are always represented with the format such as "larger than or equal to" during optimization, inversely directed edges, such as the edge from "b" to "a" with a negative weight, helps to determine the position relationships solely.

54.6 Conclusions

Floorplan representation is an important issue for floorplan optimization. We classify the representations into graph based and placement based methods based on the different strategies used in deriving the representations. The graph-based methods are based on the constraint graphs, which include the constraint graphs, the corner stitching, the twin binary tree and the O-tree. The placement-based methods are based on the local topology relations in deriving the floorplans, which include the sequence pair, the bounded sliceline grid, the corner block list and the slicing tree.

The floorplans can also be classified into different categories: general, mosaic and slicing floorplans. Slicing floorplan is a subset of mosaic floorplan, which is again a subset of the general floorplan. We have different representations for different types of floorplans. The constraint graphs, the corner stitching, the sequence pair, the bounded sliceline grid and the O-trees are for general floorplans. The twin binary tree and the corner block list are for mosaic floorplans. The slicing tree is for slicing floorplans. Different representations have different solution spaces and time complexities to derive a floorplan.

Acknowledgments

The authors are grateful to Mr. Simon Chu of Sun Microsystems for his thorough review of the chapter.

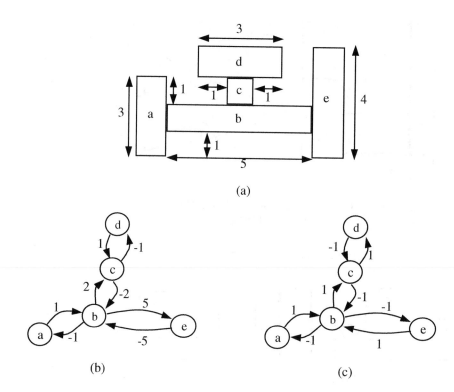

FIGURE 54.35 An example of adding extra constraints.

References

1. W. Dai and E.S. Kuh, "Hierarchical Floor Planning for Building Block Layout," in *IEEE Int. Conf. on Computer-Aided Design*, 1986, pp. 454–457.
2. B. Yao, H. Chen, C.K. Cheng, and R. Graham, "Revisiting Floorplan Representations," in *Proc. International Symposium on Physical Design*, 2001, pp. 138–143.
3. X. Tang and M. Wong, "FAST-SP A Fast Algorithm for Block Placement Based on Sequence Pair," in *Proc. of Asia and South Pacific Design Automation Conference*, pp. 521–526, Yokohama, Japan, Jan. 2001.
4. S. Nakatake, K. Fujiyoshi, H. Murata, and Y. Kajitani, "Module Packing Based on the BSG-Structure and IC Layout Applications," *IEEE Trans. on Computer-Aided Design of Integrated Circuits and Systems*, vol. 17, no. 6, June 1998, pp. 519–530.
5. X. Hong, G. Huang, Y. Cai, J. Gu, S. Dong, C.K. Cheng, and J. Gu, "Conner Block List: An Effective and Efficient Topological Representation of Non-Slicing Floorplan," in *Proceeding of ICCAD-2000*, 2000, pp. 8–12.
6. M. Abe, "Covering the Square by Squares without Overlapping (in Japanese)," *J. Japan Math. Phys.*, vol. 4, no. 4, 1930, pp. 359–366.
7. K. Sakanushi, Y. Kajitani, and D.P. Mehta, "The Quarter-State-Sequence Floorplan Representation," *IEEE Trans. on Circuits and Systems—I*, vol. 50, no. 3, 2003, pp. 376–385.
8. F.Y. Young, C.N. Chu, and Z.C. Shen, "Twin Binary Sequences: A Non-Redundant Representation for General Non-Slicing Floorplan," in *Proc International Symposium on Physical Design*, 2002, pp. 196–201.
9. K. Kozminsky and E. Kinnen, "Rectangular Duals of Planar Graph," *Networks*, vol. 15, 1985, pp. 145–157.
10. R.L. Brooks, C.A.B. Smith, A.H. Stone, and W.T. Tutte, "The Dissection of Rectangle into Squares," *Duke Math, Journal*, vol. 7, 1940, pp. 312–340.
11. J.M. Lin and Y.W. Chang, "TCG: A Transitive Closure Graph-Based Representation for Non-Slicing Floorplans," in *Proc: DAC*, 2001, pp. 764–76.
12. J. Ousterhout, "Corner Stitching: A Data-Structuring Technique for VLSI Layout Tools," in *IEEE Transactions on Computer-Aided Design*, CAD-3, January 1984, pp. 87–100.
13. P. Guo, C.K. Cheng, and T. Yoshimura, "An O-Tree Representation of Non-Slicing Floorplan and Its Applications," in *Proc: 36th DAC*, 1999, pp. 268–273.
14. Y. Pang, C.K. Cheng, K. Lampaert, and W. Xie, "Rectilinear Block Packing Using O-Tree Representation," in *Proc. ISPD*, 2001, pp. 156–161.

15. Y. Chang, G. Wu, and S. Wu, "B*-Trees: A New Representation for Non-Slicing Floorplans," in *Proc. 37th DAC*, 2000, pp. 458–463.

16. H. Murata, K. Fujiyoushi, S. Nakatake, and Y. Kajitani, "Rectangle-Packing-Based Module Placement," in *Proc. of International Conference on Computer Aided Design*, 1995, pp. 472–479.

17. J. Xu, P. Guo, and C.K. Cheng, "Sequence-Pair Approach for Rectilinear Module Placement," in *IEEE Trans. on CAD*, April 1999, pp. 484–493.

18. S. Zhou, S. Dong, X. Hong, Y. Cai, J. Gu, and C.K. Cheng, "ECBL: An Extended Corner Block List with O(n) Complexity and Solution Space Including Optimum Placement," in *Int: Symp. on Physical Design*, 2001, pp. 150–155.

19. Y. Ma, X. Hong, S. Dong, Y. Cai, C.K. Cheng, and J. Gu, "Floorplanning with Abutment Constraints and L-Shaped /T-Shaped Blocks Based on Corner Block List," in *ACM/IEEE Design Automation Conf.*, 2001, pp. 770–775.

20. R.H.J.M. Otten, "Automatic Floorplan Design," in *Proc. ACM/IEEE Design Automation Conf.*, 1982, pp. 261–267.

21. D.F. Wong and C.L. Liu, "A New Algorithm for Floorplan Design," in *ACM/IEEE Design Automation Conf.*, 1986, pp. 101–107.

22. A.A. Szepieniec and R.H.J.M. Otten, "The Genealogical Approach to the Layout Problem," in *Proc. of the 17th Design Automation Conference*, 1980, pp. 535–542.

23. D.E. Knuth, *The Art of Computer Programming*, Addison-Wesley Press, 1997.

24. R.H.J.M. Otten, "What is a Floorplan," in *Proc. of International Symposium on Physical Design*, 2000, pp. 212–217.

25. J. Grason, *A Dual Linear Graph Representation for Space-filling Location Problems of the Floor-planning Type*, MIT Press, Cambridge (Mass), USA, 1970.

26. K. Sakanushi and Y. Kajitani, "The Quarter-State Sequence (Q-Sequence) to Represent the Floorplan and Applications to Layout Optimization," in *Proc. IEEE APCCAS*, 2000, pp. 829–832.

27. J. Xu, P. Guo, and C.K. Cheng, "Cluster Refinement for Block Placement," in *Proc. ACM/IEEE Design Automation Conf.*, 1997, pp. 762–765.

28. J. Xu, P. Guo, and C.K. Cheng, "Empirical Study of Block Placement by Cluster Refinement," *VLSI Design*, vol. 10, no. 1, pp. 71–86, 1999.

29. C. Zhuang, K. Sakanushi, J. Liyan, and Y. Kajitani, "An Enhanced Q-Sequence Augmented with Empty-Room-Insertion and Parenthesis Trees," in *Design Automation and Test in Europe Conference and Exhibition*, 2002, pp. 61–68.

55

Computer Graphics*

55.1	Introduction.. 857
	Hardware and Pipeline
55.2	Basic Applications 858
	Meshes • CAD/CAM Drawings • Fonts • Bitmaps • Texture Mapping
55.3	Data Structures 862
	Vertices, Edges, and Faces • Vertex, Normal, and Face Lists • Winged Edge • Tiled, Multidimensional Array • Linear Interpolation and Bezier Curves
55.4	Applications of Previously Discussed Structures........................ 867
	Hidden Surface Removal: An Application of the BSP Tree • Proximity and Collision: Other Applications of the BSP Tree • More With Trees: CSG Modeling
	References... 869

Dale McMullin†
Colorado School of Mines

Alyn Rockwood†
Colorado School of Mines

55.1 Introduction

Like all major applications within the vast arenas of computer science and technology, the computer graphics industry depends on the efficient synergy of hardware and software to deliver to the growing demands of computer users. From computer gaming to engineering to medicine, computer graphics applications are pervasive. As hardware capabilities grow, the potential for new feasible uses for computer graphics emerge.

In recent years, the exponential growth of chipset speeds and memory capacities in personal computers have made commonplace the applications that were once only available to individuals and companies with specialized graphical needs. One excellent example is flight simulators. As recently as twenty years ago, an aviation company would have purchased a flight simulator, capable of rendering one thousand shaded polygons per second, for ten million dollars. Even with a room full of processing hardware and rendering equipment, it would be primitive by today's graphics standards. With today's graphics cards and software, one renders tens of millions of shaded polygons every second for two hundred dollars.

As the needs for graphics applications grow and change, research takes the industry in many different directions. However, even though the applications may evolve, what happens under the scenes is much more static; the way graphics primitives are represented, or stored in computer memory, have stayed relatively constant. This can be mainly attributed to the continued use of many standard, stable data structures, algorithms, and models. As this chapter will illustrate, data and algorithmic standards familiar to computer science lend themselves quite well to turning the mathematics and geometries of computer graphics into impressive images.

55.1.1 Hardware and Pipeline

Graphics hardware plays an important role in nearly all applications of computer graphics. The ability for systems to map 3–D vertices to 2–D locations on a monitor screen is critical. Once the object or "model" is interpreted by the CPU, the hardware draws and shades the object according to a user's viewpoint. The "pipeline" [1], or order in which the computer turns mathematical expressions into a graphics scene, governs this process. Several complex sub-processes define the pipeline. Figure 55.1 illustrates a typical example.

The Model View is the point where the models are created, constructed as a combination of meshes and mapped textures. The Projection point is where the models are transformed (scaled, rotated, and moved) through a series of affine transformations to

---

* This chapter has been reprinted from first edition of this Handbook, without any content updates.
† We have used this author's affiliation from the first edition of this handbook, but note that this may have changed since then.

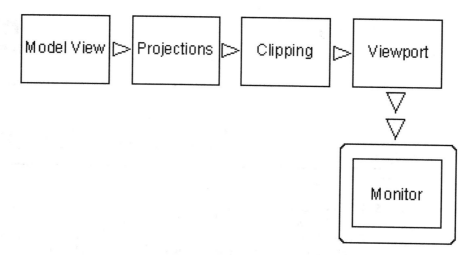

FIGURE 55.1   Graphics pipeline.

their final position. Clipping involves invoking algorithms to determine perspective, and what objects are visible in the Viewport. The Viewport readies the scene for display on the computer's monitor. The final scene is "rasterized" to the monitor [1].

Standard data structures and algorithms apply to all steps in the pipeline process. Since the speed of most rendering hardware (and hence the pipeline) is directly dependent on the number of models to be drawn, it becomes important to utilize structures that are as efficient as possible. In fact, graphics hardware is often designed and engineered to cater to a specific, standard form of data representation.

55.2 Basic Applications

55.2.1 Meshes

In both 2-D and 3-D applications, objects are modeled with polygons and polygon meshes. The basic elements of a polygon mesh include the vertex, the edge, and the face. An edge is composed of the line segment between two vertices, and a face is defined by the closed polygon of three or more edges. A mesh is formed when two or more faces are connected by shared vertices and edges. A typical polygon (triangle) mesh is shown in Figure 55.2.

Specific examples of meshes include "triangle meshes, polygon meshes, and polygon meshes with holes." [2] However, the triangle is currently the most popular face geometry in standard meshes. This is mainly because of its simplicity, direct application to trigonometric calculations, and ease of mesh storage (as we will see).

55.2.2 CAD/CAM Drawings

In Computer Aided Design (CAD), Computer Aided Manufacturing (CAM), and other channels of engineering design, we see the same basic elements. During the design process of everything from automobiles to telephones, applications are used to "see"

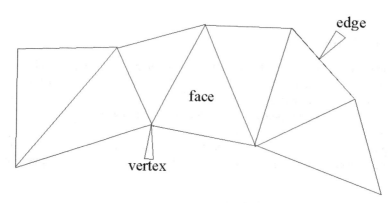

FIGURE 55.2   A triangle mesh.

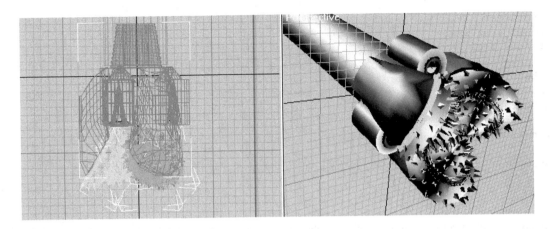

FIGURE 55.3 Models of a drill bit.

$$Ma\beta\gamma\triangle\,\delta\,\phi\,T$$

FIGURE 55.4 Fonts.

the parts before they are manufactured. The physical interactions and assemblies of parts can be tested before any steel is milled or plastic is poured.

Figure 55.3 illustrates a wire-frame (and fully rendered) model of a drill bit used in the oil and gas industry. Every modeled element of this design requires a definition of vertices and edges. These vertices define single-line edges, polyline edges, and the curved edges. The proximity of vertices to edges, edges to edges, and vertices to vertices may be tested for tolerances and potential problems. The software makes the stored model visible to the user within the current coordinate system and viewpoint. It is then up to the user to interface the visual medium with his or her engineering (or design) training to create a useful design.

Even in wire frame models, we see the application of more sophisticated graphics practices. Note how many of the contours of the drill bit parts are not connected line segments, but actual smooth curves. This is an example of where vertices have been used not to define edges, but the "control points" of a Bezier Curve. How curves like the B-Spline and Bezier utilize vertices in their structures is discussed later in this chapter. For now it is sufficient to mention that these types of curves are present in nearly all corners of the graphics industry.

55.2.3 Fonts

Fonts are another example of where the vertex is used, in two dimensions, as the basis of a data structure in order to define the appearance of each character and number in various languages. Several examples of interesting Greek characters, each defined by a different font definition file, and at different sizes, are shown in Figure 55.4.

The typical Postscript font is defined by a number of points, read in as control points for each curve (or line) in the font character. The "Postscript" driver software is then used to interpret the control points and render the characters with Bezier curves (see Section 55.3.5).

Perhaps the most important concept is that each pixel of the font character does not have to be stored, only a small number of vertex definitions. As the control points are transformed through operations like italicizing, the curves remain aligned in the proper proportion to the original character. Note the "M" character in the above figure has been italicized. The "Postscript" font representation serves as a very effective example of how a single standardized data structure of specific size can be utilized to define a potentially infinite number of characters.

55.2.4 Bitmaps

Bitmaps are a cornerstone to computer graphics. In fact, the name "bitmap" has become a commodity to most computer users in the way that it describes a computer generated picture. A bitmap is a very simple, yet versatile, way to store a graphic image as a binary file. The file header structure of a bitmap [3] in itself is a standard data structure, as the following illustrates:

```
struct _BITMAP
{
Uint8 bmType;
```

```
Uint8 bmBitsPerPixel;
Uint8 bmBytesPerPixel;
Uint8 bmReserved;

Uint32 bmColorKey;

Uint32 bmWidth;
Uint32 bmHeight;
Uint32 bmPitch;

void* bmBits;
};
```

The bitmap is initially defined by a number of parameters that define the type, width, height, etc. of the graphic image. These properties are stored in the header of the file as shown. The actual "pixels" required to fill that space is then defined in the "bmBits" pointer. As a rule, the total number of pixels in the image memory will equal the width times the height divided by the "bits per pixel" property. The "bytes per pixel" property determines how the "bits per pixel" are divided among the individual color components of each pixel. For instance, a bitmap with RGB color map is commonly defined by twenty four (24) bits per pixel and three (3) bytes per pixel. Each of the three bytes for each pixel use 8 of the twenty four bits to define red, green, and blue values respectively.

Early bitmaps, when color monitors were first being used, were defined with 4-bit color. In other words, each color was defined by a 4-bit (or half-byte) word. Since a half-byte has a maximum value of 24 or 16, 4-bit bitmaps were capable of supporting sixteen (16) different colors. The required disk space to store a 100 × 100 4-bit bitmap would then be 10,000* .5 bytes or 5,000 Bytes (5kB).

In the past fifteen years, the need for more realistic graphics has driven the increase in the memory used for color. Today 24-bit (also called true color) and 32-bit color, which both represent 16.7 million colors (32-bit adds an extra byte for the alpha channel), are the most commonly supported formats in today's hardware. The alpha channel refers to a single byte used to store the transparency of the color. Bitmap files in these formats now require 3 and 4 bytes per pixel. Additionally, current monitor resolutions require over 1,000 pixels in width or height, and thus graphic files are growing even larger. Figure 55.5 is a photograph that has been scanned in and stored as a bitmap file.

A typical 5×7 photograph, at 100 dpi (dots per inch), and 24-bit color, would require 500×700 × 3 or 1.05 megabytes (MB) to store in a conventional bitmap format. Because of the increased size requirements in bitmaps, compression algorithms have become common-place. File formats such as JPG (jpeg) and PNG (ping) are examples of widely used formats. However, there is a tradeoff. When a compression algorithm is applied to a bitmap, a degree of image quality is inevitably lost. Consequently, in applications like publishing and graphic art, uncompressed bitmaps are still required where high image quality is expected.

FIGURE 55.5   Bitmap file.

55.2.5 Texture Mapping

The processes of texture and surface mapping represent an application where bitmaps and polygonal meshes are combined to create more realistic models. Texture mapping has become a cornerstone of graphics applications in recent years because of its versatility, ease of implementation, and speed in graphical environments with a high number of objects and polygons. In fact, today's graphics hardware ships with the tools necessary to implement the texture and surface mapping processes on the local chipset, and the data structures used in rendering environments are largely standardized.

Prior to texture mapping, 3-D polygonal meshes were processed through shading models such as Gouraud and Phong to provide realism. Each shading model provides the means for providing color, shadows, and even reflection (glare) to individual components of the model. This was accomplished through a mathematical model of how light behaves within specular, diffuse, reflective, and refractive contexts. The general shading equation for light intensity, I, based on the individual ambient, diffuse, and specular components is shown below.

$$I = I_a p_a + (I_d p_d \times \text{lambert}) + (I_s p_s \times \text{phong})$$

where

$$\text{lambert} = \max\left(0, \frac{s \cdot m}{|s||m|}\right)$$
$$\text{phong} = \max\left(0, \frac{h \cdot m}{|h||m|}\right)$$
$$p_a = \text{ambient coefficient} 0 \le p_a \le 1$$
$$p_d = \text{diffuse coefficient} 0 \le p_d \le 1$$
$$p_s = \text{specular coefficient} 0 \le p_s \le 1$$

The primary difference between Gouraud and Phong is in that Phong provides the additional component for specular reflection, which gives objects a more realistic glare when desired. Also, because the Phong model requires more detail, the intensity values are calculated (interpolated) at each pixel rather than each polygon. The vectors m, s, and h represent the normal, reflected, and diffuse light from a given polygonal surface (or face). Linear interpolation, a widely used algorithm in graphics, is discussed in Section 55.3.5.

Although still used today, these shading models have limitations when more realistic results are desired. Because color and light reflection are modeled on individual model components, the model would have to be constructed as a composite of many different components, each with different color (or material) properties, in order to achieve a more realistic effect. This requires models to be much more complex, and increasingly difficult to construct. If changes in surface appearance or continuity were desired, they would have to be physically modeled in the mesh itself in order to be viewed.

Texture and surface mapping have provided a practical solution to these model complexity dilemmas. The mapping methods involve taking the input from a bitmap file, like those described previously, and "stretching" them over a polygonal mesh. The end result is a meshed object.which takes on the texture properties (or appearance) of the bitmap. The figure below illustrates a simple example of how the mesh and bitmap can be combined to create a more interesting object. Figure 55.6 shows how how Figure 55.5 has been "mapped" onto a sphere.

FIGURE 55.6 Combination of mesh and bitmap.

FIGURE 55.7   Texture mapping.

The object seems to take on a more realistic appearance even when modeled by a very simple polygonal mesh. This technology has made fast rendering of realistic environments much more feasible, especially in computer games.

Texture, or "bump," mapping utilizes a similar process as surface mapping, where a bitmap is stretched over a polygonal mesh. However, the pixels, commonly called "texels" [1], are used to alter how the light intensity interacts with the surface. Initially, the lighting model shown above would calculate an intensity, I, for a given surface. With texture mapping, individual grayscale values at each texel are used to alter the intensity vectors across each polygon in order to produce roughness effects.

Figure 55.7 illustrates a model of a sphere that has been rendered with the traditional Gouraud model, then Phong, and then finally rendered again with a texture map. This approach to improving model realism through mapping applies also to reflection, light intensity, and others.

55.3 Data Structures

55.3.1 Vertices, Edges, and Faces

As mentioned previously, vertices, edges, and faces form the most basic elements of all polygonal representations in computer graphics. The simplest point can be represented in Cartesian coordinates in two (2D) and three dimensions (3D) as (x, y) and (x, y, z) respectively (Figure 55.8).

As a simple data structure, each vertex may then be stored as a two or three-dimensional array. Edges may then be represented by two-dimensional arrays containing the indexes of two points. Further, a face may be dimensioned based on the number of desired sides per face, and contain the indices of those edges. At first, this approach may seem acceptable, and in basic applications it is common to model each vertex, edge, and face as a separate class. Relative relationships between classes are then governed by the software to build meshes. However, modeling objects in this manner does have disadvantages.

Firstly, important information becomes difficult to track, such as the normal at each vertex, adjacency of faces, and other properties required for blending and achieving realism. Furthermore, the number of intersecting edges at each vertex may vary throughout the model, and mesh sizes between objects may be unpredictable. The ability to manage this approach then becomes increasingly difficult, with the potential for unnecessary overhead and housekeeping of relationships within the model. It then becomes necessary to create higher level data structures that go beyond these basic elements, and provide the elements necessary for efficient graphics applications.

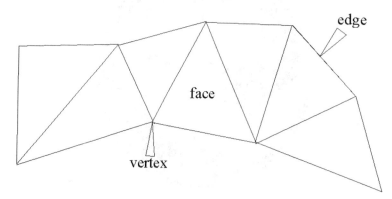

FIGURE 55.8   Vertices, edges, and faces.

55.3.2 Vertex, Normal, and Face Lists

In this storage method, list structures are used to store three, inter-dependent lists of data. The first list defines the vertexes contained in the scene as follows. Each vertex is assigned an index, and a coordinate location, or (x, y, z) point. The second list defines the normals for each vertex. Again, each normal is assigned a numbered index and a 3-D coordinate point. The final list serves to integrate the first two. Each face is identified by a numbered index, an array of vertex indexes, and an array of indexed normals for each vertex. Figure 55.9 illustrates typical examples of a similar application with vertex, face, and edge lists.

In the table, three specific lists are evident. The first list represents each vertex in the model as it is defined in 3D space. The "vertex" column defines the id, index, or label of the vertex. The x, y, and z coordinates are then defined in the adjacent columns.

In the second list, six faces are defined. Each face has a label or index similar to the vertex list. However, rather than specifying coordinates in space, the adjacent column stores the id's of the vertexes that enclose the face. Note that each face consists of four vertices, indicating that the each face will be defined by a quadrilateral.

The final list defines edges. Again, each edge definition contains a label or index column, followed by two adjacent vertex columns. The vertices of each edge define the start point and end point of each edge. In many applications of graphics, it is important to define the direction of an edge. In right handed coordinate systems, the direction of an edge will determine the direction of the normal vector which is orthogonal to the face surrounded by the edges.

55.3.3 Winged Edge

Although one of the oldest data structures relating to polygonal structures, the Winged Edge approach is very effective, and still widely used [4]. This structure is different from the wireframe model in that edges are the primary focal point of organization.

In the structure, each edge is stored in an indexed array, with its vertices, adjacent faces, previous, and successive edges. This allows the critical information for each edge to be stored in an array of eight integer indexes; it is both consistent and scalable between applications. The structure is

Figure 55.11 illustrates a typical layout for a winged edge.

An important aspect of the Winged Edge structure is the order in which entries are listed. The edge itself has a direction, from the start vertex to the end vertex. The other entries are then defined by their proximity to the edge. If the direction of the edge were reversed, the right and left faces would change accordingly, as would the previous and succeeding entries of both the left and right traverse.

There is a time/space trade-off in using this model. What is saved in storage space adds to the needed time to find previous and successive edges. See Chapter 18 for more details.

vertex	x	Y	Z	face	vertex number	edge	vertex begin	vertex end
0	0	0	0	0	4, 5, 6, 7	0	0	1
1	0	1	0	1	0, 1, 2, 3	1	1	2
2	1	1	0	2	3, 2, 6, 5	2	2	3
3	1	0	0	3	4, 7, 1, 0	3	3	0
4	0	0	1	4	2, 1, 7, 6	4	4	5
5	1	0	1	5	4, 0, 3, 5	5	5	6
6	1	1	1			6	6	7
7	0	1	1			7	7	4

FIGURE 55.9   Example of vertex, normal, and face lists.

Edge	Vertices		Faces		Left Traverse		Right Traverse	
Name	Start	End	Left	Right	Previous	Succeeding	Previous	Succeeding
a	X	Y	1	2	b	d	e	c

FIGURE 55.10   Winged edge table.

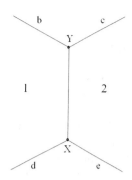

FIGURE 55.11 Winged edge.

55.3.4 Tiled, Multidimensional Array

In this section we will discuss a data structure that is important to most geometric implementations. In many ways, the tiled array behaves identically to matrix data structures. For instance, a p by q matrix is normally stored as a single dimension array. However, since the matrix has p-rows and q-columns, the array needs to be addressed by creating an "offset" of size q in order to traverse each row. The following example should illustrate this concept.

Consider a matrix with $p = 3$ rows and $q = 3$ columns:

is stored as:

$$A = [1, 3, 4, 5, 2, 7, 3, 9, 3]$$

This represents where a 3×3 matrix is stored as an array with $(3)(3) = 9$ entries. In order to find the entry at row$(i) = 3$ and column$(j) = 2$ we employ the following method on the array:

$$Entry = j + (i - 1)q$$

Or in this case, Entry $= A[2 + (3 - 1)(3)] = A[8] = 9$

We use a similar method for tiling a single bitmap into several smaller images. This is analogous to each number entry in the above array being replaced by a bitmap with n by n pixels.

"Utilizing cache hierarchy is a crucial task in designing algorithms for modern architectures." [1] In other words, tiling is a crucial step to ensuring multi-dimensional arrays are stored in an efficient, useful manner. Indexing mechanisms are used to locate data in each dimension. For example, a p by q array is stored in a one-dimensional array of length $p * q$, and indexed in row-column fashion as above.

When we wish to tile a p by q matrix into n by n tiles, the number of blocks in x is defined by q/n and the number of blocks in y is defined by p/n. Therefore, to find the "tile index" or the row and column of the tile for a value (x, y) we first calculate the tile location, or bitmap within the matrix of bitmaps. Then once the bitmap is located, the pixel location within that bitmap tile is found (sub-indexed). The entire two-step process can be simplified into a single equation that is executed once. Figure 55.12 illustrates this concept.

The final formula for locating x, y in a p by q array divided into n, n tiles is:

$$n^2 \left(\left(\frac{q}{n} \right) \left(\frac{y}{n} \right) + \frac{x}{n} \right) + (y \mod n)n + (x \mod n)$$

When dealing with matrices and combining multiple bitmaps and/or data into a Tile Multidimensional Array, performance and speed can both improve substantially.

55.3.5 Linear Interpolation and Bezier Curves

This section will introduce one of the most significant contributions to design and graphics: the interpolated curve structure.

Linear interpolation refers to the parameterization of a straight line as a function of t, or:

$$L(t) = (1 - t)a + tb$$

where a, b are points in space. This equation represents both an affine invariant and barycentric combination of the points a and b. Affine invariance means that the point $L(t)$ will always be collinear with the straight line through the point set $\{a, b\}$, regardless of the positioning of a and b. Describing this set as barycentric simply means that for t values between 0 and 1, $L(t)$ will always

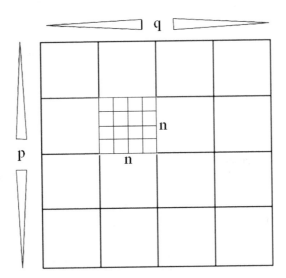

FIGURE 55.12 Tiled array.

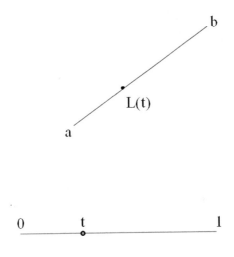

FIGURE 55.13 Linear interpolation.

occur between a and b. Another accurate description is that the equation $L(t)$ is a linear "mapping" of t into some arbitrary region of space. Figure 55.13 illustrates this concept.

Linear interpolation is an extremely powerful concept. It is not only the foundation of many curve approximation algorithms, but the method is also used in many non-geometric applications as well, including calculating individual pixel intensities in the Phong shading method mentioned previously.

The basic Bezier curve derivation is based on an expanded form of linear interpolation. This concept uses the parameterization of t to represent two connected lines. The curve is then derived by performing a linear interpolation between the points interpolated on each line; a sort of linear interpolation in n parts, where n is the number of control points. The following example should illustrate:

Given are three points in space, $\{a, b, c\}$. These three points form the two line segments ab and bc (Figure 55.14).

During the first "iteration" of the Bezier curve derivation, linear interpolation is performed on each line segment for a given t value. These points are then connected by an additional line segment. The resulting equations (illustrated in Figure 55.15) are:

$$x = (1 - t)a + tb$$
$$y = (1 - t)b + tc$$

The linear interpolation is performed one last time, with the same t value between the new points $\{x, y\}$ (Figure 55.16):

$$z = (1 - t)x + ty$$

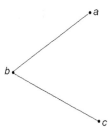

FIGURE 55.14   Expanded form of linear interpolation.

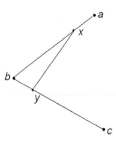

FIGURE 55.15   First iteration of Bezier curve derivation.

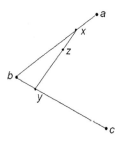

FIGURE 55.16   Result of three linear interpolations.

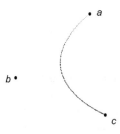

FIGURE 55.17   Smooth curve.

This final point z, after three linear interpolations, is on the curve. Following this 3-step process for several "stepped" values for t between 0 and 1 would result in a smooth curve of z-values from a to c, and is illustrated in Figure 55.17:

This is a quadratic Bezier curve, and is represented mathematically by a linear interpolation between each set of x and y points, which were also derived through linear interpolation, for every t. By substituting the equations for x and y into the basic form, we obtain:

$$z(t) = (1-t)[(1-t)a + tb] + t[(1-t)b + tc]$$

Simplified, we obtain a quadratic polynomial for a, b, and c as a function of the parameter t, or

$$z(t) = (1-t)^2 a + 2(1-t)tb + t^2 c]$$

The "string art" algorithm described previously is also referred to as the de Causteljau algorithm. Programmatically, the algorithm performs an n-step process for each value of t, where n is the number of "control points" in the curve. In this example, $\{a, b, c\}$ are the control points of the curve.

Because of their ease of implementation and versatility, Bezier curves are invaluable in CAD, CAM, and graphic design. Also, Bezier curves require only a small number of control points for an accurate definition, so the data structure is ideal for compact storage. As mentioned previously, the Bezier form is the foundation for the Postscript font definition structure.

However, the Bezier form has limitations. When several attached curves are required in a design component, it takes considerable effort, or pre-calculation, to ensure the connected curves are "fair" or continuous with each other. Merely joining the end points of the curves may not be enough, resulting in "kinks" and other undesirable effects. For this reason, the B-spline curve provides an alternative to the Bezier where continuity is important. In fact, B-spline curves have often been referred to as a "user interface" for the Bezier form.

The goal of this section was to illustrate the versatility of linear interpolation and basic iteration structures in graphics and design. If more information is desired, numerous texts are currently available which describe the properties, mathematics, and applications of Bezier and B-spline curves, including the references listed at the end of this chapter.

55.4 Applications of Previously Discussed Structures

55.4.1 Hidden Surface Removal: An Application of the BSP Tree

In addition to other algorithms, BSP (Binary Space Partitioning) trees are one example where a specific, widely used, data structure has direct application in computer graphics. Hidden surface removal is essential to realistic 3-D graphics, and a primary application of BSP trees.

Hidden surface removal is used anywhere 3-dimensional objects must be made visible from multiple, if not infinite view points within a graphics scene. Whether a scene is being viewed in a 3-D design application, or the latest sci-fi game, the problem is the same: objects furthest away from the user's viewpoint may be hidden by closer objects. Therefore, the algorithm used to determine visibility must effectively sort the objects prior to rendering. This has been a hot topic until recent years, with researchers finding new and creative ways to tackle the problem. The z-buffer is now undeniably the most widely used algorithm. Tree algorithms are, on the other hand, also widely used where time-based rendering (animation) is not an issue, especially where the object positions are static. BSP trees are discussed in greater detail in Chapter 21.

The BSP tree is an example of a "painter's algorithm." [2] The basic concept of this algorithm involves sorting the objects (polygons) from back to front "relative to [the] viewpoint" and then drawing them in order. The key to the BSP algorithm in hidden surface removal is in the pre-processing, and encoding of the objects into a data structure that is later used to determine visibility. In other words, the data structure does not change, only the way it is viewed.

For hidden surface removal, the BSP tree is built by passing a plane through each polygon in the scene. For each point p in front of the plane, $f(p) > 0$ and for each point behind the plane, $f(p) < 0$. The tree structure is created by applying this object to each polygon, and defining a "negative branch" and "positive branch" for each polygon relative to the current position in the tree. Also called the "half space" on each side of the plane, each vertex position in the tree is dictated by it position relative to the passing planes. One plane is treated as the root of the tree, and successive branches are defined from that root.

Because the relative position of vertices to each other is defined by the tree, regardless of position, the entire BSP tree can be pre-computed. Whether or not polygons in the tree are visible is then a function of their position in the viewer's plane. Figure 55.19 demonstrates a BSP tree for vertices A through E shown in Figure 55.18.

Note how each vertex can be located relative to at least two planes. For a viewpoint along h3, it is immediately apparent that vertex D is on the right, while A and E are on the left. Vertices C and B are located in the half space of h1 and h2, and are therefore of no interest. Vertices C and B will not be rendered. This relative method works for all positions in the BSP tree.

As mentioned, BSP trees are not a universal solution for hidden surface removal, mainly because of the pre-processing requirement. There is a major caveat; if the objects in the scene are moving, the pre-processing of the BSP tree is no longer valid as the polygons change relative position. The tree must be built every time the relative positions of objects within the tree change. Recalculating the tree at each time step is often too slow, especially in 3-D gaming, where several million polygons must be rendered every second.

Another problem is that the BSP tree works in hidden surface removal only when "no polygon crosses the plane defined by any other polygon." In other words, no object can be both behind and in front of another object for the BSP sorting algorithm to work. In gaming, it is common for objects to "collide" so this algorithm becomes even less desirable in these unpredictable conditions.

55.4.2 Proximity and Collision: Other Applications of the BSP Tree

Although BSP Tree structures are not as useful for determining the rendering order of moving polygons, they have other applications in graphics and gaming. For instance, trees are commonly used for collision detection, line-of sight, and other algorithms

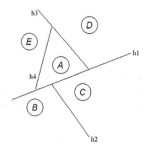

FIGURE 55.18   Objects in a plane.

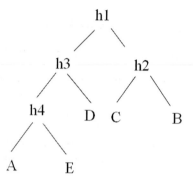

FIGURE 55.19   BSP tree for Figure 55.18.

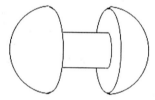

FIGURE 55.20   CSG object.

where the number of required calculations is lower. Determining between time-steps the relative positions of several (or several hundred) models, rather than several hundred million polygons, is much more feasible with today's hardware. Enhanced Tree structures are even used in many of today's more innovative artificial intelligence algorithms for gaming. These structures are used within the game space to quickly determine how an object sees, touches, and ultimately reacts with other objects.

55.4.3 More With Trees: CSG Modeling

Another widely used application of tree-based structures is application of Boolean operations to object construction. Constructive Solid Geometry, or CSG, modeling involves the construction of complex objects using only simple, primitive shapes. These shapes are added and subtracted to each other through "set operations," or "Boolean operations." The final objects are referred to as "compound," "Boolean," or "CSG" objects [5].

Figure 55.20 of a dumbbell illustrates a typical example of a CSG object.

To construct this object in a CSG environment, Boolean operations are applied to primitive objects. These operations are "intersection," "union," and "difference"[5]. Each step in the tree represents a boolean combination between two objects. The resulting object at each point in the tree is then combined again with another Boolean operation at the next step. This type of progression is continued until the finished object is created. Figure 55.21 illustrates the CSG tree used to construct this dumbbell using two spheres, a cylinder, and a half-plane.

A union operation is analogous to gluing two primitives together. An intersection operation results in only the portion (or volume) that both primitives occupy. Finally, a difference operation in effect removes the intersected section of one primitive from another. Armed with these three Boolean operations, modeling and displaying very complex shapes are possible. However,

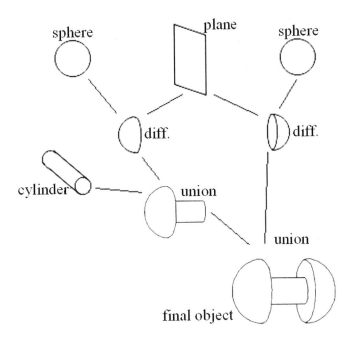

FIGURE 55.21   CSG tree corresponding to Figure 55.20.

attempting to cover the surface of a final shape with a continuous mesh is another problem entirely, and the subject of numerous texts and papers. Consequently, CSG objects are largely used for initial visualization of a complete model. This solid model is then sent to another application that converts the model to a polygonal mesh.

From a data structure and algorithm standpoint, CSG trees are quite useful. Firstly, and most obvious, the method utilizes the tree structure already mentioned throughout the text. Secondly, the final objects do not require individual mesh models to define them. Rather, each object is simply defined by its tree, where each node of the tree references primitives, such as the sphere, cone, cylinder, cube, and plane. With the basic models for the sphere, cone, etc. preprocessed and stored in memory, the overhead of CSG applications is kept to a minimum.

Many commercially available solid modeling and animation packages still provide CSG as a standard user-interface.

References

1. Hill, *Computer Graphics Using Open GL*, 1990, Macmillan Publishing Company.
2. Shirley, *Fundamentals of Computer Graphics*, 2002, School of Computing University of Utah.
3. Pappas and Murray, *Visual Basic Programming with the Windows API*, Chapter 15, 1998, Prentice Hall.
4. Arvo, *Graphics Gems II*, Chapter IV.6, 1991, Academic Press, Inc.
5. Rockwood, *Geometric Primitives*, 1997, CRC Press Inc.
6. Farin, *Curves and Surfaces for CAGD, A Practical Guide*, Fifth Edition, 2002, Academic Press.

56

Geographic Information Systems

56.1	Geographic Information Systems: What They Are All About	871
	Geometric Objects • Topological versus Metric Data • Geometric Operations • Geometric Data Structures • Applications of Geographic Information	
56.2	Space Filling Curves: Order in Many Dimensions	874
	Recursively Defined SFCs • Range Queries for SFC Data Structures • Are All SFCs Created Equal? • Many Curve Pieces for a Query Range • One Curve Piece for a Query Range	
56.3	Spatial Join	879
	External Algorithms • Advanced Issues	
56.4	Models, Toolboxes, and Systems for Geographic Information	882
	Standardized Data Models • Spatial Database Systems • Spatial Libraries	
	Acknowledgment	884
	References	884

Bernhard Seeger
University of Marburg

Peter Widmayer
ETH Zürich

56.1 Geographic Information Systems: What They Are All About

Geographic Information Systems (GISs) serve the purpose of maintaining, analyzing, and visualizing spatial data that represent geographic objects, such as mountains, lakes, houses, roads, and tunnels. For spatial data, geometric (spatial) attributes play a key role, representing, for example, points, lines, and regions in the plane or volumes in three-dimensional space. They model geographical features of the real world, such as geodesic measurement points, boundary lines between adjacent pieces of land (in a cadastral database), trajectories of mobile users, lakes or recreational park regions (in a tourist information system). In three dimensions, spatial data describe tunnels, underground pipe systems in cities, mountain ranges, or quarries. In addition, spatial data in a GIS possess nongeometric, so-called thematic attributes, such as the time when a geodesic measurement was taken, the name of the owner of a piece of land in a cadastral database, the usage history of a park.

This chapter aims to highlight some of the data structures and algorithms aspects of GISs that define challenging research problems, and some that show interesting solutions. More background information and deeper technical expositions can be found in books [1–10].

56.1.1 Geometric Objects

Our technical exposition will be limited to geometric objects with a vector representation. Here, a point is described by its coordinates in the Euclidean plane with a Cartesian coordinate system. We deliberately ignore the geographic reality that the earth is (almost) spherical, to keep things simple. A line segment is defined by its two end points. A polygonal line is a sequence of line segments with coinciding endpoints along the way. A (simple) polygonal region is described by its corner points, in clockwise (or counterclockwise) order around its interior. In contrast, in a raster representation of a region, each point in the region, discretized at some resolution, has an individual representation, just like a pixel in a raster image. Satellites take raster images at an amazing rate, and hence raster data abound in GISs, challenging current storage technology with terabytes of incoming data per day. Nevertheless, we are not concerned with raster images in this chapter, even though some of the techniques that we describe have implications for raster data [11]. The reason for our choice is the different flavor that operations with raster images have, as compared with vector data, requiring a chapter of their own.

56.1.2 Topological versus Metric Data

For some purposes, not even metric information is needed; it is sufficient to model the topology of a spatial dataset. *How many states share a border with Iowa?* is an example of a question of this topological type. In this chapter, however, we will not specifically study the implications that this limitation has. There is a risk of confusing the limitation to topological aspects only with the explicit representation of topology in the data structure. Here, the term explicit refers to the fact that a topological relationship need not be computed with substantial effort. As an example, assume that a partition of the plane into polygons is stored so that each polygon individually is a separate clockwise sequence of points around its interior. In this case, it is not easy to find the polygons that are neighbors of a given polygon, that is, share some boundary. If, however, the partition is stored so that each edge of the polygon explicitly references both adjacent polygons (just like the *doubly connected edge list* in computational geometry [12]), then a simple traversal around the given polygon will reveal its neighbors. It will always be an important design decision for a data structure which representation to pick.

56.1.3 Geometric Operations

Given this range of applications and geometric objects, it is no surprise that a GIS is expected to support a large variety of operations. We will discuss a few of them now, and then proceed to explain in detail how to perform two fundamental types of operations in the remainder of the chapter, spatial searching and spatial join. Spatial searching refers to rather elementary geometric operations without which no GIS can function. Here are a few examples, always working on a collection of geometric objects, such as points, lines, polygonal lines, or polygons. A *nearest neighbor query* for a query point asks for an object in the collection that is closest to the query point, among all objects in the collection. A *distance query* for a query point and a certain distance asks for all objects in the collection that are within the given distance from the query point. A *range query* (or window query) for a query range asks for all objects in the collection that intersect the given orthogonal window. A *ray shooting query* for a point and a direction asks for the object in the collection that is hit first if the ray starts at the given point and shoots in the given direction. A *point-in-polygon query* for a query point asks for all polygons in the collection in which the query point lies. These five types of queries are illustrations only; many more query types are just as relevant. For a more extensive discussion on geometric operations, see the chapters on geometric data structures in this Handbook. In particular, it is well understood that great care must be taken in geometric computations to avoid numeric problems, because already tiny numerical inaccuracies can have catastrophic effects on computational results. Practically all books on geometric computation devote some attention to this problem [12–14], and geometric software libraries such as CGAL [15] take care of the problem by offering exact geometric computation.

56.1.4 Geometric Data Structures

Naturally, we can only hope to respond to queries of this nature quickly, if we devise and make use of appropriate data structures. An extra complication arises here due to the fact that GISs maintain data sets too large for internal memory. So far, GIS have also not been designed for large distributed computer networks that would allow to exploit the internal memory of many nodes as it is known from the recently emerged Big Data technologies [16]. Data maintenance and analysis operations on a central system can therefore be efficient only if they take external memory properties into account, as discussed also in other chapters in this Handbook. We will limit ourselves here to external storage media with direct access to storage blocks, such as disks and flash memory. A block access to a random block on disk takes time to move the read–write–head to the proper position (the latency), and then to read or write the data in the block (the transfer). With today's disks, where block sizes are on the order of several kilobytes, latency is a few milliseconds, and transfer time is less by a factor of about 100. Therefore, it pays to read a number of blocks in consecution, because they require the head movement only once, and in this way amortize its cost over more than one block. We will discuss in detail how to make use of this cost savings possibility.

All of the above geometric operations (see Section 56.1.3) on an external memory geometric data structure follow the general *filter-refinement* pattern [17] that first, all relevant blocks are read from disk. This step is a first (potentially rough) *filter* that makes a superset of the relevant set of geometric objects available for processing in main memory. In a second step, a refinement *identifies* the exact set of relevant objects. Even though complicated geometric operators can make this refinement step quite time-consuming, in this chapter we limit our considerations to the filter step. Because queries are the dominant operations in GISs by far, we do not explicitly discuss updates (see the chapters on external memory spatial data structures in this Handbook for more information).

56.1.5 Applications of Geographic Information

Before we go into technical detail, let us mention a few of the applications that make GISs a challenging research area up until today, with more fascinating problems to expect than what we can solve.

56.1.5.1 Map Overlay

Maps are the most well-known visualizations of geographical data. In its simplest form, a map is a partition of the plane into simple polygons. Each polygon may represent, for instance, an area with a specific thematic attribute value. For the attribute *land use*, polygons can stand for *forest, savanna, lake* areas in a simplistic example, whereas for the attribute *state*, polygons represent *Arizona, New Mexico, Texas*. In a GIS, each separable aspect of the data (such as the planar polygonal partitions just mentioned) is said to define a *layer*. This makes it easy to think about certain analysis and viewing operations, by just super-imposing (overlaying) layers or, more generally, by applying Boolean operations on sets of layers. In our example, an overlay of a *land use* map with a *state* map defines a new map, where each new polygon is (some part of) the intersection of two given polygons, one from each layer. In *map overlay* in general, a Boolean combination of all involved thematic attributes together defines polygons of the resulting map, and one resulting attribute value in our example is the *savannas* of *Texas*. Map overlay has been studied in many different contexts, ranging from the special case of convex polygons in the partition and an internal memory plane-sweep computation [18] to the general case that we will describe in the context of spatial join processing later in this chapter.

56.1.5.2 Map Labeling

Map visualization is an entire field of its own (traditionally called cartography), with the general task to layout a map in such a way that it provides just the information that is desired, no more and no less; one might simply say, the map *looks right*. What that means in general is hard to say. For maps with texts that label cities, rivers, and the like, *looking right* implies that the labels are in proper position and size, that they do not run into each other or into important geometric features, and that it is obvious to which geometric object a label refers. Many simply stated problems in map labeling turn out to be NP-hard to solve exactly, and as a whole, map labeling is an active research area with a variety of unresolved questions (see Reference 19 for a tutorial introduction). In particular, the boundary-labeling problem [20] has attracted research attention assuming that labels are placed around an iso-oriented rectangle that contains a set of point objects. Dynamic map labeling [21] is another recently addressed problem of maps that are generated in an ad hoc manner using a set of operations such as zooming and panning. The underlying goal is to preserve a consistent layout of the labels among source and target maps.

56.1.5.3 Cartographic Generalization

If cartographers believe that automatically labeled maps will never look really good, they believe even more that another aspect that plays a role in map visualization will always need human intervention, namely map generalization. Generalization of a map is the process of reducing the complexity and contents of a map by discarding less important information and retaining the more essential characteristics. This is most prominently used in producing a map at a low resolution, given a map at a high resolution. Generalization ensures that the reader of the produced low resolution map is not overwhelmed with all the details from the high resolution map, displayed in small size in a densely filled area. Generalization is viewed to belong to the art of map making, with a whole body of rules of its own that can guide the artist [22,23]. The collection [24] provides a summary of the various flavors of generalization. Nevertheless, computational solutions of some subproblem help a lot, such as the simplification of a high resolution polygonal line to a polygonal line with fewer corner points that does not deviate too much from the given line. For line simplification, old algorithmic ideas [25] have seen efficient implementations [26]. A survey and an extensive experimental comparison of various algorithms is given in Reference 27 in terms of positional accuracy and processing time. Interesting extensions of the classical problem exist with the goal of preserving topological consistency [24,28].

Maps on demand, with a selected viewing window, to be shown on a screen with given resolution, imply the choice of a corresponding scale and therefore need the support of data structures that allow the retrieval up to a desired degree of detail [29]. More recently, van Oosterom and Meijers [30] have proposed a vario-scale structure that supports a continuous generalization while preserving consistency [110]. Apart from the simplest aspects, automatic map generalization, access support and labeling [21] are still open problems.

56.1.5.4 Road Maps

Maps have been used for ages to plan trips. Hence, we want networks of roads, railways, and the like to be represented in a GIS, in addition to the data described earlier. This fits naturally with the geometric objects that are present in a GIS in any case, such as polygonal lines. A point where polygonal lines meet (a node) can then represent an intersection of roads (edges), with a choice which road to take as we move along. The specialty in storing roads comes from the fact that we want to be able to find paths between nodes efficiently, for instance in car navigation systems, while we are driving. The fact that not all roads are equally important can be expressed by weights on the edges. Because a shortest path computation is carried out as a breadth first search on the edge weighted graph, in one way or another (e.g., bidirectional), it makes sense to partition the graph into pages so as to minimize the weight of edges that cross the cuts induced by the partition. Whenever we want to maintain road data together with other thematic data, such as land use data, it also makes sense to store all the data in one structure, instead of using an extra structure for the road network. It may come as no surprise that for some data structures, partitioning the graph and partitioning the other thematic aspects go together very well (compromising a little on both sides), while for others this is not easily the case. The compromise in partitioning the graph does almost no harm, because it is NP-complete to find the optimum partition, and hence a suboptimal solution of some sort is all we can get anyway. An excellent recent survey on road networks and algorithms for shortest-paths can be found in Reference 31. It does not only provide a concise summary of various methods for different problem settings but also contains a very detailed experimental evaluation in terms of multiple factors such as query time, preprocessing cost, and space usage.

56.1.5.5 Spatiotemporal Data

Just like for many other database applications, a *time* component brings a new dimension to spatial data (even in the mathematical sense of the word, if you wish). *How did the forest areas in New Mexico develop over the last 20 years?* Questions like this one demonstrate that for environmental information systems, a specific branch of GISs, keeping track of developments over time is a must. Spatiotemporal database research is concerned with all problems that the combination of space with time raises, from models and languages, all the way through data structures and query algorithms, to architectures and implementations of systems [32]. In this chapter, we refrain from the temptation to discuss spatiotemporal data structures in detail; see Chapter 23 for an introduction into this lively field.

56.1.5.6 Data Mining

The development of spatial data over time is interesting not only for explicit queries but also for data mining. Here, one tries to find relevant patterns in the data, without knowing beforehand the character of the pattern (for an introduction to the field of data mining, see Reference 4). For more advanced topics, the reader is referred to Reference 33. Let us briefly look at a historic example for spatial data mining. A London epidemiologist identified a water pump as the centroid of the locations of cholera cases, and after the water pump was shut down, the cholera subsided. This and other examples are described in Reference 9. If we want to find patterns in quite some generality, we need a large data store that keeps track of data extracted from different data sources over time, a so-called data warehouse. It remains as an important, challenging open problem to efficiently run a spatial data warehouse and mine the spatial data. The spatial nature of the data seems to add the extra complexity that comes from the high autocorrelation present in typical spatial data sets, with the effect that most knowledge discovery techniques today perform poorly. This omnipresent tendency for data to cluster in space has been stated nicely [34]: *Everything is related to everything else but nearby things are more related than distant things.*

Due to the ubiquitous availability of GPS devices and their capability of tracking arbitrary movements, mining algorithms for trajectories have emerged over the last decade [35]. In addition to clustering and classification of trajectories, one of the most challenging spatial mining tasks is to extract behavioral patterns from very large set of trajectories [36]. A typical pattern is to identify sequences of regions, frequently visited by trajectories in a specified order with similar average stays. The work on trajectories differentiates between physical and semantic ones. A physical trajectory consists of typical physical measurements, such as position, speed, and acceleration, while a semantic or conceptual trajectory [37] adds annotations to the physical trajectory representing semantically richer patterns that are more relevant to a user. For example, someone might be interested in trajectories with high frequency of stops at touristic points. We refer to a recent survey [38] that gives an excellent summary of mining techniques for semantic trajectories.

56.2 Space Filling Curves: Order in Many Dimensions

As explained above, our interest in space filling curves (SFCs) comes from two sources. The first one is the fact that we aim at exploiting the typical database support mechanisms that a conventional database management system (DBMS) offers, such as support for transactions and recovery. On the data structures level, this support is automatic if we resort to a data structure that is inherently supported by a DBMS. These days, this is the case for a number of one-dimensional data structures, such as those of the B-tree family (see Chapter 16). In addition, spatial DBMS support one of a small number of multidimensional data structures, such as those of the R-tree family (see Chapter 22) or of a grid-based structure. The second reason for our interest in SFCs is the general curiosity in the gap between one dimension and many: In what way and to what degree can we bridge this gap for the sake of supporting spatial operations of the kind described above? In our setting, the gap between one dimension and many goes back to the lack of a linear order in many dimensions that is useful for all purposes. An SFC tries to overcome this problem at least to some degree, by defining an artificial linear order on a regular grid that is as useful as possible for the intended purpose.

Limiting ourselves in this section to two-dimensional space, we define an SFC more formally as a bijective mapping p from an index pair of a grid cell to its number in the linear order:

$$p : N \times N \longrightarrow \{1, \ldots, N^2\}.$$

For the sake of simplicity, we limit ourselves to numbers $N = 2^n$ for some positive integer n.

Our main attention is on the choice of the linear order in such a way that range queries are efficient. In the GIS setting, it is not worst-case efficiency that counts, but efficiency in expectation. It is, unfortunately, hard to say what queries can be expected. Therefore, a number of criteria are conceivable according to which the quality of an SFC should be measured. Before entering the discussion about these criteria, let us present some of the most prominent SFCs that have been investigated for GIS data structures.

56.2.1 Recursively Defined SFCs

Two SFCs have been investigated most closely for the purpose of producing a linear ordering suitable for data structures for GIS, the z-curve, by some also called Peano curve or Morton encoding, and the Hilbert curve. We depict them in Figures 56.1 and 56.2. They can both be defined recursively with a simple refinement rule. To obtain a $2^{n+1} \times 2^{n+1}$ grid from a $2^n \times 2^n$ grid, replace each cell by the elementary pattern of four cells as in Figure 56.1, with the appropriate rotation for the Hilbert curve, as indicated in Figure 56.2 for the four by four grid (i.e., for $n = 2$). In the same way, a slightly less popular SFC can be defined, the Gray code curve (see Figure 56.3).

For all recursively defined SFCs, there is an obvious way to compute the mapping p (and also its inverse), namely just along the recursive definition. Any such computation therefore takes time proportional to the logarithm of the grid size. Without going to great lengths in explaining how this computation is carried out in detail (refer [39] for this and other mathematical aspects of SFCs), let us just mention that there is a particularly nice way of viewing it for the z-curve: Here, p simply alternately interleaves the bits of both of its arguments, when these are expressed in binary. This may have made the z-curve popular among geographers at an early stage, even though our subsequent discussion will reveal that it is not necessarily the best choice.

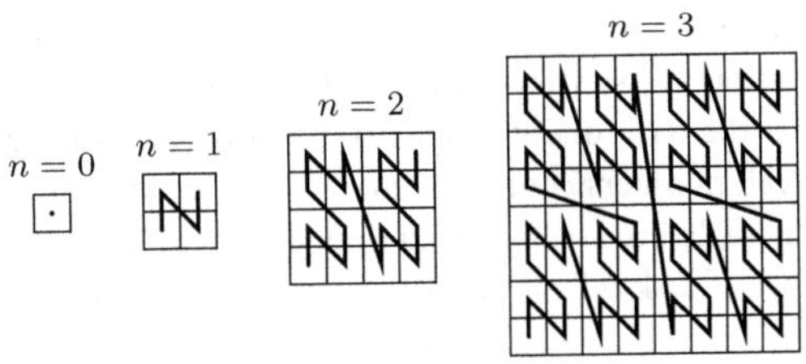

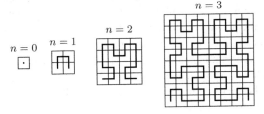

FIGURE 56.1 The z-curve for a $2^n \times 2^n$ grid, for $n = 0, \ldots, 3$.

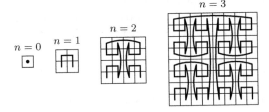

FIGURE 56.2 The Hilbert curve for a $2^n \times 2^n$ grid, for $n = 0, \ldots, 3$.

FIGURE 56.3 The Gray curve for a $2^n \times 2^n$ grid, for $n = 0, \ldots, 3$.

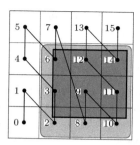

FIGURE 56.4   Range queries and disk seek operations.

56.2.2   Range Queries for SFC Data Structures

An SFC defines a linear order that is used to store the geographical data objects. We distinguish between two extreme storage paradigms that can be found in spatial data structures, namely a partition of the data space according to the objects present in the data set (such as in R-trees or in B-trees for a single dimension), or a partition of the space only, regardless of the objects (such as in regular cell partitions). The latter makes sense if the distribution of objects is somewhat uniform (see the spatial data structures chapters in this Handbook), and in this case achieves considerable efficiency. Naturally, a multitude of data structures that operate in between both extremes have been proposed, such as the grid file (see the spatial data structures chapters in this Handbook). As to the objects, let us limit ourselves to points in the plane, so that we can focus on a single partition of the plane into grid cells and need not worry about objects that cross cell boundaries (these are dealt with elsewhere in this Handbook, see, e.g., the chapter on R-trees or the survey chapter). For simplicity, let us restrict our attention to partitions of the data space into a $2^n \times 2^n$ grid, for some integer n. Partitions into other numbers of grid cells (i.e., not powers of 4) can be achieved dynamically for some data structures (see, e.g., z-Hashing [40]), but are not always easy to obtain; we will ignore that aspect for now.

Corresponding to both extreme data structuring principles, there are two ways to associate data with external storage blocks for SFCs. The simplest way is to identify a grid cell with a disk block. This is the usual setting. Grid cell numbers correspond to physical disk block addresses, and a range query translates into the need to access the corresponding set of disk blocks. In the example of Figure 56.4, for instance, the dark query range results in the need to read disk blocks corresponding to cells with numbers 2–3, 6, 8–12, 14, that is, cell numbers come in four consecutive sequences. In the best case, the cells to be read have consecutive numbers, and hence a single disk seek operation will suffice, followed by a number of successive block read operations. In the worst case, the sequence of disk cell numbers to be read breaks into a large number of consecutive pieces. It is one concern in the choice of an SFC to bound this number or some measure related to it (see the next subsection).

The association of a grid cell with exactly one disk block gives a data structure based on an SFC very little flexibility to adapt to a not so uniform data set. A different and more adaptive way of associating disk blocks with grid cells is based on a partition of the data space into cells so that for each cell, the objects in the cell fit on a disk block (as before), but a disk block stores the union of the sets of objects in a consecutive number of cells. This method has the advantage that relatively sparsely populated cells can go together in one disk block, but has the potential disadvantage that disk block maintenance becomes more complex. Not too complex, though, because a disk block can simply be maintained as a node of a B-tree (or, more specifically, a leaf, depending on the B-tree variety), with the content of a cell limiting the granularity of the data in the node. Under this assumption, the adaptation of the data structure to a dynamically changing population of data objects simply translates to split and merge operations of B-tree nodes. In this setting, the measure of efficiency for range queries may well be different from the above. One might be interested in running a range query on the B-tree representation, and ignoring (skipping) the contents of retrieved cells that do not contribute to the result. Hence, for a range query to be efficient, the consecutive single piece of the SFC that includes all cells of the query range should not run outside the query range for too long, that is, for too high a number of cells. We will discuss the effect of a requirement of this type on the design of an SFC in more detail below.

Obviously, there are many other ways in which an SFC can be the basis for a spatial data structure, but we will abstain from a more detailed discussion here and limit ourselves to the two extremes described so far.

56.2.3   Are All SFCs Created Equal?

Consider an SFC that visits the cells of a grid by visiting next an orthogonal neighbor of the cell just visited. The Hilbert curve is of this *orthogonal neighbor* type, but the z-curve is not. Now consider a square query region of some fixed (but arbitrary) size, say k by k grid cells, and study the number of consecutive pieces of the SFC that this query region defines (see Figure 56.5 for a

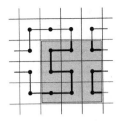

FIGURE 56.5 Disk seeks for an orthogonal curve.

range query region of 3 × 3 grid cells that defines three consecutive pieces of a Hilbert curve). We are interested in the average number of curve pieces over all locations of the query range.

For a fixed location of the query range, the number of curve pieces that the range defines is half the number of orthogonal neighbor links that cross the range boundary (ignoring the starting or ending cell of the entire curve). The reason is that when we start at the first cell of the entire curve and follow the curve, it is an orthogonal neighbor on the curve that leads into the query range and another one that leads out, repeatedly until the query range is exhausted. To obtain the average number of curve pieces per query range, we sum the number of boundary crossing orthogonal neighbors over all query range locations (and then divide that number by twice the number of locations, but this step is of no relevance in our argument).

This sum, however, amounts to the same as summing up for every orthogonal neighbor link the number of range query locations in which it is a crossing link. We content ourselves for the sake of simplicity with an approximate average, by ignoring cells close to the boundary of the data space, an assumption that will introduce only a small inaccuracy whenever k is much less than N (and this is the only interesting case). Hence, each orthogonal link will be a crossing link for $2k$ query range positions, namely k positions for the query range on each of both sides. The interesting observation now is that this summation disregards the position of the orthogonal link completely (apart from the fact that we ignore an area of size $O(kN)$ along the boundary of the universe). Hence, for any square range query, the average number of pieces of an SFC in the query range is the same across all SFCs of the orthogonal neighbor type.

In this sense, therefore, all SFCs of the orthogonal neighbor type have the same quality. This includes snake like curves such as row-major zigzag snakes and spiral snakes. Orthogonal neighbors are not a painful limitation here: If we allow for other than orthogonal neighbors, this measure of quality can become only worse, because a nonorthogonal neighbor is a crossing link for more query range locations. In rough terms, this indicates that the choice of SFC may not really matter that much. Nevertheless, it also shows that the above measure of quality is not perfectly chosen, since it captures the situation only for square query ranges, and that is not fully adequate for GISs. A recent measure of curve quality looks at a contiguous piece of the SFC and divides the area of its bounding box by the area that this piece of the curve covers. The quality of the curve is then the maximum of these ratios, taken over all contiguous pieces of the curve, appropriately called *worst-case bounding-box area ratio (WBA)* [41]. When instead of the area of the curve, its perimeter is taken as a basis, it needs to be properly normalized to be comparable to the area of the bounding box. When we normalize by pretending the perimeter belongs to a square, we get the ratio between the area of the bounding box and 1/16th of the squared perimeter. This view results in the *worst-case bounding-box squared perimeter ratio (WBP)* [41] as another quality measure, taking into account how wiggly the SFC is. For quite a few SFCs, WBA and WBP are at least 2, and the Peano curve comes very close to this bound [41]. These are not the only quality measures; for other measures and a thorough discussion of a variety of SFCs (also in higher dimension) and their qualities, see References 41 and 42. High dimension is useful when spatial objects have a high number of descriptive coordinates, such as four coordinates that describe an axis parallel rectangle, but high dimensional SFCs turn out to be far more complex to understand than those in two dimensions [43–45]. In this chapter, we will continue to discuss the basics for two dimensions only.

56.2.4 Many Curve Pieces for a Query Range

The properties that recursively definable SFCs entail for spatial data structure design have been studied from a theoretical perspective [46,47]. The performance of square range queries has been studied [46] on the basis that range queries in a GIS can have any size, but are most likely to not deviate from a square shape by too much. Based on the external storage access model in which disk seek and latency time by far dominates the time to read a block, they allow a range query to skip blocks for efficiency's sake. Skipping a block amounts to reading it in a sequence of blocks, without making use of the information thus obtained. In the example of Figure 56.4, skipping the blocks of cells 4, 5, 7, and 13 leads to a single sequence of the consecutive blocks 2–14. This can obviously be preferable to just reading consecutive sequences of relevant cell blocks only and therefore performing more disk seek operations, provided that the number of skipped blocks is not too large. In an attempt to quantify one against the other, consider a range query algorithm that is allowed to read a linear number of additional cells [46], as compared with those in the

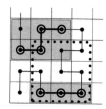

FIGURE 56.6   Bounding a piece of a curve.

query range. It turns out that for square range query regions of arbitrary size and the permission to read at most a linear number of extra cells, each recursive SFC needs at least three disk seek operations in the worst case. While no recursive SFC needs more than four disk seeks in the worst case, none of the very popular recursive SFCs, including the z-curve, the Hilbert curve, and the Gray code curve, can cope with less than four. One can define a recursive SFC that needs only three disk seeks in the worst case [46], and hence the lower and upper bounds match. This result is only a basis for data structure design, though, because its quality criterion is still too far from GIS reality to guide the design of practical SFCs. Nevertheless, this type of fragmentation has been studied in depth for a number of SFCs [48].

Along the same lines, one can refine the cost measure and account explicitly for disk seek cost [47]. For a sufficiently short sequence of cells, it is cheaper to skip them, but as soon as the sequence length exceeds a constant (that is defined by the disk seek cost, relative to the cost of reading a block), it is cheaper to stop reading and perform a disk seek. A refinement of the simple observation from above leads to performance formulas expressed in the numbers of links of the various types that the SFC uses, taking the relative disk seek cost into consideration [47]. It turns out that the local behavior of SFCs can be modeled well by random walks, again perhaps an indication that at least locally, the choice of a particular SFC is not crucial for the performance.

56.2.5 One Curve Piece for a Query Range

The second approach described above reads a single consecutive piece of the SFC to respond to a range query, from the lowest numbered cell in the range to the highest numbered. In general, this implies that a high number of irrelevant cells will be read. As we explained above, such an approach can be attractive, because it allows immediately to make use of well-established data structures such as the B-tree, including all access algorithms. We need to make sure, however, that the inefficiency that results from the extra accessed cells remains tolerable. Let us therefore now calculate this inefficiency approximately. To this end, we change our perspective and calculate for any consecutive sequence of cells along the SFC its shape in two-dimensional space. The shape itself may be quite complicated, but for our purpose a simple rectangular bounding box of all grid cells in question will suffice. The example in Figure 56.6 shows a set of 3 and a set of 4 consecutive cells along the curve, together with their bounding boxes (shaded), residing in a corner of a Hilbert curve. Given this example, a simple measure of quality suggests itself: The fewer useless cells we get in the bounding box, the better. A quick thought reveals that this measure is all too crude, because it does not take into consideration that square ranges are more important than skinny ranges. This bias can be taken into account by placing a piece of the SFC into its smallest enclosing square (see the dotted outline of a 3 × 3 square in Figure 56.6 for a bounding square of the 3 cells), and by taking the occupancy (i.e., percentage of relevant cells) of that square as the measure of quality. The question we now face is to bound the occupancy for a given SFC, across all possible pieces of the curve, and further, to find a best possible curve with respect to this bound. For simplicity's sake, let us again limit ourselves to curves with orthogonal links.

Let us first argue that the lower bound on the occupancy, for any SFC, cannot be higher than one-third. This is easy to see by contradiction. Assume to this end that there is an SFC that guarantees an occupancy of more than one-third. In particular, this implies that there cannot be two vertical links immediately after each other, nor can there be two consecutive horizontal links (the reason is that the three cells defining these two consecutive links define a smallest enclosing square of size 3 by 3 cells, with only 3 of these 9 cells being useful, and hence with an occupancy of only one-third). But this finishes the argument, because no SFC can always alternate between vertical and horizontal links (the reason is that a corner of the space that is traversed by the curve makes two consecutive links of the same type unavoidable, see Figure 56.7).

On the positive side, the Hilbert curve guarantees an occupancy of one-third. The reason is that the Hilbert curve gives this guarantee for the 4 × 4 grid, and that this property is preserved in the recursive refinement.

This leads to the conclusion that in terms of the worst-case occupancy guarantee, the Hilbert curve is a best possible basis for a spatial data structure.

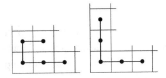

FIGURE 56.7 All possible corner cell traversals (excluding symmetry).

56.3 Spatial Join

In order to compute map overlays, a GIS provides an operator called spatial join that allows flexible combinations of multiple inputs according to a spatial predicate. The spatial join computes a subset of the Cartesian product of the inputs. Therefore, it is closely related to the join operator of a DBMS. A (binary) join on two sets of spatial objects, say R and S, according to a binary spatial predicate P is given by

$$SJ_P(R, S) = \{(r, s) \mid r \in R \wedge s \in S \wedge P(r, s)\}$$

The join is called a spatial join if the binary predicate P refers to spatial attributes of the objects. Among the most important ones are the following:

- Intersection predicate: $P_\cap(r, s) = (r \cap s \neq \emptyset)$.
- Distance predicate: Let DF be a distance function and $\epsilon > 0$. Then, $P_{DF}(r, s) = (DF(r, s) < \epsilon)$.

In the following we assume that R and S consist of N two-dimensional spatial objects and that the join returns a total of T pairs of objects. N is assumed being sufficiently large such that the problem cannot be solved entirely in main memory of size M. Therefore, we are particularly interested in external memory algorithms. Due to today's large main memory, the reader might question this assumption. However, remember that a GIS is a resource-intensive multiuser system where multiple complex queries are running concurrently. Thus the memory assigned to a specific join algorithm might be substantially lower than the total physically available. Given a block size of B, we use the following notations: $n = N/B$, $m = M/B$, and $t = T/B$. Without loss of generality, we assume that B is a divisor of N, M, and T.

A naive approach to processing a spatial join is simply to perform a so-called nested-loop algorithm, which checks the spatial predicate for each pair of objects. The nested-loop algorithm is generally applicable, but requires $O(N^2)$ time and $O(n^2)$ I/O operations. For special predicates, however, we are able to design new algorithms that provide substantial improvements in runtime compared to the naive approach. Among those is the intersection predicate, which is also the most frequently used predicate in GIS. We therefore will restrict our discussion to the intersection predicate to which the term spatial join will refer to by default. The discussion of other predicates is postponed to the end of the section. We refer to Reference 49 for an excellent survey on spatial joins.

The processing of spatial joins follows the general paradigm of multistep query processing [50] that consists at least of the following two processing steps. In the filter step, a spatial join is performed using conservative approximations of the spatial objects like their minimum bounding rectangles (MBRs). In the refinement step, the candidates of the filter step are checked against the join predicate using their exact geometry. In this section, we limit our discussion on the filter step that utilizes the MBR of the spatial objects. The reader is referred to References 51 and 52 for a detailed discussion of the refinement step and intermediate processing steps that are additionally introduced.

If main memory is large enough to keep R and S entirely in memory, the filter step is (almost) equivalent to the rectangle intersection problem, one of the elementary problems in computational geometry. The problem can be solved in $O(N \log N + T)$ runtime where T denotes the size of the output. This can be accomplished by using either a sweep-line algorithm [53] or a divide-and-conquer algorithm [54]. However, the disadvantage of these algorithms is their high overhead that results in a high constant factor. For real-life spatial data, it is generally advisable, see Reference 55 for example, to employ an algorithm that is not optimal in the worst case. The problem of computing spatial joins in main memory has attracted research attention again [56]. In comparison to external algorithms, it has been recognized that performance of main memory solutions is more sensitive to specific implementation details.

56.3.1 External Algorithms

In this subsection, we assume that the spatial relations are larger than main memory. According to the availability of indices on both relations, on one of the relations, or on none of the relations, we obtain three different classes of spatial join algorithms for the filter step. This common classification scheme also serves as a foundation in the most recent survey on spatial joins [49].

56.3.1.1 Index on Both Spatial Relations

In References 57 and 58, spatial join algorithms were presented where each of the relations is indexed by an R-tree. Starting at the roots, this algorithm synchronously traverses the trees node by node and joins all pairs of overlapping MBRs. If the nodes are leaves, the associated spatial objects are further examined in the refinement step. Otherwise, the algorithm is called recursively for each qualifying pair of entries. As pointed out in Reference 59, this algorithm is not limited to R-trees, but can be applied to a broad class of index structures. Important to both, I/O and CPU cost, is the traversal strategy. In Reference 58, a depth-first strategy was proposed where the qualifying pairs of nodes are visited according to a spatial ordering. If large buffers are available, a breadth-first traversal strategy was shown to be superior [60] in an experimental comparison. Though experimental studies have shown that these algorithms provide fast runtime in practice, the worst-case complexity of the algorithm is $O(n^2)$. The general problem of computing an optimal schedule for reading pages from disk is shown [61] to be NP-hard for spatial joins. For specific situations where two arbitrary rectangles from R and S do not share boundaries, the optimal solution can be computed in linear time. This is generally satisfied for bounding boxes of the leaves of R-trees.

56.3.1.2 Index on One Spatial Relation

Next, we assume that only one of the relations, say R, is equipped with a spatial index. We distinguish between the following three approaches. The first approach is to issue a range query against the index on R for each MBR of S. By using worst-case optimal spatial index structures, this already results in algorithms with subquadratic runtime. When a page buffer is available, it is also beneficial to sort the MBRs of S according to a criterion that preserves locality, for example, Hilbert value of the center of the MBRs. Then, two consecutive queries will access the same pages with high probability and therefore, the overall number of disk accesses decreases. The second approach as proposed in Reference 62 first creates a spatial index on S, the relation without spatial index. The basic idea for the creation of the index is to use the upper levels of the available index on R as a skeleton. Thereafter, one of the algorithms is applied that requires an index on both of the relations. In Reference 63, an improvement of the algorithm is presented that makes better use of the available main memory. A third approach is presented in References 64 and 65 where the spatial index is used for sorting the data according to one of the minimum boundaries of the rectangles. The sorted sequence then serves as input to an external plane-sweep algorithm.

56.3.1.3 Index on None of the Inputs

There are two early proposals on spatial join processing that require no index. The first one [66] suggests using an external version of a computational geometry algorithm. This is basically achieved by employing an external segment tree. The other [17] can be viewed as a sweep-line algorithm where the ordering is derived from z-order. Though the algorithm was originally not designed for rectangles, this can be accomplished in a straightforward manner [67,68]. Like any other sweep-line algorithm, R and S are first sorted, where the z-order serves as criterion in the following way. A rectangle receives the z-value of the smallest cell that still covers the entire rectangle. Let x be a z-value and $b(x) = (x_0, x_1, \ldots, x_k)$ its binary representation where $k = l(x)$ denotes the level of the corresponding cell. Then, $x <_z y$, if $b(x) <_{lexi} b(y)$ where $<_{lexi}$ denotes the lexicographical order on strings. Note that if x is a prefix of y, x will precede y in lexicographical order. After sorting the input, the processing continues while maintaining a stack for each input. The stacks satisfy the following invariant. The z-value of each element (except the last) is a prefix of its successor. The algorithm simply takes the next element from the sorted inputs, say x from input R. Before x is pushed to the stack, all elements from both stacks are removed that are not prefix of $b(x)$. Thereafter, the entire stack of S is checked for overlap. The worst case of the join occurs when each of the rectangles belongs to the cell that represents the entire data space. Then, the join runs like a nested loop and requires $O(n^2)$ I/O operations. This can practically be improved by introducing redundancy [67,69]. However, the worst-case bound remains the same.

Other methods like the Partition Based Spatial-Merge Join (PBSM) [70] and the Spatial Hash-Join [71] are based on the principles of divide-and-conquer where spatial relations are partitioned into buckets that fit into main memory and the join is computed for each pair of corresponding buckets. Let us discuss PBSM in more detail. PBSM performs in four phases. In the first phase, the number of partitions p is computed such that the join for each pair of partitions is likely to be processed in main memory. Then, a grid is introduced with g cells, $g \geq p$ and each of the cells is then associated with a partition. A rectangle is then assigned to a partition if it intersects with at least one of its cells. In the second phase, pairs of partitions have to be processed that still contain too much data. This usually requires repartitioning of the data into smaller partitions. In the third phase, the join is processed in main memory for every pair of related partitions. The fourth phase consists of sorting, in order to get rid of duplicates in the response set. This however can be replaced by applying an inexpensive check of the result whenever it is produced [72]. Overall, the worst case is $O(n^2)$ I/O operations for PBSM.

Spatial Hash-Joins [71] differ from PBSM in the following way. One of the relations is first partitioned using a spatial index structure like an R-tree which uniquely assigns rectangles to leaves, where each of them corresponds to a partition. The other relation is then partitioned using the index of the first relation. Each rectangle has to be stored in all partitions where an overlap with the MBR of the partition exists. Overall, this guarantees the avoidance of duplicate results.

At the end of the 1990s, Arge et al. [55] proposed the first spatial join algorithm that meets the lower I/O bound $O(n \log_m n + t)$. The method is an external sweep-line algorithm that is also related to the divide-and-conquer algorithm of Reference 66. Rather than partitioning the problem recursively into two, it is proposed to partition the input recursively into $k = \sqrt{m}$ strips of almost equal size until the problem is small enough for being solved in memory. This results in a recursion tree of height $O(\log_m n)$. At each level, $O(m)$ sorted lists of size $\Theta(B)$ are maintained where simultaneously an interval join is performed. Since each of the interval joins runs in $O(n' + t')$ (n' and t' denote the input size and result size of the join, respectively) and at most $O(N)$ intervals are maintained on each level of the recursion, it follows that at most $O(n)$ accesses are required for each level.

Instead of using the optimal algorithm, Arge et al. [55] employs a plane-sweep algorithm in their experiments where the sweep-line is organized by a dynamic interval tree. This is justified by the observation that the sweep-line is small as it holds only a small fraction of the entire spatial relations. If the algorithm really runs out of memory, it is recommended invoking the optimal algorithm for the entire problem. A different overflow strategy has been presented in Reference 73 where multiple iterations over the spatial relations might be necessary. The advantage of this strategy is that each result element is computed exactly once.

56.3.2 Advanced Issues

There are various extensions for spatial joins which are discussed in the following. We first address the processing of spatial joins on multiple inputs. Next, we discuss the processing of distance joins. Eventually, we conclude the section with a brief summary of requirements on the implementation of algorithms within a system.

The problem of joining more than two spatial relations according to a set of binary spatial predicates has been addressed in References 74 and 75. Such a join can be represented by a connected graph where its nodes represent the spatial relations and its edges binary join predicates. A first approach, called pairwise join method (PJM), is to decompose the multiway join into an operator tree of binary joins. The graphs with cycles therefore need to be transformed into a tree by ignoring the edges with lowest selectivity. Depending on the availability of spatial indexes, it is then advantageous to use from each class of spatial join algorithms the most efficient one. A different approach generalizes synchronous traversal to multiple inputs. The most important problem is not related to I/O, but to the processing cost of checking the join predicates. In Reference 74, different strategies are examined for an efficient processing of the join predicates. Results of an experimental study revealed that neither PJM nor synchronous traversal performs best in all situations. Therefore, an algorithm is presented for computing a hybrid of these approaches by using dynamic programming.

In addition to the intersection predicate, there are many other spatial predicates that are of practical relevance for spatial joins in a GIS, see Reference 76 for a survey. Among those, the distance predicate has received most attention. The distance join of R and S computes all pairs within a given distance. This problem has been even more extended in the following directions. First, pairs should be reported in an increasing order of the distances of the spatial objects. Second, only a fixed number of pairs should be reported. Third, answers should be produced on demand one at a time (without any limitations on the total number of answers). This problem has been addressed in References 77 and 78 where R-trees are assumed to be available on the spatial relations. The synchronized traversal is then controlled by two priority queues, where one maintains pairs of nodes according to their minimum distance and the other is primarily used for pruning irrelevant pairs of entries. In Reference 78, it was recognized that there are many overlapping nodes which are not distinguishable in the priority queues. In order to break ties, a secondary ordering has been introduced that assigns a high priority to such pairs that are likely to contribute to the final result. The problem of distance joins is not only important for spatial data in two dimensions, but it is possible to extend it to high dimensional spaces [79] and metric spaces [80]. Another member of the distance join family is the k-nearest neighbor join [81] that returns for each object of the one set the k-nearest neighbors in the other sets. This operation is of particular interest for moving object databases [82,83] where the one set refers to moving users and the other set consists of a spatial set with points of interest like restaurants.

Recently, the problem of iterated distance joins has been studied in the literature for supporting extensive simulations on a two-dimensional point set [84]. First, a distance self join returns for every point p of the input all points within distance ϵ to p. Then, these neighbors trigger a positional update of p. These two steps are recursively repeated on the updated input until the user quits the processing.

The design of algorithms for processing spatial joins largely depends on the specific system requirements. Similar to DBMSs, a complex query in a GIS is translated into an operator tree where nodes may correspond to spatial joins or other operators. There are two important issues that arise in this setting. First, an operator for processing spatial joins should have the ability to produce an estimation of the processing cost before the actual processing starts. Therefore, cost formulas [85] are required that are inexpensive to compute, only depend on a few parameters of the input and produce sufficiently accurate estimations. In general, accuracy of selectivity estimation can be largely improved using histograms [86]. Special spatial histograms have been proposed for estimating the selectivity of intersection joins [87–89] and k-nearest neighbor joins [90].

Second, it is generally advantageous to use join methods that are nonblocking, that is, first results are produced without having the entire input available. Nonblocking supports the production of early results and leads to small intermediate results in processing pipelines consisting of a large number of operators. This is a necessity for continuously arriving spatial streams [91] where the arrival of new data items triggers the processing and the corresponding results are immediately forwarded to the next consuming operator. A nonblocking operator is also beneficial for traditional demand-driven data processing engines [92] where an explicit request for the next result triggers the next processing steps of a query pipeline.

56.4 Models, Toolboxes, and Systems for Geographic Information

GISs differ substantially with respect to their specific functionality, which makes a comparison of the different systems quite difficult. We restrict our evaluation to the core functionality of a GIS related to manipulating vector data. Moreover, we stress the following three criteria in our comparison:

Data Model: The spatial data model offered by the system is very important to a user since it provides the geometric data types and the operations.

Spatial indexing: Spatial index structures are crucial for efficiently supporting the most important spatial queries. It is therefore important what kind of index structures are actually available, particularly for applications that deal with very large databases.

Spatial Join: Since spatial joins are the most meaningful operations for combining different maps, system performance depends on the efficiency of the underlying algorithm.

These criteria are among the most relevant ones for spatial query processing in a GIS [93].

Though we already have restricted our considerations to specific aspects, we limit our comparison to a few important systems and libraries. We actually start our discussion with introducing common standardized data models that are implemented by many commercial systems. Thereafter, we will discuss a few commercial systems that are used in the context of GIS in industry. Next, we present a few spatial libraries, some of them serve as plug-ins in systems.

56.4.1 Standardized Data Models

The most important standard for GIS [94,95] is published by Open Geospatial Consortium (OGC). This so-called simple feature model of OGC provides an object-oriented vector model and basic geometric data types. The actual implementations of commercial vendors like Oracle are closely related to the OGC standard. All of the geometric data types are subclasses of the class Geometry that provides an attribute that specifies the spatial reference system. One of the methods of Geometry delivers the so-called envelope of an object that is called MBR in our terminology. The data model distinguishes between atomic geometric types, such as points, curves and surfaces, and the corresponding collection types. The most complex atomic type is a polygonal surface that consists of an exterior polygonal ring and a set of internal polygonal rings where each of them represents a hole in the surface. Certain assertions have to be obeyed for such a polygonal surface to be in a consistent state.

The topological relationships of two spatial objects are expressed by using the nine-intersection model [96]. This model distinguishes between the exterior, interior, and boundary of an object. Spatial predicates like overlaps are then defined by specifying which of the assertions has to be satisfied. In addition to predicates, the OGC specification defines different constructive methods like the one for computing the convex hull of a spatial object. Another important function allows to compute the buffer object which contains those points that are within distance ε of a given object. Moreover, there are methods for computing the intersection (and other set operations) on two objects.

The OGC standard has also largely influenced other ones for geographic data like the standard for storing, retrieving, and processing spatial data using SQL [97] and the standard for the Geography Markup Language (GML) that is based on XML. The latest published version of the GML standard [98] additionally provides functionality to support three-dimensional objects and spatiotemporal applications. Because GML is often considered to be too complex, syntactically simpler markup languages, such as WKT [94] and GeoJSON [99], have gained popularity. For example, WKT serves in many database systems for supporting spatial data types. Users can express objects and functions in WKT, while the internal storage format is in a more compressed binary representation. The open standard format GeoJSON has become popular for the specification of vector data. Today, many front-end tools use GeoJSON to import and export spatial data in an easy way.

Due to the great variety of spatial data models, GDAL (Geospatial Data Abstraction Library)* becomes an indispensable C++-library today. Originally designed for the conversion of raster data, it also comes with routines to convert vector models in its package OGR. While similar libraries exists, GDAL is undoubtedly the one with the largest support of data models so far.

---

* www.gdal.org.

56.4.2 Spatial Database Systems

In this section, we give an overview on the geographic query processing features of spatial database systems. We first start our discussion with the Oracle system that is still the first choice for the management of spatial data. Then, we introduce other important database systems with spatial support.

56.4.2.1 Oracle

Among the big database vendors, Oracle currently offers the richest support for the management of spatial data. The data model of Oracle [100] is conformant to the simple feature model of OGC. Oracle additionally offers curves where the arcs might be circular. The spatial functionality is implemented on top of Oracle's DBMS and therefore, it is not fully integrated. This is most notable when SQL is used for specifying spatial queries where the declarative flavor of SQL is in contrast to the imperative procedural calls of the spatial functionality.

The processing of spatial queries is performed by using the filter-refinement approach. Moreover, intermediate filters might be employed where a kernel approximation of the object is used. This kind of processing is applied to spatial selection queries and to spatial joins.

There are R-trees and quadtrees in Oracle for indexing spatial data. In contrast to R-trees, (linear) quadtrees are built on a grid decomposition of the data space into tiles, each of them keep a list of its intersecting objects. The linear quadtree is implemented within Oracle's B+-tree. In case of fixed indexing, the tiles are all of the same size. Oracle provides a function to enable users to determine a good setting of the tile size. In the case of hybrid indexing, tiles may vary in size. This is accomplished by locally increasing the grid resolution if the number of tiles is still below a given threshold. A comparison of Oracle's spatial index structures [101] shows that the query performance of the R-tree is superior compared to the quadtree.

56.4.2.2 Open Source Systems and Research Prototypes

Today, there is a plethora of database systems coming with spatial query processing support. The most well-known open source system is probably PostGIS,[*] a spatial extension of PostgreSQL. Similar to Oracle, PostGIS implements the simple feature model of OGC and supports the common topological 9-intersection model. For spatial indexing, R-trees are offered as a specialization of a more generic indexing framework. There is a large number of GIS products, such as ArcGIS and QGIS [102], that can use PostGIS as their underlying spatial data store. Among those is also CartoDB,[†] a unique GIS system running in a cloud infrastructure and using PostGIS instances on the computing nodes. Another remarkable spatial database systems is MonetDB/GIS.[‡] The unique features of this system are its column architecture, data compression, and in-memory processing. As mentioned before, main-memory processing is only possible if memory is also sufficiently available. Note however that compression rates in column stores tend to be extremely high such that this assumption is not anymore unrealistic for specific applications. Finally, systems are becoming of interest not running in server mode, but within the application process. Examples of such systems are SQLite[§] and its spatial extender SpatiaLite. Both of them come with an R*-tree as spatial index. In a similar direction, H2GIS is the spatial extension of the open source system H2[¶] that is fully implemented in Java. Java applications can take advantage of H2GIS as the high communication and serialization overhead can be avoided.

56.4.3 Spatial Libraries

The success in rapidly developing spatial extensions of database systems is also due to the availability of powerful spatial libraries that already offer the building blocks for spatial query processing. We already mentioned the C++ library GDAL that supports a large number of spatial data models (raster and vector). However, the processing capabilities are not sufficient in GDAL. Therefore, other tools have been developed over the last decade.

56.4.3.1 JTS Topology Suite, GeoTools, and GEOS

The JTS Topology Suite (JTS) [103] is a Java class library providing fundamental geometric functions according to the geometry model defined by OGC [95]. Hence, it provides the basic spatial data types like polygonal surfaces and spatial predicates and operations like buffer and convex hull. The library also supports a user-definable precision model and contains code for robust geometric computation. There are also a few classes available for indexing MBRs (envelopes). The one structure is the MX-CIF

---

[*] postgis.net/.

[†] cartodb.com.

[‡] www.monetdb.org/Documentation/Extensions/GIS.

[§] www.sqlite.org/.

[¶] www.h2database.com.

quadtree [8] that is a specialized quadtree for organizing a dynamic set of rectangles. The other structure is a static R-tree that is created by using a bulk-loading technique [104]. So far, there has been no support of managing data on disk efficiently. The JTS library did not change for a long time until recently when the project was reactivated and a new version was published.

There are two important extensions of JTS. GEOS (Geometry Engine, Open Source) is a C++ library port of JTS Topology Suite. GEOS is a very active open source project and it can be viewed as the successor of JTS. It makes use of JTS and its mature numerically stable geometry operations, and also added extensions like support for raster data.

56.4.3.2 LEDA and CGAL

LEDA [105] and CGAL [15] are C++ libraries (see Chapter 42 for more on LEDA) that offer a rich collection of data structures and algorithms. Among the more advanced structures are spatial data types suitable for being used for the implementation of GIS. Most interesting to GIS is LEDA's and CGAL's ability to compute the geometry exactly by using a so-called rational kernel, that is, spatial data types whose coordinates are rational numbers. LEDA provides the most important two-dimensional data types such as points, iso-oriented rectangles, polygons, and planar subdivisions. Moreover, LEDA provides efficient implementations of important geometric algorithms such as convex hull, triangulations, and line intersection. AlgoComs, a companion product of LEDA, also provides a richer functionality for polygons that is closely related to a map overlay. In contrast to LEDA, the focus of CGAL is limited to computational geometry algorithms where CGAL's functionality is generally richer in comparison to LEDA. CGAL contains kd-trees for indexing multidimensional point data and supports incremental nearest neighbor queries.

Both, LEDA and CGAL, do not support external algorithms and spatial index structures. Therefore, they only partly cover the functionality required for a GIS. There has been an extension of LEDA, called LEDA-SM [106], that supports the most important external data structures.

56.4.3.3 XXL

XXL (eXtensible and fleXible Library) [107] is a pure Java library that does not originally support spatial data types, but only points and rectangles. Similar to GeoTools, this missing functionality is added by making use of JTS. XXL provides powerful support for various kinds of (spatial) indexes. There are different kinds of implementations of R-trees [108] as well as B-trees that might be combined with SFCs (e.g., z-order and Hilbert order). The concept of containers is introduced in XXL to provide an abstract view on external memory. Implementations of containers exist that are based on main memory, files, and raw disks. XXL offers a rich source of different kinds of spatial joins [72,109] that are based on using SFCs and the sort–merge paradigm. XXL is equipped with an object-relational algebra of query operators and a query optimizer that is able to rewrite Java programs. Query operators are iterators that deliver the answers of a query on demand, one by one. XXL is available under an open source licensing agreement (GNU LGPL).

Acknowledgment

We gratefully acknowledge the assistance of Michael Cammert.

References

1. S.A. Ahson and M. Ilyas, *Location-Based Services Handbook: Applications, Technologies, and Security*. CRC Press, Boca Raton, FL, 2010.

2. P.A. Burrough, R.A. McDonnell, and C.D. Lloyd, *Principles of Geographical Information Systems*, Oxford University Press, Oxford, UK, 2015.

3. R.H. Güting and M. Schneider, *Moving Objects Databases*, Morgan Kaufmann, San Francisco, CA, 2005.

4. J. Han, M. Kamber, and J. Pei, *Data Mining: Concepts and Techniques*, 3rd edition, Morgan Kaufmann, San Francisco, CA, 2011.

5. P.A. Longley, M.F. Goodchild, D.J. Maguire, and D.W. Rhind, *Geographic Information Science and Systems*, John Wiley & Sons, Hoboken, NJ, 2015.

6. H.J. Miller, and J. Han, *Geographic Data Mining and Knowledge Discovery*, CRC Press, Boca Raton, FL, 2009.

7. P. Rigaux, M.O. Scholl, and A. Voisard, *Spatial Databases: With Application to GIS*, Morgan Kaufmann, San Francisco, CA, 2001.

8. H. Samet, *Foundations of Multidimensional and Metric Data Structures*, Morgan Kaufmann, San Francisco, CA, 2006.

9. S. Shekhar and S. Chawla, *Spatial Databases: A Tour*, Prentice-Hall, Upper Saddle River, NJ, 2003.

10. M.J. van Kreveld, J. Nievergelt, and T. Roos, P. Widmayer (eds.), *Algorithmic Foundations of Geographic Information Systems*, Lecture Notes in Computer Science, vol. 1340, Springer, Heidelberg, Germany, 1997.

11. R. Pajarola and P. Widmayer. An image compression method for spatial search, *IEEE Transactions on Image Processing*, 9:357–365, 2000.

12. F.P. Preparata and M.I. Shamos, *Computational Geometry: An Introduction*, Springer, Heidelberg, Germany, 1985.

13. M.T. de Berg, M.J. van Kreveld, M.H. Overmars, and O. Schwarzkopf, *Computational Geometry: Algorithms and Applications* (second edition), Springer, Heidelberg, Germany, 2000.

14. J. Nievergelt and K. Hinrichs, *Algorithms and Data Structures: With Applications to Graphics and Geometry*, Prentice-Hall, Upper Saddle River, NJ, 1993.

15. CGAL, *CGAL 4.6—Manual, release 4.6*, http://doc.cgal.org/latest/Manual/, 2015.

16. H. Zhang, G. Chen, B.-C. Ooi, K.-L. Tan, and M. Zhang, In-memory big data management and processing: A survey, *IEEE TKDE*, 27(7):1920–1948, 2015.

17. J.A. Orenstein, Spatial query processing in an object-oriented database system, In *Proceedings of the 1986 (ACM) (SIGMOD) International Conference on Management of Data*, Washington, D.C., May 28–30, 1986, pp. 326–336.

18. J. Nievergelt and F. Preparata, Plane-sweep algorithms for intersecting geometric figures, *CACM*, 25(10):739–747, 1982.

19. G. Neyer, Map labeling with application to graph drawing, *Drawing Graphs, Methods and Models* (M. Kaufmann and D. Wagner, eds.), Lecture Notes in Computer Science, vol. 2025, Springer, pp. 247–273, 2001.

20. M.A. Bekos, M. Kaufmann, A. Symvonis, and A. Wolff, Boundary labeling: Models and efficient algorithms for rectangular maps, *Computational Geometry*, 36(3):215–236, 2007.

21. K. Been, E. Daiches, and C. Yap, Dynamic map labeling, *IEEE Transactions on Visualization and Computer Graphics*, 12(5):773–780, 2006.

22. B.P. Buttenfield and R.B. McMaster, *Map Generalization: Making Rules for Knowledge Representation*, Wiley, New York, 1991.

23. J.-C. Müller, J.-P. Lagrange, and R. Weibel, *GIS and Generalization: Methodology and Practice*, Taylor & Francis, London, 1995.

24. W.A. Mackaness, A. Ruas, and L. T. Sarjakoski, *Generalisation of Geographic Information: Cartographic Modelling and Applications*, Elsevier, Amsterdam, The Netherlands, 2007.

25. D.H. Douglas and T.K. Peucker, Algorithms for the reduction of the number of points required to represent a digitized line or its caricature, *Canadian Cartographer*, 10(2):112–122, 1973.

26. J. Hershberger and J. Snoeyink, Cartographic line simplification and polygon CSG formulæ in $O(n \log n)$ time, *Computational Geometry*, 11:175–185, 1998.

27. W. Shi and C. Cheung, Performance evaluation of line simplification algorithms for vector generalization, *The Cartographic Journal*, 43(1):27–44, 2006.

28. A. Saalfeld, Topologically consistent line simplification with the Douglas–Peucker algorithm, *Cartography and Geographic Information Science*, 26(1):7–18, 1999.

29. B. Becker, H.-W. Six, and P. Widmayer, Spatial priority search: An access technique for scaleless maps, In *Proceedings of the 1991 (ACM) (SIGMOD) International Conference on Management of Data*, Denver, Colorado, May 29–31, 1991, pp. 128–137. ACM Press.

30. P. van Oosterom and M. Meijers, Vario-scale data structures supporting smooth zoom and progressive transfer of 2D and 3D data, *International Journal of GIS*, 28(3):455–478, 2014.

31. H. Bast, D. Delling, A. Goldberg, M. Müller-Hannemann, T. Pajor, P. Sanders, D. Wagner, and R.F. Werneck, *Route planning in transportation networks*, arXiv preprint arXiv:1504.05140, 2015.

32. M. Koubarakis, T.K. Sellis, A.U. Frank, S. Grumbach, R.H. Güting, C.S. Jensen, N.A. Lorentzos et al. (eds.), *Spatio-Temporal Databases: The Chorochronos Approach*, Lecture Notes in Computer Science, vol. 2520, Springer, Heidelberg, Germany, 2003.

33. C.C. Aggarwal, *Data Mining: The Textbook*, Springer, Heidelberg, Germany, 2015.

34. W. Tobler, Cellular geography, *Philosophy in Geography*, Reidel Publishing Company, Dordrecht, Holland, 1979, pp. 379–386.

35. N. Preparata and Y. Theodoridis, *Mobility Data Management and Exploration*, Springer, Heidelberg, Germany, 2014.

36. F. Giannotti, M. Nanni, F. Pinelli, and D. Pedreschi, Trajectory pattern mining, In *Proceedings of the 13th (ACM) (SIGKDD) International Conference on Knowledge Discovery and Data Mining*, (P. Berkhin, R. Caruana, and X. Wu, eds.) San Jose, CA, August 12–15, 2007, pp. 330–339.

37. S. Spaccapietra, C. Parent, M.L. Damiani, J.A. de Macedo, F. Porto, and C. Vangenot, A conceptual view on trajectories, *Data & Knowledge Engineering*, 65(1):126–146, 2008.

38. C. Parent, S. Spaccapietra, C. Renso, G. Andrienko, N. Andrienko, V. Bogorny, M. L. Damiani et al., Semantic trajectories modeling and analysis, *ACM Computing Surveys*, 45(4):42, 2013.

39. H. Sagan, *Space Filling Curves*, Springer, Heidelberg, Germany, 1994.

40. A. Hutflesz, H.-W. Six, and P. Widmayer, Globally order preserving multidimensional linear hashing, *IEEE ICDE*, 1988, pp. 572–579, IEEE, New York.

41. H. J. Haverkort and F. van Walderveen, Locality and bounding-box quality of two-dimensional space-filling curves, *Computational Geometry*, 43(2):131–147, 2010.

42. H. J. Haverkort and F. van Walderveen, Four-dimensional Hilbert curves for r-trees, *Journal of Experimental Algorithmics (JEA)*, 16:3–4, 2011.

43. H. Haverkort, *An inventory of three-dimensional Hilbert space-filling curves*, arXiv preprint arXiv:1109.2323, 2011.

44. H. Haverkort, *Harmonious Hilbert curves and other extradimensional space-filling curves*, arXiv preprint arXiv:1211.0175, 2012.

45. M.F. Mokbel, W.G. Aref, and I. Kamel, Analysis of multi-dimensional space-filling curves, *GeoInformatica*, 7(3):179–209, 2003.

46. T. Asano, D. Ranjan, T. Roos, E. Welzl, and P. Widmayer, Space filling curves and their use in the design of geometric data structures, *Theoretical Computer Science*, 18(1):3–15, 1997.

47. E. Bugnion, T. Roos, R. Wattenhofer, and P. Widmayer, Space filling curves versus random walks, in *Proceedings of the Algorithmic Foundations of Geographic Information Systems* (M. van Kreveld, J. Nievergelt, T. Roos, and P. Widmayer, eds.), Lecture Notes in Computer Science, vol. 1340, Springer, Heidelberg, Germany, 1997.

48. H. Haverkort, Recursive tilings and space-filling curves with little fragmentation, *Journal of Computational Geometry*, 2(1):92–127, 2011.

49. E.H. Jacox and H. Samet, Spatial join techniques, *ACM TODS*, 32(1):7, 2007.

50. J.A. Orenstein, Redundancy in spatial databases, *Proceedings of the 1989 ACM SIGMOD International Conference on Management of Data*, Portland, Oregon, 1989, pp. 294–305. ACM, New York.

51. T. Brinkhoff, H.-P. Kriegel, R. Schneider, and B. Seeger, Multi-step processing of spatial joins, *Proceedings of the 1994 ACM SIGMOD International Conference on Management of ata*, Minneapolis, Minnesota, May 24–27, 1994, pp. 197–208. ACM, New York.

52. Y.-W. Huang, M.C. Jones, and E.A. Rundensteiner, Improving spatial intersect joins using symbolic intersect detection, *SSD*, Lecture Notes in Computer Science, vol. 1262, 1997, Springer, Heidelberg, Germany, pp. 165–177.

53. H.-W. Six and D. Wood, The rectangle intersection problem revisited, *BIT*, 20(4):426–433, 1980.

54. R.H. Güting and D. Wood, Finding rectangle intersections by divide-and-conquer, *IEEE Transactions on Computers*, 33(7):671–675, 1984.

55. L. Arge, O. Procopiuc, S. Ramaswamy, T. Suel, and J.S. Vitter, Scalable sweeping-based spatial join, *Proceedings of 24rd International Conference on Very Large Data Bases, New York, August 24–27*, 1998, pp. 570–581. Morgan Kaufmann.

56. D. Šidlauskas and C.S. Jensen, Spatial joins in main memory: Implementation matters!, *VLDB Conference*, 8(1):97–100, 2014.

57. N. Beckmann, H.-P. Kriegel, R. Schneider, and B. Seeger, The R*-tree: An efficient and robust access method for points and rectangles, In *Proceedings of the 1990 (ACM) (SIGMOD) International Conference on Management of Data*, Atlantic City, NJ, May 23–25, 1990, pp. 322–331. ACM Press.

58. T. Brinkhoff, H.-P. Kriegel, and B. Seeger, Efficient processing of spatial joins using R-trees, In *Proceedings of the 1993 (ACM) (SIGMOD) International Conference on Management of Data*, Washington, DC, May 26–28 (P. Buneman and S. Jajodiaeds, eds.), 1993, pp. 237–246. ACM Press.

59. O. Günther, Efficient computation of spatial joins, In *Proceedings of the Ninth International Conference on Data Engineering*, Vienna, Austria, April 19–23, 1993, pp. 50–59. IEEE Computer Society.

60. Y.-W. Huang, N. Jing, and E.A. Rundensteiner, Spatial joins using R-trees: Breadth-first traversal with global optimizations, In *Proceedings of 23rd International Conference on Very Large Data Bases*, Athens, Greece, August 25–29, (M. Jarke, M.J. Carey, K.R. Dittrich, F.H. Lochovsky, P. Loucopoulos, and M.A. Jeusfeld, eds.), 1997, pp. 396–405. Morgan Kaufmann.

61. G. Neyer and P. Widmayer, Singularities make spatial join scheduling hard, *International Symposium on Algorithms and Computation*, Lecture Notes in Computer Science, vol. 1350, Springer, Heidelberg, Germany, pp. 293–302, 1997.

62. M.-L. Lo and C.V. Ravishankar, Spatial joins using seeded trees, In *Proceedings of the 1994 (ACM) (SIGMOD) International Conference on Management of Data*, Minneapolis, Minnesota, May 24–27, (R.T. Snodgrass and M. Winslett, eds.), 1994, pp. 209–220. ACM Press.

63. N. Mamoulis and D. Papadias, Slot index spatial join, *IEEE TKDE*, 15(1):211–231, 2003.

64. L. Arge, O. Procopiuc, S. Ramaswamy, T. Suel, J. Vahrenhold, and J.S. Vitter, A unified approach for indexed and non-indexed spatial joins, In *Advances in Database Technology-EDBT 2000, 7th International Conference on Extending Database Technology*, Konstanz, Germany, March 27–31, (C. Zaniolo, P.C. Lockemann, M.H. Scholl, and T. Grust, eds.). Lecture Notes in Computer Science, 2000, pp. 413–429. Springer.

65. C. Gurret and P. Rigaux, The sort/sweep algorithm: A new method for R-Tree based spatial joins, In *Proceedings of the 12th International Conference on Scientific and Statistical Database Management*, Berlin, Germany, July 26–28, (O. Günther and H.-J. Lenz, eds.), 2000, pp. 153–165. IEEE Computer Society.

66. R.H. Güting and W. Schilling, A practical divide-and-conquer algorithm for the rectangle intersection problem, *Information Sciences*, 42:95–112, 1987.

67. J.-P. Dittrich and B. Seeger, GESS: A scalable similarity-join algorithm for mining large data sets in high dimensional spaces, In *Proceedings of the Seventh ACM SIGKDD International Conference on Knowledge Discovery and Data Mining*, San Francisco, CA, USA, August 26–29, (D. Lee, M. Schkolnick, F.J. Provost, and R. Srikant, eds.), 2001, pp. 47–56. ACM Press.

68. N. Koudas and K.C. Sevcik, Size separation spatial join, In *Proceedings ACM SIGMOD International Conference on Management of Data*, Tucson, Arizona, May 13–15, (J. Peckham, ed.), 1997, pp. 324–335. ACM Press.

69. J.A. Orenstein, An algorithm for computing the overlay of k-dimensional spaces, In *Advances in Spatial Databases, Second International Symposium, SSD'91*, Zürich, Switzerland, August 28–30, Lecture Notes in Computer Science, vol. 525, 1991, pp. 381–400. Springer.

70. J.M. Patel and D.J. DeWitt, Partition based spatial-merge join, In *Proceedings of the 1996 ACM SIGMOD International Conference on Management of Data*, Montreal, Quebec, Canada, June 4–6, (H.V. Jagadish and I.S. Mumick, eds.), 1996, pp. 259–270. ACM Press.

71. M.-L. Lo and C.V. Ravishankar, Spatial hash-joins, In *Proceedings of the 1996 ACM SIGMOD International Conference on Management of Data*, Montreal, Quebec, Canada, June 4–6, (H.V. Jagadish and I.S. Mumick, eds.), 1996, pp. 247–258. ACM Press.

72. J.-P. Dittrich and B. Seeger, Data redundancy and duplicate detection in spatial join processing, In *Proceedings of the 16th International Conference on Data Engineering*, San Diego, California, February 28–March 3, (D.B. Lomet and G. Weikum, eds.), 2000, pp. 535–546. IEEE Computer Society.

73. E.H. Jacox and H. Samet, Iterative spatial join, *ACM TODS*, 28(3):230–256, 2003.

74. N. Mamoulis and D. Papadias, Multiway spatial joins, *TODS*, 26(4):424–475, 2001.

75. D. Papadias, N. Mamoulis, and Y. Theodoridis, Constraint-based processing of multiway spatial joins, *Algorithmica*, 30(2):188–215, 2001.

76. D. Papadias, Y. Theodoridis, T. Sellis, and M. Egenhofer, Topological relations in the world of minimum bounding rectangles, In *Proceedings of the 1995 ACM SIGMOD International Conference on Management of Data*, San Jose, California, May 22–25, (M.J. Carey and D.A. Schneider, eds.), 1995, pp. 92–103. ACM Press.

77. G.R. Hjaltason an H. Samet, Incremental distance join algorithms for spatial databases, In *Proceedings ACM SIGMOD International Conference on Management of Data*, Seattle, Washington, June 2–4, (L.M. Haas and A. Tiwary, eds.), 1998, pp. 237–248. ACM Press.

78. H. Shin, B. Moon, and S. Lee, Tie-breaking strategies for fast distance join processing, *Data and Knowledge Engineering*. 41(1):67–83, 2002.

79. C. Böhm, B. Braunmüller, F. Krebs, and H.-P. Kriegel, Epsilon grid order: An algorithm for the similarity join on massive high-dimensional data, In *Proceedings of the 2001 ACM SIGMOD International Conference on Management of Data*, Santa Barbara, CA, May 21–24, (S. Mehrotra and T.K. Sellis, eds.), 2001, pp. 379–388. ACM Press.

80. E.H. Jacox and H. Samet, Metric space similarity joins, *ACM TODS*, 33(2):7, 2008.

81. C. Böhm and F. Krebs, The k-nearest neighbour join: Turbo charging the kdd process, *Knowledge and Information Systems*, 6(6):728–749, 2004.

82. S. Ilarri, E. Mena, and A. Illarramendi, Location-dependent query processing: Where we are and where we are heading, *ACM Computing Surveys*, 42(3):12, 2010.

83. X. Xiong, M.F. Mokbel, and W.G. Aref, SEA-CNN: Scalable processing of continuous k-nearest neighbor queries in spatio-temporal databases, In *Proceedings of the 21st International Conference on Data Engineering*, Tokyo, Japan, April 5–8, (K. Aberer, M.J. Franklin, and S. Nishio, eds.), 2005, pp. 643–654. IEEE Computer Society.

84. B. Sowell, M.V. Salles, T. Cao, A. Demers, and J. Gehrke, An experimental analysis of iterated spatial joins in main memory, *VLDB Conference*, 6(14):1882–1893, 2013.

85. Y. Theodoridis, E. Stefanakis, and T.K. Sellis, Efficient cost models for spatial queries using R-Trees, *IEEE TKDE*, 12(1): 19–32, 2000.

86. Y. Ioannidis, The history of histograms (abridged), In *Proceedings of 29th International Conference on Very Large Data Bases*, Berlin, Germany, September 9–12, (J.C. Freytag, P.C. Lockemann, S. Abiteboul, M.J. Carey, P.G. Selinger, and A. Heuer, eds.), 2003, pp. 19–30. Morgan Kaufmann.

87. N. An, Z.-Y. Yang, and A. Sivasubramaniam, Selectivity estimation for spatial joins, In *Proceedings of the 17th International Conference on Data Engineering*, Heidelberg, Germany, April 2–6, (D. Georgakopoulos and A. Buchmann, eds.) 2001, pp. 368–375. IEEE Computer Society.

88. R. Beigel and E. Tanin, The geometry of browsing, *LATIN'98: Theoretical Informatics, Third Latin American Symposium*, Campinas, Brazil, April, 20–24, (C.L. Lucchesi and A.V. Moura, eds.), 1998, pp. 331–340. Springer.

89. C. Sun, D. Agrawal, and A. El Abbadi, Selectivity estimation for spatial joins with geometric selections, In *Advances in Database Technology-EDBT 8th International Conference on Extending Database Technology*, Prague, Czech Republic, March 25–27, (C.S. Jensen, K.G. Jeffery, J. Pokorný, S. Saltenis, E. Bertino, K. Böhm, and M. Jarke, eds.), 2002, pp. 609–626. Lecture Notes in Computer Science. Springer.

90. A.M. Aly, W.G. Aref, and M. Ouzzani, Cost estimation of spatial k-nearest-neighbor operators, In *Proceedings of the 18th International Conference on Extending Database Technology*, Brussels, Belgium, March 23–27, (G. Alonso, F. Geerts, L. Popa, P. Barcelö, J. Teubner, M. Ugarte, J. Van den Bussche, and J. Paredaens, eds.), 2015, pp. 457–468. OpenProceedings.org

91. R. Zhang, D. Lin, K. Ramamohanarao, and E. Bertino, Continuous intersection joins over moving objects, In *Proceedings of the 24th International Conference on Data Engineering*, Cancun, Mexico, April 7–12, (G. Alonso, J.A. Blakeley, and A.L.P. Chen, eds.), 2008, pp. 863–872. IEEE Computer Society.

92. G. Graefe, Query evaluation techniques for large databases, *ACM Computing Surveys*, 25(2):73–170, 1993.

93. R.H. Güting, An introduction to spatial database systems, *VLDB Journal*, 3(4):357–399, 1994.

94. Open GIS Consortium, *OpenGIS Implementation Specification for Geographic information—Simple feature access—Part 1: Common architecture option, revision 1.2.1*, Wayland, MA, 2010.

95. Open GIS Consortium, *OpenGIS Implementation Specification for Geographic information—Simple feature access—Part 2: SQL option, revision 1.2.1*, Wayland, MA, 2010.

96. M.J. Egenhofer, Reasoning about binary topological relations, *Symposium on Spatial Databases*, Lecture Notes in Computer Science, vol. 525, Springer, Heidelberg, Germany, pp. 143–160, 1991.

97. ISO/IEC 13249-3:2011 FDIS, *Information technology—database languages—SQL multimedia and application packages—Part 3: Spatial*, 2nd edition, Geneva, Switzerland, 2011.

98. Open GIS Consortium, *OpenGIS Geography Markup Language (GML) Encoding Standard, Version 3.2.1*, Wayland, MA, 2007.

99. GeoJSON.org, *The GeoJSON Format Specification, Revison 1.0*, 2008.

100. Oracle, *Spatial and Graph Developer's Guide release 12.1*, Redwood City, CA, http://docs.oracle.com/database/121, 2014.

101. R.K.V. Kothuri, S. Ravada, and D. Abugov, Quadtree and R-tree indexes in oracle spatial: A comparison using GIS data, In *Proceedings of the 2002 ACM SIGMOD International Conference on Management of Data*, Madison, Wisconsin, June 3–6, (M.J. Franklin, B. Moon, and A. Ailamaki, eds.), 2002, pp. 546–557. ACM Press.

102. QGIS, *User Guide, Release 2.18*, 2016.

103. Vivid Solutions, *JTS topology suite technical specifications version 1.4*, Victoria, BC, Canada, http://www.vividsolutions.com/, 2003.

104. S. Leutenegger, M. Lopez, and J. Edgington, A simple and efficient algorithm for R-tree packing, In *Proceedings of the 13th International Conference on Data Engineering*, Birmingham, UK, April 7–11, (W.A. Gray and P. Larson, eds.), 1997, pp. 497–506. IEEE Computer Society.

105. K. Mehlhorn and S. Näher, *LEDA 2 Part Set: A Platform for Combinatorial and Geometric Computing*, Cambridge University Press, Cambridge, UK, 2009.

106. A. Crauser and K. Mehlhorn, LEDA-SM: Extending LEDA to secondary memory, In *Algorithm Engineering, 3rd International Workshop*, London, UK, July 19–21, (J.S. Vitter and C.D. Zaroliagis, eds.), Lecture Notes in Computer Science, 1999, vol. 1668, pp. 228–242. Springer.

107. M. Cammert, C. Heinz, J. Krämer, M. Schneider, and B. Seeger, A status report on XXL—a software infrastructure for efficient query processing, *IEEE Data Engineering Bulletin*, 26(2):12–18, 2003.

108. A. Guttman, R-trees: A dynamic index structure for spatial searching, In *Proceedings of Annual Meeting*, Boston, Massachusetts, June 18–21, (B. Yormark, ed.), 1984, pp. 47–57. ACM Press.

109. J.A. Orenstein, Strategies for optimizing the use of redundancy in spatial databases, In *Design and Implementation of Large Spatial Databases*, First Symposium, Santa Barbara, California, July 17/18, (A.P. Buchmann, O. Gnther, T.R. Smith, and Y.-F. Wang, eds.), Lecture Notes in Computer Science, vol. 409, pp. 115–136, 1989. Springer.

110. M. Meijers, Simultaneous & topologically-safe line simplification for a variable-scale planar partition, *Advancing Geoinformation Science for a Changing World*, Utrecht, The Netherlands, April 18–21, (S.C.M. Geertman, W. Reinhardt, and F. Toppen, eds.). Lecture Notes in Geoinformation and Cartography, pp. 337–358, 2011. Springer.

57

Collision Detection*

57.1	Introduction	889
57.2	Convex Polytopes	890

Linear Programming • Voronoi-Based Marching Algorithm • Minkowski Sums and Convex Optimization

| 57.3 | General Polygonal Models | 892 |

Interference Detection Using Trees of Oriented Bounding Boxes • Performance of Bounding Volume Hierarchies

| 57.4 | Penetration Depth Computation | 896 |

Convex Polytopes • Incremental Penetration Depth Computation • Non-Convex Models

| 57.5 | Large Environments | 898 |

Multiple-Object Collision Detection • Two-Dimensional Intersection Tests

References.................. 901

Ming C. Lin
University of North Carolina at Chapel Hill

Dinesh Manocha
University of North Carolina at Chapel Hill

57.1 Introduction

In a geometric context, collision detection refers to checking the relative configuration of two or more objects. The goal of collision detection, also known as interference detection or contact determination, is to automatically report a geometric contact when it is about to occur or has actually occurred. The objects may be represented as polygonal objects, spline or algebraic surfaces, deformable models, etc. Moreover, the objects may be static or dynamic.

Collision detection is a fundamental problem in computational geometry and frequently arises in different applications. These include:

1. *Physically-based Modeling and Dynamic Simulation:* The goal is to simulate dynamical systems and the physical behavior of objects subject to dynamic constraints. The mathematical model of the system is specified using geometric representations of the objects and the differential equations that govern the dynamics. The objects may undergo rigid or non-rigid motion. The contact interactions and trajectories of the objects are affected by collisions. It is important to model object interactions precisely and compute all the contacts accurately [1].
2. *Motion Planning:* The goal of motion planning is to compute a collision free path for a robot from a start configuration to a goal configuration. Motion planning is a fundamental problem in algorithmic robotics [2]. Most of the practical algorithms for motion planning compute different configurations of the robot and check whether these configurations are collision-free, that is, no collision between the robot and the objects in the environment. For example, probabilistic roadmap planners can spend up to 90% of the running time in collision checking.
3. *Virtual Environments and Walkthroughs:* A large-scale virtual environment, like a walkthrough, creates a computer-generated world, filled with real, simulated or virtual entities. Such an environment should give the user a feeling of presence, which includes making the images of both the user and the surrounding objects feel solid. For example, the objects should not pass through each other, and things should move as expected when pushed, pulled or grasped. Such actions require accurate and interactive collision detection. Moreover, runtime performance is critical in a virtual environment and all collision computations need to be performed at less than 1/30th of a second to give a sense of presence [3].
4. *Haptic Rendering:* Haptic interfaces, or force feedback devices, improve the quality of human-computer interaction by accommodating the sense of touch. In order to maintain a stable haptic system while displaying smooth and realistic forces and torques, haptic update rates must be as high as 1000 Hz. This involves accurately computing all contacts between the

---

* This chapter has been reprinted from first edition of this Handbook, without any content updates.

object attached to the probe and the simulated environment, as well as the restoring forces and torques—all in less than one millisecond [4].

In each of these applications, collision detection is one of the major computational bottlenecks.

Collision detection has been extensively studied in the literature for more than four decades. Hundreds of papers have been published on different aspects in computational geometry and related areas like robotics, computer graphics, virtual environments and computer-aided design. Most of the algorithms are designed to check whether a pair of objects collide. Some algorithms have been proposed for large environments composed of multiple objects and perform some form of culling or localize pairs of objects that are potentially colliding. At a broad level, different algorithms for collision detection can be classified based on the following characteristics:

- *Query Type:* The basic collision query checks whether two objects, described as set of points, polygons or other geometric primitives, overlap. This is the boolean form of the query. The enumerative version of the query yields some representations of the intersection set. Other queries compute the separation distance between two non-overlapping objects or the penetration distance between two overlapping objects.
- *Object Types:* Different algorithms have been proposed for convex polytopes, general polygonal models, curved objects described using parametric splines or implicit functions, set-theoretic combinations of objects, deformable models, etc.
- *Motion Formulation:* The collision query can be augmented by adding the element of time. If the trajectories of two objects are known, then we can determine when is the next time that a particular boolean query will become true or false. These queries are called *dynamic queries*, whereas the ones that do not use motion information are called *static queries*. In the case where the motion of an object can not be represented as a closed form function of time, the underlying application often performs static queries at specific time steps in the application.

In this chapter, we give a brief survey of different collision detection algorithms for convex polytopes, general polygonal models, penetration computations and large-scaled environments composed of multiple objects. In each category, we give a detailed description of one of the algorithms and the underlying data structures.

57.2 Convex Polytopes

In this section, we give a brief survey of algorithms for collision detection between a pair of convex polytopes. This problem has been extensively studied and a number of algorithms with good asymptotic performance have been proposed. The best known runtime algorithm for boolean collision queries takes $O(log^2 n)$ time, where n is the number of features [5]. It precomputes the Dobkin-Kirkpatrick hierarchy for each polytope and uses it to perform the runtime query. In practice, three classes of algorithms are commonly used for convex polytopes. These are linear programming, Minkowski sums, and tracking closest features based on Voronoi diagrams.

57.2.1 Linear Programming

The problem of checking whether two convex polytopes intersect or not can be posed as a linear programming (LP) problem. In particular, two convex polytopes do not overlap, if and only if there exists a separation plane between them. The coefficients of the separation plane equation are treated as unknowns. The linear constraints are formulated by imposing that all the vertices of the first polytope lie in one half-space of this plane and those of the other polytope lie in the other half-space. The linear programming algorithms are used to check whether there is any feasible solution to the given set of constraints. Given the fixed dimension of the problem, some of the well-known linear programming algorithms [6] can be used to perform the boolean collision query in expected linear time.

57.2.2 Voronoi-Based Marching Algorithm

An expected constant time algorithm for collision detection was proposed by Lin and Canny [7,8]. This algorithm tracks the closest features between two convex polytopes. The features may correspond to a vertex, an edge or a face of each polytope. Variants of this algorithm have also been presented in References 9,10. The original algorithm basically works by traversing the external Voronoi regions induced by the features of each convex polyhedron toward the pair of the closest features between the two given polytopes. The invariant is that at each step, either the inter-feature distance is reduced or the dimensionality of one or both of the features decreases by one, that is, a move from a face to an edge or from an edge to a vertex.

The algorithm terminates when the pair of testing features contain a pair of points that lie within the Voronoi regions of the other feature. It returns the pair of closest features and the Euclidean distance between them, as well as the contact status (i.e., colliding or not). This algorithm uses a modified boundary representation to represent convex polytopes and a data structure for describing "*Voronoi regions*" of convex polytopes.

57.2.2.1 Polytope Representation

Let A be a polytope. A is partitioned into "*features*" $f_1, \ldots, f_n$ where n is the total number of features, that is, $n = f + e + v$ where f, e, v stands for the total number of faces, edges, vertices respectively. Each feature (except vertex) is an open subset of an affine plane and does not contain its boundary.

Definition: B is in the *boundary* of F and F is in *coboundary* of B, if and only if B is in the closure of F, that is, $B \subseteq \overline{F}$ and B has one fewer dimension than F does.

For example, the coboundary of a vertex is the set of edges touching it and the coboundary of an edge are the two faces adjacent to it. The boundary of a face is the set of edges in the closure of the face. It uses winged edge representation, commonly used for boundary evaluation of boolean combinations of solid models [11]. The edge is oriented by giving two incident vertices (the head and tail). The edge points from tail to head. It has two adjacent faces cobounding it as well. Looking from the the tail end toward the head, the adjacent face lying to the right hand side is labeled as the "right face" and similarly for the "left face."

Each polytope's data structure has a field for its features (faces, edges, vertices) and Voronoi cells to be described below. Each feature is described by its geometric parameters. Its data structure also includes a list of its boundary, coboundary, and *Voronoi regions*.

Definition: A *Voronoi region* associated with a *feature* is a set of points exterior to the polyhedron which are closer to that feature than any other. The Voronoi regions form a partition of space outside the polyhedron according to the closest feature. The collection of Voronoi regions of each polyhedron is the generalized Voronoi diagram of the polyhedron. Note that the Voronoi diagram of a convex polyhedron has linear size and consists of polyhedral regions.

Definition: A Voronoi *cell* is the data structure for a Voronoi region. It has a set of constraint planes that bound its Voronoi region with pointers to the neighboring cells (each of which shares a common constraint plane with the given Voronoi cell) in its data structure.

Using the geometric properties of convex sets, "applicability criteria" are established based upon the Voronoi regions. That is, if a point P on object A lies inside the Voronoi region of f_B on object B, then f_B is a closest feature to the point P. If a point lies on a constraint plane, then it is equi-distant from the two features that share this constraint plane in their Voronoi cells.

57.2.2.2 Local Walk

The algorithm incrementally traverses the features of each polytope to compute the closest features. For example, given a pair of features, *Face 1* and *vertex* V_a on objects A and B, respectively, as the pair of initial features (Figure 57.1). The algorithm verifies if the vertex V_a lies within *Cell 1* of *Face 1*. However, V_a violates the constraint plane imposed by *CP* of *Cell 1*, that is, V_a does not line in the half-space defined by *CP* which contains *Cell 1*. The constraint plane *CP* has a pointer to its adjacent cell *Cell 2*, so the walk proceeds to test whether V_a is contained within *Cell 2*. In similar fashion, vertex V_a has a cell of its own, and the algorithm checks whether the nearest point P_a on the edge to the vertex V_a lies within V_a's Voronoi cell. Basically, the algorithm checks whether a point is contained within a Voronoi region defined by the constraint planes of the region. The constraint plane, which causes this test to fail, points to the next pair of closest features. Eventually, the algorithm computes the closest pair of features.

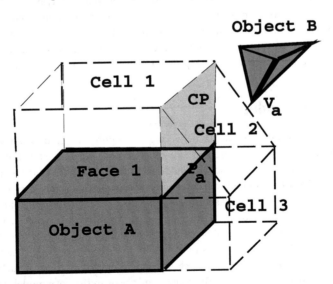

FIGURE 57.1 A walk across Voronoi cells. Initially, the algorithm checks whether the vertex V_a lies in *Cell 1*. After it fails the containment test with respect to the plane *CP*, it walks to *Cell 2* and checks for containment in *Cell 2*.

Since the polytopes and its faces are convex, the containment test involves only the neighboring features of the current candidate features. If either feature fails the test, the algorithm steps to a neighboring feature of one or both candidates, and tries again. With some simple preprocessing, the algorithm can guarantee that every feature of a polytope has a constant number of neighboring features. As a result, it takes a constant number of operations to check whether two features are the closest features.

This approach can be used in a static environment, but is especially well-suited for dynamic environments in which objects move in a sequence of small, discrete steps. The method takes advantage of coherence within two successive static queries: that is, the closest features change infrequently as the polytopes move along finely discretized paths. The closest features computed from the previous positions of the polytopes are used as the initial features for the current positions. The algorithm runs in *expected constant time* if the polytopes are not moving quickly. Even when a closest feature pair is changing rapidly, the algorithm takes only slightly longer. The running time is proportional to the number of feature pairs traversed, which is a function of the relative motion the polytopes undergo.

57.2.2.3 Implementation and Application

The Lin-Canny algorithm has been implemented as part of several public-domain libraries, include I-COLLIDE and SWIFT [12]. It has been used for different applications including dynamic simulation [13], interactive walkthrough of architectural models [3] and haptic display (including force and torque computation) of polyhedral models [14].

57.2.3 Minkowski Sums and Convex Optimization

The collision and distance queries can be performed based on the Minkowski sum of two objects. Given two sets of points, P and Q, their Minkowski sum is the set of points:

$$\{\mathbf{p} + \mathbf{q} \mid \mathbf{p} \in P, \mathbf{q} \in Q.\}$$

It has been shown in Reference 15, that the minimum separation distance between two objects is the same as the minimum distance from the origin of the Minkowski sums of A and $-B$ to the surface of the sums. The Minkowski sum is also referred to as the *translational C-space obstacle* (TCSO) If we take the TCSO of two convex polyhedra, A and B, then the TCSO is another convex polyhedra, and each vertex of the TCSO correspond to the vector difference of a vertex from A and a vertex from B. While the Minkowski sum of two convex polytopes can have $O(n^2)$ features [16], a fast algorithm for separation distance computation based on convex optimization that exhibits linear-time performance in practice has been proposed by Gilbert et al. [17]. It is also known as the GJK algorithm. It uses pairs of vertices from each object that define simplices (i.e., a point, bounded line, triangle or tetrahedron) within each polyhedra and a corresponding simplex in the TCSO. Initially the simplex is set randomly and the algorithm refines it using local optimization, till it computes the closest point on the TCSO from the origin of the Minkowski sums. The algorithm assumes that the origin is not inside the TCSO.

By taking the similar philosophy as the Lin-Canny algorithm [8], Cameron [18] presented an extension to the basic GJK algorithm by exploiting motion coherence and geometric locality in terms of connectivity between neighboring features. It keeps track of the *witness points*, a pair of points from the two objects that realize the minimum separation distance between them. As opposed to starting from a random simplex in the TCSO, the algorithm starts with the witness points from the previous iteration and performs hill climbing to compute a new set of witness points for the current configuration. The running time of this algorithm is a function of the number of refinement steps that the algorithm has to perform.

57.3 General Polygonal Models

Algorithms for collision and separation distance queries between general polygons models can be classified based on the fact whether they are closed polyhedral models or represented as a collection of polygons. The latter, also referred to as "polygon soups," make no assumption related to the connectivity among different faces or whether they represent a closed set.

Some of the commonly known algorithms for collision detection and separation distance computation use spatial partitioning or bounding volume hierarchies (BVHs) . The spatial subdivisions are a recursive partitioning of the embedding space, whereas bounding volume hierarchies are based on a recursive partitioning of the primitives of an object. These algorithms are based on the divide-and-conquer paradigm. Examples of spatial partitioning hierarchies include k-D trees and octrees [19], R-trees and their variants [20], cone trees, BSPs [21] and their extensions to multi-space partitions [22]. The BVHs use bounding volumes (BVs) to bound or contain sets of geometric primitives, such as triangles, polygons, curved surfaces, etc. In a BVH, BVs are stored at the internal nodes of a tree structure. The root BV contains all the primitives of a model, and children BVs each contain separate partitions of the primitives enclosed by the parent. Each of the leaf node BVs typically contains one primitive. In some variations, one may place several primitives at a leaf node, or use several volumes to contain a single primitive. The BVHs are used to perform collision and separation distance queries. These include sphere-trees [23,24], AABB-trees [20,25, 26], OBB-trees [27–29], spherical shell-trees [30,31], *k*-DOP-trees [32,33], SSV-trees [34], and convex hull-trees [35]. Different BVHs can be classified based on:

- *Choice of BV:* The AABB-tree uses an axis-aligned bounding box (AABB) as the underlying BV. The AABB for a set of primitives can be easily computed from the extremal points along the *X*, *Y* and *Z* direction. The sphere tree uses a sphere as the underlying BV. Algorithms to compute a minimal bounding sphere for a set of points in 3D are well known in computational geometry. The *k*-DOP-tree is an extension of the AABB-tree, where each BV is computed from extremal points along *k* fixed directions, as opposed to the 3 orthogonal axes. A spherical shell is a subset of the volume contained between two concentric spheres and is a tight fitting BV. A SSV (swept sphere volume) is defined by taking the Minkowski sum of a point, line or a rectangle in 3D with a sphere. The SSV-tree corresponds to a hybrid hierarchy, where the BVs may correspond to a point-swept sphere (PSS), a line-swept sphere (LSS) or a rectangle-swept sphere (RSS). Finally, the BV in the convex-hull tree is a convex polytope.
- *Hierarchy generation:* Most of the algorithms for building hierarchies fall into two categories: bottom-up and top-down. Bottom-up methods begin with a BV for each primitive and merge volumes into larger volumes until the tree is complete. Top-down methods begin with a group of all primitive, and recursively subdivide until all leaf nodes are indivisible. In practice, top-down algorithms are easier to implement and typically take $O(n \lg n)$ time, where n is the number of primitives. On the other hand, the bottom-up methods use clustering techniques to group the primitives at each level and can lead to tighter-fitting hierarchies.

The collision detection queries are performed by traversing the BVHs. Two models are compared by recursively traversing their BVHs in tandem. Each recursive step tests whether BVs *A* and *B*, one from each hierarchy, overlap. If *A* and *B* do not overlap, the recursion branch is terminated. But if *A* and *B* overlap, the enclosed primitives may overlap and the algorithm is applied recursively to their children. If *A* and *B* are both leaf nodes, the primitives within them are compared directly. The running time of the algorithm is dominated by the overlap tests between two BVs and a BV and a primitive. It is relatively simple to check whether two AABBs overlap or two spheres overlap or two k-DOPs overlap. Specialized algorithms have also been proposed to check whether two OBBs, two SSVs, two spherical shells or two convex polytopes overlap. Next, we described a commonly used interference detection algorithm that uses hierarchies of oriented bounding boxes.

57.3.1 Interference Detection Using Trees of Oriented Bounding Boxes

In this section we describe an algorithm for building a BVH of OBBs (called OBBTree) and using them to perform fast interference queries between polygonal models. More details about the algorithm are available in References 27,29.

The underlying algorithm is applicable to all triangulated models. They need not represent a compact set or have a manifold boundary representation. As part of a preprocess, the algorithm computes a hierarchy of oriented bounding boxes (OBBs) for each object. At runtime, it traverses the hierarchies to check whether the primitives overlap.

57.3.1.1 OBBTree Construction

An OBBTree is a bounding volume tree of OBBs. Given a collection of triangles, the algorithm initially approximates them with an OBB of similar dimensions and orientation. Next, it computes a hierarchy of OBBs.

The OBB computation algorithm makes use of first and second order statistics summarizing the vertex coordinates. They are the mean, μ, and the covariance matrix, $\mathbf{C}$, respectively [36]. If the vertices of the i'th triangle are the points $\mathbf{p}^i$, $\mathbf{q}^i$, and $\mathbf{r}^i$, then the mean and covariance matrix can be expressed in vector notation as:

$$\mu = \frac{1}{3n} \sum_{i=0}^{n} (\mathbf{p}^i + \mathbf{q}^i + \mathbf{r}^i),$$

$$\mathbf{C}_{jk} = \frac{1}{3n} \sum_{i=0}^{n} (\bar{\mathbf{p}}_j^i \bar{\mathbf{p}}_k^i + \bar{\mathbf{q}}_j^i \bar{\mathbf{q}}_k^i + \bar{\mathbf{r}}_j^i \bar{\mathbf{r}}_k^i), \quad 1 \le j, k \le 3$$

where n is the number of triangles, $\bar{\mathbf{p}}^i = \mathbf{p}^i - \mu$, $\bar{\mathbf{q}}^i = \mathbf{q}^i - \mu$, and $\bar{\mathbf{r}}^i = \mathbf{r}^i - \mu$. Each of them is a 3×1 vector, for example, $\bar{\mathbf{p}}^i = (\bar{\mathbf{p}}_1^i, \bar{\mathbf{p}}_2^i, \bar{\mathbf{p}}_3^i)^T$ and $\mathbf{C}_{jk}$ are the elements of the 3 by 3 covariance matrix.

The eigenvectors of a symmetric matrix, such as $\mathbf{C}$, are mutually orthogonal. After normalizing them, they are used as a basis. The algorithm finds the extremal vertices along each axis of this basis. Two of the three eigenvectors of the covariance matrix are the axes of maximum and of minimum variance, so they will tend to align the box with the geometry of a tube or a flat surface patch.

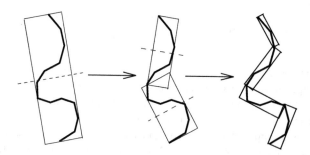

FIGURE 57.2 Building the OBBTree: recursively partition the bounded polygons and bound the resulting groups. This figure shows three levels of an OBBTree of a polygonal chain.

The algorithm's performance can be improved by using the convex hull of the vertices of the triangles. To get a better fit, we can sample the surface of the convex hull densely, taking the mean and covariance of the sample points. The uniform sampling of the convex hull surface normalizes for triangle size and distribution.

One can sample the convex hull "infinitely densely" by integrating over the surface of each triangle, and allowing each differential patch to contribute to the covariance matrix. The resulting integral has a closed form solution. Let the area of the i'th triangle in the convex hull be denoted by

$$A^i = \frac{1}{2}|(\mathbf{p}^i - \mathbf{q}^i) \times (\mathbf{p}^i - \mathbf{r}^i)|$$

Let the surface area of the entire convex hull be denoted by

$$A^H = \sum_i A^i$$

Let the centroid of the i'th convex hull triangle be denoted by

$$\mathbf{c}^i = (\mathbf{p}^i + \mathbf{q}^i + \mathbf{r}^i)/3$$

Let the centroid of the convex hull, which is a weighted average of the triangle centroids (the weights are the areas of the triangles), be denoted by

$$\mathbf{c}^H = \frac{\sum_i A^i \mathbf{c}^i}{\sum_i A^i} = \frac{\sum_i A^i \mathbf{c}^i}{A^H}$$

The elements of the covariance matrix $\mathbf{C}$ have the following closed-form,

$$\mathbf{C}_{jk} = \sum_{i=1}^{n} \frac{A^i}{12A^H} \left(9\mathbf{c}_j^i \mathbf{c}_k^i + \mathbf{p}_j^i \mathbf{p}_k^i + \mathbf{q}_j^i \mathbf{q}_k^i + \mathbf{r}_j^i \mathbf{r}_k^i\right) - \mathbf{c}_j^H \mathbf{c}_k^H$$

Given an algorithm to compute tight-fitting OBBs around a group of polygons, we need to represent them hierarchically. The simplest algorithm for OBBTree computation uses a top-down method. It is based on a subdivision rule that splits the longest axis of a box with a plane orthogonal to one of its axes, partitioning the polygons according to which side of the plane their center point lies on (a 2-D analog is shown in Figure 57.2). The subdivision coordinate along that axis is chosen to be that of the mean point, μ, of the vertices. If the longest axis cannot not be subdivided, the second longest axis is chosen. Otherwise, the shortest one is used. If the group of polygons cannot be partitioned along any axis by this criterion, then the group is considered indivisible.

Given a model with n triangles, the overall time to build the tree is $O(n \log^2 n)$ if we use convex hulls, and $O(n \log n)$ if we don't. The recursion is similar to that of quicksort. Fitting a box to a group of n triangles and partitioning them into two subgroups takes $O(n \log n)$ with a convex hull and $O(n)$ without it. Applying the process recursively creates a tree with leaf nodes $O(\log n)$ levels deep.

57.3.1.2 Interference Detection

Given OBBTrees of two objects, the interference algorithm typically spends most of its time testing pairs of OBBs for overlap. The algorithm computes axial projections of the bounding boxes and check for disjointness along those axes. Under this projection, each box forms an interval on the axis (a line in 3D). If the intervals don't overlap, then the axis is called a 'separating axis' for the boxes, and the boxes must then be disjoint.

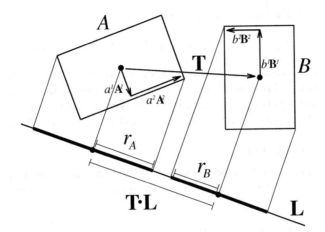

FIGURE 57.3 $\vec{L}$ is a separating axis for OBBs A and B because A and B become disjoint intervals under projection (shown as r_A and r_B, respectively) onto $\vec{L}$.

It has been shown that we need to perform at most 15 axial projections in 3D to check whether two OBBs overlap or not [29]. These 15 directions correspond to the face normals of each OBB, as well as 9 pairwise combinations obtained by taking the cross-product of the vectors representing their edges.

To perform the test, the algorithm projects the centers of the boxes onto the axis, and also to compute the radii of the intervals. If the distance between the box centers as projected onto the axis is greater than the sum of the radii, then the intervals (and the boxes as well) are disjoint. This is shown in 2D in Figure 57.3.

57.3.1.3 OBB Representation and Overlap Test

We assume we are given two OBBs, A and B, with B placed relative to A by rotation matrix $\vec{R}$ and translation vector $\vec{T}$. The half-dimensions (or "radii") of A and B are a_i and b_i, where $i = 1, 2, 3$. We will denote the axes of A and B as the unit vectors $\vec{A}^i$ and $\vec{B}^i$, for $i = 1, 2, 3$. These will be referred to as the 6 box axes. Note that if we use the box axes of A as a basis, then the three columns of $\vec{R}$ are the same as the three $\vec{B}^i$ vectors.

The centers of each box projects onto the midpoint of its interval. By projecting the box radii onto the axis, and summing the length of their images, we obtain the radius of the interval. If the axis is parallel to the unit vector $\vec{L}$, then the radius of box A's interval is

$$r_A = \sum_i |a_i \vec{A}^i \cdot \vec{L}|$$

A similar expression is used to compute r_B.

The placement of the axis is immaterial, so we assume it passes through the center of box A. The distance between the midpoints of the intervals is $|\vec{T} \cdot \vec{L}|$. So, the intervals are disjoint if and only if

$$|\vec{T} \cdot \vec{L}| > \sum_i |a_i \vec{A}^i \cdot \vec{L}| + \sum_i |b_i \vec{B}^i \cdot \vec{L}|$$

This simplifies when $\vec{L}$ is a box axis or cross product of box axes. For example, consider $\vec{L} = \vec{A}^1 \times \vec{B}^2$. The second term in the first summation is

$$|a_2 \vec{A}^2 \cdot (\vec{A}^1 \times \vec{B}^2)| = |a_2 \vec{B}^2 \cdot (\vec{A}^2 \times \vec{A}^1)|$$
$$= |a_2 \vec{B}^2 \cdot \vec{A}^3|$$
$$= |a_2 \vec{B}_3^2|$$
$$= a_2 |\vec{R}_{32}|$$

The last step is due to the fact that the columns of the rotation matrix are also the axes of the frame of B. The original term consisted of a dot product and cross product, but reduced to a multiplication and an absolute value. Some terms reduce to zero and are eliminated. After simplifying all the terms, this axis test looks like:

$$|\vec{T}_3\vec{R}_{22} - \vec{T}_2\vec{R}_{32}| > a_2|\vec{R}_{32}| + a_3|\vec{R}_{22}| + b_1|\vec{R}_{13}| + b_3|\vec{R}_{11}|$$

All 15 axis tests simplify in similar fashion. Among all the tests, the absolute value of each element of $\vec{R}$ is used four times, so those expressions can be computed once before beginning the axis tests. If any one of the expressions is satisfied, the boxes are known to be disjoint, and the remainder of the 15 axis tests are unnecessary. This permits early exit from the series of tests. In practice, it takes up to 200 arithmetic operations in the worst case to check whether two OBBs overlap.

57.3.1.4 Implementation and Application

The OBBTree interference detection algorithm has been implemented and used as part of the following packages: RAPID, V-COLLIDE and PQP [12]. These implementations have been used for robot motion planning, dynamic simulation and virtual prototyping.

57.3.2 Performance of Bounding Volume Hierarchies

The performance of BVHs on proximity queries is governed by a number of design parameters. These include techniques to build the trees, number of children each node can have, and the choice of BV type. An additional design choice is the descent rule. This is the policy for generating recursive calls when a comparison of two BVs does not prune the recursion branch. For instance, if BVs A and B failed to prune, one may recursively compare A with each of the children of B, B with each of the children of A, or each of the children of A with each the children of B. This choice does not affect the correctness of the algorithm, but may impact the performance. Some of the commonly used algorithms assume that the BVHs are binary trees and each primitive is a single triangle or a polygon. The cost of performing the collision query is given as [27,34]:

$$T = N_{bv} \times C_{bv} + N_p \times C_p,$$

where T is the total cost function for collision queries, N_{bv} is the number of bounding volume pair operations, and C_{bv} is the total cost of a BV pair operation, including the cost of transforming and updating (including resizing) each BV for use in a given configuration of the models, and other per BV-operation overhead. N_p is the number of primitive pairs tested for proximity, and C_p is the cost of testing a pair of primitives for proximity (e.g., overlaps or distance computation).

Typically for tight fitting bounding volumes, for example, oriented bounding boxes (OBBs), N_{bv} and N_p are relatively low, whereas C_{bv} is relatively high. In contrast, C_{bv} is low while N_{bv} and N_p may be higher for simple BV types like spheres and axis-aligned bounding boxes (AABBs). Due to these opposing trends, no single BV yields optimum performance for collision detection in all possible cases.

57.4 Penetration Depth Computation

In this section, we give a brief overview of penetration depth (PD) computation algorithms between convex polytopes and general polyhedral models. The PD of two inter-penetrating objects A and B is defined as the minimum translation distance that one object undergoes to make the interiors of A and B disjoint. It can be also defined in terms of the TCSO. When two objects are overlapping, the origin of the Minkowski sum of A and $-B$ is contained inside the TCSO. The penetration depth corresponds to the minimum distance from the origin to the surface of TCSO [18]. PD computation is often used in motion planning [37], contact resolution for dynamic simulation [38,39] and force computation in haptic rendering [40]. For example, computation of dynamic response in penalty-based methods often needs to perform PD queries for imposing the non-penetration constraint for rigid body simulation. In addition, many applications, such as motion planning and dynamic simulation, require a continuous distance measure when two (non-convex) objects collide, in order to have a well-posed computation.

Some of the algorithms for PD computation involve computing the Minkowski sums and computing the closest point on its surface from the origin. The worst case complexity of the overall PD algorithm is governed by the complexity of computing Minkowski sums, which can be $O(n^2)$ for convex polytopes and $O(n^6)$ for general (or non-convex) polyhedral models [16]. Given the complexity of Minkowski sums, many approximation algorithms have been proposed in the literature for fast PD estimation.

57.4.1 Convex Polytopes

Dobkin et al. [16] have proposed a hierarchical algorithm to compute the directional PD using Dobkin and Kirkpatrick polyhedral hierarchy. For any direction d, it computes the directional penetration depth in $O(\log n \log m)$ time for polytopes with m and n vertices. Agarwal et al. [41] have presented a randomized approach to compute the PD values [41]. It runs in

$O(m^{\frac{3}{4}+\epsilon}n^{\frac{3}{4}+\epsilon} + m^{1+\epsilon} + n^{1+\epsilon})$ expected time for any positive constant ϵ. Cameron [18] has presented an extension to the GJK algorithm [17] to compute upper and lower bounds on the PD between convex polytopes. Bergen has further elaborated this idea in an expanding polytope algorithm [42]. The algorithm iteratively improves the result of the PD computation by expanding a polyhedral approximation of the Minkowski sums of two polytopes.

57.4.2 Incremental Penetration Depth Computation

Kim et al. [43] have presented an incremental penetration depth (PD) algorithm that marches towards a "locally optimal" solution by walking on the surface of the Minkowski sum. The surface of the TCSO is implicitly computed by constructing a local Gauss map and performing a local walk on the polytopes.

This algorithm uses the concept of width computation from computational geometry. Given a set of points $P = \{p_1, p_2, \ldots, p_n\}$ in 3D, the *width* of P, $W(P)$, is defined as the minimum distance between parallel planes supporting P. The width $W(P)$ of convex polytopes A and B is closely related to the penetration depth $PD(A, B)$, since it is easy to show that $W(P) = PD(P, P)$. It can be shown that width and penetration depth computation can be reduced to searching only the VF and EE antipodal pairs (where V, E and F denote a vertex, edge and face, respectively, of the polytopes). This is accomplished by using the standard dual mapping on the Gauss map (or normal diagram). The mapping is defined from object space to the surface of a unit sphere $\mathbb{S}^2$ as: a vertex is mapped to a region, a face to a point, and an edge to a great arc. The algorithm finds the antipodal pairs by overlaying the upper hemisphere of the Gauss map on the lower hemisphere and computing the intersections between them.

57.4.2.1 Local Walk

The incremental PD computation algorithm does not compute the entire Gauss map for each polytope or the entire boundary of the Minkowski sum. Rather it computes them in a lazy manner based on local walking and optimization. Starting from some feature on the surface of the Minkowski sum, the algorithm computes the direction in which it can decrease the PD value and proceeds towards that direction by extending the surface of the Minkowski sum.

At each iteration of the algorithm, a vertex is chosen from each polytope to form a pair. It is called a *vertex hub pair* and the algorithm uses it as a hub for the expansion of the local Minkowski sum. The vertex hub pair is chosen in such a way that there exists a plane supporting each polytope, and is incident on each vertex. It turns out that the vertex hub pair corresponds to two intersected convex regions on a Gauss map, which later become intersecting convex polygons on the plane after *central projection*. The intersection of convex polygons corresponds to the VF or EE antipodal pairs that are used to reconstruct the local surface of the Minkowski sum around the vertex hub pair. Given these pairs, the algorithm chooses the one that corresponds to the shortest distance from the origin of the Minkowski sum to their surface. If this pair decreases the estimated PD value, the algorithm updates the current vertex hub pair to the new adjacent pair. This procedure is repeated until the algorithm can not decrease the current PD value and converges to a local minima.

57.4.2.2 Initialization and Refinement

The algorithm starts with an initial guess on the vertex hub pair. A good estimate to the penetration direction can be obtained by taking the centroid difference between the objects, and computing an extremal vertex pair for the difference direction. In other cases, the penetrating features (for overlapping polytopes) or the closest features (from non-overlapping polytopes) from the previous instance can also suggest a good initial guess.

After the algorithm obtains a initial guess for a **VV** pair, it iteratively seeks to improve the PD estimate by jumping from one **VV** pair to an adjacent **VV** pair. This is accomplished by looking around the neighborhood of the current **VV** pair and walking to a pair which provides the greatest improvement in the PD value. Let the current vertex hub pair be $v_1 v_1'$. The next vertex hub pair $v_2 v_2'$ is computed as follows:

1. Construct a local Gauss map each for v_1 and v_1',
2. Project the Gauss maps onto $z = 1$ plane, and label them as G and G', respectively. G and G' correspond to convex polygons in 2D.
3. Compute the intersection between G and G' using a linear time algorithm such as [44]. The result is a convex polygon and let u_i be a vertex of the intersection set. If u_i is an original vertex of G or G', it corresponds to the VF antipodal pair in object space. Otherwise, it corresponds to an EE antipodal pair.
4. In object space, determine which u_i corresponds to the best local improvement in PD, and set an adjacent vertex pair (adjacent to u_i) to $v_2 v_2'$.

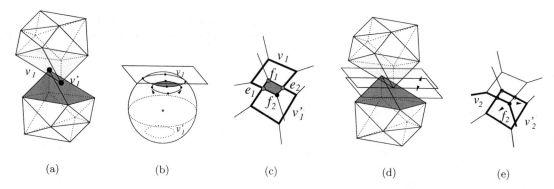

FIGURE 57.4 Iterative optimization for incremental PD computation: (a) The current **VV** pair is $v_1 v_1'$ and a shaded region represents edges and faces incident to $v_1 v_1'$. (b) shows local Gauss maps and their overlay for $v_1 v_1'$. (c) shows the result of the overlay after central projection onto a plane. Here, f_1, e_1, f_2 and e_2 comprise vertices (candidate PD features) of the overlay. (d) illustrates how to compute the PD for the candidate PD features in object space. (e) f_2 is chosen as the next PD feature, thus $v_2 v_2'$ is determined as the next vertex hub pair.

This iteration is repeated until either there is no more improvement in the PD value or number of iterations reach some maximum value. At step 4 of the iteration, the next vertex hub pair is selected in the following manner. If u_i corresponds to VF, then the algorithm chooses one of the two vertices adjacent to F assuming that the model is triangulated. The same reasoning is also applied to when u_i corresponds to EE. As a result, the algorithm needs to perform one more iteration in order to actually decide which vertex hub pair should be selected. A snapshot of a typical step during the iteration is illustrated in Figure 57.4. Eventually the algorithm computes a local minima.

57.4.2.3 Implementation and Application

The incremental algorithm is implemented as part of DEEP [12]. It works quite well in practice and is also able to compute the global penetration depth in most cases. It has been used for 6-DOF haptic rendering (force and torque display), dynamic simulation and virtual prototyping.

57.4.3 Non-Convex Models

Algorithms for penetration depth estimation between general polygonal models are based on discretization of the object space containing the objects or use of digital geometric algorithms that perform computations on a finite resolution grid. Fisher and Lin [45] have presented a PD estimation algorithm based on the distance field computation using the fast marching level-set method. It is applicable to all polyhedral objects as well as deformable models, and it can also check for self-penetration. Hoff et al. [46,47] have proposed an approach based on performing discretized computations on the graphics rasterization hardware. It uses multi-pass rendering techniques for different proximity queries between general rigid and deformable models, including penetration depth estimation. Kim et al. [43] have presented a fast approximation algorithm for general polyhedral models using a combination of object-space as well discretized computations. Given the global nature of the PD problem, it decomposes the boundary of each polyhedron into convex pieces, computes the pairwise Minkowski sums of the resulting convex polytopes and uses graphics rasterization hardware to perform the closest point query up to a given discretized resolution. The results obtained are refined using a local walking algorithm. To further speed up this computation and improve the estimate, the algorithm uses a hierarchical refinement technique that takes advantage of geometry culling, model simplification, accelerated ray-shooting, and local refinement with greedy walking. The overall approach combines discretized closest point queries with geometry culling and refinement at each level of the hierarchy. Its accuracy can vary as a function of the discretization error. It has been applied to haptic rendering and dynamic simulation.

57.5 Large Environments

Large environments are composed of multiple moving objects. Different methods have been proposed to overcome the computational bottleneck of $O(n^2)$ pairwise tests in an environment composed of n objects. The problem of performing proximity queries in large environments, is typically divided into two parts [3,23]: the *broad phase*, in which we identify the pair of objects on which we need to perform different proximity queries, and the *narrow phase*, in which we perform the exact pairwise queries. In this section, we present a brief overview of algorithms used in the broad phase.

Architecture for Multi-body Collision Detection

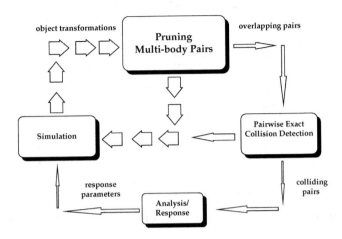

FIGURE 57.5 Architecture for multiple body collision detection algorithm.

The simplest algorithms for large environments are based on spatial subdivisions. The space is divided into cells of equal volume, and at each instance the objects are assigned to one or more cells. Collisions are checked between all object pairs belonging to each cell. In fact, Overmars has presented an efficient algorithm based on hash table to efficient perform point location queries in fat subdivisions [48]. This approach works well for sparse environments in which the objects are uniformly distributed through the space. Another approach operates directly on four-dimensional volumes swept out by object motion over time [49].

57.5.1 Multiple-Object Collision Detection

Large-scale environments consist of stationary as well as moving objects. Let there be N moving objects and M stationary objects. Each of the N moving objects can collide with the other moving objects, as well as with the stationary ones. Keeping track of $(N2) + NM$ pairs of objects at every time step can become time consuming as N and M get large. To achieve interactive rates, we must reduce this number before performing pairwise collision tests. In this section, we give an overview of sweep-and-prune algorithm used to perform multiple-object collision detection [3]. The overall architecture of the multiple object collision detection algorithm is shown in Figure 57.5.

The algorithm uses sorting to prune the number of pairs. Each object is surrounded by a 3-dimensional bounding volume. These bounding volumes are sorted in 3-space to determine which pairs are overlapping. The algorithm only needs to perform exact pairwise collision tests on these remaining pairs.

However, it is not intuitively obvious how to sort objects in 3-space. The algorithm uses a *dimension reduction* approach . If two bodies collide in a 3-dimensional space, their orthogonal projections onto the xy, yz, and xz-planes and x, y, and z-axes must overlap. Based on this observation, the algorithm uses axis-aligned bounding boxes and efficiently project them onto a lower dimension, and perform sorting on these lower-dimensional structures.

The algorithm computes a rectangular bounding box to be the tightest axis-aligned box containing each object at a particular orientation. It is defined by its minimum and maximum x, y, and z-coordinates. As an object moves, the algorithm recomputes its minima and maxima, taking into account the object's orientation.

As a precomputation, the algorithm computes each object's initial minima and maxima along each axis. It is assumed that the objects are convex. For non-convex polyhedral models, the following algorithm is applied to their convex hulls. As an object moves, its minima and maxima are recomputed in the following manner:

1. Check to see if the current minimum (or maximum) vertex for the x, y, or z-coordinate still has the smallest (or largest) value in comparison to its neighboring vertices. If so the algorithm terminates.
2. Update the vertex for that extreme direction by replacing it with the neighboring vertex with the smallest (or largest) value of all neighboring vertices. Repeat the entire process as necessary.

This algorithm recomputes the bounding boxes at an expected constant rate. It exploits the temporal and geometric coherence. The algorithm does not transform all the vertices as the objects undergo motion. As it is updating the bounding boxes, new positions are computed for current vertices using matrix-vector multiplications. This approach is optimized based on the fact that

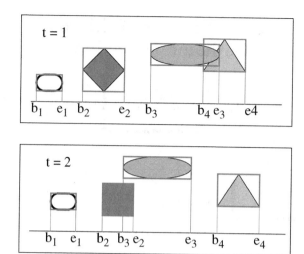

FIGURE 57.6   Bounding box behavior between successive instances. Notice the coherence between the 1D list obtained after projection onto the X-axis.

the algorithm is interested in *one* coordinate value of each extremal vertex, say the *x* coordinate while updating the minimum or maximum value along the x-axis. Therefore, there is no need to transform the other coordinates in order to compare neighboring vertices. This reduces the number of arithmetic operations by two-thirds, as we only compute a dot-product of two vectors and do not perform matrix-vector multiplication.

57.5.1.1 One-Dimensional Sweep and Prune

The one-dimensional sweep and prune algorithm begins by projecting each three-dimensional bounding box onto the *x*, *y*, and *z* axes. Because the bounding boxes are axis-aligned, projecting them onto the coordinate axes results in intervals (see Figure 57.6). The goal is to compute the overlaps among these intervals, because a pair of bounding boxes can overlap if and only if their intervals overlap in all three dimensions.

The algorithm constructs three lists, one for each dimension. Each list contains the values of the endpoints of the intervals corresponding to that dimension. By sorting these lists, it determines which intervals overlap. In the general case, such a sort would take $O(n \log n)$ time, where n is the number of objects. This time bound is reduced by keeping the sorted lists from the previous frame, changing only the values of the interval endpoints. In environments where the objects make relatively small movements between frames, the lists will be nearly sorted, so we can sort in expected $O(n)$ time, as shown in References 50,51. In practice, insertion sort works well for almost sorted lists.

In addition to sorting, the algorithm keeps track of changes in the overlap status of interval pairs (i.e., from overlapping in the last time step to non-overlapping in the current time step, and vice-versa). This can be done in $O(n + e_x + e_y + e_z)$ time, where e_x, e_y, and e_z are the number of exchanges along the x, y, and z-axes. This also runs in expected linear time due to coherence, but in the worst case e_x, e_y, and e_z can each be $O(n^2)$ with an extremely small constant.

This algorithm is suitable for dynamic environments where coherence between successive static queries is preserved. In computational geometry literature several algorithms exist that solve the static version of determining 3-D bounding box overlaps in $O(n \log^2 n + s)$ time, where s is the number of pairwise overlaps [52,53]. It has been reduced to to $O(n + s)$ by exploiting coherence.

57.5.1.2 Implementation and Application

The sweep-and-prune has been used in some of the widely uses collision detection systems, including I-COLLIDE, V-COLLIDE, SWIFT and SWIFT++ [12]. It has been used for multi-body simulations and interactive walkthroughs of complex environments.

57.5.2 Two-Dimensional Intersection Tests

The two-dimensional intersection algorithm begins by projecting each three-dimensional axis-aligned bounding box onto any two of the *x-y*, *x-z*, and *y-z* planes. Each of these projections is a rectangle in 2-space. Typically there are fewer overlaps of these 2-D rectangles than of the 1-D intervals used by the sweep and prune technique. This results in fewer swaps as the objects move.

In situations where the projections onto one-dimension result in densely clustered intervals, the two-dimensional technique is more efficient. The interval tree is a common data structure for performing such two-dimensional range queries [54].

Each query of an interval intersection takes $O(\log n + k)$ time where k is the number of reported intersections and n is the number of intervals. Therefore, reporting intersections among n rectangles can be done in $O(n \log n + K)$ where K is the total number of intersecting rectangles [55].

References

1. D. Baraff and A. Witkin, Physically-Based Modeling: ACM SIGGRAPH Course Notes, 2001.
2. J. Latombe, *Robot Motion Planning*. Kluwer Academic Publishers, Boston, MA, 1991.
3. J. Cohen, M. Lin, D. Manocha, and M. Ponamgi. I-collide: An interactive and exact collision detection system for large-scale environments. *Proc. of ACM Interactive 3D Graphics Conference*, pp. 189–196, 1995.
4. M. A. Otaduy, and M. C. Lin. Sensation preserving simplification for haptic rendering. *Proc. of ACM SIGGRAPH*, 2003.
5. D. P. Dobkin and D. G. Kirkpatrick, Determining the separation of preprocessed polyhedra—a unified approach, in *Proc. 17th Internat. Colloq. Automata Lang. Program.*, vol. 443 of Lecture Notes in Computer Science, pp. 400–413, Springer-Verlag, 1990.
6. R. Seidel, Linear programming and convex hulls made easy, *Proc. 6th Ann. ACM Conf. on Computational Geometry*, (Berkeley, California), pp. 211–215, 1990.
7. M. Lin, Efficient Collision Detection for Animation and Robotics: PhD thesis, Department of Electrical Engineering and Computer Science, University of California, Berkeley, December 1993.
8. M. Lin, and J. F. Canny, Efficient algorithms for incremental distance computation. *IEEE Conference on Robotics and Automation*, pp. 1008–1014, 1991.
9. S. Ehmann, and M. C. Lin, Accelerated proximity queries between convex polyhedra using multi-level voronoi marching. *Proc. of IEEE/RSJ International Conference on Intelligent Robots and Systems*, pp. 2101–2106, 2000.
10. B. Mirtich, V-Clip: Fast and robust polyhedral collision detection, *ACM Transactions on Graphics*, vol. 17, pp. 177–208, July 1998.
11. C. Hoffmann, *Geometric and Solid Modeling*. San Mateo, California: Morgan Kaufmann, 1989.
12. Collision detection systems. http://gamma.web.unc.edu/software/#collision, 2002.
13. B. Mirtich and J. Canny, Impulse-based simulation of rigid bodies, *Proc. of ACM Interactive 3D Graphics*, (Monterey, CA), 1995.
14. A. Gregory, A. Mascarenhas, S. Ehmann, M. C. Lin, and D. Manocha, 6-dof haptic display of polygonal models. *Proc. of IEEE Visualization Conference*, 2000.
15. S. Cameron, and R. K. Culley, Determining the minimum translational distance between two convex polyhedra. *Proceedings of International Conference on Robotics and Automation*, pp. 591–596, 1986.
16. D. Dobkin, J. Hershberger, D. Kirkpatrick, and S. Suri, Computing the intersection-depth of polyhedra, *Algorithmica*, vol. 9, pp. 518–533, 1993.
17. E. G. Gilbert, D. W. Johnson, and S. S. Keerthi, A fast procedure for computing the distance between objects in three-dimensional space, *IEEE J. Robotics and Automation*, vol. RA-4, pp. 193–203, 1988.
18. S. Cameron. Enhancing GJK: Computing minimum and penetration distance between convex polyhedra. *Proceedings of International Conference on Robotics and Automation*, pp. 3112–3117, 1997.
19. H. Samet, *Spatial Data Structures: Quadtree, Octrees and Other Hierarchical Methods*. Addison Wesley, Reading, MA, 1989.
20. M. Held, J. Klosowski, and J. Mitchell. Evaluation of collision detection methods for virtual reality fly-throughs. *Canadian Conference on Computational Geometry*, 1995.
21. B. Naylor, J. Amanatides, and W. Thibault. Merging bsp trees yield polyhedral modeling results. *Proc. of ACM SIGGRAPH*, pp. 115–124, 1990.
22. W. Bouma, and G. Vanecek. Collision detection and analysis in a physically based simulation. *Proceedings Eurographics Workshop on Animation and Simulation*, pp. 191–203, 1991.
23. P. M. Hubbard, Interactive collision detection, *Proceedings of IEEE Symposium on Research Frontiers in Virtual Reality*, October 1993.
24. S. Quinlan . Efficient distance computation between non-convex objects. *Proceedings of International Conference on Robotics and Automation*, pp. 3324–3329, 1994.
25. N. Beckmann, H. Kriegel, R. Schneider, and B. Seeger, The r*-tree: An efficient and robust access method for points and rectangles. *Proc: SIGMOD Conf. on Management of Data*, pp. 322–331, 1990.
26. M. Ponamgi, D. Manocha, and M. Lin, Incremental algorithms for collision detection between solid models, *IEEE Transactions on Visualization and Computer Graphics*, vol. 3, no. 1, pp. 51–67, 1997.

27. S. Gottschalk, M. Lin, and D. Manocha. OBB-Tree: A hierarchical structure for rapid interference detection. *Proc. of ACM Siggraph'96*, pp. 171–180, 1996.

28. G. Barequet, B. Chazelle, L. Guibas, J. Mitchell, and A. Tal, Boxtree: A hierarchical representation of surfaces in 3D. *Proc. of Eurographics'96*, 1996.

29. S. Gottschalk, Collision Queries using Oriented Bounding Boxes: PhD thesis, University of North Carolina. Department of Computer Science, 2000.

30. S. Krishnan, A. Pattekar, M. Lin, and D. Manocha. Spherical shell: A higher order bounding volume for fast proximity queries. *Proc. of Third International Workshop on Algorithmic Foundations of Robotics*, pp. 122–136, 1998.

31. S. Krishnan, M. Gopi, M. Lin, D. Manocha, and A. Pattekar, Rapid and accurate contact determination between spline models using shelltrees, *Computer Graphics Forum, Proceedings of Eurographics*, vol. 17, no. 3, pp. C315–C326, 1998.

32. M. Held, J. Klosowski, and J. S. B. Mitchell, Real-time collision detection for motion simulation within complex environments, in *Proc. ACM SIGGRAPH'96 Visual Proceedings*, p. 151, 1996.

33. J. Klosowski, M. Held, J. Mitchell, H. Sowizral, and K. Zikan, Efficient collision detection using bounding volume hierarchies of k-dops, *IEEE Trans. on Visualization and Computer Graphics*, vol. 4, no. 1, pp. 21–37, 1998.

34. E. Larsen, S. Gottschalk, M. Lin, and D. Manocha, Fast proximity queries with swept sphere volumes," Tech. Rep. TR99-018, Department of Computer Science, University of North Carolina, 1999. 32 pages.

35. S. Ehmann and M. C. Lin, Accurate and fast proximity queries between polyhedra using convex surface decomposition, *Computer Graphics Forum (Proc. of Eurographics'2001)*, vol. 20, no. 3, pp. 500–510, 2001.

36. R. Duda and P. Hart, *Pattern Classification and Scene Analysis*. John Wiley and Sons, 1973.

37. D. Hsu, L. Kavraki, J. Latombe, R. Motwani, and S. Sorkin. On finding narrow passages with probabilistic roadmap planners. *Proc. of 3rd Workshop on Algorithmic Foundations of Robotics*, pp. 25–32, 1998.

38. M. McKenna and D. Zeltzer, Dynamic simulation of autonomous legged locomotion, *Computer Graphics (SIGGRAPH '90 Proceedings)* (F. Baskett, ed.), vol. 24, pp. 29–38, Aug. 1990.

39. D. E. Stewart and J. C. Trinkle, An implicit time-stepping scheme for rigid body dynamics with inelastic collisions and coulomb friction, *International Journal of Numerical Methods in Engineering*, vol. 39, pp. 2673–2691, 1996.

40. Y. Kim, M. Otaduy, M. Lin, and D. Manocha . 6-dof haptic display using localized contact computations. *Proc. of Haptics Symposium*, pp. 209–216, 2002.

41. P. Agarwal, L. Guibas, S. Har-Peled, A. Rabinovitch, and M. Sharir, Penetration depth of two convex polytopes in 3d, *Nordic J: Computing*, vol. 7, pp. 227–240, 2000.

42. G. Bergen. Proximity queries and penetration depth computation on 3d game objects. *Game Developers Conference*, 2001.

43. Y. Kim, M. Lin, and D. Manocha. Deep: An incremental algorithm for penetration depth computation between convex polytopes. *Proc. of IEEE Conference on Robotics and Automation*, pp. 921–926, 2002.

44. J. O'Rourke, C.-B. Chien, T. Olson, and D. Naddor A new linear algorithm for intersecting convex polygons, *Comput. Graph. Image Process.*, vol. 19, pp. 384–391, 1982.

45. S. Fisher, and M. C. Lin. Deformed distance fields for simulation of non-penetrating flexible bodies. *Proc. of EG Workshop on Computer Animation and Simulation*, pp. 99–111, 2001.

46. K. Hoff, A. Zaferakis, M. Lin, and D. Manocha. Fast and simple 2d geometric proximity queries using graphics hardware. *Proc. of ACM Symposium on Interactive 3D Graphics*, pp. 145–148, 2001.

47. K. Hoff, A. Zaferakis, M. Lin, and D. Manocha, Fast 3d geometric proximity queries between rigid and deformable models using graphics hardware acceleration, Tech. Rep. TR02-004, Department of Computer Science, University of North Carolina, 2002.

48. M. H. Overmars, Point location in fat subdivisions, *Inform. Proc. Lett.*, vol. 44, pp. 261–265, 1992.

49. S. Cameron. Collision detection by four-dimensional intersection testing. *Proceedings of International Conference on Robotics and Automation*, pp. 291–302, 1990.

50. D. Baraff, Dynamic simulation of non-penetrating rigid body simulation. PhD thesis, Cornell University, 1992.

51. M. Shamos, and D. Hoey, Geometric intersection problems. *Proc. 17th An. IEEE Symp. Found. on Comput. Science*, pp. 208–215, 1976.

52. J. Hopcroft, J. Schwartz, and M. Sharir, Efficient detection of intersections among spheres, *The International Journal of Robotics Research*, vol. 2, no. 4, pp. 77–80, 1983.

53. H. Six, and D. Wood, Counting and reporting intersections of D-ranges. *IEEE Transactions on Computers*, pp. 46–55, 1982.

54. F. Preparata and M. I. Shamos, *Computational Geometry*. New York: Springer-Verlag, 1985.

55. H. Edelsbrunner, A new approach to rectangle intersections, Part I, *Internat. J. Comput. Math.*, vol. 13, pp. 209–219, 1983.

58

Image Data Structures*

58.1   Introduction.. 903
58.2   What is Image Data?.. 904
58.3   Quadtrees.. 905
        What is a Quadtree? • Variants of Quadtrees
58.4   Virtual Quadtrees 907
        Compact Quadtrees • Forest of Quadtrees (FQT)
58.5   Quadtrees and R-trees.. 910
58.6   Octrees 911
58.7   Translation Invariant Data Structure (TID)..................................... 912
58.8   Content-Based Image Retrieval System 914
        What is CBIR? • An Example of CBIR System
58.9   Summary.. 915
Acknowledgments 915
References... 916

S. S. Iyengar
Florida International University

V. K. Vaishnavi[†]
Georgia State University

S. Gunasekaran[†]
Louisiana State University

58.1 Introduction

Image has been an integral part of our communication. Visual information aids us in understanding our surroundings better. Image processing, the science of manipulating digital images, is one of the methods used for digitally interpreting images. Image processing generally comprises three main steps:

1. Image acquisition: Obtaining the image by scanning it or by capturing it through some sensors.
2. Image manipulation/analysis: Enhancing and/or compressing the image for its transfer or storage.
3. Display of the processed image.

Image processing has been classified into two levels: low-level image processing and high-level image processing. Low-level image processing needs little information about the content or the semantics of the image. It is mainly concerned with retrieving low-level descriptions of the image and processing them. Low-level data include matrix representation of the actual image. Image calibration and image enhancement are examples of low-level image processing.

High-level image processing is basically concerned with segmenting an image into objects or regions. It makes decisions according to the information contained in the image. High-level data are represented in symbolic form. The data include features of the image such as object size, shape and its relation with other objects in the image. These image-processing techniques depend significantly on the underlying image data structure. Efficient data structures for region representation are important for use in manipulating pictorial information.

Many techniques have been developed for representing pictorial information in image processing [1]. These techniques include data structures such as quadtrees, linear quadtrees, Forest of Quadtrees, and Translation Invariant Data structure (TID). We will discuss these data structures in the following sections. Research on quadtrees has produced several interesting results in different areas of image processing [2–7]. In 1981, Jones and Iyengar [8] proposed methods of refining quadtrees. A good tracing of the history of the evolution of quadtrees is provided by Klinger and Dyer [9]. See also Chapter 20 of this handbook.

---

* This chapter has been reprinted from first edition of this Handbook, without any content updates.
† We have used this author's affiliation from the first edition of this handbook, but note that this may have changed since then.

58.2 What is Image Data?

An image is a visual reproduction of an object using an optical or electronic device. Image data include pictures taken by satellites, scanned images, aerial photographs and other digital photographs. In the computer, image is represented as a data file that consists of a rectangular array of picture elements called pixels. This rectangular array of pixels is also called a raster image. The pixels are the smallest programmable visual unit. The size of the pixel depends on the resolution of the monitor. The resolution can be defined as the number of pixels present on the horizontal axis and vertical axis of a display monitor. When the resolution is set to maximum the pixel size is equal to a dot on the monitor. The pixel size increases with the decrease of the resolution. Figure 58.1 shows an example of an image.

In a monochrome image each pixel has its own brightness value ranging from 0 (black) to 255 (white). For a color image each pixel has a brightness value and a RGB color value. RGB is an additive color state that has separate values for red, green, and blue. Hence each pixel has independent values (0–255) for red, green, and blue colors. If the values for red, green, and blue components of the pixel are the same then the resulting pixel color is gray. Different shades of gray pixels constitute a gray-scale image. If pixels of an image have only two states, black or white, then the image is called a binary image.

Image data can be classified into raster graphics and vector graphics. Raster graphics, also known as bitmap graphics, represents an image using x and y coordinates (for 2d-images) of display space; this grid of x and y coordinates is called the raster. Figure 58.2 shows how a circle will appear in raster graphics. In raster graphics an image is divided up in raster. All raster dots that are more than half full are displayed as black dots and the rest as white dots. This results in step like edges as shown in the figure. The appearance of jagged edges can be minimized by reducing the size of the raster dots. Reducing the size of the dots will increase the number of pixels needed but increases the size of the storage space.

Vector graphics uses mathematical formulas to define the image in a two-dimensional or three-dimensional space. In a vector graphics file an image is represented as a sequence of vector statements instead of bits, as in bitmap files. Thus it needs just minimal amount of information to draw shapes and therefore the files are smaller in size compared to raster graphics files.

FIGURE 58.1   Example of an image.

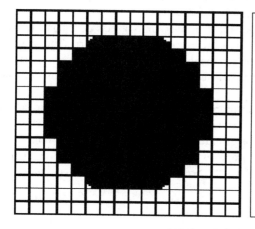

FIGURE 58.2   A circle in raster graphics.

Vector graphics does not consist of black and white pixels but is made of objects like line, circle, and rectangle. The other advantage of vector graphics is that it is flexible, so it can be resized without loss of information. Vector graphics is typically used for simple shapes. CorelDraw images, PDF, and PostScript are all in vector image format. The main drawbacks of vector graphics is that it needs longer computational time and has very limited choice of shapes compared to raster graphics.

Raster file, on the other hand, is usually larger than vector graphics file and is difficult to modify without loss of information. Scaling a raster file will result in loss of quality whereas vector graphics can be scaled to the quality of the device on which it is rendered. Raster graphics is used for representing continuous numeric values unlike vector graphics, which is used for discrete features. Examples of raster graphic formats are GIF, JPEG, and PNG.

58.3 Quadtrees

Considerable research on quadtrees has produced several interesting results in different areas of image processing. The basic relationship between a region and its quadtree representation is presented in [3]. In 1981, Jones and Iyengar [8] proposed methods for refining quadtrees. The new refinements were called virtual quadtrees, which include compact quadtrees and forests of quadtrees. We will discuss virtual quadtrees in a later section of this chapter. Much work has been done on the quadtree properties and algorithms for manipulations and translations have been developed by Samet [10,11], Dyer [2] and others [3,5,12].

58.3.1 What is a Quadtree?

A quadtree is a class of hierarchical data structures used for image representation. The fundamental principle of quadtrees is based on successive regular decomposition of the image data until each part contains data that is sufficiently simple so that it can be represented by some other simpler data structure. A quadtree is formed by dividing an image into four quadrants, each quadrant may further be divided into four sub quadrants and so on until the whole image has been broken down to small sections of uniform color.

In a quadtree the root represents the whole image and the non-root nodes represent sub quadrants of the image. Each node of the quadtree represents a distinct section of the image; no two nodes can represent the same part of the image. In other words, the sub quadrants of the image do not overlap. The node without any child is called a leaf node and it represents a region with uniform color. Each non-leaf node in the tree has four children, each of which represents one of the four sub regions, referred to as NW, NE, SW, and SE, that the region represented by the parent node is divided into.

The leaf node of a quadtree has the color of the pixel (black or white) it is representing. The nodes with uniform colored children have the color of their children and all their child nodes are removed from the tree. All the other nodes with non-uniform colored children have the gray color. Figure 58.3 shows the concept of quadtrees.

The image can be retrieved from the quadtree by using a recursive procedure that visits each leaf node of the tree and displays its color at an appropriate position. The procedure starts with visiting the root node. In general, if the visited node is not a leaf then the procedure is recursively called for each child of the node, in order from left to right.

The main advantage of the quadtree data structure is that images can be compactly represented by it. The data structure combines data having similar values and hence reduces storage size. An image having large areas of uniform color will have very small storage size. The quadtree can be used to compress bitmap images, by dividing the image until each section has the same color. It can also be used to quickly locate any object of interest.

The drawback of the quadtree is that it can have totally different representation for images that differ only in rotation or translation. Though the quadtree has better storage efficiency compared to other data structures such as the array, it also has considerable storage overheads. For a given image it stores many white leaf nodes and intermediate gray nodes, which are not required information.

58.3.2 Variants of Quadtrees

The numerous quadtree variants that have been developed so far can be differentiated by the type of data they are designed to represent [13]. The many variants of the quadtree include region quadtrees, point quadtrees, line quadtrees, and edge quadtrees. Region quadtrees are meant for representing regions in images, while point quadtrees, edge quadtrees, and line quadtrees are used to represent point features, edge features, and line features, respectively. Thus there is no single quadtree data structure which is capable of representing a mixture of features of images like regions and lines.

58.3.2.1 Region Quadtrees

In the region quadtree, a region in a binary image is a subimage that contains either all 1's or all 0's. If the given region does not consist entirely of 1's or 0's, then the region is divided into four quadrants. This process is continued until each divided section

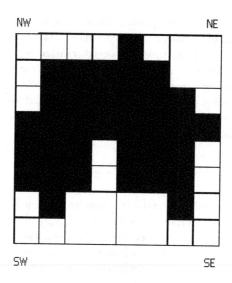

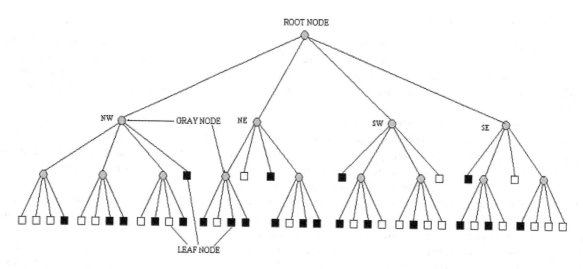

FIGURE 58.3   The concept of quadtrees.

consists entirely of 1's or 0's; such regions are called final regions. The final regions (either black or white) are represented by leaf nodes. The intermediate nodes are called gray nodes. A region quadtree is space efficient for images containing square-like regions. It is very inefficient for regions that are elongated or for representing line features.

58.3.2.2 Line Quadtrees

The line quadtree proposed by Samet [14] is used for representing the boundaries of a region. The given region is represented as in the region quadtree with additional information associated with nodes representing the boundary of the region. The data structure is used for representing curves that are closed. The boundary following algorithm using the line quadtree data structure is at least twice as good as the one using the region quadtree in terms of execution time; the map superposition algorithm has execution time proportional to number of nodes in the line quadtree [13]. The main disadvantage of the line quadtree is that it cannot represent independent linear features.

58.3.2.3 Edge Quadtrees

Shneier [15] formulated the edge quadtree data structure for storing linear features. The main principle used in the data structure is to approximate the curve being represented by a number of straight line segments. The edge quadtree is constructed using a recursive procedure. If the sub quadrant represented by a node does not contain any edge or line, it is not further subdivided and is represented by a leaf. If it does contain one then an approximate equation is fitted to the line. The error caused by approximation is calculated using a measure such as least squares. When the error is less than the predetermined value, the node becomes a leaf;

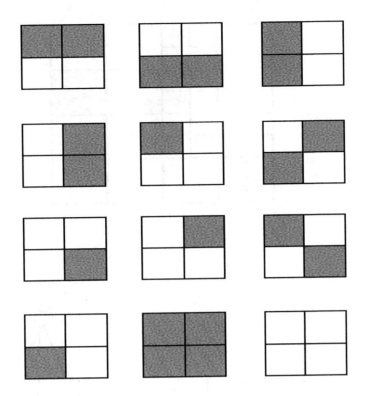

FIGURE 58.4 Template sets.

otherwise the region represented by the node is further subdivided. If an edge terminates (within the region represented by) a node then a special flag is set. Each node contains magnitude, direction, intercept, and error information about the edge passing through the node. The main drawback of the edge quadtree is that it cannot efficiently handle two or more intersecting lines.

58.3.2.4 Template quadtrees

An image can have regions and line features together. The above quadtree representational schemas are not efficient for representing such an image in one quadtree data structure. The template quadtree is an attempt toward development of such a quadtree data structure. It was proposed by Manohar, Rao, and Iyengar [13] to represent regions and curvilinear data present in images.

A template is a $2^k \times 2^k$ sub image, which contains either a region of uniform color or straight run of black pixels in horizontal, vertical, left diagonal, or right diagonal directions spanning the entire sub image. All the possible templates for a 2×2 sub image are shown in Figure 58.4.

A template quadtree can be constructed by comparing a quadrant with any one of the templates, if they match then the quadrant is represented by a node with the information about the type of template it matches. Otherwise the quadrant is further divided into four sub quadrants and each one of them is compared with any of the templates of the next lower size. This process is recursively followed until the entire image is broken down into maximal blocks corresponding to the templates. The template quadtree representation of an image is shown in Figure 58.5.

Here the leaf is defined as a template of variable size, therefore it does not need any approximation for representing curves present in the images. The advantage of template quadtree is that it is very accurate and has the capabilities for representing features like regions and curves. The main drawback is that it needs more space compared to edge quadtrees. For more information on template quadtrees see Reference 13.

58.4 Virtual Quadtrees

The quadtree has become a major data structure in image processing. Though the quadtree has better storage efficiency compared to other data structures such as the array, it also has considerable storage overhead. Jones and Iyengar [16] have proposed two ways in which quadtrees may be efficiently stored: as "forest of quadtrees" and as "compact quadtrees." They called these new data structures virtual quadtrees because the basic operations performed in quadtrees can also be performed on the new

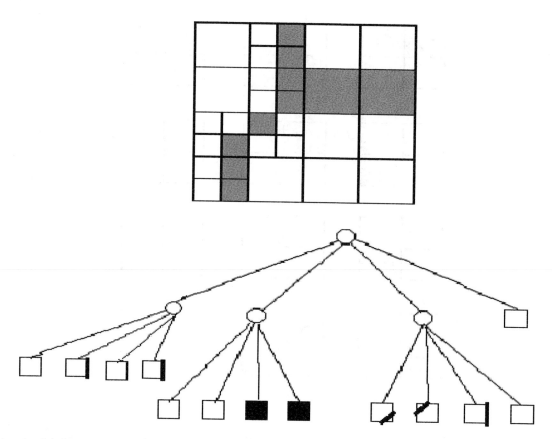

FIGURE 58.5   Template quadtree representation.

representations. The virtual quadtree is a space-efficient way of representing a quadtree. It is a structure that simulates quadtrees in such a way that we can,

1. Determine the color of any node in the quadtree.
2. Find the child node of any node, in any direction in the quadtree.
3. Find the parent node of any node in the quadtree.

The two types of virtual quadtrees, which we are going to discuss, are the compact quadtree and the forest of quadtrees.

58.4.1 Compact Quadtrees

The compact quadtree has all the information contained in a quadtree but needs less space. It is represented as C(T), where T is the quadtree it is associated with. Each set of four sibling nodes in the quadtree is represented as a single node called metanode in the corresponding compact quadtree.

The metanode M has four fields, which are explained as follows:

- MCOLOR(M, D)—The colors of the nodes included in M. Where D ∈ {NW,NE,SW,SE}.
- MSONS(M)—Points to the first metanode that represents offsprings of a node represented in M; NIL if no offspring exists.
- MFATHER(M)—Points to the metanode that holds the representation of the parent of the nodes that M represents.
- MCHAIN(M)—If there are more than one metanode that represent offspring of nodes represented by a given metanode M, then these are linked by MCHAIN field.

The compact quadtree for the quadtree shown in the Figure 58.6 is given in Figure 58.7. In Figure 58.7 downward links are MSON links, horizontal links are MCHAIN links and upward links are MFATHER links.

The compact quadtree uses the same amount of space for the storage of color but it uses very less space for storage of pointers. In Figure 58.7, the compact quadtree has 10 metanodes, whereas the quadtree has 41 nodes. Thus it saves about 85% of the storage space. Since the number of nodes in a compact quadtree is less a simple recursive tree traversal can be done more efficiently.

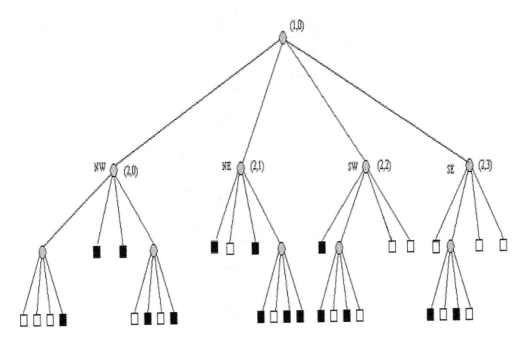

FIGURE 58.6 Quadtree with coordinates.

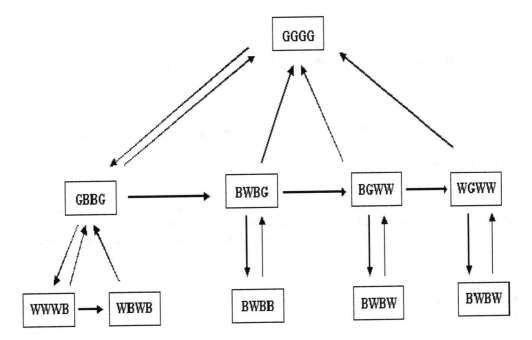

FIGURE 58.7 Compact quadtree C(T).

58.4.2 Forest of Quadtrees (FQT)

Jones and Iyengar [8] proposed a variant of quadtrees called forest of quadtrees to improve the space efficiency of quadtrees. A forest of quadtrees is represented by F(T) where T is the quadtree it represents. A forest of quadtrees, F(T) that represents T consists of a table of triples of the form (L, K, P), and a collection of quadtrees where,

1. Each triple (L, K, P) consists of the coordinates, (L, K), of a node in T, and a pointer, P, to a quadtree that is identical to the subtree rooted at position (L, K) in T.
2. If (L, K) and (M, N) are coordinates of nodes recorded in F(T), then neither node is a descendant of the other.
3. Every black leaf in T is represented by a black leaf in F(T).

Table in the FQT

2	0	Pointer to A
2	1	Pointer to B
3	8	Pointer to C
3	9	Pointer to D
4	3	Pointer to E
3	13	Pointer to F

Trees in the FQT

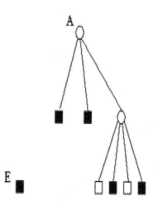
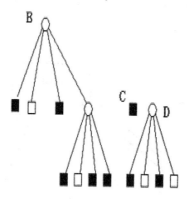

FIGURE 58.8   Forest of quadtrees F(T).

To obtain the corresponding forest of quadtrees, the nodes in a quadtree need to be classified as good and bad nodes. A leaf node with black color or an intermediate node that has two or more black child nodes are called good nodes, the rest are called bad nodes. Only the good nodes are stored in the forest of quadtrees thereby saving lot of storage space. Figure 58.8 contains a forest of quadtrees that represents the quadtree shown in Figure 58.6.

To reduce a quadtree to a forest of quadtrees, we first need to identify the good nodes and the bad nodes. We then break down the quadtree into smaller quadtrees in such a way that each of them has a good root node and none of them is a subtree of another and the bad nodes encountered by forest are freed. This collection of quadtrees is called as forest of quadtrees. The time required for the execution of the conversion is obviously linear in the number of nodes in the quadtree [16].

Theorem:

The maximum number of trees in a forest of quadtrees derived from a quadtree that represents a square of dimension $2^k \times 2^k$ is 4^{k-1}, that is, one-fourth the area of the square. For proof see Reference 16.

The corresponding quadtree can be easily reconstructed from a forest of quadtrees F. The reconstructed quadtree R(F) consists of real nodes and virtual nodes (nodes corresponding to the bad nodes that are deleted while creating the forest). Since the virtual nodes require no storage they are located by giving their coordinates. The virtual nodes are represented as v(L, K).

58.5 Quadtrees and R-trees

Quadtrees and R-trees [17] are two commonly used spatial data structures to locate or organize objects in a given image. Both quadtrees and R-trees use bounding boxes to depict the outline of the objects. Therefore, the data structure only needs to keep

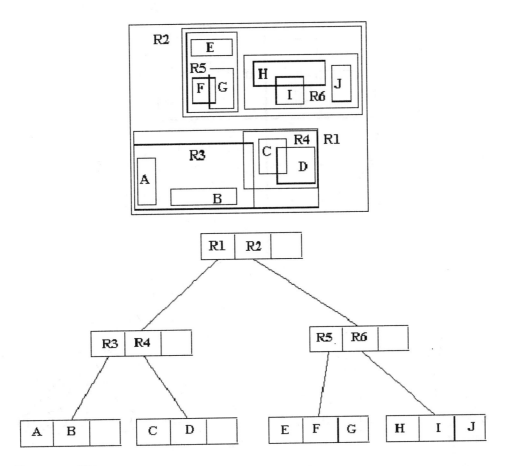

FIGURE 58.9   The concept of R-tree.

track of the boundaries instead of the actual shape and size of the objects in the image. The size of the bounding boxes usually depends on the size of the object we are trying to locate.

The R-tree (see also Chapter 22) is a hierarchical, height balanced data structure, similar to the B^+-tree that aggregates objects based on their spatial proximity. It is a multidimensional generalization of the B^+-tree. The R-tree indexes multidimensional data objects by enclosing data objects in bounding rectangles, which may further be enclosed in bounding rectangles; these bounding rectangles can overlap each other. Each node except the leaf node has a pointer and the corresponding bounding rectangle. The pointer points to the subtree with all the objects enclosed in its corresponding rectangle. The leaf node has a bounding rectangle and the pointer pointing to the actual data object. The root node has a minimum of two children unless it is a leaf node and all the leaf nodes appear at the same level. Figure 58.9 Shows the R-tree representation of an image.

The main difference between quadtrees and R-trees is that unlike R-trees the bounding rectangle of quadtrees do not overlap. In real world objects overlap; in such cases more than one node in a quadtree can point to the same object. This is a serious disadvantage for the quadtree. Figure 58.10 shows the quadtree representation of the image in Figure 58.9.

As you can see from the Figure 58.10 objects like "C" and "D" are represented by more than one node. This makes it difficult to find the origin of the quadtree.

The R-tree is a dynamic structure, so its contents can be modified without having to reconstruct the entire tree. It can be used to determine which objects intersect a given query region. The R-tree representation of an image is not unique; size and placement of rectangles in the tree depends on the sequence of insertions and deletions resulting in the tree starting from the empty R-tree.

58.6 Octrees

Octrees are 3D equivalent of quadtrees and are used for representing 3D images. An octree is formed by dividing a 3D space into 8 sub-cubes called cells. This process of dividing the cube into cells is carried on until the (image) objects lie entirely inside or outside the cells. The root node of an octree represents a cube, which encompasses the objects of interest.

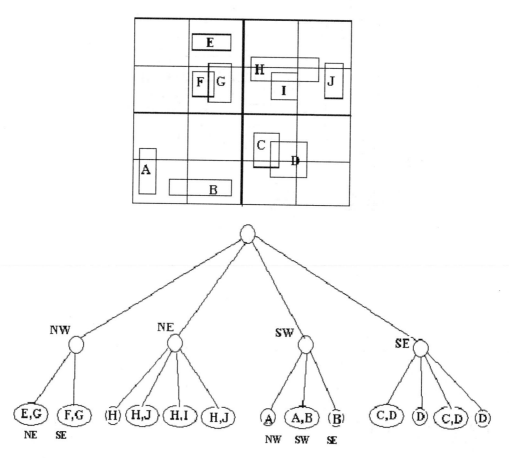

FIGURE 58.10   Quadtree representation of the image in Figure 58.9.

In the octree each node except the leaf node has eight child nodes, which represent the sub-cubes of the parent node. The node stores information like pointer to the child nodes, pointer to the parent node, and pointers to the contents of the cube. An example of an octree representation is shown in Figure 58.11.

The Octree become very ineffective if the objects in the image that it is representing are not uniformly distributed, as in that case many nodes in the octree will not contain any objects. Such an octree is also very costly to compute as a large amount of data has to be examined.

58.7 Translation Invariant Data Structure (TID)

Methods for the region representation using quadtrees exist in the literature [2,10,18]. Much work has been done on quadtree properties, and algorithms for translations and manipulations have been derived by Dyer [2], Samet [11,19], Shneier [20] and others.

Various improvements to quadtrees have been suggested including forests of quadtrees, hybrid quadtrees, linear quadtrees, and optimal quadtrees for image segments. All of these methods try to optimize quadtrees by removing some are all of the gray and white nodes. All of them maintain the same number of black nodes.

All these methods are sensitive to the placement of the origin. An image, which has been translated from its original position, can have a very different looking structure [21]. We explain this phenomenon by using the example given in Figure 58.12. In Example 1, the black square is in the upper left corner. In Example 2, it is translated down and right by one pixel. Figure 58.13 gives the quadtree representation for these two examples.

The shift sensitivity of the image data structure derives from the fact that the positions of the maximal blocks represented by leaf nodes are not explicitly represented in the data structure. Instead, these positions are determined by the paths leading to them from the root of the tree. Thus, when the image is shifted, the maximal blocks are formed in a different way.

For this reason Scott and Iyengar [21] have introduced a new data structure called Translation Invariant Data structure (TID), which is not sensitive to the placement of the region and is translation invariant.

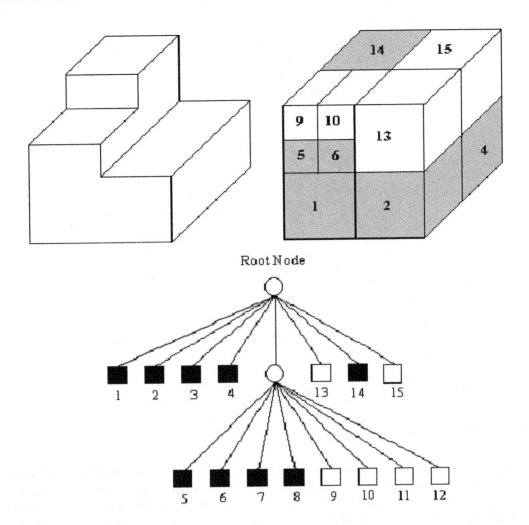

FIGURE 58.11 Octree representation.

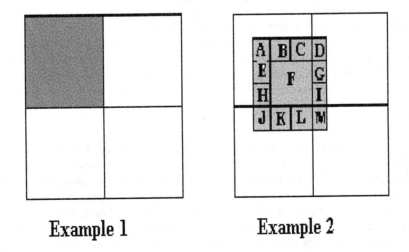

FIGURE 58.12 Sample Regions.

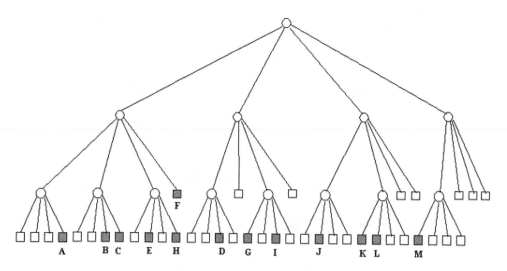

FIGURE 58.13   Quadtrees for Example 1 and 2 of Figure 58.12.

A maximal square is a black square of pixels that is not included in any large square of black pixels. TID is made of such maximal squares, which are represented as (i, j, s) where (i, j) is the coordinates of the northwest vertex of the square and "s" is the length of the square. Translation made to any image can be represented as a function of these triples. For example consider a triple (i, j, s), translating it x units to the right and y units up yields (i+x, j+y, s) [21].

The rotation of the square by $\pi/2$ is only slightly more complicated due to the fact that the NW corner of the square changes upon rotation. The $\pi/2$ rotation around the origin gives (− j, i +s, s).

58.8 Content-Based Image Retrieval System

Images have always been a part of human communication. Due to the increase in the use of Internet the interest in the potential of digital images has increased greatly. Therefore we need to store and retrieve images in an efficient way. Locating and retrieving a desired image from a large database can be a very tedious process and is still an active area of research. This problem can be reduced greatly by using Content-Based Image Retrieval (CBIR) systems, which retrieves images based only on the content of the image. This technique retrieves images on the basis of automatically-derived features such as color, texture, and shape.

58.8.1 What is CBIR?

Content-Based Image Retrieval is a process of retrieving desired images from a large database based on the internal features that can be obtained automatically from the images themselves. CBIR techniques are used to index and retrieve images from databases based on their pictorial content, typically defined by a set of features extracted from an image that describe the color, texture, and/or shape of the entire image or of specific objects in the image. This feature description is used to index a database through various means such as distance-based techniques, rule-based decision-making, and fuzzy inferencing [22–24]. Images can be matched in two ways. Firstly, an image can be compared with another image to check for similarity. Secondly, images similar to the given image can be retrieved by searching a large image database. The latter process is called content-based image retrieval.

58.8.1.1 General Structure of CBIR Systems

The general computational framework of a CBIR system as shown in Figure 58.14 was proposed in Reference 25.

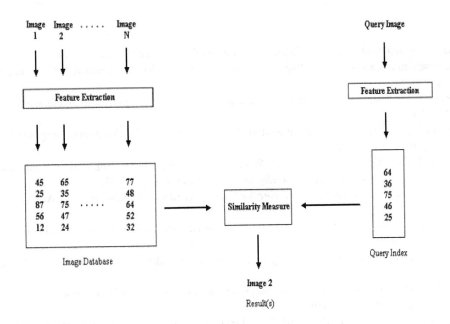

FIGURE 58.14 Computational framework of CBIR systems.

At first, the image database is created, which stores the images as numerical values supplied by the feature extraction algorithms. These values are used to locate an image similar to the query image. The query image is processed by the same feature extraction algorithm that is applied to the images stored in the database.

The similarity between the query image and the images stored in the database can be verified with the help of the similarity measure algorithm, which compares the results obtained from the feature extraction algorithms for both the query image and the images in the database. Thus after comparing the query image with all the images in the database the similarity measure algorithm gives a set as the result, which has all the images from the database that are similar to the query image.

58.8.2 An Example of CBIR System

An example of Content-Based Image Retrieval System is BlobWorld. The BlobWorld system, developed at the University of California, Berkeley, supports color, shape, spatial, and texture matching features. Blobworld is based on finding coherent image regions that roughly correspond to objects. The system automatically separates each image into homogeneous regions in order to improve content-based image retrieval. Querying is based on the user specifying attributes of one or two regions of interest, rather than a description of the entire image. For more information on Blobworld see Reference 26.

CBIR techniques are likely to be of most use in restricted subject areas, where merging with other types of data like text and sound can be achieved. Content-based image retrieval provides an efficient solution to the restrictions and the problems caused by the traditional information retrieval technique. The number of active research systems is increasing, which reflects the increasing interest in the field of content-based image retrieval.

58.9 Summary

In this chapter we have explained what an image data is and how it is stored in raster graphics and vector graphics. We then discussed some of the image representation techniques like quadtrees, virtual quadtrees, octrees, R-trees and translation invariant data structures. Finally, we have given a brief introduction to Content-Based Image Retrieval (CBIR) systems. For more information on image retrieval concepts see References 22–25.

Acknowledgments

The authors acknowledge the contributions of Sridhar Karra and all the previous collaborators in preparing this chapter.

References

1. H. Samet, *The Design and Analysis of Spatial Data Structures*. Addison-Wesley Pub. Co., 1990.

2. C. R. Dyer, A. Rosenfeld, H. Samet, Region Representation: Boundary Codes from Quadtrees, *Communication of the ACM*, 23, 1980, 171–179.

3. G. M. Hunter and K. Steiglitz, Operations on Images Using Quadtrees, *IEEE Transactions on Pattern Analysis and Machine Intelligence*, 1(2), April 1979, 145–153.

4. G. M. Hunter and K. Steiglitz, Linear Transformation of Pictures Represented By Quadtrees, *Computer, Graphics and Image Processing*, 10(3), July 1979, 289–296.

5. E. Kawaguchi and T. Endo, On a Method of Binary-Picture Representation and Its Application to Data Compression, *IEEE Transactions on Pattern Analysis and Machine Intelligence*, 2(1), January 1980, 27–35.

6. A. Klinger, *Patterns and Search Statistics, Optimizing Methods in Statistics*, J. D. Rustagi, (Ed.), Academic Press, New York, 1971.

7. A. Klinger and M. L. Rhodes, Organization and Access of Image Data by Areas, *IEEE Transactions on Pattern Analysis and Machine Intelligence*, 1(11), January 1979, 50–60.

8. L. Jones and S. S. Iyengar, Representation of a Region as a Forest of Quadtrees, *Proc. IEEE-PRIP 81 Conference*, TX, IEEE Publications, 57–59.

9. A. Klinger and C. R. Dyer, Experiments in Picture Representation Using Regular Decomposition, *Computer Graphics and Image Processing*, 5(1), March 1976, 68–105.

10. H. Samet, Region Representation: Quadtree from Boundary Codes, *Communications of the ACM*, 23(3), March 1980, 163–170.

11. H. Samet, Connected Component Labeling using Quadtrees, *JACM*, 28(3), July 1981, 487–501.

12. L. Jones and S. S. Iyengar, Virtual Quadtrees, *Proc. IEEE-CVIP 83 Conference*, Virginia, IEEE Publications101–105.

13. M. Manohar, P. Sudarsana Rao, S. Sitarama Iyengar, Template Quadtrees for Representing Region and Line Data Present in Binary Images, *NASA Goddard Flight Center*, Greenbelt, MD, 1988.

14. H. Samet and R. E. Webber, On Encoding Boundaries with Quadtrees, *IEEE Transcations on Pattern Analysis and Machine Intelligence*, 6, 1984, 365–369.

15. Edge Pyramids and Edge Quadtrees, *Computer Graphics and Image Process*, 17, 1981, 211–224.

16. L. P. Jones and S. S. Iyengar, Space and Time Efficient Virtual Quadtrees, *IEEE Transactions on Pattern Analysis and Machine Intelligence*, 6(2), March 1984, 244–247.

17. A. Guttman, R-Trees: A Dynamic Index Structure for Spatial Searching, *Proceedings of the SIGMOD Conference*, 47–57.

18. H. Samet, Region Representation: Quadtree from Binary Arrays, *Computer Graphics Image Process*, 18(1), 1980, 88–93.

19. H. Samet, An Algorithm for Converting Rasters to Quadtrees, *IEEE Transcations on Pattern Analysis and Machine Intelligence*, 3(1), January 1981, 93–95.

20. M. Shneier, Path Length Distances for Quadtrees, *Information Sciences*, 23(1), 1981, 49–67.

21. D. S. Scott and S. S. Iyengar, TID- A Translation Invariant Data Structure For Storing Images, *Communications of the ACM*, 29(5), May 1985, 418–429.

22. C. Hsu, W. W. Chu, and R. K. Taira, A Knowledge-Based Approach for Retrieving Images by Content, *IEEE Transcations on Knowledge and Data Engineering*, 8(4), August 1996.

23. P. W. Huang and Y. R. Jean, Design of Large Intelligent Image Database Systems, *International Journal of Intelligent Systems*, 11, 1996, 347.

24. B. Tao and B. Dickinson, Image Retrieval and Pattern Recognition, *The International Society for Optical Engineering*, 2916, 1996, 130.

25. J. Zachary, S. S. Iyengar, J. Barhen, Content Based Image Retrieval and Information Theory, A General Approach, *JASIS*.

26. C. Carson, M. Thomas, S. Belongie, J. M. Hellerstein, J. Malik, Blobworld: A System for Region-based Image Indexing and Retrieval, *Proc. Int'l Conf. Visual Information System*, 509–516.

59

Computational Biology

Paolo Ferragina
Università di Pisa

Stefan Kurtz
University of Hamburg

Stefano Lonardi
University of California, Riverside

Giovanni Manzini
Università del Piemonte Orientale

59.1 Introduction.. 917
59.2 Discovering Unusual Words 917
    Statistical Analysis of Words • Detecting Unusual Words
59.3 Comparing Whole Genomes 923
    Basic Definitions • Computation of *multiMEMs* • Space Efficient Computation of *MEMs* for Two Genomes
59.4 The FM-index.. 926
    Fast Rank and Select Operations Using Wavelet Trees
References... 931

59.1 Introduction

In the last 20 years, biological sequence data have been accumulating at exponential rate under continuous improvement of sequencing technology, progress in computer science, and steady increase of funding. Molecular sequence databases (e.g., EMBL, Genbank, DDJB, Entrez, Swissprot, etc.) currently collect hundreds of thousands of sequences of nucleotides and amino acids from biological laboratories all over the world, reaching into the hundreds of terabytes. Such an exponential growth makes it increasingly important to have fast and automatic methods to process, analyze, and visualize massive amounts of data.

The exploration of many computational problems arising in contemporary molecular biology has now grown to become a mature field of computer science. A coarse selection would include sequence homology and alignment, physical and genetic mapping, protein folding and structure prediction, gene-expression analysis, evolutionary tree construction, gene finding, assembly for shotgun sequencing, gene rearrangements and pattern discovery, among others.

In this chapter, we focus the attention on the applications of suffix trees and suffix arrays to computational biology. In fact, suffix trees, suffix arrays, and their variants are among the most used (and useful) data structures in computational biology. (See Chapters 30 and 31 for more on suffix trees and DAWGs.)

This chapter is divided into three sections. In Sections 59.2 and 59.3, we describe two distinct applications of suffix trees to computational biology. In the first application, the use of suffix trees and DAWGs is essential in developing a linear-time algorithm for solving a pattern discovery problem. In the second application, suffix trees are used to solve a challenging algorithmic problem: the alignment of two or more complete genomes.

In Section 59.4, we describe the FM-index which is a sort of "compressed" suffix array based on a fascinating mathematical transformation called the Burrows–Wheeler transform. Because of its reduced space occupancy, the FM-index for a mammalian genome can be stored entirely in primary memory. Due to its memory efficiency, this data structure is a cornerstone of the majority of alignment tools, as well some assembly tools.

59.2 Discovering Unusual Words

In the context of computational biology, "pattern discovery" refers to the automatic identification of biologically significant patterns (or *motifs*) by statistical methods. These methods originated from the work of R. Staden [1] and have been recently expanded in a subfield of data mining. The underlying assumption is that biologically significant words show distinctive distributional patterns within the genomes of various organisms, and therefore they can be distinguished from the others. The reality is not too far from this hypothesis [2–10], because during the evolutionary process, living organisms have accumulated certain biases toward or against some specific motifs in their genomes. For instance, highly recurring oligonucleotides are often found in correspondence to regulatory regions or protein binding sites of genes [11–14]. Vice versa, rare oligonucleotide motifs may be discriminated against due to structural constraints of genomes or specific reservations for global transcription controls [9,15]. In some prokaryotes, rare motifs can also corresponds to restriction enzyme sites—bacteria have evolved to avoid those motifs to protect themselves.

From a statistical viewpoint, over- and underrepresented words have been studied quite rigorously (see, e.g., Reference 16 for a review). A substantial corpus of works has been also produced by the scientific community which studies combinatorics on words (see, e.g., References 17 and 18, and references therein). In the application domain, however, the "success story" of pattern discovery has been its ability to find previously unknown regulatory elements in *DNA* sequences [13,19]. Regulatory elements control the expression (i.e., the amount of *mRNA* produced) of the genes over time, as the result of different external stimuli to the cell and other metabolic processes that take place internally in the cell. These regulatory elements are typically, but not always, found in the *upstream* sequence of a gene. The upstream sequence is defined as a portion of *DNA* of 1–2 k bases of length, located upstream of the site that controls the start of transcription. The regulatory elements correspond to binding sites for the factors involved in the transcriptional process. The complete characterization of these elements is a critical step in understanding the function of different genes and the complex network of interaction between them.

The use of pattern discovery to find regulatory elements relies on a conceptual hypothesis that genes which exhibit similar expression patterns are assumed to be involved in the same biological process or functions. Although many believe that this assumption is an oversimplification, it is still a good working hypothesis. Co-expressed genes are therefore expected to share common regulatory domains in the upstream regions for the coordinated control of gene expression. In order to elucidate genes which are co-expressed as a result of a specific stress or condition, a *DNA* microarray experiment is typically designed. After collecting data from the microarray, co-expressed genes are usually obtained by clustering the time series corresponding to their expression profiles.

As said, words that occur unexpectedly often or rarely in genetic sequences have been variously linked to biological meanings and functions. With increasing availability of whole genomes, exhaustive statistical tables and global detectors of unusual words on a scale of millions, even billions of bases become conceivable. It is natural to ask how large such tables may grow with increasing length of the input sequence, and how fast they can be computed. These problems need to be regarded not only from the conventional perspective of asymptotic space and time complexities, but also in terms of the volumes of data produced and ultimately, of practical accessibility and usefulness. Tables that are too large at the outset saturate the perceptual bandwidth of the user and might suggest approaches that sacrifice some modeling accuracy in exchange for an increased throughput.

The number of distinct substrings in a string is at worst quadratic in the length of that string. The situation does not improve if one only computes and displays the most *unusual* words in a given sequence. This presupposes comparing the frequency of occurrence of every word in that sequence with its expectation: a word that departs from expectation beyond some preset *threshold* will be labeled as *unusual* or *surprising*. Departure from expectation is assessed by a distance measure often called a *score* function. The typical format for a *z*-score is that of a difference between observed and expected counts, usually normalized to some suitable moment. For most *a priori* models of a source, it is not difficult to come up with extremal examples of *observed* sequences in which the number of, say, overrepresented substrings grows itself with the square of the sequence length.

An extensive study of probabilistic models and scores for which the population of potentially unusual words in a sequence can be described by tables of size at worst linear in the length of that sequence was carried out in Reference 20. That study not only leads to more palatable representations for those tables but also supports (nontrivial) linear time and space algorithms for their constructions, as described in what follows. These results do not mean that the number of unusual words must be linear in the input, but just that their representation and detection can be made such. Specifically, it is seen that it suffices to consider as candidate surprising words only the members of an *a priori* well-identified set of "representative" words, where the cardinality of that set is linear in the text length. By the representatives being identifiable *a priori*, we mean that they can be known before any score is computed. By neglecting the words other than the representatives we are not ruling out that those words might be surprising. Rather, we maintain that any such word: (i) is embedded in one of the representatives and (ii) does not have a bigger score or degree of surprise than its representative (hence, it would add no information to compute and give its score explicitly).

59.2.1 Statistical Analysis of Words

For simplicity of exposition, assume that the source can be modeled by a Bernoulli distribution, that is, symbols are generated i.i.d., and that strings are ranked based on their number of occurrences (possibly overlapping). The results reported in the rest of this section can be extended to other models and counts [20].

We use standard concepts and notation about strings, for which we refer to References 21 and 22. For a substring y of a text x over an alphabet Σ, we denote by $f_x(y)$ the number of occurrences of y in x. We have $f_x(y) = |pos_x(y)| = |endpos_x(y)|$, where $pos_x(y)$ is the *start-set* of starting positions of y in x and $endpos_x(y)$ is the similarly defined *end-set*. Clearly, for any *extension uyv* of y, $f_x(uyv) \leq f_x(y)$.

Suppose now that string $x = x[0]x[1]\ldots x[n-1]$ is a realization of a stationary ergodic random process and $y = y[0]y[1]\ldots y[m-1] = y$ is an arbitrary but fixed pattern over Σ with $m < n$. We define Z_i, for all $i \in [0\ldots n-m]$, to be

1 if y occurs in x starting at position i, 0 otherwise, so that

$$Z_y = \sum_{i=0}^{n-m} Z_i$$

is the random variable for $f_x(y)$.

Expressions for the expectation and the variance for the number of occurrences in the Bernoulli model have been given by several authors [23–27]. Here we adopt derivations in References 21 and 22. Let p_a be the probability of symbol $a \in \Sigma$ and $\hat{p} = \prod_{i=0}^{m-1} p_{y[i]}$, we have

$$E(Z_y) = (n - m + 1)\hat{p}$$

$$Var(Z_y) = \begin{cases} (1-\hat{p})E(Z_y) - \hat{p}^2(2n - 3m + 2)(m-1) + 2\hat{p}B(y) & \text{if } m \leq (n+1)/2 \\ (1-\hat{p})E(Z_y) - \hat{p}^2(n-m+1)(n-m) + 2\hat{p}B(y) & \text{otherwise} \end{cases}$$

where

$$B(y) = \sum_{d \in \mathcal{P}(y)} (n - m + 1 - d) \prod_{j=m-d}^{m-1} p_{y[j]} \tag{59.1}$$

is the *autocorrelation factor* of y that depends on the set $\mathcal{P}(y)$ of the lengths of the periods[*] of y.

Given $f_x(y)$, $E(Z_y)$, and $Var(Z_y)$, a statistical significance score that measures the degree of "unusual-ness" of a substring y must be carefully chosen. Ideally, the score function should be independent of the structure and size of the word. That would allow one to make meaningful comparisons among substrings of various compositions and lengths based on the value of the score.

There is some general consensus that z-scores may be preferred over other types of score function [28]. For any word w, a standardized frequency called z-score can be defined by

$$z(y) = \frac{f_x(y) - E(Z_y)}{\sqrt{Var(Z_y)}}$$

If $E(Z_y)$ and $Var(Z_y)$ are known, then under rather general conditions, the statistics $z(y)$ is asymptotically normally distributed with zero mean and unit variance as n tends to infinity. In practice, $E(Z_y)$ and $Var(Z_y)$ are seldom known, but are estimated from the sequence under study.

59.2.2 Detecting Unusual Words

Consider now the problem of computing exhaustive tables reporting scores for all substrings of a sequence or perhaps at least for the most surprising among them. While the complexity of the problem ultimately depends on the probabilistic model and type of count, a table for all words of any size would require at least quadratic space in the size of the input, not to mention that such a table would take at least quadratic time to be filled.

As seen towards the end of the section, such a limitation can be overcome by partitioning the set of all words into equivalence classes with the property that it suffices to account for only one or two candidate surprising words in each class, while the number of classes is linear in the textstring size. More formally, given a score function z, a set of words C, and a real positive *threshold T*, we say that a word $w \in C$ is *T-overrepresented* in C (resp., *T-underrepresented*) if $z(w) > T$ (resp., $z(w) < -T$) and for all words $y \in C$ we have $z(w) \geq z(y)$ (resp., $z(w) \leq z(y)$). We say that a word w is *T-surprising* if $z(w) > T$ or $z(w) < -T$. We also call $\max(C)$ and $\min(C)$ respectively the longest and shortest words in C, when $\max(C)$ and $\min(C)$ are unique.

Let now x be a textstring and $\{C_1, C_2, \ldots, C_l\}$ a partition of all its substrings, where $\max(C_i)$ and $\min(C_i)$ are uniquely determined for all $1 \leq i \leq l$. For a given score z and a real positive constant T, we call $\mathcal{O}_z^T$ the set of T-overrepresented words of C_i, $1 \leq i \leq l$, with respect to that score function. Similarly, we call $\mathcal{U}_z^T$ the set of T-underrepresented words of C_i, and $\mathcal{S}_z^T$ the set of all T-surprising words, $1 \leq i \leq l$.

For strings u and $v = suz$, a (u,v)-*path* is a sequence of progressively longer words $\{w_0 = u, w_1, w_2, \ldots, w_j = v\}$, $l \geq 0$, such that w_i is the extension of w_{i-1} $(1 \leq i \leq j)$ with an additional symbol to its left or its right. In general, a (u,v)-path is not unique. If all $w \in C$ belong to some $(\min(C_i), \max(C_i))$-path, we say that class C is *closed*.

A score function z is (u,v)-*increasing* (resp., *nondecreasing*) if given any two words w_1, w_2 belonging to a (u,v)-path, the condition $|w_1| < |w_2|$ implies $z(w_1) < z(w_2)$ (resp., $z(w_1) \leq z(w_2)$). The definitions of a (u,v)-*decreasing* and (u,v)-*nonincreasing* z-scores are symmetric. We also say that a score z is (u,v)-*monotone* when specifics are unneeded or understood. The following fact and its symmetric are immediate.

---

[*] String z has a *period* w if z is a nonempty prefix of w^k for some integer $k \geq 1$.

Fact 59.1

If the z-score under the chosen model is $(min(C_i), max(C_i))$-increasing, and C_i is closed, $1 \leq i \leq l$, then

$$\mathcal{O}_z^T \subseteq \bigcup_{i=1}^{l} \{max(C_i)\} \text{ and } \mathcal{U}_z^T \subseteq \bigcup_{i=1}^{l} \{min(C_i)\}$$

In Reference 20, extensive results on the monotonicity of several scores for different probabilistic models and counts are reported. For the purpose of this chapter, we just need the following result.

Theorem 59.1

Let x be a text generated by a Bernoulli process, and p_{max} be the probability of the most frequent symbol [20]. If $f_x(w) = f_x(wv)$ and $p_{max} < \min\{1/\sqrt[m]{4m}, \sqrt{2} - 1\}$ then

$$\frac{f_x(wv) - E(Z_{wv})}{\sqrt{Var(Z_{wv})}} > \frac{f_x(w) - E(Z_w)}{\sqrt{Var(Z_w)}}$$

Here, we pursue substring partitions $\{C_1, C_2, \ldots, C_l\}$ in forms which would enable us to restrict the computation of the scores to a constant number of candidates in each class C_i. Specifically, we require, (1) for all $1 \leq i \leq l$, $max(C_i)$ and $min(C_i)$ to be unique; (2) C_i to be closed, that is, all w in C_i belong to some $(min(C_i), max(C_i))$ path; (3) all w in C_i have the same count. Of course, the partition of all substrings of x into singleton classes fulfills those properties. In practice, l should be as small as possible.

We begin by recalling a few basic facts and constructs from, for example, Reference 29. We say that two strings y and w are *left-equivalent* on x if the set of starting positions of y in x matches the set of starting positions of w in x. We denote this equivalence relation by $\equiv_l$. It follows from the definition that if $y \equiv_l w$, then either y is a prefix of w, or vice versa. Therefore, each class has unique shortest and longest words. Also by definition, if $y \equiv_l w$ then $f_x(y) = f_x(w)$.

EXAMPLE 59.1

For instance, in the string `ataatataataatataatatag` the set $\{$`ataa, ataat, ataata`$\}$ is a left-equivalent class (with position set $\{1, 6, 9, 14\}$) and so are $\{$`taa, taat, taata`$\}$ and $\{$`aa, aat, aata`$\}$. There are 39 left-equivalent classes, much less than the total number of substrings, which is $22 \times 23/2 = 253$, and than the number of distinct substrings, in this case 61.

We similarly say that y and w are *right-equivalent* on x if the set of ending positions of y in x matches the set of ending positions of w in x. We denote this by $\equiv_r$. Finally, the equivalence relation $\equiv_x$ is defined in terms of the *implication* of a substring of x [29,30]. Given a substring w of x, the implication $imp_x(w)$ of w in x is the longest string uwv such that every occurrence of w in x is preceded by u and followed by v. We write $y \equiv_x w$ iff $imp_x(y) = imp_x(w)$. It is not difficult to see that

Lemma 59.1

The equivalence relation $\equiv_x$ is the transitive closure of $\equiv_l \cup \equiv_r$.

More importantly, the size l of the partition is linear in $|x| = n$ for all three equivalence relations considered. In particular, the smallest size is attained by $\equiv_x$, for which the number of equivalence classes is at most $n + 1$.

Each one of the equivalence classes discussed can be mapped to the nodes of a corresponding automaton or word graph, which becomes thereby the natural support for the statistical tables. The table takes linear space, since the number of classes is linear in $|x|$. The automata themselves are built by classical algorithms, for which we refer to, for example, References 21 and 29 with their quoted literature, or easy adaptations thereof. The graph for $\equiv_l$, for instance, is the compact subword tree T_x of x, whereas the graph for $\equiv_r$ is the DAWG, or *Directed Acyclic Word Graph* D_x, for x. The graph for $\equiv_x$ is the compact version of the DAWG.

These data structures are known to commute in simple ways, so that, say, an $\equiv_x$-class can be found on T_x as the union of some left-equivalent classes or, alternatively, as the union of some right-equivalent classes. Beginning with left-equivalent classes, that correspond one to one to the nodes of T_x, we can build some right-equivalent classes as follows. We use the elementary fact that whenever there is a branching node μ in T_x, corresponding to $w = ay$, $a \in \Sigma$, then there is also a node ν corresponding to y, and there is a special *suffix link* directed from ν to μ. Such auxiliary links induce another tree on the nodes of T_x, that we may call

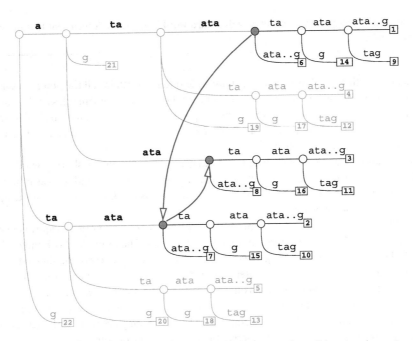

FIGURE 59.1   The tree T_x for $x=$ `ataatataataatataatatag`: subtrees rooted at solid gray nodes are isomorphic.

S_x. It is now easy to find a right-equivalent class with the help of suffix links. For this, traverse S_x bottom-up while grouping in a single class all strings such that their terminal nodes in T_x are roots of isomorphic subtrees of T_x. When a subtree that violates the isomorphism condition is encountered, we are at the end of one class and we start with a new one.

EXAMPLE 59.2

For example, the three subtrees rooted at the solid nodes in Figure 59.1 correspond to the end-sets of `ataata`, `taata`, and `aata`, which are the same, namely, $\{6, 11, 14, 19\}$. These three words define the right-equivalent class $\{$`ataata`, `taata`, `aata`$\}$. In fact, this class cannot be made larger because the two subtrees rooted at the end nodes of `ata` and `tataata` are not isomorphic to the subtree of the class. We leave it as an exercise for the reader to find *all* the right-equivalence classes on T_x. It turns out that there are 24 such classes in this example.

Subtree isomorphism can be checked by a classical linear-time algorithm by Aho et al. [31]. But on the suffix tree T_x this is done even more quickly once the f counts are available [32,33].

Lemma 59.2

Let T_1 and T_2 be two subtrees of T_x. T_1 and T_2 are isomorphic if and only if they have the same number of leaves and their roots are connected by a chain of suffix links.

If, during the bottom-up traversal of S_x, we collect in the same class strings such that their terminal *arc* leads to nodes with the same frequency counts f, then this would identify and produce the $\equiv_x$-classes, that is, the smallest substring partition.

EXAMPLE 59.3

For instance, starting from the right-equivalent class $C=\{$`ataata`, `taata`, `aata`$\}$, one can augment it with of all words which are left-equivalent to the elements of C. The result is one $\equiv_x$-class composed by $\{$`ataa`, `ataat`, `ataata`, `taa`, `taat`, `taata`, `aa`, `aat`, `aata`$\}$. Their respective *pos* sets are $\{1,6,9,14\}$, $\{1,6,9,14\}$, $\{1,6,9,14\}$, $\{2,7,10,15\}$, $\{2,7,10,15\}$, $\{2,7,10,15\}$, $\{3,8,11,16\}$, $\{3,8,11,16\}$, $\{3,8,11,16\}$. Their respective *endpos* sets are $\{4,9,12,17\}$, $\{5,10,13,18\}$, $\{6,11,14,19\}$, $\{4,9,12,17\}$, $\{5,10,13,18\}$, $\{6,11,14,19\}$, $\{4,9,12,17\}$, $\{5,10,13,18\}$, $\{6,11,14,19\}$. Because of Lemma 59.1, given two words y and w in the class, either they share the start set, or they share the end set, or they share the start set by transitivity with a third word in the class, or they share the end set by transitivity with a third word in the class. It turns out that there are only seven $\equiv_x$-classes in this example.

Note that the longest string in this $\equiv_x$-class is unique (ataata) and that it contains all the others as substrings. The shortest string is unique as well (aa). As said, the number of occurrences for all the words in the same class is the same (4 in the example). Figure 59.2 illustrates the seven equivalence classes for the running example. The words in each class have been organized in a lattice, where edges correspond to extensions (or contractions) of a single symbol. In particular, horizontal edges correspond to right extensions and vertical edges to left extensions.

While the longest word in an $\equiv_x$-class is unique, there may be in general more than one shortest word. Consider, for example, the text $x = a^k g^k$, with $k > 0$. Choosing $k = 2$ yields a class which has three words of length two as minimal elements, namely, aa, gg, and ag. (In fact, $imp_x(\text{aa}) = imp_x(\text{gg}) = imp_x(\text{ag}) = \text{aagg}$.) Taking instead $k = 1$, all three substrings of $x = $ ag coalesce into a single class which has two shortest words.

Recall that by Lemma 59.1, each $\equiv_x$-class C can be expressed as the union of one or more left-equivalent classes. Alternatively, C can be also expressed as the union of one or more right-equivalent classes. The example above shows that there are cases in which left- or right-equivalent classes *cannot* be merged without violating the uniqueness of the shortest word. Thus we may use the $\equiv_x$-classes as the C_i's in our partition only if we are interested in detecting overrepresented words. If underrepresented words are also wanted, then we must represent a same $\equiv_x$-class once for each distinct shortest word in it.

It is not difficult to accommodate this in the subtree merge procedure. Let $p(u)$ denote the parent of u in T_x. While traversing S_x bottom-up, merge two nodes u and v with the same f count if and only if both pairs (u, v) and $(p(u), p(v))$ are connected by a suffix link. This results in a substring partition slightly coarser than $\equiv_x$. It will be denoted by $\tilde{\equiv}_x$.

Fact 59.2

Let $\{C_1, C_2, \ldots, C_l\}$ be the set of equivalence classes built on the equivalence relation $\tilde{\equiv}_x$ on the substrings of text x. Then, for all $1 \le i \le l$,

1. $\max(C_i)$ and $\min(C_i)$ are unique
2. All $w \in C_i$ are on some $(\min(C_i), \max(C_i))$-path
3. All $w \in C_i$ have the same number of occurrences $f_x(w)$.

We are now ready to address the computational complexity of our constructions. In Reference 21, linear-time algorithms are given to compute and store expected value $E(Z_w)$ and variance $Var(Z_w)$ for the number of occurrences under Bernoulli model of

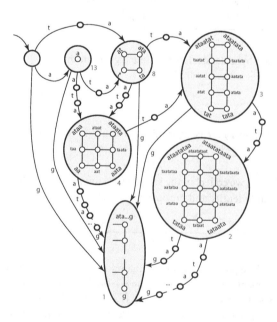

FIGURE 59.2   A representation of the seven $\equiv_x$-classes for $x = $ ataatataataatataatatag. The words in each class can be organized in a lattice. Numbers refer to the number of occurrences.

all prefixes of a given string. The crux of that construction rests on deriving an expression of the variance (see Expression 59.1) that can be cast within the classical linear time computation of the "failure function" or smallest periods for all prefixes of a string [31]. These computations are easily adapted to be carried out on the linked structure of graphs such as S_x or D_x, thereby yielding expectation and variance values at all nodes of T_x, D_x, or the compact variant of the latter. These constructions take time and space linear in the size of the graphs, hence linear in the length of x. Combined with the monotonicity results this yields immediately:

Theorem 59.2

Under the Bernoulli model [20], the sets $\mathcal{O}_z^T$ and $\mathcal{U}_z^T$ associated to the score

$$z(w) = \frac{f_x(w) - E(Z_w)}{\sqrt{Var(Z_w)}}$$

can be computed in linear time and space, provided that $p_{max} < \min\{1/\sqrt[m]{4m}, \sqrt{2} - 1\}$.

59.3 Comparing Whole Genomes

As of writing this article (March 2015), Genbank contains "complete" genomes for more than 3900 virus, over 3200 bacterial, and about 180 eukaryote species. For many of these species, there are even several different genomes available. The abundance of complete genome sequences has given an enormous boost to comparative genomics. Association studies have become a powerful tool for the functional identification of genes and molecular genetics has in many cases revealed the biological basis of diversity. Comparing the genomes of related species gives new insights into the complex structure of organisms at the *DNA* level and protein level.

The first step when comparing genomes is to produce an alignment, that is, a collinear arrangement of sequence similarities. Alignment of nucleic or amino acid sequences has been one of the most important methods in sequence analysis, with much dedicated research and now many sophisticated algorithms available for aligning sequences with similar regions. These require defining a score for all possible alignments (typically, the sum of the similarity/identity values for each aligned symbol, minus a penalty for the introduction of gaps), along with a dynamic programming method to find optimal or near-optimal alignments according to this scoring scheme [34]. These dynamic programming methods run in time proportional to the product of the length of the sequences to be aligned. Hence they are not suitable for aligning entire genomes. In the last decade, a large number of genome alignment programs have been developed, most of which use an anchor-based method to compute an alignment (for an overview, see Reference 35). An *anchor* is an exact match of some minimum length occurring in all genomes to be aligned (see Figure 59.3). The anchor-based method can roughly be divided into the following three phases:

1. Computation of all potential anchors
2. Computation of an optimal collinear sequence of nonoverlapping potential anchors: these are the anchors that form the basis of the alignment (see Figure 59.4)
3. Closure of the gaps in between the anchors.

In the following, we will focus on phase 1 and describe two algorithms to compute potential anchors. The first algorithm allows one to align more than two genomes, while the second is limited to two genomes, but uses less space than the former. Both algorithms are based on suffix trees. For phase 2 of the anchor-based method, one uses methods from computational geometry.

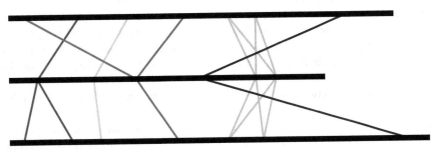

FIGURE 59.3 Three genomes represented as horizontal lines. Potential anchors of an alignment are connected by vertical lines. Each anchor is a sequence occurring at least once in all three genomes.

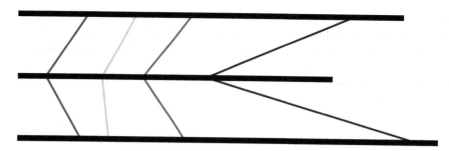

FIGURE 59.4 A collinear chain of nonoverlapping anchors.

The interested reader is referred to the algorithms described in References 36–39. In phase 3, one can apply any standard multiple or pairwise alignment methods, such as T-Coffee [40] or MAFFT [41]. Alternatively, one can even use the same anchor-based method with relaxed parameters, like it is possible in the CoCoNUT software [42].

59.3.1 Basic Definitions

We recall and extend the definition introduced in Section 59.2. A *sequence*, or a *string*, S of length n is written as $S = S[0]S[1] \ldots S[n-1] = S[0 \ldots n-1]$. A *prefix* of S is a sequence $S[0 \ldots i]$ for some i, $0 \leq i \leq n-1$. A *suffix* of S is a sequence $S[i \ldots n-1]$ for some i, $0 \leq i \leq n-1$. Consider a set $\{G_0, \ldots, G_{k-1}\}$ of $k \geq 2$ sequences (the genomes) over some alphabet Σ. Let $n_q = |G_q|$ for $0 \leq q \leq k-1$. To simplify the handling of boundary cases, assume that $G_0[-1] = \$_{-1}$ and $G_{k-1}[n_{k-1}] = \$_{k-1}$ are unique symbols not occurring in Σ. A *multiple exact match* is a $(k+1)$-tuple $(l, p_0, p_1, \ldots, p_{k-1})$ such that $l > 0$, $0 \leq p_q \leq n_q - l$, and $G_q[p_q \ldots p_q + l - 1] = G_{q'}[p_{q'} \ldots p_{q'} + l - 1]$ for all q, q', $0 \leq q, q' \leq k-1$. A multiple exact match is *left maximal* if for at least one pair (q, q'), $0 \leq q, q' \leq k-1$, we have $G_q[p_q - 1] \neq G_{q'}[p_{q'} - 1]$. A multiple exact match is *right maximal* if for at least one pair (q, q'), $0 \leq q, q' \leq k-1$, we have $G_q[p_q + l] \neq G_{q'}[p_{q'} + l]$. A multiple exact match is *maximal* if it is left maximal and right maximal. A maximal multiple exact match is also called *multiMEM*. Roughly speaking, a *multiMEM* is a sequence of length l that occurs in all sequences $G_0, \ldots, G_{k-1}$ (at positions $p_0, \ldots, p_{k-1}$), and cannot simultaneously be extended to the left or to the right in every sequence. The *ℓ-multiMEM*-problem is to enumerate all *multiMEMs* of length at least ℓ for some given length threshold $\ell \geq 1$. For $k = 2$, we use the notion *MEM* and *ℓ-MEM*-problem. The notions of matches defined here can be generalized in several ways. One important generalization is to require that for all q, the matching substring $G_q[p_q \ldots p_q + l - 1]$ is allowed to occur at most r_q times in G_q for some user-defined parameter r_q. This allows to model, for example, maximal unique matches (by defining $r_q = 1$) or excludes repeats from being part of a *multiMEM* (by defining r_q as small constant). Methods to compute such rare multiple matches are described in Reference 74. Here we however restrict to solutions of the *ℓ-multiMEM*-problem.

Let $\$_0, \ldots, \$_{k-1}$ be pairwise different symbols not occurring in any G_q. These symbols are used to separate the sequences in the concatenation $S = G_0 \$_0 G_1 \$_1 \ldots G_{k-2} \$_{k-2} G_{k-1}$. $\$_{k-1}$ will be used as a sentinel attached to the end of G_{k-1}. All these special symbols are collected in a set $Seps = \{\$_{-1}, \$_0, \ldots, \$_{k-2}, \$_{k-1}\}$.

Let $n = |S|$. For any i, $0 \leq i \leq n$, let $S_i = S[i \ldots n-1] \$_{k-1}$ denote the ith nonempty suffix of $S \$_{k-1}$. Hence $S_n = \$_{k-1}$. For all q, $0 \leq q \leq k-1$ let t_q be the start position of G_q in S. For $0 \leq i \leq n-1$ such that $S[i] \notin Seps$, let

- $\sigma(i) = q$ iff position i occurs in the qth sequence in S and
- $\rho(i) = i - t_q$ denote the relative position of i in G_q.

We consider trees whose edges are labeled by nonempty sequences. For each symbol a, every node α in these trees has at most one a-edge $\alpha \xrightarrow{av} \beta$ for some sequence v and some node β. Let $\mathcal{T}$ be a tree and α be a node in $\mathcal{T}$. A node α is denoted by $\overline{w}$ if and only if w is the concatenation of the edge labels on the path from the root of $\mathcal{T}$ to α. A sequence w *occurs* in $\mathcal{T}$ if and only if $\mathcal{T}$ contains a node $\overline{wv}$, for some (possibly empty) sequence v. The *suffix tree* for S, denoted by $\mathrm{ST}(S)$, is the tree $\mathcal{T}$ with the following properties: (i) each node is either the root, a leaf or a branching node, and (ii) a sequence w occurs in $\mathcal{T}$ if and only if w is a substring of $S \$_{k-1}$. For each branching node $\overline{au}$ in $\mathrm{ST}(S)$, where a is a symbol and u is a string, u is also a branching node, and there is a *suffix link* from $\overline{au}$ to $\overline{u}$.

There is a one-to-one correspondence between the leaves of $\mathrm{ST}(S)$ and the nonempty suffixes of $S \$_{k-1}$: leaf $\overline{S_i}$ corresponds to suffix S_i and vice versa.

For any node $\overline{u}$ of $\mathrm{ST}(S)$ (including the leaves), let $\mathcal{P}_{\overline{u}}$ be the set of positions i such that u is a prefix of S_i. In other words, $\mathcal{P}_{\overline{u}}$ is the set of positions in S where sequence u starts. $\mathcal{P}_{\overline{u}}$ is divided into disjoint and possibly empty position sets:

- For any q, $0 \leq q \leq k-1$, define $\mathcal{P}_{\overline{u}}(q) = \{i \in \mathcal{P}_{\overline{u}} \mid \sigma(i) = q\}$ that is, $\mathcal{P}_{\overline{u}}(q)$ is the set of positions i in S where u starts and i is a position in genome G_q.

- For any $a \in \Sigma \cup Seps$, define $\mathcal{P}_{\overline{u}}(a) = \{i \in \mathcal{P}_{\overline{u}} \mid S[i-1] = a\}$, that is, $\mathcal{P}_{\overline{u}}(a)$ is the set of positions i in S where u starts and the symbol to the left of this position is a. We also state that the left context of i is a.
- For any q, $0 \le q \le k-1$ and any $a \in \Sigma \cup Seps$, define $\mathcal{P}_{\overline{u}}(q, a) = \mathcal{P}_{\overline{u}}(q) \cap \mathcal{P}_{\overline{u}}(a)$.

59.3.2 Computation of *multiMEMs*

We now describe an algorithm to compute all *multiMEMs*, using the suffix tree for *S*. The algorithm is part of the the multiple genome alignment software *MGA*. It improves on the method which was described in Reference 72.

The algorithm computes position sets $\mathcal{P}_{\overline{u}}(q, a)$ by processing the edges of the suffix tree in a bottom-up traversal. That is, the edge leading to node $\overline{u}$ is processed only after all edges in the subtree below $\overline{u}$ have been processed.

If $\overline{u}$ is a leaf corresponding to, say suffix S_i, then compute $\mathcal{P}_{\overline{u}}(q, a) = \{i\}$ where $q = \sigma(i)$ and $a = S[i-1]$. For all $(q', a') \ne (q, a)$ compute $\mathcal{P}_{\overline{u}}(q', a') = \emptyset$. Now suppose that $\overline{u}$ is a branching node with r outgoing edges. These are processed in any order. Consider an edge $\overline{u} \longrightarrow \overline{w}$. Due to the bottom-up strategy, $\mathcal{P}_{\overline{w}}$ is already computed. However, only a subset of $\mathcal{P}_{\overline{u}}$ has been computed since only, say $j < r$, edges outgoing from $\overline{u}$ have been processed. The corresponding subset of $\mathcal{P}_{\overline{u}}$ is denoted by $\mathcal{P}_{\overline{u}}^j$. The edge $\overline{u} \to \overline{w}$ is processed in the following way: at first, *multiMEMs* are output by combining the positions in $\mathcal{P}_{\overline{u}}^j$ and $\mathcal{P}_{\overline{w}}$. In particular, all $(k+1)$-tuples $(l, p_0, p_1, \ldots, p_{k-1})$ satisfying the following conditions are enumerated:

1. $l = |u|$
2. $p_q \in \mathcal{P}_{\overline{u}}^j(q) \cup \mathcal{P}_{\overline{w}}(q)$ for any q, $0 \le q \le k-1$
3. $p_q \in \mathcal{P}_{\overline{u}}^j(q)$ for at least one q, $0 \le q \le k-1$
4. $p_q \in \mathcal{P}_{\overline{w}}(q)$ for at least one q, $0 \le q \le k-1$
5. $p_q \in \mathcal{P}_{\overline{w}}(q, a)$ and $p_{q'} \in \mathcal{P}_{\overline{w}}(q', b)$ for at least one pair (q, q'), $0 \le q, q' \le k-1$ and different symbols a and b.

By definition of $\mathcal{P}_{\overline{u}}$, u occurs at positions $p_0, p_1, \ldots, p_{k-1}$ in S. Moreover, for all q, $0 \le q \le k-1$, we define $\rho(p_q)$ as the relative position of u in G_q. Hence $(l, \rho(p_0), \rho(p_1), \ldots, \rho(p_{k-1}))$ is a multiple exact match. Conditions 3 and 4 guarantee that not all positions are exclusively taken from $\mathcal{P}_{\overline{u}}^j(q)$ or from $\mathcal{P}_{\overline{w}}(q)$. In other words, at least two of the positions in $\{p_0, p_1, \ldots, p_{k-1}\}$ are taken from different subtrees below $\overline{u}$. Suppose that $p_q \in \mathcal{P}_{\overline{u}}^j(q)$ and $p_{q'} \in \mathcal{P}_{\overline{w}}(q')$. Hence p_q comes from a subtree below $\overline{u}$ that can be reached by one of the previous j edges processed so far. Let c be the first character of label of this edge. $p_{q'}$ comes from the subtree reached by $\overline{u} \to \overline{w}$ and the first character of the label of this edge is different from c. As a consequence $(l, \rho(p_0), \rho(p_1), \ldots, \rho(p_{k-1}))$ is right maximal. Condition 5 means that at least two of the positions in $\{p_0, p_1, \ldots, p_{k-1}\}$ have a different left context. This implies that $(l, \rho(p_0), \rho(p_1), \ldots, \rho(p_{k-1}))$ is left maximal.

As soon as for the current edge $\overline{u} \to \overline{w}$ the *multiMEMs* are enumerated, the algorithm adds $\mathcal{P}_{\overline{w}}(q, a)$ to $\mathcal{P}_{\overline{u}}^j(q, a)$ to obtain position sets $\mathcal{P}_{\overline{u}}^{j+1}(q, a)$ for all q, $0 \le q \le k-1$ and all $a \in \Sigma \cup Seps$. That is, the position sets are inherited from node $\overline{w}$ to the parent node $\overline{u}$. Finally, $\mathcal{P}_{\overline{u}}(q, a)$ is obtained as soon as all edges outgoing from $\overline{u}$ are processed.

The algorithm performs two operations on position sets, namely enumeration of multiple exact matches by combining position sets and accumulation of position sets. A position set $\mathcal{P}_{\overline{u}}(q, a)$ is the union of position sets from the subtrees below $\overline{u}$. Recall that we consider processing an edge $\overline{u} \to \overline{w}$. If the edges to the children of $\overline{w}$ have been processed, the position sets of the children are obsolete. Hence it is not required to copy position sets. At any time of the algorithm, each position is included in exactly one position set. Thus the position sets require $O(n)$ space. For each branching node one maintains a table of $k(|\Sigma| + 1)$ references to possibly empty position sets. In particular, to achieve independence of the number of separator symbols, we store all positions from $\mathcal{P}_{\overline{u}}(q, a)$, $a \in Seps$, in a single set. Hence, the space requirement for the position sets is $O(|\Sigma|kn)$. The union operation for the position sets can be implemented in constant time using linked lists. For each node, there are $O(|\Sigma|k)$ union operations. Since there are $O(n)$ edges in the suffix tree, the union operations thus require $O(|\Sigma|kn)$ time.

Each combination of position sets requires to enumerate the following cartesian product:

$$\mathop{\times}_{q=0}^{k-1} \left(\mathcal{P}_{\overline{u}}^j(q) \cup \mathcal{P}_{\overline{w}}(q) \right) \setminus \left(\left(\mathop{\times}_{q=0}^{k-1} \mathcal{P}_{\overline{u}}^j(q) \right) \cup \left(\mathop{\times}_{q=0}^{k-1} \mathcal{P}_{\overline{w}}(q) \right) \cup \left(\mathop{\times}_{a \in \Sigma \cup Seps} (\mathcal{P}_{\overline{w}}(a) \cup \mathcal{P}_{\overline{u}}(a)) \right) \right) \quad (59.2)$$

The enumeration is done in three steps, as follows. In a first step one enumerates all possible k-tuples $(P_0, P_1, \ldots, P_{k-1})$ of nonempty sets where each P_q is either $\mathcal{P}_{\overline{u}}^j(q)$ or $\mathcal{P}_{\overline{w}}(q)$. Such a k-tuple is called father/child choice, since it specifies to either choose a position from the father $\overline{u}$ (a father choice) or from the child $\overline{w}$ (a child choice). One rejects the two k-tuples specifying only father choices or only child choices and processes the remaining father/child choices further. In the second step, for a fixed father/child choice $(P_0, P_1, \ldots, P_{k-1})$ one enumerates all possible k-tuples $(a_0, \ldots, a_{k-1})$ (called symbol choices) such that $P_q(a_q) \ne \emptyset$. At most $|\Sigma|$ symbol choices consisting of k identical symbols (this can be decided in constant time) are rejected.

The remaining symbol choices are processed further. In the third step, for a fixed symbol choice $(a_0, \ldots, a_{k-1})$ we enumerate all possible k-tuples $(p_0, \ldots, p_{k-1})$ such that $p_q \in P_q(a_q)$ for $0 \le q \le k-1$. By construction, each of these k-tuples represents a *multiMEM* of length l. The cartesian product (59.2) thus can be enumerated in $O(k)$ space and in time proportional to its size.

The suffix tree construction and the bottom-up traversal (without handling of position sets) require $O(n)$ time. Thus the algorithm described here runs in $O(|\Sigma|kn)$ space and $O(|\Sigma|kn + z)$ time where z is the number of *multiMEM*s. It is still unclear if there is an algorithm which avoids the factors $|\Sigma|$ or k in the space or time requirement.

59.3.3 Space Efficient Computation of *MEM*s for Two Genomes

The previous algorithm, when applied to two genomes, requires to construct the suffix tree of the concatenation $S = G_0 \$_0 G_1$. The algorithm described next computes all *MEM*s of length at least ℓ by constructing the suffix tree of G_0 and matches G_1 against it. Thus, for genomes of similar sizes, the space requirement is roughly halved.

The matching process delivers substrings of G_1 represented by locations in the suffix tree, where a location is defined as follows: Suppose a string u occurs in $\text{ST}(G_0)$. If there is a branching node $\bar{u}$, then $\bar{u}$ is the *location of* u. If $\bar{v}$ is the branching node of maximal depth, such that $u = vw$ for some nonempty string w, then $(\bar{v}, w)$ is the *location* of u. The location of u is denoted by $loc(u)$.

$\text{ST}(G_0)$ represents all suffixes $T_i = G_0[i \ldots n_0-1]\$_0$ of $G_0 \$_0$. The algorithm processes G_1 suffix by suffix from longest to shortest. In the jth step, the algorithm processes suffix $R_j = G_1[j \ldots n_1-1]\$_2$ and computes the locations of two prefixes p^j_{min} and p^j_{max} of R_j defined as follows:

- p^j_{max} is the longest prefix of R_j that occurs in $\text{ST}(G_0)$.
- p^j_{min} is the prefix of p^j_{max} of length $\min\{\ell, |p^j_{max}|\}$.

If $|p^j_{min}| < \ell$, then $p^j_{min} = p^j_{max}$ and the longest prefix of R_j matching any substring of G_1 is not long enough. So in this case R_j is skipped. If $|p^j_{min}| = \ell$, then at least one suffix represented in the subtree below $loc(p^j_{min})$ matches the first ℓ characters of R_j. To extract the *MEM*s, this subtree is traversed in a depth first order. The depth first traversal maintains for each visited branching node u the length of the longest common prefix of u and R_j. Each time a leaf $\bar{T}_i$ is visited, one first checks if $G_0[i-1] \ne G_1[j-1]$. If this is the case, then $(\bot, \rho(i), \rho(j))$ is a left maximal exact match and one determines the length l of the longest common prefix of T_i and R_j. By construction, $l \ge \ell$ and $G_0[i+l] \ne G_1[j+l]$. Hence $(l, \rho(i), \rho(j))$ is an *MEM*. Now consider the different steps of the algorithm in more detail.

Computation of $loc(p^j_{min})$: For $j=0$, one computes $loc(p^j_{min})$ by greedily matching $G_1[0 \ldots \ell-1]$ against $\text{ST}(G_0)$. For all j, $1 \le j \le n_1-1$, one follows the suffix link of $loc(p^{j-1}_{min})$, if this is a branching node, or of $\bar{v}$ if $loc(p^{j-1}_{min}) = (\bar{v}, w)$. This shortcut via the suffix link leads to a branching node on the path from the root to $loc(p^j_{min})$, from which one matches the next characters in G_1. The method is similar to the matching statistics computation of Reference 43, and one can show that its overall running time for the computation of all $loc(p^j_{min})$, $0 \le j \le n_1-1$, is $O(n_1)$.

Computation of $loc(p^j_{max})$: Starting from $loc(p^j_{min})$ one computes $loc(p^j_{max})$ by greedily matching $G_1[|p^j_{min}| \ldots n_j-1]$ against $\text{ST}(G_0)$. To facilitate the computation of longest common prefixes, one keeps track of the list of branching nodes on the path from $loc(p^j_{min})$ to $loc(p^j_{max})$. This list is called the *match path*. Since $|p^{j-1}_{max}| \ge 1$ implies $|p^j_{max}| \ge |p^{j-1}_{max}| - 1$, we do not always have to match the edges of the suffix tree completely against the corresponding substring of G_1. Instead, to reach $loc(p^j_{max})$, one rescans most of the edges by only looking at the first character of the edge label to determine the appropriate edge to follow. Thus the total time for this step in $O(n_1 + \alpha)$ where α is the total length of all match paths. α is upper bounded by the total size β of the subtrees below $loc(p^j_{min})$, $0 \le j \le n_1-1$. β is upper bounded by the number r of right maximal exact matches between G_0 and G_1. Hence the running time for this step of the algorithm is $O(n_1 + r)$.

The depth first traversal: This maintains an *lcp*-stack which stores for each visited branching node, say $\bar{u}$, a pair of values $(onmatchpath, lcpvalue)$, where the Boolean value $onmatchpath$ is true, if and only $\bar{u}$ is on the match path, and $lcpvalue$ stores the length of the longest common prefix of u and R_j. Given the match path, the *lcp*-stack can be maintained in constant time for each branching node visited. For each leaf $\bar{T}_i$ visited during the depth first traversal, the *lcp*-stack allows to determine in constant time the length of the longest common prefix of T_i and R_j. As a consequence, the depth first traversal requires time proportional to the size of the subtree. Thus the total time for all depth first traversals of the subtrees below $loc(p^j_{min})$, $0 \le j \le n_1-1$, is $O(r)$.

Altogether, the algorithm described here runs in $O(n_0 + n_1 + r)$ time and $O(n_0)$ space. It is implemented as part of the widely used *MUMmer* genome alignment software [44] which is available at http://mummer.sourceforge.net/.

59.4 The FM-index

In the last decade, next-generation sequencing (NGS) machines have become the main tool for genomic analysis. The widespread diffusion of NGS machines and their continuous increase in efficiency have posed new challenges to the bioinformatics

	SA		F	L
1	13	$	$ agaatatagta	g
2	3	aatatagtag$	aatatagtag$ a	g
3	11	ag$	ag$ agaatatag	t
4	1	agaatatagtag$	agaatatagtag	$
5	8	agtag$	agtag$ agaata	t
6	6	atagtag$	atagtag$ agaa	t
7	4	atatagtag$ ⟹	atatagtag$ ag	a
8	12	g$	g$ agaatatagt	a
9	2	gaatatagtag$	gaatatagtag$ a	a
10	9	gtag$	gtag$ agaatat	a
11	10	tag$	tag$ agaatata	g
12	7	tagtag$	tagtag$ agaat	a
13	5	tatagtag$	tatagtag$ aga	a

FIGURE 59.5 Suffix array and Burrows–Wheeler transform for the string s = agaatatagtag$.

community. Usually, the first step in the analysis of NGS data is the alignment of millions of short (a few hundred bases) fragments to a reference genome. Such alignment is indeed an approximate matching problem since fragments can differ from the reference because of sequencing errors or individual differences. Because of the importance of the problem it is not surprising that many different alignment tools have been proposed in the literature [45,46]. Some of the most popular alignment tools, such as Bowtie2 [47], BWA-MEM [48], GEM [49], and Soap2 [50], solve the "inexact" alignment problem by first finding exact matches between substrings of the fragment and the reference genome. The FM-index has also been used in some genome assembly tools (e.g., SGA [51]).

Because of the size of the data involved, to find such exact matches these tools do not use the suffix array or suffix tree data structures introduced in Chapters 30 and 31. Instead, they use some variant of the more recent FM-index data structure described in this section. As we will see the FM-index is strictly related to the suffix array but it is much more compact and searches for exact matchings using a completely different, and somewhat surprising, strategy in which the problem of searching substrings is reduced to the easier problem of counting single characters.

Figure 59.5 shows the suffix array for the string agaatatagtag $, where, as usual, we have introduced a unique sentinel character $ which is lexicographically smaller than all other symbols. On the right of the suffix array we show the matrix of characters obtained by transforming each suffix into a cyclic rotation of the input string. In the matrix we have highlighted the first (F) and last (L) columns. It is immediate to see that any column of this matrix, hence also F and L, is a permutation of the input string. Clearly column F contains the string characters alphabetically sorted, while the last column L has no apparent properties. The following result, due to Mike Burrows and David Wheeler, shows that indeed these two columns are related by the following nontrivial property.

Theorem 59.3

For any character α [52], the ith occurrence of α in L corresponds to the ith occurrence of α in F.

For example, in Figure 59.5 the second a in L and F are respectively L[8] and F[3]; these two characters coincide in the sense that they are both the last a of the input string agaatatagtag$. In view of the above property, it is natural to define the *Last-to-First* column mapping LF, such that $LF(i)$ is the index in the first column F of the character corresponding to L[i]. For example since L[8] corresponds to F[3] we have $LF(8) = 3$. As another example $LF(2) = 9$ since L[2], the second occurrence of g in L, corresponds to F[9], the second occurrence of g in F; indeed they are both the first g in agaatatagtag$.

Let C denote the array indexed by characters in Σ such that C[α] is the number of occurrences of text characters alphabetically smaller than α, and let $\text{Rank}_\alpha(L, k)$ denote the number of occurrences of the character α in the prefix L[1, k]. By Theorem 59.3, if L[i] = α then the character in column F corresponding to L[i] is the $\text{Rank}_\alpha(L, i)$-th occurrence of α in F. Since the characters in F are alphabetically sorted, this occurrence is in position C[α] + $\text{Rank}_\alpha(L, i)$. Summing up, we have that the LF map can be computed as

$$LF(i) = C[L[i]] + \text{Rank}_{L[i]}(L, i) \tag{59.3}$$

Given the last column L and the *LF* map, it is possible to retrieve the original string *s* because of the following observation: since each row *j* is a cyclic rotation of *s*, character L[*j*] immediately precedes F[*j*] in *s*, and L[*LF*(*j*)] immediately precedes L(*j*) = F[*LF*(*j*)]. Given L we can therefore retrieve *s* as follows: (i) compute F sorting the characters in L; (ii) by construction, the special character $ will be in F[1], hence the last character of *s* is *s*[*n*] = L[1]; (iii) by the above observation we have *s*[*n* − 1] = L[*LF*(1)], *s*[*n* − 2] = L[*LF*(*LF*(1))], and by induction *s*[*n* − *i*] = L[*LF^i*(1)], where *LF^i*(·) = *LF*(*LF^{i−1}*(·)).

The function mapping the string *s* to the last column L is therefore an invertible transformation, now called the Burrows–Wheeler Transform. In Reference 52, Burrows and Wheeler not only proved the reversibility of the transformation but also noted that column L is often easy to compress and thus originated the family of the so - called Burrows - - Wheeler compressors (for further reading, see Reference [53] and references therein).

In the context of computational biology, rather than in its compressibility, we are interested in another remarkable property of the Burrows–Wheeler Transform. As we now show, because of its relationship with the suffix array, column L can be used as a full - text index for the original string *s*. This property was first established in Reference 54 and it is based on the following result.

Theorem 59.4

Let SA(*s*) denote the suffix array of the string *s* [54]. Assume that [begin,end] denotes the range of rows of SA(*s*) prefixed by a given pattern *Q*, where *Q* is a substring of *s*. Then, for any character α, the range of rows [begin′,end′] prefixed by αQ is given by

$$\text{begin}' = C[\alpha] + \text{Rank}_\alpha(L, \text{begin} - 1) + 1 \quad \text{end}' = C[\alpha] + \text{Rank}_\alpha(L, \text{end}) \tag{59.4}$$

If begin′ > end′ then no row in SA(*s*) is prefixed by αQ, hence the pattern αQ does not appear inside *s*.

The above theorem follows from the observation that [begin′,end′] are precisely the positions in column F corresponding, via the LF-map, to the occurrences of α in L[begin,end]. This means that begin′ = *LF*(*i*) (resp. end′ = *LF*(*j*)), where *i* (resp. *j*) is the index of the first (resp. last) position in L[begin,end] containing the character α. Note that Equation 59.4 computes begin′ and end′ without actually computing the two indices *i* and *j*. From Theorem 59.4, we immediately get the procedure get_rows in Figure 59.6 for determining the rows of SA(*s*) prefixed by an arbitrary pattern *P*.

Note that Theorem 59.4 and Algorithm get_rows determine a range of rows in SA(*s*) without using SA(*s*). The algorithm relies only on the array C, which has size $|\Sigma|$, and the computation of the function Rank$_\alpha$(L, ·). Indeed, the problem of searching substrings in *s* is reduced to the problem of counting single characters in L. The FM-index is loosely defined as a data structure taking *O*(|*s*|) bits of space and supporting the Rank operation in constant time [55]; if the alphabet is not constant, both space and time can have an additional *O*(polylog(|Σ|)) factor.

Using algorithm get_rows an FM-index can report the number of occurrences of a pattern *P* in *s* in *O*(|*P*|) time using *O*(|*s*|) bits of space. The same problem can be solved by a suffix array in *O*(|*P*| log *n*) time, or in *O*(|*P*| + log *n*) time using also the LCP array (see Chapters 30 and 31). Given the simplicity of the binary search procedure, the suffix array is likely to be faster in practice. However, the suffix array uses *O*(|*s*| log |*s*|) bits of space that, both in theory and in practice, is significantly larger than the space usage of FM-index [56]. Indeed, a remarkable feature of the FM-index is that it can take advantage of the compressibility of the input string. Exploiting properties of the Burrows - Wheeler Transform, the FM-index can be stored in space asymptotically bounded by the *k*th-order entropy of the sequence *s* and still support the Rank operation in constant time [57–59]. The *k*th-order entropy is a natural lower bound for the compressibility of a sequence and it was somewhat surprising that in such a "minimal" space we can not only represent *s* but also search it.

Algorithm get_rows(*P*[1, *p*])

    1. *i* ← *p*, α ← *P*[*p*], begin ← C[α] + 1, end ← C[α + 1];
    2. **while** ((begin ≤ end) **and** (*i* ≥ 1)) **do**
    3. α ← *P*[*i*];
    4.   begin ← C[α] + Rank$_c$(L, begin − 1) + 1;
    5.   end ← C[α] + Rank$_c$(L, end);
    6.   *i* ← *i* − 1;
    7. **if** (end < begin) **return** "no rows prefixed by *P*[1, *p*]" **else return** [begin, end].

FIGURE 59.6 Algorithm getrows for finding the set of suffix array rows prefixed by *P*[1, *p*]. This procedure is known in the literature as "backward search" since the characters of *P* are considered right to left.

Algorithm get_position(i)

    1. $t \leftarrow 0$;

    2. **while** row i is not marked **do**

    3. $i \leftarrow \mathrm{C}[\mathrm{L}[i]] + \mathrm{Rank}_{L[i]}(\mathrm{L}, i)$;

    4. $t \leftarrow t + 1$;

    5. **return** SA$[i] + t$;

FIGURE 59.7   Algorithm `get_position` for the computation of SA$[i]$ on an FM-index enriched with a sample of the suffix array. Note that the value computed at line 3 is $LF(i)$ (cf. Equation 59.3).

In some applications, we are interested not only in counting the occurrences but also in finding their positions within s. For this task, the speed of the suffix array is hard to beat: once the range of rows [begin,end] is determined the positions are simply the values stored in SA[begin...end]. To efficiently solve the same problem on an FM-index it is necessary to enrich it with additional information. The common approach is to store along with the FM-index a sampling of the suffix array. For example, we can store the values SA$[i]$ which are multiples of a given value d. Then, using a hash table or a bit array (possibly compressed), we mark the rows containing the stored values.

In the example of Figure 59.5, if we take $d = 4$, we mark the rows $\{5,7,8\}$ since they contains the SA values $\{8,4,12\}$. To determine the position of the occurrences of the pattern tag we first use algorithm `get_rows` to find the SA rows prefixed by tag, which are the rows 11 and 12 of the suffix array. Then, we proceed determining the position in s of these two rows (or more precisely of the two suffixes prefixing these two rows). If we had the complete suffix array the desired values SA$[11] = 10$ and SA$[12] = 7$ would be immediately available, but we only have the SA values for the rows $\{5,7,8\}$. To determine for example SA$[11]$, we apply the *LF*-map (59.3) and compute the value $10 = LF(11)$ which is the entry in F corresponding to the character L$[11] = $ g. By construction, we have that row 10 contains the suffix which is one character longer than the suffix in row 11, hence SA$[10] = $ SA$[11]-1$. Since row 10 is not marked, we still do not know the value SA$[10]$. We thus compute $LF(10) = 5$, which is the index of the row containing the suffix one character longer than the one in row 10. Reasoning as above we have

$$\mathrm{SA}[5] = \mathrm{SA}[10] - 1 = \mathrm{SA}[11] - 2.$$

Now we find that row 5 is marked hence the value SA$[5] = 8$ is available, from which we correctly derive SA$[11] = 10$. The above procedure for retrieving the position of a single occurrence is formalized in Figure 59.7. The idea is to use the *LF*-map to move backward in the text until we reach a marked position for which the SA value is known. By construction we reach a marked position in at most $d - 1$ steps, so the cost of locating a single occurrence is bounded by d times the cost of a Rank operation. Since the cost of storing the SA values is $O(n \log n)/d$ bits, the parameter d offers a simple trade-off between memory usage and speed.

Although the above simple strategy works well in most cases, improvements or other approaches to the problem of locating the occurrences using an FM-index have been proposed in the literature. The reader should refer to References 60–62 and references therein for further information.

59.4.1 Fast Rank and Select Operations Using Wavelet Trees

Wavelet trees have been introduced in Reference 71 at roughly the same time as compressed indices and have quickly gained popularity as a very versatile data structure. They can be described as a data structure offering some degree of compression and supporting fast Rank/Select operations (see below), but over the years they have been used to provide elegant and efficient solutions for a myriad of problems in different algorithmic areas [63].

Consider again the string s of Figure 59.5 and let $s' = $ ggt$ttaaaagaa denote its Burrows--Wheeler transform, that is, column L of Figure 59.5. Figure 59.8 shows a wavelet tree for the string s'. The tree is a complete binary tree and each leaf corresponds to an alphabet character. To build the wavelet tree, we initially label the root with the string s'. Then at each internal node each character goes left or right depending on whether the leaf labeled with that character is on the left or the right subtree. So for example at the root $ and a both go left, while at the root's left child $ goes left and a goes right. Finally, we replace each string in the internal nodes with a binary string obtained replacing with 0 each character that goes left, and with 1 each character that goes right. The wavelet tree consists of the set of binary strings associated to the internal nodes; in our example the wavelet tree for $s' = $ ggt$ttaaaagaa is represented by the binary strings $w_{root} = 1110110000100$, $w_{left} = 0111111$, $w_{right} = 011110$ (we are assuming the shape of the tree and the correspondence from leaves to characters are established in advance so there is no need to encode them).

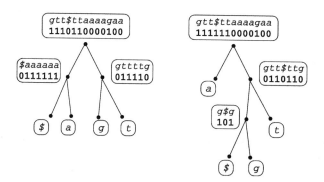

FIGURE 59.8 Two wavelet trees of different shapes for the string $s' = $ ggt\$ttaaaagaa. The tree on the left is balanced and has all leaves at depth two. The tree on the right has the leaf associated to a at depth one: as a result it uses three less bits in the internal nodes.

Suppose now that we are given the wavelet tree and we want to retrieve the character $s'[6]$. Since $w_{root}[6] = 1$ we know that it is a character on the right subtree, so either g or t. To find out which, we count how many 1's there are in w_{root} up to position 6. Using the notation of the previous section this number is $\text{Rank}_1(w_{root}, 6) = 5$. Hence, $s'[6]$ corresponds to the fifth bit of w_{right}. Since $w_{right}[5] = 1$ and w_{right}'s right child is the leaf labeled t we have $s'[6] = $ t. From the above decoding procedure, we can draw the following observations:

1. Each character in the input text is essentially encoded with one bit for each internal node in the path from the root to the corresponding leaf. Using a balanced tree each character requires at most $\lceil \log |\Sigma| \rceil$ bits so the total space usage will be similar to the space used for s' uncompressed. However, using a tree of different shape we can represent s' with less bits (see the tree on the right in Figure 59.8). For any input string, we can minimize the overall length of the internal node's binary strings using the shape of an optimal Huffman tree. Note that the resulting data structure would offer essentially the same compression as a Huffman code but it will also provide random access to the single characters, something which is not possible with Huffman codes. For further discussions on the compressibility of wavelet trees, see Reference 64 and references therein.

2. Assuming that the computation of Rank_1 on a binary string takes $O(r)$ time, decoding a single input character with the above procedure on a balanced wavelet tree takes $O(r \log |\Sigma|)$ time (notice that $\text{Rank}_0(w, k) = k - \text{Rank}_1(w, k)$). If the wavelet tree is skewed some characters could be more expensive to decode than others. For example, if we use a Huffman-shaped tree decoding will be faster for frequent characters and slower for those rarely occurring, a trade-off that could be appropriate in some situations. More in general, if we know the distribution of the accesses we can sacrifice space and choose the shape of the wavelet tree in order to minimize the average access time.

The above observations show that wavelet trees are interesting *per se*. However, their use for FM-indices is justified by another remarkable property. Recall that to efficiently search for patterns in s an FM-index need to support efficient Rank operations on s'. Consider again the example of Figures 59.5 and 59.8 and suppose we need to compute $\text{Rank}_a(s', 9)$, that is, count how many a's are there in the first nine positions of s'. In the string w_{root} a is represented by 0, so by computing $\text{Rank}_0(w_{root}, 9) = 4$ we find that in the first nine positions of s' there are four occurrences of \$'s and a's. To compute $\text{Rank}_a(s', 9)$, we need to count how many a's are there among these four occurrences. This can be done looking at w_{left}: since there a is represented with a 1, we compute $\text{Rank}_1(w_{left}, 4) = 3$ and this is precisely the value $\text{Rank}_a(s', 9)$ that we were looking for. Summing up, Rank queries on s' for any character α can be resolved by $\text{Rank}_0/\text{Rank}_1$ queries on the nodes on the path from the root to the leaf associated to α. Note that, again, the wavelet tree shape influences the cost of the Rank queries on s'.

Putting all together, we have that combining FM-indices and wavelet trees the "difficult" problem of locating patterns in a reference text is reduced to the easier problem of computing Rank queries on binary strings (see Figure 59.9). Note that with an appropriate compressed representation of the binary strings and the use of compression boosting [65], the resulting data structure takes space close to the information theoretic lower bound given by the kth-order entropy of the indexed sequence [66].

Computing Rank queries on binary strings is by no means trivial if we want to compress the binary strings involved. However, if the space usage is not a major issue, there are simple data structures for this task that even first-year students can easily implement. For example, suppose we want to support Rank_1 operations on a binary string w (recall Rank_0 values can be obtained by difference). We split the content of w into blocks of 256 bits and we store each block into four 64-bit words. We interleave these blocks with additional 64-bit words storing the number of 1's in all the preceding blocks. These interleaved values are represented in gray in Figure 59.10.

FIGURE 59.9 The problem of searching patterns in a string is reduced to the problem of counting 1's in binary strings thanks to the combined use of the FM-index and the wavelet tree data structure.

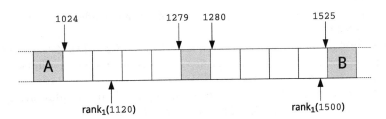

FIGURE 59.10 Simple scheme supporting Rank queries on a binary string described in Reference 67. Each cell represents a 64-bit word. Four consecutive white cells contain a 256-bit block from input string, each gray cell contains the number of 1's in the preceding blocks. The figure shows the 5th and 6th blocks of the input string containing bit ranges $[1024, 1279]$ and $[1280, 1525]$, respectively. Cell A stores the number of 1's in $[0, 1023]$, cell B stores the number of 1's in $[0, 1525]$.

To compute $\text{Rank}_1(w, k)$, we first locate the block containing $w[k]$. If $w[k]$ is in the first half of the block we sum the value in the immediately preceding gray cell with the number of 1's from said cell up to position k. Otherwise, if $w[k]$ is in the second half of a block, we subtract from the value stored in the immediately following gray cell the number of 1's from position $k + 1$ up to that cell. For example, in Figure 59.10 to compute $\text{Rank}_1(1120)$ we sum the value in A to the number of 1's in $[1024, 1120]$, while to compute $\text{Rank}_1(1500)$ we subtract from the value in B the number of 1's in $[1501, 1525]$.

Summing up, using the above data structure the computation of $\text{Rank}_1(w, k)$ only requires the access to up to three consecutive 64-bit words and counting the 1's in up to two of them. Since the latter operation is natively supported by modern Intel and AMD processors (under the name POPCNT), the whole procedure is extremely fast and simple to implement. This data structure uses one additional 64-bit word for each 256-bit block, so the total space usage in bits is $1.25|w|$. If we use this simple scheme for the binary strings of a wavelet tree, a representation of the Burrow–Wheeler Transform for the human genome (≈ 3 billion bases) would take less than 1 GB. Adding a sampling of the suffix array, say with $d = 20$, we get an FM-index of roughly 1.5 GB: this is almost 10 times smaller than the 12 GB used by the suffix array alone.

Together with the Rank operation many data structures support also its inverse, called Select. By definition $\text{Select}_\alpha(s', k)$ is the position in s' of the kth occurrence of character α, or some invalid value if α appears less than k times. As for the Rank operation, given a wavelet tree for s', we can compute $\text{Select}_\alpha(s', k)$ performing $\text{Select}_0/\text{Select}_1$ operations on the binary strings associated to the nodes in the path from the leaf associated to α up to the tree root (i.e., for Select we traverse the tree upward rather than downward). Computing Select on $s' = \text{Bwt}(s)$ allows us to compute the inverse of the LF-map in an FM-index which is useful, for example, for extracting substrings of the indexed text s [59].

For a discussion of sophisticated schemes for Rank/Select queries offering a variety of time/space trade-offs the reader should refer to Reference 59 for the theoretical results, and to References 67–69 for implementations and extensive experiments. The SDSL library [70] offers efficient and ready to use implementations of succinct (compressed) data structures including FM-indices, Suffix trees, and Wavelet trees.

References

1. R. Staden. Methods for discovering novel motifs in nucleic acid sequences. *Comput. Appl. Biosci.*, 5(5), 1989, 293–298.
2. C. Burge, A. Campbell, and S. Karlin. Over- and under-representation of short oligonucleotides in DNA sequences. *Proc. Natl. Acad. Sci. U.S.A.*, 89, 1992, 1358–1362.
3. T. Castrignano, A. Colosimo, S. Morante, V. Parisi, and G. C. Rossi. A study of oligonucleotide occurrence distributions in DNA coding segments. *J. Theor. Biol.*, 184(4), 1997, 451–469.
4. J. W. Fickett, D. C. Torney, and D. R. Wolf. Base compositional structure of genomes. *Genomics*, 13, 1992, 1056–1064.

5. S. Karlin, C. Burge, and A. M. Campbell. Statistical analyses of counts and distributions of restriction sites in DNA sequences. *Nucleic Acids Res.*, 20, 1992, 1363–1370.

6. R. Nussinov. The universal dinucleotide asymmetry rules in DNA and the amino acid codon choice. *J. Mol. Evol.*, 17, 1981, 237–244.

7. P. A. Pevzner, M. Y. Borodovsky, and A. A. Mironov. Linguistics of nucleotides sequences II: Stationary words in genetic texts and the zonal structure of DNA. *J. Biomol. Struct. Dyn.*, 6, 1989, 1027–1038.

8. G. J. Phillips, J. Arnold, and R. Ivarie. The effect of codon usage on the oligonucleotide composition of the *E. coli* genome and identification of over- and underrepresented sequences by Markov chain analysis. *Nucleic Acids Res.*, 15, 1987, 2627–2638.

9. S. Volinia, R. Gambari, F. Bernardi, and I. Barrai. Co-localization of rare oligonucleotides and regulatory elements in mammalian upstream gene regions. *J. Mol. Biol.*, 1(5), 1989, 33–40.

10. S. Volinia, R. Gambari, F. Bernardi, and I. Barrai. The frequency of oligonucleotides in mammalian genic regions. *Comput. Appl. Biosci.*, 5(1), 1989, 33–40.

11. Predicting gene regulatory elements *in silico* on a genomic scale. *Genome Res.*, 8(11), 1998, 1202–1215.

12. M. S. Gelfand, E. V. Koonin, and A. A. Mironov. Prediction of transcription regulatory sites in archaea by a comparative genomic approach. *Nucleic Acids Res.*, 28(3), 2000, 695–705.

13. J. van Helden, B. André, and J. Collado-Vides. Extracting regulatory sites from the upstream region of the yeast genes by computational analysis of oligonucleotides. *J. Mol. Biol.*, 281, 1998, 827–842.

14. J. van Helden, A. F. Rios, and J. Collado-Vides. Discovering regulatory elements in non-coding sequences by analysis of spaced dyads. *Nucleic Acids Res.*, 28(8), 2000, 1808–1818.

15. M. S. Gelfand and E. V. Koonin. Avoidance of palindromic words in bacterial and archaeal genomes: A close connection with restriction enzymes. *Nucleic Acids Res.*, 25, 1997, 2430–2439.

16. G. Reinert, S. Schbath, and M. S. Waterman. Probabilistic and statistical properties of words: An overview. *J. Comput. Biol.*, 7, 2000, 1–46.

17. M. Lothaire. Ed. *Combinatorics on Words*. second ed. Cambridge University Press, London, 1997.

18. M. Lothaire Ed. *Algebraic Combinatorics on Words*. Cambridge University Press, London, 2002.

19. M. Tompa. An exact method for finding short motifs in sequences, with application to the ribosome binding site problem. In *Proceedings of the 7th International Conference on Intelligent Systems for Molecular Biology*, 1999, AAAI Press, Menlo Park, CA, pp. 262–271.

20. A. Apostolico, M. E. Bock, and S. Lonardi. Monotony of surprise and large-scale quest for unusual words. In *Proceedings of Research in Computational Molecular Biology (RECOMB)*, Washington, DC, 2002. G. Myers, S. Hannenhalli, S. Istrail, P. Pevzner, and M. Waterman, Eds., Also in *J. Comput. Biol.*, 10:3–4, (July 2003), 283–311.

21. A. Apostolico, M. E. Bock, S. Lonardi, and X. Xu. Efficient detection of unusual words. *J. Comput. Biol.*, 7(1/2), Jan 2000, 71–94.

22. A. Apostolico, M. E. Bock, and X. Xu. Annotated statistical indices for sequence analysis. In *Sequences*, Positano, Italy, 1998, B. Carpentieri, A. De Santis, U. Vaccaro and J. Storer, Eds., IEEE Computer Society Press, pp. 215–229.

23. J. Gentleman. The distribution of the frequency of subsequences in alphabetic sequences, as exemplified by deoxyribonucleic acid. *Appl. Statist.*, 43, 1994, 404–414.

24. J. Kleffe and M. Borodovsky. First and second moment of counts of words in random texts generated by Markov chains. *Comput. Appl. Biosci.*, 8, 1992, 433–441.

25. P. A. Pevzner, M. Y. Borodovsky, and A. A. Mironov. Linguistics of nucleotides sequences I: The significance of deviations from mean statistical characteristics and prediction of the frequencies of occurrence of words. *J. Biomol. Struct. Dy.*, 6, 1989, 1013–1026.

26. M. Régnier and W. Szpankowski. On pattern frequency occurrences in a Markovian sequence. *Algorithmica*, 22, 1998, 631–649.

27. E. Stückle, C. Emmrich, U. Grob, and P. Nielsen. Statistical analysis of nucleotide sequences. *Nucleic Acids Res.*, 18(22), 1990, 6641–6647.

28. M. Y. Leung, G. M. Marsh, and T. P. Speed. Over and underrepresentation of short DNA words in herpesvirus genomes. *J. Comput. Biol.*, 3, 1996, 345–360.

29. A. Blumer, J. Blumer, A. Ehrenfeucht, D. Haussler, and R. McConnel. Complete inverted files for efficient text retrieval and analysis. *J. Assoc. Comput. Mach.*, 34(3), 1987, 578–595.

30. B. Clift, D. Haussler, R. McConnell, T. D. Schneider, and G. D. Stormo. Sequences landscapes. *Nucleic Acids Res.*, 14, 1986, 141–158.

31. A. V. Aho, J. E. Hopcroft, and J. D. Ullman. *The Design and Analysis of Computer Algorithms*. Addison-Wesley, Reading, MA, 1974.

32. A. Apostolico and S. Lonardi. A speed-up for the commute between subword trees and DAWGs. *Inf. Process. Lett.*, 83(3), 2002, 159–161.

33. D. Gusfield. *Algorithms on Strings, Trees, and Sequences: Computer Science and Computational Biology.* Cambridge University Press, 1997.

34. S. B. Needleman and C. D. Wunsch. A general method applicable to the search for similarities in the amino acid sequence of two proteins. *J. Mol. Biol.*, 48, 1970, 443–453.

35. P. Chain, S. Kurtz, E. Ohlebusch, and T. Slezak. An applications-focused review of comparative genomics tools: Capabilities, limitations and future challenges. *Brief. Bioinform*, 4(2), 2003, 105–123.

36. M. I. Abouelhoda and E. Ohlebusch. A local chaining algorithm and its applications in comparative genomics. In *Proceedings of the Third Workshop on Algorithms in Bioinformatics, WABI* 2003, vol. 2812 of *Lecture Notes in Computer Science*, Springer Verlag, Budapest, Hungary, pp. 1–16.

37. M. I. Abouelhoda and E. Ohlebusch. Multiple genome alignment: Chaining algorithms revisited. In *Proceeding of the 14th Annual Symposium on Combinatorial Pattern Matching CPM*, 2003, Springer Verlag, Morelia, Mexico, pp. 1–16.

38. E. W. Myers and W. Miller. Chaining multiple-alignment fragments in sub-quadratic time. In *Proceedings of the 6th ACM–SIAM Annual Symposium on Discrete Algorithms*, 1995, San Francisco, CA, pp. 38–47.

39. Z. Zhang, B. Raghavachari, R. Hardison, and W. Miller. Chaining multiple alignment blocks. *J. Comput. Biol.*, 1(3), 1994, 217–226.

40. C. Notredame, D. G. Higgins, and J. Heringa. T-Coffee: A novel method for fast and accurate multiple sequence alignment. *J. Mol. Biol.*, 302(1), 2000, 205–217.

41. K. Katoh and D. M. Standley. MAFFT multiple sequence alignment software version 7: Improvements in performance and usability. *Mol. Biol. Evol.*, 30(4), 2013, 772–780.

42. M. I. Abouelhoda, S. Kurtz, and E. Ohlebusch. CoCoNUT: An efficient system for the comparison and analysis of genomes. *BMC Bioinformatics*, 9, 2008, 476.

43. W. I. Chang and E. L. Lawler. Sublinear approximate string matching and biological applications. *Algorithmica*, 12(4/5), 1994, 327–344.

44. S. Kurtz, A. Phillippy, A. Delcher, M. Smoot, M. Shumway, C. Antonescu, and S. Salzberg. Versatile and open software for comparing large genomes. *Genome Biol.*, 5(R12), 2004. https://genomebiology.biomedcentral.com/articles/10.1186/gb-2004-5-2-r12

45. N. Fonseca, J. Rung, A. Brazma, and J. Marioni. Tools for mapping high-throughput sequencing data. *Bioinformatics*, 28(24), 2012, 3169–3177.

46. H. Li and N. Homer. A survey of sequence alignment algorithms for next-generation sequencing. *Brief. Bioinform.*, 11(5), 2010, 473–483.

47. B. Langmead and S. L. Salzberg. Fast gapped-read alignment with Bowtie 2. *Nature Methods*, 9(4), 2012, 357–360.

48. H. Li and R. Durbin. Fast and accurate short read alignment with Burrows–Wheeler transform. *Bioinformatics*, 25(14), 2009, 1754–1760.

49. S. Marco-Sola, M. Sammeth, R. Guigó, and P. Ribeca. The GEM mapper: Fast, accurate and versatile alignment by filtration. *Nature Methods*, 9(12), 2012, 1185–1188.

50. R. Li, C. Yu, Y. Li, T. W. Lam, S. Yiu, K. Kristiansen, and J. Wang. SOAP2: An improved ultrafast tool for short read alignment. *Bioinformatics*, 25(15), 2009, 1966–1967.

51. J. T. Simpson and R. Durbin. Efficient construction of an assembly string graph using the FM-index. *Bioinformatics*, 26(12), 2010, i367–i373.

52. M. Burrows, D. Wheeler. A block-sorting lossless data compression algorithm. Tech. Rep. 124, Digital Equipment Corporation, 1994.

53. P. Ferragina, G. Manzini, and S. Muthukrishnan Eds. Special issue on the Burrows–Wheeler transform and its applications. *Theor. Comput. Sci.*, 387(3), 2007.

54. P. Ferragina and G. Manzini. Opportunistic data structures with applications. In *Proceedings of the 41st IEEE Symposium on Foundations of Computer Science*, Redondo Beach, CA, 2000, pp. 390–398.

55. P. Ferragina and G. Manzini. An experimental study of a compressed index. *Inf. Sci.*, 135(1–2), 2001, 13–28.

56. P. Ferragina, R. González, G. Navarro, and R. Venturini Compressed text indexes: From theory to practice. *ACM J. Exp. Algorithms*, 13, 2009, article 12. 30 pages.

57. D. Belazzougui and G. Navarro. Alphabet-independent compressed text indexing. *ACM Trans. Algorithms*, 10(4), 2014, article 23.

58. P. Ferragina and G. Manzini. Indexing compressed text. *J. ACM*, 52(4), 2005, 552–581.

59. G. Navarro and V. Mäkinen. Compressed full-text indexes. *ACM Comput. Surveys*, 39(1), 2007.

60. P. Ferragina, J. Sirén, and R. Venturini. Distribution-aware compressed full-text indexes. *Algorithmica*, 67(4), 2013, 529–546.

61. S. Gog, A. Moffat, J. S. Culpepper, A. Turpin, and A. Wirth. Large-scale pattern search using reduced-space on-disk suffix arrays. *IEEE Trans. Knowl. Data Eng.*, 26(8), 2014, 1918–1931.

62. S. Gog, G. Navarro, and M. Petri. Improved and extended locating functionality on compressed suffix arrays. *J. Discrete Algorithms*, 32, 2015, 53–63.

63. G. Navarro. Wavelet trees for all. *J. Discrete Algorithms*, 25, 2014, 2–20.

64. P. Ferragina, R. Giancarlo, and G. Manzini. The myriad virtues of wavelet trees. *Inform. Comput.*, 207, 2009, 849–866.

65. P. Ferragina, R. Giancarlo, G. Manzini, and M. Sciortino. Boosting textual compression in optimal linear time. *J. ACM*, 52(4), 2005, 688–713.

66. P. Ferragina, G. Manzini, V. Mäkinen, and G. Navarro. Compressed representations of sequences and full-text indexes. *ACM Trans. Algorithms*, 3(2), 2007.

67. S. Gog and M. Petri. Optimized succinct data structures for massive data. *Softw., Pract. Exp.*, 44(11), 2014, 1287–1314.

68. G. Navarro and E. Providel. Fast, small, simple rank/select on bitmaps. In *Proceedings of the 11th International Symposium on Experimental Algorithms, SEA*, Bordeaux, France, 2012, R. Klasing, Ed., vol. 7276 of *Lecture Notes in Computer Science*, Springer, pp. 295–306.

69. S. Vigna. Broadword implementation of rank/select queries. In *Proceedings of the 7th International Workshop on Experimental Algorithms, WEA*, Provincetown, MA, 2008, C. C. McGeoch, Ed., vol. 5038 of *Lecture Notes in Computer Science*, Springer, pp. 154–168.

70. S. Gog, T. Beller, A. Moffat, and M. Petri. From theory to practice: Plug and play with succinct data structures. In *Proceedings of the 13th International Symposium on Experimental Algorithms SEA*, Copenhagen, Denmark, 2014, J. Gudmundsson, and J. Katajainen, Eds., vol. 8504 of *Lecture Notes in Computer Science*, Springer, pp. 326–337.

71. R. Grossi, A. Gupta, and J. Vitter. High-order entropy-compressed text indexes. In *Proceedings of the 14th ACM–SIAM Symposium on Discrete Algorithms (SODA)*, 2003, pp. 841–850.

72. M. Höhl, S. Kurtz, and E. Ohlebusch. Efficient multiple genome alignment. *Bioinformatics*, 18(Suppl. 1), 2002, S312–S320.

73. E. M. McCreight. A space-economical suffix tree construction algorithm. *J. Assoc. Comput. Mach.*, 23(2), 1976, 262–272.

74. E. Ohlebusch and S. Kurtz. Space efficient computation of rare maximal exact matches between multiple sequences. *J. Comp. Biol.*, 15(4), 2008, 357–377.

75. P. Weiner. Linear pattern matching algorithm. In *Proceedings of the 14th Annual IEEE Symposium on Switching and Automata Theory*, 1973, Washington, DC, pp. 1–11.

60

Data Structures for Cheminformatics

60.1 Introduction..935
60.2 Exact Searches...936
    Graph Theoretic Representations • Canonical Representations • Canonical
    Labeling Methods
60.3 Chemical Fingerprints and Similarity Search...939
    Tanimoto Similarity • *Bitbound* Algorithm • Fingerprint Compression • Hashing
    Approaches for Improved Searches • Triangle Inequality and Improvements • XOR Filter •
    Inverted Indices • LINGO • Trie-Based Approaches for Storing Fingerprints • Multibit
    and Union Tree • Combining Hashing and Trees
References..943

Dinesh P. Mehta
Colorado School of Mines

John D. Crabtree
University of North Alabama

60.1 Introduction

Cheminformatics is an interdisciplinary science that exists at the interface between chemistry and computer and information sciences. Its goal is to design new molecules that meet societal needs. Among the many fields in which it is used, the design of new drugs (medicines) is an area that has seen the greatest application of cheminformatics. The goal of drug discovery is to find the optimal molecule that binds to a biological target, typically a protein. The number of theoretical molecules (known as *chemical space*) from which to find the optimal molecule is infinite. This chemical space can be reduced to a finite *druglike* chemical space (estimated to contain between 10^{12} and 10^{180} molecules) by eliminating molecules that are unlikely to be usable as drugs. The Chemical Abstracts Service (CAS) whose objective is "to find, collect, and organize all publicly disclosed substance information" currently (as of December 29, 2015) contains only approximately 105 million molecules, a small fraction of the druglike chemical space. In practice, the quest for a new molecule starts from lists of existing molecules. Cheminformatics techniques are used to filter these lists to generate a subset of molecules that are tested experimentally against the biological target using *high-throughput screening*. Molecules that bind to the target are said to be *hits*. From the list of hits, the filtering process identifies *leads* and from these, a candidate that enters preclinical development. Perhaps because of the strong connection with medicinal chemistry, cheminformatics is sometimes seen as being related to bioinformatics. However, a key distinction in the underlying algorithmic techniques is that bioinformatics focuses on sequence data (e.g., DNA sequences), whereas cheminformatics focuses on the structure of small molecules represented as graphs. Excellent introductions and surveys of the field include References 1–4.

The science of cheminformatics is largely based on a principle known as the *similar property principle*, which states that molecules with similar structures tend to have similar properties. This principle motivates a graph-theoretic approach to several problems. Graph-theoretic representations are often referred to as topological or 2D approaches in cheminformatics terminology. It should be noted that a molecule actually has 3D geometric structure and that this captures important information about the molecule not available in a 2D representation. In most cases, however, 2D techniques outperform 3D techniques, possibly because the former are a more mature computational technology than the latter. We will accordingly focus on 2D techniques in the remainder of this chapter.

Specific cheminformatics problems that require efficient data structures are listed below.

1. *Exact searches*: Given a query molecule q, is there a molecule d in database D such that $q = d$? This is discussed in Section 60.2.
2. *Similarity searches*: Given a query molecule q, identify molecules d in database D that are similar to q. This is discussed in Section 60.3.

60.2 Exact Searches

This section first describes a graph-theoretic representation of a molecule and then describes canonical representations that are used in the exact search problem.

60.2.1 Graph Theoretic Representations

A molecule can be represented mathematically as a chemical or molecular graph: an atom of the molecule corresponds naturally to a graph vertex while (covalent) bonds connecting atoms correspond naturally to edges between graph vertices. More precisely, a molecular graph is a labeled, undirected, connected multigraph.

1. Labels denote atom type (e.g., a carbon atom is represented by a graph vertex with a label "C").
2. A pair of atoms may be connected by double- or triple-bonds. This requires multiple edges between a pair of vertices in the graph making it a *multigraph*. (Recall that a simple graph permits at most one edge between a pair of vertices.)
3. If a molecular graph is not connected and consists of multiple connected components, each component is considered to be a separate molecule.

A molecular graph could be hydrogen-included (hydrogen atoms are explicitly included) or hydrogen-excluded (hydrogen atoms are excluded from the representation). In the latter, positions of the H atoms can be inferred from the valences and bonds of the remaining atoms. The valence of an atom is the maximum number of bonds incident on it and is determined by the number of electrons in the atom's outermost shell and the octet rule. For example, carbon has a valence of 4. If a carbon atom in a hydrogen-excluded representation consists of (1) a double bond to a second carbon atom and (2) a single bond to a nitrogen atom, we can infer that it has a fourth implicit bond to a hydrogen atom. Other scenarios can also be accommodated within the graph representation.

- In conjugated systems such as benzene, bonds are not well defined and are represented by alternating single/double bonds or alternatively as bonds with a valency of 1.5.
- In a reduced graph model, each vertex represents a group of connected atoms.

Molecular graphs are typically represented using a variant of the well-known adjacency matrix. The adjacency matrix representation is known to be inefficient for sparse graphs (relative to adjacency lists), but the relatively small sizes of molecules make its use acceptable in this context.

60.2.2 Canonical Representations

Consider a data structure (database) that stores a set of molecules. A classical operation on such a data structure is to search for a particular query molecule. Underlying this is the need to solve the following problem: given two molecules M_1 and M_2, are they the same molecule? This is usually a trivial question for most data. For example, two integers can be compared for equality in $O(1)$ time; or two strings can be compared in time proportional to their lengths by a simple character-by-character comparison. For graphs, however, this check for equality reduces to the graph isomorphism problem.

A graph G is said to be isomorphic to graph H if there exists a bijection $f : V(G) \rightarrow V(H)$ such that u and v in G are adjacent if and only if $f(u)$ and $f(v)$ are adjacent in H.

The status of the isomorphism problem is undetermined for general graphs: no polynomial-time algorithm has been discovered, nor has a proof of NP-completeness been found. However, polynomial-time algorithms have been found for special classes of graphs including planar graphs [5], interval graphs [6], and trees [7]. Just as efficient graph isomorphism algorithms have been discovered for other special classes of graphs, a theoretical polynomial-time solution has been proven to exist for graphs of bounded valence [8,9]. However, a practical, polynomial-time, graph isomorphism algorithm for graphs of bounded valence has not been implemented [10].

Faulon showed that a chemical graph can be converted to a bounded degree graph [10], implying that isomorphism can be solved in polynomial time for chemical graphs. However, the underlying algorithms are highly theoretical and are considered to be unimplementable and impractical.

For these and other reasons, the cheminformatics community has considered alternative approaches to implement exact searches. The main idea is to convert a molecular graph to a unique text string known as a canonical label. We denote the canonical label of a molecular graph M by $CL(M)$. The question of whether $M_1 == M_2$ is now reduced to the question $CL(M_1) == CL(M_2)$, which is easily resolved by a simple string comparison. Note that if the canonical labeling algorithm runs in polynomial time, this approach yields a polynomial time solution to the isomorphism problem for molecular graphs. Unfortunately existing canonical labeling algorithms suffer from one of two drawbacks.

1. They run in polynomial time, but do not result in unique labels; that is, two distinct molecular graphs get the same label.
2. They generate unique labels, but run in exponential time on some inputs.

These drawbacks are mitigated in that they do not arise often in practice. A theoretical analysis based on random chemical graphs [11] appears to confirm these observations.

60.2.3 Canonical Labeling Methods

A thorough treatment of all of the individual canonical labeling techniques is beyond the scope of this article. In the following, we describe three canonical labeling methods.

60.2.3.1 Morgan's Algorithm

The first attempt at creating a canonical name for molecules in systems came from the CAS [12]. This algorithm creates a name for a given compound based on the names, connectivity, and other properties of the atoms encountered during the traversal of its graphical representation. The key to Morgan's algorithm is the calculation of the graph's extended connectivity (EC):

1. Set the EC of all atoms to their degree.
2. Count the number of different EC values in the graph (let this be the cardinality C of the EC values).
3. Calculate a new EC value for each atom by summing the old values of its connected vertices (except if the EC is 1, do not change it).
4. Determine cardinality of the new EC values C_{new} and compare to the old value C_{old}; if $C_{new} > C_{old}$, goto Step 3.
5. If $C_{new} < C_{old}$, use C_{old}.
6. If $C_{new} == C_{old}$, use C_{new}.

Many researchers have modified this algorithm to fit their particular needs or data sets. Chen et al. [13] describe a variant of Morgan's Algorithm in which each atom is assigned a code which consists of the atom's elemental symbol, its connectivity, and particle charge.

The example in Figure 60.1 (inspired by Chenet al. [13]) shows how the connectivity is calculated during each iteration of the algorithm.

Initially, there are three distinct connectivity values. After the next iteration, there are four. Since the next two iterations both produce six distinct values, the end condition is triggered and the current (last iteration) is used as the final values.

While this algorithm works for almost all chemical structures, it can fail when the graph is highly regular producing oscillatory behavior. Morgan's Algorithm has been adapted with approximation techniques to solve such problems [14]. As mentioned, many different versions of this algorithm have been produced. The variation from Chen et al. is $O(mn^2)$ where m is the number of starting points and n is the number of symmetry points [13].

60.2.3.2 Nauty

One of the fastest algorithms for isomorphism testing and canonical naming is Nauty (No AUTomorphism, Yes?) [15]. It is based on a method of finding the automorphism group for a graph [16]. An automorphism is a mapping of a graph onto itself. A graph may have many such mappings, which are collectively referred to as the automorphism group.

Given a graph $G(V, E)$, a partition of V is a set of disjoint subsets of V. The set of all partitions of V is denoted by $\Pi(V)$. The divisions of a particular partition π are called its cells. A partition with one cell is called a unit partition and a partition in which each cell contains only one vertex is called a discrete partition. Given two partitions π_1 and π_2, π_1 is finer than π_2 (and

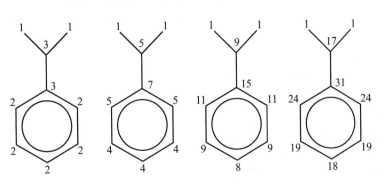

FIGURE 60.1 Connectivity graph of cumin.

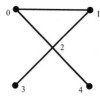

FIGURE 60.2   Graph for Nauty example.

π_2 is coarser than π_1) if every cell of π_1 is a subset of some cell in π_2. The unit partition is the coarsest partition and the discrete partition is the finest. The key to the algorithm is producing *equitable* partitions. A partition π is said to be equitable if, given a partition π of V into n cells (where a cell W is a set of vertices): $\pi = W_1, W_2, \ldots, W_n$, π is equitable if for every cell W_i there exists a d_i such that every vertex in cell W_i is adjacent to exactly d_i vertices in cell W_{i+1}. McKay [16] proves that every graph has a unique, coarsest equitable partition that is guaranteed to exist. The Nauty package implements a refinement process that takes any partition (usually a unit partition) as input and produces the unique equitable coarsest partition as output. The vertices are initially divided into cells by degree, sorted in increasing order. Each cell is then iteratively compared with subsequent cells by comparing the degrees in each cell. Whenever a discrepancy in degree is found, the cell is divided. If no discrepancies are found, the loop ends. Since a discrete partition cannot be further divided, it is also an end condition. Each cell in the partition is "colored" producing a unique name.

Given the graph in Figure 60.2, and the initial unit partition $\pi = [01234]$, the initial refinement process would create the partition $\pi = [34|01|2]$ which, for this simple example, is the coarsest equitable partition. Nodes 3 and 4 would be colored the same as would nodes 0 and 1.

The time complexity of Nauty was analyzed in Reference 17, where it is proven that the worst-case time complexity is exponential. In practice, when used on chemical graphs, Nauty appears to have a complexity that is linear.

60.2.3.3 Signature-Based Canonization

This naming algorithm produces a molecular descriptor called a signature which is based on extended valence sequences [10].

The algorithm first constructs a graph, which is similar in form to a tree, for each atom in the molecule. The graph is not formally a tree since vertices representing atoms may appear more than once in the graph. The root of the "tree" is the atom being considered. The next level is composed of nodes representing all of the atoms connected to the root atom. The next level contains all of the atoms connected to atoms in the previous level. This process continues until all bonds in the molecule have been considered. As previously mentioned, it is possible that a node representing an atom could appear multiple times in the graph. In fact, when these repeated nodes occur at the same level, entire subtree structures are repeated. For this reason, the space requirements of the algorithm are minimized by pointing to repeated subtrees, thus forming a directed acyclic graph (DAG).

Each node in the tree is then initialized with an integer invariant that is derived from the represented atom's type. Then, starting from the level furthest from the root, the nodes are sorted based on their invariants. The invariants of the nodes at the next level closer to the root are then recalculated based on their invariants and those of their neighbors. This process continues until the root is reached. The purpose of this procedure is the production of a unique ordering of the edges of the graph based on the node invariants.

The graph is then traversed in a depth-first manner, producing a signature for the atom. This signature contains enough information to completely rebuild the molecule. In fact, the molecule can be reconstructed from the signature produced for any of the atoms in the graph. The lexicographically largest signature of all of the atom signatures is used to represent the molecule. Since all signatures contain all of the needed information, the other signatures can be discarded. Pseudocode for Signature-based canonization follows.

1. For each atom in the molecule:
    a. Create the signature tree with the current atom at the root
    b. Assign initial integer invariant values
    c. For each level of the signature tree, from the leaves to the root:
        i. Sort the nodes at the current level, based on invariant values
        ii. Recalculate the invariant values for the nodes on the next level using the values of the neighboring nodes
    d. Produce a signature string for the atom based on a depth-first search for every node in the tree
    e. Associate the signature string with the current atom
2. Choose the lexicographically largest signature string to represent the molecule
3. Discard all other signature strings.

The worst-case time complexity of the Signature algorithm on pathological inputs is exponential.

60.3 Chemical Fingerprints and Similarity Search

Because of the difficulties associated with the graph isomorphism and subgraph isomorphism problems, molecules are encoded and stored in chemical databases as fingerprints. Chemical fingerprints are representations of 2D chemical structure and are used in chemical database substructure searching, similarity search, clustering, and classification. Several chemical fingerprinting schemes are described in the literature. One of these is the extended connectivity fingerprint (ECFP).

The ECFP is a circular fingerprint based on Morgan's algorithm. ECFPs are not designed for substructure searches, but are instead used for similarity searches, clustering, and virtual screening. They are generated in three steps: (1) an initial assignment stage, where each atom is assigned an integer identifier; (2) an iterative updating stage in which each atom's identifier is updated to include its neighbors; and (3) finally, a duplicate identifier removal stage. The ECFP set includes all of the identifiers generated in each iteration. For more details about ECFP and a comprehensive discussion of other fingerprint techniques, we refer the reader to Reference 18.

In the remainder of this chapter, we do not focus on a particular fingerprinting scheme, but on the underlying concept as illustrated in Figure 60.3.

60.3.1 Tanimoto Similarity

Given two sets a and b, the Tanimoto similarity $S(a, b)$ between them is defined as

$$
\begin{aligned}
S(a, b) &= \frac{|a \cap b|}{|a \cup b|} \\
&= \frac{|a \cap b|}{|a| + |b| - |a \cap b|}
\end{aligned}
$$

where $|a|$ denotes the cardinality of set a. The cardinality of a set is given by the number of "1"s in the corresponding fingerprint. The Tanimoto similarity is well defined as long as at least one set is nonempty. Note that $0 \le S(a, b) \le 1$, with $S(a, b) = 0$ when the intersection is empty and $S(a, b) = 1$ when the two sets are identical.

Tanimoto similarities are used to find all sets (represented by fingerprints) in a database that are similar to a query fingerprint as follows:

Given a database D of fingerprints and a query fingerprint q, return all fingerprints d∈D such that S(q, d) ≥ t, where t is a user-specified threshold.

Note that the time complexity of this operation is $O(|D|f)$, where f denotes the time needed to compute the Tanimoto similarity between a pair of fingerprints, which is proportional to the length of the fingerprint vector.

An alternative formulation for similarity searches does not require the user to specify an arbitrary t, but instead returns the top C hits:

Given a database D of fingerprints and a query fingerprint q, return C fingerprints d∈D with the highest S(q, d) values.

This approach requires the computation of all Tanimoto similarities as before $O(|D|f)$. A max-heap is used to keep track of the top C hits for an additional computation of $O(|D| (\log C)$. Given the large sizes of molecular databases, a linear search that

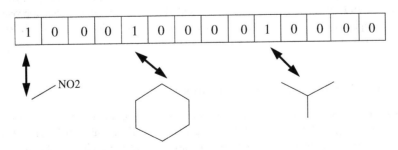

FIGURE 60.3 Illustration of fingerprints: the fingerprint vector is a bit string with each element in the vector denoting a particular molecular substructure. Examples of substructure features used in the fingerprint include all possible labeled paths or labeled trees (circular substructures) up to a certain length. A bit is set to "1" if and only if the corresponding substructure is present in the molecule. In some versions, each element of the vector is a count of the number of occurrences of the particular substructure.

compares a query fingerprint with each of the database fingerprints is unacceptably expensive. In the sections below, we describe several improvements to the linear search based on the use of innovative data structures.

60.3.2 *Bitbound* Algorithm

One strategy [19] to improve search times is to use information about the cardinalities of both sets to reduce the number of database fingerprints against whom similarities are computed. For a given fixed $|q|$ and $|d|$, $S(q, d)$ is maximized when $|q \cap d|$ is maximized and $|q \cup d|$ is minimized. This occurs when $q \subseteq d$ or $d \subseteq q$ and results in the following bound:

$$S(q, d) \leq \frac{\min(|q|, |d|)}{\max(|q|, |d|)}.$$

Consider an example of a similarity search where the size of the query fingerprint $|q| = 200$ and the user-defined similarity threshold $t = 0.8$.

Case 1 $|q| \leq |d|$: $S(q, d) \leq 200/|d|$. Any fingerprints $d \in D$ such that $|d| > 250$ result in $S(q, d) \leq 0.8$ and can be immediately discarded without further analysis.

Case 2 $|q| \geq |d|$: $S(q, d) \leq |d|/200$. Any fingerprints $d \in D$ such that $|d| < 160$ result in $S(q, d) \leq 0.8$ and can also be immediately discarded.

Only d with $160 \leq |d| \leq 250$ need to be evaluated further against query q. To exploit this observation, the database is preprocessed so that fingerprints are placed in bins indexed by size. Only bins corresponding to sizes in the range $[|q|t, |q|/t]$ need to be searched.

This organization of fingerprints by cardinality can also be used in the top C hits formulation. We compute the bound for each bin and sort the bins in decreasing order of bound. The fingerprints are explored (i.e., their Tanimoto similarity against q is computed) in this order. Once again, a max-heap is used to store the top C hits. The search is terminated when the smallest similarity value in the heap is greater than the highest bound remaining.

60.3.3 Fingerprint Compression

Fingerprints can be very long ($2^{15} - 2^{20}$ bits) and sparse. Recall that the computation time for similarity and top-C searches includes a factor f corresponding to fingerprint size. Compressing fingerprints not only reduces the size of the database but also improves computation time. Long fingerprints may be compressed by a factor of k by folding them using a simple modulo operation.

Compressed fingerprints can now be used instead of the original fingerprints for similarity searches as described earlier. This introduces an error that results in a systematic overestimate of Tanimoto similarity. (A given bit location in a pair of compressed fingerprints could both be 1 even though none of the corresponding k pairs of bits in the original fingerprints were both "1"s.) Molecules that would not meet the similarity threshold using the original fingerprints might meet them using compressed fingerprints. This overestimate results in a reduction in the quality of the similarity search. A mathematical correction of this overestimate is derived in Reference 20.

The folding scheme described above is *lossy* because it is not possible to recover the original fingerprint from the compressed fingerprint. An alternative *lossless* approach for reducing fingerprint size exploits the sparsity of fingerprints by storing the indices of "1" bits. In the example, (1 0 0 0 1 1 0 0 0 0 0 0 1 0 1) can be represented by (1 5 6 13 15). Another approach is to store the lengths of 0 runs: (0 3 6 1) in our example. These reduced vectors can be compressed optimally using Golomb-Rice or Elias Codes based on statistical assumptions [21].

60.3.4 Hashing Approaches for Improved Searches

Additional improvements over the bounds presented in Section 60.3.2 can be obtained by associating with each fingerprint a short integer signature vector of length M. When $M = 1$, the integer is simply the number of "1"s in the fingerprint, making the improvement of Section 60.3.2 a special case of this approach. The case when $M = 5$ is illustrated by a simple modification to the example in Figure 60.4: instead of taking the OR of corresponding bits, we add their values giving the vector (2 0 1 0 2).

In our explanation, we focus on the case where $M = 2$. The two integers in this case may be obtained by separately counting the number of "1"s in the first half and the second half, respectively. For fingerprint q, the two counts are denoted by q_1 and q_2, respectively.

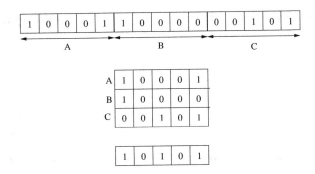

FIGURE 60.4 Illustration of folding: a long fingerprint consisting of 15 bits (1 0 0 0 1 1 0 0 0 0 0 0 1 0 1) is folded by a factor of 3 to give a fingerprint (1 0 1 0 1) of length 5. The original fingerprint is interpreted as having been subdivided into three segments of length 5 each. Bit j in the compressed fingerprint is the logical OR of bits j in the three segments.

We can now compute a bound on $S(q, d)$ as follows:

$$
\begin{aligned}
S(q, d) &\leq \frac{\min(q_1, d_1) + \min(q_2, d_2)}{\max(q_1, d_1) + \max(q_2, d_2)} \\
&= \frac{\min(q_1, d_1) + \min(q_2, d_2)}{|q| + |d| - \min(q_1, d_1) - \min(q_2, d_2)} \\
&\leq t.
\end{aligned}
$$

From, this we get $\min(q_1, d_1) + \min(q_2, d_2) \leq \frac{t}{t+1}(|q| + |d|)$. There are now four cases, depending on the outcome of the min operations:

1. $q_1 + q_2 \leq \frac{t}{t+1}(|q| + |d|)$
2. $d_1 + d_2 \leq \frac{t}{t+1}(|q| + |d|)$
3. $q_1 + d_2 \leq \frac{t}{t+1}(|q| + |d|)$
4. $d_1 + q_2 \leq \frac{t}{t+1}(|q| + |d|)$.

The first two cases give us $|q|t \leq |d| \leq |q|/t$, the inequalities corresponding to $M = 1$, identical to those obtained in Section 60.3.2. For a fixed value of $|d|$, cases 3 and 4 give us $-q_2 + t(|q| + |d|)/(t + 1) < d_1 < |d| + q_1 - t(|q| + |d|)/(1 + t)$. Consider an example with $|q| = 200$ and $t = 0.8$ as before. Recall that this already reduces the search to values of $|d|$ in [160, 250]. Suppose $q_1 = 125$ and $q_2 = 75$. Suppose also that we are considering database fingerprints with $|d| = 200$. Substituting, this gives $102 < d_1 < 147$. This allows us to further eliminate fingerprints in the $|d| = 200$ bin that do not meet this criteria. This computation is facilitated by a secondary sorting within each bin on d_1. Recall that bins are already ordered by $|d|$.

60.3.5 Triangle Inequality and Improvements

Recall that Tanimoto similarity is guaranteed to be in the range [0, 1] with a high value denoting a high similarity and vice versa. We can also define *Tanimoto distance* $D(a, b) = 1 - S(a, b)$. The Tanimoto distance is a distance metric and satisfies the triangle inequality; that is, given three fingerprints a, b, and c: $|D(b, a) - D(a, c)| \leq D(b, c) \leq D(b, a) + D(a, c)$. This inequality can be used to prune searches as follows: suppose $D(a, b)$ is small (say 0.1) and $D(a, c)$ is large (say 0.9). We have $0.8 \leq D(b, c) \leq 1.0$, which implies that $0 \leq S(b, c) \leq 0.2$. For $t > 0.2$, we are able to avoid computing $S(b, c)$. Notice that this computation does not utilize information about the precise bits that are common between a and c or a and b to compute a bound between b and c. A more sophisticated analysis that considers this information gives an *intersection bound* that is tighter than the distance bound that results in more efficient searches [22].

60.3.6 XOR Filter

Another strategy [23] for improving similarity searches is to augment the original fingerprint with a signature: a folded fingerprint based on the XOR operator unlike the example in Figure 60.4 which uses the OR operator. The XOR-based signature on the example is (0, 0, 1, 0, 0). The signature is used to prune the search space as described below.

First, note that the union and intersection used in the computation of Tanimoto's similarity can be expressed in terms of the XOR ($\oplus$) operator as follows. In set-theoretic terms, $|a \oplus b| = |a - b| + |b - a|$.

$$|a \cap b| = \frac{1}{2}[|a| + |b| - |a \oplus b|]$$

$$|a \cup b| = \frac{1}{2}[|a| + |b| + |a \oplus b|].$$

Let the XOR-folded fingerprints of a and b be α and β, respectively. We then have $|a \oplus b| \geq |\alpha \oplus \beta|$, which gives.

$$|a \cap b| \leq \frac{1}{2}[|a| + |b| - |\alpha \oplus \beta|]$$

$$|a \cup b| \geq \frac{1}{2}[|a| + |b| + |\alpha \oplus \beta|]$$

resulting in

$$S(a, b) \leq \frac{|a| + |b| - |\alpha \oplus \beta|}{|a| + |b| + |\alpha \oplus \beta|}.$$

If the quantity in the RHS is less than the similarity threshold t, we can discontinue the computation of $S(a, b)$. The XOR filter will sometimes result in better bounds than *BitBound* and can be used in addition to it.

60.3.7 Inverted Indices

Consider a database of $|D|$ molecules represented as f-bit binary fingerprints, with each bit i, $1 \leq i \leq f$, corresponding to the presence/absence of a feature. In an inverted index, a list of molecules is maintained for each of the f features. List i consists of all molecules whose ith bit is set to 1. Consider the fingerprint corresponding to a query molecule q. In order to compute its Tanimoto similarity with each fingerprint in the database, we only need to traverse the lists corresponding to features present in q. We maintain a counter for each molecule in D that is initialized to 0. When a molecule is encountered in one of the relevant lists, its counter is incremented. The final value of the counter for each molecule represents the size of its intersection with the query molecule, the numerator of the Tanimoto similarity formula.

Nasr et al. note that the similarity threshold problem can be reduced to the T occurrence problem in information retrieval (i.e., finding all fingerprints with an intersection of size at least T with the query), for which efficient solutions based on the inverted index exist. We have

$$
\begin{aligned}
S(a, b) &= \frac{|a \cap b|}{|a \cup b|} \\
&= \frac{|a \cap b|}{|a| + |b| - |a \cap b|} \\
&\geq t
\end{aligned}
$$

which gives

$$T = |a \cap b| \geq \frac{t(|a| + |b|)}{t + 1}.$$

The invertex index-based *DivideSkip* solution [24] to the T-occurrence problem is presented and applied in Reference 23.

60.3.8 LINGO

A set of LINGOs [25] is defined as all substrings of length q in the modified SMILES string for a molecule. A simplified SMILES string $S = c0ccccc0L$ for $q = 4$ gives the multiset {c0cc, 0ccc, cccc, cccc, ccc0, cc0L}. The similarity between two molecules represented by their SMILES strings can be computed as the Tanimoto similarity between their LINGO multisets (LINGOsim). The LINGOsim computation is performed by representing the LINGOs of the query by a trie/finite state machine to improve the efficiency of the computation [26]. An alternative solution stores the database of molecules (with each molecule represented by a LINGO multiset) as an inverted index [27].

60.3.9 Trie-Based Approaches for Storing Fingerprints

A database of binary fingerprints can be stored in a binary tree. Starting with the root, the complete set of fingerprints is recursively split into two sets corresponding to its left and right subtrees. Each fingerprint is assigned to one of the two sets depending on the value of its first (leftmost) bit. If the value is 0 (1), it is assigned to the left (right) subtree. At level i of the tree, bit i of the fingerprint is used to determine the assignment of the fingerprint to the appropriate subtree. For an exact search for a fingerprint, the algorithm starts at the root and follows left or right pointers that spell out the fingerprint in time proportional to f.

For a threshold similarity search, a naive solution visits all of the leaves of the trie and computes the Tanimoto similarity. This search can be pruned be employing the Tanimoto lookahead condition [28] described later. Let

1. $B(n, d)$ denote the partial bit string spelled by the path from the root to node n at depth d.
2. $Q(d)$ denote the partial bit string formed from the first d bits of the query fingerprint.
3. $QL(d)$ denote the number of "1"s in the query bit string after position d.

The Running Union Count and Running Intersection Counts are defined respectively as $RUC(n, d) = |B(n, d) \cup Q(d)|$ and $RIC(n, d) = |B(n, d) \cap Q(d)|$. The maximum value of the Tanimoto similarity between query q and *any* molecule that is reached by traversing the bit tree through node n is given by

$$Smax(n, q) = \frac{RIC(n, d) + QL(d)}{RUC(n, d) + QL(d)}.$$

The search can be terminated at node n if this value computed at node n is less than the threshold t.

A more efficient organization of the tree splits the fingerprints into two sets using a bit of maximal entropy (i.e., one that yields two subsets as much as possible of equal size) rather than picking the next bit in left-to-right order. This is the *Singlebit* tree described in Reference 29. The equation for *Smax* above applies to the modified tree, but requires that each tree node keep track of the bit used at that node as well as the bits used by its ancestors to split the fingerprint set.

60.3.10 Multibit and Union Tree

The Singlebit tree was found to be inefficient in experiments [29] because a relatively small tree depth was found to be enough to partition the fingerprints. This meant that the algorithm is only aware of a small fraction of bits, resulting in bounds that do not prune a large number of fingerprints.

The Multibit tree addresses this shortcoming by recording in each node the list of bits that have constant value 0 or 1 in all of the fingerprints stored in the node's subtree. For example, consider a node whose subtree contains two fingerprints: 10, 101, and 10110. The node would store the three common bits as (1,1) (2,0) (3,1) where the first element denotes the position of the bit and the second is the common value. Using this information results in tighter bounds and greater pruning [29,30].

An alternative strategy (UnionBit Tree) has a node store the union of the fingerprint vectors in its leaves. In the example above, the node would store 10111. A query fingerprint's intersection with the union vector has cardinality guaranteed to be greater than or equal to that of any of the individual fingerprints. The UnionBit tree uses less CPU time and memory relative to the MultiBit treei [31].

60.3.11 Combining Hashing and Trees

The *BitBound* algorithm (described in Section 60.3.2) organizes fingerprints by placing each fingerprint in a bin corresponding to its cardinality, the number of 1s in that fingerprint; that is, all fingerprints with 15 1s are placed in bin 15. The bins are sorted by cardinality; the BitBound algorithm searches bins in a specified range as described earlier. Improved solutions are obtained by storing the fingerprints in each bin using a multibit tree [30].

References

1. N. Brown, Chemoinformatics—An introduction for computer scientists, *ACM Computing Surveys*, vol. 41, no. 2, pp. 8:1–8:38, February 2009.
2. J. K. Wegner, A. Sterling, R. Guha, A. Bender, J.-L. Faulon, J. Hastings, N. O'Boyle, J. Overington, H. Van Vlijmen, and E. Willighagen, Cheminformatics, *Communications of the ACM*, vol. 55, no. 11, pp. 65–75, November 2012.
3. J.-L. Faulon, and A. Bender, *Handbook of Cheminformatics Algorithms*, Boca Raton, FL: CRC Press, 2010.
4. D. J. Wild, *Introducing Cheminformatics: An Intensive Self-Study Guide*, McGraw-Hill Open-publishing, 2013.
5. J. E. Hopcroft and J. K. Wong, Linear time algorithm for isomorphism of planar graphs (preliminary report), in *STOC'74: Proceedings of the Sixth Annual ACM Symposium on Theory of Computing*. New York, NY: ACM Press, 1974, pp. 172–184.

6. M. R. Garey and D. S. Johnson, *Computers and Intractability; A Guide to the Theory of NP-Completeness*, New York, NY: W. H. Freeman & Co., 1990.

7. A. Aho, J. Hopcroft, and J. Ullman, *The Design and Analysis of Computer Algorithms*, Reading, MA: Addison-Wesley, 1974.

8. E. Luks, Isomorphism of graphs of bounded valence can be tested in polynomial time, *Journal of Computer and System Sciences*, vol. 25, no. 1, pp. 42–65, August 1982.

9. C. M. Hoffmann, *Group-Theoretic Algorithms and Graph Isomorphism*, Springer-Verlag GmbH and Co. KG: Berlin/Heidelberg, December 1982.

10. J.-L. Faulon, M. J. Collins, and R. D. Carr, The signature molecular descriptor. 4. Canonizing molecules using extended valence sequences, *Journal of Chemical Information and Modeling*, vol. 44, no. 2, pp. 427–436, 2004. [Online]. Available: http://dx.doi.org/10.1021/ci0341823

11. T. M. Kouri, D. Pascua, and D. P. Mehta, Random models and analyses for chemical graphs, *International Journal of Foundations of Computer Science*, vol. 26, no. 2, pp. 269–291, 2015.

12. H. L. Morgan, The generation of a unique machine description for chemical structures—A technique developed at chemical abstracts service, *Journal of Chemical Documentation*, vol. 5, no. 2, pp. 107–113, 1965.

13. W. Chen, J. Huang, and M. K. Gilson, Identification of symmetries in molecules and complexes, *Journal of Chemical Information and Modeling*, vol. 44, no. 4, pp. 1301–1313, 2004.

14. T. Cieplak and J. Wisniewski, A new effective algorithm for the unambiguous identification of the stereochemical characteristics of compounds during their registration in databases, *Molecules*, vol. 6, 915–926, 2001.

15. B. McKay, No automorphisms, yes?" http://cs.anu.edu.au/ bdm/nauty/, 2004.

16. B. McKay, Practical graph isomorphism, *Congressus Numerantium*, vol. 30, 45–87, 1981.

17. T. Miyazaki, The complexity of McKay's canonical labeling algorithm, *Groups and Computation II, DIMACS Series in Discrete Mathematics and Theoretical Computer Science*, vol. 28, 239–256, 1997.

18. D. Rogers and M. Hahn, Extended-connectivity fingerprints, *Journal of Chemical Information and Modeling*, vol. 50, no. 5, pp. 742–754, 2010.

19. S. J. Swamidass and P. Baldi, Bounds and algorithms for fast exact searches of chemical fingerprints in linear and sublinear time, *Journal of Chemical Information and Modeling*, vol. 47, no. 2, pp. 302–317, 2007.

20. S. J. Swamidass and P. Baldi, Mathematical correction for fingerprint similarity measures to improve chemical retrieval, *Journal of Chemical Information and Modeling*, vol. 47, no. 3, pp. 952–964, 2007.

21. P. Baldi, R. W. Benz, D. S. Hirschberg, and S. J. Swamidass, Lossless compression of chemical fingerprints using integer entropy codes improves storage and retrieval, *Journal of Chemical Information and Modeling*, vol. 47, no. 6, pp. 2098–2109, 2007.

22. P. Baldi and D. S. Hirschberg, An intersection inequality sharper than the Tanimoto triangle inequality for efficiently searching large databases, *Journal of Chemical Information and Modeling*, vol. 49, no. 8, pp. 1866–1870, 2009.

23. P. Baldi, D. S. Hirschberg, and R. J. Nasr, Speeding up chemical database searches using a proximity filter based on the logical exclusive or, *Journal of Chemical Information and Modeling*, vol. 48, no. 7, pp. 1367–1378, 2008.

24. C. Li, J. Lu, and Y. Lu, Efficient merging and filtering algorithms for approximate string searches,' in *Proceedings of the 2008 IEEE 24th International Conference on Data Engineering*, series ICDE'08. Washington, DC: IEEE Computer Society, 2008, pp. 257–266.

25. D. Vidal, M. Thormann, and M. Pons, Lingo, an efficient holographic text based method to calculate biophysical properties and intermolecular similarities, *Journal of Chemical Information and Modeling*, vol. 45, no. 2, pp. 386–393, 2005.

26. J. A. Grant, J. A. Haigh, B. T. Pickup, A. Nicholls, and R. A. Sayle, Lingos, finite state machines, and fast similarity searching, *Journal of Chemical Information and Modeling*, vol. 46, no. 5, pp. 1912–1918, 2006.

27. T. G. Kristensen, J. Nielsen, and C. N. S. Pedersen, Using inverted indices for accelerating lingo calculations, *Journal of Chemical Information and Modeling*, vol. 51, no. 3, pp. 597–600, 2011.

28. A. Smellie, Compressed binary bit trees: A new data structure for accelerating database searching, *Journal of Chemical Information and Modeling*, vol. 49, no. 2, pp. 257–262, 2009.

29. T. Kristensen, J. Nielsen, and C. Pedersen, A tree based method for the rapid screening of chemical fingerprints, in *Algorithms in Bioinformatics*, series Lecture Notes in Computer Science, S. Salzberg and T. Warnow, Eds. Springer, Berlin/Heidelberg, 2009, vol. 5724, pp. 194–205.

30. R. Nasr, T. Kristensen, and P. Baldi, Tree and hashing data structures to speed up chemical searches: Analysis and experiments, *Molecular Informatics*, vol. 30, no. 9, pp. 791–800, 2011.

31. S. Saeedipour, D. Tai, and J. Fang, Chemcom: A software program for searching and comparing chemical libraries, *Journal of Chemical Information and Modeling*, vol. 55, no. 7, pp. 1292–1296, 2015.

61

Elimination Structures in Scientific Computing*

61.1 The Elimination Tree 946
The Elimination Game • The Elimination Tree Data Structure • An Algorithm •
A Skeleton Graph • Supernodes

61.2 Applications of Etrees... 950
Efficient Symbolic Factorization • Predicting Row and Column Nonzero Counts • Three
Classes of Factorization Algorithms • Scheduling Parallel Factorizations • Scheduling
Out-of-Core Factorizations

61.3 The Clique Tree 954
Chordal Graphs and Clique Trees • Design of Efficient Algorithms with Clique Trees •
Compact Clique Trees

61.4 Clique Covers and Quotient Graphs 957
Clique Covers • Quotient Graphs • The Problem of Degree Updates • Covering the
Column-Intersection Graph and Biclique Covers

61.5 Column Elimination Trees and Elimination DAGS 959
The Column Elimination Tree • Elimination DAGS • Elimination Structures for the
Asymmetric Multifrontal Algorithm

Acknowledgments 962
References... 962

Alex Pothen
Purdue University

Sivan Toledo
Tel Aviv University

The most fundamental computation in numerical linear algebra is the factorization of a matrix as a product of two or more matrices with simpler structure. An important example is Gaussian elimination, in which a matrix is written as a product of a lower triangular matrix and an upper triangular matrix. The factorization is accomplished by elementary operations in which two or more rows (columns) are combined together to transform the matrix to the desired form. In Gaussian elimination, the desired form is an upper triangular matrix, in which nonzero elements below the diagonal have been transformed to be equal to zero. We say that the subdiagonal elements have been eliminated. (The transformations that accomplish the elimination yield a lower triangular matrix.)

The input matrix is usually sparse, that is, only a few of the matrix elements are nonzero to begin with; in this situation, row operations constructed to eliminate nonzero elements in some locations might create new nonzero elements, called fill, in other locations, as a side-effect. Data structures that predict fill from graph models of the numerical algorithm, and algorithms that attempt to minimize fill, are key ingredients of efficient sparse matrix algorithms.

This chapter surveys these data structures, known as *elimination structures*, and the algorithms that construct and use them. We begin with the *elimination tree*, a data structure associated with symmetric Gaussian elimination, and we then describe its most important applications. Next we describe other data structures associated with symmetric Gaussian elimination, the *clique tree*, the *clique cover*, and the *quotient graph*. We then consider data structures that are associated with asymmetric Gaussian elimination, the *column elimination tree* and the *elimination directed acyclic graph*.

This survey has been written with two purposes in mind. First, we introduce the algorithms community to these data structures and algorithms from combinatorial scientific computing; the initial subsections should be accessible to the non-expert. Second, we wish to briefly survey the current state of the art, and the subsections dealing with the advanced topics move rapidly. A collection of articles describing developments in the field circa 1991 may be found in [1]; Duff provides a survey as of 1996 in [2].

---

* This chapter has been reprinted from first edition of this Handbook, without any content updates.

61.1 The Elimination Tree

61.1.1 The Elimination Game

Gaussian elimination of a symmetric positive definite matrix A, which factors the matrix A into the product of a lower triangular matrix L and its transpose L^T, $A = LL^T$, is one of the fundamental algorithms in scientific computing. It is also known as Cholesky factorization. We begin by considering the graph model of this computation performed on a symmetric matrix A that is sparse, that is, few of its matrix elements are nonzero. The number of nonzeros in L and the work needed to compute L depend strongly on the (symmetric) ordering of the rows and columns of A. The graph model of sparse Gaussian elimination was introduced by Parter [3], and has been called the *elimination game* by Tarjan [4]. The goal of the elimination game is to symmetrically order the rows and columns of A to minimize the number of nonzeros in the factor L.

We consider a sparse, symmetric positive definite matrix A with n rows and n columns, and its adjacency graph $G(A) = (V, E)$ on n vertices. Each vertex in $v \in V$ corresponds to the vth row of A (and by symmetry, the vth column); an edge $(v, w) \in E$ corresponds to the nonzero a_{vw} (and by symmetry, the nonzero a_{wv}). Since A is positive definite, its diagonal elements are positive; however, by convention, we do not explicitly represent a diagonal element a_{vv} by a loop (v, v) in the graph $G(A)$. (We use $v, w, \ldots$ to indicate unnumbered vertices, and $i, j, k, \ldots$ to indicate numbered vertices in a graph.)

We view the vertices of the graph $G(A)$ as being initially unnumbered, and number them from 1 to n, as a consequence of the elimination game. To number a vertex v with the next available number, add new *fill edges* to the current graph to make all currently unnumbered neighbors of v pairwise adjacent. (Note that the vertex v itself does not acquire any new neighbors in this step, and that v plays no further role in generating fill edges in future numbering steps.)

The graph that results at the end of the elimination game, which includes both the edges in the edge set E of the initial graph $G(A)$ and the set of fill edges, F, is called the filled graph. We denote it by $G^+(A) = (V, E \cup F)$. The numbering of the vertices is called an elimination ordering, and corresponds to the order in which the columns are factored. An example of a filled graph resulting from the elimination game on a graph is shown in Figure 61.1. We will use this graph to illustrate various concepts throughout this paper.

The goal of the elimination game is to number the vertices to minimize the fill since it would reduce the storage needed to perform the factorization, and also controls the work in the factorization. Unfortunately, this is an NP-hard problem [5]. However, for classes of graphs that have small separators, it is possible to establish upper bounds on the number of edges in the filled graph, when the graph is ordered by a nested dissection algorithm that recursively computes separators. Planar graphs, graphs of "well-shaped" finite element meshes (aspect ratios bounded away from small values), and overlap graphs possess elimination orderings with bounded fill. Conversely, the fill is large for graphs that do not have good separators.

Approximation algorithms that incur fill within a polylog factor of the optimum fill have been designed by Agrawal, Klein and Ravi [6]; but since it involves finding approximate concurrent flows with uniform capacities, it is an impractical approach for

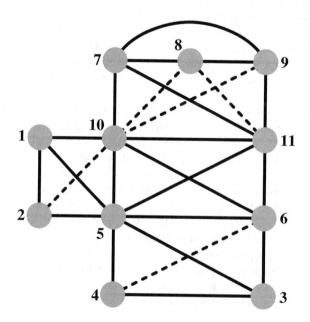

FIGURE 61.1 A filled graph $G^+(A)$ resulting from the elimination game on a graph $G(A)$. The solid edges belong to $G(A)$, and the broken edges are filled edges generated by the elimination game when vertices are eliminated in the order shown.

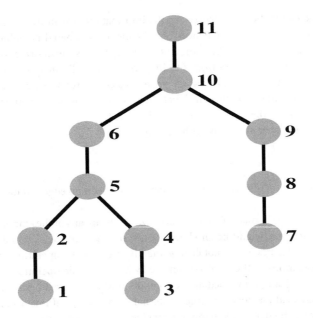

FIGURE 61.2   The elimination tree of the example graph.

large problems. A more recent approximation algorithm, due to Natanzon, Shamir and Sharan [7], limits fill to within the square of the optimal value; this approximation ratio is better than that of the former algorithm only for dense graphs.

The elimination game produces sets of cliques in the graph. Let $\text{hadj}^+(v)$ ($\text{ladj}^+(v)$) denote the higher-numbered (lower-numbered) neighbors of a vertex v in the graph $G^+(A)$; in the elimination game, $\text{hadj}^+(v)$ is the set of unnumbered neighbors of v immediately prior to the step in which v is numbered. When a vertex v is numbered, the set $\{v\} \cup \text{hadj}^+(v)$ becomes a clique by the rules of the elimination game. Future numbering steps and consequent fill edges added do not change the adjacency set (in the filled graph) of the vertex v. (We will use $\text{hadj}(v)$ and $\text{ladj}(v)$ to refer to higher and lower adjacency sets of a vertex v in the original graph $G(A)$.)

61.1.2 The Elimination Tree Data Structure

We define a forest from the filled graph by defining the parent of a vertex v to be the lowest numbered vertex in $\text{hadj}^+(v)$. It is clear that this definition of parent yields a forest since the parent of each vertex is numbered higher than itself. If the initial graph $G(A)$ is connected, then indeed we have a tree, the *elimination tree*; if not we have an *elimination forest*.

In terms of the Cholesky factor L, the elimination tree is obtained by looking down each column below the diagonal element, and choosing the row index of the first subdiagonal nonzero to be the parent of a column. It will turn out that we can compute the elimination tree corresponding to a matrix and a given ordering without first computing the filled graph or the Cholesky factor.

The elimination tree of the graph in Figure 61.1 with the elimination ordering given there is shown in Figure 61.2.

A *fill path* joining vertices i and j is a path in the original graph $G(A)$ between vertices i and j, all of whose interior vertices are numbered lower than both i and j. The following theorem offers a static characterization of what causes fill in the elimination game.

Theorem 61.1: ([8])

The edge (i, j) is an edge in the filled graph if and only if a fill path joins the vertices i and j in the original graph $G(A)$.

In the example graph in Figure 61.1, vertices 9 and 10 are joined a fill path consisting of the interior vertices 7 and 8; thus $(9, 10)$ is a fill edge. The next theorem shows that an edge in the filled graph represents a dependence relation between its end points.

Theorem 61.2: ([9])

If (i, j) is an edge in the filled graph and $i < j$, then j is an ancestor of the vertex i in the elimination tree $T(A)$.

This theorem suggests that the elimination tree represents the information flow in the elimination game (and hence sparse symmetric Gaussian elimination). Each vertex i influences only its higher numbered neighbors (the numerical values in the column i affect only those columns in $\text{hadj}^+(i)$). The elimination tree represents the information flow in a minimal way in that we need consider only how the information flows from i to its parent in the elimination tree. If j is the parent of i and ℓ is another higher neighbor of i, then since the higher neighbors of i form a clique, we have an edge (j, ℓ) that joins j and ℓ; since by Theorem 61.2, ℓ is an ancestor of j, the information from i that affects ℓ can be viewed as being passed from i first to j, and then indirectly from j through its ancestors on the path in the elimination tree to ℓ.

An immediate consequence of the Theorem 61.2 is the following result.

Corollary 61.1

If vertices i and j belong to vertex-disjoint subtrees of the elimination tree, then no edge can join i and j in the filled graph.

Viewing the dependence relationships in sparse Cholesky factorization by means of the elimination tree, we see that any topological reordering of the elimination tree would be an elimination ordering with the same fill, since it would not violate the dependence relationships. Such reorderings would not change the fill or arithmetic operations needed in the factorization, but would change the schedule of operations in the factorization (i.e., when a specific operation is performed). This observation has been used in sparse matrix factorizations to schedule the computations for optimal performance on various computational platforms: multiprocessors, hierarchical memory machines, external memory algorithms, etc. A postordering of the elimination tree is typically used to improve the spatial and temporal data locality, and thereby the cache performance of sparse matrix factorizations.

There are two other perspectives from which we can view the elimination tree.

Consider directing each edge of the filled graph from its lower numbered endpoint to its higher numbered endpoint to obtain a directed acyclic graph (DAG). Now form the transitive reduction of the directed filled graph; that is, delete an edge (i, k) whenever there is a directed path from i to k that does not use the edge (i, k) (this path necessarily consists of at least two edges since we do not admit multiple edges in the elimination game). The minimal graph that remains when all such edges have been deleted is unique, and is the elimination tree.

One could also obtain the elimination tree by performing a depth-first search (DFS) in the filled graph with the vertex numbered n as the initial vertex for the DFS, and choosing the highest numbered vertex in $\text{ladj}^+(i)$ as the next vertex to search from a vertex i.

61.1.3 An Algorithm

We begin with a consequence of the repeated application of the following fact: If a vertex i is adjacent to a higher numbered neighbor k in the filled graph, and k is not the parent of i, p_i, in the elimination tree, then i is adjacent to both k and p_i in the filled graph; when i is eliminated, by the rules of the elimination game, a fill edge joins p_i and k.

Theorem 61.3

If (i, k) is an edge in the filled graph and $i < k$, then for every vertex j on an elimination tree path from i to k, (j, k) is also an edge in the filled graph.

This theorem leads to a characterization of $\text{ladj}^+(k)$, the set of lower numbered neighbors of a vertex k in the filled graph, which will be useful in designing an efficient algorithm for computing the elimination tree. The set $\text{ladj}^+(k)$ corresponds to the column indices of nonzeros in the kth row of the Cholesky factor L, and $\text{ladj}(k)$ corresponds to the column indices of nonzeros in the lower triangle of the kth row of the initial matrix A.

Theorem 61.4: ([10])

Every vertex in the set $\text{ladj}^+(k)$ is a vertex reachable by paths in the elimination tree from a set of leaves to k; each leaf l corresponds to a vertex in the set $\text{ladj}(k)$ such that no proper descendant d of l in the elimination tree belongs to the set $\text{ladj}(k)$.

Theorem 61.4 characterizes the kth row of the Cholesky factor L as a *row subtree* $T_r(k)$ of the elimination subtree rooted at the vertex k, and pruned at each leaf l. The leaves of the pruned subtree are contained among $\text{ladj}(k)$, the column indices of the nonzeros in (the lower triangle of) the kth row of A. In the elimination tree in Figure 61.2, the pruned elimination subtree corresponding to row 11 has two leaves, vertices 5 and 7; it includes all vertices on the etree path from these leaves to the vertex 11.

```
for  k := 1 to n  →
     p_k := 0;
     for  j ∈ ladj(k) (in increasing order) →
          find the root r of the tree containing j;
          if (k ≠ r) then k := p_r; fi
     rof
rof
```

FIGURE 61.3   An algorithm for computing an elimination tree. Initially each vertex is in a subtree with it as the root.

The observation above leads to an algorithm, shown in Figure 61.3, for computing the elimination tree from the row structures of A, due to Liu [10].

This algorithm can be implemented efficiently using the union-find data structure for disjoint sets. A height compressed version of the p. array, ancestor, makes it possible to compute the root fast; and union by rank in merging subtrees helps to keep the merged tree shallow. The time complexity of the algorithm is $O(e\alpha(e, n) + n)$, where n is the number of vertices and e is the number of edges in $G(A)$, and $\alpha(e, n)$ is a functional inverse of Ackermann's function. Liu [11] shows experimentally that path compression alone is more efficient than path compression and union by rank, although the asymptotic complexity of the former is higher. Zmijewski and Gilbert [12] have designed a parallel algorithm for computing the elimination tree on distributed memory multiprocessors.

The concept of the elimination tree was implicit in many papers before it was formally identified. The term elimination tree was first used by Duff [13], although he studied a slightly different data structure; Schreiber [9] first formally defined the elimination tree, and its properties were established and used in several articles by Liu. Liu [11] also wrote an influential survey that delineated its importance in sparse matrix computations; we refer the reader to this survey for a more detailed discussion of the elimination tree current as of 1990.

61.1.4 A Skeleton Graph

The filled graph represents a supergraph of the initial graph $G(A)$, and a skeleton graph represents a subgraph of the latter. Many sparse matrix algorithms can be made more efficient by implicitly identifying the edges of a skeleton graph $G^-(A)$ from the graph $G(A)$ and an elimination ordering, and performing computations only on these edges. A skeleton graph includes only the edges that correspond to the leaves in each row subtree in Theorem 61.4. The other edges in the initial graph $G(A)$ can be discarded, since they will be generated as fill edges during the elimination game. Since each leaf of a row subtree corresponds to an edge in $G(A)$, the skeleton graph $G^-(A)$ is indeed a subgraph of the former. The skeleton graph of the example graph is shown in Figure 61.4.

The leaves in a row subtree can be identified from the set $\text{ladj}(j)$ when the elimination tree is numbered in a postordering. The subtree $T(i)$ is the subtree of the elimination tree rooted at a vertex i, and $|T(i)|$ is the number of vertices in that subtree. (It should not be confused with the row subtree $T_r(i)$, which is a pruned subtree of the elimination tree.)

Theorem 61.5: ([10])

Let $\text{ladj}(j) = \{i_1 < i_2 < \ldots < i_s\}$, and let the vertices of a filled graph be numbered in a postordering of its elimination tree T. Then vertex i_q is a leaf of the row subtree $T_r(j)$ if and only if either $q = 1$, or for $q \geq 2$, $i_{q-1} < i_q - |T(i_q)| + 1$.

61.1.5 Supernodes

A supernode is a subset of vertices S of the filled graph that form a clique and have the same higher neighbors outside S. Supernodes play an important role in numerical algorithms since loops corresponding to columns in a supernode can be blocked to obtain high performance on modern computer architectures. We now proceed to define a supernode formally.

A maximal clique in a graph is a set of vertices that induces a complete subgraph, but adding any other vertex to the set does not induce a complete subgraph. A supernode is a maximal clique $\{i_s, i_{s+1}, \ldots, i_{s+t-1}\}$ in a filled graph $G^+(A)$ such that for each $1 \leq j \leq t - 1$,

$$\text{hadj}^+(i_s) = \{i_{s+1}, \ldots, i_{s+j}\} \cup \text{hadj}^+(i_{s+j}).$$

Let $\text{hd}^+(i_s) \equiv |\text{hadj}^+(i_s)|$; since $\text{hadj}^+(i_s) \subseteq \{i_{s+1}, \ldots, i_{s+j}\} \cup \text{hadj}^+(i_{s+j})$, the relationship between the higher adjacency sets can be replaced by the equivalent test on higher degrees: $\text{hd}^+(i_s) = \text{hd}^+(i_{s+j}) + j$.

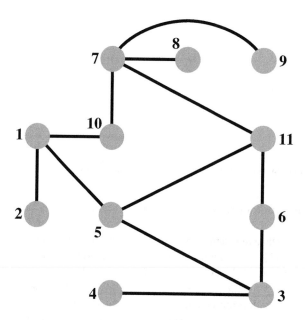

FIGURE 61.4　The skeleton graph $G^-(A)$ of the example graph.

In practice, *fundamental supernodes*, rather than the maximal supernodes defined above, are used, since the former are easier to work with in the numerical factorization. A fundamental supernode is a clique but not necessarily a maximal clique, and satisfies two additional conditions: (1) i_{s+j-1} is the only child of the vertex i_{s+j} in the elimination tree, for each $1 \leq j \leq t-1$; (2) the vertices in a supernode are ordered consecutively, usually by post-ordering the elimination tree. Thus vertices in a fundamental supernode form a path in the elimination tree; each of the non-terminal vertices in this path has only one child, and the child belongs to the supernode.

The fundamental supernodes corresponding to the example graph are: $\{1, 2\}$; $\{3, 4\}$; $\{5, 6\}$; $\{7, 8, 9\}$; and $\{10, 11\}$.

Just as we could compute the elimination tree directly from $G(A)$ without first computing $G^+(A)$, we can compute fundamental supernodes without computing the latter graph, using the theorem given below. Once the elimination tree is computed, this algorithm can be implemented in $O(n + e)$ time, where $e \equiv |E|$ is the number of edges in the original graph $G(A)$.

Theorem 61.6: ([14])

A vertex i is the first node of a fundamental supernode if and only if i has two or more children in the elimination tree T, or i is a leaf of some row subtree of T.

61.2 Applications of Etrees

61.2.1 Efficient Symbolic Factorization

Symbolic factorization (or symbolic elimination) is a process that computes the nonzero structure of the factors of a matrix without computing the numerical values of the nonzeros.

The symbolic Cholesky factor of a matrix has several uses. It is used to allocate the data structure for the numeric factor and annotate it with all the row/column indices, which enables the removal of most of the non-numeric operations from the innermost loop of the subsequent numeric factorization [15,16]. It is also used to compute relaxed supernode (or amalgamated node) partitions, which group columns into supernodes even if they only have approximately the same structure [17,18]. Symbolic factors can also be used in algorithms that construct approximate Cholesky factors by dropping nonzeros from a matrix A and factoring the resulting, sparser matrix B [19,20]. In such algorithms, elements of A that are dropped from B but which appear in the symbolic factor of B can be added to the matrix B; this improves the approximation without increasing the cost of factoring B. In all of these applications a supernodal symbolic factor (but not a relaxed one) is sufficient; there is no reason to explicitly represent columns that are known to be identical.

The following algorithm for symbolically factoring a symmetric matrix A is due to George and Liu [21] (and in a more graph-oriented form due to [8]; see also [16, Section 5.4.3] and [11, Section 8]).

The algorithm uses the elimination tree implicitly, but does not require it as input; the algorithm can actually compute the elimination tree on the fly. The algorithm uses the observation that

$$\text{hadj}^+(j) = \text{hadj}(j) \bigcup \cup_{i,p_i=j} \text{hadj}^+(i).$$

That is, the structure of a column of L is the union of the structure of its children in the elimination tree and the structure of the same column in the lower triangular part of A. Identifying the children can be done using a given elimination tree, or the elimination tree can be constructed on the fly by adding column i to the list of children of p_i when the structure of i is computed (p_i is the row index of the first subdiagonal nonzero in column i of L). The union of a set of column structures is computed using a boolean array P of size n (whose elements are all initialized to false), and an integer stack to hold the newly created structure. A row index k from a child column or from the column of A is added to the stack only if P$[k]$ = false. When row index k is added to the stack, P$[k]$ is set to true to signal that k is already in the stack. When the computation of hadj$^+(j)$ is completed, the stack is used to clear P so that it is ready for the next union operation. The total work in the algorithm is $\Theta(|L|)$, since each nonzero requires constant work to create and constant work to merge into the parent column, if there is a parent. (Here $|L|$ denotes the number of nonzeros in L, or equivalently the number of edges in the filled graph $G^+(A)$; similarly $|A|$ denotes the number of nonzeros in A, or the number of edges in the initial graph $G(A)$.)

The symbolic structure of the factor can usually be represented more compactly and computed more quickly by exploiting supernodes, since we essentially only need to represent the identity of each supernode (the constituent columns) and the structure of the first (lowest numbered) column in each supernode. The structure of any column can be computed from this information in time proportional to the size of the column. The George-Liu column-merge algorithm presented above can compute a supernodal symbolic factorization if it is given as input a supernodal elimination tree; such a tree can be computed in $O(|A|)$ time by the Liu-Ng-Peyton algorithm [14]. In practice, this approach saves a significant amount of work and storage.

Clearly, column-oriented symbolic factorization algorithms can also generate the structure of rows in the same asymptotic work and storage. But a direct symbolic factorization by rows is less obvious. Whitten [22], in an unpublished manuscript cited by Tarjan and Yannakakis [23], proposed a row-oriented symbolic factorization algorithm (see also [10] and [11, Sections 3.2 and 8.2]). The algorithm uses the characterization of the structure of row i in L as the row subtree $T_r(i)$. Given the elimination tree and the structure of A by rows, it is trivial to traverse the ith row subtree in time proportional to the number of nonzeros in row i of L. Hence, the elimination tree along with a row-oriented representation of A is an effective implicit symbolic row-oriented representation of L; an explicit representation is usually not needed, but it can be generated in work and space $O(|L|)$ from this implicit representation.

61.2.2 Predicting Row and Column Nonzero Counts

In some applications the explicit structure of columns of L is not required, only the number of nonzeros in each column or each row. Gilbert, Ng, and Peyton [24] describe an almost-linear-time algorithm for determining the number of nonzeros in each row and column of L. Applications for computing these counts fast include comparisons of fill in alternative matrix orderings, preallocation of storage for a symbolic factorization, finding relaxed supernode partitions quickly, determining the load balance in parallel factorizations, and determining synchronization events in parallel factorizations.

The algorithm to compute row counts is based on Whitten's characterization [22]. We are trying to compute $|L_{i*}| = |T_r(i)|$. The column indices $j < i$ in row i of A define a subset of the vertices in the subtree of the elimination tree rooted at the vertex i, $T[i]$. The difficulty, of course, is counting the vertices in $T_r(i)$ without enumerating them. The Gilbert-Ng-Peyton algorithm counts these vertices using three relatively simple mechanisms: (1) processing the column indices $j < i$ in row i of A in postorder of the etree, (2) computing the distance of each vertex in the etree from the root, and (3) setting up a data structure to compute the least-common ancestor (LCA) of pairs of etree vertices. It is not hard to show that the once these preprocessing steps are completed, $|T_r(i)|$ can be computed using $|A_{i*}|$ LCA computations. The total cost of the preprocessing and the LCA computations is almost linear in $|A|$.

Gilbert, Ng, and Peyton show how to further reduce the number of LCA computations. They exploit the fact that the leaves of $T_r(i)$ are exactly the indices j that cause the creation of new supernodes in the Liu-Ng-Peyton supernode-finding algorithm [14]. This observation limits the LCA computations to leaves of row subtrees, that is, edges in the skeleton graph $G^-(A)$. This significantly reduces the running time in practice.

Efficiently computing the column counts in L is more difficult. The Gilbert-Ng-Peyton algorithm assigns a weight $w(j)$ to each etree vertex j, such that $|L_{*j}| = \sum_{k \in T[j]} w(k)$. Therefore, the column-count of a vertex is the sum of the column counts of its children, plus its own weight. Hence, w_j must compensate for (1) the diagonal elements of the children, which are not included in the column count for j, (2) for rows that are nonzero in column j but not in its children, and (3) for duplicate counting stemming

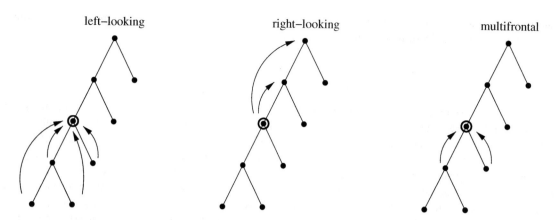

FIGURE 61.5   Patterns of data access in the left-looking, right-looking, and multifrontal algorithms. A subtree of the elimination tree is shown, and the circled node corresponds to the current submatrix being factored.

from rows that appear in more than one child. The main difficulty lies in accounting for duplicates, which is done using least-common-ancestor computations, as in the row-counts algorithm. This algorithm, too, benefits from handling only skeleton-graph edges.

Gilbert, Ng, and Peyton [24] also show in their paper how to optimize these algorithms, so that a single pass over the nonzero structure of *A* suffices to compute the row counts, the column counts, and the fundamental supernodes.

61.2.3   Three Classes of Factorization Algorithms

There are three classes of algorithms used to implement sparse direct solvers: left-looking, right-looking, and multifrontal; all of them use the elimination tree to guide the computation of the factors. The major difference between the first two of these algorithms is in how they schedule the computations they perform; the multifrontal algorithm organizes computations differently from the other two, and we explain this after introducing some concepts.

The computations on the sparse matrix are decomposed into subtasks involving computations among dense submatrices (supernodes), and the precedence relations among them are captured by the supernodal elimination tree. The computation at each node of the elimination tree (subtask) involves the partial factorization of the dense submatrix associated with it.

The right-looking algorithm is an eager updating scheme: Updates generated by the submatrix of the current subtask are applied immediately to future subtasks that it is linked to by edges in the filled graph of the sparse matrix. The left-looking algorithm is a lazy updating scheme: Updates generated by previous subtasks linked to the current subtask by edges in the filled adjacency graph of the sparse matrix are applied just prior to the factorization of the current submatrix. In both cases, updates always join a subtask to some ancestor subtask in the elimination tree. In the multifrontal scheme, updates always go from a child task to its parent in the elimination tree; an update that needs to be applied to some ancestor subtask is passed incrementally through a succession of vertices on the elimination tree path from the subtask to the ancestor.

Thus the major difference among these three algorithms is how the data accesses and the computations are organized and scheduled, while satisfying the precedence relations captured by the elimination tree. An illustration of these is shown in Figure 61.5.

61.2.4   Scheduling Parallel Factorizations

In a parallel factorization algorithm, dependences between nonzeros in *L* determine the set of admissible schedules. A diagonal nonzero can only be factored after all updates to it from previous columns have been applied, a subdiagonal nonzero can be scaled only after updates to it have been applied (and after the diagonal element has been factored), and two subdiagonal nonzeros can update elements in the reduced system only after they have been scaled.

The elimination tree represents very compactly and conveniently a superset of these dependences. More specifically, the etree represents dependences between columns of *L*. A column can be completely factored only after all its descendants have been factored, and two columns that are not in an ancestor-descendant relationship can be factored in any order. Note that this is a superset of the element dependences, since a partially factored column can already perform some of the updates to its ancestors.

But most sparse elimination algorithms treat column operations (or row operations) as atomic operations that are always performed by a single processor sequentially and with no interruption. For such algorithms, the etree represents exactly the relevant dependences.

In essence, parallel column-oriented factorizations can factor the columns associated with different children of an etree vertex simultaneously, but columns in an ancestor-descendant relationship must be processed in postorder. Different algorithms differ mainly in how updates are represented and scheduled.

By computing the number of nonzeros in each column, a parallel factorization algorithm can determine the amount of computation and storage associated with each subtree in the elimination tree. This information can be used to assign tasks to processors in a load-balanced way.

Duff was the first to observe that the column dependences represented by the elimination tree can guide a parallel factorization [25]. In that paper Duff proposed a parallel multifrontal factorization. The paper also proposed a way to deal with indefinite and asymmetric systems, similar to Duff and Reid's sequential multifrontal approach [18]. For further references up to about 1997, see Heath's survey [26]. Several implementations described in papers published after 1997 include PaStiX [27], PARADISO [28], WSSMP [29], and MUMPS [30], which also includes indefinite and unsymmetric factorizations. All four are message-passing codes.

61.2.5 Scheduling Out-of-Core Factorizations

In an out-of-core factorization at least some of the data structures are stored on external-memory devices (today almost exclusively magnetic disks). This allows such factorization algorithms to factor matrices that are too large to factor in main memory. The factor, which is usually the largest data structure in the factorization, is the most obvious candidate for storing on disks, but other data structures, for example, the stack of update matrices in a multifrontal factorization, may also be stored on disks.

Planning and optimizing out-of-core factorization schedules require information about data dependences in the factorization and about the number of nonzeros in each column of the factor. The etree describes the required dependence information, and as explained above, it is also used to compute nonzero counts.

Following Rothberg and Schreiber [31], we classify out-of-core algorithms into *robust* algorithms and *non-robust algorithms*. Robust algorithms adapt the factorization to complete with the core memory available by performing the data movement and computations at a smaller granularity when necessary. They partition the submatrices corresponding to the supernodes and stacks used in the factorization into smaller units called panels to ensure that the factorization completes with the available memory. Non-robust algorithms assume that the stack or a submatrix corresponding to a supernode fits within the core memory provided. In general, non-robust algorithms read elements of the input matrix only once, read from disk nothing else, and they only write the factor elements to disk; Dobrian and Pothen refer to such algorithms as read-once-write-once, and to robust ones as read-many-write-many [32].

Liu proposed [33] a non-robust method that works as long as for all $j = 1, \ldots, n$, all the nonzeros in the submatrix $L_{j:n,1:j}$ of the factor fit simultaneously in main memory. Liu also shows in that paper how to reduce the amount of main memory required to factor a given matrix using this technique by reordering the children of vertices in the etree.

Rothberg and Schreiber [31,34] proposed a number of robust out-of-core factorization algorithms. They proposed multifrontal, left-looking, and hybrid multifrontal/left-looking methods. Rotkin and Toledo [35] proposed two additional robust methods, a more efficient left-looking method, and a hybrid right/left-looking method. All of these methods use the etree together with column-nonzero counts to organize the out-of-core factorization process.

Dobrian and Pothen [32] analyzed the amount of main memory required for read-once-write-once factorizations of matrices with several regular etree structures, and the amount of I/O that read-many-write-many factorizations perform on these matrices. They also provided simulations on problems with irregular elimination tree structures. These studies led them to conclude that an external memory sparse solver library needs to provide at least two of the factorization methods, since each method can outperform the others on problems with different characteristics. They have provided implementations of out-of-core algorithms for all three of the multifrontal, left-looking, and right-looking factorization methods; these algorithms are included in the direct solver library OBLIO [36].

In addition to out-of-core techniques, there exist techniques that reduce the amount of main memory required to factor a matrix without using disks. Liu [37] showed how to minimize the size of the stack of update matrices in the multifrontal method by reordering the children of vertices in the etree; this method is closely related to [33]. Another approach, first proposed by Eisenstat, Schultz and Sherman [38] uses a block factorization of the coefficient matrix, but drops some of the off-diagonal blocks. Dropping these blocks reduces the amount of main memory required for storing the partial factor, but requires recomputation of these blocks when linear systems are solved using the partial factor. George and Liu [16, Chapter 6] proposed a general algorithm to partition matrices into blocks for this technique. Their algorithm uses *quotient graphs*, data structures that we describe later in this chapter.

61.3 The Clique Tree

61.3.1 Chordal Graphs and Clique Trees

The filled graph $G^+(A)$ that results from the elimination game on the matrix A (the adjacency graph of the Cholesky factor L) is a *chordal* graph, that is, a graph in which every cycle on four or more vertices has an edge joining two non-consecutive vertices on the cycle [39]. (The latter edge is called a chord, whence the name chordal graph. This class of graphs has also been called triangulated or rigid circuit graphs.)

A vertex v in a graph G is *simplicial* if its neighbors adj(v) form a clique. Every chordal graph is either a clique, or it has two non-adjacent simplicial vertices. (The simplicial vertices in the filled graph in Figure 61.1 are 1, 2, 3, 4, 7, 8, and 9.) We can eliminate a simplicial vertex v without causing any fill by the rules of the elimination game, since adj(v) is already a clique, and no fill edge needs to be added. A chordal graph from which a simplicial vertex is eliminated continues to be a chordal graph. A *perfect elimination ordering* of a chordal graph is an ordering in which simplicial vertices are eliminated successively without causing any fill during the elimination game. A graph is chordal if and only if it has a perfect elimination ordering.

Suppose that the vertices of the adjacency graph $G(A)$ of a sparse, symmetric matrix A have been re-numbered in an elimination ordering, and that $G^+(A)$ corresponds to the filled graph obtained by the elimination game on $G(A)$ with that ordering. This elimination ordering is a perfect elimination ordering of the filled graph $G^+(A)$. Many other perfect elimination orderings are possible for $G^+(A)$, since there are at least two simplicial vertices that can be chosen for elimination at each step, until the graph has one uneliminated vertex.

It is possible to design efficient algorithms on chordal graphs whose time complexity is much less than $O(|E \cup F|)$, where $E \cup F$ denotes the set of edges in the chordal filled graph. This is accomplished by representing chordal graphs by tree data structures defined on the maximal cliques of the graph. (Recall that a clique K is maximal if $K \cup \{v\}$ is not a clique for any vertex $v \notin K$.)

Theorem 61.7

Every maximal clique of a chordal filled graph $G^+(A)$ is of the form $K(v) = \{v\} \cup \text{hadj}^+(v)$, with the vertices ordered in a perfect elimination ordering.

The vertex v is the lowest-numbered vertex in the maximal clique $K(v)$, and is called the *representative vertex* of the clique. Since there can be at most $n \equiv |V|$ representative vertices, a chordal graph can have at most n maximal cliques. The maximal cliques of the filled graph in Figure 61.1 are: $K_1 = \{1, 2, 5, 10\}$; $K_2 = \{3, 4, 5, 6\}$; $K_3 = \{5, 6, 10, 11\}$; and $K_4 = \{7, 8, 9, 10, 11\}$. The lowest-numbered vertex in each maximal clique is its representative; note that in our notation $K_2 = K(3)$, $K_1 = K(1)$, $K_3 = K(5)$, and $K_4 = K(7)$.

Let $\mathcal{K}_G = \{K1, K_2, \ldots, K_m\}$ denote the set of maximal cliques of a chordal graph G. Define a clique intersection graph with the maximal cliques as its vertices, with two maximal cliques K_i and K_j joined by an edge (K_i, K_j) of weight $|K_i \cap K_j|$. A *clique tree* corresponds to a maximum weight spanning tree (MST) of the clique intersection graph. Since the MST of a weighted graph need not be unique, a clique tree of a chordal graph is not necessarily unique either.

In practice, a rooted clique tree is used in sparse matrix computations. Lewis, Peyton, and Pothen [40] and Pothen and Sun [41] have designed algorithms for computing rooted clique trees. The former algorithm uses the adjacency lists of the filled graph as input, while the latter uses the elimination tree. Both algorithms identify representative vertices by a simple degree test. We will discuss the latter algorithm.

First, to define the concepts needed for the algorithm, consider that the the maximal cliques are ordered according to their representative vertices. This ordering partitions each maximal clique $K(v)$ with representative vertex v into two subsets: new($K(v)$) consists of vertices in the clique $K(v)$ whose higher adjacency sets are contained in it but not in any earlier ordered maximal clique. The residual vertices in $K(v) \backslash \text{new}(K(v))$ form the ancestor set anc($K(v)$). If a vertex $w \in \text{anc}(K(v))$, by definition of the ancestor set, w has a higher neighbor that is not adjacent to v; then by the rules of the elimination game, any higher-numbered vertex $x \in K(v)$ also belongs to anc($K(v)$). Thus the partition of a maximal clique into new and ancestor sets is an ordered partition: vertices in new($K(v)$) are ordered before vertices in anc($K(v)$). We denote the lowest numbered vertex f in anc($K(v)$) the *first ancestor* of the clique $K(v)$. A rooted clique tree may be defined as follows: the parent of a clique $K(v)$ is the clique P in which the first ancestor vertex f of K appears as a vertex in new(P).

The reason for calling these subsets "new" and "ancestor" sets can be explained with respect to a rooted clique tree. We can build the chordal graph beginning with the root clique of the clique tree, successively adding one maximal clique at a time, proceeding down the clique tree in in-order. When a maximal clique $K(v)$ is added, vertices in anc($K(v)$) also belong to some ancestor clique(s) of $K(v)$, while vertices in new($K(v)$) appear for the first time. A rooted clique tree, with vertices in new(K) and anc(K) identified for each clique K, is shown in Figure 61.6.

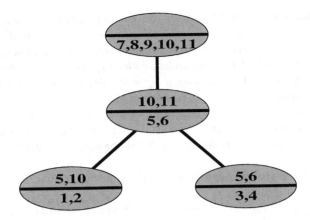

FIGURE 61.6 A clique tree of the example filled graph, computed from its elimination tree. Within each clique K in the clique tree, the vertices in new(K) are listed below the bar, and the vertices in anc(K) are listed above the bar.

```
for  v := 1 to n  →
        if v has a child u in etree with hd⁺(v) + 1 = hd⁺(u) then
                let Kᵤ be the clique in which u is a new vertex;
                add v to the set new(Kᵤ);
        else
                make v the representative vertex of a maximal clique K(v);
                add v to the set new(K(v));
        fi
        for each child s of v in etree such that v and s are new vertices in different cliques →
                let Kₛ be the clique in which s is a new vertex;
                make Kₛ a child of the clique Kᵥ in which v is a new vertex;
        rof
    rof
```

FIGURE 61.7 An algorithm for computing a clique tree from an elimination tree, whose vertices are numbered in postorder. The variable $hd^+(v)$ is the higher degree of a vertex v in the filled graph.

This clique tree algorithm can be implemented in $O(n)$ time, once the elimination tree and the higher degrees have been computed. The rooted clique tree shown in Figure 61.6, is computed from the example elimination tree and higher degrees of the vertices in the example filled graph, using the clique tree algorithm described in Figure 61.7. The clique tree obtained from this algorithm is not unique. A second clique tree that could be obtained has the clique $K(5)$ as the root clique, and the other cliques as leaves.

A comprehensive review of clique trees and chordal graphs in sparse matrix computations, current as of 1991, is provided by Blair and Peyton [42].

61.3.2 Design of Efficient Algorithms with Clique Trees

Shortest Elimination Trees. Jess and Kees [43] introduced the problem of modifying a fill-reducing elimination ordering to enhance concurrency in a parallel factorization algorithm. Their approach was to generate a chordal filled graph from the elimination ordering, and then to eliminate a maximum independent set of simplicial vertices at each step, until all the vertices are eliminated. (This is a greedy algorithm in which the largest number of pairwise independent columns that do not cause fill are eliminated in one step.) Liu and Mirzaian [44] showed that this approach computed a shortest elimination tree over all perfect elimination orderings for a chordal graph, and provided an implementation linear in the number of edges of the filled graph. Lewis, Peyton, and Pothen [44] used the clique tree to provide a faster algorithm; their algorithm runs in time proportional to the size of the clique tree: the sum of the sizes of the maximal cliques of the chordal graph.

A vertex is simplicial if and only if it belongs to exactly one maximal clique in the chordal graph; a maximum independent set of simplicial vertices is obtained by choosing one such vertex from each maximal clique that contains simplicial vertices, and thus the clique tree is a natural data structure for this problem. The challenging aspect of the algorithm is to update the rooted clique tree when simplicial vertices are eliminated and cliques that become non-maximal are *absorbed* by other maximal cliques.

Parallel Triangular Solution. In solving systems of linear equations by factorization methods, usually the work involved in the factorization step dominates the work involved in the triangular solution step (although the communication costs and synchronization overheads of both steps are comparable). However, in some situations, many linear systems with the same coefficient matrix but with different right-hand-side vectors need to be solved. In such situations, it is tempting to replace the triangular solution step involving the factor matrix L by explicitly computing an inverse L^{-1} of the factor. Unfortunately L^{-1} can be much less sparse than the factor, and so a more space efficient "product-form inverse" needs to be employed. In this latter form, the inverse is represented as a product of triangular matrices such that all the matrices in the product together require exactly as much space as the original factor.

The computation of the product form inverse leads to some interesting chordal graph partitioning problems that can be solved efficiently by using a clique tree data structure.

We begin by directing each edge in the chordal filled graph $G^{+}(A)$ from its lower to its higher numbered end point to obtain a directed acyclic graph (DAG). We will denote this DAG by $G(L)$. Given an edge (i, j) directed from i to j, we will call i the predecessor of j, and j the successor of i. The elimination ordering must eliminate vertices in a topological ordering of the DAG such that all predecessors of a vertex must be eliminated before it can be eliminated. The requirement that each matrix in the product form of the inverse must have the same nonzero structure as the corresponding columns in the factor is expressed by the fact that the subgraph corresponding to the matrix should be transitively closed. (A directed graph is *transitively closed* if whenever there is a directed path from a vertex i to a vertex j, there is an edge directed from i to j in the graph.) Given a set of vertices P_i, the *column subgraph* of P_i includes all the vertices in P_i and vertices reached by directed edges leaving vertices in P_i; the edges in this subgraph include all edges with one or both endpoints in P_i.

The simpler of the graph partitioning problems is the following:

Find an ordered partition $P_1 \prec P_2 \prec \ldots P_m$ of the vertices of a directed acyclic filled graph $G(L)$ such that

1. Every $v \in P_i$ has all of its predecessors included in $P_1, \ldots, P_i$;
2. The column subgraph of P_i is transitively closed; and
3. The number of subgraphs m is minimum over all topological orderings of $G(L)$.

Pothen and Alvarado [45] designed a greedy algorithm that runs in $O(n)$ time to solve this partitioning problem by using the elimination tree.

A more challenging variant of the problem minimizes the number of transitively closed subgraphs in $G(L)$ over all perfect elimination orderings of the undirected chordal filled graph $G^{+}(A)$. This variant could change the edges in the DAG $G(L)$, (but not the edges in $G^{+}(A)$) since the initial ordering of the vertices is changed by the perfect elimination ordering, and after the reordering, edges are directed from the lower numbered end point to its higher numbered end point.

This is quite a difficult problem, but two surprisingly simple greedy algorithms solve it. Peyton, Pothen, and Yuan provide two different algorithms for this problem; the first algorithm uses the elimination tree and runs in time linear in the number of edges in the filled graph [46]. The second makes use of the clique tree, and computes the partition in time linear in the size of the clique tree [47]. Proving the correctness of these algorithms requires a careful study of the properties of the minimal vertex separators (these are vertices in the intersections of the maximal cliques) in the chordal filled graph.

61.3.3 Compact Clique Trees

In analogy with skeleton graphs, we can define a space-efficient version of a clique tree representation of a chordal graph, called the compact clique tree. If K is the parent clique of a clique C in a clique tree, then it can be shown that $\text{anc}(C) \subset K$. Thus trading space for computation, we can delete the vertices in K that belong to the ancestor sets of its children, since we can recompute them when necessary by unioning the ancestor sets of the children. The partition into new and ancestor sets can be obtained by storing the lowest numbered ancestor vertex for each clique. A compact clique K^c corresponding to a clique K is:

$$K^c = K \setminus \cup_{C \in \text{child}(K)} \text{anc}(C).$$

Note that the compact clique depends on the specific clique tree from which it is computed.

A compact clique tree is obtained from a clique tree by replacing cliques by compact cliques for vertices. In the example clique tree, the compact cliques of the leaves are unchanged from the corresponding cliques; and the compact cliques of the interior cliques are $K^c(5) = \{11\}$, and $K^c(7) = \{7, 8, 9\}$.

The compact clique tree is potentially sparser (asymptotically $O(n)$ instead of $O(n^2)$ even) than the skeleton graph on pathological examples, but on "practical" examples, the size difference between them is small. Compact clique trees were introduced by Pothen and Sun [41].

61.4 Clique Covers and Quotient Graphs

Clique covers and quotient graphs are data structures that were developed for the efficient implementation of minimum-degree reordering heuristics for sparse matrices. In Gaussian elimination, an elimination step that uses a_{ij} as a pivot (the elimination of the jth unknown using the ith equation) modifies every coefficient a_{kl} for which $a_{kj} \neq 0$ and $a_{il} \neq 0$. Minimum-degree heuristics attempt to select pivots for which the number of modified coefficients is small.

61.4.1 Clique Covers

Recall the graph model of symmetric Gaussian elimination discussed in Section 61.1.1. The adjacency graph of the matrix to be factored is an undirected graph $G = (V, E)$, $V = \{1, 2, \ldots, n\}$, $E = \{(i,j) : a_{ij} \neq 0\}$. The elimination of a row/column j corresponds to eliminating vertex j and adding edges to the remaining graph so that the neighbors of j become a clique. If we represent the edge set E using a clique cover, a set of cliques $\mathcal{K} = \{K : K \subseteq V\}$ such that $E = \cup_{K \in \mathcal{K}} \{(i,j) : i,j \in K\}$, the vertex elimination process becomes a process of merging cliques [39]: The elimination of vertex j corresponds to merging all the cliques that j belongs to into one clique and removing j from all the cliques. Clearly, all the old cliques that j used to belong to are now covered by the new clique, so they can be removed from the cover. The clique-cover can be initialized by representing every nonzero of A by a clique of size 2. This process corresponds exactly to symbolic elimination, which we have discussed in Section 61.2, and which costs $\Theta(|L|)$ work. The cliques correspond exactly to frontal matrices in the multifrontal factorization method.

In the sparse-matrix literature, this model of Gaussian elimination has been sometimes called the *generalized-element* model or the *super-element* model, due to its relationship to finite-element models and matrices.

The significance of clique covers is due to the fact that in minimum-degree ordering codes, there is no need to store the structure of the partially computed factor, so when one clique is merged into another, it can indeed be removed from the cover. This implies that the total size $\sum_{K \in \mathcal{K}} |K|$ of the representation of the clique cover, which starts at exactly $|A| - n$, shrinks in every elimination step, so it is always bounded by $|A| - n$. Since exactly one clique is formed in every elimination step, the total number of cliques is also bounded, by $n + (|A| - n) = |A|$. In contrast, the storage required to explicitly represent the symbolic factor, or even to just explicitly represent the edges in the reduced matrix, can grow in every elimination step and is not bounded by $O(|A|)$.

Some minimum-degree codes represent cliques fully explicitly [48,49]. This representation uses an array of cliques and an array of vertices; each clique is represented by a linked list of vertex indices, and each vertex is represented by a linked list of clique indices to which it belongs. The size of this data structure never grows—linked-list elements are moved from one list to another or are deleted during elimination steps, but new elements never need to be allocated once the data structure is initialized.

Most codes, however, use a different representation for clique covers, which we describe next.

61.4.2 Quotient Graphs

Most minimum-degree codes represent the graphs of reduced matrices during the elimination process using *quotient graphs* [50]. Given a graph $G = (V, E)$ and a partition $\mathcal{S}$ of V into disjoint sets $S_j \in \mathcal{S}$, the quotient graph $G/\mathcal{S}$ is the undirected graph $(\mathcal{S}, \mathcal{E})$ where $\mathcal{E} = \{(S_i, S_j) : \text{adj}(S_i) \cap S_j \neq \emptyset\}$.

The representation of a graph G after the elimination of vertices $1, 2, \ldots, j - 1$, but before the elimination of vertex j, uses a quotient graph $G/\mathcal{S}$, where $\mathcal{S}$ consists of sets S_k of eliminated vertices that form maximal connected components in G, and sets $S_i = \{i\}$ of uneliminated vertices $i \geq j$. We denote a set S_k of eliminated vertices by the index k of the highest-numbered vertex in it.

This quotient graph representation of an elimination graph corresponds to a clique cover representation as follows. Each edge in the quotient graph between uneliminated vertices $S_{\{i_1\}}$ and $S_{\{i_2\}}$ corresponds to a clique of size 2; all the neighbors of an eliminated set S_k correspond to a clique, the clique that was created when vertex k was eliminated. Note that all the neighbors of an uneliminated set S_k are uneliminated vertices, since uneliminated sets are maximal with respect to connectivity in G.

The elimination of vertex j in the quotient-graph representation corresponds to marking S_j as eliminated and merging it with its eliminated neighbors, to maintain the maximal connectivity invariant.

Clearly, a representation of the initial graph G using adjacency lists is also a representation of the corresponding quotient graph. George and Liu [51] show how to maintain the quotient graph efficiently using this representation without allocating more storage through a series of elimination steps.

Most of the codes that implement minimum-degree ordering heuristics, such as GENMMD [52], AMD [53], and Spindle [54,55], use quotient graphs to represent elimination graphs.

It appears that the only advantage of a quotient graph over an explicit clique cover in the context of minimum-degree algorithms is a reduction by a small constant factor in the storage requirement, and possibly in the amount of work required. Quotient graphs, however, can also represent symmetric partitions of symmetric matrices in applications that are not directly related to elimination graphs. For example, George and Liu use quotient graphs to represent partitions of symmetric matrices into block matrices that can be factored without fill in blocks that only contain zeros [16, Chapter 6].

In [56], George and Liu showed how to implement the minimum degree algorithm without modifying the representation of the input graph at all. In essence, this approach represents the quotient graph implicitly using the input graph and the indices of the eliminated vertices. The obvious drawback of this approach is that vertex elimination (as well as other required operations) are expensive.

61.4.3 The Problem of Degree Updates

The minimum-degree algorithm works by repeatedly eliminating the vertex with the minimum degree and turning its neighbors into a clique. If the reduced graph is represented by a clique cover or a quotient graph, then the representation does not reveal the degree of vertices. Therefore, when a vertex is eliminated from a graph represented by a clique cover or a quotient graph, the degrees of its neighbors must be recomputed. These degree updates can consume much of the running time of minimum-degree algorithms.

Practical minimum-degree codes use several techniques to address this issue. Some techniques reduce the running time while preserving the invariant that the vertex that is eliminated always has the minimum degree. For example, mass elimination, the elimination of all the vertices of a supernode consecutively without recomputing degrees, can reduce the running time significantly without violating this invariant. Other techniques, such as multiple elimination and the use of approximate degrees, do not preserve the minimum-degree invariant. This does not imply that the elimination orderings that such technique produce are inferior to true minimum-degree orderings. They are often superior to them. This is not a contradiction since the minimum-degree rule is just a heuristic which is rarely optimal. For further details, we refer the reader to George and Liu's survey [57], to Amestoy, Davis, and Duff's paper on approximate minimum-degree rules [53], and to Kumfert and Pothen's work on minimum-degree variants [54,55]. Heggernes, Eisenstat, Kumfert and Pothen prove upper bounds on the running time of space-efficient minimum-degree variants [58].

61.4.4 Covering the Column-Intersection Graph and Biclique Covers

Column orderings for minimizing fill in Gaussian elimination with partial pivoting and in the orthogonal-triangular (QR, where Q is an orthogonal matrix, and R is an upper triangular matrix) factorization are often based on symmetric fill minimization in the symmetric factor of $A^T A$, whose graph is known as the the column intersection graph $G_\cap(A)$ (we ignore the possibility of numerical cancellation in $A^T A$). To run a minimum-degree algorithm on the column intersection graph, a clique cover or quotient graph of it must be constructed. One obvious solution is to explicitly compute the edge-set of $G_\cap(A)$, but this is inefficient, since $G_\cap(A)$ can be much denser than $G(A)$.

A better solution is to initialize the clique cover using a clique for every row of A; the vertices of the clique are the indices of the nonzeros in that row [57]. It is easy to see that each row in A indeed corresponds to a clique in $G_\cap(A)$. This approach is used in the COLMMD routine in MATLAB [49] and in COLAMD [59].

A space-efficient quotient-graph representation for $G_\cap(A)$ can be constructed by creating an adjacency-list representation of the symmetric 2-by-2 block matrix

$$\begin{pmatrix} I & A \\ A^T & 0 \end{pmatrix}$$

and eliminating vertices 1 through n. The graph of the Schur complement matrix

$$G(0 - A^T I^{(-1)} A) = G(A^T A) = G_\cap(A).$$

If we maintain a quotient-graph representation of the reduced graph through the first n elimination steps, we obtain a space-efficient quotient graph representation of the column-intersection graph. This is likely to be more expensive, however, than constructing the clique-cover representation from the rows of A. We learned of this idea from John Gilbert; we are not aware of any practical code that uses it.

The nonzero structure of the Cholesky factor of $A^T A$ is only an upper bound on the structure of the LU factors in Gaussian elimination with partial pivoting. If the identities of the pivots are known, the nonzero structure of the reduced matrices can be represented using biclique covers. The nonzero structure of A is represented by a bipartite graph $(\{1, 2, \ldots, n\} \cup \{1', 2', \ldots, n'\}, \{(i, j'): a_{ij} \neq 0\})$. A biclique is a complete bipartite graph on a subset of the vertices. Each elimination step corresponds to a removal of two connected vertices from the bipartite graph, and an addition of a new biclique. The vertices of the new biclique are the neighbors of the two eliminated vertices, but they are not the union of a set of bicliques. Hence, the storage requirement of this representation may exceed the storage required for the initial representation. Still, the storage requirement is always smaller than the storage required to represent each edge of the reduced matrix explicitly. This representation poses the same degree update problem that symmetric clique covers pose, and the same techniques can be used to address it. Version 4 of UMFPACK, an asymmetric multifrontal LU factorization code, uses this idea together with a degree approximation technique to select pivots corresponding to relatively sparse rows in the reduced matrix [60].

61.5 Column Elimination Trees and Elimination DAGS

Elimination structures for asymmetric Gaussian elimination are somewhat more complex than the equivalent structures for symmetric elimination. The additional complexity arises because of two issues. First, the factorization of a sparse asymmetric matrix A, where A is factored into a lower triangular factor L and an upper triangular factor U, $A = LU$, is less structured than the sparse symmetric factorization process. In particular, the relationship between the nonzero structure of A and the nonzero structure of the factors is much more complex. Consequently, data structures for predicting fill and representing data-flow and control-flow dependences in elimination algorithms are more complex and more diverse.

Second, factoring an asymmetric matrix often requires *pivoting*, row and/or column exchanges, to ensure existence of the factors and numerical stability. For example, the 2-by-2 matrix $A = [0\ 1; 1\ 0]$ does not have an LU factorization, because there is no way to eliminate the first variable from the first equation: that variable does not appear in the equation at all. But the permuted matrix PA does have a factorization, if P is a permutation matrix that exchanges the two rows of A. In finite precision arithmetic, row and/or column exchanges are necessary even when a nonzero but small diagonal element is encountered. Some sparse LU algorithms perform either row or column exchanges, but not both. The two cases are essentially equivalent (we can view one as a factorization of A^T), so we focus on row exchanges (partial pivoting). Other algorithms, primarily multifrontal algorithms, perform both row and column exchanges; these are discussed toward the end of this section.

For completeness, we note that pivoting is also required in the factorization of sparse symmetric indefinite matrices. Such matrices are usually factored into a product LDL^T, where L is lower triangular and D is a block diagonal matrix with 1-by-1 and 2-by-2 blocks. There has not been much research about specialized elimination structures for these factorization algorithms; such codes invariably use the symmetric elimination tree of A to represent dependences for structure prediction and for scheduling the factorization.

The complexity and diversity of asymmetric elimination arises not only due to pivoting, but also because asymmetric factorizations are less structured than symmetric ones, so a rooted tree can no longer represent the factors. Instead, directed acyclic graphs (dags) are used to represent the factors and dependences in the elimination process. We discuss *elimination dags (edags)* in Section 61.5.2.

Surprisingly, dealing with partial pivoting turns out to be simpler than dealing with the asymmetry, so we focus next on the *column elimination tree*, an elimination structure for LU factorization with partial pivoting.

61.5.1 The Column Elimination Tree

The *column elimination tree (col-etree)* is the elimination tree of $A^T A$, under the assumption that no numerical cancellation occurs in the formation of $A^T A$. The significance of this tree to LU with partial pivoting stems from a series of results that relate the structure of the LU factors of PA, where P is some permutation matrix, to the structure of the Cholesky factor of $A^T A$.

George and Ng observed that, for any permutation matrix P, the structure of the LU factors of PA is contained in the structure of the Cholesky factor of $A^T A$, as long as A does not have any zeros on its main diagonal [61]. (If there are zeros on the diagonal of a nonsingular A, the rows can always be permuted first to achieve a zero-free diagonal.) Figure 61.8 illustrates this phenomenon. Gilbert [62] strengthened this result by showing that for every nonzero R_{ij} in the Cholesky factor R of $A^T A = R^T R$, where A has a zero-free diagonal and no nontrivial block triangular form, there exists a matrix $A^{(U_{ij})}$ with the same nonzero structure as A, such that in the LU factorization of $A^{(U_{ij})} = P^T L^{(U_{ij})} U^{(U_{ij})}$ with partial pivoting, $U_{ij}^{(U_{ij})} \neq 0$. This kind of result is known as a *one-at-a-time* result, since it guarantees that every element of the predicted factor can fill for some choice of numerical values in A, but not that all the elements can fill simultaneously. Gilbert and Ng [63] later generalized this result to show that an equivalent one-at-a-time property is true for the lower-triangular factor.

These results suggest that the col-etree, which is the elimination tree of $A^T A$, can be used for scheduling and structure prediction in LU factorizations with partial pivoting. Because the characterization of the nonzero structure of the LU factors in terms of the structure of the Cholesky factor of $A^T A$ relies on one-at-a-time results, the predicted structure and predicted dependences are necessarily only loose upper bounds, but they are nonetheless useful.

The col-etree is indeed useful in LU with partial pivoting. If $U_{ij} \neq 0$, then by the results cited above $R_{ij} \neq 0$ (recall that R is the Cholesky factor of the matrix $A^T A$). This, in turn, implies that j is an ancestor of i in the col-etree. Since column i of L updates column j of L and U only if $U_{ij} \neq 0$, the col-etree can be used as a task-dependence graph in column-oriented LU factorization codes with partial pivoting. This analysis is due to Gilbert, who used it to schedule a parallel factorization code [62]. The same

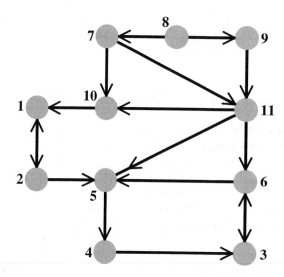

FIGURE 61.8   The directed graph $G(A)$ of an asymmetric matrix A. The column intersection graph of this graph $G_\cap(B)$ is exactly the graph shown in Figure 61.1, so the column elimination tree of A is the elimination tree shown in Figure 61.2. Note that the graph is sparser than the column intersection graph.

technique is used in several current factorization codes, including SuperLU [64], SuperLU_MT [65], UMFPACK 4 [60], and TAUCS [66,67]. Gilbert and Grigori [68] recently showed that this characterization is tight in a strong *all-at-once* sense: for every strong Hall matrix A (i.e., A has no nontrivial block-triangular form), there exists a permutation matrix P such that every edge of the col-etree corresponds to a nonzero in the upper-triangular factor of PA. This implies that the a-priori symbolic column-dependence structure predicted by the col-etree is as tight as possible.

Like the etree of a symmetric matrix, the col-etree can be computed in time almost linear in the number of nonzeros in A [69]. This is done by an adaptation of the symmetric etree algorithm, an adaptation that does not compute explicitly the structure of $A^T A$. Instead of constructing $G(A^T A)$, the algorithm constructs a much sparser graph G' with the same elimination tree. The main idea is that each row of A contributes a clique to $G(A^T A)$; this means that each nonzero index in the row must be an ancestor of the preceding nonzero index. A graph in which this row-clique is replaced by a path has the same elimination tree, and it has only as many edges as there are nonzeros in A. The same paper shows not only how to compute the col-etree in almost linear time, but also how to bound the number of nonzeros in each row and column of the factors L and U, using again an extension of the symmetric algorithm to compute the number of nonzeros in the Cholesky factor of $A^T A$. The decomposition of this Cholesky factor into fundamental supernodes, which the algorithm also computes, can be used to bound the extent of fundamental supernodes that will arise in L.

61.5.2 Elimination DAGS

The col-etree represents all possible column dependences for any sequence of pivot rows. For a specific sequence of pivots, the col-etree includes dependences that do not occur during the factorization with these pivots. There are two typical situations in which the pivoting sequence is known. The first is when the matrix is known to have a stable LU factorization without pivoting. The most common case is when A^T is strictly diagonally dominant. Even if A is not diagonally dominant, its rows can be pre-permuted to bring large elements to the diagonal. The permuted matrix, even if its transpose is not diagonally dominant, is fairly likely to have a relatively stable LU factorization that can be used to accurately solve linear systems of equations. This strategy is known as *static pivoting* [70]. The other situation in which the pivoting sequence is known is when the matrix, or part of it, has already been factored. Since virtually all sparse factorization algorithms need to collect information from the already-factored portion of the matrix before they factor the next row and column, a compact representation of the structure of this portion is useful.

Elimination dags (edags) are directed acyclic graphs that capture a minimal or near minimal set of dependences in the factors. Several edags have been proposed in the literature. There are two reasons for this diversity. First, edags are not always as sparse and easy to compute as elimination trees, so researchers have tried to find edags that are easy to compute, even if they represent a superset of the actual dependences. Second, edags often contain information only about a specific structure in the factors or a specific dependence in a specific elimination algorithm (e.g., data dependence in a multifrontal algorithm), so different edags are used for different applications. In other words, edags are not as universal as etrees in their applications.

The simplest edag is the graph $G(L^T)$ of the transpose of the lower triangular factor, if we view every edge in this graph as directed from the lower-numbered vertex to a higher-numbered vertex. This corresponds to orienting edges from a row index to a column index in L. For example, if $L_{6,3} \neq 0$, we view the edge $(6, 3)$ as a directed edge $3 \rightarrow 6$ in $G(L^T)$. Let us denote by $G((L^{(j-1)})^T)$ the partial lower triangular factor after $j - 1$ columns have been factored. Gilbert and Peierls showed that the nonzeros in the jth rows of L and U are exactly the vertices reachable, in $G((L^{(j-1)})^T)$, from the nonzero indices in the jth column of A [71]. This observation allowed them to use a depth-first search (DFS) to quickly find the columns in the already-factored part of the matrix that update the jth column before it can be factored. This resulted in the first algorithm that factored a general sparse matrix in time linear in the number of arithmetic operations (earlier algorithms sometimes performed much more work to manipulate the sparse data structure than the amount of work in the actual numerical computations).

Eisenstat and Liu showed that a simple modification of the graph $G((L^{(j-1)})^T)$ can often eliminate many of its edges without reducing its ability to predict fill [72]. They showed that if both L_{ik} and U_{ki} are nonzeros, then all the edges $i \rightarrow \ell$ for $\ell > i$ can be safely pruned from the graph. In other words, the nonzeros in column k of L below row i can be pruned. This is true since if $U_{ki} \neq 0$, then column k of L updates column i, so all the pruned nonzeros appear in column i, and since the edge $k \rightarrow i$ is in the graph, they are all reachable when k is reachable. This technique is called *symmetric pruning*. This edag is used in the SuperLU codes [64,65] to find the columns that update the next supernode (set of consecutive columns with the same nonzero structure in L). Note that the same codes use the col-etree to predict structure before the factorization begins, and an edag to compactly represent the structure of the already-factored block of A.

Gilbert and Liu went a step further and studied the *minimal* edags that preserve the reachability property that is required to predict fill [73]. These graphs, which they called the *elimination dags* are the transitive reductions of the directed graphs $G(L^T)$ and $G(U)$. (The graph of U can be used to predict the row structures in the factors, just as $G(L^T)$ can predict the column structures.) Since these graphs are acyclic, each graph has a unique transitive reduction; If A is symmetric, the transitive reduction is the symmetric elimination tree. For the graph $G(A)$ shown in Figure 61.8, the graph of the upper triangular factor $G(U)$ is depicted in Figure 61.9, and its minimal edag is shown in Figure 61.10. Gilbert and Liu also proposed an algorithm to compute these transitive reductions row by row. Their algorithm computes the next row i of the transitive reduction of L by traversing the reduction of U to compute the structure of row i of L, and then reducing it. Then the algorithm computes the structure of row i of U by combining the structures of earlier rows whose indices are the nonzeros in row i of L. In general, these minimal edags are often more expensive to compute than the symmetrically-pruned edags, due to the cost of transitively reducing each row. Gupta recently proposed a different algorithm for computing the minimal edags [74]. His algorithm computes the minimal structure of U by rows and of L by columns. His algorithm essentially applies to both L and U the rule that Gilbert and Liu apply to U. By computing the structure of U by rows and of L by columns, Gupta's algorithm can cheaply detect supernodes that are suitable for

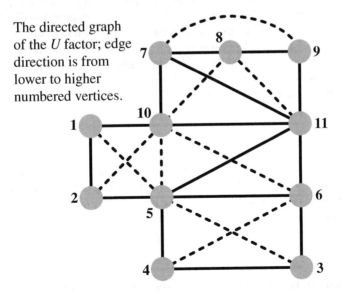

The directed graph of the U factor; edge direction is from lower to higher numbered vertices.

FIGURE 61.9 The directed graph of the U factor of the matrix whose graph is shown in Figure 61.8. In this particular case, the graph of L^T is exactly the same graph, only with the direction of the edges reversed. Fill is indicated by dashed lines. Note that the fill is indeed bounded by the fill in the column-intersection graph, which is shown in Figure 61.1. However, that upper bound is not realized in this case: the edge $(9, 10)$ fills in the column-intersection graph, but not in the LU factors.

The minimal dag of the *U* factor; edge direction is from lower to higher numbered vertices.

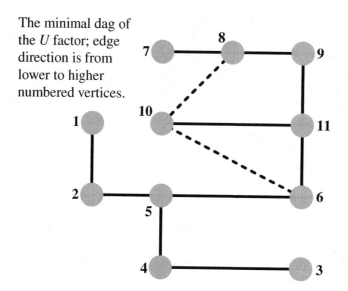

FIGURE 61.10   The minimal edag of *U*; this graph is the transitive reduction of the graph shown in Figure Figure 61.9.

asymmetric multifrontal algorithms, where a supernode consists of a set of consecutive indices for which both the rows of *U* all have the same structure and the columns of *L* have the same structure (but the rows and columns may have different structures).

61.5.3   Elimination Structures for the Asymmetric Multifrontal Algorithm

Asymmetric multifrontal *LU* factorization algorithms usually use both row and column exchanges. UMFPACK, the first such algorithm, due to Davis and Duff [75], used a pivoting strategy that factored an arbitrary row and column permutation of *A*. The algorithm tried to balance numerical and degree considerations in selecting the next row and column to be factored, but in principle, all row and column permutations were possible. Under such conditions, not much structure prediction is possible. The algorithm still used a clever elimination structure that we described earlier, a biclique cover, to represent the structure of the Schur complement (the remaining uneliminated equations and variables), but it did not use etrees or edags.

Recent unsymmetric multifrontal algorithms still use pivoting strategies that allow both row and column exchanges, but the pivoting strategies are restricted enough that structure prediction is possible. These pivoting strategies are based on *delayed pivoting*, which was originally invented for symmetric indefinite factorizations. One such code, Davis's UMFPACK 4, uses the column elimination tree to represent control-flow dependences, and a biclique cover to represent data dependences [60]. Another code, Gupta's WSMP, uses conventional minimal edags to represent control-flow dependences, and specialized dags to represent data dependences [74]. More specifically, Gupta shows how to modify the minimal edags so they exactly represent data dependences in the unsymmetric multifrontal algorithm with no pivoting, and how to modify the edags to represent dependences in an unsymmetric multifrontal algorithm that employs delayed pivoting.

Acknowledgments

Alex Pothen was supported, in part, by the National Science Foundation under grant CCR-0306334, and by the Department of Energy under subcontract DE-FC02-01ER25476. Sivan Toledo was supported in part by an IBM Faculty Partnership Award, by grant 572/00 from the Israel Science Foundation (founded by the Israel Academy of Sciences and Humanities), and by Grant 2002261 from the US-Israeli Binational Science Foundation.

References

1. A. George, J. R. Gilbert and J. W. H. Liu, editors *Graph Theory and Sparse Matrix Computation*. Springer-Verlag, New York, 1993. IMA Volumes in Applied Mathematics, Volume 56.
2. I. S. Duff. Sparse numerical linear algebra: direct methods and preconditioning. Technical Report RAL-TR-96-047, Department of Computation and Information, Rutherford Appleton Laboratory, Oxon OX11 0QX, England, 1996.
3. S. V. Parter. The use of linear graphs in Gaussian elimination. *SIAM Review*, 3:119–130, 1961.

4. R. E. Tarjan. Unpublished lecture notes. 1975.

5. M. Yannakakis. Computing the minimum fill-in is NP-complete. *SIAM Journal on Algebraic and Discrete Methods*, 2:77–79, 1981.

6. A. Agrawal, P. Klein, and R. Ravi. Cutting down on fill using nested dissection: Provably good elimination orderings. In [1], pages 31–55, 1993.

7. A. Natanzon, R. Shamir, and R. Sharan. A polynomial approximation algorithm for the minimum fill-in problem. In *Proceedings of the 30th Annual ACM Symposium on the Theory of Computing (STOC 98)*, Dallas, TX, pages 41–47, 1998.

8. D. J. Rose, R. E. Tarjan, and G. S. Lueker. Algorithmic aspects of vertex elimination on graphs. *SIAM Journal on Computing*, 5:266–283, 1976.

9. R. S. Schreiber. A new implementation of sparse Gaussian elimination. *ACM Transactions on Mathematical Software*, 8:256–276, 1982.

10. J. W. H. Liu. A compact row storage scheme for Cholesky factors using elimination trees. *ACM Transactions on Mathematical Software*, 12(2):127–148, 1986.

11. J. W. H. Liu. The role of elimination trees in sparse factorization. *SIAM Journal on Matrix Analysis and Applications*, 11:134–172, 1990.

12. E. Zmijewski and J. R. Gilbert. A parallel algorithm for sparse symbolic Cholesky factorization on a multiprocessor. *Parallel Computing*, 7:199–210, 1988.

13. I. S. Duff. Full matrix techniques in sparse matrix factorizations. In G. A. Watson, editor, *Lecture Notes in Mathematics*, pages 71–84. Springer-Verlag, 1982.

14. J. W. H. Liu, E. G. Ng, and B. W. Peyton. On finding supernodes for sparse matrix computations. *SIAM Journal on Matrix Analysis and Applications*, 14:242–252, 1993.

15. I. S. Duff, A. M. Erisman, and J. K. Reid. *Direct Methods for Sparse Matrices*, Second edition. Oxford University Press, Oxford, 2017.

16. A. George and J. W. H. Liu. *Computer Solution of Large Sparse Positive Definite Systems*. Prentice-Hall, 1981.

17. C. Ashcraft and R. Grimes. The influence of relaxed supernode partitions on the multifrontal method. *ACM Transactions on Mathematical Software*, 15(4):291–309, 1989.

18. I. S. Duff and J. K. Reid. The multifrontal solution of indefinite sparse symmetric linear equations. *ACM Transactions on Mathematical Software*, 9:302–325, 1983.

19. M. Bern, J. R. Gilbert, B. Hendrickson, N. Nguyen, and S. Toledo. Support-graph preconditioners. *SIAM Journal on Matrix Analysis and Applications*, 27(4), 930–951, 2006.

20. P. M. Vaidya. Solving linear equations with symmetric diagonally dominant matrices by constructing good preconditioners. Unpublished manuscript. *A talk based on the manuscript was presented at the IMA Workshop on Graph Theory and Sparse Matrix Computation*, Minneapolis October 1991.

21. A. George and J. W. H. Liu. An optimal algorithm for symbolic factorization of symmetric matrices. *SIAM Journal on Computing*, 9:583–593, 1980.

22. G. F. Whitten. Computation of fill-in for sparse symmetric positive definite matrices. Technical report, Computer Science Department, Vanderbilt University, Nashville, TN, 1978.

23. R. E. Tarjan and M. Yannakakis. Simple linear-time algorithms to test chordality of graphs, test acyclicity of hypergraphs and selectively reduce acyclic hypergraphs. *SIAM Journal of Computing*, 13:566–579, 1984. Addendum in 14:254–255, 1985.

24. J. R. Gilbert, E. G. Ng, and B. W. Peyton. An efficient algorithm to compute row and column counts for sparse Cholesky factorization. *SIAM Journal on Matrix Analysis and Applications*, 15(4):1075–1091, 1994.

25. I. S. Duff. Parallel implementation of multifrontal schemes. *Parallel Computing*, 3:193–204, 1986.

26. M. T. Heath. Parallel direct methods for sparse linear systems. In D. E. V. Keyes, A. Sameh, and Venkatakrishnan, editors, *Parallel Numerical Algorithms*, pages 55–90. Kluwer, 1997.

27. P. Hénon, P. Ramet, and J. Roman. PaStiX: A high-performance parallel direct solver for sparse symmetric definite systems. *Parallel Computing*, 28(2):301–321, 2002.

28. O. Schenk and K. Gärtner. Sparse factorization with two-level scheduling in PARADISO. In *Proceedings of the 10th SIAM Conference on Parallel Processing for Scientific Computing*, Portsmouth, Virginia, March 2001SEP. 10 pp. on CDROM.

29. A. Gupta, M. V. Joshi, and V. Kumar. WSSMP: A high-performance serial and parallel symmetric sparse linear solver. In B. Kågström, J. Dongarra, E. Elmroth, and J. Wasniewski, editors, *Proceedings of the 4th International Workshop on Applied Parallel Computing, Large Scale Scientific and Industrial Problems (PARA '98)*, volume 1541 of Lecture Notes in Computer Science, pages 182–194, Umeå, Sweden, June 1998. Springer.

30. P. R. Amestoy, I. S. Duff, J.-Yves L'Excellent, and J. Koster. A fully asynchronous multifrontal solver using distributed dynamic scheduling. *SIAM Journal on Matrix Analysis and Applications*, 23(1):15–41, 2001.

31. E. Rothberg and R. S. Schreiber. Efficient methods for out-of-core sparse Cholesky factorization. *SIAM Journal on Scientific Computing*, 21:129–144, 1999.

32. F. Dobrian and A. Pothen. The design of I/O-efficient sparse direct solvers. In *Proceedings of the 2001 ACM/IEEE conference on Supercomputing*. Denver, ACM Press, 2001. 21 pp. on CDROM.

33. J. W. H. Liu. An adaptive general sparse out-of-core Cholesky factorization scheme. *SIAM Journal on Scientific and Statistical Computing*, 8(4):585–599, 1987.

34. E. Rothberg and R. S. Schreiber. An alternative approach to sparse out-of-core factorization *presented at the 2nd SIAM Conference on Sparse Matrices*, Coeur d'Alene, Idaho, October 1996.

35. V. Rotkin and S. Toledo. The design and implementation of a new out-of-core sparse Cholesky factorization method. *ACM Transactions on Mathematical Software*, 30(1), pp. 19–46, 2004.

36. F. Dobrian and A. Pothen. A comparison of three external memory algorithms for factoring sparse matrices. In *Proceedings of the SIAM Conference on Applied Linear Algebra*, Williamsburg, VA, 11 pages, SIAM, 2003.

37. J. W. H. Liu. On the storage requirement in the out-of-core multifrontal method for sparse factorization. *ACM Transactions on Mathematical Software*, 12(3):249–264, 1986.

38. S. C. Eisenstat, M. H. Schultz, and A. H. Sherman. Software for sparse Gaussian elimination with limited core memory. In I. S. Duff and G. W. Stewart, editors, *Sparse Matrix Proceedings*, pages 135–153. SIAM, Philadelphia, 1978.

39. D. J. Rose. A graph-theoretic study of the numerical solution of sparse positive definite systems of linear equations. In R. C. Read, editor, *Graph Theory and Computing*, pages 183–217. Academic Press, New York, 1972.

40. J. G. Lewis, B. W. Peyton, and A. Pothen. A fast algorithm for reordering sparse matrices for parallel factorization. *SIAM Journal on Scientific Computing*, 6:1147–1173, 1989.

41. A. Pothen and C. Sun. Compact clique tree data structures for sparse matrix factorizations. In T. F. Coleman and Y. Li, editors, *Large Scale Numerical Optimization*, pages 180–204. SIAM, Philadelphia, 1990.

42. J. R. S. Blair and B. W. Peyton. An introduction to chordal graphs and clique trees. In [1], pages 1–30, 1993.

43. J. Jess and H. Kees. A data structure for parallel LU decomposition. *IEEE Transactions on Computers*, C-31:231–239, 1982.

44. J. W. H. Liu and A. Mirzaian. A linear reordering algorithm for parallel pivoting of chordal graphs. *SIAM Journal on Discrete Mathematics*, 2:100–107, 1989.

45. A. Pothen and F. Alvarado. A fast reordering algorithm for parallel sparse triangular solution. *SIAM Journal on Scientific and Statistical Computation*, 13:645–653, 1992.

46. B. W. Peyton, A. Pothen, and X. Yuan. Partitioning a chordal graph into transitive subgraphs for parallel sparse triangular solution. *Linear Algebra and its Applications*, 192:329–354, 1993.

47. B. W. Peyton, A. Pothen, and X. Yuan. A clique tree algorithm for partitioning a chordal graph into transitive subgraphs. *Linear Algebra and its Applications*, 223:553–588, 1995.

48. T.-Yi Chen, J. R. Gilbert, and S. Toledo. Toward an efficient column minimum degree code for symmetric multiprocessors. In *Proceedings of the 9th SIAM Conference on Parallel Processing for Scientific Computing*, San Antonio, Texas, 1999. 11 pp. on CDROM.

49. J. R. Gilbert, C. Moler, and R. S. Schreiber. Sparse matrices in MATLAB: Design and implementation. *SIAM Journal on Matrix Analysis and Applications*, 13(1):333–356, 1992.

50. A. George and J. W. H. Liu. A quotient graph model for symmetric factorization. In I. S. G. W. Duff and Stewart, editors, *Sparse Matrix Proceedings*, pages 154–175. SIAM, Philadelphia, 1978.

51. A. George and J. W. H. Liu. A fast implementation of the minimum degree algorithm using quotient graphs. *ACM Transactions on Mathematical Software*, 6(3):337–358, 1980.

52. J. W. H. Liu. Modification of the minimum-degree algorithm by multiple elimination. *ACM Transactions on Mathematical Software*, 11(2):141–153, 1985.

53. P. R. Amestoy, T. A. Davis, and I. S. Duff. An approximate minimum degree ordering algorithm. *SIAM Journal on Matrix Analysis and Applications*, 17(4):886–905, 1996.

54. F. Dobrian, G. Kumfert, and A. Pothen. The design of a sparse direct solver library using object-oriented techniques. In A. M. Bruaset, H. P. Langtangen, and E. Quak, editors, *Modern Software Tools in Scientific Computing*, pages 89–131. Springer-Verlag, Birkhauser, Basel, 2000.

55. G. Karl Kumfert. *An object-oriented algorithmic laboratory for ordering sparse matrices*. PhD thesis, Old Dominion University, December 2000.

56. A. George and J. W. H. Liu. A minimal storage implementation of the minimum degree algorithm. *SIAM Journal on Numerical Analysis*, 17:283–299, 1980.

57. A. George and J. W. H. Liu. The evolution of the minimum-degree ordering algorithm. *SIAM Review*, 31:1–19, 1989.

58. P. Heggernes, S. C. Eisenstat, G. Kumfert, and A. Pothen. The computational complexity of the minimum degree algorithm. In *Proceedings of the 14th Norwegian Computer Science Conference (NIK 2001)*, Tromso, Norway, November 2001, 12 pages. Also available as ICASE Report 2001-42, NASA/CR-2001-211421, NASA Langley Research Center.

59. T. A. Davis, J. R. Gilbert, S. I. Larimore, and E. G. Ng. A column approximate minimum degree ordering algorithm. Technical Report TR-00-005, Department of Computer and Information Science and Engineering, University of Florida, 2000.

60. T. A. Davis. A column pre-ordering strategy for the unsymmetric-pattern multifrontal method. Technical Report TR-03-006, Department of Computer and Information Science and Engineering, University of Florida, May 2003.

61. A. George and E. G. Ng. An implementation of Gaussian elimination with partial pivoting for sparse systems. *SIAM Journal on Scientific and Statistical Computing*, 6:390–409, 1985.

62. J. R. Gilbert. An efficient parallel sparse partial pivoting algorithm. Technical Report 88/45052-1, Christian Michelsen Institute, Bergen, Norway, 1988.

63. J. R. Gilbert and E. G. Ng. Predicting structure in nonsymmetric sparse matrix factorizations. In [1], pages 107–139, 1993.

64. J. W. Demmel, S. C. Eisenstat, J. R. Gilbert, X. S. Li, J. W. H. Liu. A supernodal approach to sparse partial pivoting. *SIAM Journal on Matrix Analysis and Applications*, 20:720–755, 1999.

65. J. W. Demmel, J. R. Gilbert, and X. S. Li. An asynchronous parallel supernodal algorithm for sparse Gaussian elimination. *SIAM Journal on Matrix Analysis and Applications*, 20:915–952, 1999.

66. J. R. Gilbert and S. Toledo. High-performance out-of-core sparse LU factorization. In *Proceedings of the 9th SIAM Conference on Parallel Processing for Scientific Computing*, San Antonio, Texas, 1999. 10 pp. on CDROM.

67. H. Avron, G. Shklarski, and S. Toledo. Parallel unsymmetric-pattern multifrontal sparse LU with column preordering, *ACM Transactions on Mathematical Software*, 34(2): 8:1–8:31, 2008.

68. J. R. Gilbert and L. Grigori. A note on the column elimination tree. *SIAM Journal on Matrix Analysis and Applications*, 25:143–151, 2003.

69. J. R. Gilbert, X. S. Li, E. G. Ng, and B. W. Peyton. Computing row and column counts for sparse QR and LU factorization. *BIT*, 41:693–710, 2001.

70. X. S. Li and J. W. Demmel. SuperLU_DIST: A scalable distributed memory sparse direct solver for unsymmetric linear systems. *ACM Transactions on Mathematical Software*, 29:110–140, 2003.

71. J. R. Gilbert and T. Peierls. Sparse partial pivoting in time proportional to arithmetic operations. *SIAM Journal on Scientific and Statistical Computing*, 9:862–874, 1988.

72. S. C. Eisenstat, J. W. H. Liu. Exploiting structural symmetry in unsymmetric sparse symbolic factorization. *SIAM Journal on Matrix Analysis and Applications*, 13:202–211, January 1992.

73. J. R. Gilbert and J. W. H. Liu. Elimination structures for unsymmetric sparse LU factors. *SIAM Journal on Matrix Analysis and Applications*, 14:334–352, 1993.

74. A. Gupta. Improved symbolic and numerical factorization algorithms for unsymmetric sparse matrices. *SIAM Journal on Matrix Analysis and Applications*, 24:529–552, 2002.

75. T. A. Davis and I. S. Duff. An unsymmetric-pattern multifrontal method for sparse LU factorization. *SIAM Journal on Matrix Analysis and Applications*, 18:140–158, 1997.

Data Structures for Databases[*]

62.1	Overview of the Functionality of a Database Management System	967
62.2	Data Structures for Query Processing	968
	Index Structures • Sorting Large Data Sets • The Parse Tree • Expression Trees • Histograms	
62.3	Data Structures for Buffer Management	974
62.4	Data Structures for Disk Space Management	976
	Record Organizations • Page Organizations • File Organization	
62.5	Conclusion	981
	References	981

Joachim Hammer[†]
University of Florida

Markus Schneider
University of Florida

62.1 Overview of the Functionality of a Database Management System

Many of the previous chapters have shown that efficient strategies for complex data-structuring problems are essential in the design of fast algorithms for a variety of applications, including combinatorial optimization, information retrieval and Web search, databases and data mining, and geometric applications. The goal of this chapter is to provide the reader with an overview of the important data structures that are used in the implementation of a modern, general-purpose database management system (DBMS). In earlier chapters of the book the reader has already been exposed to many of the data structures employed in a DBMS context (e.g., B-trees, buffer trees, quad trees, R-trees, interval trees, hashing). Hence, we will focus mainly on their application but also introduce other important data structures to solve some of the fundamental data management problems such as *query processing and optimization, efficient representation of data on disk*, as well as the *transfer of data from main memory to disk*. Due to space constraints, we cannot cover applications of data structures to manage non-standard data such as multi-dimensional data, spatial and temporal data, multimedia data, or XML.

Before we begin our treatment of how data structures are used in a DBMS, we briefly review the basic architecture, its components, and their functionality. Unless otherwise noted, our discussion applies to a class of DBMSs that are based on the relational data model. These so-called relational database management systems make up the majority of systems in use today and are offered by all major vendors including IBM, Microsoft, Oracle, and Sybase. Most of the components described here can also be found in DBMSs based on other models such as the object-based model or XML.

Figure 62.1 depicts a conceptual overview of the main components that make up a DBMS. Rectangles represent system components, double-sided arrows represent input and output, and the solid connectors indicate data as well as process flow between two components. Please note that the inner workings of a DBMS are quite complex and we are not attempting to provide a detailed discussion of its implementation. For an in-depth treatment the reader should refer to one of the many excellent database textbooks books, for example [1–6].

Starting from the top, users interact with the DBMS via commands generated from a variety of user interfaces or application programs. These commands can either retrieve or update the data that is managed by the DBMS or create or update the underlying metadata that describes the schema of the data. The former are called queries, the latter are called data definition statements. Both types of commands are processed by the *Query Evaluation Engine* which contains components for parsing the input, producing an execution plan, and executing the plan against the underlying database. In the case of queries, the parsed command is presented to a query optimizer component, which uses information about how the data is stored to produce an efficient execution plan from the possibly many alternatives. We discuss data structures that represent parsed queries, execution plans, and statistics about a database, including the data structures that are used by an external sorting algorithm in Section 62.2 when we focus on the query evaluation engine.

---

[*] This chapter has been reprinted from first edition of this Handbook, without any content updates.

[†] We have used this author's affiliation from the first edition of this handbook, but note that this may have changed since then.

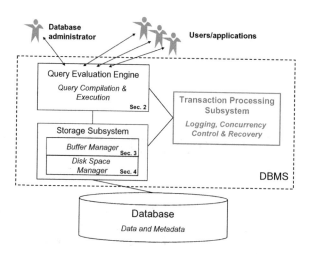

FIGURE 62.1   A simplified architecture of a database management system (DBMS).

Since databases are normally too large to fit into the main memory of a computer, the data of a database resides in secondary memory, generally on one or more magnetic disks. However, to execute queries or modifications on data, that data must first be transferred to main memory for processing and then back to disk for persistent storage. It is the job of the *Storage Subsystem* to accomplish a sophisticated placement of data on disk, to assure an efficient localization of these persistent data, to enable their bidirectional transfer between disk and main memory, and to allow direct access to these data from other DBMS subsystems. The storage subsystem consists of two components: The *Disk Space Manager* is responsible for storing physical data items on disk, managing free regions of the disk space, hiding device properties from higher architecture levels, mapping physical blocks to tracks and sectors of a disc, and controlling the transfer of data items between external and main memory. The *Buffer Manager* organizes an assigned, limited main memory area called *buffer* and may comprise several smaller buffers (buffer pool). Other subsystems may have direct access to data items in these buffers.

In Sections 62.3 and 62.4, we discuss data structures that are used to represent both data in memory as well as on disk such as fixed and variable-length records, large binary objects (LOBs), heap, sorted, and clustered files, as well as different types of index structures. Given the fact that a database management system must manage data that is both resident in main memory as well as on disk, one has to deal with the reality that the most appropriate data structure for data stored on disk is different from the data structures used for algorithms that run in main memory. Thus when implementing the storage manager, one has to pay careful attention to select not only the appropriate data structures but also to map the data between them in an efficient manner.

In addition to the above two subsystems, today's modern DBMSs include a *Transaction Management Subsystem* to support concurrent execution of queries against the database and recovery from failure. Although transaction processing is an important and complex topic, it is less interesting for our investigation of data structures and is mentioned here only for completeness.

The rest of this chapter is organized as follows. Section 62.2 describes important data structures used during query evaluation. Data structures used for buffer management are described in Section 62.3, and data structures used by the disk space manager are described in Section 62.4. Section 62.5 concludes the chapter.

62.2 Data Structures for Query Processing

Query evaluation is performed in main memory in several steps as outlined in Figure 62.2. Starting with the high-level input query expressed in a declarative language called SQL (see, e.g., [7]) the *Parser* scans, parses, and validates the query. The goal is to check whether the query is formulated according to the syntax rules of the language supported in the DBMS. The parser also validates that all attribute and relation names are part of the database schema that is being queried.

The parser produces a *parse tree* which serves as input to the *Query Translation and Rewrite* module shown underneath the parser. Here the query is translated into an internal representation, which is based on the relational algebra notation [8]. Besides its compact form, a major advantage of using relational algebra is that there exist transformations (re-write rules) between equivalent expressions to explore alternate, more efficient forms of the same query. Different algebraic expressions for a query are called *logical query plans* and are represented as *expression trees* or *operator trees*. Using the re-write rules, the initial logical query plan is transformed into an equivalent plan that is expected to execute faster. Query re-writing is guided by heuristics which help reduce the amount of intermediary work that must be performed by the query in order to arrive at the same result.

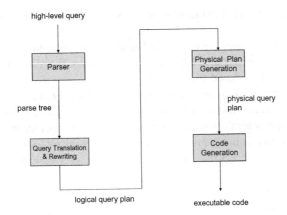

FIGURE 62.2   Outline of query evaluation.

A particularly challenging problem is the selection of the best join ordering for queries involving the join of three or more relations. The reason is that the *order* in which the input relations are presented to a join operator (or any other binary operator for that matter) tends to have an important impact on the cost of the operation. Unfortunately, the number of candidate plans grows rapidly when the number of input relations grows.

The outcome of the query translation and rewrite module is a set of "improved" logical query plans representing different execution orders or combinations of operators of the original query. The *Physical Plan Generator* converts the logical query plans into *physical query plans* which contain information about the algorithms to be used in computing the relational operators represented in the plan. In addition, physical query plans also contain information about the access methods available for each relation. Access methods are ways of retrieving tuples from a table and consist of either a file scan (i.e., a complete retrieval of all tuples) or an index plus a matching selection condition. Given the many different options for implementing relational operators and for accessing the data, each logical plan may lead to a large number of possible physical plans. Among the many possible plans, the physical plan generator evaluates the cost for each and chooses the one with the lowest overall cost.

Finally, the best physical plan is submitted to the *Code Generator* which produces the executable code that is either executed directly or is stored and executed later whenever needed. Query re-writing and physical plan generation are referred to as *query optimization*. However, the term is misleading since in most cases the chosen plan represents a reasonably efficient strategy for executing a query.

In the following paragraphs, we focus on several important data structures that are used during query evaluation, some of which have been mentioned above: The *parse tree* for storing the parsed and validated input query (Section 62.2.3), the *expression tree* for representing logical and physical query plans (Section 62.2.4), and the histogram which is used to approximate the distribution of attribute values in the input relations (Section 62.2.5). We start with a summary of the well-known *index structures* and how they are used *to speed up database operations*. Since sorting plays an important role in query processing, we include a separate description of the data structures used to *sort large data sets using external memory* (Section 62.2.2).

62.2.1 Index Structures

An important part of the work of the physical plan generator is to choose an efficient implementation for each of the operators in the query. For each relational operator (e.g., `selection`, `projection`, `join`) there are several alternative algorithms available for implementation. The best choice usually depends on factors such as size of the relation, available memory in the buffer pool, sort order of the input data, and availability of index structures. In the following, we briefly highlight some of the important index structures that are used by a modern DBMS.

62.2.1.1 One-dimensional Indexes

One-dimensional indexes contain a single search key, which may be composed of multiple attributes. The most frequently used data structures for one-dimensional database indexes are dynamic tree-structured indexes such as B/B^+-*Trees* (from now on collectively referred to as B-Trees, see also Chapter 16) and *hash-based indexes* using extendible and linear hashing (see Chapter 9). In general, hash-based indexes are especially good for equality searches. For example, in the case of an equality selection operation, one can use a one-dimensional hash-based index structure to examine just the tuples that satisfy the given condition. Consider the selection of student records having a certain grade point average (GPA). Assuming students are randomly distributed throughout the relation, an index on the GPA value could lead us to only those records satisfying the selection condition and resulting in a lot fewer data transfers than a sequential scan of the relation (if we assume the tuples satisfying the condition make up only a fraction of the entire relation).

Given their superior performance for equality searches hash-based indexes prove to be particularly useful in implementing relational operations such as joins. For example, the index-nested-loop join algorithm generates many equality selection queries, making the difference in cost between a hash-based and the slightly more expensive tree-based implementation significant.

B-Trees provide efficient support for range searches (all data items that fall within a range of values) and are almost as good as hash-based indexes for equality searches. Besides their excellent performance, B-Trees are "self-tuning," meaning they maintain as many levels of the index as is appropriate for the size of the relation being indexed. Unlike hash-based indexes, B-Trees manage the space on the disk blocks they use and do not require any overflow blocks. As we have mentioned in Section 62.1, database index structures are an example of data structures that have been designed as secondary memory structures.

62.2.1.2 Multi-dimensional Indexes

In addition to these one-dimensional index structures, many applications (e.g., geographic database, inventory and sales database for decision-support) also require data structures capable of indexing data existing in two or higher-dimensional spaces. In these domains, important database operations are selections involving partial matches (all points that match specified values in one or more dimensions), range queries (all points that fall within a range of values in one or more dimensions), nearest-neighbor queries (closest point to a given point), and so-called "where-am-I" queries (all the regions in which a given point is located).

The following are some of the most important data structures that support these types of operations.

Grid file. A multi-dimensional extension of one-dimensional hash tables. Grid files support range queries, partial-match queries, and nearest-neighbor queries well, as long as data is uniformly distributed.

Multiple-key index. The index on one attribute leads to indexes on another attribute for each value of the first. Multiple-key indexes are useful for range and nearest-neighbor queries.

R-tree. A B-Tree generalization suitable for collections of regions. R-Trees are used to represent a collection of regions by grouping them into a hierarchy of larger regions. They are well suited to support "where-am-I" queries as well as the other types of queries mentioned above if the atomic regions are individual points. (See also Chapter 22.)

Quad tree. Recursively divide a multi-dimensional data set into quadrants until each quadrant contains a minimal number of points (e.g., amount of data that can fit on a disk block). Quad trees support partial-match, range, and nearest-neighbor queries well. (See also Chapter 20)

Bitmap index. A collection of bit vectors which encode the location of records with a given value in a given field. Bitmap indexes support range, nearest-neighbor, and partial-match queries and are often employed in data warehouses and decision-support systems. Since bitmap indexes tend to get large when the underlying attributes have many values, they are often compressed using a run-length encoding.

Given the importance of database support for non-standard applications, many relational database management systems support one or more of these multi-dimensional indexes, either directly (e.g., bitmap indexes), or as part of a special-purpose extension to the core database engine (e.g., R-trees in a spatial extender).

In general, indexes are also used to answer certain types of queries without having to access the data file. For example, if we need only a few attribute values from each tuple and there is an index whose search key contains all these fields, we can choose an index scan instead of examining all data tuples. This is faster since index records are smaller (and hence fit into fewer buffer pages). Note that an index scan does not make use of the search structure of the index: for example, in a B-Tree index one would examine all leaf pages in sequence. All commercial relational database management systems support B-Trees and at least one type of hash-based index structure.

62.2.2 Sorting Large Data Sets

The need to sort large data sets arises frequently in data management. Besides outputting the result of a query in sorted order, sorting is useful for eliminating duplicate data items during the processing of queries. In addition, an efficient algorithm for performing a join operation (sort-merge join) requires the input relations to be sorted. Since the size of databases routinely exceeds the amount of available main memory, most DBMSs use an external sorting technique called *merge sort*, which is based on the main-memory version with the same name. The idea behind merge sort is that a file which does not fit into main memory can be sorted by breaking it into smaller pieces (runs), sorting the smaller runs individually, and then merging them to produce a single run that contains the original data items in sorted order. External merge sort is another example where main memory versions of algorithms and data structures need to be changed to accommodate a computing environment where all data resides on secondary and perhaps even tertiary storage. We will point out more such examples in Section 62.4 when we describe the disk space manager.

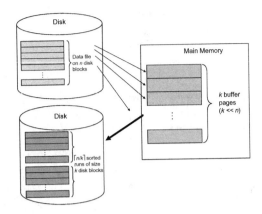

FIGURE 62.3   Arrangement of buffer pages and disk blocks during the run generation phase.

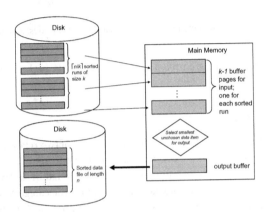

FIGURE 62.4   Arrangement of buffer pages and disk blocks during the one pass, multi-way merging phase.

During the first phase, also called the run-generation phase, merge-sort fills the available buffer pages in main memory with blocks containing the data records from a file stored on disk. We will have more to say about the management of buffer pages when we discuss data structures for buffer management in Section 62.3. Sorting is done using any of the main-memory algorithms (e.g., Heapsort, Quicksort). The sorted records are written back to new blocks on disk, forming a sorted run containing as many blocks as there are available buffer pages in main memory. This process is repeated until all records in the data file are in one of the sorted runs. The arrangement of buffer pages and disk blocks during run generation is depicted in Figure 62.3.

In the second phase, also called the merging phase, all but one of the main memory buffers are used to hold input data from one of the sorted runs. In most instances, the number of sorted runs is less than the number of buffer pages and the merging can be done in one pass. Note, this so-called *multi-way merging* is different from the main-memory version of merge sort which merges pairs of runs (two-way merge). The arrangement of buffers and disk blocks to complete this one-pass multi-way merging step is shown in Figure 62.4. Note that the two-way merge strategy results in reading data in and out of memory $2 * \log_2(n)$ times for n runs (versus reading all n runs only once for the n-way strategy).

In situations when the number of sorted runs exceeds the available buffer pages in main memory, the merging step must be performed in *several passes* as follows: assuming k buffer pages in main memory, each pass involves the repeated merging of $k - 1$ runs until all runs have been merged. At this point the number of runs has been reduced by a factor of $k - 1$. If the reduced number of sublists is still greater than k, the merging is repeated until the number of sublists is less than k. A final merge generates the sorted output. In this scenario, the number of merge passes required is $\lceil \log_{k-1}(n/k) \rceil$.

62.2.3 The Parse Tree

A *parse tree* is an m-ary tree data structure that represents the structure of a query. Each interior node of the tree is labeled with a non-terminal symbol from the grammar of the query language. The root node is labeled with the goal symbol. The query being parsed appears at the bottom with each token of the query being a leaf in the tree. In the case of SQL, leaf nodes are lexical elements such as keywords of the language (e.g., *SELECT*), names of attributes or relations, operators, and other schema elements.

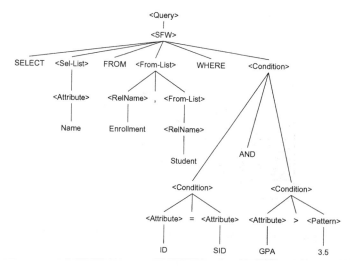

FIGURE 62.5 Sample parse tree for an SQL query showing query goal, interior and leaf nodes. Adapted from Garcia-Molina et al. [2].

The parse tree for the SQL query selecting all the enrolled students with a GPA higher than 3.5.

```
SELECT Name
FROM Enrollment, Student
WHERE ID = SID AND GPA > 3.5;
```

is shown in Figure 62.5. For this example, we are tacitly assuming the existence of two relations called `Enrollment` and `Student` which store information about enrollment records for students in a school or university.

The parse tree shown in Figure 62.5 is based on the grammar for SQL as defined in [2] (which is a subset of the full grammar for SQL). Non-terminal symbols are enclosed in angled brackets. At the root is the category $<Query>$ which forms the goal for the parsed query. Descending down the tree, we see that this query is of the form *SFW* (select-from-where). In case one of the relations in the *FROM* clause is a view, it must be replaced by its own parse tree since a view is essentially a query. A parse tree is said to be valid if it conforms to the syntactic rules of the grammar as well as the semantic rules on the use of the schema names.

62.2.4 Expression Trees

An *expression tree* is a binary tree data structure that represents a logical query plan for a query after it has been translated into a relational algebra expression. The expression tree represents the input relations of the query as *leaf nodes* of the tree, and the relational algebra operations together with estimates for result sizes as *internal nodes*.

Figure 62.6 shows an example of three expression trees representing different logical query plans for the following SQL query, which selects all the students enrolled in the course "COP 4720" during the term "Sp04" who have a grade point average of 3.5 (result sizes are omitted in the figure):

```
SELECT Name FROM Enrollment, Student
WHERE Enrollment.ID = Student.SID AND Enrollment.Course = 'COP 4720'
    AND Enrollment.TermCode = 'Sp04' AND Student.GPA = 3.5;
```

An execution of a tree consists of executing an internal node operation whenever its operands are available and then replacing that internal node by the relation that results from executing the operation. The execution terminates when the root node is executed and produces the result relation for the query.

Note that expression trees representing *physical* query plans differ in the information that is stored in the nodes. For example, internal nodes contain information such as the operation being performed, any parameters if necessary, general strategy about the algorithm that is used, whether materialization of intermediate results or pipelining is used, and the anticipated number of buffers the operation will require (rather than result size as in logical query plans). At the leaf nodes table names are replaced by scan operators such as `TableScan`, `SortScan`, `IndexScan`, etc.

There is an interesting twist to the types of expression trees that are actually considered by the query optimizer. As we have previously pointed out, the number of different query plans (both logical and physical) for a given query can be very large. This is even more so the case, when the query involves the join of two or more relations since we need to take the join order into account when choosing the best possible plan. Today's query optimizers prune a large portion of the candidate expression trees

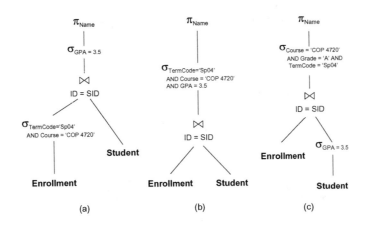

FIGURE 62.6 Three expression trees representing different logical query plans for the same query.

and concentrate only on the class of *left-deep trees*. A left-deep tree is an expression tree in which the right child of each join is a leaf (i.e., a base table). For example, in Figure 62.6, the tree labeled (a) is an example of a left-deep tree. The tree labeled (b) is an example of a *nonlinear* or bushy tree (a join node may have no leaf nodes), tree (c) is an example of a *right-deep tree* (the left child of each join node is a base table).

Besides the fact that the number of left-deep trees is smaller than the number of all trees, there is another advantage for considering only left-deep expression trees: Left-deep trees allow the query optimizer to generate more efficient plans by avoiding the intermediate storage (materialization) of join results. Note that in most join algorithms, the inner table must be materialized because we must examine the entire inner table for each tuple of the outer table. However, in a left-deep tree, all inner tables exist as base tables and are therefore already materialized. IBM DB2, Informix, Microsoft SQL Server, Oracle 8, and Sybase ASE all search for left-deep trees using a dynamic programming approach [5].

62.2.5 Histograms

Whether choosing a logical query plan or constructing a physical query plan from a logical query plan, the query evaluation engine needs to have information about the expected cost of evaluating the expressions in a query. Cost estimation is based on statistics about the database which include number of tuples in a relation, number of disk blocks used, distribution of values for each attribute, etc. Frequent computation of statistics, especially in light of many changes to the database, lead to more accurate cost estimates. However, the drawback is increased overhead since counting tuples and values is expensive.

An important data structure for cost estimation is the *histogram*, which is used by the DBMS to *approximate* the distribution of values for a given attribute. Note that in all but the smallest databases, counting the exact occurrence of values is usually not an option. Having access to accurate distributions is essential in determining how many tuples satisfy a certain selection predicate, for example, how many students there are with a GPA value of 3.5. This is especially important in the case of joins, which are among the most expensive operations. For example, if a value of the join attribute appears in the histograms for both relations, we can determine exactly how many tuples of the result will have this value.

Using a histogram, the data distribution is approximated by dividing the range of values, for example, GPA values, into subranges or *buckets*. Each bucket contains the number of tuples in the relation with GPA values within that bucket. Histograms are more accurate than assuming a uniform distribution across all values.

Depending on how one divides the range of values into the buckets, we distinguish between *equiwidth* and *equidepth* histograms [5]. In equiwidth histograms, the value range is divided into buckets of equal size. In equidepth histograms, the value range is divided so that the number of tuples in each bucket is the same (usually within a small delta). In both cases, each bucket contains the average frequency. When the number of buckets gets large, histograms can be compressed, for example, by combining buckets with similar distributions.

Consider the Students-Enrollments scenario from above. Figure 62.7 depicts two sample histograms for attribute GPA in relation Student. Values along the horizontal axis denote GPA, the vertical bars indicate the number of students that fall in each range. For this example we are assuming that GPA values are rounded to one decimal and that there are 50 students total. Histogram a) is an equiwidth histogram with bucket size = 2. Histogram b) is an equidepth histogram containing between 7 and 10 students per bucket.

Consider the selection *GPA = 3.5*. Using the equidepth histogram, we are led to bucket 3, which contains only the GPA value 3.5 and we arrive at the correct answer, 10 (vs. 1/2 of 12 = 6 in the equiwidth histogram). In general, equidepth histograms

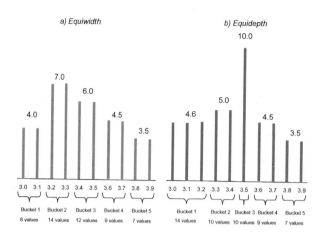

FIGURE 62.7 Two sample histograms approximating the distribution of GPA values in relation `Student`. Adapted from Ramakrishnan and Gehrke [5].

provide better estimates than equiwidth histograms. This is due to the fact that buckets with very frequently occurring values contain fewer values. Thus the uniform distribution assumption is applied to a smaller range of values, leading to a more accurate estimate. The converse is true for buckets containing infrequent values, which are better approximated by equiwidth histograms. However, in query optimization, good estimation for frequent values are more important.

Histograms are used by the query optimizers of all of the major DBMS vendors. For example, Sybase ASE, IBM DB2, Informix, and Oracle all use one-dimensional, equidepth histograms. Microsoft's SQL Server uses one-dimensional *equiarea histograms* (a combination of equiwidth and equidepth) [5].

62.3 Data Structures for Buffer Management

A *buffer* is partitioned into an array of *frames* each of which can keep a *page*. Usually a page of a buffer is mapped to a *block** of a file so that reading and writing of a page only require one disk access each. Application programs and queries make requests on the buffer manager when they need a block from disk, that contains a data item of interest. If the block is already in the buffer, the buffer manager conveys the address of the block in main memory to the requester. If the block is not in main memory, the buffer manager first allocates space in the buffer for the block, throwing out some other block if necessary, to make space for the new block. The displaced block is written back to disk if it was modified since the most recent time that it was written to the disk. Then, the buffer manager reads in the requested block from disk into the free frame of the buffer and passes the page address in main memory to the requester. A major goal of buffer management is to minimize the number of block transfers between the disk and the buffer.

Besides pages, so-called *segments* are provided as a counterpart of files in main memory. This allows one to define different segment types with additional attributes, which support varying requirements concerning data processing. A segment is organized as a contiguous subarea of the buffer in a virtual, linear address space with visible page borders. Thus, it consists of an ordered sequence of pages. Data items are managed so that page borders are respected. If a data item is required, the address of the page in the buffer containing the item is returned.

An important question now is how segments are mapped to files. An appropriate mapping enables the storage system to preserve the merits of the file concept. The distribution of a segment over several files turns out to be unfavorable in the same way as the representation of a data item over several pages. Hence, a segment S_k is assigned to exactly one file F_j, and m segments can be stored in a file. Since block size and page size are the same, each page $P_{k_i} \in S_k$ is assigned to a block $B_{j_l} \in F_j$. We distinguish four methods of realizing this mapping.

The *direct page addressing* assumes an implicitly given mapping between the pages of a segment S_k and the blocks of a file F_j. The page P_{k_i} ($1 \le i \le s_k$) is stored in the block B_{j_l} ($1 \le l \le d_j$) so that $l = K_j - 1 + i$ and $d_j \ge K_j - 1 + s_k$ holds. K_j denotes the number of the first block reserved for S_k (Figure 62.8). Frequently, we have a restriction to a 1:1-mapping, that is, $K_j = 1$ and $s_k = d_j$ hold. Only in this case, a dynamic extension of segments is possible. A drawback is that at the time of the segment creation

---

* A block is a contiguous sequence of bytes and represents the unit used for both storage allocation and data transfer. It is usually a multiple of 512 Bytes and has a typical size of 1KB to 8KB. It may contain several data items. Usually, a data item does not span two or more blocks.

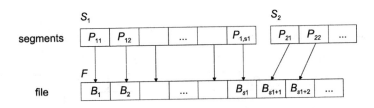

FIGURE 62.8   Direct page addressing.

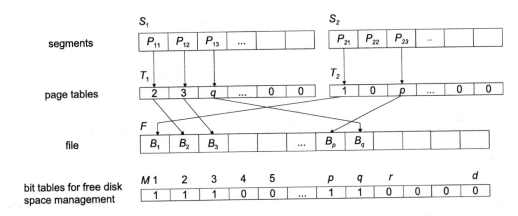

FIGURE 62.9   Indirect page addressing.

the assigned file area has to be allocated so that a block is occupied for each empty page. For segments whose data stock grows slowly, the fixed block allocation leads to a low storage utilization.

The *indirect page addressing* offers a much larger flexibility for the allocation of pages to blocks and, in addition, dynamic update and extension functionality (Figure 62.9). It requires two auxiliary data structures.

- Each segment S_k is associated with a *page table* T_k which for each page of the segment contains an entry indicating the block currently assigned to the page. Empty pages obtain a special null value in the page table.
- Each file F_j is associated with a *bit table* M_j which serves for free disk space management and quotes for each block whether currently a page is mapped to it or not. $M_j(l) = 1$ means that block B_{jl} is occupied; $M_j(l) = 0$ says that block B_{jl} is free. Hence, the bit table enables a dynamic assignment between pages and blocks.

Although this concept leads to an improved storage utilization, for large segments and files, the page tables and bit tables have to be split because of their size, transferred into main memory and managed in a special buffer. The provision of a page P_{k_i} that is not in the buffer can require two physical block accesses (and two enforced page removals), because, if necessary, the page table T_k has to be loaded first in order to find the current block address $j = T_k(i)$.

The two methods described so far assume that a modified page is written back to the block that has once been assigned to it (*update in place*). If an error occurs within a transaction, as a result of the direct placement of updates, the recovery manager must provide enough log information (*undo* information) to be able to restore the old state of a page. Since the writing of large volumes of log data leads to notable effort, it is often beneficial to perform updates in a page in a manner so that the old state of the page is available until the end of the transaction. The following two methods are based on an indirect update of changes and provide extensive support for recovery.

The *twin slot method* can be regarded as a modification of the direct page addressing. It causes very low costs for recovery but partially compensates this advantage through double disk space utilization. For a page P_{k_i} of a segment S_k, two physically consecutive blocks B_{jl-1} and B_{jl} of a file F_j with $l = K_j - 1 + 2 \cdot i$ are allocated. Alternately, at the beginning of a transaction, one of both block keeps the current state of the page whereas changes are written to the other block. In case of a page request, both blocks are read, and the block with the more recent state is provided as the current page in the buffer. The block with the older state then stores the changed page. By means of page locks, a transaction-oriented recovery concept can be realized without explicitly managing log data.

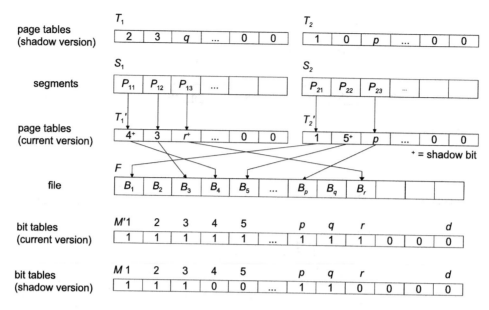

FIGURE 62.10   The shadow paging concept (segments S_1 and S_2 currently in process).

The *shadow paging concept* (Figure 62.10) represents an extension of the indirect page addressing method and also supports indirect updates of changes. Before the beginning of a new *save-interval* given by two *save-points*[*] the contents of all current pages of a segment are duplicated as so-called *shadow pages* and can thus be kept unchanged and consistent. This means, when a new save-point is created, all data structures belonging to the representation of a segment S_k (i.e., all occupied pages, the page table T_k, the bit table M) are stored as a consistent snapshot of the segment on disk. All modifications during a save-interval are performed on copies T_k' and M' of T_k and M. Changed pages are not written back to their original but to free blocks. At the creation of a new save-point, which must be an atomic operation, the tables T_k' and M' as well as all pages that belong to this state and have been changed are written back to disk. Further, all those blocks are released whose pages were subject to changes during the last save-interval. This just concerns those shadow pages for which a more recent version exists. At the beginning of the next save-interval the current contents of T_k' and M' has to be copied again to T_k and M. In case of an error within a save-interval, the DBMS can roll back to the previous consistent state represented by T_k and M.

As an example, Figure 62.10 shows several changes of pages in two segments S_1 and S_2. These changes are marked by so-called *shadow bits* in the page tables. Shadow bits are employed for the release of shadow pages at the creation time of new save-points. If a segment consists of *s* pages, the pertaining file must allocate *s* further blocks, because each changed page occupies two blocks within a save-interval.

The save-points orientate themselves to segments and not to transaction borders. Hence, in an error case, a *segment-oriented recovery* is executed. For a *transaction-oriented recovery* additional log data have to be collected.

62.4 Data Structures for Disk Space Management

Placing data items on disc is usually performed at different logical granularities. The most basic items found in relational or object-oriented database systems are the values of attributes. They consist of one or several bytes and are represented by *fields*. Fields, in turn, are put together in collections called *records*, which correspond to tuples or objects. Records need to be stored in physical *blocks* (see Section 62.3). A collection of records that forms a relation or the extent of a class is stored in some useful way as a collection of blocks, called a *file*.

62.4.1 Record Organizations

A collection of field names and their corresponding data types constitute a *record format* or *record type*. The data type of a field is usually one of the standard data types (e.g., *integer, float, bool, date, time*). If all records in a file have the same size in bytes,

---

[*] Transactions are usually considered as being atomic. But a limited concept of "subtransactions" allows one to establish intermediate *save-points* while the transaction is executing, and subsequently to roll back to a previously established save-point, if required, instead of having to roll back all the way to the beginning. Note that updates made at save-points are invisible to other transactions.

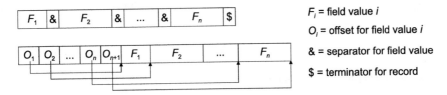

FIGURE 62.11 Organization of records with fields of fixed length.

FIGURE 62.12 Alternative organizations of records with fields of variable length.

we call them *fixed-length records*. The fields of a record all have a fixed length and are stored consecutively. If the *base address*, that is, the start position, of a record is given, the address of a specific field can be calculated as the sum of the lengths of the preceding fields. The sum assigned to each field is called the *offset* of this field. Record and field information are stored in the data dictionary. Figure 62.11 illustrates this record organization.

Fixed-length records are easy to manage and allow the use of efficient search methods. But this implies that all fields have a size so that all data items that potentially are to be stored may find space. This can lead to a waste of disk space and to more unfavorable access times.

If we assume that each record of a file has the same, fixed number of fields, a *variable-length record* can only be formed if some fields have a variable length. For example, a string representing the name of an employee can have a varying length in different records. Different data structures exist for implementing variable-length records. A first possible organization amounts to a consecutive sequence of fields which are interrupted by separators (such as ? or % or $). *Separators* are special symbols that do not occur in data items. A special *terminator* symbol indicates the end of the record. But this organization requires a pass (scan) of the record to be able to find a field of interest (Figure 62.12). Instead of separators, each field of variable length can also start with a counter that specifies the needed number of bytes of a field value.

Another alternative is that a header precedes the record. A *header* represents the "administrative" part of the record and can include information about integer offsets of the beginnings of the field values (Figure 62.12). The ith integer number is then the start address of the ith field value relatively to the beginning of the record. Also for the end of the record we must store an offset in order to know the end of the last field value. This alternative is usually the better one. Costs arise due to the header in terms of storage; the benefit is direct field access. Problems arise with changes. An update can let a field value grow which necessitates a "shift" of all consecutive fields. Besides, it can happen that a modified record does not fit any more on the page assigned to it and has to be moved to another page. If record identifiers contain a page number, on this page the new page number has to be left behind pointing to the new location of the record.

A further problem of variable-length records arises if such a record grows to such an extent that it does not fit on a page any more. For example, field values storing image data in various formats (e.g., GIF or JPEG), movies in formats such as MPEG, or spatial objects such as polygons can extend from the order of many kilobytes to the order of many megabytes or even gigabytes. Such truly large values for records or field values of records are called *large objects* (*lobs*) with the distinction of *binary large objects* (*blobs*) for large byte sequences and *character large objects!character* (*clobs*) for large strings.

Since, in general, lobs exceed page borders, only the non-lob fields are stored on the original page. Different data structures are conceivable for representing lobs. They all have in common that a lob is subdivided into a collection of linked pages. This organization is also called *spanned*, because records can span more than one page, in contrast to the *unspanned* organization where records are not allowed to cross page borders. The first alternative is to keep a pointer instead of the lob on the original page as attribute value. This pointer (also called *page reference*) points to the start of a linked page or block list keeping the lob (Figure 62.13a). Insertions, deletions, and modifications are simple but direct access to pages is impossible. The second alternative is to store a *lob directory* as attribute value (Figure 62.13b). Instead of a pointer, a directory is stored which includes the lob size, further administrative data, and a *page reference list* pointing to the single pages or blocks on a disk. The main benefit of this structure is the direct and sequential access to pages. The main drawback is the fixed and limited size of the lob directory and thus the lob. A lob directory can grow so much that it needs itself a lob for its storage.

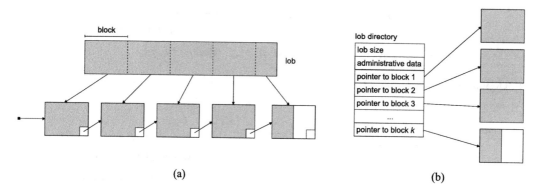

(a)   (b)

FIGURE 62.13   A lob as a linked list of pages (a), and the use of a lob directory (b).

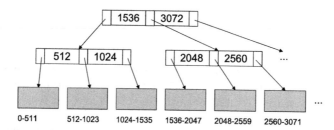

FIGURE 62.14   A lob managed by a positional B$^+$ - tree.

The third alternative is the usage of *positional B$^+$ - trees* (Figure 62.14). Such a B - tree variant stores relative byte positions in its inner nodes as separators. Its leaf nodes keep the actual data pages of the lob. The original page only stores as the field value a pointer to the root of the tree.

62.4.2 Page Organizations

Records are positioned on pages (or blocks). In order to reference a record, often a *pointer* to it suffices. Due to different requirements for storing records, the structure of pointers can vary. The most obvious pointer type is the *physical address* of a record on disk or in a virtual storage and can easily be used to compute the page to be read. The main advantage is a direct access to the searched record. But it is impossible to move a record within a page, because this requires the locating and changing of all pointers to this record. We call these pointers *physical pointers*. Due to this drawback, a pointer is often described as a pair (p, n) where p is the number of the page where the record can be found and where n is a number indicating the location of the record on the page. The parameter n can be interpreted differently, for example, as a relative byte address on a page, as a number of a slot, or as an index of a directory in the *page header*. The entry at this index position yields the relative position of the record on the page. All pointers (s, p) remain unchanged and are named *page - related pointers*. Pointers that are completely stable against movements in main memory can be achieved if a record is associated with a *logical address* that reveals nothing about its storage. The record can be moved freely in a file without changing any pointers. This can be realized by *indirect addressing*. If a record is moved, only the respective entry in a *translation table* has to be changed. All pointers remain unchanged, and we call them *logical pointers*. The main drawback is that each access to a record needs an additional access to the translation table. Further, the table can become so large that it does not fit completely in main memory.

A page can be considered as a collection of *slots*. Each slot can capture exactly one record. If all records have the same length, all slots have the same size and can be allocated consecutively on the page. Hence, a page contains so many records as slots fit on a page plus page information like directories and pointers to other pages. A first alternative for arranging a set of N fixed - length records is to place them in the first N slots (see Figure 62.15). If a record is deleted in slot $i<N$, the last record on the page in slot N is moved to the free slot i. However, this causes problems if the record to be moved is pinned[*] and the slot number has to be changed. Hence, this "packed" organization is problematic, although it allows one to easily compute the location of the ith record. A second alternative is to manage deletions of records on each page and thus information about free slots by means of a directory represented as a bitmap. The retrieval of the ith record as well as finding the next free slot on a page require a traversal

---

[*] If pointers of unknown origin reference a record, we call the record *pinned*, otherwise *unpinned*.

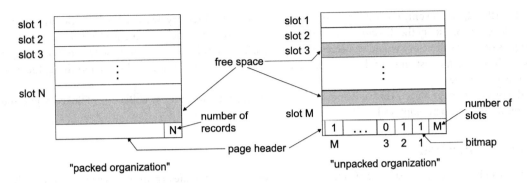

FIGURE 62.15 Alternative page organizations for fixed-length records.

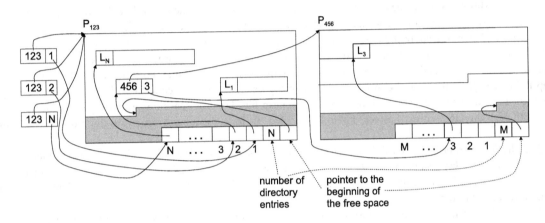

FIGURE 62.16 Page organization for variable-length records.

of the directory. The search for the next free slot can be sped up if an additional, special field stores a pointer on the first slot whose deletion flag is set. The slot itself then contains a pointer to the next free slot so that a chaining of free slots is achieved.

Also variable-length records can be positioned consecutively on a page. But deletions of records must be handled differently now because slots cannot be reused in a simple manner any more. If a new record is to be inserted, first a free slot of "the right size" has to be found. If the slot is too large, storage is wasted. If it is too small, it cannot be used. In any case, unused storage areas (*fragmentation*) at the end of slots can only be avoided if records on a page are moved and condensed. This leads to a connected, free storage area. If the records of a page are unpinned, the "packed" representation for fixed-length records can be adapted. Either a special terminator symbol marks the end of the record, or a field at the beginning of the record keeps its length. In the general case, indirect addressing is needed which permits record movements without negative effects and without further access costs. The most flexible organization of variable-length records is provided by the *tuple identifier* (*TID*) *concept* (Figure 62.16). Each record is assigned a unique, stable pointer consisting of a page number and an index into a page-internal directory. The entry at index i contains the relative position, that is, the offset, of slot i and hence a pointer to record i on the page. The length information of a record is stored either in the directory entry or at the beginning of the slot (L_i in Figure 62.16). Records which grow or shrink can be moved on the page without being forced to modify their TIDs. If a record is deleted, this is registered in the corresponding directory entry by means of a deletion flag.

Since a page cannot be subdivided into predefined slots, some kind of free disk space management is needed on each page. A pointer to the beginning of the free storage space on the page can be kept in the page header. If a record does not fit into the currently available free disk space, the page is compressed (i.e., defragmented) and all records are placed consecutively without gaps. The effect is that the maximally available free space is obtained and is located after the record representations.

If, despite defragmentation, a record does still not fit into the available free space, the record must be moved from its "home page" to an "overflow page". The respective TID can be kept stable by storing a "proxy TID" instead of the record on the home page. This proxy TID points to the record having been moved to the overflow page. An overflow record is not allowed to be moved to another, second overflow page. If an overflow record has to leave its overflow page, its placement on the home page is attempted. If this fails due to a lack of space, a new overflow page is determined and the overflow pointer is placed on the home page. This procedure assures that each record can be retrieved with a maximum of two page accesses.

If a record is deleted, we can only replace the corresponding entry in the directory by a deletion flag. But we cannot compress the directory since the indexes of the directory are used to identify records. If we deleted an entry and compress, the indexes of the subsequent slots in the directory would be decremented so that TIDs would point to wrong slots and thus wrong records. If a new record is inserted, the first entry of the directory containing a deletion flag is selected for determining the new TID and pointing to the new record.

If a record represents a large object, that is, it does not fit on a single page but requires a collection of linked pages, the different data structures for blobs can be employed.

62.4.3 File Organization

A *file* (*segment*) can be viewed as a sequence of *blocks* (*pages*). Four fundamental file organizations can be distinguished, namely files of unordered records (heap files), files of ordered records (sorted files), files with dispersed records (hash files), and tree-based files (index structures).

Heap files are the simplest file organization. Records are inserted and stored in their unordered, chronological sequence. For each heap file we have to manage their assigned pages (blocks) to support scans as well as the pages containing free space to perform insertions efficiently. Doubly-linked lists of pages or directories of pages using both page numbers for page addressing are possible alternatives. For the first alternative, the DBMS uses a *header page* which is the first page of a heap file, contains the address of the first data page, and information about available free space on the pages. For the second alternative, the DBMS must keep the first page of the heap file in mind. The directory itself represents a collection of pages and can be organized as a linked list. Each directory entry points to a page of the heap file. The free space on each page is recorded by a counter associated with each directory entry. If a record is to be inserted, its length can be compared to the number of free bytes on a page.

Sorted files physically order their records based on the values of one (or several) of their fields, called the *ordering field(s)*. If the ordering field is also a *key field* of the file, that is, a field guaranteed to have a unique value in each record, then the field is called the *ordering key* for the file. If all records have the same fixed length, binary search on the ordering key can be employed resulting in faster access to records.

Hash files are a file organization based on hashing and representing an important indexing technique. They provide very fast access to records on certain search conditions. Internal hashing techniques have been discussed in different chapters of this book; here we are dealing with their external variants and will only explain their essential features. The fundamental idea of hash files is the distribution of the records of a file into so-called *buckets*, which are organized as heaps. The distribution is performed depending on the value of the *search key*. The direct assignment of a record to a bucket is computed by a *hash function*. Each bucket consists of one or several pages of records. A *bucket directory* is used for the management of the buckets, which is an array of pointers. The entry for index i points to the first page of bucket i. All pages for bucket i are organized as a linked list. If a record has to be inserted into a bucket, this is usually done on its last page since only there space can be found. Hence, a pointer to the last page of a bucket is used to accelerate the access to this page and to avoid traversing all the pages of the bucket. If there is no space left on the last page, overflow pages are provided. This is called a *static hash file*. Unfortunately, this strategy can cause long chains of overflow pages. *Dynamic hash files* deal with this problem by allowing a variable number of buckets. *Extensible hash files* employ a directory structure in order to support insertion and deletion efficiently without the employment of overflow pages. *Linear hash files* apply an intelligent strategy to create new buckets. Insertion and deletion are efficiently realized without using a directory structure.

Index structures are a fundamental and predominantly tree-based file organization based on the search key property of values and aiming at speeding up the access to records. They have a paramount importance in query processing. Many examples of index structures are already described in detail in this handbook, for example, B-trees and variants, quad-trees and octtrees, R-trees and variants, and other multidimensional data structures. We will not discuss them further here. Instead, we mention some basic and general organization forms for index structures that can also be combined. An index structure is called a *primary organization* if it contains search key information together with an embedding of the respective records; it is named a *secondary organization* if it includes besides search key information only TIDs or TID lists to records in separate file structures (e.g., heap files or sorted files). An index is called a *dense index* if it contains (at least) one index entry for each search key value which is part of a record of the indexed file; it is named a *sparse index* (Figure 62.17) if it only contains an entry for each page of records of the indexed file. An index is called a *clustered index* (Figure 62.17) if the logical order of records is equal or almost equal to their physical order, that is, records belonging logically together are physically stored on neighbored pages. Otherwise, the index is named *non-clustered*. An index is called a *one-dimensional index* if a linear order is defined on the set of search key values used for organizing the index entries. Such an order cannot be imposed on a *multi-dimensional index* where the organization of index entries is based on spatial relationships. An index is called a *single-level index* if the index only consists of a single file; otherwise, if the index is composed of several files, it is named a *multi-level index*.

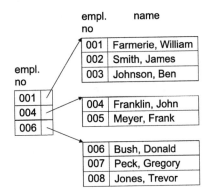

FIGURE 62.17 Example of a clustered, sparse index as a secondary organization on a sorted file.

62.5 Conclusion

A modern database management system is a complex software system that leverages many sophisticated algorithms, for example, to evaluate relational operations, to provide efficient access to data, to manage the buffer pool, and to move data between disk and main memory. In this chapter, we have shown how many of the data structures that were introduced in earlier parts of this book (e.g., B-trees, buffer trees, quad trees, R-trees, interval trees, hashing) including a few new ones such as histograms, LOBs, and disk pages, are being used in a real-world application. However, as we have noted in the introduction, our coverage of the data structures that are part of a DBMS is not meant to be exhaustive since a complete treatment would have easily exceeded the scope of this chapter. Furthermore, as the functionality of a DBMS must continuously grow in order to support new applications (e.g., GIS, federated databases, data mining), so does the set of data structures that must be designed to efficiently manage the underlying data (e.g., spatio-temporal data, XML, bio-medical data). Many of these new data structure challenges are being actively studied in the database research communities today and are likely to form a basis for tomorrow's systems.

References

1. Ramez Elmasri and Shamkant B. Navathe. *Fundamentals of Database Systems*. Addison-Wesley, fourth edition, 2003.
2. Hector Garcia-Molina, Jeffrey D. Ullman, and Jennifer Widom. *Database Systems - The Complete Book*. Prentice Hall, Upper Saddle River, New Jersey, first edition, 2002.
3. Philip M. Lewis, Arthur Bernstein, and Michael Kifer. *Databases and Transaction Processing*. Addison Wesley, first edition, 2002.
4. Patrick O'Neil and Elizabeth O'Neil: *Database: Principles, Programming, and Performance*. Morgan Kaufmann, second edition, 2000.
5. Raghu Ramakrishnan and Johannes Gehrke. *Database Management Systems*. McGraw-Hill, third edition, 2003.
6. Abraham Silberschatz, Henry F. Korth, and S. Sudharshan. *Database System Concepts*. McGraw-Hill, fourth edition, 2002.
7. Chris J. Date and Hugh Darwen. *A Guide to The SQL Standard*. Addison-Wesley Publishing Company, Inc., third edition, 1997.
8. E. Codd. A relational model of data for large shared data banks. *Communications of the ACM*, 13(6):377–387, 1970.

63

Data Structures for Big Data Stores

63.1 Introduction... 983
63.2 Data Models... 984
63.3 Partitioning.. 984
    Consistent Hashing
63.4 Replication and Consistency.. 985
    Merkle Tree
63.5 Persistence.. 987
    Performance Metrics • B+ Trees • Log Structured Merge Trees • Intuition behind LSM •
    Performance Optimizations
63.6 Concurrency... 992
63.7 Conclusion... 993
References.. 993

Arun A. Ravindran
University of North Carolina at Charlotte

Dinesh P. Mehta
Colorado School of Mines

63.1 Introduction

The emergence of large-scale web services, smartphones, social media, and high throughput scientific instruments within the last decade or so has resulted in a deluge of data. For example, Facebook has an incoming daily data rate of 600 TB with data warehouses storing upwards of 300 PB of data [1]. The amount of data generated by DNA sequencing centers is many terabytes every day with the amount doubling every 7 months [2]. The data problem is expected to worsen with the roll out of the Internet-of-Things (IoT) [3]. A key feature of IoT is equipping physical infrastructure with sensing and information transmission capabilities with the goal of realizing intelligent transportation, energy management, environmental monitoring, manufacturing, and health care systems.

In addition to the sheer *volume* of data generated by current and emerging applications, big-data is characterized by *velocity* and *variety*. Velocity pertains to the high data throughput and often real-time data processing requirements. Variety refers to the different types of data that need to be aggregated for storage, querying, and processing. File systems and databases are two ways of persisting the data on storage devices. Unlike file systems, databases allow for efficient querying of the data while supporting concurrent access and often providing ACID semantics: (1) *Atomicity*, which guarantees all-or-nothing execution of transactions, (2) *Consistency*, which ensures that relationship constraints among data are preserved, (3) *Isolation*, which allows for multiple transactions to execute without unintentionally affecting each other, and (4) *Durability*, for persistence of transactions. While traditional relational database management systems (RDBMS) allow complex querying of the data, their reliance on static schema and their inability to scale-out with additional nodes, seriously limit their utility for big-data applications [4].

The NoSQL or Not Only SQL paradigm refers to the new class of nonrelational data stores designed to address the big-data characteristics outlined above. NoSQL data stores typically employ a scale-out architecture with a distributed approach where commodity servers can be added on the fly. According to the CAP theorem [5], in distributed systems, only two out of the three *Consistency, Availability,* and *Partition tolerance* properties are achievable. The intuition behind the CAP theorem is that in the presence of partitions where one set of nodes in the distributed system is not reachable from the other, access to nodes from different partitions may result in inconsistent data. On the other hand, if the node partitions are to have consistent data, then they should not be made available. To achieve both consistency and availability, no partitions should be allowed in the distributed system. Unlike RDBMS which are designed to be consistent and available, NoSQL data stores choose partition tolerance, and either consistency or availability. Also, many NoSQL data stores do not support ACID transactions. Further, they support a variety of data models ranging from simple key-value stores to NewSQL data stores that bring a semirelational model to NoSQL data stores.

In this chapter, we present an overview of the key data structures used by NoSQL data stores, providing references to the literature for details. We first give a short summary of the different data models used by NoSQL systems. We then outline the key design features of NoSQL data stores with a brief description of the algorithms and the associated data structures. These features include partitioning, replication, consistency, persistence, and concurrency. Among these features, persistence is treated in greater detail since the choice of data structures is particularly important in the overall performance of the data store.

63.2 Data Models

NoSQL databases can be classified into five types based on the data model realized—key-value stores, column-family stores, documents stores, graph databases, and NewSQL databases [6].

Key-Value Stores: Key-value stores use a schema-free key-value data model similar to associative maps or dictionary data structures. The value data type is opaque, that is, all querying and indexing is done through keys. Key-value stores such as Memcached [7] and Redis [8] are in-memory, while others such as BerkeleyDB [9], Amazon Dynamo [10], LevelDB [11], Voldemort [12], and Riak [13] persist data on disk.

Column Family Stores: A column family store is a multidimensional sorted map where a key has multiple attributes such as row-key, column family, column, and timestamp. Examples of column family stores include Big Table [14], Cassandra [15], and HBase [16].

Document Stores: Document stores are a type of key-value store where the values are documents stored in a format such as XML or JSON that can be interpreted by the data store. Document stores are schema free and allow storing of arbitrarily complex data. They are more flexible than key-value stores since, in addition to the key, data can also be accessed by the queries supported by the document format. Examples of document stores include MongoDB [17], CouchDB [18], and RavenDB [19].

Graph Databases: Graph data stores use graph-like nodes and edges where each node maintains direct references to adjacent nodes allowing for efficient graph traversals. The different graph components such as nodes, edges, labels, and properties are stored in separate files with pointers linking the corresponding elements across files allowing for rapid traversal. The properties of the nodes and edges are stored as key-value pairs. Graph databases are well suited to represent relationships between elements (e.g., determining relations in a social network, recommendations, and route finding). Examples of graph databases include Neo4j [20] and HypergraphDB [21].

NewSQL Databases: NewSQL data stores seek to extend the NoSQL data stores by providing a semirelational model, ACID semantics, and use of SQL as their primary interface. They try to achieve scalability by using techniques such as nonlocking concurrency control, in-memory architectures, and short-lived limited scope transactions. Examples of NewSQL databases include Google Spanner [22], VoltDB [23], NuoDB [24], TokuDB [25], and ScaledDB [26].

63.3 Partitioning

The goal of partitioning is to divide the key-values among multiple nodes in a distributed system so as to realize a distributed data store. The partition scheme seeks to balance the load between the nodes while enabling quick access of key-values. Moreover, nodes should be able to join and leave the distributed system with minimum disruption. The most commonly used partition scheme is consistent hashing [27]. Among the NoSQL databases that use consistent hashing are Voldemort [12], Riak [13], Cassandra [15], DynamoDB [10], CouchDB [18], and VoltDB [23].

63.3.1 Consistent Hashing

In a standard hash table, when the number of buckets changes, the mapping from keys to buckets changes completely. In the data store context, the storage nodes represent the buckets. As a result, all the data has to be reshuffled to be consistent with the new mapping scheme. In a distributed system with many nodes, where node failures are frequent and new machines are added regularly, such reshuffling of data between nodes can be expensive. Also, consistency issues may arise due to the delays associated with notifying the nodes of the new key to node mapping, thus resulting in multiple system views. "Multiple views" refers to differing key ownership information among the nodes. To support such multiple views, a given node may have to host keys consistent with different views, leading to both an increase in the load per node, and an increase in the node-spread of keys (the number of nodes responsible for a key). Therefore, among the requirements of a good hashing scheme are—load balance, minimal data reshuffling with addition or removal of nodes, and minimal load increase and key spread in maintaining multiple views. Consistent hashing aims to achieve these goals.

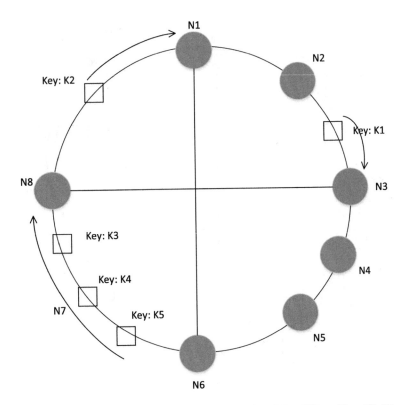

FIGURE 63.1 Consistent Hashing—Key Assignment: Key K1 is assigned to node N3, K2 to N1, and keys K3, K4, and K5 are assigned to N8.

In consistent hashing, data and nodes are mapped to points on a circle using a standard hash function. The data is then assigned to the closest node in a clockwise traversal from the point it is mapped on the circle (Figure 63.1). When a new node is inserted (Figure 63.2), only data mapped to points near the new node are copied to the new node. Since data is assigned to nodes that are at close by points on the circle, even in the presence of a large number of views, the spread of the data is limited. Also, since the data is assumed to be distributed randomly, only a limited amount of data will have their mapped points on the circle close to that of a given node. Thus the load on the node does not increase drastically with multiple views. Karger et al. [27] show that both load and the spread only increases logarithmically in the number of nodes when consistent hashing is used, as compared to linear increase with standard hashing. To ensure that the nodes are not assigned to disproportionate sections of the circle, each node is assigned m points around the circle ($m \geq 1$) where $m = k \log N$. Here, N is the number of nodes and k is a constant. A balanced binary search tree is used to store the mapping between segments of the circle and the nodes. For N nodes, there are $kN \log N$ segments since each node is virtually replicated $k \log N$ times. The search tree thus has a depth of $O(\log N)$. Since we insert or delete $k \log N$ points for each node, the addition or deletion of a node is $O(\log^2 N)$. Typical hash functions used include linear congruential function ($x \mapsto ax + b(modp)$) and cryptographic hash functions such as the secure hash algorithm [28].

63.4 Replication and Consistency

Replication not only improves fault tolerance but also increases performance by spreading data accesses among multiple nodes. A coordinator node keeps track of the location of the keys and their replicas, directing client key-value access requests to appropriate nodes. Note that separate coordinators could be maintained for different keys. Among the distributed computing primitives required to implement replication are—election of a coordinator, maintenance of a membership list, detection of node failures, and reaching consensus on the replicated value. Replication consistency involves keeping the replicated data consistent with each other. According to the CAP theorem [5], in a distributed system, it is impossible to simultaneously achieve consistency, availability, and partition tolerance. Traditional RDBMS trade-off partition tolerance for consistency and availability. Since the ability to scale out is an important goal of big data stores, NoSQL stores try to maintain partition tolerance while giving up on either consistency or availability. NoSQL stores, such as Cassandra [29] and DynamoDB [10], trade-off consistency for availability and partition tolerance. Cassandra, for example, implements an eventual consistency model [30], where typically a quorum of replicas acknowledge a read and write operation. Cassandra also supports consistency levels ranging from a single replica response, to

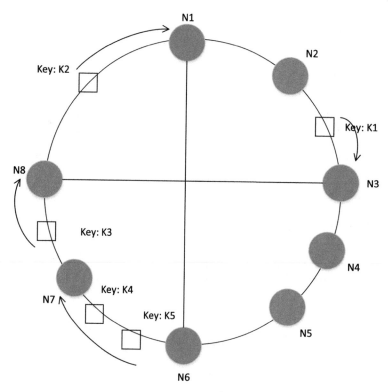

FIGURE 63.2   Consistent Hashing—Node addition: When new node N7 is added, only K4 and K5 are reassigned to N7. The mapping of the remaining keys is not affected.

all the replicas responding (strong consistency). Other NoSQL stores such as HBase [16] and NewSQL databases such as Google Spanner [22] and VoltDB [23] trade-off availability for partition tolerance and consistency.

The coordinator election and consensus on replicated values is achieved through distributed consensus algorithms such as Zab [31] and Paxos [32,33]. In order to bring the divergent replicas to a consistent state, the Merkle Tree data structure [34] is commonly used.

63.4.1 Merkle Tree

A Merkle tree is a tree of hashes where the leaf vertices are the hashes of data blocks, and the internal vertices are the hashes (actually, concatentations of the hashes) of their children. Figure 63.3 shows a binary Merkle tree. In this example, hash(3,4) is the result of concatenating hash of data block 0 and hash of data block 1. A cryptographic hash function such as the secure hash function could be used [28]. To determine if the two replicas are consistent, the root vertices of their corresponding Merkle trees are compared. If the roots are equal, the two replicas are considered to be consistent. If the root vertices do not match, the two trees are recursively searched until corresponding matching vertices are found. A matching vertex pair indicates that the data blocks in the two replicas that have the vertices as the root are consistent. For example, in Figure 63.3 if vertex hash (5,6) between the two trees match, then data blocks 2 and 3 are consistent. The search terminates when the leaf vertices are reached. Unmatched leaf vertices indicate that the corresponding data blocks between the two replicas are not consistent. The Merkle tree thus allows us to drill down the two replicas, to determine the blocks that do not match. Since only hash values are stored, the entire tree can be maintained in memory and transferred between the replica nodes with low overhead. Moreover, comparisons can proceed even with a partial tree. For example, in Figure 63.3, the consistency of data block 0 can be checked by knowing hash(data1) and hash (5,6). The root hash can then be computed and compared against the root vertex of the Merkle tree associated with the replica.

A gossip-style protocol can be used to implement both failure detection and membership lists. For example, Cassandra uses the Scuttlebutt gossip-style protocol [29] for maintaining membership lists. Examples of failure-detection protocols include SWIM [35] and adaptive-phi [36]. Apache Zookeeper [37] and Google Chubby [38] are projects used by many NoSQL stores for services such as consensus, failure detection, and distributed locks.

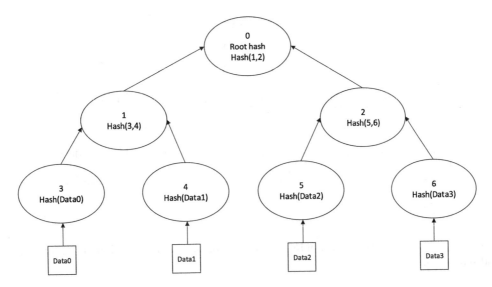

FIGURE 63.3 Merkle Tree—A binary hash tree where the leaves are hashes of the data blocks and the internal nodes are concatenation of the hashes of the children.

63.5 Persistence

NoSQL database workloads are characterized by high throughput inserts, and point or range queries [39,40]. Workloads such as interactive web and online gaming require low latency reads and writes. Analytics workloads such as predictive modeling and web analytics are characterized by sequential reads and high write throughputs. However, with the explosion of mobile devices and the emergence of the IoT paradigm, the need for integrating interactive applications with analytics is fast rising. Example use cases include gaining real-time behavioral insights from user mobile activity, and remote monitoring and control of physical infrastructure such as smart power grids. This has resulted in workloads that are both write heavy, and have stringent latency requirements. For example, as reported by Yahoo [41], the percentage of writes in their workloads has increased from 10% to 20% in 2010 to 50% in 2012, with a further increase expected.

On the storage side, hard disk seek times have improved much more slowly compared to disk bandwidths. Utilizing typical hard disk performance numbers of 3 ms access time, 2 ms rotational latency, and disk bandwidth of 200 MB/s—for a block size of 4 KB, the seek time of 5 ms is 250× compared to the 0.02 ms time required for data transfer. High I/O performance thus requires long sequential I/Os to amortize disk seek times. Another trend is the growth of popularity of solid state drives (SSDs) in cloud computing systems due to their falling costs and growing capacity. Since SSDs have no moving parts, random read is about 100× times faster than hard disks. However unlike hard disks, data cannot be directly overwritten in SSDs. Instead, data needs to be erased at a much higher level of granularity before the write operation. As a result, SSD random writes are 100× slower than random reads. Also, write bandwidths are further reduced by 10× if the data is fragmented [42]. Therefore, SSDs favor sequential writes over random writes to achieve high write performance. Moreover, SSDs are characterized by a finite number of write cycles and require data structures that reduce the number of writes. Write-optimized data structures are thus crucially important in providing data persistence in NoSQL databases.

In this section, we review the two most widely used data structures for persistence used by NoSQL data stores today—the B+ tree and the Log Structured Merge (LSM) tree. After a brief description of the LSM tree, we carry out a simplified analysis of its behavior to provide the reader some intuition about its performance. We also examine optimizations that seek to improve the performance of these data structures.

63.5.1 Performance Metrics

Traditional metrics used to evaluate data structure performance include asymptotic time complexity for random point query, scan, insert, update, and delete operations. Point query involves finding a value for the queried key, while scan involves finding all values in sorted order that come after the queried key. Insert and delete have their usual meaning and update is a read–modify–write operation. Additional performance metrics include write amplification and read amplification. Write amplification is the ratio of the amount of data written to the size of the object. A higher write amplification means worse performance especially if the extra writes are not sequential. Additionally, for SSDs, a higher writer amplification leads to a quicker drive wear out. Read

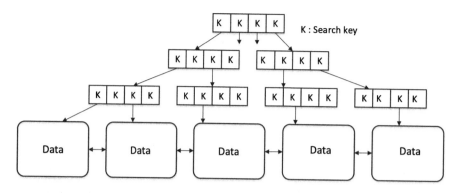

FIGURE 63.4   Structure of B+ tree.

amplification is defined as the worst-case number of I/Os for a read. Similar to write amplification, a higher read amplification results in degraded performance.

63.5.2 B+ Trees

The most widely used data structure for data indexing is the B+ tree. B+ trees are search trees with a high degree of fanout (child nodes) with data held in leaf nodes and index keys stored in interior nodes (Figure 63.4). In the analysis below, the size of the nodes (block size) of the B+ tree is denoted by B and the size of the database by N. Assuming the Disk Access Model (DAM) which counts the number of memory transfers from slow to faster memory [43], the cost of query and insertion is given by the height $O(\log_B(N/B))$ of the tree. If the keys fit in memory (i.e., the internal nodes are cached), the B+ tree takes a single I/O for a random point query. Two I/Os are required if the data to be inserted or updated is not cached in memory (one to read it to memory from disk and one to write the updated leaf node to disk). However, as mentioned earlier, random in-place writes lead to poor performance due to large disk seek times. The choice of B is impacted by the following observations: (1) reads perform better with larger block sizes (e.g., 1 MB) since a larger block size reduces the depth of the tree. The read amplification is $O(\log_B N/B)$ for uncached keys and 1 for cached keys [44]. (2) However, the worst-case write amplification is $O(B)$ since a random write requires the entire block to be written. Thus a smaller block size is preferred for write-intensive databases. Block size is a configurable parameter. For example, in MongoDB, where the blocks are referred to as buckets, the db.collection.indexStats() method has a parameter bucketBodyBytes with a default size of 8 KB.

63.5.3 Log Structured Merge Trees

The Log Structured Merge Tree [45] is a tree data structure used by a number of NoSQL databases such as Bigtable [14], Cassandra [29], and HBase [16]. While random point queries run slower than B+ trees, LSM-trees can perform random insertions and updates without incurring external I/O. This trade-off of a slight slowdown in read performance for a much greater speed up in write performance is well suited for increasingly write-intensive NoSQL workloads.

The LSM-tree is a collection of exponentially larger trees treated as a single key-value store. The individual trees in the collection can be implemented using B+ trees or Sorted String Table (SSTable) [14,29,46]. An SSTable is a simplified B+ tree which only maintains a single layer of index nodes with the leaf nodes pointed at by the entries in the index node. The leaf nodes are written to a storage device in a single contiguous allocation with the index nodes stored at the end of the leaf nodes along with the offsets of their values. In an SSTable, insertions are performed by adding elements into the last empty leaf node with a new leaf node created if required. The offset of the new leaf node is then appended to the last empty index node. New index nodes are created if required. A look up on a key K is performed by a binary search on the index nodes to determine keys K_i and K_{i+1} such that $K_i \leq K \leq K_{i+1}$. The leaf pointed to by index K_i is the leaf node with the value associated with K [46].

The basic LSM-tree design is shown in Figure 63.5. When a tree at level i is filled, it is flushed out and its contents merged, sorted, and packed with the tree at level $i+1$. This process is known as tree compaction. Since the trees are stored in sorted key order, the merges can be done in a single pass [47]. The updated tree is then written to a new part of the disk allowing sequential writes at storage bandwidth speeds. The smallest tree, that is, the tree at the lowest level is kept in memory thus requiring no external I/O access. Upon merging with higher level trees, the tree in memory is ready to accept new insertions. To perform updates, the merge process selects the most recent of the values with the same key to flush to the larger tree. Deletes are similar to updates and involves setting the key with a special tombstone flag. The tombstones are then removed during the merge process. Regarding query performance, for uncached data, in the worst case, a query involves binary search of each of the trees. Since there

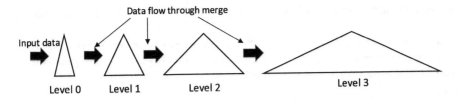

FIGURE 63.5 Basic LSM-tree design. When a tree at level i is filled, it is merged with the tree at level $i + 1$.

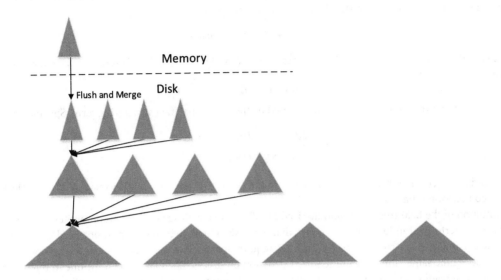

FIGURE 63.6 Size tiered LSM tree with 4 trees per level.

are $O(\log_k N/B)$ levels, where k is the tree growth factor, the query complexity is $O(\log N/B * \log_k N/B)$ or $O(\log^2 N/B)/\log k$. This is the read amplification factor as well. Similarly, since there are $O(\log_k N/B)$ levels, and on an average the data written to a level is merged again to the same level $k/2$ times, the asymptotic update I/O cost is $\Theta(k \log_k N/B)$. However, updates use sequential I/O, and hence can handle more intense update loads than B+ trees. The write amplification factor is $\Theta(k \log_k N/B)$ as well [44] which if amortized for sequential transfers, leads to an effective write amplification less than 1. Thus while LSM trees have superior insert and update performance, their random query performance is unsatisfactory. Moreover, the time required to merge two trees is unbounded since the size of the trees is unbounded. Such unbounded merges can potentially lead to sporadic write outages [41].

The data structure described above can be thought of as a level-tiered LSM tree [44]. As shown in Figure 63.6, in a size-tiered LSM-tree (also known as Stepped Merge Method and Sorted Array Merge Tree when implemented using SSTables) rather than storing one tree per level, N_T trees are stored at each level [44,46,48]. The merge is performed after N_T smaller trees are flushed to the larger tree. Cassandra [15] provides both size-tiered and level-tiered LSMs. Compared to the level-tiered approach, the size-tiered LSM has a write amplification $O(\log_k N/B)$ which is k times smaller than the level-tiered approach since the data is written to each level only once. However, read amplification is $O\left(k \frac{\log_k N/B}{\log k}\right)$ which is k times larger than the level-tiered approach. Thus level-tiered LSM trees are better suited for read-intensive workloads while size-tiered LSM trees are better suited for write-intensive workloads. NoSQL databases such as BigTable [14], Cassandra [15], and Hbase [16] maintain a commit log to disk for crash recovery. The write to the in-memory LSM-tree is performed only after a successful write to the commit log. A rolling commit log is used to purge old commit log entries.

63.5.4 Intuition behind LSM

We present a somewhat simplified analysis of the LSM tree data structure to provide the reader with some intuition of its advantages over traditional disk-based data structures. To facilitate this analysis, we make three assumptions.

- A (random) insertion into a dictionary data structure in internal memory requires $c \log n$ time, where n is the size of the data structure and c is a constant associated with writes in main memory.

- A similar insertion into a dictionary on disk requires $D \log n$ time. Here D is a constant associated with random writes to disk. Clearly $c \ll D$.
- Finally, we assume that the merging of the contents of two dictionaries of size n_1 and n_2 that results in a merged dictionary of size $n_1 + n_2$ on disk requires $d(n_1 + n_2)$ time. Here, d is a constant associated with sequential writes to a disk. Clearly $d \ll D$ (sequential writes are faster than random writes).

In both of the following scenarios, assume that a dictionary of size N is already present on disk. In Scenario 1 (traditional), n insertions are directly made into the dictionary on disk. This will require approximately $Dn \log (N + n)$ time. In Scenario 2 (essence of LSM), n insertions are first made into an initially empty in-memory dictionary ($cn \log n$) followed by a merge with the dictionary on disk ($d(n + N)$). To compare the two scenarios, we compare

$$cn \log n + d(n+N) \quad vs. \quad Dn \log(N+n).$$

Here, $cn \log n$ is negligible relative to $Dn \log (N + n)$ because $c \ll D$ and $n < N + n$. We therefore restrict our analysis of the two data structuring approaches to

$$d(n+N) \quad vs. \quad Dn \log(N+n).$$

Now, if $n = N$, we find that the LSM term is again dominated by the traditional data structuring term. Specifically,

$$2dn \quad vs. \quad Dn \log(2n)$$
$$2d \quad vs. \quad D \log(2n).$$

Here $d \ll D$ and $2 < \log (2n)$ for most n. It is instructive to note that if $N \gg n$, we would evaluate dN versus $Dn \log N$ and the outcome of the comparison is unclear.

The main intuition of the LSM tree can be summarized as follows: relatively expensive random writes that result in an initial sorted structure are performed in fast memory rather than on disk. The sorted structure so obtained can then be efficiently merged with the existing sorted structure on disk (merges require linear scans, a strength of disk technology), provided their relative sizes are about the same. If the size of the sorted structure on disk is much larger than the one in main memory, we are going to the expense of reading and writing this large structure to accommodate only a relatively small set of new insertions. This observation motivates the need for several levels of gradually increasing sized structures in an LSM tree.

63.5.5 Performance Optimizations

A number of optimizations have been proposed in the literature to improve the performance of LSM trees. Some of these techniques have been implemented in production NoSQL databases. We give a brief description of these techniques below.

Bloom Filter: Point queries can be made more efficient by adding an in-memory Bloom filter to each level, thus avoiding I/Os for levels that do not contain the key of interest. A Bloom filter is a space-efficient probabilistic data structure that is used to test whether an element is a member of a set. The basic data structure in a Bloom filter is a bit vector. For a given input, a series of independent hash functions is used to set different bits of the bit vector. If two inputs map to the same bits, a look-up for one of the inputs can result in a false positive. However, no false negatives occur. The false positive rate can be controlled through the choice of number of hash functions and the length of the bit vector. With the use of a Bloom filter, the read amplification factor can be made 1 for random point queries. Bloom filters are used by a number of NoSQL databases including Cassandra [15], HBase [16], TokuDB [25], and LevelDB [11]. In the LSM tree variant, bLSM, 10 bits are used per key leading to a false-positive rate of 1% [41]. Note that Bloom filters do not improve scan performance.

Fractional Cascading: Fractional cascading is a data structuring technique that has its origins in the computational geometry literature from the 1980s [49,50]. We first describe the general principle through an example followed by a brief description of its use in LSM trees. Consider three sorted lists:

L_1: 5, 15, 16, 30, 50, 65
L_2: 4, 6, 9, 10
L_3: 2, 12, 13, 20, 25.

Suppose we wish to search for 14 in the three lists. This is accomplished by performing three independent binary searches. Adopting the convention that a failed search returns the predecessor, this will return 5 in L_1, 10 in L_2, and 13 in L_3. Assuming k lists, each of size n, the time and space complexity of this approach are $O(k \log n)$ and $\Theta(kn)$, respectively.

Fractional cascading gives an improved run time of $O(\log n + k)$ while retaining a space complexity of $\Theta(kn)$. The individual lists are preprocessed and modified in order to achieve these bounds as follows: list F_k is identical to L_k; list F_i is obtained by merging L_i with *every second* element of F_{i+1} for all $i < k$. In our example, the F_is are shown in the order in which they are computed.

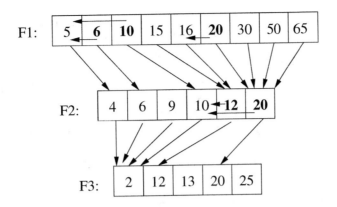

FIGURE 63.7 Fractional cascading structure.

F_3: 2, 12, 13, 20, 25
F_2: 4, 6, 9, 10, **12**, **20**
F_1: 5, **6**, **10**, 15, 16, **20**, 30, 50, 65.

Bold-face is used to denote elements in F_i that were obtained from F_{i+1} rather than from L_i. The F_i arrays are further augmented to include

1. An *internal* pointer from each bold-faced element in F_i to its largest nonbold-faced predecessor in F_i. For example, the **10** in F_1 points to 5 in the same list.
2. An *external* pointer from each element x in F_i to the largest element less than or equal to x in F_{i+1}.

The resulting data structure is shown in Figure 63.7.

Consider now a search for 14 in F_1. A binary search in F_1 gives 10, whose internal pointer yields 5 (the correct response to a query in the original list L_1). 10's external pointer points to the element 10 in F_2. A $\Theta(1)$ time scan yields 12 as being the predecessor of 14 in F_2. Since 12 is bold-faced, its internal pointer returns 10 as the response to the query in L_2. Next, 12's external pointer gives 12 in F_3. The $\Theta(1)$ local search yields 13, which is the correct response to the query in L_3. Fractional cascading does result in an increase in space because of duplicated elements, but the total space remains $\Theta(kn)$.

In an LSM tree, forwarding pointers are stored from keys at one level to the same or predecessor key at the next level. So instead of performing binary search to find a key or its predecessor at all $O(\log_k N/B)$ levels at $O(\log_k N/B * \log N/B)$ cost, a binary search need only be performed at one level, followed by constant number access down the higher levels through the forwarding pointers. In practice, only one forwarding pointer per block needs to be maintained. However, to ensure forwarding pointers to the higher level in the absence of the corresponding blocks in the lower level, ghost pointers are added to the lower level. The point query cost is thus $O(\log N/B + \log_k N/B)$. Fractional cascading has been used in LSM-tree variants such as FD-tree [42], cache oblivious lookup arrays (COLA) [51,52], write buffer tree [53], and in LevelDB [11], AsterixDB [54], and TokuDB [25] databases.

Key-Based Partitioning: In the partitioning technique, the LSM trees are divided into many partitions with each partition storing data within a range of keys [55]. Each partition is of a bounded size thus improving merge performance. Query performance is also improved since the partition key ranges are disjoint allowing for searches to be limited to a partition. Additionally, the partitioning is a natural unit of concurrency control and handles write skews well by concentrating merges on frequently updated key ranges [41]. LSM tree partitioning is used in LSM-tree variants such as write buffer tree [53], range table [56] and in LevelDB [11], AsterixDB [54], and TokuDB [25] databases.

Merge Scheduling: A merge schedule defines when and which LSM trees are merged. The merge time needs to be minimized for reducing write pauses and improving read latencies. Cassandra [15] prioritizes merging of trees with recent reads to improve read latency. The spring and gear merge scheduler used in bLSM [41] attempts to keep the size of the tree between a low and a high watermark. If the size of the tree dips below the low watermark, the scheduler pauses the merge of the higher level trees. On the other hand, if the size of the tree exceeds the high watermark, the scheduler puts back pressure on the application. However, as described by Sears and Ramakrishnan [41], the implementation is tricky due to the costs associated with acquiring coarse-grained mutex for each merged tuple-page, and the risk of deadlocks or disk idling. A theoretical analysis of compaction algorithms is presented in References 57 and 58 where compaction is presented as an optimization problem that minimizes system costs such disk I/O [58] and CPU load [57].

Fractal Tree: Fractal trees [44,59] are a variant of B+ trees in which each internal node of the tree is partitioned into two regions. The first region is used to store pointers to child nodes as before. The second region is used to maintain a buffer which is used

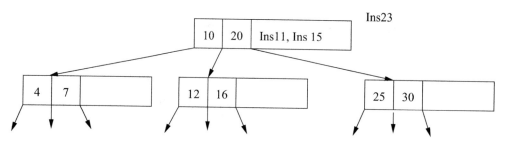

FIGURE 63.8   Fractal tree's root node's buffer region is filled to capacity with two pending operations "Insert 11" and "Insert 15." The remaining nodes' buffer regions are empty.

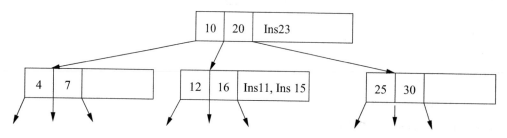

FIGURE 63.9   The addition of the "Insert 23" operation causes the "Insert 11" and "Insert 15" operations to be propagated down the middle pointer (both are in the [10, 20] range) and placed in the child node's buffer.

to store pending operations such as insert, delete, and update. When a key-value pair is inserted, it is considered to be a pending operation and is stored in the root node's *buffer* region (as opposed to being propagated down the tree to a leaf node). When the root node's buffer gets filled with pending operations, they are recursively cascaded to an appropriate child node and stored in its buffer. As buffers in the tree get filled, pending operations ultimately reach the leaves and are executed just like in a B+ tree. Figures 63.8 and 63.9, respectively, show a fractal tree before and after the execution of an "Insert 23" operation, which causes some pending buffer operations to be propagated down the tree.

Part of the block is used for implementing the buffer. The write amplification is k for each level of the tree where k is the number of children. Since there are $O(\log_k N/B)$ levels, the write amplification for the fractal tree is $O(k \log_k N/B)$. Read amplification is $O(\log_k N/B)$. Additionally write performance is improved by batching the writes of leaf nodes. Regarding the choice of parameters, Kuzmaul recommends the number of children $k = 10$, and block size of 64 KB [44].

63.6 Concurrency

The traditional approach of using read/write locks to support concurrent transactions suffers from performance issues due to readers and writers blocking each other. For example, an analytics application may execute a long read transaction, blocking out transactions that need to perform short updates on the data store. A solution adopted by data stores such as Voldemort [12], Riak [13], Hbase [16], NuoDB [24], HypergraphDB [21] is Multi Version Concurrency Control (MVCC). In MVCC, multiple versions of the key-value are maintained with writers updating a new version of the key-value. The read visibility set consists of the committed transactions. MVCC thus allows reads and writes to proceed concurrently. Locking is only required if two writes seek to update the same key-values in the same version. The versioned dictionary data structure supports implementation of MVCC. In a versioned dictionary, in addition to the key and value, a version number is also stored. The query and modify operations have the version number as an additional parameter. A significant challenge in implementing such a data structure is in realizing optimal query/update operations similar to B+ trees or LSM trees while maintaining linear space usage. The traditional Copy-On-Write (CoW) B+ trees result in excessive space usage since each update may cause an entire new path to be written. A large space usage in turn makes garbage collection expensive.

The stratified B-tree (also known as stratified doubling array) [60] supports multiversioning, optimal query/update performance ($O(\log(N_v)/B)$), where N_v is the number of keys live at version v, and optimal space ($O(N)$). The basic organization is similar to the SSTables described in Section 63.5 with the addition that the arrays are tagged with a set of version IDs. The arrays double in size going up the levels. Disjoint sets of versions are maintained in the arrays at the same level. The density of the array for a particular version is the fraction of the elements of that version that is live. A high density, where the array is dominated by elements from a single version, results in inefficient space usage and costly updates. On the other hand, low density results in poor point and range query performance. A density of 1/3 was found to be a good compromise value [60]. In the merge process, additional care has to be taken to ensure that the density is maintained for all versions. A technique known as density amplification achieves this by replacing large arrays with smaller density arrays by carefully choosing the versions to be included in the denser array. The details of the algorithm are described in Reference 61.

An additional consideration for high performance data stores is the concurrency of in-memory data structures so as to achieve vertical scaling on servers with a large number of processor cores. A very recent work in this space is the concurrent LSM (cLSM) data structure [62] which enhances Google's LevelDB [11] data store to achieve $1.5\times$ to $2.5\times$ performance. The key-ideas employed by cLSM are the use of highly concurrent in-memory data structures [63–65], careful minimization of critical sections to pointer update operations, use of time-stamp based version control to support snapshot scans and range queries, and optimistic concurrency control with efficient in-memory conflict detection to support atomic read–modify–write operations.

63.7 Conclusion

The rise of big-data has resulted in the emergence of diverse set of data store designs and products tailored to meet different application needs. A key feature of these NoSQL and NewSQL big-data stores is the distributed storage needed to accommodate massive datasets (exabytes and beyond), and replication for fault tolerant operation. Meanwhile, application requirements are rapidly changing from traditional read-heavy workloads, to both read and write-heavy workloads. Currently, considerable research and industry efforts are focused on providing ACID properties to the new generation data stores. On the hardware side, solid state storage is increasingly being adopted in large data centers. Also, the increasing memory capacity and falling prices give rise to the possibility of some databases entirely residing in memory. With the rise of cloud computing, an emerging paradigm is that of database-as-a-service. Some of the challenges associated with this paradigm are outlined in the Beckman report on database research [66]. In this chapter, we have provided a brief overview of the techniques involved in the design of the these new generation data store with an emphasis on the data structures required to persist data. As big-data applications become more ubiquitous with stringent performance and power requirements, the coming years will likely see work on data structures that are highly scalable, concurrent, and adaptive [67].

References

1. Scaling the Facebook data warehouse to 300 PB, https://code.facebook.com/posts/229861827208629/scaling-the-facebook-da ta-warehouse-to-300-pb/, accessed: 2015-10-09.
2. Genomic Data Growing Faster Than Twitter and YouTube, http://spectrum.ieee.org/tech-talk/biomedical/diagnostics/the-human-os- is-at-the-top-of-big-data, accessed: 2015-10-09.
3. L. Atzori, A. Iera, and G. Morabito, The internet of things: A survey, *Computer Networks*, vol. 54, no. 15, pp. 2787–2805, 2010.
4. R. Cattell, Scalable sql and nosql data stores, *SIGMOD Rec.*, vol. 39, no. 4, pp. 12–27, May 2011.
5. S. Gilbert and N Lynch, Perspectives on the CAP theorem, *IEEE Computer*, vol. 45, no. 2, pp. 30–36, 2012.
6. K. Grolinger, W. Higashino, A. Tiwari, and M Capretz, Data management in cloud environments: NoSQL and NewSQL data stores, *Journal of Cloud Computing*, vol. 2, no. 22, pp. 49:1–49:24, 2013.
7. Memcached, http://memcached.org/, accessed: 2015-10-09.
8. Redis, http://redis.io/, accessed: 2015-10-09.
9. Oracle Berkeley DB 12c, http://www.oracle.com/technetwork/database/database-technologies/berkeleydb/overview/index.html, accessed: 2015-10-09.
10. G. DeCandia, D. Hastorun, M. Jampani, G. Kakulapati, A. Lakshman, A. Pilchin, S. Sivasubramanian, P. Vosshall, and W. Vogels, Dynamo: Amazon's highly available key-value store, in *Proceedings of Twenty-First ACM SIGOPS Symposium on Operating Systems Principles*, Stevenson, WA, October 14–17, 2007, pp. 202–220.
11. LevelDB, http://leveldb.org/, accessed: 2015-10-09.
12. Project Voldemort, http://www.project-voldemort.com/voldemort/, accessed: 2015-10-09.
13. Riak KV, http://basho.com/products/riak-kv/, accessed: 2015-10-09.
14. F. Chang, J. Dean, S. Ghemawat, W. C. Hsieh, D. A. Wallach, M. Burrows, T. Chandra, A. Fikes, and R. E. Gruber, Bigtable: A distributed storage system for structured data, *ACM Transactions on Computer Systems*, vol. 26, no. 2, pp. 1–26, 2008.
15. Cassandra, http://cassandra.apache.org/, accessed: 2015-10-09.
16. Apache Hbase, http://hbase.apache.org/, accessed: 2015-10-09.
17. MongoDB, https://www.mongodb.org/, accessed: 2015-10-09.
18. Apache CouchDB, http://couchdb.apache.org/, accessed: 2015-10-09.

19. RavenDB, http://ravendb.net/, accessed: 2015-10-09.

20. Neo4j, http://neo4j.com/, accessed: 2015-10-09.

21. HperGraphDB, http://hypergraphdb.org/index/, accessed: 2015-10-09.

22. J. C. Corbett, J. Dean, M. Epstein, A. Fikes, C. Frost, J. J. Furman, S. Ghemawat et al., Spanner google's globally-distributed database, in *Proceedings of the 10th USENIX Conference on Operating Systems Design and Implementation*, Hollywood, California, October 8–10, 2012, pp. 251–264.

23. VoltDB, https://voltdb.com/, accessed: 2015-10-09.

24. NuoDB, http://www.nuodb.com/, accessed: 2015-10-09.

25. TokuDB introduction, https://www.percona.com/doc/percona-server/5.6/tokudb/tokudb_intro.html, accessed: 2015-10-09.

26. ScaleDB, http://www.scaledb.com/, accessed: 2015-10-09.

27. D. Karger, E. Lehman, T. Leighton, R. Panigrahy, M. Levine, and D. Lewin, Consistent hashing and random trees: Distributed caching protocols for relieving hot spots on the world wide web, in *Proceedings of the Twenty-Ninth Annual ACM Symposium on Theory of Computing*, El Paso, Texas, May 4–6, 1997, pp. 654–663.

28. SHA1 Secure Hash Algorithm—Version 1.0, http://www.w3.org/PICS/DSig/SHA1_1_0.html/, accessed: 2015-10-09.

29. A. Lakshman and P. Malik, Cassandra: A decentralized structured storage system. *SIGOPS Operating System Reviews*, vol. 44, no. 2, pp. 35–40, 2010.

30. W. Vogels, Eventually consistent. *Communications of the ACM*, vol. 52, no. 1, pp. 40–44, 2009.

31. F. P. Junqueira, B. C. Reed, and M. Serafini, Zab: High-performance broadcast for primary-backup systems, in *Proceedings of the 2011 IEEE/IFIP 41st International Conference on Dependable Systems & Networks*, Hong Kong, China, June 27–30, 2011, pp. 245–256.

32. L. Lamport, Paxos made simple, *ACM SIGACT News*, vol. 32, pp. 51–58, 2001.

33. T. D. Chandra, R. Griesemer, and J. Redstone, Paxos made live: An engineering perspective, in *Proceedings of the Twenty-Sixth Annual ACM Symposium on Principles of Distributed Computing*, Portland, Oregon, August 12–15, pp. 2007, 398–407.

34. R. Merkle, A digital signature based on a conventional encryption function, in *Conference on the Theory, Applications of Cryptographic Techniques on Advances in Cryptology*, Santa Barbara, California, August 16–20, 1988, pp. 369–378.

35. A. Das, I. Gupta, and A. Motivala, Swim: Scalable weakly-consistent infection-style process group membership protocol, in *Proceedings of the 2002 International Conference on Dependable Systems and Networks*, Bethesda, Maryland, June 23–26, 2002, pp. 303–312.

36. N. Hayashibara, X. Defago, R. Yared, and T. Katayama, The phi; accrual failure detector, in *Proceedings of the 23rd IEEE International Symposium on Reliable Distributed Systems*, Florianopolis, Brazil, October 18–20, 2004, pp. 66–78.

37. Apache Zookeeper, https://zookeeper.apache.org/, accessed: 2015-10-09.

38. M. Burrows, The Chubbylock service for loosely-coupled distributed systems, in *Proceedings of the 7th Symposium on Operating Systems Design, Implementation*, ser. SoCC'10, Seattle, Washington, November 6–8, 2006, pp. 335–350.

39. J. Schindler, I/o characteristics of NoSQL databases, *Proceedings of the VLDB Endowment*, vol. 5, no. 12, pp. 2020–2021, August 2012.

40. B. F. Cooper, A. Silberstein, E. Tam, R. Ramakrishnan, and R. Sears, Benchmarking cloud serving systems with YCSB, in *Proceedings of the First ACM Symposium on Cloud Computing*, Indianapolis, Indiana, June 10–11, 2010, pp. 143–154.

41. R. Sears and R. Ramakrishnan, bLSM: A general purpose log structured merge tree, in *Proceedings of the International Conference on Management of Data (SIGMOD)*, Scottsdale, Arizona, May 20–25, 2012, pp. 217–228.

42. Y. Li, B. He, R. J. Yang, Q. Luo, and K. Yi, Tree indexing on solid state drives, in *Proceedings of the VLDB Endowment*, 2010, pp. 1195–1206.

43. A. Aggarwal and J. Vitter, The input/output complexity of sorting and related problems, *Communations of the ACM*, vol. 31, no. 9, pp. 1116–1127, 1988.

44. B. Kuszmaul, A Comparison of Fractal Trees to Log-Structured Merge (LSM) Trees, *Tokutek Technical Report*, 2014.

45. P. O'Neil, E. Cheng, D. Gawlick, and E. O'Neil, The log-structured merge-tree (LSM-tree), *Acta Inf.*, vol. 33, no. 4, pp. 351–385, 1996.

46. R. Spillane, Efficient, Scalable, and Versatile Application and System Transaction Management for Direct Storage Layers, *Ph.D. Dissertation*, 2012.

47. J. S. Vitter, External memory algorithms and data structures: Dealing with massive data, *ACM Computing Surveys*, vol. 33, no. 2, pp. 209–271, June 2001.

48. H. V. Jagadish, P. P. S. Narayan, S. Seshadri, S. Sudarshan, and R. Kanneganti, Incremental organization for data recording and warehousing, in *Proceedings of the 23rd International Conference on Very Large Data Bases*, Athens, Greece, August 26–29, 1997, pp. 16–25.

49. B. Chazelle and L. J. Guibas, Fractional cascading. I: a data structuring technique, *Algorithmica*, vol. 1, pp. 133–162, 1986.

50. B. Chazelle and L. J. Guibas, Fractional cascading. II: a data structuring technique, *Algorithmica*, vol. 1, pp. 163–191, 1986.

51. M. A. Bender, M. Farach-Colton, J. T. Fineman, Y. R. Fogel, B. C. Kuszmaul, and J. Nelson, Cache-oblivious streaming B-trees, in *Proceedings of the Nineteenth Annual ACM Symposium on Parallel Algorithms and Architectures*, 2007, San Diego, California, June 9–11, pp. 81–92.

52. J. Wang, Y. Zhang, Y. Gao, and C. Xing, pLSM: A highly efficient LSM-tree index supporting real-time big data analysis. in *Proceedings of the IEEE Computer Society International Conference on Computers, Software and Applications*, Paris, France, June 1–2, 2013, pp. 240–245.

53. H. Amur, D. G. Andersen, M. Kaminsky, and K. Schwan, Design of a Write-Optimized Data Store, *Georgia Tech Technical Report*, vol. GIT-CERCS-13-08, 2013.

54. AsterixDB, https://asterixdb.ics.uci.edu/, accessed: 2015-10-09.

55. C. Jermaine, E. Omiecinski, and W. G. Yee, The partitioned exponential file for database storage management, *The VLDB Journal*, vol. 16, no. 4, pp. 417–437, 2007.

56. G. Margaritis and S. Anastasiadis, Efficient range-based storage management for scalable datastores, *IEEE Transactions on Parallel and Distributed Systems*, vol. 25, no. 11, pp. 2851–2866, 2014.

57. C. Mathieu, C. Staelin, and N. E. Young, K-slot SSTable stack compaction, *arXiv preprint arXiv:1407.3008*, 2014.

58. M. Ghosh, I. Gupta, S. Gupta, and N. Kumar, *IEEE 35th International Conference on Distributed Computing Systems (ICDCS)*, Columbus, Ohio, June 29–July 2, 2015.

59. L. Arge, The buffer tree: A new technique for optimal i/o-algorithms (extended abstract), in *Proceedings of the 4th International Workshop on Algorithms and Data Structures*, Kingston, Canada, August 16–18, 1995, pp. 334–345.

60. A. Twigg, A. Byde, G. Milos, T. Moreton, J. Wilkes, and T. Wilkie, Stratified b-trees and versioned dictionaries, in *Proceedings of the 3rd USENIX Conference on Hot Topics in Storage and File Systems*, ser. HotStorage'11, Portland, Oregon, June 14–17, 2011, pp. 10–10.

61. A. Byde and A. Twigg, Optimal query/update tradeoffs in versioned dictionaries, *CoRR*, vol. abs/1103.2566, 2011. [Online]. Available at: http://arxiv.org/abs/1103.2566

62. G. Golan-Gueta, E. Bortnikov, E. Hillel, and I. Keidar, Scaling concurrent log-structured data stores, in *Proceedings of the Tenth European Conference on Computer Systems*, Bordeaux, France, April 22–24, 2015, pp. 32:1–32:14.

63. N. G. Bronson, J. Casper, H. Chafi, and K. Olukotun, A practical concurrent binary search tree, in *Proceedings of the 15th ACM SIGPLAN Symposium on Principles and Practice of Parallel Programming*, ser. PPoPP'10, Bordeaux, France, April 22–24, 2010, pp. 257–268.

64. libcds: Library of lock-free and fine-grained algorithms, http://libcds.sourceforge.net/, accessed: 2015-10-22.

65. Java's ConcurrentSkipListMap, http://docs.oracle.com/javase/7/docs/api/java/util/concurrent/ConcurrentSkipListMap.html, accessed: 2015-10-22.

66. D. Abadi, R. Agrawal, A. Ailamaki, M. Balazinska, P. A. Bernstein, M. J. Carey, S. Chaudhuri et al., The Beckman report on database research, *SIGMOD Record*, vol. 43, no. 3, pp. 61–70, December 2014.

67. J. Eastep, D. Wingate, and A. Agarwal, Smart data structures, an online machine learning approach to multicore data structures, in *Proceedings of the 8th ACM International Conference on Autonomic Computing*, Karlsruhe, Germany, June 14–18, 2011, pp. 11–20.

64

Data Mining*

64.1 Introduction.. 997
    Data Mining Tasks and Techniques • Challenges of Data Mining • Data Mining and the
    Role of Data Structures and Algorithms

64.2 Classification.. 1000
    Nearest-Neighbor Classifiers • Proximity Graphs for Enhancing Nearest Neighbor Classifiers

64.3 Association Analysis... 1002
    Hash Tree Structure • FP-Tree Structure

64.4 Clustering... 1007
    Hierarchical and Partitional Clustering • Nearest Neighbor Search and Multi-Dimensional
    Access Methods

64.5 Conclusion 1010
Acknowledgments 1010
References.. 1010

Vipin Kumar
University of Minnesota

Pang-Ning Tan
Michigan State University

Michael Steinbach
University of Minnesota

64.1 Introduction

Recent years have witnessed an explosive growth in the amounts of data collected, stored, and disseminated by various organizations. Examples include (1) the large volumes of point-of-sale data amassed at the checkout counters of grocery stores, (2) the continuous streams of satellite images produced by Earth-observing satellites, and (3) the avalanche of data logged by network monitoring software. To illustrate how much the quantity of data has grown over the years, Figure 64.1 shows an example of the number of Web pages indexed by a popular Internet search engine since 1998.

In each of the domains described above, data is collected to satisfy the information needs of the various organizations: Commercial enterprises analyze point-of-sale data to learn the purchase behavior of their customers; Earth scientists use satellite image data to advance their understanding of how the Earth system is changing in response to natural and human-related factors; and system administrators employ network traffic data to detect potential network problems, including those resulting from cyber-attacks.

One immediate difficulty encountered in these domains is how to extract useful information from massive data sets. Indeed, getting information out of the data is like *drinking from a fire hose*. The sheer size of the data simply overwhelms our ability to manually sift through the data, hoping to find useful information. Fueled by the need to rapidly analyze and summarize the data, researchers have turned to *data mining* techniques [1–5]. In a nutshell, data mining is the task of *discovering interesting knowledge automatically from large data repositories*.

Interesting knowledge has different meanings to different people. From a business perspective, knowledge is interesting if it can be used by analysts or managers to make profitable business decisions. For Earth Scientists, knowledge is interesting if it reveals previously unknown information about the characteristics of the Earth system. For system administrators, knowledge is interesting if it indicates unauthorized or illegitimate use of system resources.

Data mining is often considered to be an integral part of another process, called *Knowledge Discovery in Databases* (or KDD). KDD refers to the overall process of turning raw data into interesting knowledge and consists of a series of transformation steps, including data preprocessing, data mining, and postprocessing. The objective of data preprocessing is to convert data into the right format for subsequent analysis by selecting the appropriate data segments and extracting attributes that are relevant to the data mining task (feature selection and construction). For many practical applications, more than half of the knowledge discovery efforts are devoted to data preprocessing. Postprocessing includes all additional operations performed to make the data mining

---

* This chapter has been reprinted from first edition of this Handbook, without any content updates.

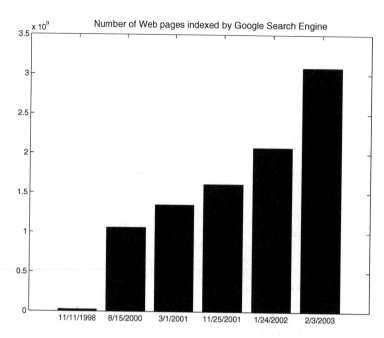

FIGURE 64.1 Number of Web pages indexed by the Google© search engine (Source: Internet Archive, http://www.archive.org).

results more accessible and easier to interpret. For example, the results can be sorted or filtered according to various *measures* to remove uninteresting patterns. In addition, *visualization* techniques can be applied to help analysts explore data mining results.

64.1.1 Data Mining Tasks and Techniques

Data mining tasks are often divided into two major categories:

Predictive The goal of predictive tasks is to use the values of some variables to predict the values of other variables. For example, in Web mining, e-tailers are interested in predicting which online users will make a purchase at their Web site. Other examples include biologists, who would like to predict the functions of proteins, and stock market analysts, who would like to forecast the future prices of various stocks.

Descriptive The goal of descriptive tasks is to find human-interpretable patterns that describe the underlying relationships in the data. For example, Earth Scientists are interested in discovering the primary forcings influencing observed climate patterns. In network intrusion detection, analysts want to know the kinds of cyber-attacks being launched against their networks. In document analysis, it is useful to find groups of documents, where the documents in each group share a common topic.

Data mining tasks can be accomplished using a variety of data mining techniques, as shown in Figure 64.2.

- **Predictive modeling** is used primarily for predictive data mining tasks. The input data for predictive modeling consists of two distinct types of variables: (1) explanatory variables, which define the essential properties of the data, and (2) one or more target variables, whose values are to be predicted. For the Web mining example given in the previous section, the input variables correspond to the demographic features of online users, such as age, gender, and salary, along with their browsing activities, for example, what pages are accessed and for how long. There is one binary target variable, Buy, which has values, Yes or No, indicating, respectively, whether the user will buy anything from the Web site or not. Predictive modeling techniques can be further divided into two categories: *classification* and *regression*. Classification techniques are used to predict the values of discrete target variables, such as the Buy variable for online users at a Web site. For example, they can be used to predict whether a customer will most likely be lost to a competitor, that is, customer churn or attrition, and to determine the category of a star or galaxy for sky survey cataloging. Regression techniques are used to predict the values of continuous target variables, for example, they can be applied to forecast the future price of a stock.
- **Association rule mining** seeks to produce a set of dependence rules that predict the occurrence of a variable given the occurrences of other variables. For example, association analysis can be used to identify products that are often purchased together by sufficiently many customers, a task that is also known as *market basket analysis*. Furthermore, given a database that records a sequence of events, for example, a sequence of successive purchases by customers, an important task is that

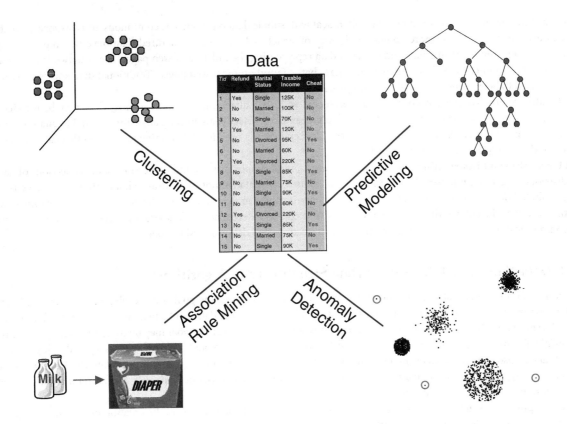

FIGURE 64.2   Data mining techniques.

of finding dependence rules that capture the temporal connections of events. This task is known as *sequential pattern analysis*.

- **Cluster analysis** finds groupings of data points so that data points that belong to one cluster are more similar to each other than to data points belonging to a different cluster, for example, clustering can be used to perform market segmentation of customers, document categorization, or land segmentation according to vegetation cover. While cluster analysis is often used to better understand or describe the data, it is also useful for summarizing a large data set. In this case, the objects belonging to a single cluster are replaced by a single *representative* object, and further data analysis is then performed using this reduced set of representative objects.

- **Anomaly detection** identifies data points that are significantly different than the rest of the points in the data set. Thus, anomaly detection techniques have been used to detect network intrusions and to predict fraudulent credit card transactions. Some approaches to anomaly detection are statistically based, while other are based on distance or graph-theoretic notions.

64.1.2 Challenges of Data Mining

There are several important challenges in applying data mining techniques to large data sets:

Scalability Scalable techniques are needed to handle the massive size of some of the datasets that are now being created. As an example, such datasets typically require the use of efficient methods for storing, indexing, and retrieving data from secondary or even tertiary storage systems. Furthermore, parallel or distributed computing approaches are often necessary if the desired data mining task is to be performed in a timely manner. While such techniques can dramatically increase the size of the datasets that can be handled, they often require the design of new algorithms and data structures.

Dimensionality In some application domains, the number of dimensions (or attributes of a record) can be very large, which makes the data difficult to analyze because of the "curse of dimensionality" [6]. For example, in bioinformatics, the development of advanced microarray technologies allows us to analyze gene expression data with thousands of attributes. The dimensionality of a data mining problem may also increase substantially due to the temporal, spatial, and sequential nature of the data.

Complex Data Traditional statistical methods often deal with simple data types such as continuous and categorical attributes. However, in recent years, more complicated types of structured and semi-structured data have become more important. One example of such data is graph-based data representing the linkages of web pages, social networks, or chemical structures. Another example is the free-form text that is found on most web pages. Traditional data analysis techniques often need to be modified to handle the complex nature of such data.

Data Quality Many data sets have one or more problems with data quality, for example, some values may be erroneous or inexact, or there may be missing values. As a result, even if a "perfect" data mining algorithm is used to analyze the data, the information discovered may still be incorrect. Hence, there is a need for data mining techniques that can perform well when the data quality is less than perfect.

Data Ownership and Distribution For a variety of reasons, for example, privacy and ownership, some collections of data are distributed across a number of sites. In many such cases, the data cannot be centralized, and thus, the choice is either distributed data mining or no data mining. Challenges involved in developing distributed data mining solutions include the need for efficient algorithms to cope with distributed and heterogeneous data sets, the need to minimize the cost of communication, and the need to accommodate data security and data ownership policies.

64.1.3 Data Mining and the Role of Data Structures and Algorithms

Research in data mining is motivated by a number of factors. In some cases, the goal is to develop an approach with greater efficiency. For example, a current technique may work well as long as all of the data can be held in main memory, but the size of data sets has grown to the point where this is no longer possible. In other cases, the goal may be to develop an approach that is more flexible. For instance, the nature of the data may be continually changing, and it may be necessary to develop a model of the data that can also change. As an example, network traffic varies in volume and kind, often over relatively short time periods. In yet other cases, the task is to obtain a more accurate model of the data, that is, one that takes into account additional factors that are common in many real world situations.

The development and success of new data mining techniques is heavily dependent on the creation of the proper algorithms and data structures to address the needs such as those just described: efficiency, flexibility, and more accurate models. (This is not to say that system or applications issues are unimportant.) Sometimes, currently existing data structures and algorithms can be directly applied, for example, data access methods can be used to efficiently organize and retrieve data. However, since currently existing data structures and algorithms were typically not designed with data mining tasks in mind, it is frequently the case that some modifications, enhancements, or completely new approaches are needed, that is, new work in data structures and algorithms is needed. We would emphasize, though, that sometimes it is the concepts and viewpoints associated with currently existing algorithms and data structures that are the most useful. Thus, the realization that a problem can be formulated as a particular type of a graph or tree may quickly lead to a solution.

In the following sections, we provide some examples of how data structures play an important role, both conceptually and practically, for classification, association analysis, and clustering.

64.2 Classification

Classification [7,8] is the task of assigning objects to their respective categories. For example, stock analysts are interested in classifying the stocks of publicly-owned companies as buy, hold, or sell, based on the financial outlook of these companies. Stocks classified as buy are expected to have stronger future revenue growth than those classified as sell. In addition to the practical uses of classification, it also helps us to understand the similarities and differences between objects that belong to different categories.

The data set in a classification problem typically consists of a collection of *records* or data objects. Each record, also known as an *instance* or *example*, is characterized by a tuple $(\mathbf{x}, y)$, where $\mathbf{x}$ is the set of explanatory variables associated with the object and y is the object's class label. A record is said to be *labeled* if the value of y is known; otherwise, the record is *unlabeled*. Each attribute $x_k \in \mathbf{x}$ can be discrete or continuous. On the other hand, the class label y must be a discrete variable whose value is chosen from a finite set $\{y_1, y_2, \ldots y_c\}$. If y is a continuous variable, then this problem is known as *regression*.

The classification problem can be stated formally as follows:

Classification is the task of learning a function, $f : \mathbf{x} \to y$, that maps the explanatory variables $\mathbf{x}$ of an object to one of the class labels for y.

f is known as the *target function* or *classification model*.

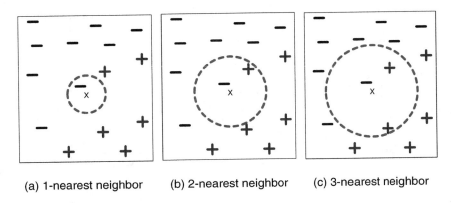

(a) 1-nearest neighbor   (b) 2-nearest neighbor   (c) 3-nearest neighbor

FIGURE 64.3   The 1-, 2- and 3-nearest neighbors of an instance.

(k: number of nearest neighbor, E: training instances, z: unlabeled instance)
1: Compute the distance or similarity of z to all the training instances
2: Let $E' \subset E$ be the set of k closest training instances to z
3: Return the predicted class label for z: *class* $\leftarrow$ *Voting*(E').

FIGURE 64.4 k-nearest neighbor classification algorithm.

64.2.1 Nearest-Neighbor Classifiers

Typically, the classification framework presented involves a two-step process: (1) an inductive step for constructing classification models from data, and (2) a deductive step for applying the derived model to previously unseen instances. For decision tree induction and rule-based learning systems, the models are constructed immediately after the training set is provided. Such techniques are known as *eager learners* because they intend to learn the model as soon as possible, once the training data is available.

An opposite strategy would be to delay the process of generalizing the training data until it is needed to classify the unseen instances. One way to do this is to find all training examples that are relatively similar to the attributes of the test instance. Such examples are known as the nearest neighbors of the test instance. The test instance can then be classified according to the class labels of its neighbors. This is the central idea behind the *nearest-neighbor classification* scheme [8–11], which is useful for classifying data sets with continuous attributes. A nearest neighbor classifier represents each instance as a data point embedded in a d-dimensional space, where d is the number of continuous attributes. Given a test instance, we can compute its distance to the rest of the data objects (data points) in the training set by using an appropriate distance or similarity measure, for example, the standard Euclidean distance measure.

The k-nearest neighbors of an instance z are defined as the data points having the k smallest distances to z. Figure 64.3 illustrates an example of the 1-, 2- and 3-nearest neighbors of an unknown instance, $\times$, located at the center of the circle. The instance can be assigned to the class label of its nearest neighbors. If the nearest neighbors contain more than one class label, then one takes a majority vote among the class labels of the nearest neighbors.

The nearest data point to the unknown instance shown in Figure 64.3a has a negative class label. Thus, in a 1-nearest neighbor classification scheme, the unknown instance would be assigned to a negative class. If we consider a larger number of nearest neighbors, such as three, the list of nearest neighbors would contain training examples from 2 positive classes and 1 negative class. Using the majority voting scheme, the instance would be classified as a positive class. If the number of instances from both classes are the same, as in the case of the 2-nearest neighbor classification scheme shown in Figure 64.3b, we could choose either one of the classes (or the default class) as the class label.

A summary of the k-nearest neighbor classification algorithm is given in Figure 64.4. Given an unlabeled instance, we need to determine its distance or similarity to all the training instances. This operation can be quite expensive and may require efficient indexing techniques to reduce the amount of computation.

While one can take a majority vote of the nearest neighbors to select the most likely class label, this approach may not be desirable because it assumes that the influence of each nearest neighbor is the same. An alternative approach is to weight the influence of each nearest neighbor according to its distance, so that the influence is weaker if the distance is too large.

64.2.2 Proximity Graphs for Enhancing Nearest Neighbor Classifiers

The nearest neighbor classification scheme, while simple, has a serious problem as currently presented: It is necessary to store all of the data points, and to compute the distance between an object to be classified and all of these stored objects. If the set of

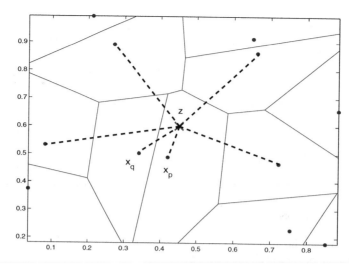

FIGURE 64.5   Voronoi diagram.

original data points is large, then this can be a significant computational burden. Hence, a considerable amount of research has been conducted into strategies to alleviate this problem.

There are two general strategies for addressing the problem just discussed:

Condensing The idea is that we can often eliminate many of the data points without affecting classification performance, or without affecting it very much. For instance, if a data object is in the "middle" of a group of other objects with the same class, then its elimination will likely have no effect on the nearest neighbor classifier.

Editing Often, the classification performance of a nearest neighbor classifier can be enhanced by deleting certain data points. More specifically, if a given object is compared to its nearest neighbors and most of them are of another class (i.e., if the points that would be used to classify the given point are of another class), then deleting the given object will often improve classifier performance.

While various approaches to condensing and editing points to improve the performance of nearest neighbor classifiers have been proposed, there has been a considerable amount of work that approaches this problem from the viewpoint of computational geometry, especially proximity graphs [12]. Proximity graphs include nearest neighbor graphs, minimum spanning trees, relative neighborhood graphs, Gabriel graphs, and the Delaunay triangulation [13]. We can only indicate briefly the usefulness of this approach, and refer the reader to [12] for an in depth discussion.

First, we consider how Voronoi diagrams can be used to eliminate points that add nothing to the classification. (The Voronoi diagram for a set of data points is the set of polygons formed by partitioning all of points in the space into a set of convex regions such that every point in a region is closer to the data point in the same region than to any other data point. Figure 64.5 shows a Voronoi diagram for a number of two-dimensional points.) Specifically, if all the Voronoi neighbors of a point, that is, those points belonging to Voronoi regions that touch the Voronoi region of the given point, have the same class as the given data point, then discarding that point cannot affect the classification performance. The reason for this is that the Voronoi regions of the neighboring points will expand to "occupy" the space once occupied by the the Voronoi region of the given point, and thus, classification behavior is unchanged. More sophisticated approaches based on proximity graphs are possible [12].

For editing, that is, discarding points to approve classification performance, proximity graphs can also be useful. In particular, instead of eliminating data points whose k nearest neighbors are of a different class, we build a proximity graph and eliminate those points where a majority of the neighbors in the proximity graph are of a different class. Of course, the results will depend on the type of proximity graph. The Gabriel graph has been found to be the best, but for further discussion, we once again refer the reader to [12], and the extensive list of references that it contains.

In summary, our goal in this section was to illustrate that—for one particular classification scheme, nearest neighbor classification—the rich set of data structures and algorithms of computational geometry, that is, proximity graphs, have made a significant contribution, both practically and theoretically.

64.3 Association Analysis

An important problem in data mining is the discovery of association patterns [14] present in large databases. This problem was originally formulated in the context of *market basket* data, where the goal is to determine whether the occurrence of certain items

TID	Items
1	Bread, Coke, Milk
2	Beer, Bread
3	Beer, Coke, Diaper, Milk
4	Beer, Bread, Diaper, Milk
5	Coke, Diaper, Milk

FIGURE 64.6 Market-basket transactions.

in a transaction can be used to infer the occurrence of other items. If such interesting relationships are found, then they can be put to various profitable uses such as marketing promotions, shelf management, inventory management, etc.

To formalize the problem, let $T = \{t_1, t_2, \ldots, t_N\}$ be the set of all transactions and $I = \{i_1, i_2, \ldots, i_d\}$ be the set of all items. Any subset of I is known as an *itemset*. The *support count* of an itemset C is defined as the number of transactions in T that contain C, that is,

$$\sigma(C) = |\{t | t \in T, C \subseteq t\}|.$$

An *association rule* is an implication of the form $X \rightarrow Y$, where X and Y are itemsets and $X \cap Y = \emptyset$. The strength of an association rule is given by its *support* (*s*) and *confidence* (*c*) measures. The support of the rule is defined as the fraction of transactions in T that contain itemset $X \cup Y$.

$$s(X \longrightarrow Y) = \frac{\sigma(X \cup Y)}{|T|}.$$

Confidence, on the other hand, provides an estimate of the conditional probability of finding items of Y in transactions that contain X.

$$c(X \longrightarrow Y) = \frac{\sigma(X \cup Y)}{\sigma(X)}$$

For example, consider the market basket transactions shown in Figure 64.6. The support for the rule {Diaper, Milk}→{Beer} is $\sigma(Diaper, Milk, Beer)/5 = 2/5 = 40\%$, whereas its confidence is $\sigma(Diaper, Milk, Beer)/\sigma(Diaper, Milk) = 2/3 = 66\%$.

Support is useful because it reflects the significance of a rule. Rules that have very low support are rarely observed, and thus, are more likely to occur by chance. Confidence is useful because it reflects the reliability of the inference made by each rule. Given an association rule $X \rightarrow Y$, the higher the confidence, the more likely it is to find Y in transactions that contain X. Thus, the goal of association analysis is to automatically discover association rules having relatively high support and high confidence. More specifically, an association rule is considered to be interesting only if its support is greater than or equal to a minimum support threshold, *minsup*, and its confidence is greater than or equal to a minimum confidence threshold, *minconf*.

The association analysis problem is far from trivial because of the exponential number of ways in which items can be grouped together to form a rule. In addition, the rules are constrained by two completely different conditions, stated in terms of the *minsup* and *minconf* thresholds. A standard way for generating association rules is to divide the process into two steps. The first step is to find all itemsets that satisfy the minimum support threshold. Such itemsets are known as *frequent itemsets* in the data mining literature. The second step is to generate high-confidence rules only from those itemsets found to be frequent. The completeness of this two-step approach is guaranteed by the fact that any association rule $X \rightarrow Y$ that satisfies the *minsup* requirement can always be generated from a frequent itemset $X \cup Y$.

Frequent itemset generation is the computationally most expensive step because there are 2^d possible ways to enumerate all itemsets from I. Much research has therefore been devoted to developing efficient algorithms for this task. A key feature of these algorithms lies in their strategy for controlling the exponential complexity of enumerating candidate itemsets. Briefly, the algorithms make use of the anti-monotone property of itemset support, which states that all subsets of a frequent itemset must be frequent. Put another way, if a candidate itemset is found to be infrequent, we can immediately prune the search space spanned by supersets of this itemset. The *Apriori* algorithm, developed by Agrawal et al. [15], pioneered the use of this property to systematically enumerate the candidate itemsets. During each iteration k, it generates only those *candidate* itemsets of length k whose $(k-1)$-subsets are found to be frequent in the previous iteration. The support counts of these candidates are then determined by scanning the transaction database. After counting their supports, candidate k-itemsets that pass the *minsup* threshold are declared to be frequent.

Well-designed data structures are central to the efficient mining of association rules. The *Apriori* algorithm, for example, employs a hash-tree structure to facilitate the support counting of candidate itemsets. On the other hand, algorithms such as FP-growth [16] and H-Miner [17] employ efficient data structures to provide a compact representation of the transaction database. A brief description of the hash tree and FP-tree data structures is presented next.

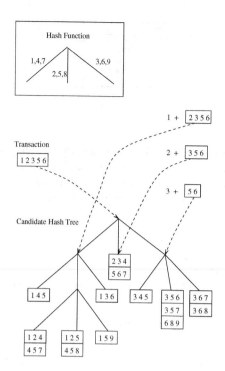

FIGURE 64.7   Hashing a transaction at the root node of a hash tree.

64.3.1 Hash Tree Structure

Apriori is a level-wise algorithm that generates frequent itemsets one level at a time, from itemsets of size-1 up to the longest frequent itemsets. At each level, candidate itemsets are generated by extending the frequent itemsets found at the previous level. Once the candidate itemsets have been enumerated, the transaction database is scanned once to determine their actual support counts. This generate-and-count procedure is repeated until no new frequent itemsets are found.

Support counting of candidate itemsets is widely recognized as the key bottleneck of frequent itemset generation. This is because one has to determine the candidate itemsets contained in each transaction of the database. A naive way for doing this is to simply match each transaction against every candidate itemset. If the candidate is a subset of the transaction, its support count is incremented. This approach can be prohibitively expensive if the number of candidates and number of transactions are large.

In the *Apriori* algorithm, candidate itemsets are hashed into different buckets and stored in a hash tree structure. During support counting, each transaction is also hashed into its appropriate buckets. This way, instead of matching a transaction against all candidate itemsets, the transaction is matched only to those candidates that belong to the same bucket.

Figure 64.7 illustrates an example of a hash tree for storing candidate itemsets of size 3. Each internal node of the hash tree contains a hash function that determines which branch of the current node is to be followed next. The hash function used by the tree is also shown in this figure. Specifically, items 1, 4 and 7 are hashed to the left child of the node; items 2, 5, 8 are hashed to the middle child; and items 3, 6, 9 are hashed to the right child. Candidate itemsets are stored at the leaf nodes of the tree. The hash tree shown in Figure 64.7 contains 15 candidate itemsets, distributed across 9 leaf nodes.

We now illustrate how to enumerate candidate itemsets contained in a transaction. Consider a transaction t that contains five items, $\{1, 2, 3, 5, 6\}$. There are $^5C_3 = 10$ distinct itemsets of size 3 contained in this transaction. Some of these itemsets may correspond to the candidate 3-itemsets under investigation, in which case, their support counts are incremented. Other subsets of t that do not correspond to any candidates can be ignored.

Figure 64.8 shows a systematic way for enumerating size-3 itemsets contained in the transaction t by specifying the items one-by-one. It is assumed that items in every 3-itemset are stored in increasing lexicographic order. Because of the ordering constraint, all itemsets of size-3 derived from t must begin with item 1, 2, or 3. No 3-itemset may begin with item 5 or 6 because there are only two items in this transaction that are greater than or equal to 5. This is illustrated by the level 1 structures depicted in Figure 64.8. For example, the structure 1 | 2 3 5 6 | represents an itemset that begins with 1, followed by two more items chosen from the set $\{2, 3, 5, 6\}$.

After identifying the first item, the structures at level 2 denote the various ways to select the second item. For example, the structure 1 2 | 3 5 6 | corresponds to itemsets with prefix (1 2), followed by either item 3, 5, or 6. Once the first two items have

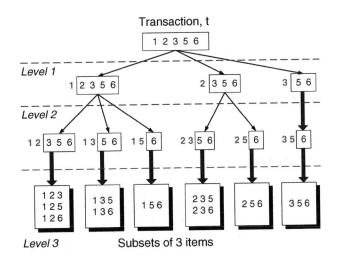

FIGURE 64.8 Enumerating subsets of three items from a transaction *t*.

been chosen, the structures at level 3 represent the complete set of 3-itemsets derived from transaction *t*. For example, the three itemsets beginning with the prefix {1 2} are shown in the leftmost box at level 3 of this figure.

The tree-like structure shown in Figure 64.8 is simply meant to demonstrate how subsets of a transaction can be enumerated, that is, by specifying the items in the 3-itemsets one-by-one, from its left-most item to its right-most item. For support counting, we still have to match each subset to its corresponding candidate. If there is a match, then the support count for the corresponding candidate is incremented.

We now describe how a hash tree can be used to determine candidate itemsets contained in the transaction $t = \{1, 2, 3, 5, 6\}$. To do this, the hash tree must be traversed in such a way that all leaf nodes containing candidate itemsets that belong to *t* are visited. As previously noted, all size-3 candidate itemsets contained in *t* must begin with item 1, 2, or 3. Therefore, at the root node of the hash tree, we must hash on items 1, 2, and 3 separately. Item 1 is hashed to the left child of the root node; item 2 is hashed to the middle child of the root node; and item 3 is hashed to the right child of the root node. Once we reach a child of the root node, we need to hash on the second item of the level 2 structures given in Figure 64.8. For example, after hashing on item 1 at the root node, we need to hash on items 2, 3, and 5 at level 2. Hashing on items 2 or 5 will lead us to the middle child node while hashing on item 3 will lead us to the right child node, as depicted in Figure 64.9. This process of hashing on items that belong to the transaction continues until we reach the leaf nodes of the hash tree. Once a leaf node is reached, all candidate itemsets stored at the leaf are compared against the transaction. If a candidate belongs to the transaction, its support count is incremented. In this example, 6 out of the 9 leaf nodes are visited and 11 out of the 15 itemsets are matched against the transaction.

64.3.2 FP-Tree Structure

Recently, an interesting algorithm called FP-growth was proposed that takes a radically different approach to discovering frequent itemsets. The algorithm does not subscribe to the generate-and-count paradigm of Apriori. It encodes the database using a compact data structure called an FP-tree and infers frequent itemsets directly from this structure.

First, the algorithm scans the database once to find the frequent singleton items. An order is then imposed on the items based on decreasing support counts. Figure 64.10 illustrates an example of how to construct an FP-tree from a transaction database that contains five items, A, B, C, D, and E. Initially, the FP-tree contains only the root node, which is represented by a null symbol. Next, each transaction is used to create a path from the root node to some node in the FP-tree.

After reading the first transaction, {A, B}, a path is formed from the root node to its child node, labeled as A, and subsequently, to another node labeled as B. Each node in the tree contains the symbol of the item along with a count of the transactions that reach the particular node. In this case, both nodes A and B would have a count equal to one. After reading the second transaction {B,C,D} a new path extending from null → B → C → D is created. Again, the nodes along this path have support counts equal to one. When the third transaction is read, the algorithm will discover that this transaction shares a common prefix A with the first transaction. As a result, the path null → A → C → D is merged to the existing path null → A → B. The support count for node A is incremented to two, while the newly-created nodes, C and D, each have a support count equal to one. This process is repeated until all the transactions have been mapped into one of the paths in the FP-tree. For example, the state of the FP-tree after reading the first ten transactions is shown at the bottom of Figure 64.10.

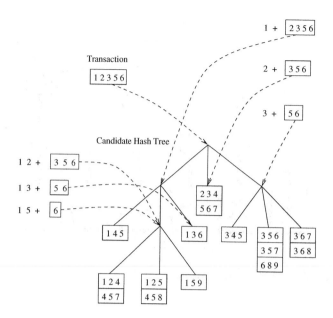

FIGURE 64.9 Subset operation on the left most subtree of the root of a candidate hash tree.

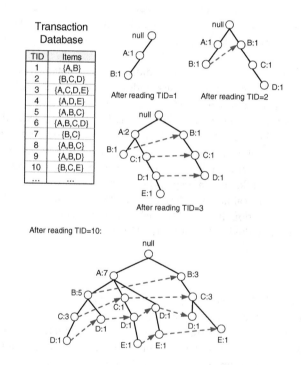

FIGURE 64.10 Construction of an FP-tree.

By looking at the way the tree is constructed, we can see why an FP-tree provides a compact representation of the database. If the database contains many transactions that share common items, then the size of an FP-tree will be considerably smaller than the size of the database. The best-case scenario would be that the database contains the same set of items for all transactions. The resulting FP-tree would contain only a single branch of nodes. The worst-case scenario would be that each transaction contains a unique set of items. In this case, there is no sharing of transactions among the nodes and the size of the FP-tree is the same as the size of the database.

During tree construction, the FP-tree structure also creates a linked-list access mechanism for reaching every individual occurrence of each frequent item used to construct the tree. In the above example, there are five such linked lists, one for each item, A, B, C, D, and E.

1: Initialization: Select K points as the initial centroids.
2: **repeat**
3:    Form K clusters by assigning all points to the closest centroid.
4:    Recompute the centroid of each cluster.
5: **until** The centroids do not change

FIGURE 64.11 Basic K-means algorithm.

The algorithm used for generating frequent itemsets from an FP-tree is known as *FP-growth*. Given the FP-tree shown in Figure 64.10, the algorithm divides the problem into several subproblems, where each subproblem involves finding frequent itemsets having a particular suffix. In this example, the algorithm initially looks for frequent itemsets that end in E by following the linked list connecting the nodes for E. After all frequent itemsets ending in E are found, the algorithm looks for frequent itemsets that end in D by following the linked list for D, and so on.

How does FP-growth find all the frequent itemsets ending in E? Recall that an FP-tree stores the support counts of every item along each path, and that these counts reflect the number of transactions that are collapsed onto that particular path. In our example, there are only three occurrences of the node E. By collecting the *prefix paths* of E, we can solve the subproblem of finding frequent itemsets ending in E. The prefix paths of E consist of all paths starting from the root node up to the parent nodes of E. These prefix paths can form a new FP-tree to which the FP-growth algorithm can be recursively applied.

Before creating a new FP-tree from the prefix paths, the support counts of items along each prefix path must be updated. This is because the initial prefix path may include several transactions that do not contain the item E. For this reason, the support count of each item along the prefix path must be adjusted to have the same count as node E for that particular path. After updating the counts along the prefix paths of E, some items may no longer be frequent, and thus, must be removed from further consideration (as far as our new subproblem is concerned). An FP-tree of the prefix paths is then constructed by removing the infrequent items. This recursive process of breaking the problem into smaller subproblems will continue until the subproblem involves only a single item. If the support count of this item is greater than the minimum support threshold, then the label of this item will be returned by the FP-growth algorithm. The returned label is appended as a prefix to the frequent itemset ending in E.

64.4 Clustering

Cluster analysis [18–21] groups data objects based on information found in the data that describes the objects and their relationships. The goal is that the objects in a group be similar (or related) to one another and different from (or unrelated to) the objects in other groups. The greater the similarity (or homogeneity) within a group, and the greater the difference between groups, the "better" or more distinct the clustering.

64.4.1 Hierarchical and Partitional Clustering

The most commonly made distinction between clustering techniques is whether the resulting clusters are nested or unnested or, in more traditional terminology, whether a set of clusters is *hierarchical* or *partitional*. A partitional or unnested set of clusters is simply a division of the set of data objects into non-overlapping subsets (clusters) such that each data object is in exactly one subset, that is, a partition of the data objects. The most common partitional clustering algorithm is K-means, whose operation is described by the psuedo-code in Figure 64.11. (K is a user specified parameter, that is, the number of clusters desired, and a centroid is typically the mean or median of the points in a cluster.)

A hierarchical or nested clustering is a set of nested clusters organized as a hierarchical tree, where the leaves of the tree are singleton clusters of individual data objects, and where the cluster associated with each interior node of the tree is the union of the clusters associated with its child nodes. Typically, hierarchical clustering proceeds in an agglomerative manner, that is, starting with each point as a cluster, we repeatedly merge the closest clusters, until only one cluster remains. A wide variety of methods can be used to define the distance between two clusters, but this distance is typically defined in terms of the distances between pairs of points in different clusters. For instance, the distance between clusters may be the minimum distance between any pair of points, the maximum distance, or the average distance. The algorithm for agglomerative clustering is described by the psuedo-code in Figure 64.12.

The tree that represents a hierarchical clustering is called a *dendrogram*, a term that comes from biological taxonomy. Figure 64.13 shows six points and the hierarchical clustering that is produced by the MIN clustering technique. This approach creates a hierarchical clustering by starting with the individual points as clusters, and then successively merges pairs of clusters with the minimum distance, that is, that have the closest pair of points.

1: Compute the pairwise distance matrix.
2: **repeat**
3:   Merge the closest two clusters.
4:   Update the distance matrix to reflect the distance between the new cluster and the original clusters.
5: **until** Only one cluster remains

FIGURE 64.12   Basic agglomerative hierarchical clustering algorithm.

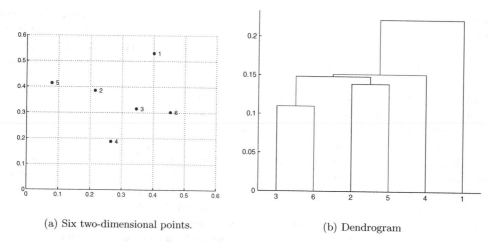

(a) Six two-dimensional points.   (b) Dendrogram

FIGURE 64.13   A hierarchical clustering of six points.

64.4.2 Nearest Neighbor Search and Multi-Dimensional Access Methods

In both the K-means and agglomerative hierarchical clustering algorithms, the time required is heavily dependent on the amount of time that it takes to find the the distance between sets of points—step 3 in both algorithms. This is true of most other clustering schemes as well, and thus, efficient implementations of clustering techniques often require considerations of nearest-neighbor search and the related area of multi-dimensional access methods. In the remainder of this section, we discuss these areas and their relevance to cluster analysis.

We begin by considering a general situation. Given a set of objects or data points, including possibly complicated data such as images, text strings, DNA sequences, polygons, etc., two issues are key to efficiently utilizing this data:

1. **How can items be located efficiently?** While a "representative" feature vector is commonly used to "index" these objects, these data points will normally be very sparse in the space, and thus, it is not feasible to use an array to store the data. Also, many sets of data do not have any features that constitute a "key" that would allow the data to be accessed using standard and efficient database techniques.

2. **How can similarity queries be efficiently conducted?** Many applications, including clustering, require the nearest neighbor (or the k nearest neighbors) of a point. For instance, the clustering techniques DBSCAN [22] and Chameleon [23] will have a time complexity of $O(n^2)$ unless they can utilize data structures and algorithms that allow the nearest neighbors of a point to be located efficiently. As a non-clustering example of an application of similarity queries, a user may want to find all the pictures similar to a particular photograph in a database of photographs.

Techniques for nearest-neighbor search are often discussed in papers describing multi-dimensional access methods or spatial access methods, although strictly speaking the topic of multi-dimensional access methods is broader than nearest-neighbor search since it addresses all of the many different types of queries and operations that a user might want to perform on multi-dimensional data. A large amount of work has been done in the area of nearest neighbor search and multi-dimensional access methods. Examples of such work include the kdb tree [24,25], the R [26] tree, the R* tree [27], the SS-tree [28], the SR-tree [29], the X-tree [30], the GNAT tree [31], the M-tree [32], the TV tree [33], the hB tree [34], the "pyramid technique" [35], and the "hybrid" tree [36]. A good survey of nearest-neighbor search, albeit from the slightly more general perspective of multi-dimensional access methods is given by [37].

As indicated by the prevalence of the word "tree" in the preceding references, a common approach for nearest neighbor search is to create tree-based structures, such that the "closeness" of the data increases as the tree is traversed from top to bottom. Thus,

1: Add the children of the root node to a search queue
2: **while** the search queue is not empty **do**
3:    Take a node off the search queue
4:    **if** that node and its descendants are not 'close enough' to be considered for the search **then**
5:       Discard the subtree represented by this node
6:    **else**
7:       **if** the node is not a leaf node, i.e., a data point **then**
8:          Add the children of this node to a search queue
9:       **else** {if the node is a leaf node}
10:          add the node to a list of possible solutions
11:       **end if**
12:    **end if**
13: **end while**
14: Sort the list of possible solutions by distance from the query point and return the k-nearest neighbors

FIGURE 64.14 Basic algorithm for a nearest neighbor query.

the nodes towards the bottom of the tree and their children can often be regarded as representing "clusters" of data that are relatively cohesive. In the reverse directions, we also view clustering as being potentially useful for finding nearest neighbors. Indeed, one of the simplest techniques for generating a nearest neighbor tree is to cluster the data into a set of clusters and then, recursively break each cluster into subclusters until the subclusters consist of individual points. The resulting cluster tree tree consists of the clusters generated along the way. Regardless of how a nearest neighbor search tree is obtained, the general approach for performing a k-nearest-neighbor query is given by the algorithm in Figure 64.14.

This seems fairly straightforward and, thus it seems as though nearest neighbor trees should useful for clustering data, or conversely, that clustering would be a practical way to find nearest neighbors based on the results of clustering. However, there are some problems.

Goal Mismatch One of the goals of many nearest-neighbor tree techniques is to serve as efficient secondary storage based access methods for non-traditional databases, for example, spatial databases, multimedia databases, document databases, etc. Because of requirements related to page size and efficient page utilization, "natural" clusters may be split across pages or nodes. Nonetheless, data is normally highly "clustered" and this can be used for actual clustering as shown in [38], which uses an R* tree to improve the efficiency of a clustering algorithm introduced in [39].

Problems with High-dimensional Data Because of the nature of nearest-neighbor trees, the tree search involved is a branch-and-bound technique and needs to search large parts of the tree, that is, at any particular level, many children and their descendants may need to be examined. To see this, consider a point and all points that are within a given distance of it. This hyper-sphere (or hyper-rectangle in the case of multi-dimensional range queries) may very well cut across a number of nodes (clusters)—particularly if the point is on the edge of a cluster and/or the query distance being considered is greater than the distance between clusters. More specifically, it is difficult for the algorithms that construct nearest neighbor trees to avoid a significant amount of overlap in the volumes represented by different nodes in the tree. In [30], it has been shown that the degree of overlap in a frequently used nearest neighbor tree, the R* tree, approaches 100% as the dimensionality of the vectors exceeds 5. Even in two dimensions the overlap was about 40%. Other nearest neighbor search techniques suffer from similar problems.

Furthermore, in [40] it was demonstrated that the concept of "nearest neighbor" is not meaningful in many situations, since the minimum and maximum distances of a point to its neighbors tend to be very similar in high dimensional space. Thus, unless there is significant clustering in the data and the query ranges stay within individual clusters, the points returned by nearest neighbor queries are not much closer to the query point than are the points that are not returned. In this latter case, the nearest neighbor query is 'unstable, to use the terminology of [40]. Recently, for example, in [35], there has been some work on developing techniques that avoid this problem. Nonetheless, in some cases, a linear scan can be more efficient at finding nearest neighbors than more sophisticated techniques.

Outliers Typically, outliers are not discarded, that is, all data points are stored. However, if some of the data points do not fit into clusters particularly well, then the presence of outliers can have deleterious effects on the lookups of other data points.

To summarize, there is significant potential for developments in the areas of nearest neighbor search and multidimensional access methods to make a contribution to cluster analysis. The contribution to efficiency is obvious, but the notions of distance (or similarity) are central to both areas, and thus, there is also the possibility of conceptual contributions as well. However, currently, most clustering methods that utilize nearest neighbor search or multidimensional access methods are interested only in the efficiency aspects [38,41,42].

64.5 Conclusion

In this chapter we have provided some examples to indicate the role that data structures play in data mining. For classification, we indicated how proximity graphs can play an important role in understanding and improving the performance of nearest neighbor classifiers. For association analysis, we showed how data structures are currently used to address the exponential complexity of the problem. For clustering, we explored its connection to nearest neighbor search and multi-dimensional access methods—a connection that has only been modestly exploited.

Data mining is a rapidly evolving field, with new problems continually arising, and old problems being looked at in the light of new developments. These developments pose new challenges in the areas of data structures and algorithms. Some of the most promising areas in current data mining research include multi-relational data mining [43–45], mining streams of data [46], privacy preserving data mining [47], and mining data with complicated structures or behaviors, for example, graphs [45,48] and link analysis [49,50].

Acknowledgments

This work was partially supported by NASA grant # NCC 2 1231 and by the Army High Performance Computing Research Center under the auspices of the Department of the Army, Army Research Laboratory cooperative agreement number DAAD19-01-2-0014. The content of this work does not necessarily reflect the position or policy of the government and no official endorsement should be inferred. Access to computing facilities was provided by the AHPCRC and the Minnesota Supercomputing Institute. Figures 64.1-64.11, 64.13 and some parts of the text were taken from *Introduction to Data Mining* by Pang-Ning Tan, MIchael Steinbach, and Vipin Kumar, published by Addison-Wesley, and are reprinted with the permission of Addison-Wesley.

References

1. J. Han and M. Kamber. *Data Mining: Concepts and Techniques.* Morgan Kaufmann Publishers, San Francisco, 2001.
2. D. Hand, H. Mannila, and P. Smyth. *Principles of Data Mining.* MIT Press, 2001.
3. I. H. Witten and E. Frank. *Data Mining: Practical Machine Learning Tools and Techniques with Java Implementations.* Morgan Kaufmann, 1999.
4. M. H. Dunham. *Data Mining: Introductory and Advanced Topics.* Prentice Hall, 2002.
5. R. L. Grossman, C. Kamath, V. K. Philip Kegelmeyer, and R. R. Namburu, editors, *Data Mining for Scientific and Engineering Applications.* Kluwer Academic Publishers, October 2001.
6. R. Bellman. *Adaptive Control Processes: A Guided Tour.* Princeton University Press, 1961.
7. T. M. Mitchell. *Machine Learning.* McGraw-Hill, March 1997.
8. R. O. Duda, P. E. Hart, and D. G. Stork. *Pattern Classification.* John Wiley & Sons, Inc., New York, second edition, 2001.
9. T. M. Cover and P. E. Hart. Nearest neighbor pattern classification. *Knowledge Based Systems,* 8(6):373–389, 1995.
10. S. Cost and S. Salzberg. A weighted nearest neighbor algorithm for learning with symbolic features. *Machine Learning,* 10:57–78, 1993.
11. D. Aha. *A study of instance-based algorithms for supervised learning tasks: mathematical, empirical, and psychological evaluations.* PhD thesis, University of California, Irvine, 1990.
12. G. T. Toussaint. Proximity graphs for nearest neighbor decision rules: recent progress. In *Interface-2002, 34th Symposium on Computing and Statistics,* Montreal, Canada, April 17–20, 2002.
13. J. W. Jaromczyk and G. T. Toussaint. Relative neighborhood graphs and their relatives. *Proceedings of the IEEE,* 80(9):1502–1517, September 1992.
14. R. Agrawal, T. Imielinski, and A. Swami. Mining association rules between sets of items in large databases. In *Proc. ACM SIGMOD Intl. Conf. Management of Data,* pages 207–216, Washington D.C., USA, 1993.
15. R. Agrawal and R. Srikant. Fast algorithms for mining association rules. In *Proc. of the 20th VLDB Conference,* pages 487–499, Santiago, Chile, 1994.
16. J. Han, J. Pei, and Y. Yin. Mining frequent patterns without candidate generation. In *Proc. 2000 ACM-SIGMOD Int'l Conf on Management of Data (SIGMOD'00),* Dallas, TX, May 2000.
17. J. Pei, J. Han, H. Lu, S. Nishio, S. Tang, and D. Yang. H-mine: Hyperstructure mining of frequent patterns in large databases. In *Proc. 2001 Int'l Conf on Data Mining (ICDM'01),* San Jose, CA, Nov 2001.

18. A. K. Jain and R. C. Dubes. *Algorithms for Clustering Data*. Prentice Hall Advanced Reference Series. Prentice Hall, Englewood Cliffs, New Jersey, March 1988.

19. L. Kaufman and P. J. Rousseeuw. *Finding Groups in Data: An Introduction to Cluster Analysis*. Wiley Series in Probability and Statistics. John Wiley and Sons, New York, 1990.

20. M. R. Anderberg. *Cluster Analysis for Applications*. Academic Press, New York, December 1973.

21. P. Arabie, L. Hubert, and G. De Soete. An overview of combinatorial data analysis. In P. Arabie, L. Hubert, and G. De Soete, editors, *Clustering and Classification*, pages 188–217. World Scientific, Singapore, January 1996.

22. M. Ester, H.-P. Kriegel, J. Sander, and X. Xu. A density-based algorithm for discovering clusters in large spatial databases with noise. In *KDD96*, pages 226–231, 1996.

23. G. Karypis, E.-H. Han, and V. Kumar. Chameleon: Hierarchical clustering using dynamic modeling. *Computer*, 32(8):68–75, 1999.

24. J. T. Robinson. The k-d-b-tree: a search structure for large multidimensional dynamic indexes. In *Proceedings of the 1981 ACM SIGMOD International Conference on Management of Data*, pages 10–18. ACM Press, 1981.

25. T. B. B. Yu, R. Orlandic, and J. Somavaram. Kdb$_{KD}$-tree: A compact kdb-tree structure for indexing multidimensional data. In *Proceedings of the 2003 IEEE International Symposium on Information Technology (ITCC 2003)*. IEEE, April, 28–30, 2003.

26. A. Guttman. R-trees: a dynamic index structure for spatial searching. In *Proceedings of the 1984 ACM SIGMOD International Conference on Management of Data*, pages 47–57. ACM Press, 1984.

27. N. Beckmann, H.-P. Kriegel, R. Schneider, and B. Seeger. The r*-tree: an efficient and robust access method for points and rectangles. In *Proceedings of the 1990 ACM SIGMOD International Conference on Management of Data*, pages 322–331. ACM Press, 1990.

28. R. Jain and D. A. White. Similarity indexing with the ss-tree. In *Proceedings of the 12th International Conference on Data Engineering*, pages 516–523, 1996.

29. N. Katayama and S. Satoh. The sr-tree: an index structure for high-dimensional nearest neighbor queries. In *Proceedings of the 1997 ACM SIGMOD International Conference on Management of Data*, pages 369–380. ACM Press, 1997.

30. S. Berchtold, D. A. Keim, and H.-P. Kriegel. The X-tree: An index structure for high-dimensional data. In T. M. Vijayaraman, A. P. Buchmann, C. Mohan, and N. L. Sarda, editors, *Proceedings of the 22nd International Conference on Very Large Databases*, pages 28–39, San Francisco, U.S.A., 1996. Morgan Kaufmann Publishers.

31. S. Brin. Near neighbor search in large metric spaces. In *The VLDB Journal*, pages 574–584, 1995.

32. P. Ciaccia, M. Patella, and P. Zezula. M-tree: An efficient access method for similarity search in metric spaces. In *The VLDB Journal*, pages 426–435, 1997.

33. K.-I. Lin, H. V. Jagadish, and C. Faloutsos. The tv-tree: An index structure for high-dimensional data. *VLDB Journal*, 3(4):517–542, 1994.

34. D. B. Lomet and B. Salzberg. The hb-tree: a multiattribute indexing method with good guaranteed performance. *ACM Transactions on Database Systems (TODS)*, 15(4):625–658, 1990.

35. S. Berchtold, C. Böhm, and H.-P. Kriegal. The pyramid-technique: towards breaking the curse of dimensionality. In *Proceedings of the 1998 ACM SIGMOD International Conference on Management of Data*, pages 142–153. ACM Press, 1998.

36. K. Chakrabarti and S. Mehrotra. The hybrid tree: An index structure for high dimensional feature spaces. In *Proceedings of the 15th International Conference on Data Engineering*, 23-26 March 1999, Sydney, Austrialia, pages 440–447. IEEE Computer Society, 1999.

37. V. Gaede and O. Günther. Multidimensional access methods. *ACM Computing Surveys (CSUR)*, 30(2):170–231, 1998.

38. M. Ester, H.-P. Kriegel, and X. Xu. Knowledge discovery in large spatial databases: focusing techniques for efficient class identification. In M. Egenhofer and J. Herring, editors, *Advances in Spatial Databases, 4th International Symposium, SSD'95*, volume 951, pages 67–82, Portland, ME, 1995. Springer.

39. R. T. Ng and J. Han. Efficient and effective clustering methods for spatial data mining. In J. Bocca, M. Jarke, and C. Zaniolo, editors, *20th International Conference on Very Large Data Bases*, September 12–15, 1994, Santiago, Chile proceedings, pages 144–155, Los Altos, CA 94022, USA, 1994. Morgan Kaufmann Publishers.

40. K. Beyer, J. Goldstein, R. Ramakrishnan, and U. Shaft. When is "nearest neighbor" meaningful? In *Proceedings 7th International Conference on Database Theory (ICDT'99)*, pages 217–235, 1999.

41. Alsabti , Ranka , and Singh . An efficient parallel algorithm for high dimensional similarity join. In *IPPS: 11th International Parallel Processing Symposium*. IEEE Computer Society Press, 1998.

42. F. Murtagh. Clustering in massive data sets. In J. Abello, P. M. Pardalos, and M. G. C. Resende, editors, *Handbook of Massive Data Sets*, pages 501–543. Kluwer Academic Publishers, Dordrecht, Netherlands, May 2002.

43. S. Dzeroski and L. D. Raedt. Multi-relational data mining: The current frontiers. *SIGKDD Explorations*, 5(1), July 2003.

44. P. Domingos. Prospects and challenges for multi-relational data mining. *SIGKDD Explorations*, 5(1), July 2003.

45. L. B. Holder and D. J. Cook. Graph-based relational learning: Current and future directions. *SIGKDD Explorations*, 5(1), July 2003.

46. P. Domingos and G. Hulten. A general framework for mining massive data streams. *Journal of Computational and Graphical Statistics*, 12, 2003.

47. R. Agrawal and R. Srikant. Privacy-preserving data mining. In *Proc. of the ACM SIGMOD Conference on Management of Data*, pages 439–450. ACM Press, May 2000.

48. M. Kuramochi and G. Karypis. Frequent subgraph discovery. In *The 2001 IEEE International Conference on Data Mining*, pages 313–320, 2001.

49. D. Jensen and J. Neville. Data mining in social networks. In *National Academy of Sciences Symposium on Dynamic Social Network Analysis*, 2002.

50. D. Mladenic, M. Grobelnik, N. Milic-Frayling, S. Donoho, and T. D., editors, Kdd 2003: Workshop on link analysis for detecting complex behavior, August 2003.

65

Computational Geometry: Fundamental Structures*

65.1 Introduction.. 1013
65.2 Arrangements 1014
      Substructures and Complexity • Decomposition • Duality
65.3 Convex Hulls.. 1017
      Complexity • Construction • Dynamic Convex Hulls
65.4 Voronoi Diagrams 1020
      Complexity • Construction • Variations
65.5 Triangulations.. 1022
      Delaunay Triangulation • Polygons • Polyhedra • Pseudo-Triangulations
References... 1025

Mark de Berg
Eindhoven University of Technology

Bettina Speckmann
Eindhoven University of Technology

65.1 Introduction

Computational geometry deals with the design and analysis of algorithms and data structures for problems involving spatial data. The questions that are studied range from basic problems such as line-segment intersection ("Compute all intersection points in a given set of line segments in the plane.") to quite involved problems such as motion-planning ("Compute a collision-free path for a robot in workspace from a given start position to a given goal position.") Because spatial data plays an important role in many areas within and outside of computer science—CAD/CAM, computer graphics and virtual reality, and geography are just a few examples—computational geometry has a broad range of applications. Computational geometry emerged from the general algorithms area in the late 1970s. It experienced a rapid growth in the 1980s and by now is a recognized discipline with its own conferences and journals and many researchers working in the area. It is a beautiful field with connections to other areas of algorithms research, to application areas like the ones mentioned earlier, and to areas of mathematics such as combinatorial geometry.

To design an efficient geometric algorithm or data structure, one usually needs two ingredients: a toolbox of algorithmic techniques and geometric data structures and a thorough understanding of the geometric properties of the problem at hand. As an example, consider the classic *post-office problem*, where we want to preprocess a set S of n points in the plane—the points in S are usually called *sites*—for the following queries: report the site in S that is closest to a query point q. A possible approach is to subdivide the plane into n regions, one for each site, such that the region of a site $s \in S$ consists of exactly those points $q \in \mathbb{R}^2$ for which s is the closest site. This subdivision is called the Voronoi diagram of S. A query with a point q can now be answered by locating the region in which q lies, and reporting the site defining that region. To make this idea work, one needs an efficient data structure for point location. But one also needs to understand the geometry: What does the Voronoi diagram look like? What is its complexity? How can we construct it efficiently?

In Part IV, many data structures for spatial data were already discussed. Hence, in this chapter we will focus on the second ingredient: we will discuss a number of basic geometric concepts. In particular, we will discuss arrangements in Section 65.2, convex hulls in Section 65.3, Voronoi diagrams in Section 65.4, and triangulations in Section 65.5.

More information on computational geometry can be found in various sources: there are several general textbooks on computational geometry [1–4], as well as more specialized books, for example, on arrangements [5,6] and Voronoi diagrams [7]. Finally, there are two handbooks that are devoted solely to (discrete and) computational geometry [8,9].

---

* This chapter has been reprinted from first edition of this Handbook, without any content updates.

65.2 Arrangements

The *arrangement* $\mathcal{A}(S)$ defined by a finite collection S of curves in the plane is the subdivision of the plane into open cells of dimensions 2 (the *faces*), 1 (the *edges*), and 0 (the *vertices*), induced by S—see Figure 65.1 for an example. This definition generalizes readily to higher dimensions: the arrangement defined by a set S of geometric objects in $\mathbb{R}^d$ such as hyperplanes or surfaces, is the decomposition of $\mathbb{R}^d$ into open cells of dimensions $0, \ldots, d$ induced by S. The cells of dimension k are usually called *k-cells*. The 0-cells are called *vertices*, the 1-cells are called *edges*, the 2-cells are called *faces*, the $(d-1)$-cells are called *facets*, and the d-cells are sometimes just called *cells*.

Arrangements have turned out to form a fundamental concept underlying many geometric problems and the efficiency of geometric algorithms is often closely related to the combinatorial complexity of (certain parts of) some arrangement. Moreover, to solve a certain problem geometric algorithms often rely on some decomposition of the arrangement underlying the problem. Hence, the following subsections give some more information on the complexity and the decomposition of arrangements.

65.2.1 Substructures and Complexity

Let H be a collection of n hyperplanes in $\mathbb{R}^d$. As stated earlier, the arrangement $\mathcal{A}(H)$ is the decomposition of $\mathbb{R}^d$ into open cells of dimensions $0, \ldots, d$ induced by H. The *combinatorial complexity* of $\mathcal{A}(H)$ is defined to be the total number of cells of the various dimensions. This definition immediately carries over to arrangements induced by other objects, such as segments in the plane, or surfaces in $\mathbb{R}^d$. For example, the complexity of the arrangement in Figure 65.1 is 58, since it consists of 27 vertices, 27 edges, and 4 faces (one of which is the unbounded face).

It is easy to see that the maximum complexity of an arrangement of n lines in the plane is $\Theta(n^2)$: there can be at most $n(n-1)/2$ vertices, at most n^2 edges, and at most $n^2/2 + n/2 + 1$ faces. Also for an arrangement of curves the maximum complexity is $\Theta(n^2)$, provided that any pair of curves intersects at most s times, for a constant s. More generally, the maximum complexity of an arrangement of n hyperplanes in $\mathbb{R}^d$ is $\Theta(n^d)$. The same bound holds for well-behaved surfaces (such as algebraic surfaces of constant maximum degree) or well-behaved surface patches in $\mathbb{R}^d$.

Single cells. It becomes more challenging to bound the complexity when we consider only a part of an arrangement. For example, what is the maximum complexity of a *single cell*, that is, the maximum number of i-cells, for $i < d$, on the boundary of any given d-cell? For lines in the plane this is still rather easy—the maximum complexity is $\Theta(n)$, since any line can contribute at most one edge to a given face—but the question is already quite hard for arrangements of line segments in the plane. Here it turns out that the maximum complexity can be $\Theta(n\alpha(n))$, where $\alpha(n)$ is the extremely slowly growing functional inverse of Ackermann's function. More generally, for Jordan curves where each pair intersects in at most s points, the maximum complexity is $\Theta(\lambda_{s+2}(n))$, where $\lambda_{s+2}(n)$ is the maximum length of a Davenport-Schinzel sequence of order $s+2$ on n symbols. The function $\lambda_{s+2}(n)$ is only slightly super-linear for any constant s. In higher dimensions, tight bounds are known for hyperplanes: the famous *Upper Bound Theorem* states that the maximum complexity of a single cell is $\Theta(n^{\lfloor d/2 \rfloor})$. For most other objects, the known upper and lower bounds are close but not tight—see Table 65.1.

Lower envelopes. Another important substructure is the *lower envelope*. Intuitively, the lower envelope of a set of segments in the plane is what one would see when looking at the segments from below—see Figure 65.2. More formally, if we view the segments as graphs of partially defined (linear) functions, then the lower envelope is the point-wise minimum of these functions. Similarly, the *upper envelope* is defined as the point-wise maximum of the functions. The definition readily extends to x-monotone curves in the plane, to planes, triangles, or xy-monotone surface patches in $\mathbb{R}^3$, etc.

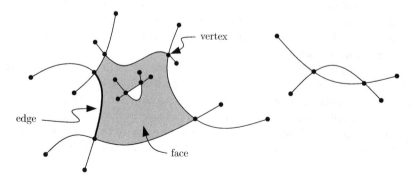

FIGURE 65.1 An arrangement of curves in the plane.

TABLE 65.1 Maximum Complexity of Single Cells and Envelopes in Arrangements

		Single cell	Upper envelope	Reference
$d=2$	Lines	$\Theta(n)$	$\Theta(n)$	Trivial
	Segments	$\Theta(n\alpha(n))$	$\Theta(n\alpha(n))$	[10]
	Circles	$\Theta(n)$	$\Theta(n)$	Linearization
	Jordan arcs	$\Theta(\lambda_{s+2}(n))$	$\Theta(\lambda_{s+2}(n))$	[11]
$d=3$	Planes	$\Theta(n)$	$\Theta(n)$	Euler's formula
	Triangles	$\Omega(n^2\alpha(n)), O(n^2\log n)$	$\Theta(n^2\alpha(n))$	[12], [13]
	Spheres	$\Theta(n^2)$	$\Theta(n^2)$	Linearization
	Surfaces	$\Omega(n\lambda_q(n)), O(n^{2+\varepsilon})$	$\Omega(n\lambda_q(n)), O(n^{2+\varepsilon})$	[14]
$d>3$	Hyperplanes	$\Theta(n^{\lfloor d/2\rfloor})$	$\Theta(n^{\lfloor d/2\rfloor})$	Upper Bound Thm [15]
	$(d-1)$-simplices	$\Omega(n^{d-1}\alpha(n)), O(n^{d-1}\log n)$	$\Theta(n^{d-1}\alpha(n))$	[12], [13]
	$(d-1)$-spheres	$\Theta(n^{\lceil d/2\rceil})$	$\Theta(n^{\lceil d/2\rceil})$	Linearization
	Surfaces	$\Omega(n^{d-2}\lambda_q(n)), O(n^{d-1+\varepsilon})$	$O(n^{d-1+\varepsilon})$	[16], [17]

The parameter s is the maximum number of points in which any two curves meet; the parameter q is a similar parameter for higher dimensional surfaces. The function $\lambda_t(n)$ is the maximum length of a Davenport-Schinzel sequence [6] of order t on n symbols, and is only slightly super-linear for any constant t. Bounds of the form $O(n^{d-1+\varepsilon})$ hold for any constant $\varepsilon > 0$.

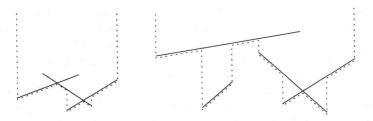

FIGURE 65.2 The lower envelope of a set of segments in the plane.

Envelopes are closely related to single cells. The lower envelope of a set of lines in the plane, for instance, is the boundary of the single cell in the arrangement that is below all the lines. The vertices of the lower envelope of a set of segments in the plane are also vertices of the unbounded cell defined by those segments, but here the reverse is not true: vertices of the unbounded cell that are above other segments are not on the lower envelope. Nevertheless, the worst-case complexities of lower envelopes and single cells are usually very similar—see Table 65.1.

Other substructures. More types of substructures have been studied than single cells and envelopes: zones, levels, multiple cells, etc. The interested reader may consult the Chapter 21 of the *CRC Handbook of Discrete and Computational Geometry* [8], or the books by Edelsbrunner [5] or Sharir and Agarwal [6].

65.2.2 Decomposition

Full arrangements, or substructures in arrangements, are by themselves not convenient to work with, because their cells can be quite complex. Thus it is useful to further decompose the cells of interest into constant-complexity subcells: triangles or trapezoids in 2D, and simplices or trapezoid-like cells in higher dimensions. There are several ways of doing this.

Bottom-vertex triangulations. For arrangements of hyperplanes, the so-called *bottom-vertex triangulation* is often used. This decomposition is obtained as follows. Consider a bounded face f in a planar arrangement of lines. We can decompose f into triangles by drawing a line segment from the bottommost vertex v of f to all other vertices of f, except the vertices that are already adjacent to v—see Figure 65.3a. Note that this easy method for triangulating f is applicable since f is always convex. To decompose the whole arrangement of lines (or some substructure in it) we simply decompose each face in this manner.[*]

To decompose a d-cell C in a higher-dimensional arrangement of hyperplanes, we proceed inductively as follows. We first decompose each $(d-1)$-cell on the boundary of C, and then extend each $(d-1)$-simplex in this boundary decomposition into

---

[*] Unbounded faces require a bit of care, but they can be handled in a similar way.

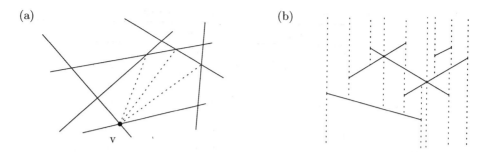

FIGURE 65.3   (a) The bottom-vertex triangulation of a face in a planar arrangement. (b) The vertical decomposition of a planar arrangement of line segments.

a d-simplex by connecting each of its vertices to the bottommost vertex of C. Again, this can readily be used to decompose any subset of cells in an arrangement of hyperplanes.

The total number of simplices in a bottom-vertex decomposition is linear in the total complexity of the cells being decomposed.

Vertical decompositions. The bottom-vertex triangulation requires the cells to be convex, so it does not work for arrangements of segments or for arrangements of surfaces. For such arrangements one often uses the *vertical decomposition* (or: *trapezoidal decomposition*). For an arrangement of line segments or curves in the plane, this decomposition is defined as follows. Each vertex of the arrangement—this can be a segment endpoint or an intersection point—has a vertical connection to the segment immediately above it, and to the segment immediately below it. If there is no segment below or above a vertex, then the connection extends to infinity. This decomposes each cell into trapezoids: subcells that are bounded by at most two vertical connections and by at most two segments—see Figure 65.3b. This definition can be generalized to higher dimensions as follows. Suppose we wish to decompose the arrangement $\mathcal{A}(S)$ induced by a collection S of surfaces in $\mathbb{R}^d$, where each surface is vertically monotone (i.e., any line parallel to the x_d-axis intersects the surface in at most one point). Each point of any $(d-2)$-dimensional cell of $\mathcal{A}(S)$ is connected by a vertical segment to the surface immediately above it and to the surface immediately below it. In other words, from each $(d-2)$-cell we extend a vertical wall upward and downward. These walls together decompose the cells into subcells bounded by vertical walls and by at most two surfaces from S—one from above and one from below. These subcells are vertically monotone, but do not yet have constant complexity. Hence, we recursively decompose the bottom of the cell, and then extend this decomposition vertically upward to obtain a decomposition of the entire cell.

The vertical decomposition can be used for most arrangements (or substructures in them). In the plane, the maximum complexity of the vertical decomposition is linear in the total complexity of the decomposed cells. However, in higher dimensions this is no longer true. In $\mathbb{R}^3$, for instance, the vertical decomposition of an arrangement of n disjoint triangles can consist of $\Theta(n^2)$ subcells, even though in this case the total complexity of the arrangement in obviously linear. Unfortunately, this is unavoidable, as there are collections of disjoint triangles in $\mathbb{R}^3$ for which any decomposition into convex subcells must have $\Omega(n^2)$ subcells. For n intersecting triangles, the vertical decomposition has complexity $O(n^2\alpha(n)\log n + K)$, where K is the complexity of the arrangement of triangles [12]. More information about the complexity of vertical decompositions in various settings can be found in Halperin's survey on arrangements [8, Chapter 21]. In many cases, vertical decompositions can be constructed in time proportional to their complexity, with perhaps a small (logarithmic or $O(n^\varepsilon)$) multiplicative factor.

65.2.3 Duality

Consider the transformation in the plane that maps the point $p = (p_x, p_y)$ to the line p^*: $y = p_x x - p_y$, and the line ℓ:$y = ax + b$ to the point $\ell^* = (a, -b)$. Such a transformation that maps points to lines and vice versa is called a *duality transform*. Often the term *primal plane* is used for the plane in which the original objects live, and the term *dual plane* is used for the plane in which their images live. The duality transform defined above has a few easy-to-verify properties:

    1. It is *incidence preserving*: if a point p lies on a line ℓ, then the point ℓ^* dual to ℓ lies on the line p^* dual to p.
    2. It is *order preserving*: if a point p lies above a line ℓ, then ℓ^* lies above p^*.

These properties imply several others. For example, three points on a line become three lines through a point under the duality transform—see Figure 65.4. Another property is that for any point p we have $(p^*)^* = p$. Notice that the duality transform above is not defined for vertical lines. This technicality is usually not a problem, as vertical lines can often be handled separately.

This duality transform is so simple that it does not seem very interesting at first sight. However, it turns out to be extremely useful. As an example, consider the following problem. We are given a set P of n points in the plane, which we wish to preprocess for *strip-emptiness* queries: given a query strip—a strip is the region between two parallel lines—decide whether that strip is empty

primal plane   dual plane

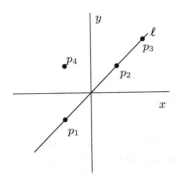   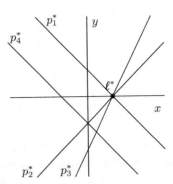

FIGURE 65.4   Illustration of the duality transform.

or if it contains one or more points from P. If we know about duality, and we know about data structures for point location, then this problem is easy to solve: we take the duals of the points in P to obtain a set P^* of lines in the plane, and we preprocess the arrangement $\mathcal{A}(P^*)$ induced by P^* for logarithmic-time point location. To decide whether the strip bounded by lines ℓ_1 and ℓ_2 is empty, we perform point locations with ℓ_1^* and ℓ_2^* in $\mathcal{A}(P^*)$; the strip is empty if and only if ℓ_1^* and ℓ_2^* lie in the same face of the arrangement. (This does not work if ℓ_1 and ℓ_2 are vertical, since then their duals are undefined. But for this case we can simply build one extra data structure, which is a balanced binary search tree on the x-coordinates of the points.)

In principle, of course, we could also have arrived at this solution without using duality. After all, duality does not add any new information: it is just a different way of looking at things. Hence, every algorithm or data structure that works in the dual plane can also be interpreted as working in the primal plane. But some things are simply more easy to see in the dual plane. For example, a face in the arrangement $\mathcal{A}(P^*)$ in the dual plane is much more visible than the collection of all lines in the primal plane dividing P into two subsets in a certain way. So without duality, we would not have realized that we could solve the strip-emptiness problem with a known data structure, and we would probably not have been able to develop that structure ourselves either.

This is just one example of the use of duality. There are many more problems where duality is quite useful. In fact, we will see another example in the next section, when we study convex hulls.

65.3 Convex Hulls

A set $A \subset \mathbb{R}^d$ is *convex* if for any two points $p, q \in A$ the segment $\overline{pq}$ is completely contained in A. The *convex hull* of a set S of objects is the smallest convex set that contains all objects in S, that is, the most tightly fitting convex bounding volume for S. For example, if S is a set of objects in the plane, we can obtain the convex hull by taking a large rubber band around the objects and then releasing the band; the band will snap around the objects and the resulting shape is the convex hull. More formally, we can define $\mathcal{CH}(S)$ as the intersection of all convex sets containing all objects in S:

$$\mathcal{CH}(S) := \bigcap \{A : A \text{ is convex, and } o \subset A \text{ for all } o \in S\}.$$

We denote the convex hull of the objects in S by $\mathcal{CH}(S)$.

It is easy to see that the convex hull of a set of line segments in the plane is the same as the convex hull of the endpoints of the segments. More generally, the convex hull of a set of bounded polygonal objects in $\mathbb{R}^d$ is the same as the convex hull of the vertices of the objects. Therefore we will restrict our discussion to convex hulls of sets of points. Table 65.2 gives an overview of the results on the complexity and construction of convex hulls discussed below.

65.3.1 Complexity

Let P be a set of n points in $\mathbb{R}^d$. The *convex hull* of P, denoted by $\mathcal{CH}(P)$, is a convex polytope whose vertices are a subset of the points in P. The complexity of a polytope is defined as the total number of k-facets[*] (i.e., k-dimensional features) on the

---

[*]   In the previous section we used the term k-cells for the k-dimensional features in an arrangement, but since we are dealing here with a single polytope we prefer the term k-facet.

TABLE 65.2   Maximum Complexity of the Convex Hull of a Set of
n Points, and the Time Needed to Construct the Convex Hull

	Complexity	Construction	Reference
$d = 2$, worst case	$\Theta(n)$	$O(n \log n)$	[1,18]
$d = 3$, worst case	$\Theta(n)$	$O(n \log n)$	[19–22]
$d > 3$, worst case	$\Theta(n^{\lfloor d/2 \rfloor})$	$O(n^{\lfloor d/2 \rfloor})$	[19,20,22,23]
$d \geq 2$, uniform distr.	$\Theta(\log^{d-1} n)$	$O(n)$	[24]

The bounds on uniform distribution refer to points drawn uniformly at
random from a hypercube or some other convex polytope.

boundary of the polytope, for $k = 0, 1, \ldots, d - 1$: the complexity of a planar polygon is the total number of vertices and edges, the complexity of a 3-dimensional polytope is the total number of vertices, edges, and faces, and so on.

Because the vertices of $\mathcal{CH}(P)$ are a subset of the points in P, the number of vertices of $\mathcal{CH}(P)$ is at most n. In the plane this means that the total complexity of the convex hull is $O(n)$, because the number of edges of a planar polygon is equal to the number of vertices. In higher dimensions this is no longer true: the number of k-facets ($k > 0$) of a polytope can be larger than the number of vertices. How large can this number be in the worst case? In $\mathbb{R}^3$, the total complexity is still $O(n)$. This follows from Euler's formula, which states that for a convex polytope in $\mathbb{R}^3$ with V vertices, E edges, and F faces it holds that $V - E + F = 2$. In higher dimensions, the complexity can be significantly higher: the worst-case complexity of a convex polytope with n vertices in $\mathbb{R}^d$ is $\Theta(n^{\lfloor d/2 \rfloor})$.

In fact, the bound on the complexity of the convex hull immediately follows from the results of the previous section if we apply duality. To see this, consider a set P of n points in the plane. For simplicity, let us suppose that $\mathcal{CH}(P)$ does not have any vertical edges and that no three points in P are collinear. Define the *upper hull* of P, denoted by $\mathcal{UH}(P)$, as the set of edges of $\mathcal{CH}(P)$ that bound $\mathcal{CH}(P)$ from above. Let P^* be the set of lines that are the duals of the points in P. A pair $p, q \in P$ defines an edge of $\mathcal{UH}(P)$ if and only if all other points $r \in P$ lie below the line through p and q. In the dual this means that all lines $r^* \in P^*$ lie above the intersection point $p^* \cap q^*$. In other words, $p^* \cap q^*$ is a vertex on the lower envelope $\mathcal{LE}(P^*)$ of the lines in P^*. Furthermore, a point $p \in P$ is a vertex of $\mathcal{UH}(P)$ if and only if its dual p^* defines an edge of $\mathcal{LE}(P^*)$. Thus there is a one-to-one correspondence between the vertices (or, edges) of $\mathcal{UH}(P)$, and the edges (or, vertices) of $\mathcal{LE}(P^*)$. In higher dimensions a similar statement is true: there is a one-to-one correspondence between the k-facets of the upper hull of P and the $(d - k - 1)$-facets of the lower envelope of P^*. The bound on the complexity of the convex hull of a set of n points in $\mathbb{R}^d$ therefore follows from the $\Theta(n^{\lfloor d/2 \rfloor})$ bound on the complexity of the lower envelope of a set of n hyperplanes $\mathbb{R}^d$.

The $\Theta(n^{\lfloor d/2 \rfloor})$ bound implies that the complexity of the convex hull can be quite high when the dimension gets large. Fortunately this is not the case if the points in P are distributed uniformly: in that case only few of the points in P are expected to show up as a vertex on the convex hull. More precisely, the expected complexity of the convex hull of a set P that is uniformly distributed in the unit hypercube is $O(\log^{d-1} n)$ [24,25].

65.3.2 Construction

We now turn our attention to algorithms for computing the convex hull of a set P of n points in $\mathbb{R}^d$. In the previous subsection we have shown that there is a correspondence between the upper hull of P and the lower envelope of P^*, where P^* is the set of hyperplanes dual to the points in P. It follows that any algorithm that can compute the convex hull of a set of points in $\mathbb{R}^d$ can also be used to compute the intersection of a set of half-spaces in $\mathbb{R}^d$, and vice versa.

First consider the planar case. By a reduction from sorting, one can show that $\Omega(n \log n)$ is a lower bound on the worst-case running time of any convex-hull algorithm. There are many different algorithms that achieve $O(n \log n)$ running time and are thus optimal in the worst case. One of the best known algorithms is called *Graham's scan* [1,18]. It treats the points from left to right, and maintains the upper hull of all the points encountered so far. To handle a point p_i, it is first added to the end of the current upper hull. The next step is to delete the points that should no longer be part of the hull. They always form a consecutive portion at the right end of the old hull, and can be identified easily—see Figure 65.5.

After these points have been deleted, the next point is handled. Graham's scan runs in linear time, after the points have been sorted from left to right. This is optimal in the worst case.

The $\Omega(n \log n)$ lower bound does not hold if only few points show up on the convex hull. Indeed, in this case it is possible to do better: Kirkpatrick and Seidel [26], and later Chan [27], gave output-sensitive algorithms that compute the convex hull in $O(n \log k)$ time, where k is the number of vertices of the hull.

In $\mathbb{R}^3$, the worst-case complexity of the convex hull is still linear, and it can be computed in $O(n \log n)$ time, either by a deterministic divide-and-conquer algorithm [21] or by a simpler randomized algorithm [19,20,22]. In dimensions $d > 3$, the convex

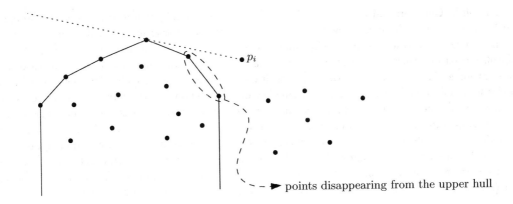

FIGURE 65.5 Constructing the upper hull.

hull can be computed in $\Theta(n^{\lfloor d/2 \rfloor})$ time [23]. This is the same as the worst-case complexity of the hull, and therefore optimal. Again, the simplest algorithms that achieve this bound are randomized [19,20,22]. There is also an output-sensitive algorithm by Chan [27], which computes the convex hull in $O(n \log k + (nk)^{1 - 1/(\lfloor d/2 \rfloor + 1)} \log^{O(1)} n)$ time, where k is its complexity.

As remarked earlier, the expected complexity of the convex hull of a set of n uniformly distributed points is much smaller than the worst-case complexity. This means that if we use an output-sensitive algorithm to compute the convex hull, its expected running time will be better than its worst-case running time. One can do even better, however, by using specialized algorithms. For example, Dwyer [24] has shown that the convex hull of points distributed uniformly in, for example, the unit hypercube can be computed in linear expected time.

65.3.3 Dynamic Convex Hulls

In some applications the set P changes over time: new points are inserted into P, and some existing points are deleted. The convex hull can change drastically after an update: the insertion of a single point can cause $\Theta(n)$ points to disappear from the convex hull, and the deletion of a single point can cause $\Theta(n)$ points to appear on the convex hull. Surprisingly, it is nevertheless possible to store the convex hull of a planar point set in such a way that any update can be processed in $O(\log^2 n)$ in the worst case, as shown by Overmars and van Leeuwen [28]. The key to their result is to not only store the convex hull of the whole set, but also information about the convex hull of certain subsets. The structure of Overmars and van Leeuwen roughly works as follows. Suppose we wish to maintain $\mathcal{UH}(P)$, the upper hull of P; maintenance of the lower hull can be done similarly. The structure to maintain $\mathcal{UH}(P)$ is a balanced binary tree $\mathcal{T}$, whose leaves store the points from P sorted by x-coordinate. The idea is that each internal node ν stores the upper hull of all the points in the subtree rooted at ν. Instead of storing the complete upper hull at each node ν, however, we only store those parts that are not already on the upper hull of nodes higher up in the tree. In other words, the point corresponding to a leaf μ is stored at the highest ancestor of μ where it is on the upper hull—see Figure 65.6 for an illustration.

Note that the root still stores the upper hull of the entire set P. Because a point is stored in only one upper hull, the structure uses $O(n)$ storage. Overmars and van Leeuwen show how to update the structure in $O(\log^2 n)$ time in the worst case.

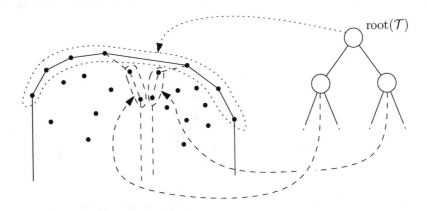

FIGURE 65.6 The Overmars-van Leeuwen structure to maintain the convex hull.

Although the result by Overmars and van Leeuwen is more than 20 years old by now, it still has not been improved in its full generality. Nevertheless, there have been several advances in special cases. For example, for the semi-dynamic case (only insertions, or only deletions), there are structures with $O(\log n)$ update time [14,29]. Furthermore, $O(\log n)$ update time can be achieved in the off-line setting, where the sequence of insertions and deletions is known in advance [30]. Improvements are also possible if one does not need to maintain the convex hull explicitly. For example, in some applications the reason for maintaining the convex hull is to be able to find the point that is extreme in a query direction, or the intersection of the convex hull with a query line. Such queries can be answered in logarithmic time if the convex hull vertices are stored in order in a balanced binary tree, but it is not necessary to know all the convex hull vertices explicitly to answer such queries. This observation was used by Chan [31], who described a fully dynamic structure that can answer such queries in $O(\log n)$ time and that can be updated in $O(\log^{1+\varepsilon} n)$ time. This result was recently improved by Brodal and Jacob [32], who announced a structure that uses $O(n)$ storage, has $O(\log n)$ update time, and can answer queries in $O(\log n)$ time. Neither Chan's structure nor the structure of Brodal and Jakob, however, can report the convex hull in time linear in its complexity, and the update times are amortized.

65.4 Voronoi Diagrams

Recall the *post-office problem* mentioned in the introduction. Here we want to preprocess a set S of n points, referred to as *sites*, in the plane such that we can answer the query: which site in S is closest to a given query point q? In order to solve this problem we can divide the plane into regions according to the nearest-neighbor rule: each site s gets assigned the region which is closest to s. This subdivision, which compactly captures the distance information inherent in a given configuration, is called the *Voronoi diagram* of S—see Figure 65.7a.

More formally, the Voronoi diagram of a set of sites $S = \{s_1, \ldots, s_n\}$ in $\mathbb{R}^d$, which we refer to as Vor(S), partitions space into n regions—one for each site—such that the region for a site s_i consists of all points that are closer to s_i than to any other site $s_j \in S$. The set of points that are closest to a particular site s_i forms the so-called *Voronoi cell* of s_i, and is denoted by $V(s_i)$. Thus, when S is a set of sites in the plane we have

$$V(s_i) = \{p \in \mathbb{R}^2 : \text{dist}(p, s_i) < \text{dist}(p, s_j) \text{ for all } j \neq i\},$$

where dist(.,.) denotes the Euclidean distance.

Now consider the dual graph of the Voronoi diagram, that is, the graph that has a node for every Voronoi cell and an arc between any two Voronoi cells that share a common edge—see Figure 65.7b. (Observe that the concept of dual graph used here has nothing to do with the duality transform discussed in Section 65.2.3.) Suppose we embed this graph in the plane, by using the site s_i to represent the node corresponding to the cell $V(s_i)$ and by drawing the edges as straight line segments, as in Figure 65.7c. Somewhat surprisingly perhaps, this graph is always planar. Moreover, it is actually a triangulation of the point set S, assuming that no four points in S are co-circular. More details on this special triangulation, which is called the *Delaunay triangulation*, will be given in Section 65.5.1.

There exists a fascinating connection between Voronoi diagrams in $\mathbb{R}^d$ and half-space intersections in $\mathbb{R}^{d+1}$. Assume for simplicity that $d = 2$, and consider the transformation that maps a site $s = (s_x, s_y)$ in $\mathbb{R}^2$ to the non-vertical plane $h(s) : z = 2s_x x + 2s_y y - \left(s_x^2 + s_y^2\right)$ in $\mathbb{R}^3$. Geometrically, $h(s)$ is the plane tangent to the unit paraboloid $z = x^2 + y^2$ at the point vertically above $(s_x, s_y, 0)$. Let $H(S)$ be the set of planes that are the image of a set of point sites S in the plane. Let $\mathcal{S}$ denote the convex polyhedron that is formed by the intersection of the positive half-spaces defined by the planes in $H(S)$, that is, $\mathcal{S} = \bigcap_{h \in H(S)} h^+$,

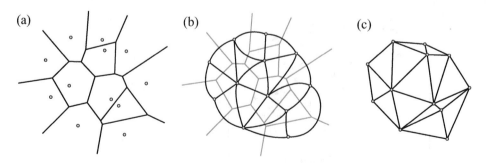

FIGURE 65.7 (a) The Voronoi diagram of a set of points. (b) The dual graph of the Voronoi diagram. (c) The Delaunay triangulation of the points.

where h^+ denotes the half-space above h. Surprisingly, the projection of the edges and vertices of S vertically downward on the xy-plane is exactly the Voronoi diagram of S.

The Voronoi diagram can be defined for various sets of sites, for example points, line segments, circles, or circular arcs, and for various metrics. Sections 65.4.1 and 65.4.2 discuss the complexity and construction algorithms of Voronoi diagrams for the usual case of point sites in $\mathbb{R}^d$, while Section 65.4.3 describes some of the possible variations. Additional details and proofs of the material presented in this section can be found in References 33–35.

65.4.1 Complexity

Let S be a set of n points in $\mathbb{R}^d$. The Voronoi diagram of S is a cell complex in $\mathbb{R}^d$. If $d = 2$ then the Voronoi cell of a site is the interior of a convex, possibly infinite polygon. Its boundary consists of *Voronoi edges*, which are equidistant from two sites, and *Voronoi vertices*, which are equidistant from at least three sites. The Voronoi diagram of $n \geq 3$ sites has at most $2n - 5$ vertices and at most $3n - 6$ edges, which implies that the Delaunay triangulation has at most $2n - 5$ triangles and $3n - 6$ edges. What is the complexity of the Voronoi diagram in $d \geq 3$? Here the connection between Voronoi diagrams in $\mathbb{R}^d$ and intersections of half-spaces in $\mathbb{R}^{d+1}$ comes in handy: we know from the Upper Bound Theorem that the intersection of n half-spaces in $\mathbb{R}^{d+1}$ has complexity $O(n^{\lfloor (d+1)/2 \rfloor}) = O(n^{\lceil d/2 \rceil})$. This bound is tight, so the maximum complexity of the Voronoi diagram—and of the Delaunay triangulation, for that matter—in $\mathbb{R}^d$ is $\Theta(n^{\lceil d/2 \rceil})$.

65.4.2 Construction

In this section we present several algorithms to compute the Voronoi diagram or its dual, the Delaunay triangulation, for point sites in $\mathbb{R}^d$. Several of these algorithms can be generalized to work with metrics other than the Euclidean, and to sites other than points. Since the Voronoi diagram in the plane has only linear complexity one might be tempted to search for a linear time construction algorithm. However the problem of sorting n real numbers is reducible to the problem of computing Voronoi diagrams and, hence, any algorithm for computing the Voronoi diagram must take $\Omega(n \log n)$ time in the worst case.

Two data structures that are particularly well suited to working with planar subdivisions like the Voronoi diagram are the *doubly-connected edge list* (DCEL) by Muller and Preparata [36] and the *quad-edge structure* by Guibas and Stolfi [37]. Both structures require $O(n)$ space, where n is the complexity of the subdivision, and allow to efficiently traverse the edges adjacent to a vertex and the edges bounding a face. In both structures, one can easily obtain in $O(n)$ time a structure for the Voronoi diagram from a structure for the Delaunay triangulation, and vice versa. In fact, the quad-edge structure, as well as a variant of the DCEL [1], are representations of a planar subdivision and its dual at the same time, so there is nothing to convert (except, perhaps, that one may wish to explicitly compute the coordinates of the vertices in the Voronoi diagram, if the quad-edge structure or DCEL describes the Delaunay triangulation).

Divide-and-conquer. The first deterministic worst-case optimal algorithm to compute the Voronoi diagram was given by Shamos and Hoey [38]. Their divide-and-conquer algorithm splits the set S of point sites by a dividing line into subsets L and R of approximately the same size. The Voronoi diagrams Vor(L) and Vor(R) are computed recursively and then merged into Vor(S) in linear time. This algorithm constructs the Voronoi diagram of a set of n points in the plane in $O(n \log n)$ time and linear space.

Plane Sweep. The strategy of a sweep line algorithm is to move a line, called the *sweep line*, from top to bottom over the plane. While the sweep is performed, information is maintained regarding the structure one wants to compute. It is tempting to apply the same approach to Voronoi diagrams, by keeping track of the Voronoi edges that are currently intersected by the sweep line. It is problematic, however, to discover a new Voronoi region in time: when the sweep line reaches a new site, then the line has actually been intersecting the Voronoi edges of its region for a while. Fortune [39] was the first to find a way around this problem. *Fortune's algorithm* applies the plane sweep paradigm in a slightly different fashion: instead of maintaining the intersection of the sweep line with the Voronoi diagram, it maintains information of the part of the Voronoi diagram of the sites above the line that can not be changed by sites below it. This algorithm provides an alternative way of computing the Voronoi diagram of n points in the plane in $O(n \log n)$ time and linear space.

Randomized incremental construction. A natural idea is to construct the Voronoi diagram by *incremental insertion*, that is, to obtain Vor(S) from Vor(S\{s}) by inserting the site s. Insertion of a site means integrating its Voronoi region into the diagram constructed so far. Unfortunately the region of s can have up to $n - 1$ edges, for $|S| = n$, which may lead to a running time of $O(n^2)$. The insertion process is probably better described and implemented in the dual environment, for the Delaunay triangulation DT: construct $DT_i = DT(\{s_1, \ldots, s_i\})$ by inserting s_i into DT_{i-1}. The advantage of this approach over a direct construction of Vor(S) is that Voronoi vertices that appear only in intermediate diagrams but not in the final one need not be computed or stored. DT_i is constructed by exchanging edges, using edge flips [40], until all edges invalidated by s_i have been removed. Still, the worst-case running time of this algorithm can be quadratic. However, if we insert the sites in *random order*, and the algorithm is implemented carefully, then one can prove that the expected running time is $O(n \log n)$, and that the expected amount of storage is $O(n)$.

Other approaches. Finally, recall the connection between Delaunay triangulations and convex hulls. Since there exist $O(n \log n)$ algorithms to compute the convex hull of points in $\mathbb{R}^3$ (see Section 65.3) we therefore have yet another optimal algorithm for the computation of Voronoi diagrams.

65.4.3 Variations

In this section we present some of the common variations on Voronoi diagrams. The first is the *order-k Voronoi diagram* of a set S of n sites, which partitions $\mathbb{R}^d$ on the basis of the first k closest point sites. In other words, each cell in the order-k Voronoi diagram of a set S of sites in the plane corresponds to a k-tuple of sites from S and consists of those points in the plane for which that k-tuple are the k closest sites. One might fear that the order-k Voronoi diagram has $\Theta(n^k)$ cells, but this is not the case. In two dimensions, for example, its complexity is $O(k(n - k))$, and it can be computed in $O(k(n - k) \log n + n \log^3 n)$ expected time [41].

The *furthest-site Voronoi diagram* partitions $\mathbb{R}^d$ according to the furthest site, or equivalently, according to the closest $n - 1$ of n sites. The furthest-site Voronoi diagram can be computed in $O(n \log n)$ time in two dimensions, and in $O(n^{\lceil d/2 \rceil})$ in dimension $d \geq 3$.

One can also consider different distance functions than the Euclidean distance. For example, one can alter the distance function by the addition of *additive* or *multiplicative weights*. In this case every point site s_i is associated with a weight w_i and the distance function $d(s_i, x)$ between a point x and a site s_i becomes $d(s_i, x) = w_i + \text{dist}(s_i, x)$ (additive weights) or $d(s_i, x) = w_i \cdot \text{dist}(s_i, x)$ where $\text{dist}(s_i, x)$ denotes the Euclidean distance between s_i and x. The Voronoi diagram for point sites in 2 dimensions with additive weights can be computed in $O(n \log n)$ time, for multiplicative weights the time increases to $O(n^2)$ time.

Finally the *power diagram*, or *Laguerre diagram*, is another Voronoi diagram for point sites s_i that are associated with weights w_i. Here the distance function is the *power distance* introduced by Aurenhammer [42], where the distance from a point x to a site s_i is measured along a line tangent to the sphere of radius $\sqrt{w_i}$ centered at s_i, that is, $d(s_i, x) = \sqrt{\text{dist}(s_i, x)^2 - w_i}$. The power diagram can be computed in $O(n \log n)$ time in two dimensions.

65.5 Triangulations

In geometric data processing, structures that partition the geometric input, as well as connectivity structures for geometric objects, play an important role. Versatile tools in this context are *triangular meshes*, often called *triangulations*. A triangulation of a geometric domain such as a polygon in $\mathbb{R}^2$ or a polyhedron in $\mathbb{R}^3$ is a partition into simplices that meet only at shared faces. A triangulation of a point set S is a triangulation of the convex hull of S, such that the vertices in the triangulation are exactly the points in S.

In the following sections, we first discuss the most famous of all triangulations, the Delaunay triangulation. We then address triangulations of polygons and polyhedra in $\mathbb{R}^3$. Finally we describe a recent generalization of triangulations: the pseudo-triangulation.

65.5.1 Delaunay Triangulation

In this section we provide additional detail on the Delaunay triangulation of a set $S = \{s_1, \ldots, s_n\}$ of points in $\mathbb{R}^d$, which was introduced in Section 65.4. There we defined the Delaunay triangulation as the dual of the Voronoi diagram. In the plane, for instance, the Delaunay triangulation has an edge between sites s_i and s_j if and only if the Voronoi cells of s_i and s_j share a boundary edge. In higher dimensions, there is an edge between s_i and s_j if their Voronoi cells share a $(d - 1)$-facet. The Delaunay triangulation can also be defined directly. If we restrict ourselves for the moment to a set S of points in $\mathbb{R}^2$ then the Delaunay triangulation of S, $DT(S)$, is defined by the *empty-circle condition*: a triangle Δ defined by three points s_i, s_j, and s_k is part of $DT(S)$ if and only if Δ's circumcircle neither encloses nor passes through any other points of S. More generally, $d + 1$ points in $\mathbb{R}^d$ define a simplex in the Delaunay triangulation if and only if its circumscribed sphere neither encloses nor passes through any other points of S. If no $d + 1$ points of S are co-spherical then $DT(S)$ is indeed a triangulation. If $d + 2$ or more points are co-spherical, then $DT(S)$ can contain cells with more than $d + 1$ sides. Fortunately, such cells can easily be further decomposed in simplices—with a bottom-vertex triangulation, for example—so such degeneracies are not a real problem. To simplify the description, we from now on assume that these degeneracies do not occur.

In the previous section we have seen a close connection between Voronoi diagrams in $\mathbb{R}^d$ and intersections of half-spaces in $\mathbb{R}^{d+1}$. Similarly, there is a close connection between the Delaunay triangulation in $\mathbb{R}^d$ and convex hulls in $\mathbb{R}^{d+1}$. Let's again restrict ourselves to the case $d = 2$. Consider the transformation that maps a site $s = (s_x, s_y)$ in $\mathbb{R}^2$ onto the point $\lambda(s_x, s_y) = \left(s_x, s_y, s_x^2 + s_y^2\right)$

in $\mathbb{R}^3$. In other words, s is "lifted" vertically onto the unit paraboloid to obtain $\lambda(s)$. Let $\lambda(S)$ be the set of lifted sites. Then if we project the lower convex hull of $\lambda(S)$—the part of the convex hull consisting of the facets facing downward—back onto the xy-plane, we get the Delaunay triangulation of S.

The Delaunay triangulation is the "best" triangulation with respect to many optimality criteria. The Delaunay triangulation:

- Minimizes the maximum radius of a circumcircle;
- Maximizes the minimum angle;
- Maximizes the sum of inscribed circle radii;
- Minimizes the "potential energy" of a piecewise-linear interpolating surface.

Also, the distance between any two vertices of the Delaunay triangulation along the triangulation edges is at most 2.42 times their Euclidean distance, that is, the *dilation* of the Delaunay triangulation is 2.42. Finally, the Delaunay triangulation contains as a subgraph many other interesting graphs:

$$EMST \subseteq RNG \subseteq GG \subseteq DT$$

where *EMST* is the Euclidean minimum spanning tree, *RNG* is the relative neighborhood graph, and *GG* is the Gabriel graph (see Reference 4 for details on these graphs).

Since the Delaunay triangulation is the dual of the Voronoi diagram, any algorithm presented in Section 65.4.2 can by used to efficiently compute the Delaunay triangulation. We therefore refrain from presenting any additional algorithms at this point.

65.5.2 Polygons

Triangulating a simple polygon P is not only an interesting problem in its own right, but it is also an important preprocessing step for many algorithms. For example, many shortest-path and visibility problems on a polygon P can be solved in linear time if a triangulation of P is given [43]. It is fairly easy to show that any polygon can indeed by decomposed into triangles by adding a number of diagonals and, moreover, that the number of triangles in any triangulation of a simple polygon with n vertices is $n - 2$.

There are many algorithms to compute a triangulation of a simple polygon. Most of them run in $O(n \log n)$ time. Whether it is possible to triangulate a polygon in linear time was a prominent open problem for several years until Chazelle [44], after a series of interim results, devised a linear-time algorithm. Unfortunately, his algorithm is more of theoretical than of practical interest, so it is probably advisable to use a deterministic algorithm with $O(n \log n)$ running time, such as the one described below, or one of the slightly faster randomized approaches with a time complexity of $O(n \log^* n)$ [20].

In the remainder of this section we sketch a deterministic algorithm that triangulates a simple polygon P with n vertices in $O(n \log n)$ time. We say that a polygon P is *monotone* with respect to a line ℓ if the intersection of any line ℓ' perpendicular to ℓ with P is connected. A polygon that is monotone with respect to the x-axis is called *x-monotone*. Now the basic idea is to decompose P into monotone polygons and then to triangulate each of these monotone polygons in linear time.

There are several methods to decompose a simple polygon into x-monotone polygons in $O(n \log n)$ time. One approach is to sweep over P twice, from left to right and then from right to left, and to add appropriate edges to vertices that did not previously have at least one edge extending to the left and at least one edge extending to the right. A more detailed description of this or related approaches can be found in References 1,8.

Triangulating a monotone polygon. Now suppose we are given an x-monotone polygon P—see Figure 65.8. We consider its vertices $p_1, \ldots, p_n$ from left to right and use a stack to store the vertices of the not-yet-triangulated part of the polygon (which necessarily form a reflex chain) to the left of our current vertex p_i. If p_i is adjacent to p_{i-1}, as in see Figure 65.8a, then we pop vertices from the stack and connect them to p_i until the stack (including p_i) forms a reflex chain again. In particular that might mean that we simply add p_i to the stack. If p_i is adjacent to the leftmost vertex on the stack, which could be p_{i-1}, as in see Figure 65.8b, then we connect p_i to each vertex of the stack and clear the stack of all vertices except p_i and p_{i-1}. This algorithm triangulates P in linear time.

65.5.3 Polyhedra

In this section we briefly discuss triangulations (or *tetrahedralizations*) of three-dimensional polyhedra. A polyhedron P is a connected solid with a piecewise linear boundary (i.e., its boundary is composed of polygons). We assume P to be non-degenerate: A sufficiently small ball around any point of the boundary of P contains a connected component of the interior as well as the exterior of P. The number of vertices, edges, and faces of a non-degenerate tetrahedron are linearly related.

Three dimensions unfortunately do not behave as nicely as two. Two triangulations of the same input data may contain quite different numbers of tetrahedra. For example, a triangulation of a convex polyhedron with n vertices may contain between $n - 3$

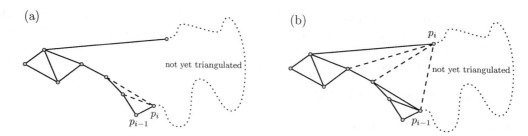

FIGURE 65.8   Triangulating an *x*-monotone polygon.

and $\binom{n-2}{2}$ tetrahedra. Furthermore, some non-convex polyhedra can not be triangulated at all without the use of additional (so-called *Steiner*) points. The most famous example is *Schönhardt's* polyhedron. Finally, it is NP-complete to determine whether a polyhedron can be triangulated without Steiner points and to test whether *k* Steiner points are sufficient [45].

Chazelle [46] constructed a polyhedron with *n* vertices that requires $\Omega(n^2)$ Steiner points. On the other hand, any polyhedron can be triangulated with $O(n^2)$ tetrahedra, matching his lower bound. If one pays special attention to the reflex vertices of the input polyhedron *P*, then it is even possible to triangulate *P* using only $O(n + r^2)$ tetrahedra, where *r* is the number of reflex edges of *P* [47].

65.5.4 Pseudo-Triangulations

In recent years a relaxation of triangulations, called *pseudo-triangulations*, has received considerable attention. Here, faces are bounded by three concave chains, rather than by three line segments. More formally, a pseudo-triangle is a planar polygon that has exactly three convex vertices with internal angles less than π, all other vertices are concave. Note that every triangle is a pseudo-triangle as well. The three convex vertices of a pseudo-triangle are called its *corners*. Three concave chains, called *sides*, join the three corners. A pseudo-triangulation for a set *S* of points in the plane is a partition of the convex hull of *S* into pseudo-triangles whose vertex set is exactly *S*—see Figure 65.9. Pseudo-triangulations, also called *geodesic triangulations*, were originally studied for convex sets and for simple polygons in the plane because of their applications to visibility [48,49] and ray shooting [50,51]. But in the last few years they also found application in robot motion planning [52] and kinetic collision detection [53,54].

Of particular interest are the so-called *minimum pseudo-triangulations*, which have the minimum number of pseudo-triangular faces among all possible pseudo-triangulations of a given domain. They were introduced by Streinu [52], who established that every minimum pseudo-triangulation of a set *S* of *n* points consists of exactly $n - 2$ pseudo-triangles (here we do not count the outer face). Note that such a statement cannot be made for ordinary triangulations: there the number of triangles depends on the number of points that show up on the convex hull of *S*. Minimum pseudo-triangulations are also referred to as *pointed pseudo-triangulations*. The name stems from the fact that every vertex *v* of a minimum pseudo-triangulation has an incident region whose angle at *v* is greater than π. The converse is also true (and can be easily established via Euler's relation): If every vertex of a pseudo-triangulation is *pointed*—it has an incident angle greater than π—then this pseudo-triangulation has exactly $n - 2$ pseudo-triangles and is therefore minimum. A pseudo-triangulation is called *minimal* (as opposed to minimum) if the union of any two faces is not a pseudo-triangle. In general, all minimum pseudo-triangulations are also minimal, but the opposite is not necessarily true—see Figure 65.9a for an example of a minimal but not minimum pseudo-triangulation.

The great variety of applications in which pseudo-triangulations are successfully employed prompted a growing interest in their geometric and combinatoric properties, which often deviate substantially from those of ordinary triangulations. An example of a nice property of pseudo-triangulations is that every point set in the plane admits a pseudo-triangulation of maximum vertex

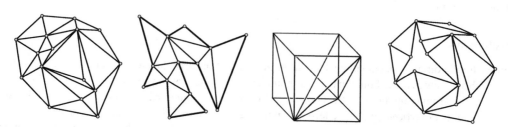

FIGURE 65.9   Triangulations of a point set, a simple polygon, and a polyhedron; a pseudotriangulation of a point set.

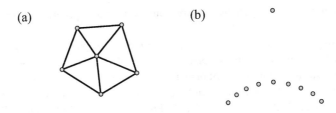

FIGURE 65.10 (a) A minimal but not minimum pseudo-triangulation. (b) A point set for which every triangulation has a vertex of high degree.

degree 5 [55]. (This bound is tight.) Again, this is not the case for ordinary triangulations: any ordinary triangulation of the point set depicted in Figure 65.10b, contains a vertex of degree $n - 1$.

Up to now there are unfortunately no extensions of pseudo-triangulations to dimensions higher than two which retain a significant subset of their useful planar properties.

References

1. M. de Berg, M. van Kreveld, M. Overmars, and O. Schwarzkopf: *Computational Geometry: Algorithms and Applications* (2nd Ed.). Springer-Verlag, 2000.
2. J.-D. Boissonnat and M. Yvinec: *Algorithmic Geometry*. Cambridge University Press, 1998.
3. J. O'Rourke: *Computational Geometry in C* (2nd Ed.). Cambridge University Press, 1998.
4. F. P. Preparata and M. I. Shamos: *Computational Geometry: An Introduction*. Springer-Verlag, 1985.
5. H. Edelsbrunner: *Algorithms in Combinatorial Geometry*. Springer-Verlag, 1987.
6. M. Sharir and P. K. Agarwal: *Davenport-Schinzel Sequences and Their Geometric Applications*. Cambridge University Press, 1995.
7. A. Okabe, B. Boots, K. Sugihara, and S. N. Chiu: *Spatial Tessellations: Concepts and Applications of Voronoi Diagrams* (2nd Ed.). Wiley, 2000.
8. J. E. Goodman and J. O'Rourke (Eds.): *Handbook of Discrete and Computational Geometry*. CRC Press LLC, 1997.
9. J.-R. Sack and J. Urrutia (Eds.): *Handbook of Computational Geometry*. Elsevier Science, 2000.
10. R. Pollack, M. Sharir, and S. Sifrony: Separating two simple polygons by a sequence of translations. *Discr. Comput. Geom.* 3:123–136, 1988.
11. L. J. Guibas, M. Sharir, and S. Sifrony: On the general motion planning problem with two degrees of freedom. *Discr. Comput. Geom.* 4:491–521, 1989.
12. B. Tagansky: A new technique for analyzing substructures in arrangements of piecewise linear surfaces. *Discr. Comput. Geom.* 16:455–479, 1996.
13. H. Edelsbrunner: The upper envelope of piecewise linear functions: Tight complexity bounds in higher dimensions. *Discr. Comput. Geom.* 4:337–343, 1989.
14. D. Halperin and M. Sharir: New bounds for lower envelopes in three dimensions, with applications to visibility in terrains. *Discr. Comput. Geom.* 12:313–326, 1994.
15. P. McMullen: The maximal number of faces of a convex polytope. *Mathematica* 17:179–184, 1970.
16. S. Basu: On the combinatorial and topological complexity of a single cell. In *Proc. 39th Annu. IEEE Sympos. Found. Comput. Sci.*, pp. 606–616, 1998.
17. M. Sharir: Almost tight upper bounds for lower envelopes in higher dimensions. *Discr. Comput. Geom.* 12:327–345, 1994.
18. R. L. Graham: An efficient algorithm for determining the convex hull of a finite planar set. *Inf. Proc. Lett.* 1:132–133, 1972.
19. K. L. Clarkson and P. W. Shor: Applications of random sampling in computational geometry, II. *Discr. Comput. Geom.* 4:387–421, 1989.
20. K. Mulmuley: *Computational Geometry: An Introduction through Randomized Algorithms*. Prentice Hall, 1994.
21. F. P. Preparata and S. J. Hong: Convex hulls of finite sets of points in two and three dimensions. *Comm. ACM* 20:87–93, 1977.
22. R. Seidel: Small-dimensional linear programming and convex hulls made easy. *Discr. Comput. Geom.* 6:423–434, 1991.
23. B. Chazelle: An optimal convex hull algorithm in any fixed dimension. *Discr. Comput. Geom.* 10:377–409, 1993.
24. R. A. Dwyer: On the convex hull of random points in a polytope. *J. Appl. Probab.* 25:688–699, 1988.
25. R. A. Dwyer: Kinder, gentler average-case analysis for convex hulls and maximal vectors. *SIGACT News* 21:64–71, 1990.
26. D. Kirkpatrick and R. Seidel: The ultimate planar convex hull algorithm? *SIAM J. Comput.* 15:287–299, 1986.

27. T. Chan: Output-sensitive results on convex hulls, extreme points, and related problems. *Discr. Comput. Geom.* 16:369–387, 1996.

28. M. H. Overmars and J. van Leeuwen: Maintenance of configurations in the plane. *J. Comput. Systs. Sci.* 23:166–204, 1981.

29. F. P. Preparata: An optimal real-time algorithm for planar convex hulls. *Comm. ACM* 22:405–408, 1979.

30. J. Hershberger and S. Suri: Off-line maintenance of planar configurations. *J. Algorithms* 21:453–475, 1996.

31. T. Chan: Dynamic planar convex hull operations in near-logarithmic amortized time. *J. ACM.* 48:1–12, 2001.

32. G. S. Brodal and R. Jacob: Dynamic Planar Convex Hull. In *Proc. 43rd Annu. IEEE Sympos. Found. Comput. Sci.*, pp. 617–626, 2002.

33. F. Aurenhammer: Voronoi diagrams—A survey of a fundamental geometric data structure. *ACM Comput. Surv.*, 23:345–405, 1991.

34. F. Aurenhammer and R. Klein: Voronoi diagrams. In: J. Sack and G. Urrutia (eds.), *Handbook of Computational Geometry*, Chapter V, pp. 201–290. Elsevier Science Publishing, 2000.

35. S. Fortune: Voronoi diagrams and delaunay triangulations. In: F. K. Hwang and D.-Z. Du (eds.), *Computing in Euclidean Geometry* (2nd Ed.). World Scientific, pp. 225–265, 1995.

36. D. E. Muller and F. P. Preparata: Finding the intersection of two convex polyhedra. *Theoret. Comput. Sci.*, 7:217–236, 1978.

37. L. J. Guibas and J. Stolfi: Primitives for the manipulation of general subdivisions and the computation of Voronoi diagrams. *ACM Trans. Graph.*, 4:74–123, 1985.

38. M. I. Shamos and D. Hoey: Closest-point problems. In *Proc. 16th IEEE Symp. Found. Comput. Science*, pp. 151–162, 1975.

39. S. Fortune: A sweepline algorithms for Voronoi diagrams. *Algorithmica*, 2:153–174, 1987.

40. C. L. Lawson: Software for C^1 surface interpolation. In: J. R. Rie (ed.), *Math. Software III*, Academic Press, New York, NY, pp. 161–194, 1977.

41. P. K. Agarwal, M. de Berg, J. Matousek, and O. Schwartzkopf: Constructing levels in arrangements and higher order Voronoi diagrams. In *Proc. 11th Symp. Comput. Geom.*, pp. 71–78, 1995.

42. F. Aurenhammer: Power diagrams: properties, algorithms, and applications. *Siam J. Comput.*, 16:78–96, 1987.

43. L. J. Guibas, J. Hershberger, D. Leven, M. Sharir, and R. E. Tarjan. Linear-time algorithms for visibility and shortest path problems inside triangulated simple polygons. *Algorithmica*, 2:209–233, 1987.

44. B. Chazelle: Triangulating a simple polygon in linear time. *Discrete Comput. Geom.*, 6:485–524, 1991.

45. J. Rupert and R. Seidel: On the difficulty of tetrahedralizing 3-dimensional non-convex polyhedra. *Discrete Comput. Geom.*, 7:227–253, 1992.

46. B. Chazelle: Convex partitions of polyhedra:A lower bound and worst-case optimal algorithm. *SIAM J. Comput.*, 13:488–507, 1984.

47. B. Chazelle and L. Palios: Triangulating a nonconvex polytope. *Discrete Comput. Geom.*, 5:505–526, 1990.

48. M. Pocchiola and G. Vegter: Minimal tangent visibility graphs. *Comput. Geom. Theory Appl.*, 6:303–314, 1996.

49. M. Pocchiola and G. Vegter: Topologically sweeping visibility complexes via pseudo-triangulations *Discrete Comp. Geom.*, 16:419–453, 1996.

50. B. Chazelle, H. Edelsbrunner, M. Grigni, L. J. Guibas, J. Hershberger, M. Sharir, and J. Snoeyink: Ray shooting in polygons using geodesic triangulations. *Algorithmica*, 12:54–68, 1994.

51. M. Goodrich and R. Tamassia: Dynamic ray shooting and shortest paths in planar subdivision via balanced geodesic triangulations. *J. Algorithms* 23:51–73, 1997.

52. I. Streinu: A combinatorial approach to planar non-colliding robot arm motion planning. In *Proc. 41st FOCS*, 2000, pp. 443–453.

53. P. K. Agarwal, J. Basch, L. J. Guibas, J. Hershberger, and L. Zhang: Deformable free space tilings for kinetic collision detection. *Int. J. Robotics Research* 21(3):179–197, 2002.

54. D. Kirkpatrick, J. Snoeyink, and B. Speckmann: Kinetic collision detection for simple polygons. *Int. J. Comput. Geom. Appl.* 12(1&2):3–27, 2002.

55. L. Kettner, D. Kirkpatrick, A. Mantler, J. Snoeyink, B. Speckmann, and F. Takeuchi: Tight degree bounds for pseudo-triangulations of points. *Comp. Geom. Theory Appl.*, 25:1–12, 2003.

66

Computational Geometry: Proximity and Location[*]

66.1 Introduction.. 1027
66.2 Point Location... 1027
    Kirkpatrick's Algorithm • Slab-Based Methods and Persistent Trees • Separating Chains and Fractional Cascading • Trapezoidal Maps and the History Graph • Worst- and Expected-Case Optimal Point Location
66.3 Proximity Structures 1032
    Voronoi Diagrams • Delaunay Triangulations • Other Geometric Proximity Structures
66.4 Nearest Neighbor Searching 1035
    Nearest Neighbor Searching Through Point Location • K-d Trees • Other Approaches to Nearest Neighbor Searching • Approximate Nearest Neighbor Searching • Approximate Voronoi Diagrams
66.5 Sources and Related Material... 1038
Acknowledgments 1039
References.. 1039

Sunil Arya
Hong Kong University of Science and Technology

David M. Mount
University of Maryland

66.1 Introduction

Proximity and location are fundamental concepts in geometric computation. The term *proximity* refers informally to the quality of being close to some point or object. Typical problems in this area involve computing geometric structures based on proximity, such as the Voronoi diagram, Delaunay triangulation and related graph structures such as the relative neighborhood graph. Another class of problems are retrieval problems based on proximity. These include nearest neighbor searching and the related concept of range searching. (See Chapter 19 for a discussion of data structures for range searching.) Instances of proximity structures and proximity searching arise in many fields of applications and in many dimensions. These applications include object classification in pattern recognition, document analysis, data compression, and data mining.

The term *location* refers to the position of a point relative to a geometric subdivision or a given set of disjoint geometric objects. The best known example is the point location problem, in which a subdivision of space into disjoint regions is given, and the problem is to identify which region contains a given query point. This problem is widely used in areas such as computer graphics, geographic information systems, and robotics. Point location is also used as a method for proximity searching, when applied in conjunction with Voronoi diagrams.

In this chapter we will present a number of geometric data structures that arise in the context of proximity and location. The area is so vast that our presentation will be limited to a relatively few relevant results. We will discuss data structures for answering point location queries first. After this we will introduce proximity structures, including Voronoi diagrams and Delaunay triangulations. Our presentation of these topics will be primarily restricted to the plane. Finally, we will present results on multidimensional nearest neighbor searching.

66.2 Point Location

The planar *point location* problem is one of the most fundamental query problems in computational geometry. Consider a *planar straight line graph S*. (See Chapter 18 for details.) This is an undirected graph, drawn in the plane, whose edges are straight

---

[*] This chapter has been reprinted from first edition of this Handbook, without any content updates.

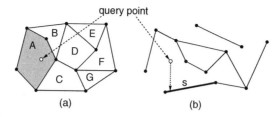

FIGURE 66.1   Illustration of (a) point location and (b) vertical ray shooting queries.

line segments that have pairwise disjoint interiors. The edges of S subdivide the plane into (possibly unbounded) polygonal regions, called *faces*. Henceforth, such a structure will be referred to as a *polygonal subdivision*. Throughout, we let n denote the *combinatorial complexity* of S, that is, the total number of vertices, edges and faces. (We shall occasionally abuse notation and use n to refer to the specific number of vertices, edges, or faces of S.) A planar subdivision is a special case of the more general topological concept of a *cell complex* [1], in which vertices, edges, and generally faces of various dimensions are joined together so that the intersection of any two faces is either empty or is a face of lower dimension.

The *point location problem* is to preprocess a polygonal subdivision S in the plane into a data structure so that, given any query point q, the polygonal face of the subdivision containing q can be reported quickly. (In Figure 66.1a, face A would be reported.) This problem is a natural generalization of the binary search problem in 1-dimensional space, where the faces of the subdivision correspond to the intervals between the 1-dimensional key values. By analogy with the 1-dimensional case, the goal is to preprocess a subdivision into a data structure of size $O(n)$ so that point location queries can be answered in $O(\log n)$ time.

A slightly more general formulation of the problem, which is applicable even when the input is not a subdivision is called *vertical ray shooting*. A set S of line segments is given with pairwise disjoint interiors. Given a query point q, the problem is to determine the line segment of S that lies vertically below q. (In Figure 66.1b, the segment s would be reported.) If the ray hits no segment, a special value is returned. When S is a polygonal subdivision, point location can be reduced to vertical ray shooting by associating each edge of S with the face that lies immediately above it.

66.2.1 Kirkpatrick's Algorithm

Kirkpatrick was the first to present a simple point location data structure that is asymptotically optimal [2]. It answers queries in $O(\log n)$ time using $O(n)$ space. Although this is not the most practical approach to point location, it is quite easy to understand.

Kirkpatrick starts with the assumption that the planar subdivision has been refined (through the addition of $O(n)$ new edges and vertices) so that it is a triangulation whose external face is a triangle. Let T_0 denote this initial triangulation subdivision. Kirkpatrick's method generates a finite sequence of increasingly coarser triangulations, $\langle T_0, T_1, T_2, \ldots, T_m \rangle$, where T_m consists of the single triangle forming the outer face of the original triangulation. This sequence satisfies the following constraints: (a) each triangle of T_{i+1} intersects a constant number of triangles of T_i, and (b) the number of vertices of T_{i+1} is smaller than the number of vertices of T_i by a constant fraction. (See Figure 66.2.)

The data structure itself is a rooted DAG (directed acyclic graph), where the root of the structure corresponds to the single triangle of T_m, and the leaves correspond to the triangles of T_0. The interior nodes of the DAG correspond to the triangles of each of the triangulations. A directed edge connects each triangle in T_{i+1} with each triangle in T_i that it overlaps.

Given a query point q, the point location query proceeds level-by-level through the DAG, visiting the nodes corresponding to the triangles that contain q. By property (a), each triangle in T_{i+1} overlaps a constant number of triangles of T_i, which implies that it is possible to descend one level in the data structure in $O(1)$ time. It follows that the running time is proportional to the

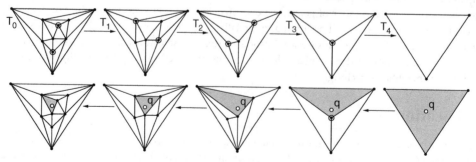

FIGURE 66.2   The sequence of triangulations generated in the construction of Kirkpatrick's structure (above) and the triangles visited in answering a point location query (below).

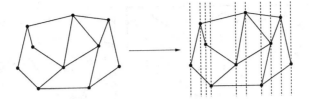

FIGURE 66.3 Slab refinement of a subdivision.

number of levels in the tree. By property (b), the number of vertices decreases at each level by a fixed constant fraction, and hence, the number of levels is $O(\log n)$. Thus the overall query time is $O(\log n)$.

Kirkpatrick showed how to build the data structure by constructing a sequence of triangulations satisfying the above properties. Kirkpatrick's approach is to compute an *independent set* of vertices (that is, a set of mutually nonadjacent vertices) in T_i where each vertex of the independent set has constant degree. (An example is shown at the top of Figure 66.2. The vertices of the independent set are highlighted.) The three vertices of the outer face are not included. Kirkpatrick showed that there exists such a set whose size is a constant fraction of the total number of vertices, and it can be computed in linear time. These vertices are removed along with any incident edges, and the resulting "holes" are then retriangulated. Kirkpatrick showed that the two properties hold for the resulting sequence of triangulations.

66.2.2 Slab-Based Methods and Persistent Trees

Many point location methods operate by refining the given subdivision to form one that is better structured, and hence, easier to search. One approach for generating such a refinement is to draw a vertical line through each vertex of the subdivision. These lines partition the plane into a collection of $O(n)$ *vertical slabs*, such that there is no vertex within each slab. As a result, the intersection of the subdivision with each slab consists of a set of line segments, which cut clear through the slab. These segments thus partition the slab into a collection of disjoint trapezoids with vertical sides. (See Figure 66.3.)

Point location queries can be answered in $O(\log n)$ time by applying two binary searches. The first search accesses the query point's x coordinate to determine the slab containing the query point. The second binary search tests whether the query point lies above or below individual lines of the slab, in order to determine which trapezoid contains the query point. Since each slab can be intersected by at most n lines, this second search can be done in $O(\log n)$ time as well.

A straightforward implementation of this method is not space efficient, since there are $\Omega(n)$ slabs,[*] each having up to $\Omega(n)$ intersecting segments, for a total of $\Omega(n^2)$ space. However, adjacent slabs are very similar, since the only segments that change are those that are incident to the vertices lying on the slab boundary. Sarnak and Tarjan [4] exploited this idea to produce an optimal point location data structure. To understand their algorithm, imagine sweeping a line segment continuously from left to right. Consider the sorted order of subdivision line segments intersecting this sweep line. Whenever the sweep line encounters a vertex of the subdivision, the edges incident to this vertex lying to the left of the vertex are removed from the sweep-line order and incident edges to the right of the vertex are inserted. Since every edge is inserted once and deleted once in this process, the total number of changes over the entire sweep process is $O(n)$.

Sarnak and Tarjan proposed maintaining a persistent variant of the search tree. A *persistent search tree* is a dynamic search tree (supporting insertion and deletion) which can answer queries not only to the current tree, but to any of the previous versions in the history of the tree's lifetime as well. (See Chapter 33.) In this context, the history of changes to the search tree is maintained in a left to right sweep of the plane. The persistent search tree supports queries to any of these trees, that is, in any of the slabs, in $O(\log n)$ time. The clever aspect of Sarnak and Tarjan's tree is that it can be stored in $O(n)$ total space (as opposed to $O(n^2)$ space, which would result by generating $O(n)$ copies of the tree). This is done by a method called *limited node copying*. Thus, this provides an asymptotically optimal point location algorithm. A similar approach was discovered independently by Cole [5].

66.2.3 Separating Chains and Fractional Cascading

Slab methods use vertical lines to help organize the search. An alternative approach, first suggested by Lee and Preparata [6], is to use a divide-and-conquer approach based on a hierarchy of monotone polygon chains, called *separating chains*. A simple polygon is said to be *x-monotone* if the intersection of the interior of the polygon with a vertical line is connected. An *x-monotone subdivision* is one in which all the faces are *x*-monotone. The separating chain method requires that the input be an *x*-monotone subdivision. Fortunately, it is possible to convert any polygonal subdivision in the plane into an *x*-monotone subdivision in

---

[*] For readers unfamiliar with this notation, $\Omega(f(n))$ is analogous to the notation $O(f(n))$, but it provides an asymptotic lower bound rather than an upper bound. The notation $\Theta(f(n))$ means that both upper and lower bounds apply [3].

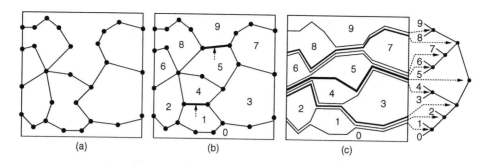

FIGURE 66.4   Point location by separating chains: (a) the original subdivision, (b) the addition of one or more edges to make the subdivision *x*-monotone, (c) decomposition of the subdivision into a hierarchy of separating chains.

$O(n \log n)$ time, through the addition of $O(n)$ new edges. (See e.g., [6–8].) For example, Figure 66.4a shows a subdivision that is not *x*-monotone, but the addition of two edges suffice to produce an *x*-monotone subdivision shown in Figure 66.4b.

Consider an *x*-monotone subdivision with n faces. It is possible to order the faces $f_0, f_1, \ldots, f_{n-1}$ such that if $i < j$, then every vertical line that intersects both of these faces intersects f_i below f_j. (See Figure 66.4b.) For each i, $0 < i < n$, define the ith separating chain to be the *x*-monotone polygonal chain separating faces whose indices are less than i from those that are greater than or equal to i.

Observe that, given a chain with m edges, it is possible to determine whether a given query point lies above or below the chain in $O(\log m)$ time, by first performing a binary search on the *x*-coordinates of the chain, in order to find which chain edge overlaps the query point, and then determining whether the query point lies above or below this edge in $O(1)$ time. The separating chain method works intuitively by performing a binary search on these chains. The binary search can be visualized as a binary tree imposed on the chains, as shown in Figure 66.4c.

Although many chains traverse the same edge, it suffices to store each edge only once in the structure, namely with the chain associated with the highest node in the binary tree. This is because once a discrimination of the query point is made with respect to such an edge, its relation is implicitly known for all other chains that share the same edge. It follows that the total space is $O(n)$.

As mentioned above, at each chain the search takes logarithmic time to determine whether the query point is above or below the chain. Since there are $\Omega(n)$ chains, this would lead to an $\Omega(\log^2 n)$ algorithm [6]. There is a clever way to reduce the search time to $O(\log n)$, through the use of a simple and powerful method called *fractional cascading* [9,10]. Intuitively, fractional cascading seeks to replace a sequence of independent binary searches with a more efficient sequence of coordinated searches. After searching through a parent's chain, it is known which edge of this chain the query point overlaps. Thus, it is not necessary to search the entire range of *x*-coordinates for the child's chain, just the sublist of *x*-coordinates that overlap this interval.

However, in general, the number of edges of the child's chain that overlaps this interval may be as large as $\Omega(n)$, and so this observation would seem to be of no help. In fractional cascading, this situation is remedied by augmenting each list. Starting with the leaf level, the *x*-coordinate of every fourth vertex is passed up from each child's sorted list of *x*-coordinates and inserted into its parent's list. This is repeated from the parent to the grandparent, and so on. After doing this, once the edge of the parent's chain that overlaps the query point has been determined, there can be at most four edges of the child's chain that overlap this interval. (e.g., in Figure 66.5 the edge $\overline{pq}$ is overlapped by eight edges at the next lower level. After cascading, it is broken into three subedges, each of which overlaps at most four edges at the next level.) Thus, the overlapping edge in the child's chain can be found in $O(1)$ time. The root requires $O(\log n)$ time, and each of the subsequent $O(\log n)$ searches can be performed in $O(1)$ additional time. It can be shown that this augmentation of the lists increases the total size of all the lists by at most a constant factor, and hence the total space is still $O(n)$.

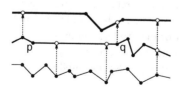

FIGURE 66.5   Example of fractional cascading. Every fourth vertex is sampled from each chain and inserted in its parent's chain.

66.2.4 Trapezoidal Maps and the History Graph

Next we describe a randomized approach for point location. It is quite simple and practical. Let us assume that the planar subdivision is presented simply as a set of n line segments $S = \{s_1, s_2, \ldots, s_n\}$ with pairwise disjoint interiors. The algorithm answers vertical ray-shooting queries as described earlier. This approach was developed by Mulmuley [11]. Also see Seidel [12].

The algorithm is based on a structure called a *trapezoidal map* (or *trapezoidal decomposition*). First, assume that the entire domain of interest is enclosed in a large rectangle. Imagine shooting a bullet vertically upwards and downwards from each vertex in the polygonal subdivision until it hits another segment of S. To simplify the presentation, we shall assume that the x-coordinates of no two vertices are identical. The segments of S together with the resulting bullet paths subdivide the plane into $O(n)$ trapezoidal cells with vertical sides, which may degenerate to triangles. (See Figure 66.6a.)

For the purposes of point location, the trapezoidal map is created by a process called a *randomized incremental construction*. The process starts with the initial bounding rectangle (i.e., one trapezoid) and then the segments of S are inserted one by one in random order. As each segment is added, the trapezoidal map is updated by "walking" the segment through the subdivision, and updating the map by shooting new bullet paths through the segments endpoints and trimming existing paths that hit the new segment. See References 7,11,12 for further details. The number of changes in the diagram with each insertion is proportional to the number of vertical segments crossed by the newly added segment, which in the worst case may be as high as $\Omega(n)$. It can be shown, however, that on average each insertion of a new segment results in $O(1)$ changes. This is true irrespective of the distribution of the segments, and the expectation is taken over all possible insertion orders.

The point location data structure is based on a rooted directed acyclic graph, or DAG, called the *history DAG*. Each node has either two outgoing edges (internal nodes) or none (leaves). Leaves correspond one-to-one with the cells of the trapezoidal map. There are two types of internal nodes, x-nodes and y-nodes. Each x-node contains the x-coordinate x_0 of an endpoint of one of the segments, and its two children correspond to the points lying to the left and to the right of the vertical line $x = x_0$. Each y-node contains a pointer to a line segment of the subdivision. The left and right children correspond to whether the query point is above or below the line containing this segment, respectively. (In Figure 66.7, x-nodes are shown as circles, y-nodes as hexagons, and leaves as squares.)

As with Kirkpatrick's algorithm, the construction of the point location data structure encodes the history of the randomized incremental construction. Let $\langle T_0, T_1, \ldots, T_n \rangle$ denote the sequence of trapezoidal maps that result through the randomized incremental process. The point location structure after insertion of the ith segment has one leaf for each trapezoid in T_i. Whenever a segment is inserted, the leaf nodes corresponding to trapezoids that were destroyed are replaced with internal x- and y-nodes that direct the search to the location of the query point in the newly created trapezoids, after the insertion. (This is illustrated

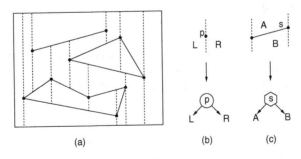

FIGURE 66.6 A trapezoidal map of a set of segments (a), and the two types of internal nodes: x-node (b) and y-node (c).

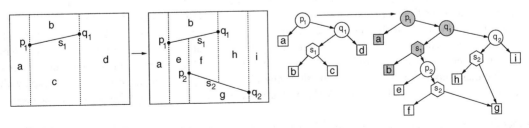

FIGURE 66.7 Example of incremental construction of a trapezoidal map and the associated history DAG. The insertion of segment s_2 replaces the leaves associated with destroyed trapezoids c and d with an appropriate search structure for the new trapezoids e–i.

in Figure 66.7.) Through the use of node sharing, the resulting data structure can be shown to have expected size $O(n)$, and its expected depth is $O(\log n)$, where the expectation is over all insertion orders. Details can be found in References 7,11,12.

66.2.5 Worst- and Expected-Case Optimal Point Location

Goodrich, Orletsky and Ramaiyer [13] posed the question of bounding the minimum number of comparisons required, in the worst case, to answer point location queries in a subdivision of n segments. Adamy and Seidel [14] provided a definitive answer by showing that point location queries can be answered with $\log_2 n + 2\sqrt{\log_2 n} + o(\sqrt{\log n})$ primitive comparisons. They also gave a similar lower bound.

Another natural question involves the expected-case complexity of point location. Given a polygonal subdivision S, assume that each cell $z \in S$ is associated with the probability p_z that a query point lies in z. The problem is to produce a point location data structure whose expected search time is as low as possible. The appropriate target bound on the number of comparisons is given by the *entropy* of the subdivision, which is denoted by H and defined:

$$\text{entropy}(S) = H = \sum_{z \in S} p_z \log_2(1/p_z).$$

In the 1-dimensional case, a classical result due to Shannon implies that the expected number of comparisons needed to answer such queries is at least as large as the entropy of the probability distribution [15,16]. Mehlhorn [17] showed that in the 1-dimensional case it is possible to build a binary search tree whose expected search time is at most $H + 2$.

Arya, Malamatos, and Mount [18,19] presented a number of results on this problem in the planar case, and among them they showed that for a polygonal subdivision of size n in which each cell has constant combinatorial complexity, it is possible to answer point location queries with $H + o(H)$ comparisons in the expected case using space that is nearly linear in n. Their results also applied to subdivisions with convex cells, assuming the query distribution is uniform within each cell. Their approach was loosely based on computing a binary space partition (BSP) tree (see Chapter 21) satisfying two properties:

1. The entropy of the subdivision defined by the leaves of the BSP should be close to the entropy of the original subdivision.
2. The depth of a leaf should be close to $\log_2(1/p)$, where p is the probability that a query point lies within the leaf.

Arya, Malamatos, and Mount [20] also presented a simple weighted variant of the randomized incremental algorithm and showed that it can answer queries in $O(H)$ expected time and $O(n)$ space. Iacono [21] presented a deterministic weighted variant based on Kirkpatrick's algorithm.

66.3 Proximity Structures

Proximity structures arise from numerous applications in science and engineering. It is a fundamental fact that nearby objects tend to exert a greater influence and have greater relevance than more distant objects. Proximity structures are discrete geometric and graph structures that encode proximity information. We discuss a number of such structures, including Voronoi diagrams, Delaunay triangulations, and various geometric graph structures, such as the relative neighborhood graph.

66.3.1 Voronoi Diagrams

The *Voronoi diagram* of a set of sites S is a partition of space into regions, one per site, where the region for site s is the set of points that are closer to s than to any other site of S. This structure has been rediscovered and applied in many different branches of science and goes by various names, including Thiessen diagrams and Dirichlet tessellations.

Henceforth, we consider the most common case in which the sites S consist of a set of n points in real d-dimensional space, $\mathbb{R}^d$, and distances are measured using the Euclidean metric. The set of points of $\mathbb{R}^d$ that are closer to some site $s \in S$ than any other site is called the *Voronoi cell* of s, or $V(s)$. (See Figure 66.8.) The union of the boundaries of the Voronoi cells is the *Voronoi diagram* of S, denoted $Vor(S)$. Observe that the set of points of $\mathbb{R}^d$ that are closer to s than some other site t consists of the points that lie in the open halfspace defined by a plane that bisects the pair (s, t). It follows that each Voronoi cell is the intersection of $n - 1$ halfspaces, and hence, it is a (possibly unbounded) convex polyhedron. A Voronoi diagram in dimension d is a cell complex whose faces of all dimensions are convex polyhedra. In the plane a Voronoi diagram is a planar straight line graph with possibly unbounded edges. It can be represented using standard methods for representing polygonal subdivisions and cell complexes (see Chapter 18).

The Voronoi diagram possesses a number of useful geometric properties. For example, for a set of points in the plane, each edge of the Voronoi diagram lies on the perpendicular bisector between two sites. The vertices of the Voronoi diagram lie at the center of an empty circle passing through the incident sites. If the points are in general position (and in particular if no four

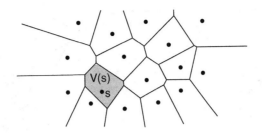

FIGURE 66.8   The Voronoi diagram and a Voronoi cell $V(s)$.

points are cocircular) then every vertex of the diagram is incident to exactly three edges. In fact, it is not hard to show that the largest empty circle whose center lies within the convex hull of a given point set will coincide with a Voronoi vertex. In higher dimensions, each face of dimension k of the Voronoi diagram consists of the points of $\mathbb{R}^d$ that are equidistant from a subset of $d-k+1$ sites, and all other sites are strictly farther away. In the plane the combinatorial complexity of the Voronoi diagram is $O(n)$, and in dimension d its complexity is $\Theta(n^{\lceil d/2 \rceil})$.

Further information on algorithms for constructing Voronoi diagrams as well as variants of the Voronoi diagram can be found in Chapter 65. Although we defined Voronoi diagrams for point sites, it is possible to define them for any type of geometric object. One such variant involves replacing point sites with line segments or generally the boundary of any region of the plane. Given a region P (e.g., a simple polygon), the *medial axis* is defined to be the set of centers of maximal balls contained in P, that is, balls contained in P that are not contained in another ball in P [22]. The medial axis is frequently used in pattern recognition and shape matching. It consists of a combination of straight-line segments and hyperbolic arcs. It can be computed in $O(n \log n)$ time by a modification of Fortune's sweepline algorithm [23]. Finally, it is possible to generalize Voronoi diagrams to other metrics, such as the L_1 and L_∞ metrics (see Section 66.4).

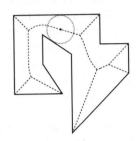

66.3.2 Delaunay Triangulations

The Delaunay triangulation is a structure that is closely related to the Voronoi diagram. The Delaunay triangulation is defined as follows for a set S of n point sites in the plane. Consider any subset $T \subseteq S$ of sites, such that there exists a circle that passes through all the points of T, and contains no point of S in its interior. Such a subset is said to satisfy the *empty circumcircle property*. For example, in Figure 66.9a, the pair $\{p, q\}$ and triple $\{r, s, t\}$ both satisfy the empty circumcircle property. The *Delaunay triangulation* is defined to be the union of the convex hulls of all such subsets. It can be shown that the result is a cell complex. Furthermore, if the points are in general position, and in particular, no four points are cocircular, then the resulting structure is a triangulation of S. (If S is not in general position, then some faces may have more than three edges, and it is common to *complete* the triangulation by triangulating each such face.) A straightforward consequence of the above definition is that the Delaunay triangulation is dual to the Voronoi diagram. For example, Figure 66.9b shows the overlay of these two structures in the plane.

Delaunay triangulations are widely used in practice, and they possess a number of useful properties. For example, among all triangulations of a planar point set the Delaunay triangulation maximizes the minimum angle. Also, in all dimensions, the Euclidean minimum spanning tree (defined below) is a subgraph of the Delaunay triangulation. Proofs of these facts can be found in Reference 7.

In the plane the Delaunay triangulation of a set of points has $O(n)$ edges and $O(n)$ faces. The above definition can be generalized to arbitrary dimensions. In dimension d, the Delaunay triangulation can have as many as $\Theta(n^{\lceil d/2 \rceil})$ faces. However, it can be much

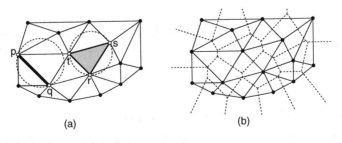

(a)   (b)

FIGURE 66.9   (a) The Delaunay triangulation of a set of points and (b) its overlay with the Voronoi diagram.

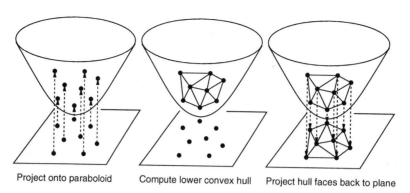

Project onto paraboloid   Compute lower convex hull   Project hull faces back to plane

FIGURE 66.10 The Delaunay triangulation can be computed by lifting the points to the paraboloid, computing the lower convex hull, and projecting back to the plane.

smaller. In particular, Dwyer [24] has shown that in any fixed dimension, if n points are drawn from a uniform distribution from within a unit ball, then the expected number of simplices is $O(n)$.

There is an interesting connection between Delaunay triangulations in dimension d and convex hulls in dimension $d+1$. Consider the *lifting map* $f : \mathbb{R}^2 \to \mathbb{R}^3$ defined $f(x, y) = (x, y, x^2 + y^2)$. This projects points in the plane onto the paraboloid $z = x^2 + y^2$. Given a planar point set S, let S' denote the set of points of $\mathbb{R}^3$ that results by applying this map to each point of S. Define the lower hull of S' to be the set of faces whose outward pointing normal has a negative z coordinate. It can be shown that, when projected back to the plane, the edges of the lower convex hull of S' are exactly the edges of the Delaunay triangulation of S. (See Figure 66.10.)

Although there exist algorithms specially designed for computing Delaunay triangulations, the above fact makes it possible to compute Delaunay triangulations in any dimension by computing convex hulls in the next higher dimension. There exist $O(n \log n)$ time algorithms for computing planar Delaunay triangulations, for example, based on divide-and-conquer [25] and plane sweep [23]. Perhaps the most popular method is based on randomized incremental point insertion [26]. In dimension $d \geq 3$, Delaunay triangulations can be computed in $O(n^{\lceil d/2 \rceil})$ time through randomized incremental point insertion [27].

66.3.3 Other Geometric Proximity Structures

The Delaunay triangulation is perhaps the best known example of a proximity structure. There are a number of related graph structures that are widely used in pattern recognition, learning, and other applications. Given a finite set S of points in d-dimensional Euclidean space, we can define a graph on these points by joining pairs of points that satisfy certain neighborhood properties. In this section we will consider a number of such neighborhood graphs.

Let us first introduce some definitions. For $p, q \in \mathbb{R}^d$ let dist(p, q) denote the Euclidean distance from p to q. Given positive $r \in \mathbb{R}$, let $B(p, r)$ be the open ball consisting of points whose distance from point p is strictly less than r. Define the *lune*, denoted $L(p, q)$, to be the intersection of two balls both of radius dist(p, q) centered at these points, that is,

$$L(p, q) = B(p, \text{dist}(p, q)) \cap B(q, \text{dist}(p, q)).$$

The following geometric graphs are defined for a set S consisting of n points in $\mathbb{R}^d$. (See Figure 66.11.)

Nearest Neighbor Graph (NNG): The directed graph containing an edge (p, q) if q is the nearest neighbor of p, that is, $B(p, \text{dist}(p, q)) \cap S = \emptyset$.

Euclidean Minimum Spanning Tree (EMST): This is an undirected spanning tree on S that minimizes the sum of the Euclidean edge lengths.

Relative Neighborhood Graph (RNG): The undirected graph containing an edge (p, q) if there is no point $r \in S$ that is simultaneously closer to p and q than dist(p, q) [28]. Equivalently, (p, q) is an edge if $L(p, q) \cap S = \emptyset$.

Gabriel Graph (GG): The undirected graph containing an edge (p, q) if the ball whose diameter is $\overline{pq}$ does not contain any other points of S [29], that is, if

$$B\left(\frac{p+q}{2}, \frac{\text{dist}(p, q)}{2}\right) \cap S = \emptyset.$$

Delaunay Graph (DT): The 1-skeleton (edges) of the Delaunay triangulation.

These graphs form an interesting hierarchical relationship. If we think of each edge of an undirected graph as consisting of two directed edges, then we have the following hierarchical relationship, which was first established in Reference 28.

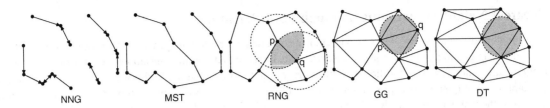

FIGURE 66.11 Common geometric graphs on a point set.

Also see Reference 30.

$$\text{NNG} \subseteq \text{MST} \subseteq \text{RNG} \subseteq \text{GG} \subseteq \text{DT}.$$

This holds in all finite dimensions and generalizes to Minkowski (L_m) metrics, as well.

66.4 Nearest Neighbor Searching

Nearest neighbor searching is an important problem in a variety of applications, including knowledge discovery and data mining, pattern recognition and classification, machine learning, data compression, multimedia databases, document retrieval, and statistics. We are given a set S of objects in some space to be preprocessed, so that given a query object q, the closest object (or objects) of S can be reported quickly.

There are many ways in which to define the notion of similarity. Because the focus of this chapter is on geometric approaches, we shall assume that proximity is defined in terms of the well known Euclidean distance. Most of the results to be presented below can be generalized to any *Minkowski* (or L_m) *metric*, in which the distance between two points $\mathbf{p}$ and $\mathbf{q}$ is defined to be

$$\text{dist}_m(\mathbf{p}, \mathbf{q}) = \left(\sum_{i=1}^{d} |p_i - q_i|^m \right)^{1/m}$$

where $m \geq 1$ is a constant. The case $m = 2$ is the Euclidean distance, the case $m = 1$ is the Manhattan distance, and the limiting case $m = \infty$ is the max distance. In typical geometric applications the dimension d is assumed to be a fixed constant. There has also been work on high dimensional proximity searching in spaces of arbitrarily high dimensions [31] and in arbitrary (nongeometric) metric spaces [32], which we shall not cover here.

There are a number of natural extensions to the nearest neighbor problem as described above. One is to report the k nearest neighbors to the query point, for some given integer k. Another is to compute all the points lying within some given distance, that is, a range query in which the range is defined by the distance function.

Obviously, without any preprocessing whatsoever, the nearest neighbor search problem can be solved in $O(n)$ time through simple brute-force search. A number of very simple methods have been proposed which assume minimal preprocessing. For example, points can be sorted according to their projection along a line, and the projected distances can be used as a method to prune points from consideration [33–35]. These methods are only marginally effective, and provide significant improvements over brute-force search only in very low dimensions.

For uniformly distributed point sets, good expected case performance can be achieved by simple decompositions of space into a regular grid of hypercubes. Rivest [36] and later Cleary [37] provided analyses of these methods. Bentley, Weide, and Yao [38] also analyzed a grid-based method for distributions satisfying certain bounded-density assumptions. Intuitively, the points are bucketed into grid cells, where the size of the grid cell is based on the expected distance to the nearest neighbor. To answer a query, the grid cell containing the query point is located, and a spiral-like search working outwards from this cell is performed to identify nearby points. Suppose for example that q is the query point and p is its closest neighbor. Then all the grid cells overlapping a ball centered at q of radius $\text{dist}(p, q)$ would be visited.

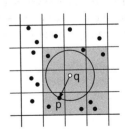

Grids are easy to implement, since each bucket can be stored as a simple list of points, and the complete set of buckets can be arranged in a multi-dimensional array. Note that this may not be space efficient, since it requires storage for empty cells. A more space-efficient method is to assign a hash code to each grid cell based on its location, and then store only the nonempty grid buckets in a hash table. In general, grid methods do not work well for nearest neighbor search unless the point distribution is roughly uniform. As will be discussed below, more sophisticated methods are needed to achieve good efficiency for nonuniformly distributed data.

66.4.1 Nearest Neighbor Searching Through Point Location

One of the original motivations for the Voronoi diagram is nearest neighbor searching. By definition, the Voronoi diagram subdivides space into cells according to which site is the closest. So, in order to determine the closest site, it suffices to compute the Voronoi diagram and generate a point location data structure for the Voronoi diagram. In this way, nearest neighbor queries are reduced to point location queries. This provides an optimal $O(n)$ space and $O(\log n)$ query time method for answering point location queries in the plane. Unfortunately, this solution does not generalize well to higher dimensions. The worst-case combinatorial complexity of the Voronoi diagram in dimension d grows as $\Theta(n^{\lceil d/2 \rceil})$, and optimal point location data structures are not known to exist in higher dimensions.

66.4.2 K-d Trees

Perhaps the most popular class of approaches to nearest neighbor searching involves some sort of hierarchical spatial subdivision. Let S denote the set of n points in $\mathbb{R}^d$ for which queries are to be answered. In such an approach, the entire space is subdivided into successively smaller regions, and the resulting hierarchy is represented by a rooted tree. Each node of the tree represents a region of space, called a *cell*. Implicitly, each node represents the subset of points of S that lie within its cell. The root of the tree is associated with the entire space and the entire point set S. For some arbitrary node u of the tree, if the number of points of S associated with u is less than some constant, then this node is declared to be a leaf of the tree. Otherwise, the cell associated with u is recursively subdivided into smaller (possibly overlapping) subcells according to some *splitting rule*. Then the associated points of S are distributed among these children according to which subcell they lie in. These subcells are then associated with the children of u in the tree.

There are many ways in which to define such a subdivision. Perhaps the earliest and best known example is that of the k-d tree data structure. Bentley [39] introduced the *k-d tree* data structure (or *kd-tree*) as a practical general-purpose data structure for many types of geometric retrieval problems. Although it is not the asymptotically most efficient solution for these problems, its flexibility makes it a popular choice for implementation. The cells of a k-d tree are axis-aligned hyperrectangles. Each internal node is associated with an axis-orthogonal splitting hyperplane. This hyperplane splits the rectangular cell into two rectangular subcells, each of which is associated with one of the two children. An example is shown in Figure 66.12.

The choice of the splitting hyperplane is an important issue in the implementation of the k-d tree. For the purpose of nearest neighbor searching, a good split is one that divides the points into subsets of similar cardinalities and which produces cells that are not too skinny, that is, the ratio between the longest and shortest sides is bounded. However, it is not always possible to achieve these goals. A simple and commonly used method is to cycle through the various coordinate axes (i.e., splitting along x, then y, then z, then back to x, and so on). Each time the split is made through the median coordinate along the splitting dimension [7,40]. Friedman, Bentley and Finkel [41] suggested the following method, which is more sensitive to the data distribution. First, compute the minimum axis-aligned bounding box for the set of points associated with the current cell. Next choose the splitting axis to be the one that is parallel to the longest side of this box. Finally, split the points by a hyperplane that is orthogonal to this axis, and which splits the points into two sets of equal size. A number of other splitting rules have been proposed for k-d trees, including the sliding midpoint rule by Arya and Fu [42] and Maneewongvatana and Mount [43], variance minimization by White and Jain [44], and methods by Silva Filho [45] and Sproull [46]. We will discuss other subdivision methods in the next section as well.

It is possible to construct the k-d tree of an n-element point set in $O(n \log n)$ time by a simple top-down recursive procedure. The process involves determining the splitting axis and the splitting coordinate along this axis, and then partitioning the point set about this coordinate. If the splitting rule partitions the point set about its median coordinate then it suffices to compute the median by any linear-time algorithm for computing medians [3]. Some splitting methods may not evenly partition the point set.

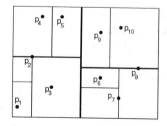 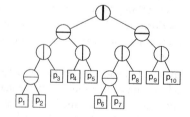

FIGURE 66.12 An example of a k-d tree of a set of points in the plane, showing both the associated spatial subdivision (left) and the binary tree structure (right).

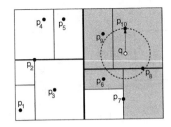 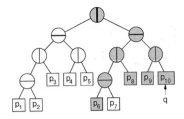

FIGURE 66.13   Nearest neighbor search in a k-d tree. The point p_{10} is the initial closest, and only the shaded cells and nodes are visited. The final answer is p_8.

In the worst case this can lead to quadratic construction time. Vaidya showed that it is possible to achieve $O(n \log n)$ construction time, even when unbalanced splitting takes place [47]. The total space requirements are $O(n)$ for the tree itself.

Given a query point q, a nearest neighbor search is performed by the following recursive algorithm [41]. Throughout, the algorithm maintains the closest point to q encountered so far in the search, and the distance to this closest point. As the nodes of the tree are traversed, the algorithm maintains the d-dimensional hyperrectangular cell associated with each node. (This is updated incrementally as the tree is traversed.) When the search arrives at a leaf node, it computes the distance from q to the associated point(s) of this node, and updates the closest point if necessary. (See Figure 66.13.) Otherwise, when it arrives at an internal node, it first computes the distance from the query point to the associated cell. If this distance is greater than the distance to the closest point so far, the search returns immediately, since the subtree rooted at this node cannot provide a closer point. Otherwise, it is determined which side of the splitting hyperplane contains the query point. First, the closer child is visited and then the farther child. A somewhat more intelligent variant of this method, called *priority search*, involves storing the unvisited nodes in a priority queue, sorted according to the distance from the query point to the associated cell, and then processes the nodes in increasing order of distance from the query point [48].

66.4.3 Other Approaches to Nearest Neighbor Searching

The k-d tree is but one example of a general class of nearest neighbor search structures that are based on hierarchical space decomposition. A good survey of methods from database literature was given by Böhm, Berchtold, and Keim [49]. These include the R-tree [50] and its variants, the R*-tree [51], the R$^+$-tree [52], and the X-tree [53], which are all based on recursively decomposing space into (possibly overlapping) hyperrectangles. (See Chapter 22 for further information.) For the cases studied, the X-tree is reported to have the best performance for nearest neighbor searching in high dimensional spaces [49]. The SS-tree [44] is based on subdividing space using (possibly overlapping) hyperspheres rather than rectangles. The SR-tree [54] uses the intersection of an enclosing rectangle and enclosing sphere to represent a cell. The TV-tree [55] applies a novel approach of considering projections of the data set onto higher dimensional subspaces at successively deeper levels in the search tree.

A number of algorithms for nearest neighbor searching have been proposed in the algorithms and computational geometry literature. Higher dimensional solutions with sublinear worst-case performance were considered by Yao and Yao [56]. Clarkson [57] showed that queries could be answered in $O(\log n)$ time with $O(n^{\lceil d/2 \rceil + \delta})$ space, for any $\delta > 0$. The O-notation hides constant factors that are exponential in d. Agarwal and Matoušek [58] generalized this by providing a tradeoff between space and query time. Meiser [59] showed that queries could be answered in $O(d^5 \log n)$ time and $O(n^{d+\delta})$ space, for any $\delta > 0$, thus showing that exponential factors in query time could be eliminated by using sufficient space.

66.4.4 Approximate Nearest Neighbor Searching

In any fixed dimensions greater than two, no method for exact nearest neighbor searching is known that achieves the simultaneous goals of roughly linear space and logarithmic query time. For methods achieving roughly linear space, the constant factors hidden in the asymptotic running time grow at least as fast as 2^d (depending on the metric). Arya et al. [60] showed that if n is not significantly larger than 2^d, then boundary effects decrease this exponential dimensional dependence. Nonetheless, the so called "curse of dimensionality" is a significant impediment to computing nearest neighbors efficiently in high dimensional spaces.

This suggests the idea of computing nearest neighbors approximately. Consider a set of points S and a query point q. For any $\epsilon > 0$, we say that a point $p \in S$ is an ϵ-*approximate nearest neighbor* of q if

$$\mathrm{dist}(p,q) \leq (1+\epsilon)\mathrm{dist}(p^*,q),$$

where p^* is the true nearest neighbor of q in S. The approximate nearest neighbor problem was first considered by Bern [61]. He proposed a data structure that achieved a fixed approximation factor depending on dimension. Arya and Mount [62] proposed a

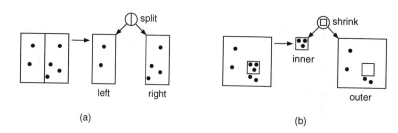

FIGURE 66.14   Splitting nodes (a) and shrinking nodes (b) in a BBD-tree.

randomized data structure that achieves polylogarithmic query time in the expected case, and nearly linear space. Their approach was based on a combination of the notion of neighborhood graphs, as described in Section 66.3.3, and skip lists. In their algorithm the approximation error factor ϵ is an arbitrary positive constant, which is given at preprocessing time.

Arya et al. [63] proposed a hierarchical spatial subdivision data structure, called the *BBD-tree*. This structure has the nice features of having $O(n)$ size, $O(\log n)$ depth, and each cell has bounded aspect ratio, that is, the ratio between its longest and shortest side is bounded. They achieved this by augmenting the axis-aligned splitting operation of the k-d tree (see Figure 66.14a) with an additional subdivision operation called *shrinking* (see Figure 66.14b). A shrinking node is associated with an axis-aligned rectangle, and the two children correspond to the portions of space lying inside and outside of this rectangle, respectively. The resulting cells are either axis-aligned hyperrectangles, or the set-theoretic difference of two axis-aligned hyperrectangles. They showed that, in all fixed dimensions d and for $\epsilon > 0$, it is possible to answer ϵ-nearest neighbor queries in $O(\log n)$ time using the BBD-tree. The hidden asymptotic constants in query time grow as $(1/\epsilon)^d$.

Duncan et al. [64] proposed an alternative structure, called the *BAR tree*, which achieves all of these combinatorial properties and has convex cells. The BAR tree achieves this by using cutting planes that are not necessarily axis-aligned. Clarkson [65] and Chan [66] presented data structures that achieved better ϵ dependency in the query time. In particular, they showed that queries could be answered in $O((1/\epsilon)^{d/2} \log n)$ time.

66.4.5 Approximate Voronoi Diagrams

As mentioned in Section 66.4.1 it is possible to answer nearest neighbor queries by applying a point location query to the Voronoi diagram. However, this approach does not generalize well to higher dimensions, because of the rapid growth rate of the Voronoi diagram and the lack of good point location structures in dimension higher than two.

Har-Peled [67] proposed a method to overcome these problems. Given an error bound $\epsilon > 0$, an *approximate Voronoi diagram* (AVD) of a point set S is defined to be a partition of space into cells, where each cell c is associated with a *representative* $r_c \in S$, such that r_c is an ϵ-nearest neighbor for all the points in c [67]. Arya and Malamatos [68] generalized this by allowing up to some given number $t \geq 1$ representatives to be stored with each cell, subject to the requirement that for any point in the cell, one of these t representatives is an ϵ-nearest neighbor. Such a decomposition is called a (t, ϵ)-AVD. (See Figure 66.15.)

Of particular interest are AVDs that are constructed from hierarchical spatial decompositions, such as quadtrees and their variants, since such structures support fast point location in all dimensions. This yields a very simple method for performing approximate nearest neighbor searching. In particular, a tree descent determines the leaf cell containing the query point and then the closest of the t representatives is reported.

Har-Peled [67] showed that it is possible to construct a $(1, \epsilon)$ AVD in which the number of leaf cells is $O((n/\epsilon^d)(\log n) \log(n/\epsilon))$. Arya and Malamatos [68] and later Arya, Malamatos, and Mount [69] improved these results by showing how to construct more space-efficient AVDs. In all constant dimensions d, their results yield a data structure of $O(n)$ space (including the space for representatives) that can answer ϵ-nearest neighbor queries in $O(\log n + (1/\epsilon)^{(d-1)/2})$ time. This is the best asymptotic result known for approximate nearest neighbor searching in fixed dimensional spaces.

66.5 Sources and Related Material

General information regarding the topics presented in the chapter can be found in standard texts on computational geometry, including those by Preparata and Shamos [8], Edelsbrunner [1], Mulmuley [70], de Berg et al. [7], and Boissonnat and Yvinec [71] as well as Samet's book on spatial data structures [40]. Further information on point location can be found in a survey paper written by Snoeyink [72]. For information on Voronoi diagrams see the book by Okabe, Boots and Sugihara [73] or surveys by Aurenhammer [74], Aurenhammer and Klein [75], and Fortune [76]. For further information on geometric graphs see the survey

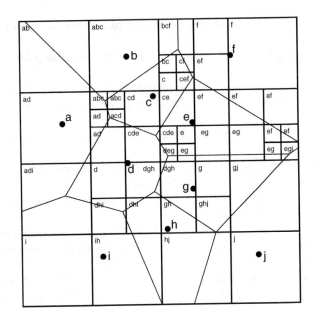

FIGURE 66.15 A $(3, 0)$-AVD implemented as a quadtree subdivision for the set $\{a, b, \ldots, j\}$. Each cell is labeled with its representatives. The Voronoi diagram is shown for reference.

by O'Rourke and Toussaint [77]. Further information on nearest neighbor searching can be found in surveys by Böhm et al. [49], Indyk [31], and Chavez et al. [32].

Acknowledgments

The work of the first author was supported in part by a grant from the Hong Kong Research Grants Council (HKUST6229/00E and HKUST6080/01E). The work of the second author was supported in part by National Science Foundation grant number CCR-0098151.

References

1. H. Edelsbrunner: *Algorithms in Combinatorial Geometry, Volume 10 of EATCS Monographs on Theoretical Computer Science.* Springer-Verlag, Heidelberg, West Germany, 1987.
2. D. G. Kirkpatrick: Optimal search in planar subdivisions. *SIAM J. Comput.*, 12(1):28–35, 1983.
3. T. H. Cormen, C. E. Leiserson, R. L. Rivest, and C. Stein: *Introduction to Algorithms.* MIT Press, Cambridge, MA, 2nd edition, 2001.
4. N. Sarnak and R. E. Tarjan: Planar point location using persistent search trees. *Commun. ACM*, 29(7):669–679, July 1986.
5. R. Cole: Searching and storing similar lists. *J. Algorithms*, 7:202–220, 1986.
6. D. T. Lee and F. P. Preparata: Location of a point in a planar subdivision and its applications. *SIAM J. Comput.*, 6(3):594–606, 1977.
7. M. de Berg, M. van Kreveld, M. Overmars, and O. Schwarzkopf: *Computational Geometry: Algorithms and Applications.* Springer-Verlag, Berlin, Germany, 2nd edition, 2000.
8. F. P. Preparata and M. I. Shamos: *Computational Geometry: An Introduction.* Springer-Verlag, New York, 3rd edition, October 1990.
9. B. Chazelle and L. J. Guibas: Fractional cascading: I. A data structuring technique. *Algorithmica*, 1(3):133–162, 1986.
10. H. Edelsbrunner, L. J. Guibas, and J. Stolfi: Optimal point location in a monotone subdivision. *SIAM J. Comput.*, 15(2):317–340, 1986.
11. Mulmuley K.: A fast planar partition algorithm, I. *J. Symbolic Comput.*, 10(3–4):253–280, 1990.
12. R. Seidel: A simple and fast incremental randomized algorithm for computing trapezoidal decompositions and for triangulating polygons. *Comput. Geom. Theory Appl.*, 1(1):51–64, 1991.
13. Goodrich M. T., M. Orletsky, and K. Ramaiyer: Methods for achieving fast query times in point location data structures. In *Proc. 8th ACM-SIAM Sympos. Discrete Algorithms*, pp. 757–766, 1997.

14. U. Adamy and R. Seidel: Planar point location close to the information-theoretic lower bound. In *Proc. 9th ACM-SIAM Sympos. Discrete Algorithms*, 1998.

15. D. E. Knuth: *Sorting and Searching, Volume 3 of The Art of Computer Programming.* Addison-Wesley, Reading, MA, second edition, 1998.

16. C. E. Shannon: A mathematical theory of communication. *Bell Sys. Tech. Journal*, 27:379–423, 623–656, 1948.

17. K. Mehlhorn: Best possible bounds on the weighted path length of optimum binary search trees. *SIAM J. Comput.*, 6:235–239, 1977.

18. S. Arya, T. Malamatos, and D. M. Mount: Nearly optimal expected-case planar point location. In *Proc. 41 Annu. IEEE Sympos. Found. Comput. Sci.*, pp. 208–218, 2000.

19. S. Arya, T. Malamatos, and D. M. Mount: Entropy-preserving cuttings and space-efficient planar point location. In *Proc. 12th ACM-SIAM Sympos. Discrete Algorithms*, pp. 256–261, 2001.

20. S. Arya, T. Malamatos, and D. M. Mount: A simple entropy-based algorithm for planar point location. In *Proc. 12th ACM-SIAM Sympos. Discrete Algorithms*, pp. 262–268, 2001.

21. J. Iacono: Optimal planar point location. In *Proc. 12th ACM-SIAM Sympos. Discrete Algorithms*, pp. 340–341, 2001.

22. Duda R. O. and P. E. Hart: *Pattern Classification and Scene Analysis.* Wiley-Interscience, New York, 1973.

23. S. J. Fortune: A sweepline algorithm for Voronoi diagrams. *Algorithmica*, 2:153–174, 1987.

24. Dwyer R. A.: Higher-dimensional Voronoi diagrams in linear expected time. *Discrete Comput. Geom.*, 6:343–367, 1991.

25. M. I. Shamos and D. Hoey: Closest-point problems. In *Proc. 16th Annu. IEEE Sympos. Found. Comput. Sci.*, pp. 151–162, 1975.

26. L. J. Guibas, D. E. Knuth, and M. Sharir: Randomized incremental construction of Delaunay and Voronoi diagrams. *Algorithmica*, 7:381–413, 1992.

27. K. L. Clarkson and P. W. Shor: Applications of random sampling in computational geometry II. *Discrete Comput. Geom.*, 4:387–421, 1989.

28. G. T. Toussaint: The relative neighbourhood graph of a finite planar set. *Pattern Recogn.*, 12:261–268, 1980.

29. K. R. Gabriel and R. R. Sokal: A new statistical approach to geographic variation analysis. *Systematic Zoology*, 18:259–278, 1969.

30. J. W. Jaromczyk and G. T. Toussaint: Relative neighborhood graphs and their relatives. Proc. IEEE, 80(9):1502–1517, September 1992.

31. A. Andoni and P. Indyk: Nearest neighbors in high-dimensional spaces. In J. E. Goodman, J. O'Rourke, and C. D. Toth, editors, *Handbook of Discrete and Computational Geometry*, 3rd edition, CRC Press LLC, Boca Raton, FL, 2017.

32. E. Chavez, G. Navarro, R. A. Baeza-Yates, and J. L. Marroquin: Searching in metric spaces. *ACM Comput. Surveys*, 33:273–321, 2001.

33. J. H. Friedman, F. Baskett, and L. J. Shustek: An algorithm for finding nearest neighbors. *IEEE Trans. Comput.*, C-24(10):1000–1006, 1975.

34. L. Guan and M. Kamel: Equal-average hyperplane partitioning method for vector quantization of image data. *Pattern Recogn. Lett.*, 13:693–699, 1992.

35. C.-H. Lee and L.-H. Chen: Fast closest codeword search algorithm for vector quantisation. *IEE Proc.-Vis. Image Signal Process.*, 141:143–148, 1994.

36. R. L. Rivest: On the optimality of Elias's algorithm for performing best-match searches. In *Information Processing*, pp. 678–681. North Holland Publishing Company, 1974.

37. Cleary J. G.: Analysis of an algorithm for finding nearest neighbors in Euclidean space. *ACM Trans. on Math. Softw.*, 5(2):183–192, 1979.

38. J. L. Bentley, B. W. Weide, and A. C. Yao: Optimal expected-time algorithms for closest point problems. *ACM Trans. on Math. Softw.*, 6(4):563–580, 1980.

39. J. L. Bentley: Multidimensional binary search trees used for associative searching. *Commun. ACM*, 18(9):509–517, September 1975.

40. H. Samet: *The Design and Analysis of Spatial Data Structures.* Addison-Wesley, Reading, MA, 1990.

41. J. H. Friedman, J. L. Bentley, and R. A. Finkel: An algorithm for finding best matches in logarithmic expected time. *ACM Trans. Math. Softw.*, 3:209–226, 1977.

42. S. Arya and H. Y. Fu: Expected-case complexity of approximate nearest neighbor searching. *SIAM J. of Comput.*, 32:793–815, 2003.

43. S. Maneewongvatana and D. M. Mount: Analysis of approximate nearest neighbor searching with clustered point sets. In M. H. Goldwasser, C. C. McGeoch, and D. S. Johnson, editors, *Data Structures, Near Neighbor Searches, and Methodology, Volume 59 of DIMACS Series in Discrete Mathematics and Theoretical Computer Science*, pp. 105–123. American Mathematics Society, Providence, Rhode Island, 2002.

44. D. A. White and R. Jain: Similarity indexing with the SS-tree. In *Proc. 12th IEEE Internat. Conf. Data Engineering*, pp. 516–523, 1996.

45. Silva Filho Y. V.: Optimal choice of discriminators in a balanced K-D binary search tree. *Inform. Proc. Lett.*, 13:67–70, 1981.

46. R. F. Sproull: Refinements to nearest-neighbor searching in k-dimensional trees. *Algorithmica*, 6(4):579–589, 1991.

47. P. M. Vaidya: An $O(n \log n)$ algorithm for the all-nearest-neighbors problem. *Discrete Comput. Geom.*, 4:101–115, 1989.

48. S. Arya and D. M. Mount: Algorithms for fast vector quantization. In *Data Compression Conference*, pp. 381–390. IEEE Press, 1993.

49. C. Böhm, S. Berchtold, and D. A. Keim: Searching in high-dimensional spaces: Index structures for improving the performance of multimedia databases. *ACM Comput. Surveys*, 33:322–373, 2001.

50. A. Guttman: R-trees: A dynamic index structure for spatial searching. In *Proc. ACM SIGMOD Conf. on Management of Data*, pp. 47–57, 1984.

51. N. Beckmann, H.-P. Kriegel, R. Schneider, and B. Seeger: The R*-tree: An efficient and robust access method for points and rectangles. In *Proc. ACM SIGMOD Conf. on Management of Data*, pp. 322–331, 1990.

52. T. Sellis, N. Roussopoulos, and C. Faloutsos: The R+ -tree: A dynamic index for multi-dimensional objects. In *Proc. 13th VLDB Conference*, pp. 507–517, 1987.

53. S. Berchtold, D. A. Keim, and H.-P. Kriegel: The X-tree: An index structure for high-dimensional data. In *Proc. 22nd VLDB Conference*, pp. 28–39, 1996.

54. N. Katayama and S. Satoh: The SR-tree: An index structure for high-dimensional nearest neighbor queries. In Proc. ACM SIGMOD Conf. on Management of Data, pp. 369–380, 1997.

55. K. I. Lin, H. V. Jagadish, and C. Faloutsos: The TV-tree: An index structure for high-dimensional data. *VLDB Journal*, 3(4):517–542, 1994.

56. A. C. Yao and F. F. Yao: A general approach to d-dimensional geometric queries. In *Proc. 17th Ann. ACM Sympos. Theory Comput.*, pp. 163–168, 1985.

57. K. L. Clarkson: A randomized algorithm for closest-point queries. *SIAM J. Comput.*, 17(4):830–847, 1988.

58. P. K. Agarwal and J. Matoušek: Ray shooting and parametric search. *SIAM J. Comput.*, 22(4):794–806, 1993.

59. S. Meiser: Point location in arrangements of hyperplanes. *Information and Computation*, 106(2):286–303, 1993.

60. S. Arya, D. M. Mount, and O. Narayan: Accounting for boundary effects in nearest-neighbor searching. *Discrete Comput. Geom.*, 16:155–176, 1996.

61. M. Bern: Approximate closest-point queries in high dimensions. *Inform. Process. Lett.*, 45:95–99, 1993.

62. S. Arya and D. M. Mount: Approximate nearest neighbor queries in fixed dimensions. In *Proc. 4th ACM-SIAM Sympos. Discrete Algorithms*, pp. 271–280, 1993.

63. S. Arya, D. M. Mount, N. S. Netanyahu, R. Silverman, and A. Wu: An optimal algorithm for approximate nearest neighbor searching in fixed dimensions. *J. ACM*, 45:891–923, 1998.

64. C. A. Duncan, M. T. Goodrich, and S. Kobourov: Balanced aspect ratio trees: Combining the advantages of k-d trees and octrees. *J. Algorithms*, 38:303–333, 2001.

65. K. L. Clarkson: An algorithm for approximate closest-point queries. In *Proc. 10th Annu. ACM Sympos. Comput. Geom.*, pp. 160–164, 1994.

66. T. Chan: Approximate nearest neighbor queries revisited. In *Proc. 13th Annu. ACM Sympos. Comput. Geom.*, pp. 352–358, 1997.

67. S. Har-Peled: A replacement for Voronoi diagrams of near linear size. In *Proc. 42 Annu. IEEE Sympos. Found. Comput. Sci.*, pp. 94–103, 2001.

68. S. Arya and T. Malamatos: Linear-size approximate Voronoi diagrams. In *Proc. 13th ACM-SIAM Sympos. Discrete Algorithms*, pp. 147–155, 2002.

69. S. Arya, T. Malamatos, and D. M. Mount: Space-efficient approximate Voronoi diagrams. In *Proc. 34th Annual ACM Sympos. Theory Comput.*, pp. 721–730, 2002.

70. K. Mulmuley: *Computational Geometry: An Introduction Through Randomized Algorithms*. Prentice Hall, Englewood Cliffs, NJ, 1993.

71. J.-D. Boissonnat and M. Yvinec: *Algorithmic Geometry*. Cambridge University Press, UK, 1998. Translated by H. Brönnimann.

72. J. Snoeyink: Point location. In Jacob E. Goodman and Joseph O'Rourke, editors, *Handbook of Discrete and Computational Geometry*, Chapter 30, pp. 559–574. CRC Press LLC, Boca Raton, FL, 1997.

73. A. Okabe, B. Boots, and K. Sugihara: *Spatial Tessellations: Concepts and Applications of Voronoi Diagrams*. John Wiley & Sons, Chichester, UK, 1992.

74. F. Aurenhammer: Voronoi diagrams: A survey of a fundamental geometric data structure. *ACM Comput. Surv.*, 23(3):345–405, September 1991.

75. F. Aurenhammer and R. Klein: Voronoi diagrams. In J.-R. Sack and J. Urrutia, editors, *Handbook of Computational Geometry*, pp. 201–290. Elsevier Science Publishers B.V. North-Holland, Amsterdam, 2000.

76. S. Fortune: Voronoi diagrams and Delaunay triangulations. In Jacob E. Goodman and Joseph O'Rourke, editors, *Handbook of Discrete and Computational Geometry*, chapter 20, pp. 377–388. CRC Press LLC, Boca Raton, FL, 1997.

77. J. O'Rourke and G. T. Toussaint: Pattern recognition. In Jacob E. Goodman and Joseph O'Rourke, editors, *Handbook of Discrete and Computational Geometry*, Chapter 43, pp. 797–814. CRC Press LLC, Boca Raton, FL, 1997.

67

Computational Geometry: Generalized (or Colored) Intersection Searching

Prosenjit Gupta
NIIT University

Ravi Janardan
University of Minnesota

Saladi Rahul
University of Illinois

Michiel Smid
Carleton University

67.1 Geometric Intersection Searching Problems 1043
    Generalized Intersection Searching
67.2 Summary of Known Results 1044
    Axes-Parallel Objects • Arbitrarily Oriented Objects • Problems on the Grid • Single-Shot
    Problems • External Memory and Word-RAM Algorithms
67.3 Techniques 1046
    A Transformation-Based Approach • A Sparsification-Based Approach •
    A Persistence-Based Approach • A General Approach for Reporting Problems •
    Adding Range Restrictions • Exploiting the Output Size • A Reverse Transformation
67.4 Conclusion and Future Directions 1056
Acknowledgments 1056
References 1056

67.1 Geometric Intersection Searching Problems

Problems arising in diverse areas, such as VLSI layout design, database query-retrieval, robotics, and computer graphics can often be formulated as *geometric intersection searching problems*. In a generic instance of such a problem, a set, S, of geometric objects is to be preprocessed into a suitable data structure so that given a query object, q, we can answer efficiently questions regarding the intersection of q with the objects in S. The problem comes in four versions, depending on whether we want to report the intersected objects or simply count their number—the *reporting* version and the *counting* version, respectively—and whether S remains fixed or changes through insertion and deletion of objects—the *static* version and the dynamic version, respectively. In the dynamic version, which arises very often owing to the highly interactive nature of the above-mentioned applications, we wish to perform the updates more efficiently than simply recomputing the data structure from scratch after each update, while simultaneously maintaining fast query response times. We call these problems *standard* intersection searching problems in order to distinguish them from the *generalized* intersection searching problems that are the focus of this chapter. Due to their numerous applications, standard intersection searching problems have been the subject of much study and efficient solutions have been devised for many of them (see, for instance, References 1,2 and the references therein).

The efficiency of a standard intersection searching algorithm is measured by the space used by the data structure, the query time, and, in the dynamic setting, the update time. In a counting problem, these are expressed as a function of the input size n (i.e., the size of S); in a reporting problem, the space and update time are expressed as a function of n, whereas the query time is expressed as a function of both n and the output size k (i.e., the number of intersected objects) and is typically of the form $O(f(n) + k)$ or $O(f(n) + k \cdot g(n))$, for some functions f and g. Such a query time is called *output-sensitive*.

67.1.1 Generalized Intersection Searching

In many applications, a more general form of intersection searching arises. Here the objects in S come aggregated in disjoint groups and of interest are questions regarding the intersection of q with the groups rather than with the objects. (q intersects a group if and only if it intersects some object in the group.) In our discussion, it will be convenient to associate with each group a different color and imagine that all the objects in the group have that color. Then, in the *generalized reporting* (resp., *generalized counting*) problem, we want to report (resp., count) the distinct colors of the objects intersected by q; in the dynamic setting, an object of some (possibly new) color is inserted in S or an object in S is deleted. Note that the generalized problem reduces to the standard one when each color class has cardinality 1. (We remark that the generalized problems discussed here are also sometimes referred to as *colored* problems; we use the two terms interchangeably.)

We give some examples of such generalized problems:

- Consider a database of mutual funds which contains for each fund its annual total return and its beta (a real number measuring the fund's volatility). Thus each fund can be represented as a point in two dimensions. Moreover, funds are aggregated into groups according to the fund family they belong to. A typical query is to determine the families that offer funds whose total return is between, say, 15% and 20%, and whose beta is between, say, 0.9 and 1.1. This is an instance of the generalized two-dimensional range searching problem. The output of this query enables a potential investor to initially narrow his/her search to a few families instead of having to plow through dozens of individual funds (all from the same small set of families) that meet these criteria. (From a database perspective, this query is similar to an SQL query with a GROUP-BY clause.)
- In the Manhattan layout of a VLSI chip, the wires (line segments) can be grouped naturally according to the circuits they belong to. A problem of interest to the designer is determining which circuits (rather than wires) become electrically connected when a new wire is added. This is an instance of the generalized orthogonal segment intersection searching problem.

One approach to solving a generalized problem is to try to take advantage of solutions known for the corresponding standard problem. For instance, we can solve a generalized reporting problem by first determining the objects intersected by q (a standard reporting problem) and then reading off the distinct colors. However, the query time can be very high since q could intersect $k = \Omega(n)$ objects but only $O(1)$ distinct colors. For a generalized reporting problem, we seek query times that are sensitive to the number, i, of distinct colors intersected, typically of the form $O(f(n) + i)$ or $O(f(n) + i \cdot g(n))$, where f and g are polylogarithmic. (This is attainable using the approach just described if each color class has cardinality $O(1)$. On the other hand, if there are only $O(1)$ different color classes, we could simply run a standard algorithm on each color class in turn, stopping as soon as an intersection is found and reporting the corresponding color. The real challenge is when the number of color classes and the cardinalities of the color classes are not constants, but rather are (unknown) functions of n; throughout, we will assume this to be the case.) For a generalized counting problem, the situation is worse; it is not even clear how one can extract the answer for such a problem from the answer (a mere count) to the corresponding standard problem. One could, of course, solve the corresponding reporting problem and then count the colors, but this is not efficient. Thus it is clear that different techniques are needed.

In this chapter, we describe the research that has been conducted over the past two decades on generalized intersection searching problems. We begin with a brief review of known results and then discuss a variety of techniques for these problems. For each technique, we give illustrative examples and provide pointers to related work. We conclude with a discussion of possible directions for further research.

67.2 Summary of Known Results

Generalized intersection searching problems were introduced by Janardan and Lopez [3]. Subsequent work in this area may be found in References 4–28, among others. In this section, we give a broad overview of the work on these problems to date; details may be found in the cited references.

67.2.1 Axes-Parallel Objects

In Reference 3, efficient solutions were given for several generalized reporting problems, where the input objects and the query were axes-parallel. Examples of such input/query pairs considered include: points/interval in $\mathbb{R}^1$; line segments/segment, points/rectangle, and rectangles/rectangle, all in $\mathbb{R}^2$; and rectangles/points in $\mathbb{R}^d$, where $d \geq 2$ is a constant. Several of these results were further extended in Reference 13 to include counting and/or dynamic reporting, and new results were presented for input/query pairs such as intervals/interval in $\mathbb{R}^1$, points/quadrant in $\mathbb{R}^2$, and points/rectangle in $\mathbb{R}^3$. Furthermore, a new type of counting problem, called a *type-2 counting problem* was also introduced, where the goal was to count for each color intersected the number of objects of that color that are intersected. In Reference 6, improved solutions were given for counting and/or reporting problems involving points/interval in $\mathbb{R}^1$, points/rectangle in $\mathbb{R}^2$, and line segments/segment in $\mathbb{R}^2$.

In References 9 and 10, variants of the reporting problem were considered for points/rectangle in $\mathbb{R}^2$, with the goal of reporting those colors that appeared "many" times in the output. Specifically, in Reference 10, an efficient dynamic algorithm was given for reporting each color for which the number of points of that color in the query rectangle was more than a user-specified fraction of the total number points in the query rectangle. On the other hand, in Reference 9, an efficient (approximation) algorithm was given for reporting each color such that the number of points of that color in the query rectangle was at least a user-specified fraction of the total number of points of that color in the input set. More recently, interesting connections have been shown between generalized counting problems and other problems. For instance, as shown in Reference 21, standard range counting involving points/rectangle in $\mathbb{R}^2$ can be solved using an algorithm for generalized range counting on points/interval in $\mathbb{R}^1$. Also, as discussed at length in Reference 18, it is possible to use an "offline" version of the generalized counting problem on points/rectangle in $\mathbb{R}^2$ to do sparse matrix multiplication.

67.2.2 Arbitrarily Oriented Objects

Efficient solutions were given in Reference 3 for generalized reporting on nonintersecting line segments using a query line segment. Special, but interesting, cases of intersecting line segments, such as when each color class forms a polygon or a connected component, were considered in Reference 5. Efficient solutions were given in Reference 14 for input/query pairs consisting of points/half-space in $\mathbb{R}^d$, points fat-triangle, and fat-triangles/point in $\mathbb{R}^2$. (A *fat-triangle* is a triangle where each internal angle is at least a user-specified constant, hence "well-shaped.") Some of these results were improved subsequently in Reference 6. In Reference 15, alternative bounds were obtained for the fat-triangle problems within the framework of a general technique for adding range restriction capability to a generalized data structure. Results were presented in Reference 8 for querying, with a polygon, a set of polygons whose sides are oriented in at most a constant number of different directions, with a polygon. In Reference 28, a general method was given for querying intersecting line segments with a segment and for querying points in $\mathbb{R}^d$ with a half-space or a simplex. Generalized problems involving various combinations of circular objects (circles, discs, annuli) and points, lines, and line segments were considered in Reference 16.

67.2.3 Problems on the Grid

Problems involving document retrieval or string manipulation can often be cast in the framework of generalized intersection searching. For example, in the context of document retrieval, the following problem (among others) was considered in Reference 22. Preprocess an array of colored nonnegative integers (i.e., points on the one-dimensional grid) such that, given two indices into the array, each distinct color for which there is a pair of points in the index range at distance less than a specified constant can be reported efficiently. In the context of substring indexing, the following problem was considered in Reference 11. Preprocess a set of colored points on the one-dimensional grid, so that given two nonoverlapping intervals, the list of distinct colors that occurs in both intervals can be reported efficiently. I/O efficient algorithms were given in the standard external memory model [29] for this problem. See References 23 and 30 and references therein for a discussion of more recent work on document retrieval and string manipulation problems. Other grid-related work in this area includes Reference 4, where efficient solutions were given for the points/rectangle and rectangles/point problems, under the condition that the input and query objects lie on a d-dimensional grid.

67.2.4 Single-Shot Problems

In this class of problems, we are given a collection of geometric objects and the goal is to report all pairs that intersect. Note that there is no query object as such here and no notion of preprocessing the input. As an example, suppose that we are given a set of convex polygons with a total of n vertices in $\mathbb{R}^2$, and we wish to report or count all pairs that intersect, with the goal of doing this in time proportional to the number of intersecting pairs (i.e., output-sensitively). If the number of polygons and their sizes are both functions of n (instead of one or the other being a constant), then, as discussed in Reference 17, standard methods (e.g., testing each pair of polygons or computing all boundary intersections and polygon containments in the input) are inefficient. In Reference 17, an efficient and output-sensitive algorithm is given for this problem. Each polygon is assigned a color and then decomposed into simpler elements, that is, trapezoids, of the same color. The problem then becomes one of reporting all distinct color pairs (c_1, c_2) such that a trapezoid of color c_1 intersects one of color c_2. An improved algorithm was given subsequently in Reference 31 for both $\mathbb{R}^2$ and $\mathbb{R}^3$. Other related work on such colored single-shot problems may be found in References 7 and 19. In Reference 19, interesting connections between sparse matrix multiplication and colored single-shot problems are established.

67.2.5 External Memory and Word-RAM Algorithms

The results discussed earlier (and in the rest of this chapter) are in the well-known RAM model or pointer machine model. Recently, generalized problems have also been considered in other machine models such as the external memory model and the word-RAM model. In the external memory model [32], the data resides primarily on disk, in blocks of some fixed size, data transfer (I/O) between disk and main memory happens in blocks, and space and query time are measured in terms of the number of blocks used and the number of I/O operations, respectively. In this model, efficient algorithms were first given in Reference 24 (see also Reference 25) for generalized range search in $\mathbb{R}^2$, where the query rectangle is grounded, that is, three-sided. These results have been improved subsequently in Reference 21. A limitation of these results is that they report each color $O(1)$ times, instead of exactly once, which results in an additional overhead in the external memory model for removal of duplicate colors via sorting. (This is in contrast to internal memory algorithms, where duplicate removal is easy.) This limitation has recently been removed in Reference 26 where each color is reported exactly once. The above papers also present results in the word-RAM model [33], where it is assumed that standard arithmetic and bitwise operations can be done in constant time on a computer word of some fixed length.

67.3 Techniques

We describe in some detail several techniques that have emerged over the past several years for generalized intersection searching. Briefly, these include: a geometric transformation-based approach, an approach based on generating a sparse representation of the input, an approach based on persistent data structures, a generic method that is applicable to any reporting problem, an approach for searching on a subset of the input satisfying a specified range restriction, and an approach that exploits the output size to gain efficiency. We illustrate each method with examples. Finally, we also discuss a "reverse" transformation method that shows an interesting connection between a standard counting and a generalized counting problem.

67.3.1 A Transformation-Based Approach

We describe an approach for certain reporting and counting problems, which transforms the original generalized reporting/counting problem to an instance of a related standard reporting/counting problem on which efficient known solutions can be brought to bear. We illustrate this approach by considering the *generalized one-dimensional range searching* problem, where the input consists of a set, S, of n colored points in $\mathbb{R}^1$ and the query, q, is an interval. Let S be a set of n colored points on the x-axis. We show how to preprocess S so that for any query interval q, we can solve efficiently the dynamic reporting problem, the static and dynamic counting problems, and the static type-2 counting problem. The solutions for the dynamic reporting problem and the static and dynamic counting problems are from Reference 13. The type-2 counting solution is from Reference 6.

We now describe the transformation. For each color c, we sort the distinct points of that color by increasing x-coordinate. For each point p of color c, let $pred(p)$ be its predecessor of color c in the sorted order; for the leftmost point of color c, we take the predecessor to be the point $-\infty$. We then map p to the point $p' = (p, pred(p))$ in the plane and associate with it the color c. Let S' be the resulting set of points. Given a query interval $q = [l, r]$, we map it to the grounded rectangle $q' = [l, r] \times (-\infty, l)$.

Lemma 67.1

There is a point of color c in S that is in $q = [l, r]$ if and only if there is a point of color c in S' that is in $q' = [l, r] \times (-\infty, l)$. Moreover, if there is a point of color c in q', then this point is unique.

Proof. Let p' be a c-colored point in q', where $p' = (p, pred(p))$ for some c-colored point $p \in S$. Since p' is in $[l, r] \times (-\infty, l)$, it is clear that $l \leq p \leq r$ and so $p \in [l, r]$.

For the converse, let p be the leftmost point of color c in $[l, r]$. Thus $l \leq p \leq r$ and since $pred(p) \notin [l, r]$, we have $l > pred(p)$. It follows that $p' = (p, pred(p))$ is in $[l, r] \times (-\infty, l)$. We prove that p' is the only point of color c in q'. Suppose for a contradiction that $t' = (t, pred(t))$ is another point of color c in q'. Thus we have $l \leq t \leq r$. Since $t > p$, we also have $pred(t) \geq p \geq l$. Thus $t' = (t, pred(t))$ cannot lie in q'—a contradiction. ∎

Lemma 67.1 implies that we can solve the generalized one-dimensional range reporting (resp., counting) problem by simply reporting the points in q' (resp., counting the number of points in q'), without regard to colors. In other words, we have reduced the generalized reporting (resp., counting) problem in $\mathbb{R}^1$ to the standard grounded range reporting (resp., counting) problem in $\mathbb{R}^2$. In the dynamic case, we also need to update S' when S is updated. We discuss these issues in more detail below.

67.3.1.1 The Dynamic Reporting Problem

Our data structure consists of the following: For each color c, we maintain a balanced binary search tree, T_c, in which the c-colored points of S are stored in increasing x-order. We maintain the colors themselves in a balanced search tree CT and store with each color c in CT a pointer to T_c. We also store the points of S' in a *balanced priority search tree (PST)* [34]. (Recall that a *PST* on m points occupies $O(m)$ space, supports insertions and deletions in $O(\log m)$ time, and can be used to report the k points lying inside a grounded query rectangle in $O(\log m + k)$ time [34]. Although this query is designed for query ranges of the form $[l, r] \times (-\infty, l]$, it can be trivially modified to ignore the points on the upper edge of the range without affecting its performance.) Clearly, the space used by the entire data structure is $O(n)$, where $n = |S|$.

To answer a query $q = [l, r]$, we simply query the *PST* with $q' = [l, r] \times (-\infty, l)$ and report the colors of the points found. Correctness follows from Lemma 67.1. The query time is $O(\log n + k)$, where k is the number of points inside q'. By Lemma 67.1, $k = i$, and so the query time is $O(\log n + i)$.

Suppose that a c-colored point p is to be inserted into S. If $c \notin CT$, then we create a tree T_c containing p, insert $p' = (p, -\infty)$ into the *PST*, and insert c, with a pointer to T_c, into CT. Suppose that $c \in CT$. Let u be the successor of p in T_c. If u exists, then we set $pred(p)$ to $pred(u)$ and $pred(u)$ to p; otherwise, we set $pred(p)$ to the rightmost point in T_c. We then insert p into T_c, $p' = (p, pred(p))$ into the *PST*, delete the old u' from the *PST*, and insert the new u' into it.

Deletion of a point p of color c is essentially the reverse. We delete p from T_c. Then we delete p' from the *PST* and if p had a successor, u, in T_c then we reset $pred(u)$ to $pred(p)$, delete the old u' from the *PST*, and insert the new one. If T_c becomes empty in the process, then we delete c from CT. Clearly, the update operations are correct and take $O(\log n)$ time.

Theorem 67.1

Let S be a set of n colored points on the real line. S can be preprocessed into a data structure of size $O(n)$ such that the i distinct colors of the points of S that are contained in any query interval can be reported in $O(\log n + i)$ time and points can be inserted and deleted online in S in $O(\log n)$ time.

For the static reporting problem, we can dispense with CT and the T_c's and simply use a static form of the *PST* to answer queries. This provides a simple $O(n)$-space, $O(\log n + i)$-query time alternative to another solution given in Reference 3.

67.3.1.2 The Static Counting Problem

We store the points of S' in nondecreasing x-order at the leaves of a balanced binary search tree, T, and store at each internal node t of T an array A_t containing the points in t's subtree in nondecreasing y-order. The total space is clearly $O(n \log n)$. To answer a query, we determine $O(\log n)$ canonical nodes v in T such that the query interval $[l, r]$ covers v's range but not the range of v's parent. Using binary search we determine in each canonical node's array the highest array position containing an entry less than l (and thus the number of points in that node's subtree that lie in q') and add up the positions thus found at all canonical nodes. The correctness of this algorithm follows from Lemma 67.1. The total query time is $O(\log^2 n)$.

We can reduce the query time to $O(\log n)$ as follows: At each node t we create a linked list, B_t, which contains the same elements as A_t and maintain a pointer from each entry of B_t to the same entry in A_t. We then apply the technique of fractional cascading [35] to the B-lists, so that after an initial $O(\log n)$-time binary search in the B-list of the root, the correct positions in the B-lists of all the canonical nodes can be found directly in $O(\log n)$ total time. (To facilitate binary search in the root's B-list, we build a balanced search tree on it after the fractional cascading step.) Once the position in a B-list is known, the appropriate position in the corresponding A-array can be found in $O(1)$ time.

It is possible to reduce the space slightly (to $O(n \log n / \log \log n)$) at the expense of a larger query time ($O(\log^2 n / \log \log n)$), by partitioning the points of S' recursively into horizontal strips of a certain size and doing binary search, augmented with fractional cascading, within the strips. Details can be found in Reference 13.

Theorem 67.2

Let S be a set of n colored points on the real line. S can be preprocessed into a data structure of size $O(n \log n)$ (resp., $O(n \log n / \log \log n)$) such that the number of distinctly colored points of S that are contained in any query interval can be determined in $O(\log n)$ (resp., $O(\log^2 n / \log \log n)$) time.

67.3.1.3 The Dynamic Counting Problem

We store the points of S' using the same basic two-level tree structure as in the first solution for the static counting problem. However, T is now a $BB(\alpha)$ tree [36] and the auxiliary structure, $D(t)$, at each node t of T is a balanced binary search tree where the points are stored at the leaves in left to right order by nondecreasing y-coordinate. To facilitate the querying, each node v of $D(t)$ stores a count of the points in its subtree. Given a real number, l, we can determine in $O(\log n)$ time the number of points in $D(t)$ that have y-coordinate less than l by searching for l in $D(t)$ and adding up the count for each node of $D(t)$ that is not on the search path but is the left child of a node on the path. It should be clear that $D(t)$ can be maintained in $O(\log n)$ time under updates.

In addition to the two-level structure, we also use the trees T_c and the tree CT, described previously, to maintain the correspondence between S and S'. We omit further discussion about the maintenance of these trees.

Queries are answered as in the static case, except that at each auxiliary structure we use the above-mentioned method to determine the number of points with y-coordinate less than l. Thus the query time is $O(\log^2 n)$. (We cannot use fractional cascading here.)

Insertion/deletion of a point is done using the worst-case updating strategy for $BB(\alpha)$ trees, and take $O(\log^2 n)$ time.

Theorem 67.3

Let S be a set of n colored points on the real line. S can be preprocessed into a data structure of size $O(n \log n)$ such that the number of distinctly colored points of S that are contained in any query interval can be determined in $O(\log^2 n)$ time and points can be inserted and deleted online in S in $O(\log^2 n)$ worst-case time.

67.3.1.4 The Static Type-2 Problem

We wish to preprocess a set S of n colored points on the x-axis, so that for each color intersected by a query interval $q = [l, r]$, the number of points of that color in q can be reported efficiently. The solution for this problem originally proposed in Reference 13 takes $O(n \log n)$ space and supports queries in $O(\log n + i)$ time. The space bound was improved to $O(n)$ in Reference 6, as follows.

The solution consists of two priority search trees, PST_1 and PST_2. PST_1 is similar to the priority search tree built on S' in the solution for the dynamic reporting problem, with an additional count stored at each node. Let $p' = (p, pred(p))$ be the point that is stored at a node in PST_1 and c the color of p. Then at this node, we store an additional number $t_1(p')$, which is the number of points of color c to the right of p.

PST_2 is based on a transformation that is symmetric to the one used for PST_1. For each color c, we sort the distinct points of that color by increasing x-coordinate. For each point p of color c, let $next(p)$ be its successor in the sorted order; for the rightmost point of color c, we take the successor to be the point $+\infty$. We then map p to the point $p'' = (p, next(p))$ in the plane and associate with it the color c. Let S'' be the resulting set of points. We build PST_2 on S'', with an additional count stored at each node. Let $p'' = (p, next(p))$ be the point that is stored at a node in PST_2 and c the color of p. Then at this node, we store an additional number $t_2(p'')$, which is the number of points of color c to the right of $next(p)$.

We also maintain an auxiliary array A of size n. Given a query $q = [l, r]$, we query PST_1 with $q' = [l, r] \times (-\infty, l)$ and for each color c found, we set $A[c] = t_1(p')$, where p' is the point stored at the node where we found c. Then we query PST_2 with $q'' = [l, r] \times (r, +\infty)$ and for each color c found, we report c and $A[c] - t_2(p'')$, where p'' is the point stored at the node where we found c. This works because the queries on PST_1 and PST_2 effectively find the leftmost and rightmost points of color c in $q = [l, r]$ (cf. proof of Lemma 67.1). Thus $A[c] - t_2(p'')$ gives the number of points of color c in q.

Theorem 67.4

A set S of n colored points on the real line can be preprocessed into a data structure of size $O(n)$ such that for any query interval, a type-2 counting query can be answered in $O(\log n + i)$ time, where i is the output size.

Finally, we note that Theorem 67.1 combined with the notion of persistence (which is discussed in detail in Section 67.3.3) yields an efficient solution to the *generalized grounded range reporting* problem in $\mathbb{R}^2$. In this problem, we wish to preprocess a set of n colored points in $\mathbb{R}^2$ so that the distinct colors of the points lying in any three-sided axes-parallel query rectangle can be reported efficiently. The solution uses $O(n \log n)$ and $O(\log n + i)$ query time. As shown in Reference 27, the space bound can be further improved to $O(n)$ by using persistence in conjunction with a transformation that maps the colored points in $\mathbb{R}^1$ to colored points in $\mathbb{R}^2$ such that the y-coordinates are integers in the range $[0 : \lceil \log n \rceil]$. (This transformation is different from the one underlying Lemma 67.1.) We refer the reader to Reference 27 for details. In Section 67.3.6, we describe another solution to this problem which has the same bounds but is based on a different technique.

67.3.2 A Sparsification-Based Approach

The idea behind this approach is to generate from the given set, S, of colored objects a colored set, S'—possibly consisting of different objects than those in S—such that a query object q intersects an object in S if and only if it intersects an object in S'. Moreover, if q intersects objects in S' then it intersects at most a constant number of them. This allows us to use a solution to a standard problem on S' to solve the generalized reporting problem on S. (In the case of a generalized counting problem, the requirement is more stringent: exactly one object in S' must be intersected.) We illustrate this method with the generalized half-space range searching problem in $\mathbb{R}^d$, $d = 2, 3$.

67.3.2.1 Generalized Half-Space Range Searching in $\mathbb{R}^2$ and $\mathbb{R}^3$

Let S be a set of n colored points in $\mathbb{R}^d$, $d = 2, 3$. In the *generalized half-space range searching* problem we wish to preprocess S so that for any query hyperplane Q, the i distinct colors of the points lying in the closed half-space Q^- (i.e., below Q) can be reported or counted efficiently. Without loss of generality, we may assume that Q is nonvertical since vertical queries are easy to handle. The approach described here is from Reference 14.

We denote the coordinate directions by $x_1, x_2, \ldots, x_d$. Let $\mathcal{F}$ denote the well-known point-hyperplane duality transform [37]: If $p = (p_1, \ldots, p_d)$ is a point in $\mathbb{R}^d$, then $\mathcal{F}(p)$ is the hyperplane $x_d = p_1 x_1 + \cdots + p_{d-1} x_{d-1} - p_d$. If $H : x_d = a_1 x_1 + \cdots + a_{d-1} x_{d-1} + a_d$ is a (nonvertical) hyperplane in $\mathbb{R}^d$, then $\mathcal{F}(H)$ is the point $(a_1, \ldots, a_{d-1}, -a_d)$. It is easily verified that p is above (resp., on, below) H, in the x_d-direction, if and only if $\mathcal{F}(p)$ is below (resp., on, above) $\mathcal{F}(H)$. Note also that $\mathcal{F}(\mathcal{F}(p)) = p$ and $\mathcal{F}(\mathcal{F}(H)) = H$.

Using $\mathcal{F}$, we map S to a set S' of hyperplanes and map Q to the point $q = \mathcal{F}(Q)$, both in $\mathbb{R}^d$. Our problem is now equivalent to: "Report or count the i distinct colors of the hyperplanes lying on or above q, that is, the hyperplanes that are intersected by the vertical ray r emanating upwards from q."

Let S_c be the set of hyperplanes of color c. For each color c, we compute the *upper envelope* E_c of the hyperplanes in S_c. E_c is the locus of the points of S_c of maximum x_d-coordinate for each point on the plane $x_d = 0$. E_c is a d-dimensional convex polytope which is unbounded in the positive x_d-direction. Its boundary is composed of j-faces, $0 \leq j \leq d - 1$, where each j-face is a j-dimensional convex polytope. Of particular interest to us are the $(d-1)$-faces of E_c, called *facets*. For instance, in $\mathbb{R}^2$, E_c is an unbounded convex chain and its facets are line segments; in $\mathbb{R}^3$, E_c is an unbounded convex polytope whose facets are convex polygons.

Let us assume that r is well behaved in the sense that for no color c does r intersect two or more facets of E_c at a common boundary—for instance, a vertex in $\mathbb{R}^2$ and an edge or a vertex in $\mathbb{R}^3$. (This assumption can be removed; details can be found in Reference 14.) Then, by definition of the upper envelope, it follows that (i) r intersects a c-colored hyperplane if and only if r intersects E_c and, moreover, (ii) if r intersects E_c, then r intersects a unique facet of E_c (in the interior of the facet). Let $\mathcal{E}$ be the collection of the envelopes of the different colors. By the above discussion, our problem is equivalent to: "Report or count the facets of $\mathcal{E}$ that are intersected by r," which is a standard intersection searching problem. We will show how to solve efficiently this ray-envelope intersection problem in $\mathbb{R}^2$ and in $\mathbb{R}^3$. This approach does not give an efficient solution to the generalized half-space searching problem in $\mathbb{R}^d$ for $d > 3$; for this case, we will give a different solution in Section 67.3.4.

To solve the ray–envelope intersection problem in $\mathbb{R}^2$, we project the endpoints of the line segments of $\mathcal{E}$ on the x-axis, thus partitioning it into $2n + 1$ *elementary intervals* (some of which may be empty). We build a *segment tree* T which stores these elementary intervals at the leaves. Let v be any node of T. We associate with v an x-interval $I(v)$, which is the union of the elementary intervals stored at the leaves in v's subtree. Let $Strip(v)$ be the vertical strip defined by $I(v)$. We say that a segment $s \in \mathcal{E}$ is *allocated* to a node $v \in T$ if and only if $I(v) \neq \emptyset$ and s crosses $Strip(v)$ but not $Strip(parent(v))$. Let $\mathcal{E}(v)$ be the set of segments allocated to v. Within $Strip(v)$, the segments of $\mathcal{E}(v)$ can be viewed as lines since they cross $Strip(v)$ completely. Let $\mathcal{E}'(v)$ be the set of points dual to these lines. We store $\mathcal{E}'(v)$ in an instance $D(v)$ of the standard half-plane reporting (resp., counting) structure for $\mathbb{R}^2$ given in Reference 38 (resp., [39]). This structure uses $O(m)$ space and has a query time of $O(\log m + k_v)$ (resp., $O(m^{1/2})$), where $m = |\mathcal{E}(v)|$ and k_v is the output size at v.

To answer a query, we search in T using q's x-coordinate. At each node v visited, we need to report or count the lines intersected by r. But, by duality, this is equivalent to answering, in $\mathbb{R}^2$, a half-plane query at v using the query $\mathcal{F}(q)^- = Q^-$, which we do using $D(v)$. For the reporting problem, we simply output what is returned by the query at each visited node; for the counting problem, we return the sum of the counts obtained at the visited nodes.

Theorem 67.5

A set S of n colored points in $\mathbb{R}^2$ can be stored in a data structure of size $O(n \log n)$ so that the i distinct colors of the points contained in any query half-plane can be reported (resp., counted) in time $O(\log^2 n + i)$ (resp., $O(n^{1/2})$).

Proof. Correctness follows from the preceding discussion. As noted earlier, there are $O(|S_c|)$ line segments (facets) in E_c; thus $|\mathcal{E}| = O(\sum_c |S_c|) = O(n)$ and so $|T| = O(n)$. Hence each segment of $\mathcal{E}$ can get allocated to $O(\log n)$ nodes of T. Since the structure $D(v)$ has size linear in $m = |\mathcal{E}(v)|$, the total space used is $O(n \log n)$. For the reporting problem, the query time at a node v is $O(\log m + k_v) = O(\log n + k_v)$. When summed over the $O(\log n)$ nodes visited, this gives $O(\log^2 n + i)$. To see this, recall that the ray r can intersect at most one envelope segment of any color; thus the terms k_v, taken over all nodes v visited, sum to i.

For the counting problem, the query time at v is $O(m^{1/2})$. It can be shown that if v has depth j in T, then $m = |\mathcal{E}(v)| = O(n/2^j)$ [40, page 675]. Thus, the overall query time is $O(\sum_{j=0}^{O(\log n)} (n/2^j)^{1/2})$, which is $O(n^{1/2})$. ∎

The (standard) problem of reporting the segments in $\mathbb{R}^2$ that are intersected by a vertical query ray has been revisited recently in Reference 41, in the context of solving a different problem defined on so-called "uncertain points." For a set of n segments, the solution in Reference 41 uses $O(n)$ space and has a query time of $O(\log n + k)$, where k is the number of reported segments. The approach is based on using a combination of a segment tree and an interval tree together with some other ideas; we refer the reader to Reference 41 for details. By the preceding discussion, this result implies an $O(n)$-space and $O(\log n + i)$-query time solution to the generalized half-space range reporting problem in $\mathbb{R}^2$.

In $\mathbb{R}^3$, the approach is similar, but somewhat more complex. Our goal is to solve the ray–envelope intersection problem in $\mathbb{R}^3$. As shown in Reference 14, this problem can be reduced to certain standard half-space range queries in $\mathbb{R}^3$ on a set of triangles (obtained by triangulating the E_c's). This problem can be solved by building a segment tree on the x-spans of the triangles projected to the xy-plane and augmenting each node of this tree with a data structure based on partition trees [42] or cutting trees [43] to answer the half-plane queries. Details may be found in Reference 14.

Theorem 67.6

The reporting version of the generalized half-space range searching problem for a set of n colored points in $\mathbb{R}^3$ can be solved in $O(n \log^2 n)$ (resp., $O(n^{2+\epsilon})$) space and $O(n^{1/2+\epsilon} + i)$ (resp., $O(\log^2 n + i)$) query time, where i is the output size and $\epsilon > 0$ is an arbitrarily small constant. The counting version is solvable in $O(n \log n)$ space and $O(n^{2/3+\epsilon})$ query time.

Additional examples of the sparsification-based approach may be found in Reference 3. (An example also appears in the next section, enroute to a persistence-based solution of a generalized problem.)

67.3.3 A Persistence-Based Approach

Roughly speaking, we use persistence as follows: To solve a given generalized problem we first identify a different, but simpler, generalized problem and devise a data structure for it that also supports updates (usually just insertions). We then make this structure partially persistent [44] and query this persistent structure appropriately to solve the original problem.

We illustrate this approach for *generalized three-dimensional range searching*, where we are required to preprocess a set, S, of n colored points in $\mathbb{R}^3$ so that for any query box $q = [a, b] \times [c, d] \times [e, f]$ the i distinct colors of the points inside q can be reported efficiently. We first show how to build a semidynamic (i.e., insertions-only) data structure for the generalized versions of the quadrant searching and two-dimensional range searching problems. These two structures will be the building blocks of our solution for the three-dimensional problem.

67.3.3.1 Generalized Semidynamic Quadrant Searching

Let S be a set of n colored points in the plane. For any point $q = (a, b)$, the *northeast quadrant* of q, denoted by $NE(q)$, is the set of all points (x, y) in the plane such that $x \geq a$ and $y \geq b$. We show how to preprocess S so that for any query point q, the distinct colors of the points of S contained in $NE(q)$ can be reported, and how points can be inserted into S. This is the *generalized two-dimensional quadrant range searching* problem supporting insertions. The data structure uses $O(n)$ space, has a query time of $O(\log^2 n + i)$, and an amortized insertion time of $O(\log n)$. This solution is based on the sparsification approach described previously.

For each color c, we determine the c-maximal points. (A point p is called c-maximal if it has color c and there are no points of color c in p's northeast quadrant.) We discard all points of color c that are not c-maximal. In the resulting set, let the predecessor, $pred(p)$, of a c-colored point p be the c-colored point that lies immediately to the left of p. (For the leftmost point of color c, the predecessor is the point $(-\infty, \infty)$.) With each point $p = (a, b)$, we associate the horizontal segment with endpoints (a', b) and (a, b), where a' is the x-coordinate of $pred(p)$. This segment gets the same color as p. Let S_c be the set of such segments of color c. The data structure consists of two parts, as follows.

The first part is a structure $\mathcal{T}$ storing the segments in the sets S_c, where c runs over all colors. $\mathcal{T}$ supports the following query: given a point q in the plane, report the segments that are intersected by the upward-vertical ray starting at q. Moreover, it allows segments to be inserted and deleted. We implement $\mathcal{T}$ as the structure given in Reference 45. This structure uses $O(n)$ space, supports insertions and deletions in $O(\log n)$ time, and has a query time of $O(\log^2 n + l)$, where l is the number of segments intersected.

The second part is a balanced search tree CT, storing all colors. For each color c, we maintain a balanced search tree, T_c, storing the segments of S_c by increasing y-coordinate. This structure allows us to dynamically maintain S_c when a new c-colored point p is inserted. The general approach (omitting some special cases; see Reference 13) is as follows: By doing a binary search in T_c we can determine whether or not p is c-maximal in the current set of c-maximal points, that is, the set of right endpoints of the segments of S_c. If p is not c-maximal, then we simply discard it. If p is c-maximal, then let $s_1, \ldots, s_k$ be the segments of S_c

whose left endpoints are in the southwest quadrant of p. We do the following: (i) delete $s_2, \ldots, s_k$ from T_c; (ii) insert into T_c the horizontal segment which starts at p and extends leftwards up to the x-coordinate of the left endpoint of s_k; and (iii) truncate the segment s_1 by keeping only the part of it that extends leftwards up to p's x-coordinate. The entire operation can be done in $O(\log n + k)$ time.

Let us now consider how to answer a quadrant query, $NE(q)$, and how to insert a point into S. To answer $NE(q)$, we query T with the upward-vertical ray from q and report the colors of the segments intersected. The correctness of this algorithm follows from the easily proved facts that (i) a c-colored point lies in $NE(q)$ if and only if a c-maximal point lies in $NE(q)$ and (ii) if a c-maximal point is in $NE(q)$, then the upward-vertical ray from q must intersect a segment of S_c. The correctness of T guarantees that only the segments intersected by this ray are reported. Since the query can intersect at most two segments in any S_c, we have $l \leq 2i$, and so the query time is $O(\log^2 n + i)$.

Let p be a c-colored point that is to be inserted into S. If c is not in CT, then we insert it into CT and insert the horizontal, leftward-directed ray emanating from p into a new structure T_c. If c is present already, then we update T_c as just described. In both cases, we then perform the same updates on T. Hence, an insertion takes $O((k+1)\log n)$ time.

What is the total time for n insertions into an initially empty set S? For each insertion, we can charge the $O(\log n)$ time to delete a segment $s_i, 2 \leq i \leq k$, to s_i itself. Notice that none of these segments will reappear. Thus each segment is charged at most once. Moreover, each of these segments has some previously inserted point as a right endpoint. It follows that the number of segments existing over the entire sequence of insertions is $O(n)$ and so the total charge to them is $O(n \log n)$. The rest of the cost for each insertion ($O(\log n)$ for the binary search plus $O(1)$ for steps (ii) and (iii)) we charge to p itself. Since any p is charged in this mode only once, the total charge incurred in this mode by all the inserted points is $O(n \log n)$. Thus the time for n insertions is $O(n \log n)$, which implies an amortized insertion time of $O(\log n)$.

Lemma 67.2

Let S be a set of n colored points in the plane. There exists a data structure of size $O(n)$ such that for any query point q, we can report the i distinct colors of the points that are contained in the northeast quadrant of q in $O(\log^2 n + i)$ time. Moreover, if we do n insertions into an initially empty set then the amortized insertion time is $O(\log n)$.

67.3.3.2 Generalized Semidynamic Two-Dimensional Range Searching

Our goal here is to preprocess a set S of n colored points in the plane so that for any axes-parallel query rectangle $q = [a, b] \times [c, d]$, we can report efficiently the distinct colors of the points in q and moreover perform insertions in S. This is the *generalized two-dimensional range reporting* problem supporting insertions.

Our solution is based on the quadrant reporting structure of Lemma 67.2. We first show how to solve the problem for query rectangles $q' = [a, b] \times [c, \infty)$. We store the points of S in sorted order by x-coordinate at the leaves of a $BB(\alpha)$ tree T'. At each internal node v, we store an instance of the structure of Lemma 67.2 for NE-queries (resp., NW-queries) built on the points in v's left (resp., right) subtree. Let $X(v)$ denote the average of the x-coordinate in the rightmost leaf in v's left subtree and the x-coordinate in the leftmost leaf of v's right subtree; for a leaf v, we take $X(v)$ to be the x-coordinate of the point stored at v.

To answer a query q', we do a binary search down T', using $[a, b]$, until either the search runs off T' or a (highest) node v is reached such that $[a, b]$ intersects $X(v)$. In the former case, we stop. In the latter case, if v is a leaf, then if v's point is in q' we report its color. If v is a nonleaf, then we query the structures at v using the NE-quadrant and the NW-quadrant derived from q' (i.e., the quadrants with corners at (a, c) and (b, c), respectively), and then combine the answers. Updates on T' are performed using the amortized-case updating strategy for $BB(\alpha)$ trees [36]. The correctness of the method should be clear. The space and query time bounds follow from Lemma 67.2. Since the amortized insertion time of the quadrant searching structure is $O(\log n)$, the insertion in the $BB(\alpha)$ tree takes amortized time $O(\log^2 n)$ [36].

To solve the problem for general query rectangles $q = [a, b] \times [c, d]$, we use the above approach again, except that we store the points in the tree by sorted y-coordinates. At each internal node v, we store an instance of the data structure above to answer queries of the form $[a, b] \times [c, \infty)$ (resp., $[a, b] \times (-\infty, d]$) on the points in v's left (resp., right) subtree. The query strategy is similar to the previous one, except that we use the interval $[c, d]$ to search in the tree. The query time is as before, while the space and update times increase by a logarithmic factor.

Lemma 67.3

Let S be a set of n colored points in the plane. There exists a data structure of size $O(n \log^2 n)$ such that for any query rectangle $[a, b] \times [c, d]$, we can report the i distinct colors of the points that are contained in it in $O(\log^2 n + i)$ time. Moreover, points can be inserted into this data structure in $O(\log^3 n)$ amortized time.

67.3.3.3 Generalized Three-Dimensional Range Searching

The semidynamic structure of Lemma 67.3 coupled with persistence allows us to go up one dimension and solve the original problem of interest: Preprocess a set S of n colored points in $\mathbb{R}^3$ so that for any query box $q = [a,b] \times [c,d] \times [e,f]$ the i distinct colors of the points inside q can be reported efficiently.

First consider queries of the form $q' = [a,b] \times [c,d] \times [e,\infty)$. We sort the points of S by nonincreasing z-coordinates, and insert them in this order into a partially persistent version of the structure of Lemma 67.3, taking only the first two coordinates into account. To answer q', we access the version corresponding to the smallest z-coordinate greater than or equal to e and query it with $[a,b] \times [c,d]$.

To see that the query algorithm is correct, observe that the version accessed contains the projections on the xy-plane of exactly those points of S whose z-coordinate is at least e. Lemma 67.3 then guarantees that among these only the distinct colors of the ones in $[a,b] \times [c,d]$ are reported. These are precisely the distinct colors of the points contained in $[a,b] \times [c,d] \times [e,\infty)$. The query time follows from Lemma 67.3. To analyze the space requirement, we note that the structure of Lemma 67.3 satisfies the conditions given in Reference 44. Specifically, it is a pointer-based structure, where each node is pointed to by only $O(1)$ other nodes. As shown in Reference 44, any modification made by a persistent update operation on such a structure adds only $O(1)$ amortized space to the resulting persistent structure. By Lemma 67.3, the total time for creating the persistent structure, via insertions, is $O(n \log^3 n)$. This implies the same bound for the number of modifications in the structure, so the total space is $O(n \log^3 n)$.

To solve the problem for general query boxes $q = [a,b] \times [c,d] \times [e,f]$, we follow an approach similar to that described for the two-dimensional case: We store the points in a balanced binary search tree, sorted by z-coordinates. We associate with each internal node v in the tree the auxiliary structure described above for answering queries of the form $[a,b] \times [c,d] \times [e,\infty)$ (resp., $[a,b] \times [c,d] \times (-\infty, f]$) on the points in v's left (resp., right) subtree. (Note that since we do not need to do updates here the tree need not be a $BB(\alpha)$ tree.) Queries are done by searching down the tree using the interval $[e,f]$. The query time is as before, but the space increases by a logarithmic factor.

Theorem 67.7

Let S be a set of n colored points in 3-space. S can be stored in a data structure of size $O(n \log^4 n)$ such that for any query box $[a,b] \times [c,d] \times [e,f]$, we can report the i distinct colors of the points that are contained in it in $O(\log^2 n + i)$ time.

Additional applications of the persistence-based approach to generalized intersection problems can be found in References 13,14,16,27.

67.3.4 A General Approach for Reporting Problems

We describe a general method from Reference 16 for solving any generalized reporting problem given a data structure for a "related" standard decision problem.

Let S be a set of n colored geometric objects and let q be any query object. In preprocessing, we store the distinct colors in S at the leaves of a balanced binary tree CT (in no particular order). For any node v of CT, let $C(v)$ be the set of colors stored in the leaves of v's subtree and let $S(v)$ be the set of those objects of S colored with the colors in $C(v)$. At v, we store a data structure $DEC(v)$ to solve the following *standard decision* problem on $S(v)$: "Decide whether or not q intersects any object of $S(v)$." $DEC(v)$ returns "true" if and only if there is an intersection.

To answer a generalized reporting query on S, we do a depth-first search in CT and query $DEC(v)$ with q at each node v visited. If v is a nonleaf node then we continue searching below v if and only if the query returns "true"; if v is a leaf, then we output the color stored there if and only if the query returns "true."

Theorem 67.8

Assume that a set of n geometric objects can be stored in a data structure of size $M(n)$ such that it can be decided in $f(n)$ time whether or not a query object intersects any of the n objects. Assume that $M(n)/n$ and $f(n)$ are nondecreasing functions for nonnegative values of n. Then a set S of n colored geometric objects can be preprocessed into a data structure of size $O(M(n) \log n)$ such that the i distinct colors of the objects in S that are intersected by a query object q can be reported in time $O(f(n) + i \cdot f(n) \log n)$.

Proof. We argue that a color c is reported if and only if there is a c-colored object in S intersecting q. Suppose that c is reported. This implies that a leaf v is reached in the search such that v stores c and the query on $DEC(v)$ returns "true." Thus, q intersects some object in $S(v)$. Since v is a leaf, all objects in $S(v)$ have the same color c and the claim follows.

For the converse, suppose that q intersects a c-colored object p. Let v be the leaf storing c. Thus $p \in S(v')$ for every node v' on the root-to-v path in CT. Thus, for each v', the query on $DEC(v')$ will return "true," which implies that v will be visited and c will be output.

If $v_1, v_2, \ldots, v_r$ are the nodes at any level, then the total space used by CT at that level is $\sum_{i=1}^{r} M(|S(v_i)|) = \sum_{i=1}^{r} |S(v_i)| \cdot (M(|S(v_i)|)/|S(v_i)|) \leq \sum_{i=1}^{r} |S(v_i)| \cdot (M(n)/n) = M(n)$, since $\sum_{i=1}^{r} |S(v_i)| = n$ and since $|S(v_i)| \leq n$ implies that $M(|S(v_i)|)/|S(v_i)| \leq M(n)/n$. Now since there are $O(\log n)$ levels, the overall space is $O(M(n) \log n)$. The query time can be upper bounded as follows: If $i = 0$, then the query on $DEC(\text{root})$ returns "false" and we abandon the search at the root itself; in this case, the query time is just $O(f(n))$. Suppose that $i \neq 0$. Call a visited node v *fruitful* if the query on $DEC(v)$ returns "true" and *fruitless* otherwise. Each fruitful node can be charged to some color in its subtree that gets reported. Since the number of times any reported color can be charged is $O(\log n)$ (the height of CT) and since i colors are reported, the number of fruitful nodes is $O(i \log n)$. Since each fruitless node has a fruitful parent and CT is a binary tree, it follows that there are only $O(i \log n)$ fruitless nodes. Hence the number of nodes visited by the search is $O(i \log n)$. At each such node, v, we spend time $f(|S(v)|)$, which is $O(f(n))$ since $|S(v)| \leq n$ and f is nondecreasing. Thus the total time spent in doing queries at the visited nodes is $O(i \cdot f(n) \log n)$. The claimed query time follows. $\blacksquare$

As an application of this method, consider the generalized half-space range searching problem in $\mathbb{R}^d$, for any fixed $d \geq 2$. For $d = 2, 3$, we discussed a solution for this problem in Section 67.3.2. For $d > 3$, the problem can be solved by extending (significantly) the ray–envelope intersection algorithm outlined in Section 67.3.2. However, the bounds are not very satisfactory—$O(n^{d\lfloor d/2 \rfloor + \epsilon})$ space and logarithmic query time or near-linear space and superlinear query time. The solution we give below has more desirable bounds.

The colored objects for this problem are points in $\mathbb{R}^d$ and the query is a closed half-space in $\mathbb{R}^d$. We store the objects in CT, as described previously. The standard decision problem that we need to solve at each node v of CT is "Does a query half-space contain any point of $S(v)$." The answer to this query is "true" if and only if the query half-space is nonempty. We take the data structure, $DEC(v)$, for this problem to be the one given in Reference 46. If $|S_v| = n_v$, then $DEC(v)$ uses $O(n_v^{\lfloor d/2 \rfloor}/(\log n_v)^{\lfloor d/2 \rfloor - \epsilon})$ space and has query time $O(\log n_v)$ [46]. The conditions in Theorem 67.8 hold, so applying it gives the following result.

Theorem 67.9

For any fixed $d \geq 2$, a set S of n colored points in $\mathbb{R}^d$ can be stored in a data structure of size $O(n^{\lfloor d/2 \rfloor}/(\log n)^{\lfloor d/2 \rfloor - 1 - \epsilon})$ such that the i distinct colors of the points contained in a query half-space Q^- can be reported in time $O(\log n + i \log^2 n)$. Here $\epsilon > 0$ is an arbitrarily small constant.

Other applications of the general method may be found in Reference 16.

67.3.5 Adding Range Restrictions

We describe the general technique of Reference 15 that adds a range restriction to a generalized intersection searching problem.

Let PR be a generalized intersection searching problem on a set S of n colored objects and query objects q belonging to a class Q. We denote the answer to a query by $PR(q, S)$. To add a *range restriction*, we associate with each element $p \in S$ a real number k_p. In a range-restricted generalized intersection searching problem, denoted by TPR, a query consists of an element $q \in Q$ and an interval $[l, r]$, and

$$TPR(q, [l, r], S) := PR(q, \{p \in S : l \leq k_p \leq r\}).$$

For example, if PR is the generalized $(d-1)$-dimensional range searching problem, then TPR is the generalized d-dimensional version of this problem, obtained by adding a range restriction to the dth dimension.

Assume that we have a data structure DS that solves PR with $O((\log n)^u + i)$ query time using $O(n^{1+\epsilon})$ space and a data structure TDS that solves TPR for generalized (semi-infinite) queries of the form $TPR(q, [l, \infty), S)$ with $O((\log n)^v + i)$ query time using $O(n^w)$ space. (Here u and v are positive constants, $w > 1$ is a constant, and $\epsilon > 0$ is an arbitrarily small constant.) We will show how to transform DS and TDS into a data structure that solves generalized queries $TPR(q, [l, r], S)$ in $O((\log n)^{\max(u,v,1)} + i)$ time, using $O(n^{1+\epsilon})$ space.

Let $S = \{p_1, p_2, \ldots, p_n\}$, where $k_{p_1} \geq k_{p_2} \geq \cdots \geq k_{p_n}$. Let m be an arbitrary parameter with $1 \leq m \leq n$. We assume for simplicity that n/m is an integer. Let $S_j = \{p_1, p_2, \ldots, p_{jm}\}$ and $S'_j = \{p_{jm+1}, p_{jm+2}, \ldots, p_{(j+1)m}\}$ for $0 \leq j < n/m$.

The transformed data structure consists of the following. For each j with $0 \le j < n/m$, there is a data structure DS_j (of type DS) storing S_j for solving generalized queries of the form $PR(q, S_j)$, and a data structure TDS_j (of type TDS) storing S'_j for solving generalized queries of the form $TPR(q, [l, \infty), S'_j)$.

To answer a query $TPR(q, [l, \infty), S)$, we do the following. Compute the index j such that $k_{p_{(j+1)m}} < l \le k_{p_{jm}}$. Solve the query $PR(q, S_j)$ using DS_j, solve the query $TPR(q, [l, \infty), S'_j)$ using TDS_j, and output the union of the colors reported by these two queries. It is easy to see that the query algorithm is correct. The following lemma gives the complexity of the transformed data structure.

Lemma 67.4

The transformed data structure uses $O(n^{2+\epsilon}/m + n\,m^{w-1})$ space and can be used to answer generalized queries $TPR(q, [l, \infty), S)$ in $O((\log n)^{\max(u,v,1)} + i)$ time.

Theorem 67.10

Let S, DS, and TDS be as above. There exists a data structure of size $O(n^{1+\epsilon})$ that solves generalized queries $TPR(q, [l, r], S)$ in $O((\log n)^{\max(u,v,1)} + i)$ time.

Proof. We will use Lemma 67.4 to establish the claimed bounds for answering generalized queries $TPR(q, [l, \infty), S)$. The result for queries $TPR(q, [l, r], S)$ then follows from a technique, based on $BB(\alpha)$ trees, that we used in Section 67.3.3.

If $w > 2$, then we apply Lemma 67.4 with $m = n^{1/w}$. This gives a data structure having size $O(n^2)$ that answers queries $TPR(q, [l, \infty), S)$ in $O((\log n)^{\max(u,v,1)} + i)$ time. Hence, we may assume that $w = 2$.

By applying Lemma 67.4 repeatedly, we obtain, for each integer constant $a \ge 1$, a data structure of size $O(n^{1+\epsilon+1/a})$ that answers queries $TPR(q, [l, \infty), S)$ in $O((\log n)^{\max(u,v,1)} + i)$ time. This claim follows by induction on a; in the inductive step from a to $a+1$, we apply Lemma 67.4 with $m = n^{a/(a+1)}$. ∎

Using Theorem 67.10, we can solve efficiently, for instance, the generalized orthogonal range searching problem in $\mathbb{R}^d$. (Examples of other problems solvable via this method may be found in Reference 15.)

Theorem 67.11

Let S be a set of n colored points in $\mathbb{R}^d$, where $d \ge 1$ is a constant. There exists a data structure of size $O(n^{1+\epsilon})$ such that for any query box in $\mathbb{R}^d$, we can report the i distinct colors of the points that are contained in it in $O(\log n + i)$ time.

Proof. The proof is by induction on d. For $d = 1$, the claim follows from Theorem 67.1. Let $d \ge 2$, and let DS be a data structure of size $O(n^{1+\epsilon})$ that answers generalized $(d-1)$-dimensional range queries in $O(\log n + i)$ time. Observe that for the generalized d-dimensional range searching problem, there are only polynomially many distinct semi-infinite queries. Hence, there exists a data structure TDS of polynomial size that answers generalized d-dimensional semi-infinite range queries in $O(\log n + i)$ time. Applying Theorem 67.10 to DS and TDS proves the claim. ∎

67.3.6 Exploiting the Output Size

In this approach, the design of the data structure and the query algorithm is based on the (unknown) size of the output (i.e., i). It involves building and querying two structures, one to handle "large" output size and the other to handle "small" output size. The definition of "large" and "small" depends on the problem at hand and will become clear in the discussion below.

Suppose that we have designed a low-space data structure, $\mathcal{D}_L$, to handle the case where the output size is large; let the query time of this be $O(f(n) + i)$. Then the crucial observation is that if i is $\Omega(f(n))$ (i.e., i is large), then the query time is $O(i)$, which is the best one can hope for since it takes this much time to merely output the query results. However, if i is $O(f(n))$ (i.e., i is small), then the query time is $O(f(n))$ which is undesirable if $f(n)$ is large. The challenge now is to design a separate data structure, $\mathcal{D}_S$, that can handle efficiently the case where the output size is small, by taking advantage of this very fact. Intuitively, this step involves "precomputing" and storing the answers to certain carefully chosen queries; the space used for this is kept small by exploiting the fact that the output size is small. Note that an additional challenge is that one does not know *a priori* whether i is small or large for a given query instance. Therefore, the overall query algorithm proceeds as follows: First, $\mathcal{D}_S$ is queried with the given query object. If this succeeds in returning all of the query results (i.e., the output size is small) then the query algorithm terminates. Otherwise, the query on $\mathcal{D}_S$ aborts and the algorithm proceeds to reissue the query on $\mathcal{D}_L$ to compute the query result.

We illustrate this approach by presenting an optimal algorithm for generalized grounded range reporting in $\mathbb{R}^2$, where the input is a set of n colored points and the query is a rectangle $q = [x_l, x_r] \times [y_0, \infty)$. This algorithm uses $O(n)$ space and answers queries in $O(\log n + i)$ time. The following description is based on ideas from References 24 and 25, where the problem is solved in external memory.

For $\mathcal{D}_L$, we use a data structure presented in Reference 3 which, for a set of n colored points in $\mathbb{R}^2$, occupies $O(n)$ space and answers a generalized grounded range query in $O(\log^2 n + i)$ time. Thus $f(n) = \log^2 n$ and we take the output size to be large if $i \geq \log^2 n$. We build an instance of $\mathcal{D}_L$ on the given set S.

We design the structure $\mathcal{D}_S$ as follows: We sort the points of S in nondecreasing order of their x-coordinates and partition them into groups consisting of $\log^3 n$ consecutive points each. For each group we build an instance of $\mathcal{D}_L$. Next, we build a balanced binary search tree, $\mathcal{T}$, based on the left-to-right ordering of the groups and associate a leaf with each group. Let v be a proper ancestor of a leaf u and let $\Pi(u, v)$ be the path from u to v (excluding u and v). Let $S_l(u, v)$ (resp., $S_r(u, v)$) be the union of the sets of points in the subtrees rooted at nodes that are left (resp., right) children of nodes on $\Pi(u, v)$ but not themselves on the path. For each distinct color in $S_l(u, v)$, we select from among the points of that color in $S_l(u, v)$ the one with the highest y-coordinate. Let $S_l'(u, v)$ be the set of such points selected. Let $S_l''(u, v)$ be the subset of $S_l'(u, v)$ consisting of the $\log^2 n$ points with largest y-coordinate (ties broken arbitrarily). If fewer than $\log^2 n$ points are in $S_l'(u, v)$, then we include all of them in $S_l''(u, v)$. A symmetric discussion for $S_r(u, v)$ yields a set $S_r''(u, v)$. For each pair (u, v), we store $S_l''(u, v)$ and $S_r''(u, v)$ in linked lists, in nonincreasing order of the y-coordinates of their points (ties broken arbitrarily). The number of (u, v)-pairs stored is $O((n/\log^3 n) \times (\log n)) = O(n/\log^2 n)$, so the space occupied by all sets $S_l''(u, v)$ and $S_r''(u, v)$ is $O(n)$. The space occupied by $\mathcal{T}$ and the $\mathcal{D}_L$ structures at all leaves is also $O(n)$.

To answer a query $q = [x_l, x_r] \times [y_0, \infty)$, we first determine the leaf u_l (resp., u_r) of $\mathcal{T}$ whose range of x-coordinates contains x_l (resp., x_r). If $u_l = u_r$, then we query the $\mathcal{D}_L$ structure of that leaf and stop. Otherwise, we find the lowest common ancestor, v, of u_l and u_r and do the following:

- First, we report the distinct colors of the points in the groups associated with u_l and u_r that lie in q. This is done by querying the $\mathcal{D}_L$ structure of u_l (resp., u_r) with $[x_l, \infty) \times [y_0, \infty)$ (resp., $(-\infty, x_r] \times [y_0, \infty)$).
- Next, we scan the list for $S_r''(u_l, v)$ and report the colors in it until either (a) we find a point with y-coordinate less than y_0 or (b) all the points in the list have been scanned. If case (a) holds, then the distinct colors of the points of $S_r(u_l, v)$ that lie in q have been reported. If case (b) holds, then we check the size of $S_r''(u_l, v)$. If $|S_r''(u_l, v)| < \log^2 n$, then again the distinct colors of the points of $S_r(u_l, v)$ that lie in q have been reported. (Recall that, by construction, if $S_r(u_l, v)$ has fewer than $\log^2 n$ distinctly colored points, then all of them are included in $S_r''(u_l, v)$.) However, if $|S_r''(u_l, v)| \geq \log^2 n$, then we conclude that i is $\Omega(\log^2 n)$. Similarly, we also scan the list for $S_l''(u_r, v)$ and either report the distinct colors of the points of $S_l(u_r, v)$ that lie in q or conclude that i is $\Omega(\log^2 n)$. If we conclude i is $\Omega(\log^2 n)$ at least once in the above process, then we discard the reported colors and proceed to query the structure $\mathcal{D}_L$ built on the entire set S with q.

The time to find u_l, u_r, and v is $O(\log n)$. If $u_l = u_r$, then querying the $\mathcal{D}_L$ structure at that leaf node takes $O((\log(\log^3 n))^2 + i) = O(\log n + i)$ time. If $u_l \neq u_r$, then, as above, querying the $\mathcal{D}_L$ structure at the two leaves takes $O(\log n + i)$ time. The time to scan $S_r''(u_l, v)$ and $S_l''(u_r, v)$ is $O(i)$. Finally, the time to query the $\mathcal{D}_L$ structure for S is $O(\log^2 n + i) = O(i)$, since at this point we know that i is $\Omega(\log^2 n)$. Therefore, the overall query time is $O(\log n + i)$.

Theorem 67.12

Let S be a set of n colored points in $\mathbb{R}^2$. S can be preprocessed into a data structure of size $O(n)$ such that the i distinct colors of the points of S lying in any grounded query rectangle can be reported in $O(\log n + i)$ time.

We remark that the idea of exploiting the output size to gain efficiency can be traced back to Reference 47, where it was used within the framework of filtering search. It has manifested itself in various forms in several subsequent papers on standard and generalized range search, including References 21,24,25,41,48–53 among others.

67.3.7 A Reverse Transformation

In Section 67.3.1, we discussed how the generalized range counting problem in $\mathbb{R}^1$ can be transformed to a standard range counting problem in $\mathbb{R}^2$ (with a grounded query rectangle). In this section, we show a reverse transformation which maps a standard range counting problem in $\mathbb{R}^2$ to a generalized range counting problem in $\mathbb{R}^1$. The approach, which is based on Reference 21, shows that the two problems (i.e., generalized range counting in $\mathbb{R}^1$ and standard range counting in $\mathbb{R}^2$) are in fact equivalent and it also yields a lower bound for the former problem based on a known lower bound for the latter (in the word-RAM model). This reverse transformation is not a solution technique for generalized problems per se, but it does reveal an interesting connection between a generalized problem and a standard problem and prompts the question of whether other pairs of generalized and standard problems might share a similar connection.

We begin by noting that a standard range counting query in $\mathbb{R}^2$ with a four-sided (i.e., nongrounded) query rectangle can be answered by doing four standard quadrant counting queries, where each quadrant is defined by one of the vertices of the rectangle, and then adding and subtracting the counts returned using the principle of inclusion/exclusion. Therefore, in what follows, it suffices to focus on standard quadrant counting queries. Specifically, we wish to preprocess a set S of n points in $\mathbb{R}^2$ so that given any query quadrant $q = (-\infty, a] \times (-\infty, b]$, we can count efficiently the number of points of S that are in q. We assume, without loss of generality, that all coordinates are positive.

We map each point $p = (x(p), y(p)) \in S$ to two points in $\mathbb{R}^1$, one with coordinate $-x(p)$ and the other with coordinate $y(p)$. We give each of these points the color p (any unique identifier associated with p). Let S_1 be the set of these newly constructed $2n$ points in $\mathbb{R}^1$. Next we make a copy, S_2, of S_1 and recolor the points so that all of them have distinct colors. For each of S_1 and S_2, we build a data structure for answering generalized range counting queries in $\mathbb{R}^1$. (Note that the structure on S_2 is actually a standard structure since all colors in S_2 are distinct, but it is convenient for our purposes to view it as a generalized structure.)

Given q, we query the two data structures with $q' = [-a, b]$ to obtain two integers t_1 and t_2, where t_1 is the the number of distinct colors among the points of S_1 in q' and t_2 is the number of distinct colors among the points of S_2 in q' (hence also the number of points of S_2 in q', since the points in S_2 have distinct colors). We report $t_2 - t_1$ as the answer to the query q on S, that is, the number of points of S that are in q.

The correctness of this can be seen as follows: If p is in q, then we have $x(p) \leq a$ and $y(p) \leq b$, that is, $-a \leq -x(p)$ and $y(p) \leq b$, that is, $-a \leq -x(p) < b$ and $-a < y(p) \leq b$ (since all coordinates are positive). Hence the points $-x(p)$ and $y(p)$ are both in $q' = [-a, b]$. Thus this pair contributes 2 to t_2 and 1 to t_1, and it follows that $t_2 - t_1$ correctly returns the count of the points of S that are in q. On the other hand, if p is not in q, then we have $-a > -x(p)$ or $y(p) > b$ (or both). In this case, the pair contributes the same amount to both t_2 and t_1 (1, if exactly one inequality holds, and 0 if both hold), so $t_2 - t_1$ again returns the correct overall count.

67.4 Conclusion and Future Directions

We have reviewed research on a class of geometric query-retrieval problems, where the objects to be queried come aggregated in disjoint groups and of interest are questions concerning the intersection of the query object with the groups (rather than with the individual objects). These problems include the well-studied standard intersection problems as a special case and have many applications. We have described several general techniques that have been identified for these problems and have illustrated them with examples.

Some potential directions for future work include: (i) extending the transformation-based approach to higher dimensions; (ii) improving the time bounds for some of the problems discussed here—for instance, can the generalized orthogonal range searching problem in $\mathbb{R}^d$, for $d \geq 4$, be solved with $O(\text{polylog}(n) + i)$ query time and $O(n(\log n)^{O(1)}n)$ space; (iii) developing general dynamization techniques for generalized problems, along the lines of, for instance, Reference 54 for standard problems; (iv) developing efficient solutions to generalized problems where the objects may be in time-dependent motion; (v) identifying other pairs of standard and generalized problems (beyond the pair discussed in Section 67.3.7) that are equivalent; and (vi) implementing and testing experimentally some of the solutions presented here.

Acknowledgments

Parts of the material in this chapter are drawn from the publications of some of the chapter's authors: References 13–15, with permission from Elsevier (http://www.elsevier.com), and Reference 16, with permission from Taylor & Francis (http://www.tandf.co.uk). The work of one of the authors (S. Rahul) is supported in part by a Doctoral Dissertation Fellowship from the Graduate School of the University of Minnesota.

References

1. P. K. Agarwal and J. Erickson. Geometric range searching and its relatives. In B. Chazelle, J. E. Goodman, and R. Pollack, editors, *Advances in Discrete and Computational Geometry*, volume 223 of *Contemporary Mathematics*, pages 1–56. American Mathematical Society, Providence, RI, 1999.

2. M. de Berg, M. van Kreveld, M. Overmars, and O. Schwarzkopf. *Computational Geometry: Algorithms and Applications*. Springer-Verlag, Berlin, Germany, 2nd edition, 2000.

3. R. Janardan and M. Lopez. Generalized intersection searching problems. *International Journal of Computational Geometry and Applications*, 3:39–69, 1993.

4. P. K. Agarwal, S. Govindarajan, and S. Muthukrishnan. Range searching in categorical data Colored range searching on grid. In *Proceedings of the 10th European Symposium on Algorithms, volume 2461 of Lecture Notes in Computer Science*, Berlin, 17–28, Springer-Verlag, 2002.

5. P. K. Agarwal and M. van Kreveld. Polygon and connected component intersection searching. *Algorithmica*, 15:626–660, 1996.

6. P. Bozanis, N. Kitsios, C. Makris, and A. Tsakalidis. New upper bounds for generalized intersection searching problems. In *Proceedings of the 22nd International Colloqium on Automata, Languages and Programming*, volume 944 of Lecture Notes in Computer Science, pages 464–475, Berlin, 1995. Springer-Verlag.

7. P. Bozanis, N. Kitsios, C. Makris, and A. Tsakalidis. Red-blue intersection reporting for objects of non-constant size. *The Computer Journal*, 39:541–546, 1996.

8. P. Bozanis, N. Kitsios, C. Makris, and A. Tsakalidis. New results on intersection query problems. *The Computer Journal*, 40:22–29, 1997.

9. M. de Berg and H. Haverkort. Significant-presence range queries in categorical data. In *Proceedings of the 8th International Workshop on Algorithms and Data Structures*, pages 462–473. Springer, Ottawa, 2003.

10. A. Elmasry, M. He, J. Munro, and P. Nicholson. Dynamic range majority data structures. In *Proceedings of the 22nd International Symposium on Algorithms and Computation*, pages 150–159. Springer, Yokohama, 2011.

11. P. Ferragina, N. Koudas, S. Muthukrishnan, and D. Srivastava Two-dimensional substring indexing. In *Proceedings of the 20th ACM Symposium on Principles of Database Systems*, Santa Barbara, pages 282–288, 2001.

12. P. Gupta, Efficient algorithms and data structures for geometric intersection problems. Ph.D. dissertation, Department of Computer Science, University of Minnesota, Minneapolis, MN, 1995.

13. P. Gupta, R. Janardan, and M. Smid. Further results on generalized intersection searching problems: Counting, reporting and dynamization. *Journal of Algorithms*, 19:282–317, 1995.

14. P. Gupta, R. Janardan, and M. Smid. Algorithms for generalized halfspace range searching and other intersection searching problems. *Computational Geometry: Theory and Applications*, 5:321–340, 1996.

15. P. Gupta, R. Janardan, and M. Smid. A technique for adding range restrictions to generalized searching problems. *Information Processing Letters*, 64:263–269, 1997.

16. P. Gupta, R. Janardan, and M. Smid. Algorithms for some intersection searching problems involving circular objects. *International Journal of Mathematical Algorithms*, 1:35–52, 1999.

17. P. Gupta, R. Janardan, and M. Smid. Efficient algorithms for counting and reporting pairwise intersections between convex polygons. *Information Processing Letters*, 69:7–13, 1999.

18. H. Kaplan, N. Rubin, M. Sharir, and E. Verbin. Counting colors in boxes. In *Proceedings of the Annual ACM–SIAM Symposium on Discrete Algorithms*, pages 785–794, 2007.

19. H. Kaplan, M. Sharir, and E. Verbin. Colored intersection searching via sparse rectangular matrix multiplication. In *Proceedings of the ACM Symposium on Computational Geometry*, Sedona, pages 52–60, 2006.

20. Y. Lai, C. Poon, and B. Shi. Approximate colored range and point enclosure queries. *Journal of Discrete Algorithms*, 6(3):420–432, 2008.

21. K. G. Larsen and F. van Walderveen. Near-optimal range reporting structures for categorical data. In *Proceedings of the Annual ACM–SIAM Symposium on Discrete Algorithms*, New Orleans, pages 265–276, 2013.

22. S. Muthukrishnan. Efficient algorithms for document retrieval problems. In *Proceedings of the 13th ACM–SIAM Symposium on Discrete Algorithms*, San Francisco, pages 657–666, 2002.

23. G. Navarro and Y. Nekrich. Top-k document retrieval in optimal time and linear space. In *Proceedings of the Annual ACM–SIAM Symposium on Discrete Algorithms*, Kyoto, pages 1066–1077, 2012.

24. Y. Nekrich. Space-efficient range reporting for categorical data. In *Proceedings of the 31st symposium on Principles of Database Systems*, Scottsdale, pages 113–120, 2012.

25. Y. Nekrich. Efficient range searching for categorical and plain data. *ACM Transactions on Database Systems*, 39(1):9, 2014.

26. M. Patil, S. V. Thankachan, R. Shah, Y. Nekrich, and J. S. Vitter. Categorical range maxima queries. In *Proceedings of the ACM Symposium on Principles of Database Systems*, Snowbird, pages 266–277, 2014.

27. Q. Shi and J. JáJá. Optimal and near-optimal algorithms for generalized intersection reporting on pointer machines. *Information Processing Letters*, 95(3):382–388, 2005.

28. M. J. van Kreveld. New Results on Data Structures in Computational Geometry. Ph.D. dissertation, Department of Computer Science, Utrecht University, The Netherlands, 1992.

29. J. S. Vitter. External memory algorithms and data structures: Dealing with massive data. *ACM Computing Surveys*, 33:209–271, 2001.

30. G. Navarro. Spaces, trees, and colors: The algorithmic landscape of document retrieval on sequences. *ACM Computing Surveys*, 46(4):52, 2014.

31. P. K. Agarwal, M. de Berg, S. Har-Peled, M. H. Overmars, M. Sharir, and J. Vahrenhold. Reporting intersecting pairs of convex polytopes in two and three dimensions. *Computational Geometry: Theory and Applications*, 23:195–207, 2002.

32. A. Aggarwal and J. Vitter. The input/output complexity of sorting and related problems. *Communications of the ACM*, 31(9):1116–1127, 1988.

33. T. Hagerup. Sorting and searching on the word RAM. In *Proceedings of the 15th Annual Symposium on Theoretical Aspects of Computer Science*, pages 366–398. Springer, Paris, 1998.

34. E. M. McCreight. Priority search trees. *SIAM Journal on Computing*, 14:257–276, 1985.

35. B. Chazelle and L. J. Guibas. Fractional cascading: I. A data structuring technique. *Algorithmica*, 1:133–162, 1986.

36. D. E. Willard and G. S. Lueker. Adding range restriction capability to dynamic data structures. *Journal of the ACM*, 32:597–617, 1985.

37. H. Edelsbrunner, *Algorithms in Combinatorial Geometry*. Springer-Verlag, New York, 1987.

38. B. Chazelle, L. J. Guibas, and D. T. Lee. The power of geometric duality. *BIT*, 25:76–90, 1985.

39. J. Matoušek. Range searching with efficient hierarchical cuttings. *Discrete & Computational Geometry*, 10:157–182, 1993.

40. S. W. Cheng and R. Janardan. Algorithms for ray-shooting and intersection searching. *Journal of Algorithms*, 13:670–692, 1992.

41. P. K. Agarwal, S. W Cheng, Y. Tao, and K Yi. Indexing uncertain data. In *Proceedings of the ACM Symposium on Principles of Database Systems*, Providence, pages 137–146, 2009.

42. J. Matoušek. Efficient partition trees. *Discrete & Computational Geometry*, 8:315–334, 1992.

43. J. Matoušek. Cutting hyperplane arrangements. *Discrete & Computational Geometry*, 6:385–406, 1991.

44. J. R. Driscoll, N. Sarnak, D. D. Sleator, and R. E. Tarjan. Making data structures persistent. *Journal of Computer and System Sciences*, 38:86–124, 1989.

45. S. W. Cheng and R. Janardan. Efficient dynamic algorithms for some geometric intersection problems. *Information Processing Letters*, 36:251–258, 1990.

46. J. Matoušek and O. Schwarzkopf. On ray shooting in convex polytopes. *Discrete & Computational Geometry*, 10:215–232, 1993.

47. B. Chazelle. Filtering search: A new approach to query-answering. *SIAM Journal on Computing*, 15(3):703–724, 1986.

48. P. Afshani, L. Arge, and K. G. Larsen. Higher-dimensional orthogonal range reporting and rectangle stabbing in the pointer machine model. In *Proceedings of the ACM Symposium on Computational Geometry*, Chapel Hill, pages 323–332, 2012.

49. P. Afshani and T. M. Chan. Optimal halfspace range reporting in three dimensions. In *Proceedings of the Annual ACM–SIAM Symposium on Discrete Algorithms*, New York, pages 180–186, 2009.

50. P. K. Agarwal, L. Arge, and K. Yi. An optimal dynamic interval stabbing-max data structure?. In *Proceedings of the Annual ACM–SIAM Symposium on Discrete Algorithms*, Vancouver, pages 803–812, 2005.

51. A. Agrawal, S. Rahul, Y. Li, and R. Janardan. Range search on tuples of points. *Journal of Discrete Algorithms*, 2014. doi:10.1016/j.jda.2014.10.006.

52. S. Rahul. Improved bounds for orthogonal point enclosure query and point location in orthogonal subdivisions in R^3. In *Proceedings of the 26th Annual ACM–SIAM Symposium on Discrete Algorithms*, San Diego, 2015.

53. Y. Tao, Stabbing horizontal segments with rays. In *Proceedings of the ACM Symposium on Computational Geometry*, Chapel Hill, pages 313–322, 2012.

54. J. L. Bentley, J. B. Saxe. Decomposable searching problems. I: Static-to-dynamic transformations. *Journal of Algorithms*, 1:301–358, 1980.

Index

Note: Page numbers followed by "*fn*" indicate footnotes.

A

AA-trees, 157
AABB-trees, 256, 893
AABB, *see* Axis-aligned bounding box
ABIT, *see* Alternative BIT structure
Abstract data types (ADT), 77, 79, 117, 197, 595, 653, 655
(a, b)-trees, 166–167
Accept nodes, 477
Access functions, 279, 282
Access operations, 182–183
Accessors, 681–682
Access structure, 260
Accounting method, 14
    maintenance contract, 15–16
    McWidget company, 17–18
    subset generation, 19
ACID semantics, *see* Atomicity, Consistency, Isolation,
        Durability semantics
Ackermann's function, 538, 1014
    functional inverse of, 95
    inverse, 322
Active pair, 225
Active vertex, 190
Actual complexity, 14
Acyclic digraphs, 58
Acyclic graph, 53
Adaptive merging algorithm, 176–177
Adaptive multidimensional histogram (AMH), 367
Adaptive-phi, 986
Adaptive properties of pairing heaps, 94
Adaptive sorting algorithm, 176–177
Additional operations, 69, 343, 389, 395, 540, 599, 668
Additive weights, 1022
Address computation methods, 253
Adjacency
    lists, 50, 51
    lists representation, 277
    matrix, 50, 51
Adjacent polygons, 872
ADT, *see* Abstract data types
Advanced geometric data structures, 656
Advanced operations, 350
    nearest neighbor queries, 350–351
    spatial joins, 351–353
Agglomerative clustering, 1007, 1008

Aggregate method, 14
    maintenance contract, 15
    McWidget company, 17
    subset generation, 19
Aggregate query, 367
Aggregation
    B+-tree, 246
    operators, 245
    query in B+-tree, 245–247
Air-traffic controls, 370
AlgoComs, 884
Algorithmic Solutions Software GmbH, 654
Algorithm(s), 656, 679, 683–684, 688, 948–949
    animation, 697
    designers, 7
    Dijkstra's, 689
    euler tour tree traversal, 688
    graph traversals, 688–689
    iterator-based tree traversals, 688
    Prim-Jarník, 689
    sequence sorting, 688
    topological numbering, 689
    visualization, 697
All-pairs approximate shortest paths (APASP), 611
All-pairs shortest path, 64–65
All-pairs shortest paths, 591
    updates in $O(n^{2.5}C\log n)$ time, 592
    updates in $O(n^2 \log^3 n)$ time, 592
AllocatePage function, 234
Almost-linear-time algorithm, 951
α-balanced region, 406
    canonical region, 413
Alphabetic tree problem, 224
Alpha channel, 859
Alternative BIT structure (ABIT), 775
Alternative node structures, 447–449
ALU, *see* Arithmetic and logic unit
Amalgamated node, 950
Ambivalent data structures, 574, 582; *see also* Cache-oblivious
        data structures; Randomized graph
        data-structure
Amethyst system, 704
AMH, *see* Adaptive multidimensional histogram
Amortization, 644–648
Amortized analysis, 78–79

Amortized complexity, 14
    maintenance contract, 15–16
    McWidget company, 16–18
    subset generation, 18–20
Amortized cost, 14–19, 79, 94, 98, 121, 160, 161, 164, 185, 427,
      532–534
    analysis of Insert or DeleteMin operation, 552, 556–557, 559–560
    of meld operation, 82–83
    of single splaying, 188
Amortized time, 69, 90, 181–182
Amortized update cost to worst-case, 633–634
Analogous methods for 3D polytopes, 383
Analysis of algorithms, 3
    amortized complexity, 14–20
    asymptotic complexity, 9–11
    counting cache misses, 7–9
    operation counts, 4–5
    practical complexities, 20–21
    recurrence equations, 12–13
    step counts, 5–7
Analytical models of R-trees, 353–356
Analytics workloads, 987
Ancestor, 714–715
    node, 36
Anchor, 923
Angular resolution, 723
Annulus wedge, 719
Anomaly detection, 999
Anti-monotone property of itemset support, 1003
Apache Zookeeper projects, 986
APASP, *see* All-pairs approximate shortest paths
Append operation, 295, 296
Applicability criteria, 891
"Apply" algorithm, 506
Apportion, 715
Approximate distance oracle, 611, 616
    computing $Ball^i(u)$ efficiently, 616–618
    computing $p^i(u)$, $\forall u \in V$, 616
Approximate geometric query structures
    approximate queries, 406–409
    BAR trees, 412–416
    BBD trees, 410–411
    general terminology, 406
    maximum-spread k-d trees, 416
    quasi-BAR bounds, 409–410
Approximate nearest neighbor searching, 1037–1038
Approximate pattern matching, 473–474
Approximate queries, 406–409
Approximate range query, 407
Approximates optimal Huffman code, 230
Approximate Voronoi diagram (AVD), 1038; *see also*
      Voronoi diagrams
Approximation algorithm, 389
Approximation error factor, 1038
Apriori algorithm, 1003, 1004, 1005
AQT, *see* Area-based Quadtree
Arbitrarily oriented objects, 1045
Arbitrarily shaped queries, 345
Arbitrary element deletion from Max HBLT, 74
Arbitrary merging order, 175–176
Arbitrary range minimum query, 472
Arbitrary space decomposition, 262

Arbitrary weights, 63–64
ArcGIS, 883
Archetypical data structure, 595
Architectural design, 817
Arc of dynamic trees, 193
Area-based Quadtree (AQT), 790–791, 792
Area Searches operation, 820
Arithmetic and logic unit (ALU), 7
Arithmetic model, 530, 538
Arithmetic operation, 7
Arrangement(s), 288, 385, 1014
    of curves in plane, 1014
    decomposition, 1015–1016
    duality, 1016–1017
    of lines, 288
    maximum complexity of single cells and envelopes in
      arrangements, 1015
    substructures and complexity, 1014–1015
Array(s), 23, 468, 607
    access structure, 253
    of arrays representation, 26
    doubling, 25
    dynamic, 607
    handling, 518
    heterogeneous arrays, 25–26
    implementation of queue, 33
    index, 24
    multidimensional, 26–27
    multiple lists in single array, 25
    operations on, 24–25
    resizable, 607
    sorted, 25
    sparse matrices, 27
ArrayHeap, 687
ArrayQuickSort, 688
ArraySequence, 686
Assignment pedigree, 519
Association analysis, 1002
    basic K-means algorithm, 1007
    enumerating subsets of three items from transaction, 1005
    FP-tree structure, 1005–1007
    hashing transaction at root node of hash tree, 1004
    hash tree structure, 1004–1005
    subset operation on left most subtree of root of candidate hash
      tree, 1006
Association rule, 1003
    mining, 998–999
Associative containers, 667, 669
    maps and multimaps, 670–672
    sets and multisets, 669–670, 671
"Assume, verify, and conquer" strategy (AVC strategy), 392
Asymmetric hashing, 125
Asymmetric matrix, 959
Asymmetric multifrontal algorithm, elimination structures
      for, 962
Asymptotic complexity
    big oh notation (O), 9–11
    little oh notation (o), 11
    omega and theta notations, 11
Asymptotic notation, 10
Atomicity, Consistency, Isolation, Durability semantics (ACID
      semantics), 983

Attribute, 278
    edge, 278, 280, 281
    face, 278, 280, 281
    geometric, 378
    geometric, 871
    primary, 252
    vertex, 278, 280, 281
Augmented ephemeral data structure, 517
Autocorrelation factor, 919
Automatic Graph Drawing, 707
Automatic translation techniques, 819
Automorphism, 937
    geometric, 728, 729
    group, 937
Auxiliary 3D R-tree, 369
AVC strategy, *see* "Assume, verify, and conquer" strategy
AVD, *see* Approximate Voronoi diagram
Average path length, 810
AVL-trees, 155–156, 159, 164, 166, 179, 180, 389, 390, 708
Axes-parallel objects, 1044–1045
Axial case, 729
Axial symmetry, 728, 730–732
Axioms of probability, 199
Axis-aligned bounding box (AABB), 256, 893, 896

B

Back-end structure, 774
Back edges, 54
Backoff technique, 743
Backward analysis, 207
Backward finger search, 174
Backward pointers, 456
Bad nodes, 910
Balanced aspect ratio tree (BAR tree), 405, 412–416, 426, 1038; *see also* Binary space partitioning trees (BSP trees); R-trees*
    construction algorithm, 414
Balanced binary search trees (BBST), 78, 151–152, 183, 233, 316, 389, 644; *see also* Finger search trees
    application to multi-dimensional search trees, 161–162
    balancing, 153–155
    binary trees as dictionaries, 152–153
    classic balancing schemes, 155–158
    general balanced trees, 160–161
    implementation of binary search trees, 153
    implicit representation of balance information, 159–160
    low height schemes, 162–164
    rebalancing tree to perfect balance, 158
    relaxed balance, 164–167
    schemes with no balance information, 158
Balanced binary tree, 574
    based on multi-way trees, 157–158
Balanced box decomposition tree (BBD tree), 405, 410–411, 412, 426, 1038; *see also* Binary space partitioning trees (BSP trees); R-trees*
Balanced BST implementation, 393–395
Balance definitions, 153, 154
    in AVL-trees, 155
    rebalancing tree to perfect, 158
Balanced parenthesis representation, 600

Balanced priority search tree, 1046
Balanced search tree, 554
Balanced structures, 390, 391, 393
Balanced tree, 331, 599
Balance information, implicit representation of, 159–160
Balance of node, 156
Balancer, 750
Balancing, 153
    balance definitions, 154
    complexity results, 155
    rebalancing algorithms, 154–155
$Ball^i(u)$ efficiently computiation, 616–618
BANG file, 258
Barriers, 748
Bars", horizontal line segments, 732
BAR tree, *see* Balanced aspect ratio tree
Bar visibility algorithm, 734–735
Bar visibility representations
    and layered drawings, 735–736
    for orthogonal drawings, 736–737
Barycenter algorithm, 726
Base repertoire of actions, 529
Base repertoire plus member, 529
BasicintervalPointer, 774
Basic Library (*libL*), 654
Basic search algorithms, 783
Basket weaving, 338
Batched dynamic operations, 427
Batched filtering, 422
Batched geometric problems, EM algorithms for, 421–424
Batched incremental construction, 422
Batched problems, 420, 421
Batched techniques, 421
Batcher's merging technique, 635
Bayes' rule, 200
BB[α]-trees, *see* Weight-balanced trees
BBD tree, *see* Balanced box decomposition tree
BBST, *see* Balanced binary search trees
BDDs, *see* Binary decision diagrams
Beam-tracing based algorithm, 334
Beam tracing, 336
Benes network, 604–605
Bernoulli distribution, 202, 918, 919
Bernoulli Trial, *see* Coin flip
Beyond planar graphs, 738
Beyond planarity, 738
Bezier curves, 864–867
BFS, *see* Breadth-first search
Bibliometrics, 805
Biclique covers, 958–959
Bidirectional iterators, 674
Big data stores, data structures for, 983
    concurrency, 992–993
    data models, 984
    partitioning, 984–985
    persistence, 987–992
    replication and consistency, 985–987
Big data technologies, 872
Big oh notation (*O*), 9–11
BigTable, 989
Binarizations of multi-way search tree schemes, 154
Binary B-tree, 157

Binary decision diagrams (BDDs), 495
    BDD-based techniques, 495
    concepts, 495–498
    construction from Boolean expressions, 499–502
    data structure, 498–499
    package, 506
    related research activities, 506–507
    shared BDD, 498
    truth table and, 496
    for typical Boolean functions, 498
    variable ordering for, 502–503
    Winter of BDDs", 507
    zero-suppressed BDD, 504–506
Binary dispatching problem, 514
Binary encoding, 800
Binary image, 904
Binary interval tree (BIT), 774
Binary large objects (blobs), 977
Binary logic operation, 500–501
Binary mergers, 548–549
Binary merge sort, 175
Binary merge trees, 548
Binary search recurrence of equation, 13
Binarysearchtree-on-Trie (BoT), 793
Binary search trees (BSTs), 42, 152, 209, 223, 253, 329,
            331–332, 421, 447, 642–644, 774–775
    algorithm, 470
    delete, 43
    implementation, 153
    implementations, 391–393
    insert, 42–43
    miscellaneous, 43–44
    search, 42
Binary space partitioning trees (BSP trees), 261, 262,
            329–330, 340, 384, 406, 416, 867–868, 1032;
            see also R-trees*
    boundary representations *vs.* BSP trees, 340
    converting B-Reps to trees, 339–340
    elementary operation, 331
    good BSP trees, 337–338
    as hierarchy of regions, 334
    of inter-object spatial relations, 330
    of intra-object spatial relations, 330
    merging BSP trees, 336
    multi-dimensional search structure, 331–332
    multiresolution representation, 335
    plane power, 330
    tree merging, 335–337
    visibility orderings, 332–334
BinaryTree, 686
Binary tree on binary tree (BOB), 777
Binary trees, 35–37, 39, 152, 470, 599, 708
    data structure, 427
    as dictionaries, 152–153
    hashing, 127
    level layout for, 709–710
    properties, 38–39
Binary tree traversals, 39
    inorder traversal, 40
    level order traversal, 41
    postorder traversal, 40–41
    preorder traversal, 40

Binary trie, 448, 449, 531, 533, 603
    compressed, 456, 457
Binary union tree, 538
Binomial coefficients, 202
Binomial distribution, 135, 137, 140, 202, 206
    negative, 202–204
Binomial heaps, 69, 85–89; *see also* Skew heaps
Binomial trees, 85–86, 87
Bintree, 254, 261
Biological sequence data, 917
Biological taxonomies, 332, 1007
Bipartite cores, 805
Bipartitions, 540–541
Bisector list implementation in HVT, 829
Bisector List Quad Trees (BLQT), 264, 884
Bisector List Quad Trees (BLQT), 826
Bisectors, 826
Bit-wise boolean operations, 595
BIT, *see* Binary interval tree
Bitbound algorithm, 940, 943
Bit collision", 146
Bitmap-intersection, 796–797
Bitmap graphics, *see* Raster graphics
Bitmap index, 970
Bits, 604
Bit string, 317
Bit table, 975
Bitvector, 596–597
Black box", 179
blobs, *see* Binary large objects
BlobWorld system, 915
Block(s), 234, 833–834, 974, 976, 980
    decomposition, 261
    size, 988
Blocked bloom filter, 136
    bloom-1 filter, 136–137
    bloom-1 *vs.* bloom with optimal k, 139–140
    bloom-1 *vs.* bloom with small k, 138–139
    bloom-g, 140–141
    bloom-g in dynamic environment, 144
    bloom-g *vs.* bloom with optimal k, 142–144
    bloom-g *vs.* bloom with small k, 141–142
    discussion, 144
    impact of word size, 137–138
Blocking phenomenon, 743
Blocking techniques, 743–744
Bloom-1 filter, 136–140
Bloom-$g$filter, 140–141
    vs. bloom with optimal k, 142–144
    vs. bloom with small k, 141–142
    in dynamic environment, 144
Bloom filter, 131, 132, 990
    black box view, 132
    blocked bloom filter, 136–144
    CBF, 135–136
    for dynamic set, 146–147

false-positive ratio and optimal *k*, 133–134
  performance metrics, 133
Bloom filter variants, 136, 144
  bloom filter for dynamic set, 146–147
  improving read/write efficiency, 145–146
  improving space efficiency, 145
  reducing false-positive ratio, 145
  reducing hash complexity, 146
BloomFlash, 146
Bloom with optimal *k*
  bloom-1 *vs.*, 139–140
  bloom-*g* *vs.*, 142–144
Bloom with small *k*
  bloom-1 *vs.*, 138–139
  bloom-*g* *vs.*, 141–142
BLQT, *see* Bisector List Quad Trees
BOB, *see* Binary tree on binary tree
Boolean" objects, 868
Boolean AND, 800
Boolean expressions, BDDs construction from, 499
  binary logic operation, 500–501
  complement edges, 501–502
  primitive operations for BDD manipulation, 499
Boolean function, 495–496
  systematic methods for, 495
Boolean NOT, 800
Boolean operations, 868, 873
Boole's Inequality, *see* Sub additivity
Bootstrapping
  paradigm, 428, 431, 432
  for three-sided orthogonal 2-D range search, 432–433
Boruvka's MST algorithm, 62
BoT, *see* Binarysearchtree-on-Trie
Bottom-up
  algorithms, 393
  construction, 316
  methods, 430, 893
  splay trees, 195
  traversal, 470
Bottom-vertex triangulations, 1015–1016
Bottom trees, 549
Boundary-based representation, 259
Boundary-labeling problem, 873
Boundary model, 266
Boundary of face, 891
Boundary representation, 339, 340
  vs. BSP trees, 340
Boundary vertices, 574
Bounded-sliceline grid (BSG), 833, 844, 847–848
Bounded aspect ratio, 434
Bounded Quad Trees (BQT), 828–829
Bounded slicing grid, 835
Bounding
  box, 437
  functions, 350
  size of $B_u^{d,S}$ under edge-deletions, 621
  sphere, 383
Bounding volume hierarchies (BVHs), 340, 383, 892, 896
Bounding volumes (BVs), 334, 892
  choice of, 893
Bounds from computational geometry, 786
Bowtie2 alignment tool, 927

BQT, *see* Bounded Quad Trees
Branch-and-bound algorithm, 351
Branch-node pointers, 456
Breadth-first search (BFS), 54, 56, 578–579
  simple applications of, 56
  topological sorting, 57–58
  tree, 618, 622
Breadth-first traversal strategy, 880
Brent's method, 124, 127
BRep, *see* winged-edge data structure
Brightness value, 904
Broad phase, 898
Brute force algorithm, 345
BSG, *see* Bounded-sliceline grid
B-spline curves, 867
BSP trees, *see* Binary space partitioning trees
BSTs, *see* Binary search trees
B+-tree, 233, 239, 245, 911, 987, 988
  aggregation query in, 245–247
  bulk-loading, 245
  copy-up and push-up, 240
  deletion, 242–244
  insertion, 241–242
  query, 240–241
B-tree based technique, 360
B-trees, 157, 166, 233, 234–235, 389, 425–426, 878, 969, 970;
      see also Splay trees
  B+-tree, 239–244
  deletion, 237–239
  disk-based environment, 233–234
  efficiency analysis, 245
  family, 874
  insertion, 236–237
  query, 235–236
  structure, 421
  variants, 360
B*-trees, 233, 425, 844
Bubble sort, 25
Bucketing, 163
  methods, 252, 256–259
Bucket(s), 252, 254, 562, 980, 988
  directory, 980
Buddy trees, 429
Buffer management, data structures for, 974
  direct page addressing, 974, 975
  indirect page addressing, 975
  shadow paging concept, 976
Buffer Manager, 968
Buffer(s), 356, 523, 557, 968
  pool, 234
  trees, 427–428
  zone, 407
Bugs, 640
Bulk dynamic operations, 430
Bulk loading, 349, 884
  B+-tree, 245
"Bump" mapping, 862
Burrows–Wheeler transform, 927, 928
Bursty access pattern, 775*fn*
BVHs, *see* Bounding volume hierarchies
BVs, *see* Bounding volumes
BWA-MEM alignment tool, 927

C

C++, data structures in
    components of STL, 676–677
      containers, 667–673
      iterators, 674–676
C++ programming language, 23, 679
Cache-coherent multiprocessor, 743
Cached assertions, 378
Cache hierarchy, utilizing, 864
Cache-oblivious algorithm, 545–546
Cache-oblivious data structures; *see also* Randomized graph
        data-structure
    cache-oblivious model, 545–546
    dynamic B-trees, 550–553
    fundamental primitives, 546
    k-merger, 546, 548–550
    priority queues, 554–560
    2d orthogonal range searching, 560–563
    van Emde Boaz layout, 546–548
Cache-oblivious kd-tree, 560
    construction, 561
    query, 560–561
    structure, 560
    updates, 561
Cache-oblivious model, 545–546
Cache-oblivious range tree, 561
    four-sided queries, 563
    three-sided queries, 562–563
Cache misses effect on run time, 7–8
Caching mechanisms, 360
CAD, *see* Computer aided design
CAD/CAM drawings, 858–859
CAGD, *see* Computer-aided geometric design
Calinescu's algorithm, 541
CAM, *see* Computer Aided Manufacturing
Candidate itemsets, 1004
Canonical bounding cuts, 412
Canonical covering, 294, 295
Canonical cuts, 412
Canonical geometric structures, 386
Canonical label, 936
Canonical labeling methods, 937
    Morgan's algorithm, 937
    Nauty, 937–938
    signature-based canonization, 938
Canonical order, 605
    algorithm using, 727–728
Canonical region, 412
Canonical representations, 936–937
Canonical spatial join algorithm, 352
Canonical triangulation, 399, 400
Capacitance, 818
Capacity of cut, 190
CAP theorem, *see* Consistency, availability, and partition tolerance
        theorem
Cardinality degree, 571
Cardinal trees, 599, 601–602
CartoDB, 883
Cartographic generalization, 873
Cartography, *see* Map visualization
CAS, *see* Chemical abstracts service; Compare-and-swap
CAS-based lock-free linked list, 753–754

Cascade hashing, 127
Cascading cuts, 90, 91
Cassandra, 986, 989
Catalan numbers, 39
Catenate, 522
Causteljau algorithm, 866
CBAC, *see* Context-based access control
CBF, *see* Counting Bloom filter
CBIR systems, *see* Content-Based Image Retrieval systems
CBL, *see* Corner block list
Cell(s), 312, 911, 1014
    complex, 1028
    deletion, 317
    insertion, 317
    orderings, 314–315
    probe model, 530
    queries, 316–317
Central projection, 897
Centroid shrink, 410
Certificate, 378
CGAL, *see* Computational geometry algorithms library
Chain codes, 265
Chaining, hashing with, 122
Chains, 28–29, 30
Characteristic functions, 504
Character large objects!character (clobs), 977
Chemical abstracts service (CAS), 935
Chemical fingerprints, 939
    bitbound algorithm, 940
    combining hashing and trees, 943
    fingerprint compression, 940
    hashing approaches for improved searches, 940–941
    inverted indices, 942
    LINGO, 942
    multibit and union tree, 943
    Tanimoto similarity, 939–940
    triangle inequality and improvements, 941
    trie-based approaches for storing fingerprints, 943
    XOR filter, 941–942
Chemical graph, 936
Chemical space, 935
Cheminformatics, 935
    bitbound algorithm, 940
    chemical fingerprints and similarity search, 939
    combining hashing and trees, 943
    exact searches, 936–938
    fingerprint compression, 940
    hashing approaches for improved searches, 940–941
    inverted indices, 942
    LINGO, 942
    multibit and union tree, 943
    tanimoto similarity, 939–940
    triangle inequality and improvements, 941
    trie-based approaches for storing fingerprints, 943
    XOR filter, 941–942
Chernoff bounds, 203
Child cache, 432
Child deque, 523
childNode, 448
Children node, 36, 92
Chip multithreading (CMT), 741
Cholesky factorization, 946

Chord, 954

Chordal graphs, 954–955

Circular deque, 607

Circular lists, 29–30

C language, 640

Classical binary merge sort algorithm, 176

Classical $O(n \log n)$ time sorting algorithm, 175

Classical union-find problem and variants, 539–540

Classic balancing schemes, 155
> AVL-trees, 155–156
> balanced binary trees based on multi-way trees, 157–158
> weight-balanced trees, 156–157

Classic search trees, 164

Classification, 1000
> k-nearest neighbor classification algorithm, 1001
> nearest-neighbor classifiers, 1001
> 1-, 2-and 3-nearest neighbors of instance, 1001
> proximity graphs for enhancing nearest neighbor classifiers, 1001–1002
> system, 541
> techniques, 998

Classifier, 783

Clique, 291

Clique covers, 945, 957
> covering column-intersection graph and biclique covers, 958–959
> problem of degree updates, 958
> quotient graphs, 957–958

Clique tree, 945, 954
> algorithm, 955
> chordal graphs and, 954–955
> compact clique trees, 956
> design of efficient algorithms with, 955–956

clobs, *see* Character large objects!character

Closely-related variables, 502

Cluster(s), 571, 574, 582
> analysis, 999
> changes, 384
> path, 574

Clustered graph drawing, 737

Clustered index, 980, 981

Clustering, 384, 582, 999, 1007
> hierarchical and partitional clustering, 1007–1008
> mobile nodes, 384
> multi-dimensional access methods, 1008–1010
> nearest neighbor search, 1008–1010
> technique, 574, 582

CLWS problem, *see* Concave Least Weight Subsequence problem

c-maximal point, 1050

CMT, *see* Chip multithreading

CNFs, *see* Conjunctive normal forms

Co-expressed genes, 918

Coarsest equitable partitions, 938

Coboundary of vertex, 891

Code fragments for sequential and lock-based fetch-and-inc operations, 742

Code Generator, 969

Coding information, 159

Cofactors, 496

Coin-collector, 215, 216

Coin collector problem, 221

> algorithm for, 221–222
> reduction to, 220–221

Coin flip, 202

Col-etree, *see* Column elimination tree

Collection of prefix trees (CPT), 774

Collision detection, 377, 383–384, 889, 893, 894–895
> convex polytopes, 890–892
> general polygonal models, 892–896
> large environments, 898–901
> PD computation, 896–898

Collisions, 532, 801, 867–868
> array, 749
> resolution strategy, 122

Color, 859

Colored intersection searching, *see* Generalized intersection searching

Colored problems, 1044

Color monitors, 859

Column-count of vertex, 951

Column-intersection graph, 958–959

Column-oriented symbolic factorization algorithms, 951

Column elimination tree (Col-etree), 945, 959–960
> asymmetric multifrontal algorithm, elimination structures for, 962
> elimination dags, 960–962

Column family stores, 984

Column indices, 950

Column intersection graph, 958, 961

Column major representation, 26

Column nonzero counts prediction, 951–952

Column orderings, 958

Column subgraph, 956

Combinatorial and geometric computing, 653, 654

Combinatorial complexity, 1014, 1028

Combining funnel, 749

Combining hashing and trees, 943

Combining nodes, 389

Combining techniques, 635

Combining tree approach, 743, 749

Command completion system, 451

Common-sense approach, 224

COMMON statements, 538

Communication systems, 359

Compact clique trees, 956

Compact DAWG, 478, 479, 486
> using compact DAWG to finding locations of string in text, 487–488
> edge labels representation, 488
> variations and applications, 488–489

Compaction, 819, 820

Compactness, 380

Compact quadtrees, 907, 908–909

Comparators, 683

Compare-and-swap (CAS), 744

Comparison-based model, 627, 634

Complementary range search problem, 104–105

Complemented edges, 501

Complement edges, 501–502

Complete binary tree, 599

Complete directed bipartite graph, 277

Complete subset graph, 536

Complex data, 704, 1000

Complexity, 679, 779
  measurement, 155, 545, 745
"Compound" objects, 868
Compressed binary trie, 456, 457
Compressed data structure, 435
Compressed path method, 522
Compressed quadtrees
  bottom-up construction, 316
  construction of, 315
  divide-and-conquer construction algorithm, 315–316
  and octrees, 313–314
Compressed tries, 452
  with digit numbers, 452–454
  with edge information, 454–456
  with skip fields, 454
  space required by compressed trie, 456
Compressed tries with digit numbers
  inserting into, 452–453
  removing element from, 453–454
  searching, 452
Compressed tries with edge information, 454
  inserting into, 455
  removing element from, 456
  searching, 455
"Compresses" multiple R-trees, 368
Compression process, 756
Computational biology, 917
  FM-index, 926–931
  unusual words, 917–923
  whole genomes comparison, 923–926
Computational geometry, 377, 1013
  arrangements, 1014–1017
  bounds from, 786
  convex hulls, 1017–1020
  motion in, 377
  triangulations, 1022–1025
  Voronoi diagrams, 1020–1022
Computational geometry algorithms library (CGAL),
    679, 872, 884
Computational platforms, 948
Computational tasks, 622
Computation model, 530, 627
Computer aided design (CAD), 343, 858
Computer-aided geometric design (CAGD), 377
Computer Aided Manufacturing (CAM), 858
Computer graphics, 857
  bitmaps, 859–860
  CAD/CAM drawings, 858–859
  CSG modeling, 868–869
  data structures, 862–867
  fonts, 859
  hardware and pipeline, 857–858
  hidden surface removal, 867
  meshes, 858
  proximity and collision, 867–868
  texture mapping, 861–862
Concave Least Weight Subsequence problem (CLWS problem), 230
Concave matrix multiplication techniques, 230
Concurrency, 992–993
Concurrent data structures
  barriers, 748
  blocking techniques, 743–744

complexity measures, 745
correctness, 745–746
designing, 741
hash tables, 754
linked lists, 753–754
locks, 747
nonblocking techniques, 744–745
performance, 742–743
pools, 752–753
priority queues, 756–757
search trees, 754–756
shared counters and fetch-and-ϕ structures,
    749–750
stacks and queues, 751–752
tools of trade, 746
transactional synchronization
    mechanisms, 748
verification techniques, 746
Concurrent queue implementation, 748–749
Condensing, 1002
Conditional probability, 200–201
Confidence bounds, 199
Conflict-free ranges, 780
Confluently persistent data structures, 511, 518–522
Conjugated systems, 936
Conjunctive normal forms (CNFs), 501, 506
Connected components, 57
  labeling, 322
Connected graph, 53
Connectivity, 51–53, 384, 587
  updates in $O(\log^2 n)$ time, 587–588
Consistency, availability, and partition tolerance theorem (CAP
    theorem), 983
Consistent hashing, 984–985
  key assignment, 985
  node addition, 986
Consolidation process, 91
Constant, 478
Constant size, 401
  alphabet, 463
Constant time, 381, 892
Constrained MST, 62
Constraint graphs, 833, 838
  generation, 838–839
  relationship between constraint graph based representation, 845
  triangulation, 839
  Tutte's duality, 839–841
Constraint plane, 891
Construction
  in RAM model, 561
  of topology trees, 572
Constructive Solid Geometry (CSG), 259, 868–869
Constructors, 641
Contact determination, *see* Collision detection
Container adapters, 667, 672
  priority_queue, 672–673
  stack and queue, 672
Containers, 667, 681–682
  associative containers, 669–672
  classes, 667
  container adapters, 672–673
  sequence containers, 668–669

Content-Based Image Retrieval systems (CBIR systems), 914, 915
    computational framework, 915
    general structure, 914–915
Context-based access control (CBAC), 131
Continuous differential equations, 806
Continuously moving items, 436–437
Continuously moving objects, indexing structures for, 370
    R^{EXP}-tree, 373
    STAR-tree, 373
    TPR-tree, 371–372
    TPR*-tree, 373
Continuous query, 367
Contract cases, 286
Conventional DBMS, 874
Converting B-reps to trees, 339–340
Convex bounding volumes, 336
Convex drawing, 726; *see also* Symmetric drawing; Visibility drawing
    algorithm using canonical ordering, 727–728
    Barycenter algorithm, 726
    divide and conquer algorithm, 726–727
Convex hulls, 378–379, 385, 402, 894, 1017
    complexity, 1017–1018
    construction, 1018–1019
    convex hull-trees, 893
    dynamic, 1019–1020
    maximum complexity, 1018
    Overmars-van Leeuwen structure, 1019
    revisited, 380–381
Convex optimization, 892
Convex-polygon clipping, 337
Convex polytopes, 382, 890, 896–897
    linear programming, 890
    Minkowski sums and convex optimization, 892
    Voronoi-based marching algorithm, 890–892
Coordinate values, 253
Copy pointer, 517
Copy-up node, 240
Corner block list (CBL), 833, 844, 848–849, 851
Corner cut canonical set, 416
Corner(s), 1024
    stitching, 833, 841
Corner stitching data structure, 819
    point finding operation, 820–821
    storage requirements of, 822
    Tile Creation, 821–822
Corner stitching extensions, 822
    curved tiles, 823
    expanded rectangles, 822–823
    L-shaped tiles, 823–826
    trapezoidal tiles, 823
Correctness, 225, 654, 745–746
Correspondence property, 101, 108, 109
Corruption, 94–95
Counter overflow, 135–136
Counter size, 135–136
Counting Bloom filter (CBF), 135
    counter size and counter overflow, 135–136
Counting cache misses, 7
    effect of cache misses on run time, 7–8
    matrix multiplication, 8–9
    simple computer model, 7
Counting networks, 750

COUNT operator, 245–246
CPT, *see* Collection of prefix trees
Crawlers fingerprint, 802
Crawling, 813–814
Credits, 17–18
    invariant, 556
Crossing minimization methods, 737
Cross-producting, 789–790
Cross tree, 429
CRUST algorithm, 661
Cryptographic hash function, 986
CSE method, 360
CSG, *see* Constructive Solid Geometry
Cuckoo hashing, 126
Current edge, 192
currentNode, 448
"Curse of dimensionality", 359, 999, 1037
Curved tiles, 823
Curve reconstruction, 661–662
Cuttings
    applications, 401
    convex hulls and Voronoi diagrams, 402
    cutting construction, 397
    geometric sampling, 398–399
    Hopcroft's problem, 401
    optimal cuttings, 399–401
    point location, 401
    range searching, 402
Cyber communities, 805
Cycle detection, 57
Cycle or circuit, 52–53
Cyclic case, 729
Cyclic ordering of characters, 489

D

DAG, *see* Directed acyclic graph
DAM, *see* Disk access model
Dangling edges, 278
Data
    compression method, 471, 800
    definition statements, 967
    element, 389
    field, 447
    flow diagrams, 707
    models, 882, 984
    ownership and distribution, 1000
    quality, 1000
    set, 349
    structuring techniques, 583
    warehouse, 874
Database management system (DBMS), 233, 245, 874, 881, 967
    functionality, 967–968
    simplified architecture, 968
Databases, data structures for, 967
    for buffer management, 974–976
    for disk space management, 976–981
    functionality of database management system, 967–968
    for query processing, 968–974
Data-driven spatial data structures, 428
Dataflow analysis, 589

Data mining, 874, 997
  application, 507, 1002–1007
  challenges, 999–1000
  classification, 1000–1002
  clustering, 1007–1010
  and role of data structures and algorithms, 1000
  tasks and techniques, 998–999
  Web pages indexed by Google© search engine, 998
Data structure(s), 23, 151, 179, 186, 197, 263, 331, 363, 419, 420, 426, 489, 551, 579, 584, 628, 639, 655, 698, 767, 787, 862, 891
  advanced geometric data structures, 656
  and algorithms, 1000
  BDDs, 498–499
  BOB, 777–778
  for buffer management, 974–976
  for disk space management, 976–981
  geometry Kernels, 656
  graph data structures, 655
  hierarchical distance maintaining, 619–621
  hierarchical tries, 787, 788
  linear interpolation and Bezier curves, 864–867
  linear search, 787
  numbers and matrices, 655
  Patricia, 456
  for query processing, 968–974
  rendering value, 698
  set-pruning tries, 787–788
  tiled, multidimensional array, 864
  vertex, normal, and face lists, 863
  vertices, edges, and faces, 862
  views, 699
  Winged Edge approach, 863–864
Data structures library in Java (JDSL), 680, 681
  algorithms, 683–684, 688–689
  architecture of, 684
  Class FlightDijkstra, 694–695
  Class IntegerDijkstraPathfinder, 693–694
  Class IntegerDijkstraTemplate, 690–693
  comparators, 683
  containers and accessors, 681–682
  decorations, 683
  design concepts in, 681
  iterators, 682–683
  key-based containers, 687–688
  minimum-time flight itineraries, 689–690
  packages, 684
  positional containers, 684–687
  sample application, 689
Data structure visualization, 697
  data structure views, 699
  DDD, 702–704
  existing research and systems, 700
  GELO, 702
  incense, 701
  interacting with system, 699–700
  purpose and environment, 698–699
  systems, 698, 704
  value of data structure rendering, 698
  VIPS, 701–702
Data types, 655
  advanced data types, 655
  basic, 655

geometry Kernels, 656
numbers and matrices, 655
DAWG, *see* Directed acyclic word graph
db.collection.indexStats() method, 988
d-box, 389
DBMS, *see* Database management system
DCAS, *see* Double-compare-and-swap
DDD system, 700, 702–704
d-dimensional array, 252
d-dimensional packet classification problem, 765
d-dimensional points, 343
Deadlocks, 744
DeallocatePage function, 234
Deaps, 98, 108
  inserting element, 109
  removing min element, 109–111
Decision nodes, *see* Nonterminal nodes
Decision trees, 332
Decomposability property, 291
Decomposition, 1015–1016
  process, 309, 310
  rules, 259, 260
Decorations, 683
Decremental APASP, randomized data-structure for, 618
  bounding size of $B_u^{d,S}$ under edge-deletions, 621–622
  hierarchical distance maintaining data-structure, 619–621
  improved decremental algorithm for APASP up to distance d, 622–624
  main idea, 618–619
  notations, 619
Decremental graph problem, 581
defaultAncestor, 714
Default prefix, 766
Deferred splitting, policy for, 347
Deformable objects, 383
Defragmentation, 979
Degree, 49
  distribution, 806
  node, 36
Degree updates, problem of, 958
Delaunay diagrams, 385
Delaunay flipping, 663–664
Delaunay graph, 1034
Delaunay triangulation (DT), 277, 289, 1020, 1021–1023, 1033–1034
  under point motion, 382
Delayed pivoting, 962
Delete algorithm, 237–238, 242–244
Delete-max operation, 77
Delete-min operation, 77, 80, 304, 554
Delete operation, 14, 85, 182, 197, 554, 987
  BSTs, 43
Deletion(s), 152–153, 160, 161
  algorithm, 317
  of arbitrary element from Max HBLT, 74
  max-heap, 45–46
  of max element from Max HBLT, 72
  in $O(\log^2 n)$ time, 589
  of object, 362
  in RBST, 211–212
Delta-encoding technique, 800
Dendrogram, 1007
Dense case, 120

Dense index, 980
Dense interconnection, 805
"Denseness" of interconnection, 805
Density based search tree, 550–552
Density of intervals set, 291
Density thresholds, 551
DEPQs, *see* Double-ended priority queues
Depth first expression (DF-expression), 362
Depth-first search (DFS), 54, 683, 688, 948, 961
    adjacency lists and depth-first traversal, 55
    algorithm for depth-first search on undirected graph, 55
    on digraph, 57
    simple applications of, 56
    topological sorting, 57–58
    traversal, 472
Depth-first strategy, 880
Depth first traversal, 55, 926
Depth-first unary degree sequence representation, 600
Depth of node, 152
Deques, *see* Double ended queues
Descent pointers, 204, 207
Descriptive tasks, 998
Designating algorithms, 402
Design Rule Checking (DRC), 263, 818, 829
Destination address, 765
Deterministic algorithms, 198, 630
    exponential search trees, 632–633
    fusion trees, 630–632
Deterministic finite automaton, 477
Deterministic sum of n random variables, 203
Dewey decimal notation, 35
DF-expression, *see* Depth first expression
DFS, *see* Depth-first search
Dictionary, 204, 687–688
    binary trees as, 152–153
    dynamic, 598–599
    problem, 529
Dictionary ADT, 197, 198, 212
Dictionary operations, 207
    analysis of dictionary operations, 207–209
Difference operation, 868–869
Diffracting balancers, 750
Diffracting trees, 750
Diffusing computation tree, 748
Diffusion layer, 817
Digital searching, *see* Radix searching
digitNumber field, 452
Digit numbers, compressed tries with, 452–454
digitValue, 448
digraph, *see* Directed graph
Digraph, DFS, 57
Dihedral case, 729
Dihedral symmetry, displaying, 732
Dijkstra's algorithm, 616, 689
Dijkstra's label-setting algorithm, 63
Dijkstra's shortest-path algorithm, 63, 659
Dimensionality, 309, 999
Dimension reduction approach, 899
Directed acyclic graph (DAG), 511, 938, 948, 956, 959
    effective depth of, 512
    of five versions, 520

Directed acyclic word graph (DAWG), 477, 479, 920; *see also* Compact DAWG
    constructing DAWG in linear time, 482
    modifying, 484–486
    property, 477–478
    simple algorithm for constructing, 481–482
Directed Area Enumeration, 820
Directed graph (digraph), 49, 50, 581, 584, 805
    Kleene closures, 584–585
    locality, 585–586
    long paths, 585
    matrices, 586–587
Direct page addressing, 974, 975
Direct preprocessing, 473
Discrete event simulation, 427
Discrete geometry, 377
Discrete partition, 937–938
Discrete random variable, expected value of, 201
Disjoint set union-find problem, 538
    classical union-find problem and variants, 539–540
Disjunctive normal forms (DNFs), 501, 506
Disk access model (DAM), 988
Disk-based environment, 233–234
Disk model, 419–420
DiskRead function, 234
Disk space management, data structures for, 976
    file organization, 980–981
    page organizations, 978–980
    record organizations, 976–978
Disk space manager, 968
DiskWrite function, 234
Dismissals of false alarms, 365
Distance, 51–53
    distance-based techniques, 914
    joins, 353
    predicate, 879
    query, 872
Distributed consensus algorithms, 986
Distributed memory multiprocessors, 949
Distributed R-trees, 343
Distribution sweep(ing), 421, 422–423
Divide-and-conquer, 880, 1021
    algorithms, 12, 13, 175, 726–727, 879
    construction algorithm, 315–316
    problem-solving paradigm, 268
    technique, 582
Division, hashing by, 118
DNA sequences, 918
DNFs, *see* Disjunctive normal forms
Document stores, 984
Dopants, 817
Double-compare-and-swap (DCAS), 752
Double-ended priority queues (DEPQs), 77, 97; *see also* Meldable priority queues
    application, 97–98
    deaps, 108–111
    generic methods for, 111–112
    interval heaps, 100–105
    MDEPQs, 113
    min-max heaps, 105–108
    SMMH, 98–100

Double ended queues (Deques), 31, 512, 607, 752
  of elements, 523
Double hashing, 124
Double rotation, 154
Doubling technique, 607, 630
Doubly connected edge list (DCEL), *see* Halfedge
    data structure
Doubly linked circular lists, 30
Down buffers, 557, 558
Downward pointers, *see* Branch-node pointers
Downward traversal, 466
Drawing algorithm input, 729
Drawing trees
    hv-layout, 719–720
    level layout for binary trees, 709–710
    level layout for *n*-ary trees, 710–718
    preliminaries, 708–709
    radial layout, 718–719
    typical layout of binary tree, 708
DRC, *see* Design Rule Checking
Drill bit models, 859
Driscoll, Sarnak, Sleator, and Tarjan (DSST),
    511, 512, 516
    limitations of transformations, 512
droppedElement, 107–108
Druglike chemical space, 935
DSST, *see* Driscoll, Sarnak, Sleator, and Tarjan
DS-Viewer system, 704
DT, *see* Delaunay graph; Delaunay
    triangulation
Dual heap, 491, 492
Duality, 1016–1017
    transform, 1016, 1017
Dual plane, 1016
Dual priority queues, 111
Dual structure method, 111
Dual transformed space, 360
Dynamic algorithms, 349, 583
Dynamic arrays, 24, 607
Dynamic B-trees, 550
    density based, 550–552
    exponential tree based, 552–553
Dynamic binary trees, 602–603
Dynamic bit vector problem, 606
Dynamic computational geometry, 377
Dynamic convex hulls, 1019–1020
Dynamic counting problem, 1047–1048
Dynamic data structures, 421, 435
    continuously moving items, 436–437
    logarithmic method for decomposable search problems, 435–436
Dynamic dictionary, 598–599
Dynamic environments, 892
    bloom-*g* in, 144
Dynamic finger search trees, 171–172; *see also* Randomized finger
    search trees
Dynamic finger theorem, 185
Dynamic graphs, 577, 581
    algorithms, 574, 581
    all-pairs shortest paths, 591–592
    connectivity, 587–588
    directed graphs, techniques for, 584–587
    minimum spanning tree, 588–589

    transitive closure, 589–591
    undirected graphs, techniques for, 582–584
Dynamic hash files, 980
Dynamic hashing, 127
Dynamic interval management, 432
Dynamic map labeling, 873
Dynamic multi-level tree algorithms, 793
"Dynamic Optimality Conjecture", 185
Dynamic perfect hashing, 120–122
Dynamic programming, 224, 881
    recurrences, 223
Dynamic queries, 890
Dynamic reporting problem, 1046–1047
Dynamic rule, 767
Dynamic set
    bloom filter for, 146–147
    intersections and subset testing, 537–538
Dynamic setting, 567
Dynamic simulation, 889, 896
Dynamic subset testing problem, 537
Dynamic tree
    clustering techniques, 574
    ET trees, 567, 577–578
    linking and cutting trees, 567, 568–571
    membership, 567
    reachability trees, 567, 578–579
    topology trees, 567, 571–574
    top trees, 567, 574–577
Dynamic variable ordering, 503
Dynamization, 291
    static algorithms, 583
DynamoDB, 986

E

Eager learners, 1001
EC, *see* Extended connectivity
ECFP, *see* Extended connectivity fingerprint
edags, *see* Elimination dags
Edge-deletions, bounding size of $B_u^{d,S}$ under, 621
    maintaining BFS tree, 622
    technical details, 622
Edge-flip, 382
Edge information, 454
    compressed tries with, 454–456
Edge(s), 576, 581, 862, 863
    algebra, 288
    attributes, 278, 280, 281
    bends, 723
    contraction, 279–280, 284–287
    crossings, 723
    deletion, 279, 283–284
    insertion, 279, 283–284, 587
    label, 461, 468
    quadtrees, 267, 906–907
    resolution, 723
    ring, 287
Editing, 1002
Efficiency, 380, 680
    efficient symbolic factorization, 950–951
Element, 131, 454, 681
Elementary operations, 331, 945

Elementary range, 294
Elimination
    directed acyclic graph, 945
    forest, 947
    game, 946–947
    ordering, 946, 954
    technique, 749, 752
Elimination dags (edags), 959, 960, 961
    column elimination tree, 959–960
    elimination structures for asymmetric multifrontal
        algorithm, 962
Elimination structures, 945
    applications of etrees, 950–953
    clique covers and quotient graphs, 957–959
    clique tree, 945, 954–956
    column elimination trees and elimination dags,
        959–962
    elimination tree, 946–950
Elimination tree, 945–947
    algorithm, 948–949
    data structure, 947–948
    elimination game, 946–947
    skeleton graph, 949
    supernode, 949–950
EM, *see* External memory
Empty-circle condition, 1022
Empty circumcircle property, 1033
Empty pointers, 160
Empty string, 477
EMST, *see* Euclidean minimum spanning tree
Enclosing interval searching problem, 293, 296
Encode set, 132
End-point array, 767–768
Entity-relationship diagrams, 707
Entropy
    Huffman codes and, 220
    of subdivision, 1032
Entry coordinate, 298
Ephemeral data structures, 511
Epochs, 533
ε-approximation, 398
ε-cutting, 397
ε-net, 398
Equal-length prefixes sets, 768–769
Equality testing, 533–535
Equidepth histograms, 973
Equivalence classes, 53, 478, 484, 535, 919, 920
EQUIVALENCE statements, 538
Equiwidth histograms, 973
Error rate, 95
Etrees applications, 950
    efficient symbolic factorization, 950–951
    predicting row and column nonzero counts, 951–952
    scheduling out-of-core factorizations, 953
    scheduling parallel factorizations, 952–953
    three classes of factorization algorithms, 952
ET trees, *see* Euler tour trees
Euclidean minimum spanning tree (EMST),
        1023, 1034
Eulerian graph, 65–66
Eulerian trail, 65
Euler's formula, 1018

Euler tour trees (ET trees), 567, 577, 582
    applications, 578
    data structure, 583
    traversal, 688
    updates, 578
Event, 199, 378
    KDS, 382
    queue, 378
Evolutionary web graph models, 808
Exact-match query, 234, 240, 241
Exact searches, 935, 936
    canonical labeling methods, 937–938
    canonical representations, 936–937
    graph theoretic representations, 936
EXCELL, 255, 256
Exclusive mode, 754, 755
Exit coordinate, 298
Expanded rectangles, 822–823
Expectation of X, 200–201
Expected mean of X, *see* Expectation of X
Expected value of discrete random variable, 201
Expected value of X, *see* Expectation of X
Explanatory variables, 998
Exponential backoff lock, 747
Exponential growth models, 808
Exponential layout, 552
Exponential level based priority queue, 557
    analysis, 559–560
    layout, 558
    operations, 558–559
    structure, 557–558
Exponential search, 171
    trees, 628, 632–633
Exponential tree based structure, 552–553
Expression trees, 40, 968, 969, 972–973
Extended binary tree, 70
Extended connectivity (EC), 937
Extended connectivity fingerprint (ECFP), 939
Extended trees, 841, 842
Extendible hashing, 127
Extending LEDA to secondary memory (LEDA-SM), 884
Extensibility, 654
eXtensible and fleXible Library (XXL), 884
Extensible hash files, 980
Extensible hash table, 754
Extent problems, 382
External algorithms, 879
    index on both spatial relations, 880
    index on none of inputs, 880–881
    index on one spatial relation, 880
External boundary vertices, 574
External degree, 571
External fractional cascading, 422
External fragmentation, 598
"Externalizing" plane sweep algorithms, 422
External marriage-before-conquest, 422
External memory (EM), 421
    algorithms, 419, 421–424, 1045–1046
    design criteria for EM data structures, 420–421
    disk model, 419–420
    dynamic and kinetic data structures, 435–437
    finger search trees, 173

External memory (*Continued*)
    online data structure, 421
    related problems, 434–435
    spatial data structures and range search, 428–434
    tree data structures, 424–428
External merge sort, 970
External nodes, 37, 70, 152
External path length, 37
Extractmin, 93
    operation, 85, 88
Extra fields, 517
Extremal sets and subset testing, 535
    dynamic set intersections and subset testing, 537–538
    static extremal sets, 536–537
Extremal sets problem, 536

F

Faces, 725, 862, 1014, 1028
    attributes, 278, 280, 281
Factorization, 953, 956
Failure-detection protocols, 986
Fair split, 411
False negative, 132
False positive, 132, 352
    probability, 133
    ratio, 133–134, 137, 139, 142, 145
False sharing phenomenon, 743
Farach's algorithm, 467
Faster randomized algorithms, 635
Fat Inverted Segment tree (FIS-tree), 791–792
Fat nodes, 519, 521
    method, 511, 516
Fat regions, 406, 410
Fat-triangle, 1045
FCFS hashing, *see* First-come first-served hashing
FD, *see* Fixed length-depth
Fetch-and-ϕ structures, 749
    combining, 749
    counting networks, 750
Fibonacci heaps, 69, 86, 89–92; *see also* Skew heaps
    data structure, 85
Fibonacci number, 155
Fibonacci trees, 708
FID, *see* Fully indexable dictionary
FIFO, *see* First in first out
File, 976, 980
    formats, 859
    index data structure, 799
    path, 947
File organization, 980–981; *see also* Page organizations; Record organizations
Filled graph, 946, 949
Filtering, 421
Filter-refinement approach, 872, 883
Fingerprints, 801, 939
    compression, 940
Finger search trees, 155, 171, 179; *see also* Balanced binary search trees (BBST)
    adaptive merging and sorting algorithm, 176–177
    applications, 175
    arbitrary merging order, 175–176

    dynamic finger search trees, 171–172
    level linked (2,4)-trees, 172–173
    list splitting problem, 176
    optimal merging and set operations, 175
    randomized finger search trees, 173–175
Fine-grained locking scheme, 743
FinishPointer, 774
Finite element methods, 323
Finite trees, 151
First-come first-served hashing (FCFS hashing), 125
First in first out (FIFO), 31, 192
    dynamic tree implementation, 193–194
First matching rule in table, 766, 783
FIS-tree, *see* Fat Inverted Segment tree
Fixed-grids, 252, 256
Fixed-length records, 977
Fixed-stride trie optimization (FSTO), 771
Fixed-stride tries (FSTs), 770–772
Fixed length (FL), 361
Fixed length-depth (FD), 361
Fixed size alphabet, 463
Fixed string of diamonds, 730
FKS-($\alpha/2$) data structure, 121
FKS-α data structure, 120–121
FKS hashing scheme, 597
FL, *see* Fixed length
Flexibility, 680
FlightDijkstra class, 694–695
Flight simulators, 857
Flip bit, 288
Float-kernel, 656
Floating point Cartesian coordinates, 656
Floats data type, 699
Floorplan representation in VLSI, 833, 834, 849
    combinations and complexities of representations, 834–835
    extra constraints, 854
    graph based representations, 837–844
    motivation of representation, 834
    placement based representations, 844–850
    rectilinear shape handling, 851–853
    relationships of representations, 851
    slicing, mosaic, LB compact, and general floorplans, 835–837, 838
    statement of floorplanning problem, 833–834
Flow, 190, 765*fn*
    across cut, 190
Flow-Excess, 190
FM-index, 917, 926
    algorithm, 928, 929
    fast rank and select operations using wavelet trees, 929–931
    problem of searching patterns in string, 931
    simple scheme supporting Rank queries on binary string, 931
    suffix array and Burrows–Wheeler transform, 927
    theorem, 927–929
Fonts, 859
Force directed methods, 737
Forced reinsertion, 429
Forest
    forest-structured Bloom filter, 146
    ordered set of trees, 37
    set of trees, 53
Forest of Quadtrees (FQT), 903, 907, 909–910
Fortune's algorithm, 1021

4-dimensional real-life classifier, 785
Four-dimensional space, 251
Four-sided queries, 563
FP-growth algorithm, 1005, 1007
FP-tree structure, 1005–1007
FQT, *see* Forest of Quadtrees
Fractal trees, 991–992
Fractional cascading, 1029–1030
Fractional cascading, 990, 991
Fragmentation, 979
Freelist, 752
Free tree, 35
Frequent itemsets, 1003
Frequently-encountered problem, 58
Frontier-based methods, 507
FSTO, *see* Fixed-stride trie optimization
FSTs, *see* Fixed-stride tries
Full path method, 519
Fully dynamic algorithms, 582, 589
Fully dynamic connectivity, 574
    problem, 587
Fully dynamic graph problem, 581
Fully dynamic minimum spanning tree, 588–589
    problem, 574
Fully dynamic tree membership, 573
Fully indexable dictionary (FID), 597, 598
Fully persistent data structures, 511
Functional data structures, 639
    binary search trees, 642–644
    data structures in functional languages, 639–640
    decreased synchronization, 640
    difficulties, 649
    fewer bugs, 640
    functional data structures in mainstream
        languages, 640
    increased sharing, 640
    in mainstream languages, 640
    skew heaps, 644–649
    stacks, 640–642
Functionality, 343, 680
Functional languages, data structures in, 639–640
Functional programming language, 641
Functions, 605–606
Fundamental primitives, 546
    k-merger, 546, 548–550
    van Emde boaz layout, 546–548
Fundamental supernodes, 950, 960
Funnel heap structure, 554, 555
Funnelsort, 548, 550, 554
Furthest-site Voronoi diagram partitions, 1022
Fusion tree, 597, 628, 630–632

G

Gabriel Graph (GG), 1023, 1034
Garbage collection, 640
Garsia-Wachs algorithm, 216, 224, 226, 227, 228
Gaussian elimination, 945, 946, 957, 958
Gauss map, 897
GB(c)-trees, *see* General balanced trees (GB-tree)
GB-tree, *see* General balanced trees
GDAL, *see* Geospatial Data Abstraction Library

GDBX system, 704
GELO system, 702, 703
GEM alignment tool, 927
Genbank, 923
Genealogical information, 35
General balanced trees (GB-tree), 160–161
General floorplans, 835–837
    constraint graphs for, 838
    example, 851
    O-tree representation, 844
General graph, 49
Generalization
    of bloom-1, 140–141
    of convex drawing, 737–738
    of map, 873
Generalized-element model, 957
Generalized grounded range reporting problem, 1048
Generalized half-space range searching, 1049–1050; *see also*
        Persistence-based approach; Sparsification-based
        approach
Generalized intersection searching; *see also* Nearest neighbor
        searching
    adding range restrictions, 1053–1054
    approach for reporting problems, 1052–1053
    arbitrarily oriented objects, 1045
    axes-parallel objects, 1044–1045
    exploiting output size, 1054–1055
    external memory and word-RAM algorithms, 1045–1046
    geometric intersection searching problems, 1043–1044
    persistence-based approach, 1050–1052
    problems on grid, 1045
    reverse transformation, 1055–1056
    single-shot problems, 1045
    sparsification-based approach, 1048–1050
    summary of known results, 1044–1046
    techniques, 1046–1056
    transformation-based approach, 1046–1048
Generalized lists, 31
Generalized one-dimensional range searching problem, 1046
Generalized reporting problem, 1043–1044
Generalized search trees (GiSTs), 756
Generalized semidynamic
    quadrant searching, 1050–1051
    two-dimensional range searching, 1051
Generalized suffix trees, 466
Generalized three-dimensional range searching, 1050, 1052
Generalized two-dimensional quadrant range searching, 1050
General orthogonal 2-D range search, 433–434
General polygonal models, 892
    interference detection using trees of oriented bounding boxes,
        893–896
    performance of bounding volume hierarchies, 896
Generating function framework, 809–810
Generic algorithms, 667
Geodesic triangulations, *see* Pseudo-triangulations
Geographic information systems (GISs), 251, 343, 359,
        419, 871, 878
    applications, 872
    cartographic generalization, 873
    data mining, 874
    geometric data structures, 872
    geometric objects, 871

Geographic information systems (*Continued*)
  geometric operations, 872
  map labeling, 873
  map overlay, 873
  models, toolboxes, and systems for geographic information, 882
  open source systems and research prototypes, 883
  road maps, 873–874
  space filling curves, 874–879
  spatial database systems, 883
  spatial join, 879–882
  spatial libraries, 883–884
  spatiotemporal data, 874
  standardized data models, 882
  topological *vs.* metric data, 872
Geography Markup Language (GML), 882
GeoJSON, 882
Geometric
  algorithms, 783, 788, 884
  AQT, 790–791, 792
  attributes, 378, 871
  automorphisms, 728
  computing, 277
  cross-producting, 789–790
  data, 331
  data structures, 872
  distribution, 202
  dynamic multi-level tree algorithms, 793
  entity, 332, 334, 338
  FIS-tree, 791–792
  graphs, 384
  grid-of-tries, 788–789
  intersection searching problems, 1043–1044
  objects, 429, 871, 872
  operations, 872
  primitives, 892
  relation, 378
  sampling, 398–399
  scene, 658
  set systems, 399
  software libraries, 872
  structures, 385
  2-dimensional classification scheme, 790, 791
Geometry, 882
Geometry Engine, Open Source (GEOS), 883–884
Geometry Kernels, 656
GEOS, *see* Geometry Engine, Open Source
Geospatial Data Abstraction Library (GDAL), 882, 883
Geotagging, 269
GeoTools, 883–884
GeoWin, 358
GG, *see* Gabriel Graph
Giant components, emergence of, 811
GISs, *see* Geographic information systems
GiSTs, *see* Generalized search trees
GJK algorithm, 892, 896–897
Global rebuilding
  approach, 535
  operation, 121
GML, *see* Geography Markup Language
Golumb code, 800
Good BSP trees, 337–338
Good nodes, 910

Google Chubby projects, 986
Gossip-style protocol, 986
Gouraud model, 861
GPA, *see* Grade point average
GPS
  devices, 874
  systems, 370
Grade point average (GPA), 969–970
Graham's scan, 1018
Graph(s), 277, 686–687, 938, 960
  based representations, 837
  connectivity, distance, and spanning trees, 51–54
  constraint graphs, 838–841
  corner stitching, 841
  databases, 984
  data structures, 655
  data type, 657
  digraph with 6 vertices and 11 edges, 50
  Eulerian and Hamiltonian graphs, 65–66
  graph-based approach, 833
  "graph-like" representations, 533
  graph-theoretic approach, 935
  graph-theoretic representations, 935
  MST, 58–62
  pseudograph of 6 vertices and 10 edges, 50
  representations, 50, 52
  representations, 603
  searching graph, 54–56
  shortest paths, 62–65
  simple applications of DFS and BFS, 56–58
  single tree representations, 842–844
  theoretic representations, 936
  traversals, 688–689
  twin binary tree, 841–842
  undirected graph with 5 vertices and 6 edges, 50
  weighted graph representation, 51, 52
Graph Drawing, 707, 723
  convex drawing, 726–728
  has-a-joint-paper-with Relation, 724
  preliminaries, 725–726
  prerequisite diagram, 725
  symmetric drawing, 728–732
  two drawings of graph, 724
  two drawings of social network, 724
  visibility drawing, 732–737
Graphical display of data structures, 698
GraphLibrary (*libG*), 655
GraphWin, 657–658
Graycode, 315
Gray code curve, 875
Gray encoding, 320–321
"Greedy" algorithms, 62, 216, 217, 955
Greengard's fast multipole method, 323
Grid
  cells, 252, 255
  directory, 255
  file, 255, 256, 429, 876, 970
  grid-based methods, 323
  grid-of-tries, 788–789
  placement, 845
  problems on, 1045
  vertex, 347

Grounded 2D-range search problem, 305–306
Grouping process, 256
Group mutual exclusion, 748
Group queries, unified algorithm for, 319
Growth model, 807
Gunther's algorithm, 352

H

H2GIS, 883
Halfedge data structure, 281; *see also* Kinetic data structures (KDS);
    Layout data structures; Online dictionary structures;
    Spatial data structures
    access functions, 282
    edge insertion and deletion, 283–284
    effects of half_splice, 283
    vertex split and edge contraction, 284–287
Halfedge data structure, 872, 1021
Half-space in d-dimensional space, 261*fn*
Halfspace range searching, 435
Hamiltonian graph, 65–66
Hand, 172
Hand-over-hand locking approach, 753
Haptic rendering, 889
Hardware, 857–858
    bitmap-intersection, 796–797
    hardware-based algorithms, 795
    ternary CAMs, 795–796
Hardware depth buffer, 384
Hash-based cache, 501
Hash-based indexes, 969
Hash-based join approach, 353
Hash complexity, reducing, 146
Hashed consing, 532
Hash files, 980
Hash functions, 117, 118, 132*fn*
Hashing, 152, 318
    approaches for improved searches, 940–941
    with chaining, 122
    by division, 118
    by multiplication, 118–119
    with open addressing, 123
HashtableDictionary, 687
Hash tables, 117, 612, 754
    developments, 127
    historical notes, 126
    for integer keys, 118–122
    operations, 506
    random probing, 122–126
Hash tree structure, 1004–1005
Hash trie, 531–533
Hash value, 117
Haskell
    binary search trees in, 643
    skew heaps in, 645
    stacks in, 641
Hasse diagram, 479
hB-tree, 258, 429
Hbase, 984, 986, 988, 989, 990, 992
HBLTs, *see* Height-biased leftist trees
HBSTR-tree, 360
Header, 977

Header page, 980
Heap-on-Trie (HoT), 793
Heap(s), 44, 77
    deletion, 45–46
    files, 980
    heap-based priority queues, 756–757
    insertion, 45
    max-heap, 44
    order, 173, 599
    priority queues, 44
Height balanced trees, 180
Height-biased leftist trees (HBLTs), 70
    deletion of arbitrary element from max HBLT, 74
    deletion of max element from max HBLT, 72
    initialization, 72–74
    insertion into max HBLT, 71
    max trees, 71
    melding two max HBLTs, 72, 73
Height limited Huffman trees, 220
    algorithm for coin collector problem, 221–222
    reduction to coin collector problem, 220–221
Height of trie, 447
Heights of subtrees, 154
Heterogeneous arrays, 25–26
Heuristic(s), 793
    algorithms, 345, 783
    HiCuts, 793–794
    optimization, 346
    optimizing algorithms, 834
    RFC, 793, 794
    tuple space search, 795
HiCuts, *see* Hierarchical intelligent cuttings
Hidden surface removal, 867
Hierarchical clustering, 1007–1008
Hierarchical information, 35
Hierarchical intelligent cuttings (HiCuts), 793–794
Hierarchical motion descriptions, 386
Hierarchical partitioning, 434
Hierarchical tries, 787, 788
Hierarchy generation, 893
High confidence bounds, 198, 199
High-dimensional data, problems with, 1009
Highest-priority matching, 777; *see also*
         Most-specific-range matching
    data structure BOB, 777–778
    search for highest-priority matching range, 779
Highest-priority rule, 766
High-level data, 903
High-level image processing, 903
Highly recurring oligonucleotides, 917
High probability bounds, 198
High-speed on-die cache memory, 132
High-throughput screening, 935
Hilbert curve, 347, 875, 876–877, 878
Hilbert R-tree, 429
Hilbert sort (HS), 349
Hilbert tree, 346–348
Hilbert value, 347
Hinted Quad Trees (HQT), 829–830
Histograms, 881, 973–974
Historical R-tree (HR-trees), 368
Historical shortest path, 586

Historical synopsis, 367
History DAG, 1031
History graph, 1031–1032
Homology, 845
Hopcroft's problem, 401
Horizontal-vertical (hv), 719
    hv-layout, 719–720
    hv-layout algorithms, 708
    HV Trees, 829
Horizontal combination, 719
Horizontal O-tree, 842, 844
Horizontal pointers, 204
HoT, *see* Heap-on-Trie
HQT, *see* Hinted Quad Trees
HR-tree, 360, 368, 370
HR-trees, *see* Historical R-tree
HS, *see* Hilbert sort
"Hubs", 805
Huffman algorithm, 215
Huffman algorithm for *t*-ary trees, 220
Huffman codes, 216
    and entropy, 220
HuffmanCost(S), 216, 217
Huffman trees, 216, 226
    Huffman algorithm for *t*-ary trees, 220
    Huffman codes and entropy, 220
    linear time algorithm for presorted
        sequence of items, 218
    $O(n \log n)$ time algorithm, 217
    relation between general uniquely decipherable codes
        and prefix-free codes, 218–219
Hydrogen-excluded representation, 936
Hydrogen-included representation, 936
Hyperoctrees, 309
Hyperplane, 329, 332, 333, 401
Hyperrectangles, 298

I

IDE, *see* Integrated development environment
Image acquisition, 903
Image array, 259
Image data structures; *see also* Kinetic data structures (KDS); Layout
        data structures; Online dictionary structures; Spatial data
        structures
    CBIR systems, 914–915
    image data, 904–905
    octrees, 911–912
    quadtrees, 905–907, 910–911
    R-trees, 910–911
    TID, 912–914
    virtual quadtrees, 907–910
Image dilation, 263
Image manipulation/analysis, 903
Image processing, 903
    applications, 320
    connected component labeling, 322
    construction of image quadtrees, 321
    gray encoding, 320–321
    rotation and scaling, 321
    union and intersection of images, 321
Image quadtrees construction, 321

Immediate subcells, 312, 313, 314, 317
Immediate supercell, 312
Immutability, 639
Immutable data structures, 639
Immutable nodes, 647–648
Implicit data structures, 98
Implicit representation of balance information, 159–160
Improved decremental algorithm for APASP up to
        distance d, 622–624
Incense system, 701
IncidenceListGraph, 687
Incidence preserving, 1016
Inclusion exclusion pruning, 814
Incorporating randomness, 197
Incremental dynamic graph problem, 581
Incremental graph problem, 581
Incremental landscape, 482, 484
Incremental penetration depth computation, 897
    implementation and application, 898
    initialization and refinement, 897–898
    local walk, 897
Independence of two events, 200
Independent set of vertices, 1029
Index
    on both spatial relations, 880
    compression, 800
    generation efficiency, 360
    granularity, 800–801
    index-nested-loop join algorithm, 970
    multi-dimensional indexes, 970
    on none of inputs, 880–881
    one-dimensional indexes, 969–970
    on one spatial relation, 880
    structure, 260, 423, 969, 980
Indexability, 434
Indexable bitvector, 597
Indexable circular deques, 607
Indexable dictionary, 597–598
Indexing
    hybrid type of indexing methods, 360
    mechanisms, 864
Indirect addressing, 978
Indirect page addressing, 975
Inductance, 818
Induction
    hypothesis, 620
    step, 485, 620
Influential variables, 502
Information processors, 679
Information retrieval, 799
Information-theoretic $O(\log n)$ barrier, 628
Information visualization, 697
Initialization, 897–898
    HBLTs, 72–74
    interval heap, 103
Inner convex drawings, 738
Inner radius, 406
In-order invariant, 152
Inorder traversal, 40, 152
    of threaded binary tree, 41–42
Input communication, 419
Input data distribution, 355

Input/output (I/O), 343
   algorithms, 419
   cost, 246, 421, 426
   cost of insertion, deletion and exact-match query, 245
   efficient algorithms, 1045
   high I/O performance, 987
   I/O-model, 545, 546
   model, 546, 548
   performance, 427
   two-level I/O-model, 545
"Insert 23" operation, 992
Insert algorithm, 236, 241–242
Inserting element, 457–458
   deaps, 109, 110
   interval heaps, 101–102, 103
   min-max heaps, 105–106
Inserting into trie, 449
Insertion, 152, 160
   algorithm, 317
   max-heap, 45
   into max HBLT, 71
   of object, 362
   in RBST, 210–211
   sort, 25
   sort algorithm, 176
   of string, 466
Insert operation, 14, 80, 85, 182, 197, 554
   BSTs, 42–43
InspectableBinaryTree, 686
InspectableGraph, 687
InspectableKeyBasedContainer, 687
InspectableTree, 686
Instance, see Records
IntegerDijkstraPathfinder class, 693–694
IntegerDijkstraTemplate class, 690–693
Integer keys, hash tables for, 118
   dynamic perfect hashing, 120–122
   hashing by division, 118
   hashing by multiplication, 118–119
   static perfect hashing, 119–120
   universal hashing, 119
Integers data type, 699
Integrated development environment (IDE), 699
Intention content, 700
Interference detection, see Collision detection
Interior-based representation, 259
Internal memory, 429, 431
Internal nodes, 37, 70, 152, 972, 1031
Internal path length, 37
Internet-of-Things (IoT), 983
Internet Protocol (IP), 765
   highest-priority matching, 777–779
   longest matching-prefix, 767–777
   most-specific-range matching, 780–780
   prefixes and ranges, 766
   router tables, 765
Internet router, 765, 783
Inter-object spatial relations, BSP tree representation of, 330
Inter-object visibility orderings, 336
Intersection
   bound, 941
   of images, 321

   model, 882
   operation, 868
   predicate, 879, 881
   queries, 344–345
   query, 425, 428
Interval graph, 291
Interval heaps, 98, 100, 102
   complementary range search problem, 104–105
   complexity of interval heap operations, 104
   initializing interval heap, 103
   inserting element, 101–102, 103
   removing min element, 102–103, 104
Interval intersection graph, 291
Interval query, 368–369
Interval tree, 263, 291, 292, 901
   construction, 292–293
   example and applications, 293–294
Intra-object visibility, 333
Invariants, 588
Inverse Ackermann's function, 322
Inverted indices, 799, 942
   index compression, 800
   index granularity, 800–801
Inverted list, 252, 799
Inverter, 817, 819
Invocation, 548
IoT, see Internet-of-Things
IP, see Internet Protocol
IPv4
   prefix sets, 771
   router tables, 770
Irregular arrays, 27
Iso-oriented rectangles, 291
Isolated vertex, 53
Itemset, 1003
Iterated logarithm function, 539
Iterator(s), 668, 674, 682–683
   basics, 674–676
   classes, 667
   iterator-based tree traversals, 688
   reverse iterators, 676

J

Jaccard coefficient of similarity, 802
Java, 26
   applications, 883
   binary search trees in, 643
   language, 640
   programming language, 679
   skew heaps in, 646
   stacks in, 641
Java collections (JC), 679, 681
Java generic libraries (JGL), 679, 681
JC, see Java collections
JDSL, see Data structures library in Java
JGL, see Java generic libraries
Join algorithms, 352
Join operations, 183
Join procedure, 211
JPG file formats, 859, 860
JTS Topology Suite (JTS), 883–884

K

Käräkkanen and Sanders' algorithm, 467–468
k-ary tree, 35–36, 253
k-cells, 1014
k-crossing filter set (k-CFS), 790
k-cut, 413
kd–B-trees, 429
KDD, *see* Knowledge discovery in databases
k-DOP-tree, 893
*k*D-range, 298, 301
KDS, *see* Kinetic data structures
k-d tree, 254, 256, 261, 311, 331, 343, 360, 412, 426,
        560, 826, 892, 1036
    nearest neighbor search in, 1037
    set of points in plane, 1036
KeyBasedContainer, 687
Keys, 152
    dictionaries, 687–688
    with different length, 446–447
    field, 445, 980
    key-based containers, 682, 687
    key-based partitioning, 991
    key-value stores, 984
    priority queues, 687
Kinetic data structures (KDS), 378, 435; *see also*
        Layout data structures; Online dictionary structures;
        Spatial data structures
    application survey, 381
    collision detection, 383–384
    connectivity and clustering, 384
    continuously moving items, 436–437
    convex hull example, 378–379
    convex hull, revisited, 380–381
    event, 382
    extent problems, 382
    logarithmic method for decomposable search problems, 435–436
    motion in computational geometry, 377
    motion models, 377–378
    open problems, 385–387
    performance measures for, 379–380
    proximity problems, 382, 384
    querying moving objects, 387
    triangulations and tilings, 382–383
    visibility, 384–385
Kinetic methods, 383
Kinetic motion-sensitive algorithms, 386
Kirchoff voltage law, 840
Kirkpatrick's algorithm, 1028–1029
Kleene closures, 584
    logarithmic decomposition, 585
    recursive decomposition, 585
Kleinberg's HITS algorithm, 805
k-merger, 546, 548
    binary mergers and merge trees, 548–549
    funnelsort, 550
k-nearest neighbors, 320
    classification algorithm, 1001
    k-nearest neighbor joins, 881
    query, 367
k-neighbor trees, 162
Knowledge discovery in databases (KDD), 997
Knowledge processing application, 507

Kraft's inequality, 219
Kruskal's MST algorithm, 59, 60

L

Label-correcting method, 63
Labeled buffer, 430
Labeled Hilbert, 431
Labeled naive, 430
Labeled universe, 531
Label of directed path, 477
Laguerre diagram, 1022
Land use, 873
Large binary objects (LOBs), 968
Large environments, 898
    algorithms for, 890
        multiple-object collision detection, 899–900
        two-dimensional intersection tests, 900–901
Large objects (lobs), 977
Largest strongly connected component, 806
Larmore-Hirschberg algorithm, 215, 216
Last-come first-served hashing (LCFS hashing), 125–126
Last in first out (LIFO), 31
Las Vegas type algorithms, 198
Las Vegas type randomized algorithm, 203
Latency, 872
Law of total probability in conditional form, 200
Laws of probability, 199
Layer, 873
Layered drawings, 735–736
Layout, 547, 555, 558
    editor, 819
    process, 819
Layout data structures, 818, 819; *see also* Halfedge data structure;
        Multidimensional spatial data structures
    corner stitching data structure, 819–822
    corner stitching extensions, 822–826
    quad trees and variants, 826–830
    VLSI technology, 817–819
Lazy evaluation, 644–648
Lazy skew heaps analysis, 648–649
Lazy strategy, 373
LB compact floorplans, 835–837, 838
LCA, *see* Least common ancestor
LCFS hashing, *see* Last-come first-served hashing
LCS data structure, *see* L-shaped corner stitching data structure
LC-tries, *see* Level-compressed tries
Leaf
    correspondence, 112
    entry updating, 363
    label, 462
    leaf-oriented trees, 165
    node, 36, 905, 972
    quads, 828
    selection, 345
Least common ancestor (LCA), 174, 951
Least Weight Subsequence problem (LWS problem), 230
LEDA, *see* Library of efficient data types and algorithms
LEDA extension packages (LEPs), 654
LEDA-SM, *see* Extending LEDA to secondary memory
Left child-right sibling representation, 37
Left contour, 709

Left-deep tree, 973
Leftist orders, 426
Leftist trees, 77; *see also* Trees
    binary tree and extended binary tree, 70
    HBLTs, 70–74
    WBLTs, 74–75
Left-justified trees property, 230–231
Left-looking algorithm, 952
Leftmost canonical ordering, 728
Left tree, 195
Length of path, 215
Length of string, 477
Lens system, 704
LEPs, *see* LEDA extension packages
Level, 153
Level-balanced B-trees, 426–427
Level-compressed tries (LC-tries), 449
Level drawings, 709
Level equivalent, 225
Leveling, 204
Level layout
    for binary trees, 709–710
    for *n*-ary trees, 710–718
Level linked (2,4)-trees, 171, 172–173, 175
Level links, 172
Level number, 361
Level-order unary degree sequence representation, 600
Level order traversal, 41
Level-pointers, 153
Level-tiered LSM trees, 989
Lexicographic depth, 464
Lexicographic ordering, 464
Library of efficient data types and algorithms (LEDA),
    653, 679, 884
    algorithms, 656
    availability and usage, 654
    combinatorial and geometric computing, 653
    correctness, 654
    curve reconstruction, 661–662
    data structures and data types, 655–656
    Delaunay flipping, 663–664
    discussion, 664
    ease of use, 653–654
    example programs, 659
    extensibility, 654
    projects enabled by, 665
    shortest paths, 659–661
    structure, 654–655
    upper convex hull, 663
    visualization, 657–658
    word count, 659
LIFO, *see* Last in first out
Lifting map, 1034
Limited node copying, 1029
Lin-Canny algorithm, 892
Line-based constraint graphs, 840
Linear growth models, 808
Linear hash files, 980
Linear hashing, 127
Linear in *d*-dimensional space, 261*fn*
Linear interpolation, 864–867
Linearizability, 746

Linearization point, 746
Linear list, 767
Linear ordering, 315
Linear probing, 123–124
Linear programming problem (LP problem), 890
Linear quadtrees, 903
Linear region quadtree, 361
Linear scales, 253, 260
Linear search, 634, 787
Linear space, 630
    exponential search trees, 632–633
    fusion trees, 630–632
    spatial structures, 429
Linear speedup, 742
Linear split algorithm, 345
Linear time
    constructing DAWG in, 482
    tree traversal method, 471
Linear time algorithm, 467, 482, 709, 728, 732, 1036
    for constructing convex drawings of planar graphs, 726
    for constructing planar convex grid drawing, 727
    for drawing binary trees, 707
    for drawing planar graphs, 728
    for presorted sequence of items, 218
Linear time construction algorithms, 463
    Käräkkanen and Sanders' algorithm, 467–468
    space issues, 468
    of suffix arrays, 467
    of suffix trees, 464–466
Line data and boundaries of regions, 265–268
Line quadtrees, 906
Line road data sets, 423
Line segments, 265, 268, 871
Line-swept sphere (LSS), 893
LINGO, 942
LINGOsim computation, 942
Linked lists, 28, 204, 753–754
    chains, 28–29, 30
    circular lists, 29–30
    doubly linked circular lists, 30
    generalized lists, 31
Linking and cutting trees, 185, 567, 568, 582, 584
    data structure, 186
    implementation of primitives for, 189
    implementation without, 192–193
    implementing operations on vertex-disjoint paths, 569–571
    operations, 185–186
    using operations on vertex-disjoint paths, 568–569
    rotation, 187–188
    solid trees, 186–187
    splay analysis in virtual tree, 188–189
    splay in virtual tree, 188
    splicing, 188
Linking operation, 85
    two heap-ordered trees and result of linking, 86
LISP functional programming languages, 532
List, 668
    ranking, 427
    representation, 37
    splitting problem, 176
    structures, 862
Little oh notation (*o*), 11

LL/SC, *see* Load-linked/store-conditional
lmp, *see* Locally minimal pair
LMPBOB, *see* Longest matching-prefix BOB
l-multiMEM-problem, 924
Load-linked/store-conditional (LL/SC), 744
Load balancing algorithm, 753
Loading algorithms, 352
Lob directory, 977, 978
LOBs, *see* Large binary objects
lobs, *see* Large objects
Locality, 380, 585–586
    geometric locality, 892
    locality-guided work stealing algorithm, 753
    locality-sensitive hashing, 435
    principle of, 338
    of reference, 819, 825–826
Locally minimal pair (lmp), 225–226
"Locally optimal" solution, 897
Local rebuilding, 155
Local retiling of space, 382
Local spinning, 743
Local walk, 891–892, 897
Location, 1027; *see also* Point location
    code, 255, 361
    planar point, 554
Locator, 682
Lock, *see* Mutual exclusion lock
Logarithmic method, 435–436, 561
Logarithmic on binary search tree, 35
Logical address, 978
Logical left spine, 648–649
Logical operation, 7
Logical pointers, 978
Logical query plans, 968
Logical right spine, 648–649
Logical view of tree, 648–649
Logic design, 817
Logic elements, 817
Logic gates, 817
Log structured merge tree (LSM tree), 987, 988–989
    intuition behind, 989–990
Long development time, 679
Longest common prefix, 462
Longest common substrings, 470–471
Longest matching-prefix
    binary search trees, 774–775
    end-point array, 767–768
    linear list, 767
    PST, 775–777
    sets of equal-length prefixes, 768–769
    tries, 769–774
Longest matching-prefix BOB (LMPBOB), 777
Longest prefix matching for routing lookups, 784
Longest-side *k*-d trees, 416
Long paths, 585
Loose quadtree, 264
Lopsided tree, 215, 226
Loser trees, 48
Lower envelope, 1014–1015
Lowest common ancestors, 472
    Bender and Farach's lca algorithm, 472–473
    suffix links from, 473

Low height schemes, 162–164
Low-level data, 903
Low-level image processing, 903
LP problem, *see* Linear programming problem
L-shaped corner stitching data structure (LCS data structure), 823–824
L-shaped tiles, 823–826
    comments on corner stitching, 825–826
    LCS data structure, 823–824
    parallel corner stitching, 825
    rectilinear polygons, 825
    space requirements of tiles, 825
LSM tree, *see* Log structured merge tree
LSS, *see* Line-swept sphere
Lune, 1034
LWS problem, *see* Least Weight Subsequence problem

M

MAGIC system, 819, 825
Mainstream languages, functional data structures in, 640
Maintenance contract, 15
    accounting method, 15–16
    aggregate method, 15
    potential method, 16
    problem definition, 15
    worst-case method, 15
Map, 670–672, 873
    labeling, 873
    overlay, 873
    visualization, 873
Mapping data to display, 704
Marker partitioning technique, 769
Market basket
    analysis, 998
    data, 1002–1003
Mark procedure, 207
Markup languages, 882
Master Equation Approach, 808
Matching basic interval, 774, 775
Match path, 926
Mathematical Graph Theory, 725
Matrices, 586–587
Matrix-vector multiplications, 899–900
Matrix multiplication, 8–9
Max-Flow Min-Cut Theorem, 190–191
Max-heap, 44
Max arrays, 101
Max HBLT
    arbitrary element deletion from, 74
    deletion of max element from, 72
Max heap, 108
Maximal clique in graph, 949
Maximal multiple exact match, 924–926
Maximal palindrome, 474
Maximum Clique Size of set of Intervals, 297
Maximum-spread k-d trees, 416
Maximum clique problem, 291, 296
Maximum elements, 43
Maximum weight spanning tree (MST), 954
MAX operator, 245–246
maxPQ, 111

Max tree, 71
Max WBLT operations, 75
MB, *see* Megabytes
MBB, *see* Minimum bounding boxes
MBRs, *see* Minimum bounding rectangles
McCreight's algorithm, 465–466, 470
McWidget company, 16
    accounting method, 17–18
    aggregate method, 17
    potential method, 18
    problem definition, 16
    worst-case method, 17
MDEPQs, *see* Meldable DEPQs
"Mean field theory" based model, 806
Medial axis, 1033
Megabytes (MB), 859
Meldable DEPQs (MDEPQs), 113
Meldable priority queues, 77, 80; *see also* Double-ended priority
        queues (DEPQs)
    amortized cost of meld operation, 82–83
    left and right children of nodes, 81
    left subtrees of nodes in merged path, 81
Melding operation, 77
Melding two max HBLTs, 72, 73
Meld operation, 77, 80, 85, 113, 153
    amortized cost of, 82–83
    max HBLT meld operation, 71
    skew heaps for, 80
    of WBLT, 75
Membership
    bits, 132
    problem, 291
    word, 137
Memoization, 647–648
Memory contention, 742–743
Memory management techniques, 506
Memory model, 595
Memory transfer, 545
MEMS, *see* Microelectromechanical systems
Merge, 153, 212
    analysis, 556–557
    based priority queue, 554
    layout, 555
    operation, 29, 555–556, 576
    scheduling, 991
    sort, 970
    sort recurrence of equation, 13
    structure, 554–555
    trees, 548–549
Merged constraint graph, 839
Mergesort, 546
Merging
    BSP trees, 336
    phase, 971
    process, 175
Merkle tree, 986, 987
Metanode, 908
Metric data, 872
Metric distance, 406
Microelectromechanical systems (MEMS), 823
Middle tree, 195
Min array, 101

minconf thresholds, 1003
Min heap, 108
Minimal edags, 961
Minimalistic approach, 166
Minimal perfect hash function, 127
Minimal pseudo-triangulations, *see* Minimum pseudo-triangulations
Minimum-degree algorithms, 957, 958
Minimum-degree ordering heuristics, 957
Minimum-time flight itineraries, 689–690
Minimum AABB, 256
Minimum bounding boxes (MBB), 365, 373
    leaf-level MBBs, 371
    MBB-based index, 365
Minimum bounding rectangles (MBRs), 343, 365, 370, 879, 882
Minimum element, 43, 111
Minimum pseudo-triangulations, 1024, 1025
Minimum separation determination, 377
Minimum spanning forest, 59
Minimum spanning tree (MST), 58, 381, 384, 588, 689; *see also*
        n–ary trees; Ordered-tree (O-tree)
    Boruvka's MST algorithm, 62
    constrained MST, 62
    deletions in $O(\log^2 n)$ time, 589
    of graph, 35
    Kruskal's MST algorithm, 59, 60
    Prim's MST algorithm, 59–61
    updates in $O(\log^4 n)$ time, 589
Minkey queries, 567
Minkowski metric (L_m metric), 1035
Minkowski sums, 892, 896
Min-max heaps, 98, 105
    inserting element, 105–106
    removing min element, 106–108
MIN operator, 245–246
minPQ, 111, 112
minsup thresholds, 1003
Min tree, 71
Model evolutionary graphs, 808
Model View, 857–858
Molecular graph, 936
Molecular sequence databases, 917
MonetDB/GIS, 883
Monge property, 216, 228
MongoDB, 988
Monoid, 530
Monotone, 1023
    monotone-matrix concepts, 228
    subdivisions, 427
Monotonicity property, 223
Monte Carlo algorithms, 198
Morgan's algorithm, 937
Morton encoding, 875
Morton order, 255
Morton ordering, 315
Mosaic floorplans, 835–837
    constraint graphs for, 839
    example, 851
    and representations, 852
Most-specific-range matching, 780; *see also* Highest-priority
        matching
    conflict-free ranges, 780
    nonintersecting ranges, 780

Most-specific-rule matching, 766
Motion, 377
    formulation, 890
    models, 377–378
    plan, 378
    planning, 889, 896
    sensitivity, 386
MPBSM, 423
MSQT, *see* Multiple Storage Quad Trees
MST, *see* Minimum spanning tree
Multibit tree, 943
Multi-dimensional access methods, 1008–1010
Multidimensional arrays, 26, 864
    array of arrays representation, 26
    irregular arrays, 27
    row-or column major representation, 26
Multidimensional balanced binary search trees, 393, 395
Multidimensional binary search trees, 311
Multidimensional data, 251
Multi-dimensional indexes, 970, 980
Multi-dimensional packet classification
    bounds from computational geometry, 786
    classification algorithms, 785
    data structures, 787–788
    geometric algorithms, 788–793
    hardware-based algorithms, 795–797
    heuristics, 793–795
    performance metrics for classification algorithms, 785
    problem statement, 783–785
    range lookups, 786–787
    taxonomy of classification algorithms, 787
Multidimensional range search, 428
Multi-dimensional search
    application to multi-dimensional search trees, 161–162
    structure, 331–332
Multidimensional spatial data structures; *see also* Kinetic data structures (KDS); Layout data structures; Online dictionary structures; Spatial data structures
    bucketing methods, 256–259
    line data and boundaries of regions, 265–268
    point data, 252–256
    rectangle data, 263–265
    region data, 259–263
    transformation approach, 251–252
Multi edges, 49
Multifrontal factorization method, 957
Multigraph, *see* Multi edges
Multi-level index, 980
Multimaps, 670–672
Multimedia systems, 359
multiMEM, *see* Maximal multiple exact match
Multiple-choice hashing method, 125, 126
Multiple-key index, 970
Multiple-object collision detection, 899
    implementation and application, 900
    one-dimensional sweep and prune, 900
Multiple elimination, 958
Multiple exact match, 924
Multiple lists in single array, 25
Multiple posting problem, 258
Multiple quad, 826, 827

Multiple Storage Quad Trees (MSQT), 827
Multiple views, 984
Multiplication, 506, 595, 627, 630
    hashing by, 118–119
    matrix, 8–9, 584
Multiplicative weights, 1022
Multi-resolution representation, 335, 337
Multirooted BDDs, 498
Multisets, 669–670
Multislab, 422, 431
Multistep query processing, 879
Multiversion 3D R-trees (MV3R-trees), 360, 370
Multiversion B-trees (MVBT), 360
Multiversion linear quadtree (MVLQ), 360
    deletion of object in, 362
    insertion of object in, 362
    updating object in, 363
Multiversion linear quadtree (MVLQ), 360, 361
    deletion of object, 362
    insertion of object, 362
    updating object, 363
Multiversion R-tree (MVR-tree), 368, 369, 370
Multi-way merging, 971
Multi-way trees, 166–167, 425, 431
    balanced binary trees based on, 157–158
Munro's equations, 591
Mutex, *see* Mutual exclusion lock
Mutual exclusion lock, 742, 747
    coupling approach, 753
    elision, 749
    granularity, 742
    lock-free counter, 745
    lock-free operation, 744
MV3R-trees, *see* Multiversion 3D R-trees
MVBT, *see* Multiversion B-trees
MVLQ, *see* Multiversion linear quadtree
MVR-tree, *see* Multiversion R-tree
MX-CIF quadtree, *see* Bisector List Quad Trees (BLQT)
MX quadtree, 266
"Mysterious" behavior, 216

N

n–ary trees, 708–709; *see also* Minimum spanning tree (MST); Ordered-tree (O-tree)
    ancestor, 714–715
    apportion, 715
    combining subtree and left subforest, 713–714
    level layout for, 710
    level layout of *PQ*-tree, 712
    PrePosition, 713
    shifting smaller subtrees, 715–718
    spacing out smaller subtrees, 711
Nahar-Sahni algorithm, 825
Naïve approach, 364
Naive scheme, 511
NAND gate, 817
NAP, *see* Network Access Point
Narrow phase, 898
Nauty, *see* No AUTomorphism, Yes?
N-body problem, 323–325
N-component, 552

Near-duplicate detection algorithm, 802

Near-duplicate documents, finding, 801–802

Nearest neighbor, 59–60

    classification scheme, 1001

    classifiers, 1001

    proximity graphs for enhancing, 1001–1002

    query, 310, 350–351, 407, 425, 428, 872

Nearest Neighbor Graph (NNG), 1034

Nearest neighbor searching, 421, 1008–1010, 1035; *see also*
        Generalized intersection searching

    approximate nearest neighbor searching, 1037–1038

    AVD, 1038

    K-d trees, 1036–1037

    other approaches to, 1037

    through point location, 1036

Nearest-X (NX), 349

Negative binomial distribution, 202–203

Neighbor(s), 574

    Finding operation, 820

    list, 323

Nested-loop algorithm, 879

Nested clustering, *see* Hierarchical clustering

Network Access Point (NAP), 783

Network flows, application to, 190–192

Networks, 50

newElement, 105, 109

NewSQL databases, 984

Next-generation sequencing (NGS), 926–927

NGS, *see* Next-generation sequencing

NNG, *see* Nearest Neighbor Graph

No AUTomorphism, Yes? (Nauty), 937–938

Node, 49, 151, 152, 344, 425, 496, 688, 826

    capacity, 345

    copying method, 512, 516–517

    deletion rule, 497

    reduction rules, 497

    sharing rule, 497

NodeBinaryTree, 686

Node splitting, 345, 517–518

    algorithm, 346

    method, 516–517

NodeTree, 686

Nonblocking

    algorithms, 743

    linearizable heap-based priority queue algorithms,
        756

    operator, 882

    techniques, 744–745

Non-canonical structures, 386–387

Non-clustered index, 980

Non-convex

    models, 898

    polyhedral models, 899

Nondestructive updates, 531

Non-indexed data set, 352

Nonintersecting ranges, 780

Nonnegative weights, single-source shortest paths and, 62–63

Nonoverlapping, 258

    collinear chain of nonoverlapping anchors, 923, 924

    intervals

    polytopes, 897

Nonpoint data, 429

Non-robust

    algorithms, 953

    method, 953

Nonroot nodes, 194

Non-slicing structure, 849

Nonterminal nodes, 496

Non-uniform memory access (NUMA), 743

Nonzero element, 27

NOR gate, 817

NoSQL paradigm, *see* Not Only SQL paradigm

NOT gate, 817

Not Only SQL paradigm (NoSQL paradigm), 983

    database workloads, 987

    data stores, 983–984, 987, 993

NP-complete problem, 502

N-type dopants, 817–818

NUMA, *see* Non-uniform memory access

Number-of-brep-faces, 340

Number-of-tree-nodes, 340

Number processors, 679

Numerical linear algebra, 945

NX, *see* Nearest-X

O

OAT problem, *see* Optimal alphabetic tree problem

OBBs, *see* Oriented bounding boxes

OBDDs, *see* Ordered BDDs

Object, 332

    hierarchy, 256

    object-oriented language, 699

    total ordering of collection of, 332–333

    types, 343, 890

    updating, 363

OBST, *see* Optimal binary search trees

Obstruction-free operation, 744

OCC, *see* Optimistic concurrency control

Occupancy of hash table, 117

OCR, *see* Optical character recognition

Octants, 262

Octrees, 260, 322, 340, 412, 892, 911–912; *see also* Quadtrees

    compressed quadtrees and, 313–314

    representation, 913

Odd-even merge sort, 635

Offset, 977

OGC, *see* Open Geospatial Consortium

$O(\log n)$ Time, searching and priority queues in, 627

    achieving sub-logarithmic time per element by simple means,
        628–630

    from amortized update cost to worst-case, 633–634

    deter*m*inistic algorithms and linear space, 630–633

    model of computation, 627

    overview, 628

    sorting and priority queues, 634–636

OM data structure, *see* Order Maintenance data structure

Omega notation (Ω), 11

One-at-a-time result, 959

1-bit tries, 769–770

One-cut, 413

One-dimension (1D)

    1D-range tree, 299

    array, 26

One-dimension (*Continued*)
  compaction, 819
  data structures, 874
  indexes, 969–970, 980
  packet classification, 765, 783
  sweep and prune, 900
1-edge, 496
One-level search tree, 800
1-terminal, 496
Online data structures, 421
Online dictionary structures; *see also* Kinetic data structures (KDS);
      Spatial data structures
  additional operations, 395
  balanced BST implementation, 393–395
  binary search tree implementations, 391–393
  discussion, 396
  trie implementations, 390–391
$O(n^2 \log n)$ time, updates in, 590
$O(n^2 \log^3 n)$ time, updates in, 592
$O(n \log n)$ planesweep algorithm, 825
$O(n \log n)$ time algorithm, 217
$O(n^2)$ time, updates in, 591
$O(n^{\beta+2})$-timedynamic programming algorithm, 227
Open addressing, hashing with, 123
Open Geospatial Consortium (OGC), 882, 883
Open source systems, 883
Open transaction times, handling query with, 367–368
Operating system, 451
Operations, 555–556, 558–559, 601
  on array, 24–25
  cache, 501
  counts, 4–5
  types, 350
  on vertex-disjoint paths, 568–569, 569–571
Operator trees, 968
Optical character recognition (OCR), 541
Optimal algorithm, 881
Optimal alphabetic tree problem (OAT problem), 216, 224–226
  computing cost, 224–226
  construction, 226
  for presorted items, 226
Optimal binary search trees (OBST), 216, 222–224
Optimal construction, 402
Optimal cuttings, 399–401
Optimal hashing, 127
Optimality of splay trees, 184–185
Optimal k, 133–134
  bloom-1 *vs.* bloom with, 139–140
  bloom-g *vs.* bloom with, 142–144
Optimal lopsided trees, 216, 226–229
Optimal merging, 175
Optimal paging strategy, 560
Optimal policy matrix, 65
Optimal static finger search algorithm, 171
Optimal structure, 432
Optimistic approach, 744
Optimistic concurrency control (OCC), 749
Oracle, 612, 882, 883
Orbit-stabilizer theorem, 728
Order, 969
  preserving, 1016
  queries, 427

Ordered BDDs (OBDDs), 496
Ordered dictionary, 151, 687
Ordered partition, 954
Ordered queries, 152
Ordered-tree (O-tree), 429, 833, 842, 844; *see also* Minimum
      spanning tree (MST); n–ary trees
  building, 842
  horizontal O-tree, 842
  with linear lists, 843
  performance, 843
  representations, 844, 851, 853
  structures, 835
Ordering, 255
  fields, 980
  invariant, 152, 165
  key for file, 980
  of polygons, 334
Order–kVoronoi diagram, 1022
Order Maintenance data structure (OM data structure), 516, 518
Order-preserving hash functions, 424, 425
Ordinal trees, 599–601
Oriented bounding boxes (OBBs), 893, 896
  implementation and application, 896
  interference detection, 894–895
  interference detection using trees of, 893
  OBB-trees, 893
  OBBtree construction, 893–894
  representation, 895–896
Original heuristics, 345
Original tree, 186
Orthogonal drawings, bar visibility representations for, 736–737
Orthogonal neighbors, 876–877
Orthogonal range
  lower bounds for orthogonal range search, 434
  queries, 561
O-tree, *see* Ordered-tree
Outer radius, 406
Outliers, 1009
Out-of-core
  algorithms, 419
  factorization, 953
Output communication, 419
Output-sensitive
  algorithms, 402, 1018
  query time, 1043
Output size, exploiting, 1054–1055
Overlap(ping)
  query, 367
  test, 895–896
  interval searching problem, 293
  polytopes, 897
Overlapping linear quadtrees, *see* Multiversion linear quadtree (MVLQ)
Overmars-van Leeuwen structure, 1019

P

P2P search problem, *see* Peer to peer search problem
Packed bucketing, 635
Packed computation, 628, 629
Packed sorting, 634–635

Packet, 787
    packet-header information, 765
Packing algorithms, 349
Packing function, 409
Packing keys, 629–630
Packing problem, 852
Page header, 978
pageID, 234
Page organizations, 978; *see also* File organization;
        Record organizations
    alternative page organizations for fixed-length records, 979
    slots, 978
    for variable-length records, 979
Page reference list, 977
Page reference pointer, 977
Page-related pointers, 978
Pages, 234
Page table, 975
Painter's Algorithm, 329, 332, 867
Pairing heaps, 69, 86, 92–94; *see also* Skew heaps
    adaptive properties, 94
    data structure, 85
    variations, 94
Pairwise independence, 200
Pairwise join method (PJM), 881
Palindrome, 474
Panning, 873
Parallel algorithms, 230
    property of left-justified trees, 230–231
Parallel column-oriented factorizations, 953
Parallel disk model (PDM), 419–420, 421
Parallel edges, 49
Parallel factorization algorithm, 952
Parallelism, 799
Parallel R-trees, 343
Parallel Random Access Machines (PRAM), 230, 745
Parallel triangular solution, 956
Parameterization, *see* Transformation approach
Parasitic extraction, 818–819
Parasitics, 818
parentElement, 105
parentNode, 106
Parent-pointers, 153, 426–427
Parse tree, 968, 969, 971–972
Partially dynamic graph problem, 581
Partially dynamic problem, 581
Partially persistent data structures, 511
Partial pivoting, 959
Partial rebuilding technique, 155, 160, 161
Partial sums, 606
Particle-based methods, 323
Partitional clustering, 1007–1008
Partition-Based Spatial Merge (QPBSM), 423
Partition Based Spatial-Merge Join (PBSM), 880
Partitioned Bloom filter, 145
Partitioning, 984
    consistent hashing, 984–985
    edges, 574
    process, 445
    scheme, 819
    tree representation of intra-object spatial relations, 330
Partition maintenance algorithms, 540–542

Partition trees, 387, 402–403, 426
    data structure, 535
Path, 152
    caching, 433
    copying technique, 642–644
    path-balancing, 157
    problems, 584
PA-tree, 368
Patricia, *see* Practical Algorithm To Retrieve Information Coded In
        Alphanumeric
Pattern discovery, 917, 918
Pattern matching, 468, 640
    pattern matching using suffix arrays, 469–470
    pattern matching using suffix trees, 469
p-biased coin, 202
PBSM, *see* Partition Based Spatial-Merge Join
PD, *see* Penetration depth
PDM, *see* Parallel disk model
Peano curve, 347, 875
Peano–Hilbert order, 255
Peano–Hilbert space-filling curve, 429
Pedigree, 519, 522
Peer to peer search problem (P2P search problem), 811
Pending node, 648–649
Penetration depth (PD), 897
    computation, 896
        convex polytopes, 896–897
        incremental penetration depth computation, 897–898
        non-convex models, 898
Perfect elimination ordering, 954
Perfect hash function, 118, 127
Performance analysis of 3D R-Trees, 367
Performance metrics, 133, 785, 987–988
Performance optimizations, 990
    fractional cascading, 990, 991
    "Insert 23" operation, 992
Permutations, 604–606
Persistence-based approach, 1050; *see also* Sparsification-based
        approach; Transformation-based approach
    generalized semidynamic quadrant searching, 1050–1051
    generalized semidynamic 2D range searching, 1051
    generalized 3D range searching, 1052
Persistence, 987
    B+ tree, 422, 553–554, 988
    intuition behind LSM, 989–990
    LSM-tree, 988–989
    performance metrics, 987–988
    performance optimizations, 990–992
Persistent data structures, 361, 511, 639; *see also* Kinetic data
        structures (KDS); Layout data structures; Online
        dictionary structures; Spatial data structures
    algorithmic applications, 513–516
    fat node method, 516
    general techniques for making, 516
    handling arrays, 518
    making data structures confluently persistent, 518–522
    making specific data structures more efficient, 522
    node copying, 517
    node splitting, 517–518
    persistent deques, 524–526
    redundant binary counters, 524
    representation of deque of elements, 523

Persistent deques, 524–526
Persistent search tree, 1029
Persistent trees, 1029
Phong model, 861
Physical address, 978
Physically-based modeling, 889
Physical Plan Generator, 969
Physical pointers, 978
Physical query plans, 969
Physical trajectory, 874
πDD, 495, 507
ping, *see* PNG
Pipeline, 857–858, 882
Pivot, 97
    element, 558
Pivoting strategy, 959, 962
Pixels, 255, 904
    two dimensional array of, 320
PJM, *see* Pairwise join method
PK-tree, 256, 258–259
Placeholders, 681
Placement based representations, 833, 844
    BSG, 847–848
    corner block list, 848–849
    sequence pair, 845–847
    slicing tree, 849–850
Planar drawings, 723
Planarity, 725
Planarization algorithm, 603
Planar point location, 554
    algorithm for, 513, 514
    problem, 1027
Planar *st*-graphs, 427, 733–734
Planar straight line graphs (PSLGs), 277, 1027
    edge deletion, 279
    edge insertion, 279
    features, 277–278
    halfedge data structure, 281–287
    operations on, 279
    quadedge data structure, 287–288
    vertex split and edge contraction, 279–280
    winged-edge data structure, 280–281
Planar subdivision, 1028
Plane-sweep algorithm, 881, 1021
Plowing, 820
pmf, *see* Probability mass function
PM quadtree family, 267
PMR quadtree, 267, 268
PNG (ping), 859, 860
Point-in-polygon query, 872
Point-swept sphere (PSS), 893
Point data, 252–256
Pointed pseudo-triangulations, *see*
        Minimum pseudo-triangulations
Pointed pseudotriangulations, 1024
Pointer(s), 23, 28, 92, 978
    based data structure, 517
    element, 454
    fields, 519–520
    machine model, 530
    triple, 449
    types, 391

Point Finding operation, 820–821, 825
Point location, 401, 421, 429, 786, 1027, 1028
    fractional cascading, 1029–1030
    Kirkpatrick's algorithm, 1028–1029
    nearest neighbor searching through, 1036
    persistent trees, 1029
    query, 425, 428
    separating chains, 1029–1030
    sequence of triangulations, 1028
    slab-based methods, 1029
    slab refinement of subdivision, 1029
    trapezoidal maps and history graph, 1031–1032
    worst-and expected-case optimal, 1032
Point quadtree, 253, 254, 310–311
Point queries, 316–317, 354, 987
Point region quadtree (PR-quadtree), 253, 254, 255–256, 312
Point search tree (PTST), 777
Polish expression, 849, 850
Polycrystalline silicon, 818
Polygon(al), 330, 343, 350, 872, 1023
    data sets, 352
    line, 871
    maps, 267
    mesh, 858
    polygon-area approaches, 338
    representations, 265, 266
    soups, 892
    subdivision, 1028
Polyhedra, 343, 1023–1024
Polyhedral, 330
Polyhedral regions, 891
Polyhedron, 891
Polyline, 265
Polynomial time algorithms, 345, 936
Polytopes, 330
    data structure, 891
    representation, 891
Pools, 752–753
Pop operation, 522, 642
Positional B$^+$-trees, 978
Positional containers, 681, 684
    graphs, 686–687
    sequences, 685
    trees, 686
Position concept, 681–682
Position heap, 478, 489
    of *aabcabcaa*, 479
    building position heap, 489–490
    construction, 489
    improvements to time bounds, 491–494
    querying, 490–491
    for string, 478
    time bounds, 491
Post-office problem, 1013, 1020
PostGIS, 883
Postorder traversal, 40–41
Postprocessing, 997
Postscript
    driver software, 859
    font, 859
Potential method, 15
    maintenance contract, 16

McWidget company, 18
    subset generation, 19–20
Power diagram, 382, 1022
Power distance, 1022
Power law distribution, 806, 808
PQ, *see* Priority queues
Practical Algorithm To Retrieve Information Coded In
        Alphanumeric (Patricia), 456
    inserting element, 457–458
    removing element, 458
    searching, 457
PRAM, *see* Parallel Random Access Machines
Predecessor, 49
    pointers, 517
    search, 152
Predecessor fields (p), 517
Predictive category, 998
Predictive modeling, 998
Preemption-safe locks, 747
Preferential attachment paradigm, 806, 807, 808, 809
Prefix(es), 523
    node, 774–775
    paths, 1007
    prefix-free codes, 216
    range, 766
    and ranges, 766
    relation between general uniquely decipherable codes and
        prefix-free codes, 218–219
    search and applications, 451–452
PreFlow, 190
Preflow-push algorithms, 191–192
Preorder traversal, 40, 262, 316, 710
PrePosition method, 713
Preprocessing time, 513
Presorted items, OAT for, 226
Presorted sequence of items, 218
Prim-Jarník algorithm, 689
Primal heap, 491
Primary attribute, 252
Primary organization, 980
Primary storage, 233–234
Prime implicant, 486, 487
Prim's MST algorithm, 59–61
Priori, 294
Priorities, 212
Priority_queue template, 672–673
Priority_Search_Tree_root(S), 304
Priority queues (PQ), 44, 69, 77, 98, 304, 407, 554, 634, 687, 756
    combining techniques, 635
    exponential level based, 557–560
    further techniques and faster randomized algorithms, 635–636
    heap-based priority queues, 756
    merge based, 554
    packed sorting, 634–635
    range reduction, 634
    tree-based priority pools, 757
Priority R-tree, 431
Priority search trees (PSTs), 263, 291, 304–306, 432,
        775–777, 1037, 1046
    construction, 304–305
    example and applications, 305–306
PR k-d tree, 254

Probabilistic priority queues, 212
Probability density function, *see* Probability mass function (pmf)
Probability distribution, 199
Probability mass function (pmf), 201
Probability theory, 199
Procedure Interval_Insertion function, 295–296
Process function, 319
Process visualization and debugging environment system
        (PROVIDE system), 704
Product-form inverse, 956
Programming languages, 23
Program visualization, 697
Projection point, 857
Projects enabled by LEDA, 665
PROVIDE system, *see* Process visualization and debugging
        environment system
Proximity, 867–868, 1027
    Delaunay triangulation, 1033–1034
    geometric graphs on point set, 1035
    geometric proximity structures, 1034–1035
    graphs for enhancing nearest neighbor classifiers, 1001–1002
    problems, 382, 384
    queries, 343
    set, 323
    structures, 1032–1035
    Voronoi diagrams, 1032–1033
"Proxy TID", 979
PR-quadtree, *see* Point region quadtree
Pseudo-algebraic motions, 380
Pseudograph, *see* General graph
Pseudo-node, 157, 160
Pseudotriangulation, 383, 385, 1024–1025
    of convex hull of point set, 382–383
    mixed, 383
    pointed pseudotriangulations, 1024
    pseudotriangulation-based methods, 384
PSLGs, *see* Planar straight line graphs
PSS, *see* Point-swept sphere
PSTs, *see* Priority search trees
PTST, *see* Point search tree
Public-domain libraries, 892
Pure Heap Model, 94
Pure silicon, 817
Push-up node, 240
Push operation, 522, 642
Pyramid technique, 269, 1008

Q

QGIS, 883
QLQT, *see* Quad List Quad Trees
QMAT, *see* Quadtree medial axis transform
QoS routers, 765
QPBSM, *see* Partition-based spatial merge
q tokens, 802
Quad-edge data structure, 287–288, 1021
Quad List Quad Trees (QLQT), 828
Quadrangle inequality, 230
Quadrants, 267–268, 309, 320–321, 361, 790, 791, 905, 907
Quadratic Bezier curve, 866
Quadratic probing, 124
Quadtree-based approach, 259, 264, 360

Quadtree Complexity Theorem, 262, 268
Quadtree medial axis transform (QMAT), 260
Quadtrees, 252, 256, 260, 309, 312, 361, 819, 883, 903, 905, 906, 910–911
   basic operations, 316
   cell deletion, 317
   cell insertion, 317
   cell orderings and space-filling curves, 314–315
   compressed quadtrees and octrees, 313–314
   constructing, 309–310
   construction of compressed quadtrees, 315–316
   with coordinates, 909
   image processing applications, 320–322
   insertions and deletions, 317
   point and cell queries, 316–317
   for point data, 310
   point quadtrees, 310–311
   practical considerations, 317–318
   region quadtrees, 311–313
   representation of image, 912
   scientific computing applications, 322–325
   spatial queries with region quadtrees, 318–320
   variants, 905–907
Quad trees, 429, 826, 970
   BLQT, 826
   BQT, 828–829
   HQT, 829–830
   HV Trees, 829
   k-d trees, 826
   QLQT, 828
   and variants, 826, 827
Quasi-bar bounds, 409–410
Queries, 560–561, 562–563, 593, 967
   algorithm, 235, 236
   curve pieces for query range, 877–878
   data structures for query processing, 968
   evaluation, 967, 968, 969
   expression trees, 972–973
   histograms, 973–974
   index structures, 969–970
   operators, 884
   optimization, 969
   optimizer component, 967
   parse tree, 971–972
   performance, 360
   re-writing, 968
   sorting large data sets, 970–971
   time, 795, 1043
   types, 343, 360, 890
"Query-reply" loop, 698
Querying position heap, 490–491
Query Translation and Rewrite module, 968
Queue(s), 31, 192, 522, 672, 751–752
   implementation, 33–34
Queuelocks, 747
Quicksort, 210, 546, 548
Quiescent consistency condition, 746
Quotient graphs, 945, 953, 957–958
   clique covers, 957
   covering column-intersection graph and biclique covers, 958–959
   problem of degree updates, 958

R

Rabin's fingerprints, 801, 802
Radial layout, 718–719
Radix priority-search tree (RPST), 775
Radix searching, 253
RAM, *see* Random Access Machine model of computation
Random-access iterators, 674
Random access, 649
Random Access Machine model of computation (RAM), 172, 530, 536, 545, 546, 561, 627
Random hash functions, 137
Randomization, 197, 583–584
   graph decomposition, 583–584
   maintaining spanning forests, 583
   random sampling, 583
Randomized algorithms, 198–199, 531
Randomized binary search tree (RBST), 171, 197, 209–210, 212
   deletion in, 211–212
   insertion in, 210–211
Randomized data-structure
   for decremental APASP, 618–624
   for static APASP, 611–618
Randomized dictionary structures; *see also* Online dictionary structures
   analysis of dictionary operations, 207–209
   Bernoulli distribution, 202
   binomial distribution, 202
   conditional probability, 200–201
   data structures, 197–198
   Dictionary ADT, 198
   dictionary operations, 207
   geometric distribution, 202
   negative binomial distribution, 202–203
   preliminaries, 198
   probability theory, 199
   randomized algorithms, 198–199
   RBST, 209–212
   skip lists, 204–205
   structural properties of skip lists, 205–206
   tail estimates, 203–204
Randomized finger search trees, 173
   skip lists, 174–175
   treaps, 173–174
Randomized graph data-structures; *see also* Cache-oblivious data structures
   (2k–1)-approximate distance oracle, 613–615
   3-approximate distance oracle, 612–613
   bounding size of $B_u^{d,S}$ under edge-deletions, 621–622
   computing approximate distance oracles, 616–618
   hierarchical distance maintaining data-structure, 619–621
   improved decremental algorithm, 622–624
   main idea, 618–619
   notations, 619
   preliminaries, 613
   randomized data-structure for decremental APASP, 618
   randomized data-structure for static APASP, 611
Randomized incremental construction, 1021–1022, 1031
Randomized search tree, 533
Randomized set representations
   hash trie, 531–533
   remarks on unique representations, 533
   simple, 530

Random probing, 122
    assumption, 122
    asymmetric hashing, 125
    Brent's method, 124
    Cuckoo hashing, 126
    double hashing, 124
    hashing with chaining, 122
    hashing with open addressing, 123
    LCFS hashing, 125–126
    linear probing, 123–124
    multiple-choice hashing, 125
    quadratic probing, 124
    Robin-Hood hashing, 126
Random sum, 203, 208
Random variables, 200–201
Range allocation rule, 777
Range lookups, 786–787
Range query, 240, 241, 310, 318–319, 367, 425, 434,
    547–548, 551, 872
    analytical models for, 353
    for SFC data structures, 876
Range reduction, 628–629, 634
Range restrictions, adding, 1053–1054
Range search(ing), 402, 428, 429, 434
    bootstrapping for 2-D diagonal corner and stabbing queries,
      431–432
    bootstrapping for three-sided orthogonal 2-D range search,
      432–433
    general orthogonal 2-D range search, 433–434
    linear-space spatial structures, 429
    lower bounds for orthogonal range search, 434
    problem, 298
    R-trees, 429–431
    structures, 387
    tools, 387
Range search-tree (RST), 778
Range trees, 291, 298, 387, 560
    construction, 299–301
    example and applications, 301–304
Rank search, 153
Rank/Select queries, 931
Raster, 904
    graphics, 904
    images, 871, 904
Rat-kernel, 656
"Rate equation" approach, 806
Rational kernel, 656, 884
Ray shooting, 786
    query, 425, 428, 872
RBPST, *see* Red-black priority-search tree
RBST, *see* Randomized binary search tree
RCS data structure, *see* Rectangular corner stitching data structure
RDBMS, *see* Relational database management systems
Reachability trees, 567, 578–579, 584
Read amplification, 987–989
Reader-writer locks, 747–748
Read/write efficiency, 145–146
Real-life classifier, 785, 795
Realism, 861, 862
Rebalancing
    algorithm, 153, 154–155, 156
    binary search tree, 159

    operations, 180
    scheme, 153, 154, 155
    technique, 155
    tree to perfect balance, 158
Rebuilding, 155, 634
Record organizations, 976; *see also* File organization; Page
    organizations
    alternative organizations of records with fields of
      variable length, 977
    organization of records with fields of fixed length, 976
    positional B^+-trees, 978
Records, 976, 1000
    format, 976–978
    type, 976–978
Rectangle-swept sphere (RSS), 893
Rectangle(s), 343, 350
    data, 263–265
    enclosure, 786
    intersection graph, 291*fn*
Rectangular corner stitching data structure (RCS data structure),
    823, 824
Rectilinear polygons, 825
Rectilinear shape handling, 851–853
Recurrence equations, 12
    substitution method, 12–13
    table-lookup method, 13
Recursion, 87, 640
Recursive algorithm, 40, 42, 222, 1037
Recursive decomposition of space, 309
Recursive flow classification (RFC), 793, 794
Recursive halving process, 253
Recursively defined SFCs, 875
Recursive process, 336, 1007
Red-black, 179, 180
    method, 162
    trees, 164, 165–166, 774–775
Red-black priority-search tree (RBPST), 775
RedBlackTree, 688
Red, green, and blue (RGB), 904
Reduced ordered BDD (ROBDD), 497, 498
Redundancy, 834, 850
Redundant binary counters, 524
Redundant binary representation, 524
Redundant counters, 522
Reference(s), 23
    counter, 499
    element, 454
Refinement, 897–898
Reflection, 709, 731
Reflectional symmetry, 728
Regional images, 361
Regional quadtrees, 361
Region data, 259–263
Region octrees, 260, 262
Region quadtrees, 260, 266, 311–313, 322, 905–906
    k-nearest neighbors, 320
    range query, 318–319
    spatial queries with, 318
    spherical region queries, 319–320
Region Search operation, 827
Regression techniques, 998, 1000
Regular decomposition, 253, 261, 264, 266

Regulatory elements, 918
Reingold-Tilford algorithm, 710, 711
Relational database management systems (RDBMS), 967, 983
Relative Neighborhood Graph (RNG), 1023, 1034
Relaxed balance, 164
    AVL-trees, 166
    multi-way trees, 166–167
    other results, 167
    red-black trees, 165–166
    update operations, 165
Relaxed depth, 167
Relaxed height of node, 166
Relaxed multi-way trees, 167
Relaxed supernode partitions, 950
Reliability, 680
removeMax operation, 103, 112
removeMin operation, 103, 112
Removing element, 449–451, 458
Removing min element
    deaps, 109–111
    interval heaps, 102–103, 104
    min-max heaps, 106–108
Repairable certification, 380
Replication consistency, 985
    Merkle tree, 986, 987
Reporting distance with stretch at-most (2k−1), 614–615
Reporting problems, general approach for, 1052–1053
Representative vertex, 954
Research prototypes, 700, 883
Resemblance, 802
Residual capacity, 190
Residual graph, 190
Resistance, 818
Resizable arrays, 607
Restricted Dijkstra's algorithm, 617, 618
Restricted multilevel partition, 572, 573
Restricted partition, 571
    update, 572–573
Restructuring primitive, 154
Reverse-nearest-neighbor query, 367
Reverse iterators, 676
Reverse transformation, 1055–1056
R^{EXP}-tree, 373
RFC, *see* Recursive flow classification
RGB, *see* Red, green, and blue
RGB color value, 904
Right-deep tree, 973
Right-looking algorithm, 952, 953
Right contour, 709, 710, 715
Rightist orders, 426
Right tree, 195
Rigid circuit graphs, 954
RNG, *see* Relative Neighborhood Graph
Road maps, 873–874
ROBDD, *see* Reduced ordered BDD
Robin-Hood hashing, 126, 127
Robust algorithms, 953
Room synchronization, 748
Root, 496, 708
    cell, 312
    level of min-max heap, 105
    node, 47, 194, 452

Root*table, 362
Rooted clique tree, 954, 955
Rooted tree, 35, 53, 708
Rotational symmetry, displaying, 729–730
Rotation(s), 154, 187–188, 204, 321, 389, 395, 426
Routers, 165
Row-counts algorithm, 952
Row-prime order method, 255
Row indices, 950
Row nonzero counts prediction, 951–952
Row order method, 255
RPST, *see* Radix priority-search tree
RSS, *see* Rectangle-swept sphere
RST, *see* Range search-tree
R*-tree, 883, 1037
R-tree(s), 233, 252, 256, 266, 343, 360, 429–431, 880, 881, 883, 884,
        910–911, 970, 1037
R-trees*, 343, 346; *see also* Binary space partitioning trees (BSP trees)
    advanced operations, 350–353
    analytical models, 353–356
    basic concepts, 343–344
    bulk loading, 349
    hilbert tree, 346–348
    improving performance, 346
    intersection queries, 344–345
    R* tree, 346
    updating tree, 345–346
R+-tree, 258, 259, 266, 1037
Rule-based decision-making, 914
Rule table, 765, 766, 767
Run-generation phase, 971
Running time, 198, 407, 545

S

Sample space, 199, 200, 338
Sarnak and Tarjan's tree, 1029
Saturating push, 191
Save-points, 976
SBB-trees, *see* Symmetric Binary B-trees
SBB/red-black trees, 157, 158
SB-tree, 425
Scalability, 136, 742, 749, 785, 999
Scalable sweeping-based spatial join (SSSJ), 423
Scalable techniques, 742, 999
Scaling, 321, 905
Scene, geometric, 658
Scheduling out-of-core factorizations, 953
Scheduling parallel factorizations, 952–953
Schönhardt's polyhedron, 1024
Scientific computing applications, 322–325
    N-body problem, 323–325
    physical system behavior, 322–323
Score function, 918, 919
Scuttlebutt gossip-style protocol, 986
SDD, *see* Sentential Decision Diagram
Searchable partial sums problem, 606
Search(ing), 457, 547, 553, 676, 677
    BFS, 54, 56
    BSTs, 42
    DFS, 54, 55
    engines, 478, 799, 801

graph, 54
for highest-priority matching range, 779
index, 365
operation, 14, 197, 204
problem, 291
procedures, 153
trees, 209, 513, 754–756
trie, 446
Secondary clustering phenomenon, 124
Secondary organization, 980, 981
Secondary storage, 234, 468
Seeded tree creation algorithm, 352, 353
Seed levels, 352
Segments, 974
segment-oriented recovery, 976
Segment tree(s), 263, 291, 294, 422, 791
construction, 294–296
example and applications, 296–298
Select-from-where (SFW), 972
Select functions, 319
Selection, 676, 677
sort, 25
Self-adjusting binary search trees, 172, 571
Self-loop edges, 49, 51
Self-tunable spatiotemporal B+-tree (SP^2 B-tree), 360
Self pointer, 457, 458, 459
"Semi-splaying" tree, 195
Sentential Decision Diagram (SDD), 507
Separability assumption, 540
Separating axis, 894, 895
Separating chains, 1029–1030
Separator decomposition tree, 603
Separators, 309, 946, 977
Sequence pair (SP), 833, 835, 844, 845–847, 851
Sequence(s), 685, 924
BDD, 495, 507
containers, 667–669
sorting algorithms, 688
Sequential bottleneck, 742
Sequential consistency, 746
Sequential greedy algorithm for Huffman coding, 230
Sequential pattern analysis, 999
Serializability, 746
Set abstract data type (Set ADT), 117, 119, 122, 127
SETI, spatial partition method, 360, 368
Set(s), 175, 669–670, 671
ADT, 117
of combinations, 504, 506
data structures for, 529
disjoint set union-find problem, 538–540
equality testing, 533–535
extremal sets and subset testing, 535–538
models of computation, 530
operations, 175
of operators, 363
partition maintenance algorithms, 540–542
set-pruning tries, 787–788
set-theoretic operations, 259
simple randomized set representations, 530–533
system, 398
SFCs, *see* Space filling curves
SFW, *see* Select-from-where

Shading model, 861
Shadow bits, 976
Shadow paging concept, 976
Shannon expansion, 496
Shape handling, 851–852
Shared-memory multiprocessor machines, 741, 751
Shared BDDs, 498, 499
Shared counters, 749
combining, 749
counting networks, 750
data structure, 742
Shared mode, 754
Shatter function, 398, 399
Shield region, 412, 413, 415
Shifting smaller subtrees, 715–718
Shingling technique, 802
Shortcut representation, 604
Shortest elimination trees, 955
Shortest paths, 62, 585, 589, 659–661
all-pairs shortest path, 64–65
single-source shortest paths, arbitrary weights, 63–64
single-source shortest paths, nonnegative weights, 62–63
Shrinking node, 1038
Shrink operation, 410, 411
Siblings, 708
Siblings node, 36, 294–295
Sides, 256, 263, 268, 1024, 1045
Sift-up operation, 86, 88, 89
Signature, 938
signature-based canonization, 938
sort, 635
Similarity, 208
detection mechanisms, 802
Jaccard coefficient, 802
Similarity search, 935, 939
bitbound algorithm, 940
combining hashing and trees, 943
fingerprint compression, 940
hashing approaches for improved searches, 940–941
inverted indices, 942
LINGO, 942
multibit and union tree, 943
tanimoto similarity, 939–940
triangle inequality and improvements, 941
trie-based approaches for storing fingerprints, 943
XOR filter, 941–942
Similar property principle, 935
"Simpath" algorithm, 495, 507
Simple computer model, 7
Simple graph, 49
Simple means, 628–630
Simple path, 65
Simple searching, 152
Simple splay trees, 195
Simple updates, 152
Simplicial partition, 402
Simulators, 817
Singlebit tree, 943
Single cells, 1014, 1015
Single-ended priority queues, 69, 98
Single-level index, 980
Single quad, 826, 827

Single-shot problems, 1045
Single Source LWS problem, 230
Single-source shortest paths
    arbitrary weights, 63–64
    nonnegative weights, 62–63
Single tree representations, 842–844
Sites, 1013, 1020, 1022, 1032
Sizes of subtrees, 154
Size-tiered LSM-tree, 989
Skeleton graph, 949, 950, 956
Sketch subset, 802
Skewed tree, 39
Skew heaps, 69, 78, 644; *see also* Binomial heaps; Fibonacci heaps;
        Pairing heaps
    amortized analysis, 78–79
    analysis of lazy, 648–649
    executing pending merge, 648
    meldable priority queues and, 80–83
    merging two, 645
    skew heaps with lazy evaluation in Java, 647
    unbalanced, 646
Skew operation, 158
Skip fields, compressed tries with, 454
Skip lists, 173, 174–175, 198, 204–205, 206, 533
    number of levels in, 205–206
    space complexity, 206
    structural properties, 205
Skip value, 455
Slabs, 349, 422, 431, 433, 1029
    slab-based methods, 1029
    x-range, 562
Slicing
    floorplans, 835–837
    structure, 849
    trees, 833, 844, 845, 849–850
Slot index spatial join, 353
"Slots", 352, 978
Smallest quad, 826, 827
"Small motions", 385
SMAWK algorithm, 228
SMMH, *see* Symmetric min-max heap
Smoothed particle hydrodynamics, 323
Smoothing strategy, 586, 592
Soap2 alignment tool, 927
Social network, 723, 724
Social security number, 445, 447
Soft heaps, 94–95
Software environment, 698
Software transactional memory, 749
Software visualization, 697, 698
Solid-state storage, 145–146
Solid path, 186, 569, 571
Solid state drives (SSDs), 987
Solid tiles, 819
Solid trees, 186–187, 569
Sort-tile-recursive (STR), 349
Sorted array merge tree, *see* Size-tiered LSM-tree
Sorted arrays, 25, 171, 597
Sorted files, 980
Sorted String Table (SSTable), 988
"Sort heap", 94
Sorting, 634, 676–677, 969

    algorithms, 634
    arrangement of buffer pages and disk blocks, 971
    combining techniques, 635
    further techniques and faster randomized algorithms, 635–636
    large data sets, 970
    by mindist, 351
    by minmaxdist, 351
    networks 634
    packed sorting, 634–635
    policy, 351
    range reduction, 634
Sort–merge paradigm, 884
SP, *see* Sequence pair
SP^2 B-tree, *see* Self-tunable spatiotemporal B+-tree
Space bounds, 435, 512, 556, 602, 603, 1048
Space complexity, 9, 206
Space decomposition process, 256, 259
Space-driven
    methods, 428
    partitioning of methods, 429
    structures, 359–360
Space efficiency, 145, 234, 909
Space efficient computation of MEMs for two genomes, 926
Space filling curves (SFCs), 314–315, 347, 429, 874, 884
    corner cell traversals, 879
    creation, 876–877
    curve pieces for query range, 877–878
    orderings, 315
    range queries for SFC data structures, 876
    recursively defined, 875
Space issues, 468
Space-ordering methods, 255
Space partitioning, binary tree representation of, 334
Space required by compressed trie, 456
Space required structures, 447–449
Spanned organization, 977
Spanning
    forest, 54, 578, 583, 584, 588
    trees, 53–54, 57, 582
Sparse
    case, 120
    direct solvers, 952
    environments, 899
    matrices, 27
    matrix factorizations, 948
    multislab lists, 432
    segment, 598
    symmetric factorization, 959
Sparsification, 582–583, 1050
Sparsification-based approach, 1048; *see also* Persistence-based
        approach; Transformation-based approach
    generalized half-space range searching, 1049–1050
Spatial 2D R-tree, 365
Spatial data, 871, 1013
    mining, 263
    models, 882
    operations, 343
    sets, 351
Spatial database systems, 883
Spatial data structures, 309, 425, 428; *see also* Kinetic data
    structures (KDS)

bootstrapping for 2-D diagonal corner and stabbing queries, 431–432

bootstrapping for three-sided orthogonal 2-D range search, 432–433

general orthogonal 2-D range search, 433–434

linear-space spatial structures, 429

lower bounds for orthogonal range search, 434

R-trees, 429–431

Spatial Hash-Join, 880–881

Spatial index(ing), 365, 882

Spatial join(s), 343, 351–353, 365, 872, 879, 882

advanced issues, 881–882

algorithms, 880, 881

external algorithms, 879–881

Spatial libraries, 883

JTS, GeoTools, and GEOS, 883–884

LEDA and CGAL, 884

XXL, 884

Spatial partitioning hierarchies, 892

Spatial partition methods, 360

Spatial queries

with region quadtrees, 318–320

types, 425, 428

Spatial searching, 872

Spatial search structures, 331

Spatial sorting, 329

Spatial–temporal queries, 364

Spatio-temporal data, 343, 359, 874

HR-trees, 368

indexing structures for continuously moving objects, 370–373

MV3R-tree, 368–370

overlapping linear quadtree, 360–363

spatiotemporal indexing techniques, 360

3D R-tree, 363

2+3 R-tree, 368

Spatiotemporal databases (STB), 359

Spatiotemporal data structures (STDS), 359

Spatiotemporal indexing techniques, 360

Spatiotemporal joins, 370

Spatiotemporal operators, 363

Spatiotemporal query types, 367

Spatio-temporal self-adjusting R-tree (STAR-tree), 373

Specialized data structures, 359

Speedup of application, 742

Sphere-trees, 893

Spherical region query, 309, 310, 319–320

Spherical shell-trees, 893

Spinlocks technique, 747

Spinning, 743, 747

Spiral storage, 127

Splay

analysis in virtual tree, 188–189

sequence of rotations, 180

Splay trees, 78, 155, 158, 172, 179, 180–181; *see also* B-tree

access and update operations, 182–183

amortized time, 181–182

analysis, 181

application to network flows, 190–192

FIFO dynamic tree implementation, 193–194

implementation without linking and cutting trees, 192–193

linking and cutting trees, 185–189

optimality, 184–185

variants and top-down splaying, 195

Splice effects, 289

Splicing, 188, 578

Split

cases, 285

operation, 153, 158, 183, 395, 410, 576

policy, 346

split-find problem, 540

Splitting, 389

nodes, 1038

rule, 1036

threshold, 267

SQL database, 882, 968, 984

SQLite, 883

SR-tree, 1008, 1037

SSDs, *see* Solid state drives

SSSJ, *see* Scalable sweeping-based spatial join

SSTable, *see* Sorted String Table

SS-tree, 1008, 1037

SSV, *see* Swept sphere volume

SSV-trees, 893

Stabbing queries, 421, 426, 431–432

Stability, 583

Stacks, 31–32, 522, 640–642, 672, 751

Standardized data models, 882

Standard range, 294, 295, 301, 1055

Standard template library (STL), 667, 679

additional components, 676

sorting, searching, and selection, 676–677

Star-shaped drawings, 738

STAR-tree, *see* Spatio-temporal self-adjusting R-tree

Star-triangulation, 729

Start node, 477, 478

StartPointer, 774, 775

Static APASP

(2k−1)-approximate distance oracle, 613–615

3-approximate distance oracle, 612–613

computing approximate distance oracles, 616–618

preliminaries, 613

randomized data-structure for, 611

Static array, 24

Static counting problem, 1047

Static data structure, 291, 633

Static extremal sets, 536–537

Static finger theorem, 184

Static hash file, 980

Static optimality, 184

Static perfect hashing, 119–120

Static pivoting, 960

Static point interval query, 373

Static queries, 890, 892

Static R-tree, 884

Static search tree, 546

Static structure, 433

Static type-2 problem, 1048

Statistical analysis of words, 918–919

Status function, 319

STB, *see* Spatiotemporal databases

STDS, *see* Spatiotemporal data structures

Steiner points, 382, 383, 1024

Step-count(s), 5–7
  analysis, 10
  method, 5
Stepped merge method, *see* Size-tiered LSM-tree
Steps per execution (s/e), 6
STL, *see* Standard template library
Storage, 348
  subsystem, 968
  systems, 419
  utilization, 360
Stout-Warren algorithm, 158
STR-tree, 360
STR, *see* Sort-tile-recursive
Stride of node, 770
String(s), 461, 924
  containment, 471
  data type, 699
"String art" algorithm, 866
String searching, 477
  compact DAWG, 479, 486–489
  DAWG, 478, 479–486
  position heap, 479, 489–494
  preliminaries, 479
Strip-emptiness queries, 1016–1017
Strip tree data structure, 266
Structural operations, 279
Structural theorem, 225
Sub additivity, 199
Subcell, 312, 314, 317, 320, 1016, 1036
Sub-circuits, 833
Subgraph of graph, 53
Sub-hyperplanes, 337
Sub-logarithmic time per element by simple means, 628
  combining, 630
  packing keys, 629–630
  range reduction, 628–629
Sub-quadratic space, 612
Subset enumerator, 18
Subset generation, 18
  accounting method, 19
  aggregate method, 19
  potential method, 19–20
  problem definition, 18–19
  worst-case method, 19
Subset testing
  dynamic set intersections and, 537–538
  extremal sets and, 535
  static extremal sets, 536–537
Substitution method, 12–13
Subtrees, 35, 72, 90, 93, 151, 154, 156, 163, 223, 301, 319,
    344–345, 501, 547, 561, 571, 605, 688, 708, 711, 713, 715,
    777, 926, 1055
Successor
  search, 152
  vertex, 49
Succinct dictionaries, 597
  dynamic dictionary, 598–599
  FID, 598
  indexable dictionary, 597–598
Succinct representation of data structures
  arrays, 607
  bitvector, 596–597

graph representations, 603
partial sums, 606
permutations and functions, 604–606
succinct dictionaries, 597–599
succinct structures for indexing, 603–604
tree representations, 599–603
Succinct structures for indexing, 603–604
Suffix array(s), 461, 462, 463, 467, 603
  advanced applications, 473
  applications, 468
  approximate pattern matching, 473–474
  construction algorithms, 463
  Käräkkanen and Sanders' algorithm, 467–468
  linear time construction algorithms, 463–468
  longest common substrings, 470–471
  lowest common ancestors, 472
  maximal palindromes, 474
  pattern matching, 468–470
  properties, 461–463
  string containment, 471
  suffix-prefix overlaps, 471–472
  suffix links from lowest common ancestors, 473
  text compression, 471
Suffix(es), 462, 523, 924
Suffix link, 462, 464, 465
Suffix tree(s), 461, 462, 463, 467, 603, 924
  advanced applications, 473
  applications, 468
  approximate pattern matching, 473–474
  construction algorithms, 463, 464
  generalized, 466
  linear time construction algorithms, 463–468
  longest common substrings, 470–471
  lowest common ancestors, 472
  maximal palindromes, 474
  McCreight's algorithm, 465–466, 470
  pattern matching, 468–470
  properties, 461–463
  string containment, 471
  suffix-prefix overlaps, 471–472
  suffix arrays vs., 464
  suffix links from lowest common ancestors, 473
  text compression, 471
SUM operator, 245–246
Super-element model, 957
Supercell, 312, 314
Supernodes, 62, 949–950
Support count of itemset, 1003
Surface mapping, 861, 862
SWAN system, 704
Swapping pointers, 160
Sweeping, 380, 427, 515
  distribution, 422
  phases, 748
  plane sweeping algorithm, 516
Sweep-line algorithm, 352, 423, 879, 880, 1021
Sweep-line order, 352, 1029
Swept sphere volume (SSV), 893
SWIM, failure-detection protocols, 986
Switch pointer, 788, 789
Symbolic Cholesky factor, 950
Symbolic elimination, *see* Symbolic factorization

Symbolic factorization, 950–951, 957

Symbolic perturbation, 654

Symmetric Binary B-trees (SBB-trees), 157

Symmetric drawing, 728; *see also* Convex drawing; Visibility drawing
- displaying axial symmetry, 730–732
- displaying dihedral symmetry, 732
- displaying rotational symmetry, 729–730

Symmetric matrix, 893, 946, 950

Symmetric min-max heap (SMMH), 98–100

Symmetric pruning, 961

Synchronous traversal method, 353

"Syntactic" duplicates, 801, 802

T

Table-lookup method, 13

Tag value, 166

Tail estimates, 203–204

"Tailor" layout system, 823

Tall cache assumption, 546, 556

Tanimoto distance, 941

Tanimoto similarity, 939–940, 943

Target function, 1000

t-ary trees, Huffman algorithm for, 220

TB-tree, 360

TBT, *see* Twin binary tree

TCAMs, *see* Ternary content-addressable memories

TCG, *see* Transitive closure graph

Template method pattern, 683, 684, 689, 692

Template quadtrees, 907, 908

Template sets, 907

Temporal 1D R-tree, 365

Temporal index, 365

Terminal nodes, 496

Ternary content-addressable memories (TCAMs), 767, 795–796

Tertiary storage, 234, 970

Test bit, 629, 630, 631

Tetrahedralizations, 1023

"Texels", 862

Text, 477
- compression, 471
- index, 603

Texture mapping, 861–862

Thematic attributes, 871, 873

Theoretical growth models, 806–809

Theta notation (Θ), 11

Threaded binary trees
- inorder traversal, 41–42
- threads, 41

Thread(s), 41, 709, 741

3-approximate distance oracle, 612–613

3D Geometry Library (*libP*), 655
- cache-oblivious kd-tree, 560–561
- cache-oblivious range tree, 561–563

Three-dimensional lines, 368

3-D polygonal meshes, 861

Three-sided queries, 433, 562
- layout, 562
- query, 562–563
- structure, 562

Three-sided range search(ing), 433

Three dimensional graph drawing, 737

3D R-tree, 363
- handling queries with open transaction times, 367–368
- performance analysis, 367
- spatiotemporal queries using unified schema, 365–367

TID, *see* Translation invariant data structure

TID concept, *see* Tuple identifier concept

Tiered vector, 607

Tiled array, 864, 865

Tile(s), 384, 819
- creation, 820, 821–822
- curved, 823
- deletion, 820
- L-shaped, 823–826
- trapezoidal, 823

Tilings, 382–383

Time bounds, 94, 491, 568, 582, 606, 1056
- of finger search trees, 176
- improvements to, 491–494
- logarithmic, 523
- run–time bounds, 468
- for updates and queries, 515

Time complexity, 3, 9, 54, 204, 224, 391, 541, 776, 836, 847, 949

Time-forward processing, 427

Time-Parameterized R*-tree (TPR*-tree), 371–373

Timeslice query, 371

Timestamp query, 368

TMS, *see* Truth Maintenance System

Token, 750, 802

Tombstones, 988

Top-down
- hierarchical methodology, 834
- methods, 893
- splaying, 195
- traversal of suffix tree, 470

Topological
- approach, 935
- data, 872

Topological number(ing), 689, 734, 735, 736

Topological sorting, 57–58

Topology trees, 567, 571, 582
- applications, 573–574
- construction, 572
- update, 573
- updates, 572–573

Toponym recognition, 269

Toponym resolution, 269

Top trees, 567, 574
- representation and applications, 576–577
- updates, 575–576

Total correspondence, 111–112

Tournament trees, 46
- loser trees, 48
- winner trees, 47

TPIE system, *see* Transparent parallel I/O environment system

TPR*-tree, *see* Time-Parameterized R*-tree

Traditional indexing
- schemas, 371
- techniques, 360

Traditional metrics, 987

Traditional spatial data structures, 359

Traditional spatial indexes, 360
Trajectory indexing techniques, 359
Transaction-oriented recovery, 976
Transactional memory mechanism, 749
Transactional synchronization mechanisms, 748
Transaction management subsystem, 968
Transaction time, 359, 360, 367–368
Transformation-based approach, 1046
    dynamic counting problem, 1047–1048
    dynamic reporting problem, 1046–1047
    static counting problem, 1047
    static type-2 problem, 1048
Transformed data structure, 1054
Transistor, 817, 818
Transition function, 745
Transitive closure, 589
    updates in $O(n^2)$ time, 591
    updates in $O(n^2 \log n)$ time, 590
Transitive closure graph (TCG), 841
Transitively closed subgraphs, 956
Transitive reduction, 479, 961
Translational C-space obstacle (TCSO), *see* Minkowski sum
Translation invariant data structure (TID), 903, 912–914
Translation table, 978
Transparent parallel I/O environment system (TPIE system), 423
Trapezoidal decomposition, *see* Trapezoidal map;
        Vertical decompositions
Trapezoidal maps, 1031–1032
Trapezoidal tiles, 823
Trashing, 391
Trawling, 813–814
Treaps, 173–174
Tree-depth, 462
TreeLayout algorithm, 711
TreeNode class, 39
Trees, 35, 49, 53, 151–152, 686; *see also* Leftist trees
    binary trees, 37–39, 599
    binary tree traversals, 39–41
    BSTs, 42–44
    cardinal trees, 601–602
    color problem, 514
    compaction, 988
    dynamic binary trees, 602–603
    heaps, 44–46
    height limited Huffman trees, 220–222
    Huffman trees, 216–220
    left child-right sibling representation, 37
    membership, 579
    merging, 335–337
    with minimum weighted path length, 215
    nodes, 601
    OAT problem, 224–226
    OBST, 222–224
    optimal lopsided trees, 226–229
    ordinal trees, 599–601
    parallel algorithms, 230–231
    rebalancing algorithm, 163
    references primitives, 869
    representations, 37, 599
    threaded binary trees, 41–42
    tournament trees, 46–48
    tree-based data structures, 425

tree-based priority pools, 757
tree-push procedure, 193
type data structure, 825
updating, 345–346
Treewise merging process, 175
Triangle inequality, 620, 941
Triangle mesh, 858
Triangulated graphs, 898, 954
Triangulations, 277, 288, 382–383, 839, 1022
    DT, 1022–1023
    of point set, simple polygon, and polyhedron, 1024
    polygons, 1023
    polyhedra, 1023–1024
    pseudo-triangulations, 1024–1025
Trickle-down process, 107, 108
Triconnected planar graph, 728
Tridiagonal matrix, 27
Trie(s), 390, 445, 769
    1-bit tries, 769–770
    compressed, 452–456
    FSTs, 770–772
    height of trie, 447
    implementations, 390–391
    inserting into, 449
    keys with different length, 446–447
    patricia, 456–458
    prefix search and applications, 451–452
    removing element, 449–451
    representation, 449
    searching, 446
    space required and alternative node structures, 447–449
    trie-array implementation, 391
    trie-based approaches for storing fingerprints, 943
    trie-bst approach, 396
    VST, 772–774
Trivial labeling, 190
Trivial rebalancing scheme, 153
True color, 859
Truth Maintenance System (TMS), 507
Tuple identifier concept (TID concept), 979
Tuple space search, 795
Tutte's duality, 839–841
TV-tree, 1037
Twin binary tree (TBT), 833, 834, 841–842, 849, 851
Twin heap, 111
Twin slot method, 975
2-3-trees, 179, 180, 389
2-3-4 tree, 234
Two-cuttable theorem, 415
Two-dimension (2D), 8
    approaches, 935
    arrays, 8, 26, 862
    bootstrapping for 2-D diagonal corner,, 431–432
    classification scheme, 790, 791
    classifier, 786
    compaction, 819
    data set, 353
    data types, 884
    homothetic range search, 435
    intersection tests, 900–901
    orthogonal range queries, 428
    points, 368

space, 251, 371
topology trees, 574, 582
2D-range search problem, 302–304
2D Geometry Library (*libP*), 655
2D orthogonal range searching, 560
2-edge connectivity, 574
Two-handed emulation technique, 754
Two-level hash function, 597
Two-pass pairing, 93
Two-step approach, 628
Two dimensional array of pixels, 320
(2k–1)-approximate distance oracle, 613
reporting distance with stretch at-most (2k–1), 614–615
size of (2k–1)-approximate distance oracle, 615
Type-2 counting problem, 1044
(t, ε)-AVD decomposition, 1038, 1039

U

UML, *see* Unified modelling language
Unary encoding, 800
Uncertain points, 1050
Uncompressed region quadtrees, 317
Undirected graph(s), 50, 51, 53, 581
clustering, 582
randomization, 583–584
sparsification, 582–583
techniques for, 582
Unified algorithm for group queries, 319
Unified modelling language (UML), 707
Unified schema, spatiotemporal query using, 365–366
Uniform distribution, 199, 338, 973, 1034
Uniform grid, 252, 253, 259, 260
Uniform hashing assumption, 122*fn*
Union
of images, 321
operation, 868–869
UnionBit Tree, 943
Union-copy problem, 540
Union tree, 943
Unique representation
property, 531
remarks on, 533
of sets, 530, 532
Unit partition, 937, 938
UnitWeightedTopologicalNumbering, 689
Universal hashing, 119
Unmatched clusters, 572
Unordered BDD, 496
Unordered dictionaries, 151
Unreliability, 679
Unrooted tree, 35
Unspanned organization, 977
Unusual words, 917
detection, 919–923
statistical analysis of words, 918–919
subtrees rooted at solid gray nodes, 921
Up buffer, 557, 558, 559
Update(s), 551, 553, 561
on graph, 581
in $O(\log^2 n)$ time, 587–588
in $O(\log^4 n)$ time, 589

in $O(n^{2.5} \sqrt{C \log n})$ time, 592
operations, 182–183, 345, 581
tradeoffs, 593
Upper bound theorem, 1014
Upper convex hull, 663
Upper envelope, 380, 381, 1014, 1049
Upper hull, 1018, 1019
Upper triangular matrix, 27, 945
Upstream sequence, 918
Upward star-shaped polytopes, 738
User-defined displays, 704
User interface, 867, 967

V

Vacant tiles, 819, 822, 823, 824
Valence of atom, 936
Valid interval, 517, 518
Valid labeling, 190
Valid time, 359
Value of flow, 190
van derWaal forces, 323
Van Emde Boaz layout, 546, 547, 560, 561
range query, 547–548
search, 547
Vapnik and Chervonenkis dimension (VC-dimension), 398, 399
Variable-Increment Counting Bloom Filter (VI-CBF), 146
Variable-length record, 968, 977
Variable-stride trie (VST), 772–774
Variable length (VL), 361, 977
encoding scheme, 800
records, 979
strings, 118
Variable length signatures (VBF), 146
Variable ordering for BDDs, 502–503
Variant(s), 425–426, 428, 819
edge quadtrees, 906–907
line quadtrees, 906
of quadtrees, 905
region quadtrees, 905–906
of splay trees, 195
template quadtrees, 907
Variety, 983
VBF, *see* Variable length signatures
VC-dimension, *see* Vapnik and Chervonenkis dimension
Vector graphics, 904, 905
Velocity vector, 371, 983
Verification techniques, 746
Version list, 516, 517, 518
Vertex, 49, 576, 859
attributes, 278, 280, 281
elimination process, 957
hub pair, 897
resolution, 723
split, 279–280, 284–287
Vertex, normal, and face lists, 863
Vertical combination, 719
Vertical decompositions, 1016
Vertical inversions of polygon, 825
Vertical O-tree, 843–844
Vertical ray shooting, 1028

Vertical slabs, 422, 432, 513, 1029
Vertices, 581, 611, 1014
Vertices, 862
Very Large Scale Integration (VLSI), 817–819
    combinations and complexities of representations, 834–835
    design process, 507
    design stages, 818
    example of extra constraints, 854
    floorplan representation in, 833
    graph based representations, 837–844
    logic design, 495
    motivation of representation, 834
    placement based representations, 844–850
    rectilinear shape handling, 851–853
    relationships of representations, 851
    slicing, mosaic, LB compact, and general floorplans, 835–837, 838
    statement of floorplanning problem, 833–834
VI-CBF, *see* Variable-Increment Counting Bloom Filter
Violations of CV property, 822
VIPS, 701–702
Virtual environments, 889
Virtual hashing, 127
Virtual quadtrees, 905, 907
    compact quadtrees, 908–909
    FQT, 909–910
Virtual trees, 186
    splay analysis in, 188–189
    splay in, 188
Visibility, 384–385
    representation, 732
Visibility drawing, 732; *see also* Convex drawing; Symmetric drawing
    bar visibility algorithm, 734–735
    bar visibility representations and layered drawings, 735–736
    bar visibility representations for orthogonal drawings, 736–737
    planar st-graphs, 733–734
Visibility orderings, 332
    binary tree representation of space partitioning, 334
    intra-object visibility, 333
    ordering of polygons, 334
    total ordering of collection of objects, 332–333
    as tree traversal, 333
    visibility ordering as tree traversal, 333
Visualization, 657
    GeoWin, 358
    GraphWin, 657–658
    systems, 700
VL, *see* Variable length
VLSI, *see* Very Large Scale Integration
Voronoi-based marching algorithm, 890
    implementation and application, 892
    local walk, 891–892
    polytope representation, 891
Voronoi cell, 891, 1020, 1032, 1033
Voronoi diagrams, 277, 288, 385, 402, 1002, 1013, 1020, 1021, 1032–1033
    complexity, 1021
    construction, 1021–1022
    variations, 1022
Voronoi edges, 1021
Voronoi regions, 890, 891, 1002
VST, *see* Variable-stride trie

W

Wait-freedom, 744
Wait-free operation, 744
Walkthroughs, 889
Warm-up filter, 146
Warshall-Floyd algorithm, 64
Wavelet trees, fast rank and select operations using, 929–931
WBA, *see* Worst-case bounding-box area ratio
WBLTs, *see* Weight-biased leftist trees
WBP, *see* Worst-case bounding-box squared perimeter ratio
Weak version condition, 369
Web algorithmics, properties of, 809
    average path length, 810
    crawling and trawling, 813–814
    emergence of giant components, 811
    generating function framework, 809–810
    search on web graphs, 811–813
Web as dynamic graph
    experimental observations, 806
    properties of web graphs and web algorithmics, 809–814
    theoretical growth models, 806–809
    web model building techniques, 805–806
Web crawl, 805, 806
Web graph(s), 806
    average path length, 810
    crawling and trawling, 813–814
    emergence of giant components, 811
    generating function framework, 809–810
    properties, 809–814
    search on, 811–813
Web information retrieval
    data structures in, 799
    finding near-duplicate documents, 801–802
    fingerprints, 801
    inverted indices, 799–801
Web model building techniques, 805
Web search engines, 799
Wedge cut canonical set, 416
Weight
    function, 230
    matrix, 51
    of node, 156, 165
Weight-balanced B-trees, 426, 429, 431, 435
Weight-balanced trees, 156–157, 389, 390
Weight-balanced trees, 426
Weight-biased leftist trees (WBLTs), 74
    max WBLT operations, 75
Weighted graph, 50, 59, 694
    connected weighted graph for MST algorithm, 60
    representation, 51, 52
    undirected, 611
Weighted path length, 215
Well-separatedness criteria, 323–324
"Where-am-I" queries, 970
Whole genomes comparison, 923
    basic definitions, 924–925
    collinear chain of nonoverlapping anchors, 923, 924
    computation of multiMEMs, 925–926
    space efficient computation of MEMs for two genomes, 926
    three genomes represented as horizontal lines, 923
Widgets, 16, 17
Window data type, 657

Window Library (*libW*), 655
Window query, *see* Range query
Winged-edge
    approach, 863–864
    data structure, 280–281
Winged-edge data structure, 266
Winner trees, 47
Witness points, 892
WKT, 882
Word count, 659
Word processor, 452
Word RAM
    algorithms, 1045–1046
    model, 530
Word size impact, 137–138
"Working-set" theorem, 94, 185
Workload(s), 370, 372, 987
    distribution algorithm, 753
    NoSQL database, 987
Work sharing, classes of algorithm, 753
Work stealing, classes of algorithm, 753
World Wide Web, 269, 805
Worst-and expected-case optimal point location, 1032
Worst-case bounding-box area ratio (WBA), 877
Worst-case bounding-box squared perimeter ratio (WBP), 877
Worst-case efficient exponential search trees, 634
Worst-case method
    maintenance contract, 15
    McWidget company, 17
    subset generation, 19
Worst-case search time, 122
Write amplification, 987, 988, 989, 992

X

x-monotone, 1023
    polygon triangulation, 1024
    subdivision, 1029–1030
x-nodes, 1031
XOR filter, 935, 941–942
X-tree, 1037
XXL, *see* eXtensible and fleXible Library

Y

y-nodes, 1031

Z

ZBDD, *see* Zero-suppressed BDD (ZDD)
z-buffers, 340
z-curve, 875
ZDD, *see* Zero-suppressed BDD
0-edge, 496
Zero-suppressed BDD (ZDD), 495, 504–506
0-terminal, 496
Zig-zig step, 180, 195
Zig step, 180
Ziv-Lempel compression, 471
Zooming, 873
Z-order curve, 347, 352
z-score, 918, 919
Z-space filling curve, *see* Morton ordering
z-values, 334, 866